COHERENT STATES

Applications in Physics and
Mathematical Physics

COHERENT STATES

Applications in Physics and Mathematical Physics

John R. Klauder
Bo-Sture Skagerstam

G. Domokos + S.K. Domokos
1986

World Scientific

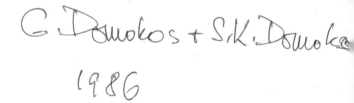

Published by

World Scientific Publishing Co. Pte. Ltd.
P. O. Box 128, Farrer Road, Singapore 9128

The authors and publisher are indebted to the authors and the following publishers of the various journals and books for their assistance and permission to reproduce the selected articles found in this volume:

Academic Press Inc. *(Ann. Phys.)*; American Institute of Physics *(Am. J. Phys., J. Math. Phys.* and *Sov. J. Nucl. Phys.)*; American Physical Society *(Phys. Rev., Phys. Rev. Lett.* and *Rev. Modern Phys.)*; Kyoto University *(Prog. Theor. Phys.* and *Prog. Theor. Phys. Suppl.)*; North-Holland Publishing Co. *(Nucl. Phys.* and *Phys. Lett)*; Plenum Publishing Co. *(Int. J. Theor. Phys.* and *Symmetries in Science)*; Physical Society of Japan *(J. Phys. Soc. of Japan)*; Springer-Verlag *(Comm. Math. Phys.* and *Z. Phys. C)*; The Institute of Physics *(J. Phys. A: Gen. Phys.* and *J. Phys. A: Math. Gen.)*.

COHERENT STATES — APPLICATIONS IN PHYSICS AND MATHEMATICAL PHYSICS

ISBN 9971-966-52-2
9971-966-53-0 pbk

Printed in Singapore by Kyodo-Shing Loong Printing Industries Pte Ltd.

To Our

Families

PREFACE

As a general concept, some notion of *coherence* applies to nearly every type of phenomenon. This is particularly so for those phenomena involving waves of one kind or another. Indeed, coherence has been recognized as an important feature of optics for more than three centuries, and it is fitting that the concept of *coherent states* arose in the modern theory of light, quantum optics. Related concepts now permeate many fields of physics and lead to a wide variety of applications.

The purpose of the present volume is to offer a representative collection of articles dealing with coherent states. The reprinted articles are loosely organized by subject matter in Chapters II-XI, and within each chapter they have been chronologically ordered. Our literature searches have revealed over 1000 articles on this subject, and clearly only a small fraction of these could be included. We apologize to those authors whose work we have overlooked and/or who feel their work should have been included.

While the applications of coherent-state methods are quite diverse, there is nevertheless a common core to the subject. In our first chapter, "A Coherent-State Primer", we have attempted to provide an introduction to the variety of definitions and the associated formalism used for coherent states. We hope the reader will find this to be a useful survey, a guide to the various reprints that follow in the remaining chapters, and even an entré to the vast literature on this subject.

The authors thank a number of their colleagues for discussions on the reprinted material and/or their comments on parts or all of the primer, specifically J. P. Blaizot, I. Daubechies, B. DeFacio, R. Gilmore, H. Kuratsuji, E. Lieb, J. Negele, M. Nieto, and T. Suzuki. We also warmly thank Ms. Terry Yulick without whose patience and expertise our task would not have been possible.

Copenhagen,
July 1984

John R. Klauder
Bo-Sture Skagerstam

CONTENTS

REPRINTED PAPERS

6. Equivalence of Semiclassical and Quantum Mechanical Descriptions of Statistical Light Beams
by E. C. G. Sudarshan, *Phys. Rev. Lett.* **10** (1963) 277 — 199

7. Coherent and Incoherent States of the Radiation Field
by R. J. Glauber, *Phys. Rev.* **131** (1963) 2766 — 202

8. Weak Correspondence Principle
by J. R. Klauder, *J. Math. Phys.* **8** (1967) 2392 — 225

9. Exponential Hilbert Space: Fock Space Revisited
by J. R. Klauder, *J. Math. Phys.* **11** (1970) 609 — 233

10. New "Coherent" States Associated with Non-Compact Groups
by A. O. Barut and L. Girardello, *Commun. Math. Phys.* **21** (1971) 41 — 255

11. Some Properties of Coherent Spin States
by J. M. Radcliffe, *J. Phys. A: Gen. Phys.* **4** (1971) 313 — 270

12. Coherent States for Arbitrary Lie Group
by A. M. Perelomov, *Commun. Math. Phys.* **26** (1972) 222 — 281

13. A Note on Coherent State Representations of Lie Groups
by E. Onofri, *J. Math. Phys.* **16** (1975) 1087 — 296

14. On Intelligent Spin States
by C. Aragone, E. Chalbaud and S. Salamó, *J. Math. Phys.* **17** (1976) 1963 — 299

15. Charged Bosons and the Coherent State
by D. Bhaumik, K. Bhaumik and B. Dutta-Roy, *J. Phys. A: Math. Gen.* **9** (1976) 1507 — 308

16. Coherent States Associated with the Continuous Spectrum of Noncompact Groups
by M. Hongoh, *J. Math. Phys.* **18** (1977) 2081 — 314

17. Quasi-Coherent States for Unitary Groups
by B. –S. Skagerstam, *J. Phys. A: Math. Gen.* **18** (1985) 1. 318

III. *Coherent States and Quantum Mechanics*

1. Coherent States and the Number-Phase Uncertainty Relation
by P. Carruthers and M. M. Nieto, *Phys. Rev. Lett.* **14** (1965) 387 333

2. Coherent States and the Forced Harmonic Oscillator
by P. Carruthers and M. M. Nieto, *Am. J. Phys.* **7** (1965) 537 336

3. Classical Behavior of Systems of Quantum Oscillators
by R. J. Glauber, *Phys. Lett.* **21** (1966) 650 344

4. The Classical Limit for Quantum Mechanical Correlation Functions
by K. Hepp, *Commun. Math. Phys.* **35** (1974) 265 347

5. Quantum Nondemolition Measurement and Coherent States
by W. G. Unruh, *Phy. Rev.* **D17** (1978) 1180 360

6. Coherent States and Quantum Nonperturbing Measurements
by V. Moncrief, *Ann. Phys.* (N. Y.) **114** (1978) 201 362

7. Coherent States for General Potentials
by M. M. Nieto and L. M. Simmons, Jr., *Phys. Rev. Lett.* **41** (1978) 207 376

8. Quantum Limits on Resonant-Mass Gravitational-Radiation Detectors
by J. N. Hollenhorst, *Phys. Rev.* **D19** (1979) 1669 380

9. Charged Schrödinger Particle in a c-Number Radiation Field
by T. Toyoda and K. Wildermuth, *Phys. Rev.* **D22** (1980) 2391 — 391

10. Large N Limits as Classical Mechanics
by L. G. Yaffe, *Rev. Mod. Phys.* **54** (1982) 407 — 400

11. Coherent States With ıassical Motion; From an Analytical Method Complementary to Group Theory
by M. M. Nieto in *Group Theoretical Methods in Physics,* Proceedings, Vol. II, of International Seminar at Zvenigorod, 1982, ed. by M. A. Markov (Nauka, Moscow, 1983), p. 174. — 429

IV. *Path Integrals and Coherent States*

1. Coherent States of Fermi Operators and the Path Integral
by Y. Ohnuki and T. Kashiwa, *Prog. Theor. Phys.* **60** (1978) 548 — 449

2. Path Integrals and Stationary-Phase Approximations
by J. R. Klauder, *Phys. Rev.* **D19** (1979) 2349 — 466

3. Path Integrals for the Nuclear Many-Body Problem
by J. P. Blaizot and H. Orland, *Phys. Rev.* **C24** (1981) 1740 — 474

4. Path-Integral Approach to Problems in Quantum Optics
by M. Hillery and M. S. Zubairy, *Phys. Rev.* **A26** (1982) 451 — 496

5. Path Integral Approach to Many-Body Systems and Classical Quantization of Time-Dependent Mean Field
by H. Kuratsuji and T. Suzuki, *Prog. Theor. Phys. Suppl.* **74 & 75** (1983) 209 — 506

6. Quantum Mechanical Path Integrals with Wiener Measures for all Polynomial Hamiltonians
by J. R. Klauder and I. Daubechies, *Phys. Rev. Lett.* **52** (1984) 1161 518

V. *Pseudospin Techniques*

1. Semi-Classical Spectrum of the Continuous Heisenberg Spin Chain
by A. Jevicki and N. Papanicolaou, *Ann. Phys.* (N. Y.) **120** (1979) 107 525

2. Bohr-Sommerfeld Quantization of Pseudospin Hamiltonians
by R. Shankar, *Phys. Rev. Lett.* **45** (1980) 1088 547

3. Holstein-Primakoff Theory for Many-Body Systems
by L. R. Mead and N. Papanicolaou, *Phys. Rev.* **B28** (1983) 1633 551

VI. *Applications in Condensed Matter Physics*

1. Coherent States and Irreversible Processes in Anharmonic Crystals
by P. Carruthers and K. S. Dy, *Phys. Rev.* **147** (1966) 214 557

2. Theory of Superfluidity
by F. W. Cummings and R. J. Johnston, *Phys. Rev.* **151** (1966) 105 566

3. Coherent States in the Theory of Superfluidity
by J. S. Langer, *Phys. Rev.* **167** (1968) 183 574

4. Landau Diamagnetism from the Coherent States of an Electron in a Uniform Magnetic Field
by A. Feldman and A. H. Kahn, *Phys. Rev.* **B1** (1970) 4584 582

5. Spin Coherent State Representation in Non-Equilibrium Statistical Mechanics
by Y. Takahashi and F. Shibata, *J. Phys. Soc. Japan* **38** (1975) 656 588

6. Coherent-State Langevin Equations for Canonical Quantum Systems with Applications to the Quantized Hall Effect
by J. R. Klauder, *Phys. Rev.* **A29** (1984) 2036 601

VII. *Applications in Thermodynamics*

1. The Classical Limit of Quantum Spin Systems
by E. H. Lieb, *Commun. Math. Phys.* **31** (1973) 327 615

2. Coherent States and Partition Function
by M. Rasetti, *Int. J. Theor. Phys.* **14** (1975) 1 629

3. Applications of Coherent States in Thermodynamics and Dynamics
by R. Gilmore, in *Symmetries in Science,* eds. B. Gruber and R. S. Millman (Plenum Publ. Corp., New York, 1980), p. 105 650

4. On the Large N_c Limit of the $SU(N_c)$ Colour Quark-Gluon Partition Function
by B. –S. Skagerstam, *Z. Phys.* **C24** (1984) 97 677

VIII. *Aplications in Atomic Physics*

1. Atomic Coherent States in Quantum Optics
by F. T. Arecchi, E. Courtens, R. Gilmore and H. Thomas, *Phys. Rev.* **A6** (1972) 2211 685

2. Phase Transition in the Dicke Model of Superradiance
by Y. K. Wang and F. T. Hioe, *Phys. Rev.* **A7** (1973) 831 712

3. Equilibrium Statistical Mechanics of Matter Interacting with the Quantized Radiation Field
by K. Hepp and E. H. Lieb, *Phys. Rev.* **A8** (1973) 2517 718

4. Multitime-Correlation Functions and the Atomic Coherent-State Representation
by L. M. Narducci, C. M. Bowden, V. Bluemel, G. P. Garrazana and R. A. Tuft, *Phys. Rev.* **A11** (1975) 973 727

IX. *Applications in Nuclear Physics*

1. Coherent States of the Quantum-Mechanical Top
by D. Janssen, *Sov. J. Nucl. Phys.* **25** (1977) 479 737

2. Studies of the Ground-State Properties of the Lipkin-Meshkov-Glick Model via the Atomic Coherent States
by R. Gilmore and D. H. Feng, *Phys. Lett.* **76B** (1978) 26 743

3. Phase Transitions in Nuclear Matter Described by Pseudospin Hamiltonians
by R. Gilmore and D. H. Feng, *Nucl. Phys.* **A301** (1978) 189 746

4. Condensates and Bose-Einstein Correlations
by G. N. Fowler, N. Stelte and R. M. Weiner, *Nucl. Phys.* **A319** (1979) 349 762

X. *Applications in Elementary Particle Physics*

1. Coherent Production of Pions
by D. Horn and R. Silver, *Ann. Phys.* (N. Y.) **66** (1971) 509 779

2. Coherent States and Particle Production
by J. C. Botke, D. J. Scalapino and R. L. Sugar, *Phys. Rev.* **D9** (1974) 813 812

3. Coherent Quark-Gluon Jets
by G. Curci, M. Greco and Y. Srivastava, *Phys. Rev. Lett.* **43** (1979) 834 ... 823

4. Coherent-State Representation of a Non-Abelian Charged Quantum Field
by K.-E. Eriksson, N. Mukunda and B.-S. Skagerstam, *Phys. Rev.* **D24** (1981) 2615 ... 827

5. Correlations and Fluctuations in Hadronic Multiplicity Distributions: The Meaning of the KNO Scaling
by P. Carruthers and C. C. Shih, *Phys. Lett.* **127B** (1983) 242 ... 838

6. Hadronic Multiplicity Distribution: Example of a Universal Stochastic Process
by P. Carruthers, in *Proceedings of the XIVth International Symposium on Multiparticle Dynamics,* eds. P. Yager and J. F. Gunion (World Scientific Publishing Co., 1984), p. 825 ... 847

XI. *Applications in Quantum Field Theory*

1. Infrared Divergences in Quantum Electrodynamics
by V. Chung, *Phys. Rev.* **140** (1965) B1110 ... 857

2. Coherent States and Indefinite Metric — Applications to the Free Electromagnetic and Gravitational Fields
by J. Gomatam, *Phys. Rev.* **D3** (1971) 1292 ... 870

3. Extended Particles and Solitons
by K. Cahill, *Phys. Lett.* **53B** (1974) 174 ... 877

4. Classical Source Emitting Self-Interacting Bosons
by K.-E. Eriksson and B.-S. Skagerstam, *Phys. Rev.* **D18** (1978) 3858 ... 880

5. Path-Integral Representation for the S Matrix
by C. L. Hammer, J. E. Shrauner and B. De Facio, *Phys. Rev.* **D18** (1978) 373 ... 885

6. Coherent State Approach to the Infra-red Behaviour of Non-Abelian Gauge Theories
by M. Greco, F. Palumbo, G. Pancheri-Srivastava and Y. Srivastava, *Phys. Lett.* **77B** (1978) 282 — 897

7. Creation and Annihilation Operators for Nielsen-Olesen Vortices in the Coherent State Approximation
by D. E. L. Pottinger, *Phys. Lett.* **78B** (1978) 476 — 902

8. Vacuum Instability and the Critical Value of the Coupling Parameter in Scalar QED
by G. -J. Ni and Y. -P. Wang, *Phys. Rev.* **D27** (1983) 969 — 905

I. A Coherent-State Primer

1. ORIGIN AND SCOPE

In a contemporary context the appellation "coherent states" was first used by Roy Glauber [1] in the field of quantum optics [2,3] in a rather specific and carefully chosen sense. However, a catchy phrase like "coherent states" is hard to resist and it was not long before it was applied to cases well outside its original meaning and to families of quantum-mechanical states quite different from the original one. Some authors have attempted to preserve a distinction between the original and expanded definitions — for example, calling the latter "generalized coherent states" — but this appears to be a losing cause. Consequently in the present review we accept the inevitable and adopt the commonly used wider meaning of the phrase coherent states, saving qualifiers to denote specific examples.

When the meaning of a concept is widened to encompass a larger spectrum of objects it also becomes possible to look backward in time to see if and when examples of the enlarged concept have been previously used. In so doing we will encounter examples of coherent states (although not called that at the time) which appear in some of the earliest literature on quantum theory. In addition, under the umbrella of an expanded definition, we can even find applications of coherent states outside of quantum theory in widely diverse fields.

The purpose of the present volume is to draw together in one place a selected collection of papers and references in which coherent states play a significant role. Our selection includes articles in the general area of physics (quantum optics, quantum mechanics, condensed matter physics, atomic physics, nuclear physics, elementary particle physics), chemistry (semiclassical behavior), mathematical physics (group theory, path integrals), and other areas (biology, plasma physics, radar theory). We could, of course, not include all the relevant reprints and references we would like, but only a representative sample chosen from our own perspective and in consultation with a number of colleagues. We apologize to those authors whose work we have overlooked and/or who feel that we should have included their work in our survey.

To gain a greater perspective of the wide variety of coherent-state applications, we offer in what follows an introduction to and a brief overview of the general subject matter and reprinted articles of this volume.

2. COHERENT STATES – THE BARE ESSENTIALS

Minimum Requirements

Although the name "coherent states" is applied to a wide class of objects, all such definitions have two basic properties in common, and which thereby may be logically termed the minimum set of requirements for a set of states to be termed coherent. In every case the set of states refers to vectors in a Hilbert space $\mathfrak{H}$ [finite or (countably) infinite dimensional]. Without prejudging the nature or specifying an application we denote the states in question, in Dirac notation, by $|\ell\rangle$ where ℓ is an element (generally multidimensional) of an appropriate label space $\mathfrak{L}$ endowed with a notion of continuity (topology). More specifically, an appropriate label space is one for which every finite dimensional subspace is locally Euclidean.

There are in essence just two properties that all coherent states share in common. The first property is

CONTINUITY: *The vector* $|\ell\rangle$ *is a strongly continuous function of the label* ℓ.

Here continuity has its typical meaning, namely, for every convergent label set such that $\ell' \to \ell$ (say) it follows that $\| |\ell'\rangle - |\ell\rangle \| \to 0$. As usual $\| |\psi\rangle \| \equiv \langle\psi|\psi\rangle^{1/2}$ defines the vector norm, which is positive save for the case $|\psi\rangle = 0$; for all $\ell \in \mathcal{L}$ we suppose $|\ell>\neq 0$. Stated otherwise the set of vectors $\{|\ell>: \ell \in \mathfrak{L}\}$ form a continuous (usually connected) submanifold in the Hilbert space $\mathfrak{H}$. Clearly such vectors are rarely if ever mutually orthogonal. In applications the continuity property will be almost self-evident.

It is important to emphasize that the continuity property already rules out as coherent states two familiar sets of vectors from quantum theory and from spectral theory, more generally. We have in mind either a set of discrete orthogonal vectors $\{ |n> : n = 0, 1, 2,... \}$, with $<n|m> = \delta_{nm}$, or a set of δ-normalized, continuum orthogonal vectors $\{ |x> : -\infty<x<\infty \}$, with $<x|x'> = \delta(x-x')$. In the discrete case the vectors exist but they do not form a continuous set; in the continuous case the labels are continuous but the vectors are not continuous in the labels (they are also, strictly speaking, not vectors as well). The conclusion is that a conventional basis set of vectors, for example, the eigenvectors, in the usual or generalized sense, of any self-adjoint operator *never* constitute a set of "coherent states".

The second and last property that all sets of coherent states share in common is

COMPLETENESS (RESOLUTION OF UNITY): *There exists a positive measure $\delta\ell$ on $\mathfrak{L}$ such that the unit operator I admits the "resolution of unity"*

$$I = \int |\ell><\ell|\delta\ell \tag{2.1}$$

when integrated over $\mathfrak{L}$.

This formal "resolution of unity" is to be interpreted in the sense of weak convergence, namely, that arbitrary matrix elements of the indicated expression converge as desired. This will generally be implicit in our discussion.

It is noteworthy that (2.1) appears exactly like a resolution of unity appropriate to a self-adjoint operator but it differs in that the one-dimensional projection operators $|\ell><\ell|$ are *not* in general mutually orthogonal. For a self-adjoint operator a resolution of unity in terms of mutually orthogonal (possibly generalized) projection operators *always* holds as a matter of course. For coherent states the existence of the "resolution of unity" must be verified and imposes severe restrictions on the choice of the set of states, as can be imagined. Here we have used quotation marks so as to distinguish the coherent-state resolution of unity from the conventional one which involves mutually orthogonal states. However, hereafter we shall drop the quotation marks thereby broadening the concept of resolution of unity to include projection onto states that typically are not mutually orthogonal.

If the resolution of unity (2.1) is the final property imposed on the coherent states then it is clear that a continuous change of phase leads to an essentially equivalent set of coherent states, and it is also clear that we can, if we like, always rescale $\delta\ell$ at each point $\ell\in\mathfrak{L}$ in such a way that we can choose $\| |\ell\rangle \| = 1$ for all $\ell\in\mathfrak{L}$. In other words, we can assume, without any loss of generality, that the vectors $|\ell\rangle$ are all of unit length. This normalization is actually commonly assumed, and for the most part we shall adopt it as well. However, there are several important exceptions to the convention of normalized coherent states, a variant made to enable other properties to hold.

There is also a weaker version of the completeness property, which has occasionally been used, and that is

COMPLETENESS (TOTAL SET OF VECTORS): *The closed linear span of* $\{|\ell\rangle : \ell\in\mathfrak{L}\}$ *is the Hilbert space* $\mathfrak{H}$.

This condition simply means that any vector in the Hilbert space may be represented as a (possibly infinite) linear sum of coherent states. We shall return to this weaker form of completeness in Secs. 5 and 7. Unless otherwise stated we shall hereafter assume the validity of the stronger form as represented by the resolution of unity.

Continuous Representation of $\mathfrak{H}$

One of the standard uses of conventional resolutions of unity, such as

$$I = \sum |n\rangle \langle n| \,,$$

$$I = \int |x\rangle \langle x| dx \,,$$

is the introduction of various functional representations of a Hilbert space in the truly elegant Dirac notation. Thus, simply by "inserting the unit operator", as in

$$\langle\psi|\phi\rangle = \sum \langle\psi|n\rangle \langle n|\phi\rangle \,,$$

$$\langle\psi|\phi\rangle = \int \langle\psi|x\rangle \langle x|\phi\rangle dx \,,$$

one is automatically led to a representation of $\mathfrak{H}$, first, by l^2, the space of square summable sequences, with $\phi_n \equiv \langle n|\phi\rangle$, etc., or, second, to a representation of $\mathfrak{H}$ by $L^2(\mathbb{R})$, the space of square integrable functions, with $\phi(x) \equiv \langle x|\phi\rangle$. (Strictly speaking the L^2 representation is one of equivalence classes of functions that differ on sets of measure zero, but such a distinction can often be safely ignored.) In like fashion operators are given a functional representation according to

$$\langle n|B|\phi\rangle = \sum \langle n|B|n'\rangle \langle n'|\phi\rangle \,,$$

$$\langle x|B|\phi\rangle = \int \langle x|B|x'\rangle \langle x'|\phi\rangle dx' \,,$$

expressions which involve the kernel representatives $B_{nn'} \equiv \langle n|B|n'\rangle$ and $B(x\,,\,x') \equiv \langle x|B|x'\rangle$.

Evidently the key to the introduction of a functional representation is the existence of a resolution of unity expressed as a positive linear superposition of one-dimensional projection operators. The coherent states and their associated resolution of unity then give rise to a completely analogous functional representation of vectors and operators. In particular we have

$$<\psi|\phi> = \int <\psi|\ell> <\ell|\phi> \delta\ell \, ,$$

$$<\ell|B|\phi> = \int <\ell|B|\ell'> <\ell'|\phi> \delta\ell' \, .$$

It is characteristic of the functional representation $\phi(\ell) \equiv <\ell|\phi>$ induced by coherent states that the vector representatives are *continuous functions,* or *bounded, continuous functions* when the coherent states are normalized (or uniformly bounded). Continuity excludes the need for an equivalence class of functions differing on sets of measure zero. We shall have frequent occasion to use, either explicitly or implicitly, the "continuous representation" $\mathfrak{C}$ of Hilbert space that coherent states induce.

The functions of the continuous representation are all square integral since

$$<\phi|\phi> = \int |<\ell|\phi>|^2 \delta\ell < \infty \, ,$$

but they constitute a *subspace* of all square-integrable functions. That this is true follows from the resolution of unity (2.1) used in the form

$$<\ell|\phi> = \int <\ell|\ell'> <\ell'|\phi> \delta\ell' .$$

Each admissable function therefore satisfies an integral equation involving the *reproducing kernel* $\mathcal{K}(\ell;\ell') \equiv <\ell|\ell'> = <\ell'|\ell>^*$, and one sometimes says that $\mathfrak{C}$ is a reproducing kernel Hilbert space. Thanks to the continuity requirement, the reproducing kernel is a jointly continuous function, nonzero for $\ell' = \ell$, and therefore nonzero for all ℓ' in an open neighborhood of ℓ. This remark simply implies that the integral equation above is a real restriction on admissable functions. On the other hand, a similar integral equation also holds for conventional bases, but in that case, due to the orthonormality of the basis vectors, the integral equation collapses to an identity and imposes no restriction whatsoever.

One further use of the resolution of unity reveals that

$$<\ell''|\ell'> = \int <\ell''|\ell> <\ell|\ell'> \delta\ell ,$$

namely that the reproducing kernel satisfies the integral equation of a *projection operator,* as it should. Linear dependency of the coherent states is made evident by the related equation

$$|\ell'> = \int |\ell> <\ell|\ell'> \delta\ell ,$$

which expresses any coherent state as a linear sum (integral) of the remaining coherent states. It is no wonder that coherent states have also been termed an "overcomplete family of states".

Lastly we note that since

$$\| \, |\ell'> - |\ell> \|^2 = <\ell'|\ell'> + <\ell|\ell> - 2 \, \mathrm{Re} <\ell'|\ell>$$

then the requisite continuity of the vectors is a simple consequence of the joint continuity of the reproducing kernel, i.e., of the continuity of the matrix element of any two coherent states in their label variables. This offers a simple way to check for strong vector continuity.

Continuity, completeness, and their consequences, as outlined above, are common features of all coherent state systems. When it comes to any specific example of coherent states, then generally a corresponding set of additional properties holds true — as we shall already see in the next section.

Additional Remarks and Introduction to Selected Reprints

The postulates for a generic class of coherent states presented in Sec. 2, and the general properties they infer, have been adapted from Klauder [4]. The notion of reproducing kernel Hilbert spaces (without the need for an integral representation of the inner product) is much older and can be traced from Aronszajn [5]. An example of a reproducing kernel Hilbert space without invoking an integral representation is given in Sec. 7. A thorough account of reproducing kernel Hilbert spaces is given by Meschkowski [6], and a recent reprint volume dealing with the theory and applications of reproducing kernel Hilbert spaces is given in Ref. [7].

3. CANONICAL COHERENT STATES

Complex Variable Characterization

We start with a brief description of the original and still most widely used of all coherent states, those generated by canonical annihilation (a) and creation ($a^\dagger$) operators. Let a and $a^\dagger$ denote an irreducible pair of operators that satisfy the basic commutation relation $[a,a^\dagger] \equiv aa^\dagger - a^\dagger a = I$. We introduce the normalized state $|0\rangle$ with the property that $a|0\rangle = 0$, and define the state space as that spanned by repeated action of $a^\dagger$ on $|0\rangle$. The uncertainty relation for the operators a and $a^\dagger$ simply reads $\langle a^\dagger a \rangle \geqslant \langle a^\dagger \rangle \langle a \rangle$, where $\langle (\cdot) \rangle \equiv \langle \psi | (\cdot) | \psi \rangle$, and it follows that the fiducial state $|0\rangle$ is a minimum uncertainty state. We shall discuss other minimum uncertainty states later.

The canonical coherent states are generated by unitary transformations of the fiducial state $|0\rangle$, and are defined for each complex number z by the relations

$$\begin{aligned}
|z\rangle &\equiv e^{(za^\dagger - z^* a)}|0\rangle \\
&= e^{-\frac{1}{2}|z|^2} e^{za^\dagger} e^{-z^* a}|0\rangle \\
&= e^{-\frac{1}{2}|z|^2} e^{za^\dagger}|0\rangle \\
&= e^{-\frac{1}{2}|z|^2} \sum_{n=0}^{\infty} \frac{1}{\sqrt{n!}} z^n |n\rangle \,.
\end{aligned} \tag{3.1}$$

Here, in obtaining the second line, we have used the elementary Baker-Campbell-Hausdorff formula

$$e^{A+B} = e^{-½[A,B]} e^A e^B$$

valid whenever $[A\ ,\ B]$ commutes with both A and B. In the last line we have introduced the orthonormal vectors $|n> \equiv \frac{1}{\sqrt{n!}}(a^\dagger)^n|0>$, which are nothing but the various eigenstates of the number operator $N \equiv a^\dagger a$. We have labelled the state by z since, apart from a normalization factor, it depends only on z and not on z^*. However, in only an apparent abuse of notation we set $<z| \equiv (|z>)^\dagger$ as the adjoint vector. It follows immediately that

$$<z_2|z_1> = e^{-½|z_2|^2+z_2^* z_1-½|z_1|^2}, \tag{3.2}$$

which is a nowhere vanishing, continuous function of the variables z_1 and z_2. As for the resolution of unity, it readily follows that $[d^2z \equiv d(\text{ Re } z)d(\text{ Im } z)]$

$$\pi^{-1}\int |z><z|d^2z$$

$$= \pi^{-1} \sum_{n,m} \frac{1}{\sqrt{n!m!}} \int e^{-|z|^2} z^{*n} z^m |m><n|\ d(\text{Re } z)d(\text{Im } z)$$

$$= \sum_n \frac{1}{n!} \int e^{-|z|^2} |z|^{2n}|n><n|d|z|^2$$

$$= \sum_n |n><n| = I\ , \tag{3.3}$$

where polar coordinates have been used at an intermediate stage of calculation.

Consequences of Analyticity

The fact that apart from an overall normalization the canonical coherent states depend only on z and not z^* has important consequences. For any $|\psi>$ it follows that

$$\psi(z) \equiv <z|\psi> = e^{-\frac{1}{2}|z|^2} f(z^*) ,$$

where f is an entire function of z^* with a growth restriction such that $|<z|\psi>| \leqslant \| \, |\psi> \|$. There is no clearer evidence of the overcompleteness of the canonical coherent states than the existence of characteristic sets $\mathfrak{B}$ of the complex plane for which the vanishing of $\psi(z)$ for all $z \in \mathfrak{B}$ implies the vanishing for all z, and hence that $|\psi> = 0$. An evident characteristic set is any curve of nonzero length (e.g., $0 \leqslant \text{Re}\, z \leqslant 1$, $\text{Im}\, z = 0$). Less evident are the characteristic sets of points for which $z = \gamma(m+in)$, $m, n = 0, \pm 1, \pm 2, \ldots$, for any γ such that $0 < \gamma < \sqrt{\pi}$; the case $\gamma = \sqrt{\pi}$ corresponds to the von Neumann lattice [8], and is also characteristic; however, for $\gamma > \sqrt{\pi}$ the set is *not* characteristic [9,10]. The points of the latter characteristic sets have a density of γ^{-2} per unit area in the complex plane, $\gamma^2 \leqslant \pi$. When this is translated into phase space coordinates (p,q) — more on this subject later — defined by $z \equiv \sqrt{\omega/2\hbar}\; q + i 1/\sqrt{2\hbar\omega}\; p$, the characteristic sets have a density of $(2\gamma^2\hbar)^{-1}$ points per unit area in phase space. The critical value $\gamma^2 = \pi$ corresponds to one point per Planck cell (phase-space area equal to $h = 2\pi\hbar$, Planck's constant) in accord with heuristic quantum arguments. Evidently any set of coherent states that entails a representation (possibly up to a common factor) by entire functions enjoys corresponding properties.

A related aspect pertains to the matrix element of operators. For convenience let $\mathscr{B}$ be a bounded operator or a polynomial in a and $a^\dagger$, and consider

$$<z|\mathscr{B}|z'> = e^{-\frac{1}{2}|z|^2-\frac{1}{2}|z'|^2} \sum_{n,m} \frac{1}{\sqrt{n!m!}} <n|\mathscr{B}|m> z^{*n} z'^{\,m} .$$

Apart from the normalization factors this series defines an entire function of two variables, z^* and z'. Now it is straightforward to see that such a function is uniquely determined by its "diagonal" expression, i.e., the one that results when $z = z'$. For example, each monomial such as $z^{*n} z^m$ may be isolated by its separate θ and r dependence ($z \equiv re^{i\theta}$) and uniquely extended to $z^{*n} z'^{\,m}$. Consequently, an operator $\mathscr{B}$ is uniquely determined by its *diagonal* matrix elements

$$B(z) \equiv <z|\mathscr{B}|z>$$

in the canonical coherent states. For conventional bases this is never the case. Evidently any other set of coherent states that entails a representation (possibly up to a common factor) by analytic functions leads to the same conclusion.

Finally we observe that the trace of $\mathscr{B}$ may be represented by $Tr(\mathscr{B}) = \pi^{-1} \int B(z) d^2z$, which is a commonly used expression.

Diagonal Representation of Operators

The two foregoing properties have been based on analyticity but have not made any specific use of the form of the coherent state overlap $<z_2|z_1>$, i.e., the reproducing kernel. A remarkable and seemingly more special property entails the diagonal representation of operators in which one represents an operator in the form (often called the P-representation)

$$\mathscr{B} = \pi^{-1} \int b(z) |z><z| d^2z \qquad (3.4)$$

as a weighted superposition of one-dimensional projection operators (as with the resolution of unity). It follows that

$$B(z') \equiv \langle z'|\mathscr{B}|z'\rangle$$

$$= \pi^{-1} \int b(z)|\langle z'|z\rangle|^2 d^2z$$

$$= \pi^{-1} \int b(z) e^{-|z'-z|^2} d^2z \, .$$

In one form this relation implies

$$b(z) = e^{-\partial^2/\partial z^* \partial z} B(z) \, ;$$

in another, which involves the Fourier transform of these quantities,

$$\tilde{b}(v) = \tilde{B}(v) e^{|v|^2} .$$

For $\mathscr{B}$ an arbitrary bounded operator or a polynomial in a and $a^\dagger$ it follows that $\tilde{b}(v)$ may be regarded as a distribution on $\mathscr{D}_2$ (associsted with C^∞ test functions of compact support), and $b(z)$ a distribution on Z_2, the Fourier transform of $\mathscr{D}_2$ (cf., Sec. 4). Apart from such technicalities the point is that an immense variety of operators admit a diagonal representation (3.4), and indeed, a general operator may be represented as the limit of a sequence of operators each of which admits a diagonal representation with a smooth weight function. Clearly this is in marked contrast to the strictly commutative family of operators that can be so represented with conventional bases.

We note that the trace of $\mathscr{B}$ may also be represented by $Tr(\mathscr{B}) = \pi^{-1} \int b(z) d^2z$, and more generally that

$$Tr(\mathscr{A}\mathscr{B}) = \pi^{-1} \int a(z) B(z) d^2z = \pi^{-1} \int A(z) b(z) d^2z \, .$$

This relation serves to make clear an evident duality between the expressions $a(z)$ and $A(z)$ associated with the operator $\mathscr{A}$.

Although our discussion of the diagonal representation and its generality has not used analyticity we shall show in Sec. 4 that analyticity leads to the same conclusion.

Eigenproperties of $|z\rangle$

It is clear that $e^{-za^\dagger} a e^{za^\dagger} = a + z$ holds in virtue of the basic commutation relation, and thus (by acting on $|0\rangle$) it follows that

$$a|z\rangle = z|z\rangle \; ; \tag{3.5}$$

namely, that for each complex z the coherent state $|z\rangle$ is an eigenstate of the annihilation operator a with eigenvalue z. It follows that $\langle z|a|z\rangle = z$, and one may now interpret the label z by saying it is the mean of a in the coherent state $|z\rangle$. Each state $|z\rangle$ satisfies $\langle z|a^\dagger a|z\rangle = |z|^2 = |\langle z|a|z\rangle|^2$, and thus $|z\rangle$ is a minimum uncertainty state for all z.

Normal-ordered operators are those with all a's standing to the right of all $a^\dagger$'s. We let : : denote the operation of normal ordering, e.g., $:aa^\dagger: \equiv a^\dagger a$. As one consequence we have

$$\langle z|: F(a^\dagger, a):|z'\rangle = F(z^*, z')\langle z|z'\rangle \; ,$$

as another consequence we have

$$\langle z|\mathcal{O}|z\rangle = F(z^*, z) \quad \text{implies} \quad \mathcal{O} = : F(a^\dagger, a): .$$

For further such properties we refer, e.g., to Refs.[1-3].

Differential Representation of Operators

The action of various operators on functions of the form $\psi(z) \equiv \langle z|\psi\rangle$ may be represented by integration kernels as discussed in Sec. 2, but they may also be represented in another way. For example, it follows from the relation $\langle z|a^\dagger|\psi\rangle = z^* \langle z|\psi\rangle$ that the operator $a^\dagger$ may be represented by multiplication by z^*. And, it follows from the functional form of $\langle z|z'\rangle$ that $\langle z|a|z'\rangle = z'\langle z|z'\rangle = (z/2 + \partial/\partial z^*)\langle z|z'\rangle$. Multiplication by $\langle z'|\psi\rangle$ and integration over z' then leads to $\langle z|a|\psi\rangle = (z/2 + \partial/\partial z^*)\langle z|\psi\rangle$, or to the fact that a may be represented as $(z/2 + \partial/\partial z^*)$. Thus we find the general differential operator representation given by

$$<z|\mathscr{W}(a^\dagger, a)|\psi> = \mathscr{W}(z^*, z/2 + \partial/\partial z^*)<z|\psi> . \qquad (3.6)$$

Stability Under Time Evolution

The elementary fact represented by

$$e^{-it(\omega a^\dagger a)}|z> = e^{-\frac{1}{2}|z|^2} \sum_n \frac{1}{\sqrt{n!}} z^n e^{-in\omega t}|n>$$

$$= |e^{-i\omega t}z>$$

implies that under the dynamical evolution of a harmonic oscillator the coherent states evolve into other coherent states by a time-dependent label change that follows the classical oscillator solution. These properties were noticed by Schrödinger [11] who in his fundamental work was implicitly the first to study properties of coherent states. The most general dynamical system under which canonical coherent states evolve into canonical coherent states involves a forced harmonic-oscillator Hamiltonian composed of a number operator plus terms linear in the creation and annihilation operators. This property is implicit in Schrödinger's work and was later expanded on by a number of workers [12-19].

Additional Remarks and Introduction to Selected Reprints

1. The first appearance of canonical coherent states to build and use a representation space in which inner products are given by integration over the complex plane is by Klauder [20]. These states were applied to the construction of path integrals. The importance and generality of coherent-state path integrals is sufficient that we shall devote a special section, Sec. 6, of this primer to that subject.

2. In Sec. 2 we indicated that it is occasionally useful not to choose the coherent state to be normalized but rather to rescale the

measure that enters the resolution of unity. Let us then absorb into the measure the common normalization factor $\exp(-\frac{1}{2}|z|^2)$ of the canonical coherent states, and to avoid confusion, as well as to come more into line with common terminology, set $\zeta = z^*$. The result is a representation by entire functions

$$\psi(\zeta) \equiv \sum_{n=0}^{\infty} \frac{1}{\sqrt{n!}} \zeta^n <n|\psi> ,$$

with an inner product given by

$$(\psi,\psi) \equiv \pi^{-1} \int |\psi(\zeta)|^2 e^{-|\zeta|^2} d^2\zeta ,$$

which is known as the Bargmann-Segal representation of Hilbert space. In this representation space $a^\dagger$ acts by multiplication by ζ, while a is represented simply by $\partial/\partial\zeta$. Note that the eigenvalue equation for the number operator $N = a^\dagger a$ simply reads $\zeta(\partial/\partial\zeta)\psi(\zeta) = \lambda\psi(\zeta)$. All solutions of this equation are of the form $\psi(\zeta) = c\zeta^\lambda$; however, analyticity of ψ restricts λ such that $\lambda = 0, 1, 2, \ldots$. What a simple and elegant derivation of the spectrum of the number operator! Additional properties of this Hilbert space representation are discussed in Refs. [21] and [22].

3. In the presence of an optical field characterized by the state $|\psi>$ let us accept the fact that the counting rate for n single-mode photons by a photodetector is proportional to $r_n \equiv <\psi| : (a^\dagger a)^n : |\psi>$. If r_n had the form $r_n = (r_1)^n$, then the counting rates would be equivalent to those of a *nonfluctuating* field intensity, i.e., of what might be termed a *coherent* field. Such power-law behavior holds whenever $|\psi> = |z>$, a canonical coherent state, for then $r_n = |z|^{2n}$. And it is just for this reason that Glauber [23] called the corresponding quantum states "coherent states". When applied to the electromagnetic field such states are sometimes referred to as Glauber states.

An important step in understanding and formulating quantum optics is provided by Sudarshan's optical equivalence theorem [24], which establishes the equivalence of quantum and suitable semiclassical approaches. Such questions are fully analyzed, e.g., in Ref. [2].

4. Coherent states can be used to solve the quantum mechanical problem of an harmonic oscillator acted on by a time dependent external force as was done in the paper by Carruthers and Nieto [25]. This fundamental problem has been considered in the literature by making use of other methods [26,27], which has been followed up in Refs. [28,29]. The use of coherent states in this case clearly exhibits the role of the classical orbits in the quantum domain, a point which is discussed in later sections. In the paper by Hepp [30] a discussion is given of the classical limit of quantum mechanical correlation functions.

5. An application of canonical coherent states defined on a von Neumann lattice is given in the paper by Toyoda and Wildermuth [31] in the context of a non-relativistic atomic electron interacting with a strong radiation field. Further applications of this formalism may be found in Refs. [32-35].

6. The canonical coherent states have been used extensively in many other areas of physics. The number-phase operator uncertainty relationship was discussed by Carruthers and Nieto [36], while related topics appear in the review article [37] and Refs. [38,39]. In the papers by Moncrief [40] and Unruh [41] there is a discussion of the sensitivity limit imposed by quantum mechanics on detectors for gravitational radiation. In the paper by Hollenhorst [42] the notion of "squeezed" coherent states was introduced to arrive at similar conclusions.

The use of the concept of "coherence" in condensed matter physics may be traced back to a famous paper by Anderson [43]. A study of irreversible processes in anharmonic crystals is given in the paper of Carruthers and Dy [44]. A coherent state approach to the theory of superfluidity was developed by Cummings and Johnston [45] and by Langer [46]. For some other applications in the context of condensed matter physics see Refs. [47-53], and for an application in plasma physics see Ref. [54].

In the context of nuclear physics coherent states were used in order to describe condensates in nuclear matter by Fowler, Stelte and Weiner [55]. Models for heavy ion collisions in terms of canonical coherent states have also been discussed in the literature [56].

Straightforward applications of coherent states to questions of elementary particle physics may be found in the papers of Carruthers [57] and Carruthers and Shih [58] where hadronic multiplicity distributions are addressed.

7. There are many other fields of application of canonical coherent states, and to illustrate the very broad range of applications we mention a few additional examples. In Ref. [59] there is an application in the context of radar theory. In chemistry, linear superpositions of coherent states, parametrized by classical trajectories, have been used [60] in order to construct multidimensional wavefunctions, and a semiclassical collision theory has been developed in a similar fashion [61]. Canonical coherent states have also been used in the field of biology in order to describe the long-range forces between human blood cells [62] and the long-range phase coherence in the bacteriorhodospin macromolecules [63].

Phase-Space Formulation

Although the analytic properties of canonical coherent states have played a prominent role in our discussion in this section there are some advantages to be gained by focussing on real labels and the associated self-adjoint generators. To this end let

$$Q \equiv \frac{\sqrt{\hbar}}{\sqrt{2\omega}} (A+A^{\dagger})$$

$$P \equiv i \frac{\sqrt{\hbar\omega}}{\sqrt{2}} (A^{\dagger}-A)$$

denote two self-adjoint canonical operators which satisfy the Heisenberg commutation relation $[Q\ , P] = i\hbar I$. As above, $\hbar$ denotes Planck's constant $(h/2\pi)$, and ω denotes an arbitrary positive scale factor. It follows that $A = \sqrt{\omega\ /2\hbar}\ Q + i\ 1/\sqrt{2\hbar\omega}\ P$, and in like fashion we decompose z according to the rule $z \equiv \sqrt{\omega/2\hbar}\ q\ + i\ 1/\sqrt{2\hbar\omega}\ p$. Here (p, q) range over all of $\mathbb{R}^2$. The fundamental unitary operator

$$U[p\ , q] \equiv e^{i(pQ-qP)/\hbar}$$

is called the Weyl operator, and it is just an alternative rewriting of the unitary operator in the first line of (3.1) which defines the canonical coherent states. Consequently,

$$|z> = |p\ , q> \equiv U[p\ , q]|0> ;$$

the overlap of two such states reads

$$<p_2\ , q_2|p_1\ , q_1>$$

$$=\exp\{\frac{i}{2\hbar}(p_1q_2-q_1p_2) - \frac{1}{4\hbar}\ [\omega^{-1}(p_2-p_1)^2 + \omega(q_2-q_1)^2]\} ;$$

and the resolution of unity takes the form

$$I = \int |p\,,q\rangle\langle p\,,q|\;(dp\,dq/2\pi\hbar)\,.$$

The Weyl operators serve to act as translation operators for Q and P in the sense that

$$U^{\dagger}[p\,,q]QU[p\,,q] = Q + qI \equiv Q + q\,,$$

$$U^{\dagger}[p\,,q]PU[p\,,q] = P + pI \equiv P + p\,, \qquad (3.7)$$

where in the last form the unit operator has been left implicit. It follows that $\langle p\,,q|Q|p\,,q\rangle = q$ and $\langle p\,,q|P|p\,,q\rangle = p$, which provide an interpretation for the labels p and q. From their construction and parameterization it is clear that the states $|p\,,q\rangle$ generate a canonical *phase-space* continuous representation according to the usual rule, $\psi(p\,,q) \equiv \langle p\,,q|\psi\rangle$. Frequently it is convenient to set $\omega = \hbar = 1$, and this choice is commonly found in the literature.

Fiducial Vector Generalization

The Weyl operators provide a representation of the two-dimensional Abelian group apart from a multiplicative phase factor as is evident in the relation

$$U[p_2\,,q_2]\,U\,[p_1\,,q_1] = e^{i(q_1p_2-p_1q_2)/2\hbar}U[p_2+p_1\,,q_2+q_1]\,, \qquad (3.8)$$

along with $U^{-1}[p\,,q] = U^{\dagger}[p\,,q] = U[-p,-q]$ and $U[0,0] = I$. From a group-theoretic viewpoint one may adopt (3.8) as the basic "commutation rule" for strongly continuous unitary Weyl operators and deduce that there is only one irreducible representation up to unitary equivalence. A fundamental property of Weyl operators, which was proved and effectively used by Moyal [64], is given by the equation

$$\int <\alpha|U[p\,,\,q]|\beta> <\gamma|U[p\,,\,q]|\delta>^* \,(dp\,dq/2\pi\hbar) = <\alpha|\gamma> <\delta|\beta> \tag{3.9}$$

for any four states $|\alpha>$, $|\beta>$, $|\gamma>$ and $|\delta>$. This equation may be proved in a straightforward fashion if one employs the Schrödinger representation for which, for example,

$$<\alpha|U[p\,,\,q]|\beta> = \int \alpha^*(x+q/2)e^{ipx}\beta(x-q/2)dx\,.$$

In essence, Eq. (3.9) is just a group orthogonality relation. However, it can be "read" in two interesting but different ways. In one way, drop $<\alpha|$ and $|\beta>$ to learn that

$$\int U\,[p\,,\,q]\, <\gamma|U[p\,,\,q]|\delta>^* \,(dp\,dq/2\pi\hbar) = |\gamma> <\delta|\,,$$

which is recognized as an elementary generating form of the Weyl representation of operators given more completely by the relation

$$\mathscr{C} = \int Tr(U^\dagger[p\,,\,q]\mathscr{C})\, U[p\,,\,q]\,(dp\,dq/2\pi\hbar)\,.$$

In the other way, drop $<\alpha|$ and $|\gamma>$ to learn that

$$\int U[p\,,\,q]|\beta> <\delta|U^\dagger[p\,,\,q]\,(dp\,dq/2\pi\hbar) = <\delta|\beta>I\,,$$

namely, the result is a multiple of the unit operator. That such an equation should hold is strongly suggested by the irreducibility of the Weyl operators and Schur's Lemma. Specifically, if $\mathcal{O}$ denotes the left-hand side of the previous equation then $U[p'\,,\,q']\mathcal{O} = \mathcal{O}\,U[p'\,,\,q']$ for all $(p'\,,\,q')$ as follows directly from the Weyl operator composition law and the translation invariance of the integration measure. Thus $\mathcal{O}$ must be a multiple of unity, but since phase space is not compact here that multiple could be infinite; that it is not infinite for the Weyl operators is equivalent to the fact that the representation is "square integrable", i.e., all its matrix elements are square integrable.

If we define $|p\,,\,q;\,\phi_o> \equiv U[p\,,\,q]|\phi_o>$ for all $(p\,,\,q)\in\mathbb{R}^2$, where $|\phi_o>$ is an arbitrary unit vector, then

$<p_2\,, q_2;\, \phi_o|p_1\,, q_1;\, \phi_o>$ is clearly a jointly continuous function of its phase-space arguments, and

$$\int |p\,, q;\, \phi_o><p\,, q;\, \phi_o|\,(dp\,dq/2\pi\hbar) = I\,.$$

Thus for *any* choice of the fiducial vector $|\phi_o>$ these vectors constitute a set of *coherent states* in our sense, namely in the sense of Sec. 2. This is a vast generalization over the states $|z>$ discussed in the earlier part of this section for which $|\phi_o> = |0>$. The states $|p\,, q;\, \phi_o>$ for an arbitrary fiducial vector are overcomplete and possess characteristic sets, but analyticity (up to a factor) of the vectors is generally not available to help determine them. It may be shown that the diagonal matrix elements of polynomials in P and Q uniquely determine the operator for any $|\phi_o>$, while this is true for a general operator provided the coherent state overlap (reproducing kernel) nowhere vanishes. A nonvanishing reproducing kernel actually arises for uncountably many fiducial vectors. By an argument discussed in Sec. 4, a nonvanishing reproducing kernel also implies that a general operator admits a diagonal representation as a superposition of projection operators. Of course, the states $|p\,, q;\, \phi_o>$ are generally not minimum uncertainty states or eigenstates of any particular operator. On the other hand, on the space of functions $\psi(p\,, q) \equiv <p\,, q;\, \phi_o|\psi>$, for an arbitrary fiducial vector $|\phi_o>$, the operator Q is universally represented by $(q/2 + i\hbar\partial/\partial p)$ while P is universally represented by $(p/2 - i\hbar\partial/\partial q)$. Thus — with the major exception of analyticity — the canonical phase-space coherent states based on a general fiducial vector enjoy all the essential properties of those based on the special choice $|\phi_o> = |0>$.

Although the states $|z>$ have the greatest range of applications the wider class of canonical phase-space coherent states do have interesting uses. Let us mention just two such applications. As a first application consider the anharmonic oscillator Hamiltonian

$$\mathscr{H}(P , Q) \equiv \tfrac{1}{2}(P^2 + Q^2) + \lambda Q^4 \;, \qquad \lambda>0 .$$

for which we seek to study the spectrum. If $|\underline{n}>$, $n = 0, 1, 2,...$, denotes the nth eigenfunction and E_n the associated eigenvalue then

$$<p , q; \phi_o|\mathscr{H}(P , Q)|\underline{n}> = E_n<p , q; \phi_o|\underline{n}>$$

However, if $|\phi_o>$ is chosen as $|\underline{m}>$, one of the (as yet unknown) eigenstates of the Hamiltonian $\mathscr{H}$, then, using (3.7), we find the relation

$$<p , q; \underline{m}|[\mathscr{H}(P , Q)- \mathscr{H}(P-p , Q-q)]|\underline{n}>$$

$$= (E_n-E_m) <p , q; \underline{m}|\underline{n}> .$$

The left side of this equation may be written as a differential operator acting on $<p , q; \underline{m}|\underline{n}>$ which is different and of lower order than the one for the preceding equation. In particular, if we let $F_{mn}(p , q) \equiv <p , q; \underline{m}|\underline{n}>$, then

$$[\mathscr{H}(p/2-i\hbar\partial/\partial q , q/2+i\hbar\partial/\partial p) - \mathscr{H}(-p/2-i\hbar\partial/\partial q,-q/2+i\hbar\partial/\partial p)]$$

$$\times F_{mn}(p , q) = (E_n-E_m)F_{mn}(p , q) ,$$

and for $\underline{m} = \underline{n}$ this is an equation independent of any energy eigenvalue that is satisfied by *all* the eigenstates of $\mathscr{H}$. These differential equations have been studied by Truong [65], and he has shown that they lead to a novel integral equation which the eigenstates of $\mathscr{H}$ must satisfy. In turn, this has led to interesting numerical work on the eigenvalues of the anharmonic oscillator. For additional studies of the anharmonic oscillator using coherent states see Refs. [66-71].

For the second application we first appeal to the Schrödinger representation formula

$$\psi(p , q) = e^{ipq/2\hbar} \int \phi_o^*(x-q)e^{-ipx/\hbar}\psi(x)dx . \qquad (3.10)$$

For $\phi_o(x) = \pi^{-1/4}\exp(-\tfrac{1}{2}x^2)$ Bargmann [72] has extended the use of such an expression to map tempered distributions $\psi(x) \in \mathscr{S}'$ into continuous functions $\psi(p, q)$. If $f(x)$ denotes an arbitrary test function in $\mathscr{S}$, and $f(p, q)$ its image under the coherent-state map (3.10), then it follows that

$$\psi(f^*) = \int f^*(p, q)\psi(p, q)(dp\, dq/2\pi\hbar). \qquad (3.11)$$

Such an expression gives a genuine integral representation for the value of a tempered distribution ψ acting on a test function f^*. We now simply observe that if, for example, we choose $\phi_o(x) = N\exp[-(1-x^2)^{-1}]$ for $|x|<1$, and zero elsewhere, where N is chosen so that $\int |\phi_o(x)|^2 dx = 1$, then it follows that the previous argument extends to the wider class of distributions $\psi(x) \in \mathscr{D}'$ and associated test functions $f(x) \in \mathscr{D}$, to their representation by continuous functions $\psi(p, q)$ and $f(p, q)$, respectively, and to the integral realization given in (3.11) of $\psi(f^*)$. Different choices of ϕ_o lead to analogous descriptions of other distribution spaces, such as Z' (the Fourier transform of $\mathscr{D}'$), etc. This extension to various sorts of distributions is based on unpublished work of Klauder and McKenna.

4. SPIN COHERENT STATES

In this section we discuss the next most important set of coherent states after the canonical coherent states, namely the spin coherent states, so named because they are based on the familiar algebra of spin operators. The properties of the spin operators that we will require are covered in any of a number of excellent textbooks on quantum theory and/or group theory. The spin coherent states that we shall describe here are similar (or perhaps identical depending on allowed variations in their definition) to what are sometimes referred to in the literature as atomic coherent states or Bloch coherent states. Their definition begins with the introduction of an irreducible representation of self-adjoint spin operators, S_1, S_2, and S_3, that satisfy the standard angular momentum commutation relation $[S_1, S_2] = iS_3$ plus cyclic permutations; note that the *physical* spin operators are $\hat{S}_j \equiv \hbar S_j$. As an irreducible representation it follows that $\vec{S}^2 \equiv \sum S_j{}^2 = s(s+1)I$, where $s = 0, 1/2, 1, 3/2, \ldots$, denotes the spin, and I denotes the unit operator in the $(2s+1)$-dimensional complex Hilbert space $\mathfrak{H}$ that carries this representation. The case $s=0$ is trivial and will be excluded in what follows. The spin operators S_j are generators of the group SU(2) according to any of several parameterizations:

$$U(\theta, \phi, \psi) \equiv e^{-i\phi S_3} e^{-i\theta S_2} e^{-i\psi S_3},$$

$$V(\vec{\alpha}) \equiv \exp[i(\vec{\alpha}\cdot\vec{S})] ,$$

$$W(z, \psi) \equiv N e^{zS_-} e^{-z^* S_+} e^{-i\psi S_3}, \tag{4.1}$$

where $S_\pm \equiv (S_1 \pm iS_2)/\sqrt{2}$ and N is a normalization factor chosen so that W is unitary (N is discussed later). The operators U are defined in terms of Euler angles for which $0 \leqslant \theta \leqslant \pi$, $0 \leqslant \phi < 2\pi$, and $0 \leqslant \psi < 2\pi$ (or $0 \leqslant \psi < 4\pi$ if s is nonintegral). The operators V are in the form of so-called canonical group coordinates (of the first kind). The operators W involve the raising (S_+) and lowering (S_-) ladder operators which satisfy $[S_3, S_\pm] = \pm S_\pm$ and $[S_+, S_-] = S_3$. The variable z is an arbitrary complex variable, and ψ is as before. Every element of SU(2) can be represented in either the U, V, or W form, but we shall primarily concentrate on the U and W forms.

Next we choose a normalized fiducial vector $|\,\rangle$, and for convenience we choose $|\,\rangle$ to be an eigenvector of S_3. Thus $S_3|\,\rangle = m|\,\rangle$ and so we may also denote $|\,\rangle$ by $|(m)\rangle$, where $m = -s, -s+1, \ldots, s$. In both vectors $U|(m)\rangle$ and $W|(m)\rangle$ the operator $\exp(-i\psi S_3)$ may be replaced by $\exp(-im\psi)$, and as a simple phase factor will be henceforth ignored (i.e., set $\psi \equiv 0$). Hence we also introduce the related expressions

$$U(\theta, \phi) \equiv e^{-i\phi S_3} e^{-i\theta S_2} ,$$

$$W(z) \equiv N e^{zS_-} e^{-z^* S_+} ,$$

and correspondingly we define

$$|\theta, \phi\rangle \equiv U(\theta, \phi)|(m)\rangle ,$$

$$|z\rangle \equiv W(z)|(m)\rangle . \tag{4.2}$$

It is sometimes convenient to set $\Omega \equiv (\theta, \phi)$, $|\Omega\rangle \equiv |\theta, \phi\rangle$, etc., and this notation often appears in the literature.

One of the uncertainty relations for spin operators is given by $|\langle S_3\rangle| \leqslant 2\langle(\Delta S_1)^2\rangle^{1/2} \langle(\Delta S_2)^2\rangle^{1/2}$, where $\langle \cdot \rangle \equiv \langle\psi| \cdot |\psi\rangle$ and $\Delta S_j \equiv S_j - \langle S_j\rangle$. For $|\psi\rangle = |(m)\rangle$ this inequality reads $|m| \leqslant s(s+1)-m^2$. Equality holds when $m = \pm s$, and thus $|(s)\rangle$ and $|(-s)\rangle$ are minimum uncertainty states.

Unlike the canonical coherent states, it is noteworthy, for any choice of fiducial vector, that the spin coherent states cannot all *simultaneously* be minimum uncertainty states. To see this in a general context consider the uncertainty relation for self-adjoint A and B given by

$$|\langle[A, B]\rangle|^2 = |\langle[\Delta A, \Delta B]\rangle|^2$$

$$\leqslant 4\,|\langle\Delta A \Delta B\rangle|^2 \leqslant 4\,\langle(\Delta A)^2\rangle\langle(\Delta B)^2\rangle ,$$

where $\langle(\cdot)\rangle \equiv \langle\psi|(\cdot)|\psi\rangle$ and $\Delta A \equiv A - \langle A\rangle$, etc. Equality in this relation demands that

$$(rA + iB)|\psi\rangle = (r\langle A\rangle + i\langle B\rangle)|\psi\rangle \equiv \gamma|\psi\rangle ,$$

where $r \equiv -(i/2)\langle[A, B]\rangle / \langle(\Delta A)^2\rangle$ is real. In a *finite*-dimensional (say d) Hilbert space A, B and I are equivalent to $d\times d$ matrices, and the characteristic equation $\det(rA + iB - \gamma I) = 0$

leads to (at most) d distinct one-parameter complex eigenvalues $\gamma_j(r)$, $j = 1, \ldots, d$. As r varies the d corresponding curves in the complex plane are insufficient to continuously parameterize the two variables (θ, ϕ) or $(\mathrm{Re}\, z, \mathrm{Im}\, z)$ of the spin coherent states, and thus they cannot all be minimum uncertainty states. In the context of the spin algebra, the corresponding set of states which do fulfill the Heisenberg equality are referred to as intelligent spin states by Aragone, Chalbaud, and Salamó [73].

Let us start our further discussion with the W-set of coherent states for which we will choose $m = s$.

Complex Variable Characterization

Since the commutation relation $S_3 S_+ = S_+(S_3+1)$ implies that $S_+|(m)\rangle = C|(m+1)\rangle$, C a normalization factor, it follows that $S_+|(s)\rangle = 0$ as s is the maximum weight allowed for m in this representation. Let us adopt $|\,\rangle = |(s)\rangle$ in which case the definition of $|z\rangle$ reduces to

$$|z\rangle \equiv N e^{zS_-}|\,\rangle\,, \tag{4.3}$$

which apart from the normalization factor N is only a function of the complex variable z. There is evidently a close parallel between (4.3) and the complex variable form of the canonical coherent states given in (3.1). Hence we may immediately anticipate that a number of features similar to those found in Sec. 3 will again follow from analyticity.

In order to determine the normalization factor N we first recall the relation $S_-|(m+1)\rangle = [(s-m)(s+m+1)]^{1/2}|(m)\rangle$, and thus that

$$N^{-1}|z\rangle = |(s)\rangle + z\sqrt{2s}\,|(s-1)\rangle$$

$$+ \tfrac{1}{2}z^2\sqrt{2s}\sqrt{2\cdot(2s-1)}\,|(s-2)\rangle + \cdots + z^{2s}|(-s)\rangle \,. \tag{4.4}$$

Consequently, we find that $N = (1+|z|^2)^{-s}$, and the overlap of two such states is given by

$$<z|z'> = (1+|z|^2)^{-s}\ (1+|z'|^2)^{-s}\ (1+z^*z')^{2s}, \qquad (4.5)$$

which is clearly a continuous function of z and z'.

For the resolution of unity we note that

$$\int |z><z| \left(\frac{2s+1}{\pi}\right) \frac{d^2z}{(1+|z|^2)^2} =$$

$$\left(\frac{2s+1}{\pi}\right) \int \sum_{n,m=0}^{2s} (n!m!)^{-1}\ z^n z^{*m} S_-^n |(s)><(s)| S_+^m \frac{d(\mathrm{Re}\ z)d(\mathrm{Im}\ z)}{(1+|z|^2)^{2s+2}}$$

$$= (2s+1) \sum_{n=0}^{2s} \binom{2s}{n} \int_0^\infty x^n\ |(s-n)><(s-n)|\ \frac{dx}{(1+x)^{2s+2}}$$

$$= \sum_{n=-s}^{s} |(n)><(n)| = I \qquad (4.6)$$

as desired, where polar coordinates were used at an intermediate stage.

Consequences of Analyticity

Let us consider the vector representatives $\psi(z) \equiv <z|\psi>$, which are continuous bounded functions, $|<z|\psi>| \leqslant \| |\psi> \|$. Furthermore, $<z|\psi> = (1+|z|^2)^{-s} f(z^*)$ where f is analytic, which together with the bound necessarily implies that $f(z^*) = \sum_{n=0}^{2s} f_n z^{*n}$, a fact already clear from (4.4). If $f(z^*)$ vanishes on a characteristic set of points it vanishes identically, and in the present case any set with at least $2s+1$ distinct points is sufficient.

An arbitrary operator matrix element

$$<z|\mathscr{B}|z'> = NN' \sum_{n,m=0}^{2s} B_{nm} z^{*n} z'^{m}$$

is determined by its diagonal matrix elements,

$$B(z) = <z|\mathscr{B}|z> = N^2 \sum_{n,m=0}^{2s} B_{nm}\, z^{*n} z^{m},$$

since each monomial $z^{*n}z^m$ may be isolated by its separate θ and r dependence ($z \equiv re^{i\theta}$) and uniquely extended to $z^{*n}z'^m$. In a similar vein, every operator admits a diagonal representation of the form

$$\mathscr{B} = \int b(z)\; |z><z| \left(\frac{2s+1}{\pi}\right) \frac{d^2z}{(1+|z|^2)^2}\,. \tag{4.7}$$

To prove such a representation holds we need only observe that, apart from fixed factors, the right-hand side of (4.7) is a sum of terms such as

$$\int b(z)\; z^n z^{*m}\; |(s-n)> <(s-m)|\; \frac{d^2z}{(1+|z|^2)^2}\,.$$

With $z \equiv re^{i\theta}$ the θ dependence of $b(z)$ can be used to project out any diagonal row $n - m = k$ while the r dependence of $b(z)$ can be used to project out any cross-diagonal row $n + m = p$; the sum of such terms leads to the diagonal representation of an arbitrary operator. It is evident that $b(z)$ is *not* unique since any term with a phase factor exceeding $\pm 2s\theta$ makes no contribution to (4.7).

The preceding argument may be used to establish a diagonal representation of operators for any coherent-state representation of a finite-dimensional Hilbert space that is composed (possibly up to a factor) of analytic functions. Such an argument also holds for infinite-dimensional Hilbert spaces, although each case may need special attention. Our argument to prove the universality of the diagonal representation is similar in spirit to that of Sudarshan [24] who was the first to show this fact for the canonical coherent states.

Euler Angle Characterization

Let us now consider the alternative set of spin coherent states defined by (4.2) for all $(\theta, \phi) \in S^2$. For any choice of fiducial vector $| > = |(m)>$ the easiest fact to ascertain about such coherent states relates to the form of the resolution of unity. Since $|\theta, \phi><\theta, \phi| = U(\theta, \phi, \psi)| > < |U(\theta, \phi, \psi)^\dagger$ for all ψ it is suggestive to first examine the group invariant measure, which for this group and these parameters is known to be (proportional to) $\sin\theta \, d\theta \, d\phi \, d\psi$. Let us therefore consider the operator

$$\mathcal{O} = \int |\theta, \phi><\theta, \phi| \sin\theta \, d\theta \, d\phi \, d\psi \, .$$

Due to the invariance of the measure (see below for further examples) it follows that $U'\mathcal{O} = \mathcal{O} U'$ for all $U' = U(\theta', \phi', \psi')$. Since the representation is irreducible we conclude from Schur's Lemma that $\mathcal{O} = cI$ for some c. With $Tr(\mathcal{O}) = c(2s+1) = 4\pi \int d\psi$, we learn that

$$I = \left(\frac{2s+1}{4\pi}\right) \int |\theta, \phi><\theta, \phi| \sin\theta \, d\theta \, d\phi \, , \tag{4.8}$$

where we have now dropped the ψ integration altogether. This resolution of unity holds for any choice of m in the fiducial vector $| > = |(m)>$. However, only for $m = s$ (or $m = -s$) is the expression of the coherent-state overlap especially simple. The coherent-state overlap

$$<\theta, \phi|\theta', \phi'> = <(m)|e^{i\theta S_2} \, e^{i(\phi-\phi')S_3} \, e^{-i\theta' S_2} \, |(m)>$$

$$= \sum_{n=-s}^{s} <(m)|e^{i\theta S_2}|(n)><(n)|e^{-i\theta' S_2}|(m)> e^{i(\phi-\phi')n}$$

is thus expressed in terms of the reduced Wigner coefficients of the spin-s representation,

$$d^s_{nm}(\theta) \equiv <(n)|e^{-i\theta S_2}|(m)> .$$

These functions are given in Chapter 15 of Wigner's book [74], for example, and we indicate here only the special case for $m = s$, namely

$$d^s_{ns}(\theta) = \binom{2s}{s-n}^{1/2} \cos^{s+n}(\tfrac{1}{2}\theta)\sin^{s-n}(\tfrac{1}{2}\theta) .$$

Insertion of this expression readily leads to the result (valid for $m=s$)

$$<\theta, \phi|\theta', \phi'> = \left[\cos(\tfrac{1}{2}\theta)\cos(\tfrac{1}{2}\theta')e^{i\frac{1}{2}(\phi-\phi')} + \sin(\tfrac{1}{2}\theta)\sin(\tfrac{1}{2}\theta')e^{-i\frac{1}{2}(\phi-\phi')}\right]^{2s}$$

Observe this is a nonvanishing expression save when the two points refer to diametrically opposite points on the unit sphere.

There is an evident similarity between (4.9) and the result (4.5) obtained previously. Clearly the vector $|z=0> = |(s)> = |\theta=0, \phi=0>$, and a further connection follows if we rewrite (4.9) according to

$$<\theta, \phi|\theta', \phi'> = \left[\cos(\tfrac{1}{2}\theta)\cos(\tfrac{1}{2}\theta')e^{i\frac{1}{2}(\phi-\phi')}\right]^{2s}$$

$$\times \left[1 + \tan(\tfrac{1}{2}\theta)\tan(\tfrac{1}{2}\theta')e^{-i(\phi-\phi')}\right]^{2s} .$$

This expression enables us to relate the two forms of spin coherent states according to

$$z = \tan(\tfrac{1}{2}\theta)e^{i\phi} ,$$

$$|\theta, \phi> = e^{-is\phi}|z> ,$$

which are thus seen to be identical apart from an (essentially irrelevant) phase factor. Indeed, if the choice $\psi = -\phi$ (rather than $\psi = 0$) was made in the Euler angle parameterization, then even that phase factor would be absent. Note that this identification (up to a phase factor) applies only to $|\theta, \phi>$ when $|(m)> = |(s)>$.

In view of our earlier remarks it is clear when $|(m)> = |(s)>$ that the diagonal matrix elements $B(\theta, \phi) \equiv <\theta, \phi|\mathscr{B}|\theta, \phi>$ uniquely determine the operator $\mathscr{B}$, and that every operator $\mathscr{B}$ admits a diagonal representation

$$\mathscr{B} = \int b(\theta, \phi)|\theta, \phi> <\theta, \phi| \left(\frac{2s+1}{4\pi}\right) \sin\theta\, d\theta\, d\phi\,.$$

An expression for $b(\theta, \phi)$ may be found as follows. Consider the relation

$$B(\theta', \phi') = \left(\frac{2s+1}{4\pi}\right) \int b(\theta, \phi)|<\theta, \phi|\theta', \phi'>|^2 \sin\theta\, d\theta\, d\phi\,.$$

From (4.9) it follows that

$$\begin{aligned}(2s+1)|<\theta, \phi|\theta', \phi'>|^2 &= (2s+1)2^{-2s}(1+\cos\beta)^{2s} \\ &\equiv \sum_{l=0}^{2s} (2l+1)r_l^{-1} P_l(\cos\beta) \\ &= 4\pi \sum_{l=0}^{2s} \sum_{m=-l}^{l} r_l^{-1} Y_{lm}(\theta', \phi') Y_{lm}^*(\theta, \phi)\,,\end{aligned}$$

where $\cos\beta \equiv \cos\theta \cos\theta' + \sin\theta \sin\theta' \cos(\phi-\phi')$, and $r_o \equiv 1$ along with

$$r_{l+1} = \frac{(2s+l+2)}{(2s-l)}\, r_l\,.$$

Consequently, we learn that

$$b(\theta, \phi) = \sum_{l=0}^{2s} \sum_{m=-l}^{l} \frac{(2s+l+1)!}{(2s+1)!} \frac{(2s-l)!}{(2s)!} Y_{lm}(\theta, \phi)$$

$$\times \int Y^*_{lm}(\theta', \phi') B(\theta', \phi') \sin\theta' d\theta' d\phi' . \qquad (4.10)$$

It should be emphasized that the expression for $b(\theta, \phi)$ is not unique. One can add any convergent series $\sum A_{\ell m} Y_{\ell m}(\theta, \phi)$ with $\ell > 2s$ to the expression in (4.10) and still obtain the same operator $\mathscr{B}$.

Choices Other Than The Maximal (or Minimal) Weight

When $|(m)\rangle \neq |(s)\rangle$ [or $|(-s)\rangle$] then a unique specification of any operator by its diagonal matrix elements and a diagonal representation for a general operator may be lacking. For example, let $s = 1$ and choose $m = 0$. In this case

$$|\theta, \phi\rangle = \begin{pmatrix} -e^{-i\phi} \sin\theta / \sqrt{2} \\ \cos\theta \\ e^{i\phi} \sin\theta / \sqrt{2} \end{pmatrix} , \qquad (4.11)$$

which leads to a real coherent-state overlap given by

$$\langle\theta, \phi|\theta', \phi'\rangle = \cos\theta \cos\theta' + \sin\theta \sin\theta' \cos(\phi - \phi') .$$

It is easy to see that any operator of the form

$$\mathscr{O} = \begin{pmatrix} a & b & 0 \\ c & 0 & b \\ 0 & c & -a \end{pmatrix} \qquad (4.12)$$

has vanishing diagonal matrix elements for arbitrary complex a, b, and c, and that it also cannot be generated by a diagonal representation. Conversely, any operator of the form

$$\mathscr{D} = \begin{pmatrix} d & h & f \\ k & e & -h \\ g & -k & d \end{pmatrix} \qquad (4.13)$$

is uniquely specified by its diagonal matrix elements, for arbitrary complex d, e, f, g, h, and k, and it can also be generated by a diagonal representation. Clearly a general matrix $\mathscr{M}$ is uniquely decomposed as $\mathscr{M} = \mathscr{D} + \mathscr{O}$. The lack of a full diagonal representation is a direct consequence of the fact that the θ dependence is identical (viz., $\sin\theta$) in two of the three rows of $|\theta, \phi>$. Nevertheless, there is a perfectly satisfactory representation of $\mathfrak{H}$ given by $\psi(\theta, \phi) \equiv <\theta, \phi|\psi>$ and of operators as integral kernels $<\theta, \phi|\mathscr{B}|\theta', \phi'>$ for any operator $\mathscr{B}$ in the manner outlined in Sec. 2.

Range of the Diagonal Representation of Operators

The example at hand enables us to stress the general fact that the set of operators uniquely specified by their diagonal elements is identical to the set of operators admitting a diagonal representation. For this discussion let us return to the notation of Sec. 2 appropriate to a general set of coherent states. For simplicity we initially assume that the dimension d of the Hilbert space is finite. We say that two operators $\mathscr{A}$ and $\mathscr{B}$ are orthogonal if $Tr(\mathscr{A}^\dagger \mathscr{B}) = 0$. Denote by $\mathfrak{D}$ the set of operators admitting a diagonal representation, and by $\mathfrak{O}$ the set of operators with vanishing diagonal coherent-state matrix elements. Clearly, both $\mathfrak{D}$ and $\mathfrak{O}$ are linear spaces closed under the adjoint operation. We now prove that $\mathfrak{O} = \mathfrak{D}^\perp$, i.e., that every $\mathscr{A} \in \mathfrak{O}$ is orthogonal to every $\mathscr{B} \in \mathfrak{D}$. If $\mathscr{A} \in \mathfrak{O}$ and $\mathscr{B}$ (hence $\mathscr{B}^\dagger$) $\in \mathfrak{D}$ then $Tr(\mathscr{B}^\dagger \mathscr{A}) = \int b^*(\ell)A(\ell)\delta\ell = 0$ since $A(\ell) = <\ell|A|\ell> \equiv 0$; hence $\mathscr{A} \in \mathfrak{D}^\perp$. Conversely, if $\mathscr{A} \in \mathfrak{D}^\perp$, then $0 = Tr(\mathscr{B}^\dagger \mathscr{A}) = \int b^*(\ell)A(\ell)\delta\ell$ holds for all integrable $b(\ell)$, which implies that $A(\ell) \equiv 0$; hence $\mathscr{A} \in \mathfrak{O}$. Therefore $\mathfrak{O} = \mathfrak{D}^\perp$. As a linear vector space, the space of all $d \times d$ matrices $\mathfrak{M} = \mathfrak{D} + \mathfrak{D}^\perp = \mathfrak{D} + \mathfrak{O}$, which implies that an arbitrary matrix has a unique decomposition $\mathscr{M} = \mathscr{D} + \mathscr{O}$, where $\mathscr{D} \in \mathfrak{D}$ and $\mathscr{O} \in \mathfrak{O}$. The results established here generalize those of the example of (4.11), (4.12) and (4.13) to the case of an arbitrary set of coherent states in a finite-dimensional Hilbert space.

To give a precise characterization when the Hilbert space is infinite dimensional attention is restricted to the space $\mathfrak{D}_2$ of Hilbert-Schmidt operators admitting a diagonal representation, namely those operators with diagonal weight functions that satisfy

$$Tr(\mathscr{B}^{\dagger}\mathscr{B}) = \int b^*(\ell)b(\ell')\,|<\ell'\,|\,\ell>|^2\,\delta\ell\,\delta\ell' < \infty\,.$$

Correspondingly, $\mathfrak{O}_2$ denotes the space of Hilbert-Schmidt operators with vanishing diagonal coherent-state matrix elements. Then the previous argument extends to $d = \infty$ and leads to the conclusion that $\mathfrak{O}_2 = \mathfrak{D}_2^{\perp}$. The linear vector space $\mathfrak{M}_2$ of all Hilbert-Schmidt operators decomposes into $\mathfrak{M}_2 = \mathfrak{D}_2 + \mathfrak{D}_2^{\perp} = \mathfrak{D}_2 + \mathfrak{O}_2$, which implies an analogous unique decomposition of an arbitrary Hilbert-Schmidt operator as $\mathscr{M} = \mathscr{D} + \mathscr{O}$, with $\mathscr{D} \in \mathfrak{D}_2$ and $\mathscr{O} \in \mathfrak{O}_2$. If the diagonal coherent-state matrix elements uniquely determine the operator, then $\mathfrak{O}_2$ consists of the zero operator, and as a consequence all Hilbert-Schmidt operators admit a diagonal representation. Since a general operator can be constructed as the strong (sequential) limit of Hilbert-Schmidt operators, then the foregoing is a precise statement of the fact alluded to in Sec. 3 that a general operator admits a diagonal representation whenever the diagonal coherent-state matrix elements uniquely determine an operator. The preceding argument is an extension of one given by Simon [75].

Group Parameterization

Finally, let us briefly consider the last choice of group parameterization given in (4.1), namely that of

$$|\vec{\alpha}> = e^{i(\vec{\alpha}\cdot\vec{S})}\,|\,>, \tag{4.14}$$

where we now suppose that $|\,>$ is an *arbitrary* unit vector. If $d\alpha$ denotes the group invariant measure in these parameters, normalized so that $\int d\alpha = (2s+1)$, then Schur's Lemma again ensures that

$$I = \int |\vec{\alpha}><\vec{\alpha}|\,d\alpha\,.$$

Thus there is a representation by $\psi(\vec{\alpha}) \equiv <\vec{\alpha}|\psi>$, and all the features outlined in Sec. 2 for such representations hold true. However, the three parameters $\vec{\alpha}$ do *not* label distinct rays, i.e., distinct vectors modulo phase vectors. This was already true in the complex variable and Euler angle parameterization, but we eliminated the problem when we set $\psi = 0$ [or any other choice $\psi = \psi(\theta,\phi)$]. Here the redundancy is made manifest in those $\vec{\beta}$ for which

$$e^{i\vec{\beta}\cdot\vec{S}}|\,\rangle = e^{i\gamma}|\,\rangle,$$

where γ is a phase factor. Clearly these $\vec{\beta}$ label a subgroup. For each allowed $\vec{\beta}$ it follows that $|\,\rangle$ is an eigenvector of $\vec{\beta}\cdot\vec{S}$ with eigenvalue γ. Since only one component of $\vec{S}$ may be diagonalized at a time it follows that the allowed $\vec{\beta}$ span a one-dimensional space, $\vec{\beta} = \beta\vec{n}$, where $\vec{n}$ is a unit vector in some direction, and thus $\gamma = \beta m$, where $m = -s, -s+1, \ldots, s$. Only the two vector directions orthogonal to $\vec{n}$ label distinct rays, and the constraint $\vec{\alpha}\cdot\vec{n} = 0$ reduces the three labels to two as was the case previously. We should emphasize that redundant labelling of coherent states arises in other contexts as well, and we will see further examples in the next section.

Additional Remarks and
Introduction to Selected Reprints

1. Spin coherent states for spin ½ and their resolution of unity appear in Ref. [20]. In this paper both the parameterization $|z\rangle = N\begin{pmatrix}1\\z\end{pmatrix}$, which is used here in our initial discussion, and the equivalent but alternative parameterization $|\chi\rangle = \begin{pmatrix}\sqrt{1-|\chi|^2}\\ \chi\end{pmatrix}$, $|\chi| \leqslant 1$, appear.

2. A rather complete discussion of spin coherent states along with several simple applications appears in the work of Radcliffe [76].

An elegant discussion of spin coherent states and their application to problems in quantum optics appears in the paper of Arecchi, Courtens, Gilmore, and Thomas [77]. Some of these topics are also well discussed in the group theory text of Gilmore [78]. For still further treatments, we cite the reviews [79,80].

3. An application of spin coherent states and the diagonal representation of operators to formulate spin coherent-state path integrals and to derive rigorous upper and lower bounds for the partition function of a statistical mechanics problem involving spin variables was given by Lieb [81]. This construction involved the first use of the diagonal representation of operators in path integrals, and we give an analogous construction in our discussion of path integrals in Sec. 6.

4. The spin coherent states have successfully been applied in many branches of physics: Hepp and Lieb [82] studied the thermodynamic phase transition in the Dicke model of resonant interaction of the radiation field with N two-level atoms, and by making use of spin coherent states simplified and improved an earlier treatment of Wang and Hioe [83], which had used canonical coherent states to treat this problem. Narducci, Bowden, Bluemel, Garrazana and Tuft [84] used spin coherent states in order to study multitime correlation functions for systems with observables satisfying an angular momentum algebra. In the field of nuclear physics Gilmore and Feng [85,86] used spin coherent states in order to study phase transitions in nuclear matter. Langevin equations associated with fermions and quantum spins can also conveniently be derived in the framework of spin coherent states [87].

5. SOME OTHER COHERENT STATES

Guided by the discussion of the canonical coherent states and spin coherent states of the preceding two sections, we now turn our attention to the characterization of additional coherent states.

Group-Related Coherent States

Let G denote a continuous (Lie) group and $U(g)$, $g \in G$, a strongly continuous irreducible unitary representation of the group on an appropriate Hilbert space $\mathfrak{H}$. Let $|0\rangle$ denote a normalized fiducial vector, and for all $g \in G$,

$$|g\rangle \equiv U(g)|0\rangle$$

are the corresponding (normalized) coherent states of the group G. The overlap of two such states, given by

$$\langle g_2|g_1\rangle = \langle 0|U(g_2)^{\dagger}U(g_1)|0\rangle = \langle 0|U(g_2^{-1}\cdot g_1)|0\rangle \ ,$$

is bounded by unity and is jointly continuous in the group parameters by definition. Here the symbol $g_2^{-1}\cdot g_1$ denotes group multiplication.

We denote by dg the left-invariant group measure, chosen so that for any fixed $g_o \in G$, $d(g_o\cdot g) = dg$. For compact groups the left-invariant and right-invariant measures are equal, but for noncompact groups they may be different (an example appears below). Attention is confined to those unitary representations of G and fiducial vectors $|0\rangle$ for which $\int|\langle 0|U(g)|0\rangle|^2 dg < \infty$; if the volume of group space $\int dg$ is finite, as for compact groups, this bound is automatic, but for noncompact groups it may impose a restriction. Such representations are termed "square integrable". If dg is both left and right invariant then all vectors lead to square-integrable representations if one vector does; this is not necessarily the case if dg is not right invariant. We now scale dg by a finite factor so that $\int|\langle 0|U(g)|0\rangle|^2 dg = 1$, which we choose hereafter to fix the normalization of dg. By invariance of the measure it follows that $\int|\langle g'|g\rangle|^2 dg = 1$ for all $g' \in G$. Next consider the operator $\mathcal{O} = \int |g\rangle\langle g|dg$, defined for a noncompact group as a weak limit. Invariance of the measure ensures that

$U(g')\mathcal{O} = \mathcal{O}U(g')$ for all $g' \in G$, and Schur's Lemma and the chosen normalization of dg then ensure that

$$I = \int_G |g\rangle\langle g| dg \,. \tag{5.1}$$

This relation establishes the desired resolution of unity for this representation of the group G.

As was the case for the group SU(2) and the spin coherent states it may happen that the points $g \in G$ do not label distinct states. Let the points $h \in H \subset G$ be defined by the property $U(h)|\,\rangle = e^{i\alpha(h)}|\,\rangle$, where α is a phase factor. Clearly, H represents a subgroup of G, and the manifold $M \equiv G/H$ labels distinct states. Denote by x a point in M and by $|x\rangle$ the associated coherent state. When it occurs, which is the case in many applications, denote by dx the G-invariant measure induced by dg on G/H [88]. In this terminology one may also show, in the same manner as above but without reference to (5.1), that

$$I = \int_{G/H} |x\rangle\langle x| dx \,, \tag{5.2}$$

with a suitable normalization of the measure dx. Both (5.1) and (5.2) are correct and it is somewhat a matter of taste or depending perhaps on the application which one is to be preferred. In either case there is a representation of the abstract Hilbert space $\mathfrak{H}$ by functions $\psi(g) \equiv \langle g|\psi\rangle$ on G, or by functions $\psi(x) \equiv \langle x|\psi\rangle$ on $M = G/H$. Clearly

$$\langle\psi|\psi\rangle = \int_G |\psi(g)|^2 dg = \int_M |\psi(x)|^2 dx \,,$$

and all the properties of such function spaces outlined in Sec. 2 hold for both formulations. We emphasize that without further specialization these examples are not necessarily analytic function representations (possibly up to a factor); nevertheless, they evidently satisfy the criteria of Sec. 2 to qualify as coherent states.

Additional Remarks and Introduction to Selected Reprints

1. Group-related coherent states, their resolution of unity, and their associated representation spaces already appear in a paper of Klauder [89]. In this paper these states were applied to analyze the relation between classical and quantum theories. It is shown that when properly interpreted the classical action functional is nothing but the quantum action functional with a restriction imposed on the allowed variations. This relation may be seen directly, but further support for it comes from the construction of coherent-state path integrals, which is discussed in the next section.

2. All non-Abelian two parameter continuous groups are equivalent to the noncompact affine group corresponding to translations (q) and dilatations ($p>0$) (without inversion) of the real line: $y \rightarrow p^{-1}y + q$. This group is an example of one for which the left- and right-group invariant measures are different. In the parameters indicated $dg \propto d_L g = dp\, dq$ while $d_R g = p^{-1} dp\, dq$. Moreover — and unlike the canonical case — not every fiducial vector leads to square-integrable representations for the affine group. Suitable affine coherent states (based on the affine group) have been discussed by Aslaksen and Klauder [90]. For certain fiducial vectors the coherent-state induced representation of Hilbert space is given (up to a common factor) by analytic functions. In the parameters indicated these functions are analytic in a *half plane* (e.g., $\mathrm{Re}\, z > 0$). The properties of such analytic function representations are briefly discussed in Ref. [91], and extensively analyzed by Paul [92].

3. An important step in pointing the way beyond the canonical coherent states was taken by Barut and Girardello [93], who presented a thorough analysis of analytic-variable coherent states for the group SU(1,1). Like the canonical coherent states, the coherent

states of Barut and Girardello are eigenfunctions of one of the ladder operators. The importance of SU(1,1) to dynamical group applications formed part of their motivation. An explicit formula was obtained by Hongoh [94] for the SU(1,1) coherent states associated with the continuous spectrum of an infinite-dimensional unitary representation of SU(1,1).

4. In the context of multilevel atomic systems Gilmore [95] was led to generalize spin coherent states and introduced coherent states for a general unitary group. The first full-scale discussion of coherent states for an arbitrary Lie group was given by Perelomov [96]. In this paper the general theory that he develops is illustrated by coherent states for the groups SU(2) and SU(1,1). However, almost no emphasis is put on the possibility of representations by spaces of analytic functions if suitable fiducial vectors are chosen.

5. In the canonical decomposition given by Cartan for semisimple Lie algebras into the commutator subgroup and the analog of raising and lowering operators one is led to the important concept of representations induced by the highest (or lowest) weight vectors. Such a fiducial vector is the analog of the one with highest (or lowest) eigenvalue of S_3 in the case of the spin algebra. The special importance of the highest (or lowest) weight vectors was recognized and emphasized by Gilmore [97]. Choosing the highest (or lowest) weight representation and acting on this fiducial vector with a suitable combination of lowering or raising operators with coefficients chosen as suitable powers of complex variables leads to the characterization of group-related coherent states by analytic expressions (possibly up to a common factor due to normalization). In this fashion it becomes possible to define group-related coherent states so that the associated Hilbert space representation is by a space of analytic functions. The generality of an analytic function representation was recognized and emphasized by Onofri [98].

6. An excellent review of coherent states for general groups is given by Perelomov [99]. Coherent states defined for various compact Lie groups have been used in the study of the large N-limit of quantum mechanical and field theoretical models by Yaffe [100], who has also provided a pedagogical discussion [101]. For various applications of these coherent states in thermodynamics and nuclear physics see the papers by Rasetti [102] and Gilmore [103].

Conserved-Charge Coherent States

In certain applications, where a conserved charge should be respected, a special kind of coherent states proves useful. We first consider the simplest case of an Abelian charge. Let $a_1^\dagger$ and $a_2^\dagger$ denote two independent boson creation operators, and take as fiducial vector the ground state $|0>$ for which $a_1|0> = a_2|0> = 0$. The conserved-charge coherent states $|z; q>$ are then defined to be such that the Abelian charge $Q \equiv a_1^\dagger a_1 - a_2^\dagger a_2$ has an eigenvalue, i.e., $Q|z; q> = q|z; q>$. However, this condition does not fully characterize the state $|z; q>$. Since the operator $a_1 a_2$ commutes with the charge operator Q, we can also impose the constraint that $a_1 a_2|z; q> = z|z; q>$ with the result that

$$|z; q> = N \sum_{n=0}^{\infty} z^n \frac{(a_1^\dagger)^{n+q}}{(n+q)!} \frac{(a_2^\dagger)^n}{n!} |0>$$

$$= N \sum_{n=0}^{\infty} \frac{z^n}{\sqrt{(n+q)!}\sqrt{n!}} |n + q\ ,\ n> \ , \tag{5.3}$$

where $q \geqslant 0$ is a fixed integer and $N = [|z|^{-q} I_q(2|z|)]^{-1/2}$ is a normalization factor (I_q is the modified Bessel function of order q). Observe that these states are just a subset of a two-variable family of states each of the form given in (3.1).

In the subspace where $Q = q$ the resolution of unity $I_{(q)}$ is given by the expression

$$I_{(q)} = \int |z;q\rangle\langle z;q|\, \Phi_q(z)\, d^2z\,, \tag{5.4}$$

where

$$\Phi_q(z) \equiv (2/\pi)\, I_q(2|z|)K_q(2|z|)\,.$$

For fixed q the coherent states and the representation of Hilbert space they induce are (up to a factor) analytic functions, and so consequences such as unique specification of an operator by its diagonal matrix elements and a diagonal representation of a general operator all hold.

Of course, if completeness for all charge sectors is desired then we have

$$I = \sum_{q=0}^{\infty} \int |z;q\rangle\langle z;q|\Phi_q(z)d^2z\,.$$

In the latter case it is clear that the label space is *not* connected, and the coherent states and the induced representation are *not* by analytic functions (but by a direct sum of such functions). Thus the usual consequences of analyticity do not automatically follow in the case of all charge sectors for this kind of coherent state parameterization.

As we observed above, the conserved-charge coherent states are closely related to the two-variable canonical coherent states $|z_1, z_2\rangle$ given by an extension of (3.1),

$$|z_1, z_2\rangle = \exp[(a^\dagger, z) - (z^*, a)]|0\rangle\,, \tag{5.5}$$

where the scalar product (x,y) is given by $x_1y_1 + x_2y_2$. The conserved-charge coherent state (5.3) can then be obtained from the state (5.5) by considering a suitable average over U(1)-phases (generated by the charge operator Q) according to the following expression

$$|z;q> = Ne^{\frac{1}{2}(|z_1|^2+|z_2|^2)} z_1^{-q} \int_{-\pi}^{\pi} (d\alpha/2\pi) e^{iq\alpha} |e^{-i\alpha} z_1, e^{i\alpha} z_2> \quad (5.6)$$

where $z = z_1 z_2$. Such a representation is very useful since the properties of the canonical coherent states now can be employed in a study of the properties of the conserved-charge coherent states.

The form of the conserved-charge coherent states given by (5.6) has a very simple group-theoretical interpretation. On the two-variable canonical coherent states one acts with a U(1) transformation, generated by the charge operator Q, on the $z_1(z_2)$ variable corresponding to a "particle" ("anti-particle") state. In (5.6) one suitably averages over the U(1)-group action on the two-variable coherent state, which then projects out the Q=q charge subspace contribution.

This procedure can easily be extended to general compact groups. Let us illustrate the construction of the corresponding non-Abelian conserved-charge coherent states by considering the group G=SU(2) and a multi-component canonical coherent state $|f>$ generated by the one-particle state $f_{k,\alpha}$ which transforms according to some real representation R of SU(2), i.e., $f_{k,\alpha} \rightarrow R_{\alpha\beta}(g) f_{k,\beta}$ under a group action (summation convention). The index k denotes any further labelling necessary to fully characterize the one-particle state (e.g., a possible momentum dependence). Equation (5.5) trivially extends to

$$|f> = \exp[(a^\dagger, f) - (f^*, a)]|0> ,$$

where now $(x,y) = x_{k,\alpha} y_{k,\alpha}$. A natural extension of (5.6) to the group G is then

$$|D^\ell_{\alpha\beta}; f> = M_{\alpha\beta}^{-\frac{1}{2}} \int_G dg D^\ell_{\alpha\beta}(g) |R(g)f> , \quad (5.7)$$

provided, of course, the representation $D^\ell_{\alpha\beta}$ [74] can be obtained from $R_{\alpha\beta}$, where $M_{\alpha\beta}$ is a normalization factor and dg is the invariant measure on the group under consideration. The non-Abelian conserved-charge coherent state (5.7) clearly transforms irreducibly under group actions according to the SU(2) representation $D^{\ell}_{\alpha\beta}$. The above construction easily extends to complex representations as well.

In the subspace given by the representation $D^{\ell}_{\alpha\beta}$, the resolution of the identity operator is an extension of (5.4), i.e.,

$$I_{(\ell)} = \int \sum_{\alpha\beta} |D^{\ell}_{\alpha\beta}; f\rangle \langle D^{\ell}_{\alpha\beta}; f| (2\ell+1) M_{\alpha\beta}\, d^2 f\,, \qquad (5.8)$$

where $d^2 f \equiv \prod_{k,\alpha} (d^2 f_{k,\alpha}/\pi)$, and the discussion below (5.4) applies also to this case. Summing over all the representations $D^{\ell}_{\alpha\beta}$ in (5.8), i.e., summing over the index ℓ, naturally leads to the unit operator I in analogy with the Abelian conserved-charge coherent states discussed above.

Additional Remarks and Introduction to Selected Reprints

1. The derivation of various properties of the conserved-charge coherent states has been given by Bhaumik, Bhaumik and Dutta-Roy [104]. The construction of these states is also discussed in the paper by Horn and Silver [105] in the context of coherent production of pions. For an application in nuclear physics of similar ideas see, e.g., Ref. [106].

2. The conserved-charge coherent state of Bhaumik ct al. were extended to the field theoretical case in Ref. [107], where the corresponding states are applied to the problem of soft emission of charged particles from a c-number current. The vacuum instability in scalar quantum electrodynamics was studied by Ni and Wang [108] in terms of these states.

3. The construction of the SU(2) non-Abelian conserved-charge coherent states was carried out in Refs. [109,110] in the quantum mechanical case, and in field theory by Eriksson, Mukunda and Skagerstam [111] extending considerations by Botke, Scalapino and Sugar [112]. Various applications of these states in the context of elementary particle physica can be found in Refs. [111-113]. Somewhat related states are used in a variational context in Ref. [114].

Non-Abelian conserved-charge coherent states for arbitrary compact Lie groups were constructed and applied to various problems by Skagerstam [115]. These coherent state techniques were used by Skagerstam [116] in order to extract the color-singlet contribution to the quark-gluon partition function in the high temperature limit. For other applications see, e.g., Refs. [117,118].

4. Other constructions of coherent states associated with non-Abelian Lie groups have been discussed in the literature. In an elegant paper of Bargmann [119], the Bargmann-Segal representation [21,22] of Hilbert space is used to construct representations of the rotation group in terms of Schwinger's [120] well-known boson representation of the spin operators. Similar realizations appear in the paper of Takahashi and Shibata [121], where an application is given in the context of spin relaxation processes, and in Refs. [122-124].

Coherent states which satisfy the Heisenberg equality relation for the angular momentum operators, the so-called intelligent spin states [125], and their relationship to the spin coherent states, are discussed in Ref. [73]. Further discussions on the properties of intelligent spin states can be found in Ref. [126].

Coherent states appropriate for the description of an electron moving in a uniform magnetic field are discussed in the papers by Man'ko and collaborators [127-129], and also in the paper by

Feldman and Kahn [130], in which some thermodynamics properties are computed.

In a study of the quantum mechanical top in nuclear physics, Janssen [131] developed a set of coherent states closely related to a direct product of two spin coherent states. The Bargmann-Segal representation [21,22] of Hilbert space was used in Ref. [132] in a treatment of the same problem.

Finally, we mention a construction of SU(3) coherent states due to Fan and Ruan [133] similar in spirit to the SU(2) conserved-charge coherent states [109].

Fermion Coherent States

The coherent states to be defined here are of a qualitatively different kind than others in that the basic label variables are *not* ordinary real or complex variables` but are rather so-called *anticommuting c-numbers* (or linear elements of a Grassmann algebra as defined below). All such variables anticommute among themselves: If ψ_1 and ψ_2 denote two such quantities then $\psi_1\psi_2+\psi_2\psi_1=0$. Consequently any such variable satisfies $\psi^2=0$, a relation of fundamental importance. It is convenient to illustrate some aspects of the following discussion for several (K) of freedom at once.

The anticommuting *c*-numbers ψ_k, $k=1, 2,..., K$, $K<\infty$, stand in one-to-one relation with elements of a corresponding set of fermion annihilation (or destruction) operators $\{b_k\}$. An associated set of anticommuting *c*-numbers $\{\overline{\psi}_k\}$ stands in relation with the associated set of fermion creation operators $\{b_k^\dagger\}$. Although the creation operators $\{b_k^\dagger\}$ are adjoint operators (in the sense of Hilbert space) to the annihilation operators $\{b_k\}$ there is no corresponding relation between $\{\overline{\psi}_k\}$ and $\{\psi_k\}$. All the elements of $\{\psi_k\}$ and $\{\overline{\psi}_k\}$ anticommute among themselves, $\psi_j\overline{\psi}_k + \overline{\psi}_k\psi_j=0$, and the two sets must be chosen *independently* of each other (for if $\overline{\psi}_j$ is a mathematical adjoint of ψ_j, then they both vanish). Furthermore, it is assumed that elements of $\{\psi_k\}$ and $\{\bar{\psi}_k\}$ anticommute with all the operators in $\{b_k\}$ and $\{b_k^\dagger\}$, namely, $\psi_j b_k + b_k \psi_j = 0$, etc. Choose $|0\rangle$ as the no-particle state such that $b_k|0\rangle = 0$ for all k, and define $|k_1, \ldots, k_m\rangle = b_{k_1}^\dagger \cdots b_{k_m}^\dagger|0\rangle$, $m \leqslant K$, which are normalized states totally antisymmetric in the variables k_1 through k_m. The linear span of such states determines the Hilbert space $\mathfrak{H}$. We assume that the anticommuting c-numbers commute with the no-particle state, $\psi_k^\#|0\rangle = |0\rangle\psi_k^\#$; here $\psi^\#$ denotes either ψ or $\bar{\psi}$. As a consequence

$$\begin{aligned}\psi_\ell^\#|k_1, \ldots, k_m\rangle &= \psi_\ell^\# b_{k_1}^\dagger \cdots b_{k_m}^\dagger|0\rangle \\ &= (-1)^m b_{k_1}^\dagger \cdots b_{k_m}^\dagger \psi_\ell^\#|0\rangle \\ &= (-1)^m|k_1, \ldots, k_m\rangle\psi_\ell^\# . \end{aligned} \tag{5.9}$$

The adjoint of (5.9) holds as well.

The elements of the Grassmann algebra are made up of complex linear sums of the nonvanishing, independent, multiple products $\psi_{\ell_1}^\# \cdots \psi_{\ell_m}^\#$, $0 \leqslant m \leqslant K$. If $m=0$ we interpret the result as the number one, 1, the unit element. Apart from the unit element all elements are generated from multiple products of the linear elements $\{\psi_\ell^\#\}$. Even elements are those unchanged if $\psi^\# \to -\psi^\#$; odd elements are those that change sign. All odd elements anticommute with each other and (odd multiples of) the annihilation and creation operators. All even elements commute with all elements of the Grassmann algebra and (all multiples of) the annihilation and creation operators. All Grassmann elements commute with all complex numbers, which is not entirely a trivial statement. Consider the example

$$\psi_\ell\langle 0|b_k^\dagger|0\rangle = \langle 0|\psi_\ell b_k^\dagger|0\rangle = -\langle 0|b_k^\dagger\psi_\ell|0\rangle = -\langle 0|b_k^\dagger|0\rangle\psi_\ell ,$$

which is no contradiction since $\langle 0|b_k^\dagger|0\rangle = 0$.

We note that a concrete realization of the anticommuting c-numbers may be given as follows. Let $\{\theta_p\}$, $\{\theta_p^\dagger\}$, $p = 1, 2, \ldots, 2K$ denote an auxiliary set of fermion annihilation and creation operators such that $\{\theta_p, \theta_q\}=0$ and $\{\theta_p, \theta_q^\dagger\}=\delta_{pq}$. Note that the operator norm for such quantities is just given by $\|\theta_p\|=\|\theta_p^\dagger\|=1$ for all p. Now if c_k, $k = 1, 2, \ldots, K$ denote arbitrary complex variables we may choose $\psi_k=c_k\theta_{2k-1}$ and $\overline{\psi}_k=c_k^*\theta_{2k}$, $k = 1, 2, \ldots, K$; the $\theta^\dagger$ operators are not used. Continuity in ψ (or $\overline{\psi}$) is interpreted as continuity in the complex coefficients c.

In an occupation number notation the basic states are represented by

$$|n_1, n_2, \ldots, n_K> \equiv (b_1^\dagger)^{n_1}(b_2^\dagger)^{n_2} \cdots (b_K^\dagger)^{n_K}|0> ,$$

where $n_k= 0$ or 1 for eack k. If the order is preserved these states may be regarded as direct-product states

$$|n_1, n_2, \ldots, n_K> = |n_1>_1 \otimes |n_2>_2 \otimes \cdots \otimes |n_K>_K .$$

In this notation the no-particle state is given by

$$|0,0, \ldots, 0> = |0>_1 \otimes |0>_2 \otimes \cdots \otimes |0>_K ,$$

and it follows that

$$\psi_\ell^\# |n_1, n_2, \ldots, n_K> = (-1)^{\Sigma n_k}|n_1, n_2, \ldots, n_K>\psi_\ell^\# .$$

The fermion coherent states for K degrees of freedom are essentially the ordered direct product of K single-degree-of-freedom fermion coherent states. Consequently we now specialize to a single degree of freedom (K=1) and drop the subscript.

For a single degree of freedom the associated coherent states are defined by

$$|\psi> \equiv M(|0> + |1>\psi)$$

where M is a normalization factor, which we shall see is an even element of the Grassmann algebra. The phase convention here is chosen so that

$$b|\psi\rangle = Mb|1\rangle\psi = M|0\rangle\psi = \psi M|0\rangle = \psi|\psi\rangle ,$$

namely, so that $|\psi\rangle$ is a right eigenvector of b with eigenvalue ψ very much as in the canonical case discussed in Sec. 3. The adjoint vector is defined by

$$\langle\psi| \equiv (\langle 0| + \bar{\psi}\langle 1|)M ;$$

this is not a proper mathematical adjoint since $\bar{\psi}$ is independent of ψ, but we nevertheless refer to it as such. Again we have

$$\langle\psi|b^\dagger = \bar{\psi}\langle 1|b^\dagger M = \bar{\psi}\langle 0|M = \langle 0|M\bar{\psi} = \langle\psi|\bar{\psi}$$

showing that $\langle\psi|$ is a left eigenvector of $b^\dagger$ with eigenvalue $\bar{\psi}$. The overlap $\langle\psi|\psi\rangle = M^2(1 + \bar{\psi}\psi)$ can be normalized if we choose $M = (1 + \bar{\psi}\psi)^{-1/2}$, or equivalently as $M = \exp(-\bar{\psi}\psi/2)$. Note this equivalence is a formal one due to the fact that $(\bar{\psi}\psi)^2 = 0$; it is not due to any "numerical equality" since ψ or $\bar{\psi}\psi$ do not have numerical values in the conventional sense. It is noteworthy that the only nonvanishing commutator among the elements $b^\dagger\psi$, $\bar{\psi}b$, and $\bar{\psi}\psi$ is given by $[\bar{\psi}b, b^\dagger\psi] = \bar{\psi}\psi$. When combined with the foregoing it follows that

$$\begin{aligned} |\psi\rangle &= e^{(b^\dagger\psi - \bar{\psi}b)}|0\rangle \\ &= e^{-\frac{1}{2}\bar{\psi}\psi}e^{b^\dagger\psi}|0\rangle \\ &= e^{-\frac{1}{2}\bar{\psi}\psi}(|0\rangle - \psi|1\rangle) , \end{aligned} \tag{5.10}$$

in essentially an exact parallel with the canonical case [cf., (3.1)]. In like fashion we have

$$\begin{aligned} <\psi| &= <0|e^{(\bar{\psi}b - b^\dagger\psi)} \\ &= e^{-\frac{1}{2}\bar{\psi}\psi}<0|e^{\bar{\psi}b} \\ &= e^{-\frac{1}{2}\bar{\psi}\psi}(<0| - <1|\bar{\psi}) . \end{aligned} \tag{5.11}$$

Such vectors were first used as generating functions for fermion states by Schwinger [134].

To develop a resolution of unity requires the concept of an integral which is evidently a problem here since ψ and $\bar{\psi}$ have no values in a conventional sense. Instead one adopts a linear functional which is then clothed in the fictitious dress of traditional integral notation. In particular, one conventionally adopts the relations

$$\begin{aligned} &\int d\psi = 0 , \quad \int d\bar{\psi} = 0 , \\ &\int \psi d\psi = 1 , \quad \int \bar{\psi} d\bar{\psi} = 1 , \end{aligned} \tag{5.12}$$

although other choices for the last line have also been made. Do not ask what the integration limits are; these are just symbolic integrals. The formal differentials $d\psi$ and $d\bar{\psi}$ are chosen to have the same commutation properties as ψ and $\bar{\psi}$, respectively. Multiple integrals then follow from the appropriate commutation properties and the basic integration rules, as in the example

$$\int (A + B\psi + C\bar{\psi} + D\bar{\psi}\psi) d\bar{\psi} d\psi = -D$$

where A, B, C, and D are arbitrary complex numbers. Note the integrand here is the most general element of the Grassmann algebra that is possible for a single ψ and $\bar{\psi}$. The indicated integral is then, in reality, a linear transformation of the Grassmann algebra elements into the complex numbers; nothing more, nothing less. The rules of integration given above are often called Grassmann integration, or Berezin [135] integration in view of the latter's extensive study of this formalism.

Next let us evaluate the integral

$$\int |\psi\rangle\langle\psi| d\bar{\psi} d\psi = \int e^{-\bar{\psi}\psi}(|0\rangle + |1\rangle\psi)(\langle 0| + \bar{\psi}\langle 1|) d\bar{\psi} d\psi$$

$$= \int [(1-\bar{\psi}\psi)|0\rangle\langle 0| + \psi\bar{\psi}|1\rangle\langle 1|] d\bar{\psi} d\psi$$

$$= |0\rangle\langle 0| + |1\rangle\langle 1| = I \,, \tag{5.13}$$

the identity operator, as desired. The rules of integration have indeed been well conceived!

Consider next the trace of an operator $\mathscr{B}$ (a two-dimensional matrix) expressed as an integral over coherent states. It is straightforward to see that the usual expression gives

$$\int \langle\psi|\mathscr{B}|\psi\rangle d\bar{\psi} d\psi = B_{00} - B_{11} \,, \tag{5.14}$$

where $B_{nm} \equiv \langle n|\mathscr{B}|m\rangle$, which is *not* the trace. To find the trace it is necessary to take $\psi_1 = -\psi$, $\bar{\psi}_1 = -\bar{\psi}$, and then

$$\int \langle\psi_1|\mathscr{B}|\psi\rangle d\bar{\psi} d\psi = B_{00} + B_{11} \,; \tag{5.15}$$

in this case one commonly writes $\langle\psi_1| = \langle -\psi|$. (This relation is responsible for the need to use antiperiodic temporal boundary conditions in path integrals to evaluate the trace for fermion variables within the Grassmann formalism.) Another matrix element is given if $\psi_2 = \psi$, $\bar{\psi}_2 = -\bar{\psi}$, for which

$$\int \langle\psi_2|\mathscr{B}|\psi\rangle d\bar{\psi} d\psi = B_{11} \,. \tag{5.16}$$

Still other relations are

$$\int \langle\psi|\mathscr{B}|\psi\rangle \psi d\bar{\psi} d\psi = -B_{10} \,, \tag{5.17}$$

$$\int \bar{\psi}\langle\psi|\mathscr{B}|\psi\rangle d\bar{\psi} d\psi = -B_{01} \,, \tag{5.18}$$

$$\int \bar{\psi} <\psi|\mathscr{B}|\psi> \psi d\bar{\psi} d\psi = -B_{00} \,, \tag{5.19}$$

$$\int <\psi_3|\mathscr{B}|\psi> d\bar{\psi}_3 d\psi_3 d\bar{\psi} d\psi = \tfrac{1}{4}\, B_{00} \,. \tag{5.20}$$

The asymmetry between $|0>$ and $|1>$ in these relations can largely be eliminated by using two complementary sets of fermion coherent states following Ohnuki and Kashiwa [136].

For the fermion coherent states it is straightforward to observe that an operator is uniquely determined by its diagonal elements,

$$B(\psi) \equiv <\psi|\mathscr{B}|\psi>$$

$$= e^{-\bar{\psi}\psi}[B_{00} + \psi B_{01} + \bar{\psi} B_{10} + \bar{\psi}\psi B_{11}] \,.$$

For example, four of the integral relations given above [(5.14) and (5.17)-(5.19)] isolate independent linear combinations of matrix elements. According to the general discussion of Sec. 4 it should follow that every operator admits a diagonal representation in the appropriate coherent states. Indeed, this holds true here as well, as follows directly from the observation that

$$\int [A + B\psi + C\bar{\psi} + (A-D)\bar{\psi}\psi]|\psi><\psi| d\bar{\psi} d\psi$$

$$= D\,|0><0| + C|1><0| + B|0><1| + A|1><1| \,, \tag{5.21}$$

for arbitrary complex A, B, C, and D. It is even tempting to say that these two results are consequences of "analyticity", for, apart from normalization, $|\psi>$ depends only on ψ.

An integral of paramount importance when dealing with Grassmann variables is

$$\int e^{-A\bar{\psi}\psi} d\bar{\psi} d\psi = A \tag{5.22}$$

valid for arbitrary complex A. This is the prototype of a basic multiple integral of the form

$$\int \dots \int e^{-\Sigma \bar{\psi}_j A_{jk} \psi_k} \Pi d\bar{\psi}_j d\psi_j = \det(A) , \qquad (5.23)$$

where A is a complex-valued matrix and the sums and products extend from 1 to K. If A were diagonal, $A_{jk}=A_j\delta_{jk}$, the result would be ΠA_k, i.e., $\det(A)$. If A is diagonalizable under the complex linear transformation $\psi_j \to \psi'^j = \sum S_{jk}\psi_k$, $\bar{\psi}_j \to \bar{\psi}'_j = \sum \overline{S_{jk}^{-1}}\bar{\psi}_k$, then it follows from a transformation of variables that the result is $\det(A)$. In any case, even if A is nondiagonalizable, it follows by expansion of the exponent and the rules for Grassmann integration that the result is $\det(A)$.

While the treatment of fermion degrees of freedom by Grassmann variables is really quite slick one should not be seduced into believing that Grassmann variables are actually required. In view of the Jordan-Wigner [137] representation of fermion operators by spin 1/2 operators it follows that the spin-coherent states for s=1/2 discussed in Sec. 4 are also suitable to discuss fermions. In the next example as well as in Sec. 7 we will discuss alternative coherent-state treatments of fermions that make no use of anticommuting c-numbers whatsoever.

Bilinear Fermion Coherent States

A convenient set of coherent states for fermion degrees of freedom may be defined without the use of Grassmann variables or anticommuting c-numbers as follows. Let b_k, $b_k^\dagger$, $k = 1, 2, \dots, K$, denote fermion annihilation and creation operators satisfying

$$b_j b_k^\dagger + b_k^\dagger b_j = \delta_{jk} ,$$

$$b_j b_k + b_k b_j = 0 .$$

If $|0>$ denotes the normalized state for which $b_k|0> = 0$ for all k, then we choose as fiducial state the state $|\psi_o>$ defined by $|\psi_o> \equiv b_J^\dagger b_{J-1}^\dagger \cdots b_1^\dagger|0>$, where $1 \leqslant J \leqslant K-1$. As coherent states we choose Slater determinants in the Thouless form [138]

$$|z> \equiv N e^{\Sigma z_{ph} b_p^\dagger b_h} |\psi_o> ,$$

where $z = \{z_{ph}\}$ is a $(K-J) \times J$-dimensional matrix array of arbitrary independent complex coefficients, N is a normalization constant, and in the sum p (for "particle") runs from $J+1$ to K while h (for "hole") runs from 1 to J. It follows that $N = \det(1+z^\dagger z)^{-1/2}$, where $z^\dagger z$ is the $J \times J$ matrix $\{\Sigma z_{ph}^* z_{ph'}\}$, and thus

$$|z> = \det(1+z^\dagger z)^{-1/2} e^{\Sigma z_{ph} b_p^\dagger b_h} |\psi_o> . \tag{5.24}$$

The overlap of two bilinear fermion coherent states is given by

$$<z|z'> = \det(1+z^\dagger z)^{-1/2} \det(1+z'^\dagger z')^{-1/2} \det(1+z^\dagger z') ,$$

which is a jointly continuous function of its arguments. The resolution of unity for the bilinear fermion coherent states can be obtained form the geometric properties of the label space [139], which in this case is the Grassmannian manifold $U(K)/U(J) \times U(K-J)$ [140], or by an explicit calculation of the appropriate Jacobians as discussed in the papers by Monastyrsky and Perelmov [141] and Blaizot and Orland [142]. The result is that

$$\eta \int |z><z| \frac{d^2 z}{\det(1+z^\dagger z)^K} = I , \tag{5.25}$$

where $d^2 z \equiv \prod_{ph} d^2 z_{ph}/\pi$ and $\eta = 1!2!...K!/1!2!...(K-J)!1!2!...J!$ is a geometrical factor which can be computed by making use of the properties of a classical domain [143].

If $J = 1$, $K = 2$ these states are seen to be equivalent to the spin coherent states for $s = 1/2$. More generally, observe that, apart form a common factor, the bilinear fermion coherent states are analytic functions of the array z of complex variables. Thus we can expect the by-now standard consequence of analyticity to hold for these coherent states as well.

A number of properties of bilinear fermion coherent states have been worked out and applied to problems in nuclear physics in Ref. [142] and also in Ref. [144], an area well-presented by Suzuki [145]. In the paper of Mead and Papanicolaou [146] one finds an application of these coherent states in many-body theory.

Minimum-Uncertainty Coherent States

In previous sections we have commented on the fact that certain families of coherent states are minimim-uncertainty states in that they achieve equality in an uncertainty inequality among operators. Let us briefly reexamine how minimum-uncertainty states arise for the harmonic oscillator. For the classical Hamiltonian $H = \frac{1}{2}(p^2+\omega^2 q^2)$, classical solutions are given by

$$q(t) = A \sin \omega t \,,$$

$$p(t) = \omega A \cos \omega t \,. \tag{5.26}$$

Upon quantization these operators are represented by Q and P satisfying $[Q, P] = i\hbar$. If $<(\cdot)> \equiv <\psi|(\cdot)|\psi>$ and $\Delta Q \equiv Q - <Q>$, etc., then the uncertainty relation $<(\Delta Q)^2><(\Delta P)^2> \geqslant \hbar^2/4$ attains equality for only those states — the minimum-uncertainty states — for which

$$(rQ + iP)|\psi> = (r<Q> + i<P>)|\psi> \tag{5.27}$$

where $r > 0$. Note that

$$\mathcal{H} = \tfrac{1}{2}(rQ - iP)(rQ + iP) = \tfrac{1}{2}\{(P^2 + r^2 Q^2) - \hbar r\} \,,$$

and if $r = \omega$, then $\mathcal{H}$ may be chosen as the quantum Hamiltonian for this oscillator. The state $|0>$, for which $(\omega Q + iP)|0> = 0$, satisfies (5.27), and is at the same time the oscillator ground state. Thus the minimum-uncertainty coherent states for this system are conveniently taken as the eigenvectors of the relation

$$(\omega Q + iP)|\beta\rangle = \beta|\beta\rangle ,$$

where we have labelled the state by the complex eigenvalue $\beta = \omega\langle Q\rangle + i\langle P\rangle$. Of course, these states are just the canonical coherent states of Sec. 3; recall that $a|z\rangle = z|z\rangle$, and that all canonical coherent states are minimum-uncertainty states.

Suppose we now consider a general classical Hamiltonian of the form $H = \tfrac{1}{2}p^2+V(q)$, restricted to have periodic orbits, at least for a certain range of energies. Let $q(t) = F(E, t)$ denote a periodic solution of the classical equations of motion, where E denotes the energy of the solution. As a periodic function of time we can find an energy-dependent transformation $\bar{q} = G(q, E) \equiv G(q)$ such that $\bar{q}(t) = A(E)\sin[\omega(E)t]$. Defining $\bar{p}(t) \equiv d\bar{q}(t)/dt = G'(q)dq/dt = G'(q)p$ it follows that $\bar{p}(t) = \omega(E)A(E)\cos[\omega(E)t]$. These solutions satisfy $\bar{p}^2+\omega^2(E)\bar{q}^2 = \omega^2(E)A^2(E) \equiv T(E)$. On quantization, $\bar{q} \to \bar{Q} \equiv G(Q,\mathcal{H})$ (with a choice being made for factor ordering ambiguity if necessary) and $\bar{p} \to \bar{P} \equiv \tfrac{1}{2}[G'(Q,\mathcal{H})P+PG'(Q,\mathcal{H})]$. The uncertainty relation for these operators reads

$$|\langle[\bar{P},\bar{Q}]\rangle|^2 = |\langle[\Delta\bar{P},\Delta\bar{Q}]\rangle|^2$$

$$\leqslant 4|\langle\Delta\bar{P}\Delta\bar{Q}\rangle|^2 \leqslant 4\langle(\Delta\bar{P})^2\rangle\langle(\Delta\bar{Q})^2\rangle .$$

Equality in this uncertainty relation is attained only for those $|\psi\rangle$ for which

$$(r\bar{Q} + i\bar{P})|\psi\rangle = (r\langle\bar{Q}\rangle + i\langle\bar{P}\rangle)|\psi\rangle , \qquad (5.28)$$

where $r \equiv \langle i[\bar{P}, \bar{Q}]\rangle/2\langle(\Delta\bar{Q})^2\rangle$ is real. This leads to a three (real) parameter solution, viz., r, $\langle\bar{Q}\rangle$, and $\langle\bar{P}\rangle$. To eliminate one parameter consider

$$\mathcal{T} \equiv (r\bar{Q} - i\bar{P})(r\bar{Q} + i\bar{P}) = (\bar{P}^2 + r^2\bar{Q}^2) - ir[\bar{P}, \bar{Q}] .$$

Since the commutator is $O(\hbar)$ we can, to leading order in $\hbar$, relate $\mathcal{T}$ to the classical expression $T = \bar{p}^2+\omega^2\bar{q}^2$ provided we identify r and ω. Since $\omega = \omega(E)$ a choice for E must be made, and it proves convenient to let $r = r_o \equiv \omega(E_o)$, where E_o is the ground-state energy of the quantum Hamiltonian $\mathcal{H}$. The function $T = T(E)$ is evidently a monotonic (increasing) function related to the classical Hamiltonian, and at least to leading order in $\hbar$, we may expect that $\mathcal{T}= \mathcal{T}(\mathcal{H})$ is a rather similar function and also is monotonic increasing. Since $\mathcal{T} \geqslant 0$ it achieves its minimum for the state $|0>$ which satisfies $(r_o\bar{Q}+i\bar{P})|0> = 0$, and therefore it is plausible that $|\psi_o> = |0>$, where $|\psi_o>$ is the ground state of $\mathcal{H}$, again at least to order $\hbar$. In their studies of these types of coherent states, Nieto and his collaborators [147,148] have shown that $|\psi_o> = |0>$ to leading order in a WKB approximation, in analogy to our discussion, and even, in a handful of analytically soluble examples, that $|\psi_o> = |0>$ exactly (to all orders in $\hbar$).

Following Nieto, we define the minimum-uncertainty coherent states as the eigensolutions of the equation

$$(r_o\bar{Q} + i\bar{P})|\beta> = \beta|\beta> , \tag{5.29}$$

where we have labelled the solution by the complex eigenvalue $\beta = r_o<\bar{Q}> + i<\bar{P}>$. These states may be assumed to be normalized. However, it is not immediately clear what is the allowed range of β values, nor as β ranges over its domain whether the states $\{|\beta>\}$ span the Hilbert space. For most sets of coherent states completeness is assured by the existence of the resolution of unity. In the present case, not only may the states $\{|\beta>\}$ be incomplete, but even if they are complete it is not certain that a resolution of unity of the standard form $\int|\beta><\beta|\rho(\beta)d^2\beta$, $\rho(\beta) \geqslant 0$, exists. On the other hand, it is certainly plausible that the states $\{|\beta>\}$ are complete, and a standard resolution of unity even exists in two special cases. For our purposes we will assume that the minimum-uncertainty coherent states are complete, and thus at least satisfy the second form of the completeness criterion of Sec. 2. In this sense the states defined above qualify as coherent states as we have defined the term.

It should be clear that the minimum-uncertainty coherent states are defined in a rather different fashion from those defined previously, and generally can be expected to yield new sets of coherent states. These states have the virtue of relating to a specific dynamical system, which in principle can be chosen quite generally. Nieto and his coworkers have introduced and championed this general approach to define general classes of minimum-uncertainty coherent states, and they have analytically and/or numerically solved several examples and made comparison with other definitions of coherent states. As a representative sample of this work we include the paper by Nieto [148], in which references to earlier work can be found. In a recent paper by Nieto [149] a connection between special minimum-uncertainty coherent states and conserved-charge coherent states is discussed.

6. COHERENT-STATE PATH INTEGRALS

Ever since Feynman [150] introduced path integrals into quantum theory they have come to play an ever-increasing role in nearly all quantum problems. Path-integral formulations in terms of coherent states follow one of two general patterns quite independently of which set of coherent states is being considered. Consequently for the purposes of the present section we revert to the generic notation of Sec. 2 assuming that we deal with a set of coherent states $\{|\ell\rangle\}$ and a positive measure $\delta\ell$ for which $\int |\ell\rangle\langle\ell|\delta\ell = I$. Without any significant loss of generality we assume that $\delta\ell$ has been chosen so that $\| |\ell\rangle \| = 1$ for all $\ell \in \mathfrak{L}$.

Path Integral — First Form

The principal object represented by a path integral is the propagator, the matrix element of the evolution operator. If the self-adjoint operator $\mathcal{H}$ denotes a time-independent Hamiltonian, then $\exp(-it\mathcal{H}/\hbar)$ is the evolution operator, and $<\ell'', t''|\ell', t'> \equiv <\ell''|\exp[-i(t''-t')\mathcal{H}/\hbar]|\ell'>$ is the propagator from t' to t''. We initially restrict our attention to Hamiltonians that satisfy

$$||\mathcal{H}^N|\ell>||^2 \leqslant ab^N(2N!) \tag{6.1}$$

for some positive (possibly ℓ dependent) constants a and b. It follows in this case that ($T \equiv t'' - t'$)

$$<\ell''|\exp(-iT\mathcal{H}/\hbar)|\ell'>$$

$$= \lim_{N\to\infty} <\ell''|[1-iT\mathcal{H}/(N+1)\hbar]^{(N+1)}|\ell'> . \tag{6.2}$$

If we now insert the coherent-state resolution of unity between each factor we find that

$$<\ell''|\exp(-iT\mathcal{H}/\hbar)|\ell'>$$

$$= \lim_{N\to\infty} \int\ldots\int \prod_{n=0}^{N} <\ell_{n+1}|(1-i\epsilon\mathcal{H}/\hbar)|\ell_n> \prod_{n=1}^{N} \delta\ell_n$$

$$\equiv \lim_{N\to\infty} \int\ldots\int \prod_{n=0}^{N} <\ell_{n+1}|\ell_n> [1-i(\epsilon/\hbar)H(\ell_{n+1},\ell_n)] \prod_{n=1}^{N} \delta\ell_n \tag{6.3}$$

where we have introduced $\epsilon \equiv T/(N+1)$, $\ell_{N+1} \equiv \ell''$, $\ell_o \equiv \ell'$, and $H(\ell,\ell') \equiv <\ell|\mathcal{H}|\ell'> / <\ell|\ell'>$. Equation (6.3) represents a valid (discrete form of) path-integral representation of the propagator. Provided the indicated integrals exist (and for some $\mathcal{H}$ they may not!) we may replace (6.3) by the more suggestive expression

$$<\ell''|\exp(-iT\mathcal{H}/\hbar)|\ell'>$$

$$= \lim_{N\to\infty} \int\ldots\int \prod_{n=0}^{N} <\ell_{n+1}|\ell_n> e^{-i(\epsilon/\hbar)H(\ell_{n+1},\ell_n)} \prod_{n=1}^{N} \delta\ell_n , \tag{6.4}$$

which is the form typically encountered in the literature.

We now return to the bound (6.1) which was necessary to ensure that (6.2) held true. However, for some problems the desired self-adjoint Hamiltonian $\mathscr{H}$ may not satisfy (6.1), although we shall still require that $\|\mathscr{H}|\ell>\| < \infty$ for all ℓ. For an alternative approach let us introduce a regularized form of $\mathscr{H}$ given, for example, by

$$\mathscr{H}_\epsilon \equiv \mathscr{H}/(1+\epsilon\mathscr{H}^2), \quad \epsilon > 0\,,$$

which on the domain of $\mathscr{H}$ has the property, as $\epsilon \rightarrow 0$, that s–lim $\mathscr{H}_\epsilon = \mathscr{H}$ ("s" means strong), and on the whole Hilbert space that

$$\text{s-}\lim_{N\rightarrow\infty} \left[1-\frac{i}{\hbar}\left(\frac{T}{N+1}\right)\mathscr{H}_{T/(N+1)}\right]^{(N+1)} = e^{-iT\mathscr{H}/\hbar}\,.$$

Consequently, with the same notation as before, we have

$$\begin{aligned}
&<\ell''|\exp(-iT\mathscr{H}/\hbar)|\ell'> \\
&= \lim_{N\rightarrow\infty} <\ell''|[1-i\epsilon\mathscr{H}_\epsilon/\hbar]^{(N+1)}|\ell'> \\
&= \lim_{N\rightarrow\infty} \int\!...\!\int \prod_{n=0}^{N} <\ell_{n+1}|(1-i\epsilon\mathscr{H}_\epsilon/\hbar)|\ell_n> \prod_{n=1}^{N} \delta\ell_n \\
&\equiv \lim_{N\rightarrow\infty} \int\!...\!\int \prod_{n=0}^{N} <\ell_{n+1}|\ell_n>[1-i(\epsilon/\hbar)H_\epsilon(\ell_{n+1},\ell_n)] \prod_{n=1}^{N} \delta\ell_n
\end{aligned} \tag{6.5}$$

where $H_\epsilon(\ell,\ell') \equiv <\ell|\mathscr{H}_\epsilon|\ell'>/<\ell|\ell'>$. Equation (6.5) represents a valid (discrete form of) path-integral representation of the propagator suitable for a general Hamiltonian. As before, provided the indicated integrals exist, (6.5) may be replaced by the expression

$$<\ell''|\exp(-iT\mathcal{H}/\hbar)|\ell'>$$

$$= \lim_{N\to\infty} \int\cdots\int \prod_{n=0}^{N} <\ell_{n+1}|\ell_n> e^{-i(\epsilon/\hbar)H_\epsilon(\ell_{n+1},\ell_n)} \prod_{n=1}^{N} \delta\ell_n \,. \quad (6.6)$$

Formal Path Integral

Although it is utterly unacceptable, the temptation is nonetheless irresistible to interchange the operations of integration and the limit $N\to\infty$ in (6.3) or (6.5). As $\epsilon\to 0$ one can imagine that the set of points ℓ_n, $n = 0, 1, 2,\ldots, N+1$, defines in the limit a (possibly generalized) function $\ell(t)$, $t' \leqslant t \leqslant t''$. In the expression obtained by interchanging these operations it is conventional to adopt for the integrand the form it assumes for *continuous and differentiable paths* $\ell(t)$. To help identify that expression first observe that

$$<\ell_{n+1}|\ell_n> = 1 - <\ell_{n+1}|(|\ell_{n+1}> - |\ell_n>)$$

$$\cong \exp[\,-<\ell_{n+1}|(|\ell_{n+1}> - |\ell_n>)]\,.$$

Therefore, as $\epsilon\to 0$ the limiting form of the integrand for continuous and differentiable paths is correctly given by

$$\exp\left[-\int_{\ell'}^{\ell''} <\ell(t)|d\ell(t)> - (i/\hbar)\int_{t'}^{t''} H(\ell(t))\,dt\right],$$

where we have introduced

$$H(\ell) \equiv H(\ell,\ell) = <\ell|\mathcal{H}|\ell>$$

and the coherent-state differential,

$$|d\ell(t)> \equiv d|\ell(t)> \equiv |\ell(t)+d\ell(t)> - |\ell(t)>\,.$$

Hence in the ubiquitously used abuse of language common in path integrals we are led to the first formal coherent-state path-integral expression

$$<\ell'',t''|\ell',t'> = \int \exp\{(i/\hbar)\int_{t'}^{t''} [i\hbar<\ell(t)|\dot{\ell}(t)> - H(\ell(t))]dt\}\ \mathcal{D}\ell \tag{6.7}$$

where we have introduced

$$|\dot{\ell}(t)> \equiv (d/dt)|\ell(t)> ,$$

$$\mathcal{D}\ell \equiv \prod_t \delta\ell(t) \equiv \lim_{N\to\infty} \prod_{n=1}^{N} \delta\ell_n .$$

Let us next examine the merits of (6.7).

What's Right with (6.7)

Although (6.7) may only be formal it does offer intuitive insight and even limited calculational possibilities. From the general path-integral prescription it is clear that

$$S \equiv \int_{t'}^{t''} [i\hbar<\ell(t)|\dot{\ell}(t)> - H(\ell(t))]dt \tag{6.8}$$

assumes the role of the classical action in this expression. Note that Planck's constant $\hbar$ is not set equal to zero in the classical action. The first term is strictly real as follows from the fact that $<\ell|\ell> \equiv 1$, and thus $<\dot{\ell}|\ell> + <\ell|\dot{\ell}> = 0$; hence $(i\hbar<\ell|\dot{\ell}>)* = -i\hbar<\dot{\ell}|\ell> = i\hbar<\ell|\dot{\ell}>$. This term necessarily vanishes if the coherent-state overlap $<\ell|\ell'>$ is real. We therefore assume that $<\ell|\ell'>$ is complex and that $i\hbar<\ell|\dot{\ell}>$ does not vanish, which is a selection of favorable coherent states for this purpose. The part of $i\hbar<\ell|\dot{\ell}>$ which is a total time derivative, as contributed, e.g., by phase factors, may be ignored as far as the classical action is concerned, and so we further assume that $i\hbar<\ell|\dot{\ell}>$ is a function of ℓ, linear in first-order time derivatives that is not (simply) a total time derivative. For example, by this argument, ℓ cannot be a scalar but must have at least two components. The linear, first-order time derivative means that the action is expressed in so-called first-order form, implying further that the variables ℓ are *phase-space variables* (at least that subset which labels distinct coherent states), and that $\mathfrak{L}$ (or a subspace) is *phase space*. For example, for the canonical coherent states of Sec. 3 it follows that $i\hbar\langle\ell|\dot{\ell}\rangle = \frac{1}{2}(p\dot{q}-q\dot{p})$. Evidently, the real expression $H(\ell) \equiv \langle\ell|\mathcal{H}|\ell\rangle$ assumes the role of the classical Hamiltonian. This interpretation, which has been referred to as the "weak correspondence principle" [151], has been adopted and exploited by many authors. Within a classical context it is, of course, completely appropriate that the paths $\ell(t)$ are continuous and differentiable. Moreover for numerous examples the explicit form of S reduces to the usual phase-space action functional (cf., Sec. 7).

It is interesting to note that the *classical* action (6.8), for which stationary variations lead to the classical equations of motion, is actually identical in form to the *quantum* action

$$S = \int [i\hbar \langle\psi(t)|\dot{\psi}(t)\rangle - \langle\psi(t)|\mathcal{H}|\psi(t)\rangle]dt \,,$$

for which stationary variations lead to the Schrödinger equation. The difference lies in the allowed range of vectors: For the classical action variations are made among only the set of coherent states $\{|\ell(t)\rangle\}$, while for the quantum action variations are made among the set of all vectors $\{|\psi(t)\rangle\}$, or equivalently the set of all unit vectors $\{|\psi(t)\rangle\}$ modulo simple phase changes. For further discussion of this kind of relation between classical and quantum theories see Refs. [89] and [142].

One common approximation scheme of evaluating integrals, including path integrals, is the stationary-phase approximation. At first sight this method can also be applied to (6.7); however, on second glance, trouble arises in that the boundary conditions

$\ell(t'') = \ell''$ and $\ell(t') = \ell'$ are generally incompatible with the *first-order* set of equations of motion that arise from (6.8). How this paradox is overcome is an important feature of coherent-state path integrals (see below).

What's Wrong with (6.7)

It is appropriate to ask why the interchange of limits is not justified and why therefore (6.7) is *not* an integral and remains only a formal expression. For (6.7) to be an integral $\mathscr{D}\ell$ should be a measure, which it decidedly is *not* whenever $\dim(\mathfrak{H}) = Tr(I) = \int \delta\ell > 1$. To convert it to a measure it must be associated with a positive weight factor to render its volume finite, say unity. For example, for $1 < d \equiv \int \delta\ell < \infty$ it follows that $D\ell \equiv \lim_{N\to\infty} d^{-N} \prod_{n=1}^{N} \delta\ell_n$ leads to a measure. Superficially, the fact that

$$<\ell''|\ell'> = \lim_{N\to\infty} \int \dots \int \prod_{n=0}^{N} <\ell_{n+1}|\ell_n> \prod_{n=1}^{N} \delta\ell_n$$

suggests that incorporating the indicated weight would also lead to a measure. However, that notion is incorrect here since the inescapable divergence,

$$\int \dots \int \prod_{n=0}^{N} |<\ell_{n+1}|\ell_n>| \prod_{n=1}^{N} \delta\ell_n \to \infty \quad \text{as } N \to \infty ,$$

implies in the limit that the finite result $<\ell''|\ell'>$ arises from a cancellation among infinite terms. The absence of a measure is not merely a technicality, it means that the coherent-state path integral is *not* an integral; instead, it is a linear functional. The difference is significant since analytical tools appropriate for integrals (e.g., dominated convergence theorem, etc.) are not applicable to linear functionals. Indeed, similar remarks apply to *all* common forms of quantum-mechanical path integrals, not only the coherent-state variety.

In discussions of formal expressions such as (6.7) one often encounters the question: What is the nature of the paths? Strictly speaking, this is a meaningless question for it presumes the existence of a path-space integral which in actual fact does not exist! (It is like asking your father to describe the experience of child birth.) Nevertheless the question is persistently asked and our dismissal will not be accepted by all. To provide *an* answer (of sorts) to this question let us specialize to a case where $1 < d = \int \delta\ell < \infty$ and set $\mathscr{H} = 0$. Then it follows that

$$d = \lim_{N\to\infty} \int \ldots \int \prod_{n=0}^{N} <\ell_{n+1}|\ell_n> \prod_{n=0}^{N} \delta\ell_n \,,$$

where now $\ell_{N+1} \equiv \ell_0$ and we have integrated over ℓ_0 as well. Now consider the expression

$$c_m \equiv \lim_{N\to\infty} \int \ldots \int <\ell_{m+1}|\ell_m> \prod_{n=0}^{N} <\ell_{n+1}|\ell_n> \prod_{n=0}^{N} \delta\ell_n \,,$$

which due to the cyclic nature of the integrand is independent of m and may simply be evaluated as

$$c_m \equiv c \equiv \int\int <\ell_2|\ell_1>|<\ell_2|\ell_1>|^2 \, \delta\ell_2\delta\ell_1 \,.$$

Since d may be expressed as

$$d = \int\int |<\ell_2|\ell_1>|^2 \, \delta\ell_2\delta\ell_1$$

it follows that $|c/d| < 1$. Consequently, in an *average sense,* $\lim_{N\to\infty} \frac{1}{N} \sum_{n=1}^{N} <\ell_{n+1}|\ell_n> = \frac{c}{d}$, which shows that an average "path" $\ell(t)$ — far from being continuous and differentiable — suffers discontinuities everywhere such that $<\ell_+|\ell_-> \approx c/d$.

Finally, if we observe that the average "paths" are nowhere continuous and differentiable, then another formal aspect of (6.7) becomes apparent in that the form adopted for the integrand is wholly unrepresentative! In other words, although (6.8) represents the classical action for the problem at hand it is not altogether evident how that expression should be extended and defined for the unruly "paths" required in quantization via the path integral. The inherent ambiguity in this extension and some of its consequences for the kind of statistics that emerge upon quantization are discussed in Refs. [20,152].

Additional Remarks and Introduction to Selected Reprints

1. The first use of coherent states for the construction of quantum-mechanical path integrals was in Ref. [20]. The inherent limitations of (6.4) and the preference for (6.3) or another related form was first recognized by Ciafaloni and Onofri [153], and is clearly presented by Onofri [154].

2. The application of coherent-state path integrals to problems of nuclear physics, and particularly an emphasis of the equivalence of the Hartree and Hartree-Fock equations with those classical equations of motion that arise from demanding stationarity of the classical action (6.8), is well treated in the article by Kuratsuji and Suzuki [155], where additional references can be found. A similar treatment is given in the work of Blaizot and Orland [142].

3. It was remarked earlier that the classical, first-order equations of motion that arise from the action (6.8) are generally incompatible with the required boundary conditions $\ell(t'') = \ell''$ and $\ell(t') = \ell'$. Just how this situation is overcome, and more precisely, what is the nature of the stationary-phase approximation in typical coherent-state path integrals is spelled out by Klauder [156]. An analogous discussion for canonical coherent states appears in the elegant and widely-quoted review of Faddeev [157]. For a related discussion see Ref. [158].

4. In the paper by Hillery and Zubairy [159] coherent-state path integrals are applied in evaluating the propagator for single-mode and multi-mode Hamiltonians which are at most quadratic in the field operators. Langevin equations for canonical quantum systems are derived in terms of a coherent-state path-integral representation by Klauder [160], where the results are applied in a study of the quantized Hall effect. In the paper by Jevicki and Papanicolaou [161], a coherent-state path integral formalism is used in order to study the continuous Heisenberg spin chain.

5. Coherent states for Fermi operators are applied in the paper by Ohnuki and Kashiwa [136] in order to construct a path-integral representation for Fermi fields. Such a formalism has been applied to various problems such as in a recent path-integral form of the Wilson-loop in gauge theories [162] or in supersymmetric quantum mechanics as discussed in Ref. [163].

Path Integral — Second Form

In deriving the second form for the path integral we need to assume the existence of a diagonal representation for the self-adjoint Hamiltonian given by

$$\mathscr{H} = \int h(\ell)\, |\ell\rangle \langle\ell|\, \delta\ell \tag{6.9}$$

for some real function $h(\ell)$. Such a representation is not required to hold for a general operator nor is it even required that $h(\ell)$ be uniquely determined by $\mathscr{H}$, although in some situations, as we have seen in earlier sections, both these conditions hold true. We also impose the requirement that

$$\int |h(\ell_1)h(\ell_2)\langle\ell|\ell_1\rangle \langle\ell_1|\ell_2\rangle \langle\ell_2|\ell\rangle|\delta\ell_1\delta\ell_2 < \infty \tag{6.10}$$

for all ℓ, which is slightly stronger than the condition $\| \mathscr{H}|\ell\rangle \|^2 < \infty$. Now consider the operator given for all real t by

$$\mathscr{F}(t) \equiv \int e^{-ith(\ell)/\hbar} |\ell\rangle \langle\ell|\delta\ell \,.$$

It follows that $|\langle\phi|\mathscr{F}(t)|\psi\rangle| \leqslant \| |\phi\rangle \| \cdot \| |\psi\rangle \|$ for any $|\phi\rangle$ and $|\psi\rangle$, and as $t \to 0$ that $(\hbar/it)[I - \mathscr{F}(t)]$ converges strongly to $\mathscr{H}$ on the coherent states. By a theorem of Chernoff [164] these conditions are sufficient to ensure as $N \to \infty$ that

$$\text{s-lim}[\mathscr{F}(T/N)]^N = e^{-iT\mathscr{H}/\hbar}$$

on all of $\mathfrak{H}$. In application we find that

$$\langle\ell''|\exp(-iT\mathscr{H}/\hbar)|\ell'\rangle$$

$$= \lim_{N\to\infty} \int\ldots\int \prod_{n=0}^{N} \langle\ell_{n+1}|\ell_n\rangle \prod_{n=1}^{N} e^{-i\epsilon h(\ell_n)/\hbar}\, \delta\ell_n \,, \qquad (6.11)$$

where here $\epsilon \equiv T/N$, $\ell_{N+1} \equiv \ell''$ and $\ell_0 \equiv \ell'$. Equation (6.11) represents a valid (discrete form of) path-integral representation of the propagator.

Once again we succumb to the irresistible urge to interchange the order of integration and the limit $N \to \infty$. As before we adopt for the integrand the form it assumes for continuous and differentiable paths. As a result we are led to the second formal coherent-state path-integral expression given by

$$\langle\ell'', t''|\ell', t'\rangle = \int \exp\{(i/\hbar) \int_{t'}^{t''} [i\hbar\langle\ell(t)|\dot{\ell}(t)\rangle - h(\ell(t))]dt\}\, \mathscr{D}\ell \qquad (6.12)$$

where the notation is the same as that used in (6.7). In (6.7) and (6.12) we have two presumably identical yet different expressions for the same quantity, a fact which further underscores the formal character of such path integrals. That these expressions are different follows from the relation

$$H(\ell) = \int h(\ell')|<\ell'|\ell>|^2 \delta\ell' \tag{6.13}$$

showing that $H(\ell)$ and $h(\ell)$ are generally not identical. Of course, the remarks (both positive and negative) made earlier regarding the formal expression (6.7) hold with equal validity for the formal expression (6.12).

On the basis of (6.12) it would seem that

$$S' \equiv \int_{t'}^{t''} [i\hbar<\ell(t)|\dot{\ell}(t)> - h(\ell(t))]dt \tag{6.14}$$

assumes the role of the classical action, which seems to contradict (6.8). In reality this conflict is no problem since (6.7) and 6.12) are both formal expressions. However, in several cases, especially in the case of the canonical coherent states of Sec. 3, it happens that

$$H(\ell) = H_c(\ell) + O(\hbar),$$

$$h(\ell) = H_c(\ell) + O(\hbar), \tag{6.15}$$

where $H_c(\ell)$ denotes the usual classical Hamiltonian. In other words, the two classical actions are identical apart from terms of order $\hbar$ and coincide in the ultimate classical limit in which $\hbar \to 0$.

Statistical Mechanics

Let us recast the second form of coherent-state path integral into a form suitable for quantum statistical mechanics problems. Assume that $\mathscr{H}$ is bounded below and that this bound has been adjusted so that $h(\ell) \geqslant 0$. If, in the present case, we define

$$\mathscr{G}(t) \equiv \int e^{-th(\ell)} \, |\ell><\ell|\delta\ell$$

for all $t \geqslant 0$, and impose similar conditions as before, it follows as $N \to \infty$ that

$$\text{s-lim}[\mathscr{G}(\beta/N)]^N = e^{-\beta\mathscr{H}} .$$

In application this relation becomes

$$\mathrm{Tr}(e^{-\beta\mathcal{H}}) = \lim_{N\to\infty} \int \ldots \int \prod_{n=1}^{N} <\ell_{n+1}|\ell_n> e^{-\epsilon h(\ell_n)} \delta\ell_n \,, \quad (6.16)$$

where $\epsilon \equiv \beta/N$, $\ell_{N+1} \equiv \ell_1$ and all ℓ-variables are integrated out. The result is a valid (discrete form of) path-integral representation of the partition function.

The orders of operation may formally be exchanged in such an expression as well leading to the formal path integral

$$\mathrm{Tr}(e^{-\beta\mathcal{H}}) = \int e^{-\int_0^\beta [<\ell(t)|\dot{\ell}(t)> + h(\ell(t))]dt} \mathcal{D}\ell \,.$$

Similar remarks apply to this formal path integral as those made earlier.

In another direction, closely following work of Lieb, we can arrange to put rigorous upper and lower bounds on the partition function as follows. First recall Jensen's inequality which states for any convex function $\phi(x)$, $x \in \mathbb{R}$, that $<\phi(X)> \geqslant \phi(<X>)$, where X denotes a real random variable and $<\cdot>$ an average over the distribution of X. An important application arises for $\phi(x) = \exp(-x)$, in which case

$$<e^{-X}> \geqslant e^{-<X>} \,. \quad (6.17)$$

Let $|p>$ and μ_p, p=1, 2,..., be a complete set of eigenvectors and eigenvalues for $\mathcal{H}$: $\mathcal{H}|p> = \mu_p|p>$. Then from (6.17) it follows that

$$<\ell|e^{-\beta\mathscr{H}}|\ell> = \sum_{p=1}^{\infty} e^{-\beta\mu_p}|<\ell|p>|^2$$

$$\geqslant \exp\left(-\beta\sum_{p=1}^{\infty}\mu_p|<\ell|p>|^2\right)$$

$$= \exp(-\beta<\ell|\mathscr{H}|\ell>) ,$$

which immediately leads to the inequality

$$\mathrm{Tr}(e^{-\beta\mathscr{H}}) = \int <\ell|e^{-\beta\mathscr{H}}|\ell> \delta\ell$$

$$\geqslant \int e^{-\beta H(\ell)}\delta\ell .$$

On the other hand,

$$<p|e^{-\beta\mathscr{H}}|p> = \exp(-\beta<p|\mathscr{H}|p>)$$

$$= \exp[-\beta\int h(\ell)|<\ell|p>|^2\delta\ell]$$

$$\leqslant \int e^{-\beta h(\ell)}|<\ell|p>|^2\delta\ell .$$

By summing over p we arrive at the relation

$$\mathrm{Tr}(e^{-\beta\mathscr{H}}) \leqslant \int e^{-\beta h(\ell)}\delta\ell ,$$

and in summary we have determined that

$$\int e^{-\beta H(\ell)}\delta\ell \leqslant \mathrm{Tr}(e^{-\beta\mathscr{H}}) \leqslant \int e^{-\beta h(\ell)}\delta\ell . \tag{6.18}$$

In addition, in those cases where (6.15) holds true the convergence of the upper and lower bounds establishes that

$$\lim_{\hbar\to 0} \mathrm{Tr}(e^{-\beta\mathscr{H}}) = \int e^{-\beta H_c(\ell)}\delta\ell .$$

Additional Remarks and
Introduction to Selected Reprints

1. The application of the diagonal representation of operators (6.9) to the construction of path integrals first appeared in the work of Berezin [165] and Lieb [81]. They both established the inequalities in (6.18), and in particular, derived the upper bound by a path-integral representation much like (6.16). Our derivation of the upper bound follows unpublished work of Lieb, as quoted in a paper by Simon [75] who generalized Lieb's argument from the case of spin coherent-states to more general classes of coherent-states based on compact groups for which diagonal representations of operators generally occur. The Berezin-Lieb inequalities, and the existence of the classical limit that follows therefrom, were generalized to other groups in Refs. [166,167,75]. These inequalities were applied by Shankar [168] to a study of the Lipkin-Meshkov-Glick Hamiltonian.

The importance of the diagonal representation of operators for the construction of quantum-mechanical path integrals was stressed by Berezin [165], Ciafaloni and Onofri [153], and Onofri [154].

2. The question frequently arises whether quantum-mechanical path integrals can, in some way, be represented as genuine path integrals with well-defined, countably additive measures. The nearest attainment of this goal has been given by Klauder and Daubechies [169] and it makes full use of the coherent state machinery as well as the representation of the Hamiltonian by the diagonal representation (6.9).

7. COHERENT STATES FOR FIELDS

The extension of coherent states from a finite number of degrees of freedom to an infinite number of degrees of freedom is basically straightforward — yet it also contains some fascinating aspects regarding field-operator representations. In this section we shall discuss some of the issues that arise for two of the most important cases: scalar fields and fermion fields. As a preliminary we discuss the case of an infinite number of canonical degrees of freedom each of which is modelled after the discussion in Sec. 3.

Fock Representation for Bosons

Let a_k, $a_k^\dagger$, $k=1, 2, \ldots$, denote an infinite set of annihilation and creation operators, satisfying the canonical commutation relation $[a_k, a_\ell^\dagger]=\delta_{k\ell}I$. Denote the no-particle state by the normalized vector $|0\rangle$, where $a_k|0\rangle=0$ for all k, and define the Hilbert space $\mathfrak{H}$ by the closed linear span of vectors of the form $a_{k_1}^\dagger \ldots a_{k_N}^\dagger|0\rangle$ for all $k_1, \ldots, k_N$, where $N = 0, 1, 2, \ldots$. The so-defined construction generates the justly-famous Fock representation of the canonical commutation relations.

Now let z_k, $k = 1, 2, \ldots$, denote an arbitrary *square-summable* sequence of complex coefficients, $\sum|z_k|^2<\infty$. This, and all sums in this section, unless stated otherwise, extend from 1 to ∞. It is noteworthy that the sequence $\{z_k\}$ defines an element in an infinite-dimensional Hilbert space $\mathfrak{Z}$ realized here as ℓ^2. To each element $\{z_k\}\in\mathfrak{Z}$ we associate a unit vector $|\{z_k\}\rangle\in\mathfrak{H}$ according to the rule

$$|\{z_k\}\rangle = e^{-\frac{1}{2}\sum|z_k|^2}\, e^{\sum z_k a_k^\dagger}|0\rangle\,; \tag{7.1}$$

these are the appropriate canonical coherent states. Alternatively formulated these states are represented by an infinite direct product in the form

$$|\{z_k\}\rangle = \bigotimes_{k=1}^{\infty} e^{-\frac{1}{2}|z_k|^2}\, e^{z_k a_k^\dagger}|0\rangle_k\,.$$

In either form it follows that the overlap of two canonical coherent states is just the infinite product over the single degree-of-freedom overlap, i.e.,

$$<\{z_k\}|\{z'_k\}> = \exp[-½\sum|z_k|^2 + \sum z_k^* z'_k - ½\sum|z'_k|^2] ,$$

an expression which never vanishes. Continuity of the canonical coherent states in the labels $\{z_k\}$ is a direct consequence of the fact that

$$\| \,|\{z_k\}> - |\{z'_k\}>\|^2 \leqslant 2\sqrt{2}\,(\|\{z_k\}\| + \|\{z'_k\}\|)\|\{z_k - z'_k\}\| ,$$

where $\|\{z_k\}\|^2 \equiv \sum|z_k|^2$. Stated otherwise we see that the coherent-state overlap is a continuous function of its labels in the Hilbert space topology of $\mathfrak{Z}$. The example under discussion is one of the simplest illustrations of a so-called exponential Hilbert space [170], a concept of remarkable generality.

For an arbitrary $|\psi> \in \mathfrak{H}$, we set

$$\psi(\{z_k\}) \equiv <\{z_k\}|\psi>$$

and take up the question of an integral representation of the inner product in $\mathfrak{H}$. Based on the discussion in Sec. 3 it is reasonable and straightforward to show [2] that

$$<\psi|\psi> = \lim_{N\to\infty} \int |\psi(\{z_k^N\})|^2 \prod_{k=1}^{N} (d^2 z_k/\pi)$$

$$\equiv \int |\psi(\{z_k\})|^2 \, d\sigma(\{z_k\}) , \qquad (7.2)$$

where $z_k^N \equiv z_k$ for $k \leqslant N$ and $z_k^N \equiv 0$ for $k > N$. This formula is entirely suitable to define an inner product, but, strictly speaking, it is not an integral representation (σ is not a true measure). For the most part this fine point is ignored and one adopts (7.2) as an integral representation. However, it is noteworthy that a true integral representation does exist for the closely associated Bargmann-Segal [21,22] functional representatives in which the normalization factor is included in the measure. In that case define

$$\psi(\{z_k\}) \equiv e^{-\frac{1}{2}\sum|z_k|^2} f(\{z_k^*\}) ,$$

where f is an entire function of infinitely many complex variables constrained so that $|\psi(\{z_k\})| \leqslant \| \, |\psi> \|$. The expression

$$d\mu(\{z_k\}) = \prod_{k=1}^{\infty} e^{-|z_k|^2} (d^2 z_k/\pi)$$

formally defines a true measure on $\mathbb{R}^\infty$, one that is more properly represented by its characteristic functional

$$C(\{s_k\}) \equiv \int e^{\sum(z_k{}^* s_k - s_k{}^* z_k)} d\mu(\{z_k\})$$

$$= e^{-\sum|s_k|^2} ,$$

which is well-defined for every $\{s_k\}$ such that $\sum k^2 |s_k|^2 < \infty$ [171]. In this notation it follows that

$$<\psi|\psi> = \int |f(\{z_k{}^*\})|^2 \, d\mu(\{z_k\}) ,$$

which provides a genuine integral representation for the infinite degree of freedom case.

We note further that the canonical coherent states are eigenstates of the annihilation operators,

$$a_k|\{z_k\}> = z_k|\{z_k\}>$$

for all k. Diagonal coherent-state matrix elements determine the operator, e.g.,

$$<\{z_k\}|\mathscr{B}(\{a_\ell^\dagger\},\{a_m\})|\{z_k\}> = B(\{z_\ell^*\}, \{z_m\})$$

implies that

$$\mathscr{B}(\{a_\ell^\dagger\}, \{a_k\}) = :B(\{a_\ell^\dagger\}, \{a_m\}): .$$

All operators are limits of those admitting a diagonal representation, e.g.,

$$\mathscr{B} = \int b(\{z_k\})|\{z_k\}><\{z_k\}|\, d\sigma(\{z_k\}) \,.$$

Clearly, if $b=1$ then $\mathscr{B}=I$, the identity operator.

Field Formulation

The foregoing may also be given a field description in an s-dimensional configuration space, $\mathbb{R}^s$. Let us adopt $\{h_k(x)\}$, $x \in \mathbb{R}^s$, as a complete orthonormal set of functions, $\int h_k^* h_\ell dx = \delta_{k\ell}$, and introduce

$$A(x) \equiv \sum a_k h_k(x) \,, \quad A^\dagger(x) \equiv \sum a_k^\dagger h_k^*(x) \,,$$

$$V(x) \equiv \sum z_k h_k(x) \,, \quad V^*(x) \equiv \sum z_k^* h_k^*(x) \,.$$

From the conditions on $\{z_k\}$ it follows that V is square integrable. We conclude that: $[A(x),A^\dagger(y)]=\delta(x-y)$; $|V>\equiv|\{z_k\}>$; $A(x)|V>=V(x)|V>$; and

$$<V|V'> = \exp(-\tfrac{1}{2}\int|V|^2dx + \int V^*V'dx - \tfrac{1}{2}\int|V'|^2dx) \,. \qquad (7.3)$$

Apart from normalization, $|V>$ is an analytic functional of the field V, and as such the diagonal matrix elements determine an operator; $B(V^*,V) \equiv <V|\mathscr{B}|V>$ implies that $\mathscr{B}=:B(A^\dagger,A):$. The resolution of unity is contained as a special case of the diagonal representation of a general operator

$$\mathscr{B} = \int b(V)\, |V><V|\, \mathscr{D}V^* \mathscr{D}V \,,$$

where

$$\int (\cdot)\mathscr{D}V^* \mathscr{D}V \equiv \lim_{N\to\infty} \int (\cdot) \prod_{k=1}^{N} (d^2z_k/\pi) \,.$$

Here if $b=1$ then $\mathscr{B}=I$, the identity operator.

As one can see from the foregoing discussion the extension of the canonical coherent states to an infinite number of degrees of freedom is relatively straightforward. However there are interesting subtleties involved with infinitely many degrees of freedom that will become apparent in the next topic of discussion.

Scalar Fields

We orient the present discussion of coherent states for fields to properties of the functional representation and their relation to field-operator representations. We deliberately forego any discussion of an integral representation of the resolution of unity so as to emphasize an alternative functional representation of the Hilbert-space inner product making direct use of the coherent-state overlap functional, the quantity which in Sec. 2 was referred to as the reproducing kernel.

We begin conventionally by considering a scalar field operator $\phi(x)$ and its canonically conjugate momentum $\pi(x)$, where $x \in \mathbb{R}^s$. We assume that the fields are local self-adjoint operators, namely that when smeared with suitable real test functions, f and g, respectively, they become self-adjoint operators. For many purposes it suffices to assume that f and g are both in Schwartz's space $\mathscr{S}$ of C^∞ functions of rapid decrease. However, we will be purposely vague on this point since we make almost no demands on the test functions besides smoothness and localizability, sufficiency in numbers, and some unspecified notion of continuity (topology). The unitary Weyl operators are defined for suitably many f and g by

$$W[f, g] \equiv \exp\{(i/\hbar)\int [f(x)\phi(x)-g(x)\pi(x)]dx\} .$$

The formal canonical commutation relation $[\phi(x),\pi(y)]=i\hbar\delta(x-y)$ is embodied in the multiplication law

$$W[f_1, g_1]W[f_2, g_2] = e^{-i(\beta/\hbar)}W[f_1+f_2, g_1+g_2] ,$$

$$\beta \equiv (1/2) \int [g_1(x)f_2(x)-f_1(x)g_2(x)]dx . \tag{7.4}$$

Now choose $|0\rangle$ as a normalized cyclic vector in the representation space. This remark simply means that the scalar field coherent states defined for suitably many f and g by

$$|f, g\rangle \equiv W[f, g]|0\rangle$$

are a total set in the Hilbert space $\mathfrak{H}$; a set of vectors is total if the closure of its linear span yields $\mathfrak{H}$. If the representation of the Weyl operators is irreducible any nonzero vector is cyclic; however, we do not need to assume the representation is irreducible. In view of the multiplication law of Weyl operators (7.4) it follows that the coherent-state overlap is given by

$$\langle f_1, g_1|f_2, g_2\rangle = e^{i(\beta/\hbar)}\langle 0|f_2-f_1, g_2-g_1\rangle ,$$

and is thus completely determined by the basic *expectation functional*

$$E(f, g) \equiv \langle 0|W[f, g]|0\rangle = \langle 0|f, g\rangle .$$

We shall always assume that the expectation functional is continuous in both f and g in a suitable test-function space topology.

The expectation functional is so basic because one may begin with *it* and determine the representation of the field operators up to unitary equivalence. For example, every continuous functional $E(f, g)$ for which

$$\mathscr{K}(f_1, g_1; f_2, g_2) \equiv e^{i(\beta/\hbar)}E(f_2-f_1, g_2-g_1) ,$$

with β as before, is a positive-definite functional, defines a representation of the canonical Weyl operators, or of the canonical commutation relations, or, briefly, of the CCR. A positive-definite functional $\mathscr{K}$ is one which for all $N<\infty$, all complex coefficients $c_1,..., c_N$, and all function pairs $(f_1, g_1),..., (f_N, g_N)$, satisfies

$$\sum_{p,n=1}^{N} c_p^* c_n \mathscr{K}(f_p, g_p; f_n, g_n) \geqslant 0 . \tag{7.5}$$

Any continuous functional $\mathcal{K}$ satisfying (7.5) serves as a reproducing kernel generating a functional Hilbert space representation which we denote by $\mathfrak{C}$. This space may have an integral representation for its inner product (as we have generally required in earlier sections) but the inner product can always be defined in an alternative fashion. Each of a dense set of vectors in $\mathfrak{C}$ is defined by the continuous functional

$$\psi(f,\, g) \equiv \sum_{n=1}^{N} c_n \mathcal{K}(f,\, g;\, f_n,\, g_n)$$

with parameter conditions as before. For any two such vectors, e.g.,

$$\phi(f,\, g) \equiv \sum_{\ell=1}^{L} a_\ell \mathcal{K}(f,\, g;\, f_\ell,\, g_\ell)\;,$$

$$\chi(f,\, g) \equiv \sum_{m=1}^{M} b_m \mathcal{K}(f,\, g;\, \hat{f}_m,\, \hat{g}_m)\;,$$

the inner product in $\mathfrak{C}$ is defined by

$$(\chi,\phi) \equiv \sum_{m=1}^{M} \sum_{\ell=1}^{L} b_m^* a_\ell \mathcal{K}(\hat{f}_m,\, \hat{g}_m;\, f_\ell,\, g_\ell)\;.$$

It follows from (7.5) that $(\psi,\psi)\geqslant 0$, with equality holding if and only if $\psi(f,\, g)\equiv 0$. The completion of the space in the norm $\sqrt{(\psi,\psi)}$ defines the continuous-representation Hilbert space $\mathfrak{C}$.

In this kind of Hilbert space the Weyl operators are universally represented by

$$(W[\hat{f},\, \hat{g}]\psi)(f,\, g) = e^{i(\gamma/\hbar)}\psi(f-\hat{f},\, g-\hat{g})\;,$$

$$\gamma \equiv (1/2)\int [\hat{f}(x)g(x) - \hat{g}(x)f(x)]dx\;.$$

By functional differentiation then it follows that

$$[\phi(x)\psi](f, g) = [i\hbar\frac{\delta}{\delta f(x)} + \frac{g(x)}{2}]\,\psi(f, g)\,,$$

$$[\pi(x)\psi](f, g) = [-i\hbar\frac{\delta}{\delta g(x)} + \frac{f(x)}{2}]\,\psi(f, g)\,.$$

Comparison with the latter half of Sec. 3 will show a close parallel with our present discussion.

We note that any realization of the Hilbert space and the action of various operators is unitarily equivalent to the one outlined above, and in particular our construction is completely equivalent to the well-known GNS (Gel'fand, Naimark, Segal) [172] construction of a representation space based on a positive-definite functional.

Inequivalent Field-Operator Representations

Let us use the scalar field example to illustrate the wealth of inequivalent field-operator representations that exist. To begin, consider the group of unitary translation operators $U(a)$, $a \in \mathbb{R}^s$, formally defined so that $U(a)\phi(x)U^{-1}(a)=\phi(x-a)$. If we assume that $|0>$ is invariant under translation, $U(a)|0>=|0>$, for all $\mathbf{a}$, then it follows that the expectation functional E is also invariant, $E(f_a, g_a)=E(f, g)$, for all f and g, where $f_a(x)\equiv f(x+a)$ and $g_a(x)\equiv g(x+a)$. Suppose further that (up to a constant factor) $|0>$ is the unique translationally invariant state, a situation which is completely equivalent to the cluster condition given by

$$E(\hat{f}+f_a, \hat{g}+g_a) \rightarrow E(\hat{f}, \hat{g})E(f, g)$$

as $|a|\rightarrow\infty$. It follows from the foregoing that ("w" means weak)

$$\text{w-}\lim_{|a|\rightarrow\infty} W[f_a, g_a] = E(f, g)I\,,$$

where I denotes the identity operator. It is evident that the complex coefficient of unity deduced here is independent of the realization of W; such a factor has been termed a *tag* in that it clearly helps to identify and classify representations (apart from unitary equivalence).

As an application of tags consider two CCR representations $W_1[f, g]$ and $W_2[f, g]$ satisfying the foregoing conditions. We ask the question when can they be unitarily equivalent? That is, when does a unitary operator V exist such that

$$W_1[f, g] = VW_2[f, g]V^{-1}$$

for all f and g. If V exists it follows that

$$\begin{aligned} E_1(f, g)I_1 &= \lim W_1[f_a, g_a] \\ &= V \lim W_2[f_a, g_a]\, V^{-1} = VE_2(f, g)I_2V^{-1} \\ &= E_2(f, g)\, I_1 \ , \end{aligned}$$

for an arbitrary f and g. In other words unitary equivalence demands complete equality of the expectation functionals; or stated differently, every distinct translationally invariant cluster-fulfilling expectation functional labels an inequivalent CCR representation.

For example, if we choose $|0>$ as the ground state of a free, relativistic scalar field of mass m then the expectation functional

$$E_o(f, g) = \exp\{-\tfrac{1}{4}\int[(k^2+m^2)^{-\frac{1}{2}}|\tilde{f}(k)|^2 + (k^2+m^2)^{\frac{1}{2}}|\tilde{g}(k)|^2]dk\} \tag{7.6}$$

where

$$\tilde{f}(k) \equiv (2\pi)^{-s/2}\int e^{-ik\cdot x} f(x)dx \ ,$$

etc. Thus each m, $0<m<\infty$, already labels an inequivalent CCR representation. In turn these free-field CCR representations all come from one Fock representation made up of annihilation operators $a(k)$, creation operators $a^\dagger(k)$, which satisfy $[a(k),a^\dagger(q)]=\delta(k-q)$, and a unique no-particle state $|0>$, such that $a(k)|0>=0$ for all $k\in\mathbb{R}^s$. The free field operators of mass m

are given by a linear superposition of the Fock operators,

$$\phi(x) = (2\pi)^{-s/2}\int [e^{ik\cdot x}a(k) + e^{-ik\cdot x}a^{\dagger}(k)]\frac{dk}{(k^2+m^2)^{1/4}} \, ,$$

$$\pi(x) = -i(2\pi)^{-s/2}\int [e^{ik\cdot x}a(k) - e^{-ik\cdot x}a^{\dagger}(k)](k^2+m^2)^{1/4}dk \, ,$$

and they depend parametrically on m in a clear fashion. Inequivalence simply means there exists no unitary transformation relating ϕ's and π's for unequal mass values. It is common to refer to a free-field CCR representation as the Fock representation as well.

For the free field, normal ordering is given by the rule

$$:W[f, g]: \equiv W[f, g]/E_o(f, g) \, ,$$

which is identical to the usual rule of all creation operators to the left, all annihilation operators to the right. For a general CCR representation normal ordering (with respect to $|0\rangle$) is defined by

$$:W[f, g]: \equiv W[f, g]/\langle 0|W[f, g]|0\rangle \, . \tag{7.7}$$

As a consequence of (7.7) and the commutation relation (7.4) it follows that

$$B(f, g) \equiv \langle f, g|\mathscr{B}|f, g\rangle$$

is true for $\mathscr{B}$=$:B(\pi, \phi):$ since

$$\begin{aligned}\langle f, g|:B(\pi, \phi):|f, g\rangle &= \langle 0|:B(\pi+f, \phi+g):|0\rangle \\ &= B(f, g) \, .\end{aligned}$$

If $\mathscr{B}$ is a polynomial then $\mathscr{B}$ is uniquely determined by this rule. For free fields, $\mathscr{B}$ is generally related to B in this way. However, it should be emphasized that not all functionals $B(f, g)$ are associated with operators by this prescription. For a proposed $\mathscr{B}$ to be an operator here it is also necessary that $\|\mathscr{B}|f, g\rangle\|<\infty$, and it is this condition that markedly restricts the possible allowed range of diagonal coherent-state matrix-element values and relates it with a specific CCR representation. For example, for the relativistic free scalar field of mass m there is a Hamiltonian operator $\mathscr{H}_o$, formally given by $\mathscr{H}_o = :H_o(\pi, \phi):$, such that $\mathscr{H}_o|0\rangle = 0$, $\mathscr{H}_o \geqslant 0$, and

$$H_o(f, g) = \langle f, g|\mathscr{H}_o|f, g\rangle$$

$$= (1/2)\int \{f^2(x) + [\nabla g(x)]^2 + m^2 g^2(x)\} dx \;, \quad (7.8)$$

if and only if the CCR representation is given by (7.6) with the same value of m. In this fashion dynamics is inextricably linked with kinematics when one deals with field theory. The restrictions imposed by this linkage are even better known under the guise of Haag's theorem [173], which, suitably paraphrased, reads: *Thou shalt not use the Fock representation for interacting, translationally invariant theories.*

Fermion Fields

Fock Representation

Consider, for simplicity, a single-component nonrelativistic canonical fermion field annihilation operator $\psi(x)$ and creation operator $\psi^\dagger(x)$, $x \in \mathbb{R}^s$, for which the only nonvanishing anticommutator is given by $\{\psi(x), \psi^\dagger(y)\} = \delta(x-y)$. We take as fiducial vector the normalized no-particle state $|0\rangle$, which satisfies $\psi(x)|0\rangle = 0$ for all $\mathbf{x}$. The Hilbert space is spanned by the vectors $\psi^\dagger(x_1)\cdots\psi^\dagger(x_N)|0\rangle$, $N = 0, 1, 2, \ldots$. This construction defines the justly-famous fermion Fock representation.

With the use of anticommuting c-number fields $\eta(x)$ and $\bar{\eta}(x)$, which are direct analogues of the discrete anticommuting c-numbers of Sec. 5, it is straightforward to introduce canonical coherent states for fermion fields. The anticommuting c-number fields η and $\bar{\eta}$ may be represented as

$$\eta(x) = \sum c_m \theta_m h_m(x) \,,$$

$$\bar{\eta}(x) = \sum c_m^* \bar{\theta}_m h_m(x) \,, \qquad (7.9)$$

where $\{h_m\}$ forms a fixed complete orthornormal system, $\{\theta_m\}$ and $\{\bar{\theta}_m\}$ are fixed, independent, infinite sets of conventional anticommuting c-numbers (as in Sec. 5), chosen so that their mathematical norm $\|\theta_m\| = \|\bar{\theta}_m\| = 1$ for all m. In this case

$$\int \bar{\eta}(x)\eta(x)dx = \sum |c_m|^2 \bar{\theta}_m \theta_m \,,$$

which is convergent provided $\sum |c_m|^2 < \infty$. In this case we say η and $\bar{\eta}$ are square integrable. This condition is analogous to the square summability or integrability of the coherent-state labels in the Bose case. The relevant Grassmann algebra is generated by arbitrary polynomials (including the unit element) involving complex sums of products of the fields η and $\bar{\eta}$, and their limits.

The canonical fermion field coherent states are then defined for all square integrable η and $\bar{\eta}$ by

$$|\eta\rangle \equiv e^{-\frac{1}{2}\int \bar{\eta}\eta dx} e^{\int \psi^\dagger \eta dx} |0\rangle \,,$$

$$\langle\eta| \equiv e^{-\frac{1}{2}\int \bar{\eta}\eta dx} \langle 0| e^{\int \bar{\eta}\psi dx} \,. \qquad (7.10)$$

The overlap of two such states is expressed by

$$\langle\eta_1|\eta_2\rangle = \exp(-\tfrac{1}{2}\int \bar{\eta}_1\eta_1 dx + \int \bar{\eta}_1\eta_2 dx - \tfrac{1}{2}\int \bar{\eta}_2\eta_2 dx) \,.$$

Continuity in the fields η and $\bar{\eta}$ is defined through continuity in the coefficients $\{c_m\}$.

A resolution of unity for the canonical fermion coherent states is formally given by

$$I = \lim_{M\to\infty} \int |\eta^M\rangle \langle\eta^M|\, d^M\bar{\eta}\, d^M\eta\,,$$

$$\equiv \int |\eta\rangle\langle\eta|\, \mathcal{D}\bar{\eta}\, \mathcal{D}\eta\,.$$

Here η^M denotes the field defined by (7.9) in which the sum includes only those $m \leqslant M$, and for each $M < \infty$ the rules of Grassmann integration discussed in Sec. 5 are implied.

The canonical fermion coherent states are eigenstates in the sense that $\psi(x)|\eta\rangle = \eta(x)|\eta\rangle$, and $\langle\eta|\psi^\dagger(x) = \langle\eta|\bar{\eta}(x)$ for all x. Operators are determined by their diagonal matrix elements; if $B(\bar{\eta}, \eta) = \langle\eta|\mathcal{B}|\eta\rangle$ then it follows that $\mathcal{B} = {:}B(\psi^\dagger, \psi){:}$, where : : orders all creation operators to the left of all annihilation operators. As usual this implies that a general operator admits a diagonal representation.

Fermi Field Coherent States without Anticommuting c-Numbers

Before leaving the fermion field Fock representation we wish to illustrate another set of coherent states that do not use anticommuting c-numbers in any way. The states defined in (7.10) may be written, apart from a normalization N, as

$$|\eta\rangle = N \sum_{n=0}^{\infty} (n!)^{-1} \int \ldots \int |x_1, x_2, \ldots, x_n\rangle_A\, \eta(x_n) \cdots \eta(x_2)\eta(x_1)\, dx_1 \cdots dx_n\,,$$

where

$$|x_1, x_2, \ldots, x_n\rangle_A \equiv \psi^\dagger(x_1)\psi^\dagger(x_2)\cdots\psi^\dagger(x_n)|0\rangle$$

and the subscript A emphasizes the complete antisymmetry in the arguments. Now, following Friedrichs [174], let us introduce an ordering ($<$) of points $x = (x^1, x^2, \ldots, x^s) \in \mathbb{R}^s$. For example, let us say that $x_1 < x_2$ if : (i) $x_1^1 < x_2^1$; or (ii) $x_1^1 = x_2^1$, $x_1^2 < x_2^2$; or (iii) ... or (s) $x_1^1 = x_2^1, \ldots, x_1^{s-1} = x_2^{s-1}$, $x_1^s < x_2^s$. For each $n \geqslant 2$ introduce the sign function $\sigma_n(x_1, x_2, \ldots, x_n)$ defined to be $+1$ if the x's are in standard order, $x_1 < x_2 < \cdots < x_n$, and continued to be totally antisymmetric; i.e., $\sigma_n=\pm 1$, depending on the permutation to restore its arguments to a standard order. If any two x's coincide then $\sigma_n\equiv 0$.

For each $n\geqslant 2$ we introduce

$$|x_1,x_2,\ldots,x_n>_S \equiv \sigma_n(x_1,x_2,\ldots,x_n)|x_1,x_2,\ldots,x_n>_A \,,$$

which are evidently fully symmetric in their arguments and which vanish if any two arguments coincide. Since $\sigma_n^2=1$ for distinct arguments we have

$$|x_1,x_2,\ldots,x_n>_A = \sigma_n(x_1,x_2,\ldots,x_n)|x_1,x_2,\ldots,x_n>_S$$

showing that the antisymmetric states can be fully recovered from the symmetric ones. We extend the notation so that $|x>_S \equiv |x>_A \equiv \psi^\dagger(x)|0>$.

We now introduce the commuting-variable fermion field coherent states according to the rule

$$|\chi> \equiv \hat{N}\sum_{n=0}^{\infty}(n!)^{-1}\int\ldots\int\chi(x_1)\chi(x_2)\cdots\chi(x_n)|x_1,x_2,\ldots,x_n>_S dx_1\cdots dx_n$$

where $\hat{N}$ is a normalization factor and $\chi(x)$ denotes an arbitrary square-integrable complex (c-number!) field. These states are defined in the very same way as in the Bose case save that no contributions are included for coinciding points. However, since the numerical value of a multidimensional Lebesgue integral of an integrable function is unchanged if sets of lower dimensionality are excluded, it follows, as in the Bose case, that $\hat{N}=\exp(-\frac{1}{2}\int|\chi(x)|^2dx)$, and that the coherent state overlap is

$$<\chi|\chi'> = \exp(-\tfrac{1}{2}\int|\chi|^2dx + \int\chi^*\chi'dx - \tfrac{1}{2}\int|\chi'|^2dx)\,,$$

which is indistinguishable from the corresponding overlap for the Bose field case (7.3)! Although the $|\chi>$ states are generally not eigenstates of the annihilation operator $\psi(x)$, the relation

$$<\chi|\psi^{\dagger}(x)\psi(x)|\chi> = \chi^{*}(x)\chi(x)$$

holds true, and this and similar kinds of relations suffice to relate quantum and classical expressions as needed. The analysis of fermion fields by coherent states without anticommuting c-numbers along these lines has been carried out in Ref. [20], and has been studied more recently by Garbaczewski [175].

Non-Fock Representations

As in the Bose case, there are uncountably many inequivalent representations of the canonical anticommutation relations (CAR) for an infinite number of degrees of freedom. The characterization of such representations by expectation functionals, and the construction of a functional Hilbert space representation based on a reproducing kernel carries through much as in the Bose case. Typically, a specific dynamical problem is linked with a given representation of the CAR, and Haag's theorem applies to fermions just as much as it does to bosons [176].

Needless to say, the topics raised in the present section inevitably go to the heart of nonlinear quantum field theory and raise difficult questions that have as yet not been satisfactorily resolved. Certainly any further discussion of these problems is outside the scope of this primer.

Additional Remarks and Introduction to Selected Reprints

1. The formulation and description of boson fields and associated models by expectation functionals is most convincingly presented in the fundamental work of Araki [177]. Many related ideas are due to Segal, which may be traced from Ref. [22] and references therein.

2. Coherent state techniques are frequently used to provide

justification of a study of the classical equation in regard to its quantum properties. With regard to field theory, we observe in (7.8) that the diagonal coherent-state matrix elements of the Hamiltonian operator has resulted in the appropriate classical Hamiltonian (with $\pi_{cl} \equiv f$ and $\phi_{cl} \equiv g$); this is a concrete example of the weak correspondence principle [151] alluded to in Sec. 6. In the paper by Cahill [178] (see also Ref. [179]) an application of the field theoretical methods discussed above is applied in a calculation of quantum corrections to the mass of a one-dimensional soliton; for alternative calculational methods see Ref. [180]. Pottinger [181] gives a similar discussion of Abelian vortices. For a general discussion and for further references we refer the reader to Ref. [182].

3. The infrared problem in quantum electrodynamics (QED) [183] can conveniently be studied by making use of coherent states. In the paper by Chung [184] coherent states for soft photons are used in such a way that matrix elements of the Feynman-Dyson S-matrix, describing scattering of charged particles, are free of infrared divergences. The soft photons correspond to long wavelength excitations and hence one may treat the charged matter currents as classical and the radiation field as quantum mechanical. One is therefore considering an external field problem in structure similar to the problem of a forced harmonic oscillator as discussed in Sec. 3. An important advance in Chung's work is the construction of appropriate asymptotic states by Kibble [185]. A proper treatment of the Coulomb phase was given in the paper by Kulish and Faddeev [186], and related work is found in Refs. [187,188]. Papanicolaou [189] has reviewed the infrared problem in QED, while for an axiomatic approach we refer the reader to the work of Strocchi and collaborators [190].

4. The infrared sector of quantum gravity can successfully be studied to all orders in perturbation theory as was shown by

Weinberg [191]. The time evolution of soft graviton emission following any scattering process is discussed in Ref. [192]. In the paper by Gomatam [193] a canonical coherent-state formulation was laid out for photons and gravitons in terms of an indefinite metric formalism.

5. With regard to the present field theoretical model of the strong interaction, quantum chromodynamics (QCD) [194], Greco, Palumbo, Pancheri-Srivastava and Srivastava [195], as well as other workers [196-198], have tried to carry over the coherent state techniques of QED to the case of non-Abelian fields. In such a context the probability for the emission of generally self-interacting, non-Abelian, soft bosons up to a certain total energy, was derived in the paper by Eriksson and Skagerstam [199].

An application of the coherent state approach to the emission of non-Abelian quanta, i.e., gluons, in the context of quark-gluon jet physics is discussed in the paper by Curci, Greco and Srivastava [200] and in Ref. [196]. A canonical coherent-state framework for the description of e^+e^- annihilation was discussed by Rayne [201].

6. When suitably extended to infinitely many degrees of freedom, the forced harmonic oscillator problem, as discussed in Sec. 3, can be used as a tool in order to derive a compact form for the S-matrix in quantum field theory. An elegant exposition of this derivation can be found in lecture notes of Faddeev [157]. Early work on transition amplitudes in terms of coherent states can be found in papers by Schweber [202] and Berezin [203], and the article by Hammer, Schrauner and DeFacio [204] assembles some of these developments.

REFERENCES
(* *denotes reprinted paper*)

[1] R. J. Glauber, in *Quantum Optics and Electronics,* eds. C. DeWitt, A. Blandin and C. Cohen-Tannoudji (Gordon and Breach, New York, 1964).

[2] J. R. Klauder and E. C. G. Sudarshan, *Fundamentals of Quantum Optics,* (W. A. Benjamin, New York, 1968).

[3] H. M. Nussenzweig, *Introduction to Quantum Optics,* (Gordon and Breach, New York, 1973).

*[4] J. R. Klauder, "Continuous Representation Theory. I. Postulates of Continuous Representation Theory", *J. Math. Phys.* **4,** 1055-1058 (1963), Paper II.2.

[5] N. Aronszajn, "La Theorie des Noyaux Reproduisants et ses Applications. Premiere Partie." *Proc. Cambridge Phil. Soc.* **39,** 133-153 (1943); "Theory of Reproducing Kernels", *Trans. Am. Math. Soc.* **68,** 337-404 (1950).

[6] H. Meschkowski, *Hilbertsche Raume mit Kernfunktion* (Springer Verlag, Berlin, 1962).

[7] *Reproducing Kernel Hilbert Spaces: Application in Statistical Signal Processing,* ed. H. L. Weinert (Hutchinson Ross, Stroudsberg, PA, 1983).

[8] J. von Neumann, *Mathematical Foundations of Quantum Mechanics,* (Princeton, 1955).

[9] V. Bargmann, P. Butera, L. Giradello and J. R. Klauder, "On The Completeness of the Coherent States", *Rep. Math. Phys.* **2,** 221-228 (1971).

[10] A. M. Perelomov, "On The Completeness of a System of Coherent States", *Teor. Mat. Fiz.* **6,** 213-224 (1971).

[11] E. Schrödinger, "Der Stetige Übergang von der Mikro-zur Makromechanik", *Naturwissenschaften* **14,** 664-666 (1926).

*[12] R. J. Glauber, "Classical Behavior of Systems of Quantum Oscillators", *Phys. Lett.* **21,** 650-652 (1966), Paper III.5.

[13] B. Crosignani, P. Di. Porto and S. Solimeno, "Evolution of a Coherent State of an Harmonic Oscillator with Time-Dependent Frequency", *Phys. Lett.* **28A,** 271-272 (1968).

[14] Y. Kano, "A Remark on the Time Evolution of Coherent States", *Phys. Lett.* **56A,** 7-8 (1976).

[15] H. Letz, "Evolution Operator and Stable Coherent States", *Phys. Lett.* **60A,** 399-400 (1977).

[16] A. Trias, "On the Time Evolution of Coherent States", *Phys. Lett.* **61A,** 149-150 (1977).

[17] P. Chand, "Necessary Condition for a Coherent State to Remain Coherent", *Phys. Lett.* **67A,** 99-100 (1978); "Time Evolution of Coherent States", *Nuovo Cimento* **50B,** 17-20 (1979).

[18] R. R. Puri and S. V. Lawande, "Time-Dependent Invariants and Stable Coherent States", *Phys. Lett.* **70A,** 69-70 (1979).

[19] A. K. Rajagopal and J. T. Marshall, "New Coherent States with Applications to Time-Dependent Systems", *Phys. Rev.* **A26**, 2977-2980 (1982).

*[20] J. R. Klauder, "The Action Option and a Feynman Quantization of Spinor Fields in Terms of Ordinary C-

Numbers", *Ann. Phys. (N.Y.)* **11,** 123-168 (1960), Paper II.1.

[21] V. Bargmann, "On a Hilbert Space of Analytic Functions and an Associated Integral Transform. Part 1", *Commun. Pure and Appl. Math.* **14,** 187-214 (1961).

[22] See, e.g., I. Segal, *Mathematical Problems of Relativistic Physics,* (Providence, Rhode Island, 1963).

*[23] R. J. Glauber, "Photon Correlations", *Phys. Rev. Lett.* **10,** 84-86 (1963), Paper II.4; "The Quantum Theory of Optical Coherence", *Phys. Rev.* **130,** 2529-2539 (1963), Paper II.5; "Coherent and Incoherent States of the Radiation Field", *Phys. Rev.* **131,** 2766-2788 (1963), Paper II.7.

*[24] E. C. G. Sudarshan, "Equivalence of Semiclassical and Quantum Mechanical Descriptions of Statistical Light Beams", *Phys. Rev. Lett.* **10,** 277-279 (1963), Paper II.6.

*[25] P. Carruthers and M. M. Nieto, "Coherent States and the Forced Quantum Oscillator", *Am. J. Phys.* **33,** 537-544 (1965), Paper III.2.

[26] R. P. Feynman, "Mathematical Formulation of the Quantum Theory of Electromagnetic Interaction", *Phys. Rev.* **80,** 440-457 (1950).

[27] J. Schwinger, "The Theory of Quantized Fields. III", *Phys. Rev.* **91,** 728-740 (1953).

[28] R. W. Fuller, S. M. Harris and E. L. Slaggie, "S-Matrix Solution for the Forced Harmonic Oscillators", *Am. J. Phys.* **31,** 431-439 (1963).

[29] L. M. Scarfone, "Transition Probabilities for the Forced Quantum Oscillator", *Am. J. Phys.* **32,** 158-162 (1964).

*[30] K. Hepp, "The Classical Limit for Quantum Mechanical Correlation Functions", *Commun. Math. Phys.* **35,** 265-277 (1966), Paper III.4.

*[31] T. Toyoda and K. Wildermuth, "Charged Schrödinger Particle in a C-Number Radiation Field", *Phys. Rev.* **D22,** 2391-2399 (1980), Paper III.9.

[32] H. Bacry, A. Grossmann and J. Zak, "Proof of Completeness of Lattice States in the KQ Representation", *Phys. Rev.* **B12,** 1118-1120 (1975).

[33] M. Boon and J. Zak, "Coherent States and Lattice Sums", *J. Math. Phys.* **19,** 2308-2311 (1978); "Discrete Coherent States on the von Neumann Lattice", *Phys. Rev.* **B18,** 6744-6751 (1978).

[34] J. Zak, "Coherent States in Solids", *Phys. Rev.* **B21,** 3345-3348 (1980).

[35] I. Dana, "Composite von Neumann Lattice", *Phys. Rev.* **A28,** 2594-2596 (1983).

*[36] P. Carruthers and M. M. Nieto, "Coherent States and the Number-Phase Uncertainty Relation", *Phys. Rev. Lett.* **14,** 387-389 (1965), Paper III.1.

[37] P. Carruthers and M. M. Nieto, "Phase and Angle Variables in Quantum Mechanics", *Rev. Mod. Phys.* **40,** 411-440 (1968).

[38] M. M. Nieto, "Quantized Phase Effect and Josephson Tunneling", *Phys Rev.* **167,** 416-418 (1968); "Quantum Phase Operators and the Josephson Plasma Resonance", *Nuovo Cimento,* **62B,** 95-102 (1969); "Phase-Difference Operator Analysis of Microscopic Radiation Field Measurements", *Phys. Lett.* **60A,** 401-403 (1977).

[39] R. Jackiw, "Minimum Uncertainty Product, Number-Phase Uncertainty Product and Coherent States", *J. Math. Phys.* **9,** 339-346 (1968).

*[40] V. Moncrief, "Coherent States and Quantum Nonperturbing Measurements", *Ann. Phys. (N.Y.)* **114,** 201-214 (1978), Paper III.6.

*[41] W. G. Unruh, "Quantum Nondemolition Measurements and Coherent States", *Phys. Rev.* **D17,** 1180-1181 (1978), Paper III.5.

*[42] J. N. Hollenhorst, "Quantum Limits on Resonant-Mass Gravitational-Radiation Detectors", *Phys. Rev.* **D19,** 1669-1679 (1979), Paper III.8.

[43] P. W. Anderson, "Coherent Excited States in the Theory of Superconductivity: Gauge Invariance and the Meissner Effect", *Phys. Rev.* **110,** 827-835 (1958).

*[44] P. Carruthers and K. S. Dy, "Coherent States and Irreversible Processes in Anharmonic Crystals", *Phys. Rev.* **147,** 214-222 (1966), Paper VI.1.

*[45] F. W. Cummings and J. R. Johnston, "Theory of Superfluidity", *Phys. Rev.* **151,** 105-112 (1966), Paper VI.2.

*[46] J. S. Langer, "Coherent States in the Theory of Superfluidity", *Phys. Rev.* **167,** 183-190 (1968), Paper VI.3.

[47] M. Rasetti and T. Regge, "Coherent States and Virial Theorem for Liquid ^4He", *J. Low Temp. Phys.* **13,** 249-259 (1973).

[48] J. S. Langer, "Coherent States in the Theory of Superfluidity. II. Fluctuations and Irreversible Processes",

Phys. Rev. **184,** 219-229 (1969).

[49] J. R. Johnston, "Coherent States in Superfluids: The Ideal Einstein-Bose Gas", *Am. J. Phys.* **38,** 516-528 (1970).

[50] G. Chapline, "Theory of the Superfluid Transition in Liquid Helium", *Phys. Rev.* **A3,** 1671-1679 (1971).

[51] Y. Kano, "Coherence Theory of the Bose-Einstein Condensation, I and II", *J. Phys. Soc. Jap.* **36,** 649-652 (1974); *ibid.* **37,** 310-321 (1974).

[52] V. Srinivasan, "On the Connection Between the Boson Transformed Ground State and the Coherent State in Superconductivity", *Prog. Theor. Phys.* **55,** 939-940 (1976).

[53] J. C. Lee, "Rescaled Coherent State Theory of the Lambda-Transition in Liquid ^4He", *Physica A* **116,** 604-611 (1982).

[54] G. Ghosh, B. Dutta-Roy and D. Bhaumik, "The Coherent State Approach to the Vlasov Equation", *Plasma Phys.* **19,** 1051-1054 (1977).

*[55] G. N. Fowler, N. Stelte and R. M. Weiner, "Condensates and Bose-Einstein Correlations", *Nucl. Phys.* **A319,** 349-363 (1979), Paper IX.4.

[56] See e.g., R. A. Broglia, C. H. Dasso and H. Esbensen, "On the Spectroscopy of Coherent States", in *Progress in Particle and Nuclear Physics,* **4,** ed. D. Wilkinson (Pergamon Press, New York, 1980), pp. 345-381, and references cited therein.

*[57] P. Carruthers, "Hadronic Multiplicity Distribution: Example of a Universal Stochastic Process", in *Proceedings of the XIVth International Symposium on*

Multiparticle Dynamics (Granlibakken, Lake Tahoe, June, 1983), Paper X.6.

*[58] P. Carruthers and C. C. Shih, "Correlations and Fluctuations in Hadronic Multiplicity Distributions: The Meaning of the KNO Scaling", *Phys. Lett.* **127B,** 242-250 (1983), Paper X.5.

[59] J. R. Klauder, "The Design of Radar Signals Having Both High Range Resolution and High Velocity Resolution", *Bell Syst. Tech. Journ.* **34,** 809-820 (1960).

[60] M. J. Davies and E. J. Heller, "Multidimensional Wave Functions from Classical Trajectories", *J. Chem. Phys.* **75,** 3919-3924 (1981).

[61] S. Shi, "A New Semiclassical Approach to the Molecular Dynamics: Label Variable Classical Mechanics", *J. Chem. Phys.* **79,** 1343-1352 (1983).

[62] R. Paul, "Production of Coherent States in Biological Systems", *Phys. Lett.* **96A,** 263-268 (1983).

[63] L. J. Dunne, A. D. Clark and E. J. Brandas, "Evidence for the Existence of Long-Range Phase Coherence in the Bacteriorhodospin Macro-molecule", *Chem. Phys. Lett.* **97,** 573-576 (1983).

[64] J. E. Moyal, "Quantum Theory as a Statistical Theory", *Proc. Camb. Phil. Soc.* **45,** 99-124 (1949); see also M. S. Bartlett and J. E. Moyal, "The Exact Transition Probabilities of Quantum-Mechanical Oscillators Calculated by the Phase-Space Method", *Proc. Camb. Phil. Soc.* **45,** 545-553 (1949).

[65] T. T. Truong, "New Integral Equation for the Quartic Anharmonic Oscillator", *Nuovo Cimento Lett.* **9,** 533-536 (1974); "Weyl Quantization of Anharmonic Oscillators",

J. Math. Phys. **16,** 1034-1043 (1975).

[66] K. Bhaumik and B. Dutta-Roy, "The Classical Nonlinear Oscillator and the Coherent State", *J. Math. Phys.* **16,** 1131-1133 (1975).

[67] R. Dutt and M. Lakshmanan, "Application of Coherent State Representation to Classical x^6 and Coupled Anharmonic Oscillators", *J. Math. Phys.* **17,** 482-484 (1976).

[68] F. T. Hioe, "Quantum Theory of Coupled Anharmonic Oscillators", *Phys. Rev.* **D15,** 488-496 (1977); F. T. Hioe and E. W. Montroll, "Quantum Theory of Anharmonic Oscillators. I. Energy Levels of Oscillators with Positive Quartic Anharmonicity", *J. Math. Phys.* **16,** 1945-1955 (1975); F. T. Hioe, D. MacMillen and E. W. Montroll, "Quantum Theory of Anharmonic Oscillators. II. Energy Levels of Oscillators with Positive $x^{2\alpha}$ Anharmonicity", *J. Math. Phys.* **17,** 1320-1337 (1976).

[69] S. K. Bose and D. N. Tripathy, "The General Coupled Anharmonic Oscillators and the Coherent State Representation", *J. Math. Phys.* **19,** 2555-2557 (1978); "Uniformly Damped General Coupled Anharmonic Oscillators and the Coherent State Representation", *Lett. Math. Phys.* **4,** 265-273 (1980); "Studies of Coupled Anharmonic Oscillator Problem Using Coherent States and Path Integral Approaches", *Fortschr. Phys.* **31,** 131-163 (1983).

[70] A. P. Shustov, "Coherent States and Energy Spectrum of the Anharmonic Oscillator", *J. Phys. A: Math. Gen.* **11,** 1771-1780 (1978).

[71] S. G. Krivoshlykov, V. I. Man'ko and I. N. Sissakian, "Coherent State Evolution for the Quantum Anharmonic

Oscillator", *Phys. Lett.* **90A,** 165-168 (1982).

[72] V. Bargmann, "On a Hilbert Space of Analytic Functions and an Associated Integral Transform. Part II. A Family of Related Function Spaces Application to Distribution Theory", *Commun. Pure and Appl. Math.* **20,** 1-101 (1967).

*[73] C. Aragone, E. Chalbaud and S. Salamó, "On Intelligent Spin States", *J. Math. Phys.* **17,** 1963-1971 (1976), Paper II.14.

[74] E. P. Wigner, *Group Theory and Its Application to the Quantum Mechanics of Atomic Spectra,* (Academic Press, New York, 1959).

[75] B. Simon, "The Classical Limit of Quantum Partition Functions", *Commun. Math. Phys.* **71,** 247-276 (1980).

*[76] J. M. Radcliffe, "Some Properties of Coherent Spin States", *J. Phys. A.: Gen. Phys.* **4,** 313-323 (1971), Paper II.11.

*[77] F. T. Arecchi, E. Courtens, R. Gilmore and H. Thomas, "Atomic Coherent States in Quantum Optics", *Phys. Rev.* **A6,** 2211-2237 (1972), Paper VIII.1.

[78] R. Gilmore, *Lie Groups, Lie Algebras, and Some of Their Applications,* (Wiley & Sons, New York, 1974).

[79] M. Ducloy, "Application of the Formalism of Angular Momentum Coherent States to Atomic Physics Problem", *J. Phys.* **36,** 927-941 (1975).

[80] F. T. Arecchi, "An Introduction to Quantum Optics", in *Interaction of Radiation with Condensed Matter,* (IAEA-publication, Vienna, 1977), pp. 93-159.

*[81] E. H. Lieb, "The Classical Limit of Quantum Spin Systems", *Commun. Math. Phys.* **31,** 327-340 (1973), Paper VII.1.

*[82] K. Hepp and E. H. Lieb, "Equilibrium Statistical Mechanics of Matter Interacting with the Quantized Radiation Field", *Phys. Rev.* **A8,** 2517-2525 (1973), Paper VIII.3.

*[83] Y. K. Wang and F. T. Hioe, "Phase Transitions in the Dicke Model of Superradiance", *Phys. Rev.* **A7,** 831-836 (1973), Paper VIII.2.

*[84] L. M. Narducci, C. M. Bowden, V. Bleumel, G. Garrazana and R. A. Tuft, "Multitime-Correlation Functions and the Atomic Coherent-State Representation", *Phys. Rev.* **A11,** 973-980 (1975), Paper VIII.4.

*[85] R. Gilmore and D. H. Feng, "Phase Transitions in Nuclear Matter Described by Pseudospin Hamiltonians", *Nucl. Phys.* **A301,** 189-204 (1978), Paper IX.3.

*[86] R. Gilmore and D. H. Feng, "Studies of the Ground-State Properties of the Lipkin-Meshkov-Glick Model via Atomic Coherent States", *Phys. Lett.* **76B,** 26-28 (1978), Paper IX.2.

[87] J. R. Klauder, "A Langevin Approach to Fermion and Quantum Spin Correlation Functions", *J. Phys. A: Math. Gen.* **16,** L317-L319 (1983).

[88] S. Helgason, *Differential Geometry and Symmetric Spaces,* (Academic Press, New York, 1962), p. 369.

*[89] J. R. Klauder, "Continuous Representation Theory. II. Generalized Relation Between Quantum and Classical Dynamics", *J. Math. Phys.* **4,** 1058-1073 (1963), Paper

II.3.

[90] E. W. Aslaksen and J. R. Klauder, "Continuous Representation Theory Using the Affine Group", *J. Math. Phys.* **10,** 2267-2275 (1969).

[91] J. R. Klauder, "Path Integrals for Affine Variables", in *Functional Integration,* eds. J.-P. Antoine and E. Tirapegui (Plenum, New York, 1980), pp. 101-119.

[92] T. Paul, "Functions Analytic on the Half-Plane as Quantum Mechanical States", *J. Math. Phys.* **25,** 3252-3263 (1984).

*[93] A. O. Barut and L. Girardello, "New 'Coherent' States Associated with Non-Compact Groups", *Commun. Math. Phys.* **21,** 41-55 (1972), Paper II.10.

*[94] M. Hongoh, "Coherent States Associated with the Continuous Spectrum of Non-Compact Groups", *J. Math. Phys.* **18,** 2081-2084 (1977), Paper II.16.

[95] R. Gilmore, "Geometry of Symmetrized States", *Ann. Phys. (N.Y.)* **74,** 391-463 (1972).

*[96] A. M. Perelomov, "Coherent States for Arbitrary Lie Group", *Commun. Math. Phys.* **26,** 222-236 (1972), Paper II.12.

[97] R. Gilmore, "On the Properties of Coherent States", *Revista Mexicana de Fisica* **23,** 143-187 (1974).

*[98] E. Onofri, "A Note on Coherent State Representation of Lie Groups", *J. Math. Phys.* **16,** 1087-1089 (1975), Paper II.13.

[99] A. M. Perelomov, "Generalized Coherent States and Some of Their Applications", *Sov. Phys. Usp.* **20,** 703-720 (1977).

*[100] L. G. Yaffe, "Large N Limits as Classical Mechanics", *Rev. Mod. Phys.* **54,** 407-435 (1982), Paper III.10.

[101] L. G. Yaffe, "Large N Quantum Mechanics and Classical Limits", *Physics Today,* August 1983, 50-57.

*[102] M. Rasetti, "Coherent States and Partition Function", *International Journ. Theor. Phys.* **14,** 1-21 (1975), Paper VII.2.

*[103] R. Gilmore, "Applications of Coherent States in Thermodynamics and Dynamics", in *Symmetries in Science,* eds. B. Gruber and S. Millman (Plenum Publ. Corp., New York, 1980), pp. 105-131; Paper VII.3.

*[104] D. Bhaumik, K. Bhaumik and B. Dutta-Roy, "Charged Bosons and the Coherent State", *J. Phys. A: Math. Gen.* **9,** 1507-1512 (1976), Paper II.15.

*[105] D. Horn and R. Silver, "Coherent Production of Pions", *Ann. Phys. (N.Y.)* **66,** 509-541 (1971), Paper X.1.

[106] P. Chattopadhyay and J. Da Providencia, "Investigations in the Problem of Pion Condensation Using Generator Co-ordinate Methods", *Nucl. Phys.* **A370,** 445-467 (1981).

[107] B.-S. Skagerstam, "Coherent State Representation of a Charged Relativistic Boson Field", *Phys. Rev.* **D19,** 2471-2476 (1979); *ibid.* **D22,** 534(E) (1980).

*[108] G.-J. Ni and Y.-P. Wang, "Vacuum Instability and the Critical Value of the Coupling Parameter in Scalar QED", *Phys. Rev.* **D27,** 969-975 (1983), Paper XI.8.

[109] K.-E. Eriksson and B.-S. Skagerstam, "Isotopic Spin and Coherent States", *J. Phys. A: Math. Gen.* **12,** 2175-2185 (1979); *ibid.* **14,** 545 (E) (1981).

[110] B.-S. Skagerstam, "Some Remarks on Coherent States and Conserved Charges", *Phys. Lett.* **A69,** 76-78 (1978).

*[111] K.-E. Eriksson, N. Mukunda and B.-S. Skagerstam, "Coherent State Representation of a Non-Abelian Charged Quantum Field", *Phys. Rev.* **D24,** 2615-2625 (1981), Paper X.4.

*[112] J. C. Botke, D. J. Scalapino and R. L. Sugar, "Coherent States and Particle Production", *Phys. Rev.* **D9,** 813-823 (1974), Paper X.2.

[113] M. Martinis and V. Mikuta, "Coherent States with Definite Charge and Isospin and Cluster Production", *Phys. Rev.* **D12,** 909-917 (1975).

[114] M. Bosterli, "Interaction of Isovector Scalar Mesons with Simple Sources", *Phys. Rev.* **D16,** 1749-1753 (1977).

*[115] B.-S. Skagerstam, "Quasi-Coherent States for Unitary Groups", *J. Phys. A: Math. Gen.* **18,** 1-13 (1985), Paper II.17.

*[116] B.-S. Skagerstam, "On the Large N_c Limit of the SU(N_c) Color Quark-Gluon Partition Function", *Z. Phys.* **C24,** 97-101 (1984), Paper VII.4.

[117] B.-S. Skagerstam, "On Finite Lattice Corrections to Gauge Field Thermodynamics", *Phys. Lett.* **133B,** 419-422 (1983).

[118] A. T. M. Hertz, T. H. Hansson and B.-S. Skagerstam, "Thermodynamics of Boson-Fermion Duality in Confined Two-Dimensional Models" *Phys. Lett.* **145B,** 123-126 (1984).

[119] V. Bargmann, "On the Representation of the Rotation Group", *Rev. Mod. Phys.* **34,** 829-845 (1962).

[120] J. Schwinger, "On Angular Momentum", in *Quantum Theory of Angular Momentum,* eds. L. C. Biederharn and H. Van Dam (Academic Press, New York, 1965), pp. 229-279.

*[121] Y. Takahashi and F. Shibata, "Spin Coherent States Representation in Non-Equilibrium Statistical Mechanics", *J. Phys. Soc. Japan* **38,** 656-668 (1975), Paper VI.5.

[122] P. W. Atkins and J. C. Dobson, "Angular Momentum Coherent States", *Proc. Roy. Soc. Lond.* **A321,** 321-340 (1971).

[123] V. V. Mikhailov, "Boson Representation of Angular Momentum", *Theor. Math. Phys.* **15,** 367-374 (1973).

[124] P. Gulshani, "Oscillator-Like Coherent States of an Asymmetric Top", *Phys. Lett.* **71A,** 13-16 (1979); "Generalized Schwinger-Boson Realizations and the Oscillator-Like Coherent States of the Rotation Groups and the Asymmetric Top", *Can. J. Phys.* **57,** 998-1021 (1979).

[125] C. Aragone, G. Guerri, S. Salamó and J. L. Tani, "Intelligent Spin States", *J. Phys. A: Math. Gen.* **7,** L149-L151 (1974).

[126] G. Vetri, "Rotations and Intelligence of Coherent Spin States", *J. Phys. A: Math. Gen.* **8,** L55-L57 (1975); L. Kolodziejczyk, "Relationship Between Coherent States and Intelligent States as Applied to Spins", *J. Phys. A: Math. Gen.* **8,** L99-L101 (1975); M. A. Rashid, "The Intelligent States. I. Group Theoretic Study and the Computation of Matrix Elements", *J. Math. Phys.* **19,** 1391-1396 (1978); "The Intelligent States. II. The Computation of the Clebsch-Gordon Coefficients", *J.*

Math. Phys. **19,** 1397-1402 (1978).

[127] I. A. Malkin and V. I. Man'ko, "Coherent States of a Charged Particle in a Magnetic Field", *Sov. Phys. JETP* **28,** 527-532 (1969); "Coherent States and Excitation of a Charged Particle in a Constant Magnetic Field by Means of an Electric Field", *Theor. Math. Phys.* **6,** 71-77 (1971); "Coherent States and Green's Function of a Charged Particle in Variable Electric and Magnetic Fields", *Sov. Phys. JETP* **32,** 949-953 (1971).

[128] I. A. Malkin, V. I. Man'ko and D. A. Trifonov, "Invariants and the Evolution of Coherent States for a Charged Particle in a Time-Dependent Magnetic Field", *Phys. Lett.* **30A,** 414 (1969); "Evolution of Coherent States of a Charged Particle in a Variable Magnetic Field", *Sov. Phys. JETP* **31,** 386-390 (1970); "Coherent States and Transition Probabilities in a Time-Dependent Electromagnetic Field", *Phys. Rev.* **D2,** 1371-1385 (1970).

[129] V. V. Dodlonov, I. A. Malkin and V. I. Man'ko, "Coherent States of a Charged Particle in a Time-Dependent Uniform Electromagnetic Field of a Plane Current", *Physica* **59,** 241-256 (1972).

*[130] A. Feldman and A. H. Kahn, "Landau Diamagnetism from the Coherent States of an Electron in a Uniform Magnetic Field", *Phys. Rev.* **B1,** 4584-4589 (1970), Paper VI.4.

*[131] D. Janssen, "Coherent States of the Quantum-Mechanical Top", *Sov. J. Nucl. Phys.* **25,** 479-484 (1978), Paper IX.1.

[132] I. M. Pavlichenkov, "Quantum Theory of the Asymmetric Top", *Sov. J. Nucl. Phys.* **33,** 52-58 (1981).

[133] H.-Y. Fan and T.-N. Ruan, "SU(3) Charged and Hypercharged Coherent State and the Color Quantum Number", *Commun. in Theor. Phys.* (Beijing, China) **2,** 1405-1417 (1983).

[134] See, e.g., J. Schwinger, *Quantum Kinematics and Dynamics,* (W. A. Benjamin, Inc., 1970). This reference also contains a great deal of other coherent-state related material.

[135] F. A. Berezin, *The Method of Second Quantization,* (Academic Press, New York, 1966).

*[136] Y. Ohnuki and T. Kashiwa, "Coherent States of Fermi Operators and the Path Integral", *Prog. Theor. Phys.* **60,** 548-568 (1978), Paper IV.1.

[137] P. Jordan and E. Wigner, "Über das Paulische Äquivalenzverbot", *Z. Physik* **47,** 631-651 (1928).

[138] D. J. Thouless, "Stability Conditions and Nuclear Rotations in the Hartree-Fock Theory", *Nucl. Phys.* **21,** 225-232 (1960).

[139] H. Kuratsuji and T. Suzuki, "Path Integral Approach to Many-Nucleon Systems and Time-Dependent Hartree-Fock", *Phys. Lett.* **92B,** 19-22 (1980).

[140] S. Kobayashi and K. Nomizu, *Foundations of Differential Geometry,* Vol. II (Wiley & Sons, New York, 1964), p. 163.

[141] M. I. Monastyrsky and A. M. Perelomov, "Coherent States and Bounded Homogeneous Domains", *Rep. Math. Phys.* **6,** 1-14 (1974).

*[142] J. P. Blaizot and H. Orland, "Path Integrals for the Nuclear Many-Body Problem", *Phys. Rev.* **C24,** 1740-

1761 (1981), Paper IV.3.

[143] L. K. Hua, *Harmonic Analysis of Functions of Several Complex Variables in the Classical Domains,* Translations of Mathematical Monographs **6,** (American Mathematical Society, Providence, 1963), p. 40.

[144] H. Kuratsuji, "Classical Quantization of Time-Dependent Hartree-Fock Solutions by Coherent-State Path Integrals", *Phys. Lett.* **103B,** 79-82 (1981); "Classical Quantization Condition in the Generalized Coherent-State Representation with Application to Many-Body Systems", *Phys. Lett.* **108B,** 367-371 (1982); U. S. Van Roosmalen and A. E. L. Diepernik, "Properties of a Generalized Pseudo-Spin System: Application of the Time-Dependent Mean Field Method to an SU(4) Invariant Hamiltonian", *Ann. Phys. (N.Y.)* **139,** 198-211 (1982); S. E. Koonin, "Functional Integrals in Nuclear Physics", in *Proceedings of the Nuclear Theory Summer Workshop,* Santa Barbara, 1981, ed. G. F. Bertsch (World Scientific Publ. Co., Singapore, 1982), pp. 183-254.

[145] T. Suzuki, "Classical and Quantum Aspects of Time-Dependent Hartree-Fock Trajectories", *Nucl. Phys.* **A398,** 557-596 (1983).

*[146] L. R. Mead and N. Papanicolaou, "Holstein-Primakoff Theory for Many-Body Systems", *Phys. Rev.* **B28,** 1633-1636 (1983), Paper V.3.

*[147] M. M. Nieto and L. M. Simmons, Jr., "Coherent States for General Potentials", *Phys. Rev. Lett.* **41,** 207-210 (1978), Paper III.7.

*[148] M. M. Nieto, "Coherent States With Classical Motion From an Analytical Method Complementary to Group Theory", in *Group Theoretical Methods in Physics,* Vol.

II, Moscow, 1983, 174-188, Paper III.11.

[149] M. M. Nieto, "Resolution of the Identity for Minimum Uncertainty Coherent States: An Example Related to Charged-Boson Coherent States", *Phys. Rev.* **D30,** 770-773 (1984).

[150] R. P. Feynman, "Space-Time Approach to Non-Relativistic Quantum Mechanics", *Rev. Mod. Phys.* **20,** 367-387 (1948); see also R. P. Feynman and A. R. Hibbs, *Quantum Mechanics and Path Integrals* (McGraw-Hill, New York, 1965).

*[151] J. R. Klauder, "Weak Correspondence Principle", *J. Math. Phys.* **8,** 2392-2399 (1967), Paper II.8.

[152] J. R. Klauder, "Path Integrals", *Acta Physica Austriaca, Suppl. XXII,* 3-49 (1980).

[153] M. Ciafaloni and E. Onofri, "Path Integral Formulation of Reggeon Quantum Mechanics", *Nucl. Phys.* **B151,** 118-146 (1979).

[154] E. Onofri, "Path Integrals Over Coherent States", in *Functional Integration,* eds. J.-P. Antoine and E. Tirapegui (Plenum, New York, 1980) pp. 121-124.

*[155] H. Kuratsuji and T. Suzuki, "Path Integral Approach to Many-Body Systems and Classical Quantization of Time-Dependent Mean Field", *Prog. Theor. Phys. Suppl.* **74&75,** 209-220 (1983), Paper IV.5.

*[156] J. R. Klauder, "Path Integrals and Stationary-Phase Approximations", *Phys. Rev.* **D19,** 2349-2356 (1979), Paper IV.2.

[157] L. D. Faddeev, "Introduction to Functional Methods", in *Methods in Field Theory,* eds. R. Bailan and J. Zinn-

Justin (North-Holland and World Scientific Publ. Co., Singapore, 1981), pp. 1-39.

[158] Y. Weissman, "On the Stationary Phase Evaluation of Path Integrals in the Coherent States Representation", *J. Phys. A: Math. Gen.* **16,** 2693-2701 (1983).

*[159] M. Hillery and M. S. Zubairy, "Path-Integral Approach to Problems in Quantum Optics", *Phys. Rev.* **A26,** 451-460 (1982), Paper IV.4.

*[160] J. R. Klauder, "Coherent-State Langevin Equations for Canonical Quantum Systems with Applications to the Quantized Hall Effect", *Phys. Rev.* **A29,** 2036-2047 (1984), Paper VI.6.

*[161] A. Jevicki and N. Papanicolaou, "Semi-Classical Spectrum of the Continuous Heisenberg Spin Chain", *Ann. Phys. (N.Y.)* **120,** 107-128 (1979), Paper V.1.

[162] J. Ishida and A. Hosoya, "Path-Integral For a Color Spin and Path-Ordered Phase Factor", *Prog. Theor. Phys.* **62,** 544-553 (1979). See also S. Samuel, "Color Zitterbewegung", *Nucl. Phys.* **B149,** 517-524 (1979) and references therein.

[163] H. Ezawa and J. R. Klauder, "Fermions Without Fermions: The Nicolai Map Revisited", University of Bielefeld preprint, 74/1984, and references cited therein.

[164] P. Chernoff, "Note on Product Formulas for Operator Semigroups, *J. Func. Analysis* **2,** 238-242 (1968).

[165] F. A. Berezin, "General Concept of Quantization", *Commun. Math. Phys.* **40,** 153-174 (1975).

[166] R. Gilmore, "The Classical Limit of Quantum Non-Spin Systems", *J. Math. Phys.* **20,** 891-893 (1979).

[167] W. Fuller and A. Lenard, "Generalized Quantum Spins, Coherent States, and Lieb Inequalities", *Commun. Math. Phys.* **67,** 69-84 (1979).

*[168] R. Shankar, "Bohr-Sommerfeld Quantization of Pseudospin Hamiltonians", *Phys. Rev. Lett.* **45,** 1088-1091 (1980), Paper V.2.

*[169] J. R. Klauder and I. Daubechies, "Quantum Mechanical Path Integrals with Wiener Measures for All Polynomial Hamiltonians", *Phys. Rev. Lett.* **52,** 1161-1164 (1984), Paper IV.6; see also I. Daubechies and J. R. Klauder, "Quantum Mechanical Path Integrals with Wiener Measures for All Polynomial Hamiltonians. II" (to appear).

*[170] J. R. Klauder, "Exponential Hilbert Space: Fock Space Revisited", *J. Math. Phys.* **11,** 609-630 (1970), Paper II.9.

[171] See, e.g., A. V. Skorohod, *Integration in Hilbert Space,* (Springer-Verlag, New York, 1974), Chapter 1.

[172] See, e.g., G. C. Emch, *Algebraic Methods in Statistical Mechanics and Quantum Field Theory,* (Wiley-Interscience, New York, 1972).

[173] R. Haag, "On Quantum Field Theories", *Kgl. Danske Videnskab. Selskab, Mat.-Fys. Medd.* **29,** No. 12, 3-36 (1955); L. Streit, "A Generalization of Haag's Theorem", *Nuovo Cimento* **62A,** 673-680 (1969).

[174] K. O. Friedrichs, *Mathematical Aspects of the Quantum Theory of Fields,* (Interscience, New York, 1953).

[175] P. Garbaczewski, "The Method of Boson Expansion in Quantum Theory", *Phys. Rep.* **36,** 65-135 (1978); *Classical and Quantum Field Theory of Exactly*

Soluable Non-Linear Systems, to be published by World Scientific, Singapore.

[176] See, e.g., R. T. Powers, "Absence of Interaction as a Consequence of Good Ultraviolet Behavior in the Case of a Local Fermi Field", *Commun. Math. Phys.* **4,** 145-156 (1967).

[177] H. Araki, "Hamiltonian Formalism and the Canonical Commutation Relations in Quantum Field Theory", *J. Math. Phys.* **1,** 492-504 (1960).

*[178] K. Cahill, "Extended Particles and Solitons", *Phys. Lett.* **53B,** 174-176 (1974), Paper XI.3.

[179] P. Vinciarelli, "Sum Over Configurations (Non-Perturbative Quantum Expansions for Solitons and Other States in Field Theory)", *Phys. Lett.* **59B,** 380-386 (1975).

[180] R. F. Dashen, B. Hasslacher and A. Neveu, "Nonperturbative Methods and Extended-Hadron Models in Field Theory. I & II & III", *Phys. Rev.* **D10,** 4114-4142 (1974); L. D. Faddeev and V. E. Korepin, "Quantum Theory of Solitons", *Phys. Rep.* **42C,** 1-87 (1978); R. U. Konoplich and S. G. Rubin, "Quantum Corrections to Nontrivial Classical Solutions in ϕ^4 Theory", *Sov. J. Nucl. Phys.* **37,** 790-793 (1983).

*[181] D. E. L. Pottinger, "Creation and Annihilation Operators for Nielsen-Olesen Vortices in the Coherent State Approximation", *Phys. Lett.* **78B,** 476-478 (1978), Paper XI.7.

[182] J. R. Klauder, "Classical Concepts in Quantum Contexts", *Acta Physica Austriaca, Suppl. XVIII,* 1-62 (1977).

[183] See, e.g., C. Itzykson and J.-B. Zuber, *Quantum Field Theory* (McGraw-Hill, New York, 1980).

*[184] V. Chung, "Infrared Divergences in Quantum Electrodynamics", *Phys. Rev.* **140,** B1110-B1122 (1965), Paper XI.1.

[185] T. W. B. Kibble, "Coherent Soft-Photon States and Infrared Divergences. I. Classical Currents", *J. Math. Phys.* **9,** 315-324 (1968); "II. Mass-Shell Singularities of Green's Functions", *Phys. Rev.* **173,** 1527-1535 (1968); "III. Asymptotic States and Reduction Formulas", *Phys. Rev.* **174,** 1882-1901 (1968); "IV. The Scattering Operator", *Phys. Rev.* **175,** 1624-1640 (1968); see also T. W. Kibble, "Some Applications of Coherent States", in *Cargèse Lectures in Physics,* 1967, ed. M. Lévy (Gordon and Breach, New York, 1968), pp. 299-345, and "Coherent States and Infrared Divergences", in *Mathematical Methods in Theoretical Physics,* eds. K. T. Mahanthappa and W. E. Brittin (Gordon and Breach, New York, 1969), pp. 387-477.

[186] P. P. Kulish and L. D. Faddeev, "Asymptotic Conditions and Infrared Divergences in Quantum Electrodynamics", *Theor. Math. Phys.* **4,** 745-757 (1970).

[187] Ph. Blanchard, "Discussion Mathematique du Modèle de Pauli et Fierz Relatif à la Catastrophe Infrarouge", *Commun. Math. Phys.* **15,** 156-172 (1969).

[188] K.-E. Eriksson, "Asymptotic States in Quantum Electrodynamics", *Physica Scripta* **1,** 3-12 (1970).

[189] N. Papanicolaou, "Infrared Problems in Quantum Electrodynamics", *Phys. Rep.* **24C,** 229-313 (1976).

[190] J. Fröhlich, G. Morchio and F. Strocchi, "Charged Sectors and Scattering States in Quantum Electrodynamics", *Ann. Phys. (N.Y.)* **119,** 241-284 (1979); G. Morchio and F. Strocchi, "A Non-Perturbative Approach to the Infrared problem in QED", *Nucl. Phys.* **B211,** 471-508 (1983).

[191] S. Weinberg, "Infrared Photons and Gravitons", *Phys. Rev.* **140,** B516-B524 (1965).

[192] C. Alvegard, K.-E. Eriksson and C. Högfors, "Soft Graviton Radiation", *Physica Scripta* **17,** 95-102 (1978).

*[193] J. Gomatam, "Coherent States and Indefinite Metric - Applications to the Free Electromagnetic and Gravitational Fields", *Phys. Rev.* **D3,** 1292-1298 (1971), Paper XI.2.

[194] For a review see, e.g., W. Marciano and H. Pagels, "Quantum Chromodynamics", *Phys. Rep.* **36C,** 137-276 (1978).

*[195] M. Grecco, F. Palumbo, G. Pancheri-Srivastava and Y. Srivastava, "Coherent State Approach to the Infrared Behavior of Non-Abelian Gauge Theories", *Phys. Lett.* **77B,** 282-286 (1978), Paper XI.6.

[196] G. Curci and M. Greco, "Mass Singularities and Coherent States in Gauge Theories", *Phys. Lett.* **79B,** 406-410 (1978); G. Curci, M. Greco and Y. Srivastava, "QCD Jets from Coherent States", *Nucl. Phys.* **B159,** 451-468 (1979).

[197] D. R. Butler and C. A. Nelson, "Non-Abelian Structure of Yang-Mills Theory and Infrared-Finite Asymptotic States", *Phys. Rev.* **D18,** 1196-1198 (1978); C. A. Nelson, "Origin of Cancellation of Infrared Divergences in

Coherent State Approach", *Nucl. Phys.* **B181,** 141-156 (1981); "Avoidance of Counter-Example to Non-Abelian Bloch-Nordsieck Conjecture by Using Coherent State Approach", *Nucl. Phys.* **B186,** 187-204 (1981).

[198] H. D. Dahmen and F. Steiner, "Asymptotic Dynamics of QCD, Coherent States, and the Quark Form Factor", *Z. Phys.* **C11,** 247-249 (1981); H. D. Dahmen, B. Scholz and F. Steiner, "First Results of a Calculation of the Long Range Quark-Antiquark Potential from Asymptotic QCD Dynamics", *Z. Phys.* **C12,** 229-234 (1982).

*[199] K.-E. Eriksson and B.-S. Skagerstam, "Classical Source Emitting Self-Interacting Bosons", *Phys. Rev.* **D18,** 3858-3862 (1978), Paper XI.4.

*[200] G. Curci, M. Greco and Y. Srivastava, "Coherent Quark-Gluon Jets", *Phys. Rev. Lett.* **43,** 834-837 (1979), Paper X.3.

[201] F. P. Rayne, "A Coherent State Model for Electron-Positron Annihilation", *Nucl. Phys.* **B159,** 244-262 (1979).

[202] S. S. Schweber, "On Feynman Quantization", *J. Math. Phys.* **3,** 831-842 (1962).

[203] F. A. Berezin, "Non-Wiener Functional Integrals", *Theor. Math. Phys.* **6,** 141-155 (1971).

*[204] C. L. Hammer, J. E. Shrauner and B. DeFacio, "Path-Integral Representation for the S-Matrix", *Phys. Rev.* **D18,** 373-384 (1978), Paper XI.5.

II. General Developments

Reprinted from ANNALS OF PHYSICS, Volume 11, No. 2, October 1960

ANNALS OF PHYSICS: **11**, 123–168 (1960)

The Action Option and a Feynman Quantization of Spinor Fields in Terms of Ordinary C-Numbers*

JOHN R. KLAUDER†

Palmer Physical Laboratories, Princeton, New Jersey

The Feynman sum represents a convenient formulation of quantum mechanics for Bose fields, but, to secure a similar formulation applicable to Fermion fields, it has been necessary to use "anticommuting c-number" field histories to insure the anticommutivity of the quantum field operators. Here, a method is presented to sum over histories for spinor fields which (1) employs the familiar classical c-number expression for the action, (2) predicts anticommutation rules and Fermi statistics, and (3) retains the invariance of the theory under a change in phase of the complex ψ field.

The Feynman procedure demands a numerical action value for histories outside the domain for which the action integral was intended, for example, for histories which are discontinuous with respect to space or time. One is therefore presented with an "action option", i.e., the action value for such "unruly" histories may be defined in various ways. Depending on the choice made, the resulting quantum theory can be made to manifest either Bose or Fermi statistics. This ambiguity is inherent in the formalism itself. However, the proper choice to extend the classical information is most readily determined by constructing the sum over histories by a summation over multiple products of matrix elements of the unitary operator which advances the state an infinitesimal time. This summation need not be limited to the familiar discrete basis vectors; instead a "generalized representation" can be employed which involves, for each Fermion degree of freedom, continuously many, nonindependent vectors. When a suitable parameterization is chosen for this "overcomplete family of states" the multiple product of matrix elements for a given history reduces to the exponential of the appropriate action functional evaluated for that history.

A unified formulation of both statistics for the Schrödinger field is presented which includes a detailed account of the necessary properties of the overcomplete family of states and a derivation of the functional measure for Fermion fields. The propagator and a functional expression for the ground state of the neutrino field are presented as applications of the method to relativistic spinor fields.

---

* Based on a thesis submitted to Princeton University, May, 1959.

† Now at Bell Telephone Laboratories, Murray Hill, New Jersey.

I. INTRODUCTION AND SURVEY OF THE PAPER; THE FEYNMAN SUM OVER HISTORIES AND SOME AMBIGUITIES IN ITS CONSTRUCTION; SUMMARY OF PRINCIPAL RESULTS

I.1 Introduction

The sum over histories, developed by Feynman (*1*, *2*), represents a functional approach to quantum theory which has been widely used to quantize fields obeying Bose statistics (*3–6*). In order to include the quantization of spinor fields into the same formalism it has been necessary to employ "classical" field histories which are constructed from anticommuting *c*-numbers (*4*). Such field histories have been used to insure the anticommutivity of the quantum field operators in order to obtain Fermi statistics. However, the end (Fermi statistics) does not necessarily justify the means (anticommuting *c*-numbers), and in this paper we investigate a purely *c*-number procedure to quantize spinor fields via a sum over histories, a procedure which also describes the desired Fermi statistics.

Not only do the anticommuting *c*-numbers represent an undesirable addition to the physical description of nature, they even appear to be dubious mathematical entities as well. For example, Symanzik (*7*) makes the following remark regarding the necessary properties they must process: "Bei der Behandlung von Fermionfeldern werden spinorielle Funktionen eingeführt, die mit sich selbst, mit ihrer Adjugierten und mit anderen spinoriellen Funktionen antikommutieren. In dieser Antivertauschbarkeit mit der konjugiert komplexen Funktion liegt eine Schwierigkeit, da hieraus das identische Verschwinden dieser spinoriellen Funktionen folgt . . .".

One may perhaps avoid the use of anticommuting *c*-numbers by considering a sum over histories which concentrates on the observable aspects of the theory. However, an undesirable feature in this approach has been pointed out by Nambu (*8*), who remarked: "Instead of allowing the coordinates q to take on all real values from $-\infty$ to $+\infty$, we could assign here two values 1 and 0 (or discrete values in general) to the quantity $\psi^*\psi$, but the formulation does not seem to be so simple and advantageous as in Bose systems". Can one not avoid *both* anticommuting *c*-numbers *and* discrete field values? Let us discuss the principal techniques we shall use by treating the simplest possible example: the case of a single harmonic oscillator quantized according to Bose and Fermi statistics.

I.2 The Harmonic Oscillator as an Example

The classical theory of a single harmonic oscillator is summarized, in complex form, by the simple equations of motion $i\dot{a}(t) = a(t)$ and $-i\dot{a}^*(t) = a^*(t)$. These equations follow, for example, from the classical action principle

$$I = \int \left[\left(\frac{i}{2}\right)\left(a^* \frac{\overleftrightarrow{\partial}}{\partial t} a\right) - a^*a \right] dt \tag{1}$$

by varying the independent coordinates a^* and a. Here, the "double arrow" derivative is a convenient shorthand defined by

$$A \overset{\leftrightarrow}{\frac{\partial}{\partial t}} B = A \frac{\partial B}{\partial t} - \frac{\partial A}{\partial t} B\,. \tag{2}$$

Upon quantization, the complex coordinate $a(t)$ is replaced by an operator $\mathbf{a}(t)$ which satisfies the same equation of motion as the classical variable: $i\dot{\mathbf{a}}(t) = \mathbf{a}(t)$ and $-i\dot{\mathbf{a}}^*(t) = \mathbf{a}^*(t)$. To obtain either (1) Bose statistics or (2) Fermi statistics one imposes the commutation rules (1) $[\mathbf{a},\mathbf{a}^*] = 1$ or (2) $\{\mathbf{a},\mathbf{a}\} = 0$ and $\{\mathbf{a},\mathbf{a}^*\} = 1$, respectively, for the operators $\mathbf{a}$ and $\mathbf{a}^*$ at a fixed time.[1] These commutation rules predict that the operator $\mathbf{N} = \mathbf{a}^*\mathbf{a}$ has the following eigenvalues (1) $N' = 0, 1, 2, \cdots$ for Bose statistics, and (2) $N' = 0, 1$ for Fermi statistics, and we let $| N'\rangle$ denote the eigenvector associated with the eigenvalue N'. The propagator for this quantum system is defined by the matrix element

$$\langle N''t'' \mid N't'\rangle = \langle N'' \mid e^{-i(t''-t')\mathbf{H}} \mid N'\rangle. \tag{3}$$

For this elementary example, where $\mathbf{H} = \mathbf{a}^*\mathbf{a} = \mathbf{N}$, the propagator is simply $\delta_{N''N'} \exp[-i(t'' - t')N']$. Physically, the propagator represents the probability amplitude for the system to undergo a transition from an initial configuration to some final configuration, and it contains all the relevant dynamical information for the system.

We first construct a Feynman sum for the propagator (3) in the case of Bose statistics. It will be very convenient to work not with the basis $| N'\rangle$ but rather to introduce an *overcomplete family of states* containing the vector $| a\rangle$ for all complex a where we define

$$| a\rangle \equiv \exp(-\tfrac{1}{2} | a |^2) \sum_{N'=0}^{\infty} (N'!)^{-1/2} a^{N'} \mid N'\rangle. \tag{4}$$

In fact, as we shall see, this choice of states is dictated if we use the form (1) for the action in the Feynman sum. It follows directly from the definition (4) that $\langle a \mid a\rangle = 1$, and $\langle a \mid \mathbf{a}^*\mathbf{a} \mid a\rangle = a^*a$. Less obvious is the fact that the set of states $| a\rangle$ forms a "basis" for the quantum mechanical Hilbert space. To see this let us calculate the following integral over the real part, a_r, and the imaginary part, a_i, of a.

$$\begin{aligned} \int_{-\infty}^{\infty}\int_{-\infty}^{\infty} | a\rangle \frac{da_r da_i}{\pi} \langle a | &\equiv \int | a\rangle \text{“}d^B a\text{”} \langle a | \\ &\doteq \int_{-\infty}^{\infty}\int_{-\infty}^{\infty} \exp(- | a |^2) \sum_{N',N''=0}^{\infty} (N'!N''!)^{-1/2} a^{*N''} a^{N'} \mid N'\rangle\langle N'' \mid \frac{da_r da_i}{\pi}\,. \end{aligned} \tag{5a}$$

[1] Throughout this paper, units are chosen so that $\hbar = c = 1$.

Transforming to polar coordinates one readily determines that (5a) reduces to

$$\int_0^\infty \exp(-\mid a \mid^2) \sum_{N'=0}^{\infty} (N'!)^{-1} \mid a \mid^{2N'} \mid N'\rangle\langle N' \mid d \mid a \mid^2 = \sum_{N'=0}^{\infty} \mid N'\rangle\langle N' \mid = 1. \tag{5b}$$

This "resolution of the identity" is a very useful tool and we shall calculate the propagator between two states, $\mid a'\rangle$ and $\mid a''\rangle$, which are members of the over-complete family:

$$\langle a''t'' \mid a't'\rangle \equiv \langle a'' \mid e^{-i(t''-t')\mathbf{H}} \mid a'\rangle. \tag{6}$$

To generate a functional expression for (6) we first subdivide the time interval, $t'' - t'$, into $N + 1$ equal intervals of length ϵ and repeatedly insert the unit operator as expressed by (5):

$$\langle a''t'' \mid a't'\rangle = \int \cdots \int \langle a'' \mid e^{-i\epsilon\mathbf{H}} \mid a_N\rangle\langle a_N \mid e^{-i\epsilon\mathbf{H}} \mid a_{N-1}\rangle \cdots \langle a_1 \mid e^{-i\epsilon\mathbf{H}} \mid a'\rangle \, d^B a_N \cdots d^B a_1 . \tag{7}$$

When ϵ is small (N large) we may evaluate, to order ϵ, each term in the integrand in (7) as follows

$$\langle a_{l+1} \mid e^{-i\epsilon\mathbf{H}} \mid a_l\rangle \approx \langle a_{l+1} \mid \mathbf{1} - i\epsilon\mathbf{H} \mid a_l\rangle = \langle a_{l+1} \mid a_l\rangle - i\epsilon \langle a_{l+1} \mid \mathbf{H} \mid a_l\rangle$$
$$= \langle a_{l+1} \mid a_l\rangle \left(1 - i\epsilon \frac{\langle a_{l+1} \mid \mathbf{H} \mid a_l\rangle}{\langle a_{l+1} \mid a_l\rangle}\right) \approx \langle a_{l+1} \mid a_l\rangle \exp\left(-i\epsilon \frac{\langle a_{l+1} \mid \mathbf{H} \mid a_l\rangle}{\langle a_{l+1} \mid a_l\rangle}\right). \tag{8}$$

If we substitute this evaluation back into (7), the resulting expression for the propagator, valid to order ϵ, takes the form

$$\langle a''t'' \mid a't'\rangle = \int \cdots \int \prod_{l=0}^{N} \left[\langle a_{l+1} \mid a_l\rangle \exp\left(- i\epsilon \frac{\langle a_{l+1} \mid \mathbf{H} \mid a_l\rangle}{\langle a_{l+1} \mid a_l\rangle}\right)\right] \prod_{m=1}^{N} d^B a_m , \tag{9}$$

where $a_0 \equiv a'$ and $a_{N+1} \equiv a''$. The ratio of the matrix elements in the exponent may be easily evaluated by using the relation $\mathbf{a} \mid a\rangle = a \mid a\rangle$, so that

$$\langle a_{l+1} \mid \mathbf{H} \mid a_l\rangle / \langle a_{l+1} \mid a_l\rangle = a^*_{l+1} a_l \,;$$

from (4) it is a simple matter to show that

$$\langle a_{l+1} \mid a_l\rangle = \exp(-\tfrac{1}{2} \mid a_{l+1} \mid^2 + a^*_{l+1} a_l - \tfrac{1}{2} \mid a_l \mid^2).$$

In terms of these results the propagator will be expressed, to order ϵ, by

$$\langle a''t'' \mid a't' \rangle = \int \cdots \int \exp\left[\sum_{l=0}^{N} (-\tfrac{1}{2} \mid a_{l+1} \mid^2 + \overset{*}{a_{l+1}} a_l - \tfrac{1}{2} \mid a_l \mid^2 - i\,\epsilon\, \overset{*}{a_{l+1}} a_l)\right] \prod_{m=0}^{N} d^B a_m . \tag{10}$$

This finite number of Gaussian integrals may be evaluated to yield the propagator

$$\begin{aligned} \langle a''t'' \mid a't' \rangle &= \exp[-\tfrac{1}{2}a^{*\prime\prime}a'' + a^{*\prime\prime}e^{-i\epsilon(N+1)}a' - \tfrac{1}{2}a^{*\prime}a'], \\ &= \exp[-\tfrac{1}{2}a^{*\prime\prime}a'' + a^{*\prime\prime}e^{-i(t''-t')}a' - \tfrac{1}{2}a^{*\prime}a'], \end{aligned} \tag{11}$$

which, as $N \to \infty$, $\epsilon \to 0$ and $(N + 1)\epsilon = t'' - t'$, constitutes an exact evaluation of (6).

The limit $N \to \infty$, $\epsilon \to 0$ will define a formal expression for the right side of (10) which represents the symbolic Feynman sum. As $\epsilon \to 0$ the exponent in (10) may be abbreviated

$$\sum \left\{i\left(\frac{i}{2}\right)[\overset{*}{a_{l+1}}(a_{l+1} - a_l) - (\overset{*}{a_{l+1}} - a_l^*)a_l] - i\,\epsilon\, \overset{*}{a_{l+1}} a_l\right\} \to i \int \left[\left(\frac{i}{2}\right)(a^*\dot{a} - \dot{a}^*a) - a^*a\right] dt,$$

and we symbolize

$$\lim_{N\to\infty} \prod_{m=1}^{N} d^B a_m = N\mathfrak{D}\, a(t),$$

where N represents a normalization factor. Consequently, we have generated the familiar Feynman rule

$$\langle a''t'' \mid a't' \rangle = N \int \exp\left\{i \int_{t'}^{t''} \left[\left(\frac{i}{2}\right)\left(a^* \overset{\leftrightarrow}{\frac{\partial}{\partial t}} a\right) - a^*a\right] dt\right\} \mathfrak{D}a(t) \tag{12}$$

for constructing a sum over histories which is based on a functional integration and the classical action (1). It is clear from the derivation that only for histories which are continuous in time can the classical action functional (1) have any validity as a weighting in the exponent in (12). Discontinuous histories are outside the domain of applicability of (1); loosely speaking, the prescription in (10) may be viewed as *defining* the action value for such unruly histories. The above procedure will be extended, in Section II.2, to quantize the Schrödinger field according to Bose statistics.

Let us follow an analogous route to a functional description for the propagator in the case of Fermi statistics. For this purpose we express the propagator as

$$\langle b''t'' \mid b't' \rangle \equiv \langle b'' \mid e^{-i(t''-t')\mathbf{H}} \mid b' \rangle \tag{13}$$

in terms of vectors in another overcomplete family of states containing $\mid b\rangle$, for all complex b for which $\mid b \mid \leqq 1$, and where

$$\mid b\rangle \equiv (1 - \mid b \mid^2)^{1/2} \mid 0\rangle + b \mid 1\rangle. \tag{14}$$

It follows immediately that $\langle b \mid b\rangle = 1$, and $\langle b \mid \mathbf{a}^*\mathbf{a} \mid b\rangle = b^*b$. The set of vectors $\mid b\rangle$ will also behave like a "basis" where the "resolution of the identity" is given by

$$\begin{aligned}\iint_{|b|\leq 1} \mid b\rangle \frac{2}{\pi}\, db_r\, db_i\, \langle b \mid &\equiv \int \mid b\rangle \text{“}db\text{''} \langle b \mid \\ = \int_0^1 [(1 - \mid b \mid^2) \mid 0\rangle\langle 0 \mid + \mid b \mid^2 \mid 1\rangle\langle 1 \mid] \, 2\, d \mid b \mid^2 &= \sum_{N'=0}^{1} \mid N'\rangle\langle N' \mid = 1.\end{aligned} \tag{15}$$

The derivation which led from (6) to (9) made no use of specific properties of the overcomplete family of states and would apply equally well if the vectors $\mid b\rangle$ were substituted for $\mid a\rangle$. Therefore, the propagator (13) may be approximated, to order ϵ, by the expression

$$\langle b''t'' \mid b't' \rangle = \int \cdots \int \prod_{l=0}^{N} \left[\langle b_{l+1} \mid b_l \rangle \exp\left(- i\,\epsilon \frac{\langle b_{l+1} \mid \mathbf{H} \mid b_l \rangle}{\langle b_{l+1} \mid b_l \rangle} \right) \right] \prod_{m=1}^{N} db_m . \tag{16}$$

Let us find the weighting given by (16), in the limit $N \to \infty$, $\epsilon \to 0$, for continuous histories $b(t)$. When the difference $b_{l+1} - b_l$ is small and of order ϵ it follows, from (14), that

$$\langle b_{l+1} \mid b_l \rangle \approx \exp\{ -\tfrac{1}{2}[b_{l+1}^*(b_{l+1} - b_l) - (b_{l+1}^* - b_l^*)b_l]\}; \tag{17}$$

in addition, to the same order, $\epsilon \langle b_{l+1} \mid \mathbf{H} \mid b_l\rangle / \langle b_{l+1} \mid b_l \rangle = \epsilon\, b_{l+1}^* b_l$. Consequently, for every continuous history the weighting given by the limit of (16) is exactly expressed by

$$\exp(-\textstyle\sum\{(\tfrac{1}{2})[b_{l+1}^*(b_{l+1} - b_l) - (b_{l+1}^* - b_l^*)b_l] + i\,\epsilon\, b_{l+1}^* b_l\}) \to$$

$$\exp\left\{ i \int_{t'}^{t''} \left[\left(\frac{i}{2}\right) \left(b^* \frac{\overleftrightarrow{\partial}}{\partial t} b \right) - b^*b \right] dt \right\}$$

and *with the same validity as* (12), we may write the Feynman sum as

$$\langle b''t'' \mid b't' \rangle = N \int \exp\left\{ i \int_{t'}^{t''} \left[\left(\frac{i}{2}\right) \left(b^* \frac{\overleftrightarrow{\partial}}{\partial t} b \right) - b^*b \right] dt \right\} \mathfrak{D}b(t) \tag{18}$$

in the case of Fermi statistics. This formula for the Feynman sum is identical with that found for Bose statistics. However, the propagators in the two cases are quite different. In the Fermi case the propagator is found by evaluating (13) and is

$$\langle b''t'' \mid b't' \rangle = (1 - |b''|^2)^{1/2}(1 - |b'|^2)^{1/2} + b^{*\prime\prime}e^{-i(t''-t')}b'. \tag{19}$$

Compare this with the Bose result in Eq. (11)!

As for Eq. (12) in the Bose case, so here Eq. (18) is a *symbolic equation*, where the exponent is defined by ordinary methods of calculus only for continuous histories but where this standard definition has been *extended* to deal with discontinuous histories. The action is extended in a different way for the Fermi case than for the Bose case.

I.3 The Case of Fields

The calculations made on the single degree of freedom form the prototype of our analysis of the Schrödinger field and the relativistic spinor fields in Section II. In the case of fields additional classes of unruly histories arise for which the classical action is inapplicable. These classes are connected with the infinite number of degrees of freedom and contain, for example, fields represented by improper functions or by functions which are not square integrable. For such histories, which are outside the "classical" domain, the "action option" applies, i.e., their action value may be variously defined. In constructing the Feynman sum we shall be formally able to give the weight, $\exp[iI(\psi)]$, to *every continuous field history in Hilbert space, independently of the resulting statistics.* Here, $I(\psi)$ represents, for example, the familiar c-number action functional for the Schrödinger field:

$$I(\psi) = \int \left[\left(\frac{i}{2}\right) \left(\psi^* \frac{\overleftrightarrow{\partial}}{\partial t} \psi\right) - \psi^* H \psi \right] d^4x, \tag{20}$$

where H is the Hamiltonian, for example $H = -\nabla^2/2m + V(x)$.

The main contentions of this paper may be summarized in the following fundamental theses:

(1) A classical action principle does not suffice to define the Feynman sum over histories uniquely. One is presented with *an action option for unruly histories*; i.e., an ambiguity in the definition of the action for those field histories outside the domain of continuity.

(2) Depending on which action option one selects for the Schrödinger field one can obtain either Bose or Fermi statistics (or one of the many intermediate cases). For relativistic spinor fields, as well, one will be presented with the customary mathematical arbitrariness in the choice of statistics.

(3) If we choose the action in the form which is invariant under the change $\psi(x) \to \exp(i\lambda)\psi(x)$ and extend its definition to unruly histories (action option) then the Feynman integral will give the propagator in a representation corresponding to an overcomplete family of states.

I.4 Survey of the Paper

In II.1, an introductory part of the next section, we present a brief outline of some customary uses of the Feynman sum in the quantization of fields. A brief illustration of the customary use of anticommuting *c*-numbers is presented by studying the quantization of the free Dirac field. In Sections II.2 and II.3 we treat the Schrödinger field quantized with Bose and Fermi statistics according to the methods developed above in this section. Propagators and the representation of operators are discussed for these two cases.

To unify the treatment of all statistics we present (Section II.4) a very general formulation which discusses the detailed properties of the overcomplete family of states. The functional measure for Fermion fields is also derived. The quantization of spinor fields is formally no more difficult that the quantization of the Schrödinger field and we conclude Section II by presenting a detailed account of the two-component neutrino field which includes an expression for the propagator and an expression for the ground state functional (Section II.5).

Also in the case of interacting fields (Section III) the Feynman sum is readily stated in formal terms even though the evaluation can no longer be done explicitly in the general case.

II. A SUM OVER HISTORIES FOR THE SCHRÖDINGER FIELD AND FOR RELATIVISTIC SPINOR FIELDS

II.1 Anticommuting *C*-Number Field Histories

Before proceeding to a discussion of the quantization of the Schrödinger field, we shall first make some general remarks regarding several previous applications of the Feynman sum to the study of fields.

In the case of a field theory, characterized by a general field variable f, the sum over histories is defined in a manner analogous to the single particle case:

$$(f_2\sigma_2 \mid f_1\sigma_1) = N \int e^{iI(f)} \mathfrak{D}f. \tag{21}$$

In this equation $I(f)$ represents the classical action functional for the theory under discussion and $\int \cdots \mathfrak{D}f$ signifies that a functional integration is to be carried out over all field histories which satisfy the prescribed boundary conditions: f_2 on surface σ_2 and f_1 on surface σ_1. The normalization factor, N, is conveniently determined by the composition law which the propagator is required to satisfy.

$$(f_3\sigma_3 \mid f_1\sigma_1) = \int (f_3\sigma_3 \mid f_2\sigma_2)\delta f_2(f_2\sigma_2 \mid f_1\sigma_1). \tag{22}$$

Here, $\int \cdots \delta f_2$ signifies that a summation is to be carried out over all field configurations, f_2, on surface σ_2. Various matrix elements of operators may be defined in a similar manner but this definition is deferred until Section II.3.

Numerous authors have used such formulas to discuss field quantization (*3, 4, 9, 10*). The reader's attention is drawn to Tobocman's discussion of the free Dirac field quantized according to Fermi statistics (*9*) and the technique of Matthews and Salam (*4*) for quantizing spinor fields. Both of these methods require the use of eigenvectors which involve *anticommuting* c-number eigenvalues to insure the anticommutivity of the field operators. Similar "entities" have been used by other authors in alternate field-theory definitions (*11–14*). As an illustration of the use of anticommuting c-numbers, we choose a "sum over histories" style of representation of the Schwinger generating functional (the T-product generating functional) for a noninteracting Dirac field. The construction is analogous to that appearing in Eq. (21), but involves independent, external sources, η and $\bar{\eta}$, which are required to be anticommuting c-number fields. The sum over histories representation then takes the form:

$$\Omega(\eta,\bar{\eta}) = \frac{1}{N}\int \exp\left\{i\int\left[\bar{\eta}\psi + \bar{\psi}\eta - \bar{\psi}\left(\frac{1}{2}\gamma^\mu\overleftrightarrow{\partial}_\mu + m\right)\psi\right]d^4x\right\}\mathfrak{D}\psi\mathfrak{D}\bar{\psi}.$$

In addition, in this functional integration, the variables $\psi(x)$ and $\bar{\psi}(x)$ are themselves required to be anticommuting c-number spinor "functions". In order to avoid the Symanzik "disaster" mentioned in Section I.4, one also chooses these two fields to be independent, i.e., the field $\bar{\psi}(x)$ is *not* taken as the adjoint of the field $\psi(x)$. The sum over histories for this example is taken over the entire four-dimensional space-time and the composition law in Eq. (22) is no longer imposed. Instead, to determine the constant N one uses the relation $\Omega(0,0) = 1$. This equality follows from the fact that $\Omega(\eta,\bar{\eta})$ is related to the Heisenberg field operators, $\boldsymbol{\psi}(x)$ and $\bar{\boldsymbol{\psi}}(x)$, according to

$$\Omega(\eta,\bar{\eta}) - \langle 0 \mid T\left\{\exp\left[i\int(\bar{\eta}\boldsymbol{\psi} + \bar{\boldsymbol{\psi}}\eta)\,d^4x\right]\right\}\mid 0\rangle,$$

where $\mid 0\rangle$ denotes the physical vacuum state and T denotes the time ordering operator of Wick (*15*).

The functional integrations over anticommuting c-number fields have been made more "precise" by Matthews and Salam by the introduction of anticommuting orthonormal basis sets. Further details of their method may be found in an excellent and extensive account given by Novozilov (*16*).

II.2 Bose Statistics for the Schrödinger Field

A. Determination of the propagator

This section concentrates on a Bose quantization of the Schrödinger field by a sum over histories. The method will bc: (1) formulate the sum indicating the necessary computations and (2) show subsequently that Bose statistics have resulted.

To formulate the sum, as in (21), one needs an action integral I and a measure $\mathfrak{D}\psi$ on the functional space of histories. The action functional was given in (20), and we rewrite the form it takes:

$$I = \int \left[\left(\frac{i}{2}\right) (\psi^*\dot{\psi} - \dot{\psi}^*\psi) - \psi^* H \psi \right] d^4x, \tag{23}$$

where H represents the Hamiltonian of the system under consideration. Following the analysis of the single harmonic oscillator in Section I, we shall define the Feynman sum by first breaking up the time integration, in the action (23), into a finite number of equal time intervals, ϵ:

$$I_s \equiv \sum_{l=0}^{N} \int \left\{ \left(\frac{i}{2}\right) [\psi_{l+1}^*(\psi_{l+1} - \psi_l) - (\psi_{l+1}^* - \psi_l^*)\psi_l] - \epsilon\, \psi_{l+1}^* H \psi_l \right\} d^3x. \tag{24}$$

This choice of a finite difference to represent the integrand in (23) has two features: (1) the value of the action will coincide with that given by (23) for all paths $\psi_l(x)$ which, in the limit $\epsilon \to 0$, are continuous in time, and (2) both variables, ψ_r and ψ_i, occur at every "time slice." If the propagator be expressed in the symbolic form $(\psi''t'' \mid \psi't')$, then the specification of the field ψ' at the initial time slice and of the field ψ'' at the final time slice demands a knowledge of both real and imaginary parts of ψ.

It is interesting to contrast this theory with another theory which would have seemed plausible. In that theory the integral I of (23), regarded as the action functional of a classical theory, demands as proper boundary value data the specification of the real (or imaginary) part of ψ at an initial and a final time, or for a single time slice the data $\psi_{r,l}$ and $\psi_{r,l+1}$. The classical history defined by this boundary value data has a well-defined action,

$$I_H = I(\psi_{r,l+1}, t_{l+1}; \psi_{r,l}, t_l).$$

This action can be used in a familiar way (*18*, *19*) to define the propagator when $t_{l+1} - t_l$ is small:

$$(\psi_{r,l+1}, t_{l+1} \mid \psi_{r,l}, t_l) \approx e^{iI_H}.$$

However, for the present Lagrangian, the action I_H as so defined happens to be identically zero! Moreover, the resulting propagator—even if it did not have the trivial value unity—would depend on only half as many variables as the propagator

$$(\psi_{r,l+1}, \psi_{i,l+1}, t_{l+1} \mid \psi_{r,l}, \psi_{i,l}, t_l)$$

of the present theory—a theory that uses the prescription (24) to calculate the action (see Ref. *20*).

When we use the extended action given by (24) then, following the example of the single harmonic oscillator, we replace the functional measure $\mathfrak{D}\psi$ for histories by a finite product of measures, $\delta\psi$, for fields, each at a fixed time. We choose this measure, in a form customary for quadratic Lagrangians (*5*), as $\delta\psi = \mathfrak{N} \prod_x d\psi_r(x) d\psi_i(x)$ where $\mathfrak{N}$ represents an appropriate normalization factor. This selection for $\delta\psi$ is invariant under unitary changes of basis (*4*), such as $\psi(x) = \sum a_n \varphi_n(x)$ where $\int \varphi_n{}^* \varphi_m \, d^3x = \delta_{nm}$. Therefore we may define $\delta\psi$ equally well by the following expression:[2]

$$\delta\psi = \prod_{n=0}^{\infty} \left(\frac{da_{n,r} \, da_{n,i}}{\pi} \right). \tag{25}$$

In the integration that defines the sum over histories each variable, $a_{n,r}$ or $a_{n,i}$, will run from $-\infty$ to $+\infty$.

Equations (24) and (25) contain the required definitions, implicitly, to construct the propagator; the details of this construction appear in Part C, below. The actual computation is straightforward and involves only repeated Gaussian integrals.[3] The resultant propagator, when reexpressed in terms of the field $\psi(x)$, becomes

$$(\psi''t'' \mid \psi't') = \exp\left(-\frac{1}{2} \int \psi^{*\prime\prime} \psi'' \, d^3x \right.$$
$$\left. + \iint \psi^{*\prime\prime}(x) \, d^3x \, \langle xt'' \mid yt' \rangle \, d^3y \, \psi'(y) - \frac{1}{2} \int \psi^{*\prime} \psi' \, d^3y \right), \tag{26}$$

where $\langle xt'' \mid yt' \rangle = \langle x \mid \exp[-i(t'' - t')\mathbf{H}] \mid y \rangle$ is the single particle propagator associated with the Hamiltonian $\mathbf{H}$.

B. Functional unit matrix

Let us examine (26) for $t'' = t'$; the result will play the role of the unit matrix and will be of central importance in interpreting the quantum theory we have obtained.

$$(\psi'' \mid \psi') = \exp\left(-\frac{1}{2} \int \psi^{*\prime\prime} \psi'' \, d^3x + \int \psi^{*\prime\prime} \psi' \, d^3x - \frac{1}{2} \int \psi^{*\prime} \psi' \, d^3x \right) \tag{27}$$

[2] For these arguments any convenient orthonormal set of functions, $\varphi_n(x)$, may be used.
[3] A detailed discussion of integration over complex function space by Gaussian integrals and "Hermite functionals" has been given by Friedrichs (*21*).

since $\langle xt \mid yt \rangle = \delta^3(x - y)$. To interpret the result (27), note that a particular value of ψ, such as $\psi = \psi'(x)$, over a 3-space at a given time defines a *configuration* of the field quantity ψ at that time. For a single harmonic oscillator degree of freedom the coordinate x' at a given time defines a configuration of the system. Likewise for the ensemble of oscillators which describe the electromagnetic field a configuration of the system is specified by choosing a value for the magnetic field, $H'(x)$. In these familiar quantum mechanical systems the basis states associated with two different configurations are orthogonal to each other. No such orthogonality shows up in the bilinear form $(\psi'' \mid \psi')$ in Eq. (27). The base states[4] $\mid \psi')$ and $\mid \psi'')$ are not orthogonal although each is a normalized vector since $(\psi' \mid \psi') = 1$ follows from Eq. (27).

Can one illustrate such an unusual situation in the quantum mechanics of a one particle system? Certainly the state vectors $\mid x' \rangle$ and $\mid x'' \rangle$ are orthogonal if $x' \neq x''$. However, $\mid x' \rangle$ and $\mid p' \rangle$ are not orthogonal. More generally, as Condon has shown, one can define observables part way in character between x and p,

$$\xi_{\theta} = x \cos \theta + p \sin \theta.$$

Then the basis vectors $\mid \xi'_{\theta'} \rangle$ and $\mid \xi''_{\theta''} \rangle$ are not in general orthogonal. As in the primitive situation, so also in second-quantized theory the states $\mid \psi')$ and $\mid \psi'')$ ordinarily are not and need not be orthogonal.

Consequently, the choice involved in (24) has *led* us to a "representation" in terms of an overcomplete family of states. To show that this overcomplete family of states actually provides the basis for a representation, let us first expand (formally![5]) the middle term in the exponent of (27).

$$(\psi'' \mid \psi') = N''N' \sum_{n=0}^{\infty} (n!)^{-1} \int \cdots \int \psi^{*''}(x_1) \cdots \psi^{*''}(x_n) \cdot \psi'(x_1) \cdots \psi'(x_n)\, d^3x_1 \cdots d^3x_n, \tag{28}$$

where $N = \exp(-\tfrac{1}{2} \int \psi^*\psi\, d^3x)$. In the form (28) we easily recognize that $(\psi'' \mid \psi')$ may be viewed as the inner product of two normed states of a Fock representation (*22*). A given one of these normed states is defined by

$$\mid \psi) = N \sum_{n=0}^{\infty} (n!)^{-1} \int \cdots \int \psi(x_1) \cdots \psi(x_n)\, d^3x_1 \cdots d^3x_n \mid x_1, \cdots x_n), \tag{29}$$

where $\mid x_1, \cdots x_n)$ are a set of basis vectors such as are commonly used for Bose statistics, symmetric and orthonormal in the sense of Dirac. To complete

[4] The symbol $\mid\,)$ stands for a vector in the Hilbert space of the second-quantized theory, in distinction to the symbol $\mid\,\rangle$ which is reserved for vectors in the Hilbert space of the one-particle system.

[5] The explanation of this remark appears in II.4.

the proof that the set of vectors $|\psi)$ forms a "basis" we may evaluate the matrix $\int |\psi)\delta\psi(\psi|$. Using $\delta\psi$ as given in (25) for example, and the definition of $|\psi)$, above, in (29) we find

$$\int |\psi)\delta\psi(\psi| = \sum_{n=0}^{\infty} (n!)^{-1} \int \cdots \int | x_1, \cdots x_n)\, d^3x_1 \cdots d^3x_n\, (x_1, \cdots x_n | = 1$$

by a repeated application of the evaluation in Eq. (5).

The use of linear combinations of orthogonal vectors such as (29) is a common artifice but their use as a "basis" is less common. Dirac (*23*) has employed a unit matrix for Fock states where the integration consists of a path in the complex plane; we require an integration over the whole plane.

C. Reconstruction of the sum over histories

As an example of the use of the states $|\psi)$ as a basis, we may fold together a great number of propagators of the form (26), each referring to a very short time interval. In this way we will obtain a description of the propagator for a finite time interval as a Feynman sum in essentially the same manner discussed for simple systems in Section I. We can then see if this Feynman sum allows itself to be described as an exponential of the form $\exp(iI_H)$. If so, we then have a way to reconstruct the classical action from the quantum propagator.

To spell out this reconstruction process, we write the elementary propagator (26) in the form

$$(\psi''t'' | \psi't') = (\psi'' | e^{-i(t''-t')\mathbf{H}} | \psi'), \tag{30}$$

using the fact that every unitary operator can be written as the exponential of an anti-Hermitian operator. In this case the anti-Hermitian operator involves the second-quantized Hamiltonian operator $\mathbf{H}$. The time, $t'' - t'$, is subdivided and the "unit operator," $\int |\psi)\delta\psi(\psi|$, is repeatedly introduced. In this manner (30) may be represented by

$$\begin{aligned}(\psi''t'' | \psi't') = \int \cdots \int (\psi'' | e^{-i\epsilon\mathbf{H}} | \psi_N)\delta\psi_N(\psi_N | e^{-i\epsilon\mathbf{H}} | \psi_{N-1})\delta\psi_{N-1} \\ \times \cdots \delta\psi_1(\psi_1 | e^{-i\epsilon\mathbf{H}} | \psi').\end{aligned} \tag{31}$$

We now approximate, to an order of ϵ, the typical term appearing in the integrand of (31). Formally

$$\begin{aligned}(\psi_{l+1} | e^{-i\epsilon\mathbf{H}} | \psi_l) &\approx (\psi_{l+1} | \mathbf{1} - i\epsilon\mathbf{H} | \psi_l) = (\psi_{l+1} | \psi_l) - i\epsilon(\psi_{l+1} | \mathbf{H} | \psi_l) \\ &= (\psi_{l+1} | \psi_l)\left(1 - i\epsilon\frac{(\psi_{l+1} | \mathbf{H} | \psi_l)}{(\psi_{l+1} | \psi_l)}\right) \approx (\psi_{l+1} | \psi_l) \exp\left(-i\epsilon\frac{(\psi_{l+1} | \mathbf{H} | \psi_l)}{(\psi_{l+1} | \psi_l)}\right).\end{aligned} \tag{32}$$

When this expression for $(\psi_{l+1} \mid \exp(-i\,\epsilon\,\mathbf{H}) \mid \psi_l)$ is substituted back into Eq. (31) the propagator becomes

$$(\psi''t'' \mid \psi't') = \int \cdots \int (\psi'' \mid \psi_N) \exp\left(-i\,\epsilon\,\frac{(\psi'' \mid \mathbf{H} \mid \psi_N)}{(\psi'' \mid \psi_N)}\right) \delta\psi_N (\psi_N \mid \psi_{N-1})$$
$$\times \cdots \delta\psi_1 (\psi_1 \mid \psi') \exp\left(-i\,\epsilon\,\frac{(\psi_1 \mid \mathbf{H} \mid \psi')}{(\psi_1 \mid \psi')}\right). \tag{33}$$

Suppose the Hamiltonian operator, $\mathbf{H}$, is given by an expression of the form $\int \boldsymbol{\psi}^*(x) H \boldsymbol{\psi}(x)\, d^3x$. Here H represents a one-particle operator, and the quantities $\boldsymbol{\psi}^*(x)$ and $\boldsymbol{\psi}(x)$ are the familiar creation and annihilation operators. The calculation of $(\psi_{l+1} \mid \mathbf{H} \mid \psi_l)/(\psi_{l+1} \mid \psi_l)$ may be readily carried out using the fact that $\boldsymbol{\psi}(x) \mid \psi) = \psi(x) \mid \psi)$. Hence, this ratio of matrix elements reduces to, $\int \psi_{l+1}^*(x) \cdot H\psi_l(x)\, d^3x$, a result which is substituted back into Eq. (33). Use is also made of the expression (27) for the inner product $(\psi'' \mid \psi')$. Then the propagator reduces to a straightforward functional integral. Does the functional thus constructed agree with the Feynman sum? Yes! The resulting exponent in (33) agrees exactly with the action defined by the sum in Eq. (24) for skeletonized histories. In other words, the classical action has been recovered from the quantum expression for the propagator. In later sections, particularly Section III, important information about the definition of the Feynman sum will be gained by requiring the classical action recovered from the quantum propagator to have the same functional form as the usual classical action.

II.3 Fermi Statistics for the Schrödinger Field

It is not difficult to develop a formal sum over histories for Fermi statistics by extending, to the case of fields, the treatment of the single oscillator given in Section I.B. One such extension was employed in the last section to obtain a Bose quantization. The other extension—to give Fermi statistics—is equally straightforward. However, there is an instructive alternative route to the Fermi quantization which asks whether the exponent in the propagator (26) is well defined for unruly field configurations.

A. Assumptions are present in the derivation of Bose statistics

To secure Bose statistics in II.2 it was essential (1) to identify (28) with the inner product of two vectors, given by (29), and (2) to note that these vectors were superpositions of base vectors of a type frequently used for Bose statistics. Let us now rewrite (28) with a slight modification, *omitting contributions to the integrals whenever any* $x_i = x_j$. Let $\mathcal{P}$ preceding the integral signs indicate this alteration, so the modified expression (28) becomes

$$(\psi'' \mid \psi')' = N''N' \sum_{n=0}^{\infty} (n!)^{-1} \mathcal{P} \underset{(\text{omit: } x_1 = x_2,\, x_1 = x_3,\, \text{etc.})}{\int \cdots \int} \psi^{*\prime\prime}(x_1) \cdots \psi^{*\prime\prime}(x_n) \cdot \psi'(x_1) \cdots \psi'(x_n)\, d^3x_1 \cdots d^3x_n . \tag{34}$$

Clearly, only a set of measure zero has been omitted from each integral in (34). Such an omission leads to numerical values for (28) and (34) which can be distinct *only* for improper configurations which are singular in x space. For the class of functions which constitute a representation of the familiar quantum mechanical separable Hilbert spaces, the numerical value of (28) and (34) are *absolutely identical.* This important class of functions would have had the same property if (34) had been written as a sum of integrals in momentum space—integrals which similarly omitted a set of points of measure zero. Furthermore, it is of utmost importance to recognize that the configurations which represent vectors in Hilbert space constitute the "classical" domain of our theory; improper and non-normalizable functions are excluded (*24*)! Therefore, if singular configurations contributed significantly to the integrations made in II.2, then the particular treatment accorded these unruly fields would constitute an unwarranted assumption beyond the information contained in the "classical" theory. It is easy to see that singular configurations were included in the integrations. Construct a typical singular function that is contained in the domain of integration of the a_n's. Recall that the field ψ was related to the a_n via $\psi(x) = \sum a_n\varphi_n(x)$. Now, select an a_n set by letting $a_n = \varphi_n^*(x')$. Because of the closure property, this a_n set will generate the field $\psi(x) = \delta^3(x - x')$. For such fields the numerical value of the integrals with and without omissions ((34) and (28)) will formally no longer agree.[6]

B. Establishing Fermi statistics; Construction of the antisymmetric states

We now show how the modified treatment of unruly fields, implicit in (34), provides a description of Fermi statistics. For this purpose we shall express the sum (34) of integrals with omissions in terms of a *new* overcomplete family of states

$$\mid \psi) = N \sum_{n=0}^{\infty} (n!)^{-1} \int \cdots \int \psi(x_1) \cdots \psi(x_n)\, d^3x_1 \cdots d^3x_n \mid x_1, \cdots x_n)_S . \tag{35}$$

These states themselves are in turn expressed by (35) in terms of a standard orthonormal set of basis vectors of the type $\mid x_1, x_2, \cdots x_n)_S$. These basis vectors are symmetric in their arguments but are defined to vanish if any two indices

[6] Actually, the number of such unruly fields may be so enormous as to make the well-behaved "classical" fields by comparison only a set of measure zero (see, for example, Ref. *25*).

coincide. They describe a field such that two of its quanta can never be located at the same position, i.e., Fermi statistics. The above kets are, by definition, not antisymmetric in their arguments, but the familiar antisymmetric kets can easily be introduced by the use of an ordering sign function $\sigma(x_1, x_2, \cdots, x_n)$ discussed, for example, by Friedrichs (*21*).

Ordering of the elements and the ordering sign function. In order to introduce the antisymmetric basis vectors it is necessary, according to Jordan and Wigner (*26*), to select an ordering for the labels of the wave function. If the label were completely characterized by one integer, one could conveniently choose the "natural" ordering of the integers. However, our wave functions are labeled by the 3-vector, x.

An ordering for the points x is determined by a relation $<$ ("less than") between points which satisfies:

(1) Either $x < x'$, $x = x'$, or $x' < x$ and, of course, only one of these is true.
(2) If $x < x'$ and $x' < x''$, then $x < x''$.

One example of an ordering for the points x would be the following: We would say $x < x'$ if (1) $x_{(1)} < x'_{(1)}$ (here we use $<$ in the usual sense), or if (2) $x_{(1)} = x'_{(1)}$, $x_{(2)} < x'_{(2)}$ or if (3) $x_{(1)} = x'_{(1)}$, $x_{(2)} = x'_{(2)}$, $x_{(3)} < x'_{(3)}$, where the $x_{(i)}$ are the coordinates of the point x. An ordering referring to the points x will be called an x-ordering.

Now, the ordering sign function, $\sigma(x_1, x_2, \cdots x_n)$, is defined to have the value of ± 1 given by the sign of the permutation, P, which is necessary to bring the arguments of σ to the "standard" order, $x_{P_1} < x_{P_2} < \cdots < x_{P_n}$; if any two x's are equal, σ is defined to be zero. Then, the antisymmetric basis vectors are fully described by

$$| x_1, x_2, \cdots, x_n)_A = \sigma(x_1, x_2, \cdots, x_n) \, | x_1, x_2, \cdots, x_n)_S . \tag{36}$$

This expression defines $| x_1, x_2, \cdots, x_n)_A$ for all x_i and n.

Generally, one uses the anticommuting field operators $\psi^*(x)$ and $\psi(x)$ to define the antisymmetric basis vectors. Here one uses the base vectors to define the anticommuting creation and destruction operators; thus

$$\psi^*(x_1)\psi^*(x_2) \cdots \psi^*(x_n) \, | 0) \equiv | x_1, x_2, \cdots, x_n)_A , \tag{37a}$$

where $| 0)$ is the state annihilated by the adjoint operator:

$$\psi(x) \, | 0) = 0. \tag{37b}$$

C. Elementary dynamical problem with Fermi statistics

Having this new overcomplete family of states at hand, we can do everything for Fermi quantization that we have previously done for Bose quantization: (1) the concept of field configuration is unchanged, (2) the method of summing over histories—of integrating over field configurations—will turn out to be a little different than it was before, (3) the propagator, which previously had the value (26), will still have the same *formal* appearance. To spell out the use of the propagator in a particular problem and to show that it describes a system truly of Fermi character, we will not find it convenient, however, to operate in the x-representation. For this reason the following example of the quantization of the free-particle Schrödinger field is considered in the p-representation. This example is essentially an extension of the single oscillator example considered in Section I.B.

In order to construct the Feynman sum according to (33) we require: (1) a Hamiltonian operator, (2) an overcomplete family of states, and (3) a measure for fields at a fixed time. The Hamiltonian operator for the free particle is given by the familiar expression

$$\mathbf{H} = \frac{1}{2m} \int \varphi^*(p) p^2 \varphi(p)\, d^3p, \tag{38}$$

where $\varphi(p)$ and $\varphi^*(p)$ are operators satisfying

$$\{\varphi(p), \varphi(p')\} = 0 \quad \text{and} \quad \{\varphi(p), \varphi^*(p')\} = \delta^3(p - p').$$

Corresponding to this description in momentum space we define the vectors $|\,\varphi)$, which constitute the members of our overcomplete family of states, by the expression

$$|\,\varphi) \equiv N \sum_{n=0}^{\infty} (n!)^{-1} \int \cdots \int \varphi(p_1) \cdots \varphi(p_n)\, d^3p_1 \cdots d^3p_n \,|\, p_1, \cdots p_n)_S, \tag{39}$$

and N is a normalization factor chosen so that $(\varphi\,|\,\varphi) = 1$. Here, the vectors $|\, p_1, p_2, \cdots, p_n)_S$ form a set of states which are symmetric and orthonormal in the sense of Dirac and which are defined to vanish if any two indices coincide. To make a connection between the operators $\varphi(p)$ and the symmetric basis states $|\, p_1, p_2, \cdots, p_n)_S$, we must first select an ordering for the points p (a p-ordering) so that we may define a p-ordering sign function, $\sigma(p_1, p_2, \cdots, p_n)$. Then, we have the relations:

$$\begin{aligned} \varphi^*(p_1)\varphi^*(p_2) \cdots \varphi^*(p_n)\,|\,0) &= |\, p_1, p_2, \cdots, p_n)_A \\ &= \sigma(p_1, p_2, \cdots, p_n)\,|\, p_1, p_2, \cdots, p_n)_S, \end{aligned} \tag{40a}$$

where, again, $|\,0)$ is defined as the state satisfying

$$\varphi(p)\,|\,0) = 0. \tag{40b}$$

The next quantities necessary for (33) are the matrix elements $(\varphi''\,|\,\varphi')$ and

$(\varphi'' \mid \mathbf{H} \mid \varphi')$ where $\mid \varphi')$ and $\mid \varphi'')$ are members of the overcomplete family. It follows directly from the definition of the states $\mid \varphi)$ that

$$\begin{aligned}(\varphi'' \mid \varphi') &= N''N' \sum_{n=0}^{\infty} (n!)^{-1} \mathcal{P} \int \cdots \int \varphi^{*''}(p_1) \cdots \varphi^{*''}(p_n)\varphi'(p_1) \\ &\quad \times \cdots \varphi'(p_n)\, d^3p_1 \cdots d^3p_n \\ &= N''N' \prod_p (1 + \varphi^{*''}(p)\varphi'(p)\, d^3p\,).\end{aligned} \tag{41}$$

This latter functional form follows from the preceding expression by formally expanding the product and regrouping the terms. Numerically, of course, for *every* field φ' and φ'' in the "classical" domain

$$\prod_p (1 + \varphi^{*''}(p)\varphi'(p)\, d^3p\,) = \exp\left(\int \varphi^{*''}(p)\varphi'(p)\, d^3p\right), \tag{42}$$

as follows from the discussion below Eq. (34). The "multiple product" form is a convenient *formal shorthand* and clearly shows the independent status of each degree of freedom. The ratio of matrix elements

$$\frac{(\varphi'' \mid \mathbf{H} \mid \varphi')}{(\varphi'' \mid \varphi')} = \frac{1}{2m} \int \frac{\varphi^{*''}(p)p^2\varphi'(p)\, d^3p}{1 + \varphi^{*''}(p)\varphi'(p)\, d^3p} \tag{43}$$

follows from (39) and the properties of the operators $\boldsymbol{\varphi}(p)$ and $\boldsymbol{\varphi}^*(p)$. To evaluate this expression one may use the formal relation between the Dirac δ function and the Kronecker δ function (*21*) expressed in momentum space by

$$\delta^3(p - p')\, d^3p = \delta_{p,p'}\,. \tag{44}$$

If the fields φ' and φ'' are improper functions such that $\varphi^{*''}(p')\varphi'(p') = C\delta^3(0)$, for some p', then $\varphi^{*''}(p')\varphi'(p')\, d^3p = C\delta^3(0)\, d^3p = C$. This evaluation is used in the denominator as well as the numerator. However, for fields φ' and φ'' in the classical domain the denominator *plays no role* and (43) becomes

$$\frac{(\varphi'' \mid \mathbf{H} \mid \varphi')}{(\varphi'' \mid \varphi')} = \frac{1}{2m} \int \varphi^{*''}(p)p^2\varphi'(p)\, d^3p. \tag{45}$$

Formal expressions like (43) have been studied in detail by Friedrichs (*21*).

The measure $\delta\varphi$, for fields at a fixed time, takes the formal form

$$\delta\varphi = \mathfrak{N} \prod_p \frac{d\varphi_r(p)\, d\varphi_i(p)}{(1 + \varphi^*(p)\varphi(p)\, d^3p)^2}, \tag{46}$$

as we shall prove in Section II.4.

For the free particle every momentum degree of freedom is independent.

Therefore the calculation of the propagator simplifies to a repeated calculation of two simple integrals:

$$\int_0^\infty \frac{x\,dx}{(1+x)^3} = \int_0^\infty \frac{dx}{(1+x)^3} = \frac{1}{2}. \tag{47}$$

The resulting propagator for the free-particle non-relativistic Schrödinger field is then formally represented by

$$\langle \varphi''t'' \mid \varphi't' \rangle = N''N' \prod_p \left\{1 + \varphi^{*\prime\prime}(p) \exp\left[-i(t'' - t')\frac{p^2}{2m}\right] \varphi'(p)\, d^3p\right\}. \tag{48}$$

For fields which are members of the original "classical" domain ($\varphi'(p)$ and $\varphi''(p)$ in Hilbert space), Eq. (48) is numerically identical to

$$\begin{aligned}\langle \varphi''t'' \mid \varphi't' \rangle = \exp\Big(& -\frac{1}{2}\int \varphi^{*\prime\prime}\varphi''\, d^3p \\ & + \iint \varphi^{*\prime\prime}(p)\, d^3p\, \langle pt'' \mid p't' \rangle\, d^3p'\varphi'(p') - \frac{1}{2}\int \varphi^{*\prime}\varphi' d^3p\Big),\end{aligned} \tag{49}$$

where we have used $\langle pt'' \mid p't' \rangle = \delta^3(p - p') \exp[-i(t'' - t')p^2/2m]$. Equation (49) may be re-expressed in terms of any $\psi(x)$ whose Fourier transform, $\varphi(p) = (2\pi)^{-3/2} \int \exp(ip\cdot x)\psi(x)\, d^3x$, is nonsingular. Thus, for Fermi statistics, Eq. (49) becomes

$$\begin{aligned}\langle \psi''t'' \mid \psi't' \rangle = \exp\Big(& -\frac{1}{2}\int \psi^{*\prime\prime}\psi''\, d^3x \\ & + \iint \psi^{*\prime\prime}(x)\, d^3x\, \langle xt'' \mid x't' \rangle\, d^3x'\, \psi'(x') - \frac{1}{2}\int \psi^{*\prime}\psi' d^3x\Big),\end{aligned} \tag{50}$$

an expression identical to (26), the expression for the Bose propagator. We conclude, in this case, that the propagation functional valid for Bose statistics is also equally valid to describe Fermi statistics for the important class of functions (in Hilbert space) which make up the classical domain of the theory.

D. Functional representation of operators

It suffices to illustrate the functional representation of operators by treating the field operators. We first calculate the matrix elements of the field operators with respect to the overcomplete family of states. For Bose statistics, it follows from II.2 that

$$\langle \psi'' \mid \boldsymbol{\psi}(x) \mid \psi' \rangle = \psi'(x)\langle \psi'' \mid \psi' \rangle, \tag{51a}$$

and

$$\langle \psi'' \mid \boldsymbol{\psi}^*(x) \mid \psi' \rangle = \psi^{*\prime\prime}(x)\langle \psi'' \mid \psi' \rangle. \tag{51b}$$

In the case of Fermi statistics it follows that

$$(\varphi'' \mid \varphi(p) \mid \varphi') = N''N' \prod_{p'<p} (1 - \varphi^{*''}(p')\varphi'(p')\, d^3p')\varphi'(p) \cdot \prod_{p''>p} (1 + \varphi^{*''}(p'')\varphi'(p'')\, d^3p''), \tag{52a}$$

and

$$(\varphi'' \mid \varphi^*(p) \mid \varphi') = N''N' \prod_{p'<p} (1 - \varphi^{*''}(p')\varphi'(p')\, d^3p')\varphi^{*''}(p) \cdot \prod_{p''>p} (1 + \varphi^{*''}(p'')\varphi'(p'')\, d^3p''). \tag{52b}$$

The matrix elements of various operator combinations can be constructed from the basic matrix elements in (51) or (52) combined with integrations over $\delta\psi$ or $\delta\varphi$. Of course, the familiar commutation and anticommutation rules emerge automatically.

In the Feynman formalism one computes operator matrix elements from the rule (*27*, *6*)

$$(f_2\sigma_2 \mid \mathcal{O}(\mathbf{f}) \mid f_1\sigma_1) = N \int \mathcal{O}_F(f) e^{iI(f)} \mathfrak{D}f, \tag{53}$$

where $\mathcal{O}_F(f)$ is the *c*-number functional which represents the abstract quantum operator $\mathcal{O}$ (compare Eqs. (21) and (22)). As customary, one writes for $\mathcal{O}_F(f)$ the form it assumes for *continuous, classical* histories; in (51) and (52), therefore, we select *expectation values* and fields ψ and φ in the *classical* domain. Thus, for example, in the case of Fermi statistics, one may say

$$(\varphi''t'' \mid \varphi(p) \mid \varphi't') = N \int \mathcal{O}_F{}^{\varphi}(\varphi) e^{iI(\varphi)} \mathfrak{D}\varphi, \tag{54}$$

where, formally,

$$\mathcal{O}_F{}^{\varphi}(\varphi) = \varphi(p) \exp\left(-2 \int_{p'<p} \varphi^*(p')\varphi(p')\, d^3p'\right). \tag{55}$$

Other functionals $\mathcal{O}_F$ may be constructed out of the matrix elements above. Unlike the action functional I, the functional $\mathcal{O}_F$ does not always coincide in algebraic form with the corresponding quantum operator; compare $\varphi(p)$ in (54) with $\mathcal{O}_F{}^{\varphi}$, its *c*-number functional, in (55)! Nevertheless, *every anticommuting operator is represented by a c-number functional.*

II.4 Unified Formulation of the Sum Over Histories for Both Statistics

The discussion in the preceding two sections demonstrates either for Bose or for Fermi statistics that the Feynman sum can be conveniently defined with the aid of the familiar operator and vector techniques of quantum mechanics. Furthermore, motivated by the results of the two preceding sections, it is highly desirable to develop in a straightforward manner the elements, such as the over-complete family of states and the measure for field configurations, necessary to construct the sum over histories. In this section we present the essentials of such a straightforward construction and are able, thereby, to achieve a more natural and unified presentation serving to emphasize the ambiguity in statistics present in quantizing the Schrödinger field. The discussion in this section will also prove useful for the study of relativistic fields presented in Sections II.5 and III. The Hilbert space employed in our discussion is the "Complete Direct Product Space" introduced by von Neumann (*17*).

A. Construction of the Complete Direct Product Space

Let us introduce a set of separable Hilbert spaces, $\mathfrak{H}_\alpha$, where α is a member of I, $\alpha \in I$, and I is some set of indices. The power of I may be unrestricted, but we shall concentrate on I's which have a continuous or denumberably infinite number of elements.[7] We wish to establish the necessary definitions to construct a Hilbert space $\mathfrak{H}$ which is the direct product of all the spaces $\mathfrak{H}_\alpha$, and is denoted by $\mathfrak{H} = \prod_{\alpha \in I} \otimes \mathfrak{H}_\alpha$ or by $\prod_\alpha \otimes \mathfrak{H}_\alpha$ for brevity. The infinite number of elements in I will make $\mathfrak{H}$ a nonseparable Hilbert space, i.e., a space whose basis elements are nondenumerable.

First, we concentrate on those elements of $\mathfrak{H}$ which we may symbolically write as $\Phi = \prod_\alpha \otimes f_\alpha$, where f_α is a vector in $\mathfrak{H}_\alpha$, $f_\alpha \in \mathfrak{H}_\alpha$. The inner product of two such vectors is defined by

$$(\prod_\alpha \otimes f_\alpha , \prod_\alpha \otimes g_\alpha) \equiv \prod_\alpha (f_\alpha , g_\alpha), \tag{56}$$

where (f_α , g_α) is the inner product in $\mathfrak{H}_\alpha$. The length of Φ is determined from (56) in terms of the length of f_α, $\| f_\alpha \| \equiv +[(f_\alpha , f_\alpha)]^{1/2}$ by

$$\| \Phi \| = +[(\prod_\alpha \otimes f_\alpha , \prod_\alpha \otimes f_\alpha)]^{1/2} = \prod_\alpha \| f_\alpha \| . \tag{57}$$

Clearly, only sequences of f_α for which $\prod_\alpha \| f_\alpha \| < \infty$ can be admitted to $\mathfrak{H}$. To interpret Eqs. (56) and (57) we need to be able to evaluate infinite products of complex numbers such as $\prod_\alpha z_\alpha$. The product $\prod_\alpha z_\alpha$ is defined to converge (*17*) to a value z if there exists, for every $\delta > 0$, a finite set $I_0(\delta) \subset I$ such that $| \prod_{\alpha \in J} z_\alpha - z | < \delta$ for every finite set J satisfying $I_0 \subset J \subset I$. This prescription is extended to give a value to all cases of $\prod_\alpha (f_\alpha , g_\alpha)$ when $\prod_\alpha \| f_\alpha \|$ and $\prod_\alpha \| g_\alpha \|$ converge. A product $\prod_\alpha z_\alpha$ is now called quasi-convergent if $\prod_\alpha | z_\alpha |$ converges; its value is (1) that given above if $\prod_\alpha z_\alpha$ itself is convergent or (2)

[7] The index α will generally refer to a momentum space label, p. When the spin, s, is added α will denote the combination of p and s. In Section III, α will describe a coordinate space label, x, combined with the spin degree of freedom. Later in this section we shall refer to $dm(\alpha)$, the Lebesque measure of a point $\alpha \in I$; this would be either d^3p or d^3x for the examples mentioned in this footnote.

defined to be zero if $\prod_\alpha z_\alpha$ is not convergent. This prescription gives zero for the cases where the phase of $\prod_\alpha z_\alpha$ is infinite.

Let us denote by $\mathfrak{T}_\alpha$ the set of all vectors $f_\alpha \in \mathfrak{H}_\alpha$ for which $\| f_\alpha \| = 1$. Then the vectors $\Phi = \prod_\alpha \otimes f_\alpha$, where $f_\alpha \in \mathfrak{H}_\alpha$ and $f_\alpha \in \mathfrak{T}_\alpha$, are certainly members of $\mathfrak{H} = \prod_\alpha \otimes \mathfrak{H}_\alpha$. Consider the set of all finite linear combinations of such vectors which contains, for example, the vector $\Phi = \sum_{\nu=1}^{p} c_\nu \prod_\alpha \otimes f_{\alpha,\nu}$, where $| c_\nu | < \infty$, p is finite, and $f_{\alpha,\nu} \in \mathfrak{H}_\alpha$ and $\mathfrak{T}_\alpha$ for each ν. These vectors span a space denoted by $\prod_\alpha' \otimes \mathfrak{H}_\alpha$. The closure of this space (inclusion of limit points defined by Cauchy sequences) defines the set of vectors which are members of the complete direct product space, $\mathfrak{H} = \prod_\alpha \otimes \mathfrak{H}_\alpha$. The inner product of vectors in $\prod_\alpha' \otimes \mathfrak{H}_\alpha$ involves a finite linear combination of terms such as appear in (56). The limit of such linear combinations defines a unique inner product for all vectors in $\mathfrak{H} = \prod_\alpha \otimes \mathfrak{H}_\alpha$ *(17)*.

The space $\mathfrak{H}_\alpha$ is given a topology by using the norm to define the "distance" between two elements as follows: distance between f_α and $g_\alpha = \| f_\alpha - g_\alpha \|$. The complete direct product space, $\mathfrak{H} = \prod_\alpha \otimes \mathfrak{H}_\alpha$, is given a topology by defining the distance between Φ and Ψ by $\| \Phi - \Psi \|$, for Φ and $\Psi \in \mathfrak{H}$.

B. Construction of the "unit operator"

We shall first construct our unit operator in terms of the vectors discussed above; subsequently we shall give it a functional description.

(1) Let f_α, g_α, and h_α be vectors in $\mathfrak{H}_\alpha$ and $\mathfrak{T}_\alpha$ (so that $\| f_\alpha \| = \| g_\alpha \| = \| h_\alpha \| = 1$) and form the following product of inner products:

$$(f_\alpha , g_\alpha)(g_\alpha , h_\alpha).$$

(2) Choose a set $\mathfrak{S}_\alpha \subset \mathfrak{T}_\alpha$ and an associated measure μ_α on the vectors $g_\alpha \in \mathfrak{S}_\alpha$, denoted by $\mu_\alpha = \mu_\alpha(g_\alpha)$, in such a way that

$$\int_{g_\alpha \in \mathfrak{S}_\alpha} (f_\alpha , g_\alpha) \, d\mu_\alpha (g_\alpha)(g_\alpha , h_\alpha) = (f_\alpha , h_\alpha); \tag{58}$$

the set $\mathfrak{S}_\alpha$ and measure μ_α must be chosen to make (58) valid for all f_α and h_α in $\mathfrak{T}_\alpha$. A minimal set, $\mathfrak{S}_\alpha$, would contain a basis set of vectors, g_α', of total number D_α, the dimension of $\mathfrak{H}_\alpha$, which are all mutually orthogonal. This need *not* be the only choice for $\mathfrak{S}_\alpha$ and even when D_α is finite we find it convenient to use an $\mathfrak{S}_\alpha$ which contains an infinite number of elements.

(3) In addition to (58) we require, when D_α is finite, that

$$\mu_\alpha(\mathfrak{S}_\alpha) = \int_{\mathfrak{S}_\alpha} d\mu_\alpha (g_\alpha) = D_\alpha ; \tag{59}$$

this corresponds to the "trace of the unit matrix" in $\mathfrak{H}_\alpha$.[8]

[8] The numbers D_α for the principal cases of interest are (1) $D_\alpha = 2$ for Fermi statistics and (2) $D_\alpha = \infty$ for Bose statistics. Here ∞ means the power of the integers.

(4) For the space $\mathfrak{H} = \prod_\alpha \otimes \mathfrak{H}_\alpha$, we let an element $\Phi \in \mathfrak{H}$ be also in $\mathfrak{T}$, $\Phi \in \mathfrak{T}$, if $\Phi = \prod_\alpha \otimes f_\alpha$ where all $f_\alpha \in \mathfrak{T}_\alpha$. We also say $\Phi \in \mathfrak{S}$, where $\mathfrak{S} \subset \mathfrak{T}$, if $\Phi = \prod_\alpha \otimes f_\alpha$ for all $f_\alpha \in \mathfrak{S}_\alpha$. In both cases $\| \Phi \| = 1$.

(5) Now let $(\Phi, \Psi)(\Psi, X) = \prod_\alpha (f_\alpha, g_\alpha)(g_\alpha, h_\alpha)$ for Φ, Ψ, X generated by f_α, g_α, h_α, respectively. The integration over $\Psi \in \mathfrak{S}$ is then defined by a multiple integration over the vectors $g_\alpha \in \mathfrak{S}$ which make up the vector Ψ:

$$\begin{aligned} \int_{\mathfrak{S}} (\Phi,\Psi)\delta\mu(\Psi)(\Psi,X) &\equiv \prod_\alpha \int_{\mathfrak{S}} (f_\alpha, g_\alpha)\, d\mu_\alpha\, (g_\alpha)(g_\alpha, h_\alpha) \\ &= \prod_\alpha (f_\alpha, h_\alpha) = (\Phi,X). \end{aligned} \tag{60}$$

It is apparent that "$\int \Psi\rangle\delta\mu(\Psi)\langle\Psi$" behaves as the unit operator, or unit matrix, in the complete direct product space $\mathfrak{H}$.

C. Preliminaries to the introduction of dynamics; Restrictions on the sets $\mathfrak{S}_\alpha$

We suppose now that the matrix element (Φ^{N+1}, Φ^0) is constructed by an N-fold repeated application of (60). This construction involves an integration over N different sets of vectors, each set comprising all the vectors in $\mathfrak{S}$. The integrand of this expression would therefore appear as follows:

$$\prod_{s=0}^{N} (\Phi^{s+1}, \Phi^s) = (\Phi^{N+1}, \Phi^N)(\Phi^N, \Phi^{N-1}) \cdots (\Phi^1, \Phi^0), \tag{61}$$

where all $\Phi^s \in \mathfrak{S}$. We wish to find an alternative expression for (61) for sequences of vectors Φ^s where "Φ^s is close to Φ^{s+1}." In order to be able to let Φ^s be close to Φ^{s+1} we select the sets $\mathfrak{S}_\alpha$ in a very special way: *Let the sets* $\mathfrak{S}_\alpha$ *be so chosen that for every* $\delta > 0$ *and* $g_\alpha \in \mathfrak{S}_\alpha$ *there exists a vector* $f_\alpha \in \mathfrak{S}_\alpha$ *such that* $\| f_\alpha - g_\alpha \| < \delta$.[9] This restriction on $\mathfrak{S}_\alpha$ permits a similar property to hold in $\mathfrak{S}$, namely, for every $\delta > 0$ and $\Phi \in \mathfrak{S}$ there exits a $\Psi \in \mathfrak{S}$ such that $\| \Phi - \Psi \| < \delta$.

Consider the vector sequences Φ^s where $\Phi^s \in \mathfrak{S}$ for which $\| \Phi^s - \Phi^{s+1} \| < M\epsilon$, where M is finite, and ϵ is a small parameter. For these restricted sequences we may approximate each matrix element in (61) as follows:

$$\begin{aligned} (\Phi^{s+1}, \Phi^s) &= 1 - \tfrac{1}{2}[(\Phi^{s+1}, \Phi^{s+1} - \Phi^s) - (\Phi^{s+1} - \Phi^s, \Phi^s)] \\ &\approx \exp\{- \tfrac{1}{2}[(\Phi^{s+1}, \Phi^{s+1} - \Phi^s) - (\Phi^{s+1} - \Phi^s, \Phi^s)]\}. \end{aligned} \tag{62}$$

The final expression is valid to order ϵ. For the restricted class of vector sequences (61) is approximated to order ϵ by

$$\prod_{s=0}^{N} (\Phi^{s+1}, \Phi^s) = \exp\left\{-\sum_{s=0}^{N} \frac{1}{2} [(\Phi^{s+1}, \Phi^{s+1} - \Phi^s) - (\Phi^{s+1} - \Phi^s, \Phi^s)]\right\}. \tag{63}$$

Let us pass to the limit where $N \to \infty$, $\epsilon \to 0$, M remains finite, and s becomes a

[9] It is this requirement which puts an infinite number of vectors in $\mathfrak{S}_\alpha$ and which overcomes the difficulty of Nambu mentioned in Section I.A.

continuum label called t (the time) such that $N\epsilon = t'' - t'$. Then the right side of (63) becomes

$$\exp\left\{-\frac{1}{2}\int_{t'}^{t''}\left[\left(\Phi(t), \frac{\partial\Phi(t)}{\partial t}\right) - \left(\frac{\partial\Phi(t)}{\partial t}, \Phi(t)\right)\right]dt\right\} \tag{64a}$$

and constitutes an exact evaluation of the limit of the left side of (63). There are, however, "histories" obtained from (61) as $N \to \infty$ which are not continuous—but only when M becomes infinite. Consider a history which in the limit possesses just one point, say at $t = \hat{t}$, where the history is discontinuous and has no derivative. The limit of (61) is no longer (64a) but is given instead by

$$\exp\left[-\frac{1}{2}\int_{\hat{t}_+}^{t''}\left(\Phi(t), \frac{\overleftrightarrow{\partial}}{\partial t}\Phi(t)\right)dt\right]\left(\Phi(\hat{t}_+), \Phi(\hat{t}_-)\right)$$
$$\cdot\exp\left[-\frac{1}{2}\int_{t'}^{\hat{t}_-}\left(\Phi(t), \frac{\overleftrightarrow{\partial}}{\partial t}\Phi(t)\right)dt\right], \tag{64b}$$

where $\hat{t}_\pm$ are, respectively, just above and below $\hat{t}$.

Functional representation of the unit operator. We now discuss the relation of the unit operator determined above with the functional unit operators discussed in Sections II.2 and II.3. Let each vector $f_\alpha \in \mathfrak{S}_\alpha$ be completely specified by a "sufficient" set of parameters. Only one complex variable $\psi(\alpha)$ will be used such that $f_\alpha(\psi(\alpha))$ describes a unique vector in $\mathfrak{S}_\alpha$. As $\psi(\alpha)$ covers its domain $f_\alpha(\psi(\alpha))$ covers all vectors in $\mathfrak{S}_\alpha$. Note that only one complex variable is used to specify f_α independently of D_α, the dimension of $\mathfrak{H}_\alpha$. Since $\psi(\alpha)$ for a fixed α describes a unique vector $f_\alpha(\psi(\alpha))$ it follows that the sequence of complex numbers $\psi(\alpha)$ for all $\alpha \in I$ describes a sequence of unique vectors, $f_\alpha(\psi(\alpha))$. This sequence in turn generates a specific vector $\Phi = \prod_\alpha \otimes f_\alpha(\psi(\alpha))$ which is a member of the complete direct product space. The specification of certain vectors in $\mathfrak{H} = \prod_\alpha \otimes \mathfrak{H}_\alpha$ by means of the sequence of complex numbers $\psi(\alpha)$ corresponds to a functional representation for those vectors. The parameterization of the vectors f_α is also chosen to be continuous, i.e., for every $\delta > 0$ there exists an $\epsilon > 0$ such that if $\| f_\alpha(\psi(\alpha)) - f_\alpha(\psi'(\alpha)) \| < \delta$, then

$$|\psi(\alpha) - \psi'(\alpha)| < \epsilon.$$

The existence of such vectors in $\mathfrak{S}_\alpha$ is insured by the very special demands placed on the subsets $\mathfrak{S}_\alpha$ above.

We now define, formally, the inner product of two vectors in $\mathfrak{S}_\alpha$ by the following expression in terms of $\psi(\alpha)$:

$$(f_\alpha(\psi''(\alpha)), f_\alpha(\psi'(\alpha))) \equiv N_{\psi''(\alpha)}N_{\psi'(\alpha)}\left\{\sum_{p=0}^{n_\alpha}(p!)^{-1}[\psi^{*\prime\prime}(\alpha)\psi'(\alpha)\,dm(\alpha)]^p\right\}, \tag{65}$$

where (1) $dm(\alpha)$ is the Lebesque measure of the point $\alpha \in I$, (2) n_α is the "maximum occupation number of the αth state" and is related to D_α by $D_\alpha = n_\alpha + 1$, and (3) $N_{\psi''(\alpha)}$ and $N_{\psi'(\alpha)}$ are normalization constants.[10]

For the Fermi case, $n_\alpha = 1$, and Eq. (65) becomes

$$(f_\alpha(\psi''(\alpha)), f_\alpha(\psi'(\alpha))) = N_{\psi''(\alpha)}N_{\psi'(\alpha)}[1 + \psi^{*\prime\prime}(\alpha)\psi'(\alpha)\, dm(\alpha)]. \quad (66)$$

For the vectors $\Phi \in \mathfrak{S}$ where $\Phi(\psi) \equiv \prod_\alpha \otimes f_\alpha(\psi(\alpha))$ we have, from (56), the inner product

$$(\Phi(\psi''), \Phi(\psi')) = N''N' \prod_\alpha [1 + \psi^{*\prime\prime}(\alpha)\psi'(\alpha)\, dm(\alpha)], \quad (67)$$

where $N \equiv \prod_\alpha N_{\psi(\alpha)}$. This formula is easily recognized to be in the form that appears in (41). It also shows that no matter how many singularities $\psi(x)$, or $\varphi(p)$, may have—even if they possess a continuous set of singular points—they still correspond to valid vectors in $\mathfrak{H}$; this is the justification for footnote 5.

Let us consider the evaluation of the exponent in Eq. (64a) when $n_\alpha = 1$ (Fermi). We imagine that Φ depends on t through a continuous dependence of $\psi(\alpha)$ on t. Thus, from (67), it follows that

$$\begin{aligned}\left(\Phi(\psi''), \frac{\partial}{\partial t'}\Phi(\psi')\right) &= \frac{\partial}{\partial t'}(\Phi(\psi''),\Phi(\psi')) \\ &= \sum_\alpha N_{\psi''(\alpha)}N_{\psi'(\alpha)}\left\{\psi^{*\prime\prime}(\alpha)\frac{\partial\psi'(\alpha)}{\partial t'}\, dm(\alpha)\right. \\ &\quad \left. - \frac{1}{2}N^2_{\psi'(\alpha)}[1 + \psi^{*\prime\prime}(\alpha)\psi'(\alpha)\, dm(\alpha)]\frac{\partial}{\partial t'}(\psi^{*\prime}(\alpha)\psi'(\alpha))\, dm(\alpha)\right\},\end{aligned} \quad (68)$$

where we have used $N_{\psi'(\alpha)} = [1 + \psi^{*\prime}(\alpha)\psi'(\alpha)\, dm(\alpha)]^{-1/2}$. When we set $\psi' = \psi'' = \psi$, (68) simplifies to

$$\left(\Phi(\psi), \frac{\partial}{\partial t}\Phi(\psi)\right) = \sum_\alpha N^2_{\psi(\alpha)}\left[\psi^*(\alpha)\frac{\partial}{\partial t}\psi(\alpha) - \frac{1}{2}\frac{\partial}{\partial t}(\psi^*(\alpha)\psi(\alpha))\right] dm(\alpha).$$

By taking the Hermitian adjoint we obtain

$$\left(\frac{\partial}{\partial t}\Phi(\psi),\Phi(\psi)\right) = \sum_\alpha N^2_{\psi(\alpha)}\left[\left(\frac{\partial}{\partial t}\psi^*(\alpha)\right)\psi(\alpha) - \frac{1}{2}\frac{\partial}{\partial t}(\psi^*(\alpha)\psi(\alpha))\right] dm(\alpha),$$

and finally the combination in the exponent of (64a) becomes

$$\frac{1}{2}\sum_\alpha N^2_{\psi(\alpha)}\left[\psi^*(\alpha)\frac{\partial}{\partial t}\psi(\alpha) - \left(\frac{\partial}{\partial t}\psi^*(\alpha)\right)\psi(\alpha)\right] dm(\alpha).$$

[10] While $n_\alpha = 1$ (Fermi) and $n_\alpha = \infty$ (Bose) are the usual cases, other statistics exist mathematically and have been studied; see, for example Ref. *28*.

When α denotes x, for example, an evaluation of this expression for fields $\psi(x)$ which are members of Hilbert space takes the following form:

$$\frac{1}{2}\sum_{\alpha}\psi^*(\alpha)\overleftrightarrow{\frac{\partial}{\partial t}}\psi(\alpha)\,dm(\alpha)=\frac{1}{2}\int\psi^*(x)\overleftrightarrow{\frac{\partial}{\partial t}}\psi(x)\,d^3x. \tag{69}$$

If one carries through a similar derivation, from (68) to (69), for another value of n_α in (65), the last formula, (69),—which is exactly valid for $\psi(x)$ in Hilbert space—will be reproduced *independent* of n_α. *Its form is clearly no indication of statistics.* Even discontinuous histories, for which (64b) "typically" applies, are insensitive to statistics so long as the fields $\psi(x)$ are in Hilbert space. It is only for improper singular fields that the evaluation of $(\Phi(\psi''), \Phi(\psi'))$ begins to depend on the statistics assumed, i.e., on the value of n_α.

D. Introduction of dynamics; Construction of the propagator

In general, the introduction of dynamics into the complete direct product space is not without problems (*29–31*)! These occur principally because the Hamiltonian can be defined only on a separable subspace of the full, nonseparable space $\mathfrak{H}$; this restriction makes it difficult to apply "unitary" operators, such as $\mathbf{U} = \exp(-it\mathbf{H})$, to all vectors in $\mathfrak{H}$.

In spite of these difficulties, there appear some compelling formal reasons for the use of the full space $\mathfrak{H}$(*32, 33*). For example, prior to the introduction of any specific dynamics how can one be sure on which separable subspace of $\mathfrak{H}$ the physical theory will depend? The action of various operators has not yet been introduced. Even for the free relativistic spinor fields it is well known that the physical subspace of $\mathfrak{H}$ is not the one most obviously connected with the classical variables (see Section II.5). This is why in a previous paragraph which dealt with the case of no dynamics it was necessary to construct the functional unit matrix by integrating over *all* possible subspaces—hence, the whole space $\mathfrak{H}$.

One may easily carry the construction of the propagator to a preliminary point. Consider those vectors $\Phi \in \mathfrak{S}$ on which the desired Hamiltonian $\mathbf{H}$ may be applied. For "histories" which involve only vectors in this class we may evaluate the corresponding term that enters into the construction of the sum over histories. For one of these restricted histories we want to evaluate the expression

$$\prod_{s=0}^{N}(\Phi^{s+1}, e^{-i\epsilon\mathbf{H}}\Phi^{s}) \tag{70}$$

in the limit where $N \to \infty$, $\epsilon \to 0$, and $N\epsilon = t'' - t'$. Each term in (70) may be approximated to an order ϵ as was done in (32):

$$(\Phi^{s+1}, e^{-i\epsilon\mathbf{H}}\Phi^{s}) \approx (\Phi^{s+1}, \Phi^{s})\exp\left[-i\,\epsilon\,\frac{(\Phi^{s+1}, \mathbf{H}\Phi^{s})}{(\Phi^{s+1}, \Phi^{s})}\right]. \tag{71}$$

This expression may now be combined with the expression for (Φ^{s+1}, Φ^s) given by (62). In the limit above in which also M remains finite (so that $\Phi(t)$ is continuous in time, see Eq. (64a)), the expression in (70) is accurately describable by the formula

$$\exp\left\{-\int_{t'}^{t''}\left[\frac{1}{2}\left(\Phi(t), \frac{\overleftrightarrow{\partial}}{\partial t}\Phi(t)\right) + i(\Phi(t), \mathbf{H}\Phi(t))\right]dt\right\}. \tag{72}$$

This result is the obvious generalization of (64a) to include dynamics for those vector histories for which $\mathbf{H}\Phi(t)$ can be defined. A rigorous treatment for the remaining histories requires further study.

It is possible to bypass the problem of histories for which $\mathbf{H}\Phi(t)$ is undefined by using a procedure of von Neumann (*34*) to determine the dynamical behavior of the system. Von Neumann takes as the basic dynamical postulate a certain automorphism of the state vectors with the passage of time. This procedure may be conveniently illustrated for the free-particle Schrödinger field. For this example, let the members, α, of the set of indices denote momentum space labels, p; the vectors $\Phi \in \mathfrak{S}$ are therefore represented by $\Phi(\psi) = \prod_p \otimes f_p(\psi(p))$. The passage of time is now defined by the automorphism $\Phi(\psi(0)) \to \Phi(\psi(t))$ where

$$\Phi(\psi(t)) = \prod_p \otimes f_p(\psi(p,t)) \equiv \prod_p \otimes f_p(z(p,t)\psi(p,0)), \tag{73}$$

and $z(p,t) = \exp(-itp^2/2m)$.[11] From these arguments we may obtain the formal expression for the free particle propagator for Fermi statistics that was given in Section II.3:

$$\begin{aligned}(\psi''t'' \mid \psi't') &\equiv (\Phi(\psi''(t'')),\Phi(\psi'(t'))) \\ &= N''N'\prod_p\left\{1 + \psi^{*\prime\prime}(p)\exp\left[-i(t''-t')\frac{p^2}{2m}\right]\psi'(p)\,d^3p\right\}.\end{aligned} \tag{74}$$

The same procedure will be used for the relativistic fields in Section II.5.

E. Determination of the functional measure

We close Section II.4 by illustrating the way the particular form of the inner product (66) can help determine the measure $\mu_\alpha(g_\alpha)$ on vectors $g_\alpha \in \mathfrak{S}_\alpha$, and consequently determine the measure $\mu(\Psi)$ on vectors $\Psi \in \mathfrak{S}$. Two separate derivations will be given for the case of Fermi statistics.

We first consider the measure that is *induced* on the Fermi case by the familiar measure assumed for Bose statistics. Choose $d\mu_\alpha(g_\alpha) \to d\mu_\alpha(\psi(\alpha)) \equiv \rho_\alpha(\psi(\alpha)) \cdot d\psi_r(\alpha)\,d\psi_i(\alpha)$ for the Fermi case; and for the Bose case, let $d\mu_\alpha(g_\alpha) \to d\nu_\alpha(\psi^B(\alpha)) = (\text{const.})\,d\psi_r^B(\alpha)\,d\psi_i^B(\alpha)$. Inspection of (65) for the two cases of

[11] This result is chosen to conform with the result that the unitary operator should predict if it existed.

statistics shows that the relation between a Bose and a Fermi parameterization may be formally expressed by

$$\exp[\psi^{*B\prime\prime}(\alpha)\psi^{B\prime}(\alpha)\, dm(\alpha)] = 1 + \psi^{*\prime\prime}(\alpha)\psi^{\prime}(\alpha)\, dm(\alpha). \tag{75}$$

This relation may be viewed as defining a transformation of variables from $\psi^B(\alpha)$ to $\psi(\alpha)$; the Jacobian of this transformation may then be used to define the Fermi weight function, $\rho_\alpha(\psi(\alpha))$. A simple calculation based on (75) shows that

$$\begin{aligned} d\psi_r^{\,B}(\alpha)\, d\psi_i^{\,B}(\alpha) &= \frac{\partial(\psi_r^{\,B}(\alpha), \psi_i^{\,B}(\alpha))}{\partial(\psi_r(\alpha), \psi_i(\alpha))}\, d\psi_r(\alpha)\, d\psi_i(\alpha) \\ &= [1 + \psi^*(\alpha)\psi(\alpha)\, dm(\alpha)]^{-2}\, d\psi_r(\alpha)\, d\psi_i(\alpha). \end{aligned} \tag{76}$$

Therefore, $\rho_\alpha(\psi(\alpha))$ is proportional to $[1 + \psi^*(\alpha)\psi(\alpha)\, dm(\alpha)]^{-2}$. A consideration of the requirements (58) and (59), expressed with the aid of (66), determines the formal constant of proportionality so that, for Fermi statistics, the measure becomes

$$d\mu_\alpha(\psi(\alpha)) = \left(\frac{2}{\pi}\, dm(\alpha)\right)[1 + \psi^*(\alpha)\psi(\alpha)\, dm(\alpha)]^{-2}\, d\psi_r(\alpha)\, d\psi_i(\alpha). \tag{77}$$

From (77) it follows that the functional measure $\delta\mu(\psi) \equiv \delta\mu(\Psi)$ is formally expressed by

$$\delta\mu(\psi) = \prod_\alpha d\mu_\alpha(\psi(\alpha)) = \prod_\alpha \left(\frac{2}{\pi}\, dm(\alpha)\right) \frac{d\psi_r(\alpha)\, d\psi_i(\alpha)}{[1 + \psi^*(\alpha)\psi(\alpha)\, dm(\alpha)]^2}. \tag{78}$$

The second method of obtaining the measure, $d\mu_\alpha(\psi(\alpha))$, follows as closely as possible the invariance arguments discussed recently by Misner (*5*). This procedure is based on finding a transitive group of permutations of the field variables which will carry any point of the domain of integration into any other. The functional measure is then assumed to be the invariant integral measure associated with the group. For example, for the complex field $\psi^B(\alpha)$ of Bose quantization, an additive group of permutations at a point is selected so that the law of group composition is $\psi^{B\prime}(\alpha) + \psi^B(\alpha) = \psi^{B\prime\prime}(\alpha)$. The desired measure is required to be invariant under the action of this group at a point:

$$d\nu_\alpha(\psi^{B\prime\prime}(\alpha)) = d\nu_\alpha(\psi^B(\alpha))$$

for a fixed $\psi^{B\prime}(\alpha)$. Clearly, the desired measure $d\nu_\alpha(\psi^B(\alpha))$ is proportional to $d\psi_r^{\,B}(\alpha)\, d\psi_i^{\,B}(\alpha)$.

In the case of Fermi statistics the elementary Hilbert space, $\mathfrak{H}_\alpha$, belonging to a single degree of freedom α is 2-dimensional, as is easily seen in the occupation-number representation. Therefore, we can conveniently represent the vectors $f_\alpha(\psi(\alpha))$, in the overcomplete family, in a 2-dimensional space as follows:

$$f_\alpha(\psi(\alpha)) = N_{\psi(\alpha)} \begin{pmatrix} 1 \\ \psi(\alpha)[dm(\alpha)]^{1/2} \end{pmatrix}. \tag{79}$$

The inner product of two such vectors, in the usual sense, reproduces the definition (66) of the inner product.[12] It is important to notice that as $\psi(\alpha)$ covers its domain *not all vectors of unit norm are reached*; in particular, the phase of the first component is always zero.

Were we looking for a transitive group to carry vectors in $\mathfrak{T}_\alpha$, the set of all unit vectors, into other vectors in $\mathfrak{T}_\alpha$, the answer would be simple: use the 2-dimensional unitary-unimodular group $\mathfrak{U}_2$. The elements of this group may be represented by[13]

$$W = \begin{pmatrix} \alpha & -\beta^* \\ \beta & \alpha^* \end{pmatrix}, \quad \text{where} \quad \alpha^*\alpha + \beta^*\beta = 1. \tag{80}$$

The operation of an element W of this group on any *fixed* 2-dimensional vector creates a unique element in $\mathfrak{T}_\alpha$. In particular, consider the vector $v = \begin{pmatrix} 1 \\ 0 \end{pmatrix}$; then $g = Wv = \begin{pmatrix} \alpha \\ \beta \end{pmatrix}$ and, because of (80), the new vector g is a vector of unit length. The product of two elements in the group $\mathfrak{U}_2$ is represented by ordinary matrix multiplication. This multiplication law in $\mathfrak{U}_2$ induces a group operation in the set of vectors in $\mathfrak{T}_\alpha$, with the aid of the isomorphic mapping from $W \to g$ provided by the definition $g = Wv$, a relation expressed in terms of the fixed vector v.

However, the set of vectors spanned by Eq. (79) is *not* $\mathfrak{T}_\alpha$ nor cannot even be generated by a subgroup of $\mathfrak{U}_2$. How do we proceed? Let us rewrite the completeness relation, (58):

$$(f_\alpha, h_\alpha) = \int_{\mathfrak{S}_\alpha} (f_\alpha, g_\alpha)\, d\mu_\alpha\, (g_\alpha)(g_\alpha, h_\alpha).$$

[12] One should not be bothered by the appearance of "$[dm(\alpha)]^{1/2}$". This stems from our postulate here that the norm of the vector $\begin{pmatrix} 0 \\ 1 \end{pmatrix}$, $\left\| \begin{pmatrix} 0 \\ 1 \end{pmatrix} \right\|$, is equal to 1. Square roots of measures have been discussed in an analogous sense by von Neumann (*35*).

[13] The matrix element α appearing in W should not be confused with the labeling element α in the set I of indices.

We now modify the right side of this equation by multiplying by the number 1, written in the following manner

$$1 = \int_0^{2\pi} e^{i\omega(\alpha)} \frac{d\omega\ (\alpha)}{2\pi} e^{-i\omega(\alpha)}, \tag{81}$$

which we distribute so that

$$(f_\alpha, h_\alpha) = \int_0^{2\pi} \int_{\mathfrak{S}_\alpha} (f_\alpha, e^{i\omega(\alpha)} g_\alpha) \frac{d\omega\ (\alpha)}{2\pi} d\mu_\alpha\ (g_\alpha)(e^{i\omega(\alpha)} g_\alpha, h_\alpha). \tag{82a}$$

Now, as $\omega(\alpha)$ varies between 0 and 2π and as g_α varies over $\mathfrak{S}_\alpha$, the vectors $g_\alpha \exp[i\omega(\alpha)]$ completely cover $\mathfrak{T}_\alpha$, the set of all normed 2-dimensional vectors. Therefore, (82a) may be rewritten as an integral over all the vectors in $\mathfrak{T}_\alpha$:

$$(f_\alpha, h_\alpha) = \int_{\mathfrak{T}_\alpha} (f_\alpha, g_\alpha)\ d\mu_\alpha'\ (g_\alpha)(g_\alpha, h_\alpha), \tag{82b}$$

where $d\mu_\alpha'(g_\alpha) \equiv d\mu_\alpha(g_\alpha)[d\omega(\alpha)/2\pi]$. If we can obtain $d\mu_\alpha'(g_\alpha)$ *in this form*, then $d\mu_\alpha(g_\alpha)$ may be obtained by "splitting off" $[d\omega(\alpha)/2\pi]$.[14]

It is Misner's contention (*5*) that the proper choice of $\mu_\alpha'(g_\alpha)$ is the invariant group measure for the transitive group under consideration. Therefore, we set $d\mu_\alpha'(g_\alpha) = d\mu_\alpha'(W)$, the invariant measure on $\mathfrak{U}_2$ (so that $d\mu_\alpha'(W) = d\mu_\alpha'(W')$, for W and $W' \in \mathfrak{U}_2$), and we use a fixed vector v, say $\begin{pmatrix}1\\0\end{pmatrix}$, to secure the isomorphism between the elements of $\mathfrak{U}_2$ and the elements of $\mathfrak{T}_\alpha$ by the previous prescription: $Wv = g_\alpha$. It is easy to check that the invariant measure is consistent with (82b). Clearly, for any measure μ_α' there exists some matrix η' for which

$$\int_{\mathfrak{T}_\alpha} (f_\alpha, g_\alpha)\ d\mu_\alpha'\ (g_\alpha)(g_\alpha, h_\alpha) = (f_\alpha, \eta' h_\alpha). \tag{83}$$

The postulated invariance of $d\mu_\alpha'$ will then require η' to commute with every element of $\mathfrak{U}_2 : W\eta' = \eta' W$. By Schur's lemma, η' must be a multiple of the 2-dimensional unit matrix as was to be proven.

The invariant measure may be expressed in terms of the variables α and β defining the matrix W which appears in Eq. (80). However, it is more convenient to introduce other variables which are chosen to represent the vectors in $\mathfrak{T}_\alpha$ in a

[14] It is clear that there exist measure functions $\mu_\alpha'(g_\alpha)$ which satisfy (82b) but which do not "factor" into the particular form we desire; one example is obtained by letting $\mu_\alpha'(g_\alpha)$ vanish except at the two points $g_\alpha = \begin{pmatrix}1\\0\end{pmatrix}$ and $g_\alpha = \begin{pmatrix}0\\1\end{pmatrix}$.

form analogous to the representation in (79). Thus, α and β of (80) are re-expressed as follows:

$$\alpha = e^{i\omega(\alpha)}[1 + \psi^*(\alpha)\psi(\alpha)\, dm(\alpha)]^{-1/2}, \tag{84a}$$

and

$$\beta = e^{i\omega(\alpha)}\psi(\alpha)[dm(\alpha)]^{1/2}[1 + \psi^*(\alpha)\psi(\alpha)\, dm(\alpha)]^{-1/2}. \tag{84b}$$

A calculation following customary procedures[15] shows that the invariant measure has the form

$$d\mu_\alpha'(W) = (\text{const.})[1 + \psi^*(\alpha)\psi(\alpha)\, dm(\alpha)]^{-2}\, d\psi_r(\alpha)\, d\psi_i(\alpha)\cdot\frac{d\omega(\alpha)}{2\pi} \tag{85}$$

which does factor into the desired form. From the preceding arguments, therefore, the measure for the vectors in $\mathfrak{S}_\alpha$ is

$$d\mu_\alpha(g_\alpha) = \left(\frac{2}{\pi}\, dm(\alpha)\right)[1 + \psi^*(\alpha)\psi(\alpha)\, dm(\alpha)]^{-2}\, d\psi_r(\alpha)\, d\psi_i(\alpha),$$

where again the formal constant in front was uniquely determined by (58) and (59). It is important to remark that it was necessary to go outside the class of variables $\psi(\alpha)$, the class on which the "classical" theory depends to construct the proper transitive group. The only transformation for which the measure on $\mathfrak{S}_\alpha$ *alone* remains invariant is the simple phase transformation

$$\psi(\alpha) \rightarrow \exp[i\lambda(\alpha)]\psi(\alpha),$$

and corresponds to the invariance of the set of vectors, $\mathfrak{S}_\alpha$, under the subgroup of unitary matrices

$$\mathcal{V}(\lambda(\alpha)) = \begin{pmatrix} 1 & 0 \\ 0 & \exp[i\lambda(\alpha)] \end{pmatrix}.$$

In conclusion, the completeness relation and considerations of group theory have led to a unique measure for integrating over the overcomplete family of states belonging to one degree of freedom, α. In this section, the abstract properties of the overcomplete family of states and the functional representation of states have been presented in detail. In addition we have discussed the construction of the propagator and the measure on field configurations. All of these general aspects of our formalism can now be applied to specific problems..

[15] See, for example, Bargmann (*36*) who calculates the invariant measure for the 2-dimensional unimodular group. The calculation procedure is very similar.

II.5 Propagator and State Functionals for Relativistic Spinor Fields

In this section we use the method of summing over histories to quantize fields which obey the familiar relativistic spinor equations. Customarily, the fields in the "classical" domain (first quantized) of such theories consist of all spinor fields $\psi(x)$ in a separable Hilbert space. We see, from our preceding results, that we shall be faced with the familiar mathematical arbitrariness in the choice of statistics. From the outset we shall consider only those "extensions" which yield Fermi statistics. To illustrate the techniques we concentrate our effort on the two-component neutrino theory; all results are easily extended to other free fields which have mass and additional spin degrees of freedom.

A. Determination of the propagator

The Hamiltonian operator for the neutrino field may be written as an integral over 3-space.

$$\mathbf{H} = -\frac{i}{2} \int \psi^* \sigma \cdot \overleftrightarrow{\nabla} \psi \, d^3x, \tag{86}$$

where the "double arrow" derivative is defined in Eq. (2). It is convenient to work in terms of operators in which (86) becomes diagonalized. For this purpose we shall introduce operators $\varphi_s(p)$ which are related to the field operators $\psi_\alpha(x)$ by the following equation[16]:

$$\psi_\alpha(x) = (2\pi)^{-3/2} \sum_s \int U_{\alpha s}(p) e^{ip\cdot x} \varphi_s(p) \, d^3p, \tag{87}$$

where the spin transformation matrix $U_{\alpha s}(p)$ is given by

$$\begin{pmatrix} U_{11} & U_{12} \\ U_{21} & U_{22} \end{pmatrix} = (2p^2 + 2\,|\,p\,|\,p_{(z)})^{-1/2} \begin{pmatrix} |\,p\,| + p_{(z)} & -(p_{(x)} - ip_{(y)}) \\ p_{(x)} + ip_{(y)} & |\,p\,| + p_{(z)} \end{pmatrix}. \tag{88}$$

The operators $\varphi_s(p)$ obey the usual anticommutation rules:

$$\{\varphi_s(p), \varphi_{s'}^*(p')\} = \delta_{ss'}\delta^3(p - p'), \quad \text{and} \quad \{\varphi_s(p), \varphi_{s'}(p')\} = 0. \tag{89}$$

When we substitute $\psi_\alpha(x)$ given by (87) into Eq. (86) we find the Hamiltonian in terms of the field operators $\varphi_s(p)$.

$$\mathbf{H} = \int \varphi^*(p) \sigma_z \,|\,p\,|\, \varphi(p) \, d^3p. \tag{90}$$

We see that (90) is expressed in the "hole theory" formulation where $\varphi_s^*(p)$ creates a neutrino of + or − energy depending on whether s is 1 or 2. This form of the theory is, of course, formally equivalent to the more modern form which is expressed in terms of neutrinos and antineutrinos.

[16] The subscript α denotes a spin degree of freedom throughout this section.

Construction of the functional unit matrix. Prior to discussing dynamics we introduce the unit matrix for Fermi statistics in terms of "basis" vectors analogous to those used in II.3. The basis vectors are formally represented by

$$| \varphi) = N \sum_{n=0}^{\infty} (n!)^{-1} \int \cdots \int \sum_{s_1,s_2,\ldots} \varphi_{s_1}(p_1) \cdots \varphi_{s_n}(p_n)\, d^3p_1 \times \cdots d^3p_n \,|\, p_1s_1, \cdots p_ns_n)_S, \tag{91}$$

where $| p_1s_1, \cdots p_ns_n)_S$ denotes a set of vectors which are orthonormal in the sense of Dirac, symmetric in the argument pairs, p_is_i, and which vanish if any pair equals any other pair; for example, if $p_1 = p_2$ *and* $s_1 = s_2$. In terms of a ps-ordering and a ps-ordering sign function the antisymmetric states are defined by $| p_1s_1, \cdots p_ns_n)_A = \sigma(p_1s_1, \cdots p_ns_n) \,|\, p_1s_1, \cdots p_ns_n)_S$ and are connected with the operators $\boldsymbol{\varphi}_s(p)$ and $\boldsymbol{\varphi}_s^*(p)$ via $\overset{*}{\boldsymbol{\varphi}}_{s_1}(p_1) \cdots \overset{*}{\boldsymbol{\varphi}}_{s_n}(p_n) \,|\, \text{“}0\text{”}) = | p_1s_1, \cdots p_ns_n)_A$ and $\boldsymbol{\varphi}_s^*(p) \,|\, \text{“}0\text{”}) = 0$. Clearly, $|\, \text{“}0\text{”})$ is the state in the overcomplete family described by $\varphi_s(p) = 0$. In the hole theory the "vacuum" state $|\, \text{“}0\text{”})$ is not identical to the physical vacuum as we shall see below (see also footnote 20).

The definition of the basis vector, (91), allows us to construct the inner product, $(\varphi'' \,|\, \varphi')$, which is an element of the unit matrix.

$$(\varphi'' \,|\, \varphi') = N''N' \sum_{n=0}^{\infty} (n!)^{-1} \mathcal{P} \int \cdots \int \sum \varphi_{s_1}^{*\prime\prime}(p_1) \cdots \varphi_{s_n}^{*\prime\prime}(p_n) \varphi_{s_1}'(p_1) \cdots \varphi_{s_n}'(p_n)\, d^3p_1 \cdots d^3p_n. \tag{92}$$

In this expression the symbol $\mathcal{P}$ instructs one to omit any contribution when simultaneously any $p_i = p_j$ and $s_i = s_j$; again, this is the omission only of a set of measure zero from each integral. Therefore, the relation in (92) evaluated for fields φ'' and φ' which are members of the "classical" domain is numerically identical to the following expression:

$$(\varphi'' \,|\, \varphi') = N''N' \exp\left[\sum_s \int \varphi_s^{*\prime\prime}(p) \varphi_s'(p)\, d^3p\right]. \tag{93}$$

The elements of the unit matrix when evaluated for "classical" fields, remain insensitive to the assumed choice of statistics, even with the introduction of the discrete spin variable s.

Introduction of dynamics. The most elegant manner in which dynamics may be introduced follows the methods discussed in part D of Section II.4. Because the Hamiltonian which appears in (90) does not mix the individual degrees of freedom, the appropriate automorphism of past states to future states due to the passage of time is similar to that in the case of the free nonrelativistic particle.

Let $\Phi(\varphi)$ be an alternative name for the vector $|\varphi)$ in (91). The time automorphism is represented by $\Phi(\varphi(0)) \to \Phi(\varphi(t))$ where

$$\Phi(\varphi(t)) = \prod_{p,s} \otimes f_{p,s}(\varphi_s(p,t)) \equiv \prod_{p,s} \otimes f_{p,s}(z_s(p,t)\varphi_s(p,0)), \tag{94}$$

and where $z_s(p,t) = \exp(-it\sigma_s \mid p \mid)$. Here $\sigma_s \equiv 1$ if $s = 1$ and $\sigma_s \equiv -1$ if $s = 2$. The complete propagator then takes the formal form:

$$\begin{aligned}(\varphi''t'' \mid \varphi't') &\equiv (\Phi(\varphi''(t'')), \Phi(\varphi'(t'))) \\ &= N''N' \prod_{p,s} \{1 + \varphi_s^{*\prime\prime}(p) \exp[-i(t'' - t')\sigma_s \mid p \mid]\varphi_s'(p)\, d^3p\}.\end{aligned} \tag{95}$$

This result, of course, takes on a simpler appearance when the fields φ'' and φ' are in the classical domain.

Alternatively, the dynamics may be calculated by summing over histories by a formula such as (33). For this purpose it is necessary to know, in addition to the inner product $(\varphi'' \mid \varphi')$, the matrix elements of the Hamiltonian (90). A simple calculation involving the states $|\varphi)$ in (91) yields the formal relation

$$\frac{(\varphi'' \mid \mathbf{H} \mid \varphi')}{(\varphi'' \mid \varphi')} = \sum_s \int \frac{\varphi_s^{*\prime\prime}(p)\sigma_s \mid p \mid \varphi_s'(p)\, d^3p}{1 + \varphi_s^{*\prime\prime}(p)\varphi_s'(p)\, d^3p}. \tag{96}$$

The functional measure for field configurations needed to complete the construction of the sum over histories is similar to that appearing in (46) with, now, an additional product over the two spin variables s:

$$\delta\varphi = \mathfrak{N} \prod_{p,s} \frac{d\varphi_{s,r}(p)\, d\varphi_{s,i}(p)}{[1 + \varphi_s^*(p)\varphi_s(p)\, d^3p]^2}, \tag{97}$$

where $\mathfrak{N}$ represents a suitable normalization constant. The Feynman sum may be formally carried out with the definitions provided by (92), (96), and (97); the resulting propagator, of course, matches that given in (95).

B. State Functionals

The vast amount of information contained in a completed propagator may be used to generate functional expressions which represent the eigenstates of the system (*37*). For example, the analytic properties of propagators under a substitution such as $t \to -i\tau$ may be used followed by passage to the limit $\tau \to \infty$. Such a procedure makes the only significant contribution to the transformed propagator come from the lowest energy state. In applying this procedure to the propagator expression (95) we find we should keep only the first term in the parenthesis when $s = 1$ (since $\sigma_1 = 1$), and we should keep only the second term when $s = 2$ (since $\sigma_2 = -1$). Consequently, with $\Psi_0(\varphi)$ denoting the vacuum state functional, we find

$$\Psi_0(\varphi'')\Psi_0^*(\varphi') = N''N' \prod_p \varphi_2^{*\prime\prime}(p)\varphi_2'(p)\, d^3p. \tag{98}$$

If we set $\varphi'' = \varphi' = \varphi$ and write out the definition of the normalization constant, N, Eq. (98) becomes

$$| \Psi_0(\varphi) |^2 = \prod_p \frac{1}{[1 + \varphi_1^*(p)\varphi_1(p)\, d^3p]} \frac{\varphi_2^*(p)\varphi_2(p)\, d^3p}{[1 + \varphi_2^*(p)\varphi_2(p)\, d^3p]}. \qquad (99)$$

The interpretation of this equation is as follows: $\Psi_0(\varphi)$ is large when $\varphi_1 \approx 0$, i.e., almost no probability of finding neutrinos; $\Psi_0(\varphi)$ is also large when $\varphi_2 \approx \infty$, i.e., there is almost certainty of finding a negative energy neutrino, or stated alternatively, there is almost no probability to find an antineutrino. The fact that the vacuum state is to be described by $\varphi_2 \approx \infty$ reflects the myriotic nature of the hole theory. The adjective myriotic was introduced by Friedrichs (*21*) to describe fields containing an infinite number of quanta.

Excited states of this field receive an equally simple expression. For example, the $|$ state $|^2$ for two positive energy neutrinos of momentum p_1 and p_2 is represented formally by the symmetric, c-number expression

$$| \Psi_{2\nu}(\varphi) |^2 = (1 - \delta_{p_1,p_2})N^2 \, | \varphi_1(p_1)\varphi_1(p_2) |^2 (d^3p)^2 \prod_p | \varphi_2(p) |^2 d^3p, \qquad (100)$$

where δ_{p_1,p_2} denotes the Kronecker δ symbol.

While we could continue to work with the field variable $\varphi_s(p)$, it is worthwhile illustrating another, more symmetric, set of field variables closely related to the "b-variables" of the single oscillator example in Section I.B. These variables are formally defined by

$$\chi_s(p) = \varphi_s(p)[1 + \varphi_s^*(p)\varphi_s(p)\, d^3p]^{-1/2}. \qquad (101)$$

Note that $\chi_s(p) = \varphi_s(p)$ for all field configurations $\varphi_s(p)$ in the "classical" domain. The new variables are limited in their variation according to

$$0 \leqq | \chi_s(p) |^2 d^3p \leqq 1,$$

in contrast to the field variables $\varphi_s(p)$, which are unrestricted. In terms of the field χ, the vacuum expression (99) becomes

$$| \Psi_0(\chi) |^2 = \prod_p \{1 - \chi_1^*(p)\chi_1(p)\, d^3p\}\{1 - [1 - \chi_2^*(p)\chi_2(p)\, d^3p]\}. \qquad (102)$$

It is instructive to regard multiple products like (102) as having been generated by the expansion of an infinite diagonal determinant, say by $\det(1 - J)$. The invariance of a determinant under a unitary transformation may then be exploited to obtain a spatial description of the exact ground-state expression (102). This procedure is schematically represented in the following manner:

$$\det(1 - J) = \det[U(1 - J)U^{-1}] = \det(1 - UJU^{-1}) \equiv \det(1 - K). \qquad (103)$$

If the unitary matrix in (103) is exactly the one that was used to diagonalize the Hamiltonian (86) then the description will be formulated in terms of the original spatial variables. The transformation (103) may be carried out by use of (87) and (88). We express the matrix K in terms of its "matrix elements" as follows:

$$K_{\alpha\alpha'}(x,x') = \frac{1}{2}\,\delta_{\alpha\alpha'}\delta^3(x - x') - \frac{i\delta_{\alpha\alpha'}}{8\pi^2 V}\iint \frac{[\chi^*(z)\cdot\sigma\overleftrightarrow{\nabla}\chi(y)]}{|x - x' + z - y|^2}\,d^3y\,d^3z$$
$$+ \frac{i\sigma_{\alpha\alpha'}\cdot\nabla_x}{4\pi^2}\frac{1}{|x - x'|^2} - \frac{i\sigma_{\alpha\alpha'}\cdot\nabla_x}{4\pi^2 V}\iint \frac{\chi^*(z)\chi(y)}{|x - x' + z - y|^2}\,d^3y\,d^3z. \quad (104)$$

V is a formal constant representing $\int d^3x$ which appears only at this intermediate stage of the calculation. To calculate the vacuum $|$ state $|^2$, write

$$|\Psi_0(\chi)|^2 = \det(1 - K) = \exp \operatorname{Tr}\ln(1 - K)$$
$$= \exp(-\operatorname{Tr}K - \tfrac{1}{2}\operatorname{Tr}K^2 - \tfrac{1}{3}\operatorname{Tr}K^3 - \cdots), \quad (105)$$

where Tr denotes a spin and space summation such that, for example,

$$\operatorname{Tr}K = \sum_{\alpha}\int K_{\alpha\alpha}(x,x)\,d^3x, \quad (106a)$$

and

$$\operatorname{Tr}K^2 = \sum_{\alpha,\alpha'}\iint K_{\alpha\alpha'}(x,x')K_{\alpha'\alpha}(x',x)\,d^3x\,d^3x'. \quad (106b)$$

Because J is a positive semidefinite matrix the first deviation of (105) from 1 will always come from the linear term in K. Therefore, we may approximate (105) by

$$|\Psi_0(\chi)|^2 \approx e^{-\operatorname{Tr}K} = N_0\exp\left\{-\frac{1}{4\pi^2}\iint \frac{[\chi^*(z)\sigma\cdot(-i)\overleftrightarrow{\nabla}\chi(y)]}{|z - y|^2}\,d^3y\,d^3z\right\}, \quad (107)$$

where N_0 is a field independent normalization factor: $N_0 = \exp[-\int \delta^3(0)\,d^3x]$. Actually, for a certain domain of the field variables χ_s, Eq (107) represents an exact evaluation of the vacuum expression $|\Psi_0(\chi)|^2$. Inspecting (102), we find that the exponential form in (107) is valid for those fields χ where $\chi_1(p)$ is finite and $\chi_2(p)$ is infinite in a manner such that $\delta^3(0) - \chi_2^*(p)\chi_2(p)$ is finite. It is interesting to compare the vacuum expression (107) for the neutrino field with the functional expression for the ground state of the electromagnetic field obtained by Wheeler (*38, 5*). The vacuum state for the electromagnetic field can be represented as a functional of the magnetic field given by

$$\Psi_{\text{vac}}(H) = N_0'\exp\left\{-\frac{1}{16\pi^3}\iint \frac{H(z)\cdot H(y)}{|z - y|^2}\,d^3y\,d^3z\right\}$$

when expressed in natural units. The similarity between the two expressions representing vacuum functionals is striking!

Further discussion of the χ field. To a large extent the field χ defined in Eq. (101) is a very natural variable to describe Fermi statistics. Consider, for example, the customary elimination of the negative energy states in quantum field theory. The Hamiltonian operator, (90), is rewritten as follows:

$$\mathbf{H} = \int [\varphi_1{}^*(p)\varphi_1(p) - \varphi_2{}^*(p)\varphi_2(p)] \, | \, p \, | \, d^3p. \tag{108}$$

Now, the creation operator of a negative energy neutrino, $\varphi_2{}^*(p)$, is reinterpreted as the destruction operator of the antineutrino and renamed, $\varphi_{-1}(p)$. Thus, Eq. (108) becomes

$$\begin{aligned} \mathbf{H} &= \int [\varphi_1{}^*(p)\varphi_1(p) - \varphi_{-1}(p)\overset{*}{\varphi}_{-1}(p)] \, | \, p \, | \, d^3p \\ &= \int [\varphi_1{}^*(p)\varphi_1(p) + \overset{*}{\varphi}_{-1}(p)\varphi_{-1}(p)] \, | \, p \, | \, d^3p - \infty, \end{aligned} \tag{109}$$

where we have used the anticommutation rules (89) to interchange the operators and have written ∞ for the following quantity:

$$\infty \equiv \int \delta^3(0) \, | \, p \, | \, d^3p. \tag{110}$$

This infinite energy is present even if the Heisenberg operator prescription is used, and it must be discarded as unobservable (or added originally to Eq. (90)). We now give a parallel classical treatment at a level extended to include functions outside the usual classical Hilbert space. Consider the expectation value of the above Hamiltonian, (108), expressed in states which are labeled by values of the field χ. If we combine (96) and (101),

$$(\chi \, | \, \mathbf{H} \, | \, \chi) = \int [\chi_1{}^*(p)\chi_1(p) - \chi_2{}^*(p)\chi_2(p)] \, | \, p \, | \, d^3p, \tag{111}$$

where $| \, \chi)$ is exactly the same state as (91), just relabeled with χ according to (101). From the formal definition of χ, follows the inequality

$$0 \leqq | \, \chi_2(p) \, |^2 \, d^3p \leqq 1.$$

We may therefore introduce an additional field $\chi_{-1}(p)$ formally satisfying

$$\overset{*}{\chi}_{-1}(p)\chi_{-1}(p) \, d^3p \equiv 1 - \chi_2{}^*(p)\chi_2(p) \, d^3p, \tag{112}$$

or, equally well, (see Eq. (44))

$$\overset{*}{\chi}_{-1}(p)\chi_{-1}(p) = \delta^3(0) - \chi_2{}^*(p)\chi_2(p). \tag{113}$$

The phase of $\chi_{-1}(p)$ may be chosen, for example, equal to the phase of $\chi_2^*(p)$. It also follows that $0 \leqq |\chi_{-1}(p)|^2 d^3p \leqq 1$. With these definitions (111) becomes

$$(\chi|\mathbf{H}|\chi) = \int [\chi_1^*(p)\chi_1(p) + \overset{*}{\chi}_{-1}(p)\chi_{-1}(p)]\,|p|\,d^3p - \infty, \tag{114}$$

where "∞" has the same meaning as in (110).

Physically, χ_{-1} represents the probability amplitude of a positive energy antineutrino since χ_2 represents the probability amplitude of a negative energy neutrino. From the construction of the sum over histories in Eq (33), it is clear that $(\chi|\mathbf{H}|\chi)$ represents the "classical" Hamiltonian functional. Equations (111) and (114) show, therefore, that this classical Hamiltonian possesses, like the quantum theory, the important feature of a lowest energy state!

It is also interesting to consider the form assumed by the functional measure (97) when it is expressed in terms of the χ field variables. This measure reduces to the more compact form

$$\delta\varphi = \delta\chi = \mathfrak{N} \prod_{p,s} d\chi_{s,r}(p)\, d\chi_{s,i}(p) \tag{115}$$

with no additional weight factors. Despite appearances, the measure $\delta\chi$ coupled with the domain of the integration of the field configuration χ is not fully invariant under a unitary change of the type $\chi(p) = \sum a_n\varphi_n(p)$ as was the measure (25) for Bose fields. Of course, this unusual feature in no way puts the correctness of the theory in question.

Nature of state functionals for spinor fields. A natural requirement for a physically acceptable state functional is that it should be invariant under a transformation by the unit matrix. In other words, it is natural to define the unit matrix as a projection operator which projects out the unacceptable parts of all functionals and which leaves the acceptable parts unaltered.[17] In this respect the unit matrix is a special case of the dynamical propagator which also projects out the unacceptable parts of state functionals. In the case of Fermi statistics, any functional containing true powers of field variables, such as $\varphi^3(p)$, will have these particular terms projected out when acted upon by the unit matrix. Consider, for example, the functional

$$\Omega(\varphi) = N \iiint \varphi^*(p_1)\varphi^*(p_2)\varphi^*(p_3)\omega(p_1,p_2,p_3)\, d^3p_1 d^3p_2 d^3p_3. \tag{116}$$

[17] An elementary analog of this projection process is provided by Fourier analysis. Let $f(x)$, for example, be a function which is continuous everywhere except at $x = 0$, and let $f(0^-) = 2$, $f(0) = 5$, and $f(0^+) = 6$. Now, let the Fourier transform of $f(x)$ be calculated; and from this Fourier transform let one attempt to recalculate $f(x)$. One finds instead a new function $f'(x)$ which agrees with $f(x)$ everywhere except at the point of discontinuity. At that point, $f'(0) = \frac{1}{2}[f(0^-) + f(0^+)] = 4$, not 5. Thus, the double Fourier transformation serves as a unit operator which projects out all parts of a function which do not satisfy a certain mean value requirement.

When the unit operator acts on Ω, it gives a projected state functional which is formally defined by

$$\Omega^P(\varphi) = N\mathcal{P} \iiint \varphi^*(p_1)\varphi^*(p_2)\varphi^*(p_3)\omega(p_1, p_2, p_3)\, d^3p_1 d^3p_2 d^3p_3, \quad (117)$$

but which remains numerically the same as (116) for all "classical" fields, $\varphi(p)$.

Apart from a fixed normalization factor, the complex field φ enters into any state functional in the form φ^*. That it enters as φ^* and not φ follows from the definition of the "basis" vector $|\,\varphi)$ and the fact that state functionals are defined by the inner product $(\varphi\,|$ state). In consequence of the appearance only of φ^* and not both φ and φ^*, one *cannot* construct state functionals which are strongly peaked about an arbitrary preselected field $\varphi^{(0)}$. The general lack of ability to construct peaked state functionals stems from the theorem that a function of a complex variable, analytic in a certain region, takes its maximum and minimum values on the boundary of that region. In order for $(\varphi\,|\,\text{state}) \equiv N\,F(\varphi^*)$ to be peaked about $\varphi = \varphi^{(0)}$ it is necessary that $F(\varphi^*)$ itself be peaked about $\varphi^{*(0)}$; but, since F is regular about $\varphi^{*(0)}$ it follows from the above theorem that $|\,F\,|$ will assume a larger value in the neighborhood of $\varphi^{*(0)}$ than it has for $\varphi^{*(0)}$. An exception to the general rule occurs in the special case when $\varphi^{(0)} = 0$. It should be noted that only the complex property of the field φ has been used. Therefore, the general lack of peaked state functionals is also a property of the Schrödinger field, for example, quantize with either Bose or Fermi statistics. On the other hand, for the electromagnetic field, one is able to construct state functionals such as

$$\Psi(H) = N^{(0)} \exp\left\{-\frac{1}{\epsilon}\int [H(x) - H^{(0)}(x)]^2\, d^3x\right\}$$

which, if $\epsilon \ll 1$, is strongly peaked about an arbitrarily prescribed magnetic field $H^{(0)}(x)$, a field which is subject only to $\nabla\cdot H^{(0)} = 0$.

III. A SUM OVER HISTORIES FOR SPINOR FIELDS IN THE PRESENCE OF AN INTERACTION

The functional quantization for relativistic spinor fields presented in Section II.5 was especially adapted to treat free fields. In this section we shall illustrate a *c*-number quantization by the method of summing over histories that may be carried out when local interactions are present; a sum over histories which leads to a description in terms of quanta localized in position space.

A typical local interaction, for example, is afforded by the electron field in the presence of an external electromagnetic field. The action functional for this familiar example is simply

$$I^A(\psi) = -\int \left[\bar{\psi}\left(\frac{1}{2}\gamma^\mu \overleftrightarrow{\partial}_\mu + m\right)\psi + ie\bar{\psi}\gamma^\mu\psi A_\mu\right] d^4x. \quad (118)$$

Quantization of the spinor field, $\psi(x)$, in the presence of the prescribed external field, A_μ, would be schematically carried out as follows: (1) Substitute the action (118) into the exponent of the expression (21) which defines the sum over histories, (2) exercise the action option to select an appropriate action value for unruly histories required in constructing the propagator, and (3) calculate! In this section we illustrate a formal procedure to select the action value for unruly histories and thereby establish for interacting spinor fields the existence of a c-number sum over histories which leads to Fermi statistics.

A. Outline of the sum over histories construction

We shall establish the appropriate sum over histories by reconstructing it from the operator formalism in a manner similar to that employed in Section II.2. The necessary formalities are briefly sketched in outline form in the following steps:

(1) It will be sufficient to illustrate the construction of the sum for cases in which the spinor fields are in the presence of external unquantized sources; such a case is illustrated in (118). As long as this external source is maintained as an arbitrary space-time function, one can always invoke the familiar Feynman quantization for Bose fields to obtain a completely quantized interacting system (4). This procedure formally encompasses all the familiar Fermi interactions as well.

(2) A very general Hamiltonian density operator for relativistic spinor fields can be represented (apart from additive constants) by a finite linear combination of local operators typically represented by the two types:

$$c_{\beta\beta'}(x)\psi_\beta^*(x)\psi_{\beta'}(x) + \text{H.c.} \tag{119a}$$

and

$$d_{\beta\beta'}(x)\left(\psi_\beta^*(x)\,\overleftrightarrow{\frac{\partial}{\partial x_i}}\,\psi_{\beta'}(x)\right) + H.c. \tag{119b}$$

In these two expressions: (a) β and β' are discrete indices signifying different spin components or, perhaps, labeling distinct particles (such as neutron and protron), (b) the spinor operators obey the following anticommutation rules

$$\{\psi_\beta(x), \psi_{\beta'}(x')\} = \{\psi_\beta^*(x), \psi_{\beta'}^*(x')\} = 0, \tag{120a}$$

and

$$\{\psi_\beta(x), \psi_{\beta'}^*(x')\} = \delta_{\beta\beta'}\delta^3(x - x'), \tag{120b}$$

(c) the operation $\overleftrightarrow{\partial/\partial x_i}$ denotes a particular spatial derivative applied to the operators $\psi_{\beta'}$ and ψ_β^*, (d) H.c. denotes the Hermitian conjugate expression, and

(e) $c_{\beta\beta'}(x)$ and $d_{\beta\beta'}(x)$ are arbitrary, complex c-number functions of space and time which represent the external sources.[18]

(3) The local nature of the interaction invites us to consider a quantum mechanical representation expressed in terms of *localized quanta.*[19] To discuss a description in terms of localized quanta the following machinery is useful:

(a) An overcomplete family of states is introduced whose typical member is formally defined by

$$| \psi) \equiv N \sum_{n=0}^{\infty} (n!)^{-1} \int \cdots \int \sum_{\beta_1, \beta_2, \cdots} \psi_{\beta_1}(x_1) \cdots \psi_{\beta_n}(x_n) \cdot d^3x_1 \cdots d^3x_n \mid x_1\beta_1, \cdots x_n\beta_n)_S. \tag{112}$$

In this expression the vectors $| x_1\beta_1 , \cdots x_n\beta_n)_S$ are symmetric in their index pairs and vanish if any pair equals any other pair.

(b) An ordering relation $<$ ("less than") between the label pairs $x\beta$ and $x'\beta'$ is introduced and an ordering sign function, $\sigma(x_1\beta_1 , x_2\beta_2 , \cdots x_n\beta_n)$, is defined. Some particular ordering relations will be discussed below in part B.

(c) The antisymmetric basis vectors are introduced. They are formally connected with the symmetric states and with the localized operator "vacuum" state,[20] denoted by $|$ "0"), by the following relations:

$$\begin{aligned} | x_1\beta_1 , x_2\beta_2 , \cdots x_n\beta_n)_A &= \sigma(x_1\beta_1 , x_2\beta_2 , \cdots x_n\beta_n) \mid x_1\beta_1 , x_2\beta_2 , \cdots x_n\beta_n)_S \\ &= \psi^*_{\beta_1}(x_1)\psi^*_{\beta_2}(x_2) \cdots \psi^*_{\beta_n}(x_n) \mid \text{“0”}), \end{aligned} \tag{122a}$$

where $|$ "0") is defined as the state satisfying

$$\psi_\beta(x) \mid \text{“0”}) = 0. \tag{122b}$$

(4) The propagator between two states $| \psi')$ and $| \psi'')$ in the overcomplete family is formally defined by the following expression involving the time-ordering operator, T, *(15)*:

$$(\psi''t'' \mid \psi't') = (\psi'' \mid T \left\{ \exp\left[-i \int_{t'}^{t''} \mathbf{H}(t)\, dt \right] \right\} \mid \psi'), \tag{123}$$

where $\mathbf{H}(t)$ represents the time dependent Hamiltonian operator. Let us assume that the integral in (123) is approximated by a finite number of terms and that the "unit operator" is repeatedly inserted expressed in the form $\int | \psi)\delta\psi(\psi |$,

[18] The factor $d_{\beta\beta'}(x)$ is necessary, for example, when the spinor field is coupled to the gravitational field; the details of a gravitational coupling are contained in the paper by Brill and Wheeler *(39)*.

[19] Localized "number" operators have been discussed, for example, by Heisenberg and Pauli *(40)*.

[20] The localized operator "vacuum" state has been recently considered by Thirring *(41)*; this state does not coincide with the physical vacuum state for relativistic fields.

where the vector $|\psi)$ is given in (121) and $\delta\psi$ is the functional measure for fields at a fixed time:

$$\delta\psi = \mathfrak{N} \prod_{x\beta} \frac{d\psi_{\beta,r}(x)\, d\psi_{\beta,i}(x)}{[1 + \psi_\beta^*(x)\psi_\beta(x)\, d^3x]^2}. \tag{124}$$

Each infinitesimal propagator may be formally manipulated, as illustrated in Eq. (32), to obtain a sum over histories which is defined by the following limiting operation:

$$\begin{aligned}(\psi''t'' \mid \psi't') = \lim \int \cdots \int (\psi'' \mid \psi_N) \exp\left[-i\,\epsilon \frac{(\psi'' \mid \mathbf{H}(t_N) \mid \psi_N)}{(\psi'' \mid \psi_N)}\right] \delta\psi_N \\ \times \cdots \delta\psi_1 (\psi_1 \mid \psi') \exp\left[-i\,\epsilon \frac{(\psi_1 \mid \mathbf{H}(t') \mid \psi')}{(\psi_1 \mid \psi')}\right],\end{aligned} \tag{125}$$

where "lim" denotes passage to the limit in which $N \to \infty$, $\epsilon \to 0$, and $N\epsilon = t'' - t'$.

The limiting expression on the right side of (125) will certainly generate *some* description of the propagator, $(\psi''t'' \mid \psi't')$ as a functional sum over histories. However, this description need not be based on an extension of the corresponding classical action principle. For example, if $\mathbf{H}(t)$ represents the Hamiltonian for the electron field in the presence of an external potential A_μ, there is no guarantee that the action functional (118) will be contained in the expression (125) after passage to the limit.

(5) We need to establish the necessary criteria in order that the limiting form of (125) shall represent a functional sum over histories which *is* based on an extension of the familiar classical action expression, such as (118).

We are already aware, for continuous histories $\psi_\beta(x,t)$ which are in the "classical" Hilbert space, that a portion of the integrand in (125) is exactly represented by the following expression:

$$\lim \prod_{m=0}^{N} (\psi_{m+1} \mid \psi_m) = \exp\left[i \sum_\beta \int_{t'}^{t''} \left(\frac{i}{2}\right)\left(\psi_\beta^* \overleftrightarrow{\frac{\partial}{\partial t}} \psi_\beta\right) d^4x\right]. \tag{126}$$

This part of the multiple product appearing in (125) establishes the desired time-derivative term in the classical action functional (see, for example, the first term in (118)).

The remaining product in the integrand of (125) yields an exponent which may be described, for continuous histories, by the expression

$$-i \int_{t'}^{t''} (\psi(t) \mid \mathbf{H}(t) \mid \psi(t))\, dt. \tag{127}$$

Our criteria will be fulfilled if, for those continuous histories $\psi_\beta(x,t)$ which are also in Hilbert space, the expression (127) reproduces the algebraic form of the

Hamiltonian operator, **H**. This phrase is best explained by an example: Suppose

$$\mathbf{H} = \int \left[d_{\beta\beta'} \boldsymbol{\psi}_\beta^* \frac{\overleftrightarrow{\partial}}{\partial x_i} \boldsymbol{\psi}_{\beta'} + c_{\beta\beta'} \boldsymbol{\psi}_\beta^* \boldsymbol{\psi}_{\beta'} + \text{H.c.} \right] d^3x ,$$

then, for all continuous histories $\psi_\beta(x,t)$ in Hilbert space, it is necessary that

$$\begin{aligned} \int_{t'}^{t''} \int (\psi(t) \mid d_{\beta\beta'} \boldsymbol{\psi}_\beta^* \frac{\overleftrightarrow{\partial}}{\partial x_i} \boldsymbol{\psi}_{\beta'} + c_{\beta\beta'} \boldsymbol{\psi}_\beta^* \boldsymbol{\psi}_{\beta'} + \text{H.c.} \mid \psi(t)) \, d^4x \\ = \int_{t'}^{t''} \int \left[d_{\beta\beta'} \psi_\beta^* \frac{\overleftrightarrow{\partial}}{\partial x_i} \psi_{\beta'} + c_{\beta\beta'} \psi_\beta^* \psi_{\beta'} + \text{H.c.} \right] d^4x . \end{aligned} \tag{128}$$

Since $c_{\beta\beta'}$ and $d_{\beta\beta'}$ are independent arbitrary functions, Eq. (128) must hold, for each term, at every space-time point. The validity of (128), under the restrictions placed on the history $\psi_\beta(x,t)$, is by no means assured. In order to secure (128) we now investigate the only remaining freedom available in our quantum representation: the particular selection of the ordering relation, $<$.

B. Selection of the ordering relation

A primary objective in selecting a particular ordering relation $<$ is to insure that the expression

$$(\psi \mid \boldsymbol{\psi}_\beta^*(x) \boldsymbol{\psi}_{\beta'}(x) \mid \psi) = \psi_\beta^*(x) \psi_{\beta'}(x) \tag{129}$$

is valid for all fields $\psi_\beta(x)$ in Hilbert space where the state $\mid \psi)$ is given by (121). Equation (129) has essentially been established already for the case $\beta = \beta'$ by the work in Section II; we concentrate here on the case where $\beta \neq \beta'$.

With the aid of the states $\mid \psi)$, the information contained in the relations (122), and the as yet unselected ordering relation, we may establish that the following formula for the ratio of matrix elements is valid when $\beta \neq \beta'$, for all fields $\psi_\beta''(x)$ and $\psi_\beta'(x)$:

$$\begin{aligned} \frac{(\psi'' \mid \boldsymbol{\psi}_\beta^*(x) \boldsymbol{\psi}_{\beta'}(x) \mid \psi')}{(\psi'' \mid \psi')} &= \frac{\psi_\beta^{*\prime\prime}(x) \psi_{\beta'}'(x)}{[1 + \psi_\beta^{*\prime\prime} \psi_\beta' \, d^3x][1 + \psi_{\beta'}^{*\prime\prime} \psi_{\beta'}' \, d^3x]} \\ &\times \prod_{y\gamma \in \mathfrak{A}} \frac{[1 - \psi_\gamma^{*\prime\prime}(y) \psi_\gamma'(y) \, d^3y]}{[1 + \psi_\gamma^{*\prime\prime}(y) \psi_\gamma'(y) \, d^3y]} . \end{aligned} \tag{130}$$

Here, $\mathfrak{A}$ represents the set of points "between" $x\beta$ and $x\beta'$ and is defined by one of the following conditions: $y\gamma \in \mathfrak{A}$ if (1) $x\beta < y\gamma < x\beta'$ (if $x\beta < x\beta'$) or (2) $x\beta' < y\gamma < x\beta$ (if $x\beta' < x\beta$). It is clear that we may satisfy (129) identically for all fields, $\psi_\beta(x)$, in Hilbert space if the measure of the set $\mathfrak{A}$ is selected to be zero for all combinations of β and β'.[21] We therefore divide the possible relations

[21] The measure of the set $\mathfrak{A}$, $m(\mathfrak{A})$, may be defined by an integral over all $\alpha \in \mathfrak{A}$: $m(\mathfrak{A}) = \int dm(\alpha)$. This expression employs the notation of Section II.4 where α denotes a member of the set of indices I; here, α stands for the pair $y\gamma$.

into two classes: (I) A relation, $<$, is called *physical* if the set $\mathcal{A}$, defined above, is a set of measure zero for all β and β', and (II) a relation is called *unphysical* if $\mathcal{A}$ is a set of nonzero measure for some choice of β and β'. The following examples will illustrate these two types of orderings:

(I) *A physical ordering example.* We call the pair $x\beta < x'\beta'$ if (1) $x < x'$[22] or if (2) $x = x'$, $\beta < \beta'$, where $\beta < \beta'$ employs some ordering relation among the finite number of indices β. For this physical ordering example the number of elements in the set $\mathcal{A}$ cannot exceed the total finite number of indices and $\mathcal{A}$ clearly constitutes a set of measure zero.

(II) *An unphysical ordering example.* For this example we call the pair $x\beta < x'\beta'$ if (1) $\beta < \beta'$ or if (2) $\beta = \beta'$ and $x < x'$; it is easily established that $\mathcal{A}$ is no longer a set of measure zero for any choice of β and β' such that $\beta \neq \beta'$.

Therefore, by selecting a physical ordering relation, $<$, for the set of points $x\beta$ we fully satisfy the requirement expressed in (128) for all the operators of the form given in (119a). The physical orderings are also valuable for the second form of operators illustrated in (119b), which involve spatial derivatives.

For all fields $\psi_\beta(x)$ in Hilbert space we need to establish that

$$(\psi \mid \boldsymbol{\psi}_\beta{}^*(x) \overleftrightarrow{\frac{\partial}{\partial x_i}} \boldsymbol{\psi}_{\beta'}(x) \mid \psi) = \psi_\beta{}^*(x) \overleftrightarrow{\frac{\partial}{\partial x_i}} \psi_{\beta'}(x), \tag{131}$$

where the state $\mid \psi)$ is given in (121); establishing (131) will constitute the final step in proving that a functional sum over histories can be constructed based on an extension of the familiar classical action.

In the case $\beta = \beta'$, we may formally compute the following expression

$$\frac{(\psi'' \mid \boldsymbol{\psi}_\beta{}^*(x) \overleftrightarrow{\frac{\partial}{\partial x_i}} \boldsymbol{\psi}_\beta(x) \mid \psi')}{(\psi'' \mid \psi')} = \frac{\psi_\beta^{*\prime\prime}(x) \overleftrightarrow{\frac{\partial}{\partial x_i}} \psi_\beta{}'(x)}{[1 + \psi_\beta^{*\prime\prime} \psi_\beta{}' \, d^3x]} \tag{132}$$

by a direct use of the states $\mid \psi)$ and the relations (122). Equation (132) therefore provides a verification of (131) for $\beta = \beta'$. A direct calculation of similar matrix elements when $\beta \neq \beta'$ is complicated by the need to take spatial derivatives of the ordering sign function. However, it can be shown, for $\beta \neq \beta'$, that

$$\begin{aligned} &\frac{(\psi'' \mid \boldsymbol{\psi}_\beta{}^*(x) \overleftrightarrow{\frac{\partial}{\partial x_i}} \boldsymbol{\psi}_{\beta'}(x) \psi')}{(\psi'' \mid \psi')} \\ &\quad = \frac{\psi_\beta^{*\prime\prime}(x) \overleftrightarrow{\frac{\partial}{\partial x_i}} \psi_{\beta'}'(x)}{[1 + \psi_\beta^{*\prime\prime} \psi_\beta{}' \, d^3x][1 + \psi_{\beta'}^{*\prime\prime} \psi_{\beta'}' \, d^3x]} \prod_{y\gamma \in \mathcal{A}} \frac{[1 - \psi_\gamma^{*\prime\prime}(y) \psi_\gamma{}'(y) \, d^3y]}{[1 + \psi_\gamma^{*\prime\prime}(y) \psi_\gamma{}'(y) \, d^3y]}. \end{aligned} \tag{133}$$

[22] $x < x'$ may be described by any ordering relation for 3-space (see Section II.3).

Here, again, $\mathfrak{A}$ is the set containing the points $y\gamma$ "between" $x\beta$ and $x\beta'$. The expression in (133) is very similar to that in (130) and, by selecting a physical ordering of the points $x\beta$, we can fully satisfy the requirement expressed by (131). This completes the necessary requirements and establishes that *locally interacting spinor fields can be described by a c-number sum over histories.*

It should be pointed out that the unusual extensions for singular fields, given in (130), (132), and (133), are related, by a canonical transformation, to the momentum space extensions given in Section II and therefore the new extensions contain the same physics as the earlier extensions.

C. Conclusion

We have partially clarified Feynman's symbolic solution of the quantum dynamical problem,

$$(f_2\sigma_2 \mid f_1\sigma_1) = N \int e^{iI(f)} \mathfrak{D}f,$$

by emphasizing the inherent ambiguity, as well as the importance, in the action assigned to unruly histories outside the classical domain. Our results do not always have the simplicity available in the solution above, but they clearly define a sum over histories which (1) in the case of a well-defined c-number Schrödinger theory leads either to Bose or to Fermi statistics, according as the unruly histories receive one or another kind of treatment, and (2) in the case of familiar relativistic fields leads to Fermi statistics. Anticommuting c-numbers are an unnecessary addition to mathematical physics.

Acknowledgments

The author greatly appreciates the guidance and continued interest of Professor J. A. Wheeler in this work as well as the invaluable assistance given him by Professor R. Haag and Professor Wheeler in clarifying the presentation of this paper. Several discussions with Professor C. C. Grosjean and Dr. J. G. Fletcher proved valuable.

It is a pleasure, also, to acknowledge the generous assistance given the author from 1956 to 1959 under a C.D.T. Fellowship awarded by Bell Telephone Laboratories.

Received: December 3, 1959

REFERENCES

1. R. P. Feynman, thesis, Princeton University, 1942.
2. R. P. Feynman, *Revs. Modern Phys.* **20,** 367 (1948).
3. J. G. Polkinghorne, *Proc. Roy. Soc.* **A230,** 272 (1955).
4. P. T. Matthews and A. Salam, *Nuovo cimento* [10], **2,** 120 (1955).
5. C. W. Misner, *Revs. Modern Phys.* **29,** 497 (1957).
6. G. Rosen, thesis, Princeton University, 1959.
7. K. Symanzik, *Z. Naturforsch.* **9a,** 809 (1954).
8. Y. Nambu, *Progr. Theoret. Phys.* (*Japan*) **7,** 131 (1952).
9. W. Tobocman, *Nuovo cimento* [10], **3,** 1213 (1956).

10. K. Goto, *Nuovo cimento* [10], **3,** 533 (1956).
11. J. Schwinger, *Proc. Nat. Acad. Sci.* **37,** 452, 455 (1951); J. Schwinger, *Phys. Rev.* **92,** 1283 (1953).
12. F. Coester, *Phys. Rev.* **95,** 1318 (1954).
13. J. M. Jauch, *Helv. Phys. Acta* **29,** 287 (1954).
14. J. G. Valatin, *Nuovo cimento* [10], **4,** 726 (1956).
15. G. Wick, *Phys. Rev.* **80,** 268 (1950).
16. J. V. Novozilov and A. V. Tulub, *Fortschr. Physik* **6,** 50 (1958).
17. J. von Neumann, *Compositio Math.* **6,** 1 (1938).
18. C. Morette, *Phys. Rev.* **81,** 848 (1951); W. Pauli, "Ausgewählte Kapitel aus der Feldquantisierung," Zurich, 1951.
19. P. Choquard, *Helv. Phys. Acta* **28,** 89 (1955).
20. W. K. Burton and A. H. deBorde, *Nuovo cimento* [10], **2,** 197 (1955).
21. K. O. Friedrichs, "Mathematical Aspects of the Quantum Theory of Fields." Interscience, New York, 1953.
22. V. Fock, *Physik. Z. Sowjetunion* **6,** 425 (1934).
23. P. A. M. Dirac, *Dublin Institute Ser.* A # 1, 1943.
24. J. von Neumann, "Mathematical Foundations of Quantum Mechanics." Princeton University Press, Princeton, 1953.
25. J. Doob, *Trans. Am. Math. Soc.* **42,** 107 (1937); J. Doob, *Annals of Math.* **41,** 737 (1940); M. Kac, "Probability and Related Topics in Physical Sciences." Interscience, New York, 1959.
26. P. Jordan and E. Wigner, *Z. Physik* **47,** 631 (1928).
27. F. J. Dyson, "Lecture Notes on Quantum Electrodynamics," Cornell University, 1954.
28. T. Okayama, *Progr. Theoret. Phys. (Japan)* **7,** 517 (1952); H. S. Green, *Phys. Rev.* **90,** 270 (1953).
29. L. van Hove, *Physica* **18,** 145 (1952).
30. R. Haag, *Kgl. Danske Videnskab. Seskab, Mat.-fys. Medd.* **29,** 12 (1955).
31. S. Schweber and A. Wightman, *Phys. Rev.* **98,** 812 (1955).
32. S. Albertoni and F. Duimio, *Nuovo cimento* [10], **6,** 1193 (1957).
33. M. Schwartz, *Bull. Am. Phys. Soc.* [2], **4,** 30 (1959).
34. J. von Neumann, "The Theory of the Positron," Institute for Advanced Study lecture notes, 1936.
35. J. von Neumann, *Annals of Math.* **50,** 401 (1949).
36. V. Bargmann, *Annals of Math.* **48,** 568 (1947).
37. R. P. Feynman, *Phys. Rev.* **97,** 660 (1955); J. M. Gelfand and A. M. Jaglom, *Fortschr. Physik* **5,** 517 (1957).
38. J. A. Wheeler, "Fields and Particles" (unpublished). Lectures given at Princeton (1954–1955 and 1956–1957) and at Leiden (1956).
39. D. Brill and J. Wheeler, *Revs. Modern Phys.* **29,** 465 (1957).
40. W. Heisenberg and W. Pauli, *Z. Physik* **59,** 168 (1929).
41. W. Thirring, *Annals of Physics* **3,** 91 (1958).

Continuous-Representation Theory. I. Postulates of Continuous-Representation Theory

John R. Klauder
Bell Telephone Laboratories, Murray Hill, New Jersey
(Received 26 December 1962)

In a continuous representation of Hilbert space, each vector Ψ is represented by a complex, continuous, bounded function $\psi(\Phi) \equiv (\Phi, \Psi)$ defined on a set $\mathfrak{S}$ of continuously many, nonindependent unit vectors Φ having rather special properties: Each vector in $\mathfrak{S}$ possesses an arbitrarily close neighboring vector, and the identity operator is expressable as an integral over projections onto individual vectors in $\mathfrak{S}$. In particular cases it is convenient to introduce labels for the vectors in $\mathfrak{S}$ whereupon each Ψ is represented by a complex, continuous, bounded, label-space function. Basic properties common to all continuous representations are presented, and some applications of the general formalism are indicated.

INTRODUCTION

The essential ingredients in the basic structure of quantum mechanics are surprisingly few: (i) unit vectors in a Hilbert space $\mathfrak{H}$ correspond to states of a system; (ii) dynamics or scattering involves an automorphism among unit vectors; and (iii) the natural inner products in $\mathfrak{H}$ are interpreted as probability amplitudes. Physics enters, as well as being read out of the formalism, by means of one or more mappings $\mathfrak{M}$ from appropriate label sets characterizing the physical problem into unit vectors in $\mathfrak{H}$. For example, the "in" and "out" states of scattering theory are manifestations of two different mappings into $\mathfrak{H}$ of a parameter set including four-momentum, spin, baryon number, etc. However, independent of the particular form a mapping takes, it provides the "bridge" between the abstract quantum formalism and the label-space framework in which stochastic statements pertinent to a particular system are made.

Now dynamics, or any other Lie group of automorphisms, entails a continuous permutation among unit vectors in $\mathfrak{H}$. It would frequently be desirable to have Hilbert-space representations expressed as label-space functions, admitting direct parameterization of such continuous transformations simply by means of label-space transformations. That is, the labels by themselves are rich enough to parameterize continuous permutations among unit vectors in $\mathfrak{H}$. This clearly requires that the labels must in part assume values in the continuum. For present purposes, such a requirement rules out representations wherein vectors are functions of discrete variables, e.g., the eigenvalues of a complete set of commuting observables with discrete spectrum. The Dirac prescription to generate representations with continuous labels (e.g., the Schrödinger representation) cannot be regarded as a continuous mapping of the label space into unit vectors in $\mathfrak{H}$. Instead, continuity demands that the image set of unit vectors $\mathfrak{S}$ cannot be an orthonormal set, but, in contrast, $\mathfrak{S}$ *must contain vectors arbitrarily close to one another*. This basic property is common to all the image sets $\mathfrak{S}$ we consider, and is fundamental for the definition of a continuous representation. Accounts of specific label sets, mappings, and continuous representations will be treated in the following paper,[1] and in subsequent papers in connection with various applications. Here it is our purpose to present the requirements on the mapping $\mathfrak{M}$ and the image sets $\mathfrak{S}$ that are necessary for a continuous representation to exist, and to discuss some of the basic properties of such representations common to all systems.

POSTULATES OF CONTINUOUS-REPRESENTATION THEORY

We choose a Hilbert space $\mathfrak{H}$, finite- or infinite-dimensional, with positive-definite metric. Among all the vectors in $\mathfrak{H}$, let us focus our attention on *unit* vectors, to be universally denoted by Φ with an arbitrary array of sub- or superscripts, etc. Thus

$$(\Phi, \Phi)^{\frac{1}{2}} \equiv ||\Phi|| = 1 = ||\Phi_1|| = ||\Phi'|| = \cdots .$$

Let $\mathfrak{T}$ denote the set of all unit vectors, and $\mathfrak{S}$ denote a subset of $\mathfrak{T}$: $\mathfrak{S} \subset \mathfrak{T}$. It is the set $\mathfrak{S}$ that will be the image set of the mapping $\mathfrak{M}$. Then, we assert

Postulate 1. (Local density and continuity). For each $\Phi \in \mathfrak{S}$ and every $\delta > 0$, there exists a vector $\Phi' \in \mathfrak{S}$, different from Φ, for which $||\Phi - \Phi'|| < \delta$. The set $\mathfrak{S}$ is an arcwise connected subset of $\mathfrak{H}$ or a union thereof.

[1] J. R. Klauder, J. Math. Phys. 4, 1058 (1963) (following paper).

Clearly, the conventional orthonormal basis set fails to satisfy Postulate 1.

Further, let $\mathfrak{L}$ denote the "label" space whose points l may be correlated with vectors in $\mathfrak{S}$ by the mapping $\mathfrak{M}$ of $\mathfrak{L}$ onto $\mathfrak{S}$. Regarding this correlation, we require

Postulate 2. (Label continuity). The mapping $\mathfrak{M}$: $l \rightarrow \Phi[l]$ is a many-one continuous map of a Hausdorff, i.e., separable topological space $\mathfrak{L}$ onto $\mathfrak{S}$. By continuity in $\mathfrak{S}$, we mean the usual weak continuity in $\mathfrak{H}$. Thus if $l_n \rightarrow l$, then $(\Phi[l_n], \Psi) \rightarrow (\Phi[l], \Psi)$ for all $\Psi \in \mathfrak{H}$.

It is often the case that the topology for $\mathfrak{L}$ is the usual topology where open sets are identified as open intervals. The basic purpose of this postulate is to provide a parameterization, i.e., a "handle" for the states in $\mathfrak{S}$.

For purposes of using the vectors in $\mathfrak{S}$ to define a representation with as familiar a form as possible, we assume

Postulate 3. (Completeness and resolution). The set $\mathfrak{S}$ spans the space $\mathfrak{H}$, i.e., completion in norm of the set of all linear combinations of elements in $\mathfrak{S}$ yields $\mathfrak{H}$. The identity operator in $\mathfrak{H}$ can be resolved into an integral over projection operators onto individual vectors in $\mathfrak{S}$.

When $\mathfrak{S}$ is locally compact, then the last postulate means that some additive real measure μ on elementary sets in $\mathfrak{S}$ exists such that

$$\Psi = \int_{\mathfrak{S}} \Phi \, d\mu\,(\Phi)(\Phi, \Psi),$$

or

$$(\Psi', \Psi) = \int_{\mathfrak{S}} (\Psi', \Phi) \, d\mu\,(\Phi)(\Phi, \Psi), \tag{1}$$

for arbitrary vectors Ψ and Ψ'. If $\mathfrak{S}$ is not locally compact, the concept of integral can be generalized to give meaning to (1), e.g., in the manner of Friedrichs of Shapiro.[2] The only role of μ shall be to generate the expansion (1), a restriction which, in general, is insufficient to fix μ uniquely. It is certainly plausible, therefore, and we shall provide a simple proof in the Appendix, for a compact set $\mathfrak{S}$, that (i) because of invariance under unitary transformations, and (ii) the existence of (1), we have

[2] K. O. Friedricks and H. N. Shapiro, Proc. Natl. Acad. Sci. U. S. **43**, 336 (1957); *Integration of Functionals,* New York University, Institute of Mathematical Sciences (1957). Here we treat such cases formally.

Theorem 1. (Invariant measure.) There is no loss of generality in assuming μ invariant under any and all unitary transformations U that leave the compact set $\mathfrak{S}$ invariant: $U\mathfrak{S} = \mathfrak{S}$. The stated invariance takes the form $\mu(U\mathfrak{R}) = \mu(\mathfrak{R})$ for all $\mathfrak{R} \subset \mathfrak{S}$. A suitable choice for μ is one for which $d\mu(U\Phi) = d\mu(\Phi)$.

Consequently, an important application of Theorem 1 arises if the group $\mathcal{G}$ of such unitary transformations U forms a transitive permutation group on vectors in $\mathfrak{S}$ (i.e., if $\Phi \in \mathfrak{S}$, then $\{U\Phi : U \in \mathcal{G}\} = \mathfrak{S}$), for then μ can be chosen without loss of generality as the invariant group measure. Examples of this approach to find μ are given in the following paper.[1]

Form of the Continuous Representation

Equation (1) provides the basis for a continuous representation of Hilbert space. In such a representation, the vector Ψ is represented by the complex, bounded, continuous function

$$\psi(\Phi) \equiv (\Phi, \Psi); \qquad \psi^*(\Phi) \equiv (\Psi, \Phi), \tag{2}$$

[or by $\psi(l) \equiv (\Phi[l], \Psi)$ if specific labels are introduced]. The inner product of two vectors is then a restatement of (1):

$$(\Psi', \Psi) = \int_{\mathfrak{S}} \psi'^*(\Phi) \, d\mu\,(\Phi)\psi(\Phi). \tag{3}$$

Not all functions on $\mathfrak{S}$ represent vectors, but only those which satisfy

$$\psi(\Phi') = \int_{\mathfrak{S}} \mathcal{K}(\Phi'; \Phi) \, d\mu\,(\Phi)\psi(\Phi), \tag{4}$$

where the reproducing kernel

$$\mathcal{K}(\Phi'; \Phi) \equiv (\Phi', \Phi), \tag{5}$$

as follows from Eqs. (3) and (2). Furthermore, $\mathcal{K}$ fulfills the indempotent relation

$$\mathcal{K}(\Phi'; \Phi'') = \int_{\mathfrak{S}} \mathcal{K}(\Phi'; \Phi) \, d\mu\,(\Phi)\mathcal{K}(\Phi; \Phi''). \tag{6}$$

An operator $\mathfrak{B}$ defined on $\mathfrak{S}$ is represented as a function of two points in $\mathfrak{S}$ by

$$\mathfrak{B}(\Phi'; \Phi) \equiv (\Phi', \mathfrak{B}\Phi) \tag{7}$$

[or by $\mathfrak{B}(l'; l)$ if specific labels are introduced], which is separately continuous in each argument and bounded if $\mathfrak{B}$ is a bounded operator. The representation of $\mathfrak{B}\Psi$ is clearly

$$(\mathfrak{B}\psi)(\Phi') = \int_{\mathfrak{S}} \mathfrak{B}(\Phi'; \Phi) \, d\mu\,(\Phi)\psi(\Phi). \tag{8}$$

Not every function of two points in $\mathfrak{S}$ represents an operator, but only those which satisfy

$$\mathcal{B}(\Phi'; \Phi'') = \int_{\mathfrak{S}} \mathcal{B}(\Phi'; \Phi)\, d\mu\, (\Phi) \mathcal{K}(\Phi; \Phi'')$$
$$= \int_{\mathfrak{S}} \mathcal{K}(\Phi'; \Phi)\, d\mu\, (\Phi) \mathcal{B}(\Phi; \Phi'')$$
$$= \iint_{\mathfrak{S}} \mathcal{K}(\Phi'; \Phi_1)\, d\mu\, (\Phi_1)$$
$$\times\, \mathcal{B}(\Phi_1; \Phi_2)\, d\mu\, (\Phi_2) \mathcal{K}(\Phi_2; \Phi''), \quad (9)$$

as follows from (7) and (8) and their adjoints.

It is clear that the set $\mathfrak{S}$ plays an important role in establishing a continuous representation and determining its properties: Postulate 1 leads, with the aid of continuity in the inner product, to continuous functions ψ defined on unit vectors; Postulate 2 leads, in turn, to continuous functions ψ defined on points of label space; and Postulate 3 ensures that the functions ψ provide a representation of $\mathfrak{H}$. As a matter of nomenclature, we call an abstract set of unit vectors $\mathfrak{S}$ that satisfies Postulates 1 and 3 an *overcomplete family of states* (OFS). As is common in quantum mechanics, we shall often refer to the continuous representation generated (in the manner described above) by an OFS simply by reference to the particular OFS itself.

Some Applications of Overcomplete Families of States

An OFS may be used as any other representation would be used in quantum or quantum statistical mechanics. As such, the OFS appears generally to be a more proper way to introduce vectors with continuous labels than the conventionally used but nonexistent eigenstates for operators with continuous spectra. In particular, there exists one special choice of an OFS that is closely related to the Fock representation of an infinite-dimensional Hilbert space by entire analytic functions, which has recently seen a renewed interest.[3] For applications to field theories, one natural set of "labels" for the OFS are the well-defined test functions of Distribution Theory.[4]

In the following paper[1] we shall establish a generalized form of "classical" dynamics expressed in terms of the continuously variable c-number labels that characterize vectors of the OFS. This analysis shows, for example, that, for certain classes of Hamiltonians, we can exactly describe quantum mechanics by the classical dynamical formalism merely by reinterpreting the classical dynamical variables as c-number labels belonging to Hilbert space vectors. In addition, it is possible that overcomplete families of states may prove useful in placing the sum-over-histories on a more sound mathematical footing; already in formal studies they have permitted the usual Fresnel integrals to be replaced by absolutely convergent Gaussian integrals.[5] On the other hand, an OFS may be used for a Hamiltonian-*less* approach to dynamics directly through a postulated evolutionary[6] or in–out scattering automorphism.

Finally, it should be remarked that a natural choice of an OFS for fermion degrees of freedom[7] corresponds to the set of states generated by all Bogoliubov transformations. Thus, this and other OFS may be directly relevant in making interesting approximations.

The author thanks J. McKenna for several discussions.

APPENDIX

We wish to prove for compact sets $\mathfrak{S}$ that the measure on vectors in the resolution

$$(\Psi', \Psi) = \int_{\mathfrak{S}} (\Psi', \Phi)\, d\mu\, (\Phi)(\Phi, \Psi) \quad (A1)$$

may be chosen invariant under all unitary transformations leaving $\mathfrak{S}$ unchanged. Let U be such a transformation:

$$U\mathfrak{S} = \mathfrak{S}. \quad (A2)$$

Then unitary invariance of (A1) combined with (A2) yields

$$(\Psi', \Psi) = \int_{\mathfrak{S}} (\Psi', U\Phi)\, d\mu\, (\Phi)(U\Phi, \Psi)$$
$$= \int_{\mathfrak{S}} (\Psi', \Phi)\, d\mu\, (U^{-1}\Phi)(\Phi, \Psi). \quad (A3)$$

Namely, if $d\mu(\Phi)$ was a suitable weighting, we see that an equally good weighting on vectors Φ is $d\mu(U^{-1}\Phi)$. If $d\mu(U^{-1}\Phi) = d\mu(\Phi)$, then the measure already possesses the desired invariance. If μ is not invariant we may proceed as follows:

It follows from (A2) that $U^p\mathfrak{S} = \mathfrak{S}$, where p is an arbitrary integer. The generalization of (A3) would imply that $d\mu(U^p\Phi)$ is as good a weighting

[3] V. Bargmann, Commun. Pure Appl. Math. **14**, 187 (1961); Proc. Natl. Acad. Sci. U. S. **48**, 199 (1962). For the closely related formalism, see reference 1, Sec. 2.C.

[4] One type of labeling for boson fields is discussed by H. Araki, J. Math. Phys. **1**, 492 (1960).

[5] J. R. Klauder, Ann. Phys. (NY) **11**, 123 (1960), especially p. 127, and pp. 142–153; S. S. Schweber, J. Math. Phys. **3**, 831 (1962).

[6] Such an approach to dynamics was discussed by J. von Neumann, "The Theory of the Positron," Lecture Notes, Institute for Advanced Study, Princeton, New Jersey, 1936.

[7] See reference 1, Sec. 4.

as $d\mu(\Phi)$. If U is a cyclic element of order P, i.e., $U^P = 1$, then the combination

$$d\mu_P(\Phi) \equiv \frac{1}{P} \sum_{p=0}^{P-1} d\mu\,(U^p\Phi) \tag{A4}$$

is invariant under U, $d\mu_P(U^{-1}\Phi) = d\mu_P(\Phi)$. Henceforth, we would use only the invariant form and rename the measure $d\mu(\Phi)$. This procedure can be extended to include invariance under all cyclic elements that leave $\mathfrak{S}$ invariant.

On the other hand, if U is not a cyclic element, then we shall regard it as an element of a one-parameter Lie group whose elements $U[\alpha]$ satisfy $U[\beta]U[\alpha] = U[\beta + \alpha]$. Invariance of the compact set $\mathfrak{S}$ under all powers of U leads by continuity and Postulate 1 to invariance of $\mathfrak{S}$ under the entire one-parameter subgroup containing U. We now can form a quantity analogous to (A4), namely,

$$d\mu_c(\Phi) = \frac{1}{\int d\alpha} \int d\mu\,(U[\alpha]\Phi)\,d\alpha, \tag{A5}$$

an expression which exists in virtue of the finite parameter range in compact groups. Clearly, $d\mu_c$ is invariant under $U[\beta]$, i.e., $d\mu_c(U[\beta]\Phi) = d\mu_c(\Phi)$. Again we rename this invariant form $d\mu(\Phi)$. By extending the preceding techniques to all invariant transformations of $\mathfrak{S}$, we establish Theorem 1.

While the above proof holds only for compact spaces, the left-invariant group measure as suggested by Theorem 1 may always be examined for individual noncompact spaces to see whether or not it satisfies (A1).

Continuous-Representation Theory. II. Generalized Relation between Quantum and Classical Dynamics

JOHN R. KLAUDER
Bell Telephone Laboratories, Murray Hill, New Jersey
(Received 26 December 1962)

This paper discusses an application to the study of dynamics of the typical overcomplete, nonindependent sets of unit vectors that characterize continuous-representation theory. It is shown in particular that the conventional, classical Hamiltonian dynamical formalism arises from an analysis of quantum dynamics restricted to an overcomplete, nonindependent set of vectors which lie in one-to-one correspondence with, and are labeled by, points in phase space. A generalized "classical" mechanics is then defined by the extremal of the quantum-mechanical action functional with respect to a restricted set of unit vectors whose c-number labels become the dynamical variables. This kind of "classical" formalism is discussed in some generality, and is applied not only to simple single-particle problems, but also to finite-spin degrees of freedom and to fermion field oscillators. These latter cases are examples of an important class of problems called exact, for which a study of the classical dynamics alone is sufficient to infer the correct quantum dynamics.

1. INTRODUCTION

THE general postulates of continuous representations of Hilbert space have been stated elsewhere.[1] The overcomplete family of states, hereafter abbreviated to OFS and denoted by $\mathfrak{S}$, that is involved in such a representation may be visualized as forming one or more arcwise, connected "patches" on the unit sphere in Hilbert space defined by $||\Phi|| = 1$. In order to discuss dynamics and time evolution we shall define a *path* to be a continuous, unit-vector-valued time function $\Phi(t)$. Now, a general variation of the path $\Phi(t)$, apart from simple ray rotations (e.g., $\Phi(t) = \exp[i\lambda(t)]\Phi$), in the action functional

$$I = \int [i\hbar(\Phi, d\Phi/dt) - (\Phi, \mathcal{H}\Phi)]\, dt, \qquad (1)$$

yields as the Euler–Lagrange equations, the Schrödinger equation of motion. However, we may ask what are the dynamical consequences if Eq. (1) is extremized over only a *restricted* set of paths, such as those constrained to lie in $\mathfrak{S}$: $\Phi(t) \in \mathfrak{S}$?

Through the study of a one-dimensional, single-particle problem in Sec. 2,[2] we conclude that when

[1] J. R. Klauder, J. Math. Phys. 4, 1055 (1963) (preceding paper) hereafter referred to as I.

[2] A brief discussion based on this example appears in J. R. Klauder, Helv. Phys. Acta 35, 333 (1962).

evaluated for a particular restricted set of unit vectors, parameterized or labeled by phase-space points p, q [the vectors of this set being given by (3)], the quantum action functional, Eq. (1), reduces in form to the *classical* action functional. That is, Eq. (1) reduces to

$$I = \int [p\dot{q} - H(p, q)]\, dt. \qquad (2)$$

Consequently, by extremizing (2) with respect to just this special set of states, we would obtain the classical equations and not the quantum equations. While the action functional (2) and its extremal equations thus have the classical form, these dynamical variables are still to be interpreted as labels for Hilbert-space vectors. Likewise, the physical interpretation of the theory still follows the stochastic quantum prescription, i.e., vector inner products are transition amplitudes. Thus, the distinction between classical mechanics based on (2), and (an approximate form of) quantum mechanics also based on (2), lies in the *interpretation* of p and q and in how results are read out of the formalism; the quantum interpretation generally leads to an approximate form of quantum mechanics since (1) is described by (2) for only a *restricted* set of paths. We shall refer to an action functional of the form (2)—wherein the variables p and q are ordinary c-number time functions, but which are interpreted as labels for Hilbert-space vectors—as a "classical" action functional and the Euler–Lagrange equations derived therefrom as "classical" equations of motion. We observe, therefore, that merely by reinterpreting the c-number variables of the classical theory, we can view the classical action functional as a restricted evaluation of the true quantum mechanical action.

The above example, suggested by the study of a particular restricted set $\mathfrak{S}$, permits an obvious abstract extension to an arbitrary set $\mathfrak{S}$. Namely, *the "classical" equations of motion relative to* $\mathfrak{S}$ *arise as a result of extremizing* (1) *over just those vector functions for which* $\Phi(t) \in \mathfrak{S}$. Here we have a *relative* definition of the attribute "classical"—its relative nature depending on the size of the set $\mathfrak{S}$—that is generally applicable to any system with arbitrary statistics.

As an extreme situation, suppose $\mathfrak{S}$ were so large as to equal $\mathfrak{T}$, the set of all unit vectors. In that case, the resultant "classical" equations would be physically equivalent to the usual quantum-mechanical equations. It must be stressed, however, that it is not always necessary that the set $\mathfrak{S}$ be as large as $\mathfrak{T}$ in order that the "classical" equations be physically equivalent to the quantum equations. For example, if $\mathcal{H} = 0$, then extremizing (1) over any (complete) set $\mathfrak{S}$ would lead to $\Phi(t) = \Phi(0)$ for each vector $\Phi(0) \in \mathfrak{S}$; the completeness of $\mathfrak{S}$ would then correctly imply the simple evolutionary behavior for any state vector in the Hilbert space $\mathfrak{H}$. This trivial example has the property that the *exact* solution to the quantum-mechanical equations, $\Phi(t)$, obeys $\Phi(t) \in \mathfrak{S}$ if only $\Phi(0) \in \mathfrak{S}$; namely, that the true extremal to (1) is a path that remains in $\mathfrak{S}$ if only it started in $\mathfrak{S}$. This possibility is by no means confined to the case $\mathcal{H} = 0$, and for the general case we introduce the

Definition: An *exact* "classical" action functional $I_{\mathfrak{S}}\{\Phi(t)\}$ relative to the set $\mathfrak{S}$ is one whose extremal solutions correctly correspond to true extremal vector-valued time functions. In other words, if the true quantum solution $\Phi(t) = e^{-it\mathcal{H}/\hbar}\Phi(0)$ lies within the complete set $\mathfrak{S}$, $\Phi(t) \in \mathfrak{S}$, assuming only that it initially lay within $\mathfrak{S}$, then the "classical" theory is exact. If a "classical" action functional is not exact, then we shall call it inexact.

Clearly the existence of an exact "classical" action functional depends strongly on both the set $\mathfrak{S}$ and the Hamiltonian operator $\mathcal{H}$.

It follows as a corollary to the preceding definition that the *value* of the action evaluated for an extremal path will vanish for an exact "classical" action principle. Therefore, for an action functional to be exact, it is necessary but not sufficient (this latter aspect is further discussed in Sec. 2) that

$$I_{\mathfrak{S}}\{\Phi_{\text{extremal}}(t)\} = 0, \qquad (E)$$

when $I_{\mathfrak{S}}$ is extremized with respect to all $\Phi(t) \in \mathfrak{S}$. We shall refer to this vanishing of I for exact action principles as "criterion E." If criterion E is not obeyed then, of course, the action principle is inexact.

It is the subject of this paper to study some properties of our generalized "classical" formalism and learn some of the consequences that arise from restricted variations of quantum-mechanical action functionals. In Sec. 2 we study in some detail the properties of a single-particle, nonrelativistic, one-dimensional example. For Hamiltonians linear in the momentum and position operators, an exact "classical" action functional arises. Oscillator Hamiltonian operators can also lead to exact cases but only for a unique choice of the set $\mathfrak{S}$. Other Hamiltonians lead to inexact equations of motion (unless $\mathfrak{S}$ becomes significantly enlarged). Our study of canonical transformations suggests that partial physical significance of the dynamical variables is contained in the important *canonical kinematical form*, $i\hbar(\Phi, d\Phi)$, expressed as a function of the labels. A "resolution of unity" and the associated continuous representation is briefly discussed.

Generalization of the preceding analysis to an abstract N-dimensional Lie group of unitary transformations is the subject of Sec. 3. Various formulas are given for exact and inexact "classical" action functionals to discuss the "classical" formalism under certain special circumstances. Of special note, a fact which is stressed here possibly for the first time, is that classical theories with nonunique solutions may nevertheless have distinguished solutions that exhibit the exact character of their equations of motion, and therefore a study of these solutions may in turn shed light on the quantum analysis of classical theories with symmetries or with gauge freedoms. As an example, the "classical" action for a two-component-spin degree of freedom is presented.

In Sec. 4 we discuss a particularly simple labeling for a finite-dimensional Hilbert space where the labels are simply the vector components themselves. Application is made to a single-fermion oscillator, and generalization to a fermion field shows that Dirac action functionals for c-number spinor fields may be considered exact even in the presence of external sources.

Our viewpoint of "classical" mechanics is simply as quantum mechanics, evaluated for a restricted class of paths. Quantization is, therefore, already accomplished in part merely by the reinterpretation of classical variables as vector labels. The continuity of the labels and the corresponding continuity of the associated vectors plays a fundamental role in our dynamical viewpoint. It leads, for example, to an essential difference between our formalism and that of Schwinger[3] who considers only orthogonal vector or operator sets, and thus must exclude finite-dimensional Hilbert spaces. Other recent abstract dynamical studies include those of Sudarshan and coworkers,[4] which focus almost exclusively on operators forming an orthogonal operator basis and their time evolution, generalizing earlier work of Moyal.[5]

[3] J. Schwinger, Proc. Natl. Acad. Sci. U. S. **46**, 883, 1401 (1960).

[4] E. C. G. Sudarshan, *Brandeis Summer Institute Lecture Notes*," (W. A. Benjamin, Inc., New York, 1961), Vol. 2, p. 143. T. F. Jordan and E. C. G. Sudarshan, Rev. Mod. Phys. **33**, 515 (1961). Related ideas appear in: E. H. Wichmann, J. Math. Phys. **2**, 876 (1961); F. Bopp, *Heisenberg-Festschrift* (Frederick Vieweg und Sohn, Braunschweig, Germany, 1961), p. 128.

[5] J. E. Moyal, Proc. Cambridge Phil. Soc. **45**, 99 (1949).

2. AN ELEMENTARY EXAMPLE

A. An OFS Parameterized by Phase Space Points

Let us consider a single, nonrelativistic particle free to move in only one dimension. We denote Hermitian position and momentum operators by Q and P, respectively; these operators satisfy

$$[Q, P] = i\hbar.$$

Along with these operators, let us introduce two c numbers q and p with the dimensions of coordinate and momentum, respectively. We now build the unitary operator

$$U[p, q] \equiv \exp(-iqP/\hbar) \exp(ipQ/\hbar). \quad (3a)$$

With the help of a fiducial unit vector Φ_0, we define

$$\Phi[p, q] \equiv U[p, q]\Phi_0; \quad (3b)$$

each $\Phi[p, q]$ is a unit vector and the set of these vectors for all p and q define $\mathfrak{S}$. Thus, in the present example, the label space consists of all points in phase space, and the mapping from label points to Hilbert-space vectors $p, q \to \Phi[p, q]$, is a homeomorphism explicitly displayed in Eqs. (3a) and (3b). We wish to emphasize that $\Phi[p, q]$ is not a particular *representation* of a Hilbert-space vector, but it is a vector in its own right.

In order for the set of states $\mathfrak{S}$ to be an OFS suitable for a continuous representation, it is necessary that $\mathfrak{S}$ and the labeling of the vectors therein satisfy three postulates.[1] The verification of Postulate 1 of I regarding local density and continuity of the vectors in $\mathfrak{S}$, as well as Postulate 2 of I regarding labeling continuity may be established by a study of the quantity

$$\begin{aligned}\mathcal{K}(p', q'; p, q) &\equiv (\Phi[p', q'], \Phi[p, q]) \\ &= e^{-ip'(q-q')/\hbar}(\Phi_0, e^{-i(q-q')P/\hbar}e^{i(p-p')Q/\hbar}\Phi_0). \quad (4)\end{aligned}$$

$\mathcal{K}$ may be brought arbitrarily close to the value one for fixed p and q, and $\mathcal{K}$ is in addition separately continuous in p and q. Part of Postulate 3 of I, the completeness of the set $\mathfrak{S}$, has been shown by Moyal.[5] Implicit in his work is the fulfillment of the remaining condition of Postulate 3 of I, which will be discussed in part C below. Hence, the vectors, defined in Eq. (3) form an OFS. Rigorous proof of this fact will be given in Part IV in collaboration with J. KcKenna.

The preceding remarks are valid for any choice of Φ_0, independent of the fact that the set $\mathfrak{S}$ clearly depends on Φ_0. We wish now to eliminate some of the arbitrariness of Φ_0 so as to simplify the interpretation of p and q. In particular, we ask that

$$(\Phi[p, q], P\Phi[p, q]) = p, \quad (5a)$$

$$(\Phi[p, q], Q\Phi[p, q]) = q, \quad (5b)$$

two relations which impose on Φ_0 the modest restrictions

$$(\Phi_0, P\Phi_0) = 0, \quad (6a)$$

$$(\Phi_0, Q\Phi_0) = 0. \quad (6b)$$

Equation (6) follows as a consequence of (5) directly, with the help of the familiar equality

$$U^{-1}[p, q](\alpha Q + \beta P)U[p, q] = \alpha(Q + q) + \beta(P + p),$$

where α and β are arbitrary c numbers. Henceforth, we shall assume Eqs. (5) and (6) to be satisfied.

The adoption of (5) and (6) narrows the possible forms that can be taken by $\mathcal{K}$ in Eq. (4). In particular, Eqs. (5) and (6) lead to the following canonical kinematical form:

$$i\hbar(\Phi[p, q], d\Phi[p, q]) = p\, dq, \quad (7)$$

for a kind of differential form for $\mathcal{K}$. We now take up several aspects of this example in more detail.

Relation of Quantum and "Classical" Dynamics[2]

In order to discuss restricted dynamical equations with the aid of the action functional (1), we let the unit vectors $\Phi[p, q] \in \mathfrak{S}$ be functions of time. This we do simply by permitting p and q to be arbitrary, independent c-number time functions $p(t)$ and $q(t)$, and define

$$\Phi(t) \equiv \Phi[p(t), q(t)]. \quad (8)$$

Note that the operators P and Q remain unchanged here. Combining (7) and (8) we find

$$i\hbar(\Phi, d\Phi/dt) = p\dot{q};$$

here, and elsewhere, the dot signifies a time derivative, $\dot{q} \equiv dq/dt$.

With the help of the preceding expression, Eq. (1) reduces to

$$I_{\mathfrak{S}} = \int [p\dot{q} - H(p, q)]\, dt, \quad (9)$$

where

$$\begin{aligned} H(p, q) &\equiv (\Phi[p, q], \mathcal{H}(P, Q)\Phi[p, q]) \\ &= (\Phi_0, \mathcal{H}(P + p, Q + q)\Phi_0) \\ &= \mathcal{H}(p, q) + \mathcal{O}(\hbar; \Phi_0; p, q). \quad (10) \end{aligned}$$

Equation (9) has the form of a *classical* action functional where the classical Hamiltonian is $H(p, q)$. According to (10), we see that $H(p, q)$ has the functional form of the quantum mechanical Hamiltonian with explicit c-number substitution, i.e., $\mathcal{H}(p, q)$ plus an additional term $\mathcal{O}$ depending on $\hbar$, the fiducial vector Φ_0 as well as on the momentum and coordinate. For nonpathological Hamiltonian operators, $\mathcal{O}$ depends only on positive powers of $\hbar$: hence in this case,

$$\lim_{\hbar \to 0} \mathcal{O}(\hbar; \Phi_0; p, q) = 0.$$

Thus as $\hbar \to 0$ we obtain $H(p, q) = \mathcal{H}(p, q)$, which is just the conventional relation in order that $H(p, q)$ be the appropriate classical Hamiltonian for the system under discussion. In this same limit, p and q achieve their conventional, classical sharp physical significance since, e.g.,

$$\lim_{\hbar \to 0} (\Phi[p, q], (P - p)^2\Phi[p, q]) = 0.$$

When $\hbar \to 0$, it is clear that a stationary variation of (9) yields the conventional classical equations of motion, and not the quantum equations. For a macroscopic system, where $\mathcal{H}(p, q) \gg \mathcal{O}$, we expect the classical equations and interpretation to be very accurate. Hence we have established that a restricted variation of I over just those vectors $\Phi(t)$ which obey (8) yields essentially the classical equations of motion, for macroscopic systems, and in the limit $\hbar \to 0$, it yields precisely the entire classical "picture."

But such limiting cases are not the only ones in which the conventional classical equations can arise. We shall shortly prove that, for any Hamiltonian operator of the form $\mathcal{H} = \frac{1}{2}P^2 + V(Q)$, we can always choose Φ_0, consistent with (6), so that $\mathcal{O}$ *is as small as desired* for any system and not only for macroscopic systems. Thus, with $\mathcal{O}$ small, we again recover the conventional classical equations of motion, but since $\hbar \neq 0$, p and q do not have a sharp physical meaning, and these "classical" variables must be correctly interpreted for what they are: labels for Hilbert-space vectors as in (3). Thus, simply by a reinterpretation of the classical variables, we can regard the classical action functional as a restricted evaluation of the quantum action. With this wider viewpoint understood, $\mathcal{O}$ ceases to be conceptually bothersome, and we can just as well regard $H(p, q)$ itself as the "classical" Hamiltonian; indeed, the harmonic oscillator is a very important case where $\mathcal{O}$ is best chosen not to vanish.

Certain classical statements contained in the present formalism may be compared with those predicted by the conventional viewpoint based on Ehrenfest's theorem. This theorem states[6] that the expectation values

$$\bar{P} \equiv (\Phi(t), P\Phi(t)), \tag{11a}$$

$$\bar{Q} \equiv (\Phi(t), Q\Phi(t)), \tag{11b}$$

satisfy Hamilton's equations when $\Phi(t)$ is the true quantum solution. Clearly, for exact action principles when the true solution remains within $\mathfrak{S}$ the quantities defined in (11) are just those defined in (5). For inexact action principles, however, (5) differs from (11). From the point of view of this paper, the equations of motion that follow from (9) are of precisely the conventional form, e.g.,

$$dp/dt = -\partial H(p, q)/\partial q.$$

the other hand, Ehrenfest's theorem leads to

$$lt = i(\Phi(t), [\mathcal{H}, P]\Phi(t))$$
$$= \overline{i[\mathcal{H}, P]} \neq -\partial\mathcal{H}(\bar{P}, \bar{Q})/\partial\bar{Q},$$

inequality holding as a general statement. herefore, while the label point of view for position momentum leads to Hamilton's equations, the xpectation point of view in (11) for such variables leads to Hamilton's equations only in the limit $\hbar \to 0$.

Additional information regarding the "classical" namics of our one-dimensional examples may more sily be found if we first modify the set $\mathfrak{S}$ in a very simple manner.

B. Modification of the OFS to Include Phase Factors

Instead of the set of vectors defined in Eq. (3), let us choose

$$\Phi[p, q, \alpha] \equiv U[p, q, \alpha]\Phi_0 \equiv \exp(-i\alpha/\hbar) \times \exp(-iqP/\hbar)\exp(ipQ/\hbar)\Phi_0 \tag{12}$$

to be members of $\mathfrak{S}$, where α, $0 \leq \alpha < 2\pi\hbar$ is a new c-number label. This expanded set of vectors may be treated exactly as in part A above. In particular, if p, q, and α become functions of time, the action functional for this restricted set of unit vectors has the form

$$I_{\mathfrak{S}} = \int [p\dot{q} + \dot{\alpha} - H(p, q)]\, dt, \tag{13}$$

where $H(p, q)$ is the same as in (10). Clearly α in no way alters the dynamics of p and q, nor is the evolution of α determined by extremizing the action (13). If $\alpha(t)$ remains a free and undetermined function, then Eq. (13) evaluated for "an extremal path" has not one but many values, which is why arbitrary ray rotations were excluded in the consideration of Eq. (1). Instead of regarding α as an independent "classical" dynamical variable having undetermined behavior, we shall elaborate a different viewpoint.

Along with the unrestricted set $\mathfrak{S}$ defined by Eq. (12), let us also consider a family of its subsets restricted so that $\alpha = \alpha(p, q, t)$ for various functions α. Each of these restricted sets remains a valid OFS; e.g., if we put $\alpha = 0$, we recover the set $\mathfrak{S}$ discussed in part A. We now consider several applications of various OFS restricted by $\alpha = \alpha(p, q, t)$ where the form of α is chosen to accomplish one or another specific purpose, e.g., to share with the restricted set a certain desirable property exhibited in the unrestricted set.

Exact "Classical" Action Functions

We wish to demonstrate that for linear and quadratic Hamiltonians it is possible to choose $\alpha = \alpha(p, q)$ in (12) so as to obtain a determinate, exact "classical" system. This is possible in cases where the unrestricted set (12) contains the evolution of its own members.

As a first example let us choose $\mathcal{H} = AP$, where A is a constant. It is clear that

$$\Phi(t) = e^{-itAP/\hbar}\Phi(0) = \Phi[p(0), q(0) + tA, \alpha(0)],$$

so that $\Phi(t) \in \mathfrak{S}$. The phase α does not change at all here so we may set $\alpha = 0$ for this example, picking out the OFS discussed in A. Since $H(p, q) = Ap$, it is clear that criterion E is satisfied.

As a second example, let $\mathcal{H} = BQ$, where B is a constant. Then

$$\Phi(t) = e^{-itBQ/\hbar}\Phi(0) = \Phi[p(0) - tB, q(0), \alpha(0) + tBq(0)], \tag{14}$$

a vector which is of the general form of (12) for all t. If we evaluate (13) for the indicated functions p, q, and α, we see that criterion E is obeyed. An alternate way to express $\alpha(t)$ is $\alpha = -pq$, where we arbitrarily choose $\alpha(0) = -p(0)q(0)$. Thus, the set of vectors of the form (12) restricted so that $\alpha = -pq$ also contains the evolution of its members for the Hamiltonian $\mathcal{H} = BQ$. The "classical" action assumes the form $-\int q(\dot{p} + B)\, dt$ which clearly satisfies criterion E. Other choices of $\alpha = \alpha(p, q, t)$ are appropriate for the general linear case $\mathcal{H}_{\text{lin}} =$

[6] See, e.g., L. I. Schiff, *Quantum Mechanics* (McGraw-Hill Book Company, Inc., New York, 1955), 2nd ed., p. 25.

$AP + BQ + C$, to exhibit the exact character of $I_{\mathfrak{S}}$. For any linear Hamiltonian, $\mathcal{O}$ in Eq. (10) vanishes.

The fact that (13) may be exact for linear Hamiltonians can be seen as follows. The unitary transformations $U[p, q, \alpha]$ acting on Φ_0 in (12) form a group in virtue of the closed Lie algebra of 1, P, and Q; specifically,

$$U[p, q, \alpha]U[p', q', \alpha'] = U[p + p', q + q', \alpha + \alpha' - pq'], \quad (15a)$$

which, on introducing Φ_0, states that

$$U[p, q, \alpha]\Phi[p', q', \alpha'] = \Phi[p + p', q + q', \alpha + \alpha' - pq']. \quad (15b)$$

Hence, for any evolution operator $\exp(-it\mathcal{H}/\hbar)$ that can be expressed in the form $U[p, q, \alpha]$, a choice for $\alpha(t)$ [that implicit in (15b)] exists so that the associated OFS leads to an exact action principle satisfying criterion E. Such $\mathcal{H}$ are of the linear form $AP + BQ + C$. A generalization to coefficients A, B, and C that are functions of time offers no difficulty.

To illustrate the analysis of quadratic Hamiltonians, we consider only the particular quadratic form $\mathcal{H}_{\text{h.o.}} = \frac{1}{2}(P^2 + \omega^2Q^2 - \hbar\omega)$ and we choose $\alpha = -\frac{1}{2}pq$ in (12). [When linear driving terms are present other choices for $\alpha(t)$ are needed to exhibit the exact nature of a subset of the unrestricted set $\mathfrak{S}$.] With this choice for α we are effectively considering the set $\mathfrak{S}$ whose vectors are

$$\Phi[p, q] = \exp[-i(qP - pQ)/\hbar]\Phi_0, \quad (16)$$

and the corresponding classical action functional

$$I_{\mathfrak{S}} = \int [\tfrac{1}{2}(p\dot{q} - q\dot{p}) - H(p, q)]\, dt. \quad (17)$$

Consider now the exactness of this action principle. The quantum evolution of the states in (16) is

$$\Phi(t) = e^{-it\mathcal{H}_{\text{h.o.}}/\hbar}\Phi(0) = \exp\{-i[q(t)P - p(t)Q]/\hbar\}e^{-it\mathcal{H}_{\text{h.o.}}/\hbar}\Phi_0, \quad (18)$$

where

$$q(t) = q(0)\cos\omega t + \omega^{-1}p(0)\sin\omega t,$$

$$p(t) = -\omega q(0)\sin\omega t + p(0)\cos\omega t.$$

It follows from (18) that $\Phi(t) \in \mathfrak{S}$, i.e., it is of the form (16), if and only if Φ_0 is the ground state of $\mathcal{H}_{\text{h.o.}}$, although other eigenstates of $\mathcal{H}_{\text{h.o.}}$ would only give phase factors that could easily be absorbed into $\alpha(t)$. By choosing Φ_0 as an eigenstate, the addition of $\mathcal{H}_{\text{h.o.}}$ to the Lie algebra of 1, P, and Q can thus be *effected* without its actual *inclusion*, again giving rise to an exact "classical" action functional.

Armed with this result, we now observe the interesting conclusion for oscillator Hamiltonians that an exact "classical" theory arises by extremizing the "classical" action function's dependence on the *fiducial vector* Φ_0 and thus ensuring the eigenstate property of Φ_0; a global extremization further narrows Φ_0 to be the ground state of the oscillator. With this choice, $\mathcal{O} = \frac{1}{2}\hbar\omega$. The "classical" Hamilton Eq. (10) then becomes $H_{\text{h.o.}} = \frac{1}{2}(p^2 + \omega^2q^2)$ corresponding to the preceding form for the harmonic-oscillator Hamiltonian, $\mathcal{H}_{\text{h.o.}}$, in which the zero-point energy was subtracted off (a physically very attractive correspondence indeed when generalized to a Bose field!). Thus, (i) the classical equations for a harmonic oscillator imply the correct quantum-mechanical time automorphism, (ii) the spectrum of the corresponding quantum Hamiltonian begins at zero, without a zero-point subtraction being necessary, and (iii) criterion E is evidently satisfied in (17) by virtue of Euler's theorem on homogeneous forms.

The success of the analysis in (18) mainly lay in being able to commute the evolutionary operator with the unitary operator in (16) with the only "cost" being a change of the labels. This situation will prevail if the evolutionary operator generates a family of outer automorphisms of the Lie algebra of 1, P, and Q, which is only true for a general quadratic Hamiltonian.[7]

Canonical Transformations and Inexact "Classical" Action Functionals

While criterion E is necessary to have an exact "classical" action functional it is not sufficient for by a suitable choice of $\alpha = \alpha(p, q)$ in (12) and in (13) we can always satisfy criterion E. In particular, we could choose α equal to $F(\mathfrak{q}, q)$, a function related to Hamilton's characteristic function, chosen such that $p \equiv -\partial F/\partial q$ and $\mathfrak{p} \equiv \partial F/\partial \mathfrak{q} = H(-\partial F/\partial q, q)$. After such a canonical transformation, the "classical" action function reads[8]

$$I_{\mathfrak{S}} = \int (\mathfrak{p}\dot{\mathfrak{q}} - \mathfrak{p})\, dt,$$

which clearly fulfills criterion E. The restricted subset of (12) determined by the above rule is an OFS conveniently labeled by $\mathfrak{p}$, $\mathfrak{q}$, $\Phi[\mathfrak{p}, \mathfrak{q}]$, in terms of

[7] J. E. Moyal and M. S. Bartlett, Proc. Cambridge Phil. Soc. **45**, 545 (1949).

[8] See, e.g., H. Goldstein, *Classical Mechanics* (Addison-Wesley Publishing Company, Inc., Reading, Massachusetts, 1950), Chaps. 8 and 9.

which $i\hbar(\Phi, d\Phi) = \mathfrak{p}\, d\mathfrak{q}$ consistent with (7). But the functional form of Eq. (5)—let alone that of the more involved relation (4)—would in general be very different when expressed in terms of the variables $\mathfrak{p}$, $\mathfrak{q}$. This difference reflects the fact that Eqs. (4) and (5) give to the canonical pair p, q a certain physical significance that may not be shared by the canonical pair $\mathfrak{p}$, $\mathfrak{q}$. With this physical difference recognized, it is clear that the Hamiltonian $\mathfrak{p}$ would no longer be the expectation value of an infinitesimal element that generates an Abelian subgroup of the set $U[p, q, \alpha]$ as (5a) represents. Instead, $\mathfrak{p}$ is the expectation value of the Hamiltonian $\mathcal{H}(P, Q)$ in the state $\Phi[\mathfrak{p}, \mathfrak{q}]$. If the Hamiltonian were not one of the linear or quadratic ones we discussed above, then the action functional would be inexact even though criterion E were fulfilled. These results in no way prohibit $\mathfrak{p}$ and $\mathfrak{q}$ from being "good" labels; they simply call our attention to the fact that (7) has many solutions [including all sets arising from (12) by the restriction $\alpha = \alpha(p, q)$], and among these is the solution (3) for variables p and q having a physical, translational invariance as is implied by (4).

It is worth emphasizing the basic relation of a classical canonical transformation to an associated quantum transformation from our point of view. To carry out any canonical transformation, we merely pass from one OFS to another OFS differing trivially from the first by having different phase factors; no involved unitary transformation acting both on operators and vectors is coupled to the classical canonical transformation. Such unitary transformations are a separate invariance group of the quantum theory.

By our definition, an inexact "classical" action functional is one whose associated OFS does not contain the true quantum-dynamical evolution of its members. The example above involving $\mathfrak{p}$ and $\mathfrak{q}$ implies that merely changing the labels of the vectors and introducing phase factors cannot, in general, make an inexact action become exact. In order to conclude that a given "classical" action functional satisfying criterion E is exact or not requires some additional information regarding the physical significance of the variables in which it is expressed. One convenient way to analyze "classical" action functionals—and that which we follow in this and in subsequent sections of this paper—is, by means of Eq. (1), to express the action functionals directly in terms of *specific* labels whose physical significance is implicitly contained in inner products such as (4).

The discussion associated with Eq. (18) has shown us that an inexact "classical" theory will result when $\mathcal{H}$ is neither linear nor quadratic in P and Q. There is even no choice of Φ_0 that will lead to an exact theory. However, in the case

$$\mathcal{H} = \tfrac{1}{2}P^2 + V(Q),$$

a suggestive choice for Φ_0 can be put forward that gives to the "classical" theory its conventional form. For this class of Hamiltonians we find from (10) that

$$\mathcal{O} = \tfrac{1}{2}(\Phi_0, P^2\Phi_0) + (\Phi_0, [V(Q + q) - V(q)]\Phi_0) \equiv c + v(q).$$

Thus $q(t)$ is the only dynamical variable on which $\mathcal{O}$ depends. By choosing Φ_0 sharp in Q space about zero we can make $v(q)$ arbitrarily small. In addition to Eq. (6), such a Φ_0 satisfies the relation $(\Phi_0, Q^2\Phi_0) =$ arbitrarily small. The price for reducing $v(q)$ to a negligible quantity is that now $c \equiv \frac{1}{2}(\Phi_0, P^2\Phi_0)$ becomes arbitrarily large. But we can cancel this constant by the choice of phase $\alpha = ct$ in (13), thus eliminating $\mathcal{O}$ altogether. In summary, if we (i) choose Φ_0 arbitrarily sharp in Q space about zero, and (ii) use a set of states including $\Phi[p, q, ct]$ as defined in (12), then we can bring the "classical" Hamiltonian $H(p, q)$ arbitrarily close to the conventional form $\frac{1}{2}p^2 + V(q)$ even when $\hbar \neq 0$.

Operationally we can argue that the choice of Φ_0 to make $v(q)$ negligible is a result of extremizing the "classical" action functional with respect to Φ_0, as was the case for the harmonic oscillator. In the present case we simply give priority to those parts of $\mathcal{O}$ that do not lead to surface terms.

A further investigation of inexact "classical" action functionals will be the subject of a separate study. There we shall consider the relative accuracy of the approximate vector solution $\Phi[p(t), q(t)]$, where p and q are solutions of the extremal equations based on (13), as compared to the true quantum-mechanical solution $\Phi(t) = \exp(-it\mathcal{H}/\hbar)\Phi(0)$. We anticipate that the approximate solutions will possess some form of "maximum accuracy" compared to the true solutions when we choose Φ_0 to extremize the "classical" action functional.

C. Resolution of Unity and Continuous Representations

While the phase variable α is itself eliminated in favor of some specific functional form $\alpha = \alpha(p, q, t)$, in general the precise form of the elimination cannot be made *a priori* until the Hamiltonian is selected. It is fitting, therefore, that the resolution of unity in Eq. (1) of I is most directly found when all α are included. The completeness of the vectors in $\mathfrak{S}$ (α plays no role here, of course) has been demonstrated by Moyal, and is easily proved by taking recourse to a Schrödinger representation of Hilbert space. We wish rather to illustrate the utility of Theorem 1, which is discussed in I, in deriving the measure on vectors in $\mathfrak{S}$ in a "resolution of unity."

Quite generally, the resolution of unity in terms of such vectors will have the form

$$1 = \int \Phi[p, q, \alpha]\Delta(p, q, \alpha)\, dp\, dq\, d\alpha\, \Phi^\dagger[p, q, \alpha]. \quad (19a)$$

Because of the group property in (15a) it is clear that the $U[p, q, \alpha]$ are also the elements of the unitary transformation group that leave $\mathfrak{S}$ invariant. It is further clear that they form a transitive permutation group acting in $\mathfrak{S}$ and in view of Theorem 1 of I, if Δ exists, it can be determined everywhere up to a constant directly from the left-invariant group measure.[9] A simple calculation shows that Δ is constant, i.e., the weighting in (19a) is independent of p, q, and α. That the weighting would be independent of α could have been anticipated since the elements $U[0, 0, \alpha]$ form an invariant subgroup of the set $U[p, q, \alpha]$ with an additive law of combination. Further, it is clear that α also disappears from the special integrand $\Phi\Phi^\dagger$ in (19a), so that the integral over α simply multiplies Δ by a factor. It is important that this scaling of Δ is by a *finite* factor, which follows from the periodic nature of the parameter α. Consequently, the resolution of unity assumes the form

$$1 = \int \Phi[p, q]\, \frac{dp\, dq}{2\pi\hbar}\, \Phi^\dagger[p, q], \quad (19b)$$

the over-all constant being determined, for example, by the single requirement that the expectation value of (19b) with Φ_0 is one. Although obtained and used in different ways, Eq. (19b) is a result which agrees with the solution in reference 5. Thus, while α has no fundamental dynamical role, it is extremely useful in deducing the resolution of unity (19b) expressed as an integration over the true dynamical variables. It is observed that the functional form of the resolution in (19b) is invariant under canonical transformations.[10]

The existence of (19b) as a valid "resolution of unity" shows that a representation of Hilbert space can be realized by a suitable class of phase space functions. In particular,

$$(\Psi_1, \Psi_2) = \int \psi_1^*(p, q)\, \frac{dp\, dq}{2\pi\hbar}\, \psi_2(p, q), \quad (20a)$$

where, according to (19b),

$$\psi(p, q) \equiv (\Phi[p, q], \Psi).$$

This definition for $\psi(p, q)$ does not lead to a vector space of arbitrary functions but rather to one composed of continuous functions that fulfill the relation

$$\psi(p', q') = (2\pi\hbar)^{-1} \int \mathcal{K}(p', q'; p, q)\psi(p, q)\, dp\, dq,$$

where $\mathcal{K}$ is defined in (4). This restriction on ψ is implied by (20a) when we set $\Psi_1 = \Phi[p', q']$. Among the properties required of ψ is the bound

$$|\psi(p, q)| \leq ||\Psi||,$$

since Φ is a unit vector.

The preceding representation reduces essentially to the Fock representation by entire analytic functions[11] when Φ_0 is chosen as the ground state of an harmonic oscillator. To see this, let us first reexpress the vectors in (3) in an equivalent form:

$$\Phi[p, q] = e^{-\frac{1}{2}(x^2+y^2)+ixy} e^{(x-iy)A^*} e^{-(x+iy)A} \Phi_0,$$

where

$$x \equiv (\tfrac{1}{2}\omega/\hbar)^{\frac{1}{2}} q, \qquad y \equiv -(\tfrac{1}{2}\omega/\hbar)^{\frac{1}{2}} p,$$
$$A \equiv (\tfrac{1}{2}\omega/\hbar)^{\frac{1}{2}} Q + i(\tfrac{1}{2}\omega\hbar)^{\frac{1}{2}} P,$$

and A^* is the adjoint of A. The operator A is the usual annihilation operator for an oscillator. If Φ_0 is the oscillator ground state, then $A\Phi_0 = 0$, and the related term in the expression for Φ will vanish. Apart from multiplicative normalization factors, therefore, $\Phi[p, q]$ depends only on $x - iy$.[12] To

[9] The invariant group measure is discussed, e.g., by E. P. Wigner, *Group Theory* (Academic Press Inc., New York, 1959), Chap. 10.

[10] It is worth remarking at this point why we do not consider a set $\mathfrak{S}$ that includes $\Phi[p] = \exp(ipQ)\Phi^0$ for all p. If this set $\mathfrak{S}$ is not complete then its utility is severely impaired; if it is considered complete then a resolution of unity in terms of these states should exist. Recourse to a Schrödinger representation shows that any kernel $K(x', x'')$ proposed as a matrix representation of unity fails to satisfy translational invariance unless Φ_0 is an eigenvector of the momentum operator P. Since such eigenvectors do not exist, one would be forced into resolutions of unity having physically undesirable characteristics.

[11] V. Bargmann, Commun. Pure Appl. Math. **14**, 187 (1961).

[12] A related set of normalized states, which depends essentially only on one complex variable, is discussed by J. R. Klauder, Ann. Phys. (NY) **11**, 123 (1960), p. 125.

eliminate the normalization scale factors, let us introduce

$$w \equiv (\Phi[p, q], \Phi_0);$$

in view of our choice for Φ_0, w never vanishes. Then

$$f(p, q) \equiv w^{-1}(\Phi[p, q], \Psi) = f(z),$$

i.e., f is a function of $z = x + iy$, or stated otherwise, f is analytic, which furthermore is defined everywhere. To account for the weight factor w we have introduced, we must redefine the inner product (20a) as

$$(\Psi_1, \Psi_2) = \int f_1^*(z) \, |w|^2 \frac{dx\, dy}{\pi} f_2(z), \qquad (20b)$$

where $|w|^2 = \exp[-(x^2 + y^2)]$. These results conform with those given by Bargmann.[11] In the representation associated with (20b), the functions $f(z)$ need only be entire; in a manner of speaking, the measure now contains the boundedness property required of ψ.

Whether or not we choose Φ_0 as above, we wish to emphasize that Eq. (20) entails a representation of *vectors* by phase-space functions as contrasted with the more conventional representation of *operators* by phase-space functions.[13] In the present formalism, operators are continuous functions of two phase-space points, e.g., for (20a)

$$\mathcal{B}(p', q'; p, q) \equiv (\Phi[p', q'], \mathcal{B}\Phi[p, q]),$$

and their operation on vectors is effected by integrating $\mathcal{B}\psi$ over both p and q:

$$(\mathcal{B}\psi)(p', q') = (2\pi\hbar)^{-1} \int \mathcal{B}(p', q'; p, q) \, dp\, dq\, \psi(p, q)$$

[cf. Eqs. (7) and (8) of I]. Equation (9) of I ensures that the representation of $\mathcal{B}$ is both continuous and unique.

3. LABELING BY PARAMETERS OF UNITARY LIE GROUPS

Let us consider a generalization of the examples discussed in the last section to the case of an N-dimensional Lie group of unitary transformations acting in an n-dimensional Hilbert space. An N-dimensional Lie group element is characterized by N parameters, l^a, $a = 1, 2, \cdots, N$. Elements near to unity may be generated from N skew-Hermitian infinitesimal elements L_a. The L_a are assumed to be elements of a Lie algebra whose commutator product satisfies the well-known conditions

$$[L_a, L_b] = c_{ab}^d L_d, \qquad (21)$$

wherein the summation convention for label indices has been adopted. For the sake of clarity and to facilitate the comparison with the preceding section, we shall assume our labels to be the so-called "canonical coordinates."[14] In terms of these labels, the finite unitary transformation

$$V[l^a] = \exp(l^a L_a).$$

The set of states $\mathfrak{S}$ is now defined to contain all vectors of the form

$$\Phi[l^a] = V[l^a]\Phi_0 = \exp(l^a L_a)\Phi_0. \qquad (22)$$

Postulates 1 and 2 of I may be verified by a study of the quantity

$$\mathcal{K}(l'^a; l^a) \equiv (\Phi[l'^a], \Phi[l^a]), \qquad (23)$$

a continuous function of the single-parameter set $(l'^{-1} \cdot l)^a$, where the dot denotes group multiplication in label space. We remark on Postulate 3 of I below.

Considered as a function of time, we let

$$\Phi(t) = \Phi[l^a(t)] = \exp[l^a(t)L_a]\Phi_0.$$

By making use of the general operator rule

$$e^{(A+B)} = e^A + \int_0^1 e^{sA} B e^{(1-s)A} \, ds,$$

valid to first order in B, we find that

$$d\Phi = dl^b(t) M_b^c(t) L_c \Phi(t), \qquad (24)$$

where the numerical coefficients M_b^c are defined through the relation

$$M_b^c(t) L_c = \int_0^1 \exp[s l^a(t) L_a] L_b \exp[-s l^c(t) L_c] \, ds. \qquad (25)$$

Now let us introduce additional numerical coefficients U_c^d by

$$U_c^d(t) L_d = \exp[-l^a(t) L_a] L_c \exp[l^b(t) L_b]. \qquad (26a)$$

An implicit expression for the label space matrix $U = \{U_c^d\}$ is given by

$$U(t) = \exp[-l^a(t) c_a], \qquad (26b)$$

where c_a is the matrix formed from the structure constants whose bd element is c_{ab}^d. In terms of the coefficients U_c^d, we have

$$(\Phi(t), L_c\Phi(t)) = U_c^d(t)(\Phi_0, L_d\Phi_0) \equiv -iU_c^d(t) v_d/\hbar. \qquad (27)$$

The constants v_d are real and characterize the expectation value of $i\hbar L_d$ in the fiducial state Φ_0.

[13] E. P. Wigner, Phys. Rev. 40, 749 (1932); see also references 3–5, and additional references therein.

[14] C. Chevalley, *Theory of Lie Groups I* (Princeton University Press, Princeton, New Jersey, 1946), Chaps. IV and V.

Depending on the relative dimensions of N and n, the constants v_d may or may not determine Φ_0 up to a phase factor given an explicit representation of the infinitesimal elements. At any rate, in terms of the quantities defined above, the canonical kinematical form is

$$i\hbar(\Phi, d\Phi) \equiv y_b(l^a)\, dl^b, \tag{28}$$

where

$$y_b \equiv M_b^c U_c^d v_d. \tag{29a}$$

Adopting a matrix notation once again, we may express the "vector" y in terms of the "vector" v as

$$y = [1 - \exp(-l^a c_a)](l^a c_a)^{-1} v, \tag{29b}$$

where again c_a is the matrix of structure constants.[14]

The "classical" Hamiltonian is defined by

$$H(l^a) \equiv (\Phi[l^a], \mathcal{H}\Phi[l^a]). \tag{30a}$$

In the event that $\mathcal{H}$ is a linear sum of infinitesimal elements, i.e., $\mathcal{H} = i\hbar h^e L_e$, then from (27) we have

$$H(l^a) = h^e U_e^d(l^a) v_d. \tag{30b}$$

In either case, the "classical" action functional assumes the form

$$I_{\mathfrak{S}} = \int [y_b(l^a)\dot{l}^b - H(l^a)]\, dt. \tag{31}$$

The "classical" equations of motion follow from extremizing (31) with respect to independent variations in l^b. These equations are

$$A_{cb}\dot{l}^b \equiv \left(\frac{\partial y_b}{\partial l^c} - \frac{\partial y_c}{\partial l^b}\right)\dot{l}^b = \frac{\partial H}{\partial l^c}. \tag{32}$$

A more explicit form for A_{cb} may be found as follows. The time derivative of (26a) may be expressed with the aid of (25):

$$\dot{U}_e^e(t) L_e = \dot{l}^b M_b^d \exp(-l^a L_a)[L_e, L_d] \exp(l^e L_e)$$
$$= \dot{l}^b M_b^d c_{ed}^f U_f^e L_e.$$

This relation is true for all time functions $l^b(t)$, and due to their linear independence and to the linear independence of the infinitesimal elements L_e, we have

$$\partial U_c^e/\partial l^b = M_b^d c_{cd}^f U_f^e. \tag{33}$$

In view of (29a), the partial derivative $\partial y_b/\partial l^c$ is given by

$$\frac{\partial y_b}{\partial l^c} = \frac{\partial M_b^f}{\partial l^c} U_f^e v_e + M_b^a \frac{\partial U_a^e}{\partial l^c} v_e.$$

Substituting from (33) and antisymmetrizing in c and b we finally obtain

$$A_{cb} = [\partial M_b^f/\partial l^c - \partial M_c^f/\partial l^b$$
$$+ M_b^a M_c^d c_{ad}^f - M_c^a M_b^d c_{ad}^f] U_f^e v_e. \tag{34}$$

Returning to Eq. (32), we see from the antisymmetry of A_{cb} that

$$\dot{l}^c A_{cb} \dot{l}^b = \dot{l}^c \frac{\partial H}{\partial l^c} = \frac{dH}{dt} - \frac{\partial H}{\partial t} = 0,$$

which expresses the constancy of H if H is not an explicit function of t. If A_{cb} is nonsingular, Eq. (32) fully determines the solution. Conversely, if A_{cb} is singular, the equations of motion do not determine the solution $l^b(t)$ uniquely; furthermore, if the dimension N of the Lie group is odd, A is necessarily singular. Such was the case for the dynamical example involving p, q, and α in Sec. 2. But just as there was a distinguished choice for $\alpha(t)$ for linear Hamiltonians to make manifest the exact nature of their action functionals, we find an analogous distinguished solution for (32) whenever H has the form (30b), even if A_{cb} is singular. While an explicit form for this solution $l^a(t)$ is difficult to write down, it is clearly defined for any Φ_0 through the relation

$$\Phi(t) = \exp(th^b L_b) \exp[l^a(0) L_a]\Phi_0 \equiv \exp[l^a(t) L_a]\Phi_0. \tag{35}$$

That the time evolution in (35) remains a vector in $\mathfrak{S}$ is a consequence of the Baker–Hausdorff theorem, but closed-form solutions are available for only a few algebras.[15] The exact quantum-mechanical solution (35) is also the extremal solution for (31), and the evaluation of $I_{\mathfrak{S}}$ for this solution vanishes, thus satisfying criterion E. If in addition the set $\mathfrak{S}$, i.e., the set of vectors $\Phi[l^a]$ at any one time, is *complete*, then the "classical" action principle (31) is exact.

For exact action functionals, the physical transition matrix element S to go from one state $\Phi[l^a] \in \mathfrak{S}$ at time 0 to another state $\Phi[l'^a] \in \mathfrak{S}$ at time t, has a simple appearance. In particular, from (35), we see that

$$S_{l',l} = (\Phi[l'^a], \Phi[l^a(t)])$$
$$= \mathcal{K}(l'^a; l^a(t)).$$

Thus for exact action functionals, dynamical transition amplitudes may be read directly out of (23), which, in turn, involves only the projection of the vectors in $\mathfrak{S}$ on the fiducial state Φ_0.

When $\mathcal{H}$ lies outside the Lie algebra, we are generally led to an inexact "classical" action prin-

[15] For a recent discussion of this theorem, see: G. H. Weiss and A. A. Maradudin, J. Math. Phys. 3, 771 (1962).

ciple. Although the equations derived from such an action principle have the form shown in (32), they generally remain inexact for any choice of Φ_0; the exceptional cases, analogous to $\mathcal{H}_{\mathrm{h.o.}}$ treated in Sec. 2, are discussed below. For reasons similar to those presented in Sec. 2, we may expect to secure classical equations of "maximum accuracy" if we choose Φ_0 so as to extremize $I_{\mathfrak{S}}$. These topics will be discussed elsewhere.

Consideration of Simplifiable, Enlargable, and Special Cases

Suppose now that the Lie algebra is simplifiable in the sense that there exists a choice of infinitesimal elements such that two or more subsets of elements are totally unconnected with one another by the structure constants c^d_{ab}. Then, according to (25) and (26), both M and U provide admixtures only within each individual subset. The canonical kinematical form becomes a sum of terms, each similar to that appearing in (28). It is clear that the Hilbert space in such a case may conveniently be chosen as a product space, a product over as many spaces as there are disconnected subsets of the Lie algebra. If the Hamiltonian is a linear sum of infinitesimal elements, then the Hamiltonian part also breaks up into a sum of terms like (30b), each depending on the parameters within a subset. Such a Hamiltonian, therefore, does not mix the dynamics in one product space with the dynamics in another. The complete problem is a sum of noninteracting smaller problems, one for each of the disconnected subsets of the original Lie algebra.

However, if $\mathcal{H}$ is *not* simply a linear sum of infinitesimal elements, then the Hamiltonian defined by Eq. (30a) will possess interaction terms, terms in which the labels from two or more subsets may be involved. In principle, the Lie algebra could be enlarged so as to include $\mathcal{H}$ (and generally other elements as well). In this enlarged algebra, with additional parameters, $\mathcal{H}$ is now an infinitesimal element and the form in (30b) prevails. Disconnected subsets may be sought in the enlarged algebra. If they are found the problem can be reduced to a sum of simpler noninteracting problems. More specific statements can be made if the Lie algebra were semisimple, for then the disconnected subsets would be a direct sum over simple algebras whose properties are well known.

A particularly simple dynamics arises for those labels belonging to elements in the center $\mathfrak{C}$ of a Lie algebra, i.e., those infinitesimal elements that commute with all other infinitesimal elements. If L_b is such an element, then (25) and (26) state, respectively, that $M^c_b = U^c_b = \delta^c_b$. From (29a) we see that the contribution to (28) of such elements is simply a total differential, $v_b\, dl^b$, summed only over the elements in $\mathfrak{C}$. The contribution of these terms to the Hamiltonian part is trivial in the case (30b), i.e., $H = h^b v_b +$ (terms involving noncentral labels). Thus, the complete dynamics is not fully determined, the evolution of the parameters of the elements in $\mathfrak{C}$ being arbitrary; e.g., if we were to set $\dot{l}^b = h^b$, as suggested by (35), then the appearance of these variables as well as their energy shifts would disappear completely. The basic dynamical elements in this case lie outside the center $\mathfrak{C}$.

It would be possible to use this information regarding the time behavior of the parameters of the elements in $\mathfrak{C}$ to simplify the "classical" action functional. Thus, of the possible paths $\Phi[l^a(t)]$, we might consider only those for which the central element parameters equal specific time functions which satisfy their elementary equations of motion $\dot{l}^b = h^b$; in the derivation of the "classical" equations only the remaining variables would be varied. Care should be taken, however, lest the restricted set of states with only noncentral element parameters free to vary fails to span $\mathfrak{H}$.

If, on the other hand, the Hamiltonian is *not* simply a linear sum of infinitesimal elements, then $H(l^a)$ defined by (30a) may very well depend on the parameters of those elements in the center. Equations generated by extremal conditions for central element parameters are then constraints, $0 = \partial H/\partial l^b$, i.e., the parameters relating to the elements in $\mathfrak{C}$ enter the Lagrangian at most only in the form of Lagrange multipliers.

After establishing that the Hamiltonian $\mathcal{H}$ is a member of the Lie algebra, suppose we further find that $\mathcal{H}$ lies in a *sub*algebra. Then it may be desirable to simplify the action functional by simply setting $l^a(t) = 0$ (or more generally their values for the identity element) for those infinitesimal elements *outside* the subalgebra. Thus, we are restricting our algebra in such a way that only the subalgebra containing $\mathcal{H}$ appears. Such a restriction should be carried out only if (or should be carried out only to an extent that) the vectors remaining in $\mathfrak{S}$ form a complete set. Otherwise the exact "classical" action principle possible in such a case would be restricted to apply to an incomplete set, and the dynamics for an arbitrary state vector could not be predicted.

A special case arises if the Hamiltonian $\mathcal{H}$ is an element of a Lie algebra, the remaining elements of which form an invariant subalgebra, and if $\mathcal{H}$

has at least one discrete eigenvalue. In such a case, the solution (35) is applicable and we express it as

$$\Phi(t) = \exp(tL_h) \exp[l^a(0)L_a]\Phi_0;$$

the summation over "a" also includes the element L_h, the infinitesimal element representing $\mathcal{H}$. This solution may be written in the form

$$\Phi(t) = \exp[\bar{l}^a(t)L_a] \exp(tL_h)\Phi_0, \qquad (36)$$

where

$$\bar{l}^a(t)L_a \equiv l^a(0)e^{tL_h}L_a e^{-tL_h}.$$

From the definition of $\bar{l}^a$, it is clear that

$$\bar{l}^h(t) = l^h(0), \qquad (37a)$$

$$\bar{l}^b(t) = \bar{l}^b[l^c(0), t]; \qquad b, c \neq h. \qquad (37b)$$

Now choose Φ_0 to be one of the eigenvectors of $\mathcal{H}$, hence also of L_h, such that $\exp(tL_h)\Phi_0 = \exp(-i\omega t)\Phi_0$. Then, apart from a trivial phase factor (which may be absorbed if one of the elements L_a is, or effectively acts as unity), the solution (36) has the special form

$$\Phi(t) = \exp[\bar{l}^a(t)L_a]\Phi_0.$$

According to (37), if we choose $l^h(0) = 0$, then $\bar{l}^h(t) = 0$, and furthermore, there is no disturbance to the remaining labels $\bar{l}^b(t)$, $b \neq h$. Thus by this choice, all appearance of the Hamiltonian label l^h can be eliminated from $\mathfrak{S}$, and if the remaining vectors in $\mathfrak{S}$ are a complete set, then the action functional remains exact. It is just this situation that occurred for the harmonic-oscillator example discussed in Sec. 2.

There also may be a simplification in the parameterization when certain of the constants v_d in (27) vanish. From Eq. (29a) it is clear that y_b will in general be simpler if some of the v_d vanish. We shall exclude cases where all the constants v_d vanish, since then the canonical kinematical form itself vanishes. In the elementary example of Sec. 2, for instance, we chose Φ_0 in Eq. (6) so that two out of three such terms would vanish. The vanishing of these expectation values was extended even further to include the Hamiltonian in the case of the harmonic oscillator.

An even greater simplification of the parameterization may take place if the stronger conditions $L_{m_1}\Phi_0 = L_{m_2}\Phi_0 = \cdots = 0$ hold true for a set of elements $\{L_{m_1}, L_{m_2}, \cdots\}$ which form a subalgebra. Let us order our labeling so that L_m, $m = 1, \cdots, M$ denotes the elements in such a subalgebra, and L_p, $p = M + 1, \cdots, N$ denotes the remainder of the elements in the algebra. Then $\Phi[l^a]$ in (22) has the special property that if $l^p = 0$, $p = M + 1, \cdots, N$, then $\Phi[l^m] = \Phi_0$, *independent* of the values of the remaining parameters. This suggests that $\Phi[l^a]$ really depends not on N variables but only on $N - M$ variables. Call these independent variables r^q, $q = 1, \cdots, N - M$. Then in general we expect that there is a many-one mapping of points l^a to points r^q such that $\Phi[l^a] = \Phi[r^q]$. The set of points l^a mapped onto one point r^q may be found from

$$\exp(l^a L_a)\Phi_0 = \exp(l^a L_a)\exp(s^m L_m)\Phi_0$$
$$\equiv \exp[\bar{l}^a(l^b, s^m)L_a]\Phi_0.$$

For fixed l^b the set

$$\mathcal{W} \equiv \{\bar{l}^a \mid \bar{l}^a = \bar{l}^a(l^b, s^m),\ s^m \text{ arbitrary}\}$$

is mapped onto a single point whose coordinate values r^q are determined from $N - M$ continuous, linearly independent set functions: $r^q = f^q(\mathcal{W})$. It is important that the L_m form a subalgebra in the above analysis.

The possibility that $L_m\Phi_0 = 0$ seems to be not an uncommon circumstance as the following example shows. Consider the Lie group $SU(n)$, the n-dimensional unitary–unimodular group, acting on an n-dimensional Hilbert space. Without loss of generality, we can take a representation in which Φ_0 is represented by one in the first row and zero in the remaining rows. Now there exists a subgroup of $SU(n)$ that leaves Φ_0 invariant, and this subgroup is clearly isomorphic to $SU(n - 1)$. Thus of the $n^2 - 1$ parameters in $SU(n)$, a number corresponding to $SU(n - 1)$, i.e., $(n - 1)^2 - 1$ are totally arbitrary. This leaves $(n^2 - 1) - (n - 1)^2 + 1 = 2n - 1$ effective parameters to describe Φ. While an expression of these effective parameters in terms of those of $SU(n)$ is, in general, very complicated, it is easy to see that the *number* $2n - 1$ is correct, since, in a complex n-dimensional space, there are $2n$ real variables needed to describe a vector, less one to account for normalization.

It is worth speculating at this point that an extension of the analysis of those cases where several L_m annihilate Φ_0 may shed some light on the form taken in quantum mechanics by classical guage groups. Although our present analysis is basically relevant to a finite number of degrees of freedom, it certainly contains non-Abelian classical c-number symmetries. Thus, the introduction of infinitely many similar spaces to describe a field, and the enlargement of the symmetries to describe locally variable guages may well clarify the quantum treatment of such questions. We hope to comment on this possibility in subsequent work.

An Example

It is clear that one example of the formalism developed in this section is the one-dimensional problem treated in Sec. 2. We quote without proof some results of another application of our general formalism to a two-dimensional Hilbert space. We choose as infinitesimal elements of our Lie algebra $\frac{1}{2}i\boldsymbol{\sigma}$, where $\boldsymbol{\sigma}$ are the three Pauli spin matrices. A vector notation will be used throughout to treat the label indices. The members of the OFS are $\Phi[\mathbf{l}] \equiv \exp(\frac{1}{2}i\mathbf{l}\cdot\boldsymbol{\sigma})\Phi_0$. The fiducial state is characterized as in (27) by giving $\mathbf{v} \equiv (\frac{1}{2}\hbar)(\Phi_0, \boldsymbol{\sigma}\Phi_0)$. In terms of these quantities, the canonical kinematical form reads

$$i\hbar(\Phi, d\Phi) = \mathbf{y}\cdot d\mathbf{l},$$

where

$$\mathbf{y} \equiv (l^{-3}\sin l - l^{-2})(\mathbf{l}\cdot\mathbf{v})\mathbf{l} - l^{-1}\sin l\mathbf{v} + l^{-2}(1 - \cos l)\mathbf{l}\times\mathbf{v}.$$

Here l denotes the magnitude of $\mathbf{l}$. If we choose a Hamiltonian of the form $(\frac{1}{2}\hbar)\mathbf{h}\cdot\boldsymbol{\sigma}$ [cf. Eq. (30b)], then an exact "classical" action principle results whose Hamiltonian is

$$H(\mathbf{l}) = \cos l(\mathbf{h}\cdot\mathbf{v}) + l^{-2}(1 - \cos l)(\mathbf{h}\cdot\mathbf{l})(\mathbf{v}\cdot\mathbf{l}) + l^{-1}\sin l(\mathbf{l}\times\mathbf{h}\cdot\mathbf{v}).$$

We shall not pursue the resulting equations of motion, save to remark that the evolution of $\mathbf{l}$ is necessarily nonunique since the Lie algebra has odd dimensionality.

The introduction of new labels other than the "canonical coordinates" we have been using cannot change the physics of a given problem but only its description. We now wish to point out that quite different labels give to the preceding example a much simpler appearance. For this purpose we choose to label an equivalent OFS by Eulerian angles:

$$\Phi[\theta, \varphi, \psi] = \begin{bmatrix} e^{-\frac{1}{2}i(\psi+\varphi)}\cos\frac{1}{2}\theta \\ e^{-\frac{1}{2}i(\psi-\varphi)}\sin\frac{1}{2}\theta \end{bmatrix}.$$

In these variables, the canonical kinematical form is expressed by

$$i\hbar(\Phi, d\Phi) = (\tfrac{1}{2}\hbar)\cos\theta\, d\varphi + (\tfrac{1}{2}\hbar)\, d\psi,$$

and the "classical" Hamiltonian becomes $(\frac{1}{2}\hbar)\mathbf{h}\cdot\mathbf{w}$, where

$$\mathbf{w} \equiv (\sin\theta\cos\varphi, \sin\theta\sin\varphi, \cos\theta).$$

While the unit vector $\mathbf{w}$ appears to be an ordinary three-vector, the unusual role of $(\frac{1}{2}\hbar)\cos\theta$ as a momentum conjugate to φ can be shown to infer the Poisson bracket relation $[w_x, w_y]_{\text{P.b.}} = -2w_z/\hbar$ characteristic of an angular momentum. Thus, spin degrees of freedom in our formalism have a "classical" description of the *form* but not the *interpretation* extensively discussed by Bohm and coworkers.[16]

Resolutions of Unity and Continuous Representations

For completeness, let us make some remarks regarding resolutions of unity expressed in terms of the vectors $\Phi[l^a]$, which we assume to satisfy Postulate 3 of I, i.e., so that the form

$$1 = \int \Phi[l^a]\delta l^a \Phi^\dagger[l^a] \tag{38}$$

is true. Since the $V[l^a]$ coefficients in (22) form a group, they are also the elements of the invariance group $\mathcal{G}$ of $\mathfrak{S}$. As such they form a transitive group in which any vector can be transformed into any other vector in $\mathfrak{S}$. When $\mathcal{G}$ is compact, the invariant measure theorem, Theorem 1 of I, assures us that if (38) is true, we may without loss of generality choose δl^a as the invariant group measure. Indeed, when the representation of the $V[l^a]$ is irreducible, Schur's lemma guarantees (38) for any Φ_0. If the group $\mathcal{G}$ is not compact, a possible candidate for δl^a is still the (left-) invariant group measure. If the Lie algebra is simplifiable in our earlier sense, then the resolution of unity is a product resolution over each of the product spaces that make up the Hilbert space $\mathfrak{H}$.

The assumed validity of (38) gives rise to a representation of $\mathfrak{H}$ by means of continuous label-space functions. In particular such functions are

$$\psi(l^a) \equiv (\Phi[l^a], \Psi),$$

while from (38) the inner product takes the form

$$(\Psi_1, \Psi_2) = \int \psi_1^*(l^a)\delta l^a \psi_2(l^a). \tag{39}$$

The functions $\psi(l^a)$ representing vectors in Hilbert space are not arbitrary but must satisfy the projection identity

$$\psi(l''^a) = \int \mathcal{K}(l''^a; l^a)\psi(l^a)\delta l^a, \tag{40}$$

where $\mathcal{K}$ is defined in (23). The solutions to (40) form a linear vector space and include as special cases those functions which represent vectors in $\mathfrak{S}$. That is, when $\Psi = \Phi[l''^b]$, Eq. (40) holds for the special case $\psi(l^a) = \mathcal{K}(l^a; l''^a)$ [cf. Eq. (6) of I].

[16] See, e.g., D. Bohm, R. Schiller, and J. Tiomno, Nuovo Cimento Suppl. 1, 48 (1955).

Such a constraint on $\mathcal{K}$ used in conjunction with the canonical kinematical form (28) could be helpful in *deriving* $\mathcal{K}$ starting purely from the "classical" theory. Two examples of this are given in Part III.

4. LABELING BY VECTOR COMPONENTS

Let us confine our attention initially to an n-dimensional Hilbert space, but allow any Hermitian operator to be a potential candidate for the Hamiltonian. To ensure that we obtain an exact "classical" action principle, we shall parameterize and include all unit vectors in our overcomplete family of states. For example, the parameters of the Lie group $SU(n)$ could be used and even arranged so that just one such parameter was associated with the Hamiltonian. However, many of the remaining parameters would be superfluous, and the vectors would be independent of them. A symmetric and virtually nonredundant set of parameters may be introduced in the following manner.

Let $\psi = \psi_1, \cdots, \psi_n$ be an n-tuple of complex numbers lying on the complex n-dimensional unit sphere S_c^n;

$$|\psi|^2 \equiv \sum_{k=1}^{n} |\psi_k|^2 = 1. \tag{41}$$

Then each point in S_c^n corresponds to a unit vector in Hilbert space and, by means of a suitable mapping $\mathfrak{M}: \psi \rightarrow \Phi[\psi]$, we can characterize each unit vector by a "label" ψ. The inner product of two such vectors can be expressed as a function of their labels, which we shall define as

$$(\Phi[\psi'], \Phi[\psi]) \equiv \psi'^*\psi \equiv \sum_k \psi_k'^*\psi_k. \tag{42}$$

Postulate 1 of I, regarding the local density of the vectors in $\mathfrak{S}$ as well as the completeness aspect of Postulate 3 of I are trivially fulfilled. The continuity of the labeling, Postulate 2 of I, is satisfied in virtue of the form adopted in (42).

If we now consider vector-valued time functions we put as before $\Phi(t) = \Phi[\psi(t)]$. The canonical kinematical form follows from the differential of (42) as

$$i\hbar(\Phi, d\Phi) = i\hbar\psi^*\, d\psi. \tag{43}$$

The "classical" Hamiltonian must be a bilinear functional in ψ^* and ψ and is of the general form

$$H = (\Phi[\psi], \mathcal{K}\Phi[\psi]) \equiv \psi^*\tilde{\mathcal{K}}\psi. \tag{44}$$

The equations of motion that follow from an action principle based on (43) and (44) are, as expected, $i\hbar\, \partial\psi/\partial t = \tilde{\mathcal{K}}\psi$. Since the solution must remain a unit vector in the Hilbert space, it is characterized by some label ψ, and consequently the "classical" action functional is exact. This example is one where the generalized "classical" theory, as we have defined the term, will contain the same physics and very nearly the same formalism as the quantum mechanics. Here the relative term "classical" is interchangeable with the term "quantum."

As regards the "resolution of unity," we can expect the following form to hold:

$$1 = \int \Phi[\psi]\, d\bar{\mu}\,(\psi)\Phi^\dagger[\psi]. \tag{45}$$

The set $\mathfrak{S}$ is invariant under any and all unitary transformations. If we invoke the invariant measure theorem, Theorem 1 of I, then Schur's lemma insures that the integral in (45) is in fact necessarily proportional to the unit matrix. The appropriate measure on vectors has the symmetric form

$$d\bar{\mu}(\psi) = \left(\frac{n!}{\pi^n}\right)\delta(1 - \sum_k |\psi_k|^2) \prod_k d\psi_{kr}\, d\psi_{ki}, \tag{46}$$

where r and i denote the real and imaginary parts, respectively.

From (45) there arises a representation of Hilbert-space vectors by functions of ψ^* homogeneous in the first degree. The inner product of two vectors is then expressed by

$$(\Psi_1, \Psi_2) = \int \omega_1^*(\psi)\, d\bar{\mu}\,(\psi)\omega_2(\psi),$$

where

$$\omega(\psi) \equiv (\Phi[\psi], \Psi) \equiv \sum_{k=1}^{n} \psi_k^*\nu_k,$$

and ν_k are n complex coefficients.

Two-Dimensional Space

Consider the case $n = 2$. Of the four real parameters in ψ_1 and ψ_2, one may be eliminated by means of the constraint (41) and another represents an over-all phase factor which can not be a true dynamical variable. [The over all phase was useful in establishing the weight factor (46) in the resolution of unity, but may now be eliminated.] Thus there are only two dynamical degrees of freedom. In order to more clearly display these two degrees of freedom, we proceed as follows.

Let N be a projection operator with eigenvectors $\Phi^{(0)}$, and $\Phi^{(1)}$ such that $N\Phi^{(r)} = r\Phi^{(r)}$. We now define

$$N\Phi[\psi] = (1 - |\chi|^2)^{\frac{1}{2}}e^{-i\alpha/\hbar}\Phi^{(1)}, \tag{47a}$$

where χ is a complex variable restricted so that $0 \leq |\chi| \leq 1$, and $-\alpha/\hbar$ is the phase of the projected component. To fix the phase of χ, we set

$$(1 - N)\Phi[\psi] = \chi e^{-i\alpha/\hbar}\Phi^{(0)}. \tag{47b}$$

When expressed in these variables, the canonical kinematical form becomes

$$i\hbar(\Phi, d\Phi) = (i\hbar/2)(\chi^* \overleftrightarrow{\partial}_t \chi)\, dt + d\alpha, \tag{48}$$

where $A \overleftrightarrow{\partial}_t B \equiv A(\partial B/\partial t) - (\partial A/\partial t)B$. The Hamiltonian is clearly a function only of χ and χ^* defined by

$$H(\chi) = (\Phi[\psi], \mathcal{H}\Phi[\psi]). \tag{49}$$

The total differential $d\alpha$ will not effect the dynamics, and may be arbitrarily specified. As an example, therefore, let us restrict $\mathfrak{S}$ so as to include only those vectors of the form (47) *for which* $\alpha = 0$. Now that we have restricted our OFS to only those vectors parameterized by true dynamical degrees of freedom, the question arises whether there remain any Hamiltonians for which the "classical" action principle will be exact.

To answer this question we observe that any vector that we choose as an initial vector will have a real coefficient of $\Phi^{(1)}$. As time progresses, this property must be maintained for any choice of χ at $t = 0$; hence there can be no mixing of (47a) and (47b) as time passes. The state $\Phi^{(1)}$ must be an eigenvector of the Hamiltonian with eigenvalue zero. Thus the most general solution is $\mathcal{H} = \hbar\omega(1 - N)$, which in turn leads to

$$H(\chi) = \hbar\omega\chi^*\chi. \tag{50}$$

We see that the "classical" action principle

$$I = \int (\tfrac{1}{2}i\hbar\chi^* \overleftrightarrow{\partial}_t \chi - \hbar\omega\chi^*\chi)\, dt \tag{51}$$

for a single fermion oscillator is an exact action principle. This conclusion remains true even if ω is an explicit function of t. Furthermore, the interpretation of χ follows from (50); as usual, it is simply a probability amplitude for oscillator excitation.

Generalization to a Fermion Field

An infinite linear sum of action functionals of the type in (51), each characterizing an independent fermion oscillator, can be used to describe a fermion field. So long as the oscillators remain independent, the over-all action functional will be exact. It follows, for example, that the conventional Dirac c-number action functional in the presence of an external, fixed c-number source may be considered as an exact "classical" action principle since it may always be resolved into independent noninteracting oscillators. If the source is also allowed to respond dynamically it means, in general, that the "classical" action is no longer exact. In either case the "classical" action functional arises as a restricted evaluation of that action, of the general form in Eq. (1), which leads to the so-called second quantized Schrödinger equations.[17] Thus we find the satisfying result that the Dirac equations are simply "classical" equations relative to the second quantized formalism for fermion fields, exact in the absence of dynamical interactions, with all the Fermi–Dirac statistics being correctly included by the limitation $0 \leq |\chi| \leq 1$ placed on the "classical" amplitude of each independent oscillator.

5. SUMMARY

We have focused our attention in this paper on the relation of quantum and classical dynamics from the standpoint of continuous-representation theory and its associated overcomplete families of states. A study of the elementary examples in Sec. 2 suggested that classical mechanics can already be viewed as the study of quantum mechanics for a certain restricted class of vectors if only we reinterpret the classical variables as labels for those vectors. The generalization of this result led to our concept of "classical" dynamics relative to a set $\mathfrak{S}$ as the study of quantum dynamics for unit vectors restricted to the subset $\mathfrak{S}$. Such a definition for the restricted dynamics merits the name "classical" since it deals with c-number variables capable of continuous variation with the aid of conventional action principle techniques.

In a larger and more abstract sense it should be recognized that the dynamical variables are invariantly characterized as the Hilbert space vectors themselves, it being expedient to discuss these vectors by the labels we introduce.

There are several aspects of our formalism and viewpoint worth noting. Firstly, the construction and analysis of classical theories becomes at the same time a partial study of quantum-mechanical theories. For simple enough systems we have learned that the classical dynamics is sufficient to infer the correct quantum dynamics.

[17] For further details relating to the evaluation of the action principle for the relevant restricted set of states, see J. R. Klauder, Ann. Phys. (NY) **11**, 123 (1960), pp. 159 and 160. Many of the formal manipulations in that reference regarding the measure on label-space points may be eliminated by the conventional device of first working in a "box" of finite volume and later passing to the limit.

Secondly, we have seen that the appropriate "classical" theory is dictated once we are given the subset $\mathfrak{S}$ of Hilbert space vectors, the labels for these vectors, and the Hamiltonian. Thus the possible forms of "classical" action functionals can be classified and catalogued, e.g., in the manner discussed in Sec. 3. Then, were we confronted with a specific classical theory, the possible associated quantum theories and their properties could be readily determined, at least in principle.

Thirdly, let us reconsider the "process of quantization" from our viewpoint. Initially suppose we are given a classical theory in the form of an action functional expressed in terms of c-number dynamical variables. The first step is to reinterpret the classical variables as vector labels and to view the action functional as a restricted evaluation of the true quantum action functional. It is in this step that the conventional factor-ordering ambiguity, if any, would show up. For example, given only $\mathcal{H}(p, q)$ in Eq. (10) the term $\mathcal{O}$ must first be chosen before the "classical" Hamiltonian $H(p, q)$ is determined. While some of the freedom in $\mathcal{O}$ stems from the arbitrariness in Φ_0, some may also lie in the factor ordering in $\mathcal{H}$. It is our contention that the proper choice of the classical Hamiltonian should already coincide with one of the possible expectation values H; the general separation of the classical Hamiltonian into $\mathcal{O}$ and $H - \mathcal{O}$ is to be regarded as heuristic and not of fundamental significance. Adopting this point of view, our first step then neither changes the form of the classical action functional nor alters the mathematical properties of the dynamical variables (e.g., c-number $\rightarrow$ operator); it is strictly a *reinterpretation* of the physical meaning of the old c-number variables.

The second step of quantization involves an *enlargement,* in one way or another, of the *domain* of the action functional so as to infer the true quantum dynamics. One approach to this domain enlargement is discussed in Part III.[18] As we have seen, this enlargement need only proceed to a point where an exact "classical" action functional arises. Independent of just how far this domain enlargement proceeds, the process of quantization, which—as far as the dynamical formalism is concerned—is entirely contained in this domain expansion, is seen to involve a smooth and continuous transition. This desirable conceptual feature, coupled with the universal applicability of our approach to any dynamical system, and coupled with the physically desirable correspondence that eliminates the zero-point field energies, all provide strong reasons to favor our view of "classical" theories as simply being restricted quantum theories.

---

[18] An alternate means to pass from a classical to a quantum theory is by means of the Feynman sum-over-histories. The analogue of this technique in our formalism has a somewhat different form than the usual one; it is discussed formally in general terms similar to those of the present paper in J. R. Klauder, Ann. Phys. (NY) **11**, 123 (1960), pp. 142–149, and in unpublished lecture notes "The Sum-Over-Histories: Formalism and Some Applications," University of Bern, Switzerland, 1962.

A related formulation of the Schwinger Action Principle approach to quantum mechanics that is suitable only for infinite-dimensional Hilbert spaces is discussed in reference 3. A more general statement of the Schwinger Action Principle is implicitly contained in our formalism. For example, the basic kinematical effects are contained in an expression of $\delta(\Phi[l^{a\prime}], \Phi[l^a])$ in terms of label differentials with the aid of the formulas in Sec. 3.

VOLUME 10, NUMBER 3 PHYSICAL REVIEW LETTERS 1 FEBRUARY 1963

PHOTON CORRELATIONS*

Roy J. Glauber
Lyman Laboratory, Harvard University, Cambridge, Massachusetts
(Received 27 December 1962).

In 1956 Hanbury Brown and Twiss[1] reported that the photons of a light beam of narrow spectral width have a tendency to arrive in correlated pairs. We have developed general quantum mechanical methods for the investigation of such correlation effects and shall present here results for the distribution of the number of photons counted in an incoherent beam. The fact that photon correlations are enhanced by narrowing the spectral bandwidth has led to a prediction[2] of large-scale correlations to be observed in the beam of an optical maser. We shall indicate that this prediction is misleading and follows from an inappropriate model of the maser beam. In considering these problems we shall outline a method of describing the photon field which appears particularly well suited to the discussion of experiments performed with light beams, whether coherent or incoherent.

The correlations observed in the photoionization processes induced by a light beam were given a simple semiclassical explanation by Purcell,[3] who made use of the methods of microwave noise theory. More recently, a number of papers have been written examining the correlations in considerably greater detail. These papers[2,4-6] retain the assumption that the electric field in a light beam can be described as a classical Gaussian stochastic process. In actuality, the behavior of the photon field is considerably more varied than such an assumption would indicate. Whereas a stationary Gaussian stochastic process is described completely by its frequency-dependent power spectrum, a great deal more information in the form of amplitude and phase relations between differing quantum states may be required to describe a steady light beam. Beams of identical spectral distributions may exhibit altogether different photon correlations or, alternatively, none at all. There is ultimately no substitute for the quantum theory in describing quanta.

We assume, for convenience, that the field has discrete propagation modes labeled by an index k (which in free space, for example, specifies propagation vector and polarization). To describe the quantum state of the kth mode we must specify an infinite set of complex amplitudes, one for each quantum occupation state $|n_k\rangle$, $n_k = 0, 1, 2, \cdots$. Since the states we wish to describe include ones in which the phase of the kth mode is fairly well defined, and a large number of states $|n_k\rangle$ must then be superposed, it is preferable to use an altogether different set of basis states. We take these to be of the form

$$|\alpha_k\rangle = \exp(-\tfrac{1}{2}|\alpha_k|^2)\sum_n [\alpha_k^{\,n}/(n!)^{1/2}]|n\rangle, \qquad (1)$$

where α_k is an arbitrary complex amplitude. We shall call the $|\alpha_k\rangle$ coherent states; their use is well known in discussions of the harmonic oscillator in the classical limit. The expectation value in the state $|\alpha_k\rangle$ of the contribution of the kth mode to the total field is a monochromatic wave with complex amplitude proportional to α_k. The coherent states $|\alpha_k\rangle$, for all complex α_k, form a complete set in a sense best expressed by the relation

$$(1/\pi)\int |\alpha_k\rangle\langle\alpha_k| d^{(2)}\alpha_k = 1, \qquad (2)$$

where $d^{(2)}\alpha_k$ is a real element of area of the complex α_k plane. It follows that any state may be expanded linearly in terms of coherent states. The most general light beam can thus be described by a density operator of the form

$$\rho \doteq \int \mathcal{P}(\{\alpha_k, \alpha_k'\})\prod_k |\alpha_k\rangle\langle\alpha_k'| d^{(2)}\alpha_k d^{(2)}\alpha_k', \qquad (3)$$

which deals with all the modes of the field at once.

An incoherent light beam must be described as a statistical mixture of all the excitation states available for each mode excited. For the kth mode, the probability to be associated with the state $|n_k\rangle$ is proportional to $\{\langle N_k\rangle/(1+\langle N_k\rangle)\}^{n_k}$, where $\langle N_k\rangle$ is the mean number of photons occupying the mode. A simple theorem expresses this mixture in terms of the coherent states defined earlier: The density operator (3) reduces to a product of operators of the form

$$\rho_k = \int p(\alpha_k)|\alpha_k\rangle\langle\alpha_k| d^{(2)}\alpha_k, \qquad (4)$$

where the probability $p(\alpha_k)$ is a Gaussian function,

$$p(\alpha_k) = \{\pi\langle N_k\rangle\}^{-1}\exp(-|\alpha_k|^2/\langle N_k\rangle). \qquad (5)$$

In particular, blackbody radiation may be described as a mixture of coherent waves by substituting for $\langle N_k\rangle$ the familiar value for thermal excitation of a field oscillator.

To discuss photon correlations we examine the photoionization probability of a pair of atoms, labeled 1 and 2, which lie at $\vec{r}_1$ and $\vec{r}_2$ within the light beam. We assume that the incident beam is of narrow enough spectral bandwidth that any variation of frequency-dependent parameters entering the photoionization probabilities may be neglected. Then, if we sum the transition probabilities over final electron energies, there is no difficulty in defining a time at which each electron emission takes place. The probability density for ionization of atom 1 at time t_1 and for atom 2 at t_2 may be written as

$$w(t_1t_2) = w_1w_2C(\vec{r}_1t_1\vec{r}_2t_2), \qquad (6)$$

where w_1 and w_2 are the constant transition probabilities for each atom placed individually in the beam, and C is the function whose departure from unity expresses a tendency for the two events to be correlated.

We assume, to simplify notation, that photons of only one polarization are present. The appropriate vector component of the electric field operator has a positive-frequency part $\mathcal{E}^{(+)}(\vec{r}, t)$ and a negative-frequency part $\mathcal{E}^{(-)}(\vec{r}, t)$. The correlation function may be expressed in terms of these operators as

$$C(\vec{r}_1t_1\vec{r}_2t_2) = \frac{\mathrm{tr}\{\rho\, \mathcal{E}^{(-)}(\vec{r}_1t_1)\, \mathcal{E}^{(-)}(\vec{r}_2t_2)\, \mathcal{E}^{(+)}(\vec{r}_1t_1)\, \mathcal{E}^{(+)}(\vec{r}_2t_2)\}}{\mathrm{tr}\{\rho\, \mathcal{E}^{(-)}(\vec{r}_1t_1)\, \mathcal{E}^{(+)}(\vec{r}_1t_1)\}\mathrm{tr}\{\rho\, \mathcal{E}^{(-)}(\vec{r}_2t_2)\, \mathcal{E}^{(+)}(\vec{r}_2t_2)\}}, \qquad (7)$$

where tr stands for trace, and ρ is a density operator of the general form (3).

It is easily shown that coherent states of the field lead to no photoionization correlations at all. If the state of the field is specified by any density operator of the form $\prod_k |\alpha_k\rangle\langle\alpha_k|$, the correlation function C reduces to unity. A correlation between photons only appears when incoherent mixtures or superpositions of the coherent states are present. For collimated, completely incoherent beams of the type described earlier (e.g., filtered thermal radiation), we find

$$C = 1 + |R(t_1 - t_2 - c^{-1}(x_1 - x_2))|^2, \tag{8}$$

where the coordinates x_j are the components of $\vec{r}_j$ in the propagation direction, and the function R is given by

$$R(t) = \sum_k \langle N_k\rangle \exp(-i\omega_k t) / \sum_k \langle N_k\rangle. \tag{9}$$

This result, for incoherent beams, corresponds in the classical limit, when the modes are treated as forming a continuum, to that derived using stochastic models.[2-6] It may be associated, in this limit, with a tendency of the complex total field strength of the beam to fluctuate in modulus with time.

The density operator which represents an actual maser beam is not yet known. It is clear that such a beam cannot be represented by a product of individual coherent states, $\prod_k |\alpha_k\rangle\langle\alpha_k|$, unless the phase and amplitude stability of the device is perfect. On the other hand, a maser beam is not at all likely to be described by the ideally incoherent classical model which underlies the calculation of Mandel and Wolf,[2] and leads them to results corresponding to Eqs. (8) and (9). More plausible models for a steady maser beam are much closer in behavior to the ideal coherent states. They may be shown to lead to photon correlations only to the extent that random amplitude modulation is present in the statistically averaged beam.

If photoionization processes tend to be correlated in time, the distribution of the number of photons recorded by a counter in a fixed interval of time should differ from the Poisson distribution. We have developed a general technique for finding this distribution for incoherent light beams.[7] The result for the important case in which the spectral distribution has the Lorentz line shape, $\langle N_k\rangle \sim \{(\omega_k - \omega)^2 + \gamma^2\}^{-1}$, with central frequency ω, may be stated as follows: Let W be the average rate at which photons are recorded; then the probability that n photons are recorded in a time t, long compared to $1/\gamma$, is

$$P(n,t) = \frac{1}{n!}\frac{(\gamma W t)^n}{(\gamma^2 + 2\gamma W)^{\frac{1}{2}n}} S_n[t(\gamma^2 + 2\gamma W)^{1/2}] \times \exp\{-t[(\gamma^2 + 2\gamma W)^{1/2} - \gamma]\}. \tag{10}$$

The functions $S_n(x)$ are nth order polynomials in x^{-1}, familiar in the theory of modified Bessel functions of half-integral order. They may be found from the relations $S_0 = S_1 = 1$ and $S_{n+1} = -S_n' + (1 + nx^{-1})S_n$. The full set of moments of the distribution is given by the averages of products of the form $n!/(n-j)!$. These are

$$\langle n!/(n-j)!\rangle_{\text{av}} = (Wt)^j S_j(\gamma t). \tag{11}$$

When the number of photons recorded during a relaxation time is small, $W \ll \gamma$, e.g., when the linewidth becomes appreciable, the distribution (10) approaches the Poisson distribution as a limit.

The author is grateful to Dr. Saul Bergmann for bringing these problems to his attention, and to Dr. S. M. MacNeille and the American Optical Company for partial support of their solution.

*Work supported in part by the Air Force Office of Scientific Research.

[1]R. Hanbury Brown and R. Q. Twiss, Nature 177, 27 (1956); G. A. Rebka and R. V. Pound, Nature 180, 1035 (1957).

[2]L. Mandel and E. Wolf, Phys. Rev. 124, 1696 (1961).

[3]E. M. Purcell, Nature 178, 1449 (1956).

[4]F. D. Kahn, Optica Acta 5, 93 (1958).

[5]L. Mandel, Proc. Phys. Soc. (London) 71, 1037 (1958).

[6]L. Mandel, Proc. Phys. Soc. (London) 74, 233 (1959).

[7]Incoherent light beams of exceedingly narrow bandwidth may, in principle, be formed by superposing the outputs of many identical but independent masers.

PHYSICAL REVIEW VOLUME 130, NUMBER 6 15 JUNE 1963

The Quantum Theory of Optical Coherence*

ROY J. GLAUBER
Lyman Laboratory of Physics, Harvard University, Cambridge, Massachusetts
(Received 11 February 1963)

The concept of coherence which has conventionally been used in optics is found to be inadequate to the needs of recently opened areas of experiment. To provide a fuller discussion of coherence, a succession of correlation functions for the complex field strengths is defined. The nth order function expresses the correlation of values of the fields at $2n$ different points of space and time. Certain values of these functions are measurable by means of n-fold delayed coincidence detection of photons. A fully coherent field is defined as one whose correlation functions satisfy an infinite succession of stated conditions. Various orders of incomplete coherence are distinguished, according to the number of coherence conditions actually satisfied. It is noted that the fields historically described as coherent in optics have only first-order coherence. On the other hand, the existence, in principle, of fields coherent to all orders is shown both in quantum theory and classical theory. The methods used in these discussions apply to fields of arbitrary time dependence. It is shown, as a result, that coherence does not require monochromaticity. Coherent fields can be generated with arbitrary spectra.

I. INTRODUCTION

CORRELATION, it has long been recognized, plays a fundamental role in the concept of optical coherence. Techniques for both the generation and detection of various types of correlations in optical fields have advanced rapidly in recent years. The development of the optical maser, in particular, has led to the generation of fields with a range of correlation unprecedented at optical frequencies. The use of techniques of coincidence detection of photons[1,2] has, in the same period, shown the existence of unanticipated correlations in the arrival times of light quanta. The new approaches to optics, which such developments will allow us to explore, suggest the need for a fundamental discussion of the meaning of coherence.

The present paper, which is the first of a series on fundamental problems of optics, is devoted largely to defining the concept of coherence. We do this by constructing a sequence of correlation functions for the field vectors, and by discussing the consequences of certain assumptions about their properties. The definition of coherence which we reach differs from earlier ones in several significant ways. The most important difference, perhaps, is that complete coherence, as we define it, requires that the field correlation functions satisfy an infinite succession of coherence conditions. We are led then to distinguish among various orders of incomplete coherence, according to the number of conditions satisfied. The fields traditionally described as coherent in optics are shown to have only first-order coherence. The fields generated by the optical maser, on the other hand, may have a considerably higher order of coherence. A further difference between our approach and previous ones is that it is constructed to apply to fields of arbitrary time dependence, rather than just to those which are, on the average, stationary in time. We have also attempted to develop the discussion in a fully quantum theoretical way.

It would hardly seem that any justification is necessary for discussing the theory of light quanta in quantum theoretical terms. Yet, as we all know, the successes of classical theory in dealing with optical experiments have been so great that we feel no hesitation in introducing optics as a sophomore course. The quantum theory, in other words, has had only a fraction of the influence upon optics that optics has historically had upon quantum theory. The explanation, no doubt, lies in the fact that optical experiments to date have paid very little attention to individual photons. To the extent that observations in optics have been confined to the measurement of ordinary light intensities, it is not surprising that classical theory has offered simple and essentially correct insights.

Experiments such as those on quantum correlations suggest, on the other hand, the growing importance of studies of photon statistics. Such studies lie largely outside the grasp of classical theory. To observe that the quantum theory is fundamentally necessary to the treatment of these problems is not to say that the semiclassical approach always yields incorrect results. On the contrary, correct answers to certain classes of problems of photon statistics[3] may be found through adaptations of classical methods. There are, however, distinct virtues to knowing where such methods succeed and where they do not. For that reason, as well as for its intrinsic interest, we shall formulate the theory in quantum theoretical terms from the outset. Quite a few of our arguments can easily be paraphrased in classical terms. Several seem to be new in the context of classical theory.

We shall try to construct this paper so that it can be followed with little more than a knowledge of elementary quantum mechanics. Since its subject matter is, in the deepest sense, quantum electrodynamics, we begin with

* Supported in part by the U. S. Air Force Office of Scientific Research.

[1] R. Hanbury Brown and R. Q. Twiss, Nature **177**, 27 (1956); Proc. Roy. Soc. (London) **A242**, 300 (1957); **A243**, 291 (1957).

[2] G. A. Rebka and R. V. Pound, Nature **180**, 1035 (1957).

[3] E. M. Purcell, Nature **178**, 1449 (1956).

a section which describes the few simple aspects of that subject which are referred to later.

II. ELEMENTS OF FIELD THEORY

The observable quantities of the electromagnetic field will be taken to be the electric and magnetic fields which are represented by a pair of Hermitian operators, $\mathbf{E}(\mathbf{r}t)$ and $\mathbf{B}(\mathbf{r}t)$. The state of the field will be described by means of a state vector, $|\rangle$, on which the fields operate from the left, or by means of its adjoint, $\langle|$, on which they operate from the right. Since we shall use the Heisenberg representation, the choice of a fixed state vector specifies the properties of the field at all times. The theory is constructed, by whatever formal means, so that in a vacuum the field operators $\mathbf{E}(\mathbf{r}t)$ and $\mathbf{B}(\mathbf{r}t)$ satisfy the Maxwell equations

$$\begin{aligned} \nabla\cdot\mathbf{E}&=0,\\ \nabla\times\mathbf{E}&=-\frac{1}{c}\frac{\partial\mathbf{B}}{\partial t},\\ \nabla\times\mathbf{B}&=\frac{1}{c}\frac{\partial\mathbf{E}}{\partial t},\\ \nabla\cdot\mathbf{B}&=0. \end{aligned} \tag{2.1}$$

We omit the source terms in the equations since, for the present, we are more interested in the fields themselves than the explicit way in which they are generated or detected. It follows from the Maxwell equations that the electric field operator obeys the wave equation

$$\left(\nabla^2-\frac{1}{c^2}\frac{\partial^2}{\partial t^2}\right)\mathbf{E}(\mathbf{r}t)=0, \tag{2.2}$$

and the magnetic field operator does likewise.

One of the essential respects in which quantum field theory differs from classical theory is that two values of the field operators taken at different space-time points do not, in general, commute with one another. The components of the electric field, which is the only field we shall discuss at length, obey a commutation relation of the general form

$$[E_\mu(\mathbf{r}t),E_\nu(\mathbf{r}'t')]=D_{\mu\nu}(\mathbf{r}-\mathbf{r}',t-t'). \tag{2.3}$$

That the tensor function $D_{\mu\nu}$ has as arguments the coordinate differences $\mathbf{r}-\mathbf{r}'$ and $t-t'$ follows from the invariance of the theory under translations in space and time. We shall not need any further details of the function $D_{\mu\nu}$, but may mention that it vanishes when the four-vector $(\mathbf{r}-\mathbf{r}', t-t')$ lies outside the light cone, i.e., for $(\mathbf{r}-\mathbf{r}')^2>c^2(t-t')^2$. The vanishing of the commutator, for points with spacelike separations, corresponds to the fact that measurements of the stated field components at such points can be carried out to arbitrary accuracy. Such accuracy is attainable since no disturbances can propagate through the field rapidly enough to reach one point from the other.

An important element of the discussion in this paper will be the separation of the electric field operator $\mathbf{E}(\mathbf{r}t)$ into its positive and negative frequency parts. The separation is most easily accomplished when the time dependence of the operator is represented by a Fourier integral. If, for example, the field operator has a representation

$$\mathbf{E}(\mathbf{r}t)=\int_{-\infty}^{\infty}\mathfrak{e}(\omega,\mathbf{r})e^{-i\omega t}d\omega, \tag{2.4}$$

where the Hermitian property is secured by the relation $\mathfrak{e}(-\omega,\mathbf{r})=\mathfrak{e}^\dagger(\omega,\mathbf{r})$, then we define the positive frequency part of $\mathbf{E}$ as

$$\mathbf{E}^{(+)}(\mathbf{r}t)=\int_0^{\infty}\mathfrak{e}(\omega,\mathbf{r})e^{-i\omega t}d\omega, \tag{2.5}$$

and the negative frequency part as

$$\mathbf{E}^{(-)}(\mathbf{r}t)=\int_{-\infty}^{0}\mathfrak{e}(\omega,\mathbf{r})e^{-i\omega t}d\omega, \tag{2.6}$$

$$=\int_0^{\infty}\mathfrak{e}^\dagger(\omega,\mathbf{r})e^{-i\omega t}d\omega. \tag{2.7}$$

It is evident from these definitions that the field is the sum of its positive and negative frequency parts,

$$\mathbf{E}(\mathbf{r}t)=\mathbf{E}^{(+)}(\mathbf{r}t)+\mathbf{E}^{(-)}(\mathbf{r}t). \tag{2.8}$$

The two parts, regarded separately, are not Hermitian operators; the fields they represent are intrinsically complex, and mutually adjoint,

$$\mathbf{E}^{(-)}(\mathbf{r}t)=\mathbf{E}^{(+)\dagger}(\mathbf{r}t). \tag{2.9}$$

In the absence of a Fourier integral representation of $\mathbf{E}(\mathbf{r}t)$, the positive and negative frequency parts of the field may be defined more formally as the limits of the integrals,

$$\mathbf{E}^{(+)}(\mathbf{r}t)=\lim_{\eta\to+0}\frac{1}{2\pi i}\int_{-\infty}^{\infty}\frac{\mathbf{E}(\mathbf{r},t-\tau)}{\tau-i\eta}d\tau, \tag{2.10}$$

$$\mathbf{E}^{(-)}(\mathbf{r}t)=-\lim_{\eta\to+0}\frac{1}{2\pi i}\int_{-\infty}^{\infty}\frac{\mathbf{E}(\mathbf{r},t-\tau)}{\tau+i\eta}d\tau. \tag{2.11}$$

It follows from the intrinsically different time dependences of $\mathbf{E}^{(+)}(\mathbf{r}t)$ and $\mathbf{E}^{(-)}(\mathbf{r}t)$ that they act to change the state of the field in altogether different ways, one associated with photon absorption, the other with photon emission. In particular, the positive frequency part, $\mathbf{E}^{(+)}(\mathbf{r}t)$, may be shown[4] to be a photon annihilation operator. Applied to an n-photon state it produces an $(n-1)$-photon state. Further applications of $\mathbf{E}^{(+)}(\mathbf{r}t)$

[4] See, for example, P. A. M. Dirac, *The Principles of Quantum Mechanics* (Oxford University Press, New York, 1947), 3rd ed., pp. 239–242.

reduce the number of photons present still further, but the regression must end with the state in which the field is empty of all photons. It is part of the definition of this state, which we represent as $|\text{vac}\rangle$, that

$$\mathbf{E}^{(+)}(\mathbf{r}t)|\text{vac}\rangle=0. \tag{2.12}$$

The adjoint relation is

$$\langle\text{vac}|\mathbf{E}^{(-)}(\mathbf{r}t)=0. \tag{2.13}$$

Since the operator $\mathbf{E}^{(+)}(\mathbf{r}t)$, annihilates photons, its Hermitian adjoint, $\mathbf{E}^{(-)}(\mathbf{r}t)$, must create them; applied to an n-photon state it produces an $(n+1)$-photon state. In particular, the state

$$\mathbf{E}^{(-)}(\mathbf{r}t)|\text{vac}\rangle$$

is a one-photon state.

It has become customary, in discussions of classical theory, to regard the electric field $\mathbf{E}(\mathbf{r}t)$ as the quantity one measures experimentally, and to think of the complex fields $\mathbf{E}^{(\pm)}(\mathbf{r}t)$ as convenient, but fictitious, mathematical constructions. Such an attitude can only be held be held in the classical domain, where quantum phenomena play no essential role. The frequency ω of a classical field must be so low that the quantum energy $\hbar\omega$ is negligible. In such a case, we can not tell whether a classical test charge emits or absorbs quanta. In measuring a classical field strength, $\mathbf{E}(\mathbf{r}t)$, we implicitly sum the effects of photon absorption and emission which are described individually by the fields $\mathbf{E}^{(+)}(\mathbf{r}t)$ and $\mathbf{E}^{(-)}(\mathbf{r}t)$.

Where quantum phenomena are important the situation is usually quite different. Experiments which detect photons ordinarily do so by absorbing them in one or another way. The use of any absorption process, such as photoionization, means in effect that the field we are measuring is the one associated with photon annihilation, the complex field $\mathbf{E}^{(+)}(\mathbf{r}t)$. We need not discuss the details of the photoabsorption process to find the appropriate matrix element of the field operator. If the field makes a transition from the initial state $|i\rangle$ to a final state $|f\rangle$ in which one photon, polarized in the μ direction, has been absorbed, the matrix element takes the form

$$\langle f|E_\mu^{(+)}(\mathbf{r}t)|i\rangle. \tag{2.14}$$

We shall define an ideal photon detector as a system of negligible size (e.g., of atomic or subatomic dimensions) which has a frequency-independent photoabsorption probability. The advantage of imagining such a detector, as we shall show more explicitly in a later paper, is that the rate at which it records photons is proportional to the sum over all final states $|f\rangle$ of the squared absolute values of the matrix elements (2.14). In other words, the probability per unit time that a photon be absorbed by an ideal detector at point $\mathbf{r}$ at time t is proportional to

$$\begin{aligned}\textstyle\sum_f|\langle f|E_\mu^{(+)}(\mathbf{r}t)|i\rangle|^2 &=\textstyle\sum_f\langle i|E_\mu^{(-)}(\mathbf{r}t)|f\rangle\langle f|E_\mu^{(+)}(\mathbf{r}t)|i\rangle\\ &=\langle i|E_\mu^{(-)}(\mathbf{r}t)E_\mu^{(+)}(\mathbf{r}t)|i\rangle.\end{aligned} \tag{2.15}$$

We may verify immediately from (2.12) that the rate at which photons are detected in the empty, or vacuum, state vanishes.

The photodetector we have described is the quantum-mechanical analog of what, in classical experiments, has been called a square-law detector. It is important to bear in mind that such a detector for quanta measures the average value of the product $E_\mu^{(-)}E_\mu^{(+)}$, and not that of the square of the real field $E_\mu(\mathbf{r}t)$. Indeed, it is easily seen from the foregoing work that the average value of $E_\mu^2(\mathbf{r}t)$ does not vanish in the vacuum state;

$$\langle\text{vac}|E_\mu^2(\mathbf{r}t)|\text{vac}\rangle>0.$$

The electric field in the vacuum undergoes zero-point oscillations which, in the correctly formulated theory, have nothing to do with the detection of photons.

Recording photon intensities with a single detector does not exhaust the measurements we can make upon the field, though it does characterize, in principle, virtually all the classic experiments of optics. A second type of measurement we may make consists of the use of two detectors situated at different points $\mathbf{r}$ and $\mathbf{r}'$ to detect photon coincidences or, more generally, delayed coincidences. The field matrix element for such transitions takes the form

$$\langle f|E_\mu^{(+)}(\mathbf{r}'t')E_\mu^{(+)}(\mathbf{r}t)|i\rangle, \tag{2.16}$$

if both photons are required to be polarized along the μ axis. The total rate at which such transitions occur is proportional to

$$\begin{aligned}&\textstyle\sum_f|\langle f|E_\mu^{(+)}(\mathbf{r}'t')E_\mu^{(+)}(\mathbf{r}t)|i\rangle|^2\\ &\quad=\langle i|E_\mu^{(-)}(\mathbf{r}t)E_\mu^{(-)}(\mathbf{r}'t')E_\mu^{(+)}(\mathbf{r}'t')E_\mu^{(+)}(\mathbf{r}t)|i\rangle.\end{aligned} \tag{2.17}$$

Such a total rate is to be interpreted as a probability per unit $(\text{time})^2$ that one photon is recorded at $\mathbf{r}$ at time t and another at $\mathbf{r}'$ at time t'. Photon correlation experiments of essentially the type we are describing were performed by Hanbury Brown and Twiss[1] in 1955 and have, subsequently, been performed by others.[2]

Whatever may be the practical difficulties of more elaborate experiments, we may at least imagine the possibility of detecting n-fold delayed coincidences of photons for arbitrary n. The total rate per unit $(\text{time})^n$ for such coincidences will be proportional to

$$\langle i|E_\mu^{(-)}(\mathbf{r}_1t_1)\cdots E_\mu^{(-)}(\mathbf{r}_nt_n)E_\mu^{(+)}(\mathbf{r}_nt_n)\cdots E_\mu^{(+)}(\mathbf{r}_1t_1)|i\rangle,\quad n=1,2,3,\cdots. \tag{2.18}$$

The entire succession of such expectation values, therefore, possesses a simple physical interpretation.

In closing this survey we add a note on the commutation rules obeyed by the fields $\mathbf{E}^{(+)}$ and $\mathbf{E}^{(-)}$. It is easy to find these rules from the relation (2.3) for the real field $\mathbf{E}$, by decomposing its dependence on the two variables t and t' into positive and negative frequency parts. If the function $D_{\mu\nu}$ has the Fourier transform

$$D_{\mu\nu}(\mathbf{r}-\mathbf{r}', t-t')=\int_{-\infty}^{\infty} \mathfrak{D}_{\mu\nu}(\omega, \mathbf{r}-\mathbf{r}')e^{-i\omega(t-t')}d\omega, \quad (2.19)$$

we see immediately that the commutator (2.3) has no part which is of positive frequency in both its t and t' dependences. Neither does it have any part of negative frequency in both its time dependences. It follows that all values of the field $\mathbf{E}^{(+)}(\mathbf{r}t)$ commute with one another. and so too do those of $\mathbf{E}^{(-)}(\mathbf{r}t)$, i.e., we have

$$[E_\mu^{(+)}(\mathbf{r}t),E_\nu^{(+)}(\mathbf{r}'t')]=0, \quad (2.20)$$

$$[E_\mu^{(-)}(\mathbf{r}t),E_\nu^{(-)}(\mathbf{r}'t')]=0, \quad (2.21)$$

for all points $\mathbf{r}t$ and $\mathbf{r}'t'$, and all μ and ν. Products of the $\mathbf{E}^{(+)}$ operators or products of the $\mathbf{E}^{(-)}$ operators such as occur in (2.18) may, therefore, be freely rearranged, but the operators $\mathbf{E}^{(+)}$ and $\mathbf{E}^{(-)}$ do not, in general, commute.

III. FIELD CORRELATIONS

The electromagnetic field may be regarded as a dynamical system with an infinite number of degrees of freedom. Our knowledge of the condition of such a system is virtually never so complete or so precise in practice as to justify the use of a particular quantum state $|\rangle$ in its description. In the most accurate preparation of the state of a field which we can actually accomplish some parameters, usually an indefinitely large number of them, must be regarded as random variables. Since there is no possibility in practice of controlling these parameters, we can only hope ultimately to compare with experiment quantities which are averages over the distributions of the unknown parameters.

Our actual knowledge of the state of the field is specified fully by means of a density operator ρ which is constructed as an average, over the uncontrollable parameters, of an expression bilinear in the state vector. If $|\rangle$ is a precisely defined state of the field corresponding to a particular set of random parameters, the density operator is defined as the averaged outer product of state vectors

$$\rho=\{|\rangle\langle|\}_{\text{av}}. \quad (3.1)$$

The weightings to be used in the averaging are the ones which best describe the actual preparation of the fields. It is clear from the definition that ρ is Hermitian, $\rho^\dagger=\rho$.

The average of an observable $\mathcal{O}$ in the quantum state $|\rangle$ is the expectation value, $\langle|\mathcal{O}|\rangle$. It is the average of this quantity over the randomly prepared states which we compare with experiment. The average taken in this twofold sense may be written as

$$\{\langle|\mathcal{O}|\rangle\}_{\text{av}}=\text{tr}\{\rho\mathcal{O}\}, \quad (3.2)$$

where the symbol tr stands for the trace, or the sum of the diagonal matrix elements. Since we require the average of the unit operator to be one, we must have $\text{tr}\rho=1$. These considerations show that the average counting rate of an ideal photodetector, which is proportional to (2.15) in a completely specified quantum state of the field, is more generally proportional to

$$\text{tr}\{\rho E_\mu^{(-)}(\mathbf{r}t)E_\mu^{(+)}(\mathbf{r}t)\} \quad (3.3)$$

when the state is less completely specified.

It is convenient at this point, as a simplification of notation, to confine our attention to a single vector component of the electric field. We suppose, for the present, that all of our detectors are fitted with polarizers and record only photons polarized parallel to an arbitrary unit vector $\mathbf{e}$. (If $\mathbf{e}$ is chosen as a complex unit vector, $\mathbf{e}^*\cdot\mathbf{e}=1$, the photons detected may have arbitrary elliptical polarization.) We then introduce the symbols $E^{(+)}$ and $E^{(-)}$ for the projections of the complex fields in the direction $\mathbf{e}$ and $\mathbf{e}^*$,

$$E^{(+)}(\mathbf{r}t)=\mathbf{e}^*\cdot\mathbf{E}^{(+)}(\mathbf{r}t) \quad (3.4)$$

$$E^{(-)}(\mathbf{r}t)=\mathbf{e}\cdot\mathbf{E}^{(-)}(\mathbf{r}t). \quad (3.5)$$

We resume a fully general treatment of photon polarizations in Sec. V.

The field average (3.3) which determines the counting rate of an ideal photodetector is a particular form of a more general type of expression whose properties are of considerable interest. In the more general expression, the fields $E^{(-)}$ and $E^{(+)}$ are evaluated at different space-time points. Statistical averages of the latter type furnish a measure of the correlations of the complex fields at separated positions and times. We shall define such a correlation function, $G^{(1)}$, for the $\mathbf{e}$ components of the complex fields as

$$G^{(1)}(\mathbf{r}t,\mathbf{r}'t')=\text{tr}\{\rho E^{(-)}(\mathbf{r}t)E^{(+)}(\mathbf{r}'t')\}. \quad (3.6)$$

Only the values of this function at $\mathbf{r}=\mathbf{r}'$ and $t=t'$ are needed to predict the counting rate of an ideal photodetector. However, other values of the function become necessary, quite generally, when we use as detectors less ideal systems such as actual photo-ionizable atoms. In actual photodetectors the absorption of photons can not be localized too closely, either in space or in time. Atomic photo ionization rates must be written, in general, as double integrals, over a microscopic range, of all the variables in $G^{(1)}(\mathbf{r}t,\mathbf{r}'t')$. Our interest in the function $G^{(1)}$ extends to widely spaced values of its variables as well. That field correlations may extend over considerable intervals of distance and time is essential to the idea of coherence, which we shall shortly discuss.

As we have noted earlier, our interest in averages of the field operators extends beyond quadratic ones. Just as we generalized the expression for the photon detection rate to define $G^{(1)}$, we may generalize the expression

(2.17) for the photon coincidence rate and thereby define a second-order correlation function,

$$G^{(2)}(\mathbf{r}_1t_1\mathbf{r}_2t_2,\mathbf{r}_3t_3\mathbf{r}_4t_4)$$
$$=\mathrm{tr}\{\rho E^{(-)}(\mathbf{r}_1t_1)E^{(-)}(\mathbf{r}_2t_2)E^{(+)}(\mathbf{r}_3t_3)E^{(+)}(\mathbf{r}_4t_4)\}. \quad (3.7)$$

This too is a function whose values, even at widely separated arguments, interest us.

In view of the possibility of discussing n-photon coincidence experiments for arbitrary n it is natural to define an infinite succession of correlation functions $G^{(n)}$. It is convenient in writing these to abbreviate a set of coordinates $(\mathbf{r}_j,t_j)$ by a single symbol, x_j. We then define the nth-order correlation function as

$$G^{(n)}(x_1\cdots x_n,x_{n+1}\cdots x_{2n})$$
$$=\mathrm{tr}\{\rho E^{(-)}(x_1\cdots E^{(-)}x_n)E^{(+)}(x_{n+1})\cdots E^{(+)}(x_{2n})\}. \quad (3.8)$$

The correlation functions have a number of simple properties. It is easily verified that interchanging the arguments in $G^{(1)}$ leads to the complex conjugate function

$$G^{(1)}(\mathbf{r}'t',\mathbf{r}t)=\{G^{(1)}(\mathbf{r}t,\mathbf{r}'t')\}^*. \quad (3.9)$$

The same type of relation holds for all of the higher order functions

$$G^{(n)}(x_{2n}\cdots x_1)=\{G^{(n)}(x_1\cdots x_{2n})\}^*. \quad (3.10)$$

Furthermore, the commutation relations (2.20) and (2.21) show us that $G^{(n)}$ is unchanged by any permutation of its arguments $(x_1\cdots x_n)$, or its arguments $(x_{n+1}\cdots x_{2n})$. The fact that the complex fields $E^{(\pm)}$ individually satisfy the wave equation (2.2) leads to another useful property of the $G^{(n)}$. The nth-order function satisfies $2n$ different wave equations, one for each of its arguments x_j, $(j=1,\cdots,2n)$.

A large number of inequalities satisfied by the functions $G^{(n)}$ may be derived from the positive definite character of the density operator ρ. Derivations of several classes of these are presented in the Appendix. We confine ourselves, in this section, to mentioning some of the simpler and more useful inequalities, those which are linear or quadratic in the correlation functions. It is clear from (3.10) that all of the functions $G^{(n)}(x_1\cdots x_n,x_n\cdots x_1)$ are real. The linear inequalities assert that these functions are positive definite as well. We have then, in particular for $n=1$, the self-evident relation

$$G^{(1)}(x_1,x_1)\geq 0, \quad (3.11)$$

and for arbitrary n

$$G^{(n)}(x_1\cdots x_n,x_n\cdots x_1)\geq 0. \quad (3.12)$$

These relations simply affirm that the average photon intensity of a field and the average coincidence counting rates are all intrinsically positive.

The simplest of the quadratic inequalities takes the form

$$G^{(1)}(x_1,x_1)G^{(1)}(x_2,x_2)\geq |G^{(1)}(x_1,x_2)|^2. \quad (3.13)$$

Higher order inequalities of this type are given by

$$G^{(n)}(x_1\cdots x_n,x_n\cdots x_1)G^{(n)}(x_{n+1}\cdots x_{2n},x_{2n}\cdots x_{n+1})$$
$$\geq |G^{(n)}(x_1\cdots x_n,x_{n+1}\cdots x_{2n})|^2, \quad (3.14)$$

which holds for arbitrary n. Different forms of these relations are obtained by permuting or equating coordinates. Various other inequalities are proved in the Appendix along with those noted.

It is interesting to note that when the number of quanta present in the field is bounded, the sequence of functions $G^{(n)}$ terminates. If the density operator restricts the number of photons present to be smaller than or equal to some value M, the properties of $E^{(\pm)}$ as annihilation and creation operators show that $G^{(n)}=0$ for $n>M$.

Classical correlation functions bearing some analogy to $G^{(1)}$ have received a great deal of discussion in recent years, mainly in connection with the theory of noise in radio waves. A detailed application of the classical correlation theory to optics has been made by Wolf.[5] At the core of Wolf's analysis is a single correlation function Γ, defined as an average over an infinite time span of the product of two fields, evaluated at times separated by a fixed interval. The procedure of time averaging restricts the application of such an approach to the treatment of field distributions which are statistically stationary in time.

If we were to restrict the character of our density operator ρ to describe only stationary field distributions (e.g., by choosing ρ to commute with the field Hamiltonian) our function $G^{(1)}(\mathbf{r}t,\mathbf{r}'t')$ would depend only on the difference of the two times, $t-t'$. In that case the function $G^{(1)}$ would, in the classical limit (strong, low-frequency fields), agree numerically[6] with Wolf's function Γ. It should be clear, however, that the concepts of correlation and ultimately of coherence are quite useful in the discussion of nonstationary field distributions. The correlation functions $G^{(n)}$ which we have defined are ensemble averages rather than time averages and hence remain well-defined in fields of arbitrary time dependence.

IV. COHERENCE

The term "coherence" has had long if somewhat varied use in areas of physics concerned with the electromagnetic field. In physical optics the term is used to denote a tendency of two values of the field at distantly separated points or at greatly separated times to take on correlated values. When optical means are used to superpose the fields at such points (e.g., as in Young's two-slit experiment) intensity fringes result. The possibility of producing such fringes in hypothetical superposition experiments epitomizes the optical definition of

[5] M. Born and E. Wolf, *Principles of Optics* (Pergamon Press, Inc., London, 1959), Chap. X. An extensive bibliography is given there.

[6] This is true provided Wolf's "disturbance" field V behaves ergodically and is identified with $E^{(+)}$.

coherence. The definition has remained a satisfactorily explicit one only as long as optical experiments were confined to measuring field intensities, or more generally quantities quadratic in the field strengths. We have already noted that the photon correlation experiment of Hanbury Brown and Twiss,[1] performed in 1955, is of an altogether new type and measures the average of a quartic expression.[7] The study of quantities of fourth and higher powers in the field strengths is the basis of all work in the recently developed area of nonlinear optics. It appears safe to assume that the number of such experiments will increase in the future, and that the concept of coherence should be extended to apply to them.

Another pressing reason for sharpening the meaning of coherence is provided by the recent development of the optical maser. The maser produces light beams of narrow spectral bandwidth which are characterized by field correlations extending over quite long ranges. Such light is inevitably described as coherent, but the sense in which the term is used has not been made adequately clear. If the sense is simply the optical one then, as we shall see, it may scarcely do justice to the potentialities of the device. The optical definition does not at all distinguish among the many ways in which fields may vary while remaining equally correlated at all pairs of points. That much greater regularities may exist in the field variations of a maser beam than are required by the optical definition of coherence may be seen by comparing the maser beam with the carrier wave of a radio transmitter. The latter type of wave ideally possesses a stability of amplitude which optically coherent fields need not have.[8] Furthermore, the field values of such a wave possess correlations of a much more detailed sort than the optical definition requires. These are properties best expressed in terms of the higher order correlation functions $G^{(n)}$, for $n>1$.

To discuss coherence in quantitative terms it is convenient to introduce normalized forms of the correlation functions. Corresponding to the first-order function $G^{(1)}$ we define

$$g^{(1)}(\mathbf{r}t,\mathbf{r}'t')=\frac{G^{(1)}(\mathbf{r}t,\mathbf{r}'t')}{\{G^{(1)}(\mathbf{r}t,\mathbf{r}t)G^{(1)}(\mathbf{r}'t',\mathbf{r}'t')\}^{1/2}}. \tag{4.1}$$

It is immediately seen from (3.13) that $g^{(1)}$ obeys the inequality

$$|g^{(1)}(\mathbf{r}t,\mathbf{r}'t')|\leq 1. \tag{4.2}$$

For $\mathbf{r}=\mathbf{r}'$, $t=t'$ we have, of course, $g^{(1)}\equiv 1$.

The normalized forms of the higher order correlation functions are defined as

$$g^{(n)}(x_1\cdots x_{2n})=G^{(n)}(x_1\cdots x_{2n})/\prod_{j=1}^{2n}\{G^{(1)}(x_j,x_j)\}^{1/2}. \tag{4.3}$$

These functions, for $n>1$, are not, in general, restricted in absolute value as is $g^{(1)}$.

We shall try in this paper to give the concept of coherence as precise a definition as is both realizable in physical terms, and useful as well.[9] We, therefore, begin by stating an infinite sequence of conditions on the functions $g^{(n)}$ which are to be satisfied by a fully coherent field. These necessary conditions for coherence are that the normalized correlation functions all have unit absolute magnitude,

$$|g^{(n)}(x_1\cdots x_{2n})|=1, \quad n=1,2\cdots. \tag{4.4}$$

That there exist at least some states which meet these conditions at all points of space and time is immediately clear from the example of a classical plane wave, $E^{(+)}\sim\exp[i(\mathbf{k}\cdot\mathbf{r}-\omega t)]$. We shall presently show that the class of coherent fields is vastly larger than that of individual plane waves.

The conditions (4.4) on the functions $g^{(n)}$ are stated only as necessary ones and need not be construed as defining coherence completely. We shall shortly, in fact, sharpen the definition somewhat further. It is worth noting at this point, however, that not all of the fields which have been described as "coherent" in the past meet the set of conditions (4.4) even approximately. There may be some virtue, therefore, in constructing a hierarchy of orders of coherence to discuss fields which do not have that property in its fullest sense. We shall state as a condition necessary for first-order coherence that $|g^{(1)}(\mathbf{r}t,\mathbf{r}'t')|=1$. More generally, for a field to be characterized by nth order coherence we shall require $|g^{(j)}|=1$ for $j\leq n$. For fields which occur in practice, one can not expect relations such as these to hold exactly for all points in space and time. We shall, therefore, often employ the term nth order coherence more loosely to mean that the first n coherence conditions are fairly accurately satisfied over appreciable intervals of the variables surrounding all points $x_1=x_2=\cdots=x_{2n}$.

The definition of coherence which has been used to date in all studies of physical optics corresponds only to first-order coherence. The most coherent fields which have been generated by optical means prior to the development of the maser, in fact, lack second and higher order coherence. On the other hand, the optical maser, functioning with ideal stability, may produce fields which are coherent to all orders.

The various orders of coherence may, in principle, be distinguished fairly directly in experimental terms. The inequality (3.12), which states that the n-fold coincidence counting rate is positive, requires that $g^{(n)}(x_1\cdots x_n,x_n\cdots x_1)$ be positive. If the field in question possesses nth-order coherence, it must, therefore, have

$$g^{(j)}(x_1\cdots x_j,x_j\cdots x_1)=1, \tag{4.5}$$

[7] R. J. Glauber, Phys. Rev. Letters **10**, 84 (1963). The particular field referred to as incoherent in that note may have first-order coherence if it is monochromatic, but not second- or higher order coherence.

[8] This point has been noted with particular clarity by M. J. E. Golay, Proc. IRE **49**, 959 (1961); also **50**, 223 (1962).

[9] A brief account of this work was presented by R. J. Glauber, in Proceedings of the Third International Conference on Quantum Electronics, Paris, France, 1963 (to be published).

for $j \leq n$. It follows from the definitions of the $g^{(j)}$ that the corresponding values of the correlation functions $G^{(j)}$ factorize, i.e.,

$$G^{(j)}(x_1 \cdots x_j, x_j \cdots x_1) = \prod_{i=1}^{j} G^{(1)}(x_i, x_i), \qquad (4.6)$$

for $j \leq n$. These relations mean, in observational terms, that the rate at which j-fold delayed coincidences are detected by our ideal photon counters, reduces to a product of the detection rates of the individual counters.[7] In photon coincidence experiments of multiplicity up to and including n, the photon counts registered by the individual counters may then be regarded as statistically independent events. No tendency of photon counts to be statistically correlated will be evident in j-fold coincidence experiments for $j \leq n$.

The experiments of Hanbury Brown and Twiss[1] were designed to detect correlations in the fluctuating outputs of two photomultipliers. These detectors were placed in fields made coherent with one another (in the optical sense) through the use of monochromatic, pinhole illumination and a semitransparent mirror. The photocurrents of the two detectors were observed to show a positive correlation for small delay times, rather than independent fluctuations. A similar experiment has been performed by Rebka and Pound,[2] using coincidence counting equipment. Their experiment, performed with a more monochromatic beam and better geometrical definition, shows an explicit correlation in the counting probabilities of the two detectors. These observations verify that light beams from ordinary sources such as discharge tubes, when made optimally coherent in the first-order sense, still lack second-order coherence.

The coherence conditions (4.4) can also be stated as a requirement that the functions $|G^{(n)}(x_1 \cdots x_{2n})|$ factorize into a product of $2n$ functions of the same form, each dependent on a single space-time variable,

$$|G^{(n)}(x_1 \cdots x_{2n})| = \prod_{j=1}^{2n} \{G^{(1)}(x_j, x_j)\}^{1/2}. \qquad (4.7)$$

This statement of the necessary conditions for coherence suggests that it may be convenient to give a stronger definition to coherence by regarding it as a factorization property of the correlation functions,

Let us suppose that there exists a function $\mathcal{E}(x)$, independent of n, such that the correlation functions for all n may be expressed as the products

$$G^{(n)}(x_1 \cdots x_n, x_{n+1} \cdots x_{2n}) = \mathcal{E}^*(x_1) \cdots \mathcal{E}^*(x_n) \mathcal{E}(x_{n+1}) \cdots \mathcal{E}(x_{2n}). \qquad (4.8)$$

It is immediately clear that these functions satisfy the conditions (4.4) and (4.7). To show that fields with such correlations exist we need only refer again to the case of a classical plane wave. In fact, any classical field of predetermined (i.e., nonrandom) behavior has correlation functions which fall into this form, and such fields are at times called coherent in communication theory. We shall, therefore, adopt the factorization conditions (4.8) as the definition of a coherent field and turn next to the question of how they may be satisfied in the quantum domain.

If it were possible for the field to be in an eigenstate of the operators $E^{(+)}$ and $E^{(-)}$, the correlation functions for such states would factorize immediately to the desired form. The operators $E^{(+)}(\mathbf{r}t)$ and $E^{(-)}(\mathbf{r}'t')$ do not commute, however, so no state can be an eigenstate of both in the usual sense. Not only are these operators non-Hermitian, but the failure of each to commute with its adjoint shows that $E^{(+)}$ and $E^{(-)}$ are non-normal as well. Operators of this type can not, as a rule, be diagonalized at all, but may nonetheless have eigenstates. In general, we must distinguish between their left and right eigenstates; the two types need not occur in mutually adjoint pairs. The operator $E^{(+)}(\mathbf{r}t)$, in particular, has no left eigenstates, but does have right eigenstates[10] corresponding to complex eigenvalues for the field, which are functions of position and time. We shall suppose that $|\rangle$ is a right eigenstate of $E^{(+)}$ and that the equation it satisfies takes the form

$$E^{(+)}(\mathbf{r}t)|\rangle = \mathcal{E}(\mathbf{r}t)|\rangle, \qquad (4.9)$$

in which the function $\mathcal{E}(\mathbf{r}t)$ is to be interpreted as the complex eigenvalue. The Hermitian adjoint of this relation shows us that the conjugate state, $\langle|$, is a left eigenstate of $E^{(-)}(\mathbf{r}t)$,

$$\langle|E^{(-)}(\mathbf{r}t) = \langle|\mathcal{E}^*(\mathbf{r}t). \qquad (4.10)$$

The density operator for such states is simply the projection operator, $\rho = |\rangle\langle|$. It follows immediately from these relations that the correlation functions $G^{(n)}$ all factorize into the form of Eq. (4.8). In other words, the state of the field defined by Eqs. (4.9) or (4.10) meets our definition precisely and is fully coherent. We shall discuss the properties of such states[11] at length in the paper to follow. For the present it may suffice to say that we can find an eigenstate $|\rangle$ which corresponds to the choice, as an eigenvalue, of any function $\mathcal{E}(\mathbf{r}t)$ which satisfies certain conditions. One condition, which is clear from Eq. (4.9), is that $\mathcal{E}(\mathbf{r}t)$ must satisfy the wave equation. The other, which corresponds to the positive frequency character of $E^{(+)}$, is that $\mathcal{E}(\mathbf{r}t)$, when regarded as a function of a complex time variable, be analytic in the lower half-plane. The eigenstates which correspond to different fields $\mathcal{E}(\mathbf{r}t)$ are not mutually orthogonal, but nontheless form a natural basis for the discussion of photon detection problems. We have introduced them

[10] States of the harmonic oscillator which have an analogous property were introduced in a slightly different but related connection by E. Schrödinger, Naturwiss. **14**, 664 (1926). The electromagnetic field, as is well known, may be treated as an assembly of oscillators.

[11] Some of the properties of these states have already been noted in references 7 and 9.

here only to demonstrate the possibility of satisfying the coherence conditions in quantum theory. Such quantum states do not exhaust the possibility of describing coherent fields. Statistical mixtures, for example, of the states for which the eigenvalues $\mathcal{E}(\mathbf{r}t)$ differ by constant phase factors satisfy the coherence conditions equally well.

The fields which have been described as most coherent in optical contexts have tended to be those of the narrowest spectral bandwidth. If coherent fields in optics have necessarily been chosen as monochromatic ones, it is because that has been virtually the only means of securing appreciably correlated fields from intrinsically chaotic sources. For this reason, perhaps, there has been a natural tendency to associate the concept of coherence with monochromaticity. The association was, in fact, made an implicitly rigid one by earlier discussions[5] of optical (i.e., first-order) coherence which were applicable only to statistically stationary fields. By extending the definition of coherence to nonstationary fields we see that it places no constraint on the frequency spectrum. Coherent fields exist corresponding to eigenvalues $\mathcal{E}(\mathbf{r}t)$ with arbitrary spectra. The coherence conditions restrict randomness of the fields rather than their bandwidth.

Having defined full coherence by means of the factorization conditions (4.8), we may now use them in defining the various orders of coherence. We shall speak of mth-order coherent fields when the conditions (4.8) are satisfied for $n\leq m$, a definition which accords with our earlier conditions on $|g^{(j)}|$.

Photon correlation experiments have shown the importance of distinguishing between the first two orders of coherence. At the other end of the scale, we have shown that there exist, in principle at least, states which are fully coherent. We are entitled to ask, therefore, whether the intermediate orders of coherence will also be useful classifications. In the absence of any experimental information, we can only guess that they may be useful, though perhaps not in the sharp sense in which we we have defined them. One may easily imagine the possibility that, for light sources such as the maser, the correlation functions $G^{(n)}$ show gradually increasing departure from the factored forms (4.8) as n increases, even when the variables $x_1\cdots x_{2n}$ are not too widely separated. In such contexts the order of coherence can only be defined approximately.[12] Something of the same approximate character must be present in all applications of the definitions we have given. The field correlations we have discussed can extend over great intervals of distance and time, though never infinite ones in practice. Coherence conditions, such as $|g^{(n)}|=1$, can only be met within a finite range of relative values of the coordinates $x_1\cdots x_{2n}$. It is only within such ranges, and therefore as an approximation, that we can speak of coherence at all.

[12] The characterization we have given the nth-order coherent fields is, in principle, an accurately realizable one, however. States with such properties may be constructed in a variety of ways. The factorization conditions can be met for $j\leq n$, for example, by suitably chosen states in which the number of photons present may take on any value up to n. The correlation functions of order $j>n$ then vanish, as we have noted earlier. The vanishing of these correlation functions for states with bounded numbers of quanta shows, incidentally, that no bound can be placed on the photon number in a fully coherent field.

V. COHERENCE AND POLARIZATION

We have to this point, in the interest of simplicity, dealt only with the projections of the fields along a single (possibly complex) unit vector $\mathbf{e}$. To take fuller account of the vector nature of the fields we must define tensor rather than scalar correlation functions. The first-order function is taken to be

$$G_{\mu\nu}^{(1)}(x,x')=\mathrm{tr}\{\rho E_\mu^{(-)}(x)E_\nu^{(+)}(x')\},\tag{5.1}$$

in which the indices μ and ν label Cartesian components. This function satisfies the symmetry relation

$$G_{\nu\mu}^{(1)}(x',x)=\{G_{\mu\nu}^{(1)}(x,x')\}^*,\tag{5.2}$$

and is shown in the appendix to obey the inequalities,

$$G_{\mu\mu}^{(1)}(x,x)\geq 0\tag{5.3}$$

and

$$G_{\mu\mu}^{(1)}(x,x)G_{\nu\nu}^{(1)}(x',x')\geq|G_{\mu\nu}^{(1)}(x,x')|^2.\tag{5.4}$$

The photon intensities which can be detected at the space-time point x are found from $G_{\mu\nu}^{(1)}(x,x')$ for $x'=x$. We shall abbreviate this 3×3 matrix as $\mathbf{G}^{(1)}(x)$, and use it as the basis of a brief discussion of polarization correlations in three dimensions, a subject which seems to have received little attention in comparison to plane polarizations. The symmetry relation (5.2) for $x'=x$ shows that the intensity matrix $\mathbf{G}^{(1)}(x)$ is Hermitian; an argument given in the Appendix shows it to be a positive definite matrix as well. It follows that $\mathbf{G}^{(1)}(x)$ has positive real eigenvalues, $\lambda_p(x)$, $(p=1,2,3)$, which correspond to a set of (generally, complex) eigenvectors. The eigenvectors, which we write as $\mathbf{e}^{(p)}$ satisfy

$$\mathbf{G}^{(1)}(x)\cdot\mathbf{e}^{(p)*}=\lambda_p\mathbf{e}^{(p)*},$$
$$\mathbf{e}^{(p)}\cdot\mathbf{G}^{(1)}(x)=\lambda_p\mathbf{e}^{(p)}.\tag{5.5}$$

If the three eigenvalues $\lambda_p(x)$ are all different, it is clear that the three eigenvectors must be orthogonal; if not they may be chosen so. If the eigenvectors are normalized to obey the relations

$$\mathbf{e}^{(p)}\cdot\mathbf{e}^{(q)*}=\delta_{pq},\tag{5.6}$$

their components form the unitary matrix which diagonalizes $\mathbf{G}^{(1)}(x)$. The eigenvectors, or equivalently the unitary matrix, are determined by a set of eight independent real parameters.

A tensor product, such as

$$\mathbf{e}^{(p)}\cdot\mathbf{G}^{(1)}(x)\cdot\mathbf{e}^{(q)*}=\lambda_p\delta_{pq},\tag{5.7}$$

expresses the correlation, at the point x, of the field components in the $\mathbf{e}^{(p)}$ and $\mathbf{e}^{(q)}$ directions. It is clear, then, that there always exist a set of three (complex) orthogonal polarization vectors such that the field components in these directions are statistically uncorrelated. The eigenvalues λ_p correspond to the intensities for these polarizations. For quantitative discussions of polarization it is convenient to define the normalized intensities $I_p=\lambda_p/\sum_q\lambda_q$, $(p=1,2,3)$, which sum to unity, $\sum_p I_p=1$. When the normalized intensities are all equal to $\frac{1}{3}$ we have the case of an isotropic field, as in a hohlraum filled with thermal radiation.

The triad of eigenvectors at a point in an arbitrary field depends, in general, on time as well as position. If the density operator, ρ, represents a stationary ensemble, however, the triad becomes fixed. A particular example which has been studied in minute detail in optics is that of a beam of plane waves.[5,13] In that case, since the fields are transverse, one of the eigenvectors may be chosen as the beam direction and obviously corresponds to the eigenvalue zero. The net polarization of the beam is usually defined as the magnitude of the difference of the normalized intensities, $|I_1-I_2|$, which correspond to the remaining two eigenvalues.

We next define the higher order correlation functions as

$$G^{(n)}{}_{\mu_1\cdots\mu_{2n}}(x_1\cdots x_n,x_{n+1}\cdots x_{2n})=\mathrm{tr}\{\rho E_{\mu_1}{}^{(-)}(x_1)\cdots \times E_{\mu_n}{}^{(-)}(x_n)E_{\mu_{n+1}}{}^{(+)}(x_{n+1})\cdots E_{\mu_{2n}}{}^{(+)}(x_{2n})\}. \quad (5.8)$$

These functions are unchanged by simultaneous permutations of the coordinates $(x_1\cdots x_n)$ and the indices $(\mu_1\cdots\mu_n)$; they are likewise invariant under permutations of the $(x_{n+1}\cdots x_{2n})$ and $(\mu_{n+1}\cdots\mu_{2n})$. They satisfy the symmetry relation

$$G^{(n)}{}_{\mu_{2n}\cdots\mu_1}(x_{2n}\cdots x_1)=\{G^{(n)}{}_{\mu_1\cdots\mu_{2n}}(x_1\cdots x_{2n})\}^* \quad (5.9)$$

and are shown, in the Appendix, to obey the inequalities

$$G^{(n)}{}_{\mu_1\cdots\mu_n\mu_n\cdots\mu_1}(x_1\cdots x_n,x_n\cdots x_1)\geq 0 \quad (5.10)$$

and

$$G^{(n)}{}_{\mu_1\cdots\mu_n\mu_n\cdots\mu_1}(x_1\cdots x_n,x_n\cdots x_1) \times G^{(n)}{}_{\mu_{n+1}\cdots\mu_{2n}\mu_{2n}\cdots\mu_{n+1}}(x_{n+1}\cdots x_{2n},x_{2n}\cdots x_{n+1}) \geq |G^{(n)}{}_{\mu_1\cdots\mu_n\mu_{n+1}\cdots\mu_{2n}}(x_1\cdots x_n,x_{n+1}\cdots x_{2n})|^2. \quad (5.11)$$

As in our earlier discussion of coherence, it is convenient to make use of the normalized correlation functions

$$g^{(n)}{}_{\mu_1\cdots\mu_{2n}}(x_1\cdots x_{2n}) =G^{(n)}{}_{\mu_1\cdots\mu_{2n}}(x_1\cdots x_{2n})/\prod_{j=1}^{2n}\{G^{(1)}{}_{\mu_j\mu_j}(x_j,x_j)\}^{1/2}. \quad (5.12)$$

The necessary conditions for full coherence are

$$|g^{(n)}{}_{\mu_1\cdots\mu_{2n}}(x_1\cdots x_{2n})|=1, \quad (5.13)$$

which must hold for all components $\mu_1\cdots\mu_{2n}$, as well as all n. It is clear, however, that these conditions do not constitute an adequate definition of coherence, since they are not, in general, invariant under rotations of the coordinate axes. We therefore turn once again to a definition of coherence as a factorization property of the correlation functions.

We define full coherence to hold when the set of correlation functions $G^{(n)}$ may be expressed as products of the components of a vector field $\mathcal{E}_\mu(x)$, $(\mu=1,2,3)$, i.e.,

$$G^{(n)}{}_{\mu_1\cdots\mu_{2n}}(x_1\cdots x_n,x_{n+1}\cdots x_{2n}) =\prod_{j=1}^{n}\mathcal{E}^*{}_{\mu_j}(x_j)\prod_{l=n+1}^{2n}\mathcal{E}_{\mu_l}(x_l), \quad (5.14)$$

where it is understood that the vector field $\mathcal{E}_\mu(x)$ is independent of n. It is immediately clear, from the transformation properties of the definition, that a field coherent in one coordinate frame is equally coherent in any rotated frame. Furthermore, all of the normalized correlation functions $g^{(n)}$, which follow from the definition, satisfy the conditions (5.13).

The coherence conditions (5.14) imply that the field is fully polarized in the direction of the vector $\boldsymbol{\mathcal{E}}(x)$ at each point x. The formal way of seeing this is to note that the intensity matrix $G_{\mu\nu}{}^{(1)}(x,x)$, which we discussed earlier in general terms, reduces for a coherent field to,

$$G_{\mu\nu}{}^{(1)}(x,x)=\mathcal{E}_\mu{}^*(x)\mathcal{E}_\nu(x). \quad (5.15)$$

Such a matrix represents an unnormalized projection operator for the direction of $\boldsymbol{\mathcal{E}}(x)$. It obviously has, as an eigenvector in the sense of Eq. (5.5), the vector $\mathcal{E}_\mu(x)$ itself. The corresponding eigenvalue is the full intensity $\sum_\mu|\mathcal{E}_\mu(x)|^2$. The two remaining eigenvalues, which correspond to orthogonal directions, clearly vanish.

It is interesting to note that for coherent fields many of the inequalities stated earlier, e.g., (3.13), (3.14), (5.4), (5.11), reduce to statements of equality. This reduction holds quite generally, as is shown in the Appendix, for those inequalities of quadratic and higher degree in the correlation functions.

The arguments by which we exhibit fields satisfying the coherence conditions, are essentially unchanged from the previous section. In particular, as we shall discuss in the next paper, there exist states which are simultaneously right eigenstates of all three components of $E_\mu{}^{(+)}(\mathbf{r}t)$ and correspond to a set of three complex eigenvalues $\mathcal{E}_\mu(\mathbf{r}t)$. Such states satisfy the coherence conditions (5.14) precisely.

If we have chosen to discuss only the correlations of the electric field in this paper, it is because that field plays the dominant role in all detection mechanisms for photons of lower frequency than x rays. It is not difficult to construct correlation functions which involve the magnetic field as well as the electric field, and perhaps these too will someday prove useful. One method is to use the relativistic field tensor, $F_{\mu\nu}$, in precisely the way we have used the field E_μ. The field tensor may be written as a 4×4 antisymmetric matrix, made up of the

[13] Most of these studies have been confined to stationary, quasimonochromatic beams. See, for example, G. B. Parrent, Jr., and P. Roman, Nuovo Cimento 15, 370 (1960).

components of both the electric and magnetic fields. The nth-order correlation function for the complex components of those fields would have $4n$ four-valued indices. Coherence may then be defined as a requirement that the correlation functions all be separable into the the products of 4×4 antisymmetric fields, just as Eq. (5.14) requires a separation into products of three-vector field components. The advantage of such a definition is to make it clear that coherence is a relativistically invariant concept; that a field which is coherent in any one Lorentz frame is coherent in any other. Fields which are coherent in this relativistic sense are automatically coherent in the more limited senses we have described earlier.

ACKNOWLEDGMENTS

The author is grateful to the Research Laboratory of the American Optical Company and its director, Dr. S. M. MacNeille, for partial support of this work.

APPENDIX

In this section we derive a number of inequalities obeyed by the correlation functions defined in the paper. Fundamentally, these relations are all consequences of a single inequality

$$\mathrm{tr}\{\rho A^\dagger A\}\geq 0, \tag{A1}$$

which holds for arbitrary choice of the operator A. To prove this inequality, we note that the density operator ρ is Hermitian and can always be diagonalized, i.e., we can find a set of basis states such that the matrix representation of ρ is

$$\langle k|\rho|l\rangle=\delta_{kl}p_k. \tag{A2}$$

The numbers p_k may be interpreted as probabilities associated with the states $|k\rangle$. They are, therefore, non-negative, $p_k\geq 0$; which is to say that ρ is a positive definite operator. The normalization condition on the density operator, $\mathrm{tr}\rho=\sum p_k=1$, shows that not all the p_k vanish. The trace (A1) may be reduced, in the representation defined by (A2), to the form

$$\mathrm{tr}\{\rho A^\dagger A\}=\textstyle\sum_k p_k\langle k|A^\dagger A|k\rangle. \tag{A3}$$

The diagonal matrix elements on the right of (A3) are all non-negative since they may be expressed as a sum of squared absolute values,

$$\begin{aligned}\langle k|A^\dagger A|k\rangle&=\textstyle\sum_l\langle k|A^\dagger|l\rangle\langle l|A|k\rangle\\&=\textstyle\sum_l|\langle l|A|k\rangle|^2.\end{aligned} \tag{A4}$$

This statement completes the proof of (A1), since the trace is invariant under unitary transformations of the basis states.

The trace which occurs in the inequality (A1) has the same basic structure as all of the correlation functions $G^{(n)}$. Various inequalities relating the correlation functions follow, more or less directly, from different choices of the operator A. If, for example, we choose A to be $E^{(+)}(x)$, as defined by (3.4), we find the inequality (3.11),

$$G^{(1)}(x,x)\geq 0. \tag{A5}$$

If we choose A to be the n-fold product $E^{(+)}(x_1)\cdots E^{(+)}(x_n)$ we find the inequality (3.12),

$$G^{(n)}(x_1\cdots x_n,x_n\cdots x_1)\geq 0. \tag{A6}$$

The proofs are no different if the components of the three-dimensional field are used in place of $E^{(+)}$, i.e., if a component index μ_j is associated with each coordinate x_j. Hence, we have also derived (5.3) and (5.10).

The remaining inequalities are of second and higher degree in the correlation functions. Those obeyed by the first-order function, $G^{(1)}$, may be found as follows: We choose at random a set of m space-time points $x_1\cdots x_m$, and consider as the operator A,

$$A=\sum_{j=1}^{m}\lambda_j E^{(+)}(x_j), \tag{A7}$$

where the superposition coefficients $\lambda_1\cdots\lambda_m$ are an arbitrary set of complex numbers. When we substitute (A7) into the basic inequality, (A1), we find

$$\sum_{i,j}\lambda_i^*\lambda_j G^{(1)}(x_i,x_j)\geq 0. \tag{A8}$$

In other words, the set of correlation functions $G^{(1)}(x_i,x_j)$ for $i,\ j=1,\ \cdots,\ m$ forms the matrix of coefficients of a positive definite quadratic form. It follows, in particular, that the determinant of the matrix is non-negative,

$$\det[G^{(1)}(x_i,x_j)]\geq 0\quad i,\ j=1,\ \cdots,\ m. \tag{A9}$$

For $m=1$ this inequality is simply (A5). For $m=2$ it becomes the one noted in the text as (3.13),

$$G^{(1)}(x_1,x_1)G^{(1)}(x_2,x_2)\geq|G^{(1)}(x_1,x_2)|^2. \tag{A10}$$

For larger values of m the inequalities are perhaps best left in the form (A9). When tensor components are introduced, we have only to replace the coordinate x_j in the proofs by the combination of x_j and a tensor index μ_j. The relation (5.4) thereby follows from the form of (A10). If, in particular for $m=3$, we choose the three coordinates to be the same and the tensor indices all different, i.e., we choose

$$A=\sum_{\nu=1}^{3}\lambda_\nu E_\nu^{(+)}(x), \tag{A11}$$

we find that the 3×3 matrix $G_{\mu\nu}^{(1)}(x,x)$ is positive definite, a property used in the text in the discussion of polarizations.

Since the succession of inequalities which follows from (A1) is endless, we only mention the quadratic ones for the higher order functions. To find these, we choose a set of $2n$ coordinates at random and let A be any operator of the form

$$A=\lambda_1 E^{(+)}(x_1)\cdots E^{(+)}(x_n)$$
$$+\lambda_2 E^{(+)}(x_{n+1})\cdots E^{(+)}(x_{2n}). \quad \text{(A12)}$$

The positive definiteness of the quadratic form which results from substituting this expression in (A1) shows that the inequality (3.14) must hold. When vector indices are attached to the operators $E^{(+)}$, the same proof leads to (5.11).

We have noted in the text that, for the particular case of coherent fields, the inequalities of second degree in the correlation functions reduce to equalities. The reason for the reduction lies in the way the correlation functions factorize. The factorization causes all of the second and higher order determinants involved in the statement of positive definiteness conditions [e.g., (A9)] to vanish.

VOLUME 10, NUMBER 7 PHYSICAL REVIEW LETTERS 1 APRIL 1963

EQUIVALENCE OF SEMICLASSICAL AND QUANTUM MECHANICAL DESCRIPTIONS OF STATISTICAL LIGHT BEAMS

E. C. G. Sudarshan
Department of Physics and Astronomy, University of Rochester, Rochester, New York
(Received 1 March 1963)

With the advent of the laser, attention has been focused on the problem of the complete description of the electromagnetic field associated with arbitrary light beams. The classical theory of optical coherence[1] works almost exclusively with two-point correlations; and this theory is adequate for the description of the classical optical phenomena of interference and diffraction in general. More sophisticated experiments on intensity interferometry and photoelectric counting statistics necessitated special higher order correlations. Most of this work[2] was done using a classical or a semiclassical formulation of the problem. On the other hand, statistical states of a quantized (electromagnetic) field have been considered recently,[3] and a quantum mechanical definition of coherence functions of arbitrary order presented. It is the aim of this note to elaborate on this definition and to demonstrate its complete equivalence to the classical description as long as no nonlinear effects are considered.

We begin with an outline of the analytic function representation[4] of canonical creation and destruction operators. If a and $a^\dagger$ satisfy the relations

$$[a, a^\dagger] = 1,$$

every irreducible representation is equivalent to the Fock representation in terms of the states $\psi(n)$, satisfying

$$a^\dagger a\psi(n) = n\psi(n); \quad (\psi(m), \psi(n)) = \delta_{mn}.$$

The matrix elements of a and $a^\dagger$ in this representation are

$$(\psi(m), a\psi(n)) = \sqrt{n}\,\delta_{m,n-1},$$

$$(\psi(m), a^\dagger\psi(n)) = (n+1)^{1/2}\delta_{m,n+1}.$$

One could, however, introduce an overcomplete set of eigenstates of the destruction operator given by

$$|re^{i\theta}\rangle \equiv |z\rangle = \exp(-\tfrac{1}{2}|z|^2)\sum_{n=0}^{\infty}\frac{z^n}{(n!)^{1/2}}\psi(n), \tag{1}$$

satisfying

$$a|z\rangle = z|z\rangle; \quad \langle z|a^\dagger = z^*\langle z|; \quad \langle z|z\rangle = 1,$$

for every complex number z. These states are all normalized but not orthogonal[5]; they are complete in the sense that they furnish a resolution of the identity

$$1 = (1/\pi)\iint r\,dr\,d\theta\,|re^{i\theta}\rangle\langle re^{i\theta}|.$$

More generally,

$$\int\frac{d\theta}{2\pi}|re^{i\theta}\rangle\langle re^{i\theta}| = e^{-r^2}\sum_{n=0}^{\infty}\frac{r^{2n}}{n!}\psi(n)\psi^\dagger(n). \tag{2}$$

We can make use of the overcompleteness[6] of the states to represent every density matrix,

$$\rho = \sum_{n=0}^{\infty}\sum_{n'=0}^{\infty}\rho(n,n')\psi(n)\psi^\dagger(n'),$$

VOLUME 10, NUMBER 7 PHYSICAL REVIEW LETTERS 1 APRIL 1963

in the "diagonal" form

$$\rho = \sum_{n=0}^{\infty}\sum_{n'=0}^{\infty} \rho(n,n') \frac{(n!n'!)^{1/2}}{(n+n')!} \left\{ \left(\frac{\partial}{\partial r}\right)^{n+n'} \int \frac{d\theta}{2\pi} \exp[r^2 + i(n'-n)\theta] |re^{i\theta}\rangle\langle re^{i\theta}| \right\}\Big|_{r=0}. \tag{3}$$

This form is particularly interesting since if $O = (a^\dagger)^\lambda a^\mu$ be any normal ordered operator (i.e., all creation operators to the left of all annihilation operators), its expectation value in the statistical state represented by the density matrix in the "diagonal" form

$$\rho = \int d^2z\, \phi(z) |z\rangle\langle z| \tag{4}$$

is given by

$$\mathrm{tr}\{\rho O\} = \mathrm{tr}\{\rho (a^\dagger)^\lambda a^\mu\} = \int d^2z\, \phi(z) (z^*)^\lambda z^\mu. \tag{5}$$

This is the same as the expectation value of the complex classical function $(z^*)^\lambda z^\mu$ for a probability distribution $\phi(z)$ over the complex plane. The demonstration above shows that <u>any</u> statistical state of the quantum mechanical system may be described by a classical probability distribution over a complex plane, provided all operators are written in the normal ordered form. In other words, the classical complex representations[1] can be put in one-to-one correspondence with quantum mechanical density matrices. Hermiticity of ρ implies that $\phi(z)$ is a "real" function in the sense that $\phi^*(z^*) = \phi(z)$, but not necessarily positive definite.

These considerations generalize in a straightforward manner to an arbitrary (countable) number of degrees of freedom, finite or infinite.[7] The states are now represented by a sequence of complex numbers $\{z\}$; and the Fock representation basis is labeled by a sequence of non-negative integers $\{n\}$ and density matrices by functions of two such sequences $\rho(\{n\},\{n'\})$. Any such state can be put into one-to-one correspondence with classical probability distributions in a sequence of complex variables $\phi(\{z\})$ such that the expectation value of any normal ordered operator $O(\{a^\dagger\},\{a\})$ is given by

$$\mathrm{tr}\left\{O(\{a^\dagger\},\{a\})\rho\right\} = \prod_\lambda \int d^2z_\lambda\, O(\{z^*\},\{z\})\phi(\{z\}),$$

where the "real" function $\phi(\{z\})$ is given by

$$\phi(\{z\}) = \prod_\lambda \left[\sum_{n_\lambda=0}^{\infty}\sum_{n_\lambda'=0}^{\infty} \frac{\rho(\{n\},\{n'\})(n_\lambda! n_\lambda'!)^{1/2}}{(n_\lambda+n_\lambda')!(2\pi r_\lambda)} \exp[r_\lambda^2 + i(n_\lambda'-n_\lambda)\theta_\lambda] \left\{ \left(-\frac{\partial}{\partial r_\lambda}\right)^{n_\lambda+n_\lambda'} \delta(r_\lambda) \right\} \right]. \tag{6}$$

Consequently the description of statistical states of a quantum mechanical system with an arbitrary (countably infinite) number of degrees of freedom is completely equivalent to the description in terms of classical probability distributions in the same (countably infinite) number of complex variables. In particular, the statistical states of the quantized electromagnetic field may be described uniquely by classical complex linear functions on the classical electromagnetic field. This functional will be "real" reflecting the Hermiticity of the density matrix; and leads in either version to real expectation values for Hermitian (real) dynamical variables.

Several additional remarks are in order. Firstly, since the states $|\{z\}\rangle$ are eigenfunctions of the annihilation operators, the analogous states for a quantized field are eigenfunctions of the positive-frequency (annihilation) part of the field. The corresponding classical theory should then work with positive-frequency parts of the classical field; but this is precisely what is involved in the concept of the (classical) analytic signal.[1] Secondly, while thermal beams are usually represented by Gaussian classical probability functions corresponding to a density matrix diagonal in the occupation numbers $\{n\}$ given by the grand canonical ensemble for the blackbody radiation, there are other probability functions! A particular one may not be diagonal in the occupation number sequence $\{n\}$ and this implies, in accordance with Eq. (6), that not all phase-angle sequences $\{\theta\}$ have equal weight. In such a case the expectation values of operators with unequal number of creation and destruction operators need not all vanish. We note in passing, that Eq. (6) for $\phi(\{z\})$ in terms of $\rho(\{n\},\{n'\})$ can be inverted to yield

$$\rho(\{n\},\{n'\}) = \prod_\lambda \int d^2z_\lambda\, \phi(\{z\}) e^{-|z_\lambda|^2} \frac{z_\lambda^{n_\lambda}(z_\lambda^*)^{n_\lambda'}}{(n_\lambda! n_\lambda')^{1/2}}. \tag{7}$$

If we do this for the Gaussian functions, we obtain the Bose-Einstein distribution, diagonal in the occupation number sequences corresponding to the equal weightage of all phase angles. It is worth pointing out that this result reproduces the Purcell-Mandel derivation[8] for photoelectric counting statistics. The method of inverting the expectation values to obtain the probability distribution as the Fourier transform of the characteristic function can be generalized in the present case, using probability functions and characteristic functionals.[9] The methods developed here are thus adequate to determine the quantum mechanical density matrix, provided all the correlation functions are given.[10]

It is a pleasure to thank Professor Emil Wolf for introducing me to the subject and for his interest in this work. Mr. C. L. Mehta and Mr. N. Mukunda made several helpful suggestions.

---

[1]For a comprehensive review, see M. Born and E. Wolf, Principles of Optics (Pergamon Press, New York, 1959), Chap. X.

[2]See L. Mandel (to be published), for a systematic account.

[3]R. J. Glauber, Phys. Rev. Letters 10, 84 (1963).

[4]V. Bargmann, Commun. Pure Appl. Math. 14, 187 (1961); I. E. Segal (to be published); J. R. Klauder, Ann. Phys. (N.Y.) 11, 123 (1960); S. S. Schweber, J. Math. Phys. 3, 831 (1962).

[5]These states are easily represented in a Schrödinger coordinate representation

$$|0\rangle \equiv \psi(0) \to \pi^{-1/4} \exp(-\tfrac{1}{2}x^2),$$

$$|z\rangle = \exp(za^\dagger)\psi(0) \to \pi^{-1/4} \exp(-\tfrac{1}{2}[x+z]^2),$$

as pointed out by Dr. C. Ryan.

[6]For a general theory of representation by an overcomplete family of states, see J. R. Klauder (to be published).

[7]All strange representations of the infinite canonical ring are deliberately ignored here.

[8]E. M. Purcell, Nature 178, 1449 (1956). L. Mandel, Proc. Phys. Soc. (London) 71, 1037 (1958); 74, 233 (1959). See also reference 2.

[9]I. E. Segal, Com. J. Math. 13, 1 (1961). See also E. Hopf, J. Rat. Mech. Anal. 2, 587 (1953). These have to be generalized here by admitting indefinite linear functionals.

[10]The only case where all correlation functions are known is for the important but familiar example of the blackbody radiation. We hope that this circumstance is not time independent!

PHYSICAL REVIEW VOLUME 131, NUMBER 6 15 SEPTEMBER 1963

Coherent and Incoherent States of the Radiation Field*

ROY J. GLAUBER
Lyman Laboratory of Physics, Harvard University, Cambridge, Massachusetts
(Received 29 April 1963)

Methods are developed for discussing the photon statistics of arbitrary radiation fields in fully quantum-mechanical terms. In order to keep the classical limit of quantum electrodynamics plainly in view, extensive use is made of the coherent states of the field. These states, which reduce the field correlation functions to factorized forms, are shown to offer a convenient basis for the description of fields of all types. Although they are not orthogonal to one another, the coherent states form a complete set. It is shown that any quantum state of the field may be expanded in terms of them in a unique way. Expansions are also developed for arbitrary operators in terms of products of the coherent state vectors. These expansions are discussed as a general method of representing the density operator for the field. A particular form is exhibited for the density operator which makes it possible to carry out many quantum-mechanical calculations by methods resembling those of classical theory. This representation permits clear insights into the essential distinction between the quantum and classical descriptions of the field. It leads, in addition, to a simple formulation of a superposition law for photon fields. Detailed discussions are given of the incoherent fields which are generated by superposing the outputs of many stationary sources. These fields are all shown to have intimately related properties, some of which have been known for the particular case of blackbody radiation.

I. INTRODUCTION

FEW problems of physics have received more attention in the past than those posed by the dual wave-particle properties of light. The story of the solution of these problems is a familiar one. It has culminated in the development of a remarkably versatile quantum theory of the electromagnetic field. Yet, for reasons which are partly mathematical and partly, perhaps, the accident of history, very little of the insight of quantum electrodynamics has been brought to bear on the problems of optics. The statistical properties of photon beams, for example, have been discussed to date almost exclusively in classical or semiclassical terms. Such discussions may indeed be informative, but they inevitably leave open serious questions of self-consistency, and risk overlooking quantum phenomena which have no classical analogs. The wave-particle duality, which should be central to any correct treatment of photon statistics, does not survive the transition to the classical limit. The need for a more consistent theory has led us to begin the development of a fully quantum-mechanical approach to the problems of photon statistics. We have quoted several of the results of this work in a recent note,[1] and shall devote much of the present paper to explaining the background of the material reported there.

Most of the mathematical development of quantum electrodynamics to date has been carried out through the use of a particular set of quantum states for the field. These are the stationary states of the noninteracting field, which corresponds to the presence of a precisely defined number of photons. The need to use these states has seemed almost axiomatic inasmuch as nearly all quantum electrodynamical calculations have been carried out by means of perturbation theory. It is characteristic of electrodynamical perturbation theory that in each successive order of approximation it describes processes which either increase or decrease the number of photons present by one. Calculations performed by such methods have only rarely been able to deal with more than a few photons at a time. The

* Supported in part by the U. S. Air Force Office of Scientific Research under Contract No. AF 49(638)-589.

[1] R. J. Glauber, Phys. Rev. Letters **10**, 84 (1963).

description of the light beams which occur in optics, on the other hand, may require that we deal with states in which the number of photons present is large and intrinsically uncertain. It has long been clear that the use of the usual set of photon states as a basis offers at best only an awkward way of approaching such problems.

We have found that the use of a rather different set of states, one which arises in a natural way in the discussion of correlation and coherence[2,3] properties of fields, offers much more penetrating insights into the role played by photons in the description of light beams. These states, which we have called coherent ones, are of a type that has long been used to illustrate the time-dependent behavior of harmonic oscillators. Since they lack the convenient property of forming an orthogonal set, very little attention has been paid them as a set of basis states for the description of fields. We shall show that these states, though not orthogonal, do form a complete set and that any state of the field may be represented simply and uniquely in terms of them. By suitably extending the methods used to express arbitrary states in terms of the coherent states, we may express arbitrary operators in terms of products of the corresponding state vectors. It is particularly convenient to express the density operator for the field in an expansion of this type. Such expansions have the property that whenever the field possesses a classical limit, they render that limit evident while at the same time preserving an intrinsically quantum-mechanical description of the field.

The earlier sections of the paper are devoted to a detailed introduction of the coherent states and a survey of some of their properties. We then undertake in Secs. IV and V the expansion of arbitrary states and operators in terms of the coherent states. Section VI is devoted to a discussion of the particular properties of density operators and the way these properties are represented in the new scheme. The application of the formalism to physical problems is begun in Sec. VII, where we introduce a particular form for the density operator which seems especially suited to the treatment of radiation by macroscopic sources. This form for the density operator leads to a particularly simple way of describing the superposition of radiation fields. A form of the density operator which corresponds to a very commonly occurring form of incoherence is then discussed in Sec. VIII and shown to be closely related to the density operator for blackbody radiation. In Sec. IX the results established earlier for the treatment of single modes of the radiation field are generalized to treat the entire field. The photon fields generated by arbitrary distributions of classical currents are shown to have an especially simple description in terms of coherent states. Finally, in Sec. X the methods of the preceding sections are illustrated in a discussion of certain forms of coherent and incoherent fields and of their spectra and correlation functions.

II. FIELD-THEORETICAL BACKGROUND

We have, in an earlier paper,[3] discussed the separation of the electric field operator $\mathbf{E}(\mathbf{r}t)$ into its positive-frequency part $\mathbf{E}^{(+)}(\mathbf{r}t)$ and its negative-frequency part $\mathbf{E}^{(-)}(\mathbf{r}t)$. These individual fields were then used to define a succession of correlation functions $G^{(n)}$, the simplest of which takes the form

$$G_{\mu\nu}^{(1)}(\mathbf{r}t,\mathbf{r}'t')=\mathrm{tr}\{\rho E_\mu^{(-)}(\mathbf{r}t)E_\nu^{(+)}(\mathbf{r}'t')\}\,, \qquad (2.1)$$

where ρ is the density operator which describes the field and the symbol tr stands for the trace. We noted, in discussing these functions, that there exist quantum-mechanical states which are eigenstates of the positive- and negative-frequency parts of the fields in the senses indicated by the relations

$$E_\mu^{(+)}(\mathbf{r}t)|\ \rangle=\mathcal{E}_\mu(\mathbf{r}t)|\ \rangle\,, \qquad (2.2)$$

$$\langle\ |E_\mu^{(-)}(\mathbf{r}t)=\mathcal{E}_\mu^*(\mathbf{r}t)\langle\ |\,, \qquad (2.3)$$

in which the function $\mathcal{E}_\mu(\mathbf{r}t)$ plays the role of an eigenvalue. It is possible, as we shall note, to find eigenstates $|\ \rangle$ which correspond to arbitrary choices of the eigenvalue function $\mathcal{E}_\mu(\mathbf{r}t)$, provided they obey the Maxwell equations satisfied by the field operator $E_\mu(\mathbf{r}t)$ and contain only positive frequency terms in their Fourier resolutions.

The importance of the eigenstates defined by Eqs. (2.2) and (2.3) is indicated by the fact that they cause the correlation functions to factorize. If the field is in an eigenstate of this type we have $\rho=|\ \rangle\langle\ |$, and the first-order correlation function therefore reduces to

$$G_{\mu\nu}^{(1)}(\mathbf{r}t,\mathbf{r}'t')=\mathcal{E}_\mu^*(\mathbf{r}t)\mathcal{E}_\nu(\mathbf{r}'t')\,. \qquad (2.4)$$

An analogous separation into a product of $2n$ factors takes place in the nth- order correlation function. The existence of such factorized forms for the correlation functions is the condition we have used to define fully coherent fields. The eigenstates $|\ \rangle$, which we have therefore called the coherent states, have many properties which it will be interesting to study in detail. For this purpose, it will be useful to introduce some of the more directly related elements of quantum electrodynamics.

The electric and magnetic field operators $\mathbf{E}(\mathbf{r}t)$ and $\mathbf{B}(\mathbf{r}t)$ may be derived from the operator $\mathbf{A}(\mathbf{r}t)$, which represents the vector potential, via the relations

$$\mathbf{E}=-\frac{1}{c}\frac{\partial\mathbf{A}}{\partial t},\quad \mathbf{B}=\nabla\times\mathbf{A}. \qquad (2.5)$$

We shall find it convenient, in discussing the quantum states of the field, to describe the field by means of a discrete succession of dynamical variables rather than

[2] R. J. Glauber, in Proceedings of the Third International Conference on Quantum Electronics, Paris, France, 1963 (to be published).

[3] R. J. Glauber, Phys. Rev. **130**, 2529 (1963).

a continuum of them. For this reason we assume that the field we are discussing is confined within a spatial volume of finite size, and expand the vector potential within that volume in an appropriate set of vector mode functions. The amplitudes associated with these oscillation modes then form a discrete set of variables whose dynamical behavior is easily discussed.

The most convenient choice of a set of mode functions, $\mathbf{u}_k(\mathbf{r})$, is usually determined by physical considerations which have little direct bearing on our present work. In particular, we need not specify the nature of the boundary conditions for the volume under study; they may be either the periodic boundary conditions which lead to traveling wave modes, or the conditions appropriate to reflecting surfaces which lead to standing waves. If the volume contains no refracting materials, the mode function $\mathbf{u}_k(\mathbf{r})$, which corresponds to frequency ω_k, may be taken to satisfy the wave equation

$$\nabla^2\mathbf{u}_k+\frac{\omega_k^2}{c^2}\mathbf{u}_k=0 \qquad (2.6)$$

at interior points. More generally, whatever the form of the wave equation or the boundary conditions may be, we shall assume that the mode functions form a complete set which satisfies the orthonormality condition

$$\int \mathbf{u}_k^*(\mathbf{r})\cdot\mathbf{u}_l(\mathbf{r})d\mathbf{r}=\delta_{kl}, \qquad (2.7)$$

and the transversality condition

$$\nabla\cdot\mathbf{u}_k(\mathbf{r})=0. \qquad (2.8)$$

The plane-wave mode functions appropriate to a cubical volume of side L may be written as

$$\mathbf{u}_k(\mathbf{r})=L^{-3/2}\hat{e}^{(\lambda)}\exp(i\mathbf{k}\cdot\mathbf{r}), \qquad (2.9)$$

where $\hat{e}^{(\lambda)}$ is a unit polarization vector. This example illustrates the way in which the mode index k may represent an abbreviation for several discrete variables, i.e., in this case the polarization index ($\lambda=1,2$) and the three Cartesian components of the propagation vector $\mathbf{k}$. The polarization vector $\hat{e}^{(\lambda)}$ is required to be perpendicular to $\mathbf{k}$ by the condition (2.8), and the permissible values of $\mathbf{k}$ are determined in a familiar way by means of periodic boundary conditions.

The expansion we shall use for the vector potential takes the form

$$\mathbf{A}(\mathbf{r}t)=c\sum_k\left(\frac{\hbar}{2\omega_k}\right)^{1/2} \times(a_k\mathbf{u}_k(\mathbf{r})e^{-i\omega_kt}+a_k^\dagger\mathbf{u}_k^*(\mathbf{r})e^{i\omega_kt}), \qquad (2.10)$$

in which the normalization factors have been chosen to render dimensionless the pair of complex-conjugate amplitudes a_k and $a_k^\dagger$. In the classical form of electromagnetic theory these Fourier amplitudes are complex numbers which may be chosen arbitrarily but remain constant in time when no charges or currents are present. In quantum electrodynamics, on the other hand, these amplitudes must be regarded as mutually adjoint operators. The amplitude operators, as we have defined them, will likewise remain constant when no field sources are active in the system studied.

The dynamical behavior of the field amplitudes is governed by the electromagnetic Hamiltonian which, in rationalized units, takes the form

$$H=\tfrac{1}{2}\int(\mathbf{E}^2+\mathbf{B}^2)d\mathbf{r}. \qquad (2.11)$$

With the use of Eqs. (2.7,8) and of a suitable set of boundary conditions on the mode functions, the Hamiltonian may be reduced to the form

$$H=\tfrac{1}{2}\sum_k\hbar\omega_k(a_k^\dagger a_k+a_ka_k^\dagger). \qquad (2.12)$$

This expression is the source of a well-known and extremely fruitful analogy between the mode amplitudes of the field and the coordinates of an assembly of one-dimensional harmonic oscillators. The quantum mechanical properties of the amplitude operators a_k and $a_k^\dagger$ may be described completely by adopting for them the commutation relations familiar from the example of independent harmonic oscillators:

$$[a_k,a_{k'}]=[a_k^\dagger,a_{k'}^\dagger]=0, \qquad (2.13a)$$

$$[a_k,a_{k'}^\dagger]=\delta_{kk'}. \qquad (2.13b)$$

Having thus separated the dynamical variables of the different modes, we are now free to discuss the quantum states of the modes independently of one another. Our knowledge of the state of each mode may be described by a state vector $|\ \rangle_k$ in a Hilbert space appropriate to that mode. The states of the entire field are then defined in the product space of the Hilbert spaces for all of the modes.

To discuss the quantum states of the individual modes we need only be familiar with the most elementary aspects of the treatment of a single harmonic oscillator. The Hamiltonian $\frac{1}{2}\hbar\omega_k(a_k^\dagger a_k+a_ka_k^\dagger)$ has eigenvalues $\hbar\omega_k(n_k+\frac{1}{2})$, where n_k is an integer ($n_k=0,1,2\cdots$). The state vector for the ground state of the oscillator will be written as $|\ \rangle_k$. It is defined by the condition

$$a_k|0\rangle_k=0. \qquad (2.14)$$

The state vectors for the excited states of the oscillator may be obtained by applying integral powers of the operator $a_k^\dagger$ to $|0\rangle_k$. These states are written in normalized form as

$$|n_k\rangle_k=\frac{(a_k^\dagger)^{n_k}}{(n_k!)^{1/2}}|0\rangle_k, \quad (n_k=0,1,2\cdots). \qquad (2.15)$$

The way in which the operators a_k and $a_k^\dagger$ act upon these states is indicated by the relations

$$a_k|n_k\rangle_k = n_k^{1/2}|n_k-1\rangle_k, \tag{2.16}$$

$$a_k^\dagger|n_k\rangle = (n_k+1)^{1/2}|n_k+1\rangle_k, \tag{2.17}$$

$$a_k^\dagger a_k|n_k\rangle = n_k|n_k\rangle. \tag{2.18}$$

With these preliminaries completed we are now ready to discuss the coherent states of the field in greater detail. The expansion (2.10) for the vector potential exhibits its positive frequency part as the sum containing the photon annihilation operators a_k and its negative frequency part as that involving the creation operators $a_k^\dagger$. The positive frequency part of the electric field operator is thus given, according to (2.10), by

$$\mathbf{E}^{(+)}(\mathbf{r}t) = i\sum_k (\tfrac{1}{2}\hbar\omega_k)^{1/2} a_k \mathbf{u}_k(\mathbf{r}) e^{-i\omega_k t}. \tag{2.19}$$

The eigenvalue functions $\boldsymbol{\mathcal{E}}(\mathbf{r}t)$ defined by Eq. (2.2) must clearly satisfy the Maxwell equations, just as the operator $\mathbf{E}^{(+)}(\mathbf{r}t)$ does. They therefore possess an expansion in normal modes similar to Eq. (2.19). In other words we may introduce a set of c-number Fourier coefficients α_k which permit us to write the eigenvalue function as

$$\boldsymbol{\mathcal{E}}(\mathbf{r}t) = i\sum_k (\tfrac{1}{2}\hbar\omega_k)^{1/2} \alpha_k \mathbf{u}_k(\mathbf{r}) e^{-i\omega_k t}. \tag{2.20}$$

Since the mode functions $\mathbf{u}_k(\mathbf{r})$ form an orthogonal set, it then follows that the eigenstate $|\ \rangle$ for the field obeys the infinite succession of relations

$$a_k|\ \rangle = \alpha_k|\ \rangle, \tag{2.21}$$

for all modes k. To find the states which satisfy these relations we seek states, $|\alpha_k\rangle_k$, of the individual modes which individually obey the relations

$$a_k|\alpha_k\rangle_k = \alpha_k|\alpha_k\rangle_k. \tag{2.22}$$

The coherent states $|\ \rangle$ of the field, considered as a whole, are then seen to be direct products of the individual states $|\alpha_k\rangle$,

$$|\ \rangle = \prod_k |\alpha_k\rangle_k. \tag{2.23}$$

III. COHERENT STATES OF A SINGLE MODE

The next few sections will be devoted to discussing the description of a single mode oscillator. We may therefore simplify the notation a bit by dropping the mode index k as a subscript to the state vector and to the amplitude parameters and operators. To find the oscillator state $|\alpha\rangle$ which satisfies

$$a|\alpha\rangle = \alpha|\alpha\rangle, \tag{3.1}$$

we begin by taking the scalar product of both sides of the equation with the nth excited state, $\langle n|$. By using the Hermitian adjoint form of the relation (2.17), we find the recursion relation

$$(n+1)^{1/2}\langle n+1|\alpha\rangle = \alpha\langle n|\alpha\rangle \tag{3.2}$$

for the scalar products $\langle n|\alpha\rangle$. We immediately find from the recursion relation that

$$\langle n|\alpha\rangle = \frac{\alpha^n}{(n!)^{1/2}}\langle 0|\alpha\rangle. \tag{3.3}$$

These scalar products are the expansion coefficients of the state $|\alpha\rangle$ in terms of the complete orthonormal set $|n\rangle$ $(n=0, 1, \cdots)$. We thus have

$$|\alpha\rangle = \sum_n |n\rangle\langle n|\alpha\rangle = \langle 0|\alpha\rangle \sum_n \frac{\alpha^n}{(n!)^{1/2}}|n\rangle. \tag{3.4}$$

The squared length of the vector $|\alpha\rangle$ is thus

$$\langle\alpha|\alpha\rangle = |\langle 0|\alpha\rangle|^2 \sum_n \frac{|\alpha|^{2n}}{n!} = |\langle 0|\alpha\rangle|^2 e^{|\alpha|^2}. \tag{3.5}$$

If the state $|\alpha\rangle$ is normalized so that $\langle\alpha|\alpha\rangle = 1$ we may evidently define its phase by choosing

$$\langle 0|\alpha\rangle = e^{-\frac{1}{2}|\alpha|^2}. \tag{3.6}$$

The coherent states of the oscillator therefore take the forms

$$|\alpha\rangle = e^{-\frac{1}{2}|\alpha|^2}\sum_n \frac{\alpha^n}{(n!)^{1/2}}|n\rangle \tag{3.7}$$

and

$$\langle\alpha| = e^{-\frac{1}{2}|\alpha|^2}\sum_n \frac{(\alpha^*)^n}{(n!)^{1/2}}\langle n|. \tag{3.8}$$

These forms show that the average occupation number of the nth state is given by a Poisson distribution with mean value $|\alpha|^2$,

$$|\langle n|\alpha\rangle|^2 = \frac{|\alpha|^{2n}}{n!}e^{-|\alpha|^2}. \tag{3.9}$$

They also show that the coherent state $|\alpha\rangle$ corresponding to $\alpha=0$ is the unique ground state of the oscillator, i.e., the state $|n\rangle$ for $n=0$.

An alternative approach to the coherent states will also prove quite useful in the work to follow. For this purpose we assume that there exists a unitary operator D which acts as a displacement operator upon the amplitudes $a^\dagger$ and a. We let D be a function of a complex parameter β, and require that it displace the amplitude operators according to the scheme

$$D^{-1}(\beta)aD(\beta) = a+\beta, \tag{3.10}$$

$$D^{-1}(\beta)a^\dagger D(\beta) = a^\dagger+\beta^*. \tag{3.11}$$

Then if $|\alpha\rangle$ obeys Eq. (3.1), it follows that $D^{-1}(\beta)|\alpha\rangle$ is an eigenstate of a corresponding to the eigenvalue $\alpha-\beta$,

$$aD^{-1}(\beta)|\alpha\rangle=(\alpha-\beta)D^{-1}(\beta)|\alpha\rangle. \quad (3.12)$$

In particular, if we choose $\beta=\alpha$, we find

$$aD^{-1}(\alpha)|\alpha\rangle=0.$$

Since the ground state of the oscillator is uniquely defined by the relation (2.14), it follows that $D^{-1}(\alpha)|\alpha\rangle$ is just the ground state, $|0\rangle$. The coherent states, in other words, are just displaced forms of the ground state of the oscillator,

$$|\alpha\rangle=D(\alpha)|0\rangle. \quad (3.13)$$

To find an explicit form for the displacement operator $D(\alpha)$, we begin by considering infinitesimal displacements in the neighborhood of $D(0)=1$. For arbitrary displacements $d\alpha$, we see easily from the commutation rules (2.13) that $D(d\alpha)$ may be chosen to have the form

$$D(d\alpha)=1+a^\dagger d\alpha-a d\alpha^*, \quad (3.14)$$

which holds to first order in $d\alpha$. To formulate a simple differential equation obeyed by the unknown operator we consider increments of α of the form $d\alpha=\alpha d\lambda$ where λ is a real parameter. Then if we assume the operators D to possess the group multiplication property

$$D(\alpha(\lambda+d\lambda))=D(\alpha d\lambda)D(\alpha\lambda), \quad (3.15)$$

we find the differential equation

$$\frac{d}{d\lambda}D(\alpha\lambda)=(\alpha a^\dagger-\alpha^* a)D(\alpha\lambda), \quad (3.16)$$

whose solution, evaluated for $\lambda=1$, is the unitary operator

$$D(\alpha)=e^{\alpha a^\dagger-\alpha^* a}. \quad (3.17)$$

The coherent states $|\alpha\rangle$ may therefore be written in the form

$$|\alpha\rangle=e^{\alpha a^\dagger-\alpha^* a}|0\rangle \quad (3.18)$$

which is correctly normalized since $D(\alpha)$ is unitary.

It is interesting to discuss the relationship between the two forms we have derived for the coherent states. For this purpose we invoke a simple theorem on the multiplication of exponential functions of operators. If $\mathcal{A}$ and $\mathcal{B}$ are any two operators, whose commutator $[\mathcal{A},\mathcal{B}]$ commutes with each of them,

$$[[\mathcal{A},\mathcal{B}],\mathcal{A}]=[[\mathcal{A},\mathcal{B}],\mathcal{B}]=0, \quad (3.19)$$

it may be shown[4] that

$$\exp(\mathcal{A})\exp(\mathcal{B})=\exp\{\mathcal{A}+\mathcal{B}+\tfrac{1}{2}[\mathcal{A},\mathcal{B}]\}. \quad (3.20)$$

If we write $\mathcal{A}=a^\dagger$ and $\mathcal{B}=a$, this theorem permits us to resolve the exponential $D(\alpha)$ given by Eq. (3.17) into the product

$$D(\alpha)=e^{-\frac{1}{2}|\alpha|^2}e^{\alpha a^\dagger}e^{-\alpha^* a}. \quad (3.21)$$

Products of this type, which have been ordered so that the annihilation operators all stand to the right of the creation operators, will be said to be in normal form. Their convenience is indicated by the fact that the exponential $\exp[-\alpha^* a]$, when applied to the ground state $|0\rangle$, reduces in effect to unity, i.e., we have

$$e^{-\alpha^* a}|0\rangle=|0\rangle, \quad (3.22)$$

since the exponential may be expanded in series and the definition (2.14) of the ground state applied. It follows then that the coherent states may be written as

$$|\alpha\rangle=D(\alpha)|0$$

$$=e^{-\frac{1}{2}|\alpha|^2}e^{\alpha a^\dagger}|0\rangle \quad (3.23)$$

$$=e^{-\frac{1}{2}|\alpha|^2}\sum_n \frac{(\alpha a^\dagger)^n}{n!}|0\rangle. \quad (3.24)$$

Since the excited states of the oscillator are given by $|n\rangle=(n!)^{-1/2}(a^\dagger)^n|0\rangle$, we have once again derived the expression

$$|\alpha\rangle=e^{-\frac{1}{2}|\alpha|^2}\sum_n \frac{\alpha^n}{n!}|n\rangle.$$

It may help in visualizing the coherent states if we discuss the form they take in coordinate space and in momentum space. We therefore introduce a pair of Hermitian operators q and p to represent, respectively, the coordinate of the mode oscillator and its momentum. These operators, which must satisfy the canonical commutation relation, $[q,p]=i\hbar$, may be defined for our purposes by the familiar expressions

$$q=(\hbar/2\omega)^{1/2}(a^\dagger+a), \quad (3.25a)$$

$$p=i(\hbar\omega/2)^{1/2}(a^\dagger-a). \quad (3.25b)$$

To find the expectation value of q and p in the coherent states we need only use Eq. (3.1), which defines these states, and its corresponding Hermitian adjoint form. We have then

$$\langle\alpha|q|\alpha\rangle=(2\hbar/\omega)^{1/2}\,\mathrm{Re}\,\alpha, \quad (3.26a)$$

$$\langle\alpha|p|\alpha\rangle=(2\hbar\omega)^{1/2}\,\mathrm{Im}\,\alpha, \quad (3.26b)$$

where $\mathrm{Re}\,\alpha$ and $\mathrm{Im}\,\alpha$ stand for the real and imaginary parts of α.

To find the wave functions for the coherent states, we write the defining equation (3.1) in the form

$$(2\hbar\omega)^{-1/2}(\omega q+ip)|\alpha\rangle=\alpha|\alpha\rangle, \quad (3.27)$$

and take the scalar product of both members with the conjugate state $\langle q'|$, which corresponds to the eigenvalue q' for q. Since the momentum may be represented by a derivative operator, i.e., $\langle q'|p=-i\hbar(d/dq')\langle q'|$, we find that the coordinate space wave function, $\langle q'|\alpha\rangle$,

[4] A. Messiah, *Quantum Mechanics* (North-Holland Publishing Company, Amsterdam, 1961), Vol. I, p. 442.

obeys the differential equation

$$\frac{d}{dq'}\langle q'|\alpha\rangle = -2\left(\frac{\omega}{2\hbar}\right)^{1/2}\left\{\left(\frac{\omega}{2\hbar}\right)^{1/2} q' - \alpha\right\}\langle q'|\alpha\rangle. \quad (3.28)$$

The equation may be integrated immediately to yield a solution for the wave function which, in normalized form, is

$$\langle q'|\alpha\rangle = (\omega/\pi\hbar)^{1/4}\exp\{-[(\omega/2\hbar)^{1/2}q' - \alpha]^2\}. \quad (3.29)$$

An analogous argument furnishes the momentum space wave function. If we take the scalar product of Eq. (3.27) with a momentum eigenstate $\langle p'|$, and use the relation $\langle p'|q = i\hbar(\partial/\partial p')\langle p'|$, we reach a differential equation whose normalized solution is

$$\langle p'|\alpha\rangle = (\pi\hbar\omega)^{-1/4}\exp\{-[(2\hbar\omega)^{-1/2}p' + i\alpha]^2\}. \quad (3.30)$$

Both of these wave functions are simply displaced forms of the ground-state wave function of the oscillator. The parameters $(\hbar/\omega)^{1/2}$ and $(\hbar\omega)^{1/2}$ correspond to the amplitudes of the zero-point fluctuations of the coordinate and momentum, respectively, for an oscillator of unit mass. The fact that the wave functions for the coherent states have this elementary structure should be no surprise in view of the way they are generated in Eq. (3.13), by means of displacements in the complex α plane.

The time-independent states $|\alpha\rangle$ which we have been describing are those characteristic of the Heisenberg picture of quantum mechanics. The Schrödinger picture, alternatively, would make use of the time-dependent states $\exp(-iHt/\hbar)|\alpha\rangle$. If we omit the zero-point energy $\frac{1}{2}\hbar\omega$ from the oscillator Hamiltonian and write $H = \hbar\omega a^\dagger a$, it is then clear from the expansion (3.7) for $|\alpha\rangle$ that the corresponding Schrödinger state takes the same form with α replaced by $\alpha e^{-i\omega t}$. We may thus write the Schrödinger state as $|\alpha e^{-i\omega t}\rangle$. With the substitution of $\alpha e^{-i\omega t}$ for α in Eqs. (3.26a) and (3.26b), we see that the expectation values of the coordinate and momentum carry out a simple harmonic motion with coordinate amplitude $(2\hbar/\omega)^{1/2}|\alpha|$. The same substitutions in the wave functions (3.29) and (3.30) show that the Gaussian probability densities characteristic of the ground state of the oscillator are simply carried back and forth in the same motion as the expectation values. Such wave packets are, of course, quite familiar; they were introduced to quantum mechanics at a very early stage by Schrödinger,[5] and have often been used to illustrate the way in which the behavior of the oscillator approaches the classical limit.

Another connection in which the wave packets (3.29) and (3.30) have been discussed in the past has to do with the particular way in which they localize the coordinate q' and the momentum p'. Wave packets can, of course, be found which localize either variable more sharply, but only at the expense of the localization of the other. There is a sense in which the wave packets (3.29) and (3.30) furnish a unique compromise; they minimize the product of the uncertainties of the variables q' and p'. If we represent expectation values by means of the angular brackets $\langle\ \rangle$ and define the variances

$$(\Delta q)^2 = \langle q^2\rangle - \langle q\rangle^2, \quad (3.31a)$$

$$(\Delta p)^2 = \langle p^2\rangle - \langle p\rangle^2, \quad (3.31b)$$

we find, for the wave functions (3.29) and (3.30), that the product of the variances is

$$(\Delta p)^2(\Delta q)^2 = \tfrac{1}{4}\hbar^2.$$

According to the uncertainty principle, this is the minimum value such a product can have.[6] There thus exists a particular sense in which the description of an oscillator by means of the wave functions (3.29) and (3.30) represents as close an approach to classical localization as is possible.

The uses we shall make of the coherent states in quantum electrodynamics will not, in fact, require the explicit introduction of coordinate or momentum variables. We have reviewed the familiar representations of the coherent states in terms of these variables in the hope that they may be of some help in understanding the various applications of the states which we shall shortly undertake.

One property of the states $|\alpha\rangle$ which is made clear by the wave-function representations is that two such states are not, in general, orthogonal to one another. If we consider, for example, the wave functions $\langle q'|\alpha\rangle$ and $\langle q'|\alpha'\rangle$ for values of α' close to α, it is evident that the functions are similar in form and overlap one another appreciably. For values of α' quite different from α, however, the overlap is at most quite small. We may therefore expect that the scalar product $\langle\alpha|\alpha'\rangle$, which is unity for $\alpha' = \alpha$, will tend to decrease in absolute magnitude as α' and α recede from one another in the complex plane. The scalar product may, in fact, be calculated more simply than by using wave functions if we employ the representations (3.7) and (3.8). We then find

$$\langle\alpha|\beta\rangle = e^{-\frac{1}{2}|\alpha|^2 - \frac{1}{2}|\beta|^2}\sum_{n,m}\frac{(\alpha^*)^n\beta^m}{(n!m!)^{1/2}}\langle n|m\rangle,$$

which, in view of the orthonormality of the $|n\rangle$ states, reduces to

$$\langle\alpha|\beta\rangle = \exp\{\alpha^*\beta - \tfrac{1}{2}|\alpha|^2 - \tfrac{1}{2}|\beta|^2\}. \quad (3.32)$$

The absolute magnitude of the scalar product is given by

$$|\langle\alpha|\beta\rangle|^2 = \exp\{-|\alpha-\beta|^2\}, \quad (3.33)$$

[5] E. Schrödinger, Naturwissenschaften **14**, 664 (1926). For a more recent treatment see L. I. Schiff, *Quantum Mechanics* (McGraw-Hill Book Company, Inc., New York, 1955), 2nd ed., p. 67.

[6] W. Heisenberg, *The Physical Principles of the Quantum Theory* (University of Chicago Press, Chicago, 1930, reprinted by Dover Publications, Inc., New York, 1930), pp. 16–19.

which shows that the coherent states tend to become approximately orthogonal for values of α and β which are sufficiently different. The fact that these states are not even approximately orthogonal for $|\alpha-\beta|$ of order unity may be regarded as an expression of the overlap caused by the presence of the displaced zero-point fluctuations.

Since the coherent states do not form an orthogonal set, they appear to have received little attention as a possible system of basis vectors for the expansion of arbitrary states.[7] We shall show in the following section that such expansions can be carried out conveniently and uniquely and that they possess exceedingly useful properties. In later sections we shall, by generalizing the procedure to deal with bilinear combinations of states $|\alpha\rangle$ and $\langle\beta|$, develop analogous expansions for operators[1] as well.

IV. EXPANSION OF ARBITRARY STATES IN TERMS OF COHERENT STATES

While orthogonality is a convenient property for a set of basis states it is not a necessary one. The essential property of such a set is that it be complete. The set of coherent states $|\alpha\rangle$ for a mode oscillator can be shown without difficulty to form a complete set. To give a proof we need only demonstrate that the unit operator may be expressed as a suitable sum or an integral, over the complex α plane, of projection operators of the form $|\alpha\rangle\langle\alpha|$. In order to describe such integrals we introduce the differential element of area in the α plane

$$d^2\alpha = d(\mathrm{Re}\,\alpha)d(\mathrm{Im}\,\alpha) \tag{4.1}$$

(i.e., $d^2\alpha$ is real). If we write $\alpha=|\alpha|e^{i\vartheta}$, we may easily prove the integral identity

$$\begin{aligned}\int (\alpha^*)^n\alpha^m e^{-|\alpha|^2}d^2\alpha &= \int_0^\infty |\alpha|^{n+m+1}e^{-|\alpha|^2}d|\alpha| \int_0^{2\pi} e^{i(m-n)\vartheta}d\vartheta \\ &= \pi n!\delta_{nm},\end{aligned} \tag{4.2}$$

in which the integration is carried out, as indicated, over the entire area of the complex plane. With the aid of this identity and the expansions (3.7,8) for the coherent states, we may immediately show

$$\int |\alpha\rangle\langle\alpha|d^2\alpha = \pi\sum_n |n\rangle\langle n|\,.$$

Since the n-quantum states are known to form a complete orthonormal set, the indicated sum over n is simply the unit operator. We have thus shown[1]

$$\frac{1}{\pi}\int |\alpha\rangle\langle\alpha|d^2\alpha = 1\,, \tag{4.3}$$

which is a completeness relation for the coherent states of precisely the type desired.

An arbitrary state of an oscillator must possess an expansion in terms of the n-quantum states of the form

$$\begin{aligned}|\ \rangle &= \sum_n c_n|n\rangle\,, \\ &= \sum_n c_n\frac{(a^\dagger)^n}{(n!)^{1/2}}|0\rangle\,,\end{aligned} \tag{4.4}$$

where $\sum|c_n|^2=1$. The series which occurs in Eq. (4.4) may be used to define a function f of a complex variable z,

$$f(z) = \sum c_n\frac{z^n}{(n!)^{1/2}}\,. \tag{4.5}$$

It is clear from the normalization condition on the c_n that this series converges for all finite z, and thus represents a function which is analytic throughout the finite complex plane. We shall speak of the functions $f(z)$ for which $\sum|c_n|^2=1$ as the set of normalized entire functions. There is evidently a one-to-one correspondence which exists between such entire functions and the states of the oscillator. One way of approaching the description of the oscillator is to regard the functions $f(z)$ themselves as the elements of a Hilbert space. The properties of this space and of expansions carried out in it have been studied in some detail by Segal[8] and Bargmann.[9] The method we shall use for expanding arbitrary states in terms of the coherent states has been developed as a simple generalization of the usual method for carrying out changes of basis states in quantum mechanics. It is evidently equivalent, however, to one of the expansions stated by Bargmann.

If we designate the arbitrary state which corresponds to the function $f(z)$ by $|f\rangle$, then we may rewrite Eq. (4.4) as

$$|f\rangle = f(a^\dagger)|0\rangle\,. \tag{4.6}$$

To secure the expansion of $|f\rangle$ in terms of the states $|\alpha\rangle$, we multiply $|f\rangle$ by the representation (4.3) of the unit operator. We then find

$$|f\rangle = \frac{1}{\pi}\int |\alpha\rangle\langle\alpha|f(a^\dagger)|0\rangle d^2\alpha\,,$$

---

[7] Uses of these states as generating functions for the n-quantum states have, however, been made by J. Schwinger, Phys. Rev. **91**, 728 (1953).

[8] I. E. Segal, Illinois J. Math. **6**, 520 (1962).

[9] V. Bargmann, Commun. Pure and Appl. Math. **14**, 187 (1961); Proc. Natl. Acad. Sci. U. S. **48**, 199 (1962).

which reduces, since $\langle\alpha|f(a^\dagger)=\langle\alpha|f(\alpha^*)$, to

$$|f\rangle=\frac{1}{\pi}\int|\alpha\rangle f(\alpha^*)e^{-\frac{1}{2}|\alpha|^2}d^2\alpha\,, \qquad (4.7)$$

which is an expansion of the desired type.

It is worth noting that the expansion (4.7) can easily be inverted to furnish an explicit form for the function $f(\alpha^*)$ which corresponds to any vector $|f\rangle$. For this purpose we take the scalar product of both sides of Eq. (4.7) with the coherent state $\langle\beta|$, and then, using Eq. (3.32), evaluate the scalar product $\langle\beta|\alpha\rangle$ to find

$$\langle\beta|f\rangle=\frac{1}{\pi}e^{-\frac{1}{2}|\beta|^2}\int e^{\beta^*\alpha-|\alpha|^2}f(\alpha^*)d^2\alpha. \qquad (4.8)$$

Since $f(\alpha^*)$ may be expanded in a convergent power series we note the relation

$$\frac{1}{\pi}\int e^{\beta^*\alpha-|\alpha|^2}(\alpha^*)^n d^2\alpha=(\beta^*)^n\,, \qquad (4.9)$$

from which we may derive the more general identity

$$\frac{1}{\pi}\int e^{\beta^*\alpha-|\alpha|^2}f(\alpha^*)d^2\alpha=f(\beta^*). \qquad (4.10)$$

On substituting the latter identity in Eq. (4.8) we find

$$f(\beta^*)=e^{\frac{1}{2}|\beta|^2}\langle\beta|f\rangle. \qquad (4.11)$$

There is thus a unique correspondence between functions $f(\alpha^*)$ which play the role of expansion amplitudes in Eq. (4.7) and the vectors $|f\rangle$ which describe the state of the oscillator.

An expansion analogous to Eq. (4.7) also exists for the adjoint state vectors. If we let $g(\alpha^*)$ be an entire function of α^* we may construct for the state $\langle g|$ the expansion

$$\langle g|=\frac{1}{\pi}\int[g(\beta^*)]^*\langle\beta|e^{-\frac{1}{2}|\beta|^2}d^2\beta. \qquad (4.12)$$

The scalar product of the two states $\langle g|$ and $|f\rangle$ may then be expressed as

$$\langle g|f\rangle=\pi^{-2}\int[g(\beta^*)]^*f(\alpha^*)\exp\{\beta^*\alpha-|\alpha|^2-|\beta|^2\}d^2\alpha d^2\beta.$$

The identity (4.10) permits us to carry out the integration over the variable α to find

$$\langle g|f\rangle=\frac{1}{\pi}\int[g(\beta^*)]^*f(\beta^*)e^{-|\beta|^2}d^2\beta. \qquad (4.13)$$

This expression for the scalar product of two vectors is, in essence, the starting point used by Bargmann in his discussion[10] of the Hilbert space of functions $f(z)$.

It may be worth noting, for its mathematical interest, that the coherent states $|\alpha\rangle$ are not linearly independent of one another, as the members of a complete orthogonal set would be. Thus, for example, the expansion (4.7) may be used to express any given coherent state linearly in terms of all of the others, i.e., in view of Eqs. (4.11) and (3.32) we may write

$$|\alpha\rangle=\frac{1}{\pi}\int|\beta\rangle e^{\beta^*\alpha-\frac{1}{2}|\alpha|^2-\frac{1}{2}|\beta|^2}d^2\beta. \qquad (4.14)$$

There exist many other types of linear dependence among the states $|\alpha\rangle$. We may, for example, note the identity

$$\int|\alpha\rangle\alpha^n e^{-\frac{1}{2}|\alpha|^2}d^2\alpha=0\,, \qquad (4.15)$$

which holds for all integral $n>0$. It is clear from the latter result that if we admitted as expansion coefficients in Eq. (4.7) more general functions than $f(\alpha^*)$, say functions $F(\alpha,\alpha^*)$, there would be many additional ways of expanding any state in terms of coherent states. The constraint implicit in Eq. (4.7), that the expansion function must depend analytically upon the variable α^* is what renders the expansion unique. The virtue of an expansion scheme in which the coefficients are uniquely determined is evident. It becomes possible, by inverting the expansion as in Eq. (4.11), to construct an explicit solution for the expansion coefficient of any state, no matter what representation it was expressed in initially.

V. EXPANSION OF OPERATORS IN TERMS OF COHERENT STATE VECTORS

Our knowledge of the condition of an oscillator mode is rarely explicit enough in practice to permit the specification of its quantum state. Instead, we must describe it in terms of a mixture of states which is expressed by means of a density operator. The same reasons that lead us to express arbitrary states in terms of the coherent states, therefore, suggest that we develop an expansion for the density operator in terms of these states as well. We shall begin by considering in the present section a rather more general class of operators and then specialize to the case of the density operator in the section which follows.

A general quantum mechanical operator T may be expressed in terms of its matrix elements connecting states with fixed numbers of quanta as

$$T=\sum_{n,m}|n\rangle T_{nm}\langle m|\,, \qquad (5.1)$$

$$=\sum T_{nm}(n!m!)^{-1/2}(a^\dagger)^n|0\rangle\langle 0|a^m. \qquad (5.2)$$

[10] Some of Bargmann's arguments are summarized by S. Schweber, J. Math. Phys. 3, 831 (1962), who has used them in connection with the formulation of quantum mechanics in terms of Feynman amplitudes. We are indebted to Dr. S. Bergmann for calling this reference to our attention.

If we use this expression for T to calculate the matrix element which connects the two coherent states $\langle\alpha|$ and $\langle\beta|$ we find

$$\langle\alpha|T|\beta\rangle=\sum_{n,m} T_{nm}(n!m!)^{-1/2}(\alpha^*)^n\beta^m\langle\alpha|0\rangle\langle 0|\beta\rangle. \quad (5.3)$$

It is evidently convenient to define a function $\mathcal{T}(\alpha^*,\beta)$ as

$$\mathcal{T}(\alpha^*,\beta)=\sum_{n,m} T_{nm}(n!m!)^{-1/2}(\alpha^*)^n\beta^m. \quad (5.4)$$

The operators which occur in quantum mechanics are often unbounded ones such as those of Eqs. (2.16)–(2.18). Those operators and the others we are apt to encounter have the property that the magnitudes of the matrix elements T_{nm} are dominated by an expression of the form Mn^jm^k for some fixed positive values of M, j, and k. It then follows that the double series (5.4) converges throughout the finite α^* and β planes and represents an entire function of both variables.

To secure the expansion of the operator T in terms of the coherent states, we may use the representation (4.3) of the unit operator to write

$$T=\frac{1}{\pi^2}\int|\alpha\rangle\langle\alpha|T|\beta\rangle\langle\beta|d^2\alpha d^2\beta, \quad (5.5)$$

$$=\frac{1}{\pi^2}\int|\alpha\rangle\mathcal{T}(\alpha^*,\beta)\langle\beta|\langle\alpha|0\rangle\langle 0|\beta\rangle d^2\alpha d^2\beta,$$

$$=\frac{1}{\pi^2}\int|\alpha\rangle\mathcal{T}(\alpha^*,\beta)\langle\beta|\exp\{-\tfrac{1}{2}|\alpha|^2-\tfrac{1}{2}|\beta|^2\}d^2\alpha d^2\beta. \quad (5.6)$$

The inversion of this expansion, or the solution for $\mathcal{T}(\alpha^*,\beta)$, is accomplished by the same method we used to invert Eq. (4.7) and secure the amplitude function (4.11). The result of the inversion is

$$\mathcal{T}(\alpha^*,\beta)=\langle\alpha|T|\beta\rangle\exp\{\tfrac{1}{2}|\alpha|^2+\tfrac{1}{2}|\beta|^2\}. \quad (5.7)$$

We see, thus, that the expansion of operators, as well as of arbitrary quantum states, in terms of the coherent states is a unique one.

The law of operator multiplication is easily expressed in terms of the functions $\mathcal{T}$. If $T=T_1T_2$ and $\mathcal{T}_1$ and $\mathcal{T}_2$ are the functions appropriate to the latter two operators, we note that

$$\langle\alpha|T|\beta\rangle=\langle\alpha|T_1T_2|\beta\rangle$$

$$=\frac{1}{\pi}\int\langle\alpha|T_1|\gamma\rangle\langle\gamma|T_2|\beta\rangle d^2\gamma. \quad (5.8)$$

The function $\mathcal{T}$ which represents the product is therefore given by

$$\mathcal{T}(\alpha^*,\beta)=\frac{1}{\pi}\int\mathcal{T}_1(\alpha^*,\gamma)\mathcal{T}_2(\gamma^*,\beta)e^{-|\gamma|^2}d^2\gamma. \quad (5.9)$$

The expansion function for the operator $T^\dagger$, the Hermitian adjoint of T, is obtained by substituting T_{mn}^* for T_{nm} in Eq. (5.4). It is given by $[\mathcal{T}(\beta^*,\alpha)]^*$. If the operator T is Hermitian the function $\mathcal{T}$ must satisfy the identity

$$\mathcal{T}(\alpha^*,\beta)=[\mathcal{T}(\beta^*,\alpha)]^*, \quad (5.10)$$

since the expansions of T and $T^\dagger$ are unique.

The functions $\mathcal{T}(\alpha^*,\beta)$ which represent normal products of the operators $a^\dagger$ and a such as $(a^\dagger)^n a^m$ are immediately seen from Eqs. (5.7) and (3.32) to be

$$\mathcal{T}(\alpha^*,\beta)=(\alpha^*)^n\beta^m\exp[\alpha^*\beta]. \quad (5.11)$$

In particular, the unit operator corresponds to $n=m=0$.

It may be worth noting at this point that many of the foregoing formulas can be abbreviated somewhat by adopting a normalization different from the conventional one for the coherent states. If we introduce the symbol $\|\alpha\rangle$ for the states normalized in the new way and define these as

$$\|\alpha\rangle=|\alpha\rangle e^{\frac{1}{2}|\alpha|^2}, \quad (5.12)$$

then we may write the scalar product of two such states as $\langle\alpha\|\beta\rangle$. We see from Eq. (3.32) that this scalar product is

$$\langle\alpha\|\beta\rangle=\exp[\alpha^*\beta]. \quad (5.13)$$

We may next, following Bargmann,[9] introduce an element of measure $d\mu(\alpha)$ which is defined as

$$d\mu(\alpha)=\frac{1}{\pi}e^{-|\alpha|^2}d^2\alpha. \quad (5.14)$$

With these alterations, all of the Gaussian functions, and factors of π, in the preceding formulas become absorbed, as it were, into the notation. The Eqs. (5.6) and (5.7), for example, reduce to the briefer forms

$$T=\int\|\alpha\rangle\mathcal{T}(\alpha^*,\beta)\langle\beta\|d\mu(\alpha)d\mu(\beta) \quad (5.15)$$

and

$$\mathcal{T}(\alpha^*,\beta)=\langle\alpha\|T\|\beta\rangle. \quad (5.16)$$

A more significant property of the states $\|\alpha\rangle$ is that they are given by the expansion

$$\|\alpha\rangle=\sum_n\frac{\alpha^n}{(n!)^{1/2}}|n\rangle \quad (5.17)$$

and thus obey the relation

$$a^\dagger\|\alpha\rangle=\frac{\partial}{\partial\alpha}\|\alpha\rangle. \quad (5.18)$$

While the properties of the alternatively normalized states $\|\alpha\rangle$ are worth bearing in mind, we have chosen not to adopt this normalization in the present paper in order to retain the more conventional interpretation of scalar products as probability amplitudes. The advantage afforded by the relation (5.18) is not a great one since all of the operators we shall have to deal with are either already in normally ordered form, or easily so ordered.

VI. GENERAL PROPERTIES OF THE DENSITY OPERATOR

The formalism we have developed in the two preceding sections has been intended to provide a background for the expression of the density operator of a mode in terms of the vectors that represent coherent states. Viewed in mathematical terms, the use of the coherent state vectors in this way leads to considerable simplification in the calculation of statistical averages. The fact that these states are eigenstates of the field operators $\mathbf{E}^{(\pm)}(\mathbf{r}t)$ means that normally ordered products of the field operators, when they are to be averaged, may be replaced by the products of their eigenvalues, i.e., treated not as operators, but as numbers. The field correlation functions such as $G^{(1)}$ given by Eq. (2.1) are averages of just such operator products. Their evaluation may be carried out quite conveniently through use of the representations we shall discuss.

Any density operator ρ may, according to the methods of the preceding section, be represented in a unique way by means of a function of two complex variables, $R(\alpha^*,\beta)$, which is analytic throughout the finite α^* and β planes. The function R is given explicitly, by means of Eq. (5.7), as

$$R(\alpha^*,\beta)=\langle\alpha|\rho|\beta\rangle\exp[\tfrac{1}{2}|\alpha|^2+\tfrac{1}{2}|\beta|^2]. \quad (6.1)$$

If we happen to know the matrix representation of ρ in the basis formed by the n-quantum states, the function R is evidently given by

$$R(\alpha^*,\beta)=\sum_{n,m}\langle n|\rho|m\rangle(n!m!)^{-1/2}(\alpha^*)^n\beta^m. \quad (6.2)$$

If we do not know the matrix elements $\langle n|\rho|m\rangle$ they may be found quite simply from a knowledge of $R(\alpha^*,\beta)$. One method for finding them is to consider $R(\alpha^*,\beta)$ as a generating function and identify its Taylor series with the series (6.2). A second method is to note that if we multiply Eq. (6.2) by $\alpha^i(\beta^*)^j\exp[-(|\alpha|^2+|\beta|^2)]$ and integrate over the α and β planes, then all terms save that for $n=i$ and $m=j$ vanish in the sum on the right and we have

$$\langle i|\rho|j\rangle=\frac{1}{\pi^2}\int R(\alpha^*,\beta)(i!j!)^{-1/2}\alpha^i(\beta^*)^je^{-(|\alpha|^2+|\beta|^2)}d^2\alpha d^2\beta. \quad (6.3)$$

Given the knowledge of $R(\alpha^*,\beta)$, we may write the density operator as

$$\rho=\frac{1}{\pi^2}\int|\alpha\rangle R(\alpha^*,\beta)\langle\beta|e^{-\frac{1}{2}(|\alpha|^2+|\beta|^2)}d^2\alpha d^2\beta. \quad (6.4)$$

The statistical average of an operator T is given by the trace of the product ρT. If we calculate this average by using the representation (6.4) for ρ we must note that the trace of the expression $|\alpha\rangle\langle\beta|T$, regarded as an operator, is the matrix element $\langle\beta|T|\alpha\rangle$. Then, if we express the matrix element in terms of the function $\mathcal{T}(\alpha^*,\beta)$ defined by Eq. (5.7) we find

$$\mathrm{tr}\,\{\rho T\}=\frac{1}{\pi^2}\int R(\alpha^*,\beta)\mathcal{T}(\beta^*,\alpha)e^{-|\alpha|^2-|\beta|^2}d^2\alpha d^2\beta. \quad (6.5)$$

If T is any operator of the form $(a^\dagger)^na^m$, its representation $\mathcal{T}(\beta^*,\alpha)$ is given by Eq. (5.11). In particular for $n=m=0$, we have the unit operator $T=1$ which is represented by $\mathcal{T}(\beta^*,\alpha)=\exp[\beta^*\alpha]$. Hence, the trace of ρ itself, which must be normalized to unity, is

$$\mathrm{tr}\rho=1$$

$$=\frac{1}{\pi^2}\int R(\alpha^*,\beta)\exp[\beta^*\alpha-|\alpha|^2-|\beta|^2]d^2\alpha d^2\beta.$$

Since $R(\alpha^*,\beta)$ is an entire function of α^*, we may use Eq. (4.10) to carry out the integration over the α plane. In this way we see that the normalization condition on R is

$$\frac{1}{\pi}\int R(\beta^*,\beta)e^{-|\beta|^2}d^2\beta=1. \quad (6.6)$$

The density operator is Hermitian and hence has real eigenvalues. These eigenvalues may be interpreted as probabilities and so must be positive numbers. Since ρ is thus a positive definite operator, its expectation value in any state, e.g., the state $|f\rangle$ defined by Eq. (4.6), must be non-negative,

$$\langle f|\rho|f\rangle\geq 0. \quad (6.7)$$

If, for example, we choose the state $|f\rangle$ to be a coherent state $|\alpha\rangle$ we find that the function R, which is given by Eq. (6.1), satisfies the inequality

$$R(\alpha^*,\alpha)\geq 0. \quad (6.8)$$

If we let the state $|f\rangle$ be specified as in Eq. (4.7) by an entire function $f(\alpha^*)$, then we find from the inequality (6.7) the more general condition for positive definiteness

$$\int[f(\alpha^*)]^*f(\beta^*)R(\alpha^*,\beta)e^{-|\alpha|^2-|\beta|^2}d^2\alpha d^2\beta\geq 0, \quad (6.9)$$

which must hold for all entire functions f.

In many types of physical experiments, particularly those dealing with fields which oscillate at extremely high frequencies, we cannot be said to have any *a priori* knowledge of the time-dependent parameters. The predictions we make in such circumstances are unchanged by displacements in time. They may be derived from a density operator which is stationary, that is, one which commutes with the Hamiltonian operator or, more simply, with $a^\dagger a$. The necessary and sufficient condition that a function $R(\alpha^*,\beta)$ correspond to a stationary density operator is that it depend only on the product of its two variables, $\alpha^*\beta$. There must, in other words, exist an analytic function $\mathcal{S}$ such that

$$R(\alpha^*,\beta)=\mathcal{S}(\alpha^*\beta). \tag{6.10}$$

That this condition is a sufficient one is clear from the invariance of R under the multiplication of both α and β by a phase factor, $e^{i\varphi}$. The condition may be derived as a necessary one directly from the vanishing of the commutator of ρ with $a^\dagger a$. An alternative and perhaps simpler way of seeing the result depends on noting that a stationary ρ can only be a function of the Hamiltonian for the mode, or of $a^\dagger a$. It is therefore diagonal in the basis formed by the n-quantum states, i.e., $\langle n|\rho|m\rangle=\delta_{nm}\langle n|\rho|n\rangle$. Examination of the series expansion (6.2) for R then shows that it then takes the form of Eq. (6.10).

VII. THE P REPRESENTATION OF THE DENSITY OPERATOR

In the preceding sections we have demonstrated the generality of the use of the coherent states as a basis. Not all fields require for their description density operators of quite so general a form. Indeed for a broad class of radiation fields which includes, as we shall see, virtually all of those studied in optics, it becomes possible to reduce the density operator to a considerably simpler form. This form is one which brings to light many similarities between quantum electrodynamical calculations and the corresponding classical ones. Its use offers deep insights into the reasons why some of the fundamental laws of optics, such as those for superposition of fields and calculation of the resulting intensities, are the same as in classical theory, even when very few quanta are involved. We shall continue, for the present, to limit consideration to a single mode of the field.

One type of oscillator state which interests us particularly is, of course, a coherent state. The density operator for a pure state $|\alpha\rangle$ is just the projection operator

$$\rho=|\alpha\rangle\langle\alpha|. \tag{7.1}$$

The unique representation of this operator as a function $R(\beta^*,\gamma)$ is easily shown, from Eq. (6.1), to be

$$R(\beta^*,\gamma)=\exp[\beta^*\alpha+\gamma\alpha^*-|\alpha|^2]. \tag{7.2}$$

Other functions $R(\beta^*,\gamma)$, which satisfy the analyticity requirements necessary for the representations of density operators, may be constructed by forming linear combinations of exponentials such as (7.2) for various values of the complex parameter α. The functions R, which we form in this way, represent statistical mixtures of the coherent states. The most general such function R may be written as

$$R(\beta^*,\gamma)=\int P(\alpha)\exp[\beta^*\alpha+\gamma\alpha^*-|\alpha|^2]d^2\alpha, \tag{7.3}$$

where $P(\alpha)$ is a weight function defined at all points of the complex α plane. Since $R(\beta^*,\gamma)$ must satisfy the Hermiticity condition, Eq. (5.10), we require that the weight function be real-valued, i.e., $[P(\alpha)]^*=P(\alpha)$. The function $P(\alpha)$ need not be subject to any regularity conditions, but its singularities must be integrable ones.[11] It is convenient to allow $P(\alpha)$ to have delta-function singularities so that we may think of a pure coherent state as represented by a special case of Eq. (7.3). A real-valued two-dimensional delta function which is suited to this purpose may be defined as

$$\delta^{(2)}(\alpha)=\delta(\mathrm{Re}\,\alpha)\delta(\mathrm{Im}\,\alpha). \tag{7.4}$$

The pure coherent state $|\beta\rangle$ is then evidently described by

$$P(\alpha)=\delta^{(2)}(\alpha-\beta), \tag{7.5}$$

and the ground state of the oscillator is specified by setting $\beta=0$.

The density operator ρ which corresponds to Eq. (7.3) is just a superposition of the projection operators (7.1),

$$\rho=\int P(\alpha)|\alpha\rangle\langle\alpha|d^2\alpha. \tag{7.6}$$

It is the kind of operator we might naturally be led to if we were given knowledge that the oscillator is in a coherent state, but one which corresponds to an unknown eigenvalue α. The function $P(\alpha)$ might then be thought of as playing a role analogous to a probability density for the distribution of values of α over the complex plane.[12] Such an interpretation may, as we shall see, be justified at times. In general, however, it is not possible to interpret the function $P(\alpha)$ as a probability distribution in any precise way since the projection operators $|\alpha\rangle\langle\alpha|$ with which it is associated are not orthogonal to one another for different values of α. There is an approximate sense, as we have noted in connection with Eq. (3.33), in which two states $|\alpha\rangle$ and $|\alpha'\rangle$ may be said to become orthogonal to one another for $|\alpha-\alpha'|\gg 1$, i.e., when their wave packets (3.29) and those of the form (3.30) do not appreciably overlap. When the function $P(\alpha)$ tends to vary little over such large ranges of the parameter α, the non-orthogonality of the coherent states will make little difference, and $P(\alpha)$ will then be interpretable approximately as a probability density. The functions $P(\alpha)$

[11] If the singularities of $P(\alpha)$ are of types stronger than those of delta functions, e.g., derivatives of delta functions, the field represented will have no classical analog.

[12] The existence of this form for the density operator has also been observed by E. C. G. Sudarshan, Phys. Rev. Letters **10**, 277 (1963). His note is discussed briefly at the end of Sec. X.

which vary this slowly will, in general, be associated with strong fields, ones which may be described approximately in classical terms.

We shall call the expression (7.6) for the density operator the P representation in order to distinguish it from the more general form based on the functions R discussed earlier. The normalization property of the density operator requires that $P(\alpha)$ obey the normalization condition

$$\text{tr}\rho=\int P(\alpha)d^2\alpha=1. \tag{7.7}$$

It is interesting to examine the conditions that the positive definiteness of ρ places upon $P(\alpha)$. If we apply the condition (6.9) to the function $R(\beta^*,\gamma)$ given by Eq. (7.3) we find

$$\int [f(\beta^*)]^* f(\gamma^*)P(\alpha)\exp[\beta^*\alpha+\gamma\alpha^*-|\alpha|^2-|\beta|^2-|\gamma|^2] \times d^2\alpha d^2\beta d^2\gamma \geq 0. \tag{7.8}$$

The γ integration may be carried out via Eq. (4.10) and the β integration by means of its complex conjugate. We then have the condition that

$$\int |f(\alpha^*)|^2 P(\alpha)e^{-|\alpha|^2}d^2\alpha \geq 0 \tag{7.9}$$

must hold for all entire functions $f(\alpha^*)$. In particular, the choice $f(\alpha^*)=\exp[\beta\alpha^*-\frac{1}{2}|\beta|^2]$ leads to the simple condition

$$\int P(\alpha)e^{-|\alpha-\beta|^2}d^2\alpha \geq 0, \tag{7.10}$$

which must hold for all complex values of β. It corresponds to the requirement $\langle\beta|\rho|\beta\rangle\geq 0$. These conditions are immediately satisfied if $P(\alpha)$ is positive valued as it would be, were it a probability density. They are not strong enough, however, to exclude the possibility that $P(\alpha)$ takes on negative values over some suitably restricted regions of the plane.[13] This result serves to underscore the fact that the weight function $P(\alpha)$ cannot, in general, be interpreted as a probability density.[14]

If a density operator is specified by means of the P representation, its matrix elements connecting the n-quantum states are given by

$$\langle n|\rho|m\rangle=\int P(\alpha)\langle n|\alpha\rangle\langle\alpha|m\rangle d^2\alpha. \tag{7.11}$$

When Eqs. (3.3) and (3.6) are used to evaluate the scalar products in the integrand we find

$$\langle n|\rho|m\rangle=(n!m!)^{-1/2}\int P(\alpha)\alpha^n(\alpha^*)^m e^{-|\alpha|^2}d^2\alpha. \tag{7.12}$$

This form for the density matrix indicates a fundamental property of the fields which are most naturally described by means of the P representation. If $P(\alpha)$ is a weight function with singularities no stronger than those of delta function type, it will, in general, possess nonvanishing complex moments of arbitrarily high order. [The unique exception is the choice $P(\alpha)=\delta^{(2)}(\alpha)$ which corresponds to the ground state of the mode.] It follows then that the diagonal matrix elements $\langle n|\rho|n\rangle$, which represent the probabilities for the presence of n photons in the mode, take on nonvanishing values for arbitrarily large n. There is thus no upper bound to the number of photons present when the function P is well behaved in the sense we have noted.[15]

Stationary density operators correspond in the P representation to functions $P(\alpha)$ which depend only on $|\alpha|$. This correspondence is made clear by Eq. (7.2) which shows that such $P(\alpha)$ lead to functions $R(\beta^*,\gamma)$ which are unaltered by a common phase change of β and γ. It is seen equally well through Eq. (7.12) which shows that $\langle n|\rho|m\rangle$ reduces to diagonal form when the weight function $P(\alpha)$ is circularly symmetric.

Some indication of the importance, in practical terms, of the P representation for the density operator can be found by considering the way in which photon fields produced by different sources become superposed. Since we are only discussing the behavior of one mode of the field for the present, we are only dealing with a fragment of the full problem, but all the modes may eventually be treated similarly. We shall illustrate the superposition law by assuming there are two different transient radiation sources coupled to the field mode and that they may be switched on and off separately. The first source will be assumed, when it is turned on alone at time t_1, to excite the mode from its ground state $|0\rangle$ to the coherent state $|\alpha_1\rangle$. If we assume that the source has ceased radiating by a time t_2, the state of the field remains $|\alpha_1\rangle$ for all later times. We may alternatively consider the case in which the first source remains inactive and the second one is switched on at

---

[13] An example of a weight function $P(\alpha)$ which takes on negative values but leads to a positive-definite density operator is given by the form

$$P(\alpha)=(1+\lambda)(\pi n)^{-1}\exp[-|\alpha|^2/n]-\lambda\delta^{(2)}(\alpha)$$

for $n>0$ and $0<\lambda<n^{-1}$. The matrix representation of the corresponding density operator, which is given by Eq. (7.12), is seen to be diagonal and to have only positive eigenvalues.

[14] A familiar example of a function which plays a role analogous to that of a probability density, but may take on negative values in quantum-mechanical contexts is the Wigner distribution function, E. P. Wigner, Phys. Rev. **40**, 749 (1932).

[15] Density operators for fields in which the number of photons present possesses an upper bound N are represented by functions $R(\beta^*,\gamma)$ which are polynomials of Nth degree in β^* and in γ. It is evident from the behavior of such polynomials for large $|\beta|$ and $|\gamma|$ that any weight function $P(\alpha)$ which corresponds to $R(\beta^*,\gamma)$ through Eq. (7.2) would have to have singularities much stronger than those of a delta function. Such fields are probably represented more conveniently by means of the R function.

time t_2. The second source will then be assumed to bring the mode from its ground state to the coherent state $|\alpha_2\rangle$. We now ask what state the mode will be brought to if the two sources are allowed to act in succession, the first at t_1 and the second at t_2.

The answer for this simple case may be seen without performing any detailed calculations by making use of the unitary displacement operators described in Sec. III. The action of the first source is represented by the unitary operator $D(\alpha_1)$ which displaces the oscillator state from the ground state to the coherent state $|\alpha_1\rangle=D(\alpha_1)|0\rangle$. The action of the second source is evidently represented by the displacement operator $D(\alpha_2)$, so that when it is turned on after the first source, it brings the oscillator to the superposed state

$$|\;\rangle=D(\alpha_2)D(\alpha_1)|0\rangle. \qquad (7.13)$$

Since the displacement operators are of the exponential form (3.17), their multiplication law is given by Eq. (3.20). We thus find

$$D(\alpha_2)D(\alpha_1)=D(\alpha_1+\alpha_2)\exp[\tfrac{1}{2}(\alpha_2\alpha_1^*-\alpha_2^*\alpha_1)]. \qquad (7.14)$$

The exponential which has been separated from the D operators in this relation has a purely imaginary argument and, hence, corresponds to a phase factor. The superposed state, (7.13), in other words, is just the coherent state $|\alpha_1+\alpha_2\rangle$ multiplied by a phase factor. The phase factor has no influence upon the density operator for the superposed state, which is

$$\rho=|\alpha_1+\alpha_2\rangle\langle\alpha_1+\alpha_2|. \qquad (7.15)$$

To vary the way in which the sources are turned on in the imaginary experiment we have described, e.g., to turn the two sources on at other times or in the reverse order, would only alter the final state through a phase factor and would thus lead to the same final density operator. The amplitudes of successive coherent excitations of the mode add as complex numbers in quantum theory, just as they do in classical theory.

Let us suppose next that the sources in the same experiment are somewhat less ideal and that, instead of exciting the mode to pure coherent states, they excite it to conditions described by mixtures of coherent states of the form (7.6). The first source acting alone, we assume, brings the field to a condition described by the density operator

$$\rho_1=\int P_1(\alpha_1)|\alpha_1\rangle\langle\alpha_1|d^2\alpha_1. \qquad (7.16)$$

The condition produced by the second source, when it acts alone, is assumed to be represented by

$$\rho=\int P_2(\alpha_2)|\alpha_2\rangle\langle\alpha_2|d^2\alpha_2,$$

$$=\int P_2(\alpha_2)D(\alpha_2)|0\rangle\langle 0|D^{-1}(\alpha_2)d^2\alpha_2.$$

If the second source is turned on after the first, it brings the field to a condition described by the density operator

$$\rho=\int P_2(\alpha_2)D(\alpha_2)\rho_1 D^{-1}(\alpha_2)d^2\alpha_2,$$

$$=\int P_2(\alpha_2)P_1(\alpha_1)|\alpha_1+\alpha_2\rangle\langle\alpha_1+\alpha_2|d^2\alpha_1 d^2\alpha_2. \qquad (7.17)$$

The latter density operator may be written in the general form

$$\rho=\int P(\alpha)|\alpha\rangle\langle\alpha|d^2\alpha,$$

if we define the weight function $P(\alpha)$ for the superposed excitations to be

$$P(\alpha)=\int \delta^{(2)}(\alpha-\alpha_1-\alpha_2)P_1(\alpha_1)P_2(\alpha_2)d^2\alpha_1 d^2\alpha_2, \qquad (7.18)$$

$$=\int P_1(\alpha-\alpha')P_2(\alpha')d^2\alpha'. \qquad (7.19)$$

We see immediately from Eq. (7.18) that P is correctly normalized if P_1 and P_2 are. The simple convolution law for combining the weight functions is one of the unique features of the description of fields by means of the P representation. It is quite analogous to the law we would use in classical theory to describe the probability distribution of the sum of two uncertain Fourier amplitudes for a mode.

The convolution theorem can often be used to separate fields into component fields with simpler properties. Suppose we have a field described by a weight function $P(\alpha)$ which has a mean value of α given by

$$\bar{\alpha}=\int \alpha P(\alpha)d^2\alpha. \qquad (7.20)$$

It is clear from Eq. (7.19) that any such field may be regarded as the sum of a pure coherent field which corresponds to the weight function $\delta^{(2)}(\alpha-\bar{\alpha})$ and an additional field represented by $P(\alpha+\bar{\alpha})$ for which the mean value of α vanishes. Fields with vanishing mean values of α will be referred to as unphased fields.

The use of the P representation of the density operator, where it is not too singular, leads to simplifications in the calculation of statistical averages which go somewhat beyond those discussed in the last section. Thus, for example, the statistical average of any normally ordered product of the creation and annihilation operators, such as $(a^\dagger)^n a^m$, reduces to a simple average of $(\alpha^*)^n\alpha^m$ taken with respect to the weight function $P(\alpha)$, i.e., we have

$$\operatorname{tr}\{\rho(a^\dagger)^n a^m\} = \int P(\alpha)\langle\alpha|(a^\dagger)^n a^m|\alpha\rangle d^2\alpha,$$

$$= \int P(\alpha)(\alpha^*)^n \alpha^m d^2\alpha. \qquad (7.21)$$

This identity means, in practice, that many quantum-mechanical calculations can be carried out by means which are analogous to those already familiar from classical theory.

The mean number of photons which are present in a mode is the most elementary measure of the intensity of its excitation. The operator which represents the number of photons present is seen from Eq. (2.18) to be $a^\dagger a$. The average photon number, written as $\langle n\rangle$, is therefore given by

$$\langle n\rangle = \operatorname{tr}\{\rho a^\dagger a\}. \qquad (7.22)$$

According to Eq. (7.21), with its two exponents set equal to unity, we have

$$\langle n\rangle = \int P(\alpha)|\alpha|^2 d^2\alpha, \qquad (7.23)$$

i.e., the average photon number is just the mean squared absolute value of the amplitude α. When two fields described by distributions P_1 and P_2 are superposed, the resulting intensities are found from rules of the form which have always been used in classical electromagnetic theory. For unphased fields the intensities add "incoherently"; for coherent states the amplitudes add "coherently."

The use of the P representation of the density operator in describing fields brings many of the results of quantum electrodynamics into forms similar to those of classical theory. While these similarities make applications of the correspondence principle particularly clear, they must not be interpreted as indicating that classical theory is any sort of adequate substitute for the quantum theory. The weight functions $P(\alpha)$ which occur in quantum theoretical applications are not accurately interpretable as probability distributions, nor are they derivable as a rule from classical treatments of the radiation sources. They depend upon Planck's constant, in general, in ways that are unfathomable by classical or semiclassical analysis.

Since a number of calculations having to do with photon statistics have been carried out in the past by essentially classical methods, it may be helpful to discuss the relation between the P representation and the classical theory a bit further. It is worth noting in particular that the definition we have given the amplitude α as an eigenvalue of the annihilation operator is an intrinsically quantum-mechanical one. If we wish to represent a given classical field amplitude for the mode as an eigenvalue, then we see from Eq. (2.20) that the appropriate value of α has a magnitude which is proportional to $\hbar^{-1/2}$. In the dimensionless terms in which α is defined, the classical description of the mode only applies to the region $|\alpha|\gg 1$ of the complex α plane, i.e., to amplitudes of oscillation which are large compared with the range of the zero-point fluctuations present in the wave packet (3.29) and (3.30). Classical theory can therefore, in principle, only furnish us with the grossest sort of information about the weight function $P(\alpha)$. When the weight function extends appreciably into the classical regions of the plane, classical theory can only be relied upon, crudely speaking, to tell us average values of the function $P(\alpha)$ over areas whose dimensions, $|\Delta\alpha|$, are of order unity or larger. From Eq. (7.10) we see that such average values will always be positive; in the classical limit they may always be interpreted as probabilities.

VIII. THE GAUSSIAN DENSITY OPERATOR

The Gaussian function is a venerable statistical distribution, familiar from countless occurrences in classical statistics. We shall indicate in this section that it has its place in quantum field theory as well, where it furnishes the natural description of the most commonly occurring type of incoherence.[1]

Let us assume that the field mode we are studying is coupled to a number of sources which are essentially similar but are statistically independent of one another in their behavior. Such sources might, in practice, simply be several hypothetical subdivisions of one large source. If we may represent the contribution of each source (numbered $j=1,\cdots N$) to the excitation of the mode by means of a weight function $p(\alpha_j)$, we may then construct the weight function $P(\alpha)$ which describes the superposed fields by means of the generalized form of the convolution theorem

$$P(\alpha) = \int \delta^{(2)}\Big(\alpha - \sum_{j=1}^{N}\alpha_j\Big)\prod_{j=1}^{N} p(\alpha_j)d^2\alpha_j. \qquad (8.1)$$

Since the weight functions which appear in this expression are all real valued, it is sometimes convenient to think of the amplitudes α in their arguments not as complex numbers, but as two-dimensional real vectors $\boldsymbol{\alpha}$ (i.e., $\boldsymbol{\alpha}_x = \operatorname{Re}\alpha$, $\boldsymbol{\alpha}_y = \operatorname{Im}\alpha$). Then if λ is an arbitrary complex number represented by the vector $\boldsymbol{\lambda}$, we may use a two-dimensional scalar product for the abbreviation

$$\operatorname{Re}\lambda\operatorname{Re}\alpha + \operatorname{Im}\lambda\operatorname{Im}\alpha = \boldsymbol{\alpha}\cdot\boldsymbol{\lambda}. \qquad (8.2)$$

Using this notation, we may define the two-dimensional Fourier transform of the weight function $p(\boldsymbol{\alpha})$ as

$$\xi(\boldsymbol{\lambda}) = \int \exp(i\boldsymbol{\lambda}\cdot\boldsymbol{\alpha})p(\boldsymbol{\alpha})d^2\alpha. \qquad (8.3)$$

The superposition law (8.1) then shows that the Fourier transform of the weight function $P(\boldsymbol{\alpha})$ is given by

$$\Xi(\boldsymbol{\lambda}) = \int \exp(i\boldsymbol{\lambda}\cdot\boldsymbol{\alpha})P(\boldsymbol{\alpha})d^2\alpha\,,$$
$$= [\xi(\boldsymbol{\lambda})]^N. \tag{8.4}$$

If the individual sources are stationary ones their weight function $p(\boldsymbol{\alpha})$ depends only on $|\boldsymbol{\alpha}|$. The transform $\xi(\boldsymbol{\lambda})$ may then be approximated for small values of $|\boldsymbol{\lambda}|$ by

$$\xi(\boldsymbol{\lambda}) = 1 - \tfrac{1}{4}\lambda^2 \int |\boldsymbol{\alpha}|^2 p(\boldsymbol{\alpha})d^2\alpha\,,$$
$$= 1 - \tfrac{1}{4}\lambda^2\langle|\boldsymbol{\alpha}|^2\rangle. \tag{8.5}$$

For values of $|\boldsymbol{\lambda}|$ which are smaller still (i.e., $|\boldsymbol{\lambda}|^2 < N^{-1/2}(\langle|\boldsymbol{\alpha}|^2\rangle^{-1})$, the transform Ξ for the superposed field may be approximated by

$$\Xi(\boldsymbol{\lambda}) \approx \exp\{-\tfrac{1}{4}\lambda^2 N\langle|\boldsymbol{\alpha}|^2\rangle\}. \tag{8.6}$$

Since the weight function $p(\boldsymbol{\alpha})$ may take on negative values it is necessary at this point to verify that the second moment $\langle|\boldsymbol{\alpha}|^2\rangle$ is positive. That it is indeed positive is indicated by Eqs. (7.22) and (7.23) which show that $\langle|\boldsymbol{\alpha}|^2\rangle$ is the mean number of photons which would be radiated by each source in the absence of the others. For large values of N the transform $\Xi(\boldsymbol{\lambda})$ therefore decreases rapidly as $|\boldsymbol{\lambda}|$ increases. Since the function becomes vanishingly small for $|\boldsymbol{\lambda}|$ lying outside the range of approximation noted earlier, we may use (8.6) more generally as an asymptotic approximation to $\Xi(\boldsymbol{\lambda})$ for large N. When we calculate the transform of this asymptotic expression for $\Xi(\boldsymbol{\lambda})$ we find

$$P(\boldsymbol{\alpha}) = (2\pi)^{-2}\int \exp(-i\boldsymbol{\alpha}\cdot\boldsymbol{\lambda})\Xi(\boldsymbol{\lambda})d^2\lambda\,,$$
$$= \frac{1}{\pi N\langle|\boldsymbol{\alpha}|^2\rangle}\exp(-\alpha^2/N\langle|\boldsymbol{\alpha}|^2\rangle). \tag{8.7}$$

The mean value of $|\boldsymbol{\alpha}|^2$ for such a weight function is evidently $N\langle|\boldsymbol{\alpha}|^2\rangle$, but by the general theorem expressed in Eq. (7.23), this mean value is just the average of the total number of quanta present in the mode. If we write the latter average as $\langle n\rangle$, and resume the use of the complex notation for the variable α, the weight function (8.7) may be written as

$$P(\alpha) = \frac{1}{\pi\langle n\rangle}e^{-|\alpha|^2/\langle n\rangle}. \tag{8.8}$$

The weight function $P(\alpha)$ is positive everywhere and takes the same form as the probability distribution for the total displacement which results from a random walk in the complex plane. However, because the coherent states $|\alpha\rangle$ are not an orthogonal set, $P(\alpha)$ can only be accurately interpreted as a probability distribution for $\langle n\rangle \gg 1$. We may note that it is not ultimately necessary, in order to derive Eq. (8.8), to assume that the weight functions corresponding to the individual sources are all the same. All that is required to carry out the proof is that the moments of the individual functions be of comparable magnitudes. The mean squared value of $|\alpha|$ is then given more generally by $\sum_j\langle|\alpha_j|^2\rangle$, rather than the value in Eq. (8.7), but this value is still the mean number of quanta in the mode, as indicated in Eq. (8.8).

It should be clear from the conditions of the derivation that the Gaussian distribution $P(\alpha)$ for the excitation of a mode possesses extremely wide applicability. The random or chaotic sort of excitation it describes is presumably characteristic of most of the familiar types of noncoherent macroscopic light sources, such as gas discharges, incandesant radiators, etc.

The Gaussian density operator

$$\rho = \frac{1}{\pi\langle n\rangle}\int e^{-|\alpha|^2/\langle n\rangle}|\alpha\rangle\langle\alpha|d^2\alpha \tag{8.9}$$

may be seen to take on a very simple form as well in the basis which specifies the photon numbers. To find this form we substitute in Eq. (8.9) the expansions (3.7) and (3.8) for the coherent states and note the identity

$$\pi^{-1}(l!m!)^{-1/2}\int \exp[-C|\alpha|^2]\alpha^l(\alpha^*)^m d^2\alpha = \delta_{lm}C^{-(m+1)},$$

which holds for $C>0$. If we write $C = (1+\langle n\rangle)/\langle n\rangle$ we then find

$$\rho = \frac{1}{1+\langle n\rangle}\sum_m\left\{\frac{\langle n\rangle}{1+\langle n\rangle}\right\}^m |m\rangle\langle m|. \tag{8.10}$$

In other words, the number of quanta in the mode is distributed according to the powers of the parameter $\langle n\rangle/(1+\langle n\rangle)$. The Planck distribution for blackbody radiation furnishes an illustration of a density operator which has long been known to take the form of Eq. (8.10). The thermal excitation which leads to the blackbody distribution is an ideal example of the random type we have described earlier, and so it should not be surprising that this distribution is one of the class we have derived. It is worth noting, in particular, that while the Planck distribution is characteristic of thermal equilibrium, no such limitation is implicit in the general form of the density operator (8.9). It will apply whenever the excitation has an appropriately random quality, no matter how far the radiator is from thermal equilibrium.

The Gaussian distribution function $\exp[-|\alpha|^2/\langle n\rangle]$ is phrased in terms which are explicitly quantum mechanical. In the limit which would represent a classical field both $|\alpha|^2$ and the average quantum number $\langle n\rangle$ become infinite as $\hbar^{-1}$, but their quotient, which is the argument of the Gaussian function, remains well defined. The form which the distribution takes in the classical limit is a familiar one. Historically, one of the origins of the random walk problem is to be found in the discussion of a classical harmonic oscillator which is subject to random excitations.[16] Such oscillators have complex amplitudes which are described under quite general conditions by a Gaussian distribution. If we were armed with this knowledge, and lacked the quantum-mechanical analysis given earlier, we might be tempted to assume that a Gaussian distribution derived in this way from classical theory can describe the photon distribution. To demonstrate the fallacy of this view we must examine more closely the nature of the parameter $\langle n\rangle$ which is, after all, the only physical constant involved in the distribution. We may take, as a simple illustration, the case of thermal excitation corresponding to temperature T. Then the mean photon number is given by $\langle n\rangle=[\exp(\hbar\omega/\kappa T)-1]^{-1}$, where κ is Boltzmann's constant, and the distribution $P(\alpha)$ takes the form

$$P(\alpha)=\frac{1}{\pi}[e^{\hbar\omega/\kappa T}-1]\exp[-(e^{\hbar\omega/\kappa T}-1)|\alpha|^2]. \quad (8.11)$$

To reach the classical analog of this distribution we would assume that the classical field energy in the mode, $H=\frac{1}{2}\int(\mathbf{E}^2+\mathbf{B}^2)d\mathbf{r}$, is distributed with a probability proportional to $\exp[-H/\kappa T]$. The distribution for the amplitude α that results is

$$P_{cl}(\alpha)=(\hbar\omega/\pi\kappa T)\exp[-\hbar\omega|\alpha|^2/\kappa T], \quad (8.12)$$

which is seen to be a first approximation in powers of $\hbar$ to the correct distribution. (Again, we must remember that the quantity $\hbar|\alpha|^2$ is to be construed as a classical parameter.) The distribution $P_{cl}(\alpha)$ only extends into the classical region of the plane, $|\alpha|\gg 1$, for low-frequency modes, that is, only for $(\hbar\omega/\kappa T)\ll 1$ are the modes sufficiently excited to be accurately described by classical theory. For higher frequencies the two distributions differ greatly in nature even though both are Gaussian. The classical distribution retains much too large a radius in the α plane as $\hbar\omega$ increases beyond κT, rather than narrowing extremely rapidly as the correct distribution does.[17] That error, in fact, epitomizes the ultraviolet catastrophe of the classical radiation theory. The example we have discussed is, of course, an elementary one, but it should serve to illustrate some of the points noted in the preceding section regarding the limitations of the classical distribution function.

The expression for the thermal density operator of an oscillator in terms of coherent quantum states appears to offer new and instructive approaches to many familiar problems. It permits us, for example, to derive the thermal averages of exponential functions of the operators a and $a^\dagger$ in an elementary way. The thermal average of the operator $D(\beta)$ defined by Eq. (3.17) is an illustration. It is given by

$$\mathrm{tr}\{\rho D(\beta)\}=\frac{1}{\pi\langle n\rangle}\int e^{-|\alpha|^2/\langle n\rangle}\langle\alpha|D(\beta)|\alpha\rangle d^2\alpha. \quad (8.13)$$

The expectation value in the integrand is, in this case

$$\begin{aligned}\langle\alpha|D(\beta)|\alpha\rangle&=\langle 0|D^{-1}(\alpha)D(\beta)D(\alpha)|0\rangle,\\&=\exp[\beta\alpha^*-\beta^*\alpha]\langle 0|D(\beta)|0\rangle,\\&=\exp[\beta\alpha^*-\beta^*\alpha]\langle 0|\beta\rangle,\\&=\exp[\beta\alpha^*-\beta^*\alpha-\tfrac{1}{2}|\beta|^2],\end{aligned} \quad (8.14)$$

where the properties of $D(\alpha)$ as a displacement operator have been used in the intermediate steps. When the integration indicated in Eq. (8.13) is carried out, we find

$$\mathrm{tr}\{\rho D(\beta)\}=\exp[-|\beta|^2(\langle n\rangle+\tfrac{1}{2})], \quad (8.15)$$

which is a frequently used corollary of Bloch's theorem on the distribution function of an oscillator coordinate.[18]

IX. DENSITY OPERATORS FOR THE FIELD

The developments introduced in Secs. III–VIII have all concerned the description of the quantum state of a single mode of the electromagnetic field. We may describe the field as a whole by constructing analogous methods to deal with all its modes at once. For this purpose we introduce a basic set of coherent states for the entire field and write them as

$$|\{\alpha_k\}\rangle\equiv\prod_k|\alpha_k\rangle_k, \quad (9.1)$$

where the notation $\{\alpha_k\}$, which will be used in several other connections, stands for the set of all amplitudes α_k. It is clear then, from the arguments of Sec. IV, that any state of the field determines uniquely a function $f(\{\alpha_k^*\})$ which is an entire function of each of the variables α_k^*. If the Hilbert space vector which represents the state is known and designated as $|f\rangle$, the function f is given by

$$f(\{\alpha_k^*\})=\langle\{\alpha_k\}|f\rangle\exp(\tfrac{1}{2}\textstyle\sum_k|\alpha_k|^2), \quad (9.2)$$

which is the direct generalization of Eq. (4.11). The expansion for the state $|f\rangle$ in terms of coherent states is then

$$|f\rangle=\int|\{\alpha_k\}\rangle f(\{\alpha_k^*\})\prod_k\pi^{-1}e^{-\frac{1}{2}|\alpha_k|^2}d^2\alpha_k, \quad (9.3)$$

which generalizes Eq. (4.7).

All of the operators which occur in field theory possess expansions in terms of the vectors $|\{\alpha_k\}\rangle$ and their

---

[16] Lord Rayleigh, *The Theory of Sound*, (MacMillan and Company Ltd., London, 1894), 2nd ed., Vol. I, p. 35; *Scientific Papers* (Cambridge University Press, Cambridge, England, 1899–1920), Vol. I, p. 491, Vol. IV, p. 370.

[17] For frequencies in the middle of the visible spectrum and temperatures under 3000°K the quantum mechanical distribution (8.11) will have a radius which corresponds to $|\alpha|^2\ll 10^{-3}$, i.e., the distribution is far from classical in nature. Comparable radii characterize the distributions for nonthermal incoherent sources.

[18] F. Bloch, Z. Physik 74, 295 (1932).

adjoints. To construct such representations is simply a matter of generalizing the formulas of Sec. V to deal with an infinite set of amplitude variables. We therefore proceed directly to a discussion of the density operator. For any density operator ρ we may define a function $R(\{\alpha_k^*\},\{\beta_k\})$ which is an entire function of each of the variables α_k^* and β_k for all modes k. This function, as may be seen from Eq. (6.1), is given by

$$R(\{\alpha_k^*\},\{\beta_k\})=\langle\{\alpha_k\}|\rho|\{\beta_k\}\rangle \times\exp[\tfrac{1}{2}\sum(|\alpha_k|^2+|\beta_k|^2)]. \quad (9.4)$$

The corresponding representation of the density operator is

$$\rho=\int|\{\alpha_k\}\rangle R(\{\alpha_k^*\},\{\beta_k\})\langle\{\beta_k\}|\prod_k \pi^{-2} \times e^{-\frac{1}{2}(|\alpha_k|^2+|\beta_k|^2)}d^2\alpha_k d^2\beta_k. \quad (9.5)$$

If the set of integers $\{n_k\}$ is used to specify the familiar stationary states which have n_k photons in the kth mode, we may regard R as a generating function for the matrix elements of ρ connecting these states, i.e., as a generalization of Eq. (6.2) we have

$$R(\{\alpha_k^*\},\{\beta_k\})=\sum_{\{n_k\},\{m_k\}}\langle\{n_k\}|\rho|\{m_k\}\rangle \times\prod_k (n_k!m_k!)^{-1/2}(\alpha_k^*)^{n_k}\beta_k^{m_k}. \quad (9.6)$$

The matrix elements of ρ in the stationary basis are then given by

$$\langle\{n_k\}|\rho|\{m_k\}\rangle =\int R(\{\alpha_k^*\},\{\beta_k\})\prod_k \pi^{-2}(n_k!m_k!)^{-1/2}\alpha_k^{n_k}(\beta_k^*)^{m_k} \times e^{-(|\alpha_k|^2+|\beta_k|^2)}d^2\alpha_k d^2\beta_k. \quad (9.7)$$

The normalization condition on R is clearly

$$\int R(\{\beta_k^*\},\{\beta_k\})\prod_k \pi^{-1}e^{-|\beta_k|^2}d^2\beta_k=1. \quad (9.8)$$

The positive definiteness condition, Eq. (6.9), may also be generalized in an evident way to deal with the full set of amplitude variables.

It may help as a simple illustration of the foregoing formulae to consider the representation of a single-photon wave packet. The state which is empty of all photons is the one for which the amplitudes α_k all vanish. If we write that state as $|\text{vac}\rangle$, then we may write the most general one-photon state as $\sum_k q(k)a_k^\dagger|\text{vac}\rangle$, where the function $q(k)$ plays the role of a packet amplitude. The function f which represents this state is then

$$f(\{\alpha_k^*\})=\sum_k q(k)\alpha_k^*, \quad (9.9)$$

and the corresponding function R which determines the density operator is

$$R(\{\alpha_k^*\},\{\beta_k\})=\sum_k q(k)\alpha_k^*\sum_{k'} q^*(k')\beta_{k'}. \quad (9.10)$$

The normalization condition (9.8) corresponds to the requirement $\sum|q(k)|^2=1$. Since the state we have considered is a pure one, the function R factorizes into the product of two functions, one having the form of f and the other of its complex conjugate. If the packet amplitudes $q(k)$ were in some degree unpredictable, as they usually are, the packet could no longer be represented by a pure state. The function R would then be an average taken over the distribution of the amplitudes $q(k)$ and hence would lose its factorizable form in general. Whenever an upper bound exists for the number of photons present, i.e., the number of photons is required to be less than or equal to some integer N, we will find that R is a polynomial of at most Nth degree in the variables $\{\alpha_k^*\}$ and of the same degree in the $\{\beta_k\}$.

There will, of course, exist many types of excitation for which the photon numbers are unbounded. Among these are the ones which are more conveniently described by means of a generalized P distribution, i.e., the excitations for which there exists a reasonably well-behaved real-valued function $P(\{\alpha_k\})$ such that

$$R(\{\beta_k^*\},\{\gamma_k\})=\int P(\{\alpha_k\}) \times\exp\left[\sum_k(\beta_k^*\alpha_k+\gamma_k\alpha_k^*-|\alpha_k|^2)\right]\prod_k d^2\alpha_k. \quad (9.11)$$

When R possesses a representation of this type the density operator (9.5) may be reduced by means of Eq. (4.14) and its complex conjugate to the simple form

$$\rho=\int P(\{\alpha_k\})|\{\alpha_k\}\rangle\langle\{\alpha_k\}|\prod_k d^2\alpha_k, \quad (9.12)$$

which is the many-mode form of the P representation given by Eq. (7.6). The function P must satisfy the positive definiteness condition

$$\int|f(\{\alpha_k^*\})|^2P(\{\alpha_k\})\prod_k e^{-|\alpha_k|^2}d^2\alpha_k\geq 0 \quad (9.13)$$

for all possible choices of entire functions $f(\{\alpha_k^*\})$. The matrix elements of the density operator in the representation based on the n-photon states are

$$\langle\{n_k\}|\rho|\{m_k\}\rangle=\int P(\{\alpha_k\}) \times\prod_k (n_k!m_k!)^{-1/2}\alpha_k^{n_k}(\alpha_k^*)^{m_k}e^{-|\alpha_k|^2}d^2\alpha_k. \quad (9.14)$$

Stationary density operators, i.e., ones which commute with the Hamiltonian correspond to functions $P(\{\alpha_k\})$ which depend on the amplitude variables only through their magnitudes $\{|\alpha_k|\}$.

The superposition of two fields is described by forming the convolution integral of their distribution functions, much as in the case of a single mode. Thus, if two fields, described by $P_1(\{\beta_k\})$ and $P_2(\{\gamma_k\})$, respectively, are superposed, the resulting field has a distribution function

$$P(\{\alpha_k\}) = \int \prod_k \delta^{(2)}(\alpha_k - \beta_k - \gamma_k) \times P_1(\{\beta_k\}) P_2(\{\gamma_k\}) \prod_k d^2\beta_k d^2\gamma_k. \quad (9.15)$$

For fields which are represented by means of the density operator (9.12) all of the averages of normally ordered operator products can be calculated by means of formulas which, as in the case of a single mode, greatly resemble those of classical theory. Thus, the parameters $\{\alpha_k\}$ play much the same role in these calculations as the random Fourier amplitudes of the field do in the familiar classical theory of microwave noise.[19] Furthermore, the weight function $P(\{\alpha_k\})$ plays a role similar to that of the probability distribution for the Fourier amplitudes. Although this resemblance is extremely convenient in calculations, and offers immediate insight into the application of the correspondence principle, we must not lose sight of the fact that the function $P(\{\alpha_k\})$ is, in general, an explicitly quantum-mechanical structure. It may assume negative values, and is not accurately interpretable as a probability distribution except in the classical limit of strongly excited or low frequency fields.

In the foregoing discussions we have freely assumed that the density operator which describes the field is known and that it may, therefore, be expressed either in the representation of Eq. (9.5) or in the P representation of Eq. (9.12). For certain types of incoherent sources which we have discussed in Sec. VIII and will mention again in Sec. X, the explicit construction of these density operators is not at all difficult. But to find accurate density operators for other types of sources, including the recently developed coherent ones, will require a good deal of physical insight. The general problem of treating quantum mechanically the interaction of a many-atom source both with the radiation field and with an excitation mechanism of some sort promises to be a complicated one. It will have to be approached, no doubt, through greatly simplified models.

Since very little is known about the density operator for radiation fields, some insight may be gained by examining the form it takes on in one of the few completely soluble problems of quantum electrodynamics. We shall study the photon field radiated by an electric current distribution which is essentially classical in nature, one that does not suffer any noticeable reaction from the process of radiation. We may then represent the radiating current by a prescribed vector function of space and time $\mathbf{j}(\mathbf{r},t)$. The Hamiltonian which describes the coupling of the quantized electromagnetic field to the current distribution takes the form

$$H_1(t) = -\frac{1}{c}\int \mathbf{j}(\mathbf{r},t)\cdot\mathbf{A}(\mathbf{r},t)d\mathbf{r}. \quad (9.16)$$

The introduction of an explicitly time-dependent interaction of this type means that the state vector for the field, $|\ \rangle$, which previously was fixed (corresponding to the Heisenberg picture) will begin to change with time in accordance with the Schrödinger equation

$$i\hbar\frac{\partial}{\partial t}|\ \rangle = H_1(t)|\ \rangle, \quad (9.17)$$

which is the one appropriate to the interaction representation. The solution of this equation is easily found.[20] If we assume that the initial state of the field at time $t=-\infty$ is one empty of all photons, then the state of the field at time t may be written in the form

$$|t\rangle = \exp\left\{\frac{i}{\hbar c}\int_{-\infty}^{t} dt' \int \mathbf{j}(\mathbf{r},t')\cdot\mathbf{A}(\mathbf{r},t')d\mathbf{r} + i\varphi(t)\right\}|\mathrm{vac}\rangle. \quad (9.18)$$

The function $\varphi(t)$ which occurs in the exponent is a real-valued c-number phase function. It is easily evaluated, but cancels out of the product $|t\rangle\langle t|$ and so has no bearing on the construction of the density operator. The exponential operator which occurs in Eq. (9.18) may be expressed quite simply in terms of the displacement operators we discussed in Sec. III. For this purpose we define a displacement operator D_k for the kth mode as

$$D_k(\beta_k) = \exp[\beta_k a_k^\dagger - \beta_k^* a_k]. \quad (9.19)$$

Then it is clear from the expansion (2.10) for the vector potential that we may write

$$\exp\left\{\frac{i}{\hbar c}\int_{-\infty}^{t} dt' \int \mathbf{j}(\mathbf{r},t')\cdot\mathbf{A}(\mathbf{r},t')d\mathbf{r}\right\} = \prod_k D_k[\alpha_k(t)], \quad (9.20)$$

where the time-dependent amplitudes $\alpha_k(t)$ are given by

$$\alpha_k(t) = \frac{i}{(2\hbar\omega)^{1/2}}\int_{-\infty}^{t} dt' \int d\mathbf{r}\, \mathbf{u}_k^*(\mathbf{r})\cdot\mathbf{j}(\mathbf{r},t')e^{i\omega t'}. \quad (9.21)$$

The density operator at time t may therefore be written

[19] J. Lawson and G. E. Uhlenbeck, *Threshold Noise Signals* (McGraw-Hill Book Company, Inc., New York, 1950), pp. 33–56.

[20] R. J. Glauber, Phys. Rev. **84**, 395 (1951).

as

$$|t\rangle\langle t| = \prod_k D_k[\alpha_k(t)]|\text{vac}\rangle\langle\text{vac}|\prod_k D_k^{-1}[\alpha_k(t)] \quad (9.22)$$

$$= |\{\alpha_k(t)\}\rangle\langle\{\alpha_k(t)\}|. \quad (9.23)$$

The radiation by any prescribed current distribution, in other words, always leads to a pure coherent state.

It is only a slight generalization of the model we have just considered to imagine that the current distribution $\mathbf{j}(\mathbf{r},t)$ is not wholly predictable. In that case the amplitudes $\alpha_k(t)$ defined by Eq. (9.21) become random variables which possess collectively a probability distribution function which we may write as $p(\{\alpha_k\},t)$. The density operator for the field radiated by such a random current then becomes

$$\rho(t) = \int p(\{\alpha_k\},t)|\{\alpha_k\}\rangle\langle\{\alpha_k\}|\prod_k d^2\alpha_k. \quad (9.24)$$

We see that the density operator for a field radiated by a random current which suffers no recoil in the radiation process always takes the form of the P representation of Eq. (9.12). The weight function in this case does admit interpretation as a probability distribution, but it has a classical structure associated directly with the properties of the radiating current rather than with particular (nonorthogonal) states of the field. The assumption we have made in defining the model, that the current suffers negligible reaction, is a strong one but is fairly well fulfilled in radiating systems operated at radio or microwave frequencies. The fields produced by such systems should be accurately described by density operators of the form (9.24).

X. CORRELATION AND COHERENCE PROPERTIES OF THE FIELD

Any eigenvalue function $\boldsymbol{\mathcal{E}}(\mathbf{r}t)$ which satisfies the appropriate field equations and contains only positive frequency terms determines a set of mode amplitudes $\{\alpha_k\}$ uniquely through the expansion (2.20). This set of mode amplitudes then determines a coherent state of the field, $|\{\alpha_k\}\rangle$, such that

$$\mathbf{E}^{(+)}(\mathbf{r}t)|\{\alpha_k\}\rangle = \boldsymbol{\mathcal{E}}(\mathbf{r}t)|\{\alpha_k\}\rangle. \quad (10.1)$$

To discuss the general form which the field correlation functions take in such states it is convenient to abbreviate a set of coordinates $(\mathbf{r}_j,t_j)$ by a single symbol x_j. The nth-order correlation function is then defined as[3]

$$G_{\mu_1\cdots\mu_{2n}}^{(n)}(x_1\cdots x_{2n}) = \text{tr}\{\rho E_{\mu_1}^{(-)}(x_1)\cdots$$
$$\times E_{\mu_n}^{(-)}(x_n)E_{\mu_{n+1}}^{(+)}(x_{n+1})\cdots E_{\mu_{2n}}^{(+)}(x_{2n})\}. \quad (10.2)$$

The density operator for the coherent state defined by Eq. (10.1) is the projection operator

$$\rho = |\{\alpha_k\}\rangle\langle\{\alpha_k\}|. \quad (10.3)$$

For this operator it follows from Eq. (10.1) and its Hermitian adjoint that the correlation functions reduce to the factorized form

$$G_{\mu_1\cdots\mu_{2n}}^{(n)}(x_1\cdots x_{2n}) = \prod_{j=1}^{n}\mathcal{E}_{\mu_j}^*(x_j)\prod_{l=n+1}^{2n}\mathcal{E}_{\mu_l}(x_l). \quad (10.4)$$

In other words, the field which corresponds to the state $|\{\alpha_k\}\rangle$ satisfies the conditions for full coherence according to the definition[3] given earlier.

It is worth noting that the state $|\{\alpha_k\}\rangle$ is not the only one which leads to the set of correlation functions (10.4). Indeed, let us consider a state which corresponds not to the amplitudes $\{\alpha_k\}$, but to a set $\{e^{i\varphi}\alpha_k\}$ which differs by a common phase factor (i.e., φ is real and independent of k). Then the corresponding eigenvalue function becomes $e^{i\varphi}\boldsymbol{\mathcal{E}}(\mathbf{r}t)$, but such a change leaves the correlation functions (10.4) unaltered. It is clear from this invariance property of the correlation functions that certain mixtures of the coherent states also lead to the same set of functions. Thus, if $|\{\alpha_k\}\rangle$ is the state defined by Eq. (10.1), and $\mathcal{L}(\varphi)$ is any real-valued function of φ normalized in the sense

$$\int_0^{2\pi}\mathcal{L}(\varphi)d\varphi = 1, \quad (10.5)$$

we see that the density operator

$$\rho = \int_0^{2\pi}\mathcal{L}(\varphi)|\{e^{i\varphi}\alpha_k\}\rangle\langle\{e^{i\varphi}\alpha_k\}|d\varphi \quad (10.6)$$

leads for all choices of $\mathcal{L}(\varphi)$ to the set of correlation functions (10.4). Such a density operator is, of course, a special case of the general form (9.12), one which corresponds to an over-all uncertainty in the phase of the $\{\alpha_k\}$. The particular choice $\mathcal{L}(\varphi) = (2\pi)^{-1}$, which corresponds to complete ignorance of the phase, represents the usual state of our knowledge about high-frequency fields. We have been careful, therefore, to define coherence in terms of a set of correlation functions which are independent of the over-all phase.

Since nonstationary fields of many sorts can be represented by means of eigenvalue functions, it becomes a simple matter to construct corresponding quantum states. As an illustration we may consider the example of an amplitude-modulated plane wave. For this purpose we make use of the particular set of mode functions defined by Eq. (2.9). Then if the carrier wave has frequency ω and the modulation is periodic and has frequency $\zeta\omega$ where $0<\zeta<1$, we may write an appropriate eigenvalue function as

$$\mathcal{E}(\mathbf{r}t) = i\left(\frac{\hbar\omega}{2L^3}\right)^{1/2}\hat{e}^{(\lambda)}\alpha_{\mathbf{k}}$$
$$\times\{1+M\cos[\zeta(\mathbf{k}\cdot\mathbf{r}-\omega t)-\delta]\}e^{i(\mathbf{k}\cdot\mathbf{r}-\omega t)}. \quad (10.7)$$

When this expression is expanded in plane-wave modes it has only three nonvanishing amplitude coefficients. These are $\alpha_{\mathbf{k}}$ itself and the two sideband amplitudes

$$\alpha_{\mathbf{k}(1-\zeta)}=\tfrac{1}{2}M(1-\zeta)^{-1/2}e^{i\delta}\alpha_{\mathbf{k}}\,,\qquad \alpha_{\mathbf{k}(1+\zeta)}=\tfrac{1}{2}M(1+\zeta)^{-1/2}e^{-i\delta}\alpha_{\mathbf{k}}\,. \tag{10.8}$$

The coherent state which corresponds to the modulated wave may be constructed immediately from the knowledge of these amplitudes. In practice, of course, we will not often know the phase of $\alpha_{\mathbf{k}}$, and so the wave should be represented not by a single coherent state, but by a mixture of the form (10.6). Representations of other forms of modulated waves may be constructed similarly.

Incoherent fields, or the broad class of fields for which the correlation functions do not factorize, must be described by means of density operators which are more general in their structure than those of Eqs. (10.3) or (10.6). To illustrate the form taken by the correlation functions for such cases we may suppose the field to be described by the P representation of the density operator. Then the first-order correlation function is given by

$$G_{\mu\nu}^{(1)}(\mathbf{r}t,\mathbf{r}'t')=\int P(\{\alpha_k\})\sum_{k,k'}\tfrac{1}{2}\hbar(\omega\omega')^{1/2}u_{k\mu}^{*}(\mathbf{r})u_{k'\nu}(\mathbf{r}')\times\alpha_k^{*}\alpha_{k'}e^{i(\omega t-\omega' t')}\prod_l d^2\alpha_l\,. \tag{10.9}$$

Fields for which the P representation is inconveniently singular may, as we have noted earlier, always be described by means of analytic functions $R(\{\alpha_k^*\},\{\beta_k\})$ and corresponding density operators of the form (9.5). When that form of density operator is used to evaluate the first-order correlation function we find

$$G_{\mu\nu}^{(1)}(\mathbf{r}t,\mathbf{r}'t')=\int R(\{\alpha_k^*\},\{\beta_k\})\sum_{k',k''}\tfrac{1}{2}\hbar(\omega'\omega'')^{1/2}\times u_{k'\mu}^{*}(\mathbf{r})u_{k''\nu}(\mathbf{r}')\beta_{k'}^{*}\alpha_{k''}e^{i(\omega' t-\omega'' t')}\times\prod_l e^{\beta_l^*\alpha_l}d\mu(\alpha_l)d\mu(\beta_l)\,, \tag{10.10}$$

where the differentials $d\mu(\alpha_l)$ and $d\mu(\beta_l)$ are those defined by Eq. (5.14). The higher order correlation functions are given by integrals analogous to (10.9) and (10.10). Their integrands contain polynomials of the $2n$th degree in the amplitude variables α_k and β_k^* in place of the quadratic forms which occur in the first-order functions.

The energy spectrum of a radiation field is easily derived from a knowledge of its first-order correlation function. If we return for a moment to the expansion (2.19) for the positive-frequency field operator, and write the negative-frequency field as its Hermitian adjoint, we see that these operators obey the identity

$$2\int \mathbf{E}^{(-)}(\mathbf{r}t)\cdot\mathbf{E}^{(+)}(\mathbf{r}t')d\mathbf{r}=\sum_k \hbar\omega a_k^{\dagger}a_k\exp[i\omega(t-t')]. \tag{10.11}$$

If we take the statistical average of both sides of this equation we may write the resulting relation as

$$\sum_\mu \int G_{\mu\mu}^{(1)}(\mathbf{r}t,\mathbf{r}t')d\mathbf{r}=\tfrac{1}{2}\sum_k \hbar\omega\langle n_k\rangle\exp[i\omega(t-t')], \tag{10.12}$$

where $\langle n_k\rangle$ is the average number of photons in the kth mode. The Fourier representation of the volume integral of $\sum_\mu G_{\mu\mu}^{(1)}$ therefore identifies the energy spectrum $\hbar\omega\langle n_k\rangle$ quite generally.

For fields which may be represented by stationary density operators, it becomes still simpler to extract the energy spectrum from the correlation function. For such fields the weight function $P(\{\alpha_k\})$ depends only on the absolute values of the α_k, so that we have

$$\int P(\{\alpha_k\})\alpha_{k'}^{*}\alpha_{k''}\prod_l d^2\alpha_l=\langle|\alpha_{k'}|^2\rangle\delta_{k'k''}=\langle n_{k'}\rangle\delta_{k'k''}\,. \tag{10.13}$$

By using Eq. (10.9) to evaluate the correlation function, and specializing to the case of plane-wave modes, we then find

$$\sum_\mu G_{\mu\mu}^{(1)}(\mathbf{r}t,\mathbf{r}t')=\tfrac{1}{2}L^{-3}\sum_{\mathbf{k},\lambda}\hbar\omega\langle n_{\mathbf{k},\lambda}\rangle e^{i\omega(t-t')}\,, \tag{10.14}$$

in which we have explicitly indicated the role of the polarization index λ. If the volume which contains the field is sufficiently large in comparison to the wavelengths of the excited modes, the sum over the modes in Eq. (10.14) may be expressed as an integral over $\mathbf{k}$ space $[\sum_k\to\int L^3(2\pi)^{-3}d\mathbf{k}]$. By defining an energy spectrum for the quanta present (i.e., an energy per unit interval of ω) as

$$w(\omega)=(2\pi)^{-3}\hbar k^3\sum_\lambda\int\langle n_{\mathbf{k},\lambda}\rangle d\Omega_{\mathbf{k}}\,, \tag{10.15}$$

where $d\Omega_{\mathbf{k}}$ is an element of solid angle in $\mathbf{k}$ space, we may then rewrite Eq. (10.14) in the form

$$\sum_\mu G_{\mu\mu}^{(1)}(\mathbf{r}t,\mathbf{r}t')=\tfrac{1}{2}\int_0^\infty w(\omega)e^{i\omega(t-t')}d\omega\,. \tag{10.16}$$

With the understanding that $w(\omega)=0$ for $\omega<0$, we may extend the integral over ω from $-\infty$ to ∞. It is then clear that the relation (10.16) may be inverted to express the energy spectrum as the Fourier transform of the time-dependent correlation function,

$$w(\omega)=\frac{1}{\pi}\int_{-\infty}^{\infty}\sum_{\mu} G_{\mu\mu}^{(1)}(\mathbf{r}0,\mathbf{r}t)e^{i\omega t}dt. \qquad (10.17)$$

A pair of relations analogous to Eqs. (10.16) and (10.17), and together called the Wiener-Khintchine theorem, has long been of use in the classical theory of random fields.[21] The relations we have derived are, in a sense, the natural quantum mechanical generalization of the Wiener-Khintchine theorem. All we have assumed is that the field is describable by a stationary form of the P representation of the density operator. The proof need not, in fact, rest upon the use of the P representation since we can construct a corresponding statement in terms of the more general representation (9.5).

Stationary fields, according to Eq. (6.10), are represented by entire functions $R=\mathcal{S}(\{\alpha_k^*\beta_k\})$, i.e., functions which depend only on the set of products $\alpha_k^*\beta_k$. For such fields, then, the integral over the α and β planes which is required in Eq. (10.10) takes the form

$$\langle\beta_{k'}^*\alpha_{k''}\rangle=\int \mathcal{S}(\{\alpha_k^*\beta_k\})\beta_{k'}^*\alpha_{k''}\prod_l e^{\beta_l^*\alpha_l}d\mu(\alpha_l)d\mu(\beta_l). \qquad (10.18)$$

Since the range of integration of each of the α and β variables covers the entire complex plane, this integral cannot be altered if we change the signs of any of the variables. If, however, we replace the particular variables $\alpha_{k''}$ and $\beta_{k''}$ by $-\alpha_{k''}$ and $-\beta_{k''}$ the integral is seen to reverse in sign, unless we have

$$\langle\beta_{k'}^*\alpha_{k''}\rangle=\delta_{k'k''}\langle\beta_{k'}^*\alpha_{k'}\rangle. \qquad (10.19)$$

The average $\langle\beta_k^*\alpha_k\rangle$, we may note from Eqs. (5.11) and (6.5), is just the mean number of quanta in the kth mode,

$$\langle\beta_k^*\alpha_k\rangle=\mathrm{tr}\{\rho a^\dagger{}_k a_k\}=\langle n_k\rangle. \qquad (10.20)$$

We have thus shown that the general expression (10.10) for the first-order correlation function always satisfies Eq. (10.14) when the field is described by a stationary density operator. The derivation of the equations relating the energy spectrum to the time-dependent correlation function then proceeds as before.

The simplest and most universal example of an incoherent field is the type generated by superposing the outputs of stationary sources. We have shown in some detail in Sec. VIII that as the number of sources which contribute to the excitation of a single mode increases, the density operator for the mode takes on a Gaussian form in the P representation. It is not difficult to derive an analogous result for the case of sources which excite many modes at once. We shall suppose that the sources ($j=1\cdots N$) are essentially identical, and that their contributions to the excitation are described by a weight function $p(\{\alpha_{jk}\})$. The weight function $P(\{\alpha_k\})$ for the superposed fields is then given by the convolution theorem as

$$P(\{\alpha_k\})=\int\prod_k\delta^{(2)}\left(\alpha_k-\sum_{j=1}^{N}\alpha_{jk}\right)\prod_{j=1}^{N}p(\{\alpha_{jk}\})\prod_k d^2\alpha_{jk}. \qquad (10.21)$$

Since the individual sources are assumed to be stationary, the function $p(\{\alpha_{jk}\})$ will only depend on the variables α_{jk} through their absolute magnitudes, $|\alpha_{jk}|$.

The derivation which leads from Eq. (10.21) to a Gaussian asymptotic form for $P(\{\alpha_k\})$ is so closely parallel to that of Eqs. (8.1)–(8.8) that there is no need to write it out in detail. The argument makes use of second-order moments of the function p which may, with the same type of vector notation used previously, be written as

$$\langle\boldsymbol{\alpha}_k\boldsymbol{\alpha}_{k'}\rangle=\int\boldsymbol{\alpha}_k\boldsymbol{\alpha}_{k'}p(\{\boldsymbol{\alpha}_k\})\prod_l d^2\alpha_l. \qquad (10.22)$$

The stationary character of the function p implies that such moments vanish for $k\neq k'$. With this observation, we may retrace our earlier steps to show that the many-dimensional Fourier transform of P takes the form of a product of Gaussians, one for each mode and each similar in form to that of Eq. (8.6). It then follows immediately that the weight function P for the field as a whole is given by a product of Gaussian factors each of the form of Eq. (8.8). We thus have

$$P(\{\alpha_k\})=\prod_k\frac{1}{\pi\langle n_k\rangle}e^{-|\alpha_k|^2/\langle n_k\rangle}, \qquad (10.23)$$

where $\langle n_k\rangle$ is the average number of photons present in the kth mode when the fields are fully superposed. One of the striking features of this weight function is its factorized form. It is interesting to remember, therefore, that no assumption of factorizability has been made regarding the weight functions p which describe the individual sources. These sources may, indeed, be ones for which the various mode amplitudes are strongly coupled in magnitude. It is the stationary property of the sources which leads, because of the vanishing of the moments (10.22) for $k\neq k'$, to the factorized form for the weight function (10.23).

The density operator which corresponds to the Gaussian weight function (10.23) evidently describes an ideally random sort of excitation of the field modes. We may reasonably surmise that it applies, at least as a good approximation, to all of the familiar sorts of incoherent sources in laboratory use. It is clear, in particular, from the arguments of Sec. VII that the Gaussian weight function describes thermal sources

[21] The Wiener-Khintchine theorem is usually expressed in terms of cosine transforms since it deals with a real-valued correlation function for the classical field $\mathbf{E}$, rather than a complex one for the fields $\mathbf{E}^{(\pm)}$. The complex correlation functions are considerably more convenient to use for quantum mechanical purposes, as is shown in Ref. 3.

correctly. The substitution of the Planck distribution $\langle n_k\rangle=[\exp(\hbar\omega_k/\kappa T)-1]^{-1}$ into Eq. (10.23) leads to the density operator for the entire thermal radiation field. To the extent that the Gaussian weight function (10.23) may describe radiation by a great variety of incoherent sources there will be certain deep-seated similarities in the photon fields generated by all of them. One may, for example, think of these sources all as resembling thermal ones and differing from them only in the spectral distributions of their outputs. As a way of illustrating these similarities we might imagine passing blackbody radiation through a filter which is designed to give the spectral distribution of the emerging light a particular line profile. We may choose this artificial line profile to be the same as that of some true emission line radiated, say, by a discharge tube. We then ask whether measurements carried out upon the photon field can distinguish the true emission-line source from the artificial one. If the radiation by the discharge tube is described, as we presume, by a Gaussian weight function, it is clear that the two sources will be indistinguishable from the standpoint of any photon counting experiments. They are equivalent sorts of narrow-band, quantum-mechanical noise generators.

It is a simple matter to find the correlation functions for the incoherent fields[2] described by the Gaussian weight function (10.23). If we substitute this weight function into the expansion (10.9) for the first-order correlation function we find

$$G_{\mu\nu}^{(1)}(\mathbf{r}t,\mathbf{r}'t')=\tfrac{1}{2}\sum_k \hbar\omega u_{k\mu}^*(\mathbf{r})u_{k\nu}(\mathbf{r}')\langle n_k\rangle e^{i\omega(t-t')}. \quad (10.24)$$

When the mode functions $\mathbf{u}_k(\mathbf{r})$ are the plane waves of Eq. (2.9), and the volume of the system is sufficiently large, we may write the correlation function as the integral

$$G_{\mu\nu}^{(1)}(\mathbf{r}t,\mathbf{r}'t')=\frac{\hbar c}{2(2\pi)^3}\int\sum_\lambda e_\mu^{(\lambda)*}e_\nu^{(\lambda)}\langle n_{\mathbf{k},\lambda}\rangle k$$
$$\times\exp\{-i[\mathbf{k}\cdot(\mathbf{r}-\mathbf{r}')-\omega(t-t')]\}d\mathbf{k}, \quad (10.25)$$

in which the index λ again labels polarizations. To find the second-order correlation function defined by Eq. (10.2) we may write it likewise as an expansion in terms of mode functions. The only new moments of the weight function which we need to know are those given by $\langle|\alpha_k|^4\rangle=2\langle|\alpha_k|^2\rangle^2=2\langle n_k\rangle^2$. We then find that the second-order correlation function may be expressed in terms of the first-order function as

$$G_{\mu_1\mu_2\mu_3\mu_4}^{(2)}(x_1x_2,x_3x_4)=G_{\mu_1\mu_3}^{(1)}(x_1,x_3)G_{\mu_2\mu_4}^{(1)}(x_2,x_4)$$
$$+G_{\mu_1\mu_4}^{(1)}(x_1,x_4)G_{\mu_2\mu_3}^{(1)}(x_2,x_3). \quad (10.26)$$

It is easily shown that all of the higher order correlation functions as well reduce to sums of products of the first-order function. The nth-order correlation function may be written as

$$G_{\mu_1\cdots\mu_{2n}}^{(n)}(x_1\cdots x_n,x_{n+1}\cdots x_{2n})=\sum_{\mathcal{P}}\prod_{j=1}^{n}G_{\mu_j\nu_j}^{(1)}(x_j,y_j), \quad (10.27)$$

where the indices ν_j and the coordinates y_j for $j=1\cdots n$ are a permutation of the two sets $\mu_{n+1}\cdots\mu_{2n}$ and $x_{n+1}\cdots x_{2n}$, and the sum is carried out over all of the $n!$ permutations. One of the family resemblances which links all fields represented by the weight function (10.23) is that their properties may be fully described through knowledge of the first-order correlation function.

The fields which have traditionally been called coherent ones in optical terminology are easily described in terms of the first-order correlation function given by Eq. (10.25). Since the light in such fields is accurately collimated and nearly monochromatic, the mean occupation number $\langle n_{\mathbf{k},\lambda}\rangle$ vanishes outside a small volume of **k**-space. The criterion for accurate coherence is ordinarily that the dimensions of this volume be extremely small in comparison to the magnitude of **k**. It is easily verified, if the field is fully polarized, and the two points $(\mathbf{r},t)$ and $(\mathbf{r}',t')$ are not too distantly separated, that the correlation function (10.25) falls approximately into the factorized form of Eq. (2.4). That is to say, fields of the type we have described approximately fulfill the condition for first-order coherence.[3] It is easily seen, however, from the structure of the higher order correlation functions that these fields can never have second or higher order coherence. In fact, if we evaluate the function $G^{(n)}$ given by Eq. (10.27) for the particular case in which all of the coordinates are set equal, $x_1=\cdots=x_{2n}=x$, and all of the indices as well, $\mu_1=\cdots=\mu_{2n}=\mu$, we find the result

$$G_{\mu\cdots\mu}^{(n)}(x\cdots x,x\cdots x)=n![G_{\mu\mu}^{(1)}(x,x)]^n. \quad (10.28)$$

The presence of the coefficient $n!$ in this expression is incompatible with the factorization condition (10.4) for the correlation functions of order n greater than one. The absence of second or higher order coherence is thus a general feature of stationary fields described by the Gaussian weight function (10.23). There exists, in other words, a fundamental sense in which these fields remain incoherent no matter how monochromatic or accurately collimated they are. We need hardly add that other types of fields such as those generated by radio transmitters or masers may possess arbitrarily high orders of coherence.

During the completion of the present paper a note by Sudarshan[12] has appeared which deals with some of the problems of photon statistics that have been treated here.[22] Sudarshan has observed the existence of what

[22] In an accompanying note, L. Mandel and E. Wolf [Phys. Rev. Letters **10**, 276 (1963)] warmly defend the classical approach to photon problems. Some of the possibilities and fundamental limitations of this approach should be evident from our earlier work. We may mention that the "implication" they draw from Ref. 1 and disagree with cannot be validly inferred from any reading of that paper.

we have called the P representation of the density operator and has stated its connection with the representation based on the n-quantum states. To that extent, his work agrees with ours in Secs. VII and IX. He has, however, made a number of statements which appear to attach an altogether different interpretation to the P representation. In particular, he regards its existence as demonstrating the "complete equivalence" of the classical and quantum mechanical approaches to photon statistics. He states further that there is a "one-to-one correspondence" between the weight functions P and the probability distributions for the field amplitudes of classical theory.

The relation between the P representation and classical theory has already been discussed at some length in Secs. VII–IX. We have shown there that the weight function $P(\alpha)$ is, in general, an intrinsically quantum-mechanical structure and not derivable from classical arguments. In the limit $\hbar \rightarrow 0$, which corresponds to large amplitudes of excitation for the modes, the weight functions $P(\alpha)$ may approach classical probability functions as asymptotic forms. Since infinitely many quantum states of the field may approach the same asymptotic form, it is clear that the correspondence between the weight functions $P(\alpha)$ and classical probability distributions is not at all one-to-one.

ACKNOWLEDGMENTS

The author is grateful to the Research Laboratory of the American Optical Company and its director, Dr. S. M. MacNeille, for partial support of this work.

Reprinted from
JOURNAL OF MATHEMATICAL PHYSICS VOLUME 8, NUMBER 12 DECEMBER 1967

Weak Correspondence Principle

JOHN R. KLAUDER*
Bell Telephone Laboratories, Incorporated Murray Hill, New Jersey

(Received 5 April 1967)

The weak correspondence principle (WCP) for a scalar field states that the diagonal matrix elements

$$G(f,g) \equiv \langle f,g|\, \mathcal{G}\, |f,g\rangle$$

of a quantum generator $\mathcal{G}$ necessarily have the form of the appropriate classical generator G in which $f(\mathbf{x})$ and $g(\mathbf{x})$ are interpreted as the classical momentum and field, respectively. For a field operator $\varphi(\mathbf{x})$ and its canonically conjugate momentum $\pi(\mathbf{x})$ the states in question are given by

$$|f,g\rangle \equiv \exp\,\{i\int[\varphi(\mathbf{x})f(\mathbf{x}) - \pi(\mathbf{x})g(\mathbf{x})]\,d\mathbf{x}\}\,|0\rangle,$$

where $|0\rangle$ denotes the vacuum. The validity of the WCP is established for the six Euclidean generators (plus the Hamiltonian) of a Euclidean-invariant theory, and for the ten Poincaré generators of a Lorentz-invariant theory. Only general properties and certain operator domain conditions are essential to our argument. The WCP holds whether the representation of π and φ is irreducible or reducible; in the latter case, the WCP holds even if the vectors $|f,g\rangle$ do not span the Hilbert space, or even if the generator $\mathcal{G}$ is not a function solely of π and φ. Thus, the WCP is an exceedingly general and completely representation-independent connection between a classical theory and its quantum generators which is especially useful in the formulation of nontrivial, Euclidean-invariant quantum field theories.

1. INTRODUCTION

BASIC to any quantization procedure is a prescription for relating the quantum problem to its classical counterpart. The traditional guide in this respect has been a prescription which involves, essentially, a straightforward operator substitution for "coordinates" and "momenta" in the classical generators. In field theory, a normal ordering and subsequent renormalization is often necessary to make any sense of such a prescription. In this paper we shall establish a more general correspondence rule—the weak correspondence principle (WCP)—and show how it accounts for normal ordering and certain renormalizations in field theory. More significantly, however, we show that the WCP is far more general and that it applies in cases where the traditional prescription is manifestly incorrect. This latter aspect has been concretely demonstrated in recent analyses of the soluble "rotationally-symmetric" models.[1,2]

We confine our attention to a neutral scalar quantum field $\varphi(\mathbf{x}, t)$ in the presence of Euclidean-invariant self-interactions. In Sec. 2 some basic definitions and the general statement of the WCP for

* Present address: Department of Physics, Syracuse University, Syracuse, N.Y.

[1] J. R. Klauder, J. Math. Phys. **6**, 1666 (1965).
[2] H. D. I. Abarbanel, J. R. Klauder, and J. G. Taylor, Phys. Rev. **152**, 1198 (1966).

such fields are formulated. Applications to Euclidean- and Lorentz-invariant interactions are carried out in Secs. 3 and 4, respectively. Finally, in Sec. 5, we discuss the significance and utility of the WCP. In order to ensure clarity of the concepts involved we adopt a semiheuristic approach in this paper ignoring questions of operator domains, test-function smearing, and the like.

2. BASIC DEFINITIONS AND GENERAL STATEMENT OF WEAK CORRESPONDENCE PRINCIPLE

We deal with a neutral scalar field $\varphi(\mathbf{x})$ and its conjugate momentum $\pi(\mathbf{x})$, which at a common time satisfy the usual canonical commutation relations (CCR),

$$[\varphi(\mathbf{x}), \pi(\mathbf{x}')] = i\delta(\mathbf{x} - \mathbf{x}'), \tag{1}$$

where $\mathbf{x}$, $\mathbf{x}'$ are points in a three-dimensional Euclidean space. Unless otherwise stated, all operators are to be evaluated at time t, a variable which is generally suppressed. We especially use the unitary Weyl operators

$$U[f, g] \equiv \exp\left\{i\int [\varphi(\mathbf{x})f(\mathbf{x}) - \pi(\mathbf{x})g(\mathbf{x})]\, d\mathbf{x}\right\}, \tag{2}$$

which are defined for sufficiently many well-behaved functions $f(\mathbf{x})$ and $g(\mathbf{x})$. To fix the idea, it may be assumed that $f(\mathbf{x})$ and $g(\mathbf{x})$ are infinitely differentiable and fall off at infinity faster than any inverse power of $|\mathbf{x}|$. Evidently, we have $U[f, g]^\dagger = U[-f, -g]$. Some straightforward consequences of the CCR which we frequently use are

$$U[f, g] = \exp\left\{\tfrac{1}{2}i\int f(\mathbf{x})g(\mathbf{x})\, d\mathbf{x}\right\}U[0, g]U[f, 0] \tag{3a}$$

$$= \exp\left\{-\tfrac{1}{2}i\int f(\mathbf{x})g(\mathbf{x})\, d\mathbf{x}\right\}U[f, 0]U[0, g], \tag{3b}$$

and the "translation property"

$$U[f, g]^\dagger\{\alpha\varphi(\mathbf{x}) + \beta\pi(\mathbf{x})\}U[f, g] = \alpha\{\varphi(\mathbf{x}) + g(\mathbf{x})\} + \beta\{\pi(\mathbf{x}) + f(\mathbf{x})\}, \tag{4}$$

where α and β are complex numbers.

We introduce a distinguished set of unit vectors according to the definition

$$|f, g\rangle \equiv U[f, g]\,|0\rangle \tag{5}$$

for sufficiently many f and g. In our subsequent applications we regard $|0\rangle$ as the unique translationally invariant state, or equivalently as the unique ground state for the problem at hand. Generally, the vectors $|f, g\rangle$ are not mutually orthogonal for different arguments; on the contrary, telling information regarding the problem and the CCR representation leaves its imprint in the overlap $\langle f, g \mid f', g'\rangle$ between such vectors. While the states $|f, g\rangle$ are most useful when they span the Hilbert space $\mathfrak{H}$, the basic statements embodied in the weak correspondence principle do *not* require that they span $\mathfrak{H}$; the WCP remains valid in the subspace that these vectors do span. For notational convenience we shall not distinguish these cases and we refer simply to the collection of states $|f, g\rangle$ as the overcomplete family of states (OFS).

We are primarily interested in the *diagonal* OFS matrix elements

$$G(f, g) \equiv \langle f, g|\, \mathcal{G}\, |f, g\rangle$$

for various operators interpreted as quantum generators. Additionally, from (3), we may deduce that

$$G(f, g) = \langle 0|\, U[0, -g]U[-f, 0]\mathcal{G}\, U[f, 0]U[0, g]\,|0\rangle, \tag{6a}$$

$$= \langle 0|\, U[-f, 0]U[0, -g]\mathcal{G}\, U[0, g]U[f, 0]\,|0\rangle, \tag{6b}$$

where we note that

$$U[f, 0] = \exp\left\{i\int \varphi(\mathbf{x})f(\mathbf{x})\, d\mathbf{x}\right\}, \tag{7a}$$

$$U[0, g] = \exp\left\{-i\int \pi(\mathbf{x})g(\mathbf{x})\, d\mathbf{x}\right\}. \tag{7b}$$

Suppose, now, one seeks to choose an operator $\mathcal{G}$ to associate with some classical generator (e.g., Hamiltonian) $G(\pi_{cl}, \varphi_{cl})$ depending on the classical field $\varphi_{cl}(\mathbf{x})$ and its conjugate momentum $\pi_{cl}(\mathbf{x})$. In the traditional approach one adopts

$$\mathcal{G} = :G(\pi, \varphi): \tag{8}$$

for the quantum generator where the colons denote some sort of normal ordering. With this choice we see from (4) that

$$\langle 0|\, U[f, g]^\dagger :G(\pi, \varphi): U[f, g]\,|0\rangle = \langle 0|\, :G(\pi + f, \varphi + g):\,|0\rangle = G(f, g), \tag{9}$$

with all other terms vanishing. Thus one consequence of the traditional identification (8) is that diagonal OFS matrix elements of the quantum generator yield the classical generator with the understanding that

$$f(\mathbf{x}) \equiv \pi_{cl}(\mathbf{x}), \quad g(\mathbf{x}) \equiv \varphi_{cl}(\mathbf{x}). \tag{10}$$

This property is not limited to the usual definition of normal ordering (all creation operators to the left of all annihilation operators), but applies to the generalized normal ordering implicit in the generating functional

$$:U[f, g]: \equiv U[f, g]/\langle 0|\, U[f, g]\,|0\rangle,$$

whatever form the c-number denominator may take.

If the CCR representation is irreducible, then the prescription (8) is generally correct. However, if the CCR representation is *reducible*, then for many important operators the preceding prescription *necessarily fails*. This failure is often so complete that $\mathcal{G}$ cannot be expressed as *any* function solely of π and φ. Such is the case, for example, for the three generators of space translations $\mathcal{T}_k$, $k = 1, 2, 3$, and the Hamiltonian $\mathcal{H}$, if we assume that $|0\rangle$ is a nondegenerate eigenstate of these operators.[3] The space–time translation generators for a generalized free field are a case in point.

It is the purpose of the WCP to shed light on the situation even in those cases where (8) breaks down. The WCP states that the diagonal OFS matrix elements of quantum generators have the proper classical functional form independent of whether Eq. (8) holds or not. In symbols, if $\mathcal{G}$ is the quantum generator associated with the classical generator G, then the WCP states that these quantities are connected by the relation

$$G(f, g) = \langle f, g|\, \mathcal{G}\, |f, g\rangle \tag{11}$$

not only when (8) is true but even in cases where $\mathcal{G}$ is not any function solely of the π and φ.[4]

Roughly speaking, the WCP does not precast the CCR representation into the narrow confinements of an "irreducible mold"; on the contrary, the WCP leaves completely open the ultimate nature of the CCR representation. As was shown elsewhere,[1] such liberalism is necessary to achieve any solution at all in the case of the "rotationally-symmetric" models.

The remainder of this paper is devoted to verifying (11) for the principal generators in Euclidean- and Lorentz-invariant problems. We treat the simpler Euclidean-invariant cases first as preliminaries to the more interesting Lorentz-invariant cases.

3. WEAK CORRESPONDENCE PRINCIPLE IN EUCLIDEAN-INVARIANT THEORIES

For a given Euclidean-invariant theory we assume there exist three space-translation generators $\mathcal{T}_k$, three infinitesimal rotation generators $\mathcal{J}_k$, $k = 1, 2, 3$, having traditional commutation properties, and a positive Hamiltonian operator $\mathfrak{H}$, which commutes with the generators of the Euclidean group. We assume the state $|0\rangle$ is a simultaneous eigenstate such that (for all k)

$$\mathcal{T}_k\, |0\rangle = \mathcal{J}_k\, |0\rangle = \mathcal{H}\, |0\rangle = 0.$$

From Euclidean invariance of $|0\rangle$ it follows that

$$\nabla_k \langle 0|\, \varphi(\mathbf{x})\, |0\rangle = \nabla_k \langle 0|\, \pi(\mathbf{x})\, |0\rangle = 0. \tag{12}$$

Momentum Operators

To illustrate our basic idea most simply let us first consider the space-translation generators $\mathcal{T}_k$. Almost instinctively these operators are identified with the quantities $\int \pi \nabla_k \varphi \, d\mathbf{x}$ suitably normal ordered. Yet, as we have noted,[3] for a reducible CCR representation and unique translationally invariant state, the above identification is manifestly false.

We proceed to determine the three functionals

$$P_k(f, g) = \langle f, g|\, \mathcal{T}_k\, |f, g\rangle. \tag{13}$$

From the condition $\mathcal{T}_k\, |0\rangle = 0$, we conclude that $P_k(0, 0) = 0$. From $[\mathcal{T}_k, \varphi(\mathbf{x})] = -i\nabla_k\, \varphi(\mathbf{x})$ coupled with (6a), (4), and (12), we learn that

$$\begin{aligned}[\delta/\delta f(\mathbf{x})]P_k(f, g) &= i\langle f, g|\, [\mathcal{T}_k, \varphi(\mathbf{x})]\, |f, g\rangle \\ &= \langle f, g|\, \nabla_k \varphi(\mathbf{x})\, |f, g\rangle = \nabla_k g(\mathbf{x}).\end{aligned}$$

In similar fashion, $[\mathcal{T}_k, \pi(\mathbf{x})] = -i\nabla_k \pi(\mathbf{x})$ coupled with (6b), (4), and (12) leads to

$$\begin{aligned}[\delta/\delta g(\mathbf{x})]P_k(f, g) &= -i\langle f, g|\, [\mathcal{T}_k, \pi(\mathbf{x})]\, |f, g\rangle \\ &= -\langle f, g|\, \nabla_k \pi(\mathbf{x})\, |f, g\rangle \\ &= -\nabla_k f(\mathbf{x}).\end{aligned}$$

The only functionals consistent with these three conditions are easily seen to be

$$\langle f, g|\, \mathcal{T}_k\, |f, g\rangle = P_k(f, g) = \int f(\mathbf{x}) \nabla_k g(\mathbf{x})\, d\mathbf{x}. \tag{14}$$

Clearly if we interpret $g(\mathbf{x})$ as a *classical* c-number field and $f(\mathbf{x})$ as its conjugate momentum—as we hereafter shall—then (14) states that the diagonal expectation value of the quantum generator of space translations in the states $|f, g\rangle$ yields the classical generator of space translations. This is just the WCP as applied to the space-translation generators. In obtaining this result we note that no functional form for $\mathcal{T}_k$ was assumed, nor was it assumed that the states $|f, g\rangle$ span $\mathfrak{H}$.

[3] The field and momentum operators are reducible if they admit a common decomposition $\varphi = \varphi_1 \oplus \varphi_2$, $\pi = \pi_1 \oplus \pi_2$. If $\mathcal{H}$ (say) were a function solely of π and φ, then $\mathcal{H} = \mathcal{H}_1 \oplus \mathcal{H}_2$ which would violate the assumption that $|0\rangle$ is a nondegenerate eigenstate of $\mathcal{H}$. Consequently, $\mathcal{H}$ cannot be a function solely of π and φ. A similar argument applies to $\mathcal{T}_k$.

[4] The word "classical" as used here need only be taken to infer that the diagonal OFS matrix elements have a functional dependence on the fields f and g similar to that of the appropriate generator of the classical theory. No deterministic interpretation of the fields f and g is implied or intended; indeed, $\hbar$ is not set to zero but remains unaltered in the WCP (having the value one in the units chosen here). Further discussion of the WCP including the $\hbar$ dependence is contained in a related analysis for particle mechanics by J. R. Klauder, J. Math. Phys. **4**, 1058 (1963); **5**, 177 (1964).

Angular Momentum Operators

A similar computation determines the diagonal matrix elements

$$J_k(f, g) \equiv \langle f, g|\, \mathfrak{J}_k \,|f, g\rangle$$

for the rotation group generators. From $\mathfrak{J}_k\,|0\rangle = 0$ it follows that $J_k(0, 0) = 0$. Since

$$[\mathfrak{J}_k, \varphi(\mathbf{x})] = -i\epsilon_{klm}x_l\,\nabla_m\varphi(\mathbf{x}),$$

and similarly for $\pi(\mathbf{x})$, it follows that

$$[\delta/\delta f(\mathbf{x})]J_k(f, g) = i\langle f, g|\,[\mathfrak{J}_k, \varphi(\mathbf{x})]\,|f, g\rangle = \epsilon_{klm}x_l\nabla_m g(\mathbf{x}),$$

and

$$[\delta/\delta g(\mathbf{x})]J_k(f, g) = -i\langle f, g|\,[\mathfrak{J}_k, \pi(\mathbf{x})]\,|f, g\rangle = -\epsilon_{klm}x_l\nabla_m f(\mathbf{x}).$$

These conditions uniquely fix J_k and we find that

$$\langle f, g|\,\mathfrak{J}_k\,|f, g\rangle = J_k(f, g) = \int \epsilon_{klm}f(\mathbf{x})x_l\nabla_m g(\mathbf{x})\,d\mathbf{x}. \tag{15}$$

As is readily seen, the J_k are just the classical generators of infinitesimal rotations as required by the WCP.[5]

Hamiltonian

We now turn our attention to the generator of time translations, the Hamiltonian $\mathcal{H}$. We seek to study the functional

$$H(f, g) \equiv \langle f, g|\,\mathcal{H}\,|f, g\rangle \tag{16}$$

which is evidently real and (granted suitable spectral conditions on $\mathcal{H}$) is positive subject to the single exception $H(0, 0) = 0$ since $\mathcal{H}\,|0\rangle = 0$. From the condition $[\mathcal{H}, \varphi(\mathbf{x})] = -i\pi(\mathbf{x})$ coupled with (6a) and (4), it follows that

$$[\delta/\delta f(\mathbf{x})]H(f, g) = i\langle f, g|\,[\mathcal{H}, \varphi(\mathbf{x})]\,|f, g\rangle = \langle f, g|\,\pi(\mathbf{x})\,|f, g\rangle = f(\mathbf{x}),$$

since $\langle 0|\,\pi(\mathbf{x})\,|0\rangle = i\langle 0|\,[\mathcal{H}, \varphi(\mathbf{x})]\,|0\rangle \equiv 0$. Consequently we can put

$$H(f, g) = \frac{1}{2}\int f^2(\mathbf{x})\,d\mathbf{x} + W(g), \tag{17}$$

with $W(g) > 0$ excepting the case $g = 0$, where $W(0) = 0$. Euclidean invariance of $\mathcal{H}$ implies that $W(g)$ is invariant under spatial rotations and translations of $g(\mathbf{x})$. Further information about W is of course difficult to deduce in general but some features may be inferred from plausible requirements on $\mathcal{H}$.

If we introduce the notation

$$\pi(g) = \int \pi(\mathbf{x})g(\mathbf{x})d\mathbf{x},$$

then we may set

$$\begin{aligned} W(g) &= H(0, g) = \langle 0|\, e^{i\pi(g)}\mathcal{H}e^{-i\pi(g)}\,|0\rangle \\ &= \sum_{n=2}^{\infty}(n!)^{-1}i^n\langle 0|\,[\pi(g), [\pi(g), \cdots, [\pi(g), \mathcal{H}]\cdots]\,|0\rangle \\ &\equiv \sum_{n=2}^{\infty}(n!)^{-1}W_n(g) \\ &\equiv \sum_{n=2}^{\infty}(n!)^{-1}\int\cdots\int w_n(\mathbf{x}_1, \cdots, \mathbf{x}_n) \\ &\qquad \times g(\mathbf{x}_1)\cdots g(\mathbf{x}_n)\,d\mathbf{x}_1\cdots d\mathbf{x}_n, \end{aligned} \tag{18}$$

where $W_n(g)$ is homogeneous of degree n. If $W(-g) = W(g)$ is a plausible *even symmetry*, then $W_n(g) = 0$ for all odd integral n. For a *polynomial-type interaction* the series in (18) should terminate; indeed, for example, a general φ^4-type interaction Hamiltonian should at least be sensitive to the power *four* in such a way that a *five* fold multiple commutator of $\pi(g)$ with $\mathcal{H}$ vanishes identically. If we combine these arguments, then a φ^4-type theory should involve only $W_2(g)$ and $W_4(g)$.

In addition to limiting the number of terms which contribute to $W(g)$, some general properties of $w_n(\mathbf{x}_1, \cdots, \mathbf{x}_n)$ may be postulated in special cases. In particular, besides Euclidean invariance of these quantities, an *essentially local* theory may be defined as one for which all nonvanishing $w_n(\mathbf{x}_1, \cdots, \mathbf{x}_n)$ are distributions with but a single point of support at $\mathbf{x}_1 = \mathbf{x}_2 = \cdots = \mathbf{x}_n$ for all n. This is an important classification and, as we note below, it includes the relativistic interactions.

Let us examine the specific term $W_2(g)$ in somewhat greater detail. Evidently we have the several relations

$$\begin{aligned} W_2(g) &= -\langle 0|\,[\pi(g), [\pi(g), \mathcal{H}]]\,|0\rangle \\ &= 2\langle 0|\,\pi(g)\mathcal{H}\pi(g)\,|0\rangle \\ &= \iint w_2(\mathbf{x}_1, \mathbf{x}_2)g(\mathbf{x}_1)g(\mathbf{x}_2)\,d\mathbf{x}_1\,d\mathbf{x}_2 \\ &\equiv \int \tilde{w}_2(k)\,|\tilde{g}(\mathbf{k})|^2\,d\mathbf{k}. \end{aligned} \tag{19}$$

In this expression,

$$w_2(\mathbf{x}_1, \mathbf{x}_2) = 2\langle 0|\,\pi(\mathbf{x}_1)\mathcal{H}\pi(\mathbf{x}_2)\,|0\rangle,$$

$$\tilde{g}(\mathbf{k}) = (2\pi)^{-\frac{3}{2}}\int e^{-i\mathbf{k}\cdot\mathbf{x}}g(\mathbf{x})\,d\mathbf{x};$$

[5] It is evident from the preceding examples that any "kinematic" generator (i.e., one whose commutator with an arbitrary linear sum of φ and π is again a linear sum of φ and π) which annihilates the state $|0\rangle$ will fulfill the WCP.

also

$$\tilde{w}_2(k) = \int e^{-i\mathbf{k}\cdot\mathbf{x}} w_2(\mathbf{x}, 0)\, d\mathbf{x},$$

which by rotational invariance is a function only of $k = |\mathbf{k}|$. In the case of a unique ground state, then $\tilde{w}_2(k) > 0$ (almost everywhere). For an essentially local theory, $\tilde{w}_2$ is a polynomial in k^2; for a relativistic theory we show below that $\tilde{w}_2(k) = k^2 + m_0^2$.

In the simplest of examples, $W_2(g)$ is the only nonvanishing term in (18), and a free-field Hamiltonian results. As a "collection of independent oscillators" such a free-field theory admits a straightforward irreducible quantization whose CCR representation is characterized by the fact that

$$\langle f, g \mid 0\rangle = \exp\left\{-\frac{1}{4}\int [\tilde{w}_2(k)^{-\frac{1}{2}} |\tilde{f}(\mathbf{k})|^2 + \tilde{w}_2(k)^{\frac{1}{2}} |\tilde{g}(\mathbf{k})|^2]\, d\mathbf{k}\right\}.$$

In the present case, $\mathcal{H} = :H(\pi, \varphi):$ so that the validity of the WCP follows from Eq. (9). It may also be worth noting in this case that each distinct $\tilde{w}_2(k)$ corresponds to a unitarily inequivalent CCR representation.

Concrete examples of Euclidean-invariant models with nonvanishing interaction are the "rotationally symmetric" models.[1,2] Only discrete energy levels arise in these models so no scattering takes place. Elsewhere, we will present other Euclidean-invariant models with nonvanishing interaction [e.g., $W_4(g) \neq 0$] which exhibit both scattering and production. Some of these latter examples are essentially local in the sense discussed above. Each of these models exploits a reducible CCR representation, and although they satisfy no conventional prescription like Eq. (8), they nonetheless all fulfill the WCP.

4. WEAK CORRESPONDENCE PRINCIPLE IN RELATIVISTIC INVARIANT THEORIES

We divide our relativistic discussion into two parts: the first part is an extension of the arguments in the preceding section, while the second part is a covariant reformulation of the results of the first part.

Quite clearly the previous discussion of the momentum and angular momentum generators applies in the present case. We begin by examining the Hamiltonian somewhat further.

Hamiltonian

We note initially that local commutativity on a single spacelike surface (t = const) requires that

$$[\pi(\mathbf{x}_1), [\pi(\mathbf{x}_2), \mathcal{H}]] = 0; \quad \mathbf{x}_1 \neq \mathbf{x}_2,$$

and similarly for additional commutators with $\pi(\mathbf{x})$. Thus relativistic fields are essentially local fields in the previous sense. We next show in a two-stage analysis that covariance arguments enable us to determine the functional form of

$$W(g) = \sum_{n=2}^{\infty} (n!)^{-1} W_n(g)$$

in Eq. (18).

We may determine the second-order term $W_2(g)$ with the aid of the Lehmann representation for the two-point function. In particular, relativistic invariance leads to the well-known expression[6]

$$\langle 0|\, \varphi(\mathbf{x}) e^{-i\mathcal{H}t} \varphi(\mathbf{y})\, |0\rangle = \frac{1}{2(2\pi)^3} \iint e^{i\mathbf{k}\cdot(\mathbf{x}-\mathbf{y}) - i\omega t} \times \frac{d\mathbf{k}}{\omega} \rho(m^2)\, dm^2, \quad (20)$$

where $\omega^2 = k^2 + m^2$, and (since we assume CCR)

$$\int \rho(m^2)\, dm^2 = 1; \quad \rho(m^2) \geq 0.$$

Combining the relation

$$\begin{aligned}\langle 0|\, \varphi(\mathbf{x})\mathcal{H}^3\varphi(\mathbf{y})\, |0\rangle &= \langle 0|\, [\varphi(\mathbf{x}), \mathcal{H}]\mathcal{H}[\mathcal{H}, \varphi(\mathbf{y})]\, |0\rangle \\ &= \langle 0|\, \pi(\mathbf{x})\mathcal{H}\pi(\mathbf{y})\, |0\rangle = \tfrac{1}{2} w_2(\mathbf{x}, \mathbf{y})\end{aligned}$$

with the result of three time derivatives of (20) at the origin, we find that

$$w_2(\mathbf{x}, \mathbf{y}) = \frac{1}{(2\pi)^3} \iint e^{i\mathbf{k}\cdot(\mathbf{x}-\mathbf{y})} (k^2 + m^2)\, d\mathbf{k} \rho(m^2)\, dm^2.$$

Evidently, in the relativistic case, we may conclude that

$$\tilde{w}_2(k) = k^2 + m_0^2, \quad (21)$$

where

$$m_0^2 \equiv \int m^2 \rho(m^2)\, dm^2. \quad (22)$$

The latter expression is recognized as the usual definition of the bare mass, although it is deduced by different arguments.[7] In the present formulation, Eq. (22) becomes a consequence of the WCP. By way of summary at this point we note that (19) and (21) lead to

$$W_2(g) = \int [(\nabla g)^2 + m_0^2 g^2]\, d\mathbf{x}, \quad (23)$$

as befits the second-order contribution to a relativistic Hamiltonian.

The remainder of $W(g)$ is best treated as a unit.

---

[6] See, e.g., S. S. Schweber, *An Introduction to Relativistic Quantum Field Theory* (Harper & Row, Publishers, Inc., New York, 1962), p. 659.

[7] S. S. Schweber, Ref. 6, p. 667.

Let us introduce

$$V(g) \equiv W(g) - \tfrac{1}{2}W_2(g) = \sum_{n=3}^{\infty} (n!)^{-1}\, W_n(g) \quad (24)$$

as the classical interaction potential. When we discuss the relativity generators below we shall be able to show that $V(g)$ has a form given by

$$V(g) = \int V[g(\mathbf{x})]\, d\mathbf{x} = \sum_{n=3}^{\infty} (n!)^{-1} c_n \int g^n(\mathbf{x})\, d\mathbf{x}, \quad (25)$$

where c_n are constants.

Accepting this result for $V(g)$ temporarily, we learn in the relativistic case that the diagonal OFS matrix elements of the Hamiltonian have the form

$$\begin{aligned} H(f, g) &= \langle f, g|\, \mathcal{H}\, |f, g\rangle \\ &= \tfrac{1}{2}\int \{f^2(\mathbf{x}) + [\nabla g(\mathbf{x})]^2 + m_0^2 g^2(\mathbf{x})\}\, d\mathbf{x} \\ &\qquad + \int V[g(\mathbf{x})]\, d\mathbf{x}, \\ &\equiv \int H(\mathbf{x})\, d\mathbf{x}. \end{aligned} \quad (26)$$

Here, $H(\mathbf{x})$ is a nonnegative density defined in obvious fashion from $f(\mathbf{x})$, $\nabla g(\mathbf{x})$, and $g(\mathbf{x})$. If in addition we assume the conditions of (say) a φ^4-type theory as discussed in Sec. 3, then $H(f, g)$ would have the form (26) in which $V[g(\mathbf{x})] = \lambda g^4(\mathbf{x})$ for some positive constant λ. Evidently Eq. (26) gives just the desired family of classical relativistic Hamiltonians, and so fulfills the WCP in splendid fashion.

Relativity Transformations

We may adjoin three relativity-transformation generators $\mathcal{K}_k$, $k = 1, 2, 3$ (all defined at $t = 0$) to our Euclidean generators and Hamiltonian discussed above and make a set of Poincaré generators. The three functionals

$$K_k(f, g) = \langle f, g|\, \mathcal{K}_k\, |f, g\rangle$$

may be found as before. From $\mathcal{K}_k|0\rangle = 0$ there follows the condition $K_k(0, 0) = 0$. From

$$[\mathcal{K}_k, \varphi(\mathbf{x})] = -i[t\nabla_k\varphi(\mathbf{x}) + x_k\dot{\varphi}(\mathbf{x})],$$

we learn that

$$\begin{aligned} \frac{\delta}{\delta f(\mathbf{x})} K_k(f, g) &= i\langle f, g|\, [\mathcal{K}_k, \varphi(\mathbf{x})]\, |f, g\rangle \\ &= \langle f, g|\, t\nabla_k\varphi(\mathbf{x}) + ix_k[\mathcal{H}, \varphi(\mathbf{x})]\, |f, g\rangle \\ &= t\nabla_k g(\mathbf{x}) + x_k \frac{\delta}{\delta f(\mathbf{x})} H(f, g). \end{aligned}$$

Likewise,

$$[\mathcal{K}_k, \pi(\mathbf{x})] = -i[t\nabla_k\pi(\mathbf{x}) + \nabla_k\varphi(\mathbf{x}) + x_k\dot{\pi}(\mathbf{x})],$$

which is just the time derivative of the previous commutator, leads to

$$\begin{aligned} \frac{\delta}{\delta g(\mathbf{x})} K_k(f, g) &= -i\langle f, g|\, [\mathcal{K}_k, \pi(\mathbf{x})]\, |f, g\rangle \\ &= \langle f, g|\, -t\nabla_k\pi(\mathbf{x}) - \nabla_k\varphi(\mathbf{x}) \\ &\qquad - ix_k[\mathcal{H}, \pi(\mathbf{x})]\, |f, g\rangle \\ &= -t\nabla_k f(\mathbf{x}) - \nabla_k g(\mathbf{x}) \\ &\qquad + x_k \frac{\delta}{\delta g(\mathbf{x})} H(f, g). \end{aligned} \quad (27)$$

In terms of $H(\mathbf{x})$ [implicitly defined in (26)], the solution for the functionals K_k is readily seen to be

$$\begin{aligned} \langle f, g|\, \mathcal{K}_k\, |f, g\rangle &= K_k(f, g) \\ &= \int [f(\mathbf{x}) t\nabla_k g(\mathbf{x}) + x_k H(\mathbf{x})]\, d\mathbf{x}. \end{aligned} \quad (28)$$

This expression makes explicit use of the $(\nabla g)^2$ term in $H(\mathbf{x})$ [to win the term $-\nabla_k g(\mathbf{x})$ in (27)], and makes implicit use of the absence of any other field gradient in $H(\mathbf{x})$. Stated otherwise, if some term in $V(g)$ contained a field gradient, then the solution for $K_k(f, g)$ would differ from that in (28) over and above a mere redefinition of $H(\mathbf{x})$ to reflect the gradients in $V(g)$. However, with a standard argument we next show that only (28) can be correct thereby clinching the fact that $V(g)$ has no gradients and hence has the form stated in (25).

We recall the basic commutation relation

$$[\mathcal{K}_k, \mathcal{P}_l] = -i\delta_{kl}\mathcal{H}$$

among the Poincaré generators, which we can immediately restate as

$$e^{-ia_l\mathcal{P}_l}\mathcal{K}_k e^{ia_l\mathcal{P}_l} = \mathcal{K}_k + a_k\mathcal{H},$$

where a_k are the components of a three vector $\mathbf{a}$. Since

$$e^{ia_l\mathcal{P}_l}\, |f(\mathbf{x}), g(\mathbf{x})\rangle = |f(\mathbf{x} - \mathbf{a}), g(\mathbf{x} - \mathbf{a})\rangle,$$

it follows that $[f = f(\mathbf{x}), g = g(\mathbf{x})]$

$$\begin{aligned} K_k(f(\mathbf{x} - \mathbf{a}), g(\mathbf{x} - \mathbf{a})) &= \langle f, g|\, e^{-ia_l\mathcal{P}_l}\mathcal{K}_k e^{ia_l\mathcal{P}_l}\, |f, g\rangle \\ &= \langle f, g|\, (\mathcal{K}_k + a_k\mathcal{H})\, |f, g\rangle \\ &= K_k(f, g) + a_k H(f, g). \end{aligned}$$

This relation is fulfilled by (28) since $H(\mathbf{x})$ is the same density involved in defining $H(f, g)$. In turn, in order for this to be true, it is required that Eq. (25) hold, thus establishing its validity. This is the same argument used to prove Eq. (25) in a strictly classical theory.

In summary, we may combine the present results with those of Sec. 3 to conclude that in a relativistic theory all ten generators of the Poincaré group fulfill the WCP for some local interaction density $V[g(\mathbf{x})]$.

Additional conditions—such as those corresponding to a φ^4-type theory—may be imposed to restrict the form of the interaction potential.

Covariant Formulation of the Weak Correspondence Principle

In this section we recast the previous results into a Lorentz covariant form. We employ a metric with signature -2, and a unit surface-normal vector n^μ and three-volume element $d\sigma^\mu$, which in an appropriate frame have vanishing spacelike components and timelike components of 1 and $d\mathbf{x}$, respectively. Let σ denote the spacelike surface which in the same appropriate Lorentz frame is just the surface $t =$ const. Then we may relate the function pair $f(\mathbf{x})$ and $g(\mathbf{x})$ used previously to a space–time function $g(x)$ through the relations

$$g(\mathbf{x}) = g(x); \quad x \in \sigma,$$
$$f(\mathbf{x}) = n^\mu \partial_\mu g(x); \quad x \in \sigma.$$

Note by this that all previously used spatial coordinates are reinterpreted as contravariant components x^k as usual. With this notation, the covariant definition of the vectors in the OFS is given by

$$|f, g\rangle \equiv \exp\left\{i\int \varphi \overleftrightarrow{\partial}_\mu g \, d\sigma^\mu\right\} |0\rangle, \qquad (29)$$

which in the appropriate frame just reduces to (5).

The ten generators of the Poincaré group are given by the skew tensor $\mathcal{M}^{\alpha\beta}$ and four-momentum $\mathcal{P}^\mu$ which fulfill the standard commutation rules

$$[\mathcal{P}^\mu, \mathcal{M}^{\alpha\beta}] = -i(g^{\beta\mu}\mathcal{P}^\alpha - g^{\alpha\mu}\mathcal{P}^\beta),$$
$$[\mathcal{M}^{\alpha\beta}, \mathcal{M}^{\sigma\tau}] = -i(g^{\alpha\sigma}\mathcal{M}^{\beta\tau} - g^{\beta\sigma}\mathcal{M}^{\alpha\tau} + g^{\beta\tau}\mathcal{M}^{\alpha\sigma} - g^{\alpha\tau}\mathcal{M}^{\beta\sigma}).$$

The connection of these expressions with our previous generators is given by

$$\mathcal{P}_k = -\mathcal{P}^k, \quad \mathcal{H} = \mathcal{P}_0 = (\mathcal{P}^0), \quad \mathcal{K}_k = \mathcal{M}_{0k} \; (= \mathcal{M}^{0k}),$$
$$\mathcal{J}_k = \mathcal{M}_{lm} \; (= \mathcal{M}^{lm}); \; k, l, m = 1, 2, 3 \text{ cyclic}.$$

We may immediately generalize the diagonal OFS matrix elements to covariant form from their values in a specific frame. In particular, the covariant WCP for the ten Poincaré generators reads

$$\langle f, g| \mathcal{P}^\mu |f, g\rangle = P^\mu(f, g) = \int T^{\mu\nu}(x)\, d\sigma_\nu \qquad (30)$$

and

$$\langle f, g| \mathcal{M}^{\alpha\beta} |f, g\rangle = M^{\alpha\beta}(f, g) = \int [x^\alpha T^{\beta\nu}(x) - x^\beta T^{\alpha\nu}(x)]\, d\sigma_\nu. \qquad (31)$$

In these expressions, $T^{\mu\nu}(x)$ is the classical stress–energy tensor given by

$$T^{\mu\nu}(x) = \partial^\mu g(x)\partial^\nu g(x) - g^{\mu\nu}L(x),$$

where

$$L(x) \equiv \tfrac{1}{2}\{[\partial_\alpha g(x)]^2 - m_0^2 g^2(x)\} - V[g(x)]$$

as follows from our analysis in a specific frame coupled with Lorentz covariance.

In summary, the diagonal OFS matrix elements of the ten Poincaré generators yield the ten generators of a classical covariant theory expressed in the traditional form with the aid of a stress–energy tensor.

5. SUMMARY AND DISCUSSION

In the previous sections, we have shown that diagonal matrix elements of quantum generators in the states $|f, g\rangle$ yield the appropriate classical generators as required by the weak correspondence principle. Basically, these results are only contingent on certain operator domain conditions, and are not wedded to a three-dimensionality of space nor to a complete expandability of $W(g)$ in a power series. If the canonical operators π and φ are irreducible, then the WCP is equivalent to normal ordering. However, if π and φ are reducible—independent of whether the states $|f, g\rangle$ span $\mathcal{H}$ or not—the WCP is far more general. As already noted in such a case,[3] $\mathcal{H}$ cannot be solely a function of π and φ. It must clearly be understood that, in general, this property implies that no conventional field equations (i.e., $\ddot{\varphi} = F\{\varphi\}$ for some functional F of φ at a fixed time) can hold; hence no conventional τ-function equations, nor conventional Schwinger equations can hold. These features are not related to renormalizations and are easily illustrated for generalized free fields.

The utility of the WCP stems from the constraints it places on the CCR representation and the quantum generators, constraints which can serve as a guide in formulating the quantum theory. For example, given a representation of the operators π and φ, the WCP can help to test which operators if any can serve as generators appropriate to a specified classical theory. Generally, for a given CCR representation, it happens that *no* operator can be found which fulfills the WCP for a prescribed classical Hamiltonian and different CCR representations must be tried. Once a representation and generators compatible with the WCP are found, it follows that a consistent quantum theory exists. Sometimes it ends up that there are several quantum theories compatible with the WCP as, e.g., happens with a generalized free field in which only the first moment of the spectral weight is specified classically [cf., Eq. (22)]. For the "rotationally symmetric" models there are a two-parameter family of quantum solutions compatible with the WCP (these solutions may be labeled by the mass values of the two asymptotic "one-particle" states that arise in these models). Such nonuniqueness is by no means unexpected since the WCP, by itself, generally provides only a *partial* constraint on the quantum generator, especially if the states $|f, g\rangle$ do not even span $\mathfrak{H}$.

Because the reducible representations for π and φ which we advocate are somewhat unconventional, let us note here several features in their favor. For example, reducible representations readily take care of Haag's theorem which requires that the CCR representation of a nontrivial Euclidean-invariant theory be unitarily inequivalent to an irreducible, free-field Fock representation.[8] One fashionable irreducibility axiom is the so-called time-slice axiom, which roughly states that the operators $\varphi(\mathbf{x}, t)$ for all $\mathbf{x}$ and $|t| < \epsilon$, $\epsilon > 0$, are irreducible. From this axiom it need not follow that just $\varphi(\mathbf{x})$ and $\dot{\varphi}(\mathbf{x}) = \pi(\mathbf{x})$ at $t = 0$ (say) are irreducible. Also, since the commutation relations for fields are known to have uncountably many inequivalent irreducible representations, it is conceivable that a reducible representation could arise, loosely speaking, so as to be able to display some of this variety as so often occurs, e.g., with the rotation group.

Finally, although not directly related to this paper, it is interesting to note that recent results for relativistic Fermi fields (depending likewise on a few domain conditions) imply that a nontrivial relativistic theory fulfilling traditional anticommutation rules must necessarily employ reducible representations of the spinor field and its adjoint.[9] We hope to develop a weak correspondence principle for Euclidean- and Lorentz-invariant Fermi fields in a subsequent paper.

[8] A. S. Wightman, *Lecture Notes at the French Summer School of Theoretical Physics, Cargese Corsica, July*, 1964 (Gordon & Breach, to be published)

[9] R. T. Powers, Commun. Math. Phys. **4**, 145 (1967).

Reprinted from:

JOURNAL OF MATHEMATICAL PHYSICS VOLUME 11, NUMBER 2 FEBRUARY 1970

Exponential Hilbert Space: Fock Space Revisited

JOHN R. KLAUDER
Bell Telephone Laboratories, Murray Hill, New Jersey 07974

(Received 8 May 1969)

An exponential Hilbert space, which is an abstraction of the familiar Fock space for bosons, provides a natural framework to discuss a wide class of field-operator representations. This framework is especially convenient when wide invariance groups, such as a unique translationally invariant state, are involved. In this paper, we develop the theory of exponential Hilbert spaces in a functional fashion suitable to discuss representations of field operators enjoying such invariance features. Representations of both current algebras and canonical field operators are discussed, and it is shown that these representations are natural generalizations of those characterizing infinitely divisible random processes. Questions of reducibility and equivalence are treated, and we prove that our construction gives rise to infinitely many unitarily inequivalent representations. Nevertheless, an extremely simple expression, bilinear in annihilation and creation operators, abstractly characterizes the operators of both the current algebras and canonical fields. Dynamical applications to quantum field theory will be treated in subsequent papers.

1. INTRODUCTION

In numerous applications, and especially for dynamical considerations in quantum field theory, the representation of basic field operators becomes important. In a canonically formulated theory, for example, a commonly used representation is the familiar Fock representation, although it is known to be inapplicable for interacting theories possessing wide invariance groups.[1] Compatibility with the invariance requirements is the minimum demand we can impose on a representation for it to be relevant; and for our purposes, we demand that there exists a unique, translationally invariant state in the representation space. The general representation of the appropriate algebra, such as the commutation relations or a current algebra, lacks a translationally invariant state or at least a unique one. Thus, it is of some interest that numerous field-operator representations, consistent with the invariance conditions, can be constructed in the framework of exponential Hilbert spaces.[2] Such representations—which may

[1] R. Streater and A. S. Wightman, *PCT, Spin and Statistics, and All That* (W. A. Benjamin, New York, 1964); A. S. Wightman, *Lecture Notes at the French Summer School of Theoretical Physics, Cargese Corsica, July, 1964* (Gordon & Breach, New York, 1967); L. Streit, Bull. Am. Phys. Soc. **14,** 86 (1969).

[2] Rudimentary ideas regarding exponential Hilbert spaces appear in K. O. Friedrichs, *Mathematical Aspects of the Quantum Theory of Fields* (Interscience, New York, 1953), and in J. R. Klauder, Ann. Phys. (N.Y.) **11,** 123 (1960), pp. 133 and 134; for a more explicit construction, see H. Araki and E. J. Woods, Publ. Res. Inst. Math. Sci. (Kyoto), Ser.A, **2,** 157 (1966).

be called exponential representations—are the subject of this paper, along with the intimately related theory of exponential Hilbert spaces that "support" them. Sections 2 and 3 are devoted to our formulation of the exponential Hilbert space by appealing to familiar and intuitive properties of the usual Fock representation.[3] In Sec. 4 we develop the exponential representations of current algebras, while in Sec. 5 we discuss exponential solutions of the canonical commutation relations. In each case, questions of reducibility and equivalence of various representations are treated.

The analysis of these representations within the context of exponential Hilbert spaces has several advantages. In the first place, questions of invariance and uniqueness of a state in the representation space are easily dealt with. Secondly, it provides a natural class of representations with which to deal, a class which is a natural generalization of those characterizing the so-called infinitely divisible random variables.[4] It lends itself to a natural and unified approach to the representation of diverse algebras like current algebras and canonical operators. Moreover, it is difficult to envisage a representation consistent with having a unique translationally invariant state that does not fit into our framework, although, of course, such a wild conjecture is undoubtedly false.

In spite of the fact that there are infinitely many inequivalent representations of operator algebras—and, indeed, infinitely many inequivalent exponential representations—there is, nevertheless, a comparatively simple abstract operator solution that covers all our cases. To present this operator solution, suppose we consider as an example the canonical field operators $\varphi(\mathbf{x})$ and $\pi(\mathbf{y})$ for which

$$[\varphi(\mathbf{x}), \pi(\mathbf{y})] = i\delta(\mathbf{x} - \mathbf{y}). \tag{1.1}$$

The general form of exponential representation we are led to for these operators may be given as follows: Initially, let $\varphi_1(\mathbf{x})$, $\pi_1(\mathbf{y})$ denote a Fock representation (or a two-fold direct sum of such representations) which is obtained from a suitable linear combination of annihilation and creation operators, $A_1(\mathbf{x})$ and $A_1^\dagger(\mathbf{x})$. In addition, let us introduce an auxiliary, independent set of annihilation and creation operators, A_{2r} and $A_{2r}^\dagger$, $r = 1, 2, \cdots$. Then the class of solutions we are led to has the basic form given by

$$\varphi(\mathbf{x}) = \varphi_1(\mathbf{x}) + A_2^\dagger\tilde{\varphi}(\mathbf{x})A_2, \tag{1.2a}$$

$$\pi(\mathbf{x}) = \pi_1(\mathbf{x}) + A_2^\dagger\tilde{\pi}(\mathbf{x})A_2. \tag{1.2b}$$

Here $\tilde{\varphi}(\mathbf{x}) = \{\tilde{\varphi}_{rs}(\mathbf{x})\}$ and $\tilde{\pi}(\mathbf{x}) = \{\tilde{\pi}_{rs}(\mathbf{x})\}$ are general, commuting, formally self-adjoint operators defined on the index space of A_2, and

$$A_2^\dagger\tilde{\varphi}(\mathbf{x})A_2 \equiv \sum_{r,s} A_{2r}^\dagger\tilde{\varphi}_{rs}(\mathbf{x})A_{2s}, \quad \text{etc.} \tag{1.3}$$

Provided that the operators A_2 differ from a Fock representation by a suitable additive multiple of the identity operator, the resultant operators φ and π will have all the desired features. Suitable solutions to a current algebra have essentially the same structure in which the first (linear) terms are generally absent, and the analogs of the operators $\tilde{\varphi}(\mathbf{x})$ and $\tilde{\pi}(\mathbf{x})$ also fulfill the current algebra. To achieve the desired invariance, it is important that the representation of the operators A_2 is unitarily inequivalent to a Fock representation, but differs from one only by an additive multiple of unity. Such operators give rise to "translated Fock representations," which were among the earliest non-Fock representations to be studied.[5] Utilizing this fact, we may summarize by noting that all the basic field operators we consider admit a bilinear expansion in terms of annihilation and creation operators of an embedding, or "parental" Fock representation.

This comparatively simple and completely general form of solution makes it practical to study such representations for possible application to quantum dynamical models. Elsewhere, we shall study our solutions from this point of view and show that nontrivial model field theories having wide invariance groups can be constructed with their help.[6] Here we content ourselves with the formulation of exponential Hilbert spaces and the exponential representations they so naturally contain.

Earlier studies along these lines have been directed at continuous tensor product representations,[7] which are extensions of the notion of direct product representations. The form of inner product adopted therein did not always generate a Hilbert space with positive-definite metric and gave rise to only a limited class of solutions of the basic field algebras. Our formulation and results are more general in character, always yielding proper Hilbert spaces and yielding far larger classes of solutions of the basic equations. In turn, we feel, this has been achieved with considerable

[3] For an introductory treatment, see J. R. Klauder, *Coherence, Correlation and Quantum Field Theory, Brandeis Summer School, 1967* (Gordon & Breach, New York, to be published), Sec. 4.3.

[4] E. Lukacs, *Characteristic Functions* (Charles Griffin, London, 1960), p. 79; I. M. Gel'fand and N. Y. Vilenkin, *Generalized Functions, Vol. 4: Applications of Harmonic Analysis*, translated by A. Feinstein (Academic Press, Inc., New York, 1964), Chap. III.

[5] L. van Hove, Physica **18**, 145 (1952); K. O. Friedrichs, Ref. 2.

[6] J. R. Klauder, See also Ref. 3, Sec. 6.

[7] R. F. Streater, Nuovo Cimento **53A**, 487 (1968); D. A. Dubin and R. F. Streater, Nuovo Cimento **50**, 154 (1967); R. F. Streater, *Lectures at 1968 Karpacz Winter School, Karpacz, Poland* (to be published); R. F. Streater and A. Wulfsohn, Nuovo Cimento **57B**, 330 (1968); R. F. Streater, *Lectures at Varenna Summer School, 1968* (to be published). See also H. Araki and E. J. Woods, Publ. Res. Inst. Math. Sci. (Kyoto), Ser. A, **2**, 157 (1966), and A. Guichardet, Commun. Math. Phys. **5**, 262 (1967), for the introduction of continuous tensor products.

simplification in the mathematical prerequisites, since, in essence, we merely exploit Fock-space methods to the hilt. This does not mean that our notation remains uncomplicated; simpler notational choices could be made, but, in the interests of maintaining some over-all unity in our presentation, we felt it necessary to resist that temptation.

2. EXPONENTIAL HILBERT SPACE: MOTIVATION AND ABSTRACTION

A. Fock Representation and Coherent States

We begin with a brief review of well-known properties of the Fock representation of a countable collection of annihilation and creation operators, A_l, $A_m^\dagger$, respectively, which fulfill the commutation relations

$$[A_l, A_m^\dagger] = \delta_{lm}. \quad (2.1)$$

The Fock representation is singled out by the requirement that the collection of operators A_l and $A_m^\dagger$ is irreducible and that there is a normed state $|0\rangle$, the no-particle state, for which

$$A_l\,|0\rangle = 0 \quad (2.2)$$

for all l. States in which n excitations are present are defined by

$$|l_1, l_2, \cdots, l_n\rangle \equiv A_{l_1}^\dagger A_{l_2}^\dagger \cdots A_{l_n}^\dagger\,|0\rangle, \quad (2.3)$$

and the collection of such states for all l_j and all n span the Hilbert space $\mathfrak{H}$. If $\{z_l\}$ is a square-summable sequence of complex numbers, then the state

$$|\{z_l\}\rangle \equiv \exp(-\tfrac{1}{2}\sum|z_l|^2)\exp(\sum z_l A_l^\dagger)\,|0\rangle \quad (2.4)$$

defines a normalized vector in $\mathfrak{H}$ depending on $\{z_l\}$. These vectors constitute a specific overcomplete family of states (OFS), the so-called coherent states, and their properties are well known.[8] In particular, we note that

$$A_m\,|\{z_l\}\rangle = z_m\,|\{z_l\}\rangle, \quad (2.5)$$

i.e., that these states are eigenstates of the annihilation operators with eigenvalues given by the parameters z_l. The overlap of two such states is given by the formula

$$\begin{aligned}\langle\{z_k'\}\mid\{z_k\}\rangle &= \exp(-\tfrac{1}{2}\sum|z_k'|^2 - \tfrac{1}{2}\sum|z_k|^2 + \sum z_k^{*\prime} z_k)\\ &= \exp(-\tfrac{1}{2}\sum|z_k' - z_k|^2 + i\,\mathrm{Im}\sum z_k^{*\prime} z_k),\end{aligned} \quad (2.6)$$

which never vanishes. Although the vectors $|\{z_k\}\rangle$ are never orthogonal pairwise, they span the Hilbert space $\mathfrak{H}$, or, as we shall say, they constitute a total set $\mathfrak{T}$. A total set $\mathfrak{T}$ is characterized by the equivalent properties that every vector may be given as a linear combination (possibly not unique) of the vectors in $\mathfrak{T}$, or that a vector is the zero vector if its inner product vanishes with every element of $\mathfrak{T}$. In fact, various subsets of the coherent states, which we may call characteristic sets, also yield total sets. Analyticity arguments[9] indicate, for example, that as far as, say, the parameter z_1 goes, either a line segment or a set with a finite accumulation point are sufficient to generate a total set. Analyticity combined with growth restrictions[10] lead to the sufficiency of a lattice in Re z_1 and Im z_1, with density greater than one point per unit circle (i.e., an area of π). Square-integrability leads to the sufficiency of such a lattice with density equal to one point per unit circle,[11] which just turns out to be one point in phase space per Planck cell (i.e., an area $h = 2\pi\hbar$). Lattices with a density less than this value do not yield a total set for $\mathfrak{H}$.[12] From these examples it should be apparent that there is a wide variety of characteristic subsets of coherent states each of which yields a total set for $\mathfrak{H}$.

Coherent-state matrix elements of normally ordered operators are particularly simple. If $:B(A_l^\dagger, A_m):$ denotes a normally ordered operator—all creation operators to the left of all destruction operators—then it follows from (2.5) that

$$\langle\{z_k'\}|\,:B(A_l^\dagger, A_m):\,|\{z_k\}\rangle = B(z_l^{*\prime}, z_m)\langle\{z_k'\}\mid\{z_k\}\rangle. \quad (2.7)$$

The diagonal matrix elements become

$$\langle\{z_k\}|\,:B(A_l^\dagger, A_m):\,|\{z_k\}\rangle = B(z_l^*, z_m). \quad (2.8)$$

From this expression it is clear that the diagonal matrix elements actually determine the operator, since we may regard z_l^* and z_m as independent variables in the argument of B. Subsets of diagonal coherent state matrix elements may also suffice to uniquely determine the operator. As far as the variable z_1 goes, for example, any open interval in the complex plane suffices, or a set of points of the form $z_1 \equiv x_{1n} + iy_{1m}$, where $\{x_{1n}\}$ and $\{y_{1m}\}$ are two sets with a finite accumulation point.[13] On the other hand, the dependence of just $x_1 = \mathrm{Re}\,z_1$ does not suffice to determine B; information from both the real and imaginary parts must be provided.

---

[8] J. R. Klauder and E. C. G. Sudarshan, *Fundamentals of Quantum Optics* (W. A. Benjamin, New York, 1968), Chap. 7.

[9] V. Bargmann, Commun. Pure Appl. Math. **14**, 187 (1961).

[10] P. Butera and L. Girardello, "On the Completeness of the Coherent States" (University of Milan, preprint No. IFUM 084/FT); V. Bargmann (private communication).

[11] J. von Neumann, *Mathematical Foundations of Quantum Mechanics* (Princeton University Press, Princeton, N.J., 1955), p. 406; J. R. Klauder (unpublished).

[12] V. Bargmann (private communication).

[13] K. E. Cahill, Phys. Rev. **138**, B1566 (1965).

We conclude this elementary review with a discussion of Hermitian operators of the form

$$W = \sum_{l,m} A_l^\dagger w_{lm} A_m . \tag{2.9a}$$

For simplicity, we confine our initial remarks to the specialized operators for which

$$W = \sum_l w_l A_l^\dagger A_l . \tag{2.9b}$$

Since these involve a weighted sum of number operators $N_l = A_l^\dagger A_l$, we can directly determine that

$$\begin{aligned} e^{-iWt} |\{z_k\}\rangle &= \exp(-\tfrac{1}{2}\sum |z_k|^2) \exp(-iWt) \exp(\sum z_k A_k^\dagger) |0\rangle \\ &= \exp(-\tfrac{1}{2}\sum |z_k|^2) \exp(\sum z_k e^{-iWt} A_k^\dagger e^{iWt}) |0\rangle \\ &= \exp(-\tfrac{1}{2}\sum |z_k|^2) \exp(\sum z_k e^{-iw_k t} A_k^\dagger) |0\rangle \\ &\equiv |\{e^{-iw_k t} z_k\}\rangle. \end{aligned} \tag{2.10}$$

Consequently, we learn that

$$\langle\{z_k'\}| e^{-iWt} |\{z_k\}\rangle = \exp(-\tfrac{1}{2}\sum |z_k'|^2 - \tfrac{1}{2}\sum |z_k|^2 + \sum z_k^{*\prime} e^{-iw_k t} z_k). \tag{2.11}$$

Evidently, if we had considered the more general operator (2.9a), then the last sum in the exponent of (2.11) would be replaced by

$$\sum z_k^{*\prime} (e^{-iwt})_{kl} z_l . \tag{2.12}$$

If an operator of the form (2.9) represented the Hamiltonian $\mathcal{H}$, then the determination of the dynamics is simple and straightforward. More interesting, however, is the dynamics based on Hamiltonians of the form

$$\begin{aligned} \mathcal{H} = \sum A_k^\dagger \omega_{kl} A_l & \\ + \sum g_{klm} A_k^\dagger A_l^\dagger A_m + \text{h.c.} & \\ + \sum v_{klmp} A_k^\dagger A_l^\dagger A_m A_p , & \end{aligned} \tag{2.13}$$

including terms representing production and decay, scattering, etc. As we shall subsequently see, the exploitation of Fock space for "embedding" non-Fock representations of the basic field operators will permit the study of Hamiltonians of the complexity of (2.13) which exhibit nontrivial invariance groups, such as the Euclidean group of rotations and translations.

B. Abstraction to General Hilbert Space

The next stage in our development amounts to a relatively simple abstraction of the foregoing presentation. We note that square-summable, complex-valued sequences $\{z_l\}$ themselves constitute a Hilbert space, the so-called l^2. The vectors $|\{z_l\}\rangle \in \mathfrak{H}$ are then images of vectors in l^2 as elements in $\mathfrak{H}$. We may make these notions more general in character in the following way: Rather than restrict ourselves to l^2, let us consider a general Hilbert space $\mathfrak{h}$ with elements ψ, etc., denoted by lower case Greek letters. The inner product in this space will be denoted by (ψ', ψ) and takes the place of the expression $\sum z_k^{*\prime} z_k$ used earlier. The image vectors, previously denoted by $|\{z_k\}\rangle$, may now be called Ψ, the capital Greek letter associated with the corresponding element in $\mathfrak{h}$. The appropriate inner product in $\mathfrak{H}$ will be denoted by (Ψ', Ψ). This is, of course, just one of many possible notational conventions characterizing our abstraction. The essential point is that there are two Hilbert spaces involved, $\mathfrak{h}$ and $\mathfrak{H}$, and that there is a correspondence, i.e., a map, between the elements $\psi \in \mathfrak{h}$ and a *subset*, a *total set* $\mathfrak{T}$ of unit vectors $\Psi \in \mathfrak{H}$. To make the correspondence precise, we may restate the overlap (2.6) in the form

$$\begin{aligned} (\Psi', \Psi) &= e^{-\frac{1}{2}\|\psi'\|^2 - \frac{1}{2}\|\psi\|^2 + (\psi', \psi)} \\ &= N'N e^{(\psi', \psi)}. \end{aligned} \tag{2.14}$$

Here

$$N \equiv e^{-\frac{1}{2}\|\psi\|^2}, \tag{2.15a}$$

$$N' \equiv e^{-\frac{1}{2}\|\psi'\|^2} \tag{2.15b}$$

are normalization factors to ensure that

$$(\Psi, \Psi) = 1 \tag{2.16}$$

for all $\Psi \in \mathfrak{T}$.

Since $\mathfrak{T}$ is a total set, we may represent an arbitrary vector $X \in \mathfrak{H}$ as a linear sum of elements in $\mathfrak{T}$, such as

$$X = \sum_{n=1}^{\infty} c_n \Psi_n , \quad \Psi_n \in \mathfrak{T}. \tag{2.17}$$

The inner product of two such elements follows from (2.14), namely,

$$(X', X) = \sum c_m^{*\prime} c_n (\Psi'_m, \Psi_n), \tag{2.18}$$

or the matrix elements of an arbitrary operator $\mathcal{O}$ as

$$(X', \mathcal{O}X) = \sum c_m^{*\prime} c_n (\Psi'_m, \mathcal{O}\Psi_n). \tag{2.19}$$

For simplicity, we shall generally confine our attention to matrix elements between the special vectors in the total set $\mathfrak{T}$.

We may express the correspondence between ψ and Ψ in other ways as well. Let

$$\mathfrak{H}_{(n)} \equiv \bigotimes_{s=1}^{n} \mathfrak{h}_s \Big|_S \tag{2.20}$$

denote the symmetrized (hence the subscript S), n-fold tensor product of $\mathfrak{h}$ with itself. For $n = 0$, $\mathfrak{H}_{(0)} \equiv C$, the set of complex numbers. Then the connection between $\mathfrak{H}$ and $\mathfrak{h}$ may be stated essentially as

$$\mathfrak{H} = \bigoplus_{n=0}^{\infty} \mathfrak{H}_{(n)} = \bigoplus_{n=0}^{\infty} \left(\bigotimes_{s=1}^{n} \mathfrak{h}_s \right) \Big|_S . \tag{2.21}$$

In particular, the vectors $\Psi \in \mathfrak{T}$ and the vectors $\psi \in \mathfrak{h}$ are related as follows: For $\psi \in \mathfrak{h}$, we first set

$$\Psi'_{(n)} = \bigotimes_{s=1}^{n} \psi_s|_S \in \bigotimes_{s=1}^{n} \mathfrak{h}_s|_S = \mathfrak{H}_{(n)}, \tag{2.22}$$

where by definition, if $n = 0$, we choose $\Psi'_{(0)} = 1$. Then the appropriate Ψ is given by

$$\Psi = N \bigoplus_{n=0}^{\infty} \frac{1}{\sqrt{n!}} \Psi'_{(n)}$$
$$= N \bigoplus_{n=0}^{\infty} \frac{1}{\sqrt{n!}} \left(\bigotimes_{s=1}^{n} \psi_s \right) \Big|_S . \tag{2.23}$$

Here, as before, N is a normalization factor given by $\exp(-\frac{1}{2}\|\psi\|^2)$. It is evident that the inner product of two such vectors has the form given in (2.14). While all the vectors $\Psi \in \mathfrak{T}$ imaged by all vectors $\psi \in \mathfrak{h}$ constitute a total set, there are, according to our earlier discussion, "characteristic" subsets of $\mathfrak{h}$ for which the image vectors would still constitute a total set in $\mathfrak{H}$. We suggestively set $\mathfrak{H} = \exp \mathfrak{h}$ and refer to $\mathfrak{H}$ as an exponential Hilbert space.

Along with our abstraction of the Hilbert spaces involved we shall generalize the operators. We take Eq. (2.5) as an essential defining property of the annihilation operators and abstract that relation as follows: To every $\lambda \in \mathfrak{h}$ we associate an operator $A(\lambda)$, such that

$$A(\lambda)\Psi = (\lambda, \psi)\Psi \tag{2.24}$$

for every $\Psi \in \mathfrak{T}$ imaged by $\psi \in \mathfrak{h}$. We note that $A(\lambda)$ is *antilinear* in λ, and, permitting ourselves a notational abuse, we could suggestively write (λ, A) for $A(\lambda)$. It is clear that, as straightforward abstractions of earlier relations,

$$(\Psi', A(\lambda)\Psi) = (\lambda, \psi)(\Psi', \Psi), \tag{2.25a}$$
$$(\Psi', A(\lambda)^\dagger\Psi) = (\psi', \lambda)(\Psi', \Psi), \tag{2.25b}$$
$$(\Psi', A(\lambda_1)^\dagger \cdots A(\lambda_n)^\dagger A(\mu_1) \cdots A(\mu_m)\Psi)$$
$$= (\psi', \lambda_1) \cdots (\psi', \lambda_n)(\mu_1, \psi) \cdots (\mu_m, \psi) \times (\Psi', \Psi), \tag{2.25c}$$

as well as

$$[A(\lambda), A(\lambda')^\dagger] = (\lambda, \lambda') \tag{2.26}$$

for any two elements $\lambda, \lambda' \in \mathfrak{h}$.

To every bounded operator b on $\mathfrak{h}$ we can associate a particular operator B on $\mathfrak{H}$ defined as

$$B = \bigoplus_{n=0}^{\infty} B_{(n)} = \bigoplus_{n=0}^{\infty} \bigotimes_{s=1}^{n} b, \tag{2.27}$$

where $B_{(0)} \equiv 1$, which maps a vector $\Psi \in \mathfrak{T}$ into another such vector apart from normalization. Specifically,

$$(\Psi', B\Psi) = N'Ne^{(\psi', b\psi)}, \tag{2.28}$$

from which we see that $B\Psi/\|B\Psi\| \in \mathfrak{T}$. If $b \equiv e^{-iwt}$, where w is a self-adjoint operator on $\mathfrak{h}$, then b is unitary. As the direct sum of direct products of unitary operators,

$$B = \bigoplus_{n=0}^{\infty} \bigotimes_{s=1}^{n} e^{-iwt} \equiv e^{-iWt} \tag{2.29}$$

defines a unitary operator on $\mathfrak{H}$ mapping $\mathfrak{T}$ onto $\mathfrak{T}$. Specifically,

$$(\Psi', e^{-iWt}\Psi) = N'N \exp(\psi', e^{-iwt}\psi), \tag{2.30}$$

which associates a self-adjoint operator W on $\mathfrak{H}$ for every self-adjoint operator w on $\mathfrak{h}$. From the relation

$$(\Psi', W\Psi) = (\psi', w\psi)(\Psi', \Psi), \tag{2.31}$$

which follows from (2.30), it is clear on comparison with (2.25c) that W is bilinear in $A^\dagger$ and A. Specifically, if $\{\lambda_n\}$ constitutes a complete orthonormal set in $\mathfrak{h}$, then

$$W = \sum_{n,m} A^\dagger(\lambda_n)(\lambda_n, w\lambda_m)A(\lambda_m), \tag{2.32}$$

which we may suggestively abbreviate by

$$W = (A, wA) \tag{2.33a}$$

or just by

$$W = A^\dagger wA, \tag{2.33b}$$

as we did in Eq. (1.3). We note that the diagonal matrix elements for such operators are given simply by

$$(\Psi, W\Psi) = (\psi, w\psi), \tag{2.34}$$

since $\|\Psi\| = 1$. As noted in Sec. 2A, the diagonal elements $W(\Psi) \equiv (\Psi, W\Psi)$ for all $\Psi \in \mathfrak{T}$ uniquely determine W. In the present special example this is evident from the fact that we know $w(\psi) \equiv (\psi, w\psi)$ for all $\psi \in \mathfrak{h}$, and by the polarization identity[14] we therefore know $(\psi', w\psi)$ for all $\psi', \psi \in \mathfrak{h}$ (strictly speaking in the domain of w). According to (2.31), this is sufficient to fix $(\Psi', W\Psi)$.

Let us briefly examine the diagonal elements of (2.30) further. Since $N = \exp(-\frac{1}{2}\|\psi\|^2)$, it follows that

$$(\Psi, e^{-iWt}\Psi) = e^{(\psi, (e^{-iwt}-1)\psi)}. \tag{2.35}$$

The left-hand side is the characteristic function for the distribution of W in the state Ψ and acts both as a generator of the moments and of the linked moments (subscript L, sometimes called the cumulants, which

[14] F. Riesz and B. Sz.-Nagy, *Functional Analysis* (Fredrick Ungar, New York, 1955), p. 211.

are related to truncated functions and connected diagrams in field theory). In particular,

$$(\Psi, e^{-iWt}\Psi) \equiv \langle e^{-iWt}\rangle = \sum_{m=0}^{\infty} \frac{(-it)^m}{m!} \langle W^m\rangle$$
$$\equiv \exp\{\langle e^{-iWt} - 1\rangle_L\}$$
$$= \exp\left\{\sum_{m=1}^{\infty} \frac{(-it)^m}{m!} \langle W^m\rangle_L\right\}. \quad (2.36)$$

On identification with (2.35), we see that

$$(\psi, (e^{-iwt} - 1)\psi) \equiv \langle e^{-iwt} - 1\rangle = \sum_{m=1}^{\infty} \frac{(-it)^m}{m!} \langle w^m\rangle$$
$$= \langle e^{-iWt} - 1\rangle_L = \sum_{m=1}^{\infty} \frac{(-it)^m}{m!} \langle W^m\rangle_L, \quad (2.37)$$

namely, that the mth *linked* moment $\langle W^m\rangle_L$ equals the mth *ordinary* moment $\langle w^m\rangle$ as calculated in $\mathfrak{h}$.

To make these notions precise, we need to make some remarks regarding operator domains. We recall that in order for $\psi \in \mathfrak{d}_{w^p}$, i.e., for $w^p\psi \in \mathfrak{h}$, it is necessary and sufficient that

$$\frac{d^{2p}}{dt^{2p}} (\psi, e^{-iwt}\psi) \quad (2.38)$$

exist at $t = 0$.[15] A parallel criterion establishes the conditions for $W^p\Psi \in \mathfrak{H}$. In view of the direct connection between the characteristic functions afforded by (2.35), we see that $\psi \in \mathfrak{d}_{w^p}$ implies that $\Psi \in \mathfrak{D}_{W^p}$, and conversely.

From the prescribed form of the inner product (2.14), it is not difficult to establish the following inequalities:

$$2[1 - \exp(-\tfrac{1}{2}\|\psi_1 - \psi_2\|^2)] \leq \|\Psi_1 - \Psi_2\|^2$$
$$\leq 2(\|\psi_1\| + \|\psi_2\|)\,\|\psi_1 - \psi_2\|, \quad (2.39)$$

for any two elements $\psi_1, \psi_2 \in \mathfrak{h}$ and their exponential images.[16] It follows that a stongly convergent sequence of vectors in $\mathfrak{h}$ is imaged into a strongly convergent sequence of vectors in $\mathfrak{T}$, and conversely. Another convergence condition follows from the relation

$$(\Psi', N_1^{-1}\Psi_1 - N_2^{-1}\Psi_2) = N'[e^{(\psi',\psi_1)} - e^{(\psi',\psi_2)}]. \quad (2.40)$$

From this relation we conclude that a weakly convergent sequence of vectors in $\mathfrak{h}$ is imaged into a weakly convergent sequence of vectors in $\mathfrak{H}$ [of the special form $N^{-1}\Psi$, $\Psi \in \mathfrak{T}$, where $N^{-1} = \exp(\tfrac{1}{2}\|\psi\|^2)$, $\psi \in \mathfrak{h}$], and conversely.

[15] E. Lukacs, *Characteristic Functions* (Charles Griffin, London, 1960), p. 29, Theorem 2.3.1, Corollaries 1 and 2.

[16] The first inequality arises from the fact that $\cos\theta \leq 1$; the second inequality is derived in Ref. 8, p. 113.

3. EXPONENTIAL HILBERT SPACE: LABEL SPACES AND REDUCED PARAMETERIZATION

We develop the formalism of Sec. 2 one further stage by considering the consequences of a reduced parameterization. Rather than considering every $\psi \in \mathfrak{h}$ and its image $\Psi \in \mathfrak{T} \subset \mathfrak{H}$, let us consider the vectors, say, $\varphi \in \mathfrak{s} \subset \mathfrak{h}$ and their images $\Phi \in \mathfrak{S} \subset \mathfrak{T} \subset \mathfrak{H}$. It is especially convenient at this point to introduce a label space $\mathfrak{L}$ and to regard φ as a map from points $l \in \mathfrak{L}$ to vectors $\varphi[l] \in \mathfrak{s}$, a subset of $\mathfrak{h}$. That is, we imagine labeling the vectors in $\mathfrak{s}$ by points $l \in \mathfrak{L}$. Clearly, we can use the same label to denote the image vector $\Phi = \Phi[l] \in \mathfrak{S}$ in the exponential Hilbert space $\mathfrak{H} = \exp \mathfrak{h}$. In this notation we have

$$(\Phi[l_1], \Phi[l_2]) = N_1 N_2 e^{(\varphi[l_1],\varphi[l_2])}, \quad (3.1a)$$

where

$$N = e^{-\frac{1}{2}\|\varphi[l]\|^2}. \quad (3.1b)$$

We emphasize that this relation is nothing but an alternative parameterization for a subset of the vectors introduced in Sec. 2B. The question of whether or not

$$\mathfrak{S} \equiv \{\Phi[l] : l \in \mathfrak{L}\} \quad (3.2a)$$

is a total set for $\mathfrak{H}$ depends both on $\mathfrak{L}$ and on the particular labeling. Obviously, the question can be turned around to be one for

$$\mathfrak{s} \equiv \{\varphi[l] : l \in \mathfrak{L}\}. \quad (3.2b)$$

In a manner of speaking, in analogy with the discussion of the coherent states in Sec. 2A, it is sufficient if $\mathfrak{s}$ is a "characteristic set" in some sense. While we shall tacitly assume such to be the case, our main conclusions often do not hinge on this assumption, and, occasionally, it is useful to regard the total set spanned by $\mathfrak{S}$ to be a proper subset of $\mathfrak{H}$.

All of the operator equations of Sec. 2B have immediate applicability. For example, we have

$$A(\lambda)\Phi[l] = (\lambda, \varphi[l])\Phi[l], \quad (3.3)$$

$$(\Phi[l'], e^{-iWt}\Phi[l]) = N'N e^{(\varphi[l'], e^{-iwt}\varphi[l])}, \quad \text{etc.} \quad (3.4)$$

Let us temporarily digress at this point to present an example which should help clarify some of the concepts we have introduced as well as point the way to the next stage of formal development.

Example: In the light of the formalism developed, suppose we wish to determine unitary representations of the one-parameter group

$$U(t) = e^{-iWt}, \quad (3.5)$$

which fulfills the combination law

$$U(t)U(t') = U(t + t'), \quad (3.6)$$

for all $t, t' \in R$, the real line. For the label space $\mathfrak{L}$, we choose the real line R, and, in fact, we choose the naturally induced parameterization given by

$$\Phi[t] \equiv U(t)\Phi_0, \tag{3.7}$$

where Φ_0 is some fixed fiducial vector and $U(t)$ is the representation under study. If we now insist that this representation be realized through an exponential construction, then we require that

$$(\Phi[t'], \Phi[t]) = \exp\{-\tfrac{1}{2}\|\varphi[t']\|^2 - \tfrac{1}{2}\|\varphi[t]\|^2 + (\varphi[t'], \varphi[t])\}. \tag{3.8a}$$

According to (3.6), this expression also equals

$$(\Phi_0, \Phi[t-t']) = \exp\{-\tfrac{1}{2}\|\varphi[0]\|^2 - \tfrac{1}{2}\|\varphi[t-t']\|^2 + (\varphi[0], \varphi[t-t'])\}. \tag{3.8b}$$

Equating real and imaginary parts in the exponent, we are led to the two conditions [cf. Eq. (2.6)]

$$\|\varphi[t] - \varphi[t']\|^2 = \|\varphi[0] - \varphi[t-t']\|^2, \tag{3.9a}$$

$$\operatorname{Im}(\varphi[t'], \varphi[t]) = \operatorname{Im}(\varphi[0], \varphi[t-t']). \tag{3.9b}$$

As a solution to these relations, we choose

$$\varphi[t] = a + bt \oplus \int^{\oplus} e^{-ixt}\, d\mu(x), \tag{3.10}$$

where a and b are fixed complex numbers and where the inner product is defined by

$$(\varphi[t'], \varphi[t]) = (a^* + b^*t')(a + bt) + \int e^{ix(t'-t)}\, d\mu(x). \tag{3.11}$$

Observe that

$$\|\varphi[t]\|^2 \geq \int d\mu(x) \equiv \mu(R), \tag{3.12}$$

which means [if $\mu(R) > 0$] that $\varphi[t]$ is never the zero vector. In this realization,

$$\mathfrak{h} = C \oplus \int^{\oplus} C\, d\mu(x), \tag{3.13}$$

for some measure μ, where C denotes the space of complex numbers. If we set $c \equiv \operatorname{Im} ba^*$, then the representations we have found have characteristic functions of the form

$$\begin{aligned}(\Phi_0, e^{-iWt}\Phi_0) &= (\Phi_0, \Phi[t]) \\ &= \exp\left[ict - \tfrac{1}{2}|b|^2 t^2 - \int (1 - e^{-ixt})\, d\mu(x)\right],\end{aligned} \tag{3.14}$$

which may be recognized as the characteristic function of an infinitely divisible random variable.[4] However, our derivation has not yielded the most general such characteristic function, since we necessarily have $\mu(R) < \infty$. In essence, the formal development we next take up is aimed at rectifying this deficiency and its analog in related examples.

A. Translated Parameterization

We now take up, in the abstract framework, the consequences of a translated parameterization and, subsequently, return to our example for illustration. Let us assume we are given a set $\mathfrak{s} \subset \mathfrak{h}$ and have constructed the exponential map from $\varphi[l] \in \mathfrak{s}$ to $\Phi[l] \in \mathfrak{S}$. We now introduce a family of phase-related image vectors according to the rule

$$\Phi'[l] \equiv e^{-i \operatorname{Im}(\xi, \varphi[l])}\Phi[l], \tag{3.15}$$

where ξ is some fixed element of $\mathfrak{h}$. Clearly,

$$(\Phi'[l_1], \Phi'[l_2]) = (\Phi[l_1], \Phi[l_2])e^{i \operatorname{Im}\{(\xi, \varphi[l_1]) - (\xi, \varphi[l_2])\}}, \tag{3.16}$$

which can be manipulated to become

$$\begin{aligned}(\Phi'[l_1], \Phi'[l_2]) &= \exp\{-\tfrac{1}{2}\|\varphi[l_1] - \xi\|^2 - \tfrac{1}{2}\|\varphi[l_2] - \xi\|^2 \\ &\quad + (\varphi[l_1] - \xi, \varphi[l_2] - \xi)\} \\ &\equiv N_1' N_2' e^{(\varphi'[l_1], \varphi'[l_2])},\end{aligned} \tag{3.17}$$

where

$$\varphi'[l] \equiv \varphi[l] - \xi. \tag{3.18}$$

From this equation we can read that a uniform translation of the set $\mathfrak{s}$, i.e., $\varphi[l] \to \varphi'[l] = \varphi[l] - \xi$, leads to an associated phase change in the image vectors in $\mathfrak{H}$. Note that

$$\operatorname{Im}(\xi, \varphi[l]) = \operatorname{Im}(\xi, \varphi[l] - \xi) = \operatorname{Im}(\xi, \varphi'[l]), \tag{3.19}$$

so that the phase change may be given by (3.15) or by

$$\Phi'[l] = e^{-i \operatorname{Im}(\xi, \varphi'[l])}\Phi[l]. \tag{3.20}$$

Certain matrix elements are insensitive to such phase changes. In particular, for any operator B it follows that

$$\frac{(\Phi'[l_1], B\Phi'[l_2])}{(\Phi'[l_1], \Phi'[l_2])} = \frac{(\Phi[l_1], B\Phi[l_2])}{(\Phi[l_1], \Phi[l_2])} \tag{3.21}$$

and, thus, for diagonal matrix elements, that

$$(\Phi'[l], B\Phi'[l]) = (\Phi[l], B\Phi[l]). \tag{3.22}$$

The annihilation operator associated with the translated parameterization is defined by the property that

$$A'(\lambda)\Phi'[l] = (\lambda, \varphi'[l])\Phi'[l]. \tag{3.23a}$$

Since a phase factor cannot affect this, we must also have

$$A'(\lambda)\Phi[l] = (\lambda, \varphi'[l])\Phi[l] = (\lambda, \varphi[l] - \xi)\Phi[l]. \tag{3.23b}$$

It follows that

$$A'(\lambda) = A(\lambda) - (\lambda, \xi) \tag{3.24a}$$

and, consequently,

$$A'(\lambda)^\dagger = A(\lambda)^\dagger - (\xi, \lambda). \tag{3.24b}$$

By direct computation we see that

$$[A'(\lambda_1), A'(\lambda_2)^\dagger] = [A(\lambda_1), A(\lambda_2)^\dagger] = (\lambda_1, \lambda_2). \tag{3.25}$$

The representation A' of the commutation relations is, in fact, unitarily equivalent to the representation A, since we have assumed $\xi \in \mathfrak{h}$.[17] Specifically,

$$V^{-1}A'(\lambda)V = A(\lambda), \quad \text{for all} \quad \lambda \in \mathfrak{h}, \tag{3.26}$$

where

$$V = \exp\,[A(\xi)^\dagger - A(\xi)]. \tag{3.27}$$

(Shortly we shall break the assumption that $\xi \in \mathfrak{h}$ and actually be dealing with unitarily inequivalent representations A' and A.)

The kind of change involved in the annihilation and creation operators carries over to many other operators as well. Suppose we consider

$$(\Phi'[l_1], e^{-iW't}\Phi'[l_2]) = N_1'N_2' \exp\{\varphi'[l_1], e^{-iwt}\varphi'[l_2])\}. \tag{3.28}$$

Then the same argument as before shows that

$$W' = (A', wA') \tag{3.29}$$

in keeping with the fact that

$$\frac{(\Phi'[l_1], W'\Phi'[l_2])}{(\Phi'[l_1], \Phi'[l_2])} = (\varphi'[l_1], w\varphi'[l_2]). \tag{3.30}$$

In addition,

$$(\Phi'[l], W'\,\Phi'[l]) = (\Phi[l], W'\Phi[l]) = (\varphi'[l], w\varphi'[l]), \tag{3.31}$$

which is different from the expression that would follow from (3.4), in which

$$W = (A, wA) = (A' + \xi, w[A' + \xi]). \tag{3.32}$$

Among all the possible choices of translation vectors ξ we can always choose $\xi \equiv \varphi[l_0] \in \mathfrak{s}$ for some $l_0 \in \mathfrak{L}$, so that

$$\varphi'[l_0] = \varphi[l_0] - \xi = 0. \tag{3.33}$$

[17] K. O. Friedrichs, Ref. 2, p. 79.

That is, even if $\varphi[l] \neq 0$ for all $l \in \mathfrak{L}$, we can always choose ξ so that in the *translated* set $\mathfrak{s}' = \mathfrak{s} - \xi$ the zero vector appears. A parameterization, such that $\varphi'[l_0] = 0$, exhibits certain additional properties. For example, we note first that

$$(\Phi'[l], \Phi'[l_0]) = e^{-\frac{1}{2}\|\varphi'[l]\|^2} = N', \tag{3.34}$$

which is a *real* expression for all $l \in \mathfrak{L}$. To make the next point, let us assume that w is an arbitrary self-adjoint operator on $\mathfrak{h}$. In general, the only eigenvectors of w which are imaged directly into eigenvectors of W by the exponential map $\psi \to \Psi$ are those with eigenvalue zero. However, when $\varphi'[l_0] = 0$ is an element of $\mathfrak{s}$, it follows from (3.28) that its image $\Phi'[l_0]$ is an eigenvector with eigenvalue zero for *every* self-adjoint operator $W' = (A', wA')$. This is a particularly important property. Moreover, if w has no eigenvectors with eigenvalue zero, then $\Phi'[l_0]$ is a nondegenerate eigenvector with eigenvalue zero for the operator $W' = (A', wA')$. In summary, we note that if $\mathfrak{s}$ does not already contain the zero vector a translated set $\mathfrak{s}' = \mathfrak{s} - \xi$ may be considered which, by suitable choice of ξ, does contain the zero vector.

1. *Improper Translations*

The transition to a translated set assumes its primary importance when the transformation is an improper one. Suppose that $\varphi'[l] \in \mathfrak{s}' \subset \mathfrak{h}$ constitutes a valid initial set of vectors. Let us consider, in addition, the relation

$$\varphi'[l] = \hat{\varphi}[l] - \hat{\xi}, \tag{3.35}$$

where in the present case we admit the possibility that $\hat{\xi} \notin \mathfrak{h}$, i.e., $\|\hat{\xi}\| = \infty$, in which case $\hat{\varphi}[l] \notin \mathfrak{h}$ for all $l \in \mathfrak{L}$. To distinguish this possibility, we have resorted to the carat over the usual vector symbol. Note that although $\hat{\varphi}$ and $\hat{\xi}$ may not be elements of $\mathfrak{h}$, their formal difference lies in $\mathfrak{h}$ by assumption. Such a relation can always be viewed as the limit of valid vectors in $\mathfrak{h}$, such as

$$\varphi'[l] = \hat{\varphi}[l] - \hat{\xi} \equiv \lim_{n\to\infty} \{\varphi_n[l] - \xi_n\}, \tag{3.36}$$

where $\varphi_n, \xi_n \in \mathfrak{h}$, for all n. Abstractly, we may regard $\hat{\varphi} = \lim \varphi_n$ and $\hat{\xi} = \lim \xi_n$ as elements of the distribution space (dual to the test function, or nuclear space) in a rigged Hilbert space triplet.[18] Indeed we shall loosely use $\hat{\varphi}$ and $\hat{\xi}$ in inner products in this way. In a representation on some L^2 space, it generally suffices that $\hat{\varphi}$ and $\hat{\xi}$ are functions which in the present case need not be square integrable. To make this notion clearer, let us reconsider the example of unitary

[18] I. M. Gel'fand and N. Y. Vilenkin, Ref. 4, Chap. III, Sec. 4.

representations for the operators $U(t) = \exp(-iWt)$, treated earlier.

Example revisited: In the present formulation of the example we seek a set of vectors of the form

$$\Phi'[t] = e^{-i\,\mathrm{Im}(\hat{\xi},\varphi'[t])}U(t)\Phi_0, \tag{3.37}$$

for which an exponential construction exists, i.e.,

$$(\Phi'[t_1], \Phi'[t_2]) = N_1'N_2'e^{(\varphi'[t_1],\varphi'[t_2])}. \tag{3.38}$$

We assume $\mathrm{Im}(\xi, \varphi'[0]) = 0$, and, thus, $\Phi'[0] = \Phi_0$. The combination law $U(t)U(t') = U(t + t')$ now leads to

$$(\Phi'[t_1], \Phi'[t_2]) = (\Phi_0, \Phi'[t_2 - t_1])e^{i\alpha}, \tag{3.39}$$

where

$$\alpha \equiv \mathrm{Im}(\xi, \varphi'[t_2 - t_1] - \varphi'[t_2] + \varphi'[t_1]). \tag{3.40}$$

Equating real and imaginary terms in the exponent, we find that

$$\|\varphi'[t_1] - \varphi'[t_2]\|^2 = \|\varphi'[0] - \varphi'[t_2 - t_1]\|^2, \tag{3.41a}$$

while

$$\mathrm{Im}(\varphi'[t_1], \varphi'[t_2]) = \mathrm{Im}(\varphi'[0], \varphi'[t_2 - t_1]) + \mathrm{Im}(\xi, \varphi'[t_2 - t_1] - \varphi'[t_2] + \varphi'[t_1]). \tag{3.41b}$$

Apart from the last term in ξ, these relations are identical to those found earlier in (3.9). As a solution to the first relation, we now choose

$$\varphi'[t] = a + bt \oplus \int^{\oplus} (e^{-itx} - 1)\, d\hat{\mu}(x), \tag{3.42}$$

instead of (3.10), which leads to the expression

$$\|\varphi'[t]\|^2 = |a + bt|^2 + \int |e^{-itx} - 1|^2\, d\hat{\mu}(x). \tag{3.43}$$

Note, in the present case, that a finite norm requires only that

$$\int \left(\frac{x^2}{1 + x^2}\right) d\hat{\mu}(x) < \infty, \tag{3.44}$$

while it is possible that $\hat{\mu}(R) = \int d\hat{\mu}(x) = \infty$, in contrast with the requirement that $\mu(R) < \infty$. This is just the type of generalization in this class of representations which we sought. Let us adopt

$$\xi = \frac{a}{c} Y \oplus \int^{\oplus} d\hat{\mu}(x), \tag{3.45a}$$

where $c = \mathrm{Im}\, ba^*$ as before and

$$Y \equiv \int \frac{x}{1 + x^2}\, d\hat{\mu}(x). \tag{3.45b}$$

With this choice we find that

$$\mathrm{Im}(\xi, \varphi'[t]) = \int \left(\frac{xt}{1 + x^2} - \sin xt\right) d\hat{\mu}(x), \tag{3.46}$$

which is well defined in virtue of (3.44); moreover, we find that the second relation (3.41b) is satisfied.

In summary, we have determined that

$$\begin{aligned}(\Phi_0, \Phi'[t]) &= \exp\left(ict - \tfrac{1}{2}|b|^2 t^2 - \tfrac{1}{2}\int |e^{-itx} - 1|^2\, d\hat{\mu}(x)\right)\\ &= e^{-i\,\mathrm{Im}(\hat{\xi},\varphi'[t])}(\Phi_0, U(t)\Phi_0),\end{aligned} \tag{3.47}$$

from which it follows that

$$(\Phi_0, e^{-iWt}\Phi_0) = \exp\left[ict - \tfrac{1}{2}|b|^2 t^2 - \int \left(1 - \frac{itx}{1 + x^2} - e^{-itx}\right) d\hat{\mu}(x)\right]. \tag{3.48}$$

In this final expression we have arrived at the most general characteristic function for an infinitely divisible probability distribution (compatible with an infinite-dimensional $\mathfrak{H}$); in fact, Eq. (3.48), coupled with (3.44), is just the Lévy canonical representation for such distributions.[19] The translations ξ are not restricted to elements of $\mathfrak{h}$. This is clear since

$$\|\xi\|^2 = |a/c|^2\, Y^2 + \hat{\mu}(R), \tag{3.49}$$

neither term of which need be finite. In this particular example the translated set $\mathfrak{s}'$ does not contain the zero vector unless $a = 0$, as can be seen from (3.43).

2. *Improper Translations and Unitary Inequivalence*

In our initial discussion of those representations of $U(t) = \exp(-iWt)$ admitting an exponential construction, our parameterization was directly associated with the group combination law. To achieve the utmost generality in such representations, it was expedient to consider a translated parameterization, a translation which may well be an improper one. Nevertheless, we have shown that an exponential construction encompasses the set of infinitely divisible probability distributions, and that any unitary, one-parameter group of operators, realized in an exponential way, is such a distribution. Of primary importance, of course, is the additional machinery we have introduced to realize and represent various associated operators.

When the translation vector ξ is improper, the operators $A'(\lambda)$ and $A(\lambda)$ are unitarily inequivalent.[17]

[19] Ref. 15, p. 90.

Heuristically, this is almost evident from the connecting relation

$$A(\lambda) = A'(\lambda) + (\lambda, \xi). \tag{3.50}$$

For if ξ is improper, i.e., $\xi \notin \mathfrak{h}$, then, for some $\lambda = \lambda_0 \in \mathfrak{h}$, the expression $(\lambda_0, \xi) = \infty$ and thus, although $A'(\lambda_0)$ is an acceptable operator, $A(\lambda_0)$ is not.[20] This could not happen if the two operator sets were unitarily equivalent. The operator $A(\lambda)$ is defined now for those $\lambda \in \mathfrak{h}$ such that $|(\lambda, \xi)| < \infty$. If $\mathfrak{h}$ has infinite dimensionality, then $A(\lambda)$ is defined for a dense set of λ. It follows that the operators $A(\lambda_1)$ and $A(\lambda_2)^\dagger$ constitute an inequivalent, irreducible representation of the canonical commutation relations:

$$\begin{aligned}[A(\lambda_1), A(\lambda_2)^\dagger] &= [A'(\lambda_1), A'(\lambda_2)^\dagger] \\ &= (\lambda_1, \lambda_2). \end{aligned} \tag{3.51}$$

For completeness we quote the complete theorem on equivalence of "translated Fock representations": Given a Fock representation $A'(\lambda)$ and two translated Fock representations

$$A_1(\lambda) = A'(\lambda) + (\lambda, \xi_1), \tag{3.52a}$$

$$A_2(\lambda) = A'(\lambda) + (\lambda, \xi_2), \tag{3.52b}$$

defined for a set of λ dense in $\mathfrak{h}$, then $A_1(\lambda)$ is unitarily equivalent to $A_2(\lambda)$ if and only if $(\xi_1 - \xi_2) \in \mathfrak{h}$.[17,21] This may be made concrete by choosing an orthonormal sequence λ_m in the common dense domain of definition and requiring that

$$\sum |\xi_{1m} - \xi_{2m}|^2 < \infty, \tag{3.53}$$

where $\xi_{1m} = (\lambda_m, \xi_1)$, etc.

B. Operator-Field Representations: Space-Translation Invariance and Cluster Decomposition

Consider a field operator $W(\mathbf{x})$, where for concreteness we may take $\mathbf{x}$ as a point in Euclidean three-dimensional space R^3, which satisfies the commutation relation

$$[W(\mathbf{x}), W(\mathbf{y})] = 0. \tag{3.54}$$

We assume W to be formally self-adjoint so that the smeared operators

$$W(f) = \int W(\mathbf{x}) f(\mathbf{x})\, d\mathbf{x} \tag{3.55}$$

are, for suitable real $f(\mathbf{x})$, self-adjoint operators. As a suitable test function space we adopt Schwartz's space $\mathfrak{D}$ composed of real C^∞ functions $f(\mathbf{x})$ which have compact support.

In many applications a central question concerns the representation of the field operators $W(f)$. We approach this question, as before, by considering the expectation functional

$$E(f) \equiv (\Phi_0, \exp[-iW(f)]\Phi_0) \quad . \tag{3.56}$$

and imposing an exponential construction. Both the natural ($\xi = 0$) and translated parameterizations ($\xi \neq 0$) are important, and we shall illustrate them both. We adopt $\mathfrak{D}$ as the label space $\mathfrak{L}$ and $f(\mathbf{x})$ as the labels l.

If we set

$$\Phi'[f] = e^{-i\,\mathrm{Im}\,(\xi, \varphi'[f])} e^{-iW(f)}\Phi_0, \tag{3.57}$$

then we seek representations consistent with the relation

$$(\Phi'[f_1], \Phi'[f_2]) = N_1' N_2' e^{(\varphi'[f_1], \varphi'[f_2])}, \tag{3.58a}$$

where

$$N' = e^{-\frac{1}{2}\|\varphi'[f]\|^2}. \tag{3.58b}$$

As before we assume that $\mathrm{Im}\,(\xi, \varphi'[0]) = 0$. The combination law

$$e^{-iW(f_1)} e^{-iW(f_2)} = e^{-iW(f_1+f_2)} \tag{3.59}$$

leads, just as in the elementary case, to the two relations

$$\|\varphi'[0] - \varphi'[f_1 - f_2]\|^2 = \|\varphi'[f_2] - \varphi'[f_1]\|^2, \tag{3.60a}$$

$$\begin{aligned}\mathrm{Im}\,(\varphi'[0], \varphi'[f_1 - f_2]) &= \mathrm{Im}\,(\varphi'[f_2], \varphi'[f_1]) \\ &\quad - \mathrm{Im}\,(\xi, \varphi'[f_1 - f_2] - \varphi'[f_1] + \varphi'[f_2]). \end{aligned} \tag{3.60b}$$

There are various solutions to these relations, a few of which we shall illustrate. In the natural or group-oriented parameterization, we set $\xi = 0$, and, for example, we may adopt

$$\varphi'[f] = \varphi[f] = (bf)(\mathbf{x}) \oplus \int^{(\oplus)} e^{-i(w,f)}\, d\mu(w). \tag{3.61}$$

Here b is a general linear operator from $\mathfrak{D}$ into L^2. The expression (w, f) is a real number for all f which implies that w is a (real) distribution in $\mathfrak{D}'$. Consequently, $\mu(w)$ is a measure on $\mathfrak{D}'$, which being the dual of a nuclear space is well defined.[22] We have arbitrarily chosen only the linear term $(bf)(\mathbf{x})$ in the present solution [cf. Eq. (3.10)]. In order for $\varphi[f]$ to have finite norm, it is clear that $\mu(\mathfrak{D}') \equiv \int d\mu(w) < \infty$. It follows that

$$(\Phi_0, e^{-iW(f)}\Phi_0) = \exp\left\{-\tfrac{1}{2}\int |bf(\mathbf{x})|^2\, d\mathbf{x} - \int [1 - e^{-i(w,f)}]\, d\mu(w)\right\}. \tag{3.62}$$

This expression characterizes a vast number of field-operator distributions. As a very simple example,

[20] Ref. 14, p. 78.

[21] J. R. Klauder, J. McKenna, and E. J. Woods, J. Math. Phys. 7, 822 (1966).

[22] I. M. Gel'fand and N. Y. Vilenkin, Ref. 4, Chap. IV, Sec. 2.

suppose that μ is concentrated on Dirac δ functions at the point $\mathbf{x} = \mathbf{z}$. In other words, let

$$\int e^{-i(w,f)}\, d\mu(w) = \int e^{-i(\lambda\delta_z, f)}\, d\sigma(\lambda)$$
$$= \int e^{-i\lambda f(\mathbf{z})}\, d\sigma(\lambda), \qquad (3.63a)$$

where λ is a single real variable and

$$\mu(\mathfrak{D}') = \int d\mu(w) = \int d\sigma(\lambda) < \infty. \qquad (3.63b)$$

The inclusion of several derivatives of the δ function is an obvious generalization of this example—and many more come to mind. However, more important representations may be obtained if we exploit a translated parameterization with an improper translation.

For the second class of examples we choose as solution to (3.60) the vectors

$$\xi = 0 \oplus \int^{\oplus} d\hat{\mu}(w), \qquad (3.64a)$$

$$\varphi'[f] = (bf)(\mathbf{x}) \oplus \int^{\oplus} [e^{-i(w,f)} - 1]\, d\hat{\mu}(w). \qquad (3.64b)$$

Note that this is a different type of translation vector from the one used before, since we principally wish to emphasize a different type of improper translation. Because $\varphi'[0] = 0$, it follows that

$$(\Phi_0, \Phi'[f])$$
$$= e^{-\frac{1}{2}\|\varphi'[f]\|^2}$$
$$= \exp\left[-\tfrac{1}{2}\int |bf(\mathbf{x})|^2\, d\mathbf{x} - \tfrac{1}{2}\int |e^{-i(w,f)} - 1|^2\, d\hat{\mu}(w)\right]. \qquad (3.65)$$

From this expression we learn that

$$(\Phi_0, e^{-iW(f)}\Phi_0)$$
$$= \exp\left\{-\tfrac{1}{2}\int |bf(\mathbf{x})|^2\, d\mathbf{x} - \int [1 - e^{-i(w,f)}]\, d\hat{\mu}(w)\right\}. \qquad (3.66)$$

These relations do not require that

$$\hat{\mu}(\mathfrak{D}') \equiv \int d\hat{\mu}(w) = \|\xi\|^2 \qquad (3.67)$$

be finite; in fact, we may choose many improper translation vectors ξ.

As a very simple example, suppose $\hat{\mu}$ is again concentrated on δ functions at $\mathbf{x} = \mathbf{z}$, but now with a uniform distribution of $\mathbf{z}$ values. In particular, we assume that

$$\int [1 - e^{-i(w,f)}]\, d\hat{\mu}(w) = \int [1 - e^{-i(\lambda\delta_z, f)}]\, d\hat{\mu}_0(\lambda)\, d\mathbf{z}$$
$$= \int [1 - e^{-i\lambda f(\mathbf{z})}]\, d\hat{\mu}_0(\lambda)\, d\mathbf{z}. \qquad (3.68)$$

Although this corresponds to an improper translation ξ, since

$$\|\xi\|^2 = \int d\hat{\mu}(w) = \int d\hat{\mu}_0(\lambda)\, d\mathbf{z} = \infty, \qquad (3.69)$$

the desired relation (3.68) is finite since $f(\mathbf{z}) = 0$ outside a compact set in R^3. As it stands, this relation is also finite so long as $\int [|\lambda|/(1 + |\lambda|)]\, d\hat{\mu}_0(\lambda) < \infty$. If $\hat{\mu}_0(-\lambda) = -\hat{\mu}_0(\lambda)$, then only the even part of (3.68) is nonzero and the condition for existence is the finiteness of

$$\int \frac{\lambda^2}{(1 + \lambda^2)}\, d\hat{\mu}_0(\lambda). \qquad (3.70)$$

Even if such symmetry is not present, a different translation vector, patterned after the earlier example [cf. Eq. (3.45)], gives a more general expression. For the most part we are content to assume such symmetry for the purpose of illustration.

1. Space-Translation Invariance

In order that the vector Φ_0 possess space-translation invariance, it is necessary that

$$(\Phi_0, e^{-iW(f_a)}\Phi_0) = (\Phi_0, e^{-iW(f)}\Phi_0) \qquad (3.71)$$

for all $f(\mathbf{x}) \in \mathfrak{D}$ and all $\mathbf{a} \in R^3$, where

$$f_a(\mathbf{x}) \equiv f(\mathbf{x} + \mathbf{a}). \qquad (3.72)$$

An example of such behavior is given by (3.65), where b is simply a real constant and where (3.68) is adopted. An extension of this example is given by

$$(\Phi_0, e^{-iW(f)}\Phi_0)$$
$$= \exp\left\{-\tfrac{1}{2}b^2\int f^2(\mathbf{x})\, d\mathbf{x} - \int [1 - e^{-i(w,f_z)}]\, d\hat{\mu}_0(w)\, d\mathbf{z}\right\}, \qquad (3.73)$$

where f_z is the translated test function (3.72) and where $\hat{\mu}_0(w)$ is concentrated on distributions "at or near" $\mathbf{x} = 0$. For example, we could assume that, for each $f(\mathbf{x}) \in \mathfrak{D}$ such that $f(\mathbf{x}) = 0$ for $|\mathbf{x}| < R_0$, the distributions of interest would fulfill $(w, f) = 0$. The translation invariance of the expression (3.73) [in the sense of (3.71)] is evident.

We may profitably write $\mathfrak{h}$ as a direct integral space over R^3 when representations such as (3.73) apply.

That is, we set

$$\mathfrak{h} = \int^{\oplus} \mathfrak{h}_z \, d\mathbf{z} \tag{3.74}$$

and

$$\varphi'[f] = \int^{\oplus} \varphi'_z[f] \, d\mathbf{z}, \tag{3.75}$$

so that

$$\|\varphi'[f]\|^2 = \int (\varphi'_z[f], \varphi'_z[f])_z \, d\mathbf{z}. \tag{3.76}$$

In this language, the example of (3.73) has a direct integral entry of the form

$$\varphi'_z[f] = bf(\mathbf{z}) \oplus \int^{\oplus} [e^{-i(w, f_z)} - 1] \, d\hat{\mu}_0(w). \tag{3.77}$$

Among the possible representations of this particular form are those for which

$$\varphi'_z[f] = \varphi'_z[f(\mathbf{z})], \tag{3.78}$$

i.e., only the specific value $f(\mathbf{z})$ enters. Equation (3.68) has this "ultralocal" characteristic which arises when $\hat{\mu}_0$ is concentrated on δ functions at $\mathbf{x} = 0$, and these are essentially the only such representations. This class of representations has the property that

$$E(f_1 + f_2) = E(f_1)E(f_2) \tag{3.79}$$

whenever

$$f_1(\mathbf{x})f_2(\mathbf{x}) = 0, \tag{3.80}$$

where $E(f)$ is given by (3.56). Special representations of this form have been formulated as continuous-tensor-product representations, and have been studied by several authors.[7,23]

2. *Cluster Decomposition*

The physical idea behind cluster decomposition is that equal-time field operators localized about remote regions of space are statistically independent. It is a familiar property for statistically independent variables that the probability (amplitude) and the related characteristic function factorize. Such factorization is just the statement of (3.79) for the special class of ultralocal representations. More generally, if we assume translation invariance, $E(f_a) = E(f)$, then, to ensure cluster decomposition, it is necessary that

$$\lim_{|\mathbf{a}|\to\infty} E(f' + f_a) = E(f')E(f), \quad \text{for all } f', f \in \mathfrak{D}. \tag{3.81}$$

Examples of translationally invariant distributions that exhibit cluster decomposition are conveniently formulated in an exponential Hilbert space. If we assume $\varphi'[0] = 0$, then we have the relation

$$E(f) = (\Phi_0, \Phi'[f])e^{i\,\mathrm{Im}\,(\hat{\xi}, \varphi'[f])} = e^{-\frac{1}{2}\|\varphi'[f]\|^2 + i\,\mathrm{Im}\,(\xi, \varphi'[f])}. \tag{3.82}$$

To win the desired features, we must have invariance:

$$\|\varphi'[f_a]\|^2 = \|\varphi'[f]\|^2, \tag{3.83a}$$

$$\mathrm{Im}\,(\hat{\xi}, \varphi'[f_a]) = \mathrm{Im}\,(\hat{\xi}, \varphi'[f]); \tag{3.83b}$$

and asymptotic independence:

$$\lim_{|\mathbf{a}|\to\infty} \|\varphi'[f' + f_a]\|^2 = \|\varphi'[f']\|^2 + \|\varphi[f]\|^2, \tag{3.84a}$$

$$\lim_{|\mathbf{a}|\to\infty} \mathrm{Im}\,(\hat{\xi}, \varphi'[f' + f_a]) = \mathrm{Im}\,(\hat{\xi}, \varphi'[f']) + \mathrm{Im}\,(\hat{\xi}, \varphi'[f]). \tag{3.84b}$$

In the direct integral form of (3.74)–(3.76), we have

$$\|\varphi'[f]\|^2 = \int \|\varphi'_z[f]\|_z^2 \, d\mathbf{z}, \tag{3.85}$$

which to exhibit invariance need only fulfill

$$\|\varphi'_{z+a}[f_a]\|_{z+a}^2 = \|\varphi'_z[f]\|_z^2. \tag{3.86a}$$

Asymptotic independence is fulfilled if

$$\lim_{|\mathbf{a}|\to\infty} \|\varphi'_z[f' + f_a]\|_z^2 = \|\varphi'_z[f']\|_z^2. \tag{3.86b}$$

These relations are consistent with the intuitive notion that $\varphi'_z[f]$ should depend only on the values of $f(\mathbf{x})$ when $\mathbf{x}$ is "near" $\mathbf{z}$. Equation (3.73) gives an example of this type.

These examples, as well as our general discussion, should serve to demonstrate how ideal the exponential Hilbert space construction really is for representing field operators and invariant vectors exhibiting space-translation invariance and cluster decomposition. In the next two sections we discuss exponential representations of current algebras and of the familiar canonical commutation relations for scalar fields.

4. EXPONENTIAL REPRESENTATIONS OF FIELD-OPERATOR ALGEBRAS

We consider an equal-time field algebra (or "current" algebra of particle physics) which is characterized by a family of formally self-adjoint field operators $W_l(\mathbf{x})$, $l = 1, 2, \cdots, L$, with the commutation rule

$$[W_l(\mathbf{x}), W_m(\mathbf{y})] = ic_{lmn}W_n(\mathbf{x})\delta(\mathbf{x} - \mathbf{y}). \tag{4.1}$$

Here c_{lmn} are the structure constants of a Lie group, and summation over repeated indices is understood. If $f(\mathbf{x})$, $g(\mathbf{x})$ are suitable real test functions, such as elements of $\mathfrak{D}$, then

$$[W_l(f), W_m(g)] = ic_{lmn}W_n(fg) \tag{4.2}$$

---

[23] Such representations for canonical operators were effectively first treated by H. Araki, thesis, Princeton University, Princeton, N.J., 1960.

is the proper statement of (4.1) for self-adjoint operators. The unitary group elements in canonical coordinates are given by

$$U[f_l] = e^{iW_l(f_l)} \tag{4.3}$$

(summation understood in the exponent on the right-hand side), where $f_l(\mathbf{x}) \in \mathfrak{D}$ for all l; it is representations of these field operators that we seek.

Guided by our formulation in Sec. 3, we introduce the overcomplete family of states

$$\Phi'[f_l] \equiv e^{-i\,\mathrm{Im}\,(\hat{\xi},\varphi'[f_l])}U[f_l]\Phi_0, \tag{4.4}$$

for some choice of $\hat{\xi}$ and of $\varphi'[f_l] \in \mathfrak{h}$. We insist on an exponential construction such that

$$(\Phi'[f_l'], \Phi'[f_l]) = N'Ne^{(\varphi'[f_l'],\varphi'[f_l])} \tag{4.5}$$

The group law reads

$$U[f_l]U[\bar{f}_l] \equiv U[(f\cdot\bar{f})_l], \tag{4.6}$$

where $f\cdot\bar{f}$ symbolizes the parametric combination law characterizing the group. This law implies a functional identity within the exponential form that leads to the two relations

$$\|\varphi'[f_l'] - \varphi'[f_l]\|^2 = \|\varphi'[0] - \varphi'[(-f'\cdot f)_l]\|^2, \tag{4.7a}$$

$$\begin{aligned}\mathrm{Im}\,(\varphi'[f_l'], \varphi'[f_l]) &= \mathrm{Im}\,(\varphi'[0], \varphi'[(-f'\cdot f)_l]) \\ &\quad - \mathrm{Im}\,(\hat{\xi}, \varphi'[(-f'\cdot f)_l] - \varphi'[f_l] + \varphi'[f_l']).\end{aligned} \tag{4.7b}$$

Solutions to these equations generate the representations we seek.

As our first solution, we adopt

$$\hat{\xi} = 0, \tag{4.8a}$$

$$\varphi'[f_l] \equiv \varphi[f_l] = u[f_l]\varphi_0, \tag{4.8b}$$

where the $u[f_l]$ are unitary operators which satisfy

$$u[f_l']u[f_l] = u[(f'\cdot f)_l]; \tag{4.9}$$

i.e., $u[f_l]$ also forms a unitary representation of the sought-for group. Indeed, $U[f_l]$ is related to $u[f_l]$ in the fashion that B is related to b in Eq. (2.27), namely,

$$U[f_l] = \bigoplus_{n=0}^{\infty} \bigotimes_{s=1}^{n} u[f_l]. \tag{4.10}$$

Moreover, in canonical coordinates,

$$U[f_l] = e^{iW_l(f_l)}, \tag{4.11a}$$

$$u[f_l] = e^{iw_l(f_l)}, \tag{4.11b}$$

so that an equation like (2.29) holds as well. Thus, in this solution it is clear that the representation $U[f_l]$ is always *reducible*, whether $u[f_l]$ is irreducible or not. Equivalent (inequivalent) representations of $u[f_l]$ lead to equivalent (inequivalent) representations of $U[f_l]$. In order that $U[f_l]$ be a cyclic representation with Φ_0 a cyclic vector, it is necessary that $u[f_l]$ be a cyclic representation (on $\mathfrak{h}$) with φ_0 a cyclic vector. However, it is not *a priori* clear that this is sufficient in the general case.

With the solution (4.8) the expectation functional takes the form

$$\begin{aligned}E(f_l) &\equiv (\Phi_0, e^{iW_l(f_l)}\Phi_0) \\ &= \exp\{(\varphi_0, [e^{iw_l(f_l)} - 1]\varphi_0)\},\end{aligned} \tag{4.12}$$

as follows from (2.35). In the present parameterization $A(\lambda)\Phi_0 = (\lambda, \varphi_0)\Phi_0$ and thus, according to (2.32) and (2.33), we find

$$W_l(f_l) = (A, w_l(f_l)A). \tag{4.13a}$$

Hence, for each l we have

$$W_l(f) = (A, w_l(f)A). \tag{4.13b}$$

It is a well-known property of creation and annihilation operators that

$$\begin{aligned}[W_l(f), W_m(g)] &= [(A, w_l(f)A), (A, w_m(g)A)] \\ &= (A, [w_l(f), w_m(g)]A).\end{aligned} \tag{4.14}$$

Thus, the validity of the commutation relations (4.2) follows from (4.14) and the fact that the $w_l(f)$ fulfill the same algebra on $\mathfrak{h}$.

The reducible nature of the representations given by $W_l(f)$ may be seen another way. Let

$$N = (A, A) \tag{4.15}$$

denote the total number operator in the Fock representation. If

$$\mathrm{X} = \bigoplus_{n=0}^{\infty} \mathrm{X}_{(n)} \in \mathfrak{H}, \tag{4.16}$$

then

$$N\mathrm{X} \equiv \bigoplus_{n=0}^{\infty} n\mathrm{X}_{(n)}, \tag{4.17}$$

so that N has zero and the positive integers for eigenvalues. Now, no matter what the operators w_l are, it follows that

$$\begin{aligned}[W_l(f), N] &= [(A, w_l(f)A), (A, A)] \\ &= (A, [w_l(f), I]A) = 0.\end{aligned} \tag{4.18}$$

Implicitly, we have demonstrated reducibility by showing the existence of an operator N different from the identity operator, which commutes with all the generators.

As a second solution to relations (4.7), let us select

$$\xi = \hat{\varphi}_0, \tag{4.19a}$$

$$\varphi'[f_l] = (u[f_l] - 1)\hat{\varphi}_0, \tag{4.19b}$$

where again $u[f_l]$ is a unitary group representation on $\mathfrak{h}$. Although we may give up the finiteness of $\|\xi\| = \|\hat{\varphi}_0\|$, we assume that $\varphi'[f_l] \in \mathfrak{h}$ and that $(\hat{\varphi}_0, \varphi'[f_l])$ is well defined. In particular, we are immediately led to representations which are essentially characterized by

$$(\Phi_0, e^{iW_l(f_l)}\Phi_0) = \exp\{(\hat{\varphi}_0, [e^{iw_l(f_l)} - 1]\hat{\varphi}_0)\}. \tag{4.20}$$

These representations, of course, are not fundamentally different than those given above unless $\|\hat{\varphi}_0\| = \infty$. As an important example of the difference that this condition makes, let us suppose that $w \equiv w_l(f_l)$, for some choice of f_l, only has an absolutely continuous spectrum. Then

$$\operatorname*{w-lim}_{t\to\infty} e^{iwt} = 0, \tag{4.21}$$

where the weak operator limit is meant. If $\|\hat{\varphi}_0\| < \infty$, then

$$\operatorname*{w-lim}_{t\to\infty} e^{iWt} = P_0, \tag{4.22a}$$

where $W = W_l(f_l)$ and where P_0 is a projection operator onto the subspace of zero-W eigenvalue. This subspace always includes the vector $\hat{\Phi}_0$, which is characterized by the fact that $A(\lambda)\hat{\Phi}_0 = 0$ for all λ. On the other hand, for suitable choices of $\hat{\varphi}_0 \notin \mathfrak{h}$, it follows that

$$\operatorname*{w-lim}_{t\to\infty} e^{iWt} = 0, \tag{4.22b}$$

so that, like w, the associated operator W has only an absolutely continuous spectrum. This is a common and frequently desirable property of field operators and should be incorporated whenever suitable. An example of the two cases in (4.22) is given later in this section.

In the translated form appropriate to (4.19), we have, according to (3.23a), the basic annihilation operators $A'(\lambda)$ such that

$$A'(\lambda)\Phi'[f_l] = (\lambda, \varphi'[f_l])\Phi'[f_l].$$

Since $\varphi'[f_l] \in \mathfrak{h}$, this relation is defined for all $\lambda \in \mathfrak{h}$. Indeed, $\varphi'[0] = 0$, so that $A'(\lambda)\Phi'[0] = 0$, which shows that $A'(\lambda)$ is surely the Fock representation. On the other hand, the group generators W_l are still given by (4.13) in terms of the operators A. The relation between the two, as in (3.50), reads $A(\lambda) = A'(\lambda) + (\lambda, \hat{\varphi}_0)$, which, if $\hat{\varphi}_0 \notin \mathfrak{h}$, is not defined for all λ. By hypothesis, however, $\varphi'[f_l] = (e^{iw_l(f_l)} - 1)\hat{\varphi}_0 \in \mathfrak{h}$ and is, in fact, "enough" within $\mathfrak{h}$, we assume, so that $(\hat{\varphi}_0, \varphi'[f_l])$ makes sense; this notion will be made clearer below through an example. We assume for simplicity that even $w_l(f_l)\hat{\varphi}_0 \in \mathfrak{h}$, and, moreover, that $(\hat{\varphi}_0, w_l(f_l)\hat{\varphi}_0)$ makes sense. With this simplification we have

$$\begin{aligned} W_l(f) &= (A, w_l(f)A) \\ &= (A' + \hat{\varphi}_0, w_l(f)[A' + \hat{\varphi}_0]) \\ &= (A', w_l(f)A') + (A', w_l(f)\hat{\varphi}_0) \\ &\quad + (w_l(f)\hat{\varphi}_0, A') + (\hat{\varphi}_0, w_l(f)\hat{\varphi}_0). \end{aligned} \tag{4.23}$$

While this equation still has the same appearance as (4.13b), it is, of course, quite different, since A is no longer a Fock representation. This inequivalence in representations suggests that, when $\hat{\varphi}_0 \notin \mathfrak{h}$, we are dealing with an *inequivalent* representation of $U[f_l]$ (with no change of the representation $u[f_l]$). More generally, consider two such representations

$$W_{l1}(f) = (A_1, w_l(f)A_1), \tag{4.24a}$$

$$W_{l2}(f) = (A_2, w_l(f)A_2), \tag{4.24b}$$

where A_1 and A_2 are related to a standard Fock representation A', as in Eq. (3.52). The first representation W_{l1} is unitarily equivalent to the second W_{l2}, if A_1 is unitarily equivalent to A_2; but this is not the only possibility. Unitary equivalence of W_{l1} and W_{l2} is also assured, if A_1 is unitarily equivalent to $e^{i\gamma}A_2$; here, γ is any over-all phase factor, since such a factor obviously drops out of (4.24b). In turn, this holds if and only if $\xi_1 - e^{i\gamma}\xi_2 \in \mathfrak{h}$, for some real γ. This exhausts the equivalence class of unitarily equivalent representations, unless there are additional unitary operators v on $\mathfrak{h}$ which commute with all the $w_l(f)$. For any v such that

$$v^\dagger w_l(f)v = w_l(f), \tag{4.25}$$

the representations W_{l1} and W_{l2} are unitarily equivalent, provided that

$$\xi_1 - v\xi_2 \in \mathfrak{h}. \tag{4.26}$$

The existence of such v (not simply $e^{i\gamma}$) depends on whether or not the $w_l(f)$ representation is irreducible.

According to (4.26), whenever $\xi_1 = v\xi_2$ where the unitary operator v fulfills (4.25), the two representations $W_{l1}(f)$ and $W_{l2}(f)$ are unitarily equivalent. This is also evident from the equality of their expectation functionals, which follows from the form of (4.20) for $\hat{\varphi}_0 = \xi_1 = v\xi_2$ and from the commutation property (4.25). It is well known that the expectation functional only determines a cyclic representation up to unitary equivalence and that equal expectation functionals correspond to unitarily equivalent representations.[24]

[24] M. A. Naimark, *Normed Rings*, translated by L. F. Boron (P. Noordhoff Ltd., Groningen, The Netherlands, 1964), p. 242.

Conversely, unequal expectation functionals may or may not correspond to equivalent representations, depending on whether (4.26) is fulfilled or not.

Whenever *no* unitary v exists such that (4.26) holds, the two representations W_{l1} and W_{l2} are necessarily inequivalent. In particular, for the cases illustrated above where $\xi_1 = 0$ and $\xi_2 = \xi = \hat{\varphi}_0$, unitary inequivalence arises whenever $\hat{\varphi}_0 \notin \mathfrak{h}$. Moreover, any time $\xi \notin \mathfrak{h}$, the decomposition of $U[f_l]$ given in (4.10) does not hold.

To examine the reducibility of a given representation $W_l(f)$, we seek an operator different from the unit operator, which commutes with the generators $W_l(f)$. It is not possible to use $N = (A, A)$, for in a non-Fock representation N is not an operator.[25] What is needed is another operator y for which $[w_l(f), y] = 0$ and for which

$$Y = (A, yA) \tag{4.27}$$

is a *bona fide* operator. In that case,

$$[W_l(f), Y] = (A, [w_l(f), y]A) = 0, \tag{4.28}$$

and reducibility would be established. Such a y can exist only if the representation w_l on $\mathfrak{h}$ is itself reducible. Inspection of the Weyl operators shows that operators of the form (4.27) generate the subalgebra of operators which commute with the $W_l(f)$. Hence, if no such y exists, then the representation of $W_l(f)$ is *irreducible*, in marked contrast to the situation when A is a Fock representation.

To illustrate some of these aspects, let us discuss a particularly simple example of a current algebra, that appropriate to the "affine fields." The field algebra we wish to examine is given formally by the commutation relation

$$[\kappa(\mathbf{x}), \pi(\mathbf{y})] = i\delta(\mathbf{x} - \mathbf{y})\pi(\mathbf{y}), \tag{4.29}$$

and we take the unitary operators as

$$U[f, g] = e^{-i\pi(g)}e^{i\kappa(f)}. \tag{4.30}$$

These are also canonical coordinates (of the so-called second kind[26]) and are chosen for reasons of convenience. The group combination law reads

$$U[f, g]U[f', g'] = U[f + f', g + e^{-f}g'], \tag{4.31}$$

where multiplication is pointwise.

This algebra is the local field analog of that based on the two-parameter Lie algebra,[27] whose commutator reads

$$[B, P] = iP. \tag{4.32}$$

If we imagine that $B = \frac{1}{2}(QP + PQ)$ and heuristically regard Q and P as Heisenberg operators, then a reasonable feel for this algebra is obtained. However, unlike the Heisenberg algebra, there are two unitarily inequivalent, irreducible representations of (4.33), one for which $P > 0$ and the other for which $P < 0$.[27] The unitary operators of the affine group,

$$u_0[r, s] = e^{-isP}e^{irB}, \tag{4.33}$$

are the analogs of $U[f, g]$. If we diagonalize P, then we have the representation

$$(u_0[r, s]\varphi)(k) = e^{-\frac{1}{2}r}e^{-isk}\varphi(e^{-r}k), \tag{4.34}$$

where $\varphi(k) \in L^2(R) = L^2(-\infty, \infty)$. Note that $L^2(0, \infty)$ and $L^2(-\infty, 0)$ form invariant subspaces; in the former $P > 0$, while in the latter $P < 0$. The irreducible representations are given by restricting attention to $L^2(0, \infty)$ [or $L^2(-\infty, 0)$] as representation spaces. For our purposes, we choose $L^2(0, \infty)$ and, thereby, obtain an irreducible representation space for u_0.

As our example of a solution of the affine field algebra, we adopt an exponential construction and choose

$$\varphi'[f, g] = \int^{\oplus} \{u_0[f(\mathbf{z}), g(\mathbf{z})] - 1\}\tilde{\varphi}_0\, d\mathbf{z}. \tag{4.35}$$

Here we have taken

$$\mathfrak{h} = \int^{\oplus} \mathfrak{h}_{\mathbf{z}}\, d\mathbf{z} \tag{4.36}$$

and may identify $\mathfrak{h}_z = L^2(0, \infty)$. Thus, $\tilde{\varphi}_0$ is (presently!) an element of $L^2(0, \infty)$, and is chosen the same for all $\mathbf{z}$. Note that

$$\xi = \int^{\oplus} \tilde{\varphi}_0\, d\mathbf{z}, \tag{4.37}$$

and, therefore,

$$\|\xi\|^2 = \int \|\tilde{\varphi}_0\|_z^2\, d\mathbf{z} = \|\tilde{\varphi}_0\|_0^2 \int d\mathbf{z}. \tag{4.38}$$

In the infinite-configuration space which we consider, $\|\xi\| = \infty$ and corresponds to an improper translation. Moreover, $\tilde{\varphi}_0$ vectors with unequal norms manifestly lead to unitarily inequivalent representations of $U[f, g]$. As we shall see, thanks to the irreducibility assumed for u_0, each class of unitarily equivalent representations is labeled by the rays $[\tilde{\varphi}_0]$, i.e., by

$$[\tilde{\varphi}_0] \equiv \{e^{i\gamma}\tilde{\varphi}_0 : 0 \le \gamma < 2\pi\}. \tag{4.39}$$

---

[25] Ref. 17, p. 141; L. Gärding and A. S. Wightman, Proc. Nat. Acad. Sci. U.S. **40**, 622 (1954); J. M. Chaiken, Ann. Phys. (N.Y.) **42**, 23 (1967).

[26] P. M. Cohn, *Lie Groups* (Cambridge University Press, Cambridge, England, 1961), p. 110.

[27] I. M. Gel'fand and M. A. Naimark, Dokl. Akad. Nauk SSSR **55**, 570 (1947); E. W. Aslaksen and J. R. Klauder, J. Math. Phys. **9**, 206 (1968).

The representation induced by (4.35) has an expectation functional given by

$$E_A(f, g) \equiv (\Phi_0, U[f, g]\Phi_0)$$
$$= \exp\left[\int (\tilde{\varphi}_0, \{u_0[f(\mathbf{z}), g(\mathbf{z})] - 1\}\tilde{\varphi}_0)_z\, d\mathbf{z}\right]$$
$$= \exp\left(\iint_0^\infty \{\tilde{\varphi}_0^*(k)[e^{-\frac{1}{2}f(\mathbf{z})-ikg(\mathbf{z})}\tilde{\varphi}_0(e^{-f(\mathbf{z})}k) - \tilde{\varphi}_0(k)]\}\, dk\, d\mathbf{z}\right). \quad (4.40)$$

Observe that the vector Φ_0 is translation invariant and that the representation fulfills cluster decomposition. Indeed, this example is ultralocal in the sense of (3.79) and (3.80), although not every solution of (4.31) need be ultralocal by any means. Note also that if $f(\mathbf{z}) \equiv 0$, $U[0, g]U[0, g'] = U[0, g + g']$ and we are necessarily led to an exponential representation for the single commuting field operator $\pi(\mathbf{x})$. Clearly, the resultant representation (4.40) that arises when $f = 0$ has the general form derived in Sec. 3. This property is also fulfilled if we instead set $g = 0$, although the form of (4.40) does not make that result immediately obvious. Clearly, this is a general feature of exponential group representations in canonical coordinates.

The presence or absence of a linear term in the exponent of the expectation functional [such as appears in (3.66)] is dictated by requirements of the group representation. For example, in the case of the affine field we could legitimately adopt

$$\varphi'[f, g] = \int^{\oplus} (bf(\mathbf{z}) \oplus \{u_0[f(\mathbf{z}), g(\mathbf{z})] - 1\}\tilde{\varphi}_0)\, d\mathbf{z}, \quad (4.41)$$

where, say, b is a real constant. The modified expectation functional is

$$E'_A(f, g) = \exp\left\{-\tfrac{1}{2}b^2 \int f^2(\mathbf{z})\, d\mathbf{z}\right\} E_A(f, g), \quad (4.42)$$

where $E_A(f, g)$ is given by (4.40). Again, this representation fulfills cluster decomposition and has a translationally invariant Φ_0. In addition, distinct b values lead to inequivalent representations. However, there is an unfaithful subrepresentation of the field operator $\pi(\mathbf{x})$, which may or may not be important in applications.

To see this feature most simply, let us return to our single degree of freedom case where $[B, P] = iP$. One set of solutions is to assume $P = 0$! In that case

$$u_0[r, s] = e^{-isP}e^{irB} = e^{irB}. \quad (4.43)$$

In the subspace where this occurs, B can be diagonalized and each distinct eigenvalue b corresponds to a unitarily inequivalent representation of u_0. Each such representation is one-dimensional. For our purposes, we choose not to consider such representations either in the one-dimensional case or in the field analog in which Eq. (4.29) is satisfied by assuming a subspace where $\pi(\mathbf{y}) \equiv 0$ while $\kappa(\mathbf{x}) \neq 0$. Thus, we ignore the possibility of a linear term in $\varphi'[f, g]$. In other group representations it may be appropriate to include such terms. When we discuss the canonical commutation relations in the next section, we shall see, in a way, that linear terms are required.

Recall that the annihilation operator $A'(\lambda)$ is defined by

$$A'(\lambda)\Phi'[f, g] = (\lambda, \varphi'[f, g])\Phi'[f, g] \quad (4.44)$$

and satisfies

$$[A'(\lambda_1), A'(\lambda_2)^\dagger] = (\lambda_1, \lambda_2). \quad (4.45)$$

Since in this example $\mathfrak{h} = \int^{\oplus} \mathfrak{h}_z\, d\mathbf{z}$, it is convenient to introduce the operators $\mathcal{A}'(\lambda_z)$ and their formal (and nonexistent!) adjoints $\mathcal{A}'(\lambda_z)^\dagger$ defined as follows: For each $\lambda_z \in \mathfrak{h}_z$, we set

$$\mathcal{A}'(\lambda_z)\Phi'[f, g] = (\lambda_z, \varphi'_z[f, g])_z\Phi'[f, g] \quad (4.46)$$

and

$$[\mathcal{A}'(\lambda_{1z}), \mathcal{A}'(\lambda_{2z'})^\dagger] = \delta(\mathbf{z} - \mathbf{z}')(\lambda_{1z}, \lambda_{2z})_z. \quad (4.47)$$

In the manner of (2.32) and (2.33a), we let

$$(\mathcal{A}', w\mathcal{A}')_z \equiv \sum_{n,m} \mathcal{A}'(\lambda_{nz})^\dagger(\lambda_{nz}, w\lambda_{mz})_z\mathcal{A}'(\lambda_{mz}), \quad (4.48)$$

where $\{\lambda_{nz}\}$ is a complete orthonormal set in $\mathfrak{h}_z$. This form is not an operator but becomes one on integration, rather like the number-density operator in usual quantum field theory.

Along with $\mathcal{A}'$ let us introduce

$$\mathcal{A}(\lambda_z) = \mathcal{A}'(\lambda_z) + (\lambda_z, \xi_z)_z, \quad (4.49a)$$

which in our example becomes

$$\mathcal{A}(\lambda_z) = \mathcal{A}'(\lambda_z) + (\lambda_z, \tilde{\varphi}_0)_z. \quad (4.49b)$$

Since A' is the Fock representation, so too, we may say, is $\mathcal{A}'$. Although A is not the Fock representation because $\|\xi\| = \infty$, we may say that A is "locally Fock" whenever $\mathcal{A}(\lambda_z)$ is unitarily equivalent to $\mathcal{A}'(\lambda_z)$. This occurs in our case, provided that $\tilde{\varphi}_0 \in \mathfrak{h}_z$. In other words, the representations would be equivalent if we dealt with a finite-configuration-space volume (quantization in a box). However, we can also take $\tilde{\varphi}_0 \notin \mathfrak{h}_z$, as we shall see, in which case the representation A is not even locally Fock.

In essentially the same manner as (4.13b), we may realize the group generators for ultralocal representations as

$$W_l(f) = \int f(\mathbf{z})(\mathcal{A}, w_l\mathcal{A})_z\, d\mathbf{z}. \quad (4.50)$$

Note that on $\mathfrak{h}_z$ the operators $\mathscr{w}_l$ simply satisfy

$$[\mathscr{w}_l, \mathscr{w}_m] = ic_{lmn}\mathscr{w}_n, \tag{4.51}$$

rather like the "algebra of charges" associated with the "algebra of currents." If the representation A is locally Fock, then, for $g(\mathbf{z}) \in \mathfrak{D}$,

$$N(g) \equiv \int g(\mathbf{z})(\mathscr{A}, \mathscr{A})_z \, d\mathbf{z} \tag{4.52}$$

is a meaningful operator, although

$$N = N(1) = \int (\mathscr{A}, \mathscr{A})_z \, d\mathbf{z} = (A, A) \tag{4.53}$$

is not defined. However, it is clear that

$$\begin{aligned}[W_l(f), N(g)] &= \iint f(\mathbf{z})g(\mathbf{x}) \\ &\quad \times [(\mathscr{A}, \mathscr{w}_l\mathscr{A})_z, (\mathscr{A}, \mathscr{A})_x] \, d\mathbf{z}\, d\mathbf{x} \\ &= \int f(\mathbf{z})g(\mathbf{z})(\mathscr{A}, [\mathscr{w}_l, I]\mathscr{A})_z \, d\mathbf{z} \\ &= 0. \end{aligned} \tag{4.54}$$

Thus, the operator $N(g)$ plays the role of Y in Eq. (4.27). Since $N(g)$, for any $g(\mathbf{z}) \in \mathfrak{D}$, is not a multiple of the identity, we have established the reducibility of ultralocal exponential representations of $U[f_l]$ which are based on locally Fock representations of the operators A and $A^\dagger$.

Finally, we consider those representations for which $\tilde{\varphi}_0 \notin \mathfrak{h}_z$. To help visualize this situation, we appeal to our example of the affine fields. [Not all groups admit a generalization to $\tilde{\varphi}_0 \notin \mathfrak{h}_z$; in particular, the "algebra of charges" (4.51) cannot correspond to a compact group.] On reference to (4.40), the quantity which must be well defined is obviously

$$\int_0^\infty \tilde{\varphi}_0^*(k)[e^{-\frac{1}{2}r-isk}\tilde{\varphi}_0(e^{-r}k) - \tilde{\varphi}_0(k)]\, dk. \tag{4.55}$$

Since this must be meaningful for all r and s, it is clear that $\tilde{\varphi}_0$ can fail to be L^2 only near $k = 0$. For example, let us assume that $\tilde{\varphi}_0(k)$ is square integrable for $K \le k < \infty$, for some $K > 0$. To complete the specification of $\tilde{\varphi}_0(k)$, we assume that

$$\tilde{\varphi}_0(k) \equiv k^{-\frac{1}{2}}e(k), \tag{4.56}$$

where $e(k)$ fulfills the Lipschitz condition of order δ,

$$|e(k') - e(k)| \le C\,|k' - k|^\delta, \tag{4.57}$$

for $0 \le k, k' \le K$, and for some C and $\delta > 0$. From this condition it readily follows that (4.55) is well defined for all r and s. Two examples of suitable $\tilde{\varphi}_0(k) \notin L^2(0, \infty)$ are given by $(k + k^4)^{-\frac{1}{2}}$ and $k^{-\frac{1}{2}}e^{-k}$.

With such a choice for $\tilde{\varphi}_0$, the representation A is not even locally Fock, so that although $W_l(f)$ is still given by (4.50), the reducing operator $N(g)$ is no longer an operator. In fact, there is no operator different from unity which commutes with all the $W_l(f)$. As a consequence, the ultralocal exponential representations based on nonlocally Fock representations of A and $A^\dagger$ are irreducible representations of $U[f_l]$. In symbols, if $\tilde{\varphi}_0 \notin \mathfrak{h}_z$, the representation $U[f_l]$ is irreducible. In our example, since $\mathscr{w}_l$ $(= B, P)$ form an irreducible representation in $\mathfrak{h}_z$, only those representations for which $\tilde{\varphi}_{01} - e^{i\gamma}\tilde{\varphi}_{02} \in L^2(0, \infty)$, for some real γ, are locally equivalent. Full equivalence of two representations of the form (4.40) requires not only local equivalence but the condition that $\hat{\varphi}_{01} - e^{i\gamma}\hat{\varphi}_{02} \in \mathfrak{h}$, where $\hat{\varphi}_{0l} = \int \tilde{\varphi}_{0l}\, d\mathbf{z}$, $l = 1, 2$. The infinite volume of configuration space requires that $\tilde{\varphi}_{01} = e^{i\gamma}\tilde{\varphi}_{02}$ for equivalence, which establishes that the rays (4.39) label inequivalent representations of (4.40). The validity of this result may also be seen by the "tag test."[28] In this approach one first shows for (4.40) that

$$\operatorname*{w-lim}_{|\mathbf{a}| \to \infty} U[f_a, g_a] = E_A(f, g). \tag{4.58}$$

That is, these operators have a weak limit which is a multiple of the identity, and that multiple—the "tag"—is simply $E_A(f, g)$. As is easily seen, distinct tags label inequivalent representations. Unless $\tilde{\varphi}_{01} = e^{i\gamma}\tilde{\varphi}_{02}$, the tags are evidently unequal for some choice of $f, g \in \mathfrak{D}$. Clearly, this kind of technique applies directly to exponential representations fulfilling translation invariance and cluster decomposition for general current algebras.

Let us briefly discuss some spectral properties of the field operator $\pi(\mathbf{x})$ in the locally and nonlocally Fock-representation cases. We define[29]

$$P_\Delta \equiv \int_\Delta \pi(\mathbf{x})\, d\mathbf{x}, \tag{4.59a}$$

$$B_\Delta \equiv \int_\Delta \kappa(\mathbf{x})\, d\mathbf{x} \tag{4.59b}$$

for some compact set $\Delta \subset R^3$. It follows from (4.29) that

$$[B_\Delta, P_\Delta] = iP_\Delta, \tag{4.60}$$

which is just the commutation relation (4.32). The representation of P_Δ is determined by our expectation functional; it will be highly reducible in $\mathfrak{H}$, but we may

[28] J. R. Klauder and J. McKenna, J. Math. Phys. 6, 68 (1965), Sec. 4.C.

[29] Although a characteristic function is not an element of the test function space $\mathfrak{D}$, it is a limit of such functions. Its appropriateness as a smearing function can be determined directly from the expectation functional for ultralocal representations.

well ask the question whether P_Δ only has an absolutely continuous spectrum, $P_\Delta > 0$ in our case, or whether there is also a subspace in which $P_\Delta = 0$. We may examine this question most simply in the context of $(\Phi_0, e^{-isP_\Delta}\Phi_0)$.

$$= \exp\left[\int_\Delta\int_0^\infty (e^{-iks} - 1)\,|\tilde{\varphi}_0(k)|^2\, dk\, d\mathbf{z}\right]$$

$$= \exp\left[-\Delta\int_0^\infty (1 - e^{-iks})\,|\tilde{\varphi}_0(k)|^2\, dk\right]. \quad (4.61)$$

If $\tilde{\varphi}_0(k) \in L^2(0, \infty)$, then the Riemann–Lebesque lemma insures that

$$\lim_{s\to\infty} (\Phi_0, e^{-isP_\Delta}\Phi_0) = \exp(-\Delta\,\|\tilde{\varphi}_0\|^2) \neq 0, \quad (4.62a)$$

and, consequently, there is necessarily a subspace where $P_\Delta = 0$. The bilinear form of W_l assures us that in this subspace $B_\Delta = 0$ as well. Thus, we are confronted with a representation for B_Δ and P_Δ, a portion of which is unfaithful, and in which every group element is represented by unity.[30] However, if $\tilde{\varphi}_0(k) \notin L^2(0, \infty)$, e.g., in the manner described previously, then clearly

$$\lim_{s\to\infty} (\Phi_0, e^{-isP_\Delta}\Phi_0) = 0. \quad (4.62b)$$

In fact, it can be easily shown that

$$\lim_{s\to\infty} (\Phi[f_1, g_1], e^{-isP_\Delta}\Phi[f_2, g_2]) = 0, \quad (4.63)$$

from which it follows that

$$\operatorname{w-lim}_{s\to\infty} e^{-isP_\Delta} = 0. \quad (4.64)$$

This is just the condition that P_Δ only has an absolutely continuous spectrum. Depending on the proposed application of the representation, the spectral properties of a given smeared field (P_Δ, say) may determine that certain field representations are appropriate. Elsewhere, in an application of the local affine fields to physically motivated model problems, we shall discuss this question further.[31]

The non-Abelian nature of the combination law (4.31) imposes restrictions on the allowed representations. To demonstrate that there are representations, other than the ultralocal ones, exhibiting translation invariance and cluster decomposition, it suffices to give a class of different examples. Let $0 \leq k_q < \infty$, $q = 1, \cdots, Q$, denote Q integration variables; $f(\mathbf{z}_q) \equiv f(\mathbf{a}_q + \mathbf{z})$, $g(\mathbf{z}_q) \equiv g(\mathbf{a}_q + \mathbf{z})$, where $\mathbf{a}_q$ denote Q fixed, distinct points in R^3 and where $\tilde{\varphi}_0(k_q) = \tilde{\varphi}_0(k_1, \cdots, k_Q)$ is an element of L^2 or possibly more general, roughly in the manner of (4.57). Then, it is essentially clear that a representation of (4.31) exists such that

$$E_A(f, g) = \exp\Big(\int d\mathbf{z}\int_0^\infty\cdots\int_0^\infty \tilde{\varphi}_0^*(k_q) \times \{\exp[-\textstyle\sum \frac{1}{2}f(\mathbf{z}_q) - i\sum k_q g(\mathbf{z}_q)] \times \tilde{\varphi}_0(e^{-f(\mathbf{z}_q)}k_q) - \tilde{\varphi}_0(k_q)\}\, dk^Q\Big), \quad (4.65)$$

which is inequivalent to all those discussed previously.

5. EXPONENTIAL REPRESENTATIONS OF CANONICAL FIELD OPERATORS

In this section we consider exponential representations of scalar field and momentum operators which fulfill the equal-time-commutation relation

$$[\varphi(\mathbf{x}), \pi(\mathbf{y})] = i\delta(\mathbf{x} - \mathbf{y}). \quad (5.1)$$

Unlike (4.1), the canonical operators do not form a closed algebra, so that our analysis will necessarily be somewhat different than in Sec. 4. If $f(\mathbf{x})$ and $g(\mathbf{x})$ denote suitable real test functions, such as elements of $\mathcal{D}$, then

$$[\varphi(f), \pi(g)] = i(f, g) \quad (5.2)$$

is the proper statement of (5.1) for self-adjoint operators on a suitable domain. We shall be particularly interested in the unitary Weyl operators

$$U[f, g] = e^{i[\varphi(f) - \pi(g)]}, \quad (5.3)$$

where we have chosen our sign convention for convenience. These operators obey the relation

$$U[f', g']U[f, g] = e^{(\frac{1}{2}i)[(f', g) - (g', f)]}U[f' + f, g' + g], \quad (5.4)$$

which is the Weyl form of the commutation relations.

We introduce the overcomplete family of states

$$\Phi'[f, g] \equiv e^{-i\,\mathrm{Im}(\hat{\xi}, \varphi'[f, g])}U[f, g]\Phi_0 \quad (5.5)$$

for some choice of ξ and $\varphi'[f, g] \in \mathfrak{h}$. We assume that $\mathrm{Im}(\xi, \varphi'[0, 0]) = 0$. Once again we insist on an exponential construction such that

$$(\Phi'[f', g'], \Phi'[f, g]) = N'Ne^{(\varphi'[f', g'], \varphi'[f, g])} = e^{-\frac{1}{2}\|\varphi'[f', g'] - \varphi'[f, g]\|^2} \times e^{i\,\mathrm{Im}(\varphi'[f', g'], \varphi'[f, g])}. \quad (5.6)$$

[30] For example, see E. P. Wigner, *Group Theory*, translated by J. J. Griffin (Academic Press, Inc., New York, 1959), p. 72. We use this term to denote representations having subrepresentations in which one or more generators are represented by the zero operator.

[31] E. W. Aslaksen, thesis, Lehigh University, Bethlehem, Pa., 1968; J. R. Klauder, *5th International Conference on Gravitation and the Theory of Relativity* (Tbilisi University, Tbilisi, USSR, to be published); J. R. Klauder, *Proceedings of the Relativity Conference in the Midwest* (Plenum Press, New York, to be published).

If we employ the combination law (5.4), then we learn [cf. Eq. (4.7)] that

$$\|\varphi'[f', g'] - \varphi'[f, g]\|^2 = \|\varphi'[0, 0] - \varphi'[f - f', g - g']\|^2, \quad (5.7a)$$

$$\begin{aligned}\operatorname{Im}(\varphi'[f', g'], \varphi'[f, g]) &= \operatorname{Im}(\varphi'[0, 0], \varphi'[f - f', g - g']) \\ &\quad - \operatorname{Im}(\hat{\xi}, \varphi'[f - f', g - g'] \\ &\quad - \varphi'[f, g] + \varphi'[f', g']) \\ &\quad - \tfrac{1}{2}[(f', g) - (g', f)]. \end{aligned} \quad (5.7b)$$

The basic solution to these relations gives rise to the Fock representation of Sec. 2A, which motivated our study in the first place. Somewhat more generally, we adopt as our first solution $\mathfrak{h} \equiv \mathfrak{h}_1$, $\xi = 0$, and

$$\begin{aligned}\varphi'[f, g] &\equiv \varphi_1[f, g] \\ &= 2^{-\frac{1}{2}}\{[(\Omega^{-\frac{1}{2}}\tilde{f})(\mathbf{k}) - i(\Omega^{\frac{1}{2}}\tilde{g})(\mathbf{k})] \oplus (\Gamma^{\frac{1}{2}}\tilde{f})(\mathbf{k})\}. \end{aligned} \quad (5.8)$$

Here Ω and Γ are suitable linear operators, such as $\Omega = \Omega(k)$, $\Gamma = \Gamma(k)$, i.e., multiplication operators in momentum space such that $\Omega(k) > 0$ almost everywhere. On the other hand, $\Gamma(k) \geq 0$, and, in fact, Γ may be chosen identically zero. While (5.8) gives a convenient specific realization, we shall also treat $\mathfrak{h}_1$ abstractly. It is clear that $\varphi_1[f, g]$ already fulfills (5.7) and thus generates an exponential representation of the canonical commutation relations. The resultant expectation functional is given by

$$\begin{aligned}E_1(f, g) &= (\Phi_0, \Phi'[f, g]) = e^{-\frac{1}{2}\|\varphi'[f,g]\|^2} \\ &= e^{-\frac{1}{4}[(\tilde{f}, \Lambda\tilde{f}) + (\tilde{g}, \Omega\tilde{g})]}, \end{aligned} \quad (5.9a)$$

where

$$\Lambda \equiv \Omega^{-1} + \Gamma. \quad (5.9b)$$

This solution is translationally invariant and satisfies cluster decomposition whenever $\Lambda = \Lambda(k)$ and $\Omega = \Omega(k)$ are polynomially bounded. In this case, distinct function pairs $\Lambda(k)$ and $\Omega(k)$ lead to unitarily inequivalent representations.

The representation characterized by $E_1(f, g)$ is just the direct product of two independent Fock representations; it reduces to only one such representation if $\Gamma \equiv 0$.[32] Whether one or two Fock representations are involved, the field and momentum operators are given by suitable linear combinations of annihilation and creation operators. According to (3.3), we set

$$A_1(\lambda)\Phi[f, g] = (\lambda, \varphi_1[f, g])\Phi[f, g]. \quad (5.10)$$

[32] H. Araki and E. J. Woods, J. Math. Phys. **4**, 637 (1965); J. R. Klauder and L. Streit, J. Math. Phys. **10**, 1661 (1969).

It can be shown[32] that the smeared momentum operator $\pi_1(e)$ is given by

$$\pi_1(e) = (i/2^{\frac{1}{2}})[A_1(\lambda_{\pi(e)})^\dagger - A_1(\lambda_{\pi(e)})], \quad (5.11a)$$

where

$$\lambda_{\pi(e)} \equiv -i(\Omega^{\frac{1}{2}}\tilde{e})(\mathbf{k}) \oplus 0. \quad (5.11b)$$

In like manner, the smeared field operator $\varphi_1(e)$ is given by

$$\varphi_1(e) = 2^{-\frac{1}{2}}[A_1(\lambda_{\varphi(e)})^\dagger + A_1(\lambda_{\varphi(e)})], \quad (5.12a)$$

where

$$\lambda_{\varphi(e)} = -i(\Omega^{-\frac{1}{2}}\tilde{e})(\mathbf{k}) \oplus -i(\Gamma^{\frac{1}{2}}\tilde{e})(\mathbf{k}). \quad (5.12b)$$

Consequently, we find that

$$[\varphi_1(e), \pi_1(e)] = i(\lambda_{\varphi(e)}, \lambda_{\pi(e)}) = i(e, e), \quad (5.13)$$

as desired. The representation of φ_1 and π_1 determined by the above construction is reducible whenever $\Gamma \not\equiv 0$.

Although the solution presented above is rather special, all the other solutions build on it. For our second solution, we still choose $\xi = 0$, but set

$$\mathfrak{h} = \mathfrak{h}_1 \oplus \mathfrak{h}_2, \quad (5.14a)$$

$$\varphi'[f, g] \equiv \varphi[f, g] = \varphi_1[f, g] \oplus \varphi_2[f, g], \quad (5.14b)$$

where $\mathfrak{h}_1$ and $\varphi_1[f, g]$ are as before. Since φ_1 already fulfills Eq. (5.7), it follows that $\varphi_2[f, g]$ satisfies those same relations, with the last term in (5.7) absent. Consequently, we adopt as our solution

$$\varphi_2[f, g] = \tilde{u}_0[f, g]\varphi_0, \quad (5.15a)$$

where the unitary operators $\tilde{u}_0$ fulfill

$$\tilde{u}_0[f', g']\tilde{u}_0[f, g] = \tilde{u}_0[f' + f, g' + g], \quad (5.15b)$$

a strictly Abelian combination law. Accordingly, we may set

$$\tilde{u}_0[f, g] = e^{i[\tilde{\varphi}(f) - \tilde{\pi}(g)]}, \quad (5.15c)$$

where $\tilde{\varphi}(f)$ and $\tilde{\pi}(g)$ are *fully commuting* self-adjoint generators. Moreover, these operators act on $\mathfrak{h}_2$.

In terms of these expressions the resulting expectation functional has the form

$$\begin{aligned}E(f, g) &= E_1(f, g)e^{(\varphi_0, \{\tilde{u}_0[f,g] - 1\}\varphi_0)} \\ &= E_1(f, g)\exp(\varphi_0, \{e^{i[\tilde{\varphi}(f) - \tilde{\pi}(g)]} - 1\}\varphi_0), \end{aligned} \quad (5.16)$$

where E_1 is still given by (5.9). So long as $\varphi_0 \in \mathfrak{h}_2$, as is presently the case, none of these combined solutions can exhibit translational invariance and cluster decomposition. This defect will be remedied later when we treat improper translations.

The Weyl operators for this solution may be given as follows: Let $U_1[f, g]$ denote the Weyl operators characterized by Eq. (5.9), and define

$$\tilde{U}_0[f, g] = \bigoplus_{n=0}^{\infty} \bigotimes_{s=1}^{n} \tilde{u}_0[f, g]. \quad (5.17)$$

Then the solution characterized by Eq. (5.16) is given by

$$U[f, g] = U_1[f, g] \otimes \tilde{U}_0[f, g]. \quad (5.18)$$

If we extend the operators U_1 and $\tilde{U}_0$ in obvious fashion to the direct product space, then we may drop the direct product and observe that

$$U[f, g] = U_1[f, g]\tilde{U}_0[f, g] \equiv e^{i[\varphi(f)-\pi(g)]}. \quad (5.19)$$

Since $\tilde{U}_0[f, g]$ obeys an Abelian combination law, in virtue of (5.15b), it is clear that $U[f, g]$ will satisfy the Weyl form of the commutation relations (5.4) and that the generators $\varphi(f)$ and $\pi(g)$ will obey the usual Heisenberg rules.

In the present solution, the annihilation operators $A(\lambda)$, $\lambda \in \mathfrak{h}_1 \oplus \mathfrak{h}_2$, fulfill the relation

$$A(\lambda)\Phi[f, g] = (\lambda, \varphi[f, g])\Phi[f, g]. \quad (5.20)$$

If we set $\lambda = \lambda_1 \oplus \lambda_2$, $\lambda_j \in \mathfrak{h}_j$, then we define

$$A(\lambda) = A(\lambda_1 \oplus \lambda_2) = A(\lambda_1 \oplus 0) + A(0 \oplus \lambda_2) \equiv A_1(\lambda) + A_2(\lambda), \quad (5.21)$$

which may be seen to be in accord with (5.10). Note that A_1 commutes with A_2 and $A_2^\dagger$. Equation (5.20) leads to the relations

$$A_j(\lambda)\Phi[f, g] = (\lambda_j, \varphi_j[f, g])\Phi[f, g], \quad (5.22)$$

where we have put $\varphi_1 = \varphi_1 \oplus 0$ and $\varphi_2 = 0 \oplus \varphi_2$. The basic field and momentum operators in this representation are given as a combination of two different terms. On comparison with (4.13b), we determine that

$$\varphi(f) = \varphi_1(f) + (A_2, \tilde{\varphi}(f)A_2), \quad (5.23a)$$

$$\pi(g) = \pi_1(g) + (A_2, \tilde{\pi}(g)A_2), \quad (5.23b)$$

where φ_1 and π_1 are given by Eqs. (5.11a) and (5.12a). Note that the field operators involve one term which is *linear* in A and $A^\dagger$ and another term which is *quadratic*. This is a characteristic feature of such representations. Since

$$[(A_2, \tilde{\varphi}(f)A_2), (A_2, \tilde{\pi}(g)A_2)] = (A_2, [\tilde{\varphi}(f), \tilde{\pi}(g)]A_2) = 0, \quad (5.24)$$

it follows that

$$[\varphi(f), \pi(g)] = [\varphi_1(f), \pi_1(g)] = i(f, g), \quad (5.25)$$

as desired. The reducibility of this representation is evident, assuming that $\tilde{\varphi}$ and $\tilde{\pi}$ are not both identically zero.

Before we pass to another solution of the basic relations (5.7), we note the fact that since $\tilde{\varphi}$ and $\tilde{\pi}$ commute, they may both be simultaneously diagonalized. Alternatively stated, we may write

$$\mathfrak{h}_2 = \int^{\oplus} C \, d\mu(a, b), \quad (5.26a)$$

$$e^{i[\tilde{\varphi}(f)-\tilde{\pi}(g)]} = \int^{\oplus} e^{i[(a,f)-(b,g)]} \, d\mu(a, b), \quad (5.26b)$$

such that

$$\varphi_0 = \int^{\oplus} d\mu(a, b). \quad (5.26c)$$

Here $a = a(\mathbf{x})$ and $b = b(\mathbf{x})$ are elements of $\mathfrak{D}'$ and the integration is over the space $\mathfrak{D}' \times \mathfrak{D}'$. Note that

$$\|\varphi_0\|^2 = \int d\mu(a, b) = \mu(\mathfrak{D}' \times \mathfrak{D}'), \quad (5.27)$$

which must be finite in the present case. In this language, the expectation functional (5.16), for example, takes the form

$$E(f, g) = E_1(f, g) \exp\left[\int (e^{i[(a,f)-(b,g)]} - 1) \, d\mu(a, b)\right]. \quad (5.28)$$

Evidently, other expressions may be "diagonalized" in $\mathfrak{h}_2$ and given a concrete representation in similar fashion. As a specific example, it may be assumed that μ is concentrated on δ functions at a specific point $\mathbf{z}$, just as was assumed for Eq. (3.63).

We turn now to another set of solutions of the basic relations (5.7) in which the translation vector $\xi \neq 0$. In essence, we will retain the second form of our solution, in which $\mathfrak{h} = \mathfrak{h}_1 \oplus \mathfrak{h}_2$, and consider $\xi = \xi_1 \oplus \xi_2$. The role of ξ_1 is especially simple. From the form of $\varphi_1[f, g]$ in (5.8), it is clear that ξ_1 does not influence Eq. (5.7) in any way. The only appearance ξ_1 makes is in the phase factor which accompanies (5.5). We can relate any solution with $\xi_1 \neq 0$ to the corresponding solution with $\xi_1 = 0$ by appending the phase factor $e^{-i\,\mathrm{Im}\,(\xi_1, \varphi_1[f,g])}$, where $\varphi_1[f, g]$ is given in (5.8), to the expectation functional for the case when $\xi_1 = 0$. The effect of ξ_1, stated otherwise, is to add to φ_1 and π_1 constant multiples of the identity. For convenience, we shall confine our attention to the consequences of ξ_2, and we set $\xi_1 = 0$.

With these remarks in mind we choose as our solution to (5.7) the relations

$$\xi = 0 \oplus \xi_2 \equiv 0 \oplus \hat{\varphi}_0, \quad (5.29a)$$

$$\varphi'[f, g] \equiv \varphi_1[f, g] \oplus \varphi_2'[f, g], \quad (5.29b)$$

where

$$\varphi_2'[f, g] = (\tilde{u}_0[f, g] - 1)\hat{\varphi}_0. \quad (5.30)$$

Here $\varphi_1[f, g] \in \mathfrak{h}_1$ has the same form as before, and likewise for the Abelian group $\tilde{u}_0[f, g]$. The expectation functional to which this solution leads,

$$\begin{aligned} E(f, g) &= E_1(f, g) \exp(\hat{\varphi}_0, \{\tilde{u}_0[f, g] - 1\}\hat{\varphi}_0) \\ &= E_1(f, g) \exp(\hat{\varphi}_0, \{e^{i[\tilde{\varphi}(f) - \tilde{\pi}(g)]} - 1\}\hat{\varphi}_0), \end{aligned} \tag{5.31}$$

is similar to the one in (5.16). A real difference arises only when $\|\hat{\varphi}_0\| = \infty$, corresponding to an improper translation. Since this is the most interesting case, we shall concentrate on improper translations.

Although $\hat{\varphi}_0 \notin \mathfrak{h}_2$, we assume that $\varphi_2'[f, g] \in \mathfrak{h}_2$ and that $(\hat{\varphi}_0, \varphi_2'[f, g])$ makes sense. Moreover, for convenience, let us assume that $\tilde{\varphi}(f)\hat{\varphi}_0$ and $\tilde{\pi}(g)\hat{\varphi}_0$ are in $\mathfrak{h}_2$, and that $(\hat{\varphi}_0, \tilde{\varphi}(f)\hat{\varphi}_0)$ and $(\hat{\varphi}_0, \tilde{\pi}(g)\hat{\varphi}_0)$ make sense. With these assumptions, which are like those made in conjunction with Eq. (4.23), we can determine expressions for the field operators $\varphi(f)$ and $\pi(g)$.

We define the operators $A'(\lambda)$, $\lambda \in \mathfrak{h}_1 \oplus \mathfrak{h}_2$, according to the relation

$$A'(\lambda)\Phi'[f, g] = (\lambda, \varphi'[f, g])\Phi'[f, g]. \tag{5.32}$$

If we set $\lambda = \lambda_1 \oplus \lambda_2$, then we define

$$\begin{aligned} A'(\lambda) = A'(\lambda_1 \oplus \lambda_2) &= A'(\lambda_1 \oplus 0) + A'(0 \oplus \lambda_2) \\ &\equiv A_1(\lambda) + A_2'(\lambda). \end{aligned} \tag{5.33}$$

Since

$$\begin{aligned} A_2'(\lambda)\Phi'[f, g] &= (\lambda_2, \varphi_2'[f, g])\Phi'[f, g] \\ &= (\lambda_2, \{\tilde{u}_0[f, g] - 1\}\hat{\varphi}_0)\Phi'[f, g], \end{aligned} \tag{5.34}$$

it is clear that

$$A_2'(\lambda) = A_2(\lambda) - (\lambda, \hat{\varphi}_0), \tag{5.35}$$

where here $\hat{\varphi}_0 = 0 \oplus \hat{\varphi}_0$ is understood. Since $\|\hat{\varphi}_0\| = \infty$, it follows that A_2 and A_2' are unitarily inequivalent and, in particular, that A_2 is (now) inequivalent to the Fock representation A_2'.

The basic field and momentum operators of this solution have the *same form* as in (5.23), namely,

$$\varphi(f) = \varphi_1(f) + (A_2, \tilde{\varphi}(f)A_2), \tag{5.36a}$$

$$\pi(g) = \pi_1(g) + (A_2, \tilde{\pi}(g)A_2), \tag{5.36b}$$

but are inequivalent to those previous solutions. The latter terms in each expression may be interpreted in the manner of (4.23). General criteria for equivalence of two such representations follow the discussion pertaining to Eq. (4.24) (in addition to taking into account the equivalence of the first term). In particular (assuming Ω and Γ remain fixed), two such representations are equivalent, provided that there exists a unitary v which commutes with $\tilde{\varphi}$ and $\tilde{\pi}$ such that

$$\hat{\varphi}_{01} - v\hat{\varphi}_{02} \in \mathfrak{h}_2. \tag{5.37}$$

To demonstrate the reducibility of these representations, we need only exhibit one operator, different from unity, which commutes with φ and π. Although (A_2, A_2) fails to be an operator when A_2 is not equivalent to the Fock representation, we have, by assumption, the fact that

$$Y \equiv (A_2, \tilde{\varphi}(e)A_2), \tag{5.38}$$

for some fixed but arbitrary test function e, is a meaningful operator. Since $\tilde{\varphi}$ and $\tilde{\pi}$ commute, it is clear that

$$[Y, \varphi(f)] = 0 = [Y, \pi(g)], \tag{5.39}$$

which demonstrates reducibility.

The commutativity of $\tilde{\varphi}$ and $\tilde{\pi}$ permits us again to diagonalize them both simultaneously. In particular, we let

$$\mathfrak{h}_2 = \int^{\oplus} C\, d\hat{\mu}(a, b), \tag{5.40a}$$

$$\hat{\varphi}_0 = \int^{\oplus} d\hat{\mu}(a, b), \tag{5.40b}$$

$$\varphi_2'[f, g] = \int^{\oplus} (e^{i[(a,f)-(b,g)]} - 1)\, d\hat{\mu}(a, b). \tag{5.40c}$$

As before, the integration is over $\mathfrak{D}' \times \mathfrak{D}'$, but now we assume that

$$\|\hat{\varphi}_0\|^2 = \int d\hat{\mu}(a, b) = \hat{\mu}(\mathfrak{D}' \times \mathfrak{D}') = \infty. \tag{5.41}$$

With this expression, it follows that

$$E(f, g) = E_1(f, g) \exp\left[\int (e^{i[(a,f)-(b,g)]} - 1)\, d\hat{\mu}(a, b)\right]. \tag{5.42}$$

Among the many possible ways in which improper translations can be used, we shall consider only two basic examples. Initially, let us create a translationally invariant expectation functional exhibiting cluster decomposition. This we may do, following the lead of Eq. (3.73), by assuming that

$$\begin{aligned} L(f, g) &\equiv \int (e^{i[(a,f)-(b,g)]} - 1)\, d\hat{\mu}(a, b) \\ &= \int (e^{i[(a,f_{\mathbf{z}})-(b,g_{\mathbf{z}})]} - 1)\, d\hat{\mu}_0(a, b)\, d\mathbf{z}, \end{aligned} \tag{5.43}$$

where $\hat{\mu}_0$ is concentrated on distributions "at or near" the origin. As a further specialization, we may let $\hat{\mu}_0$ be concentrated on δ functions at the origin so that

$$L(f, g) = \int (e^{i[\lambda f(\mathbf{z}) - \nu g(\mathbf{z})]} - 1)\, d\hat{\sigma}_0(\lambda, \nu)\, d\mathbf{z}. \tag{5.44}$$

Such a specialization would be an ingredient in leading to an ultralocal representation [in the sense of (3.78)–(3.80)]. Note that it is not necessary that $\hat{\sigma}_0$ have a finite measure. For example, a possible choice for $\hat{\sigma}_0$

would be

$$d\hat{\sigma}_0(\lambda, \nu) = (\lambda^2 + \nu^2)^{-1} e^{-(\lambda^2 + \nu^2)}\, d\lambda\, d\nu. \quad (5.45)$$

We have already noted that when $\|\hat{\varphi}_0\| = \infty$, the representation of φ and π is inequivalent to that based on a Fock representation for A_2. In light of the discussion in Sec. 4, it should be clear that in the example discussed above, if $\hat{\sigma}_0$ has infinite measure, the representation of φ and π is not even "locally Fock." The issues of unfaithful representations which arose in the group case do not arise for the canonical operators, for it is impossible to fulfill the Heisenberg commutation relation, if there is a subspace where φ or π or both act as zero. This is reflected in the fact that all solutions have the property that $\varphi(f)$ and $\pi(g)$ have a strictly absolutely continuous spectrum, as may be seen by applying the test of Eq. (4.22). In all such tests the properties of φ_1 and π_1 are controlling.[33]

6. NONTRIVIAL NATURE OF HAMILTONIAN

As our concluding remarks we want only to indicate, in a simple and heuristic fashion, that the Hamiltonians associated with exponential representations are generally far from trivial and contain terms for production, annihilation, and scattering. Elsewhere[6] we shall study these models in their own right for their physical content and predictions.

A very common feature of a canonical theory is the identity between the momentum operator $\pi(\mathbf{x})$ and the first time derivative of the field $\dot{\varphi}(\mathbf{x})$. If $\mathcal{H}$ denotes the Hamiltonian for the problem, then we require that

$$i[\mathcal{H}, \varphi(\mathbf{x})] = \pi(\mathbf{x}). \quad (6.1)$$

Roughly speaking, this means that

$$\mathcal{H} = \tfrac{1}{2}\int \pi^2(\mathbf{x})\, d\mathbf{x} + \mathcal{W}, \quad (6.2)$$

where $[\mathcal{W}, \varphi(\mathbf{x})] = 0$. This form holds true whether or not the canonical operators are given by an irreducible representation. In an exponential representation of the canonical variables, as we considered in Sec. 5, the momentum operator $\pi(\mathbf{x})$ is a bilinear expression in annihilation and creation operators, as is made explicit in Eq. (5.36b). Consequently, $\mathcal{H}$, being quadratic in $\pi(\mathbf{x})$, is (at least) a quartic in these operators, much as in Eq. (2.13). Thus, it is clear that theories exhibiting production, annihilation, and scattering—as these terms are conventionally understood—can be constructed with the aid of exponential representations, since such terms appear in the Hamiltonian. It is noteworthy that these terms already appear in the kinetic energy factor, usually treated as part of the free theory. This is, of course, a consequence of the uncommon representations which we have employed.

[33] H. G. Tucker, Pacific J. Math. 12, 1125 (1962).

Commun. math. Phys. 21, 41—55 (1971)

New "Coherent" States Associated with Non-Compact Groups*

A. O. BARUT and L. GIRARDELLO**

Institute of Theoretical Physics

and

Department of Physics, University of Colorado, Boulder, Colorado

Received November 2, 1970

Abstract. Generalized "Coherent" States are the eigenstates of the lowering and raising operators of non-compact groups. In particular the discrete series of representations of $SO(2, 1)$ are studied in detail: the resolution of the identity and the connection with the Hilbert spaces of entire functions of growth (1, 1). Also discussed are the application to the evaluation of matrix elements of finite group elements and the contraction to the usual coherent states.

I. Introduction

The definition and use of coherent states associated with the Heisenberg algebra is well known (Section II). The purpose of this paper is to generalize this notion to the Lie algebra of non compact groups. In particular, we deal with the simplest semi-simple Lie algebra of $SO(2,1)$ isomorphic to the algebra of $SU(1,1)$ and $SL(2,R)$. We call generalized "coherent" states the eigenstates of the ladder operators in the discrete series of representations. Generalizations of these continuous bases will be indicated. The new "coherent" states are useful mathematically, aside from their intrinsic interest, in the evaluation of matrix elements of the finite transformations of the group and will have physical applications as the ordinary coherent states have.

II. Coherent States Associated with the Heisenberg Algebra

In this Section we review briefly, for reference purposes, some important properties of the usual coherent states[1] which are introduced

* Supported in part by the Air Force Office of Scientific Research under Grant AF-AFOSR-30-67.

** On leave from Istituto di Fisica dell'Universitá, Parma and Istituto Nazionale di Fisica Nucleare, Sezione di Milano (Italy).

[1] For more details see, for example, Klauder, J. R.: Ann. Phys. (N. Y.) **11**, 123 (1960); Glauber, J.: Phys. Rev. **131**, 2766 (1963), and Klauder, J., Sudarshan, E. C. G.: Quantum Optics. New York: Benjamin 1968.

as the eigenstates in an Hilbert space of the boson annihilation operator

$$a|z\rangle = z|z\rangle, \quad [a, a^+] = \mathbb{1}, \tag{2.1}$$

where z is a complex eigenvalue. In terms of the eigenstates $|n\rangle$ of a^+a (or the Hamiltonian of the linear harmonic oscillator) one obtains

$$|z\rangle = e^{-\frac{1}{2}|z|^2} \sum_{n=0}^{\infty} \frac{z^n}{\sqrt{n!}} |n\rangle. \tag{2.2}$$

The factor in front of the sum is so chosen that the states are normalized

$$\langle z|z\rangle = 1. \tag{2.3}$$

But they are not orthogonal to each other

$$\langle z'|z\rangle = \exp[-\tfrac{1}{2}|z|^2 - \tfrac{1}{2}|z'|^2 + z'^* z] \tag{2.4}$$

so that they form an over-complete linearly dependent set. The resolution of the identity holds in the form

$$\frac{1}{\pi}\int d^2z|z\rangle\langle z| = \sum_{n=0}^{\infty} |n\rangle\langle n| = 1; \; d^2z = d(\operatorname{Re} z)\, d(\operatorname{Im} z). \tag{2.5}$$

It has been recently shown [1] under which conditions countable subsets are complete. In particular one can take lattice points $z_{mn} = \gamma(m + in)$, $m, n = 0, \pm 1, \pm 2, \ldots$. For $0 < \gamma < \sqrt{\pi}$ the set of states $\{|z_{m,n}\rangle\}$ is still overcomplete; for $\gamma > \sqrt{\pi}$ it is not complete. For $\gamma = \sqrt{\pi}$ one obtains a complete set (von Neumann case).

An arbitrary vector f can be expanded

$$|f\rangle = \frac{1}{\pi}\int d^2z|z\rangle\langle z|f\rangle. \tag{2.6}$$

The coefficients satisfy the equation

$$\langle z|f\rangle = \frac{1}{\pi}\int d^2z'\langle z|z'\rangle\langle z'|f\rangle \tag{2.7}$$

so that $\langle z|z'\rangle$, Eq. (2.4), acts as a reproducing kernel.

The coherent states are special quantum states most closely approximating classical states in the sense that for them the uncertainty relation $\Delta p \Delta q \geqq \hbar/2$ has its minimum value. This can be seen by comparing the ground state wave function of the oscillator with the relation

$$\langle z|n\rangle = e^{-\frac{1}{2}|z|^2} \frac{z^{*n}}{\sqrt{n!}}.$$

III. Coherent States Associated with the Lie-Algebra of $SU(1, 1)$

$SU(1,1) \sim SO(2,1) \sim SL(2,R)$.

The Lie-algebra is defined by the commutation relations [2]

$$[L^+, L^-] = -L_{12}, [L_{12}, L^{\pm}] = \pm L^{\pm}. \tag{3.1}$$

The Casimir operator is given by

$$Q = -L_{12}(L_{12}-1) + 2L^+L^- = -L_{12}(L_{12}+1) + 2L^-L^+. \tag{3.2}$$

Note that $L^{\pm} = \frac{1}{\sqrt{2}}(L_{13} \pm i L_{23})$. The fundamental spinor-representation of the algebra is

$$L_{12} = \tfrac{1}{2}\sigma_3, \quad L^{\pm} = \frac{i}{2\sqrt{2}}(\sigma_1 \pm i\sigma_2). \tag{3.3}$$

The corresponding parametrization of the group is defined by the Euler angles such that the group element is

$$\begin{aligned} W &= e^{i\mu L_{12}} e^{i\xi L_{23}} e^{i\nu L_{12}} \\ &= \begin{pmatrix} e^{i\frac{\mu}{2}} & \\ & e^{-i\frac{\mu}{2}} \end{pmatrix} \begin{pmatrix} \cos\left(\varepsilon\frac{\xi}{2}\right) & \sin\left(\varepsilon\frac{\xi}{2}\right) \\ -\sin\left(\varepsilon\frac{\xi}{2}\right) & \cos\left(\varepsilon\frac{\xi}{2}\right) \end{pmatrix} \begin{pmatrix} e^{i\nu/2} & \\ & e^{-i\nu/2} \end{pmatrix} \\ &= \begin{pmatrix} \alpha & \beta \\ \bar{\beta} & \bar{\alpha} \end{pmatrix}, \quad \alpha = e^{i(\mu+\nu)/2}\cos\left(\varepsilon\frac{\xi}{2}\right), \quad \beta = e^{i(\mu-\nu)/2}\sin\left(\varepsilon\frac{\xi}{2}\right) \end{aligned} \tag{3.4}$$

$\det W = 1$, $\varepsilon = i$ for $SU(1,1)$, $\varepsilon = 1$ for $SO(3)$.

1. *Discrete Representations $D^+(\Phi)$*

The discrete class of unitary representations in the Hilbert space with the basis vectors $|\Phi, m\rangle$ can be defined by the following relations [2]

$$\begin{aligned} L_{12}|\Phi, m\rangle &= (E_0 + m)\,|\Phi, m\rangle \\ L^+|\Phi, m\rangle &= \frac{1}{\sqrt{2}}[(\Phi + E_0 + m + 1)(E_0 - \Phi + m)]^{\frac{1}{2}}\,|\Phi, m+1\rangle \\ L^-|\Phi, m\rangle &= \frac{1}{\sqrt{2}}[(\Phi + E_0 + m)(E_0 - \Phi + m - 1)]^{\frac{1}{2}}\,|\Phi, m-1\rangle \\ Q|\Phi, m\rangle &= \Phi(\Phi+1)\,|\Phi, m\rangle. \end{aligned} \tag{3.5}$$

Here E_0 is an arbitrary invariant (coming from the universal covering group). For discrete representations E_0 and the Casimir invariant are not independent

$$\Phi + E_0 = 0 . \tag{3.6}$$

The spectrum of L_{12} is discrete and bounded below:

$$L_{12} - E_0 = L_{12} + \Phi = 0, 1, 2, 3 \ldots . \tag{3.7}$$

For unitary representations

$$\begin{gathered} \operatorname{Im} E_0 = 0, \Phi < 0, \langle \Phi, m | L^- L^+ | \Phi, m \rangle = \text{real} > 0 \\ -2\Phi = 1, 2, 3 \ldots . \end{gathered} \tag{3.8}$$

The representations with Φ and $-\Phi - 1$ are equivalent. The inner product in terms of the basis states is

$$\langle \Phi, m | \Phi, m' \rangle = \delta_{mm'} . \tag{3.9}$$

2. Diagonalization of L^- in $D^+(\varphi)$: "Coherent" States

We introduce the generalized coherent states as the eigenvectors of L^-. From (3.5)

$$L^- |\Phi, m\rangle = \frac{1}{\sqrt{2}} [(m(-2\Phi + m - 1)]^{\frac{1}{2}} |\Phi, m - 1\rangle . \tag{3.10}$$

We define the eigenvectors $|z\rangle$ of L^- as linear combinations of the basis vectors $\{|\Phi, m\rangle\}$ which is a complete orthonormal set in the Hilbert space:

$$L^- |z\rangle = z |z\rangle, \quad z = \text{any complex number}, \tag{3.11}$$

$$|z\rangle = \left(\sqrt{\Gamma(-2\Phi)}\right) \sum_{n=0}^{\infty} \frac{(\sqrt{2}\, z)^n}{[n!\, \Gamma(-2\Phi + n)]^{\frac{1}{2}}} |\Phi, n\rangle . \tag{3.12}$$

The adjoint states are given by

$$\langle z| = [\Gamma(-2\Phi)]^{\frac{1}{2}} \sum_{n=0}^{\infty} \frac{(\sqrt{2}\, z^*)^n}{[n!\, \Gamma(-2\Phi + n)]^{\frac{1}{2}}} \langle \Phi, n| . \tag{3.13}$$

Eq. (3.11) can directly be verified using (3.10).

The inner product of the new "coherent" states is

$$\langle z'|z\rangle = \Gamma(-2\Phi) \sum_{n=0}^{\infty} \frac{(2z'^* z)^n}{n!\, \Gamma(-2\Phi + n)} = {}_0F_1(-2\Phi;\ 2z'^* z) . \tag{3.14}$$

Hence the norm is given by

$$\| |z\rangle \|^2 = \langle z|z\rangle = \Gamma(-2\Phi) \sum_{n=0}^{\infty} \frac{|\sqrt{2}\, z|^{2n}}{n!\, \Gamma(-2\Phi + n)} = {}_0F_1(-2\Phi; 2|z|^2) \tag{3.15}$$

where

$$ {}_0F_1(c;z)=1+\frac{1}{c!}\,\frac{1}{1!}+\frac{1}{c(c+1)}\,\frac{z^2}{2!}+\frac{1}{c(c+1)(c+2)}\,\frac{z^3}{3!}+\cdots \quad (3.16) $$

is a confluent hypergeometric function (entire function in z).

The norm can also be written in terms of the Bessel functions of integer order (for $2\Phi=$ integer)

$$ \langle z|z\rangle = \Gamma(-2\Phi)\,(i\sqrt{2}|z|)^{2\Phi+1}J_{-2\Phi-1}(2\sqrt{2}\,i|z|) \quad (3.15') $$

because of the relation

$$ J_\nu(z)=\frac{(z/2)^\nu}{\Gamma(\nu+1)}\;{}_0F_1(\nu+1;\;-z^2/4)\,. \quad (3.17) $$

Again we see that the coherent states are overcomplete, and do not form an orthonormal set. Two vectors $|z'\rangle$ and $|z\rangle$ are orthogonal if the entire function ${}_1F_0(-2\Phi, 2z'^*z)$ has a zero at the point $2z'^*z$. There is only qualitative information about the location of such zeros. Clearly, if z'^*z is a positive number the entire function has no zeros. Also it is not very useful to normalize the vectors $|z\rangle$ by dividing it with the square root of the norm (3.15).

The connection to the Hilbert spaces of entire functions of the exponential type of growth (1, 1) will be treated in Section VI.

3. *The Adjoint Operator* L^+

In the *unitary* discrete representation $D^+(\Phi)$, we have immediately from $L^+=(L^-)^\dagger$

$$ \langle z|\,L^+ = z^*\langle z|\,. \quad (3.18) $$

Hence

$$ \langle z'|\,L^+L^-\,|z\rangle = z'^*z\langle z'|z\rangle\,. \quad (3.19) $$

Consequently, we can directly verify that the Casimir operator acts as

$$ Q|z\rangle = (-L_{12}(L_{12}-1)+2L^+L^-)\,|z\rangle=\Phi(\Phi+1)\,|z\rangle\,, \quad (3.20) $$

as it should.

We also easily obtain the coefficients $\langle\Phi,m|z\rangle$

$$ \langle\Phi,m|z\rangle = [\Gamma(-2\Phi)]^{\frac{1}{2}}\,\frac{\sqrt{2}\,z^m}{[m!\,\Gamma(-2\Phi+m)]^{\frac{1}{2}}}\,. \quad (3.21) $$

4. Resolution of the Identity

The problem here consists in finding a weight function $\sigma(z)$ such that

$$\int d\sigma(z)\,|z\rangle\langle z| = \sum_{m=0}^{\infty} |\Phi, m\rangle\langle m, \Phi| = \mathbb{1}. \tag{3.22}$$

Let $|f\rangle$ and $|g\rangle$ be two arbitrary vectors in $\mathscr{H}$; then Eq. (3.22) means that

$$\langle f|g\rangle = \int d\sigma(z)\,\langle f|z\rangle\langle z|g\rangle. \tag{3.23}$$

We shall now determine $\sigma(z)$. Let

$$d\sigma(z) = \sigma(r)\, r\, dr\, d\theta, \quad r = |z|.$$

Then

$$\begin{aligned}
&\langle f|g\rangle \\
&= \int_0^\infty r\sigma(r)dr \int_0^{2\pi} d\theta \sum_{m=0}^{\infty}\sum_{n=0}^{\infty} \frac{(\sqrt{2}z^*)^m(\sqrt{2}z)^n\,\Gamma(-2\Phi)}{(m!\,n!\,\Gamma(-2\Phi+n)\,\Gamma(-2\Phi+m))^{\frac{1}{2}}}\langle f|n\rangle\langle m|g\rangle \\
&= \frac{\Gamma(-2\Phi)}{\sqrt{2}} \sum_{m=0}^{\infty}\sum_{n=0}^{\infty} \frac{\langle f|n\rangle\langle m|g\rangle}{(m!\,n!\,\Gamma(-2\Phi+n)\,\Gamma(-2\Phi+m))^{\frac{1}{2}}} \\
&\qquad \times \int_0^\infty \int_0^{2\pi} (\sqrt{2}r)^{m+n+1} e^{i\theta(n-m)}\sigma(r)\,dr\,d\theta \\
&= \frac{2\pi}{\sqrt{2}}\Gamma(-2\Phi)\sum_{n=0}^{\infty} \frac{\langle f|n\rangle\langle n|g\rangle}{n!\,\Gamma(-2\Phi+n)} \int_0^\infty (\sqrt{2}r)^{2n+1}\sigma(r)\,dr.
\end{aligned} \tag{3.24}$$

Hence we must have

$$\frac{2\pi\Gamma(-2\Phi)}{\sqrt{2}} \int_0^\infty (\sqrt{2}r)^{2n+1}\sigma(r)\,dr = \Gamma(n+1)\,\Gamma(-2\Phi+n). \tag{3.25}$$

Eq. (3.25) is a Mellin transform. We start from the formula [3]:

$$\int_0^\infty 2x^{\alpha+\beta} K_{\alpha-\beta}(2x^{\frac{1}{2}})\, x^{s-1}\, dx = \Gamma(2\alpha+s)\,\Gamma(2\beta+s) \tag{3.26}$$

where

$$K_\nu(z) = \frac{\pi}{2}\,\frac{I_{-\nu}(z) - I_\nu(z)}{\sin \pi\nu} \tag{3.27}$$

is the modified Bessel function of the third kind and

$$I_\nu(z) = \sum_0^\infty n\, \frac{(\frac{1}{2}z)^{\nu+2n}}{n!\,\Gamma(\nu+n+1)} \tag{3.28}$$

is the modified Bessel function of the first kind. Substituting $x^{\frac{1}{2}}=\sqrt{2}r$, $\alpha=\frac{1}{2}, 2\beta=-2\Phi$ in (3.26) and rearranging terms we obtain

$$4\sqrt{2}\int_0^\infty (\sqrt{2}r)^{-2\Phi-1}K_{\frac{1}{2}+\Phi}(2\sqrt{2}r)(\sqrt{2}r)^{2n+1}dr=\Gamma(n+1)\Gamma(-2\Phi+n)\,.$$

Thus, comparing with (3.25) we obtain finally the desired weight function

$$\sigma(r)=\frac{4}{\pi\Gamma(-2\Phi)}(\sqrt{2}r)^{-2\Phi-1}K_{\frac{1}{2}+\Phi}(2\sqrt{2}r),$$

$$\sigma(r)>0, \qquad r>0\,.$$

There is no problem of convergence of the integrals for $r\to 0$. The change of summation and integration in (3.24) may be rigorously justified by taking finite limit on the integration and going to the limit.

5. *Diagonalization of L^+ in $D^-(\Phi)$*

The unitary discrete series of representations $D^-(\Phi)$ are bounded above. Instead of (3.6), (3.7), and (3.10) we have

$$\begin{aligned} D^-(\Phi):\Phi-E_0&=0\\ \operatorname{Im}E_0&=0, \Phi<0\ \ 2\Phi=-1,-2,\dots \\ L_{12}-E_0&=0,-1,-2,\dots, \end{aligned} \tag{3.29}$$

$$L^+|\Phi,-m\rangle=\frac{1}{\sqrt{2}}[m(-2\Phi+m-1)]^{\frac{1}{2}}|\Phi,-m+1\rangle,\ m>0.$$

Consequently we can find eigenstates of L^+ analogous to Section II. It is easy to verify that

$$L^+|z\rangle=z|z\rangle$$

$$|z\rangle=[\Gamma(-2\Phi)]^{\frac{1}{2}}\sum_{m=0}^{\infty}\frac{(\sqrt{2}z)^m}{(m!\,\Gamma(-2\Phi+m))^{\frac{1}{2}}}|\Phi,-m\rangle \tag{3.30}$$

and that these states have all the properties of the eigenstates of L^- in $D^+(\Phi)$.

6. *Remark*

Continuous basis for $SO(2,1)\simeq SU(1,1)\simeq SL(2,R)$ have been studied before by a number of authors [4]. With the exception of Vilenkin, who considered this problem in a different context, in all these studies a non-compact, self-adjoint generator has been diagonalized. What we did here amounts to a diagonalization of a non-compact, non self-adjoint generator (lowering or raising operators). This was easy in the case of the discrete series D^+ or D^-. In the case of the principal or the supplementary series, however, this cannot be done in the Hilbert space: these representations are unbounded both below and above, hence the corresponding eigenvectors of L^- and L^+ would have infinite norm. In these cases appeal must be made to more general spaces, such as the rigged Hilbert spaces. (This problem will be studied elsewhere.)

IV. Matrix Elements of Finite Group Transformations

1. First we calculate the matrix elements of L_{12} in the coherent state basis [L^- is diagonal in $D^+(\Phi)$]. We have immediately, using (3.21),

$$\begin{aligned}\langle z'|L_{12}|z\rangle &= \sum_{n=0}^{\infty} \langle z'|L_{12}|\Phi, n\rangle \langle \Phi, n|z\rangle \\ &= \sum_{n=0}^{\infty} (-\Phi+n)\langle z'|\Phi, n\rangle \langle \Phi, n|z\rangle \\ &= -\Phi\langle z'|z\rangle + \Gamma(-2\Phi)\sum_{n=0}^{\infty} \frac{n(2z'^*z)^n}{n!\,\Gamma(-2\Phi+n)} \\ &= -\Phi\, {}_0F_1(-2\Phi, 2z'^*z) - \frac{z'^*z}{\Phi}\, {}_0F_1(-2\Phi+1;\, 2z'^*z)\end{aligned} \tag{4.1}$$

2. In this Section we evaluate the matrix elements of a general group element (3.4) in the continuous "coherent" basis $|z\rangle$. We can change the parameters such that

$$\begin{aligned}W &= \begin{pmatrix}1 & b\\ 0 & 1\end{pmatrix}\begin{pmatrix}e^{\frac{a}{2}} & 0\\ 0 & e^{-\frac{a}{2}}\end{pmatrix}\begin{pmatrix}1 & 0\\ c & 1\end{pmatrix}, \quad a, b, c \text{ real. } (SL(2R)) \\ &= e^{-ibL^+}e^{aL_{12}}e^{-icL^-}\end{aligned} \tag{4.2}$$

det $W = 1$ is satisfied.

Then

$$\begin{aligned}&\langle z'|e^{-ibL^+}e^{aL_{12}}e^{-icL^-}|z\rangle \\ &\qquad = e^{i(bz'^*-cz)}\sum_{n=0}^{\infty}\langle z'|e^{aL_{12}}|\Phi, m\rangle\langle\Phi, m|z\rangle \\ &\qquad = e^{i(bz'^*-cz)}e^{-a\Phi}\Gamma(-2\Phi)\sum_{m=0}^{\infty} e^{am}\frac{(2z'^*z)^m}{m!\,\Gamma(-2\Phi+m)}.\end{aligned}$$

Or, with $e^a = \delta^{-2}$,

$$= (\delta)^{2\Phi}e^{i(bz'^*-cz)}\,{}_0F_1\left(-2\Phi; \frac{2z'^*z}{\delta^2}\right). \tag{4.3}$$

V. Contraction to the Usual Coherent States

Referring to Eq. (3.1) we define ($L_3 \equiv L_{12}$)

$$L^{+\prime} = \sqrt{\varepsilon}\, L^+,\ L^{-\prime} = \sqrt{\varepsilon}\, L^-,\ L_3' = \varepsilon L_3,\ \varepsilon > 0. \tag{5.1}$$

Then

$$[L^{+\prime}, L^{-\prime}] = -L_3',\ [L_3', L^{\pm\prime}] = \pm\varepsilon L^{\pm\prime}. \tag{5.2}$$

Hence in the limit $\varepsilon \to 0$,

$$[L^{+\prime}, L^{-\prime}] = -L_3',\ [L_3', L^{\pm\prime}] = 0. \tag{5.3}$$

This contracted algebra is isomorphic to the Heisenberg algebra: We have the correspondence with the boson creation and annihilation operators:

$$\begin{gathered} L^{-\prime} \to a,\ L^{+\prime} \to a^+,\ L_3' \to I \\ [a, a^+] = 1,\ [I, a^+] = [I, a] = 0. \end{gathered} \tag{5.4}$$

Next we consider the matrix elements. From (3.5) for $D^+(\Phi)$:

$$\begin{aligned} \langle \Phi, m' | L_3 | \Phi, m \rangle &= (-\Phi + m)\, \delta_{m'm} \\ \langle \Phi, m' | L^+ | \Phi, m \rangle &= \frac{1}{\sqrt{2}} [(m+1)(-2\Phi + m)]^{\frac{1}{2}} \delta_{m'+1,m} \\ \langle \Phi, m' | L^- | \Phi, m \rangle &= \frac{1}{\sqrt{2}} [m(-2\Phi + m - 1)]^{\frac{1}{2}} \delta_{m'-1,m}. \end{aligned} \tag{5.5}$$

We evaluate the limit of these matrix elements as [5]

$$\varepsilon \to 0,\ \Phi \to -\infty, \quad \text{but} \quad \varepsilon\Phi \to -1, \tag{5.6}$$

i.e. through a sequence of representations, and obtain

$$\begin{aligned} \langle \Phi, m' | L_3' | \Phi, m \rangle &= \varepsilon(-\Phi + m)\, \delta_{m'm} \to \delta_{m'm} \\ \langle \Phi, m' | L^{+\prime} | \Phi, m \rangle &= \frac{1}{\sqrt{2}} (\varepsilon(m+1)(-2\Phi + m))^{\frac{1}{2}} \to (m+1)^{\frac{1}{2}} \delta_{m'-1,m} \\ \langle \Phi, m' | L^{-\prime} | \Phi, m \rangle &= \frac{1}{\sqrt{2}} [\varepsilon m(-2\Phi + m - 1)]^{\frac{1}{2}} \to m^{\frac{1}{2}} \delta_{m'+1,m}. \end{aligned} \tag{5.7}$$

These limits are precisely the matrix elements of the boson operators

$$\begin{gathered} \langle m' | I | m \rangle = \delta_{m',m};\ \langle m' | a^+ | m \rangle = (m+1)^{\frac{1}{2}} \delta_{m'-1,m'} \\ \langle m' | a | m \rangle = m^{\frac{1}{2}} \delta_{m'+1,m}. \end{gathered} \tag{5.8}$$

Now we can evaluate the limit of our coherent states. Because the limit of an infinite linear combinations of states $|\Phi, m\rangle$ is not defined, we evaluate the limit of the norm. The eigenstates of $L^{-\prime}=\sqrt{\varepsilon}\,L^-$ are

$$|z'\rangle=\sum_{m=0}^{\infty}\frac{(\sqrt{2}\,z')^m(\Gamma(-2\Phi))^{\frac{1}{2}}}{(\varepsilon^m m!\,\Gamma(-2\Phi+m))^{\frac{1}{2}}}\,|\phi,m\rangle \tag{5.9}$$

with the norm

$$\langle z'|z'\rangle_\Phi=\Gamma(-2\Phi)\sum_{n=0}^{\infty}\frac{(2|z|^2)^n}{n!\,\varepsilon^n\,\Gamma(-2\Phi+n)}={}_0F_1\left(-2\Phi,\frac{2|z|^2}{\varepsilon}\right).$$

Now because

$$\lim_{\substack{\varepsilon\to 0\\ \Phi\to-\infty\\ \varepsilon\Phi\to -1}}\frac{\Gamma(-2\Phi)}{\varepsilon^n\,\Gamma(-2\Phi+n)}=\frac{1}{2^n},$$

$$\lim_{\substack{\varepsilon\to 0\\ \Phi\to-\infty}}\langle z'|z'\rangle_\Phi=\sum_n\frac{|z|^{2n}}{n!}=e^{|z|^2} \tag{5.10}$$

which is exactly the square of the norm of the usual coherent states (see Eqs. (2.2) and (2.4)). Or in terms of the Bessel functions we have the new relation

$$\lim_{\substack{\varepsilon\to 0\\ \Phi\to-\infty}}{}_0F_1\left(-2\Phi,\frac{2}{\varepsilon}|z|^2\right)=\lim_{\substack{\varepsilon\to 0\\ \Phi\to-\infty}}\Gamma(-2\Phi)\left[\sqrt{\frac{2}{\varepsilon}}\,i|z|\right]^{2\Phi+1}J_{-2\Phi-1}\left(\frac{2\sqrt{2}}{\sqrt{\varepsilon}}\,i|z|\right)$$
$$=e^{|z|^2}. \tag{5.11}$$

VI. The Connection with the Hilbert Spaces of Entire Functions of the Exponential Type

1.

Let $|f\rangle$ denote an arbitrary vector of the Hilbert space and let us consider the function[2]

$$f_\Phi(z)\equiv\langle f|z\rangle_\Phi=\sqrt{\Gamma(-2\Phi)}\sum_{n=0}^{\infty}\frac{(z)^n}{\sqrt{n!\,\Gamma(-2\Phi+n)}}\langle f|\Phi,n\rangle \tag{6.1}$$

with

$$\sum_{n=0}^{\infty}|\langle f|\Phi,n\rangle_\Phi|^2<\infty.$$

[2] In this section we prefer to write z instead of $\sqrt{2}z$ for $|z\rangle$. Consequently the measure changes slightly as indicated in (6.3). Furthermore we shall not write the label ϕ when no confusion arises.

As it can be easily seen, $f_\Phi(z)$ is an entire analytic function of order 1 and type 1 (exponential type), i.e., growth (1, 1). It is clear that $f_\Phi(z)$ uniquely determines $|f\rangle$ and *vice versa.* It is obvious at this point that we can state a connection between the eigenstates of L^- in $D^+(\Phi)$ (or of L^+ in $D^-(\Phi)$) and the Hilbert spaces of entire analytic functions or growth (1, 1), in the same way as the usual coherent states are connected to the Segal-Bargmann [6, 7] space of entire functions of growth ($\frac{1}{2}$, 2).

We introduce the countable set of Hilbert spaces $\mathscr{F}_\Phi$, whose elements are entire analytic functions. For each Φ, the inner product is defined by[2]

$$(f, g)_\Phi = \int \bar{f}(z)\, g(z)\, d\sigma_\Phi(z)\,, \tag{6.2}$$

$$d\sigma_\Phi(z) = \frac{4}{2\pi\Gamma(-2\Phi)}\, r^{-2\Phi-1} K_{\frac{1}{2}+\Phi}(2r)\, r\, d\theta\, dr\,. \tag{6.3}$$

f belongs to $\mathscr{F}_\Phi$ if and only if $(f, f)_\Phi < \infty$; its norm is $\|f\|_\Phi = \sqrt{(f, f)_\Phi}$. Let $f(z)$ be an entire function with the power series $\sum_n c_n z^n$. The norm in terms of the expansion coefficients is given by

$$(f, f)_\Phi^{\frac{1}{2}} = \left[[\Gamma(-2\Phi)]^{-1} \sum_0^\infty |c_n|^2\, n!\, \Gamma(-2\Phi+n) \right]^{\frac{1}{2}}. \tag{6.4}$$

Every set of coefficients c_n for which the sum in (6.4) converges defines an entire function $f \in \mathscr{F}_\Phi$. From the linearity we get the inner product of two functions f, g:

$$(f, g)_\Phi = [\Gamma(-2\Phi)]^{-1} \sum_{n=0}^\infty \bar{c}_n b_n\, n!\, \Gamma(-2\Phi+n)$$
$$g(z) = \sum_{n=0}^\infty b_n z^n\,. \tag{6.5}$$

An orthonormal set of vectors in $\mathscr{F}_\Phi$ is given by

$$u_{n,\Phi}(z) = [\Gamma(-2\Phi)]^{\frac{1}{2}} \frac{z^n}{\sqrt{n!\, \Gamma(-2\Phi+n)}}\,. \tag{6.6}$$

For any function $f \in \mathscr{F}_\Phi$,

$$(u_n, f)_\Phi = c_n \frac{\sqrt{n!\, \Gamma(-2\Phi+n)}}{\sqrt{\Gamma(-2\Phi)}}\,.$$

Eq. (6.4) expresses the completeness of the system $u_m(\Phi, z)$. The Schwarz inequality gives

$$|f(z)|^2 \leqq \left(\sum_{n=0}^{\infty} |c_n z^n|\right)^2 \leqq \left(\sum_{n=0}^{\infty} |c_n|^2 n!\,\Gamma(-2\Phi+n)\,[\Gamma(-2\Phi)]^{-1}\right)$$

$$\left(\sum_{n=0}^{\infty} \frac{|z|^2}{n!\,\Gamma(-2\Phi+n)}\,\Gamma(-2\Phi)\right)$$

from which

$$|f(z)|^2 \leqq \|f\|_\Phi^2\ {}_0F_1(-2\Phi;|z|^2)$$

or

$$|f(z)| \leqq \|f\|_\Phi ({}_0F_1(-2\Phi;|z|^2))^{\frac{1}{2}}. \tag{6.7}$$

As a consequence, strong convergence in $\mathscr{F}_\Phi$ implies pointwise convergence, because

$$|f(z)-g(z)| \leqq [{}_0F_1(-2\Phi;|z|^2)]^{\frac{1}{2}}\,\|f-g\|_\Phi \tag{6.8}$$

for any $f, g \in \mathscr{F}_\Phi$ and from (6.7) we see that the convergence is uniform on any compact set.

2. *The Principal Vector and the Reproducing Kernel*

Following Bargmann [6] we can introduce the *principal vectors* and the reproducing kernel for each $\mathscr{F}_\Phi$. For a fixed complex number a, the mapping $f \to f(a)$ defines a bounded linear functional. Because f is an element of an Hilbert space $\mathscr{F}_\Phi$, the functional is of the form

$$f(a) = (e_a, f)_\Phi, \tag{6.9}$$

where e_a is uniquely defined in $\mathscr{F}_\Phi$. The vectors e_a are called the principal vectors of $\mathscr{F}_\Phi$ and they behave like a continuous set of orthonormal vectors. In particular

$$(f, g)_\Phi = \int (f, e_a)_\Phi (e_a, g)_\Phi\, d\sigma_\Phi(a). \tag{6.10}$$

This expression corresponds to the expression for the resolution of the identity in terms of the eigenstates $|a\rangle_\Phi$ of L^- in $D^+(\Phi)$ (or L^+ in $D^-(\Phi)$) (see (3.23)). Thus $(f, e_a)_\Phi$ corresponds to $\langle f|a\rangle_\Phi$ which is the functional representative of the abstract vector $|f\rangle$. The vectors e_a are complete: i.e. their finite linear combinations are dense in $\mathscr{F}_\Phi$, because the only vector orthogonal to all of them is $f \equiv 0$. This is due to the fact that $(e_a, f)_\Phi$ is an entire function of a. In integral form, (6.10) reads

$$f(a) = \int W_\Phi(a, z)\, f(z)\, d\sigma_\Phi(z) \tag{6.11}$$

and $W_\Phi(a, z)$ is the "reproducing kernel". Because $e_a(z) = (e_z, e_a)_\Phi$, we have

$$W(a, z)_\Phi = \overline{W(z, a)_\Phi} = (e_a, e_z)_\Phi \tag{6.12}$$

and $W(a, z)$ is analytic in a and $\bar{z}$. It is the analog of the delta function in the usual Hilbert space of quantum mechanics. In terms of any complete orthonormal discrete set $v_1, v_2 \ldots$, we have, by (6.9):

$$e_a = \sum_{n=0}^{\infty} (v_n, e_a)\, v_n = \sum_{n=0}^{\infty} \overline{v_n(a)}\, v_n \tag{6.13}$$

and since strong convergence implies pointwise convergence, we have

$$e_a(z) = \sum_{n=0}^{\infty} \overline{v_n(a)}\, v_n(z)\,, \tag{6.14}$$

irrespective of the choice of the system $\{v_n\}$. Using the set $u_{n,\Phi}(z)$ (6), we find

$$\begin{aligned} e_a(z) &= \sum_{n=0}^{\infty} u_n(a)\, u_n(z) \\ &= \Gamma(-2\Phi) \sum_{o}^{\infty} \frac{(\bar{a}z)^n}{n!\,\Gamma(-2\Phi+n)} = {}_0F_1(-2\Phi; \bar{a}z) \end{aligned} \tag{6.15}$$

or

$$W(a, z)_\Phi = {}_0F_1(-2\Phi; a\bar{z})\,.$$

Likewise from (9) and (7) we find again

$$\|e_a\|_\Phi = ({}_0F_1(-2\Phi; a\bar{a}))^{\frac{1}{2}}\,. \tag{6.16}$$

We have thus the reproducing formula

$$\begin{aligned} f(a) &= \int W_\Phi(a, z)\, f(z)\, d\sigma_\Phi(z) \\ &= \int {}_0F_1(-2\Phi; a\bar{z})\, f(z)\, d\sigma_\Phi(z)\,, \qquad \forall f \in \mathscr{F}_\Phi. \end{aligned} \tag{6.17}$$

This could have been established also by looking at the corresponding formula in terms of the "coherent" states:

$$f(a) = \langle f | a\rangle = \int d\sigma_\Phi(z) \langle f | z\rangle \langle z | a\rangle\,, \tag{6.18}$$

from which we have the formal equality between the reproducing kernel and the inner product of two "coherent" states $|a\rangle_\Phi$ and $|z\rangle_\Phi$. Any bounded linear operator may be represented by means of the principal vectors as an integral transform. This may be adopted directly from Bargmann's case [6] and we shall not discuss here. Furthermore one can also carry an analysis similar to the one used in Ref. [1] concerning the problem of the „characteristic" sets or, in other words, the problem of the completeness of a countable subset of „coherent" states.

3. Realization of the Algebra $SO(2,1) \simeq SU(1,1) \simeq SL(2,R)$ in the Hilbert Spaces $\mathcal{F}_\Phi$ of Entire Functions

In the last section we have equipped the class of entire functions of growth (1,1) with a countable set of inner products $(f,f)_\Phi$ making them into a countable set of Hilbert spaces. We want to study in this section a realization of the generators of our algebra in these spaces. We consider here the case $D^+(\Phi)$. A similar analysis applies to the case $D^-(\Phi)$.

For each ϕ, we introduce the linear operators, acting on $\mathcal{F}_\Phi$:

$$\mathcal{L}_{12}^{(\Phi)} \equiv z\frac{d}{dz} - \Phi\,,$$

$$\mathcal{L}_{+}^{(\Phi)} \equiv \frac{1}{\sqrt{2}} z\,, \tag{6.19}$$

$$\mathcal{L}_{-}^{(\Phi)} \equiv \frac{1}{\sqrt{2}}\left(-2\Phi\frac{d}{dz} + z\frac{d^2}{dz^2}\right).$$

They satisfy the required commutation rules. In each $\mathcal{F}_\Phi$, $\mathcal{L}_{12}^{(\Phi)}$ is automatically diagonal in the orthonormal basis (6.6)

$$u_n^\Phi(z) = \sqrt{\Gamma(-2\Phi)}\;\frac{z^n}{\sqrt{n!\,\Gamma(-2\Phi+n)}}\,.$$

Furthermore, we have

$$\begin{aligned}
\mathcal{L}_+^\Phi u_n^\Phi(z) &= \frac{1}{\sqrt{2}}\sqrt{(n+1)(-2\Phi+n)}\,u_{n+1}^\Phi(z)\\
\mathcal{L}_-^\Phi u_n^\Phi(z) &= \frac{1}{\sqrt{2}}\sqrt{n(-2\Phi+n-1)}\,u_{n-1}(z)\,.
\end{aligned} \tag{6.20}$$

It is easy to verify the unitarity of the realization:

$$(f,\mathcal{L}_-^\Phi g)_\Phi = (\mathcal{L}_+^\Phi f, g) \qquad \forall f,g \in \mathcal{F}_\Phi\,, \tag{6.21}$$

i.e. $(\mathcal{L}_+^\Phi)^\dagger = \mathcal{L}_-^\Phi$.

By inspection of the confluent hypergeometric differential equation

$$z\frac{d^2}{dz^2}\varphi(z) - 2\Phi\frac{d}{dz}\varphi(z) - \sqrt{2}\lambda\,\varphi(z) = 0 \tag{6.22}$$

we find the eigenvectors of $\mathcal{L}_-^\Phi$, in $\mathcal{F}_\Phi$, to be

$$\varphi_\lambda(z) = {}_0F_1(-2\Phi, \sqrt{2}\lambda z) \tag{6.23}$$

with eigenvalues $\sqrt{2}\lambda$ where λ is any complex number.

We can understand now the meaning of $e_a^\Phi(z)$ in $\mathcal{F}_\lambda$: as it was to be expected $e_{\sqrt{2}\lambda}(z)$ with eigenvalue $\sqrt{2}\lambda$ are just the eigenvectors of $\mathcal{L}_-^\Phi$ or, in other words, they are the "coherent" states in $\mathcal{F}_\Phi$.

References

1. Bargmann, V., Butera, P., Girardello, L., Klauder, J. R.: On the Completeness of the Coherent States (to be published).
2. We use the notation in Barut, A. O., Fronsdal, C.: Proc. Roy. Soc. (London) Ser. A **287**, 532 (1965). This paper also contains the representations of the universal covering group. For a unified treatment of $SU(1, 1)$ and $SU(2)$ see also Barut, A. O., Phillips, C.: Commun. Math. Phys. **8**, 52 (1968).
3. Bateman Project: Vol. I. Integral transformations, p. 349. Erdelyi (editor). New York: McGraw-Hill 1954.
4. Barut, A. O., Phillips, C.: Cited in Ref. 2; Mukunda, M.: J. Math. Phys. **8**, 2210 (1967); Lindblad, G., Nagel, B.: Stockholm preprint, April 1969. See also Vilenkin, N. J.: Special functions and the theory of group representations, Chapt. VII. Providence, Rhode Island: American Mathematical Society 1968.
5. Inönü, E., Wigner, E. P.: Proc. Natl. Acad. Sci. U.S. **39**, 510 (1953). Inönü, E.: In: Group theoretical concepts and methods in elementary particle physics, ed. by F. Gürsey. New York: Gordon and Breach 1964.
6. Bargmann, V.: Commun. Pure Appl. Math. **14**, 187 (1961).
7. Segal, I. E.: Illinois J. Math. **6**, 500 (1962).

A. O. Barut and L. Girardello
University of Colorado
Dept. of Physics and Astrophysics
Boulder, Col. 80302, U.S.A.

J. Phys. A: Gen. Phys., 1971, Vol. 4.

Some properties of coherent spin states

J. M. RADCLIFFE†

School of Mathematical and Physical Sciences, University of Sussex, Falmer, Brighton, Sussex, England

MS. received 6*th November* 1970

Abstract. Spin states analogous to the coherent states of the linear harmonic oscillator are defined and their properties discussed. They are used to discuss some simple problems (a single spin in a field, a spin wave, two spin $\frac{1}{2}$ particles with Heisenberg coupling) and it is shown that their use may often give increased physical insight.

1. Introduction

The point of this paper is to show that there exist spin states analogous to the 'coherent' states of the harmonic oscillator. The latter have been studied extensively in recent years (see for example Carruthers and Nieto 1968) and appear to be useful in discussing the statistical mechanics and superfluid properties of boson fluids (Langer 1968); they also give a convenient description of the radiation from lasers. It is still an open question as to whether the spin states defined here will prove useful. They may, at the very least, give some physical insight into problems involving spins and their correlations.

2. Coherent states of the harmonic oscillator

Before defining the spin states it will be useful to look briefly at the problem of the one-dimensional harmonic oscillator.

In this case the coherent states are functions of a variable α which runs over the entire complex plane, and are given explicitly by

$$\begin{aligned}|\alpha\rangle &= \pi^{-1/2}\exp(-\tfrac{1}{2}|\alpha|^2)\sum_{n=0}^{\infty}\frac{\alpha^n}{(n!)^{1/2}}|n\rangle \\ &= \pi^{-1/2}\exp(-\tfrac{1}{2}|\alpha|^2)\sum_{n=0}^{\infty}\frac{\alpha^n(\alpha^+)^n}{n!}|0\rangle \\ &= \pi^{-1/2}\exp(-\tfrac{1}{2}|\alpha|^2)\exp(\alpha a^+)|0\rangle \qquad (2.1)\end{aligned}$$

where $|n\rangle$ is the nth energy eigenstate of the oscillator and a^+ the usual creation operator. These states form a complete set, in the sense that

$$\int d^2\alpha|\alpha\rangle\langle\alpha| = \sum_{n=0}^{\infty}|n\rangle\langle n| = \mathbf{1} \qquad (2.2)$$

where the right hand side is the unit matrix. However, they are neither normalized nor orthogonal. In fact, from the definition (2.1), the overlap of two states $|\alpha\rangle$, $|\beta\rangle$ is given by

$$\begin{aligned}\langle\beta|\alpha\rangle &= \pi^{-1}\exp(-\tfrac{1}{2}|\alpha|^2-\tfrac{1}{2}|\beta|^2)\langle 0|\exp(\beta^* a)\exp(\alpha a^+)|0\rangle \\ &= \pi^{-1}\exp(\alpha\beta^*-\tfrac{1}{2}|\alpha|^2-\tfrac{1}{2}|\beta|^2). \qquad (2.3)\end{aligned}$$

Of course, these states $|\alpha\rangle$ do not span Hilbert space in the two-dimensional α plane.

† See *Note added in proof*, p. 323.

To see this explicitly, write $\alpha = \rho e^{i\phi}$. Then

$$|\alpha\rangle = |\rho;\phi\rangle = \pi^{-1/2}\exp(-\tfrac{1}{2}\rho^2)\sum_{n=0}^{\infty}\frac{\rho^n e^{in\phi}}{(n!)^{1/2}}|n\rangle. \tag{2.4}$$

The functions

$$f_n(\rho;\phi) \equiv \pi^{-1/2}\exp(-\tfrac{1}{2}\rho^2)\frac{\rho^n e^{in\phi}}{(n!)^{1/2}} \tag{2.5}$$

are just a subset of the eigenfunctions of the two-dimensional harmonic oscillator. In fact, a possible classification of the states of this system is by the energy $\epsilon = (m+1)$ and the angular momentum $L_z = l$ (l, m integers); for any given value of m, l can take the values $-m$, $-m+2 \ldots m-2$, m. The functions $f_n(\phi)$ are clearly the subset of states corresponding to $l = m = n$.

It is clear that, quite generally, one can construct functions $|\xi\rangle$ by making any (enumerably infinite) selection from any set of functions $\phi_n(\xi)$ which are orthonormal and complete in ξ space. Let the chosen subset be denoted by $\{n\}$, where the association of functions in this subset with a particular one-dimensional oscillator state is quite arbitrary. Now define

$$|\xi\rangle = \sum_{\{n\}}\phi_n(\xi)|n\rangle. \tag{2.6}$$

These sets are complete in the oscillator Hilbert space:

$$\begin{aligned}\int d\xi|\xi\rangle\langle\xi| &= \sum_{nn'}|n\rangle\langle n'|\int d\xi\,\phi_{n'}^*(\xi)\phi_n(\xi)\\ &= \sum_{n}|n\rangle\langle n| = \mathbf{1}.\end{aligned} \tag{2.7}$$

However, the states $|\xi\rangle$ are not orthogonal, and cannot be, since

$$\langle\xi'|\xi\rangle = \sum_{\{n\}}\phi_n^*(\xi')\,\phi_n(\xi). \tag{2.8}$$

Only if the subset $\{n\}$ runs over a complete set in ξ space will the right hand side be equal to $\delta(\xi-\xi')$. If the space ξ has two or more dimensions the subset is certainly not complete.

The states $|\xi\rangle$ are not normalized, but it may happen that the set $\{n\}$ can be chosen so that

$$\langle\xi|\xi\rangle = \sum_{n}|\phi_n(\xi)|^2 \tag{2.9}$$

is a constant. For example, in the case of the coherent states of the harmonic oscillator, the normalized wavefunctions in the set $\{n\}$ are

$$f_n(\rho;\phi) = \pi^{-1/2}\exp(-\tfrac{1}{2}\rho^2)\frac{\rho^n e^{in\phi}}{(n!)^{1/2}} \tag{2.5}$$

and so

$$\langle\alpha|\alpha\rangle = \sum_{n=0}^{\infty}|f_n(\rho;\phi)|^2 = \pi^{-1}\exp(-\rho^2)\sum_{n=0}^{\infty}\frac{\rho^{2n}}{n!} = \frac{1}{\pi}. \tag{2.10}$$

In practice it does not appear to matter whether or not the states do normalize to a constant.†

† It is of course equally possible to choose the states $|\xi\rangle$ to be normalized and put in a weighting factor in the left hand side of the completeness relation (2.7). This is in fact the alternative we shall choose in the next section.

To conclude this section, we emphasize that the point of introducing states such as $|\xi\rangle$ is that, being complete, they can be used perfectly well in the evaluation of such quantities as the partition function of the harmonic oscillator problem:

$$Z = \sum_n \langle n|e^{-\beta\mathscr{H}}|n\rangle = \int d\xi \langle\xi|e^{-\beta\mathscr{H}}|\xi\rangle.$$

Such states $|\xi\rangle$ may well be better starting functions in 'perturbation' expansions of Z than the original oscillator states $|n\rangle$. For more details of applications and tricks in evaluating Z, see Carruthers and Nieto (1968) and Langer (1968).

3. Analogous spin states

We consider a single particle of spin S. Define the ground state $|0\rangle$ as the state such that $\hat{S}_z|0\rangle = S|0\rangle$, where $\hat{S}_z$ is the operator of the z component of spin. Then the operator $\hat{S}_- \equiv \hat{S}_x - i\hat{S}_y$ creates spin deviations. In fact we have

$$(\hat{S}_-)^p|0\rangle = \left(\frac{p!\,2S!}{(2S-p)!}\right)^{1/2} |p\rangle \qquad 0 \leqslant p \leqslant 2S \tag{3.1}$$

where $|p\rangle$ is the eigenstate of $\hat{S}_z$ such that

$$\hat{S}_z|p\rangle = (S-p)|p\rangle. \tag{3.2}$$

Consider the state

$$|\mu\rangle \equiv N^{-1/2}\exp(\mu\hat{S}_-)|0\rangle = N^{-1/2}\sum_{p=0}^{2S}\left(\frac{2S!}{p!(2S-p)!}\right)^{1/2}\mu^p|p\rangle \tag{3.3}$$

where μ runs over the complex plane and N is a normalization factor. We have

$$\langle\mu|\mu\rangle = N^{-1}\sum_{p=0}^{2S}\frac{(2S)!}{p!(2S-p)!}|\mu|^{2p} = N^{-1}(1+|\mu|^2)^{2S} \tag{3.4}$$

and hence the normalized state is

$$|\mu\rangle = (1+|\mu|^2)^{-S}\exp(\mu\hat{S}_-)|0\rangle. \tag{3.5}$$

The overlap integral between two states $|\lambda\rangle$, $|\mu\rangle$ is

$$\langle\lambda|\mu\rangle = \frac{(1+\lambda^*\mu)^{2S}}{(1+|\lambda|^2)^S(1+|\mu|^2)^S} \tag{3.6}$$

and so

$$|\langle\lambda|\mu\rangle|^2 = \left(1-\frac{|\lambda-\mu|^2}{(1+|\lambda|^2)(1+|\mu|^2)}\right)^{2S}. \tag{3.7}$$

The states $|\mu\rangle$ defined by (3.5) do form a complete set, although it is necessary to include a weight function $m(|\mu|^2) \geqslant 0$ in the integral. We require

$$\int d^2\mu|\mu\rangle m(|\mu|^2)\langle\mu| = \sum_{p=0}^{2S}|p\rangle\langle p| = \mathbf{1}. \tag{3.8}$$

By doing the angular integration and putting $|\mu| = \rho$, one finds

$$\begin{aligned}\int d^2\mu|\mu\rangle m(|\mu|^2)\langle\mu| &= 2\pi\sum_{p=0}^{2S}|p\rangle\langle p|\frac{(2S)!}{p!(2S-p)!}\int_0^\infty d\rho\frac{\rho^{2p}}{(1+\rho^2)}2Sm(\rho^2)\\ &= \sum_{p=0}^{2S}|p\rangle\langle p|\frac{(2S)!}{p!(2S-p)!}I(p,S)\end{aligned} \tag{3.9}$$

where

$$I(p;S) \equiv \pi \int_0^\infty \mathrm{d}\sigma \frac{\sigma^p}{(1+\sigma)^{2S}} m(\sigma). \tag{3.10}$$

Now one seeks a form for $m(\sigma)$ such that $I(p;S) = \{p!(2S-p)!/(2S)!\}$. A little thought shows that a suitable choice is

$$m(\sigma) = \frac{2S+1}{\pi}\frac{1}{1+\sigma^2}. \tag{3.11}$$

So, finally, the completeness relation is

$$\frac{2S+1}{\pi}\int \frac{\mathrm{d}^2\mu}{(1+|\mu|^2)^2}|\mu\rangle\langle\mu| = \sum_{p=0}^{2S}|p\rangle\langle p| = \mathbf{1}. \tag{3.12}$$

This result can be obtained more neatly by transforming back from a different parametrization of the states, namely $\mu = \tan(\theta/2)\mathrm{e}^{\mathrm{i}\phi}$, which is used later on. It is convenient to work with μ while drawing analogies with the harmonic oscillator. The case of the oscillator is obtained in the limit $S \gg 1$. To see this, write

$$\hat{S}_- \to (2S)^{1/2}a^+ \tag{3.13}$$

(which is the high-spin limit of the Holstein–Primakoff transformation) and

$$\mu \to \alpha/(2S)^{1/2}. \tag{3.14}$$

The normalized states $|\alpha\rangle_{(S)}$ are then

$$|\alpha\rangle_{(S)} = \left(1+\frac{|\alpha|^2}{2S}\right)^{-S}\exp(\alpha a^+)|0\rangle. \tag{3.15}$$

But we have

$$\lim_{S\to\infty}\left(1+\frac{|\alpha|^2}{2S}\right)^{S} = \exp(\tfrac{1}{2}|\alpha|^2) \tag{3.16}$$

and so

$$\lim_{S\to\infty}|\alpha\rangle_{(S)} = \exp(-\tfrac{1}{2}|\alpha|^2)\exp(a\alpha^+)|0\rangle \tag{3.17}$$

which apart from normalization is precisely a coherent state of the harmonic oscillator (cf. equation (2.1) above). It is easy to show that, for example, the spin state overlap integrals go to the correct limit.

We conclude this section with a remark on the completeness of the spin states. For consistency, we must have

$$\frac{2S+1}{\pi}\int\frac{\mathrm{d}^2\mu}{(1+|\mu|^2)}f(\mu)\langle\mu|\lambda\rangle = f(\lambda) \tag{3.18}$$

where in general the $f(\mu)$ is an overlap of the state $|\mu\rangle$ on some (arbitrary) spin state. This relation does hold so long as $f(\mu)$ is of the general form $P(\mu)/(1+|\mu|^2)^S$, where $P(\mu)$ is an arbitrary polynomial in μ with terms up to μ^{2S}. Now, in fact, only functions of precisely this form can occur in calculations if one stays within the Hilbert space appropriate to a particle of spin S, so in all such cases (3.18) is valid.

4. Some typical matrix elements

Define

$$\hat{p} \equiv S-\hat{S}_z \qquad \hat{S}_+ \equiv \hat{S}_x+i\hat{S}_y. \tag{4.1}$$

Then we have the following relations:

(i)

$$\langle\mu|\hat{p}|\mu\rangle = (1+|\mu|^2)^{-2S}\sum_{p=0}^{2S}\frac{(2S)!}{p!(2S-p)!}|\mu|^{2p}p$$
$$= \frac{2S|\mu|^2}{1+|\mu|^2}. \tag{4.2}$$

The second equality can be derived either by direct computation or by the observation that the sum can be written in the form

$$\rho\frac{\partial}{\partial\rho}(1+\rho)^{2S} \qquad \rho \equiv |\mu|^2.$$

(ii)

$$\langle\mu|\hat{S}_+|\mu\rangle = (1+|\mu|^2)^{-2S}\frac{\partial}{\partial\mu^*}(1+|\mu|^2)^{2S}$$
$$= \frac{2S\mu}{1+|\mu|^2}. \tag{4.3}$$

(iii) Hence

$$\langle\mu|\hat{S}_-|\mu\rangle = \frac{2S\mu^*}{(1+|\mu|^2)^{2S}} \tag{4.4}$$

(iv)

$$\langle\lambda|\hat{p}|\mu\rangle = \frac{2S\lambda^*\mu}{1+\lambda^*\mu}\langle\lambda|\mu\rangle \tag{4.5}$$

(v)

$$\langle\lambda|\hat{S}_+|\mu\rangle = \frac{2S\mu}{1+\lambda^*\mu}\langle\lambda|\mu\rangle \tag{4.6}$$

(vi)

$$\langle\lambda|\hat{S}_-|\mu\rangle = \frac{2S\lambda^*}{1+\lambda^*\mu}\langle\lambda|\mu\rangle. \tag{4.7}$$

Note that from (4.5–4.7)

$$\langle\lambda|\hat{S}_+|\mu\rangle = \frac{1}{\lambda^*}\langle\lambda|\hat{p}|\mu\rangle \tag{4.8}$$

$$\langle\lambda|\hat{S}_-|\mu\rangle = \frac{1}{\mu}\langle\lambda|\hat{p}|\mu\rangle. \tag{4.9}$$

5. An alternative parametrization

Let us write

$$\mu = \tan(\tfrac{1}{2}\theta)e^{i\phi} \quad 0 \leqslant \theta < \pi \quad 0 \leqslant \phi < 2\pi. \tag{5.1}$$

Then the normalized states can be written

$$|\theta,\phi\rangle \equiv |\Omega\rangle = (\cos\tfrac{1}{2}\theta)^{2S}\exp\{\tan(\tfrac{1}{2}\theta)e^{i\phi}\hat{S}_-\}|0\rangle \tag{5.2}$$

and the completeness relation is

$$\frac{2S+1}{4\pi}\int d\phi\, d\theta \sin\theta|\Omega\rangle\langle\Omega| \equiv (2S+1)\int\frac{d\Omega}{4\pi}|\Omega\rangle\langle\Omega| = \mathbf{1}. \tag{5.3}$$

There is a simple geometrical construction relating μ and the variables θ, ϕ. In fact, if we write $\mu = \rho e^{i\phi'}$ and draw the μ plane as tangent plane to a sphere of unit diameter where the z axis meets the sphere, then the point μ is the projection onto the μ plane of the point (θ, ϕ) on the sphere from the opposite pole. Clearly $\phi' = \phi$, $\rho = \tan(\theta/2)$.

From equation (3.6) we find for the overlap integral between states $|\Omega\rangle$, $|\Omega'\rangle$:

$$\langle\Omega'|\Omega\rangle = \{\cos\tfrac{1}{2}\theta\cos\tfrac{1}{2}\theta' + \sin\tfrac{1}{2}\theta\sin\tfrac{1}{2}\theta'\, e^{i(\phi-\phi')}\}^{2S} \tag{5.4}$$

and so

$$|\langle\Omega'|\Omega\rangle| = \left(\frac{1+\boldsymbol{n}\,.\,\boldsymbol{n}'}{2}\right)^S$$

where $\boldsymbol{n}$ and $\boldsymbol{n}'$ are unit vectors in the directions specified by (θ, ϕ) and (θ', ϕ') respectively.

Finally, we calculate the expectation values of the spin components in the state $|\Omega\rangle$. From equations (4.2–4.7) we have

$$\langle\Omega|\hat{p}|\Omega\rangle = S(1-\cos\theta) \tag{5.5}$$

$$\langle\Omega|\hat{S}_+|\Omega\rangle = S\sin\theta\, e^{i\phi} \tag{5.6}$$

$$\langle\Omega|\hat{S}_-|\Omega\rangle = S\sin\theta\, e^{-i\phi} \tag{5.7}$$

from which we get the result for the expectation value of the spin vector

$$\langle\Omega|\hat{\boldsymbol{S}}|\Omega\rangle = S\boldsymbol{n} \tag{5.8}$$

where $\boldsymbol{n}$ is the unit vector specified by Ω.

6. The effect of changing the ground state

At the moment we are describing spins by states of the form

$$|\mu\rangle = \hat{A}(\mu)|0\rangle \tag{6.1}$$

where

$$\hat{A}(\mu) \equiv (1+|\mu|^2)^{-S}\exp(\mu\hat{S}_-)|0\rangle \tag{6.2}$$

and $|0\rangle$ is the state such that $\hat{S}_z|0\rangle = S|0\rangle$. Consider now making a rotation to a new axis of quantization z' and write

$$|\lambda'\rangle' = \hat{A}'(\lambda')|0'\rangle \tag{6.3}$$

where $|0'\rangle$ is the state such that $\hat{S}_{z'}|0'\rangle = S|0'\rangle$ and

$$\hat{A}'(\lambda') \equiv (1+|\lambda'|^2)^{-S}\exp(\lambda'\hat{S}'_-) \tag{6.4}$$

$(\hat{S}'_- \equiv \hat{S}_{x'} - i\hat{S}_{y'})$. The problem now is to express the states $|\lambda'\rangle'$ in terms of the states $|\mu\rangle$. This question is relevant, for example, to a discussion of the structure of the density matrix (and mean values) for a pair of spins coupled by the Heisenberg interaction (which is invariant under rotations) and the Ising coupling (which is not). Explicitly, we seek the amplitude $\langle\mu|\lambda'\rangle'$ in the expansion

$$|\lambda'\rangle' = \left(\frac{2S+1}{\pi}\right)\int\frac{d^2\mu}{(1+|\mu|^2)^2}|\mu\rangle\langle\mu|\lambda'\rangle'. \tag{6.5}$$

Let a unitary rotation operator which carries $|0\rangle$ to $|0\rangle'$ be denoted by $\hat{R}$, so that

$$|0\rangle' = \hat{R}|0\rangle. \tag{6.6}$$

Then we have

$$\hat{S}_{z'} = \hat{R}\hat{S}_z\hat{R}^\dagger \tag{6.7}$$

$$(\hat{S}_\pm)' = \hat{R}\hat{S}_\pm\hat{R}^\dagger. \tag{6.8}$$

Hence we get simply

$$\begin{aligned}|\lambda'\rangle' &= (1+|\lambda'|^2)^{-S}\exp(\lambda'\hat{S}'_-)|0\rangle' \\ &= (1+|\lambda'|^2)^{-S}\hat{R}\exp(\lambda'\hat{S}_-)R^\dagger R|0\rangle \\ &= (1+|\lambda'|^2)^{-S}\hat{R}\exp(\lambda'\hat{S}_-)|0\rangle \\ &= \hat{R}|\lambda'\rangle.\end{aligned} \tag{6.9}$$

That is, one needs to evaluate

$$\langle\mu|\lambda'\rangle' = \langle\mu|R|\lambda'\rangle \tag{6.10}$$

for any calculations in which this amplitude is required explicitly. But we have (see Brink and Satchler 1968)

$$\hat{R} = \exp(-i\alpha\hat{S}_z)\exp(-i\beta\hat{S}_y)\exp(-i\gamma\hat{S}_z) \tag{6.11}$$

where α, β, γ are the Euler angles describing the rotation. Expressing the states $\langle\mu|$, $|\lambda'\rangle$ in terms of states $\langle p|$, $|p'\rangle$, we find that

$$\begin{aligned}\langle\mu|R|\lambda'\rangle = (1+|\mu|^2)^{-S}(1+|\lambda'|^2)^{-S}\sum_{p,p'=0}^{2S}\left(\frac{(2S)!}{p!(2S-p)!}\right)^{1/2}\left(\frac{(2S)!}{p'!(2S-p')!}\right)^{1/2} \\ \times(\mu^*)^p(\lambda')^{p'}\langle p|\hat{R}|p'\rangle.\end{aligned} \tag{6.12}$$

Now consider the amplitude

$$\begin{aligned}\langle p|\hat{R}|p'\rangle &= \langle p|\exp(-i\alpha\hat{S}_z)\exp(-i\beta\hat{S}_y)\exp(-i\gamma\hat{S}_z)|p'\rangle \\ &= \exp\{-i\alpha(S-p)\}\exp\{-i\gamma(S-p')\}\langle p|\exp(-i\beta\hat{S}_y)|p'\rangle.\end{aligned} \tag{6.13}$$

We shall use the explicit expression for

$$\langle p|\exp(-i\beta\hat{S}_y)|p'\rangle \equiv d^S_{S-p,S-p'}(\beta)$$

given in Brink and Satchler (1968 p. 22), namely

$$\begin{aligned}\langle p|\exp(-i\beta\hat{S}_y)|p'\rangle = \sum_t(-1)^t\frac{\{(2S-p)!p!(2S-p')!p'!\}^{1/2}}{(2S-p-t)!(p'-t)!t!(t+p-p')!} \\ \times(\cos\tfrac{1}{2}\beta)^{2S+p'-p-2t}(\sin\tfrac{1}{2}\beta)^{2t+p-p'}.\end{aligned} \tag{6.14}$$

To evaluate $\langle\mu|\hat{R}|\lambda'\rangle$ we use the alternative parametrization, writing

$$\mu \equiv \tan(\tfrac{1}{2}\theta)e^{i\phi} \qquad \lambda' \equiv \tan(\tfrac{1}{2}\theta')e^{i\phi'}.$$

Then we have, after some cancellations,

$$\begin{aligned}\langle\theta,\phi|\hat{R}|\theta',\phi'\rangle &= (\cos\tfrac{1}{2}\beta\cos\tfrac{1}{2}\theta\cos\tfrac{1}{2}\theta')^{2S}\exp\{-i(\alpha+\gamma)S\} \\ &\times\sum_{p,p'=0}^{2S}\sum_t(-1)^t\frac{2S!}{(2S-p-t)!(p'-t)!t!(t+p-p')!} \\ &\times(\tan\tfrac{1}{2}\beta)^{2t+p-p'}(\tan\tfrac{1}{2}\theta)^p(\tan\tfrac{1}{2}\theta')^{p'}\exp\{-ip(\phi-\alpha)\} \\ &\times\exp\{ip'(\phi'+\gamma)\}.\end{aligned} \tag{6.15}$$

We write the sum in the form

$$(2S)! \sum_{p,p',t} \frac{(-1)^t x^p y^{p'} z^{2t+p-p'}}{(2S-p-t)!(p'-t)!t!(t+p-p')!} \tag{6.16}$$

where

$$x \equiv \tan(\tfrac{1}{2}\theta)\exp\{-i(\phi-\alpha)\} \qquad y \equiv \tan(\tfrac{1}{2}\theta')\exp\{i(\phi'+\gamma)\} \qquad z \equiv \tan(\tfrac{1}{2}\beta). \tag{6.17}$$

The limits on the sum over t are such as to ensure that no factorials shall have negative arguments, while p and p' run from 0 to $2S$. That is,

$$p' \geqslant t \qquad p+t \geqslant p' \qquad 2S-p \geqslant t \geqslant 0 \qquad 2S-p \geqslant 0. \tag{6.18}$$

We take the sums in the following sequence:

(I) sum over p' from t to $p+t$
(II) sum over t from 0 to $2S-p$
(III) sum over p from 0 to $2S$.

Sum (I):

$$\sum_{p'=t}^{p+t} \frac{y^{p'} z^{2t+p-p'}}{(p'-t)!(t+p-p')!} = y^t z^t \sum_{q=0}^{p} \frac{y^q z^{p-q}}{q!(p-q)!} = \frac{y^t z^t (y+z)^p}{p!}. \tag{6.19}$$

Sum (II):

$$\sum_{t=0}^{2S-p} (-1)^t \frac{y^t z^t}{t!(2S-p-t)!} = \frac{(1-yz)^{2S-p}}{(2S-p)!} \tag{6.20}$$

Sum (III):

$$\begin{aligned}
&\sum_{p=0}^{2S} \frac{(2S)!}{p!(2S-p)!} x^p (y+z)^p (1-yz)^{2S-p} \\
&= (1+xy-yz+zx)^{2S} \\
&= [1+\tan(\tfrac{1}{2}\theta)\exp\{-i(\phi-\alpha)\}\tan(\tfrac{1}{2}\theta')\exp\{i(\phi'-\gamma)\} - \tan(\tfrac{1}{2}\beta)\tan(\tfrac{1}{2}\theta')\exp\{i(\phi'-\gamma)\} \\
&\quad + \tan(\tfrac{1}{2}\beta)\tan(\tfrac{1}{2}\theta)\exp\{-i(\phi-\alpha)\}]^{2S}.
\end{aligned} \tag{6.21}$$

So, finally:

$$\begin{aligned}
\langle\theta,\phi|R|\theta',\phi'\rangle &= \exp(-i\alpha S)\exp(-i\gamma S)(\cos\tfrac{1}{2}\beta\cos\tfrac{1}{2}\theta\cos\tfrac{1}{2}\theta')^{2S} \\
&\quad \times [1+\tan(\tfrac{1}{2}\theta)\exp\{-i(\phi-\alpha)\}\tan(\tfrac{1}{2}\theta')\exp\{i(\phi'-\gamma)\} - \tan(\tfrac{1}{2}\theta') \\
&\quad \times \exp\{i(\phi'+\gamma)\}\tan(\tfrac{1}{2}\beta) + \tan(\tfrac{1}{2}\theta)\exp\{-i(\phi-\alpha)\}\tan(\tfrac{1}{2}\beta)]^{2S}.
\end{aligned} \tag{6.22}$$

As a simple check on this expression, take $\hat{R} = 1$, that is, $\alpha = \beta = \gamma = 0$. Then we find

$$\langle\theta,\phi|\theta',\phi'\rangle = [\cos\tfrac{1}{2}\theta\cos\tfrac{1}{2}\theta' + \sin\tfrac{1}{2}\theta\sin\tfrac{1}{2}\theta'\exp\{i(\phi'-\phi)\}]^{2S} \tag{6.23}$$

in agreement with (5.4). Another property which the amplitude $\langle\Omega|R|\Omega'\rangle$ must satisfy follows from the unitary property $\hat{R}(\hat{R}^{\dagger} = \hat{R}^{-1})$:

$$(\langle A|\hat{R}|B\rangle)^* = \langle B|\hat{R}^{\dagger}|A\rangle = \langle B|\hat{R}^{-1}|A\rangle. \tag{6.24}$$

Since $\hat{R}^{-1}$ is the rotation specified by Euler angles $(-\gamma, -\beta, -\alpha)$, we expect

$$\{\langle\theta,\phi|\hat{R}(\alpha,\beta,\gamma)|\theta',\phi'\rangle\}^* = \langle\theta',\phi'|\hat{R}(-\gamma,-\beta,-\alpha)|\theta,\phi\rangle. \tag{6.25}$$

We easily check that the condition (6.25) is indeed satisfied by the expression (6.22).

7. Some simple applications

The results obtained here with the help of the coherent-state formalism are of course well known; the point of doing the problems by this method is simply to give some extra physical insight. In particular, the connection with the classical limit comes out very clearly.

7.1. *Partition function for a single spin in a magnetic field*

With a suitable choice of the zero of energy, we can write the partition function in the form

$$Z = \mathrm{Tr}\{\exp(-\hat{p}h)\} \equiv \sum_{p=0}^{2S} \exp(-ph) \tag{7.1}$$

(where $h = \beta\gamma H$; γ is the particle's magnetic moment, H the external field and β is $1/k_B T$ as usual). It is straightforward to verify that we can write this in the form

$$\begin{aligned} Z &= (2S+1) \int \frac{\mathrm{d}\Omega}{4\pi} \{\tfrac{1}{2}(1+\mathrm{e}^{-h}) + \tfrac{1}{2}(1-\mathrm{e}^{-h}) \cos\theta\}^{2S} \\ &= (2S+1) \int \frac{\mathrm{d}\Omega}{4\pi} \langle \Omega | \exp(-\beta\hat{\mathscr{H}}) | \Omega \rangle. \end{aligned} \tag{7.2}$$

If we calculate the mean value of the operator $\hat{p}$ (equation (4.1)) by the relation $\langle\hat{p}\rangle = -Z^{-1}\partial Z/\partial H$, we find

$$\langle\hat{p}\rangle = \frac{2S(2S+1)}{Z} \int \frac{\mathrm{d}\Omega}{4\pi} \left(\frac{\sin^2(\tfrac{1}{2}\theta)\mathrm{e}^{-h}}{\cos^2(\tfrac{1}{2}\theta) + \sin^2(\tfrac{1}{2}\theta)\mathrm{e}^{-h}}\right) \{\tfrac{1}{2}(1+\mathrm{e}^{-h}) + \tfrac{1}{2}(1-\mathrm{e}^{-h})\cos\theta\}^{2S} \tag{7.3}$$

so that in this particular case $\langle\hat{p}\rangle$ can be written in the form

$$\langle\hat{p}\rangle = (2S+1) \int \frac{\mathrm{d}\Omega}{4\pi} \langle p(\Omega)\rangle \langle \Omega|\hat{\rho}|\Omega\rangle \tag{7.4}$$

where $\hat{\rho} \equiv \exp(-\beta\hat{\mathscr{H}})/Z$ is the density matrix and

$$p(\Omega) = \left(\frac{2S\sin^2(\tfrac{1}{2}\theta)\mathrm{e}^{-h}}{\cos^2(\tfrac{1}{2}\theta) + \sin^2(\tfrac{1}{2}\theta)\mathrm{e}^{-h}}\right). \tag{7.5}$$

In the limits $h \to 0$ and $h \gg 1$ we get, respectively, $\langle\hat{p}\rangle \to S$ and $\langle\hat{p}\rangle \to \mathrm{e}^{-h}$, as of course we must.

7.2. *Ferromagnetic spin wave*

The ground state $|0\rangle$ of the ferromagnet has $p_i = 0$ for all spins i, $(i = 1, 2, \ldots N)$. In terms of the μ representation we can write

$$|0\rangle = \int \mathrm{d}M(\mu)\, |\mu\rangle\langle\mu|0\rangle \tag{7.6}$$

where $|\mu\rangle$ is shorthand for $|\mu_1, \mu_2 \ldots \mu_N\rangle$ and $\int \mathrm{d}M(\mu)$ for the expression (cf. equation (3.2))

$$\left(\frac{2S+1}{\pi}\right)^N \prod_{i=1}^{N} \int \mathrm{d}^2\mu_i (1+|\mu_i|^2)^{-2}. \tag{7.7}$$

The amplitude or 'wavefunction' of the ground state in the μ representation is

$$\langle\mu|0\rangle \equiv \Phi_0(\mu) = \prod_i (1+|\mu_i|^2)^{-S}. \tag{7.8}$$

In the p representation a state containing a single spin wave of wavevector $|\boldsymbol{k}\rangle$ is given by

$$|\boldsymbol{k}\rangle = N^{-1/2}\sum_i \exp(\mathrm{i}\boldsymbol{k}\,.\,\boldsymbol{R}_i)|0,0,\dots p_i = 1, 0, \dots\rangle. \tag{7.9}$$

Therefore in μ space we have

$$|\boldsymbol{k}\rangle = \int \mathrm{d}M(\mu)|\mu\rangle\langle\mu|\boldsymbol{k}\rangle$$

$$= (2S)^{1/2}\int \mathrm{d}M(\mu)|\mu\rangle\left\{N^{-1/2}\sum_i \exp(\mathrm{i}\boldsymbol{k}\,.\,\boldsymbol{R}_i)\mu_i^*\right\}\langle\mu|0\rangle. \tag{7.10}$$

That is, the amplitude of the spin-wave state in μ space is

$$\Phi_{\boldsymbol{k}}(\mu) = (2S)^{1/2}N^{-1/2}\sum_i \exp(\mathrm{i}\boldsymbol{k}\,.\,\boldsymbol{R}_i)\mu_i^*\Phi_0(\mu) = \mu_k^*\Phi_0(\mu) \tag{7.11}$$

(where $\mu_k^* \equiv (2S/N)^{1/2}\sum_i \exp(\mathrm{i}\boldsymbol{k}\,.\,\boldsymbol{R}_i)\mu_i^*$). Thus, $\Phi_{\boldsymbol{k}}(\mu)$ is a simple algebraic multiple of $\Phi_0(\mu)$.

7.3. *Two spin $\frac{1}{2}$ particles interacting via the Heisenberg Hamiltonian*

Here we have

$$\mathscr{H}' = -2J\hat{S}_1\,.\,\hat{S}_2. \tag{7.12}$$

It is straightforward to show that the diagonal elements of the density matrix $\hat{\rho} \equiv \exp(-\beta\mathscr{H})/\mathrm{Tr}(\exp-\beta\mathscr{H})$ are given by ($j \equiv \beta J$)

$$\langle\mu_1\mu_2|\hat{\rho}|\mu_1\mu_2\rangle = \frac{1}{3\mathrm{e}^{2j}+1}\,\frac{1}{1+|\mu_1|^2}\,\frac{1}{1+|\mu_2|^2}$$
$$\times\{\mathrm{e}^{2j}(1+\tfrac{1}{2}|\mu_1+\mu_2|^2+|\mu_1|^2|\mu_2|^2)+\tfrac{1}{2}|\mu_1-\mu_2|^2\}. \tag{7.13}$$

We notice that this expression satisfies the conditions

(i) when $j = 0$,

$$\langle\mu_1\mu_2|\hat{\rho}|\mu_1\mu_2\rangle = \tfrac{1}{4} = \frac{1}{(2S+1)^2}$$

(ii) an integration over the coordinate μ_1 of spin 1 gives

$$\int \mathrm{d}M(\mu_1)\langle\mu_1\mu_2|\hat{\rho}|\mu_1\mu_2\rangle = \tfrac{1}{2}.$$

Correlations between the two spins come out most clearly if we write equation (7.13) in the form

$$\langle\mu_1\mu_2|\hat{\rho}|\mu_1\mu_2\rangle = \frac{1}{3+\mathrm{e}^{-2j}}\left(1-\frac{(1-\mathrm{e}^{-2j})|\mu_1-\mu_2|^2}{2(1+|\mu_1|^2)(1+|\mu_2|^2)}\right). \tag{7.14}$$

This in turn can be written in terms of the angular variables (θ_1, ϕ_1) and (θ_2, ϕ_2). One finds

$$\langle\Omega_1\Omega_2|\hat{\rho}|\Omega_1\Omega_2\rangle = \tfrac{1}{4}\left(1+\frac{1-\mathrm{e}^{-2j}}{3+\mathrm{e}^{-2j}}\boldsymbol{n}_1\,.\,\boldsymbol{n}_2\right) \tag{7.15}$$

where $\boldsymbol{n}_1$, $\boldsymbol{n}_2$ are unit vectors in the directions specified by (θ_1, ϕ_1) and (θ_2, ϕ_2) respectively.

This shows absolutely transparently that

(i) for $j > 0$, that is, ferromagnetic coupling, the spins are correlated and tend to align parallel (i.e. with $\boldsymbol{n}_1 \cdot \boldsymbol{n}_2 > 0$).

(ii) for $j < 0$, that is, antiferromagnetic coupling, the spins tend to align antiparallel (i.e. with $\boldsymbol{n}_1 . \boldsymbol{n}_2 < 0$).

(iii) the density matrix is clearly invariant under rotations as it should be (it contains no reference to any particular axis.)

In conclusion, although the problems treated here are basically trivial, we may hope that there are also some nontrivial problems for which the point of view developed here may be illuminating.

Note added in proof. This paper was completed in draft form by D Radcliffe before his death and was edited for publication by A. J. Leggett.

References

Brink, D. M., and Satchler, G. R., 1968, *Angular Momentum* (Oxford: Clarendon Press).
Carruthers, P., and Nieto, M. M., 1968, *Rev. mod. Phys.*, **40**, 411.
Langer, J. S., 1968, *Phys. Rev.*, **167**, 183–90.

Commun. math. Phys. 26, 222—236 (1972)

Coherent States for Arbitrary Lie Group

A. M. Perelomov

Institute for Theoretical and Experimental Physics, Moscow, USSR

Received November 11, 1971

Abstract. The concept of coherent states originally closely related to the nilpotent group of Weyl is generalized to arbitrary Lie group. For the simplest Lie groups the system of coherent states is constructed and its features are investigated.

1. Introduction

In a number of fields of quantum theory, and especially in quantum optics and radiophysics, it is convenient to use the system of so called coherent states [1–3].

These states are in close connection with the nilpotent group first considered by Weyl [4].

A question arises: are there exist analogous systems of states for other Lie groups?

The recent paper [5] generalizes the concept of coherent states to some Lie groups. However, the method proposed in this paper cannot be applied to all Lie groups and, in particular, it is inapplicable to compact groups. Besides, with this approach the set of coherent states is noninvariant relative to the action of the group representation operators.

The present paper proposes another method to extend the concept of coherent state[1]. This method can be applied to any Lie group and is consistent with the action of the group on the set of coherent states (see Section 2). Sections 3–5 of the paper deal with construction of the system of coherent states and with the investigation of its features for the simplest Lie groups.

---

[1] Note in this connection that although some states of such type were considered previously, the properties of the system of states as a whole do not appear to have been investigated (except for the Weyl group).

2. General Properties of Coherent States

Let G be an arbitrary Lie group and T be its irreducible unitary representation acting in the Hilbert space $\mathscr{H}$. A vector of this space is denoted by the symbol $|\psi\rangle$, the scalar product of the vectors $|\psi\rangle$ and $|\varphi\rangle$, linear on $|\psi\rangle$ and antilinear on $|\varphi\rangle$, by the symbol $\langle\varphi|\psi\rangle$, and the projection operator on the vector $|\psi\rangle$ by $|\psi\rangle\langle\psi|$.

Let $|\psi_0\rangle$ be some fixed vector in the space $\mathscr{H}$. Consider the set of vectors $\{|\psi_g\rangle\}$, where $|\psi_g\rangle = T(g)|\psi_0\rangle$ and g goes over all the group G. It is easy to see that two vectors $|\psi_{g_1}\rangle$ and $|\psi_{g_2}\rangle$ differ from one another only by a phase factor ($|\psi_{g_1}\rangle = e^{i\alpha}|\psi_{g_2}\rangle$, $|e^{i\alpha}| = 1$), or in other words determine the same state only if $T(g_2^{-1}\cdot g_1)|\psi_0\rangle = e^{i\alpha}|\psi_0\rangle$.

Let $H = \{h\}$ be the set of elements of the group G such that $T(h)|\psi_0\rangle = e^{i\alpha(h)}|\psi_0\rangle$. It is evident that H is a subgroup of the group G and we denote it as the stationary group of the state $|\psi_0\rangle$.

From these construction we see that the vectors $|\psi_g\rangle$ for all g which belong to one left coset G on H differ from one another only by a phase factor and that these vectors determine the same state.

Selecting in each coset x one representative $g(x)$ of the group G[2] we get the set of states $\{|\psi_{g(x)}\rangle\}$, or in an abridged form of writing $\{|x\rangle\}$ where $|x\rangle \in \mathscr{H}$, $x \in M = G/H$.

Now we may give the definition of generalized coherent states.

The system of coherent states of the type $(T, |\psi_0\rangle)$ (T is the representation of the group G acting in the some space $\mathscr{H}$ and $|\psi_0\rangle$ is a fixed vector of this space) is called a set of states $\{|\psi_g\rangle\}$, $|\psi_g\rangle = T(g)|\psi_0\rangle$ where g runs over all the group G. Let H be a stationary subgroup of the state $|\psi_0\rangle$. Then the coherent state $|\psi_g\rangle$ is determined by the point $x = x(g)$ of the factor space G/H corresponding to the element g:

$$|\psi_g\rangle = e^{i\alpha}|x\rangle, \ |\psi_0\rangle = |0\rangle.$$

Let us consider some general properties of the coherent states. It is easy to see that $e^{i\alpha(h_2h_1)} = e^{i\alpha(h_2)}e^{i\alpha(h_1)}$, i.e. $e^{i\alpha(h)}$ is a one-dimensional unitary representation of the group H.

If this representation is not identity, i.e. if $\alpha(h) \not\equiv 0$, then the factor group A of the group H on its commutant H'[3] is not trivial, i.e. it contains elements different from unity and the character of the group A determines completely the representation of the group H.

[2] The group G can be considered as a fiber bundle with the base $M = G/H$, and the fiber H. Then the choice of $g(x)$ is the choice of a certain cross section of this fiber bundle.

[3] Remind that the commutant H' of the group H consists of the elements h' of the type of $h' = h_1 h_2 h_1^{-1} h_2^{-2}$. The commutant is an invariant subgroup of the group H and the factor group H/H' is the Abelian group.

If $\alpha(h) \equiv 0$, then H is the stationary subgroup of the vector $|\psi_0\rangle$ in the usual sense. In the first (second) case the representation T of the group G being restricted on the subgroup H has to contain the one-dimensional (identity) representation of the group H[4].

Note that if the subgroup H is connected then the vector $|\psi_0\rangle$ is the eigenvector of the infinitesimal operators of the representation of the subgroup itself.

Let us consider now the action of the operator $T(g)$ on the state $|\psi_0\rangle = |0\rangle$

$$T(g)\,|0\rangle = e^{i\alpha(g)}|x(g)\rangle\,. \tag{1}$$

Here the function $\alpha(g)$ is determined on the whole group G and at $g \in H$ it coincides with the previously considered function $\alpha(h)$[5]. Substituting in Eq. (1) g by gh we get

$$\alpha(gh) = \alpha(g) + \alpha(h)\,. \tag{2}$$

Let us now act by the operator $T(g)$ on an arbitrary coherent state $|x\rangle$

$$T(g_1)\,|x\rangle = e^{-i\alpha(g)}T(g_1)\,T(g)\,|0\rangle = e^{i\beta(g_1,g)}|g_1 \cdot x\rangle\,. \tag{3}$$

Here $\beta(g_1, g) = \alpha(g_1 \cdot g) - \alpha(g)$; $x = x(g)$, $g_1 \cdot x = x_1 \in M$ and the element x_1 is determined by the action of the group G on the homogeneous space $M = G/H$. Note that due to (2) Eq. (3) is correct, i.e. the right-hand side of equality depends not on g but only on the cosets of $x(g)$: $\beta(g_1, g) = \beta(g_1, x)$.

It can be easily seen that the scalar product of two coherent states $|x_1\rangle = |x(g_1)\rangle$ and $|x_2\rangle = |x(g_2)\rangle$ is of the form

$$\langle x_1\,|\,x_2\rangle = e^{i[\alpha(g_1)-\alpha(g_2)]}\langle 0|\,T(g_1^{-1}g_2)|0\rangle \tag{4}$$

and due to (2) it is independent of the choice of the representatives g_1 and g_2. But due to unitarity of the representations $T(g)\,|\langle x_1\,|\,x_2\rangle| < 1$ at $x_1 \neq x_2$ and the following equalities take place

$$\langle x_1\,|\,x_2\rangle = \overline{\langle x_2\,|\,x_1\rangle}\,, \tag{5}$$

$$\langle g \cdot x_1\,|\,g \cdot x_2\rangle = e^{i[\beta(g,x_1)-\beta(g,x_2)]}\langle x_1\,|\,x_2\rangle\,. \tag{6}$$

Turning to the problem of completeness, note first of all that the completeness of the system follows immediately from the irreducibility

[4] In many cases a useful information on possible representations $T(g)$ may be obtained from the reciprocity theorem of Frobenius. Stating that if T_α is the representation of the group G induced by the character $e^{i\alpha}$ of the group H, then the representation T has to be contained in the decomposition of representation T_α into irreducible representations [6].

[5] Formula (1) defines the mapping π: $G \to \tilde{M}$ where $\tilde{M}$ is the fiber bundle the base of which is $M - G/H$ and the fiber is a circle.

of the representation T. Let it exist the invariant measure dg on the group G. In many cases it induces the invariant measure dx on the homogeneous space $M = G/H$. Supposing the convergence conditions to be fulfilled let us consider the operator

$$B = \int dx|x\rangle\langle x| . \tag{7}$$

From definition of B, the invariance of measure dx and from formula (3) it immediately follows that

$$T(g)\,BT(g)^{-1} = B . \tag{8}$$

Thus B commutes with all the operators $T(g)$ and so due to irreducibility of the representation T, the operator B is multiple of the identity operator

$$\frac{1}{d}B = I . \tag{9}$$

To find the constant d let us calculate the average value of the operator B in the state $|y\rangle$ ($\langle y|y\rangle = 1$)

$$\langle y|B|y\rangle = \int |\langle y|x\rangle|^2\,dx = \int |\langle 0|x\rangle|^2\,dx = d . \tag{10}$$

Hence it is, in particular, seen that a necessary condition for the existence of the operator B is the convergence of the integral (10). In this case, which we call the case of the square-integrable system of coherent states, an important identity holds

$$\frac{1}{d}\int dx|x\rangle\langle x| = I . \tag{11}$$

Making use of this one may expand the arbitrary state in coherent states

$$|\psi\rangle = \frac{1}{d}\int dx\,c(x)\,|x\rangle, \qquad c(x) = \langle x|\psi\rangle . \tag{12}$$

Here

$$\langle\psi|\psi\rangle = \frac{1}{d}\int dx|c(x)|^2 \tag{13}$$

and the function $c(x)$ is not arbitrary but it must satisfy the condition

$$c(x) = \frac{1}{d}\int\langle x|y\rangle\,c(y)\,dy . \tag{14}$$

Thus the kernel $K(x, y) = \frac{1}{d}\langle x|y\rangle$ is the reproducing one

$$K(x, z) = \int dy\,K(x, y)\,K(y, z) \tag{15}$$

and the function $f(x) = \int K(x, y)\,f(y)dy$ satisfies Eq. (14) for an arbitrarily chosen function $f(x)$.

It can be also easily seen that between the coherent states there are "linear dependences". Indeed, from (12) it follows that

$$|x\rangle = \frac{1}{d}\int \langle y|x\rangle\, |y\rangle\, dy\,. \tag{16}$$

It means that the system of coherent states is overcomplete, i.e. it contains subsystems of coherent states which are complete systems.

The simplest subsets arise from consideration of the discrete subgroups of the group G. Let Γ be a discrete subgroup of the group G such that the volume V_Γ of the factor space M/Γ is finite. Let us consider the subsystem of the coherent states

$$\{|x_l\rangle\}, \qquad x_l = x(\gamma_l), \qquad \gamma_l \in \Gamma\,. \tag{17}$$

A question arises concerning the completeness of such subsystems. It would be interesting to know whether the following statement is valid: at $V_\Gamma > d$ the system of states $\{|x_l\rangle\}$ is not complete but at $V_\Gamma < d$ this system is complete and remains complete even after eliminating any finite number of states. The most interesting case is characterized by $V_\Gamma = d$ (if such a condition can be fulfilled) and it requires a separate, more detailed consideration. Note that for the simplest nilpotent group this problem has been solved in the paper [7].

Let us now illustrate the concept of generalized coherent states using concrete examples.

3. Case of the Special Nilpotent Group[6]

Let us first review some well known facts. The Lie algebra of this group is isomorphic to the Lie algebra produced by the annihilation operators $a_1, \ldots, a_N$, Hermitian-conjugate creation operators $a_1^+, \ldots, a_N^+$ and the identity operator I. The commutation relations between these operators are of the form

$$[a_i, a_j] = [a_i^+, a_j^+] = [a_i, I] = [a_j^+, I] = 0, \qquad [a_i, a_j^+] = \delta_{ij} I\,. \tag{18}$$

A general element of the Lie algebra can be written as

$$tI + i(\bar{\alpha} a - \alpha a^+) \tag{19}$$

where t is a real number, $\alpha = (\alpha_1, \ldots, \alpha_N)$, $a = (a_1, \ldots, a_N)$ are N-dimensional vectors, α_i and $\bar{\alpha}_i$ are complex conjugate to each other. Here and in the following we use an abbreviated notation for the scalar product of

[6] This group appears if one writes the Heisenberg commutation relations in the Weyl form [4]. Its properties are considered in detail in the paper [8].

two such vectors:

$$\bar{\alpha}a = \sum_{i=1}^{N} \bar{\alpha}_i a_i, \qquad \alpha a^+ = \sum_{i=1}^{N} \alpha_i a_i^+ .$$

The Lie group W_N is obtained from the Lie algebra by means of the exponential mapping. Thus to the element of the algebra (19) corresponds the element g of the group which is denoted by (t, α). Then the multiplication law in W_N is given by the formula

$$(s, \alpha)(t, \beta) = (s + t + \mathrm{Im}(\alpha\bar{\beta}), \alpha + \beta) . \tag{20}$$

The operators of the irreducible unitary representation of the group W_N are of the form

$$T(g) = T(t, \alpha) = e^{it} D(\alpha) . \tag{21}$$

From (21) it follows in particular that to the set of elements $G_0 = \{g_k = (2\pi k, 0)\}$ (k is integer) corresponds the identity operator, i.e. the representation under consideration is not a faithful one.

Making use of the commutation relations (18) we can easily obtain the multiplication law of the operators

$$D(\alpha) D(\beta) = e^{i\,\mathrm{Im}(\alpha\bar{\beta})} D(\alpha + \beta) . \tag{22}$$

Lět us now take some vector $|\psi_0\rangle$ in the representation space. We denote the stationary subgroup of this vector as H. Consider two different cases.

I. Let $|\psi_0\rangle$ be an arbitrary vector of the Hilbert space $\mathscr{H}$. In this case the subgroup H consists of the elements of type $(t, 0)$ and the factor space $M = W_N/H$ is the N-dimensional complex space $\boldsymbol{C}^N$. In correspondence with Section 2 of this paper, the system of generalized coherent states is the set of vectors where

$$|\alpha\rangle = D(\alpha)\,|\psi_0\rangle . \tag{23}$$

Note that the usual system of coherent states whose properties have been considered in detail for instance in the papers [1–3] corresponds to the choice in (23) of the so called "vacuum" vector $|0\rangle$[7] as initial vector $|\psi_0\rangle$. However, it can be readily shown that the system of generalized coherent states possesses the same properties as the usual system of coherent states. In particular the main identity

$$\frac{1}{\pi^N} \int d^{2N}\alpha |\alpha\rangle \langle\alpha| = I, \qquad d^{2N}\alpha = \prod_{i=1}^{N} d\,\mathrm{Re}\alpha_i \cdot d\,\mathrm{Im}\alpha_i \tag{24}$$

remains valid for it.

[7] The vacuum vector $|0\rangle$ is determined by the equations $a_i|0\rangle = 0$.

II. Let us extend the Hilbert space $\mathscr{H}$ up to the space of generalized functions $\mathscr{H}_{-\infty}$ [8] and try to find the eigenvector $|\theta_0\rangle \in \mathscr{H}_{-\infty}$ of the operators $T(h)$, where $h = (t, \alpha_n)$ is the element of the subgroup H, $\alpha_n = \sum_{j=1}^{2N} n_j \omega_j$, n_j are integer numbers. The set of vectors α_n form a lattice L in the space C^N and we suppose the periods ω_j of this lattice to be really linearly independent. It can be easily seen that the commutant H' of the group H consists of the elements $h' = (t_n, 0)$ where $t_n = 2\pi \sum_{i,j=1}^{2N} B_{ij} n_i n_j$ and the $2N \times 2N$ matrix B has the form $B_{ij} = \frac{1}{\pi} \operatorname{Im}(\omega_i \bar{\omega}_j)$. But according to Section 2 the operator $T(h')$ must be equal to the identity operator, and from this follows that the elements of the matrix B are integral numbers:

$$B_{ij} = \frac{1}{\pi} \operatorname{Im}(\omega_i \bar{\omega}_j) \equiv 0 (\operatorname{mod} 1). \tag{25}$$

We call such lattice L admissible [9]. The conditions $T(h)\,|\theta_0\rangle = e^{i\alpha(h)}|\theta_0\rangle$ are in this case equivalent to the equations

$$D(\omega_i)\,|\theta_0\rangle = e^{i\pi\varepsilon_i}|\theta_0\rangle. \tag{26}$$

The state $|\theta_0\rangle$ is determined by the real numbers ε_i, $i = 1, \ldots 2N$. Acting with the operator $D(\alpha)$ on $|\theta_0\rangle$ we get the system of generalized coherent states

$$|\theta_\alpha\rangle = D(\alpha)\,|\theta_0\rangle \tag{27}$$

where α runs over the complex torus $M = G/H$. It appears that the states $|\theta_\alpha\rangle$ under the choice of a definite realization of space $\mathscr{H}$ coincide in the essence with the theta functions. Some properties of the theta functions in frame of such approach are considered in the papers [7, 8].

4. Case of the Simplest Compact Group-$SU(2)$ Group

The $SU(2)$ group is the unitary matrix group of the second order with unity determinant. It is locally isomorphic to the $SO(3)$ group – the rotation group of three-dimensional space and it is the most investigated group among all the non-Abelian Lie groups [10]. Nevertheless, the coherent states for this group as a special system do not appear to have been considered so far.

[8] The definition of space $\mathscr{H}_{-\infty}$ and the consideration of some of its properties is given in paper [8].

[9] Note that the factor space $M = G/H$ is a complex torus and in the case of admissible lattice it can be considered as an Abelian variety.

[10] The properties of this group are considered in detail e.g. in the books [9, 10].

Let us first review some well known facts. The T^j representation of this group is determined by the nonnegative number j, integer or half-integer. The basis vectors $|j, \mu\rangle$ of the space in which the representation acts are labelled by a number μ, that takes the $2j+1$ integer (if j is integer) or half-integer (if j is half-integer) values from $-j$ up to j. The vectors $|j, \mu\rangle$ satisfy the condition

$$T^j(h)\,|j, \mu\rangle = e^{i\mu\varphi}|j, \mu\rangle, \qquad h = \begin{pmatrix} e^{i\varphi/2} & 0 \\ 0 & e^{-i\varphi/2} \end{pmatrix}. \tag{28}$$

Here h is the element of the rotation subgroup H around the axis x_3. Note that vector $|j, \mu\rangle$ is the eigenvector of the infinitesimal operator J_3 of the representation T^j corresponding to the subgroup H

$$J_3\,|j, \mu\rangle = \mu|j, \mu\rangle\,. \tag{29}$$

The factor space G/H is the two-dimensional sphere S^2, the point of this sphere is determined by the unit vector $\boldsymbol{n}$, $\boldsymbol{n}^2 = 1$. Let $g(\boldsymbol{n})$ be the element of the group G which transforms the vector $\boldsymbol{n}_0 = (0, 0, 1)$ into the vector $\boldsymbol{n} = (\sin\theta\cos\varphi, \sin\theta\sin\varphi, \cos\theta)$. As $g(\boldsymbol{n})$ we may choose, for example, the element $g^3_\varphi g^2_\theta$ where g^3_φ corresponds to rotation around the axis x_3 by the angle φ and g^2_θ to rotation around the axis x_2 by the angle θ. We thus come to the system of coherent states $\{|\mu, \boldsymbol{n}\rangle\}$:

$$|\mu, \boldsymbol{n}\rangle = T(g(\boldsymbol{n}))\,|\mu\rangle = T(g^3_\varphi)\,T(g^2_\theta)\,|\mu\rangle\,. \tag{30}$$

(Here we have to omit for simplicity index j to abbreviate the notations.)

The system of coherent states can be also determined up to the phase factor $e^{i\alpha(\boldsymbol{n})}$ using the equation

$$(n_i J_i)\,|\mu, \boldsymbol{n}\rangle = \mu|\mu, \boldsymbol{n}\rangle \tag{31}$$

where J_i is the infinitesimal operator of the representation which corresponds to a rotation around the axis x_i. Here μ should be considered as a fixed parameter and the vector $\boldsymbol{n}$ as a variable quantity.

Let us consider the properties of this system. The scalar product of two coherent states is generally speaking non-zero and it equals

$$\langle\mu, \boldsymbol{n}'\,|\,\mu, \boldsymbol{n}\rangle = e^{i\phi(\boldsymbol{n}', \boldsymbol{n})}\,d^j_{\mu\mu}(\theta) = e^{i\phi}\left(\cos\frac{\theta}{2}\right)^{2|\mu|} P^{(0,2|\mu|)}_{j-|\mu|}(\cos\theta) \tag{32}$$

where $\cos\theta = \boldsymbol{n}'\boldsymbol{n}$, $d^j_{\mu\nu}(\theta)$ are standard matrix elements of the $SU(2)$ group [9, 10] and $P^{(a,b)}_n$ are Jacobi polynomials. Hence we find that

$$d = \int |\langle\mu, \boldsymbol{n}'\,|\,\mu, \boldsymbol{n}\rangle|^2\, d\boldsymbol{n} = \frac{4\pi}{2j+1} \tag{33}$$

and correspondingly

$$\frac{2j+1}{4\pi}\int d\boldsymbol{n}|\mu,\boldsymbol{n}\rangle\langle\mu,\boldsymbol{n}| = I\,. \tag{34}$$

Especially simple is the system of coherent states for $\mu = j$. Then, for instance, we get from Eq. (32)

$$|\langle j,\boldsymbol{n}'|j,\boldsymbol{n}\rangle|^2 = \left(\frac{1+\boldsymbol{n}'\boldsymbol{n}}{2}\right)^{2j}. \tag{35}$$

Let us also give the expression for the coherent states in the so-called z-representation [9, 10]. In this case the representation T acts in the space of polynomials of $2j$ degree and the operator of the representation is given by the formula

$$T^j(g)f(z) = (\beta z+\bar{\alpha})^{2j} f\left(\frac{\alpha z-\bar{\beta}}{\beta z+\bar{\alpha}}\right),\quad g=\begin{pmatrix}\alpha & \beta\\ -\bar{\beta} & \bar{\alpha}\end{pmatrix},\quad |\alpha|^2+|\beta|^2=1\,. \tag{36}$$

The basis vectors $|j,\mu\rangle$ in this representation are of the form

$$\langle z|j,\mu\rangle = \sqrt{\frac{(2j)!}{(j+\mu)!\,(j-\mu)!}}\, z^{j+\mu}\,. \tag{37}$$

In this case from (30) and (36) it immediately follows

$$\langle z|\mu,\boldsymbol{n}\rangle = \sqrt{\frac{(2j)!}{(j+\mu)!\,(j-\mu)!}}\,(\beta z+\bar{\alpha})^{j-\mu}(\alpha z-\bar{\beta})^{j+\mu} \tag{38}$$

where

$$\alpha = \cos\frac{\theta}{2}\, e^{i\varphi/2},\qquad \beta = \sin\frac{\theta}{2}\, e^{i\varphi/2}.$$

Let us map the sphere S^2 onto the plane of the complex variable ζ using the stereographic projection

$$\zeta = \operatorname{ctg}\frac{\theta}{2}\, e^{i\varphi}\,. \tag{39}$$

Expression (38) takes now the form

$$\langle z|\mu,\boldsymbol{n}\rangle = e^{i\phi}\langle z|\mu,\zeta\rangle \quad \text{where} \quad e^{i\phi} = (-1)^{j+\mu}e^{-i\mu\varphi} \tag{40}$$

$$\langle z|\mu,\zeta\rangle = \sqrt{\frac{(2j)!}{(j+\mu)!\,(j-\mu)!}}\,(1+|\zeta|^2)^{-j}(z+\bar{\zeta})^{j-\mu}(1-\zeta z)^{j+\mu}\,. \tag{41}$$

Acting on the function $|\mu, \zeta\rangle$ by the operator $T(g)$ we get $T(g)|\mu, \zeta\rangle = e^{i\phi}|\mu, g \cdot \zeta\rangle$, where

$$e^{i\phi} = \left(\frac{\bar{\beta}\zeta + \bar{\alpha}}{\beta\bar{\zeta} + \alpha}\right)^{\mu}, \qquad g \cdot \zeta = \frac{\alpha\zeta - \beta}{\bar{\beta}\zeta + \bar{\alpha}}. \tag{42}$$

Let us now give the formula for the scalar product of coherent states in this representation

$$\langle j, \zeta' | j, \zeta\rangle = (1 + |\zeta'|^2)^{-j}(1 + |\zeta|^2)^{-j}(1 + \bar{\zeta}'\zeta)^{2j} \tag{43}$$

Note that just as the usual coherent states appear naturally in the problem of an oscillator which is under the action of an external time-dependent force [1, 2] the states $|\mu, \boldsymbol{n}\rangle$ appear naturally when one considers the problem of the spin motion in a time-dependent magnetic field. In this case the variation of the state over time is determined by the Schrödinger equation

$$i\frac{\partial}{\partial t}|\psi(t)\rangle = -\boldsymbol{aJ}|\psi(t)\rangle \tag{44}$$

where $\boldsymbol{a} = \mu \boldsymbol{H}$, μ is the magnetic moment of the particle, $\boldsymbol{H}(t)$ is the magnetic field, $\boldsymbol{J} = (J_1, J_2, J_3)$, J_i is the operator of infinitesimally small rotations around the axis x_i.

It can be easily seen that if at initial time we have the coherent state $|\psi(0)\rangle = |\boldsymbol{n}_0\rangle$, then at any following time this state remains coherent, i.e.

$$|\psi(t)\rangle = e^{i\alpha(t)}|\boldsymbol{n}(t)\rangle \tag{45}$$

where the vector $\boldsymbol{n}(t)$ is determined by the classical equation of motion

$$\dot{\boldsymbol{n}}(t) = -[\boldsymbol{a}(t), \boldsymbol{n}(t)]. \tag{46}$$

The coherent states may be also used to describe the density matrix ϱ of a particle with spin[11]. Namely, the density matrix ϱ is completely determined either by the function $P(\boldsymbol{n})$ or $Q(\boldsymbol{n})$, according to the formulae

$$\varrho = \int d\boldsymbol{n}\, P(\boldsymbol{n})\, |\boldsymbol{n}\rangle\langle \boldsymbol{n}|, \tag{47}$$

$$Q(\boldsymbol{n}) = \langle \boldsymbol{n}|\varrho|\boldsymbol{n}\rangle. \tag{48}$$

Note one more useful identity

$$|\boldsymbol{n}\rangle\langle \boldsymbol{n}| = \frac{2j+1}{16\pi^2}\int c(\boldsymbol{n}, g^{-1})\, T(g) dg, \tag{49}$$

$$c(\boldsymbol{n}, g) = \langle \boldsymbol{n}|T(g)|\boldsymbol{n}\rangle \tag{50}$$

which follows from the orthogonality of the matrix elements $T^j_{\mu\nu}(g)$.

[11] The description of the oscillator density matrix using the usual coherent states can be find in the papers [1–3].

Note that if in (47) $P(\boldsymbol{n})$ is expanded in a series of spherical functions

$$P(\boldsymbol{n}) = \sum_{l,m} C_{lm} Y_{lm}(\boldsymbol{n}) \tag{51}$$

one gets the expansion of the density matrix

$$\varrho = \sum_{l,m} C_{lm} \hat{P}_{lm} \tag{52}$$

in the operators

$$\hat{P}_{lm} = \int d\boldsymbol{n}\, Y_{lm}(\boldsymbol{n})\, |\boldsymbol{n}\rangle \langle \boldsymbol{n}| \,. \tag{53}$$

Calculating the integral entering (53) we find

$$\langle v' | \hat{P}_{lm} | v \rangle = \frac{\sqrt{4\pi(2l+1)}}{2j+1} (j, v'; l, m \,|\, j, v)\, (j, \mu; l, 0 \,|\, j, \mu) \tag{54}$$

where $(j, v'; l, m \,|\, j, v)$ is the Clebsch-Gordan coefficient.

In conclusion note that the formulae obtained in this section carry over to the case arbitrary compact Lie group. In order to do this it is only necessary to replace the group H by the Cartan subgroup and to take into consideration that $2j+1$ is the dimension of the representation $T(g)$ and 4π is the volume of the factor space $M = G/H$.

5. Case of the Simplest Noncompact Group-$SU(1,1)$ Group

The $SU(1,1)$ group is the group of unimodular matrices that leave the form $|z_1|^2 - |z_2|^2$ invariant. The element g of this group has the form

$$g = \begin{pmatrix} \alpha & \beta \\ \bar{\beta} & \bar{\alpha} \end{pmatrix}, \qquad |\alpha|^2 - |\beta|^2 = 1 \,. \tag{55}$$

The $SU(1,1)$ group is isomorphic to the $Sp(2, R)$ group (to the group of real simplectic second order matrices) and it is locally isomorphic to the $SO(2,1)$ group, the group of "rotations" of the three-dimensional pseudoeuclidian space. It has several series of unitary irreducible representations and, in particular, two discrete series T^+ and T^-. It is sufficient to consider only one of these, e.g. T^+, because all the results are automatically carried over to the other case.

The representation of the series T^+ is characterized by the positive integer or half-integer numbers k. It may be realized in the space of the function $\mathscr{F}_k$ that are analytic in the unit disk $|\zeta|<1$ and satisfy the condition

$$\|f\|^2 = \int d\mu_k(z)\,|f(z)|^2 < \infty \tag{56}$$

where

$$d\mu_k(z) = \frac{2k-1}{\pi}(1-|z|^2)^{2k-2}\,d^2 z\,. \tag{57}$$

It is not difficult to see that if $f(z) = \sum_{n=0}^{\infty} c_n z^n$ then

$$\|f\|^2 = \sum_{n=0}^{\infty} \frac{\Gamma(n+1)\,\Gamma(2k)}{\Gamma(n+2k)}|c_n|^2\,. \tag{58}$$

The space $\mathscr{F}_k$ becomes a Hilbert space if the scalar product of two vectors is defined in it according to the formula

$$\langle f|g\rangle = \int d\mu_k(z)\,\bar{f}(z)\,g(z)\,. \tag{59}$$

Now it can be easily checked that the functions

$$|n\rangle = f_n(z) = \sqrt{\frac{\Gamma(n+2k)}{\Gamma(n+1)\,\Gamma(2k)}}\,z^n \tag{60}$$

form an orthonormal basis in the space $\mathscr{F}_k$ [12].

Let us define the action of the operators $T(g)$ in the space $\mathscr{F}_k$

$$T^k(g)\,f(z) = (\beta z+\bar{\alpha})^{-2k} f\left(\frac{\alpha z+\bar{\beta}}{\beta z+\bar{\alpha}}\right). \tag{61}$$

It can be shown [10–12] that the operators $T^k(g)$ determine the unitary irreducible representation of the $SU(1,1)$ group.

Let us choose in $\mathscr{F}_k$ the vector of the lowest weight $|0\rangle$ as the fixed vector $|\psi_0\rangle$ (the corresponding function $f_0(z)\equiv 1$). Acting on it by the operators $T^k(g)$ we get the system of states

$$|g\rangle = T(g)\,|0\rangle = (\beta z+\bar{\alpha})^{-2k}\,. \tag{62}$$

This expression can be easily transformed to the form

$$|g\rangle = e^{i\phi}|\zeta\rangle; \quad |\zeta\rangle = (1-|\zeta|^2)^k(1-\zeta z)^{-2k}, \quad |\zeta|<1\,. \tag{63}$$

The set $\{|\zeta\rangle\}$ is just the system of coherent states.

[12] Note that if k tends to infinity and z to zero so that kz = const then as the functions $f_n(z)$, and all other quantities go over into corresponding quantities for a special nilpotent group (see Section 3).

Note that the group of matrices of type $h = \begin{pmatrix} e^{i\varphi/2} & 0 \\ 0 & e^{-i\varphi/2} \end{pmatrix}$ is the stationary subgroup H of the vector $|0\rangle$ and the factor space G/H is the unit disk. We see that according to Section 2, the coherent state $|\zeta\rangle$ is completely determined by the point ζ of the factor-space G/H. Note that this space could be also realized as an upper sheet of the hyperboloid $n_0^2 - n_1^2 - n_2^2 = 1$.

Expanding the state $|\zeta\rangle$ in states $|n\rangle$ we get

$$|\zeta\rangle = (1 - |\zeta|^2)^k \sum_{n=0}^{\infty} \sqrt{\frac{\Gamma(n+2k)}{\Gamma(n+1)\,\Gamma(2k)}}\, \zeta^n |n\rangle\,. \tag{64}$$

Hence for the scalar product of two coherent states the following formula obtains

$$\langle \zeta' | \zeta \rangle = (1 - |\zeta'|^2)^k (1 - |\zeta|^2)^k (1 - \bar{\zeta}'\zeta)^{-2k}\,. \tag{65}$$

Correspondingly

$$d = \int d\mu(\zeta)\, |\langle 0 | \zeta \rangle|^2 = \frac{\pi}{2k-1} \tag{66}$$

and the condition of completeness takes the form

$$\frac{2k-1}{\pi} \int d\mu(\zeta)\, |\zeta\rangle \langle \zeta| = I\,. \tag{67}$$

Here $d\mu(\zeta) = \dfrac{d^2\zeta}{(1-|\zeta|^2)^2}$ is the invariant measure on the disk $|\zeta| < 1$.

Let us now consider an arbitrary normalized vector $|\psi\rangle$ belonging to the Hilbert space $\mathscr{H}$. To this vector the function $\langle \zeta | \psi \rangle$ may be taken into correspondence and if $|\psi\rangle = \Sigma\, c_n |n\rangle$, then

$$\langle \zeta | \psi \rangle = (1 - |\zeta|^2)^k\, \psi(\bar{\zeta}) \tag{68}$$

where

$$\psi(\zeta) = \sum_{n=0}^{\infty} \sqrt{\frac{\Gamma(n+2k)}{\Gamma(n+1)\,\Gamma(2k)}}\, c_n \zeta^n\,. \tag{69}$$

Note that in this case

$$\|\psi\|^2 = \langle \psi | \psi \rangle = \int d\mu_k(\zeta)\, |\psi(\zeta)|^2\,, \tag{70}$$

i.e. $\psi(\zeta) \in \mathscr{F}_k$. The formula (69) establishes the isomorphism between the spaces $\mathscr{H}$ and $\mathscr{F}_k$.

Moreover, from the inequality $|\langle \zeta | \psi \rangle|^2 \leqq \|\psi\|^2$ follows a restriction on the growth of function $\psi(\zeta)$

$$|\psi(\zeta)|^2 \leq (1 - |\zeta|^2)^{-2k}\, \|\psi\|^2\,. \tag{71}$$

From (71) we immediately obtain that strong convergence of the sequence $|\psi_n\rangle$ implies pointwise convergence $\psi_n(\zeta)$ uniform on any compact subset of the plane ζ.

A characteristic feature of spaces of the type $\mathscr{F}_k$ are the so called "reproducing kernels" which play the role of usual δ-functions. Such kernel can be found in the usual way. Namely

$$\delta_{z'}(z) = \sum_{n=0}^{\infty} \overline{f_n(z')}\, f_n(z) = (1 - \bar{z}' z)^{-2k}. \tag{72}$$

At fixed $z' \delta_{z'}(z)$ is a function of z and its norm is equal to

$$\|\delta_{z'}\|^2 = (1 - |z'|^2)^{-2k}. \tag{73}$$

It can be easily checked by direct calculations that $\delta_{z'}(z)$ is the analog of δ-function, i.e. the equality

$$\langle \delta_z | f \rangle = \int d\mu(z')\, \bar{\delta}_z(z')\, f(z') \equiv f(z), \qquad (f(z) \in \mathscr{F}_k) \tag{74}$$

holds.

Up to now we considered the representations of the $SU(1,1)$ group. One can however consider also the representations of its universal covering group, namely the group $\overline{SU(1,1)}$ [13] which as is well known, covers the group $SU(1,1)$ an infinite number of times. It can be easily seen that this results only in replacing the number k, that was previously restricted to non-negative integer or half-integer values, by an arbitrary non-negative number.

Note, that analogous results obtain also for other semi-simple Lie groups, having a discrete series of representations.

In this paper we have briefly considered the simplest systems of coherent states. It would be interesting to consider other systems of such states, in particular, the systems related to the continuous spectrum.

Thanks are due to F. A. Berezin for the discussion of these results and F. Calogero for his help in translating the paper into English.

References

1. Glauber, R. J.: Phys. Rev. **130**, 2529 (1963); **131**, 2766 (1963).
2. Klauder, J. R., Sudarshan, E. C. G.: Fundamentals of quantum optics. New York: Benjamin 1968.
3. Zeldovich, B. Ya., Perelomov, A. M., Popov, V. S.: JETP **55**, 589 (1968); **57**, 196 (1969); Preprints ITEP No. **612**, **618** (1968).
4. Weyl, H.: Gruppentheorie und Quantenmechanik. Leipzig: S. Hirzel 1928.
5. Barut, A. O., Girardello, L.: Commun. math. Phys. **21**, 41 (1971).

[13] Note that the group $\overline{SU(1,1)}$ is a dynamical symmetry group in some model many body problems [13].

6. Mackey, G. W.: Bull. Am. Math. Soc. **69**, 628 (1963).
7. Perelomov, A. M.: Theoret. Math. Phys. **6**, 213 (1971).
8. Cartier, P.: Proc. Symp. Pure Math., v. 9, Algebraic groups and discontinuous subgroups, p. 361. Providence, R.I.: Amer. Math. Soc. 1966.
9. Gelfand, I. M., Minlos, R. A., Shapiro, Z. Ya.: Representations of the rotation group and the Lorentz group. Oxford: Pergamon Press 1963.
10. Vilenkin, N. Ya.: Special functions and the theory of group representations. Providence, R.I.: Amer. Math. Soc. 1968.
11. Bargmann, V.: Ann. Math. **48**, 568 (1947).
12. — Comm. Pure Appl. Math. **14**, 187 (1961).
13. Perelomov, A. M.: Theoret. Math. Phys. **6**, 368 (1971).

A. M. Perelomov
Institute for Theoretical
and Experimental Physics
Bolshaya Cheremushkinskaya 89
Moscow, USSR

A note on coherent state representations of Lie groups

Enrico Onofri

Istituto di Fisica dell'Università, Parma, Italy

Istituto Nazionale di Fisica Nucleare, Sezione di Milano, Milano, Italy

(Received 25 January 1974)

The analyticity properties of coherent states for a semisimple Lie group are discussed. It is shown that they lead naturally to a classical "phase space realization" of the group.

In a recent article Peremolov[1] introduced the concept of "coherent states" for unitary irreducible representations (UIR) of any Lie group G. The idea to consider the translates $T_g|0\rangle$ of a fixed cyclic vector $|0\rangle$ under the group action is as old as the celebrated Gel'fand–Raikov theorem on locally bicompact groups.[2] The contribution of Ref. 1 was then to show how the concept of coherent state, first introduced in the case of Heisenberg–Weyl group, fitted in the general theory of group representations. Let us note that no mention to analyticity question is made in the very general approach given in Ref. 1. Yet all the examples [Heisenberg–Weyl group, $SU(2)$, $SU(1,1)$] have in common the following properties: (a) the homogeneous space G/H has a complex homogeneous structure, i.e., G acts on G/H by means of holomorphic transformations [in the above examples G/H is given by the complex plane, by $S^2 \approx$ complex projective space $P_1(\mathbb{C})$ and by the unit disc $D_1 \subset \mathbb{C}$, respectively]; (b) the Hilbert space of the UIR is identified, in the coherent state basis, with a space of holomorphic functions on G/H [namely, $\exp(\frac{1}{2}|z|^2)$-bounded entire functions, polynomials of degree $2l$ in z and $(1-|z|^2)^{-l}$-bounded analytic functions in D_1, respectively).

Purpose of the present note is to show that

(1) *a homogeneous complex structure is actually present quite generally*, and

(2) that on the basis of the homogeneous complex structure *the manifolds G/H are just the classical phase spaces on which G acts through canonical transformations.*

From this point of view, coherent states appear just as probability wave packets over the classical phase space, a well-known result for the harmonic oscillator coherent states. The converse problem, i.e., the construction of a UIR of G starting from a phase space realization of G. (*quantization of a dynamical group*[3]) was considered in Ref. 4 and found a definite mathematical setting in Refs. 5, 6; here, however, we shall adhere to Ref. 1 scheme and deduce the phase space structure from the UIR.

First of all we restrict our attention to *compact semisimple Lie groups*. Let us fix the notation: $\mathbf{g}$ is the real Lie algebra of the group G, $\mathbf{g}_c = \mathbf{g} \oplus i\mathbf{g}$ its complexification, H a Cartan subgroup of G, $\mathbf{g}_c = \mathbf{h}_c \oplus \sum_{\alpha\in\Delta}\mathbf{g}_\alpha$ the corresponding Cartan decomposition, Δ_+ the set of positive roots (chosen once for all), $\rho = \frac{1}{2}\sum_{\alpha\in\Delta_+}\alpha$. Let $g\in G \to U(g)$ be a UIR of G in a (finite-dimensional) Hilbert space $\mathcal{H}$. Let λ be the highest weight and let us suppose that it is nonsingular. Then there exists a vector $|0\rangle \in \mathcal{H}$ such that

$$U(h)|0\rangle = e^{\lambda}(h)|0\rangle, \quad h\in H$$
$$\{e^{\lambda}\exp(X) = \exp[\lambda(X)], \quad X\in\mathbf{g}, \quad \lambda\in i\mathbf{g}^*\}, \tag{1}$$
$$X|0\rangle = 0, \quad X\in \mathbf{n}_+ = \sum_{\alpha\in\Delta_+}\mathbf{g}_\alpha,$$

or, alternatively,

$$\langle 0|U(h) = e^{\lambda}(h)\langle 0|, \quad h\in H,$$
$$\langle 0|X = 0, \quad X\in\mathbf{n}_- = \sum_{\alpha\in\Delta_+}\mathbf{g}_{-\alpha}. \tag{2}$$

Given $U(g)$ there is no problem in extending it to a holomorphic representation $T(g)$ of G_c in $\mathcal{H}$ (it is sufficient to exponentiate the finite-dimensional matrices which represent the elements of $\mathbf{g}_c$; for noncompact groups the situation is different). Equation (2) shows that the stability subgroup of $\langle 0|$ under $T(g)$ is just the *Borel subgroup B*, which is the complex Lie subgroup of G_c with Lie algebra $\mathbf{b} = \mathbf{h}_c \oplus \mathbf{n}_-$. Let us call $\pi(b)$ the *holomorphic character* of B defined by

$$\langle 0|T(b) = \pi(b)\langle 0|. \tag{3}$$

It is wellknown that the quotient manifolds G/H and G_c/B coincide[7]; then G/H inherits a complex homogeneous structure from G_c, being B a complex subgroup. Let us now consider the family of coherent states $T(g^{-1})^\dagger|0\rangle$: the representatives

$$\Psi(g) = \langle 0|T(g^{-1})|\Psi\rangle \tag{4}$$

are holomorphic on G_c; moreover,

$$\Psi(gb) = \langle 0|T(b^{-1})T(g^{-1})|\Psi\rangle$$
$$= \pi(b^{-1})\Psi(g), \quad g\in G_c, \; b\in B. \tag{5}$$

This means that $\Psi(g)$ defines a *holomorphic section of the homogeneous line bundle $E_\pi(G/H, \mathbb{C})$ associated to the principal fibre bundle $B\to G_c\to G_c/B$ by the holomorphic character π* (see Ref. 8 for the definition of "associated" fibre bundle). The local trivializations of the line bundle E_π associates to every such Ψ a holomorphic function on an open domain in G/H. For instance, let $\mathcal{U} = \{g\in G_c | \langle 0|T(g^{-1})|0\rangle \neq 0\}$; then the function

$$\Psi(g) = \frac{\langle 0|T(g^{-1})|\Psi\rangle}{\langle 0|T(g^{-1})|0\rangle} \tag{6}$$

is actually a function only of the projection $g \xrightarrow{p} g\{H\}\in G/H$ and is holomorphic in the domain $p(\mathcal{U})\subset G/H$. Let $z = (z_1, \ldots, z_n)$ be a local chart $p(\mathcal{U})\to\mathbb{C}^n$ such that $0 = (0, \ldots, 0)$ correspond to $\{H\}$. We can, therefore, introduce a new set of states in a one-to-one correspondence with the points in $p(\mathcal{U})$:

$$|g\cdot 0\rangle = \frac{T(g^{-1})^\dagger|0\rangle}{\langle 0|T(g^{-1})^\dagger|0\rangle}, \quad g\in G_c, \tag{7}$$

or, restricting to G,

$$|g\cdot 0\rangle = \frac{U(g)|0\rangle}{\langle 0|U(g)|0\rangle}, \quad g\in G. \tag{8}$$

Owing to the orthogonality relations

$$\int_G \overline{\langle\Psi_1|U(g)|\Psi_2\rangle}\langle\Psi_3|U(g)|\Psi_4\rangle dg = d_\lambda^{-1}\langle\Psi_2|\Psi_4\rangle\langle\Psi_3|\Psi_1\rangle$$

(being $d_\lambda = \Pi_{\alpha\in\Delta_+}(\langle\lambda+\rho,\alpha\rangle/\langle\rho,\alpha\rangle)$ = dimension of U), the scalar product $\langle\Psi_1|\Psi_2\rangle$ can be written as follows (completeness relation for the coherent states $|z\rangle = |g\cdot 0\rangle$):

$$\begin{aligned}\langle\Psi_1|\Psi_2\rangle &= d_\lambda\int_G \overline{\langle\Psi_2|U(g)|0\rangle}\langle\Psi_1|U(g)|0\rangle dg\\ &= d_\lambda\int_G \overline{\langle 0|U(g)^\dagger|\Psi_1\rangle}\langle 0|U(g)^\dagger|\Psi_2\rangle dg\\ &= d_\lambda\int_G \overline{\langle g\cdot 0|\Psi_1\rangle}\langle g\cdot 0|\Psi_2\rangle|\langle 0|U(g)|0\rangle|^2 dg\\ &= d_\lambda\int_{G/H}\overline{\psi_1(z)}\psi_2(z)\exp[-f(z,\bar z)]\mathring{z},\end{aligned} \tag{9}$$

with

(i) $z = g\cdot 0$,

(ii) $f(z,\bar z) = \log|\langle 0|U(g)|0\rangle|^{-2} \geq 0 \quad (f=0 \Rightarrow z=0)$,

(iii) $\mathring{z}$ = invariant volume density on G/H induced by dg.

Thus $\mathcal{H}$ is isomorphic to a Hilbert space of holomorphic functions bounded by $\exp[\frac{1}{2}f(z,\bar z)]$. In this derivation we tacitly assumed that $\{g\in G|\langle 0|U(g)|0\rangle = 0\}$ is of measure zero in G so that integrals over G/H can be restricted to integrals over $p(\mathcal{U})$.

The representation of G in the basis of the coherent states takes the form of a "multiplier" representation:

$$\begin{aligned}[U(g')\psi](z) &= \frac{\langle 0|U(g)^\dagger U(g')|\Psi\rangle}{\langle 0|U(g)^\dagger|0\rangle}\\ &= \frac{\langle 0|U(g'^{-1}g)^\dagger|0\rangle}{\langle 0|U(g)^\dagger|0\rangle}\psi(g'^{-1}z)\\ &= \mu(g',z)\psi(g'^{-1}z).\end{aligned} \tag{10}$$

Let us now construct the symplectic (Kaehler) structure on G/H. Let

$$\omega = i\partial\bar\partial f(z,\bar z), \quad z\in p(\mathcal{U}), \tag{11}$$

with ∂ and $\bar\partial$ the exterior differentiation operators with respect to z and $\bar z$, respectively $(\partial\bar\partial + \bar\partial\partial = \partial^2 = \bar\partial^2 = 0)$;

explicitly, $\omega = i\sum_{j,k}\frac{\partial^2 f}{\partial z^j\partial\bar z^k}dz^j\wedge d\bar z^k$.

It can be shown that $\omega(X,Y)$ coincides at $z=0$ with the bilinear functional $-i\lambda([X,Y])$, with $X,Y\in\mathbf{g}$ and λ identified with its image in $i\mathbf{g}^*$. Then ω is nonsingular at $z=0$. Moreover, since ω is G-invariant, it is nonsingular everywhere. Let us prove the G-invariance of ω:

$$\begin{aligned}g'_*\omega &= i\partial\bar\partial f(g'^{-1}z, \overline{g'^{-1}z})\\ &= i\partial\bar\partial\log|\langle 0|U(g'^{-1}g)|0\rangle|^{-2}\\ &= i\partial\bar\partial f(z,\bar z) - i\partial\bar\partial\log|\mu(g',z)|^2\end{aligned} \tag{12}$$

and the invariance follows, being $\mu(g',z)$ holomorphic in z.

Up to now we have made only local considerations. Actually, it can be easily shown that a Kaehler form ω_α can be given in a whole covering $\mathcal{O}_\alpha$ of G/H and $\omega_\alpha = \omega_\beta$ on the overlapping $\mathcal{O}_\alpha\cap\mathcal{O}_\beta$. Then we have a homogeneous Kaehler manifold G/H associated to the UIR, i.e., a *G-homogeneous Hamiltonian dynamical system*. Given $X\in\mathbf{g}$, let $\hat X$ be its self-adjoint representation in $\mathcal{H}$, $\mathcal{X}$ the holomorphic vectorfield on G/H, $H(z,\bar z)$ the "classical generating function," defined by $i_{\mathcal{X}+\bar{\mathcal{X}}}\omega = dH(z,\bar z)$. For every ψ in the domain of $\hat X$ it holds that

$$[\hat X\psi](z) = [H(z,\bar z) + i\mathcal{X}(f)]\psi(z) - i(\mathcal{X}\psi)(z). \tag{13}$$

From this it follows that

$$H(z,\bar z) = \langle z|\hat X|z\rangle/\langle z|z\rangle; \tag{14}$$

i.e., *the value of the classical generating function at a point z coincides with the expectation value of the corresponding self-adjoint operator on the coherent state* $|z\rangle$. In general, for a normalized $|\Psi\rangle$ it holds that

$$\begin{aligned}\langle\Psi|\hat X|\Psi\rangle &= \gamma\int_{G/H}\exp[-f(z,\bar z)]|\psi(z)|^2 H(z,\bar z)\mathring{z}\\ &= \int_{G/H}\rho(z,\bar z)H(z,\bar z)\mathring{z},\end{aligned} \tag{15}$$

with γ a factor depending on the UIR (but not on X!); i.e., *the expectation values in the* UIR *coincide with a phase space average of the classical generating function with a probability distribution*

$$\rho(z,\bar z) = \gamma|\psi(z)|^2\exp[-f(z,\bar z)] = \gamma\frac{\langle z|\hat\rho|z\rangle}{\langle z|z\rangle}, \quad \hat\rho \equiv |\Psi\rangle\langle\Psi|. \tag{16}$$

$\rho(z,\bar z)$ is then proportional to the diagonal matrix element of the density matrix in the coherent states representation. Equations (15), (16) hold for a generic density matrix $\hat\rho$. Let us note that $\rho(z,\bar z)$ is positive definite, unlike other correspondential statistical distributions, such as Wigner's function.

A simple relation exists between $\mu(g,z)$ and the classical generating function of the (finite) canonical transformations in G. Let us define $\vartheta = -i\partial f$, then

$$\omega = \mathbf{d}\vartheta \quad (\mathbf{d} = \partial + \bar\partial \Rightarrow \mathbf{d}\partial = \bar\partial\partial = -\partial\bar\partial)$$

and

$$S(g,z) = \int_0^z (\vartheta - g_*\vartheta) + S(g,0)$$

is the (holomorphic) generating function, analogous to the usual $S = \int(pdq - p'\,dq')$. Then we have

$$\mu(g,z) = \exp[iS(g,z)] \tag{17}$$

and we recover an expression first introduced by Van Hove[9] for a Euclidean phase space:

$$[U(g)\psi](z) = \exp[iS(g,z)]\psi(g^{-1}z). \tag{18}$$

Let us now consider the case of a *singular* λ. In this case the stability subgroup of the highest weight vector $\langle 0|$ is bigger than H; by complexifying the parameters, one finds a complex subgroup $P\subset G_c$ (called a *parabolic* subgroup) which includes B. Yet the same statements hold as in the nonsingular case; we simply get Kaehler manifolds with dimension smaller than $\dim(G/H)$, i.e., the so-called singular realizations.[10] It is actually known that every G-homogeneous symplectic manifold is Kaehler,[11] a result which does not generalize to noncompact groups.

We give now some hints about the question of the extension of the above results to (a) noncompact semisimple Lie groups and (b) to a wider class of Lie groups. In Ref. 6 we have reviewed the relevant results for the noncompact semisimple case. Obviously we must restrict ourselves to discrete series, in order to maintain orthogonality relations; yet the simple results of the compact case hold only in the "Hermitian symmetric" case, while in general we must consider Hilbert spaces of *harmonic differential forms* rather than holomorphic functions. Details can be found in Ref. 6.

As for point (b), a similar treatment as presented here can be given for nilpotent and, more generally, *solvable* Lie groups, on the basis of the UIR theory developed by Kostant, Auslander, Moore and others.[12] For Lie groups having an intermediate structure (between solvable and semisimple) results are yet incomplete. Let us conclude by remarking that the theory as it stands can already deal with groups such as $SU(n,m)$, $SO(2n,1)$, and $SO(n,2)$, which have attracted physicists' attention as good candidates for a fundamental dynamical group.

ACKNOWLEDGMENTS

The ideas presented in this paper arose during a collaboration with Professor Massimo Pauri. I am indebted to Professor Fiorenzo Duimio for reading the manuscript.

[1]A. M. Peremolov, Commun. Math. Phys. **26**, 222 (1972); see also in this connection the article by A.O. Barut and L. Girardello cited by Peremolov.
[2]I. M. Gel'fand and D. Raikov, Mat. Sb. (N.S.) **13**(55), 301 (1942) [Am. Math. Soc. Transl. **2**(36), 1 (1964)].
[3]J. M. Souriau, *Structure des systèmes dynamiques* (Dunod, Paris, 1970).
[4]E. Onofri and M. Pauri, J. Math. Phys. **13**, 533 (1972).
[5]E. Onofri and M. Pauri, Lett. Nuovo Cimento 2(3), 35 (1972).
[6]E. Onofri, "Quantization theory for homogeneous Kaehler manifolds," Università di Parma, Preprint IFPR-T-038 (1974).
[7]R. Bott, Ann. Math. **66**, 203 (1957); P. Griffiths and W. Schmid, Acta Math. **123**, 253 (1969).
[8]S. Kobayashi and K. Nomizu, *Foundations of Differential Geometry* (Interscience, New York, 1963), Vol. 1.
[9]L. Van Hove, Acad. Roy. Belg. Bull. Classe Sci. Mem. 37, 610 (1951).
[10]M. Pauri and G. M. Prosperi, J. Math. Phys. **9**, 1146 (1968).
[11]J. P. Serre, "Représentation linéaires et espaces homogènes Kaehleriens des groupes de Lie semisimples compacts," Sém. Bourbaki, Exp. 100 (1954).
[12]L. Auslander and C. C. Moore, Amer. Math. Soc. Mem. **62**, (1966); R. F. Streater, Commun. Math. Phys. **4**, 217 (1967); L. Auslander and B. Kostant, Bull. Amer. Math. Soc. **73**, 692 (1967); J. Brezin, Amer. Math. Soc. Mem. **79**, (1968); B. Kostant, in *Lecture Notes in Mathematics*, edited by C. C. Taam (Springer, New York, 1970).

On intelligent spin states

C. Aragone, E. Chalbaud, and S. Salamó

Departamento de Física, Universidad Simón Bolívar, Apartado Postal 80569, Caracas 108, Venezuela
(Received 14 July 1975; revised manuscript received 9 April 1976)

In this paper we give a more compact representation of the intelligent spin states defined by Aragone, Guerri, Salamó, and Tani. Using this new representation, we discuss the differences between minimum uncertainty states, coherent Bloch spin states and intelligent states. The evolution of these states under a particular time dependent Hamiltonian is studied, showing the relevance of the noncompact subgroup K of the Lorentz group. Finally we analyze the radiative properties connected with the intelligent states for a pointlike medium. The main results are: (I) they have a nonvanishing dipole moment (as the Bloch states) and (II) the proper intelligent states give a spontaneous emission intensity which is different from the one provided by the Bloch states.

1. INTRODUCTION

In a recent paper, Aragone, Guerri, Salamó and Tani, constructed the intelligent spin states as those which satisfy the Heisenberg equality for the angular momentum operators. Many questions of physical interest were not discussed there.

The purpose of this work is threefold: (a) to give a clear distinction between intelligent states, minimum uncertainty states, and Bloch states; (b) to show a more compact representation of intelligent states; and (c) to determine the time evolution and some radiative properties of two different systems initially set in an intelligent state.

This article is organized as follows: In the next section we give a more compact expression for the intelligent states than the original, and we discuss the connection between intelligent states and coherent spin states.[2-4] We will show the difference between the $2j+1$ intelligent states and the $2j+1$ states obtained by applying the two-parameter rotation $R(\tau)$, defined by Arecchi, Courtens, Gilmore, and Thomas (ACGT),[4] to the standard Wigner states $|j,m\rangle$.

Section 3 is devoted to analyzing the difference between minimum uncertainty states, atomic coherent spin states, and intelligent states. We calculate the expectation values of J_x, J_y, J_z and their quadratic deviations for intelligent states, using the technique of generating functionals, whose details are presented in Appendix A.

In Sec. 4 we present the explicit evolution of a nonrelativistic high spin system, initially set in an intelligent state, immersed in a magnetic atmosphere.

We also estimate the macroscopic dipole and emission rates of a pointlike laser.

In the last section we make some comments and remarks.

2. COHERENT SPIN STATES AND INTELLIGENT STATES

The SU(2) algebra is defined by the usual commutation relations,

$$[J_i, J_j] = i\epsilon_{ijk}J_k, \quad i,j,k=1,2,3, \tag{1a}$$

or, in terms of the ladder operators $J_\epsilon \equiv J_1 + i\epsilon J_2$ ($\epsilon = +1, -1$) and J_3, by

$$[J_\epsilon, J_{-\epsilon}] = 2\epsilon J_3, \quad [J_3, J_\epsilon] = \epsilon J_\epsilon. \tag{1b}$$

The $(2j+1)$-dimensional Hilbert space spanned by the eigenvectors of J^2 and J_3 (labeled by $|j,m\rangle$ or by $|m\rangle$)

$$J^2|j,m\rangle = j(j+1)|j,m\rangle, \quad J_3|j,m\rangle = m|j,m\rangle, \tag{2}$$

is denoted by H_j.

A useful formula for computation is

$$(j+\epsilon m)!\,|m\rangle = \begin{pmatrix} 2j \\ j+\epsilon m \end{pmatrix}^{-1/2} J_\epsilon^{j+\epsilon m}|-\epsilon j\rangle. \tag{3}$$

The ladder operators are useful in order to construct[4] the atomic coherent spin states or Bloch states $|\tau\rangle$,

$$\begin{aligned} |\tau\rangle &\equiv (1+|\tau|^2)^{-j}\exp(\tau J_+)|-j\rangle \\ &= \exp(\tau J_+)\exp[\ln(1+|\tau|^2)J_3]\exp(-\tau^* J_-|-j\rangle \\ &\equiv R(\tau)|-j\rangle, \end{aligned} \tag{4}$$

where $\tau \equiv \tan\frac{1}{2}\theta\exp(-i\varphi)$, $\theta \in [0, 2\pi)$. $R(\tau)$ represents a rotation through an angle θ about the axis $\hat{a} \equiv \sin\varphi\,\hat{e}_1 - \cos\varphi\,\hat{e}_2$.

Two different Bloch states are not necessarily orthogonal. In fact their inner product is

$$\langle\tau_1|\tau_2\rangle = (1+|\tau_1|^2)^{-j}(1+|\tau_2|^2)^{-j}(1+\tau_1^*\tau_2)^{2j}. \tag{5}$$

The expression of the atomic coherent spin given in Eq. (4) is analogous to that for Glauber states, $|z\rangle = N(z)\exp(za+)|0\rangle$, where the operator $\exp(za^+)$ is applied to the ground state of the harmonic oscillator.[5,6]

The Glauber states satisfy the Heisenberg equality $\Delta x\,\Delta p = \frac{1}{2}$. Therefore, one could also enquire whether the states $|\tau\rangle$ satisfy the Heisenberg equality for the SU(2) algebra,

$$\Delta J_1^2\,\Delta J_2^2 = \tfrac{1}{4}\langle J_3\rangle^2 \tag{6}$$

or, what are all the states $|w\rangle$ which verify Eq. (6)?

For a careful analysis of Eq. (6), let us define two homogeneous functionals of zeroth order, the uncertainty functional $I(\psi)$,

$$I(\psi) \equiv \langle\psi|\Delta J_1^2|\psi\rangle\langle\psi|\Delta J_2^2|\psi\rangle\langle\psi|\psi\rangle^{-2}, \tag{7a}$$

and the half-commutator squared functional $C(\psi)$,

$$C(\psi) \equiv 4^{-1}|\langle\psi|[J_1, J_2]|\psi\rangle|^2\langle\psi|\psi\rangle^{-2}. \tag{7b}$$

In terms of these functionals the Heisenberg equality looks like

$$I(\psi)=C(\psi). \tag{6'}$$

We shall refer to $|u\rangle$ as a minimum (maximum, stationary) uncertainty state if $I(\psi)$ has a local minimum (maximum, stationary point) at $|\psi\rangle=|u\rangle$. Moreover, $|w\rangle$ shall be called an intelligent state if $I(w)=C(w)$.

Therefore, in principle we have three different kind of states related to the angular momentum algebra: the Bloch states $|\tau\rangle$, the intelligent states $|w\rangle$, and the minimum uncertainty states $|u\rangle$.

It is worthwhile to stress that, in the case of the Heisenberg algebra $\{x,p,[x,p]=i\}$, the corresponding functional $C(\psi)=4^{-1}\langle\psi|[x,p]|\psi\rangle^2\cdot\langle\psi|\psi\rangle^{-2}$ has a constant value: $\frac{1}{4}$. Therefore, any intelligent state of this algebra must be a minimum uncertainty state too.

However, this property does not necessarily hold for other algebras where $C(\psi)$ is not a constant number, as in the case of SU(2).

It is a well established property of quantum mechanics[7] that all the intelligent spin states are given by the set of all the states that satisfy the linear equation,

$$J_\alpha|w\rangle\equiv(J_1-i\alpha J_2)|w\rangle=(\langle J_1\rangle_w-i\alpha\langle J_2\rangle_w)|w\rangle\equiv w|w\rangle, \tag{8a}$$

where α is a real number. Defining $\gamma_\epsilon\equiv\frac{1}{2}(1-\epsilon\alpha)$, $\epsilon=\pm1$, J_α can also be written as a linear combination of the ladder operators,

$$J_\alpha=\gamma_+J_++\gamma_-J_-=\gamma_\epsilon J_\epsilon, \tag{8b}$$

leading to the explicit expression of the intelligent spin states shown in Ref. 1. With the present notation they can be written as

$$|w_N(\tau_\alpha)\rangle=\hat{a}_N\sum_{l=0}^{N}\binom{N}{l}(2j-l)!(-2\tau_\alpha J_+)^l|\tau_\alpha\rangle,\quad 0\leq N\leq 2j,$$
$$\tau_\alpha^2=\gamma_+\gamma_-^{-1},\quad w_N\equiv 2\gamma_+\tau_\alpha^{-1}(j-N), \tag{9}$$

where $\hat{a}_N$ is a normalizing factor which shall be determined later on.

We note that for a given τ_α we have $2j+1$ different eigenvalues w_N, as we see from the explicit form of w_N. Therefore, the set $\{|w_N(\tau_\alpha)\rangle\}$ is for a given α, $|\alpha|\neq1$, a basis of H_j.[8]

It is also worthwhile to point out that, due to the fact that α must be real (therefore $\gamma_+\gamma_-^{-1}$ is real too), $\tau_\alpha=\pm(\gamma_+/\gamma_-)^{1/2}$ can only be real or pure imaginary.[9]

However, we could think of enlarging the definition (9) for $|w_N(\tau)\rangle$ to any complex number without giving raise to any mathematical inconsistency. In this case one has to stress that for complex τ not on the real or imaginary axis, $|w_N(\tau)\rangle$ does not represent a solution of the Heisenberg equation anymore. We shall call these states the generalized intelligent states.

There are two special cases of N, the extremes 0 and $2j$. In fact $|w_0(\tau)\rangle=|\tau\rangle$ and (it shall be shown in this section) $|w_{2j}(\tau)\rangle=|-\tau\rangle$. Actually these are the simpler cases of the general law relating intelligent states corresponding to opposite complex numbers,

$$|w_{N_1}(\tau_1)\rangle=|w_{N_2}(\tau_2)\rangle,\quad N_1+N_2=2j,\quad \tau_1+\tau_2=0. \tag{10}$$

This relation is easily seen after having established the value of the inner product $\langle\rho|w_{N_1}(\tau)\rangle=\langle w_0(\rho)|w_{N_1}(\tau)\rangle$ given in Appendix A.[10]

In order to perform calculations of physical interest, it is convenient to have a simpler expression than Eq. (9) to describe the intelligent states. Fortunately this can be done just by ordinary straightforward algebra. It turns out that $|w_N(\tau)\rangle$ can be written as

$$|w_n(\tau)\rangle=a_nY_1\partial_y^n y^{2j}\exp(\tau_yJ_+)|-j\rangle,$$
$$n=0,\ldots,2j, \tag{11}$$

where

$$a_n\equiv\hat{a}_NN!(1+|\tau|^2)^{-j},\quad n,\ Y_1,\ \tau_y \text{ given by}$$
$$n\equiv 2j-N,\quad Y_1f(y)\equiv f(1),$$
$$(\partial_y^n)f(y)\equiv\partial^nf/\partial_y^n,\quad \tau_y\equiv\tau(1-2y^{-1}), \tag{12}$$

and the corresponding eigenvalue w_n is given by $w_n\equiv2\gamma_+\tau^{-1}(j-n)$. Taking into account definition (4) and introducing the auxiliary polynomials $p_j(y,z,|\tau|)$,

$$p_j(y,z,\tau)\equiv(yz+\tau\tau^*(y-2)(z-2))^j, \tag{13}$$

one can write down the intelligent states as

$$|w_n(\tau)\rangle=a_nY_1\partial_y^n\exp(\tau_yJ_+)\exp(-2\ln yJ_3)|-j\rangle$$
$$=a_nY_1\partial_y^np_j(y,y,\tau)|\tau_y\rangle, \tag{14a}$$

where the normalizing factor a_n is shown to be (see Appendix A)

$$a_n=\{Z_1Y_1\partial_y^n\partial_z^np_{2j}(y,z,\tau)\}^{-1/2}\equiv(p_{2j}^{nn})^{-1/2} \tag{14b}$$

and $|\tau_y\rangle$ means the Bloch state corresponding to the complex number $\tau(1-2y^{-1})=\tau_y$.

We note that in the expression given in Eq. (14a) for the intelligent spin states the operator $Y_1\partial_y^n$ occurs. Therefore, one has to know the behavior of $p_j(y,y,\tau)|\tau_y\rangle$ in a neighborhood of $y=1$, in order to obtain the corresponding derivatives.

States having the structure $p_j(y,y,\tau)|\tau_y\rangle=\exp(\tau_yJ_+)\times\exp(-2\ln yJ_3)|-j\rangle$ are not atomic coherent, since the group parameters τ_y, y do not verify the condition for a Bloch-type rotation $R(\tau)$ $(y\neq1+|\tau_y|^2)$.

However, the structure (14a) proves to be very useful in order to deduce many properties of intelligent states from the corresponding properties of the associated Bloch states $|\tau_y\rangle$.

One can also ask if an intelligent state $|w_n(\tau)\rangle$ coincides with some Bloch state $|\mu\rangle$. In order to answer this question, one can prove that[11]

$$|w_n(\tau)\rangle=|\mu\rangle\rightleftarrows n=0,\quad \tau=-\mu \text{ or } n=2j,\quad \tau=\mu, \tag{15}$$

which shows that proper intelligent states ($|w_n(\tau)\rangle$, $n\neq0,2j$) are not Bloch states, but a refinement of them.

Moreover, since for each τ we have $2j+1$ different intelligent states, it is natural to enquire whether they could be obtained through some operation applied to the

Wigner basis $|m\rangle$. In other words: Are the $2j+1$ states $|\tau, m\rangle \equiv R(\tau)|m\rangle$ $(m=-j+1,\ldots,j-1,j)$ intelligent?

Straightforward calculation yields (notice that $\tau = \tan\frac{1}{2}\theta \exp(-il\pi/2)$, l integer)

$$J_\alpha|\tau,m\rangle = -m\sin\theta|\tau,m\rangle + \cos\theta(j+m)^{1/2} \times (j-m+1)^{1/2}|\tau, m-1\rangle. \quad (16)$$

The second term in the right-hand side shows that $|\tau, m\rangle$ is not an eigenvector of J_α, unless $\cos\theta(j+m)=0$.

As in general $|\tau|\neq 1$, the only possibility we are left with is $m=-j$, which means that in the set $\{|\tau m\rangle\}$, only $|\tau,-j\rangle = |\tau\rangle$ is intelligent. In the particular case where $\cos\theta = 0$ $(\theta = \pi/2 + k\pi$, k integer), it is immediate to see that such a situation corresponds to $\alpha = 0$, ∞, i.e., $J_\alpha = J_1$ or J_2, respectively. In that case it is easy to understand why $|\tau = \exp(-il\pi/2), m\rangle$ is an eigenstate of J_1 (or J_2): $R(\tau = \exp(-il\pi/2)$ corresponds to $\pi/2$ rotations about J_2 (or J_1), therefore the states $|\tau\rangle = \exp(-il\pi/2), m\rangle$ are nothing else but the Wigner basis with respect to the x (or y) axis.

3. EXPECTATION VALUES FOR INTELLIGENT STATES AND MINIMUM UNCERTAINTY STATES

In order to define calculations of physical quantities for systems prepared in an intelligent state, one has to develop a suitable technique to handle the corresponding matrix elements. As ACGT have shown for the Bloch states, the technique of the generating functions has been proved to be very useful. In Appendix A we present with some details how the technique due to ACGT is extended to deal with intelligent spin states.

If we define the operators $(\cdot)^{n_1 n_2}$ as

$$f^{n_1 n_2} \equiv Y_1 Z_1 \partial_y^{n_1}\partial_z^{n_2} f(y,z) \equiv \left[\frac{\partial^{n_1}}{\partial_y^{n_1}}\frac{\partial^{n_2}}{\partial_z^{n_2}} f(y,z)\right]_{y=z=1}, \quad (17)$$

one finds (see Appendix A) for the expectation values of J_i for a system in an intelligent state,

$$\langle w_n(\tau)|J_1|w_n(\tau)\rangle \equiv \langle J_1\rangle_{n\tau} = 2j\,\mathrm{Re}\tau[y(z-2)p_{2j-1}(y,z,\tau)]^{nn}(p_{2j}^{nn})^{-1}$$

$$\langle w_n(\tau)|J_2|w_n(\tau)\rangle \equiv \langle J_2\rangle_{n\tau} = -2j\,\mathrm{Im}\tau[y(z-2)p_{2j-1}(y,z,\tau)]^{nn}(p_{2j}^{nn})^{-1} \quad (18)$$

$$\langle w_n(\tau)|J_3|w_n(\tau)\rangle \equiv \langle J_3\rangle_{n\tau} = j[(\tau\tau^*(y-2)(z-2) - zy)p_{2j-1}]^{nn}(p_{2j}^{nn})^{-1}$$

Further on, by taking second-order derivatives of the generating function X_A, defined in Eq. (A8), we evaluate the quadratic deviations $\langle\Delta J_i^2\rangle_{n\tau}$,

$$\langle\Delta J_1^2\rangle_{n\tau} = \tfrac{1}{2}j(2j-1)(\tau^2+\tau^{*2})[y^2(z-2)^2p_{2j-2}]^{nn}(p_{2j}^{nn})^{-1} + j[(y^2z^2 + 2j\tau\tau^* yz(y-2)(z-2))p_{2j-2}]^{nn}(p_{2j}^{nn})^{-1} + \tfrac{1}{2}j[(\tau\tau^*(y-2)(z-2) - zy)p_{2j-1}]^{nn} - 4j^2(\mathrm{Re}\tau)^2\{[y(z-2)p_{2j-1}]^{nn}\}^2(p_{2j}^{nn})^{-2}, \quad (19)$$

$$\langle\Delta J_2^2\rangle_{n\tau} = -\tfrac{1}{2}j(2j-1)(\tau^2+\tau^{*2})[y^2(z-2)^2p_{2j-2}]^{nn}(p_{2j}^{nn})^{-1} + j[(y^2z^2 + 2j\tau\tau^* yz(y-2)(z-2))p_{2j-2}]^{nn}(p_{2j}^{nn})^{-1} + \tfrac{1}{2}j[((y-2)(z-2)\tau\tau^* - zy)P_{2j-1}]^{nn}(p_{2j}^{nn})^{-1} - 4j^2(\mathrm{Im}\tau)^2\{[y(z-2)P_{2j-1}]^{nn}\}^2(p_{2j}^{nn})^{-2},$$

$$\langle\Delta J_3^2\rangle_{n\tau} = -4j^2\{[zyp_{2j-1}]^{nn}\}^2(p_{2j}^{nn})^{-2} + 2j[zyp_{2j-1}]^{nn} \times (p_{2j}^{nn})^{-1} + 2j(2j-1)[y^2z^2p_{2j-2}]^{nn}(p_{2j}^{nn})^{-1}.$$

In a similar way, the mean values of monomials of the type $J_{i_1}^{a_1}J_{i_2}^{a_2}J_{i_3}^{a_3}$ can also be calculated by an appropriate number of derivatives of the generating functions, one of which is $X_A(\alpha\beta\gamma)$, defined in Eq. (A8).

Once we have obtained the values of $\langle\Delta J_{1,2}^2\rangle_{n\tau}$ and $\langle J_3\rangle_{n\tau}$ we are in a position to discuss more precisely what are the differences between minimum uncertainty states $|\mu\rangle$, and intelligent states $|w\rangle$ of the SU(2) algebra. As we know this algebra has commutators which are not numbers, it is a good candidate to find out explicit examples of intelligent states which are not minimum uncertainty states.

Actually, in order to determine all the minimum uncertainty states, one should have to parametrize H_j and thereafter calculate $I(\psi)$ and $C(\psi)$ for this H_j parametrization. Proceeding in that way, one obtains two functions depending upon $4j+1$ independent real parameters and it is a standard task to find both the local minimums of $I(\psi)$ and the subvariety where $I(\psi) = C(\psi)$.

If we restrict ourselves to a subset B of H_j, we can explore what happens on B. Evidently, any intelligent state that belongs to B is an intelligent state in H_j. On the contrary, that $|u_B\rangle$ is a minimum uncertainty state on B does not necessarily imply that $|u_B\rangle$ shall be a minimum uncertainty state on the large variety H_j.

For $B \equiv \{|\tau\rangle$, $\tau = \tan\frac{1}{2}\theta\exp(-i\varphi)\}$, the uncertainty functional $I(\psi)$ has, on B, the value[12]

$$I(\tau) = \tfrac{1}{4}j^2(1-\sin^2\theta\sin^2\varphi)(1-\sin^2\theta\cos^2\varphi), \quad (20)$$

while for $C(\psi)$, we have

$$C(\tau) = 4^{-1}j^2\cos^2\theta. \quad (21)$$

Due to the simplicity of both $I(\tau)$ and $C(\tau)$, it is immediate to solve the Heisenberg equation $I(\tau) = C(\tau)$. That gives

$$j^2\sin^4\theta\sin^2 2\varphi = 0, \quad (22)$$

or equivalently

$\theta = 0$, φ arbitrary, θ arbitrary, $\varphi = n\pi/2$ (n integer).

Because of the degeneracy at the origin in the polar representation (θ, φ) of the complex plane, the solution given in Eq. (22) is exactly the set of the two axes of the complex plane. That corresponds to the fact already mentioned: The only intelligent Bloch states are those contained in the two axes. Of course, as we have shown before, there are intelligent states which are not Bloch states.

In connection with the possible minimum uncertainty states located on B, one has to find the local minimums of $I(\tau)$. $I(\tau)$ has nine stationary points τ_s,

$$\tau_s = \tan(m\pi/4)\exp(-in\pi/4), \quad m = 0, 1, \quad n = 0, 1, \ldots, 6, 7. \quad (23)$$

It is straightforward to verify that $\tau_s = 0$ gives a maximum of $I(\tau)$, and that $\tau_s = \exp(-in\pi/2)$ give the four minimums while the remaining four points $\tau_s = \exp[-i(\pi/4$

$+n\pi/2)]$ give saddle points of $I(\tau)$ in the subset B. That means that only the four points of $B(\tau_s=\exp(-in\pi/2))$ can be minimum uncertainty states on H_j.

Nevertheless we have a lot of intelligent states defined on B (τ any real or pure imaginary number) which shall proceed to be intelligent states when we enlarge the calculations to the whole H_j.

4. DYNAMICAL PROPERTIES OF THE INTELLIGENT STATES

The first situation that we want to consider is the time evolution of a nonrelativistic spin j system (of magnetic moment γ), in a magnetic environment $\mathbf{B}(t)$ of the type considered by Gilmore[13]:

$$\mathbf{B}(t)\equiv 2B_\perp(\cos 2\omega_1 t\hat{x}+\sin 2\omega_1 t\hat{y})+2B_\parallel\hat{z}, \tag{24}$$

where $2B_\parallel\hat{z}$ is a constant magnetic field along a fixed direction and B is the strength of a perpendicular field of proper frequency $2\omega_1$.

The corresponding time-dependent Hamiltonian is

$$H(t)=-\hbar\gamma\mathbf{J}\cdot\mathbf{B}(t)=-\hbar\gamma(B_\perp\exp(-2i\omega_1 t)J_+ + B_\perp\exp(2i\omega_1 t)J_- + 2B_\parallel J_3), \tag{25}$$

with $\mathbf{J}$ represented in the $(2j+1)$-dimensional space H_j. By going to the two-dimensional representation of SU(2), Gilmore has evaluated the time evolution operator $U(t)$ which satisfies the Schrödinger equation $i\hbar\dot{U}=HU$,

$$U(t)=\begin{Bmatrix}\cos^2\psi\exp(i\omega_- t)+\sin^2\psi\exp(-i\omega_+ t) & i\sin 2\psi\sin\omega_2 t\exp(-i\omega_1 t)\\ i\sin 2\psi\sin\omega_2 t\exp(i\omega_1 t) & \cos^2\psi\exp(-i\omega_- t)+\sin^2\psi\exp(i\omega_+ t)\end{Bmatrix}, \tag{26a}$$

where ω_+, ω_-, and ψ are given by

$$\omega_\pm\equiv\omega_2\pm\omega_1,\quad \omega_2\equiv[\gamma^2B_\perp^2+(\gamma B_\parallel+\omega_1)^2]^{1/2},$$
$$\sin 2\psi\equiv\gamma B_\perp\omega_2^{-1},\quad \cos 2\psi\equiv(\gamma B_\parallel+\omega_1)\omega_2^{-1}. \tag{26}$$

Let us assume that our system has been initially prepared in an intelligent state $|w_n(\tau)\rangle$. Therefore, in any other subsequent instant t, the system shall be in a certain state $|w_n(t,\tau)\rangle$ determined by the evaluation operator $U(t)$; namely, $|w_n(t,\tau)\rangle=U(t)|w_n(\tau)\rangle$. We want to investigate whether $|w_n(t,\tau)\rangle$ is an intelligent state or, at least, how close to an intelligent state it is while it evolves. We know, after ACGT, that a Bloch state remains a Bloch state along its evolutions under the Hamiltonian (25).

Moreover, as both $|w_0(\tau)\rangle$ and $|w_{2j}(\tau)\rangle$ are Bloch states, it might happen that any proper intelligent state could evolve remaining in the subset of the intelligent states too.

In order to give an answer to this question, let us briefly mention some useful facts concerning SU(2) and $|w_n(\tau)\rangle$, as has been given in Eq. (14a).

The first property we want to point out concerns the structure of $|w_n(\tau)\rangle$ itself; $|w_n(\tau)\rangle$ can be written

$$|w_n(\tau)\rangle=a_nY_1\partial_y^n k(y,\tau)|-j\rangle,$$
$$k(y,\tau)\equiv\exp(\tau_y J_+)\exp(-J_3\ln y^2), \tag{27}$$

where $k(y,\tau)$ belongs to SL(2, C),[1] the analytic continuation of SU(2).[14] In the two-dimensional representation of SL(2, C), $k(y,\tau)$ has the form

$$k(y,\tau)=\exp(\tau_y J_+)\exp(-(\ln y^2)J_3)$$
$$=\begin{Bmatrix}1&\tau_y\\ \cdot&1\end{Bmatrix}\begin{Bmatrix}y^{-1}&\cdot\\ \cdot&y\end{Bmatrix}=\begin{Bmatrix}y^{-1}&\tau(y-2)\\ \cdot&y\end{Bmatrix}, \tag{28}$$

showing that it belongs to the well-known four parameter subgroup K of SL(2, C),[15] as reviewed in Appendix B. We prove in this appendix that for $y\neq 1$, $k(y,\tau)$ contains a Lorentz boost and, therefore, $k(y,\tau)$ does not represent a proper rotation.

The operator $U(t)k(y,\tau)\equiv\hat{l}(t,y,\tau)$ has also been explicitly evaluated in Appendix B, Eq. (B8). This allows us to write the state $|w_n(t,\tau)\rangle$ as follows:

$$|w_n(t,\tau)\rangle=a_n(\tau)Y_1\partial_y^n\hat{l}_4^{2j}\exp(\hat{l}_2\hat{l}_4^{-1}J_+)|-j\rangle, \tag{29a}$$

where

$$\hat{l}_2\equiv[\tau(y-2)\cos^2\psi+y\sin\psi\cos\psi]\exp(i\omega_- t)+[\tau(y-2)\sin^2\psi-y\sin\psi\cos\psi]\exp(-i\omega_+ t),$$
$$\hat{l}_4\equiv[\tau(y-2)\sin\psi\cos\psi+y\sin^2\psi]\exp(i\omega_+ t)+[y\cos^2\psi-\tau(y-2)\sin\psi\cos\psi]\exp(-i\omega_- t).$$

Although the structure of the state $|w_n(t,\tau)\rangle$ seems complicated, it is proved in Appendix B that this state becomes, up to a phase factor, an intelligent state if the transverse magnetic field vanishes, i.e., $B_\perp=0$. Only in this case the evolution of an intelligent state of order n determined by the complex number τ is a generalized intelligent state, of the same order n, corresponding to the complex t-dependent number $\tau'=\tau\exp(2i\gamma B_\parallel t)$. If $n=0$ we recover the result of ACGT: $|w_0(t,\tau)\rangle=\exp(2i\times\arg\hat{l}_4)|\hat{l}_2\hat{l}_4^{-1}\rangle$. That is, the evolution of a Bloch state keeps being a Bloch state, up to a phase factor.

The second situation we want to treat here is the relevance of the intelligent states in connection with the pointlike laser,[16] either with a semiclassical or a fully quantized representation of the laser field.

By a semiclassical pointlike laser we mean a collection of identical atoms, each with two effective energy levels (with $\hbar\omega$ the energy gap) interacting with a classical field $\mathbf{E}(t)=2\mathrm{Re}\{\mathbf{E}_0\exp(i\omega t)\}$, which has the resonant mode of frequency ω.

The Hamiltonian corresponding to this system is, following ACGT,

$$H=H_A+H_{AF}\equiv\hbar\omega J_3-(\mathbf{p}\cdot\mathbf{E}_0^*)J_+\exp(-i\omega t)-(\mathbf{p}^*\cdot\mathbf{E}_0)J_-\exp(i\omega t), \tag{30}$$

where the vector $\mathbf{p}$ is the complex dipole moment associated to each atom giving rise to the total dipole moment

$$\mathbf{D} \equiv \mathbf{p}J_+ + \mathbf{p}^* J_- . \tag{31}$$

For a system of N_0 atoms, the cooperation number j must satisfy the inequality

$$j \leq \tfrac{1}{2} N_0 . \tag{32}$$

We are assuming either that $\mathbf{p}$ verifies $\mathbf{p} \cdot \mathbf{E}_0 = 0 = \mathbf{p}^* \cdot \mathbf{E}_0^*$ or, if this selection rule does not apply, that we are working in the rotating wave approximation.

If one neglects the interaction term H_{AF} between matter and the electromagnetic field, namely $H_{AF} = 0$ in Eq. (30), it is possible to give an estimate of the expectation value of $\mathbf{D}$. For the system initially in an intelligent state $|w_n(\tau)\rangle$, the state $|w_n(t,\tau)\rangle$ becomes $|w_n(t,\tau)\rangle = \exp(-ij\omega t)\,|w_n(\tau \exp(-ij\omega t))\rangle$. Therefore,

$$\begin{aligned} \langle w_n(t,\tau)\,|\mathbf{D}|\,w_n(t,\tau)\rangle &= \mathbf{p}\langle J_+\rangle_{n\tau}(t) + \mathbf{p}^* \langle J_-\rangle_{n\tau}(t) \\ &= 2j(\mathbf{p}\tau^* \exp(i\omega t) + \mathbf{p}^* \tau \exp(-i\omega t)) \\ &\quad \times [y(z-2)p_{2j-1}]^{nn} (p_{2j}^{nn})^{-1} . \end{aligned} \tag{33}$$

This result is a refinement of the corresponding one for the Bloch state, which is reobtained here by taking $n=0$. (It is worthwhile to remind the reader that for the Wigner–Dicke states the expectation value of $\mathbf{D}$ vanishes.)

As the macroscopic dipole of the system does not vanish, there exists a nonvanishing classical radiation intensity I_c generated by this oscillating dipole, which in the wave zone is

$$I_c = I_0 \cdot 4j^2 \tau\tau^* \{[y(z-2)p_{2j-1}]^{nn}\}^2 (p_{2j}^{nn})^{-2} . \tag{34}$$

Introducing the fully quantized Hamiltonian

$$H = H_A + H_F + H_{AF} \equiv \hbar\omega J_3 + \hbar\omega a^+ a + \gamma a J_+ + \gamma a^+ J_- , \tag{35}$$

we can calculate the emission rate for the pointlike laser.[17]

The spontaneous emission intensity can be calculated for an initial intelligent state $|w_n(\tau)\rangle$, in a way similar to what ACGT did for this model,

$$\begin{aligned} I_n^{sp} &= I_0 \sum_{m=-j}^{j} |\langle m\,|J_-|\,w_n(\tau)\rangle|^2 \\ &= I_0 \langle w_n(\tau)\,|J_+ J_-|\,w_n(\tau)\rangle \\ &= I_0 \langle J_- J_+\rangle_{n\tau} + 2 I_0 \langle J_3\rangle_{n\tau} . \end{aligned} \tag{36a}$$

The matrix elements occurring in this relation are easily evaluated by means of the generating function $X_A(\alpha, \beta, \gamma)$, given by Eq. (A9),

$$\begin{aligned} I_{n\tau}^{sp} = I_0 2j\tau\tau^* \{ &(2j-1)[yz(y-2)(z-2)p_{2j-2}]^{nn} \\ &\times (p_{2j}^{nn})^{-1} + [(y-2)(z-2)p_{2j-1}]^{nn} (p_{2j}^{nn})^{-1} \}, \end{aligned} \tag{36b}$$

an expression which reduces for $n=0$ to the results found by ACGT for Bloch states.

In the case of a Dicke–Wigner initial state $|m\rangle$, the spontaneous emission intensity is $I_D^{sp} = I_0 (j+m)(j-m+1)$. In order to compare the spontaneous emission intensities between intelligent states and Dicke–Wigner states we have to evaluate I_D^{sp} for a Dicke–Wigner state having the same energy expectation value that $|w_n(\tau)\rangle$.[18] Therefore, introducing $m = \langle J_3\rangle_{n\tau}$ in I_D^{sp}, we get

$$\begin{aligned} [I_D^{sp}]_{m=\langle J_3\rangle_{n\tau}} &= I_0 \cdot 2j[1 - (yzp_{2j-1})^{nn} (p_{2j}^{nn})^{-1}] \\ &\quad \times [1 + 2j(yzp_{2j-1})^{nn} (p_{2j}^{nn})^{-1}] \neq I_{n\tau}^{sp} . \end{aligned} \tag{37}$$

A similar calculation for the stimulated intensity I^{st} leads to:

$$\begin{aligned} I_{n\tau}^{st} &= I_0 \sum_m \{ |\langle m\,|J_-|\,w_n(\tau)\rangle|^2 - |\langle m\,|J_+|\,w_n(\tau)\rangle|^2 \} \\ &= 2\langle J_3\rangle_{n\tau} \cdot I_0 . \end{aligned} \tag{38}$$

Consequently, using the value given in Eq. (18) of $\langle J_3\rangle_{n\tau}$ we have

$$I_{n\tau}^{st} = I_0 \cdot 2j[1 - 2(yzp_{2j-1})^{nn} (p_{2j}^{nn})^{-1}] = I_D^{st} , \tag{39}$$

which is identical to the stimulated intensity emitted for an initial Dicke state with quantum number $m = \langle J_3\rangle_{n\tau}$.

Just for completeness, one can explicitly calculate $\langle J_3\rangle_{0\tau}$, $\langle J_3\rangle_{1\tau}$, and $\langle J_3\rangle_{2\tau}$. It happens that, for $j \geq 3$, the three values decrease for $0 \leq \theta < \pi/2$, and increase for $\pi/2 < \theta < \pi$,

$$\langle J_3\rangle_{2\tau} < \langle J_3\rangle_{1\tau} < \langle J_3\rangle_{0\tau} \quad (j \geq 3), \tag{40a}$$

$$\langle J_3\rangle_{0\tau} = -2j\cos\theta, \quad \langle J_3\rangle_{1\tau} = -2j\cos\theta \left[1 + \frac{(2-j^{-1})\sin^2\theta}{2j\cos^2\theta + \sin^2\theta} \right], \tag{40b}$$

$$\langle J_3\rangle_{2\tau} = -2\cos\theta \, \frac{[(j-1)(j-2)(2j-3)\cos^4\theta + 2(j-1)(4j-5)\cos^2\theta + (5j-4)]}{[(j-1)(2j-3)\cos^4\theta + 4(j-1)\cos^2\theta + 1]} \tag{40c}$$

However, as we have not been able to proceed a step further we are not allowed to claim a general property from Eqs. (40). The only statement we are making is that the stimulated emission intensity (and also the energy expectation value) of the proper intelligent states ($n = 1, 2$) is greater than the stimulated emission intensity arising from the Bloch state corresponding to the same value of the parameter τ.

The last point we want to mention concerning the different behavior of intelligent states in comparison with Bloch states is the following: Suppose we have initially prepared a system of spin j in an intelligent state $|w_n(\tau)\rangle$ and we want to know what is the probability that, under the magnetic Hamiltonian (25), the system could be found in $t > 0$ in a Wigner state $|m\rangle$. Making use of the results of Appendices A and B, we obtain the transition probabilities

$$p_{(n,\tau)\to|m\rangle} = |\langle m|w_n(t\tau)\rangle|^2 = (n!)^2 a_n^2(\tau)\binom{2j}{j+m}$$

$$\times\left\{\sum_{(l_1,l_2)=(0,0)}^{(n,n)} a_2^{l_1} a_2^{*l_2} a_4^{n-l_1} a_4^{*n-l_2}\right.$$

$$\times c_2^{j+m-l_1} c_2^{*j+m-n+l_2} c_4^{j-m-n+l_1} c_4^{*j-m-n+l_2}$$

$$\times\binom{j+m}{l_1}\binom{j+m}{l_2}\binom{j-m}{n-l_1}\binom{j-m}{n-l_2}, \quad (41a)$$

where a_i and c_i, $i=2,4$ are

$$a_2 \equiv \exp(-i\omega_1 t)[\tau\cos\omega_2 t + i\sin\omega_2 t(\tau\cos 2\psi + \sin 2\psi)],$$

$$a_4 \equiv \exp(i\omega_1 t)[\cos\omega_2 t + i\sin\omega_2 t(\tau\sin 2\psi - \cos 2\psi)], \quad (41b)$$

$$c_2 \equiv \exp(-i\omega_1 t)[(\tau-2)\cos\omega_2 t + i\sin\omega_2 t((\tau-2)\cos 2\psi + \sin 2\psi)],$$

$$c_4 \equiv \exp(i\omega_1 t)[\cos\omega_2 t - i\sin\omega_2 t(\tau\sin 2\psi + \cos 2\psi)].$$

In order to see how a pure intelligent state behaves, one can take a particular case of Eqs. (41). For instance, let us choose $|m\rangle = |-j\rangle$. Making use of the above result, it turns out that ($\tau = \tan\frac{1}{2}\theta\exp(+in\pi/2)$)

$$\Gamma \equiv \frac{p_{(1,\tau)\to|-j\rangle}}{p_{(0,\tau)\to|-j\rangle}} = (2j\cos^2\theta + \sin^2\theta)^{-1}$$

$$\times\frac{1+\sin^2\omega_2 t\cdot[\sin^2 2\psi(\tau^2-1) - \tau\sin 4\psi]}{1+\sin^2\omega_2 t\cdot[\sin^2 2\psi(\tau^2-1) + \tau\sin 4\psi]}. \quad (42)$$

This ratio Γ is finite for any ψ, τ, and t unless τ takes the value $\tau'_\psi = -\cot\text{an}2\psi$. In that case, Eq. (42) becomes

$$\Gamma \equiv \frac{p_{(1,\tau'_\psi)\to|-j\rangle}}{p_{(0,\tau'_\psi)\to|-j\rangle}} = [1 + (2j-1)\cos^2 4\psi]^{-1}$$

$$\times(1 + \sin 2\psi^{-1}\sin 6\psi\sin^2\omega_2 t)\cos^{-2}\omega_2 t, \quad (43)$$

showing that, for $t_n = (n+\frac{1}{2})\pi/\omega_2$ the value of Γ is infinite. Consequently we see that the behavior of the proper intelligent state $|w_1(t,\tau'_\psi)\rangle$ is qualitatively different from the behavior of the Bloch state $|w_0(t,\tau'_\psi)\rangle$.

Further, as for τ'_ψ, the function $c_4(t)$ appearing in Eq. (41) has the value

$$c_4(\tau'_\psi) = \exp(i\omega_1 t)\cos\omega_2 t. \quad (44)$$

It is clear that for instants $t_n = (2n+1)\pi/2\omega_2$ and for numbers n, m (which have to verify $n+m \leq j-1$)[19] the transition probability $p_{(n,\tau'_\psi)\to|m\rangle}$ vanishes with period $T_2 - \pi/\omega_2$.

Looking at the structure of the probability $p_{(n,\tau)\to|m\rangle}$, one gets two other special values of τ,

$$\tau''_\psi = 2 - \tan 2\psi, \quad \tau'''_\psi = 2. \quad (45)$$

These values cause the periodic vanishing of $p_{(n,\tau)\to|m\rangle}$ too, now because $c_2(t)$ vanishes with the same period as above, for each instant $t'_n = n\pi/\omega_2$ and for quantum numbers n, m such that $n+1 \leq j+m$.

5. DISCUSSION AND COMMENTS

We have been able to establish a clear distinction between intelligent spin states, minimum uncertainty states, and Bloch states. We have shown that the generalized intelligen states constitute a refinement of the Bloch states containing them as extreme cases.

We also pointed out in Eq. (10) the symmetry in the definition of intelligent states which allow us to restrict the analysis of $|w_n(\tau)\rangle$ to any half-plane containing the origin of the whole complex plane.

Thereafter we evaluated, through the technique of the generating functions, the expectation values of both the components of the angular momentum vector and of their mean square deviations. They turned out to be rational functions of $\tau\tau^* = \tan^2\frac{1}{2}\theta$.

Moreover, by making use of some algebraic properties of the noncompact subgroup K of SL(2, C) we studied some dynamical properties of the intelligent states valid both for a reasonable time dependent model of a spin-j particle in a magnetic atmosphere and for a pointlike laser.

One important result found is that for a permanent magnetic field $\mathbf{B} = 2B_\parallel\hat{e}_3$, proper intelligent states evolve continuously in the set of generalized intelligent states. Of course, the two extreme states ($n=0, 2j$) which are Bloch and intelligent evolve in the assembly of the complex Bloch states.

The transition probabilities, for a system prepared in an intelligent state, of becoming in time t a Wigner–Dicke state, have been computed. It turned out that there exist three values of the real parameter τ defining an intelligent state for which $p_{(n,\tau)\to|m\rangle}$ vanishes periodically.

In the case of the pointlike laser, the spontaneous and stimulated emission intensities and the macroscopic dipole of the system have also been evaluated showing again a refinement of the results obtained using Bloch states.

We have also proved that, in general, an intelligent state is not a minimum uncertainty state and we pointed out where the noncoincidence of both kind of states stems.

It is also worthwhile to note that, contrary to what has recently been asserted by Kolodziejczyk,[20] the coherent states defined by Mikhailov[21] cannot be used to explain the relationship between coherent and intelligent states, essentially because the only Mikhailov coherent state which is intelligent is, trivially, the ground state.

Finally, let us remark that Vetri's comment[22] that Radcliffe states which do not point in the z direction and are labeled "intelligent" in Ref. 1 are actually those oriented in such a way that the $\hat{n}$ axis is along $\hat{x}$ or $\hat{y}$ is precisely what Aragone, Guerri, Salamó, and Tani meant when they said that "only those Radcliffe states located on the real line or the imaginary axis are intelligent states."

APPENDIX A

In this Appendix we are going to show the details concerning some of the calculations whose results have been used in the text.

Let us recall that the states we are dealing with have been written in the form [Eqs. (13)].

$$|w_n\rangle = a_{n_1}\partial_y^{n_1}\{p_j(y,y,|\tau|)|\tau_y\rangle\}_{y=1}, \tag{A1}$$

where

$$p_j(y,z,|\tau|) \equiv [yz + |\tau|^2(y-2)(z-2)]^j. \tag{A2}$$

Suppose we are interested in computing the value of $\langle w_{n_2}|w_{n_1}\rangle$, where $|w_{n_1}\rangle$ remains as in (A1). $|w_{n_2}\rangle$ may be written

$$|w_{n_2}\rangle = a_{n_2}\partial_z^{n_2}\{p_j(z,z,|\tau|)|\tau_z\rangle\}_{z=1}, \tag{A3}$$

where instead of y we use a different variable z, in order to avoid confusion. Making the scalar product we have (p_j is real for y, z real numbers)

$$\langle w_{n_2}|w_{n_1}\rangle = a_{n_2}^* a_{n_1}\partial_z^{n_2}\partial_y^{n_1}\{p_j(y,y,\tau) \times p_j(z,z,\tau)\langle\tau_z|\tau_y\rangle\}_{z=1} \tag{A4}$$

but, by virtue of Eq. (5),

$$\begin{aligned}\langle\tau_z|\tau_y\rangle &= (1+|\tau_y|^2)^{-j}(1+|\tau_z|^2)^{-j}(1+\tau_z^*\tau_y)^{2j}\\ &\equiv y^{2j}p_j(y,y,|\tau|)^{-1}z^{2j}p_j(z,z,|\tau|)^{-1}\\ &\quad\times[1+|\tau|^2(1-2/y)(1-2/z)]^{2j}\\ &= p_j(y,y,|\tau|)^{-1}p_j(z,z,|\tau|)^{-1}p_{2j}(y,z,|\tau|).\end{aligned} \tag{A5}$$

Introducing this value of $\langle\tau_{z_2}|\tau_y\rangle$ into Eq. (A4) we get the final value of $\langle w_{n_2}|w_{n_1}\rangle$,

$$\begin{aligned}\langle w_{n_2}|w_{n_1}\rangle &= a_{n_2}^* a_{n_1}\{\partial_z^{n_2}\partial_y^{n_1}p_{2j}(y,z,|\tau|)\}_{y=1=z}\\ &= a_{n_2}^* a_{n_1}p_{2j}^{n_1 n_2}.\end{aligned} \tag{A6}$$

If we take here $n_2 = n_1$ and impose that the result found must be 1, we get the modulus of the normalizing factor a_n, as was mentioned in Eq. (14). Once we get the value of the a_n, the scalar product (A6) is completely defined,

$$\langle w_{n_2}|w_{n_1}\rangle = \frac{p_{2j}^{n_1 n_2}}{(p_{2j}^{n_1 n_1})^{1/2}(p_{2j}^{n_2 n_2})^{1/2}}. \tag{A7}$$

In order to calculate expected values of observables contained in the SO(3) algebra, it is of crucial importance to evaluate the generator function $X_A(\alpha,\beta,\gamma)$, defined in the ACGT paper as

$$X_A(\alpha,\beta,\gamma) \equiv \langle w_n|\exp(\gamma J_-)\exp(\beta J_3)\exp(\alpha J_+)|w_n\rangle. \tag{A8}$$

Introducing the form (A3) of $|w_n\rangle$ and applying the Baker–Campbell–Haussdorff formula we have that

$$\begin{aligned}X_A(\alpha,\beta,\gamma) = |a_n|^2\partial_z^n\partial_y^n\{&zy\exp(-\beta/2)\\ &+[\tau(y-2)+\alpha y][\tau^*(z-2)+\gamma z]\exp(\beta/2\}^{2j}\end{aligned} \tag{A9}$$

which, if we define the auxiliary function q in y, z, α, β, γ, by

$$\begin{aligned}q_{2j}(\alpha,\beta,\gamma,y,z,\tau) \equiv \{&zy\exp(-\beta/2)+\exp(\beta/2)\\ &\times[\tau(y-2)+\alpha y][\tau^*(z-2)+\gamma z]\}^{2j},\end{aligned} \tag{A10}$$

can be rewritten in the shorter form

$$X_A(\alpha\beta\gamma) = |a_n|^2 q_{2j}^{nn}(\alpha,\beta,\gamma). \tag{A11}$$

Once we have evaluated X_A, it is very simple to estimate the expected values of, for instance, J_1, J_2, J_3, and $(\Delta J_1)^2$, $(\Delta J_2)^2$ for the intelligent states $|w_n\rangle$.

In fact,

$$\begin{aligned}\langle w_n|J_1|w_n\rangle &= \tfrac{1}{2}\langle w_n|J_+|w_n\rangle + \tfrac{1}{2}\langle w_n|J_-|w_n\rangle\\ &= \tfrac{1}{2}(\partial_\alpha X_A)_{\alpha=\beta=\gamma=0} + \tfrac{1}{2}(\partial_\gamma X_A)_{\alpha=\beta=\gamma=0},\end{aligned} \tag{A12}$$

$$\langle w_n|J_3|w_n\rangle = (\partial_\beta X_A)_{\alpha=\beta=\gamma=0}, \tag{A13}$$

and

$$4(\Delta J_1)^2 = \langle J_+^2\rangle + \langle J_-^2\rangle + 2\langle J_- J_+\rangle + 2\langle J_3\rangle - 4\langle J_1\rangle^2.$$

Consequently,

$$\begin{aligned}4(\Delta J_1)^2 = (&\partial_{\alpha\alpha}^2 X_A)_{\alpha=\beta=\gamma=0} + (\partial_{\gamma\gamma}^2 X_A)_{\alpha=\beta=\gamma=0}\\ &+ 2(\partial_{\alpha\gamma}^2 X_A)_{\alpha=\beta=\gamma=0} + 2(\partial_\beta X_A)_{\alpha=\beta=\gamma=0}\\ &- [(\partial_\alpha X_A)_{\alpha=\beta=\gamma=0} + (\partial_\gamma X_A)_{\alpha=\beta=\gamma=0}]^2,\end{aligned} \tag{A14}$$

and in the same way the value of $4(\Delta J_2)^2$ can be given,

$$\begin{aligned}4(\Delta J_2)^2 = &-(\partial_{\alpha\alpha}^2 X_A)_0 - (\partial_{\gamma\gamma}^2 X_A)_0 + 2(\partial_{\alpha\gamma}^2 X_A)_0\\ &+ 2(\partial_\beta X_\alpha)_0 + [(\partial_\alpha X_A)_0 - (\partial_\gamma X_A)_0]^2.\end{aligned} \tag{A15}$$

It is interesting to observe that $q_{2j}(0,0,0) = p_{2j}(y,z,|\tau|)$.

APPENDIX B

In this Appendix we shall give some group results concerning SL(2, C) and its subgroup K.

The four-parameter subgroup K has been extensively used in connection with the irreducible representations of the Lorentz group (see for instance Ref. 15). K is defined as the set of all the elements k of SL(2, C) of the form

$$K = k(p,q) = \begin{pmatrix} p^{-1} & q \\ 0 & p\end{pmatrix}, \quad p, q \text{ complex numbers.} \tag{B1}$$

The importance of K lies in the fact that any element l of SL(2, C) can uniquely be decomposed in the form

$$l = kz, \quad k = \begin{pmatrix} p^{-1} & q \\ 0 & p\end{pmatrix}, \quad z = \begin{pmatrix} 1 & \cdot \\ z & 1\end{pmatrix}. \tag{B2}$$

Moreover, as any $k(pq)$ can uniquely be factorized in the form

$$\begin{aligned}k &= \begin{pmatrix} 1 & qp^{-1} \\ \cdot & 1\end{pmatrix}\begin{pmatrix} p^{-1} & \cdot \\ \cdot & p\end{pmatrix} = \begin{pmatrix} p^{-1} & q \\ \cdot & p\end{pmatrix}\\ &= \exp(qp^{-1}J_+)\exp(-2\ln p J_3),\end{aligned} \tag{B3}$$

l can be uniquely decomposed as a product of three exponentials,

$$l = \exp(qp^{-1}J_+)\exp(-2\ln p\cdot J_3)\exp(zJ_-).$$

Let an arbitrary $l \in$ SL(2, C) be given,

$$l = \begin{pmatrix} l_1 & l_2 \\ l_3 & l_4\end{pmatrix}. \tag{B4}$$

It is easy to check that

$$l = \exp(l_2 l_4^{-1}J_+)\exp(-2\ln l_4 J_3)\exp(l_3 l_4^{-1}J_-). \tag{B5}$$

The elements $k(y, \tau)$ defined in Eq. (28) have the structure (B1), therefore the convenience of dealing with K (even if the restriction one could make of keeping p real could suggest that the three-dimensional subgroup $K' \equiv \{k \in K : p \text{ real}\}$ should play some specific role, more centrally than K itself).

Just for the sake of completeness it is possible to write down the four-dimensional Lorentz transformation $\Lambda(k)$ represented by $k(y, \tau)$. Following Gel'fand, Graev, and Vilenkin[23] it is straightforward to prove that $\Lambda(k) = \Lambda_1\Lambda_2$, where Λ_1 is the standard Lorentz boost ($|y| \neq 1$) and Λ_2 is a distortion of the $\{x^2, x^-\}$ two-dimensional plane (or of the $\{x^1, x^-\}$ two-plane accordingly to whether τ is a real or an imaginary number, respectively). The distortion Λ_2 turns out to be

$$
\begin{aligned}
(\Lambda_2 x)^+ &= x^+,\\
(\Lambda_2 x)^- &= x^- + \tau_y\tau_y^* x^+ + 2^{1/2}\mathrm{Re}\,\tau_y x^2 - 2^{1/2}\mathrm{Im}\,\tau_y x^1, \qquad \text{(B6)}\\
(\Lambda_2 x)^1 &= x^1 - 2^{1/2}\mathrm{Im}\,\tau_y x^+, \quad (\Lambda_2 x)^2 = x^2 + 2^{1/2}\mathrm{Re}\,\tau_y x^+,
\end{aligned}
$$

while the boost Λ_1 applied to $\hat{x} \equiv \Lambda_2 x$ gives

$$
\begin{aligned}
(\Lambda_1\hat{x})^+ &= y^{-2}\hat{x}^+, \quad (\Lambda_1\hat{x})^- = y^2\hat{x}^-,\\
(\Lambda_1\hat{x})_1 &= \hat{x}_1, \quad (\Lambda_1\hat{x})_2 = \hat{x}_2, \qquad \text{(B7)}
\end{aligned}
$$

where we denoted by $x^{\pm} \equiv 2^{-1/2}(x^0 \mp x^3)$ the usual two null coordinates.

We are interested in the decomposition (B5) for the operator $U(t)k(y, \tau)$ in order to have $|w_n(t, \tau)\rangle$ written in a way resembling an intelligent state. Calculating the matrix product, we get

$$
\hat{l} \equiv U(t)k(y,\tau) \equiv \begin{pmatrix} \hat{l}_1 & \hat{l}_2 \\ \hat{l}_3 & \hat{l}_4 \end{pmatrix} \equiv \begin{cases} y^{-1}\cos^2\psi \exp(i\omega_- t) + y^{-1}\sin^2\psi \exp(-i\omega_+ t), & [\tau(y-2)\cos^2\psi + y\sin\psi\cos\psi]\exp(i\omega_- t) \\ & \quad + [\tau(y-2)\cos^2\psi - y\sin\psi\cos\psi]\exp(i\omega_+ t), \\ y^{-1}\sin\psi\cos\psi[\exp(i\omega_+ t) - \exp(-i\omega_- t)], & [y\sin^2\psi + \tau(y-2)\sin\psi\cos\psi]\exp(i\omega_+ t) \\ & \quad + [y\cos^2\psi - \tau(y-2)\sin\psi\cos\psi]\exp(-i\omega_- t). \end{cases} \qquad \text{(B8)}
$$

With this result, one obtains for $|w_n(t, \tau)\rangle = U(t)|w_n(\tau)\rangle$,

$$
\begin{aligned}
|w_n(t,\tau)\rangle &= a_n(\tau)Y_1\partial_y^n\hat{l}(y,t,\tau)|-j\rangle\\
&= a_n Y_1\partial_y^n \hat{l}_4^{2j}\exp(\hat{l}_2\hat{l}_4^{-1}J_+|-j\rangle, \qquad \text{(B9)}
\end{aligned}
$$

or what is the same,

$$
\begin{aligned}
|w_n(t,\tau)\rangle &= a_n(\tau)Y_1\partial_y^n\{\exp(2ij\arg\hat{l}_4)\\
&\quad \times(|\hat{l}_2|^2 + |\hat{l}_4|^2)^j|\hat{l}_2\hat{l}_4^{-1}\rangle\}, \qquad \text{(B10)}
\end{aligned}
$$

in terms of the Bloch state $|\hat{\tau}\rangle = |\hat{l}_2\hat{l}_4^{-1}\rangle$. In the case where $n=0$ (and consequently the term has been prepared in a Bloch state), we have for $|w_n(t, \tau)\rangle$,

$$
|w_0(t,\tau)\rangle = \exp(2ijY_1\arg\hat{l}_4)\cdot Y_1|\hat{l}_2\hat{l}_4^{-1}\rangle, \qquad \text{(B11)}
$$

a state which differs by a phase factor $2jY_1\arg l_4$ from the standard Bloch state corresponding to the complex number $\tau(t) = Y_1(\hat{l}_2\hat{l}_4^{-1})$.

The explicit expression shown in Eq. (B10) for $|w_n(t, \tau)\rangle$ allows an easy calculation of the transition number $\langle\mu|w_n(t, \tau)\rangle$ for an arbitrary coherent spin state $|\mu\rangle$,

$$
\begin{aligned}
\langle\mu|w_n(t,\tau)\rangle &= a_n(\tau)a_0(\mu)Y_1\partial_y^n\{(\hat{l}_4 + \mu^*\hat{l}_2)^{2j}\}\\
&= a_n(\tau)a_0(\mu)(!n)\binom{2j}{n}[\sin\psi(\sin\psi + \tau\cos\psi)\\
&\quad \times\exp(i\omega_+ t) + \cos\psi(\cos\psi - \tau\sin\psi)\exp(-i\omega_- t)\\
&\quad + \mu^*\sin\psi(\sin\psi\tau - \cos\psi)\exp(-i\omega_+ t)\\
&\quad + \mu^*\cos\psi(\tau\cos\psi + \sin\psi)\exp(i\omega_- t)]^n\\
&\quad \times[\sin\psi(\sin\psi - \tau\cos\psi)\exp(i\omega_+ t)\\
&\quad + \cos\psi(\cos\psi + \tau\sin\psi)\exp(-i\omega_- t)\\
&\quad - \mu^*\sin\psi(\tau\sin\psi + \cos\psi)\exp(-i\omega_+ t)\\
&\quad - \mu^*\cos\psi(\tau\cos\psi - \sin\psi)\exp(i\omega_- t)]^{2j-n'}. \qquad \text{(B12)}
\end{aligned}
$$

This expression is very useful in order to investigate under what conditions $|w_n(t, \tau)\rangle$ could be an intelligent state. It is sufficient to calculate $\langle\mu|w_n'(\tau')\rangle$ and to compare its value with (B12). If we prove that there exists (n', τ') such that for any complex μ, $\langle\mu|w_{n'}(\tau')\rangle = \langle\mu|w_n(t, \tau)\rangle$, then the state $|w_n(t, \tau)\rangle$ keeps being intelligent along its evolution under the influence of the Hamiltonian given in Eq. (26). Since

$$
\begin{aligned}
&\langle\mu|w_{n'}(\tau')\rangle\\
&\quad = a_{n'}(\tau')a_0(\mu)(!n')\binom{2j}{n''}(1+\mu^*\tau')^{n'}(1-\mu^*\tau')^{2j-n'}, \qquad \text{(B13)}
\end{aligned}
$$

and both polynomials in the variable μ^* (B12) and (B13) must be identical, they have to contain the same roots with the same multiplicity. Therefore, n' has to be equal to n. Moreover, if we proceed with the analysis, one can immediately recognize that they are going to coincide iff $\sin 2\psi = 0$. That implies $\cos 2\psi = (-1)^p$ or, equivalently, $\psi = n\pi/2$. The condition $\psi = n\pi/2$ [see Eq. (26b)] is equivalent to saying that $B_\perp = 0$. Thus, after Eq. (B8), we have

$$
|w_n(t,\tau)\rangle = \exp(-2ij\gamma B_\parallel t)|w_n(\tau\exp(2i\gamma B_\parallel t))\rangle. \qquad \text{(B14)}
$$

Of course, if $\tau = |\tau|\exp(in\pi/2)$, $\tau' = \tau\exp(2i\gamma B_\parallel t) = |\tau|\exp(i(n\pi/2 + 2\gamma B_\parallel t))$ we get a generalized intelligent state, which is strictly intelligent for t such that $2\gamma B_\parallel t = m\pi/2$, i.e., it is periodically intelligent.

[1] C. Aragone, G. Guerri, S. Salamó, and J.L. Tani, J. Phys. A: Math. Nucl. Gen. 7, L149 (1974).
[2] F. Bloch, Phys. Rev. 70, 460 (1946).
[3] J.M. Radcliffe, J. Phys. A: Gen. Phys. 4, 313 (1971).
[4] F. Arecchi, E. Courtens, R. Gilmore, and H. Thomas, Phys. Rev. A 6, 2211 (1972).
[5] J.R. Klauder and E.C.G. Sudarshan, *Fundamentals of Quantum Optics* (Benjamin, New York, 1968).

[6]W. H. Louisell, *Quantum Statistical Properties of Radiation* (Wiley, New York, 1973).

[7]K. Gottfreid, *Quantum Mechanics, Vol. I: Fundamentals* (Benjamin, New York, 1966).

[8]In the sense that they are $2j+1$ nonvanishing linearly independent vectors. Of course we do not know whether they are orthogonal. Their inner product is given in Appendix A.

[9]Actually it can be easily seen that τ_α ranges over ail the points different from the origin of the two axes of the complex plane while α takes any real value $\alpha : |\alpha| \neq 1$. It is easy to see that $|\alpha| = 1$ only gives a trivial solution to Eq. (8a): If $\alpha = +1$, $w = 0$, $|w_N\rangle = |-\hat{j}\rangle$, and if $\alpha = -1$, $w = 0$, $|w_N\rangle = |\hat{j}\rangle$. If one wants to extend the definition of $|w_N(\tau)\rangle$ to $\tau = 0$, it turns out that $|w_N(0)\rangle = |-\hat{j}\rangle$ for all N.

[10]Even directly, it is enough to realize that $|w_{N_1}(\tau_1)\rangle$ and $|w_{N_2}(\tau_2)\rangle$ are normalized eigenvectors corresponding to the same eigenvalue of $J\alpha$ and that both of them have the same signature for their projection along $|-\hat{j}\rangle$.

[11]In fact, take from Eq. (9), $|w_n(\tau)\rangle = a_n(2j-n)!^{-1}(1+\tau\tau^*)^j \times\sum_{l=0}^{2j-n}\binom{2j-n}{l}(2j-l)!(-2\tau J_+)^l|\tau|\rangle = |\mu\rangle = (1+\mu\mu^*)^{-j}\exp(\mu J_+)|-\hat{j}\rangle$. Then, if we multiply both sides times $\exp(\tau J_+)$ and compare them, we see that (n,τ) has to be either $(0,-\mu)$ or $(2j,\mu)$.

[12]E. Lieb, Commun. Math. Phys. **31**, 327 (1973).

[13]R. Gilmore, *Lie Groups, Lie Algebras, and Some of Their Applications* (Wiley, New York, 1974).

[14]Actually, one could think, by comparison with what happens in the case of the Bloch states, that if $k(y,\tau)$ does not belong to SU(2), at least $k(y,\tau)\exp(-\tau_y^* J_-)$ belongs to this subgroup. However, it can be proved that for any $|\tau| > 0$ there always exists a neighborhood of the complex point $y = 1 + i0$ where none of the elements $\exp(\tau_y J_+)\exp(-2\ln y J_3)\exp(-\tau_y^* J_-)$ belong to SU(2).

[15]N. Sciarrino and M. Toller, J. Math. Phys. **8**, 1252 (1967).

[16]Especially see Secs. VI and VII of Ref. 4.

[17]Recently, M. E. Smithers and E. C. Lu, Phys. Rev. A **9**, 790 (1974), have given a beautiful treatment of the time evolution of both spontaneous and stimulated omission in this model using both the Wigner—Dicke states and the Bloch states. We conjecture that these results can be extended when the matter-initial state is an intelligent one.

[18]I. R. Senitzky, Phys. Rev. **111**, 3 (1958).

[19]That condition easily follows from the structure of $p_{\langle n\tau\rangle\to|m\rangle}$, given in the first of Eqs. (41) taking into account that C_4 contains all the powers between $j-m-n$ and $j-m$.

[20]I. Kolodziejczyk, J. Phys. A: Math. Nucl. Gen. **8**, L99 (1975).

[21]V. V. Mikhailov, Teor. Math. Fiz. **15**, 367 (1973).

[22]G. Vetri, J. Phys. A: Math. Nucl. Gen. **8**, L55 (1975).

[23]I. M. Gel'fand, M. I. Graev, and N. Ya. Vilenkin, *Generalized Functions* (Academic, New York, 1964), Vol. 5.

J. Phys. A: Math. Gen., Vol. 9, No. 9, 1976.

Charged bosons and the coherent state

Debajyoti Bhaumik†, Kamales Bhaumik‡ and Binayak Dutta-Roy‡
† Bose Institute, Calcutta, India
‡ Saha Institute of Nuclear Physics, Calcutta, India

Received 4 May 1976

Abstract. The coherent state for charged bosons is constructed, its properties are investigated and the corresponding classical model is discussed.

1. Introduction

The coherent state was first constructed (Schrödinger 1926, Glauber 1963) for the simple harmonic oscillator. The Hamiltonian of the system

$$H = p^2/2m + \tfrac{1}{2}m\omega^2x^2 \tag{1}$$

may be rewritten as

$$H = \hbar\omega(a^\dagger a + \tfrac{1}{2}) \tag{2}$$

by defining annihilation and creation operators

$$a = (p - \mathrm{i}m\omega x)/(2m\omega\hbar)^{1/2} \qquad a^\dagger = (p + \mathrm{i}m\omega x)/(2m\omega\hbar)^{1/2}. \tag{3}$$

The eigenstates of the Hamiltonian, $|n\rangle$, belonging to the energy eigenvalue $E_n = \hbar\omega(n + \frac{1}{2})$, where n is a non-negative integer, may then be obtained with the properties

$$a^\dagger a|n\rangle = n|n\rangle \qquad a^\dagger|n\rangle = (n+1)^{1/2}|n+1\rangle \qquad a|n\rangle = n^{1/2}|n-1\rangle. \tag{4}$$

The coherent state may then be constructed out of these states thus

$$|\alpha\rangle = \exp(-\tfrac{1}{2}|\alpha|^2)\sum_{n=0}^{\infty}\frac{\alpha^n}{(n!)^{1/2}}|n\rangle \tag{5}$$

where α is a complex number and the factor outside the summation is the normalization constant. The coherent state $|\alpha\rangle$ is an eigenstate of the annihilation operator a, namely

$$a|\alpha\rangle = \alpha|\alpha\rangle. \tag{6}$$

The coherent state may also be written in the form

$$|\alpha\rangle = \exp(-\alpha^* a + \alpha a^\dagger)|0\rangle \tag{7}$$

and is thus 'a displacement of the vacuum'. The coherent states form a complete (albeit an overcomplete) set in the sense that

$$\int \frac{\mathrm{d}^2\alpha}{\pi}|\alpha\rangle\langle\alpha| = 1 \tag{8}$$

where the integration is over the whole complex α plane. The coherent state constitutes a state of minimum uncertainty, namely

$$\Delta p \Delta X = \tfrac{1}{2}\hbar. \tag{9}$$

Since the coherent state is a non-stationary state it develops with time in an interesting manner and, taking $\alpha(t=0) = \lambda\, e^{-i\theta}$, it follows that

$$\langle \alpha, t|X|\alpha, t\rangle = \left[2\lambda\left(\frac{\hbar}{2m\omega}\right)^{1/2}\right]\sin(\omega t + \theta). \tag{10}$$

Identifying the constant in square brackets with the amplitude, the expectation value of the displacement in the coherent state behaves like the displacement of a classical oscillator. In this sense the coherent state is called a 'classical state'.

The coherent state has found widespread applications (Glauber 1969, Klauder and Sudarshan 1968, Perina 1971) in nonlinear optics and laser physics, in the discussion of the superfluid state (Langer 1969) and in nuclear physics (Bhaumik *et al* 1975). However, in all these cases the quanta involved are uncharged and the type of coherent state discussed above is adequate. Recent attempts (Botke *et al* 1974) to use the coherent state basis for the description of pion production in high-energy collisions necessitate the present discussion of the coherent state for charged bosons. We therefore discuss in the present paper the construction, properties and corresponding classical model of the coherent state for bosons possessing some 'charge' which is absolutely conserved.

2. Coherent state for charged bosons

The coherent state is a superposition of states containing different numbers of quanta phase-locked in the manner depicted in equation (5). However, if these quanta possess some absolutely conserved 'charge' Q it is impossible to construct coherent superpositions of states with different values of Q or to measure the corresponding phases. This is the content of the superselection rule (Wick *et al* 1952). Thus coherent states for charged quanta need careful consideration.

Let us introduce 'charge' by defining two types of quanta possessing 'charge' +1 and −1 with corresponding annihilation operators a and b. Thus

$$[a, a^\dagger] = 1 = [b, b^\dagger] \qquad [a, a] = 0 = [b, b] \qquad [a, b] = 0 = [a^\dagger, b] \tag{11}$$

and the charge operator is given by

$$Q = a^\dagger a - b^\dagger b. \tag{12}$$

Clearly the charge operator Q does not commute with a or with b. Therefore we cannot demand that the coherent state be simultaneously an eigenstate of the charge and the annihilation operators a or b. Nevertheless, in view of the fact that

$$[Q, ab] = 0 \tag{13}$$

we may define the modified coherent state $|\xi, q\rangle$ for charged quanta to be simultaneously an eigenstate of Q and ab belonging to the eigenvalues q and ξ (a complex number) respectively; thus

$$Q|\xi, q\rangle = q|\xi, q\rangle \qquad ab|\xi, q\rangle = \xi|\xi, q\rangle. \tag{14}$$

This state can easily be constructed out of eigenstates $|n, m\rangle$ where

$$a^\dagger a|n, m\rangle = n|n, m\rangle \qquad b^\dagger b|n, m\rangle = m|n, m\rangle \tag{15}$$

to yield

$$|\xi, q\rangle = N_q \sum_{n=0}^{\infty} \frac{\xi^n}{[n!(n+q)!]^{1/2}}|n+q, n\rangle \tag{16}$$

where the normalization constant N_q is given by

$$N_q = \left(\sum_n \frac{(|\xi|^2)^n}{n!(n+q)!}\right)^{-1/2} = [(-\mathrm{i}|\xi|)^q J_q(2\mathrm{i}|\xi|)]^{-1/2} \tag{17}$$

where J_q is the Bessel function of order q. This coherent state may also be generated from the vacuum state (Schwinger 1965). Thus

$$F_q(a^\dagger b^\dagger \xi)a^{\dagger q}|0\rangle \tag{18a}$$

is the coherent state except for a normalization factor. The function F_q is given by

$$F_q(z) = q!(-z)^{-q/2} J_q(2\mathrm{i}z^{1/2}) = \sum_{n=0}^{\infty} \frac{q!}{n!(n+q)!} z^q. \tag{18b}$$

The above expressions are for $q>0$ and analogous expressions for $q<0$ are obtained by replacing a by b. These states constitute a complete set of states in the sense that

$$\sum_{q=0}^{\infty} \int \frac{\mathrm{d}^2\xi}{\pi} \phi_q(\xi)|\xi, q\rangle\langle\xi, q| = 1 \tag{19}$$

where

$$\phi_q(\xi) = 2(-\mathrm{i})^q J_q(2\mathrm{i}|\xi|)K_q(2|\xi|) \tag{20a}$$

with

$$K_q(z) = \tfrac{1}{2}\pi\mathrm{i}\,\exp(\tfrac{1}{2}\mathrm{i}\pi q)(J_q(\mathrm{i}z) + \mathrm{i}N_q(\mathrm{i}z)). \tag{20b}$$

The coherence of the conventional coherent states (discussed in § 1) is realized through the correlation of signals from photon counters. For the coherent states defined here for charged quanta, the coherence is manifested by replacing the photon counters by detectors which respond to the simultaneous detection of positive and negative quanta.

The coherent state for charged quanta may also be obtained by projecting out a state of definite charge from the two-mode coherent state

$$|\alpha\beta\rangle = \exp(\alpha a^\dagger + \beta b^\dagger)|0\rangle = \sum_{n,m=0}^{\infty} \frac{\alpha^n \beta^m}{n!m!} a^{\dagger n} b^{\dagger m}|0\rangle \tag{21}$$

which does not have a definite charge. This is accomplished by putting

$$\alpha = \lambda\,\mathrm{e}^{-\mathrm{i}(\theta+\varphi)}, \qquad \beta = \mu\,\mathrm{e}^{-\mathrm{i}(\theta-\varphi)} \tag{22}$$

and observing that the state

$$|\lambda\mu\theta; q\rangle = \frac{1}{2\pi}\int_0^{2\pi} \mathrm{d}\varphi\,\mathrm{e}^{\mathrm{i}q\varphi}|\alpha\beta\rangle \tag{23}$$

is identical to the charged coherent state, equation (16), if we make the identification $\lambda\mu\,\mathrm{e}^{-2\mathrm{i}\theta} = \xi$.

Introducing the 'coordinates' corresponding to the a and b quanta

$$x_1=-\mathrm{i}\left(\frac{\hbar}{2m\omega}\right)^{1/2}(a^\dagger-a) \qquad x_2=-\mathrm{i}\left(\frac{\hbar}{2m\omega}\right)^{1/2}(b^\dagger-b) \tag{24}$$

it is easily seen that the average displacements of the two oscillators vanish and that

$$\langle\xi;q|x_1^2|\xi;q\rangle=\frac{\hbar}{2m\omega}\left(-2\mathrm{i}|\xi|\frac{J_{q+1}(2\mathrm{i}|\xi|)}{J_q(2\mathrm{i}|\xi|)}+1\right) \tag{25a}$$

$$\langle\xi;q|x_2^2|\xi;q\rangle=\frac{\hbar}{2m\omega}\left(2\mathrm{i}|\xi|\frac{J_{q-1}(2\mathrm{i}|\xi|)}{J_q(2\mathrm{i}|\xi|)}+1\right). \tag{25b}$$

In order to investigate in what sense these states are 'classical' it is useful to consider the classical limit ($\hbar\to 0$, $|\xi|\to\infty$: $\hbar\xi\to$ finite limit) of the above expectation values. Thus the relevant expectation values are given by

$$\langle\xi;q|x_1|\xi;q\rangle=0=\langle\xi;q|x_2|\xi;q\rangle \tag{26a}$$

$$\langle\xi;q|x_1^2|\xi;q\rangle=\frac{\hbar}{2m\omega}(2|\xi|+q+\tfrac{3}{2})+O(\hbar/|\xi|) \tag{26b}$$

$$\langle\xi;q|x_2^2|\xi,q\rangle=\frac{\hbar}{2m\omega}(2|\xi|-q+\tfrac{3}{2})+O(\hbar/|\xi|) \tag{26c}$$

$$\langle\xi;q|x_1x_2|\xi;q\rangle=\frac{\hbar}{m\omega}|\xi|\cos(2\theta+2\omega t). \tag{26d}$$

It may be observed that

$$\langle\xi;q|(x_1^2-x_2^2)|\xi;q\rangle=\frac{\hbar}{m\omega}q+O(\hbar/|\xi|). \tag{27}$$

Equation (27), expressing the charge as the semiclassical approximation to the expectation value of $(x_1^2-x_2^2)$ provides the clue to the construction of the classical analogue. Since the energy of a classical oscillator is proportional to the square of the amplitude, the classical analogue of a state with definite charge is obtained by constraining the two oscillators described by the Hamiltonian

$$H=\frac{p_1^2}{2m}+\tfrac{1}{2}m\omega^2x_1^2+\frac{p_2^2}{2m}+\tfrac{1}{2}m\omega^2x_2^2 \tag{28}$$

to oscillate, keeping the difference of their action functions

$$\frac{E_1}{\omega}-\frac{E_2}{\omega}=\text{fixed}. \tag{29}$$

This quantity is taken to be proportional to the 'charge'. The role of this 'charge' in the motion of the classical oscillators is further clarified by performing a canonical transformation (Goldstein 1950), from (x_1, x_2, p_1, p_2) to (X_1, X_2, P_1, P_2), through the generating function

$$F(x_i, X_i)=\tfrac{1}{2}m\omega(x_1^2\cot X_1+x_2^2\cot X_2) \tag{30}$$

whence the transformed Hamiltonian becomes

$$K=H=\omega(P_1+P_2). \tag{31}$$

Thus X_i and P_i take the role of (phase) angle and action variables, and

$$P_1 - P_2 = \frac{E_1}{\omega} - \frac{E_2}{\omega} \tag{32}$$

is canonically conjugate to $\frac{1}{2}(X_1 - X_2)$ which is the relative phase between the two oscillators. Thus if one fixes the 'charge', namely the quantity $(P_1 - P_2)$, the various possible motions differ from each other in the relative phase φ of the two motions given by

$$x_1 = A \sin(\omega t + \theta + \varphi) \tag{33a}$$

$$x_2 = B \sin(\omega t + \theta - \varphi). \tag{33b}$$

If F_{cl} is a classical dynamical variable of this system with the 'charge' held fixed, the average, $\bar{F}_{\text{cl}}$, of this quantity over all motions of the system in the state of given 'charge' (here $(P_1 - P_2)$) is obtained by taking an average (Messiah 1967) over the relative phase of the two oscillators (which is canonically conjugate to the 'charge'). Thus

$$\overline{x_1} = 0 = \overline{x_2} \tag{34a}$$

$$\overline{x_1^2} = \tfrac{1}{2}A^2 \tag{34b}$$

$$\overline{x_2^2} = \tfrac{1}{2}B^2 \tag{34c}$$

$$\overline{x_1 x_2} = AB \cos(2\theta + 2\omega t). \tag{34d}$$

Comparing these results with the classical limits of the quantal expectation values (equations (26)) we see that these are in agreement provided we identify $A^2 = B^2 = 2\hbar|\xi|/m\omega$. It may also be observed that $\langle X_1^2\rangle - \langle X_2^2\rangle \to 0$ in the classical limit and the 'charge' is thus a semiclassical quantity. It is in the sense discussed above that the modified coherent states for charged quanta discussed in this paper are 'classical'.

The coherent states for charged bosons have thus been constructed, their properties investigated and the corresponding classical model has been discussed.

Acknowledgments

One of us (BDR) would like to thank Binayak Basu for helpful discussions. One of us (DB) is grateful to the Council of Scientific and Industrial Research, India for financial support, to the Director, Bose Institute for his hospitality and to Professor A M Ghose for his constant encouragement.

References

Bhaumik D, Dasgupta B and Dutta-Roy B 1975 *Phys. Lett.* **56B** 119
Botke J C, Scalapino D J and Sugar R L 1974 *Phys. Rev.* D **9** 813
Glauber R J 1963 *Phys. Rev.* **131** 2766
—— 1969 *Quantum Optics* ed R J Glauber (New York: Academic Press) pp 15–56
Goldstein H 1950 *Classical Mechanics* (New York: Addison-Wesley)
Klauder J R and Sudarshan E C G 1968 *Fundamentals of Quantum Optics* (New York: Benjamin)

Langer J S 1969 *Phys. Rev.* **184** 219–29
Messiah A 1967 *Quantum Mechanics* vol 1 (Amsterdam: North-Holland) p 444
Perina J 1971 *Coherence of Light* (London: Van Nostrand)
Schrödinger E 1926 *Naturwiss.* **14** 664
Schwinger J 1965 *Quantum Theory of Angular Momentum* ed L C Biedenharn and H Van Dam (New York: Academic Press) p 229
Wick G C, Wightman A S and Wigner E P 1952 *Phys. Rev.* **88** 101

Coherent states associated with the continuous spectrum of noncompact groups*

M. Hongoh

Department of Physics, University of Montreal, Montréal, Québec, Canada
(Received 12 July 1976)

A new, explicit formula is obtained for the coherent states associated with a continuous spectrum of the noncompact group SU(1,1). The method is based on a simple and unified algebraic approach. We briefly discuss its relations to the generalized coherent states of Barut and Girardello, and of Perelomov.

I. INTRODUCTION

The notion of coherent states was first introduced by Weyl for the nilpotent group.[1] Recently it has been extended to any Lie group.[2,3] Barut and Girardello[4] have constructed in particular the coherent states associated with the discrete series of representations of SO(2, 1). They are the eigenstates of the lowering and raising operators of noncompact groups. On the other hand, Perelomov has generalized the coherent states in such a way that a set of coherent states is invariant relative to the action of the group representation operators.

It has been suggested since,[2,4] that one extend the notion of coherent states for the continuous spectrum corresponding to the infinite-dimensional unitary representations of the noncompact groups. But so far no explicit construction seems to have appeared in the literature. The purpose of this paper is to present a general method for constructing such states. In particular we deal with the continuous basis of the simplest semisimple Lie algebra of SU(1, 1). Explicit coherent states are then constructed on this basis, which are somewhat analogous to the Bloch coherent states of SU(2).

In atomic physics, coherent states associated with a system of spins are known as the Bloch coherent states.[5,6] They can be obtained by rotating the lowest angular momentum state (called the lowest Dicke state) in the manifold of the group of the angular momentum; they provide natural and useful bases for various calculations of superradiance and superconductivity.[7,8] Moreover they possess a unique property in that they represent special quantum states most closely approximating classical states, e.g., the uncertainty relation takes its minimum value for these states. The Bloch coherent states of SU(2) form a subsystem of the systems of generalized coherent states of Perelomov. Thus, they are invariant relative to the action of the group representation operators. We might call the generalization of this particular system of coherent states a *Bloch-type* system, and in the text we shall restrict ourselves to this particular system of coherent states. In Sec. II we demonstrate our method by deriving the coherent states for the discrete spectrum of SU(1, 1), and we compare the results with those of Barut and Girardello. In Sec. III, we construct an explicit formula for the coherent states associated with the continuous spectrum, and Sec. IV is devoted to discussion and to some further remarks.

II. COHERENT STATES ASSOCIATED WITH THE DISCRETE SPECTRUM OF SU(1, 1)[17]

In connection with the general scalar coefficients, Van der Waerden[9,10] introduced an invariant form for the compact SU(2) group,

$$W=\prod_i \Lambda_i^{a_i} \tag{2.1}$$

where $\Lambda_i=(\eta_j\xi_k-\eta_k\xi_j)$, $i,j,k=1,2,3$ cyclic. a_i are nonnegative integers. η and ξ are the two fundamental complex variables of the group, and i refers to the three distinct representation spaces. Λ_i are called elementary scalars.

The skew-symmetric form of Λ_i reflects the following theorem[11-14]:

Malcev-Dynkin Theorem: The self contragradient IR (SCIR) of the connected semisimple Lie groups leaves the nondegenerate bilinear form invariant which is symmetric (skew-symmetric) according to whether the SCIR is orthogonal (symplectic).

All the IR's of SU(2) are SC[15] and the smallest SCIR is even-dimensional. Letting $a_1=a_2=0$ and $a_3=2S$, the SU(2) Bloch coherent states can be derived.[16]

We shall now extend the notion of the Van der Waerden invariant for noncompact groups, i.e., let

$$\Lambda_{\text{n.c.}}\equiv(\eta_1\xi_2-\eta_2\xi_1)^{2S}, \quad 2S\text{: real number}. \tag{2.2}$$

The Bargmann–Schwinger realization of the SU(1, 1) Lie algebra is well known,

$$\begin{aligned} &L_+=\epsilon\eta\partial_\xi, \quad L_-=\epsilon\xi\partial_\eta, \\ &L_3=\tfrac{1}{2}(\eta\partial_\eta-\xi\partial_\xi), \quad \epsilon=+\sqrt{-g_{33}}, \end{aligned} \tag{2.3}$$

and the basis functions are,[17]

$$g_m(Z)=\frac{(-g_{33})^{\phi-m}\Gamma(2\phi+1)}{\Gamma(\phi+m+1)\Gamma(\phi-m+1)}Z_1^{\phi+m}Z_2^{\phi-m}, \tag{2.4}$$

where $g_{33}=-1$ for SU(2) ~ O(3), and +1 for SU(1, 1) ~ O(2, 1). One can easily see that $\{Z,Z\}\equiv|Z_1|^2-g_{33}|Z_2|^2$ is preserved by the fundamental representation of the group. Let us consider the following function,

$$F(W,Z)=\sum_{m=0}^{\infty} g_m^*(W)g_m(Z). \tag{2.5}$$

F can be viewed as a scalar product of two basis vectors which belong to the two distinct Hilbert spaces of IR characterized by 2ϕ. In fact,

$$F(W,Z)=\{W,Z\}^{2\phi}. \tag{2.6}$$

Comparing Eq. (2.6) and Eq. (2.2) with $2S \to 2\phi$, we establish the correspondences

$$\begin{pmatrix}\eta\\ \xi\end{pmatrix}_1 \to \begin{pmatrix}Z_1\\ Z_2\end{pmatrix}, \quad \begin{pmatrix}\xi\\ \eta\end{pmatrix}_2 \to \begin{pmatrix}W_1\\ -g_{33}W_2\end{pmatrix}^*. \tag{2.7}$$

Let us rewrite $W_1^*, W_2^* \to \nu, \mu$ respectively. Then Eq. (2.7) means that the operators in Eq. (2.3), once operated upon $W(\nu,\mu)$ space, would be replaced by,[18]

$$L_+ \to \epsilon\mu\partial_\nu, \quad L_- \to \epsilon\nu\partial_\mu, \qquad L_3 \to \tfrac{1}{2}(\mu\partial\mu - \nu\partial\nu). \tag{2.8}$$

We are interested in the coherent states defined over the homogeneous factor space $G/H \sim SU(1,1)/SO(2)$. Let Z be a vector belonging to the Hilbert space of the set of pure states. This Hilbert space is isomorphic to an upper sheet of the hyperboloid. Without a loss of generality we can set,

$$\{Z, Z\}^{2\phi} = 1. \tag{2.9}$$

The completeness relation for the basis function $g_n(Z) \equiv |\phi n\rangle$ is written

$$F(Z,Z) = \sum_{n=0}^{\infty} |\phi n\rangle\langle\phi n| = 1. \tag{2.10}$$

The homogeneous factor space is a unit disc that can be parametrized by a single complex variable, say ω.[19] We fix our parametrization by letting $w \in G/H$, and $\nu \to \omega$, $\mu \to 1$, respectively. We have

$$F(W,W) = (1 - g_{33}|\omega|^2)^{2\phi}. \tag{2.11}$$

Note that the fundamental length $(1 - g_{33}|\omega|^2)$ is preserved in the operation of the group. Substituting (2.4) for (2.6) with $\phi + m \to N$, $\phi - m \to 2\phi - N$, respectively, and using the binomial expansion,[20] we obtain

$$|\omega\rangle = \sum_{N=0}^{\infty}\left[\frac{(-g_{33})^{2\phi-N}\Gamma(2\phi+1)}{\Gamma(N+1)\Gamma(2\phi-N+1)}\right]^{1/2}\omega^N|\phi N\rangle, \tag{2.12}$$

where the normalization factor is ignored. Note that $|\omega\rangle$ is not identical with the SU(1,1) coherent states given by Barut and Girardello,[4] in which L_- (L_+) is diagonalized. In fact, their coherent states are

$$|Z\rangle_{B.G.} = [\Gamma(-2\phi)]^{1/2}\sum_{n=0}^{\infty}\frac{(\sqrt{2}\,Z)^n}{[\Gamma(n+1)\Gamma(-2\phi+n)]^{1/2}}|\phi n\rangle. \tag{2.13a}$$

It can be seen easily that the following holds for $|Z\rangle_{B.G.}$,

$$L_-|Z\rangle_{B.G.} = Z|Z\rangle_{B.G.}. \tag{2.13b}$$

On the other hand, we have

$$L_-|\omega\rangle = \sum_N\left[\frac{(-g_{33})^{2\phi-N}\Gamma(2\phi+1)}{\Gamma(N+1)\Gamma(2\phi-N+1)}\right]^{1/2}\omega^{N+1}(2\phi-N)|\phi N\rangle, \tag{2.14}$$

i.e., $|\omega\rangle$ fully retains the property of the Bloch coherent states. Equation (2.11) gives the norm of $|\omega\rangle$, and the nonorthogonality is written

$$\langle\omega'|\omega\rangle = (1 - g_{33}\omega'^*\omega)^{2\phi}. \tag{2.15}$$

With the resolution of unity,

$$\int d\sigma(\omega)|\omega\rangle\langle\omega| = \sum_{N=0}^{\infty}|\phi N\rangle\langle\phi N| = 1, \tag{2.16}$$

one can write

$$\langle f|i\rangle = \int d\sigma(\omega)\langle f|\omega\rangle\langle\omega|i\rangle, \tag{2.17}$$

where $|f\rangle$ and $|i\rangle$ belong to the Hilbert space of IR's of the vectors of G/H. Letting $d\sigma(\omega) = \sigma(r)r\,dr\,d\phi$, $|\omega| = r$, the rhs is,

$$2\pi\int_0^\infty dr\,\sigma(r)r^{2N+1} \times \sum_N \frac{\Gamma(-2\phi+1)}{\Gamma(N+1)\Gamma(-2\phi+N+1)}\langle f|\phi N\rangle\langle\phi N|i\rangle. \tag{2.18}$$

Due to Eq. (2.16), $\sigma(r)$ satisfies

$$\int_0^\infty dr\,\sigma(r)r^{2N+1} = \frac{\Gamma(N+1)\Gamma(-2\phi+N+1)}{2\pi\Gamma(-2\phi+1)}, \tag{2.19}$$

and via the Mellin transform, we obtain

$$\sigma(r) = \frac{2r^{-2\phi}K_\phi(2r)}{\pi\Gamma(-2\phi+1)} > 0, \quad r > 0, \tag{2.20a}$$

where

$$K_\phi(2r) = \frac{\pi}{2}\circ\frac{I_{-\phi}(2r) - I_{+\phi}(2r)}{\sin(\phi\pi)} \tag{2.20b}$$

and

$$I_\phi(2r) = \sum_{m=0}^{\infty}\frac{r^{\phi+2m}}{\Gamma(m+1)\Gamma(\phi+m+1)}. \tag{2.20c}$$

$K_\phi(2r)$ is the modified Bessel function of the third kind, and $I_\phi(2r)$ is the modified Bessel function of the first kind. Equations (2.20) are the counterparts of the formulas given in Ref. 4. Equations (2.12), (2.16), and (2.15) completely specify the system of the Bloch-type coherent states.

III. COHERENT STATES ASSOCIATED WITH THE CONTINUOUS SPECTRUM OF SU(1, 1)

Let us consider the SU(1,1) algebra in which L_{23} is diagonalized.[21]

$$L_{12} = (i/\sqrt{2})(\partial_d - d\partial_u), \quad L_{13} = (\epsilon/\sqrt{2})(u\partial_d + d\partial_u), \quad L_{23} = (\epsilon/\sqrt{2})(u\partial_u - d\partial_d), \tag{3.1}$$

where $u = (1/\sqrt{2})(\eta + i\xi)$, $d = (1/\sqrt{2})(\xi + i\eta)$. The basis functions are,

$$f^\alpha(Z) = A_\alpha u^{\phi+\epsilon\alpha}d^{\phi-\epsilon\alpha}, \quad Z(u,d), \tag{3.2}$$

where α is in general a complex number. Our task is to construct the coherent states defined over $G/H \sim SU(1,1)/O(1,1)$ as a certain linear combination of $f^\alpha(Z)$. Clearly it suffices to consider rotations around the third axis,

$$R = e^{i\theta L_{12}} = \begin{pmatrix} e^{i\theta/2} & \\ & e^{-i\theta/2}\end{pmatrix} \tag{3.3}$$

which preserve the form $\{|u|^2 + |d|^2\}$ for both compact and noncompact cases. Defining the state conjugate to $|\phi\alpha\rangle \equiv f^\alpha(Z)$,

$$\langle\phi\alpha| = A_\alpha(u^*)^{\phi+\epsilon\alpha}(d^*)^{\phi-\epsilon\alpha}, \tag{3.4}$$

the orthogonality of the basis function can be expressed as follows:

$$\langle \phi\alpha' | \phi\alpha \rangle = \delta(\alpha' - \alpha) . \quad (3.5)$$

The normalization constant A_α can be obtained by using the complex binomial expansion of the invariant form,

$$|A_\alpha|^2 = \frac{1}{2\pi} \frac{\Gamma(-\phi + \epsilon\alpha)\Gamma(-\phi - \epsilon\alpha)}{\Gamma(-2\phi)} \geq 0 . \quad (3.6)$$

Let us introduce a function

$$\mathcal{J}(W, Z) = \int_{-(i/\epsilon)\infty}^{+(i/\epsilon)\infty} d\alpha f^\alpha(W)^* f^\alpha(Z) , \quad (3.7)$$

which is the continuum version of Eq. (2.5). One can see immediately that

$$\mathcal{J}(Z, Z) = \{|u|^2 + |d|^2\}^{2\phi} . \quad (3.8)$$

We now consider the Hilbert space which corresponds to an upper half of the hyperboloid lying along the first axis. Let $Z(u, d)$ belong to this Hilbert space, and we set

$$\{|u|^2 + |d|^2\} = 1 . \quad (3.9)$$

Then from Eqs. (3.2) and (3.7) we obtain,

$$\mathcal{J}(W, Z) = \int d\alpha |A_\alpha|^2 (W_1^*)^{\phi + \epsilon\alpha} (W_2)^{\phi - \epsilon\alpha} | \phi\alpha \rangle . \quad (3.10)$$

Due to the correspondences (2.6), we replace w_1^*, w_2^* $\to \theta, \rho$ respectively; where ρ, θ are complex variables similar to u, d. The factor space may be projected onto a unit disc perpendicular to the first axis, which can be parametrized by a single complex variable. This amounts to having $\theta \to \theta$, $\rho \to 1$, and the coherent states are,

$$|\theta\rangle = \int_{-(i/\epsilon)\infty}^{+(i/\epsilon)\infty} d\alpha \left[\frac{\Gamma(-\phi + \epsilon\alpha)\Gamma(-\phi - \epsilon\alpha)}{2\pi\Gamma(-2\phi)} \right]^{1/2} \theta^{\phi + \epsilon\alpha} | \phi\alpha \rangle . \quad (3.11)$$

To the best of our knowledge, this expression has not appeared in the literature before. The nonorthogonality of $|\theta\rangle$ is,

$$\langle \theta' | \theta \rangle = \frac{1}{2\pi} \int d\alpha \frac{\Gamma(-\phi + \epsilon\alpha)\Gamma(-\phi - \epsilon\alpha)}{\Gamma(-2\phi)} (\theta'^* \theta)^{\phi + \epsilon\alpha} . \quad (3.12)$$

$\Gamma(-\phi + \epsilon\alpha)$ and $\Gamma(-\phi - \epsilon\alpha)$ have poles at $\alpha = y + i(N - x)$ and at $-y + i(x - N)$ respectively (x and y are the real and imaginary parts of ϕ).

Integrating clockwise along the contour and summing contributions from poles of $\Gamma(-\phi - \epsilon\alpha)$, we obtain

$$\langle \theta' | \theta \rangle = \frac{1}{\Gamma(-2\phi)} \sum_{N=0}^{\infty} \frac{(-)^N \Gamma(-2\phi + N)}{\Gamma(N+1)} (\theta'^* \theta)^N (\rho'^* \rho)^{2\phi - N} \Big|_{\rho, \rho' = 1} . \quad (3.13)$$

Note the rhs is the complex binomial expansion. Thus

$$\langle \theta' | \theta \rangle = (1 + \theta'^* \theta)^{2\phi} , \quad |\theta'^* \theta| < 1 , \quad |\arg(-\theta'^* \theta)| < \pi . \quad (3.14)$$

In particular the norm of $|\theta\rangle$ is,

$$\| |\theta\rangle \|^2 = (1 + |\theta|^2)^{2\phi} . \quad (3.15)$$

Dividing $|\theta\rangle$ by the square root of its norm, we define

$$|\bar{\theta}\rangle = (1 + |\theta|^2)^{-\phi} |\theta\rangle .$$

Then the completeness relation can be written as

$$\langle f | i \rangle = \int d\sigma(\theta) \langle f | \bar{\theta} \rangle \langle \bar{\theta} | i \rangle . \quad (3.16)$$

With $d\sigma(\theta) = \sigma(r) r dr d\phi$ we have,

$$\langle f | i \rangle = \int d\alpha \langle f | \phi\alpha \rangle \langle \phi\alpha | i \rangle$$
$$= \frac{2\pi}{(1 + |\theta|^2)^{2\phi}} \int_0^\infty \sigma(r) dr$$
$$\times \int_{-\infty}^{+\infty} d\alpha |A_\alpha|^2 r^{2(\phi + \epsilon\alpha)} \langle f | \phi\alpha \rangle \langle \phi\alpha | i \rangle . \quad (3.17)$$

Inserting the beta function for $|A_\alpha|^2$

$$B(-\phi + \epsilon\alpha, -\phi - \epsilon\alpha) = \int_0^\infty dr \frac{2r^{-2\phi - 2\epsilon\alpha - 1}}{(1 + r^2)^{-2\phi}} , \quad (3.18)$$

we obtain

$$2 \int_0^\infty dr \, \sigma(r) \frac{1}{r} \int_{-\infty}^{+\infty} d\alpha \langle f | \phi\alpha \rangle \langle \phi\alpha | i \rangle = \langle f | i \rangle , \quad (3.19)$$

where $\sigma(r)$ is,

$$\sigma(r) = (r/2)\delta(r) , \quad r \geq 0 , \quad (3.20a)$$

or

$$= \frac{r}{\sqrt{2}} \exp(-r^2) , \quad r \geq 0 . \quad (3.20b)$$

Equation (3.20) together with Eqs. (3.11) and (3.16) completely specify the system of the coherent states associated with the continuous spectrum.

IV. CONCLUDING REMARKS

A simple method is introduced for constructing the explicit *Bloch-type* coherent states for the UIR of noncompact Lie groups. In particular we have studied the coherent states associated with both the discrete and the continuous spectra of SU(1, 1). The method might be useful in constructing the general coherent state representations for other noncompact groups which appear frequently in the applications to physical problems. This will be discussed elsewhere.

ACKNOWLEDGMENTS

The author wishes to thank Professor J. Patera for drawing his attention to the Malcev-Dynkin theorem, and Professor J. Destry for reading the manuscript. It is also a pleasure to express his gratitude to the Centre de Recherches Mathématiques, Université de Montreal for their hospitality, where the present work was partially completed.

*Work supported in part by the National Research Council of Canada.

[1]H. Weyl, *Gruppentheorie und Quantenmechanik* (Leipzig, S. Hirzel, 1928).
[2]A.M. Perelomov, Commun. Math. Phys. 26, 222 (1972).
[3]E. Onofri, J. Math. Phys. 16, 1078 (1975).
[4]A.O. Barut and L. Girardello, Commun. Math. Phys. 21, 41 (1971).
[5]J.M. Radcliffe, J. Phys. A (London) 4, 313 (1971).
[6]F.T. Arecchi, E. Courtens, R. Gilmore, and H. Thomas, Phys. Rev. A 6, 2211 (1972).
[7]F.T. Hioe, Phys. Rev. A 8, 1440 (1973).

[8]F. T. Hioe, J. Math. Phys. 15, 445 (1974).
[9]B. L. Van der Waerden, *Die Gruppentheoretische Methode in der Quantenmechanik* (Berlin, Springer, 1932).
[10]M. Hongoh, R. T. Sharp, and D. E. Tilley, J. Math. Phys. 15, 782 (1974).
[11]A. I. Malcev, Am. Math. Soc. Transl. Ser. 1, 9, 172 (1962).
[12]E. B. Dynkin, Dokl. Akad. Nauk SSSR (N.S.) 71, 221 (1950).
[13]E. B. Dynkin, Dokl. Akad. Nauk SSSR (N.S.) 76, 629 (1951).
[14]A. K. Bose and J. Patera, J. Math. Phys. 11, 2231 (1970).
[15]The two vectors of Λ_i can be non-SCIR, since they belong to two distinct representation spaces. For example, the smallest nontrivial orthogonal SCIR of SU(3) is $D(1,1)$ (octet) and the symmetric Van der Waerden invariant consists of vectors of $D(10)$ and $D(01)$, which separately are not SCIR.
[16]M. Hongoh, notes on the SU(2) Bloch coherent states, Univ. of Montreal preprint (unpublished).
[17]Throughout the work we shall adopt the notation of A. O. Barut, *Lectures in Theorectical Physics* (Gordon and Breach, New York, 1967), Vol. IXA, pp. 125–71.
[18]To adopt the phase convention used in Z space, the operators acting in w space must be of this form.
[19]For the lowest state, $|\phi n=0\rangle$, the stabilizer is SO(2) and the corresponding orbit is SU(1,1)/SO(2).
[20]The use of the (complex) binomial expansion imposes a certain restriction on the values that ϕ and E_0 can take. While this causes no trouble for discrete cases $D^{\pm}(\phi)$, the general cases can be handled by going to the covering group SU(1,1) $\otimes Z$. See Ref. 9.
[21]A. O. Barut and E. C. Phillips, Commun. Math. Phys. 8, 52 (1968).

J. Phys. A: Math. Gen. **18** (1985) 1–13.

Quasi-coherent states for unitary groups

B-S Skagerstam
Institute of Theoretical Physics, S-41296 Göteborg, Sweden and NORDITA, Blegdamsvej 17, DK-2100 Copenhagen, Denmark

Received 13 February 1984, in final form 6 August 1984

Abstract. We extend a recent construction of isotopic spin-coherent states to unitary groups. Expectation values in such a quasi-coherent state can be obtained from a generating functional which is given explicitly for the fundamental and the adjoint representations of the unitary groups SU(N) and U(N). The close connection between the corresponding generating functional and the external field problem in QCD is pointed out. Free quantum fields in such a condensed, colour singlet, coherent state are formally related to two-dimensional lattice gauge theories with, in general, a mixed action. In the large-N limit, we exhibit a phase transition for a free quantum field in a singlet quasi-coherent state. A path integral representation of the transition amplitude in terms of quasi-coherent states is also given. The corresponding effective action describes dynamics on a complex Kähler manifold.

1. Introduction

The concept of coherent states plays an important role in various branches of physics (we refer to Klauder and Sudarshan (1968), Glauber (1963, 1964) and Klauder and Skagerstam (1984) for a general introduction to coherent states). Coherent state techniques can for example be used in order to justify the study of classical equations of motion in quantum mechanics or in quantum field theory, i.e. semi-classical methods (Klauder 1977), and in this context we also mention a recent application to Yang–Mills instanton gauge field configuration considerations (Duff and Isham 1980).

Due to their very interesting properties various attempts have been made to generalise the concept of coherent states. One such very successful generalisation to arbitrary Lie groups is due to Perelemov (1972) (see also Gilmore 1972), who constructed what is referred to in the literature as generalised coherent states (see Perelemov (1977) for a review and applications of these states). This construction extends other work on SU(2) spin-coherent states (Radcliffe 1971, Arechi *et al* 1972, Klauder 1963, 1979, 1982, 1983).

Recently an alternative generalisation has been put forward (Skagerstam 1978, 1979, 1980, Eriksson and Skagerstam 1979, 1981, Mukunda *et al* 1981). The corresponding over-complete set of quasi-coherent states transforms according to a given irreducible representation of a compact Lie group under consideration. The construction of these quasi-coherent states is based on the considerations by Bhaumik *et al* (1976), where it was noted that in the case when an Abelian conserved charge is present a complete set of 'coherent states', in which the charge is diagonal, can easily be constructed. In the papers by Skagerstam (1979, 1980) their construction was extended to the field-theoretical situation (for a related discussion see Horn and Silver 1971).

These U(1)-charged coherent states have recently been applied to a study of the vacuum state in some field-theoretical models by Ni and Wang (1983). An extension of the work by Bhaumik *et al* (1976) to the case of the non-Abelian group SU(2)/Z(2) was given by Skagerstam (1978), Eriksson and Skagerstam (1979, 1981), which recently was also extended to the field-theoretical situation by Mukunda *et al* (1981).

In the present paper we will extend the analysis of Skagerstam (1978), Eriksson and Skagerstam (1979, 1981) and Mukunda *et al* (1981) to higher-dimensional compact Lie groups. We will specialise in the unitary groups U(N) and SU(N), but, as will be clear from our presentation below, the present work can easily be extended to any compact Lie group. We will mainly consider one degree of freedom, as in Skagerstam (1978), Eriksson and Skagerstam (1979, 1981), although most of our results can be extended directly to an arbitrary number of degrees of freedom and will briefly be discussed in the text. Concerning applications of the quasi-coherent states of the form we are considering in the present paper we mention a recent study of the thermodynamics of the non-Abelian, ideal and colourless quark-gluon gas (Skagerstam 1983, 1984). Elsewhere we will return to applications in other areas of physics.

The paper is organised as follows: in § 2 we construct the quasi-coherent states for real representations. We give some specific results for the adjoint representation, which are relevant for the study of condensed gauge fields (Mukunda *et al* 1981). In § 3 we consider complex representations. Some field-theoretical considerations are presented in § 4, where we also include a study of the large-N limit of the colour singlet quasi-coherent states. It is shown that a free quantum field, in a condensed colour singlet state, may exhibit a first-order phase transition if the one-particle state transforms according to the adjoint representation. The origin of this phase transition has a counterpart in the study of some two-dimensional lattice gauge theories (Chen and Zheng 1982, Makeenko and Polikarpov 1982, Samuel 1982, Ogilvie and Horowitz 1983, Jurkiewicz and Zalewski 1983a, b). In § 5 we give some general conclusions of our work including a discussion of the transition amplitude in terms of a quasi-coherent state representation. The effective Lagrangian in the corresponding path integral representation of the transition amplitude describes classical dynamics on a complex Kähler phase space manifold, where the Kähler metric is determined by the generating functional.

2. Quasi-coherent states with real representations

We follow the construction given by Mukunda *et al* (1981) by making use of conventional coherent states. Let D be the dimension of the real representation, $M(g)$, of the group G under consideration and consider D creation (annihilation) operators $a^{\dagger}_{k,\alpha}(a_{k,\alpha})$, $\alpha = 1, \ldots, D$, in Fock space. We introduce in the usual manner the unitary operator

$$U(f) = \exp[(a^{\dagger}, f) - (f^{*}, a)], \tag{2.1}$$

where the D-dimensional vector f describes a one-particle state which transforms according to the $M(g)$ representation and

$$(a^{\dagger}, f) = a^{\dagger}_{k,\alpha} f_{k,\alpha} \tag{2.2}$$

defines a scalar product. The index k can be thought of as a labelling of, for example, the momentum degrees of freedom of the one-particle state. The conventional coherent state $|f\rangle$ is then given by

$$|f\rangle = U(f)|0\rangle, \tag{2.3}$$

where $|0\rangle$ is the Fock-space vacuum and, furthermore,

$$a_{k,\alpha}|f\rangle = f_{k,\alpha}|f\rangle. \tag{2.4}$$

For later purposes we also recall the expression for the overlap between two coherent states (2.3) i.e.

$$\langle f|g\rangle = \exp[(f^*, g) - (f^*, f)/2 - (g^*, g)/2]. \tag{2.5}$$

In the following we will suppress the index k but, when appropriate, it is straightforward to make it explicit.

In order to extract from the coherent state (2.3) the component which transforms according to a given representation $D^{\{n\}}_{ab}$ of the group G, we construct the following quasi-coherent state:

$$|D^{\{n\}}_{ab}; f\rangle = M^{\{n\}}_{ab}(f) \int_G \mathrm{d}g\, D^{\{n\}}_{ab}(g)|M(g)f\rangle \tag{2.6}$$

where, of course, only those representations can occur which can be generated by the representation $M(g)$. In (2.6) $\mathrm{d}g$ stands for the invariant Haar-measure on the group (see, for example, Talman 1968) and $M^{\{n\}}_{ab}(f)$ is a normalisation factor which, in general, depends on the representation chosen. $\{n\}$ stands for the sequence of integers, $n_0 \leq n_1 \leq \ldots \leq n_{N-1}$, which characterise the irreducible representation $D^{\{n\}}_{ab}$ (see, for example, Weyl 1949). By its very construction it is now obvious that the state (2.6) transforms irreducibly under group actions which, of course, can be verified by an explicit calculation. The normalisation factor $M^{\{n\}}_{ab}(f)$ can be evaluated by making use of (2.5) i.e.

$$\begin{aligned}\langle f; D^{\{n\}}_{ab}|D^{\{n\}}_{cd}; f\rangle \\ &= |M^{\{n\}}_{ab}(f)|^2 \exp[-(f^*, f)] \int_G \mathrm{d}g\, \mathrm{d}h\, D^{\{n\}}_{ab}(g)^* D^{\{n\}}_{cd}(h) \\ &\quad \times \exp[(f^*, M(g^{-1}h)f)].\end{aligned} \tag{2.7}$$

The invariance of the group measure and the orthogonality relation

$$\int_G \mathrm{d}g\, D^{\{n\}}_{ab}(g) D^{\{m\}}_{cd}(g)^* = \delta(\{n\}, \{m\})\delta_{ac}\delta_{bd}/d_{\{n\}}, \tag{2.8}$$

where $d_{\{n\}}$ is the dimension of the representation $D^{\{n\}}_{ab}$, then leads to the following definition of normalised quasi-coherent states:

$$|D^{\{n\}}_{ab}; f\rangle = \exp[(f^*, f)/2] M^{\{n\}}_{b}(f)^{-1/2} \int_G \mathrm{d}g\, D^{\{n\}}_{ab}(g)|M(g)\rangle. \tag{2.9}$$

Here $M^{\{n\}}_{b}(f) \equiv M^{\{n\}}_{bb}(f^*, f)/d_{\{n\}}$, where $M^{\{n\}}_{ab}(h^*, f)$ is given by

$$M^{\{n\}}_{ab}(h^*, f) = \int_G \mathrm{d}g\, D^{\{n\}}_{ab}(g) \exp[(h^*, M(g)f)]. \tag{2.10}$$

$M^{\{n\}}_{b}(f)$ is a real quantity as expected.

The overlap between two quasi-coherent states (2.9) can be evaluated in an analogous manner with the following result

$$\langle h; D^{\{n\}}_{ab}|D^{\{m\}}_{cd}; f\rangle = \delta(\{m\},\{n\})\delta_{ac}\delta_{bd}M^{\{n\}}_{bb}(h^*, f)[M^{\{n\}}_b(h)M^{\{n\}}_b(f)]^{-1/2}. \tag{2.11}$$

Expectation values of normal ordered operators, invariant under group transformations, can easily be obtained from the generating functional (2.10). Let us illustrate the procedure by considering the number operator

$$N = a^\dagger_{k,\alpha}a_{k,\alpha}. \tag{2.12}$$

Proceeding as in the calculation presented above, we find that

$$\langle f; D^{\{n\}}_{ab}|N|D^{\{n\}}_{ab}; f\rangle = (M^{\{n\}}_b(f)d_{\{n\}})^{-1}\int_G \mathrm{d}g\, D^{\{n\}}_{bb}(g)(f^*, M(g)f)\exp[(f^*, M(g)f)]. \tag{2.13}$$

The equations (2.10) and (2.13) can then be combined to give

$$\langle f; D^{\{n\}}_{ab}|N|D^{\{n\}}_{ab}; f\rangle = \partial/\partial\lambda \ln[M^{\{n\}}_{bb}(f^*, \lambda f)]|_{\lambda=1}. \tag{2.14}$$

The corresponding expression for the dispersion of the number operator, $[\langle\Delta N\rangle_{\{n\},a,b,f}]^2$, has the same formal structure as in the case of quasi-coherent SU(2) states discussed in the literature (Mukunda *et al* 1981) and reads

$$[\langle\Delta N\rangle_{\{n\},a,b,f}]^2 = (\partial^2/\partial\lambda^2 + \partial/\partial\lambda)\ln[M^{\{n\}}_{bb}(f^*, \lambda f)]|_{\lambda=1}. \tag{2.15}$$

By making use of the completeness relation

$$\sum_{\{n\},a,b} d_{\{n\}}D^{\{n\}}_{ab}(g)D^{\{n\}}_{ab}(g')^* = \delta(g, g'), \tag{2.16}$$

we can easily verify (compare the discussion by Skagerstam (1979, 1980), Mukunda *et al* (1982)) the following completeness relation for the quasi-coherent states (2.9)

$$1 = \sum_{\{n\},a,b}\int \mathrm{d}f \exp[-(f^*, f)]M^{\{n\}}_b(f)|D^{\{n\}}_{ab}; f\rangle\langle D^{\{n\}}_{ab}; f|. \tag{2.17}$$

As is the case for the SU(2) quasi-coherent states constructed in Mukunda *et al* (1981), a further representation reduced set of quasi-coherent states can be constructed in a straightforward manner which are labelled only by the Casimir invariants of the group G

$$|\{n\}; f\rangle = \{d_{\{n\}}\exp[(f^*, f)]/M^{\{n\}}(f)\}^{1/2}\int_G \mathrm{d}g\,\chi_{\{n\}}(g)|M(g)f\rangle, \tag{2.18}$$

where $M^{\{n\}}(f) \equiv M^{\{n\}}(f^*, f)$ and

$$M^{\{n\}}(f^*, h) = \int_G \mathrm{d}g\,\chi_{\{n\}}(g)\exp[(f^*, M(g)h)]. \tag{2.19}$$

The overlap between two quasi-coherent states (2.18) can easily be computed to be

$$\langle f; \{n\}|\{m\}; h\rangle = \delta(\{n\},\{m\})M^{\{n\}}(f^*, h)/[M^{\{n\}}(f)M^{\{n\}}(h)]^{1/2} \tag{2.20}$$

and the following completeness relation holds

$$1 = \sum_{\{n\}} d_{\{n\}}\int \mathrm{d}f \exp[-(f^*, f)]M^{\{n\}}(f)|\{n\}; f\rangle\langle f; \{n\}|. \tag{2.21}$$

Expectation values of G-invariant operators can then be derived in terms of the generating functional (2.19) in a way similar to the one used in the discussion of the quasi-coherent states (2.9).

We realise that in any practical application of the quasi-coherent states (2.9) (some applications have already been indicated by Mukunda *et al* 1981) the analytical structure of the generating functional (2.10) is essential. It turns out however to be a rather difficult problem in itself to evaluate the integral over the group. For reasons of simplicity and because of its practical importance, we now consider the adjoint representation of the group SU(N) (or U(N)) in which case (2.10) takes the following form

$$M(h^*,f)=\int_G \exp[\mathrm{Tr}(h^\dagger g f g^\dagger)/2], \tag{2.22}$$

where

$$f=f_\alpha\lambda_\alpha, \qquad h=h_\beta\lambda_\beta, \tag{2.23}$$

and where λ_α, normalised in such a way that $\mathrm{Tr}(\lambda_\alpha\lambda_\beta)=2\delta_{\alpha\beta}$, generates the fundamental representation of the SU(N) groups (or U(N)). Integrals of the form (2.22) have actually been considered in the literature in the context of the planar approximation (Itzykson and Zuber 1980) and in the study of chiral U(N)⊗U(N) models (Bars *et al* 1983, Brihaye and Rossi 1983). In terms of a character expansion, (2.22) takes the form ($n_0=0$ for $G=\mathrm{SU}(N)$)

$$M(h^*,f)=\sum_{\{n\}} \sigma_{\{n\}}/|n|!\,d_{\{n\}}\chi_{\{n\}}(h^\dagger)\chi_{\{n\}}(f), \tag{2.24}$$

where $\sigma_{\{n\}}$ is the number of times the representation $D^{\{n\}}(U)$ occurs in the tensor product $\otimes^{|n|}U$ and $|n|=\Sigma_0^{N-1}n_i$. If we consider diagonal elements only, (2.24) can be written in a more explicit form. In this case the matrix f can be diagonalised i.e. it is sufficient to consider f in a diagonal form $f=\mathrm{diag}(\lambda_1,\ldots,\lambda_N)$. (2.22) can then be evaluated explicitly with the result (Itzykson and Zuber 1980)

$$M(f^*,f)=\left(\prod_{p=0}^{N-1} p!\right)\det[\exp(\lambda_i\lambda_j^*)]\Big/\left|\prod_{i>j}(\lambda_i-\lambda_j)\right|^2. \tag{2.25}$$

For the group SU(2), (2.25) reduces to

$$M(f^*,f)=\exp(\lambda_i^*\lambda_i)[1-\exp(-|\lambda_1-\lambda_2|^2)]/|\lambda_1-\lambda_2|^2. \tag{2.26}$$

An analogous expression can be similarly derived for the group SU(3) which for two eigenvalues equal ($\lambda_2=\lambda_3$) reduces to

$$M(f^*,f)=2\exp(\lambda_i^*\lambda_i)[1-|\lambda_1-\lambda_2|^2\exp(-|\lambda_1-\lambda_2|^2)-\exp(-|\lambda_1-\lambda_2|^2)]/|\lambda_1-\lambda_2|^4. \tag{2.27}$$

Similar expressions can be derived for higher-dimensional unitary groups.

3. Quasi-coherent states for complex representations

In the present section we study the construction of quasi-coherent states for one-particle states which transform according to a complex representation, $C(g)$, of dimension d. Such a situation occurs when one is considering particles and anti-particles which transform differently under group actions. In practical applications complex representations may for example be associated with fermions (see e.g. Skagerstam 1983, 1984).

In order to take the Dirac–Fermi statistics properly into account one can make use of a coherent state representation of fermionic operators (Klauder 1977, Ohnuki 1977) and a corresponding integration over Grassmannian variables (Berezin 1966). Here we will restrict ourselves to bosonic variables but, as will be clear from the presentation below, it is straightforward to extend our discussion to fermionic variables.

When we consider a complex representation $C(g)$ we must explicitly distinguish between particles and anti-particles. We introduce creation and annihilation operators for particles (anti-particles) $a^\dagger_{k,\alpha}(b^\dagger_{k,\alpha})$ and $a_{k,\alpha}(b_{k,\alpha})$ where $\alpha = 1, \ldots, d$. Proceeding as in the construction of U(1)-charged quasi-coherent states (Skagerstam 1979, 1980) we then consider the coherent state

$$|f, h\rangle = U_1(f)U_2(h)|0\rangle, \tag{3.1}$$

where U_1 (U_2) corresponds to the U-operator defined by the equation (2.1) for particles (anti-particles). A quasi-coherent state which transforms according to the representation $D^{\{n\}}_{ab}$ is then defined by

$$|D^{\{n\}}_{ab}, f, h\rangle = \exp[(f^*, f)/2 + (h^*, h)/2]C^{\{n\}}_b(f, h)^{-1/2} \int_G \mathrm{d}g\, D^{\{n\}}_{ab}(g)|C(g)f, C(g)^*h\rangle, \tag{3.2}$$

where the normalisation constant $C^{\{n\}}_b(f, h)$ can be evaluated by making use of the procedure given in § 2 with the result $C^{\{n\}}_b(f, h) \equiv C^{\{n\}}_{bb}(h^*, f^*|f, h)/d_{\{n\}}$. Here we have introduced the generating functional

$$C^{\{n\}}_{ab}(h^*_1, f^*_1|f_2, h_2) = \int_G \mathrm{d}g\, D^{\{n\}}_{ab}(g) \exp[\mathrm{Tr}(CF^\dagger_{12} + C^\dagger H_{12})]. \tag{3.3}$$

In (3.3) the matrix F_{12} has the matrix elements $[F_{12}]_{\alpha\beta} = f^*_{1\alpha} f_{2\beta}$ and similarly for the matrix H_{12}. A completeness relationship can, furthermore, be derived for the quasi-coherent states (3.2) with the following result

$$1 = \sum_{\{n\}} \int \mathrm{d}f\, \mathrm{d}h \exp[-(f^*, f) - (h^*, h)]C^{\{n\}}_b(f, h)|D^{\{n\}}_{ab}; f, h\rangle\langle f, h; D^{\{n\}}_{ab}|. \tag{3.4}$$

Reduction of coherent states in terms of characters can also be carried out in complete analogy with the construction of the quasi-coherent states in terms of characters as discussed in § 2 (cf (2.18)).

Expectation values of operators in the quasi-coherent state (3.2) can be obtained from the generating functional (3.3) in a fashion similar to the discussion of quasi-coherent states with real representations. For the number operator

$$N = a^\dagger_{k,\alpha} a_{k,\alpha} + b^\dagger_{k,\alpha} b_{k,\alpha}, \tag{3.5}$$

the expectation value becomes

$$\langle f, h; D^{\{n\}}_{ab}|N|D^{\{n\}}_{ab}; f, h\rangle = \partial/\partial\lambda \ln[C^{\{n\}}_{bb}(h^*, f^*|\lambda f, \lambda h)]|_{\lambda=1}. \tag{3.6}$$

For the dispersion of the number operator, $[\langle \Delta N\rangle_{\{n\},a,b,f,h}]^2$, we obtain in analogy with the equation (2.15) the following result

$$[\langle \Delta N\rangle_{\{n\},a,b,f,h}]^2 = [\partial^2/\partial\lambda^2 + \partial/\partial\lambda]\ln[C_{ab}^{\{n\}}(h^*, f^*|\lambda f, \lambda h)], \tag{3.7}$$

evaluated for $\lambda = 1$.

In the case of the fundamental representation of the groups SU(N) and U(N), and for $F \equiv F_{12} = H_{12}$, the generating functional (3.3) has been studied in detail for the singlet representation in the literature either in the context of the external field problem in QCD (Brezin and Gross 1980) or in the context of one-link integrals in the lattice regularisation of QCD (Brower and Nauenberg 1981, Bars 1981, Brower *et al* 1981, Eriksson *et al* 1981, Fateev and Onofri 1981). For the U(N) group the following expression has been derived (Brower *et al* 1981) in terms of the eigenvalues x_i of the matrix $FF^\dagger$ for the singlet generating functional $C(FF^\dagger) \equiv C(f^*, f^*|f, f)$

$$C(FF^\dagger) = \left(2^{N(N-1)/2} \prod_{p=0}^{N-1} p!\right) \det[\lambda_j^{i-1} I_{i-1}(\lambda_j)]/\det[(\lambda_j^2)^{i-1}], \tag{3.8}$$

where $\lambda_i = 2\sqrt{x_i}$. For the groups SU(N) and U(N) with $N \leqslant 3$ there exists explicit expressions for $C(FF^\dagger)$ in terms of the invariants of the matrix $FF^\dagger$ (Eriksson *et al* 1981).

4. Field-theoretical considerations

In order to be explicit, we consider free quantum fields which transform according to the fundamental or the adjoint representation. We furthermore restrict ourselves to the singlet representation quasi-coherent states. Remarkably enough, a rather rich structure is exhibited in these rather trivial examples as will be clear from the presentation below.

Let us first consider the fundamental representation. The free field Hamiltonian then reads

$$H_0 = \int d\mu(k)\omega(k)[a_\alpha^\dagger(k)a_\alpha(k) + b_\alpha^\dagger(k)b_\alpha(k)], \tag{4.1}$$

where $d\mu(k) \equiv d^3k/2\omega(k)$ is the Lorentz-invariant phase-space measure. The canonical commutation relations have been normalised in such a way that

$$[a_\alpha(k), a_\beta^\dagger(k')] = 2\omega(k)\delta^3(k-k')\delta_{\alpha\beta}, \tag{4.2}$$

and similarly for the anti-particles. $\omega(k)$ is, of course, the energy of the one-particle state. The scalar product as defined by (2.2) is in terms of the invariant normalisation

$$(a^\dagger, f) = \int d\mu(k)a_\alpha^\dagger(k)f_\alpha(k) \tag{4.3}$$

and the Fock-space states are obtained by the successive actions of the operators $(a^\dagger, f)$ and $(b^\dagger, h)$ on the vacuum state. Let us now consider the singlet quasi-coherent state $|D_{ab}^{\{n\}} = 1; f, f\rangle \equiv |f\rangle_s$ and the expectation value of the free field Hamiltonian H_0 defined above divided by the number of internal degrees of freedom (N^2) i.e.

$$\langle H_0\rangle \equiv {}_s\langle f|H_0|f\rangle_s/N^2 = E_0/2N^2[\lambda^{-1}\partial/\partial\lambda \ln Z_\lambda(FF^\dagger)]_{\lambda=1}, \tag{4.4}$$

where $Z_\lambda(FF^\dagger)$ is the (one-link) integral

$$Z_\lambda(FF^\dagger)=\int_G \mathrm{d}g\ \exp[\lambda\ \mathrm{Tr}(F^\dagger g+Fg^\dagger)]. \tag{4.5}$$

Here we have for reasons of simplicity assumed that the one-particle overlap integral matrix F in (4.4) satisfies

$$[F]_{\alpha\beta}=\int \mathrm{d}\mu\ (k) f^*_\alpha(k) f_\beta(k)=2\lambda/E_0\int \mathrm{d}\mu\ (k)\omega(k) f^*_\alpha(k) f_\beta(k). \tag{4.6}$$

If the matrix F is diagonal, i.e. $[F]_{\alpha\beta}=\lambda N\delta_{\alpha\beta}$, then $Z_\lambda(FF^\dagger)$ takes the following form

$$Z(\lambda)\equiv Z_\lambda(FF^\dagger)=\int_G \mathrm{d}g\ \exp\{\lambda N[\chi_F(g)+\chi_F(g)^*]\}, \tag{4.7}$$

and (4.3) should be evaluated for an arbitrary λ. E_0 is, for a diagonal F, the expectation value of H_0 in the conventional coherent states $|f,f\rangle$ divided by the number of internal degrees of freedom (N^2). Similarly λ is related to the mean value of the number operator i.e. $2\lambda=\langle f,f|N|f,f\rangle/N^2$. χ_F is the character of the fundamental representation. We recognise in (4.7) the one-plaquette partition function for the two-dimensional lattice gauge theory with Wilson's action and with the gauge group G (Gross and Witten 1980. See also Wadia 1979). In the large-N limit, $Z(\lambda)$ has been computed exactly by means of steepest descent methods with the following results for $\langle H_0\rangle$

$$\langle H_0\rangle=\begin{cases} E_0\lambda, & \lambda\leqslant 0.5\\ E_0(1-\frac{1}{4}\lambda), & \lambda\geqslant 0.5.\end{cases} \tag{4.8}$$

The 'free energy', $-\ln Z(\lambda)$, and its first and second derivatives with respect to λ are continuous at $\lambda=0.5$. Equation (4.8) exhibits a third-order phase transition since the third derivative of the free energy is discontinuous at $\lambda=0.5$. The existence of this third-order phase transition actually persists in the case of a general one-particle matrix $[F]_{\alpha\beta}$ (Brezin and Gross 1980). $\langle H_0\rangle$ as a function of λ is exhibited in figure 1. For finite λ the singlet quasi-coherent state is less condensed as compared to a conventional coherent state $|f,f\rangle$ (cf the discussion by Mukunda *et al* 1981).

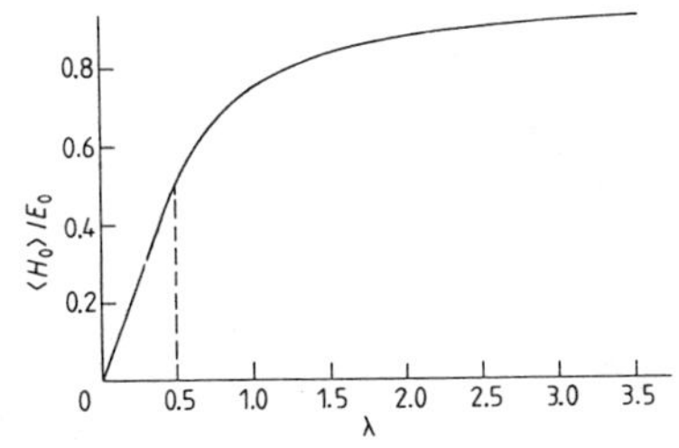

Figure 1. The large-N limit of the expectation value of the free field Hamiltonian, with fields transforming according to the fundamental representation, in a singlet quasi-coherent state. E_0 is defined in the text. $\lambda=0.5$ corresponds to a third-order phase transition.

The discussion above can easily be carried over to the adjoint representation in which case the free field Hamiltonian reads

$$H_0 = \int d\mu\,(k)\omega(k)a_\alpha^\dagger(k)a_\alpha(k). \tag{4.9}$$

The corresponding expectation value in a singlet quasi-coherent state then becomes

$$\langle H_0\rangle = E_0/N^2[\lambda^{-1}\partial/\partial\lambda\ \ln Z_\lambda(FF^\dagger)]_{\lambda=1} \tag{4.10}$$

where $Z_\lambda(FF^\dagger)$ is now given by

$$Z_\lambda(FF^\dagger) = \int_G dg\ \exp\{\tfrac{1}{2}[F]_{\alpha\beta}\ \mathrm{Tr}(g\lambda_\alpha g^\dagger\lambda_\beta)\}, \tag{4.11}$$

and where we have assumed a relation similar to (4.6). If the matrix F is diagonal, i.e. $[F]_{\alpha\beta} = \lambda\delta_{\alpha\beta}$, equation (4.10) should be evaluated for an arbitrary λ. As can easily be verified, E_0 is then the expectation value of H_0 in the coherent state $|f\rangle$ divided by the number of internal degrees of freedom (N^2) and $\lambda = \langle f|N|f\rangle/N^2$. $Z_\lambda(FF^\dagger)$ then becomes the one-plaquette partition function for two-dimensional lattice gauge theory with Wilson's action in the adjoint representation (Chen and Zheng 1982, Makeenko and Polikarpov 1982, Samuel 1982, Ogilvie and Horowitz 1983, Jurkiewicz and Zalewski 1983a, b) i.e.

$$Z(\lambda) \equiv Z_\lambda(FF^\dagger) = \int_G dg\ \exp(\lambda\chi_A), \tag{4.12}$$

where χ_A is the character of the adjoint representation. In the large-N limit, $Z(\lambda)$ can be computed exactly with the following result

$$\langle H_0\rangle = \begin{cases} 0, & \lambda < 1 \\ E_0/4[1+(1-1/\lambda)^{1/2}]^2, & \lambda \geq 1. \end{cases} \tag{4.13}$$

In figure 2 we exhibit $\langle H_0\rangle$ as a function of λ. In the large-N limit we therefore obtain a first-order phase transition for one-particle states transforming according to the adjoint representation. For $\lambda < 1$ there are not sufficiently many states available to form a non-trivial singlet state and, hence, the only attainable state is the vacuum state.

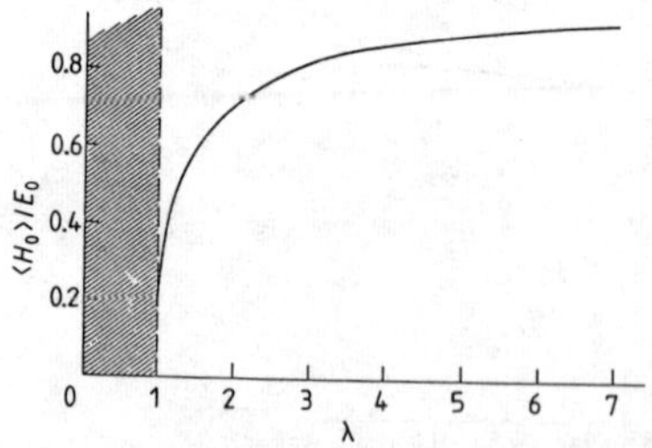

Figure 2. The large-N limit of the expectation value of the free field Hamiltonian, where the fields transform according to the adjoint representation, in a singlet quasi-coherent state. E_0 is defined in the text. For $\lambda < 1$, the shaded area in the figure, the only attainable state is the zero-energy state due to the existence of a first-order phase transition at $\lambda = 1$.

One can trace the presence of the first-order phase transition back to the spontaneous breaking of the global symmetry under the action of the central elements of SU(N). In the large-N limit the central elements, Z_N, form a U(1) sub-group which, for $\lambda \geqslant 1$, is spontaneously broken. This breaking of the central U(1) symmetry is similar to the breaking of the $U(1)^d$-symmetry breaking in the quenched Eguchi–Kawai model (Eguchi and Kawai 1982, Bhanot *et al* 1982). It is amazing to notice that the existence of a first-order phase transition for an ideal, colourless gluon gas in the large-N limit (Skagerstam 1983, 1984) has a formal counterpart in the first-order phase transition exhibited in (4.13).

More generally we can, of course, consider a free field Hamiltonian with, for example, one set of fields transforming according to the fundamental representation and another set of fields which transforms according to the adjoint representation. The generating functionals (4.7) and (4.11) will then combine to yield the one-plaquette partition function for the two-dimensional lattice gauge theory with the mixed action (Chen and Zheng 1982, Makeenko and Polikarpov 1982, Samuel 1982, Ogilvie and Horowitz 1983, Jurkiewicz and Zalewski 1983a, b). Clearly, the analysis can be extended to any representation. The large-N limit of the corresponding generating functional has been studied to some extent in the literature (Yee 1983). The general structure which emerges from such a study is a phase-diagram with, in general, first and higher-order phase transitions in the sense defined above.

For interacting fields a perturbative scheme can, of course, be developed in terms of quasi-coherent states. We also mention the possibility of performing a variational calculation for interacting fields along the lines suggested by Ni and Wang (1983).

5. Final remarks

Coherent states have the property that they closely describe the classical dynamics of a given dynamical system under consideration (Klauder and Sudarshan 1968, Glauber 1963, 1964). In the present paper we have considered quasi-coherent states appropriate for dynamical systems where an additional constraint on the dynamics is present, namely the existence of a conserved and, in general, non-Abelian charge.

It has been pointed out by Klauder (1978, 1979, 1982, 1983) and Kuratsuji and Suzuki (1980a, b, 1981, 1983) (see also in this context Shankar 1980) that a path integral representation of the transition amplitude in terms of generalised coherent states may be useful in order to study the relationship between the quantum and classical aspects of a dynamical system with a 'curved phase space' as in the case of a spin system. In such a context it is natural to extend these considerations by making use of quasi-coherent states. Let H be a Hamiltonian which commutes with the generators of a Lie group G and let $|\psi\rangle$ denote a corresponding quasi-coherent state with the following decomposition of the unity operator

$$1 = \int d\psi |\psi\rangle\langle\psi|, \tag{5.1}$$

where we have suppressed all group-theoretical indices. A path integral representation of the transition amplitude

$$T_{12} \equiv T(\psi_2, t_2; \psi_1, t_1) = \langle\psi_2| \exp[-i/\hbar H(t_2 - t_1)]|\psi_1\rangle \tag{5.2}$$

can then easily be written down

$$T_{12}=\lim_{n\to\infty}\int\ldots\int\prod_{k=1}^{n-1}\mathrm{d}\psi_k\prod_{l=1}^{n}\langle\psi_l|\psi_{l-1}\rangle\prod_{m=1}^{n}[1-\mathrm{i}\varepsilon/\hbar\langle\psi_m|H|\psi_{m-1}\rangle/\langle\psi_m|\psi_{m-1}\rangle] \tag{5.3}$$

where $\varepsilon=(t_2-t_1)/n$. We now specifically consider a real representation, $M(g)$, and quasi-coherent states of the form (2.18) i.e. states which are diagonal in terms of the Casimir invariants. Other cases can be treated similarly. By making use of (2.19) and (2.20) we derive

$$\langle f+\Delta f;\{n\}|\{n\};f\rangle=\exp\{\tfrac{1}{2}[(\Delta f^*,g^{\{n\}}f)-(f^*,g^{\{n\}}\Delta f)]\}[1+\mathrm{O}((\Delta f)^2)] \tag{5.4}$$

where we have defined a metric

$$g_{\alpha\beta}^{\{n\}}(f^*,f)=M^{\{n\}}(f^*,f)^{-1}\int_G \mathrm{d}g\chi_{\{n\}}(g)M_{\alpha\beta}(g)\exp[(f^*,M(g)f)]. \tag{5.5}$$

The transition amplitude T_{12} can then be written as follows

$$\begin{aligned}T_{12}&\equiv\langle f_2;\{n\};t_2|t_1;\{m\};f_1\rangle\\&=\delta(\{n\},\{m\})\int D[f(t)]\exp\{\mathrm{i}/\hbar S[f^*(f),\dot f^*(t),f(t),\dot f(t)]\},\end{aligned} \tag{5.6}$$

where the action S is defined by

$$S=\int_1^2 \mathrm{d}t\,\mathscr{L}[f^*(t),\dot f^*(t),f(t),\dot f(t)]. \tag{5.7}$$

The Lagrangian density $\mathscr{L}$ is defined by the Legendre transformation

$$\mathscr{L}=\mathrm{i}\hbar/2[(f^*,g^{\{n\}}\dot f)-(\dot f^*,g^{\{n\}}f)]-\mathscr{H}(f^*,f) \tag{5.8}$$

and

$$\mathscr{H}(f^*,f)=\delta(\{n\},\{m\})\langle f;\{n\}|H|\{m\};f\rangle. \tag{5.9}$$

The functional integration measure $D[f(t)]$ is derived by making use of the decomposition of the unity operator (2.21) i.e. formally the measure has the form

$$D[f(t)]=\prod_t\sum_{\{n\}}\mathrm{d}f(t)\exp[-(f^*(t),f(t))]M^{\{n\}}(f^*(t),f(t))\,d_{\{n\}}. \tag{5.10}$$

The presence of the metric $g_{\alpha\beta}^{\{n\}}(f^*,f)$ in the 'kinetic' part of $\mathscr{L}$ is related to the geometrical structure of the phase space in a fashion similar to the complex Kähler manifold derived by making use of the SU(2) generalised spin coherent states (Klauder 1978, 1979, 1982, 1983, Kuratsuji and Suzuki 1980a, b, 1981, 1983). The equations of motion derived from the action (5.7) in general take the following form

$$\mathrm{i}\hbar G_{\alpha\beta}^{\{n\}}(f^*,f)\dot f_\beta^*=\partial\mathscr{H}/\partial f_\alpha,\qquad \mathrm{i}\hbar G_{\alpha\beta}^{\{n\}}(f^*,f)\dot f_\beta=-\partial\mathscr{H}/\partial f_\alpha^* \tag{5.11}$$

where

$$G_{\alpha\beta}^{\{n\}}(f^*,f)=\partial^2\ln M^{\{n\}}(f^*,f)/\partial f_\alpha\partial f_\beta^* \tag{5.12}$$

defines a metric on a Kähler manifold (Kobayashi and Nomizu 1969).

Semi-classical aspects on the dynamical system defined by the Langrangian (5.8) have been discussed in the literature (Kuratsiju and Suzuki 1980a, b, 1981, 1983) and we will not dwell further on this question in the present paper. Here we only make the following observations. If H describes the interaction of a Yang–Mills particle with an external gauge field (Balachandran *et al* 1983), a path integral representation of T_{12} in terms of conventional coherent states yields an effective action of the form (5.8) with $g^{\{n\}}_{\alpha\beta} = \delta_{\alpha\beta}$. The corresponding c-number Lagrangian has been studied in detail in the literature (Balachandran *et al* 1977), where it was observed that the corresponding canonical quantisation in general leads to reducible representations of the group G. Allowing for only one irreducible representation to contribute in the derivation of (5.6) therefore corresponds to the non-trivial form of the metric $G^{\{n\}}_{\alpha\beta}$ (or $g^{\{n\}}_{\alpha\beta}$). The canonical quantisation of the dynamical system described by the Lagrangian (5.8) must, of course, yield only one irreducible representation of the group G.

As a final remark we notice that the probability, $P(\Delta E)$, for the emission of soft non-Abelian massless vector bosons up to a certain total energy ΔE from a classical c-number source, neglecting self-interactions among the vector bosons, can be computed along the lines of Skagerstam (1979, 1980). By making use of the asymptotic form of equation (2.25) it can then be verified that one obtains an infrared finite result for $P(\Delta E)$ with a structure similar to the corresponding expression in quantum electrodynamics (cf Mukunda *et al* 1981 and references cited therein).

Acknowledgments

The author wishes to thank M Jacob and the CERN Theory Division for hospitality for the period when the present paper was completed and J R Klauder for his encouragement. Guang-Jiong Ni is also acknowledged for valuable discussions.

References

Arecchi F T, Courtens E, Gilmore R and Thomas H 1972 *Phys. Rev.* A **6** 2211
Balachandran A P, Marmo G, Skagerstam B-S and Stern A 1983 *Gauge Symmetries and Fibre Bundles Applications to Particle Dynamics, Lecture Notes in Physics* (Berlin: Springer)
Balachandran A P, Salomonson P, Skagerstam B-S and Winnberg J-O 1977 *Phys. Rev.* D **15** 2308
Bars I 1981 *Phys. Scr.* **23** 983
Bars I, Günaydin M and Yankielowicz S 1983 *Nucl. Phys.* B **219** 81
Berezin F A 1966 *The Method of Second Quantization* (New York: Academic)
Bhanot G, Heller U and Neuberger H 1982 *Phys. Lett.* **113B** 47
Bhaumik D, Bhaumik K and Dutta-Roy B 1976 *J. Phys. A: Math. Gen.* **9** 1507
Brezin E and Gross D J 1980 *Phys. Lett.* **97B** 120
Brihaye Y and Rossi P 1983 *Nucl. Phys.* B **235** 226
Brower R and Nauenberg M 1981 *Nucl. Phys.* B **180** 221
Brower R, Rossi P and Tan C 1981 *Phys. Rev.* D **23** 942
Chen T and Zheng Z 1982 *Phys. Lett.* **109B** 383
Duff M J and Isham C J 1980 *Nucl. Phys.* B **162** 271
Eguchi T and Kawai H 1982 *Phys. Rev. Lett.* **48** 1063
Eriksson K-E and Skagerstam B-S 1979 *J. Phys. A: Math. Gen.* **12** 2175
—— 1981 *J. Phys. A: Math. Gen.* **14** 545
Eriksson K-E, Skagerstam B-S and Svartholm N 1981 *J. Math. Phys.* **22** 2276
Fateev V A and Onofri E 1981 *Lett. Math. Phys.* **5** 367
Gilmore R 1972 *Ann. Phys., NY* **74** 391
Glauber R J 1963 *Phys. Rev.* **131** 2766
—— 1964 in *Quantum Optics and Electronics* ed C Dewitt and C Cohen-Tannoudji (New York: Gordon and Breach)
Gross D and Witten E 1980 *Phys. Rev.* D **21** 446

Horn D and Silver R 1971 *Ann. Phys., NY* **66** 509
Itzykson C and Zuber J-B 1980 *J. Math. Phys.* **21** 411
Jurkiewicz J and Zalewski K 1983a *Nucl. Phys.* B **210** 167
—— 1983b *Acta Phys. Polon.* B **14** 517
Klauder J R 1963 *J. Math. Phys.* **4** 1058
—— 1977 *Acta Phys. Austriaca* suppl. XVIII 1
—— 1978 in *Path Integrals, Proceedings of the NATO Advanced Summer Institute*, ed G J Papadopoulos and J T Devreese (New York: Plenum)
—— 1979 *Phys. Rev.* D **19** 2349
—— 1982 *J. Math. Phys.* **23** 1797
—— 1983 *Acta Phys. Austriaca* suppl. XXV 251
Klauder J R and Skagerstam B-S 1984 *Coherent States—Applications to Physics and Mathematical Physics* (Singapore: World Scientific), to appear
Klauder J R and Sudarshan E C G 1968 *Fundamentals of Quantum Optics* (New York: Benjamin)
Kobayashi S and Nomizu K 1969 *Foundations of Differential Geometry*, vol 2 (New York: Wiley)
Kuratsuji H and Suzuki T 1980a *Phys. Lett.* **92B** 19
—— 1980b *J. Math. Phys.* **21** 472
—— 1981 *J. Math. Phys.* **22** 757
—— 1983 *Suppl. Prog. Theor. Phys.* **74/75** 209
Makeenko Y M and Polikarpov M I 1982 *Nucl. Phys.* B **205** 386
Mukunda N, Eriksson K-E and Skagerstam B-S 1981 *Phys. Rev.* D **24** 2615
Ni G-J and Wang Y-P 1983 *Phys. Rev.* D **27** 969
Ogilvie M C and Horowitz H 1983 *Nucl. Phys.* B **215** 249
Ohnuki Y 1977 *Lectures given at the National Laboratory for High Energy Physics (KEK), Tsukuba* (unpublished)
Perelemov A M 1972 *Commun. Math. Phys.* **26** 391
—— 1977 *Usp. Fiz. Nauk.* **123** 23 (Engl. Transl. 1978 *Sov. Phys. Usp.* **20** 703)
Radcliffe J M 1971 *J. Phys. A: Math. Gen.* **4** 313
Samuel S 1982 *Phys. Lett.* **112B** 237
Shankar R 1980 *Phys. Rev. Lett.* **45** 1088
Skagerstam B-S 1978 *Phys. Lett.* **69A** 76
—— 1979 *Phys. Rev.* D **19** 2471
—— 1980 *Phys. Rev.* D **22** 534
—— 1983 *Phys. Lett.* **133B** 419
—— 1984 *Z. Phys.* C **24** 97
Talman J D 1968 *Special Functions—A Group Theoretical Approach* (New York: Benjamin)
Wadia S 1979 *preprint EFI-79/44, University of Chicago* (unpublished)
Weyl H 1949 *The Classical Groups* (Princeton NJ: University Press)
Yee F G 1983 *Phys. Lett.* **124B** 225

III. Coherent States and Quantum Mechanics

PHYSICAL REVIEW LETTERS

VOLUME 14 15 MARCH 1965 NUMBER 11

COHERENT STATES AND THE NUMBER-PHASE UNCERTAINTY RELATION*

P. Carruthers†
Laboratory of Nuclear Studies and Laboratory of Atomic and Solid State Physics, Cornell University, Ithaca, New York

and

M. M. Nieto‡
Laboratory of Nuclear Studies, Cornell University, Ithaca, New York

(Received 25 January 1965)

Attention has recently been called to the fact that the traditional[1] number-phase "uncertainty relation" connecting the excitation number of an oscillator to the phase angle,

$$\Delta N \Delta\varphi \geq \tfrac{1}{2}, \qquad (1)$$

lacks precise meaning for small quantum numbers, since an appropriate Hermitian phase operator $\varphi_{\rm op}$ does not exist.[2,3] Our purpose here is (1) to propose a substitute for Eq. (1) in terms of the (well-defined) operators S and C, and (2) to evaluate the new expression for the so-called coherent states, whose importance has been stressed by Glauber[4] and Sudarshan.[5] It will be shown that for large N the coherent states are minimum-uncertainty number-phase states as well as minimum-uncertainty position-momentum states.[4] Even for very small N the uncertainty product is very small.

We consider a single mode of the radiation field, described by the usual harmonic-oscillator variables a, a^* obeying $[a, a^*]=1$. The operators S and C are defined[2] in terms of the operators $E_\pm$, whose classical analogs are $\exp(\mp i\psi)$, ψ being the classical phase. Denoting the number operator a^*a by $N_{\rm op}$, we have

$$E_- \equiv (N_{\rm op}+1)^{-1/2}a, \quad E_+ \equiv a^*(N_{\rm op}+1)^{-1/2}. \qquad (2)$$

The $E_\pm$ are one-sided unitary, as is shown by the following matrix elements in the number basis:

$$(E_-E_+)_{mn} = \delta_{mn}, \quad (E_+E_-)_{mn} = \delta_{mn} - \delta_{m0}\delta_{n0}. \qquad (3)$$

Here the integers m, n, denote the usual number states. The $E_\pm$ are raising and lowering operators:

$$E_\pm |m\rangle = |m \pm 1\rangle. \qquad (4)$$

Despite the nonunitary nature of $E_\pm$, we can define Hermitian "sine and cosine" operators[6]

$$S \equiv (1/2i)(E_- - E_+),$$
$$C \equiv \tfrac{1}{2}(E_- + E_+). \qquad (5)$$

From the commutation rules

$$[N_{\rm op}, S] = iC,$$
$$[N_{\rm op}, C] = -iS, \qquad (6)$$

one can deduce the uncertainty relations $[(\Delta x)^2 = \langle x^2\rangle - \langle x\rangle^2)]$

$$(\Delta N)^2(\Delta S)^2 \geq \tfrac{1}{4}\langle C\rangle^2,$$
$$(\Delta N)^2(\Delta C)^2 \geq \tfrac{1}{4}\langle S\rangle^2. \qquad (7)$$

The proposed relation, which treats the fluctuations in S and C symmetrically, and reduces to (1) in the appropriate limit, is deduced from (7):

$$U \equiv (\Delta N)^2 \frac{[(\Delta S)^2 + (\Delta C)^2]}{[\langle S\rangle^2 + \langle C\rangle^2]} \geq \frac{1}{4}. \qquad (8)$$

VOLUME 14, NUMBER 11 PHYSICAL REVIEW LETTERS 15 MARCH 1965

The coherent states, well known in the study of radiation emitted by a classical current source,[7,8] are defined by[9]

$$a|\alpha\rangle = \alpha|\alpha\rangle;$$

$$|\alpha\rangle = \exp[-(\tfrac{1}{2})|\alpha|^2]\sum_{n=0}^{\infty}\frac{\alpha^n}{(n!)^{1/2}}|n\rangle. \qquad (9)$$

The average number N is $|\alpha|^2$; the states $|n\rangle$ are Poisson distributed and $(\Delta N)^2 = N$. The complex number α has magnitude $N^{1/2}$ and its phase corresponds to the classical phase angle.[10]

The foregoing equations can now be used to give

$$\langle\alpha|S^2|\alpha\rangle = \tfrac{1}{2} - \tfrac{1}{4}e^{-N} - \tfrac{1}{2}e^{-N}N(1-2\xi) \times\sum_{n=0}^{\infty}\frac{N^n}{n![(n+1)(n+2)]^{1/2}}. \qquad (10)$$

In Eq. (10) the parameter ξ is defined by

$$\xi = (\mathrm{Im}\alpha)^2/[(\mathrm{Re}\alpha)^2 + (\mathrm{Im}\alpha)^2], \qquad (11)$$

and lies in the range $0 \leq \xi \leq 1$. Also,

$$\langle\alpha|S|\alpha\rangle = e^{-N}\,\mathrm{Im}\alpha\sum_{n=0}^{\infty}\frac{N^n}{n!(n+1)^{1/2}}. \qquad (12)$$

The corresponding results for C follow by substituting $1-\xi$ for ξ in (10), (11), and (12). [In (12) this means $\mathrm{Im}\alpha \rightarrow \mathrm{Re}\alpha$.]

It is interesting to note that

$$\langle\alpha|C^2 + S^2|\alpha\rangle = 1 - \tfrac{1}{2}e^{-N}, \qquad (13)$$

which has limits $\frac{1}{2}$ and 1 for N small and large, respectively. The asymptotic expressions (for large N)[11]

$$\sum_{n=0}^{\infty}\frac{N^n}{n![(n+1)(n+2)]^{1/2}} \sim \frac{e^N}{N}\left[1 - \frac{1}{2N} - \frac{1}{8N^2} + \cdots\right], \qquad (14)$$

$$\sum_{n=0}^{\infty}\frac{N^n}{n!(n+1)^{1/2}} \sim \frac{e^N}{N^{1/2}}\left[1 - \frac{1}{8N} + \cdots\right], \qquad (15)$$

are also useful. In particular, we note the result

$$(\langle\alpha|C|\alpha\rangle)^2 + (\langle\alpha|S|\alpha\rangle)^2 = Ne^{-2N}\left[\sum_{n=0}^{\infty}\frac{N^n}{n!(n+1)^{1/2}}\right]^2 \sim 1 - \frac{1}{4N}. \qquad (16)$$

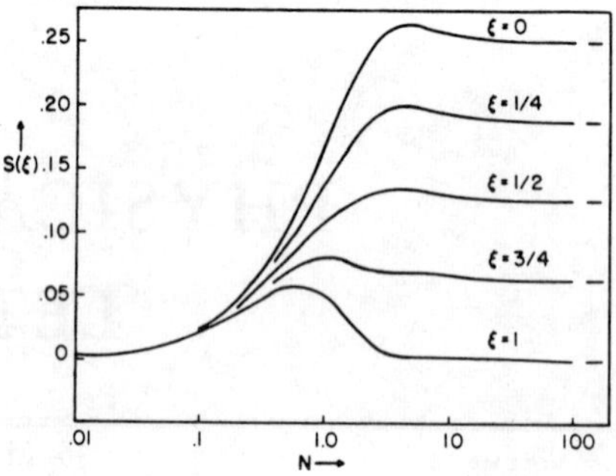

FIG. 1. The uncertainty product $S(\xi) = (\Delta N)^2(\Delta S^2)$ is shown as a function of $(\Delta N)^2 = N$ for various values of the parameter ξ defined in Eq. (11). All expectation values refer to coherent states.

We have evaluated expressions entering in Eqs. (7) and (8) for coherent states having a wide range of mean excitation N. The cumbersome summations were dealt with by means of a CDC 1604 computer, and the results were checked for large and small N by means of the limits discussed above. Figure 1 shows the quantity

$$S(\xi) \equiv (\Delta N)^2(\Delta S)^2 \qquad (17)$$

evaluated for coherent states with parameters $\xi = 0, \frac{1}{4}, \frac{1}{2}, \frac{3}{4}, 1$. Figure 2 shows explicitly that the first uncertainty relation in Eq. (7) is satisfied and closely approached for all N when $\xi = 0$ (real α). Corresponding results

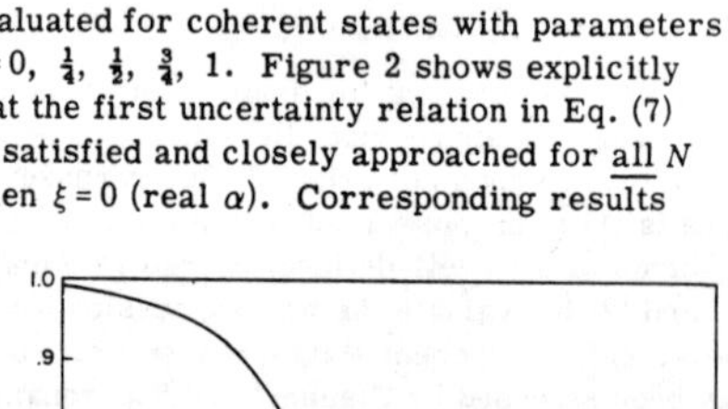

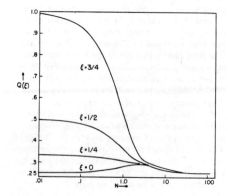

FIG. 2. The quantity $Q(\xi) \equiv S(\xi)/\langle C\rangle^2$ is shown as a function of $(\Delta N)^2 = N$ for various ξ, for the coherent states. According to Eq. (7), $Q(\xi)$ must be larger than $\frac{1}{4}$.

for the second relation in Eq. (7) follow on using the symmetry mentioned after Eq. (12). The maximum is 0.2922 and occurs near $N = 2.4$.

Figure 3 shows the uncertainty product U of Eq. (8), which does not discriminate between S and C (U is independent of ξ). The coherent states are seen to be nearly as classical as permitted for all values of mean excitation. In particular, we see that $\frac{1}{2} \geq U \geq \frac{1}{4}$. Brunet[12] has also proposed states having a small uncertainty product. It appears that his definition of phase uncertainty relies on the classical limit and so is only valid for large N. Moreover, the uncertainty product is larger than that which we have found for the coherent states. (Even more important for application is the direct physical meaning of the coherent states.)

Finally we consider the limitation on simultaneous measurements of S and C. From the commutation relation

$$[S, C] = (1/2i)(1 - E_+ E_-), \tag{18}$$

one finds the relation

$$(\Delta S)(\Delta C) \geq \tfrac{1}{4} e^{-N}. \tag{19}$$

If $N \leq 1$, the equality in Eq. (19) is almost exactly satisfied, for all ξ. This is seen to be in agreement with the fact that as N approaches zero, both ΔS and ΔC go to $\frac{1}{2}$. For large N, the left-hand side decreases as

$$\Delta S \Delta C \sim \left[\frac{1}{16N^2}\xi(1-\xi) + O\left(\frac{1}{N^3}\right) + \cdots\right]^{1/2}. \tag{20}$$

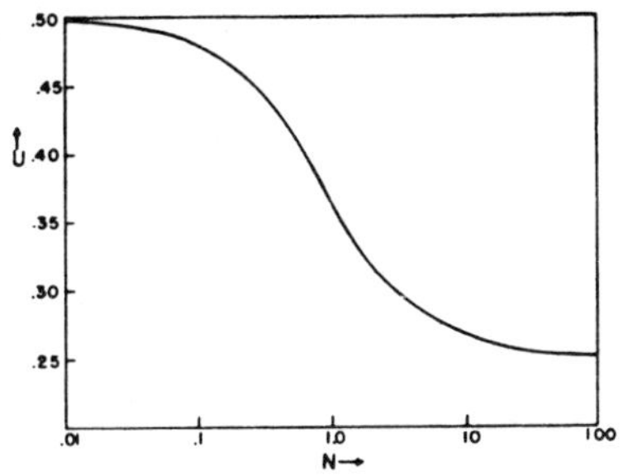

FIG. 3. The dependence of the uncertainty product U [defined in Eq. (8)] on $(\Delta N)^2 = N$ is shown for the coherent states. Note that U is independent of ξ.

*Work supported in part by the Office of Naval Research and the U. S. Atomic Energy Commission.

†Alfred P. Sloan Foundation Fellow.

‡National Science Foundation Predoctoral Fellow.

[1]See, e.g., W. Heitler, Quantum Theory of Radiation (Oxford University Press, London, 1954), Chap. 2.

[2]L. Susskind and J. Glowgower, Physics 1, 49 (1964).

[3]W. H. Louisell, Phys. Letters 7, 60 (1963).

[4]R. J. Glauber, Phys. Rev. 131, 2766 (1963).

[5]E. C. G. Sudarshan, Phys. Rev. Letters 10, 277 (1963).

[6]The operators S and C have a continuous eigenvalue spectrum in the interval -1 to $+1$. The eigenvalue can be labeled by $\sin\varphi$ and $\cos\varphi$, respectively, for S and C; however, the sum of the squares does not equal unity except in the limit of large N. In the classical limit the quantity φ corresponds to the classical phase angle. Although $S^2 + C^2$ is not equal to the unit operator [cf. Eq. (13)] the quantities S and C bear such a close relation to the classical quantities that we have retained a notation suggestive of this correspondence.

[7]R. J. Glauber, Phys. Rev. 84, 395 (1951).

[8]W. Thirring, Principles of Quantum Electrodynamics (Academic Press, Inc., New York, 1958), Chap. 9.

[9]Another useful form is $|\alpha\rangle = A(\alpha)|0\rangle$, where $|0\rangle$ is the ground state and $A(\alpha)$ is the unitary operator $\exp(\alpha a^* - \alpha^* a)$. The $A(\alpha)$ give a non-Abelian ray representation of the Abelian group of phase translations in the variable α of the coherent states $|\alpha\rangle$, where α ranges over the complex α plane. The multiplication law is

$$A(\alpha_2)A(\alpha_1) = \exp[\tfrac{1}{2}(\alpha_2\alpha_1^* - \alpha_1\alpha_2^*)]A(\alpha_2 + \alpha_1).$$

[10]P. Carruthers and M. M. Nieto (to be published) discuss the problem of the forced quantum oscillator from this point of view.

[11]In calculating (15) the fact that

$$1/(n+1)^{z+1} = \int_0^\infty t^z e^{-(n+1)t} dt/\Gamma(z+1)$$

is useful. Here $z = -\frac{1}{2}$.

[12]H. Brunet, Phys. Letters 10, 172 (1964).

Reprinted from AMERICAN JOURNAL OF PHYSICS, Vol. 33, No. 7, 537–544, July 1965

Coherent States and the Forced Quantum Oscillator*

P. CARRUTHERS†
Laboratory of Nuclear Studies and Laboratory of Atomic and Solid-State Physics, Cornell University, Ithaca, New York
M. M. NIETO‡
Laboratory of Nuclear Studies, Cornell University, Ithaca, New York
(Received 27 January 1965)

It is shown that the forced quantum oscillator subject to a transient classical force is easily described in terms of the "coherent states" recently found useful in the description of light from optical masers. The natural role these states play in understanding the notions of the phase of a quantum oscillator and the transition to the classical limit is also explained. Very simple derivations of the state vectors, energy transfer, and various transition probabilities are given.

I. INTRODUCTION

RECENT papers in this journal have applied powerful formal methods having their origin in quantum field theory to the study of the one-dimensional quantum mechanical oscillator subject to a transient classical force.[1,2] Although this problem arises in many contexts[3–5] and has been solved in many ways,[1–6] we wish to present a discussion which is not only extremely simple (on the level of a first year graduate course in quantum mechanics), but also provides insight into the concept of the phase of a quantum oscillator and the classical limit. The simplicity of our discussion rests on the use of the "coherent states" of an oscillator.[7] These states, which go over into coherent classical states for large quantum numbers, have recently been recognized to be of great utility in the quantum mechanical description of coherence of light from optical masers.

In Sec. II a summary of the harmonic oscillator variables and other conventions required is given. Section III is largely a summary of the properties of the coherent states, following Glauber.[7] In Sec. IV the forced oscillator problem is formulated in the Heisenberg picture in order to keep closely to the classical interpretation of the result. A simple Green's function method is used to construct the states of the system. In Sec. V the transition probability between arbitrary number states of the oscillator is computed in a completely elementary way, in contrast to the intricate methods used in Refs. 1 and 2.

II. HARMONIC OSCILLATOR IN QUANTUM MECHANICS

The position x of a classical oscillator of mass m, spring constant k subject to a uniform driving force $F(t)$ obeys Newton's equation of motion

$$m\ddot{x}+kx=F(t). \tag{2.1}$$

Until Sec. IV we shall be concerned only with the free oscillator, in which case $F(t)\equiv 0$. The solution of (2.1) is then

$$x=|A|\cos(\omega t-\phi), \tag{2.2}$$

where $\omega=(k/m)^{\frac{1}{2}}$ is the (circular) frequency, $|A|$ the amplitude and ϕ the phase angle of the oscillator. The point to be observed here is that the description of motion of a classical oscillator requires the specification of both amplitude and phase. It is not possible to ascribe such detailed information to the quantum oscillator.

To discuss the quantum oscillator we write

---

* Supported in part by the U. S. Office of Naval Research and the United States Atomic Energy Commission.
† Alfred P. Sloan Foundation Fellow.
‡ National Science Foundation Predoctoral Fellow.
[1] R. W. Fuller, S. M. Harris, and E. L. Slaggie, Am. J. Phys. **31**, 431 (1963).
[2] L. M. Scarfone, Am. J. Phys. **32**, 158 (1964).
[3] The most important example arises in the study of the infrared divergence problem, reviewed in Ref. 4.
[4] J. M. Jauch and F. Rohrlich, *Theory of Photons and Electrons* (Addison-Wesley Publishing Company, Reading, Massachusetts, 1955), Chap. 16.
[5] R. J. Glauber, Phys. Rev. **84**, 395 (1951).
[6] R. P. Feynman, Rev. Mod. Phys. **20**, 267 (1948).
[7] See, especially, R. J. Glauber, Phys. Rev. **131**, 2766 (1963).

down the classical energy, or Hamiltonian,

$$H=\frac{p^2}{2m}+\tfrac{1}{2}m\omega^2x^2, \tag{2.3}$$

where $p=m\dot{x}$ is the momentum conjugate to x,

$$xp-px\equiv[x,p]=i\hbar. \tag{2.4}$$

Let us work in the Heisenberg picture,[8] wherein state vectors are constants in time, the operators $\mathcal{O}(t)$ varying according to

$$i\hbar\dot{\mathcal{O}}(t)=[\mathcal{O}(t),H]. \tag{2.5}$$

From Eqs. (2.3)–(2.5) one finds

$$\dot{x}(t)=p(t)/m,$$
$$\dot{p}(t)=-m\omega^2x(t), \tag{2.6}$$

which when combined show that the harmonic oscillator equation

$$\ddot{x}(t)+\omega^2x(t)=0 \tag{2.7}$$

is an operator identity satisfied by $x(t)$.

In order to discuss the eigenvalue spectrum and the relations among the eigenfunctions it is convenient to introduce two new dynamical variables, to be used in place of the old ones, p and x. To discover these let us note that the classical solution (2.2) can be written as

$$x_{\text{classical}}=\tfrac{1}{2}Ae^{-i\omega t}+\tfrac{1}{2}A^*e^{+i\omega t}, \tag{2.8}$$

where $A=|A|e^{i\phi}$. Here the asterisk denotes complex conjugate. In analogy we invent the time-independent dimensionless operators a and $a^\dagger$ ($a^\dagger$ is the Hermitian adjoint of a) by

$$x(t)=x_0[ae^{-i\omega t}+a^\dagger e^{i\omega t}], \tag{2.9}$$

where for convenience the length

$$x_0=(\hbar/2m\omega)^{\frac{1}{2}} \tag{2.10}$$

has been factored out. x_0 turns out to be the root-mean-square zero-point displacement. The momentum operator is accordingly

$$p(t)=-im\omega x_0[ae^{-i\omega t}-a^\dagger e^{i\omega t}]. \tag{2.11}$$

Alternatively, we can invert (2.10) and (2.11) to obtain a definition of a and $a^\dagger$ in terms of x and p

$$a(t)=\frac{i}{(2m\hbar\omega)^{\frac{1}{2}}}[p(t)-im\omega x(t)],$$
$$a^\dagger(t)=\frac{-i}{(2m\hbar\omega)^{\frac{1}{2}}}[p(t)+im\omega x(t)], \tag{2.12}$$

where $a(t)$ and $a^\dagger(t)$ are given by

$$a(t)=ae^{-i\omega t},\quad a^\dagger(t)=a^\dagger e^{i\omega t}. \tag{2.13}$$

One notes the basic commutation rule

$$[a,a^\dagger]=1, \tag{2.14}$$

as follows from (2.12) and (2.4). In addition to (2.14) one has the trivial relations

$$[a,a]=[a^\dagger,a^\dagger]=0. \tag{2.15}$$

The Hamiltonian can now be expressed in the form

$$H=\tfrac{1}{2}\hbar\omega(a^\dagger a+aa^\dagger)=(a^\dagger a+\tfrac{1}{2})\hbar\omega. \tag{2.16}$$

Now we may verify from (2.14) and (2.5) that (2.13) is valid. $a(t)$ and $a^\dagger(t)$ are *normal mode* operators

$$i\hbar\dot{a}(t)=[a(t),H]=\hbar\omega a(t),\ \ddot{a}(t)+\omega^2a(t)=0. \tag{2.17}$$

As everybody knows, the eigenvalues of (2.16) are

$$E_n=(n+\tfrac{1}{2})\hbar\omega\quad n=0,1,2,\ldots, \tag{2.18}$$

so that the *number operator*

$$N_{\text{op}}\equiv a^\dagger a \tag{2.19}$$

has eigenvalues 0, 1, 2, ... The orthonormal eigenfunctions ψ_n belonging to E_n are given by

$$\psi_n=\frac{1}{(n!)^{\frac{1}{2}}}(a^\dagger)^n\psi_0, \tag{2.20}$$

where the ground state ψ_0 of the oscillator is defined by

$$a\psi_0=0. \tag{2.21}$$

The $a^\dagger$ and a operators raise or lower [with the exception of (2.21)] the degree of excitation of the oscillator by $\hbar\omega$

$$a^\dagger\psi_n=(n+1)^{\frac{1}{2}}\psi_{n+1}, \tag{2.22}$$

$$a\psi_n=(n)^{\frac{1}{2}}\psi_{n-1}, \tag{2.23}$$

[8] Heisenberg picture operators are labeled explicitly by the time variable t; otherwise the Schrödinger picture operator is meant.

from which we obtain the matrix elements

$$\langle m|a^\dagger|n\rangle=(n+1)^{\frac{1}{2}}\delta_{m,n+1},$$
$$\langle m|a|n\rangle=(n)^{\frac{1}{2}}\delta_{m,n-1}. \quad (2.24)$$

An oscillator excited to its nth quantum state behaves just as a collection of n indistinguishable Bose particles. One conventionally says that $a^\dagger$ creates (a destroys) a quantum $\hbar\omega$. Thus, the a's are called destruction, annihilation, or absorption operators according to the author's taste. The $a^\dagger$'s are accordingly called creation, or emission operators.

Let us now verify the claim that x_0 defined in Eq. (2.10) is the rms zero-point displacement, as claimed. We have

$$\langle x^2\rangle_0\equiv\langle 0|x^2|0\rangle=x_0^2\langle(a+a^\dagger)^2\rangle_0$$
$$=x_0^2=\hbar/2m\omega \quad (2.25)$$

on using the commutation rules and (2.21).

However, the mean position and momentum vanish in *any* number eigenstate, regardless of the size of n

$$\langle n|x|n\rangle=\langle n|p|n\rangle=0, \quad (2.26)$$

since x and p are nondiagonal operators in this so-called number representation in which we are working. Thus, the number states are definitely not the appropriate ones for a transition to the classical limit. The physical reason for this result is that the phase is completely undefined once the excitation number is specified, leading to the result (2.25). This "complementarity" of number and phase requires a careful discussion in order to make precise the meaning of the idea of the phase of a quantum oscillator.[9] The present problem is closely related to the subtle interpretation required of the energy–time uncertainty relation.

III. COHERENT STATES OF AN OSCILLATOR

What quantum mechanical states correspond to the kinds of motion encountered in "classical" circumstances? Besides having a high degree of mean excitation, they must satisfy, at least to a good approximation,

$$\langle\Psi_{\text{class}}|x(t)|\Psi_{\text{class}}\rangle=\text{const}\cos(\omega t-\phi). \quad (3.1)$$

As a matter of fact, we can find states satisfying (3.1) exactly, regardless of the number of oscillator quanta in the state.[7] Of course, from this set of states only those having $\langle\Psi|N_{\text{op}}|\Psi\rangle\gg1$ are necessary for the particular job of satisfying the correspondence principle. However, we wish to emphasize the utility of these states for a wide variety of problems not necessarily involving large quantum numbers.

Comparing Eq. (2.9) with (3.1) shows immediately that if we can find states $|\alpha\rangle$ which are eigenfunctions of the annihilation operator with complex eigenvalue α

$$a|\alpha\rangle=\alpha|\alpha\rangle, \quad (3.2)$$

then the mean value of $x(t)$ in these states

$$\langle\alpha|x(t)|\alpha\rangle=x_0(\alpha e^{-i\omega t}+\alpha^*e^{+i\omega t})\langle\alpha|\alpha\rangle \quad (3.3)$$

does indeed agree with (3.1) where the phase angle ϕ is given by

$$\alpha=|\alpha|e^{i\phi}. \quad (3.4)$$

The modulus of α is related to the mean excitation of the oscillator by

$$N=\frac{\langle\alpha|a^\dagger a|\alpha\rangle}{\langle\alpha|\alpha\rangle}=|\alpha|^2. \quad (3.5)$$

Thus, we can write the *eigenvalue* of a in the unsurprising form $(N)^{\frac{1}{2}}e^{i\phi}$. However, the corresponding decomposition of the *operator* a into $(N_{\text{op}})^{\frac{1}{2}}\exp(i\phi_{\text{op}})$ where ϕ_{op} is a Hermitian operator, is impossible.[9]

The states $|\alpha\rangle$ can be constructed out of number eigenstates

$$|\alpha\rangle=\sum_{n=0}^{\infty}|n\rangle\langle n|\alpha\rangle. \quad (3.6)$$

From (3.2) and (2.24) we find the expansion coefficient

$$\langle n|\alpha\rangle=\frac{\alpha}{(n)^{\frac{1}{2}}}\langle n-1|\alpha\rangle=\cdots=\frac{\alpha^n}{(n!)^{\frac{1}{2}}}\langle 0|\alpha\rangle, \quad (3.7)$$

whence

$$|\alpha\rangle=\langle 0|\alpha\rangle\sum_{n=0}^{\infty}\frac{\alpha^n}{(n!)^{\frac{1}{2}}}|n\rangle. \quad (3.8)$$

[9] L. Susskind and J. Glogower, Physics 1, 49 (1964) have given a careful and mathematically precise discussion of this question.

Normalizing $|\alpha\rangle$ determines $\langle 0|\alpha\rangle$ up to an arbitrary phase chosen as follows

$$1=\langle\alpha|\alpha\rangle=|\langle 0|\alpha\rangle|^2\sum_{n=0}^{\infty}\frac{|\alpha|^{2n}}{n!}$$

$$=\exp(|\alpha|^2)|\langle 0|\alpha\rangle|^2,$$

$$\langle 0|\alpha\rangle=\exp(-\tfrac{1}{2}|\alpha|^2). \qquad (3.9)$$

Finally,

$$|\alpha\rangle=\exp(-\tfrac{1}{2}|\alpha|^2)\sum_{n=0}^{\infty}\frac{\alpha^n}{(n!)^{\frac{1}{2}}}|n\rangle. \qquad (3.10)$$

The probability of finding the oscillator in the nth level in the state $|\alpha\rangle$ is

$$P_n(\alpha)=e^{-|\alpha|^2}\frac{|\alpha|^{2n}}{n!}=\frac{e^{-N}N^n}{n!}. \qquad (3.11)$$

We have therefore derived the familiar Poisson distribution which also expresses the photon distribution in classical waves.

The expression (3.10) for $|\alpha\rangle$ can be simplified further by introducing (2.20)

$$|\alpha\rangle=e^{-\frac{1}{2}|\alpha|^2}\sum_{n=0}^{\infty}\frac{(\alpha a^\dagger)^n}{n!}|0\rangle$$

$$\equiv\exp[\alpha a^\dagger-\tfrac{1}{2}|\alpha|^2]|0\rangle. \qquad (3.12)$$

Thus, by applying a very simple operator to the ground state, one generates a *coherent state* $|\alpha\rangle$. These states are the well-known minimum wave packets of a displaced harmonic-oscillator ground state, discussed very clearly in the book by Henley and Thirring.[10] In the Schrödinger picture the states evolve in time according to

$$|\alpha(t)\rangle\equiv e^{-iHt}|\alpha\rangle=\exp[\alpha a^\dagger e^{i\omega t}-\tfrac{1}{2}|\alpha|^2]|0\rangle. \qquad (3.13)$$

This state describes a displaced ground-state wavefunction which vibrates back and forth with frequency ω without any change of shape or spreading.

The fact that $|\alpha\rangle$ and $|0\rangle$ are both normalized suggests that $\exp[\alpha a^\dagger-\frac{1}{2}|\alpha|^2]$ is equivalent to a unitary operator. Such an operator is expected to be of the form exp(ih) with h Hermitian. This suggests that we replace (3.12) by

$$|\alpha\rangle\equiv A(\alpha)|0\rangle,\quad A(\alpha)\equiv\exp(\alpha a^\dagger-\alpha^*a). \qquad (3.14)$$

That this is correct follows from the identity[11,12]

$$\exp A\,\exp B=\exp(A+B+\tfrac{1}{2}[A,B]), \qquad (3.15)$$

valid when A and B commute with their commutator $[A,B]$. Specifically we get

$$\exp(\alpha a^\dagger)\exp(-\alpha a)$$
$$=\exp[\alpha a^\dagger-\alpha^*a]\exp(\tfrac{1}{2}|\alpha|^2), \qquad (3.16)$$

which establishes the equivalence of (3.12) and (3.14).

The virtues of the operator $A(\alpha)$ are exhibited by the following properties. First, we note the relation

$$A(\alpha)=A^\dagger(-\alpha), \qquad (3.17)$$

by means of which the unitarity statement

$$A^\dagger(\alpha)A(\alpha)=A(\alpha)A^\dagger(\alpha)=1 \qquad (3.18)$$

takes the form[13]

$$A(-\alpha)A(\alpha)=A(\alpha)A(-\alpha)=1. \qquad (3.19)$$

From these relations we see that $A(\alpha)$ and $A(-\alpha)$ are a kind of creation and annihilation operator for the coherent states. In particular,

$$|\alpha\rangle=A(\alpha)|0\rangle;\quad A(-\alpha)|\alpha\rangle=|0\rangle. \qquad (3.20)$$

The basic origin of these properties lies in the fact that the $A(\alpha)$ are displacement operators of the normal coordinates a and $a^\dagger$. To show this we need to use the identity

$$[a,(a^\dagger)^n]=n(a^\dagger)^{n-1}, \qquad (3.21)$$

which gives us

$$[a,\exp(\alpha a^\dagger)]=\sum_{n=0}^{\infty}[a,(a^\dagger)^n]\frac{\alpha^n}{n!}$$

$$=\alpha\exp(\alpha a^\dagger). \qquad (3.22)$$

It follows that

$$[a,A(\alpha)]=e^{-\frac{1}{2}|\alpha|^2}[a,e^{\alpha a^\dagger}]e^{\alpha a}$$
$$=\alpha A(\alpha),$$

[10] E. M. Henley and W. Thirring, *Elementary Quantum Field Theory* (McGraw-Hill Book Company, Inc., New York, 1962), Chap. 2.

[11] This identity (Baker–Hausdorff) is proved in most modern books on quantum mechanics, for instance, Ref. 12.

[12] A. Messiah, *Quantum Mechanics* (North-Holland Publishing Company, Amsterdam, 1964), Vol. I, p. 442.

[13] The mathematically inclined reader can note that the unitary operators $A(\alpha)$ give a non-Abelian ray representation of the Abelian group of phase translations in the variable α of the coherent states $|\alpha\rangle$, where α may range over the complex α plane. The multiplication law is $A(\alpha_2)A(\alpha_1)=\exp[\frac{1}{2}(\alpha_2\alpha_1^*-\alpha_1\alpha_2^*)]A(\alpha_1+\alpha_2)$. The unimodular exponential factor vanishes only for real α_1, α_2 or for the special case $\alpha_2=\pm\alpha_1$.

so that rearrangement yields

$$A^\dagger(\alpha)aA(\alpha)=A^\dagger(\alpha)A(\alpha)a+A^\dagger(\alpha)[a,A(\alpha)]$$
$$=a+\alpha. \quad (3.23)$$

From this we find

$$A^\dagger(\alpha)a^\dagger A(\alpha)=a^\dagger+\alpha^*. \quad (3.24)$$

Let us now prove that the (nonspreading) states $|\alpha\rangle$ give rise to the minimum uncertainty product. First note that

$$\begin{aligned}(\bar{x})_\alpha{}^2 &\equiv (\langle\alpha|x|\alpha\rangle)^2=x_0{}^2(\alpha+\alpha^*)^2,\\ (\overline{x^2})_\alpha &\equiv \langle\alpha|x^2|\alpha\rangle=x_0{}^2\langle\alpha|(a+a^\dagger)^2|\alpha\rangle\\ &=(\bar{x})_\alpha{}^2+x_0{}^2;\\ (\Delta x)^2 &= \overline{x^2}-\bar{x}^2=x_0{}^2=\hbar/2m\omega.\end{aligned} \quad (3.25)$$

Hence, the rms displacement is the same as that in the ground state. A similar calculation shows that

$$(\Delta p)^2=m^2\omega^2x_0{}^2=\tfrac{1}{2}m\hbar\omega, \quad (3.26)$$

whence

$$\Delta p\Delta x=\tfrac{1}{2}\hbar. \quad (3.27)$$

Thus, the states $|\alpha\rangle$ are as "classical as possible" according to the principles of quantum mechanics, in the sense of (3.27).

Next, let us consider questions of orthogonality and completeness of the coherent states $|\alpha\rangle$. As these are eigenfunctions of a non-Hermitian operator there is no guarantee that states with differing phase are orthogonal. From Eq. (3.10) we compute the inner product $\langle\alpha|\beta\rangle$:

$$\begin{aligned}\langle\alpha|\beta\rangle &=\exp[-\tfrac{1}{2}(|\alpha|^2+|\beta|^2)]\sum_{mn}\frac{(\alpha^*)^m\beta^n}{(m!n!)^{\frac{1}{2}}}\langle n|m\rangle\\ &=\exp(\alpha^*\beta-\tfrac{1}{2}|\alpha|^2-\tfrac{1}{2}|\beta|^2);\\ |\langle\alpha|\beta\rangle|^2 &=\exp(-|\alpha-\beta|^2).\end{aligned} \quad (3.28)$$

Despite this lack of orthogonality, the $|\alpha\rangle$ states are complete. More precisely, we prove the following identity[7,14]

$$\frac{1}{\pi}\int|\alpha\rangle\langle\alpha|d^2\alpha=1, \quad (3.29)$$

where $d^2\alpha=d\,\mathrm{Re}\,\alpha\,d\,\mathrm{Im}\,\alpha$ and the integration is to be taken over the whole complex α plane.

[14] E. C. G. Sudarshan, Phys. Rev. Letters 10, 277 (1963).

Transforming to polar coordinates $\alpha=re^{i\theta}$ and using (3.10) to express (3.29) in terms of number states converts Eq. (3.29) to

$$\begin{aligned}&\sum_{mn}\frac{|m\rangle\langle n|}{(m!n!)^{\frac{1}{2}}}\frac{1}{\pi}\int_0^\infty dre^{-r^2}r^{m+n+1}\int_0^{2\pi}d\theta e^{i(m-n)\theta}\\ &=\sum_m|m\rangle\langle m|\frac{1}{m!}\int_0^\infty dr^2e^{-r^2}r^{2m}\\ &=\sum_m|m\rangle\langle m|=1.\end{aligned} \quad (3.30)$$

By means of Eq. (3.29) one can expand state vectors and matrix elements in terms of the coherent states. For more details on such procedures, and in particular for a discussion of special density matrices (coherent states, black-body radiation, etc.) the papers by Glauber[7] and Sudarshan[14] should be consulted.

IV. THE FORCED QUANTUM OSCILLATOR

We now show that the coherent states have a wider range of utility than simply to describe a highly excited oscillator. We show that the application of a "classical" driving force (i.e., one which is unaffected by the motion of the oscillator) generates coherent states regardless of the size of the average excitation of the oscillator.

We wish to solve the "scattering" problem of a quantum oscillator subject to a position independent classical force $F(t)$. The interaction energy of the oscillator with this force may be taken to be[1]

$$V=-xF(t)=-x_0(a+a^\dagger)F(t). \quad (4.1)$$

For orientation let us first consider the trivial problem of a constant force F_0. Then the oscillator is simply displaced to a different equilibrium position. The Hamiltonian

$$H=\hbar\omega(a^\dagger a+\tfrac{1}{2})-x_0(a+a^\dagger)F(t), \quad (4.2)$$

specialized to $F(t)=F_0$, can be diagonalized by "completing the square",

$$H=\hbar\omega(b^\dagger b+\tfrac{1}{2})-\frac{(x_0F_0)^2}{\hbar\omega}, \quad (4.3)$$

where the new normal coordinates b are

$$b=a-x_0F_0/\hbar\omega. \quad (4.4)$$

The b operators obey $[b,b^\dagger]=1$ and the eigenvalue spectrum is changed only by an over-all downward shift by an amount $-x_0{}^2F_0{}^2/\hbar\omega = -F_0{}^2/2m\omega^2$. The new ground state $|0\rangle'$ obeys

$$b|0\rangle'=0, \quad \text{or} \quad a|0\rangle'=(x_0F_0/\hbar\omega)|0\rangle', \qquad (4.5)$$

so that in terms of the original coordinates and ground state $|0\rangle$

$$|0\rangle'=A\left(\frac{x_0F_0}{\hbar\omega}\right)|0\rangle=\exp\left[\frac{x_0F_0}{\hbar\omega}(a^\dagger-a)\right]|0\rangle. \qquad (4.6)$$

The excited states of the displaced oscillator are found by applying $b^\dagger=a^\dagger-x_0F_0/\hbar\omega$ to (4.6).

Next, let us permit a time-dependent force to act on the oscillator. For simplicity suppose $F(\pm\infty)=0$, so that the oscillator is free at early and late times. As we wish to emphasize the similarities between the classical and quantum aspects, we use the Heisenberg picture, in which the displacement obeys the equation

$$\ddot{x}(t)+\omega_0{}^2x(t)=F(t)/m. \qquad (4.7)$$

We solve (4.7) by means of a Green's function $G(t-t')$, defined by the differential equation

$$\left(\frac{d^2}{dt^2}+\omega_0{}^2\right)G(t-t')=\omega_0\delta(t-t') \qquad (4.8)$$

and supplemented by appropriate boundary conditions. The delta function corresponds to an impulsive force applied at $t=t'$, and ω_0 has been included to make $G(t)$ dimensionless. Eq. (4.8) is solved in the conventional way by expanding $G(t)$ in a Fourier intergral. The solutions needed here are:

$$G_R(t)=\begin{cases}0 & t<0,\\ \sin\omega_0t & t>0,\end{cases} \qquad (4.9)$$

$$G_A(t)=\begin{cases}-\sin\omega_0t & t<0,\\ 0 & t>0.\end{cases} \qquad (4.10)$$

We may now write down solutions of Eq. (4.7) according to the alternative boundary conditions

$$\begin{aligned}x(t)&\to x_{\text{in}}(t), & t&\to-\infty,\\ x(t)&\to x_{\text{out}}(t), & t&\to+\infty,\end{aligned} \qquad (4.11)$$

where x_{in} and x_{out} obey the free oscillator equation

$$\ddot{x}_{\text{in}}+\omega_0{}^2x_{\text{in}}=0, \quad \ddot{x}_{\text{out}}+\omega_0{}^2x_{\text{out}}=0. \qquad (4.12)$$

The solutions clearly are

$$\begin{aligned}x(t)&=x_{\text{in}}(t)+\frac{1}{m}\int_{-\infty}^{\infty}G_R(t-t')F(t')dt',\\ x(t)&=x_{\text{out}}(t)+\frac{1}{m}\int_{-\infty}^{\infty}G_A(t-t')F(t')dt'.\end{aligned} \qquad (4.13)$$

As a simple illustration let us find the state of the oscillator after the force has ceased to act. For this we need to express x_{out} in terms of x_{in}

$$x_{\text{out}}=x_{\text{in}}+\frac{1}{m}\int_{-\infty}^{\infty}G(t-t')F(t')dt', \qquad (4.14)$$

$$G(t)=G_A(t)-G_R(t)=\sin\omega_0t. \qquad (4.15)$$

Introducing the normal mode coordinates a and b by (cf. Eq. 2.9)

$$\begin{aligned}x_{\text{in}}(t)&=x_0(ae^{-i\omega_0t}+a^\dagger e^{i\omega_0t}),\\ x_{\text{out}}(t)&=x_0(be^{-i\omega_0t}+b^\dagger e^{i\omega_0t}),\end{aligned} \qquad (4.16)$$

one finds from (4.14) that

$$b=a+iF(\omega_0)/(2m\hbar\omega_0)^{\frac{1}{2}}=a+i\alpha_0, \qquad (4.17)$$

where $F(\omega)$ is defined by

$$F(\omega)=\int_{-\infty}^{\infty}e^{i\omega t}F(t)dt. \qquad (4.15)$$

Hence, the total effect of the transient force is to displace the incoming normal mode a by an amount depending on that Fourier coefficient of $F(t)$ with the fundamental frequency of the oscillator. From the discussion of Sec. III we know that a unitary operator S exists which transforms a to b

$$a\to b=S^\dagger aS \qquad (4.16)$$

and relates the in and out state vectors[15]

$$\Psi_{\text{in}}\to\Psi_{\text{out}}=S^\dagger\Psi_{\text{in}}. \qquad (4.17)$$

(The operator S is often called the S matrix.)

---

[15] For a discussion of the significance of these (Heisenberg) state vectors and the S matrix, see Ref. 10, Chaps. 8–10, or Ref. 2. It is important to remember that Ψ_{out} and Ψ_{in} are constant state vectors; they are not $\psi(\pm\infty)$ as in the interaction picture.

Comparing Eqs. (3.23) and (4.16) indicates that

$$\exp\left\{i\frac{a^\dagger F(\omega_0)+aF^*(\omega_0)}{(2m\hbar\omega_0)^{\frac{1}{2}}}.\right\} \tag{4.18}$$

An interesting alternate way of writing (4.18) results on expressing the argument of the exponential as a time integral

$$\Psi_{\text{out}}=\exp\left\{\frac{-i}{(2m\hbar\omega_0)^{\frac{1}{2}}}\right.$$
$$\left.\times\int_{-\infty}^{\infty}(a^\dagger e^{i\omega_0 t}+ae^{-i\omega_0 t})F(t)dt\right\}\Psi_{\text{in}};$$

$$\Psi_{\text{out}}=\exp\left[\frac{-i}{\hbar}\int_{-\infty}^{\infty}x_{\text{in}}(t)F(t)dt\right]\Psi_{\text{in}}. \tag{4.19}$$

These formulas express Ψ_{out} in terms of the "incoming" variables a and $a^\dagger$. Using (4.16) one can easily verify that S has the same form expressed in terms of b and $b^\dagger$ as it does in terms of a and $a^\dagger$.

After the force has ceased, the motion of the system is described by x_{out}, and measurements on the oscillator are appropriately described in terms of Ψ_{out}. In order to find how often a given state $\Psi_{\text{out},\nu}$ appears, we have to expand the state vector $\Psi_{\text{in},\mu}$ (which contains information about the preparation of the state μ of the oscillator) in terms of the $\Psi_{\text{out},\nu}$. Using the unitarity of S and labeling the states in (4.17) appropriately we have

$$\Psi_{\text{in},\mu}=\sum_\nu \Psi_{\text{out},\nu}\langle\Psi_{\text{out},\nu}|S|\Psi_{\text{out},\mu}\rangle\equiv\sum_\nu S_{\nu\mu}\Psi_{\text{out},\nu},$$

$$S_{\nu\mu}\equiv\langle\Psi_{\text{out},\nu}|S|\Psi_{\text{out},\mu}\rangle=\langle\Psi_{\text{in},\nu}|S|\Psi_{\text{in},\mu}\rangle. \tag{4.20}$$

In Eq. (4.20) the completeness of the Ψ_{out} was utilized. Eq. (4.17) was used again to establish the last equality.

The explicit form of the scattering matrix Eq. (4.18) completes the solution of the problem.

V. TRANSITION PROBABILITIES BETWEEN NUMBER EIGENSTATES

An interesting and important special case of Eq. (4.20) occurs when the μ, ν labels refer to number states. For example, the model of Sec. IV can with little effort be extended to describe the emission of photons by a classical current source.[5,7] Although it is the coherent states which are radiated by such a source, one often uses photon counters in conducting interesting experiments on the emitted radiation.

If the oscillator was initially in the number state n, then the probability amplitude that it is finally in the number state m is

$$S_{mn}=\langle m|S|n\rangle=\langle m|A(i\alpha_0)|n\rangle, \tag{5.1}$$

[see Eq. (4.18)]. In the simplest case of an oscillator initially in its ground state, the result follows directly from the analysis of Sec. II [see Eqs. (3.10) and (3.14)]

$$S_{m0}=\langle m|i\alpha_0\rangle=\frac{(i\alpha_0)^m}{(m!)^{\frac{1}{2}}}\exp(-\tfrac{1}{2}|\alpha_0|^2). \tag{5.2}$$

The probability of the transition $0\rightarrow m$ is therefore Poisson

$$P_{m0}=|S_{m0}|^2=\frac{|\alpha_0|^{2m}}{m!}\exp(-|\alpha_0|^2), \tag{5.3}$$

with a peak at $N=|\alpha_0|^2$

$$N=|\alpha_0|^2=|F(\omega_0)|^2/(2m\hbar\omega_0). \tag{5.4}$$

Hence, the most probable energy transferred to the oscillator,

$$\Delta E=N\hbar\omega_0=|F(\omega_0)|^2/2m, \tag{5.5}$$

also coincides with the average energy transfer

$$\sum_{m=0}^{\infty}P_{m0}(m\hbar\omega_0)=N\hbar\omega_0, \tag{5.6}$$

as well as with the expression obtained for a classical oscillator initially at rest.[1,16]

The general expression (5.1) for S_{mn} can also be evaluated explicitly in terms of known functions, as was shown by Fuller *et al.*[1] Our method is considerably shorter than theirs. First, we use Eq. (2.20) to write (in the following $m\geq n$)

$$(n!)^{\frac{1}{2}}S_{mn}=\langle m|A^\dagger(-i\alpha_0)(a^\dagger)^n|0\rangle. \tag{5.7}$$

Here we have written $A(\alpha)=A^\dagger(-\alpha)$. This allows us to exploit Eq. (3.24) and write the

---

[16] Note that in lowest order perturbation theory the energy transfer, $\hbar\omega|S_{10}^{(1)}|^2$ is also given by Eq. (5.5) (Here $S_{10}^{(1)}$ is the lowest order contribution to S_{10}).

right-hand side of (5.7) in the form

$$\langle m|(A^\dagger(-i\alpha_0)a^\dagger A(-i\alpha_0))^n A^\dagger(-i\alpha_0)|0\rangle$$
$$=\langle m|(a^\dagger+i\alpha_0^*)^n A(i\alpha_0)|0\rangle. \quad (5.8)$$

The binomial theorem is now employed to obtain

$$S_{mn}=\sum_{j=0}^{n}\frac{(n!)^{\frac{1}{2}}}{(n-j)!j!}(i\alpha_0^*)^{n-j}$$
$$\times\langle m|(a^\dagger)^j A(i\alpha_0)|0\rangle. \quad (5.9)$$

Combining Eq. (3.21) with Eq. (2.20), one easily shows that

$$(a)^j|m\rangle=\left(\frac{m!}{(m-j)!}\right)^{\frac{1}{2}}|m-j\rangle, \quad m\geqslant j, \quad (5.10)$$

so that the matrix element in Eq. (5.9) is simply $S_{m-j,0}$. Inserting the explicit form of $S_{m-j,0}$ from Eq. (5.2) gives

$$S_{mn}=(m!n!)^{\frac{1}{2}}e^{-\frac{1}{2}|\alpha_0|^2}(i\alpha_0)^{m-n}$$
$$\times\sum_{k=0}^{n}\frac{(-|\alpha_0|^2)^k}{k!(n-k)!(m+k-n)!}. \quad (5.11)$$

In obtaining the latter form the substitution $k=n-j$ has been made. It only remains to recognize the definition of the associated Laguerre polynomials[17]

$$L_m^{m-n}(x)=\sum_{k=0}^{n}\frac{(-x)^k m!}{k!(n-k)!(m-n+k)!} \quad (5.12)$$

to write (5.11) in the succinct form

$$S_{mn}=(n!/m!)^{\frac{1}{2}}e^{-\frac{1}{2}|\alpha_0|^2}(i\alpha_0)^{m-n}$$
$$\times L_m^{m-n}(|\alpha_0|^2), \quad m\geq n, \quad (5.13)$$

where α_0 is $F(\omega_0)/(2mh\omega_0)^{\frac{1}{2}}$ (see Eq. 4.17). One thus obtains for the transition probability $(m\geq n)$

$$P_{mn}=|S_{mn}|^2=\frac{n!}{m!}e^{-x}x^{m-n}[L_m^{m-n}(x)]^2;$$
$$x=|\alpha_0|^2. \quad (5.14)$$

If $m<n$ the sum in Eq. (5.9) actually cuts off at $j=m$. Repeating the calculation shows that

$$S_{mn}=(m!/n!)^{\frac{1}{2}}e^{-\frac{1}{2}|\alpha_0|^2}(i\alpha_0^*)^{n-m}L_n^{n-m}(|\alpha_0|^2),$$
$$m\leq n, \quad (5.15)$$

so that P_{mn} can be obtained from (5.14) by interchanging m and n.

In Ref. 1 it is shown by direct summation that the mean energy transferred to an oscillator initially in an arbitrary number state n is the same as for $n=0$. This result, too, can be obtained in a completely elementary manner. The energy shift is simply (the state Ψ_{in} is fixed)

$$\Delta E=\langle\Psi_{\text{in}}|(H_{\text{out}}-H_{\text{in}})|\Psi_{\text{in}}\rangle$$
$$=h\omega[|\alpha_0|^2+i(\langle\Psi_{\text{in}}|a^\dagger|\Psi_{\text{in}}\rangle\alpha_0-\text{c.c.})], \quad (5.16)$$

using Eq. (4.17) to express b in terms of a. For number states the term in parentheses vanishes and (5.16) simplifies to

$$\Delta E=h\omega|\alpha_0|^2=|F(\omega_0)|^2/2m. \quad (5.17)$$

The vanishing of the last terms in (5.16) for the number states can be regarded as due to the complete uncertainty in the phase of such states. If Ψ_{in} is a coherent state with phase parameter β, for instance, then a term $2h\omega\,\text{Im}(\beta\alpha_0^*)$ has to be added to (5.17).

Finally, we mention another way that coherent states can be used to illuminate the concept of phase of a quantum mechanical oscillator. Using the operators appropriate[9] to describe the phase variable ϕ one can introduce a suitable "number-phase" uncertainty relation which reduces to the familiar $\Delta N\Delta\phi\geq\frac{1}{2}$ in the classical limit but is still meaningful for small quantum numbers. The coherent states are found to be very good minimum-uncertainty-product states. For further details the reader is referred to Ref. 18.

[17] *Higher Transcendental Functions*, edited by A. Erdelyi (McGraw-Hill Book Company, Inc., New York, 1953), Vol. II, p. 188.

[18] P. Carruthers and M. M. Nieto, Phys. Rev. Letters **14**, 387 (1965).

Volume 21, number 6 PHYSICS LETTERS 1 July 1966

CLASSICAL BEHAVIOR OF SYSTEMS OF QUANTUM OSCILLATORS *

R. J. GLAUBER
Lyman Laboratory of Physics, Harvard University,
Cambridge Massachusetts

Received 1 June 1966

We show that if the equations of motion of a system of coupled oscillators assume a certain general form, states of the system which are intitially coherent remain coherent at all times. It follows that the motion of the system is nearly classical in nature.

There are many contexts in which the quantum mechanical behavior of a harmonic oscillator or of a set of coupled oscillators is found to be quite similar to the classical behavior of the same systems. A considerable number of instances in which this similarity exists may be explained by means of an elementary theorem which we shall prove in the present note.

It has recently been observed that a number of problems involving harmonic oscillator degrees of freedom may be simplified by making systematic use of the coherent states of the oscillators in describing their motion. For a single oscillator these states are the right eigenstates [1] of the annihilation operator a. For any complex number α there exists a coherent state $|\alpha\rangle$ of the oscillator with the property $a|\alpha\rangle = \alpha|\alpha\rangle$. Such states have the same uncertainty of position and momentum as the ground state of the oscillator; they may be regarded simply as forms of the ground state which have been displaced both in coordinate and momentum space. While the states $|\alpha\rangle$ do not form an orthogonal set, they are complete (in fact overcomplete). Simple means are available, notwithstanding the non-orthogonality of the states, for expanding any state in terms of them.

There exists a well-defined sense in which the coherent states of an oscillator are as nearly classical in character as it is possible for a quantum state to be. If $(\delta q)^2$ and $(\delta p)^2$ are the mean variances of the coordinate and momentum of an oscillator respectively, and ω and m its angular frequency and mass, the coherent states

* Supported in part by the Air Force Office of Scientific Research.

are uniquely determined by the condition [2] that $(2m)^{-1}\{(\delta p)^2 + m^2\omega^2(\delta q)^2\}$ assume its absolute minimum value *, which is $\frac{1}{2}\hbar\omega$. It was first noted by Schrödinger [3] that if a single free oscillator is initially in such a state then its Gaussian wave packet undergoes no spreading with time, and the center of the packet moves both in coordinate and momentum space precisely as a classical oscillator. This observation corresponds to the statement that an initially coherent state of the oscillator remains coherent at all later times. We ask, therefore, whether more general oscillator systems do not show analogous behavior in preserving the coherent character of their quantum states as a function of time.

We consider systems of oscillators for which the Heisenberg equations of motion can be written in the form

$$\dot{a}_j(t) = F_j(\{a_k(t)\}, t), \qquad j, k = 1 \ldots n \; , \qquad (1)$$

where the functions F_j may, as indicated, depend explicitly on time. For such systems we shall show that if the Schrödinger state is initially a coherent state it remains a coherent state at all times. To prove the theorem we define the Heisenberg and Schrödinger pictures of the motion of the system so that they coincide at time $t = 0$. The initial Schrödinger state of the system, which we take to be the coherent state $|\{\alpha_k\}\rangle$, is therefore the constant Heisenberg state vector. This state is an eigenstate of the initial values $a_k(0) \equiv a_k$ of the amplitude operators,

$$a_k|\{\alpha_k\}\rangle = \alpha_k|\{\alpha_k\}\rangle \; . \qquad (2)$$

We now form the variances of the complex amplitude operators, which are defined as

$$V_j(t) = \langle\{a^\dagger_j(t) - \langle a^\dagger_j(t)\rangle\}\{a_j(t) - \langle a_j(t)\rangle\}\rangle$$

$$= \langle a^\dagger_j(t)\, a_j(t)\rangle - \langle a^\dagger_j(t)\rangle\langle a_j(t)\rangle \; , \qquad (3)$$

where the state for which the expectation values are taken is given by eq. (2). It is clear from eq. (2) that the initial values of the variances $V_j(t)$ vanish. If we evaluate the time derivative of $V_j(t)$ by making use of eq. (1) and its Hermitian adjoint, we see that the initial values of the time derivatives also vanish. More generally, the n-th time derivative of $V_j(t)$ may be constructed by making repeated use of the equations of motion, and it is clear from eq. (2) that the initial values of all of these derivatives vanish as well. Since $V_j(t)$ vanishes at $t = 0$ together with its derivatives of all orders we infer that $V_j(t)$ vanishes at times t different from zero.

Let us write the state of the system at time t in the Schrödinger picture as $|t\rangle$. Then the variances $V_j(t)$ are given in the Schrödinger picture by

$$V_j(t) = \langle t|\{a^\dagger_j - \langle a^\dagger_j(t)\rangle\}\{a_j - \langle a_j(t)\rangle\}|t\rangle \; . \qquad (4)$$

These variances are non-negative, according to the Schwarz inequality, which tells us that the values $V_j(t) = 0$ can only be attained if

$$a_j|t\rangle = \langle a_j(t)\rangle|t\rangle, \qquad j = 1 \ldots n \; . \qquad (5)$$

The state $|t\rangle$, in other words, is always an eigenstate of all the operators a_j. It may be written as the coherent state $|\{\alpha_k(t)\}\rangle$, where the eigenvalues of the complex amplitude operators are just

$$\alpha_j(t) = \langle a_j(t)\rangle \; , \qquad (6)$$

i.e. the time-dependent eigenvalues are just the expectation values of the amplitudes.

The same result may be seen somewhat more directly if we solve the differential eqs. (1) and express the solutions in the form

$$a_j(t) = L_j(\{a_k\}, t) \; . \qquad (7)$$

Then it is clear that the state $|\{\alpha_k\}\rangle$ remains an eigenstate of $a_j(t)$ for all times t, i.e. we have

$$a_j(t)|\{\alpha_k\}\rangle = L_j(\{\alpha_k\}, t)|\{\alpha_k\}\rangle$$

$$= \langle a_j(t)\rangle|\{\alpha_k\}\rangle \; . \qquad (8)$$

Now the time dependent operators $a_j(t)$ may be written in terms of their initial values as

$$a_j(t) = U^{-1}(t)\, a_j U(t) \; , \qquad (9)$$

where $U(t)$ is the unitary operator which relates the Heisenberg and Schrödinger pictures. It follows then that

$$a_j U(t)|\{\alpha_k\}\rangle = \langle a_j(t)\rangle\, U(t)|\{\alpha_k\}\rangle \; . \qquad (10)$$

But $U(t)|\{\alpha_k\}\rangle$ is the Schrödinger state of the system at time t and we see once again that it is coherent in character for all t. This property of the state permits us to solve for it in a particularly simple way. By writing the state as $|\{\alpha_k(t)\}\rangle$ and making use of eqs. (6) and (7) we see that the complex amplitudes $\alpha_k(t)$ satisfy the set of c-number equations of motion

$$\dot{\alpha}_j(t) = F_j(\{\alpha_k(t)\}\, t) \; , \qquad (11)$$

which show that the motion of the system is nearly classical in character. The wave packets which represent all of the oscillators in configuration or momentum space retain their minimum uncertain-

* An additional sense in which the coherent states have minimum uncertainty is that they minimize the product $(\delta q)^2(\delta p)^2$. This condition, however, does not characterize them uniquely.

Volume 21, number 6 PHYSICS LETTERS 1 July 1966

ty character at all times and move along trajectories which are precisely those of classical theory.

A simple example of a system exhibiting this behavior is an oscillator whose motion is forced by an arbitrary external field [4]. Other examples are the phase diffusion model of a laser beam [5] and the parametric frequency converter [6]. More generally, any coupling of the oscillators for which the interaction Hamiltonian takes the form

$$H' = \sum_{jk} f_{jk}(t) a^\dagger_j a_k + \sum_j \{g_j(t) a^\dagger_j + g^*_j(t) a_j\} , \quad (12)$$

where $f_{jk} = f^*_{kj}$ and the f_{jk} and g_j are arbitrary functions of time, leads to states which remain coherent at all times [8].

The theorem we have demonstrated does not extend in general to cases in which the equations of motion express the time derivatives $\dot{a}_j(t)$ in terms of the adjoint operators $\{a^\dagger_k(t)\}$ as well as the $\{a_k(t)\}$. When the equations of motion take this more general form, as they do for example in the case of the parametric amplifier [7], the behavior of the system is typically less susceptible to classical description.

References

1. R.J.Glauber, Proc.3rd Intern.Congr.on Quantum electronics, Paris 1963 (Dunod, Paris, 1964) Vol.I, p.111; and Phys.Rev.131 (1963) 2766.
2. U.M.Titulaer and R.J.Glauber, Phys.Rev., to be published.
3. E.Schrödinger, Naturwissenschaften 14 (1926) 664.
4. P.Carruthers and M.M.Nieto, Am.J.Phys.33 (1965) 537.
5. R.J.Glauber, 1964 Summer School for Theoretical Physics, Les Houches, France (Gordon and Breach Science Publishers, Inc., New York 1965) p.165.
6. W.H.Louisell, Radiation and noise in quantum electronics (McGraw-Hill Book Company, New York 1964) p.274.
7. The case of $g_j = 0$ and f_{jk} time independent has been discussed by C.W.Helstrom, J.Math.Phys., to be published.

* * * * *

Commun. math. Phys. 35, 265—277 (1974)

The Classical Limit for Quantum Mechanical Correlation Functions

Klaus Hepp

Physics Department, ETH, Zürich, Schweiz

Received November 5, 1973

Abstract. For quantum systems of finitely many particles as well as for boson quantum field theories, the classical limit of the expectation values of products of Weyl operators, translated in time by the quantum mechanical Hamiltonian and taken in coherent states centered in x- and p-space around $\hbar^{-1/2}$ (coordinates of a point in classical phase space) are shown to become the exponentials of coordinate functions of the classical orbit in phase space. In the same sense, $\hbar^{-1/2}$ [(quantum operator) (t) – (classical function) (t)] converges to the solution of the linear quantum mechanical system, which is obtained by linearizing the non-linear Heisenberg equations of motion around the classical orbit.

§ 1. Introduction

Consider the canonical system with the real Hamilton function

$$\mathscr{H}(\pi, \xi) = \pi^2/2m + V(\xi) \tag{1.1}$$

in the $2f$-dimensional phase space $\mathbb{R}^{2f} \ni (\pi, \xi)$. If $\operatorname{grad} V = \nabla V$ is Lipschitz around ξ, then the canonical equations

$$m\dot{\xi}(t) = \pi(t), \qquad \dot{\pi}(t) = -\operatorname{grad} V(\xi(t)) \tag{1.2}$$

have a unique solution $(\xi(\alpha, t), \pi(\alpha, t))$ for times $|t| < T(\alpha)$ (possibly $0 < T(\alpha) \leqq \infty$) with the initial data

$$\xi(\alpha, 0) = \xi, \qquad \pi(\alpha, 0) = \pi, \qquad \alpha = (\xi + i\pi)/\sqrt{2}. \tag{1.3}$$

While the classical equations (1.2) have locally unique but globally possibly nonexistent solutions (escape to infinity in finite times or collisions in the N-body problem), the corresponding quantum mechanical problem

$$i\hbar \frac{\partial \psi}{\partial t}(x, t) = -\frac{\hbar^2}{2m} \Delta \psi(x, t) + V(x)\,\psi(x, t) \tag{1.4}$$

in $L^2(\mathbb{R}^f)$ has always global solutions, if $p_\hbar^2/2m$ and $V_\hbar$ have a common dense domain $\mathscr{D}$ and if $\psi = \psi(\cdot, 0) \in \mathscr{D}$, by taking any selfadjoint extension $H_\hbar$ of the real and symmetric operator $p_\hbar^2/2m + V_\hbar$, $U_\hbar(t) = \exp(-iH_\hbar t/\hbar)$

and $\psi_t = U_\hbar(t)\,\psi$. However, these global solutions are not unique, if $p_\hbar^2/2m + V_\hbar$ is not essentially self-adjoint on $\mathscr{D}$.

The discussion of the connection between (1.2) and (1.4) is as old as quantum mechanics (see e.g. [1–3]). The WKB method relates an asymptotic expansion of solutions of (1.4) for $\hbar \to 0$ to solutions of the Hamilton Jacobi equations for (1.2) [4]. For more than one degree of freedom, the mathematical difficulties of this approach are considerable [5]. The Feynman integral approach [6] is very suggestive, but also difficult in rigorous mathematical terms [7]. The simplest connection between quantum and classical mechanics, however, goes back to Ehrenfest [8]: For every $\psi \in \mathscr{D}$ and V sufficiently regular,

$$\frac{d}{dt}(\psi_t, q_\hbar \psi_t) = (\psi_t, p_\hbar \psi_t)/m$$
$$\frac{d}{dt}(\psi_t, p_\hbar \psi_t) = -(\psi_t, \nabla V_\hbar \psi_t)\,. \tag{1.5}$$

However (1.5) does not define a solution of (1.2) since $(\psi_t, \nabla V_\hbar \psi_t) \neq \nabla V((\psi_t, q_\hbar \psi_t))$, unless ∇V is linear, and even if the error is small for some t, it need not be controllable for all t, if $\hbar > 0$.

It is a folk-theorem (see [9, 36]) that (1.5) establishes a rigorous transition to (1.2), when $\hbar \to 0$ in minimal uncertainty states for $p_\hbar$ and $q_\hbar$, i.e. in coherent states [10] centered around large mean values $\hbar^{-1/2}\pi$, $\hbar^{-1/2}\xi$. This becomes apparent in the following symmetric representation of the CCR:

$$p_\hbar = \sqrt{\hbar}\,p\,, \qquad q_\hbar = \sqrt{\hbar}\,q\,, \tag{1.6}$$

where $p = -i\,d/dx$, $q = x$ and $a = (q + ip)/\sqrt{2}$ are $\hbar$-independent. Let $\alpha \in \mathbb{C}$ and

$$U(\alpha) = \exp(\alpha a^* - \alpha^* a) = \exp i(\pi q - \xi p)\,. \tag{1.7}$$

Because of $U(\alpha)\,a\,U(\alpha)^* = a - \alpha$, one has in the coherent state $|\alpha\rangle = U(\alpha)|0\rangle$ (where $a|0\rangle = 0$) for an arbitrary monomial in the p's and q's:

$$\langle \hbar^{-1/2}\alpha|(q - \hbar^{-1/2}\xi)\dots(p - \hbar^{-1/2}\pi)|\hbar^{-1/2}\alpha\rangle = \langle 0|q\dots p|0\rangle\,, \tag{1.8}$$

and hence

$$\lim_{\hbar \to 0} \langle \hbar^{-1/2}\alpha|q_\hbar \dots p_\hbar|\hbar^{-1/2}\alpha\rangle = \xi \dots \pi\,. \tag{1.9}$$

We shall show that (1.9) (in Weyl form) is preserved under the time evolution $U_\hbar(t)$ of any selfadjoint extension $H_\hbar$ of $p_\hbar^2/2m + V_\hbar$:

$$\lim_{\hbar \to 0} \langle \hbar^{-1/2}\alpha|q_\hbar(s)\dots p_\hbar(t)|\hbar^{-1/2}\alpha\rangle = \xi(\alpha, s)\dots \pi(\alpha, t)\,, \tag{1.10}$$

as long as the classical orbit exists. The fact, that along coherent states the quantum mechanical evolution $\langle \hbar^{-1/2}\alpha | a_\hbar(t) | \hbar^{-1/2}\alpha \rangle$ and the classical evolution $\xi(\alpha, t) = \langle \hbar^{-1/2}\alpha(t) | a_\hbar | \hbar^{-1/2}\alpha(t) \rangle$ are in "weak correspondence" (which becomes exact for $\hbar \to 0$) has been analyzed by Klauder [9]. But to the best of the present author's knowledge no general proof has been given of (1.10), nor has it been recognized that also (1.8) is preserved under time-evolution (for the technical details, see Theorem 2.1 and [37] for a probabilistic setting):

$$\lim_{\hbar \to 0} \langle \hbar^{-1/2}\alpha | \hbar^{-1/2}(q_\hbar(s) - \xi(\alpha, s)) \ldots \hbar^{-1/2}(p_\hbar(t) - \pi(\alpha, t)) | \hbar^{-1/2}\alpha \rangle = \langle 0 | q(\alpha, s) \ldots p(\alpha, t) | 0 \rangle . \tag{1.11}$$

Here the $q(\alpha, t)$ and $p(\alpha, t)$ are solutions of the linearized classical equations (1.2) around $\xi(\alpha, t)$:

$$\dot{q}(\alpha, t) = p(\alpha, t)/m, \qquad \dot{p}(\alpha, t) = -\nabla V(\xi(\alpha, t))\, q(\alpha, t), \tag{1.12}$$

with initial conditions $q(\alpha, 0) = q$, $p(\alpha, 0) = p$. Both, (1.10) and (1.11) have an easy generalization to more complicated Hamiltonians and to relativistic and non-relativistic infinite boson systems. In the latter case the compensation of singularities for $\hbar \to 0$, when expanding the quantum dynamics around a classical solution, is implicit in the work of Goldstone [11] and Gross [12], but again a mathematical proof is desirable.

Our work has been most strongly influenced by the findings of Lieb and the author [13] in mean field models, as lasers and strongly coupled superconductors, that "intensive" quantities $a_N(t)$, i.e. space averages $N^{-1} \sum_{n=1}^{N} A_n$ of local observables translated in time by mean field Hamiltonians, become classical $\alpha(t)$ in the limit $N \to \infty$ along classical states, while the "fluctuations" $\sqrt{N}(a_N(t) - \alpha(t))$ become boson operators $a(\alpha, t)$, which follow linearized equations of motion, if in the classical states the fluctuations at $t = 0$ have a limit. We think that the analogy between $N \to \infty$ and $\hbar \to 0$ is significant for the understanding of classical operations within the framework of quantum mechanics [14]. It is the pedagogical goal of this paper to elaborate a unified picture of the classical limit in quantum mechanical correlation functions, which is so simple that it could belong into an elementary course on quantum mechanics.

The author is indebted to M. Fierz, J. Glimm, A. M. Jaffe, J. R. Klauder, B. Kostant and J. Lascoux for helpful discussions and bibliographical information and, last but not least, to E. H. Lieb whithout whom this paper would never have been written.

§ 2. Finitely Many Degrees of Freedom

The passage to the classical limit in quantum mechanical correlation functions can be completely illustrated for the Hamiltonian $\mathscr{H}(\pi, \xi) = \pi^2/2m + V(\xi)$ with only one degree of freedom:

Theorem 2.1. *Let $V(\xi)$ be real and $\xi(\alpha, t)$ a solution of (1.2) for $|t| < T > 0$ and initial data α. Let V be $C^{2+\delta}$, $\delta > 0$, in a neighborhood of $\xi(\alpha, t)$ and assume that $\int |V(x)|^2 \exp(-\varrho x^2)\, dx < \infty$ for some $\varrho < \infty$. Let $H_\hbar$ be any selfadjoint extension of*

$$-\frac{\hbar}{2m} d^2/dx^2 + V(\sqrt{\hbar}\, x) \quad in \quad L^2(\mathbb{R}^1) \quad and \quad U_\hbar(t) = \exp - iH_\hbar t/\hbar .$$

Then for all $(r, s) \in \mathbb{R}^2$ and uniformly on compacts in $\{|t| < T\}$:

$$\begin{aligned} \text{s-}\lim_{\hbar \to 0} U(\hbar^{-1/2}\alpha)^* U_\hbar(t)^* \exp i[r(q - \hbar^{-1/2}\xi(\alpha, t)) + s(p - \hbar^{-1/2}\pi(\alpha, t))] \\ \cdot U_\hbar(t)\, U(\hbar^{-1/2}\alpha) = \exp i[rq(\alpha, t) + sp(\alpha, t)] , \end{aligned} \tag{2.1}$$

and

$$\begin{aligned} \text{s-}\lim_{\hbar \to 0} U(\hbar^{-1/2}\alpha)^* U_\hbar(t)^* \exp i[rq_\hbar + sp_\hbar]\, U_\hbar(t)\, U(\hbar^{-1/2}\alpha) \\ = \exp i[r\xi(\alpha, t) + s\pi(\alpha, t)] . \end{aligned} \tag{2.2}$$

Here $(p(\alpha, t), q(\alpha, t))$ are the solutions of (1.2) linearized around $\xi(\alpha, t)$ with initial data (p, q), which arise from the selfadjoint Hamiltonian

$$H(t) = p^2/2m + V''(\xi(\alpha, t))\, q^2/2 . \tag{2.3}$$

Proof. One expands $H_\hbar/\hbar$ around the classical orbit $\xi(\alpha, t) = \xi_t$:

$$H_\hbar/\hbar = H_\hbar^0(t) + H_\hbar^1(t) + H_\hbar^2(t) + H_\hbar^3(t) , \tag{2.4}$$

$$H_\hbar^0(t) = \mathscr{H}(\pi, \xi)/\hbar , \tag{2.5}$$

$$H_\hbar^1(t) = \pi_t(p - \hbar^{-1/2}\pi_t)\, \hbar^{-1/2} + V'(\xi_t)\,(q - \hbar^{-1/2}\xi_t)\, \hbar^{-1/2} , \tag{2.6}$$

$$H_\hbar^2(t) = (p - \hbar^{-1/2}\pi_t)^2/2 + V''(\xi_t)\,(q - \hbar^{-1/2}\xi_t)^2/2 . \tag{2.7}$$

The propagator $U_\hbar^1(t) = T \exp - i \int_0^t ds\, H_\hbar^1(s)$ exists for all $|t| < T$ (by the unitary extension of its strongly convergent Dyson series on the linear hull of all Hermite functions) and defines an automorphism of the Weyl algebra:

$$U_\hbar^1(t)^*(a^\# - \hbar^{-1/2}\alpha_t^\#)\, U_\hbar^1(t) = a^\# - \hbar^{-1/2}\alpha^\# . \tag{2.8}$$

Hence the l.h.s. of (2.1) can be written as

$$W_\hbar(t, 0)^* \exp i[rq + sp]\, W_\hbar(t, 0) , \tag{2.9}$$

where

$$W_\hbar(t,s)=U(\hbar^{-1/2}\alpha)^* U_\hbar^1(t)^* U_\hbar(t-s)\,U_\hbar^1(s)\,U(\hbar^{-1/2}\alpha) \cdot \exp i\int_s^t dr\, H_\hbar^0(r)\,. \tag{2.10}$$

Hence (2.1) is proved, if on a dense subspace $s\text{-lim}\, W_\hbar(t,s)=W(t,s) = T\exp-\int_s^t dr\, H(r)$ holds.

The normalized states $\{\psi_a(x)=\pi^{-1/4}\exp-(x-a)^2/2\,|\,a\in\mathbb{R}\}$ span $L^2(\mathbb{R})$. We claim that for every $0<k<T$ there exists some $\hbar_k>0$, such that for all $\hbar<\hbar_k$ and all $|s|\leqq k$, the total set of states

$$\{\psi_a^{\hbar s}=U_\hbar^1(s)\,U(\hbar^{-1/2}\alpha)\,W(s,0)\,\psi_a\}\subset D(p^2)\cap D(\hbar^{-1}V(\sqrt{\hbar}q))\,. \tag{2.11}$$

For, $H(r)$ is quadratic with $V''(\xi_r)$ continuous in r. Hence the Dyson series for $W(t,s)$ converges for small $|t-s|$ and

$$\begin{aligned} W(s,0)\,q\,W(s,0)^* &= \alpha q+\beta p\,,\\ W(s,0)\,p\,W(s,0)^* &= \gamma q+\delta p\,,\\ A=A(s) &= \begin{pmatrix}\alpha & \beta\\ \gamma & \delta\end{pmatrix}\in \mathrm{Sp}(2,\mathbb{R})\,, \end{aligned} \tag{2.12}$$

with continuous dependence on s. Since ψ_a satisfies $[q-a+ip]\,\psi_a=0$,

$$\begin{aligned} 0 &= U_\hbar^1(s)\,U(\hbar^{-1/2}\alpha)\,W(s,0)\,[q-a+ip]\,\psi_a\\ &= [(\alpha+i\gamma)(q-\hbar^{-1/2}\xi_s)-a+i(\delta-i\beta)(p-\hbar^{-1/2}\pi_s)]\,\psi_a^{\hbar s}\,, \end{aligned} \tag{2.13}$$

or

$$\psi_a^{\hbar s}(x)=\text{const}\exp\left[-\frac{(\alpha+i\gamma)}{2(\delta-i\beta)}\left(x-\hbar^{-1/2}\xi_s-\frac{a}{(\alpha+i\gamma)}\right)^2+i\pi_s\hbar^{-1/2}x\right]. \tag{2.14}$$

Since $\mathrm{Re}(\alpha+i\gamma)/(\delta-i\beta)\,2=1/2(\delta^2+\beta^2)>\eta_k>0$ for all $|s|\leqq k$, and since $\int dx\,|V(x)|^2\exp-\varrho x^2<\infty$ for some $\varrho<\infty$, one obtains (2.11) for $\hbar_k=2\eta_k/\varrho$.

Therefore $W_\hbar(t,s)\,W(s,r)\,\psi_a$ is strongly differentiable with respect to s, if $0<k<T$, $|s|$, $|t|\leqq k$ and if $\hbar<\hbar_k$, for any selfadjoint extension $H_\hbar$ of $p_\hbar^2/2m+V_\hbar$. We obtain the Duhamel formula

$$W(t,0)\,\psi_a-W_\hbar(t,0)\,\psi_a=\int_0^t ds\,\frac{d}{ds}\,W_\hbar(t,s)\,W(s,0)\,\psi_a\,, \tag{2.15}$$

$$\begin{aligned} \frac{d}{ds}\,W_\hbar(t,s)\,W(s,0)\,\psi_a &= iW_\hbar(t,s)\,\{\hbar^{-1}V(\xi_s+\sqrt{\hbar}q)\\ &\quad -\hbar^{-1}V(\xi_s)-\hbar^{-1/2}V'(\xi_s)\,q-V''(\xi_s)\,q^2/2\}\,W(s,0)\,\psi_a\,. \end{aligned} \tag{2.16}$$

The norm of (2.16) will be estimated as follows: There exists some $\sigma > 0$, such that $V(\xi_s + x)$ is $C^{2+\delta}$ for all $|s| \leqq k$ and $|x| \leqq \sigma$. We consider

$$\int dx |\hbar^{-1} V(\xi_s + \sqrt{\hbar} x) - \hbar^{-1} V(\xi_s) - \hbar^{-1/2} V'(\xi_s)\, x - V''(\xi_s)\, x^2/2|^2 \cdot |(W(s,0)\, \psi_a)(x)|^2 . \tag{2.17}$$

In $\{|x| \geqq \hbar^{-1/2}\sigma\}$, each term is $O(\hbar^N)$ for every N, since $|V(\xi_s + \sqrt{\hbar} x)|^2$ increases at infinity at most as $\exp \hbar \varrho x^2$, while $|(W(s,0)\, \psi_a)(x)|^2$ decreases as $\exp - 2\eta_k x^2$. On the other hand, for $|x| \leqq \hbar^{-1/2}\sigma$, one uses the Hölder continuity of V'' $(\delta \leqq 1)$:

$$\begin{aligned} &|\hbar^{-1} V(\xi_s + \sqrt{\hbar} x) - \hbar^{-1} V(\xi_s) - \hbar^{-1/2} V'(\xi_s)\, x - V''(\xi_s)\, x^2/2| \\ &\leqq x^2 \int_0^1 dy (1-y) |V''(\xi_s + \sqrt{\hbar} xy) - V''(\xi_s)| \leqq \text{const}\, x^{2+\delta} \hbar^{\delta/2} . \end{aligned} \tag{2.18}$$

Hence $\|W(t,0)\, \psi_a - W_\hbar(t,0)\, \psi_a\| = O(\hbar^{\delta/2})$ leads to (2.1). By the same argument

$$\begin{aligned} &\|U(\hbar^{-1/2}\alpha)^* U_\hbar(t)^* e^{i[rq_\hbar + sp_\hbar]} U_\hbar(t)\, U(\hbar^{-1/2}\alpha)\, \psi - e^{i[r\xi_t + s\pi_t]} \psi\| \\ &= \|W_\hbar(t,0)^* \exp i \sqrt{\hbar}(rq + sp)\, W_\hbar(t,0)\, \psi - \psi\| . \end{aligned} \tag{2.19}$$

Since $s\text{-lim}\, W_\hbar(t,0) = W(t,0)$ and $s\text{-lim} \exp i\sqrt{\hbar}(rq + sp) = \mathbb{1}$, (2.2) follows.

Q.E.D.

Remark. It is helpful for the interpretation of Theorem 2.1 to note the analogy to time dependent scattering theory [15] between

$$\lim_{\hbar \to 0} \|U_\hbar(t)\, U(\hbar^{-1/2}\alpha)\, \psi - U(\hbar^{-1/2}\alpha_t)\, W(t,0)\, \psi\| = 0 \tag{2.20}$$

and

$$\lim_{t \to \infty} \|e^{iHt} \Omega_- \psi - e^{iH_0 t} \psi\| = 0 . \tag{2.21}$$

If ψ is Gaussian, then also $U(\hbar^{-1/2}\alpha)\, \psi$ and $U(\hbar^{-1/2}\alpha_t)\, W(t,0)\, \psi$, and (2.20) shows that under the time evolution $U_\hbar(t)$ the difference of $U_\hbar(t)\, U(\hbar^{-1/2}\alpha)\, \psi$ from a Gaussian wave packet centered around the classical orbit and with the shape wobbling according to the quadratic Hamiltonian of the linearized theory goes to zero, as $\hbar \to 0$.

The error in (2.2) between Gaussian wave packets can be reduced to $O(\sqrt{\hbar})$ uniformly for bounded time intervals, if $\xi(\alpha, t)$ exists for all t and if V is C^3 in a neighborhood of this orbit. Hence the Ehrenfest theorem describes well the classical aspects of the motion of wave packets for finite times, but not for $t \to \infty$. Our method is complementary to the WKB-method, which is successful for describing the stationary states in quantum mechanics.

One learns from Theorem 2.1 that equilibrium points (π_0, ξ_0) of the classical motion, $\pi_0 = 0$ and $V'(\xi_0) = 0$, are driven by the quantum fluctuations in $O(\sqrt{\hbar})$: If $V''(\xi_0)>0$, then (2.3) leads to an oscillatory behavior of Δp^2 and Δq^2 in any wave packet ψ_t. For $V''(\xi_0)\leqq 0$ the spectrum of (2.3) is purely continuous and the wave packets spread, for $V''(\xi_0)=0$ with a power law and exponentially fast for $V''(\xi_0)<0$.

A slight modification of the kinematics leads to the classical limit for heavy particles, if $\lambda=\hbar/m\to 0$ in Hamiltonians

$$\mathscr{H}(\pi,\xi)=\pi^2/2m+mV(\xi)\,. \tag{2.22}$$

Corollary 2.2. *Consider* (2.22) *(under the same assumptions on V as in Theorem 2.1) around the solution $\xi(t)$ of the classical equation $\ddot{\xi}(t)=-V'(\xi(t))$ with initial data $\alpha=(\xi(0)+i\dot{\xi}(0))/\sqrt{2}$. Let*

$$p_\lambda=m\sqrt{\lambda}\,p,\qquad q_\lambda=\sqrt{\lambda}\,q\,, \tag{2.23}$$

and let $H_\lambda\hbar^{-1}$ be any selfadjoint extension of $p^2/2+\lambda^{-1}V(\sqrt{\lambda}q)$ with $U_\lambda(t)=\exp-iH_\lambda t/\hbar$. Then

$$\begin{aligned}\operatorname*{s-lim}_{\lambda\to 0}\,& U^*(\lambda^{-1/2}\alpha)\,U_\lambda(t)^*\exp i[r(q-\lambda^{-1/2}\xi_t)+s(p-\lambda^{-1/2}\dot{\xi}_t)]\\ &\cdot U_\lambda(t)\,U(\lambda^{-1/2}\alpha)=\exp i[rq(t)+sp(t)]\,,\end{aligned} \tag{2.24}$$

$$\begin{aligned}\operatorname*{s-lim}_{\lambda\to 0}\,& U(\lambda^{-1/2}\alpha)^*\,U_\lambda(t)^*\exp i[rq_\lambda+sp_\lambda/m]\,U_\lambda(t)\,U(\lambda^{-1/2}\alpha)\\ &=\exp i[r\xi_t+s\dot{\xi}_t]\,,\end{aligned} \tag{2.25}$$

where $\dot{q}(t)=p(t)$, $\dot{p}(t)=-V''(\xi_t)\,q(t)$, $q(0)=q$, $p(0)=p$.

For N-particle Hamiltonians of the type

$$\mathscr{H}(\pi,\xi)=\sum_{n=1}^{N}\left(\pi_n-\frac{e}{c}A(\xi_n,t)\right)^2/2M_n+V(\xi,t) \tag{2.26}$$

with nontrivial time-dependence, we have to assume the following regularity property:

(R): There exists a propagator $U_\hbar(t,s)$ which is strongly continuous for $-\infty<s,t<+\infty$ with $U_\hbar(t,s)\,U_\hbar(s,r)=U_\hbar(t,r)$, $U_\hbar(t,s)^*=U_\hbar(s,t)$ and $U_\hbar(t,t)=\mathbb{1}$. For some $\varrho<\infty$ and all Gauss packets with

$$\begin{gathered}\sup|\psi(x)\exp\varrho\|x\|^2|<\infty\,,\\ \operatorname*{s-lim}_{r\to 0}\hbar r^{-1}[U_\hbar(t,s+r)-U_\hbar(t,s)]\,\psi=iU_\hbar(t,s)\,H_\hbar(s)\,\psi\,,\end{gathered} \tag{2.27}$$

where $H_\hbar(s)\,\psi$ is naturally defined as partial differential operator.

Theorem 2.3. *For N-body systems* (2.26) *with A and V satisfying* (R), *a generalized Ehrenfest theorem of the type* (2.1), (2.2) *holds along every classical orbit $\xi(t)$, in the neighborhood of which A and V are $C^{2+\delta}$, $\delta>0$.*

In the classical limit, there is no difference between the coherent and incoherent superposition of states of the type $U(\hbar^{-1/2}\alpha_n)\,\varphi_n$ [with $\alpha_n \in \mathbb{C}^f$ and $\varphi_n \in L^2(\mathbb{R}^f)$], if $\Sigma\|\varphi_n\|^2 = 1$ and $\alpha_m \neq \alpha_n$ for $m \neq n$: For all $(r,s) \in \mathbb{R}^2$ and $|t| < \min T(\alpha_n)$ and

$$\psi_\hbar = \Sigma\, U(\hbar^{-1/2}\alpha_n)\,\varphi_n, \quad P_\hbar = \Sigma\, U(\hbar^{-1/2}\alpha_n)|\varphi_n\rangle\langle\varphi_n|\,U(\hbar^{-1/2}\alpha_n)^*,$$

$$\begin{aligned}&\lim_{\hbar\to 0}(\psi_\hbar, U_\hbar(t)^* e^{i[rq_\hbar + sp_\hbar]} U_\hbar(t)\,\psi_\hbar)\\ &\quad = \lim_{\hbar\to 0}\mathrm{Tr}(P_\hbar U_\hbar(t)^* e^{i[rq_\hbar + sp_\hbar]} U_\hbar(t)) \qquad (2.28)\\ &\quad = \sum_n \|\varphi_n\|^2 \exp i[r\xi(\alpha_n, t) + s\pi(\alpha_n, t)]\,.\end{aligned}$$

This is important for fermions or bosons, where the (anti-)symmetrization of spatial wave functions of the type $U(\hbar^{-1/2}\alpha)\,\varphi$ usually leads to a classical ensemble in phase space with discrete density. Classical ensembles with continuous densities $\varrho(\pi,\xi) \geqq 0$, $\int d\pi\, d\xi\, \varrho(\pi,\xi) = 1$ can obviously be reached from any density matrix P, by forming

$$P_\hbar = \int d\pi\, d\xi\, \varrho(\pi,\xi)\, U(\hbar^{-1/2}\alpha)\, P\, U(\hbar^{-1/2}\alpha)^*$$

and by passing to the limit as in (2.27) for $|t| < \min T(\alpha)$.

A classical problem is the limit $\hbar \to 0$ and the related high temperature expansion in statistical mechanics [16, 17]:

$$\lim_{\hbar\to 0} h^f\, \mathrm{Tr}\, e^{-\beta H_\hbar} = \int d\pi\, d\xi\, e^{-\beta\mathscr{H}(\pi,\xi)}\,, \qquad (2.29)$$

$$\begin{aligned}&\lim_{\hbar\to 0}\mathrm{Tr}(e^{-\beta H_\hbar} p_\hbar^m q_\hbar^n)/\mathrm{Tr}\, e^{-\beta H_\hbar}\\ &\quad = \int d\pi\, d\xi\, e^{-\beta\mathscr{H}(\pi,\xi)}\pi^m\xi^n \Big/ \int d\pi\, d\xi\, e^{-\beta\mathscr{H}(\pi,\xi)}\,. \qquad (2.30)\end{aligned}$$

For finite f, (2.29) has been proved by Berezin [18] for a large class of Hamiltonians. In [19] the limit (2.30) of the correlation functions was considered for finitely many harmonic oscillators compled linearly to large systems of multilevel atoms. This method can be generalized to N-particle systems in an anharmonic oszillator well and interacting via regular short range two-body potentials.

Finally let us remark that there exist coherent states on a large class of Lie groups [20], which allow the passage to the classical limit for dynamical systems with more exotic phase spaces than $\mathbb{R}^{2f}$ [21–23]. One example, $\mathbb{R}^2 \times SU_2$, with "atomic" coherent states [24, 25] for SU_2, is important in the thermodynamic limit of the laser [13].

§ 3. Boson Systems of Infinitely Many Degrees of Freedom

Some of the results of the preceeding section can be generalized to systems with infinitely many degrees of freedom.

The best understood models for a relativistic quantum dynamics are the scalar boson theories in two dimensional space time with polynomial interaction (see e.g. [26]). Let $\Phi(x)$, $x \in \mathbb{R}^1$, be the free boson field of mass $m>0$ at $t=0$, $H_0(m^2)$ the corresponding free Hamiltonian and $\Pi(x)=i[H_0, \Phi(x)]$ in Fock space $\mathscr{F}$ with $[\Phi(x), \Phi(y)]=[\Pi(x), \Pi(y)]=0$ and $[\Phi(x), \Pi(y)]=i\,\delta(x-y)$. Let $\Phi(f)=\int dx\, f(x)\,\Phi(x)$ for $f \in \mathscr{D}(\mathbb{R}^1)$. For every $\alpha \in \mathscr{D}(\mathbb{R}^1)$ with the decomposition $\alpha(x)=(\varphi(x)+i\pi(x))/\sqrt{2}$ into real and imaginary part, the shift operator $U(\alpha)$ satisfies

$$U(\alpha)=\exp i[\Phi(\pi)-\Pi(\varphi)]\,, \tag{3.1}$$

$$\begin{aligned} U(\alpha)^* :\Phi(x)^m: U(\alpha) &= :(\Phi(x)+\varphi(x))^m: \\ U(\alpha)^* :\Pi(x)^m: U(\alpha) &= :(\Pi(x)+\pi(x))^m: \\ U(\alpha)^* H_0\, U(\alpha) &= H_0 + \int dx\, \{\pi(x)\,\Pi(x) + \nabla\varphi\ \nabla\Phi(x) \\ &\quad + m^2\,\varphi(x)\,\Phi(x)\} + \mathscr{H}_0(\alpha) \end{aligned} \tag{3.2}$$

where $\mathscr{H}_0(\alpha)$ is the classical energy of the free field $(\Box+m^2)\,\varphi(x,t)=0$ with Cauchy data α, and $:\ :$ is the Wick ordering w.r.t. the free vacuum.

For Cauchy data $\alpha \in \mathscr{D}(\mathbb{R}^1)$, the classical nonlinear real wave equation

$$(\Box+m^2)\,\varphi(\alpha,t,x)+\sum_{n=1}^{N} n a_n \varphi(\alpha,t,x)^{n-1}=0 \tag{3.3}$$

has for finite times, $|t|<T(\alpha)>0$, a unique smooth solution with propagation speed 1 (see [27–29]), where $T(\alpha)=\infty$ for N even and $a_N>0$. These solutions will be compared with the quantum solutions of

$$(\Box+m^2)\,\Phi_\hbar(t,x)+\sum_{n=1}^{N} n a_n :\Phi_\hbar(t,x)^{n-1}: \tag{3.4}$$

with $\Phi_\hbar(0,x)=\Phi_\hbar(x)=\sqrt{\hbar}\,\Phi(x)$, $\Pi_\hbar(0,x)=\Pi_\hbar(x)=\sqrt{\hbar}\,\Pi(x)$, which have been constructed by Glimm and Jaffe [26] for N even and $a_N>0$.

Let $r>0$ and $0 \leqq g_r \in \mathscr{D}(\mathbb{R}^1)$ with $g_r(x)=1$ for $|x| \leqq r$. Let $H_{\hbar r}$ be any self-adjoint extension of $\hbar H_0 + V_\hbar(g_r)$ from $D(H_0) \cap D(V_\hbar(g_r))$, where

$$V_\hbar(g_r)=\sum_{n=1}^{N} a_n \int dx\, g_r(x) :\Phi_\hbar^n(x):. \tag{3.5}$$

Let $U_{\hbar r}(t)=\exp - itH_{\hbar r}/\hbar$. On $D(H_0) \cap D(V_\hbar(g_r))$ the generator of $U_{\hbar r}(t)$ is

$$H_0+\sum_{n=1}^{N} a_n \hbar^{\frac{n}{2}-1} \int dx\, g_r(x) :\Phi(x)^n:. \tag{3.6}$$

Theorem 3.1. *Let $\alpha \in \mathscr{D}(\mathbb{R}^1)$ and $\varphi(\alpha,t,x)$ exist for $|t|<T$. Let $f,g \in \mathscr{D}(\mathbb{R}^1)$ be real and r be sufficiently large. Then*

$$\begin{aligned}
&\operatorname*{s-lim}_{\hbar\to 0} U(\hbar^{-1/2}\alpha)^* U_{\hbar r}(t)^* \exp i[(\Phi(f)-\hbar^{-1/2}\varphi(\alpha,t,f)) \\
&\qquad +(\Pi(g)-\hbar^{-1/2}\pi(\alpha,t,g))]\, U_{\hbar r}(t)\, U(\hbar^{-1/2}\alpha) \qquad (3.7)\\
&\quad = \exp i[\Phi(\alpha,t,f)+\Pi(\alpha,t,g)]\,,
\end{aligned}$$

$$\begin{aligned}
&\operatorname*{s-lim}_{\hbar\to 0} U(\hbar^{-1/2}\alpha)^* U_{\hbar r}(t)^* \exp i[\Phi_\hbar(f)+\Pi_\hbar(g)]\, U_{\hbar r}(t)\, U(\hbar^{-1/2}\alpha) \\
&\quad = \exp i[\varphi(\alpha,t,f)+\pi(\alpha,t,g)]\,, \qquad (3.8)
\end{aligned}$$

where $\varphi(\alpha,t,f)=\int dx\, f(x)\,\varphi(\alpha,t,x)$, $\Phi(\alpha,t,f)=\int dx\, f(x)\,\Phi(\alpha,t,x)$, *and where the* $\Phi(\alpha,t,x)$ *and* $\Pi(\alpha,t,x)=\dot{\Phi}(\alpha,t,x)$ *are the unique global solutions of* (3.3) *linearized around* $\varphi(\alpha,t,x)$:

$$0=(\Box+m^2)\,\Phi(\alpha,t,x)+\sum_{n=2}^{N} n(n-1)\,a_n\varphi(\alpha,t,x)^{n-2}\,\Phi(\alpha,t,x) \qquad (3.9)$$

with initial conditions $\Phi(x)$, $\Pi(x)$ *at* $t=0$.

Proof. The proof of Theorem 2.1 applies with few changes. (3.6) is developed around $\varphi(\alpha,t,x)$ in an obvious way. Again

$$H_r(t)=H_0+\sum_{n=2}^{N}\binom{n}{2}a_n\int dx\, g_r(x)\,\varphi(\alpha,t,x)^{n-2}:\Phi(x)^2: \qquad (3.10)$$

has a propagator $W_r(t,s)$, which generates (3.9) for r sufficiently large. $W_r(t,s)$ has for small $|t-s|$ a controllable action on $\mathscr{F}_0$, the subspace of finite particle states with momentum space wave functions of compact support, by its convergent Dyson expansion. Here one easily sees that s-lim $W_{\hbar r}(t,s)\,\psi=W_r(t,s)\,\psi$ for $\hbar\to 0$, $|t-s|$ small, using the Duhamel formula. Hence, using unitarity and the composition law,

$$W_r(t,u)\,W_r(u,s)=W_r(t,s)\,, \qquad \text{s-lim}\, W_{\hbar r}(t,s)=W_r(t,s)$$

for all $|s|,|t|<T$. Q.E.D.

Remark that by the Wick reordering automorphism one can transfer mass from H_0 to V. In the translation to the classical limit the coefficients in (3.6) are $\hbar$-dependent and make that only the unique highest order term in the transition from $\Phi(x_1)\ldots\Phi(x_n)$ to a Wick product $:\Phi(x)^n:$ contributes in the limit $\hbar\to 0$.

The classical limit in Theorem 3.1, which in perturbation theory corresponds to the sum over all tree graphs (see e.g. [30]), gives a rigorous meaning to the Goldstone picture [11] as the leading asymptotic term in an expansion in $\sqrt{\hbar}$. The $O(\sqrt{\hbar})$-correction gives an interesting instability, whenever the classical field equations have a non-zero stationary solution φ_0, e.g. for $2a_2 < -m^2$, $a_4 > 0$ and $a_n = 0$ for $n \neq 2, 4$. In this case, (3.10) with $\alpha(x) = \varphi_0/\sqrt{2} = \pm(-2a_2 - m^2)^{1/2}(24a_4)^{-1/2}$ for $|x| \leqq r$ is locally equivalent to $H_0(-4a_2 - 2m^2)$ with positive mass elementary excitations, while for the unstable stationary state $\varphi(x, t) \equiv 0$ one has local equivalence to $H_0(m - 2a_2)$ with purely imaginary mass.

In more than 2-dimensional space-time, the renormalized local Hamiltonians for Φ_3^4 [31] and Φ_4^3 [32] are for $\hbar > 0$ defined in non-Fock representations of the CCR. Perturbation theory indicates that the classical limit is again of the structure of Theorem 3.1. In Φ_4^4 one can introduce an ultraviolet cut-off at $|\boldsymbol{k}| \leqq \kappa = \text{const}\,\hbar^{-1/3}$ and obtain the classical limit without any renormalization.

In non-relativistic many-body theory, the classical limit for bosons with the second quantized Hamiltonian (in Fock space over $L^2(\mathbb{R}^3)$)

$$\begin{aligned} H_\hbar &= -\frac{\hbar^2}{2m} \int dx\, a^*(x)\, \Delta a(x) \\ &\quad + \tfrac{1}{2} \int dx\, dy\, a^*(x)\, a^*(y)\, V(x-y)\, a(x)\, a(y) \qquad (3.11) \\ [a(x), a(y)] &= 0, \qquad [a(x), a^*(y)] = \delta(x-y), \end{aligned}$$

has been discussed by Gross [12] as the first step in a series of canonical transformations for diagonalizing $H_\hbar$ in the thermodynamic limit.

We shall assume $V(x) = V(-x) = V(x)^*$ to be a Kato potential [15] and $\partial V/\partial x_i$, $\partial^2 V/\partial x_i \partial x_j$ to be $-\Delta$-bounded. By a fixed-point argument one can show that for every initial condition $\beta \in D(\Delta)$ there exists a unique solution of the classical non-linear wave equation

$$\frac{\partial \alpha}{\partial t}(\beta, t, x) = \frac{i}{2\mu} \Delta \alpha(\beta, t, x) + i \int dy\, V(x-y)\, |\alpha(\beta, t, y)|^2\, \alpha(\beta, t, x), \qquad (3.12)$$

with $\alpha(\beta, t, \cdot) \in D(\Delta)$ for $|t| < T > 0$ and $\alpha(\beta, 0, \cdot) = \beta$. Furthermore, $H_\hbar$ is essentially self-adjoint on $\mathscr{F}_0$. Let $\beta \in D(\Delta)$, $a^{\#}(\beta) = \int dx\, \beta(x) a^{\#}(x)$ and $U(\beta) = \exp[a^*(\beta) - a(\beta^*)]$. In an almost coherent state $U(\hbar^{-1/2}\beta)\varphi$, $\varphi \in \mathscr{F}_0$, the particle number is $O(\hbar^{-1})$ for $\hbar \to 0$. However, $\hbar^{-1} H_\hbar$ is not extensive, as in Theorem 3.1, since $\hbar^{-1} H_{0\hbar} = O(1)$ and $\hbar^{-1} V_\hbar = O(\hbar^{-3})$ for $\hbar \to 0$. A non-trivial classical limit can be obtained by setting $m = \hbar^3 \mu$, $t = \hbar^2 \tau$ and by keeping $\mu > 0$ and τ fixed. This leads to $U_\hbar(\tau) = \exp - iK_\hbar \tau = \exp - iH_\hbar t/\hbar$, where

$$K_\hbar = -\frac{1}{2\mu} \int dx\, a^*(x)\, \Delta a(x) + \frac{\hbar}{2} \int dx\, dy\, a^*(x)\, a^*(y)\, V(x-y)\, a(x)\, a(y). \quad (3.13)$$

For $\gamma \in L^2(\mathbb{R}^3)$ we set $\alpha^{\#}(\beta, t, \gamma) = \int dx\, \gamma(x) \alpha^{\#}(\beta, t, x)$ and $a^{\#}(\beta, t, \gamma) = \int dx\, \gamma(x) a^{\#}(\beta, t, x)$, where $a(\beta, t, x)$ are the solutions of the linearization of (3.12) around $\alpha(t, \beta, x)$ with initial data $a(x)$. Furthermore, let $a_\hbar(x) = \sqrt{\hbar}\, a(x)$. Then

Theorem 3.2. *Under the above assumptions one has for* $|\tau| < T$:

$$\begin{aligned} \text{s-}\lim_{\hbar\to 0} U(\hbar^{-1/2}\beta)^* U_\hbar(\tau)^* \exp[(a^*(\gamma) - \hbar^{-1/2}\alpha^*(\beta,\tau,\gamma)) - \text{h.c.}] \\ \cdot U_\hbar(\tau)\, U(\hbar^{-1/2}\beta) \\ = \exp[a^*(\beta,\tau,\gamma) - a(\beta,\tau,\gamma^*)]\,, \end{aligned} \tag{3.14}$$

$$\begin{aligned} \text{s-}\lim_{\hbar\to 0} U(\hbar^{-1/2}\beta)^* U_\hbar(\tau)^* \exp[a_\hbar^*(\gamma) - a_\hbar(\gamma^*)]\, U_\hbar(\tau)\, U(\hbar^{-1/2}\beta) \\ = \exp[\alpha^*(\beta,\tau,\gamma) - \alpha(\beta,\tau,\gamma^*)]\,. \end{aligned} \tag{3.15}$$

Proof. As in Theorem 3.1 one proves $\text{s-}\lim W_\hbar(t,s)\,\psi = W(t,s)\,\psi$ first for $\psi \in \mathscr{F}_0$, by using the Dyson series for $W(t,s)$ and the Duhamel formula for small $|t-s|$.

It is amusing but not surprising that the classical limit is not unique: in coherent states centered around $\hbar^{-1/2}(\boldsymbol{\alpha}_1, \ldots, \boldsymbol{\alpha}_N)$ with fixed N, one obtains the classical mechanics of N mass points by Theorem 2.1, while in boson coherent states centered in Fock space around a classical field $\hbar^{-1/2}\alpha(x)$ one obtains a classical field theory, if $m \sim \hbar^3$ and $t \sim \hbar^2$.

The transition from the quantum to the classical correlation functions in Gibbs states in the thermodynamical limit is presently only understood for small activities, where the Kirkwood Salsburg equations have a unique solution (see [33–34], and [35] for the diagrammatic analysis).

§ 4. Conclusion

The main objective of this paper was to give a simple and mathematically rigorous discussion of the classical limit in quantum mechanics. We hope that our construction can sometimes be used as a reliable starting point for understanding some of the intriguing features of infinite quantum systems, as for the boson condensation and the appearance of broken symmetries.

References

1. Schrödinger, E.: Ann. d. Phys. **79**, 489 (1926)
2. Heisenberg, W.: Die physikalischen Prinzipien der Quantentheorie. Leipzig: Hirzel 1930
3. Pauli, W.: Die allgemeinen Prinzipien der Wellenmechanik, Handbuch der Physik, V. 1, Berlin-Göttingen-Heidelberg: Springer 1958
4. Andrié, M.: Comment. Phys. Math. **41**, 333 (1971)
5. Maslov, V.P.: Uspekhi Mat. Nauk, **15**, 213 (1960); Théorie des perturbations et méthodes asymptotiques. Paris: Dunod 1972
6. Feynman, R.P.: Rev. Mod. Phys. **20**, 367 (1948)
7. Nelson, E.: J. Math. Phys. **5**, 332 (1964)
   Berezin, F.A., Šubin, M.A.: Coll. Math. Soc. J. Bolyai, **5** (1970)

8. Ehrenfest, P.: Z. Physik **45**, 455 (1927)
9. Klauder, J. R.: J. Math. Phys. **4**, 1058 (1963); **5**, 177 (1964); **8**, 2392 (1967)
10. Glauber, R. J.: Phys. Rev. **131**, 2766 (1963)
11. Goldstone, J.: Nuovo Cim. **19**, 154 (1961)
12. Gross, E. P.: Phys. Rev. **100**, 1571 (1955); **106**, 161 (1957); Ann. Phys. **4**, 57 (1958); **9**, 292 (1960)
13. Hepp, K., Lieb, E. H.: Ann. Phys. **76**, 360 (1973); Helv. Phys. Acta **46** (1973). — Constructive quantum field theory. Velo, G., Wightman, A. S., eds., Lecture Notes in Physics, Berlin-Heidelberg-New York: Springer 1973
14. Hepp, K.: Helv. Phys. Acta **45**, 237 (1972)
15. Kato, T.: Perturbation theory of linear operators. Berlin-Heidelberg-New York: Springer 1966
16. Wigner, E. P.: Phys. Rev. **40**, 749 (1932)
17. Kirkwood, J. G.: Phys. Rev. **44**, 31 (1933); **45**, 116 (1934)
18. Berezin, F. A.: Math. USSR Sbornik **15**, 577 (1971); **17**, 269 (1972); Izv. Akad. Nauk Ser. Mat. **37**, 1134 (1972)
19. Hepp, K., Lieb, E. H.: Phys. Rev. **A 8**, 2517 (1973)
20. Perelomov, A. M.: Commun. math. Phys. **26**, 222 (1972)
21. Kostant, B.: In: Group representations in mathematics and physics, Bargmann, V., ed., Berlin-Heidelberg-New York: Springer 1970
22. Souriau, J. M.: Structure des systèmes dynamiques. Paris: Dunod 1970
23. Kirillov, A. A.: Elements of representation theory. Moscow: Nauka 1972
24. Arecchi, F. T., Courtens, E., Gilmore, R., Thomas, H.: Phys. Rev. **A 6**, 2211 (1972)
25. Lieb, E. H.: Commun. math. Phys. **31**, 327 (1973)
26. Glimm, J., Jaffe, A. M.: In: Mathematics of contemporary physics, Streater, R. F., ed., London: Academic P. 1972
27. Jörgens, K.: Math. Z. **77**, 295 (1961)
28. Browder, F. E.: Math. Z. **80**, 249 (1962)
29. Segal, I. E.: Ann. Math. **78**, 339 (1963)
30. Coleman, S., Weinberg, E.: Phys. Rev. D **7**, 1888 (1973)
31. Glimm, J.: Commun. math. Phys. **10**, 1 (1968)
32. Osterwalder, K.: Fortschr. Physik **19**, 43 (1971)
33. Ruelle, D.: Statistical mechanics: Rigorous results, New York: Benjamin 1969
34. Ginibre, J.: In: Statistical mechanics and quantum field theory, de Witt, C., Stora, R., eds. Paris: Gordon & Breach 1971
35. Bloch, C., de Dominicis, C.: Nucl. Phys. **10**, 181 (1959)
36. Bialynicki-Birula, I.: Ann. Phys. **67**, 252 (1971)
37. Martin-Löf, A.: Skand. Aktuarietidskr. **1967**, 70

Communicated by W. Hunziker

Klaus Hepp
Physics Department
E.T.H.
Hönggerberg
CH-8049 Zürich
Switzerland

PHYSICAL REVIEW D VOLUME 17, NUMBER 4 15 FEBRUARY 1978

Quantum nondemolition measurement and coherent states

W. G. Unruh
Department of Physics, University of British Columbia, Vancouver, B.C. V6T 1W5, Canada
(Received 1 July 1977)

This note investigates the possibility of quantum nondemolition measurements on an oscillator in a coherent state by means of first-order interactions with a probe particle.

Work by Braginsky and co-workers[1] has aroused interest in the possibility of experimentally measuring the state of an oscillator without altering the state of the oscillator. In a recent paper[2] I showed that the interaction must be of second order in the generalized coordinate (e.g., electric field in an electromagnetic oscillator, displacement of the ends in a bar, etc.) if such a scheme is to work. That work was done assuming that the oscillator is an energy eigenstate. Moncrief[3] has suggested that the oscillator is more likely to be in what is called a coherent state, and that a first-order interaction might work in that case. This note will examine the effect of an interaction linear in the generalized coordinate on such a coherent state.

I shall use a similar model system to that proposed in my original paper with a Lagrangian action given by

$$I=\int\left[\tfrac{1}{2}(\dot{x}^2+\dot{y}^2+\dot{z}^2)+\tfrac{1}{2}(\dot{q}^2-q^2)\right.$$
$$\left.-\alpha V(x,y,z)q\right]dt\,.$$

Here, x,y,z are the coordinates of the particle, and q is the generalized coordinate of the oscillator. The system can be quantized in the usual way.

The initial state of the oscillator is assumed to be the coherent state

$$\phi_c(q,t)=\exp(ce^{-it}a^\dagger)\phi_0(q,t)/e^{-|c|^2/2},$$

where ϕ_0 is the ground state of the oscillator, c is a constant, and $a^\dagger$ is the creation operator

$$a^\dagger=\frac{1}{\sqrt{2}}\left(q-\frac{d}{dq}\right)$$

(I will work throughout in the Schrödinger representation).

The wave function ϕ_c obeys the equation

$$a\phi_c=(ce^{-it})\phi_c,$$

where

$$a=\frac{1}{\sqrt{2}}\left(q+\frac{d}{dq}\right)$$

and is a solution to the Schrödinger equation for the free oscillator. Another normalized solution to the free oscillator equation is

$$\phi_p=(a^\dagger e^{-it}-c^*)\phi_c.$$

This function has unit norm and is orthogonal to ϕ_c. ϕ_p is chosen in this way because

$$q\phi_c=\frac{1}{\sqrt{2}}(ce^{-it}+c^*e^{it})\phi_c+\frac{1}{\sqrt{2}}e^{it}\phi_p.$$

Let us assume that the particle is initially in some state ψ_0 (assumed to be a normalized wave packet).

Now, because of the interaction, the particle can be scattered into some orthogonal state ψ_1 either leaving the oscillator in the state ϕ_c or by altering that state. To lowest order in α, that altered state must be ϕ_p. The amplitude for a nonperturbative scatter is therefore

$$A_n=\int_{-\infty}^{\infty}\frac{dt}{\sqrt{2}}\left[(ce^{-it}+c^*e^{it})\right.$$
$$\times\int dx\,dy\,dz\,\psi_1^*(x,y,z,t)V(x,y,z)$$
$$\left.\times\psi_0(x,y,z,t)\right].$$

The first term represents an interaction in which the particle has gained energy from the oscillator, while the second represents an energy loss to the oscillator. The perturbative scattering amplitude is given by

$$A_p=\int_{-\infty}^{\infty}\frac{dt}{\sqrt{2}}e^{it}\left[\int dx\,dy\,dz\,\psi_1^*(x,y,z,t)\right.$$
$$\left.\times V(x,y,z)\psi_0(x,y,z,t)\right].$$

Note that A_p is equal to $1/c^*$ times the second term in A_n. If c^* is large, the probability of a perturbative scatter is much less than that of an energy-emitting nonperturbative scatter.[4] As $|c|^2$ is just the mean number of particles in the state ϕ_c, we find that in the region of interest, i.e., small values of $|c|^2$, the ratio of perturbative to nonperturbative energy-emitting scatterings is approximately equal to one.

The energy-absorbing scatterings can only be nonperturbative. If the energy of the particle is much greater than that of the oscillator, and if $V(x,y,z)$ is reasonable, for any final state ψ_1, which corresponds to an energy-emitting interaction, there will be a similar final state ψ_1' corresponding to energy absorption. In other words, the particle will have roughly equal probability of scattering nonperturbatively through an energy emission or absorption, and for small $|c|^2$, also approximately equal probability of scattering perturbatively. Using high-energy particles to attempt to measure the state of the oscillator will result in a large probability of altering the coherent state of the system (if the detection of scattered particles is inefficient, as it will be, the probability of altering the state will increase).

If the energy of the particle is less than the energy of one quantum of the oscillator, both the energy-emitting nonperturbative scattering and the perturbative scattering have zero probability. In this case the oscillator is left in its coherent state by any scattering.

However, this does not seem too helpful for any practical scheme for measuring the state. Any mechanical oscillator has a low frequency, and even the electromagnetic oscillator used would probably have frequencies less than 10^{12} Hz which would imply electron energies of less than about 10^{-4} eV, a rather difficult requirement.

Furthermore, the above analysis is true *only* if the oscillator is in a coherent state. It is not at all clear to me that this is a reasonable expectation for a real oscillator with very few quanta. The effects of thermal fluctuations will be to destroy the coherence of the state.

One comment is appropriate here regarding the situation investigated by Moncrief. He investigated the use of electrons to study a coherence state of a free electromagnetic field. For this situation momentum conservation completely suppresses the interaction to first order in the field. The interaction is thus effectively of second order, which can allow nonperturbative measurement even of energy eigenstates.

For a unidirectional plane wave or wave packet, this does not occur, again because of momentum and energy conservation. However, the Kapitza-Dirac[5] effect can be regarded as such a measurement for two oppositely traveling plane waves.

---

[1]V. B. Braginsky, Yu. I. Vorontsov, and V. D. Krivchenkov, Zh. Eksp. Teor. Fiz. 68, 55 (1975) [Sov. Phys.—JETP 41, 28 (1975)].
[2]W. G. Unruh, Univ. of British Columbia report, 1977 (unpublished).
[3]V. Moncrief, Yale Univ. report, 1977 (unpublished).
[4]Energy absorbing and emitting will refer always to the particle.
[5]P. Kapitza and P. A. M. Dirac, Proc. Camb. Philos. Soc. 29, 297 (1933).

Reprinted from ANNALS OF PHYSICS
All Rights Reserved by Academic Press, New York and London
Vol. 114, Nos. 1-2, September 1, 1978

Coherent States and Quantum Nonperturbing Measurements*

VINCENT MONCRIEF

Department of Physics, Yale University, New Haven, Connecticut 06520

Received August 10, 1977

Spurred by the recent proposal of Braginskii and co-workers, we study the possibility of making quantum nonperturbing (qnp) measurements of coherent states of the electromagnetic field. As a model problem we consider the Compton scattering of electrons from a coherent photon beam. We find (to second order) that the electrons can scatter with only a small probability (per scattered electron) of disturbing the coherence of the ongoing incident beam. Measurement of the scattering rate for a known density of target electrons gives direct information about the absolute square of the coherence eigenvalue of the incident photon beam. We also consider the (somewhat academic) process of pair production by colliding coherent photon beams. In this case coherence is completely preserved at second order, and only higher-order amplitudes (e.g., Delbrück scattering) contribute to its disturbance. Our results, though not directly related to Braginskii's proposed experiment, suggest the potential importance of coherent states for the design of a quantum nonperturbing measuring device.

1. INTRODUCTION

Braginskii *et al.* [1, 2] have recently studied the possibility of measuring the n quantum state of an electromagnetic oscillator without disturbing the quantum state of the oscillator. They argue that with a suitable experimental design one can measure the photon number of an excited mode of a resonant cavity with only a small probability of changing the quantum state of the mode. Their design involves passing a beam of electrons through the cavity and measuring a diffraction pattern of the emerging beam. If most of the electrons are freely allowed to reenter the cavity (which consists of two separated components) after inversion of the beam through a magnetic lens, the effect of second passage through the cavity can be arranged to cancel the perturbing influence of the first passage, leaving the oscillator in its original state. They have emphasized the importance of such quantum nonperturbing (qnp) measurements for the design of gravitational wave detectors. In fact, it now seems that the possibility of detecting gravitational radiation from known astrophysical sources (such as supernova explosions in nearby galaxies) may ultimately depend upon reaching the quantum limit of allowed measurements.

A specific argument for the quantum nonperturbing properties of their design is given by Braginskii *et al.* only for the special case of an initial $n = 0$ quantum state

* Research supported in part by NSF Grant PHY76-82353.

of the oscillator. In fact it seems possible that their argument may be limited in application to only this single value of n, since the $n = 0$ state has a special property (the coherence property discussed below) which is not shared by the $n \neq 0$ energy eigenstates of the oscillator. On the other hand it seems that their argument does generalize to measurements of the other coherent states of the oscillator provided one reinterprets the aim of the experiment accordingly. If one expected the oscillator to be in a coherent state rather than an energy eigenstate then it would be natural to try to measure the coherence eigenvalue of the state without disturbing its quantum properties. In fact if the oscillator were initially in a coherent state it would be inappropriate to measure its energy eigenvalue, since, according to standard measurement theory, such a measurement would project the system into a state of definite photon number, thus perturbing its quantum properties.

Thus it seems important to determine whether the quantum oscillations to be measured will be in energy eigenstates, coherent states, or something else. If, for example, the oscillator can be regarded as excited from its ground state by specified classical currents then it will certainly be driven into a coherent state rather than a state of specific photon number [3]. Of the usual n quantum states only the special $n = 0$ state is also a coherent state. Intrinsically quantum mechanical processes such as the stimulated emission responsible for laser action can also excite (approximately) coherent photon states [4].

Whether Braginskii's oscillator would actually be driven up to an n quantum state, a coherent state, or something else is a complicated question. We discuss it briefly in Section 3. For the moment we ignore that question and focus on a simpler one: Can one perform quantum nonperturbing measurements of the coherent states of a quantized oscillator. As we have mentioned it seems that the semiclassical Braginskii argument (which treats the electrons as applying a specified classical force) does extend to arbitrary coherent states. Instead of pursuing it, however, we consider a completely quantum mechanical model problem. A beam of photons with wave vector $\mathbf{k}$ (discrete normalization, for convenience) is initially in a coherent state specified by the (complex) number $\alpha(\mathbf{k})$. The photons Compton scatter from electrons with initial momentum $\mathbf{p}_i$, and one measures the distribution of scattered electrons in an effort to determine $\alpha(\mathbf{k})$.

At first sight it would seem that we shall definitely disturb the quantum state of the photon beam by such a process. An electron is scattered to a new momentum $\mathbf{p}_f \neq \mathbf{p}_i$ by absorbing a photon of momentum $\mathbf{k}$ from the incident beam and emitting a photon of momentum $\mathbf{k}' \neq \mathbf{k}$ in the process. However, the important feature of the coherence of the incident beam is that one can absorb a photon (or any number of photons) from it without disturbing its coherence properties and without changing the value of $\alpha(\mathbf{k})$. Only forward scattering in which a new photon is emitted directly into the ongoing incident beam can disturb the coherence of the beam. We derive this result for second-order process in Section 2. A brief discussion of possible higher-order effects is given there also. A Feynman diagram is given to show how the (second-order) Compton scattering from an initially coherent beam can leave undisturbed the quantum state of the incident beam.

The actual process we compute involves a coherent photon beam in empty space interacting with an electron taken (for simplicity) to be initially at rest. We adopt the usual convenient mixture of discrete and continuum normalizations (the latter being more convenient for evaluation of integrals). A more accurate model for Braginskii's experiment would involve only the discrete modes of a cavity oscillator. Our computation is more appropriate for, say, electrons scattered by a laser beam (which at a high laser pumping rate approximates a coherent beam [4]). We show also that our results extend to the more general coherent states in which a spectrum of momentum modes is coherently excited. Again the scattering gives information about the coherent state without disturbing the coherence of the incident beam (except by forward scattering). By considering a pencil beam of some small angular width we can estimate the probability of quantum nonperturbing scattering. We find that qnp measurements do seem to be possible for coherent oscillator states and argue that a thorough understanding of the expected quantum state (coherent or otherwise) may be necessary for the design of a qnp measuring device.

In a recent preprint Unruh [5] analyzed the Braginskii experiment using a fully quantum mechanical model. He argues that measuring the diffraction pattern of the emerging electron beam will unavoidably affect the quantum state of the cavity oscillator. His analysis is made in the spirit of the original Braginskii proposal in which the oscillator is assumed to be initially in an n quantum state. Since we do not consider an actual resonant cavity our results do not bear directly upon the controversy. They do, however, suggest that coherent states offer greater promise for qnp measurements than the more familiar n quantum oscillator states.

A different example allowing for possible qnp measurements is the (somewhat academic) process of pair production by colliding, coherent photon beams. We study this process briefly in Section 2 and show that the second-order amplitudes are completely coherence nonperturbing. Only higher-order processes (such as Delbrück scattering) can contribute to disturbing the coherence of the ongoing beams. Some other processes for which the coherence disturbance is suppressed until a higher order are discussed in Section 3.

2. Compton Scattering from Coherent States

The importance of coherent photon states for quantum optics was first emphasized by Glauber [3], who derived many of their properties. To describe these states let us fix a photon four-momentum k ($k \cdot k = 0$) and a polarization state s. The usual n quantum (Fock) states are written $|n_{(k,s)}\rangle$. The normalized elementary coherent states (which have only a single mode (k, s) excited) are given by

$$|\alpha(k, s)\rangle = \exp\left(-\frac{1}{2}|\alpha(k, s)|^2\right) \sum_{n_{(k,s)}=0}^{\infty} \frac{(\alpha(k, s))^{n_{(k,s)}}}{(n_{(k,s)}!)^{1/2}} |n_{(k,s)}\rangle, \tag{2.1}$$

where $\alpha(k, s)$ is an arbitrary complex number. These states are eigenstates of the annihilation operator $\hat{\mathbf{a}}(k, s)$ with eigenvalues $\alpha(k, s)$:

$$\hat{\mathbf{a}}(k, s) \mid \alpha(k, s)\rangle = \alpha(k, s) \mid \alpha(k, s)\rangle. \tag{2.2}$$

One can define more general coherent states in which a family of modes is excited, each with its coherence eigenvalue $\alpha(k, s)$. Thus a state $\mid \{\alpha(k, s)\}\rangle$ is coherent if it is normalizable and obeys

$$\hat{\mathbf{a}}(k', s') \mid \{\alpha(k, s)\}\rangle = \alpha(k', s') \mid \{\alpha(k, s)\}\rangle \tag{2.3}$$

for all (k', s'). If $\hat{\mathbf{A}}_\mu^{(+)}(x)$ represents the positive frequency (annihilation) part of the photon field operator then Eq. (2.3) is equivalent to

$$\hat{\mathbf{A}}_\mu^{(+)}(x) \mid\{\alpha(k, s)\}\rangle = a_\mu(x) \mid\{\alpha(k, s)\}\rangle, \tag{2.4}$$

where $a_\mu(x)$ is the vector potential field with Fourier components $\{\alpha(k, s)\}$.

A. *Elementary Coherent States*

We can build up the amplitude for Compton scattering from an elementary coherent photon state from the amplitudes for scattering from n quantum states. The non-vanishing second-order scattering amplitudes are

$$\begin{aligned}&\langle 1_{(k',s')}, n'_{(k,s)}; (p_f, \sigma_f)| \mathscr{S}^{(2)} |(p_i, \sigma_i); n_{(k,s)}\rangle \\ &= \langle 1_{(k',s')}, n'_{(k,s)}; (p_f, \sigma_f)| \left\{-\frac{1}{2!}\int_{-\infty}^{\infty} dt \int_{-\infty}^{\infty} dt'\, T[\hat{\mathbf{H}}_I(t)\, \hat{\mathbf{H}}_I(t')]\right\} |(p_i, \sigma_i); n_{(k,s)}\rangle,\end{aligned} \tag{2.5}$$

where (p_i, σ_i) and (p_f, σ_f) are the initial and final electron states (with $p \cdot p = m^2$) and where $\hat{H}_I(t)$ is the interaction Hamiltonian

$$\hat{\mathbf{H}}_I(t) = e \int d^3x : \bar{\psi}(x)\, \gamma^\mu \psi(x)\, \hat{\mathbf{A}}_\mu(x):. \tag{2.6}$$

The first-order amplitudes all vanish (by energy momentum conservation), and the second-order amplitudes vanish except for out states with only a single scattered photon (k', s') satisfying $p_f + k' = p_i + k$. In fact this amplitude vanishes unless $n'_{(k,s)} = n_{(k,s)} - 1$, since a photon is absorbed from the incident beam. The amplitude with n photons in the incident beam can be expressed simply in terms of that for a single photon in the incident beam as

$$\begin{aligned}&\langle 1_{(k',s')}, n'_{(k,s)}; (p_f, \sigma_f)| \mathscr{S}^{(2)} |(p_i, \sigma_i); n_{(k,s)}\rangle \\ &= [n_{(k,s)}]^{1/2}\, \delta_{n'_{(k,s)}+1, n_{(k,s)}} \langle 1_{(k',s')}; (p_f, \sigma_f)| \mathscr{S}^{(2)} |(p_i, \sigma_i); 1_{(k,s)}\rangle.\end{aligned} \tag{2.7}$$

This follows from making the usual Dyson–Wick reduction of (2.5) and observing that

$$\hat{\mathbf{a}}_{(k,s)} \mid (p_i, \sigma_i); n_{(k,s)}\rangle = [n_{(k,s)}]^{1/2} \mid (p_i, \sigma_i); n_{(k,s)} - 1\rangle, \tag{2.8}$$

whereas the other details of the computation remain the same as those in the one-photon case. A more complete derivation (for the generalized coherent states) is outlined below.

The matrix element

$$\mathscr{S}_{fi}^{(2)} \equiv \langle 1_{(k',s')}; (p_f, \sigma_f)| \mathscr{S}^{(2)} |(p_i, \sigma_i); 1_{(k,s)}\rangle \tag{2.9}$$

is given by [6]

$$\begin{aligned}\mathscr{S}_{fi}^{(2)} = {} & \frac{e^2}{V^2}\left(\frac{m^2}{E_f E_i}\right)^{1/2} \frac{1}{(2k)^{1/2}} \frac{1}{(2k')^{1/2}} (2\pi)^4 \delta^{(4)}(p_f + k' - p_i - k) \\ & \times \bar{u}(p_f, \sigma_f)\left[(-i\not{\epsilon}')\frac{i}{\not{p}_i + \not{k} - m}(-i\not{\epsilon}) \right. \\ & \left. + (-i\not{\epsilon})\frac{i}{\not{p}_i - \not{k}' - m}(-i\not{\epsilon}')\right] u(p_i, \sigma_i),\end{aligned} \tag{2.10}$$

where V is a normalization volume E_i and E_f are the initial and final electron energies, ϵ_μ and ϵ'_μ are the photon polarizations, and the $u(p, \sigma)$ are plane wave (positive energy) spinor solution of the Dirac equation. Our notation is essentially that of Bjorken and Drell [6]. (In particular, $\hbar = c = 1$ and $\alpha = e^2/4\pi \simeq 1/137$.)

Combining Eqs. (2.1), (2.7), and (2.9), we obtain after a short computation

$$\begin{aligned}&\langle 1_{(k',s')}, n'_{(k,s)}; (p_f, \sigma_f)| \mathscr{S}^{(2)} |(p_i, \sigma_i); \alpha(k, s)\rangle \\ &\quad = \alpha(k, s)\, \mathscr{S}_{fi}^{(2)} \left[\exp\left(-\frac{1}{2} \mid \alpha(k, s)|^2\right) \frac{(\alpha(k, s))^{n'_{(k,s)}}}{(n'_{(k,s)}!)^{1/2}}\right].\end{aligned} \tag{2.11}$$

Comparing Eq. (2.11) with Eq. (2.1) we see that the amplitudes for fixed (p_f, σ_f) and (k', s') are precisely those for a coherent state of the mode (k, s) with eigenvalue $\alpha(k, s)$. The overall ($n'_{(k,s)}$-independent) factor $\alpha(k, s)\, \mathscr{S}_{fi}^{(2)}$ determines the relative amplitude for scattering of the electron into its final state (p_f, σ_f) with an associated scattered photon in the state (k', s'). Since the δ-function in Eq. (2.10) assures conservation of four-momentum we get a nonzero amplitude for $p_f \neq p_i$ only if $k' \neq k$.

The foregoing argument requires modification for the special case of forward scattering ($(k, s) = (k', s')$). In this case the coherence of the forward beam is disturbed, since the initial coherent state is not an eigenstate of the creation operator $\hat{\mathbf{a}}^\dagger(k, s)$. Thus purely forward scattering, for which $p_f = p_i$, does disturb the coherence of the incident beam. The amplitude for this coherence disturbing process must be distinguished from the (zeroth-order) identity operator contribution to the $\mathscr{S}$ matrix, which obviously also gives $p_f = p_i$ but does not disturb the coherence of the forward beam. Only the forward scattering amplitude (proportional to e^2) contributes to disturbing the coherence by allowing for the emission of a new photon into the beam. We thus have the unusual result that (to second order) the scattering of an electron out of its initial state ($p_f \neq p_i$) leaves the ongoing photon beam in its original coherent state, whereas the forward scattering ($p_f = p_i$) is accompanied by an amplitude for disturbing the coherence of the beam. In the former case a new photon is scattered out of the incident beam ($k' \neq k$) and may be detected along with the scattered electron to gain information about the coherence eigenvalue $\alpha(k, s)$ of the undisturbed forward beam. A Feynman diagram illustrating this process is given in Fig. 1.

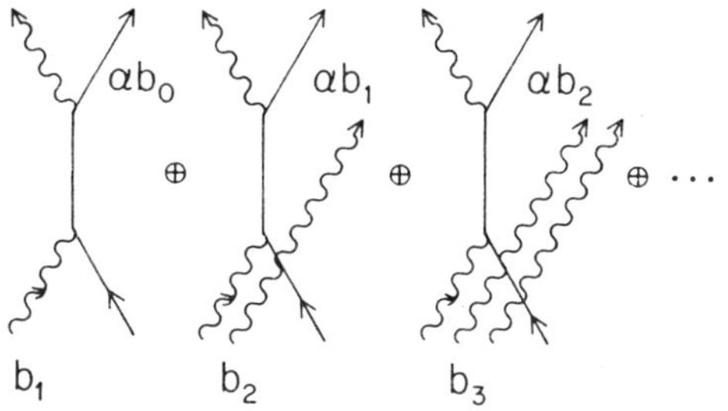

FIG. 1. The incident photon state is a superposition of n-quantum states with the relative amplitudes $b_n = \alpha^n/(n!)^{1/2}$, where α is the coherence eigenvalue. These relative amplitudes are undisturbed in the ongoing forward beam unless the new photon is emitted into the same momentum state. The complete amplitude includes (symmetrized) scattering from each incident photon and emission and absorbtion in reverse temporal order. These contributions are not shown.

To estimate the probability for such a qnp scattering of the electron let us suppose that the incident photon beam's momentum $\mathbf{k}$ is only known to be directed within a cone of (solid) angular subtension $\delta\Omega_0$ about some specified direction. This would be true if, for example (and more realistically), the beam actually contained a distribution of nonvanishing Fourier components. We study this more general case (which allows for a spatially localized beam) below. For the moment we regard the coherence of the quantum stated of the incident beam as being disturbed if a photon is scattered into this forward cone.

The transition rate per unit volume for photon scattering into a solid angle $d\Omega_{k'}$ is obtained by squaring the amplitude (2.11), dividing by $(2\pi)^4\delta^{(4)}(0)$, dividing by the number of target particles per unit volume $1/V$, summing over phase space $[V^2/(2\pi)^6]\, d^3p_f\, d^3k'$, and summing over the unobserved photon numbers $\{n'_{(k,s)}\}$ of the final state:

$$dw = |\alpha(k, s)|^2 \int \frac{|\mathscr{S}_{fi}^{(2)}|^2}{(2\pi)^{10}\,\delta^{(4)}(0)} V^3\, d^3p_f\, d^3k' \exp(-|\alpha(k, s)|^2) \sum_{n'=0}^{\infty} \frac{(|\alpha(k, s)|^2)^{n'}}{n'!}$$

$$= \frac{|\alpha(k, s)|^2\, e^4 m^2}{(2k)\, E_i V (2\pi)^2} \int \frac{d^3p_f\, d^3k'}{E_f(2k')}\, \delta^{(4)}(p_f + k' - p_i - k)\, |\mathscr{M}|^2, \tag{2.12}$$

where

$$\mathscr{M} = \bar{u}(p_f, \sigma_f) \left[(-i\not{\epsilon}') \frac{i}{\not{p}_i + \not{k} - m} (-i\not{\epsilon}) + (-i\not{\epsilon}) \frac{i}{\not{p}_i - \not{k}' - m} (-i\not{\epsilon}')\right] u(p_i, \sigma_i) \tag{2.13}$$

and where the integral is over the momenta of the unobserved scattered particles. This rate dw is simply $|\alpha(k, s)|^2/V$ times the usual differential cross section for Compton scattering [6]. Thus $|\alpha(k, s)|^2/V$ is the effective flux of the incident coherent beam, and the angular distribution of scattered electrons and photons is exactly the same as that for an n quantum (Fock) state incident beam. By passing an electron beam of known intensity through the photon beam and measuring the rate of scattered electrons one can thus estimate $|\alpha(k, s)|^2/V$, which is equal to the expectation value of the photon number density of the incident coherent beam. The probability that we have disturbed the coherence of the ongoing photon beam can also be estimated by computing the probability per scattered electron of emitting a photon into the forward beam.

For simplicity we take the electrons to be initially at rest ($E_i = m$) and unpolarized. If we sum over final spin states and average over initial spin states the result is [6]

$$d\bar{w} = \frac{|\alpha(k, s)|^2 e^4 m}{4kV(2\pi)^2} \int \frac{d^3p_f\, d^3k'}{E_f 2k'} \delta^{(4)}(p_f + k' - p_i - k)\, |\bar{\mathscr{M}}|^2, \tag{2.14}$$

where

$$|\bar{\mathscr{M}}|^2 = \frac{1}{2m^2}\left[\frac{k'}{k} + \frac{k}{k'} + 4(\epsilon' \cdot \epsilon)^2 - 2\right], \tag{2.15}$$

in which $(\epsilon^\mu) = (0, \boldsymbol{\epsilon})$ and $(\epsilon^{\mu\prime}) = (0, \boldsymbol{\epsilon}')$ are transverse polarizations obeying $\boldsymbol{\epsilon} \cdot \mathbf{k} = \boldsymbol{\epsilon}' \cdot \mathbf{k}' = 0$. Energy momentum conservation, arising via the δ-function, implies the Compton relationship

$$k' = \frac{k}{1 + (k/m)(1 - \cos\theta)}, \tag{2.16}$$

where θ is the photon scattering angle

$$\mathbf{k} \cdot \mathbf{k}' = kk' \cos\theta. \tag{2.17}$$

The angular distribution of scattered photons is obtained in the usual way by integrating over the final electron momentum and over the photon energy k' (noting that $d^3k' = d\Omega_{k'} k'^2\, dk'$). If we sum over final photon polarizations and average over initial polarizations to get a rate for unpolarized photons, the result is [6]

$$\left.\frac{d\bar{w}}{d\Omega_\gamma}\right|_{\text{unpol}} = \frac{|\alpha(k, s)|^2}{V}\left\{\frac{\alpha^2}{2m^2}\left(\frac{k'}{k}\right)^2\left[\frac{k'}{k} + \frac{k}{k'} - \sin^2\theta\right]\right\}$$
$$= \frac{|\alpha(k, s)|^2}{V}\frac{d\bar{\sigma}}{d\Omega_\gamma}, \tag{2.18}$$

where $\alpha = e^2/4\pi$ is the fine structure constant and where $d\bar{\sigma}/d\Omega_\gamma$ is the usual (unpolarized) Klein–Nishina cross section.

The probability per scattered electron $p(k, \delta\Omega_0)$ of emitting a (coherence disturbing) photon into the forward beam is thus

$$p(k, \delta\Omega_0) \approx \frac{d\bar{\sigma}/d\Omega_\gamma|_{\theta=0}\, \delta\Omega_0}{\bar{\sigma}} = \frac{(\alpha^2/m^2)\, \delta\Omega_0}{\bar{\sigma}}, \tag{2.19}$$

where $\bar{\sigma}$ is the total unpolarized Compton cross section. At low energies ($k/m \to 0$) this result becomes

$$p(k, \delta\Omega_0) \xrightarrow[k/m \to 0]{} \frac{3}{8\pi} \delta\Omega_0 , \tag{2.20}$$

where, as defined before, $\delta\Omega_0$ is the solid angle subtended by the cone defining the forward beam directions.

To measure $|\alpha(k, s)|^2/V$, it may be more practical to observe the scattered electrons. Their angular distribution is obtained from Eq. (2.14) by integrating over the (unobserved) photon momentum $\mathbf{k}'$ and over $p_f = |\mathbf{p}_f|$ (after setting $d^3p_f = p_f^2\, dp_f\, d\Omega_f$). The result (again for unpolarized photons) is

$$\left.\frac{d\bar{w}}{d\Omega_f}\right|_{\text{unpol}} = 2\,\frac{|\alpha(k, s)|^2}{V}\,\alpha^2 \left[\frac{k'}{k} + \frac{k}{k'} - \sin^2\theta\right] \frac{(m + k)^2 \cos\psi}{[(m + k)^2 - k^2 \cos^2\psi]^2}, \tag{2.21}$$

where ψ is the electron scattering angle defined by

$$\mathbf{p}_f \cdot \mathbf{k} = p_f k \cos\psi \tag{2.22}$$

and related to θ, through energy momentum conservation, by

$$\cos^2\psi = \frac{(1 + m/k)^2(1 - \cos\theta)}{[2(m/k)^2 + (1 + 2m/k)(1 - \cos\theta)]}. \tag{2.23}$$

As θ ranges from 0 to π, ψ ranges from 0 to $\pi/2$, since the electron does not scatter backward in the lab frame. This result may be written

$$\left.\frac{d\bar{w}}{d\Omega_f}\right|_{\text{unpol}} = \frac{|\alpha(k, s)|^2}{V}\,\frac{d\bar{\sigma}}{d\Omega_f}, \tag{2.24}$$

where $d\bar{\sigma}/d\Omega_f$ is the usual unpolarized differential cross section for the scattered electrons. Thus by measuring the rate of scattered electrons, for a fixed density $1/V$ of target electrons, we obtain a measure of the expected density

$$\eta(k, s) \equiv |\alpha(k, s)|^2/V \tag{2.25}$$

of photons of the incident beam. Fixing the normalization volume we can thus measure the expectation value $|\alpha(k, s)|^2$ of the number of photons in the volume V. Since the angular distribution is precisely the same as that for an n quantum (Fock) state incident beam, no information is lost by measuring just the total scattering rate into all angles.

$$\bar{w}\,|_{\text{unpol}} = \frac{|\alpha(k, s)|^2}{V}\,\bar{\sigma}, \tag{2.26}$$

where $\bar{\sigma}$ is, as before, the total (unpolarized) Compton cross section.

We wish to scatter a sufficient number of electrons to gain an accurate estimate of $|\alpha(k, s)|^2$, but not so many as to give a high probability for disturbing the coherence of the forward photon beam. In practice one could pass a low-energy beam of electrons (of known intensity) through the photon beam and measure the scattering rate of the emerging electrons. Suppose that during some fixed time interval, N electrons pass through the interaction region and n of these are observed to scatter. The expected ratio is

$$n/N \approx |\alpha(k, s)|^2 \Gamma, \tag{2.27}$$

where Γ is an ($\alpha(k, s)$-independent) factor which depends upon the details of the experimental arrangement. In particular, $\Gamma \propto \bar{\sigma}$ and Γ could be calculated explicitly for any specified experimental conditions. Assuming that each scattering event is independent of the others one can apply the binomial distribution to estimate the expected error in the value(2.27) for $|\alpha(k, s)|^2$. Recalling that the variance is given by

$$\sigma^2 = Npq, \tag{2.28}$$

where p is the probability that a single electron will scatter and $q = 1 - p$, we have estimate

$$\Delta n \approx \sigma \approx \left[N \cdot \left(\frac{n}{N}\right)\left(1 - \frac{n}{N}\right)\right]^{1/2} \approx n^{1/2}, \tag{2.29}$$

where we have assumed that $n \ll N$. Therefore

$$\Delta n/N \approx \Delta |\alpha(k, s)|^2 \Gamma \approx n^{1/2}/N, \tag{2.30}$$

and we get

$$\Delta |\alpha(k, s)|^2 / |\alpha(k, s)|^2 \approx 1/n^{1/2}. \tag{2.31}$$

Since $|\alpha(k, s)|^2$ is the expected number $\bar{n}$ of photons in the volume V we can write this as

$$\Delta\bar{n}/\bar{n} \approx 1/n^{1/2}. \tag{2.32}$$

The probability that no coherence distributing photons have been emitted into the forward beam is, from Eq. (2.19), $(1 - p(k, \delta\Omega_0))^n$, which, if $p(k, \delta\Omega_0) \ll 1$ and $n \cdot p(k, \delta\Omega_0) \ll 1$, will remain reasonably close to unity.

B. *Generalized Coherent States*

To complete this section we sketch the extension of our basic result (2.11) to the case of the generalized coherent states of Eq. (2.4). These generalized coherent states can be used to describe a localized photon beam and thus to alleviate the need for idealizing the beam as a single coherently excited mode (k, s). Given this generalization we can make more precise our notion of the cone of directions defined by the beam and its corresponding angular width $\delta\Omega_0$. We suppose that the state is chosen so that its associated vector potential $a_\mu(x)$ (cf. Eq. (2.4)) describes a narrow beam passing through an interaction region through which the electrons will pass. Such a beam will contain nonvanishing Fourier amplitudes $\{\alpha(k, s)\}$ for $\mathbf{k}$ directed within some cone of (solid) angular width $\delta\Omega_0$. We wish to demonstrate that photons scattered out of this beam do not disturb the coherence properties of this generalized coherent state.

The nonvanishing (second-order) scattering amplitudes are given by

$$\begin{aligned} &\langle 1_{(k',s')}, \{n'_{(k,s)}\}; (p_f, \sigma_f)| \mathscr{S}^{(2)} |(p_i, \sigma_i); \{\alpha(k, s)\}\rangle \\ &= \langle 1_{(k',s')}, \{n'_{(k,s)}\}; (p_f, \sigma_f)| \\ &\quad \times \left\{-\frac{1}{2!}\int_{-\infty}^{\infty} dt \int_{-\infty}^{\infty} dt'\, T[\hat{\mathbf{H}}_I(t)\, \hat{\mathbf{H}}_I(t')]\right\} |(p_i, \sigma_i); \{\alpha(k, s)\}\rangle, \end{aligned} \tag{2.33}$$

where $\{\alpha(k, s)\}$ and $\{n'_{(k,s)}\}$ designate coherence eigenvalues and occupation numbers for the states $\{(k, s)\}$ which are excited in the incident photon beam. When the scattered photon (k', s') lies outside the incident beam, $(k', s') \notin \{(k, s)\}$, the only nonvanishing contribution to the amplitude arises from the absorbtion of a photon from the incident beam via the operator $\hat{\mathbf{A}}_\mu^{(+)}(x)$ and the emission of the scattered photon via the operator $\hat{\mathbf{a}}^\dagger(k', s')\, \epsilon'_\mu e^{ik'\cdot y'}$. These events are linked by the Feynman propagator

$$S_F(x, y) = -i\langle 0 \mid T(\psi(x), \bar{\psi}(y)) \mid 0\rangle \tag{2.34}$$

through the usual Dyson–Wick reduction procedure. The main point is simply that $|(p_i, \sigma_i); \{\alpha(k, s)\}\rangle$ is an eigenvector of $\hat{\mathbf{A}}_\mu^{(+)}(x)$, whereas $\hat{\mathbf{a}}^\dagger(k', s')$ produces a photon outside the set of occupied $\{(k, s)\}$. Neither of these operators affects the coherence of the forward beam.

The full (second-order) amplitude for $(k', s') \notin \{(k, s)\}$ is given by

$$\begin{aligned} &\langle 1_{(k',s')}, \{n'_{(k,s)}\}; (p_f, \sigma_f)| \mathscr{S}^{(2)} |(p_i, \sigma_i); \{\alpha(k, s)\}\rangle \\ &\quad = \langle \{n'_{(k,s)}\}; 0 \mid 0; \{\alpha(k, s)\}\rangle\, \mathscr{S}_{fi}(\{\alpha(k, s)\}), \end{aligned} \tag{2.35}$$

where

$$\begin{aligned} \mathscr{S}_{fi}(\{\alpha(k, s)\}) &= -e^2 \left[\frac{m^2}{E_f E_i (2\pi)^6}\right]^{1/2} \left[\frac{1}{2k'(2\pi)^3}\right]^{1/2} \\ &\quad \times \iint d^4x_1\, d^4x_2 [\bar{u}(p_f, \sigma_f)\, e^{i(p_f+k')\cdot x_1} \not{\epsilon}' [iS_F(x_1, x_2)]\, \not{a}(x_2)\, e^{-ip_i\cdot x_2} u(p_i, \sigma_i) \\ &\quad + \bar{u}(p_f, \sigma_f)\, e^{ip_f\cdot x_1} \not{a}(x_1) [iS_F(x_1, x_2)]\, \not{\epsilon}' e^{i(k'-p_i)\cdot x_2} u(p_i, \sigma_i)]. \end{aligned} \tag{2.36}$$

Here $\langle \{n'_{(k,s)}\}; 0 \mid 0; \{\alpha(k, s)\}\rangle$ are simply the expansion coefficients of the coherent state $|\{\alpha(k, s)\}\rangle$ in the usual (Fock) basis and $\mathscr{S}_{fi}(\{\alpha(k, s)\})$ is an $\{n'(k, s)\}$-independent amplitude which determines the angular distribution of the scattered particles. Thus, just as in Eq. (2.11), the amplitudes are precisely those for a scattered electron and photon accompanied by an undisturbed coherent forward beam.

As before, the argument breaks down for photons scattering into the forward beam ($(k', s') \in \{(k, s)\}$), since the coherent states are not eigenstates of the creation operators. One can recover Eq. (2.11) from Eqs. (2.35) and (2.36) by making the substitution

$$a_\mu(x) \to \frac{\epsilon_\mu(k, s)}{[2k(2\pi)^3]^{1/2}} e^{-ik\cdot x} \alpha(k, s), \tag{2.37}$$

Fourier transforming the Feynman propagator

$$S_F(x_1, x_2) = \int \frac{d^4q}{(2\pi)^4} \frac{e^{-iq\cdot(x_1 - x_2)}}{\not{q} - m}, \tag{2.38}$$

and reverting to discrete normalization

$$\int d^3k \frac{1}{(2\pi)^{3/2}} \to \frac{1}{V^{1/2}} \sum_k . \tag{2.39}$$

In all of the above we have considered only second-order processes. For a weak incident beam one would expect the second order calculation to be an excellent approximation. In any case it seems that higher-order amplitudes would be rather similar to those we have computed. Since $|\{\alpha(k, s)\}\rangle$ is an eigenvector of $\hat{A}_\mu^{(+)}(x)$, we can absorb as many photons as we please without altering the coherence of the state. Only the emission of one or more photons into the forward beam can disturb this coherence and, as in the second-order computation, the amplitudes for these processes will always be limited by phase space considerations.

C. *Pair Production*

A closely related (though rather academic) process is pair production by colliding, coherent photon beams. At second order, only photon annihilations contribute, so there is no disturbance of the coherence of the two beams. In this case the amplitudes for disturbing the coherence arise only at a higher order (through, e.g., forward Delbrück scattering). In the following we simplify the notation slightly by suppressing the electron spin and photon polarization labels.

Let the incident photon beams be represented by the coherent state

$$|\alpha_{k_1}, \alpha_{k_2}\rangle = \exp\left(-\frac{1}{2}|\alpha_{k_1}|^2 - \frac{1}{2}|\alpha_{k_2}|^2\right) \sum_{n_{k_1}} \sum_{n_{k_2}} \frac{(\alpha_{k_1})^{n_{k_1}}}{(n_{k_1}!)^{1/2}} \frac{(\alpha_{k_2})^{n_{k_2}}}{(n_{k_2}!)^{1/2}} |n_{k_1}, n_{k_2}\rangle, \tag{2.40}$$

where α_{k_1}, α_{k_2} are the coherence eigenvalues for the momenta k_1, k_2. By arguments similar to those given above we obtain the second-order amplitudes

$$\langle n'_{k_1}, n'_{k_2}; p_-, p_+ | \mathscr{S}^{(2)} | \alpha_{k_1}, \alpha_{k_2}\rangle \tag{2.41}$$

$$= \alpha_{k_1}\alpha_{k_2} \mathscr{S}_{\rm fi}^{(\rm pair)} \left[\exp\left(-\frac{1}{2}|\alpha_{k_1}|^2 - \frac{1}{2}|\alpha_{k_2}|^2\right) \frac{(\alpha_{k_1})^{n'_{k_1}}}{(n'_{k_1}!)^{1/2}} \frac{(\alpha_{k_2})^{n'_{k_2}}}{(n'_{k_2}!)^{1/2}}\right],$$

where

$$\mathscr{S}_{\mathrm{fi}}^{(\mathrm{pair})} \equiv \langle p_-, p_+ \mid \mathscr{S}^{(2)} \mid 1_{k_1}, 1_{k_2} \rangle \tag{2.42}$$

is the usual Feynman amplitude for pair production by two colliding photons:

$$\begin{aligned} \mathscr{S}_{\mathrm{fi}}^{(\mathrm{pair})} = & \frac{e^2}{V^2}\left(\frac{m^2}{E_+E_-2k_12k_2}\right)^{1/2}(2\pi)^4\,\delta^{(4)}(p_+ + p_- - k_1 - k_2) \\ & \times \bar{u}(p_-, \sigma_-)\Big[(-i\not{\epsilon}_1)\frac{i}{\not{p}_- - \not{k}_1 - m}(-i\not{\epsilon}_2) \\ & + (-i\not{\epsilon}_2)\frac{i}{\not{p}_- - \not{k}_2 - m}(-i\not{\epsilon}_1)\Big]\, v(p_+, \sigma_+). \end{aligned} \tag{2.43}$$

Here E_+, E_- and $v(p_+, \sigma_+)$, $u(p_-, \sigma_-)$ are the positron and electron energies and Dirac wavefunctions, respectively.

Comparing Eqs. (2.40) and (2.41) we see that the amplitudes for fixed p_-, p_+ are precisely those of two coherent photon beams with the eigenvalues α_{k_1} and α_{k_2}. The multiplicative factor $\alpha_{k_1}\alpha_{k_2}\mathscr{S}_{\mathrm{fi}}^{(\mathrm{pair})}$ determines the angular dependence of the created pairs which is evidently equivalent to that for the usual two photon processes. In this case a measurement of the production rate would give information only about the product $|\alpha_{k_1}|^2 |\alpha_{k_2}|^2$ and not about the individual coherence eigenvalues.

The calculations in this section treat only single electron (or single pair) processes and do not consider the possible cumulative effects which may arise when many electrons interact with the coherent photon beam (a corresponding limitation applies to the studies by Braginskii *et al* [1, 2] and Unruh [5, 7]). A more sophisticated study would be needed to decide whether such cumulative effects contribute disturbing the beam's coherence.

3. Discussion

We have shown that quantum nonperturbing (qnp) measurements are in fact possible for coherent states of the electromagnetic field. Our model assumed a coherent wave in empty space, but it seems likely that the results obtained could be extended to the discrete modes of a resonant cavity. The main idea is rather simple. A coherent state is an eigenvector of the photon annihilation operators so that the absorption of one or more photons does not alter the coherence of the state. Only the emission (through, e.g., forward scattering) of a photon into the coherent mode can disturb its coherence, and the amplitudes for such processes are limited by phase space considerations.

Coherent states are the most natural quantum analog of classical oscillations. In both the configuration and the momentum space of the oscillator these states are described by (minimum uncertainty) Gaussian wave packets which oscillate (in the Schrodinger picture) without change of shape even when an external force is applied. The centers of the packets follow exactly the classical motion.

A resonant cavity coupled to a Weber detector would be driven by currents which signal the arrival of an incident gravitational wave. It would also be excited by the thermal noise of the cavity and fluctuations of the signal due to thermal fluctuations of the Weber detector itself. Except at 0°K, the quantum state of the quiescent detector would surely not be a pure coherent state. However, if the temperature T of the cavity and the angular frequency ω of the particular mode of interest obey $\hbar\omega \gg kT$ then the expectation value $\bar{n}$ of the photon number of the mode will satisfy $\bar{n} \ll 1$. In this case the most probable state of the (thermally excited) mode is the $n = 0$, coherent state. For the frequencies considered by Braginskii *et al.*, however ($\omega \approx 2 \times 10^{10}$ sec^{-1}), this requires a temperature $T \lesssim 10^{-3}$°K (for which $\bar{n} \approx 0.01$).

At less extreme temperatures the thermal noise of the cavity will require a more detailed study. In this connection the quantum theory of laser light may provide an instructive example. The spontaneous emission of photons from a laser active medium disturbs the coherence which the stimulated emission is trying to reinforce. The result is a statistical mixture of coherent and chaotic light [4].

An important difference between our model problem and the Braginskii proposal is that the presence of the microwave cavity destroys translational invariance of the basic equations (unless the cavity's own degrees of freedom are included in the system). The total momentum of electrons and photons need no longer be conserved, since momentum can be absorbed by the cavity itself. This difference would seem to allow first-order transitions of the electron–photon quantum state, whereas, in our model problem, energy momentum conservation excluded such transitions until second order. As in our problem, however, the absorbtion of a photon from a coherent state of the cavity mode will leave the state undisturbed. Only the amplitude for emission of a photon into the given mode can disturb its coherence, and one expects this process to be limited by phase space considerations. However, the emission of a new photon into even a different mode would seem to interfere with the aim of Braginskii's proposal, since subsequent electrons would interact with it as well as with those of the coherent mode. One would like to have a process in which the amplitude for emission of a new photon is suppressed until a higher order than that of the absorbtion necessary for qnp measurements.

One possible design would involve the use of atomic or molecular beams chosen so that a suitable internal transition is resonant with the frequency of the cavity mode of interest. If the constituents of the beam are initially in their ground states (to a sufficiently high probability) then the only first-order processes would be coherence nonperturbing excitations to the resonant excited states. In this case only higher-order processes could disturb the quantum states of the cavity's electromagnetic field. Unruh [5] has proposed another model in which qnp measurements seem to be possible even for the n quantum states of the field.

In a recent note Unruh [7] also considered the scattering of electrons by coherent electromagnetic states. In his model the electrons interact with a single mode of the electromagnetic field (e.g., a cavity oscillator). In this model the scattering and coherence disturbing probabilities are almost always of equal magnitude. Thus qnp measurements are almost impossible (unless the electrons have energy less than the energy of one quantum of the oscillator). There is, however, a qualitative difference between Unruh's model and that considered here. Since Unruh treats only a single electromagnetic mode, all emission or scattering processes necessarily contaminate the main beam (and usually disturb its coherence). In our model scattering is almost invariably out of the main beam so that, whereas the full electromagnetic state is altered (by the scattered photon), the main beam's coherence is left intact.

We have noted that the quantum theory of measurement imposes important limitations upon the design of a qnp measuring device. A properly designed experiment to measure the photon number of a mode would be inappropriate unless the state of the mode were truly one of definite photon number (i.e., a Fock state). Any other state would, by the process of measurement, be projected to an eigenstate of the photon number operator with a probability given by the square of its expansion coefficient in the Fock basis. Thus it seems important to predict the likely quantum states of any proposed detector before one can accurately estimate its quantum nonperturbing measurement properties.

Acknowledgments

I am grateful to Beverly Berger, Douglas Eardley, Charles Sommerfield, Kip Thorne, and William Unruh for helpful conversations.

References

1. V. B. Braginskii and Yu. I. Vorontsov, *Sov. Phys. Usp.* **17** (1975), 644.
2. V. B. Braginskii, Yu. I. Vorontsov, and V. D. Krivchenkov, *Sov. Phys. JETP* **41** (1975), 28.
3. R. J. Glauber, *Phys. Rev.* **131** (1963), 2766.
4. R. Loudon, "The Quantum Theory of Light," Oxford Univ. Press, London/New York, 1973. See esp. Chapter 10 for a discussion of the coherence properties oflaser light.
5. W. Unruh, An analysis of quantum-nondemolition measurement, University of British Columbia preprint, 1977.
6. J. Bjorken and S. Drell, "Relativistic Quantum Mechanics," McGraw–Hill, New York, 1964. See Section (7.7) on Compton scattering.
7. W. Unruh, *Phys. Rev. D* **17** (1978), 1180.

VOLUME 41, NUMBER 4 PHYSICAL REVIEW LETTERS 24 JULY 1978

Coherent States for General Potentials

Michael Martin Nieto and L. M. Simmons, Jr.
Theoretical Division, Los Alamos Scientific Laboratory, University of California, Los Alamos, New Mexico 87545
(Received 15 February 1978)

We define coherent states for general potentials, requiring that they have the physically interesting properties of the harmonic-oscillator coherent states. We exhibit these states for several solvable examples and show that they obey a quantum approximation to the classical motion.

The well-known[1-3] coherent states $|\alpha\rangle$ for the simple harmonic oscillator were originally obtained by Schrödinger[1] as those quantum states which obey the classical motion: $\langle\alpha|x(t)|\alpha\rangle = A\sin(\omega t+\varphi)$. These states have a number of other interesting properties including the following: (1) They minimize the uncertainty relation $(\Delta x)^2(\Delta p)^2 \geq \hbar^2/4$, and have $\Delta x = (\hbar/2m\omega)^{1/2}$. (2) They are eigenstates of the destruction operator: $a^-|\alpha\rangle = \alpha|\alpha\rangle$. (3) They are created from the ground state by a unitary displacement operator: $\{\exp[\alpha a^+ - \alpha^* a^-]\}|0\rangle = |\alpha\rangle$. These properties are all equivalent. In fact, usually one of them is adopted as the definition of the harmonic-oscillator coherent states.

For systems other than the harmonic oscillator, definitions of coherent states have been proposed based[4] on property (2) and[5] on property (3), but have been applied to systems having equal level spacing. In a general system the discrete energy levels are not equally spaced, there may be a continuous part of the spectrum, and the problem may not be solvable in closed form. In such cases these definitions are difficult to apply. Moreover, they represent a departure from the original motivation[1] for studying coherent states; namely, that they obey the classical motion. We seek a definition which retains this property and which is generally applicable.

Consider a one-dimensional, single-particle, quantum-mechanical system described by a local potential with one confining region.[6] Classically, the bound-state motion is periodic and by a suitable mapping one can find a function $X_c(x)$ which varies sinusoidally (as will its associated momentum $P_c = m\dot{X}_c$).

Specifically, if $m\ddot{x} = -dV(x)/dx$, then $X_c(x) \equiv A\sin(\omega_c t+\varphi)$ and $P_c = pX_c'(x)$ obey

$$m\dot{X}_c = P_c, \quad \dot{P}_c = -m\omega_c^2 X_c, \tag{1}$$

where

$$X_c'(x) \equiv \frac{dX_c}{dx} = \omega_c\left[\frac{\frac{1}{2}m(A^2 - X_c^2)}{E - V(x)}\right]^{1/2} \tag{2}$$

and

$$(2m)^{-1}P_c^2 + \tfrac{1}{2}m\,\omega_c^2 X_c^2 = \tfrac{1}{2}m\omega_c^2 A^2. \tag{3}$$

Note that, in general, ω_c and A depend upon the total energy E.

The corresponding quantum-mechanical operators are (to within overall normalizations which can be arranged for convenience)

$$X \equiv X_c, \quad P \equiv \frac{-i\hbar}{2}\left[\frac{d}{dx}X' + X'\frac{d}{dx}\right]. \tag{4}$$

They obey

$$[X, P] = i\hbar(X')^2 \tag{5}$$

and

$$(\Delta X)^2(\Delta P)^2/\langle(X')^2\rangle^2 \geq \tfrac{1}{4}\hbar^2. \tag{6}$$

The states which minimize this uncertainty relation satisfy[7]

$$A^-\psi_\alpha(x) \equiv \tfrac{1}{2}[X/\Delta X + iP/\Delta P]\psi_\alpha(x) = \alpha\psi_\alpha(x), \tag{7}$$

where $\alpha = \frac{1}{2}[\langle X\rangle/\Delta X + i\langle P\rangle/\Delta P]$. [For simplicity, we have assumed that X_c is independent of E. If not, then to obtain X, one must make the replacement $E \to H$, add a possible zero-point energy, and symmetrize. Such a more complicated $X_c(x, E)$ occurs for the Morse potential, which system will be discussed elsewhere.]

We claim that a subset of these minimum-uncertainty states, labeled by a particular value of $\Delta X/\Delta P$, are the appropriate generalization of the harmonic-oscillator coherent states. We shall now demonstrate that they obey the classical motion and suitably generalize other coherent-state properties.

Equations (4) imply that $\dot{X} = -i\hbar^{-1}[X, H] = P/m$, so that the first classical equation of motion (1) is obeyed. The second of the classical equations of motion (1) cannot be obeyed precisely because ψ_α is a superposition of energy eigenstates and in general ω_c depends upon E (the harmonic oscillator is exceptional in this regard). We shall demonstrate in our examples, however, that if $\dot{P}_c = -(K_1E + K_2)X_c$, then quantum mechanically $\dot{P}$

$\equiv -i\hbar^{-1}[P,H] = -\frac{1}{2}\{K_1H + K_2 + Z, X\}$, where Z is a quantum correction and $\{,\}$ denotes the anticommutator. One obtains a quantum approximation to the classical motion, with correct amplitude and frequency, by specifying a particular value for $\Delta X/\Delta P$.

By construction the $\psi_\alpha(x)$ satisfy a generalized version of property (1). They are also [see Eq. (7)] eigenstates of the generalized annihilation operator A^- [property (2)]. For the required value of $\Delta X/\Delta P$, A^- will turn out to be the ground-state destruction operator, A_0^-. In general this relationship is not obvious, for the following reason. For a general potential, in contrast to the harmonic oscillator, the raising and lowering operators $A_n^\pm$ for the energy eigenstates E_n depend explicitly upon the state label n. Moreover, it is not generally true that $(A_n^-)^\dagger = A_n^+$. Nevertheless, it will turn out in our examples that the "natural" position and momentum variables are expressible in the form [$K(n)$ a c-number]

$$X = \tfrac{1}{4}K(n)\{[A_n^- + (A_n^+)^\dagger] + [A_n^+ + (A_n^-)^\dagger]\},$$
$$P = \frac{-i}{4}\{[A_n^- + (A_n^+)^\dagger] - [A_n^+ + (A_n^-)^\dagger]\}, \tag{8}$$

and these operators do not depend explicitly upon n. Equation (7) is therefore seen to be a generalization of property (2), with $a^- = (2m\hbar\omega)^{-1/2}[m\omega x + ip]$ replaced by A^-. The n dependence of $A_n^\pm$ makes the connection of our coherent states to property (3) more difficult to establish.

Note also, that in principle our definition of the coherent states can be implemented even for systems which cannot be solved in closed form. Equation (2) can be solved for X_c by either analytic or numerical approximation methods. Such a solution then suffices for the calculation of X and P via (4), and Eq. (7) can be solved for the coherent states $\psi_\alpha(x)$.

We now summarize the results of applying our method to several solvable examples.

(A) Harmonic oscillator.—For the case of the harmonic oscillator, the natural variables are the usual x and p, and our generalized coherent states reduce, as already stated, precisely to the familar coherent states,[1-3] with $\Delta x/\Delta p = 1/m\omega$.

(B) Symmetric Rosen-Morse potential.[8]—It is convenient to add U_0 to the usual $U_0\cosh^{-2}ax$ and deal with $V(x) = U_0\tanh^2 z$, $z \equiv ax$. Making the convenient choice $\varphi = 0$, the classical bound state solutions are $X_c = \sinh z = [E/(U_0 - E)]^{1/2}\sin\omega_c t$, where $\omega_c = [2a^2(U_0 - E)/m]^{1/2}$ and $U_0 > E$. For free particles, $U_0 < E$, the circular functions become hyperbolic functions. In either case, the equations of motion are $\dot{X}_c = P_c/m$, $\dot{P}_c = -2a^2(U_0 - E)X_c$.

The quantum operators $X = \sinh z$ and $P = -\frac{1}{2}i\hbar a^2 \times[\cosh z(d/dz) + (d/dz)\cosh z]$ obey $[X,P] = i\hbar a^2 \times\cosh^2 z$, and the normalized minimum-uncertainty states are

$$\psi = \left[\frac{a\,\Gamma(B+\frac{1}{2}+iu)\,\Gamma(B+\frac{1}{2}-iu)}{\pi^{1/2}\,\Gamma(B)\,\Gamma(B+\frac{1}{2})}\right]^{1/2}(\cosh z)^{-B}\exp[C\sin^{-1}(\tanh z)], \tag{9}$$

where $B = \frac{1}{2}[\langle\cosh^2 z\rangle/(\Delta\sinh z)^2 + 1]$ and $C = u + iv = B\langle\sinh z\rangle + \langle\cosh z(d/dz)\rangle$. One can verify that (9) satisfies (7) and that $(\Delta X)^2(\Delta P)^2 = \frac{1}{4}\hbar^2a^4\langle\cosh^2 z\rangle^2$. The second equation of motion is $\dot{P} = -a\{U_0 - H - \frac{1}{4}E_0, X\}$, where $E_0 = a^2\hbar^2/2m$. These allow one to calculate $X(t)$ exactly. A quantum approximation to the classical motion follows if $B = s$, where $U_0 = E_0 s(s+1)$.

The normalized energy eigenfunctions[9] are $[a(s-n)\,\Gamma(2s-n+1)/\Gamma(n+1)]^{1/2}P_s^{(n-s)}(\tanh z)$. It follows that $A_n^\pm = (s-n)\sinh z \mp \cosh z(d/dz)$. Therefore, Eq. (8) yields X and P in agreement with the forms obtained from X_c and P_c.

It is interesting to note that in the limit $a \to 0$, $a^2 s \to m\omega/\hbar$, the potential, eigenfunctions, eigenvalues, X, P, and the coherent states as defined here all approach their counterparts for the harmonic oscillator.

(C) Symmetric Pöschl-Teller potential.[10]—Here it is convenient to subtract $U_0 \equiv \lambda(\lambda-1)E_0$ from the usual form $U_0\cos^{-2}z$ and deal with $V(x) = U_0\tan^2 z$, where $z = ax$. The problem is mathematically similar to that for the Rosen-Morse potential (except that there is no continuum). The "natural" quantum operators are $X = \sin z$ and $P = -\frac{1}{2}i\hbar a^2[\cos z(d/dz) + (d/dz)\cos z]$, which obey $[X,P] = i\hbar a^2\cos^2 z$. The normalized minimum-uncertainty states are found to be

$$\psi = \left[\frac{a}{\pi^{1/2}}\,\frac{\Gamma(B+\frac{1}{2})\,\Gamma(B+1)}{\Gamma(B+\frac{1}{2}+u)\,\Gamma(B+\frac{1}{2}-u)}\right]^{1/2}(\cos z)^B\left[\frac{1+\sin z}{1-\sin z}\right]^{C/2}, \tag{10}$$

where $B=\frac{1}{2}[-1+\langle\cos^2 z\rangle/(\Delta\sin z)^2]$, $C=u+iv=\langle\cos z(d/dz)\rangle+B\langle\sin z\rangle$. Again, except for quantum corrections, the second classical equation of motion is obtained: $\dot{P}=-a^2\{U_0+H-\frac{1}{4}E_0,X\}$, and one can claculate $X(t)$ exactly. The classical motion follows when $B=\lambda$.

The normalized energy eigenfunctions[9] can be expressed in terms of Legendre functions and the raising and lowering operators are analogous to those for case B. All of these results approach those for the harmonic oscillator in the limit $a\to 0$, $\lambda a^2\to m\omega/\hbar$.

(D) Infinite square well.—In the limit $\lambda\to 1$ the symmetric Pöschl-Teller potential approaches the infinite square well, with walls at $\pm d=\pm\pi/2a$. Notice, however, that a minimum-uncertainty state for the potential $V(x)=U_0 u(ax)$ is also[11] a minimum-uncertainty state for the potential $U_0'\times u(ax)$. It follows that any minimum-uncertainty state for a symmetric Pöschl-Teller potential is also a minimum-uncertainty state for the corresponding infinite square well (with $d=\pi/2a$). In particular, even though any state in a flat well has difficulty producing the classical motion, Eq. (10), with $B=1$, is a coherent state for the infinite square well.

Observe that the first classical equation of motion for the Pöschl-Teller potential becomes $\sin(\pi x/2d)=\sin\omega t$. This is indeed a correct expression for the behavior of a classical particle confined by rigid walls.

Finally, we briefly consider problems in three dimensions. As in the one-dimensional case, the coherent states should reproduce, as closely as possible, the classical motion. We therefore use the classical problem as a guide. A problem which is separable, such as the three-dimensional harmonic oscillator, can be treated trivally by analogy to the one-dimensional case. More generally, one must deal with non-Cartesian coordinate systems.

In the following example, we limit ourselves to a spherically symmetric potential and treat the radial portion of the problem.[12] One must realize that the natural angular variable is no longer the time but a generalized variable $\varphi(t)$ which varies between successive apsidal distances: $\dot{\varphi}(t)=L/mr^2(t)$.

(E) Coulomb potential.—The classical Kepler solution is $X_c=(1/r-me^2/L^2)=A\sin\varphi(t)$, where $A=[(m^2e^4/L^4)+(2mE/L^2)]^{1/2}$. $P_c=(p_r)_c=-AL\times\cos\varphi(t)$ and one has $[me^4/(2L^2)+E]=\frac{1}{2}P_c{}^2/m+\frac{1}{2}L^2X_c{}^2/m$.

The quantum operators are $X=X_c$, with $L^2=\hbar^2 l(l+1)$, and $P=p_r\equiv-i\hbar[(d/dr)+1/r]$. The resulting minimum-uncertainty states are

$$\psi=(2u)^{B+1/2}[\Gamma(2B+1)]^{-1/2}r^{B-1}e^{-Cr}, \tag{11}$$

with $B=\frac{1}{2}\langle 1/r^2\rangle/[\Delta(1/r)]^2$, and $C=u+iv=B\langle 1/r\rangle-i\langle P\rangle/\hbar$. One finds also that X and P are related as in Eq. (8) to the operators $A_l^{\pm}$ which raise and lower l for the radial eigenfunctions $R_{nl}(r)$. Here the "ground-state" annihilation operator is $A_{n-1}{}^{+}$, which indicates that $B=n$ for the coherent states.[13]

Further details and other results, including numerical studies of the time evolution of our coherent states, will appear elsewhere.

This work was initiated due to a suggestion by P. Carruthers that generalized coherent states may be useful for describing the interactions of molecules and coherent radiation. We thank R. S. Berry, D. Campbell, T. Cotter, L. Durand, M. Feigenbaum, D. Finley, T. Goldman, J. Hamilton, R. Haymaker, J. Louck, E. Schrauner, W. Wilson, and especially K. Macrae for a number of useful conversations. One of us (L.M.S.) is grateful for the hospitality of the Aspen Center for Physics where part of this work was done. This work was supported by the U. S. Department of Energy.

[1]E. Schrödinger, Naturwissenschaften 14, 664 (1926).

[2]R. J. Glauber, Phys. Rev. Lett. 10, 84 (1963), and Phys. Rev. 130, 2529 (1963), and 131, 2766 (1963); E. C. G. Sudarshan, Phys. Rev. Lett. 10, 227 (1963).

[3]P. Carruthers and M. M. Nieto, Am. J. Phys. 33, 537 (1965), and Rev. Mod. Phys. 40, 411 (1968).

[4]A. O. Barut and L. Girardello, Commun. Math. Phys. 21, 41 (1971).

[5]M. Perelemov, Commun. Math. Phys. 26, 222 (1972).

[6]For potentials with more than one minimum, the classical motion will have more than one period in certain energy regions. Therefore, the coherent states must be able to describe simultaneously the various allowed classical motions.

[7]K. Gottfried, *Quantum Mechanics* (Benjamin, New York, 1966), Vol. I, pp. 213–215; R. Jackiw, J. Math. Phys. (N.Y.) 9, 339 (1968).

[8]N. Rosen and P. M. Morse, Phys. Rev. 42, 210 (1932).

[9]M. M. Nieto, Phys. Rev. A 17, 1273 (1978); also see M. Bauhain, Lett. Nuovo Cimento 14, 475 (1975).

[10]G. Pöschl and E. Teller, Z. Phys. 83, 143 (1933).

[11]This is the generalization of the (trivial) statement that a minimum-uncertainty state (namely a Gaussian) for a harmonic oscillator with given m, ω is also a coherent state for a harmonic oscillator with any other m', ω'.

[12]For the azimuthal portion of the problem, the opera-

tors $L_{\pm}$ Eqs. (7) and (8) would lead to the "intelligent spin states" of C. Aragone, G. Guerri, S. Salamó, and J. L. Tani, J. Phys. A 7, L149 (1974); C. Aragone, E. Chalbaud, and S. Salamó, J. Math. Phys. (N.Y.) 17, 1963 (1976).

[13]Observe that in the special case Im$C=0$, our coherent states include the circular-motion "classical wave packets" which L. S. Brown, Am. J. Phys. 41, 525 (1973), obtained on physical grounds, for the large-n case. Also see J. Mostowski, Lett. Math. Phys. 2, 1 (1977), who, using the Perelomov formulation, has obtained wave packets which, for the case of circular motion, are "similar to the wave packets discussed by Brown."

PHYSICAL REVIEW D VOLUME 19, NUMBER 6 15 MARCH 1979

Quantum limits on resonant-mass gravitational-radiation detectors

James N. Hollenhorst
Department of Physics, Stanford University, Stanford, California 94305
(Received 21 November 1978)

The methods of quantum detection theory are applied to a resonant-mass gravitational-radiation antenna. Quantum sensitivity limits are found which depend strongly on the quantum state in which the antenna is prepared. Optimum decision strategies and their corresponding sensitivities are derived for some important initial states. The linear detection limit ($E_{\min} \sim \hbar\omega$) is shown to apply when the antenna is prepared in a coherent state. Preparation of the antenna in an excited energy eigenstate or in a state highly localized in position or momentum space leads to increased sensitivity. A set of minimum-uncertainty states for phase-sensitive detection is introduced.

I. INTRODUCTION

The problem of detection of gravitational radiation is a difficult one. Even optimistic calculations of possible sources suggest extreme sensitivity requirements for resonant detectors.[1] The state of the experimental science is now beginning to approach fundamental limits imposed by quantum-mechanical effects. These considerations have led to considerable interest in methods for optimizing the measurement strategy in order to obtain the highest sensitivity within the quantum limits. In view of this it is useful to perform explicit calculations of the quantum limits under various conditions.

All past resonant antenna systems and apparently all those presently under development use linear amplification to detect the state of the antenna. It is well known that linear amplifiers have a sensitivity limit imposed by quantum mechanics.[2-5] We will refer to this limit as the linear detection limit. Braginskiĭ and co-workers[6-9] have shown by uncertainty-principle arguments that the linear detection limit may be surpassed by performing measurements of the energy eigenstate of an antenna and have suggested devices which might perform the desired measurement. These and other energy measuring devices have been analyzed further by Unruh.[10] Moncrief[11] and Unruh[12] have suggested the possible usefulness of coherent states in evading the linear limit. Thorne and co-workers[13] have described conceptually a phase-sensitive device for improving on the linear limit. Braginskiĭ and co-workers[14] have also proposed a phase-sensitive or "stroboscopic" technique.

In this paper we do not attempt to describe any specific device for measuring the state of an antenna. Rather, we attempt to consider all possible measurements in order to find the fundamental limits which arise once an initial state for the antenna has been chosen, using the techniques of quantum detection theory. This theory has been developed along the lines of classical detection theory and a large body of literature exists on the subject.[15-17] The results we obtain are consistent with those obtained, in specific cases, by the previously cited authors, but are of general applicability.

A resonant-mass gravitational-radiation antenna is a damped harmonic oscillator which couples to the Riemann tensor via the nonvanishing mass quadrupole moment of a vibrational eigenmode. As a simple model of such an antenna one may consider two point particles each of mass $m/2$ connected by a spring of length l along the x axis. The classical equation of motion is[18,19]

$$m\frac{d^2x}{dt^2}+\frac{m\omega}{Q}\frac{dx}{dt}+m\omega^2x=-c^2lmR_{x0x0}(t)\equiv-F(t), \qquad (1)$$

where x is the change in separation of the masses, R_{x0x0} is a component of the Riemann tensor, c is the speed of light, and Q is the quality factor which characterizes the damping of the oscillator. The motion induced by a burst of gravitational radiation which occurs in the interval $-\tau \leqslant t \leqslant 0$ is

$$x(t)=\mathrm{Re}\left[(U_i+U_s)\exp\left(-\frac{\omega t}{2Q}-i\omega t\right)\right] \text{ for } t\geqslant 0, \qquad (2)$$

where U_s is a complex amplitude given by

$$U_s\equiv-\frac{i}{m\omega}\int_{-\tau}^{0}F(t')e^{i\omega t'}e^{\omega t'/2Q}dt' \qquad (3)$$

and U_i is the complex amplitude of the antenna before the pulse arrives. We have assumed that the Q is sufficiently high that the frequency shift due to finite Q may be neglected. We see that the pulse merely displaces the complex amplitude of the antenna by the amount U_s. A convenient measure of the signal energy available from the antenna is the quantity E_s given by

$$E_s = \tfrac{1}{2} m\omega^2 |U_s|^2$$
$$= \frac{1}{2m} \left| \int_{-\tau}^{0} F(t') e^{i\omega t'} e^{\omega t' / 2Q} dt' \right|^2 . \qquad (4)$$

E_s will be called the signal energy and is the energy that would be deposited in an antenna originally at rest. It is important to realize that the energy change imparted to an excited antenna ($U_i \neq 0$) may be much greater than E_s and may be either positive or negative. The signal causes a displacement $U_s = U_f - U_i$ while the energy change is

$$\tfrac{1}{2} m\omega^2 (|U_f|^2 - |U_i|^2) = E_s + m\omega^2 \mathrm{Re}(U_s U_i^*).$$

The interference term may easily be larger than E_s if U_i is sufficiently large and has the proper phase.

The quantum limit for linear amplifiers has been well studied. Specific models for linear amplifiers have been given a full quantum treatment.[2] Model-independent calculations which assume only linearity have been studied with both the uncertainty principle[3] and with a full quantum treatment.[4] The application of these results to resonant gravitational-radiation detectors has also been studied.[5] The result is that in order to detect a pulse with an antenna whose state is measured with a linear amplifier the signal energy must satisfy $E_s \gtrsim \hbar\omega$. This is what we have called the linear detection limit.

II. QUANTUM DETECTION THEORY

We will consider a gravity-wave antenna to consist of a single-mode harmonic oscillator with no coupling to any other modes. We are thus assuming an infinite mechanical Q and no thermal noise. We treat the gravitational-radiation field as a classical force which couples to the oscillator. This assumption is very well justified since the coupling between the radiation field and the antenna is so weak. We are interested in detecting short pulses of gravitational radiation with a large spectral density at the oscillator frequency.

The technique is as follows. The antenna is prepared in some initial state. Next the antenna is allowed to interact with the radiation field for some length of time. A measurement is then performed on the antenna. Finally a detection algorithm is used to make one of two conclusions. One conclusion is that no gravity wave pulse has arrived; we will refer to this as the null conclusion. Alternatively the conclusion will be that a pulse has arrived; this will be called the alternative or positive conclusion. In general, conclusions as to the size of the pulse, its time of arrival, and so forth may be desired. We will restrict our attention to the more primitive question of whether or not a pulse has arrived at all. The combination of measurement and detection algorithm will be referred to as a decision strategy. Our task is to determine what the optimum decision strategy is and what sensitivity limit it leads to. We will find that this will depend on the initial state of the antenna.

In order to make analytical progress a criterion must be found for assessing the value of a given decision strategy. To facilitate this we define two probabilities. The detection probability Q_D is the probability that a given decision strategy will result in the positive conclusion under the hypothesis that a pulse has in fact arrived. Q_D is thus the "efficiency" of the detector. The false-alarm probability Q_0 is the probability that the decision strategy will result in the positive conclusion under the hypothesis that no pulse has arrived. Q_0 is thus the probability of "accidentals." Clearly it is desirable to make Q_D high and Q_0 low.

The optimum strategy will be found in the following way. A tolerable false-alarm probability Q_0 is prescribed. The decision strategy which maximizes Q_D is then found. A decision strategy which maximizes Q_D for a prescribed Q_0 is said to satisfy the Neyman-Pearson criterion.

Binary decision theory has been studied for quantum-mechanical systems by several groups[15,16,17]; we follow closely the book by Helstrom.[15] In order to optimize the decision strategy one must define a set of possible measurements. The set of measurements considered in the theory is the set of "probability operator measures."[15,20] This set includes not only all conventional "projection valued measures" such as energy and position measurements, but also more general types of measurements. For example, we can imagine a measurement which is made by allowing a second quantum system to interact with a primary system. A "projection valued" measurement on the second system will not in general be describable as a "projection valued" measurement on the first. Such a measurement will be a member of the class of "probability operator measures" on the primary system, however, and thus is considered in our optimization.

To set up the problem we must first compute the density operator which describes the state of the system at the time of measurement under the hypothesis that no gravity wave pulse has been received. This density operator will be labeled ρ_0. ρ_1 will refer to the density operator under the alternative hypothesis. We wish to find the set of Neyman-Pearson decision strategies for distinguishing between ρ_0 and ρ_1. This complex problem has been reduced to the following eigenvalue problem[15]:

$$(\rho_1 - \lambda\rho_0)|\eta_i\rangle = \eta_i|\eta_i\rangle . \quad (5)$$

We must solve for the eigenvalues η_i and eigenstates $|\eta_i\rangle$ of the operator $\rho_1 - \lambda\rho_0$ where λ is an arbitrary, real Lagrange multiplier. Each value of λ will correspond to a different decision strategy which satisfies the Neyman-Pearson criterion. Thus, λ will parametrize a curve of optimal strategies in the Q_0, Q_D domain. Once the eigenvalue problem is solved, Q_0 and Q_D may be computed according to

$$Q_0 = \sum_{\eta_i>0} \langle\eta_i|\rho_0|\eta_i\rangle ,$$
$$Q_D = \sum_{\eta_i>0} \langle\eta_i|\rho_1|\eta_i\rangle . \quad (6)$$

(Here we have assumed for simplicity that $\rho_1 - \lambda\rho_0$ has no zero eigenvalues.) The decision strategy to be followed for each λ is to measure the operator $\rho_1 - \lambda\rho_0$; the measurement will give one of the eigenvalues η_i, if η_i is negative the null conclusion is made, if η_i is positive the alternative conclusion is made.

The eigenvalue problem is easily solved if ρ_0 and ρ_1 represent pure states. Suppose $\rho_0 = |\psi_0\rangle\langle\psi_0|$ and $\rho_1 = |\psi_1\rangle\langle\psi_1|$. We try to find states $|\eta_i\rangle$ which are linear combinations of $|\psi_0\rangle$ and $|\psi_1\rangle$. After diagonalizing the two by two matrix representing $|\psi_1\rangle\langle\psi_1| - \lambda|\psi_0\rangle\langle\psi_0|$ we solve for $Q_0(\lambda)$ and $Q_D(\lambda)$. Eliminating λ we obtain[21]

$$Q_D = \begin{cases} [\sqrt{Q_0}\,|\gamma| + (1-Q_0)^{1/2}(1-|\gamma|^2)^{1/2}]^2, & 0 \le Q_0 \le |\gamma|^2 \\ 1, \quad |\gamma|^2 \le Q_0 \le 1 , & \end{cases} \quad (7)$$

where $\gamma \equiv \langle\psi_0|\psi_1\rangle$ is just the overlap between the two possible states. Figure 1 is a plot of equation (7) for several values of the overlap. Note that if the overlap is small it is easy to distinguish states and we can find decision strategies with small Q_0 and large Q_D. For large overlaps $|\gamma| \lesssim 1$ it is hard to distinguish the two states and the locus approaches a straight line which characterizes a decision strategy based on random guessing. In the case of gravity wave reception we expect a very low event rate. It will thus be necessary to maintain a very low false-alarm probability. Consequently the case where $Q_0 = 0$ will be of special interest to us.

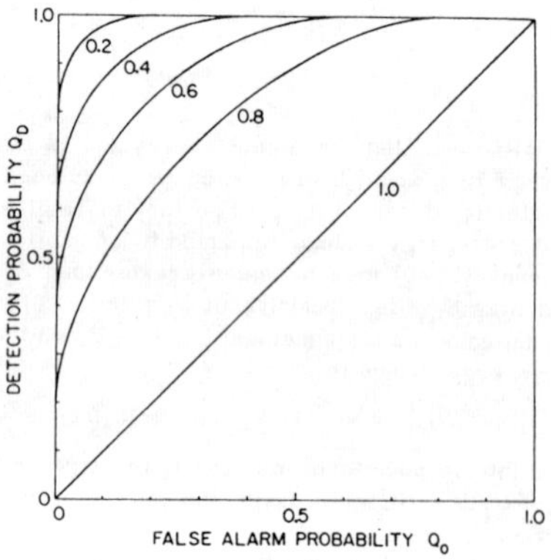

FIG. 1. Loci of Neyman-Pearson strategies in Q_0, Q_D domain for binary decisions between pure states $|\psi_0\rangle$ and $|\psi_1\rangle$. The curves are labeled by $|\gamma|^2 = |\langle\psi_0|\psi_1\rangle|^2$.

III. COMPUTATION OF DENSITY OPERATORS

The Hamiltonian describing the system is

$$H = \tfrac{1}{2}m\omega^2\hat{x}^2 + \frac{\hat{p}^2}{2m} + F(t)\hat{x} - \tfrac{1}{2}\hbar\omega$$

or (8)

$$H = \hbar\omega a^\dagger a + F(t)(\hbar/2m\omega)^{1/2}(a^\dagger + a) ,$$

where $a^\dagger = (m\omega/2\hbar)^{1/2}\hat{x} - i(2m\hbar\omega)^{-1/2}\hat{p}$. The term $(\hbar/2m\omega)^{1/2}F(t)(a^\dagger + a)$ is a classical driving term representing the interaction with the gravitational-radiation field. For convenience we have subtracted away the zero-point energy.

Suppose the oscillator is prepared at $t = -\tau$ in a state described by the density operator ρ_i. Further, suppose the oscillator is allowed to interact with the radiation field for a time τ. If $F(t)$ is zero in this interval then at $t = 0$ the state of the oscillator will be

$$\rho_0 = e^{-i\omega\tau a^\dagger a}\rho_i e^{i\omega\tau a^\dagger a} . \quad (9)$$

Alternatively, if $F(t) \neq 0$ in the interaction interval then the state will evolve differently. Fortunately, the quantum harmonic oscillator with classical driving force is exactly soluble. The density operator at $t = 0$ is given by[22]

$$\rho_1 = e^{-i\omega\tau a^\dagger a}D(\mu e^{i\omega\tau})\rho_i D^\dagger(\mu e^{i\omega\tau})e^{i\omega\tau a^\dagger a} , \quad (10)$$

where $D(\mu)$ is referred to as the displacement operator and is given by

$$D(\mu) \equiv \exp(\mu a^\dagger - \mu^* a) \quad (11)$$

and μ is a normalized complex amplitude given by

$$\mu = -\frac{i}{(2m\hbar\omega)^{1/2}} \int_{-T}^{0} F(t')e^{i\omega t'}dt' . \quad (12)$$

By comparison with Eq. (3) we see that $\mu = (m\omega/2\hbar)^{1/2}U_s$ where U_s is the classical displacement amplitude for $Q\to\infty$. The signal energy E_s is given by

$$E_s = \tfrac{1}{2}m\omega^2|U_s|^2 = \hbar\omega|\mu|^2 . \quad (13)$$

In the Appendix it is shown that

$$e^{-i\omega\tau a^\dagger a}D(\mu)e^{i\omega\tau a^\dagger a} = D(\mu e^{-i\omega\tau}) , \quad (14)$$

so we may rewrite Eq. (10) as

$$\rho_1 = D(\mu)e^{-i\omega\tau a^\dagger a}\rho_i e^{i\omega\tau a^\dagger a}D^\dagger(\mu) \quad (15)$$

and recalling Eq. (9) we have

$$\rho_0 = e^{-i\omega\tau a^\dagger a}\rho_i e^{i\omega\tau a^\dagger a} ,$$
$$\rho_1 = D(\mu)\rho_0 D^\dagger(\mu) . \quad (16)$$

IV. COHERENT STATES

An important set of initial states is the set of coherent states.[23] These are minimum-uncertainty states for the position and momentum operators. They are the closest analogs of classical oscillator states. In the position representation the coherent states are Gaussian wave packets with the same width as the ground-state wave function moving about in the oscillator along classical trajectories. We may generate the set of coherent states by displacing the ground state by the complex amplitude β. We define the coherent state $|\beta\rangle$ by

$$|\beta\rangle \equiv D(\beta)|0\rangle . \quad (17)$$

An oscillator could be prepared in a coherent state by starting in the ground state and driving it with a classical source. In the energy representation the coherent state $|\beta\rangle$ is given by

$$|\beta\rangle = \exp(-\tfrac{1}{2}|\beta|^2)\sum_{n=0}^{\infty}\frac{\beta^n}{\sqrt{n!}}|n\rangle .$$

Suppose the oscillator has been prepared in an initial state $|\psi_i\rangle = |\beta'\rangle$, a coherent state. Under the null hypothesis we will have

$$|\psi_0\rangle = e^{-i\omega\tau a^\dagger a}|\beta'\rangle = e^{-i\omega\tau a^\dagger a}D(\beta')e^{i\omega\tau a^\dagger a}|0\rangle$$
$$= D(\beta' e^{-i\omega\tau})|0\rangle$$
$$= |\beta' e^{-i\omega\tau}\rangle .$$

Thus the state evolves into a new coherent state $|\beta\rangle$ where $\beta \equiv \beta' e^{-i\omega\tau}$. Under the null hypothesis we will have $\rho_0 = |\beta\rangle\langle\beta|$. Under the alternative hypothesis we will have $\rho_1 = D(\mu)|\beta\rangle\langle\beta|D^\dagger(\mu)$. In this example both ρ_0 and ρ_1 are pure states. To solve the problem we need only compute the overlap between the two states. We have

$$\gamma \equiv \langle\psi_0|\psi_1\rangle = \langle\beta|D(\mu)|\beta\rangle$$
$$= \langle 0|D^\dagger(\beta)D(\mu)D(\beta)|0\rangle .$$

But two important properties of the displacement operator are

$$D(\mu)D(\beta) = D(\mu+\beta)\exp[\tfrac{1}{2}(\beta^*\mu - \beta\mu^*)] \quad (18)$$

and

$$D^\dagger(\beta) = D(-\beta) . \quad (19)$$

Using these we find

$$D^\dagger(\beta)D(\mu)D(\beta) = D(\mu)\exp(\beta^*\mu - \beta\mu^*) , \quad (20)$$

so $\gamma = \langle 0|D(\mu)|0\rangle \exp[2i\,\mathrm{Im}(\beta^*\mu)]$. In the Appendix it is shown that $\langle 0|D(\mu)|0\rangle = \exp(-\tfrac{1}{2}|\mu|^2)$ so we conclude $\gamma = \exp[-\tfrac{1}{2}|\mu|^2 + 2i\,\mathrm{Im}(\beta^*\mu)]$ and $|\gamma|^2 = \exp(-|\mu|^2) = \exp(-E_s/\hbar\omega)$. Now Eq. (7) may be used to compute Q_D in terms of Q_0. For the case $Q_0 = 0$ we have

$$Q_D = 1 - \exp(-E_s/\hbar\omega) \quad \text{(coherent states, } Q_0 = 0). \quad (21)$$

We note that the expression for Q_D is independent of β. This immediately tells us that if the oscillator is prepared in a coherent state, there is no improvement in sensitivity obtained by preparing it in a highly excited state. In Fig. 2 we have plotted Eq. (21). We will define a minimum detectable pulse energy E_{min} as follows: For $Q_0 = 0$ we will maximize Q_D and find the minimum signal energy such that the detection probability is at least 50% for all $E_s \geq E_{min}$. For coherent initial states the result is

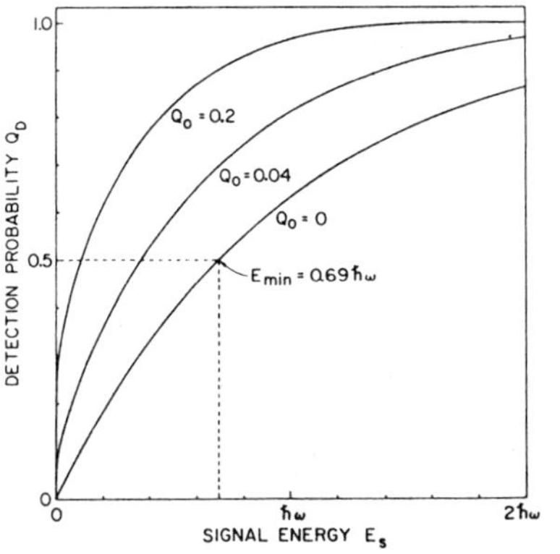

FIG. 2. Detection probability Q_D vs signal energy E_s for an oscillator prepared in a coherent state. Curves are shown for several values of the false-alarm probability Q_0.

$$E_{min} = \hbar\omega \ln 2 \text{ (coherent states)}. \qquad (22)$$

This is very close to the quantum limit for linear amplification. We conclude that the linear detection limit cannot be improved on by using coherent initial states.

V. ENERGY EIGENSTATES

Several schemes have been proposed for gravity-wave detection using energy measurements.[6-10] It is therefore natural to consider the case in which the oscillator is prepared in an energy eigenstate. Assume $\rho_i = |n\rangle\langle n|$, where $|n\rangle$ is the energy eigenstate with eigenvalue $n\hbar\omega$. We have

$$\rho_0 = e^{-i\omega\tau a^\dagger a}|n\rangle\langle n|e^{i\omega\tau a^\dagger a} = |n\rangle\langle n|$$

and

$$\rho_1 = D(\mu)|n\rangle\langle n|D^\dagger(\mu).$$

Again the problem involves only pure states and we need $\gamma = \langle n|D(\mu)|n\rangle$. It is shown in the Appendix that

$$\langle n|D(\mu)|n\rangle = L_n(|\mu|^2)e^{-|\mu|^2/2},$$

where $L_n(x)$ is the Laguerre polynomial of order n. For the case $Q_0 = 0$ we have

$$Q_D = 1 - e^{-E_s/\hbar\omega}[L_n(E_s/\hbar\omega)]^2 \quad (n \text{ states, } Q_0 = 0). \qquad (23)$$

Recalling that $L_0 = 1$ we see that this result reduces to Eq. (21) in the ground state. This is as it must be since the ground state is a coherent state. In order to do better than the linear detection limit one must prepare the oscillator in an excited state. Figure 3 is a plot of Q_D vs E_s for the cases $n = 10$ and $n = 0$. For large n and small argument we may approximate the Laguerre polynomial by

$$L_n(x) \approx e^x J_0(2\sqrt{nx}), \quad 0 \leq x \ll n.$$

To find E_{min} we set $1 - e^{-E_{min}/\hbar\omega}[L_n(E_{min}/\hbar\omega)]^2 = \frac{1}{2}$. Using the above approximation we get

$$E_{min} \approx \frac{0.32}{n}\hbar\omega \quad (n \text{ states}). \qquad (24)$$

We see that it is possible to reduce the quantum limit as far as we like by preparing the oscillator in a highly excited energy eigenstate. The increased sensitivity is due to the classical interference term mentioned in the introduction. In quantum language we say that the incident radiation field has stimulated emission or absorption of quanta by the antenna. This result is in agreement with the result obtained by Braginskiĭ using uncertainty-principle arguments.[6]

Now we wish to examine what the decision strategy is for this case, that is, how does one actually

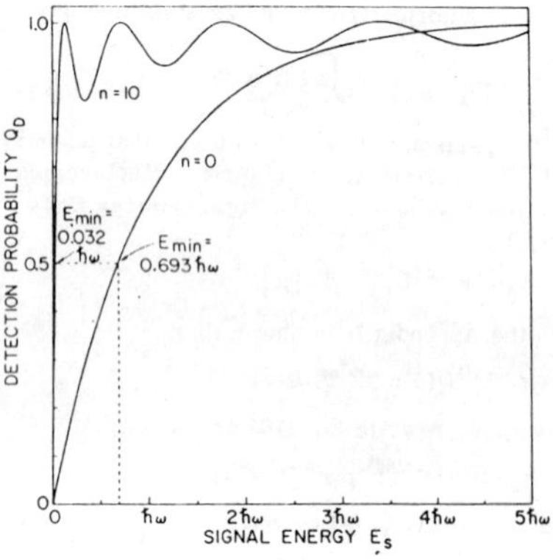

FIG. 3. Detection probability vs signal energy E_s for an oscillator prepared in the energy states with $n = 10$ and $n = 0$.

achieve the optimum sensitivity. As discussed in Sec. II one measures the operator

$$\rho_1 - \lambda\rho_0 = D(\mu)|n\rangle\langle n|D^\dagger(\mu) - \lambda|n\rangle\langle n|.$$

Let us examine this for the case $Q_0 = 0$. The appropriate value of λ for $Q_0 = 0$ is $\lambda \to \infty$. As $\lambda \to \infty$, ρ_1 may be ignored in the operator $\rho_1 - \lambda\rho_0$. The optimum measurement for $Q_0 = 0$ is a measurement of the projection operator $\rho_0 = |n\rangle\langle n|$. This may be accomplished by performing an energy measurement. If the measured eigenvalue is different from n (the initial value) then the positive conclusion is made, otherwise the null conclusion is made. It is clear that the false-alarm probability is zero since if no pulse occurs the measurement will always give the eigenvalue n. On the other hand, the probability that the measurement will yield the value n after a pulse has arrived will be $p = |\langle n|\psi_1\rangle|^2$ or $p = |\langle n|D(\mu)|n\rangle|^2 = |\gamma|^2$ leading to a detection probability $Q_D = 1 - p = 1 - |\gamma|^2$ in agreement with the above result.

VI. WAVE-PACKET STATES

Perhaps the most ubiquitous nonlinear detection scheme used in laboratory practice is synchronous or phase-sensitive detection. In classical linear detection one measures the displacement of the oscillator $x(t)$. In phase-sensitive detection one measures $X_1(t)$ or $X_2(t)$, where $x(t) = X_1(t)\cos\omega t + X_2(t)\sin\omega t$. It has been realized for some time that it is possible to escape the linear detection limit by doing a "quantum-mechanical" phase-sensitive measurement.[24]

To examine this possibility we introduce operators which correspond to the classical variables $X_1(t)$ and $X_2(t)$. We follow Thorne *et al.*[13] by defining

$$\begin{aligned}\hat{X}_1 &\equiv a^\dagger e^{-i\omega t} + a e^{i\omega t}\\ &= (2m\omega/\hbar)^{1/2}[\hat{x}\cos\omega t - (\hat{p}/m\omega)\sin\omega t],\\ \hat{X}_2 &\equiv i(a^\dagger e^{-i\omega t} - a e^{i\omega t})\\ &= (2m\omega/\hbar)^{1/2}[\hat{x}\sin\omega t + (\hat{p}/m\omega)\cos\omega t].\end{aligned} \tag{25}$$

Note that $\hat{X}_1\cos\omega t + \hat{X}_2\sin\omega t = (2m\omega/\hbar)^{1/2}\hat{x}$. These operators are explicitly time dependent and it must be remembered that we are working in the Shrödinger picture. A measurement of the $\hat{X}_1$ operator corresponds to a position measurement at $\omega t = 0$ and to a momentum measurement at $\omega t = \pi/2$. Since $[\hat{X}_1, \hat{X}_2] = 2i$ there is an uncertainty principle which reads $\Delta\hat{X}_1\Delta\hat{X}_2 \geq 1$ where $\Delta\hat{X}_1 \equiv (\langle\hat{X}_1^2\rangle - \langle\hat{X}_1\rangle^2)^{1/2}$.

Now let us introduce a set of states which will turn out to be minimum-uncertainty states for $\hat{X}_1$ and $\hat{X}_2$. First, let us consider the unitary operator

$$S(z) \equiv \exp[\tfrac{1}{2}z(a^\dagger)^2 - \tfrac{1}{2}z^* a^2].$$

In the Appendix it is shown that if $|\psi\rangle$ is the state of a system and r is a real number, then $S(r)|\psi\rangle$ represents the same system compressed in position space by the factor $\alpha = e^{-r}$ and expanded in momentum space by the factor $1/\alpha = e^r$. For this reason we call $S(z)$ the "squeeze" operator. Let us define the state $|0, r\rangle$ according to

$$|0, r\rangle \equiv S(r)|0\rangle. \tag{26}$$

Since the ground state $|0\rangle$ is a Gaussian wave packet with position spread $\Delta x = (\hbar/2m\omega)^{1/2}$, we know that $|0, r\rangle$ is a Gaussian packet with $\Delta x = e^r(\hbar/2m\omega)^{1/2} = (1/\alpha)(\hbar/2m\omega)^{1/2}$. For very-large positive values of r, the state $|0, r\rangle$ is highly localized in momentum space. For very-large negative values of r $|0, r\rangle$ is highly localized in position space. The state $|0, 0\rangle$ with $r = 0$ is the ground state.

We may generalize this set of wave-packet states by defining

$$|\beta, z\rangle \equiv D(\beta)S(z)|0\rangle, \tag{27}$$

where β is a complex displacement and z is a complex squeeze factor. The state $|\beta, r\rangle$ is a Gaussian packet with the same shape as $|0, r\rangle$ but displaced from the origin in position and momentum space. In the Appendix we show that these states develop in time according to

$$e^{-i\omega t a^\dagger a}|\beta, z\rangle = |\beta e^{-i\omega t}, z e^{-2i\omega t}\rangle. \tag{28}$$

That is, they remain wave-packet states with the complex amplitudes following the classical trajectory and the complex squeeze factor z rotating at twice the resonant frequency.

Since the set of wave-packet states is unitarily equivalent to the set of coherent states, many of the useful properties of the coherent states such as the overcompleteness relation may be generalized for these states. It is not surprising that these states and the techniques we have developed for dealing with them have appeared in the literature many times before. The first use of these states for their low noise properties is, to our knowledge, in the paper by Takahasi[24] in which he discusses the degenerate parametric amplifier, which is a type of phase-sensitive amplifier. A very clear presentation of a unitary transformation technique just like that used in this paper may be found in the papers by Stoler.[25,26] These states are again introduced in a paper by Yuen[27] in which a more complete set of references may be found.

To get a better understanding of the wave-packet states we quote some expectation values for the state $|\beta e^{-i\omega t}, re^{-2i\omega t}\rangle$ with $\alpha = e^{-r}$ and $\beta = \beta_1 + i\beta_2$:

$$\begin{aligned}\langle\hat{x}\rangle &= (2\hbar/m\omega)^{1/2}(\beta_1\cos\omega t + \beta_2\sin\omega t),\\ \langle\hat{p}\rangle &= (2m\hbar\omega)^{1/2}(-\beta_1\sin\omega t + \beta_2\cos\omega t),\\ \Delta\hat{x} &= (\langle\hat{x}^2\rangle - \langle\hat{x}\rangle^2)^{1/2}\\ &= (\hbar/2m\omega)^{1/2}[\alpha^2\sin^2\omega t + (1/\alpha^2)\cos^2\omega t]^{1/2},\\ \Delta\hat{p} &= (m\hbar\omega/2)^{1/2}[(1/\alpha^2)\sin^2\omega t + \alpha^2\cos^2\omega t]^{1/2},\\ \Delta\hat{x}\Delta\hat{p} &= \tfrac{1}{2}\hbar\{1 + [\tfrac{1}{2}(\alpha^2 - 1/\alpha^2)\sin 2\omega t]^2\},\\ \langle\hat{E}\rangle &= \hbar\omega\{|\beta|^2 + [(1-\alpha^2)/2\alpha]^2\},\\ \langle\hat{X}_1\rangle &= 2\beta_1, \quad \langle\hat{X}_2\rangle = 2\beta_2,\\ \Delta\hat{X}_1 &= 1/\alpha, \quad \Delta\hat{X}_2 = \alpha,\\ \Delta\hat{X}_1\Delta\hat{X}_2 &= 1.\end{aligned} \tag{29}$$

Recall that for $r = 0$ $(\alpha = 1)$ the states $|\beta, 0\rangle$ are just the coherent states. The last five equations make clear the connection to the $\hat{X}_1$ and $\hat{X}_2$ operators. We see that the states $|\beta e^{-i\omega t}, re^{-2i\omega t}\rangle$ are minimum-uncertainty states $(\Delta\hat{X}_1\Delta\hat{X}_2 = 1)$ for $\hat{X}_1$ and $\hat{X}_2$, that α gives the spread in $\hat{X}_2$, $1/\alpha$ gives the spread in $\hat{X}_1$, and $\mathrm{Re}\{\beta\}$ and $\mathrm{Im}\{\beta\}$ give the time-independent expectation values of $\hat{X}_1$ and $\hat{X}_2$, respectively.

Now let us consider preparing the system initially in a wave-packet state. For notational convenience we will suppose that the initial state is $|\psi_i\rangle = e^{i\omega\tau a^\dagger a}|\beta, r\rangle$. At $t = 0$ we have $|\psi_0\rangle = e^{-i\omega\tau a^\dagger a}|\psi_i\rangle = |\beta, r\rangle$, thus $\rho_0 = |\beta, r\rangle\langle\beta, r|$ and $\rho_1 = D(\mu)|\beta, r\rangle\langle\beta, r|D^\dagger(\mu)$. We need to compute $\gamma = \langle\beta, r|D(\mu)|\beta, r\rangle = \langle 0, r|D^\dagger(\beta)D(\mu)D(\beta)|0, r\rangle$. Recalling Eq. (20) we may write this as $\gamma = \langle 0, r|D(\mu)|0, r\rangle\exp[2i\,\mathrm{Im}(\beta^*\mu)]$.

The value of this matrix element is shown in the Appendix to be

$$\gamma = \exp\{-\tfrac{1}{2}|\mu|^2[\alpha^2\cos^2\phi + (1/\alpha^2)\sin^2\phi] + 2i\,\mathrm{Im}(\beta^*\mu)\}$$

and

$$|\gamma|^2 = \exp\{-(E_s/\hbar\omega)[\alpha^2\cos^2\phi + (1/\alpha^2)\sin^2\phi]\},$$

where ϕ is defined by $\mu = |\mu|e^{i\phi}$. Note that for $\alpha = 1$ the result reduces to the coherent state result as it must. As in the coherent state result we find that $|\gamma|^2$ is independent of the initial displacement β. A new feature has appeared in this result, namely phase dependence. The overlap depends on the phase ϕ which is a function of the arrival time and shape of the gravity wave pulse. To this point we have assumed that this phase is known. In practice this assumption is incorrect and we must in principle do a more difficult calculation as described in Sec. VII. However, let us examine the present result in more detail. For $Q_0 = 0$ we have

$$Q_D = 1 - \exp\{-(E_s/\hbar\omega)[\alpha^2\cos^2 + (1/\alpha^2)\sin^2\phi]\}$$

$$\text{(wave packets, } Q_0 = 0, \text{ known phase)} \qquad (30)$$

and

$$E_{\min} = \frac{\hbar\omega \ln 2}{[\alpha^2\cos^2\phi + (1/\alpha^2)\sin^2\phi]}$$

$$\text{(wave packets, known phase).} \qquad (31)$$

Figure 4 is a plot of Eq. (31) vs α for various

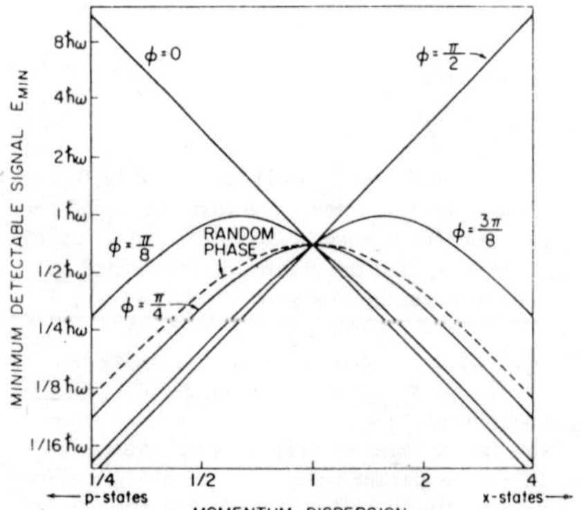

FIG. 4. Minimum detectable signal for detection of gravity waves of known phase for an oscillator prepared in the initial state $|\beta, r\rangle$ vs the momentum dispersion $\alpha = e^{-r}$. Also shown is the curve for the case of random phase.

values of the phase. We see that for favorable values of ϕ we can make the minimum detectable energy very small either by making $\alpha \ll 1$ or $\alpha \gg 1$. It turns out that for the case $Q_0 = 0$, the optimum decision strategy for pulses of unknown phase is identical to the strategy for signals of known phase. In this case Q_D is gotten by integrating over the phase, that is

$$Q_D = \int_0^{2\pi} (d\phi/2\pi)(1 - \exp\{-(E_s/\hbar\omega) \times [\alpha^2\cos^2\phi + (1/\alpha^2)\sin^2\phi]\})$$

and we obtain

$$Q_D = 1 - \exp[-(E_s/2\hbar\omega)(\alpha^2 + 1/\alpha^2)] \times I_0[(E_s/2\hbar\omega)(\alpha^2 - 1/\alpha^2)]$$

$$\text{(wave packets, } Q_0 = 0, \text{ random phase),} \qquad (32)$$

where $I_0(x)$ is the modified Bessel function of order zero. We can approximate Q_D in two limits: $Q_D \approx 1 - (\hbar\omega/\pi E_s\alpha^2)^{1/2}$ for $\hbar\omega/E_s\alpha^2 \ll 1$ and $Q_D \approx 1 - (\hbar\omega\alpha^2/\pi E_s)^{1/2}$ for $\hbar\omega\alpha^2/E_s \ll 1$. We also conclude

$$E_{\min} \approx 1.8\hbar\omega/\alpha^2, \quad \text{for } \alpha^2 \gg 1,$$

$$E_{\min} \approx 1.8\alpha^2\hbar\omega, \quad \text{for } \alpha^2 \ll 1 \qquad (33)$$

$$\text{(wave packets, } Q_0 = 0, \text{ random phase).}$$

We see that as in the case of n states we can obtain high sensitivity by preparing the system in a highly excited state. We may choose to localize the oscillator either in position space ($\alpha \gg 1$) or in momentum space ($\alpha \ll 1$).

We now consider another scheme which is analogous to a two-channel phase-sensitive detector.[28] One builds two gravity-wave antennas. One of them is prepared in a state $|0, r\rangle$ with $\alpha = e^{-r} \gg 1$ which is highly localized in position space. The other is prepared in the state $|0, -r\rangle$ which is highly localized in momentum space. The optimum decision strategy is used separately on each antenna. If both antennas give null results we make the null conclusion, otherwise we make the alternative conclusion. With this strategy we still have $Q_0 = 0$, but now,

$$Q_D = \int_0^{2\pi} \frac{d\phi}{2\pi}\left\{1 - \exp\left[-\frac{E_s}{\hbar\omega}\left(\alpha^2\cos^2\phi + \frac{1}{\alpha^2}\sin^2\phi\right)\right] \times \exp\left[-\frac{E_s}{\hbar\omega}\left(\frac{1}{\alpha^2}\cos^2\phi + \alpha^2\sin^2\phi\right)\right]\right\}$$

$$= \int_0^{2\pi} \frac{d\phi}{2\pi}\left\{1 - \exp\left[-\frac{E_s}{\hbar\omega}\left(\alpha^2 + \frac{1}{\alpha^2}\right)\right]\right\},$$

so

$$Q_D = 1 - \exp[-(E_s/\hbar\omega)(\alpha^2 + 1/\alpha^2)]$$
$$\text{(two-phase detector, } Q_0 = 0). \quad (34)$$

This implies

$$E_{\min} = \frac{\hbar\omega \ln 2}{(\alpha^2 + 1/\alpha^2)}$$
$$\text{(two-phase detector, } Q_0 = 0). \quad (35)$$

The advantage of this technique is that the detection efficiency depends exponentially on the signal strength. With a single antenna the detection efficiency is a weak function of the signal energy for random phase signals. This can be seen in Fig. 5.

The optimum measurement for $Q_0 = 0$ is a measurement of the projection operator $\rho_0 = |\beta, r\rangle\langle r, \beta|$. For large positive r the optimum measurement can be approximated by measuring $\hat{X}_1$. The almost optimum decision strategy (for a single antenna) is as follows. The oscillator is prepared in the state that evolves to $|0, r\rangle$ at $t = 0$ with $\alpha \gg 1$. This state is highly localized in $\hat{X}_1$ since $\Delta\hat{X}_1 = (1/\alpha) \ll 1$. The expectation value of $\hat{X}_1$ is $\langle\hat{X}_1\rangle = 0$. If a gravity wave pulse arrives, the state at $t = 0$ will be $D(\mu)|0, r\rangle = |\mu, r\rangle$ which is also highly localized in $\hat{X}_1$ but at a shifted value $\langle\hat{X}_1\rangle = 2\mu$, with $\Delta\hat{X}_1 = (1/\alpha) \ll 1$. A measurement of $\hat{X}_1$ is performed and compared to zero, if a shift much bigger than $\Delta\hat{X}_1 = 1/\alpha$ is observed we conclude that a gravity wave has been received.

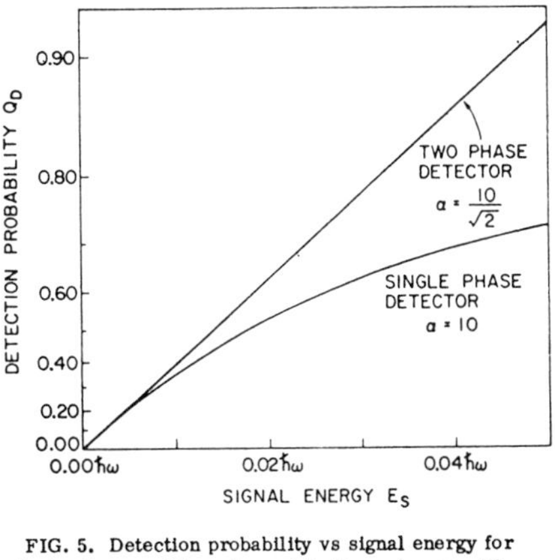

FIG. 5. Detection probability vs signal energy for single-phase and two-phase detection schemes with $\alpha = 10$ and $\alpha = 10/\sqrt{2}$, respectively.

VII. RANDOM PHASE

In order to treat the more realistic problem of gravity waves of unknown phase we no longer have a simple pure-state problem. The density matrix ρ_1 is given by

$$\rho_1 = \int_0^{2\pi} \frac{d\phi}{2\pi} D(|\mu| e^{i\phi}) \rho_0 D^\dagger(|\mu| e^{i\phi}),$$

where we have assigned equal probability to each value of phase. Although we do not present the results here we have solved this problem for coherent states and n states. In the case $Q_0 = 0$ the results are exactly the same as we have calculated for the assumption of known phase. However, as Q_0 is allowed to differ from zero the value of Q_D does not improve as fast for random phase as it does for known phase. For the wave-packet states the result with $Q_0 = 0$ is obtained by integrating over the phase as we have done in Sec. VI.

VIII. CONCLUSION

We have seen that for an infinite-Q gravity-wave antenna interacting with a classical radiation field quantum mechanics imposes no ultimate sensitivity limit. However, for any given initial state of the antenna there is a sensitivity limit. We have found this limit for several important classes of initial states by finding the optimum detection strategy in each case. When the antenna is prepared in a coherent state the sensitivity limit is of the same order as the limit arising in a detection scheme employing a linear amplifier. Higher sensitivity may be obtained by preparing the antenna initially in an excited energy eigenstate or in a highly localized wave-packet state. The wave-packet states look particularly interesting since they are minimum-uncertainty states and it is not hard to imagine devices which will prepare an antenna in such a state and subsequently read it out. An example of such a device is the degenerate parametric amplifier.

Although the state of the art of gravity-wave astronomy has not yet reached the linear detection limit, we may hope that a future generation of detectors will reach and surpass it.

ACKNOWLEDGMENTS

It is a pleasure to acknowledge R. C. Taber, P. F. Michelson, and especially S. P. Boughn and R. P. Giffard for fruitful discussions and support in the course of this work. This work was supported by the National Science Foundation under Grant No. PHY76-80105.

APPENDIX

In this appendix we will establish some results which are used in the body of the paper. First we recall the operator identity

$$e^{\xi A}Be^{-\xi A}=B+\xi[A,B]+\frac{\xi^2}{2!}[A,[A,B]]+\cdots . \quad \text{(A1)}$$

Letting $A=a^\dagger a$ and $B=a^\dagger$ we find $e^{\xi a^\dagger a}a^\dagger e^{-\xi a^\dagger a}=a^\dagger e^{\xi}$. Similarly we find $e^{\xi a^\dagger a}ae^{-\xi a^\dagger a}=ae^{-\xi}$. If $f(a,a^\dagger)$ is any function of a and $a^\dagger$ which may be written as a power series we have

$$e^{\xi a^\dagger a}f(a,a^\dagger)e^{-\xi a^\dagger a}=f(e^{\xi a^\dagger a}ae^{-\xi a^\dagger a},\ e^{\xi a^\dagger a}a^\dagger e^{-\xi a^\dagger a})$$

$$=f(ae^{-\xi},a^\dagger e^{\xi}). \quad \text{(A2)}$$

Recalling the definition of the displacement operator $D(\mu)=\exp(\mu a^\dagger-\mu^* a)$ we use (A2) to obtain

$$e^{-i\omega t a^\dagger a}D(\mu)e^{i\omega t a^\dagger a}$$

$$=\exp[\mu e^{-i\omega t}a^\dagger-\mu^* e^{i\omega t}a]=D(\mu e^{-i\omega t}). \quad \text{(A3)}$$

Another important operator identity is the Baker-Hausdorff formula,[29]

$$e^{A+B}=e^Ae^Be^{-[A,B]/2}, \quad \text{(A4)}$$

for any operators A and B which commute with their commutator $[A,B]$. Applying this formula to the displacement operator we can easily show

$$D(\mu)D(\beta)=D(\mu+\beta)\exp[\tfrac{1}{2}(\beta^*\mu-\beta\mu^*)]. \quad \text{(A5)}$$

We may also use this identity to rewrite $D(\mu)$ in normal-ordered form. We get

$$D(\mu)=\exp(-\tfrac{1}{2}|\mu|^2)\exp(\mu a^\dagger)\exp(-\mu^* a). \quad \text{(A6)}$$

It follows that

$$\langle 0|D(\mu)|0\rangle=\exp(-\tfrac{1}{2}|\mu|^2). \quad \text{(A7)}$$

Equation (A6) may be used to obtain the number-state representation of the coherent state $|\alpha\rangle$. We have $|\alpha\rangle=D(\alpha)|0\rangle=\exp(-\frac{1}{2}|\alpha|^2)\exp(\mu a^\dagger)|0\rangle$ and thus

$$|\alpha\rangle=\exp(-\tfrac{1}{2}|\alpha|^2)\sum_{n=0}^{\infty}\frac{\alpha^n}{\sqrt{n!}}|n\rangle. \quad \text{(A8)}$$

If B is any operator, we define its coherent-state representation $B(\alpha^*,\beta)$ by $B(\alpha^*,\beta)\equiv\langle\alpha|B|\beta\rangle\times\exp(\frac{1}{2}|\alpha|^2+\frac{1}{2}|\beta|^2)$. Using Eq. (A8) we can write this as

$$B(\alpha^*,\beta)=\sum_{m=0}^{\infty}\sum_{n=0}^{\infty}\frac{\alpha^{*m}\beta^n}{(m!n!)^{1/2}}\langle m|B|n\rangle. \quad \text{(A9)}$$

We see that $B(\alpha^*,\beta)$ is a generating function for the matrix elements of B in the number representation. The coherent-state representation of $D(\mu)$ is

$$D(\alpha^*,\beta;\mu)=\langle\alpha|D(\mu)|\beta\rangle\exp(\tfrac{1}{2}|\alpha|^2+\tfrac{1}{2}|\beta|^2)$$

$$=\langle 0|D^\dagger(\alpha)D(\mu)D(\beta)|0\rangle\exp(\tfrac{1}{2}|\alpha|^2+\tfrac{1}{2}|\beta|^2)$$

$$=\exp(-\tfrac{1}{2}|\mu|^2+\alpha^*\beta+\alpha^*\mu-\mu^*\beta).$$

We may therefore write

$$\exp(-\tfrac{1}{2}|\mu|^2)\exp(\alpha^*\beta+\alpha^*\mu-\mu^*\beta)$$

$$=\sum_{m=0}^{\infty}\sum_{n=0}^{\infty}\frac{\alpha^{*m}\beta^n}{(m!n!)^{1/2}}\langle m|D(\mu)|n\rangle. \quad \text{(A10)}$$

Now a generating function for the associated Laguerre polynomials is[30]

$$\exp(\lambda w+\kappa z+\nu wz)=\sum_{m=0}^{\infty}\sum_{n=0}^{\infty}\frac{w^m}{m!}\frac{z^n}{n!}D_{mn}(\lambda,\kappa,\nu), \quad \text{(A11)}$$

with

$$D_{mn}(\lambda,\kappa,\nu)=\begin{cases} m!\,\nu^m\kappa^{n-m}L_m^{(n-m)}\left(-\dfrac{\lambda\kappa}{\nu}\right), & \text{for } m\le n\\ n!\,\nu^n\lambda^{m-n}L_n^{(m-n)}\left(-\dfrac{\lambda\kappa}{\nu}\right), & \text{for } m\ge n. \end{cases}$$

Setting $\lambda=\mu$, $\kappa=-\mu^*$, $w=\alpha^*$, $z=\beta$, and $\nu=1$ we obtain from Eq. (A10)

$$\langle m|D(\mu)|n\rangle=\exp(-\tfrac{1}{2}|\mu|^2)\left(\frac{m!}{n!}\right)^{1/2}(-\mu^*)^{n-m}L_m^{(n-m)}(|\mu|^2),\quad \text{for } m\le n. \quad \text{(A12)}$$

This gives the matrix element for a transition from the state $|n\rangle$ to the state $|m\rangle$ under the influence of a gravity wave. For the case $m=n$ we have

$$\langle n|D(\mu)|n\rangle=\exp(-\tfrac{1}{2}|\mu|^2)L_n(|\mu|^2). \quad \text{(A13)}$$

To facilitate calculations with the wave-packet states let us introduce the "squeeze" operator

$$S(z)\equiv\exp\left[\frac{z}{2}(a^\dagger)^2-\frac{z^*}{2}a^2\right].$$

Now

$$S^\dagger(z)=\exp\left[-\frac{z}{2}(a^\dagger)^2+\frac{z^*}{2}a^2\right]=S^{-1}(z)=S(-z).$$

Thus $S^\dagger(z)S(z)=S(z)S^\dagger(z)=1$ so $S(z)$ is a unitary operator. S induces a canonical transformation of the creation and annihilation operators. Using Eq. (A1) we may show

$$b^\dagger(z)\equiv S^\dagger a^\dagger S=\cosh|z|a^\dagger+\frac{z^*}{|z|}\sinh|z|a\,,$$
$$b(z)\equiv S^\dagger aS=\cosh|z|a+\frac{z}{|z|}\sinh|z|a^\dagger\,. \tag{A14}$$

Since the transformation is canonical we have $[b,b^\dagger]=[a,a^\dagger]=1$. If $f(a,a^\dagger)$ is any power-series function of a and $a^\dagger$, then

$$S^\dagger f(a,a^\dagger)S=f(b,b^\dagger)\,.$$

Let us examine the effect of this transformation on the position and momentum operators for the case where z is a real number r:

$$S^\dagger(r)\hat{x}S(r)=\left(\frac{2\hbar}{m\omega}\right)^{1/2}(b^\dagger+b)$$
$$=\left(\frac{2\hbar}{m\omega}\right)^{1/2}(\cosh r+\sinh r)(a^\dagger+a)\,,$$

so $S^\dagger(r)\hat{x}S(r)=\hat{x}e^{-r}=(1/\alpha)\hat{x}$, where we have defined $\alpha=e^{-r}$. Similarly we have $S^\dagger(r)\hat{p}S(r)=e^{r}\hat{p}=\alpha\hat{p}$.

If $|\psi\rangle$ is the state vector of a system then $S(r)|\psi\rangle$ represents the same system compressed in position space by the factor α and expanded in momentum space by the factor $1/\alpha$. This is seen since for any power k we have

$$\langle\psi|S^\dagger(\hat{x})^kS|\psi\rangle=\langle\psi|(\hat{x}/\alpha)^k|\psi\rangle$$

and

$$\langle\psi|S^\dagger(\hat{p})^kS|\psi\rangle=\langle\psi|(\alpha\hat{p})^k|\psi\rangle\,.$$

We define the class of states $|\beta,z\rangle$ where both z and β may be complex by

$$|\beta,z\rangle\equiv D(\beta)S(z)|0\rangle,\quad\text{with }\alpha=e^{-z}\,. \tag{A15}$$

$S(z)$ propagates in time according to

$$e^{-i\omega ta^\dagger a}\exp\left(\frac{z}{2}a^{\dagger 2}-\frac{z^*}{2}a^2\right)e^{i\omega ta^\dagger a}$$
$$=\exp\left(\frac{z}{2}e^{-2i\omega t}a^{\dagger 2}-\frac{z^*}{2}e^{2i\omega t}a^2\right)$$
$$=S(ze^{-2i\omega t})\,.$$

Thus the state $|\beta,z\rangle$ develops in time according to

$$e^{-i\omega ta^\dagger a}|\beta,z\rangle=D(\beta e^{-i\omega t})S(ze^{-2i\omega t})|0\rangle$$
$$=|\beta e^{-i\omega t},ze^{-2i\omega t}\rangle\,.$$

The displacement β has the classical time dependence, while the complex spreading factor z rotates in the complex plane at twice the resonant frequency.

Expectation values in the wave-packet states are easily calculated using the transformation equations (A14). We will illustrate this by computing $\langle 0,r|D(\mu)|0,r\rangle$ for real r. We have

$$\langle 0,r|D(\mu)|0,r\rangle=\langle 0|S^\dagger(r)D(\mu)S(r)|0\rangle$$
$$=\langle 0|\exp(\mu b^\dagger-\mu^*b)|0\rangle$$
$$=\langle 0|\exp(\eta a^\dagger-\eta^*a)|0\rangle\,,$$

where $\eta=(\mu\cosh r-\mu^*\sinh r)$. Thus

$$\langle 0,r|D(\mu)|0,r\rangle=\langle 0|D(\eta)|0\rangle=\exp(-\tfrac{1}{2}|\eta|^2)$$
$$=\exp[-\tfrac{1}{2}|\mu|^2$$
$$\times(\cosh 2r-\sinh 2r\cos^2\phi)]\,,$$

where $\mu=|\mu|e^{i\phi}$. Recalling that $\alpha=e^{-r}$ we may write this as

$$\langle 0,r|D(\mu)|0,r\rangle=\exp\left[-\tfrac{1}{2}|\mu|^2\left(\alpha^2\cos^2\phi+\frac{1}{\alpha^2}\sin^2\phi\right)\right]. \tag{A16}$$

Although we do not need it here, it is useful to have an expression for $S(z)$ which is in normal-ordered form. We find that

$$S(z)=(\cosh|z|)^{1/2}\exp\left(\frac{z}{2|z|}\tanh|z|a^{\dagger 2}\right)$$
$$\times\sum_{n=0}^{\infty}\frac{(\operatorname{sech}|z|-1)^n}{n!}(a^\dagger)^na^n$$
$$\times\exp\left(-\frac{z}{2|z|}\tanh|z|a^2\right). \tag{A17}$$

This form may be used to find the $|n\rangle$ state representation of the state $|0,z\rangle$:

$$|0,z\rangle=S(z)|0\rangle$$
$$=(\cosh|z|)^{1/2}$$
$$\times\sum_{n=0}^{\infty}\left(\frac{z}{2|z|}\tanh|z|\right)^n\frac{[(2n)!]^{1/2}}{n!}|2n\rangle\,.$$

---

[1]K. S. Thorne, in *Theoretical Principles in Astrophysics and Relativity*, edited by N. R. Lebovitz, W. H. Reid, and P. O. Vandervoort (Univ. of Chicago Press, Chicago, 1978), Chap. 8.

[2]K. Shimoda, H. Takahasi, and C. H. Townes, J. Phys. Soc. Japan 12, 686 (1957).

[3]H. Heffner, Proc. IRE 50, 1604 (1962).

[4]H. A. Haus and J. A. Mullen, Phys. Rev. 128, 2407 (1962).

[5]R. P. Giffard, Phys. Rev. D 14, 4278 (1976).

[6]V. B. Braginskiĭ, in *Topics in Theoretical and Experimental Gravitation Physics*, edited by V. de Sabbata and J. Weber (Plenum, New York, 1977).
[7]V. B. Braginskiĭ and Y. I. Vorontsov, Usp. Fiz. Nauk 114, 41 (1974) [Sov. Phys.—Usp. 17, 644 (1975)].
[8]V. B. Braginskiĭ, Y. I. Vorontsov, and V. D. Krivchenkov, Zh. Eksp. Teor. Fiz. 68, 55 (1975) [Sov. Phys.—JETP 41, 28 (1975)].
[9]V. B. Braginskiĭ, Y. I. Vorontsov, and F. Y. Khalili Zh. Eksp. Teor. Fiz. 73, 1340 (1977) [Sov. Phys.—JETP 46, 705 (1977)].
[10]W. G. Unruh, Phys. Rev. D 18, 1764 (1978).
[11]V. Moncrief. Ann. Phys. (N. Y.) 114, 201 (1978).
[12]W. G. Unruh, Phys. Rev. D 17, 1180 (1978).
[13]K. S. Thorne, R. P. Drever, C. M. Caves, M. Zimmerman, and V. D. Sandberg, Phys. Rev. Lett. 40, 667 (1978).
[14]V. B. Braginskiĭ, Y. I. Vorontsov, and F. Y. Khalili Pis'ma Zh. Eksp. Teor. Fiz. 27, 296 (1978) [JETP Lett. 27, 277 (1978)].
[15]C. W. Helstrom, *Quantum Detection and Estimation Theory* (Academic, New York, 1976).
[16]A. S. Holevo, J. Multivar. Anal. 3, 337 (1973).
[17]H. P. Yuen, R. S. Kennedy, and M. Lax, IEEE Trans. Inf. Theory IT-21, 125 (1975).
[18]J. Weber, Phys. Rev. 117, 306 (1960).
[19]G. W. Gibbons and S. W. Hawking, Phys. Rev. D 4, 2191 (1971).
[20]E. P. Davies and J. T. Lewis, Commun. Math. Phys. 17, 239 (1970).
[21]C. W. Helstrom, Inf. Control 13, 156 (1968).
[22]W. H. Louisell, *Quantum Statistical Properties of Radiation* (Wiley, New York, 1973), pp. 203 ff.
[23]R. J. Glauber, Phys. Rev. 131, 2766 (1963).
[24]H. Takahasi, in *Advances in Communications Systems*, edited by A. Balakrishnan (Academic, New York, 1965), Vol. 1, p. 227–310.
[25]D. Stoler, Phys. Rev. D 1, 3217 (1970).
[26]D. Stoler, Phys. Rev. D 4, 1925 (1971).
[27]H. P. Yuen, Phys. Rev. A 13, 2226 (1976).
[28]Essentially the same technique is proposed in Ref. 13.
[29]A. Messiah, *Quantum Mechanics* (Wiley, New York, 1968), Vol. 1, p. 442.
[30]B. Mollow and R. J. Glauber, Phys. Rev. 160, 1076 (1967).

PHYSICAL REVIEW D VOLUME 22, NUMBER 10 15 NOVEMBER 1980

Charged Schrödinger particle in a *c*-number radiation field

Tadashi Toyoda and Karl Wildermuth
Institut für Theoretische Physik der Universität Tübingen, Auf der Morgenstelle 14, D-7400 Tübingen 1, Federal Republic of Germany

(Received 27 May 1980)

We formulate a method to derive the time-dependent Schrödinger equation which describes the interaction of an electron with a *c*-number radiation field. To obtain this semiclassical approximation from the complete quantum-mechanical formulation we use the corresponding Schrödinger equation in its projection form, which is an extension of the Hill-Wheeler generator coordinate method. For the basis states of the radiation field we introduce a complete, but not overcomplete, subset of coherent states, which were found by von Neumann and whose proof of completeness was given by Bargmann *et al*. and Perelomov. We find that the conventional overcomplete continuous coherent states are not suitable for the projection form of the Schrödinger equation. The method also allows one to calculate quantum-mechanical correction terms systematically.

I. INTRODUCTION

Since the very beginning of the development of quantum electrodynamics and other field theories, their classical limits have been attracting much attention.[1] It is well known that coherent states are useful for studying such classical or semiclassical limits.[2]

Very recently, several authors have used coherent states to study various field theories and have shown their powerfulness. It has been shown that the use of coherent states with a path integral for elements of the S matrix gives a different treatment of the Callan-Coleman vacuum tunneling.[3] Coherent states combined with the variational method have been used to treat an isovector meson field interacting with a static source.[4] It has also been shown that coherent states can be used to describe equilibrium states of boson fields.[5]

Apart from their applications, coherent states themselves have been intensively studied since the pioneering work of Bargmann and Segal.[6] In 1971, Bargmann, Butera, Ghiraradello, Klauder, and Perelomov proved that a certain subset of coherent states form a complete, but not overcomplete, set.[7] This will be important for our considerations.

In the present paper we shall study the semiclassical treatment of a charged Schrödinger particle interacting with a c-number radiation field. We shall derive the corresponding Schrödinger equation from the fully quantized theory, in which the radiation field is also quantized, by using coherent states to describe the radiation field. To avoid trouble due to the above-mentioned overcompleteness of coherent states we shall use the complete subset of coherent states of Bargmann *et al.* and Perelomov, which we shall call VNLCS (see Sec. III). In order to introduce such a subset of coherent states (VNLCS) in a consistent manner into the quantized theory, we shall use the time-dependent Schrödinger equation written in projection form,[8] which is a generalization of the Hill-Wheeler method.[9] This method has been used to obtain a microscopic nuclear theory for low-energy phenomena from a unified point of view.[8]

As a charged Schrödinger particle we shall consider a nonrelativistic atomic electron interacting with a strong radiation field. In Sec. II we shall formulate the problem quantitatively by defining the semiclassical treatment of the system. In Sec. III we shall introduce coherent states, especially VNLCS, as the basis states for the radiation field. In Sec. IV we shall rewrite the Schrödinger equation for the system in the projection form. In Sec. V we shall introduce a classical approximation to the projection form of the Schrödinger equation to obtain the semiclassical Schrödinger equation. In Sec. VI we shall give some concluding remarks.

II. QUANTITATIVE FORMULATION OF THE PROBLEM

We shall consider the system of an atomic electron interacting with a strong radiation field in the nonrelativistic case. Such a system can be described semiclassically by the time-dependent Schrödinger equation

$$[\hat{H}_e - (e/mc)\vec{A}(\vec{r},t)\cdot(\hbar/i)\vec{\nabla}_r]\,|\psi(\vec{r},t)\rangle = i\hbar\frac{\partial}{\partial t}\,|\psi(\vec{r},t)\rangle\,, \qquad (2.1)$$

where $\vec{A}(\vec{r},t)$ is the classical time-dependent vector potential and $|\psi\rangle$ is the wave function of the atomic electron. For simplicitly, we consider a one-electron atom. Then, $\hat{H}_e$ has the form

$$\hat{H}_e = -(\hbar/2m)\nabla_r^{\,2} + V(r)\,, \qquad (2.2)$$

where $V(r)$ is the electrostatic potential[10] in which

the electron moves. We shall show under what conditions Eq. (2.1) can be a very good approximation to the fully quantized description, where the transverse radiation field $\vec{A}(\vec{r},t)$ is also quantized. The corresponding Schrödinger equation is

$$\hat{H}|\Psi\rangle=[\hat{H}_e+\hat{H}_A-(e/mc)\hat{\vec{A}}(\vec{r})\cdot\hat{\vec{p}}]|\Psi\rangle=i\hbar\frac{\partial}{\partial t}|\Psi\rangle. \quad (2.3)$$

Here, $|\Psi\rangle$ is the state vector which describes the quantum-mechanical state of the electron and the light-quanta states of the transverse radiation field. $\hat{H}_A$ is the Hamiltonian of the light quanta, which can be written as

$$\hat{H}_A=\sum_{\vec{k},\lambda}\hbar\omega_k a^\dagger_{\vec{k}\lambda}a_{\vec{k}\lambda}=\sum_{\vec{k},\lambda}\hat{H}_{\vec{k}\lambda},\quad |\vec{k}|=\omega_k/c, \quad (2.4a)$$

with the commutation relation for $a^\dagger_{\vec{k}\lambda}$ and $a_{\vec{k}\lambda}$

$$[a_{\vec{k}\lambda},a^\dagger_{\vec{q}\eta}]=\delta_{\vec{k},\lambda;\vec{q},\eta}, \quad (2.4b)$$

where $a_{\vec{k}\lambda}$ and $a^\dagger_{\vec{k}\lambda}$ are the annihilation and the creation operators of the light quanta with wave vector $\vec{k}$ and polarization λ. The operator $\hat{\vec{A}}(\vec{r})$ has the form

$$\hat{\vec{A}}(\vec{r})=(1/V)^{1/2}\sum_{\vec{k},\eta}c(\hbar/2\omega_k)^{1/2}(a_{\vec{k}\eta}e^{i\vec{k}\cdot\vec{r}}+a^\dagger_{\vec{k}\eta}e^{-i\vec{k}\cdot\vec{r}})\vec{\epsilon}_{\vec{k}\eta}, \quad (2.5)$$

where $\epsilon_{\vec{k}\eta}$ is the polarization vector and V is the normalization volume.

III. COHERENT LIGHT-QUANTA STATES

In this section we shall discuss the basis states for the radiation field which are especially suited for the transition from the completely quantum-mechanical Schrödinger equation (2.3) to the semiclassical equation (2.1). The most suited basis states of the radiation field for this purpose are the so-called coherent states which were introduced first by Schrödinger.[11] Coherent states are minimum wave-packet states[12] and are labeled by complex eigenvalues, whose real and imaginary parts correspond to the expectation values of two canonically conjugate observables, respectively [see Eqs. (3.5e) and (3.5f)]. This is one of the reasons why the coherent states are extremely useful for the discussion of the classical limit of the quantum-mechanical treatment. Indeed, Klauder has shown that the classical-mechanical description can be obtained formally from the quantum theory by using the continuous representation, which is a generalization of the coherent states.[2]

The coherent states were discussed by many authors during the last 20 years.[6] Here we list the main properties of these states and refer to the literature[12] for their derivations.

With the operators $a^\dagger$ and a (for simplicity from now on we write the indices $\vec{k},\eta$ explicitly only when it is necessary for discussion), which have been introduced by Eq. (2.4b), we can define the coherent state $|\alpha\rangle$, where α is a complex number to label the state, as

$$|\alpha\rangle=\exp(-|\alpha|^2/2)\sum_{n=0}^{\infty}\frac{\alpha^n}{\sqrt{n!}}|n\rangle. \quad (3.1)$$

Here,

$$|n\rangle=\frac{1}{\sqrt{n!}}(a^\dagger)^n|\varphi_0\rangle,\quad |\varphi_0\rangle=|\text{vacuum state}\rangle \quad (3.2)$$

is the eigenstate of the number operator

$$\hat{N}=a^\dagger a, \quad (3.3)$$

containing n light quanta of $(\vec{k},\eta)$. For our later considerations it is useful to split $a^\dagger$ and a into a sum of Hermitian operators, i.e.,

$$a=(\hat{u}+i\hat{p})/(2\hbar)^{1/2},\quad a^\dagger=(\hat{u}-i\hat{p})/(2\hbar)^{1/2}. \quad (3.4)$$

The coherent states $|\alpha\rangle$ have the following properties:

$$\langle\alpha|\alpha\rangle=1, \quad (3.5a)$$

$$a|\alpha\rangle=\alpha|\alpha\rangle=(1/2\hbar)^{1/2}(u+ip)|\alpha\rangle, \quad (3.5b)$$

$$\langle\alpha|a^\dagger=\langle\alpha|\alpha^*=(1/2\hbar)^{1/2}(u-ip)\langle\alpha|, \quad (3.5c)$$

$$\langle\hat{N}\rangle=\langle\alpha|\hat{N}|\alpha\rangle=\langle\alpha|a^\dagger a|\alpha\rangle=\alpha^*\alpha, \quad (3.5d)$$

$$\langle\hat{u}\rangle=\langle\alpha|\hat{u}|\alpha\rangle=(2\hbar)^{1/2}\,\mathrm{Re}(\alpha)=u, \quad (3.5e)$$

$$\langle\hat{p}\rangle+\langle\alpha|\hat{p}|\alpha\rangle=(2\hbar)^{1/2}\,\mathrm{Im}(\alpha)=p, \quad (3.5f)$$

$$|\langle u|\alpha\rangle|^2=\text{const}\times\exp[-(u-\langle\hat{u}\rangle)^2/2\hbar], \quad (3.5g)$$

$$|\langle p|\alpha\rangle|^2=\text{const}\times\exp[-(p-\langle\hat{p}\rangle)^2/2\hbar], \quad (3.5h)$$

where $\langle u|\alpha\rangle$ and $\langle p|\alpha\rangle$ are the u and p representations of the state $|\alpha\rangle$, respectively. The eigenvalues of $\hat{u}$ and $\hat{p}$ are written as u and p, respectively. Equation (3.5b) shows that the coherent state $|\alpha\rangle$ is the eigenstate of the non-Hermitian annihilation operator a with the complex eigenvalue $\alpha=(u+ip)/(2\hbar)^{1/2}$.

If this complex eigenvalue α, which labels the coherent states as already stated, runs over the whole complex plane, the coherent states become overcomplete for the Hilbert space. Such an overcomplete set of coherent states cannot be used for our considerations because they are linearly dependent (see Sec. IV and Appendix B). However, Bargmann *et al.* and Perelomov[7] proved that a subset of the overcomplete coherent states forms a complete set. This subset is given by

$$\{|\alpha\rangle:\ \alpha=(\pi)^{1/2}(l+im);\ l=0,\pm1,\pm2,\ldots;\ m=0,\pm1,\pm2,\ldots\}. \quad (3.6)$$

This fact was originally stated by von Neumann without proof. Therefore, these states are called von Neumann lattice coherent states (VNLCS). For later use we shall list some properties of VNLCS:

$$\langle\beta\,|\,\alpha\rangle = \exp[-|\alpha-\beta|^2/2 + i\,\mathrm{Im}(\beta^*\alpha)]\,,$$
$$= \exp[-(l-s)^2/2 - i\pi tl]$$
$$\times\exp[-(m-t)^2/2 + i\pi sm]\,, \qquad (3.7\mathrm{a})$$

with

$$\beta = (\pi)^{1/2}(s+it), \quad s,t = 0,\pm1,\pm2,\ldots\,, \qquad (3.7\mathrm{b})$$

$$\sum_\alpha \langle\beta\,|\,\alpha\rangle = \sum_{l=-\infty}^{\infty}\sum_{m=-\infty}^{\infty}\langle\beta\,|\,\alpha\rangle$$
$$= \vartheta_3(-\pi t/2, e^{-\pi/2})\vartheta_3(\pi s/2, e^{-\pi/2})\,, \qquad (3.7\mathrm{c})$$

$$\sum_\alpha \langle\beta\,|\,\alpha\rangle\alpha = \beta\vartheta_3(-\pi t/2, e^{-\pi/2})\vartheta_3(\pi s/2, e^{-\pi/2})\,, \qquad (3.7\mathrm{d})$$

where

$$\vartheta_3(z,q) = \sum_{n=-\infty}^{\infty} q^{n^2} e^{i2nz}\,.$$

The derivations of Eqs. (3.7c) and (3.7d) are given in Appendix A. It is important to note that the subset states of (3.6), which form a complete set of states, are not mutually orthogonal, as can be seen from Eq. (3.7a). On the other hand, the states $|\alpha_{\vec{k}\lambda}\rangle$, with different $\vec{k},\lambda$, are mutually orthogonal.

With the properties of the coherent states $|\alpha\rangle$ given in Eqs. (3.5) we can now discuss their physical qualities by considering the expectation value of the radiation field operator $\hat{\vec{A}}(\vec{r})$. For the moment we consider only coherent states belonging to a given wave vector $\vec{k}$ and polarization λ. That means the coherent states with $(\vec{k}',\lambda')\neq(\vec{k},\lambda)$ have the complex eigenvalues $\alpha_{\vec{k}'\lambda'}=0$ (i.e., these modes are in their ground states). In the interaction picture, with respect to the Hamiltonian $\hat{H}_A$ given in Eq. (2.4), the radiation field operator has the form

$$\hat{\vec{A}}(\vec{r},t) = V^{-1/2}\sum_{\vec{k},\eta} c(\hbar/2\omega_k)^{1/2}$$
$$\times(a_{\vec{k}\eta}e^{i\vec{k}\cdot\vec{r}-i\omega_k t} + a^\dagger_{\vec{k}\eta}e^{-i\vec{k}\cdot\vec{r}+i\omega_k t})\vec{\epsilon}_{\vec{k}\eta}\,, \qquad (3.8\mathrm{a})$$

where

$$|\vec{k}| = \omega_k/c\,.$$

Using the formulas (3.5), its expectation value can be obtained:

$$\langle\alpha\,|\,\hat{\vec{A}}\,|\,\alpha\rangle = c(\hbar/2\omega_k V)^{1/2}(\alpha_{\vec{k}\eta}e^{i\vec{k}\cdot\vec{r}-i\omega_k t} + \mathrm{c.c.})\,\vec{\epsilon}_{\vec{k}\eta}$$
$$= \vec{A}^{(0)}_{\vec{k}\eta}\cos(\vec{k}\cdot\vec{r} - \omega_k t + \phi_{\vec{k}\eta})\,. \qquad (3.8\mathrm{b})$$

Thus, the above coherent state $|\alpha\rangle$ corresponds to the classical radiation field $A^{(0)}_{\vec{k}\eta}\cos(\vec{k}\cdot\vec{r} - \omega_k t + \phi_{\vec{k}\eta})$. However, it remains to be examined whether the above quantum-mechanical expectation value can be approximately substituted by the corresponding classical value in any functions of A_k. For this, it must be shown that the quantum-mechanical root-mean-square deviations from the expectation values of the radiation field are relatively small. By using the formulas (3.5) we can obtain for these deviations

$$\Delta u_{\mathrm{rel}} = \Delta u(\langle\hat{u}\rangle^2 + \langle\hat{p}\rangle^2)^{-1/2}$$
$$= [\langle(\hat{u}-\langle\hat{u}\rangle)^2\rangle]^{1/2}(\langle\hat{u}\rangle^2 + \langle\hat{p}\rangle^2)^{-1/2}$$
$$= 2^{-1}(|\alpha|^2)^{-1/2} = 2^{-1}(\langle\hat{N}\rangle)^{-1/2} \qquad (3.9\mathrm{a})$$

and

$$\Delta p_{\mathrm{rel}} = \Delta p(\langle\hat{u}\rangle^2 + \langle\hat{p}\rangle^2)^{-1/2}$$
$$= [\langle(\hat{p}-\langle\hat{p}\rangle)^2\rangle]^{1/2}(\langle\hat{u}\rangle^2 + \langle\hat{p}\rangle^2)^{-1/2}$$
$$= 2^{-1}(\langle\hat{N}\rangle)^{-1/2}\,, \qquad (3.9\mathrm{b})$$

where $\langle\hat{N}\rangle$ is the number operator of the light quanta in the normalization volume V [see Eq. (2.5)]. Similarly, we obtain

$$\Delta N_{\mathrm{rel}} = \Delta N/\langle\hat{N}\rangle = (\langle\hat{N}\rangle)^{-1/2}\,. \qquad (3.9\mathrm{c})$$

Hence, we can see that if the expectation values of the light-quanta number operator for the coherent states are very large, the classical approximation can be a good approximate substitute to the quantum-mechanical description.

To obtain an impression for the magnitude of these deviations for macroscopic dimensions, let us consider the radiation field that corresponds to the classical field of the wavelength 1 cm having the volume of 10^3 cm^3 and the energy density 1 erg/cm^3—the belonging absolute value of the field strength is 1 G. For such a state $\langle\hat{N}\rangle\approx$ energy/$\hbar\omega$ becomes about 5×10^{18} and therefore Δu_{rel}, Δp_{rel}, and ΔN_{rel} become about 5×10^{-10}, which is indeed very small. It should be noted that ΔN itself is proportional to $(\langle\hat{N}\rangle)^{1/2}$ and therefore also becomes large in the classical case, i.e., only the relative deviations Δu_{rel}, Δp_{rel}, and ΔN_{rel} become small in this limit case.

Up to now we have discussed the physical properties of the single-mode radiation-field coherent state specified by the complex $\alpha = \alpha_{\vec{k}\eta}$. We can straightforwardly generalize the above discussion to arbitrary radiation fields by taking into account all modes. If we form the direct product of single-mode coherent states with a set of $\alpha_{\vec{k}\eta}$, which we write as $\{\alpha\}$,

$$|\{\alpha\}\rangle = \prod_{\vec{k},\eta}|\alpha_{\vec{k}\eta}\rangle\,, \qquad (3.10)$$

then such a state gives the expectation value

$$\langle\{\alpha\}|\hat{\vec{A}}(\vec{r},t)|\{\alpha\}\rangle = V^{-1/2}\sum_{\vec{k},\eta} c(\hbar/2\omega_k)^{1/2}(\alpha_{\vec{k}\eta}e^{i\vec{k}\cdot\vec{r}-i\omega_k t}+\text{c.c.})\vec{\epsilon}_{\vec{k}\eta}\langle\{\alpha\}|\{\alpha\}\rangle$$
$$=\sum_{\vec{k},\eta} A^{(0)}_{\vec{k}\eta}\cos(\vec{k}\cdot\vec{r}-\omega_k t+\phi_{\vec{k}\eta}) . \qquad (3.11)$$

Here we have used Eqs. (3.5b), (3.5c), and (3.8b). Thus, by choosing appropriate $\{\alpha\}$ we can construct a coherent state $|\{\alpha\}\rangle$ which corresponds to a given classical field, which is expressed in general as a linear superposition of electromagnetic plane waves.

One interesting fact should be noted: To construct a quantum-mechanical wave-packet state for a free massive particle one has to superpose linearly plane-wave states with different wave vectors $\vec{k}$, as is well known. On the contrary, to construct a quantum-mechanical state of the radiation field which corresponds to a classical electromagnetic wave packet one has to make a direct product of many single-mode states and not a superposition of such states.

IV. REFORMULATION OF THE SCHRÖDINGER EQUATION FOR THE ELECTRON-LIGHT-QUANTA SYSTEM

In this section we shall reformulate the electron–light-quanta Schrödinger equation using the projection equation method[8] with the VNLCS for the light quanta as basis states. With respect to the atomic electron we remain in the usual x representation, i.e., we use as basis states the continuous electron coordinate eigenstates. First we write down the Schrödinger equation (2.3) in the mixed picture, which is the interaction picture with regard to the radiation field and is the Schrödinger picture in the x representation with regard to the atomic electron:

$$[\hat{H}_e+\hat{H}_{\text{int}}(t)]|\Psi(t)\rangle = i\hbar\frac{\partial}{\partial t}|\Psi(t)\rangle \qquad (4.1a)$$

with

$$\hat{H}_{\text{int}}(t) = -(e/mc)\exp(i\hat{H}_A t/\hbar)\hat{\vec{A}}\exp(-i\hat{H}_A t/\hbar)\cdot\hat{\vec{p}}_e . \qquad (4.1b)$$

The Schrödinger equation (3.1a) can now be rewritten in the projection form

$$\left\langle\delta\Psi(t)\middle|\left[\hat{H}_e+\hat{H}_{\text{int}}(t)-i\hbar\frac{\partial}{\partial t}\right]\middle|\Psi(t)\right\rangle = 0 . \qquad (4.2)$$

If $\langle\delta\psi(t)|$ represents an arbitrary variation at every time t in the whole Hilbert space, then Eq. (4.2) is equivalent to Eq. (4.1a).

The great advantage of the Schrödinger equation written in the projection form is its flexibility to allow various possibilities of basis states. Thus we can introduce a desirable physical ansatz into the formalism from the very beginning in a quite natural manner. That is, we can introduce a superposition of basis states with linear variational amplitudes, which can be adapted to the physical conditions, i.e., the boundary conditions, of the system considered. It is important to note that those states do not have to be mutually orthogonal but only have to be linearly independent.

The nondegenerate stationary solutions of Eq. (4.2) are mutually orthogonal, whatever basis states are chosen. This is so because if the basis states form a complete set of states in the belonging Hilbert space, the exact solution of the Schrödinger equation (4.1a) can be obtained from Eq. (4.2). This orthogonality of the solutions of Eq. (4.2) also remains valid when one constructs approximate solutions of Eq. (4.1a) by restricting the basis states to be used in Eq. (4.2). For details see Ref. 8.

After these general remarks about the Schrödinger equation written in the projection form, let us now return to the discussion of the electron–light-quanta system. Since our purpose is to obtain a classical description of the radiation field as an approximation to the quantized field, the coherent states are suitable as basis states for the radiation field. As already mentioned in Sec. III, in order to use the projection form of the Schrödinger equation we introduce the VNLCS as basis states for the radiation field:

$$|\{\alpha\}\rangle = \prod_{\vec{k},\eta}|\alpha_{\vec{k}\eta}\rangle , \qquad (4.3a)$$

where $|\alpha_{\vec{k}\eta}\rangle$ is the VNLCS for the mode $(\vec{k},\eta)$ with

$$\alpha_{\vec{k}\eta} = (\pi)^{1/2}(l_{\vec{k}\eta}+im_{\vec{k}\eta}),\quad l_{\vec{k}\eta}, m_{\vec{k}\eta} = \text{integers}. \qquad (4.3b)$$

For the electron we use the coordinate eigenstate $|\vec{r}'\rangle = \delta(\vec{r}-\vec{r}')$ with the orthonormal relation

$$\langle\vec{r}''|\vec{r}'\rangle = \int\delta(\vec{r}''-\vec{r})\delta(\vec{r}-\vec{r}')d\vec{r}$$
$$=\delta(\vec{r}''-\vec{r}') . \qquad (4.4)$$

Then, the basis states for the whole system are the direct products

$$|\{\alpha\},\vec{r}'\rangle = |\{\alpha\}\rangle\otimes|\vec{r}'\rangle , \qquad (4.5)$$

which span the whole Hilbert space. Using these basis states we now make the following generalized

Hill-Wheeler ansatz[8] for the states of the whole system:

$$|\Psi(t)\rangle = \sum_{\{\alpha\}} \int d\vec{r}' \, |\{\alpha\}, \vec{r}'\rangle f(\{\alpha\}, \vec{r}'; t), \tag{4.6}$$

where the notation is defined as

$$\sum_{\{\alpha\}} = \sum_{\alpha_{\vec{k}_1\eta_1}} \sum_{\alpha_{\vec{k}_2\eta_2}} \cdots = \prod_{\vec{k},\eta}\Big(\sum_{\alpha_{\vec{k}\eta}}\Big)$$

$$= \prod_{\vec{k},\eta}\Big(\sum_{l=-\infty}^{\infty}\sum_{m=-\infty}^{\infty}\Big), \tag{4.7a}$$

with

$$\alpha_{\vec{k}\eta} = (\pi)^{1/2}(l + im). \tag{4.7b}$$

The time-dependent linear amplitude $f(\{\alpha\}, \vec{r}'; t)$, which has to be varied arbitrarily at every time t and at every point $\vec{r}'$ for every multiple $\{\alpha\}$, is a continuous function of the parameter coordinate $\vec{r}'$ and a discrete function of the discrete values of the multiple $\{\alpha\}$.

If we introduce the ansatz Eq. (4.6) into the projection equation (4.2), we then obtain after integrating over $\vec{r}'$

$$\sum_{\{\alpha\}} \int d\vec{r}' \langle\{\beta\}, \vec{r}''|[\hat{H}_e + \hat{H}_{\text{int}}(t)]|\{\alpha\}, \vec{r}'\rangle f(\{\alpha\}, \vec{r}'; t) = i\hbar\frac{\partial}{\partial t}\sum_{\{\alpha\}} \int d\vec{r}' \langle\{\beta\}, \vec{r}''|\{\alpha\}, \vec{r}'\rangle f(\{\alpha\}, \vec{r}'; t). \tag{4.8}$$

Using Eqs. (2.2), (4.1b), and (4.7a) we can write down the three kernels appearing in Eq. (4.8) explicitly:

$$\langle\{\beta\}, \vec{r}''|\{\alpha\}, \vec{r}'\rangle = \langle\{\beta\}|\{\alpha\}\rangle\langle\vec{r}''|\vec{r}'\rangle = \langle\{\beta\}|\{\alpha\}\rangle\delta(\vec{r}'' - \vec{r}'), \tag{4.9}$$

$$\langle\{\beta\}, \vec{r}''|\hat{H}_e|\{\alpha\}, \vec{r}'\rangle = \langle\{\beta\}|\{\alpha\}\rangle\langle\vec{r}''|\hat{H}_e|\vec{r}'\rangle = \langle\{\beta\}|\{\alpha\}\rangle\delta(\vec{r}'' - \vec{r}')[(-\hbar^2/2m)\vec{\nabla}_{r'}^2 + V(\vec{r}')], \tag{4.10}$$

$$\begin{aligned}\langle\{\beta\}, \vec{r}''|\hat{H}_{\text{int}}(t)|\{\alpha\}, \vec{r}'\rangle = V^{-1/2}\sum_{\vec{k},\eta} c(\hbar/2\omega_k)^{1/2}(\langle\{\beta\}|a_{\vec{k}\eta}|\{\alpha\}\rangle e^{i\vec{k}\cdot\vec{r}' - i\omega_k t}\\ + \langle\{\beta\}|a^\dagger_{\vec{k}\eta}|\{\alpha\}\rangle e^{-i\vec{k}\cdot\vec{r}' + i\omega_k t})\vec{\epsilon}_{\vec{k}\eta}\langle\vec{r}''|\vec{p}_e|\vec{r}'\rangle(-e/mc)\\ = V^{-1/2}\sum_{\vec{k},\eta} c(\hbar/2\omega_k)^{1/2}(\alpha_{\vec{k}\eta} e^{i\vec{k}\cdot\vec{r}' - i\omega_k t} + \beta^*_{\vec{k}\eta} e^{-i\vec{k}\cdot\vec{r}' + i\omega_k t})\vec{\epsilon}_{\vec{k}\eta}\\ \times(-e/mc)\langle\{\beta\}|\{\alpha\}\rangle\delta(\vec{r}'' - \vec{r}')(-i\hbar)\vec{\nabla}_{r'}\\ = \langle\{\beta\}|\{\alpha\}\rangle\vec{A}(\{\beta\}, \{\alpha\}, \vec{r}'; t)(-e/mc)\cdot\delta(\vec{r}'' - \vec{r}')(-i\hbar)\vec{\nabla}_{r'},\end{aligned} \tag{4.11}$$

where $\vec{A}(\{\beta\}, \{\alpha\}, \vec{r}'; t)$ is defined as

$$\vec{A}(\{\beta\}, \{\alpha\}, \vec{r}'; t) = V^{-1/2}\sum_{\vec{k},\eta} c(\hbar/2\omega_k)^{1/2}(\alpha_{\vec{k}\eta} e^{i\vec{k}\cdot\vec{r}' - i\omega_k t} + \beta^*_{\vec{k}\eta} e^{-i\vec{k}\cdot\vec{r}' + i\omega_k t})\vec{\epsilon}_{\vec{k}\eta}. \tag{4.12}$$

Introducing Eqs. (4.9), (4.10), and (4.11) into Eq. (4.8) and integrating over r' we obtain

$$\sum_{\{\alpha\}} \langle\{\beta\}|\{\alpha\}\rangle\left[-(\hbar^2/2m)\nabla_{r''}^2 + V(r'') + (-e/mc)\vec{A}(\{\beta\}, \{\alpha\}, \vec{r}''; t)\cdot\frac{\hbar}{i}\vec{\nabla}_{r''} - i\hbar\frac{\partial}{\partial t}\right] f(\{\alpha\}, \vec{r}''; t) = 0. \tag{4.13}$$

Since the basis states (4.5) are complete, Eq. (4.13) is equivalent to the Schrödinger equation (4.1a). That means until now we have not introduced any approximation.

V. CLASSICAL APPROXIMATION TO THE SCHRÖDINGER EQUATION

In the previous section we have obtained the Schrödinger equation in the projection form (4.13) using the coherent states given by Eq. (4.3a) for the radiation field. In the projection equation our coherent-states basis effectively resulted in the kernel, which is an inner product of two coherent states. In this section we shall develop a systematic way of obtaining a classical approximation to the projection equation (4.13) based on the properties of this kernel.

First we assume that the amplitude $f(\{\alpha\}, \vec{r}'; t)$ given by Eq. (4.6) varies very slowly with regard to the change of the variable $\{\alpha\}$ over the range of order $(\pi)^{1/2}$. This means if the square root of the corresponding light-quanta-number expectation value $\sum\langle\hat{N}_{\vec{k}\lambda}\rangle = \sum\alpha^*_{\vec{k}\lambda}\alpha_{\vec{k}\lambda}$ varies by a few units $f(\{\alpha\}, \vec{r}'; t)$ practically does not change [see Eq. (3.5d)]. In the classical limit region this is certainly the case (see the discussion in the second part of Sec. III). For the amplitudes which satisfy this assumption we can introduce a classical approximation into the Schrödinger equation (4.13) written in the projection form by using the property of the kernel.

The essential feature of the projection equation

(4.13) for the present purpose can be represented by the sum

$$\sum_{\{\alpha\}} \langle\{\beta\}\,|\{\alpha\}\rangle\vec{A}(\{\beta\},\{\alpha\})f(\{\alpha\}), \tag{5.1}$$

where the electron variables have been dropped, since we are interested in the classical approximation to the radiation field and the electron state is not subjected to such an approximation. The kernel $\langle\{\beta\}\,|\{\alpha\}\rangle$ is just the product of single-mode kernels $\langle\beta_{\vec{k}\eta}|\alpha_{\vec{k}\eta}\rangle$, which have Gaussian forms centered at $\alpha_{\vec{k}\eta}=\beta_{\vec{k}\eta}$. The widths of these Gaussian forms are of the order of $(\pi)^{1/2}$. Thus the kernel $\langle\{\beta\}|\{\alpha\}\rangle$ has a strongly localized peak at the point $\{\alpha\}=\{\beta\}$ in the $\{\alpha\}$ space. Therefore, for the microscopically slow varying $f(\{\alpha\})$, the above sum can be approximated as

$$\begin{aligned}\sum_{\{\alpha\}} \langle\{\beta\}\,|\{\alpha\}\rangle\vec{A}(\{\beta\},\{\alpha\})f(\{\alpha\}) &= \sum_{\alpha_{\vec{k}\eta}}\cdots\sum_{\alpha_{\vec{k}'\eta'}}\langle\beta_{\vec{k}\eta}|\alpha_{\vec{k}\eta}\rangle\cdots\langle\beta_{\vec{k}'\eta'}|\alpha_{\vec{k}'\eta'}\rangle\cdots\vec{A}(\{\beta\},\{\alpha\})f(\{\alpha\})\\ &\approx f(\{\beta\})\sum_{\alpha_{\vec{k}\eta}}\cdots\sum_{\alpha_{\vec{k}'\eta'}}\cdots\langle\beta_{\vec{k}\eta}|\alpha_{\vec{k}\eta}\rangle\cdots\langle\beta_{\vec{k}'\eta'}|\alpha_{\vec{k}'\eta'}\rangle\cdots\vec{A}(\{\beta\},\{\alpha\})\\ &\approx f(\{\beta\})\sum_{\{\alpha\}}\langle\{\beta\}\,|\{\alpha\}\rangle\vec{A}(\{\beta\},\{\alpha\}).\end{aligned} \tag{5.2}$$

It should be noted here that the correction to this approximation can be calculated by representing $f(\{\alpha\})$ in the form of the Taylor expansion around $\{\alpha\}=\{\beta\}$.

Applying this approximation to the left-hand side of the projection equation (4.13), we obtain

$$\begin{aligned}&\sum_{\{\alpha\}}\langle\{\beta\}\,|\{\alpha\}\rangle\left[-(\hbar^2/2m)\nabla_{r'}^2+V(r')+(-e/mc)\vec{A}(\{\beta\},\{\alpha\},\vec{r}';t)\cdot(-i\hbar)\vec{\nabla}_{r'}-i\hbar\frac{\partial}{\partial t}\right]f(\{\alpha\},\vec{r}';t)\\ &\approx\left[-(\hbar^2/2m)\nabla_{r'}^2+V(r')-i\hbar\frac{\partial}{\partial t}\right]f(\{\beta\},r';t)\sum_{\{\alpha\}}\langle\{\beta\}\,|\{\alpha\}\rangle\\ &\qquad+[(-e/mc)(-i\hbar)\vec{\nabla}_{r'}f(\{\beta\},\vec{r}';t)]\sum_{\{\alpha\}}\langle\{\beta\}\,|\{\alpha\}\rangle\vec{A}(\{\beta\},\{\alpha\},\vec{r}';t).\end{aligned} \tag{5.3}$$

In order to calculate the two summations on the right-hand side of Eq. (5.3) let us generalize the formulas (3.7c) and (3.7d) to the many-mode case:

$$\begin{aligned}\sum_{\{\alpha\}}\langle\{\beta\}\,|\{\alpha\}\rangle &= \prod_{\vec{k},\eta}\sum_{\alpha_{\vec{k}\eta}}\langle\beta_{\vec{k}\eta}|\alpha_{\vec{k}\eta}\rangle\\ &=\prod_{\vec{k},\eta}\vartheta_3(-\pi t_{\vec{k}\eta}/2,e^{-\pi/2})\vartheta_3(\pi s_{\vec{k}\eta}/2,e^{-\pi/2})\\ &\equiv C(\{\beta\}),\end{aligned} \tag{5.4}$$

where

$$\beta_{\vec{k},\eta}\equiv(\pi)^{1/2}(s_{\vec{k}\eta}+it_{\vec{k}\eta}) \tag{5.5}$$

and

$$\begin{aligned}\sum_{\{\alpha\}}\langle\{\beta\}\,|\{\alpha\}\rangle\alpha_{\vec{p}\xi} &= \left[\prod_{\vec{k},\eta\neq\vec{p},\xi}\sum_{\alpha_{\vec{k}\eta}}\langle\beta_{\vec{k}\eta}|\alpha_{\vec{k}\eta}\rangle\right]\sum_{\alpha_{\vec{p}\xi}}\langle\beta_{\vec{p}\xi}|\alpha_{\vec{p}\xi}\rangle\alpha_{\vec{p}\xi}\\ &=\beta_{\vec{p}\xi}\prod_{\vec{k},\eta}\vartheta_3(-\pi t_{\vec{k}\eta}/2,e^{-\pi/2})\vartheta_3(\pi s_{\vec{k}\eta}/2,e^{-\pi/2})\\ &=\beta_{\vec{p}\xi}C(\{\beta\}).\end{aligned} \tag{5.6}$$

Using Eq. (5.6), the summation in the second term on the right-hand side of Eq. (5.3) can be calculated as

$$\begin{aligned}\sum_{\{\alpha\}}\langle\{\beta\}\,|\{\alpha\}\rangle\vec{A}(\{\beta\},\{\alpha\},\vec{r}';t) &= V^{-1/2}\sum_{\vec{k},\eta}c(\hbar/2\omega_k)^{1/2}\left(e^{i\vec{k}\cdot\vec{r}'-i\omega_k t}\sum_{\{\alpha\}}\langle\{\beta\}\,|\{\alpha\}\rangle\alpha_{\vec{k}\eta}+e^{-i\vec{k}\cdot\vec{r}'+i\omega_k t}\beta^*_{\vec{k}\eta}\sum_{\{\alpha\}}\langle\{\beta\}\,|\{\alpha\}\rangle\right)\vec{\epsilon}_{\vec{k}\eta}\\ &=V^{-1/2}\sum_{\vec{k},\eta}c(\hbar/2\omega_k)^{1/2}(\beta_{\vec{k}\eta}e^{i\vec{k}\cdot\vec{r}'-i\omega_k t}+\beta^*_{\vec{k}\eta}e^{-i\vec{k}\cdot\vec{r}'+i\omega_k t})\vec{\epsilon}_{\vec{k}\eta}C(\{\beta\})\\ &=\vec{A}(\{\beta\},\{\beta\},\vec{r}';t)C(\{\beta\}).\end{aligned} \tag{5.7}$$

Introducing Eqs. (5.3), (5.4), and (5.7) into Eq. (4.13) we obtain the classical approximation to the projection equation (4.13):

$$\left[-(\hbar^2/2m)\nabla_{r'}^2+V(r')+(-e/mc)\vec{A}(\{\beta\},\{\beta\},\vec{r}';t)\cdot(-i\hbar)\vec{\nabla}_{r'}-i\hbar\frac{\partial}{\partial t}\right]f(\{\beta\},\vec{r}';t)C(\{\beta\})=0\,. \tag{5.8}$$

This equation must be fulfilled for all values of the multiple $\{\beta\}$. This comes from the approximation Eq. (5.2) and means physically that the reaction of the electron on the radiation field is neglected, because coherent states of different $\{\beta\}$ values are not coupled with each other. Because the operator on the right-hand side of Eq. (5.8) is a self-adjoint operator, the conservation law

$$\int d\vec{r}''\,|f(\{\beta\},\vec{r}'';t)|^2=\text{const in time} \tag{5.9}$$

is valid.

To connect the above equation [(5.8)] to the corresponding classical description we must introduce another requirement for $f(\{\beta\},\vec{r}';t)$. So far, $f(\{\beta\},\vec{r}';t)$ has been required to be only microscopically slow varying. Now we further assume that $f(\{\beta\},\vec{r}';t)$ is a macroscopically well-localized wave packet in the $\{\beta\}$ space, for which the absolute spreads $\Delta\beta_{\vec{k}\eta}$ around $\beta_{\vec{k}\eta}=\beta^{cl}_{\vec{k}\eta}$ are much larger than $(\pi)^{1/2}$—see the discussion to Eq. (3.9c)—but their relative spreads $\Delta\beta_{\vec{k}\eta}/|\beta^{cl}_{\vec{k}\eta}|$ are much smaller than 1. The second condition demands that the root-mean-square deviations of $\vec{A}$ from $\langle\vec{A}\rangle$ are very small. In the classical limit certainly this must be the case.

As already mentioned, Eq. (5.8) must be fulfilled for all values of the multiple $\{\beta\}$. However, the second requirement on $f(\{\beta\},\vec{r}';t)$ means that it is a macroscopically well-localized wave packet in the $\{\beta\}$ space around the point $\{\beta\}=\{\beta^{cl}\}$. Therefore, we have to consider only the case $\{\beta\}\approx\{\beta^{cl}\}$. Because $\Delta\beta_{\vec{k}\eta}/|\beta^{cl}_{\vec{k}\eta}|$ is much smaller than one, for $\{\beta\}\approx\{\beta^{cl}\}$ we may approximate $A(\{\beta\},\{\beta\},\vec{r}';t)$ by $A(\{\beta^{cl}\},\{\beta^{cl}\},\vec{r}';t)$. Then, from Eq. (5.8) we obtain for $\{\beta\}\approx\{\beta^{cl}\}$

$$\left[-(\hbar^2/2m)\nabla_{r'}^2+V(r')+(-e/mc)\vec{A}(\{\beta^{cl}\},\{\beta^{cl}\},\vec{r}',t)\cdot(-i\hbar)\vec{\nabla}_{r'}-i\hbar\frac{\partial}{\partial t}\right]f(\{\beta\},\vec{r}';t)C(\{\beta\})=0\,. \tag{5.10}$$

Here, the operator $[\cdots]$ no longer depends on $\{\beta\}$. Therefore, for $f(\{\beta\},\vec{r}';t)$ we can make the following ansatz:

$$f(\{\beta\},\vec{r}';t)=g(\{\beta\}-\{\beta^{cl}\})\psi'(\{\beta^{cl}\},\vec{r}',t)\,, \tag{5.11}$$

where $g(\{\beta\}-\{\beta^{cl}\})$ is nonvanishing only for $\{\beta\}\approx\{\beta^{cl}\}$, reflecting the macroscopic requirement on $f(\{\beta\},\vec{r}';t)$. We can also simply drop the nonvanishing factor $C(\{\beta\})$ from Eq. (5.10). In order to connect the above introduced ψ' to the electron wave function we must introduce the proper normalization. From Eq. (4.6), using the same approximation as Eq. (5.2), we get

$$1=\langle\Psi(t)\,|\,\Psi(t)\rangle$$
$$\approx\sum_{\{\beta\}}\int d\vec{r}''\,|f(\{\beta\},r';t)|^2C(\{\beta\})\,. \tag{5.12}$$

Introducing the ansatz (5.11), this can be written as

$$1\approx N(\{\beta^{cl}\})\int d\vec{r}'\,|\psi'(\{\beta^{cl}\},\vec{r}';t)|^2\,, \tag{5.13}$$

where

$$N(\{\beta^{cl}\})=\sum_{\{\beta\}}|g(\{\beta\}-\{\beta^{cl}\})|^2C(\{\beta\})\,. \tag{5.14}$$

Using this $N(\{\beta^{cl}\})$ we can define the properly normalized electron wave function ψ:

$$\psi(\{\beta^{cl}\},\vec{r}',t)=[N(\{\beta^{cl}\})]^{-1/2}\psi'(\{\beta^{cl}\},\vec{r}',t)\,. \tag{5.15}$$

From Eqs. (5.10), (5.11), and (5.15) it is obvious that this ψ obeys the equation

$$\left[-(\hbar^2/2m)\nabla_{r'}^2+V(r')+(-e/mc)\vec{A}(\{\beta^{cl}\},\{\beta^{cl}\},\vec{r}';t)\cdot(-i\hbar)\vec{\nabla}_{r'}-i\hbar\frac{\partial}{\partial t}\right]\psi(\{\beta^{cl}\},\vec{r}',t)=0\,, \tag{5.16}$$

which is the Schrödinger equation for the electron wave function ψ with the classical radiation field $\vec{A}(\{\beta^{cl}\},\{\beta^{cl}\},\vec{r}_A;t)$.

Now let us show that $\vec{A}(\{\beta^{cl}\},\{\beta^{cl}\},\vec{r}_A;t)$ is nothing else but the expectation value of the radiation field operator at the position $\vec{r}_A$ for the wave packet in the $\{\beta\}$ space, which is given by Eqs. (4.6) and (5.11). Using Eqs. (4.6), (4.11), (5.4), (5.11), (5.14), and (5.15) and introducing the same approximations that are used to obtain Eqs. (5.2) and (5.10), we can readily show

$$\langle\hat{\vec{A}}(\vec{r}_A,t)\rangle \approx \vec{A}(\{\beta^{cl}\},\{\beta^{cl}\},\vec{r}_A;t)N(\{\beta^{cl}\}) \times\int d\vec{r}'\,|\psi'(\{\beta^{cl}\},\vec{r}',t)|^2$$
$$\approx \vec{A}(\{\beta^{cl}\},\{\beta^{cl}\},\vec{r}_A;t)\int d\vec{r}'\,|\psi(\{\beta^{cl}\},\vec{r}',t)|^2$$
$$\approx \vec{A}(\{\beta^{cl}\},\{\beta^{cl}\},\vec{r}_A;t)\,. \qquad (5.17)$$

Now we have completed the derivation of the semiclassical time-dependent Schrödinger equation, because $A(\{\beta^{cl}\},\{\beta^{cl}\},\vec{r}_A,t)$ describes the classical radiation field, as can be seen from Eq. (5.17), and ψ is the proper normalized electron wave function due to Eq. (5.15).

It should be noted that the classical approximation scheme discussed here has been carried out for a group of wave-packet states for the radiation field in the $\{\beta\}$ space and not for a specific wave-packet state. Every wave-packet state which belongs to this group is characterized to be microscopically slow varying and macroscopically well localized in the $\{\beta\}$ space.

VI. CONCLUDING REMARKS

One essential feature of our approach is the use of the projection form of the Schrödinger equation (4.13), because then one can straightforwardly introduce many kinds of basis states which are appropriate to the problem. This can be done in any kind of particle quantum mechanics and relativistic or nonrelativistic quantum field theories. It is important to note that these basis states have to be linearly independent.[13] The nondegenerate solutions of the Schrödinger equation (4.13) are then automatically orthogonal. This remains true even if one uses only a subspace of the complete Hilbert space.

Another essential feature of our approach is the use of the VNLCS, which represents a nonorthogonal complete set of coherent states, as basis states for the description of the radiation field. If we use the conventional overcomplete continuous coherent states, then the linear dependences between them destroy the application of the projection form of the Schrödinger equation (see Appendix B).

Next, we shall derive from the fully quantized theory the Abraham-Lorentz equation with radiation damping of the charged Schrödinger particle including quantum corrections. As already mentioned in Sec. V, for this purpose one has to expand $f(\{\alpha\},r;t)$ into a Taylor series around $\{\beta\}$ [see Eqs. (5.2) and (5.3)].

ACKNOWLEDGMENTS

We wish to thank Professor W. Dittrich and Dr. B. Englert for bringing Ref. 7 to our attention. We also wish to thank Dr. R. Sartor, Dr. K. Shimizu, and Dr. E. J. Kannelopoulos for helpful discussions. This work was supported by Deutsche Forschungsgemeinschaft.

APPENDIX A

Equation (3.7c) is obtained as follows:

$$\sum_\alpha\langle\beta|\alpha\rangle=\sum_{l=-\infty}^{\infty}\sum_{m=-\infty}^{\infty}\exp[-\pi(l-s)^2/2-i\pi tl]\exp[-\pi(m-t)^2/2+i\pi sm]$$
$$=\sum_{l=-\infty}^{\infty}\exp[-\pi l^2/2-i\pi t(l+s)]\sum_{m=-\infty}^{\infty}\exp[-\pi m^2/2+i\pi s(m+t)]$$
$$=e^{-i\pi st}\vartheta_3(-\pi t/2,e^{-\pi/2})e^{i\pi ts}\vartheta_3(\pi s/2,e^{-\pi/2})$$
$$=\vartheta_3(-\pi t/2,e^{-\pi/2})\vartheta_3(\pi s/2,e^{\pi/2})\,, \qquad (A1)$$

where the ϑ_3 function is defined as

$$\vartheta_3(z,q)=\sum_{n=-\infty}^{\infty}q^{n^2}e^{i2nz}\,.$$

The second sum, Eq. (3.7d), is

$$\sum_\alpha\langle\beta|\alpha\rangle\alpha=\sum_{l=-\infty}^{\infty}\sum_{m=-\infty}^{\infty}(\pi)^{1/2}(l+im)\exp[-\pi(l-s)^2/2-i\pi tl]\exp[\pi(m-t)^2/2+i\pi sm]$$
$$=(\pi)^{1/2}\sum_{l=-\infty}^{\infty}l\exp[-\pi(l-s)^2/2-i\pi tl]\sum_{m=-\infty}^{\infty}\exp[-\pi(m-t)^2/2+i\pi sm]$$
$$+i(\pi)^{1/2}\sum_{l=-\infty}^{\infty}\exp[-\pi(l-s)^2/2-i\pi tl]\sum_{m=-\infty}^{\infty}m\exp[-\pi(m-t)^2/2+i\pi sm]\,. \qquad (A2)$$

The two summations which contain the linear factors l and m can be calculated as follows:

$$\sum_{l=-\infty}^{\infty} l\exp[-\pi(l-s)^2/2-i\pi tl]=\sum_{l=-\infty}^{\infty}(l+s)\exp[-\pi l^2/2-i\pi t(l+s)]$$

$$=\sum_{l=-\infty}^{\infty} l\exp[-\pi l^2/2-i\pi t(l+s)]+se^{-i\pi ts}\sum_{l=-\infty}^{\infty}\exp(-\pi l^2/2-i\pi tl). \quad (A3)$$

Since tl is an integer, the first term vanishes. Then,

$$\sum_{l=-\infty}^{\infty} l\exp[-\pi(l-s)^2/2-i\pi tl]=se^{-its}\vartheta_3(-\pi t/2,e^{-\pi/2}). \quad (A4)$$

Similarly, we can obtain

$$\sum_{m=-\infty}^{\infty} m\exp[-\pi(m-t)^2/2+i\pi sm]=te^{i\pi st}\vartheta_3(\pi s/2,e^{-\pi/2}). \quad (A5)$$

Using Eqs. (A1), (A4), and (A5) we find from Eq. (A2)

$$\sum_{\alpha}\langle\beta|\alpha\rangle\,\alpha=(\pi)^{1/2}se^{-\pi ts}\vartheta_3(-\pi t/2,e^{-\pi/2})e^{i\pi ts}\vartheta_3(\pi s/2,e^{-\pi/2})$$

$$+i(\pi)^{1/2}e^{-i\pi ts}\vartheta_3(-\pi t/2,e^{-\pi/2})te^{i\pi st}\vartheta_3(\pi s/2,e^{-\pi/2})$$

$$=(\pi)^{1/2}(s+it)\vartheta_3(-\pi t/2,e^{-\pi/2})\vartheta_3(\pi s/2,e^{-\pi/2}) \quad (A6)$$

$$=\beta\sum_{\alpha}\langle\beta|\alpha\rangle. \quad (A7)$$

APPENDIX B

If we use the conventional overcomplete continuous coherent states, we encounter the integral

$$\int d^2\alpha\langle\beta|\alpha\rangle\alpha=(2\pi)^{-1}\int_{-\infty}^{\infty}d(\mathrm{Re}\alpha)\int_{-\infty}^{\infty}d(\mathrm{Im}\alpha)\exp[-|\beta-\alpha|^2/2+i\,\mathrm{Im}(\beta^*\alpha)]\alpha \quad (B1)$$

instead of the discrete summation appearing in Eq. (5.1) for the VNLCS. Unfortunately, the above integral vanishes:

$$\int d^2\alpha\langle\beta|\alpha\rangle\alpha=(2\pi)^{-1}\int_{-\infty}^{\infty}ds\int_{-\infty}^{\infty}dt\exp[-(s-u)^2/2-iws]\exp[-(t-w)^2/2+iut](s+it)$$

$$=2\pi[(u-iw)+i(w+iu)]\exp[-(w^2+u^2)]=0, \quad (B2)$$

where we have set $\alpha=s+it$ and $\beta=u+iw$. Obviously, this leads to an unphysical approximation. In contrast to this, if we adopt the VNLCS, we get the nonvanishing discrete summations (5.4) and (5.6).

---

[1] See, for example, R. Jackiw, Rev. Mod. Phys. 49, 681 (1977); A. O. Barut, *Foundations of Radiation Theory and Quantum Electrodynamics* (Plenum, New York, 1980).

[2] See, for example, J. R. Klauder, in *Path Integrals*, edited by G. J. Papadopoulos and J. T. Devreese (Plenum, New York, 1978), p. 5; see also references cited therein; J. R. Klauder, Acta Phys. Austriaca Suppl. XVIII, 1 (1977).

[3] C. L. Hammer *et al.*, Phys. Rev. D 18, 373 (1978); 19, 667 (1979).

[4] M. Bolsterli, Phys. Rev. D 13, 1727 (1976); 16, 1749 (1977).

[5] T. Toyoda, Phys. Lett 71A, 399 (1979); 74A, 167 (1979).

[6] See, for example, Chap. 7 of Ref. 12; V. Bargmann, Commun. Pure Appl. Math. 14, 187 (1961); A. O. Barut and L. Ghiraradello, Commun. Math. Phys. 21, 41 (1971); M. M. Nieto and L. M. Simmons, Jr., Phys. Rev. D 20, 1321 (1979).

[7] V. Bargmann *et al.*, Rep. Math. Phys. 2, 221 (1971); A. M. Perelomov, Theor. Math. Phys. (USSR) 6, 156 (1971).

[8] K. Wildermuth and V. C. Tang, *A Unified Theory of the Nucleus* (Vieweg, Braunschweig, 1977).

[9] D. L. Hill and J. A. Wheeler, Phys. Rev. 89, 1102 (1953); J. J. Griffin and J. A. Wheeler, *ibid.* 108, 311 (1957).

[10] The fact that the longitudinal electrostatic potential $V(r)$ can always be treated as a c number is shown, for example, in J. J. Sakurai, *Advanced Quantum Mechanics* (Addison-Wesley, Reading, Massachusetts, 1967).

[11] E. Schrödinger, Naturwissenschaften 14, 664 (1926).

[12] See, for example, J. R. Klauder and E. C. G. Sudarshan, *Quantum Optics* (Benjamin, New York, 1967).

[13] See Secs. 8.3 and 9.2c of Ref. 8.

Large N limits as classical mechanics

Laurence G. Yaffe*

California Institute of Technology, Pasadena, California 91125

This paper discusses the sense in which the large N limits of various quantum theories are equivalent to classical limits. A general method for finding classical limits in arbitrary quantum theories is developed. The method is based on certain assumptions which isolate the minimal structure any quantum theory should possess if it is to have a classical limit. In any theory satisfying these assumptions, one can generate a natural set of generalized coherent states. These coherent states may then be used to construct a classical phase space, derive a classical Hamiltonian, and show that the resulting classical dynamics is equivalent to the limiting form of the original quantum dynamics. This formalism is shown to be applicable to the large N limits of vector models, matrix models, and gauge theories. In every case, one can explicitly derive a classical action which contains the complete physics of the $N = \infty$ theory. "Solving" the $N = \infty$ theory requires minimizing the classical Hamiltonian, and this has been possible only in simple theories. The relation between this approach and other methods which have been proposed for deriving large N limits is discussed in detail.

CONTENTS

I. INTRODUCTION

Many quantum theories possess natural generalizations in which the number of degrees of freedom is a free parameter. If N is some measure of the number of dynamical variables, then for a wide class of these theories the $N\to\infty$ limit is known to simplify the dynamics dramatically. This is true in theories ranging from the quantum mechanics of a point particle moving, in an N-dimensional central potential, to quantum spin models with spin N quantum spins, to quantum field theories containing SU(N) gauge fields. If the $N\to\infty$ limit of such a theory can be explicitly solved, then a systematic expansion in powers of $1/N$ can provide a very useful approach for studying the original finite N theory.[1]

Much of the recent interest in large N expansions is motivated by the desire to find reliable methods for analyzing the dynamics of quantum chromodynamics (QCD). One may introduce a parameter N into QCD by replacing the SU(3) gauge group with SU(N) ('t Hooft, 1974). $1/N$ then provides the only known expansion parameter which can be used in calculations of hadronic properties. Qualitative arguments suggest that the $N=\infty$ theory is surprisingly similar to the real world (Veneziano, 1976; Witten, 1979a). (For example, for large N one expects to see infinitely many narrow resonances that are purely composed of valence quarks. Exotics are absent, Zweig's rule is satisfied, and one-meson exchange dominates scattering amplitudes.) Therefore a $1/N$ expansion might be very reliable even at $N=3$. Unfortunately, the $N=\infty$ theory has not yet been explicitly solved, and for this reason quantitative predictions are totally lacking.

Because brute force methods for solving the $N\to\infty$ limit (such as summing the appropriate class of Feynman diagrams) appear to be totally hopeless in theories like QCD, there has been considerable effort directed toward finding useful ways to reformulate the large N limit of various theories. By now, quite a few different methods have been proposed [under trade names such as "collective field methods" (Jevicki and Sakita, 1980a), "string equations" (Makeenko and Migdal, 1979), "master fields" (Witten, 1979), "constrained classical solutions" (Jevicki and Papanicolaou, 1980; Halpern, 1981a), etc.]. Each of these methods is known to work in at least some specific set of models. However, questions such as "Why does the method work?" or "For what class of theories does the method work?" have not been fully answered. This paper represents an attempt to answer some of these questions.,

Essentially every method developed for studying large N limits has been based on the following fact. In every theory known to have a sensible large N limit, the vacuum expectation of any product of (reasonable) operators,[2] $\hat{A}\hat{B}$, satisfies the factorization relation,

*Richard Chace Tolman Fellow in Theoretical Physics

[1]For a review of some of the applications of $1/N$-expansions to field theories, see, for example, Coleman (1980).

[2]What constitutes a "reasonable" operator will be discussed in Sec. III.

$$\langle \hat{A}\hat{B} \rangle = \langle \hat{A} \rangle \langle \hat{B} \rangle + O(1/N) . \tag{1.1}$$

Therefore, the variance of any (reasonable) operator vanishes as $N \to \infty$,

$$\lim_{N\to\infty} (\langle \hat{A}^2 \rangle - \langle \hat{A} \rangle^2) = 0 . \tag{1.2}$$

[One way to verify these statements, at least perturbatively, is to examine the set of Feynman diagrams which survive in the $N\to\infty$ limit. (1.1) is equivalent to the statement that disconnected graphs always dominate, and this may be verified on a graph-by-graph basis without having to sum the whole series.]

Equation (1.2) shows that fluctuations become irrelevant, at least for some set of operators, when N tends to infinity. Therefore in some sense quantum theories with large N behave like classical theories. It then seems natural to ask the question, "Is the $N\to\infty$ limit a classical limit?" By this we mean the following. Can one find a classical system (i.e., a classical phase space, Poisson bracket, and classical Hamiltonian) whose dynamics is equivalent to the $N\to\infty$ limit of a given quantum theory? In this paper the following strategy will be used to explore this question. First I present a general scheme for finding classical limits in arbitrary quantum theories. This formalism is based on a small set of assumptions which explicitly isolate the minimal structure any quantum theory should possess if it is to have a classical limit. Given any quantum theory satisfying these assumptions, an explicit algorithm may then be used to construct the classical phase space, define a consistent Poisson bracket, and find a classical Hamiltonian, such that the resulting classical dynamics agrees with the limiting form of the original quantum dynamics. I then examine various theories in order to see if their large N limits can be understood as examples of this general formalism. Specifically, I consider vector models, matrix models, and gauge theories, and in every case find that the large N limit is a classical limit in the sense described above. In fact, this method for deriving classical limits is applicable to every quantum theory known to me which, in some limit, satisfies factorization (Eq. 1.1). Besides trivial $\hbar\to 0$ limits, this includes all large N limits of the type discussed here, where the invariance group of the theory grows with N, as well as limits where the underlying symmetry group is fixed, but where quantum operators in larger and larger representations of the group appear. This latter type of limit describes, for example, the large spin limit of quantum spin models. In somewhat greater detail, the outline of this paper is as follows.

Section II contains a brief discussion of the $\hbar\to 0$ limit in quantum mechanics of point particles. I review a few of the standard properties of Gaussian coherent states, and show how they may be used to construct a very simple derivation of the $\hbar\to 0$ limit. The basic purpose of this section is to provide a concrete example which will be used to illustrate many of the features of the following rather abstract discussion.

Section III presents a general formalism for finding classical limits in arbitrary quantum theories. To apply this method one must (a) choose an appropriate group of unitary transformations, (b) generate a set of coherent states by applying elements of this group to a suitable initial state, and (c) show that in some limit of the theory different coherent states become orthogonal. This structure then automatically allows one to construct a unique classical phase space, define classical dynamics on this space, and show that the limit of the quantum dynamics is equivalent to this classical dynamics. In particular, one can form a classical action which contains all the physics of the original quantum theory that survives in the classical limit.

This formalism is applied to the large N limit of vector models in Sec. IV. This class of models includes theories ranging from quantum mechanics of point particles in N dimensions to N component ϕ^4 field theories. I show that all the assumptions of the general formalism are valid for these models, and explicitly derive the classical limit. All of the standard results concerning the large N limit of these models (such as the ground state energy, spectrum, etc.) may be obtained by minimizing the classical Hamiltonian and expanding the classical action about the minimum.

Section V discusses the large N limit of matrix models. Such theories are much more complicated than simple vector models. (For example, the large N limit of the perturbation series contains all planar diagrams.) Nevertheless, the general formalism is shown to be applicable to these models, and the classical limit is derived. However, only in the case of a single matrix has it been possible to explicitly minimize the resulting classical Hamiltonian and thereby "solve" the $N=\infty$ theory.

Lattice gauge theories are the subject of Sec. VI. The analysis is essentially identical to the preceding treatment of matrix models. The large N limit may be shown to be a classical limit, and the classical Hamiltonian may be derived. The resulting classical phase space is sufficiently complicated that only the one plaquette model has been explicitly solved.

Section VII discusses the relation between this approach for understanding the large N limit, and previously proposed methods. The advantages and disadvantages of each approach, as well as their interrelationships, are considered at some length. Various open problems are mentioned.

Finally, the appendix contains a brief discussion of several topics which are related to the general formalism of Sec. III.

A few "historical" remarks are appropriate to end this introduction. The method presented in Sec. III for deriving classical limits of general quantum theories was motivated by two recent papers, one by Berezin (1978) and the other by Simon (1980). Berezin considered the large N limit of vector models and showed that the limiting theory is a classical theory. In fact, he found that vector models with finite N may be regarded as the quantization of classical mechanics on Kähler manifolds. Berezin used coherent state methods which are very similar to those employed in Sec. IV. The treatment here is somewhat simpler, but the results are equivalent. Further comments on Berezin's work will be found in Sec. VII. Simon discussed the classical limit of quantum spin systems, extending previous work by Lieb (1973). His work includes a derivation of the classical phase space for spin models whose spins represent the generators of any compact Lie group. This paper's treatment of generalized coherent states, as well as the identification of the classical phase space, is essentially patterned after Simon's discussion. I do not discuss quantum spin models in any detail in this paper simply because there already exists an extensive literature on the application of coherent state methods to spin models. [See Simon (1980), Lieb (1973), Fuller and Lenard (1979), Gilmore and Feng (1978), Gilmore (1979), Shankar (1980), and references therein.] However, it should be noted that the classical limits of quantum spin models provide beautiful examples of the general formalism described in Sec. III.

II. THE $\hbar \to 0$ LIMIT

Quantum mechanics is generally said to reduce to classical mechanics in the $\hbar \to 0$ limit. However, this statement really requires some qualification. The crucial fact is that one may form states whose uncertainty in both position and momentum vanishes as $\hbar$ tends to zero. If (and only if) the quantum system is prepared in such a state, then the quantum dynamics will reduce to classical dynamics when $\hbar \to 0$. A simple way formally to derive this result is as follows.

Consider a quantum theory describing n degrees of freedom, with basic position $\{\hat{x}_i\}$ and momentum $\{\hat{p}_i\}$ operators obeying the canonical commutation relations,

$$i[\hat{p}_i,\hat{x}_j]=\hbar\delta_{ij}\ , \quad i,j=1,\ldots,n\ . \tag{2.1}$$

Introduce a set of Gaussian coherent states,[3] $\{|p,q\rangle\}$, with wave functions given by

$$\langle x\,|p,q\rangle=(\pi\hbar)^{-n/4}\exp\{(1/\hbar)[ip\cdot x-\tfrac{1}{2}(x-q)^2]\}\ . \tag{2.2}$$

Note that different coherent states are not orthogonal; their overlaps are given by

$$\langle p,q\,|p',q'\rangle=\exp\{-(1/4\hbar)[(p-p')^2+(q-q')^2+2i(p-p')\cdot(q+q')]\}\ . \tag{2.3}$$

This set of coherent states forms an overcomplete basis for the full Hilbert space. This is expressed by the completeness relation,

$$\hat{1}=\int (dp\,dq/2\pi\hbar)\,|p,q\rangle\langle p,q\,| \tag{2.4}$$

$[(dp\,dq/2\pi\hbar)\equiv\prod_{i=1}^{n}dp_i dq_i/2\pi\hbar)]$. Consequently, any quantum state $|\psi\rangle$ may be represented by its projections onto the different coherent states, $\psi(p,q)=\langle p,q\,|\,\psi\rangle$. This is convenient representation regardless of the choice of the Hamiltonian; the quantum dynamics is not required to preserve the form of the coherent states.

Similarly, any operator, $\hat{A}$, may be represented by its coherent state matrix elements, $\langle p,q|\hat{A}|p',q'\rangle$. However, to specify an operator uniquely it is not necessary to give all possible matrix elements. Due to the overcompleteness of the coherent state basis, an arbitrary operator $\hat{A}$ may be reconstructed from just the diagonal expectation values,

$$A(p,q)\equiv\langle p,q\,|\,\hat{A}\,|p,q\rangle\ . \tag{2.5}$$

[One way to see this is based on the observation that $\langle x|p,q\rangle e^{q^2/2\hbar}$ is an analytic function of $(p-iq)$. Therefore, for any (reasonable) operator $\hat{A}$, $\langle p,q|\hat{A}|p',q'\rangle/\langle p,q\,|p',q'\rangle$ is an analytic function of $(p+iq)$ and $(p'-iq')$. Consequently, one may recover arbitrary coherent state matrix elements of $\hat{A}$ by analytic continuation from the diagonal expectation values.]

Now consider the diagonal matrix elements of a product of two operators, $\hat{A}\hat{B}$. The completeness relation (2.4) may be used to write this as

$$(AB)(p,q)=\int (dp'dq'/2\pi\hbar)\,|\langle p,q\,|p',q'\rangle|^2\times\frac{\langle p,q\,|\hat{A}\,|p',q'\rangle}{\langle p,q\,|p',q'\rangle}\,\frac{\langle p',q'\,|\hat{B}\,|p,q\rangle}{\langle p',q'\,|p,q\rangle}\ . \tag{2.6}$$

To study the classical limit, one must compute the small $\hbar$ asymptotics of this integral. The first factor,

$$|\langle p,q\,|p',q'\rangle|^2=\exp\{-(1/2\hbar)[(p-p')^2+(q-q')^2]\}$$

becomes arbitrarily highly peaked about $p=p'$ and $q=q'$ as $\hbar\to 0$. However, the remaining factors have a smooth limit when $\hbar\to 0$.[4] Therefore one finds

$$\lim_{\hbar\to 0}(AB)(p,q)=a(p,q)b(p,q)\ , \tag{2.7}$$

where $a(p,q)\equiv\lim_{\hbar\to 0}A(p,q)$, etc. Similarly, one may expand the integral (2.6) about $p=p'$ and $q=q'$ and easily find that

$$\lim_{\hbar\to 0}\frac{i}{\hbar}[A,B](p,q)=\left[\frac{\partial a(p,q)}{\partial p}\frac{\partial b(p,q)}{\partial q}-\frac{\partial b(p,q)}{\partial p}\frac{\partial a(p,q)}{\partial q}\right]=\{a(p,q),b(p,q)\}_{PB}\ . \tag{2.8}$$

[3]The properties of Gaussian coherent states are discussed in many textbooks; Klauder and Sudarshan (1968), for example, contains a good discussion.

[4]$\hat{A}$ and $\hat{B}$ must be "classical operators" as defined in Sec. III. Such operators include arbitrary polynomials in $\hat{x}$ and $\hat{p}$ with no explicit $\hbar$ dependence.

These results show how the quantum theory reduces to classical mechanics when $\hbar \to 0$. Equation (2.7) implies that quantum operators become ordinary functions on the classical phase space, while Eq. (2.8) shows that quantum commutators become classical Poisson brackets. This implies that the quantum equations of motion, $\partial\hat{A}/\partial t = (i/\hbar)[\hat{H},\hat{A}]$, reduce to the classical Hamiltonian equations, $\partial a(p,q)/\partial t = \{h(p,q),a(p,q)\}_{PB}$. If the quantum Hamiltonian is some given functions of $\hat{x}$ and $\hat{p}$, $\hat{H}=f(\hat{p},\hat{x})$, then the classical Hamiltonian $h(p,q)$ equals $f(p,q)$ regardless of the ordering of the original quantum operators.[5]

The classical equations of motion follow from the classical action,

$$S_{cl}[p(t),q(t)] = \int dt\,[p(t)\dot{q}(t) - h(p(t),q(t))] \ . \quad (2.9)$$

One may regard the classical action as containing the complete physics of the theory in the $\hbar \to 0$ limit. For example, the limiting behavior as $\hbar \to 0$ of the ground-state energy, the spectrum, or any correlation function may be obtained from the classical action. This will be discussed in more detail at the end of the next section.

III. COHERENT STATES AND COADJOINT ORBITS

The key ingredient in the preceding discussion of the $\hbar \to 0$ limit was obviously the choice of coherent states. They provided

(i) a convenient partition of unity requiring only diagonal projections onto the coherent states,

(ii) a basis sufficiently overcomplete that any operator could be completely represented by its diagonal matrix elements alone,

(iii) a simple derivation of factorization based on the fact that different coherent states become orthogonal in the classical limit, and

(iv) an identification of the classical phase space as the manifold whose coordinates could be used to label different coherent states.

We will now see how each of these features may be naturally incorporated in a more general framework. The resulting abstract formalism will be applicable to every known theory possessing a factorizing limit.

Consider a family of quantum theories labeled by some parameter χ (such as $\hbar$ or $1/N$). χ is assumed to take values in some set of positive real numbers whose limit points include zero. We are intereseted in studying the limit of these theories as χ tends to zero. Each theory is defined on some Hilbert space $\mathbf{H}_\chi$ with some Hamiltonian $\hat{H}_\chi$, etc. [We speak of a one-parameter family of theories, as opposed to a single theory depending on the parameter χ, in order to emphasize that the basic structure of the theory (such as the Hilbert space, commutation relations, etc.) may vary with χ. For example, the number of basic degrees of freedom will differ in theories with different values of N.]

Let there be given some Lie group g which, within each theory, may be represented by a set of unitary operators. In other words, acting on each Hilbert space $\mathbf{H}_\chi$ is a group of unitary operators, $G_\chi \equiv \{\hat{U} | \hat{U} = D_\chi(u),\ u \in g\}$, which provides a representation of the abstract group g. We will refer to g as the *coherence group*. This coherence group has a Lie algebra $\mathbf{g}$ which may be represented within each theory by a set of anithermitian operators, $\mathbf{G}_\chi \equiv \{\hat{\Lambda} | \hat{\Lambda} = D_\chi(\lambda),\ \lambda \in \mathbf{g}\}$, which generate one-parameter subgroups of G_χ, $\exp t\hat{\Lambda} \in G_\chi$. Note that the abstract group g and its algebra $\mathbf{g}$ do not depend on the parameter χ.

Furthermore, within each theory let there be given some chosen normalized state, $|0\rangle_\chi \in \mathbf{H}_\chi$, which we will call the *base state*. Consider the states which are generated by applying elements of the coherence group to the base state,

$$|u\rangle_\chi \equiv \hat{U}\,|0\rangle_\chi\ , \quad \hat{U} \in G_\chi \ . \quad (3.1)$$

These are precisely the coherent states we will use.[6] Henceforth, we will occasionally drop the explicit subscript χ if we do not need to emphasize which particular theory we are working in.

We will use the previous $\hbar \to 0$ example to illustrate each feature of the general discussion. For convenience, I describe the case of just one degreee of freedom. In this example χ equals $\hbar$, and the coherence group g may be chosen to be the Heisenberg group. This may be regarded as a three-parameter group whose elements, $u(p,q,\alpha)$, obey the muliplication rule

$$u(p,q,\alpha)\,u(p',q',\alpha') = u(p+p',q+q',\alpha+\alpha'-qp') \ .$$

This group may be represented by the set of operators

$$G_\hbar = \{\hat{U}(p,q,\alpha) \equiv \exp(i\alpha/\hbar)\,\exp(ip\hat{x}/\hbar)\exp(-iq\hat{p}/\hbar)\} \ .$$

Note that $\hat{U}(p,q,\alpha)$ simply translates positions by q, and momenta by p. If we choose the base state to be a simple Gaussian,

$$\langle x\,|\,0\rangle_\hbar = (\pi\hbar)^{-1/4}\exp-(x^2/2\hbar) \ ,$$

then elements of the Heisenberg group precisely generate our previous coherent states up to an overall phase factor,

$$\hat{U}(p,q,\alpha)\,|\,0\rangle_\hbar = e^{i\alpha/\hbar}\,|\,p,q\rangle \ .$$

---

[5]Note that if the Hamiltonian $\hat{H}(\hat{p},\hat{x})$ is replaced by $(1/g^2)\hat{H}(g\hat{p},g\hat{x})$, and the rescaled operators $\hat{p}' \equiv g\hat{p}$ and $\hat{x}' \equiv g\hat{x}$ are used in place of $\hat{p}$ and $\hat{x}$, then every occurrence of $\hbar$ becomes $(g^2\hbar)$. Therefore the classical $(\hbar \to 0)$ limit is equivalent to the weak coupling $(g^2 \to 0)$ limit.

[6]This method for generating generalized coherent states based on arbitrary Lie groups was first discussed by Klauder (1963) and Perelomov (1972).

The choice of the coherence group is restricted by the following assumption. We require

Assumption 1. Each representation of the coherence group, G_χ, acts irreducibly on the corresponding Hilbert space $\mathbf{H}_\chi$.

In other words, there must be no nontrivial subspace of $\mathbf{H}_\chi$ which is left invariant under the action of all elements of the coherence group G_χ. To test for irreducibility one may use Schur's lemma, which states that a group acts irreducibly if and only if the only operators which commute with all elements of the group are proportional to the identity—i.e., G_χ acts irreducibly iff $\hat{U}\hat{A}\hat{U}^{-1}=\hat{A}$ for all $\hat{U}\in G_\chi$ implies $\hat{A}\propto\hat{1}$.

This assumption has the following consequences. Consider the operator $\hat{J}\equiv\int d\mu_L(u)|u\rangle\langle u|$, where $d\mu_L(u)$ is the (left) invariant measure on the coherence group—i.e., $d\mu_L(u'u)=d\mu_L(u)$ for any fixed $u'\in g$.[7] Note that $\hat{J}$ commutes with all elements of G_χ,

$$\hat{U}'\hat{J}\hat{U}'^{-1}=\int d\mu_L(u)\,|u'u\rangle\langle u'u|=\hat{J}$$

for all $\hat{U}'\in G_\chi$, so that by Schur's lemma $\hat{J}$ is proportional to $\hat{1}$. Therefore the irreducibility of G_χ automatically provides us with a natural completeness relation,

$$\hat{1}=c_\chi\int d\mu_L(u)\,|u\rangle\langle u|\ . \tag{3.2}$$

The constant c_χ depends on the normalization of the group measure and must be computed explicitly.

Assumption 1 has another important consequence. If a group acts irreducibly, then any operator may be expressed as a linear combination of elements of the group.[8] Therefore any operator $\hat{A}$ acting in $\mathbf{H}_\chi$ may be written in the form

$$\hat{A}=\int (d\lambda)f(\lambda)\exp(\chi\hat{\Lambda}) \tag{3.3}$$

for some weight function $f(\lambda)$. [Here $(d\lambda)$ indicates uniform measure on the Lie algebra $\mathbf{g}$. The factor of χ is inserted for later convenience.]

Assumption 1 may be easily verified in our $\hbar\to 0$ example. To do so, first consider the subgroup of the Heisenberg group which is generated by $(i\hat{x}/\hbar)$, $\{\hat{U}(p,0,0)=\exp(ip\hat{x}/\hbar)\}$. One may easily see that the only operators which commute with all elements of this subgroup are of the form $a(\hat{x})$, that is, which are solely constructed from $\hat{x}$. However, any operator of this form which also commutes with the subgroup generated by $(i\hat{p}/\hbar)$, $\{\hat{U}(0,q,0)=\exp(-iq\hat{p}/\hbar)\}$, must simply be a constant. Therefore the Heisenberg group acts irreducibly. Invariant measure on the Heisenberg group is given by $d\mu(u(p,q,\alpha))\equiv dp\,dq\,d\alpha$ ($d\mu$ is both left and right invariant). Therefore (3.2) agrees with our previous completeness relation (2.4). In this example, the operator representation (3.3) becomes

$$\hat{A}=\int dp\,dq\,d\alpha\,f(p,q,\alpha)\exp i(p\hat{x}-q\hat{p}+\alpha)$$
$$=\int dp\,dq\,\bar{f}(p,q)\exp i(p\hat{x}-q\hat{p})$$

[$\bar{f}(p,q)\equiv\int d\alpha\,f(p,q,\alpha)e^{i\alpha}$]. This is simply the well-known Weyl representation. Equation (3.3) is the natural generalization of this representation to any group which acts irreducibly.

Assumtpion 1 involves only the choice of the coherence group g. It places no restriction on the base state $|0\rangle_\chi$. In fact, (3.2) shows that a complete set of states may be generated by applying elements of the coherence group to any initial state in $\mathbf{H}_\chi$. However, the choice of base state will be restricted by our next assumption.

For any operator $\hat{A}$ acting in $\mathbf{H}_\chi$, let us define the *symbol* $A_\chi(u)$ as the set of coherent-state expectation values,[9]

$$A_\chi(u)=\langle u\,|\hat{A}\,|u\rangle_\chi\ ,\quad u\in g\ . \tag{3.4}$$

For each value of χ, we require

Assumption 2. Zero is the only operator whose symbol identically vanishes.

In other words, if $A_\chi(u)=0$ for all $u\in g$, then $\hat{A}$ must equal zero. This assumption implies that two different operators cannot have the same symbol. (Otherwise, the difference of the two operators would violate the assumption.) Therefore any operator may be uniquely recovered from its symbol. This means that it is sufficient to study the behavior of the symbols of various operators in order to characterize the theory completely.

Assumption 2 may be easily verified for the $\hbar\to 0$ example. One method, based on the analyticity of $\langle p,q|\hat{A}|p',q'\rangle/\langle p,q\,|p',q'\rangle$ in $(p+iq)$ and $(p'-iq')$, was discussed in the last section. The assumption may also be proven in a more direct fashion using an argument due to B. Simon. We will present this argument in some detail, since the method naturally generalizes to later examples.

Suppose the symbol of some operator, $A_\chi(u)$, vanishes for all $u\in g$. We may choose $u=e^{t_1\lambda_1}e^{t_2\lambda_2}\cdots e^{t_n\lambda_n}$ for

[7]All integrals over the coherence group need only be taken over the coset space $g_\chi\equiv g/h_\chi$, where h_χ is the isotropy subgroup of $|0\rangle\langle 0|$, $h_\chi\equiv\{u\in g\mid |u\rangle\langle u|=|0\rangle\langle 0|\}$. I will not bother to indicate this explicitly in the notation here. The integral $\int d\mu_L(u)\,|\langle u|0\rangle|^2$ is assumed to converge.

[8]This is a sloppy version of the Von Neumann density theorem, which implies that a group acts irreducibly if and only if the algebra of operators generated by the group is strongly dense in the set of all bounded operators. For a rigorous proof see, for example, Bratteli and Robinson (1979), Sec. 2.4.2.

[9]What I am calling the *symbol* of an operator is elsewhere referred to as the *lower symbol* (Simon, 1980) or the *covariant symbol* (Berezin, 1972). There is another natural association of operators with functions on the coherence group which may be used to define *upper* or *contravariant* symbols. These are described in the appendix, but are not used in the bulk of this paper.

arbitrary $\lambda_i \in \mathbf{g}$, differentiate with respect to each t_i, and find

$$0=\frac{\partial}{\partial t_1}\cdots\frac{\partial}{\partial t_n}A_\chi(u)\,|_{t_i=0}$$

$$=\langle 0|[\cdots[[\hat{A},\hat{\Lambda}_1],\hat{\Lambda}_2],\ldots,\hat{\Lambda}_n]|0\rangle_\chi . \qquad (3.5)$$

Therefore expectations in the base state of multiple commutators of $\hat{A}$ with arbitrary generators of the coherence group vanish. Now for the Heisenberg group each $\hat{\Lambda}_i$ is some linear combination of $i\hat{p}/\hbar$, $i\hat{x}/\hbar$, and $i\hat{1}/\hbar$. Therefore we may choose each $\hat{\Lambda}_i$ to be either a creation, $\hat{a}^\dagger=(\hat{p}+i\hat{x})/\sqrt{2\hbar}$, or annihilation, $\hat{a}=(\hat{p}-i\hat{x})/\sqrt{2\hbar}$, operator. We will prove inductively that

$$\langle 0|\hat{a}\cdots\hat{a}\hat{A}\hat{a}^\dagger\cdots\hat{a}^\dagger|0\rangle=0 \qquad (3.6)$$

for any number of creation or annihilation operators. Suppose (3.6) has been verified whenever the total number of creation plus annihilation operators is less than some number L. Consider a multiple commutator of the form (3.5) with L creation or annihilation operators. Expand the multiple commutator. One term has all annihilation operators to the left, and all creation operators to the right of $\hat{A}$. Every other term contains at least one annihilation operator which may be pushed right until it annihilates the base state, or one creation operator which may be pushed left. This process may produce commutator terms (since $[\hat{a},\hat{a}^\dagger]=1$); however, each commutator reduces the number of $\hat{a}$'s plus $\hat{a}^\dagger$'s by two. Therefore the expectation of the multiple commutator contains one new term of the form in (3.6) with L creation and annihilation operators, plus lower-order terms which have already been shown to vanish. Therefore (3.6) holds for L creation and annihilation operators, and by induction holds for any number. This shows that matrix elements of $\hat{A}$ between any two states formed by applying polynomials of creation operators to the base state vanish. But such states are known to be dense, thereby implying that $\hat{A}=0$. This proves the assumption.

Henceforth, whenever I speak of an "operator," I will actually mean some given family of operators consisting of one operator acting in each Hilbert space $\mathbf{H}_\chi$. χ obviously determines which operator in a given family is appropriate; normally I will not bother to add it as an explicit label.

Clearly, an arbitrary operator need not have a sensible limit as $\chi\to 0$. In order to have some control over this limit, we will introduce a restricted class of operators, $\mathbf{K}$, consisting of operators $\hat{A}$ whose coherent state matrix elements, $\langle u|\hat{A}|u'\rangle_\chi/\langle u|u'\rangle_\chi$, have finite limits as $\chi\to 0$ for all $u,u'\in g$. We will refer to such operators as *classical operators.*

Since classical operators form only a subset of all operators, it is possible that measurements with any classical operator will fail to distinguish between different coherent states. Therefore we will call two coherent states, $|u\rangle$ and $|u'\rangle$, *classically equivalent* if

$$\lim_{\chi\to 0}A_\chi(u)=\lim_{\chi\to 0}A_\chi(u') \qquad (3.7)$$

for all $\hat{A}\in\mathbf{K}$. We will write $u\sim u'$ if $|u\rangle$ and $|u'\rangle$ are classically equivalent.

We may now formulate our third assumption, which states that classically inequivalent coherent states become orthogonal as $\chi\to 0$. Specifically, we require

Assumption 3. $\phi(u,u')\equiv -\lim_{\chi\to 0}\chi\ln\langle u|u'\rangle_\chi$ exists for all $u,u'\in g$, and satisfies

(i) if $u\nsim u'$, $\mathrm{Re}\,\phi(u,u')>0$, and

(ii) if $u\sim u'$, $\mathrm{Re}\,\phi(u,u')=0$, and

$$\frac{\partial}{\partial t}[\phi(u,e^{t\lambda}u)-\phi(u,e^{t\lambda}u')]\,|_{t=0}=0 \quad \text{for all } \lambda\in\mathbf{g} .$$

This shows that if $|u\rangle$ and $|u'\rangle$ are classically inequivalent, then their overlap $\langle u|u'\rangle$ decreases exponentially as $\chi\to 0$. Therefore for any classical operator $\hat{A}\in\mathbf{K}$, $\langle u|\hat{A}|u'\rangle_\chi$ must become highly peaked about $u\sim u'$ as $\chi\to 0$ (otherwise, $\langle u|\hat{A}|u'\rangle_\chi/\langle u|u'\rangle_\chi$ will have no limit). In other words, classical operators cannot "move" the coherent states. This excludes any fixed element $\hat{U}\in G_\chi$ of the coherence group (except $\hat{1}$). However, Assumption 3 shows that $(\chi\hat{\Lambda})$ and $\exp(\chi\hat{\Lambda})$ are acceptable classical operators, for any $\hat{\Lambda}\in\mathbf{G}_\chi$, since

$$\lim_{\chi\to 0}\frac{\langle u|\chi\hat{\Lambda}|u'\rangle_\chi}{\langle u|u'\rangle_\chi}=\frac{\partial}{\partial t}\phi(u,e^{t\lambda}u')\,|_{t=0} \qquad (3.8)$$

and

$$\lim_{\chi\to 0}\frac{\langle u|e^{\chi\hat{\Lambda}}|u'\rangle_\chi}{\langle u|u'\rangle_\chi}=\exp\left[-\frac{\partial}{\partial t}\phi(u,e^{t\lambda}u')\,|_{t=0}\right] . \qquad (3.9)$$

Consequently, a general classical operator, $\hat{A}\in\mathbf{K}$, may be represented in the form (3.3) for some weight function $f(\lambda)$ with compact support [and for which $\lim_{\chi\to 0}f(\lambda)$ exists as a distribution]. Note that

$$\lim_{\chi\to 0}A_\chi(u)=\int(d\lambda)\lim_{\chi\to 0}f(\lambda)\exp\left[-\frac{\partial}{\partial t}\phi(u,e^{t\lambda}u)\,|_{t=0}\right] .$$

This shows that if $\lim_{\chi\to 0}\chi\Lambda_\chi(u)=\lim_{\chi\to 0}\chi\Lambda_\chi(u')$ for all $\hat{\Lambda}\in\mathbf{G}_\chi$, then $|u\rangle$ and $|u'\rangle$ are classically equivalent. In other words, expectation values of the set of operators $\{\chi\hat{\Lambda}|\hat{\Lambda}\in\mathbf{G}_\chi\}$ are sufficient to distinguish classically inequivalent coherent states. Finally, note that (3.9) plus the last part of Assumption 3 implies that if $|u\rangle$ and $|u'\rangle$ are classically equivalent, then

$$\lim_{\chi\to 0}\frac{\langle u|\hat{A}|u'\rangle_\chi}{\langle u|u'\rangle_\chi}=\lim_{\chi\to 0}A_\chi(u) \quad \text{for all } \hat{A}\in\mathbf{K} . \qquad (3.10)$$

Assumption 3 allows us to prove factorization for any pair of operators $\hat{A},\hat{B}\in\mathbf{K}$. We may use the completeness relation (3.2) to write the symbol of the product $\hat{A}\hat{B}$ as

$$(AB)_\chi(u)=c_\chi \int d\mu_L(u')\,|\langle u\,|\,u'\rangle_\chi|^2 \times \frac{\langle u\,|\hat{A}\,|\,u'\rangle_\chi}{\langle u\,|\,u'\rangle_\chi}\frac{\langle u'\,|\,\hat{B}\,|\,u\rangle_\chi}{\langle u'\,|\,u\rangle_\chi} . \quad (3.11)$$

As $\chi\to 0$ the integral becomes highly peaked about points which are equivalent to u. If we define the region $R_\chi \equiv [u' \in g | \mathrm{Re}\,\phi(u,u') \le \sqrt{\chi}]$ consisting of a small neighborhood around each point $u' \sim u$, then Assumption 3 implies that the contribution from the region outside R_χ is exponentially small. However, inside R_χ, (3.10) implies that

$$\langle u\,|\hat{A}\,|\,u'\rangle_\chi / \langle u\,|\,u'\rangle_\chi = A_\chi(u)+o(1) .$$

Therefore

$$\begin{aligned}(AB)_\chi(u)&=c_\chi \int_{R_\chi} d\mu_L(u')\,|\langle u\,|\,u'\rangle_\chi|^2 \times \frac{\langle u\,|\hat{A}\,|\,u'\rangle_\chi}{\langle u\,|\,u'\rangle_\chi}\frac{\langle u'\,|\,\hat{B}\,|\,u\rangle_\chi}{\langle u'\,|\,u\rangle_\chi}+o(1)\\ &=A_\chi(u)B_\chi(u)\cdot c_\chi \int_{R_\chi} d\mu_L(u')\,|\langle u\,|\,u'\rangle_\chi|^2+o(1)\\ &=A_\chi(u)B_\chi(u)+o(1) ,\end{aligned}$$

since $c_\chi \int d\mu_L(u')|\langle u\,|\,u'\rangle_\chi|^2=1$. Consequently, the factorization

$$\lim_{\chi\to 0}[(AB)_\chi(u)-A_\chi(u)B_\chi(u)]=0 \quad (3.12)$$

holds for any pair of classical operators.

This whole discussion is a simple generalization of the $\hbar\to 0$ example. In that case classical operators obviously include arbitrary $\hbar$ independent polynomials in $\hat{x}$ and $\hat{p}$. Since measurements with classical operators can determine the mean position and momentum of a state, classically equivalent coherent states can differ only in their overall phase. To verify Assumption 3, one must simply compute the overlap $\langle u\,|\,u'\rangle_\hbar$, and check that

$$\begin{aligned}&\phi(u(p,q,\alpha),u'(p',q',\alpha'))\\ &=\tfrac{1}{4}[(p-p')^2+(q-q')^2+2i(p-p')(q+q')+4i(\alpha-\alpha')]\end{aligned}$$

satisfies the stated conditions. The operators $\{\hat{U}(p,q,\alpha)=e^{i\alpha/\hbar}e^{ip\hat{x}/\hbar}e^{-iq\hat{p}/\hbar}\}$ are not classical operators and do not obey factorization. However, the rescaled operators $\{\exp(\hbar\hat{\Lambda})=\exp i(\alpha+p\hat{x}-q\hat{p})\}$ translate positions and momenta only by $O(\hbar)$ and are prefectly acceptable classical operators.

The previous assumptions have given us some control over the structure of the theory as $\chi\to 0$, but have left the quantum dynamics completely unrestricted. To control the $\chi\to 0$ limit completely we need to place one condition on the quantum Hamiltonian $\hat{H}_\chi$. We require[10]

[10]The quantum equations of motion are taken to be $\partial\hat{A}/\partial t=i[\hat{H},\hat{A}]$, with no explicit factors of χ appearing.

Assumption 4. $(\chi\hat{H}_\chi)$ is a classical operator.

This condition will ensure that the coupling constants in the Hamiltonian are scaled in a manner that maintains sensible dynamics as $\chi\to 0$.

These assumptions will suffice to show that the complete quantum theory reduces to classical mechanics as $\chi\to 0$. However, before this can be demonstrated the appropriate classical phase space must first be described. To facilitate this, we first review certain aspects of the structure of the coherence group.[11]

Consider the Lie algebra $\mathbf{g}$. It is a linear space, and therefore has a *dual space* $\mathbf{g}^*$ consisting of linear functionals acting on $\mathbf{g}$. If we introduce a basis in $\mathbf{g}$, $\{e_i\}$, and the corresponding biorthogonal basis in $\mathbf{g}^*$, $\{e'^i\}$, then the application of any element of the dual space, $\zeta=\zeta_i e'^i\in\mathbf{g}^*$, on any element of the Lie algebra, $\lambda=\lambda^i e_i\in\mathbf{g}$, is given by

$$\langle\zeta,\lambda\rangle=\zeta_i\lambda^i . \quad (3.13)$$

The communicator of any two elements, $\lambda,\mu\in\mathbf{g}$, of the Lie algebra may be represented as

$$[\lambda,\mu]=\lambda^i\mu^j c_{ij}^k e_k ,$$

where $\{c_{ij}^k\}$ are the structure constants of $\mathbf{g}$. There is a natural action of elements of the coherence group on elements of the algebra $\mathbf{g}$, given by

$$Ad[u](\lambda)\equiv u\lambda u^{-1} \quad (3.14)$$

for any $u\in g,\lambda\in\mathbf{g}$. This is simply the *adjoint representation* of g. There is a corresponding action of the group on elements of the dual space given by

$$Ad^*[u]\equiv Ad[u^{-1}]^* . \quad (3.15)$$

This is the *coadjoint representation*.[12] It is defined so that (3.13) is invariant,

$$\begin{aligned}\langle Ad^*[u](\zeta),Ad[u](\lambda)\rangle&=\langle\zeta,Ad[u^{-1}]Ad[u](\lambda)\rangle\\ &=\langle\zeta,\lambda\rangle .\end{aligned}$$

One may now consider the set of points generated by the action of $Ad^*[u]$ applied to any given element $\zeta_o\in\mathbf{g}^*$. This is a *coadjoint orbit,*

[11]For more information on the following material see, for example, Kirillov (1976).

[12]In any semisimple Lie group the Cartan-Killing form, $(\lambda,\mu)\equiv\mathrm{tr}([\lambda,e_i][e_i,\mu])$, provides a nondegenerate invariant scalar product for the Lie algebra. This then generates a natural isomorphism between the algebra $\mathbf{g}$ and its dual space $\mathbf{g}^*$. Given any element of the dual space, $\zeta\in\mathbf{g}^*$, the corresponding element of the Lie algebra, $\bar{\zeta}\in\mathbf{g}$, is uniquely defined by the requirement that $\langle\zeta,\lambda\rangle=(\bar{\zeta},\lambda)$ for all $\lambda\in\mathbf{g}$. Consequently for semisimple groups the adjoint and coadjoint representations are equivalent. Unfortunately, none of the coherence groups considered in this paper is semisimple, and we are therefore forced to distinguish between working on the Lie algebra $\mathbf{g}$ and on its dual space $\mathbf{g}^*$.

$$\Gamma=\{Ad^*[u](\zeta_o)\,|\,u\in g\}\ . \tag{3.16}$$

We will later see that each set of classically equivalent coherent states may be naturally associated with some point on a single coadjoint orbit. In other words, the set of these equivalence classes, $[u]\equiv\{u'\in g|u\sim u'\}$, is isomorphic to a particular coadjoint orbit $\Gamma\in\mathbf{g}^*$. This is important for the following reason. To define a classical phase space one must not only specify the classical manifold, but must also give a consistent definition of a Poisson bracket. This requires the existence of an invariant symplectic structure on the classical manifold. Coadjoint orbits provide a particularly natural setting for classical mechanics, because they automatically possess the required symplectic structure.[13] This may be described as follows.

Consider an arbitrary point on some coadjoint orbit, $\zeta\in\Gamma$. The tangent space to Γ at ζ, $T_\zeta(\Gamma)$, is a subspace of $\mathbf{g}^*$ given by

$$T_\zeta(\Gamma)=\{\xi\in\mathbf{g}^*\,|\,\langle\xi,\cdot\rangle=\langle\zeta,[\lambda,\cdot]\rangle \text{ for some } \lambda\in\mathbf{g}\}\ . \tag{3.17}$$

These are simply the tangent vectors to the curves $Ad^*[e^{t\lambda}](\zeta)$ on Γ passing through ζ. We may regard elements of the Lie algebra $\mathbf{g}$ as linear functionals acting on the tangent space $T_\zeta(\Gamma)$. However, we should then identify different elements of $\mathbf{g}$ if they yield the same value when applied to any vector in $T_\zeta(\Gamma)$. Therefore elements of the cotangent space $T_\zeta^*(\Gamma)$ ("one-forms") may be regarded as equivalence classes of $\mathbf{g}$. For each $\lambda\in\mathbf{g}$, the equivalence class $[\lambda]$ is given by

$$[\lambda]=\{\lambda'\in\mathbf{g}\,|\,\langle\zeta,[\lambda',\mu]\rangle=\langle\zeta,[\lambda,\mu]\rangle \text{ for all } \mu\in\mathbf{g}\}\ .$$

Explicitly, the cotangent space $T_\zeta^*(\Gamma)$ is given by the quotient space

$$T_\zeta^*(\Gamma)=\mathbf{g}/\mathbf{h}_\zeta\ , \tag{3.18}$$

where $\mathbf{h}_\zeta\equiv\{\lambda\in\mathbf{g}\,|\,\langle\zeta,[\lambda,\mu]\rangle=0 \text{ for all } \mu\in\mathbf{g}\}$ is the Lie algebra of the isotropy subgroup of ζ,

$$h_\zeta\equiv\{u\in g\,|\,Ad^*[u](\zeta)=\zeta\}\ .$$

Note that the coadjoint orbit Γ is equivalent to the coset space g/h_ζ. We may now define a linear functional ω_ζ acting on $T_\zeta^*(\Gamma)\otimes T_\zeta^*(\Gamma)$ by

$$\omega_\zeta([\lambda],[\mu])\equiv\langle\zeta,[\lambda,\mu]\rangle\ , \tag{3.19}$$

or in components, $\omega_\zeta([\lambda],[\mu])=\lambda^i\mu^j c_{ij}^k\zeta_k$. The value of $\omega_\zeta([\lambda],[\mu])$ is clearly independent of which representative λ of the equivalence class $[\lambda]$ is used. ω_ζ is obviously antisymmetric and so may be regarded as a bivector on Γ. Furthermore, ω_ζ is nondegenerate, since if $\omega_\zeta([\lambda],[\mu])=0$ for all $[\mu]$, then $[\lambda]=0$ by our definition of equivalence.[14] Therefore ω_ζ provides an invertible mapping of the cotangent space $T_\zeta^*(\Gamma)$ onto the tangent space $T_\zeta(\Gamma)$. The inverse mapping, $\tilde{\omega}_\zeta$, may be regarded as a two-form on Γ. $\tilde{\omega}_\zeta$ is automatically closed, $d\omega_\zeta=0$, simply as a consequence of the definition (3.19). Consequently $\tilde{\omega}_\zeta$ is the exterior derivative of some one-form on Γ, $\tilde{\omega}_\zeta=d\vartheta_\zeta$, $\vartheta_\zeta\in T_\zeta^*(\Gamma)$.[15] (This construction will be described in more explicit terms later).

The bivector ω_ζ may be used to define a Poisson bracket as follows. Given any function $f(\zeta)$ on the coadjoint orbit, its gradient, $df(\zeta)\in T_\zeta^*(\Gamma)$, is one-form on Γ. We define the Poisson bracket of two such functions $f(\zeta)$ and $g(\zeta)$ by

$$\{f(\zeta),g(\zeta)\}_{PB}\equiv\omega_\zeta(df(\zeta),dg(\zeta))\ . \tag{3.20}$$

If $f(\zeta)$ and $g(\zeta)$ are defined in a neighborhood of Γ, then one may use the component form

$$\{f(\zeta),g(\zeta)\}_{PB}=\frac{\partial f(\zeta)}{\partial\zeta_i}\frac{\partial g(\zeta)}{\partial\zeta_j}c_{ij}^k\zeta_k\ . \tag{3.21}$$

One may easily verify that this definition satisfies the Jacobi identity and therefore yields a consistent definition of the Poisson bracket.

We may now define classical dynamics on the coadjoint orbit Γ. The classical Hamiltonian, $h_{cl}(\zeta)$, is some given function on Γ. The classical equations of motion are simply

$$\frac{d}{dt}f(\zeta)=\{h_{cl}(\zeta),f(\zeta)\}_{PB} \tag{3.22}$$

for any function $f(\zeta)$ defined on Γ. These equations of motion may be derived from the classical action,

$$S_{cl}[\zeta(t)]=\int dt\{-\langle\dot{\zeta}(t),\vartheta_\zeta\rangle-h_{cl}(\zeta(t))\}\ . \tag{3.23}$$

So far, we have not bothered to introduce independent coordinates on Γ. In practice it will be useful to do so. Let $\{z_\alpha\}$ be an arbitrary set of coordinates on Γ, defined as explicit functions of the natural coordinates in $\mathbf{g}^*$, $z_\alpha\equiv z_\alpha(\zeta_i)$. The components of an arbitrary vector $\xi\in T_\zeta(\Gamma)$, in the coordinate basis defined by $\{z_\alpha\}$, are $\xi_\alpha\equiv(\partial z_\alpha/\partial\zeta_i)\xi_i$. Similarly, an arbitrary one-form, $\sigma\in T_\zeta^*(\Gamma)$, may be expressed as $\sigma=\sigma^\alpha dz_\alpha$. The components of the bivector ω_ζ are

$$\omega_{\alpha\beta}=\frac{\partial z_\alpha}{\partial\zeta_i}\frac{\partial z_\beta}{\partial\zeta_j}c_{ij}^k\zeta_k(z)\ ,$$

and the Poisson bracket becomes

$$\{f(z),g(z)\}_{PB}=\frac{\partial f}{\partial z_\alpha}\frac{\partial g}{\partial z_\beta}\omega^{\alpha\beta}\ . \tag{3.24}$$

---

[13]This fact has been extensively used in the theory of group representations (Kirillov, 1962, 1976; Auslander and Kostant, 1967). It also underlies the method of "geometric quantization" developed by Kostant (1970) and Souriau (1970).

[14]This shows that coadjoint orbits are always even dimensional.

[15]ϑ_ζ need not be defined globally. This does not affect our applications. Note that ϑ_ζ is defined only up to the addition of an arbitrary gradient $df(\zeta)$.

The two-form $\tilde{\omega}_\zeta$ equals $\frac{1}{2}\tilde{\omega}^{\alpha\beta}(z)dz_\alpha\wedge dz_\beta$, where $(\tilde{\omega}^{\alpha\beta})=(\omega_{\alpha\beta})^{-1}$. (Hence, $\tilde{\omega}^{\alpha\beta}\omega_{\beta\gamma}=\delta^\alpha_\gamma$.) The closure of $\tilde{\omega}_\zeta$ reads

$$\tilde{\omega}^{[\alpha\beta,\gamma]}\equiv\frac{1}{3}\left\{\frac{\partial\tilde{\omega}^{\alpha\beta}}{\partial z_\gamma}+\frac{\partial\tilde{\omega}^{\beta\gamma}}{\partial z_\alpha}+\frac{\partial\tilde{\omega}^{\gamma\alpha}}{\partial z_\beta}\right\}=0\ .$$

Consequently,

$$\tilde{\omega}^{\alpha\beta}=\frac{\partial\vartheta^\beta}{\partial z_\alpha}-\frac{\partial\vartheta^\alpha}{\partial z_\beta}$$

for some one-form $\vartheta(z)=\vartheta^\alpha(z)\,dz_\alpha$. The classical action becomes

$$S_{cl}[z(t)]=\int dt\{-z_\alpha(t)\,\vartheta^\alpha(z(t))-h_{cl}(z(t))\}\ . \tag{3.25}$$

This construction of classical mechanics on coadjoint orbits may be illustrated by considering the Heisenberg group. The Lie algebra $\mathbf{g}$ is three dimensional, and we may choose a basis $\{e_i\}$ where the only nonzero structure constants are $c^3_{12}=-c^3_{21}=1$ (e_1, e_2, and e_3 may be represented by $i\hat{p}/\hbar$, $i\hat{x}/\hbar$, and $i\hat{1}/\hbar$, respectively). The action of the adjoint representation is given by

$$Ad[u(p,q,\alpha)](\lambda^i e_i)=\bar{\lambda}^i e_i\ ,$$

where $\bar{\lambda}^1=\bar{\lambda}^1$, $\bar{\lambda}^2=\lambda^2$, and $\bar{\lambda}^3=\lambda^3-p\lambda^1-q\lambda^2$. Similarly, the action of the coadjoint representation is

$$Ad^*[u(p,q,\alpha)](\zeta_i e'^i)=\bar{\zeta}_i e'^i\ ,$$

where $\bar{\zeta}_i=\langle\zeta,Ad[u^{-1}](e_i)\rangle$, or, explicitly, $\bar{\zeta}_1=\zeta_1+p\zeta_3$, $\bar{\zeta}_2=\zeta_2+q\zeta_3$, and $\bar{\zeta}_3=\zeta_3$. Therefore coadjoint orbits are simply two-dimensional planes specified by $\zeta_3=$constant (provided ζ_3 is nonzero; the $\zeta_3=0$ orbits are single points—we ignore this dull possibility in the following discussion.) Since the orbits are flat, all tangent spaces equal to the $\zeta_3=0$ subspace, $T_\zeta(\Gamma)=\{\zeta_1 e'^1+\zeta_2 e'^2\mid\zeta_1,\zeta_2\in\mathscr{R}\}$. The cotangent spaces are given by $T^*_\zeta(\Gamma)=\mathbf{g}/e_3$, since $\langle\xi,e_3\rangle=0$ for all $\xi\in T_\zeta(\Gamma)$. Let us relabel ζ_1 as p and ζ_2 as q; (p,q) obviously provide natural coordinates on a given orbit Γ. The bivector ω_ζ is given by $\omega_\zeta([\xi],[\mu])=(\lambda^1\mu^2-\lambda^2\mu^1)\zeta_3$, its inverse is $\tilde{\omega}_\zeta=\zeta_3^{-1}dq\wedge dp$, and $\vartheta=-\zeta_3^{-1}p\,dq$ satisfies $d\vartheta=\tilde{\omega}_\zeta$. Therefore (3.21) yields the standard Poisson bracket,

$$\{f(p,q),g(p,q)\}_{PB}=\zeta_3\left[\frac{\partial f}{\partial p}\frac{\partial g}{\partial q}-\frac{\partial f}{\partial q}\frac{\partial g}{\partial p}\right]\ , \tag{3.26}$$

up to a constant overall factor of ζ_3. Finally, the classical action (3.23), is simply

$$S_{cl}[p(t),q(t)]$$

$$=\int dt\{\zeta_3^{-1}p(t)\dot{q}(t)-h_{cl}(p(t),q(t))\}\ . \tag{3.27}$$

This shows that classical mechanics on the coadjoint orbits of the Heisenberg group essentially reproduces the standard classical dynamics of a point particle.

We must now show what this discussion of coadjoint orbits has to do with the original quantum theory we wished to study. The basic connection is provided by the following observation. For any $\lambda\in\mathbf{g}$, consider $\lim_{\chi\to 0}(-i)\langle 0|\chi\hat{\Lambda}|0\rangle_\chi$. (Since $\chi\hat{\Lambda}$ is a classical operator, the limit exists.) This expectation is a linear functional of λ and therefore equals $\langle\zeta_o,\lambda\rangle$ for some element $\zeta_o\in\mathbf{g}^*$ of the dual space. Arbitrary coherent state expectations are given by

$$\lim_{\chi\to 0}(\chi/i)\Lambda_\chi(u)=\lim_{\chi\to 0}(\chi/i)\langle 0\,|\,\hat{U}^{-1}\hat{\Lambda}\hat{U}\,|\,0\rangle$$

$$=\langle\zeta_o,Ad[u^{-1}](\lambda)\rangle$$

$$=\langle Ad^*[u](\zeta_o),\lambda\rangle\ . \tag{3.28}$$

Therefore each coherent state, $|u\rangle$, may be associated with some point, $Ad^*[u](\zeta_o)$, on a single coadjoint orbit $\Gamma\in\mathbf{g}^*$. Since expectation values of $\{\chi\hat{\Lambda}\}$ distinguish classically inequivalent states, the coherent states which are mapped onto any given point $\zeta\in\Gamma$ are all classically equivalent. In other words, each point on the coadjoint orbit may be regarded as uniquely labeling a set of classically equivalent coherent states.

We will now show that for any classical operators, $\hat{A},\hat{B}\in\mathbf{K}$, the following statements hold for all $u\in g$:

$$\lim_{\chi\to 0}A_\chi(u)=a(\zeta)\ , \tag{3.29}$$

$$\lim_{\chi\to 0}(AB)_\chi(u)=a(\zeta)b(\zeta)\ , \tag{3.30}$$

$$\lim_{\chi\to 0}\frac{i}{\chi}[A,B]_\chi(u)=\{a(\zeta),b(\zeta)\}_{PB}\ . \tag{3.31}$$

Here $\zeta=Ad^*[u](\zeta_o)\in\Gamma$.

The first relation (3.29) simply expresses the fact that in the $\chi\to 0$ limit, the symbol of a classical operator may be regarded as a function on the coadjoint orbit Γ. This is just a restatement of the fact that points on Γ label equivalence classes of coherent states.

The second relation (3.30) is a restatement of factorization (3.12) and has already been established.

The last relation (3.31) could be verified by computing the subleading terms in (3.11). Fortunately, we may avoid this tedious computation by arguing as follows. First, let $\hat{A}=(\chi/i)\hat{\Lambda}$ and $\hat{B}=(\chi/i)\hat{\Lambda}'$ be arbitrary elements of the Lie algebra (times χ/i). Then $a(\zeta)=\langle\zeta,\lambda\rangle$, $b(\zeta)=\langle\zeta,\lambda'\rangle$ and, from (3.21), $\{a(\zeta),b(\zeta)\}_{PB}=\langle\zeta,[\lambda,\lambda']\rangle$. To evaluate the left-hand side of (3.31), note that $(i/\chi)[\hat{A},\hat{B}]=(\chi/i)[\hat{\Lambda},\hat{\Lambda}']$ is again an element of $\mathbf{G}_\chi$ times (χ/i). Therefore $\lim_{\chi\to 0}(i/\chi)[A,B]_\chi(u)$ also equals $\langle\zeta,[\lambda,\lambda']\rangle$. Hence the set of operators $\{(\chi/i)\hat{\Lambda}\}$ satisfies (3.31). Next, let $\hat{A}=\exp(\chi\hat{\Lambda})$ and $\hat{B}=\exp(\chi\hat{\Lambda}')$. Then from equation (3.9) we find $a(\zeta)=\exp i\langle\zeta,\lambda\rangle$, $b(\zeta)=\exp i\langle\zeta,\lambda'\rangle$, and

$$\{a(\zeta),b(\zeta)\}_{PB}=\langle\zeta,[\lambda',\lambda]\rangle\exp i\langle\zeta,\lambda+\lambda'\rangle\ .$$

We may now evaluate $(i/\chi)[\hat{A},\hat{B}]$, dropping all terms whose expectation values vanish as $\chi\to 0$. Using factorization (3.30), one finds

$$\frac{i}{\chi}[A,B]_\chi(u) = -i\chi([\Lambda,\Lambda']\exp\chi(\Lambda+\Lambda'))_\chi(u)+o(1)$$

$$=\langle \zeta,[\lambda',\lambda]\rangle \exp i\langle \zeta,\lambda+\lambda'\rangle + o(1)$$

in agreement with (3.31). The representation (3.3) may then be used to extend this result to all classical operators. This proves the stated relation.

Finally, note that Assumption 4 plus equation (3.31) implies that the quantum equations of motion, $\partial\hat{A}/\partial t = i[\hat{H},\hat{A}]$, reduce to the classical Hamilton equations, $a(\zeta)=\{h_{cl}(\zeta),a(\zeta)\}_{PB}$, where the classical Hamiltonian is given by

$$h_{cl}(\zeta)=\lim_{\chi\to 0}\chi H_\chi(u) . \qquad (3.32)$$

These results show that in the $\chi\to 0$ limit, the complete quantum theory reduces to classical mechanics on the coadjoint orbit Γ. We emphasize that all information about the $\chi\to 0$ theory is contained in the classical action $S_{cl}[\zeta(t)]$, (3.23). For example, the limiting behavior of the ground state energy is given by $E_0 \sim (1/\chi)\varepsilon_0$, where ε_0 is the minimum of the classical Hamiltonian $h_{cl}(\zeta)$ over the coadjoint orbit Γ. Expanding the classical action about this minimum and diagonalizing the quadratic terms will yield the set of small oscillation frequencies, $\{\omega_i\}$. These frequencies give the $\chi\to 0$ limit of the spectrum, in the sense that the excitation energy to a particular excited state is given by $\Delta E \sim \sum_i n_i\omega_i$, for some set of non-negative integers $\{n_i\}$. Furthermore, the limiting behavior of the connected part of any correlation function of time-ordered products of classical operators may be computed from tree diagrams generated by $(1/\chi)S_{cl}[\zeta(t)]$. [This implies that the connected part of the vacuum expectation of any product of n classical operators vanishes as $(\chi)^{n-1}$.] Finally, the behavior of $S_{cl}[\zeta(t)]$ away from the minimum determines the dynamics of highly excited $[\Delta E \sim O(1/\chi)]$ collective excitations. These statements may be easily derived using functional integrals based on the coherent state completeness relation (3.2). This is discussed in further detail in the appendix.

Naturally, the above results reproduce the expected $\hbar\to 0$ behavior. The particular coadjoint orbit which emerges in that case is the one with

$$\zeta_3=\frac{\hbar}{i}\langle 0|\frac{i}{\hbar}\hat{1}|0\rangle = 1 .$$

On this orbit (3.27) exactly reproduces the standard classical dynamics. The ground state corresponds to the minimum of the Hamiltonian; expanding the classical action about this point yields free propagators and bare vertices, from which one can construct tree diagrams giving the leading $\hbar\to 0$ behavior of any observable.

In the next few sections this formalism will be used to study the $N\to\infty$ limit of various theories. In every case the procedure will be the same. One must

(i) Choose the coherence group g and the base state $|0\rangle_\chi$, and construct the coherent states $\{|u\rangle\}$.

(ii) Verify Assumptions 1–4 thereby deriving the classical phase space.

(iii) Compute the classical Hamiltonian $h_{cl}(\zeta)$ and minimize it, thereby "solving" the $N=\infty$ theory.

IV. VECTOR MODELS

In this section we shall examine the large N limit of O(N) vector models.[16] These are O(N) invariant theories whose fundamental degrees of freedom form O(N) vectors. For convenience, we will look at only linear bosonic models [i.e., models where the O(N) symmetry is realized linearly]; nonlinear models may always be reached as limits of linear models. Identical methods may also be applied to O(N) fermionic theories; see Berezin (1978) and Papanicolaou (1981) for explicit discussions using similar methods.

The basic operators in this set of theories are the position, $\{\hat{x}_i(\alpha)\}$, and conjugate momentum, $\{\hat{p}_i(\alpha)\}$, operators normalized so that

$$i[\hat{p}_i(\alpha),\hat{x}_j(\beta)]=\frac{1}{N}\delta_{ij}\delta_{\alpha\beta},\; i,j=1,\ldots,N$$
$$\alpha,\beta=1,\ldots,n . \qquad (4.1)$$

Here i and j are O(N) vector indices, α and β label the different O(N) vectors, and n is the total number of vectors. [I have included an unconventional factor of $1/\sqrt{N}$ in the definition of $\hat{x}_i(\alpha)$ and $\hat{p}_i(\alpha)$. This choice will allow us to avoid explicit rescalings of the coupling constants in the Hamiltonian as $N\to\infty$.]

The Hamiltonian is assumed to be O(N) invariant. Consequently we may completely restrict our attention to the O(N) invariant sector of the theory. The Hilbert space $\mathbf{H}_N$ is the space of O(N) invariant wave functions, and all physical operators may be constructed from the following basic invariants:

$$\hat{A}(\alpha,\beta)\equiv\tfrac{1}{2}\sum_{i=1}^{N}[\hat{x}_i(\alpha)\hat{x}_i(\beta)] , \qquad (4.2a)$$

$$\hat{B}(\alpha,\beta)\equiv\tfrac{1}{2}\sum_{i=1}^{N}[\hat{x}_i(\alpha)\hat{p}_i(\beta)+\hat{p}_i(\beta)\hat{x}_i(\alpha)] , \qquad (4.2b)$$

$$\hat{C}(\alpha,\beta)\equiv\tfrac{1}{2}\sum_{i=1}^{N}[\hat{p}_i(\alpha)\hat{p}_i(\beta)] . \qquad (4.2c)$$

The Hamiltonian $\hat{H}_N$ will be taken to be N times an arbitrary polynomial in $\hat{A}(\alpha,\beta)$, $\hat{B}(\alpha,\beta)$, and $\hat{C}(\alpha,\beta)$ with no explicit N dependence,

$$\hat{H}_N\equiv N\,h[\hat{A}(\alpha,\beta),\hat{B}(\alpha,\beta),\hat{C}(\alpha,\beta)] . \qquad (4.3)$$

To apply the previous formalism we choose $\chi=1/N$. The coherence group G_N is defined as the group generated by the operators $\hat{A}(\alpha,\beta)$ and $\hat{B}(\alpha,\beta)$. In other words, the Lie algebra $\mathbf{G}_N$ is given by

[16] All results may be easily extended to U(N) or Sp(N) vector models.

$$\mathbf{G}_N=\{\hat{\Lambda}(a,b)\equiv iN\sum_{\alpha,\beta}[a(\alpha,\beta)\hat{A}(\beta,\alpha)+b(\alpha,\beta)\hat{B}(\beta,\alpha)]\}\ . \tag{4.4}$$

$a=||a(\alpha,\beta)||$ is an arbitrary n-dimensional real symmetric matrix, and $b=||b(\alpha,\beta)||$ is an arbitrary $n\times n$ real matrix. These operators satisfy the commutation relations

$$[\hat{\Lambda}(a_1,b_1),\hat{\Lambda}(a_2,b_2)]=\hat{\Lambda}(a_{12},b_{12})\ , \tag{4.5}$$

where $a_{12}=(a_2b_1-a_1b_2)+(b_1^t a_2-b_2^t a_1)$ and $b_{12}=(b_2b_1-b_1b_2)$. (b^t is the transpose of b.) $\mathbf{G}_N$ may be regarded as a representation of the $2n$-dimensional matrix algebra,

$$\mathbf{g}=\{\lambda(a,b)\equiv\begin{bmatrix}-b & 0\\ a & b^t\end{bmatrix}\mid a=a^t\}\ , \tag{4.6}$$

since $\lambda(a,b)\in\mathbf{g}$ and $\hat{\Lambda}(a,b)\in\mathbf{G}_N$ obey identical commutation relations. Note that

$$\hat{\Lambda}(a,b)=iN\tfrac{1}{2}(-\hat{p},\hat{x})\,\lambda(a,b)\begin{bmatrix}\hat{x}\\ \hat{p}\end{bmatrix}.$$

The algebra $\mathbf{g}$ generates the group g given by

$$g=\{u(\psi,\phi)\equiv\begin{bmatrix}\phi^{-1} & 0\\ \psi\phi^{-1} & \phi^t\end{bmatrix}\mid\psi=\psi^t\}\ . \tag{4.7}$$

Note that $u(\psi,\phi)^{-1}=u(-(\phi^t)^{-1}\psi\phi^{-1},\phi^{-1})$. If $u(\psi,\phi)=\exp\lambda(a,b)$, then $\phi=\exp b$, and $\psi=\int_0^1 d\tau\times e^{\tau b^t}ae^{\tau b}$. The algebra G_N generates a group of unitary transformations, $G_N=\{\hat{U}(\psi,\phi)\equiv\exp\hat{\Lambda}(a,b)\}$, which provides a faithful representation of g.

Elements of G_N act on $\hat{x}$ and $\hat{p}$ as

$$[\hat{\Lambda}(a,b),\hat{x}_i(\alpha)]=\sum_\beta b(\alpha,\beta)\hat{x}_i(\beta)\ , \tag{4.8a}$$

and

$$[\hat{\Lambda}(a,b),\hat{p}_i(\alpha)]=-\sum_\beta[b(\beta,\alpha)\hat{p}_i(\beta)+a(\alpha,\beta)\hat{x}_i(\beta)]\ , \tag{4.8b}$$

or, equivalently,

$$[\hat{\Lambda},\begin{bmatrix}\hat{x}\\ \hat{p}\end{bmatrix}]=-\lambda\begin{bmatrix}\hat{x}\\ \hat{p}\end{bmatrix}.$$

Finite transformations, $\hat{U}\in G_N$, act as follows:

$$\hat{U}\begin{bmatrix}\hat{x}\\ \hat{p}\end{bmatrix}\hat{U}^{-1}=u^{-1}\begin{bmatrix}\hat{x}\\ \hat{p}\end{bmatrix},$$

or

$$\hat{U}(-\hat{p},\hat{x})\hat{U}^{-1}=(-\hat{p},\hat{x})u\ . \tag{4.9}$$

This shows that the momenta transform into linear combinations of momentum and position vectors, but position vectors only mix among themselves. The motivation behind this choice for the coherence group will be discussed in Sec. VII.

We will choose the base state, $|0\rangle_N$, to be the standard Gaussian, given by

$$\Psi_0(x)\equiv\langle x\mid 0\rangle=C\exp-\tfrac{1}{2}N\sum_{\alpha=1}^{n}x_i(\alpha)x_i(\alpha) \tag{4.10}$$

$[C\equiv(\pi/N)^{-nN/4}]$. Note that $(\hat{p}-i\hat{x})|0\rangle=0$. The coherent states $|u\rangle\equiv\hat{U}\,|\,0\rangle$ satisfy $\hat{U}(\hat{p}-i\hat{x})\hat{U}^{-1}|u\rangle=\{(\phi^t)^{-1}(\hat{p}-\psi\hat{x})-i\phi\hat{x}\}|u\rangle=0$. Therefore the coherent state wave functions are given by

$$\Psi_u(x)\equiv\langle x\mid u\rangle=C(z)\exp-\tfrac{1}{2}N\sum_{\alpha,\beta}x_i(\alpha)z(\alpha,\beta)x_i(\beta)\ , \tag{4.11}$$

where $C(z)=\det[N(z+\bar{z})/2\pi]^{N/4}$ and

$$z=||z(\alpha,\beta)||\equiv\phi^t\phi-i\psi\ . \tag{4.12}$$

z is a complex symmetric matrix which may be used uniquely to label the coherent states. Under the action of the group, $|z\rangle\to\hat{U}|z\rangle=|ad^*[u](z)\rangle$, where

$$ad^*[u(\phi,\psi)](z)\equiv\phi^t z\phi-i\psi\ . \tag{4.13}$$

For future convenience, we define $w\equiv(z+\bar{z})^{-1}$, and $v\equiv i(z-\bar{z})/2$; v is an arbitrary real symmetric matrix, while w must also be positive definite.

Since the coherent states are all Gaussian, computing their overlaps is trivial. One finds

$$\langle u_1\mid u_2\rangle=\exp\tfrac{1}{4}N\,\mathrm{tr}\,[\ln(\bar{z}_1+z_1)+\ln(\bar{z}_2+z_2)-2\ln(\bar{z}_1+z_2)]\ . \tag{4.14}$$

Finally, note that $\hat{A}$, $\hat{B}$, and $\hat{C}$ are all classical operators. Let us compute their symbols. Consider the $2n\times 2n$ matrix,

$$\hat{J}=\sum_{i=1}^{N}\begin{bmatrix}\hat{x}_i(\alpha)\\ \hat{p}_i(\alpha)\end{bmatrix}\otimes[-\hat{p}_i(\beta),\hat{x}_i(\beta)]+\frac{i}{2}\begin{bmatrix}\delta_{\alpha\beta} & 0\\ 0 & \delta_{\alpha\beta}\end{bmatrix}-\begin{bmatrix}-\hat{B} & 2\hat{A}\\ -2\hat{C} & \hat{B}^t\end{bmatrix}.$$

The symbol of $\hat{J}$ is given by

$$J(u)=\langle 0|\,\hat{U}^{-1}\begin{bmatrix}\hat{x}\\ \hat{p}\end{bmatrix}\otimes(-\hat{p},\hat{x})U\,|\,0\rangle+\frac{i}{2}\mathbb{1}$$

$$=\langle 0|\,u\begin{bmatrix}\hat{x}\\ \hat{p}\end{bmatrix}\otimes(-\hat{p},\hat{x})u^{-1}\,|\,0\rangle+\frac{i}{2}\mathbb{1}$$

$$=\tfrac{1}{2}uIu^{-1}\ ,$$

where

$$I=\begin{bmatrix}0 & 1\\ -1 & 0\end{bmatrix}.$$

Using the definitions of $u(\phi,\psi)$, $w=\frac{1}{2}(\phi^t\phi)^{-1}$, and $v=\psi$, we find

$$A(u)=\tfrac{1}{2}w\ , \tag{4.15a}$$

$$B(u)=v\,w\ , \tag{4.15b}$$

and

$$C(u)=\tfrac{1}{2}v\,w\,v+\tfrac{1}{8}w^{-1}\ . \tag{4.15c}$$

These results show that one may reconstruct $z=\frac{1}{2}w^{-1}-iv$ from the expectations of $\hat{A}$ and $\hat{B}$ in a given coherent state. Since z uniquely labels the coherent states, this shows that classically equivalent states must in fact be equal.

The preliminaries are all we will need to derive the $N\to\infty$ limit. To verify Assumption 1 (irreducibility of G_N), we first note that the only O(N) invariant operators which commute with the set of transformations $\{\hat{U}=\exp iN\,\mathrm{tr}(a\hat{A})\in G_N\}$ are those solely constructed from the basic operators $\{\hat{A}(\alpha,\beta)\}$. However, the set of transformations, $\{\hat{U}=\exp iN\,\mathrm{tr}(b\hat{B})\in G_N\}$ dilate and rotate the set of vectors $\{\hat{x}(\alpha)\}$ by arbitrary amounts. Therefore only constant operators commute with the full group G_N. This proves Assumption 1.

Assumption 2 ($Z(u)=0$ implies $\hat{Z}=0$) may be verified either by using analyticity in $z=\phi^t\phi-i\psi$ or by a direct argument analogous to the discussion following (3.5). Suppose there exists some operator $\hat{Z}$ whose symbol $Z(u)$ is identically zero. Then expectations in the base state of arbitrary multiple commutators of generators of $\mathbf{G}_N$ with $\hat{Z}$ vanish, that is, Eq. (3.5) holds. By taking linear combinations we may choose each generator to be (any component of)

$$\hat{L}\equiv\hat{B}-2i\hat{A}=\hat{x}\otimes(\hat{p}-i\hat{x})-i/2\ ,$$

or $\hat{L}^\dagger$. Using the commutation relations plus the fact that $\hat{L}\,|0\rangle=(1/2i)\mathbb{1}|0\rangle$, one may show by a simple induction that

$$\langle 0|\hat{L}\cdots\hat{L}\hat{Z}\hat{L}^\dagger\cdots\hat{L}^\dagger|0\rangle=0$$

for any number of $\hat{L}$'s or $\hat{L}^\dagger$'s. However, polynomials in $\hat{L}^\dagger$ applied to the base state $|0\rangle$ clearly form a dense set of O(N) invariant states. Therefore $\hat{Z}$ equals zero.

Assumption 3 (coherent states orthogonal in the $\chi\to 0$ limit) follows from examining Eq. (4.14). $\phi(u_1,u_2)\equiv -\lim_{N\to\infty}(1/N)\ln\langle u_1\,|\,u_2\rangle$ is given by

$$\phi(u_1,u_2)=-\tfrac{1}{4}\mathrm{tr}\,[\ln(\bar{z}_1+z_1)+\ln(\bar{z}_2+z_2)-2\ln(\bar{z}_1+z_2)]$$

and satisfies $\mathrm{Re}\,\phi(u_1,u_2)>0$ if and only if $z_1\neq z_2$.

Last, any N-independent polynomial in $\hat{A}$, $\hat{B}$, and $\hat{C}$ is clearly a classical operator. Therefore any Hamiltonian of the form (4.3) satisfies assumption 4 ($\chi\hat{H}_\chi$ classical).

Thus, all the assumptions of the general formalism are satisfied, and consequently in this set of models the $N\to\infty$ limit is a classical limit.

The classical phase space is a coadjoint orbit of the coherence group g. The orbit, Γ, is specified by (3.28), which yields

$$\langle\zeta(z),\lambda\rangle\equiv\lim_{N\to\infty}\frac{i}{N}\Lambda_N(u)$$
$$=\tfrac{1}{2}\mathrm{tr}\,(w\,a+2w\,v\,b)\ , \tag{4.16}$$

where $\hat{\Lambda}=\hat{\Lambda}(a,b)\in G_N$, $u=u(\psi,\phi)\in g$, and $z=\frac{1}{2}w^{-1}-iv\ =\phi^t\phi-i\psi$. The matrices v and w provide convenient independent coordinates on Γ. Equation (4.13) shows how these coordinates transform under the action of the coadjoint representation. The gradient of any function $f(v,w)$ on Γ may be given by[17]

$$\boldsymbol{d}f=\lambda(f_{,w}-\tfrac{1}{2}(vf_{,v}w^{-1}+w^{-1}f_{,v}v),\ \tfrac{1}{2}f_{,v}w^{-1})\ . \tag{4.17}$$

[We define the derivative with respect to any symmetric matrix s by

$$(f_{,s})_{\alpha\beta}=[\partial f/\partial s(\alpha,\beta)\ \text{if}\ \alpha\leq\beta]+[\partial f/\partial s(\beta,\alpha)\ \text{if}\ \beta\leq\alpha]\ .$$

Therefore

$$\tfrac{1}{2}\mathrm{tr}(\delta s\,f_{,s})=\sum_{\alpha\leq\beta}\delta s(\alpha,\beta)[\partial f/\partial s(\alpha,\beta)].$$

If $z\to z+\delta z$, then $f(z)\to f(z)+\delta f(z)$, where

$$\delta f(z)=\tfrac{1}{2}\mathrm{tr}\,(\delta v f_{,v}+\delta w f_{,w})=\langle\delta\zeta(z),\boldsymbol{d}f\rangle\ .$$

The Poisson bracket may now be computed from the definition (3.20). One finds that v and w are naturally canonically conjugate,

$$\{f(z),g(z)\}_{PB}=\tfrac{1}{2}\mathrm{tr}\,(f_{,v}g_{,w}-f_{,w}g_{,v})\ . \tag{4.18}$$

The classical Hamiltonian follows from Eqs. (3.32) and (4.3). It is given by

$$h_{cl}(v,w)=h\,(\tfrac{1}{2}v,\ vw,\ \tfrac{1}{2}vwv+\tfrac{1}{8}w^{-1})\ . \tag{4.19}$$

Finally, the classical action (3.23) becomes

$$S_{cl}[v(t),w(t)]=\int dt\{\ \tfrac{1}{2}\mathrm{tr}[v(t)\dot{w}(t)-h_{cl}(v(t),w(t))]\ \}\ . \tag{4.20}$$

This contains the complete physics of the theory in the $N\to\infty$ limit.

Let us apply these formulas to two simple examples. First, consider a theory describing a single-point particle moving in an N-dimensional spherically symmetric potential. The quantum Hamiltonian is $\hat{H}=N[\frac{1}{2}\hat{p}^2+V(\hat{x}^2)]$. [Remember that $\hat{p}$ and $\hat{x}$ have been scaled by $1/\sqrt{N}$. In terms of unscaled operators, $\hat{H}=\frac{1}{2}\hat{p}^2+NV(\hat{x}^2/N)$. This shows that the coefficient of a term $(\hat{x}^2)^n$ in the potential must be scaled by $N^{-(n-1)}$ in order to obtain a smooth large N limit.] The classical phase space has just two coordinates, (v,w) and the classical Hamiltonian describing the $N\to\infty$ limit is simply

[17] $\boldsymbol{d}f(z)$ is an element of the cotangent space $T_z^*(\Gamma)$, which consists of equivalence classes of elements of the Lie algebra $\mathbf{g}$. Equation (4.17) gives one representative of the appropriate equivalence class.

$$h_{cl}(v,w)=\tfrac{1}{2}v^2w+\tfrac{1}{8}w^{-1}+V(w)\ .$$

If we relabel w as r^2 and v as p/r, then $\{p,r\}_{PB}=1$, and the Hamiltonian becomes

$$h_{cl}(p,r)=\tfrac{1}{2}p^2+\tfrac{1}{8}r^{-2}+V(r^2)\ . \tag{4.21}$$

This is just the classical Hamiltonian of a particle with angular momentum $L^2=\frac{1}{4}$. The minimum of the Hamiltonian is at (p_0,r_0), where $p_0=0$ and $8r_0^4V'(r_0^2)=1$. The ground-state energy is $E_0=N\varepsilon_0+O(1)$, where $\varepsilon_0\equiv h_{cl}(p_0,r_0)=r_0^2V'(r_0^2)+V(r_0^2)$. Expanding the classical action (4.20) around the point (p_0,r_0) yields

$$S_{cl}=\int dt\,[-\varepsilon_0+p\delta-\tfrac{1}{2}(p^2+\omega_s^2\delta^2)+O(\delta^3,p^2\delta)]\ ,$$

where $\delta\equiv r-r_0$ and $\omega_s^2=8V'(r_0^2)+4r_0^2V''(r_0^2)$. ω_s is the energy gap to the lowest $O(N)$-invariant excited state. This is reflected, for example, in the ground-state expectation

$$\langle[\hat{x}(t)^2,\hat{x}(0)^2]\rangle=-\frac{4i}{N}\frac{r_0^2}{\omega_s}\sin\omega_s t+O(1/N^2)\ ,$$

which reveals the presence of an intermediate state with energy ω_s above the ground state. One may also probe the $O(N)$ noninvariant spectrum, even though we have restricted the Hilbert space to be $O(N)$ invariant. For example, consider the vacuum expectation of the operator $\hat{x}(t)\cdot\hat{x}(0)$,

$$G(t)\equiv\lim_{N\to\infty}\langle\hat{x}(t)\cdot\hat{x}(0)\rangle\ .$$

Using the quantum equation of motion, $\ddot{\hat{x}}+2V'(\hat{x}(t)^2)\hat{x}(t)=0$, plus factorization, one sees that $\ddot{G}(t)+\omega_v^2G(t)=0$, where $\omega_v^2=2V'(r_v^2)$. Since $G(0)=1/(2\omega_v)$, and $\dot{G}(0)=-i/2$, one finds $G(t)=(1/2\omega_v)\times\exp-i\omega_v t$. This reveals the presence of an $O(N)$-vector multiplet of excited states with energy ω_v above the ground state. The large N limit of any other observable may be computed in an analogous manner.

Finally, consider an N-component ϕ^4 field theory in d dimensions. The quantum Hamiltonian is

$$\hat{H}=N\int d^dx\{\tfrac{1}{2}\hat{\pi}(x)^2+\tfrac{1}{2}[\nabla\hat{\phi}(x)]^2+\tfrac{1}{2}\mu^2\hat{\phi}(x)^2+\tfrac{1}{4}\lambda(\hat{\phi}(x)^2)^2\}\ ,$$

and

$$i[\hat{\pi}_i(x),\hat{\phi}_j(x')]=\frac{1}{N}\delta_{ij}\delta^d(x-x')\ .$$

One may immediately apply the previous formula to find the $N\to\infty$ limit of the theory. The "matrices" $v(x,x')$ and $w(x,x')$ provide natural coordinates on the coadjoint orbit Γ. Note that $w(x,x')$ is simply the equal time expectation $\langle z|\hat{\phi}(x)\cdot\phi(x')|z\rangle$. The classical Hamiltonian which generates the $N=\infty$ dynamics is

$$h_{cl}=\int d^dx\,d^dx'd^dx''[\tfrac{1}{2}v(x,x')w(x',x'')v(x'',x)]+\int d^dx[\tfrac{1}{8}w^{-1}(x,x)+\tfrac{1}{2}(-\nabla_x^2+\mu^2)w(x,x')+\tfrac{1}{4}\lambda(w(x,x))^2]\,|_{x'=x}\ . \tag{4.22}$$

The Hamiltonian is minimized when $v=v_0$ and $w=w_0$, where

$$v_0(x,x')=0\ , \tag{4.23}$$

$$w_0(x,x')=\tfrac{1}{2}\int\frac{d^dk}{(2\pi)^d}(k^2+\mu^2+\lambda\sigma)^{-1/2}\exp ik\cdot(x-x')\ ,$$

and $\sigma\equiv w_0(x,x)$ satisfies the standard gap equation,

$$2\sigma=\int\frac{d^dk}{(2\pi)^d}(k^2+\mu^2+\lambda\sigma)^{-1/2}\ . \tag{4.24}$$

The vacuum expectation

$$\langle\hat{\phi}(x)\cdot\hat{\phi}(x')\rangle=w_0(x,x')\sim\exp-m\,|x-x'|$$

as $|x-x'|\to\infty$, where $m^2=\mu^2+\lambda\sigma$. This shows that the theory contains an $O(N)$-vector multiplet of physical particles with mass m. Expanding the classical action about (v_0,w_0) and inverting the quadratic terms yields a Green's function which describes the propagation of a pair of physical particles, including their mutual interactions. Higher terms in the expansion describe further multibody interactions. In general, the connected part of the scattering amplitude of $2n$ particles is of order $N^{-(n-1)}$.[18]

V. MATRIX MODELS

The next set of models we consider is matrix models. These are theories where the number of degrees of freedom grows like N^2 as N tends to infinity. Specifically, I shall discuss $U(N)$-invariant models of Hermitian matrices. All results may be easily extended to, for example, $O(N)$-, $Sp(N)$-, or $U(N)\times U(N)$-invariant models which describe real symmetric, Hermitian self-dual quaternionic, or arbitrary complex matrices, respectively.

Consider a theory where the basic degrees of freedom are described by the set of operators, $\{\hat{M}_{ij}^{a}\}$, and their conjugates, $\{\hat{E}_{ij}^{a}\}$, normalized so that

[18]All these results for simple vector models have been previously derived using many different methods. See, for example, Coleman, Jackiw, and Politzer (1974), Cornwall, Jackiw, and Tomboulis (1974), Halpern (1980a), Jevicki and Papanicolaou (1980), and Mlodinow and Papanicolaou (1980).

$$i[\hat{E}^{\alpha}_{ij},\hat{M}^{\beta}_{kl}]=\frac{1}{N}\delta_{il}\delta_{jk}\delta^{\alpha\beta}\ ,\quad i,j,k,l=1,\ldots,N \qquad \alpha,\beta=1,\ldots n\ . \tag{5.1}$$

(ij) and (kl) are to be regarded as matrix indices, while α and β simply label the different matrices. We assume that $(\hat{M}^{\alpha}_{ij})^{\dagger}=\hat{M}^{\alpha}_{ji}$ and $(\hat{E}^{\alpha}_{ij})^{\dagger}=\hat{E}^{\alpha}_{ji}$; in other words, $\hat{M}^{\alpha}=||\hat{M}^{\alpha}_{ij}||$ and $\hat{E}^{\alpha}\equiv||\hat{E}^{\alpha}_{ij}||$ form N-dimensional Hermitian matrices.

The Hamiltonian is assumed to be invariant under the transformation $\hat{M}^{\alpha}\to V\hat{M}^{\alpha}V^{\dagger}$, $\hat{E}^{\alpha}\to V\hat{E}^{\alpha}V^{\dagger}$, for any $V\in U(N)$. Hence, we may restrict our attention to the $U(N)$-invariant sector of the theory. Physical operators must have all the $U(N)$ indices contracted. Typical examples are $\widetilde{\mathrm{tr}}(\hat{E}^{\alpha})^2$ and $\widetilde{\mathrm{tr}}(\hat{M}^{\alpha_1}\hat{M}^{\alpha_2}\cdots\hat{M}^{\alpha_k})$. $\widetilde{\mathrm{tr}}$ stands for a normalized trace over $U(N)$-indices, $\widetilde{\mathrm{tr}}\,\hat{Q}\equiv(1/N)\sum_{i=1}^{N}Q_{ii}$. For any finite sequence of integers between one and n, $\Gamma\equiv\{\alpha_1,\alpha_2,\ldots,\alpha_k\}$, let us define $\hat{M}^{\Gamma}\equiv\hat{M}^{\alpha_1}\hat{M}^{\alpha_2}\cdots\hat{M}^{\alpha_k}$. Note that $(\hat{M}^{\Gamma})^{\dagger}=\hat{M}^{\bar{\Gamma}}$ where $\bar{\Gamma}\equiv\{\alpha_k,\ldots,\alpha_2,\alpha_1\}$ is the reversed sequence.

The Hamiltonian will be taken to be N^2 times an arbitrary $U(N)$-invariant polynomial in $\hat{E}^{\alpha}$ and $\hat{M}^{\alpha}$ with no explicit N depenence. For example,

$$\hat{H}=N^2\sum_{\alpha=1}^{n}\tfrac{1}{2}\widetilde{\mathrm{tr}}[(\hat{E}^{\alpha})^2+(\hat{M}^{\alpha}-\hat{M}^{\alpha-1})^2 +\omega^2(\hat{M}^{\alpha})^2+g(\hat{M}^{\alpha})^4] \tag{5.2}$$

is an acceptable choice which we will use as an explicit example in the following dicussion. ($\widetilde{\mathrm{tr}}$ is considered to have no N dependence).

To apply the previous formalism we must find a group of unitary transformations which acts irreducibly. This necessarily requires an infinite dimensional group. We shall choose a group which involves arbitrary products of the matrices $\{\hat{M}^{\alpha}\}$. Specifically, the Lie algebra $\mathbf{G}_N$ is given by[19]

$$\mathbf{G}_N=\{\ \hat{\Lambda}(a,b)\equiv iN^2\,\widetilde{\mathrm{tr}}(a[\hat{M}]+\tfrac{1}{2}(\hat{E}^{\alpha}b^{\alpha}[\hat{M}]+b^{\alpha}[\hat{M}]\hat{E}^{\alpha}))\ \}\ , \tag{5.3}$$

where $a[\hat{M}]=\sum_{\Gamma}a^{\Gamma}\hat{M}^{\Gamma}$, $b^{\alpha}[\hat{M}]=\sum_{\Gamma}b^{\alpha,\Gamma}\hat{M}^{\Gamma}$, and $\sum_{\Gamma}$ indicates for a sum over all sequences, $\sum_{\Gamma}\equiv\sum_{k=0}^{\infty}\left[\prod_{i=1}^{k}\sum_{\alpha_i=1}^{n}\right]\delta_{\Gamma}^{\alpha_1\cdots\alpha_k}$. The coefficients $\{a^{\Gamma}\}$ and $\{b^{\alpha,\Gamma}\}$ must satisfy $a^{\Gamma}=(a^{\bar{\Gamma}})^{*}$ and $b^{\alpha,\Gamma}=(b^{\alpha,\bar{\Gamma}})^{*}$ so that $\alpha[\hat{M}]$ and $b^{\alpha}[\hat{M}]$ are Hermitian. Furthermore, $\{a^{\Gamma}\}$ may be chosen to be cyclically symmetric, so that $a^{\Gamma_1\Gamma_2}=a^{\Gamma_2\Gamma_1}$ for all sequences Γ_1,Γ_2. Henceforth, to avoid a proliferation of $\sum_{\Gamma}$ signs, we will treat Γ like any other index and automatically sum over all sequences whenever repeated Γ's appear. For example, $(\partial a[\hat{M}]_{ij}/\partial\hat{M}^{\alpha}_{kl})=a^{\Gamma_1\alpha\Gamma_2}(\hat{M}^{\Gamma_1})_{ik}(\hat{M}^{\Gamma_2})_{lj}$. The commutation relations are

$$[\hat{\Lambda}(a_1,b_1),\hat{\Lambda}(a_2,b_2)]=\hat{\Lambda}(a_{12},b_{12})\ , \tag{5.4}$$

where

$$a_{12}[\hat{M}]_{ij}=b_1^{\alpha}[\hat{M}]_{kl}\frac{\partial a_2[\hat{M}]_{ij}}{\partial\hat{M}^{\alpha}_{kl}}-b_2^{\alpha}[\hat{M}]_{kl}\frac{\partial a_1[\hat{M}]_{ij}}{\partial\hat{M}^{\alpha}_{kl}}\ ,$$

and

$$b_{12}^{\alpha}[\hat{M}]_{ij}=b_1^{\beta}[\hat{M}]_{kl}\frac{\partial b_2^{\alpha}[\hat{M}]_{ij}}{\partial\hat{M}^{\beta}_{kl}}-b_2^{\beta}[\hat{M}]_{kl}\frac{\partial b_1^{\alpha}[\hat{M}]_{ij}}{\partial\hat{M}^{\beta}_{kl}}\ .$$

Equivalently,

$$a_{12}^{\Gamma}=(b_1^{\alpha,\Gamma'}a_2^{\Gamma_1\alpha\Gamma_2}-b_2^{\alpha,\Gamma'}a_1^{\Gamma_1\alpha\Gamma_2})\,\delta_{\Gamma}^{\Gamma_1\Gamma'\Gamma_2}$$

and

$$b_{12}^{\alpha,\Gamma}=(b_1^{\beta,\Gamma'}b_2^{\alpha,\Gamma_1\beta\Gamma_2}-b_2^{\beta,\Gamma'}b_1^{\alpha,\Gamma_1\beta\Gamma_2})\,\delta_{\Gamma}^{\Gamma_1\Gamma'\Gamma_2}\ .$$

Elements of $\mathbf{G}_N$ act on $\hat{E}^{\alpha}$ and $\hat{M}^{\alpha}$ as follows:

$$[\hat{\Lambda}(a,b),\hat{M}^{\alpha}_{ij}]=b^{\alpha}[\hat{M}]_{ij} \tag{5.5}$$

$$[\hat{\Lambda}g(a,b),\hat{E}^{\alpha}_{ij}]=-\tfrac{1}{2}\left\{\hat{E}^{\beta}_{lk}\frac{\partial b^{\beta}[\hat{M}]_{kl}}{\partial\hat{M}^{\alpha}_{ji}}+\frac{\partial b^{\beta}[\hat{M}]_{kl}}{\partial\hat{M}^{\alpha}_{ji}}\hat{E}^{\beta}_{lk}\right\} -\frac{\partial\,\mathrm{tr}\,(a[\hat{M}])}{\partial\hat{M}^{\alpha}_{ji}}\ . \tag{5.6}$$

Finite transformations, $\hat{U}(\psi,\phi)\in G_N$, may be labeled by the functionals $\psi[\hat{M}]\equiv\psi^{\Gamma}\hat{M}^{\Gamma}$ and $\phi^{\alpha}[\hat{M}]\equiv\phi^{\alpha,\Gamma}\hat{M}^{\Gamma}$. If $\hat{U}(\psi,\phi)$ equals $\exp\hat{\Lambda}(a,0)\exp\hat{\Lambda}(0,b)$, then $\psi[M]=a[M]$, and

$$\begin{aligned}\phi^{\alpha}[M]&=M^{\alpha}+b^{\alpha}[M]+\tfrac{1}{2}b^{\beta}[M]_{ij}(\partial b^{\alpha}[M]/\partial M^{\beta}_{ij})+\cdots\\&=\exp\{\ \mathrm{tr}\,(b^{\beta}[M]\partial/\partial M^{\beta})\ \}M^{\alpha}\ .\end{aligned}$$

$\phi^{\alpha}[M]$ has an inverse,

$$\phi^{\alpha}_{-1}[M]\equiv\exp\{\ -\mathrm{tr}\,(b^{\beta}[M]\partial/\partial M^{\beta})\ \}M^{\alpha}\ ,$$

such that $\phi^{\alpha}_{-1}[\phi[M]]=M^{\alpha}$. Elements of G_N obey the multiplication rule

$$\hat{U}(\psi_1,\phi_1)\hat{U}(\psi_2,\phi_2)=\hat{U}(\psi_{12},\phi_{12})\ , \tag{5.7}$$

where $\phi^{\alpha}_{12}[M]=\phi^{\alpha}_2[\phi_1[M]]$ and $\psi_{12}[M]=\psi_2[\phi_1[M]]+\psi_1[M]$. Consequently,

$$\hat{U}(\psi,\phi)^{-1}=\hat{U}(-\psi[\phi_{-1}],\phi_{-1})\ . \tag{5.8}$$

These transformations act on $\hat{E}^{\alpha}$ and $\hat{M}^{\alpha}$ as follows:

$$\hat{U}\hat{M}^{\alpha}_{ij}\hat{U}^{-1}=\phi^{\alpha}[\hat{M}]_{ij}\ , \tag{5.9}$$

$$\begin{aligned}\hat{U}\hat{E}^{\alpha}_{ij}\hat{U}^{-1}&=\left[\frac{\partial\hat{M}^{\beta}_{kl}}{\partial\phi^{\alpha}[\hat{M}]_{ji}}\right]\left[\hat{E}^{\beta}_{lk}-\frac{\partial\,\mathrm{tr}\,\psi[\hat{M}]}{\partial\hat{M}^{\beta}_{kl}}\right]-\frac{i}{2N}\frac{\partial}{\partial\hat{M}^{\beta}_{kl}}\left[\frac{\partial\hat{M}^{\beta}_{kl}}{\partial\phi^{\alpha}[\hat{M}]_{ji}}\right]\\&=\left[\frac{\partial\hat{M}^{\beta}_{kl}}{\partial\phi^{\alpha}[\hat{M}]_{ji}}\right]\hat{E}^{\beta}_{lk}+\frac{\partial}{\partial\phi^{\alpha}[\hat{M}]_{ji}}(-\mathrm{tr}\,\psi[\hat{M}]+\frac{i}{2N}\ln J[\hat{M}])\ .\end{aligned} \tag{5.10}$$

[19]A precise definition of this infinite-dimensional Lie algebra requires more mathematical sophistication than I possess. Following established tradition, we will dispense with excessive mathematical rigor and blindly proceed until confronted by obvious problems. Questions concerning, for example, appropriate growth conditions for the coefficients $\{a^{\Gamma}\}$ and $\{b^{\alpha,\Gamma}\}$ will simply be ignored. If only a single matrix is present, then a more precise definition may be given. See footnote 22.

Here, $J[M]\equiv\det[\partial\phi^\alpha[M]_{ij}/\partial M^\beta_{kl}]$ is the Jacobian for the change of variables from $\{M^\alpha_{ij}\}$ to $\{\phi^\alpha[M]_{ij}\}$.

Once again, the most convenient choice for the base state, $|0\rangle_N$, is given by a simple Gaussian,

$$\Psi_0[M]\equiv\langle M\,|\,0\rangle=C\exp-\{\tfrac{1}{2}N^2\sum_\alpha \widetilde{\mathrm{tr}}\,(M^\alpha)^2\} \quad (5.11)$$

$[C\equiv(\pi/N)^{nN^2/4}]$. The coherent states $|u\rangle\equiv\hat{U}|0\rangle$ satisfy $\hat{U}(\hat{E}^\alpha-i\hat{M}^\alpha)\hat{U}^{-1}|u\rangle=0$. Using (5.8) and (5.9), one finds that the coherent state wave functions are given by

$$\Psi_u[M]\equiv\langle M\,|\,u\rangle \quad (5.12)$$
$$=C(J[M])^{1/2}\exp-\{N^2\widetilde{\mathrm{tr}}\,(\tfrac{1}{2}\phi^\alpha[M]^2-i\psi[M])\}\ .$$

We shall need to compute the symbols, $\Lambda(a,b)(u)$, of elements of the Lie algebra. Using (5.7) and (5.8), one may evaluate $\hat{U}^\dagger\hat{\Lambda}(a,b)\hat{U}$, and find

$$\langle u\,|\,\hat{\Lambda}(a,b)\,|\,u\rangle=\langle 0\,|\,\hat{\Lambda}(a[\phi_{-1}]+\mathrm{tr}\left[b^\alpha[\phi_{-1}]\frac{\partial}{\partial\phi^\alpha_{-1}}\right]\psi[\phi_{-1}],\ \mathrm{tr}\left[b^\alpha[\phi_{-1}]\frac{\partial}{\partial\phi^\alpha_{-1}}\right]\phi^\beta[\phi_{-1}])\,|\,0\rangle$$
$$=iN^2\langle 0\,|\,\widetilde{\mathrm{tr}}\,a[\phi_{-1}[\hat{M}]]+b^\alpha[\phi_{-1}[\hat{M}]]_{ij}\frac{\partial\,\widetilde{\mathrm{tr}}\,\psi[\phi_{-1}[\hat{M}]]}{\partial\phi^\alpha_{-1}[\hat{M}]_{j}}\,|\,0\rangle\ . \quad (5.13)$$

This expresses the symbol $\Lambda(a,b)(u)$ as a Gaussian average of a functional of $\{M^\alpha\}$. Except in the special case of a single matrix (discussed below) it does not appear possible to express the expectation in any more explicit form. Fortunately, the representation above will be sufficient for our purposes. Note that (5.13) implies that classically equivalent states must give identical expectations (as $N\to\infty$) to the operators $\{\widetilde{\mathrm{tr}}\,\hat{M}^\Gamma\}$, for all sequences Γ. Since Gaussian expectations factorize,

$$\lim_{N\to\infty}\langle 0\,|\,(\widetilde{\mathrm{tr}}\,\hat{M}^{\Gamma_1})(\widetilde{\mathrm{tr}}\,\hat{M}^{\Gamma_2})\,|\,0\rangle$$
$$=\lim_{N\to\infty}\langle 0\,|\,\widetilde{\mathrm{tr}}\,\hat{M}^{\Gamma_1}\,|\,0\rangle\langle 0\,|\,\widetilde{\mathrm{tr}}\,\hat{M}^{\Gamma_2}\,|\,0\rangle\ ,$$

this implies that expectations of any product of such operators, $(\widetilde{\mathrm{tr}}\,\hat{M}^{\Gamma_1})\cdots(\widetilde{\mathrm{tr}}\,\hat{M}^{\Gamma_2})$, must also be equal. Furthermore, $\psi[M]$ can at most differ by a constant between any two classically equivalent states.

We may now examine the various assumptions made in Sec. III in order to see if the general formalism described there is applicable to the $N\to\infty$ limit of these models. We set $\chi=1/N^2$.

Assumption 1 (irreducibility of G_N) follows from an argument analogous to those used previously. Consider the subgroup of G_N which is generated by $\{\hat{\Lambda}(a,0)\in\mathbf{G}_N\}$. Any U($N$)-invariant operator which commutes with this subgroup must be solely constructed from the matrices $\{\hat{M}^\alpha\}$, i.e., must be a multiplication operator in the representation where all $\{\hat{M}^\alpha_{ij}\}$ are diagonal. However, the subgroup of G_N generated by $\{\hat{\Lambda}(0,b)\in\mathbf{G}_N\}$ contains transformations which independently translate each eigenvalue of each matrix M^α, $M^\alpha\to f_\alpha(M^\alpha)$, where each $f_\alpha(z)$ is an arbitrary monotonically increasing function of a single variable. Consequently, operators commuting with G_N cannot depend on the eigenvalues of the matrices $\{M^\alpha\}$. Furthermore, G_N also contains transformations which mix the different matrices, such as $M^\alpha\to c^{\alpha\beta}M^\beta$, where $||c^{\alpha\beta}||$ is an arbitrary invertible matrix. Such transformations completely change the inner products between eigenvectors of the different matrices $\{M^\alpha\}$. Therefore operators commuting with G_N can depend on neither the eigenvalues nor the eigenvectors of the matrices $\{M^\alpha\}$, and so must be constants. This proves Assumption 1.

Unlike our previous examples, simple analyticity arguments are not sufficient to verify Assumption 2 (operators uniquely determined by their symbols). However, an inductive proof similar to those used previously may be constructed. Let

$$\hat{L}^{\alpha,\Gamma}_\pm=N\,\mathrm{tr}[\hat{M}^{\alpha\Gamma}\pm\frac{i}{2}(\hat{E}^\alpha\hat{M}^\Gamma+\hat{M}^\Gamma\hat{E}^\alpha)]\ .$$

The following relations may be easily verified:

$$[\hat{L}^{\alpha,\Gamma}_+,\hat{L}^{\beta,\Gamma'}_+]=\delta^{\Gamma_1\alpha\Gamma_2}_{\Gamma'}\hat{L}^{\beta,\Gamma_1\Gamma\Gamma_2}_+-\delta^{\Gamma_1\beta\Gamma_2}_{\Gamma}\hat{L}^{\alpha,\Gamma_1\Gamma'\Gamma_2}_+\ , \quad (5.14a)$$

$$[\hat{L}^{\alpha,\Gamma}_-,\hat{L}^{\beta,\Gamma'}_-]=-\delta^{\Gamma_1\alpha\Gamma_2}_{\Gamma'}\hat{L}^{\beta,\Gamma_1\Gamma\Gamma_2}_-+\delta^{\Gamma_1\beta\Gamma_2}_{\Gamma}\hat{L}^{\alpha,\Gamma_1\Gamma'\Gamma_2}_-\ , \quad (5.14b)$$

$$[\hat{L}^{\alpha,\Gamma}_+,\hat{L}^{\beta,\Gamma'}_-]=\delta^{\Gamma_1\alpha\Gamma_2}_{\Gamma'}\hat{L}^{\beta,\Gamma_1\Gamma\Gamma_2}_-+\delta^{\Gamma_1\beta\Gamma_2}_{\Gamma}\hat{L}^{\alpha,\Gamma_1\Gamma'\Gamma_2}_+\ , \quad (5.14c)$$

and

$$\hat{L}^{\alpha,\Gamma}_+\,|\,0\rangle=\frac{1}{8N^2}\delta^{\gamma_1\Gamma_1\alpha\gamma_2\Gamma_2}_{\Gamma}(\hat{L}^{\gamma_1,\Gamma_1}_++\hat{L}^{\gamma_1,\Gamma_1}_-)$$
$$\times(\hat{L}^{\gamma_2,\Gamma_2}_++\hat{L}^{\gamma_2,\Gamma_2}_-)\,|\,0\rangle\ . \quad (5.15)$$

Now suppose that $\hat{Z}$ is an operator whose symbol $Z(u)$ vanishes for all $u\in g$. Then (3.5) holds, and this implies that expectations in the base state of multiple commutators of any number of $\hat{L}^{\alpha,\Gamma}_\pm$'s vanish,

$$\langle 0\,|\,[\cdots[[\hat{Z},\hat{L}^{\alpha_1,\Gamma_1}_\pm],\hat{L}^{\alpha_2,\Gamma_2}_\pm],\cdots,\hat{L}^{\alpha_k,\Gamma_k}_\pm]\,|\,0\rangle=0\ . \quad (5.16)$$

Consider any particular multiple commutator of this form, and let S be the length of the sequence $\alpha_1\Gamma_1\alpha_2\Gamma_2\cdots\alpha_k\Gamma_k$ (in other words, S equals the total number of matrices $\{M^\alpha\}$ or $\{E^\alpha\}$ in any term of the multiple commutator). Apply the obvious procedure; expand the multiple commutator, push all $\hat{L}_+$'s to the right of $\hat{Z}$ right, and all $\hat{L}_-$'s to the left of $\hat{Z}$ left. Equation (5.14) shows that any commutator terms which arise reduce S by two. Furthermore, (5.15) shows that any $\hat{L}_+$ which is applied to $|0\rangle$ (or $\hat{L}_-$ applied to $\langle 0|$) also drops S by two. This implies (by a trivial induction) that

$$\langle 0|\hat{L}_+^{\alpha_1,\Gamma_1}\cdots\hat{L}_+^{\alpha_k,\Gamma_k}\,\hat{Z}\,\hat{L}_-^{\beta_1,\Gamma_1'}\cdots\hat{L}_-^{\beta_l,\Gamma_l'}|0\rangle=0 \tag{5.17}$$

for all sets of operators $\{\hat{L}_+^{\alpha_i,\Gamma_i}\}$ and $\{\hat{L}_-^{\beta_j,\Gamma_j'}\}$. However, arbitrary products of $\hat{L}_-$'s applied to $|0\rangle$ clearly form a dense set of U(N)-invariant states. Consequently (5.17) implies that $\hat{Z}=0$, and this proves Assumption 2.

Assumption 3 requires that classically inequivalent coherent states become orthogonal as $N\to\infty$. Classically inequivalent states must be sufficiently different so as to give different expectation values for operators such as $\widetilde{\operatorname{tr}}\,\hat{M}^\Gamma$ or $\widetilde{\operatorname{tr}}\,\hat{E}^\alpha\hat{M}^\Gamma$. That such states become orthogonal as $N\to\infty$ seems almost obvious due to the rapid growth of the number of degrees of freedom. Unfortunately, I have been unable to express the overlap, $\langle 0|u\rangle=\langle 0|\hat{U}|0\rangle$, in any form which is sufficiently explicit to allow a completely rigorous proof of Assumption 3.[20] This appears to be a purely technical problem which we will simply ignore.

Finally, matrix elements of operators such as $\widetilde{\operatorname{tr}}(\hat{M}^\Gamma)$ or $\widetilde{\operatorname{tr}}(\hat{E}^\alpha)^2$ are easily seen to have finite limits as $N\to\infty$. Therefore, Hamiltonians such as (5.2) satisfy Assumption 4 [$(1/N^2)\hat{H}_N$ classical] and generate sensible dynamics as $N\to\infty$.

We may now use the general formalism of Sec. III to find the classical dynamics which describes the $N\to\infty$ limit of these models. The particular coadjoint orbit of g which will provide the classical phase space is defined by

$$\langle\zeta,\lambda(a,b)\rangle=\lim_{n\to\infty}(1/iN^2)\Lambda(a,b)(u)\ ,$$

where $\zeta\equiv Ad^*[u](\zeta_o)$. Equation (5.13) provides the "explicit" expression. We need to find a reasonably convenient set of coordinates for the coadjoint orbit; however, expressing $\langle\zeta,\lambda(a,b)\rangle$ in terms of the original labels (that is, $\{\phi^{\alpha\Gamma}\}$ and $\{\psi^\Gamma\}$) is very difficult. We may avoid this problem by simply choosing the set of expectations

$$W^\Gamma=\lim_{N\to\infty}\langle u|\widetilde{\operatorname{tr}}\,\hat{M}^\Gamma|u\rangle \tag{5.18}$$

plus $\{\psi^\Gamma\}$ as coordinates on the coadjoint orbit. Note that $\{\psi^\Gamma\}$ and $\{W^\Gamma\}$ are both cyclically symmetric (so that $\psi^{\Gamma_1\Gamma_2}=\psi^{\Gamma_2\Gamma_1}$ and $W^{\Gamma_1\Gamma_2}=W^{\Gamma_2\Gamma_1}$ for all sequences Γ_1,Γ_2). Therefore the classical phase space has one pair of coordinates for each cyclically identified sequence. We will refer to these equivalence classes of sequences as "loops" and will indicate the identification by enclosing the label for the sequence in parentheses, as in $\psi^{(\Gamma)}$. Let

$$\Omega^{(\Gamma)(\Gamma')}\equiv\lim_{N\to\infty}\frac{1}{N}\langle u|\frac{\partial\operatorname{tr}\hat{M}^\Gamma}{\partial\hat{M}^\alpha_{ij}}\frac{\partial\operatorname{tr}\hat{M}^{\Gamma'}}{\partial\hat{M}^\alpha_{ji}}|u\rangle$$
$$=\delta_\Gamma^{\Gamma_1\alpha\Gamma_2}W^{(\bar{\Gamma}_1\bar{\Gamma}_2\Gamma_2'\Gamma_1')}\delta_{\Gamma'}^{\Gamma_1'\alpha\Gamma_2'}\ . \tag{5.19}$$

Ω should be regarded as a matrix in the space of all loops. Ω is Hermitian and positive definite and so has an inverse, Ω_{-1}, such that

$$\sum_{(\Gamma')}\Omega_{-1}^{(\Gamma)(\Gamma')}\Omega^{(\Gamma')(\Gamma'')}=\delta^{(\Gamma)(\Gamma'')}\ . \tag{5.20}$$

The gradient of any function on the coadjoint orbit, $f(\{\psi^{(\Gamma)}\},\{W^{(\Gamma)}\})$, may be represented by

$$\boldsymbol{d}f(\{\psi^{(\Gamma)}\},\{W^{(\Gamma)}\})=\lambda(a,b)\ , \tag{5.21}$$

where

$$b^{\alpha,\Gamma}=\delta_\Gamma^{\Gamma_2\Gamma_1}\Omega_{-1}^{(\Gamma')(\Gamma_1\alpha\Gamma_2)}(\partial f/\partial\psi^{(\Gamma')})$$

and

$$a^{(\Gamma)}=(\partial f/\partial W^{(\Gamma)})-b^{\alpha,\Gamma'}\psi^{\Gamma_1\alpha\Gamma_2}\delta_\Gamma^{\Gamma_1\Gamma'\Gamma_2}\ .$$

Equivalently,

$$b^\alpha[M]=(\partial f/\partial\psi^{(\Gamma')})\Omega_{-1}^{(\Gamma')(\Gamma)}(\partial\operatorname{tr}M^\Gamma/\partial M^\alpha)\ ,$$

and

$$a[M]=(\partial f/\partial W^{(\Gamma)})M^\Gamma-b^\alpha[M]_{ij}\psi^\Gamma(\partial M^\Gamma/\partial M^\alpha_{ij})\ .$$

Inserting these expressions into the definition of the Poisson bracket, (3.20), and using the commutation relations, (5.4), we find

$$\{f,g\}_{PB}=\left\{\frac{\partial f}{\partial\psi^{(\Gamma)}}\frac{\partial g}{\partial W^{(\Gamma')}}-\frac{\partial g}{\partial\psi^{(\Gamma)}}\frac{\partial f}{\partial\omega^{(\Gamma')}}\right\}\ .$$
$$\times\lim_{N\to\infty}\frac{1}{N}\langle u|\frac{\partial\operatorname{tr}\hat{M}^\Gamma}{\partial\hat{M}^\alpha_{ij}}\frac{\partial\operatorname{tr}\hat{M}^{\Gamma''}}{\partial\hat{M}^\alpha_{ji}}\Omega_{-1}^{(\Gamma'')(\Gamma')}|u\rangle$$
$$=\sum_{(\Gamma)}\left\{\frac{\partial f}{\partial\psi^{(\Gamma)}}\frac{\partial g}{\partial W^{(\Gamma)}}-\frac{\partial g}{\partial\psi^{(\Gamma)}}\frac{\partial f}{\partial W^{(\Gamma)}}\right\}\ . \tag{5.22}$$

Thus $\{\psi^{(\Gamma)}\}$ and $\{W^{(\Gamma)}\}$ are naturally canonically conjugate.

Following the general formalism, the classical Hamiltonian, $h_{cl}(\{\psi^{(\Gamma)}\},\{W^{(\Gamma)}\})$, is given by the $N\to\infty$ limit of the coherent state expectation of the quantum Hamiltonian. For the choice of Hamiltonian in (5.2) this is given by

$$h_{cl}=\lim_{N\to\infty}\frac{1}{N}\sum_{\alpha=1}^{n}\frac{1}{2}\left\{\langle u|\left|\frac{\partial}{\partial\hat{M}^\alpha_{ij}}\left\{\operatorname{tr}\psi[\hat{M}]+\frac{i}{2}\operatorname{tr}(\phi^\beta[\hat{M}])^2-\frac{i}{2N}\ln J[M]\right\}\right|^2|u\rangle\right.$$
$$\left.+\langle u|\operatorname{tr}\{(\hat{M}^\alpha-\hat{M}^{\alpha-1})^2+\omega^2(\hat{M}^\alpha)^2+g(\hat{M}^\alpha)^4\}|u\rangle\right\}\ . \tag{5.23}$$

[20]One may verify that $\phi(0,u)\equiv-\lim_{N\to\infty}(1/N^2)\ln\langle 0|u\rangle$ has strictly positive curvature about the point $u=0$ in all directions other than the one corresponding to constant phase rotations.

We must express this expectation in terms of the classical coordinates $\{\psi^{(\Gamma)}\}$ and $\{W^{(\Gamma)}\}$. The following observation will enable us to accomplish this. If $f[\widetilde{\mathrm{tr}}(\hat{M}^{\Gamma})]$ is any functional of $\{\widetilde{\mathrm{tr}}(\hat{M}^{\Gamma})\}$, then

$$\lim_{N\to\infty} N\langle u\,|\left|\frac{\partial}{\partial M_{ij}^{\alpha}}f[\widetilde{\mathrm{tr}}(\hat{M}^{\Gamma})]\right|^{2}|\,u\rangle=\langle u\,|\frac{\partial\bar{f}}{\partial\hat{M}_{ji}^{\alpha}}\frac{\partial\,\mathrm{tr}\hat{M}^{\Gamma'}}{\partial\hat{M}_{ij}^{\alpha}}|\,u\rangle\Omega_{-1}^{(\Gamma')(\Gamma)}\langle u\,|\frac{\partial\,\mathrm{tr}\hat{M}^{\bar{\Gamma}}}{\partial\hat{M}_{kl}^{\beta}}\frac{\partial f}{\partial\hat{M}_{kl}^{\beta}}|\,u\rangle\,. \tag{5.24}$$

This follows from factorization plus the definition of $\Omega^{(\Gamma)(\Gamma')}$ in Eq. (5.19). Using this relation, one finds

$$h_{cl}(\{\psi^{(\Gamma)}\},\{W^{(\Gamma)}\})=\tfrac{1}{2}\psi^{(\bar{\Gamma})}\Omega^{(\Gamma)(\Gamma')}\psi^{(\Gamma')}+\tfrac{1}{2}\omega^{(\bar{\Gamma})}\Omega_{-1}^{(\Gamma)(\Gamma')}\omega^{(\Gamma')}+V[W^{(\Gamma)}]\,, \tag{5.25}$$

where

$$\begin{aligned}\omega^{(\Gamma)}&\equiv\lim_{N\to\infty}\langle u\,|\frac{\partial\,\widetilde{\mathrm{tr}}\hat{M}^{\bar{\Gamma}}}{\partial\hat{M}_{ij}^{\alpha}}\frac{\partial}{\partial\hat{M}_{ji}^{\alpha}}\left\{\frac{1}{2}\mathrm{tr}\,(\phi^{\beta}[\hat{M}])^{2}-\frac{1}{2N}\ln J[\hat{M}]\right\}|\,u\rangle\\&=\lim_{N\to\infty}\frac{1}{2}\langle u\,|\left[\frac{\partial}{\partial\hat{M}_{ij}^{\alpha}}\frac{\partial}{\partial\hat{M}_{ji}^{\alpha}}\widetilde{\mathrm{tr}}(\hat{M}^{\bar{\Gamma}})\right]|\,u\rangle\\&=W^{(\Gamma_2)}W^{(\Gamma_3\Gamma_1)}\delta_{\bar{\Gamma}}^{\Gamma_1\alpha\Gamma_2\alpha\Gamma_3}\,,\end{aligned} \tag{5.26}$$

and

$$\begin{aligned}V[W^{(\Gamma)}]&=\lim_{N\to\infty}\tfrac{1}{2}\sum_{\alpha=1}^{n}\langle u\,|\,\widetilde{\mathrm{tr}}\{(\hat{M}^{\alpha}-\hat{M}^{\alpha-1})^{2}+\omega^{2}(\hat{M}^{\alpha})^{2}+g(\hat{M}^{\alpha})^{4}\}\,|\,u\rangle\\&=\sum_{\alpha=1}^{n}\{(1+\tfrac{1}{2}\omega^{2})W^{(\alpha\alpha)}-W^{(\alpha(\alpha-1))}+\tfrac{1}{2}gW^{(\alpha\alpha\alpha\alpha)}\}\,.\end{aligned} \tag{5.27}$$

This is about as far as we can go in the discussion of general matrix models. Equation (5.25) gives the explicit form of the classical Hamiltonian. To "solve" the $N=\infty$ theory completely, one must find the minimum of the $h_{cl}(\{\psi^{(\Gamma)}\},\{W^{(\Gamma)}\})$. This appears to be extremely difficult.

In the special case of a single matrix, one may completely solve the $N=\infty$ theory.[21] Let us briefly see how the preceding discussion simplifies in this case.

If only a single matrix is present, then the set of sequential products of matrices, $\{\hat{M}^{\Gamma}\}$, reduces to a single set of integer powers, $\{\hat{M}^{k}\}$, $k=0,1,2,\ldots,\infty$. Therefore elements of the Lie algebra, $\mathbf{G}_N=\{\hat{\Lambda}(a,b)\}$, may be labeled by two "arbitrary" real functions of a single variable, $a(z)\equiv\sum_{k=0}^{\infty}a_k z^k$ and $b(z)\equiv\sum_{k=0}^{\infty}b_k z^k$.[22] This algebra generates a group of finite transformations $G_N=\{\hat{U}(\psi,\phi)\}$, where $\psi(z)$ is an arbitrary function, and $\phi(z)$ is necessarily monotonically increasing. If $\hat{U}(\psi,\phi)=\exp\hat{\Lambda}(a,0)\exp\hat{\Lambda}(0,b)$, then the formal expression $\phi(z)=\exp[b(z)\partial/\partial z]z$ may be solved by the implicit definition,

$$\int_{z}^{\phi(z)}dz'/b(z')=1\,.$$

The action of the coherence group is given by (5.9) and (5.10). When only a single matrix is present, the Jacobian $J(M)\equiv\det[\partial\phi(M)_{ij}/\partial M_{kl}]$ may be directly expressed in terms of the eigenvalues $\{\mu_i\}$ of the matrix M. One finds (Mehta, 1967),

$$J(M)=\prod_{i=1}^{N}\phi'(\mu_i)\prod_{i\neq j}\left\{\frac{[\phi(\mu_i)-\phi(\mu_j)]}{(\mu_i-\mu_j)}\right\}. \tag{5.28}$$

All expectations in the coherent states may now be reduced to averages over a single Gaussian random matrix. Such averages may be computed explicitly for large N by writing the integration measure in terms of the eigenvalues $\{\mu_i\}$ and by using saddlepoint methods to evaluate the resulting integrals. One finds that the Gaussian measure $dM\exp-N\,\mathrm{tr}\,M^2$ leads to a density of eigenvalues given by Wigner's semicircle distribution (Wigner, 1959),

$$d\rho(\mu)=\frac{1}{\pi}(2-\mu^2)^{1/2}\,\Theta(2-\mu^2)\,d\mu \tag{5.29}$$

(Θ is the step function; $\Theta(x)=1$ if $x\geq 0$, 0 otherwise). This allows us to express the symbols of elements of the Lie algebra, (5.13), in the form

$$\Lambda(a,b)(u)=iN^2\int d\rho(\phi(\mu))\{a(\mu)+b(\mu)\psi'(\mu)\}+O(1)\,. \tag{5.30}$$

[21]In fact, the theory is exactly soluble for arbitrary N. See Brezin *et al.* (1978) and Marchesini and Onofri (1980).

[22]Requiring $a(z)$ and $b(z)$ to be entire functions which are bounded on the real axis, and whose derivatives (to all orders) are bounded on the real axis, appears to be a sensible definition. Such functions include, for example, functions with smooth Fourier transforms which decrease faster than any exponential (on the real axis). This set of functions is closed under the commutation relations, generates a well-defined set of one-parameter subgroups, $\exp t\lambda$, $\lambda\in\mathbf{g}$, and leads to a coherence group which acts irreducibly.

As usual, these expectations determine the appropriate coadjoint orbit for the classical phase space. As shown previously,

$$W^k \equiv \lim_{N\to\infty} \langle u \mid \widetilde{\operatorname{tr}} M^k \mid u \rangle = \int d\rho(\phi(\mu))\mu^k$$

and

$$\psi^k \equiv \frac{1}{k!}\frac{\partial^k \psi(z)}{(\partial z)^k}\bigg|_{z=0} \quad (\text{for } k=1,2,\dots)$$

provide canonically conjugate coordinates on the classical phase space. However, these are not the most convenient set of coordinates. If we define $w(\mu) \equiv d\rho(\phi(\mu))/d\mu$ and $v(\mu) \equiv \psi(\mu)$, then one can show that $v(\mu)$ and $w(\mu)$ are canonically conjugate, $\{v(\mu), w(\mu')\}_{PB} = \delta(\mu-\mu')$.[23] [Alternatively, $\mu(x)$, defined on the interval $0<x<1$ by $\int_{-\infty}^{\mu(x)} d\rho(\phi(\mu)) = x$, and $v(x) \equiv \psi'(\mu(x))$, are also conjugate, $\{v(x), \mu(x')\}_{PB} = \delta(x-x')$.]

Finally, the classical Hamiltonian may be expressed in a very simple form. Using (5.28), the expectation of $\widetilde{\operatorname{tr}}\,\hat{E}^2$ may be computed as follows:

$$\begin{aligned}
\langle u \mid \widetilde{\operatorname{tr}}\hat{E}^2 \mid u \rangle &= \frac{1}{N}\langle u \mid \left| \frac{\partial}{\partial M_{ij}}\left\{ \operatorname{tr}\psi(M) + \frac{i}{2}\operatorname{tr}\phi(M)^2 - \frac{i}{2N}\ln J(M)\right\}\right|^2 \mid u \rangle \\
&= \int d\rho(\phi(\mu)) \left| \frac{\partial}{\partial \mu}\left\{ \psi(\mu) + \frac{i}{2}\phi(\mu)^2 - i\int d\rho(\phi(\mu'))\ln\frac{\phi(\mu)-\phi(\mu')}{\mu-\mu'} \right\}\right|^2 + O(1/N^2) \\
&= \int d\rho(\phi(\mu))\{\psi'(\mu)^2 + [\textstyle{⨍} d\rho(\phi(\mu'))/\mu-\mu']^2\} + O(1/N^2) \\
&= \int d\mu\, w(\mu)\{v'(\mu)^2 + \frac{\pi^2}{3}w(\mu)^2\} + O(1/N^2)\,.
\end{aligned}$$

Here we have used the defining relation for the Wigner distribution, $\mu = ⨍ d\rho(\mu')/\mu-\mu'$, and the fact that

$$\int d\mu\, w(\mu)(⨍ d\mu' w(\mu')/\mu-\mu')^2 = \frac{\pi^2}{3}w(\mu)^3\,.$$

This latter relation may be easily derived by writing the principal-value integrals as averages of contour integrals passing above and below the pole, and then symmetrizing over the ordering of the contours [see Mondello and Onofri (1981) or Shapiro (1981) for more detailed discussion]. Therefore the final classical Hamiltonian is given by

$$h_{cl}[v(\mu), w(\mu)] = \tfrac{1}{2}\int d\mu\, w(\mu)\{v'(\mu)^2 + \frac{\pi^2}{3}w(\mu)^2 + \omega^2\mu^2 + g\mu^4\}\,, \tag{5.32}$$

Minimizing this Hamiltonian, subject to the constraint $\int d\mu\, w(\mu) = 1$, yields

$$v_0(\mu) = 0$$

and

$$w_0(\mu) = \frac{1}{\pi}(2e - \omega^2\mu^2 - g\mu^4)^{1/2}\,\Theta(2e - \omega^2\mu^2 - g\mu^4)\,, \tag{5.33}$$

where the Lagrange multiplier, e, is determined by the condition $1 = \int d\mu\, w(\mu)$. The $N\to\infty$ limit of the ground-state energy is given by

$$\begin{aligned}
\lim_{N\to\infty} E_0/N^2 &= h_{cl}[v_0(\mu), w_0(\mu)] \\
&= \tfrac{1}{3}\int d\mu\, w(\mu)(\omega^2\mu^2 + g\mu^4 + e)\,.
\end{aligned} \tag{5.34}$$

The classical action is given by

$$S_{cl}[v,w] = \int d\mu\{v(\mu)\dot{w}(\mu) - \tfrac{1}{2}w(\mu)[v'(\mu)^2 + \frac{\pi^2}{3}w(\mu)^2 + \omega^2\mu^2 + g\mu^4]\}\,. \tag{5.35}$$

One may expand the classical action about the minimum to find the small oscillation frequencies. One finds

$$\omega_j = j\omega_s \quad \text{for } j = 1,2,3,\dots, \tag{5.36}$$

[23] $v(\mu)$ and $w(\mu)$ are not strictly independent coordinates. $w(\mu)$ must satisfy the constraint $\int d\mu\, w(\mu) = 1$, and adding a constant to $v(\mu)$ does not affect the dynamics.

where

$$\omega_s^{-1} = \int d\mu\, w(\mu)/(2e - \omega^2\mu^2 - g\mu^4) .$$

This is known to reproduce correctly the $N \to \infty$ limit of the U(N)-invariant spectrum of the theory (Shapiro, 1981; Mondello and Onofri, 1981).

VI. GAUGE THEORIES

As a final example, we will examine the $N \to \infty$ limit of U(N)-lattice gauge theories. [I choose to work on a lattice in order to make the theory well defined. The choice of a U(N) instead of SU(N)-gauge theory is made for convenience; the difference between the groups is irrelevant for large N. Furthermore, one may show that U(N), O(N), and Sp(N) lattice gauge theories all have equivalent large N limits (Lovelace, 1982). I will not include fermions in the following discussion, although they may be easily inserted (Yaffe, 1982).] The analysis is essentially identical to that presented in the preceeding section on matrix models. Consequently, this treatment will be as brief as possible.

Consider a U(N)-Hamiltonian lattice gauge theory. The basic operators are the link variables, $\{\hat{V}^\alpha_{ij}\}$, and their conjugate momenta, $\{\hat{E}^\alpha_{ij}\}$. α labels the links of lattice. For each link α, $\hat{V}^\alpha \equiv \|\hat{V}^\alpha_{ij}\|$ is an N-dimensional unitary matrix, $(\hat{V}^\alpha)^\dagger = (\hat{V}^\alpha)^{-1}$, and $\hat{E}^\alpha \equiv \|\hat{E}^\alpha_{ij}\|$ is Hermitian, $(\hat{E}^\alpha)^\dagger = \hat{E}^\alpha$. Each link α is assumed to have a standard orientation. Links with orientation opposite to the standard will be denoted by $\bar{\alpha}$, and $\hat{V}^{\bar{\alpha}} \equiv (\hat{V}^\alpha)^\dagger$. The commutation relations are

$$[\hat{E}^\alpha_{ij}, V^\beta_{kl}] = \frac{1}{2N}\delta^{\alpha\beta}\delta_{kj}\hat{V}^\alpha_{il} , \tag{6.1a}$$

$$[\hat{E}^\alpha_{ij}, \hat{V}^{\bar{\beta}}_{kl}] = \frac{-1}{2N}\delta^{\alpha\beta}\delta_{il}\hat{V}^{\bar{\alpha}}_{kj} , \tag{6.1b}$$

and

$$[\hat{E}^\alpha_{ij}, \hat{E}^\beta_{kl}] = \frac{1}{2N}\delta^{\alpha\beta}(\delta_{kj}\hat{E}^\alpha_{il} - \delta_{il}\hat{E}^\alpha_{kj}) . \tag{6.1c}$$

$\hat{E}^\alpha_{ij}$ may be represented by $(1/2N)\hat{V}^\alpha_{ik}(\partial/\partial\hat{V}^\alpha_{jk})$. [Our conjugate momenta, $\{\hat{E}_{ij}\}$, are related to the more conventional generators $\{\hat{E}^a\}$ (which satisfy $[\hat{E}^a, \hat{E}^b] = -if^{abc}\hat{E}^c$) by $\hat{E}_{ij} = (1/N)t^a_{ij}\hat{E}^a$. Here $\{f^{abc}\}$ are the structure constants of U(N), and the matrices $\{t^a\}$ represent the generators of U(N). They satisfy $[t^a, t^b] = if^{abc}t^c$, $\mathrm{tr}(t^a t^b) = \frac{1}{2}\delta^{ab}$, and $\sum_{a=0}^{N^2-1} t^a_{ij} t^a_{kl} = \frac{1}{2}\delta_{il}\delta_{kj}$.]

Gauge transformations are specified by giving an arbitrary element of U(N) for every site of the lattice, $\Omega^s \in$ U(N). If α denotes the link running from site s to site s', then under a gauge transformation $\hat{V}^\alpha \to (\Omega^s)\hat{V}^\alpha(\Omega^{s'})^\dagger$ and $\hat{E}^\alpha \to (\Omega^s)\hat{E}^\alpha(\Omega^s)^\dagger$. For any ordered set of links, $\Gamma = \{\alpha_1, \alpha_2, \ldots, \alpha_k\}$, which forms a single closed curve beginning and ending at some site s, let us define $\hat{V}^\Gamma \equiv \hat{V}^{\alpha_1}\hat{V}^{\alpha_2}\cdots\hat{V}^{\alpha_k}$. Note that $\hat{V}^\Gamma$ transforms covariantly under a gauge transformation, $\hat{V}^\Gamma \to (\Omega^s)\hat{V}^\Gamma(\Omega^s)^\dagger$. Henceforth, if α labels a particular link of the lattice, then we will use Γ^α to denote an arbitrary closed curve which begins with the link α. (Γ) will denote closed curves irrespective of their starting point; in other words, (Γ) labels loops.

We will assume that the lattice is cubic and choose the standard Kogut-Susskind Hamiltonian (Kogut and Susskind, 1975),

$$\hat{H} \equiv N^2\, \widetilde{\mathrm{tr}}\{\lambda \sum_\alpha (\hat{E}^\alpha)^2 - \lambda^{-1}\sum_p (\hat{V}^{\partial p} + \hat{V}^{\partial\bar{p}})\} . \tag{6.2}$$

Here $\lambda \equiv g^2 N$ and $p(\partial p)$ indicates (the boundary of) an arbitrary plaquette. $\hat{H}$ is, of course, gauge invariant.

We will choose the Lie algebra of the coherence group to be given by

$$\mathbf{G}_N = \{\, \hat{\Lambda}(a,b) \equiv iN^2\, \widetilde{\mathrm{tr}}\,(a[\hat{V}] + E^\alpha b^\alpha[\hat{V}] + b^\alpha[\hat{V}]E^\alpha)\,\} , \tag{6.3}$$

where $a[\hat{V}] \equiv \sum_{(\Gamma)} a^{(\Gamma)}\hat{V}^\Gamma$ and $b^\alpha[\hat{V}] \equiv \sum_{\Gamma^\alpha} b^{\alpha,\Gamma^\alpha}\hat{V}^{\Gamma^\alpha}$. $a[\hat{V}]$ and $b^\alpha[\hat{V}]$ must be Hermitian. The commutation relations are

$$[\hat{\Lambda}(a_1,b_1), \hat{\Lambda}(a_2,b_2)] = \hat{\Lambda}(a_{12},b_{12}) , \tag{6.4}$$

where

$$a_{12}[\hat{V}]_{ij} = i(b^\alpha_1[\hat{V}]\hat{V}^\alpha)_{kl}\frac{\partial a_2[\hat{V}]_{ij}}{\partial \hat{V}^\alpha_{kl}} - i(b^\alpha_2[\hat{V}]\hat{V}^\alpha)_{kl}\frac{\partial a_1[\hat{V}]_{ij}}{\partial \hat{V}^\alpha_{kl}} ,$$

and

$$b^\alpha_{12}[\hat{V}]_{ij} = i(b^\beta_1[\hat{V}]\hat{V}^\beta)_{kl}\frac{\partial b^\alpha_2[\hat{V}]_{ij}}{\partial \hat{V}^\beta_{kl}} - i(b^\beta_2[\hat{V}]\hat{V}^\beta)_{kl}\frac{\partial b^\alpha_1[\hat{V}]_{ij}}{\partial \hat{V}^\beta_{kl}} + i\delta^{\alpha\beta}[b^\alpha_2[\hat{V}], b^\alpha_1[\hat{V}]]_{ij} .$$

This algebra generates a coherence group which is very similar to the group used for treating matrix models. Elements of the group, $\hat{U}[\psi,\phi] \in G_N$, are labeled by the functionals $\psi[\hat{V}] \equiv \psi^{(\Gamma)}\hat{V}^\Gamma$ and $\phi^\alpha[\hat{V}] \equiv \phi^{\alpha,\Gamma^\alpha}\hat{V}^{\Gamma^\alpha}$. $\psi[\hat{V}]$ must be Hermitian and $\phi^\alpha[\hat{V}]$ unitary. The action of the coherence group is given by

$$\hat{U}\hat{V}^\alpha_{ij}\hat{U}^{-1} = (\phi^\alpha[\hat{V}]\hat{V}^\alpha)_{ij} \tag{6.5}$$

and

$$\hat{U}\hat{E}^{\alpha}_{ij}\hat{U}^{-1} = (\phi^{\alpha}[\hat{V}]\hat{V}^{\alpha})_{ik}\left[\frac{\partial \hat{V}^{\beta}_{mn}}{\partial(\phi^{\alpha}[\hat{V}]\hat{V}^{\alpha})_{jk}}\right]\left[(\hat{V}^{\beta}\hat{E}^{\beta})_{nm} - \frac{i}{2}\frac{\partial \operatorname{tr}\psi[\hat{V}]}{\partial \hat{V}^{\beta}_{mn}}\right] + \frac{1}{4N}(\phi^{\alpha}[\hat{V}]\hat{V}^{\alpha})_{ik}\frac{\partial}{\partial \hat{V}^{\beta}_{mn}}\left[\frac{\partial \hat{V}_{mn}}{\partial(\phi^{\alpha}[\hat{V}]\hat{V}^{\alpha})_{jk}}\right]$$

$$= (\phi^{\alpha}[\hat{V}]\hat{V}^{\alpha})_{ik}\left[\left[\frac{\partial \hat{V}^{\beta}_{mn}}{\partial(\phi^{\alpha}[\hat{V}]\hat{V})_{jk}}\right](\hat{V}^{\bar{\beta}}\hat{E}^{\beta})_{nm} - \frac{i}{2}\frac{\partial}{\partial(\phi^{\alpha}[\hat{V}]\hat{V})_{jk}}\left[\operatorname{tr}\psi[\hat{V}] - \frac{i}{2N}\ln J[\hat{V}]\right]\right] . \tag{6.6}$$

$J[V] \equiv \det\,[\partial(\phi^{\alpha}[V]V^{\alpha})_{ij}/\partial V^{\beta}_{kl}]$ is the Jacobian for the change of variables from $\{V^{\alpha}_{ij}\}$ to $\{(\phi^{\alpha}[V]V^{\alpha})_{ij}\}$. Note that

$$J[V]^{*} = J[V]\prod_{\alpha}\{\det(V^{\alpha})/\det(\phi^{\alpha}[V]V^{\alpha})\}^{2} .$$

The base state, $|0\rangle$, will be chosen to be the state which is annihilated by all conjugate momenta, $\hat{E}^{\alpha}|0\rangle = 0$, for all links α. Its wave function is simply a constant,

$$\Psi_0[V] \equiv \langle V\,|\,0\rangle = 1 .$$

This base state generates a set of coherent states whose wave functions are given by

$$\Psi_u[V] \equiv \langle V\,|\,u\rangle = (J[V])^{1/2}\exp iN^{2}\,\widetilde{\operatorname{tr}}(\psi[V]) . \tag{6.7}$$

We may now apply the general formalism developed earlier. The arguments needed to verify the assumptions made in Sec. III are essentially identical to those presented in the last section.[24] I will not bother to repeat that discussion here. Note that the coupling constant λ in the Hamiltonian (6.2) must remain fixed in order for the dynamics to have a sensible limit as $N \to \infty$.

The coadjoint orbit which provides the appropriate classical phase space is specified by

$$\langle \zeta,\, \lambda(a,b)\rangle = \lim_{N\to\infty}(1/iN^{2})\langle u\,|\,\hat{\Lambda}(a,b)\,|\,u\rangle$$

$$= \lim_{N\to\infty}\langle u\,|\left[\widetilde{\operatorname{tr}}\,a[\hat{V}] + i(b^{\alpha}[\hat{V}]\hat{V}^{\alpha})_{ij}\frac{\partial\,\widetilde{\operatorname{tr}}\,\psi[\hat{V}]}{\partial \hat{V}^{\alpha}_{ij}}\right]|\,u\rangle . \tag{6.8}$$

Reasonably convenient coordinates on this coadjoint orbit are provided by the coefficients $\{\psi^{(\Gamma)}\}$ plus the expectation values of Wilson loops,

$$W^{(\Gamma)} \equiv \lim_{N\to\infty}\langle u\,|\,\widetilde{\operatorname{tr}}V^{(\Gamma)}\,|\,u\rangle . \tag{6.9}$$

By applying exactly the same procedure that was used in the last section, one may derive the Poisson bracket for the classical phase space and compute the classical Hamiltonian. One finds that the coordinates $\{\psi^{(\Gamma)}\}$ and $\{W^{(\Gamma)}\}$ are naturally canonically conjugate,

$$\{f,\,g\}_{PB} = \sum_{(\Gamma)}\left[\frac{\partial f}{\partial\psi^{(\Gamma)}}\frac{\partial g}{\partial W^{(\Gamma)}} - \frac{\partial f}{\partial W^{(\Gamma)}}\frac{\partial g}{\partial\psi^{(\Gamma)}}\right] . \tag{6.10}$$

The classical Hamiltonian, $h_{cl}(\{\psi^{(\Gamma)}\},\{W^{(\Gamma)}\})$, is given by the expectation of (6.2),

$$h_{cl} = \lim_{N\to\infty}(1/N^{2})\langle u\,|\,\hat{H}_N\,|\,u\rangle$$

$$= \lim_{N\to\infty}\frac{1}{N}\left\{\lambda\sum_{\alpha}\frac{1}{4}\langle u\,|\left|\hat{V}^{\alpha}_{ij}\frac{\partial}{\partial\hat{V}^{\alpha}_{kj}}\left\{\operatorname{tr}\psi[\hat{V}] - \frac{i}{2N}\ln J[\hat{V}]\right\}\right|^{2}|\,u\rangle - \frac{1}{\lambda}\sum_{p}\langle u\,|\,\operatorname{tr}\{\hat{V}^{\partial p} + \hat{V}^{\partial\bar{p}}\}\,|\,u\rangle\right\} .$$

Let us define the following Hermitian, positive definite matrix in "loop space,"

---

[24]The easiest way to verify Assumption 2 is to regard the gauge theory as the $\varepsilon \to 0$ limit of a theory in which the matrices $\{V^{\alpha}\}$ are arbitrary complex matrices and a term $(1/2\varepsilon)\operatorname{tr}(V^{\alpha}V^{\bar{\alpha}} - 1)$ is added to the Hamiltonian. (This is completely analogous to the construction of nonlinear sigma models from limits of linear sigma models.) One may then use exactly the same procedure developed earlier for matrix models.

$$\Omega^{(\Gamma)(\Gamma')} \equiv \lim_{N\to\infty} \frac{1}{N} \langle u \,|\, \frac{\partial\, \mathrm{tr} \hat{V}^{(\bar{\Gamma})}}{\partial \hat{V}^{\bar{\alpha}}_{ji}} \frac{\partial\, \mathrm{tr} \hat{V}^{(\Gamma')}}{\partial \hat{V}^{\alpha}_{ij}} \,|\, u \rangle$$

$$= \Delta_{\Gamma}^{\Gamma_1 \alpha \Gamma_2} W^{(\bar{\Gamma}_1 \bar{\Gamma}_2 \Gamma_2' \Gamma_1')} \Delta_{\Gamma'}^{\Gamma_1' \alpha \Gamma_2'} \,. \tag{6.11}$$

$\Delta_{\Gamma}^{\Gamma_1\alpha\Gamma_2}$ is a "signed" delta function, defined to equal $\pm\delta_{\Gamma}^{\Gamma_1\alpha\Gamma_2}$, depending on the orientation of Γ ($+$ if Γ traverses the link α in the standard orientation, $-$ otherwise). The classical Hamiltonian may then be expressed in the form,

$$h_{cl}(\{\psi^{(\Gamma)}\},\{W^{(\Gamma)}\}) = \tfrac{1}{4}\lambda \psi^{(\bar{\Gamma})} \Omega^{(\Gamma)(\Gamma')} \psi^{(\Gamma')} + \lambda\, \omega^{(\bar{\Gamma})} \Omega_{-1}^{(\Gamma)(\Gamma')} \omega^{(\Gamma')} - \frac{1}{\lambda} \sum_{p} (W^{(\partial p)} + W^{(\partial \bar{p})}) \,, \tag{6.12}$$

where $\Omega_{-1}^{(\Gamma)(\Gamma')}\, \Omega^{(\Gamma')(\Gamma'')} = \delta^{(\Gamma)(\Gamma'')}$ and

$$\omega^{(\Gamma)} \equiv \lim_{N\to\infty} \langle u \,|\, \frac{\partial\, \mathrm{tr} \hat{V}^{(\bar{\Gamma})}}{\partial \hat{V}^{\bar{\alpha}}_{ji}} \frac{\partial}{\partial \hat{V}^{\alpha}_{ij}} \left[\frac{1}{4N^2} \ln J[\hat{V}] \right] |\, u \rangle$$

$$= \lim_{N\to\infty} \langle u \,|\, [\hat{E}^{\alpha}_{ij}, [\hat{E}^{\alpha}_{ji}, \mathrm{tr} \hat{V}^{(\bar{\Gamma})}]] \,|\, u \rangle$$

$$= \lim_{N\to\infty} \frac{1}{4N} \langle u \,|\, \left[\hat{V}^{\alpha}_{ik} \frac{\partial}{\partial \hat{V}^{\alpha}_{jk}} \hat{V}^{\alpha}_{jl} \frac{\partial}{\partial \hat{V}^{\alpha}_{il}} \widetilde{\mathrm{tr}}\, (\hat{V}^{(\bar{\Gamma})}) \right] |\, u \rangle$$

$$= \tfrac{1}{4} l(\Gamma) W^{(\bar{\Gamma})} + \tfrac{1}{2} W^{(\bar{\Gamma}_3 \bar{\alpha} \bar{\Gamma}_1)} W^{(\bar{\Gamma}_2 \bar{\alpha})} \Delta_{\Gamma}^{\Gamma_1 \alpha \Gamma'} \Delta_{\Gamma'}^{\Gamma_2 \alpha \Gamma_3} \,. \tag{6.13}$$

Here $l(\Gamma) \equiv V^{\alpha}_{ij} (\partial\, \mathrm{tr} V^{(\Gamma)} / \partial V^{\alpha}_{ij}) / (\mathrm{tr} V^{(\Gamma)})$ is the signed "length" of the loop (Γ). [In deriving (6.13) we have used the fact that $(1/N^2)(\ln J[V] - \ln J[V]^{\dagger}) = O(1/N)$.]

Equation (6.12) gives the explicit form of the classical Hamiltonian.[25] Solving the $N=\infty$ theory requires finding the minimum of $h_{cl}(\{\psi^{(\Gamma)}\},\{W^{(\Gamma)}\})$. Regrettably, if the lattice contains more than one plaquette, this minimization appears to be very difficult. The one plaquette theory, however, may be completely solved. I will briefly summarize the explicit results which may be obtained in that case.

If only a single plaquette is present, then the theory may be completely expressed in terms of a single unitary matrix, given by the product of the link variables around the plaquette, $\hat{V} \equiv \hat{V}^1 \hat{V}^2 \hat{V}^3 \hat{V}^4$, and its conjugate momentum $\hat{E} \equiv \hat{E}^1$. Consequently, elements of the coherence group may be labeled by two functions of a single variable, $\psi(z)$ and $\phi(z)$, defined on the unit circle [$\psi(z)$ must be real and $\phi(z)$ must provide an invertible mapping of the unit circle onto itself]. The Jacobian $J[V] = \det[\partial(\phi[V]V)_{ij} / \partial V_{kl}]$ may then be explicitly computed in terms of the eigenvalues $\{v_i\}$ of the matrix V. One finds (Mehta, 1967)

$$J[V] = \prod_{i} (\psi(\nu_i) + \nu_i \phi'(\nu_i)) \prod_{i \neq j} |\, [\nu_i \phi(\nu_i) - \nu_j \phi(\nu_j)] / (\nu_i - \nu_j) \,| \,. \tag{6.14}$$

Expectations such as (6.8) may be expressed in the form

$$\langle u \,|\, \hat{\Lambda}(a,b) \,|\, u \rangle = i N^2 \oint d\rho[\nu\phi(\nu)] \{ a(\nu) + i \nu b(\nu) \psi'(\nu) \} \,, \tag{6.15}$$

where $d\rho(z) \equiv dz/(2\pi i z)$ is the density of eigenvalues for the base state and the integration is over the unit circle. The set of Wilson loops reduces to simple powers of the matrix V, $W^k \equiv \lim_{N\to\infty} \langle u | \widetilde{\mathrm{tr}}\, (\hat{V})^k \,|\, u \rangle$, $k = \pm 1, \pm 2, \ldots$. Instead of the coordinates $\{\psi^k\}$ and $\{W^k\}$ we may use the density of the eigenvalues, $w(\vartheta) \equiv \{ d\rho(e^{i\vartheta} \phi(e^{i\vartheta})) / d\vartheta \}$, and $v(\vartheta) \equiv \psi(e^{i\vartheta})$.[26] One may easily show that $v(\vartheta)$ and $w(\vartheta)$ are canonically conjugate, $\{ v(\vartheta), w(\vartheta') \} = \delta(\vartheta - \vartheta')$. Finally, the classical Hamiltonian may be expressed in terms of these variables. One finds

$$h_{cl} = \lambda \oint d\rho(\nu\phi(\nu)) \left| \frac{\partial}{\partial \nu} \left\{ \psi(\nu) - i \oint d\rho(\nu'\phi(\nu')) \ln \left| \frac{\nu\phi(\nu) - \nu'\phi(\nu')}{\nu - \nu'} \right| \right\} \right|^2 - \lambda^{-1} \oint d\rho(\nu\phi(\nu)) (\nu + \nu^{-1})$$

$$= \oint d\rho(\nu\phi(\nu)) \left\{ \lambda \,|\, \psi'(\nu) \,|^2 + \tfrac{1}{4}\lambda \left| \oint d\rho(\nu'\phi(\nu')) (\nu + \nu') / \nu - \nu' \right|^2 - \lambda^{-1} (\nu + \nu^{-1}) \right\}$$

$$= \int_{-\pi}^{\pi} d\vartheta\, w(\vartheta) \left\{ \lambda\, v'(\vartheta)^2 + \frac{\pi^2}{3} \lambda\, w(\vartheta)^2 - 2\lambda^{-1} \cos\vartheta - \frac{\lambda}{12} \right\} . \tag{6.16}$$

[25] This form of the classical Hamiltonian has been previously derived by Sakita (1980).

[26] $v(\vartheta)$ is real and $w(\vartheta)$ is real and positive. $v(\vartheta)$ and $w(\vartheta)$ do not quite represent independent dynamical variables; $w(\vartheta)$ must satisfy $\int d\vartheta\, w(\vartheta) = 1$, and adding a constant $v(\vartheta)$ does not affect the dynamics.

(Exactly the same contour integral tricks that were used in the last section have been used here.) Minimizing (6.16) subject to the constraint $\int_{-\pi}^{\pi} d\vartheta\, w(\vartheta)=1$ yields[27]

$$v_0(\vartheta)=0$$

and (6.17)

$$w_0(\vartheta)=\frac{1}{\pi}(e+2\lambda^{-2}\cos\vartheta)^{1/2}\,\Theta(e-2\lambda^{-2}\cos\vartheta)\ ,$$

where the constant e is determined by the condition $1=\int_{-\pi}^{\pi} d\vartheta\, w_0(\vartheta)$. Finally, the ground-state energy is given by

$$\lim_{N\to\infty}(E_0/N^2)=h_{cl}[v_0,w_0]$$

$$=\lambda\left[e-\tfrac{2}{3}\pi^2\int_{-\pi}^{\pi} d\vartheta\, w_0(\vartheta)^3\right]. \qquad (6.18)$$

Note that the structure of the original lattice appears in the classical Hamiltonian (6.12) only indirectly through the set of possible loops $\{\Gamma\}$. Suppose the original lattice, Λ, is invariant under some translation T. If one identifies all loops in Λ which are equivalent under T, then the resulting set of equivalence classes of loops is isomorphic to the set of all topologically trivial loops on a smaller periodic lattice $\tilde{\Lambda}$, formed by identifying all sites of Λ equivalent under T. Comparing the classical Hamiltonians for the lattices Λ and $\tilde{\Lambda}$, one finds that the two will be identical if (a) the expectations (and conjugate momenta) of all Wilson loops in Λ which are equivalent under T are equal, and (b) the expectations (and conjugate momenta) of all topologically nontrivial loops in $\tilde{\Lambda}$ vanish. The minimum of the classical Hamiltonian will automatically satisfy these conditions if appropriate global symmetries remain unbroken. [Condition (a) requires unbroken translation invariance under T, while (b) requires unbroken invariance under gauge transformations which are only periodic up to elements of the center of the U(N) gauge group.] Repeating this argument allows one to show that the large N limits of gauge theoreis on all periodic sublattice of Λ are equivalent, provided that all global symmetries remain unbroken. In particular, the large N limit of a theory on a d-dimensional cubic lattice should be equivalent to the limit of a theory on a periodic lattice containing just one site (i.e., a matrix model of d matrices). [Eguchi and Kawai (1982) have recently discussed an analogous result for Euclidean lattice gauge theories.]

VII. DISCUSSION

We have presented a general formalism for finding classical limits in arbitrary quantum theories, based on certain assumptions shown to be sufficient to construct a classical phase space and derive the appropriate classical dynamics. These assumptions appear to isolate cleanly the minimal structure required for any classical limit; however, proving any form of necessity appears to be very difficult. Using this formalism, it has been shown that for a large class of theories the $N\to\infty$ limit is a classical limit. This class of theories (plus their obvious generalizations) includes essentially all known theories with sensible large N limits. In every case considered, exactly the same procedure has worked. The only input required is a suitable choice for the coherence group and an appropriate base state.

In any theory where the fundamental quantum operators can be divided into "coordinates," $\{\hat{x}(\alpha)\}$, and conjugate "momenta," $\{\hat{p}(\alpha)\}$, there is a natural choice for the coherence group. If $\{f_i(\hat{x}(\alpha))\}$ is a minimal set of physical operators such that every physical operator constructed from the "coordinates" $\{\hat{x}(\alpha)\}$ can be expressed as a function of the f's and if $\{g_j(\hat{x}(\alpha),p(\beta))\}$ is the set of operators obtained from the set $\{f_i\}$ by replacing any single coordinate, $\hat{x}(\alpha)$, by the corresponding momentum, $\hat{p}(\alpha)$, then the group generated by the operators $\{f_i(\hat{x}(\alpha))\}$ and $\{g_j(\hat{x}(\alpha),\hat{p}(\beta))\}$ will act irreducibly on the physical Hilbert space. Every one of the coherence groups we have considered earlier may be regarded as an example of this prescription. Note that the generators of the coherence groups have always been at most linear in the conjugate momenta. This feature ensures that one can always exponentiate the generators, and thereby construct the group of finite transformations, in a reasonably explicit manner.

We have always used the simplest possible choice for the base state. It turns out that only for these simple choices is it easy to prove Assumption 2 (operators uniquely specified by their symbols). It is not known whether, in an arbitrary quantum theory, there necessarily exists any choice for the base state which will satisfy this assumption.[28] Similarly, questions of uniqueness (such as whether or not using a different coherence group for gauge theories might allow one to avoid loop spaces) have not been adequately answered. Different choices for the coherence group or base state must give equivalent results, and it seems very doubtful that there exist any choices satisfying the required assumptions which are more convenient than the choices we have made.

Next, we will discuss how the approach we have used to derive large N limits relates to previously proposed methods. We will begin by considering various methods which have had limited applicability.

Large N limits were first studied in vector models, originally in the context of statistical mechanics (Stanley,

[27]The following results have been previously derived using different methods. See Jevicki and Sakita (1980b) and Wadia (1980a).

[28]Note that Assumption 2 was used only to justify restricting attention to the symbols of physical operators. It is conceivable that there exist theories for which a weaker form of the assumption may be appropriate. See the appendix for further consequences of this assumption.

1968), and later from the viewpoint of particle physics (Wilson, 1973). Certain specific models, such as ϕ^4 field theories, are sufficiently tractable that one may simply sum all Feynman diagrams that survive in the $N \rightarrow \infty$ limit (Dolan and Jackiw, 1974; Schnitzer, 1974; Coleman, Jackiw, and Politzer, 1974; Gross and Neveu, 1974). Equivalently, functional integral methods may be used to compute directly the $N = \infty$ limit of the effective action (Halpern, 1980). One finds an effective action which is nonlocal in both space and time and not obviously equivalent to any classical action which is local in time. However, minimizing the effective action in, for examples, ϕ^4 theories, leads to exactly the same gap equation as (4.24), and one may easily see that all other results also agree. The classical action, (4.20), is in fact closely related to the effective action one obtains from the second Legendre transform of the generating functional (Cornwall, Jackiw, and Tomboulis, 1974).

More recently, Berezin (1978) studied vector models with the specific intention of understanding the classical nature of the $N \rightarrow \infty$ limit. His paper is somewhat obscure and relies heavily on his earlier work on quantization on Kähler manifolds (Berezin, 1974, 1975). Presumably for this reason, it has received less attention than it deserved. Berezin used coherent state methods similar to those employed in Sec. IV. The major difference is that he chose to include the operators $\hat{C}(\alpha,\beta) \equiv \frac{1}{2}\hat{p}_i(\alpha)\hat{p}_i(\beta)$ among the generators of the coherence group. This enlarged group actually leads to exactly the same set of coherent states. However, including all bilinear operators in the Lie algebra **G** conveniently allows one to express the action of the coherence group in Fock space.[29] This makes it very easy to carry out the discussion for both Bose and Fermi theories in parallel. Berezin expresses the resulting classical mechanics in a form which appears quite different from (4.19) and (4.20); however, this is simply a consequence of his choice of coordinates on the classical phase space. [He uses complex coordinates which reflect the curvature of the Kähler manifold and are not naturally canonically conjugate. A stereographic projection linearizes the phase space (Jevicki and Papanicolaou, 1980) and relates his coordinates to those used in Sec. IV.]

Mlodinow and Papanicolaou (1980, 1981) have also studied certain vector models using related techniques termed "pseudospin" methods. Instead of employing coherent states, they choose to work directly at the operator level and study the algebra of the O(N)-invariant bilinear operators, $\hat{A}(\alpha,\beta)$, $\hat{B}(\alpha,\beta)$, and $\hat{C}(\alpha,\beta)$. This algebra reflects the structure of the canonical commutation relations and for Bose theories is equivalent to $\mathrm{Sp}(2n,\mathscr{R})$ [$n=1$ or 2 is the number of O(N)-vectors in their work]. They rewrite the operators $\hat{A}$, $\hat{B}$, and $\hat{C}$ in terms of a new set of elementary Bose creation and annihilation operators (a generalized Holstein-Primakoff representation) and show that in the large N limit all of these new operators may be treated classically. This then leads to the same classical Hamiltonian as in (4.19). In my opinion, this operator level approach is less convenient than the coherent state approach for deriving the $N \rightarrow \infty$ limit. However, it appears to be more convenient for deriving systematic corrections in $1/N$, since one need deal with only a finite number of O(N)-invariant basis states, instead of with an overcomplete set of coherent states. Mlodinow and Papanicolaou have computed the first three terms in the $1/N$ expansion for the ground-state energy of systems such as helium, and hydrogen in a magnetic field, and obtained surprisingly good results (N equals the dimension of space here). Recently, Papanicolaou (1981) has also used this "pseudospin" method to discuss the large N limit of the two-dimensional $(\bar{\psi}\psi)^2$ model.

Berezin, Mlodinow, and Papanicolaou have all relied heavily on the fact that for any vector model one can immediately rewrite any physical operator, such as the Hamiltonian, in terms of the bilocal "pseudospin" operators $\hat{A}$, $\hat{B}$, and $\hat{C}$. However, in order to understand the generalization of these methods to more complicated theories, it is important to realize that one need not include all three sets of operators in the Lie algebra of the coherence group ($\equiv$"pseudospin" algebra). As shown in Sec. IV, including only $\hat{A}$ and $\hat{B}$ (i.e., operators at most linear in momenta) still produces a coherence group which acts irreducibly. Furthermore, any physical operator can be expressed as a linear combination of elements of the coherence group, or equivalently, as a function of just $\hat{A}(\alpha,\beta)$ and $\hat{B}(\alpha,\beta)$. This was stated in equation (3.3), which follows from the Von Neumann density theorem (see footnote 8). However, finding this representation can be very difficult. This problem is neatly avoided in our discussion of the $N \rightarrow \infty$ limit. All we ever require is the symbol of an operator, and computing this is a deductive operation.

Next, we turn to a discussion of methods which are, or claim to be, applicable to a wider class of theories than just vector models. We begin with the method which is, in some ways, closest to the approach used in this paper. This is the "collective field method" of Jevicki and Sakita (1980a).

The collective field method is based on the idea of directly rewriting a quantum theory in terms of an overcomplete set of commuting physical operators. Specifically, all wave functions are taken to be functionals of an overcomplete set of physical variables, such as the set of bilinears $\{x(\alpha)\cdot x(\beta)\}$ for vector models or the set of Wilson loops $\{\widetilde{\mathrm{tr}}\,V^{\Gamma}\}$ for gauge theories; this set of variables is the "collective field." Using the chain rule, one then expresses the kinetic terms in the Schrödinger equation in terms of derivatives with respect to the collective field variables. Massaging the resulting expression by means of a similarity transformation then produces the "collective field Hamiltonian." Finally, it is argued that

[29]This also produces a coherence group which is semisimple, unlike all of the groups used in this paper. Consequently for this group one need not distinguish between the Lie algebra and its dual space. (See footnote 12 for further discussion.)

in the large N limit one may treat the collective field variables and their conjugates as independent classical variables. Therefore one finds a classical Hamiltonian appropriate for studying the large N limit. (Infinite terms formally suppressed by powers of $1/N$ are typcially discarded at this stage.) The ground-state energy, for example, may then be computed by minimizing this Hamiltonian.

In every example we have considered it turns out that the coherent-state method used in this paper produces exactly the same classical Hamiltonian as does the collective field method. This was basically inevitable, since both methods work entirely within the physical Hilbert space and produce classical Hamiltonians which are expressed in terms of physical variables. Despite this equivalence of the final results, the two methods of derivation differ considerably. We would like to discuss this difference in somewhat greater detail because it will bring up an important point.

Consider, for example, a one-matrix model such as the one described in Sec. V. The U(N)-invariant spectrum consists of a set of modes whose frequencies increase linearly, $\omega_j = j\omega_s$. For finite N, there are N such modes. Therefore the zero-point energy of these modes is of order N^2 and contributes to the leading large N behavior of the ground-state energy. Consequently, any method for deriving the large N limit of the model must correctly account for this zero-point energy. If one simply rewrites the Schrödinger equation for this model in terms of the eigenvalues of the original matrix and tries to neglect the gradient terms, then this zero-point energy will be missed and one will obtain the wrong answer.[30] The collective field method manages to avoid this problem. Instead of writing the theory in terms of the complete set of N eigenvalues, $\{\lambda_i\}$, it uses a continuous function with a smooth large N limit, the density of eigenvalues $\rho(\lambda)$. One might think that this would only make the problem worse, since the collective field Hamiltonian has no cutoff on the number of modes and the zero-point energy is now $\infty \times N$ instead of $O(N^2)$. However, it turns out that the correct answer is obtained by simply dropping all such terms which are (formally) suppressed by powers of $1/N$. Understanding why this is true, within the collective field approach, is not easy.

The coherent-state formalism provides a much cleaner method for deriving the large N classical Hamiltonian. Instead of requiring operator level manipulations, one simply computes the expectation of the quantum Hamiltonian in a specified set of (normalizable) coherent states. The zero-point energy is automatically included correctly.

Along this same line, I should mention the work of Lovelace (1981). He considered the general problem of changing variables to an independent set of gauge-invariant coordinates and claimed that adding a simple term based on the volume of a gauge orbit to the potential energy would reproduce the results of the collective field method. This prescription amounts to simply neglecting certain portions of the kinetic energy and does not appear correctly to include zero-point energy contributions.

The large N limit of many simple models can be solved by formulating Schwinger-Dyson equations for correlation functions of time-ordered products of physical operators. Normally, such equations generate an infinite hierarchy of relations involving arbitrarily complicated correlation functions. However, in the large N limit factorization [Eq. (1.1)] can be used to simplify the equations, and one can derive a closed set of nonlinear equations which specify the behavior of a minimal set of physical observables. In a variety of simple models, one can explicitly solve these equations, and thereby derive the leading $N\rightarrow\infty$ behavior of physical correlation functions (Paffuta and Rossi, 1980; Friedan, 1981; Brower, Rossi, and Tan, 1981a, 1981b; Kazakov and Kostov, 1980; Wadia, 1981).

Many people have attempted to study gauge theories by formulating Dyson equations for the vacuum expectations of Wilson loops (Gervais and Neveu, 1979; Nambu, 1979; Polyakov, 1979; Foerster, 1979; Eguchi, 1979; Migdal, 1980). Makeenko and Migdal (1979, 1980) were the first to emphasize that in the large N limit one can derive a closed set of equations involving the expectations of single Wilson loops. Inevitably, these equations are extremely difficult to solve.

A Hamiltonian approach, such as we have used leads to an explicit minimization problem, and varying the classical Hamiltonian (Eq. 6.12) generates a closed set of equations for the expectations of equal-time Wilson loops. In contrast, the Migdal-Makeenko equations require expectations of arbitrary space-time Wilson loops and do not follow from any explicit minimization problem.[31] It is not clear if the equations have a unique solution. Furthermore, these equations actually contain much less information than the large N classical Hamiltonian (6.12). An explict solution would at most determine the large N limit of the expectation of any single Wilson loop. However, this is not sufficient information to determine the large N behavior of all physical observables. In order to compute, for example, the gauge-invariant spectrum (i.e., glueball states), one must be able to determine the connected part of the expectation of a product of two Wilson loops. This information is not contained in the Migdal-Makeenko equations.

When applied to gauge theories, any method which

---

[30]This has been discussed in detail in the context of a self-dual quaternionic matrix model by Aragao de Carvalho and Fateev (1981).

[31]Jevicki and Sakita (1981) have shown that the Migdal-Makeenko equations follow from an effective action for loops derived from the Euclidean functional integral. However, evaluating the action requires computing a Jacobian which is only implicitly defined through a functional differential equation.

works entirely within the physical (i.e., gauge-invariant) Hilbert space seems inevitably to lead to some sort of loop space. Because analysis in such a space is intractable, a number of people have recently investigated alternative approaches which might avoid this problem.

Witten (1979) [see also Coleman (1979)] argued on the basis of factorization (Eq. 1.1) that the support of the Euclidean functional integral must reduce to a single gauge orbit when $N\to\infty$. This would imply that the large N limit of the expectation of any time-ordered product of physical operators could be computed by simply replacing all quantum gauge fields by a single classical field (unique up to gauge transformations). Such a field configuration has been termed a "master field". The major problem with this approach is that no constructive method for finding a suitable master field is known. The basic idea of a "unique" gauge orbit at $N=\infty$ is somewhat ill-defined, and it is not always clear how to define the infinite-dimensional space in which the master field is supposed to live. In some sense, one can always package an arbitrarily large amount of information into the "sum" over the group indices of the master field. Consider, for example, a ϕ^4 field theory. The original configuration space consists of real fields, $\phi_i(x)$, where the index i runs from 1 to N. The large N limit of the two-point function is given by

$$\lim_{N\to\infty} \langle T\{\hat{\phi}_i(x)\hat{\phi}_i(x')\}\rangle = \int \frac{d^{d+1}k}{(2\pi)^{d+1}} \frac{e^{i(k\cdot(x-x'))}}{(k^2+m^2)}\,, \tag{7.1}$$

where $m^2=\mu^2+\lambda\sigma$ and σ satisfies the gap equation (4.24). Construction of a master field for this theory is straightforward. One simply (a) allows the master field to be complex instead of real, (b) interprets the index i as the label for an arbitrary $(d+1)$-dimensional momentum vector, and (c) chooses the master field $\Phi_k(x)$ to be given by (Levine, 1980; Halpern, 1981b)

$$\Phi_k(x)=\frac{\exp\{-i(k\cdot x)\}}{(2\pi)^{(d+1)/2}(k^2+m^2)^{1/2}}\,. \tag{7.2}$$

Replacing every occurrence of $\hat{\phi}(x)\cdot\hat{\phi}(x')$ in any time-ordered correlation function with $\int d^{d+1}k\,\Phi_k^*(x)\,\Phi_k(x')$ will then yield the correct large N limit. Obviously, this construction simply reflects the fact that the large N limit of the spectrum consists of a single particle of mass m. Note that the master field does not allow one to compute the leading $N\to\infty$ behavior of one-particle irreducible correlation functions, and that therefore one cannot extract, for example, the large N behavior of physical scattering amplitudes from the master field. Because its derivation was solely based on factorization, the master field actually contains much less information than the classical Hamiltonian (4.22)

Since there is no constructive approach for finding such "master fields" without having first solved the theory, it is obviously desirable to find an alternative approach which will have more predictive power. Such an approach is potentially provided by the recently developed method of "constrained classical solutions." (Unlike the usage in the rest of this paper, in the following discussion "classical solutions," "classical hamiltonians," etc., will be understood to refer specifically to the $\hbar\to 0$ limit.) The first step in the development of this method was provided by the observation by Jevicki and Papanicolaou (1980) that the effective Hamiltonian describing the $N=\infty$ dynamics of a point particle, (4.21), is identical to the classical Hamiltonian of a point particle whose angular momentum (squared) is equal to 1/4. (If we had not scaled out factors of $\hbar$ and N, L^2 would equal $\hbar^2N^2/4$). Minimizing the original classical Hamiltonian subject to this constraint does in fact lead to the correct large N limit of the ground-state energy. A similar result was shown to hold in linear σ models. Subsequent work by Jevicki and Levine (1980, 1981), and Kessler and Levine (1981) has extended this idea to general vector models and to the single Hermitian matrix model. In every case, the equations of motion which follow from the large N effective Hamiltonian may be shown to be equivalent to the original classical equations of motion subject to a suitable constraint. In the one-matrix model, for example, the constraint may be written as $J_{ab} \equiv i[M,\dot{M}]_{ab}=\hbar(1-\delta_{ab})$. Bardakci (1981a) has also discussed one-vector and one-matrix models from a somewhat different viewpoint and obtained equivalent results. The clearest explanation about what's going on, and why this approach works, has been given by Halpern (1981a). A brief sketch of his argument follows.

Consider a theory which is known to satisfy factorization in the large N limit, such as a one-vector model. Imagine computing the vacuum expectation of any "index-ordered" product of field operators. ["Index-ordered" means that the quantum operators are ordered in a way that allows a natural contraction of the group indices. In vector models this means that neighboring pairs of vectors are contracted, as in $\langle 0|\hat{x}(t)\cdot\hat{x}(t')\,\hat{x}(t)\cdot\hat{x}(t'')|0\rangle$.] Consider inserting a complete set of quantum eigenstates after each field operator. This obviously requires working in the full Hilbert space, which contains states transforming under all possible irreducible representations of the symmetry group. However, the restriction to index-ordered products of field operators significantly restricts the types of intermediate states which can contribute. In vector models each intermediate state must transform either as an O(N) vector or O(N) singlet. Furthermore, factorization implies that the only O(N)-invariant state which can contribute to the leading $N\to\infty$ behavior is the ground state. Therefore in order to compute any ordered expectation in, for example, the one-vector model, one requires only the following matrix elements,

$$\langle n,i\,|\,\hat{x}_j\,|\,0\rangle=\delta_{ij}\,q_n/\sqrt{N}\,. \tag{7.3}$$

Here, n labels the (unknown) number of O(N)-vector eigenstates, i is an O(N) vector index labeling the states within a multiplet, and q_n is a reduced matrix element. Because we have taken matrix elements between quantum eigenstates, the reduced matrix elements have a simple time dependence,

$$q_n(t)=e^{i\omega_n t}q_n(0)\ , \tag{7.4}$$

where $\omega_n \equiv E_n - E_0$ is the (unknown) excitation energy of the nth eigenstate. The quantum equation of motion is simply $\ddot{\hat{x}}_i+2V'(\hat{x}^2)\hat{x}_i=0$ [for the usual Hamiltonian $\hat{H}=N\{\frac{1}{2}\hat{p}^2+V(\hat{x}^2)\}$]. Taking matrix elements and using factorization yields

$$\ddot{q}_n+2V'(q\cdot q^*)q_n=0\ . \tag{7.5}$$

Therefore q_n may be regarded as a complex vector satisfying the original classical equations of motion. However, the "index" n has nothing to do with the original O(N)-vector index; rather, it labels the set of vector eigenstates which can couple to the ground state. The relevant constraint arises from the vacuum expectation of the commutation relations and reads

$$\sum_n (q_n^*\dot{q}_n-\dot{q}_n^*q_n)=i\ . \tag{7.6}$$

Assuming that the frequencies $\{\omega_n\}$ are not degenerate (7.5)–(7.6) have a unique solution (up to an overall phase). This solution allows one to calculate the large N limit of any ordered correlation functions in the one-vector model. (It should be emphasized that the leading behavior always comes from the maximally disconnected part of the correlation function; the constrained classical solution does not retain enough information to compute the connected part).

Halpern has shown that the same approach works in any vector model and in the one-matrix model. Extending the analysis to gauge theories appears to be straightforward (Bardakci, 1981b). In each case one isolates the relevant set of transition matrix elements which can contribute to the large N-limit of any ordered correlation function. Next, one defines reduced matrix elements in such a way that the quantum equations of motion are equivalent to classical equations of motion for the reduced matrix elements. Matrix elements of the canonical commutation relations then generate a set of constraints which the reduced matrix elements must satisfy.

This approach of solving for constrained classical solutions is complementary to the coherent-state or collective field methods in the following sense. The latter methods work entirely within the invariant sector of the Hilbert space and generate an effective Hamiltonian describing the $N=\infty$ dynamics. Examining small oscillations about the minimum of the effective Hamiltonian allows one to compute the $N\to\infty$ limit of the spectrum of gauge- [or O(N)-] invariant states. The constrained classical solution does not contain this information. Rather, the time dependence of the constrained solution contains information about the spectrum of noninvariant states. (In gauge theories these would be static quark antiquark states.) In general, it is not at all clear that solving for the appropriate constrained classical solution is any easier than minimizing the large N effective Hamiltonian. [In the one-matrix model, solving for the constrained solution has only recently been accomplished (Halpern and Schwartz, 1981).]

Note that methods which use factorization from the outset, such as the Migdal-Makeenko derivation of loop equations or the preceding constrained classical solution approach, at most allow one to compute the vacuum expectation of any single physical operator. These methods do not, in general, produce enough information to determine the large N limit of the invariant spectrum, scattering amplitudes, or other quantities which require knowledge of the connected part of correlation functions of products of physical operators. Only methods leading to the complete large N effective action retain enough information to allow one to compute such quantities.

The above arguments summarize why, at least in my opinion, methods which derive the effective $N=\infty$ classical Hamiltonian appear to provide the most useful approach known for studying large N limits. It seems doubtful that any significantly more convenient approach can be developed. The problem of solving the large N limit of gauge theories, for example, is reduced to a minimization problem of an explicitly known functional (6.12). The fact that the relevant variables are defined on loops appears to be unavoidable. Although we lack an analytic solution to this problem, minimizing (6.12) numerically should be perfectly feasible. This appears to be well worth the effort, since an explicit solution would allow one directly to compute the large N spectrum of QCD.

Finally, we should say a few words about $1/N$ corrections. In general, it is not sufficient simply to turn around and try to quantize the large N classical Hamiltonian (Mondello and Onofri, 1981). Besides obvious factor ordering problems, there can be explicit corrections of order $1/N$ in the Hamiltonian. Furthermore, there can also be corrections coming from a nontrivial measure for the classical variables. One can study these problems by deriving functional integral representations based on the overcomplete set of coherent states. This is briefly discussed in the appendix. In principle, this representation can be used to derive systematic corrections in powers of $1/N$; however, only in simple theories has it been possible to carry out explicit calculations.

ACKNOWLEDGMENTS

Barry Simon is gratefully thanked for many enlightening discussions. This work was supported in part by the U. S. Department of Energy under Contract No. DE-AC-03-76ER00068.

APPENDIX

This appendix discusses some further applications of coherent states, such as the representation of operators in terms of diagonal projections onto coherent states and functional integral representations based on coherent states.

In Sec. III we defined the symbol of any operator $\hat{A}$ as the set of coherent state expectation values,

$$A(u)=\langle u\,|\hat{A}\,|\,u\rangle\;,\quad u\in g\;. \tag{A1}$$

Henceforth we will call $A(u)$ the "lower symbol" (or "covariant" symbol) of $\hat{A}$. Lower symbols provide a natural mapping of quantum operators into functions on the coherence group. One may also define a natural mapping of functions into operators. If $\mathring{A}(u)$ is some function on the coherence group such that

$$\int d\mu_L(u)\mathring{A}(u)\,|\,u\rangle\langle u\,|=\hat{A}\;, \tag{A2}$$

then $\mathring{A}(u)$ will be called an "upper symbol" (or "contravariant" symbol) of $\hat{A}$. Upper symbols of a given operator need not be unique, and in general may not even exist. Only operators which can be expressed as weighted sums of diagonal projections onto the coherent states will have upper symbols.

The irreducibility of the coherence group led to the completeness relation (3.2), and this implies that

$$\int d\mu_L(u)A(u)=\mathrm{Tr}(\hat{A})=\int d\mu_L(u)\mathring{A}(u)\;. \tag{A3}$$

[We have absorbed the constant c_X in the completeness relation into the measure $d\mu$. The second part of (A3) obviously requires that $\hat{A}$ possess an upper symbol.] One may easily show that the norm of an operator is bounded below by the maximum of its lower symbol and bounded above by the maximum of any upper symbol (Simon, 1980; Berezin, 1972). Furthermore, if $\Phi(x)$ is any convex function of a real variable [such as $\exp(x)$] and if $\hat{A}$ is a self-adjoint operator, then (Berezin, 1972; Lieb, 1973),

$$\int d\mu_L(u)\;\Phi(A(u))\le\mathrm{Tr}[\Phi(\hat{A})]$$
$$\le\int d\mu_L(u)\;\Phi(\mathring{A}(u))\;. \tag{A4}$$

[These bounds follow from a simple argument based on Jensen's inequality. See Simon (1980).] These relations motivate the names "upper" and "lower" symbols.

A natural question to ask in a particular theory is whether all operators have upper symbols. If they do, then the set of coherent projections, $\{\,|\,u\rangle\langle u\,|\;\;|\,u\in g\,\}$, will be said to be *complete*. (Completeness of the coherent projections in the space of all operators should not be confused with completeness of the set of coherent states in the Hilbert space. Completeness of the coherent states is an immediate consequence of Assumption 1.) It turns out that Assumption 2 [$Z(u)=0$ implies $\hat{Z}=0$] is equivalent to the requirement that the coherent projections be complete (Simon, 1980). The basic argument is surprisingly simple.[32] Suppose that Assumption 2 were false. Then there would exist some nonzero operator $\hat{Z}$ whose lower symbol $Z(u)$ was identically zero. This would imply that

$$\int d\mu_L(u)\mathring{A}(u)Z(u)=0 \tag{A5}$$

for any function $\mathring{A}(u)$. Since $Z(u)=\langle u\,|\,\hat{Z}\,|\,u\rangle$, (A5) may be rewritten as

$$\mathrm{Tr}(\hat{A}\hat{Z})=0\;, \tag{A6}$$

where $\hat{A}\equiv\int d\mu_L(u)\mathring{A}(u)\,|u\rangle\langle u\,|$. Hence, the operator $\hat{Z}$ would be orthogonal to all operators which possess upper symbols. In other words, the set of operators with upper symbols would have a nontrivial orthogonal complement, thereby implying that the set of coherent projections was not complete. Therefore completeness of the coherent projections implies the validity of Assumption 2. To prove the converse, one simply inverts the argument above.

This shows that in any theory satisfying Assumption 2, every operator has an upper symbol. However, the preceding argument is about as nonconstructive as one can get. Regrettably, no general constructive procedure for finding the upper symbol of an arbitrary operator is known. In specific cases, if an explicit upper symbol for the Hamiltonian can be found, then the inequalities (A4) can provide matched upper and lower bounds on the partition function. This can yield detailed information on the rate at which the classical limit is approached [in contrast to Eq. (3.29)−(3.31), which simply show that the limit exists.] This procedure has been successfully applied in a variety of quantum spin systems (Lieb, 1973; Fuller and Lenard, 1979; Gilmore and Feng, 1978; Gilmore, 1979; Shankar, 1980) and even in atomic physics (Thirring, 1981). Unfortunately, constructing upper symbols in more complicated theories appears to be very difficult.

Next, we discuss functional integrals based on coherent states. The coherent-state representation will provide a simple and direct method for showing that the tree diagrams generated from the large N classical action (3.23) correctly reproduce the large N limit of any connected correlation function of physical operators.

Consider, for example, computing the partition function $Z=\mathrm{Tr}(\exp-\beta\hat{H})$. The basic ingredient needed to derive any functional integral is a convenient completeness relation. We will use the coherent-state completeness relation, (3.2). Repeatedly inserting this into the trace leads to

[32]The following sketch is quite sloppy. In particular, the phrase *all operators* actually means a dense set of bounded operators in a particular topology. See Simon (1980) for details.

$$Z=\lim_{k\to\infty}\int\prod_{i=1}^{k}d\mu_L(u_i)\langle u_k|(1-\varepsilon\hat{H})|u_{k-1}\rangle\cdots\langle u_2|(1-\varepsilon\hat{H})|u_1\rangle\langle u_1|(1-\varepsilon\hat{H})|u_k\rangle \tag{A7}$$

($\varepsilon\equiv\beta/k$). Assumption 3 [$\langle u\,|\,u'\rangle\sim\exp-(1/\chi)\phi(u,u')$] and Assumption 4 [$(1/\chi)\hat{H}$ classical] imply that this integral is highly peaked about $u_i\sim u_{i-1}$. Therefore [using (3.10)] we may write (A7) as

$$Z\sim\lim_{k\to\infty}\int\prod_{i=1}^{k}d\mu_L(u_i)\exp-\frac{1}{\chi}\sum_{i=1}^{k}\{\phi(u_i,u_{i-1})+\varepsilon\chi H_\chi(u_i)\}\ . \tag{A8}$$

Next, we may split each integral over the coherence group g into an integral over the coset space g/h times an integral over the subgroup h, where h generates the set of coherent states which are classically equivalent to the base state. (In other words, $h=\{u\in g\,|\,u\sim 1\}$). Recall that the coset space g/h is equivalent to the coadjoint orbit Γ which provides the classical phase space. If we write $u_i=e^{\lambda_i}u_{i-1}$, then it may be shown that

$$\sum_{i=1}^{k}\phi(u_i,u_{i-1})=\sum_{i=1}^{k}i\langle\delta\zeta_i,\vartheta_{\zeta_i}\rangle+O(\delta\zeta)^2\ ,$$

where $\zeta_i=Ad^*[u_i](\zeta_0)$ and $\delta\zeta_i=(Ad^*[e^{\lambda_i}]-1)(\zeta_i)$. Finally, using the definition of the classical Hamiltonian, $\lim_{\chi\to\infty}\chi H_\chi(u)=h_{cl}(\zeta)$, we find

$$Z\sim\lim_{k\to\infty}\int\prod_{i=1}^{k}d^\Gamma\mu(\zeta_i)\exp-\frac{1}{\chi}\sum_{i=1}^{k}\{i\langle\delta\zeta_i,\vartheta_{\zeta_i}\rangle+\varepsilon\, h_{cl}(\zeta)\}\equiv\int D\mu[\zeta(t)]\exp-\frac{1}{\chi}S^E_{cl}[\zeta(t)]\ , \tag{A9}$$

where

$$S^E_{cl}[\zeta(t)]\equiv\int_0^\beta dt\{\,i\langle\dot{\zeta}(t),\vartheta_\zeta\rangle+h_{cl}(\zeta(t))\,\}$$

is the Euclidean classical action, and

$$d^\Gamma\mu(\zeta)\equiv\int d\mu_L(u)\,\delta(\zeta-Ad^*[u](\zeta_0))$$

is the invariant measure on the coadjoint orbit Γ. In a similar fashion, one may consider the expectation of any time-ordered product of classical operators, and find

$$\langle A(t_1)\cdots B(t_2)\rangle\equiv Z^{-1}\mathrm{Tr}[T\{\hat{A}(t_1)\cdots\hat{B}(t_2)\exp-\beta\hat{H}\}]$$
$$\sim Z^{-1}\int D\mu[\zeta(t)]\,a(\zeta(t_1))\cdots b(\zeta(t_2))\exp-\frac{1}{\chi}S^E_{cl}[\zeta(t)]\ . \tag{A10}$$

Equations (A9) and (A10) have the standard form of any semiclassical functional integral; all dependence on χ is isolated in a single factor of $(1/\chi)$ multiplying the classical action. We emphasize that these expressions are valid only as $\chi\to 0$. Small χ is the only justification for expanding $\phi(u_i,u_{i-1})$ and $\langle u_i|\hat{H}|u_{i-1}\rangle/\langle u_i|u_{i-1}\rangle$ about $u_i\sim u_{i-1}$. Nevertheless, one may see that the terms we have neglected to not contribute to the leading $\chi\to 0$ behavior of the connected part of correlation functions such as (A10). This limiting behavior may be computed simply by expanding the classical action $S^E_{cl}[\zeta(t)]$ about the minimum of the classical Hamiltonian and evaluating the lowest-order connected tree diagrams. Corrections to this leading behavior can come from several sources; explicit higher-order loop diagrams generated from the classical action, the nontrivial measure $D\mu[\zeta(t)]$, plus all the terms we have neglected in deriving equations (A9) and (A10).[33]

[33]See Klauder (1979) plus references therein for careful discussion of some of the subtleties involved in deriving exact functional integrals based on ordinary Gaussian coherent states.

REFERENCES

Aragao de Carvalho, C., and V. A. Fateev, 1981, J. Phys. A **14**, 1925.

Auslander, L., and B. Kostant, 1967, Bull. Am. Math. Soc. **73**, 692.

Bardakci, K., 1981a, Nucl. Phys. B **178**, 263.

Bardakci, K., 1981b, "Classical Equations for Non-Abelian Gauge Theories in the Large N Limit," Princeton, Institute for Advanced Study preprint 81-0136.

Berezin, F. A., 1972, Izv. Akad. Nauk SSR Ser. Mat. **36**, 1134 [Math. USSR Izv. **6**, 1117, (1972)].

Berezin, F. A., 1974, Izv. Akad. Nauk SSR Ser. Mat. **38**, 1116 [Math. USSR Izv. **8**, 1109, (1974)].

Berezin, F. A., 1975, Izv. Akad. Nauk SSR Ser. Mat. **39**, 363 [Math. USSR Izv. **9**, 341 (1975)].

Berezin, F. A., 1978, Commun. Math. Phys. **63** 131.

Bratteli, O., and D. W. Robinson, 1979, *Operator Algebras and Quantum Statistical Mechanics 1,* (Springer, New York).

Berezin, E., C. Itzykson, G. Parisi, and J. B. Zuber, 1978, Commun. Math. Phys. **59**, 35.

Brower, R. C., P. Rossi, and C.-I. Tan, 1981a, Phys. Rev. D **23**, 942.

Brower, R. C., P. Rossi, and C.-I. Tan, 1981b, Phys. Rev. D **23**, 953.

Coleman, S., 1980, "$1/N$", SLAC preprint Pub-2484 (Erice lectures, 1979).
Coleman, S., R. Jackiw, and H. D. Politzer, 1974, Phys. Rev. D **10**, 2491.
Cornwall, J. M., R. Jackiw, and E. Tomboulis, 1974, Phys. Rev. D **10**, 2428.
Dolan, L., and R. Jackiw, 1974, Phys. Rev. D **9**, 2904.
Eguchi, T., 1979, Phys. Lett. B **87**, 91.
Eguchi, T. and H. Kawai, 1982, "Reduction for Dynamical Degrees of Freedom in the Large N Gauge Theory," Tokyo Univ. preprint UT-378.
Foerster, D., 1979, Phys. Lett. B **87**, 87.
Friedan, D., 1981, Commun. Math. Phys. **78**, 353.
Fuller, W., and A. Lenard, 1979, Commun. Math. Phys. **67**, 69; **69**, 99.
Gervais, J. L., and A. Neveu, 1979, Phys. Lett. B **80**, 255.
Gilmore, R., 1979, J. Math. Phys. **20**, 891.
Gilmore, R., and D. H. Feng, 1978, Phys. Lett. B **76**, 26.
Gross, D. J., and A. Neveu, 1974, Phys. Rev. D **10**, 3235.
Halpern, M. B., 1980, Nucl. Phys. B **173**, 504.
Halpern, M. B., 1981a, Nucl. Phys. B **188**, 61.
Halpern, M. B., 1981b, Nuovo Cimento A **61**, 207.
Halpern, M. B., and C. Schwartz, 1981, Phys. Rev. D **24**, 2146.
Hooft, G. 't., 1974, Nucl. Phys. B **72**, 461.
Jevicki, A., and H. Levine, 1980, Phys. Rev. Lett. **44**, 1443.
Jevicki, A., and H. Levine, 1981, Ann. Phys. **136**, 113.
Jevicki, A., and N. Papanicolaou, 1980, Nucl. Phys. B **171**, 362.
Jevicki, A., and B. Sakita, 1980a, Nucl. Phys. B **165**, 511.
Jevicki, A., and B. Sakita, 1980b, Phys. Rev. D **22**, 467.
Jevicki, A., and B. Sakita, 1981, Nucl. Phys. B **185**, 89.
Kazakov, V. A., and I. Kostov, 1980, Nucl. Phys. B **176**, 199.
Kessler, D. A., and H. Levine, 1981, Ann. Phys. **133**, 13.
Kirillov, A. A., 1962, Usp. Mat. Nauk **17**, (4), 57 [1962, Russian Math. Surveys **17**, (4), 53 (1962)].
Kirillov, A. A., 1976, *Elements of the Theory of Representations,* (Springer, Berlin).
Klauder, J. R., 1963, J. Math. Phys. **4**, 1055; 1058.
Klauder, J. R., 1979, Phys. Rev. D **19**, 2349.
Klauder, J. R., and E. C. G. Sudarshan, 1968, *Fundamentals of Quantum Optics* (Benjamin, New York).
Kogut, J., and L. Susskind, 1975, Phys. Rev. D **11**, 395.
Kostant, B., 1970, in "Quantization and Unitary Representations," *Lectures in Modern Analysis and Applications III,* edited by C. T. Taam, Lecture Notes in Mathematics, (Springer, Berlin), Vol. 170, p. 87.
Levine, H., 1980, "Comments of the Large N Master Field Approach," Harvard University preprint 80/A076.
Lieb, E., 1973, Commun. Math. Phys. **31**, 327.
Lovelace, C., 1981, Nucl. Phys. B **190**, 45.
Lovelace, C. 1982, "Universality at Large N", Rutgers Univ. preprint 82-01.
Makeenko, Yu. M., and A. A. Migdal, 1979, Phys. Lett. B **88**, 135; 1980, **89**, 437(E).
Makeenko, Yu. M., and A. A. Migdal, 1980, Phys. Lett. B **97**, 253.
Marchesini, G., and E. Onofri, 1980, J. Math. Phys. **21**, 1103.
Mehta, M. L., 1967, *Random Matrices,* (Academic, New York).
Migdal, A. A., 1980, Ann. Phys. (N.Y.) **126**, 279.
Mlodinow, L. D., and N. Papanicolaou, 1980, Ann. Phys. (N.Y.) **128**, 314.
Mlodinow, L. D., and N. Papanicolaou, 1981, Ann. Phys. **131**, 1.
Mondello, M., and E. Onofri, 1981, Phys. Lett. B **98**, 277.
Nambu, Y., 1979, Phys. Lett. B **80**, 372.
Paffuta, G., and P. Rossi, 1980, Phys. Lett. B **92**, 321.
Papanicolaou, N., 1981, Ann. Phys. **136**, 210.
Perelomov, A. M., 1972, Commun. Math. Phys. **26**, 222.
Polyakov, A. M., 1979, Phys. Lett. B **82**, 247.
Sakita, B., 1980, Phys. Rev. D **21**, 1067.
Schnitzer, H., 1974, Phys. Rev. D **10**, 1800.
Shankar, R., 1980, Phys. Rev. Lett. **45**, 1088.
Shapiro, J., 1981, Nucl. Phys. B **184**, 218.
Simon, B., 1980, Commun. Math. Phys. **71**, 247.
Souriau, J. M., 1970, *Structures des systèmes dynamiques,* (Dunod, Paris).
Stanley, H. E., 1968, Phys. Rev. **176**, 718.
Thirring, W., 1981, Commun. Math. Phys. **79**, 1.
Veneziano, G., 1976, Nucl. Phys. B **117**, 519.
Wadia, S. R., 1980a, Phys. Lett. B **93**, 403.
Wadia, S. R., 1981, Phys. Rev. D **24**, 970.
Wigner, E. P. 1959, *Proceedings of the 4th Canadian Mathematical Congress, Banff 1957,* (University of Toronto, Toronto), p. 174.
Wilson, K., 1973, Phys. Rev. D **7**, 2911.
Witten, E., 1979a, Nucl. Phys. B **160**, 57.
Witten, E., 1979b, "The $1/N$ Expansion in Atomic and Particle Physics," Harvard University preprint 79/A078; Cargese Summer Institute 1979 (QCD 161:S77:1979), p. 403.
Yaffe, L. G. 1982 (unpublished).

COHERENT STATES WITH CLASSICAL MOTION; FROM AN ANALYTIC METHOD COMPLEMENTARY TO GROUP THEORY

Michael Martin Nieto
Theoretical Division, Los Alamos National Laboratory, University of California
Los Alamos, New Mexico 87545, USA

ABSTRACT

From the motivation of Schrödinger, that of finding states which follow the motion which a classical particle would in a given potential, we discuss generalizations of the coherent states of the harmonic oscillator. We focus on a method which is the analytic complement to the group theory point of view. It uses a minimum uncertainty formalism as its basis. We discuss the properties and time evolution of these states, always keeping in mind the desire to find quantum states which follow the classical motion.

1. INTRODUCTION

In 1926 Schrödinger[1] discovered what have come to be known as the coherent states of the harmónic oscillator. His motivation was to find states which follow the motion that a classical particle would in an harmonic oscillator potential. He succeeded by finding a restricted class of Gaussian wave packets which can analytically be shown to have a shape independent of time, and whose centroid oscillates back and forth in the potential the same as a classical particle with an energy ($\langle H \rangle - E_0$) would, E_0 being the ground state energy.

Amusingly, at the end of his paper[1] Schrödinger commented that, "We can definitely foresee that, in a similar way, wave groups can be constructed which move round highly quantised Kepler ellipses and are the representation by wave mechanics of the hydrogen electron. But the technical difficulties in the calculation are greater than in the especially simple case which we have treated here." Schrödinger did not pursue the problem.

In the 1960's the coherent states came into wide usage through the new field of quantum optics, and many authors popularized their use.[2-13] However, despite the many advances, generalizations concentrated on systems whose eigenspectra are equally spaced. Thus, the idea of generalizing coherent states to more general types of potentials, as Schrödinger envisioned, did not reach full fruition.

This fundamental question of Schrödinger fascinated me and my coworkers (L. M. Simmons, Jr. and V. P. Gutschick). Is it possible to find quantum states which follow the motion that a classical particle would in any given potential?

Giving the intuitive answer now, I will show that it is, "Yes, up to a point, and depending..." Effectively, the more highly bound a particle state is and, given that, the greater number of eigenstates with which the "coherent state" has significant overlap and, finally, the closer to equally spaced (i.e., harmonic) these eigenstates are, then the better and longer the "coherent state" will follow the classical motion without dispersing.

The harmonic oscillator is, of course, that system which is the best of all possible worlds. It's coherent states never change their shape and follow the classical motion forever. When we began this problem we knew the harmonic oscillator is very special. However, its very special properties became clearer and clearer to us as we understood the problem better.

We wanted to find out how important it is to have systems with equally spaced eignevalues. We also wanted to know if coherent states could be found for systems which had non-equally spaced eigenvalues and/or had continuums and/or could not be solved analytically. We found a general method to handle such systems,[14-19] and at the end of the study[14-24] we came to the intuitive answer I have just given above.

What I intend to do here is first to review the harmonic oscillator and the three standard equivalent methods for defining its coherent states. In Sec. 3 I will show that systems with equally spaced eigenvalues which are not the harmonic oscillator do not in general have coherent states which follow the classical motion forever.[22-24] They almost do, but not exactly.

We then go to general potentials. It turns out that for non-harmonic systems, the generalizations of the three equivalent methods of defining harmonic oscillator coherent states lead to states which are no longer equivalent. In particular, although one definition of coherent states is easy to generalize from the group theory point of view, the method which we have most closely investigated (and which follows Schrödinger's philosophy) is best dealt with from an analytic point of view. Thus, what I will discuss can be viewed as the complementary viewpoint to group theory, just as Schrödinger's wave mechanics is the complementary viewpoint to Heisenberg's matrix mechanics. (I shall also mention further generalizations to multidimensional systems and time-dependent potentials.)

To explicitly show how the formalism works, I will go through in detail its application to the $\cosh^{-2}$ potential. Finally, I will show how well these states do follow the classical motion by reviewing the results of a computer generated movie which displays the time evolution of these states for many potentials.

2. THE HARMONIC OSCILLATOR AND ITS COHERENT STATES

Let us begin by recalling the properties of a bound classical system of total energy E. The Hamiltonian equation is

$$E = \frac{1}{2m} p^2 + V(x) = \frac{1}{2} m\dot{x}^2 + V(x) \quad , \tag{2.1}$$

or

$$\dot{x} = (2/m)^{\frac{1}{2}}[E - V(x)]^{\frac{1}{2}} \quad . \tag{2.2}$$

For the simple harmonic oscillator with

$$V(x) = \tfrac{1}{2}kx^2 = \tfrac{1}{2}m\omega^2 x^2 \quad , \tag{2.3}$$

the solution for x(t) is

$$x(t) = (2E/m\omega^2)^{\frac{1}{2}}\sin \omega t \quad , \tag{2.4}$$

so that

$$p(t) = m\dot{x} = m(2E/m)^{\frac{1}{2}}\cos \omega t \quad . \tag{2.5}$$

For the quantum problem, with the Hamiltonian equation becoming the Schrödinger equation, $p \to (\hbar/i)d/dx$,

$$H\psi = [- \frac{\hbar^2}{2m}\frac{d^2}{dx^2} + \frac{1}{2} m\omega^2 x^2]\psi = \hbar\omega[a^+a^- + \frac{1}{2}]\psi = E\psi \quad , \tag{2.6}$$

the eigensolutions and eigenvalues are

$$\psi_n = \left(\frac{a_o}{\pi^{\frac{1}{2}}2^n n!}\right)^{\frac{1}{2}} \exp(-\tfrac{1}{2}a_o^2 x^2)H_n(a_o x) \quad , \tag{2.7}$$

$$a_o \equiv (m\omega/\hbar)^{\frac{1}{2}} \equiv 1/(2^{\frac{1}{2}}x_o) \quad , \tag{2.8}$$

$$E_n = \hbar\omega(n + \tfrac{1}{2}) \quad . \tag{2.9}$$

Two of the most important properties of this system are that the energy levels are <u>independent</u> of n (equally spaced) and that the raising and lowering operators are <u>independent</u> of n,

$$a^{\pm}|n\rangle = (n + \tfrac{1}{2} \pm \tfrac{1}{2})^{\frac{1}{2}}|n \pm 1\rangle \quad , \tag{2.10}$$

$$x = [\hbar/(2m\omega)]^{\frac{1}{2}}(a^{-} + a^{+}) \quad , \tag{2.11}$$

$$p = [m\hbar\omega/2]^{\frac{1}{2}}(a^{-} - a^{+})/i \quad , \tag{2.12}$$

$$a^{\pm} - (2m\hbar\omega)^{-\frac{1}{2}}(m\omega x \mp ip) \quad , \tag{2.13}$$

$$[a^{-},a^{+}] = 1 \quad . \tag{2.14}$$

The implications of the above are many. But for us what is foremost is that any wave packet, no matter what its shape, will return to its original shape after one classical period of oscillation

$$\tau = 2\pi/\omega \quad . \tag{2.15}$$

This can be seen by decomposing any time-dependent state into eigenstates,

$$\Psi(x,t) = \sum_{n} a_n\psi_n(x)\exp[-i\omega t(n + \tfrac{1}{2})] \quad , \tag{2.16}$$

and observing that the equal spacing of the levels means that

$$\Psi^{*}(x,t_0)\Psi(x,t_0) = \Psi^{*}(x,t = 2\pi j/\omega + t_0)\Psi(x,t = 2\pi j/\omega + t_0) \quad , \tag{2.17}$$

where j is any integer.

This property will be useful in the next section. But for now we mention that for the harmonic oscillator the coherent states not only return to their original shape after one period of oscillation, they <u>retain</u> their original shape for all time and have a centroid which follows the classical motion.

How are these coherent states defined? In modern language, they are standardly defined in three equivalent ways. The first way is essentially what Schrödinger discovered.[1]

<u>i. Minimum-Uncertainty Coherent States (MUCS)</u>. From Eq. (2.6), the Hamiltonian of the harmonic oscillator is quadratic in p and x. From the commutation relation

$$[x,p] = i\hbar \quad , \tag{2.18}$$

there is an uncertainty relation

$$(\Delta x)^2(\Delta p)^2 \geq \tfrac{1}{4}\hbar^2 \quad . \tag{2.19}$$

Now, any commutation relation

$$[A,B] = iG \tag{2.20}$$

has an uncertainty relation

$$(\Delta A)^2(\Delta B)^2 \geq \tfrac{1}{4}\langle G\rangle^2 \tag{2.21}$$

whose equality can be satisfied by that three-parameter set of states which satisfies the eigenvalue equation[5,25,26]

$$\left(A + \frac{i\langle G\rangle}{2(\Delta B)^2}\, B\right)\psi = \left(\langle A\rangle + \frac{i\langle G\rangle}{2(\Delta B)^2}\,\langle B\rangle\right)\psi \quad . \tag{2.22}$$

Note that the four parameters $\langle A\rangle$, $\langle B\rangle$, $\langle B^2\rangle$, and $\langle G\rangle$ are not independent because they satisfy the equality in Eq. (2.21).

Applying this to x and p yields

$$\psi_{cs}(x) = [2\pi(\Delta x)^2]^{-\frac{1}{4}} \exp\left\{-\left[\frac{x - \langle x\rangle}{2(\Delta x)}\right]^2 + \frac{i}{\hbar}\langle p\rangle x\right\} \quad . \tag{2.23}$$

Now demanding that the ground-state (n = 0) wave function be a special case of (2.23) for $\langle x\rangle = \langle p\rangle = 0$ gives the additional restriction that

$$(\Delta x/\Delta p)^2 = 1/(m\omega)^2 \quad , \tag{2.24}$$

or

$$(\Delta x)^2 = (2a_0^2)^{-1} = x_0^2 \quad , \tag{2.25}$$

yielding a two-parameter set of states. This last restriction is physically necessary because it corresponds to a classical particle at rest. With this restriction the coherent-state wave packets will follow the motion of a classical particle and retain their shape. If a different value of $(\Delta x/\Delta p)$ were to be chosen, the packet would not keep its shape.

It will turn out that the generalization of this method is the one we will concentrate on when we generalize from our classical motion point of view.

ii. Annihilation-Operator Coherent States (AOCS). These states are defined as the eigenvalues of the destruction operator with complex eigenvalue α:

$$a^{-}|\alpha\rangle = \alpha|\alpha\rangle \quad . \tag{2.26}$$

They can be written in terms of the number states as

$$|\alpha\rangle = \exp(-\tfrac{1}{2}|\alpha|^2) \sum_{n=0}^{\infty} \frac{\alpha^n}{(n!)^{\frac{1}{2}}} |n\rangle \quad , \tag{2.27}$$

and further expressed as the set of Gaussians,

$$|\alpha\rangle = [2\pi(\Delta x)^2]^{-\frac{1}{4}} \exp\left[-\frac{x^2}{4(\Delta x)^2} + \frac{x\alpha}{\Delta x} - \frac{1}{2}(\alpha^2 + |\alpha|^2)\right] \quad . \tag{2.28}$$

With the physical restrictions above, α can be shown to be

$$\alpha = \frac{\langle x\rangle}{2\Delta x} + \frac{i}{\hbar}\langle p\rangle\Delta x = \frac{1}{2}\left[\frac{\langle x\rangle}{\Delta x} + i\frac{\langle p\rangle}{\Delta p}\right] , \tag{2.29}$$

so that

$$|\alpha\rangle = \exp[-i\langle p\rangle\langle x\rangle/(2\hbar)]\psi_{cs} \quad . \tag{2.30}$$

Thus the AOCS are the MUCS up to an irrelevant phase factor.

iii. Displacement-Operator Coherent States (AOCS). These states are defined as those states which are created by the unitary displacement operator

$$D(\alpha) = \exp(\alpha a^{+} - \alpha^{*} a^{-}) \tag{2.31}$$

acting on the ground state. With the aid of the Baker-Campbell-Hausdorff identity, one can show that

$$D(\alpha)|0\rangle = \exp(-\tfrac{1}{2}|\alpha|^2)\exp(\alpha a^+)\exp(-\alpha a)|0\rangle$$

$$= \exp(-\tfrac{1}{2}|\alpha|^2)\exp(\alpha a^+)|0\rangle$$

$$= \exp(-\tfrac{1}{2}|\alpha|^2) \sum_{n=0}^{\infty} \frac{\alpha^n (a^+)^n}{n!} |0\rangle$$

$$= |\alpha\rangle \quad . \tag{2.32}$$

Thus, the DOCS are equivalent to the AOCS and MUCS.

This method is the one which is most easily generalized from the group theory point of view, since it involves the exponential of the anti-Hermitian operator $(\alpha a^+ - \alpha^* a^-)$. Indeed, Perelomov[11] has championed and greatly explored this point of view.

It is from these MUCS and DOCS starting points that the complementarity of the analytic vs. group theory points of view is seen. As always, an advantage of the group theory point of view is that the symmetry is a useful tool to simplify the concepts. An advantage of the analytic point of view will be that one can construct states and numerically see how they evolve in time.

3. QUANTUM AND CLASSICAL HARMONIC POTENTIALS

As mentioned in Sec. 2, two of the most striking properties of the harmonic oscillator potential are that (i) in the classical system, the classical period of oscillation $\tau = 2\pi/\omega$ is independent of energy and that (ii) in the quantum system, the eigenvalues are equally spaced by $\hbar\omega$. Any potential which satisfies property (i) I call a "classical harmonic potential," and any potential which satisfies property (ii) I call a "quantum harmonic potential."[22-24]

As is discussed in the standard work of Landau and Lifshitz,[27] there are many potentials which are classical harmonic potentials; in fact, an uncountable number. In Refs. 22-24 it was also shown that the set of quantum harmonic potentials is not the same as the set of classical harmonic potentials, even though they are the same in WKB approximation.[22]

What was found is that there are three distinct classes: a) potentials which are classical harmonic but not quantum harmonic, b) those which are quantum harmonic but not classical harmonic, and c) those which are both. Further, each class contains an uncountable number. I refer people to Refs. 22-24 but the example of Ref. 22 gives the idea.

The demonstration that quantum harmonic and classical harmonic potentials are not equivalent classes was a proof by example. What was used was an idea of Abraham and Moses[28] (AM) based on the Gel'fand-Levitan formalism.[29] One can take an exactly solvable potential with a known spectra and generate an exactly solvable potential with the same spectra, but with one or, indeed, any number of eigenstates removed.

In particular, when applied to the harmonic oscillator, one can show that[22,28] if the harmonic oscillator ground state is removed, one has the exactly solvable AM[28] potential v (in dimensionless units)

$$z = a_o x \quad , \quad a_o = (m\omega/\hbar)^{\frac{1}{2}} \quad , \tag{3.1}$$

$$v = V/\hbar\omega \quad , \quad \varepsilon_n = E_n/\hbar\omega \ , \tag{3.2}$$

$$v = v_o + v_1 \quad , \tag{3.3}$$

$$v_o = \tfrac{1}{2}z^2 \quad , \quad v_1 = 4\phi(\phi - z) \quad , \tag{3.4}$$

$$\phi(z) = \frac{e^{-z^2}}{z^{\frac{1}{2}}\mathrm{erfc}(z)} \quad , \tag{3.5}$$

$$\mathrm{erfc}(z) = \frac{2}{\pi^{\frac{1}{2}}} \int_z^\infty e^{-t^2} dt \ . \tag{3.6}$$

The solutions are

$$\varepsilon_n = n + \tfrac{1}{2} \quad , \qquad n = 1,2,\ldots \tag{3.7}$$

$$\chi_n(z) = \psi_n(z) - \left(\frac{2}{n}\right)^{\frac{1}{2}} \phi(z)\psi_{n-1}(z) \quad , \tag{3.8}$$

where the ψ_n are the harmonic oscillator wave functions of Eq. (2.7). The shape of the potential is shown in Fig. 1.

The AM potential is thus a quantum harmonic potential. However, by numerically calculating the classical period $\tau(\varepsilon)$ as a function of dimensionless energy ε, one finds that the dimensionless quantity

$$P(\varepsilon) = \tau(\varepsilon)/\tau = \tau(\varepsilon)\omega/2\pi \tag{3.9}$$

varies with energy. This is shown in Fig. 2, which has two scales on it.

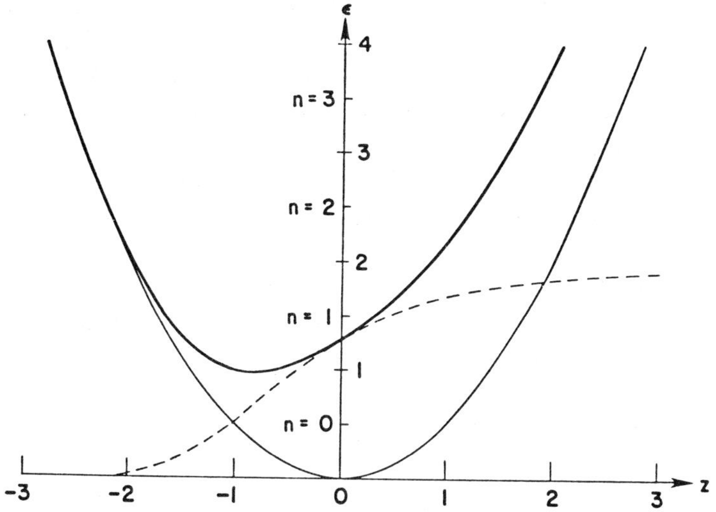

Figure 1.

The harmonic oscillator potential $v_0(z)$ is a light curve, the contribution $v_1(z)$ is a light dashed curve, and the complete AM potential v(z) is a heavy curve. The number eigenstates are also indicated.

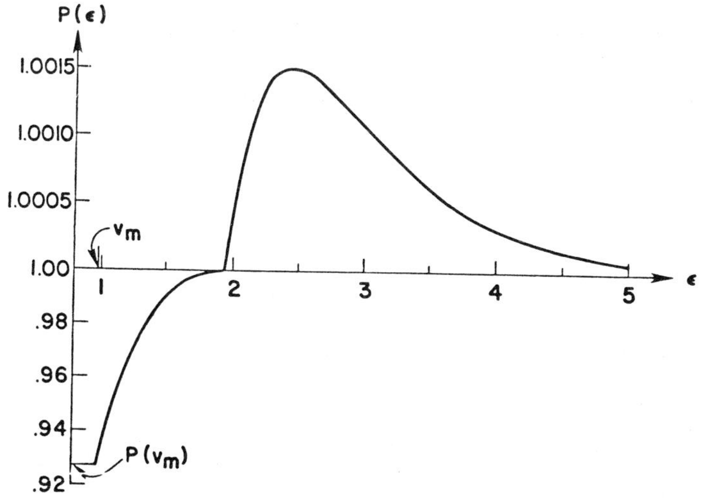

Figure 2.

P(ε) plotted as a functon of ε. There are two scales for P(ε). The scale for P(ε) > 1 is greatly expanded.

Although the variation of the classical period with energy is very small, it does exist, and the AM potential is not a classical harmonic potential.

The above means a "coherent state" will not follow the classical motion for all time. As we mentioned at Eq. (2.17), any quantum system with equally spaced eigenvalues will have any wave packet return to its original shape after every time $\tau = 2\pi/\omega$. But for the AM potential the classical period of oscillation is not independent of energy. Therefore, the quantum and classical systems will diverge; slowly, but they will diverge.

Thus, we find that a potential with a little variation from the harmonic oscillator, i.e., one which is quantum harmonic but not classical harmonic, will no longer yield perfect classical motion in the quantum system.

4. ANALYTIC COHERENT STATES FOR GENERAL POTENTIALS

We have just seen that the more "harmonic oscillator-like" a potential is, the more nearly there is a quantum state that can follow the classical motion. The thing that produces the special features of the harmonic oscillator is that its Hamiltonian is quadratic in the variables x and p, which vary sinusoidally with time. We call these the "natural classical variables" for the harmonic oscillator. What are the "natural classical variables" for other potentials?

My first physical ansatz is that the "natural classical variables" X_c and P_c for any bound system are the variables which change the total energy equation into a quadratic equation in these two variables and which vary sinusoidally with time. This makes the equation "harmonic oscillator-like" in these variables. It changes the x-p phase space plot into an ellipse in an X_c-P_c plot. Specifically, there exist variables

$$X_c = A(E)\sin[\omega_c(E)t] \quad , \tag{4.1}$$

$$P_c = m\dot{X}_c = mX_c'(x)\dot{x} \tag{4.2}$$

$$= mA(E)\omega_c(E)\cos[\omega_c(E)t] \ . \tag{4.3}$$

Because

$$\dot{x}^2 = 2(E - V)/m \quad , \tag{4.4}$$

$X_c(x)$ is the solution of the equation

$$X_c' = \frac{d}{dx} X_c(x) = \omega_c \left[\frac{m(A^2 - X_c^2)}{2[E - V(x)]}\right]^{\frac{1}{2}} \quad . \tag{4.5}$$

Equations (4.2) and (4.3) imply that the classical equations of motion are

$$\dot{X}_c = P_c/m \quad , \tag{4.6}$$

$$\ddot{X}_c = \frac{\dot{P}_c}{m} = -\omega_c^2(E)X_c \quad . \tag{4.7}$$

Thus, Eq. (2.1) is replaced by a form which is similar to the harmonic-oscillator equation for a given energy. That is, this transformation is equivalent to rewriting (2.1) as

$$\tfrac{1}{2}m\omega_c^2(E)A^2(E) = \frac{1}{2m} P_c^2 + \tfrac{1}{2}m\omega_c^2 X_c^2 \quad . \tag{4.8}$$

I now define the analogous "natural quantum operators," $p = (\hbar/i)d/dx$,

$$X \equiv X_c(x) \quad , \tag{4.9}$$

$$P \equiv \tfrac{1}{2}(X_c'p + pX_c') \quad . \tag{4.10}$$

Now obtain those states which minimize the uncertainty relation associated with the commutator

$$[X,P] = iG \ . \tag{4.11}$$

My second physical <u>ansatz</u> is that the coherent states are a two-parameter subset of the above minimum uncertainty states defined by restricting $\Delta X/\Delta P$ so that the ground state of the potential is a member of the set. (<u>A priori</u> it was not clear at first that this would be true.) Finally, observe that just as x and p can be written as sums and differences of raising and lowering operators for the harmonic oscillator, so too X and P can be written as the Hermitian sums and differences of the in-general n-<u>dependent</u> raising and lowering operators of the system $A_n^{\pm}$,

$$X = K_1(n)\{[A_n^- + (A_n^+)^\dagger] + [A_n^+ + (A_n^-)^\dagger]\} \quad , \tag{4.12}$$

$$P = \frac{1}{i} K_2(n)\{[A_n^- + (A_n^+)^\dagger] - [A_n^+ + (A_n^-)^\dagger]\} \quad , \tag{4.13}$$

where $K_1(n)$ and $K_2(n)$ are n-dependent c-numbers.

What we have is a chain in which you can go in either direction. If you can solve the classical problem, you will obtain the natural classical variables. From these you obtain the natural quantum operators, and find the coherent states as those which minimize the associated uncertainty relation subject to the ground state being a member of the set. Finally, these natural quantum operators can be written as the Hermitian sums and differences of the quantum raising and lowering operators.

Similarly, you can go in the reverse direction if you can solve the quantum problem. As a matter of fact, when we were discovering this method it was this reverse procedure which led us to the solution. Now it appears to us that for most examples it is simpler to proceed starting from the classical problem.

In our series[14-19] we also discussed analytic generalizations of annihilation-operator and displacement-operator coherent states. In general these are not the same as the generalized minimum-uncertainty coherent states. For the annihilation-operator coherent states, one uses the fact that just as a^- can be written in terms of x and p for the harmonic oscillator, so too A_n^- can be written in terms of X and P. This operator is used. Further, in general these states are equivalent to states from a displacement-operator which is <u>not</u> an exponential. One can understand this by realizing that for very confining systems, like a square well, you <u>cannot</u> displace the ground state unitarily to the side. It has nowhere to go.

I refer the reader to our series[14-19] for more details. I will now concentrate on the minimum-uncertainty coherent states, which we found to be the most physical and enlightening to study.

5. AN EXAMPLE

To show how the procedure works, I will use what I find a very illuminating example, the $\cosh^{-2}$ potential. Normalizing so that the potential's minimum is at the origin, the potential is

$$V(x) = U_0 \tanh^2 ax$$

$$\equiv \mathcal{E}_0 s(s+1) \tanh^2 z \ , \tag{5.1}$$

$$\mathcal{E}_0 = \hbar^2 a^2/2m \ . \tag{5.2}$$

This potential has a finite number of bound states, goes to zero at the origin, and goes to U_0 at $x = \pm\infty$.

The solutions for our natural classical variables X_c and P_c are

$$X_c = \sinh z = \left(\frac{E}{U_0 - E}\right)^{\frac{1}{2}} \sin \omega_c t \quad , \tag{5.3}$$

$$P_c = p \cosh z = (2ma^2E)^{\frac{1}{2}} \cos \omega_c t \quad , \tag{5.4}$$

$$\omega_c(E) = \left[\frac{2a^2(U_0 - E)}{m}\right]^{\frac{1}{2}} \quad . \tag{5.5}$$

One can verify that these variables obey classical equations of motion analogous to those for x and p in the harmonic oscillator,

$$\dot{X}_c = P_c/m \quad , \tag{5.6}$$

$$\dot{P}_c = -m\omega_c^2 X_c \quad . \tag{5.7}$$

Now we do our coherent-states procedure. The natural quantum operators are

$$X = X_c = \sinh z \quad , \tag{5.8}$$

$$P = \frac{\hbar}{2i} a^2 \left[\frac{d}{dz} \cosh z + \cosh z \frac{d}{dz}\right] , \tag{5.9}$$

so that we are looking for states which satisfy the equality in the uncertainty relation

$$(\Delta X)^2(\Delta P)^2 \geq (\hbar^2/4)a^4\langle\cosh^2 z\rangle^2 \quad . \tag{5.10}$$

These normalized states are

$$\psi \equiv \left[\frac{a\Gamma(B + \frac{1}{2} + iu)\Gamma(B + \frac{1}{2} - iu)}{\pi^{\frac{1}{2}}\Gamma(B)\Gamma(B + \frac{1}{2})}\right]^{\frac{1}{2}} (\cosh z)^{-B}$$
$$\times \exp[C \sin^{-1}(\tanh z)] , \tag{5.11}$$

$$B \equiv \frac{1}{2} \frac{\langle\cosh^2 z\rangle}{(\Delta \sinh z)^2} + \frac{1}{2} \quad , \tag{5.12}$$

$$C \equiv u + iv = B\langle \sinh z \rangle + \langle \cosh z \frac{d}{dz} \rangle \ . \tag{5.13}$$

Finally, if

$$B \equiv s \tag{5.14}$$

this set of states includes the ground state. Therefore, imposing (5.14) gives us our coherent states.

Now the quantum problem has exact eigenstates and eigenvalues

$$\psi_n \equiv \left[\frac{a(s - n)\Gamma(2s - n + 1)}{\Gamma(n + 1)}\right]^{\frac{1}{2}} P_s^{n-s}(\tanh z) \quad , \tag{5.15}$$

$$E_n = \mathcal{E}_0(2ns - n^2 + s) \ , \tag{5.16}$$

where the P_n^{n-s} are associated Legendre functions. Because of that one can figure out what the raising and lowering operators are for these eigenstates. They are

$$A_n^{\pm} = (s - n)\sinh z \mp \cosh z \frac{d}{dz} \ . \tag{5.17}$$

This verifies that indeed our natural quantum operators X and P can be expressed as the Hermitian sums and differences of the quantum raising and lowering operators, just as in Eqs. (4.12) and (4.13).

6. DISCUSSION

Given our minimum-uncertainty coherent states (MUCS), just how well do they follow the classical motion? At the beginning I gave the intuitive answer, and I now repeat it. Effectively, "The more highly bound a particle state is and, given that, the greater number of eigenstates with which a coherent state has a significant overlap and, finally, the closer to equally spaced (i.e., harmonic) these eigenstates are, then the better and longer a coherent state will follow the classical motion without dispersing."

In the 5th article of our series,[18] we showed in detail photographs displaying the time-evolution of MUCS in differing potentials and situations. I refer people to that for the detailed discussion which shows how we came to the above intuitive answer. This article[18] discusses work which was the basis for a computer generated film[30] which graphically displays the time evolution of our states (this film to be shown at the present seminar).

In this printed version I wish to include one display of the time evolution of an MUCS, shown in Fig. 3. It is for the $\cosh^{-2}$ potential of the last section. s = 399.5 (400 bound states), the coherent state starts in the middle and has an $\langle H \rangle$ that is 1/10 the way up to the continuum at U_0.

Note that each time the wave packet hits a wall, part of it begins to disperse until, at the end, after 8-5/8ths classical oscillations, there is significant spread in the packet from the main hump and the packet's $\langle x \rangle$ has deviated from the classical position.

From this and other runs like this[18,30] we came to the intuitive description of the time evolution I have given.

However, we also considered the time evolution of other types of coherent states.[14-18,31] Our conclusion there was that our MUCS always do as well as or better than other states investigated. Specifically, the less confining a potential is, the more the various methods numerically yield about the same coherent states and time evolution. However, for highly confined systems (like the infinite square-well), our MUCS method still works well whereas some others can run into trouble.[18]

One can, by a slight generalization, apply this method to multidimensional systems[17] and systems where the commutation relations involved are complicated.[16] One can also apply this method to systems which cannot be solved analytically, either by analytic approximations or by numerically solving the differential equations involved. (In fact, for the Morse potential[16] our work involved analytic approximation techniques.) Finally, the method has also been applied to time-dependent potentials.[32] Thus, I feel that this method is a general way of finding classical motion states in arbitrary potentials.

ACKNOWLEDGMENTS

I first wish to thank my partners in this project, L. M. Simmons, Jr. and V. P. Gutschick. Also, I thank the many colleagues who have helped us in our work. This is especially true of P. A. Carruthers, whose insightful question satarted me on this project, and J. R. Klauder, whose question on the comparison of different kinds of coherent states led to insightful numerical work. This work was supported by the United States Department of Energy.

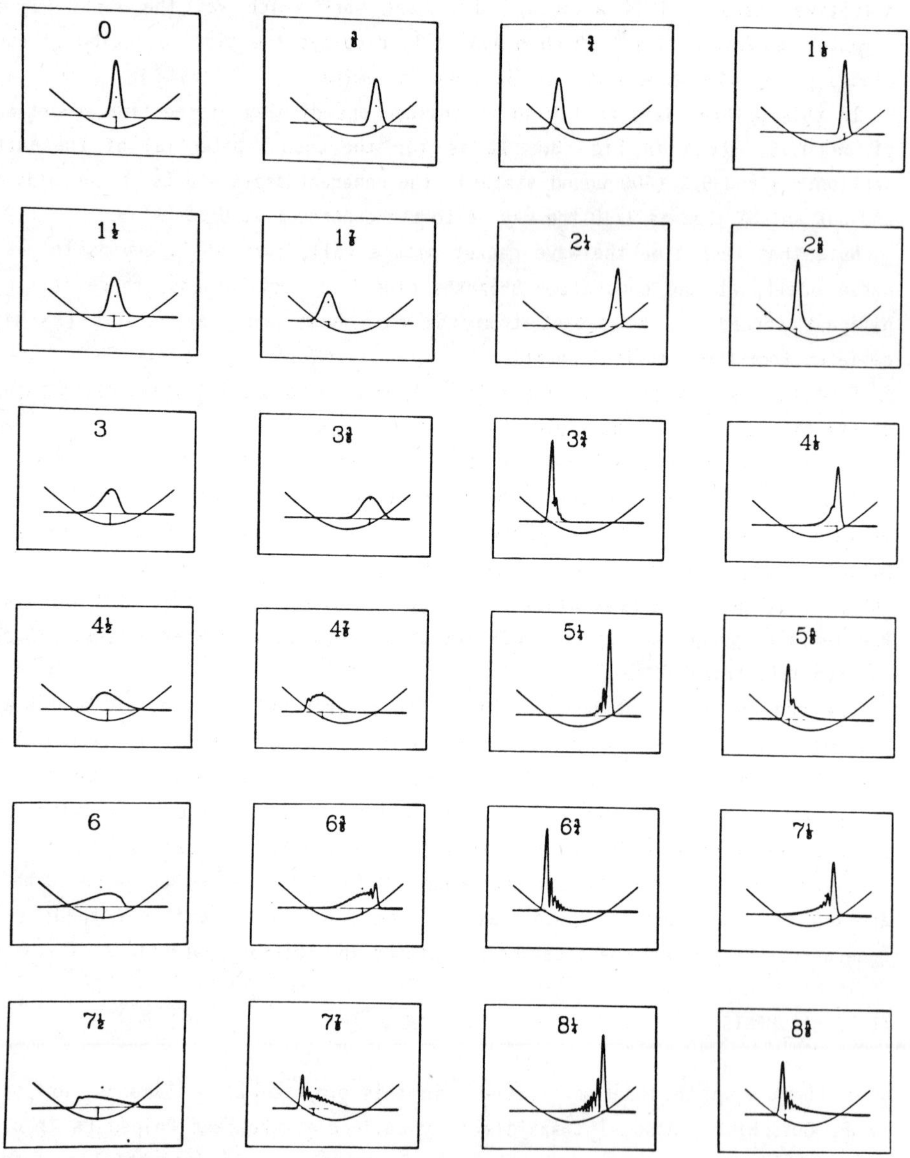

Figure 3.

Time evolution of a coherent state wave packet in the $\cosh^{-2}$ potential with s = 399.5 (400 bound state). The state has an <H> that is 1/10 the way up to the continuum at U_0. Shown is the potential, the wave packet at a height corresponding to <H>, a vertical bar that shows the classical position, and a dot that represents <x>. The numbers are the number of classical oscillations for each frame.

REFERENCES

1. E. Schrödinger, Naturwissenschaften 14, 664 (1926). Schrödinger's nine basic papers on quantum mechanics have been collected and translated into English in E. Schrödinger, Collected Papers on Wave Mechanics (Blackie & Son, London, 1928). This paper is found on p. 41.
2. R. J. Glauber, Phys. Rev. Lett. 10, 84 (1963); Phys. Rev. 130, 2529 (1963); 131, 2766 (1963).
3. E. C. G. Sudarshan, Phys. Rev. Lett. 10, 227 (1963).
4. P. Carruthers and M. M. Nieto, Am. J. Phys. 33, 537 (1965).
5. P. Carruthers and M. M. Nieto, Rev. Mod. Phys. 40, 411 (1968).
6. I. A. Malkin and V. I. Man'ko, Zh. Eksp. Toer. Fiz. 55, 1014 (1968) [Sov. Phys. JETP 28, 527 (1969)]; V. V. Dodonov, I. A. Malkin, and V. I. Man'ko, Physica (Utrecht) 59, 241 (1972).
7. J. M. Radcliffe, J. Phys. A 4, 313 (1971).
8. P. W. Atkins and J. C. Dobson, Proc. R. Sco. London A321, 321 (1971).
9. D. Stoller, Phys. Rev. D 4, 2309 (1971).
10. A. O. Barut and L. Girardello, Commun. Math. Phys. 21, 41 (1971).
11. M. Perelomov, Commun. Math. Phys. 26, 222 (1972); Usp. Fiz. Nauk 123, 23 (1977) [Sov. Phys. Usp. 20, 703 (1977)].
12. F. T. Arecchi, E. Courtens, R. Gilmore, and H. Thomas, Phys. Rev. A 6, 2211 (1972).
13. C. Aragone, C. Guerri, S. Salamó, and J. L. Tani, J. Phys. A 7, L149 (1974); A. Aragone, E. Chalbaud, and S. Salamó, J. Math. Phys. 17, 1963 (1976); M. A. Rashid, ibid. 19, 1391 (1978); 19, 1397 (1978); H. Bacry, Phys. Rev. A 18, 617 (1978).
14. M. M. Nieto and L. M. Simmons, Jr., Phys. Rev. D 20, 1321 (1979). References 14-19 comprise a series of 6 papers on our formalism.
15. M. M. Nieto and L. M. Simmons, Jr., Phys. Rev. D 20, 1332 (1979).
16. M. M. Nieto and L. M. Simmons, Jr., Phys. Rev. D 20, 1342 (1979).
17. M. M. Nieto, Phys. Rev. D 22, 391 (1980).
18. V. P. Gutschick and M. M. Nieto, Phys. Rev. D 22, 403 (1980).
19. M. M. Nieto, L. M. Simmons, Jr., and V. P. Gutschick, Phys. Rev. D 23, 927 (1981).
20. Other papers we have published on the project of Refs. 14-19 are M. M. Nieto and L. M. Simmons, Jr., Phys. Rev. Lett. 41, 207 (1978); V. P. Gutschick, M. M. Nieto, and L. M. Simmons, Jr., Phys. Lett. 76A, 15 (1980); M. M. Nieto and L. M. Simmons, Jr., in Foundations of Radiation Theory and Quantum Electrodynamics, edited by A. O. Barut (Plenum, New York, 1980), pp. 199-224.

21. Related papers on quantum mechanics which we have published are M. M. Nieto, Phys. Rev. A 17, 1273 (1978); M. M. Nieto and L. M. Simmons, Jr., Am. J. Phys. 47, 634 (1979); M. M. Nieto, Phys. Rev. A 20, 700 (1979); Am. J. Phys. 47, 1067 (1979).

22. M. M. Nieto and V. P. Gutschick, Phys. Rev. D 23, 922 (1981).

23. G. Ghosh and R. W. Hasse, Phys. Rev. D 24, 1027 (1981).

24. M. M. Nieto, Phys. Rev. D 24, 1030 (1981).

25. K. Gottfried, Quantum Mechanics (Benjamin, New York, 1966), Vol. 1, pp. 213-215.

26. R. Jackiw, J. Math. Phys. 9, 339 (1968).

27. L. D. Landau and E. M. Lifshitz, Mechanics (Pergamon, Oxford, 1960), pp. 25-29.

28. P. B. Abraham and H. E. Moses, Phys. Rev. A 22, 1333 (1980).

29. I. M. Gel'fand and B. M. Levitan, Am. Math. Soc. Transl. 1, 253 (1955) [Izv. Adad. Nauk SSSR, Ser, Mat. 15, 309 (1951)].

30. V. P. Gutschick, M. M. Nieto, and F. Baker, Time-Evolution of Coherent States for General Potentials, 13 min., 16 mm, color, sound movie, available from Cinesound Co., 915 N. Highland Ave., Hollywood California 90038. Copies are on deposit at the AAPT and Los Alamos National Laboratory film libraries. A review of the film is given in C. A. Nelson, Am. J. Phys. 47, 755 (1979).

31. J. R. Klauder, J. Math. Phys. 5, 177 (1964); in Path Integrals and Their Applications in Quantum, Statistical and Solid State Physics, edited by G. J. Papadopoulos and J. T. Devreese (Plenum, New York, 1978), pp. 5-38.

32. J. R. Ray, Phys. Rev. D 25, 3417 (1982).

IV. Path Integrals and Coherent States

548

Progress of Theoretical Physics, Vol. 60, No. 2, August 1978

Coherent States of Fermi Operators and the Path Integral*)

Yoshio OHNUKI and Taro KASHIWA**)
Department of Physics, Nagoya University, Nagoya 464

(Received March 1, 1978)

Coherent states of Fermi operators are explicitly constructed with the aid of Grassmann numbers, and their properties are discussed in detail. A rigorous form of the path integral for Fermi fields is formulated by means of the coherent states. A simple example is also discussed.

§ 1. Introduction

It is well known that the classical picture of quantum-mechanical amplitudes of Bose oscillators is given by a set of complex eigenvalues of the annihilation operators. The eigenstate corresponding to it is called the coherent state. In this connection, if one wishes to get a classical analog of Fermi operators, it seems reasonable to consider their coherent states. In this case, however, even if a coherent state exists, corresponding eigenvalues are no longer made complex numbers because of the anticommutation property of Fermi operators. Thus to be consistent the eigenvalues must be anticommutable with each other. Such peculiar anticommuting c-numbers are known as the Grassmann numbers or the Grassmann variables.[2] Schwinger[3] is perhaps the first who introduced the Grassmann variable in field theory. He used it as a source of fermion field in his theory of Green's function. Some systematic and elegant studies on applications of the Grassmann variables have been made by Berezin.[4),5),6]

The purpose of the present paper is to construct coherent states of fermions***) with the aid of the Grassmann numbers and to apply them for a description of fermion systems.

Now suppose that $\xi_1, \xi_2, \cdots, \xi_n$ are n independent Grassmann numbers. They are defined by

$$\{\xi_i, \xi_j\} = 0 \quad (i, j = 1, 2, \cdots, n) \tag{1·1}$$

and

*) Lectures[1] based on a preliminary version of this work were given by one of the authors (Y. O.) at National Laboratory of High Energy Physics (KEK), Tsukuba, on 7th and 8th of July, 1977.

**) Present Address: Research Institute for Fundamental Physics, Kyoto University, Kyoto 606.

***) After completion of this work the authors knew that almost the same idea had already been proposed by Martin.[7] Though the formalism is not complete, his paper may be the first in which coherent states of fermions are explicitly discussed. Montonen[8] also introduced the coherent states which correspond to our $|(\xi)_n\rangle$ and $\langle(\xi)_n{}^*|$.

$$\xi_1\xi_2\cdots\xi_n \neq 0 \,. \tag{1·2}$$

On the basis of these relations we shall in § 2 study some general properties of the Grassmann variables, which will play fundamental roles in our later discussion. With these preparations we shall in § 3 explicitly construct coherent states of Fermi operators. It is shown that they satisfy an orthogonality relation which is written in terms of δ-functions defined on the Grassmann variables. Several types of the completeness relation are also given by means of the coherent states. In § 4, as representatives of the ordinary state vector of fermions we shall introduce some kinds of *wavefunctions* whose arguments are Grassmann variables. All formalisms for any fermion system are shown to be rewritten in terms of these wavefunctions. Especially a rigorous expression of the path integral for time development operator will be given. We shall also write down the trace of $e^{-\beta H}$ in a form of the path integral of the Grassmann variables, in which the antiperiodic boundary condition usually imposed on the initial and final variables is shown to be an automatic consequence of our theory.

§ 2. Integrals, δ-functions and transformations of variables

By analogy with the usual theory we define a δ-function $\delta(\xi, \xi')$ for Grassmann variables ξ and ξ' by the relation

$$\int f(\xi)\delta(\xi,\xi')d\xi = f(\xi'), \tag{2·1}$$

with $f(\xi)$ being an arbitrary function of ξ. As usual the $\delta(\xi, \xi')$ is assumed to be a function of $\xi-\xi'$. The integral is also defined by (2·1). Since no infinitesimal quantity is definable for Grassmann variables, the symbol $\int\cdots d\xi$ is to be understood as a kind of mapping, which is assumed to satisfy the linearity

$$\int\{\alpha_1 f_1(\xi) + \alpha_2 f_2(\xi)\}d\xi = \alpha_1\int f_1(\xi)d\xi + \alpha_2\int f_2(\xi)d\xi \,, \tag{2·2}$$

where α_1 and α_2 are, in general, functions of the Grassmann numbers other than ξ. Though the notation $d\xi$ can have a significance as a mapping only when combined with the notation $\int$, we shall, for simplicity, sometimes drop out the latter notation in discussing the properties of the mapping and deal with $d\xi$ as if it were a kind of number.

By virtue of $\xi^2=0$ we may write the general form of $f(\xi)$ as $\alpha_1+\alpha_2\xi$. So Eq. (2·1) is equivalent to the set of equations

$$\int\delta(\xi,\xi')d\xi = 1\,, \quad \int\xi\delta(\xi,\xi')d\xi = \xi' \,. \tag{2·3}$$

Now writing the δ-function as $\beta_1+\beta_2(\xi-\xi')$, we can easily solve Eqs. (2·3) and obtain $\delta(\xi,\xi') = (\xi-\xi')/\alpha$, $\int\xi d\xi=\alpha$ and $\int d\xi=0$. Here α is a *constant* which has its own inverse and is commutable with ξ and ξ'. Although we can develop a consistent theory with any given α, we shall for later convenience put $\alpha=i$ and then we have

$$\delta(\xi,\xi')=\frac{1}{i}(\xi-\xi'), \tag{2·4}$$

$$\int \xi d\xi = i \quad \text{and} \quad \int d\xi = 0\,. \tag{2·5}$$ *)

We note that, when use is made of these relations and $\exp\{-(\xi-\xi')\xi''\}=1-(\xi-\xi')\xi''$, the δ-function is expressed in such a form of *Fourier transformation*, that is,

$$\delta(\xi,\xi')=\int e^{-(\xi-\xi')\xi''}d\xi''\,. \tag{2·6}$$

It is not a difficult task to generalize the above discussion to the case of multi-variable. Applying ξ' to both the sides of Eqs. (2·5) from left or right, we get $\xi' d\xi = -d\xi\cdot\xi'$. Thus we may write consistently

$$\{\xi_i, d\xi_j\}=0\,, \tag{2·7}$$ **)

and so from (2·4) we have

$$\iint \xi_1\xi_2\cdots\xi_n d\xi_n d\xi_{n-1}\cdots d\xi_1=(i)^n$$

and (2·8)

$$\iint \xi_{j_1}\xi_{j_2}\cdots\xi_{j_r} d\xi_n d\xi_{n-1}\cdots d\xi_1=0 \quad \text{for } 0\le r<n\,,$$

which also imply

$$\{d\xi_i, d\xi_j\}=0\,. \quad (i\neq j) \tag{2·9}$$

The notation $\iint$ stands for the multiple integral.

Now let us consider a general transformation of the set of the Grassmann variables $\{\xi_1, \xi_2, \cdots, \xi_n\}$ into a new set $\{\xi_1', \xi_2', \cdots, \xi_n'\}$. We may write it down in the form

$$\xi_i'=\alpha_i+\sum_j \alpha_i^{j}\xi_j+\sum_{j_1<j_2}\alpha_i^{j_1j_2}\xi_{j_1}\xi_{j_2}$$
$$+\sum_{j_1<j_2<j_3}\alpha_i^{j_1j_2j_3}\xi_{j_1}\xi_{j_2}\xi_{j_3}+\cdots+\alpha_i^{12\cdots n}\xi_1\xi_2\cdots\xi_n\,. \tag{2·10}$$

---

*) Berezin[4] defined the integral with $\alpha=1$.

**) This can be shown only for $i\neq j$. But for the sake of simplicity we have assumed here the anticommutability of ξ_i and $d\xi_i$, which gives a definition of integral of the form $\int d\xi\cdot\xi$.

The coefficients $\alpha_i^{\cdots}$ are, in general, functions of Grassmann numbers independent of $\xi_i (i=1, 2, \cdots, n)$,*) and in order that the new variables also obey the Grassmann algebra $\{\xi_i', \xi_j'\}=0$, they are assumed to satisfy

$$[\alpha_i^{\text{odd}}, \alpha_j^{\text{odd}}]=[\alpha_i^{\text{odd}}, \alpha_j^{\text{even}}]=[\alpha_i^{\text{odd}}, \xi_j]=0$$

and

$$\{\alpha_i^{\text{even}}, \alpha_j^{\text{even}}\}=\{\alpha_i^{\text{even}}, \xi_j\}=0\,, \tag{2·11}$$

where $\alpha_i^{\text{even}}(\alpha_i^{\text{odd}})$ stands for $\alpha_i^{\cdots}$ with an even (odd) number of superscripts. Then from (2·10) and (2·11) we obtain

$$(\xi_1'-\alpha_1)(\xi_2'-\alpha_2)\cdots(\xi_n'-\alpha_n)=\det(\alpha_i^j)\cdot\xi_1\xi_2\cdots\xi_n\,. \tag{2·12}$$

This implies that, if $\det(\alpha_i^j)^{-1}$ exists, non-vanishingness of the product $\xi_1\xi_2\cdots\xi_n$ is equivalent to that of $(\xi_1'-\alpha_1)(\xi_2'-\alpha_2)\cdots(\xi_n'-\alpha_n)$, and hence of $\xi_1'\xi_2'\cdots\xi_n'$, since α_i is independent of ξ_j's. In other words, if $\xi_1', \xi_2', \cdots, \xi_n'$ are independent of each other, so too are $\xi_1, \xi_2, \cdots, \xi_n$ under the condition that $\det(\alpha_i^j)^{-1}$ exists, and vice versa. Thus under this condition, which will be assumed in the following, ξ_i $(i=1, 2, \cdots, n)$ are inversely expressed as functions of ξ_j''s in a unique way, and each of the sets $\{\xi_1, \xi_2, \cdots, \xi_n\}$ and $\{\xi_1', \xi_2', \cdots, \xi_n'\}$ satisfies the relations (1·1) and (1·2). Consequently, all formulas are required to take the same forms for both sets of the Grassmann variables connected by transformations thus defined. In this respect we note that the following relation must hold true to keep the integral formulas (2·8) invariant:

$$d\xi_n' d\xi_{n-1}' \cdots d\xi_1' = \frac{1}{J} d\xi_n d\xi_{n-1} \cdots d\xi_1 \tag{2·13}$$

with the Jacobian

$$J \equiv \det\left(\frac{\partial \xi_i'}{\partial \xi_j}\right). \tag{2·14}$$ **)

By definition of $\alpha_i^{\cdots}$ it is evident that all elements of the determinant J are commutable with $\alpha_j^{\cdots}$ and ξ_j $(j=1, 2, \cdots, n)$. The existence of J^{-1} is assured by the existence of $\det(\alpha_i^j)^{-1}$ (see Lemma 1 in the Appendix). Unlike the case of c-number variables the Jacobian now appears as a denominator on the right-hand side of (2·13).***) The proof of (2·13) is somewhat lengthy, so we shall give it in the Appendix.

*) We consider here that $\xi_1, \xi_2, \cdots, \xi_n$ are chosen from a set of more than n independent Grassmann variables. If necessary we shall in the following regard the number of elements of this set sufficiently large.

**) For the definition of derivatives see Ref. 4). In the present paper we shall use, for definiteness, only left derivatives.

***) The simplest case, where $\xi_i'=\alpha_i+\sum_j \alpha_i^j \xi_j$, was examined by Berezin.[4]

§ 3. An extension of the Hilbert space and coherent states of Fermi operators

Let $a_i (i=1, 2, \cdots, n)$ be Fermi operators satisfying the ordinary anticommutation relations. The vacuum state is defined by

$$a_i|0\rangle = \langle 0|a_i^\dagger = 0\,. \tag{3·1}$$

The Hilbert space $\mathfrak{H}$ is spanned by linear combinations of those vectors with c-number coefficients which are obtained by acting the $a_i^\dagger$'s on $|0\rangle$. We will extend $\mathfrak{H}$ by introducing functions of Grassmann numbers as the coefficients of state vectors. The extended space thus defined will be denoted as $\mathfrak{H}_G$. In the same way the dual space of $\mathfrak{H}$ can also be extended to $\mathfrak{H}_G^*$ which consists of vectors $\langle 0|a_i \cdots a_j$ with coefficients of functions of Grassmann numbers. For definiteness we assume that any Grassmann number is anticommutable with fermion operators and is commutable with the vacuum states $|0\rangle$ and $\langle 0|$;

$$\{\xi, a_i\} = \{\xi, a_i^\dagger\} = 0 \tag{3·2}$$

and

$$\xi|0\rangle = |0\rangle\xi\,, \quad \xi\langle 0| = \langle 0|\xi\,. \tag{3·3}$$

By the use of these relations we can make a product of any element of $\mathfrak{H}_G^*$ and that of $\mathfrak{H}_G$. For instance the product of $\langle 0|a_i$ and $\xi a_j^\dagger|0\rangle$ is given by $\langle 0|a_i \xi a_j^\dagger|0\rangle = -\delta_{ij}\xi$, for $\xi a_i^\dagger|0\rangle$ is equal to $-a_j^\dagger|0\rangle\xi$ by virtue of (3·2) and (3·3). Furthermore, remembering that the right-hand sides of Eqs. (2·5) are c-numbers, we btain from (3·2) and (3·3)

$$\{d\xi, a_i\} = \{d\xi, a_i^\dagger\} = 0 \tag{3·4}$$

and

$$d\xi|0\rangle = |0\rangle d\xi\,, \quad d\xi\langle 0| = \langle 0|d\xi\,. \tag{3·5}$$

With these preparations we define the two vectors $|(\xi)_n\rangle$ and $\langle(\xi)_n|$ by

$$|(\xi)_n\rangle \equiv \exp\left(-\sum_{j=1}^{n} \xi_j a_j^\dagger\right)|0\rangle \tag{3·6}$$

and

$$\langle(\xi)_n| \equiv \langle 0|\delta(\xi_1, a_1)\,\delta(\xi_2, a_2)\cdots\delta(\xi_n, a_n) \tag{3·7}$$

with

$$\delta(\xi_j, a_j) \equiv \frac{1}{i}(\xi_j - a_j)\,. \tag{3·8}$$

Taking account of the relations

$$e^{\xi a_j^\dagger} a_i\, e^{-\xi a_j^\dagger} = a_i + \delta_{ij}\xi \tag{3·9}$$

and

$$\delta(\xi, a_j)\, a_j = \delta(\xi, a_j)\, \xi\,, \tag{3·10}$$

one easily finds that the states $|(\xi)_n\rangle$ and $\langle(\xi)_n|$ are coherent states of the Fermi operators a_j, though the eigenvalues are now the Grassmann numbers:

$$a_j|(\xi)_n\rangle = \xi_j|(\xi)_n\rangle, \tag{3·11}$$

$$\langle(\xi)_n|a_j = \langle(\xi)_n|\xi_j\,. \tag{3·12}$$

It is noted that the coherent states have the *orthogonality* relation

$$\langle(\xi)_n|(\xi')_n\rangle = \delta(\xi_1, \xi_1')\,\delta(\xi_2, \xi_2')\cdots\delta(\xi_n, \xi_n')\,, \tag{3·13}$$

which can be shown by using Eqs. (3·8) and (3·9).

We next show the completeness of these states:

$$\int\!\!\int |(\xi)_n\rangle\langle(\xi)_n|\,(d\xi)_n = 1\,, \tag{3·14}$$

where

$$(d\xi)_n \equiv d\xi_n d\xi_{n-1}\cdots d\xi_1\,. \tag{3·15}$$

Proof:

Defining the state $|j\rangle$ by

$$|j\rangle \equiv a_j^\dagger a_{j-1}^\dagger \cdots a_1^\dagger|0\rangle, \tag{3·16}$$

we have

$$\int e^{-\xi_j a_j^\dagger}|j-1\rangle\langle j-1|\delta(\xi_j, a_j)\, d\xi_i = |j\rangle\langle j| + |j-1\rangle\langle j-1|. \tag{3·17}$$

This is because the integrand is explicitly calculated as

$$\begin{aligned} e^{-\xi_j a_j^\dagger}|j-1\rangle\langle j-1|\delta(\xi_j, a_j) &= (1-\xi_j a_j^\dagger)\,|j-1\rangle\langle j-1|(-i)\,(\xi_j - a_j) \\ &= (-i)\,(|j-1\rangle\langle j-1|\xi_j + a_j^\dagger|j-1\rangle\langle j-1|a_j\xi_j - |j-1\rangle\langle j-1|a_j) \end{aligned}$$

and the integration is performed by the use of Eqs. (2·5). By mathematical induction Eq. (3·17) can be generalized to

$$\int\!\!\int |(\xi)_n\rangle\langle(\xi)_n|\,(d\xi)_n = \sum_{p;\,\text{all possible basis vectors}} |p\rangle\langle p|\,,$$

which completes the proof of the completeness. q.e.d.

We note here that the following relations are derived for any Grassmann variable ξ owing to (3·2) and (3·3):

$$\xi|(\xi)_n\rangle = |(\xi)_n\rangle\xi \tag{3·18}$$

and

$$\xi\langle(\xi)_n| = (-1)^n\langle(\xi)_n|\xi\,. \tag{3·19}$$

The similar relations evidently hold true for $d\xi$.

We next obtain coherent states of $a_j^\dagger$. To this end, following Berezin[4] we shall make use of the adjoint-operation denoted by $*$, which is a generalization of the ordinary adjoint $a_j \overset{*}{\leftrightarrow} a_j^\dagger$ and $|\rangle \overset{*}{\leftrightarrow} \langle|$ in the Fock space. It is rather trivial and we have only to keep the following rules:

$$\begin{aligned}
&(\xi)^* = \xi^*\,, \quad (\xi_i\xi_j)^* = \xi_j^*\xi_i^*\,,\\
&(\alpha_1\xi_i + \alpha_2\xi_j)^* = \xi_i^*\alpha_1^* + \xi_j^*\alpha_2^*\,,\\
&(d\xi_i d\xi_j)^* = d\xi_j^* d\xi_i^*\,, \quad (\xi_i a_j)^* = a_j^\dagger \xi_i^*\,.
\end{aligned} \tag{3·20}$$ *)

Especially we have

$$(\delta(\xi, \xi'))^* = \delta(\xi'^*, \xi^*)$$

and

$$\left(\int \xi d\xi\right)^* = \int d\xi^* \xi^* = -\int \xi^* d\xi^*\,. \tag{3·21}$$

With these rules we can see that all fundamental relations that hold for (a_i, ξ) also hold for $(a_i^\dagger, \xi^*)$. Thus defining $|(\xi)_n{}^*\rangle$ and $\langle(\xi)_n{}^*|$ by

$$|(\xi)_n{}^*\rangle \equiv \delta(a_n^\dagger, \xi_n^*)\,\delta(a_{n-1}^\dagger, \xi_{n-1}^*)\cdots\delta(a_1^\dagger, \xi_1^*)\,|0\rangle \tag{3·22}$$

and

$$\langle(\xi)_n{}^*| \equiv \langle 0|\exp\Big(-\sum_{j=1}^{n} a_j\xi_j^*\Big), \tag{3·23}$$

which are the adjoints of $\langle(\xi)_n|$ and $|(\xi)_n\rangle$ respectively, we find they are coherent states of $a_j^\dagger$, that is,

$$a_j^\dagger|(\xi)_n{}^*\rangle = \xi_j^*|(\xi)_n{}^*\rangle \tag{3·24}$$

and

$$\langle(\xi)_n{}^*|a_j^\dagger = \langle(\xi)_n{}^*|\xi_j^*\,. \tag{3·25}$$

In the same way, taking the adjoints of Eqs. (3·13) and (3·14), we obtain the following orthogonality and completeness relations:

$$\langle(\xi)_n{}^*|(\xi')_n{}^*\rangle = \delta(\xi_n^*, \xi_n'^*)\,\delta(\xi_{n-1}^*, \xi_{n-1}'^*)\cdots\delta(\xi_1^*, \xi_1'^*) \tag{3·26}$$

---

*) The adjoint given here may be regarded as a very formal operation introduced merely for convenience, because unlike the ordinary hermitian conjugation we cannot actually discriminate whether or not two Grassmann numbers arbitrarily given are related by the adjoint. Without use of it we can build the same theory in a consistent way.

and

$$\iint (d\xi)_n{}^* |(\xi)_n{}^*\rangle\langle(\xi)_n{}^*| = 1 \tag{3·27}$$

with

$$(d\xi)_n{}^* \equiv d\xi_1{}^* d\xi_2{}^* \cdots d\xi_n{}^* . \tag{3·28}$$

Moreover the products of (3·14) and (3·27) give us other forms of the completeness:

$$\iint \exp(-\sum_j \xi_j \xi_j{}^*)|(\xi)_n{}^*\rangle\langle(\xi)_n|(d\xi)_n{}^*(d\xi)_n = 1 \tag{3·29}$$

and

$$\iint \exp(-\sum_j \xi_j{}^* \xi_j)|(\xi)_n\rangle\langle(\xi)_n{}^*|(d\xi)_n(d\xi)_n{}^* = 1 . \tag{3·30}$$ [8]

Here we have used the relations

$$\langle(\xi)_n{}^*|(\xi')_n\rangle = \exp(\sum_j \xi_j{}^* \xi_j{}'),$$

$$\langle(\xi')_n|(\xi)_n{}^*\rangle = \exp(\sum_j \xi_j{}' \xi_j{}^*), \tag{3·31}$$

and the anticommutability of ξ' and ξ^*.

Before closing this section we remark that there holds the following relation:

$$\delta_{ll'} = \iint \exp(\sum_j \xi_j{}^* \xi_j)\langle(\xi)_n{}^*|l\rangle\langle l'|(\xi)_n\rangle(d\xi)_n{}^*(d\xi)_n \tag{3·32a}$$

$$= \iint (d\xi)_n \langle(-\xi)_n|l\rangle\langle l'|(\xi)_n\rangle , \tag{3·32b}$$

where $\{|l\rangle; l=1, 2, \cdots\}$ is a complete set of basis vectors of the Hilbert space $\mathfrak{H}$, with the orthonormality

$$\langle l|l'\rangle = \delta_{ll'} . \tag{3·33}$$

Proof:

Let us define a complete and orthonormal set of state vectors by

$$|N, r\rangle \equiv a_{j_N}^\dagger a_{j_{N-1}}^\dagger \cdots a_{j_1}^\dagger |0\rangle$$

with

$$j_1 < j_2 < \cdots < j_N \quad \text{and} \quad N \leq n ,$$

and

$$\langle N, r|N', r'\rangle = \delta_{NN'}\delta_{rr'} .$$

Here $r(=1, 2, \cdots, {}_NC_r)$ stands for a choice of the ${}_NC_r$ combinations of N fermion operators out of the n fermion operators $a_1, a_2, \cdots, a_n$. Without loss of generality we can use $|N, r\rangle$ in place of $|l\rangle$. Owing to Eq. (3·11) and the relation $\langle 0|(\xi)_n\rangle = 1$, we have

$$\langle N, r|(\xi)_n\rangle = \xi_{j_1}\xi_{j_2}\cdots\xi_{j_N},$$

which leads us to

$$\langle(\xi)_n^*|N', r'\rangle\langle N, r|(\xi)_n\rangle = (-1)^{NN'}\langle N, r|(\xi)_n\rangle\langle(\xi)_n^*|N', r'\rangle, \quad (3\cdot 34)$$

since ξ_j is commutable with $\langle(\xi)_n^*|$ by virtue of the adjoint relation of Eq. (3·18). Thus taking account of Eq. (3·34) as well as the relation

$$(d\xi)_n^*(d\xi)_n = (-1)^n(d\xi)_n(d\xi)_n^*,$$

we are led to

$$\begin{aligned}&\iint \exp(-\sum_j \xi_j^*\xi_j)\langle(\xi)_n^*|N', r'\rangle\langle N, r|(\xi)_n\rangle(d\xi)_n^*(d\xi)_n\\ &\quad = (-1)^{n+NN'}\iint \exp(-\sum_j \xi_j^*\xi_j)\langle N, r|(\xi)_n\rangle\langle(\xi)_n^*|N', r'\rangle(d\xi)_n(d\xi)_n^*\\ &\quad = (-1)^{n+N}\delta_{NN'}\delta_{rr'}. \qquad (3\cdot 35)\end{aligned}$$

Here we have used the completeness relation (3·30). Remembering that the inner product is invariant under the unitary transformation $a_i \to -a_i$ and $a_i^\dagger \to -a_i^\dagger$, we get

$$\langle N, r|(\xi)_n\rangle = (-1)^N\langle N, r|(-\xi)_n\rangle.$$

By this equation the left-hand side of Eq. (3·35) becomes as follows:

$$\begin{aligned}&\text{l.h.s. of } (3\cdot 35)\\ &\quad = (-1)^N\iint \exp(-\sum_j \xi_j^*\xi_j)\langle(\xi)_n^*|N', r'\rangle\langle N, r|(-\xi)_n\rangle(d\xi)_n^*(d\xi)_n. \qquad (3\cdot 36)\end{aligned}$$

Then with the change of variables $\xi_i \to -\xi_i$ we get from (3·35) and (3·36)

$$\iint \exp(\sum_j \xi_j^*\xi_j)\langle(\xi)_n^*|N', r'\rangle\langle N, r|(\xi)_n\rangle(d\xi)_n^*(d\xi)_n = \delta_{NN'}\delta_{rr'}.$$

Since the state vectors $|N, r\rangle$ span the complete set of the Hilbert space $\mathfrak{H}$, this accomplishes the proof of (3·32a).

Next we shall prove (3·32b). To do this we shall make use of the relation

$$\iint \exp(\sum_j \xi_j^*\xi_j)\langle(\xi)_n^*|(d\xi)_n^* = (-1)^n\langle(-\xi)|,$$

that can easily be derived from the definition of $\langle(\xi)_n^*|$ and Eq. (2·6) with replacement of ξ' by $-a$. With the help of this relation we can explicitly perform the $(\xi)_n{}^*$-integration in (3·32a) and then arrive at (3·32b). q.e.d.

From (3·32a) and (3·32b) we obtain the following trace formula for an operator F, which is defined on the Hilbert space $\mathfrak{H}$, by taking the product of (3·32a) or (3·32b) with $\langle l|F|l'\rangle$ (being a c-number) and summing up it over l and l':

$$\mathrm{Tr}\,F=\iint \exp(\sum_j \xi_j{}^*\xi_j)\langle(\xi)_n{}^*|F|(\xi)_n\rangle(d\xi)_n{}^*(d\xi)_n \qquad (3\cdot 37\mathrm{a})^{8)}$$

$$=\iint (d\xi)_n\langle(-\xi)_n|F|(\xi)_n\rangle\,. \qquad (3\cdot 37\mathrm{b})$$

The minus sign appearing in the bra vector is considered to be characteristic of fermions, since in the case of bosons the trace is expressed as $\int dq\langle q|F|q\rangle$. As will be seen in the next section the formula (3·37b) is powerful to represent the statistical partition function $\mathrm{Tr}e^{-\beta H}$ in a form of path integral.

§4. Grassmann representatives of state vectors and the path integral

For any state vector in $\mathfrak{H}$ we can define the Grassmann representatives $\phi_A((\xi)_n)$ etc. by

$$\phi_A((\xi)_n)\equiv\langle(\xi)_n|A\rangle,\quad \phi_A((\xi)_n{}^*)\equiv\langle(\xi)_n{}^*|A\rangle, \qquad (4\cdot 1)$$

and their adjoints

$$\tilde\phi_A((\xi)_n{}^*)\equiv\langle A|(\xi)_n{}^*\rangle,\quad \tilde\phi_A((\xi)_n)\equiv\langle A|(\xi)_n\rangle. \qquad (4\cdot 2)$$

These are analogous to wavefunctions in the ordinary theory. There is a one to one correspondence between $\phi_A((\xi)_n)$ and $|A\rangle$ in virtue of the completeness condition. Namely, when $\phi_A((\xi)_n)$ is given, we can uniquely define $|A\rangle$ corresponding to it. It is evident that the same is true for other representatives. By definition the vacuum state $|0\rangle$ corresponds to the following representatives:

$$\phi_0((\xi)_n)=(-i)^n\xi_1\xi_2\cdots\xi_n\,,\qquad \phi_0((\xi)_n{}^*)=1 \qquad (4\cdot 3)$$

and

$$\tilde\phi_0((\xi)_n{}^*)=i^n\xi_n{}^*\cdots\xi_2{}^*\xi_1{}^*\,,\qquad \tilde\phi_0((\xi)_n)=1\,. \qquad (4\cdot 4)$$

The inner product of two state vectors $|A\rangle$ and $|B\rangle$ in $\mathfrak{H}$ is expressed in terms of the Grassmann representatives by sandwiching the completeness condition (3·30) between $\langle A|$ and $|B\rangle$:

$$\langle A|B\rangle=\iint\exp(-\sum_j\xi_j{}^*\xi_j)\tilde\phi_A((\xi)_n)\phi_B((\xi)_n{}^*)\,(d\xi)_n(d\xi)_n{}^*\,. \qquad (4\cdot 5)$$

Of course there are many other ways of expressing the inner products in terms of other pairs of the Grassmann representatives. For instance, if use is made of (3·32b), one finds

$$\langle A|B\rangle = \iint (d\xi)_n \phi_B((-\xi)_n)\tilde{\phi}_A((\xi)_n). \tag{4·6}$$

Now let F be an operator, that is the product of the normal-ordered operators $F_k(a^\dagger, a)$ $(k=1, 2, \cdots, N)$, namely,

$$F = F_N(a^\dagger, a)\, F_{N-1}(a^\dagger, a) \cdots F_1(a^\dagger, a), \tag{4·7}$$

where

$$F_k(a^\dagger, a) = :F_k(a^\dagger, a):. \tag{4·8}$$

We insert the completeness condition (3·30) between F_k and F_{k-1} and rewrite $\langle(\xi_k)_n{}^*|F_k(a^\dagger, a)|(\xi_{k-1})_n\rangle$ by using the relation

$$\langle(\xi_k)_n{}^*|F_k(a^\dagger, a)|(\xi_{k-1})_n\rangle = \exp\left(\sum_j \xi^*_{k,j}\xi_{k-1,j}\right) F_k(\xi_k{}^*, \xi_{k-1}), \tag{4·9}$$

which is obtained from Eqs. (3·11), (3·25), (3·18) and the first equation of (3·31). Then we can express the operator F in the form

$$F = \iint |(\xi_N)_n\rangle \exp\left\{-\sum_{k=0}^{N}\sum_{j=1}^{n} \xi^*_{k,j}(\xi_{k,j}-\xi_{k-1,j})\right\} F_N(\xi_N{}^*, \xi_{N-1})$$
$$\times F_{N-1}(\xi^*_{N-1}, \xi_{N-2})\cdots F_1(\xi_1{}^*, \xi_0)\langle(\xi_0)_n{}^*|\prod_{k=0}^{N}(d\xi_k)_n(d\xi_k)_n{}^* \tag{4·10}$$

with $\xi_{-1,j}\equiv 0$. It is straightforward to get the path integral for the time development operator $U(t_F, t_I)$ with the aid of this relation. For this purpose, let us make the Hamiltonian normal-ordered,

$$H(a^\dagger, a, t) = :H(a^\dagger, a, t):. \tag{4·11}$$

The operator $U(t_F, t_I)$ is regarded as the limit of the product of infinitesimal time development operators:

$$U(t_F, t_I) = \lim_{N\to\infty}\{1 - i\Delta t\cdot H(a^\dagger, a, t_{N-1})\}\{1 - i\Delta t\cdot H(a^\dagger, a, t_{N-2})\}$$
$$\times\cdots\{1 - i\Delta t\cdot H(a^\dagger, a, t_0)\}, \tag{4·12}$$

where the time interval Δt is given by $(t_F - t_I)/N = t_k - t_{k-1}$ with $t_F = t_N$ and $t_I = t_0$ $(k=1, 2, \cdots, N)$, so we can use $\{1 - i\Delta t\cdot H(a^\dagger, a, t_{k-1})\}$ in place of $F_k(a^\dagger, a)$ of Eq. (4·7). Thus from (4·10) we obtain

$$\langle A|U(t_F, t_I)|B\rangle$$
$$= \lim_{N\to\infty}\iint \tilde{\phi}_A((\xi_N)_n)\exp\left[\sum_{k=0}^{N} i\Delta t\cdot \mathcal{L}_k\right]\phi_B((\xi_0)_n{}^*)\prod_{k=0}^{N}(d\xi_k)_n(d\xi_k)_n{}^* \tag{4·13}$$

with

$$\mathcal{L}_k = i\sum_{j=1}^{n}\xi^*_{k,j}(\xi_{k,j}-\xi_{k-1,j})/\Delta t - H(\xi_k{}^*, \xi_{k-1}, t_{k-1}) \tag{4·14}$$

and

$$H(\xi_0^*, \xi_{-1}, t_{-1}) \equiv 0\,.$$

Here we have equated $\{1-i\Delta t\cdot H(\xi_k^*, \xi_{k-1}, t_{k-1})\}$ to $\exp[-i\Delta t\cdot H(\xi_k^*, \xi_{k-1}, t_{k-1})]$ for infinitesimal Δt. Equation (4·13) is a correct form of the path integral, in which the initial and final states are given by the Grassmann representatives.

As a simple example let us consider a free Hamiltonian described in terms of the fermion operator a and the anti-fermion operator b together with their conjugates $a^\dagger$ and $b^\dagger$. In order to derive the Green's functions, we add source terms of Grassmann numbers to the Hamiltonian. We write such a Hamiltonian as

$$H(t) = \frac{m}{2}\{[\psi^\dagger, \sigma_3\psi] + 2\} + \psi^\dagger\eta(t) + \eta^*(t)\psi \tag{4·15}$$

with

$$\psi = \begin{pmatrix} a \\ b^\dagger \end{pmatrix}, \quad \eta(t) = \begin{pmatrix} \eta^{(1)}(t) \\ \eta^{(2)}(t) \end{pmatrix}, \quad \sigma_3 = \begin{pmatrix} 1 & 0 \\ 0 & -1 \end{pmatrix}. \tag{4·16}$$

Since the Hamiltonian (4·15) is rewritten in the normal-ordered form

$$\begin{aligned} &H(a^\dagger, b^\dagger, a, b, \eta^*(t), \eta(t)) \\ &\quad = m(a^\dagger a + b^\dagger b) + a^\dagger\eta^{(1)}(t) + b\eta^{(2)}(t) + \eta^{(1)*}(t)a + \eta^{(2)*}(t)b^\dagger\,, \end{aligned} \tag{4·17}$$

we obtain the following equation:

$$\begin{aligned} Z &= \langle 0|U(t_F, t_I)|0\rangle \\ &= \lim_{N\to\infty}\iint \exp[i\sum_{k=0}^{N}\Delta t\cdot \mathcal{L}_k]\prod_{k=0}^{N}(d\xi_k^{(2)}d\xi_k^{(1)}d\xi_k^{(1)*}d\xi_k^{(2)*}) \end{aligned} \tag{4·18}$$

with

$$\mathcal{L}_k = i\sum_{j=1,2}\xi_k^{(j)*}(\xi_k^{(j)} - \xi_{k-1}^{(j)})/\Delta t - H(\xi_k^{(1)*}, \xi_k^{(2)*}, \xi_{k-1}^{(1)}, \xi_{k-1}^{(2)}, \eta_k^*, \eta_{k-1}), \tag{4·19}$$

where the Grassmann numbers $\xi^{(1)}$ and $\xi^{(2)}$ correspond to the operators a and b respectively. In deriving (4·18) we have used the relation $\phi_0(\xi^{(1)*}, \xi^{(2)*}) = \tilde{\phi}_0(\xi^{(1)}, \xi^{(2)}) = 1$ (cf., the second equations in (4·3) and (4·4)). It may be more convenient to express the path integral (4·18) in terms of the Grassmann numbers corresponding to the operators ψ and $\psi^\dagger$. Then writing them as

$$\begin{pmatrix} \chi_k^{(1)} \\ \chi_k^{(2)} \end{pmatrix} = \begin{pmatrix} \xi_k^{(1)} \\ \xi_k^{(2)*} \end{pmatrix} \quad \text{and} \quad \begin{pmatrix} \chi_k^{(1)*} \\ \chi_k^{(2)*} \end{pmatrix} = \begin{pmatrix} \xi_k^{(1)*} \\ \xi_k^{(2)} \end{pmatrix}, \tag{4·20}$$

we find that Eq. (4·18) turns out to be

$$Z = \lim_{N\to\infty}\iint \exp[i\sum_{k=0}^{N}\Delta t\cdot \mathcal{L}_k]\prod_{k=0}^{N}(d\chi_k^{(1)}d\bar{\chi}_k^{(1)}d\chi_k^{(2)}d\bar{\chi}_k^{(2)}) \tag{4·21}$$

with

$$\Delta t\cdot\mathcal{L}_k=i\Big[\bar{\chi}_k\chi_k-\frac{1}{2}\bar{\chi}_k\{(1+\sigma_3)(1-i\Delta tm)\chi_{k-1}+(1-\sigma_3)(1-i\Delta tm)\chi_{k+1}\}$$
$$+\frac{i\Delta t}{2}\bar{\chi}_k\{(1+\sigma_3)\eta_{k-1}+(1-\sigma_3)\eta_{k+1}\}$$
$$+\frac{i\Delta t}{2}\bar{\eta}_k\{(1+\sigma_3)\chi_{k-1}+(1-\sigma_3)\chi_{k+1}\}\Big], \tag{4·22}$$

$\bar{\chi}_k=\chi_k{}^*\sigma_3$ and $\chi_{N+1}=\eta_{N+1}=0$.

It is interesting to note that the Lagrangian given here is different from the one usually taken. Evidently it is essentially due to the fact that the fermion annihilation and the anti-fermion creation are both described by the single operator ψ. Indeed, if the operator ψ would annihilate the vacuum state, we were led to the usual form of Lagrangian.

The same situation also occurs in the path integral for the relativistic Dirac fields. In this case, however, it is shown that in addition to the above feature the Lagrangian becomes inevitably non-local because of the non-locality existing in relativistic creation and annihilation operators in the configuration space.*)

Finally we try to get the path integral representation for the statistical partition function $\mathrm{Tr}\, e^{-\beta H}$. Again we assume the Hamiltonian to be of the normal order. Then the following formula is derived:

$$\mathrm{Tr}\, e^{-\beta H}=\lim_{N\to\infty}\int\!\!\int \exp[-\sum_{k=1}^{N}\{\sum_{j=1}^{n}\xi^*_{k,j}(\xi_{k+1,j}-\xi_{k,j})+\Delta\beta H(\xi_k{}^*,\xi_k)\}]|_{\xi_{N+1}=-\xi_1}$$
$$\times\prod_{k=1}^{N}(d\xi_k)_n{}^*(d\xi_k)_n \tag{4·23}$$

with

$$\Delta\beta=\beta/N.$$

Proof:

Let us substitute each of the factors $F_k(a^\dagger,a)$ of (4·7) by the operator $(1-\Delta\beta H(a^\dagger,a))$. Then from Eqs. (4·9) and (3·37b) we obtain

$$\mathrm{Tr}\, e^{-\beta H}=\lim_{N\to\infty}\int\!\!\int(d\zeta)_n\langle(-\zeta)_n|(\xi_N)_n\rangle K(\xi^*,\xi)\exp(-\sum_{j=1}^{n}\xi^*_{0,j}\xi_{0,j})$$
$$\times\langle(\xi_0)_n{}^*|(\zeta)_n\rangle\prod_{k=0}^{N}(d\xi_k)(d\xi_k)_n{}^* \tag{4·24}$$

with

$$K(\xi^*,\xi)=\exp[-\sum_{k=1}^{N}\{\sum_{j=1}^{n}\xi^*_{k,j}(\xi_{k,j}-\xi_{k-1,j})+\Delta\beta H(\xi_k{}^*,\xi_{k-1})\}].$$

---

*) Details of this problem will be discussed elsewhere together with other related problems.

Since $\langle(\xi_0)_n^*|(\zeta)_n\rangle$ is commutable with any Grassmann number, we can shift it to the leftmost in the integrand of (4·24). Furthermore, owing to Eqs. (2·7) and (3·13) the term $(d\zeta)_n\langle(-\zeta)_n|(\xi_N)_n\rangle$ in (4·24) is rewritten as $\langle(\zeta)_n|(-\xi_N)_n\rangle(d\zeta)_n$. With these arrangements of factors in the integrand, we can perform the $(\zeta)_n$-integration as follows:

$$\iint\langle(\xi_0)_n^*|(\zeta)_n\rangle\langle(\zeta)_n|(-\xi_N)_n\rangle(d\zeta)_n=\langle(\xi_0)_n^*|(-\xi_N)_n\rangle$$

$$=\exp(-\sum_{j=1}^{n}\xi_{0,j}^*\xi_{N,j}).$$

Inserting this into (4·24) and taking account of the relation

$$\prod_{k=0}^{N}(d\xi_k)_n(d\xi_k)_n^*=(-1)^n(d\xi_0)_n^*(d\xi_N)_n\prod_{k=1}^{N}(d\xi_k)_n^*(d\xi_{k-1})_n,$$

we are led to

$$\mathrm{Tr}\,e^{-\beta H}=(-1)^n\lim_{N\to\infty}\iint K(\xi^*,\xi)\exp\{\sum_{j=1}^{n}(\xi_{N,j}+\xi_{0,j})\xi_{0,j}^*\}$$

$$\times(d\xi_0)_n^*(d\xi_N)_n\prod_{k=1}^{N}(d\xi_k)_n^*(d\xi_{k-1})_n.$$

Since no $\xi_{0,j}^*$ is contained in $K(\xi^*,\xi)$, the $(\xi_0)_n^*$- and $(\xi_N)_n$-integrations in the above are easily performed with the aid of Eqs. (2·6) and (2·1). Consequently we obtain

$$\mathrm{Tr}\,e^{-\beta H}=\lim_{N\to\infty}\iint K(\xi^*,\xi)\,|_{\xi_N=-\xi_0}\prod_{k=1}^{N}(d\xi_k)_n^*(d\xi_{k-1})_n,$$

which becomes Eq. (4·23) by the change of variables $\xi_{k,j}\to\xi_{k+1,j}$ $(k=0,1,\cdots,N)$. q.e.d.

It is to be noted that the indices of ξ and the order of $(d\xi_k)_n$ and $(d\xi_k)_n^*$ in (4·23) are not the same as those of (4·13) and (4·14). The condition $\xi_{N+1}=-\xi_1$ in the above is the antiperiodic boundary condition which was pointed out by Berezin[5] and has been checked so far only in some simple cases,[9] whereas according to our formalism it is an automatic consequence which generally holds true for the trace formula in any fermion system.*)

Appendix

—*Proof of* (2·13)—

In order to show the validity of Eq. (2·13) we begin with proving the following:

*) Very recently the authors have been informed by Soper[10] that he has also derived the antiperiodic boundary condition in his formalism of the fermion path integral.

Lemma 1. If $det(\alpha_i^j)^{-1}$ exists, then there also exists the inverse of the Jacobian.

Proof:

Let $\|\alpha\|$ and $\|J\|$ be the matrices whose i-j elements are α_i^j and $\partial\xi_i'/\partial\xi_j$ respectively. Writing the transformation (2·10) as

$$\xi_i' = \alpha_i + \sum_{j=1}^{n} \alpha_i{}^j \xi_j + R_i(\xi) \tag{A·1}$$

with

$$R_i(\xi) \equiv \sum_{j_1<j_2} \alpha_i{}^{j_1 j_2} \xi_{j_1} \xi_{j_2} + \cdots + \alpha_i{}^{12\cdots n} \xi_1 \xi_2 \cdots \xi_n ,$$

we get

$$\|J\| = \|\alpha\| (1 + \|\alpha\|^{-1} \|\partial R_i(\xi)/\partial\xi_j\|) .$$

By definition, $\partial R_i(\xi)/\partial\xi_j$ is either vanishing or a polynomial in ξ whose lowest power is at least of first order. Hence we have

$$\|J\|^{-1} = \sum_{p=0}^{n} \{-\|\alpha\|^{-1} \|\partial R_i(\xi)/\partial\xi_j\|\}^p \|\alpha\|^{-1} ,$$

which completes the proof. q.e.d.

We write the transformation $\{\xi_1, \xi_2, \cdots, \xi_n\} \to \{\xi_1', \xi_2', \cdots, \xi_n'\}$ simply as $\{\xi'\} = f\cdot\{\xi\}$ or $\xi_i' = f_i\cdot\{\xi\}$, where the mapping f is completely given corresponding to a given set of transformation coefficients $\alpha_i^{\cdots}$. The composition of two successive transformations, say, $\{\xi\} \xrightarrow{f} \{\xi'\} \xrightarrow{g} \{\xi''\}$ is denoted by the product gf; $\{\xi''\} = gf\cdot\{\xi\} \equiv g\cdot\{\xi'\} = g\cdot\{f\cdot\{\xi\}\}$.

Consider the two transformations $\{\xi\} \to \{\xi'\}$ and $\{\xi'\} \to \{\xi''\}$. Owing to the assumption (2·11) we can easily show

$$\frac{\partial\xi_i''}{\partial\xi_j} = \sum_{k=1}^{n} \frac{\partial\xi_i''}{\partial\xi_k'} \frac{\partial\xi_k'}{\partial\xi_j} ,$$

from which we obtain:

Lemma 2. If each of transformations f and g satisfies Eq. (2·13), so does the product gf as well.

It is noted that general transformation (A·2) is expressed by the product of $(n+1)$ transformations, such as $f^{(n+1)} f^{(n)} f^{(n-1)} \cdots f^{(1)}$. Here the f's are defined by

$$f_i^{(k)} \cdot \{\xi\} \equiv \xi_i + \delta_{ki} \tilde{R}_i(\xi) , \quad (k = 1, 2, \cdots, n)$$

$$f_i^{(n+1)} \cdot \{\xi\} \equiv \alpha_i + \sum_{j=1}^{n} \alpha_i{}^j \xi_j , \tag{A·2}$$

where $\tilde{R}_i(\xi)$ are given by

$$\tilde{R}_1(\xi) = S_1(\xi) \quad \text{and} \quad \tilde{R}_i(\xi) = S_i([f^{(i)} f^{(i-1)} \cdots f^{(1)}]^{-1}\{\xi\}) \quad (n \geq i \geq 2)$$

with

$$S_i(\xi) = \sum_{j=1}^{n} \beta_i{}^j R_j(\xi) \quad \text{and} \quad \sum_{k=1}^{n} \alpha_k{}^j \beta_i{}^k = \delta_{ij} . \tag{A·3}$$

Thus, according to Lemma 2 it is sufficient for us to show the validity of (2·13) for every transformation given in (A·2). To this end we will show that the integral formulas (2·8) with replacement of $\{\xi\}$ by $\{\xi'\}=f\cdot\{\xi\}$ are derived from (2·8) only when use is made of (2·13).

As for the transformation $f^{(n+1)}$ it is rather trivial, since in this case $J=\det(\alpha_i^j)$, $\xi_1'\xi_2'\cdots\xi_n'=Q^{(n-1)}(\xi)+\det(\alpha_i^j)\xi_1\xi_2\cdots\xi_n$, and $\xi'_{j_1}\xi'_{j_2}\cdots\xi'_{j_r}=Q^{(r)}(\xi)\ (r<n)$, where $Q^{(s)}(\xi)$ stands for a polynomial in ξ whose highest power is equal to or less than s. So we shall examine the transformation $f^{(k)}$ $(k=1, 2, \cdots, n)$ in the following.

We write it as $\xi_i'=f_i^{(k)}\cdot\{\xi\}$, corresponding to which the Jacobian is denoted by $J^{(k)}$. Then from (A·2) we get $J^{(k)}=1+\partial\tilde{R}_k(\xi)/\partial\xi_k$. The $\tilde{R}_k(\xi)$ is the sum of two parts; the one contains ξ_k and the other does not. We factorize ξ_k to the left in the former and write $\tilde{R}_k(\xi)$ as

$$\tilde{R}_k(\xi)=\xi_k A^{(k)}(\xi)+B^{(k)}(\xi). \tag{A·4}$$

Notice that no ξ_k exists in $A^{(k)}(\xi)$ and $B^{(k)}(\xi)$. Hence the Jacobian takes the form

$$J^{(k)}=1+A^{(k)}(\xi). \tag{A·5}$$

On other hand, since $\xi_i'=\xi_i$ for $i\neq k$ and $\xi_k'=\xi_k(1+A^{(k)})+B^{(k)}$, we have

$$\begin{aligned}\xi_1'\xi_2'\cdots\xi_n'=&\{1+A^{(k)}(\xi)\}\xi_1\xi_2\cdots\xi_n\\&+B^{(k)}(\xi)\xi_1\xi_2\cdots\xi_{k-1}\xi_{k+1}\cdots\xi_n.\end{aligned} \tag{A·6}$$

No ξ_k is contained in the second term. Thus we find that Eq. (2·13) is a sufficient condition*) for the first formula of (2·8) to be invariant under the transformation $f^{(k)}$.

We next examine the second formula of (2·8). There are two cases: (i) $\xi'_{j_1}\xi'_{j_2}\cdots\xi'_{j_r}$ $(r<n)$ containing ξ_k' in itself and (ii) otherwise. In case (i) we put $\xi'_{j_1}=\xi_k'$ for simplicity. Then we obtain

$$\xi'_{j_1}\xi'_{j_2}\cdots\xi'_{j_r}=\begin{cases}\{1+A^{(k)}(\xi)\}\xi_k\xi_{j_2}\xi_{j_3}\cdots\xi_{j_r}+B^{(k)}(\xi)\xi_{j_2}\xi_{j_3}\cdots\xi_{j_r} & \text{for case (i)},\\ \xi_{j_1}\xi_{j_2}\cdots\xi_{j_r}\quad(\xi_{j_i}\neq\xi_k;\ i-1, 2, \cdots, r<n) & \text{for case (ii)}.\end{cases}$$

Consequently, if taking account of (A·5), we easily find that Eq. (2·13) is a necessary and sufficient condition for the second formula of (2·8) to remain invariant under any transformation $f^{(k)}$ given in (A·2).

Thus we have proved Eq. (2·13).

References

1) Y. Ohnuki, *KEK lecture note,* KEK-77-11 (1977).
2) *Hermann Grassmanns Gesammelte Mathematische und Physikalische Werke* Bds. I1, I2,

*) This is not a necessary condition since $A^{(k)}(\xi)\xi_1\xi_2\cdots\xi_n=0$.

Teubner (Leibzig) (1894).
3) J. Schwinger, Proc. Natl. Acad. Sci. **17** (1951), 452.
4) F. A. Berezin, *The Method of Second Quantization* (Academic Press Inc., N.Y., 1966).
5) F. A. Berezin, Theor. Math. Phys. **6** (1971), 194.
6) F. A. Berezin and M. S. Marinov, Ann. of Phys. **104** (1977), 336.
7) J. L. Martin, Proc. Roy. Soc. **A251** (1959), 543.
8) C. Montonen, Nuovo Cim. **19A** (1974), 69.
9) R. Dashen, B. Hasslachen and A. Neveu, Phys. Rev. **D12** (1975), 2443.
10) D. E. Soper, University of Oregon Preprint (1978).

PHYSICAL REVIEW D VOLUME 19, NUMBER 8 15 APRIL 1979

Path integrals and stationary-phase approximations

John R. Klauder
Bell Laboratories, Murray Hill, New Jersey 07974
(Received 26 December 1978)

The general formalism for path integrals expressed in terms of an arbitrary continuous representation (generalized coherent states) is applied to give c-number formulations for the canonical algebra and for the spin algebra, and is used to derive meaningful stationary-phase approximations for these two cases. Some clarification of a recent discussion by Jevicki and Papanicolaou for the spin case is given.

I. INTRODUCTION

Transition matrix elements expressed in the form of path integrals provide a natural scheme to obtain stationary-phase approximations for propagators.[1] Conventionally such analyses have considered transitions between initial and final coordinate eigenstates and have expressed the path integral either in the pure Lagrangian form or in the traditional phase-space form. However, there is an alternative formulation of the path integral which leads directly to transitions between initial and final states that are coherent states.[2,3] Moreover, entirely analogous formulations exist for other kinematical systems, such as spin systems, that lead to strictly c-number path integrals for a wide variety of problems.[4,5] Such an alternative path-integral formulation also leads to an unconventional form for the stationary-phase approximation. In this paper we apply the general formalism developed elsewhere to formulate and analyze path integrals[4] to the special cases of the canonical algebra and the spin algebra. The latter system has relevance for static colored quarks coupled to Yang-Mills fields, and it is pedagogically useful to discuss this case carefully. To set the stage properly we first make some general remarks.

II. PATH INTEGRALS FOR GENERAL SYSTEMS

Elsewhere we have derived an expression for the path integral in terms of an arbitrary continuous representation (generalized coherent states) suitable for analyzing stationary-phase approximations.[4] In particular, this expression is formally given by

$$\langle l'',t''|l',t'\rangle \equiv \langle l''|e^{-i(t''-t')\mathcal{H}}|l'\rangle$$

$$= \lim_{\epsilon\to 0} \mathfrak{N}\int \exp\left\{i\int_{t'}^{t''}[i\langle l|\dot{l}\rangle + \tfrac{1}{2}i\epsilon\langle \dot{l}|(1-|l\rangle\langle l|)|\dot{l}\rangle - H(l)]dt\right\}\mathfrak{D}l\,. \tag{1}$$

Here $l=l(t)$ is an L-component path connecting $l'\equiv l(t')$ and $l''\equiv l(t'')$; $|l\rangle$ denotes one vector of an overcomplete family of states (generalized coherent states) which admits a resolution of unity in the form

$$1=\int |l\rangle\langle l|\,\delta l\,, \tag{2}$$

where δl is a suitable measure, $|\dot{l}\rangle \equiv d|l\rangle/dt$, $\mathfrak{D}l\equiv\prod_t \delta l(t)$, $\mathfrak{N}$ is a normalization fixed by the composition law

$$\langle l''',t'''|l',t'\rangle = \int \langle l''',t'''|l'',t''\rangle\langle l'',t''|l',t'\rangle\,\delta l''\,, \tag{3}$$

and the classical Hamiltonian H is related to the quantum Hamiltonian $\mathcal{H}$ by

$$H(l)\equiv\langle l|\mathcal{H}|l\rangle\,. \tag{4}$$

The first and third term in the Lagrangian lead to a conventional first-order action; the unusual additional term—which incidentally implies a characteristic geometry on the Hilbert space hypersurface of vectors $\{|l\rangle\}$—has a special role to play in stationary-phase approximations. Note that the first-order extremal equations of the conventional action are generally incompatible with the required boundary conditions $l'\equiv l(t')$ and $l''\equiv l(t'')$. On the other hand, the additional term in the action changes the extremal equations to second-order ones consistent with the boundary conditions. The action evaluated for the extremal solution (or solutions) leads to the dominant stationary-phase approximation. The derivation[4] of this form of the path integral shows that the extremal solution is correctly given by this procedure and, furthermore, indicates the validity of the dominant stationary-phase approximation provided the extra (ϵ) term in the Lagrangian makes no explicit contribution to the extremal action in the limit $\epsilon \to 0$; this is actually the case for the examples of the canonical and spin algebras that we consider. On the other hand, if the extra term did not make a vanishing contribution, then recourse to the proper lattice-space formulation for the path integral would always provide a correct prescription to obtain the dominant stationary-phase approximation.[4] As usual, the first-order correction to the dominant approximation due to fluctuations is obtained most simply by applying the unitarity condition

$$\langle l'''|l'\rangle = \int \langle l'',t''|l''',t'\rangle^* \times \langle l'',t''|l',t'\rangle \delta l'' \tag{5}$$

to the needed degree of accuracy.

The examples that follow serve to illustrate and explain the general scheme. They should suggest a variety of other systems with a few or an infinite number of degrees of freedom for which parallel methods are applicable.

III. PATH INTEGRALS FOR COHERENT STATES

In the present example the states $\{|l\rangle\}$ are taken as the set of states of the form

$$|p,q\rangle \equiv \exp[i(pQ - qP)]|0\rangle , \tag{6}$$

for all real p and q, where P and Q form an irreducible Heisenberg canonical operator pair, with $[Q,P]=i$, and $|0\rangle$ denotes the (unit frequency) harmonic-oscillator ground state, a state for which $(Q+iP)|0\rangle = 0$. Of course, the states in question are just the familiar coherent states

$$|p,q\rangle \equiv |z\rangle \equiv e^{-|z|^2/2} \sum_{n=0}^{\infty} (n!)^{-1/2} z^n |n\rangle , \tag{7}$$

where $z \equiv (q+ip)/\sqrt{2}$ and $|n\rangle$ denotes the nth excited harmonic-oscillator eigenstate.[6,7] It has been established elsewhere,[8,9] in this case, that a resolution of unity is based on $\delta l = dpdq/2\pi$, and in particular

$$1 = \int |p,q\rangle\langle p,q| (dpdq/2\pi) \tag{8}$$

when integrated over all phase space.

A. Path integral

In terms of these states the formal expression for the path integral reads[4]

$$\langle p'',q'',t''|p',q',t'\rangle \equiv \langle p'',q''|e^{-i(t''-t')\mathcal{H}}|p',q'\rangle$$
$$= \lim_{\epsilon\to 0} \mathfrak{N} \int \exp\left\{ i\int_{t'}^{t''} [\tfrac{1}{2}(p\dot{q} - \dot{p}q) + \tfrac{1}{4} i\epsilon(\dot{p}^2 + \dot{q}^2) - H(p,q)]dt \right\} \mathfrak{D}p\,\mathfrak{D}q , \tag{9}$$

where $\mathfrak{D}p\mathfrak{D}q \equiv \prod_t dp(t)dq(t)$, $H(p,q) \equiv \langle p,q|\mathcal{H}|p,q\rangle$, and $\mathfrak{N}$ is a normalization fixed by the composition law (3). Observe that the real part of the action in (9) has the usual phase-space form, while the imaginary part, the term involving the ϵ factor, has the canonical form $dp^2 + dq^2$ characteristic of a flat two-dimensional space. Even under arbitrary transformations of the p,q variables, which include conventional canonical transformations, the flat-space nature implied by this term remains unchanged. (The appearance of a flat metric can be traced to the fact that the Heisenberg group is a factor representation of an Abelian group.)

B. Stationary-phase approximation

The stationary-phase approximation to the path integral in (9) leads[4] to the two extremal equations

$$\dot{q} - \partial H(p,q)/\partial p = \tfrac{1}{2} i\epsilon\ddot{p} , \tag{10a}$$

$$\dot{p} + \partial H(p,q)/\partial q = -\tfrac{1}{2} i\epsilon\ddot{q} . \tag{10b}$$

For very small ϵ, the solution to such equations generally separates into three temporal regions: Near one end, $t-t' = O(\epsilon)$, away from either end, $t-t' = O(1)$, $t''-t = O(1)$, and near the other end, $t''-t = O(\epsilon)$.

Vanishing Hamiltonian

As a first example, assume for simplicity that $H=0$, for which the solution to (10) for small ϵ reads

$$q(t) = \bar{q} + (q'-\bar{q})e^{-2(t-t')/\epsilon} + (q''-\bar{q})e^{-2(t''-t)/\epsilon} , \tag{11a}$$

$$p(t) = \bar{p} + (p'-\bar{p})e^{-2(t-t')/\epsilon} + (p''-\bar{p})e^{-2(t''-t)/\epsilon} , \tag{11b}$$

where

$$\bar{q} \equiv \tfrac{1}{2}(q''+q') - \tfrac{1}{2}i(p''-p'), \quad (12a)$$

$$\bar{p} \equiv \tfrac{1}{2}(p''+p') + \tfrac{1}{2}i(q''-q'). \quad (12b)$$

A useful rewriting of Eq. (12) is given by

$$q' + ip' = \bar{q} + i\bar{p}, \quad (13a)$$

$$q'' - ip'' = \bar{q} - i\bar{p}. \quad (13b)$$

The solution for $q(t)$ (say) clearly illustrates the three temporal regions: an initial, rapid change from q' to $\bar{q}$, a long, intermediate duration as $\bar{q}$, and a final, rapid change from $\bar{q}$ to q''. The intermediate value $\bar{q}$ is in general complex which reflects the fact that in performing the stationary-phase (or saddle-point) approximation the integration variables generally become complexified in the search for extremals of the action.

When the action is evaluated for this solution (with $H=0$ still) the result for the path integral is given by

$$\exp\{\tfrac{1}{2}i(q''p' - p''q') - \tfrac{1}{4}[(p''-p')^2 + (q''-q')^2]\}. \quad (14)$$

The ϵ term makes no contribution to this result thanks to the boundary conditions, and, moreover, this expression exactly equals $\langle p'', q'' | p', q' \rangle$. In other words, the dominant stationary-phase approximation (without fluctuations) for vanishing Hamiltonian yields the correct answer for the coherent-state path-integral (propagator).

Nonvanishing Hamiltonian

When the Hamiltonian is nonvanishing, the solution to the equations of motion (10) for small ϵ reads

$$q(t) = \bar{q}(t) + (q' - \bar{q}')e^{-2(t-t')/\epsilon} + (q'' - \bar{q}'')e^{-2(t''-t)/\epsilon}, \quad (15a)$$

$$p(t) = \bar{p}(t) + (p' - \bar{p}')e^{-2(t-t')/\epsilon} + (p'' - \bar{p}'')e^{-2(t''-t)/\epsilon}, \quad (15b)$$

where

$$\bar{q}' \equiv \bar{q}(t'), \quad \bar{p}' \equiv \bar{p}(t'), \quad (16a)$$

$$\bar{q}'' \equiv \bar{q}(t''), \quad \bar{p}'' \equiv \bar{p}(t''), \quad (16b)$$

and $\bar{q}(t)$ and $\bar{p}(t)$ are (generally) complex solutions of the usual classical equations

$$\dot{q} - \partial H(p,q)/\partial p = 0, \quad (17a)$$

$$\dot{p} + \partial H(p,q)/\partial q = 0. \quad (17b)$$

The boundary conditions appropriate to these classical equations are given by

$$q' + ip' = \bar{q}' + i\bar{p}', \quad (18a)$$

$$q'' - ip'' = \bar{q}'' - i\bar{p}''. \quad (18b)$$

In particular, for the initial conditions one can set

$$\bar{q}' = q' + w, \quad (19a)$$

$$\bar{p}' = p' + iw, \quad (19b)$$

where w is a complex parameter which is adjusted to fit the complex final condition $\bar{q}'' - i\bar{p}'' = q'' - ip''$, and where $\bar{p}'', \bar{q}''$ is the time-evolved solution of the classical equations with initial conditions $\bar{p}', \bar{q}'$.

When the action is evaluated for this extremal solution the ϵ term again vanishes due to the boundary conditions and the dominant stationary-phase approximation to the propagator (10) is given by

$$\langle p'', q'', t'' | p', q', t' \rangle \approx \exp\left\{\tfrac{1}{2}i(q''\bar{p}'' - p''\bar{q}'' + \bar{q}'p' - \bar{p}'q') + i\int_{t'}^{t''} [\tfrac{1}{2}(\bar{p}\dot{\bar{q}} - \dot{\bar{p}}\bar{q}) - H(\bar{p},\bar{q})]dt\right\}. \quad (20)$$

If H is quadratic then this expression actually gives the correct result, but otherwise not.

Discussion

In general, the extremal solution $p(t), q(t)$ is complex, but it need not be. In particular, if the solution of the classical equations starting with p', q' already evolves to p'', q'', then the solution remains real. Periodic solutions fulfill $p''=p'$, $q''=q'$, and they may either be complex or real. Real, periodic solutions are relevant for evaluating the energy spectrum through the trace of the evolution operator. It is noteworthy in this formulation that periodic solutions require equal momenta (p) as well as equal coordinates (q), rather than just equal coordinates as conventionally assumed with path integrals.

We recall that in the studies of bound-state spectra of various model problems by Dashen, Hasslacher, and Neveu,[10] additional stationary-phase arguments were used that ensured equal momenta along with equal coordinates, both of which were necessary to obtain their principal results.[11] The present work and that of Ref. 4 lead to the requirement of both equal momenta and equal coordinates in a direct and natural fashion.

IV. PATH INTEGRALS FOR SPIN STATES

In the example discussed in this section the states $\{|l\rangle\}$ are taken as the set of states of the form

$$|\theta, \phi\rangle \equiv e^{-i\phi\hat{S}_3} e^{-i\theta\hat{S}_2} |0\rangle \quad (21)$$

for all θ, ϕ, with $0 \leq \theta < \pi$ and $0 \leq \phi < 2\pi$, where $\hat{S}_j$, $j=1,2,3$, form an irreducible representation of the spin algebra,

$$[\hat{S}_j, \hat{S}_k] = i\epsilon_{jkl}\hat{S}_l\,, \qquad (22a)$$

$$\hat{\vec{S}}^2 \equiv \sum \hat{S}_j{}^2 = s(s+1)\,, \qquad (22b)$$

with $s = \frac{1}{2}, 1, \frac{3}{2}, \ldots$, and where $|0\rangle$ denotes the eigenstate of $\hat{S}_3$ with maximal eigenvalue

$$\hat{S}_3|0\rangle = s|0\rangle\,. \qquad (23)$$

Clearly the states $|\theta, \phi\rangle$ implicitly depend on s, but we suppress that particular dependence.

For $s = \frac{1}{2}$ and $\hat{\vec{S}}$ the usual spin-$\frac{1}{2}$ representation (half the standard Pauli matrices),

$$|\theta, \phi\rangle = \begin{pmatrix} e^{-i\phi/2}\cos\frac{1}{2}\theta \\ e^{i\phi/2}\sin\frac{1}{2}\theta \end{pmatrix}; \qquad (24a)$$

for $s = 1$ and $\hat{\vec{S}}$ the usual spin-1 representation,

$$|\theta, \phi\rangle = \frac{1}{\sqrt{2}} \begin{pmatrix} \cos\theta\cos\phi - i\sin\phi \\ \cos\theta\sin\phi + i\cos\phi \\ -\sin\theta \end{pmatrix}. \qquad (24b)$$

A similar construction applies for still higher spins as well. Moreover, a resolution of unity holds for each s in the form[12]

$$1 = \int |\theta, \phi\rangle\langle\theta, \phi|[(2s+1)/4\pi]\sin\theta\, d\theta\, d\phi \qquad (25)$$

integrated over the unit sphere, where 1 denotes the $(2s+1)\times(2s+1)$ unit matrix.

A. Path integral

In terms of such states the formal expression for the path integral reads

$$\langle\theta'', \phi'', t''|\theta', \phi', t'\rangle \equiv \langle\theta'', \phi''|e^{-i(t''-t')\mathcal{H}}|\theta', \phi'\rangle$$

$$= \lim_{\epsilon\to 0} \mathfrak{N} \int \exp\left\{ i \int_{t'}^{t''} [s\cos\theta\dot{\phi} + \tfrac{1}{4} is\epsilon(\dot{\theta}^2 + \sin^2\theta\dot{\phi}^2) - H(\theta, \phi)]dt \right\} \mathfrak{D}\Omega\,, \qquad (26)$$

where

$$\mathfrak{D}\Omega \equiv \prod_t d\Omega(t) \equiv \prod_t [\sin\theta(t)d\theta(t)d\phi(t)]\,,$$

$$H(\theta, \phi) \equiv \langle\theta, \phi|\mathcal{H}|\theta, \phi\rangle\,,$$

and $\mathfrak{N}$ is a normalization fixed by the composition law (3).

An alternative expression for the path integral (26) also holds. It may be determined that

$$\vec{S} \equiv \vec{S}(\theta, \phi) \equiv \langle\theta, \phi|\hat{\vec{S}}|\theta, \phi\rangle \qquad (27)$$

is given by the triplet

$$S_1 = s\sin\theta\cos\phi\,, \qquad (28a)$$

$$S_2 = s\sin\theta\sin\phi\,, \qquad (28b)$$

$$S_3 = s\cos\theta\,. \qquad (28c)$$

Furthermore, for $\mathcal{H} \equiv \mathcal{H}(\hat{\vec{S}})$ it is also possible to redefine the lattice-space path integral in such a way that the expression which enters the path integral as the classical Hamiltonian is given instead by

$$H'(\theta, \phi) \equiv \mathcal{H}(\vec{S}(\theta, \phi))\,. \qquad (29)$$

This expression possibly differs from the previous one by terms of order $\hbar$.

Note also that the term in the path integral (26) involving the ϵ factor has the canonical form $d\theta^2 + \sin^2\theta d\phi^2$ characteristic of the two-dimensional surface of the unit sphere. This geometry directly reflects the fact that we are dealing with states defined with the aid of the rotation group, and it clearly is a coordinate-invariant property.

Canonical formulation

An important and characteristic feature of the path integral (26) for spin stems from the particular form of the kinematic term in the action, namely

$$\int s\cos\theta\dot{\phi}\,dt\,. \qquad (30)$$

It is evident from this expression that $s\cos\theta$ plays the role of momentum conjugate to ϕ, i.e.,

$$p \equiv s\cos\theta\,, \qquad (31)$$

and the Hamiltonian $H(\theta, \phi)$ may be unambiguously reexpressed in terms of p and ϕ. Indeed, since (up to an irrelevant sign)

$$s\sin\theta d\theta d\phi = s d\cos\theta d\phi = dp\, d\phi\,, \qquad (32)$$

the resultant path integral expressed in p and ϕ variables has just the standard form of a phase-space integral for a single degree of freedom. Note that Dirac brackets[13] or related techniques are not required in our discussion of classical (or quantum) spin formulations, unlike the treatment elsewhere.[14] In particular, the classical variables $S_j = S_j(\theta, \phi)$, $j = 1, 2, 3$, defined in (28), exactly fulfill the Poisson brackets appropriate to spin variables,

$$\{S_j, S_k\}_{\mathrm{Pb}} = \epsilon_{jkl}\hat{S}_l\,, \tag{33}$$

where

$$\{A, B\}_{\mathrm{Pb}} \equiv \frac{\partial A}{\partial \phi}\frac{\partial B}{\partial p} - \frac{\partial A}{\partial p}\frac{\partial B}{\partial \phi}\,. \tag{34}$$

This relation shows the particular interplay of $p = s\cos\theta$ and ϕ as proper conjugate variables.

Spin variable formulation

Another way to rewrite the classical action and the path integral for spin is in terms of the variables $\vec{S}$, which is a form sometimes encountered in the literature. It is straightforward to see that

$$\frac{S_3(S_1\dot{S}_2 - \dot{S}_1 S_2)}{s^2 - S_3^{\,2}} = s\cos\theta\dot{\phi} \tag{35}$$

and that

$$\dot{\vec{S}}^2 \equiv \dot{S}_1^{\,2} + \dot{S}_2^{\,2} + \dot{S}_3^{\,2} = s^2(\dot{\theta}^2 + \sin^2\theta\dot{\phi}^2)\,. \tag{36}$$

These relations mean that the path integral (26) may alternatively be formulated as

$$\langle \vec{S}'', t'' | \vec{S}', t'\rangle \equiv \langle \theta'', \phi'', t'' | \theta', \phi', t'\rangle$$

$$= \lim_{\epsilon\to 0} \mathfrak{N} \int \exp\left\{ i\int_{t'}^{t''} \left[\frac{S_3(S_1\dot{S}_2 - \dot{S}_1 S_2)}{s^2 - S_3^{\,2}} + \tfrac{1}{4} i s^{-1}\epsilon\dot{\vec{S}}^2 - H(\vec{S})\right] dt\right\} \delta(\vec{S}^2 - s^2)\mathfrak{D}\vec{S}\,, \tag{37}$$

where the notation is as before and where the δ-functional ensures that $\vec{S}^2(t) = s^2$ for all t. The first-order action that appears here is the one studied by Doering and by Gilbert.[14] The objection to this form raised by Jevicki and Papanicolaou[15] and the preference they express for an alternative first-order Lagrangian that differs by a total derivative is difficult to understand (see below), especially when concentrating on diagonal matrix elements as they do.

Alternative for spin 1/2

Yet another approach suitable for spin $\frac{1}{2}$ is evidently provided by states of the form

$$|\chi\rangle \equiv \begin{pmatrix} (1 - |\chi|^2)^{1/2} \\ \chi \end{pmatrix}, \tag{38}$$

where the complex variable χ obeys $0 \leq |\chi| \leq 1$. Up to an overall phase, this is just the same vector introduced in (24a) for spin $\frac{1}{2}$ as is clear from the identification

$$\chi \equiv \sin\tfrac{1}{2}\theta e^{i\phi}\,. \tag{39}$$

It follows that

$$i\langle\chi|\dot{\chi}\rangle = \tfrac{1}{2} i\chi^*\overleftrightarrow{\partial}_t\chi = \tfrac{1}{2}\cos\theta\dot{\phi} - \tfrac{1}{2}\dot{\phi}\,, \tag{40}$$

which differs from the previous kinematic expression $\frac{1}{2}\cos\theta\dot{\phi}$ for $s = \frac{1}{2}$, owing to the aforementioned difference in phase factor, but such a distinction is of no physical importance whatsoever. Additionally it follows that

$$1 = \int |\chi\rangle\langle\chi| (2/\pi) d\chi_r d\chi_i\,, \tag{41}$$

where $\chi \equiv \chi_r + i\chi_i$ and the integration is over the entire unit disc, $0 \leq |\chi| \leq 1$. Consequently the formal path integral for spin $\frac{1}{2}$ (omitting the ϵ factor) becomes

$$\langle\chi'', t''|\chi', t'\rangle \equiv e^{-i\phi''/2}\langle\theta'', \phi'', t''|\theta', \phi', t'\rangle e^{i\phi'/2}$$

$$= \mathfrak{N}\int \exp\left\{ i\int_{t'}^{t''} [\tfrac{1}{2} i\chi^*\overleftrightarrow{\partial}_t\chi - H(\chi)]\, dt\right\} \mathfrak{D}\chi_r\mathfrak{D}\chi_i\,, \tag{42}$$

where $H(\chi) \equiv \langle\chi|\mathcal{H}|\chi\rangle$. Coincidentally, this formulation is identical to the one proposed by the author in 1960 for a c-number path integral for a single Fermi degree of freedom,[16] which is of course precisely that suitable for a single spin-$\frac{1}{2}$ system as well. (The ϵ factor can easily be worked out in this language, and is essential to make sense of the stationary-phase approximation, but it is omitted here for convenience.)

The difference in overall phase between the two states $|\chi\rangle$ and $|\theta, \phi\rangle$ is responsible for the difference in the kinematical term in the classical action for the two cases. If we reexpress the χ-form in terms of the spin variables $\vec{S}$, it follows that

$$\tfrac{1}{2} i\chi^*\overleftrightarrow{\partial}_t\chi = \frac{S_2\dot{S}_1 - S_1\dot{S}_2}{\frac{1}{2} + S_3}\,, \tag{43}$$

which coincides with the kinematical expression (for $s = \frac{1}{2}$) advocated by Jevicki and Papanicolaou.[15] It is now easy to see the origin of their particular form of the Lagrangian. They based their derivation of the path integral for spin variables on a reduction of states within the context of a coherent-state formulation. Since coherent states conventionally are defined with a real "first" term, i.e., $\langle 0|z\rangle$ [cf., Eq. (7)], then the overall phase is like that for the $|\chi\rangle$ states (since $\langle 0|\chi\rangle$ is also real) rather than like the $|\theta,\phi\rangle$ states as we have defined them.

Relation to Euler angles

It is also worth noting that the definition of the $|\theta,\phi\rangle$ states in (21) is clearly connected to the description of rotations in terms of Euler angles. In one such parametrization one can introduce three-parameter vectors of the form

$$|\theta,\phi,\psi\rangle \equiv e^{-i\phi\hat{S}_3}e^{-i\theta\hat{S}_2}e^{-i\psi\hat{S}_3}|0\rangle\,. \tag{44}$$

But if $\hat{S}_3|0\rangle = s|0\rangle$, as we have assumed, then

$$|\theta,\phi,\psi\rangle = e^{-is\psi}|\theta,\phi\rangle\,, \tag{45}$$

and ψ enters simply as an overall phase factor, devoid of any physical significance. On the other hand, in terms of the full set of Euler angles the condition that $|0\rangle$ be an eigenvector of $\hat{S}_3$ may be relaxed and a seemingly covariant formulation may be given. This formulation entails one too many coordinates and some form of gauge condition needs to be imposed (our implicit gauge condition was $\psi\equiv 0$). Moreover, the resultant action is only superficially covariant since there is always a direction implicit in any acceptable choice of vector $|0\rangle$, namely the direction of the vector $\vec{v}\equiv\langle 0|\hat{\vec{S}}|0\rangle$, a vector which must not vanish if the kinematical term is to make any sense at all.[17]

Our further analysis of the spin model is based on the initial formulation as expressed in terms of the variables θ and ϕ.

B. Stationary-phase approximation

The stationary-phase approximation to the path integral for spin variables (26) leads to the two extremal equations

$$s\sin\theta\dot{\phi} + \partial H(\theta,\phi)/\partial\theta = -\tfrac{1}{2}is\epsilon(\ddot{\theta} - \sin\theta\cos\theta\dot{\phi}^2)\,, \tag{46a}$$

$$s\sin\theta\dot{\theta} - \partial H(\theta,\phi)/\partial\phi = \tfrac{1}{2}is\epsilon(\sin^2\theta\ddot{\phi} + 2\sin\theta\cos\theta\dot{\theta}\dot{\phi})\,. \tag{46b}$$

As before, for very small ϵ, the solution to such equations is rapidly varying near the boundaries t' and t'', and slowly varying in between.

Vanishing Hamiltonian

As a first example, assume for simplicity that $H\equiv 0$, for which the solution to (46) for small ϵ reads

$$\cos\theta(t) = \overline{c} + (\cos\theta' - \overline{c})e^{-2(t-t')/\epsilon} + (\cos\theta'' - \overline{c})e^{-2(t''-t)/\epsilon}\,, \tag{47}$$

while

$$\phi(t) = \phi' + \tfrac{1}{2}i\ln\left(\frac{1+\cos\theta'}{1-\cos\theta'}\right) - \tfrac{1}{2}i\ln\left(\frac{1+\overline{c}+(\cos\theta'-\overline{c})e^{-2(t-t')/\epsilon}}{1-\overline{c}-(\cos\theta'-\overline{c})e^{-2(t-t')/\epsilon}}\right) \tag{48a}$$

for t near t' or intermediate times, and

$$\phi(t) = \phi'' - \tfrac{1}{2}i\ln\left(\frac{1+\cos\theta''}{1-\cos\theta''}\right) + \tfrac{1}{2}i\ln\left(\frac{1+\overline{c}+(\cos\theta''-\overline{c})e^{-2(t''-t)/\epsilon}}{1-\overline{c}-(\cos\theta''-\overline{c})e^{-2(t''-t)/\epsilon}}\right) \tag{48b}$$

for t near t'' or intermediate times. Equality of these two expressions for $\phi(t)$ at intermediate times requires that

$$\frac{1-\overline{c}}{1+\overline{c}} = \tan\tfrac{1}{2}\theta''\tan\tfrac{1}{2}\theta' e^{-i(\phi''-\phi')}\,, \tag{49}$$

a relation that fixes the (generally) complex constant $\overline{c}$.

When the action is evaluated for this solution (with $H=0$ still) the dominant stationary-phase approximation is given by

$$[\cos\tfrac{1}{2}\theta''\cos\tfrac{1}{2}\theta' e^{i(\phi''-\phi')/2} + \sin\tfrac{1}{2}\theta''\sin\tfrac{1}{2}\theta' e^{-i(\phi''-\phi')/2}]^{2s}\,. \tag{50}$$

Thanks to the boundary condition the ϵ term again makes no contribution to this result validating this expression. Moreover, this expression equals[18] $\langle\phi'',\phi''|\theta',\phi'\rangle$ for all s, $s=\frac{1}{2}, 1, \frac{3}{2}, \ldots$. In other words, the dominant stationary-phase approximation (without fluctuations) for vanishing Hamiltonian yields the correct answer for the path-integral (propagator) expressed in terms of spin continuous representations for *all* s, $s=\frac{1}{2}, 1, \frac{3}{2}, \ldots$. Since the integrals involved are all non-Gaussian this is rather a remarkable result.

Nonvanishing Hamiltonian

When the Hamiltonian is nonvanishing, the solution to the equations of motion (46) for small ϵ again has three characteristic temporal regions. In particular, the solution is given by

$$\cos\theta(t) = \cos\overline{\theta}(t) + (\cos\theta' - \cos\overline{\theta}')e^{-2(t-t')/\epsilon} + (\cos\theta'' - \cos\overline{\theta}'')e^{-2(t''-t)/\epsilon}\,, \tag{51}$$

while

$$\phi(t)=\overline{\phi}(t)-\overline{\phi}'+\phi'+\tfrac{1}{2}i\ln\left(\frac{1+\cos\theta'}{1-\cos\theta'}\right)$$
$$-\tfrac{1}{2}i\ln\left(\frac{1+\cos\overline{\theta}'+(\cos\theta'-\cos\overline{\theta}')e^{-2(t-t')/\epsilon}}{1-\cos\overline{\theta}'-(\cos\theta'-\cos\overline{\theta}')e^{-2(t-t')/\epsilon}}\right) \tag{52a}$$

for t near t' or intermediate times, and

$$\phi(t)=\overline{\phi}(t)-\overline{\phi}''+\phi''-\tfrac{1}{2}i\ln\left(\frac{1+\cos\theta''}{1-\cos\theta''}\right)$$
$$+\tfrac{1}{2}i\ln\left(\frac{1+\cos\overline{\theta}''+(\cos\theta''-\cos\overline{\theta}'')e^{-2(t''-t)/\epsilon}}{1-\cos\overline{\theta}''-(\cos\theta''-\cos\overline{\theta}'')e^{-2(t''-t)/\epsilon}}\right) \tag{52b}$$

for t near t'' or intermediate times. In these expressions

$$\overline{\theta}'\equiv\overline{\theta}(t'),\quad \overline{\phi}'\equiv\overline{\phi}(t'), \tag{53a}$$
$$\overline{\theta}''\equiv\overline{\theta}(t''),\quad \overline{\phi}''\equiv\overline{\phi}(t''), \tag{53b}$$

where $\overline{\theta}(t)$ and $\overline{\phi}(t)$ are (generally) complex solutions of the classical equations

$$s\sin\theta\dot{\phi}+\partial H(\theta,\phi)/\partial\theta=0, \tag{54a}$$
$$s\sin\theta\dot{\theta}-\partial H(\theta,\phi)/\partial\phi=0. \tag{54b}$$

The appropriate boundary conditions for these classical equations are found by insisting that $\phi(t)$ changes from ϕ' to $\overline{\phi}'$ during the rapid temporal change near t', and that $\phi(t)$ changes from $\overline{\phi}''$ to ϕ'' during the rapid temporal change near t''. This requirement leads to the boundary conditions

$$\tan\tfrac{1}{2}\theta' e^{i\phi'}=\tan\tfrac{1}{2}\overline{\theta}' e^{i\overline{\phi}'}, \tag{55a}$$
$$\tan\tfrac{1}{2}\theta'' e^{-i\phi''}=\tan\tfrac{1}{2}\overline{\theta}'' e^{-i\overline{\phi}''}. \tag{55b}$$

In particular, for the initial conditions one can set

$$\tan\tfrac{1}{2}\overline{\theta}'=e^{w}\tan\tfrac{1}{2}\theta', \tag{56a}$$
$$\overline{\phi}'=\phi'+iw, \tag{56b}$$

where w is a complex parameter which is adjusted to fit the complex final condition

$$\tan\tfrac{1}{2}\overline{\theta}'' e^{-i\overline{\phi}''}=\tan\tfrac{1}{2}\theta'' e^{-i\phi''}, \tag{57}$$

and where $\overline{\theta}'',\overline{\phi}''$ is the time evolved solution of the classical equations with initial conditions $\overline{\theta}'$, $\overline{\phi}'$.

When the action is evaluated for this extremal solution the ϵ term again vanishes due to the boundary conditions and the dominant stationary-phase approximation to the propagator (26) is given by

$$\langle\theta'',\phi'',t''|\theta',\phi',t'\rangle\approx\left(\frac{\sin\theta''\sin\theta'}{\sin\overline{\theta}''\sin\overline{\theta}'}\right)^{s}\exp\left\{i\int_{t'}^{t''}[s\cos\overline{\theta}\,\dot{\overline{\phi}}-H(\overline{\theta},\overline{\phi})]dt\right\}. \tag{58}$$

For a general, nonvanishing Hamiltonian, of course, this approximation cannot be expected to provide the correct result by itself.

Discussion

In general, the extremal solution $\theta(t)$, $\phi(t)$ is complex, but it need not be. In particular, if the solution of the classical equations starting with θ', ϕ' already evolves to θ'', ϕ'', then the solution remains real. Periodic solutions fulfill $\theta''=\theta'$, $\phi''=\phi'$, and they may either be complex or real. Real, periodic solutions are relevant for evaluating the energy spectrum through the trace of the evolution operator. Once again, it is noteworthy in this formulation that periodic solutions require equal momenta (θ) as well as equal coordinates (ϕ), rather than just equal coordinates as conventionally assumed with path integrals.

We remark that in the study of the spectra of the one-dimensional continuous Heisenberg spin chain by Jevicki and Papanicolaou[15] periodic solutions having both equal momenta and equal coordinates were assumed in order to obtain their principal results. The present work provides a clear demonstration of the necessity of that assumption, as well as clarifying the entire question of path integrals for spin variables and the associated stationary-phase approximation.

[1]See, e.g., R. P. Feynman and A. R. Hibbs, *Quantum Mechanics and Path Integrals* (McGraw-Hill, New York, 1965); M. C. Gutzwiller, J. Math. Phys. 8, 1979 (1967); 10, 1004 (1969); 11, 1791 (1970); 12, 343 (1971); R. F. Dashen, B. Hasslacher, and A. Neveu, Phys. Rev. D 10, 4114, (1974); 10, 4130 (1974); 10, 4138 (1974).

[2]J. R. Klauder, Ann. Phys. (N. Y.) 11, 123 (1960), particularly Eq. (12).

[3]S. S. Schweber, J. Math. Phys. 3, 831 (1962); F. A. Berezin, Teor. Mat. Fiz. 6, 194 (1971) [Theor. Math. Phys. 6, 141 (1971)]. See also L. D. Faddeev, *Methods in Field Theory*, 1975 Les Houches Lectures, edited

by R. Balian and J. Zinn-Justin (North-Holland, Amsterdam, 1976), p. 1.
[4]J. R. Klauder, in *Path Integrals*, Proceedings of the NATO Advanced Summer Institute, edited by G. J. Papadopoulos and J. T. Devreese (Plenum, New York, 1978), p. 5.
[5]Reference 2, cf. Eq. (65).
[6]J. R. Klauder and E. C. G. Sudarshan, *Fundamentals of Quantum Optics* (Benjamin, New York, 1968), Chap. 7.
[7]See also V. Bargmann, Commun. Pure Appl. Math. 14, 187 (1961); R. J. Glauber, Phys. Rev. 131, 2766 (1963).
[8]Reference 2, Eq. (5).
[9]See, e.g., J. Schwinger, *Quantum Kinematics and Dynamics* (Benjamin, New York, 1970), especially Secs. 4.6 and 4.7.
[10]R. F. Dashen, B. Hasslacher, and A. Neveu, Phys. Rev. D 10, 4114 (1974); see, however, K. F. Freed, Faraday Disc. Chem. Soc. 55, 68 (1973).
[11]Thanks are expressed to H. Neuberger for a discussion of this point.
[12]See, e.g., Ref. 2, Eq. (83); J. R. Klauder, J. Math. Phys. 4, 1058 (1963), Eq. (38). Compare J. M. Radcliffe, J. Phys. A 4, 313 (1971) and F. T. Arecchi, E. Courtens, R. Gilmore, and H. Thomas, Phys. Rev. A 6, 2211 (1972) for a related discussion (making allowance for a different phase convention).
[13]See, e.g., P. A. M. Dirac, Can. J. Math. 2, 129 (1950); 3, 1 (1951).
[14]See, e.g., C. F. Valenti and M. Lax, Phys. Rev. B 16, 4936 (1977).
[15]A. Jevicki and N. Papanicolaou, Ann. Phys. (N. Y.) (to be published).
[16]Reference 2, Eq. (18).
[17]J. R. Klauder, J. Math. Phys. 4, 1058 (1963), Sec. 3.
[18]See, e.g., E. U. Condon and G. H. Shortley, *The Theory of Atomic Spectra* (Cambridge Univ. Press, Cambridge, England, 1953), Chap. III; A. S. Davydov, *Quantum Mechanics*, translated by D. ter Harr (Addison-Wesley, Reading, Mass, 1965), Chap. VI.

PHYSICAL REVIEW C VOLUME 24, NUMBER 4 OCTOBER 1981

Path integrals for the nuclear many-body problem

J. P. Blaizot*
Department of Physics, University of Illinois at Urbana-Champaign, Urbana, Illinois 61801

H. Orland
Service de Physique Theorique, Saclay, 02, 91190 Gif-sur-Yvette, France
(Received 22 December 1980)

We present a general method for constructing path intergrals for the nuclear many-body problem. This method uses continuous and overcomplete sets of vectors in the Hilbert space. The state labels play the role of classical coordinates which are quantized as bosons. The equations of motion for the classical coordinates are obtained by calculating the functional integral in the saddle point approximation. In the particular case where the overcomplete set considered is the set of all Slater determinants, the classical equations of motion are the time-dependent Hartree-Fock equations. The functional integral provides a way of requantizing these classical equations. This quantization involves boson degrees of freedom and is in some cases very similar to the method of boson expansion. It is shown that the functional integral formalism provides a unifying framework to describe various approaches to the nuclear many-body problem.

[NUCLEAR STRUCTURE Functional integrals on continuous overcomplete sets. Time-dependent Hartree and Hartree-Fock theories. Boson representations for fermion systems.]

I. INTRODUCTION

The present work examines the application of path integrals to the nuclear many-body problem. It has been motivated partly by the recent developments in the time-dependent mean field theories which have been applied to the description of large amplitude collective motion or heavy ions reactions.[1–4] One of such theories is the time-dependent Hartree-Fock theory hereafter referred to as TDHF. As is well known, the mean field approximation to the many-body problem leaves out definite effects which are usually interpreted in terms of quantum mechanics. For example the vibrational and rotational modes are not quantized in TDHF. To make connection with quantum spectra, a "requantization" is obviously required. The procedure followed for this requantization is often empirical and mostly unjustified. This originates from the fact that most of the derivations of the time-dependent mean field equations do not allow for a systematic expansion beyond the mean field level. An exception to this criticism is the boson expansion method.[5–7] In this method, the time-dependent mean field equations arise from the replacement of the boson operators by c numbers. Moreover, and this is a major point, it can be shown that the boson expansion retrieves exactly the original many-body problem of interacting fermions. In other words boson expansions provide an exact quantization scheme for the time-dependent mean field equations.

Path integrals provide other possible quantization schemes. The standard procedure is to calculate first the functional integral using the saddle-point approximation. This provides the "classical" approximation of the theory. Knowing the classical solution, one can get a semiclassical expression for the transition amplitudes and apply a generalization of the Wentzel-Kramers-Brillouin (WKB) method to quantize periodic motions. Further quantum effects are recovered by calculating the successive corrections to the saddle-point approximation. In the calculation of these corrections, boson degrees of freedom appear naturally. Actually, as we shall see, the boson expansion method is closely related to the quantization through path integrals.

Path integrals have been extensively used in many different areas of physics, in particular in quantum field theory and statistical mechanics. As is well known they are mathematically ill-defined objects and some of the manipulations one usually performs on ordinary integrals are not necessarily allowed, since they may lead to completely erroneous results. This is an important point which must always be kept in mind. To circumvent part of the mathematical difficulties associated with the definition of the functional integral, one usually identifies the functional integral with the formal perturbation expansion. All the manipulations on the integral which can be interpreted as manipulations on the perturbation expansion are then allowed. Other manipulations should be examined with great care. This does not imply that the use of the functional integral is restricted to perturbative approximations. It only guarantees that the properties of the integral are identical to those of the perturbation expansion.

In this paper, we discuss a general method for constructing path integrals for the nuclear many-body problem. This method, due to Klauder,[8,9] makes use of continuous and overcomplete sets of vectors of the Hilbert space. Among those, coherent states or generalized coherent states are particularly important sets. Thus the vectors of the Hilbert space are parametrized by a set of complex numbers which play the role of classical coordinates in a generalized phase space. According to the choice of the overcomplete set, different "classical approximations" are generated from the functional integral. In the present context one should remember that the word classical does not imply that something is small compared to $\hbar$. For example, choosing the set of all the vectors in the Hilbert space as the overcomplete set, one gets as the classical approximation to the Schrödinger equation, the Schrödinger equation itself. This is certainly an extreme case and most of the interesting approximations leave out genuine quantum effects. One of the purposes of the present work is to analyze these effects in the case of the time-dependent mean field approximations.

Functional integrals have been used recently in nuclear physics by several authors.[10–15] It will be seen that all the methods used by these authors are actually particular cases of the general method presented here which has much more flexibility.

This work is organized as follows. In Sec. II of this paper we discuss the properties of some overcomplete sets which are relevant to the discussion of the nuclear many-body problem. In Sec. III we construct the functional integral. Several specific forms of the functional integral are explicitly given. In Sec. IV we discuss the link between the path integral and the formal perturbation expansion. We analyze the difficulties associated with the quantization of the time-dependent Hartree-Fock theory.

In Sec. V, we analyze the successive corrections to the mean field approximation and discuss the physical nature of the quantum effects which are left out in this approximation. We also briefly discuss the connection between path integrals and the boson expansion methods. Section VI summarizes the conclusions. Let us finally mention that a partial account of this work can be found in Refs. 16–19.

II. OVERCOMPLETE SETS

Let $\{\,|\psi(z)\rangle\,\}$ be an overcomplete set of vectors in the Hilbert space $\mathscr{H}$, depending upon a family of parameters which we denote collectively by z. We shall call the parameters z classical coordinates, and the space of variations of z the generalized phase space. The justification for this will appear in Sec. III. The overcompleteness means that any vector of $\mathscr{H}$ can be expanded on the states $|\psi(z)\rangle$ and that the states $|\psi(z)\rangle$ are linearly dependent. We assume that the parameters z vary continuously and that there exists a measure $\mu(z)$ on the space where z is defined, such that

$$\int d\mu(z)\,|z\rangle\langle z| = 1 \ , \tag{2.1}$$

where 1 denotes the unit operator in $\mathscr{H}$. In (2.1) as well as in the following, we use the abridged notation $|z\rangle$ for the state $|\psi(z)\rangle$. We give below examples of overcomplete sets which are useful in our discussion.

A. Coherent states of the harmonic oscillator

This is a well known and typical example of an overcomplete set. We recall briefly its properties. The coherent states are thus defined:

$$|z\rangle = e^{zc^\dagger}|0\rangle = \sum_n \frac{z^n}{\sqrt{n\,!}}|n\rangle \ , \tag{2.2}$$

where $|0\rangle$ is the oscillator ground state and $|n\rangle$ the state with n quanta. $c^\dagger$ is the raising operator. The closure relation is written in terms of the states

$|z\rangle$ using Bargman's measure[20]:

$$\int \frac{dz\,dz^*}{2\pi i} e^{-z^* z} |z\rangle\langle z| = 1 , \tag{2.3}$$

where

$$\frac{dz\,dz^*}{2\pi i} = \frac{d\mathrm{Re}z\, d\mathrm{Im}z}{\pi}$$

and the integration is carried over the whole complex plane. The overlap of two coherent states is given by

$$\langle z|z'\rangle = e^{z^* z'} . \tag{2.4}$$

The coherent state $|z\rangle$ is an eigenstate of the lowering operator c with eigenvalue z,

$$c|z\rangle = z|z\rangle . \tag{2.5a}$$

The matrix element of an operator $A(c^\dagger, c)$, in which the operators $c^\dagger$ and c are written in normal order (the $c^\dagger$ on the left of the c's) is therefore given by

$$\langle z|A(c^\dagger,c)|z'\rangle = e^{z^* z'} A(z^*, z') . \tag{2.5b}$$

B. Bosons coherent states

Let us consider the boson Fock space generated by the repeated action of the creation operators $c_\alpha^\dagger$ on the vacuum $|0\rangle$, the index α running over a complete set of single particle states. The operators $c_\alpha^\dagger$ and their Hermitian conjugates c_α obey boson commutation rules

$$[c_\alpha, c_\beta] = 0,\ [c_\alpha^\dagger, c_\beta^\dagger] = 0,\ [c_\alpha, c_\beta^\dagger] = \delta_{\alpha\beta} . \tag{2.6}$$

Boson coherent states are defined by

$$|Z\rangle = \exp\left[\sum_\alpha z_\alpha c_\alpha^\dagger\right] |0\rangle . \tag{2.7}$$

The properties of these coherent states generalize those of the preceding section. The closure relation in Fock space can be written

$$\int \prod_\alpha \frac{dz_\alpha^* dz_\alpha}{2\pi i} \exp\left[-\sum_\alpha z_\alpha^* z_\alpha\right] |Z\rangle\langle Z| = 1 . \tag{2.8}$$

The state (2.7) is an eigenstate of the destruction operator c_α with the eigenvalue z_α

$$c_\alpha |Z\rangle = z_\alpha |Z\rangle . \tag{2.9}$$

The overlap of two coherent states (2.7) is,

$$\langle Z|Z'\rangle = \exp\left[\sum_\alpha z_\alpha^* z'_\alpha\right] . \tag{2.10}$$

Therefore the matrix element of a normal ordered operator $A(c^\dagger, c)$ is

$$\langle Z|A(c^\dagger,c)|Z'\rangle = A(Z^*,Z') \exp\left[\sum_\alpha z_\alpha^* z'_\alpha\right] . \tag{2.11}$$

C. Fermions coherent states

The coherent states of fermions are defined by analogy with the coherent states of boson.[21] Let $a_\alpha^\dagger$, a_α be the fermion creation and destruction operators. They satisfy the anticommutation relations

$$[a_\alpha, a_\beta]_+ = 0,\ [a_\alpha^\dagger, a_\beta^\dagger]_+ = 0,\ [a_\alpha, a_\beta^\dagger]_+ = \delta_{\alpha\beta} . \tag{2.12}$$

Let us consider the state

$$|Z\rangle = \exp\left[\sum_\alpha z_\alpha a_\alpha^\dagger\right] |0\rangle , \tag{2.13}$$

where $|0\rangle$ is the vacuum of the fermion Fock space. Since $a_\alpha{}^2 = 0$, $|z\rangle$ can be an eigenstate of a_α only if $z_\alpha{}^2 = 0$. This can be realized using anticommuting Grassman variables. The rules for calculating with these objects have been widely discussed in the literature. All the formulas of Sec. (II B) hold for the fermion coherent states, provided z is understood as a Grassman variable. Note that fermion coherent states do not belong to the Fock space. However, they allow for a decomposition of the identity in Fock space

$$\int \prod_\alpha dz_\alpha^* dz_\alpha \exp\left[-\sum_\alpha z_\alpha^* z_\alpha\right] |Z\rangle\langle Z| = 1 . \tag{2.14}$$

D. Boson representation for fermions

Fermion states are usually represented by vectors of a fermion Fock space, constructed from a complete set of single particle states $\{ |\alpha\rangle \}$. It is also possible to represent fermion states as vectors belonging to a subspace (called the physical subspace) of a large boson Fock space. We consider in this section the representation which has been described in Ref. 17, and which is a generalization of that introduced in Ref. 7.

To construct the boson image of an N-fermion state, we consider a large space G, product of N boson Fock spaces B_i, associated with each of the particles i

$$G = \mathscr{B}_1 \otimes \mathscr{B}_2 \otimes \cdots \otimes \mathscr{B}_N . \tag{2.15}$$

We call $C_i^\dagger(\alpha)$, $C_i(\alpha)$ the creation and annihilation operators acting in $\mathscr{B}_i$. These operators satisfy the commutation relations

$$[C_i(\alpha),C_j^\dagger(\beta)] = \delta_{ij}\delta_{\alpha\beta} ,$$
$$[C_i(\alpha),C_j(\beta)] = [C_i^\dagger(\alpha),C_j^\dagger(\beta)] = 0 . \tag{2.16}$$

The following states

$$|\psi\rangle = \sum_P (-)^P C_1^\dagger(\alpha_{P_1})\ldots C_N^\dagger(\alpha_{P_N})|0\rangle_B , \tag{2.17}$$

where $\sum_P$ is a sum over all the possible permutations of the indices $\alpha_1\ldots\alpha_N$, and $|0\rangle_B$ is the boson vacuum, are in one-to-one correspondence with the N-fermion states of the fermion Fock space. They span the physical subspace. They are characterized by two properties. There is one and only one particle per subspace $\mathscr{B}_i$,

$$\sum_\alpha C_i^\dagger(\alpha)C_i(\alpha)|\psi\rangle = |\psi\rangle \quad (i=1,\ldots,N) . \tag{2.18}$$

The state changes sign in any transposition of the particle indices. The operator which realizes such a transposition is

$$P_{ij} = \sum_{\alpha\beta} C_i^\dagger(\alpha)C_j^\dagger(\beta)C_j(\alpha)C_i(\beta)$$
$$= -\sum_\alpha C_i^\dagger(\alpha)C_i(\alpha)$$
$$+ \sum_\alpha C_i^\dagger(\alpha)C_j(\alpha)\sum_\beta C_j^\dagger(\beta)C_i(\beta) . \tag{2.19}$$

In view of Eqs. (2.18) and (2.19), the condition

$$P_{ij}|\psi\rangle = -|\psi\rangle \tag{2.20}$$

is equivalent to the condition

$$\sum_\alpha C_i^\dagger(\alpha)C_j(\alpha)|\psi\rangle = 0 \quad (i \neq j) \tag{2.21}$$

Thus the states of the physical subspace are characterized by the following set of equations

$$\sum_\alpha C_i^\dagger(\alpha)C_j(\alpha)|\psi\rangle = \delta_{ij}|\psi\rangle \begin{bmatrix} i=1,\ldots,N \\ j=1,\ldots,N \end{bmatrix} . \tag{2.22}$$

The operators $d_{ij} = \sum_\alpha C_i^\dagger(\alpha)C_j(\alpha)$ satisfy the U(N) algebra,

$$[d_{ij},d_{kl}] = d_{il}\delta_{jk} - d_{kj}\delta_{li} . \tag{2.23}$$

They are the generators of the transformation which mixes the various components of a state vector in G. These operators can be used to construct explicitly a projector onto the physical subspace

$$P = \int \prod_{i,j} dA_{ij} e^{-i\,\mathrm{Tr}A} \exp\left[i\sum_{ij} A_{ij}d_{ij}\right] , \tag{2.24}$$

where the matrix A may be chosen to be a real matrix and the integration carried from $-\pi$ to π. Other forms are of course possible for P. The Hamiltonian in G takes the following form

$$H_B = \sum_{i=1}^N \sum_{\alpha\beta} T_{\alpha\beta} C_i^\dagger(\alpha)C_i(\beta)$$
$$+ \tfrac{1}{2}\sum_{i,j}\sum_{\alpha\beta\gamma\delta} (\alpha\beta|V|\gamma\delta) C_i^\dagger(\alpha)C_j^\dagger(\beta)C_j(\delta)C_i(\gamma) , \tag{2.25}$$

where $(\alpha\beta|V|\gamma\delta)$ denotes the nonantisymmetrized matrix element of the two-body interaction V. It is easily verified that H_B has the same matrix element within the physical subspace as the Hamiltonian

$$H = \sum_{\alpha\beta} T_{\alpha\beta} a_\alpha^\dagger a_\beta + \tfrac{1}{2}\sum_{\alpha\beta\gamma\delta} (\alpha\beta|V|\gamma\delta) a_\alpha^\dagger a_\beta^\dagger a_\delta a_\gamma \tag{2.26}$$

has in the Fermion Fock space. It is also easily checked that H_B commutes with the projector P given by (2.24), that is, H_B has no matrix elements between physical and unphysical states. This follows from the fact that physical and unphysical states belong to different representations of the unitary group, and H_B commutes with the generators d_{ij}.

The closure relation in G is conveniently written

with the help of the coherent states:

$$|Z\rangle = \exp\left[\sum_{k=1}^{N}\sum_{\alpha} Z_k(\alpha)C_k^\dagger(\alpha)\right]|0\rangle_B , \qquad (2.27a)$$

where the index α runs over all the single particle states and $|0\rangle_B$ is the boson vacuum. We shall also use continuous representation with the notation

$$|\varphi\rangle = \exp\left[\sum_{k=1}^{N}\int dx\, \varphi_k(x)\psi_k^\dagger(x)\right]|0\rangle_B . \qquad (2.27b)$$

The closure relation in G reads [see Eq. (2.8)]

$$1_G = \int \prod_{k=1}^{N}\prod_{\alpha} \frac{dZ_k^*(\alpha)dZ_k(\alpha)}{2\pi i} \times \exp\left[-\sum_k (Z_k\,|\,Z_k)\right]|Z\rangle\langle Z|$$

$$= \int \prod_{k=1}^{N}\prod_{x} \frac{d\varphi_k^*(x)d\varphi_k(x)}{2\pi i} \times \exp\left[-\sum_k (\varphi_k\,|\,\varphi_k)\right]|\varphi\rangle\langle\varphi| , \qquad (2.28)$$

where we have used the abridged notations

$$(Z_k\,|\,Z_k) = \sum_\alpha Z_k^*(\alpha)Z_k(\alpha) , \qquad (2.29)$$

$$(\varphi_k\,|\,\varphi_k) = \int dx\, \varphi_k^*(x)\varphi_k(x) .$$

The closure relation (2.28) induces a closure relation in the physical subspace of G, obtained by applying the projector P onto the physical subspace on both sides of (2.28),

$$P = \int \prod_{k=1}^{N}\prod_{\alpha} \frac{dZ_k^*(\alpha)dZ_k(\alpha)}{2\pi i} \times \exp\left[-\sum_k (Z_k\,|\,Z_k)\right]P|Z\rangle\langle Z|P . \qquad (2.30)$$

Now we note that

$$P|Z\rangle = \det[Z_k(\alpha_i)]C_1^\dagger(\alpha_1)\cdots C_N^\dagger(\alpha_N)|0\rangle . \qquad (2.31)$$

Thus, in a scale transformation

$$Z(\alpha) = \Lambda Z'(\alpha), \qquad (2.32)$$

$P|Z\rangle$ scales as $\det\Lambda$. This property can be verified using the explicit form of P given by Eq. (2.26). One has indeed

$$P|\Lambda Z\rangle = \int d\mu(Z')|Z'\rangle\langle Z'|P|\Lambda Z\rangle \qquad (2.33)$$

and

$$\langle Z'|P|\Lambda Z\rangle = \int dA e^{-i\,\mathrm{Tr}A}\prod_\alpha e^{Z'^\dagger(e^{iA}\Lambda)Z} , \qquad (2.34)$$

where we have used the property

$$e^{C_k^\dagger A_{kl}C_l} = :e^{(e^A-1)_{kl}C_k^\dagger C_l}: . \qquad (2.35)$$

Changing the integration variable A into A-$i\ln\Lambda$ and using the property $\det\Lambda = \exp\mathrm{tr}\ln\Lambda$, one obtains the desired equation

$$P|\Lambda Z\rangle = (\det\Lambda)P|Z\rangle . \qquad (2.36)$$

Equation (2.30) may then be written as follows

$$P = \int \prod_{k=1}^{N}\prod_{\alpha} \frac{dZ_k^*(\alpha)dZ_k(\alpha)}{2\pi i} \times \int d\Lambda \prod_{k,l}\delta[\Lambda_{kl} - (Z_k^*\,|\,Z_l)] \times e^{-\mathrm{Tr}\Lambda}P|Z\rangle\langle Z|P , \qquad (2.37)$$

where the integration over Λ runs over all the positive definite Hermitian matrices. Making the change of variable

$$Z(\alpha) = \Lambda^{1/2}Z'(\alpha),\quad Z^*(\alpha) = Z'^*(\alpha)\tilde{\Lambda}^{1/2} , \qquad (2.38)$$

where $\tilde{\Lambda}$ denotes the transpose of the matrix Λ one gets

$$P = \left[\int d\Lambda(\det\Lambda)^n e^{-\mathrm{Tr}\Lambda}\right] \times \int \prod_{k=1}^{N}\prod_{\alpha} \frac{dZ_k'^*(\alpha)dZ_k'(\alpha)}{2\pi i} \times \prod_{l=1}^{N}\delta[(Z_k'\,|\,Z_l') - \delta_{kl}] \times P|Z'\rangle\langle Z'|P , \qquad (2.39)$$

where n is the total number of single particle states. The integral over Λ is just a normalization constant

We thus arrive at the result

$$P=\mathcal{N}\int\prod_{k=1}^{N}\prod_{\alpha}\frac{dZ_k^*(\alpha)dZ_k(\alpha)}{2\pi i}\prod_{l=1}^{N}\delta[(Z_k\,|\,Z_l)-\delta_{kl}]P\,|\,Z\rangle\langle Z\,|\,P\ . \tag{2.40}$$

This result will be rederived in a different way in the next section.

E. Independent particle states

In this section we consider the overcomplete set formed by all the Slater determinants describing systems with a fixed number of particles N. This set can be parametrized in many ways. We give below some parametrizations which are useful in practice, together with the corresponding closure relations. The derivations are reported in the Appendix.

Let $|\phi_0\rangle$ be a particular Slater determinant. $|\phi_0\rangle$ is composed of N orthonormalized single particle orbitals, which we call "hole" states,

$$|\phi_0\rangle=\prod_h a_h^\dagger|0\rangle\ . \tag{2.41}$$

We call "particle" states the states such that

$$a_p\,|\,\phi_0\rangle=0\ . \tag{2.42}$$

We assume that the number of single particle states is finite. We call n_h the number of hole states and n_p the number of particle states. It is known that any Slater determinant nonorthogonal to $|\phi_0\rangle$ can be written[22]

$$|\,Z\,\rangle=\exp\sum_{ph}(Z_{ph}a_p^\dagger a_h)\,|\,\phi_0\rangle\ . \tag{2.43}$$

The states (2.43) are not normalized. The overlap between two of them is

$$\langle Z\,|\,Z'\rangle=\det(1+Z^\dagger Z')\ , \tag{2.44}$$

where Z denotes the complex $n_p\times n_h$ matrix made out of the Z_{ph} amplitudes. In terms of the states (2.43) the closure relation in the Hilbert space of N-fermions states takes the following form (see the Appendix),

$$\int\prod_{ph}\frac{dZ_{ph}^*\,dZ_{ph}}{2\pi i}[\det(1+Z^\dagger Z)]^{-(n_p+n_h+1)}\times|\,Z\rangle\langle Z\,|=1\ . \tag{2.45}$$

Using Wick's theorem one can express the matrix elements of any operator between two states Z and Z' in terms of the one-body density matrix defined thus

$$\rho_{\beta\alpha}(Z^\dagger,Z')=\frac{\langle Z\,|\,a_\alpha^\dagger a_\beta\,|\,Z'\rangle}{\langle Z\,|\,Z'\rangle}\ . \tag{2.46}$$

Thus, for example, one has

$$\frac{\langle Z\,|\,a_\alpha^\dagger a_\beta^\dagger a_\gamma a_\delta\,|\,Z'\rangle}{\langle Z\,|\,Z'\rangle}=\rho_{\delta\alpha}(Z^\dagger,Z')\rho_{\gamma\beta}(Z^\dagger,Z')-\rho_{\delta\beta}(Z^\dagger,Z')\rho_{\gamma\alpha}(Z^\dagger,Z')\ . \tag{2.47}$$

The matrix elements $\rho_{\alpha\beta}(Z^\dagger,Z')$ have the following expressions[6]

$$\begin{aligned}\rho_{ph}&=[Z'(1+Z^\dagger Z')^{-1}]_{ph}\ ,\\ \rho_{hp}&=[(1+Z^\dagger Z')^{-1}Z^\dagger]_{hp}\ ,\\ \rho_{pp'}&=[Z'(1+Z^\dagger Z')^{-1}Z^\dagger]_{pp'}\ ,\\ \rho_{hh'}&=[(1+Z^\dagger Z')^{-1}]_{hh'}\ .\end{aligned} \tag{2.48}$$

Performing the change of variable,

$$\beta_{ph}=Z_{ph'}[(1+Z^\dagger Z)^{-1/2}]_{h'h}\ , \tag{2.49}$$

one can simplify (2.45). In the variables β, the closure relation takes the form

$$\int\prod_{ph}\frac{d\beta_{ph}^*\,d\beta_{ph}}{2\pi i}\,|\,\widetilde{\beta}\rangle\langle\widetilde{\beta}\,|=1\ , \tag{2.50}$$

where the states $|\widetilde{\beta}\rangle$ are obtained from (2.43) by expressing Z in terms of β and normalizing. This parametrization has been used in works on boson expansions.[5] The density matrix elements have the following expressions in terms of the β's

$$\begin{aligned}\rho_{\alpha\beta}&=\langle\widetilde{\beta}\,|\,a_\beta^\dagger a_\alpha\,|\,\widetilde{\beta}\rangle\ ,\\ \rho_{ph}&=[\beta(1-\beta^\dagger\beta)^{1/2}]_{ph},\ \ \rho_{hp}=\rho_{ph}^*\ ,\\ \rho_{pp'}&=(\beta\beta^\dagger)_{pp'}\ ,\\ \rho_{hh'}&=\delta_{hh'}-(\beta^\dagger\beta)_{hh'}\ .\end{aligned} \tag{2.51}$$

Note that the measure in (2.50) is extremely simple. This results from the fact that the β_{ph} are the coefficients of the unitary transformation which carries $|\phi_0\rangle$ into the state $|\widetilde{\beta}\rangle$. The domain of integration is complicated, however, since the matrix β must satisfy

$$1-\beta^\dagger\beta\rangle 0 \ , \qquad (2.52)$$

while in the parametrization (2.43) the parameters Z_{ph} vary over the whole complex plane. Note also that the expression of the density matrix (2.48) is formally the same, whether Z' differs from Z or not. The density matrix (2.51) has a simple expression only if the bra and the ket in (2.51) are Hermitian conjugates of one another.

Using a further change of variable, one arrives at a parametrization in which the coordinates are the single particle wave functions which build up the determinant. The closure relation can be written (see Appendix), with respect to a normalization constant,

$$\int \prod_{k=1}^{N}\prod_{x} \frac{d\varphi_k^*(x)d\varphi_k(x)}{2\pi i} \prod_{l=1}^{N} \delta[\langle\varphi_k\,|\,\varphi_l\rangle-\delta_{kl}] \times |\varphi\rangle\langle\varphi| \sim 1 \ , \qquad (2.53)$$

which is identical to the relation (2.40).

III. PATH INTEGRALS

In this section we give functional integral representations for the matrix elements of the evolution operator e^{-iHt} between some initial state $|Z_i\rangle$ and some final state $\langle Z_f|$,

$$\langle Z_f\,|\,e^{-iH(t_f-t_i)}\,|\,Z_i\rangle \ , \qquad (3.1)$$

where $|Z_f\rangle$ and $|Z_i\rangle$ belong to the class of states described in the previous section. The general procedure for constructing path integrals is quite standard. First one factorizes the operator $e^{-iH(t_f-t_i)}$ into N terms $e^{-i\epsilon H}$, where $\epsilon=(t_f-t_i)/N$. Then one inserts the closure relation (2.1) between each of the factors and gets

$$\langle Z_f\,|\,e^{-iH(t_f-t_i)}\,|\,Z_i\rangle = \int \prod_{k=1}^{N} d\mu(Z_k)\langle Z_f\,|\,Z_N\rangle\langle Z_N\,|\,e^{-i\epsilon H}\,|\,Z_{N-1}\rangle \cdots \langle Z_{k+1}\,|\,e^{-i\epsilon H}\,|\,Z_k\rangle \cdots \langle Z_1\,|\,e^{-i\epsilon H}\,|\,Z_i\rangle \, . \qquad (3.2)$$

In the limit $N\to\infty$, $\epsilon\to 0$ and

$$\frac{\langle Z_{k+1}\,|\,e^{-i\epsilon H}\,|\,Z_k\rangle}{\langle Z_{k+1}\,|\,Z_k\rangle} = \exp\left[-i\epsilon\frac{\langle Z_{k+1}\,|\,H\,|\,Z_k\rangle}{\langle Z_{k+1}\,|\,Z_k\rangle}\right] + 0(\epsilon^2) \ . \qquad (3.3)$$

One then arrives at

$$\langle Z_f\,|\,e^{-iH(t_f-t_i)}\,|\,Z_i\rangle = \lim_{N\to\infty} \int \prod_{k=1}^{N} d\mu(Z_k) \prod_{k=0}^{N} \langle Z_{k+1}\,|\,Z_k\rangle \exp\left[-i\epsilon \sum_{k=0}^{N-1} \frac{\langle Z_{k+1}\,|\,H\,|\,Z_k\rangle}{\langle Z_{k+1}\,|\,Z_k\rangle}\right] \ , \qquad (3.4)$$

where $|Z_0\rangle \equiv |Z_i\rangle$ and $\langle Z_{N+1}| \equiv \langle Z_f|$. One defines

$$|\delta Z_{k+1}\rangle = |Z_{k+1}\rangle - |Z_k\rangle \ , \qquad (3.5)$$

so that (3.3) may be rewritten as follows:

$$\langle Z_f\,|\,e^{-iH(t_f-t_i)}\,|\,Z_i\rangle = \lim_{N\to\infty} \int \prod_{k=1}^{N} [d\mu(Z_k)\langle Z_k\,|\,Z_k\rangle]\langle Z_f\,|\,Z_N\rangle \times \exp\left\{\sum_{k=1}^{N}\left[\ln\left[1-\frac{\langle Z_k\,|\,\delta Z_k\rangle}{\langle Z_k\,|\,Z_k\rangle}\right] - i\epsilon\frac{\langle Z_k\,|\,H\,|\,Z_{k-1}\rangle}{\langle Z_k\,|\,Z_{k-1}\rangle}\right]\right\} \ . \qquad (3.6)$$

A further simplification is achieved if one admits that the major contribution to the integral comes from those "paths" for which $|\delta Z_k\rangle$ is of order ϵ for almost all k, that is, assuming that only piecewise continuous paths contribute in (3.6). Setting $d\,|Z\rangle/dt = |\delta Z\rangle/\epsilon$ and keeping only lowest order terms in ϵ, one finally ends up with the continuous expression

$$\langle Z_f | e^{-iH(t_f-t_i)} | Z_i \rangle = \int_{|Z(t_i)\rangle = |Z_i\rangle}^{\langle Z(t_f)| = \langle Z_f|} \mathscr{D}(Z^*(t),Z(t)) e^{iS[Z^*,Z]} , \quad (3.7)$$

where the action S is given by

$$S[Z^*,Z] = \int_{t_i}^{t_f} dt \frac{\langle Z(t) | i\partial_t - H | Z(t)\rangle}{\langle Z(t) | Z(t)\rangle} - i \ln\langle Z_f | Z(t_f)\rangle , \quad (3.8)$$

and the integration measure is

$$\mathscr{D}(Z^*,Z) = \prod_{t_i < t < t_f} d\mu[Z^*(t),Z(t)]\langle Z(t) | Z(t)\rangle . \quad (3.9)$$

The integration in (3.7) is carried over all the paths $\langle Z(t)|$ and $|Z(t)\rangle$ in the overcomplete set, subject to the boundary conditions

$$|Z(t_i)\rangle = |Z_i\rangle, \ \langle Z(t_f)| = \langle Z_f| . \quad (3.10)$$

Note that in this formulation $\langle Z(t)|$ and $|Z(t)\rangle$ have to be considered as independent variables, e.g., there are no constraints on $|Z(t_f)\rangle$ and $\langle Z(t_i)|$. It is important to keep this point in mind when applying the saddle-point approximation. (See Sec. V and Refs. 9 and 23.)

The action (3.8) may be given a more symmetrical form with respect to the boundary conditions by an integration by parts

$$S[Z^*,Z] = \int_{t_i}^{t_f} dt \frac{\langle Z(t) | i\overleftrightarrow{\partial}_t - H | Z(t)\rangle}{\langle Z(t) | Z(t)\rangle} - \frac{i}{2} \ln\langle Z(t_i) | Z_i\rangle\langle Z_f | Z(t_f)\rangle , \quad (3.8')$$

where we have used the notation

$$\langle Z(t) | \overleftrightarrow{\partial}_t | Z(t)\rangle = \tfrac{1}{2}\left[\left\langle Z(t) \middle| \frac{dZ}{dt}\right\rangle - \left\langle \frac{dZ}{dt} \middle| Z\right\rangle\right] .$$

It is worth emphasizing that the expression (3.7) has gotten no rigorous mathematical meaning from its derivation. This is known to lead to difficulties when some "unallowed" manipulations are performed on the functional integral. An example of such difficulties will be encountered in Sec. IV.

We examine now various explicit forms of the functional integral (3.8) obtained with some of the overcomplete sets described in Sec. II.

Let us first consider the form of the functional integral obtained when one uses coherent states as an overcomplete set. Due to the special form of the overlap (2.10), the integration measure simplifies into

$$\mathscr{D}(Z^*,Z) = \prod_t \frac{dZ^*(t) dZ(t)}{2\pi i} . \quad (3.11)$$

The action reads

$$S[Z^*,Z] = \int_{t_i}^{t_f} dt \left\{ \frac{i}{2}(Z^*\dot{Z} - \dot{Z}^* Z) - H(Z^*,Z) - \frac{i}{2}[Z_f^* Z(t_f) + Z^*(t_i) Z_i] \right\} , \quad (3.12)$$

where $H(Z^*,Z)$ is the normal form of the second quantized Hamiltonian, with the creation and annihilation operators $a^\dagger$ and a replaced by Z^* and Z, respectively. The formulas above hold for bosons and fermions. In the latter case, the variable Z has to be understood as a Grassman variable. This formulation has been used in Refs. 11 and 13. The formulas (3.11) and (3.12) hold also for the coherent states (2.27) described in Sec. II D. However, in this latter case, special attention must be given to the boundary conditions. Indeed, one is not interested in the matrix element of the evolution operator between two coherent states (2.27), but rather in this matrix element between two physical states. Let $|\phi\rangle$ and $\langle\psi|$ be two coherent states (2.27) and $|\Phi\rangle$ and $\langle\Psi|$ the Slater determinants built from the same single particle orbitals; that is,

$$|\Phi\rangle = \sum_P (-)^r |\phi_{P_1}\phi_{P_2} \cdots \phi_{P_N}\rangle = P|\phi) ,$$
$$\langle\Psi| = \sum_P (-)^r \langle\psi_{P_1}\psi_{P_2} \cdots \psi_{P_N}| = (\psi|P , \quad (3.13)$$

where P denotes the projector on the physical subspace (see Sec. II D). We are interested in the matrix elements

$$\langle\Psi | e^{-iH(t_f-t_i)} | \Phi\rangle , \quad (3.14)$$

which can be written

$$\int d\mu(\phi^*,\phi) d\mu(\psi^*,\psi) \times \langle\Psi|\psi)(\psi| e^{-iH_B(t_f-t_i)} |\phi)(\phi|\Phi\rangle , \quad (3.15)$$

where H_B is the boson image of H, given by Eq. (2.25). $(\psi| e^{-iH_B(t_f-t_i)} |\phi)$ may be represented by the functional integral

$$\int \mathscr{D}(\varphi^*,\varphi)e^{iS[\varphi^*,\varphi]} \ ,$$

where the measure and the action are given, respectively, by (3.11) and (3.12), except for an obvious change of notation. The overlaps $\langle \Psi \mid \psi)$ and $(\phi \mid \Phi\rangle$ determine the boundary condition

$$\varphi_k(t_i) = \phi_{P(k)}, \ \varphi_k^*(t_f) = \psi^*_{P'(k)} \ ,$$

where $p(k)$ and $p'(k)$ denote two permutations of the particle indices $1,2,\ldots,N$. The expression (3.15) thus contains an obvious summation over all such permutations. An alternative way of calculating (3.14) is to use the explicit form (2.24) for the projector P onto the physical subspace. Furthermore, since P commutes with H, it needs to be inserted only once. One then arrives at the expression

$$\begin{aligned}\langle \Psi \mid e^{-iH(t_f-t_i)} \mid \Phi\rangle &= (\psi \mid Pe^{-iH_B(t_f-t_i)} \mid \phi) \\ &= \int dAe^{-i\mathrm{Tr}A}(\psi \mid e^{-i(t_f-t_i)\{H_B-[A/(t_f-t_i)]\}} \mid \phi) \ ,\end{aligned} \tag{3.16}$$

where A is the following operator

$$A = \sum_{kl}\sum_{\alpha} C_k^\dagger(\alpha)A_{kl}C_l(\alpha) \ . \tag{3.17}$$

The matrix element $(\psi \mid e^{-it[H_B-(A/t)]} \mid \varphi)$ has the following functional integral representation

$$(\psi \mid e^{-it[H_B-(A/t)]} \mid \phi) = \int \mathscr{D}(\varphi^*,\varphi)e^{iS[\varphi^*,\varphi;A]} \tag{3.18}$$

with

$$S[\varphi^*,\varphi;A] = S[\varphi^*,\varphi] - \frac{i}{t}\sum_{kl} A_{kl}(\varphi_k \mid \varphi_l) \ , \tag{3.19}$$

where $S[\varphi^*,\varphi]$ is the action (3.12), except for an obvious change of notations. The expression (3.18) describes the evolution of a system of bosons subject to special constraints represented by the "external" field A. When Fourier transformed [see Eq. (3.16)] with respect to A this expression retrieves the original fermion dynamics.

In the two formulations above [cf. Eqs. (3.15) and (3.18)], the paths are allowed to lie outside the physical subspace; the projection onto the physical subspace is done by the overall integral over A in the case of (3.18), or by the summation over specific boundary conditions in the case of (3.15). Now it is possible to constrain the path at each time t so that it lies entirely within the physical subspace. This is achieved by inserting the projector P at each time step in the construction of the path integral. One then arrives at the following expression

$$\langle \Psi \mid e^{-iH(t_f-t_i)} \mid \Phi\rangle = \int \mathscr{D}(\varphi^*,\varphi)\prod_t \langle \varphi \mid P \mid \varphi\rangle e^{iS[\varphi^*,\varphi]} \tag{3.20}$$

with

$$S[\varphi^*,\varphi] = \int_{t_i}^{t_f} \frac{\langle \varphi \mid (i\overleftrightarrow{\partial}_t - H_B)P \mid \varphi\rangle}{\langle \varphi \mid P \mid \varphi\rangle} - \frac{i}{2}\mathrm{Tr}\ln\langle \Psi \mid \varphi(t_f)\rangle\langle \varphi(t_i) \mid \Phi\rangle \ . \tag{3.21}$$

Now let us perform the same change of variable as in Sec. II D, namely, $\varphi = \Lambda^{1/2}\varphi'$ [see Eq. (2.38)]. In this change of variable, $\langle \varphi \mid P \mid \varphi\rangle$ scales as $\det \Lambda$, as $\langle \varphi \mid HP \mid \varphi\rangle$ and $\langle \varphi \mid \overleftrightarrow{\partial}_t P \mid \varphi\rangle$ do. In this later case, it is easily verified that the possible time derivative of Λ cancel. Thus Eq. (3.21) can be rewritten as follows (with respect to an overall constant, namely, the integral over Λ; see Sec. II D)

$$\langle \Psi \mid e^{-iH(t_f-t_i)} \mid \Phi\rangle = \int \mathscr{D}(\varphi^*,\varphi)\delta[(\varphi_k \mid \varphi_l) - \delta_{kl}]e^{iS[\varphi^*,\varphi]} \ , \tag{3.22}$$

where, ignoring the boundary term

$$S[\varphi^*,\varphi] = \sum_k (\varphi_k \mid i\overleftrightarrow{\partial}_t \mid \varphi_k) - \sum_k (\varphi_k \mid T \mid \varphi_k) - \tfrac{1}{2}\sum_{kl}\langle \varphi_k\varphi_l \mid V \mid \varphi_k\varphi_l\rangle \ , \tag{3.23}$$

where now the *antisymmetrized* matrix element of the two-body interaction occurs. In contrast, the action (3.19) involves only the direct matrix elements of the two-body interaction. One recognizes in the expression (3.22) the functional integral one would have obtained working directly with Slater determinants and the measure (2.53).

The functional integrals (3.18) and (3.22) are *a priori* equivalent, i.e., they correspond to the same Schrödinger equation. However, we shall see in the next section that they have actually very different structures. Let us remark here that they differ essentially by the way the constraints are handled. In (3.18) the constraints are imposed in a global way while in (3.22) they are imposed locally (in time). One may also notice that the constraints

$(\varphi_k | \varphi_l)$ are constants of motion for the classical equations of motion. This situation is very much reminiscent of what happens in gauge theory; here the gauge group is the group U(N) which mixes the single particle orbitals. We shall not further develop this point of view here. There is still another way to take care of the constraints, namely, choose a system of coordinates in which the constraints are automatically satisfied. This is realized by the parametrizations (2.43) and (2.49) of Slater determinants. We give below the explicit form of the functional integrals in these two representations.

For the representation (2.43) the integration measure reads

$$\mathscr{D}(Z^\dagger,Z) = \prod_t \prod_{ph} \frac{dZ^*_{ph}(t)\, dZ_{ph}(t)}{2\pi i} \times \det[1+Z^\dagger(t)Z(t)]^{-(n_p+n_h)} , \quad (3.24)$$

and the action is

$$S[Z^\dagger,Z] = \int_{t_i}^{t_f} dt \frac{i}{2} \mathrm{Tr}(1+Z^\dagger Z)^{-1}(Z^\dagger \dot{Z} - \dot{Z}^\dagger Z) - E[\rho(Z^\dagger,Z)] - \frac{1}{2} \mathrm{Tr} \ln[1+Z_f^\dagger Z(t_f)][1+Z^\dagger(t_i)Z_i] , \quad (3.25)$$

where ρ is the density matrix (2.48) and $E[\rho]$ is the HF energy calculated with this density matrix

$$E[\rho] = \sum_{\alpha\beta} T_{\alpha\beta}\rho_{\beta\alpha} + \tfrac{1}{2} \sum_{\alpha\beta\gamma\delta} \langle \alpha\beta | V | \gamma\delta \rangle \rho_{\gamma\alpha}\rho_{\delta\beta} \quad (3.26)$$

and $\langle \alpha\beta | V | \gamma\delta \rangle$ is the antisymmetrized matrix element of the two-body interaction V. For later purposes, we write $E[\rho]$ using the following matrix notation

$$E[\rho] = T\cdot\rho + \tfrac{1}{2}\rho\cdot V\cdot\rho , \quad (3.27)$$

where V is the (symmetrical) matrix,

$$V_{\alpha\gamma,\beta\delta} = \langle \alpha\beta | V | \gamma\delta \rangle = V_{\beta\delta,\alpha\gamma} . \quad (3.28)$$

The action (3.25) is the one used in Ref. 14, except for the boundary term.

For the representation (2.49) the measure is simply

$$\mathscr{D}(\beta^\dagger,\beta) = \prod_t \prod_{ph} \frac{d\beta^*_{ph}(t)\, d\beta_{ph}(t)}{2\pi i} \quad (3.29)$$

and the action reads

$$S[\beta^\dagger,\beta] = \int_{t_i}^{t_f} dt \left\{ \frac{i}{2}(\beta^\dagger\dot{\beta} - \dot{\beta}^\dagger\beta) - E[\rho(\beta^\dagger,\beta)] \right\} - \tfrac{1}{2} \mathrm{tr} \ln(1-\beta_f^\dagger\beta(t_f))(1-\beta^\dagger(t_i)\beta_i) \quad (3.30)$$

Almost all the functional integrals described in this section describe boson theories with particular constraints. Indeed, the elementary fields, or coordinates, are represented by complex numbers which are quantized as boson. This boson structure has been explicitly analyzed in Sec. (II D) for the representation (3.18). The representations underlying (3.25) and (3.30) are familiar in nuclear physics for their intimate connection with perturbative boson expansions.[24,7] The method we have used to generate path integrals clearly generate at the same time boson expansions, or more precisely boson representations. In these representations, the bosons are just the quantum version of the classical parameters which label the quantum states of the overcomplete set used in the functional integral. The role of the bosons in the functional integrals, and in particular of coherent state of bosons, will be seen in the next sections.

IV. PERTURBATION EXPANSION

In this section we compare the structure of the path integrals described in the preceding section with that of the formal perturbation expansion. Let us consider the expression

$$\langle Z_f | e^{-iH(t_f-t_i)} | Z_i \rangle = \int_{\substack{Z^*(t_f)=Z_f^* \\ Z(t_i)=Z_i}} \mathscr{D}(Z^*,Z) \exp\left\{ i \int_{t_i}^{t_f} L[Z^*,Z]dt + \ln\langle Z_f | Z(t_f) \rangle \right\} , \quad (4.1)$$

where

$$L[Z^*,Z]=\frac{\langle Z\,|\,i\partial_t-H\,|\,Z\rangle}{\langle Z\,|\,Z\rangle}\ . \tag{4.2}$$

We can rewrite L as follows

$$L[Z^*,Z]=\frac{\langle Z\,|\,i\partial_t-H_0\,|\,Z\rangle}{\langle Z\,|\,Z\rangle}-\frac{\langle Z\,|\,V\,|\,Z\rangle}{\langle Z\,|\,Z\rangle}$$

$$\equiv L_0[Z^*,Z]-\frac{\langle Z\,|\,V\,|\,Z\rangle}{\langle Z\,|\,Z\rangle}$$

and expand $\exp[-i\int\langle Z\,|\,V\,|\,Z\rangle/\langle Z\,|\,Z\rangle]$ in (4.1) in powers of V. One gets

$$\langle Z_f\,|\,e^{-iH(t_f-t_i)}\,|\,Z_i\rangle=\sum_n\frac{(-i)^n}{n!}\int_{t_i}^{t_f}dt_1\ldots dt_n\int\mathcal{D}(Z^*,Z)\exp\left\{i\int_{t_i}^{t_f}L_0[Z^*,Z]+\ln\langle Z_f\,|\,Z(t_f)\rangle\right\}$$

$$\times\frac{\langle Z(t_n)\,|\,V\,|\,Z(t_n)\rangle}{\langle Z(t_n)\,|\,Z(t_n)\rangle}\cdots\frac{\langle Z(t_1)\,|\,V\,|\,Z(t_1)\rangle}{\langle Z(t_1)\,|\,Z(t_1)\rangle}$$

$$=\sum_n\frac{(-i)^n}{n!}\int_{t_i}^{t_f}dt_1\ldots dt_n\int\mathcal{D}(Z^*,Z)\langle Z_f\,|\,Z(t_f)\rangle$$

$$\times e^{i\int_{t_n}^{t_f}L_0}\frac{\langle Z(t_n)\,|\,V\,|\,Z(t_n)\rangle}{\langle Z(t_n)\,|\,Z(t_n)\rangle}e^{i\int_{t_{n-1}}^{t_n}L_0}\frac{\langle Z(t_{n-1}\,|\,V\,|\,Z(t_{n-1})\rangle}{\langle Z(t_{n-1})\,|\,Z(t_{n-1})\rangle}\cdots$$

$$\times e^{i\int_{t_1}^{t_2}L_0}\frac{\langle Z(t_1)\,|\,V\,|\,Z(t_1)\rangle}{\langle Z(t_1)\,|\,Z(t_1)\rangle}e^{i\int_{t_i}^{t_1}L_0}\ . \tag{4.3}$$

By going back to the discretized form of the functional integral (Sec. III), one easily shows that (4.3) can be rewritten as follows

$$\langle Z_f\,|\,e^{-iH(t_f-t_i)}\,|\,Z_i\rangle=\sum_n\frac{(-i)^n}{n!}\int_{t_i}^{t_f}dt_1\ldots dt_n\int d\mu(Z_n)\ldots d\mu(Z_1)d\mu(Z_n')\ldots d\mu(Z_1')$$

$$\times\langle Z_f\,|\,e^{-iH_0(t_f-t_n)}\,|\,Z_n\rangle\langle Z_n\,|\,V\,|\,Z_n'\rangle$$

$$\times\langle Z_n'\,|\,e^{-iH_0(t_n-t_{n-1})}\,|\,Z_{n-1}\rangle\ldots$$

$$\times\langle Z_1'\,|\,e^{-iH_0(t_1-t_i)}\,|\,Z_i\rangle\ , \tag{4.4}$$

where we have used the expression

$$\langle Z_n\,|\,e^{-iH_0(t_n-t_{n-1})}\,|\,Z_{n-1}\rangle=\int_{\substack{Z^*(t_n)=Z_n^*\\ Z(t_{n-1})=Z_{n-1}}}\mathcal{D}(Z^*,Z)\exp\left[i\int_{t_{n-1}}^{t_n}L_0+\ln\langle Z_n\,|\,Z(t_n)\rangle\right]\ . \tag{4.5}$$

The closure relations over $|\,Z_n\rangle\langle Z_n\,|$ can now be removed. One then ends up with

$$\langle Z_f\,|\,e^{-iH(t_f-t_i)}\,|\,Z_i\rangle=\sum_n\frac{(-i)^n}{n!}\int_{t_i}^{t_f}dt_1\ldots dt_n\langle Z_f\,|\,e^{-iH_0t_f}T[V(t_n)\ldots V(t_1)]e^{iH_0t_i}\,|\,Z_i\rangle$$

$$=\langle Z_f\,|\,e^{-iH_0t_f}T\exp\left[-\int_{t_i}^{t_f}V(t)dt\right]e^{iH_0t_i}\,|\,Z_i\rangle\ , \tag{4.6}$$

where $V(t)$ is the interaction representation of V,

$$V(t)=e^{iH_0t}Ve^{-iH_0t} . \tag{4.7}$$

One recognizes in the expression (4.6) the standard perturbation expansion in powers of V. This shows that the functional integral preserves the structure of the formal perturbation expansion. This follows from the fact that the functional integral, by construction, preserves the structure of the T product, and that we have the following identity[25]:

$$e^{-iHt}=\lim_{N\to\infty}(e^{-i(t/N)H_0}e^{-i(t/N)V})^N , \tag{4.8}$$

that is, in the continuous limit, one can neglect the noncommutation of the operators V and H_0. Had one started from the expression (4.8) instead of using

$$e^{-iHt}=\lim_{N\to\infty}(e^{-iHt/N})^N$$

for constructing the path integral, one would have obtained directly (4.6).

It should be stressed that the identification of the perturbation series obtained with the functional integral and operator methods has made explicit reference to the discretized form, which was needed to disentangle the integration over Z and Z' at different times. This is therefore not a check of the continuous limit.

In the remaining part of this section, we are going to rearrange the perturbation expansion using operator identities. The rearrangement which will be performed can be interpreted as a change of variable in the functional integral. We shall see that this change of variable is not always allowed.

Let us first notice that the T exponential may be written

$$T\exp-i\int_{t_i}^{t_f}V(t)dt$$
$$=\lim_{N\to\infty}T\prod_{k=1}^{N}[1-i\epsilon V(t_k)]$$
$$=\lim_{N\to\infty}T\prod_{k=1}^{N}:e^{-i\epsilon V(t_k)}: \left[\epsilon=\frac{t_i-t_f}{N}\right] . \tag{4.9}$$

The second line differs from the first one by terms which are negligible in the limit $\epsilon\to 0$. We shall keep them, however, for reasons which will become clear soon. Note that the second line defines the T product of two operators at equal times as their normal product:

$$T[A(t)B(t)]=:A(t)B(t): . \tag{4.10}$$

This refinement clearly does not affect the preceding discussion. But it is going to be of crucial importance in the following.

We now consider an alternative form of the perturbation expansion which relies on the following identity

$$T\exp-\frac{i}{2}\int_{t_i}^{t_f}V(t)dt=N\int\mathscr{D}W\exp\left[\frac{i}{2}\int_{t_i}^{t_f}W(t)\cdot V^{-1}\cdot W(t)dt\right]T\exp\left[-i\int_{t_i}^{t_f}W_{\alpha\beta}(t)a_\alpha^\dagger(t)a_\beta(t)dt\right] , \tag{4.11}$$

where the normalization constant N is given by

$$N^{-1}=\int\mathscr{D}W\exp\left[\frac{i}{2}\int_{t_i}^{t_f}dt\,W(t)\cdot V^{-1}\cdot W(t)\right] \tag{4.12}$$

and V^{-1} is the inverse of the matrix

$$V_{\alpha\gamma,\beta\delta}=(\alpha\beta\,|\,V\,|\,\gamma\delta) . \tag{4.13}$$

$(\alpha\beta\,|\,V\,|\,\gamma\delta)$ is the nonantisymmetrized matrix element of V and in (4.11) $\frac{1}{2}V(t)$ stands for $\frac{1}{2}\sum_{\alpha\beta\gamma\delta}(\alpha\beta\,|\,V\,|\,\gamma\delta)a_\alpha^\dagger(t)a_\beta^\dagger(t)a_\delta(t)a_\gamma(t)$. The identity (4.11) is easily proved. It is very similar to the identity used in Ref. 26. Let us simply remark here that the Gaussian integration over W operates like a Wick's theorem, the elementary contraction being

$$\langle W_{\alpha\beta}(t_1)W_{\gamma\delta}(t_2)\rangle=N\int\mathscr{D}W\exp\left[\frac{i}{2}\int_{t_i}^{t_f}W(t)\cdot V^{-1}\cdot W(t)dt\right]W_{\alpha\beta}(t_1)W_{\gamma\delta}(t_2)$$
$$=-\delta(t_1-t_2)V_{\alpha\beta,\gamma\delta} . \tag{4.14}$$

Thus the integration over W reconstructs the original two-body potential. Now it is important to realize that the integration over W involves two W at the same time, and therefore, depends crucially upon the way the T product at equal time has been defined. The formula (4.4) follows then from the application of the two identities

$$T\exp -i\int_{t_i}^{t_f} V(t)dt = \lim_{N\to\infty} T\prod_{k=1}^{N} :e^{-i\epsilon V(t_k)}: , \tag{4.15}$$

$$:e^{-i\epsilon V}: = N\int \mathscr{D}w e^{i/2\epsilon W V^{-1} W} :e^{-i\epsilon W a^\dagger a}: . \tag{4.16}$$

Let us now consider the functional integral representation of $\langle Z_f | e^{-iH(t_f-t_i)} | Z_i\rangle$ in the overcomplete set of Slater determinants. To avoid complications with the constraints, let us use for example the parametrization (3.25) or (3.30). Using the identity

$$\exp\left[-\frac{i}{2}\int \rho\cdot V\cdot\rho\, dt\right] = N\int \mathscr{D}W \exp\left[\frac{i}{2}\int W\cdot V^{-1}\cdot W\, dt - i\int W\cdot\rho\, dt\right]$$

one obtains the following expression

$$\langle Z_f | e^{-iH(t_f-t_i)} | Z_i\rangle = \int_{\substack{Z^*(t_f)=Z_f^* \\ Z(t_i)=Z_i}} \int \mathscr{D}W \exp\left[\frac{i}{2}\int W(t)\cdot V^{-1}\cdot W(t) dt\right] \times \exp\left[i\int_{t_i}^{t_f} \frac{\langle Z | i\partial_t - H_0 - W | Z\rangle}{\langle Z | Z\rangle} + \ln\langle Z_f | Z(t_f)\rangle\right] . \tag{4.17}$$

Note that in the above formula, the matrix V is constructed with antisymmetrized matrix elements of the two-body interaction. It is extremely tempting at this stage to interchange the orders of the integrations over W and Z. Writing

$$\langle Z_f | e^{-iH(t_f-t_i)} | Z_i\rangle = \int \mathscr{D}W \exp\left[\frac{i}{2}\int W(t)\cdot V^{-1}\cdot W(t) dt\right] \times \int \mathscr{D}(Z^*,Z) \exp\left[i\int_{t_i}^{t_f} \frac{\langle Z | i\partial_t - H_0 - W | Z\rangle}{\langle Z | Z\rangle} + \ln\langle Z_f | Z(t_f)\rangle\right] . \tag{4.18}$$

Now the integral over Z is the matrix element between $|Z_i\rangle$ and $\langle Z_f|$ of the evolution operator for noninteracting particles in the fluctuating field $W(t)$. It is, therefore, equal to

$$\langle Z_f | e^{-iH_0 t_f}\left[T\exp -i\int_{t_i}^{t_f} W(t)\right] e^{iH_0 t_i} | Z_i\rangle . \tag{4.19}$$

However, a careful analysis of the first terms of the perturbation expansion reveals overcounting, a signal that nonallowed manipulations have been performed. The origin of the trouble can be traced back to the fact that the contributions to the integral over W in (4.8) comes from terms which are of order $(dt)^2$, or ϵ^2 in the discretized version. Thus the integral over Z in (4.18) contains terms like

$$e^{-\epsilon W\langle a^\dagger a\rangle} \sim 1 - \epsilon W\langle a^\dagger a\rangle + \frac{\epsilon^2}{2} W\langle a^\dagger a\rangle W\langle a^\dagger a\rangle . \tag{4.20}$$

If one replaces this integral by the expression (4.19) one gets instead terms of the form

$$\langle e^{-\epsilon W a^{\dagger} a}\rangle \sim 1-\epsilon W\langle a^{\dagger} a\rangle + \frac{\epsilon^2}{2} W_{\alpha\beta} W_{\gamma\delta}\langle a_{\alpha}^{\dagger} a_{\beta}\rangle\langle a_{\gamma}^{\dagger} a_{\delta}\rangle + \frac{\epsilon^2}{2} W_{\alpha\beta} W_{\gamma\delta}\langle a_{\alpha}^{\dagger} a_{\delta}\rangle\langle a_{\beta} a_{\gamma}^{\dagger}\rangle , \tag{4.21}$$

that is, one obtains two terms corresponding to the two possible contractions, and this is the origin of the overcounting. Thus one cannot replace the integral over Z by the expression (4.19) which is really troublesome, since the integral over Z in (4.18) is really the one which in our formalism represents (4.19). Another way of stating the difficulty is to consider that the change of variable involved in (4.17) is not allowed for the integral over Slater determinants, or more precisely that one is not allowed to interchange the order of integration over Z and W in (4.17).

It is easily seen that all these difficulties disappear when one is working with a path integral constructed with coherent states. Indeed taking the matrix element of Eq. (4.16) between two coherent states yields

$$\langle :e^{-i\epsilon V}:\rangle = e^{-i\epsilon\langle V\rangle} = N\int \mathcal{D}W e^{i/2\epsilon W V^{-1} W} e^{-i\epsilon W\langle a^{\dagger} a\rangle} , \tag{4.22}$$

that is, the functional integral preserves exactly the operator identities. In this particular case, the change of variable involved in (4.17) is therefore perfectly allowed. Note that this holds for any kind of coherent states, of boson or fermions. It holds in particular for the coherent states (2.27).

V. MEAN FIELD THEORIES AND BEYOND

In the preceding section, we made explicit the similarities in the structures of the functional integral and the formal perturbation expansion. However, the most interesting feature of the functional integral is to suggest approximation schemes which are different from those of conventional perturbation theory. In this section, we examine in particular the saddle-point approximation and its successive corrections. As well known, this approximation, when performed on the standard Feynman path integral, retrieves classical mechanics. In the many-body problem, the classical approximation obtained depends on the choice of the overcomplete set of states which have been chosen to construct the functional integral. If independent particle wave functions are used, the classical equations are the time-dependent mean field equations. It turns out that these nonlinear equations have definite classical features which we analyze. We also discuss in this section the connection between path integrals and perturbative boson expansions.

Let us then apply the saddle-point approximation and its successive corrections to the calculation of the functional integral

$$\int_{\substack{Z^*(t_f)=Z_f^* \\ Z(t_i)=Z_i}} \mathcal{D}(Z^*,Z) e^{iS[Z^*,Z]} = \langle Z_f | e^{-iH(t_f-t_i)} | Z_i\rangle . \tag{5.1}$$

The saddle points are given by the following equations, with their boundary condition

$$\frac{\delta S}{\delta Z}=0, \quad Z^*(t_f)=Z_f^* , \tag{5.2a}$$

$$\frac{\delta S}{\delta Z^*}=0, \quad Z(t_i)=Z_i . \tag{5.2b}$$

We call $Z_C^{(+)*}$ and $Z_C^{(-)}$ the solutions of Eqs. (5.2a) and (5.2b), respectively. Note that $Z_C^{(+)*}(t)$ and $Z_C^{(-)}(t)$ are not, in general, complex conjugates of each other. We then expand the action $S[Z^*,Z]$ around the classical solution $(Z_C^{(+)*}, Z_C^{(-)})$:

$$S = S_C + \sum_{n\geq 2} S_n[Z^*,Z] , \tag{5.3}$$

where we have set

$$S_C \equiv S[Z_C^{(+)*}, Z_C^{(-)}] ,$$

$$S_n[Z^*,Z] = \frac{1}{n!}\sum_{p=1}^{N} C_N^p (Z^*)^p \left.\frac{\delta^n S}{\delta Z^{*p}\delta Z^{n-p}}\right|_C Z^{n-p} , \tag{5.4}$$

and the functional derivatives are evaluated for $Z^* = Z_C^{(+)*}$, $Z=Z_C^{(-)}$. The functional integral (5.1) then takes the form

$$\langle Z_f | e^{-iH(t_f-t_i)} | Z_i\rangle = e^{iS_C}\int_{\substack{Z^*(t_f)=0 \\ Z(t_i)=0}} \mathcal{D}(Z^*,Z) \exp\left[i\sum_{n\geq 2} S_n(Z^*,Z)\right] . \tag{5.5}$$

Note that the boundary conditions are now independent of Z_i and Z_f^*. That is, all the dependence of the expression (5.5) on Z_i and Z_f^* is contained in S_C and the possible boundary terms which subsist in S_n.

It is interesting to notice that Eqs. (5.2a) and (5.2b) correspond to the time-dependent variational principle (very similar to the ones developed in Ref. 27):

$$\delta S[Z^*,Z]=0 \ , \qquad (5.6)$$

where $S[Z^*,Z]$ reads explicitly:

$$S[Z^*,Z]=\int_{t_i}^{t_f}\frac{\langle Z\,|\,i\partial_t-H\,|\,Z\rangle}{\langle Z\,|\,Z\rangle} - i\ln\langle Z_f\,|\,Z(t_f)\rangle \ . \qquad (5.7)$$

It is easily verified that Eq. (5.6) leads back to the time-dependent Schrödinger equation if $|Z\rangle$ is assumed to represent any state of the Hilbert space, i.e., in the case of unrestricted variations. When $|Z\rangle$ is chosen in a given class of states, the solution of the Eq. (5.6) provides an approximation for the transition amplitude $\langle Z_f\,|\,e^{-iH(t_f-t_i)}\,|\,Z_i\rangle$. This is given by e^{iS_C}. The usefulness of this expression lies in the fact that it is a stationary quantity. It appears then clearly that the choice of an overcomplete set for the construction of the functional integral is equivalent to the choice of a class of trial states in the use of the time-dependent variational principle. Therefore the separation into a classical motion and quantum corrections, implied by Eq. (5.3) does not require that some quantity is small compared to $\hbar$. The nature of the classical approximation discussed here, or the type of quantum effects which are left out in this approximation, are entirely determined by the specific choice of an overcomplete set in the Hilbert space. In particular, if the overcomplete set is the Hilbert space itself, the classical equations of motion are identical with the Schrödinger equation.

The limitation of the time-dependent variational principle (5.6) is that it does not provide a way of estimating the error associated with a given choice of trial states. This is precisely what the functional integral (5.5) does. Although the corrections to the classical approximation would be in most cases hard to evaluate, the functional integral provides the possibility of analyzing them, and therefore, allows for a better understanding of the classical approximation itself. We shall illustrate these considerations in the case of the mean field approximations to the many-body problem.

Let us then consider that the coordinates $\{Z\}$ represent a Slater determinant, that is, $\{Z\}$ denotes any of the sets of coordinates discussed in Sec. II E. [Actually the equations of motion given below only hold if the action (3.30) or (3.23) are used. If the action (3.25) is used, extra kinematical terms appears in front of the time derivatives.] The classical equations of motion are the time-dependent Hartree-Fock equations

$$i\dot{Z}-\frac{\delta H(Z^*,Z)}{\delta Z^*}=0 \ , \qquad (5.8)$$

$$i\dot{Z}^*+\frac{\delta H(Z^*,Z)}{\delta Z}=0 \ , \qquad (5.9)$$

where $H(Z^*,Z)=\langle Z\,|\,H\,|\,Z\rangle/\langle Z\,|\,Z\rangle$ is given explicitly in Sec. II E. The state vectors $|Z(t)\rangle$ which make the action (5.7) stationary are of the form

$$|Z(t)\rangle = |Z_0(t)\rangle \exp\left[-i\int_{t_i}^{t} f(t')dt'\right] \ , \qquad (5.10)$$

where $f(t)$ is an arbitrary function of time and $Z_o(t)$ is a solution of the Eq. (5.8). This arbitrariness in the phase of $|Z(t)\rangle$ reflects the invariance of the action (5.7) with respect to the choice of phase of the state vectors. Equations (5.8) and (5.9) can be easily transformed into an equation for the one-body density matrix

$$i\dot{\rho}=[h,\rho] \qquad (5.11)$$

with

$$\rho_{\alpha\beta}(t)=\frac{\langle Z_C^{(+)}(t)\,|\,a_\alpha^\dagger a_\beta\,|\,Z_C^{(-)}(t)\rangle}{\langle Z_C^{(+)}(t)\,|\,Z_C^{(-)}(t)\rangle} \ , \qquad (5.12)$$

and $h=\delta E/\delta\rho$ is the usual Hartree-Fock Hamiltonian calculated with the density matrix (5.12). Note that the density matrix is not Hermitian, so that the Hartree-Fock Hamiltonian is in general not real. Equation (5.11) is a generalization of the ordinary time-dependent Hartree-Fock equation, appropriate to the calculation of scattering amplitudes. This equation has already been considered in Ref. 12. Note that the standard TDHF equations are recovered if one chooses the boundary conditions such that $Z(t_f)=Z_f$. Then $Z_C^{(-)}(t)$ and $Z_C^{(+)*}(t)$ are complex conjugates of each other and the density matrix, as well as the Hartree-Fock Hamiltonian are Hermitian. The classical solutions $[Z_C^{(+)}(t), Z_C^{(-)}(t)]$ can be used to get a semiclassical approximation to the transition amplitude $\langle Z_f t_f | Z_i t_i \rangle$. It can also be used to obtain semiclassical approximations to the bound state energies of the system, applying a generalization of the WKB method developed in Ref. 28. Typically one arrives at semiclassical quantization rules for the periodic trajectories of the time-dependent mean-field equations. This method has already been applied in different ways to several simple cases.[13,12,29–31]

Let us now calculate the corrections to the mean field theory. This is obtained by expanding the action S around the classical solution $[Z_C^{(+)*}(t), Z_C^{(-)}(t)]$, as indicated by Eq. (5.3). We shall limit ourselves first to the quadratic corrections, and to simplify the discussion, we shall consider the fluctuations around a static solution of Eq. (5.8). We call $|\phi_0\rangle$ the corresponding state and we calculate

$$\langle \phi_0 | e^{-\beta H} | \phi_0 \rangle \approx e^{-\beta E_{\rm HF}} \int_{\substack{Z^*(\beta)=0 \\ Z(0)=0}} \mathscr{D}(Z^*,Z) \exp\left\{ - \int_0^\beta [Z^* \dot{Z} + H_2(Z^*,Z)] dt \right\} , \tag{5.13}$$

where $E_{\rm HF}$ is the Hartree-Fock energy of the state $|\phi_0\rangle$ and we have assumed $\langle \phi_0 | \phi_0 \rangle = 1$. $H_2(Z^*,Z)$ is the quadratic form obtained by expanding $H(Z^*,Z)$ around $Z=0$ ($\equiv |\phi_0\rangle$). In terms of the amplitudes Z_{ph}, $H_2(Z^*,Z)$ have the explicit form

$$H_2(Z^*,Z) = \tfrac{1}{2}(Z_{ph}^*, Z_{ph}) \begin{bmatrix} A & B \\ B^* & A^* \end{bmatrix} \begin{bmatrix} Z_{ph} \\ Z_{ph}^* \end{bmatrix} , \tag{5.14}$$

where the matrices A and B are the usual matrices of the random phase approximation. Note that at this level of approximation, all the parametrizations considered in Sec. II E, with proper inclusion of the constraints when necessary, yield the same result, Eq. (5.14). Now the functional integral (5.13) is identical to that of a system of coupled harmonic oscillators; more precisely it can be written

$$\int_{\substack{Z^*(\beta)=0 \\ Z(0)=0}} \mathscr{D}(Z^*,Z) \exp\left\{ - \int_0^\beta [Z^* \dot{Z} + H_2(Z^*,Z)] dt \right\}$$

$$= {}_B\langle 0 | \exp -\beta(C^\dagger \cdot A \cdot C + \tfrac{1}{2} C^\dagger \cdot B \cdot C^\dagger + \tfrac{1}{2} C \cdot B \cdot C) | 0 \rangle_B , \tag{5.15}$$

where we have used the matrix notation

$$C^\dagger \cdot A \cdot C = \sum_{\substack{ph \\ p'h'}} C_{ph}^\dagger A_{ph,p'h'} C_{p'h'} .$$

$C_{ph}^\dagger$ and C_{ph} denote boson creation and annihilation operators and $|0\rangle_B$ is the boson vacuum. By using the canonical form which diagonalizes the quadratic form in (5.15) one easily obtains:

$${}_B\langle 0 | \exp(C^\dagger A C + \tfrac{1}{2} C^\dagger B C^\dagger + \tfrac{1}{2} C B^\dagger C) | 0 \rangle_B = e^{\Delta E_0} , \tag{5.16}$$

where ΔE_0 is the correlation energy associated with the random phase approximation (RPA) vibrations, that is,

$$\Delta E_0 = \tfrac{1}{2} \sum_N \omega_N - \tfrac{1}{2} {\rm Tr} A . \tag{5.17}$$

This expression is easily shown to be equal to the sum of all the ring diagrams calculated with antisymmetrized matrix element and including the well known double counting of the second order term.

The boson degrees of freedom which appear naturally in the calculation of the integral (5.15) are the usual RPA phonons. The successive corrections to the expression (5.15) represent the various couplings between these RPA phonons. A systematic expansion can be derived in the following way. We first expand $H(Z^*,Z)$ to all order in Z^* and Z. Since we have treated explicitly the terms of order 2, this expansion starts at third order. These higher order terms can be treated in perturbation, which leads to the expression

$$\langle \phi_0 | e^{-\beta H} | \phi_0 \rangle = e^{-\beta E_{\rm HF}} \exp\left[-\int_0^\beta dt \sum_{n>3} H_n \left(\frac{\delta}{\delta j}, \frac{\delta}{\delta j^*} \right) \right]$$

$$\times \int_{\substack{Z^*(\beta)=0 \\ Z(0)=0}} \mathscr{D}(Z^*,Z) \exp\left\{ -\int_0^\beta [Z^*\dot{Z} + H_2(Z^*,Z) + j^*\cdot Z + Z^*\cdot j] \right\} \Big|_{\substack{j=0 \\ j^*=0}} . \quad (5.18)$$

The expression (5.18) is very reminiscent of the familiar perturbative boson expansion. The unperturbed propagator for the bosons is the RPA propagator and the term H_n describes a coupling between n RPA bosons. The occurrence of n-body interactions between the RPA bosons arises from the Pauli principle, or in other words from the constraints necessary to project onto the physical subspace. It must be kept in mind that we are not making here an exact connection between our formalism and a perturbative boson expansion. Indeed, when going beyond the quadratic approximation, technical problems arise with the treatment of the constraints, the integration measure or the domain of integration, depending upon whether one chooses, respectively, the parametrization (2.53), (2.45), or (2.50) for the Slater determinant. In the absence of a careful treatment of these points, we consider the expression (5.18) as approximate. It is clear, however, that the physical content of (5.18) will not be very much altered by a more rigorous derivation. This physical content is indeed quite transparent. The functional integral "quantizes" as bosons the coordinates which were introduced to parametrize the states of the overcomplete set. Inversely, the classical limit obtained in the saddle-point approximation is achieved by replacing the boson operators by c numbers (see Ref. 18).

The technical difficulties mentioned above do not show up when one considers the expansion around a solution of the Hartree equation. In this case, the expansion can then be given easily a diagrammatic interpretation, using the standard technics of perturbation theory. We shall again restrict ourselves to a time-independent problem and consider the expression

$$\langle \phi_0 | e^{\beta H_0} e^{-\beta H} | \phi_0 \rangle = N \int \mathscr{D} W \exp\left[\tfrac{1}{2} \int_0^\beta W(t)\cdot V^{-1}\cdot W(t) dt \right] e^{\operatorname{Tr}\ln(1-WG_0)} , \quad (5.19)$$

which follows trivially from (4.11) and the identity

$$\langle \phi_0 | T \exp \int_0^\beta W(u) a^\dagger(u) a(u) du \, | \phi_0 \rangle = \exp \operatorname{Tr}\ln(1 - WG_0) , \quad (5.20)$$

where G_0 is the single particle Green's function:

$$G^0_{\alpha\beta}(u_1 - u_2) = \langle \phi_0 | T a_\alpha(u_1) a^\dagger_\beta(u_2) | \phi_0 \rangle . \quad (5.21)$$

Equation (5.19) can also be derived from (3.16) (see Ref. 18). Application of the saddle-point approximation on the integral over W leads to the equation

$$W\cdot V^{-1} = V^{-1}\cdot W = G_0(1 - WG_0)^{-1} = G[W] , \quad (5.22)$$

where $G[W]$ is the single particle Green's function in presence of the external field W:

$$G^{-1}[W] = G_0^{-1} - W . \quad (5.23)$$

The density matrix is related to G by

$$\rho(t) = \lim_{\tau \to 0_+} G\left(t - \frac{\tau}{2}, t + \frac{\tau}{2} \right) . \quad (5.24)$$

It satisfies the equation of motion

$$\partial_t \rho + [H_0 - W, \rho] = 0 , \quad (5.25)$$

which is the time-dependent Hartree equation written in imaginary time. The expansion around a static solution is obtained easily. Let W_0 a static field, solution of

$$[H_0 - W_0, \rho_0] = 0 . \quad (5.26)$$

The expansion of (5.19) in powers of $W' = W - W_0$ reads

$$\langle \phi_0 | e^{\beta H_0} e^{-\beta H} | \phi_0 \rangle = N e^{(\beta/2)\rho_0\cdot V\cdot \rho_0} \int \mathscr{D} W \exp\left[\int_0^\beta \tfrac{1}{2} W'\cdot V^{-1}\cdot W' - \tfrac{1}{2} \int_0^\beta \operatorname{Tr} W'\cdot G(W_0)\cdot W'\cdot G(W_0) \right]$$

$$\times \exp\left\{ \int_0^\beta \operatorname{Tr} \sum_{n>2} -\frac{1}{n} [G(W_0)\cdot W']^n \right\} . \quad (5.27)$$

In order to calculate the remaining integral, we first regroup the two quadratic terms defining

$$\Gamma^{-1} = V^{-1} - Q \ , \tag{5.28}$$

where

$$Q_{\alpha\beta,\gamma\delta}(u_1 - u_2) = G_{\beta\gamma}[W_0;u_1 - u_2]G_{\gamma\alpha}[W_0;u_2 - u_1] \ . \tag{5.29}$$

Using a standard procedure, one introduces a source term for the field W' and treats in perturbation the terms of order higher than 2 in W', in the exponent of (5.27). One then gets

$$\langle \phi_0 | e^{\beta H_0} e^{-\beta H} | \phi_0 \rangle = e^{-(\beta/2)\rho_0 \cdot V \cdot \rho_0} e^{-1/2\,\mathrm{Tr}\ln(1 - VQ_0)}$$
$$\times \exp\left\{ - \sum_{n>2} \int_0^\beta \mathrm{tr} \frac{1}{n} \left[G(W_0) \cdot \frac{\delta}{\delta j} \right]^n \right\} e^{-(1/2) j \cdot \underline{\Gamma} \cdot j} \Big|_{j=0} \ , \tag{5.30}$$

where the factor $e^{-1/2 j \Gamma j}$ comes from the Gaussian integral over W'

$$e^{-(1/2) j \cdot \Gamma \cdot j} = \frac{\int \mathscr{D} W e^{(1/2) W \cdot \Gamma^{-1} \cdot W + j \cdot W}}{\int \mathscr{D} W e^{(1/2) W \cdot \Gamma^{-1} \cdot W}} \ . \tag{5.31}$$

The diagrammatic interpretation of the formula (5.30) is very simple (we consider vacuum-vacuum diagrams corresponding to the ground state energy). The first term is the Hartree energy

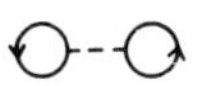 (5.32)

The second term is the sum of all ring diagrams (calculated here with direct matrix elements), plus actually the exchange counterpart of (5.32)

$\mathrm{tr}\ \ln\,(1 - VQ_0) =$ 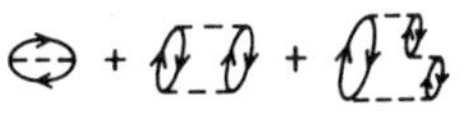 . (5.33)

To pursue the analysis we give the following representation of Γ

$\Gamma \equiv$ 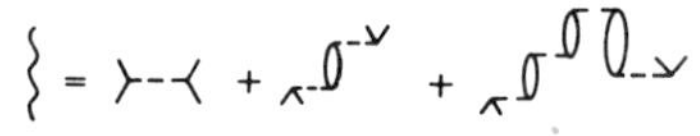 . (5.34)

Thus (5.33) can be represented by

 (5.35)

and

$$\sum_{n>2} \frac{1}{n} \mathrm{Tr} \left[G(W_0) \cdot \frac{\delta}{\delta j} \right]^n e^{-(1/2) j \cdot \Gamma \cdot j} \Big|_{j=0} \tag{5.36}$$

is the sum of all diagrams with one closed fermion loop and an arbitrary number of Γ lines:

 . (5.37)

The single particle Green's function can be written as follows:

$$G_{\alpha\beta}(u_1 - u_2) = \frac{\int \mathscr{D} W \exp\left[\frac{1}{2} \int_0^\beta W \cdot V^{-1} \cdot W + \mathrm{Tr}\ln(1 - WG_0) \right] G_{\alpha\beta}[W; u_1 - u_2]}{\int \mathscr{D} W \exp\left[\frac{1}{2} \int_0^\beta W \cdot V^{-1} \cdot W + \mathrm{Tr}\ln(1 - WG_0) \right]} \ . \tag{5.38}$$

Following a derivation similar to the one which leads to (5.30), one obtains the following expression:

$$G_{\alpha\beta}(u_1-u_2)=\left\{\exp\left\{-\int_0^\beta \operatorname{tr}\sum_{n>2}\frac{1}{n}\left[G(W_0)\cdot\frac{\delta}{\delta j}\right]^n\right\}G_{\alpha\beta}\left[W_0+\frac{\delta}{\delta j}\right]e^{-(1/2)j\cdot\Gamma\cdot j}\bigg|_{j=0}\right\}_L\ , \tag{5.39}$$

where the symbol $\{\ \}_L$ means that we have to consider only the linked diagrams. G has the following diagrammatic representation:

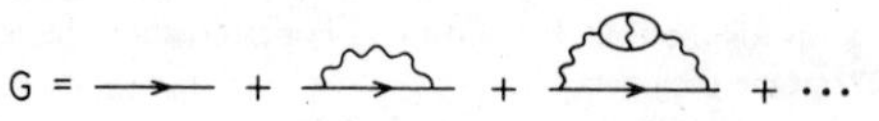

The first term may be veiwed as the classical propagator. It describes the motion of a particle in the field W_0. The other terms which describes the coupling of a particle to a vibration, *with propagation of the vibration,* are the quantum effects which are left out in the classical approximation.

VI. CONCLUSIONS

The functional integrals built on overcomplete sets of the Hilbert space provide a unifying understanding of different approaches to the nuclear many body problem. The role and the significance of the overcomplete set are best understood when calculating the functional integral using the saddle-point approximation, and its successive corrections. Then, it can be seen that the parameters which are used to label the states of the overcomplete set obey classical equations of motion. The state labels may then be viewed as classical coordinates in a generalized phase space. The classical equations of motion are identical to those obtained applying a time-dependent variational principle, using as trial states the states of the overcomplete set. But in contrast to the variational principle, the functional integral does provide a way of calculating corrections to the variational solution. A proper treatment of the fluctuations around the classical path introduces a quantization of the classical coordinates in terms of boson degrees of freedom.

As we have seen throughout this paper bosons play an important role in the functional integral formalism. In particular boson coherent states appear to be very useful because they are eigenstates of the destruction operators. This greatly facilitates the calculation of matrix elements. But more than that, it makes the structure of the functional integral simpler. Also we have seen that some changes of variables are allowed only if the overcomplete set is a set of coherent states. We have also shown that the functional integral transforms a fermion theory into a boson theory in very much the same way as the usual boson expansions do.

We have also obtained a clear physical interpretation of the classical features of the mean field approximations. In the language of boson representations, this approximation is obtained by replacing the boson propagators, e.g., the propagators corresponding to the RPA vibrations, by their classical approximation. This implies that only the static part of the particle-vibration interactions are taken into account in the mean field approximation. This point is further illustrated by diagrammatic expansion around the mean field. The processes involving a real propagation of a phonon between the time when it is emitted and the time when it is absorbed appear as quantum corrections to the mean field. Another equivalent statement, also suggested by the functional integral formalism, is that the mean field has at each time a given classical value. The functional integral allows for possible approximate schemes for calculating the "quantum" fluctuations around this value.

ACKNOWLEDGMENTS

One of the authors (J.P.B.) gratefully acknowledges the warm hospitality of the Physics Department of the University of Illinois at Urbana-Champaign where this work was completed.

APPENDIX

We construct explicitly the measures which have been used in Sec. II E to construct closure relations. The general idea underlying the method is to associate the parameters Z with some group operation and to construct the invariant measure over the group. In the case of the Slater determinants the group to be considered is the group of unitary transformations in the space of single particle states. A general element of the group is represented by the matrix

$$U=\begin{bmatrix}A & B\\ C & D\end{bmatrix},\quad UU^\dagger=U^\dagger U=1\ , \tag{A1}$$

where A, B, C, and D are $n_h \times n_h$, $n_h \times n_p$, $n_p \times n_h$, and $n_p \times n_p$ matrices, respectively. These matrices satisfy:

$$\begin{aligned} &AA^\dagger + BB^\dagger = 1, \quad CA^\dagger + DB^\dagger = 0 \ , \\ &AC^\dagger + BD^\dagger = 0, \quad CC^\dagger + DD^\dagger = 1 \ , \\ &A^\dagger A + C^\dagger C = 1, \quad A^\dagger B + C^\dagger D = 0 \ , \\ &B^\dagger A + D^\dagger C = 0, \quad B^\dagger B + D^\dagger D = 1 \ . \end{aligned} \tag{A2}$$

Let us now consider the states (2.43), normalized:

$$|\widetilde{Z}\rangle = N e^{Z_{ph} a_p^\dagger a_h} |\phi_0\rangle \ , \tag{A3}$$

where N is a normalization constant. Let S be the unitary transformation which carries $|\widetilde{Z}\rangle$ into $|\widetilde{Z}'\rangle$:

$$|\widetilde{Z}'\rangle = S|\widetilde{Z}\rangle \ . \tag{A4}$$

We look for an invariant measure $\mu(Z)$ such that

$$\begin{aligned} \mu(Z) &= \mu(Z')|J(Z',Z)| \\ &= \mu(0)|J(0,Z)| \ , \end{aligned} \tag{A5}$$

where $J(Z',Z)$ is the Jacobian of the transformation which transforms Z into Z'.

The law of transformation of the coordinates Z the transformation (A4) is easily derived. Indeed $|\widetilde{Z}\rangle$ can be written

$$|\widetilde{Z}\rangle = N \prod_h (a_h^\dagger + Z_{ph} a_p^\dagger)|0\rangle \ . \tag{A6}$$

Under the unitary transformation (A4), this becomes

$$|\widetilde{Z}\rangle = N \prod_h (a_h^\dagger + Z_{ph} a_p^\dagger)|0\rangle \ . \tag{A7}$$

where

$$b_h^\dagger = S a_h^\dagger S^\dagger, \quad b_p^\dagger = S a_p^\dagger S^\dagger, \quad S|0\rangle = |0\rangle \ . \tag{A8}$$

To the operator S is associated a matrix U of the form (A1) which realizes the linear transformation of the creation operators

$$(b_h^\dagger b_p^\dagger) = (a_h^\dagger a_p^\dagger) \begin{bmatrix} A & B \\ C & D \end{bmatrix} \ . \tag{A9}$$

Replacing $b_h^\dagger$ and $b_p^\dagger$ in the equation (A7) by their expression in terms of $a_h^\dagger$ and $a_p^\dagger$ given above, one gets:

$$\begin{aligned} |\widetilde{Z}'\rangle &= N \prod_h [a_{h'}^\dagger (A + BZ)_{h'h} \\ &\qquad + b_{p'}^\dagger (C + DZ)_{p'h}]|0\rangle \\ &= N' \prod_h (a_h^\dagger + Z'_{ph} a_p^\dagger)|0\rangle \ , \end{aligned} \tag{A10}$$

where

$$Z' = (C + DZ)(A + BZ)^{-1} \ . \tag{A11}$$

This is the desired transformation law. From this it is easy to evaluate the Jacobian which appears in (A5). First we write

$$Z'(A + BZ) = C + DZ$$

then differentiate,

$$dZ'(A + BZ) + Z'BdZ = DdZ \ .$$

We replace Z by its expression in terms of Z' by inverting the equation (A11) and finally put $Z' = 0$. We then get

$$|J(0,Z)| = |(\det D)^{2n_h} \det(A - BD^{-1}C)^{-2n_p}| \ . \tag{A12}$$

Using the relations (A2) one easily shows that

$$|\det(A - BD^{-1}C)| = |\det A|^{-1} = |\det D|^{-1} \ , \tag{A13}$$

so that the Jacobian takes the form

$$|J(0,Z)| = |\det A|^{2(n_p + n_h)} \ . \tag{A14}$$

It remains to relate the matrix A to the matrix Z. For that purpose one can use the following coset decomposition:

$$\begin{bmatrix} A & B \\ C & D \end{bmatrix} = \begin{bmatrix} 1 & -Z^\dagger \\ Z & 1 \end{bmatrix} \begin{bmatrix} U & 0 \\ 0 & U' \end{bmatrix} \begin{bmatrix} A_1 & 0 \\ 0 & D_1 \end{bmatrix} \ , \tag{A15}$$

where U and U' are, respectively, $n_h \times n_h$ and $n_p \times n_p$ arbitrary unitary matrices. A_1 and D_1 are, respectively, $n_h \times n_h$ and $n_p \times n_p$ matrices to be determined so that the matrix

$$\begin{bmatrix} A & B \\ C & D \end{bmatrix}$$

satisfies the conditions (A2). One solution is

$$A_1 = (1 + Z^\dagger Z)^{-1/2}, \quad D_1 = (1 + Z^\dagger Z)^{-1/2} \ . \tag{A16}$$

These equations define the matrices A and D, with respect to an arbitrary transformation of the form

$$\begin{bmatrix} U & 0 \\ 0 & U' \end{bmatrix}$$

which does not change the state of the system and which can be ignored. It is easily checked that the transformation thus defined carries the state $|Z\rangle$ into $|\phi_0\rangle$. Therefore, the Jacobian (A14) can be written

$$|J(0,Z)| = [\det(1+Z^\dagger Z)]^{-(n_p+n_h)}$$
$$= [\det(1+ZZ^\dagger)]^{-(n_p+n_h)} . \quad \text{(A17)}$$

The expression of the measure used in (2.24) follows trivially. This measure can also be obtained by identifying the set of Slater determinants with a complex Grassman manifold.[32] This method was used in Ref. 14. The method presented here is more elementary and similar to the methods used in Ref. 33 and Ref. 34. (See also Ref. 35.) We consider now the change of variables (2.27)

$$\beta_{ph} = \sum_{h'} Z_{ph'}[(1+Z^\dagger Z)^{-1/2}]_{h'h} . \quad \text{(A18)}$$

The expression of Z_{ph} in terms of β_{ph} is

$$Z_{ph} = \sum_{h'} \beta_{ph'}[(1-\beta^\dagger\beta)^{-1/2}]_{h'h} . \quad \text{(A19)}$$

In terms of these new variables, the matrix

$$\begin{bmatrix} A & B \\ C & D \end{bmatrix}$$

of Eq. (A15) takes the form

$$\begin{bmatrix} (1-\beta^\dagger\beta)^{1/2} & -\beta^\dagger \\ \beta & (1-\beta\beta^\dagger)^{1/2} \end{bmatrix} . \quad \text{(A20)}$$

It is easily seen that the Jacobian of the transformation (A18) is precisely given by (A17). When the β are chosen as coordinates, the measure is, therefore, extremely simple. The domain of integration is complicated, however. The volume θ of this domain can be calculated. This fixes the arbitrary constant in the measure. One has[34]

$$\theta = \frac{1!2!\ldots(n_h-1)!1!2!\ldots(n_p-1)!}{1!2!\ldots(n_p+n_h-1)!}\pi^{n_p n_h} . \quad \text{(A21)}$$

Finally it is convenient to introduce new variables $\tilde{\alpha}$ and $\tilde{\beta}$ defined as follows:

$$\tilde{\alpha} = (1-\beta^\dagger\beta)^{1/2}U, \quad \tilde{\beta} = \beta U , \quad \text{(A22)}$$

where U is a $n_h \times n_h$ unitary matrix. It is easily seen that the integral over β transforms into

$$\int d\beta d\beta^* = \int d\tilde{\beta} d\tilde{\beta}^* d\tilde{\alpha} d\tilde{\alpha}^* \delta(\tilde{\alpha}^\dagger\tilde{\alpha} + \tilde{\beta}^\dagger\tilde{\beta} - 1) . \quad \text{(A.23)}$$

But $\tilde{\alpha}$ and $\tilde{\beta}$ are the expansion coefficients of a set of N single particle states on a fixed basis. Any basis may be used to write (A23). In particular, we can choose a wave function representation, in which case we shall write the integration measure (A23) as follows:

$$\int \prod_{k=1}^{N} \prod_{x} d\varphi_k^*(x) d\varphi_k(x) \prod_{l=1}^{N} \delta(\langle\varphi_k|\varphi_l\rangle - \delta_{kl}) . \quad \text{(A24)}$$

---

*On leave from Service de Physique Theorique, CEN Saclay, France.

[1] A. K. Kerman and S. E. Koonin, Ann. Phys. (N.Y.) 100, 332 (1976).
[2] F. Villars, Nucl. Phys. A285, 269 (1977).
[3] K. Goeke and P. G. Reinhard, Ann. Phys. (N.Y.) 112, 328 (1978).
[4] M. Baranger and M. Veneroni, Ann. Phys. (N.Y.) 114, 123 (1978).
[5] E. R. Marshalek and J. Weneser, Phys. Rev. C 2, 1682 (1970).
[6] D. Janssen, F. Dönau, S. Frauendorf, and R. V. Jolos, Nucl. Phys. A172, 145 (1971).
[7] J. P. Blaizot and E. R. Marshalek, Nucl. Phys. A309, 422 (1978); A309, 453 (1978).
[8] J. R. Klauder, Ann. Phys. (N.Y.) 11, 123 (1960); J. Math. Phys. 4, 1055 (1963); 4, 1058 (1963).
[9] J. R Klauder, in *Path Integrals and their Application in Quantum, Statistical and Solid State Physics,* edited by G. Papadopoulos and J. Devrese (Plenum, New York, 1977).
[10] H. Kleinert, Phys. Lett. 69B, 9 (1977).
[11] H. Reinhardt, Nucl. Phys. A251, 317 (1975).
[12] S Levit, Phys. Rev. C 21, 1594 (1980); S. Levit, J. W. Negele, and Z. Paltiel, *ibid.* 21, 1603 (1980).
[13] H. Kleinert and H. Reinhardt, Nucl. Phys. A332, 331 (1979).
[14] H. Kuratsuji and T. Suzuki, Phys. Lett. 92B, 19 (1980).
[15] A. Kerman and T. Troudet (unpublished).

[16]J. P. Blaizot and H. Orland, J. Phys. Lett. 41, 53 (1980); 41, 401 (1980).
[17]J. P. Blaizot and H. Orland, J. Phys. Lett. 41, 523 (1980).
[18]J. P. Blaizot and H. Orland, Phys. Lett. 100B, 195 (1981).
[19]H. Orland, in *Méthodes Mathématiques de la Physique Nucléaire,* edited by B. Giraud and P. Quentin (Collége de France, Paris, 1980).
[20]V. Bargman, Commun. Pure Appl. Math. 14, 187 (1961).
[21]Y. Ohnuki and T. Kashiwa, Prog. Theor. Phys. 60, 548 (1978).
[22]D. J. Thouless, Nucl. Phys. 21, 225 (1960).
[23]L. D. Faddeev, in *Les Houches 1975,* Proceedings of the Methods in Field Theory, edited by R. Balian and J. Zinn-Justin (North-Holland, Amsterdam, 1970), p. 1.
[24]E. R. Marshalek and G. Holzwarth, Nucl. Phys. A191, 438 (1972).
[25]H. Trotter, Proc. Amer. Math. Soc. 10, 541 (1959).
[26]J. Hubbard, Phys. Rev. Lett. 3, 77 (1959).
[27]B. A. Lippman and J. Schwinger, Phys. Rev. 79, 469 (1950).
[28]R. F. Dashen, B. Hasslacher, and A. Neveu, Phys. Rev. 12D, 2443 (1975).
[29]R. Shankar, Phys. Rev. Lett. 45, 1088 (1980).
[30]K. K. Kan, J. J. Griffin, P. C. Lichtner, and M. Dworzecka, Nucl. Phys. A332, 109 (1979).
[31]H. Reinhardt, Nucl. Phys. A346, 1 (1980).
[32]S. Kobayashi and K. Nomizu, *Foundations of Differential Geometry* (Interscience, New York, 1969), Vol. II.
[33]M. I. Monastyrsky and A. M. Perelomov, Rep. Math. Phys. 6, 1 (1974).
[34]L. K. Hua, *Harmonic Analysis of Functions of Several Complex Variables in Classical Domains,* translation of Mathematical Monographs (American Mathematical Society, Providence, 1963), Vol. 6, p.46.
[35]The parametrization (A3) of Slater determinants and the derivation of the corresponding invariant measure has been extensively studied by H. Kuratsuji and T. Suzuki. These authors have also derived the invariant measure following arguments similar to those used in the first part of this appendix (private communication of T. Suzuki to one of us).

PHYSICAL REVIEW A VOLUME 26, NUMBER 1 JULY 1982

Path-integral approach to problems in quantum optics

Mark Hillery and M. S. Zubairy
Institute for Modern Optics, Department of Physics and Astronomy, University of New Mexico, Albuquerque, New Mexico 87131 and Max-Planck-Institut für Quantenoptik, D-8046 Garching bei München, West Germany
(Received 11 December 1981)

A formalism for applying path integrals to certain problems in nonlinear optics is considered. The properties of a coherent-state propagator are discussed and a path-integral representation for the propagator is presented. This representation is then employed in evaluating the propagator for general single-mode and multimode Hamiltonians which are at most quadratic in the creation and destruction operators of the field. Some examples involving parametric processes are given.

I. INTRODUCTION

Path integrals and the approximations to which they have led have been used very much in quantum field theory in recent years. The path-integral representation of the propagator allows one to see more clearly than the standard operator approach, the connection between the classical and quantum dynamics of a system. Semiclassical approximations can then be derived in a natural way.[1] So far, however, these techniques have not found much use in quantum optics.[2] In this paper we will develop some of the formalism which will be of use in applying path-integral techniques to certain problems in nonlinear optics.

The types of problems to which we would like to apply these techniques are those in which the medium with which the light interacts can be described by a nonlinear susceptibility tensor.[3] These include such processes as parametric amplification and harmonic generation. The interaction between the different modes is then described by products of various powers (depending upon the specific process) of the creation and destruction operators of the modes involved.

The type of path integral which we will consider is not the one usually used in quantum field theory in which one makes use of a coordinate representation of the field. We will be interested in problems in which only a few of the modes of the field are important and we will use a path integral which makes use of a representation of these modes in terms of coherent states. Because the Hamiltonians which we will consider will be expressed in terms of creation and destruction operators, and not the corresponding position and momentum operators, coherent states, which are eigenstates of the destruction operator, are natural objects to use. The coherent-state path integral can be used to calculate the matrix element of the time development transformation between two coherent states. This matrix element can be regarded as a type of propagator. This form of the path integral was first discussed by Klauder[4] and was subsequently examined by Schweber[5] in the context of Bargmann spaces. Klauder[6] in later work showed that the coherent-state path integral is but one example of a more general class of objects known as continuous representation path integrals.

In Sec. II, we discuss some properties of the propagator and show how it can be used to calculate various quantities of interest in quantum optics. In Sec. III, we derive formulas which can be used to calculate the propagator for single-mode systems with Hamiltonians at most quadratic in the creation and destruction operators. These are then used to calculate the propagator for the case of second subharmonic generation when the pump field is classical. In Sec. IV, we generalize our results and calculate the propagator for an N-mode system whose Hamiltonian is quadratic. This result is then used to calculate the propagator for a parametric amplifier with a classical pump field.

II. COHERENT-STATE PROPAGATOR

We consider a system which consists of one mode of the radiation field. Let the corresponding time-evolution operator be $U(t_2,t_1)$, i.e., if $|\psi(t_1)\rangle$ is the state of the system at time t_1 then the state at time t_2 is

$$|\psi(t_2)\rangle = U(t_2,t_1)|\psi(t_1)\rangle \ . \tag{1}$$

If the Hamiltonian governing the system is given by $H(t)$ then the time-evolution operator is (where we have chosen units such that $\hbar=1$)

$$U(t_2,t_1) = T\exp\left[-i\int_{t_1}^{t_2} H(t')dt'\right] , \tag{2}$$

where T is the Dyson time-ordering operator.

We will consider the propagator

$$K(\alpha_2,t_2;\alpha_1,t_1) = \langle \alpha_2 | U(t_2,t_1) | \alpha_1 \rangle \ , \tag{3}$$

where the coherent states $|\alpha_i\rangle$ are the eigenstates of the destruction operator a with eigenvalue α_i, at time $t=0$. Another expression for the propagator $K(\alpha_2,t_2;\alpha_1,t_1)$ can be derived by noting that the coherent state, at time t [i.e., the eigenstate of $a(t)$] is given by

$$|\alpha,t\rangle = U(t,0)^{-1}|\alpha\rangle \ . \tag{4}$$

We then obtain

$$\begin{aligned} K(\alpha_2,t_2;\alpha_1,t_1) &= \langle \alpha_2,t_2 | \alpha_1,t_1 \rangle \\ &= \langle \alpha_2 | U(t_2,0)U(t_1,0)^{-1} | \alpha_1 \rangle \ . \end{aligned} \tag{5}$$

In quantum optics, one is usually interested in evaluating certain correlation functions of the field. For a one-mode field these are proportional to the expectation values of products of the creation and destruction operators. These correlation functions can be expressed in terms of the propagator $K(\alpha_2,t_2;\alpha_1,t_1)$. We assume that, at $t=0$, the density matrix has a P representation, i.e.,

$$\rho = \int d^2\alpha \, P(\alpha) |\alpha\rangle\langle\alpha| \ , \tag{6}$$

so that the expectation value of any operator, $O(t)$, in the Heisenberg picture is given by

$$\begin{aligned} \langle O(t)\rangle &= \mathrm{Tr}[\rho O(t)] \\ &= \int d^2\alpha \, P(\alpha)\langle \alpha | O(t) | \alpha\rangle \ . \end{aligned} \tag{7}$$

On using the completeness property of the coherent states, namely,

$$\frac{1}{\pi}\int d^2\alpha \, |\alpha,t\rangle\langle\alpha,t| = 1 \ , \tag{8}$$

it can be easily shown that

$$\langle a(t)\rangle = \frac{1}{\pi}\int\int d^2\alpha_1 d^2\alpha_2 P(\alpha_2)|K(\alpha_1,t_1;\alpha_2,0)|^2 \alpha_1 \ , \tag{9}$$

$$\langle a^\dagger(t_1)a(t_2)\rangle = \frac{1}{\pi^2}\int\int\int d^2\alpha_1 d^2\alpha_2 d^2\alpha_3 P(\alpha_3) K(\alpha_1,t_1;\alpha_2,t_2) K(\alpha_2,t_2;\alpha_3,0) K(\alpha_3,0;\alpha_1,t_1)\alpha_1^*\alpha_2 \ , \tag{10}$$

$$\langle a^\dagger(0)a^\dagger(t)a(t)a(0)\rangle = \frac{1}{\pi^2}\int\int\int d^2\alpha_1 d^2\alpha_2 d^2\alpha_3 P(\alpha_3) K^*(\alpha_2,t;\alpha_3,0) K(\alpha_1,t;\alpha_3,0) \, |\alpha_3|^2 \alpha_2^*\alpha_1 \ . \tag{11}$$

The determination of the propagator thus enables us to calculate any correlation function of the field operators.

The propagator $K(\alpha_2,t_2;\alpha_1,t_1)$ is related to the Q representation of the radiation field, i.e.,

$$Q(\alpha,t) = \frac{1}{\pi}\langle \alpha,t|\rho|\alpha,t\rangle \ , \tag{12}$$

in a natural way. On substituting for ρ from Eq. (6), we obtain

$$Q(\alpha,t) = \frac{1}{\pi}\int d^2\alpha_1 P(\alpha_1)|K(\alpha,t;\alpha_1,0)|^2 \ . \tag{13}$$

In particular, for an initial coherent state, $P(\alpha_1)=\delta^2(\alpha_1-\alpha_0)$, and it follows from Eq. (13) that

$$Q(\alpha,t) = \frac{1}{\pi}|K(\alpha,t;\alpha_0,0)|^2 \ . \tag{14}$$

The Q representation has the property that the expectation value, at time t, of any antinormally ordered function $O_A(a,a^\dagger)$ of a and $a^\dagger$ may be determined via the relation

$$\langle O_A(a,a^\dagger)\rangle = \int d^2\alpha \, O_A(\alpha,\alpha^*)Q(\alpha,t) \ . \tag{15}$$

The close relation of propagator to the Q representation makes it easier to evaluate the expectation values of antinormally ordered products than the normally ordered products. For example, the mean number of photons at time t is most easily evaluated by using the commutation relation $[a,a^\dagger]=1$, as follows:

$$\langle a^\dagger(t)a(t)\rangle=\langle a(t)a^\dagger(t)\rangle-1=\frac{1}{\pi}\int d^2\alpha_1\int d^2\alpha_2 P(\alpha_2)\,|K(\alpha_1,t;\alpha_2,0)|^2|\alpha_1|^2-1\ . \tag{16}$$

Finally, we note that the Q and P representations are related to each other via the following relationship[7]:

$$Q(\alpha,t)=\int d^2\alpha_1 P(\alpha_1,t)\,|K(\alpha,0;\alpha_1,0)|^2\ . \tag{17}$$

We now turn to the calculation of the propagator itself for a particular set of systems.

III. REPRESENTATION OF THE PROPAGATOR

A. Path integral for the propagator

It is possible to express the coherent-state propagator in terms of a path integral. Here we outline the derivation of the path-integral representation which was first obtained by Klauder.[4]

We consider a system which is described by a Hamiltonian, $H(a^\dagger,a;t)$, which is expressed in terms of the creation and destruction operators $a^\dagger$ and a. We suppose further that $H(a^\dagger,a;t)$ is normally ordered. By inserting n resolutions of the identity into Eq. (5) we find that

$$K(\alpha_f,t_f;\alpha_i,t_i)=\left[\frac{1}{\pi}\right]^n\int d^2\alpha_1\cdots\int d^2\alpha_n\langle\alpha_f,t_f|\alpha_n,t_n\rangle\langle\alpha_n,t_n|\alpha_{n-1},t_{n-1}\rangle\cdots\langle\alpha_1,t_1|\alpha_i,t_i\rangle\ . \tag{18}$$

We also have that

$$\begin{aligned}\langle\alpha_j,t_j|\alpha_{j-1},t_{j-1}\rangle&=\left\langle\alpha_j\left|T\exp\left[-i\int_{t_{j-1}}^{t_j}d\tau H(\tau)\right]\right|\alpha_{j-1}\right\rangle\\&\cong\left\langle\alpha_j\left|\left[1-i\int_{t_{j-1}}^{t_j}d\tau H(a^\dagger,a;\tau)\right]\right|\alpha_{j-1}\right\rangle\\&\cong\langle\alpha_j|\alpha_{j-1}\rangle[1-i\epsilon H(\alpha_j^*,\alpha_{j-1};t_{j-1})]\\&\cong\exp[-\tfrac{1}{2}(|\alpha_j|^2+|\alpha_{j-1}|^2)+\alpha_j^*\alpha_{j-1}-i\epsilon H(\alpha_j^*,\alpha_{j-1};t_{j-1})]\ ,\end{aligned} \tag{19}$$

where $\epsilon=(t_f-t_i)/n+1$, $t_j=t_i+j\epsilon$, and the function $H(\alpha''^*,\alpha',t)$ is defined as

$$H(\alpha''^*,\alpha';t)=\frac{\langle\alpha''|H(a^\dagger,a;t)|\alpha'\rangle}{\langle\alpha''|\alpha'\rangle}\ . \tag{20}$$

Inserting Eq. (19) into Eq. (18) immediately yields

$$K(\alpha_f,t_f;\alpha_i,t_i)=\lim_{n\to\infty}\left[\frac{1}{\pi}\right]^n\int d^2\alpha_1\cdots\int d^2\alpha_n\exp\left[\sum_{j=1}^{n+1}[-\tfrac{1}{2}(|\alpha_j|^2+|\alpha_{j-1}|^2)+\alpha_j^*\alpha_{j-1}-i\epsilon H(\alpha_j^*,\alpha_{j-1};t_{j-1})]\right]\ . \tag{21}$$

We note that

$$\begin{aligned}&\sum_{j=1}^{n+1}[-\tfrac{1}{2}(|\alpha_j|^2+|\alpha_{j-1}|^2)+\alpha_j^*\alpha_{j-1}-i\epsilon H(\alpha_j^*,\alpha_{j-1};t_{j-1})]\\&\quad=\sum_{j=1}^{n+1}\left[-\tfrac{1}{2}\alpha_j^*\left[\frac{\alpha_j-\alpha_{j-1}}{\epsilon}\right]\epsilon+\tfrac{1}{2}\alpha_{j-1}\left[\frac{\alpha_j^*-\alpha_{j-1}^*}{\epsilon}\right]\epsilon-i\epsilon H(\alpha_j^*,\alpha_{j-1};t_{j-1})\right]\\&\quad\to\int_{t_i}^{t_f}d\tau[\tfrac{1}{2}(\alpha\dot{\alpha}^*-\alpha^*\dot{\alpha})-iH(\alpha^*,\alpha;\tau)]\ ,\end{aligned} \tag{22}$$

as $\epsilon\to 0$. It then follows that

$$K(\alpha_f,t_f;\alpha_i,t_i)=\int \mathscr{D}[\alpha(\tau)]\exp\left[\int_{t_i}^{t_f} d\tau[\tfrac{1}{2}(\alpha\dot{\alpha}^*-\alpha^*\dot{\alpha})-iH(\alpha^*,\alpha;\tau)]\right], \quad (23)$$

where $\int \mathscr{D}[\alpha(\tau)]$ designates the integration over all paths $\alpha(\tau)$, such that $\alpha(t_i)=\alpha_i$ and $\alpha(t_f)=\alpha_f$.

B. Quadratic Hamiltonian

If the Hamiltonian is at most quadratic in a and $a^\dagger$, it is possible to evaluate the path integral explicitly (Yuen[8] has calculated this propagator using a different method). The most general quadratic Hamiltonian is given by

$$H(a^\dagger,a;t)=\omega(t)a^\dagger a+f(t)a^2+f^*(t)a^{\dagger 2}+g(t)a+g^*(t)a^\dagger, \quad (24)$$

where $f(t)$ and $g(t)$ are arbitrary time-dependent functions. The evaluation of the path integral (21) corresponding to this Hamiltonian is outlined in Appendix A. The resulting expression for the propagator is

$$\begin{aligned}K(\alpha_f,t_f;\alpha_i,t_i)=\exp\Big[&-i\int_{t_i}^{t_f} d\tau[2f(\tau)X(\tau)+f(\tau)Z^2(\tau)+g(\tau)Z(\tau)]\\ &-\tfrac{1}{2}(|\alpha_f|^2+|\alpha_i|^2)+Y(t_f)\alpha_f^*\alpha_i+X(t_f)(\alpha_f^*)^2-i\alpha_i^2\int_{t_i}^{t_f} d\tau f(\tau)Y^2(\tau)+Z(t_f)\alpha_f^*\\ &-i\alpha_i\int_{t_i}^{t_f} d\tau[g(\tau)+2f(\tau)Z(\tau)]Y(\tau)\Big], \end{aligned}\quad (25)$$

where $X(t)$ satisfies the differential equation

$$\frac{dX}{dt}=-2i\omega(t)X-4if(t)X^2-if^*(t), \quad (26)$$

with $X(t_i)=0$ and

$$Y(t)=\exp\left[-i\int_{t_i}^{t} d\tau[\omega(\tau)+4f(\tau)X(\tau)]\right], \quad (27)$$

$$Z(t)=-i\int_{t_i}^{t} d\tau[g^*(\tau)+2g(\tau)X(\tau)]\exp\left[-i\int_{\tau}^{t} d\tau'[\omega(\tau')+4f(\tau')X(\tau')]\right]. \quad (28)$$

The nonlinear differential Eq. (26) for $X(t)$ can be solved if we can express $f(t)$ as

$$f(t)=\tilde{f}(t)\exp\left[2i\int_{t_i}^{t} d\tau\,\omega(\tau)\right], \quad (29)$$

where $\tilde{f}(t)$ is real or imaginary. We now consider a simple example where this condition is satisfied.

C. Degenerate parametric amplifier

The quantum statistical properties of the degenerate parametric amplifier have received considerable attention in recent years.[9] This nonlinear device is predicted to exhibit photon antibunching[10] which is a strictly quantum-mechanical effect. Squeezed states, which could prove to be useful in the efforts to detect gravitational waves, are also predicted to be generated in a degenerate parametric amplifier.[8,11]

The Hamiltonian that governs this nonlinear optical device is given by

$$H(t)=\omega a^\dagger a+\kappa(e^{2i\omega t}a^2+e^{-2i\omega t}a^{\dagger 2}), \quad (30)$$

where κ is a coupling constant and ω is the mode frequency. The Hamiltonian (30) is the same as that given by Eq. (24) if we make the following identifications:

$$\omega(t)=\omega,\quad f(t)=\kappa e^{2i\omega t},\quad g(t)=0. \quad (31)$$

Under these conditions Eq. (26) can be solved and we obtain

$$X(t)=\frac{1}{2i}e^{-2i\omega t}\tanh[2\kappa(t-t_i)], \quad (32a)$$

$$Y(t)=e^{-i\omega(t-t_i)}\operatorname{sech}[2\kappa(t-t_i)], \quad (32b)$$

$$Z(t)=0. \quad (32c)$$

On substituting from Eqs. (32a)–(32c) into Eq. (25) we obtain

$$K(\alpha_f,t_f;\alpha_i,t_i)=\{\text{sech}[2\kappa(t_f-t_i)]\}^{1/2}$$
$$\times\exp\{-\tfrac{1}{2}(|\alpha_f|^2+|\alpha_i|^2)+\alpha_f^*\alpha_i e^{-i\omega(t_f-t_i)}\text{sech}[2\kappa(t_f-t_i)]$$
$$-\tfrac{1}{2}i(\alpha_f^*)^2e^{-2i\omega t_f}\tanh[2\kappa(t_f-t_i)]-\tfrac{1}{2}i\alpha_i^2e^{2i\omega t_i}\tanh[2\kappa(t_f-t_i)]\} . \quad (33)$$

This expression for the propagator which we have derived using a path-integral approach can also be derived using a more conventional approach.[10]

IV. MULTIMODE PROBLEMS

A. Path integral

It is also possible to apply these techniques to problems involving more than one mode. If one is dealing with N modes the propagator becomes a function of $2N$ complex variables. In particular we have

$$K(\vec{\alpha}_f,t_f;\vec{\alpha}_i,t_i)=\langle\vec{\alpha}_f\,|\,U(t_f,t_i)\,|\,\vec{\alpha}_i\rangle , \quad (34)$$

where $\vec{\alpha}_i$ and $\vec{\alpha}_f$ are N-component vectors with components denoted by $\alpha_1^{(i)}, \alpha_2^{(i)}, \ldots, \alpha_N^{(i)}$ (similarly for $\vec{\alpha}_f$), and

$$|\,\vec{\alpha}_i\rangle=|\,\alpha_1^{(i)}\rangle\otimes|\,\alpha_2^{(i)}\rangle\otimes\cdots\otimes|\,\alpha_N^{(i)}\rangle .$$

Correlation functions can be computed from this propagator in ways similar to those used in the one-mode case. One must simply evaluate more integrals.

There is also a path-integral representation for the N-mode propagator. One has

$$K(\vec{\alpha}_f,t_f;\vec{\alpha}_i,t_i)=\int\mathscr{D}[\vec{\alpha}(\tau)]e^{iS}$$
$$=\int\mathscr{D}[\alpha_1(\tau)]\cdots\int\mathscr{D}[\alpha_N(\tau)]e^{iS} , \quad (35)$$

where

$$iS=\int_{t_i}^{t_f}d\tau\left[\sum_{n=1}^{N}\tfrac{1}{2}(\dot{\alpha}_n^*\alpha_n-\alpha_n^*\dot{\alpha}_n)-iH(\vec{\alpha}^*,\vec{\alpha};\tau)\right] , \quad (36)$$

$\vec{\alpha}(t_i)=\vec{\alpha}_i$, $\vec{\alpha}(t_f)=\vec{\alpha}_f$, and if $H(a_1^\dagger,\ldots,a_n^\dagger, a_1,\ldots,a_n;\tau)$ is the normally ordered Hamiltonian for the system

$$H(\vec{\alpha}''^*,\vec{\alpha}';\tau)=\langle\vec{\alpha}''\,|\,H(a_1^\dagger,\ldots,a_n^\dagger,a_1,\ldots,a_n;\tau)\,|\,\vec{\alpha}'\rangle/\langle\vec{\alpha}''\,|\,\vec{\alpha}'\rangle . \quad (37)$$

B. Quadratic Hamiltonian

If the Hamiltonian is quadratic in $a_1,\ldots,a_N$ and $a_1^\dagger,\ldots,a_N^\dagger$ one can again explicitly evaluate the path integral. We express the Hamiltonian as

$$H=\sum_{i=1}^{N}\sum_{j=1}^{N}[\omega_{ij}(t)a_i^\dagger a_j+f_{ij}(t)a_ia_j+f_{ij}^*(t)a_i^\dagger a_j^\dagger] \quad (38)$$

and we assume that f has been chosen so that $f_{ij}(t)=f_{ji}(t)$. The detailed calculation of the propagator for this Hamiltonian is performed in Appendix B. We find that

$$K(\vec{\alpha}_f,t_f;\vec{\alpha}_i,t_i)=\exp\left[-2i\int_{t_i}^{t_f}d\tau\,\text{Tr}[X(\tau)f(\tau)]-\tfrac{1}{2}[(\vec{\alpha}_f^*)\cdot\vec{\alpha}_f+(\vec{\alpha}_i^*)\cdot\vec{\alpha}_i]+(\vec{\alpha}_f^*)^TY(t_f)\vec{\alpha}_i\right.$$
$$\left.+(\vec{\alpha}_f^*)^TX(t_f)\vec{\alpha}_f^*-i\int_{t_i}^{t_f}d\tau\,\vec{\alpha}_i^TY^T(\tau)f(\tau)Y(\tau)\vec{\alpha}_i\right] . \quad (39)$$

In the above equation $X(t)$ and $f(\tau)$ are $N\times N$ symmetric matrices. The elements of $f(t)$ are simply the functions $f_{ij}(t)$ which appear in the Hamiltonian. The matrix $X(t)$ satisfies the equation

$$\frac{dX}{dt}=-i(\omega X+X\omega+f^*+4XfX) , \quad (40)$$

where $\omega(t)$ is an $N\times N$ matrix whose elements are $\omega_{ij}(t)$, and $X(t_i)=0$. The $N\times N$ matrix $Y(t)$ is given by

$$Y(t)=T\exp\left[-i\int_{t_i}^{t_f}d\tau[\omega(\tau)+4X(\tau)f(\tau)]\right] . \quad (41)$$

The superscript T appearing on some of the vectors and matrices in Eq. (39) denotes transpose.

C. Parametric amplifier

The parametric amplifier with a classical pump field is a system which has been much studied in quantum optics.[12] Here we would like to use the formulas developed in the preceding section to find the propagator for this system.

The Hamiltonian we wish to consider is

$$H=\omega_1 a_1^\dagger a_1+\omega_2 a_2^\dagger a_2 +\kappa(e^{i\omega_3 t}a_1a_2+e^{-i\omega_3 t}a_1^\dagger a_2^\dagger)\ , \quad (42)$$

where $\omega_3=\omega_1+\omega_2$. The matrices $\omega(t)$ and $f(t)$ are

$$\omega(t)=\begin{bmatrix}\omega_1 & 0\\ 0 & \omega_2\end{bmatrix}\ ,$$
$$f(t)=\tfrac{1}{2}\kappa e^{i\omega_3 t}\begin{bmatrix}0 & 1\\ 1 & 0\end{bmatrix}\ . \quad (43)$$

Considering first the equation for $X(t)$, Eq. (40), we find that

$$X(t)=-\tfrac{1}{2}ie^{-i\omega_3 t}\tanh[\kappa(t-t_i)]\begin{bmatrix}0 & 1\\ 1 & 0\end{bmatrix}\ . \quad (44)$$

Rather than solve for $Y(t)$, we instead solve for the vector

$$\tilde{u}(t)=Y(t)\vec{\alpha}_i\ . \quad (45)$$

The vector $\tilde{u}$ satisfies the equation

$$\frac{d\tilde{u}}{dt}=-i(\omega\tilde{u}+4Xf\tilde{u})\ , \quad (46)$$

where $\tilde{u}(t_i)=\vec{\alpha}_i$. One finds that

$$\tilde{u}(t)=\mathrm{sech}[\kappa(t-t_i)]\begin{bmatrix}e^{-i\omega_1(t-t_i)}\alpha_1^{(i)}\\ e^{-i\omega_2(t-t_i)}\alpha_2^{(i)}\end{bmatrix}\ . \quad (47)$$

The final result for the propagator is then

$$K(\vec{\alpha}_f,t_f;\vec{\alpha}_i,t_i)=[\mathrm{sech}\kappa(t_f-t_i)]\exp\Bigg[-\tfrac{1}{2}[(\vec{\alpha}_f^*)\cdot\vec{\alpha}_f+(\vec{\alpha}_i^*)\cdot\vec{\alpha}_i]-\tfrac{1}{2}ie^{-i\omega_3 t_f}\tanh[\kappa(t_f-t_i)](\vec{\alpha}_f^*)^T\sigma_1\vec{\alpha}_f^*$$
$$+\mathrm{sech}[\kappa(t_f-t_i)](\vec{\alpha}_f^*)^T\begin{bmatrix}e^{-i\omega_1(t_f-t_i)} & 0\\ 0 & e^{-i\omega_2(t_f-t_i)}\end{bmatrix}\vec{\alpha}_i$$
$$-\tfrac{1}{2}ie^{i\omega_3 t_i}\tanh[\kappa(t_f-t_i)]\vec{\alpha}_i^T\sigma_1\vec{\alpha}_i\Bigg]\ , \quad (48)$$

where $\sigma_1=\left(\begin{smallmatrix}0 & 1\\ 1 & 0\end{smallmatrix}\right)$.

V. CONCLUSION

We have shown how a formalism incorporating coherent-state propagators and path integrals can be of use in the consideration of certain problems in nonlinear optics. Here we concentrated on the formalism itself and certain basic results for the path integrals. These are necessary steps toward the development of approximation schemes for more complicated systems. It is in these approximations that the promise of these techniques lies.

ACKNOWLEDGMENT

Research for this work was supported by Kirtland Air Force Base under Contract No. F29601-82-K-0017.

APPENDIX A

According to Eq. (21) the propagator $K(\alpha_f,t_f;\alpha_i,t_i)$ corresponding to the Hamiltonian (24) is given by

$$K(\alpha_f,t_f;\alpha_i,t_i)=\lim_{n\to\infty}\left[\frac{1}{\pi}\right]^n\int\cdots\int\left[\prod_{j=1}^{n}d^2\alpha_j\right]e^{iS_n}\,, \tag{A1}$$

where

$$iS_n=\sum_{j=1}^{n+1}[-\tfrac{1}{2}(|\alpha_j|^2+|\alpha_{j-1}|^2)+(1-i\epsilon\omega_j)\alpha_j^*\alpha_{j-1}-i\epsilon f_{j-1}\alpha_{j-1}^2-i\epsilon f_j^*\alpha_j^{*2}-i\epsilon g_{j-1}\alpha_{j-1}-i\epsilon g_j^*\alpha_j]\,. \tag{A2}$$

The α_i integrations in Eq. (A1) are lengthy but straightforward. The resulting equation is

$$K(\alpha_f,t_f;\alpha_i,t_i)=\lim_{n\to\infty}\frac{1}{\left[\prod_{i=1}^{n}(1+4i\epsilon f_iX_i)^{1/2}\right]}$$
$$\times\exp\left[\sum_{j=0}^{n}\left\{i\epsilon\left[\left[\frac{f_jZ_j^2+g_jZ_j-i\epsilon g_j^2X_j}{1+4i\epsilon f_jX_j}\right]+\left[\frac{f_jY_j^2}{1+4i\epsilon f_jX_j}\right]\alpha_i^2\right.\right.\right.$$
$$\left.\left.\left.+\left[\frac{2f_jY_jZ_j+g_jY_j}{1+4i\epsilon f_jX_j}\right]\alpha_i\right]\right\}+X_{n+1}\alpha_f^{*2}+Y_{n+1}\alpha_i\alpha_f^*+Z_{n+1}\alpha_f^*\right]\,, \tag{A3}$$

where X_j, Y_j, and Z_j satisfy the following recursion relations:

$$Z_j=-i\epsilon f_j^*+\frac{(1-i\epsilon\omega_j)^2X_{j-1}}{1+4i\epsilon f_{j-1}X_{j-1}}\,, \tag{A4}$$

$$Y_j=\frac{(1-i\epsilon\omega_j)Y_{j-1}}{1+4i\epsilon f_{j-1}X_{j-1}}\,, \tag{A5}$$

$$Z_j=-i\epsilon g_j^*+\frac{(1-i\epsilon\omega_j)(Z_{j-1}-2i\epsilon g_jX_{j-1})}{1+4i\epsilon f_{j-1}X_{j-1}}\,, \tag{A6}$$

with $X_0=Z_0=0$ and $Y_0=1$. On taking the limit $n\to\infty$, we obtain

$$\prod_{i=1}^{n}(1+4i\epsilon f_iX_i)^{1/2}\to\exp\left[2i\int_{t_i}^{t_f}d\tau f(\tau)X(\tau)\right]\,, \tag{A7}$$

$$\sum_{j=0}^{n}\left[\frac{-i\epsilon(f_jZ_j^2+g_jZ_j-i\epsilon g_j^2X_j)}{1+4i\epsilon f_jX_j}\right]\to-i\int_{t_i}^{t_f}d\tau Z(\tau)[f(\tau)Z(\tau)+g(\tau)]\,, \tag{A8}$$

$$\sum_{j=0}^{n}\left[\frac{-i\epsilon f_jY_j^2}{1+4i\epsilon f_jX_j}\right]\to-i\int_{t_i}^{t_f}d\tau f(\tau)Y^2(\tau)\,, \tag{A9}$$

$$\sum_{j=0}^{n}\left[\frac{-i\epsilon(2f_jY_jZ_j+g_jY_j)}{1+4i\epsilon f_jX_j}\right]\to-i\int_{t_i}^{t_f}d\tau[2f(\tau)Y(\tau)Z(\tau)+g(\tau)Y(\tau)]\,, \tag{A10}$$

$$X_{n+1},Y_{n+1},Z_{n+1}\to X(t_f),Y(t_f),Z(t_f)\,, \tag{A11}$$

and, in view of the recursion relations (A4)–(A6), the functions $X(t)$, $Y(t)$, and $Z(t)$ satisfy the differential equation

$$\frac{dX}{dt}=-2i\omega(t)X-4if(t)X^2-if^*(t)\ , \tag{A12}$$

$$\frac{dY}{dt}=-i[\omega(t)+4f(t)X(t)]Y\ , \tag{A13}$$

$$\frac{dZ}{dt}=-i[\omega(t)+4f(t)X(t)]Z-i[g^*(t)+2g(t)X(t)]\ , \tag{A14}$$

where $X(t_i)=Z(t_i)=0$ and $Y(t_i)=1$.

On substituting from Eqs. (A7)–(A11) into Eq. (A3), we obtain

$$\begin{aligned}K(\alpha_f,t_f;\alpha_i,t_i)=\exp\Big[&-i\int_{t_i}^{t_f}d\tau[2f(\tau)X(\tau)+f(\tau)Z^2(\tau)+g(\tau)Z(\tau)]-\tfrac{1}{2}(|\alpha_f|^2+|\alpha_i|^2)\\&+Y(t_f)\alpha_f^*\alpha_i+X(t_f)(\alpha_f^*)^2-i\alpha_i^2\int_{t_i}^{t_f}d\tau f(\tau)Y^2(\tau)\\&-i\alpha_i\int_{t_i}^{t_f}d\tau[g(\tau)+2f(\tau)Z(\tau)]Y(\tau)+Z(t_f)\alpha_f^*\Big]\ .\end{aligned} \tag{A15}$$

Equations (A13) and (A14) can be integrated and the resulting solutions for $Y(t)$ and $Z(t)$ are

$$Y(t)=\exp\left[-i\int_{t_i}^{t}d\tau[\omega(\tau)+4f(\tau)X(\tau)]\right]\ , \tag{A16}$$

$$Z(t)=-i\int_{t_i}^{t}d\tau g^*(\tau)[1+2X(\tau)]\exp\left[-i\int_{\tau}^{t}d\tau'[\omega(\tau')+4f(\tau')X(\tau')]\right]\ , \tag{A17}$$

where $X(t)$ is determined by solving Eq. (A12) subject to $X(t_i)=0$.

APPENDIX B

We would like to compute the propagator for the system governed by the Hamiltonian given by Eq. (38). As in the one-mode case we have that the propagator is given by

$$K(\vec{\alpha}_f,t_f;\vec{\alpha}_i,t_i)\ \lim_{n\to\infty}\left[\frac{1}{\pi^N}\right]^n\int d\vec{\alpha}_1\cdots d\vec{\alpha}_n e^{iS_n}\ , \tag{B1}$$

where $d\vec{\alpha}_j=d^2\alpha_1^{(j)}d^2\alpha_2^{(j)}\cdots d^2\alpha_N^{(j)}$ and

$$iS_n=\sum_{l=1}^{n+1}\{-\tfrac{1}{2}[(\vec{\alpha}_l^*)\cdot\vec{\alpha}_l+(\vec{\alpha}_{l-1}^*)\cdot\vec{\alpha}_{l-1}]+(\vec{\alpha}_l^*)\cdot\vec{\alpha}_{l-1}-i\epsilon[(\vec{\alpha}_l^*)^T\omega_l\vec{\alpha}_{l-1}+\vec{\alpha}_{l-1}^Tf_{l-1}\vec{\alpha}_{l-1}+(\vec{\alpha}_l^*)^Tf_l^*(\vec{\alpha}_l^*)]\}\ . \tag{B2}$$

In the above equation $\vec{\alpha}^T$ designates the transpose of $\vec{\alpha}$ and $f_l=f(t_l)$ is an $N\times N$ matrix where $t_l=t_i+l\epsilon$.

To perform the integrations it is necessary to split each $\alpha_j^{(l)}$ into real and imaginary parts. That is, for each l we must go from a N-dimensional space, C^N (of which $\vec{\alpha}_l$ is a member), to a $2N$-dimensional space. It is best to view this space as a tensor product space $C^N\otimes C^2$. If $\eta_i\in C^N$ is the vector whose ith component is 1 and whose other components are 0, and $v_j\in C^2$ is the vector whose jth component is 1 and whose other component is 0, then $\vec{\alpha}\in C^N\to z\in C^N\otimes C^2$, where

$$z=\sum_{j=1}^{N}(x_j\eta_j\otimes v_1+y_j\eta_j\otimes v_2) \tag{B3}$$

and the components of $\vec{\alpha}$ are $\alpha_j=x_j+iy_j$. It is then possible to express the action as

$$iS_n=-\sum_{l=1}^{n}z_l^TM_lz_l+\sum_{l=1}^{n+1}z_l^TL_lz_{l-1}-\tfrac{1}{2}[(\vec{\alpha}_f^*)\cdot\vec{\alpha}_f+(\vec{\alpha}_i^*)\cdot\vec{\alpha}_i]-i\epsilon[\vec{\alpha}_i^Tf_i\vec{\alpha}_i+(\vec{\alpha}_f^*)^Tf_f^*\vec{\alpha}_f^*]\ , \tag{B4}$$

where $M_l = I + i\epsilon(f_l \otimes \gamma_1 + f_l^* \otimes \gamma_2)$, $L_l = (I_N - i\epsilon\omega_l)\otimes\mu$, I is the identity on $C^N \otimes C^2$, I_N is the identity on C^N, and

$$\mu = \begin{bmatrix} 1 & i \\ -i & 1 \end{bmatrix}, \quad \gamma_1 = \begin{bmatrix} 1 & i \\ i & -1 \end{bmatrix}, \quad \gamma_2 = \begin{bmatrix} 1 & -i \\ -i & -1 \end{bmatrix}. \tag{B5}$$

We now want to do the integrations starting with $l=1$, then going to $l=2$ and so on. To do this we make use of the formula for the integral (assuming that it exists)

$$\int_{-\infty}^{\infty} dx_1 \cdots \int_{-\infty}^{\infty} dx_n e^{-\vec{x}^T A \vec{x} + \vec{y}^T \vec{x}} = \frac{\pi^{n/2}}{(\det A)^{1/2}} e^{(1/4)\vec{y}^T A^{-1} \vec{y}}, \tag{B6}$$

where A is a symmetric $n\times n$ matrix and $\vec{y}$ is an n-component vector. Using this formula to do the $l=1$ integration we pick up a factor of

$$\pi^N (\det M_1)^{-1/2} \exp(\tfrac{1}{4} z_0^T L_1^T M_1^{-1} L_1 z_0)$$

and terms in the exponent which are linear and quadratic in z_2. We can express the part of the action containing z_2 (after having done the $l=1$ integration) as

$$-z_2^T M_2' z_2 + z_3^T L_2 z_2 + v_2^T z_2, \tag{B7}$$

where

$$M_2' = M_2 - \tfrac{1}{4} L_2 M_1^{-1} L_2^T \tag{B8}$$

and

$$v_2 = \tfrac{1}{4}[L_2 M_1^{-1} L_1 z_0 + L_2 (M_1^{-1})^T L_1 z_0]. \tag{B9}$$

In general, if one has done $l-1$ of the integrations the part of the action containing z_l can be expressed as

$$-z_l^T M_l' z_l + z_{l+1}^T L_l z_l + v_l^T z_l, \tag{B10}$$

where M_l' and v_l obey the recurrence relations

$$M_{l+1}' = M_{l+1} - \tfrac{1}{4} L_{l+1} (M_l')^{-1} L_{l+1}^T, \tag{B11}$$

$$v_l = \tfrac{1}{4}[L_{l+1}(M_l'^{-1})^T + L_{l+1}(M_l'^{-1})] v_l. \tag{B12}$$

Note also that each integration contributes a factor of

$$\pi^N (\det M_l')^{-1/2} \exp(\tfrac{1}{4} v_l^T M_l'^{-1} v_l).$$

One can show from the above recursion relations that it is possible to express M_l' and v_l in the form

$$M_l' = M_l - X_l \otimes \gamma_2, \quad v_l = u_l \otimes \hat{e}_1,$$

where $\hat{e}_1 = (1/\sqrt{2})(\nu_1 - i\nu_2)$ and, to first order in ϵ, X_l, and u_l obey the recursion relations

$$X_{l+1} = X_l - i\epsilon(\omega_{l+2} X_l + X_l \omega_{l+2} + f_{l+1}^* + 4X_l f_{l+1} X_l), \tag{B13}$$

$$u_{l+1} = u_l - i\epsilon(\omega_{l+1} u_l + 4X_{l-1} f_l u_l). \tag{B14}$$

Upon taking the $\epsilon \to 0$ limit these equations become

$$\frac{dX}{dt} = -i(\omega X + X\omega + f^* + 4XfX), \tag{B15}$$

$$\frac{du}{dt} = -i(\omega u + 4Xfu), \tag{B16}$$

where $X(t_i)=0$ and $u(t_i) = \sqrt{2}\,\vec{\alpha}_i$.

Upon performing all n integrations we find that

$$K(\vec{\alpha}_f, t_f; \vec{\alpha}_i, t_i) = \lim_{n\to\infty} \prod_{j=1}^{n} \frac{1}{(\det M_j')^{1/2}} \times \exp\left[-\tfrac{1}{2}[(\vec{\alpha}_i^*)\cdot\vec{\alpha}_i + (\vec{\alpha}_f^*)\cdot\vec{\alpha}_f] - z_f^T (M_{n+1}' - I) z_f + v_{n+1}^T z_f + \tfrac{1}{4}\sum_{l=1}^{n} v_l^T (M_l'^{-1}) v_l - i\epsilon[\vec{\alpha}_i^T f_i \vec{\alpha}_i + (\vec{\alpha}_f^*)^T f_f^* \vec{\alpha}_f^*]\right]. \tag{B17}$$

We now take the limit $n\to\infty$ and find that

$$\prod_{l=1}^{n} \frac{1}{(\det M')^{1/2}} \to \exp\left[-2i \int_{t_i}^{t_f} d\tau \operatorname{Tr}[X(\tau) f(\tau)]\right], \tag{B18}$$

$$-z_f^T (M_{n+1}' - I) z_f \to (\vec{\alpha}_f^*)^T X(t_f) \vec{\alpha}_f^*, \tag{B19}$$

$$v_{n+1}^T z_f \to \frac{1}{\sqrt{2}} (\vec{\alpha}_f^*)^T u(t_f), \tag{B20}$$

$$\frac{1}{4}\sum_{l=1}^{n} v_l^T(M_l'^{-1})v_l \to -\frac{1}{2}i\int_{t_i}^{t_f} d\tau\, u^T(\tau)f(\tau)u(\tau)\,. \tag{B21}$$

We can reexpress the terms involving $u(\tau)$ by defining a matrix

$$Y(t)=T\exp\left[-i\int_{t_i}^{t} d\tau[\omega(\tau)+4X(\tau)f(\tau)]\right] \tag{B22}$$

and noting that

$$u(t)=\sqrt{2}Y(t)\vec{\alpha}_i\,, \tag{B23}$$

so that $K(\vec{\alpha}_f,t_f;\vec{\alpha}_i,t_i)$ is given by the expression in Eq. (48).

One can check that this expression is correct by observing that $K(\vec{\alpha}_f,t_f;\vec{\alpha}_i,t_i)$ satisfies the equation

$$\begin{aligned} i\frac{\partial}{\partial t}K(\vec{\alpha},t;\vec{\beta},t_i) &= \langle\vec{\alpha}\,|\,H(t)U(t,t_i)\,|\,\vec{\beta}\rangle \\ &= \sum_{i=1}^{N}\sum_{j=1}^{N}\left[\omega_{ij}\alpha_i^*\left[\frac{\partial}{\partial\alpha_j^*}+\frac{1}{2}\alpha_j\right]+f_{ij}\left[\frac{\partial}{\partial\alpha_i^*}+\frac{1}{2}\alpha_i\right]\left[\frac{\partial}{\partial\alpha_j^*}+\frac{1}{2}\alpha_j\right]+f_{ij}^*\alpha_i^*\alpha_j^*\right]K(\vec{\alpha},t;\vec{\beta},t_i) \end{aligned} \tag{B24}$$

and verifying that, indeed, the expression given by Eq. (48) does satisfy this equation.

[1] R. Rajaraman, Phys. Rep. 21C, (1975).
[2] A. Zardecki, Phys. Rev. A 22, 1664 (1980).
[3] N. Bloembergen, *Nonlinear Optics* (Benjamin, New York, 1965).
[4] J. R. Klauder, Ann. Phys. (N.Y.) 11, 123 (1960).
[5] S. S. Schweber, J. Math. Phys. (N.Y.) 3, 831 (1962).
[6] J. R. Klauder, in *Path Integrals,* proceedings of the NATO Advanced Summer Institute, edited by G. J. Papadopoulos and J. T. Devreese (Plenum, New York, 1978), p. 5.
[7] M. Sargent III, M. O. Scully, and W. E. Lamb, Jr., *Laser Physics* (Addison-Wesley, Reading, Mass., 1974).
[8] Horace P. Yuen, Phys. Rev. A 13, 2226 (1976).
[9] M. Schubert and P. Wilhelmi, in *Progress in Optics,* edited by E. Wolf (North-Holland, Amsterdam, 1980), Vol. 17, p. 163; J. Perina, in *Progress in Optics,* edited by E. Wolf (North-Holland, Amsterdam, 1980), Vol. 18, p. 127.
[10] D. Stoler, Phys. Rev. Lett. 33, 1397 (1974).
[11] C. Caves, Phys. Rev. D 23, 1693 (1981).
[12] B. R. Mollow and R. J. Glauber, Phys. Rev. 160, 1076 (1967); 160, 1097 (1967).

Supplement of the Progress of Theoretical Physics, Nos. 74 & 75, 1983 209

Path Integral Approach to Many-Body Systems and Classical Quantization of Time-Dependent Mean Field

Hiroshi KURATSUJI and Tōru SUZUKI*

Department of Physics, Kyoto University, Kyoto 606
**Niels Bohr Institute, University of Copenhagen, Copenhagen ϕ*

(Received July 11, 1982)

We present the path integral formalism of many-body systems based on the representation of generalized coherent states. We give the path integral representation for the transition amplitude by utilizing the characteristic property of the coherent state and discuss its classical limit through the method of stationary phase. Applying to the bound state problem, we arrive at the classical quantization rule for the periodic mean field solutions, the utility of which is briefly discussed.

§ 1. Introduction

The time-dependent Hartree-Fock (TDHF) or mean field method has been known to provide with a basic formalism in the investigation of the large amplitude collective phenomena which is closely connected with the non-linear nature inherent to strongly interacting many-body systems. The framework of the mean field theory is, as is often recognized, essentially of classical nature, namely, the quantum nature originally inherent to many-body systems is smoothed by averaging process through the trial wave packet *labeled by the parameters of classical nature*. Hence the naive mean field theory does not give a prescription for extracting the quantum mechanical informations such as energy spectra of bound states for many-body systems. Thus it becomes a crucial problem to construct a proper quantum mechanical treatment of the time-dependent mean field so as to deduce the informations of physical importance.

In the quantum mechanical treatment of mean fields one encounters a serious difficulty of handling the non-linearity of the mean-field equations. For such a system that the non-linearity plays a dominant role, the conventional perturbative treatment does not work well and one is forced to rely upon an essentially non-perturbative approach. Path integral method is considered to provide a promising device for such a non-perturbative approach, the utility of which has been gradually recognized in nuclear many-body problems.[1)~8)]

In this report, we present the path integral formalism of many-body systems based on the representation of "generalized coherent state", which is a slightly extended version of the procedure recently developed by us.[4)~7)] The main consequence of this formalism is twofold: The first is that the time-dependent mean field theory is naturally obtained as a classical limit through the method of stationary phase. The second is that applying to the bound state problem, under the same spirit of the stationary phase, we arrive at the classical quantization rule for the periodic mean field solutions. This rule may be regarded as a many-body counterpart of the Bohr-Sommerfeld rule in old quantum mechanics and provide a useful device for evaluating the bound state spectra for many-body systems exhibiting non-linearity which cannot be treated within a perturbative approach.

An alternative path integral approach to many-body systems has recently been developed based on the functional integral over mean field.[1)~3)] The essential point of this formalism is to utilize the auxiliary field σ through the trick of the Gaussian quadratic completion for fermion pairs. On the other hand, the coherent-state path integral presented here would, without recourse to the concept of auxiliary field, enable us to *directly* describe the quantum mechanical time evolution of mean field in terms of the integration over paths in the parameter space labeling the coherent state.

§ 2. Basis Hilbert space

In order to achieve the quantum dynamical description of mean field, we first need set a basis Hilbert space of many-body system. The crucial property to be required for this Hilbert space is "continuous nature"; the "continuous nature" implies that the Hilbert space can be labeled by the point of a manifold possessing with a certain continuous property, thereby we can safely define the paths over which path integration should be performed. Such a Hilbert space can be indeed realized as a generalized coherent state (GCS), which is most readily constructed on the basis of the representation theory of Lie groups.[10)] Let $T(g)$ be the irreducible unitary representation of the Lie group G inherent to the many-body system which is concretely given by the exponential map of the Lie algebra of fermion pairs (typical example will be given below). We consider the set $\{|g\rangle = T(g)|\Phi_0\rangle\}$ where $|\Phi_0\rangle$ is a fixed state vector in the many-body Hilbert space. $|\Phi_0\rangle$ is assumed to be invariant up to a phase under the action of the so-called "stationary subgroup" H; $T(h)|\Phi_0\rangle = e^{i\alpha}|\Phi_0\rangle$. By this invariance, we get $T(gh)|\Phi_0\rangle = T(g)T(h)|\Phi_0\rangle = e^{i\alpha}T(g)|\Phi_0\rangle$. This relation, using the language of differential geometry, shows that the set $\{|g\rangle\}$ is parametrized by the point of the factor space (or quotient space) G/H. In many cases we encounter in many-body theory, the factor space has a structure of complex manifold and the point can be coordinated by the complex vector $Z = (z_1, \cdots, z_n)$. Hence we define the GCS

$$|Z\rangle = (\text{phase}) \times |g\rangle . \tag{2·1}$$

The characteristic property of the GCS thus defined is the "partition of unity"

$$\int |Z\rangle d\mu(Z) \langle Z| = 1 \tag{2·2}$$

with

$$d\mu(Z) = \text{const } \det(g_{i\bar{j}}) \prod_{i=1}^{n} d \operatorname{Re} z_i d \operatorname{Im} z_i , \tag{2·3}$$

which is the invariant measure induced by the metric tensor $g_{i\bar{j}}$ of G/H. The relation (2·2) can be verified by the use of the transformation property of $|Z\rangle$ under the action of G and Schur's lemma characterizing the irreducibility of the representation $T(g)$. We consider the overlap kernel between a couple of GCS; $\langle Z'|Z\rangle$, which is shown to be bounded as $|\langle Z'|Z\rangle| \leqq 1$ where the equality holds only for $Z'=Z$ and this just reflects the "overcompleteness" of the GCS. The overlap kernel is alternatively written by using the unnormalized Bergman kernel $F(Z, Z'^*)$:

$$\langle Z'|Z\rangle = F(Z, Z'^*) [F(Z, Z^*) F(Z', Z'^*)]^{-1/2} . \tag{2·4}$$

$F(Z, Z'^*)$ can be defined by the overlap of the unnormalized GCS $|\tilde{Z}\rangle$:

$$F(Z, Z'^*) = \langle \tilde{Z}'|\tilde{Z}\rangle . \tag{2·5}$$

The above defined GCS is applicable to a wider class of the many-body systems possessing with the Lie group structure. Here, as a specific example, the space of determinantal state can be realized as the coherent state for the unitary group, which is just relevant for the description of the TDHF. Namely, in this case, the group G becomes $U(m+n)$ which is just the transformation group acting on the fermion operators (a_μ, a_i) with $\mu=1\sim m$, $i=1\sim n$ denoting the unoccupied and occupied states with respect to the initial determinantal state $|\Phi_0\rangle = \prod_{i=1}^{n} a_i^\dagger|0\rangle$ and the stationary subgroup becomes $H = U(m) \times U(n)$ which leaves $|\Phi_0\rangle$ invariant. The GCS is thus reduced to the Thouless form:

$$|Z\rangle = [\det(I^{(m)} + ZZ^\dagger)]^{-1/2} \exp[\sum_{\mu i} Z_{\mu i} a_\mu^\dagger a_i] |\Phi_0\rangle. \tag{2·6}$$

The label space becomes the factor space $U(m+n)/U(m) \times U(n)$ ($\simeq$complex Grassmann manifold) the point of which is coordinated by an $m \times n$ complex matrix and the invariant measure is given by

$$d\mu(Z) = \text{const}[\det(I^{(m)} + ZZ^\dagger)]^{-(m+n)} \prod_{\mu i} d \operatorname{Re} Z_{\mu i} d \operatorname{Im} Z_{\mu i} . \tag{2·7}$$

The overlap kernel is calculated as

$$\langle Z'|Z\rangle = [\det(I^{(m)}+ZZ^{\dagger})\det(I^{(m)}+Z'Z'^{\dagger})]^{-1/2}$$
$$\times\det(I^{(m)}+ZZ'^{\dagger}). \tag{2·8}$$

§ 3. The path integral representation of the transition amplitude and its classical limit

We now consider the quantum dynamical time evolution of the many-body system in the Hilbert space spanned by the GCS. This is described by the transition amplitude (or propagator) joining the end points Z' and Z'' in the parameter space of the GCS,

$$K(Z''t''|Z't') = \langle Z''|\exp[-i\hat{H}(t''-t')/\hbar]|Z'\rangle, \tag{3·1}$$

where the Hamiltonian $\hat{H}$ is given by the algebraic function of the generators of the Lie group under consideration. As is usually performed, we divide the time interval into N equal segments $\varepsilon=(t''-t')/N$ with $N\to\infty$ and insert the partition of unity (2·2) at each division point, then the propagator is written as the continual integral form

$$K(Z''t''|Z't') = \lim_{N\to\infty}\int\cdots\int\prod_{k=1}^{N-1}d\mu(Z_k)$$
$$\times\prod_{k=1}^{N}\langle Z_k|\exp[-i\hat{H}\varepsilon/\hbar]|Z_{k-1}\rangle\,. \tag{3·2}$$

Each factor in the integrand is approximated as

$$\langle Z_k|\exp[-i\hat{H}\varepsilon/\hbar]|Z_{k-1}\rangle \simeq \langle Z_k|\left(1-\frac{i\varepsilon}{\hbar}\hat{H}\right)|Z_{k-1}\rangle$$
$$= \langle Z_k|Z_{k-1}\rangle\left(1-\frac{i}{\hbar}\varepsilon\langle Z_k|\hat{H}|Z_{k-1}\rangle/\langle Z_k|Z_{k-1}\rangle\right)$$
$$\simeq \langle Z_k|Z_{k-1}\rangle\exp\left[-\frac{i}{\hbar}\varepsilon\langle Z_k|\hat{H}|Z_{k-1}\rangle/\langle Z_k|Z_{k-1}\rangle\right] \tag{3·3}$$

and further using the identity $\langle Z_k|Z_{k-1}\rangle=\exp[\log\langle Z_k|Z_{k-1}\rangle]$, we get the expression

$$K(Z''t''|Z't') = \lim_{N\to\infty}\int\cdots\int\prod_{k=1}^{N-1}d\mu(Z_k)$$
$$\times\exp\left[\frac{i}{\hbar}\varepsilon\left(\sum_{k=1}^{N}-\frac{i\hbar}{\varepsilon}\log\langle Z_k|Z_{k-1}\rangle-\frac{\langle Z_k|\hat{H}|Z_{k-1}\rangle}{\langle Z_k|Z_{k-1}\rangle}\right)\right]$$
$$\Rightarrow\lim_{N\to\infty}\int\cdots\int\prod_{k=1}^{N-1}d\mu(Z_k)$$
$$\times\exp\left[\frac{i}{\hbar}\sum_{k=1}^{N}\varepsilon\left(\frac{i\hbar}{\varepsilon}\langle Z_k|\Delta Z_k\rangle-H(Z_k,\,Z_{k-1}^{*})\right)\right] \tag{3·4}$$

with $|\Delta Z_k\rangle \equiv |Z_k\rangle - |Z_{k-1}\rangle$. Thus the propagator is cast into the formal path integral

$$K(Z''t''|Z't') = \int \exp[iS/\hbar]\,\mathcal{D}\mu[Z(t)], \tag{3·5}$$

where the action functional is given as

$$S[Z(t)] = \int_{t'}^{t''} \langle Z(t)|i\hbar\,\partial/\partial t - \hat{H}|Z(t)\rangle dt \equiv \int_{t'}^{t''} L dt\,. \tag{3·6}$$

Using the (unnormalized) kernel function $F(Z, Z^*)$, the Lagrangian is calculated as

$$L = \frac{i\hbar}{2}\sum_{k=1}^{n}\left(\frac{\partial \log F}{\partial z_i}\dot{z}_i - \frac{\partial \log F}{\partial z_i^*}\dot{z}_i^*\right) - H(Z, Z^*). \tag{3·7}$$

With the aid of the representation of (3·5), we can also get the path integral representation for the transition amplitude connecting some initial and final states $|\Phi_i\rangle$ and $|\Phi_f\rangle$

$$\begin{aligned} K_{fi}(t'', t') &= \langle \Phi_f|\exp[-i\hat{H}(t''-t')/\hbar]|\Phi_i\rangle \\ &= \iint d\mu(Z'')\,d\mu(Z')\,\Phi_f^*(Z'')\,\Phi_i(Z')\,K(Z''t''|Z't'), \end{aligned} \tag{3·8}$$

where $\Phi_i(Z)$ ($\Phi_f(Z)$) may represent the probability amplitude for the initial (final) state to be in a state with the argument Z.

Now we examine the classical approximation of the propagator (3·5) in which Planck's quantum can be regarded to be extremely small in comparison with the value of the action function. The application of the method of stationary phase yields the classical propagator.

$$K^{\rm cl} \sim \sum \exp[iS^{\rm cl}/\hbar]. \tag{3·9}$$

The classical paths giving the extreme value $S^{\rm cl}$ obey the variational equation

$$\delta S = \delta \int_{t'}^{t''} \langle Z(t)|i\hbar\,\partial/\partial t - \hat{H}|Z(t)\rangle dt = 0\,. \tag{3·10}$$

If we impose the fixed end point conditions $Z(t') = Z'$, $Z(t'') = Z''$ together with the complex conjugate, Eq. (3·10) is just the time-dependent variational equation leading to the time-dependent mean field equation. In this way, the present path integral formalism naturally involves the time-dependent mean field theory as a classical limit.

With the aid of (3·7), the variational equation yields the Euler-Lagrange equations

$$\frac{d}{dt}\left(\frac{\partial L}{\partial \dot{Z}}\right) - \frac{\partial L}{\partial Z} = 0, \quad \frac{d}{dt}\left(\frac{\partial L}{\partial \dot{Z}^*}\right) - \frac{\partial L}{\partial Z^*} = 0\,, \tag{3·11}$$

which are brought to the equations of motion

$$
\begin{aligned}
i\hbar \sum_j g_{ij}\dot{z}_j &= \frac{\partial H}{\partial z_i{}^*}, \\
-i\hbar \sum_j g_{ij}\dot{z}_j{}^* &= \frac{\partial H}{\partial z_i}
\end{aligned} \tag{3·12}
$$

with

$$g_{ij} = \partial^2 \log F/\partial Z_i \partial Z_j{}^* , \tag{3·13}$$

which defines the metric of the manifold G/H. Equations of motion (3·12) can be regarded as the canonical equations in the "generalized phase space" G/H and the canonical nature has been discussed elsewhere.[4),7)]

In the above argument of the classical limit we notice that there is a problem of the overspecification of the boundary conditions. Namely, equations of motion, which are of first order, are not compatible with two complex end point condition. This suggests that if one chooses *arbitrary* fixed end points, there do not in general exist the classical paths joining these points and the corresponding classical propagator vanishes. The classical propagator takes non-vanishing value only if the classical path starting with the initial point Z' passes through the final point Z''. This feature may be alternatively stated by using the transition amplitude (3·8), namely, the classical approximation of (3·8) is given by

$$K_{fi}^{\mathrm{cl}} = \iint d\mu(Z'')d\mu(Z')\Phi_f{}^*(Z'')\Phi_i(Z')\exp[iS^{\mathrm{cl}}/\hbar]$$

and this amplitude always takes the non-vanishing value by virtue of the fact that the integration over the initial and final points Z' and Z'' guarantees the existence of at least one classical orbit which actually satisfies the equations of motion.

As a specific case of the general formalism, we here consider the time-dependent Hartree-Fock. For this case, we take as the basis space the determinantal state realized as the coherent-state for the unitary group (§ 2). By using the kernel function (2·8), the Lagrangian is given as

$$L = \frac{i\hbar}{2}\mathrm{Tr}[(I^{(m)} + ZZ^\dagger)^{-1}(Z^\dagger\dot{Z} - \dot{Z}^\dagger Z)] - H(Z, Z^\dagger), \tag{3·14}$$

where we use the formula det $A = \exp[\mathrm{Tr}\log A]$. Thus, the canonical equations (3·12) turn out to be

$$\dot{Z} = -\frac{i}{\hbar}(I^{(m)} + ZZ^\dagger)\frac{\partial H}{\partial Z^\dagger}(I^{(n)} + Z^\dagger Z),$$

$$\dot{Z}^\dagger = \frac{i}{\hbar}(I^{(n)} + Z^\dagger Z)\frac{\partial H}{\partial Z}(I^{(m)} + ZZ^\dagger), \tag{3·15}$$

where the derivative with respect to matrices Z, $Z^\dagger$ are defined as

$$\left(\frac{\partial H}{\partial Z}\right)_{\mu i} = \frac{\partial H}{\partial Z_{\mu i}}, \quad \left(\frac{\partial H}{\partial Z^\dagger}\right)_{i\mu} = \frac{\partial H}{\partial Z^*_{\mu i}}.$$

As is seen from (3·15), the use of the complex variable Z seems rather complicated due to the non-linearity of the manifold and it is desirable to write in a more tractable form. This can be achieved by using the orbital functions constituting determinantal state. Namely, expressing the Thouless form in the coordinate representation

$$\langle x_1 \cdots x_n | Z \rangle = \langle x_1 \cdots x_n | \prod_{k=1}^{n} c_k^\dagger |0\rangle = \det\{\phi_k(x_j)\} \tag{3·16}$$

with ϕ_k's are assumed to be orthonormalized, and utilizing this for Eq. (3·6), we get the action functional

$$S = \int dt \left[\int \frac{i\hbar}{2} \sum_{k=1}^{n} (\phi_k^* \dot{\phi}_k - \text{c.c.})\, dx - H(\{\phi_k\}, \{\phi_k^*\}) \right] \tag{3·17}$$

with the Hamiltonian (which consists of the kinetic and two-body interaction) being

$$H(\{\phi_k\}, \{\phi_k^*\}) = \frac{\hbar^2}{2m} \int \sum_{k=1}^{n} \nabla\phi_k^* \nabla\phi_k dx + 1/2 \int\int \sum_{kl} \{|\phi_k(x)|^2 \times V(x, y)|\phi_l(y)|^2 - \text{exchange}\}\, dxdy\,. \tag{3·18}$$

The variational equation (3·10) thus yields

$$i\hbar\dot{\phi}_k = \frac{\delta H}{\delta \phi_k^*}, \quad i\hbar\dot{\phi}_k^* = -\frac{\delta H}{\delta \phi_k}, \tag{3·19}$$

which is just brought to the familiar TDHF equations in a form of the coupled non-linear Schrödinger like equations,

$$i\hbar\dot{\phi}_k = -\frac{\hbar^2}{2m}\nabla^2\phi_k + \int \sum_j |\phi_j(y)|^2 V(x, y)\phi_k(x)\, dy - \text{exchange} \tag{3·20}$$

together with the complex conjugate. Here we note that the above derivation of TDHF is similar with the method of functional integral over mean field[1] under the spirit of stationary phase. However the use of the auxiliary field σ in Ref. 1) enevitably leads to the time-dependent Hartree (TDH) without an exchange term and the exchange term is recovered as a correction coming from the next order quantum fluctuation. Contrary to this, in our coherent-state formalism, the exchange term is naturally included even though one is restricted to the lowest order stationary phase.

§ 4. Classical quantization condition for mean field solutions

We apply the general scheme in the previous section to the bound state problem, thereby we deduce the classical quantization rule for the periodic time-dependent mean field.[6a)] For this purpose we extend the idea of the quantized periodic orbit theory which was first suggested by Gutzwiller.[11)] We start with the Green function defined by

$$K(E) = i\int_0^{\infty} \exp[iET/\hbar]\mathrm{Tr}[\exp(-i\hat{H}T/\hbar)]dT$$
$$= \mathrm{Tr}\left(\frac{1}{E-\hat{H}}\right), \qquad (4\cdot 1)^{*)}$$

the pole position of which just gives bound state spectra. The trace of the evolution operator is expressed through the coherent-state representation

$$\mathrm{Tr}[\exp(-i\hat{H}T/\hbar)] \equiv K(T) = \int d\mu(Z^0)\langle Z^0|\exp(-i\hat{H}T/\hbar)|Z^0\rangle . \qquad (4\cdot 2)$$

According to the recipe of the previous section, this is cast into the path integral form

$$K(T) = \int d\mu(Z^0)\int \mathcal{D}\mu[Z(t)]\exp[iS/\hbar]. \qquad (4\cdot 3)$$

The trace operation indicates that all the paths contributing to the path integral are closed, namely, the initial point coincides with the final point to giving $Z^0 = (X^0, Y^0)$ (X^0 and Y^0 denote the real and imaginary parts of Z^0).

We now examine the classical approximation of the Green function (4·1). Hereafter we restrict the argument to the lowest order stationary phase. To take account of the quantum fluctuation of subsequent order may be straightforward,[12)] though complicated. The procedure is carried out by the following steps: (i) As a consequence of the stationary phase, $K(T)$ is dominated by the contribution from the classical orbits,

$$K^{\mathrm{cl}}(T) \sim \sum_{\mathrm{p.o}} \int d\mu(Z^0)\exp[iS^{\mathrm{cl}}(T)/\hbar]. \qquad (4\cdot 4)$$

Here $\sum_{\mathrm{p.o}}$ indicates a contribution from several classical orbits which are assumed to be far enough apart from each other in order to avoid the inter-

---

*) Here we assume that E has a small positive imaginary part so as to ensure the convergence of integral.

ference between them. The classical orbits becomes *periodic* with the period T; $Z_{cl}(t+T)=Z_{cl}(t)$. This feature is a natural consequence of the trace operation and in this way we can safely avoid the overspecification of the boundary condition. The integral over the end point in (4·4) can be further reduced to the form

$$K^{cl}(T)\sim\sum_{p.o}\exp[iS^{cl}(T)/\hbar]\oint d\mu(Z^0), \tag{4·5}$$

where the last integral is performed along the periodic orbits and depends only on the geometric structure of the orbits. This separated form comes from the fact that the integral takes a value only on the periodic orbits as a consequence of the stationary phase and the classical action (and hence the integrand) does not depend on the end point Z^0.

(ii) Next, substituting (4·5) into (4·1) and evaluating the integration over T by the lowest order stationary phase, we get

$$K^{cl}(E)\sim\sum_{p.o}\exp[iW(E)/\hbar]\oint d\mu(Z^0). \tag{4·6}$$

Here $W(E)$ is the action integral

$$W(E)=S^{cl}(T(E))+ET(E) \tag{4·7}$$

and the period $T(E)$ is given by the solution of the extreme equation

$$\frac{\partial}{\partial T}(S^{cl}(T)+ET)=\frac{\partial S^{cl}}{\partial T}+E=0, \tag{4·8}$$

which reduces to the energy surface $H(Z,Z^*)=E$. Thus, $W(E)$ becomes the following form, in which the integral is taken along the periodic orbits on this energy surface,

$$W(E)=\oint_{H=E}\langle Z(t)|i\hbar\,\partial/\partial t|Z(t)\rangle dt. \tag{4·9}$$

(iii) Finally, the correct form of the Green function is obtained by taking account of the contribution from the *multiple traversals*[11] of the basic orbits with the basic period $T(E)$. That is, nothing that the $m\,(m=1,2,\cdots)$ times classical action mS^{cl} satisfies (4·8) for the m-times period $mT(E)$, we put $mW(E)$ for the m-times traversals and sum up over m, then $K^{cl}(E)$ turns out to be the geometric series

$$\begin{aligned}K^{cl}(E)&\propto\sum_{p.o}\sum_{m=1}^{\infty}\exp[imW(E)/\hbar]\\&=\sum_{p.o}\exp[iW(E)/\hbar]\{1-\exp[iW(E)/\hbar]\}^{-1}.\end{aligned} \tag{4·10}$$

From the pole position of (4·10) we get the rule selecting the quantized orbits

$$\oint_0^{T(E)} \langle Z(t)|i\hbar\,\partial/\partial t|Z(t)\rangle dt = 2n\pi\hbar, \quad n=\text{integers}. \tag{4·11}$$

This equation is just the Bohr-Sommerfeld like condition for the time-dependent mean field described by the general class of coherent states. Note that the condition (4·11) consists in general of several equations corresponding to the existence of the several quantizable orbits for a given energy which are mutually independent. This is due to the fact that the classical Green function consists of the sum over several independent periodic orbits. The quantization rule (4·11) is of a similar form to the one derived within the alternative methods; the functional integral over mean field[1] and the gauge invariant periodic quantization by Kan and Griffin et al.[13]

By using the kernel function $F(Z, Z^*)$, the formula (4·11) can be written as

$$\oint \frac{i\hbar}{2}\sum_{i=1}^{n}\left(\frac{\partial \log F}{\partial z_i}dz_i - \text{c.c.}\right) \equiv \oint \omega = 2n\pi\hbar \tag{4·12}$$

which is also converted to the surface integral by using the Stokes theorem $\oint \omega = \int d\omega$,

$$\int \Omega \equiv \int \sum_{ij} g_{i\bar{j}}\, dz_i \wedge dz_j^* = 2n\pi\hbar\,, \tag{4·13}$$

where Ω is the differential 2-form associated with the metric $ds^2 = \sum g_{i\bar{j}} dz_i dz_j^*$ and the integration is performed over the surface enclosed by the closed orbits.

Quantization of TDHF solutions

We can derive the quantization rule for the periodic TDHF orbits as a particular case of (4·11) and/or (4·12). If we use the Thouless form (2·6), we get

$$\oint \frac{i\hbar}{2}\mathrm{Tr}[(I^{(m)} + ZZ^\dagger)^{-1}(Z^\dagger\cdot dZ - dZ^\dagger\cdot Z)] = 2n\pi\hbar\,. \tag{4·14}$$

This form, which is rather complicated due to the non-linearity of the manifold, is converted to the familiar form by using the coordinate representation (3·16),

$$\frac{i\hbar}{2}\int_0^T dt \int \sum_{k=1}^{n}(\phi_k^*\dot{\phi}_k - \text{c.c.})\,dx = 2n\pi\hbar\,, \tag{4·15}$$

where ϕ_k's are of course the solutions of the non-linear TDHF equations (3·20) and obey the periodic conditions $\phi_k(x, t+T) = \phi_k(x, t)$ through the periodicity of $Z(t)$. Thus, (4·15) can be naively regarded as a quantization rule for the coupled "classical fields".

Outlook

We give a few remarks on the utility of the classical quantization rule to specific problems. One class of the problems is the algebraic model such as pseudo-spin[14] and its generalization.[15),16] For the $SU(2)$ pseudo-spin system, which is described by the $SU(2)$ coherent-state and parametrized by the factor space $SU(2)/U(1) \simeq$ Bloch sphere, the formula (4·13) yields

$$\frac{iN\hbar}{2}\oint\frac{z^*dz - \text{c.c.}}{(1+|z|^2)} = Ni\int\frac{dz\wedge dz^*}{(1+|z|^2)^2} = 2n\pi\hbar \qquad (4\cdot 16)$$

with N being the magnitude of pseudo-spin or using the stereo-graphic projection $z = \tan\theta/2\cdot e^{-i\phi}$,

$$\frac{N}{2}\int\sin\theta d\theta d\phi = 2n\pi\hbar \qquad (4\cdot 17)$$

the integral of the l.h.s becomes the area encircled by the closed orbit. Thus, finding out the closed orbit in the parameter space, (4·17) gives the bound state spectra of pseudo-spin system. To investigate the algebraic model may deserve a theoretical merit, since the orbits exhibit in general a peculiar feature such as the structure change, e.g., the appearance of the "separatrix"[14),16] and it would be an interesting question how the energy spectra corresponding to such a specific orbit behave.

The other class of problems is to treat the collective motion, which is closely related to the current interest in the investigation of large amplitude collective motion where it is a crucial problem to extract the collective degrees of freedom from the mean field and how to quantize it.[17] As an analytic solvable system, we have examined the one-dimensional boson system interacting with the attractive delta-function interaction:[18] For this system, the mean field equation is given by a non-linear Schrödinger equation for a single classical field ϕ and admits the exact solution of "soliton type". It has proved that through the specific feature of the soliton solution the classical quantization rule exactly yields the kinetic energy of center of mass motion as well as the internal spectrum which reproduces the exact quantum mechanical spectrum in good accuracy. The idea used in this simplified system may be promising for the TDHF, for which it may be useful to start with the formula (4·15) written in terms of the classical fields. In applying this, there is a serious difficulty of finding the solutions of the non-linear TDHF equations (3·20) in an analytic way. However, even if the analytic solution is not feasible, the quantization method based on (4·15) may be useful if one can construct the solution such that structure of collective motion is involved in a parametric form.

Acknowledgements

One of the authors (H.K.) would like to thank the organizing committe of the Kyoto Summer Institute 1982 for giving him an opportunity to talk about the authors' recent works.

References

1) S. Levit, J. W. Negele and Z. Paltiel, Phys. Rev. **21C** (1980), 1603.
2) H. Kleinert, Phys. Letters **69B** (1977), 9.
3) H. Reinhardt, Nucl. Phys. **A346** (1980), 1.
4) H. Kuratsuji and T. Suzuki, Phys. Letters **92B** (1980), 19.
5) H. Kuratsuji, Prog. Theor. Phys. **65** (1981), 224.
6a) H. Kuratsuji, Phys. Letters **103B** (1981), 79.
b) H. Kuratsuji, Phys. Letters **108B** (1982), 367.
7) T. Suzuki, Niels Bohr Institute Preprint, to be published.
8) J. P. Blaizot and H. Orland, Phys. Letters **100B** (1981), 195.
9) J. R. Klauder, Phys. Rev. **D19** (1979), 2349.
H. Kuratsuji and T. Suzuki, J. Math. Phys. **21** (1980), 472.
10) J. R. Klauder, J. Math. Phys. **4** (1963), 1055, 1058.
A. M. Perelomov, Comm. Math. Phys. **26** (1972), 222.
R. Gilmore, Ann. of Phys. **74** (1972), 391.
11) M. C. Gutzwiller, J. Math. Phys. **12** (1971), 343.
12) For example, H. Kuratsuji and Y. Mizobuchi, Phys. Letters **82A** (1981), 279; J. Math. Phys. **22** (1981), 757.
13) K. K. Kan, J. J. Griffin, P. C. Lichtner and M. Dworzecka, Nucl. Phys. **A332** (1979), 109.
14) D. H. Feng and R. Gilmore, in *Interacting Bose-Fermi System*, ed. F. Iachello (Plenum, N. Y., 1981).
R. Shankar, Phys. Rev. Letters **45** (1980), 1088.
15) O. S. van Roosmalen and A. E. L. Dieperink, Ann. of Phys. **139** (1982), 198.
16) Y. Mizobuchi, Prog. Theor. Phys. **65** (1981), 1450.
T, Suzuki, Talk presented at the symposium on "TDHF and Beyond" held at Bad Honnef, June, 1982, to be published in *Lecture Notes in Physics*.
17) See, e.g., K. Goeke and P. G. Reinhard, Ann. of Phys. **112** (1978), 328.
18) H. Kuratsuji, Kyoto Univ. Preprint KUNS 606.

PHYSICAL REVIEW LETTERS

VOLUME 52 | 2 APRIL 1984 | NUMBER 14

Quantum Mechanical Path Integrals with Wiener Measures for all Polynomial Hamiltonians

John R. Klauder
AT&T Bell Laboratories, Murray Hill, New Jersey 07974

and

Ingrid Daubechies
Theoretische Natuurkunde, Vrije Universiteit Brussel, B-1050 Brussels, Belgium

(Received 8 December 1983)

We construct arbitrary matrix elements of the quantum evolution operator for a wide class of self-adjoint canonical Hamiltonians, including those which are polynomial in the Heisenberg operators, as the limit of well defined path integrals involving Wiener measure on phase space, as the diffusion constant diverges. A related construction achieves a similar result for an arbitrary spin Hamiltonian.

PACS numbers: 03.65.Ca

Path integrals for evolution operators of quantum mechanical systems are almost always defined as the limits of expressions involving finitely many integrals.[1] Efforts to define them as integrals involving genuine measures on path spaces of continuous paths, or as limits of such integrals, have been largely unrewarding.[2] In our earlier work on this subject we have succeeded in establishing quantum mechanical path integrals with genuine measures for the limited class of quadratic Hamiltonians.[3]

In this paper, taking an alternative but closely related approach, we succeed in constructing arbitrary matrix elements of the quantum evolution operator as limits of well defined path integrals involving Wiener measure on phase space as the diffusion constant diverges. Our construction works for any self-adjoint Hamiltonian of a wide but special class (defined below) which includes all Hamiltonians polynomial in the canonical (Heisenberg) operators. A similar construction leads to an analogous description of arbitrary matrix elements of the quantum evolution operator for an arbitrary Hamiltonian composed of spin operators for any fixed spin $s > 0$, extending earlier work.[4] As above, these matrix elements are defined as limits of well defined path integrals involving Wiener measure defined here on the unit sphere, again as the diffusion constant diverges. Finally we comment on how the spin path-integral expression passes to the canonical one as $s \to \infty$. We content ourselves here with a statement of our principal results, reserving a precise formulation and detailed proofs to a separate article.[5] For clarity we confine our discussion to a single degree of freedom. The notation is that of our earlier papers.

Canonical case.— For all $(p,q) \in R^2$ let

$$|p,q\rangle = \exp[i(pQ - qP)]|0\rangle, \quad [Q,P] = i,$$

denote the canonical coherent states, where $|0\rangle$ is a normalized vector that satisfies $(Q + iP)|0\rangle = 0$. It follows[6] that a general operator H can be expressed as

$$H = \int h(p,q)|p,q\rangle\langle p,q|\, dp\, dq/2\pi,$$

where

$$h(p,q) = \exp[-(\partial^2/\partial p^2 + \partial^2/\partial q^2)/2]\langle p,q|H|p,q\rangle.$$

For H the unit operator I, this expression yields $h = 1$; for H a polynomial in P and Q, it follows that h is a polynomial in p and q. The special class of Hamiltonians we are able to discuss includes all those for which $h(p,q)$ is polynomially bounded. We suppose hereafter that H denotes the self-adjoint Hamiltonian of interest, and thus that h is real.

Now we introduce additional canonical coherent states

$$|p,q,n\rangle = \exp[i(pQ-qP)]|n\rangle,\quad n=0,1,2,\ldots,$$

where $|n\rangle$ denotes the normalized nth excited state of a harmonic oscillator of unit frequency; clearly $|p,q,0\rangle = |p,q\rangle$. For any β with $|\beta|<1$ it follows that[7]

$$|p,q\rangle\rangle = \sum_{n=0}^{\infty} \oplus\, \beta^{n/2}|p,q,n\rangle$$

defines a vector in an associated direct-sum Hilbert space $\mathscr{H}$. We define four operators in $\mathscr{H}$ as follows:

$$E_\beta = \int |p,q\rangle\rangle\langle\langle p,q|\,dp\,dq/2\pi = \sum_{n=0}^{\infty}\oplus\,|\beta|^n I_n,\quad A=\sum_{n=0}^{\infty}\oplus\, nI_n,\quad B_\beta=\int h(p,q)|p,q\rangle\rangle\langle\langle p,q|\,dp\,dq/2\pi,$$

$$C_\beta = \int \exp[-i\epsilon h(p,q)]|p,q\rangle\rangle\langle\langle p,q|\,dp\,dq/2\pi.$$

In arriving at the second form for E_β we have used the basic fact that

$$\int |p,q,n\rangle\langle p,q,m|\,dp\,dq/2\pi = \langle m|n\rangle I_n = \delta_{mn}I_n,$$

where I_n denotes the unit operator in the nth direct-sum subspace. Observe that E_1 is the identity operator in $\mathscr{H}$, while $P=E_0$ is the projection onto the 0th subspace $\mathscr{H}$. A real h implies that B_β is a symmetric operator in $\mathscr{H}$.

Now choose the following parameters: N is a variable positive integer, T a fixed positive time interval, $\epsilon = T/(N+1)$, and $\beta=(1-\epsilon\nu/2)/(1+\epsilon\nu/2)$, ν fixed and positive (ν is the diffusion constant, as will become apparent). With these identifications, and for polynomially bounded h, we are able to prove[5,8] ("s" means strong) *Lemma 1:*

$$\operatorname*{s-lim}_{N\to\infty} C_\beta^N = \exp(-\nu TA - iTB_1). \tag{1}$$

Furthermore, under the same conditions, we can prove *Lemma 2:*

$$\operatorname*{s-lim}_{\nu\to\infty} \exp(-\nu TA - iTB_1) = P\exp(-iTPB_1P)P. \tag{2}$$

Observe that PB_1P restricted to the 0th subspace is just the self-adjoint Hamiltonian H. Consequently, for any $|\phi\rangle$, $|\psi\rangle \in \mathscr{H}$, the matrix element of the evolution operator is given by

$$\langle\phi|e^{-iTH}|\psi\rangle = \lim_{\nu\to\infty}\lim_{N\to\infty}\langle\langle\phi|C_\beta^N|\psi\rangle\rangle, \tag{3}$$

where $|\phi\rangle\rangle$, $|\psi\rangle\rangle \in \mathscr{H}$ are vectors with 0th entry $|\phi\rangle$, $|\psi\rangle$, respectively, and zero in all remaining entries, $n>0$.

We now proceed to give a path-integral expression for (3). With the parameters chosen as above it follows that

$$\langle\langle p'',q''|C_\beta^N|p',q'\rangle\rangle = \int\cdots\int \prod_{l=0}^{N}\langle\langle p_{l+1},q_{l+1}|p_l,q_l\rangle\rangle \prod_{l=1}^{N}\{\exp[-i\epsilon h(p_l,q_l)]dp_l\,dq_l/2\pi\},$$

where $(p',q')=(p_0,q_0)$ and $(p'',q'')=(p_{N+1},q_{N+1})$. We observe that[3b]

$$\langle\langle p_2,q_2|p_1,q_1\rangle\rangle = \frac{1+\epsilon\nu/2}{\epsilon\nu}\exp\left\{\frac{i}{2}(p_1q_2-q_1p_2) - \frac{1}{2\epsilon\nu}[(p_2-p_1)^2+(q_2-q_1)^2]\right\}.$$

The form of this expression makes evident the result of the limit $N\to\infty$ ($\epsilon\to0$) as

$$\lim_{N\to\infty}\langle\langle p'',q''|C_\beta^N|p',q'\rangle\rangle = 2\pi e^{\nu T/2}\int \exp\{i\int_0^T[\tfrac{1}{2}(p\dot q - q\dot p) - h(p,q)]dt\}\,d\mu_W^\nu(p,q),$$

where μ_W^ν is a product of two pinned Wiener measures concentrated on continuous paths with a normalized connected covariance given for $t_1 \le t_2$ by $\langle x(t_1)x(t_2)\rangle^c = \nu t_1(1-t_2/T)$ for $x=p$ or q; here the role of ν as diffusion constant is apparent. Note that $\int(p\dot q - q\dot p)dt$ interpreted as $\int(p\,dq - q\,dp)$ involves well-defined stochastic integrals in any[9] (Itô or Stratonovich) sense. Finally, we obtain the following *Theorem:*

$$\langle\phi|e^{-iTH}|\psi\rangle = \lim_{\nu\to\infty}\int\langle\phi|p'',q''\rangle[2\pi e^{\nu T/2}\int\exp(iS)\,d\mu_W^\nu]\langle p',q'|\psi\rangle\left(\frac{dp''\,dq''}{2\pi}\right)\left(\frac{dp'\,dq'}{2\pi}\right).$$

where S denotes the "classical action,"

$$S=\int_0^T[\tfrac{1}{2}(p\dot q-q\dot p)-h(p,q)]dt.$$

This result achieves our stated goal of a path-integral representation.

It is important to compare our results here with those obtained earlier. In Ref. 3 we were able to find an expression for quadratic Hamiltonians which involves the function $H(p,q)=\langle p,q|H|p,q\rangle$ as the Hamiltonian function in S (rather than h), at the expense of introducing Wiener measures with nonvanishing drift terms determined by the usual Hamilton equations of motion.[3b] Thus, at least for the limited set of Hamiltonians in common, we gain a clearer understanding of the true underlying distinction between two formal but otherwise identical path-integral expressions given by[10]

$$\mathcal{N}\int\exp(iS)\prod_t dp(t)\,dq(t),$$

where $S=\int[\frac{1}{2}(p\dot q-q\dot p)-h(p,q)]dt$ or $S=\int[\frac{1}{2}(p\dot q-q\dot p)-H(p,q)]dt$. If $\hbar\to0$ recall that $h(p,q)\to H_c(p,q)$, $H(p,q)\to H_c(p,q)$, where $H_c(p,q)$ denotes the usual classical Hamiltonian. This fact explains how yet another expression, involving Wiener measures without drift and the function $H(p,q)$ representing the Hamiltonian, can be valid insofar as the leading term of the stationary-phase approximation is concerned, since in that approximation both H and h are equivalent to H_c [neglecting terms $O(\hbar)$].[11]

The rigorous path-integral definition described in this paper enables variable transformations (e.g., canonical transformations) to be examined much more critically than in the usual formal formulation. Such a possibility provides just one motivation for our seeking to define quantum mechanical path integrals in terms of genuine measures on continuous paths.

Spin case.—With regard to a path integral for spin s we can proceed analogously. Fix $s>0$, and for $(\theta,\phi)\in S^2$ let

$$|\theta,\phi\rangle=\exp(-i\phi S_3)\exp(-i\theta S_2)|s_s\rangle\in\mathcal{H}_s$$

denote normalized spin-coherent states, where $S_3|s_s\rangle=s|s_s\rangle$, and $\vec S^2=s(s+1)I_s$. It follows that

$$H=N_s\int h(\theta,\phi)|\theta,\phi\rangle\langle\theta,\phi|d\Omega,$$

where $N_s=(2s+1)/4\pi$, $d\Omega=\sin\theta\,d\theta\,d\phi$, represents any operator in $\mathcal{H}_s$. Here the relation between h and H is expressed[12] in terms of the usual spherical harmonics Y_{lm} by

$$h(\theta,\phi)=\sum_{l=0}^{2s}\sum_{m=-l}^{l}\frac{(2s+l+1)!}{(2s+1)!}\frac{(2s-l)!}{(2s)!}Y_{lm}(\theta,\phi)\int Y^*_{lm}(\theta',\phi')\langle\theta',\phi'|H|\theta',\phi'\rangle d\Omega'.$$

Evidently for H the identity operator I_s, $h=1$; while for any operator H, h is well defined. Hereafter we assume that H is the self-adjoint spin-operator Hamiltonian of interest, and thus h is real. Next we introduce additional normalized spin-coherent states

$$|\theta,\phi,j_m\rangle=\exp(i-\phi S_3)\exp(-i\theta S_2)|j_m\rangle$$

appropriate to spin j and magnetic quantum number m, where $S_3|j_m\rangle=m|j_m\rangle$; clearly $|\theta,\phi,s_s\rangle=|\theta,\phi\rangle$. For any β with $|\beta|<1$ it follows that

$$|\theta,\phi\rangle\rangle=\sum_{l=0}^{\infty}\oplus\,(2l+2s+1)^{1/2}\beta^{l(l+2s+1)/4}|\theta,\phi,l+s_s\rangle$$

defines a vector in an associated direct-sum Hilbert space $\mathcal{H}_s$. Four operators are defined on $\mathcal{H}_s$ as follows:

$$E_\beta=\int|\theta,\phi\rangle\rangle\langle\langle\theta,\phi|(d\Omega/4\pi)=\sum_{l=0}^{\infty}\oplus\,|\beta|^{l(l+2s+1)/2}I_{l+s},\quad A=\sum_{l=0}^{\infty}\oplus\,\tfrac{1}{2}l(l+2s+1)I_{l+s},$$

$$B_\beta=\int h(\theta,\phi)|\theta,\phi\rangle\rangle\langle\langle\theta,\phi|(d\Omega/4\pi),\quad C_\beta=\int\exp[-i\epsilon h(\theta,\phi)]|\theta,\phi\rangle\rangle\langle\langle\theta,\phi|(d\Omega/4\pi).$$

Here we have used the fact that

$$(2j+1)\int|\theta,\phi,j_m\rangle\langle\theta,\phi,j'_m|(d\Omega/4\pi)=\delta_{jj'}I_j$$

for all $j,j'\geqslant m$. We choose N a variable positive integer, T a fixed positive time integral, $\epsilon=T/(N+1)$, and $\beta=1-\epsilon\nu$, ν fixed and positive. Then again we are able to prove, *mutatis mutandis,* (1) and (2), with (3) as a

VOLUME 52, NUMBER 14 PHYSICAL REVIEW LETTERS 2 APRIL 1984

consequence. As for the path-integral expression it follows that

$$\langle\langle\theta'',\phi''|C_\beta^N|\theta',\phi'\rangle\rangle=\int\cdots\int\prod_{p=0}^{N}\langle\langle\theta_{p+1},\phi_{p+1}|\theta_p,\phi_p\rangle\rangle\prod_{p=1}^{N}\exp[-i\epsilon h(\theta_p,\phi_p)](d\Omega_p/4\pi),$$

where, for $0<\epsilon<<1$, and up to $O(\epsilon^2)$ terms,

$$\langle\langle\theta_2,\phi_2|\theta_1,\phi_1\rangle\rangle=\sum_{l=0}^{\infty}(2l+2s+1)[1-\tfrac{1}{2}\epsilon\nu l(l+2s+1)]\langle\theta_2,\phi_2,l+s_s|\theta_1,\phi_1,l+s_s\rangle$$

$$=(1+s\epsilon\nu)\langle\theta_2,\phi_2|\theta_1,\phi_1\rangle\sum_{l=0}^{\infty}\sum_{m=-l}^{l}(4\pi)[1-\tfrac{1}{2}\epsilon\nu l(l+1)]Y_{lm}(\theta_2,\phi_2)Y_{lm}^*(\theta_1,\phi_1).$$

Note here that

$$\langle\theta_2,\phi_2|\theta_1,\phi_1\rangle=\left[\cos\frac{\theta_2-\theta_1}{2}\cos\frac{\phi_2-\phi_1}{2}+i\cos\frac{\theta_2+\theta_1}{2}\sin\frac{\phi_2-\phi_1}{2}\right]^{2s} \tag{4}$$

while

$$(e^{t\nu\Delta/2})(\theta_2,\phi_2;\theta_1,\phi_1)=\sum_{l=0}^{\infty}\sum_{m=-l}^{l}\exp[-\tfrac{1}{2}t\nu l(l+1)]Y_{lm}(\theta_2,\phi_2)Y_{lm}^*(\theta_1,\phi_1), \tag{5}$$

with Δ the Laplacian on the unit sphere, is the Markov transition element for Brownian motion on the sphere with diffusion constant ν. Consequently,

$$\lim_{N\to\infty}\langle\langle\theta'',\phi''|C_\beta^N|\theta',\phi'\rangle\rangle=4\pi e^{s\nu T/2}\int\exp\{i\int_0^T[s\cos\theta\dot\phi-h(\theta,\phi)]dt\}d\mu_W^\nu\theta,\phi). \tag{6}$$

Here μ_W^ν denotes a pinned Wiener measure on the unit sphere with diffusion constant ν and weight given by (5) for $t=T$. To obtain (6) it is necessary to expand (4) to second-order differentials and use an appropriate form of the Itô calculus.[4] Here $\int\cos\theta\dot\phi\,dt=\int\cos\theta\,d\phi$ represents a well defined stochastic integral (in any sense). Finally we observe that

$$\langle\phi|e^{-iTH}|\psi\rangle=\lim_{\nu\to\infty}\int\langle\phi|\theta'',\phi''\rangle(N_s e^{s\nu T/2}\int e^{iS}d\mu_W^\nu)\langle\theta',\phi'|\psi\rangle d\Omega''d\Omega',$$

with $S=\int_0^T[s\cos\theta\dot\phi-h(\theta,\phi)]dt$, represents the desired path-integral expression.

In the spin case remarks entirely similar to those of the canonical case apply to an alternative path-integral definition in which Brownian motion on the sphere in the presence of drift and alternative expressions for the classical Hamiltonians arise. See Ref. 4.[10]

Lastly we remark that if we rescale ν in the spin case to ν/s, set $p=s^{1/2}\cos\theta$, $q=s^{1/2}\phi$ $(-\pi<\phi\leq\pi)$, and formally take the limit $s\to\infty$, then it follows that the spin path integral becomes the canonical path integral (modulo a trivial phase change).

It is a pleasure for us to thank Professor Ludwig Streit for his hospitality at the Zentrum für Interdisziplinäre Forschung, Bielefeld, Federal Republic of Germany, where this work was carried out. One of us (I.D.) acknowledges appointment as a Wetenschappelijk Medewerker at the Interuniversitair Instituut voor Kernwetenschappen, Belgium.

[1]See, e.g., E. Nelson, J. Math. Phys. (N.Y.) 5, 332 (1964).

[2]I. M. Gel'fand and A. M. Yaglom, J. Math. Phys. (N.Y.) 1, 48 (1960); R. H. Cameron, J. Analyse Math. 10, 287 (1962/1963).

[3a]J. R. Klauder and I. Daubechies, Phys. Rev. Lett. 48, 117 (1982).

[3b]I. Daubechies and J. R. Klauder, J. Math. Phys. (N.Y.) 23, 1806 (1982).

[3c]I. Daubechies and J. R. Klauder, Lett. Math. Phys. 7, 229 (1983).

[4]J. R. Klauder, J. Math. Phys. (N.Y.) 23, 1797 (1982).

[5]I. Daubechies and J. R. Klauder, to be published.

[6]J. R. Klauder and E. C. G. Sudarshan *Fundamentals of Quantum Optics* (Benjamin, New York, 1968), Chaps. 7 and 8.

[7]The normalization of the vectors $|p,q\rangle\rangle$ differs by a factor $(1-|\beta|)^{1/2}$ from that in Ref. 3b.

[8]One can prove that $\nu A+iB_1$, where $B_1=\lim_{\beta\to1}B_\beta$, generates a strongly continuous contraction semigroup which we denote by $\exp[-\nu TA-iTB_1]$.

[9]See, e.g., K. Itô, in *Mathematical Problems in Theoretical Physics*, edited by H. Araki (Springer-Verlag, New York, 1975), p. 218.

[10]Compare, in this regard, R. Shankar, Phys. Rev. Lett. 45, 1088 (1980).

[11]J. R. Klauder, in *Path Integrals*, edited G. J. Papadopoulos and J. T. Devreese (Plenum, New York, 1978), p. 5, and Phys. Rev. D 19, 2349 (1979).

[12]See, e.g., F. T. Arecchi, E. Courtens, R. Gilmore, and H. Thomas, Phys. Rev. A 6, 2211 (1972).

V. Pseudospin Techniques

Reprinted from ANNALS OF PHYSICS
All Rights Reserved by Academic Press, New York and London

Vol. 120, No. 1, July 1979

Semi-Classical Spectrum of the Continuous Heisenberg Spin Chain*

A. JEVICKI AND N. PAPANICOLAOU†

The Institute for Advanced Study, Princeton, New Jersey 08540

Received August 29, 1978

We develop a path integral formalism that allows for semi-classical quantization of systems with spin degrees of freedom. We apply it to study the continuous Heisenberg spin chain, which has been known to possess interesting classical solutions. The calculated semi-classical spectrum turns out to be essentially exact. We also construct a new infinite series of conservation laws that are nonlocal generalizations of the spin.

I. INTRODUCTION

Semi-classical methods have been vigorously investigated in recent years and applied to a number of field theories. They consist of a stationary phase approximation of the functional integral around non-trivial extrema of the action. These are classical solutions of the underlying non-linear field equations. Barring difficulties in actually solving the classical equations, the method is relatively straightforward for systems with Bose degrees of freedom. Important limitations arise for systems with Fermi degrees of freedom, which are formulated in terms of anti-commuting fields. Semi-classical methods have been successfully used for fermions only in isolated cases, where an effective Bose field may be introduced. Similar problem arise for spin degrees of freedom, since functional integral representations for spin systems have been so far given in terms of anti-commuting fields. It is the purpose of the present article to construct a functional integral representation for spin systems that involves ordinary (commuting) integration variables and, thereby, provides a scheme in which semi-classical quantization follows the standard technique.

Interesting physical systems with spin degrees of freedom are widely known and there is little need to give here a detailed enumeration. In this paper we shall concentrate on the continuous Heisenberg spin chain. The theory is defined by the Hamiltonian

$$\hat{H} = \frac{1}{2} \int dx \sum_{a=1}^{3} \left[\frac{d}{dx} \hat{S}^a(x) \right]^2 \tag{1.1}$$

* Research supported by the Department of Energy under Grant EY-76-S-02-2220.
† Present address: Department of Physics, University of California, Berkeley, Calif. 94720.

which together with the commutation relations $[\hat{S}^a(x), \hat{S}^b(x')] = i\epsilon^{abc}\hat{S}^c(x)\,\delta(x - x')$ yields the equation of motion

$$\frac{d}{dt}\hat{S}^a = \frac{1}{i}[\hat{S}^a, \hat{H}] = \epsilon^{abc}\hat{S}^b\hat{S}^c_{xx}. \tag{1.2}$$

$\hat{S}^2(x)$ is fixed to be $s(s+1)$. The exact spectrum of the discrete isotropic chain was obtained long ago by Bethe [1]. His Ansatz for the spin eigenstates (known as Bethe's hypothesis) was subsequently applied to obtain the exact solution for a class of interesting generalizations of the original Heisenberg chain. The literature in this direction is too extensive to be cited here. At any rate, these generalizations are not directly relevant to our present work.

The remarkable structure of Bethe's solution can be presently understood as deriving from an interesting property of the Heisenberg chain, namely, its complete integrability. This is most directly implied by the existence of higher conservation laws constructed in [2]. More complete results are available for the classical theory of the continuous chain, Eq. (1.2). Besides the spin-wave solution, localized soluton solutions were recently obtained in [3, 4, 5] and subsequently rederived by systematic inverse scattering methods in [6]. However, the relevance of the above classical solutions to the quantum theory has been as yet unclear. This problem is solved in the present article. Semi-classical quantization of the soliton solution leads, in fact, to the exact magnon spectrum.

Aside from providing an interesting illustration of the method for semi-classical quantization of spin, the continuous ferromagnetic chain appears to be an interesting example in which to test inverse scattering methods. Thus we present here some results concerning the inverse problem of the continuous spin chain. We construct a new infinite series of higher conservation laws that are non-local generalizations of the spin. This establishes a close connection with a large class of two-dimensional field theories that have been solved by inverse scattering methods. These are the non-linear σ-models of [7] and the classical Fermi interactions of [8]. The interesting common feature of these theories is that the origin of the Lax representation is understood to a considerable extent. Augmenting the above class by new systems, such as the ferromagnetic chain, should certainly prove useful in trying to identify the origin of the frequent occurrence of integrable systems.

The plan of the article is as follows: Section II provides the necessary background from the underlying classical theory and presents the new series of non-local conservation laws. Our path integral representation for a general spin system is given in Section III. It is used in Section IV in order to carry out a DHN [9] semi-classical calculation of the spectrum of the continuous ferromagnetic chain. The problem of small oscillations around the onesoliton solution is solved in an Appendix.

II. Classical Theory

A. *The reduction technique.* It was already mentioned in the introduction that the purpose of this section is to provide the necessary calculational background for our subsequent semi-classical discussion, and to present some new results concerning the mathematical structure of the theory. In particular, we construct a new series of conservation laws, which are non-local generalizations of the spin. This and previous results on the subject suggest a strong analogy with the situation in the non-linear σ-models and the classical Fermi interactions analyzed at length in [7, 8]. We should add, however, that our present discussion of the classical theory is not meant to be complete.

We begin with a brief outline of the reduction technique originally employed in [4] for the construction of the one-soliton solution of the continuous Heisenberg chain. Let S^a, $a = 1, 2, 3$ denote the continuous spin variable, a function of space and time, such that $S^2 = S^a S^a = 1$. We form the following set of orthonormal vectors, to be referred to as the moving trihedral:

$$e_1{}^a = \frac{S_x{}^a}{(2H)^{1/2}}, \qquad e_2{}^a = \frac{\epsilon^{abc} S^b S_x{}^c}{(2H)^{1/2}}, \qquad e_3{}^a = S^a,$$
$$H \equiv \tfrac{1}{2} S_x{}^a S_x{}^a, \qquad e_i{}^a e_j{}^a = \delta_{ij}, \tag{2.1}$$

where ϵ^{abc} is the usual antisymmetric tensor and the summation convention is implied. We have also used an obvious notation for the differentiation symbol. An arbitrary vector formed out of S^a and its derivatives may be expanded in the above complete basis with components that are invariant under $O(3)$ rotations. By a simple induction, it can be shown that all such invariants may be expressed in terms of the fundamental set of invariants

$$S^2 = 1, \qquad H \equiv \tfrac{1}{2} S_x{}^a S_x{}^a, \qquad Q \equiv \epsilon^{abc} S^a S_x{}^b S_{xx}^c \tag{2.2}$$

and their derivatives. In this notation, the following expansion will be useful for our calculations:

$$S_{xx}^a = \frac{H_x}{(2H)^{1/2}} e_1{}^a + \frac{Q}{(2H)^{1/2}} e_2{}^a - 2He_3{}^a. \tag{2.3}$$

We now examine the variation of the trihedral under space displacements. Using elementary completeness arguments, one finds:

$$e_{i,x}^a = C_{1,ik} e_k{}^a, \qquad a = 1, 2, 3, \tag{2.4}$$

where the matrix C_1 is a linear superposition of the standard anti-hermitean generators of $O(3)$, with real coefficients that are functions of the fundamental invariants H and Q:

$$C_1 = \omega_1{}^a I^a, \qquad [I^a, I^b] = \epsilon^{abc} I^c,$$
$$\{\omega_1{}^a\} = \{0, -(2H)^{1/2}, -Q/2H\}. \tag{2.5}$$

Passing to the spinor representation and using the τ-matrices defined by $t^a = (1/2i)\sigma^a$, where σ^a are the Pauli matrices, C_1 reads

$$C_1 = \frac{1}{2}\begin{bmatrix} i\,Q/2H & (2H)^{1/2} \\ -(2H)^{1/2} & -i\,Q/2H \end{bmatrix}. \tag{2.6}$$

Our statements so far have been independent of the time evolution of the spin variable. The continuous Heisenberg spin chain is described by the evolution equation

$$S_t^{\,a} = \epsilon^{abc} S^b S^c_{xx} = [\epsilon^{abc} S^b S_x^{\,c}]_x\,, \tag{2.7}$$

which is in the form of a conservation law implying the conservation of spin. Using (2.7) and the reduction technique outlined above, it is easy to find that the time displacement of the trihedral is described by

$$e^a_{i,t} = C_{2,ik} e_k^{\,a}. \tag{2.8}$$

The matrix C_2 is given, in the spinor representation, by

$$C_2 = \frac{1}{2i}\begin{bmatrix} \dfrac{Q^2}{4H^2} - \dfrac{((2H)^{1/2})_{xx}}{(2H)^{1/2}}\,, & \dfrac{H_x - iQ}{(2H)^{1/2}} \\ \dfrac{H_x + iQ}{(2H)^{1/2}}\,, & -\dfrac{Q^2}{4H^2} + \dfrac{((2H)^{1/2})_{xx}}{(2H)^{1/2}} \end{bmatrix}. \tag{2.9}$$

The integrability condition for the system of Eqs. (2.4), (2.8) results into:

$$\begin{gathered} H_t + Q_x = 0, \\ \left(\frac{Q}{2H}\right)_t = \left[H - \frac{Q^2}{4H^2} + \frac{((2H)^{1/2})_{xx}}{(2H)^{1/2}}\right]_x. \end{gathered} \tag{2.10}$$

or, in an equivalent form:

$$\begin{gathered} H_t + Q_x = 0, \\ Q_t = \left[H^2 + \frac{HH_{xx} - H_x^{\,2} - Q^2}{H}\right]_x. \end{gathered} \tag{2.11}$$

Since all invariants *can* be expressed in terms of H, Q and their space derivatives, Eqs. (2.11) provide a complete description of the dynamics of invariants. They constitute the "reduced" system associated with the continuous spin chain. They are found in the form of local conservation laws, implying the conservation of the energy $\int dx\, H$ and revealing a "higher" conserved charge, namely $\int dx\, Q$. The latter hasbeen incorrectly identified in [4] as the momentum of the system. A correct definition of a momentum integral was given in [5]. We shall return to it in Section III. We emphasize here, however, that it does not belong to the set of conservation laws following from Eqs. (2.11). For example, the next conserved charge is given by

$$Q' = \int_{-\infty}^{+\infty} dx[S^a_{xx}S^a_{xx} + \tfrac{5}{4}(S_x{}^aS_x{}^a)^2]. \tag{2.12}$$

The charge density appearing in this equation can be reduced in terms of H and Q. One obtains:

$$Q' = \int_{-\infty}^{+\infty} dx\left[9H^2 + \frac{H_x{}^2 + Q^2}{2H}\right]. \tag{3.12}$$

It is interesting to note that in its reduced form the above density is non-polynomial.

The characteristic feature of the present series of charge densities is that they are invariant under $O(3)$ rotations. Roughly speaking, they are generalizations of the energy density. An independent series of conservation laws generalizing the spin will be constructed in the following. For the moment, however, we turn our attention to an interesting identification of the system (2.11) made in [4]. It is shown by direct substitution that given a complex field $\psi = \alpha e^{i\beta}$ satisfying the non-linear Schrödinger equation:

$$i\psi_t + \psi_{xx} + \tfrac{1}{2}(\psi^*\psi)\psi = 0 \tag{2.14}$$

the quantities $H = \alpha^2/2$ and $Q = \alpha^2\beta_x$ satisfy the system of Eqs. (2.11). Hence, the "reduced system" of the continuous Heisenberg chain is the non-linear Schrödinger theory, in a manner analogous to the situation in the non-linear σ-models and the Gross–Neveu models, where the "reduced systems" are the sine- and sinh-Gordon equations and their appropriate generalizations. Correspondingly, it is natural to expect that the system of Eqs. (2.4, 9) provide the associated "linear" problem (Lax representation). However, the above line of argument is at this point incomplete, as no principle for the introduction of an eigenvalue parameter has been identified. We shall return to this question in later stages of our discussion.

B. *Non-local conservation laws.* An independent method was proposed by Takhtajan in [6]. For this discussion, it is somewhat more convenient to use matrix notation specified by

$$S = S^a\sigma^a,\ S^2 = 1. \tag{2.15}$$

The equation of motion (2.7) is transctibed into:

$$S_t = \frac{1}{2i}[S, S_{xx}] = \frac{1}{2i}[S, S_x]_x. \tag{2.16}$$

It is easy to show that the following "linear" equations:

$$R_x = -i\lambda SR, \qquad R_t = \left\{-\frac{\lambda}{2}[S, S_x] + 2i\lambda^2 S\right\} R \tag{2.17}$$

are compatible in virtue of (2.16), for arbitrary constant λ. The system (2.17) is the Lax representation used in [6] for a systematic solution of the inverse problem.

Given the reduction of the present theory to the non-linear Schrödinger theory, a procedure that involves $O(3)$ invariant quantities, it is somewhat surprising that a linear problem such as (2.17) exists—its distinguishing feature being that the Lax matrices are realized in terms of $O(3)$ variant spin variables. However, such phenomenon is not unfamiliar. It systematically occurs in the systems analyzed in [7, 8]. An equivalen way of stating Eqs. (2.17) is the following: given a solution S of (2.16), the linear system (2.17) may be solved for R, which can be chosen to be a (space-time dependent) unitary matrix. Furthermore, one can show by direct substitution that $[R^{+}SR]$ $(x + 4\lambda t, t)$ is also a solution. It may be said that the above construction provides a non-linear implementation of Galilean boosts.

Previous experience suggests that a "linear" system of the above nature generates a new series of conservation laws—the non-local conservation laws. They were first constructed in [10] for the non-linear σ-model and can be easily shown to exist in the classical Fermi interactions of [8]. Their significance for the quantum theory was studied in [11]. We briefly outline the construction of such conservation laws in the present theory and give a few examples.

For a concise derivation of the conservation laws, we introduce dual potentials as follows: Denote by $J \equiv (1/2i)[S, S_x]$ so that the equation of motion (2.16) reads $S_t = J_x$. The current densities S and J also satisfy the "pure gauge" condition

$$iS_x + \tfrac{1}{2}[S, J] = 0 \tag{2.18}$$

following merely from $S^2 = 1$. Introduce the dual potential $\Omega = \Omega^a \tau^a$ satisfying the compatible equations

$$\Omega_x = S, \Omega_t = J. \tag{2.19}$$

With this variable the current conservation reads trivially $\Omega_{xt} = \Omega_{tx}$, wheras Eq. (2.18) becomes the equation of motion for Ω:

$$i\Omega_{xx} + \tfrac{1}{2}[\Omega_x, \Omega_t] = 0. \tag{2.20}$$

Correspondingly, Eqs. (2.17) are transcribed into

$$R_x = -i\lambda\Omega_x R, R_t - [-i\lambda\Omega_t + 2i\lambda^2\Omega_x]R, \tag{2.21}$$

which are compatible in virtue of (2.20). Expanding R as $R = 1 + \lambda R_1 + \lambda^2 R_2$,..., Eqs. (2.21) may be used for a recursive determination of the coefficients R_1, R_2 .... At each stage of this recursion, the integrability condition (2.20) is iterated to yield a sequence of conservation laws, the first few of which are:

$$\Omega_{xt} = \Omega_{tx}, (\Omega_x\Omega)_t = (\Omega_t\Omega - 2\Omega_x)_x .... \tag{2.22a}$$

Expressed in the original variables, they lead to both local and non-local conservation laws. For instance, the conserved charges corresponding to the conservation laws (2.22a) read

$$
\begin{aligned}
Q^a &= \int_{-\infty}^{+\infty} dx\, S^a(x), \\
Q_1{}^0 &= \int_{-\infty}^{+\infty} dx\, dx'\, \theta(x-x')\, S^a(x)\, S^a(x') \\
Q_1{}^a &= \epsilon^{abc} \int_{-\infty}^{+\infty} dx\, dx'\, \theta(x-x')\, S^b(x)\, S^c(x'),\dots .
\end{aligned}
\tag{2.22b}
$$

$\theta(x - x')$ is the step function. As expected the first charge of this series is the integrated spin. We finally note that for ferromagnetic boundary conditions the above integrals diverge in general. They become finite after subtracting their values in the ferromagnetic ground state. This complication will cause no difficulty in our subsequent considerations.

The remarkably similar structure of such diverse systems as the non-linear σ-models, classical Fermi interactions and the present continuous ferromagnetic chain suggests that there exists a governing fundamental principle, whose full extent remains to be identified. Conserning the present theory, one more step has to be taken, namely, to explain the origin of the eigenvalue parameter in (2.17) and, consequently, introduce an eigenvalue parameter in the linear system of Section A. We shall have additional comments to make in the end of Section III.

C. *Calculational details.* In the remainder of this section, we turn our attention to more pragmatic considerations, in order to prepare our semi-classical calculation. We summarize here the necessary information concerning the one-soliton solution. Our computations are based on Takhtajan's inverse problem, Eqs. (2.17).

For later purposes, we shall need the explicit form of the kernel of the Gel'fand–Levitan-Marchenko (GLM) equation; we therefore, repeat the calculation of [6] in some more detail. The relevant GLM equation reads

$$K(x, y; t) + \Phi_1(x + y; t) + \int_x^{\infty} K(x, z; t)\, \Phi_2(z + y; t)\, dz = 0, \tag{2.23}$$

for $x \leqslant y$. The matrices Φ_1 and Φ_2 are determined from

$$F(x, t) = \frac{1}{2\pi} \int_{-\infty}^{\infty} \frac{b(\lambda, t)}{\lambda a(\lambda)}\, e^{i\lambda x}\, d\lambda + \frac{m(t)}{\zeta}\, e^{i\zeta x} \tag{2.24}$$

by the relations

$$\Phi_1 = \begin{bmatrix} 0 & -F^* \\ F & 0 \end{bmatrix}, \qquad \Phi_2 = -i \begin{bmatrix} 0 & F_x \\ F_x & 0 \end{bmatrix}. \tag{2.25}$$

The form (2.24) for F is sufficiently general to cover the one-soliton solution and small fluctuations around it. The time evolution of the scattering data is given by

$$\begin{aligned} a(\lambda) &= \text{const}, \qquad \zeta = \text{const}, \\ b(\lambda, t) &= e^{-4i\lambda^2 t} b(\lambda, 0), \qquad m(t) = e^{-4i\zeta^2 t} m, \end{aligned} \tag{2.26}$$

ζ is in general complex, Im $\zeta \geqslant 0$. In terms of the kernel K, the solution of Eq. (2.16) is given from

$$S(x, t) = [iK(x, x; t) - \sigma_3]\, \sigma_3 [iK(x, x; t) - \sigma_3]^{-1}, \tag{2.27}$$

where σ_3 is the usual Pauli matrix. The one-soliton solution is obtained by solving the GLM equation with $b = 0$. The kernel that solves Eq. (2.23) is then

$$K_1(x, y; t) = M(x; t) \exp[i\zeta y \sigma_3],$$

$$M(x; t) = \frac{\mu}{1 + \mu^2} \begin{bmatrix} \dfrac{\zeta^* - \zeta}{i\zeta^*} \mu e^{-i\zeta x} & \dfrac{|\zeta - \zeta^*|}{\zeta^*} e^{i\zeta^* x - i\phi_1} \\ -\dfrac{|\zeta - \zeta^*|}{\zeta} e^{i\phi_1 - i\zeta x} & \dfrac{\zeta^* - \zeta}{i\zeta} \mu e^{i\zeta^* x} \end{bmatrix}. \tag{2.28}$$

Here,

$$\begin{aligned} \mu^2 &= \frac{mm^*}{|\zeta - \zeta^*|^2} \exp[2i(\zeta - \zeta^*)x - 4i(\zeta^2 - \zeta^{*2})t], \\ \phi_1 &= \text{Re}[2\zeta x - 4\zeta^2 t] + \arg m. \end{aligned} \tag{2.29}$$

Inserting these values into (2.27) and taking $\zeta = v/4 + i(\epsilon/2)$, we obtain the one-soliton solution

$$\begin{aligned} S_3 &= \cos\theta = 1 - \frac{2\epsilon^2}{\epsilon^2 + v^2/4} \frac{1}{\cosh^2 \epsilon(x - vt - x_0)}, \\ S_+ &= S_1 + iS_2 = \sin\theta e^{i\phi}, \\ \phi &= \left(\epsilon^2 + \frac{v^2}{4}\right) t + \frac{v}{2}(x - vt) + \text{arctg}\left[\frac{2\epsilon}{v} \text{th}\epsilon(x - vt - x_0)\right] + \varphi_0 \end{aligned} \tag{2.30}$$

with appropriate identification of the trivial constants x_0, φ_0.

Arbitrary value s for the total spin ($S^2 = s^2$) is achieved by the substitutions $S \to sS$ and $t \to st$. By an additional rescaling $sv \to v$, we get

$$\begin{aligned} S &= s\{\sin\theta \cos\phi, \sin\theta \sin\phi, \cos\theta\}; \; S^2 = s^2; \\ \cos\theta &= 1 - \frac{2\epsilon^2}{\epsilon^2 + v^2/4s^2} \cdot \frac{1}{\cosh^2 \epsilon(x - vt - x_0)}; \\ \phi &= s\left(\epsilon^2 + \frac{v^2}{4s^2}\right) t + \frac{v}{2s}(x - vt) + \text{arctg}\left[\frac{2\epsilon s}{v} \text{th}\epsilon(x - vt - x_0)\right] + \varphi_0. \end{aligned} \tag{2.31}$$

The physical content of this solution is not difficult to identify. In the rest frame ($v = 0$), we obtain a quasi-static configuration for which the third component of the spin is time-independent, whereas the total spin undergoes a precession around the third axis with frequency $\omega = s\epsilon^2$. Concerning the space distribution of the spin density, this solution describes a localized excitation from the ferromagnetic ground state. In fact, everywhere but in a small region, the spin points on the (time) average along a fixed direction and possesses the vacuum value $S_3 \simeq 1$, which characterizes the ferromagnetic ground state configuration. For a moving soliton ($v \neq 0$), this picture remains essentially the same, except for an interesting Doppler, velocity dependent, shift of the frequency: $\omega = s(\epsilon^2 + v^2/4s^2)$.

The above intuitive picture will be gradually sharpened, as we will proceed with the semi-classical computation. To conclude this section, we calculate some relevant "observables." Thus the energy, the total spin along the third direction and the first "higher" charge are given by:

$$\begin{aligned} \mathscr{E} &= \int_{-\infty}^{+\infty} dx\, H(x) = 4\epsilon s^2, \\ Q_3 &= \int_{-\infty}^{+\infty} dx[s - S_3(x)] = \frac{4\epsilon s}{\epsilon^2 + v^2/4s^2}, \\ Q &= \int_{-\infty}^{+\infty} dx\, Q(x) = 2\epsilon v s^2. \end{aligned} \tag{2.32}$$

The above values refer to the excited spin; the contributions from the ferromagnetic ground state have been subtracted. An important omission in (2.32) is the momentum integral. This should wait for the considerations of the following section.

III. Path Integral for Spin

It is well known that the path integral method provides the most appropriate framework for semi-classical quantization, [12–14]. In practice, however, important limitations of the applicability of the method may occur: whereas for Bose fields classical solutions can be used directly to dominate the path integral, in the case of Fermi or spin degrees of freedom additional effort is required. This is mostly due to the fact that the corresponding path integral is usually defined over anti-commuting numbers. Consequently, semi-classical calculations involving fermions have been performed only in a few isolated cases, where it was possible to introduce an effective Bose field (see [15]).

We shall be interested in quantum systems that contain spin degrees of freedom. At the operator level, the procedure is standard. Let $\hat{S}^a$, $a = 1, 2, 3$ be the spin operators satisfying the commutation relations:

$$\begin{aligned} [\hat{S}^a, \hat{S}^b] &= i\epsilon^{abc}\hat{S}^c, \\ \hat{S}^2 &= s(s+1),\ s = \tfrac{1}{2}, 1\ldots . \end{aligned} \tag{3.1}$$

Given a Hamiltonian $H = H(\hat{S})$, the time evolution is dictated by the Heisenberg equation of motion

$$\frac{d}{dt}\hat{S}^a(t) = \frac{1}{i}[\hat{S}^a, H(\hat{S})]. \tag{3.2}$$

The standard functional integral representation for this system is given in terms of anti-commuting numbers. For example, for spin one-half one has, [16]

$$Z = \int \mathscr{D}\vec{\eta} \exp\left\{i \int dt \left[\frac{i}{2}\vec{\eta}\cdot\dot{\vec{\eta}} - H\left(\frac{1}{2i}\vec{\eta}\times\vec{\eta}\right)\right]\right\}, \tag{3.3}$$

where η_1, η_2 and η_3 represent generators of a Grassman algebra. Clearly, this representation is not very useful for semi-classical quantization, especially in case of many degrees of freedom. It is difficult to find classical solutions in terms of anti-commuting numbers [28].

In what follows, we describe an alternative path integral for spin that involves ordinary c-numbers and, consequently, is more appropriate for our purpose. We start from the well known harmonic oscillator representation of angular momentum [7],

$$\hat{S}^a = \hat{A}^+ \frac{\sigma^a}{2} \hat{A} \equiv \hat{A}^+ \tau^a \hat{A}. \tag{3.4}$$

Here, σ^a are the standard Pauli matrices and we have used the notation

$$\hat{A} = \begin{bmatrix}\hat{A}_1\\ \hat{A}_2\end{bmatrix}, \qquad \hat{A}^+ = [\hat{A}_1{}^+, \hat{A}_2{}^+]. \tag{3.5}$$

$\hat{A}_i$ and $\hat{A}_i{}^+$ are Bose creation-annihilation operators:

$$[\hat{A}_i, \hat{A}_j{}^+] = \delta_{ij}, \qquad i, j = 1, 2. \tag{3.6}$$

Introducing the total occupation number operator $\hat{N} = \hat{A}^+\hat{A}$, one obtains the relation

$$\hat{S}^2 = \frac{\hat{N}}{2}\left(\frac{\hat{N}}{2} + 1\right). \tag{3.7}$$

Consequently, this description allows for arbitrary value of the total spin. In order to account for a fixed value of the angular momentum, denoted by s, one is forced to work in the subspace characterized by the constraint:

$$[\hat{N} - 2s]|\rangle = 0. \tag{3.8}$$

A functional integral on this restricted subspace is obtained by using the modified Hamiltonian

$$H_\lambda = H(A^+\vec{\tau}A) + \lambda(A^+A - 2s), \tag{3.9}$$

where the Lagrange multiplier λ is introduced to enforce the constraint. The presence of a constraint necessitates a gauge condition $\phi(A)$. In choosing $\phi(A)$, the only restriction is that the Poisson bracket

$$\{\phi(A),\ A^+A\}_{P.B.} \tag{3.10}$$

be different from zero. Following the general formalism of Faddeev [18], the path integral representation of the transition amplitude reads

$$Z = \int \prod_{i=1}^{2} \mathscr{D}A_i^*\mathscr{D}A_i \prod_t \delta[A^+(t)\,A(t) - 2s] \cdot \delta[\phi(A)]$$
$$\times \det\{\phi(A),\ N\}_{\text{P.B.}} \exp\left[i \int dt\, L\right], \tag{3.11}$$

where the Lagrangian is given by

$$L = \frac{i}{2} A^+ \frac{\overleftrightarrow{d}}{dt} A - H(A^+\vec{\tau}A). \tag{3.12}$$

Next, we introduce the spin integration variables through the identity

$$\int \mathscr{D}\vec{S} \prod_{t,a} \delta[S^a(t) - A^+\tau^a A] = 1, \tag{3.13}$$

and totally eliminate the harmonic oscillator variables A and A^*. This is possible, because the δ-conditions (3.13) together with the gauge condition present in (3.11) provide four equations, which allow for the inversion.

We use the following simple gauge condition:

$$\phi(A) \equiv A_1 + A_1^* = 0. \tag{3.14}$$

It then follows that

$$A_1 = i(s + S_3)^{1/2},$$
$$A_2 = i\frac{S_1 + iS_2}{(s + S_3)^{1/2}}. \tag{3.15}$$

The Lagrangian (3.12) now becomes

$$L(\dot{S}, S) = \frac{S_2\dot{S}_1 - S_1\dot{S}_2}{s + S_3} - H(\vec{S}), \tag{3.16}$$

whereas the functional integral reduces to

$$Z = \int \mathscr{D}\vec{S} \prod_t \delta[\vec{S}^2(t) - s^2] \exp\left[i \int dt\, L(\dot{S}, \vec{S})\right], \tag{3.17}$$

which is our final form. Here, $\vec{S}(t)$ is a c-number integration variable and, consequently, such a representation allows for semi-classical approximations.

The classical equations of motion obtained by varying our Lagrangian (3.16), with the constraint $\vec{S}^2 = s^2$, are identical in form to the Heisenberg equations satisfied by the spin operators. Conserning the form of this Lagrangian, we mention the following. Classical spin Lagrangians have been proposed by solid state physicists, who have extensively studied the classical theory of rigid magnetic continuum. They all differ from (3.16). It now becomes clear, however, that the form of the Lagrangian depends on the gauge condition. If, instead of (3.14), we choose Re $A_2 = 0$ one gets

$$L' = -\frac{S_2\dot{S}_1 - S_1\dot{S}_2}{s - S_3} - H(\vec{S}). \tag{3.18}$$

The well-known Doering–Gilbert Lagrangian [9]

$$L_{DG} = \frac{S_3(S_1\dot{S}_2 - S_2\dot{S}_1)}{s^2 - S_3{}^2} - H(\vec{S})$$

is now given by $\frac{1}{2}(L + L')$. All these Lagrangians lead to the same equations of motion, as they differ by a total divergence. Nevertheless, surface terms are important in semi-classical quantization and, for such a purpose, (3.16) will be more convenient. We finally mention that the lack of manifest $O(3)$ invariance in the above expressions is quite analogous to the familiar situation in Quantum Electrodynamics, formulated in a non-covariant gauge.

The formalism described above obviously generalizes to systems with many degrees of freedom, like for example, the continuous Heisenberg ferromagnet. Having a Lagrangian, we may now deduce the generator of space translations, namely, the momentum integral. Under the variation $\vec{S}(x, t) \to \vec{S}(x + \delta X(t), t)$, the action over a finite time interval $[t_i, t_f]$ varies according to

$$\delta A(t_f, t_i) = -\delta X(t) \int dx \frac{S_1(\partial/\partial x)\, S_2}{s + S_3}(x, t)\Big|_{t_i}^{t_f}. \tag{3.20}$$

Hence, the momentum is given by

$$P = \int_{-\infty}^{+\infty} dx[s + S_3]^{-1}[S_1\, \partial_x S_2], \tag{3.21}$$

which agrees with the form derived previously in Ref. [5]. It is of particular importance to observe that in contrast to the energy and higher conserved charges, the momentum density is *locally* gauge dependent. We shall return to this point in the end of this section. We first comment on the validity of our canonical procedure. In fact, in intermediate steps of our derivation, canonical transformations were made, which are usually beset by ordering problems. For a more precise specification of the functional integral, one would need to consider the short-time definition of (3.17). Alternatively, it is more efficient to start from the formalism of geometric quantization [20–22] and derive the precise short-time expression for (3.17). These finer details are not relevant for our subsequent use of (3.17), as we will not be interested in higher order computations.

As promised earlier, we return to the classical theory, in view of the canonical procedure developed in this section. At a purely classical level, our preceding discussion may be summarized, or rephrased, as follows: given a solution A of the classical field equations

$$\begin{aligned} iA_t + [A^+\tau^a A]_{xx}\, \tau^a A + \lambda A &= 0, \\ A^+ A &= 2s \end{aligned} \tag{3.22}$$

the spin density $S^a = A^+\tau^a A$ satisfies the conservation equation (Noether's equation associated with $O(3)$ invariance)

$$S_t^a = [\epsilon^{abc} S^b S_x{}^c]_x\,. \tag{3.23}$$

In fact, this equation can be derived from (3.22) for λ chosen to be an arbitrary real function $\lambda(x, t)$, since the constraint $A^+A = 2s$ may be enforced by appropriately choosing the initial data. Therefore, the equations of motion are form invariant under the Abelian local gauge transformation:

$$A \to A e^{if(x,t)},\ \lambda \to \lambda + f_t\,, \tag{3.24}$$

with $f(x, t)$ arbitrary. The spin density and, consequently, the energy and higher conservation laws are locally gauge invariant, whereas the momentum density $P = (i/2)\, A^+\overleftrightarrow{\partial}_x A$ transforms as $P \to P - f_x$. This explains the absence of the momentum density in the reduction of Sec. IIA. A reduction in terms of the field A, however, necessitates three fundamental invariants, namely P, H and Q. This and the fact that P is locally gauge dependent is, we believe, the origin of the eigenvalue parameter in the inverse problem. In practice the eigenvalue parameter is introduced by the Galilean transformation described in section II.B.

IV. Semi-Classical Quantization

In what follows, we will quantize the classical soliton solution described in Section II and derive the associated energy spectrum. This semi-classical quantization is based on the path integral representation introduced in the preceding section and the general method of Dashen, Hasslacher and Neveu [9]. Consider

$$G(E) = i \int_0^\infty dT\, e^{iET}\, \mathrm{Tr}[e^{-i\hat{H}T}]. \tag{4.1}$$

The trace involved here has the path integral representation (3.17), with the action given by

$$A(\vec{S}) = \int_0^T dt \int_{-\infty}^{+\infty} dx \left[\frac{S_2 \overleftrightarrow{\partial}_t S_1}{s + S_3} - \frac{1}{2}\vec{S}_x{}^2\right]. \tag{4.2}$$

Summation over all periodic orbits is implied.

The general features of the following calculation are similar to those of [23], where the semi-classical spectrum of the non-linear Schrödinger theory was studied. In fact, as was explained in Sec II at the classical level our present theory is related to the Schrödinger theory. However, the spectrum is totally different. The classical solution (2.31) provides infinitely many periodic orbits, if the following requirements are imposed on the parameters:

$$\frac{2\pi}{s(\epsilon^2 + v^2/4s^2)} = \frac{T}{l} \equiv \tau, \qquad l = 1, 2, \ldots, \tag{4.3}$$

$$v = n\frac{L}{T}, \qquad n = 0, \pm 1, \ldots . \tag{4.4}$$

L represents the length of the "space box." Calculating the action for the configuration (2.31) we find

$$A_{cl} = T[-8\epsilon s^2 + 4sv \sin^{-1}\beta], \tag{4.5}$$

where

$$\beta \equiv \frac{\epsilon}{(\epsilon^2 + v^2/4s^2)^{1/2}} . \tag{4.6}$$

In terms of the parameter β, the relation (4.3) reads

$$\frac{nL}{l}\frac{1}{(1-\beta^2)^{1/2}} - 2(2\pi s\tau)^{1/2} = 0. \tag{4.7}$$

This will be enforced by an appropriate δ-function identity. Hence, the approximate form for the trace (4.1) reads:

$$\begin{aligned} G(E) = i \sum_{n,l} \int_0^\infty d(l\tau) \int_{-\infty}^{\infty} d\beta \left(\frac{2nL}{l}\right)^{1/2} \frac{\beta(2\pi s\tau)^{1/4}}{(1-\beta^2)^{5/4}} \\ \times \delta\left[\frac{nL}{l}\frac{1}{(1-\beta^2)^{1/2}} - 2(2\pi s\tau)^{1/2}\right] \Delta_1 e^{i\tau l E} \\ \times \exp\left\{i4snL\left[-\frac{\beta}{(1-\beta^2)^{1/2}} + \sin^{-1}\beta\right]\right\}. \end{aligned} \tag{4.8}$$

The δ-function has been introduced in a manner that will be convenient for subsequent calculations. Δ_1 summarizes the contribution of symmetry modes and will be evaluated explicitly in later stages. We have, finally, ignored quantum fluctuations in writing (4.8).

In order to perform the τ- and β-integrations, we introduce the following representation of the δ-function occurring in (4.8):

$$\delta[\cdots] = \frac{4ls}{2\pi}\int d\mu \exp\left\{4ils\mu\left[\frac{nL}{l}\frac{1}{(1-\beta^2)^{1/2}} - 2(2\pi s\tau)^{1/2}\right]\right\}. \tag{4.9}$$

The effective action separates into τ- and β-dependent parts. We first perform the τ-integration by stationary phase approximation. The stationary point of the corresponding action

$$A_\tau = l[-8s\mu(2\pi s\tau)^{1/2} + E\tau] \tag{4.10}$$

occurs at

$$\tau_0^{1/2} = \frac{4s\mu(2\pi s)^{1/2}}{E}, \tag{4.11}$$

and then

$$A_\tau = A_\tau{}^0 = -2\pi l\frac{16s^3\mu^2}{E}. \tag{4.12}$$

Furthermore, the Gaussian integration yields the contribution

$$8\pi\left(\frac{2\mu^2}{lE^3}\right)^{1/2}. \tag{4.13}$$

Next, we perform the β-integration again by stationary phase. The stationary point occurs at

$$\beta_0 = \mu \tag{4.14}$$

and the corresponding value of the action is

$$A_\zeta = A_\zeta{}^0 = 4nsL\sin^{-1}\mu. \tag{4.15}$$

The corresponding Gaussian integration provides the contribution

$$\left[\frac{\mu(1-\mu^2)^{3/2}}{2nsL\mu}\right]^{1/2}. \tag{4.16}$$

Putting everything together, (4.8) simplifies to

$$G(E) = i\,\text{const}\cdot\sum_{n,l}\int d\mu\frac{\mu^2}{(1-\mu^2)^{1/2}}\cdot\frac{1}{E^2}\cdot l\Delta_1\exp[i(A_\tau{}^0 + A_\beta{}^0)]. \tag{4.17}$$

We have omitted the explicit specification of the overall constant in the above expression, as it is irrelevant for the evaluation of the spectrum.

We now elaborate on the computation of Δ_1, which comes from the symmetry modes. Since there is a preferred macroscopic direction, namely, the direction of the spin in the ferromagnetic ground state, implied here by the boundary condition

$$\lim_{|x|\to\infty} S_3(x, t) = s, \tag{4.18}$$

there exist only two collective coordinates: the phase φ_0 and the position parameter x_0. In computing Δ_1, we have to consider a slightly more general situation described by the boundary conditions

$$\phi(0) = \varphi', \ \phi(T) = \varphi'' + 2\pi l \tag{4.19}$$

and

$$x(0) = x', \ x(T) = x'' + nL. \tag{4.20}$$

The relations (4.3) and (4.4) are then replaced by

$$\frac{2\pi}{s(\epsilon^2 + v^2/4s^2)} = \frac{\varphi'' - \varphi' + 2\pi l}{T}, \qquad v = \frac{x'' - x' + nL}{T} \tag{4.21}$$

which in turn imply an explicit dependence of the classical action (4.5) on the boundary values $\{\varphi'', \varphi'\}$ and $\{x'', x'\}$. Δ_1 is then defined by

$$2\pi i\,\Delta_1 = 2\pi L \det \begin{bmatrix} \dfrac{\partial^2 A_{cl}}{\partial\varphi'\,\partial\varphi''} & \dfrac{\partial^2 A_{cl}}{\partial\varphi'\,\partial x''} \\ \dfrac{\partial^2 A_{cl}}{\partial x'\,\partial\varphi''} & \dfrac{\partial^2 A_{cl}}{\partial x'\,\partial x''} \end{bmatrix}, \qquad \begin{matrix} x' = x'' = x_0, \\ \varphi' = \varphi'' = \varphi_0. \end{matrix} \tag{4.22}$$

After some algebra, one finds:

$$\Delta_1 = \frac{sL}{\pi l}. \tag{4.23}$$

Inserting this into (4.17), we obtain

$$G(E) = i \operatorname{const} L \sum_{n=-\infty}^{\infty} \sum_{l=1}^{\infty} \int d\mu \frac{\mu^2}{(1-\mu^2)^{1/2}} \frac{1}{E^2} \times \exp\left\{ i\left[-2\pi l\left(\frac{16\mu^2 s^3}{E}\right) + 4nsL \sin^{-1}\mu \right]\right\}. \tag{4.24}$$

The sum over l is a geometric series and for the summation over n we use the Poisson summation formula

$$\sum_{n=-\infty}^{\infty} e^{inx} = 2\pi \sum_{n=-\infty}^{\infty} \delta(x - 2\pi n). \tag{4.25}$$

Consequently, (4.24) becomes

$$G(E) = 2\pi i \operatorname{const} \sum_{n} \int d\mu \frac{\mu^2}{(1-\mu^2)^{1/2}} \frac{1}{E^2} \frac{e^{-2\pi(16s^3\mu^2/E)}}{1 - e^{-2\pi i(16s^3\mu^2/E)}}$$
$$\times\, \delta\left(4s \sin^{-1}\mu - \frac{2\pi n}{L}\right). \tag{4.26}$$

This is our final formula for $G(E)$. The spectrum is now easily obtained by looking at the poles of $G(E)$. They are given by

$$\frac{16s^3\mu^2}{E} = m, \qquad m = 1, 2, \ldots, \tag{4.27}$$

with

$$4s \sin^{-1}\mu = \frac{2\pi n}{L} \equiv p_n, \qquad n = 0, \pm 1, \ldots . \tag{4.28}$$

Thus we obtain the following dispersion curve ($p_n \to p$)

$$E_m(p) = 8s^3 \frac{1 - \cos(p/2s)}{m}, \qquad m = 1, 2, \ldots, \tag{4.29}$$

which provides the complete excitation spectrum for the continuous Heisenberg chain.

The classical version of this result is the dispersion relating the basic observables of the theory, the total energy momentum and spin [5]. Calculating the momentum (3.21) over the one-soliton solution (2.31), the table (2.32) is completed to be

$$\begin{aligned} \mathscr{E} &= 4\epsilon s^2, & Q_3 &= \frac{4\epsilon s}{\epsilon^2 + v^2/4s^2}, \\ p &= 4s \tan^{-1}\left(\frac{2\epsilon s}{v}\right), & Q &= 2\epsilon v s^2. \end{aligned} \tag{4.30}$$

For fixed s, the three observables ϵ, Q_3 and p are expressed in terms of two independent parameters, ϵ and v. Therefore, there exists a relation among them, which is easily found to be

$$\mathscr{E} = 8s^3 \frac{1 - \cos(p/2s)}{Q_3}. \tag{4.31}$$

Although this procedure does not explain the quantization of Q_3, it essentially reproduces Eq. (4.29).

As will be shown below, the leasing semi-classical approximation given by (4.29) appears to yield essentially the exact spectrum of the theory.
For this reason, any further considerations concerning the quantum corrections will be omitted from our discussion. However, the reader, who might be interested in this issue, will find the solution of the problem of small oscillations in our appendix.

It was mentioned already that this semi-classical spectrum agrees with Bethe's exact solution, which for a discrete spin one-half chain reads

$$\epsilon_m(K_n) = 2J\frac{1-\cos K_n}{m}\,. \tag{4.32}$$

Here, $K_n = 2\pi n/N$, $n = 0, \pm 1, \ldots N$, N is the total number of sites taken to be large, and $m = 1, 2\ldots$. To compare with our result for the continuous chain, we first write the classical Lagrangian for the discrete case

$$L = \sum_{i=1}^{N}\left[\frac{S_{2,i}\,\partial_t S_{1,i}}{\frac{1}{2}+S_{3,i}} + 2J\vec{S}_i\cdot\vec{S}_{i+1}\right], \tag{4.33}$$

where the constraint $\vec{S}_i^{\,2} = \frac{1}{4}$ is assumed and J is a coupling constant. To take the continuum limit we define

$$\vec{S}(ai) = \frac{1}{a}\vec{S}_i\,, \tag{4.34}$$

where a is the lattice spacing. With the additional specification

$$J = \frac{1}{2a^3}\,. \tag{4.35}$$

The Lagrangian (4.33) reads

$$L = a\sum_{i=1}^{N}\left\{\frac{S_2(ai)\,\partial_t S_1(ai)}{1/2a + S_3(ai)} - \frac{1}{2}\left[\frac{\Delta\vec{S}(ai)}{a}\right]^2\right\}. \tag{4.36}$$

In the continuum limit, (4.36) leads to our previous Lagrangian (4.2). For the special value of the coupling constant (4.35), and writing $K_n = 2\pi n/N = (2\pi n/L)\,a = pa$, Bethe's formula (4.32) reads

$$\epsilon_m(p) = \frac{1}{a^3}\frac{1-\cos(pa)}{m}\,, \qquad m = 1, 2, \ldots, \tag{4.37}$$

whereas the semi-classical spectrum for spin one-half is obtained from (4.29) by setting $s = 1/2a$. It is thus found to coincide with the exact spectrum.

V. Conclusion

The successful treatment of the Heisenberg spin model, with the present semi-classical method, encourages one to study other field theories, with internal degrees of freedom, in a similar fashion. Of current interest is the system of static colored quarks coupled to the Yang-Mills field. The color degrees of freedom may be treated using the first functional integral representation (3.3) which then leads to a classical mechanics over a Grassman algebra. In some sense Adler's recent proposal [24] may have some connection with such an approach.

However, it appears much simpler to work with the alternative c-number representation (3.17). A semi-classical approach would then require the solution of ordinary classical equations, in fact those constructed in [25]. Khriplovich [26] has recently discussed a particular Ansatz for the solution of this complex system of equations. An attempt to understand the quantum meaning of his Ansatz was made in [27]. In our opinion, the interesting question is to search for more general, non-perturbative solutions of the classical equations including static quark sources. The quantization can then be done in the manner discussed in the present paper. Such solutions do not exist at present. However, we have analyzed in this way the old static strong coupling theories and in that case one easily rederives the well-known isobar energy levels.

Appendix: Small Fluctuations

The Gaussian integration requires the solution of the linearized equations, around the one-soliton solution. The construction of the eigenfunctions and, thereby, the identification of the stability angles and phase shifts [9] is performed systematically through the inverse scattering formalism. The method used in the following was implied by the considerations of [29] and later amplified in [30, 31].

Specifically, a solution of the non-linear equations is parametrized by $\rho(\lambda) = b(\lambda, 0)/a(\lambda)$ and the discrete parameters m and ζ (the notation is explained in Section IIC). We denote it by $S^a(; \rho(\lambda), m, \zeta)$. The one-soliton solution is given by S^a ; $\rho(\lambda) = 0$, m, ζ). The linearized equations are solved by the eigenfunctions $\delta_\rho S^a$, $\partial_m S^a$ and $\partial_\zeta S^a$ calculated at $\rho(\lambda) = 0$. The variation with respect to the discrete parameters may be performed directly by using the explicit expressions for the one-soliton solution. A less trivial calculation is required for $\delta_\rho S^a |_{\rho=0}$ and is described below.

Consider the GLM equation (2.23), for the more general kernel (2.24) which is written as

$$F(x, t) = \frac{1}{2\pi} \int_{-\infty}^{+\infty} \frac{d\lambda}{\lambda} \rho(\lambda)\, e^{i(\lambda x - 4\lambda^2 t)} + \frac{m(t)}{\zeta} e^{i\zeta x}. \tag{A.1}$$

Varying (2.23) with respect to $\rho(\lambda)$ and setting $\rho(\lambda) = 0$ (this is implied in the following) we obtain the equation

$$\delta K(x, y) + \delta\Phi_1(x + y) + \int_x^\infty dz\{\delta K(x, z)\, \Phi_2(x + z) + K_1(x, z)\, \delta\Phi_2(z + y)\} = 0. \tag{A.2}$$

Here, K_1 is the one-soliton kernel of Eq. (2.28), Φ_2 is given by (2.25) and (A.1) with $\rho = 0$, whereas $\delta\Phi_1$ and $\delta\Phi_2$ are easily calculated to be

$$\delta\Phi_1(x+y) = \delta_\rho\Phi_1(x+y)\,|_{\rho=0} = \frac{1}{2\pi\lambda}\begin{bmatrix}0 & 0\\ e^{i(\lambda x-4\lambda^2 t)} & 0\end{bmatrix}\begin{bmatrix}e^{i\lambda y} & 0\\ 0 & 0\end{bmatrix},$$

$$\delta\Phi_2(x+y) = \delta_\rho\Phi_2(x+y)\,|_{\rho=0} = \frac{1}{2\pi}\begin{bmatrix}0 & 0\\ e^{i(\lambda x-4\lambda^2 t)} & 0\end{bmatrix}\begin{bmatrix}e^{i\lambda y} & 0\\ 0 & 0\end{bmatrix}. \qquad \text{(A.3)}$$

Hence, Eq. (A.2) is a linear GLM equation for the unknown kernel $\delta K(x, y)$. It is solved in the standard manner. From this point on the computations are straightforward, but somewhat lengthy. We only state the result. Denote by $\Lambda(x, x) = \delta K(x, x)$:

$$\Lambda(x,x) = \frac{e^{i\phi_2}}{2\pi\lambda(\lambda-\zeta^*)^2(1+\mu^2)^2} \times \begin{bmatrix} -i\mu\dfrac{\lambda}{\zeta^*}\,|\zeta-\zeta^*|\,A_1e^{-i\phi_1} & \mu^2\dfrac{\lambda}{\zeta^*}\,|\zeta-\zeta^*|^2\,e^{-2i\phi_1} \\ -\dfrac{A_1A_2}{\zeta} & -i\,\dfrac{|\zeta-\zeta^*|}{\zeta}\,\mu A_2 e^{-i\phi_1}\end{bmatrix},$$

where

$$\begin{aligned}\phi_2 &\equiv 2\lambda x - 4\lambda^2 t,\\ A_1 &\equiv (\lambda-\zeta^*) + \mu^2(\lambda-\zeta),\\ A_2 &\equiv \zeta(\lambda-\zeta^*) + \mu^2\zeta^*(\lambda-\zeta).\end{aligned} \qquad \text{(A.4)}$$

The rest of the notation is that of Sec. IIC. The final result is obtained by varying Eq. (2.27) with respect to ρ and using $\delta K(x, x) = \Lambda(x, x)$ from (A.4). One finds:

$$\begin{aligned}\delta_\rho S_3\,|_{\rho=0} &= \frac{e^{i(\phi_2-\phi_1)}}{\pi\lambda(\lambda-\zeta^*)^2}\cdot\frac{|\zeta-\zeta^*|}{\zeta\zeta^*}\cdot\frac{\mu}{(1+\mu^2)^3}\\ &\quad\times\{\lambda a a^* - (1+\mu^2)[\zeta\zeta^* a + \lambda^2 a^*] + \lambda\zeta\zeta^*(1+\mu^2)^2 - \lambda\mu^2\,|\zeta-\zeta^*|^2\},\\ \delta_\rho S_+\,|_{\rho=0} &= \frac{ie^{i\phi_2}}{\pi\lambda(\lambda-\zeta^*)^2}\,\frac{1}{\zeta^2}\,\frac{1}{(1+\mu^2)^3}\,[\lambda a^* - (1+\mu^2)\,\zeta\zeta^*]\\ &\quad\times[-aa^* + \lambda(1+\mu^2)\,a^* + \mu^2\,|\zeta-\zeta^*|^2],\\ \delta_\rho S_-\,|_{\rho=0} &= \frac{ie^{i(\phi_2-2\phi_1)}}{\pi\lambda(\lambda-\zeta^*)^2}\,\frac{\lambda\,|\zeta-\zeta^*|^2}{\zeta^{*2}}\,\frac{\mu^2}{(1+\mu^2)^3}\,[\lambda(1+\mu^2) - 2a],\\ S_\pm &= S_1 \pm iS_2\,, \qquad a \equiv \zeta^* + \mu^2\zeta.\end{aligned} \qquad \text{(A.5)}$$

As a check of consistency, one may verify that

$$\tfrac{1}{2}\delta_\rho S^2 = S_3\delta_\rho S_3 + \tfrac{1}{2}[S_-\delta_\rho S_+ + S_+\delta_\rho S_-] = 0. \qquad \text{(A.6)}$$

For the identification of the stability angles and phase shifts, it is more convenient to use a representation for the spin variable that was previously employed for discussions at the operator level [32]. Introduce a complex field ϕ through

$$S_3 = 1 - \phi^*\phi, \qquad S_+ = (2 - \phi^*\phi)^{1/2}\,\phi \tag{A.7}$$

hence,

$$\phi = \frac{S_+}{(1 + S_3)^{1/2}}. \tag{A.8}$$

Notice that our present computation is done for $S^2 = 1$. In terms of ϕ, the Lagrangian may be written as

$$L = \frac{i}{2}\,\phi^*\,\overleftrightarrow{\partial_t}\phi - H(S(\phi)). \tag{A.9}$$

The analogy with our considerations in Section III is, of course, evident. For a symmetric treatment we further use real spinors $\{{\alpha \atop \beta}\}$ defined from

$$\alpha = \frac{\phi + \phi^*}{2} = \frac{1}{2}\frac{S_+ + S_-}{(1 + S_3)^{1/2}}, \qquad \beta = \frac{\phi - \phi^*}{2i} = \frac{S_+ - S_-}{2i(1 + S_3)^{1/2}}. \tag{A.10}$$

The solution of the linearized problem is now given by

$$\chi^{(+)} = \delta_\rho \begin{pmatrix}\alpha \\ \beta\end{pmatrix} \qquad \text{and} \qquad \chi^{(-)} = \delta_{\rho^*} \begin{pmatrix}\alpha \\ \beta\end{pmatrix} = [\chi^{(+)}]^*. \tag{A.11}$$

One finds:

$$\chi^{(+)} = \frac{1}{4(1 + S_3)^{3/2}} \left\{ \begin{matrix} 2(1 + S_3)\,\delta_\rho(S_+ + S_-) - (S_+ + S_-)\,\delta_\rho S_3 \\ -2i(1 + S_3)\,\delta_\rho(S_+ - S_-) + i(S_+ - S_-)\,\delta_\rho S_3 \end{matrix} \right\}_{\rho=0}. \tag{A.12}$$

Inserting in (A.12) the value for S given in (2.30) and $\delta_\rho S$ from (A.5), the solution of the small oscillations problem is complete.

Restoring the value s for the spin, $S^2 = s^2$, the stability angles are given by

$$\nu = \pm 4s\lambda^2\tau, \qquad \tau \equiv \frac{2\pi}{s[\epsilon^2 + v^2/4s^2]} \tag{A.13}$$

and the corresponding phase shift by

$$\delta = 2\arg\left[\frac{\zeta^*(\lambda - \zeta)}{\zeta(\lambda - \zeta^*)}\right] = 4\left\{\operatorname{arctg}\left(\frac{2\epsilon s}{v}\right) + \operatorname{arctg}\left[\frac{\epsilon}{2(\lambda - v/4s)}\right]\right\}. \tag{A.14}$$

As usual, the sum over the stability angles contains divergences originating from the vacuum fluctuations as well as the short-distance singularities of the present continuous theory. The former are trivial to identify, whereas the latter require a more careful study of the renormalization structure of the theory. We have not investigated this question in detail, but it seems that these renormalization counter-terms essentially cancel the above contribution and consequently, one obtains no effect from the first quantum correction as was the case in Ref. [23].

Note: We have recently learned about the work of J. Klaudeı on C-number path integrals for spin. He used coherent states to obtain a path integral representation which is equal to our exp. (3.17). For a careful discussion and earlier references the reader should consult Ref. [33].

ACKNOWLEDGMENT

We are grateful to L. D. Faddeev for informative conversations on this subject.

REFERENCES

1. H. A. BETHE, *Z. Phys.* **71** (1931), 205.
2. M. LÜSCHER, *Nucl. Phys. B* **117** (1976), 475.
3. K. NAKAMURA AND T. SAWADA, *Phys. Lett. A* **48** (1974), 321.
4. M. LAKSHMANAN, TH. W. RUIJGROK, AND C. J. THOMPSON, *Physica The Hague A* **84** (1976), 577; AND M. LAKSHMANAN, *Phys. Lett. A* **61** (1977), 53.
5. J. TJON AND J. WRIGHT, *Phys. Rev. B* **15** (1977), 3470.
6. L. A. TAKHTAJAN, *Phys. Lett. A* **64** (1977), 235.
7. K. POHLMEYER, *Comm. Math. Phys.* **46** (1976), 207.
8. A. NEVEU AND N. PAPANICOLAOU, *Comm. Math. Phys.* **58** (1978), 31.
9. R. DASHEN, B. HASSLACHER, AND A. NEVEU, *Phys. Rev. D* **11** (1975), 3424.
10. M. LÜSCHER AND K. POHLMEYER, *Nucl. Phys.*, in press.
11. M. LÜSCHER, *Nucl. Phys.*, in press.
12. R. RAJARAMAN, *Phys. Rep. C* **21** (1975), 227.
13. A. JEVICKI, *in* "The Significance of Non-linearity in the Natural Sciences," Orbis Scientiae Plenum, New York, 1977.
14. L. D. FADDEEV AND V. E. KOREPIN, *Phys. Rep. C* **42** No. 1, 1978.
15. R. DASHEN, B. HASSLACHER, AND A. NEVEU, *Phys. Rev. D* **12** (1975), 2443.
16. F. A. BEREZIN AND M. S. MARINOV, Particle spin dynamics as the Grassman variant of classical mechanics, ITEP Preprint, 1976.
17. J. SCHWINGER, "On Angular Momentum," AEG Report NYO-3071, 1952.
18. L. D. FADDEEV, *Theor. Math. Phys. (USSR)* **1** (1969), 3.
19. See, e.g., G. F. VALENTI AND M. LAX, *Phys. Rev. B* **16** (1977), 4936.
20. A. A. KIRILLOV, *Usp. Mat. Nauk.* **17** (1962), 57.
21. B. KOSTANT, *in* Lecture Notes in Mathematics No. 170, Springer–Verlag, New York/Berlin, 1970.
22. J. M. SOURIAU, "Structure des Systèmes Dynamiques," Dunod, Paris, 1970.
23. C. NOHL, *Ann. Physics (N.Y.)* **96** (1976), 237.
24. S. L. ADLER, *Phys. Rev. D* **17** (1978), 3212.
25. S. K. WONG, *Nuovo Cimento A* **55** (1970), 689.
26. I. B. KHRIPLOVICH, *Sov. JETP* **74** (1978), 1.
27. R. GILES AND L. MCLERRAN, MIT Preprint, 1978.
28. See however: P. SENJANOVIČ, *Nucl. Phys. B* **116** (1976), 365, and Ref. [16].
29. L. D. FADDEEV AND V. E. ZAKHAROV, *Functional Anal. Appl.* **5** (1972), 280.
30. H. FLASCHKA AND A. C. NEWELL, *in* "Dynamical Systems, Theory and Applications" (Battele Seattle 1974 Rencontres) (J. Moser, Ed.), p. 441, Springer–Verlag, New York, 1975 AND A. G. NEWELL, *in* "Proceedings, N. S. F. Conference on Solitons, Tucson, Arizona" (H. Flaschka and D. W. McLaughlin, Eds.), 1975.
31. J. P. KEENER AND D. W. MCLAUGHLIN, Arizona preprint, 1976.
32. T. HOLSTEIN AND H. PRIMAKOFF, *Phys. Rev.* **58** (1940), 1098.
33. J. KLAUDER, Path integrals and stationary phase approximations, Bell Lab., preprint, 1978.

VOLUME 45, NUMBER 13 PHYSICAL REVIEW LETTERS 29 SEPTEMBER 1980

Bohr-Sommerfeld Quantization of Pseudospin Hamiltonians

R. Shankar
J. Willard Gibbs Laboratory, Yale University, New Haven, Connecticut 06520
(Received 4 June 1980)

It is shown here how to map the problem with pseudospin J into an equivalent one in which $1/J$ plays the role of $\hbar$ and canonical variables exist at the classical level. Bohr-Sommerfeld quantization of the equivalent theory is found to produce a spectrum in very good agreement with the exact results for the Lipkin-Meshkov-Glick model at $J = 15$ and 25. The method readily extends to the SU(n) case.

PACS numbers: 03.65.Ca

Consider the eigenvalues of a Hamiltonian expressed in terms of the generators of SU(n) in the symmetrized N-fold tensor product of the fundamental representation. (Such a problem arises in describing a system of N identical particles, each of which may be in one of n states.) I develop here a Bohr-Sommerfeld quantization procedure in which $1/N$ plays the role of $\hbar$ and is thus complementary to numerical methods which are good for small N. To illustrate the idea, I will work with the Lipkin-Meshkov-Glick (LMG) Hamiltonian[1]

$$H = \epsilon[J_z + (r/2J)(J_x^{\,2} - J_y^{\,2})], \tag{1}$$

where J_i are SU(2) generators of dimensionality $2J+1$, with $J = N/2$, N being the conserved particle number. Exact solution is possible if J is modest (by explicit matrix diagonalization[1]) or if $r = 0$, when it becomes a trivial example of a class of models solvable by group-theoretic methods alone.[2] Here we are concerned with a reliable approximation scheme for J large, $r \neq 0$. Using the Bloch coherent states[3]

$$|\Omega\rangle = \exp[\tfrac{1}{2}\theta(J_- e^{i\Phi} - J_+ e^{-i\Phi})]\,|JJ\rangle, \tag{2}$$

Lieb[4] showed that in general $Z_q = (2J+1)^{-1} \times \mathrm{Tr}\exp(-\beta H)$, the quantum partition function, can be bracketed by two classical partition functions:

$$\int \frac{d\Omega}{4\pi} \exp[-\beta H_Q(\Omega)] \leq Z_q \leq \int \frac{d\Omega}{4\pi} \exp[-\beta H_P(\Omega)], \tag{3}$$

where $H_Q(\Omega) = \langle\Omega|H|\Omega\rangle$ and $H_P(\Omega)$ is defined by

$$H = [(2J+1)/4\pi]\int d\Omega\,|\Omega\rangle H_P(\Omega)\langle\Omega|. \tag{4}$$

In our case we get, from Ref. 4,

$$H_P = \epsilon(J+1)[\cos\theta + \tfrac{1}{2}r(1+3/2J)\sin^2\theta\cos 2\Phi], \tag{5a}$$

$$H_Q = \epsilon J[\cos\theta + \tfrac{1}{2}r(1-1/2J)\sin^2\theta\cos 2\Phi]. \tag{5b}$$

The $\beta \to \infty$ limit of Eq. (3) brackets E_g, the ground-state energy:

$$\min_\Omega\{H_P\} \leq E_g \leq \min_\Omega\{H_Q\}. \tag{6}$$

Fend and Gilmore[5] studied this inequality numerically and found that $\min\{H_Q\}$ is a very good approximation to E_g for large J. As $J \to \infty$, $H_P/J = \hat{H}_P$ and $H_Q/J = \hat{H}_Q$ approach a common limit $\hat{H}$ and

$$\lim_{J\to\infty}(E_g/J) = \min_\Omega\{\hat{H}(\Omega)\}. \tag{7}$$

That the exact quantum ground state can be found by minimizing a classical $\hat{H}$ in the limit $1/J \to 0$ is reminiscent of the way any quantum problem becomes classical as $\hbar \to 0$. It is then natural to ask the following question: *Is there some equivalent quantum theory in which 1/J plays the role of* $\hbar$? Finding such a theory would help us move off the $J = \infty$ limit to a region of small but nonzero $1/J$. And if the equivalent theory were described by canonical variables at the classical level, Bohr-Sommerfeld (BS) quantization would give a good estimate for *all* the levels, not just the ground state. Here is how we find that theory.

We first write Z_q as a path integral:

$$Z_q = \frac{1}{2J+1}\mathrm{Tr}\,e^{-\beta H} = \int \frac{d\Omega}{4\pi}\langle\Omega|e^{-\beta H}|\Omega\rangle = \lim_{n\to\infty}(2J+1)^n \int \frac{d\Omega}{4\pi}\frac{d\Omega_1}{4\pi}\cdots\frac{d\Omega_n}{4\pi}\langle\Omega|(1-\epsilon H)|\Omega_n\rangle \times\langle\Omega_n|(1-\epsilon H)|\Omega_{n-1}\rangle\cdots\langle\Omega_1|(1-\epsilon H)|\Omega\rangle, \tag{8}$$

where $\epsilon = \beta/(n+1)$ and the expansion of the identity I [Eq. (4), with H set equal to I] has been used n times. The multiple integral expresses the Euclidean transition amplitude $\langle\Omega|e^{-\beta H}|\Omega\rangle$ as a sum over discretized paths which leave Ω at Euclidean time $\tau = 0$, pass Ω_i at τ_i, and return to Ω at time $\tau = \beta$.

(Recall $\tau = it$, and t is the Minkowski time). *Although these paths are generally nondifferentiable, let us write Z_q in a form that is appropriate to differentiable paths.* The reason will be clear in a moment. To order ϵ, we get

$$\langle \Omega_{i+1}|(1-\epsilon H)|\Omega_i\rangle = \langle \Omega_{i+1}|\Omega_i\rangle - \epsilon\langle \Omega_{i+1}|H|\Omega_i\rangle \tag{9a}$$

$$= 1 - i\epsilon J(1-\cos\theta)\dot{\Phi} - \epsilon H_Q(\Omega_i)$$

$$\cong \exp\{J[-i(1-\cos\theta)\dot{\Phi} - \hat{H}_Q(\Omega)]\epsilon\}, \tag{9b}$$

where $\langle \Omega_{i+1}|\Omega_i\rangle$ is calculated with use of

$$\langle \Omega'|\Omega\rangle = (\cos\tfrac{1}{2}\theta'\cos\tfrac{1}{2}\theta + e^{i(\Phi-\Phi')}\sin\tfrac{1}{2}\theta'\sin\tfrac{1}{2}\theta)^{2J}. \tag{10}$$

Upon dropping the $\dot{\Phi}$ term, because it is a total derivative and will be irrelevant in a classical Lagrangian, we get, in obvious notation,

$$Z_q = \int (d\Omega/4\pi)\int \mathfrak{D}\Omega \exp\{J\int_0^\beta [(i\cos\theta)\dot{\Phi} - \hat{H}_Q]\,d\tau\}. \tag{11}$$

It is clear from the path integral that the original theory has been transformed into one in which $1/J$ plays the role of $\hbar$ and for which the Minkowski-space Lagrangian is

$$L = (\cos\theta)\dot{\Phi} - \hat{H}_Q. \tag{12}$$

Knowing the effective Planck's constant $1/J$, and the action functional for continuous paths, we are ready to do Bohr-Sommerfeld quantization, which should be good when $1/J$ is small. This is why we wanted Z_q for continuous paths, and not because only continuous paths contribute to the integral. (In fact, such paths have zero measure. But they are all important for the classical limit as well as Bohr-Sommerfeld quantization.)

Proceeding along, we see that L has no kinetic term in θ. *In fact, Eq. (12) is just the Legendre transform from $\hat{H}_Q$ to L; Φ and $p \equiv \cos\theta$ are canonical variables obeying Hamilton's equations*

$$\dot{p} = -\partial\hat{H}_Q/\partial\Phi, \quad \dot{\Phi} = \partial\hat{H}_Q/\partial p. \tag{13}$$

The Bloch sphere is thus the *phase space* for this theory. Notice that it is *compact*. In this compact space, the BS condition is

$$\oint p\,d\Phi = 2\pi n/J, \quad n = 0, \pm 1, \pm 2, \ldots, \pm J, \tag{14}$$

where the upper bound on $|n|$ comes from the fact that $|p| \leq 1$. Notice also that in Eq. (11), $\int\mathfrak{D}\Omega \sim \int\mathfrak{D}p\,\mathfrak{D}\Phi$ is a sum over paths in *phase space*. This is in fact the form of the path integral in general. Only for the case where $H(p,q) = \frac{1}{2}p^2 + V(q)$ can one do the Gaussian functional integration over $\mathfrak{D}p$ explicitly and be left with the familiar Feynman sum over paths in configuration space, i.e., an integral over $\mathfrak{D}q$ with $(i/\hbar)\int dt \times L(q;\dot{q})$ in the exponential.

Here are the results of BS quantization of the LMG Hamiltonian [Eq. (5b)]. On a trajectory labeled by $\hat{H}_Q = \hat{E}_Q$,

$$p = \frac{\bar{r}^{-1} \pm [\bar{r}^{-2} + \cos 2\Phi(\cos 2\Phi - 2\hat{E}_Q/\epsilon\bar{r})]^{1/2}}{\cos 2\Phi}, \tag{15}$$

where

$$\bar{r} = r(1 - 1/2J). \tag{16}$$

Since $\hat{E}_Q \to -\hat{E}_Q$ under $p \to -p$ and $\Phi \to \Phi + \pi/2$, it is clear that the BS levels will have mirror symmetry about $\hat{E}_Q = 0$. Notice also that $\Phi \to \Phi + \pi$ does not change $\hat{E}_Q$.

In the trivial case $r = \bar{r} = 0$, it is easy to see that

$$\hat{E}_Q = n\epsilon/J, \quad n = 0, \pm 1, \pm 2, \ldots, \pm J, \tag{17}$$

which agrees with exact values of E/J. The BS orbits in this case are $2J+1$ equally spaced (in $p = \cos\theta$) latitudes on the Bloch sphere, with $\hat{E}_Q = \pm\epsilon$ being represented by $p = \pm 1$ (the poles). We shall call these *global* orbits because they go around the north pole or, equivalently, because Φ grows monotonically. For $0 < \bar{r} \leq 1$, only the minus sign in Eq. (15) satisfies $|p| \leq 1$. The orbits are again global. But now $p(\Phi)$ oscillates with maxima at $\Phi = \pm\pi/2$ and minima at $\Phi = 0$ and π, except for the $\hat{E}_Q = \pm\epsilon$ orbits which stay at the poles. For $\bar{r} > 1$, these too begin to oscillate. Four extra sets of orbits are formed in the void so created (Fig. 1). Their entry at $\bar{r} = 1$ signifies the phase transition noted by Feng and Gilmore.[5] There are two degenerate families in the upper hemisphere which circulate around the points ($p = 1/r$, $\Phi = 0$ or π) and have $\epsilon < \hat{E}_Q \leq \frac{1}{2}\epsilon(\bar{r} + 1/\bar{r})$ and two other degenerate families in the southern hemisphere which circulate around ($p = -1/\bar{r}$, $\Phi = \pm\pi/2$) with $-\frac{1}{2}\epsilon(r + 1/r) \leq \hat{E}_Q < -\epsilon$. Such *local* orbits are possible in this case because both

VOLUME 45, NUMBER 13 PHYSICAL REVIEW LETTERS 29 SEPTEMBER 1980

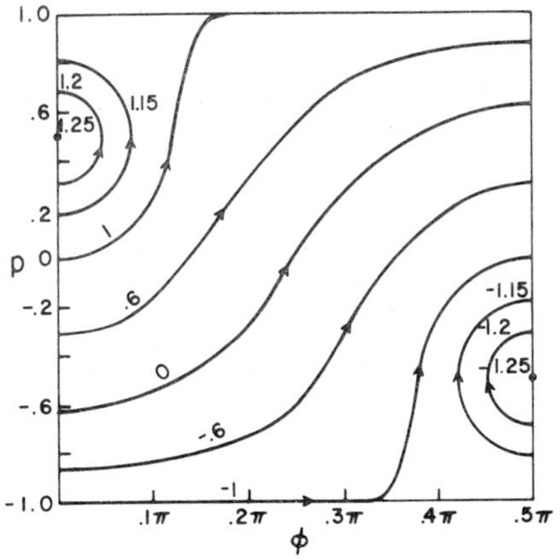

FIG. 1. A few orbits for $\bar{r}=2$, $J=25$ labeled by $\hat{E}=E/J\epsilon$. At $|\hat{E}|=1$, global orbits stop and local orbits begin. Only the region $0\leqslant\Phi\leqslant\pi/2$ is shown. Reflection on the line $\Phi=\pi/2$ gives the curves for $\pi/2\leqslant\Phi\leqslant\pi$, while invariance under $\Phi\to-\Phi$ gives curves for $-\pi\leqslant\Phi\leqslant 0$.

TABLE I. Positive energy eigenvalues (in units of ϵ) of the LMG Hamiltonian for $r=0.6$, 1, and 5 and $J=15$, compared with results of BS quantization.

$r=0.6$		$r=1$		$r=5$	
BS	LMG	BS	LMG	BS	LMG
15	15.1	15	15.3	37.8	38.0
14.2	14.3	14.5	14.8	37.8	38.0
13.3	13.4	13.8	14.1	31.3	31.4
12.4	12.5	13.0	13.3	31.3	31.4
11.4	11.5	12.1	12.4	25.3	25.4
10.5	10.5	11.2	11.4	25.3	25.4
9.5	9.5	10.2	10.4	20.0	20.1
8.5	8.5	9.1	9.3	20.0	20.0
7.5	7.5	8.1	8.3	16.1[a]	16.1
6.4	6.5	6.9	7.1	14.3	15.2
5.4	5.4	5.8	6.0	12.2	12.6
4.3	4.3	4.7	4.8	10.0	10.4
3.2	3.3	3.5	3.6	7.6	7.9
2.2	2.2	2.3	2.4	5.2	5.3
1.1	1.1	1.2	1.2	2.6	2.7
0.0	0.0	0.0	0.0	0.0	0.0

[a]Degenerate partner missing, presumed to be unbound by tunneling.

signs are allowed in Eq. (15). (The distinction between local and global orbits is not absolute and depends on our coordinate system.) Figure 1 shows some of the orbits for the case $J=25$, $\bar{r}=2$. (The Bloch sphere has been opened into a rectangle.)

Table I compares the positive energy eigenvalues from the BS calculation with the exact results from LMG[1] for $E=J\hat{E}_Q$ at $J=15$ and $r=0.6$, 1 and 5. In the first two cases the agreement is uniformly good. Notice that states with small quantum numbers are not any worse predicted. This is because small n does not mean a small global orbit. In the third case there are the following problems in the transition region ($\hat{E}_Q\simeq\epsilon$) between local and global orbits: (i) The level at $E/\epsilon=20.0$ is not exactly degenerate, (ii) a few global orbits below the transition point seem to be raised upwards in the exact answer, and, more seriously, (iii) the degenerate partner of the BS level at $E/\epsilon=16.1$ is missing in the exact answer. All these problems are presumably due to tunneling and mixing of orbits in this energy region.

The calculations were repeated at $J=25$. Not surprisingly, the same general features prevailed, with uniformly smaller errors. (Analysis at $J=25$, $\bar{r}=2$ showed that the BS analysis gives one state *less* unless I accept a BS orbit with $n=18.8$ instead of 19. Some improvisation seems inevitable in the transition region.)

This would be the end of the present discussions were it not for the fact that *there exists another continuum theory that follows from the same Z_q, which differs only in that $\hat{H}_Q\to\hat{H}_P$ in Eq. (11)*. To show this, we must, instead of inserting the identity between the factors $(1-\epsilon H)$, write each one as

$$1-\epsilon H = [(2J+1)/4\pi]\int d\Omega_i|\Omega_i\rangle[1-\epsilon H_P(\Omega_i)]\langle\Omega_i| \quad (18)$$

and work to order ϵ. That we have two continuum representations for the same theory is not a contradiction, since continuum formulas like Eq. (11) are merely formal and need to be defined by a discretization procedure. This procedure will differ in the two cases in such a way as to lead back to the same Z_q. On the other hand, with respect to BS quantization, which is not exact, $\hat{H}_P$ and $\hat{H}_Q$ offer two inequivalent possibilities. What would happen if the above analysis were repeated with $\hat{H}_P$? As for the ground state, which will be given by a pointlike BS orbit at the minimum on the Bloch sphere, we have from Lieb, Eq. (6),

$$\hat{E}_g^{\,P}\leqslant E_g/J\leqslant\hat{E}_g^{\,Q}. \quad (19)$$

As for the highest state, we have, either from mirror symmetry in this problem, or in general, from the $\beta \to -\infty$ limit of Eq. (3) (which exists for a system with an upper bound on the energy),

$$\hat{E}_h^Q \leq E_h/J \leq \hat{E}_h^P. \tag{20}$$

Mirror symmetry, plus the fact that $2J+1$ is odd implies that there is a level at $E/J=0$ in all three cases. Given the above, I conjecture that every exact level is bracketed by the two BS levels, with $\hat{H}_P$ generating the lower bound for negative energies and $\hat{H}_Q$ generating the lower bound for positive energies. Explicit computation shows that such is the case. I also find that $\hat{H}_Q$ generates values much closer to the exact ones (a feature noted by Feng and Gilmore[5] for the ground state) which is why these values are listed in Table I. (From the derivation, it must be clear that $\hat{H}_P$ and $\hat{H}_Q$ do not exhaust the possible classical Hamiltonians; that they merely bracket the continuum of possibilities.)

Since completing this work, I have learned of several pieces of interesting and related work, which will not be discussed here because of lack of space: that of Jevicki and Papanicolaou[6] on path integrals for spin, that of Levit, Negels, and Paltiel[7] on a BS calculation with use of mean-field variables, that of Kuratsuji and Mizobuchi[8] on the semiclassical treatment of spin path integrals, and the pioneering work of Klauder[9] on coherent-state path integrals. Professor F. Iachello informs me that Gilmore[10] has found the classical Lagrangian that governs the evolution of the coherent states in the time-dependent Hartree-Fock approximation for SU(2) and other groups. His L coincides with mine, as it should. Of special significance is his finding that in all cases $L = \sum_i p_i \dot{q}_i - \hat{H}_Q$, which means that the Block sphere is a classical phase space. However, BS quantization in the cases with more degrees of freedom will be harder.[11] Lastly, there is the work of Kan *et al.*,[12] which reaches many similar conclusions from the time-dependent Hartree-Fock approach and also clarified the revision between the two approaches.

I am grateful to Itzhak Bars for useful conversations, to Chia Tze for telling me about Ref. 8, and to Francesco Iachello, whose seminar stimulated this work, and whose continued enthusiasm was indispensable. This work was supported in part by the U. S. Department of Energy under Contract No. EY-76-C-02-3075.

[1] H. J. Lipkin, N. Meshkov, and A. J. Glick, Nucl. Phys. 62, 188 (1965).
[2] F. Iachello, Phys. Rev. Lett. 44, 772 (1980), and references therein.
[3] J. R. Klauder, J. Math. Phys. (N. Y.) 1, 1058 (1963); J. M. Radcliffe, J. Phys. A 4, 313 (1971); F. T. Arecchi, E. Courtens, R. Gilmore, and H. Thomas, Phys. Rev. A 6, 2211 (1972).
[4] E. Lieb, Commun. Math. Phys. 31, 327 (1973).
[5] R. Gilmore and D. H. Feng, Phys. Lett. 76B, 26 (1978).
[6] A. Jevicki and N. Papanicolaou, Ann. Phys. (N. Y.) 120, 107 (1979).
[7] S. Levit, J. W. Negele, and Z. Paltiel, Phys. Rev. C 21, 1603 (1980).
[8] H. Kuratsuji and Y. Mizobuchi, Kyoto University Report No. KUNS-519 (to be published).
[9] J. R. Klauder, Phys. Rev. D 19, 2349 (1979). Consult references therein for his earlier work.
[10] R. Gilmore, unpublished work, as communicated by F. Iachello.
[11] M. C. Gutzwiller, in *Path Integrals*, edited by G. J. Papadopoulos and J. T. Devreese (Plenum, New York, 1978).
[12] K. K. Kan, J. J. Griffin, P. C. Lichtner, and M. Dworzecka, Nucl. Phys. A332, 109 (1979), and Phys. Rev. C 21, 1098 (1979).

PHYSICAL REVIEW B VOLUME 28, NUMBER 3 1 AUGUST 1983

Holstein-Primakoff theory for many-body systems

Lawrence R. Mead and N. Papanicolaou
Department of Physics, Washington University, St. Louis, Missouri 63130
(Received 9 August 1982; revised manuscript received 24 January 1983)

We discuss the relationship between generalized coherent states and suitable extensions of the Holstein-Primakoff theory of quantum spin systems, illustrating the common origin of a variety of semiclassical approximation schemes encountered in many-body theory.

A diversity of apparently disconnected approximation schemes [the semiclassical $1/s$ expansion for magnetic systems, the Bogoliubov theory of the Bose gas, the generalized random-phase approximation (RPA) for Fermi systems, mean-field approaches of the BCS type, the $1/N$ expansion in quantum mechanics and field theory] may all be based on the theory of generalized coherent states. While the details of an overcomplete set of coherent states denoted by $|z\rangle$ depend on the particular system under consideration, the associated dynamics is governed by a Lagrangian of the general form

$$L=\left\langle z\left|\left(\frac{i}{2}\frac{\overleftrightarrow{d}}{dt}-H\right)\right|z\right\rangle \;, \tag{1}$$

where H is the Hamiltonian. Lagrangian (1) has been the starting point for the derivation of the time-dependent Hartree-Fock approximation (TDHF), recently reviewed by Kramer and Saraceno,[1] which is but the Gaussian approximation based on Eq. (1). On the other hand, the same Lagrangian occurs in a phase-space path integral in the sense of Klauder[2] suggesting that Eq. (1) is an essentially exact statement. In fact, an exact operator formalism may be abstracted from Eq. (1) which is a generalization of the Holstein-Primakoff (HP) theory[3] developed some 40 years ago for the study of spin systems.

The HP theory has already received considerable attention in its original context as well as in the study of nuclear models in terms of pseudospin algebras.[4,5] Applications have also included the $1/N$ expansion in quantum mechanics and field theory.[6–8] Although the relevance of generalized coherent states is implicit in the above work, the precise relationship is often obscured or ignored in the literature. The purpose of this Brief Report is to shed light on the precise connection in the context of many-body theory.

Thus we consider systems whose Hamiltonian $H=H(A)$ may be expressed entirely in terms of the bilinear operators

$$A_{ij}=a_i^{\sigma *}a_j^{\sigma},\quad \{a_i^{\rho},a_j^{\sigma *}\}=\delta_{ij}\delta_{\sigma\rho}\;,$$
$$\rho,\sigma=1,2,\ldots,n,\quad n=2s+1\;, \tag{2}$$

where enclosure by the curly brackets denotes the usual anticommutator and summation over the repeated spin index σ is assumed. The spin s of the particles involved is taken to be arbitrary, whereas $n=2s+1$ stands for the spin multiplicity. For convenience we assume that the indices $i,j,\ldots$ take over discrete values the total number of which may be infinite. It is a simple exercise to show that the operators A_{ij} close the unitary pseudospin algebra

$$[A_{ij},A_{kl}]=\delta_{jk}A_{il}-\delta_{il}A_{kj}\;, \tag{3}$$

which is formally identical to the pseudospin algebra occurring in the description of the Bose gas. However, the representations of the pseudospin algebra that are relevant for Fermi systems are different.

Strictly speaking, an infinite number of irreducible representations of (3) are relevant for the description of Fermi systems, in analogy with the ordinary Schrödinger equation where an infinite number of angular-momentum sectors is necessary for the description of an atom. To make the analogy more complete, we note that the symmetry group of a Hamiltonian of the form $H=H(A)$ defined in terms of operators (2) is the unitary group $U(n)\sim U(1)\times SU(n)$, with $n=2s+1$, which should be clearly distinguished from the "pseudosymmetry" (3). The U(1) component is the usual number symmetry, whereas the non-Abelian component $SU(n)$ pertains to the spin degeneracy. It should be noted that the $SU(n)$ "flavor" group is larger than the normally expected SU(2) group associated with spin rotations, except for the special case of spin-$\frac{1}{2}$ particles. This situation is somewhat analogous to the n-dimensional harmonic oscillator whose symmetry group is $U(n)$ that contains the group of $O(n)$ rotations as a subgroup.

The fact that both the algebra of the symmetry group and the pseudospin algebra (3) are unitary is merely an accident. Nevertheless, an important link between the above algebras exists, in general, when the question of specific representations is addressed, because the generators of the pseudospin algebra are invariant under transformations of the symmetry group. The Fock states associated with the original Fermi operators may be classified according to their $U(n)$ quantum numbers that reflect the underlying symmetry, in particular, according to the total particle number N and the total spin, which should be distinguished from the spin multiplicity. The essence of pseudospin algebra (3) is that it allows classification of states with definite $U(n)$ transformation properties but varying "radial" quantum numbers within the same irreducible representation.

Our intention here is certainly not to provide a complete reduction of the Fock space according to the irreducible representations of the pseudospin algebra, but to concretely implement the above general discussion in specific examples of low-lying sectors (with small total spin) which are the most relevant for physical applications. In fact, explicit results are derived only for the singlet sector that encompasses N-particale states with vanishing total spin. We further restrict ourselves to systems with a total number of particles N that is an integer multiple of the spin multiplicity:

$$N = nN_0 = (2s+1)N_0 \ , \tag{4}$$

where N_0 is an integer. [For an electron system $(s=\frac{1}{2})N$ is taken to be even.] Hence the filled Fermi sea described by the N-particle state,

$$|\Omega_0\rangle = \prod_{j=1}^{N_0}\prod_{\sigma=1}^{n} a_j^{\sigma *}|\text{vac}\rangle \ , \tag{5}$$

carries vanishing total spin (singlet state).

A complete set of N-particle singlet states may be obtained by repeated application of the operators $A_{ij}, A_{kl}, \ldots$ defined from (2) on the state $|\Omega_0\rangle$. However, we shall prefer to describe the singlet sector through the covercomplete set of generalized coherent states

$$|z\rangle = \text{const}\exp\left[\sum_{i=N_0+1}^{\infty}\sum_{j=1}^{N_0} z_{ij}A_{ij}\right]|\Omega_0\rangle \ , \tag{6}$$

which are essentially the coherent states introduced by Thouless,[9] except for the ramifications concerning the spin multiplicity discussed above. The same author pointed out the relevance of the states (6) for the derivation of the RPA as a time-dependent Hartree-Fock approximation. Those and related results on quasiboson representations will be strengthened here to obtain a complete HP theory that may be used for the derivation of systematic corrections to the RPA by means of ordinary perturbation theory in inverse powers of the spin multiplicity.

To motivate the HP theory a more detailed study of the coherent states is required than that given by Thouless. Fortunately, coherent states analogous to (6) have been studied by Perelomov in his work on pair creation of Fermi particles in an external field,[10] which applies to the current problem with suitable adaptations. Indices taking values outside the Fermi sea will now be denoted by $\mu, \nu, \ldots \geq N_0+1$, whereas indices inside the sea will be taken to be $\alpha, \beta, \ldots \leq N_0$.

The (normalized) coherent states (6) read

$$|z\rangle = [\det(\underline{I}+\underline{z}^\dagger\underline{z})]^{-n/2}\exp\left[\sum_{\mu,\alpha} z_{\mu\alpha}A_{\mu\alpha}\right]|\Omega_0\rangle \ ,$$
$$\langle z|z\rangle = 1, \quad n = 2\underline{s}+1 \ . \tag{7}$$

The matrix $\underline{z}^\dagger\underline{z}$ is the $N_0\times N_0$ matrix defined from

$$(\underline{z}^\dagger\underline{z})_{\alpha\beta} = \sum_{\mu} z^*_{\mu\alpha}z_{\mu\beta} \ , \tag{8}$$

where the asterisk denotes ordinary complex conjugation whereas the dagger also implies transposition of indices $[(\underline{z}^\dagger)_{\alpha\mu} \equiv (\underline{z}^*)_{\mu\alpha}]$.

The time-dependent dynamics associated with the overcomplete set of states (7) is governed by the Lagrangian (1) written as

$$L = \left\langle z\left|\left(\frac{i}{2}\frac{\overleftrightarrow{d}}{dt} - H\right)\right|z\right\rangle = L_0 - \langle z|H|z\rangle \ ,$$
$$L_0 = \frac{i}{2}\sum_{\mu,\alpha}\dot{z}_{\mu\alpha}\langle z|A_{\mu\alpha}|z\rangle + \text{H.c.} \ . \tag{9}$$

The evaluation of useful matrix elements in the basis of coherent states is facilitated by various identities derived by Perelomov[10] (currently adapted to account for the spin multiplicity):

$$\begin{aligned}
A_{\mu\nu}(z,z^*) &= n\{\underline{z}(\underline{I}+\underline{z}^\dagger\underline{z})^{-1}\underline{z}^\dagger\}_{\nu\mu} \ ,\\
A_{\alpha\beta}(z,z^*) &= n\{(\underline{I}+\underline{z}^\dagger\underline{z})^{-1}\}_{\beta\alpha} \ ,\\
A_{\mu\alpha}(z,z^*) &= n\{(\underline{I}+\underline{z}^\dagger\underline{z})^{-1}\underline{z}^\dagger\}_{\alpha\mu} \ ,\\
A_{\alpha\mu}(z,z^*) &= n\{\underline{z}(\underline{I}+\underline{z}^\dagger\underline{z})^{-1}\}_{\mu\alpha} \ .
\end{aligned} \tag{10}$$

We are now able to derive an explicit form for the Lagrangian L_0 of Eq. (9), namely,

$$L_0 = \frac{i}{2}n\,\text{tr}\{\dot{\underline{z}}(\underline{I}+\underline{z}^\dagger\underline{z})^{-1}\underline{z}^\dagger\} + \text{H.c.} \ , \tag{11}$$

where tr stands for the usual trace of matrices and involves summation over all particle states, including the summation over hole states implied by the matrix multiplication in (11).

The information summarized in Eqs. (10) and (11) will be sufficient for the derivation of the HP theory. As expected, the phase space implied by (11) is nonlinear; namely, the canonical momenta associated with the dynamical variables $z_{\mu\alpha}$ are nonlinear functions of the latter. We thus seek a stereographic projection that linearizes the phase space:

$$\underline{\xi} = \underline{z}[n^{1/2}(\underline{I}+\underline{z}^\dagger\underline{z})^{-1/2}] \ . \tag{12}$$

It is a simple exercise to show that

$$L_0 = \frac{1}{2}\text{tr}(\dot{\underline{\xi}}\underline{\xi}^\dagger - \underline{\xi}\dot{\underline{\xi}}^\dagger) = \frac{1}{2}i\sum_{\mu,\alpha}(\dot{\xi}_{\mu\alpha}\xi^*_{\mu\alpha} - \xi_{\mu\alpha}\dot{\xi}^*_{\mu\alpha}) \ , \tag{13}$$

whereas the diagonal matrix elements (10) transform

into

$$A_{\mu\nu}=\sum_{\alpha}\xi^*_{\mu\alpha}\xi_{\nu\alpha}\ ,$$
$$A_{\alpha\beta}=n\delta_{\alpha\beta}-\sum_{\mu}\xi^*_{\mu\beta}\xi_{\mu\alpha}\ ,$$
$$A_{\mu\alpha}=\sum_{\beta}\xi^*_{\mu\beta}R_{\alpha\beta}\ ,\qquad(14a)$$
$$A_{\alpha\mu}=\sum_{\beta}R_{\beta\alpha}\xi_{\mu\beta}\ ,$$

where $\underline{R}=(R_{\alpha\beta})$ is the $N_0\times N_0$ matrix

$$\underline{R}=(n\underline{I}-\underline{\xi}^{\dagger}\underline{\xi})^{1/2}\ .\qquad(14b)$$

Furthermore, the Lagrangian (13) implies the (Bose) commutation relations

$$[\xi_{\mu\alpha},\xi^*_{\nu\beta}]=\delta_{\mu\nu}\delta_{\alpha\beta}\ .\qquad(14c)$$

Equations (14a), (14b), and (14c) provide the HP representation suitable for the description of the singlet sector of a Fermi system interacting through a spin-independent potential.

However, the preceding derivation hides an important fact associated with the ordering of operators in (14a). In view of nonlinear coordinate transformation such as (12) one would normally expect that ordering difficulties beset the validity of (14). While the expressions (14a) close to the unitary algebra at the level of Poisson brackets essentially by construction, the transition to operators implied by (14c) may not preserve the correct commutation relations. Nevertheless, the ordering of operators judiciously chosen in (14a) can be shown to provide the correct HP representation, whose validity must be established by an independent proof.

Because of the matrix stucture associated with (14) such a proof is not straightforward. Since nontrivial examples have already been worked out in a related context,[6–8] we shall restrict ourselves to the description of some important checks of consistency of (14). Thus the matrix in (14b) may be approximated by a series expansion in inverse powers of the spin multiplicity:

$$R_{\alpha\beta}\simeq\sqrt{n}\left[\delta_{\alpha\beta}-\frac{1}{2n}\sum_{\mu}\xi^*_{\mu\alpha}\xi_{\mu\beta}+\cdots\right]\ .\qquad(15)$$

Substitution of (15) in (14a) allows for an explicit (albeit tedious) verification of the commutation relations (3) to any desired accuracy, for the various partitions of indices shown in Eq. (14a). In fact, Eqs. (14) become useful for practical purposes through an expansion of the form (15).

Other checks of consistency involve calculating the Casimir invariants from (14) and showing that they are identically equal to their eigenvalues characterizing the N-body singlet subspace. Thus the first two invariants of the pseudospin algebra (3) read

$$C_1=\sum_{i}A_{ii}\ ,\quad C_2=\sum_{i,j}A_{ij}A_{ji}\ .\qquad(16)$$

It is not difficult to rearrange the above definitions to obtain

$$C_1=\sum_{\sigma}Q^{\sigma\sigma}\ ,$$
$$C_2=-\sum_{\sigma,\rho}Q^{\sigma\rho}Q^{\rho\sigma}+(n+N_0+\Lambda)\sum_{\sigma}Q^{\sigma\sigma}\ ,\qquad(17)$$

where $N_0+\Lambda$ is the total number of available levels (which may be infinite) and $Q^{\sigma\rho}$ are the generators of the U(n) symmetry:

$$Q^{\sigma\rho}=\sum_{i}a_i^{\sigma *}a_i^{\rho}\ .\qquad(18)$$

We thus find that the Casimir invariants of the pseudospin algebra may be expressed in terms of the Casimir invariants of the underlying symmetry. Furthermore, the eigenvalues of C_1 and C_2 that characterize the N-body singlet subspace may be identified from

$$C_1|\Omega_0\rangle=N|\Omega_0\rangle,\quad C_2|\Omega_0\rangle=(n+\Lambda)N|\Omega_0\rangle\ ,\qquad(19)$$

which may be established by a simple calculation.

Equations (16)–(18) were derived using properties of the original Fermi operators alone. On the other hand, one may write

$$C_1=\sum_{i}A_{ii}=\sum_{\mu}A_{\mu\mu}+\sum_{\alpha}A_{\alpha\alpha}\ ,\quad C_2=\sum_{i,j}A_{ij}A_{ji}=\sum_{\mu,\nu}A_{\mu\nu}A_{\nu\mu}+\sum_{\alpha,\beta}A_{\alpha\beta}A_{\beta\alpha}+\sum_{\mu,\alpha}(A_{\alpha\mu}A_{\mu\alpha}+A_{\mu\alpha}A_{\alpha\mu})\ ,\qquad(20)$$

and substitute the HP representation (14) for $A_{\mu\nu}$, $A_{\alpha\beta}$, $A_{\mu\alpha}$, and $A_{\alpha\mu}$. A relatively simple calculation employing only the commutation relations (14c) shows that C_1 and C_2 are *identically* equal to the c-numbers

$$C_1=N,\quad C_2=(n+\Lambda)N\ ,\qquad(21)$$

which coincide with the eigenvalues of the Casimir invariants found in Eq. (19). The preceding calculation confirms our earlier assertion that the HP representation (14) is a restriction of the pseudospin algebra to the N-body singlet subspace.

The appearance of the parameter Λ ($\rightarrow\infty$) in (21) is an artifact associated with the definition of the Casimir invariants of the pseudospin algebra which bear no direct physical significance. The parameter Λ never appears explicitly in calculations involving the HP representation (14). Furthermore, it was noted earlier [cf. Eq. (17)] that $C_1,C_2,\ldots$ are related to the Casimir invariants of the underlying symmetry group which do possess direct physical significance. Thus the total number operator is equal to $C_1(=N)$, whereas C_2 contains information about the total spin. We illustrate this situation for the simplest spin-$\frac{1}{2}$

case for which $n=2s+1=2$, and the spin generators are defined from

$$S^a=\frac{1}{2}\sum_{i,\lambda,\rho} a_i^{\lambda *}\sigma^a{}_{\lambda\rho}a_i^\rho,\quad a=1,2,3\ , \tag{22}$$

where $\underline{\sigma}^a=(\sigma^a{}_{\lambda\rho})$ are the familiar Pauli matrices. A simple calculation shows that the total spin S^aS^a may be expressed in terms of C_1 and C_2 according to

$$S^aS^a=\frac{1}{2}(2+N_0+\Lambda)C_1-\frac{1}{4}C_1^2-\frac{1}{2}C_2\ , \tag{23}$$

a relation that is valid for all sectors. On the other hand, if the eigenvalues (21) with $n=2$ and $N=2N_0$ are inserted in (23), one finds that $S^aS^a=0$, which is characteristic of the singlet sector.

The HP representation (14) is the main result of this Brief Report. Given a Hamiltonian $H=H(A)$, its restriction to the N-body subspace with vanishing total spin is obtained by the simple substitution $A\rightarrow A(\xi,\xi^*)$ from Eq. (14). This results in an exact effective Hamiltonian for the description of the singlet sector. The ensuing method of calculation is briefly illustrated here in the case of the interacting Bose gas for which the HP theory may be obtained as a special case of (14). Hence a Bose system may be viewed as a degenerate Fermi system with a number of flavors n that is equal to the particle number N. Then $N=nN_0=n \twoheadrightarrow N_0=1$, and the $N_0\times N_0$ matrix $\xi^\dagger\xi$ in Eq. (14) becomes a scalar. Using an obvious change in notation Eqs. (14) reduce to

$$A_{00}=N-\xi^*\xi,\quad A_{pq}=\xi_p^*\xi_q \text{ for } pq\neq 0\ ,$$
$$A_{0p}=(N-\xi^*\xi)^{1/2}\xi_p,\quad A_{p0}=\xi_p^*(N-\xi^*\xi)^{1/2} \text{ for } p\neq 0\ . \tag{24}$$

The operators ξ_p and ξ_p^* are defined only for nonvanishing momenta and satisfy the usual Bose commutation relations

$$[\xi_p,\xi_q^*]=\delta_{pq}\ . \tag{25}$$

We have also used the abbreviation $\xi^*\xi=\sum_{p\neq 0}\xi_p^*\xi_p$. Roughly speaking, the operator ξ_p^* excites a particle with momentum $p\neq 0$ from the N-body condensate. The HP representation (24) coincides with that derived earlier by Okubo[11] for the symmetric representation of the unitary algebra.

The effective N-body Hamiltonian for the interacting Bose gas may then be written as

$$H=\frac{N^2}{V}V_0-\frac{N}{V}\sum_k V_k+\sum_{k\neq 0}\left[k^2\xi_k^*\xi_k+\frac{V_k}{V}\rho_k\rho_{-k}\right],$$
$$\rho_k=(N-\xi^*\xi)^{1/2}\xi_k+\xi_{-k}^*(N-\xi^*\xi)^{1/2}$$
$$+\sum_{\substack{p\neq 0\\ p+k\neq 0}}\xi_p^*\xi_{p+k}\ , \tag{26}$$

where V_k is the Fourier transform of the potential and V is the volume of the system. A systematic formal expansion in inverse powers of N and ordinary perturbation theory yield a method of successive approximations in complete analogy with the $1/N$ expansion studied in Refs. 6–8. The leading approximation coincides with the familiar Bogoliubov theory, whereas higher-order corrections are free of ordering difficulties that have been known to occur in the closely related hydrodynamical approach of Bogoliubov and Zubarev.[12,13] We have checked the current procedure with a detailed calculation of the ground-state energy in the Lieb-Liniger model[14] for which an exact solution was obtained through the Bethe ansatz. Taking $V_k=g^2$ and restricting (26) to one dimension yields, after a long calculation,

$$E_{\text{g.s.}}=N\rho^2\left[\gamma-\frac{4}{3\pi}\gamma^{3/2}+\left(\frac{1}{6}-\frac{1}{\pi^2}\right)\gamma^2+\cdots\right]\ , \tag{27}$$

where $\rho=N/V$ is the density of the system and γ is the dimensionless coupling constant $\gamma=g^2/\rho$. The result (27) agrees with the numerical solution of the Lieb-Liniger equations.[12]

ACKNOWLEDGMENTS

We wish to thank John W. Clark for informative conversations. This work was supported in part by the Department of Energy under Contract No. DEAC02-78ER04915.

[1] P. Kramer and M. Saraceno, *Geometry of the Time-Dependent Variational Principle in Quantum Mechanics*, Lecture Notes in Physics, Vol. 140 (Springer-Verlag, Berlin, 1981).
[2] J. R. Klauder, Ann. Phys. (N.Y.) 11, 123 (1960).
[3] T. Holstein and H. Primakoff, Phys. Rev. 58, 1098 (1940).
[4] P. Garbaczewski, Phys. Rep. 36C, 65 (1978).
[5] E. R. Marshalek, Nucl. Phys. A347, 253 (1980).
[6] A. Jevicki and N. Papanicolaou, Nucl. Phys. B171, 362 (1980).
[7] L. D. Mlodinow and N. Papanicolaou, Ann. Phys. (N.Y.) 128, 314 (1980); 131, 1 (1981).
[8] N. Papanicolaou, Ann. Phys. (N.Y.) 136, 310 (1981).
[9] D. J. Thouless, *The Quantum Mechanics of Many-Body Systems* (Academic, New York, 1961).
[10] A. M. Perelomov, Teor. Mat. Fiz. 19, 83 (1974) [Sov. Phys. Theor. Math. Phys. 19, 368 (1974)].
[11] S. Okubo, J. Math. Phys. 16, 528 (1975).
[12] M. Takahashi, Prog. Theor. Phys. 53, 386 (1975).
[13] A. Jevicki, Nucl. Phys. B146, 77 (1978).
[14] E. H. Lieb and W. Liniger, Phys. Rev. 130, 1605 (1963).

VI. Applications in Condensed Matter Physics

PHYSICAL REVIEW VOLUME 147, NUMBER 1 8 JULY 1966

Coherent States and Irreversible Processes in Anharmonic Crystals*

P. CARRUTHERS
Laboratory of Atomic and Solid State Physics and Laboratory of Nuclear Studies, Cornell University, Ithaca, New York

AND

K. S. DY
Laboratory of Atomic and Solid State Physics, Cornell University, Ithaca, New York

(Received 15 February 1966)

Irreversible phenomena involving phonons in a slightly anharmonic crystal are investigated using coherent states as a basis. This basis retains the advantages of the classical theory, in which phase relations are clearly exhibited, for a quantum-mechanical system. In the coherent-state basis the equation of motion for the density matrix has an obvious correspondence with the classical Liouville equation. In particular, the connection with the action-angle variables of Brout and Prigogine is elucidated. A new, rapid proof of the Brout-Prigogine equation is given using the method of semi-invariants. This technique exhibits higher-order corrections in a usable form. The quantum corrections to the classical equation of motion for the density function are shown to be due to extra terms involving second derivatives in the action-angle variables. The Peierls master equation is then derived from the Brout-Prigogine equation. The important problem of elastic scattering of phonons is treated by our method in an appendix.

I. INTRODUCTION

THE main purpose of the present work is to exhibit the utility of the coherent-state basis[1] for the description of irreversible processes. Although similar techniques may be applied to other systems of interacting bosons, we restrict our attention to the problem of phonons in a slightly anharmonic crystal. Elastic scattering by defects is discussed in the Appendix. In the coherent-state representation, the equation of motion for the density matrix although fully quantum-mechanical, has an obvious correspondence with the classical Liouville equation. The correspondence with the action-angle variables used by Brout and Prigogine[2,3] is elucidated. (For a coherent state $|\alpha\rangle$, the parameter α is related to the action J and the angle ϕ by $\alpha=J^{1/2}\exp(i\phi)$.) The coherent states, which in the classical limit describe classical harmonic-oscillator motion,[1,4] have other advantages when compared to the traditional number states. The phase relations so interesting in a discussion of irreversibility are at all times clearly exhibited. Number eigenstates have completely indeterminate phase.[5] Moreover, the coherent states correspond closely to nearly all intuitive thinking on our subject, based as it is on wave propagation. For the number states $|n\rangle$, the mean oscillator position $\langle n|x(t)|n\rangle$ always vanishes, no matter how large n. In contrast,[4] coherent states give a mean $\langle\alpha|x(t)|\alpha\rangle$ of $\cos(\omega t-\phi)$. Finally, we note that a classical driving force excites the coherent state of an oscillator.[4] A careful treatment of the coupling of crystals to massive heat reservoirs should take this into account.

In Sec. II, the necessary mathematical techniques are developed. These techniques are applied to the evolution of the density matrix in an anharmonic solid in Sec. III. We shall see that the general quantum Liouville equation derived differs from that of Brout and Prigogine by the addition of terms involving second derivatives in the action-angle variables. The latter terms are of order N^{-1} relative to the "classical terms," where N is the mean occupation of an oscillator, and hence establish the classical limit and corrections thereto. In Sec. IV a short derivation of the Brout-Prigogine master equation[2,3] is given. Higher order corrections are exhibited in a useful form. In Sec. V, we use the Brout-Prigogine master equation to derive the Peierls equation[6,7] for the time rate-of-change of the mean number of phonons in a given mode of vibration. Finally, some remarks are made on the derivation of the Peierls equation.

II. MATHEMATICAL PRELIMINARIES

The reader is referred to Refs. 1 and 4 for a full discussion of coherent states. Normalizing the creation and destruction operators for an oscillator, $a^\dagger$ and a, so that $a^\dagger a=N_{\rm op}$, $N_{\rm op}$ having eigenvalues 0, 1, 2, $\cdots$, the coherent state $|\alpha\rangle$ (α is an arbitrary complex number) is given by

$$a|\alpha\rangle=\alpha|\alpha\rangle$$

$$|\alpha\rangle=\exp(-\tfrac{1}{2}|\alpha|^2)\sum_{n=0}^{\infty}\frac{\alpha^n}{\sqrt{n!}}|n\rangle \qquad (2.1)$$

$$=\exp(\alpha a^\dagger-\alpha^* a)|0\rangle,$$

* Research supported in part by the U. S. Atomic Energy Commission and the U. S. Office of Naval Research.

[1] R. J. Glauber, Phys. Rev. **131**, 2766 (1963).

[2] R. Brout and I. Prigogine, Physica **22**, 621 (1956).

[3] I. Prigogine, *Non-Equilibrium Statistical Mechanics* (Interscience Publishers, New York, 1962).

[4] P. Carruthers and M. M. Nieto, Am. J. Phys. **33**, 537 (1965).

[5] P. Carruthers and M. M. Nieto, Phys. Rev. Letters **14**, 387 (1965).

[6] R. E. Peierls, Ann. Physik **3**, 1055 (1929).

[7] R. E. Peierls, *Quantum Theory of Solids* (Oxford University Press, London, 1955), Chap. 2.

where $|n\rangle$ denotes the normalized number state. The coherent states are not orthogonal:

$$\langle\alpha|\beta\rangle=\exp(\alpha^*\beta-\tfrac{1}{2}|\alpha|^2-\tfrac{1}{2}|\beta|^2), \qquad (2.2)$$

but they are complete, as indicated by the resolution of the identity,

$$\frac{1}{\pi}\int|\alpha\rangle\langle\alpha|d^2\alpha=I. \qquad (2.3)$$

The integration in (2.3) is to be taken over the two-dimensional α plane; $d^2\alpha=d(\mathrm{Re}\alpha)d(\mathrm{Im}\alpha)$.

For notational simplicity, let us first consider a single oscillator. In the coherent state representation, the quantum Liouville equation is

$$i\langle\alpha|\frac{\partial\rho}{\partial t}|\alpha\rangle=\frac{1}{\pi}\int d^2\beta$$
$$\times[\langle\alpha|H|\beta\rangle\langle\beta|\rho|\alpha\rangle-\langle\alpha|\rho|\beta\rangle\langle\beta|H|\alpha\rangle]. \qquad (2.4)$$

If the Hamiltonian H is expressed in terms of creation and annihilation operators, we see that Eq. (2.4) involves, in general, terms of the form [H is assumed to be expressible as an ordered polynomial as in Eq. (2.12) below]:

$$\frac{1}{\pi}\int d^2\beta(\alpha^*)^m\beta^n\langle\alpha|\beta\rangle\langle\beta|\rho|\alpha\rangle, \qquad (2.5)$$

where m, n are integers. This integral can be worked out by using the following formulas:

$$\text{(a)}\quad \frac{1}{\pi}\int d^2\beta e^{\alpha^*\beta-|\beta|^2}f(\beta^*)=f(\alpha^*),$$

$$\text{(b)}\quad \frac{1}{\pi}\int d^2\beta e^{\alpha^*\beta-|\beta|^2}\beta^n f(\beta^*)=\left(\frac{\partial}{\partial\alpha^*}\right)^n f(\alpha^*),$$

$$\text{(c)}\quad \frac{1}{\pi}\int d^2\beta e^{\alpha\beta^*-|\beta|^2}f(\beta)=f(\alpha), \qquad (2.6)$$

$$\text{(d)}\quad \frac{1}{\pi}\int d^2\beta e^{\alpha\beta^*-|\beta|^2}(\beta^*)^n f(\beta)=\left(\frac{\partial}{\partial\alpha}\right)^n f(\alpha).$$

The formulas (a)–(d) can be derived by expanding f in the integrand in a Taylor series and using the following formula[1]:

$$\frac{1}{\pi}\int d^2\beta e^{-|\beta|^2}\beta^l(\beta^*)^m d^2\beta=\delta_{lm}(l!m!)^{1/2}. \qquad (2.7)$$

To illustrate the application of these formulas, we shall evaluate the integral

$$\frac{1}{\pi}\int d^2\beta\beta\langle\alpha|\beta\rangle\langle\beta|\rho|\alpha\rangle. \qquad (2.8)$$

The expression $\langle\beta|\rho|\alpha\rangle$ in the integrand involves a factor $\langle\beta|\alpha|\rangle$, so we shall factor it out and write

$$\langle\beta|\rho|\alpha\rangle\equiv\langle\beta|\alpha\rangle\rho(\beta^*,\alpha). \qquad (2.9)$$

The functional dependence of $\rho(\beta^*,\alpha)$ on β^* and α is a result of the left- and right-handed properties of the states $\langle\beta|$ and $|\alpha\rangle$. Notice that

$$\langle\alpha|\rho|\alpha\rangle=\rho(\alpha^*,\alpha). \qquad (2.10)$$

Using these definitions and Eq. (2.6b) to do the integration, we get:

$$\frac{1}{\pi}\int d^2\beta\beta\langle\alpha|\beta\rangle\langle\beta|\rho|\alpha\rangle$$
$$=e^{-|\alpha|^2}\frac{1}{\pi}\int d^2\beta e^{\alpha^*\beta-|\beta|^2}\beta[e^{\beta^*\alpha}\rho(\beta^*,\alpha)]$$
$$=e^{-|\alpha|^2}\frac{\partial}{\partial\alpha^*}[e^{\alpha^*\alpha}\rho(\alpha^*,\alpha)]$$
$$=\left(\alpha+\frac{\partial}{\partial\alpha^*}\right)\rho(\alpha^*,\alpha). \qquad (2.11)$$

In such calculations, it is essential that α^* be considered independent of α. Note that $\exp(\beta^*\alpha)\rho(\beta^*,\alpha)$ is identical to the function $R(\beta^*,\alpha)$ defined by Glauber.[1]

The latter result can be generalized, and the form (2.9) clarified, as follows. Consider an operator A expandable in terms of the ordered series:

$$A=\sum_{mn}A_{mn}(a^\dagger)^m(a)^n\equiv A(a^\dagger,a). \qquad (2.12)$$

Then we find

$$\frac{\langle\alpha|A(a^\dagger,a)|\beta\rangle}{\langle\alpha|\beta\rangle}=\sum_{mn}A_{mn}(\alpha^*)^m\beta^n\equiv A(\alpha^*,\beta). \qquad (2.13)$$

We shall call the quantity $A(\alpha^*,\beta)$ the *reduced matrix element* of the operator A. Given sufficiently well-behaved expansion coefficients A_{mn}, the function $A(\alpha^*,\beta)$ is an analytic function of two complex variables.[8] It is therefore sufficient to calculate $A(\alpha^*,\alpha)$, which is a boundary value ($\beta=\alpha$) of $A(\alpha^*,\beta)$.[9] In fact, Eq. (2.11) can be generalized to the matrix element $\langle\alpha|A(a^\dagger,a)\rho|\alpha\rangle$.

$$\langle\alpha|A(a^\dagger,a)\rho|\alpha\rangle$$
$$=\frac{1}{\pi}\int d^2\beta|\langle\alpha|\beta\rangle|^2A(\alpha^*,\beta)\rho(\beta^*,\alpha)$$
$$=e^{-|\alpha|^2}\sum_{mn}A_{mn}(\alpha^*)^m\int\frac{d^2\beta}{\pi}e^{\alpha^*\beta-|\beta|^2}\beta^n[e^{\beta^*\alpha}\rho(\beta^*,\alpha)]$$
$$=e^{-|\alpha|^2}\sum_{mn}A_{mn}(\alpha^*)^m\left(\frac{\partial}{\partial\alpha^*}\right)^n[e^{\beta^*\alpha}\rho(\beta^*,\alpha)]_{\beta^*=\alpha^*},$$

[8] S. Bochner and W. T. Martin, *Several Complex Variables* (Princeton University Press, Princeton, New Jersey, 1948).

[9] An important consequence is therefore the reduction of the number of descriptive real variables from four (α,β) to two (α).

so

$$\langle\alpha|A(a^\dagger,a)\rho|\alpha\rangle=e^{-|\alpha|^2}A\left(\alpha^*,\frac{\partial}{\partial\alpha^*}\right)[e^{\alpha^*\alpha}\rho(\alpha^*,\alpha)]. \quad (2.14)$$

From Eq. (2.14), we can also obtain the expectation value of the operator A immediately

$$\begin{aligned}\langle A\rangle&=\mathrm{Tr}(A\rho)\\&=\frac{1}{\pi}\int d^2\alpha e^{-|\alpha|^2}A\left(\alpha^*,\frac{\partial}{\partial\alpha^*}\right)[e^{\alpha^*\alpha}\rho(\alpha^*,\alpha)].\end{aligned} \quad (2.15)$$

The method of integration used in deriving Eq. (2.11) and Eq. (2.14) will be used repeatedly when we consider specific models. It is important to note in Eq. (2.15) that all averages can be expressed in terms of the diagonal elements of the density matrix. This is an attractive feature of this representation. According to Eq. (2.15) the average of the number operator is

$$\langle N\rangle=\frac{1}{\pi}\int d^2\alpha\alpha^*\left(\alpha+\frac{\partial}{\partial\alpha^*}\right)\rho(\alpha^*,\alpha). \quad (2.16)$$

Further simplifications occur when ρ depends only on the number operator $a^\dagger a$. In this case $\rho(\alpha^*,\alpha)$ is a function of the product $\alpha^*\alpha$. Suppose $B(a^\dagger,a)$ depends on $a^\dagger a$; as a special case of Eq. (2.12) we write

$$B=\sum_n b_n(a^\dagger a)^n. \quad (2.17)$$

Since B is diagonal in the number basis we use Eq. (2.1) to obtain

$$\begin{aligned}\langle\alpha|B|\beta\rangle&=\sum_n b_n\sum_k\frac{k^n(\alpha^*\beta)^k}{k!}\exp[-\tfrac{1}{2}|\alpha|^2-\tfrac{1}{2}|\beta|^2]\\&=B_1(\alpha^*\beta)\langle\alpha|\beta\rangle,\end{aligned} \quad (2.18)$$

where

$$\begin{aligned}B_1(\alpha^*\beta)&=\sum_n b_nF_n(\alpha^*\beta),\\F_n(z)&=e^{-z}\sum_k\frac{k^nz^k}{k!}.\end{aligned} \quad (2.19)$$

Thus, if an operator B is a function of $a^\dagger a$, its reduced matrix element $\langle\alpha|B|\beta\rangle/\langle\alpha|\beta\rangle$ is a function of the product $\alpha^*\beta$. Trivial examples show that this is not the case for all operators, e.g. $(a^\dagger a^2)^2$ has reduced matrix element $(\alpha^*\beta^2)^2+2\alpha^*\beta^3$ which is not a function of $\alpha^*\beta$ alone.

Next we describe the transition to action-angle variables. The action J is defined to be the mean excitation number of the state $|\alpha\rangle$, and ϕ the corresponding phase:

$$\alpha=J^{1/2}e^{i\phi};\quad J=\alpha^*\alpha,\quad \phi=\frac{1}{2i}\ln\left(\frac{\alpha}{\alpha^*}\right). \quad (2.20)$$

Noting that α^* should be considered independent of α (see derivation of Eq. (2.11) and Eq. (2.14), for example), we find

$$\begin{aligned}\frac{\partial}{\partial\alpha}&=J^{1/2}e^{-i\phi}\left(\frac{\partial}{\partial J}-\frac{i}{2J}\frac{\partial}{\partial\phi}\right),\\\frac{\partial}{\partial\alpha^*}&=J^{1/2}e^{i\phi}\left(\frac{\partial}{\partial J}+\frac{i}{2J}\frac{\partial}{\partial\phi}\right).\end{aligned} \quad (2.21)$$

Integrations over the α plane become, in terms of the action-angle variables

$$\int d^2\alpha=\frac{1}{2}\int_0^\infty dJ\int_0^{2\pi}d\phi. \quad (2.22)$$

To illustrate this transformation, consider the Liouville equation for a free harmonic oscillator:

$$i\frac{\partial\rho}{\partial t}=[H_0,\rho]\equiv L_0\rho, \quad (2.23)$$

where $H=\omega a^\dagger a$. A short calculation shows that $\rho(\beta^*,\alpha)$ evolves as

$$\begin{aligned}i\frac{\partial\rho(\beta^*,\alpha)}{\partial t}&=L_0(\beta^*,\alpha)\rho(\beta^*,\alpha),\\L_0(\beta^*,\alpha)&=\omega\left(\beta^*\frac{\partial}{\partial\beta^*}-\alpha\frac{\partial}{\partial\alpha}\right).\end{aligned} \quad (2.24)$$

Note that if at $t=0$, $\rho(\beta^*,\alpha)$ is a function of $\beta^*\alpha$, then ρ is independent of time. This corresponds to the trivial remark that $[\rho(N),N]$ vanishes.

For $\alpha=\beta$, Eq. (2.24) reduces to

$$L_0(\alpha^*,\alpha)=i\omega\frac{\partial}{\partial\phi} \quad (2.25)$$

independent of the action. The analogy to the classical oscillator[2,3] should now be obvious. Generalizing to an arbitrary number of independent oscillators of frequency ω_k, we have

$$L_0=\sum_k i\omega_k\frac{\partial}{\partial\phi_k}. \quad (2.26)$$

The eigenfunctions of L_0 play an important role in the perturbation expansion for the density matrix. These eigenfunctions are the same as in the classical analysis:

$$\begin{aligned}f\{\nu\}&=(2\pi)^{-(N/2)}\exp[-i\sum_k\nu_k\phi_k],\\L_0f\{\nu\}&=(\sum_k\nu_k\omega_k)f\{\nu\},\end{aligned} \quad (2.27)$$

where the ν_k vary over all the positive and negative integers and the sum over k covers all N normal modes.

The set $\{\nu\}$ should never be mistaken for the occupation numbers; rather it specifies the phase of the oscillator.

The normalization in Eq. (2.27) has been chosen according to the inner product

$$(g|h)=\int g^*\{\phi\}h\{\phi\}\prod_k d\phi_k, \quad (2.28)$$

so that the eigenfunctions $f\{n\}$ are normalized as follows:

$$(f\{\nu\}|f\{\nu'\})=\delta_{\{\nu\}\{\nu\}}. \quad (2.29)$$

The Kronecker delta in Eq. (2.29) means that $\nu_k=\nu_k'$ for all $k=1, \cdots N$. We have used round brackets to denote inner products in the space of the eigenfunctions (2.27) to avoid confusion with the usual quantum mechanical inner product, which will be written with sharp brackets.

To conclude this section, we calculate some averages of functions of the number operator, for density matrices which depends only on N (or $\alpha^*\alpha$). In this case $\rho(\alpha^*,\alpha)$ depends on J alone and Eqs. (2.16), (2.21), (2.22) show that

$$\langle N\rangle=\int dJJ\rho(J)+\int dJJ\frac{\partial}{\partial J}\rho(J). \quad (2.30)$$

Integrating the second term by parts and applying the normalization condition for $\rho(J)$, we get

$$\langle N\rangle=\int dJJ\rho(J)-1$$

or

$$\langle\langle J\rangle\rangle\equiv\int dJJ\rho(J)=\langle N\rangle+1. \quad (2.31)$$

A useful generalization is easily proved

$$\langle(a^\dagger)^n a^n\rangle=(-1)^n\int_0^\infty \rho(J)L_n(J)dJ \equiv(-1)^n\langle\langle L_n(J)\rangle\rangle, \quad (2.32)$$

where L_n is the Laguerre polynomial.[10] Many interesting relations of this sort hold. For a collection of oscillators with a density matrix depending only on $\{N_k\}$

$$[\rho(\{\alpha_k^*\},\{\alpha_k\})=\rho(\{J_k\})]$$

where N_k is the occupation number of the kth mode, one finds

$$\begin{aligned}\langle N_{k1}\rangle&=(-1)\langle\langle L_1(J_{k1})\rangle\rangle=\langle\langle J_{k1}\rangle\rangle-1\\ \langle N_{k1}N_{k2}\rangle&=(-1)^2\langle\langle L_1(J_{k1})L_1(J_{k2})\rangle\rangle=\langle\langle J_{k1}J_{k2}\rangle\rangle\\ &\quad-\langle\langle J_{k1}\rangle\rangle-\langle\langle J_{k2}\rangle\rangle+1 \quad (2.33)\\ &\cdots\\ \langle N_{k1}N_{k2}\cdots N_{kn}\rangle&=(-1)^n\langle\langle L_1(J_{k1})\cdots L_1(J_{kn})\rangle\rangle\\ &=(-1)^n\langle\langle\prod_{i=1}^n(1-J_{k_i})\rangle\rangle.\end{aligned}$$

[10] *Higher Transcendental Functions*, edited by A. Erdelyi (McGraw-Hill Book Company, Inc., New York, 1953), Vol. II, p. 188.

To give an example, consider an oscillator in thermal equilibrium, for which

$$\rho=\frac{e^{-\beta\omega a^\dagger a}}{\mathrm{Tr}e^{-\beta\omega a^\dagger a}}, \quad (2.34)$$

where $\beta=(kT)^{-1}$;

$$\rho(J)=\frac{\langle\alpha|e^{-\beta\omega a^\dagger a}|\alpha\rangle}{\mathrm{Tr}e^{-\beta\omega a^\dagger a}}. \quad (2.35)$$

Using Eqs. (2.18), (2.19), we find

$$\rho(J)=\frac{\exp[-J(1-e^{-\beta\omega})]}{(1-e^{-\beta\omega})^{-1}}. \quad (2.36)$$

Then we obtain the expected results

$$\begin{aligned}\langle\langle J\rangle\rangle&=\int dJJ\rho(J)=\frac{1}{1-e^{-\beta\omega}},\\ \langle N\rangle&=\langle\langle J\rangle\rangle-1=\frac{1}{e^{\beta\omega}-1}.\end{aligned} \quad (2.37)$$

III. EVOLUTION OF THE DENSITY MATRIX IN ANHARMONIC SOLIDS; ACTION-ANGLE VARIABLES

If we denote the displacement of the atoms from the mth equilibrium lattice site by $\mathbf{u}_m$, the Hamiltonian is[11]

$$\begin{aligned}H&=H_0+V_3\\ H_0&=\tfrac{1}{2}\sum_m M\dot{\mathbf{u}}_m^2+\tfrac{1}{2}\sum_{mn}A_{mn}{}^{ij}u_m{}^iu_n{}^j\\ V_3&=\frac{1}{3!}\sum_{\substack{lmn\\ ijk}}B_{lmn}{}^{ijk}u_l{}^iu_m{}^ju_n{}^k,\end{aligned} \quad (3.1)$$

where M is the mass of the atom and i, j, k denotes the x, y, z components. In normal mode coordinates[12]

$$\dot{\mathbf{u}}_m=\sum_k\left(\frac{1}{2MN\omega_k}\right)^{1/2}(a_ke^{i\mathbf{k}\cdot\mathbf{m}}+a_k{}^\dagger e^{-i\mathbf{k}\cdot\mathbf{m}})\mathbf{e}_k, \quad (3.2)$$

where k denotes the wave vector $\mathbf{k}$ and the polarization index; $\mathbf{e}_k$ is the polarization vector, and we adopt the convention $\mathbf{e}_{-k}=-\mathbf{e}_k$. Hence,

$$\begin{aligned}H_0&=\sum_k\omega_k(a_k{}^\dagger a_k+\tfrac{1}{2})\\ V_3&=-\sum_{kk'k''}(\omega_k\omega_{k'}\omega_{k''})^{-1/2}\\ &\quad\times(V_{kk'-k''}a_{k''}{}^\dagger a_{k'}a_k+\mathrm{H.c.}), \quad (3.3)\end{aligned}$$

[11] Notation is the same as given by P. Carruthers, Rev. Mod. Phys. 33, 92 (1961).

[12] The restriction to crystals having real polarization vectors is of no concern to the purpose of this paper. Similarly, the optical modes are never considered explicitly.

where

$$V_{kk'k''}=3\times\frac{1}{3!}\left(\frac{1}{2MN}\right)^{3/2}b_{kk'k''}$$

$$b_{kk'k''}=\sum_{ijk}e_k{}^i e_{k'}{}^j e_{k''}{}^k\sum_{lmn}B_{lmn}{}^{ijk}e^{i(\mathbf{k}\cdot\mathbf{l}+\mathbf{k}'\cdot\mathbf{m}+\mathbf{k}''\cdot\mathbf{n})}. \tag{3.4}$$

The factor of three in $V_{kk'k''}$ comes from cyclic permutation of k, k', k''. We have omitted the terms which create and destroy three phonons, since they do not conserve energy. They can and must be considered when higher order terms are considered.

If we consider the quantum Liouville equation between states $|\{\alpha_k\}\rangle$, where $\{\alpha_k\}$ denotes a set of amplitudes α_k for all the modes, we get

$$i\langle\{\alpha_k\}|\frac{\partial\rho}{\partial t}|\{\alpha_k\}\rangle=\frac{1}{\pi^N}\int\prod_k d^2\beta_k[\langle\{\alpha_k\}|H|\{\beta_k\}\rangle\langle\{\beta_k\}|\rho|\{\alpha_k\}\rangle-\langle\{\alpha_k\}|\rho|\{\beta_k\}\rangle\langle\{\beta_k\}|H|\{\alpha_k\}\rangle]. \tag{3.5}$$

As remarked in Sec. II, it is sufficient to consider the diagonal elements of ρ in this basis. Off-diagonal elements are found by analytic continuation of α^* to β^*. The right hand side of Eq. (3.5) contains the following terms:

(i) From H_0:

$$\sum_q\omega_q\frac{1}{\pi^N}\int\prod_k d^2\beta_k[\alpha_q{}^*\beta_q\langle\{\alpha_k\}|\{\beta_k\}\rangle\langle\{\beta_k\}|\rho|\{\alpha_k\}\rangle-\beta_q{}^*\alpha_q\langle\{\alpha_k\}|\rho|\{\beta_k\}\rangle\langle\{\beta_k\}|\{\alpha_k\}\rangle] \tag{3.6}$$

which upon integration by the methods outlined in Sec. II becomes

$$\sum_k\omega_k\left(\alpha_k{}^*\frac{\partial}{\partial\alpha_k{}^*}-\alpha_k\frac{\partial}{\partial\alpha_k}\right)\langle\{\alpha_k\}|\rho|\{\alpha_k\}\rangle; \tag{3.7}$$

(ii) From V_3:

$$-\sum_{qq'q''}(\omega_q\omega_{q'}\omega_{q''})^{-1/2}\Big\{V_{qq'-q''}\frac{1}{\pi^N}\int\prod_k d^2\beta_k[\alpha_{q''}{}^*\beta_{q'}\beta_q\langle\{\alpha_k\}|\{\beta_k\}\rangle\langle\{\beta_k\}|\rho|\{\alpha_k\}\rangle-\beta_{q''}{}^*\alpha_{q'}\alpha_q\langle\{\alpha_k\}|\rho|\{\beta_k\}\rangle\langle\{\beta_k\}|\{\alpha_k\}\rangle]+\text{H.c.}\Big\}$$

$$=-\sum_{kk'k''}(\omega_k\omega_{k'}\omega_{k''})^{-1/2}\Big\{V_{kk'-k''}\Big[\alpha_{k''}{}^*\Big(\alpha_{k'}+\frac{\partial}{\partial\alpha_{k'}{}^*}\Big)\Big(\alpha_k+\frac{\partial}{\partial\alpha_k{}^*}\Big)-\Big(\alpha_{k''}{}^*+\frac{\partial}{\partial\alpha_{k''}}\Big)\alpha_{k'}\alpha_k\Big]+\text{H.c.}\Big\}\times\langle\{\alpha_k\}|\rho|\{\alpha_k\}\rangle. \tag{3.8}$$

Therefore, the Liouville equation in the coherent-state representation becomes

$$i\frac{\partial}{\partial t}\langle\{\alpha_k\}|\rho|\{\alpha_k\}\rangle=\sum_k\omega_k\left(\alpha_k{}^*\frac{\partial}{\partial\alpha_k{}^*}-\alpha_k\frac{\partial}{\partial\alpha_k}\right)\langle\{\alpha_k\}|\rho|\{\alpha_k\}\rangle$$
$$-\sum_{kk'k''}(\omega_k\omega_{k'}\omega_{k''})^{-1/2}\Big\{V_{kk'-k''}\Big[\alpha_{k''}{}^*\Big(\alpha_{k'}+\frac{\partial}{\partial\alpha_{k'}{}^*}\Big)\Big(\alpha_k+\frac{\partial}{\partial\alpha_k{}^*}\Big)-\Big(\alpha_{k''}{}^*+\frac{\partial}{\partial\alpha_{k''}}\Big)\alpha_{k'}\alpha_k\Big]+\text{c.c.}\Big\}\langle\{\alpha_k\}|\rho|\{\alpha_k\}\rangle. \tag{3.9}$$

Using Eqs. (2.20)–(2.21) to transform to the action-angle variables, we obtain:

$$i\frac{\partial}{\partial t}\rho(\{J_k,\phi_k\},t)=i\sum_k\omega_k\frac{\partial}{\partial\phi_k}\rho(\{J_k,\phi_k\},t)$$
$$+i\sum_{kk'k''}\Big\{\Big(\frac{J_kJ_{k'}J_{k''}}{\omega_k\omega_{k'}\omega_{k''}}\Big)^{1/2}V_{kk'-k''}e^{i(\phi_k+\phi_{k'}-\phi_{k''})}\Big[i\Big(\frac{\partial}{\partial J_k}+\frac{\partial}{\partial J_{k'}}-\frac{\partial}{\partial J_{k''}}\Big)-\frac{1}{2}\Big(\frac{1}{J_k}\frac{\partial}{\partial\phi_k}+\frac{1}{J_{k'}}\frac{\partial}{\partial\phi_{k'}}+\frac{1}{J_{k''}}\frac{\partial}{\partial\phi_{k''}}\Big)$$
$$+\Big(i\frac{\partial^2}{\partial J_k\partial J_{k'}}-\frac{1}{2J_{k'}}\frac{\partial^2}{\partial J_k\partial\phi_{k'}}-\frac{1}{2J_k}\frac{\partial^2}{\partial J_{k'}\partial\phi_k}-i\frac{1}{4J_{k'}J_k}\frac{\partial^2}{\partial\phi_{k'}\partial\phi_k}\Big)\Big]+\text{c.c.}\Big\}\rho(\{J_k,\phi_k\},t). \tag{3.10}$$

Apart from a difference in sign convention, and the addition of terms involving second derivatives in J and ϕ, Eq. (3.10) is identical to that given by Brout and Prigogine. The "quantum corrections" are clearly of order $1/J$ relative to the "classical" terms and hence vanish in the classical limit. We thus obtain the advantages of physical interpretation of the classical approach while keeping the quantum-mechanical accuracy.

In writing Eq. (3.9) and Eq. (3.10) we have omitted those contributions which do not conserve energy in the dominant approximation (see next section). A similar calculation shows that these contribute the following terms to Eq. (3.9)

$$\sum_{kk'k''}(\omega_k\omega_{k'}\omega_{k''})^{-1/2}\left\{V_{kk'k''}\left[\left(\alpha_k+\frac{\partial}{\partial\alpha_k^*}\right)\left(\alpha_{k'}+\frac{\partial}{\partial\alpha_{k'}^*}\right)\left(\alpha_{k''}+\frac{\partial}{\partial\alpha_{k''}^*}\right)-\alpha_k\alpha_{k'}\alpha_{k''}\right]+\text{c.c.}\right\}\langle\{\alpha_k\}|\rho|\{\alpha_k\}\rangle. \quad (3.11)$$

In action-angle variables, this contributes the following terms to (Eq. 3.10):

$$\sum_{kk'k''}\left\{\left(\frac{J_kJ_{k'}J_{k''}}{\omega_k\omega_{k'}\omega_{k''}}\right)^{1/2}V_{kk'k''}e^{i(\phi_k+\phi_{k'}+\phi_{k''})}\left[\left(1+\frac{\partial}{\partial J_k}+\frac{i}{2J_k}\frac{\partial}{\partial\phi_k}\right)\right.\right.$$

$$\left.\left.\times\left(1+\frac{\partial}{\partial J_{k'}}+\frac{i}{2J_{k'}}\frac{\partial}{\partial\phi_{k'}}\right)\left(1+\frac{\partial}{\partial J_{k''}}+\frac{i}{2J_{k''}}\frac{\partial}{\partial\phi_{k''}}\right)-1\right]+\text{c.c.}\right\}\rho(\{J_k,\phi_k\},t). \quad (3.12)$$

IV. THE BROUT-PRIGOGINE EQUATION

The development of the preceding section has shown that the time evolution of the ensemble is given by the solution of the differential equation for the reduced matrix element $\rho(\{J_k,\phi_k\},t)$.

$$i\frac{\partial}{\partial t}\rho(\{J_k,\phi_k\},t)=(L_0+L_1)\rho(\{J_k,\phi_k\},t), \quad (4.1)$$

where the unperturbed Liouville operator L_0 is given by Eq. (2.26) and L_1 by Eq. (3.10)

$$L_1=i\sum_{kk'k''}\left\{\left(\frac{J_kJ_{k'}J_{k''}}{\omega_k\omega_{k'}\omega_{k''}}\right)^{1/2}V_{kk'-k''}e^{i(\phi_k+\phi_{k'}-\phi_{k''})}\left[i\left(\frac{\partial}{\partial J_k}+\frac{\partial}{\partial J_{k'}}-\frac{\partial}{\partial J_{k''}}\right)-\frac{1}{2}\left(\frac{1}{J_k}\frac{\partial}{\partial\phi_k}+\frac{1}{J_{k'}}\frac{\partial}{\partial\phi_{k'}}+\frac{1}{J_{k''}}\frac{\partial}{\partial\phi_{k''}}\right)\right.\right.$$

$$\left.\left.+\left(i\frac{\partial^2}{\partial J_k\partial J_{k'}}-\frac{1}{2J_{k'}}\frac{\partial^2}{\partial J_k\partial\phi_{k'}}-\frac{1}{2J_k}\frac{\partial^2}{\partial J_{k'}\partial\phi_k}-i\frac{1}{4J_{k'}J_k}\frac{\partial^2}{\partial\phi_{k'}\partial\phi_k}\right)\right]+\text{c.c.}\right\}. \quad (4.2)$$

In the present paper, we shall solve (4.1) by well known perturbation techniques. First define the interaction-picture density matrix

$$\rho_I(\{J_k,\phi_k\},t)\equiv\exp(-iL_0t)\rho(\{J_k,\phi_k\},t). \quad (4.3)$$

ρ_I satisfies the equation

$$i\frac{\partial\rho_I}{\partial t}=L_I(t)\rho_I, \quad (4.4)$$

where

$$L_I(t)=e^{-iL_0t}L_1e^{iL_0t}. \quad (4.5)$$

The formal solution of Eq. (4.4) is

$$\rho_I(t)=T\exp\left(-i\int_{t_0}^{t}L_1(t')dt'\right)\rho_I(t_0), \quad (4.6)$$

where T is the usual time-ordering operator. We shall generally take t_0 to be zero.

$\rho_I(t)$ may be expanded in terms of the eigenfunctions of L_0, the $f\{\nu\}$ from Eq. (2.27):

$$\rho_I(t)=\sum_{\{\nu\}}\rho_I(\{J_k\}\{\nu\},t)f\{\nu\}, \quad (4.7)$$

where the expansion coefficients are

$$\rho_I(\{J_k\}\{\nu\},t)=[f\{\nu\}|\rho_I(\{J_k,\phi_k\},t)]$$

$$=\int f^*\{\nu\}\rho_I(\{J_k,\phi_k\},t)\prod_k d\phi_k. \quad (4.8)$$

The coefficients $\rho_I(\{J_k\}\{\nu\},t)$ resolve the density matrix into the phase functions $f\{\nu\}$. Inserting Eq. (4.6) into Eq. (4.8) yields (using the completeness of $f\{\nu\}$)

$$\rho_I(\{J_k\}\{\nu\},t)=\sum_{\{\nu'\}}\left(f\{\nu\}|T\exp\left(-i\int_0^t L_I(t')dt'\right)f\{\nu'\}\right)$$

$$\times(f\{\nu'\}|\rho_I(\{J_k,\phi_k\},0)). \quad (4.9)$$

This relation is independent of $\{\phi\}$ but still contains derivatives $\partial/\partial J_k$.

We next examine the consequences of the random-phase initial condition:

$$\rho_I(\{J_k\}\{\nu\},0)=0, \quad \{\nu\}\neq\{0\}. \quad (4.10)$$

As explained in Sec. II, the independence of ρ from the phase angles only occurs if ρ is a function of the number

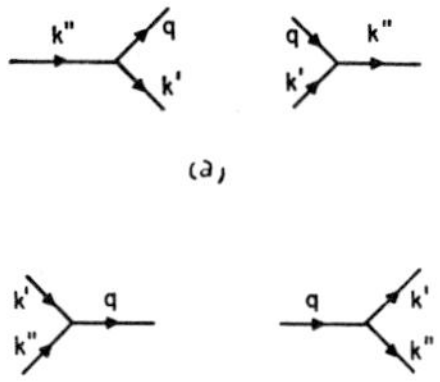

FIG. 1. The three-phonon collision terms entering in the Peierls equation are illustrated graphically. Figure 1(a) corresponds to Eq. (5.9), in which phonon **q** is emitted or absorbed. Figure 1(b), which corresponds to Eq. (5.10) shows processes in which phonon **q** is produced by the coalescence of the phonons or destroyed by splitting into two phonons.

operators $\{N_k\}$. Hence the coefficient

$$\rho_I(\{J_k\}\{0\},t)\equiv\rho_E(\{J_k\},t)$$

describes the number of phonons, or the energy distribution among the normal modes.

The energy-distribution function at time t is then given by

$$\rho_E(\{J_k\},t)=\left(f\{0\}\,|\,T\exp\left(-i\int_0^t L_I(t')dt'\right)f\{0\}\right)\times\rho_E(\{J_k\},0). \quad (4.11)$$

Quantities of the type

$$\left\langle T\exp\left(-i\int_0^t L_I(t')dt'\right)\right\rangle_0=\left(f\{0\}\,|\,T\exp\left(-i\int_0^t L_I(t')dt'\right)f\{0\}\right) \quad (4.12)$$

often arise in statistical and quantum mechanics. A systematic way exists for the evaluation of such expressions.[13,14] In Eq. (4.9) we have to calculate the average of an exponential over the phase angles. As is well-known, such an average can be expressed as the exponential of a modified series.

The method of semi-invariants (or cumulants), explained in Ref. 13, 'offers a rapid and powerful way of exhibiting this exponentiation. Complicated summations are thereby avoided. We state a few relations for clarity. Consider the moment generating function $\langle\exp(ixt)\rangle$, where the average is taken over some probability distribution $P(x)$. The logarithm of this quantity expands as

$$\ln\langle\exp(ixt)\rangle=\sum_{n=1}^{\infty}\frac{(it)^n}{n!}M_n, \quad (4.13)$$

where the first few semi-invariants M_n are given by

$$\begin{aligned} M_1&=\langle x\rangle,\\ M_2&=\langle x^2\rangle-\langle x\rangle^2,\\ M_3&=\langle x^3\rangle-2\langle x^2\rangle\langle x\rangle+2\langle x\rangle^3,\\ M_4&=\langle x^4\rangle-4\langle x^3\rangle\langle x\rangle-3\langle x^2\rangle^2+12\langle x^2\rangle\langle x\rangle-6\langle x\rangle^4. \end{aligned} \quad (4.14)$$

The case of a time-ordered exponential of an operator is not much more complicated.[13]

For the problem at hand (cubic anharmonicities) all odd M_n vanish.

$$\ln\left\langle T\exp\left(-i\int_0^t L_I(t')dt'\right)\right\rangle_0=\sum_{n=1}^{\infty}\frac{(-1)^n}{n!}M_n(t), \quad (4.15)$$

$$M_1=\int_0^t\langle L_I(t')\rangle_0 dt'=0,$$

$$M_2=\int_0^t dt_1\int_0^t dt_2[\langle TL_I(t_1)L_I(t_2)\rangle_0-\langle L_I(t_1)\rangle_0\langle L_I(t_2)\rangle_0]$$

$$=\int_0^t dt_1\int_0^t dt_2\langle TL_I(t_1)L_I(t_2)\rangle_0,$$

$$M_4=\int_0^t dt_1\cdots\int_0^t dt_4[\langle TL_I(t_1)\cdots L_I(t_4)\rangle_0-3\langle TL_I(t_1)L_I(t_2)\rangle_0\langle TL_I(t_3)L_I(t_4)\rangle_0]. \quad (4.16)$$

For long times, the M_n are proportional to t. Measuring the perturbation by a parameter λ, we see that the contributions are of order $M_n\sim\lambda^n t$ in this limit. Our procedure provides a simple method of estimating higher order corrections to transport quantities.[15,16]

Thus we find $\rho_E(\{J_k\},t)$ to be given by

$$\rho_E(\{J_k\},t)=\exp\left[\sum_n\frac{(-i)^n}{n!}M_n(t)\right]\rho_E(\{J_k\},0), \quad (4.17)$$

$$\rho_E(\{J_k\},t)\approx\exp[-\tfrac{1}{2}M_2(t)]\rho_E(\{J_k\},0). \quad (4.18)$$

In the last expression we have kept only the term of order $\lambda^2 t$. For times substantially longer than that in which the system adjusts to the perturbation (of order $1/\omega_D$ where ω_D is the Debye frequency) $-\frac{1}{2}M_2(t)$ goes over into the Brout-Prigogine operator $\lambda^2 t\mathcal{O}_0$ where $\mathcal{O}_0$ is given by

$$\mathcal{O}_0=-\sum_{\{\nu\}}(f\{0\}\,|\,L_I f\{\nu'\})\times\delta_+(\sum\nu_k'\omega_k)(f\{\nu'\}\,|\,L_I f\{0\}) \quad (4.19)$$

$$\pi\delta_+(x)=\pi\delta(x)+i\frac{P}{x},$$

P being the principal value.

[13] R. Brout and P. Carruthers, *Lectures on the Many-Electron Problem* (Interscience Publishing Company, New York, 1962).
[14] R. Brout, Phys. Rev. **115**, 824 (1959).

[15] The extension to more general interactions should be obvious.
[16] P. Carruthers, Phys. Rev. **126**, 1448 (1962). In this work the possible importance of higher order corrections is stressed. Less convincing techniques were used in that work.

In differential form, Eq. (4.16) becomes

$$\frac{\partial}{\partial t}\rho_E(\{J_k\},t)=\Theta_0\rho_E(\{J_k\},t) \quad (4.20)$$

which is the Brout-Prigogine master equation.

The invalidity of this equation for $0\leqslant t\leqslant 1/\omega_D$ is of no physical consequence, as the assumed initial condition is unphysical, involving as it does the introduction of bare phonons.

V. DERIVATION OF THE PEIERLS MASTER EQUATION

Our expression for $\partial\rho_E/\partial t$ differs from the classical one only by the addition of second derivative terms. These give rise to the spontaneous decay of the phonons, a process not present in the classical theory. To see this, we shall derive the Peierls equation from the Brout-Prigogine equation.

Explicitly, the matrix elements $(f\{\nu\}|L_If\{\nu'\})$ are

$$(f\{\nu\}|L_If\{\nu'\})=\frac{1}{(2\pi)^N}\int_0^{2\pi}\cdots\int_0^{2\pi}\prod_k d\phi_k \times\exp(-i\sum_k \nu_k\phi_k)L_I\exp(i\sum_k \nu_k'\phi_k). \quad (5.1)$$

Using Eq. (4.2), we see that the nonvanishing elements of $(f\{0\}|L_If\{\nu'\})$ and $(f\{\nu'\}|L_If\{0\})$ are

(i) $$(f\{0\}|L_I|f\{-1_k-1_{k'}1_{k''};0\})=-V_{kk'-k''}\left(\frac{J_kJ_{k'}J_{k''}}{\omega_k\omega_{k'}\omega_{k''}}\right)^{1/2}\left(\frac{\partial}{\partial J_k}+\frac{\partial}{\partial J_{k'}}-\frac{\partial}{\partial J_{k''}}+\frac{1}{2J_k}+\frac{1}{2J_{k'}}-\frac{1}{2J_{k''}}+\frac{\partial^2}{\partial J_k\partial J_{k'}}+\frac{1}{2J_k}\frac{\partial}{\partial J_{k'}}+\frac{1}{2J_{k'}}\frac{\partial}{\partial J_k}+\frac{1}{4J_kJ_{k'}}\right)e^{-i(\omega_k+\omega_{k'}-\omega_{k''})t} \quad (5.2)$$

which can be written as

$$(f\{0\}|L_If\{-1_k-1_{k'}1_{k''};0\})=-V_{kk'k''}\left(\frac{\partial}{\partial J_k}+\frac{\partial}{\partial J_{k'}}-\frac{\partial}{\partial J_{k''}}+\frac{\partial^2}{\partial J_k\partial J_{k'}}\right)\left(\frac{J_kJ_{k'}J_{k''}}{\omega_k\omega_{k'}\omega_{k''}}\right)^{1/2}e^{-i(\omega_k+\omega_{k'}-\omega_{k''})t} \quad (5.3)$$

(ii) $$(f\{-1_k-1_{k'}1_{k''};0\}|L_If\{0\})=V_{kk'-k''}^*\left(\frac{J_kJ_{k'}J_{k''}}{\omega_k\omega_{k'}\omega_{k''}}\right)^{1/2}\left(\frac{\partial}{\partial J_k}+\frac{\partial}{\partial J_{k'}}-\frac{\partial}{\partial J_{k''}}+\frac{\partial^2}{\partial J_k\partial J_{k'}}\right)e^{i(\omega_k+\omega_{k'}-\omega_{k''})t} \quad (5.4)$$

so the Brout-Prigogine equation becomes

$$\frac{\partial}{\partial t}\rho_E(t)=2\pi\sum_{kk'k''}\delta(\omega_k+\omega_{k'}-\omega_{k''})\frac{|V_{kk'-k''}|^2}{\omega_k\omega_{k'}\omega_{k''}}\times\left[\left(\frac{\partial}{\partial J_k}+\frac{\partial}{\partial J_{k'}}-\frac{\partial}{\partial J_{k''}}+\frac{\partial^2}{\partial J_k\partial J_{k'}}\right)J_kJ_{k'}J_{k''}\left(\frac{\partial}{\partial J_k}+\frac{\partial}{\partial J_{k'}}-\frac{\partial}{\partial J_{k''}}+\frac{\partial^2}{\partial J_k\partial J_{k'}}\right)\right]\rho_E(t). \quad (5.5)$$

The factor of 2 on the right hand side comes from adding the complex conjugate term and the principal part of δ_+ does not contribute because it is an odd function of $\{\nu_k'\}$.

We can now calculate the time rate of change of the average number of phonons in a certain mode q.

$$\frac{d}{dt}\langle N_q\rangle=\int(\prod_k dJ_k)J_q\frac{\partial\rho_E}{\partial t}. \quad (5.6)$$

When we substitute $\partial\rho_E/\partial t$ from Eq. (5.5) into Eq. (5.6), the sum over k, k', k'' will contain four types of terms:

$$\begin{aligned}&(a) && k,\,k',\,k''\neq q,\\ &(b) && k=q;\quad k',\,k''\neq q,\\ &(c) && k'=q;\quad k,\,k''\neq q,\\ &(d) && k''=q;\quad k,\,k'\neq q.\end{aligned} \quad (5.7)$$

Doing the integration of each of these terms by parts, we see that the first type does not contribute. The second and third types give

$$\langle\langle J_{k'}J_{k''}\rangle\rangle+\langle\langle J_qJ_{k''}\rangle\rangle-\langle\langle J_qJ_{k'}\rangle\rangle-\langle\langle J_{k''}\rangle\rangle. \quad (5.8)$$

Note that the last term of Eq. (5.8), which comes from the second-derivative term, does not appear in a classical theory and is what gives rise to the spontaneous decay process. Transforming by Eq. (2.33) to the number representation, Eq. (5.8) becomes

$$\begin{aligned}&\langle N_{k'}N_{k''}\rangle+\langle N_qN_{k''}\rangle-\langle N_qN_{k'}\rangle+\langle N_{k''}\rangle\\ &=\langle(N_q+1)(N_{k'}+1)N_{k''}\rangle-\langle N_qN_{k'}(N_{k''}+1)\rangle\end{aligned} \quad (5.9)$$

which corresponds to the processes of Fig. 1(a). The fourth type gives

$$\begin{aligned}&\langle\langle J_{k'}J_{k''}\rangle\rangle-\langle\langle J_qJ_{k'}\rangle\rangle-\langle\langle J_qJ_{k''}\rangle\rangle+\langle\langle J_q\rangle\rangle\\ &=\langle N_{k'}N_{k''}\rangle-\langle N_qN_{k'}\rangle-\langle N_qN_{k''}\rangle-\langle N_q\rangle\\ &=\langle N_{k'}N_{k''}(N_q+1)\rangle-\langle N_q(N_{k'}+1)(N_{k''}+1)\rangle,\end{aligned} \quad (5.10)$$

which corresponds to the processes of Fig. 1(b).

Therefore, changing the index q into k, we have derived the Peierls equation[6,7]:

$$\frac{d}{dt}\langle N_k\rangle = 2\pi \sum_{k'k''} \frac{|V_{kk'-k''}|^2}{\omega_k\omega_{k'}\omega_{k''}} \times\{2\delta(\omega_k+\omega_{k'}-\omega_{k''})[\langle(N_k+1)(N_{k'}+1)N_{k''}\rangle - \langle N_kN_{k'}(N_{k''}+1)\rangle]+\delta(\omega_k-\omega_{k'}-\omega_{k''}) \times[\langle N_{k'}N_{k''}(N_k+1)\rangle-\langle N_k(N_{k'}+1)(N_{k''}+1)\rangle]\}. \quad (5.11)$$

This result is not sufficient for the rigorous derivation of a transport equation since the density matrix corresponds to a homogeneous (but nonequilibrium) spatial distribution. When we use the Wigner distribution function formalism for transport processes, other components of the density matrix besides $\rho_E(t)$ have to be considered. This problem will be dealt with in a subsequent paper.

ACKNOWLEDGMENTS

The authors are indebted to Professor J. A. Krumhansl and Professor G. V. Chester for many interesting and helpful discussions.

APPENDIX

Consider the following interaction Hamiltonian for scattering by defects:

$$V_3=\frac{1}{4MN}\sum_{qq'}\frac{1}{(\omega_q\omega_{q'})^{1/2}}C_{qq'}a_q^\dagger a_{q'}+\text{H.c.} \equiv\sum_{qq'}U_{qq'}a_q^\dagger a_{q'}+\text{H.c.} \quad (A1)$$

For example, expressions of this form describe strain field or isotope scattering.[11] The perturbation L_1 is

$$L_1=\sum_{kk'}U_{kk'}(J_kJ_{k'})^{1/2}e^{i(\phi_{k'}-\phi_k)} \times\left(\frac{\partial}{\partial J_{k'}}+i\frac{1}{2J_{k'}}\frac{\partial}{\partial\phi_{k'}}-\frac{\partial}{\partial J_k}+i\frac{1}{2J_k}\frac{\partial}{\partial\phi_k}+\text{c.c.}\right). \quad (A2)$$

The nonvanishing elements of $(f\{0\}|L_If\{\nu'\})$ and $(f\{\nu'\}|L_I|f\{0\})$ are:

$$(f\{0\}|L_I|f\{1_k-1_{k'};0\}) =U_{kk'}\left(\frac{\partial}{\partial J_{k'}}-\frac{\partial}{\partial J_k}\right)(J_kJ_{k'})^{1/2}e^{-i(\omega_k-\omega_{k'})t}$$

$$(f\{1_k-1_{k'};0\}|L_I|f\{0\}) =U_{kk'}^*(J_kJ_{k'})^{1/2}\left(\frac{\partial}{\partial J_{k'}}-\frac{\partial}{\partial J_k}\right)e^{i(\omega_k-\omega_{k'})t} \quad (A3)$$

so

$$\mathfrak{O}_0=2\pi\sum_{kk'}\delta(\omega_k-\omega_{k'})|U_{kk'}|^2 \times\left(\frac{\partial}{\partial J_{k'}}-\frac{\partial}{\partial J_k}\right)J_kJ_{k'}\left(\frac{\partial}{\partial J_{k'}}-\frac{\partial}{\partial J_k}\right). \quad (A4)$$

From this result, we get

$$\frac{d\langle N_k\rangle}{dt}=2\pi\sum_{k'}|U_{kk'}|^2(\langle N_{k'}\rangle-\langle N_k\rangle). \quad (A5)$$

This is simply the "Golden rule" result; see for example Eq. (4.14) of Ref. 11.

PHYSICAL REVIEW VOLUME 151, NUMBER 1 4 NOVEMBER 1966

Theory of Superfluidity

FREDERICK W. CUMMINGS AND JAMES R. JOHNSTON
Department of Physics, University of California, Riverside, California
(Received 21 March 1966; revised manuscript received 19 May 1966)

A new structure for the microscopic theory of superfluid helium is presented which is motivated by analogy with coherent radiation of ideal lasers and with the BCS theory of superconductivity. The concept of off-diagonal long-range order is employed to put these three phenomena on a common basis. The unifying concept is that the ground state is that which completely factors the reduced density matrix appropriate to the "basic group." It is shown how the model predicted by this motivation can also be derived from the microscopic Hamiltonian by a quasiparticle self-consistent-field approximation. The superfluid ground state can be viewed as energy-lowering "coherent" excitations of the normal fluid. The normal and superfluid components are the direct analog of "noise" and "signal" in partially coherent radiation fields.

I. INTRODUCTION

THIS paper is concerned with a model for superfluid helium (He II) which is motivated by analogy with superconductivity and the coherent radiation of lasers. The concept of "off-diagonal long-range order" (ODLRO) plays a central role in this motivation.

The concept of ODLRO was originally introduced by Penrose[1] and Penrose and Onsager[2] to generalize the criterion of momentum space condensation (as occurs in an ideal Einstein-Bose gas) for superfluidity so as to be applicable to interacting systems (He II). Yang[3] later generalized the concept so as to encompass superconductivity as well.

ODLRO is present whenever one of the reduced density matrixes exhibits some degree of factorization (which may be due to the existence of a macroscopic eigenvalue). Yang[3] showed that the BCS[4] superconducting ground state was unique in that it is the only state which gives rise to the maximum possible eigenvalue in the reduced density matrix $\rho^{(2)}$, i.e., $\rho^{(2)}$ is completely factorable only for the BCS ground state. Since the large eigenvalue ($\lambda_1^{(2)}$) of $\rho^{(2)}$ is the maximum possible value at $T=0°$ and is microscopic above the superconducting transition temperature, it seems natural to associate the ratio $\lambda_1^{(2)}(T)/\mathrm{Tr}\rho^{(2)}$ with the ratio of superconducting densities $\rho_s(T)/\rho_s(0)$.

The BCS ground state can be viewed as a condensation of the Bose-like Cooper pairs into a zero (pair) momentum state analogous to the ideal Einstein-Bose gas. We prefer not to invoke this analogy for two reasons: (1) The Cooper pairs are not truly Bose particles as evidenced by the commutation rules for the Cooper pair operators,[5] and (2) the condensate is not a collection of indistinguishable particles in a single mode in that each Cooper pair is labeled with a momentum (and spin) index. Yang[3] has emphasized the point that superconductivity and superfluidity are each a special case of ODLRO.

The current application of the concept of ODLRO to superfluid helium seems to be primarily centered about the idea of a generalized condensation into some single momentum mode—which is only one of the ways in which ODLRO can occur.[6] The amount of condensate in the zero momentum mode in He II is believed to be of the order of 10%.[2,7] If the reduced density matrix $\rho^{(1)}$ for helium is defined in terms of particle operators, the factorization due to condensation into the zero-momentum mode can at most be of the order of 10%. Since the ground state of helium (100% superfluid) is a single pure quantum state, one would expect there to be a more appropriate definition of $\rho^{(1)}$ which would give complete factorization at zero temperature.

One reason for the current emphasis on condensation into a given momentum mode is the speculation of London[8] that the mechanism responsible for superfluidity should be analogous to the condensation in an ideal Bose gas. We prefer the point of view that ODLRO in superfluid helium is not as intimately connected with a condensation into any single momentum mode as it is with "coherent" excitations above the normal-fluid ground state in such a way as to lower the ground-state energy—as in superconductivity. We will show later that in the model proposed herein the occurrence of ODLRO is completely independent of the existence or nonexistence of a condensation into a single-momentum mode.

We wish at this point to comment briefly on the coherent radiation states believed to be produced by an ideal laser. If a Hamiltonian for the interaction of (two-level, say) atoms with a radiation field[9] is "averaged" over the atomic operators (the atomic system is treated "classically"), then it may be written as

$$\bar{H}=\sum_{\mathbf{k}} \omega_{\mathbf{k}}a_{\mathbf{k}}^{\dagger}a_{\mathbf{k}}-\sum_{\mathbf{k}}(\omega_{\mathbf{k}}\alpha_{\mathbf{k}}a_{\mathbf{k}}^{\dagger}+\omega_{\mathbf{k}}\alpha_{\mathbf{k}}^{*}a_{\mathbf{k}}). \qquad (I.1)$$

[1] O. Penrose, Phil. Mag. **42**, 1373 (1951).
[2] O. Penrose and L. Onsager, Phys. Rev. **104**, 576 (1956).
[3] C. N. Yang, Rev. Mod. Phys. **34**, 694 (1962).
[4] L. N. Cooper, Phys. Rev. **104**, 1189 (1956); J. Bardeen, L. N. Cooper, and J. R. Schrieffer, *ibid.* **108**, 1175 (1957).
[5] Reference 4, (BCS) Eqs. (2.9)–(2.13).
[6] M. D. Girardeau, J. Math. Phys. **6**, 1083 (1965).
[7] A. Miller, D. Pines, and P. Nozieres, Phys. Rev. **127**, 1452 (1962).
[8] F. London, *Superfluids* (John Wiley & Sons, Inc., New York, 1954).
[9] W. Louiselle, *Radiation and Noise in Quantum Electronics* (McGraw-Hill, Book Company, Inc., New York, 1964).

This can then be rewritten as

$$\bar{H}=\sum_{\mathbf{k}}\omega_{\mathbf{k}}(a_{\mathbf{k}}^{\dagger}-\alpha_{\mathbf{k}}^{*})(a_{\mathbf{k}}-\alpha_{\mathbf{k}})-\sum_{\mathbf{k}}\omega_{\mathbf{k}}|\alpha_{\mathbf{k}}|^{2}. \quad (I.2)$$

(The $\alpha_{\mathbf{k}}^*$, $\alpha_{\mathbf{k}}$ represent averages over the atom raising or lowering operators.) This Hamiltonian is now diagonal in the operators $b_{\mathbf{k}}=a_{\mathbf{k}}-\alpha_{\mathbf{k}}$, $b_{\mathbf{k}}^{\dagger}=a_{\mathbf{k}}^{\dagger}-\alpha_{\mathbf{k}}^*$ and the ground state is the vacuum state for the $b_{\mathbf{k}}$'s,

$$b_{\mathbf{k}}|\text{vac}\rangle_b=0, \quad (I.3)$$

and is given by

$$|\{\alpha_k\}\rangle=\prod_{\mathbf{k}} e^{-|\alpha_{\mathbf{k}}|^2/2}e^{\alpha_{\mathbf{k}}a_{\mathbf{k}}^{\dagger}}|\text{vac}\rangle_a, \quad (I.4)$$

where

$$|\text{vac}\rangle_b\equiv|\{\alpha_{\mathbf{k}}\}\rangle_a \quad (1.5)$$

is the vacuum state for the $b_{\mathbf{k}}$'s and $|\text{vac}\rangle_a$ is the vacuum state for the $a_{\mathbf{k}}$'s. The states (4) are the coherent radiation states discussed in detail by Glauber.[10,11]

Glauber[10] postulated that the condition for *complete* coherence of a radiation field is that all of the correlation functions

$$G^{(n)}(x,x')\equiv\text{Tr}\{\rho\psi^{\dagger}(x_1)\psi^{\dagger}(x_2)\cdots \psi^{\dagger}(x_n)\psi(x_n')\cdots\psi(x_1')\} \quad (I.6)$$

factor completely. Here, the x's indicate space-time points (r,t), $\psi^{\dagger}(x)$, and $\psi(x)$ are operators that annihilate and create photons at the space-time point x, and ρ is as usual the density operator for the system. Glauber showed[11] that one set of states which gives rise to complete factorization of all orders of the correlation functions are the states discussed above. This is due to the fact that these states are right eigenstates of the (non-normal) annihilation operator $a_{\mathbf{k}}$ and left eigenstates of the creation operator $a_{\mathbf{k}}^{\dagger}$. Titulaer and Glauber[12] have shown that, in general, the coherent modes need not be monochromatic. If the normal modes are some linear combination of monochromatic modes, say

$$b_{\mathbf{k}}=\sum_{\mathbf{p}}u_{\mathbf{k}\mathbf{p}}a_{\mathbf{p}}, \quad (I.7)$$

where u is unitary, then the coherent states are

$$e^{-1/2|\beta_{\mathbf{k}}|^2\beta_{\mathbf{k}}b_{\mathbf{k}}^{\dagger}}|\text{vac}\rangle. \quad (I.8)$$

The point to be emphasized here is that the coherent states are defined in terms of the normal modes of the normal system.

The correlation functions $G^{(n)}$ described above are formally the same as the nth order reduced density matrices in the coordinate representation. If we take the fundamental definition of ODLRO to be the existence of some finite degree of factorization of an nth order reduced density matrix in the coordinate representation, then the definitions of ODLRO and coherence are identical.[13] The coherent states described above give rise to complete factorization of all orders of correlation functions and concommitantly to all reduced density matrices.

The nature of ODLRO in these coherent states is considerably different from that due to a condensation into a given mode. For these states the first reduced density matrix is given by

$$\rho_{\mathbf{p}\mathbf{k}}^{(1)}=Tr\langle(a_{\mathbf{p}}\rho a_{\mathbf{k}}^{\dagger})=\langle\{\alpha_{\lambda}\}|a_{\mathbf{k}}^{\dagger}a_{\mathbf{p}}|\{\alpha_{\lambda}\}\rangle, \quad (I.9)$$

$$|\{\alpha_{\lambda}\}\rangle\equiv\prod_{\lambda}|\alpha_{\lambda}\rangle, \quad (I.10)$$

and $\rho_{\mathbf{p}\mathbf{k}}^{(1)}$ is simply

$$\rho_{\mathbf{p}\mathbf{k}}^{(1)}=\alpha_{\mathbf{k}}^*\alpha_{\mathbf{k}}. \quad (I.11)$$

The matrix $\rho^{(1)}$ is then the outer product of the vectors ν and $\nu^{\dagger}$ where

$$\nu\equiv\frac{1}{N^{1/2}}\begin{bmatrix}\alpha_1\\ \alpha_2\\ \vdots\end{bmatrix}, \quad \nu^{\dagger}\nu=1, \quad (I.12)$$

and N is the total average occupation of all modes. Now

$$\rho^{(1)}=N\nu\nu^{\dagger}, \quad (I.13)$$

and $\rho^{(1)}$ factors completely—regardless of the degree of occupation of the individual modes.

Yang[3] has shown that for fermions the lowest order reduced density matrix exhibiting ODLRO must be of even order. Consequently, the "basic group" must be an even number of fermions. The appropriate reduced density matrix is defined in terms of the operators which create and annihilate the normal mode "particles" of the normal fluid—the renormalized Bloch wave operators $c_{\mathbf{k}}^{\dagger}$ and $c_{\mathbf{k}}$. That is,

$$\rho_{\mathbf{p}\mathbf{k}}^{(2)}=\text{Tr}(c_{\mathbf{p}}c_{\mathbf{p}}\rho c_{\mathbf{k}}^{\dagger}c_{\mathbf{k}}^{\dagger}). \quad (I.14)$$

This is not the most general form of $\rho^{(2)}$, but it is appropriate to superconductivity since the ground-state BCS Hamiltonian can be written solely in terms of the operators $c_{\mathbf{k}}^{\dagger}c_{-\mathbf{k}}^{\dagger}$ and $c_{\mathbf{k}}c_{-\mathbf{k}}$. [Spin is included by the usual convention $\mathbf{k}=(\mathbf{k},\uparrow)$, $-\mathbf{k}=(-\mathbf{k},\downarrow)$.] As pointed out earlier, the BCS ground state[14]

$$|\Phi_0\rangle=C\prod_{\mathbf{k}} e^{g_{\mathbf{k}}c_{\mathbf{k}}^{\dagger}c_{\mathbf{k}}^{\dagger}}|\text{vac}\rangle \quad (I.15)$$

[10] R. J. Glauber, Phys. Rev. **130**, 2529 (1963).
[11] R. J. Glauber, Phys. Rev. **131**, 2766 (1963).
[12] U. M. Titulaer and R. J. Glauber, Phys. Rev. **145**, 1041 (1966).

[13] Yang (Ref. 3) showed that the existence of a large eigenvalue (and therefore factorization) in the nth-order reduced density matrix implies the existence of large eigenvalues (and factorization) in all mth-order reduced density matrices $m>n$. At first glance, this does not seem to be compatible with the fact that one can have a first-order coherent field (e.g., a highly filtered thermal radiation field) which does not have higher order coherence. However, the existence of a large eigenvalue in $\rho^{(n)}$ is a sufficient but not necessary condition for ODLRO (Ref. 6). For a filtered thermal radiation field with a narrow linewidth ($\Delta\omega\ll\omega_0$), all the terms in the spectral expansion of $G^{(1)}(t,t')(\sum_{\omega}f_{\omega}(t)f_{\omega}^*(t'))$ for which $|\omega-\omega_0|<(t-t')^{-1}$ give rise to a factored term. (The factorization is only approximate, since there is no factorization for $|t-t'|\to\infty$.) The first-order coherence is only approximate and is not due to a large eigenvalue. Therefore, the theorems of Yang do not apply to this case, and hence there is no implication of higher order coherence.
[14] J. G. Valatin, Nuovo Cimento **7**, 843 (1958).

was shown[3] to be the only state for which $\rho^{(2)}$ factors completely.

The argument given by BCS for their ground state was the lowering of energy over the normal state due to "coherent" excitations above the Fermi level which maximized the effect of the negative (effective) potential due to electron-phonon interactions.

Thus, we see that superconductivity and coherent radiation (from lasers, say) are quite analogous with respect to the points we have emphasized. We further point out that in lasers ODLRO first occurs in $\rho^{(1)}$ and the generator for the coherent radiation states is in terms of a single-particle operator $e^{\alpha_k a_k{}^\dagger}$, while for superconductivity ODLRO first occurs in $\rho^{(2)}$ and the generator for the BCS ground state is in terms of a pair operator $e^{g_k c_k{}^\dagger c_k{}^\dagger}$.

Valatin and Butler[15] in their theory of He II used a ground state similar to that in superconductivity and showed that it was the ground state for the helium "pairing" Hamiltonian. They used

$$|\Phi_0\rangle=\prod_{\mathbf{k}}(1-|g_k|^2)^{1/2}e^{g_k a_k{}^\dagger a_k{}^\dagger}|\mathrm{vac}\rangle. \qquad (I.16)$$

In this state, $\rho^{(1)}$ is diagonal in the momentum representation with the quantities $|g_{\mathbf{k}}|^2/(1-|g_{\mathbf{k}}|^2)=\langle n_{\mathbf{k}}\rangle$ on the diagonal. Thus, there is ODLRO only insofar as one of the momentum modes is macroscopically occupied—the same condition for ODLRO as in a free Bose gas. Also, the generator for the ground state is in terms of a pair operator—not an operator characteristic of the basic group for helium.

In a later article in which the ground state above was modified in an *ad hoc* manner so that $a_{\mathbf{k}}$ and $a_{\mathbf{k}}{}^\dagger$ have nonzero averages, Valatin[16] comments that the reason for this approach is that the analogies between superconductivity and superfluidity are "not merely formal and that the concepts used, perhaps with some further modifications, will prove to be relevant in a description of superfluid boson systems." Our point of view is that indeed this is the case, but that the analogy with superconductivity does not mean that the superfluid ground state should have the same form as for superconductivity. We propose that the analogy between the two "superfluids" (and with lasers) should be with respect to the following points:

(1) The appropriate reduced density matrices are defined in terms of the operators which create and destroy the normal (quasiparticle) modes of the normal fluid.

(2) The superfluid ground state is that which gives rise to complete factorization of the appropriate reduced density matrix.

(3) One expects the superfluid ground state to be generated by an operator characteristic of the basic group operating on the normal fluid ground state, and that these excitations above the normal fluid ground state would involve a lowering of the energy with respect to the normal ground state.

The first statement above is meaningful only if the fluid in the normal state can be described by a quasiparticle picture. It is not at all clear that liquid helium in the normal state can be so described. In order for this model to be applicable to liquid helium, either it must be shown that normal helium can be described in terms of quasiparticles, or the statements above must be generalized in such a way as to be meaningful in terms of the model used to describe the normal fluid. We are developing an analogy with coherent radiation and superconductors. Until it is known how to better describe normal helium, we can only present this analogy in terms of quasiparticles.

The excitation spectrum for liquid helium has a phonon character at low momenta below T_λ, and the Woods[17] experiment strongly indicates that the low-momenta spectrum does not change drastically from low temperatures through temperatures above T_λ. This indicates that the normal fluid cannot be completely described by free-particle-like quasiparticles in any temperature region. It is for this reason that we expect there to be a definition of the reduced density matrices which is more meaningful than the usual one[3] in terms of free-particle operators. As an example, we might imagine describing the coherence properties of a (hypothetical) phonon maser. Such a system would be described in terms of the coherence properties of the phonons in the "masing" mode, and not in terms of the largely irrelevant atomic coordinates.

We will now discuss some of the reasons for our belief that a quasiparticle description of the normal fluid may be possible. Quite often the quasiparticle modes and excitation spectrum for an interacting system are determined by some approximation at absolute zero, and the effect of elevated temperature is described in terms of interactions between the quasiparticles. We could, instead, imagine finding the best quasiparticle description at each value of temperature. This leads to a description of the system in terms of elementary excitations where the quasiparticle operators are temperature-dependent functions of the free-particle operators and the excitation spectrum is also temperature-dependent. (Emch[18] has recently given an interesting discussion, based on the work of Haag[19] and others, of the validity and general applicability of the temperature-dependent quasiparticle description to the BCS model of superconductivity and to certain spin systems.) Yarnell *et al.*[20] have measured the excitation spectrum

[15] J. G. Valatin and D. Butler, Nuovo Cimento **10**, 37 (1958).
[16] J. G. Valatin, in *Lectures in Theoretical Physics 1963* (University of Colorado Press, Boulder, Colorado, 1964).

[17] A. D. B. Woods, Phys. Rev. Letters **14**, 355 (1965).
[18] G. G. Emch, University of Maryland Tech. Note BN-433, 1966 (unpublished).
[19] R. Haag, Nuovo Cimento **25**, 287 (1962).
[20] J. L. Yarnell *et al.*, Phys. Rev. **113**, 1379 (1959); P. J. Bendt *et al.*, *ibid.* **113**, 1386 (1959).

of helium at temperatures between 1.1 and 1.8°K, and extrapolated their data to 2°K. They used their temperature-dependent spectrum to calculate specific heat, entropy, and fractional normal fluid density on the basis of a noninteracting quasiparticle model and found excellent agreement with experiment. It is noteworthy that such good agreement was obtained at this high temperature, where the normal fluid density ratio is ~50%. [Failure of a quasiparticle model based on neutron scattering data at these temperatures (~2°K) would not necessarily imply lack of validity of the quasiparticle description, but could possibly be attributed to the inability of neutrons to excite single "elementary excitations" at higher momenta and temperatures.]

As mentioned earlier, the quasiparticles ($\eta_{\mathbf{k}},\eta_{\mathbf{k}}^\dagger$ say) used to describe the normal fluid are not "dressed" free particles. They most likely will correspond to some combination of collective and free-particle-like motion. This will make the physical interpretation of our model somewhat more difficult in that the trace of the reduced density matrix

$$\rho_{\mathbf{pk}}^{(1)}=\mathrm{Tr}(\eta_{\mathbf{p}}\rho\eta_{\mathbf{k}}^\dagger) \tag{I.17}$$

is not necessarily equal to the total number of particles. If $|\mathrm{vac}\rangle_\eta$ is the vacuum state of the operators $\eta_{\mathbf{k}}^\dagger$, $\eta_{\mathbf{k}}$,

$$\eta_{\mathbf{k}}|\mathrm{vac}\rangle_\eta=0\,, \tag{I.18}$$

then from the consideration above, one would expect the superfluid ground state to be of the form

$$|\Phi_0\rangle=\prod_{\mathbf{k}} e^{-1/2|\alpha_{\mathbf{k}}|^2}e^{\alpha_{\mathbf{k}}\eta_{\mathbf{k}}^\dagger}|\mathrm{vac}\rangle_\eta \tag{I.19}$$

and the existence of ODLRO is completely independent of the degree of excitation of any given mode. (This property may be especially relevant in a description of superfluids bounded by finite volume.)

In the next section we obtain this result by a quasiparticle, self-consistent field approximation starting from the microscopic Hamiltonian.

II. CALCULATIONS

In this section, we show how the ground state predicted in the introduction can be obtained from a quasiparticle calculation. We have proposed that the most meaningful definition of the reduced density matrix is in terms of the quasiparticle operators which create and destroy elementary excitations in the normal fluid. In order to find the ground state, according to our prescription, one must first determine the quasiparticle modes of the normal fluid, assuming these modes exist. We do not yet know how to calculate these modes. The calculations described below are presented as an example to clarify the concepts presented in the Introduction, and hopefully to indicate how a more correct theory might proceed.

The Hamiltonian which describes a system of bosons interacting with two-body forces is, in second quantized notation,

$$H=\sum_{\mathbf{k}}(k^2/2m)a_{\mathbf{k}}^\dagger a_{\mathbf{k}}+\tfrac{1}{2}\sum_{\mathbf{k},\mathbf{p},\mathbf{r}}V(\mathbf{k})a_{\mathbf{p}-\mathbf{k}}^\dagger a_{\mathbf{r}+\mathbf{k}}^\dagger a_{\mathbf{p}}a_{\mathbf{r}}\,, \tag{II.1}$$

where the $a_{\mathbf{k}}$'s are the usual operators which destroy a helium particle in the plane-wave momentum state $\mathbf{k}$, $[a_{\mathbf{k}}, a_{\mathbf{k}'}^\dagger]=\delta_{\mathbf{kk}'}$, and $V(\mathbf{k})=(V-\mathbf{k})$.

The crudest theory might proceed by keeping only terms involving number operators, $\hat{n}_{\mathbf{k}}=a_{\mathbf{k}}^\dagger a_{\mathbf{k}}$ or their averages. Thus, one would write

$$H_{\mathrm{H.F.}}=\sum_{\mathbf{k}}E_{\mathbf{k}}a_{\mathbf{k}}^\dagger a_{\mathbf{k}}\,, \tag{II.2}$$

$$E_{\mathbf{k}}=NV(0)-\mu+(k^2/2m)+\sum_{\mathbf{p}\neq\mathbf{k}}V(\mathbf{k}-\mathbf{p})\langle\hat{n}_{\mathbf{p}}\rangle\,, \tag{II.3}$$

where the chemical potential μ is introduced in the usual way to require the conservation of total particle number $N=\sum_{\mathbf{k}}\langle a_{\mathbf{k}}^\dagger a_{\mathbf{k}}\rangle$ only in an average sense. The ground-state energy in this "Hartree-Fock" approximation is given by $E_0=N(NV(0)-\mu)=0$. One certainly expects that a lower ground-state energy can be obtained by considering further terms in the Hamiltonian. That this is the case can readily be seen by considering the "pair" Hamiltonian, as has been done by several authors,[21–23]

$$\mathcal{H}_p=\sum_{\mathbf{k}}(k^2/2m-\mu+NV(0))a_{\mathbf{k}}^\dagger a_{\mathbf{k}}+\tfrac{1}{2}\sum_{\mathbf{k}}\hat{K}(\mathbf{k})a_{\mathbf{k}}^\dagger a_{\mathbf{k}}$$
$$+\tfrac{1}{4}\sum_{\mathbf{k}}\hat{P}^\dagger(\mathbf{k})\hat{Q}_{\mathbf{k}}+\tfrac{1}{4}\sum_{\mathbf{k}}\hat{P}(\mathbf{k})\hat{Q}_{\mathbf{k}}^\dagger\,, \tag{II.4}$$

where we have defined the operators

$$\hat{K}(k)=\sum_{\mathbf{p}\neq\mathbf{k}}\nu(\mathbf{k}-\mathbf{p})a_{\mathbf{p}}^\dagger a_{\mathbf{p}}\,, \tag{II.5a}$$

$$\hat{P}(\mathbf{k})=\sum_{\mathbf{p}\neq\pm k}\nu(\mathbf{k}-\mathbf{p})\hat{Q}_{\mathbf{p}}\,, \tag{II.5b}$$

$$\hat{Q}_{\mathbf{k}}=a_{\mathbf{k}}a_{\mathbf{k}}\,. \tag{II.5c}$$

The results of Valatin and Butler[15] (V.B.) may be obtained by taking averages so that the Hamiltonian (4) becomes quadratic in the creation and destruction operators $a_{\mathbf{k}}$, $a_{\mathbf{k}}^\dagger$, and can be diagonalized by a canonical transformation of the usual sort. That is, (4) is approximated by

$$H_p=\sum_{\mathbf{k}}E_{\mathbf{k}}a_{\mathbf{k}}^\dagger a_{\mathbf{k}}+\tfrac{1}{2}\sum_{\mathbf{k}}P_{\mathbf{k}}^*\hat{Q}_{\mathbf{k}}+\tfrac{1}{2}\sum_{\mathbf{k}}P_{\mathbf{k}}\hat{Q}_{\mathbf{k}}^\dagger\,, \tag{II.6}$$

with

$$P_{\mathbf{k}}\equiv\langle\hat{P}_{\mathbf{k}}\rangle\,.$$

The canonical transformation

$$\eta_{\mathbf{k}}=\frac{a_{\mathbf{k}}-g_{\mathbf{k}}a_{-\mathbf{k}}^\dagger}{(1-|g_{\mathbf{k}}|^2)^{1/2}}\,,\quad g_{\mathbf{k}}=g_{-\mathbf{k}} \tag{II.7}$$

[21] M. Girardeau and R. Arnowitt, Phys. Rev. 113, 755 (1959).
[22] G. Wentzel, Phys. Rev. 120, 1579 (1960).
[23] M. Luban, Phys. Rev. 128, 965 (1962).

then gives

$$\tilde{H}_p = -\tfrac{1}{2}\sum_k \frac{\Omega_k |g_k|^2}{1-|g_k|^2} + \sum_k \Omega_k \eta_k^\dagger \eta_k$$
$$= -\tfrac{1}{2}\sum_k (E_k - \Omega_k) + \sum_k \Omega_k \eta_k^\dagger \eta_k , \qquad \text{(II.8)}$$

where the spectrum of elementary excitations Ω_k is given by

$$\Omega_k = (E_k^2 - |P_k|^2)^{1/2}, \qquad \text{(II.9)}$$

and

$$g_k = -(E_k - \Omega_k)/P_k^*. \qquad \text{(II.10)}$$

This is also a self-consistent-field approximation, differing from the Hartree-Fock calculation above only in that pair terms are kept in addition to number operators—which, of course, requires a cannonical transformation to diagonalize the Hamiltonian. The self-consistency equations at $T=0°$ are coupled integral equations of the form

$$n_k \equiv \langle n_k \rangle = (E_k - \Omega_k)/2\Omega_k , \qquad \text{(II.11a)}$$

$$P_k = -\tfrac{1}{2} \sum_p (V(\mathbf{k}-\mathbf{p}) P_p/\Omega_p), \qquad \text{(II.11b)}$$

which, with the condition

$$N = \sum_k n_k \qquad \text{(II.11c)}$$

are sufficient to determine the parameters μ, n_k, and P_k. In the weakly interacting limit this reduces to the results of Bogoliubov.[24] (Note that we have not taken g_k to be real as was done by V. B. The assumption of real g_k is more than a mathematical convenience in that it restricts the average of the non-Hermitian operator $\hat{P}_k$ to being real.)

In addition to the shortcomings of the V.B. theory mentioned in the Introduction, we might also point out that it gives large (Bose) fluctuations in particle number. Also, the ground state of the V.B. theory, while giving a finite value for the pair average, gives a zero expectation value for a_k.

The BCS theory was constructed to give a finite value to the pair average, the argument being that this results in a lowering of the ground-state energy over the normal state. One would expect, in analogy with BCS, that constructing a theory for superfluidity in which the average of the operator characteristic of the basic group ($\langle a_k \rangle$) is nonzero would also lower the ground-state energy. Such a model occurs in the following way. We now include the terms from the full Hamiltonian (1) which were ignored in the pairing Hamiltonian (4) and take averages in such a way as to give a linear equation of motion for a_k. The full Hamiltonian can be written

$$\mathcal{H} = \tfrac{1}{2}\hat{N}(\hat{N}-1)V(0) + \sum_k (k^2/2m - \mu) a_k^\dagger a_k$$
$$+ \tfrac{1}{2}\sum_{\mathbf{k}\neq\mathbf{p}} \nu(\mathbf{k}-\mathbf{p}) a_p^\dagger a_p a_k^\dagger a_k + \tfrac{1}{2}\sum_{\mathbf{k}\neq\pm\mathbf{p}} \nu(\mathbf{k}-\mathbf{p}) a_p^\dagger a_{-p}^\dagger a_k a_{-k}$$
$$+ \tfrac{1}{2} \sum_{\substack{\mathbf{k},\mathbf{p},\mathbf{r} \\ \{\mathbf{k}\neq 0,\ \mathbf{k}\neq -\mathbf{r},\ \mathbf{p}\neq \mathbf{r}-\mathbf{k}\}}} \nu(\mathbf{p}) a_{k+p}^\dagger a_{r-p}^\dagger a_r a_k . \qquad \text{(II.12)}$$

The averaged version of the full Hamiltonian (12) is now the averaged version of the pair Hamiltonian plus additional terms,

$$H = \sum_k E_k a_k^\dagger a_k + \tfrac{1}{2}\sum_k P_k^* \hat{Q}_k + \tfrac{1}{2}\sum_k P_k \hat{Q}_k^\dagger$$
$$- \sum_k (A_k^* a_k + A_k a_k^\dagger), \qquad \text{(II.13)}$$

where we have defined

$$A_k = -\sum_{\substack{\mathbf{p},\mathbf{r} \\ \{\mathbf{p}\neq 0,\ \mathbf{k}\neq -\mathbf{r},\ \mathbf{p}\neq \mathbf{r}-\mathbf{k}\}}} V(\mathbf{p}) \langle a_r^\dagger a_{r-p} a_{k+p} \rangle . \qquad \text{(II.14)}$$

It will be shown later that A_k is directly related to the $\langle a_k \rangle$ and $\langle a_k^\dagger \rangle$ which we assume to be nonvanishing only below the λ transition. As was mentioned in the introduction, Wood's experiment[17] gives strong evidence that the low momentum (linear) part of the spectrum does not change appreciably in going through the λ transition. Sufficient difficulty has been found with this *particular* quasiparticle model[21,23] that it cannot be taken too seriously. However, if it is to give even a rough approximation to the excitation spectrum, P_k must be nonzero even *above* the transition temperature T_λ. (Note that the Hartree-Fock energy E_k cannot be linear in k.) This is not possible unless the Fourier transformed potential $V(\mathbf{k})$ has some negative portion. We do not believe this to be a problem, since we do not believe that the qualitative aspects of a liquid [e.g., the peaked structure factor $S(k)$] can be explained by an entirely positive $V(\mathbf{k})$.[25]

We first perform the canonical transformation which diagonalizes the averaged pairing Hamiltonian, i.e., let

$$\eta_k = \mathcal{T}_k a_k \mathcal{T}_k^\dagger = \frac{a_k - g_k a_{-k}^\dagger}{(1-|g_k|^2)^{1/2}}, \qquad \text{(II.15)}$$

where

$$\mathcal{T}_k = e^{\gamma_k a_k^\dagger a_{-k}^\dagger - \gamma_k^* a_k a_{-k}} = (\mathcal{T}_k^\dagger)^{-1}, \qquad \text{(II.16a)}$$

$$g_k = \frac{\gamma_k}{|\gamma_k|} \tanh|\gamma_k| = g_{-k}. \qquad \text{(II.16b)}$$

[24] N. N. Bogoliubov, J. Phys. USSR **9**, 23 (1947).

[25] It may be relevant to point out that T. Lee, C. Yang, and K. Huang [Phys. Rev. **106**, 1135 (1957)] in their study of a hard-core, repulsive boson system obtain an $S(k) = k/(k^2 + 16\pi a\rho)^{1/2}$, a monotone function, which seems to be more appropriate to a gas rather than a liquid.

The transformed Hamiltonian is then

$$\tilde{H}=-\sum_k \tfrac{1}{2}(E_k-\Omega_k)+\sum_k \Omega_k \eta_k^\dagger \eta_k -\sum_k \Omega_k(\alpha_k\eta_k^\dagger+\alpha_k^*\eta_k)$$
$$=-\sum_k (\tfrac{1}{2}(E_k-\Omega_k)+\Omega_k|\alpha_k|^2) +\sum_k \Omega_k(\eta_k^\dagger-\alpha_k^*)(\eta_k-\alpha_k), \quad \text{(II.17)}$$

where

$$\alpha_k=\frac{1}{\Omega_k}\frac{A_k+g_kA_{-k}^*}{(1-|g_k|^2)^{1/2}}, \quad \text{(II.18)}$$

and Ω_k is again given by

$$\Omega_k=(E_k^2-|P_k|^2)^{1/2}. \quad \text{(II.19)}$$

The ground-state energy is thus apparently reduced over the V.B. expression by the term $-\sum_k \Omega_k|\alpha_k|^2$. The Hamiltonian (17)

$$\tilde{H}=E_0+\sum_k \Omega_k(\eta_k^\dagger-\alpha_k^*)(\eta_k-\alpha_k) \quad \text{(II.20)}$$

has the exact same form as the "toy" laser Hamiltonian (I.2). This Hamiltonian is taken to complete diagonal form by the transformation

$$b_k=\mathfrak{D}(\alpha_k)\eta_k\mathfrak{D}^\dagger(\alpha_k)=\eta_k-\alpha_k, \quad \text{(II.21a)}$$

where

$$\mathfrak{D}(\alpha_k)\equiv e^{\alpha_k\eta_k^\dagger-\alpha_k^*\eta_k} \quad \text{(II.21b)}$$

is the generator of the coherent radiation states.[11] The ground state in the η representation is given by

$$|\Phi_0\rangle=|\{\alpha_k\}\rangle=\prod_k \mathfrak{D}(\alpha_k)|\text{vac}\rangle_\eta =\prod_k e^{-1/2|\alpha_k|^2}e^{\alpha_k\eta_k^\dagger}|\text{vac}\rangle_\eta. \quad \text{(II.22)}$$

This is the result (I.19) predicted in the Introduction. The excited states of the system are given by

$$|\{m_k\}\rangle=\prod_k\frac{(b_k^\dagger)^{m_k}}{(m_k!)^{1/2}}|\Phi_0\rangle=\prod_k\frac{(\eta_k^\dagger-\alpha_k^*)^{m_k}}{(m_k!)^{1/2}}|\Phi_0\rangle, \quad \text{(II.23)}$$

and the temperature dependence of the momentum distribution in the b_k representation is given by

$$\langle b_k^\dagger b_k\rangle=(e^{\beta\Omega_k}-1)^{-1} =\langle\eta_k^\dagger\eta_k-\alpha_k\eta_k^\dagger-\alpha_k^*\eta_k+|\alpha_k|^2\rangle, \quad \text{(II.24a)}$$

or in the η_k representation by

$$\langle\eta_k^\dagger\eta_k\rangle=|\alpha_k|^2+(e^{\Omega_k\beta}-1)^{-1}, \quad \text{(II.24b)}$$

where we have used $\langle\eta_k\rangle=\alpha_k$ and $\beta=(k_BT)^{-1}$. In the helium particle representation, the ground state (22) takes the form[26]

$$|\Phi_0\rangle=\prod_\lambda \mathfrak{D}(\alpha_\lambda)\prod_k \mathcal{T}_k|0\rangle_a =\exp[\sum_\lambda(\alpha_\lambda a_\lambda^\dagger-\alpha_\lambda^*a_\lambda)] \times\exp[\sum_k(\gamma_k a_k^\dagger a_{-k}^\dagger-\gamma_k^*a_ka_{-k})]|0\rangle_a, \quad \text{(II.25)}$$

where $|0\rangle_a$ is the particle vacuum. To conclude the calculational details, we list the pertinent equations (for $T=0°$K only):

$$g_k=-(E_k-\Omega_k)/\langle P_k\rangle_0^*, \quad \text{(II.26a)}$$

$$\langle a_k\rangle_0=\frac{\alpha_k+g_k\alpha_{-k}^*}{(1-|g_k|^2)^{1/2}}=\frac{E_kA_k-\langle P_k\rangle_0A_{-k}^*}{\Omega_k^2}, \quad \text{(II.26b)}$$

$$\langle n_k\rangle_0=\langle a_k^\dagger\rangle_0\langle a_k\rangle_0+\frac{|g_k|^2}{(1-|g_k|^2)} =|\langle a_k\rangle_0|^2+\frac{E_k-\Omega_k}{2\Omega_k}, \quad \text{(II.26c)}$$

$$\langle Q_k\rangle_0=\langle a_k\rangle_0\langle a_k\rangle_0+\frac{g_k}{1-|g_k|^2} =\langle a_k\rangle_0\langle a_{-k}\rangle_0-\frac{\langle P_k\rangle_0}{2\Omega_k}, \quad \text{(II.26d)}$$

or using (5b),

$$\langle P_k\rangle_0=\sum_p V(\mathbf{k}-\mathbf{p})\langle a_p\rangle_0\langle a_{-p}\rangle_0 -\sum_p(\nu(\mathbf{k}-\mathbf{p})\langle P_p\rangle_0/2\Omega_p), \quad \text{(II.26e)}$$

where $\langle\ \rangle_0$ denotes ground-state averages. These coupled integral equations are even less tractable than the results of previous theories.

III. SOME QUALITATIVE ASPECTS

The model presented here is quite different from models based on an analogy with the ideal Bose gas. We will now touch on some of these differences.

The important point of departure from previous theories is that we believe the concept of ODLRO to be most transparent when the quasiparticle modes of the normal fluid are taken as a basis and that the condensation associated with superfluidity is not fundamentally related to the macroscopic occupation of a single momentum mode. We are not implying that a single-momentum mode condensation does not occur, but that it is not the fundamental source of ODLRO.

Consider the single-particle reduced density matrix

$$\rho_{pk}^{(1)}(a)\equiv\text{Tr}(a_p\rho a_k^\dagger), \quad \text{(III.1)}$$

which in the ground state (II.25) is

$$\rho_{pk}^{(1)}(a)=\langle a_k^\dagger a_p\rangle_0=\sigma_k^*\sigma_p+\mu_k\delta_{kp}, \quad \text{(III.2)}$$

where

$$\sigma_k\equiv\langle a_k\rangle_0 \quad\text{and}\quad \mu_k=|g_k|^2/(1-g_k|^2). \quad \text{(III.3)}$$

Going over to the coordinate representation,

$$\rho_a^{(1)}(\mathbf{x},\mathbf{x}')=(\sum_k e^{-i\mathbf{k}\cdot\mathbf{x}}\sigma_k^*)(\sum_p e^{+i\mathbf{p}\cdot\mathbf{x}}\sigma_p) +\sum_k\mu_k e^{i\mathbf{k}\cdot(\mathbf{x}-\mathbf{x}')},$$

[26] E. P. Gross, Ann. Phys. (N. Y.) 9, 242 (1960). Gross obtains states similar in form to ours; he was the first to introduce the states (I-4) into the study of liquid helium. His calculational procedure (and motivation) is considerably different and his states are in fact not the same as ours.

or

$$\rho_a^{(1)}(\mathbf{x},\mathbf{x}')=[n_c(\mathbf{x})]^{1/2}e^{i\varphi(\mathbf{x})}[n_c(\mathbf{x}')]^{1/2}e^{-i\varphi(\mathbf{x}')}+\bar{\rho}_a^{(1)}(\mathbf{x}-\mathbf{x}'). \quad \text{(III.4)}$$

This is the *form* postulated by Penrose and Onsager[2] and discussed more recently by Hohenberg and Martin[27] and Anderson.[28] Here, however, the factored term (the "condensate") does not pertain only to the zero-momentum mode. The condensate density $n_c(\mathbf{x})$ is related to the degree of excitation of all the coherent modes.

The ground state (II.25) is infinitely degenerate, since the transformation $\alpha_{\mathbf{k}}\to\alpha_{\mathbf{k}}e^{i\phi}$ and $g_{\mathbf{k}}\to g_{\mathbf{k}}e^{2i\phi}$ leaves the ground-state energy unchanged. We can then write the ground state as $|\Phi_0(\phi)\rangle$ and by forming a linear combination of these states with equal weights for all $\phi(0\to 2\pi)$ we may construct a state of definite N. Under this transformation, the average of the field operator

$$\langle\psi(\mathbf{x})\rangle=\sum_{\mathbf{k}}\langle a_{\mathbf{k}}\rangle e^{i\mathbf{k}\cdot\mathbf{x}}$$

undergoes the transformation $\langle\psi\rangle\to\langle\psi\rangle e^{i\phi}$. This "broken symmetry" has been recently discussed by Anderson.[28]

Hohenberg and Martin[27] introduce a "restricted ensemble" in order that the thermodynamics be consistent with the fact that part of the system (the condensate) is described by a wave function which is *per se* a macroscopic variable. We next show that this kind of an ensemble follows in a very natural way from the microscopic model of the previous section.

At all temperatures, the superfluid component is described by a pure coherent state $|\{\alpha_{\mathbf{k}}\}\rangle$. The reduction in the superfluid density at nonzero temperature is due to the reduction of the magnitudes of the $\alpha_{\mathbf{k}}$ by thermal excitations. The superfluid component is described in the η representation by the density operator

$$\rho^{(s)}=|\{\alpha_{\mathbf{k}}\}\rangle\langle\{\alpha_{\mathbf{k}}\}|. \quad \text{(III.5)}$$

This is consistent with the fact—as pointed out by London[8]—that the superfluid component is in a pure quantum state. The density operator for the normal fluid is the usual density operator appropriate for a system of noninteracting (Boson) elementary excitations,

$$\rho^{(n)}-\prod_{\mathbf{k}}e^{-\beta\Omega_{\mathbf{k}}b_{\mathbf{k}}^\dagger b_{\mathbf{k}}}\Big/\prod_{\mathbf{k}}\sum_{m_{\mathbf{k}}}e^{-\beta\Omega_{\mathbf{k}}m_{\mathbf{k}}}, \quad \text{(III.6)}$$

where the $m_{\mathbf{k}}$ are the "thermal" occupation numbers

$$m_{\mathbf{k}}\equiv\mathrm{Tr}(\rho^{(n)}b_{\mathbf{k}}^\dagger b_{\mathbf{k}}). \quad \text{(III.7)}$$

In order to construct the total density operator, we express $\rho^{(s)}$ and $\rho^{(n)}$ in the "P" representation introduced by Glauber.[11] The $P(\alpha)$ is defined so that

$$\rho=\int P(\{\alpha_{\mathbf{k}}\})|\{\alpha_{\mathbf{k}}\}\rangle\langle\{\alpha_{\mathbf{k}}\}|d^2\{\alpha_{\mathbf{k}}\}, \quad \text{(III.8)}$$

where the integration is over the entire complex plane of each $\alpha_{\mathbf{k}}$. The P representation for $\rho^{(s)}$ and $\rho^{(n)}$ separately is

$$P^{(s)}(\{\alpha_{\mathbf{k}}'\})=\prod_{\mathbf{k}}\delta^2(\alpha_{\mathbf{k}}'-\alpha_{\mathbf{k}}), \quad \text{(III.9a)}$$

$$P^{(n)}(\{\alpha_{\mathbf{k}}'\})=\prod_{\mathbf{k}}(1/\pi m_{\mathbf{k}})e^{-|\alpha_{\mathbf{k}}'|^2/m_{\mathbf{k}}}. \quad \text{(III.9b)}$$

The density operators can be combined in the P representation to obtain the P function for the combined system by a convolution theorem developed by Glauber[11] appropriate for two *noninteracting* fields

$$P(\{\alpha_{\mathbf{k}}'\})=\int P^{(s)}(\{\alpha_{\mathbf{k}}''-\alpha_{\mathbf{k}}'\})P^{(n)}(\{\alpha_{\mathbf{k}}''\})d^2\{\alpha_{\mathbf{k}}''\}. \quad \text{(III.10)}$$

Thus, the density operator for the total system is given by (8) where, using (9a) and (9b).

$$P(\{\alpha_{\mathbf{k}}'\})=\prod_{\mathbf{k}}(1/\pi m_{\mathbf{k}})e^{-|\alpha_{\mathbf{k}}'-\alpha_{\mathbf{k}}|^2/m_{\mathbf{k}}}. \quad \text{(III.11)}$$

This is the same P representation appropriate to a radiation field composed of noninteracting "signal" plus "noise"[9,29]; this P representation follows directly as a consequence of the model of Sec. II for a system of interacting bosons, and is an equivalent way of expressing those results, including thermal effects. The two-fluid aspects of the results are most apparent in this formulation in terms of the P representation. At temperatures below T_λ, the superfluid ("signal") component is macroscopic and is described by a pure quantum state. A restricted ensemble is a natural *consequence* of this theory.

The coherent states used to describe the superfluid component exhibit classical (small) fluctuations in number.[11] To the extent that the phase operator is canonically conjugate to the number operator, it follows that the phase also exhibits low fluctuations. In fact, for these states, the fluctuations in number and phase are such as to give the minimum uncertainty product. However, above T_λ, the "order parameter" is zero ($\alpha_{\mathbf{k}}\to 0$) (let us emphasize again that this in no way implies the vanishing of the *pair* averages above T_λ), and the system is described entirely by "thermal" ("noise") states, and the fluctuations are large (Bose) quantum fluctuations as expected.

[27] P. C. Hohenberg and P. C. Martin, Ann. Phys. (N. Y.) **34**, 291 (1965).

[28] P. W. Anderson, in *Lectures on the Many-Body Problem*, edited by E. R. Caianiello (Academic Press Inc., New York, 1964), Vol. 2: P. W. Anderson, Rev. Mod. Phys. **38**, 298 (1966).

[29] F. T. Arecchi, A. Berne, and P. Bulamacchi, Phys. Rev. Letters **16**, 32 (1966).

IV. SUMMARY

We have proposed that coherent radiation, superconductivity, and superfluidity are analogous phenomena in that the pure "superfluid" state of each gives rise to *complete* factorization of an appropriately defined reduced density matrix, and the coherent states for each have the form

$$\prod_{\mathbf{k}} e^{\alpha_{\mathbf{k}}\theta_{\mathbf{k}}}|0\rangle,$$

where $\theta_{\mathbf{k}}$ is an operator characteristic of the "basic group" and $|0\rangle$ is the ground state of the normal fluid. The reduced density matrix for helium is defined in terms of the (temperature-dependent) operators which create and destroy excitations of the normal fluid.

An improved version of a previous quasiparticle calculation has been described as an example. However, the concepts presented in the introduction are independent of this particular quasiparticle model, as are the qualitative aspects discussed in Sec. III which follow from the model proposed in the Introduction. A more complete theory should show the instability of the normal fluid against the formation of the coherent state below a critical temperature.

ACKNOWLEDGMENTS

It is a pleasure to express our thanks to Dr. C. A. Roberts, Dr. R. Vasudevan, and especially to Dr. A. K. Rajagopal for several stimulating conversations. One of us (J. R. J.) thanks Douglas Aircraft Company Inc., for financial support through a graduate scholarship.

Reprinted from THE PHYSICAL REVIEW, Vol. 167, No. 1, 183–190, 5 March 1968

Coherent States in the Theory of Superfluidity*

J. S. LANGER
Carnegie-Mellon University, Pittsburgh, Pennsylvania

(Received 29 September 1967; revised manuscript received 29 November 1967)

It is shown that the coherent-state representation of a many-boson wave function may be identified with the order-parameter function conventionally used to describe a superfluid. The statistical mechanics of the many-boson system is reformulated in terms of the coherent states, and a theory of the Ginzburg-Landau form is recovered in an obvious approximation. The formalism is particularly useful for describing metastable states of finite superflow and the fluctuations which may cause spontaneous decay of such states.

I. INTRODUCTION: THE PHENOMENOLOGICAL MODEL

A SLIGHTLY generalized interpretation of the Ginzburg-Landau phenomenological model of superfluidity leads to several novel conclusions regarding the nature of superfluid flow. These conclusions have been published in two recent papers,[1,2] and are summarized in this introductory section. The remainder of this paper is devoted to a systematic derivation of the Ginzburg-Landau model in the form in which we wish to use it for many-boson systems. The derivation makes use of the coherent-state representation of boson fields, a formalism which turns out to be very appropriate for the discussion of superfluidity.

In the Ginzburg-Landau model,[3,4] the states of the system are described by a complex-valued order parameter $\psi(\mathbf{r})$, which is a function of position $\mathbf{r}$. The function $\psi(\mathbf{r})$ ordinarily is interpreted as a wave function for the superfluid component of the system. In particular,

$$n_s(\mathbf{r})=|\psi(\mathbf{r})|^2 \tag{1.1}$$

is the superfluid number density, and the corresponding current density is

$$\mathbf{j}_s(\mathbf{r})=(1/2i)(\psi^*\nabla\psi-\psi\nabla\psi^*). \tag{1.2}$$

(We shall work always in units $\hbar=1$.) ψ is supposed to have the time dependence $\exp(i\mu t)$ [or $\exp(2i\mu t)$ for a superconductor], where μ is the chemical potential. More generally, if we choose two separate points in the superfluid, we may write

$$\Delta\mu=(\partial/\partial t)\Delta(\arg\psi), \tag{1.3}$$

where $\Delta\mu$ is the difference in chemical potential between the two points and $\Delta(\arg\psi)$ the corresponding difference in phase of ψ. Equation (1.3), in combination with (1.2), describes the acceleration of the supercurrent by a potential gradient, and thus characterizes the intrinsically superfluid properties of the system.[5,6]

Our specific phenomenological picture is based on the assumption that the space of all functions $\psi(\mathbf{r})$ satisfying suitable boundary conditions is appropriate for the representation of an isothermal canonical ensemble. That is, statistical fluctuations of the system, caused by interactions with a constant-temperature bath, are to be visualized as a continuous random motion of the system point $\psi(\mathbf{r})$ in the function space, the neighborhood of each point being visited by the system with a frequency proportional to the Boltzmann factor $\exp(-F\{\psi\}/k_BT)$. Here $F\{\psi\}$ will be taken to be of the form of the usual Ginzburg-Landau free-energy functional:

$$F\{\psi\}=\int d\mathbf{r}[|\nabla\psi|^2-A|\psi|^2+\tfrac{1}{2}B|\psi|^4+\cdots], \tag{1.4}$$

where A and B are temperature-dependent constants. In particular, A passes through zero at some T_0, being positive for $T<T_0$.

The fact that A and B are temperature-dependent implies that $F\{\psi\}$ is already some sort of coarse-grained free energy, i.e., that a partial partition sum has been performed in order to obtain $F\{\psi\}$. It is just this point that will be amplified in the following analysis. Here, however, it is important to emphasize that, although the state ψ_0 which minimizes $F\{\psi\}$ is the most probable state of the system, $F\{\psi_0\}$ is not the correct free energy. Rather, the true free energy is

$$-k_BT\ln\left\{\int\delta\psi(\mathbf{r})\exp(-F\{\psi\}/k_BT)\right\}, \tag{1.5}$$

where the symbol $\int\delta\psi(\mathbf{r})\cdots$ denotes an integration over the space of functions $\psi(\mathbf{r})$. It is only when this integral may be approximated by the largest value of the integrand that the usual Ginzburg-Landau (mean field) theory is valid.

The above formulation is particularly well suited to

* Supported in part by the National Science Foundation.

[1] J. S. Langer and M. E. Fisher, Phys. Rev. Letters **19**, 560 (1967), hereafter referred to as I.

[2] J. S. Langer and V. Ambegaokar, Phys. Rev. **165**, 498 (1967), hereafter referred to as II.

[3] V. L. Ginzburg and L. D. Landau, Zh. Eksperim. i Teor. Fiz. **20**, 1064 (1950). For a more recent review of the Ginzburg-Landau theory as applied to metallic superconductors, see P. G. de Gennes, *Superconductivity of Metals and Alloys* (W. A. Benjamin, Inc., New York, 1966), Chaps. 6 and 7.

[4] The use of a Ginzburg-Landau equation for the description of superfluid helium was proposed by E. P. Gross [Nuovo Cimento **20**, 454 (1961)] and by V. L. Ginsburg and L. P. Pitaevskii, Zh. Eksperim. i Teor. Fiz. **34**, 1240 (1958) [English transl.: Soviet Phys.—JETP **7**, 858 (1958)]. See also L. P. Pitaevskii, Zh. Eksperim. i Teor. Fiz. **40**, 646 (1961) [English transl.: Soviet Phys.—JETP **13**, 451 (1961)].

[5] B. D. Josephson, Advan. Phys. **14**, 419 (1965).

[6] P. W. Anderson, Rev. Mod. Phys. **38**, 298 (1966).

the discussion of metastable states. If only continuous deformations of ψ are permitted—the perturbations are small in some sense—then any local minimum of $F\{\psi\}$ locates a stable or metastable state. The stability condition is therefore

$$\delta F/\delta\psi(\mathbf{r})=0\,, \tag{1.6}$$

to which we add the requirement that the matrix

$$\delta^2 F/\delta\psi(\mathbf{r})\delta\psi(\mathbf{r}') \tag{1.7}$$

must be positive definite. Equation (1.6) is just the Ginzburg-Landau equation for general F.

The following properties of metastable states have been discussed in Refs. 1 and 2.

A. Constant-Current States

We choose F in the form (1.4) and impose periodic boundary conditions. Then the solutions of Eq. (1.6) are

$$\psi_{\mathbf{k}}(\mathbf{r})=f_k e^{i\mathbf{k}\cdot\mathbf{r}}\,,\quad f_k^2=(A-k^2)/B\,, \tag{1.8}$$

and carry current

$$\mathbf{j}_{\mathbf{k}}=\mathbf{k}f_k^2=\mathbf{k}(A-k^2)/B\,. \tag{1.9}$$

For $k^2<\frac{1}{3}A$ (the conventional Ginzburg-Landau critical wave vector[7]), each of the states (1.8) locates an isolated local minimum of F. For $\frac{1}{3}\leq k^2<A$, however, the matrix (1.7) is no longer positive definite.[8] In fact, these states are unstable against deformations of the form

$$\psi_{\mathbf{k}}\rightarrow\psi_{\mathbf{k}}+\delta_{\mathbf{q}}e^{i(\mathbf{k}-\mathbf{q})\cdot\mathbf{r}}\,, \tag{1.10}$$

where $\mathbf{q}$ is small and preferably parallel to $\mathbf{k}$ and $\delta_{\mathbf{q}}$ is an infinitesimal amplitude whose phase depends on $\mathbf{q}$.

B. Current-Reducing Fluctuations

An important feature of any fluctuation which carries the system from one constant-current state to another is that the wave function $\psi(\mathbf{r})$ must pass through zero somewhere during the transition.[9] For example, consider a ring of circumference L in which the superflow is characterized by the state $\psi_{\mathbf{k}}$. The states ψ in the neighborhood of ψ_k all will have the property that $\arg\psi$ increases by exactly kL around the ring. But if the fluctuation away from $\psi_{\mathbf{k}}$ is large enough that ψ vanishes somewhere, then the total change in $\arg\psi$ around the ring is indefinite for that particular wave function, and the system point may pass continuously from the region of states with total phase change kL to those with, say, $k'L$.

This point is particularly significant in view of the fact that it is the phase fluctuations which preclude off-diagonal long-range order in one- or two-dimensional superfluids.[10,11] Clearly, if the amplitude of ψ never vanishes, no fluctuations in the way the phase varies from point to point can possibly modify the total change of phase around a ring. In fact, if the amplitude of ψ remains constant, then the dc component of the current also must remain constant. It follows that we might expect a system to behave like a superfluid whenever there exists an energy barrier to inhibit fluctuations in the amplitude of ψ, whether or not there occurs a true phase transition with long-range order.

C. The Free-Energy Barrier

It is obvious topologically that, in order for the system point to pass from one local minimum of $F\{\psi\}$ to another, it must overcome a free-energy barrier. The lowest barrier, and thus the least improbable fluctuation which will effect the required transition, occurs at a saddle point between the two minima. Thus the current-reducing fluctuation, say $\bar{\psi}(\mathbf{r})$, satisfies the Ginzburg-Landau equation (1.6); but the matrix (1.7) must have a single negative eigenvalue at $\bar{\psi}$ determining the direction in ψ space along which the fluctuation is most likely to progress.

The properties of the saddle-point fluctuation $\bar{\psi}$ have been studied recently in several different connections. In general, $\bar{\psi}$ describes a state which is almost everywhere the same as the metastable state ($\psi_{\mathbf{k}}$ in our case), but which contains a single localized fluctuation. The simplest example of such a fluctuation is the well-known critical droplet which nucleates the condensation of a supersaturated vapor.[12,13] In the case of liquid helium, we have argued in I that $\bar{\psi}$ must describe a vortex ring of a critical size determined by the velocity of the superfluid.[14] Because ψ vanishes at the vortex core, an expanding, singly quantized, vortex ring will eventually subtract 2π from the total phase change across the system. Paper II was devoted to the study of an effectively one-dimensional superconductor, in which case the Ginzburg-Landau equation is exactly soluble and the relevant amplitude fluctuations can be examined in detail.

The point to be emphasized is that a state of nonzero superflow is truly metastable as opposed to stable in the sense that there always exists a nonzero probability for transitions to states of lower current and lower free energy. This transition probability in all cases depends on the frequency of nucleation of certain localized amplitude fluctuations, and is therefore independent of the size of the system. Rather than trying to compute this transition probability directly,[15] in I and II we

[7] J. Bardeen, Rev. Mod. Phys. **34**, 667 (1962).
[8] See II, Appendix C.
[9] The following argument owes much to the work of W. A. Little, Phys. Rev. **156**, 396 (1967).
[10] T. M. Rice, Phys. Rev. **140**, A1889 (1965).
[11] P. Hohenberg, Phys. Rev. **158**, 383 (1967).

[12] J. Frenkel, *Kinetic Theory of Liquids* (Dover Publications Inc., New York, 1955), Chap. 7.
[13] J. S. Langer, Ann. Phys. (N. Y.) **41**, 108 (1967).
[14] See also S. V. Iordanskii, Zh. Eksperim. i Teor. Fiz. **48**, 708 (1965) [English transl.: Soviet Phys.—JETP **21**, 467 (1965)].
[15] A brief discussion of the difficulties encountered in trying to make a direct calculation of the transition probability for creation of a vortex ring has been given by W. F. Vinen, in *Proceedings of the International School of Physics, "Enrico Fermi," Course XXI*, edited by G. Careri (Academic Press Inc., New York, 1963), p. 336. Our point of view differs from Vinen's in that we imagine the critical ring to be created in a very large number of small steps rather than in a single quantum jump.

have simply invoked an ergodic hypothesis and have said that the frequency of transitions has the form

$$\tau_0^{-1}\exp(-\Delta F/k_BT), \tag{1.11}$$

where ΔF is the height of the free-energy barrier and τ_0 is some sort of characteristic time for microscopic processes. (Actual calculations of, say, critical currents or resistivities turn out to be very insensitive to the choice of τ_0.) The results of these calculations are relaxation rates for states of finite superflow which are always extremely rapidly varying functions of temperature and current, being unobservably small throughout most of the conventional superfluid region. For example, in II we estimated that the width of the observable resistive transition in a tin wire about 1 μ^2 in cross section would be of the order of 10^{-4}°K. In principle, however, the wire would have a finite resistivity below this transition; but one would have to make measurements over cosmologically long times in order to observe it.

II. COHERENT STATES

The preceding discussion has been based on a phenomenological model which is slightly more general than the conventional Ginzburg-Landau theory and therefore requires additional justification. In particular, we must question whether it is possible to formulate a completely general characterization of a many-body system, including both normal and superfluid phases, in terms of an order-parameter function $\psi(\mathbf{r})$. In the following we shall attempt to show that a very simple such characterization can be constructed for many-boson systems. Hopefully, an equally simple formulation can be found for superconducting many-fermion systems.[16]

The order parameter ψ for a many-boson system is conventionally defined to be the thermodynamic expectation value of the boson field operator, the superfluid phase transition usually being associated with the loss of symmetry which allows this expectation value to be nonzero. In the preceding development, however, ψ somehow characterized the pure quantum states of the system, and was not itself a thermodynamic quantity. Our assertion is that the order parameter ψ, as we have used it above, may conveniently be chosen to be the coherent-state representation of the pure quantum states.

Consider the boson field operator, $\Psi_{\text{op}}(\mathbf{r})$, which satisfies the commutation relation[17]

$$[\Psi_{\text{op}}(\mathbf{r}),\Psi_{\text{op}}^\dagger(\mathbf{r}')]=\delta(\mathbf{r}-\mathbf{r}'). \tag{2.1}$$

[16] In an earlier paper [J. S. Langer, Phys. Rev. **134**, A553 (1964)], the problem of superconductivity has been formulated in terms of a functional integral which resembles that derived in Sec. III of the present paper. The main difference is that, for the superconductor, ψ depends on an extra timelike variable.

[17] We shall use the subscript "op" to denote second-quantized operators.

We shall often work in the Fourier representation:

$$\Psi_{\text{op}}(\mathbf{r})=\frac{1}{\sqrt{V}}\sum_{\mathbf{k}} a_{\mathbf{k}}e^{i\mathbf{k}\cdot\mathbf{r}}, \tag{2.2}$$

where

$$[a_{\mathbf{k}},a_{\mathbf{k}'}^\dagger]=\delta_{\mathbf{k},\mathbf{k}'},\quad [a_{\mathbf{k}},a_{\mathbf{k}'}]=[a_{\mathbf{k}}^\dagger,a_{\mathbf{k}'}^\dagger]=0, \tag{2.3}$$

and V is the quantization volume. The right eigenstates of the annihilation operator $a_{\mathbf{k}}$,

$$a_{\mathbf{k}}|\alpha_{\mathbf{k}}\rangle=\alpha_{\mathbf{k}}|\alpha_{\mathbf{k}}\rangle, \tag{2.4}$$

are the so-called "coherent states"[18–21]:

$$|\alpha_{\mathbf{k}}\rangle=\exp(-\tfrac{1}{2}|\alpha_{\mathbf{k}}|^2)\sum_{n_{\mathbf{k}}=0}^{\infty}\frac{\alpha_{\mathbf{k}}^{n}}{\sqrt{n_{\mathbf{k}}!}}|n_{\mathbf{k}}\rangle, \tag{2.5}$$

where the $|n_{\mathbf{k}}\rangle$ are the number states for the $\mathbf{k}$th mode of the Bose field, and $\alpha_{\mathbf{k}}$ can be any complex number. Then the order parameter $\psi(\mathbf{r})$ turns out to be

$$\psi(\mathbf{r})=\frac{1}{\sqrt{V}}\sum_{\mathbf{k}}\alpha_{\mathbf{k}}e^{i\mathbf{k}\cdot\mathbf{r}}. \tag{2.6}$$

The states given by Eq. (2.5) are normalized but not orthogonal. In fact,

$$\langle\alpha_{\mathbf{k}}|\beta_{\mathbf{k}}\rangle=\exp(\alpha_{\mathbf{k}}^*\beta_{\mathbf{k}}-\tfrac{1}{2}|\alpha_{\mathbf{k}}|^2-\tfrac{1}{2}|\beta_{\mathbf{k}}|^2). \tag{2.7}$$

They do, however, form a complete set:

$$\frac{1}{\pi}\int d^2\alpha_{\mathbf{k}}|\alpha_{\mathbf{k}}\rangle\langle\alpha_{\mathbf{k}}|=1_{\text{op}}, \tag{2.8}$$

where the integration is performed over the two-dimensional complex α plane. Thus the set of all states of the form

$$\prod_{\mathbf{k}}|\alpha_{\mathbf{k}}\rangle\equiv|\{\alpha\}\rangle\equiv|\{\psi\}\rangle,$$
$$\Psi_{\text{op}}(\mathbf{r})|\{\psi\}\rangle=\psi(\mathbf{r})|\{\psi\}\rangle \tag{2.9}$$

is a complete set for the many-boson system. We shall use the notations introduced in (2.9) interchangeably throughout the rest of the paper.

The coherent states are most useful for dealing with many-body systems which behave in some sense classically, that is, systems in which the boson modes are highly occupied. When this is true, the function $\psi(\mathbf{r})$ becomes a classical Schrödinger field which describes the complete many-boson system in just the

[18] R. J. Glauber, Phys. Rev. **131**, 2766 (1963).

[19] P. Carruthers and M. M. Nieto, Am. J. Phys. **33**, 537 (1965).

[20] P. Carruthers and K. S. Dy, Phys. Rev. **147**, 214 (1966). The problem of the anharmonic crystal discussed in this reference is closely related mathematically to the interacting-Boson problem discussed in the present paper.

[21] The relevance of the coherent states to the theory of superfluidity has previously been pointed out by F. W. Cummings and J. R. Johnston, Phys. Rev. **151**, 105 (1966).

same way that the Maxwell field describes the classical limit of quantum electrodynamics. Indeed, the most fruitful application of the coherent states has been in the systematic quantum-mechanical description of intense radiation fields for which classical electrodynamics provides a valid, although incomplete, description.[18] Our point is that, for many-particle Bose systems as opposed to many-photon systems, the validity of the classical description implies superfluidity.

It will be useful to discuss briefly the properties of the coherent states in the classical limit. The discussion is slightly complicated by the fact that the $|\alpha_{\mathbf{k}}\rangle$ are eigenstates of the annihilation operator $a_{\mathbf{k}}$, which is not Hermitian. It is a bit simpler to pursue the classical limit by introducing a coordinate q and its conjugate momentum p (we drop the subscript $\mathbf{k}$ for the moment):

$$q_{\text{op}}=\frac{1}{\sqrt{2}}(a^\dagger+a)\,,\quad p_{\text{op}}=\frac{i}{\sqrt{2}}(a^\dagger-a)\,. \tag{2.10}$$

Let the eigenstates of q_{op} be denoted $|q'\rangle$. Then it is easy to show that[18]

$$\langle q'|\alpha\rangle=\pi^{-1/4}\exp\{-\tfrac{1}{2}(q'-q)^2+ip(q'-q)+\tfrac{1}{2}p^2\}\,, \tag{2.11}$$

where

$$\alpha\equiv\frac{1}{\sqrt{2}}(q+ip)\,. \tag{2.12}$$

Thus the coherent state is a Gaussian wave packet centered at the coordinate q. Similarly, in the momentum representation, it is a wave packet centered at p. This wave packet, in either representation, becomes extremely narrow in the classical limit. To see this, note that

$$\langle n\rangle=\langle\alpha|a^\dagger a|\alpha\rangle=|\alpha|^2\gg1\ \text{(classical)}\,. \tag{2.13}$$

That is, $\hbar|\alpha|^2$ is a quantity which remains finite as $\hbar\to0$, so that (2.11) becomes a δ function. It follows that the coherent states provide an acceptable (but by no means unique) representation for the quantum-mechanical analysis of classical or semiclassical systems. The representation is particularly convenient from an analytic point of view because of the direct relationship between α and the annihilation operator, and also because of the simplicity of the canonical transformation (2.12).

Now consider boson modes $\mathbf{k}$ with occupations $|\alpha_{\mathbf{k}}|^2$ large enough that a classical approximation has some validity. The Hamiltonian equations of motion are

$$dp_{\mathbf{k}}/dt=-\partial H/\partial q_{\mathbf{k}}\,,\quad dq_{\mathbf{k}}/dt=\partial H/\partial p_{\mathbf{k}}\,. \tag{2.14}$$

Starting with (2.14) we may make a sequence of canonical transformations to derive the equations of motion for the complex field $\psi(\mathbf{r})$ which describes the classical limit of the many-boson system. From (2.12) and its complex conjugate, we have

$$d\alpha_{\mathbf{k}}/dt=-i\partial H/\partial\alpha_{\mathbf{k}}^*\,,\quad d\alpha_{\mathbf{k}}^*/dt=i\partial H/\partial\alpha_{\mathbf{k}}\,. \tag{2.15}$$

Equivalently,[22]

$$\begin{aligned} d\psi(\mathbf{r})/dt&=-i\delta H/\delta\psi^*(\mathbf{r})\,,\\ d\psi^*(\mathbf{r})/dt&=i\delta H/\delta\psi(\mathbf{r})\,. \end{aligned} \tag{2.16}$$

Finally, it is conventional to use the notation

$$\psi(\mathbf{r})=f(\mathbf{r})\exp[i\phi(\mathbf{r})]\,, \tag{2.17}$$

where f and ϕ are real functions of $\mathbf{r}$. In a pure coherent state, or in the classical limit, f^2 is just the local number density $n(\mathbf{r})$:

$$n(\mathbf{r})=\langle\Psi_{\text{op}}^\dagger(\mathbf{r})\Psi_{\text{op}}(\mathbf{r})\rangle=|\psi(\mathbf{r})|^2=f^2(\mathbf{r})\,. \tag{2.18}$$

If we transform to the conjugate variables n and ϕ (action and angle variables), we obtain

$$dn(\mathbf{r})/dt=\delta H/\delta\phi(\mathbf{r})\,,\quad d\phi(\mathbf{r})/dt=-\delta H/\delta n(\mathbf{r})\,. \tag{2.19}$$

These are exactly the superfluid equations of motion discussed by Anderson.[23] It must be emphasized, however, that Eqs. (2.14), (2.15), (2.16), or (2.19) represent, at best, a semiclassical approximation to the equations of motion for the quantum-mechanical many-body system.

III. FREE-ENERGY FUNCTIONAL

We turn now to the statistical mechanics of the many-boson system as described by the coherent states. Our entire analysis hinges on the fact that the states (2.9) form a complete set, so that we can evaluate the grand-canonical partition function as follows:

$$\begin{aligned} Z&=\mathrm{Tr}\exp\left(-\frac{H_{\text{op}}-\mu N_{\text{op}}}{k_BT}\right)\\ &=\prod_{\mathbf{k}}\left(\int\frac{d^2\alpha_{\mathbf{k}}}{\pi}\right)\langle\{\alpha\}|\exp\left(-\frac{H_{\text{op}}-\mu N_{\text{op}}}{k_BT}\right)|\{\alpha\}\rangle\\ &\equiv\int\delta\psi(\mathbf{r})\exp(-F\{\psi\}/k_BT)\,. \end{aligned} \tag{3.1}$$

In the final form of (3.1) we have defined again the functional-integral notation first introduced in (1.5), and also have made the important identification

$$F\{\psi\}=-k_BT\ln\left[\langle\{\psi\}|\exp\left(-\frac{H_{\text{op}}-\mu N_{\text{op}}}{k_BT}\right)|\{\psi\}\rangle\right]. \tag{3.2}$$

In this section we shall consider the evaluation of the free-energy functional $F\{\psi\}$.

Consider first the case of noninteracting bosons. We

---

[22] In this form, the equations of motion reduce to the nonlinear differential equation studied by E. P. Gross, Ann. Phys. (N. Y.) 4, 57 (1958).

[23] P. W. Anderson, Ref. 6, p. 300. The same equations are mentioned in a coherent-state formalism by P. Carruthers and M. M. Nieto (unpublished).

have (in units $\hbar=m=1$)

$$H_{op}^{(0)}=\sum_{\mathbf{k}} \tfrac{1}{2}k^2 a_{\mathbf{k}}^{\dagger}a_{\mathbf{k}}, \tag{3.3}$$

so that

$$\langle\{\alpha\}|\exp\left(-\frac{H_{op}^{(0)}-\mu N_{op}}{k_BT}\right)|\{\alpha\}\rangle$$

$$=\prod_{\mathbf{k}}\langle\alpha_{\mathbf{k}}|\exp\left(-\frac{\bar{\epsilon}_{\mathbf{k}}}{k_BT}a_{\mathbf{k}}^{\dagger}a_{\mathbf{k}}\right)|\alpha_{\mathbf{k}}\rangle, \tag{3.4}$$

where

$$\bar{\epsilon}_{\mathbf{k}}\equiv\tfrac{1}{2}k^2-\mu. \tag{3.5}$$

Inserting the representation (2.5) for the coherent states, we obtain

$$\langle\alpha_{\mathbf{k}}|\exp\left(-\frac{\bar{\epsilon}_{\mathbf{k}}}{k_BT}a_{\mathbf{k}}^{\dagger}a_{\mathbf{k}}\right)|\alpha_{\mathbf{k}}\rangle$$

$$=\exp[(e^{-\bar{\epsilon}_{\mathbf{k}}/k_BT}-1)|\alpha_{\mathbf{k}}|^2]. \tag{3.6}$$

Finally,

$$F^{(0)}\{\psi\}=-k_BT\sum_{\mathbf{k}}(e^{-\bar{\epsilon}_{\mathbf{k}}/k_BT}-1)|\alpha_{\mathbf{k}}|^2$$

$$=-k_BT\int d\mathbf{r}\,\psi^*(\mathbf{r})\times\left[\exp\left(\frac{-\frac{1}{2}\nabla^2+\mu}{k_BT}\right)-1\right]\psi(\mathbf{r}). \tag{3.7}$$

Equation (3.7) assumes a more familiar form if we restrict our attention to slowly varying functions ψ and remember that, near the Bose condensation, μ is small and negative. Then we may keep only the first term in an expansion of the exponent:

$$F^{(0)}\{\psi\}\cong\int d\mathbf{r}[\tfrac{1}{2}|\nabla\psi|^2+|\mu|\,|\psi|^2]. \tag{3.8}$$

The functional $F^{(0)}\{\psi\}$ given by (3.7) is a quadratic form in ψ. This means, first, that the partition function Z is a product of Gaussian integrals which may be evaluated easily and which give exactly the correct result for noninteracting bosons. More important, however, is the fact that (3.7) predicts no metastable states; that is, there are no isolated minima of $F^{(0)}\{\psi\}$ except the one at $\psi=0$ and, therefore, no superfluidity according to our criteria. The mathematical mechanism of the Bose condensation, as obtained via the coherent-state formulation, is very reminiscent of the spherical model of a ferromagnet[24]; and it is interesting to note that the spherical model is also unrealistic in its description of the ferromagnetic phase transition because it predicts no metastable states.[25]

[24] T. H. Berlin and M. Kac, Phys. Rev. 86, 821 (1952).
[25] J. S. Langer, Phys. Rev. 137, A1531 (1965).

What is needed in order to make (3.7) or (3.8) describe a system which can support metastable superflow is a higher-order term in the integrand of $F\{\psi\}$ of the form, say, $|\psi|^4$. Such a term, with the correct sign, will be generated by a repulsive two-body interaction in the original Hamiltonian. If we add a term like (2.25) to H, however, we no longer can evaluate $F\{\psi\}$ exactly. What we shall do in the following is evaluate $F\{\psi\}$ as a power series in ψ, the term of order $|\psi|^4$ being the lowest-order term involving the two-body interaction.

The technique required for evaluating $F\{\psi\}$ as a power series in ψ is the same as that used in most field-theoretic or many-body perturbation expansions. We shall use the notation

$$H_{op}-\mu N_{op}=K_{op}+H_{op}',$$

$$K_{op}=\sum_{\mathbf{k}}\bar{\epsilon}_{\mathbf{k}}a_{\mathbf{k}}^{\dagger}a_{\mathbf{k}},$$

$$H_{op}'=\frac{1}{2V}\sum_{\mathbf{k},\mathbf{k}',\mathbf{q}}\bar{v}(\mathbf{q})a_{\mathbf{k}+\mathbf{q}}^{\dagger}a_{\mathbf{k}'-\mathbf{q}}^{\dagger}a_{\mathbf{k}'}a_{\mathbf{k}}, \tag{3.9}$$

where $v(\mathbf{q})$ is the Fourier transform of the two-body interaction potential. The relevant expansion is

$$\exp\left(-\frac{H_{op}-\mu N_{op}}{k_BT}\right)=\exp\left(-\frac{K_{op}}{k_BT}\right)\times\sum_{n=0}^{\infty}(-1)^n\int_0^{(k_BT)^{-1}}d\lambda_1\int_0^{\lambda_1}d\lambda_2\cdots\int_0^{\lambda_{n-1}}d\lambda_n\times H_{op}'(\lambda_1)\cdots H_{op}'(\lambda_n), \tag{3.10}$$

$H_{op}'(\lambda)$ being the interaction representation of H_{op}':

$$H_{op}'(\lambda)\equiv e^{\lambda K_{op}}H_{op}'e^{-\lambda K_{op}}. \tag{3.11}$$

Our procedure is to evaluate, term by term, the diagonal matrix element of (3.10) for the state $|\{\alpha\}\rangle$ and then to exponentiate the resulting series to obtain a free energy F which is proportional to a single factor of the volume V. This exponentiation amounts to a linked-cluster expansion.

In evaluating the matrix element of a term in (3.10), we first make use of the fact that the interaction representation of the annihilation or creation operators always may be written in the form

$$a_{\mathbf{k}}(\lambda)=e^{-\lambda\bar{\epsilon}_{\mathbf{k}}}a_{\mathbf{k}},\quad a_{\mathbf{k}}^{\dagger}(\lambda)=e^{\lambda\bar{\epsilon}_{\mathbf{k}}}a_{\mathbf{k}}. \tag{3.12}$$

All the λ-dependent quantities now appear simply as numerical factors, and may be brought outside of the matrix element.

The next step is to bring the annihilation operators $a_{\mathbf{k}}$ to the right and the creation operators $a_{\mathbf{k}}^{\dagger}$ to the left. This is done by using the Bose commutation relations (2.3). The result is a product of matrix

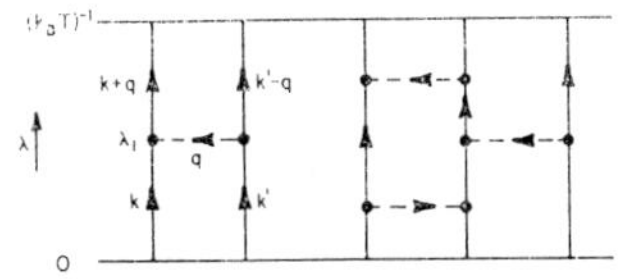

FIG. 1. A typical diagram occurring in the expansion of the partition function. The diagram contains five particle lines and four interaction lines. There are two disconnected parts, only the left-hand one of which is shown with complete momentum and λ labeling.

elements of the form

$$\langle\alpha_{\mathbf{k}}|\exp\left(-\frac{\bar{\epsilon}_{\mathbf{k}}}{k_BT}a_{\mathbf{k}}^{\dagger}a_{\mathbf{k}}\right)(a_{\mathbf{k}}^{\dagger})^s(a_{\mathbf{k}})^t|\alpha_{\mathbf{k}}\rangle, \quad (3.13)$$

one such factor for each mode $\mathbf{k}$. This expression may be evaluated as follows. Use the completeness relation (2.8) and the number-state representation (2.5) to write (3.13) in the form

$$\int\frac{d^2\beta_{\mathbf{k}}}{\pi}|\langle\alpha_{\mathbf{k}}|\beta_{\mathbf{k}}\rangle|^2(\beta_{\mathbf{k}}^*)^s(\alpha_{\mathbf{k}})^t \times\exp[(e^{-\bar{\epsilon}_{\mathbf{k}}/k_BT}-1)\alpha_{\mathbf{k}}^*\beta_{\mathbf{k}}]. \quad (3.14)$$

The integration over $\beta_{\mathbf{k}}$ can be performed using the second of the following integration formulas, valid for any function $f(\alpha)$ which has a power series in α [see Ref. 20, Eqs. (2.6)]:

$$\int\frac{d^2\beta}{\pi}e^{\alpha^*\beta-|\beta|^2}\beta^nf(\beta^*)=\left(\frac{\partial}{\partial\alpha^*}\right)^nf(\alpha^*),$$
$$\int\frac{d^2\beta}{\pi}e^{\alpha\beta^*-|\beta|^2}(\beta^*)^nf(\beta)=\left(\frac{\partial}{\partial\alpha}\right)^nf(\alpha). \quad (3.15)$$

The final form of (3.13) is then

$$\exp[(e^{-\bar{\epsilon}_{\mathbf{k}}/k_BT}-1)|\alpha_{\mathbf{k}}|^2] \times\exp(-s\bar{\epsilon}_{\mathbf{k}}/k_BT)(\alpha_{\mathbf{k}}^*)^s(\alpha_{\mathbf{k}})^t. \quad (3.16)$$

The left-hand factor in (3.16) is just the quantity we computed in Eq. (3.6); therefore the quantity $\exp(-F^{(0)}\{\psi\}/k_BT)$ must factor out of each term in the perturbation expansion.

The remaining contributions are conveniently denoted by diagrams of the kind shown in Fig. 1. The variable λ increases upward from 0 to $(k_BT)^{-1}$. Vertical solid lines denote particles; horizontal dotted lines denote interactions. Both particle lines and interaction lines are labeled by momenta, and momentum is conserved at each interaction.

The rules for evaluation of these diagrams are the following:

(1) For each particle line of momentum $\mathbf{k}$ starting at $\lambda=0$, write a factor $\alpha_{\mathbf{k}}$. For each line ending at $\lambda=(k_BT)^{-1}$, write a factor $\alpha_{\mathbf{k}}^*$.

(2) For each particle line of momentum $\mathbf{k}$ starting at λ' and ending at λ, write a factor

$$\exp[-(\lambda-\lambda')\bar{\epsilon}_{\mathbf{k}}].$$

The points λ and λ' may be either interaction vertices or the end points 0 and $(k_BT)^{-1}$.

(3) For an interaction line carrying momentum $\mathbf{q}$, write a factor

$$(1/V)\bar{v}(\mathbf{q}).$$

(4) Sum over all momenta subject to momentum conservation at each vertex. Integrate over the λ's, observing the limits of integration indicated in (3.10).

(5) Multiply by the standard symmetry factors to avoid overcounting diagrams when performing summations over momentum variables.

The above procedure yields

$$\langle\{\alpha\}|\exp\left(-\frac{H_{\rm op}-\mu N_{\rm op}}{k_BT}\right)|\{\alpha\}\rangle =\exp(-F^{(0)}\{\alpha\}/k_BT)\sum_{\Gamma}W_{\Gamma}\{\alpha\}, \quad (3.17)$$

where $W_\Gamma\{\alpha\}$ denotes the numerical contribution of the diagram Γ as determined by the preceding rules. The conventional linked-cluster analysis then tells us that

$$F\{\alpha\}-F^{(0)}\{\alpha\}=k_BT\sum_{\Gamma^{(c)}}W_{\Gamma^{(c)}}\{\alpha\}, \quad (3.18)$$

where now only connected diagrams, $\Gamma^{(c)}$, are included in the sum. By "connected," we mean that every particle line is connected to the rest of the diagram by at least one interaction line.

According to rule (1) above, a diagram with l particle lines is formally of order $|\alpha|^{2l}$; that is, it contains l factors $\alpha_{\mathbf{k}}$, $\alpha_{\mathbf{k}'}$, etc., and l factors $\alpha_{\mathbf{p}}^*$, $\alpha_{\mathbf{p}'}^*$, etc. To obtain the entire contribution to $F\{\alpha\}$ of order $|\alpha|^4$, we must sum all the two-body diagrams like those shown in Fig. 2. The three-body diagrams will make contributions of order $|\alpha|^6$, and so forth.

For completeness, we quote the numerical contribution to F of the first diagram shown in Fig. 2:

$$\frac{k_BT}{2V}\sum_{\mathbf{k},\mathbf{k}',\mathbf{q}}\bar{v}(\mathbf{q})\left(\frac{e^{-E/k_BT}-e^{-E'/k_BT}}{E'-E}\right) \times\alpha_{\mathbf{k}+\mathbf{q}}^*\alpha_{\mathbf{k}'-\mathbf{q}}^*\alpha_{\mathbf{k}'}\alpha_{\mathbf{k}}, \quad (3.19)$$

where $E=\bar{\epsilon}_{\mathbf{k}}+\bar{\epsilon}_{\mathbf{k}'}$, $E'=\bar{\epsilon}_{\mathbf{k}+\mathbf{q}}+\bar{\epsilon}_{\mathbf{k}'-\mathbf{q}}$. If the two-body interaction contains a hard core or is otherwise too strong to permit the use of (3.19), then the entire series of ladder diagrams must be summed. At this point our formalism is very similar to that developed by Lee and Yang,[26] and the reader is referred to their papers for the details of such calculations.

[26] T. D. Lee and C. N. Yang, Phys. Rev. **113**, 1165 (1959).

In conclusion, we note that we have constructed a free-energy functional $F\{\psi\}$ of basically the Ginzburg-Landau form, Eq. (1.4), the only real difference being that the derivatives of ψ enter in a rather more complicated manner when ψ varies rapidly with position. The simple form of the quartic term in (1.4) will be correct for ψ's which vary slowly over distances of the order of the two-body scattering length. Finally, the crucial temperature dependence of A, the coefficient of $|\psi|^2$, is essentially the same as the temperature dependence of the chemical potential μ, which must be chosen to fix the number of particles. For noninteracting particles, we know that μ is negative at high temperatures and goes to zero at the Bose-Einstein condensation point. When we add the effect of repulsive interactions, i.e., the term $B|\psi|^4$ with positive B, then μ may be expected to become positive below some critical temperature, as required by the phenomenological model.

IV. THERMODYNAMIC AVERAGES AND THE TWO-FLUID MODEL

The functional

$$\rho(\{\alpha^*\},\{\alpha\})=\frac{1}{Z}\langle\{\alpha\}|\exp\left(-\frac{H_{op}-\mu N_{op}}{k_BT}\right)|\{\alpha\}\rangle$$
$$=(1/Z)\exp(-F\{\psi\}/k_BT)\quad(4.1)$$

is the diagonal element of the density matrix in the coherent-state representation. We should like to interpret (4.1) as simply the probability that the system will be found in state ψ, but it must be recognized that the density matrix is not quite diagonal in this representation. For example, consider the expectation value of the number density, which is given for a pure state ψ by Eq. (2.18). We have [see Ref. 20, Eq. (2.16)]

$$n_{\mathbf{k}}=\mathrm{Tr}(a_{\mathbf{k}}^\dagger a_{\mathbf{k}}\rho_{op})$$
$$=\prod_{\mathbf{k}'}\left(\int\frac{d^2\alpha_{\mathbf{k}'}}{\pi}\right)\alpha_{\mathbf{k}}^*\left(\alpha_{\mathbf{k}}+\frac{\partial}{\partial\alpha_{\mathbf{k}}^*}\right)\rho(\{\alpha^*\},\{\alpha\}),\quad(4.2)$$

which is derived by means of the integration formulas (3.15). An integration by parts turns out to be legal, so that

$$n_{\mathbf{k}}=\prod_{\mathbf{k}'}\left(\int\frac{d^2\alpha_{\mathbf{k}'}}{\pi}\right)(|\alpha_{\mathbf{k}}|^2-1)\rho(\{\alpha^*\},\{\alpha\})$$
$$=\langle|\alpha_{\mathbf{k}}|^2\rangle_\psi-1,\quad(4.3)$$

where the angular brackets denote an average over ψ or α space with (4.1) as the statistical weight.

It is important to note that the expected identity between $n_{\mathbf{k}}$ and $\langle|\alpha_{\mathbf{k}}|^2\rangle_\psi$ is a good approximation only in the classical limit, $n_k\gg1$. The identity is exactly correct only when the mode is occupied macroscopically, i.e., n_k is of order N.

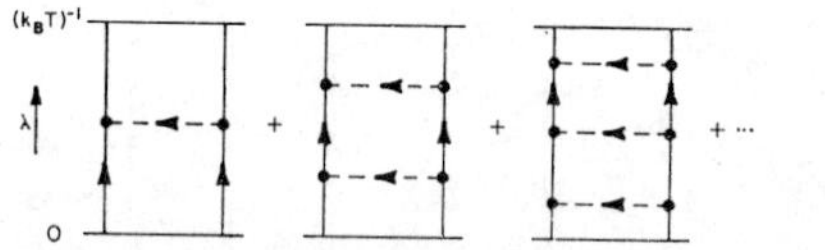

FIG. 2. The sequence of two-body ladder diagrams which must be summed to obtain the entire contribution to F of order $|\alpha|^4$.

The total density can be written in the form

$$n=\frac{1}{V}\sum_{\mathbf{k}}n_{\mathbf{k}}=\frac{1}{V}\int d\mathbf{r}\,\langle|\psi(\mathbf{r})|^2\rangle_\psi-\frac{1}{V}\sum_{\mathbf{k}}1.\quad(4.4)$$

Here it is obvious that we must expect a short-wavelength (large $\mathbf{k}$) divergence in $\langle|\psi(\mathbf{r})|^2\rangle_\psi$. In fact, this divergence is the same one which appears in similar field-theoretic calculations, and it is reassuring to see how it is subtracted out in Eq. (4.4). This divergence disappears in the formula for the current density:

$$\mathbf{j}=\frac{1}{V}\sum_{\mathbf{k}}\mathbf{k}n_{\mathbf{k}}=\frac{1}{V}\int d\mathbf{r}\left\langle\frac{1}{2i}(\psi^*\nabla\psi-\psi\nabla\psi^*)\right\rangle_\psi.\quad(4.5)$$

The Ginzburg-Landau theory, and the two-fluid model, emerge only when it is possible to evaluate the above formulas by what amounts to a mean-field calculation, augmented by a random-phase approximation. Suppose that $\rho\{\psi\}$, Eq. (4.1), is sharply peaked at some ψ, say ψ_s. In fact, let us assume that ψ_s is overwhelmingly the most probable state in its part of the function space, so that it must describe a stable or metastable state of the system. Then we can associate with ψ_s a superfluid density n_s and a supercurrent $\mathbf{j}_s$:

$$n_s=|\psi_s|^2,\quad\mathbf{j}_s=\frac{1}{2i}(\psi_s^*\nabla\psi_s-\psi_s\nabla\psi_s^*).\quad(4.6)$$

In order for the quantities defined in (4.6) to be meaningful, ψ_s must be of order unity (not, say, order $N^{-1/2}$) throughout the system. That is, ψ_s must have the properties of the classical field discussed in Sec. II, and, in particular, must obey the superfluid equations of motion, Eqs. (2.19). Thus, the most probable ψ appears to describe a superfluid component of the system.

It should be recognized, however, that the most probable ψ is not the same thing as the average ψ, nor are n_s and $\mathbf{j}_s$ the same as the average or expected values of n and $\mathbf{j}$ as defined in Eqs. (4.4) and (4.5). In order to evaluate the correct density or current, one must take proper account of the width of the peak in $\rho\{\psi\}$ near ψ_s. That is, one must include the fluctuations. There is no known way of doing this rigorously (except in one dimension), but the following procedure probably gives qualitatively correct results at temperatures far enough away from the critical point that the fluctuations are small.

In principle, we want to expand $F\{\psi\}$ out to terms quadratic in $\psi-\psi_s$ and then perform the resulting Gaussian integrals.[27] This procedure is slightly complicated by the fact that $F\{\psi\}$ is independent of the phase of ψ, so that the phase fluctuations always will be appreciable and must be handled separately. Suppose that ψ_s is one of the uniform current-carrying states given by Eq. (1.8):

$$\psi_s = f_k e^{i\mathbf{k}\cdot\mathbf{r}}. \tag{4.7}$$

Then we follow essentially Rice's procedure[10] and write

$$\psi(\mathbf{r}) = (f_k + \nu(\mathbf{r}))\exp[i\mathbf{k}\cdot\mathbf{r} + i\phi(\mathbf{r})], \tag{4.8}$$

where ν and ϕ are real functions of $\mathbf{r}$. It then makes sense to expand $F\{\psi\}$ out to terms quadratic in ν and ϕ. For the $F\{\psi\}$ given in Eq. (1.4),

$$F\{\psi\} \cong F\{\psi_s\} + \int d\mathbf{r}\,[(\nabla\nu)^2 + f_k^2(\nabla\phi)^2 + 4f_k\nu\mathbf{k}\cdot\nabla\phi + \kappa^2\nu^2], \tag{4.9}$$

where

$$\kappa^2 = 2Bf_k^2 = 2(A-k^2). \tag{4.10}$$

If we use the more general $F\{\psi\}$ computed in Sec. III, especially with the quadratic term $F^{(0)}\{\psi\}$ given by (3.7) which is necessary in order to allow for rapidly varying ψ's, then the relevant quadratic form becomes more complicated but remains qualitatively similar to (4.9). We arrive at expressions of the form

$$n = n_s + \langle\nu^2\rangle - \frac{1}{V}\sum_{\mathbf{k}'} 1 \tag{4.11}$$

and

$$\mathbf{j} = \mathbf{j}_s + 2f_k\langle\nu\nabla\phi\rangle + \mathbf{k}\langle\nu^2\rangle, \tag{4.12}$$

[27] The mathematical significance of performing these integrations in the neighborhood of a ψ_s which is not the absolute minimum of $F\{\psi\}$ is discussed in Ref. 13.

where the angular brackets here represent averages with respect to the above Gaussian approximation for the weight factor $\rho\{\psi\}$. In Eqs. (4.11) and (4.12), the fluctuations contribute additively to n and $\mathbf{j}$, and appear to describe a normal component of the fluid.

Special calculations of this kind have been published by Rice,[10] who has emphasized the fact that the phase fluctuations preclude off-diagonal long-range order in one or two dimensions. The reader may check from Rice's results that the formulas for n and $\mathbf{j}$ remain well defined in all cases except where the fluctuations become anomalously large near an apparent critical point. The fact that the method breaks down in some temperature range, especially in one or two dimensions, says nothing at all about whether a phase transition occurs, nor does it necessarily invalidate the qualitative results in regions where the fluctuations are small enough that the method seems self-consistent. In fact, the known features of the soluble one-dimensional model of the kind discussed here, using the simplest Ginzburg-Landau form for $F\{\psi\}$, are also the most striking features of the above approximation. That is, there is no long-range order, but the amplitude of ψ has a nonzero most probable value and, for low enough temperatures, is very unlikely to vanish. Thus the one-dimensional model may, in some sense, be a superfluid.

Both this one-dimensional model and further analysis concerning the behavior of the normal component as described by fluctuations will, hopefully, be the subjects of later communications.

ACKNOWLEDGMENTS

Most of this paper was written while the author was in residence at the Physics Division of the Aspen Institute, Aspen, Colorado. Stimulating discussions with many people at the Institute were most helpful in the completion of this work. The author also wishes to thank Dr. N. D. Mermin for helpful criticism of an earlier version of this paper.

PHYSICAL REVIEW B VOLUME 1, NUMBER 12 15 JUNE 1970

Landau Diamagnetism from the Coherent States of an Electron in a Uniform Magnetic Field

Albert Feldman and Arnold H. Kahn
National Bureau of Standards, Washington, D. C. 20234

(Received 29 December 1969)

A complete set of coherent-state wave packets has been constructed for an electron in a uniform magnetic field. These states are nonspreading packets of minimum uncertainty that follow the classical motion. Use was made of the ladder operators that generate all the eigen states of the Hamiltonian from any one energy eigenstate. The coherent states are the eigenstates of the two ladder operators that annihilate the zero-angular-momentum ground state. We have calculated the partition function, exploiting advantages of the coherent-state basis. The Landau diamagnetism and the de Haas–van Alphen oscillations are contained in the coherent-state framework.

I. INTRODUCTION

Coherent-state wave packets have received renewed attention since the recent article of Glauber.[1] Much of this attention is due to the recognition of their usefulness as a set of basis states for the calculation of observable physical quantities.[1,2] In addition, they have been of value in the theoretical problem of the quantum-mechanical definition of the phase of an oscillator.[3] The coherent state is a wave packet whose probability distribution is invariant in time except for a displacive translation which obeys the classical equations of motion. Louisell[2] has summarized the physical and mathematical properties of the coherent states of the linear oscillator.

In this paper we have generated the complete set of coherent states of a charged particle in a magnetic field by use of step-ladder operators.[4] Some of these states were originally found by Darwin[5] in his examination of the classical action. In order to obtain all the states, we constructed ladder operators $X_\pm$ in addition to the ladder operators $\pi_\pm$ constructed by Johnson and Lippmann.[6] Spin was neglected throughout.

As an application, we have used the coherent states to calculate the magnetization and magnetic susceptibility of the free-electron gas and have thus rederived the Landau diamagnetism. The computation is aided by simplifications that occur when using the coherent states as a basis.

II. ENERGY EIGENVALUES AND EIGENSTATES

The energy eigenvalues and eigenvectors of a charged particle in a magnetic field were first found by Landau.[7] We shall review the problem in order to obtain the operators necessary for generating the coherent states. The notation of Kubo *et al.*[8] will be used. Two constants which characterize the problem are $\Omega = eH/\mu c$, the cyclotron frequency; and $l = (\hbar/\mu\Omega)^{1/2}$, the classical radius of the ground-state Landau orbit.

The Hamiltonian for a free electron in a magnetic field, neglecting spin, is

$$\mathcal{H} = \pi^2/2\mu \, , \tag{2.1}$$

where μ is the electron mass,

$$\vec{\pi} = \vec{p} + e\vec{A}/c \tag{2.2}$$

and

$$\vec{H} = \nabla \times \vec{A} \, . \tag{2.3}$$

We choose the vector potential to be

$$\vec{A} = (-\tfrac{1}{2}Hy, \ \tfrac{1}{2}Hx, \ 0), \tag{2.4}$$

which corresponds to a uniform magnetic field parallel to the z direction. Thus, since the motion along the z direction is that of a free particle, it is necessary to solve only the two-dimensional problem in the xy plane. The appropriate Hamiltonian for the transverse motion is

$$\mathcal{H}_t = [(p_x - \tfrac{1}{2}\mu\Omega y)^2 + (p_y + \tfrac{1}{2}\mu\Omega x)^2]/2\mu \, . \tag{2.5}$$

If we introduce the operators[6]

$$\pi_\pm = p_x \pm i p_y \pm (i\hbar/2l^2)(x \pm i y) \tag{2.6}$$

which obey the commutation relation

$$[\pi_-, \ \pi_+] = 2\mu\hbar\Omega \, , \tag{2.7}$$

then

$$\mathcal{H}_t = [(\pi_+\pi_-)/2\mu] + \tfrac{1}{2}\hbar\Omega \, . \tag{2.8}$$

This is mathematically equivalent to the linear oscillator Hamiltonian with the number operator being $\pi_+\pi_-/2\mu\hbar\Omega$. In addition, the angular momentum operator

$$L_z = x p_y - y p_x , \tag{2.9}$$

which has integer eigenvalues (in units of $\hbar$), commutes with $\mathcal{H}_t$. We select energy eigenstates that are simultaneously eigenstates of L_z. Labeling them $|Nm\rangle$, one obtains the eigenvalue equations

$$\mathcal{H}_t|N, m\rangle = (N + \tfrac{1}{2})\hbar\Omega|N, m\rangle \tag{2.10}$$

and

$$L_z|N, m\rangle = m\hbar|N, m\rangle . \tag{2.11}$$

From the commutation relations of $\pi_\pm$ with $\mathcal{H}$ and L_z,

$$[\mathcal{H}, \ \pi_\pm] = \pm\hbar\Omega\pi_\pm \tag{2.12}$$

and

$$[L_z, \ \pi_\pm] = \pm\hbar\,\pi_\pm , \tag{2.13}$$

we see the operators π_+ and π_- are the raising and lowering operators, respectively, for energy and angular momentum simultaneously. Thus

$$\pi_+|N, m\rangle = (2\mu\hbar\Omega)^{1/2}(N+1)^{1/2}|N+1, \ m+1\rangle \tag{2.14}$$

and

$$\pi_-|N, m\rangle = (2\mu\hbar\Omega)^{1/2}(N)^{1/2}|N-1, \ m-1\rangle \, . \tag{2.15}$$

Figure 1 demonstrates the properties of $\pi_\pm$. In the array of states $|N, m\rangle$ the $\pi_\pm$ operator can carry us along any one diagonal. It is evident that to generate all states from any particular state we must find an operator that will move us from one diagonal to another. An operator with this property can be constructed from the orbit center-coordinate operators[8]

$$X = x - (\pi_y/\mu\Omega) \tag{2.16}$$

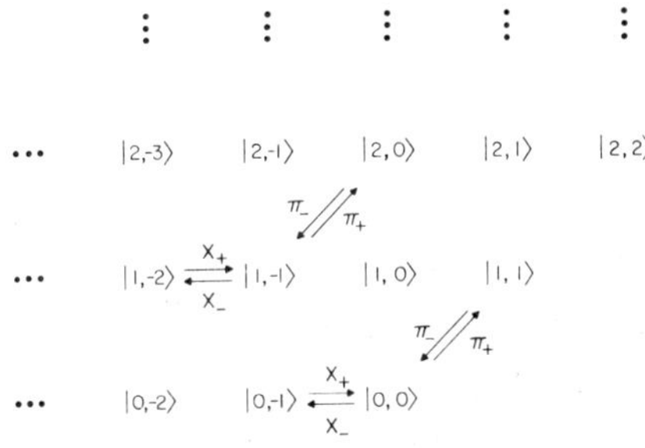

FIG. 1. Energy eigenstates $|N, m\rangle$ for the transverse motion of an electron in a magnetic field. The energy eigenvalue is $(N+\frac{1}{2})\hbar\Omega$ and the angular momentum eigenvalue is $m\hbar$. The stepping of the ladder operators operators $\pi_\pm$ and $X_\pm$ is indicated by arrows.

and $Y = y + (\pi_x/\mu\Omega)$. (2.17)

The operators

$$X_\pm = X \pm i\, Y \qquad (2.18)$$

have the desired properties since they step only the angular momentum and not the energy, as shown by the commutation relations

$$[L_z, X_\pm] = \pm\hbar X_\pm \qquad (2.19)$$

and $[\mathcal{H}, X_\pm] = 0$. (2.20)

In addition, they commute with $\pi_\pm$. We can therefore show that

$$X_+|N, m\rangle = (\sqrt{2})\, l\, (N-m)^{1/2}\, |N, m+1\rangle$$

and $X_-|N, m\rangle = \sqrt{2}\, l\, (N-m+1)^{1/2}\, |N, m-1\rangle$.

It is clear that for a given energy $(N+\frac{1}{2})\hbar\Omega$, m can take on values N, $N-1$, . . . , 0, -1, . . . , $-\infty$, because the application of X_+ to the state $|N, N\rangle$ yields zero. All energy eigenstates can now be generated from the ground state $|0, 0\rangle$ by successive applications of π_+ and X_-; thus

$$|N, m\rangle = [(2\mu\hbar\Omega)^N\, (2l^2)^{N-m}\, N!\, (N-m)!\,]^{-1/2} \times X_-^{N-m}\, \pi_+^N\, |0, 0\rangle. \qquad (2.21)$$

This is shown schematically in Fig. 1.

The ground-state zero-angular-momentum wave function, in coordinate representation, is obtained from the two conditions

$$\langle r|X_+|0,0\rangle = \left[l^2\left(\frac{\partial}{\partial x} + i\frac{\partial}{\partial y}\right) + \tfrac{1}{2}(x+iy)\right]\psi_{00} = 0,$$

$$\langle r|\pi_-|0,0\rangle = -i\hbar\left[\left(\frac{\partial}{\partial x} - i\frac{\partial}{\partial y}\right) + \frac{1}{2l^2}\,(x-iy)\right]\psi_{00} = 0 .$$

The substitution $\rho_\pm = x \pm iy$ simplifies the solution. The result is

$$\psi_{00}(x, y) = [1/(\sqrt{2\pi})l]\, e^{-(x^2+y^2)/4l^2}$$
$$= [1/(\sqrt{2\pi})l]\, e^{-\rho_+\rho_-/4l^2}$$

in which the state is normalized to unity over the xy plane.

III. COHERENT STATES

The coherent states of a harmonic oscillator have been discussed in detail by Glauber.[1] We follow his procedures for the problem of an electron in a magnetic field. We consider the transverse motion only. Let the coherent state $|\alpha, \xi\rangle$ be defined as the simultaneous eigenstate of the two commuting non-Hermitian operators which annihilate the ground state:

$$\pi_-|\alpha, \xi\rangle = (\hbar/i)\,(\alpha/l^2)|\alpha, \xi\rangle, \qquad (3.1)$$

$$X_+|\alpha, \xi\rangle = \xi|\alpha, \xi\rangle. \qquad (3.2)$$

The complex eigenvalues α and ξ have dimensions of length. Their physical significance and the coherence properties will follow.

We construct the coherent state by expanding in the complete set of energy eigenfunctions; thus

$$|\alpha, \xi\rangle = \Sigma|N, m\rangle\,\langle N, m|\alpha, \xi\rangle. \qquad (3.3)$$

We can obtain the expansion coefficients $\langle N, m|\alpha, \xi\rangle$ in terms of the single coefficient $\langle 0, 0|\alpha, \xi\rangle$ by use of the Hermitian of Eq. (2.21) and Eqs. (3.1) and (3.2). Hence,

$$\langle N, m|\alpha, \xi\rangle = \frac{(\alpha/i)^N\,(\xi)^{N-m}}{(2l^2)^{N/2}(N!)^{1/2}(2l^2)^{(N-m)/2}[(N-m)!]^{1/2}} \times\langle 0,0|\alpha, \xi\rangle. \qquad (3.4)$$

From the normalization condition

$$\langle\alpha, \xi|\alpha, \xi\rangle = 1\,, \qquad (3.5)$$

we obtain the result

$$\langle 0,0|\alpha, \xi\rangle = \exp[-(|\alpha|^2 + |\xi|^2)/4l^2], \qquad (3.6)$$

where an arbitrary phase factor has been set equal to unity.

By an additional use of Eq. (2.21), the coherent state can be put in the form

$$|\alpha, \xi\rangle = \exp\left[-\frac{1}{4l^2}\,(|\alpha|^2 + |\xi|^2) + \frac{\alpha\pi_+}{2i\hbar} + \frac{\xi X_-}{2l^2}\right]|0,0\rangle. \qquad (3.7)$$

The time dependence of the coherent state is obtained by deriving the time dependence of X_+ and π_- in the Heisenberg picture. X_+ is a constant of the motion since it commutes with $\mathcal{H}$. On the other hand,

$$\pi_-(t) = e^{-i\Omega t}\,\pi_-(0), \qquad (3.8)$$

which follows from the equation of motion

$$d\pi_-/dt = (1/i\hbar)\,[\pi_-, \mathcal{H}] = -i\Omega\pi_-. \qquad (3.9)$$

Hence

$$\pi_-(t)|\alpha\xi\rangle = e^{-i\Omega t}\,\pi_-(0)|\alpha\xi\rangle = \alpha\, e^{-i\Omega t}\,|\alpha, \xi\rangle. \qquad (3.10)$$

On transforming to the Schrödinger picture, the time-dependent coherent state becomes $|\alpha e^{-i\Omega t}, \xi\rangle$.

The Schrödinger representation of the coherent state $\psi_{\alpha\xi}(r)$ may be obtained in several ways. One method is to solve Eqs. (3.1) and (3.2) in the coordinate representation. Using $\rho_\pm = x \pm iy$, we obtain the differential equations

$$\left(\frac{\partial}{\partial\rho_+} + \frac{1}{4l^2}\,\rho_-\right)\psi_{\alpha\xi} = \frac{\alpha}{2l^2}\,\psi_{\alpha\xi}\,, \qquad (3.11a)$$

$$\left(\frac{\partial}{\partial\rho_-} + \frac{1}{4l^2}\,\rho_+\right)\psi_{\alpha\xi} = \frac{\xi}{2l^2}\,\psi_{\alpha\xi}\,. \qquad (3.11b)$$

Since the solution for $\alpha = \xi = 0$ is $e^{-\rho_+\rho_-/4l^2}$, we are led to try simple translations of the variables ρ_+ and ρ_-. Thus we find

$$\psi_{\alpha\xi} \propto \exp[-(\rho_+ - 2\xi)(\rho_- - 2\alpha)/4l^2]\, e^{f(t)}, \quad (3.12)$$

where $f(t)$ is an undetermined function of time. We obtain $f(t)$ by requiring that $\psi_{\alpha\xi}$ satisfy the time-dependent Schrödinger equation with $\alpha(t)$ having the time dependence $e^{-i\Omega t}$. We find $f(t) = [\alpha(t)\xi/2l^2]$. Thus, the wave function, normalized to unity, is

$$\psi_{\alpha\xi} = \frac{1}{l(2\pi)^{1/2}} \exp\left[-\frac{1}{4l^2}(|\alpha|^2 + |\xi|^2) - \frac{1}{4l^2}(\rho_+ - 2\xi)(\rho_- - 2\alpha) + \frac{\alpha\xi}{2l^2}\right]. \quad (3.13)$$

Equation (3.13) can be obtained directly by finding the coordinate representation of Eq. (3.7). Use is made of the operator identities

$$\exp\left(x + c\frac{\partial}{\partial x}\right) = \exp(x)\exp\left(c\frac{\partial}{\partial x}\right)\exp(\tfrac{1}{2}c)$$

and $\left[\exp\left(c\frac{\partial}{\partial x}\right)\right] f(x) = f(x+c)$.

The probability density in space is

$$|\psi_{\alpha\xi}|^2 = \frac{1}{2\pi l^2} \times \exp\left[-\left(\frac{1}{2l^2}\right)[(x-\alpha_1-\xi_1)^2 + (y+\alpha_2-\xi_2)^2]\right], \quad (3.14)$$

where $\alpha = \alpha_1 + i\alpha_2$, $\xi = \xi + i\xi_2$, and α_1, α_2, ξ_1, and ξ_2 are all real. Thus the coherent wave packet has the form of the ground state, but is displaced in space to a moving center. The mean coordinates are

$$\bar{x}(t) = \mathrm{Re}[\xi + \alpha(0)e^{-i\Omega t}],$$
$$\bar{y}(t) = \mathrm{Im}[\xi - \alpha(0)e^{-i\Omega t}]. \quad (3.15)$$

Thus the centroid of the packet follows a classical orbit of radius $|\alpha|$ around a center at (ξ_1, ξ_2). The shape of the packet is independent of time. The motion is depicted in Fig. 2. The use of the X_- operator enables us to place the center anywhere. These same techniques may be used to find the coherent states for crossed electric and magnetic fields, and for the harmonic oscillator in a magnetic field. These cases are discussed in Appendixes A and B.

The coherent states form a complete basis. Following Glauber,[1] we find that the closure relation may be expressed as

$$(1/4\pi^2 l^4)\int |\alpha, \xi\rangle\langle\alpha, \xi|\, d^2\alpha\, d^2\xi = 1, \quad (3.16)$$

where $d^2\alpha = d\alpha_1 d\alpha_2$; $d^2\xi = d\xi_1 d\xi_2$. As in the harmonic-oscillator case, these coherent states are states of minimum uncertainty, i.e., they satisfy uncertainty relations

$$\Delta x\, \Delta p_x = \Delta y\, \Delta p_y = \tfrac{1}{2}\hbar\,. \quad (3.17)$$

IV. DIAMAGNETIC SUSCEPTIBILITY

The coherent-state formulation permits us to use classical concepts for describing electron orbits, yet contains all quantum effects. We use this approach to calculate the partition function. The Landau diamagnetism is then obtained.

The partition function is given by

$$Z = \mathrm{Tr}\, e^{-\mathcal{H}/kT} = \sum \langle\alpha, \xi, k_z|\, e^{-\mathcal{H}/kT}|\alpha, \xi, k_z\rangle$$
$$= \sum_{k_z} e^{-\hbar^2 k_z^2/2\mu kT} \int \frac{d^2\xi\, d^2\alpha}{4\pi^2 l^4} \langle\alpha, \xi| \times \left[\exp\left(-\frac{\pi_+\pi_-}{2\mu kT} - \frac{1}{2}\frac{\hbar\Omega}{kT}\right)\right]|\alpha, \xi\rangle. \quad (4.1)$$

The sum on k_z gives the usual partition function for one-dimensional free motion. We evaluate Z for a cylindrical body of length L, radius R, oriented along the magnetic field. Thus we have

$$Z = Z_\parallel Z_\perp, \quad (4.2)$$

where $\quad Z_\parallel = (L/h)(2\pi\mu kT)^{1/2}. \quad (4.3)$

The transverse part may be simplified through the properties of the coherent states. We use the boson-operator identity[2]

$$e^{xa^\dagger a} = \sum_{n=0}^{\infty} \frac{(e^x - 1)^n}{n!} a^{\dagger n} a^n, \quad (4.4)$$

where a and $a^\dagger$ satisfy $[a, a^\dagger] = 1$, to evaluate Eq. (4.1). Thus the factor

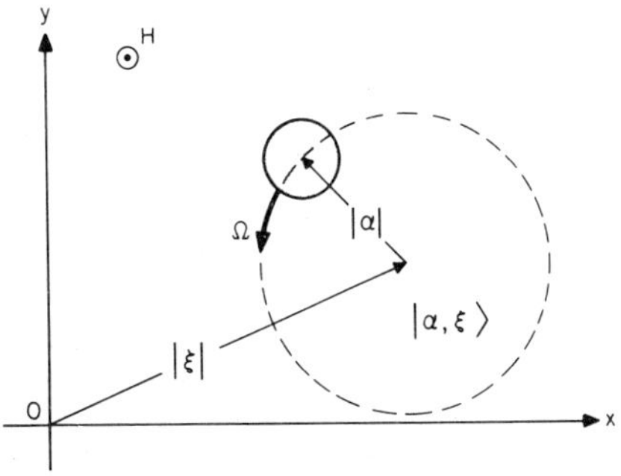

FIG. 2. Motion of the coherent state $|\alpha, \xi\rangle$. The shaded region represents the Gaussian packet which follows the classical motion of cyclotron frequency Ω.

$$\left\langle\alpha\xi\left|\left[\exp\left(-\frac{\pi_+\pi_-}{2\mu kT}\right)\right]\right|\alpha\xi\right\rangle = \exp\left(\frac{|\alpha|^2}{2l^2}\,(e^{-\hbar\Omega/kT}-1)\right) \qquad (4.5)$$

and $Z_\perp = e^{-\hbar\Omega/2kT}$

$$\times\int\frac{d^2\xi\,d^2\alpha}{4\pi^2 l^4}\exp\left(-\frac{|\alpha|^2}{2l^2}\,(1-e^{-\hbar\Omega/kT})\right). \qquad (4.6)$$

To perform the integration over ξ and α, we exclude all coherent states with $|\alpha+\xi|>R$, i.e., we sum over all orbits lying within the cylinder. The limitation on the terms to be included in the sum over coherent states is equivalent to the cutoff of energy eigenstates used by Landau.[7] However, since $R \gg l$, for our application, and the exponential falls off rapidly with $|\alpha|^2$, we can safely extend the α integration to infinity. The results are

$$Z_\perp = \frac{e^{-\hbar\Omega/2kT}}{4\pi^2 l^4}\int_0^R 2\pi|\xi|d|\xi|\int_0^\infty 2\pi|\alpha|d|\alpha| \times\exp\left(-\frac{|\alpha|^2}{2l^2}\,(1-e^{-\hbar\Omega/kT})\right) \qquad (4.7)$$

and $$Z = V\frac{(2\pi\mu kT)^{1/2}}{h}\,\frac{\mu\Omega}{4\pi\hbar}\,\frac{1}{\sinh(\hbar\Omega/2kT)}. \qquad (4.8)$$

The magnetization M is obtained from the free energy F by

$$F = -nkT\ln Z \qquad (4.9)$$

and $$M = -\frac{\partial F}{\partial H} = \frac{ne\hbar}{\mu c}\left(\frac{kT}{\hbar\Omega}-\frac{1}{2}\coth\frac{\hbar\Omega}{2kT}\right). \qquad (4.10)$$

The susceptibility χ, per electron, in the high-temperature limit is the correct Landau diamagnetism

$$\chi = -\frac{1}{n}\frac{\partial M}{\partial H} = -\frac{1}{3}\left(\frac{e\hbar}{2\mu c}\right)^2\frac{1}{kT}. \qquad (4.11)$$

Once the partition function is known for the nondegenerate case, the free energy for the free-electron gas with Fermi statistics may be found directly, as shown by Wilson.[9] This is because the density of states is given by the inverse Laplace transform of the partition function.[10] Hence all quantum effects, including the de Haas–van Alphen effect, are contained in the coherent-state treatment.

In Appendix B we obtain Z for an oscillator in a magnetic field.[11] In this case there is no need for a cutoff at radius R. In the limit of small spring constant we obtain the same M and χ as obtained above.

In this section the coherent state enabled us to use classical concepts of orbit center and radius. The quantum effects arose through the noncommutability of π_+ and π_-. This led to a distribution of orbit radii of the same form that Bloch[12] first derived for the amplitude distribution of a harmonic oscillator in thermal equilibrium.

APPENDIX A: CROSSED ELECTRIC AND MAGNETIC FIELDS

In crossed electric and magnetic fields the Hamiltonian is augmented by a term

$$V = e\vec{E}\cdot\vec{r} = e(E_x x + E_y y) = \tfrac{1}{2}e\,(E_+\rho_- + E_-\rho_+),$$

where $E_\pm = E_x \pm iE_y$. The Heisenberg equations of motion for the operators π_- and X_+ now become

$$d\pi_-/dt = -(i/\hbar)[\pi_-,\ \mathcal{H}_t + V] = -i\Omega\pi_- - eE_-$$

and $$dX_+/dt = -(i/\hbar)[X_+,\ V] = -ieE_+/\mu\Omega.$$

The solutions are

$$\pi_-(t) = e^{-i\Omega t}\pi_-(0) - eE_-/i\Omega,$$

$$X_+(t) = X_+(0) - (ieE_+/\mu\Omega)t.$$

Thus the coherent state $|\alpha_0,\ \xi_0\rangle$ at $t=0$ will have evolved at time t to the state

$$\left|\alpha_0 e^{-i\Omega t} - \frac{eE_-}{\mu\Omega^2},\ \xi_0 - \frac{ieE_+}{\mu\Omega}\,t\right\rangle.$$

The location of the centroid at time t is given by

$$\bar{x}(t) = \mathrm{Re}\left(\xi_0 - \frac{ieE_+}{\mu\Omega}\,t + \alpha_0 e^{-i\Omega t} - \frac{eE_-}{\mu\Omega^2}\right),$$

$$\bar{y}(t) = \mathrm{Im}\left(\xi_0 - \frac{ieE_+}{\mu\Omega}\,t - \alpha_0 e^{-i\Omega t} - \frac{eE_-}{\mu\Omega^2}\right).$$

This motion is the cycloidal path followed by a classical electron in crossed fields. The coherent state has the same charge distribution as before, but follows this new path.

APPENDIX B: ELECTRON IN A UNIFORM MAGNETIC FIELD AND A HARMONIC POTENTIAL

The energy eigenvalues and eigenfunctions for this problem were originally found by Fock.[11] The Hamiltonian $\mathcal{H}_t$ is

$$\mathcal{H}_t = [(p_x - \tfrac{1}{2}\mu\Omega y)^2 + (p_y + \tfrac{1}{2}\mu\Omega x)^2]/2\mu + \tfrac{1}{2}\mu\,\omega_0^2(x^2+y^2).$$

If we make the substitution

$$\Omega' = (\Omega^2 + 4\,\omega_0^2)^{1/2},$$

we then obtain

$$\mathcal{H}_t = \frac{1}{2}\left[\frac{\pi_+\pi_-}{2\mu}\left(1+\frac{\Omega}{\Omega'}\right) + \frac{1}{2}\mu\Omega'^2 X_- X_+\left(1-\frac{\Omega}{\Omega'}\right) + \hbar\Omega'\right], \qquad (B1)$$

where Ω' is substituted for Ω in the definitions of l, $X_\pm$, and $\pi_\pm$. The energy eigenstates of Eq.

(B1) are the same as before, but the eigenvalues are now $\{(N+\frac{1}{2})\hbar\Omega' - m\hbar[\frac{1}{2}(\Omega'-\Omega)]\}$. The coherent states are also the same as before except that both α and ξ will depend on time. Thus

$$\alpha(t)=\alpha(0)\exp\{-i[\tfrac{1}{2}(\Omega'+\Omega)]t\},$$

$$\xi(t)=\xi(0)\exp\{-i[\tfrac{1}{2}(\Omega'-\Omega)]t\}.$$

The wave packet will therefore follow the classical motion.

To calculate $Z_\perp$ in this case, one may take all the limits of integration from 0 to ∞. The partition function is then

$$Z_\perp=\left(4\sinh\frac{\hbar(\Omega'-\Omega)}{4kT}\sinh\frac{\hbar(\Omega'+\Omega)}{4kT}\right)^{-1} .$$

For $\omega_0 \ll \Omega$, $Z_\perp$ becomes

$$Z_\perp=\left(\frac{2\hbar\omega_0^2}{kT\,\Omega}\sinh\frac{\hbar\Omega}{2kT}\right)^{-1} .$$

Except for a proportionality constant, this equation is the same as Eq. (4.8) and therefore leads to the same magnetization and susceptibility.

---

[1]R. J. Glauber, Phys. Rev. 131, 2766 (1963).

[2]W. Louisell, *Radiation and Noise in Quantum Electronics* (McGraw-Hill, New York, 1964).

[3]P. Carruthers and M. N. Nieto, Rev. Mod. Phys. 40, 411 (1968).

[4]After the completion of this manuscript, it came to the attention of the authors that coherent states of the same type have been constructed by I. A. Malkin and V. I. Man'ko, Zh. Eksperim. i Teor. Fiz. 55, 1014 (1968) [Soviet Phys. JETP 28, 527 (1969)].

[5]G. C. Darwin, Proc. Roy. Soc. (London) 117, 258 (1928).

[6]M. H. Johnson and B. A. Lippmann, Phys. Rev. 76, 828 (1949).

[7]L. D. Landau, Z. Physik 64, 629 (1930).

[8]R. Kubo, S. J. Miyake, and N. Hashitsume, Solid State Phys. 17, 269 (1965).

[9]A. H. Wilson, *The Theory of Metals* (Cambridge U. P., Cambridge, England, 1953), pp. 160–172.

[10]C. Kittel, *Elementary Statistical Physics* (Wiley, New York, 1958), p. 56.

[11]V. Fock, Z. Physik 47, 446 (1928).

[12]F. Bloch, Z. Physik 74, 295 (1932).

Journal of the Physical Society of Japan, Vol. 38, No. 3, March, 1975

Spin Coherent State Representation in Non-Equilibrium Statistical Mechanics

Yoshinori Takahashi and Fumiaki Shibata

Department of Physics, University of Tokyo, Hongo, Bunkyo-ku, Tokyo

(Received September 19, 1974)

A new method of treating problems of spin systems is developed, in which spin operators are transformed into two kinds of boson operators according to Schwinger's method. Usual treatment of coherent state representation is generalized to spin systems with a fixed magnitude S. Some basic properties and the relations to the atomic coherent state are investigated rather in details.

The method is applied to spin relaxation process and a corresponding Fokker-Planck type equation is obtained and solved to give a multiple relaxation process due to nonlinearities. It is shown that our formulation has several advantages over other ones in some respects.

§ 1. Introduction

The coherent state representation of a bose field plays an important role in the theory of statistical mechanics and solid state physics in non-equilibrium as well as equilibrium cases. Indeed, after the work of Glauber,[1] many attempts have been done mainly in the field of quantum optics[2] and in many body problem.[3] Usefulness of the coherent state is recognized more clearly when we treat problems in non-equilibrium bose system: for example, a damped harmonic oscillator,[4] photons in laser[5] and anharmonic crystals.[6] In these problems, the Liouville equation for a density matrix is transformed into the corresponding equation for a quasi-probability function which describes the temporal evolution of the system. Sometimes such an equation is reduced to the Fokker-Planck type and becomes accordingly more tractable than the original operator equation for the density matrix.

On the other hand, in spite of its long history of spin system, the corresponding coherent state representation has not been considered explicitly until the appearance of Radcliffe's paper.[7] Some properties of this coherent state are examined by Arecchi *et al.*[8] in more details. It is called the atomic coherent state or Bloch state. Although there is a great similarity between the atomic coherent state and the usual coherent state representation of bosons, one notices some complexities in the atomic coherent state representation in contrast to the remarkable simplicity of the other.

The purpose of this paper is to present a new method of coherent state representation for spin systems which is simpler and more transparent than that of the atomic coherent state. This has also the advantage that known properties of the boson coherent state can be used in the formulation. We employ a boson-representation of spins introduced by Schwinger[9] where spin operators are replaced by two kinds of bosons and magnitude of a spin is determined by fixing total number of bosons. A coherent state is defined as a direct product of each coherent state for the two kinds of bosons. We call hereafter this representation a *spin coherent state representation*. Atkins and Dobson[10] have made a similar attempt but their spin space is too vast to be related to the atomic coherent states.

An important problem in non-equilibrium statistical mechanics is the spin relaxation; many efforts have been devoted to this area of investigation.[11] It is an important phenomenon because we can draw many informations about a physical system by observing the spin relaxation process, and also because it is essentially homologous to some other interesting physical systems, for example, superradiance[12] and a matter system of laser.

The delicate problem of equilibrium destination in spin relaxation was investigated by Kubo and Hashitsume[13] phenomenologically and its microscopic justification was given by Kawabata[14] in classical cases and by Shibata and Saito[15] in quantal cases. The same problem will be examined from a somewhat different point of view in this paper.

Starting with a formulation and investigation of properties of the spin coherent state representation in terms of bosons in § 2, its relation to the atomic coherent state is discussed in §3. In § 4, after a brief derivation of an equation of motion for a damped spin, the equation for the density matrix is transformed into that of a quasi-probability function and the non-linear relaxation processes are treated. In the final section some discussions are presented in comparison with a method which is based on the atomic coherent state representation (Appendix).

§ 2. Basic Formulation

As is seen later in this section, some analogy exists between the coherent state representation and the spin coherent state representation. It is convenient to summarize first some properties of the former.

2.1 Boson coherent state representation

A coherent state is defined as an eigenstate of the boson annihilation operator a:[1)]

$$a|\alpha\rangle=\alpha|\alpha\rangle\ , \tag{2.1}$$

where α is a complex eigenvalue; a is conjugate to the creation operator $a^\dagger$,

$$[a,\ a^\dagger]=1\ .$$

The state $|\alpha\rangle$ is expressed in terms of number states, $|n\rangle$, as

$$|\alpha\rangle=\sum_n\frac{\alpha^n}{\sqrt{n!}}e^{-|\alpha|^2/2}|n\rangle\ , \tag{2.2}$$

which can be rewritten in the form

$$|\alpha\rangle=e^{\alpha a^\dagger-\alpha^* a}|0\rangle\ . \tag{2.3}$$

These coherent states are not orthogonal; but satisfy the condition;

$$\langle\alpha|\beta\rangle=e^{(\alpha^*\beta-|\alpha|^2/2-|\beta|^2/2)}\ . \tag{2.4}$$

Instead, we have a completeness relation represented by

$$\int\frac{d^2\alpha}{\pi}|\alpha\rangle\langle\alpha|=1\ , \tag{2.5}$$

where

$$d^2\alpha=d(\mathscr{Re}\,\alpha)d(\mathscr{Im}\,\alpha)\ .$$

An arbitrary state $|f\rangle$ is expanded in the form,

$$|f\rangle=\sum_n C_n|n\rangle=\int\frac{d^2\alpha}{\pi}e^{-(|\alpha|^2/2)}f(\alpha^*)|\alpha\rangle\ , \tag{2.6}$$

where C_n's are complex numbers given by $\langle n|f\rangle$ and $f(\alpha^*)$ represented by

$$f(\alpha^*)=\sum_n C_n\frac{(\alpha^*)^n}{\sqrt{n!}}=e^{|\alpha|^2/2}\langle\alpha|f\rangle\ ,$$

is an analytic function in the α^*-plane. Using (2.6), one obtains for an overlap of two arbitrary states $|f\rangle$, and $|g\rangle$ as

$$\langle g|f\rangle=\int\frac{d^2\alpha}{\pi}e^{-|\alpha|^2}g^*(\alpha^*)f(\alpha^*)\ , \tag{2.7}$$

where $f(\alpha^*)$ and $g(\alpha^*)$ are equal to

$$e^{|\alpha|^2/2}\langle\alpha|f\rangle\ ,$$

and

$$e^{|\alpha|^2/2}\langle\alpha|g\rangle\ ,$$

respectively.

An arbitrary operator A can be expressed as follows,

$$\begin{aligned}A&=\int\frac{d^2\alpha}{\pi}\frac{d^2\beta}{\pi}|\alpha\rangle e^{\alpha^*\beta-|\alpha|^2/2-|\beta|^2/2}A(\alpha^*,\beta)\langle\beta|\\&=\int\frac{d^2\alpha}{\pi}\frac{d^2\beta}{\pi}|\alpha\rangle e^{-|\alpha|^2/2-|\beta|^2/2}\bar{A}(\alpha^*,\beta)\langle\beta|\ ,\end{aligned} \tag{2.8}$$

where we have defined $A(\alpha^*,\beta)$ and $\bar{A}(\alpha^*,\beta)$ by

$$\begin{aligned}\langle\alpha|A|\beta\rangle&=e^{\alpha^*\beta-|\alpha|^2/2-|\beta|^2/2}A(\alpha^*,\beta)\\&=e^{-|\alpha|^2/2-|\beta|^2/2}\bar{A}(\alpha^*,\beta)\ .\end{aligned} \tag{2.9}$$

Moreover, a matrix element of a product of two operators A and B is given by

$$\begin{aligned}&\langle\alpha|AB|\beta\rangle\\&=e^{\alpha^*\beta-|\alpha|^2/2-|\beta|^2/2}A\left(\alpha^*,\frac{\partial}{\partial\gamma^*}+\beta\right)B(\gamma^*,\beta)|_{\gamma^*=\alpha^*}\\&=e^{-|\alpha|^2/2-|\beta|^2/2}A\left(\alpha^*,\frac{\partial}{\partial\gamma^*}\right)\bar{B}(\gamma^*,\beta)|_{\gamma^*=\alpha^*}\\&=e^{\alpha^*\beta-|\alpha|^2/2-|\beta|^2/2}B\left(\frac{\partial}{\partial\gamma}+\alpha^*,\beta\right)A(\alpha^*,\gamma)|_{\gamma=\beta}\\&=e^{-|\alpha|^2/2-|\beta|^2/2}B\left(\frac{\partial}{\partial\gamma},\beta\right)\bar{A}(\alpha^*,\gamma)|_{\gamma=\beta}\ .\end{aligned} \tag{2.10}$$

To derive these expressions, we insert the unit matrix

$$1=\int\frac{d^2\gamma}{\pi}|\gamma\rangle\langle\gamma|\ ,$$

between the operators A and B and used the following integrals;

$$\begin{aligned}&\int\frac{d^2\beta}{\pi}e^{\alpha^*\beta-|\beta|^2}f(\beta^*)=f(\alpha^*)\ ,\\&\int\frac{d^2\beta}{\pi}e^{\alpha\beta^*-|\beta|^2}f(\beta)=f(\alpha)\ ,\end{aligned} \tag{2.11}$$

which hold for an arbitrary analytic function $f(\alpha)$. Using the properties mentioned above, an equation of motion for an arbitrary operator is obtained.

Let the Hamiltonian of a system be $\mathscr{H}$. Then the Heisenberg equation of motion,

$$i\hbar\dot{A}=[A,\ \mathscr{H}]\ , \tag{2.12}$$

is read as

$$i\hbar\dot{A}(\alpha^*,\beta)=\left\{\mathscr{H}\left(\frac{\partial}{\partial\gamma}+\alpha^*,\beta\right)-\mathscr{H}\left(\alpha^*,\frac{\partial}{\partial\delta^*}+\beta\right)\right\}A(\delta^*,\gamma)|_{\substack{\gamma=\beta\\ \delta^*=\alpha^*}},$$

or

$$i\hbar\dot{\bar{A}}(\alpha^*,\beta)=\left\{\mathscr{H}\left(\frac{\partial}{\partial\gamma},\beta\right)-\mathscr{H}\left(\alpha^*,\frac{\partial}{\partial\delta^*}\right)\right\}\bar{A}(\delta^*,\gamma)|_{\substack{\gamma=\beta\\ \delta^*=\alpha^*}}, \tag{2.13}$$

and for the Liouville equation,

$$i\hbar\dot{\rho}=[\mathscr{H},\rho],$$

we have

$$i\hbar\dot{\rho}(\alpha^*,\beta)=\left\{\mathscr{H}\left(\alpha^*,\frac{\partial}{\partial\delta^*}+\beta\right)-\mathscr{H}\left(\frac{\partial}{\partial\gamma}+\alpha^*,\beta\right)\right\}\rho(\delta^*,\gamma)|_{\substack{\delta^*=\alpha^*\\ \gamma=\beta}},$$

or

$$i\hbar\dot{\bar{\rho}}(\alpha^*,\beta)=\left\{\mathscr{H}\left(\alpha^*,\frac{\partial}{\partial\delta^*}\right)-\mathscr{H}\left(\frac{\partial}{\partial\gamma},\beta\right)\right\}\bar{\rho}(\delta^*,\gamma)|_{\substack{\delta^*=\alpha^*\\ \gamma=\beta}}. \tag{2.14}$$

A quasi-probability function $f(\alpha^*,\alpha,t)$ is defined by the diagonal part of $\rho(\alpha^*,\beta,t)$, i.e.,

$$f(\alpha^*,\alpha,t)=\rho(\alpha^*,\alpha,t), \tag{2.15}$$

which is normalized as

$$\int\frac{d^2\alpha}{\pi}f(\alpha^*,\alpha,t)=1. \tag{2.16}$$

A statistical average of an operator A is expressed as

$$\langle A(t)\rangle=\int\frac{d^2\alpha}{\pi}A\left(\alpha^*,\frac{\partial}{\partial\gamma^*}+\alpha\right)f(\gamma^*,\alpha,t)|_{\gamma^*=\alpha^*}. \tag{2.17}$$

Defining two real quantities, q and p through a relation,

$$\alpha=\frac{1}{\sqrt{2\hbar}}(q+ip), \tag{2.18}$$

we obtain

$$\frac{\partial}{\partial\alpha}=\sqrt{\frac{\hbar}{2}}\left(\frac{\partial}{\partial q}-i\frac{\partial}{\partial p}\right), \tag{2.19}$$

and

$$\frac{d^2\alpha}{\pi}=\frac{1}{2\pi\hbar}dqdp. \tag{2.20}$$

It should be remarked that the function $f(q,p,t)$ defined from $f(\alpha^*,\alpha,t)$ using (2.18) is analogous to the Winger distribution function fully discussed by Kubo.[16)]

2.2 Spin coherent state representation

Now let us introduce a coherent state for the spin operators by using Schwinger's theory of angular momentum, in which spin operators are represented by two kinds of bose operators $(a_\uparrow, a_\uparrow^\dagger)$ and $(a_\downarrow, a_\downarrow^\dagger)$:

$$S_+=a_\uparrow^\dagger a_\downarrow,\quad S_-=a_\downarrow^\dagger a_\uparrow,\quad S_0=\frac{1}{2}(a_\uparrow^\dagger a_\uparrow-a_\downarrow^\dagger a_\downarrow), \tag{2.21}$$

which satisfy the usual commutation relations of the form

$$[S_0,S_\pm]=\pm S_\pm,\quad [S_+,S_-]=2S_0. \tag{2.22}$$

A spin coherent state $|\alpha_\uparrow,\alpha_\downarrow\rangle$ in our case is defined as a direct product of two coherent states $|\alpha_\uparrow\rangle$ and $|\alpha_\downarrow\rangle$ which are eigenstates of the annihilation operators $a_\uparrow$ and $a_\downarrow$ respectively. For simplicity's sake, we designate a set of two complex numbers $\{\alpha_\uparrow,\alpha_\downarrow\}$ by $\boldsymbol{\alpha}$ which is often recognized as a vector, and $|\alpha_\uparrow|^2+|\alpha_\downarrow|^2$ by $|\boldsymbol{\alpha}|^2$; by definition one has

$$a_\uparrow|\boldsymbol{\alpha}\rangle=\alpha_\uparrow|\boldsymbol{\alpha}\rangle,\quad a_\downarrow|\boldsymbol{\alpha}\rangle=\alpha_\downarrow|\boldsymbol{\alpha}\rangle. \tag{2.23}$$

We have also

$$\boldsymbol{S}^2=\frac{n}{2}\left(\frac{n}{2}+1\right),\quad S=\frac{n}{2}=\frac{1}{2}(n_\uparrow+n_\downarrow),$$

and

$$S_0=\frac{1}{2}(n_\uparrow-n_\downarrow), \tag{2.24}$$

where

$$n_\uparrow=a_\uparrow^\dagger a_\uparrow,\quad n_\downarrow=a_\downarrow^\dagger a_\downarrow.$$

A simultaneous eigenstate of $\boldsymbol{S}^2$ and S_0, $|Sm)$, is expressed in terms of an eigenstate $|n_\uparrow,n_\downarrow\rangle$ of the number operators $n_\uparrow$ and $n_\downarrow$ as

$$|S, m) = |S+m, S-m\rangle .$$

The coherent state $|\boldsymbol{\alpha}\rangle$ can be expressed as a superposition of the states $|Sm)$:

$$|\boldsymbol{\alpha}\rangle = \sum_{S,m} \exp\left(-\frac{|\boldsymbol{\alpha}|^2}{2}\right) \frac{(\alpha_\uparrow)^{S+m}(\alpha_\downarrow)^{S-m}}{\sqrt{(S+m)!(S-m)!}} |S, m) , \tag{2.25}$$

where the summation is carried out for all the value of S, and m takes $2S+1$ values from $-S$ to S.

The magnitude of spin S is equal to the sum of two kinds of bosons $(n_\uparrow + n_\downarrow)/2$. Thus we have to fix the total number of bosons in order to treat problems in a subspace defined by a given S. For this purpose, a generating operator $P(\zeta)$ defined by

$$P(\zeta) = \sum_S \zeta^{2S} P_S , \tag{2.26}$$

is introduced in which P_S's are projection operators whose matrix elements are represented by

$$\langle n_\uparrow, n_\downarrow | P_S | n'_\uparrow, n'_\downarrow \rangle = \delta_{n_\uparrow + n_\downarrow, 2S} \delta_{n_\uparrow, n'_\uparrow} \delta_{n_\downarrow, n'_\downarrow} . \tag{2.27}$$

A matrix element of the operator $P(\zeta)$ between the coherent states is given by

$$\langle \boldsymbol{\alpha} | P(\zeta) | \boldsymbol{\beta} \rangle = \exp\left\{ \zeta \boldsymbol{\alpha}^* \cdot \boldsymbol{\beta} - \frac{|\boldsymbol{\alpha}|^2}{2} - \frac{|\boldsymbol{\beta}|^2}{2} \right\} . \tag{2.28}$$

For a product of an arbitrary operator A and $P(\zeta)$, we have

$$\langle \boldsymbol{\alpha} | P(\zeta) A | \boldsymbol{\beta} \rangle = \exp\left(-\frac{|\boldsymbol{\alpha}|^2}{2} - \frac{|\boldsymbol{\beta}|^2}{2} \right) \bar{A}(\zeta \boldsymbol{\alpha}^*, \boldsymbol{\beta}) . \tag{2.29}$$

For an operator A, the functions $A(\boldsymbol{\alpha}^*, \boldsymbol{\beta})$ and $\bar{A}(\boldsymbol{\alpha}^*, \boldsymbol{\beta})$ are defined in a similar way as in (2.9); one can also verify that all the formulae for the boson coherent states are available for spin coherent states if a complex number α is replaced by a vector $\boldsymbol{\alpha}$ and therefore, we obtain the equation of motion of a dynamical variable A:

$$i\hbar \dot{A}(\boldsymbol{\alpha}^*, \boldsymbol{\alpha}) = \left\{ \mathscr{H}\left(\frac{\partial}{\partial \boldsymbol{\beta}} + \boldsymbol{\alpha}^*, \boldsymbol{\alpha} \right) - \mathscr{H}\left(\boldsymbol{\alpha}^*, \frac{\partial}{\partial \boldsymbol{\gamma}^*} + \boldsymbol{\alpha} \right) \right\} A(\boldsymbol{\gamma}^*, \boldsymbol{\beta})|_{\substack{\boldsymbol{\gamma}^* = \boldsymbol{\alpha}^* \\ \boldsymbol{\beta} = \boldsymbol{\alpha}}} , \tag{2.30}$$

and for the density matrix,

$$i\hbar \dot{\rho}(\boldsymbol{\alpha}^*, \boldsymbol{\alpha}) = \left\{ \mathscr{H}\left(\boldsymbol{\alpha}^*, \frac{\partial}{\partial \boldsymbol{\gamma}^*} + \boldsymbol{\alpha} \right) - \mathscr{H}\left(\frac{\partial}{\partial \boldsymbol{\beta}} + \boldsymbol{\alpha}^*, \boldsymbol{\alpha} \right) \right\} \rho(\boldsymbol{\gamma}^*, \boldsymbol{\beta})|_{\substack{\boldsymbol{\gamma}^* = \boldsymbol{\alpha}^* \\ \boldsymbol{\beta} = \boldsymbol{\alpha}}} . \tag{2.31}$$

Statistical average is given by

$$\langle A(t) \rangle = \int \frac{d^2 \boldsymbol{\alpha}}{\pi^2} A\left(\boldsymbol{\alpha}^*, \frac{\partial}{\partial \boldsymbol{\beta}^*} + \boldsymbol{\alpha} \right) f(\boldsymbol{\beta}^*, \boldsymbol{\alpha})|_{\boldsymbol{\beta}^* = \boldsymbol{\alpha}^*} . \tag{2.32}$$

We can draw informations in the space of fixed S by extracting a term with a coefficient ζ^{2S} of $\langle P(\zeta) A \rangle$ with the aid of the formula (2.32).

Before concluding this section, transformation properties under three dimensional rotations of the state $|\boldsymbol{\alpha}\rangle$ are briefly remarked. It can be verified that the quantity $\boldsymbol{\alpha}$ transforms as a spinor under the rotational operation: in other words, one has

$$e^{i\theta S_\mu} \| \boldsymbol{\alpha} \rangle = \| e^{i\theta(\sigma_\mu/2)} \boldsymbol{\alpha} \rangle , \qquad \| \boldsymbol{\alpha} \rangle = e^{|\alpha|^2/2} | \boldsymbol{\alpha} \rangle , \tag{2.33}$$

where σ_μ is the Pauli matrix corresponding to the component of spin operator S_μ. An arbitrary rotation, R, is composed of successive rotations characterized by Eulerian angles ϕ, θ and ψ:

$$R = e^{-i\phi S_0} e^{-i\theta S_y} e^{-i\psi S_0} , \tag{2.34}$$

which transforms $|\boldsymbol{\alpha}\rangle$ according to a relation of the form

$$R \| \boldsymbol{\alpha} \rangle = \| \mathscr{R} \boldsymbol{\alpha} \rangle , \tag{2.35}$$

where

$$\mathscr{R} = e^{-i\phi(\sigma_z/2)} e^{-i\theta(\sigma_y/2)} e^{-i\psi(\sigma_z/2)} = \begin{pmatrix} e^{-i(\phi+\psi)/2} \cos(\theta/2), & -e^{-i(\phi-\psi)/2} \sin(\theta/2) \\ e^{i(\phi-\psi)/2} \sin(\theta/2), & e^{i(\phi+\psi)/2} \cos(\theta/2) \end{pmatrix} .$$

One can calculate, by using (2.35), a character of the representation of the rotation operator $R(\phi, \theta, \psi)$:

$$\sum \zeta^{2S}\chi^{(S)}(\phi,\theta,\psi)=\int\frac{d^2\boldsymbol{\alpha}}{\pi^2}\langle\boldsymbol{\alpha}|P(\zeta)R|\boldsymbol{\alpha}\rangle=\int\frac{d^2\boldsymbol{\alpha}}{\pi^2}\exp\{-\boldsymbol{\alpha}^*(1-\zeta\mathscr{R})\boldsymbol{\alpha}\}$$

$$=\frac{1}{\det(1-\zeta\mathscr{R})}=\frac{1}{1-2\zeta\cos\frac{1}{2}\theta\cos\frac{1}{2}(\phi+\psi)+\zeta^2}, \tag{2.36}$$

expanding the last expression in the power series of ζ, we have

$$\chi^{(S)}(\phi,\theta,\psi)=\frac{\sin\frac{2S+1}{2}\Omega}{\sin\frac{1}{2}\Omega}, \tag{2.37}$$

where

$$\cos\frac{1}{2}\Omega=\cos\frac{1}{2}\theta\cos\frac{1}{2}(\phi+\psi)\,.$$

§ 3. Relations to the Atomic Coherent State

In this section relations between our spin coherent state $|\boldsymbol{\alpha}\rangle$ and the atomic coherent state are investigated.

The atomic coherent state is defined by[7,8]

$$|\tau\rangle=\frac{1}{(1+|\tau|^2)^S}e^{\tau S_-}|S)$$

$$=\sum_{m=-S}^{S}\frac{1}{(1+|\tau|^2)^S}\left[\frac{(2S)!}{(S+m)!(S-m)!}\right]^{1/2}\tau^{S-m}|m)\,, \tag{3.1}$$

where $|m)$ is an eigenstate of $\boldsymbol{S}^2$ and S_0 with eigenvalues $S(S+1)$ and m respectively, and τ is a complex number related to the polar and azimuthal angles θ and ϕ through the relation:

$$\tau=e^{i\phi}\tan\frac{\theta}{2}\,.$$

The completeness relation in this case is represented by

$$\frac{2S+1}{\pi}\int\frac{d^2\tau}{(1+|\tau|^2)^2}|\tau\rangle\langle\tau|=1\,, \tag{3.2}$$

and therefore, an arbitrary state $|f\rangle$, given by

$$|f\rangle=\sum_m C_m|m)\,,$$

where C_m's are complex numbers $(m|f\rangle$, is written with the use of the atomic coherent states $|\tau\rangle$ as

$$|f\rangle=\frac{2S+1}{\pi}\int\frac{d^2\tau}{(1+|\tau|^2)^2}\frac{f_a(\tau^*)}{(1+|\tau|^2)^S}|\tau\rangle\,, \tag{3.3}$$

or in terms of the spin coherent state $|\boldsymbol{\alpha}\rangle$ as

$$|f\rangle=\int\frac{d^2\boldsymbol{\alpha}}{\pi^2}\exp\left(-\frac{|\boldsymbol{\alpha}|^2}{2}\right)f_s(\boldsymbol{\alpha}^*)|\boldsymbol{\alpha}\rangle\,. \tag{3.4}$$

In the above expressions we have defined the functions $f_a(\tau^*)$ and $f_s(\boldsymbol{\alpha}^*)$ by

$$f_a(\tau^*)=\sum_m C_m\left[\frac{(2S)!}{(S+m)!(S-m)!}\right]^{1/2}(\tau^*)^{S-m}\,, \tag{3.5}$$

and

$$f_s(\boldsymbol{\alpha}^*)=\sum_m C_m\left[\frac{1}{(S+m)!(S-m)!}\right]^{1/2}(\alpha_\uparrow^*)^{S+m}(\alpha_\downarrow^*)^{S-m}\,. \tag{3.6}$$

Comparing (3.5) and (3.6), one can immediately find a relation between the two functions:

$$f_s(\boldsymbol{\alpha}^*)=\frac{(\alpha_\uparrow^*)^{2S}}{[(2S)!]^{1/2}}f_a(\alpha_\downarrow^*/\alpha_\uparrow^*)\,. \tag{3.7}$$

Matrix elements of an arbitrary operator A

$$A=\sum_{m,m'=-S}^{S}C_{m,m'}|m)(m'|\,,$$

are

$$\langle\tau|A|\tau'\rangle=\frac{1}{(1+|\tau|^2)^S(1+|\tau'|^2)^S}\bar{A}_a(\tau^*,\tau')\,, \tag{3.8}$$

in the atomic coherent state representation, and

$$\langle\boldsymbol{\alpha}|A|\boldsymbol{\alpha}'\rangle=\mathrm{e}^{-|\alpha|^2/2-|\alpha'|^2/2}\bar{A}_s(\boldsymbol{\alpha}^*,\boldsymbol{\alpha}')\,, \tag{3.9}$$

in the spin coherent state representation, where we have defined the functions $\bar{A}_a(\tau^*,\tau')$ and $\bar{A}_s(\boldsymbol{\alpha}^*,\boldsymbol{\alpha}')$ by

$$\bar{A}_a(\tau^*,\tau')=\sum_{m,m'}(2S)!C_{m,m'}\left[\frac{1}{(S+m)!(S-m)!(S+m')!(S-m')!}\right]^{1/2}(\tau^*)^{S-m}(\tau')^{S-m'}\,,$$

$$\bar{A}_s(\boldsymbol{\alpha}^*,\boldsymbol{\alpha}')=\sum_{m,m'}C_{m,m'}\left[\frac{1}{(S+m)!(S-m)!(S+m')!(S-m')!}\right]^{1/2}(\alpha_\uparrow^*)^{S+m}(\alpha_\downarrow^*)^{S-m}(\alpha_\uparrow')^{S+m'}(\alpha_\downarrow')^{S-m'}\,. \tag{3.10}$$

Thus they are related each other through the relation,

$$\bar{A}_s(\boldsymbol{\alpha}^*,\boldsymbol{\alpha}')=\frac{(\alpha_\uparrow^*)^{2S}(\alpha_\uparrow')^{2S}}{(2S)!}\bar{A}_a(\alpha_\downarrow^*/\alpha_\uparrow^*,\alpha_\downarrow'/\alpha_\uparrow')\,. \tag{3.11}$$

From (3.7) and (3.11), it is seen that the $(2S+1)$ dimensional space formed by polynomials of degree $2S$ with respect to τ^* in the atomic coherent state representation corresponds to the space of homogeneous functions of degree $2S$ with respect to $\alpha_\uparrow^*$ and $\alpha_\downarrow^*$ in the spin coherent state representation.

In order to show these relations more clearly, we introduce the action and phase variables I, θ, ϕ and ψ:

$$\alpha_\uparrow=\sqrt{I}\cos\frac{\theta}{2}\,\mathrm{e}^{-i(\phi+\psi)/2}$$

$$\alpha_\downarrow=\sqrt{I}\sin\frac{\theta}{2}\,\mathrm{e}^{i(\phi-\psi)/2}\,. \tag{3.12}$$

The factor $\mathrm{e}^{-i\psi/2}$, however, plays no role in the following so that it may be neglected. A dynamical variable $A(\boldsymbol{\alpha}^*,\boldsymbol{\alpha})$ and a density matrix $\rho(\boldsymbol{\alpha}^*,\boldsymbol{\alpha})$ are then expressed in terms of I, θ and ϕ. The volume element is transformed into

$$\frac{\mathrm{d}^2\boldsymbol{\alpha}}{\pi^2}=\frac{1}{8\pi^2}I\mathrm{d}I\sin\theta\mathrm{d}\theta\mathrm{d}\phi\mathrm{d}\psi\,,$$

which after integration with respect to ψ reduces to

$$\frac{\mathrm{d}^2\boldsymbol{\alpha}}{\pi^2}=I\mathrm{d}I\frac{\mathrm{d}\Omega}{4\pi}\,, \tag{3.13}$$

where

$$\mathrm{d}\Omega=\sin\theta\mathrm{d}\theta\mathrm{d}\phi\,.$$

Using a generalized version of (2.7), an overlap integral of two arbitrary states $|f\rangle$ and $|g\rangle$ with spin of magnitude S is given by

$$\langle g|f\rangle=\int\frac{\mathrm{d}^2\boldsymbol{\alpha}}{\pi^2}\mathrm{e}^{-|\alpha|^2}g_s^*(\boldsymbol{\alpha}^*)f_s(\boldsymbol{\alpha}^*)\,, \tag{3.14}$$

where $g_s^*(\boldsymbol{\alpha}^*)f_s(\boldsymbol{\alpha}^*)$ is related to the atomic coherent state representation by

$$g_s^*(\boldsymbol{\alpha}^*)f_s(\boldsymbol{\alpha}^*)=\frac{I^{2S}}{(2S)!}\frac{g_a^*(\tau^*)f_a(\tau^*)}{(1+|\tau|^2)^{2S}}\,.$$

Carrying out the integration with respect to I, (3.14) reduces to

$$\langle g|f\rangle=(2S+1)\int\frac{\mathrm{d}\Omega}{4\pi}\frac{g_a^*(\tau^*)f_a(\tau^*)}{(1+|\tau|^2)^{2S}}\,, \tag{3.15}$$

which is the formula corresponding to (2.7) in the atomic coherent state representation first derived by Radcliffe.[7]

For the later purpose, we now transform the derivatives with respect to $\alpha_\uparrow$ and $\alpha_\downarrow$ into those with respect to I, θ and ϕ:

$$\frac{\partial}{\partial\alpha_\uparrow}=\frac{1}{\sqrt{I}}e^{i\phi/2}\sin\frac{\theta}{2}\left[\cot\frac{\theta}{2}\,I\frac{\partial}{\partial I}-\frac{\partial}{\partial\theta}+i\,\frac{1}{\sin\theta}\frac{\partial}{\partial\phi}\right],$$

$$\frac{\partial}{\partial\alpha_\downarrow}=\frac{1}{\sqrt{I}}e^{-i\phi/2}\cos\frac{\theta}{2}\left[\tan\frac{\theta}{2}\,I\frac{\partial}{\partial I}+\frac{\partial}{\partial\theta}-i\,\frac{1}{\sin\theta}\frac{\partial}{\partial\phi}\right]. \tag{3.16}$$

From (3.12) and (3.16) we obtain

$$\alpha_\uparrow\frac{\partial}{\partial\alpha_\uparrow}=\frac{1}{2}\sin\theta\left[\cot\frac{\theta}{2}\,I\frac{\partial}{\partial I}-\frac{\partial}{\partial\theta}+i\,\frac{1}{\sin\theta}\frac{\partial}{\partial\phi}\right],$$

$$\alpha_\downarrow\frac{\partial}{\partial\alpha_\downarrow}=\frac{1}{2}\sin\theta\left[\tan\frac{\theta}{2}\,I\frac{\partial}{\partial I}+\frac{\partial}{\partial\theta}-i\,\frac{1}{\sin\theta}\frac{\partial}{\partial\phi}\right],$$

$$\alpha_\uparrow\frac{\partial}{\partial\alpha_\downarrow}=e^{-i\phi}\cos^2\frac{\theta}{2}\left[\tan\frac{\theta}{2}\,I\frac{\partial}{\partial I}+\frac{\partial}{\partial\theta}-i\,\frac{1}{\sin\theta}\frac{\partial}{\partial\phi}\right],$$

$$\alpha_\downarrow\frac{\partial}{\partial\alpha_\uparrow}=e^{i\phi}\sin^2\frac{\theta}{2}\left[\cot\frac{\theta}{2}\,I\frac{\partial}{\partial I}-\frac{\partial}{\partial\theta}+i\,\frac{1}{\sin\theta}\frac{\partial}{\partial\phi}\right]. \tag{3.17}$$

As far as the space of fixed value of S is concerned, $I(\partial/\partial I)$ can be replaced by $2S$, since any operator in this space is proportional to I^{2S} as is seen from (3.11); for example, we have

$$\alpha_\uparrow\frac{\partial}{\partial\alpha_\uparrow}=\frac{1}{2}\sin\theta\left[2S\cot\frac{\theta}{2}-\frac{\partial}{\partial\theta}+i\,\frac{1}{\sin\theta}\frac{\partial}{\partial\phi}\right];$$

$$\alpha_\uparrow^*\frac{\partial}{\partial\alpha_\downarrow^*}-\alpha_\downarrow\frac{\partial}{\partial\alpha_\uparrow}=e^{i\phi}\left[\frac{\partial}{\partial\theta}+i\cot\theta\frac{\partial}{\partial\phi}\right],$$

$$\alpha_\downarrow^*\frac{\partial}{\partial\alpha_\uparrow^*}-\alpha_\uparrow\frac{\partial}{\partial\alpha_\downarrow}=-e^{-i\phi}\left[\frac{\partial}{\partial\theta}-i\cot\theta\frac{\partial}{\partial\phi}\right],$$

$$\frac{1}{2}\left(\alpha_\uparrow^*\frac{\partial}{\partial\alpha_\uparrow^*}-\alpha_\downarrow^*\frac{\partial}{\partial\alpha_\downarrow^*}-\alpha_\uparrow\frac{\partial}{\partial\alpha_\uparrow}+\alpha_\downarrow\frac{\partial}{\partial\alpha_\downarrow}\right)=\frac{1}{i}\frac{\partial}{\partial\phi}, \tag{3.18}$$

which are usual angular momentum operators L_+, L_- and L_z, respectively. Thus we can also relate the coherent state representation to the usual representation represented by polar angle θ and azimuthal angle ϕ.

Especially for the spin operators, $A(\boldsymbol{\alpha}^*, \boldsymbol{\alpha})$ can be written

$$S_0(\boldsymbol{\alpha}^*, \boldsymbol{\alpha})=\frac{I}{2}\cos\theta, \tag{3.19a}$$

$$S_x(\boldsymbol{\alpha}^*, \boldsymbol{\alpha})=\frac{I}{2}\sin\theta\cos\phi, \tag{3.19b}$$

$$S_y(\boldsymbol{\alpha}^*, \boldsymbol{\alpha})=\frac{I}{2}\sin\theta\sin\phi, \tag{3.19c}$$

and

$$\rho(\boldsymbol{\alpha}^*, \boldsymbol{\alpha})=\frac{I^{2S}}{(2S)!}e^{-I}\bar{f}(\theta, \phi). \tag{3.20}$$

Thus we obtain, for example,

$$\langle S_0\rangle=\frac{2S+1}{4\pi}\int(S+1)\cos\theta\bar{f}(\theta,\phi,t)d\Omega, \tag{3.21}$$

where function $\bar{f}(\theta, \phi)$ satisfies the normalization condition:

$$(2S+1)\int\frac{d\Omega}{4\pi}\bar{f}(\theta,\phi)=1. \tag{3.22}$$

§4. Application to Spin Relaxation Phenomena

We have developed in the preceding sections a theory of spin coherent state representation in terms of boson representation. Let us proceed to apply this formulation to the problem of spin relaxation phenomena idealized to a Brownian motion. As was mentioned in the introduction, a simple stochastic theory can not give the correct answer so that nonlinear friction terms were introduced phenomenologically[13] and their validity has been justified microscopically.[14,15] The same problem is reexamined in this section from a somewhat different point of view.

An equation of motion for the density matrix is represented by the Liouville equation of the form ($\hbar=1$)

$$\dot{\rho}=-i[\mathscr{H}, \rho(t)] \equiv -iL\rho(t), \tag{4.1}$$

which can be decomposed into two parts by the method of damping theory:[17]

$$\mathscr{P}\dot{\rho}(t)=-i\mathscr{P}L\mathscr{P}\rho(t)-i\mathscr{P}L(1-\mathscr{P})\rho(t),$$

and

$$(1-\mathscr{P})\dot{\rho}(t)=-i(1-\mathscr{P})L\mathscr{P}\rho(t)$$
$$-i(1-\mathscr{P})L(1-\mathscr{P})\rho(t)\ ,$$

where the projection operator $\mathscr{P}$ is defined as

$$\mathscr{P}X=\rho_B^{eq}\cdot Z_B^{-1}\mathrm{tr}_B X\ , \tag{4.2}$$

in which we have separated the total system into the relevant spin system (S) and the heat bath (B); ρ_B^{eq} is the equilibrium density matrix for the reservoir, Z_B the corresponding partition function.

Then the reduced density matrix $\sigma(t)$ defined by

$$\sigma(t)=\mathrm{tr}_B\rho(t)\ ,$$

obeys the following equation of motion:

$$\dot{\sigma}(t)=-iL_s'\sigma(t)+\int_0^t d\tau\Phi(t-\tau)\sigma(\tau)+I(t)\ , \tag{4.3}$$

where

$$L_s'=\mathrm{tr}_B L\rho_B^{eq}\cdot Z_B^{-1}\ ,$$
$$\Phi(t)=-\mathrm{tr}_B L\exp[-i(1-\mathscr{P})Lt](1-\mathscr{P})L\rho_B^{eq}\cdot Z_B^{-1},$$

and

$$I(t)=-i\,\mathrm{tr}_B L\exp[-i(1-\mathscr{P})Lt](1-\mathscr{P})\rho(0)\ .$$

These formulae are further simplified in the case where interactions between the system and the reservoir are weak enough so as to allow the Born approximation in evaluating the quantity $\Phi(t)$. In this situation we have

$$\dot{\sigma}(t)=-iL_s'\sigma(t)$$
$$-\int_0^\infty d\tau\mathrm{tr}_B L_{SB}e^{-i(L_S+L_B)\tau}L_{SB}\rho_B^{eq}\cdot Z_B^{-1}\sigma(t)\ , \tag{4.4}$$

where we have also assumed that the correlation time characterizing the heat bath is so short that we can regard the stochastic process as Markoffian.

To apply eq. (4.4) to the system composed of a giant spin under a magnetic field with the heat bath (see ref. 15 for more details) we assume the Hamiltonian

$$\mathscr{H}=\mathscr{H}_S+\mathscr{H}_B+\mathscr{H}_{SB}\ ,$$
$$\mathscr{H}_S=-\omega_0 S_0\ ,\quad (\omega_0=\gamma H_0)$$
$$\mathscr{H}_{SB}=\boldsymbol{R}\cdot\boldsymbol{S}\ ,$$

where $\boldsymbol{R}$ is the effective field acted by the reservoir on the spin. Equation (4.4) is now explicitly written as

$$\begin{aligned}\dot{\sigma}(t)=&i\omega_0[S_0,\sigma(t)]+A^*[S_+,\sigma(t)S_-]\\&+A[S_+\sigma(t),S_-]+B^*[S_-,\sigma(t)S_+]\\&+B[S_-\sigma(t),S_+]+C^*[S_0,\sigma(t)S_0]\\&+C[S_0\sigma(t),S_0]\ ,\end{aligned} \tag{4.5}$$

where we have defined (assuming axial symmetry around z-axis and $\langle\boldsymbol{R}\rangle_B=0$)

$$A=\frac{1}{4}\int_0^\infty\langle R^+(\tau)R^-(0)\rangle_B e^{i\omega_0\tau}d\tau\ , \tag{4.6a}$$
$$B=\frac{1}{4}\int_0^\infty\langle R^-(\tau)R^+(0)\rangle_B e^{-i\omega_0\tau}d\tau\ , \tag{4.6b}$$

and

$$C=\int_0^\infty\langle R^0(\tau)R^0(0)\rangle_B d\tau\ , \tag{4.6c}$$

in which

$$\boldsymbol{R}(\tau)=e^{iL_B\tau}\boldsymbol{R}\ ,$$

and the average is taken over ρ_B^{eq}; the quantities A and B are related each other through the relation ($\rho_B=e^{-\beta\mathscr{H}_B}$)

$$A=e^{\beta\omega_0}B^*\ ,$$

which can be verified by direct calculations of (4.6a) and (4.6b), and represents the detailed balance.

The imaginary part of A, B and C contribute only to shift the frequency ω_0 so that we may forget them hereafter and write real quantities a, b and c for A, B and C.

Now we study some consequence from eq. (4.5), especially the equilibrium solution and the relaxation process, using the corresponding coherent state representation:

$$\begin{aligned}\dot{\bar{f}}=&\frac{i\omega_0}{2}\left[\boldsymbol{\alpha}^*\sigma_z\frac{\partial}{\partial\boldsymbol{\alpha}^*}-\boldsymbol{\alpha}\sigma_z\frac{\partial}{\partial\boldsymbol{\alpha}}\right]\bar{f}+\frac{b}{4}\left[\left(\boldsymbol{\alpha}^*\sigma_+\frac{\partial}{\partial\boldsymbol{\alpha}^*}-\boldsymbol{\alpha}\sigma_-\frac{\partial}{\partial\boldsymbol{\alpha}}\right)\right.\\&\left.\times\left(e^{\beta\omega_0}\boldsymbol{\alpha}\sigma_+\frac{\partial}{\partial\boldsymbol{\alpha}}-\boldsymbol{\alpha}^*\sigma_-\frac{\partial}{\partial\boldsymbol{\alpha}^*}\right)+\mathrm{c.c.}\right]\bar{f}-\frac{c}{4}\left(\boldsymbol{\alpha}^*\sigma_z\frac{\partial}{\partial\boldsymbol{\alpha}^*}-\boldsymbol{\alpha}\sigma_z\frac{\partial}{\partial\boldsymbol{\alpha}}\right)^2\bar{f}\ ,\end{aligned} \tag{4.7}$$

where an unnormalized distribution $\bar{f}(\boldsymbol{\alpha}^*,\boldsymbol{\alpha},t)$ is defined by

$$f(\boldsymbol{\alpha}^*,\boldsymbol{\alpha})=e^{-|\boldsymbol{\alpha}|^2}\bar{f}(\boldsymbol{\alpha}^*,\boldsymbol{\alpha})=\langle\boldsymbol{\alpha}|\sigma(t)|\boldsymbol{\alpha}\rangle\ ,$$

and

$$\sigma_+=\begin{pmatrix}0, & 2\\0, & 0\end{pmatrix},\qquad \sigma_-=\begin{pmatrix}0, & 0\\2, & 0\end{pmatrix}.$$

If we transform variables from $(\boldsymbol{\alpha},t)$ to new ones $(\boldsymbol{\alpha}'=e^{-(i\omega_0/2)\sigma_z t}\boldsymbol{\alpha},t)$ then eq. (4.7) takes the form

$$\dot{g}=\frac{b}{4}\left[\left(\boldsymbol{\alpha}^*\sigma_+\frac{\partial}{\partial\boldsymbol{\alpha}^*}-\boldsymbol{\alpha}\sigma_-\frac{\partial}{\partial\boldsymbol{\alpha}}\right)\left(e^{\beta\omega_0}\boldsymbol{\alpha}\sigma_+\frac{\partial}{\partial\boldsymbol{\alpha}}-\boldsymbol{\alpha}^*\sigma_-\frac{\partial}{\partial\boldsymbol{\alpha}^*}\right)+\text{c.c.}\right]g$$
$$-\frac{c}{4}\left(\boldsymbol{\alpha}^*\sigma_z\frac{\partial}{\partial\boldsymbol{\alpha}^*}-\boldsymbol{\alpha}\sigma_z\frac{\partial}{\partial\boldsymbol{\alpha}}\right)^2 g\,, \tag{4.8}$$

where $g(\boldsymbol{\alpha}'^*, \boldsymbol{\alpha}', t)=\bar{f}(\boldsymbol{\alpha}^*, \boldsymbol{\alpha}, t)$, and we have rewritten $\boldsymbol{\alpha}'$ as $\boldsymbol{\alpha}$.

An equilibrium solution, $g_{eq.}$, satisfies the following conditions because $g_{eq.}$ must be independent of b and c:

$$\left(e^{\beta\omega_0}\boldsymbol{\alpha}\sigma_+\frac{\partial}{\partial\boldsymbol{\alpha}}-\boldsymbol{\alpha}^*\sigma_-\frac{\partial}{\partial\boldsymbol{\alpha}^*}\right)g_{eq.}=0\,,$$

and

$$\left(\boldsymbol{\alpha}^*\sigma_z\frac{\partial}{\partial\boldsymbol{\alpha}^*}-\boldsymbol{\alpha}\sigma_z\frac{\partial}{\partial\boldsymbol{\alpha}}\right)g_{eq.}=0\,, \tag{4.9}$$

which can be solved easily: The solution subject to the boundary condition,

$$[g_{eq.}(\boldsymbol{\alpha}^*, \boldsymbol{\alpha})]_{\beta=0}=\exp(|\boldsymbol{\alpha}|^2)\,,$$

is given by

$$g_{eq.}(\boldsymbol{\alpha}^*, \boldsymbol{\alpha})=\exp(e^{(\beta\omega_0/2)}|\alpha_\uparrow|^2+e^{-\beta\omega_0/2}|\alpha_\downarrow|^2)\,,$$

or

$$f_{eq.}(\boldsymbol{\alpha}^*, \boldsymbol{\alpha})=\exp[-(1-e^{\beta\omega_0/2})|\alpha_\uparrow|^2 -(1-e^{-\beta\omega_0/2})|\alpha_\downarrow|^2]\,. \tag{4.10}$$

A generating function of the partition function is calculated as

$$Z(\zeta)=\int e^{-|\boldsymbol{\alpha}|^2}g(\zeta\boldsymbol{\alpha}^*, \boldsymbol{\alpha})\frac{d^2\boldsymbol{\alpha}}{\pi^2}\,,$$
$$=[(1-\zeta e^{\beta\omega_0/2})(1-\zeta e^{-\beta\omega_0/2})]^{-1}\,,$$
$$=\sum_{2S}\zeta^{2S}\left(\sinh\frac{2S+1}{2}\beta\omega_0\right)/\sinh\frac{\beta\omega_0}{2}\,. \tag{4.11}$$

Picking up the coefficient of ζ^{2S}, we have the partition function of a spin with a fixed value of S;

$$Z_S=\left(\sinh\frac{2S+1}{2}\beta\omega_0\right)/\sinh\frac{\beta\omega_0}{2}\,,$$

thereby obtaining magnetization in the form

$$\langle S_0\rangle=\frac{\partial}{\partial\beta\omega_0}\ln Z_S=\frac{2S+1}{2}\coth\left(\frac{2S+1}{2}\beta\omega_0\right)-\frac{1}{2}\coth\frac{\beta\omega_0}{2}$$
$$\equiv SB_S(\beta\omega_0 S)\,, \tag{4.12}$$

which is a well known result.

Time dependent solutions of eq. (4.8) give us informations about relaxation process. Equation (4.8) can be solved as follows. We put

$$g=\exp h\,, \tag{4.13}$$

where h is an analytic function with respect to its variables. Then eq. (4.8) is read in the form

$$\dot{h}=b\left[\left(\alpha_\uparrow^*\frac{\partial}{\partial\alpha_\uparrow^*}-\alpha_\downarrow\frac{\partial}{\partial\alpha_\uparrow}\right)\left(e^{\beta\omega_0}\alpha_\uparrow\frac{\partial}{\partial\alpha_\downarrow}-\alpha_\uparrow^*\frac{\partial}{\partial\alpha_\uparrow^*}\right)+\text{c.c.}\right]h$$
$$-\frac{c}{4}\left(\boldsymbol{\alpha}^*\sigma_z\frac{\partial}{\partial\boldsymbol{\alpha}^*}-\boldsymbol{\alpha}\sigma_z\frac{\partial}{\partial\boldsymbol{\alpha}}\right)^2 h+b\left[\left(\alpha_\uparrow^*\frac{\partial h}{\partial\alpha_\uparrow^*}-\alpha_\downarrow\frac{\partial h}{\partial\alpha_\uparrow}\right)\right.$$
$$\left.\times\left(e^{\beta\omega_0}\alpha_\uparrow\frac{\partial h}{\partial\alpha_\downarrow}-\alpha_\downarrow^*\frac{\partial h}{\partial\alpha_\uparrow^*}\right)+\text{c.c.}\right]-\frac{c}{4}\left(\boldsymbol{\alpha}^*\sigma_z\frac{\partial h}{\partial\boldsymbol{\alpha}^*}-\boldsymbol{\alpha}\sigma_z\frac{\partial h}{\partial\boldsymbol{\alpha}}\right)^2. \tag{4.14}$$

Because h can be expanded as a power series in $\boldsymbol{\alpha}$ and $\boldsymbol{\alpha}^*$;

$$h=A_1^1(t)\alpha_\uparrow^*\alpha_\uparrow+(A_0^1\alpha_\uparrow^*\alpha_\downarrow+\text{c.c.})+A_{-1}^1\alpha_\uparrow^*\alpha_\downarrow+A^2\alpha^*\alpha^*\alpha\alpha+\cdots\,, \tag{4.15}$$

we obtain equations for $A(t)$'s by substituting this into eq. (4.14). These equations for A's can be solved iteratively from lower A's to higher A's. Equations for the lowest $A(t)$'s are given by

$$\dot{A}_0^1=-[b(1+e^{\beta\omega_0})+c]A_0^1\,, \tag{4.16a}$$

$$\dot{A}_1^1=2b(e^{\beta\omega_0}A_{-1}^1-A_1^1)\,, \tag{4.16b}$$

and

$$\dot{A}_{-1}^1=2b(A_1^1-e^{\beta\omega_0}A_{-1}^1)\,. \tag{4.16c}$$

We see that the quantity $A_1{}^1+A_{-\frac{1}{1}}$ remains constant in time. It must be equal to $2\cosh(\beta\omega_0/2)$ by the requirement that, as $t\to\infty$, the difference $A_1{}^1-A_{-\frac{1}{1}}$ must tend to the equilibrium value $e^{\beta\omega_0/2}-e^{-\beta\omega_0/2}$. Now the solutions are derived as

$$A_0{}^1=de^{-\gamma_2 t}\ ,$$

$$A_1{}^1=e^{\beta\omega_0/2}-\varepsilon e^{-\gamma_1 t}\ ,$$

and

$$A_{-\frac{1}{1}}=e^{-\beta\omega_0/2}+\varepsilon e^{-\gamma_1 t}\ , \tag{4.17}$$

where

$$\gamma_1=2b(e^{\beta\omega_0}+1)\ ,$$

$$\gamma_2=b(e^{\beta\omega_0}+1)+c\ ,$$

and d and ε represent initial conditions. Within the approximation which retain A^{1}'s, averages of the component of spins can be calculated by introducing a generating function defined by

$$Z(i\xi,\zeta)=\mathrm{Tr}P(\zeta)e^{i\xi S_\mu}\sigma(t)\ ,$$
$$=\sum_S \zeta^{2S}Z_S(i\xi)\ , \tag{4.18}$$

from which we have

$$\langle S_\mu(t)\rangle=\frac{\partial}{\partial i\xi}\ln Z_S(i\xi)|_{i\xi=0}\ . \tag{4.19}$$

A generating function can be calculated as

$$Z(i\xi,\zeta)=\int\frac{d^2\boldsymbol{\alpha}}{\pi^2}\exp[\zeta\boldsymbol{\alpha}^* e^{(i\xi/2)\sigma_\mu}A\boldsymbol{\alpha}-|\boldsymbol{\alpha}|^2]\ ,$$
$$=[\det(1-\zeta e^{(i\xi/2)\sigma_\mu}A)]^{-1}\ , \tag{4.20}$$

where matrix A is defined by

$$A=\begin{pmatrix} e^{\beta\omega_0}-\varepsilon e^{-\gamma_1 t}\ , & de^{-\gamma_2 t+i\omega_0 t} \\ d^* e^{-\gamma_2 t-i\omega_0 t}\ , & e^{-\beta\omega_0/2}+\varepsilon e^{-\gamma_1 t}\end{pmatrix}. \tag{4.21}$$

We can evaluate the partition function (4.20) in a more explicit form in the case of $S_\mu=S_0$:

$$Z(i\xi,\zeta)=[(1-\zeta x)(1-\zeta y)]^{-1}=\sum_S \zeta^{2S}\frac{x^{2S+1}-y^{2S+1}}{x-y}\ , \tag{4.22}$$

where

$$x+y=e^{i\xi/2}(e^{\beta\omega_0/2}-\varepsilon e^{-\gamma_1 t})+e^{-i\xi/2}(e^{-\beta\omega_0/2}+\varepsilon e^{-\gamma_1 t})\ ,$$

and

$$xy=(e^{\beta\omega_0/2}-\varepsilon e^{-\gamma_1 t})(e^{-\beta\omega_0/2}+\varepsilon e^{-\gamma_1 t})-|d|^2e^{-2\gamma_2 t}\ . \tag{4.23}$$

Use of (4.19) yields

$$\langle S_0\rangle=\frac{\partial}{\partial i\xi}\ln\frac{x^{2S+1}-y^{2S+1}}{x-y}\bigg|_{i\xi=0}=(2S+1)\frac{x^{2S}\frac{\partial x}{\partial i\xi}-y^{2S}\frac{\partial y}{\partial i\xi}}{x^{2S+1}-y^{2S+1}}-\frac{\frac{\partial x}{\partial i\xi}-\frac{\partial y}{\partial i\xi}}{x-y}\bigg|_{i\xi=0}\ , \tag{4.24}$$

where we find by differenciating eqs. (4.23) the following equations to hold:

$$\frac{\partial x}{\partial i\xi}+\frac{\partial y}{\partial i\xi}\bigg|_{i\xi=0}=\sinh\frac{\beta\omega_0}{2}-\varepsilon e^{-\gamma_1 t}\ ,$$

and

$$x\frac{\partial y}{\partial i\xi}+y\frac{\partial x}{\partial i\xi}=0\ , \tag{4.25}$$

and finally we have

$$\frac{\partial x}{\partial i\xi}\bigg|_{i\xi=0}=\frac{x}{x-y}\bigg|_{i\xi=0}\left[\sinh\frac{\beta\omega_0}{2}-\varepsilon e^{-\gamma_1 t}\right],\quad \frac{\partial y}{\partial i\xi}\bigg|_{i\xi=0}=-\frac{y}{x-y}\bigg|_{i\xi=0}\left[\sinh\frac{\beta\omega_0}{2}-\varepsilon e^{-\gamma_1 t}\right],$$

giving the longitudinal relaxation process in the form

$$\langle S_0(t)\rangle=\frac{1}{x-y}\bigg|_{i\xi=0}\left(\sinh\frac{\beta\omega_0}{2}-\varepsilon e^{-\gamma_1 t}\right)\left[(2S+1)\frac{x^{2S+1}+y^{2S+1}}{x^{2S+1}-y^{2S+1}}-\frac{x+y}{x-y}\right]_{i\xi=0}, \tag{4.26a}$$

where

$$x+y|_{i\xi=0}=\mathrm{Tr}A=2\cosh\frac{\beta\omega_0}{2}\ ,$$

$$xy|_{i\xi=0}=\det A=1+2\varepsilon e^{-\gamma_1 t}\sinh\frac{\beta\omega_0}{2}-\varepsilon^2e^{-2\gamma_1 t}-|d|^2e^{-2\gamma_2 t}\ . \tag{4.26b}$$

In an analogous way we can find the corresponding quantities when $S_\mu=S_+$:

$$Z=[(1-\zeta x)(1-\zeta y)]^{-1}\,, \tag{4.27}$$

but in this case x and y are given by

$$x+y=e^{\beta\omega_0/2}+e^{-\beta\omega_0/2}+i\xi d^* e^{-\gamma_2 t-i\omega_0 t}\,,$$

$$xy=1+2\varepsilon e^{-\gamma_1 t}\sinh\frac{\beta\omega_0}{2}-\varepsilon^2 e^{-2\gamma_1 t}-|d|^2 e^{-2\gamma_2 t}\,,$$

and the transverse relaxation in the form

$$\langle S_+(t)\rangle=\frac{1}{x-y}\bigg|_{i\xi=0} d^* e^{-\gamma_2 t-i\omega_0 t}\left[(2S+1)\frac{x^{2S+1}+y^{2S+1}}{x^{2S+1}-y^{2S+1}}-\frac{x+y}{x-y}\right]_{i\xi=0}, \tag{4.28}$$

where x and y are the same as (4.26b). In the final section we shall discuss some features of these results.

Before concluding this section we transform eq. (4.7) in the angle variable representation:

$$\dot{\bar f}=\left[\omega_0\frac{\partial}{\partial\phi}+\frac{1}{\sin\theta}\frac{\partial}{\partial\theta}\left\{\sin\theta(b(e^{\beta\omega_0}+1)+b(e^{\beta\omega_0}-1)\cos\theta)\frac{\partial}{\partial\theta}+2bS(e^{\beta\omega_0}-1)\sin^2\theta\right\}\right.$$
$$\left.+\frac{1}{\sin^2\theta}\{c\sin^2\theta+b(e^{\beta\omega_0}+1)\cos^2\theta+b(e^{\beta\omega_0}-1)\cos\theta\}\frac{\partial^2}{\partial\phi^2}\right]\bar f\,, \tag{4.29}$$

where $\bar f$ is the quasi-probability function and has a property

$$\mathrm{tr}\,A\sigma(t)=\int\frac{d\Omega}{4\pi}A\left(\boldsymbol{\alpha}^*,\frac{\partial}{\partial\boldsymbol{\alpha}^*}\right)\bigg|_{I(\partial|\partial I)\to 2S}\bar f(\theta,\phi)\bigg/\int\bar f(\theta,\phi)\frac{d\Omega}{4\pi}\,.$$

An equilibrium solution of eq. (4.29) is shown to be

$$\bar f_{\mathrm{eq.}}=\left(\cosh\frac{\beta\omega_0}{2}\right)^{2S}\left[1+\frac{1}{2S}2S\tanh\frac{\beta\omega_0}{2}\cos\theta\right]^{2S},$$

which tends to a classical distribution function as $S\to\infty$, $\beta\to 0$ (βS=finite):

$$\bar f_{\mathrm{eq.}}\to\exp\left(\frac{S\omega_0\cos\theta}{kT}\right).$$

In the classical limit, eq. (4.29) takes the form

$$\left(\frac{\partial}{\partial t}-\omega_0\frac{\partial}{\partial\phi}\right)\bar f=\left[\frac{1}{\sin\theta}\frac{\partial}{\partial\theta}\left\{D_\perp\sin\theta\left(\frac{\partial}{\partial\theta}+\beta\omega_0 S\sin\theta\right)\right\}\right.$$
$$\left.+\frac{1}{\sin^2\theta}\frac{\partial}{\partial\phi}(D_\parallel\sin^2\theta+D_\perp\cos^2\theta)\frac{\partial}{\partial\phi}\right]\bar f\,, \tag{4.30}$$

where

$$D_\perp=2b\,,$$
$$D_\parallel=c\,.$$

It can be seen that eq. (4.30) is precisely the same as that obtained by Kubo and Hashitsume[13,18] phenomenologically.

§ 5. Concluding Remarks

In the foregoing sections a new method of spin coherent state representation has been developed and its various properties, relations to the atomic coherent state representation and consequences of its application to spin relaxation process have been investigated in some details. Our method gives a new way of approach to equilibrium as well as non-equilibrium phenomena in spin system and atomic systems.

Our method seems to have some advantages: it is a natural extension of the boson coherent representation to the spin systems and therefore most of familiar formulae of it are generalized to our case. As is demonstrated in the Appendix, the method of the atomic coherent state representation has some complicated features which make it very difficult to obtain, in treating actual problems, an explicit non-equilibrium solution. On the other hand, by our method, we were able to find a solution for relaxation process in which it is shown that both the longitudinal and the transverse relaxations are characterized as multiple relaxation process: This is due to the nonlinearities which are represented explicitly in eq. (4.29). We could, moreover, justify the phenomenolological theory of Kubo and Hashitsume by taking the classical limit of our quantal treatment without introducing any additional approximations.

In the forthcoming papers, we want to apply the method of spin coherent representation to the superradiance phenomena and systems having many variables such as a Heisenberg ferromagnet.

Acknowledgement

The authors are indebted to Professor R. Kubo for valuable discussions and careful reading of the manuscript.

Appendix

In this appendix the method of atomic coherent state, first put forward by Radcliffe, is extended to the non-equilibrium statistical mechanics of the spin relaxation; this problem, set forth by eq. (4.5), generalizes and includes the phenomenon of superradiance whose treatments[19] are free from somewhat complicated our equilibrjum destination problem. It is convenient to start from a Dicke state having $m=S$ when $\omega_0>0$;

$$|\tau\rangle=\frac{1}{(1+|\tau|^2)^S}e^{\tau S_-}|S), \tag{A·1}$$

and hence one has

$$|\tau\rangle\langle\tau|=\frac{1}{(1+|\tau|^2)^{2S}}e^{\tau S_-}|S)(S|e^{\tau^* S_+}, \tag{A·2}$$

from which the following relations are obtained:

$$S_-|\tau\rangle\langle\tau|=\left(\frac{2S\tau^*}{1+|\tau|^2}+\frac{\partial}{\partial\tau}\right)|\tau\rangle\langle\tau| =(|\tau\rangle\langle\tau|S_+)^\dagger . \tag{A·3}$$

By using (A·2) and the relation of the form

$$e^{\tau S_-}S_+e^{-\tau S_-}=S_+-2\tau S_0-\tau^2 S_- ,$$

we have another identity:

$$S_+|\tau\rangle\langle\tau|=\left(\frac{2S\tau}{1+|\tau|^2}-\tau^2\frac{\partial}{\partial\tau}\right)|\tau\rangle\langle\tau|=(|\tau\rangle\langle\tau|S_-)^\dagger . \tag{A·4}$$

In an analogous way, using the relation

$$e^{\tau S_-}S_0e^{-\tau S_-}=S_0+\tau S_- ,$$

we get

$$S_0|\tau\rangle\langle\tau|=\left\{S\frac{1-|\tau|^2}{1+|\tau|^2}-\tau\frac{\partial}{\partial\tau}\right\}|\tau\rangle\langle\tau|=(|\tau\rangle\langle\tau|S_0)^\dagger . \tag{A·5}$$

Introducing diagonal representation of the density matrix,

$$\sigma(t)=\int d^2\tau P(\tau,\tau^*,t)|\tau\rangle\langle\tau| ,$$

we can transform eq. (4.5), with the aid of eqs. (A·3)~(A·5), into the function $P(\tau,\tau^*,t)$ after some lengthy but straightforward calculations

$$\dot{P}(\tau,\tau^*,t)=\left\{[-i\omega_0+2aS-2b(S+1)+c]\frac{\partial}{\partial\tau^*}\tau^*+[i\omega_0+2aS-2b(S+1)+c]\frac{\partial}{\partial\tau}\tau \right.$$

$$\left.+(a+b-c)\left(\frac{\partial^2}{\partial\tau^{*2}}\tau^{*2}+\frac{\partial^2}{\partial\tau^2}\tau^2\right)+2\frac{\partial^2}{\partial\tau^*\partial\tau}(b+c|\tau|^2+a|\tau|^4)\right\}P(\tau,\tau^*,t) , \tag{A·6}$$

which is a Fokker-Planck equation with τ-dependent diffusion terms.

Transforming the variable τ into

$$\tau=re^{i\phi} \quad \left(r=\tan\frac{\theta}{2}\right),$$

eq. (A·6) takes the form

$$\dot{P}(r,\phi,t)=\left\{\omega_0\frac{\partial}{\partial\phi}+\frac{1}{2}\left(ar^2+2c-a-b+\frac{b}{r^2}\right)\frac{\partial^2}{\partial\phi^2}\right.$$
$$+\frac{1}{r}\frac{\partial}{\partial r}\,[2ar^4+(2aS+2a-2bS)r^2]$$
$$\left.+\frac{1}{r}\frac{\partial}{\partial r}\frac{1}{2}\,[ar^5+(a+b)r^3+br]\frac{\partial}{\partial r}\right\}P(r,\phi,t)\,, \qquad (A\cdot7)$$

whose equilibrium solution for r is shown to be

$$P_{eq.}(r)=\int_0^{2\pi}d\phi P_{eq.}(r,\phi)$$
$$=\frac{2(2S+1)e^{\beta\omega_0(2S+1)}(e^{\beta\omega_0}-1)}{e^{\beta\omega_0(2S+1)}-1}\frac{(r^2+1)^{2S}}{(e^{\beta\omega_0}r^2+1)^{2S+2}}\,. \qquad (A\cdot8)$$

This distribution function has a maximum at $r=0$ ($\theta=0$) and monotonously decreases with r^2 irrespective of values of S and $\beta\omega_0$; a distribution of thermal and quantal fluctuations around $\theta=0$ is expressed by (A·8).

Equilibrium properties of physical quantities can be calculated by using (A·8). In particular we have the correct value of $\langle S_0\rangle_{eq.}$:

$$\langle S_0\rangle_{eq.}=\int d^2\tau P_{eq.}(\tau,\tau^*)\langle\tau|S_0|\tau\rangle$$
$$=\frac{S}{2}\int_0^\infty dxP_{eq.}(x)\frac{1-x}{1+x}\qquad(x\equiv r^2)$$
$$=SB_S(\beta\omega_0S)\,.$$

$B_S(x)$ being the Brillouin function as it should be.

This shows that a Brownian spin relaxes in somewhat complicated way towards the correct equilibrium value.

References

1) R.J. Glauber: Phys. Rev. **131** (1963) 2766.
2) P. Carruthers and M. M. Nieto: Rev. mod. Phys. **40** (1968) 411.
3) J.S. Langer: Phys. Rev. **167** (1968) 183; K.H. Douglass: Ann. Phys. **62** (1971) 383.
4) R. Bonifacio and F. Haake: Z. Phys. **200** (1967) 526.
5) H. Haken: *Handbuch der Physik*, XXV/2c, Light and Matter, L. Genzel (Ed.), 1970, Springer Verlag; W. H. Louisell: *Quantum Statistical Properties of Radiation* (John Wiley and Sons, 1973).
6) P. Carruthers and K. S. Dy: Phys. Rev. **147** (1966) 214.
7) J.M. Radcliffe: J. Phys. **A4** (1971) 313.
8) F. T. Arecchi, E. Courtens, R. Gilmore and H. Thomas: Phys. Rev. **A6** (1972) 2211.
9) J. Schwinger: in *Quantum Theory of Angular Momentum*, ed. L. Biedenharn and H. van Dam (Academic Press, New York, 1965).
10) P. W. Atkins and J. C. Dobson: Proc. Roy. Soc. (GB) **A321** (1971) 321.
11) F. Bloch: Phys. Rev. **70** (1946) 460; N. Blömbergen, E.M. Purcell and R.V. Pound: Phys. Rev. **73** (1948) 679; R. Kubo and K. Tomita: J. Phys. Soc. Japan **9** (1954) 888.
12) R. M. Dicke: Phys. Rev. **93** (1954) 493; R. Bonifacio, P. Schwendimann and F. Haake: Phys. Rev. **A4** (1971) 302, **A4** (1971) 854.
13) R. Kubo and N. Hashitsume: Progr. theor. Phys. Suppl. No. 46 (1970) 210.
14) A. Kawabata: Progr. theor. Phys. **48** (1972) 2237.
15) F. Shibata and Y. Saito: preprint.
16) R. Kubo: J. Phys. Soc. Japan **19** (1964) 2127.
17) R. Zwanzig: Physica **30** (1964) 1109; F. Haake: *Springer Tracts in Modern Physics* **66** (1973) 98; R. Kubo: to be published in "*Lectures at Sitges International School of Statistical Mechanics, 1974*," ed. L. Garrido (Springer Verlag).
18) The corrected expression as represented by eq. (4.30) is found in the following literature: N. Hashitsume: in *Tokei Butsurigaku* (Statistical Physics), ed. R. Kubo and M. Toda (Iwanami Pub. Tokyo, 1973) Chap. 7 [in Japanese]
19) R.J. Glauber and F. Haake: preprint; K. Ikeda: private communication.

Note added in proof: After the completion of this work, the following paper came into our attention; G. S. Agarwal: Phys. Rev. **178** (1969) 2025, in which relaxation process of a spin with $S=1/2$ is treated. In this case, nonlinearities disapper.

PHYSICAL REVIEW A VOLUME 29, NUMBER 4 APRIL 1984

Coherent-state Langevin equations for canonical quantum systems with applications to the quantized Hall effect

John R. Klauder
AT&T Bell Laboratories, Murray Hill, New Jersey 07974
(Received 6 October 1983)

The formulation of quantum-statistical-mechanical expectation values as long-time averages involving the solution of associated Langevin equations with complex drift terms is developed. As an example, some applications of this method to study the quantized Hall effect are presented.

I. INTRODUCTION

Statistical methods to study classical and quantum systems are extensively used at the present time with Monte Carlo and Langevin techniques among those more commonly employed.[1] In this paper we elaborate on a method to analyze canonical quantum systems through the study of associated Langevin equations the form of which has been previously announced.[2] As an application we examine some aspects of the two-dimensional (anomolous) quantized Hall effect.[3]

Section II is devoted to a careful and rather complete derivation of the Langevin equation for canonical quantum systems, including a treatment of the exclusion principle for fermions. Essentially all quantum-statistical problems may be formulated as path integrals, most generally expressed in a phase-space formulation. To relate this formulation to Langevin equations it is necessary that all the variables in the path integral assume continuous values, and this is ensured only for coherent-state, phase-space path integrals. Consequently, the first part of Sec. II is devoted to a review and discussion of canonical coherent-state path integrals. Given a path-integral representation of a problem, the introduction of an associated Fokker-Planck equation and a corresponding system of Langevin equations then follows rather standard lines here extended to the case of complex expressions.

In Sec. III the application to the two-dimensional quantized Hall effect is discussed, along with results obtained by a computer solution of the corresponding Langevin equations. While these numerical results generally support the applicability of these methods to complex systems they, unfortunately, do not shed any special new light on the physics of the anamolous quantized Hall effect. Interpretation of our results is aided by an analogous study of a harmonic oscillator, in which the concept of improved approximations naturally arises, and in the study of two uncoupled harmonic oscillators that obey the exclusion principle.

II. DERIVATION OF THE LANGEVIN EQUATIONS

A. Coherent-state and operator properties

For a single degree of freedom, let

$$|p,q\rangle \equiv e^{i(pQ-qP)}|0\rangle\ ,\quad (Q+iP)|0\rangle=0\ ,\quad \langle 0|0\rangle=1\ , \tag{2.1}$$

for all $p,q\in\mathbb{R}^2$ denote the canonical coherent states, where Q and P constitute an irreducible Heisenberg pair, $[Q,P]=i$. These states enjoy a number of interesting and useful properties,[4] the most important of which is the resolution of unity given by

$$1=\int |p,q\rangle\langle p,q|\,\frac{dp\,dq}{2\pi}\ . \tag{2.2}$$

The overlap of two such states reads

$$\langle p_2,q_2|p_1,q_1\rangle=\exp\{\tfrac{1}{2}i(q_2p_1-p_2q_1)-\tfrac{1}{4}[(q_2-q_1)^2+(p_2-p_1)^2]\}\ ,$$

an expression which never vanishes, and as a consequence an operator $\mathscr{H}$ is uniquely determined by its diagonal coherent-state matrix elements.[5] To see this connection assume that $\mathscr{H}$ is expressed in Weyl form,

$$\mathscr{H}=\int \tilde{h}_w(x,k)e^{i(kQ-xP)}\frac{dx\,dk}{2\pi}\ ,$$

where $\tilde{h}_w$ is a uniquely defined distribution associated with $\mathscr{H}$. It follows that

$$\begin{aligned}H(p,q)&\equiv\langle p,q|\mathscr{H}|p,q\rangle\\&=\int \tilde{h}_w(x,k)\langle p,q|e^{i(kQ-xP)}|p,q\rangle\frac{dx\,dk}{2\pi}\\&=\int \tilde{h}_w(x,k)e^{i(kq-xp)}\langle 0|k,x\rangle\frac{dx\,dk}{2\pi}\\&=\int \tilde{h}_w(x,k)e^{i(kq-xp)}e^{-(x^2+k^2)/4}\frac{dx\,dk}{2\pi}\ ,\end{aligned}$$

which shows that H and $\tilde{h}_w$ (hence $\mathscr{H}$) are uniquely correlated. If $h_w(p,q)$ denotes the Fourier transform of $\tilde{h}_w(x,k)$, then it follows that

$$H(p,q)=\int e^{-(p-r)^2-(q-s)^2}h_w(r,s)\frac{dr\,ds}{2\pi}\ ,$$

or stated alternatively,

$$H(p,q)=\exp\left[\frac{1}{4}\left[\frac{\partial^2}{\partial p^2}+\frac{\partial^2}{\partial q^2}\right]\right]h_w(p,q)\ .$$

The operator $\mathscr{H}$ also admits the representation[4]

$$\mathscr{H}=\int h(r,s)|r,s\rangle\langle r,s|\frac{dr\,ds}{2\pi}\ , \tag{2.3}$$

where the weight h may be determined from the relation

$$H(p,q)=\langle p,q\,|\mathscr{H}|\,p,q\rangle$$
$$=\int h(r,s)\,|\langle p,q\,|\,r,s\rangle|^2\frac{dr\,ds}{2\pi}$$
$$=\int e^{-[(p-r)^2+(q-s)^2]/2}h(r,s)\frac{dr\,ds}{2\pi}\,.$$

Thus we find that

$$H(p,q)=\exp\left[\frac{1}{2}\left[\frac{\partial^2}{\partial p^2}+\frac{\partial^2}{\partial q^2}\right]\right]h(p,q)\,,$$

which leads to the relations

$$h(p,q)=\exp\left[-\frac{1}{2}\left[\frac{\partial^2}{\partial p^2}+\frac{\partial^2}{\partial q^2}\right]\right]H(p,q)\,, \quad (2.4a)$$

$$h(p,q)=\exp\left[-\frac{1}{4}\left[\frac{\partial^2}{\partial p^2}+\frac{\partial^2}{\partial q^2}\right]\right]h_w(p,q)\,. \quad (2.4b)$$

For later use these relations are conveniently summarized with respect to the basic Weyl representation

$$\mathscr{H}=\frac{1}{2\pi}\int \tilde{h}_w(x,k)e^{i(kQ-xP)}dx\,dk\,, \quad (2.5a)$$

in the form

$$h_w(p,q)=\frac{1}{2\pi}\int \tilde{h}_w(x,k)e^{i(kq-xp)}dx\,dk\,, \quad (2.5b)$$

$$H(p,q)=\frac{1}{2\pi}\int \tilde{h}_w(x,k)e^{i(kq-xp)-1/4(x^2+k^2)}dx\,dk\,, \quad (2.5c)$$

$$h(p,q)=\frac{1}{2\pi}\int \tilde{h}_w(x,k)e^{i(kq-xp)+1/4(x^2+k^2)}dx\,dk\,. \quad (2.5d)$$

B. Path-integral representation

If we combine (2.2) and (2.3) we learn that

$$1-\epsilon\mathscr{H}=\int[1-\epsilon h(p,q)]\,|p,q\rangle\langle p,q\,|\,\frac{dp\,dq}{2\pi}$$

and thus, with an error which is $O(\epsilon^2)$, that

$$e^{-\epsilon\mathscr{H}}=\int e^{-\epsilon h(p,q)}\,|p,q\rangle\langle p,q\,|\,\frac{dp\,dq}{2\pi}\,. \quad (2.6)$$

If N is an integer chosen such that $\beta\equiv N\epsilon$, then by multiplying (2.6) by itself N times and taking the trace we find that

$$Z\equiv\mathrm{Tr}(e^{-\beta\mathscr{H}})$$
$$=\lim_{N\to\infty}\int\cdots\int\prod_{l=1}^{N}\langle p_{l+1},q_{l+1}\,|\,p_l,q_l\rangle e^{-\epsilon h(p_l,q_l)}$$
$$\times\frac{dp_l dq_l}{2\pi}\,,$$

where $p_{N+1},q_{N+1}\equiv p_1,q_1$, and we have taken the $N\to\infty$ limit to eliminate the $O(\epsilon^2)$ terms. This expression yields a (discrete form of) path-integral representation of the partition function Z, which we may approximate by keeping N large but fixed as

$$Z=\int\cdots\int e^{S(p,q)}d\mu\,,$$

where

$$d\mu\equiv\prod_{l=1}^{N}\frac{dp_l dq_l}{2\pi}\,,$$

and

$$S(p,q)\equiv\sum_{l=1}^{N}[\ln\langle p_{l+1},q_{l+1}\,|\,p_l,q_l\rangle-\epsilon h(p_l,q_l)]$$
$$=\sum_{l=1}^{N}\{\tfrac{1}{2}i(q_{l+1}p_l-p_{l+1}q_l)$$
$$-\tfrac{1}{4}[(q_{l+1}-q_l)^2+(p_{l+1}-p_l)^2]$$
$$-\epsilon h(p_l,q_l)\}\,. \quad (2.7)$$

Improved actions: With some extra effort one may improve the accuracy of (2.6) [and hence of (2.7)] by using the analog of "improved actions."[6] The idea is to replace (2.6) with the expression

$$e^{-\epsilon\mathscr{H}}=\int e^{-\epsilon h_{(n)}(p,q,\epsilon)}\,|p,q\rangle\langle p,q\,|\,\frac{dp\,dq}{2\pi}$$

which is designed to be valid apart from terms of order $O(\epsilon^{n+1})$ for some $n>1$. It is straightforward to determine an expression, at least formally, to give our version of an improved action to any order. First recall the Moyal product formula,

$$h_w^{AB}(p,q)\equiv h_w^A(p,q)e^{D}h_w^B(p,q)\,,$$

to find the Weyl representation for the operator product AB in terms of those for A and B separately. Here

$$D\equiv\tfrac{1}{2}i\left[\frac{\overleftarrow{\partial}}{\partial q}\frac{\overrightarrow{\partial}}{\partial p}-\frac{\overleftarrow{\partial}}{\partial p}\frac{\overrightarrow{\partial}}{\partial q}\right],$$

with the arrows signifying an operation on either the left or right factor. Then it follows that

$$e^{-\epsilon\mathscr{H}}=\sum_{m=0}^{\infty}(m!)^{-1}(-\epsilon)^m\mathscr{H}^m$$
$$\equiv\int\sum_{m=0}^{\infty}(m!)^{-1}(-\epsilon)^m h_{[m]}(p,q)\,|p,q\rangle\langle p,q\,|$$
$$\times\frac{dp\,dq}{2\pi}\,,$$

where

$$h_{[0]}(p,q)\equiv 1\,,$$
$$h_{[1]}(p,q)\equiv h(p,q)\,,$$
$$h_{[m]}(p,q)\equiv\exp\left[-\frac{1}{4}\left[\frac{\partial^2}{\partial p^2}+\frac{\partial^2}{\partial q^2}\right]\right]h_{w[m]}(p,q)\,,$$

and

$$h_{w[m+1]}(p,q)=h_w(p,q)e^{D}h_{w[m]}(p,q)$$

for $m=1,2,\ldots$. Thus to determine an expression correct to $O(\epsilon^n)$, i.e., with an error which is $O(\epsilon^{n+1})$, it suffices to choose

$$-\epsilon h_{(n)}(p,q,\epsilon)\equiv\ln\left[\sum_{m=0}^{n}(m!)^{-1}(-\epsilon)^m h_{[m]}(p,q)\right],$$

or any equivalent expression that differs by terms $O(\epsilon^{n+1})$. By using $h_{(n)}$ rather than h alone in numerical studies, a comparable accuracy should be obtained with fewer thermal-time steps (of order $N^{1/n}$ rather than N). In what follows we shall generally use h, but the formulas hold as well if one substitutes $h_{(n)}$ in place of h.

The usual (Metropolis) importance sampling Monte Carlo procedure to estimate Z is inapplicable since S is not real, i.e., e^S is not everywhere positive as needed if it is to be interpreted as a probability density.

C. Fokker-Planck equation

With the eventual goal of circumventing the nonpositivity of e^S in mind, we next introduce a function $G(p,q,\tau)$ that satisfies a Fokker-Planck-type equation[7,2] given by

$$\frac{\partial G(p,q,\tau)}{\partial\tau}=\frac{1}{2}\sum_{l=1}^{N}\left[\frac{\partial}{\partial q_l}\left[-\frac{\partial S}{\partial q_l}+\frac{\partial}{\partial q_l}\right]+\frac{\partial}{\partial p_l}\left[-\frac{\partial S}{\partial p_l}+\frac{\partial}{\partial p_l}\right]\right]G(p,q,\tau)\,. \tag{2.8}$$

This equation has been chosen so that

$$G(p,q,\tau)=Ce^{S(p,q)}$$

is a stationary solution for any C. Moreover, if $G_0(p,q)=G(p,q,0)$ denotes a smooth, general initial condition for (2.8) normalized so that

$$\int G_0(p,q)d\mu\equiv Z_0\neq 0\,,$$

then we require that the solution to (2.8) satisfies

$$G(p,q,\tau)\to Ce^{S(p,q)}\ \text{as}\ \tau\to\infty\,, \tag{2.9}$$

where $C=Z_0/Z$ is a finite, nonzero proportionality factor. This asymptotic criterion is of central importance to our approach, and we now embark on a detailed analysis of it.

If S were real and locally bounded, then the validity of (2.9) follows from the fact that $\exp[S(p,q)/2]$ is the nondegenerate, zero-energy ground state of a Hamiltonian (see below).[7,2] When S is complex there simply is no general argument of this nature to draw on. Since the harmonic-oscillator Hamiltonian is explicitly soluble, we shall initially discuss this case fully. At the end of this analysis we shall see how the harmonic-oscillator results can be extended to a general Hamiltonian.

Harmonic oscillator: For the oscillator example we choose $\mathscr{H}=\frac{1}{2}(P^2+Q^2-1)$, and thus $h(p,q)=\frac{1}{2}(p^2+q^2-2)$. We initially adopt

$$G_0(p,q)=(2\pi)^N\delta^{(N)}(p-p_0)\delta^{(N)}(q-q_0)\,, \tag{2.10}$$

for arbitrary $p_0=(p_{01},\ldots,p_{0N})$ and $q_0=(q_{01},\ldots,q_{0N})$, as the boundary condition at $\tau=0$. More specifically we choose

$$G_0(p,q,0^+)=\tau^{-N}\exp[-(\underline{r}-\underline{r}_0)^2/2\tau]$$

for $0<\tau\ll 1$, where $\underline{r}\equiv(q,p)$ denotes a $2N$-component vector. Given this initial condition, then the solution to (2.8) appropriate to the harmonic oscillator is necessarily of the form

$$G(p,q,\tau)=D(\tau)\exp[-\tfrac{1}{2}\underline{r}^T\underline{C}(\tau)\underline{r}-\underline{r}^T\underline{B}(\tau)]\,,$$

where $\underline{C}$ is a symmetric matrix, $\underline{B}$ a vector, and D a scalar factor. To satisfy (2.8) these quantities obey the equations of motion

$$\dot{\underline{C}}(\tau)=-\underline{C}^2(\tau)+\tfrac{1}{2}[\underline{A}\,\underline{C}(\tau)+\underline{C}(\tau)\underline{A}]\,, \tag{2.11a}$$

$$\dot{\underline{B}}(\tau)=[\tfrac{1}{2}\underline{A}-\underline{C}(\tau)]\underline{B}(\tau)\,, \tag{2.11b}$$

$$\dot{D}(\tau)=\tfrac{1}{2}\mathrm{Tr}[\underline{A}-\underline{C}(\tau)+\underline{B}(\tau)\underline{B}^T(\tau)]D(\tau)\,, \tag{2.11c}$$

where $\underline{A}$ is a $2N\times 2N$ symmetric matrix determined by the relation

$$S=S_0\equiv-\tfrac{1}{2}\underline{r}^T\underline{A}\,\underline{r}+\text{const}\,.$$

Thus $\underline{A}=\frac{1}{4}\underline{A}_0+\epsilon\underline{I}$, which apart from a multiple (ϵ) of the identity $\underline{I}$, is determined by $\underline{A}_0$, where

$$\underline{A}_0=\left[\begin{array}{ccccc|ccccc}
2 & -1 & & & -1 & 0 & i & & & -i\\
-1 & 2 & -1 & & & -i & 0 & i & & \\
 & -1 & 2 & \ddots & & & -i & 0 & \ddots & \\
 & & \ddots & \ddots & -1 & & & \ddots & \ddots & i\\
-1 & & & -1 & 2 & i & & & -i & 0\\
\hline
0 & -i & & & i & 2 & -1 & & & -1\\
i & 0 & -i & & & -1 & 2 & -1 & & \\
 & i & 0 & \ddots & & & -1 & 2 & \ddots & \\
 & & \ddots & \ddots & -i & & & \ddots & \ddots & -1\\
-i & & & i & 0 & -1 & & & -1 & 2
\end{array}\right].$$

The solution to the equations (2.11) is given by

$$\underline{C}(\tau)=\underline{A}(\underline{I}-e^{-\underline{A}\tau})^{-1}\,,$$

$$\underline{B}(\tau)=\exp\left[\int_1^\tau [\tfrac{1}{2}\underline{A}-\underline{C}(\sigma)]d\sigma\right]\underline{B}_1\,,$$

$$D(\tau)=\exp\left[\tfrac{1}{2}\int_1^\tau \mathrm{Tr}[\underline{A}-\underline{C}(\sigma)+\underline{B}(\sigma)\underline{B}^T(\sigma)]d\sigma\right]D_1\,.$$

As $\tau\to 0^+$ it follows that

$$\underline{C}(0^+)=\underline{I}/\tau\,,$$

$$\underline{B}(0^+)=\underline{B}_0/\tau\,,$$

$$D(0^+)=D_0\tau^{-N}\exp(-\underline{B}_0^T\underline{B}_0/2\tau)\,.$$

On comparison with the initial condition it is clear that $\underline{B}_1$ and D_1 should be chosen so that $\underline{B}_0=\underline{r}_0$ and $D_0=1$; this is certainly possible. On the other hand, as $\tau\to\infty$ the solution converges to a finite, nonzero multiple of $\exp(S_0)$ if and only if

$$e^{-\underline{A}\tau}\to\underline{0}\text{ as }\tau\to\infty\,. \qquad (2.12)$$

Like S_0, the scale factor is independent of $\underline{r}_0$ as determined by the fact that $\int G(p,q,\tau)d\mu$ is independent of τ and has the value unity, its value as $\tau\to 0^+$, independently of $\underline{r}_0$.

Although $\underline{A}$ is not a real symmetric matrix it does have a complete set of eigenvectors and eigenvalues. Specifically, for $\underline{A}_0$, the eigenvalues are given by

$$\lambda_n=1-e^{2\pi i(n-1)/N}\,,\quad n=1,2,\ldots,N\,,$$

each of which is doubly degenerate. Since $\mathrm{Re}\lambda_n\geq 0$, the convergence criterion (2.12) is in fact satisfied, although the rate of convergence is as slow as $e^{-\epsilon\tau}$.

The preceding argument can be extended to any smooth initial condition $G_0(p,q)$ simply by superimposing the results for sharp initial conditions. Consequently, the convergence of a general solution $G(p,q,\tau)$ to a multiple of $\exp[S(p,q)]$ has been established in the case of the harmonic oscillator. We now deduce the same conclusion for a general Hamiltonian.

General Hamiltonians: If we introduce the quantity

$$Y(p,q,\tau)\equiv G(p,q,\tau)e^{-S(p,q)/2}\,,$$

then it follows from (2.8) that Y satisfies the equation

$$\frac{\partial Y(p,q,\tau)}{\partial\tau}=\tfrac{1}{2}(\vec{\nabla}+\tfrac{1}{2}\vec{\nabla}S)\cdot(\vec{\nabla}-\tfrac{1}{2}\vec{\nabla}S)Y(p,q,\tau)\,, \qquad (2.13)$$

where $\vec{\nabla}\equiv(\partial/\partial q,\,\partial/\partial p)$. Stated otherwise, we find that

$$\frac{\partial Y(p,q,\tau)}{\partial\tau}=-\mathsf{H}Y(p,q,\tau)\,,$$

$$\mathsf{H}\equiv-\tfrac{1}{2}\nabla^2+\mathsf{V}\,,$$

where

$$\mathsf{V}=\tfrac{1}{8}(\vec{\nabla}S)^2+\tfrac{1}{4}\nabla^2 S\,.$$

It follows from (2.13) that

$$Y_0(p,q)\equiv e^{S(p,q)/2}$$

is a zero-energy eigenstate of the Hamiltonian H, i.e., $\mathsf{H}Y_0=E_0Y_0$, with $E_0=0$. If S were real and $Y_0\in L^2$, then it follows that Y_0 is the ground state (since it is nowhere vanishing), and the solutions to

$$\mathsf{H}Y_n(p,q)=E_nY_n(p,q)\,,$$

assumed discrete for simplicity, have the property that $E_n>0$, for $n>0$. Thus the solution to (2.13) given by

$$Y(p,q,\tau)=\sum_{n=0}^{\infty}a_nY_n(p,q)e^{-E_n\tau}$$

satisfies the limiting relation

$$\lim_{\tau\to\infty}Y(p,q,\tau)=a_0Y_0(p,q)\,,$$

or alternatively stated, that

$$\lim_{\tau\to\infty}G(p,q,\tau)=a_0e^{S(p,q)}\,,$$

as desired.

When S is complex, as is the case of interest, then V is complex and the general theory of self-adjoint operators is not available to us to reach the desired conclusion. The spectrum of $A+iB$ when A and B are both self-adjoint operators is a nontrivial question with diverse answers. For instance, consider the operator $rQ+iP$, which in the Schrödinger representation is given by $rx+\partial/\partial x$. The solution of the eigenvector equation

$$\left[rx+\frac{\partial}{\partial x}\right]\psi(x)=\mu\psi(x)$$

is given by

$$\psi(x)=Ne^{-rx^2/2+\mu x}\,.$$

For $r>0$, the solutions are all normalizable and yield an eigenfunction for all complex μ (these are just coherent states!), which are not mutually orthogonal. However, if $r<0$, then there are *no* eigenstates, i.e., there is *no* spectrum whatsoever.[8]

In the present case we have [cf. (2.7)]

$$S(p,q)=S_1(p,q)+iS_2(p,q)\,,$$

where S_1 and S_2 are real. But instead let us consider

$$S_\sigma(p,q)\equiv S_1(p,q)+\sigma S_2(p,q)\,,$$

where σ is a complex number. Correspondingly, we introduce

$$\mathsf{H}_\sigma\equiv-\tfrac{1}{2}\nabla^2+\mathsf{V}_\sigma\,,$$

$$\mathsf{V}_\sigma\equiv\tfrac{1}{8}(\vec{\nabla}S_\sigma)^2+\tfrac{1}{4}\nabla^2S_\sigma\,.$$

It is easy to convince oneself that H_σ does have eigenfunctions and eigenvalues (unlike the case $r<0$ above), and that both are analytic functions of σ in a strip near the real axis that includes $\sigma=i$. Moreover, if σ is real, then S_σ is real and the asymptotic condition (2.9) is fulfilled. That means, for σ real, that $E_0=0$ and $E_1\geq E_{1,\min}>0$ for σ bounded, say $|\mathrm{Re}\sigma|\leq K<\infty$; all other eigenvalues are even higher. Now as σ becomes complex, on its way toward $\sigma=i$, E_0 remains zero, while all E_n, $n>0$, generally become complex. Violations of the desired asymptotic condition (2.9) occurs whenever any of the inequalities $\mathrm{Re}E_n > 0$, $n > 0$ is not satisfied. Evidently there is some interval $J = J(K)$, where $|\mathrm{Im}\sigma| \leq J$, for which, by continuity, $\mathrm{Re}E_n > 0$, $n > 0$; whether this interval extends to $J = \infty$ is unknown, but this is not really important. What is surely true and has been borne out in numerical eigenvalue studies carried out by Petersen,[9] is that if the inequalities $\mathrm{Re}E_n > 0$, $n > 0$ are valid for some $\bar{S} \equiv \bar{S}_1 + iS_2$, then these inequalities remain valid for $S = S_1 + iS_2$ whenever $S_1 \leq \bar{S}_1$. Since we have already established the convergence criteria for the harmonic oscillator (or implicitly for a positive multiple thereof), we can choose $\bar{S}_1 = \mathrm{Re}S_0$. Consequently, whenever

$$h(p,q) \geq \alpha(p^2+q^2) + \text{const} ,$$

for some $\alpha > 0$, it follows that $S_1 \leq \bar{S}_1$ and thus $\mathrm{Re}E_n > 0$, for all $n > 0$. This result establishes the desired convergence criterion (2.9) for a general Hamiltonian and concludes our discussion of this point.

D. Langevin equations

Associated with every Fokker-Planck equation is a set of Langevin equations,[10] which in the present case are given by

$$\dot{q}_l(\tau) = \frac{1}{2}\frac{\partial S}{\partial q_l(\tau)} + \xi_l(\tau) , \qquad (2.14a)$$

$$\dot{p}_l(\tau) = \frac{1}{2}\frac{\partial S}{\partial p_l(\tau)} + \eta_l(\tau) , \qquad (2.14b)$$

where $1 \leq l \leq N$, and ξ and η denote two independent sets of standard Gaussian white-noise sources determined by their mean

$$\langle \xi_l(\tau) \rangle = \langle \eta_l(\tau) \rangle = 0$$

and by their variance

$$\langle \xi_l(\tau)\xi_m(\sigma) \rangle = \delta_{lm}\delta(\tau-\sigma) ,$$

$$\langle \eta_l(\tau)\eta_m(\sigma) \rangle = \delta_{lm}\delta(\tau-\sigma) ,$$

$$\langle \xi_l(\tau)\eta_m(\sigma) \rangle = 0 .$$

Here and elsewhere the average $\langle \cdot \rangle$ is with respect to the noise ensemble. We are interested in the solution to these equations for $\tau \geq 0$ subject to the initial conditions

$$q_l(0) = q_{0l} , \quad p_l(0) = p_{0l} ,$$

for $1 \leq l \leq N$. The white-noise sources as well as the initial conditions q_0, p_0 are chosen real, but since S is complex the solution $q(\tau), p(\tau)$ will, in general, be complex, too. For fixed (nonrandom) initial conditions q_0, p_0, the solution of the Langevin equation is related to the solution of the Fokker-Planck equation by the expression

$$\langle F(p(\tau),q(\tau)) \rangle = \int F(p,q)G(p,q,\tau)d\mu ,$$

where $G(p,q,\tau)$ is the solution subject to the initial condition (2.10). Linearity of this expression in the distribution of initial values then extends its validity to any smooth distribution $G_0(p,q)$ normalized so that $\int G_0 d\mu = 1$. Moreover, since G satisfies (2.9) it follows that

$$\lim_{\tau\to\infty} \langle F(p(\tau),q(\tau)) \rangle = \frac{\int F(p,q)e^{S(p,q)}d\mu}{\int e^{S(p,q)}d\mu} . \qquad (2.15)$$

In addition, the convergence criterion (2.9) ensures that the ensemble is also ergodic. Ergodicity means that for almost all solutions of the Langevin equations we have the relation

$$\lim_{T\to\infty} \frac{1}{T}\int_0^T F(p(\tau),q(\tau))d\tau = \frac{\int F(p,q)e^{S(p,q)}d\mu}{\int e^{S(p,q)}d\mu} . \qquad (2.16)$$

If C denotes the right side of this equation and

$$A_T \equiv \frac{1}{T}\int_0^T F(p(\tau),q(\tau))d\tau ,$$

then this limiting behavior may be understood as

$$\lim_{T\to\infty} \left\langle \exp\left[i\int_0^\infty (up - rq)dt \right](A_T - C) \right\rangle = 0$$

for all smooth functions u and r of compact support.

One degree of freedom: For the specific case at hand, where S is given by (2.7), the Langevin equations (2.14) become

$$\dot{q}_l = \frac{i}{4}(p_{l-1} - p_{l+1}) - \frac{1}{4}(2q_l - q_{l+1} - q_{l-1}) - \frac{\epsilon}{2}\frac{\partial h}{\partial q_l} + \xi_l , \qquad (2.17a)$$

$$\dot{p}_l = \frac{i}{4}(q_{l+1} - q_{l-1}) - \frac{1}{4}(2p_l - p_{l+1} - p_{l-1}) - \frac{\epsilon}{2}\frac{\partial h}{\partial p_l} + \eta_l , \qquad (2.17b)$$

for $1 \leq l \leq N$. It is important to assess the stability and sensitivity to the initial conditions of these equations.

Stability is guaranteed whenever the forces $\partial h/\partial q$ and $\partial h/\partial p$ all act attractively toward the origin, at least for large enough arguments. To study sensitivity to the initial conditions let us introduce the quantity

$$R(\tau) \equiv \sum_{l=1}^{N} [\,|q_l(\tau) - \bar{q}_l(\tau)|^2 + |p_l(\tau) - \bar{p}_l(\tau)|^2] ,$$

where q, p, and $\bar{q}, \bar{p}$ denote two solutions of (2.17) determined by different initial conditions, q_0, p_0, and $\bar{q}_0, \bar{p}_0$, but with the *same* noise histories for each solution. It follows from the Langevin equations that

$$\dot{R} = -\tfrac{1}{2}\sum_{l=1}^{N} [\,|(q_l - \bar{q}_l) - (q_{l+1} - \bar{q}_{l+1})|^2 + |(p_l - \bar{p}_l) - (p_{l+1} - \bar{p}_{l+1})|^2] - \epsilon\sum_{l=1}^{N} \mathrm{Re}\left[(q_l - \bar{q}_l)^*\left[\frac{\partial h}{\partial q_l} - \frac{\partial h}{\partial \bar{q}_l}\right] + (p_l - \bar{p}_l)^*\left[\frac{\partial h}{\partial p_l} - \frac{\partial h}{\partial \bar{p}_l}\right]\right] . \qquad (2.18)$$

Like the large-time behavior of the solution of the

Fokker-Planck equation it is difficult to make definitive statements regarding (2.18) for a general Hamiltonian. However, in the case of the harmonic oscillator, where $h=\frac{1}{2}(p^2+q^2-2)$, it follows that

$$\dot{R}=-\tfrac{1}{2}\sum_l[\,|(q_l-\bar{q}_l)-(q_{l+1}-\bar{q}_{l+1})|^2+|(p_l-\bar{p}_l)-(p_{l+1}-\bar{p}_{l+1})|^2]-\epsilon\sum_l[\,|q_l-\bar{q}_l|^2+|p_l-\bar{p}_l|^2]\,.$$

From the general fact

$$0\le|A-B|^2\le 2(|A|^2+|B|^2)\,,$$

we learn that

$$-\epsilon R\ge\dot{R}\ge-(2+\epsilon)R\,,$$

with the solution

$$e^{-\epsilon\tau}R(0)\ge R(\tau)\ge e^{-(2+\epsilon)\tau}R(0)\,.$$

Thus we see that R decays, $R(\tau)\to 0$, establishing the desired asymptotic independence on initial condition, but as noted previously this decay can be rather slow.

Several degrees of freedom: Up to this point we have largely worked as if there was only a single degree of freedom under consideration. If instead there are M distinguishable degrees of freedom, then we need only add a "particle" label and interpret the coherent states as direct products over the M variables as in

$$|p,q\rangle\equiv\bigotimes_{m=1}^{M}|p^m,q^m\rangle\,.$$

It follows directly that the modified form of the Langevin equations (2.17) is given by

$$\dot{q}_l^m=\frac{i}{4}(p_{l-1}^m-p_{l+1}^m)-\frac{1}{4}(2q_l^m-q_{l+1}^m-q_{l-1}^m)-\frac{\epsilon}{2}\frac{\partial h}{\partial q_l^m}+\xi_l^m\,, \tag{2.19a}$$

$$\dot{p}_l^m=\frac{i}{4}(q_{l+1}^m-q_{l-1}^m)-\frac{1}{4}(2p_l^m-p_{l+1}^m-p_{l-1}^m)-\frac{\epsilon}{2}\frac{\partial h}{\partial p_l^m}+\eta_l^m\,, \tag{2.19b}$$

where $1\le l\le N$ and $1\le m\le M$, and all white noises are mutually independent of one another. Here $h=h(p,q)$, which is given by an evident generalization of (2.4) as

$$h(p,q)=\exp\left[-\frac{1}{2}\sum_{m=1}^{M}\left[\frac{\partial^2}{\partial(p^m)^2}+\frac{\partial^2}{\partial(q^m)^2}\right]\right]H(p,q)\,,$$

$$=\exp\left[-\frac{1}{4}\sum_{m=1}^{M}\left[\frac{\partial^2}{\partial(p^m)^2}+\frac{\partial^2}{\partial(q^m)^2}\right]\right]h_w(p,q)\,.$$

Finally we must discuss the case of M nonrelativistic electrons, indistinguishable particles that obey the exclusion principle. A description of electrons may be accomplished by the replacement of the distinguishable particle coherent-state overlap factor

$$\langle p_2,q_2|p_1,q_1\rangle=\prod_{m=1}^{M}\langle p_2^m,q_2^m|p_1^m,q_1^m\rangle$$

by the associated Slater determinant

$$(M!)^{-1}\det(\langle p_2^m,q_2^m|p_1^n,q_1^n\rangle)\,, \tag{2.20}$$

where the determinant is of the $M\times M$ matrix of conventional coherent-state inner products. The factor $(M!)^{-1}$ ensures that this kernel satisfies the integral equation of a projection operator onto the space of antisymmetric functions. To incorporate the exclusion principle it is adequate to replace just one overlap factor in the multiparticle path integral. However, it is analytically equivalent to replace all overlap factors by determinants, and this is the approach we have generally followed in our numerical studies. Consequently the revised form of the function S is given by

$$S=\sum_{l=1}^{N}\{\ln[(M!)^{-1}\det(\langle p_{l+1},q_{l+1}|p_l,q_l\rangle)]-\epsilon h(p_l,q_l)\}\,. \tag{2.21}$$

Guided by our previous discussion we are led to reconsider the set of Langevin equations (2.14) based on the new choice for S, which then becomes

$$\dot{q}_l^m=\frac{1}{2}\frac{\partial}{\partial q_l^m}\ln(\det\langle p_{l+1},q_{l+1}|p_l,q_l\rangle\det\langle p_l,q_l|p_{l-1},q_{l-1}\rangle)-\frac{\epsilon}{2}\frac{\partial h(p_l,q_l)}{\partial q_l^m}+\xi_l^m\,, \tag{2.22a}$$

$$\dot{p}_l^m=\frac{1}{2}\frac{\partial}{\partial p_l^m}\ln(\det\langle p_{l+1},q_{l+1}|p_l,q_l\rangle\det\langle p_l,q_l|p_{l-1},q_{l-1}\rangle)-\frac{\epsilon}{2}\frac{\partial h(p_l,q_l)}{\partial p_l^m}+\eta_l^m\,, \tag{2.22b}$$

for $1\le l\le N$ and $1\le m\le M$. Although our intuition suggests that the long-time average of functions of the solution of these Langevin equations is equivalent to an average of such functions in the normalized distribution based on S as given by (2.21), we have not succeeded in directly proving this result as we did for the case of distinguishable particles. However, a model problem of two uncoupled harmonic oscillators that obey the exclusion principle does give the right results (see Sec. III) and supports the validity of these equations for more general problems.

A new feature appears in the Langevin equations (2.21) that we have not previously encountered. Since the determinant (2.20) can vanish when two (or more) electrons are in the same state, this can lead to singularities in the drift terms of the Langevin equations. However, the nature of these singularities is such as to drive the electrons apart.

An algorithm explaining how to deal with such singularities which is also suitable for numerical studies, has been discussed elsewhere[2] and will not be repeated here. In practice, for the examples studied in Sec. III, no incidence of a singularity or of a near singularity occurred and no special algorithm was required.

III. STUDY OF TWO-DIMENSIONAL QUANTIZED HALL EFFECT

A fascinating development of the past several years has been the discovery and explanation of the ordinary and anamolous two-dimensional quantized Hall effects.[11] The explanation of these effects involves the deformation of the free-electron energy levels into highly degenerate Landau levels as modified by the Coulomb interaction between electrons. We shall (i) concentrate on the low temperature, large magnetic field case in which only the first Landau level need be considered, (ii) eliminate the fast component of electronic motion, and (iii) retain only the Coulomb interaction between electrons within the first Landau level. As shown by Fukuyama and Yoshioka[12] the Hamiltonian operator under these circumstances reduces to

$$\mathscr{H}=\frac{e^2}{2\pi}\sum_{\alpha>\beta}\int\frac{d^2k}{|k|}\exp\left[ilk_1(Q^\alpha-Q^\beta)+ilk_2(P^\alpha-P^\beta)-\frac{k^2l^2}{2}\right].$$

Here $l=\sqrt{c/eH}$ denotes the Larmor radius and α,β refer to particle labels $1\le\alpha,\beta\le M$. By definition of the Weyl representation we have

$$h_w(p,q)=\frac{e^2}{2\pi}\sum_{\alpha>\beta}\int\frac{d^2k}{|k|}\exp\left[ilk_1(q^\alpha-q^\beta)+ilk_2(p^\alpha-p^\beta)-\frac{k^2l^2}{2}\right].$$

Consequently the function h needed for the functional description of Sec. II is given by [cf. (2.4b) and (2.5d)]

$$\begin{aligned}h(p,q)&=\exp\left[-\frac{1}{4}\sum_\gamma\left[\frac{\partial^2}{\partial(p^\gamma)^2}+\frac{\partial^2}{\partial(q^\gamma)^2}\right]\right]h_w(p,q)\\&=\frac{e^2}{2\pi}\sum_{\alpha>\beta}\int\frac{d^2k}{|k|}\exp[ilk_1(q^\alpha-q^\beta)+ilk_2(p^\alpha-p^\beta)]\\&=\frac{e^2}{l}\sum_{\alpha>\beta}\frac{1}{[(q^\alpha-q^\beta)^2+(p^\alpha-p^\beta)^2]^{1/2}},\end{aligned}$$

which is just the Coulomb potential again.

So as to deal with only a finite number of electrons as well as a finite electron density, we replace this potential by a periodic one

$$h(p,q)=\frac{e^2}{l}\left[\sum_{\alpha>\beta}\sum_m\frac{1}{[(q^\alpha-q^\beta+m_1)^2+(p^\alpha-p^\beta+m_2)^2]^{1/2}}+\sum_\alpha w(p^\alpha,q^\alpha)+c\right].\tag{3.1}$$

Here m_i is an integral multiple of B,

$$m_i=\ldots,-2B,-B,0,B,2B,\ldots,$$

where B is the periodic size in both the q and p directions, and w and c denote contributions from the jellium background of positive charge necessary to ensure that the sum defining the periodic potential converges. It may seem natural in this case to replace the sum defining the periodic potential by an Ewald sum,[13] however, since our coordinates q and p can become *complex* (as they appear in the Langevin equation) this procedure is not advisable. Instead we truncate the sum at some reasonable value $\bar{M}$ to approximate the results of an infinite sum so that in practice

$$m_i=-\bar{M}B,-(\bar{M}-1)B,\ldots,\bar{M}B.$$

For most of the Langevin calculations the value $\bar{M}=4$ has been chosen; this choice leads to 81 terms in the "periodic" sum and yields energies within 1% of those given by the full sum.

Closed form expressions for $w(p,q)$ and c are given by $[L\equiv(2\bar{M}+1)B]$

$$\begin{aligned}w(p,q)=\left[\frac{M}{B}\right]\Bigg[&(L+p)\sinh^{-1}\left[\frac{L+q}{L+p}\right]+(L+q)\sinh^{-1}\left[\frac{L+p}{L+q}\right]+(L+p)\sinh^{-1}\left[\frac{L-q}{L+p}\right]\\&+(L-q)\sinh^{-1}\left[\frac{L+p}{L-q}\right]+(L-p)\sinh^{-1}\left[\frac{L+q}{L-p}\right]+(L+q)\sinh^{-1}\left[\frac{L-p}{L+q}\right]\\&+(L-p)\sinh^{-1}\left[\frac{L-q}{L-p}\right]+(L-q)\sinh^{-1}\left[\frac{L-p}{L-q}\right]\Bigg],\end{aligned}$$

$$c=\frac{4M^2L}{B^2}\left[\sinh^{-1}(1)-\frac{\sqrt{2}-1}{3}\right].$$

In practice the equation for w was expanded out to terms of order L^{-3}, an approximation which contributed no more than a 1% error. The so-expanded relations complete the characterization of the energy expression entering the Langevin equations (2.19) and (2.22) used to study this problem. Most of our effort has been directed toward the four-electron problem ($M=4$), while we have also made a few studies for nine electrons ($M=9$, no data presented). We principally have calculated the mean energy

$$\langle\langle \mathscr{H} \rangle\rangle \equiv \mathrm{Tr}(\mathscr{H} e^{-\beta\mathscr{H}})/\mathrm{Tr}(e^{-\beta\mathscr{H}}) \tag{3.2a}$$

according to the rule [cf. (2.16)]

$$\frac{1}{T}\int_0^T h(p(\tau),q(\tau))d\tau = \langle\langle \mathscr{H} \rangle\rangle \tag{3.2b}$$

which holds for large T. Here $p(\tau),q(\tau)$ denote solutions of the $2NM$ coupled, complex Langevin equations (2.19) or (2.22).

The numerical solution of coupled Langevin equations (i.e., of coupled stochastic differential equations) is a problem that has been studied for some time. We have utilized the second-order, Runge-Kutta scheme systematically derived by Helfand and by Greenside and Helfand.[14] This approach has the virtue that not only is the deterministic part of the solution given correctly apart from terms $O(h^3)$, but the first two moments of the stochastic part of the solution are correct save for terms $O(h^3)$ as well. Here h denotes the discrete time step $\Delta\tau$, and we have used both $h=0.1$ and 0.01 in our work. The value of β chosen has varied between 1 and 16, while ϵ has generally been chosen as $\frac{1}{16}$. This means that N, the number of thermal-time steps, has been taken as large as 256. For four particles this has resulted in as many as 2048 coupled, complex Langevin equations to deal with in a single run. The value of T, the upper limit of the τ variable, was frequently taken as large as 100. The most involved runs took up to 40 min of Cray-1 time. A good fraction of this time was spent in evaluating the determinants and the lattice sums involved in computing the forces and energy pertinent to this problem.

We have studied the quantized Hall effect for both distinguishable and indistinguishable electrons. The energy levels are typically lower for the former case, as expected. The energy has been studied as a function of the filling factor ν, which in our notation is given by $\nu=2\pi M/B^2$; when $0\lesssim\nu\lesssim 1$ it represents the fraction of the first Landau level that is filled. Experimentally[3,11] one observes a plateau in the Hall conductivity at $\nu=\frac{1}{3}$ (and certain other fractional values) due to a special ground state and enhanced excitation gap for that value of ν. This behavior is now interpreted as the consequence of a novel quantum liquid which is analytically described by an approximate ground state proposed by Laughlin.[11] At the level our studies were made we are unable to shed any interesting light on the special phenomena that take place at $\nu=\frac{1}{3}$. We shall comment further on this point below.

As a preliminary to a full presentation of our data we first list some selected data for distinguishable particles and discuss the results. In Table I we show typical results of an evaluation (here and elsewhere in units of e^2/l) of

TABLE I. Energy per particle for various averaging times T for the distinguishable particle case. (Parameters chosen so that $\beta=8$, $\epsilon=\frac{1}{16}$, $\nu=\frac{1}{3}$, $h=0.01$, along with an ordered start.)

Averaging time T	Energy per particle
10	−0.354
11	−0.351
12	−0.348
13	−0.345
14	−0.346
15	−0.347
16	−0.349
17	−0.351
18	−0.355
19	−0.359
20	−0.362

the energy per particle

$$\frac{1}{MT}\int_0^T h(p(\tau),q(\tau))d\tau$$

based on (2.19) for several T values. Over the time interval shown there is remarkably little statistical fluctuation in the data. This is probably to be interpreted as due to some very slow relaxation times in the problem. As noted in Sec. II the time constant for the decay of correlations may be as long as ϵ^{-1}, or 16 in our case. From this point of view 20 is not a very large value for T. Hence the resultant value at 20 is still influenced by the initial value at $\tau=0$. Recognizing this fact we have opted to choose our initial configuration at or near to minimum energy classical configurations thus avoiding the prohibitively long relaxation time that starting from a random configuration entails. In Table II we show the results for a run similar to that in Table I except for a random choice of the initial conditions. Here the results are dominated by a large amplitude, under damped component, which is largely absent in the case of an ordered start as in Table I.

TABLE II. Energy per particle for various averaging times T for the distinguishable particle case. (Parameters chosen as in Table I save for a random start rather than an ordered start.)

Averaging time T	Energy per particle
10	−0.324
11	−0.328
12	−0.333
13	−0.340
14	−0.346
15	−0.352
16	−0.358
17	−0.365
18	−0.371
19	−0.378
20	−0.384

TABLE III. Energy per particle for different β and averaging times T for the distinguishable particle case. Compare also Table I. (Parameters as in Table I save for change of thermal time β.)

Averaging time T	Energy per particle $\beta=1$	$\beta=4$	$\beta=12$	$\beta=16$
10	−0.353	−0.354	−0.354	−0.353
11	−0.350	−0.350	−0.351	−0.350
12	−0.347	−0.347	−0.348	−0.347
13	−0.345	−0.345	−0.345	−0.344
14	−0.345	−0.345	−0.346	−0.344
15	−0.346	−0.346	−0.347	−0.344
16	−0.348	−0.348	−0.349	−0.346
17	−0.351	−0.351	−0.351	−0.348
18	−0.355	−0.354	−0.355	−0.352
19	−0.358	−0.358	−0.358	−0.356
20	−0.361	−0.362	−0.362	−0.360

TABLE IV. Average energy for harmonic oscillator for several different β and averaging times T. Correct values are also listed. (Parameters chosen so that $\epsilon=\beta$, $h=0.001$, and vanishing initial values.)

Averaging time T	Energy $\beta=1$	$\beta=2$	$\beta=4$
25.0	0.565	0.106	0.0121
27.5	0.472	0.0978	0.0188
30.0	0.434	0.0879	0.0273
32.5	0.521	0.109	0.0274
35.0	0.612	0.108	0.0285
37.5	0.546	0.0818	0.0196
40.0	0.476	0.0878	0.0282
42.5	0.491	0.0971	0.0280
45.0	0.460	0.111	0.0338
47.5	0.428	0.107	0.0334
50.0	0.379	0.0791	0.0275
Correct value	0.582	0.157	0.0187

In all cases the long time average excludes about 25% of the run in an effort to minimize the effects of the initial conditions.

Another and more striking feature of our results for the average energy is their dependence, or better, their *lack* of dependence on the parameter β in (3.2). Table III shows the results of a run similar in all respects to that of Table I except that the values of β are 1, 4, 12, and 16 (rather than 8 as in Table I). This remarkable insensitivity to changes in β over the range studied could arise for a suitably special density of states. In fact it was hoped to probe the density of states by studying the β dependence of the average energy (for another example, see below). However, it is unlikely that the density of states for the problem at hand is so special as to cause the insensitivity to β that was observed.[15] A more plausible explanation is that the averaging time T was simply inadequate to permit the various features of the Hamiltonian to leave an imprint on the solution to the coupled Langevin equations. Moreover, for small ϵ (large N) the effect of the Hamiltonian may simply get lost in the numerical solution at our level of precision.

Harmonic oscillator: To study further the question of β dependence we have also investigated the elementary example of a single harmonic oscillator for which $M=1$ and $h=\frac{1}{2}(p^2+q^2-2)$. In Table IV we show the average energy at $\beta=1$, 2, and 4 as a function of the averaging time T. The correct value of $1/(e^{\beta}-1)$ is indicated for each case. While these numbers show significant fluctuations they do illustrate that the method we propose here is indeed sensitive to the variable β, at least in principle. In order to obtain more accurate results much longer averaging times are needed. Table V lists the average energy at $\beta=1$ over much longer averaging times and shows a tendency for the values to settle down to the correct value of 0.582. For a harmonic oscillator with unit frequency the relaxation time is approximately unity. In order to achieve an accuracy of 1% in the average energy it is necessary to average over roughly 10^4 independent time units. Thus it is not surprising that it takes $T\simeq 5000$ in order for the average energy to be given to about two significant figures.

It is important to remark that the harmonic-oscillator examples were studied with an improved action. Since it

TABLE V. Average energy for harmonic oscillator for a variety of averaging times T for $\beta=1$ (correct answer equals 0.582). Compare also the first column of Table IV. (Parameters chosen so that $\epsilon=\beta=1$, $h=0.001$ for $T\leq 300$ while $h=0.01$ for $T>300$, and vanishing initial values.)

Averaging time T	Energy	Averaging time T	Energy	Averaging time T	Energy
150	0.514	500	0.562	2500	0.592
165	0.520	550	0.566	2750	0.591
180	0.549	600	0.563	3000	0.592
195	0.527	650	0.555	3250	0.586
210	0.523	700	0.541	3500	0.581
225	0.509	750	0.574	3750	0.583
240	0.526	800	0.594	4000	0.588
255	0.550	850	0.587	4250	0.591
270	0.552	900	0.601	4500	0.589
285	0.559	950	0.593	4750	0.586
300	0.537	1000	0.584	5000	0.582

TABLE VI. Average energy for harmonic oscillator for various thermal-time steps N and averaging times T for $\beta=1$ (correct answer equals 0.582). Compare also the last column of Table V. (Parameters chosen so that $\epsilon=\beta/N=1/N$, $h=0.01$ and vanishing initial values.)

Averaging time T	Energy $N=2$	$N=3$	$N=4$	$N=5$	$N=25$
2500	0.586	0.560	0.589	0.504	0.402
2750	0.593	0.553	0.555	0.551	0.416
3000	0.606	0.553	0.545	0.572	0.446
3250	0.614	0.530	0.553	0.602	0.432
3500	0.611	0.538	0.562	0.597	0.422
3750	0.609	0.527	0.576	0.603	0.412
4000	0.606	0.540	0.584	0.602	0.444
4250	0.607	0.525	0.575	0.592	0.457
4500	0.600	0.540	0.560	0.589	0.456
4750	0.592	0.559	0.567	0.584	0.448
5000	0.596	0.565	0.573	0.575	0.438

follows that

$$e^{-\beta\mathscr{H}}=\int e^{\beta}e^{-(e^{\beta}-1)(p^2+q^2)/2}\,|p,q\rangle\langle p,q|\,\frac{dp\,dq}{2\pi} \qquad (3.3)$$

holds as an exact relation[16] when $\mathscr{H}=\frac{1}{2}(P^2+Q^2-1)$, it follows if we use (3.3) in place of (2.6) that there are no $O(\epsilon^2)$ errors and consequently no need to decompose the thermal time β into a large number N of small steps ϵ. The results quoted above for the harmonic oscillator were all obtained for $N=1$. Table VI illustrates the N dependence of the average energy where (3.3) has been used N times for a β value of β/N. These results tend to show decreasing accuracy as N increases, as could be expected. In each solution the initial values were taken as $q_0=p_0=0$, which may help explain why the average energy for $N=25$ lags well behind that for $N\leq 5$, not only because the relaxation time is longer but because the Hamiltonian has a harder time making itself felt.

Indistinguishable harmonic oscillators: As another test case we have examined the example of two uncoupled harmonic oscillators ($M=2$), $h=\frac{1}{2}(p_1^2+q_1^2+p_2^2+q_2^2-4)$, that obey the exclusion principle. This example permits us to study the validity of the set of equations (2.22) for a soluble problem. The exact average energy in this case is given by

$$1+\frac{1}{e^{\beta}-1}+\frac{2}{e^{2\beta}-1}\ .$$

In our numerical study of this problem we have again used the exact relation (3.3), and therefore all choices of thermal-time steps N should be exact, at least in principle. In Table VII we present the resulting average energy obtained for two β values and two N values as a function of averaging time T along with the correct values. The approximate validity of the results for this model lend credence to the applicability of (2.22) to more general problems.

Quantized Hall effect: Based on the results of our study of the harmonic-oscillator test cases we are forced to conclude that our present study of the two-dimensional quantized Hall effect suffers from too short an averaging time, a problem aggrevated even further by the need to divide the thermal time β into many N factors of short duration ϵ. This latter feature could be helped by an improved action, but this has yet not been done. Also additional computer time could be devoted to this problem, but we have chosen for the present not to do so. The need for extensive computation to achieve reliable answers is well recognized in usual Monte Carlo studies as well.[17]

Having stated the shortcomings in our results, we proceed to enumerate in Table VIII a list of the energy values for different ν values for both the distinguishable and indistinguishable cases. For the distinguishable case the data all refer to an averaging time $T=80$, $\beta=2$, $\epsilon=\frac{1}{16}$, and $h=0.01$. For the indistinguishable case the data all refer to an averaging time $T=100$, $\beta=8$, $\epsilon=\frac{1}{16}$, and $h=0.1$. In each run the same set of random numbers was used (results were not too sensitive to this choice). The overall statistical accuracy of these data is less than

TABLE VII. Average energy for two uncoupled harmonic oscillators that satisfy the exclusion principle for various thermal-time steps N and averaging times T. (Parameters chosen so that $N\epsilon=\beta=0.5$ or $N\epsilon=\beta=0.3$ and $h=0.1$.)

Averaging time T	Energy ($\beta=0.5$) $N=1$	$N=4$	Energy ($\beta=0.3$) $N=1$	$N=4$
5000	3.65	3.94	6.06	6.53
5500	3.64	3.88	6.09	6.44
6000	3.60	3.74	6.03	6.26
6500	3.57	3.81	5.98	6.26
7000	3.59	3.86	6.00	6.28
7500	3.62	3.80	6.05	6.17
8000	3.60	3.73	6.00	6.02
8500	3.61	3.73	6.02	6.03
9000	3.63	3.74	6.06	6.04
9500	3.64	3.75	6.09	6.00
10000	3.64	3.76	6.10	6.00
Correct value	3.71		6.29	

TABLE VIII. Energy per particle for various filling factors ν for the classical, distinguishable (dis), exact, and indistinguishable (indis) cases. (Parameters for the dis case are $T=80$, $\beta=2$, $\epsilon=\frac{1}{16}$, $h=0.01$, and an ordered start; for the indis case the parameters are $T=100$, $\beta=8$, $\epsilon=\frac{1}{16}$, $h=0.1$, and an ordered start.)

Filling factor ν	Energy per particle Classical	dis	exact	indis
0.3125	−0.435	−0.408	−0.398	−0.283
0.3333	−0.450	−0.430	−0.413	−0.299
0.3542	−0.463	−0.445	−0.420	−0.309
0.4167	−0.503	−0.487	−0.442	−0.341
0.5000	−0.550	−0.530	−0.468	−0.379

2%, a fortuitous feature of this model. For comparison we have also included the minimum classical energy per particle values as well as the exact quantum-mechanical ground-state energy per particle values (including the exclusion principle) obtained by Yoshioka *et al.*[18]

Several remarks regarding the data presented are in order. First we observe that the energy results obtained in the distinguishable particle case lie above the classical minimum energy of (3.1) and below the exact quantum-mechanical ground-state results. For example, at $\nu=\frac{1}{3}$ the minimum classical energy per particle is -0.45, our result is -0.43, while the exact quantum-mechanical ground state is given by -0.41. However, our indistinguishable particle data all lie well above the exact ground-state results. This behavior could represent the result of an average over a suitable density of states, however, the lack of dependence of such data on β tends to belie this fact. We believe the discrepancy of our indistinguishable data arises because of two factors: (i) an insufficient averaging time T, and, perhaps more importantly, (ii) the need to use a step size $h=0.1$ in the numerical solution of the coupled Langevin equations. Note that for the distinguishable particle case we were able to choose $h=0.01$, a factor of ten better. It is interesting to add that for relatively short averaging times T ($T \lesssim 10$), the energy per particle values for the distinguishable and indistinguishable particle cases more nearly agree with each other as is to be expected for this example, particularly near $\nu=\frac{1}{3}$. This result suggests that the error in the indistinguishable particle case accumulates, more or less coherently, for larger averaging times T, indeed, due to the larger time step h. This situation could be improved, of course, by using significantly more computer time, but we have chosen not to do so.

To study the behavior at and near $\nu=\frac{1}{3}$, the average energy per particle was measured at $\nu=\frac{15}{48}$, $\frac{16}{48}$, and $\frac{17}{48}$. It is amusing to note that in both the distinguishable and indistinguishable cases the values at $\nu=\frac{1}{3}$ are slightly lower than the average of the two neighboring values (relative energies may possibly be more accurate here than absolute ones). The exact results listed in Table VIII also show a slight dip at $\nu=\frac{1}{3}$. Such a cusp in energies at $\nu=\frac{1}{3}$, not present classically, is exactly what one anticipates on the basis of the Laughlin quantum-mechanical ground state and which provides the energy gap needed to have a plateau in the Hall conductivity. At the very least our data is not inconsistent with that picture.

IV. CONCLUSION

We have presented a first-effort study of the application of coupled, complex Langevin equations to evaluate average energies of an involved system, namely, that of the two-dimensional quantized Hall effect. Our results are not unreasonable given the modest averaging times proffered to the seemingly insatiable beast that is at the heart of all statistical approaches. In the future we intend to apply the experience gained here to the study of quantum spin systems by means of associated Langevin equations, the form of which has already been presented elsewhere.[19,2]

ACKNOWLEDGMENTS

In addition to those acknowledgments already given, thanks are extended to H. S. Greenside, E. Helfand, and A. T. Ogielski for their advice and comments, and to W. P. Petersen for his help in achieving a twofold increase in program computational speed. It is especially a pleasure to thank W. F. Brinkman for his enthusiastic support of this work and for a number of discussions pertaining to it.

[1]See, e.g., M. Creutz, L. Jacobs, and C. Rebbi, Phys. Rep. 95, 201 (1983); K. Binder, J. L. Lebowitz, M. K. Phani, and M. H. Kalos, Acta Metall. 29, 1655 (1981); C. M. Bender, F. Cooper, and B. Freedman, Nucl. Phys. B 219, 61 (1983).

[2]J. R. Klauder, in *Recent Developments in High-Energy Physics*, edited by H. Mitter and C. B. Lang (Springer, Vienna, 1983), p. 251.

[3]D. C. Tsui, H. L. Stormer, and A. C. Gossard, Phys. Rev. Lett. 48, 1559 (1982); H. L. Stormer, A. Chang, D. C. Tsui, J. Hwang, A. C. Gossard, and W. Wiegmann, *ibid.* 50, 1953 (1983); D. Yoshioka, B. I. Halperin, and P. A. Lee, *ibid.* 50, 1219 (1983); R. B. Laughlin, *ibid.* 50, 1395 (1983).

[4]See, e.g., J. R. Klauder and E. C. G. Sudarshan, *Fundamentals of Quantum Optics* (Benjamin, New York, 1968), Chap. 7.

[5]J. R. Klauder, J. Math. Phys. 5, 177 (1964).

[6]See, e.g., K. G. Wilson, *Cargese Lecture Notes, 1979,* edited by G. 't Hooft (Plenum, New York, 1979); K. Symanzik, in *Mathematical Problems in Theoretical Physics,* Vol. 153 of *Lecture Notes in Physics,* edited by R. Schrader *et al.* (Springer, Berlin, 1982); K. Symanzik, Nucl. Phys. B 226, 187 (1983); B 226, 205 (1983); B. Berg, S. Meyer, I. Montvay, and K. Symanzik, Phys. Lett. B 126A, 467 (1983). It is a pleasure to thank K. Symanzik for a discussion on this subject.

[7]See, e.g., G. Parisi and Y.-S. Wu, Sci. Sin. 24, 483 (1981).

[8]See, e.g., Ref. 4, p. 113.

[9]W. P. Petersen (private communication); J. R. Klauder and W. P. Petersen (unpublished).

[10]See, e.g., N. T. J. Bailey, *The Elements of Stochastic Processes* (Wiley, New York, 1964); D. Kannan, *An Introduction to Stochastic Processes* (North-Holland, Amsterdam, 1979).

[11]See, e.g., H. L. Stormer and D. C. Tsui, Science 230, 1241 (1983); Phys. Today 36, 19 (1983).

[12]H. Fukuyama and D. Yoshioka, J. Phys. Soc. Jpn. 48, 1853 (1980).

[13]See, e.g., C. Kittel, *Introduction to Solid State Physics* (Wiley, New York, 1953), p. 347. It is a pleasure to thank W. F. Brinkman for a discussion on this point.

[14]E. Helfand, Bell Syst. Tech. J. 58, 2289 (1979); H. S. Greenside and E. Helfand, Bell Syst. Tech. J. 60, 1927 (1981).

[15]P. A. Lee (private communication).

[16]R. J. Glauber, Phys. Rev. Lett. 10, 84 (1963); J. R. Klauder and E. C. G. Sudarshan, *Fundamentals of Quantum Optics* (Benjamin, New York, 1968), p. 208.

[17]For example, in the recent work of A. T. Ogielski, Phys. Rev. D 28, 1461 (1983), it was occasionally necessary to run through as many as 400 000–600 000 Monte-Carlo updates to achieve reliable answers.

[18]D. Yoshioka, B. I. Halperin, and P. A. Lee, Phys. Rev. Lett. 50, 1219 (1983).

[19]J. R. Klauder, J. Phys. A 16, L317 (1983).

VII. Applications in Thermodynamics

Commun. math. Phys. 31, 327—340 (1973)

The Classical Limit of Quantum Spin Systems

Elliott H. Lieb*

Institut des Hautes Etudes Scientifiques, Bures-sur-Yvette, France

Received February 28, 1973

Abstract. We derive a classical integral representation for the partition function, Z^Q, of a quantum spin system. With it we can obtain upper and lower bounds to the quantum free energy (or ground state energy) in terms of two classical free energies (or ground state energies). These bounds permit us to prove that when the spin angular momentum $J \to \infty$ (but after the thermodynamic limit) the quantum free energy (or ground state energy) is equal to the classical value. In normal cases, our inequality is $Z^C(J) \leqq Z^Q(J) \leqq Z^C(J+1)$.

I. Introduction

It is generally believed in statistical mechanics that if one takes a quantum spin system of N spins, each having angular momentum J, normalizes the spin operators by dividing by J, and takes the limit $J \to \infty$, then one obtains the corresponding classical spin system wherein the spin variables are replaced by classical vectors and the trace is replaced by an integration over the unit sphere. Indeed, Millard and Leff [1] have shown this to be true for the Heisenberg model *when N is held fixed*. Their proof is quite complicated and it is therefore not surprising that this goal was not achieved before 1971. Despite that success, however, the problem is not finished. One wants to show that one can interchange the limit $N \to \infty$ with the limit $J \to \infty$, i.e. is the classical system obtained if we first let $N \to \infty$ and then let $J \to \infty$? In the Millard-Leff proof the control over the N dependence of the error is not good enough to achieve this desideratum.

A more useful result, and one which would include the above, would be to obtain, for each J, upper and lower bounds to the quantum free energy in terms of the free energies of two classical systems such that those two bounds have a common classical limit as $J \to \infty$. In this paper we do just that, and the result is surprisingly simple: In most cases of interest (including the Heisenberg model), the classical upper bound is

* On leave from the Department of Mathematics, M.I.T., Cambridge, Mass. 02139, USA. Work partially supported by National Science Foundation Grant GP-31674X and by a Guggenheim Memorial Foundation Fellowship.

obtained by replacing the quantum spin by $(J+1)$ times the classical unit vector, while the lower bound is obtained by using J instead of $(J+1)$. Symbolically,

$$Z^C(J) \leqq Z^Q(J) \leqq Z^C(J+1). \tag{1.1}$$

In other cases the result is a little more complicated to state, but it is of the same nature. With an upper and lower bound in hand, it is then possible to derive rigorous bounds on expectation values, as we shall describe in Sections V and VI.

The main tool in our derivation will be what has been termed by Arrechi *et al.* [2] the Bloch coherent state representation. These states and some of their properties were obtained earlier [3, 4], but the most complete account is in Ref. [2]. Our lower bound is obtained by a variational calculation, while the upper bound is obtained from a representation of the quantum partition function that bears some similarity to the Wiener (or path) integral. Apart from its use in deriving the upper bound, the representation may be of theoretical value in proving other properties of quantum spin systems. In particular, it provides a sensible definition of the quantum partition function for all complex J, not just when J is half an integer, and one may discuss the existence or non-existence of a phase transition as a function of the continuous parameter J.

In a forthcoming paper [7] it will be shown how to apply the methods and bounds developed herein (using not only the Bloch states but the Glauber coherent photon states as well) to certain models of the interaction of atoms with a quantized radiation field, for example the Dicke Maser model.

II. Bloch Coherent States

In this section we recapitulate results derived in Refs. [2] and [3]. We consider a single quantum spin of fixed total angular-momentum and shall denote by $\boldsymbol{S} \equiv (S_x, S_y, S_z)$ the usual angular momentum operators:

$$[S_x, S_y] = iS_z, \quad \text{and cyclically,} \quad S_\pm = S_x \pm iS_y. \tag{2.1}$$

We denote by J the total angular momentum, i.e.

$$S^2 = S_x{}^2 + S_y{}^2 + S_z{}^2 = J(J+1). \tag{2.2}$$

The Hilbert space on which these operators act has dimension $2J+1$, i.e. it is $\mathbb{C}^{2J+1}$.

On the classical side, we denote by $\mathscr{S}$ the unit sphere in three dimensions:

$$\mathscr{S} = \{(x, y, z) \mid x^2 + y^2 + z^2 = 1\}, \tag{2.3}$$

and by $L^2(\mathscr{S})$ the space of square integrable functions on $\mathscr{S}$ with the usual measure

$$\Omega = (\theta, \varphi),\ 0 \leqq \theta \leqq \pi,\ 0 \leqq \varphi < 2\pi, \tag{2.4}$$

$$d\Omega = \sin\theta \, d\theta \, d\varphi, \tag{2.5}$$

$$x = \sin\theta \cos\varphi,\ y = \sin\theta \sin\varphi,\ z = \cos\theta. \tag{2.6}$$

(Note: In Ref. [2], but not Ref. [3] the "south pole", instead of the customary "north pole" corresponds to $\theta = 0$. Hence our formulas will differ from Ref. [2] by the replacement $\theta \to \pi - \theta$).

With $|J\rangle \in \mathbb{C}^{2J+1}$ being a normalized "spin up" state, $S_z|J\rangle = J|J\rangle$, one defines the Bloch state $|\Omega\rangle \in \mathbb{C}^{2J+1}$ by

$$\begin{aligned} |\Omega\rangle &= \exp\{\tfrac{1}{2}\theta[S_- e^{i\varphi} - S_+ e^{-i\varphi}]\} \, |J\rangle \\ &= [\cos\tfrac{1}{2}\theta]^{2J} \exp\{(\tan\tfrac{1}{2}\theta) \, e^{i\varphi} S_-\} \, |J\rangle \\ &= \sum_{M=-J}^{J} \binom{2J}{M+J}^{1/2} (\cos\tfrac{1}{2}\theta)^{J+M} (\sin\tfrac{1}{2}\theta)^{J-M} \exp[i(J-M)\varphi] \, |M\rangle \end{aligned} \tag{2.7}$$

where $|M\rangle$ is the normalized state

$$|M\rangle = \binom{2J}{M+J}^{-1/2} [(J-M)!]^{-1} (S_-)^{J-M} |J\rangle \tag{2.8}$$

such that

$$S_z|M\rangle = M|M\rangle. \tag{2.9}$$

It is clear from (2.7) that the set of states $|\Omega\rangle$ are complete in $\mathbb{C}^{2J+1}$. Their overlap is given by

$$\begin{aligned} K_J(\Omega', \Omega) &\equiv \langle \Omega' | \Omega \rangle \\ &= \{\cos\tfrac{1}{2}\theta \cos\tfrac{1}{2}\theta' + e^{i(\varphi-\varphi')} \sin\tfrac{1}{2}\theta \sin\tfrac{1}{2}\theta'\}^{2J} \end{aligned} \tag{2.10}$$

so that if we think of $K_J(\Omega', \Omega)$ as the kernel of a linear transformation on $L^2(\mathscr{S})$ it is selfadjoint and compact. In fact, it is positive semidefinite. We also have

$$|K_J(\Omega', \Omega)|^2 = [\cos\tfrac{1}{2}\Theta]^{4J}, \tag{2.11}$$

where

$$\cos\Theta = \cos\theta \cos\theta' + \sin\theta \sin\theta' \cos(\varphi - \varphi') \tag{2.12}$$

is the cosine of the angle between Ω and Ω'. In particular $|\Omega\rangle$ is normalized since $K_J(\Omega, \Omega) = 1$.

Now let $\mathcal{M}^{2J+1}$ be the set of linear transformations on $\mathbb{C}^{2J+1}$ (i.e. operators on the spin space) and, for a given $G \in L^1(\mathcal{S})$, define $A_G \in \mathcal{M}^{2J+1}$ by

$$A_G = \frac{2J+1}{4\pi} \int d\Omega\, G(\Omega)\, |\Omega\rangle \langle\Omega| . \tag{2.13}$$

$\left(\text{Note: } \int d\Omega \text{ always means } \int_{\mathcal{S}} d\Omega\right)$. Since the Hilbert space is finite dimensional there is no problem in giving a meaning to (2.13). It is a remarkable fact that every operator in $\mathcal{M}^{2J+1}$ can be written in the form (2.13). In particular,

$$\mathbf{1} = \frac{2J+1}{4\pi} \int d\Omega\, |\Omega\rangle \langle\Omega| . \tag{2.14}$$

Thus, to every operator $A \in \mathcal{M}^{2J+1}$ there correspond two functions:

$$g(\Omega) = \langle\Omega| A |\Omega\rangle , \tag{2.15}$$

and the $G(\Omega)$ of (2.13). The former is, of course, unique, but the latter is not. However, it is always possible to choose $G(\Omega)$ to be infinitely differentiable. In Table 1 we list some function pairs for operators of common interest and useful formulas for calculation are given in Appendix A.

Table 1. Expectation values, $g(\Omega)$, and operator kernels, $G(\Omega)$, [cf. (2.13), (2.15)] for various operators commonly appearing in quantum spin Hamiltonians

Operator	$g(\Omega)$, (2.15)	$G(\Omega)$, (2.13)
S_z	$J\cos\theta$	$(J+1)\cos\theta$
S_x	$J\sin\theta\cos\varphi$	$(J+1)\sin\theta\cos\varphi$
S_y	$J\sin\theta\sin\varphi$	$(J+1)\sin\theta\sin\varphi$
S_z^2	$J(J-\frac{1}{2})(\cos\theta)^2 + J/2$	$(J+1)(J+3/2)(\cos\theta)^2 - \frac{1}{2}(J+1)$
S_x^2	$J(J-\frac{1}{2})(\sin\theta\cos\varphi)^2 + J/2$	$(J+1)(J+3/2)(\sin\theta\cos\varphi)^2 - \frac{1}{2}(J+1)$
S_y^2	$J(J-\frac{1}{2})(\sin\theta\cos\varphi)^2 + J/2$	$(J+1)(J+3/2)(\sin\theta\cos\varphi)^2 - \frac{1}{2}(J+1)$

We need three final remarks. The first is that if we consider $|\Omega\rangle\langle\Omega'| \in \mathcal{M}^{2J+1}$ then

$$\mathrm{Tr}\, |\Omega\rangle \langle\Omega'| = K_J(\Omega', \Omega) \tag{2.16}$$

(where Tr means Trace) as may be seen from (2.7). Hence, from (2.13)

$$\operatorname{Tr} A_G = \frac{2J+1}{4\pi} \int d\Omega\, G(\Omega)\,. \tag{2.17}$$

The second is that

$$\frac{2J+1}{4\pi} \int d\Omega\, K_J(\Omega', \Omega)\, K_J(\Omega, \Omega'') = K_J(\Omega', \Omega'')\,, \tag{2.18}$$

as may be seen from (2.14). Thus, K_J reproduces itself under convolution.

The third remark is that for any $A \in \mathcal{M}^{2J+1}$ we can use (2.14) to obtain

$$\begin{aligned} \operatorname{Tr} A &= \frac{2J+1}{4\pi} \int d\Omega \operatorname{Tr} |\Omega\rangle \langle\Omega| A \\ &= \frac{2J+1}{4\pi} \int d\Omega \sum_{M=-J}^{J} \langle M|\Omega\rangle \langle\Omega| A |M\rangle \\ &= \frac{2J+1}{4\pi} \int d\Omega\, \langle\Omega| A |\Omega\rangle\,. \end{aligned} \tag{2.19}$$

III. Lower Bound to the Quantum Partition Function

We consider a system of N quantum spins and shall label the operators and the angular momenta (which need not all be the same) by a superscript i, $i = 1, \ldots, N$. The Hamiltonian, H, can be completely general but, in any event, it can always be written as a polynomial in the $3N$ spin operators. The partition function is

$$Z^Q = \alpha_N \operatorname{Tr} \exp(-\beta H)\,, \tag{3.1}$$

where

$$\alpha_N = \prod_{i=1}^{N} (2J^i + 1)^{-1}\,. \tag{3.2}$$

[The normalization factor α_N is inessential; it is chosen to agree with the classical partition function when $\beta = 0$]. The Hilbert space is

$$\mathcal{H}_N = \bigotimes_{i=1}^{N} \mathcal{H}^i = \bigotimes_{i=1}^{N} \mathbb{C}^{2J^i+1}\,. \tag{3.3}$$

We denote by $|\Omega_N\rangle$ the complete, normalized set of states on $\mathcal{H}_N$ defined by

$$|\Omega_N\rangle = \bigotimes_{i=1}^{N} |\Omega^i\rangle\,, \tag{3.4}$$

by $\mathscr{S}_N$ the Cartesian product of N copies of the unit sphere, and by $d\Omega_N$ the product measure (2.4), (2.5) and (2.6) on $\mathscr{S}_N$. Using (2.19),

$$Z^Q = (4\pi)^{-N} \int d\Omega_N \langle \Omega_N | e^{-\beta H} | \Omega_N \rangle . \tag{3.5}$$

By the Peierls-Bogoliubov inequality, $\langle \psi | e^X | \psi \rangle \geqq \exp \langle \psi | X | \psi \rangle$ for any normalized $\psi \in \mathscr{H}_N$ and X selfadjoint. Thus,

$$Z^Q \geqq (4\pi)^{-N} \int d\Omega_N \exp\{-\beta \langle \Omega_N | H | \Omega_N \rangle\} . \tag{3.6}$$

Suppose, at first, that the polynomial, H, is linear in the operators S^i of each spin. That is we allow multiple site interactions of arbitrary complexity such as $S_x{}^1 S_y{}^2 S_y{}^3 S_z{}^4$, but do not allow monomials such as $(S_x{}^1)^2$ or $S_x{}^1 S_y{}^1$. In this case, which we shall refer to as *the normal case*, we see from (2.15) and Table 1 that the right side of (3.6) is precisely the classical partition function in which each S^i is replaced by J^i times a vector in $\mathscr{S}$. I.e.

$$S^i \to J^i (\sin\theta^i \cos\varphi^i, \sin\theta^i \sin\varphi^i, \cos\theta^i) . \tag{3.7}$$

Thus, in the normal case,

$$Z^Q \geqq Z^C (J^1, \ldots, J^N) , \tag{3.8}$$

where Z^C means the classical partition function (with the normalization $(4\pi)^{-N}$).

In more complicated cases, (3.7) is not correct and $S_z{}^1$, for example, has to be replaced by $J^1 \cos\theta^1$ if it appears linearly in H, $(S_z{}^1)^2$ has to be replaced by $[J^1 \cos\theta^1]^2 + J^1 (\sin\theta^1)^2/2$ and so forth (see Table 1). However, to leading order in J^i, (3.7) is correct.

We note in passing that it is not necessary to use the Peierls-Bogoliubov inequality for all operators appearing in H. Thus, suppose the whole Hilbert space is $\mathscr{H}' = \mathscr{H} \otimes \mathscr{H}_N$ where $\mathscr{H}$ is the Hilbert space of some additional degrees of freedom (which may or may not themselves be spins) and H is selfadjoint on $\mathscr{H}'$. Then (by a generalized Peierls-Bogoliubov inequality)

$$\begin{aligned} Z^Q &= \alpha_N \operatorname{Tr}_{\mathscr{H}'} \operatorname{Tr}_{\mathscr{H}} \exp(-\beta H) \\ &\geqq \operatorname{Tr}_{\mathscr{H}} (4\pi)^{-N} \int d\Omega_N \exp\{-\beta \langle \Omega_N | H | \Omega_N \rangle\} \end{aligned} \tag{3.9}$$

where $\langle \Omega_N | H | \Omega_N \rangle$ is a partial expectation value and defines a selfadjoint operator on $\mathscr{H}$. We shall give an example of (3.9) in Appendix B. It is clear that if $\mathscr{H}$ is itself a spin space, then (3.9) gives a better bound than (3.6) applied to the full space $\mathscr{H}'$.

IV. Upper Bound to the Quantum Partition Function

Returning to the definitions (3.1) and (3.3) we note that

$$Z^Q = \lim_{n\to\infty} Z(n)\,, \tag{4.1}$$

where

$$Z(n) = \alpha_N \operatorname{Tr}(\mathbf{1} - \beta n^{-1} H)^n\,. \tag{4.2}$$

Now, let H be represented by some $G(\Omega_N)$ as in (2.13), whence $\mathbf{1} - \beta n^{-1} H$ is represented by

$$F_n(\Omega_N) = 1 - \beta n^{-1} G(\Omega_N)\,. \tag{4.3}$$

Using (2.10), (2.13) and (2.16), we can represent Z_n as an nN fold integral:

$$Z(n) = \alpha_N \int d\Omega_N{}^1 \cdots \int d\Omega_N{}^n \prod_{j=1}^{n} F_n(\Omega_N{}^j)\, L_J(\Omega_N{}^j, \Omega_N{}^{j+1}) \tag{4.4}$$

with $n+1 \equiv 1$ in the last factor, and where

$$L_J(\Omega_N', \Omega_N) \equiv (4\pi)^{-N} \alpha_N{}^{-1} \prod_{i=1}^{N} K_{J^i}(\Omega'^i, \Omega^i)\,. \tag{4.5}$$

Thus

$$L_J(\Omega_N, \Omega_N) = (4\pi)^{-N} a_N{}^{-1}\,. \tag{4.6}$$

$$\int d\Omega_N\, L_J(\Omega_N', \Omega_N)\, L_J(\Omega_N, \Omega_N'') = L_J(\Omega_N', \Omega_N'')\,. \tag{4.7}$$

Equations (4.1) and (4.4) are our desired integral representation for Z^Q. To use them to obtain a bound, we think of F_n as a multiplication operator and of L_J as the kernel of a compact, selfadjoint operator on $L^2(\mathscr{S}_N)$. If $B(\Omega_N', \Omega_N)$ is such a kernel, then

$$\operatorname{Tr} B = \int d\Omega_N\, B(\Omega_N, \Omega_N) \tag{4.8}$$

is the trace on $L^2(\mathscr{S}_N)$. Thus,

$$Z(n) = \alpha_N \operatorname{Tr}(F_n L_J)^n\,. \tag{4.9}$$

In general, if $m = 2^j$, $j = 0, 1, 2, 3, \ldots,$

$$|\operatorname{Tr}(AB)^{2m}| \leqq \operatorname{Tr}(A^2 B^2)^m \leqq \operatorname{Tr} A^{2m} B^{2m}\,, \tag{4.10}$$

whenever A and B are selfadjoint. This follows from the Schwarz inequality (see Ref. [5] for details). Hence, if we take a sequence $n = 2^j$, $j = 1, 2, \ldots$ in (4.2) and use (4.7) n times and (4.6) we obtain in the limit

$n \to \infty$,

$$Z^Q \leqq (4\pi)^{-N} \int d\Omega_N \exp[-\beta G(\Omega_N)] \,. \tag{4.11}$$

(4.11) is our desired classical upper bound. It is just like (3.6). In the normal case we see from Table 1 that S^i is replaced by $(J^i + 1)$ times a classical unit vector. In other cases, $G(\Omega_N)$ is a bit more complicated, but the same remarks as in Section III apply. Thus, in the normal case

$$Z^C(J^1, \ldots, J^N) \leqq Z^Q \leqq Z^C(J^1 + 1, \ldots, J^N + 1) \tag{4.12}$$

This inequality says that as J increases the quantum and classical free energies form two decreasing, *interlacing* sequences.

As in Section III, if $\mathscr{H}' = \mathscr{H} \otimes \mathscr{H}_N$ an inequality similar to (4.11) can be shown to hold, i.e.

$$Z^Q \leqq \mathrm{Tr}_{\mathscr{H}} (4\pi)^{-N} \int d\Omega_N \exp[-\beta H(\cdot, \Omega_N)] \,, \tag{4.13}$$

where $H(\cdot, \Omega_N)$ is a selfadjoint operator on $\mathscr{H}$ obtained by replacing each monomial in the spin operators in H by the appropriate $G(\Omega_N)$ function found in Table 1. We shall illustrate (4.13) in Appendix B. If $\mathscr{H}$ is a spin space then (4.13) gives a better bound than (4.11) applied to the full $\mathscr{H}'$.

V. Bounds on Expectation Values and the Ground State Energy

The expectation value of a quantum operator (observable), A, is

$$\langle A \rangle^Q = \mathrm{Tr}\, A \exp(-\beta H) / \mathrm{Tr} \exp(-\beta H) \,. \tag{5.1}$$

We can always assume A is selfadjoint (otherwise consider $A + A^\dagger$ and $iA - iA^\dagger$), in which case the Peierls-Bogoliubov inequality reads, for λ real,

$$\lambda \langle A \rangle^Q \geqq f(\lambda) - f(0) \,, \tag{5.2}$$

where

$$f(\lambda) = -\beta^{-1} \ln \mathrm{Tr} \exp[-\beta(H + \lambda A)] \,, \tag{5.3}$$

is a free energy. Hence, with $\lambda > 0$,

$$[f(0) - f(-\lambda)]/\lambda \geqq \langle A \rangle^Q \geqq [f(\lambda) - f(0)]/\lambda \,. \tag{5.4}$$

The upper and lower bounds to $f(\lambda)$ derived in the preceding two sections can be used to advantage in (5.4). In particular, we use (5.4) in the next section to derive $J \to \infty$ limits of quantum expectation values.

If we take the limit $\beta \to \infty$ in (3.1) we obtain bounds on the quantum ground state energy:

$$E_-^C \leqq E^Q \leqq E_+^C, \tag{5.5}$$

where E^C is the classical ground state energy (i.e. the minimum of the classical Hamiltonian over $\mathscr{S}_N$) and the + (resp. −) refers to the substitution of the appropriate $G(\Omega_N)$ (resp. $g(\Omega_N)$) functions from Table 1. In the normal case

$$E^C(J^1, \ldots, J^N) \geqq E^Q \geqq E^C(J^1+1, \ldots, J^N+1). \tag{5.6}$$

As ground state expectation values obey an inequality similar to (5.2), with f replaced by E, a bound similar to (5.4) holds for E. This is merely the variational principle.

The upper bound in (5.6) is easy to obtain directly by a variational calculation, but the lower bound is not. It is not easy to find a direct proof of it in a system consisting of three spins antiferromagnetically coupled to each other.

VI. The Thermodynamic Limit

A. The Free Energy

We shall, for simplicity, consider only the normal case here. The general case can be handled in a similar manner.

Let H_N be a Hamiltonian (polynomial) of N spins in which each spin has angular momentum one. Replace each spin operator $\mathbf{S}^i$ by $(J)^{-1}\mathbf{S}^i$ and let $\mathbf{S}^i$ now have angular momentum J. We shall denote this symbolically by $H_N^Q(J)$ and the partition function, (3.1), by $Z_N^Q(J)$. [It would equally be possible to allow different J values for different spins, but that is a needless complication. Also, the factor J^{-1} is not crucial. One could as well use $J^{-1/2}(J+1)^{-1/2}$]. Denoting the free energy per spin by $f_N(J) = -(N\beta)^{-1}\ln Z_N(J)$, the theorem to be proved is that

$$\lim_{J\to\infty}\lim_{N\to\infty} f_N^Q(J) = f^C \equiv \lim_{N\to\infty} f_N^C, \tag{6.1}$$

where f_N^C is the free energy per spin of the classical partition function in which each $\mathbf{S}^i$ is replaced by a classical unit vector. It is assumed that H_N is known to have a thermodynamic limit for the free energy per spin. We also want to prove an analogous formula for the ground state energy per spin. Our bounds are

$$f_N^C \geqq f_N^Q(J) \geqq f_N^C(\delta_J), \tag{6.2}$$

where the right side is the classical free energy per spin in which each vector is multiplied by $\delta_J \equiv (J+1)/J$.

If we think of δ_J as a variable, δ, then $H_N^C(\delta)$, the classical Hamiltonian as a function of δ, is continuous in δ. Moreover, $N^{-1} H_N^C(\delta)$ is equicontinuous in N, i.e. given any $\varepsilon > 0$ it is possible to find a $\gamma > 0$ such that $\|N^{-1}[H_N^C(\delta+x) - H_N^C(\delta)]\| \leqq \varepsilon$ for $|x| < \gamma$, independent of N, where $\| \; \|$ means the uniform on $\mathscr{S}_N$. Hence, the limit function

$$f^C(\delta) \equiv \lim_{N\to\infty} f_N^C(\delta) \tag{6.3}$$

is continuous in δ. This, together with (6.2), proves (6.1).

The same equicontinuity holds for the classical ground state energy. Thus, the analogue of (6.1) is also true for the ground state energy per spin:

$$\lim_{J\to\infty} \lim_{N\to\infty} N^{-1} E_N^Q(J) = \lim_{N\to\infty} E_N^C . \tag{6.4}$$

B. Expectation Values

We consider expectation values of intensive observables $N^{-1} A_N$. For example, A_N might be the Hamiltonian itself, in which case $\langle N^{-1} A_N \rangle$ is the energy per spin. Alternatively, A_N could be $\sum_{i=1}^{N} S_z^i$ so that $\langle N^{-1} A_N \rangle$ is the magnetization per spin. As before, we replace each $\boldsymbol{S}^i$ by $(J)^{-1}$ times a quantum spin of angular momentum J, *both* in the Hamiltonian and in A_N. Then, using inequality (5.4) and the bounds (6.2) we have, for each positive λ, fixed N and fixed J,

$$\begin{aligned} \lambda^{-1}[f_N^C(0;1) - f_N^C(-\lambda;\delta_J)] &\geqq N^{-1}\langle A_N \rangle^Q \\ &\geqq \lambda^{-1}[f_N^C(\lambda;\delta_J) - f_N^C(0;1)] , \end{aligned} \tag{6.5}$$

where $f_N^C(\lambda;\delta)$ is the classical free energy per spin when the Hamiltonian is $H_N^C + \lambda A_N^C$ and where each classical spin unit vector in H_N^C and A_N^C is multiplied by δ. We are interested in $\delta_J = (J+1)/J$.

Now take the limit $N \to \infty$ and then the limit $J \to \infty$ in (6.5). By the same equicontinuity remark as in Section VI.A, for each $\lambda > 0$,

$$\limsup_{J\to\infty} \limsup_{N\to\infty} N^{-1}\langle A_N \rangle^Q \leqq \lambda^{-1}[f^C(0) - f^C(-\lambda)] ,$$

$$\liminf_{J\to\infty} \liminf_{N\to\infty} N^{-1}\langle A_N \rangle^Q \geqq \lambda^{-1}[f^C(\lambda) - f^C(0)] . \tag{6.6}$$

In (6.5), $f^C(\lambda)$ is the limiting classical free energy per spin for the Hamiltonian $H_N^C + \lambda A_N^C$ (with $\delta = 1$). It is easy to see that $f^C(\lambda)$ is concave in λ

and hence $\lim_{\lambda\downarrow 0}\lambda^{-1}[f^C(\lambda)-f^C(0)]\equiv G^+$ and $\lim_{\lambda\downarrow 0}\lambda^{-1}[f^C(0)-f^C(-\lambda)]$ $\equiv G^-$ exist everywhere. If $G^+=G^-$ (i.e. the right derivative equals the left derivative) then by a theorem of Griffiths (6)

$$\lim_{N\to\infty}\frac{d}{d\lambda}f_N^C(\lambda)=\frac{d}{d\lambda}f^C(\lambda). \tag{6.7}$$

This is the case in which the classical expectation value $N^{-1}\langle A_N\rangle^C$ has a well defined limit. Call it α. Then

$$\lim_{J\to\infty}\lim_{N\to\infty}N^{-1}\langle A_N\rangle^Q=\alpha, \tag{6.8}$$

as one sees by taking the limit $\lambda\to 0$ in (6.6). In other words, we have proved that *for intensive observables, as defined above, the quantum expectation value equals the classical expectation value after first taking the thermodynamic limit and then taking the classical limit* $J\to\infty$. If one takes the limits in the opposite order the theorem is trivially true and uninteresting. Note that we have not proved that the quantum thermodynamic limit, $\lim_{N\to\infty}N^{-1}\langle A_N\rangle^Q$ exists. It may not.

The same proof obviously goes through for ground state expectation values, as in Section VI.A, because the ground state energy is also concave in λ.

Acknowledgements. The author thanks the Institut des Hautes Etudes Scientifiques for its hospitality, as well as the Chemistry Laboratory III, University of Copenhagen where part of this work was done. The financial assistance of the Guggenheim Memorial Foundation is gratefully acknowleged. The author also acknowledges his gratitude to Dr. N. W. Dalton who suggested the problem to him in 1967.

Appendix A: Some Useful Formulas

The algebra $\mathcal{M}^{2J+1}$ has S_+, S_- and S_z as generators. Hence, the following generating function permits, by differentiation, easy calculation of $g(\Omega)$ in (2.15) or Table 1 for any operator. It is to be found, with appropriate modifications, in Ref. [2].

$$\begin{aligned}&\langle\Omega|\exp(\gamma S_+)\exp(\beta S_z)\exp(\alpha S_-)|\Omega\rangle \qquad \text{(A.1)}\\ &=\{[e^{-i\varphi}\sin\tfrac{1}{2}\theta+\gamma\cos\tfrac{1}{2}\theta]\,[e^{i\varphi}\sin\tfrac{1}{2}\theta+\alpha\cos\tfrac{1}{2}\theta]\,e^{-\beta/2}+e^{\beta/2}[\cos\tfrac{1}{2}\theta]^2\}^{2J}.\end{aligned}$$

Turning to (2.13), we calculate A_G for a sufficiently large class of functions $G(\Omega)$. Let

$$G(\Omega)=e^{im\varphi}(\cos\tfrac{1}{2}\theta)^p(\sin\tfrac{1}{2}\theta)^q \tag{A.2}$$

where m is an integer and p and q are complex numbers. Defining $A(m,p,q) \equiv A_G$, the matrix elements of this operator can be calculated using (2.7) to be

$$\begin{aligned} A(m,p,q;M,M') = {} & \delta(M-M'-m)\,\Gamma(J+\alpha+1+p/2)\,\Gamma(J-\alpha+1+q/2) \\ & \cdot [(J+\alpha+m/2)!\,(J+\alpha-m/2)!\,(J-\alpha-m/2)!\,(J-\alpha+m/2)!]^{-1/2} \quad \text{(A.3)} \\ & \cdot (2J+1)!/\Gamma(2J+2+p/2+q/2)\,, \end{aligned}$$

where δ is the Kroenecker delta function, Γ is the gamma function and $\alpha = (M+M')/2$. This formula has been used to calculate Table 1.

Appendix B: Application to the One Dimensional Heisenberg Chain

To illustrate the methods of this paper, we derive bounds for the free energy of a Heisenberg chain whose Hamiltonian is

$$H = -\sum_{i=1}^{N-1} \mathbf{S}^i \cdot \mathbf{S}^{i+1}\,. \quad \text{(B.1)}$$

Each spin is assumed to have angular momentum J. We have chosen the isotropic case for simplicity, but one could equally well handle the anisotropic Hamiltonian with a magnetic field. Note that $\beta > 0$ is the ferromagnetic case while $\beta < 0$ is the antiferromagnetic case.

The classical partition function is

$$Z_N^C(\beta,x) = (4\pi)^{-N} \int d\Omega_N \exp\left\{\beta x^2 \sum_{i=1}^{N-1} \boldsymbol{\Omega}^i \cdot \boldsymbol{\Omega}^{i+1}\right\} \quad \text{(B.2)}$$

with free energy per spin

$$f^C(\beta,x) = -\lim_{N\to\infty} (N|\beta|)^{-1} \ln Z_N^C(\beta,x)\,. \quad \text{(B.3)}$$

Our bounds are that

$$f^C(\beta,J) \geqq f^Q(\beta,J) \geqq f^C(\beta,J+1)\,. \quad \text{(B.4)}$$

It is easy to evaluate (B.2) by the transfer matrix method. The normalized eigenfunction (of Ω) giving the largest eigenvalue is obviously the constant function $(4\pi)^{-1/2}$. Thus,

$$f^C(\beta,x) = -|\beta|^{-1} \ln A(\beta,x)\,, \quad \text{(B.5)}$$

where

$$A(\beta, x) = (4\pi)^{-1} \int d\Omega \exp\{\beta x^2 \boldsymbol{\Omega} \cdot \boldsymbol{\Omega}'\}$$
$$= (\beta x^2)^{-1} \sinh(\beta x^2), \tag{B.6}$$

and $A(\beta, x)$ is independent of Ω' as it should be. In this approximation, (B.4), one cannot distinguish between the ferro- and antiferromagnetic cases as far as the free energy is concerned.

To illustrate the idea mentioned at the ends of Sections III and IV, we suppose that the chain has $2N+1$ spins and we let $\mathscr{H}_N$ (resp. $\mathscr{H}$) be the Hilbert space for the odd (resp. even) numbered spins. $\mathscr{H}' = \mathscr{H} \otimes \mathscr{H}_N$ is the whole space. Our bounds are

$$g(\beta, J) \geqq f^Q(\beta, J) \geqq g(\beta, J+1), \tag{B.7}$$

where

$$g(\beta, x) = -\lim_{N\to\infty} (2N|\beta|)^{-1} \ln\{(2J+1)^{-N} \tilde{Z}_N(\beta, x)\}, \tag{B.8}$$

$$\tilde{Z}_N(\beta, x) = (4\pi)^{-N} \int d\Omega_N \operatorname{Tr} \exp\left\{\beta x \sum_{i=1}^{N} \boldsymbol{S}^{2i} \cdot (\boldsymbol{\Omega}^{2i-1} + \boldsymbol{\Omega}^{2i+1})\right\} \tag{B.9}$$

and where $d\Omega_N = d\Omega^1 \, d\Omega^3 \ldots d\Omega^{2N+1}$ and the trace is over the Hilbert space of $\boldsymbol{S}^2, \boldsymbol{S}^4, \ldots, \boldsymbol{S}^{2N}$.

Since the remaining spin operators no longer interact, it is easy to calculate the trace. For a single spin:

$$\operatorname{Tr} \exp[b\boldsymbol{S} \cdot \boldsymbol{v}] = \sum_{M=-J}^{J} \exp[bMv] \tag{B.10}$$

where b is a constant and $\boldsymbol{v}$ is a vector of length v. Now we can do the integration over $\mathscr{S}_N$ by the transfer matrix method (with the same eigenvector $(4\pi)^{-1/2}$) and obtain

$$g(\beta, x) = -\tfrac{1}{2}|\beta|^{-1} \ln[A(\beta, x)/(2J+1)], \tag{B.11}$$

where

$$A(\beta, x) = (4\pi)^{-1} \int d\Omega \sum_{M=-J}^{J} \exp\{\beta x M |\boldsymbol{\Omega} + \boldsymbol{\Omega}'|\}$$
$$= 2\int_0^1 y\,dy \sinh[(2J+1)\beta x y]/\sinh[\beta x y]. \tag{B.12}$$

Again no distinction between the ferro- and antiferromagnetic cases appears.

References

1. Millard, K., Leff, H.: J. Math. Phys. **12**, 1000—1005 (1971).
2. Arecchi, F. T., Courtens, E., Gilmore, R., Thomas, H.: Phys. Rev. A **6**, 2211—2237 (1972).
3. Radcliffe, J. M.: J. Phys. A **4**, 313—323 (1971).
4. Kutzner, J.: Phys. Lett. A **41**, 475—476 (1972).
   Atkins, P. W., Dobson, J. C.: Proc. Roy. Soc. (London) A, A **321**, 321—340 (1971).
5. Golden, S.: Phys. Rev. B **137**, 1127—1128 (1965).
6. Griffiths, R. B.: J. Math. Phys. **5**, 1215—1222 (1964).
7. Hepp, K., Lieb, E. H.: The equilibrium statistical mechanics of matter interacting with the quantized radiation field. Preprint.

E. H. Lieb
I.H.E.S.
F-91440 Bures-sur-Yvette, France

International Journal of Theoretical Physics, Vol. 14, No. 1 (1975), pp. 1–21

Coherent States and Partition Function

MARIO RASETTI

Institute for Advanced Study, Princeton, New Jersey, USA

Received: 29 *November* 1974

Abstract

The concept of coherent states for arbitrary Lie group is suggested as a tool for explicitly obtaining an integral representation of the partition function, whenever the Hamiltonian has a dynamical group. Two examples are thoroughly discussed: the case of the nilpotent group of Weyl related to a generic many-body problem with two-body interactions, and the case of $\prod_{k\otimes} SU(1, 1)(k)$ relevant for a superfluid system.

1. *Introduction*

Several important properties of coherent states make them ideal for the description of a system with infinitely many degrees of freedom, in which quantum features are macroscopically relevant. They evolve according to classic equations of motion and are, therefore, the most suitable ground for the picture of a system in which low energy excitations are superimposed on a macroscopically occupied ground state which exploits a sort of quasi-classic behavior. Moreover, they constitute a set of functional representatives of the abstract state vector of the system such that every member of this set—translationally invariant in the representation space—is an entire analytic function.

These states are in close connection through Bose statistics and, through the usual commutation relations of second-quantized field theoretical creation and annihilation operators, to the well-known nilpotent Weyl group.

A. M. Perelomov (1972) and M. Rasetti (1973) generalize the concept to different groups defining a set of coherent states for any Lie group, invariant relative to the action of the group generators. These states are determined by a set of points in a suitably defined homogeneous space, and form an over-complete system which contains subsystems of complete coherent states.

As pointed out in Rasetti (1973), the requirement that there be a Lie algebra commuting with the Hamiltonian and that the Lie algebra be integrable to a Lie group, enables us to define and use the coherent states as "spectrum generating" states. It is just this feature to make coherent states interesting, whenever one is in possession of information on the Hamiltonian formulation of a theory, in the sense of its connection with the structure of the underlying dynamical group. In particular, these states may be used as the basis for obtaining a simple representation of the partition function for the given Hamiltonian, as an integral over the group manifold.

Thermodynamic properties of the system, and possible singularities or mathematical pathologies of the partition function, can then be viewed in the perspective of the group manifold topology. In Section 2, we explicit the calculation in the case of nilpotent Weyl group coherent states, essentially on the lines of J. S. Langer (1968). The dynamical group is in this case a broken symmetry, but a perturbative procedure is yet possible. In Section 3, the generalized procedure for both compact and noncompact Lie group coherent states is developed. Section 4 is devoted to the discussion of another case of particular interest: namely $G \equiv \prod_{k \otimes} SU(1,1)_{(k)}$ suitable (Solomon, 1971) for multilevel superfluid Bose system. Section 5 finally concludes with a brief discussion about the analiticity properties of the partition function Z over the symmetric space associated with the coherent state.

2. *The Special Nilpotent Weyl Group*

The Lie algebra of this group is isomorphic to the Lie algebra produced by the usual Bose creation and annihilation operators $a_k^\dagger, a_k$ through the commutation relations

$$[a_k, a_{k'}] = [a_k^\dagger, a_{k'}^\dagger] = 0 \tag{2.1}$$

$$[a_k, a_{k'}^\dagger] = \delta_{k,k'} \tag{2.2}$$

The usual Hamiltonian with two-body interactions can be used without loss of generality (we label the single particle states with the same symbol as for the momentum)

$$H = \sum_k \tilde{\epsilon}_k a_k^\dagger + \frac{1}{2\Omega} \sum_{k,p,q} V(k)\, a_{p+k}^\dagger a_{q-k}^\dagger a_p a_q \tag{2.3}$$

where the single particle energy ϵ_k has to be modified in order to account for the lack of the number of particles conservation by the introduction of the chemical potential μ:

$$\tilde{\epsilon}_k = \epsilon_k - \mu \tag{2.4}$$

and Ω is the volume of the system.

The operators of the irreducible unitary representation of the Weyl group are of the form

$$T = e^{i\phi} Д(z) \tag{2.5}$$

where ϕ is a real number, $z \equiv (z_1, z_2, \ldots z_N)$ is an N-dimensional complex vector—N is the number of levels of the system—and

$$Д(z) Д(z') = e^{i\,\mathrm{Im}(z\bar{z}')} Д(z + z') \tag{2.6}$$

If $| 0 \rangle$ is the "vacuum" vector defined by the set of equations

$$a_k \mid 0 \rangle = 0 \tag{2.7}$$

and it is chosen to be the stationary point in the Hilbert space of the system relative the action of the subgroup generated by the exponential mapping of the identity operator, the system of coherent states is the set of vectors (Klauder, 1970)

$$| \{z\} \rangle = \prod_k | z_k \rangle = e^{\sum_k z_k a_k^\dagger} | 0 \rangle \tag{2.8}$$

It possesses the completeness property

$$\int \prod_k \frac{1}{\pi} d\,\mathrm{Re}\, z_k d\,\mathrm{Im}\, z_k \mid \{z\} \rangle \langle \{z\} \mid = I \tag{2.9}$$

The partition function for H may then be written

$$Z = \mathrm{Tr}[e^{-\beta H}] = \int d\mu\{z\} \langle \{z\} \mid e^{-\beta H} \mid \{z\} \rangle \tag{2.10}$$

where

$$d\mu\{z\} = \prod_k \frac{1}{\pi} d\,\mathrm{Re}\, z_k d\,\mathrm{Im}\, z_k \tag{2.11}$$

is the invariant measure over the homogeneous space associated with the coherent state system. Explicit evaluation of the integrand is possible, recalling that

$$a_k \mid z_k \rangle = z_k \mid z_k \rangle; \quad \langle z_k \mid a_k^\dagger = \bar{z}_k \langle z_k \mid \tag{2.12}$$

In general the matrix element $\langle \{z\} \mid e^{-\beta H} \mid \{z'\} \rangle$ can be in fact expanded diagrammatically.

Introducing a free-energy functional $F\{\bar{z}, z'\}$:

$$\langle \{z\} \mid e^{-\beta H} \mid \{z'\} \rangle = \langle \{z\} \mid \{z'\} \rangle e^{-\beta F\{\bar{z}, z'\}} \tag{2.13}$$

the expansion is equivalent to giving F as a power series of $z_k, z'_{k'}$ (Dyson, 1956):

$$F\{\bar{z}, z'\} = F^{(0)}\{\bar{z}, z'\} + \sum_{n=2}^{\infty} F^{(n)}\{\bar{z}, z'\} \tag{2.14}$$

where

$$F^{(0)}\{\bar{z}, z'\} = -\frac{1}{\beta}\sum_k (e^{-\beta\tilde{\epsilon}_k} - 1)\,\bar{z}_k z'_k \tag{2.15}$$

and

$$F^{(n)}\{\bar{z}, z'\} = \frac{1}{\beta}\sum_{\{k\},\,\{k'\}} W_n(k_1, k_2, \ldots, k_n; k'_1, k'_2, \ldots, k'_n)$$

$$\bar{z}_{k_1}\bar{z}_{k_2}\cdots\bar{z}_{k_n} z'_{k'_1} z'_{k'_2}\cdots z'_{k'_n}\,\delta\Big(\sum_{i=1}^{n}\vec{k}_i - \sum_{i=1}^{n}\vec{k}'_i\Big) \tag{2.16}$$

The delta function is explicitly written because the potential conserves the total momentum. Each W_n is given by the sum of all diagrams with n particle lines.

In the present case, the nilpotent algebra, strictly speaking, does not correspond to a dynamical group of the Hamiltonian, but–according to the prescription of Rasetti (1973)–to a fixed order in $|z|$ it commutes with H, and may be integrated to form the group.

There is a small breaking of the translational invariance in the representation space, which–due to the mentioned global property–can be handled in a perturbative way.

For our purposes of exemplification the only term we are to be concerned with in the following is the two body term $W_2(k_1, k_2; k'_1, k'_2)$. Its diagrammatic expansion is obviously given by the series in Figure 1 or the equivalent integral equation in Figure 2.

The equation of the partition function is now simplified by noting that the total number of excitations, at low temperatures, is small compared to the number of particles; thus for $k_B T \ll \epsilon_k$, $\langle a_k^\dagger a_k\rangle$ must be small, and since $\langle a_k^\dagger a_k\rangle \sim |z_k|^2$ the main correction to $Z^{(0)}$ (the partition function in absence of interaction) comes from small values of z_k.

Consequently one expands the interaction part of the integrand about the origin in z-space.

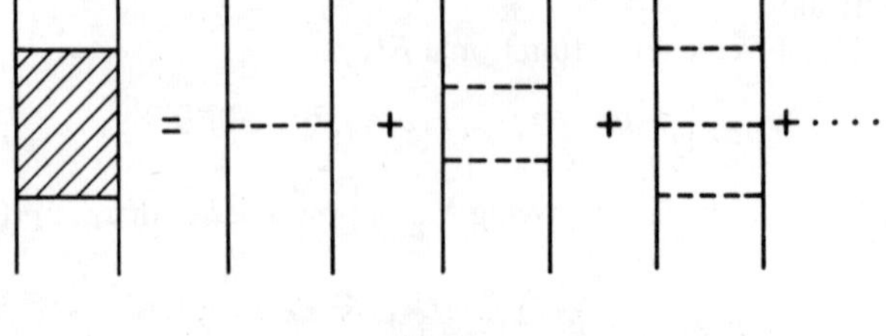

Figure 1

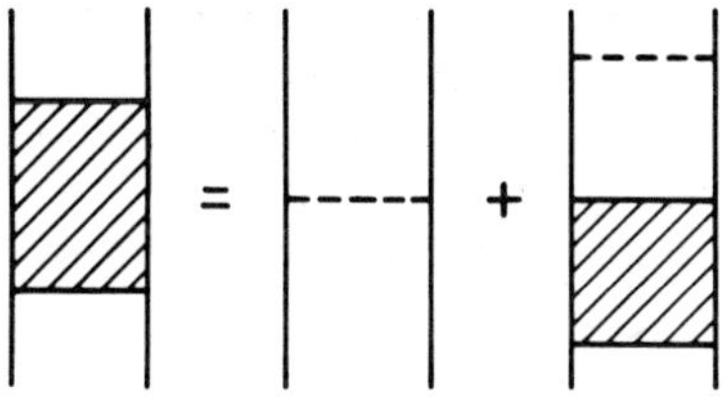

Figure 2

Denoting by $F\{z\}$ the diagonal part of the functional, $F\{z, z\}$, one gets:

$$\begin{aligned} Z &= \int d\mu\{z\}\, e^{-\beta F\{z\}} = \int d\mu\{z\}\, e^{-\beta F^{(0)}\{z\}}\, e^{-\beta F^{(2)}\{z\}} \ldots \\ &= \int d\mu\{z\}\, e^{-\beta F^{(0)}\{z\}} \Big[1 - \sum_{\{k\}} W_2(k_1, k_2; k_1', k_2')\, \bar{z}_{k_1} \bar{z}_{k_2} \\ &\qquad z_{k_1'} z_{k_2'}\, \delta(\vec{k}_1 + \vec{k}_2 - \vec{k}_1' - \vec{k}_2') + \cdots \Big] \\ &= Z^{(0)} - \sum_{\{k\}} W_2(k_1, k_2; k_1', k_2')\, \delta(\vec{k}_1 + \vec{k}_2 - \vec{k}_1' - \vec{k}_2') \\ &\qquad \int d\mu\{z\}\, e^{-\beta F^{(0)}\{z\}}\, \bar{z}_{k_1} \bar{z}_{k_2} z_{k_1'} z_{k_2'} + \cdots \end{aligned} \tag{2.17}$$

where use has been made of the overlap integral

$$\langle \{z\} \mid \{z'\} \rangle = \exp\{-\sum_k [\tfrac{1}{2} | z_k - z_k' |^2 - i\, \mathrm{Im}(\bar{z}_k z_k')]\} \tag{2.18}$$

and of course only terms involving just two particle diagrams have been kept. It is obviously straightforward to extend the procedure to include higher-order terms, but this does not affect the generality of the description of the method.

It is convenient to introduce "polar" coordinates

$$z_k = | z_k | \exp(i\theta_k) \tag{2.19}$$

and one has

$$d\, \mathrm{Re}\, z_k d\, \mathrm{Im}\, z_k = | z_k | d\, | z_k |\, d\theta_k = \tfrac{1}{2} dr_k d\theta_k \tag{2.20}$$

where $r_k = | z_k |^2$

The integral part of equation (2.17) becomes then

$$\begin{aligned} &\int d\mu\{z\}\, e^{-\beta F^{(0)}\{z\}}\, \bar{z}_{k_1} \bar{z}_{k_2} z_{k_3} z_{k_4} \\ &= \left[\prod_{i=1}^{4} \frac{1}{2\pi} \int_0^\infty r_{k_i}^{\frac{1}{2}} dr_{k_i} \int_0^{2\pi} d\theta_{k_i} \right] \int d\mu\{z'\}\, e^{-\beta F^{(0)}\{z\}}\, e^{i\Phi\{\theta_{k_i}\}} \end{aligned} \tag{2.21}$$

where $\{z'\}$ is the complementary set to $\{z_{k_1}, z_{k_2}, z_{k_3}, z_{k_4}\}$ in $\{z\}$, and $\Phi\{\theta_{k_i}\} = \theta_{k_3} + \theta_{k_4} - \theta_{k_1} - \theta_{k_2}$. Note that because of the factor $e^{i\Phi\{\theta_{k_i}\}}$ the integral will vanish unless the total phase $\Phi\{\theta_{k_i}\}$ vanishes. Together with the momentum conservation delta function in front, this leaves only two choices; either $\vec{k}_1 = \vec{k}_3$ or $\vec{k}_2 = \vec{k}_4$, giving obviously identical contributions to W_2. For higher orders the same mechanism would originate a sum over the permutations of the external particle lines.

There are, therefore, only two independent momenta left in the definition of the partition function

$$Z = Z^{(0)} - 2 \sum_{k_1, k_2} W_2(k_1, k_2; k_1, k_2) \int d\mu\{z''\} \left[\prod_{i=1}^{2} \int_0^\infty r_{k_i} dr_{k_i} \right] e^{-\beta F^{(0)}\{z\}} \tag{2.22}$$

where the phase integrals have been executed, each of them giving a factor 2π, and $\{z''\} \cup \{z_{k_1}, z_{k_2}\} = \{z\}$. Recalling the form of $F^{(0)}\{z\}$, integrals over $d\mu\{z''\}$ give each a factor

$$\frac{1}{\pi} \int_0^\infty e^{-\bar{z}_{k_i} z_{k_i}(1-e^{-\beta\tilde{\epsilon}_{k_i}})} d\,\mathrm{Re}\, z_{k_i} d\,\mathrm{Im}\, z_{k_i} = \frac{1}{1-e^{-\beta\tilde{\epsilon}_{k_i}}}, \qquad i \neq 1, 2 \tag{2.23}$$

The remaining two integrals are

$$\int_0^\infty e^{-r_{k_i}(1-e^{-\beta\tilde{\epsilon}_{k_i}})} r_{k_i} dr_{k_i} = \frac{1}{(1-e^{-\beta\tilde{\epsilon}_{k_i}})^2}, \qquad i = 1, 2 \tag{2.24}$$

Thus, there is a factor of $(1 - e^{-\beta\tilde{\epsilon}_{k_i}})^{-1}$ for all k_i, and the product of all of these is just the partition function for non interacting particles $Z^{(0)}$. Z becomes then

$$Z = Z^{(0)} \left[1 - 2 \sum_{k_1, k_2} \frac{1}{(1-e^{-\beta\tilde{\epsilon}_{k_1}})(1-e^{-\beta\tilde{\epsilon}_{k_2}})} W_2(k_1, k_2; k_1, k_2) + \cdots \right] \tag{2.25}$$

And the free energy

$$F = -\frac{1}{\beta} \lg Z \approx F^{(0)} + \frac{2}{\beta} \sum_{k_1, k_2} \frac{1}{(1-e^{-\beta\tilde{\epsilon}_{k_1}})(1-e^{-\beta\tilde{\epsilon}_{k_2}})} W_2(k_1, k_2; k_1, k_2) + \cdots \tag{2.26}$$

The second factor is the leading dynamical correction to the free energy. Other factors would of course be recovered by expanding $F\{z\}$ to higher orders about the origin in z-space. Now only the explicit calculation of W_2 is left over in order to evaluate both the partition function and the free energy dynamically corrected. The calculation is standard, and it is shortly discussed in the following for sake of completeness, even if it somewhat exorbits in its spirit the content of the present paper.

First an integral equation for W_2 should be written consisting of the formal sum of all the diagrams. Such an equation, however, depends on the temperature in a complicated way which makes it rather different from the ordinary scattering equation. Observing that temperature enters each diagram only through integrals over Boltzmann factors, which are indeed convolution integral with finite upper limit (the energy is the sum of all unperturbed single particle energies), the use of the "temperature" Laplace transform allows us to relate the equation for W_2 to that of the usual t-matrix (Bloch and De Dominicis, 1958; Abrikosov, Gorkov, and Dzialoshinski, 1963).

The transformation can be performed over each diagram separately, and this amounts to a change in the rules for evaluating the diagram itself.

Consider Figure 3. Its value, by the usual rules is

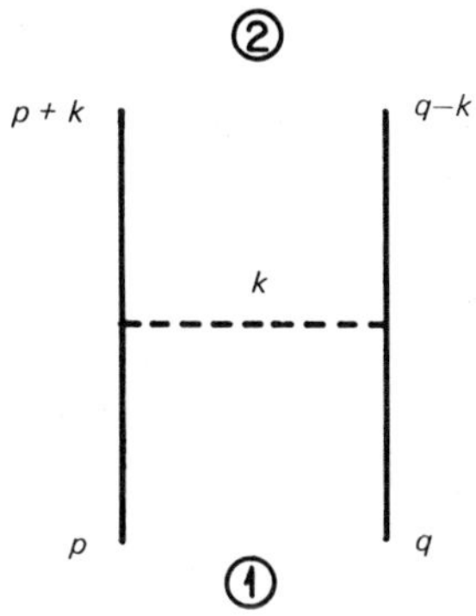

Figure 3

$$\frac{1}{\Omega} V(k) \int_0^{\beta} e^{-\beta' E_1} e^{-(\beta - \beta') E_2} \, d\beta' = \frac{1}{\Omega} V(k) \frac{e^{-\beta E_1} - e^{-\beta E_2}}{E_1 - E_2} \tag{2.27}$$

where the energy of the two free particles is, e.g.,

$$E_1 = \tilde{\epsilon}_p + \tilde{\epsilon}_q = \epsilon_p + \epsilon_q - 2\mu \tag{2.28}$$

$$E_2 = \epsilon_{p+k} + \epsilon_{q-k} - 2\mu \tag{2.29}$$

Since the total momentum of the pair $\vec{P} = \vec{p} + \vec{q}$ is a constant of the motion and will be the same in each term, it is convenient to label the two configurations ① and ② by half the relative momentum

$$\vec{Q} = \tfrac{1}{2}(\vec{p} - \vec{q}) \qquad \vec{Q}' = \tfrac{1}{2}[(\vec{p} + \vec{k}) - (\vec{q} - \vec{k})] = \vec{Q} + \vec{k} \tag{2.30}$$

The Laplace transform of the diagram then reads

$$\int_0^{\infty} e^{-\beta s} \frac{1}{\Omega} V(k) \frac{e^{-\beta E_Q} - e^{-\beta E_{Q'}}}{E_Q - E_{Q'}} \, d\beta = \frac{1}{E_{Q'} + s} \frac{V(k)}{\Omega} \frac{1}{E_Q + s} \tag{2.31}$$

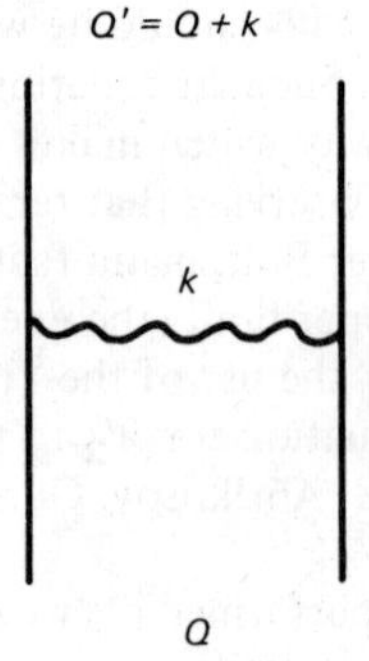

Figure 4

The Laplace transform of a convolution is just the product of individual Laplace transforms.

By comparison of this result with the diagram, one extracts the following set of rules (in the graphical representation we will use wavy lines for the interaction, in order to distinguish the two).

Given Figure 4, one has

(a) a factor $V(k)/\Omega$ for each interaction line;
(b) a factor $1/(E+s) = \int_0^\infty e^{-\beta(E+s)}\, d\beta$ for each horizontal line which may be drawn between successive interactions;
(c) symmetry factors and sum over internal momenta as usual.

In this way one may define the Laplace transform of W_2 as

$$\int_0^\infty e^{-\beta s} W_2(\vec{p},\vec{q};\vec{k}+\vec{p},\vec{q}-\vec{k})d\beta = \frac{1}{E_{Q'}+s}\,\frac{t(Q,Q')}{\Omega}\,\frac{1}{E_Q+s} \tag{2.32}$$

and the matrix $t(Q, Q')$ turns out to be the usual t-matrix for the scattering of two bosons of total energy $E = -s$. Indeed, it satisfies the integral equation (Figure 5) which, written explicitly reads:

$$t(Q,Q') = V(k) - \frac{2}{\Omega}\sum_{Q''} V(Q'-Q'')\,\frac{1}{E_{Q''}+s}\,t(Q,Q'') \tag{2.33}$$

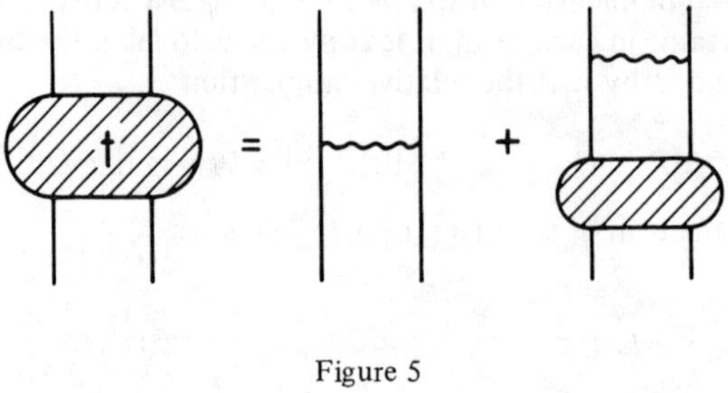

Figure 5

Once such an equation is solved, W_2 is simply determined by computing the inverse transform

$$W_2(\vec{p}, \vec{q}; \vec{p} + \vec{k}, \vec{q} - \vec{k}) = \frac{1}{2\pi i \Omega} \int_{s_0 - i\infty}^{s_0 + i\infty} \frac{1}{E_{Q'} + s} t(Q, Q') \frac{1}{E_Q + s} e^{\beta s} ds \qquad (2.34)$$

where the integration is over the usual Brömwich contour defined by s_0 to the right of any singularities in the s-plane of the integrand.

In other words, the leading dynamical correction to the partition function is related to the inverse Laplace transform of $t(Q, Q)$.

Now it is reasonable to assume that $t(Q, Q')$ is analytic in s except for the poles corresponding to the bound states of two bosons, and possibly along a branch cut, i.e., a densely discrete set of poles corresponding to a band.

In such a situation, the contour of integration for the integral defining W_2 may be deformed and the integral evaluated as the sum over the residues plus the integral around the branch cut.

One gets

$$W_2(\vec{p}, \vec{q}; \vec{p} + \vec{k}, \vec{q} - \vec{k}) = \frac{t(Q, Q')\,|_{s=-E_Q}\, e^{-\beta E_Q} - t(Q, Q')\,|_{s=-E_Q}\, e^{-\beta E_{Q'}}}{E_Q - E_{Q'}}$$

$$+ \frac{1}{\pi} \int_{s_-}^{s_+} \frac{1}{E_{Q'} + s} \operatorname{Im} t\,(Q, Q')\,|_{s - i\epsilon} \frac{1}{E_Q + s} e^{\beta s}\, ds \qquad (2.35)$$

where the branch cut lies along the axis between s_- and s_+ and Im t is to be evaluated below the cut.

Generally, the branch cut integration leads to a negligible factor in the sense that it is of higher order in the temperature. The diagonal element of W_2, which enters the correction to Z can therefore be approximated as

$$W_2(p, q; p, q) \approx \frac{1}{\Omega} \beta\, e^{-\beta E_Q}\, t(Q, Q)\,|_{s=-E_Q} \qquad (2.36)$$

3. *The General Case*

The generalized definition for coherent states of an arbitrary Lie group G is the following. Let T be the irreducible unitary representation of G acting in the Hilbert space $\mathscr{H}$; $|\Phi_0\rangle$ some fixed vector in $\mathscr{H}$; h the stationary subgroup of G with respect to $|\Phi_0\rangle$ and $M = G/h$ the homogeneous factor space. Then a system of coherent states is the set $\{|\Phi_g\rangle\}$

$$|\Phi_g\rangle = \mathrm{T}(g)\,|\Phi_0\rangle, \qquad g \in G \qquad (3.1)$$

where g runs over the whole group G. The coherent state $|\Phi_g\rangle$ is then uniquely determined by the point $x = x(g)$, $x \in M$

$$|\Phi_g\rangle = e^{i\alpha(g)}\,|x\rangle; \,|\Phi_0\rangle = |0\rangle \qquad (3.2)$$

In (3.2) $e^{i\alpha(h)}$ is a one dimensional unitary representation of $\mathscr{h}$, for $h \in \mathscr{h}$; the function $\alpha(g)$ on the other hand is determined on the entire group G and for $g \in \mathscr{h} \subset G$, it coincides with $\alpha(h)$. It can be easily seen that the set of coherent states is invariant relative to the action of the group representation operators:

$$\mathrm{T}(g') \mid \Phi_g \rangle = \mathrm{T}(g')\, \mathrm{T}(g) \mid \Phi_0 \rangle = \mathrm{T}(g'') \mid \Phi_0 \rangle = \mid \Phi_{g''} \rangle \tag{3.3}$$

where

$$g'' = g'g \in G \tag{3.4}$$

This formula can be rewritten

$$\begin{aligned} \mathrm{T}(g') \mid x \rangle &= e^{-i\alpha(g)}\, \mathrm{T}(g') \mid \Phi_g \rangle = e^{-i\alpha(g)} \mid \Phi_{g''} \rangle \\ &= e^{-i\alpha(g)}\, e^{i\alpha(g'')} \mid x'' \rangle = e^{i\beta(g,g')} \mid g'x \rangle \end{aligned} \tag{3.5}$$

with

$$\beta(g, g') = \alpha(g'g) - \alpha(g) \tag{3.6}$$

showing that the complete set of coherent states is actually generated by the action of the group G on the homogeneous space M. This isomorphism between $\mathscr{H}$, and a set $\{f\}$ of analytic functions over the manifold M, is of special physical interest in the case when the functions are square-integrable. The system of coherent states can then be defined whenever the integral

$$\int |\langle 0 \mid x \rangle|^2\, d_\mu x = \Gamma \tag{3.7}$$

where $d_\mu x$ denotes a "measure" on M, converges, i.e., Γ is a finite constant. In such a case

$$\frac{1}{\Gamma} \int d_\mu x \mid x \rangle \langle x \mid = I \tag{3.8}$$

and one can expand any arbitrary state in coherent states.

Moreover on M one has the important set of two points amplitudes ("reproducing kernels")

$$K(x, y) = \frac{1}{\Gamma} \langle x \mid y \rangle \tag{3.9}$$

which satisfy the integral equation

$$K(x, z) = \int d_\mu y\, K(x, y)\, K(y, z) \tag{3.10}$$

i.e., K reproduces itself under convolution, and for an arbitrarily chosen function $f \in \{f\}$, gives rise to the integral identity

$$f(x) = \frac{1}{\Gamma} \int d_\mu y \langle x \mid y \rangle f(y) \tag{3.11}$$

This is equivalent to performing a harmonic analysis on M in terms of irreducible representations of G (compare for example the solution of Helmholtz equation in terms of prolate spheroidal coordinates in R^3). So we may think of K as the kernel of a linear transformation in M. Due to the general differential geometric versus algebraic theoretical approach, the method is valid for both compact and noncompact groups.

The partition of identity, which depends on the overcompleteness of the set of coherent states system (this means that there are subsystems still being coherent and complete states), allows us to write the partition function Z as

$$Z = \int_{M^{\#}} d_\mu x \langle x \mid e^{-\beta H} \mid x \rangle = \lim_{n' \to \infty} \int_{M^{\#}} d_\mu x \langle x \mid (I - \frac{\beta}{n'} H)^{n'} \mid x \rangle \tag{3.12}$$

$| x \rangle$ being the coherent states for some Lie group G, and $M^{\#} \subseteq M$.

The group G is now assumed not to be a symmetry group of the Hamiltonian H, but its dynamical group, i.e., a group whose algebra may be used to generate the spectrum of H. In this case the dynamical problem involves a Hamiltonian which may in general be expressed in terms of a set of operators which are generators of the algebra g of G. In the simplest case H is a linear combination of the dynamical group generators (possibly a direct sum if the spectrum generating group is a direct product of subgroups $G^{(k)}$)

$$H = \sum_{j=1}^{n} \oplus_k \, \omega_j J_j^{(k)} \tag{3.13}$$

where the index k runs over the set of labels of $G^{(k)}$, and denoting by α_{ijl} the structural constant of G, g is defined by

$$[J_i^{(k)}, J_j^{(k')}] = \alpha_{ijl} J_l^{(k)} \delta_{k', k} \tag{3.14}$$

(summation is implied on the dummy indexes).

A general element of the Lie algebra can be written as

$$S = i(\bar{g} J - g J^\dagger) \tag{3.15}$$

where

$$\bar{g}J = \bar{g}_i J_i, \qquad gJ^\dagger = g_i J_i^\dagger \tag{3.16}$$

$g \equiv (g_1, \ldots, g_n)$ being an n-dimensional complex vector. The Lie group is obtained from the algebra by the usual exponential mapping: the operator in (3.12) is just a representative $T(g_0)$ of a selected $g_0 \in G$ (in general $g_0 = \prod_k \otimes g_0^{(k)}$).

Differential operators on $\mathscr{H}$ corresponding to the infinitesimal generators $J_i^{(k)}$ may be obtained by standardized procedure, and finally all the irreducible representations can be found by considering the action of such generators on the monomials in $\mathscr{H}$. Unitary irreducible representations are constructed by suitably normalizing the basis vectors, introducing an inner product and imposing the proper hermiticity condition on the generators.

The orthonormal basis vectors may typically be labelled by the eigenvalues of the Casimir operators $C^{(k)}$ and of one of the infinitesimal generators, say $J_\alpha^{(k)}$. (Indeed they are eigenvectors of $J_\alpha^{(k)}$ in the representation corresponding to the subgroup $\hbar$.)

In principle a rotation may be performed in the space of the algebra about some $J_\beta^{(k)}$ axs–which is just a generalization of the Bogolubov transformation (Solomon, 1971)

$$R(\theta_k) = \exp[-iJ_\beta^{(k)}\theta_k] \tag{3.17}$$

$$R = \prod_{k\otimes} R^{(k)}(\theta_k) \tag{3.18}$$

and the set $\{\theta_k\}$ be chosen in such a way that the rotated Hamiltonian

$$H_R = RHR^{-1} = \sum_{k\otimes} \tilde{\omega}_k J_\alpha^{(k)} \tag{3.19}$$

depends–linearly–only on the diagonalized generator.

The eigenvalues of the Hamiltonian are then trivially dependent on the eigenvalues of $J_\alpha^{(k)}$. That is the reason why the group G is referred to as energy spectrum generating group, even though it is not itself a symmetry group of the Hamiltonian.

According to the general definition the set of coherent states are points of the manifold M, on which the action of the group is given by a differentiable mapping

$$f : G \times M \to M : (gx) \to f(x, g) \tag{3.20}$$

For a fixed $x_0 \in M$, the corresponding coherent state is determined by the mapping

$$\underline{x}(g) = f(x, g) \tag{3.21}$$

such that $\underline{x}^{-1}(x_0) = \hbar$; where $\hbar$ is the closed subgroup of G which leaves x_0 fixed. Since $\hbar$ is closed it is again a Lie group (Rasetti, 1973) and $G/\hbar$ is an analytic manifold.

The map $\underline{x}$ induces a map $\overline{\underline{x}} : G/\hbar \to M : (g, \hbar) \to f(g, x)$. This mapping is well defined since $\hbar$ leaves the point x_0 fixed and the following diagram commutes

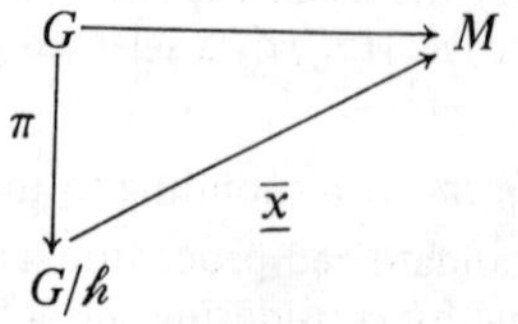

where π denotes the natural projection $g \to g\hbar$ of G onto $G/\hbar$.

It follows that the orbits of G on M are submanifolds of M. Complete coherent states constitute therefore a set $M^{\#}$ of orbits in M. On the other hand, in our case H_R is nothing but J_α and its action over $|x\rangle$ induces a flow on M.

$$\langle x' \mid e^{-\beta H_R} \mid x\rangle = \prod_{k\otimes} \langle x' \mid \mathrm{T}(g_0^{(k)}) \mid x\rangle = \sum_k \langle x' \mid g_0^{(k)} x\rangle \tag{3.22}$$

where $\mathrm{T}(g_0^{(k)})$ is the representative of the group translation induced on M by the rotated Hamiltonian H_R.

Note that since the integral in the definition of Z is over the manifold M, it is invariant under the substitution of H with H_R.

It was already pointed out that

$$\langle x' \mid x''\rangle = e^{i[\alpha(g') - \alpha(g'')]} \langle 0 \mid \mathrm{T}(g'^{-1} g'') \mid 0\rangle = \overline{\langle x'' \mid x'\rangle} \tag{3.23}$$

and therefore

$$\langle x' \mid g_0^{(k)} x''\rangle = e^{i[\alpha(g') - \alpha(g'') - \beta(g_0^{(k)}, g'')]} \langle 0 \mid \mathrm{T}(g'^{-1} g_0^{(k)} g'') \mid 0\rangle \tag{3.24}$$

It follows that evaluation of Z is nothing but an integration (indeed over the submanifold $M^{\#}$ of all the group orbits) of the scalar product of the generic coherent state with the fixed vector $|0\rangle$.

It is to be noted how, even in the case when H is not simply linear in the J_i's, the application of Baker-Campbell-Haussdorff formula, which is at the basis of our generalized Bogolubov rotation, together with the commutation relations (3.14) lead to the same structure when G is the spectrum generating group of H. Moreover, the exponential mapping does not have a vanishing Jacobian at the origin (i.e., it is a diffeomorphism of an open neighborhood of zero in g ($g_1 = \cdots = g_n = 0$) onto an open neighborhood of the identity in G) and therefore, any analytic function at the identity can be expanded in some neighborhood of $\mathbf{0} \in g$.

This leads to a diagrammatic expansion very similar to the case of the previous section when the group G is not itself a spectrum generating group (as the Weyl group) but the terms breaking the symmetry are "small." This is very similar to a random phase approximation.

Before closing this section, it is worth observing that the proposed method can be considered as a sort of generalization of the Feynman's path-integral approach to statistical mechanics.

4. *Superfluid Example*

It has been shown by A. I. Solomon (1971) that in the framework of the Foldy model approximation (Bassichis and Foldy, 1964), a many level superfluid system has a spectrum generating group which is a direct product $\prod_{k\otimes} SU(1,1)_{(k)}$ (where as usual the index k labels both the momenta and the levels).

The Hamiltonian in such a case may be written

$$H = \sum_{k\otimes} NV_k(-J_1^{(k)} + \mu_k J_2^{(k)} - \tfrac{1}{2}\mu_k) + \tfrac{1}{2}N^2 V_0 \tag{4.1}$$

where N is the total number of bosons, ϵ_k the energy of level kth, V_k the Fourier transform of the interaction potential and

$$\mu_k = 1 + \frac{\epsilon_k}{NV_k} \tag{4.2}$$

$J_i^{(k)}$ $(i = 1, 2, 3)$ are the generators of $SU(1,1)_{(k)}$, such that

$$[J_1^{(k)}, J_2^{(k')}] = -iJ_3^{(k)}\delta_{kk'}; \qquad [J_2^{(k)}, J_3^{(k')}] = iJ_1^{(k)}\delta_{kk'};$$
$$[J_3^{(k)}, J_1^{(k')}] = iJ_2^{(k)}\delta_{kk'} \tag{4.3}$$

By the hyper-rotation

$$R = \prod_{k\otimes} R^{(k)}(\theta_k), \qquad R^{(k)}(\theta_k) = \exp(-J_2^{(k)}\theta_k) \tag{4.4}$$

with

$$\theta_k = \coth^{-1}\mu_k \tag{4.5}$$

one gets

$$H_R = RHR^{-1} = \sum_{k\otimes} (\operatorname{csch}\theta_k J_3^{(k)} - \tfrac{1}{2}\mu_k)NV_k + \tfrac{1}{2}N^2 V_0 \tag{4.6}$$

The system of coherent states $\{|\xi\rangle\}$ for the universal covering group of $SU(1,1)$ has been thoroughly discussed by Perelomov (1973). Its relevant properties are briefly reconsidered hereafter.

The factor space M, isomorphic to the upper sheet of a three-dimensional hyperboloid, is the unit disk $|\xi| < 1$, whose invariant Riemannian measure is

$$d\mu(\xi) = \frac{d\,\mathrm{Re}\,\xi\, d\,\mathrm{Im}\,\xi}{(1 - |\xi|^2)^2} \tag{4.7}$$

The scalar product is given by

$$\langle\xi|\xi'\rangle = (1 - |\xi'|^2)^{\kappa}(1 - |\xi|^2)^{\kappa}(1 - \bar{\xi}'\xi)^{-2\kappa} \tag{4.8}$$

where κ is an arbitrary nonnegative number, and the condition of completeness has the form

$$\frac{2\kappa - 1}{\pi}\int d\mu(\xi)\,|\xi\rangle\langle\xi| = I \tag{4.9}$$

Finally, the action of a group representation operator $T^{(\kappa)}(g)$ on $|\xi\rangle$ results in

$$T^{(\kappa)}(g)|\xi\rangle = e^{i\phi}|\xi'\rangle \tag{4.10}$$

where, parametrizing the group by the set of 2×2 complex matrices

$$g = \begin{Vmatrix} a & b \\ \bar{b} & \bar{a} \end{Vmatrix}, \qquad \det g = |a|^2 - |b|^2 = 1 \tag{4.11}$$

one has

$$\xi' = \frac{a\xi - b}{-\bar{b}\xi + \bar{a}}; \qquad \phi = 2\kappa \arg(a - b\xi) \tag{4.12}$$

so that

$$\langle \eta | T^{(\kappa)}(g) | \zeta \rangle = e^{i\phi} \langle \eta | g\zeta \rangle = (1 - |\eta|^2)^\kappa (1 - |\zeta|^2)^\kappa (\bar{a} - \bar{b}\zeta + b\bar{\eta} - a\bar{\eta}\zeta)^{-2\kappa} \tag{4.13}$$

and

$$\langle \xi | T^{(\kappa)}(g) | \xi \rangle = (1 - |\xi|^2)^{2\kappa} (\bar{a} - a|\xi|^2 - 2i \operatorname{Im} \bar{b}\xi)^{-2\kappa} \tag{4.14}$$

If the above-mentioned procedure of rotation is performed, the exponentiation of the Hamiltonian leads simply to the irreducible representation of an element h of the subgroup $\mathcal{h}$;

$$h \equiv \begin{Vmatrix} e^{i\psi/2} & 0 \\ 0 & e^{-i\psi/2} \end{Vmatrix}$$

and the representation T of G being restricted on the subgroup has to contain, by the Fröbenius reciprocity theorem, the one-dimensional abelian representation of $\mathcal{h}$ itself. Except for a phase factor

$$\frac{a}{\bar{a}} = e^{i\psi}$$

ξ' is identical to ξ:

$$\langle \xi | T^{(\kappa)}(h) | \xi \rangle = (1 - |\xi|^2)^{2\kappa} (1 - e^{i\psi} |\xi|^2)^{-2\kappa} e^{i\kappa\psi} \tag{4.15}$$

Substituting back into (3.12) one gets

$$Z = \sum_k \left\{ e^{\frac{1}{2}\beta N(\mu_k V_k - NV_0)} \frac{2\kappa - 1}{\pi} e^{\kappa\beta \operatorname{csch} \theta_k} \int_{|\xi| < 1} \frac{d \operatorname{Re} \xi \, d \operatorname{Im} \xi (1 - |\xi|^2)^{2\kappa - 2}}{(1 - e^{\beta \operatorname{csch} \theta_k} |\xi|^2)^{2\kappa}} \right\} \tag{4.16}$$

The analysis of Solomon shows that the Casimir operators

$$C_k = -(J_1^{(k)})^2 - (J_2^{(k)})^2 + (J_3^{(k)})^2 \tag{4.17}$$

can be written

$$C_k = j_k(j_k + 1) = \tfrac{1}{4}(\Delta_k^2 - 1) \tag{4.18}$$

where

$$\Delta_k = a_k^\dagger a_k - a_{-k}^\dagger a_{-k} \tag{4.19}$$

are the differences between number of particles in opposite momentum states, so that

$$j_k = -\tfrac{1}{2} - \tfrac{1}{2} \mid \Delta_k \mid \tag{4.20}$$

On the other hand, the only allowed representation is given by the positive discrete series (Vilenkin, 1968)

$$\prod_k {}_{\otimes}\mathscr{D}^+(j_k), \qquad j_k = -\kappa \tag{4.21}$$

This suggests the choice $\kappa = \frac{1}{2}$.

Such a choice may appear singular at first sight, because of equation (4.9), but the limit $\kappa \to \frac{1}{2}$ indeed exists and it is finite.

Any integral over the unit disk can, in general, be reduced in the form of ordinary integrals performing harmonic analysis (Helgason, 1962, 1965) over the non-Euclidean manifold $D \equiv \{\xi : |\xi| \langle 1\}$. Denote in the following by $B = \partial D$ the boundary of D, and define

$$\langle \xi, b \rangle = \frac{(\xi, b)}{(1 - |\xi|^2)^2}, \qquad \xi \in D, \qquad b \in B \tag{4.22}$$

as the non-Euclidean distance from 0 to the orthogonal trajectory to the family of all parallel geodesics corresponding to $b = e^{i\phi}$, passing through $\xi((\xi, b)$ is the usual inner product in R^2). Consider moreover the Hilbert space

$$\mathscr{H}_\lambda = \{h_\lambda(\xi) = \int_B e^{(i\lambda + 1)\langle \xi, b \rangle} h(b) db \mid h \in \mathscr{L}^2(B)\} \tag{4.23}$$

Then

$$[T_\lambda(g) h_\lambda](\xi) = h_\lambda(g^{-1}\xi) \tag{4.24}$$

define the unitary irreducible representation of G in $\mathscr{H}_\lambda$. Moreover

$$h_\lambda(g^{-1}\xi) = \int_B e^{(i\lambda + 1)\langle \xi, b \rangle} e^{(-i\lambda + 1)\langle g . 0, b \rangle} h(g^{-1}b) db \tag{4.25}$$

On the other hand

$$P(\xi, b) = e^{2\langle \xi, b \rangle} = \frac{1 - |\xi|^2}{1 - 2|\xi| \cos(\theta - \phi) + |\xi|^2}, \qquad \xi = |\xi| e^{i\theta} \tag{4.26}$$

is just the Poisson Kernel expressed in non-Euclidean terms. Therefore recalling the Laplace–Beltrami operator (Karpelevic, 1965) on D

$$\Delta : f \to \frac{1}{\sqrt{\det(g_{ij})}} \sum_k \frac{\partial}{\partial x_k} \left(\sum_i g^{ik} \sqrt{\det(g_{ij})} \frac{\partial}{\partial x_i} \right) f \tag{4.27}$$

with

$$g_{ij} = [1 - |\xi|^2]^{-2} \delta_{ij} \tag{4.28}$$

$$g^{ij} = (g_{ij})^{-1}; \qquad \det(g_{ij}) = (1 - |\xi|^2)^{-4} \tag{4.29}$$

where δ_{ij} is the Kronecker delta symbol

$$\Delta_\xi = [1 - |\xi|^2]^2 \left[\frac{\partial^2}{\partial[\mathrm{Re}\,\xi]^2} + \frac{\partial^2}{\partial[\mathrm{Im}\,\xi]^2}\right] \qquad (4.30)$$

one easily finds that any power of the Poisson Kernel gives an eigenfunction of the non-Euclidean Laplacian

$$\Delta_\xi [P(\xi, b)]^\mu = 4\mu(\mu - 1)\, P(\xi, b); \qquad \mu \in \mathbf{C} \qquad (4.31)$$

with eigenvalue independent on b.

If $M(B)$ denotes the set of analytic functions on B, which is considered as an analytic manifold (observe that M carries a natural topology), the continuous linear functions $\nu : M(B) \to \mathbf{C}$ constitute a space $M^d(B)$ dual of $M(B)$; ν are called analytic functionals on B.

A theorem by S. Helgason proves that

$$F(\xi) = \int_B P(\xi, b)^\zeta\, d\nu(b); \qquad \nu \in M^d \qquad (4.32)$$

are eigenfunctions of Δ_ξ, with real eigenvalues if $\zeta \in R$.

Hence integration over ζ amounts to performing harmonic analysis over D with respect to G. One may check there exists a measure

$$d\mu(\lambda) = \frac{1}{2\pi^2} \lambda\, th\left(\frac{\pi}{2}\lambda\right) d\lambda$$

such that

$$\int_{R/Z_2} \mathcal{H}_\lambda d\mu(\lambda) = \mathcal{L}^2(D) \qquad (4.33)$$

(the integration is over R/Z_2, because of definition (4.23) and equation (4.24) which imply that T_{λ_1} is equivalent to T_{λ_2} only if $\lambda_1 = -\lambda_2$). One has therefore:

$$\int_D f(\xi) d\mu(\xi) = \int_{R \times B} d\mu(\lambda)\, db\, \tilde{f}(\lambda, b)\, \tilde{\Delta}(\lambda, b) \qquad (4.34)$$

where

$$\tilde{\Delta}(\lambda, b) = \frac{1}{(2\pi)^2} \int_D e^{(-i\lambda + 1)\langle \xi, b\rangle} d\mu(\xi) \qquad (4.35)$$

and

$$\tilde{f}(\lambda, b) = \int_D f(\xi)\, e^{(-i\lambda + 1)\langle \xi, b\rangle} d\mu(\xi) \qquad (4.36)$$

being

$$f(\xi) = \frac{1}{(2\pi)^2} \int_{R \times B} \tilde{f}(\lambda, b)\, e^{(i\lambda + 1)\langle \xi, b\rangle} d\mu(\lambda)\, db \qquad (4.37)$$

In the present case the calculation is much easier because $f(\xi) = f^{\#}(r)$ is only function of the distance r from 0 to the geodesics through ξ

$$r = \frac{1}{2} \ln \frac{1 + |\xi|}{1 - |\xi|}, \qquad |\xi| = \tanh \tag{4.38}$$

and

$$d\mu(\xi) = \pi \sinh 2r \cdot dr; \qquad 0 \leqslant r < \infty \tag{4.39}$$

Both $\tilde{f}$ and $\tilde{\Delta}$ are function of λ alone

$$\int_B e^{(i\lambda + 1)\langle \xi, b \rangle} db = P_{-\frac{1}{2}-\frac{1}{2}i\lambda}(\cosh r) \cosh (2r) \tag{4.40}$$

and

$$\tilde{\Delta}(\lambda) = \pi \int_0^\infty P_{-\frac{1}{2}-\frac{1}{2}i\lambda}(\cosh r) \cosh (2r) \sinh (2r)\, dr \tag{4.41}$$

while $f^{\#}(r)$ is simply the Mehler transform of $f(\lambda)$; $P_\eta(\cosh r)$ being the Legendre function. A simple calculation and changing of variables allows writing

$$Z = \frac{1}{2\pi} \sum_k \left\{ e^{\frac{1}{2}\beta E_k} \frac{A_k}{\cosh \theta_k} \int_{e^{-2\theta_k}}^{e^{2\theta_k}} \frac{dx}{x} \frac{(x+1)(1 - e^{-\beta E_k} x)}{x^2 + 2A_k x + e^{\beta E_k}} \right. \tag{4.42}$$

where

$$A_k = \frac{\sinh^2 \theta_k}{2 \cosh^2 \theta_k - e^{-\beta E_k}} \Bigg\} \tag{4.43}$$

θ_k being given by equation (4.5) and (4.2), and E_k by the known Bogolubov formula

$$E_k = (\epsilon_k{}^2 + 2\epsilon_k N V_k)^{\frac{1}{2}} \tag{4.44}$$

Use now the integral representation (Vilenkin, 1968)

$$\mathscr{B}^l_{mn}(\cosh \tau) = \frac{1}{2\pi i} \oint_{\Gamma'} w^{l-n} \left(\cosh \frac{\tau}{2} + z \sinh \frac{\tau}{2} \right)^{2n} \frac{z^{m-n}}{\sqrt{w^2 - 2w \cosh \tau + 1}}\, dw \tag{4.45}$$

where the contour Γ' is denoted in Figure 6, and

$$z = \frac{w - \cosh \tau \pm \sqrt{w^2 - 2w \cosh \tau + 1}}{\sinh \tau}$$

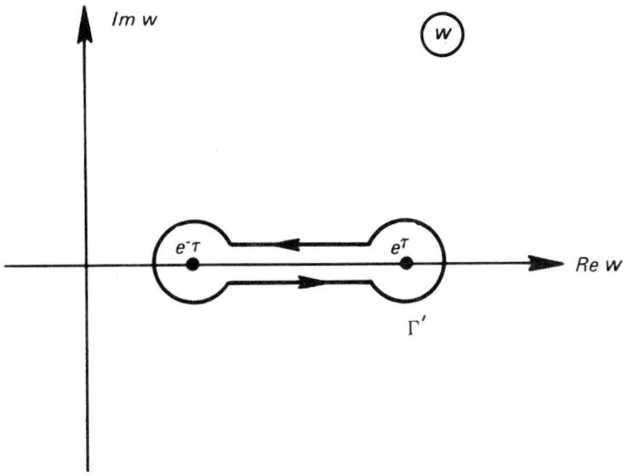

Figure 6

the radical being chosen such that $1 \leqslant |z| \leqslant \coth(\tau/2)$; for the Jacobi function $\mathscr{B}^{l}_{mn}(\cosh\tau)$, by means of which the matrix elements of the representation

$$T^{(\kappa)}(g_\tau), \quad g_\tau \equiv \begin{Vmatrix} \cosh\frac{\tau}{2} & \sinh\frac{\tau}{2} \\ \sinh\frac{\tau}{2} & \cosh\frac{\tau}{2} \end{Vmatrix} \in \Omega_1 \subset SU(1,1)$$

can be expressed as

$$t^{(j,\sigma)}_{m,n}(g_\tau) = \mathscr{B}^{j}_{m+\sigma,n+\sigma}(\cosh\tau) \tag{4.46}$$

Contracting the integration contour to the segment which it encloses, assuming $j = -\sigma = -\frac{1}{2}$, one gets

$$\mathscr{B}^{-\frac{1}{2}}_{n+\frac{1}{2},n+\frac{1}{2}}(\cosh\tau) = \frac{1}{\pi}\int_0^{\tau} \frac{\cosh[(n+\frac{1}{2})t]\,\cos[(2n+1)\alpha]}{\sqrt{\cosh^2\frac{\tau}{2} - \cosh^2\frac{t}{2}}} \tag{4.47}$$

with

$$\cos\alpha = \frac{\cosh\frac{\tau}{2}}{\cosh\frac{\tau}{2}} \tag{4.48}$$

and finally

$$Z = \sum_k e^{-\frac{1}{2}\beta E_k} \sum_{n_k} e^{-\beta n_k E_k} t^{(-\frac{1}{2},\frac{1}{2})}_{n_k, n_k} (g_0^{(k)}) \tag{4.49}$$

$$g_0^{(k)} \equiv \begin{Vmatrix} \cosh \theta_k & \sinh \theta_k \\ \sinh \theta_k & \cosh \theta_k \end{Vmatrix} \tag{4.50}$$

The use of coherent states gives at once the result which, by use of the explicitly calculated eigenstates of the Hamiltonian (Solomon, 1971)

$$| \psi\{n_k\} \rangle = \prod_{k \otimes} \sum_{m_k} t^{(-\frac{1}{2},\frac{1}{2})}_{m_k n_k} (g^{(k)}) \mid m_k \rangle \tag{4.51}$$

$$g_0^{(k)} = \begin{Vmatrix} \cosh \frac{\theta_k}{2} & \sinh \frac{\theta_k}{2} \\ \sinh \frac{\theta_k}{2} & \cosh \frac{\theta_k}{2} \end{Vmatrix} \tag{4.52}$$

could have been obtained—in the relatively simple example here examined—by direct computation.

The above discussed method is, however, more powerful. Indeed, recalling the general relation between H and G (Rasetti, 1973), and the characteristic property of coherent states (Perelomov, 1973)

$$| \xi \rangle = (1 - | \xi |^2)^{\kappa} e^{\xi \sum_{k \oplus} J_+^{(k)}} | 0 \rangle \tag{4.53}$$

and

$$\xi J_+^{(k)} | \xi \rangle = (J_3^{(k)} - \kappa) | \xi \rangle \tag{4.54}$$

$$J_-^{(k)} | \xi \rangle = \xi (J_3^{(k)} + \kappa) | \xi \rangle \tag{4.55}$$

even in the case when the Hamiltonian is not simply linear in the group generators one can—exactly in the same way as in Section 2—obtain a diagrammatic expansion of Z to any order. The calculation of the dynamical correction—not explicitly reported here—is formally identical in the case when H is bilinear, and can be straightforwardly extended to higher order interactions.

5. *Conclusion*

It is worth noticing, in going through the previous example, that the sum over k in equation (4.42) is convergent only if

$$\beta \rangle \max_k \left[\frac{2 \coth^{-1} \mu_k}{E_k} \right] \tag{5.1}$$

This points out how the topology of the homogeneous space of the group is relevant in creating the singularities of Z. The particularly simple expression of Z as an integral over a manifold whose geometry is generally known from the group structure, should allow an *a priori* discussion of such singularities, without explicitly performing the calculation.

References

Abrikosov, A. A., Gorkov, L. P., and Dzialoshinski, J. E. (1963). *Methods of Quantum Field Theory in Statistical Physics.* Prentice-Hall, New Jersey.
Bassichis, W. H., and Foldy, L. L. (1964). *Physical Review*, **133**, A935–A943.
Bloch, C., and De Dominicis, C. T. (1958). *Nuclear Physics*, 7, 459–479.
Bloch, C., and De Dominicis, C. T. (1959). *Nuclear Physics*, **10**, 181–196.
Dyson, F. J. (1956). *Physical Review*, **102**, 1217–1230.
Helgason, S. (1962). *Differential Geometry and Symmetric Spaces.* Academic Press, New York.
Helgason, S. (1965). *Bulletin of the American Mathematical Society*, 71, 757–763.
Karpelevic, F. I. (1965). *Trudy Moskovskava Matematicheskava Obsčestva*, **14**, 48–185.
Klauder, J. R. (1970). *Journal of Mathematical Physics*, **11**, 609–630.
Langer, J. S. (1968). *Physical Review*, **167**, 183–190.
Perelomov, A. M. (1972). *Communications in Mathematical Physics*, **26**, 222–236.
Perelomov, A. M. (1973). *Theoretical Mathematical Physics*, **16**, 303–314.
Rasetti, M. (1973). *International Journal of Theoretical Physics*, **13**, 425–430.
Solomon, A. I. (1971). *Journal of Mathematical Physics*, **12**, 390–394.
Vilenkin, N. J. (1968). *American Mathematical Society Translations*, **22**.

APPLICATIONS OF COHERENT STATES IN THERMODYNAMICS AND DYNAMICS

R. Gilmore

Physics Department
University of South Florida
Tampa, Florida 33620

Coherent states for an arbitrary Lie group G are defined and their properties enumerated. These properties lead to a simple algorithm for studying the thermodynamic critical properties of Hamiltonians constructed from the generators of G. The algorithm is:

1. Replace H by its "classical limit" and add an entropy term $-kTs(r)$.
2. Determine how the minima of the resulting function change as a function of T.

Step 1 is implemented using Group Theory. The classical limit is taken using coherent states, and s(r) is a group multiplicity factor. Step 2 is implemented using Catastrophe Theory. This algorithm is used to study the thermodynamics of extended Dicke models, general bialgebraic models, and nuclear models. Ground state energy phase transitions are determined by setting T = 0 and studying the bifurcations of the classical limit of H as a function of the interaction parameters. The "crossover theorem" is discussed. The dynamical properties of H are also described simply in the coherent state representation. Dynamical equations of motion are derived when $\hat{H} \in \mathfrak{g}$ and TDHF equations are derived in the general case. Phase transitions for nonequilibrium steady state systems are determined by a "dissipation-transformation" theorem, analog of the "fluctuation-transformation" theorem for equilibrium phase transitions. The identification of the equilibrium and nonequilibrium steady state equations of state (for the laser) with a (cusp) catastrophe manifold is shown.

1. History of Coherent States

Coherent states were first discussed by Schrödinger[1] in connection with the classical limit of the quantum-mechanical harmonic oscillator. They were subsequently used by Bloch and Nordsieck[2] to treat the "infrared catastrophe." The properties of these coherent states were then expounded formally by Schwinger,[3] who emphasized the relation between the properties of these states and the properties of the underlying Lie algebra. Finally, they were used by Glauber[4] in a very beautiful formulation of Quantum Optics. The harmonic oscillator coherent states are now a standard tool in any field in which the harmonic oscillator is a useful model for physical processes.

The realization came slowly that coherent states were more intimately associated with a Lie group rather than a Lie algebra, and that they could be associated with Lie groups more complicated than H(4) (harmonic oscillator group). Coherent states associated with the Lie algebra su(2) were discussed by Atkins and Dobson,[5] by Kutzner,[6] and by Radcliffe,[7] but the deep connection with the associated Lie group SU(2) was made by Arecchi, Courtens, Gilmore, and Thomas,[8] who showed that all of the properties of the "atomic coherent states" could be derived from the group SU(2) and its representations. The extension of coherent states to Lie groups more complicated than H(4) and SU(2) was made by Gilmore[9] for SU(r) and by Perelemov[10] for general Lie groups. A detailed study of the properties of coherent states for arbitrary groups, including disentangling theorems, semiclassical theorems, P- and Q- representations, operator bounds, and D-algebra mappings, was made by Gilmore.[11]

2. Definition of Coherent States

Coherent states may be defined in terms of the following mathematical structures:[11]

1. a Lie group G with Lie algebra $\mathfrak{g}$;
2. an invariant subspace V^Λ which carries an irreducible square-integrable representation Γ^Λ;
3. a highest (or extremal) weight Λ with corresponding basis vector $|\Lambda,\Lambda\rangle$ (or $|ext\rangle$) $\varepsilon\ V^\Lambda$;
4. a maximal stability subgroup $H \subset G$ with the property $h \varepsilon H \Leftrightarrow h|\Lambda,\Lambda\rangle = |\Lambda,\Lambda\rangle e^{i\Phi(h)}$.

Every element $g\ \varepsilon\ G$ can be written uniquely in the form

$$g = \Omega h \qquad \begin{array}{l} g\ \varepsilon\ G \\ h\ \varepsilon\ H \\ \Omega\ \varepsilon\ G/H \end{array} \tag{2.1}$$

The effect of g on the vector $|\Lambda,\Lambda\rangle$ is

$$g\left|{}^{\Lambda}_{\Lambda}\right\rangle = \Omega h\left|{}^{\Lambda}_{\Lambda}\right\rangle = \Omega\left|{}^{\Lambda}_{\Lambda}\right\rangle e^{i\Phi(h)} = \left|{}^{\Lambda}_{\Omega}\right\rangle e^{i\Phi(h)} \tag{2.2}$$

Definition: The states $|\Lambda,\Omega\rangle$

$$\left|{}^{\Lambda}_{\Omega}\right\rangle = \Omega\left|{}^{\Lambda}_{\Lambda}\right\rangle = \left|{}^{\Lambda}_{\text{"M"}}\right\rangle \Gamma^{\Lambda}_{\text{"M"},\Lambda}(\Omega) \tag{2.3}$$

are called coherent states of G in the invariant space V^{Λ} with respect to the state $|\Lambda,\Lambda\rangle$, or simply coherent states. Here "M" is a Gel'fand-Tsetlein pattern which indexes all the orthonormal basis states in V^{Λ}.[12,13]

Coherent states exist in 1-1 correspondence with the coset representatives $\Omega \in G/H$.[14] Since G/H has a natural geometric strucuture, including Haar metric and measure, the coherent states come endowed with a natural geometric interpretation. The coherent states $|\Lambda,\Omega\rangle$ are nondenumerable in the Hilbert space over V^{Λ}, so they must be nonorthogonal and over-complete. Many useful properties of coherent states derive from their geometric interpretation and their over-completeness.

Example 1. $G = H(4)$, $|\Lambda,\Lambda\rangle = |0\rangle$ $(a|0\rangle = 0,\ [a,a^{+}] = I)$, $H = U(1) \times U(1)$, $G/H = H(4)/U(1) \times U(1) \simeq \mathbb{R}^2$. The familiar harmonic oscillator coherent states[1-4] exist in 1-1 correspondence with points in the plane.

Example 2. $G = SU(2)$, $V^{\Lambda} \to V^{j}$, $\Gamma^{\Lambda} \to D^{j}$, $|\Lambda,\Lambda\rangle \to |j,j\rangle$, $H = U(1)$, $G/H = SU(2)/U(1) \cong S^2$ (sphere).[8]

Example 3. $G = SU(3)$, $V^{\Lambda} \to V^{(\lambda_1,\lambda_2)}$, $\Lambda = (\lambda_1-\lambda_2)(2/3,-1/3,-1/3) + \lambda_2(1/3,1/3,-2/3)$, $H = U(2)$ for $\lambda_2 = 0$ and $H = U(1) \times U(1)$ for $\lambda_2 \neq 0$, so $G/H = SU(3)/U(2)$ for symmetric representations and $G/H = SU(3)/U(1) \times U(1)$ for other representations.[9,11]

The definition of coherent states given above is not standard among all authors. For example, Perelomov[10] does not require the state $|\psi\rangle$ appearing in (2.3) to be an extremal state nor the representations Γ^{Λ} to be square-integrable, while Barut and Girardello[15] define coherent states as eigenstates of a maximal solvable subalgebra of $\mathfrak{g}$. We have adopted the above definition because it appears to be the most useful for physical applications.

3. Properties of Coherent States[11]

1. For any unitary irreducible representation Γ^{Λ} and extremal state $|ext\rangle \in V^{\Lambda}$, $\mathfrak{g}$ has a decomposition

$$\mathfrak{g} = \mathfrak{g}_{+} + \mathfrak{g}_{0} + \mathfrak{g}_{-}$$

where $\quad S_{+}|ext\rangle = 0 \qquad S_{+} \in \mathfrak{g}_{+}$

and $S_0|\text{ext}\rangle = (\text{multiple})|\text{ext}\rangle \qquad S_0 \in \mathfrak{g}_0$ (3.1)

2. All Gel'fand-Tsetlein states $|\Lambda,\text{"M"}\rangle$ in V^Λ can be written as products of shift-down operators in $\mathfrak{g}_-$ applied to $|\Lambda,\Lambda\rangle$

$$|\Lambda,\text{"M"}\rangle = E_\gamma \ldots E_\beta E_\alpha |\Lambda,\Lambda\rangle \tag{3.2}$$

3. The coset representatives can be written as exponentials

$$\Omega = \text{EXP}(S_+ + S_-) \tag{3.3}$$

where $S_+ \in \mathfrak{g}_+$, $S_- = -S_+^\dagger \in \mathfrak{g}_-$. The coordinates of the basis vectors in $S_\pm$ may be used to parameterize the space G/H.[14]

4. Baker-Campbell-Hausdorff formulas[14,16] can be developed

$$e^{(S_+ + S_-)} = e^{S'_-} e^{S'_0} e^{S'_+} \tag{3.4}$$

($S'_i \in \mathfrak{g}_i$, $i = \pm,0$) and used to simplify calculations.

5. Coherent states may be expanded in terms of Gel'fand-Tsetlein states

$$\left|{\Lambda \atop \Omega}\right\rangle = \sum_{k=0}^{\infty} \frac{(S_+ + S_-)^k}{k!} \left|{\Lambda \atop \Lambda}\right\rangle$$

$$= e^{S'_-} \left|{\Lambda \atop \Lambda}\right\rangle \times \left\langle{\Lambda \atop \Lambda}\right| e^{S'_0} \left|{\Lambda \atop \Lambda}\right\rangle \tag{3.5}$$

6. Three types of "eigenvalue" equations can be constructed

$$\Omega \text{ Operator } \Omega^{-1} \left|{\Lambda \atop \Omega}\right\rangle = \lambda \left|{\Lambda \atop \Omega}\right\rangle \tag{3.6}$$

where Operator $|\Lambda,\Lambda\rangle = \lambda|\Lambda,\Lambda\rangle$. The three useful types of eigenvalue equation correspond to the following choices

(a) Operator belongs to the universal enveloping algebra of $\mathfrak{g}$, but is not necessarily a Casimir operator;

(b) Operator $\in \mathfrak{g}_0$;

(c) Operator $\in \mathfrak{g}_+$, in which case $\lambda = 0$.

7. Uncertainty relations can be constructed for the hermitian operators Re $S'_+ = (S'_+ + S'^\dagger_+)/2$, Im $S'_+ = (S'_+ - S'^\dagger)/2i$, where $S'_+ = \Omega S_+ \Omega^{-1}$ and $S_+ \in \mathfrak{g}_+$:

$$\langle\Delta(\text{Re } S'_+)^2\rangle \, \langle\Delta(\text{Im } S'_+)^2\rangle \geq |\langle\tfrac{1}{2}[S'_+,S'^\dagger_+]\rangle|^2 \tag{3.7}$$

This uncertainty is minimized in the coherent state $|\Lambda,\Omega\rangle$, but states which minimize (3.7) are not necessarily coherent states.

8. The inner product of two coherent states is

$$<{}^{\Lambda}_{\Omega}|{}^{\Lambda}_{\Omega'}> = <{}^{\Lambda}_{\Lambda}|\Omega^{-1}\Omega'|{}^{\Lambda}_{\Lambda}> = \Gamma^{\Lambda}_{\Lambda\Lambda}(c)\, e^{i\Phi(h)}$$

where $\Omega^{-1}\,\Omega' = ch$ (3.8)

9. The overcompleteness of the coherent states allows the following resolution of the identity

$$I = \frac{\dim(\Lambda)}{\mathrm{Vol}(G/H)} \int |{}^{\Lambda}_{\Omega}> d\Omega <{}^{\Lambda}_{\Omega}| \tag{3.9}$$

Here $d\Omega$ is the Haar measure on G/H and, if G is compact, Vol (G/H) = Vol(G)/Vol(H) is the invariant volume of G/H and dim(Λ) is the dimension of the representation Γ^{Λ}.[14] If G is not compact, dim(Λ)/Vol(G/H) is the Plancherel measure.

10. Generating functions of the form

$$f(A,\Omega) = <{}^{\Lambda}_{\Omega}|\, e^{A_i X_i}\, |{}^{\Lambda}_{\Omega}> , \tag{3.10}$$

where X_i span $\mathfrak{g}$, can easily be constructed using disentangling theorems.

4. Operator Mappings and Bounds

Every linear operator $\hat{\theta}$ on V^{Λ} has a Q-representative defined by

$$Q_{\Lambda}(\hat{\theta};\Omega) = <{}^{\Lambda}_{\Omega}|\,\hat{\theta}\,|{}^{\Lambda}_{\Omega}> \tag{4.1Q}$$

If G is compact, $\hat{\theta}$ also has[8] a P-representative defined by

$$\hat{\theta} = \frac{\dim(\Lambda)}{\mathrm{Vol}(G/H)} \int |{}^{\Lambda}_{\Omega}>\ P_{\Lambda}(\hat{\theta};\Omega) <{}^{\Lambda}_{\Omega}|\, d\Omega \tag{4.1P}$$

The P- and Q-representatives are related by a group covolution integral[9,17]

$$Q_{\Lambda}(\hat{\theta};\Omega) = \frac{\dim(\Lambda)}{\mathrm{Vol}(G/H)} \int |<{}^{\Lambda}_{\Omega'}|{}^{\Lambda}_{\Omega}>|^2\, P_{\Lambda}(\hat{\theta};\Omega')\, d\Omega' \tag{4.2}$$

If $\hat{\theta}$ is an irreducible tensor operator $T^{\Lambda'}_{"M"}$ so that

$$g\, T^{\Lambda'}_{"M"}\, g^{-1} = T^{\Lambda'}_{"N"}\Gamma^{\Lambda'}_{"N","M"}(g) \tag{4.3}$$

then

a) $Q_{\Lambda}(T^{\Lambda'}_{"M"};\Omega) = q(\Lambda,\Lambda')\ y^{\Lambda'}_{"M"}(\Omega)$

b) $P_{\Lambda}(T^{\Lambda'}_{"M"};\Omega) = p(\Lambda,\Lambda')\ y^{\Lambda'}_{"M"}(\Omega)$ (4.4)

Here $\mathcal{Y}^{\Lambda'}_{\text{"M"}}(\Omega)$ is the "M"th spherical function of $\Gamma^{\Lambda'}$ on G/H.[18,19] The proportionality factors $q(\Lambda,\Lambda')$ can be computed from generating functions of type (3.10) and the ratio $p(\Lambda,\Lambda')/q(\Lambda,\Lambda')$ determined from the kernel appearing in the convolution in (4.2), which reduces to the square of a Clebsch-Gordan coefficient.[8] For SU(2)[20]

$$Q_J(\mathcal{Y}^L_M(\underset{\sim}{J});\Omega) = \frac{(2J)!}{(2J-L)!2^L} Y^L_M(\Omega)$$

$$P_J(\mathcal{Y}^L_M(\underset{\sim}{J});\Omega) = \frac{(2J+1+L)!}{(2J+1)!2^L} Y^L_M(\Omega) \quad . \tag{4.5}$$

If $\hat{\theta}$ is a hermitian operator, lower and upper bounds may be placed on the trace of EXP $\hat{\theta}$ using the Q- and P- representatives of $\hat{\theta}$

$$\frac{\dim(\Lambda)}{\mathrm{Vol}(G/H)} \int e^{Q_\Lambda(\hat{\theta};\Omega)} d\Omega \leq \mathrm{Tr}_\Lambda e^{\hat{\theta}} \leq Q \to P \quad . \tag{4.6}$$

The lower bound is due to Bogoliubov and is valid for noncompact as well as compact G. Lieb's proof of the upper bound for SU(2) extends without difficulty to other compact G.[21]

Mappings $\hat{X} \to \mathcal{D}(\hat{X})$ of operators $\hat{X} \in \mathfrak{g}$ into first order linear differential operators $\mathcal{D}(X)$ on G/H can also be constructed and are useful when G is compact. In this case any operator, for example, the density operator $\hat{\rho}$, has a diagonal P- representation. Then Lie algebraic operator products of the type $\hat{X}_i\hat{\rho}\hat{X}_j$ can be replaced by ordinary differential operators of type $\mathcal{D}(\hat{X}_i)\mathcal{D}(\hat{X}_j)\ P(\hat{\rho};\Omega)$. $\mathcal{D}$-operator algebras have been constructed for H(4),[4] SU(2),[22] and SU(r).[17]

5. Classical Limits

It is well known[23] that when the SU(2) representation label J is large the angular momentum operators $\underset{\sim}{J}$ can be replaced by c-numbers according to $J_z \to J\cos\theta$, $J_x = J\sin\theta\cos\phi$, $J_y = J\sin\theta\sin\phi$, and that this approximation becomes better as J becomes larger. The construction of this classical limit can be carried out by using atomic coherent states, for example, by taking a limit in (4.5). Classical limits analogous to the angular momentum limit can be constructed for all compact groups.[24] For $X \in \mathfrak{g}$

$$\lim_{N\to\infty} \Gamma^\Lambda(X/N) = \sum_{i=1}^{r} s_i g(\underset{\sim}{f}_i, X, \Omega) \tag{5.1}$$

where r is the rank[14] of $\mathfrak{g}$, $\underset{\sim}{f}_i$ is the highest weight of the ith fundamental irreducible representation of $\mathfrak{g}$,

$$\underset{\sim}{\Lambda} = \sum_{i=1}^{r} \mu_i \underset{\sim}{f}_i$$

$$s_i = \lim_{N\to\infty} \mu_i/N$$

$$g(\underset{\sim}{f}_i, X, \Omega) = \langle {}^{\underset{\sim}{f}_i}_{\Omega} | \, X \, | {}^{\underset{\sim}{f}_i}_{\Omega} \rangle . \qquad (5.2)$$

The proof of this result involves the construction of generating functions (3.10) and the application of Baker-Campbell-Hausdorff formulas (3.4).

6. Thermodynamics of Dicke Models

The Dicke model[25] has long been a useful tool in Quantum Optics. The Dicke model Hamiltonian describing the interaction of a single mode of the radiation field of energy $\hbar\omega$ with an ensemble of N-identical 2-level atoms is, in the long wavelength approximation,

$$\hat{H} = \hbar\omega a^\dagger a + \frac{\varepsilon}{2} \sum_{i=1}^{N} \sigma_i^z + \frac{\lambda}{\sqrt{N}} \sum_{i=1}^{N} a^\dagger \sigma_i^- + a\sigma_i^+ \qquad (6.1)$$

where $\varepsilon = \varepsilon_2 - \varepsilon_1$ is the atomic energy level spacing, λ is the coupling constant, essentially an electric dipole matrix element, and σ_i^z, $\sigma_i^\pm$ are the Pauli spin operators for the ith atom.

The thermodynamic properties of (6.1) were first discussed by Hepp and Lieb,[26] who used involved estimates to prove that a second order phase transition would occur if $|\lambda|^2 > \varepsilon\hbar\omega$ at a critical temperature $T_c = 1/k\beta_c$ defined by

$$\frac{|\lambda|^2}{\varepsilon\hbar\omega} \tanh \tfrac{1}{2} \beta_c \varepsilon = 1 . \qquad (6.2)$$

The description of this phase transition was greatly simplified by Wang and Hioe, who replaced the field operators $a^\dagger a \to \alpha^*\alpha$, $a \to \alpha$, $a^\dagger \to \alpha^*$ by their Q-representatives in the field coherent state representation, took the trace of the resulting function

$$\mathrm{Tr}\, e^{-\beta\hat{H}} \to e^{-\beta\hbar\omega\alpha^*\alpha} \left(\mathrm{Tr\ EXP} - \beta \begin{bmatrix} \varepsilon/2 & \lambda\alpha/\sqrt{N} \\ \lambda\alpha^*/\sqrt{N} & -\varepsilon/2 \end{bmatrix} \right)^N \qquad (6.3)$$

and minimized with respect to α, α^* to obtain an estimate for F/N, the free energy per particle. This estimate was not rigorous, although it was simple and heuristically useful.

A rigorous treatment using atomic coherent states was subsequently given by Hepp and Lieb.[28] They first decomposed the atomic Hilbert space of dimension 2^N into SU(2) irreducibles

$$(\mathbb{C}^2)^{\otimes N} \to \sum_{J=0 \text{ or } \frac{1}{2}}^{N/2} Y(N,J)\ V^J \qquad (6.4)$$

where the multiplicity factor $Y(N,J) = N!(2J+1)/(\frac{1}{2}N+J+1)!(\frac{1}{2}N-J)!$. Bounds on the partition function $Z_J = \mathrm{Tr}_J\, e^{-\beta\hat{H}}$ in the subspace V^J were computed using the inequalities (4.6) and the sums estimated using Stirlings approximation for

$$Y(N,J) = e^{Ns(r)}$$

$$s(r) = -\{(\tfrac{1}{2}+r)\ln(\tfrac{1}{2}+r) + (\tfrac{1}{2}-r)\ln(\tfrac{1}{2}-r)\} \qquad (6.5)$$

$$\lim_{N\to\infty} r = J/N$$

and Laplace's method[29] for estimating integrals.

These two dual methods lead to algorithms, the latter rigorous, the former not, for discussing the thermodynamics of Dicke Hamiltonians. These two approaches were synthesized[30] into a single simple algorithm for computing the ground state energy per particle E_g/N and the free energy per particle F/N in the thermodynamic limit $N \to \infty$. First, define $h = h(u_3,u_+,u_-;v_3,v_+,v_-)$ and obtain from it the operator h_Q and the function h_C through the substitutions

	$h \to h_Q$	$h \to h_C$
u_3	$a^\dagger a/N$	$\mu^*\mu = \alpha^*\alpha/N$
u_+	$a^\dagger/\sqrt{N}$	$\mu^* = \alpha^*/\sqrt{N}$
u_-	$a/\sqrt{N}$	$\mu = \alpha/\sqrt{N}$
v_3	$\frac{1}{2}\left(\sum_{i=1}^{N} \sigma_i^z\right)/N$	$r\cos\theta$
v_+	$\left(\sum_{i=1}^{N} \sigma_i^+\right)/N$	$\nu^* = r\sin\theta\, e^{i\phi}$
v_-	$\left(\sum_{i=1}^{N} \sigma_i^-\right)/N$	$\nu = r\sin\theta\, e^{-\phi}$

where μ, α are complex, $r \in [0,\frac{1}{2}]$, and $(\theta,\phi) = \Omega$ parameterize the sphere surface $SU(2)/U(1) \simeq S^2$. Then

1. If $\hat{H}/N = h_Q$

2. $\lim_{N\to\infty} E_g/N = \min_{\theta,\phi} h_C$

$\lim_{N\to\infty} F/N = \min_{r,\theta,\phi} \Phi$

where $\Phi(r,\theta,\phi;\beta) = h_C(r,\theta,\phi) - kT\, s(r)$.

This algorithm was used[30] to determine the critical properties of the following model systems

a. $h_D = u_3 + \varepsilon v_3 + \lambda(u_+ v_- + u_- v_+)$ [25-28]

b. $h_{CR} = h_D + \lambda'(u_+ v_+ + u_- v_-)$ [27,31]

c. $h_{CR+A} = h_{CR} + \kappa(u_+ + u_-)^2$ [32]

d. $h = h_D + Q(\tfrac{1}{4} - v_3^{\,2}) + K(\tfrac{1}{2} + v_3)^2$ [33]

e. $h = u_3 + f[\varepsilon v_3 + \lambda(u_+ v_- + u_- v_+)]$ [34]

f. $h = u_3 + \varepsilon v_3 + g[\lambda(u_+ v_- + u_- v_+)]$ [34]

g. $h = h_D + c^* u_+ + c^* u_-$ [35,36]

h. $h = u_3 + \varepsilon v_3 + \lambda(u_+^2 v_- + u_-^2 v_+)$ [30]

The existence of a second order phase transition is determined by the vanishing of the determinant of a 2 × 2 matrix, which is essentially the quadratic form obtained from $\Phi(r,\theta,\phi;\beta)$ along $\theta = \pi$.

7. Thermodynamics of Bialgebraic Models

The photon operators $a^\dagger a$, $a^\dagger$, a, I span the harmonic oscillator algebra $h(4)$ and the collective atomic operators $J_z = \sum_{i=1}^{N} \frac{1}{2} \sigma_i^z$, $J_\pm = \sum_{i=1}^{N} \sigma_i^\pm$ span the algebra su(2). The Dicke model (6.1) is therefore a particularly simple example of a "bialgebraic Hamiltonian"[37]

$$\hat{H} = u_i H_i + v_j K_j + \frac{1}{n} \sum_{\alpha,\gamma} E_\alpha C_{\alpha\gamma} F_\gamma \; . \tag{7.1}$$

Here H_i span the Cartan subalgebra of a Lie algebra $\mathfrak{g}_1$ and E_α are the shift operators in this algebra.[14] The operators K_j, F_β span $\mathfrak{g}_2$ with a similar Cartan decomposition. More general Hamiltonians of bialgebraic type can be written

$$\hat{H}/N = h(X_i/N, Y_j/N) \qquad (7.2)$$

where the operators X_i span $\mathcal{G}_1$ and Y_j span $\mathcal{G}_2$, and the factors 1/N are present for thermodynamic purposes.[34,38]

The form (7.1) of the bialgebraic Hamiltonian is particularly suited to a bifurcation analysis of the type originally carried out for Dicke Hamiltonians.[32] A second order phase transition can occur when a nontrivial primary branch bifurcates from the trivial branch $\langle E_\alpha \rangle = \langle F_\gamma \rangle = 0$. This in turn can occur when the matrix

$$\begin{bmatrix} u\cdot\alpha\ \delta_{\alpha\alpha'} & M_{\alpha\gamma'} \\ N_{\gamma\alpha'} & v\cdot\gamma\ \delta_{\gamma\gamma'} \end{bmatrix} \qquad (7.3a)$$

becomes singular. The eigenvector col($\langle E_{\alpha'} \rangle$, $\langle F_{\gamma'} \rangle$) of (7.3a) with zero eigenvalue indicates the initial direction in order-parameter space of the nontrivial bifurcating solution. The matrices M, N are given by

$$M_{\alpha\gamma'}(\beta) = \langle \alpha\cdot H \rangle_{1,0}\ C_{-\alpha,\gamma'}$$

$$N_{\gamma\alpha'}(\beta) = \langle \gamma\cdot K \rangle_{2,0}\ C_{\alpha',-\gamma} \qquad (7.3b)$$

where the thermodynamic expectation value $\langle\cdot\rangle_{1,0}$ is taken with respect to $H_1 = u\cdot H$ and $\langle\cdot\rangle_{2,0}$ with respect to $H_2 = v\cdot K$. Since the matrices (7.3b) are closely related to fluctuations

$$\langle \alpha\cdot H \rangle_{1,0} = \langle [E_\alpha, E_{-\alpha}] \rangle_{1,0} = \langle \{E_\alpha, E_{-\alpha}\} \rangle \tanh \frac{1}{2} \beta u\cdot\alpha$$

$$\langle \gamma\cdot K \rangle_{2,0} = \langle [F_\gamma, F_{-\gamma}] \rangle_{2,0} = \langle \{F_\gamma, F_{-\gamma}\} \rangle \tanh \frac{1}{2} \beta v\cdot\gamma \qquad (7.4)$$

the result (7.3) indicates that a transformation from disordered to ordered state ($\langle E_\alpha \rangle = 0 \to \langle E_\alpha \rangle \neq 0$ for some α, etc.) occurs when the temperature-weighted fluctuations become sufficiently large. The result (7.3) is therefore called the fluctuation-transformation theorem.[39]

More general bialgebraic Hamiltonians (7.2) can be treated following the steps indicated in §6:

1. Decompose the Hilbert space for G_1 into its invariant subspaces

$$V^{(1)} \to \sum Y_1(N,\Lambda)\ V^{\Lambda}\ . \qquad (7.5)$$

2. Use inequalities (4.6) to put upper and lower bounds on $Tr_\Lambda e^{-\beta\hat{H}}$.
3. Introduce an entropy function

$$s_1(r) = \lim_{N\to\infty} \frac{1}{N} \ell n\, Y_1(N,\Lambda) \ . \tag{7.6}$$

4. Ditto for G_2.
5. Estimate the resulting upper and lower bounds by Laplace's method.

The difference between these bounds is of order 1/N, so that

$$\lim_{N\to\infty} F/N = \min_{r_1,\Omega_1;r_2,\Omega_2} \Phi(r_1,\Omega_1,r_2,\Omega_2;\beta)$$

$$\Phi(r_1,\Omega_1,r_2,\Omega_2;\beta) = h(\langle X_i/N\rangle_1,\ \langle Y_j/N\rangle_2) - kT(s_1(r_1) + s_2(r_2)) \ . \tag{7.7}$$

Here $\langle X_i/N\rangle_1$ are the classical limits for G_1, as given by (5.1). When evaluated at the minimum, $s_1 + s_2$ gives the intensive entropy and, if T = 0, h gives the intensive ground state energy.

Under rather general conditions, involving a minimal symmetry assumption on h (7.2),[30,39] a "crossover theorem" can be proved.[40] This relates the existence of a ground state energy phase transition as a function of increasing interaction parameters ($C_{\alpha\gamma}$ in (7.1), λ in (5.1)) to the occurrence of a thermodynamic phase transition as a function of decreasing temperature.

Example 1. If we consider r-level Dicke models[32] with only one nonzero coupling constant between a single pair of levels and a resonant field mode, only one bifurcation can occur. If one of the two levels is the ground state, a second order thermodynamic phase transition will occur if the coupling constant λ exceeds a critical value λ_{cr}.[41] The bifurcation diagram is shown in Fig. 1a.

If the coupling does not involve the ground state there are four critical values of λ (Fig. 1b) with behavior as follows[41]

$\lambda < \lambda_1$. $\langle a/\sqrt{N}\rangle = 0$ at all temperatures.

$\lambda_1 < \lambda < \lambda_2$ (curve A, $\lambda = \lambda_A$). At sufficiently low temperatures a metastable ordered state exists.

$\lambda_2 < \lambda < \lambda_3$ (curves B, C, D, $\lambda_B < \lambda_C < \lambda_D$). At low temperatures an ordered state exists. For λ_B the ordered state is metastable to the right of b and stable to the left. A first order phase transition occurs at b. These first-order phase transitions are not surrounded by spinodal lines.

$\lambda_3 < \lambda < \lambda_4$ (curve E, $\lambda = \lambda_E$). Two primary branches bifurcate from the disordered branch at temperatures determined by the fluctuation-transformation theorem.[39] The lower primary branch E_2 is

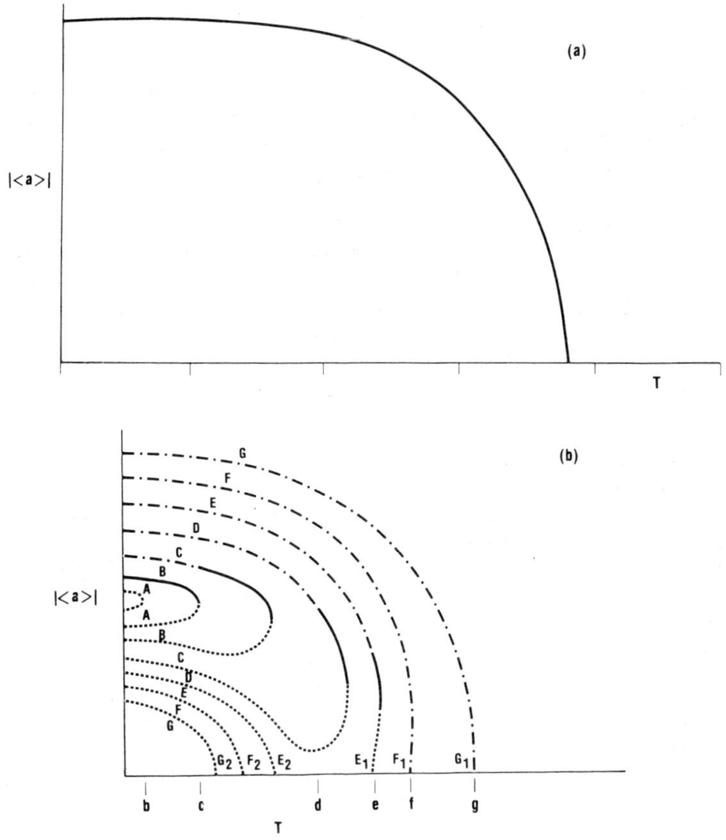

Fig. 1. The critical properties of the r-level extension of the Dicke model depend on the pair of levels between which the strong interaction occurs. 1a. If the ground state is involved a second order thermodynamic phase transition can occur. 1b. If two excited states are involved, the ground state energy critical properties depend on the value of the coupling constant λ in relation to four critical values λ_i, i = 1,2,3,4.

unstable. The upper branch E_1 is initially unstable, becomes locally stable at the point of vertical tangency, and globally stable at e. The disordered branch is stable down to e, metastable between e and the E_1 bifurcation point, unstable between E_1 and E_2, and metastable below E_2. The first order transition at e is surrounded by spinodal lines.

$\lambda = \lambda_4$ (curve F, $\lambda = \lambda_F$). The F_1 bifurcation point is a tricritical point.

$\lambda_4 < \lambda$ (curve G, $\lambda = \lambda_G$). The bifurcating branch G_1 is stable below g, and G_2 is always unstable. The disordered branch is stable above g, unstable between the G_1 and G_2 bifurcation points, and metastable below the G_2 bifurcation point.

Example 2. Primary, secondary, ..., k^{th} order branches can occur in bialgebraic models (7.1), (7.2), where $k = \min(B_1, B_2)$ and B_i is the bifurcation index[39] of $\mathfrak{g}_i$. The bifurcation index of SU(r) is r - 1, so r-level Dicke models can exhibit primary, ..., $(r-1)^{ary}$ branches, but none higher. For 3-level Dicke models only primary and secondary branches can occur. Typical temperature dependences for the order parameters on the globally stable branches are shown in Fig. 2.[42]

8. Thermodynamic Algorithm

The problem of studying the thermodynamic critical properties associated with a Hamiltonian constructed from operators belonging to Lie algebras has now been reduced to a simple algorithm for estimating the free energy per particle. This algorithm becomes increasingly accurate as N becomes large. The algorithm involves two steps:

1. Replace the Hamiltonian $\hat{H}/N$ by its classical limit, and add an entropy term $-kT\, s(r)$;
2. Determine how the minimum of Φ changes as a function of changing temperature.

The first step is implemented using the machinery of coherent states. The second step is implemented using the machinery of catastrophe theory.[43-46]

To be more specific, assume that $f(x;c)$ is a k-parameter family of (potential) functions depending on n state-variables or order-parameters $x \in \mathbb{R}^n$ and k control parameters $c \in \mathbb{R}^k$. Then under suitable conditions $\mathbb{R}^k$ is partitioned into disjoint open sets by the sets S_B (bifurcation set) and S_M (Maxwell set) defined by

$$S_B: \quad \begin{aligned} \nabla f &= 0 \\ \det \frac{\partial^2 f}{\partial x_i \partial x_j} &= 0 \end{aligned} \tag{8.1}$$

$$S_M: \quad \begin{aligned} \nabla f &= 0 \\ f(x;c) - f(x';c) &= 0 \, . \end{aligned} \tag{8.2}$$

On the bifurcation set an equilibrium becomes degenerate. On the Maxwell set two or more equilibria assume the same values. The Maxwell set is determined by equations of Clausius-Clapeyron type.[46]

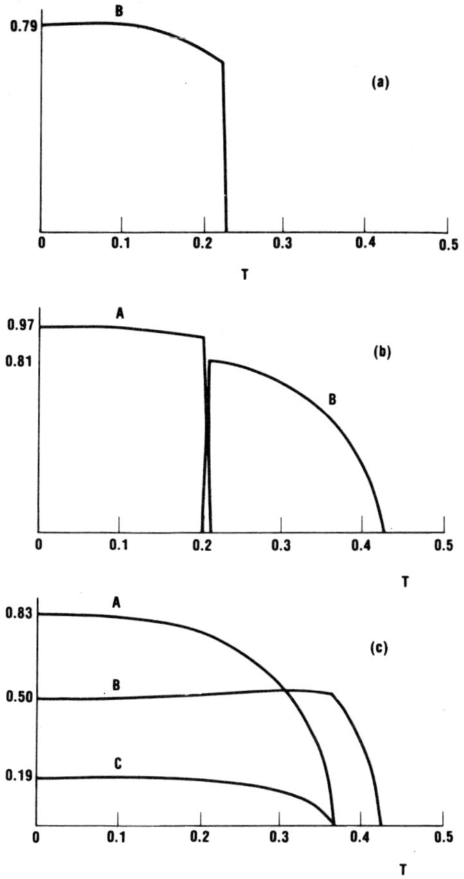

Fig. 2. The thermodynamic critical properties of the 3-level Dicke model with resonant interaction $\hbar\omega_{ji} = \varepsilon_j - \varepsilon_i$ have been studied as a function of the dimensionless coupling constants $\Lambda_{ji} = \lambda_{ji}/(\varepsilon_j - \varepsilon_i)$. Here $r = \varepsilon_2 - \varepsilon_1/\varepsilon_3 - \varepsilon_1 = .2$ and $A = \langle E_{21}\rangle$, $B = \langle E_{32}\rangle$, $C = \langle E_{31}\rangle$, for the following values of $(\Lambda_{12}, \Lambda_{23}, \Lambda_{13})$. a) (0.0,1.7,0.0) a first order phase transition occurs. b) (2.0,1.8,0.7) a second order phase transition occurs at high temperature and a first order phase transition occurs at lower temperature. c) (2.0,1.8,0.8) two phase transitions occur, both of second order.

As we follow some path $c(s) \in \mathbb{R}^k$, a second (first) order phase transition will occur if the path crosses a component of S_B (S_M) describing the global minimum.

For discussing ground state energy phase transitions, we identify the control parameters $c \in \mathbb{R}^k$ with the interaction parameters, such as $C_{\alpha\gamma}$ in (7.1). The order parameters can be chosen as the expectation values of the shift operators spanning $\mathfrak{g}_+ + \mathfrak{g}_-$. For discussing thermodynamic phase transitions we use the same order parameters but the single control variable T.

9. Thermodynamics of Nuclear Models

Nuclear systems are particularly difficult to treat because they contain too many particles for a useful description in terms of the individual particle coordinates and too few particles for a useful description in terms of statistical methods. Two general methods have emerged for the description of such systems: the group-theoretical and the liquid drop models.

Wigner initially exploited the approximate invariance of the nucleon-nucleon interaction under spin and isospin to provide the SU(4) model of nuclear Hamiltonians applicable to light ($A \leq 16$) nuclei.[47] Although this model loses its usefulness when the protons and neutrons occupy different orbitals, or when spin-orbit interactions become important, it nevertheless retains some predictive capabilities for much heavier nuclei ($30 \leq A \leq 110$).[48] The SU(3) model of Elliott[49] is also an extremely useful approximation in the range $16 \leq A \leq 24$. Recently serious attempts have been initiated[50] to understand heavy ($A > 100$) even-even nuclei on the basis of the dynamical symmetry group $\widetilde{SU}(6)$, which arises naturally under the assumptions that nucleons have a strong attractive interaction in the $L = 0$ channel, a somewhat weaker atrractive interaction in the $L = 2$ channel, and other channels can be neglected.

These approaches have been fairly successful at phenomenologically reproducing energy level spectra and some transition rates within their various regions of applicability. However, the methods described in the previous Section allow us to study the ground state energy and the thermodynamic critical properties of model Hamiltonians constructed from operators belonging to a Lie algebra.

We first tested the value of the coherent state methods in the laboratory called the Meshkov-Glick-Lipkin (MGL) pseudospin (= SU(2)) Hamiltonian.[51] The minimum value of the Q-representative of the MGL Hamiltonian gave a good approximation to the exact ground state energy computed by matrix diagonalization.[52] The error is about 4% for $N = 14$ and decreases as N increases. The ground state energy phase transition inherent in this model is particularly easy to visualize using coherent states. In the weak interaction regime the coherent state which is a best approximation to the exact ground state is represented by a point at the south pole of the sphere $S^2 \simeq SU(2)/U(1)$. When the quadrupole interaction strength exceeds a critical value the point moves off the south pole in a standard Ginzburg-Landau second-order phase transition.

The thermodynamic critical properties of the MGL Hamiltonian have also been studied.[53] The upper bound on the free energy per nucleon is a good approximation to F/N, which was computed numerically for N = 30, 50, 70. The approximation becomes better as either T decreases or N increases. Although the second order thermodynamic phase transition is not clearly signalled by matrix diagonalization, it is clearly indicated in the coherent state treatment, with a deformed state represented by a point off the south-polar axis and a spherical state by a point on the south-polar axis.

The phase transitions can also be studied by variational methods.[53,54] For the ground state case, coherent states are used as trial states to minimize the Hamiltonian operator $\hat{H}$. In the finite-temperature case Gibbs states, constructed as outer products of coherent states using (4.1P), are used as trial states to minimize the free energy operator $\hat{F} = \hat{H} - T\hat{S}$. In the former case the variational calculation leads to an upper bound identical to that constructed in Section 7, while the finite temperature variational calculation does not.

These methods were extended by Gilmore and Feng to any compact group.[55] For the r-level extension of the MGL model (G = SU(r)) with a quadrupole interaction between a single pair of levels, we find[56] a ground state energy phase transition as a function of increasing interaction strength. The transition is second order if the interaction involves the ground state, first order if it does not. The classical limit $\Phi(\theta_2,\theta_3)$ of this Hamiltonian is shown in Fig. 3 as a function of increasing quadrupole strength Q, where θ_2,θ_3 (r = 3) are the important SU(3) order parameters. We anticipate that the finite temperature critical properties of this model will be closely analogous to the finite-temperature critical properties of the r-level extension of the Dicke model.[41]

A second approach to the description of nuclear properties is based on the liquid drop model. Whereas the former approach is quantum-mechanical in spirit, this approach is primarily classical in spirit. The former approach is discrete in the sense that only a finite number of parameters are fitted in the Hamiltonian; the latter approach involves a nondenumerable number of degrees of freedom. It is a hope[57] that coherent states can bridge the gap between these two approaches. These states appear to have one foot in each school. They are constructed from a finite-dimensional Lie algebra and live in a finite dimensional space (G compact), but there is a nondenumerable number of such states.

10. Dynamics

Coherent states also provide a useful tool for treating problems of quantum dynamics. If $\hat{H}$ is constructed from operators belonging to a Lie algebra $\mathfrak{g}$ and if the quantum state is initially in the

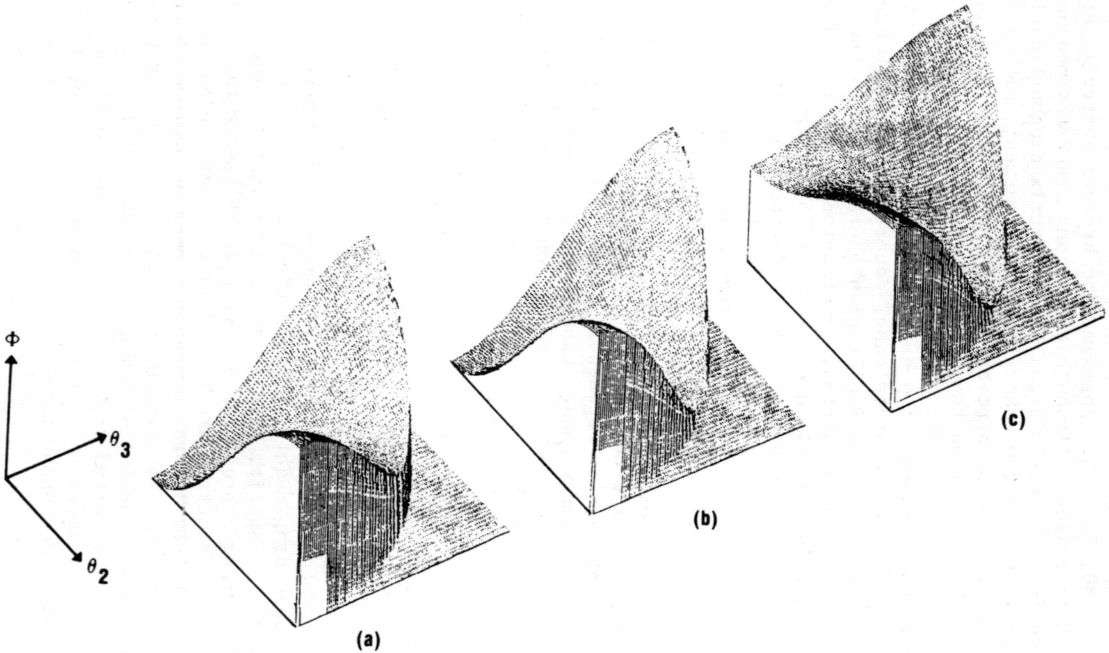

Fig. 3. The shape of $\Phi(\theta_2,\theta_3)$ at $T = 0$ is shown for the 3-level extension of the MGL Hamiltonian. Here θ_2, θ_3 are the relevant SU(3) order parameters, $r = \varepsilon_2-\varepsilon_1/\varepsilon_3-\varepsilon_1 = 0.8$, and the quadrupole interaction Q couples the two excited levels. The quadrant $0 \leq \theta \leq \pi$ is shown. A first order phase transition occurs at $Q = 3.59$. a) $Q = 2.5$; b) $Q = 3.59$; c) $Q = 6.0$.

invariant subspace V^{Λ}, it will remain in V^{Λ} for all later times. To be specific, we can write

$$|\psi(t)\rangle = \int |{}^{\Lambda}_{\Omega}\rangle\, \psi(\Omega,t)\,d\Omega \tag{10.1}$$

where $\psi(\Omega,t) = \langle\Lambda,\Omega|\psi(t)\rangle$. Then the Hamiltonian operator acting on $|\psi\rangle$ can be replaced by its $\mathcal{D}$ operator algebra equivalent to give a c-number equation of the form

$$\mathcal{D}(\hat{H})\ \psi(\Omega,t) = i\hbar \frac{\partial}{\partial t}\ \psi(\Omega,t)\ . \tag{10.2}$$

If $\hat{H}$ contains no higher than second degree operator products of the basis vectors of $\mathfrak{g}$, (10.2) is an equation of Fokker-Planck type on the (curved) surface G/H.

If the quantum system cannot be represented by a pure state $|\psi\rangle$ but is rather a statistical mixture of states, then it can be represented by a density operator

$$\hat{\rho}(t) = \int |{}^{\Lambda}_{\Omega}\rangle\, \rho(\Omega,t)\, \langle{}^{\Lambda}_{\Omega}|\ d\Omega\ . \tag{10.3}$$

A c-number equation for $\rho(\Omega,t)$ can be constructed using an appropriate $\mathcal{D}$-algebra mapping. Such equations have been particularly useful in the description of equilibrium properties ($i\hbar t \to kT$, Schrödinger equation $\to$ Bloch equation).

In the event that $\hat{H}$ is a linear superposition of the operators spanning $\mathfrak{g}$, then $\hat{H} \in \mathfrak{g}$ and more powerful statements can be made. For example, there is a powerful selection theorem: "once a coherent state, always a coherent state."[4,8,11] The proof is simple. Since $H \in \mathfrak{g}$, EXP $- i\hat{H}\Delta t/\hbar \in G$, so that the S matrix is a group element in G. Then

$$\begin{aligned} S(t_2,t_1)\ |{}^{\Lambda}_{\Omega_0}\rangle &= S(t_2,t_1)\Omega_0|{}^{\Lambda}_{\Lambda}\rangle \\ &= \Omega(t_2,t_1)\ h(t_2,t_1)|{}^{\Lambda}_{\Lambda}\rangle \\ &= |{}^{\Lambda}_{\Omega(t_2,t_1)}\rangle\ e^{i\Phi[h(t_2,t_1)]} \end{aligned} \tag{10.4}$$

where $\Omega(t_1,t_1) = \Omega_0$.

The equations of motion for the coherent state parameters can be derived quite easily using Schur's formula (Appendix).[58,59] If we write the S-matrix in a coset decomposition

$$S(t) = e^{X(t)}\ e^{Y(t)}$$

where

$$X(t) \in \mathfrak{g}_+ + \mathfrak{g}_-$$
$$Y(t) \in \mathfrak{g}_0 \tag{10.5}$$

then the equations of motion are determined from

$$\underbrace{e^{X(t + \Delta t)}\, e^{Y(t + \Delta t)}}_{S(t + \Delta t)} = e^{-i\hat{H}\Delta t/\hbar}\, e^{X(t)}\, e^{Y(t)} . \tag{10.6}$$

If we write $X(t + \Delta t) = X + \Delta X$, $X = X(t)$ and $\Delta X = \Delta t\,(dX/dt)$, and similarly for $Y(t + \Delta t)$ then (10.6) can be written

$$e^X \left\{I + \frac{I-e^{-AdX}}{AdX} \Delta X\right\}\left\{I + \frac{e^{AdY}-I}{AdY} \Delta Y\right\} e^Y$$
$$= e^X e^{-X} \left\{I - \frac{i\Delta t}{h} H\right\} e^X e^Y . \tag{10.7}$$

The equations of motion are determined from the first order terms in (10.7). In particular, if we set

$$X(t) = b_i(t)\, X_i$$
$$Y(t) = \omega_\alpha(t)\, Y_\alpha$$
$$H(t) = h_i(t)\, X_i + h_\alpha(t)\, Y_\alpha ,$$

where the X_i span $(\mathfrak{g}_+ + \mathfrak{g}_-)$ and the Y span $\mathfrak{g}_0$, then

$$i\hbar \begin{bmatrix} M_{ij}(t) & M_{i\beta}(t) \\ M_{\alpha j}(t) & M_{\alpha\beta}(t) \end{bmatrix} \begin{bmatrix} \dfrac{db_j}{dt} \\ \dfrac{d\omega_\beta}{dt} \end{bmatrix} = \begin{bmatrix} h_i(t) \\ h_\alpha(t) \end{bmatrix} \tag{10.8}$$

where

$$M_{ij} = \left(\frac{e^{AdX} - I}{AdX}\right)_{ij} , \qquad M_{i\beta} = \left(e^{AdX}\, \frac{e^{AdY} - I}{AdY}\right)_{i\beta}$$
$$M_{\alpha j} = \left(\frac{e^{AdX} - I}{AdX}\right)_{\alpha j} , \qquad M_{\alpha\beta} = \left(e^{AdX}\, \frac{e^{AdY} - I}{AdY}\right)_{\alpha\beta} . \tag{10.9}$$

In all physical cases that have arisen so far, the decomposition of $\mathfrak{g}$ has the special property

$$[\mathfrak{g}_+ + \mathfrak{g}_-, \mathfrak{g}_+ + \mathfrak{g}_-] \subseteq \mathfrak{g}_0$$
$$[\mathfrak{g}_0, \mathfrak{g}_+ + \mathfrak{g}_-] \subseteq \mathfrak{g}_+ + \mathfrak{g}_-$$
$$[\mathfrak{g}_0, \mathfrak{g}_0] \subseteq \mathfrak{g}_0 . \qquad (10.10)$$

In this case AdY is a block diagonal matrix and AdX is block off-diagonal, so that the matrices (10.9) simplify somewhat

$$M_{ij} = \left(\frac{\sinh \mathrm{AdX}}{\mathrm{AdX}}\right)_{ij}, \qquad M_{i\beta} = (\sinh \mathrm{AdX})_{i\beta'}\left(\frac{e^{\mathrm{AdY}} - I}{\mathrm{AdY}}\right)_{\beta'\beta}$$
$$M_{\alpha j} = \left(\frac{\cosh \mathrm{AdX} - I}{\mathrm{AdX}}\right)_{\alpha j}, \quad M_{\alpha\beta} = (\cosh \mathrm{AdX})_{\alpha\beta'}\left(\frac{e^{\mathrm{AdY}} - I}{\mathrm{AdY}}\right)_{\beta'\beta}. \qquad (10.11)$$

These equations have been integrated explicitly in the cases $G = H(4)$[4] and $G = SU(2)$ where $(\mathrm{AdY})_{\alpha\beta} = 0$ because $\mathfrak{g}_0$ is commutative. Special cases of the equations of motion (10.8), well-known in nuclear magnetic resonance,[60] are the Bloch equations and the spin-echo equations [$G = SU(2)$]. For $G/H = SO(3,1)/SO(3)$, the equations for $d\omega/dt$ reduce to the Thomas precession equations[14,61] when $h_i(t)$ = constant, $h_\alpha(t) = 0$.

Equations (10.8) represent a (classical) dynamical system on the Lie group G, or a flow on the direct product space $G/H \otimes H$. If the time dependence of the source terms $h_i(t)$, $h_\alpha(t)$ dies out sufficiently rapidly as $t \to \pm\infty$, then the S matrix

$$S = \lim_{\substack{t_1 \to -\infty \\ t_2 \to \infty}} S(t_2, t_1) \qquad (10.12)$$

can be determined by analyzing suitable $\alpha \to \omega$ flows. The S-matrices for several systems have been calculated by Perelomov.[62]

11. TDHF Calculations

The Time Dependent Hartree Fock equations of motion are[63,64]

$$\delta < \psi| \, i\hbar \frac{\partial}{\partial t} - \hat{H}|\psi> = 0 . \qquad (11.1)$$

If the Hamiltonian $\hat{H}$ is constructed from operators belonging to Lie algebra $\mathfrak{g}$, the most general state $|\psi>$ that can be obtained by applying a dynamical group transformation to the ground or other extremal state is a coherent state, up to phase factor. We can therefore study (11.1) by choosing $|\psi> = g|\Lambda,\Lambda>$, $g = \Omega h$. In this case

$$<\psi| \hat{H} |\psi> = Q_\Lambda(H;\Omega) \ . \tag{11.2}$$

For finite N the Q-representative of $\hat{H}$ can be written down once $\hat{H}$ has been expressed in terms of irreducible tensor operators. For N large, the Q-representative is obtained to a good approximation by taking the classical limit of $\hat{H}$.

The time derivative may be taken exactly as in the previous Section. The result is

$$<\psi| i\hbar \frac{\partial}{\partial t} |\psi> = (<X_i> \ , \ <Y_\alpha>) \begin{pmatrix} M_{ij} & M_{i\beta} \\ M_{\alpha j} & M_{\alpha\beta} \end{pmatrix} \begin{pmatrix} \dot{b}_j \\ \dot{\omega}_\beta \end{pmatrix} \tag{11.3}$$

where

$$<X_i> = <{}^{\Lambda}_{\Omega}| \ X_i |{}^{\Lambda}_{\Omega}>$$

$$<Y_\alpha> = <{}^{\Lambda}_{\Omega}| \ Y_\alpha |{}^{\Lambda}_{\Omega}> \tag{11.4}$$

are simple to compute and the notation of the previous section has been used.

The TDHF equations are therefore

$$\delta \ (\ (11.3) \ - \ Q_\Lambda(\hat{H};\Omega) \ ) \ = 0 \tag{11.5}$$

where the variation is over the coordinates in the group G, or the coordinates b_i, ω_α, where $b_i X_i \in (\mathfrak{g}_+ + \mathfrak{g}_-)$ and $\omega_\alpha Y_\alpha \in \mathfrak{g}_0$. Since $Q_\Lambda(\hat{H};\Omega)$ and Ad X are independent of the coordinates ω_α, these coordinates are easily seen to be cyclic. The equations (11.6) were explicitly derived for MGL models[55] using the coherent states $| \ J; \ \theta\phi >$. The azimuthal coordinate ϕ is also cyclic.

12. Nonequilibrium Steady State Systems

The laser is an important example of an open physical system. Energy is pumped through this system. If the rate of dissipation is sufficiently great, a phase transition from disordered to ordered state ("off" to "on") can occur.[65]

We can describe the dynamics of open quantum-mechanical systems by computing the equations of motion

$$i\hbar \frac{d}{dt} <X_i> = <[X_i,\hat{H}]> + \text{more} \ . \tag{12.1}$$

The additional terms model the dissipative process. They can be taken into account by replacing the term on the left by

$$\frac{d}{dt} <X> \rightarrow (\frac{d}{dt} + \gamma_i)(<X_i> - <X_i>_c) \ . \tag{12.2}$$

Here γ_i describes the dissipation process, i.e., how fast the expectation value $\langle X_i \rangle$ will relax to the value $\langle X_i \rangle_c$ which is "clamped" by forces causing the flow through the system.

It is useful to consider the nonequilibrium steady state ($d/dt \to 0$) properties of the bialgebraic Hamiltonian (7.1). If $\langle E_\alpha \rangle_c = \langle F_\gamma \rangle_c = 0$, then the coupled nonlinear equations are

$$i\hbar\, \gamma_i(\langle H_i \rangle_s - \langle H_i \rangle_c) = \langle [H_i, E_{\alpha'}] \rangle_s C_{\alpha'\gamma'} \langle F_{\gamma'} \rangle_s \qquad \text{(a)}$$

$$i\hbar\, \gamma_\alpha \langle E_\alpha \rangle_s = \langle [E_\alpha, E_{\alpha'}] \rangle_s\, C_{\alpha'\gamma'} \langle F_{\gamma'} \rangle_s \qquad \text{(b)}$$

$$i\hbar\, \gamma_j(\langle K_j \rangle_s - \langle K_j \rangle_c) = \langle E_{\alpha'} \rangle_s\, C_{\alpha'\gamma'} \langle [K_j, F_{\gamma'}] \rangle_s \qquad \text{(c)}$$

$$i\hbar\, \gamma_\gamma \langle F_\gamma \rangle_s = \langle E_{\alpha'} \rangle_s\, C_{\alpha'\gamma'} \langle [F_\gamma, F_\gamma'] \rangle_s \,. \qquad \text{(d)}$$

(12.3)

The subscript s indicates the steady state expectation value.

Equations (12.3) may be studied by eliminating $\langle H_i \rangle_s$ between a and b and $\langle K_j \rangle_s$ between c and d. In the absence of dissipation and for low flow rates the only solution to these coupled nonlinear equations is $\langle E_\alpha \rangle_s = \langle F_\gamma \rangle_s = 0$. Bifurcations of nontrivial solutions from the trivial solution can be determined by the standard linearization method. The bifurcation condition is given by

$$\det \begin{bmatrix} -i\hbar\gamma_\alpha\, \delta_{\alpha\alpha'} & M_{\alpha\gamma'} \\ N_{\gamma\alpha'} & -i\hbar\gamma_\gamma\, \delta_{\gamma\gamma'} \end{bmatrix} = 0 \qquad (12.4)$$

$$M_{\alpha\gamma'} = C_{-\alpha\gamma'} \langle \alpha \cdot H \rangle_c$$

$$N_{\gamma\alpha'} = C_{\alpha',-\gamma} \langle \gamma \cdot K \rangle_c \,. \qquad (12.5)$$

Because of its close relationship with the fluctuation-transformation theorem (7.3), this result is called the dissipation-transformation theorem.

When applied to the Dicke model,[36] this calculation shows that there is a formal analogy between the thermodynamic phase transition (discussed earlier) and the nonequilibrium steady state phase transition. But is does more. Part of the thermodynamic algorithm involves the study of bifurcation properties using the methods of catastrophe theory. The Ginzburg-Landau phase transition responsible for both phase transitions is structurally unstable,[42-46] and requires one additional unfolding parameter for a universal unfolding. This may be introduced by adding a classical near-resonant coherent field to the system.[66] When this is done the laser equation of state is diffeomorphic with the cusp catastrophe manifold. Increasing the amplitude of this field when the pump rate is held below threshold can lead to a first order phase transition. Such transitions have been observed by Gibbs, McCall, and Venkatesan,[67] and have been investigated extensively.[68,69]

For the laser:[45,66]
1. the nonequilibrium equations of (steady) state are ("typically") manifolds in the direct product group parameter space;
2. the density operator factors ($\hat{\rho} = \hat{\rho}_1 \otimes \hat{\rho}_2$) where $\hat{\rho}_1(\hat{\rho}_2)$ is the reduced density operator for the subsystem with group $G_1(G_2)$;
3. the reduced density operators can be written as exponentials: $\hat{\rho}_1 \simeq \mathrm{EXP}(r_i X_i)$, $\hat{\rho}_2 \simeq \mathrm{EXP}(s_j Y_j)$;
4. the expectation values $\langle X_i \rangle$, $\langle Y_j \rangle$ determine the coefficients r_i, s_j up to overall multiplicative factor;
5. this factor is determined by the variances $\langle X_i^\dagger X_i \rangle - \langle X_i^\dagger \rangle \langle X_i \rangle$, etc;
6. the reduced density operators factor

$$\rho = [\hat{\rho}\ (\text{Geometry})]^{\text{Physics}}$$

in the spirit of the Wigner-Eckart theorem. Here $\hat{\rho}$ (Geometry) is an operator defined by a point on some (catastrophe) manifold and "Physics" is a number characterizing noise;
7. the density operator $\hat{\rho} = \hat{\rho}_1 \otimes \hat{\rho}_2 = \mathrm{EXP}(r_i X_i + s_j Y_j)$ can be obtained by linearizing a Hamiltonian H'. This means the steady state dynamics of H is identifiable with the thermodynamics of H'.

It is our hope that this relation between dynamics and thermodynamics can be extended further.

13. Summary and Conclusions

Coherent states may be defined with respect to the following structures: 1) a Lie group G with Lie algebra $\mathfrak{g}$; 2) a square-integrable unitary irreducible representation Γ^Λ on an invariant Hilbert space V^Λ; 3) an extremal state $|\text{ext}\rangle$ in V^Λ annihilated by a maximal solvable subalgebra in $\mathfrak{g}$; 4) a closed subgroup $H \subset G$ fixing $|\text{ext}\rangle$ up to phase. Coherent states are "the orbit of $|\text{ext}\rangle$ under G" and are parameterized by the projective space G/H. Properties derived from their rich geometric structure are summarized in Section 3.

They provide a particularly convenient set of states for constructing operator mappings (Section 4) and for constructing classical limits for Lie algebras more general than SU(2) (Section 5). The P- and Q-representatives can be used to put upper and lower bounds on partition functions. In the thermodynamic limit the bounds on F/N converge to the value obtained by taking the classical limit of $\hat{H}/N$, adding an entropy term $-kT\, N^{-1} \ln Y(N,\Lambda)$, and minimizing over (Ω, r), $\Omega \in G/H$, $r \in$ dual to H. This two-step algorithm exploits Lie Group Theory in the first step and Elementary Catastrophe Theory in the second.

This algorithm was used to study the ground state energy phase transition and thermodynamic phase transitions in models of Dicke type (Section 6), generalized bialgebraic models (Section 7), and nuclear models (Section 9). A bifurcation analysis leads to a "fluctuation-transformation" theorem, which determines bifurcations of ordered branches from the disordered branch.

The dynamical properties of systems described by Hamiltonian $\hat{H}$ can also be treated using coherent states. If $\hat{H} \in \mathfrak{g}$, the "semi-classical" theorem can be written explicitly as a set of first order ordinary coupled nonlinear differential equations over G/H (Section 10). If $\hat{H} \notin \mathfrak{g}$ but $\hat{H}$ is constructed from operators in $\mathfrak{g}$ ($H \in U(\mathfrak{g})$) then the semiclassical theorem fails, wave packets on G/H spread, but the TDHF equations (Section 11) are closely analogous to the semiclassical equations.

Nonequilibrium steady state systems were studied in the same way that equilibrium equations can be studied, through their equations of motion. These sets of equations are closely comparable. A bifurcation analysis leads in this case to a "dissipation-transformation" theorem (Section 12). These identifications are made explicit for Dicke models, where the "unfolded" Dicke Hamiltonian gives an equation of state diffeomorphic to the cusp catastrophe manifold under both equilibrium and nonequilibrium mappings. This suggests that it may be possible to study the steady state dynamics of H variationally by studying the thermodynamics of some associated Hamiltonian H' variationally.

APPENDIX

Schur's formula[58,59] for computing operator differentials may conveniently be derived by solving the differential equation[32]

$$\frac{d}{dt} e^{-tX} e^{t(X+\varepsilon Y)} = e^{-tX} \varepsilon Y e^{t(X+\varepsilon Y)} \tag{A.1}$$

formally by iteration in powers of the small parameter ε

$$e^{-tX} e^{t(X+\varepsilon Y)} - I = \varepsilon \int_0^t dt' \, e^{-t'X} Y e^{t'X}\{I + \varepsilon \int_0^{t'} dt'' \ldots \tag{A.2}$$

By truncating beyond linear terms in ε, setting $t = 1$, and carrying out the integration formally, we obtain

$$e^{X+\varepsilon Y} = e^X \{I + \frac{I - e^{-AdX}}{AdX} \varepsilon Y\} \tag{A.3}$$

where AdX is the regular representation of X, defined by

$\mathrm{Ad}X\ Y = [X,Y]$, $e^{\mathrm{Ad}X}\ Y = e^{X}\ Ye^{-X}$. The latter relation provides the alternative decomposition

$$e^{X+\varepsilon Y} = \{I + \frac{e^{\mathrm{Ad}X} - I}{\mathrm{Ad}X}\ \varepsilon\ Y\}\ e^{X} \ . \tag{A.4}$$

REFERENCES

1. E. Schrödinger, Naturwiss. 14, 644 (1927).
2. F. Bloch and A. Nordsieck, Phys. Rev. 52, 54 (1937).
3. J. Schwinger, Phys. Rev. 91, 728 (1953).
4. R. J. Glauber, Phys. Rev. 130, 2529 (1963), 131, 2766 (1963).
5. P. W. Atkins and J. C. Dobson, Proc. Roy. Soc. (London) A321, 321 (1971).
6. J. Kutzner, Phys. Lett. 41A, 475 (1972).
7. J. M. Radcliffe, J. Phys. A4, 313 (1971).
8. F. T. Arecchi, E. Courtens, R. Gilmore, and H. Thomas, Phys. Rev. A6, 2211 (1972).
9. R. Gilmore, Ann. Phys. (NY) 74, 391 (1972).
10. A. M. Perelomov, Commun. Math. Phys. 26, 222 (1972).
11. R. Gilmore, Rev. Mex. de Fisica 23, 143 (1974).
12. I. M. Gel'fand and M. L. Tsetlein, Dokl. Akad. Nauk SSSR 71, 825 (1950), 71, 1017 (1950).
13. R. Gilmore, J. Math. Phys. 11, 3420 (1970).
14. R. Gilmore, Lie Groups, Lie Algebras, and Some of Their Applications, NY: Wiley, 1974.
15. A. O. Barut and L. Girardello, Commun. Math. Phys. 21, 41 (1971).
16. R. Gilmore, J. Math. Phys. 15, 2090 (1974).
17. R. Gilmore, C. M. Bowden, and L. M. Narducci, Phys. Rev. A12, 1019 (1975).
18. W. Miller, Jr., Lie Theory and Special Functions, NY: Academic, 1968.
19. N. Ja. Vilenkin, Special Functions and the Theory of Group Representations, Providence: American Mathematical Society, 1968.
20. R. Gilmore, J. Phys. A9, L65 (1976).
21. E. H. Lieb, Commun. Math. Phys. 31, 327 (1973).
22. L. M. Narducci, C. M. Bowden, V. Bluemel, G. P. Carrazana, and R. A. Tuft, Phys. Rev. A11, 973 (1975).
23. G. Herzberg, Atomic Spectra and Atomic Structure, NY: Dover, 1944.
24. R. Gilmore, J. Math. Phys. 20, 891 (1979).
25. R. H. Dicke, Phys. Rev. 93, 99 (1954).
26. K. Hepp and E. H. Lieb, Ann. Phys. (NY) 76, 360 (1973).
27. Y. K. Wang and F. T. Hioe, Phys. Rev. A7, 831 (1973).
28. K. Hepp and E. H. Lieb, Phys. Rev. A8, 2517 (1973).
29. D. V. Widder, An Introduction to Transform Theory, NY: Academic (1971).
30. R. Gilmore, J. Math. Phys. 18, 17 (1977).

31. H. J. Carmichael, C. W. Gardiner, and D. F. Walls, Phys. Lett. A46, 47 (1973).
32. R. Gilmore and C. M. Bowden, J. Math. Phys. 17, 1617 (1976), Phys. Rev. 13, 1898 (1976).
33. Y. A. Kudenko, A. P. Slivinsky, and G. M. Zaslavsky, Phys. Lett. A50, 411 (1975).
34. R. Gilmore, Physica 86A, 137 (1977).
35. J. P. Provost, F. Rocca, G. Vallee, and M. Sirugue, Physica 85A, 202 (1976).
36. R. Gilmore, Phys. Lett. A60, 387 (1977).
37. R. Gilmore, in: Journees Relativistes 1976, M. Cahan, R. Debener, and J. Geheniau, Eds., Brussels: Université Libre, 1976, p. 71.
38. D. Ruelle, Statistical Mechanics, NY: Benjamin, (1969).
39. R. Gilmore, Bialgebraic Models (unpublished).
40. R. Gilmore and C. M. Bowden, in Proceedings of the First Army Conference on High Energy Lasers, C. M. Bowden, D. W. Howgate, and H. R. Robl, Eds., NY: Plenum, 1978, p. 335.
41. R. Gilmore, J. Phys. A10, L131 (1977).
42. R. Gilmore, S. R. Deans, and D. H. Feng, to be published.
43. R. Thom, Structural Stability and Morphogenesis, Reading: Benjamin, 1975.
44. E. C. Zeeman, Catastrophe Theory, Selected Papers 1972-1977, Reading: Addison-Wesley, 1977.
45. T. Poston and I. N. Stewart, Catastrophe Theory and its Applications, London: Pitman, 1978.
46. R. Gilmore, Catastrophe Theory for Scientists and Engineers, NY: Wiley, 1980 (to appear).
47. E. P. Wigner, Phys. Rev. 51, 106 (1937).
48. P. Franzini and L. A. Radicati, Phys. Lett. 6, 322 (1963).
49. J. P. Elliott, Proc. Roy. Soc. (London) A245, 128 (1958).
50. A. Arima and F. Iachello, Phys. Lett. 53B, 309 (1974), Phys. Rev. Lett. 35, 1069 (1975) and 40, 385 (1978), Ann. Phys. (NY) 99, 253 (1976) and 111, 201 (1978).
51. H. J. Lipkin, N. Meshkov, and A. J. Glick, Nucl. Phys. 62, 188, 199, 211 (1965).
52. R. Gilmore and D. H. Feng, Phys. Lett. 76B, 26 (1978).
53. R. Gilmore and D. H. Feng, Nucl. Phys. A301, 189 (1978).
54. R. Gilmore and D. H. Feng, Phys. Rev. C19, 1119 (1978).
55. R. Gilmore and D. H. Feng, to be published.
56. R. Gilmore and D. H. Feng, Phys. Lett. (submitted).
57. F. Iachello, private communication.
58. I. Schur, Math. Ann. 35, 161 (1890); Leipz. Ber. 42, 1 (1890).
59. S. Helgason, Differential Geometry and Symmetry Spaces, NY: Academic, 1962.
60. A. Abragam, The Principles of Nuclear Magnetism, Oxford: Clarendon Press, 1961.
61. L. H. Thomas, Nature 117, 514 (1926).
62. A. M. Perelomov, Sov. Phys. Usp. 20, 703 (1977).

63. P. A. M. Dirac, Proc. Camb. Phil. Soc. 26, 376 (1930).
64. D. J. Rowe, Nuclear Collective Motion - Models and Theory, London: Methuen, 1970.
65. H. Haken, Rev. Mod. Phys. 47, 67 (1975).
66. R. Gilmore and L. M. Narducci, A17, 1747 (1978).
67. H. M. Gibbs, S. L. McCall, and T. N. C. Venkatesan, Phys. Rev. Lett. 36, 1135 (1976).
68. R. Bonifacio and L. A. Lagiato, Opt. Commun. 19, 1972 (1976), Phys. Rev. Lett. 40, 1023 (1978).
69. G. S. Agarwal, L. M. Narducci, R. Gilmore, and D. H. Feng, Optics Letters 2, 88 (1978), Phys. Rev. A18, 620 (1978).

Z. Phys. C – Particles and Fields 24, 97–101 (1984)

On the Large N_c Limit of the $SU(N_c)$ Colour Quark–Gluon Partition Function

B.-S. Skagerstam[1]
Nordita, Blegdamsvej 17, DK-2100 Copenhagen, Denmark

Received 7 January 1984

Abstract. The ideal, $SU(N_c)$ coloured, quark–gluon gas partition function is considered, taking the global colour-singlet condition into account. The colour-singlet condition leads to finite volume corrections to the thermodynamical quantities of the gas. Possible effects of these finite volume corrections on lattice QCD at a finite temperature are discussed. Recent Monte Carlo evaluations of the energy density in $SU(2)$ and $SU(3)$ Yang–Mills lattice gauge theories are consistent with such corrections. In the large N_c-limit, a phase transition is exhibited which, formally, has the same origin as a large N_c-limit phase transition in the two-dimensional $SU(N_c)$ lattice gauge theory.

1. Introduction

The influence of conserved, internal degrees of freedom on the state of a physical system has been studied in detail in the literature. Isospin conservation can e.g. impose constraints on the thermodynamical description of proton–antiproton annihilation (see e.g. [1]) as well as restrictions on the description of coherent pionization in hadronic collisions [2, 7].

The interesting possibility of having a phase transition from a highly compressed or excited state of hadronic matter to a quark–gluon plasma (for an incomplete list of references, see [3]) enforces a study of an ideal (due to asymptotic freedom) quark–gluon gas. Due to the confinement mechanism of QCD, one must impose the constraint of having a colour singlet on the partition function of the system*. Recently, such a study of internal degrees of freedom and the constraints they may impose on the partition function has been considered [4–6].

In the present paper, we will make use of an over-complete set of quasi-coherent states [7–8], where singlets (or any given representation) automatically can be projected out, in order to compute the partition function of an ideal, massless gas with an internal $SU(N_c)$ symmetry. In the large N_c limit, we will exhibit a phase transition which has the same origin as the third-order Gross–Witten phase transition in the two dimensional lattice $SU(N_c)$ gauge theory [9]. For the gluons we must, of course, consider the adjoint representation of $SU(N_c)$. In the large N_c-limit, the steepest descent approximation technique used in [9] for the fundamental representation can, however, straightforwardly be extended to any representation. A first-order phase transition at large N_c for an ideal gluon gas is the result of such an analysis, as will be shown below.

One consequence of our considerations is that the, naive, Stefan–Boltzmann limit of a gluon gas cannot be reached when performing Monte Carlo simulations of lattice QCD due to finite lattice effects. Such a conclusion seems to be in accordance with presently available Monte Carlo evaluations of the energy density in $SU(2)$ and $SU(3)$ Yang–Mills lattice gauge theories [10]. As a matter of fact, by taking the finite volume correction into account the approach at high temperatures, to the Stefan–Boltzmann limit in the Monte Carlo calculations of [10] is substantially improved.

2. The Non-Abelian Ideal Boson Gas

Before we discuss the boson gas with internal, non-Abelian degrees of freedom, let us compute the partition function of a photon gas making use of coherent states (see e.g. [11] for a general review of coherent states). The use of coherent states automatically takes

1 On leave of absence from the Institute of Theoretical Physics, S-41296 Göteborg, Sweden

* The colour singlet condition can, formally, be derived by considering the lattice gauge theory and imposing periodic boundary conditions. In the continuum limit the constraint then corresponds to imposing the Gauss law in the Hamiltonian formulation. For a rigorous discussion see e.g. [20]

the Bose-Einstein statistics into account. In terms of coherent states, the partition function, z, becomes ($\beta = 1/T$)

$$Z = \mathrm{Tr}\exp(-\beta H) = \int \mathrm{df} \langle f|\exp(-\beta H)|f\rangle. \quad (1)$$

Here

$$|f\rangle = \prod_k \exp\left(\frac{-f_k^* f_k}{2} + a_k^\dagger f_k\right)|0\rangle \quad (2)$$

is a coherent state constructed out of the one-particle state f_k (the index k runs over all attainable three-momenta) and

$$\int \mathrm{df} = \prod_k \int \frac{d^2 f_k}{\pi} = \prod_k \int \left(\frac{1}{\pi} d\,\mathrm{Re}\, f_k\, d\,\mathrm{Im}\, f_k\right). \quad (3)$$

With the free field hamiltonian

$$H = \sum_k \omega_k a_k^\dagger a_k, \quad (4)$$

where $\omega_k = |\mathbf{k}|$, we obtain

$$Z = \prod_k \int \frac{d^2 f_k}{\pi} \exp(-|f_k|^2(1 - \exp(-\beta\omega_k)))$$

$$= \prod_k \frac{1}{1 - \exp(-\beta\omega_k)}, \quad (5)$$

which, of course, is a well known result. The reason for presenting this elementary derivation of (5) is that we can make use of the coherent state representation to project out, say, the singlet representation of a non-Abelian Bose gas [7–8]. Let $f_{k,\alpha}$ denote the one-particle state which transforms according to a real representation R of the group G. The singlet partition function Z_s then becomes

$$Z_s = \int \mathrm{df} \int_G dg_1 dg_2 \langle R(g_1) f|\exp(-\beta H)|R(g_2) f\rangle. \quad (6)$$

By combining (2), (3) and (6) and making use of the invariance of the group measure, we obtain

$$Z_s = \int_G \mathrm{dg} \prod_{k,\alpha} \int \frac{d^2 f_{k,\alpha}}{\pi}$$

$$\cdot \exp(-|f_k|^2 + \exp(-\beta\omega_k) f_{k,\alpha}^* R_{\alpha\beta}(g) f_{k,\beta})$$

$$= \int_G \mathrm{dg} \prod_k \exp(-\mathrm{Tr}\ln(1 - \exp(-\beta\omega_k) R(g))). \quad (7)$$

Z_s can be recast into the following form

$$Z_s = \int_G \mathrm{dg} \exp\left(\sum_k \sum_{n=1}^{\infty} \frac{\exp(-\beta n \omega_k)}{n} \chi_R(g^n)\right), \quad (8)$$

where χ_R is the character in the R-representation. Equation (8) is the completely general result under the assumptions given above. Let us now consider $G = SU(N_c)$ and let R be the adjoint representation, i.e.

$$\chi_R(g) = \chi_F(g)\chi_F^*(g) - 1, \quad (9)$$

where χ_F is the character of the fundamental representation. It is clear that the integrand in (8) only depends on the eigenvalues of g. We can therefore make use of Weyl's parametrization of the reduced group measure [12] to obtain

$$Z_s = \frac{1}{(N_c)!} \int \left(\prod_i \frac{d\alpha_i}{2\pi}\right)\left(\prod_{i \neq j} \left|2\sin\left(\frac{\alpha_i - \alpha_j}{2}\right)\right|\right)$$

$$\cdot \delta\left(\sum_i \alpha_i\right) \cdot \exp((N_c - 1) B_g \xi(4))$$

$$\cdot \exp\left(2 B_g \sum_{i>j} \sum_{n=1}^{\infty} \frac{1}{n^4} \cos n(\alpha_i - \alpha_j)\right). \quad (10)$$

Here $\xi(4) = \pi^4/90$ and $B_g = 2VT^3/\pi^2$ for a massless spin one-Bose gas. We have not, in general, found a closed analytical expression for Z_s as a function of R, but (10) can of course be evaluated numerically. Here we notice that (10) can be recast into a form suitable for a steepest-descent approximation procedure. The series in the exponential function in (10) can be evaluated with the result

$$Z_s = \frac{1}{2\pi (N_c)!} \int \prod_i d\alpha_i \left(\prod_{i \neq j} |2\sin\pi(\alpha_i - \alpha_j)|\right)$$

$$\cdot \delta\left(\sum_i \alpha_i\right) \exp((N_c^2 - 1) B_g \xi(4))$$

$$\exp\left(-\frac{2\pi^4}{3} B_g \sum_{i>j} (|\alpha_i - \alpha_j|^4 - 2|\alpha_i - \alpha_j|^3 + |\alpha_i - \alpha_j|^2)\right). \quad (11)$$

For large B_g, i.e. for large T and/or V, we can apply a steepest-descent method and we obtain

$$Z_s \approx G \exp((N_c^2 - 1) B_g \xi(4)) \cdot B_g^{-(N_c^2 - 1)\, 2}, \quad (12)$$

where G is an irrelevant constant. For $N_c = 2$, this reduces to a result recently obtained by Gorenstein et al. [6].

By making use of (12) we now obtain for the energy density

$$\varepsilon = \frac{T^2}{V} \frac{\partial}{\partial T} \ln Z_c = \varepsilon_{SB}\left(1 - \frac{45}{2 \cdot \pi^2} \cdot \frac{1}{VT^3}\right), \quad (13)$$

where ε_{SB} is the naive Stefan–Boltzmann expression for the energy density, i.e.

$$\varepsilon_{SE} = (N_c^2 - 1) \cdot \frac{\pi^2}{15} T^4. \quad (14)$$

Let us comment on the size of the *universal* correction factor in (13) to the Stefan–Boltzmann distribution. In Monte Carlo simulations of pure Yang–Mills gauge theories one is, of course, forced to work on a finite lattice. The factor VT^3 is then determined by the geometrical size of the lattice only [10] and is equal to $(N/N_t)^3$, where $N(N_t)$ is the space (time) extension of the lattice. For a $10^3 \times 3$ lattice, the correction factor then is $\varepsilon/\varepsilon_{SB} \approx 0.939$ and for a $8^3 \times 3$ lattice $\varepsilon/\varepsilon_{SB} \approx 0.881$. These fairly large corrections agree very well with the $SU(2)$ and $SU(3)$ Monte Carlo data of [10], respectively.

3. The Large N_c-Limit

At very low temperatures, (8) can be approximated by (Boltzmann statistics):

$$Z_s = \int_G \mathrm{dh}\exp(B_g \chi_R(h)), \tag{15}$$

which is nothing else than the one-plaquette partition function of a lattice gauge theory with the R-representation instead of the fundamental representation for the link variables [9]. Here we would like to study the large N_c limit of Z_s. The Gross–Witten analysis [9] of the large N_c limit of two-dimensional lattice $SU(N_c)$ Yang–Mills gauge theory can be extended to our case (see in this case also [13]). In terms of a spectral density function $\rho(\alpha)$, $\alpha\varepsilon[-\pi,\pi]$, Z_s can, in the large N_c limit, be rewritten as follows

$$Z_s = \frac{1}{(N_c)!}\int D\rho\,\delta\left(\int \rho\, d\alpha - 1\right)\exp(-N_c^2 s), \tag{16}$$

where the "effective action" S is given by

$$S = -\int d\alpha\, d\beta\, \rho(\alpha)\rho(\beta)\left\{\ln\left|2\sin\left(\frac{\alpha-\beta}{2}\right)\right| + B_g\cos(\alpha-\beta)\right\} + B_g/N_c^2. \tag{17}$$

The steepest-descent approximation, which is exact for large N_c, then leads to the following integral equation

$$P\int_{-\alpha_c}^{\alpha_c} d\beta\,\rho(\beta)\left\{\tfrac{1}{2}\cot\left(\frac{\alpha-\beta}{2}\right) - B_g\sin(\alpha-\beta)\right\} = 0, \tag{18}$$

which can be solved by using standard techniques [9, 13]. A phase transition will now occur when α_c reaches π as B_g is varied (see [8] for further details). The solution of (18) has the following properties. In Z_s is zero for $B_g < B_{cr}$, where $B_{cr} = 1.0$. For $B_g \geqq B_{cr}$

$$\frac{1}{N_c^2-1}\ln Z_s = \tfrac{1}{2}(B_g + \sqrt{B_g^2 + B_g} - \ln(B_g + \sqrt{B_g^2 - B_g}) - 1), \tag{19}$$

i.e. at $B_g = B_{cr} = 1$, or at $VT^3 = \pi^2/2$, we have a first-order phase transition in the limit when the number of colour degrees of freedom goes to infinity. The energy density per degree of freedom ($\varepsilon_{\mathrm{df}} = \varepsilon/(N_c^2-1)$), as defined through (13), has the following form

$$\varepsilon_{\mathrm{df}} = 3T/4V\cdot B_g(1 + \sqrt{1 - 1/B_g})^2, \tag{20}$$

which is exhibited in Fig. 1. It is amazing to note that $\varepsilon_{\mathrm{df}}$ is zero for $B_g < B_{cr}$, i.e. all gluon degrees of freedom condense into a zero energy state. For comparison, we have included in Fig. 1 the energy density per degree of freedom for an $SU(2)$ ideal bose gas. In this case we obtain (see e.g. [7] and [6])

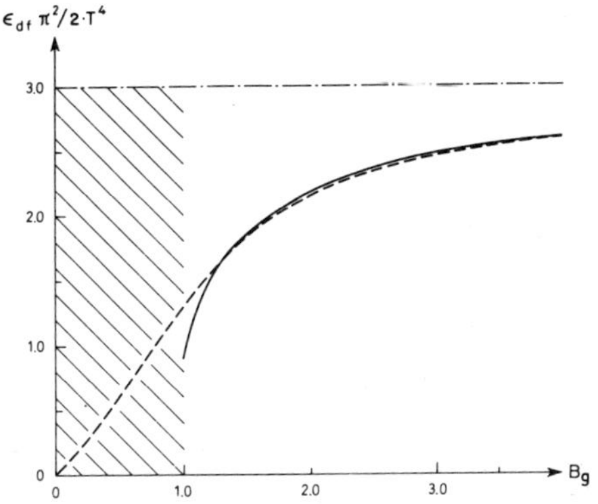

Fig. 1. The energy density per degree of freedom $\varepsilon_{\mathrm{df}}$ as a function of B_g for an ideal gluon gas in the large N_c-limit (solid curve). In the region where $B < B_{cr} = 1.0$, $\varepsilon_{\mathrm{df}}$ is zero. For comparison, we have included the corresponding function for the case where $N_c = 2$ (dashed curve). The dashed-dotted line corresponds to the Stefan–Boltzmann limit of an ideal bose gas

$$\varepsilon_{\mathrm{df}} = TB_g/V\cdot\frac{I_1(2B_g) - I_2(2B_g)}{I_0(2B_g) - I_1(2B_g)}. \tag{21}$$

It is only close to B_{cr} the energy densities per degree of freedom for the ideal $SU(2)$ and $SU(\infty)$ gluon gases differ drastically in their thermodynamical behaviour.

4. Inclusion of Fermions

The presence of non-interacting fermions can be included in a straightforward manner. Here we only consider the case with zero chemical potentials, but it is not very difficult to extend our discussion to the general situation of non-zero chemical potentials. In order to incorporate the Fermi-Dirac statistics of the fermions, we use the technique of coherent states of fermion operators ([14]; see in this context also [15] and the associated fermionic integration [16]). The difference between particles and anti-particles can, furthermore, be taken into account by a doubling of the degrees of freedom in analogy with the treatment of $U(1)$ charged coherent states [17]. The singlet partition function for an ideal quark/anti-quark gas can then be written

$$Z_s^{(q,\bar{q})} = \int_G dg\left[\prod_k \exp(\mathrm{Tr}\ln(1+\exp(-\beta\omega_k)g))\right]\left[\prod_l \exp(\mathrm{Tr}\ln(1+\exp(-\beta\omega_l)g^\dagger))\right]. \tag{22}$$

By combining (10) and (22), we now obtain the singlet

partition function for the ideal, massless quark–gluon gas

$$Z_s = \frac{1}{(N_c)!}\int\left(\prod_i \frac{d\alpha_i}{2\pi}\right)\left(\prod_{i\neq j}\left|2\sin\left(\frac{\alpha_i-\alpha_j}{2}\right)\right|\right)$$
$$\cdot\delta(\alpha_i)\exp((N_c-1)B_g\xi(4))$$
$$\cdot\exp\left(2B_g\sum_{i>j}\sum_{n=1}^{\infty}\frac{1}{n^4}\cos n(\alpha_i-\alpha_j)\right)$$
$$\cdot\exp\left(2\cdot B_q\sum_i\sum_{n=1}^{\infty}\frac{(-1)^{n-1}}{n^4}\cos n\alpha_i\right), \qquad (23)$$

where B_q is given by

$$B_q = 2\cdot n_f V T^3/\pi^2, \qquad (24)$$

and n_f is the number of quark flavours. As in the case for the ideal gluon gas, (23) can be recast into a form suitable for a steepest-descent approximation for large B_g and B_q. The fermionic contribution in (23) is then rewritten using

$$\sum_i\sum_{n=1}\frac{(-1)^{n-1}}{n^4}\cos n\alpha_i = N_c\cdot\tfrac{7}{8}\xi(4)$$
$$+\tfrac{1}{24}\sum_i(\alpha_i^4/2-\pi^2\alpha_i^2). \qquad (25)$$

The steepest-descent method then leads to

$$Z_s \approx C\exp((N_c^2-1)B_g\xi(4))\cdot B_g^{-(N_c^2-1)/2}$$
$$\cdot\exp(N_c\cdot\tfrac{7}{4}B_q\xi(4))\cdot f_{N_c}\left(\frac{B_q}{2B_g}\right), \qquad (26)$$

where

$$f_n(z) = \int_{-\infty}^{\infty} ds\left(\prod_{k=1}^{N}\int_{-\infty}^{\infty} dx_k \exp(isx_k - zx_k^2)\right)$$
$$\prod_{i>j}(|x_i-x_j|^2\exp(-|x_i-x_j|^2)). \qquad (27)$$

For a pure quark gas we obtain for the energy density ε:

$$\varepsilon = \frac{T^2}{V}\frac{\partial}{\partial T}\ln Z_c = \varepsilon_q\cdot\left(1-\frac{(N_c^2-1)}{n_f\cdot N_c}\cdot\frac{180}{7\cdot\pi^2}\cdot\frac{1}{T^3 V}\right), \qquad (28)$$

where

$$\varepsilon_q = n_f\cdot N_c\cdot\frac{7\cdot\pi^2}{60}T^4, \qquad (29)$$

i.e. the finite volume correction to the energy density is larger for a pure quark gas as compared to the pure gluon gas (13). For the quark–gluon gas, we obtain

$$\varepsilon = \frac{T^2}{V}\frac{\partial}{\partial T}\ln Z_s = \varepsilon_{SB}\left(1-\frac{45}{2\cdot\pi^2}\cdot\frac{1}{VT^3}\right)+\varepsilon_q, \qquad (30)$$

since $f(B_q/2B_g)$ is independent of T.

At very high temperatures, i.e. where the Boltzmann statistics is valid, the singlet partition function for the quark–gluon gas takes the form

$$Z_s = \int_G dh\exp(B_g\chi_R(h) + B_q(\chi_F(h)+\chi_F^*(h))), \qquad (31)$$

which we would now like to study in the large N_c-limit. Z_s is formally the one-plaquette partition function for a two-dimensional lattice $SU(N_c)$ gauge theory with a mixed action. This model has been solved exactly in the large N_c-limit [18]. If we define $w_{1,2}$ by

$$w_1 = \bar{B}_q/(1-B_g) \qquad (32)$$

and

$$w_2 = \frac{B_g - \bar{B}_q + \sqrt{(\bar{B}_q+B_g)^2 - B_g}}{2B_g^2}, \qquad (33)$$

where $\bar{B}_q = B_q/N_c$, then the correct expression for the partition function is

$$\frac{1}{N_c^2-1}\ln Z_s = w_1^2(1-B_g) \qquad (34)$$

for $B_g + 2\bar{B}_q < 1$, and

$$\frac{1}{N_c^2-1}\ln Z_s = \frac{1}{2(1-w_2)} + \tfrac{1}{2}\ln 2(1-w_2) - \tfrac{3}{4} - B_g w_2^2 \qquad (35)$$

for $B_g + 2\bar{B}_q > 1$. The energy density per degree of freedom ε_{df} takes the form

$$\varepsilon_{df} = 3T/V\cdot(2\bar{B}_q w + B_g w^2), \qquad (36)$$

where $w = w_1(w_2)$ in the phase for which $B_g + 2\bar{B}_q < 1$ ($B_g + 2\bar{B}_q > 1$). For non-zero values of $\bar{B}_q$ the phase transition is now of the Gross–Witten type [9], i.e. third order. In Fig. 2 we exhibit the energy density for various values of $\bar{B}_q = xB_g$, where $x = n_f/N_c$. It is clear that for a fixed number of quark flavours and in the large N_c-limit, the system is driven towards the first-order pahse transition situation of the pure gluon system.

5. Conclusions

In the present paper we have discussed how the global colour singlet condition affects the thermodynamical properties of an ideal quark–gluon gas. As has been pointed out, the effects of the colour singlet condition disappear in the thermodynamical limit, i.e. for infinite volume. In some practical situations, finite volume corrections may, however, be of importance. In Monte Carlo studies of finite temperatures QCD, we have seen that the naive Stefen–Boltzmann limit for the energy density acquires a correction factor which is in agreement with presently available Monte Carlo data*.

* The ideal gas limit of the $SU(2)$ Yang–Mills gauge theory with fermions and with the colour singlet condition imposed has also been considered and compared with Monte-Carlo data. In the quenched approximation, quarks and gluons seem to form colour singlets by themselves [21]

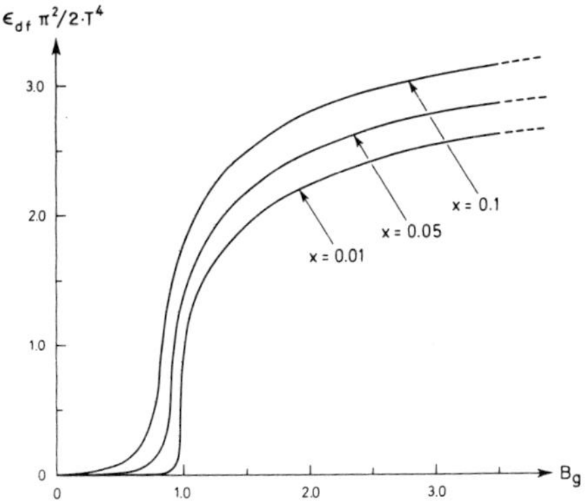

Fig. 2. The energy density per degree of freedom ε_{df} as a function of B_g for an ideal quark–gluon gas at large N_c for various values of $x = n_f/N_c$. For finite x there is a third-order phase transition below $B_{cr} = 1.0$. At $x = 0$, the phase transition becomes first order

Further-more, if the quark–gluon phase appears in nature, finite volume corrections may turn out to be important [6, 19] in studies of the actual formation of the quark–gluon plasma.

In the large N_c-limit, we have exhibited the presence of a third-order Gross–Witten phase transition in the case were n_f/N_c is finite. Otherwise the system is that of a pure gluon gas with a first-order phase transition. These results were obtained by making use of Boltzmann statistics only. In principle, the large N_c-limit can be studied by taking the appropriate boson–fermion statistics into account. A study of this problem is in progress.

Acknowledgement. The author wishes to thank C. Bernard and J. Rafelski for helpful remarks during the course of the present work. Discussions with P. Salomonson on unpublished work by him and the present author on two-dimensional lattice gauge theory models is also gratefully acknowledged. The author is also grateful to H. Satz for informing him about the work by M.I. Gorenstein et al.

References

1. H.A. Bethe: Phys. Rev. **50**, 331 (1936); E. Fermi: Prog. Theor. Phys. (Japan) **1**, 510 (1950); Z. Koba: Nuovo Cimento **18**, 608 (1961); T. Ericson: Nuovo Cimento **21**, 605 (1961); F. Cerulus: Nuovo Cimento **22**, 958 (1961); H. Satz: Fortschr. Phys. **11**, 445 (1963); H. Joos, H. Satz: Nuovo Cimento **34**, 619 (1964); K.Z. Zalewski: Acta Phys. Polon. **28**, 207 (1965); J.S. Bell, G. Karl, C.L. Llewellyn Smith: Phys. Lett. **52**, 363 (1974); J. van der Meulen, in: Proc. Symp. on Nucleon–antinucleon annihilations, ed. L. Montanet (Chexbres, 1972) CERN 72–10, 1972; B. Müller, J. Rafelski: Phys. Lett. **116B**, 274 (1982)
2. D. Horn, R. Silver: Ann. Phys. (N.Y.) **66**, 509 (1971); D. Horn, F. Zachariasen: Hadron physics at very high energies, New York: Benjamin, 1973; J.C. Botke, D.J. Scalapino, R.L. Sugar: Phys. Rev. **D9**, 813 (1974)
3. B.A. Freedman, L.D. McLerran, Phys. Rev. **D16**, 1169 (1977); S.A. Chin: Phys. Lett. **78B**, 552 (1978); B.A. Freedman, L.D. McLerran: Phys. Rev. **D16**, 1169 (1977); S.A. Chin: Phys. Lett. **78B**, 552 (1978); P.D. Morley, M.B. Kislinger: Phys. Rep. **51**, 63 (1979); J.I. Kapusta: Nucl. Phys. **B148**, 461 (1979); O.K. Kalashnikov, V.V. Kilmov: Phys. Lett. **88B**, 328 (1979); E.V. Shuryak: Phys. Lett. **81B**, 65 (1979); Phys. Rep. **61**, 71 (1980); R. Anishetty, P. Koehler, L. McLerran: Phys. Rev. **D22**, 2793 (1980); J. Rafelski, R. Hagedorn, in: Thermodynamics of quarks and hadrons, ed. H. Satz Amsterdam: North-Holland (1981); and in: Proc. of 5th European Symposium on Nucleon–antinucleon interactions, Padova, 1980 ; G. Domokos, J.I. Goldman: Phys. Rev. **D23**, 203 (1981); K. Kajantie, H.I. Mietinnen: Z. Phys. C—Particles and Fields **9**, 341 (1981); ibid. **14**, 357 (1982)
4. L. Turko: Phys. Lett. **104B**, 153 (1981); K. Redlich, L. Turko: Z. Phys. C—Particles and Fields **5**, 201 (1980)
5. H.-Th. Elze, W. Greiner, J. Rafelski: Phys. Lett. **124B**, 515 (1983)
6. M.I. Gorenstein, O.A. Mogilevsky, V.K. Petrov G.M. Zinovjev: Z. Phys. C—Particles and Fields 13 (1983); M.I. Gorenstein, S.I. Lipskikh, V.K. Petrov and G.M. Zinovjev: Phys. Lett. **123B**, 437 (1983)
7. K.-E. Eriksson, N. Mukunda, B.-S. Skagerstam: Phys. Rev. **D24**, 2615 (1981)
8. B.-S. Skagerstam: Quasicoherent states for unitary groups, Göteborg preprint (83–38) 1983, to be published.
9. D.J. Gross, E. Witten: Phys. Rev. **D21**, 446 (1980)
10. J. Engels, F. Karsch, I. Montvay, H. Satz: Nucl. Phys. **B205** [FS5], 545 (1982); T. Celik, J. Engels, H. Satz: Phys. Lett. **129B**, 419 (1983)
11. J.R. Klauder, E.C.G. Sudarshan: Fundamentals of quantum optics. New York Benjamin 1968
12. H. Weyl: The classical groups. Princeton: Princeton University Press 1946
13. S. Wadia: University of Chicago preprint EFI 79/44 (1979) unpublished; C.B. Lang, P. Salomonson, B.-S. Skagerstam: Nucl. Phys. **B190**, [FS3] 337 (1981)
14. J.R. Klauder: Acta Phys. Austr. suppl. **18**, 1 1977
15. Y. Ohnuki: lectures given at National Laboratory for High Energy Physics (KEK), Tsukuba (1977)
16. F.A. Berezin: The method of second quantization. New York: Academic Press, 1966
17. B.-S. Skagerstam: Phys. Rev. **D19**, 2471 (1979); ibid. **D22**, 534(E) (1980)
18. T. Chen, C.I. Tan, X. Zheng: Phys. Lett. **109B**, 383 (1982); Yu. M. Makeenko, M.I. Polikatpov: Nucl. Phys. **B205**, [FS5] 386 (1982); S. Samuel: Phys. Lett. **112B**, 237 (1982); M.C. Ogilvie, A. Horowitz: Nucl. Phys. **B215**, [FS7], 249 (1983)
19. J. Rafelski, M. Danos: to appear.
20. C. Borgs, E. Seiler: Nucl. Phys. **B215**, [FS7], 125 (1983); Commun. Math. Phys. **91**, 329 (1983)
21. B.-S. Skagerstam: Phys. Lett. **133B**, 419 (1983)

VIII. Applications in Atomic Physics

PHYSICAL REVIEW A VOLUME 6, NUMBER 6 DECEMBER 1972

Atomic Coherent States in Quantum Optics

F. T. Arecchi
Pavia University, and Centro Informazioni Studi Esperienze Laboratory, Milan, Italy

and

Eric Courtens
IBM Zurich Research Laboratory, 8803 Rüschlikon, Switzerland

and

Robert Gilmore*
Physics Department, Massachusetts Institute of Technology, Cambridge, Massachusetts 02139

and

Harry Thomas
Institut für Theoretische Physik, J. W. Goethe Universität, Frankfurt a. Main, Germany
(Received 22 May 1972)

For the description of an assembly of two-level atoms, atomic coherent states can be defined which have properties analogous to those of the field coherent states. The analogy is not fortuitous, but is shown to be related to the group contraction of exponential operators based on the angular momentum algebra to exponential operators based on the harmonic-oscillator algebra. The derivation of the properties of the atomic coherent states is made easier by the use of a powerful disentangling theorem for exponential angular momentum operators. A complete labeling of the atomic states is developed and many of their properties are studied. In particular it is shown that the atomic coherent states are the quantum analogs of classical dipoles, and that they can be produced by classical fields.

I. INTRODUCTION

Many problems in quantum optics can be dealt with in terms of the interaction of an assembly of two-level atoms with a transverse electromagnetic field. In these problems a particular set of quantum states has to be selected for the description of both field and atoms. The choice of a particular representation is always motivated by convenience rather than by necessity. A good example is given by the free field. Early treatments have made large use of Fock states, i.e., photon number states which are eigenstates of the free-field Hamiltonian. Although they form a perfectly valid basis for the corresponding Hilbert space, these states are poorly suited for the description of laser fields which contain a large and intrinsically uncertain number of photons. In this case the reaction of the field on the radiating atoms can be approximated by a mean field, as in classical radiation problems. The field states generated from vacuum by classical currents are well known,[1] and happen to be eigenstates of the annihilation operator. The coordinate representation is the minimum-uncertainty packet of harmonic oscillators.[2] These so-called coherent states, whose usage in atom-field interaction problems was introduced by Senitzky,[3] have now been extensively studied and applied to quantum-optical problems,[4] and will be called here Glauber states.

The coherent states of the radiation field have attractive properties. They are obtained from the vacuum state by a unitary shift operator, and are minimum-uncertainty states, i.e., products of mean-square deviations of conjugated variables are minimum in these states, e.g., $\langle \Delta p^2 \rangle \langle \Delta q^2 \rangle = \frac{1}{4}\hbar^2$. Though not orthogonal, they obey a completeness relation, and hence form a good set of basis states. In fact, the overcompleteness of coherent states allows the expansion of many important field operators as a single integral over projectors on these states. Finally, these states correspond to the field radiated by classical currents, i.e., currents produced by moving charges for which the field reaction is neglected. In this sense these states provide a quantum description of classical fields.

One purpose of the present paper is to show that states with completely analogous properties can be defined for the free-atom assembly. In fact, to each property of the atomic coherent states there exists a corresponding property of the field coherent states. This duality, far from being accidental, will be shown to be deeply rooted and related to the contraction of the rotation group describing motions on a sphere, onto a translation group describing motions in the harmonic-oscillator phase space.

For a single two-level system, that of atom n, the ground-state ket will be labeled $|\psi_2^n\rangle$ and the upper-state ket $|\psi_1^n\rangle$. Any operator acting on this system can be expanded in the set of Pauli matrices σ_x^n, σ_y^n, σ_z^n, plus the identity matrix I_2^n, associated with this particular atom. The two-level system is thus identical to a spin-$\frac{1}{2}$ system for which spin-up and spin-down operators are defined by

$$\sigma_\pm^n \equiv \tfrac{1}{2}(\sigma_x^n \pm i\sigma_y^n) \; . \qquad (1.1)$$

For recollection, the commutation rules of these operators are

$$[\sigma_z, \sigma_\pm] = \pm 2\sigma_\pm \, , \quad [\sigma_+, \sigma_-] = \sigma_z \; . \qquad (1.2)$$

The states $|\psi_1^n\rangle$ and $|\psi_2^n\rangle$ are eigenstates of σ_z^n. Such a choice of basis is convenient but by no means unique. Any other linear combination

$$|\psi_i'^n\rangle \equiv \sum_{j=1}^{2} U_{ij} |\psi_j^n\rangle \quad (i = 1, 2) \qquad (1.3)$$

may be chosen which preserves orthogonality and normalization. The most general transformations U_{ij} with these properties are the collection of 2 ×2 unitary matrices which form the group U(2).[5,6] The subgroup of transformations with determinant +1 forms the group SU(2), familiar from angular momentum analysis. U(2) and SU(2) differ by a trivial phase factor.

Turning to the assembly of N atoms, the corresponding Hilbert space is spanned by the set of 2^N product states

$$|\phi_{i_1 i_2 \cdots i_N}\rangle \equiv \prod_{n=1}^{N} |\psi_{i_n}^n\rangle \quad (i_n = 1, 2) \; . \qquad (1.4)$$

Collective angular momentum operators are defined by

$$J_\mu = \tfrac{1}{2}\textstyle\sum_n \sigma_\mu^n \quad (\mu = x, y, z) \, , \qquad (1.5a)$$

$$J_\pm = \textstyle\sum_n \sigma_\pm^n \, , \qquad (1.5b)$$

$$J^2 = J_x^2 + J_y^2 + J_z^2 \; . \qquad (1.5c)$$

For the moment the effect of the different spatial positions of atoms 1, 2, ..., N is ignored. This effect is important for the atom-field interaction and will be discussed in Sec. VI where more appropriate collective angular momentum operators are defined.

Following the historical development of quantum mechanics one could choose as another suitable basis, in place of (1.4), the set of eigenstates of the energy operator J_z. In this case, symmetry requirements actually indicate an appropriate complete set of commuting observables to which J_z belongs and whose simultaneous eigenstates form the basis. In analogy to angular momentum eigenstates these orthonormal states will be labeled

$$\left| {J \atop M} ; {\vec{\lambda} \atop i} \right\rangle , \qquad (1.6)$$

where $J(J+1)$ and M are the eigenvalues of J^2 and J_z, respectively. The quantum numbers $\vec{\lambda}$ and i are those additional eigenvalues which are required to provide a complete set of labels. They are related to the permutation properties of the free-atom Hamiltonian and will be explained in Sec. V. The energy eigenstates (1.6) have been used in the study of superradiance[7] and will be called Dicke states. They will be shown to have a close relationship to the Fock states of the free-field problem.

Another natural way to describe the N atoms is through the overcomplete set of product states

$$|\phi(a_1, b_1; \ldots ; a_N, b_N)\rangle \equiv \prod_{n=1}^{N} (a_n|\psi_1^n\rangle + b_n|\psi_2^n\rangle), \tag{1.7}$$

with $|a_n|^2+|b_n|^2=1$. These states display no correlations between different atoms.[3] For any normalized state $|\psi\rangle$ of the N-atom assembly, a degree of correlation can be defined in the following manner: One forms the overlap integral $|\langle\psi|\phi(a_1, b_1; \ldots ; a_N, b_N)\rangle|^2$ and maximizes the result with respect to the set (a_i, b_i), $i=1$ to N. The complement to one of this maximized overlap integral is defined as the degree of atomic correlation of the state $|\psi\rangle$. It can easily be seen that all states of the form (1.7) have zero correlation, whereas the Dicke states (1.6) of maximum J ($J=\frac{1}{2}N$) and small M ($M\simeq 0$) have a correlation which approaches unity for large-N values.

Another set of overcomplete states can be obtained by rotating the Dicke states $|{}^{J}_{-J}; {}^{\vec\lambda}_{i}\rangle$ through an angle (θ, φ) in angular momentum space. These states, which can be labeled

$$|{}^{J}_{\theta,\varphi}; {}^{\vec\lambda}_{i}\rangle \tag{1.8}$$

are the atomic coherent states. They will be named Bloch states in view of their resemblance to the spin states common in nuclear-induction problems.[8] The profound difference between states of type (1.6) and those of type (1.7) has already been discussed by Senitzky.[3] The Bloch states (1.8) should not be mistaken for uncorrelated coherent states of type (1.7). Only for $J=\frac{1}{2}N$ are the Bloch states a subset of (1.7).

The remainder of the paper is subdivided as follows: Sections II and III give a parallel treatment of the field and atomic states, respectively. For simplicity, Sec. II deals with a single-field mode, and Sec. III with a single member of the set $(\vec\lambda, i)$. The notation in Sec. III is therefore simplified, $|J, M\rangle$ or $|M\rangle$ replacing (1.6), and $|J, \theta\varphi\rangle$ or $|\theta, \varphi\rangle$ replacing (1.8). Section IV explains the group-contraction procedure which allows derivation of all the properties of the field states in Sec. II from the corresponding properties of the atomic states in Sec. III. Section V describes the symmetry properties of the atomic states and, in particular, explains the full notation (1.6) and (1.8). In Secs. VI and VII some aspects of the atom-field interaction are considered. In Sec. VI the spatial dependence of atomic states is introduced. In the case of a single-field mode, operators replacing the set (1.5) can be defined such that the interaction preserves the symmetry properties of the states. Various approximations are presented and radiation rates calculated. Some aspects of the interaction with a classical field are discussed in Sec. VII. Section VIII shows how the disentangling and contraction procedures can be applied to the calculation of thermal averages. Appendix A gives a disentangling theorem for exponential angular momentum operators, and some resulting properties, such as formulas for the coupling of rotations. The disentangling properties should find great use in many other fields of physics where rotations are considered and expectation values have to be calculated. Appendix B shows an example of the application of the disentangling theorem to the calculation of generating functions for expectation values of any product of angular momentum operators in Bloch states. Appendix C shows that the contraction of the rotation group onto the oscillator group can also be used to derive the Hermite polynomials and their properties from the spherical harmonics and their properties. Appendix D gives some useful formulas relating Bloch states, spherical harmonics, and irreducible representations of the full rotation group.

II. DESCRIPTION OF THE FREE FIELD

A. Harmonic-Oscillator States

In order to point out with maximum clarity the analogies between the free-field description and the free-atom description, we start by listing here, in simple terms, the properties of the single harmonic oscillator. The equation numbering here and in Sec. III is done in parallel.

The single harmonic oscillator is described by its canonically conjugated coordinates (q, p) with the commutation relation

$$[q, p] = i\hbar . \tag{2.1}$$

One forms the usual lowering and raising operators

$$a = (2\hbar\omega m)^{-1/2}(\omega m q + ip), \tag{2.2a}$$

$$a^\dagger = (2\hbar\omega m)^{-1/2}(\omega m q - ip), \tag{2.2b}$$

where $\omega m > 0$ is characteristic of the oscillator. These operators satisfy

$$[a, a^\dagger] = 1 \tag{2.3a}$$

from which one obtains

$$[a, a^\dagger a] = a , \tag{2.3b}$$

$$[a^\dagger, a^\dagger a] = -a^\dagger . \tag{2.3c}$$

The harmonic-oscillator states, or Fock states, are the eigenstates of

$$N = a^\dagger a \tag{2.4}$$

and are given by[9]

$$|n\rangle = (n!)^{-1/2}(a^\dagger)^n|0\rangle \quad (n = 0, 1, 2 \ldots) \tag{2.5}$$

with eigenvalue n. The vacuum state $|0\rangle$ is the harmonic-oscillator ground state defined by

$$a|0\rangle = 0 . \tag{2.6}$$

B. Coherent States of the Field

Let us consider the translation operator which produces a shift ξ in q and η in p:

$$T_\alpha = e^{(-i/\hbar)(\xi p - \eta q)} = e^{\alpha a^\dagger - \alpha^* a} , \tag{2.7a}$$

where

$$\alpha = (2\hbar\omega m)^{-1/2}(\omega m \xi + i\eta) . \tag{2.7b}$$

A coherent state $|\alpha\rangle$ is obtained by translation of the ground state

$$|\alpha\rangle \equiv T_\alpha |0\rangle . \tag{2.8}$$

We shall name these states Glauber states, since they have been used extensively by Glauber in quantum optics.[4] Since

$$T_\alpha a T_\alpha^{-1} = a - \alpha , \tag{2.9}$$

the state $|\alpha\rangle$ satisfies the eigenvalue equation

$$(a - \alpha)|\alpha\rangle = 0 . \tag{2.10}$$

Using a Baker–Campbell–Hausdorff formula[10, 11] or Feynman's disentangling techniques,[12] the translation operator T_α can be written in the following forms:

$$T_\alpha = e^{|\alpha|^2/2} e^{-\alpha^* a} e^{\alpha a^\dagger} = e^{-|\alpha|^2/2} e^{\alpha a^\dagger} e^{-\alpha^* a}. \tag{2.11}$$

The second of these forms, which is known as the normally ordered form, immediately gives the expansion of $|\alpha\rangle$ in terms of Fock states,

$$|\alpha\rangle = T_\alpha |0\rangle = e^{-|\alpha|^2/2} e^{\alpha a^\dagger}|0\rangle , \tag{2.12}$$

whence, expanding the exponential and using (2.5),

$$\langle n|\alpha\rangle = e^{-|\alpha|^2/2}\alpha^n/(n!)^{1/2} . \tag{2.13}$$

The scalar product of Glauber states can be obtained either from (2.12), using the disentangling theorem (2.11), or from (2.13), using the completeness property of Fock states $\sum|n\rangle\langle n| = 1$. One gets

$$\langle\alpha|\beta\rangle = e^{-[|\alpha|^2 - 2\alpha^*\beta + |\beta|^2]/2} , \tag{2.14a}$$

whence,

$$|\langle\alpha|\beta\rangle|^2 = e^{-|\alpha-\beta|^2} . \tag{2.14b}$$

The coherent states are minimum-uncertainty packets. For three observables A, B, C, which obey a commutation relation $[A, B] = iC$, it is easy to show[9] that $\langle A^2\rangle\langle B^2\rangle \geq \frac{1}{4}\langle C\rangle^2$. In particular, with $A = q - \xi$, $B = p - \eta$, and $C = \hbar$, one has

$$\langle(q-\xi)^2\rangle\langle(p-\eta)^2\rangle \geq \tfrac{1}{4}\hbar^2 \tag{2.15}$$

for any state. It is easy to show that the equality sign holds for the coherent state $|\alpha\rangle$, where α is related to ξ and η by (2.7b). This establishes the minimum-uncertainty property.

C. Coherent States as Basis

We now consider the completeness properties of the coherent states. Using (2.13), and the completeness of Fock states $\sum_n |n\rangle\langle n| = 1$, one obtains straightforwardly

$$\int \frac{d^2\alpha}{\pi} |\alpha\rangle\langle\alpha| = 1 . \tag{2.16}$$

The expansion of an arbitrary state in Glauber states follows:

$$|c\rangle \equiv \sum_n c_n|n\rangle = \int \frac{d^2\alpha}{\pi} \sum_n c_n |\alpha\rangle\langle\alpha|n\rangle$$

$$= \int \frac{d^2\alpha}{\pi} e^{-|\alpha|^2/2} f(\alpha^*)|\alpha\rangle , \tag{2.17a}$$

where

$$f(\alpha^*) \equiv \sum_n c_n(\alpha^*)^n/(n!)^{1/2} = e^{|\alpha|^2/2}\langle\alpha|c\rangle . \tag{2.17b}$$

Using (2.5), it is seen that $|c\rangle$ can also be written as

$$|c\rangle = f(a^\dagger)|0\rangle , \tag{2.18}$$

where $f(a^\dagger)$ is defined by its expansion (2.17b). The scalar product of any two states $|c'\rangle$ and $|c\rangle$ is obtained from (2.16) and (2.17b):

$$\langle c'|c\rangle = \int \frac{d^2\alpha}{\pi} \langle c'|\alpha\rangle\langle\alpha|c\rangle$$

$$= \int \frac{d^2\alpha}{\pi} e^{-|\alpha|^2} [f'(\alpha^*)]^* f(\alpha^*) . \tag{2.19}$$

In view of the completeness relations, operators F acting on this Hilbert space can be expanded as

$$F = \sum_{m,n} |m\rangle\langle m|F|n\rangle\langle n| \tag{2.20a}$$

or

$$F = \iint \frac{d^2\alpha\, d^2\beta}{\pi^2} |\beta\rangle\langle\beta|F|\alpha\rangle\langle\alpha| . \tag{2.20b}$$

Owing to the overcompleteness of the $|\alpha\rangle$ states, the expansion (2.20b) is in general not unique. This expansion is especially useful if it can be written in the diagonal form

$$F = \int d^2\alpha\, f(\alpha)|\alpha\rangle\langle\alpha| . \tag{2.20c}$$

This will be discussed further for the case of the density matrix.

D. Statistical Operator for the Field

Up to here we have considered pure quantum states. Since a field in thermal equilibrium with matter at ordinary temperatures is essentially in the ground state ($\hbar\omega \gg kT$), this is an adequate description for any field obtained from the thermal equilibrium in response to a classical current. However, the field radiated by an incoherently pumped medium is a statistical mixture described by a statistical operator ρ, which we assume normalized to unity,

$$\mathrm{Tr}\rho = 1 \ . \tag{2.21}$$

With the help of this operator, the statistical average of any observable $F(a, a^\dagger)$ is obtained as

$$\langle F \rangle = \mathrm{Tr}\rho F \ . \tag{2.22}$$

Of particular interest are statistical ensembles described by a statistical operator which is diagonal in the Glauber representation,[4]

$$\rho = \int P(\alpha) | \alpha \rangle \langle \alpha | \, d^2\alpha \ , \tag{2.23}$$

where the normalization (2.21) requires

$$\int P(\alpha) d^2\alpha = 1 \ . \tag{2.24}$$

The statistical average of an observable F is then given by an average over the diagonal elements $\langle \alpha | F | \alpha \rangle$:

$$\langle F \rangle = \int P(\alpha) \langle \alpha | F | \alpha \rangle d^2\alpha \ . \tag{2.25}$$

The weight function $P(\alpha)$ has thus the properties of a distribution function in α space, except that it is not necessarily positive.

Let us define a set of operators $\hat{X}(\lambda)$ such that their expectation values for coherent states

$$b^\alpha(\lambda) = \langle \alpha | \hat{X}(\lambda) | \alpha \rangle \tag{2.26}$$

form a basis in the function space of functions of α. If the statistical ensemble has a diagonal representation (2.23), then the statistical averages of the operators $\hat{X}(\lambda)$ form a kind of characteristic function of $P(\alpha)$:

$$X(\lambda) \equiv \langle \hat{X}(\lambda) \rangle = \int d^2\alpha \, P(\alpha) b^\alpha(\lambda) \ . \tag{2.27}$$

The weight function $P(\alpha)$ can be expressed in terms of $X(\lambda)$ with the help of the reciprocal basis $\bar{b}^\lambda(\alpha)$,

$$P(\alpha) = \int d^2\lambda \, X(\lambda) \bar{b}^\lambda(\alpha) \ . \tag{2.28}$$

A convenient basis is the Fourier basis

$$b^\alpha(\lambda) = e^{\lambda\alpha^* - \lambda^*\alpha} \ , \tag{2.29a}$$

$$\bar{b}^\lambda(\alpha) = \frac{1}{(2\pi)^2} e^{-\lambda\alpha^* + \lambda^*\alpha} \ , \tag{2.29b}$$

which is generated by the normally ordered operators

$$\hat{X}_N(\lambda) = e^{\lambda a^\dagger} e^{-\lambda^* a} \ . \tag{2.29c}$$

The question of the existence of the P representation is a complicated one.[4,13] Using the Fourier basis (2.29) it can be shown, however, that the mere existence of the inverse transformation (2.28) guarantees that the resulting function $P(\alpha)$ can be used to calculate the statistical average of any product $a^{\dagger m} a^n$ as if $P(\alpha)$ were the weight function defined in (2.23). This is due to the fact that the characteristic function $X_N(\lambda)$ plays the role of a generating function for $\langle a^{\dagger m} a^n \rangle$:

$$\langle a^{\dagger m} a^n \rangle = \left(\frac{\partial}{\partial\lambda}\right)^m \left(-\frac{\partial}{\partial\lambda^*}\right)^n X_N(\lambda)\Big|_{\lambda=0} \ ,$$

whence, by derivation of (2.27), one obtains

$$\langle a^{\dagger m} a^n \rangle = \int d^2\alpha \, P(\alpha) \langle \alpha | a^{\dagger m} a^n | \alpha \rangle \ ,$$

which is a particular case of (2.25) and proves the above statement. One could, moreover, introduce, in addition to (2.29c), symmetrically ordered $\hat{X}_S(\lambda)$ and antinormally ordered $\hat{X}_A(\lambda)$ exponential operators.[4,13] The Fourier transform of their statistical averages are the Wigner distribution and the matrix element $(1/\pi)\langle \alpha | \rho | \alpha \rangle$, respectively. We shall not develop these aspects further as the corresponding expressions for atomic coherent states are rather involved, and of no clear use as yet.

III. DESCRIPTION OF FREE ATOMS

A. Angular Momentum States

As shown in the Introduction, angular momentum operators can be defined which act on the N-atom Hilbert space. In particular we can consider a subspace of degenerate eigenstates of J^2 with eigenvalues $J(J+1)$. Since J^2 commutes with J_x, J_y, J_z, these operators only connect states within the same subspace. In general J^2 and J_z do not form a complete set of commuting observables. As explained in Sec. V, such a complete set is formed by adding to J^2 and J_z some operators of the permutation group of N objects P_N. These operators play, with respect to P_N, the same role as J^2 and J_z with respect to the three-dimensional rotation group. We shall assume that the subspace considered here has also been made invariant under these permutation operations, but for simplicity we shall omit, for the time being, to indicate this in the labeling of the states. The subspace we are dealing with is identical to a constant angular momentum Hilbert space. The Dicke states, which are the analog of the Fock states (2.5), and the Bloch states, which correspond to the Glauber states (2.8), are most easily defined within such a subspace. The equation numbering is in parallel with that of Sec. II. From the angular momentum operators J_x and J_y, which satisfy the commutation relation

$$[J_x, J_y] = iJ_z \ , \tag{3.1}$$

the lowering and raising operators are formed,

$$J_- \equiv J_x - iJ_y \ , \tag{3.2a}$$

$$J_+ \equiv J_x + iJ_y \ , \tag{3.2b}$$

which obey

$$[J_-, J_+] = -2J_z , \quad (3.3a)$$

$$[J_-, J_z] = J_- , \quad (3.3b)$$

$$[J_+, J_z] = -J_+ . \quad (3.3c)$$

The Dicke states, which are simply the usual angular momentum states,[7] are defined as the eigenstates of

$$J_z = \tfrac{1}{2}(J_+ J_- - J_- J_+) . \quad (3.4)$$

They are given by[9]

$$|M\rangle = \frac{1}{(M+J)!}\binom{2J}{M+J}^{-1/2} J_+^{M+J} |-J\rangle$$

$$(M = -J, -J+1, \ldots, J) , \quad (3.5)$$

with eigenvalue M. They span the space of angular momentum quantum number J. The ground state $|-J\rangle$ is defined by

$$J_-|-J\rangle = 0 . \quad (3.6)$$

B. Coherent Atomic States

Let us consider the rotation operator which produces a rotation through an angle θ about an axis $\hat{n} = (\sin\varphi, -\cos\varphi, 0)$ as shown in Fig. 1:

$$R_{\theta,\varphi} = e^{-i\theta J_n} = e^{-i\theta(J_x \sin\varphi - J_y \cos\varphi)} = e^{\zeta J_+ - \zeta^* J_-} , \quad (3.7a)$$

where

$$\zeta = \tfrac{1}{2}\theta\, e^{-i\varphi} . \quad (3.7b)$$

A coherent atomic state, or Bloch state, $|\theta, \varphi\rangle$ is obtained by rotation of the ground state $|-J\rangle$:

$$|\theta, \varphi\rangle \equiv R_{\theta,\varphi} |-J\rangle . \quad (3.8)$$

Referring to Fig. 1, it is seen that

$$R_{\theta,\varphi} J_n R_{\theta,\varphi}^{-1} = J_n ,$$

$$R_{\theta,\varphi} J_k R_{\theta,\varphi}^{-1} = J_k \cos\theta + J_z \sin\theta ,$$

$$R_{\theta,\varphi} J_z R_{\theta,\varphi}^{-1} = -J_k \cos\theta + J_z \sin\theta ,$$

where

$$J_n = J_x \sin\varphi - J_y \cos\varphi , \quad J_k = J_x \cos\varphi + J_y \sin\varphi ,$$

which gives

$$J_+ = (J_k - iJ_n)\, e^{i\varphi} , \quad J_- = (J_k + iJ_n)\, e^{-i\varphi} .$$

Using these relations one obtains

$$R_{\theta,\varphi} J_- R_{\theta,\varphi}^{-1} = e^{-i\varphi} [J_- e^{i\varphi} \cos^2(\tfrac{1}{2}\theta) - J_+ e^{-i\varphi} \sin^2(\tfrac{1}{2}\theta) + J_z \sin\theta] , \quad (3.9a)$$

and similar relations for J_+ and J_z:

$$R_{\theta,\varphi} J_+ R_{\theta,\varphi}^{-1} = e^{i\varphi} [J_+ e^{-i\varphi} \cos^2(\tfrac{1}{2}\theta) - J_- e^{i\varphi} \sin^2(\tfrac{1}{2}\theta) + J_z \sin\theta] , \quad (3.9b)$$

$$R_{\theta,\varphi} J_z R_{\theta,\varphi}^{-1} = J_z \cos\theta - J_- e^{i\varphi} \sin\tfrac{1}{2}\theta \cos\tfrac{1}{2}\theta - J_+ e^{-i\varphi} \sin\tfrac{1}{2}\theta \cos\tfrac{1}{2}\theta . \quad (3.9c)$$

From (3.9a) and the definition (3.8), one obtains the eigenvalue equation

$$[J_- e^{i\varphi} \cos^2(\tfrac{1}{2}\theta) - J_+ e^{-i\varphi} \sin^2(\tfrac{1}{2}\theta) + J_z \sin\theta] \,|\theta, \varphi\rangle = 0 . \quad (3.10a)$$

This equation, together with

$$J^2 |\theta, \varphi\rangle = J(J+1) |\theta, \varphi\rangle , \quad (3.10b)$$

specifies uniquely the Bloch state $|\theta, \varphi\rangle$. Note that the harmonic-oscillator analog of (3.10b) would have been the trivial relation $(a^\dagger - \alpha^*)(a - \alpha)|\alpha\rangle = 0$. Other forms of the eigenvalue equation can be obtained using the relation

$$R_{\theta,\varphi} J_z R_{\theta,\varphi}^{-1} |\theta, \varphi\rangle = -J |\theta, \varphi\rangle$$

and (3.9c). The resulting equation can be combined with (3.10a) to eliminate one of the operators J_z, J_+, or J_-, giving

$$[J_- e^{i\varphi} \cos^2(\tfrac{1}{2}\theta) + J_+ e^{-i\varphi} \sin^2(\tfrac{1}{2}\theta)] \,|\theta, \varphi\rangle = J \sin\theta \,|\theta, \varphi\rangle , \quad (3.10c)$$

$$[J_- e^{i\varphi} \cos\tfrac{1}{2}\theta + J_z \sin\tfrac{1}{2}\theta] \,|\theta, \varphi\rangle = J \sin\tfrac{1}{2}\theta \,|\theta, \varphi\rangle , \quad (3.10d)$$

$$[J_+ e^{-i\varphi} \sin\tfrac{1}{2}\theta - J_z \cos\tfrac{1}{2}\theta] \,|\theta, \varphi\rangle = J \cos\tfrac{1}{2}\theta \,|\theta, \varphi\rangle . \quad (3.10e)$$

These additional relations are not independent of (3.10a) and (3.10b). It is noted that these eigenvalue equations are more complicated than their counterpart (2.10). In particular they involve at least two of the three operators J_-, J_+, J_z. This feature is required by the more complicated commutation relation (3.1) which applies here.

Using the disentangling theorem for angular momentum operators (Appendix A), the rotation $R_{\theta,\varphi}$ given by (3.7a) becomes

$$R_{\theta,\varphi} = e^{-\tau^* J_-} e^{-\ln(1+|\tau|^2) J_z} e^{\tau J_+}$$

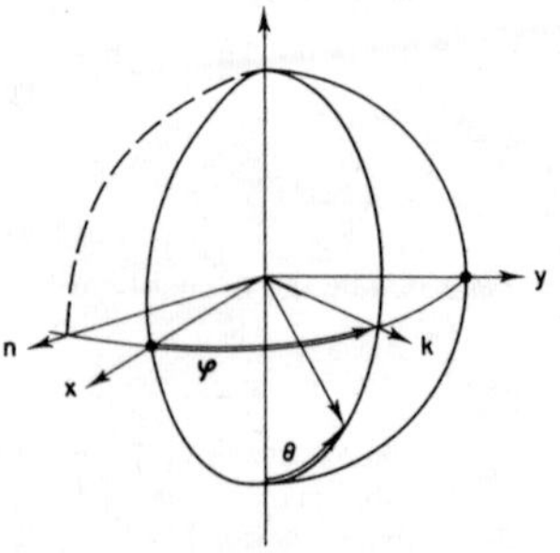

FIG. 1. Definition of the rotation $R_{\theta,\varphi}$ in angular momentum space.

$$= e^{\tau J_+} e^{\ln(1+|\tau|^2) J_z} e^{-\tau^* J_-}, \quad (3.11a)$$

where

$$\tau = e^{-i\varphi} \tan\tfrac{1}{2}\theta . \quad (3.11b)$$

Let us point out that these expressions are singular for $\theta=\pi$, i.e., for the uppermost state. We may have to exclude from some of the following consideration the states contained within an infinitesimally small circle around $\theta=\pi$. The validity of expressions such as (3.13) for $\theta=\pi$ is usually not affected and can be checked directly. The last form of (3.11a), which we call the normally ordered form, immediately gives the expansion of $|\theta, \varphi\rangle$ in terms of Dicke states:

$$|\theta,\varphi\rangle = R_{\theta,\varphi}|-J\rangle = \left(\frac{1}{1+|\tau|^2}\right)^J e^{\tau J_+}|-J\rangle , \quad (3.12)$$

whence, expanding the exponential and using (3.5),

$$\langle M|\theta,\varphi\rangle = \binom{2J}{M+J}^{1/2} \frac{\tau^{M+J}}{[1+|\tau|^2]^J}$$

$$= \binom{2J}{M+J}^{1/2} \sin^{J+M}(\tfrac{1}{2}\theta)\cos^{J-M}(\tfrac{1}{2}\theta)\, e^{-i(J+M)\varphi}. \quad (3.13)$$

Since the Dicke states form a basis for a well-known irreducible representation of the rotation group, these results could have been derived using the appropriate Wigner $\mathfrak{D}^{(J)}$ matrix.[14] The same remark applies to Eqs. (2.12) and (2.13): These could have been obtained without using a Baker–Campbell–Hausdorff formula, from the transformation properties of an irreducible representation of the group of operations T_α.[15–17]

The overlap of two Bloch states is obtained either from (3.12), using the disentangling theorem for exponential angular momentum operators, or from (3.13), using the completeness property of Dicke states $\sum_M |M\rangle\langle M| = 1$. One obtains

$$\langle\theta,\varphi|\theta',\varphi'\rangle = \left(\frac{(1+\tau^*\tau')^2}{(1+|\tau|^2)(1+|\tau'|^2)}\right)^J$$

$$= e^{iJ(\varphi-\varphi')}[\cos\tfrac{1}{2}(\theta-\theta')\cos\tfrac{1}{2}(\varphi-\varphi') - i\cos\tfrac{1}{2}(\theta+\theta')\sin\tfrac{1}{2}(\varphi-\varphi')]^{2J} , \quad (3.14a)$$

whence

$$|\langle\theta,\varphi|\theta',\varphi'\rangle|^2 = \cos^{4J}\tfrac{1}{2}\Theta , \quad (3.14b)$$

where τ is given by (3.11b), τ' is given by the same equation written with the primed quantities, and Θ is the angle between the (θ, φ) and (θ', φ') directions, as given by

$$\cos\Theta = \cos\theta\cos\theta' + \sin\theta\sin\theta'\cos(\varphi-\varphi') .$$

The Bloch states form minimum-uncertainty packets. The uncertainty relation can be defined in terms of the set of rotated operators $(J_\xi, J_\eta, J_\zeta) = R_{\theta,\varphi}(J_x, J_y, J_z)R^{-1}_{\theta,\varphi}$. These three observables obey a commutation relation of the type $[A, B] = iC$ with $A=J_\xi$, $B=J_\eta$, $C=J_\zeta$, whence they have the uncertainty property

$$\langle J_\xi^2\rangle\langle J_\eta^2\rangle \geq \tfrac{1}{4}\langle J_\zeta\rangle^2 \quad (3.15)$$

for any states. It is easy to show that the equality sign holds for the Bloch state $|\theta,\varphi\rangle$, which is therefore a minimum-uncertainty state.

C. Bloch States as Basis

Let us now consider the completeness properties of the Bloch states. Using (3.13) and the completeness of Dicke states $\sum_M |M\rangle\langle M| = 1$, one obtains

$$(2J+1)\int \frac{d\Omega}{4\pi}|\theta,\varphi\rangle\langle\theta,\varphi|$$

$$= (2J+1)\int\frac{d\Omega}{4\pi}\sum_{M,M'}\binom{2J}{M+J}^{1/2}\binom{2J}{M'+J}^{1/2} e^{i(M'-M)\varphi}(\cos\tfrac{1}{2}\theta)^{2J-M-M'}(\sin\tfrac{1}{2}\theta)^{2J+M+M'}|M\rangle\langle M'|$$

$$= (2J+1)\int_0^\pi \tfrac{1}{2}(\sin\theta\, d\theta)\sum_M\binom{2J}{M+J}(\cos\tfrac{1}{2}\theta)^{2J-2M}(\sin\tfrac{1}{2}\theta)^{2J+2M}|M\rangle\langle M| = \sum_M |M\rangle\langle M| = 1 . \quad (3.16)$$

The expansion of an arbitrary state in Bloch states follows:

$$|c\rangle = \sum_M c_M|M\rangle = (2J+1)\int\frac{d\Omega}{4\pi}\sum_M c_M|\theta,\varphi\rangle\langle\theta,\varphi|M\rangle$$

$$= (2J+1)\int\frac{d\Omega}{4\pi}\frac{f(\tau^*)}{[1+|\tau|^2]^J}|\theta,\varphi\rangle , \quad (3.17a)$$

where

$$f(\tau^*) \equiv \sum_M c_M\binom{2J}{J+M}^{1/2}(\tau^*)^{J+M} = [1+|\tau|^2]^J\langle\theta,\varphi|c\rangle . \quad (3.17b)$$

Using (3.5) it is seen that $|c\rangle$ can also be written as

$$|c\rangle = f\left(\frac{1}{J+1-J_z}J_+\right)|-J\rangle . \quad (3.18)$$

The amplitude function $f(\tau^*)$ is, by its definition

(3.17b), a polynomial of degree $2J$. However any function which has a Maclaurin expansion can be taken as a suitable amplitude function in (3.17a) or (3.18). Indeed the powers of τ^* higher than $2J$ give zero contribution in (3.17a) and (3.18). The coefficients c_M are then obtained from the first $(2J+1)$ terms of the Maclaurin series, using (3.17b).

The scalar product of two states characterized by their amplitude function is, from (3.16) and (3.17b),

$$\langle c' | c \rangle = (2J+1) \int \frac{d\Omega}{4\pi} \langle c' | \theta, \varphi \rangle \langle \theta, \varphi | c \rangle$$
$$= (2J+1) \int \frac{d\Omega}{4\pi} \frac{1}{(1+|\tau|^2)^{2J}} [f'(\tau^*)]^* f(\tau^*) . \quad (3.19)$$

Since (3.17b) was used to derive this equation, its validity is restricted to amplitude functions which are polynomials of degree $2J$.

In view of the completeness relations, operators G acting on this Hilbert space can be expanded as

$$G = \sum_{M, M'} |M\rangle \langle M | G | M' \rangle \langle M' | \quad (3.20a)$$

or

$$G = \frac{(2J+1)^2}{(4\pi)^2} \iint d\Omega \, d\Omega' \, |\theta, \varphi\rangle \langle \theta, \varphi | G | \theta', \varphi' \rangle \langle \theta', \varphi' | . \quad (3.20b)$$

However, G is completely defined by the $(2J+1)^2$ matrix elements $\langle M | G | M' \rangle$, with the result that, except for pathological cases, an operator can always be written in the diagonal form

$$G = \int d\Omega \, g(\theta, \varphi) | \theta, \varphi \rangle \langle \theta, \varphi | , \quad (3.20c)$$

where $g(\theta, \varphi)$ is given by a series expansion

$$g(\theta, \varphi) = \sum_{l, m} G_{lm} Y_{lm}(\theta, \varphi) . \quad (3.20d)$$

In accordance with the property (D21) only the $(2J+1)^2$ first terms of this sum contribute to (3.20c). These are the terms for which $0 \leq l \leq 2J$. The corresponding coefficients G_{lm} are expressed as a function of the matrix elements $\langle M | G | M' \rangle$ by Eq. (D26).

D. Statistical Operators for Atoms

In order to describe an incoherently pumped system of atoms we introduce a statistical operator ρ with the properties

$$\mathrm{Tr}\rho = 1 , \quad (3.21)$$

$$\langle G \rangle = \mathrm{Tr}\rho G . \quad (3.22)$$

As before, the considerations are restricted to states belonging to a single constant-angular-momentum subspace and, therefore, the statistical operator described here does not allow for the most general mixing of atomic states. Of particular interest is the expression of ρ in a diagonal Bloch representation

$$\rho = \int P(\theta, \varphi) | \theta, \varphi \rangle \langle \theta, \varphi | \, d\Omega , \quad (3.23)$$

with the normalization

$$\int P(\theta, \varphi) d\Omega = 1 . \quad (3.24)$$

The statistical average of an observable G is then given by

$$\langle G \rangle = \int P(\theta, \varphi) \langle \theta, \varphi | G | \theta, \varphi \rangle d\Omega . \quad (3.25)$$

The weight function $P(\theta, \varphi)$ has thus the properties of a distribution function on the unit sphere, except that it is not necessarily positive.

Let us define a set of operators $\hat{X}_\lambda$ such that their expectation values for Bloch states

$$b_\lambda^{(\theta, \varphi)} = \langle \theta, \varphi | \hat{X}_\lambda | \theta, \varphi \rangle \quad (3.26)$$

form a basis in the space of functions on the unit sphere. Since in this space a discrete basis can be chosen, the parameter λ can be restricted to discrete values $\lambda = 1, 2, \ldots$. For a statistical ensemble described by (3.23), the statistical averages of the operators $\hat{X}_\lambda$ form a set of characteristic coefficients of $P(\theta, \varphi)$:

$$X_\lambda \equiv \mathrm{Tr}(\rho \hat{X}_\lambda) = \int P(\theta, \varphi) b_\lambda^{(\theta, \varphi)} d\Omega . \quad (3.27)$$

The weight function can be expressed as a series with the help of the reciprocal basis $\bar{b}^\lambda(\theta, \varphi)$,

$$P(\theta, \varphi) = \sum_\lambda X_\lambda \bar{b}^\lambda(\theta, \varphi) . \quad (3.28)$$

A convenient basis is given by the spherical harmonics

$$b_\lambda^{(\theta, \varphi)} = Y_l^m(\theta, \varphi) , \quad \lambda \equiv (l, m) \quad (3.29a)$$

$$\bar{b}^\lambda(\theta, \varphi) = Y_l^{-m}(\theta, \varphi) , \quad (3.29b)$$

which are generated by the spherical harmonic operators[18]

$$\hat{X}_\lambda = \mathcal{Y}_l^m(\vec{J}) . \quad (3.29c)$$

The fact already mentioned that a diagonal representation always exists in the atomic case also corresponds to the fact that for a given J only the $(2J+1)^2$ operators $\mathcal{Y}_l^m$ with $l \leq 2J$ are different from zero. The finite dimensionality of the basis is required, since ρ is completely determined by its $(2J+1)^2$ matrix elements $\langle M | \rho | M' \rangle$ in the Dicke representation. An illustration is given in Appendix D, where the statistical operator corresponding to a pure Bloch state is derived [Eq. (D29)]. Other differences with the field case should also be noted. Firstly, the spherical harmonic operators are usually written in a fully symmetrized form, whereas the operators (2.29c) are normally ordered. This is only a formal difficulty as it should be possible to write normally ordered and antinormally ordered multipole operators[19] with properties similar to the $\mathcal{Y}_l^m(\vec{J})$. A second, and more fundamental difference, is that the expectation values X_λ are not generating functions for products of the type $\langle J_+^m J_z^n J_-^p \rangle$, in view of the discreteness of the set. This does not cause much difficulty, as generating functions can be defined from exponential operators (Appendix B) whose expectation values can be calculated with the help of the disentangling theorem (Appendix A). It is tempting to use for the $\hat{X}_\lambda$'s of Eq. (3.29c) these exponential operators themselves. Though the parallel with the field case then seems more transparent, the use of the discrete set $\hat{X}_\lambda$ may be of more fundamental significance as it takes into account symmetry properties of the states.

A final comment should be made about the difficulty of dealing with creation and annihilation operations in a finite Hilbert space. The existence of two terminal states, $|J\rangle$ and $|-J\rangle$, requires the presence of a third operator with the properties of J_z, and prevents the writing of an eigenvalue equation in terms of one compound operator alone. For instance, comparison of (3.18) and (2.18) suggests that $J_-(J+1-J_z)^{-1}$ could be a "good" annihilation operator. Using (3.13) one finds immediately

$$J_-\frac{1}{J+1-J_z}\,|\,\theta,\varphi\rangle=\tau\,|\,\theta,\varphi\rangle-\tau\sin^{2J}(\tfrac{1}{2}\theta)e^{-2iJ\varphi}\,|\,J\rangle\,,$$

which for small θ and large J is "almost" an eigenvalue equation, i.e., the application of the operator reproduces $|\,\theta,\varphi\rangle$ except for the uppermost Dicke state. There is no doubt that a theory could be developed in terms of more complicated annihilation and creation operators of such type, but the advantage is not clear.

IV. CONTRACTION, OR THE RELATION BETWEEN ATOMIC STATES AND FIELD STATES

The extreme similarity between the treatments of Secs. III and II suggests a close connection between atomic and field states. This connection is made here through a process known as group contraction.[20–22]

The time evolution of a single two-level atom is governed by a 2×2 unitary transformation matrix. The commutation relation for the generators of the group U(2) are rewritten here:

$$[J_z, J_\pm]=\pm J_\pm\,,\quad [J_+, J_-]=2J_z\,,\quad [\vec{J}, J_0]=0\,, \tag{4.1}$$

where J_0 in the third relation is essentially the identity. An arbitrary 2×2 unitary transformation matrix is given by

$$\mathrm{U}(2)=\exp(i\textstyle\sum_\mu \lambda_\mu J_\mu)\,, \tag{4.2}$$

where the summation is over all four indices and the λ_μ's are c-number parameters which characterize the group operation.

If another set of generators h_+, h_-, h_z, h_0 is related to J_+, J_-, J_z, J_0 by a nonsingular transformation $A_{\nu\mu}$,

$$h_\nu=\textstyle\sum_\mu A_{\nu\mu}J_\mu\,, \tag{4.3}$$

then the group operation (4.2) may be written

$$\exp(i\textstyle\sum_\mu \lambda_\mu J_\mu)=\exp(i\sum \alpha_\nu h_\nu)\,, \tag{4.4}$$

with

$$\alpha_\nu=\textstyle\sum\lambda_\nu(A^{-1})_{\mu\nu}\,. \tag{4.3'}$$

We select the following transformation A, which depends on a real parameter c:

$$\begin{bmatrix} h_+\\ h_-\\ h_z\\ h_0\end{bmatrix}=\begin{bmatrix} c & 0 & 0 & 0\\ 0 & c & 0 & 0\\ 0 & 0 & 1 & 1/2c^2\\ 0 & 0 & 0 & 1\end{bmatrix}\begin{bmatrix} J_+\\ J_-\\ J_z\\ J_0\end{bmatrix}\,. \tag{4.5}$$

It is easily verified that the h_ν's satisfy the commutation relations

$$[h_z, h_\pm]=\pm h_\pm\,,\quad [h_+, h_-]=2c^2h_z-h_0\,,\quad [\vec{h}, h_0]=0\,. \tag{4.6}$$

In the limit $c\to 0$ the transformation A becomes singular, and A^{-1} fails to exist. Nevertheless, the commutation relations (4.6) are well defined and, in fact, identical to the commutation relations (2.3) under the identification

$$\lim_{c\to 0}h_z=n=a^\dagger a\,,\quad \lim_{c\to 0}h_+=a^\dagger\,,\quad \lim_{c\to 0}h_-=a\,. \tag{4.7}$$

Although the inverse A^{-1} (4.3b) does not exist as $c\to 0$, the parameter α_ν may approach a well-defined limit if we demand all the parameters λ_μ to shrink ("contract") to zero in the limit $c\to 0$, in such a way that the following ratios are well defined:

$$\begin{aligned}
&\lim_{c\to 0}\frac{i\lambda_+}{c}=\lim_{c\to 0}\frac{e^{-i\varphi}\theta}{2c}=\alpha\,,\\
&\lim_{c\to 0}\frac{i\lambda_-}{c}=\lim_{c\to 0}-\frac{e^{i\varphi}\theta}{2c}=-\alpha^*\,,\\
&\lim_{c\to 0}\frac{\lambda_z}{c}=\lim_{c\to 0}c\,\frac{\lambda_z}{c^2}=0\,.
\end{aligned} \tag{4.8}$$

Within any $(2J+1)$-dimensional representation of the group U(2) the eigenvalue of the diagonal operator h_z is

$$h_z\,|\,J,M\rangle=(J_z+1/2c^2)\,|\,J,M\rangle=|\,J,M\rangle(M+1/2c^2)\,. \tag{4.9}$$

We demand that this have a definite limit as $c\to 0$. Physically, for both Fock and Dicke states we progress upwards from the ground or vacuum state. It is convenient to demand that the (energy) eigenvalue in (4.9) be zero for the ground state $M=-J$;

$$\lim_{c\to 0}(-J+1/2c^2)=0\ . \tag{4.10}$$

In the limit $c\to 0$, $2Jc^2=1$, the unitary irreducible representations $\mathfrak{D}^J$ [U(2)] go over into unitary irreducible representations for the contracted group with generators (4.6). In simple words, the contraction procedure amounts to letting the radius of the Bloch sphere tend to infinity as $1/c^2$, while considering smaller and smaller rotations on the sphere. The motion on the sphere then becomes identical to the motion on the bottom tangent plane which goes over into the phase plane of the harmonic oscillator.

This procedure for contracting groups, commutation relations, and representations will now be used to show the similarity between Dicke and Fock states. We define

$$|\infty, n\rangle \equiv \lim_{c\to 0}|J,M\rangle \quad (J+M=n,\ \text{fixed})\ . \tag{4.11}$$

Then

$$\begin{aligned} a^\dagger a|\infty,n\rangle &= \lim_{c\to 0}(J_z+1/c^2)|J,M\rangle \\ &= \lim_{c\to 0}|J,M\rangle[(J+M)+(-J+1/2c^2)] \\ &= n|\infty,n\rangle\ . \end{aligned} \tag{4.12a}$$

The computations for $a^\dagger$ and a are handled in an entirely analogous way:

$$a^\dagger|\infty,n\rangle=\lim_{c\to 0}h_+|J,M\rangle=(n+1)^{1/2}|\infty,n+1\rangle\ , \tag{4.12b}$$

$$a|\infty,n\rangle=\lim_{c\to 0}h_-|J,M\rangle=n^{1/2}|\infty,n-1\rangle\ . \tag{4.12c}$$

These equations provide a straightforward connection between Dicke and Fock states. The operators h_+, h_-, and h_z contract to $a^\dagger$, a, and $a^\dagger a$ with the proper commutation relations (2.3), and with the proper matrix elements between contracted Dicke states as shown in (4.12). The contracted Dicke states (4.11) can thus be identified with the Fock states, and we conclude that every property of Dicke and Bloch states listed in Sec. III must contract to a corresponding property of Fock and Glauber states listed in Sec. II. The contraction procedure is summarized in Table I.

We demonstrate this correspondence in three particular cases.

Example 1. Just as the angular momentum eigenstates $|J,M\rangle$ are obtained from the ground state $|J,-J\rangle$ by $(J+M)$ successive applications of the shift-up operator J_+, the Fock state $|n\rangle$ is obtained by n successive applications of $a^\dagger$. By contraction of (3.5) we get

$$\begin{aligned} |\infty,n\rangle &= \lim_{c\to 0}\frac{(J_+)^n|J,-J\rangle}{[(2J!/(2J-n)!)n!]^{1/2}} \\ &= \lim_{c\to 0}\frac{(cJ_+)^n|J,-J\rangle}{[(2Jc^2)^n n!]^{1/2}}=\frac{(a^\dagger)^n|\infty,0\rangle}{n!^{1/2}}\ , \end{aligned} \tag{4.13}$$

which is nothing but (2.5).

Example 2. Let us now contract Bloch states to Glauber states, using Eq. (3.12) and Table I:

$$\begin{aligned} |\alpha\rangle &= \lim_{c\to 0}|\theta,\varphi\rangle=\lim_{c\to 0}\left(\frac{1}{1+|\tau|^2}\right)^J e^{\tau J_+}|-J\rangle \\ &= \lim_{c\to 0}(1-2c^2\tfrac{1}{2}\alpha\alpha^*)^{1/2c^2}e^{\alpha a^\dagger}|0\rangle \\ &= e^{-\alpha\alpha^*/2}e^{\alpha a^\dagger}|0\rangle\ , \end{aligned} \tag{4.14}$$

which is nothing but (2.12).

We leave it to the reader to verify that every equation of Sec. III goes over to the corresponding equation of Sec. II under contraction. This is true in particular of the disentangling theorem (3.11) whose contracted limit is the Baker–Campbell–Hausdorff formula (2.11) as shown in (A5) and (A6). In general, all properties related to angular momentum have a harmonic-oscillator property as counterpart. Thus, the total momentum contracts to the harmonic-oscillator Hamiltonian, while the spherical harmonics (and their properties) contract to the harmonic-oscillator eigenfunction (and corresponding properties) (Appendix C).

Example 3. As a final case of special interest we contract the uncertainty relations (3.15). Multiplying (3.15) by c^4 on both sides, and using (3.9), we obtain the limits

$$\lim_{c\to 0}cJ_\xi=\frac{a^\dagger+a}{2}-\frac{\alpha^*+\alpha}{2}=(\omega m/2\hbar)^{1/2}(q-\xi)\ ,$$

TABLE I. Rules for the contraction of the angular momentum algebra to the harmonic-oscillator algebra. [The limit of the angular momentum quantities (first line) for $c\to 0$ are the corresponding harmonic-oscillator quantities (second line).]

Operators	Coordinates	Eigenvalues	Eigenstates	Coherent states
		Angular momentum		
cJ_+, cJ_-, $J_z+1/2c^2$	$\dfrac{\theta}{2c}e^{-i\varphi}$	$2c^2J$, $J+M$	$\|J,M\rangle$ (Dicke)	$\|\theta,\varphi\rangle$ (Bloch)
		Harmonic oscillator		
$a^\dagger$, a, $a^\dagger a$	α	$1, n$	$\|\infty,n\rangle$ (Fock)	$\|\alpha\rangle$ (Glauber)

$$\lim_{c\to 0} cJ_\eta = \frac{a^\dagger - a}{2i} - \frac{\alpha^* - \alpha}{2i} = -(2\hbar\omega m)^{-1/2}(p-\eta),$$

$$\lim_{c\to 0} c^2 J_\zeta = \tfrac{1}{2}. \qquad (4.15)$$

Introducing these relations in (3.15) we reproduce (2.15), and since the equality sign in (3.15) holds for the coherent state $|\theta, \varphi\rangle$ corresponding to the rotation $R_{\theta,\varphi}$, it follows that the equality sign in (2.15) holds for the coherent state α corresponding to the translation (ξ, η). This proves the minimum-uncertainty property of Glauber states, being given the minimum uncertainty of Bloch states.

V. SYMMETRY PROPERTIES OF ATOMIC STATES

A. Problem

We have seen in Sec. III that the symmetrized states arising in the description of N identical two-level atoms can be labeled by the quantum number J, M. These are the so-called Dicke states. The quantum numbers J, M are familiar from the study of the *angular momentum* group SU(2).

The group SU(2) arises in the description of symmetrized atomic states in the following way: The time evolution of each single two-level atom is governed by a 2×2 unitary transformation matrix. The evolution of N identical two-level atoms is described by a direct product of N such matrices when all atoms evolve in time in the same way. All transformation matrices are identical and in fact are a direct product representation of the group SU(2).

The quantum numbers J, M do not provide a large enough set of quantum numbers for a complete labeling of symmetrized atomic states. That is, many distinct symmetrized atomic states may be labeled by the same quantum number J, M. Since the study of symmetries leads to a generation of quantum numbers, we are led to a study of an additional symmetry group P_N—the *permutation* group on N atoms.[6]

The group P_N arises in the description of symmetrized atomic states in the following way: The time evolution for N indistinguishable atoms is governed by a Hamiltonian which is left unchanged under any permutation of the atomic labels

$$\mathcal{H}(1, \ldots, i, \ldots, j, \ldots, N) = +\mathcal{H}(1, \ldots, j, \ldots, i, \ldots, N). \qquad (5.1)$$

This is exactly what it means for particles to be identical.

The quantum numbers of P_N, which are in exact analogy with the quantum numbers J, M of SU(2), then provide additional labels for symmetrized atomic states. With these additional quantum numbers, all symmetrized atomic states are uniquely labeled.

B. Subspaces Invariant under Rotation Operations

The properties of such spaces are well known from the study of angular momentum in quantum theory.[23] The total angular momentum operator J^2 commutes with the operators $J_\pm$, J_z and therefore has the same eigenvalue $J(J+1)$ on all bases for an irreducible representation $\mathfrak{D}^J$.

The number of distinct basis vectors within any irreducible representation is equal to

$$\dim \mathfrak{D}^J = 2J+1. \qquad (5.2a)$$

These bases are conveniently labeled by the eigenvalues M of the diagonal operator J_z,

$$M = -J, -J+1, \ldots, +J. \qquad (5.3a)$$

Bases are therefore labeled $|{}^J_M; \alpha'\rangle$. Here α' is the set of all other quantum numbers necessary to distinguish different copies of the space with the same J label. An arbitrary element R of the group SU(2) maps a state in any J, α' subspace into a linear combination of states *within the same subspace*;

$$R|{}^J_M;\alpha'\rangle = \sum_{M'} |{}^J_{M'};\alpha'\rangle\langle{}^J_{M'}|R|{}^J_M\rangle = \sum_{M'} |{}^J_{M'};\alpha'\rangle \mathfrak{D}^J_{M'M}(R). \qquad (5.4a)$$

Referring to Fig. 2, we say that the group SU(2) "acts vertically."

C. Subspace Invariant under Permutation Operations

Operators which commute with all elements of P_N have the same eigenvalues on all bases within an irreducible representation. These eigenvalues can then be used as labels for the irreducible representations. The operators in P_N analogous to J^2 in SU(2) are the class sums[5]; their eigenvalues are the non-negative integers $\lambda_1 \ge \lambda_2$ which obey $\lambda_1 + \lambda_2 = N$.

The number of distinct basis vectors within any irreducible representation is equal to[5]

$$\dim \Gamma^{\lambda_1,\lambda_2} = \binom{\lambda_1+\lambda_2}{\lambda_2} - \binom{\lambda_1+\lambda_2}{\lambda_2 - 1}. \qquad (5.2b)$$

These bases can be labeled by eigenvalues of operators analogous to J_z: These operators are the $\frac{1}{2}N$ mutually commuting adjacent interchanges P_{12}, P_{34}, P_{56}, ... However, it is more convenient in this case to label the bases simply by the number i,

$$i = 1, 2, \ldots, \dim \Gamma^{\lambda_1,\lambda_2}. \qquad (5.3b)$$

Bases are therefore labeled $|\alpha; {}^{\lambda_1,\lambda_2}_{i}\rangle$. Here α is the set of all other quantum numbers necessary to distinguish different copies of the space with the same λ_1, λ_2 label. An arbitrary element P of the group P_N maps a state in any λ_1, λ_2; α subspace into a linear combination of states *within the same subspace*,

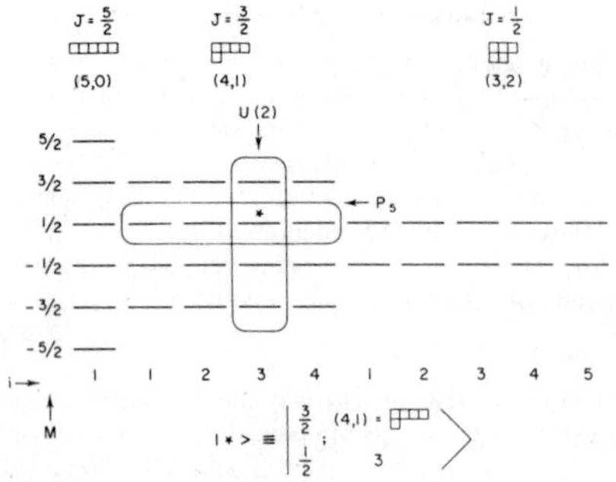

FIG. 2. An arbitrary state in the symmetry-adapted basis is labeled by (i) the U(2) invariant subspace J in which it lies, and its position M within that space, and by (ii) the P_N invariant subspace $\vec{\lambda}$ in which it lies, and its position i within that space. Moreover, the intersection of any P_N invariant subspace with any U(2) invariant subspace is at most one dimensional, so the quantum numbers J, M and $\vec{\lambda}$, i are sufficient for a complete labeling of states. The intersection is exactly one dimensional when the partitions describing the U(2) and P_N invariant subspace are identical: $2J=\lambda_1-\lambda_2$. The state marked ★ in the figure is labeled $\left|{J=3/2 \atop M=1/2};{\vec{\lambda}=(4,1) \atop i=3}\right\rangle$. Under a U(2) operation R, this state is mapped into a linear combination of states within the same vertical box,

$$R\left|{3/2 \atop 1/2};{(4,1) \atop 3}\right\rangle=\sum_M\left|{3/2 \atop M};{(4,1) \atop 3}\right\rangle\mathfrak{D}^{3/2}_{M,1/2}(R).$$

Under a P_5 operation p, it is mapped into a linear combination of states within the horizontal box,

$$p\left|{3/2 \atop 1/2};{(4,1) \atop 3}\right\rangle=\sum_j\left|{3/2 \atop 1/2};{(4,1) \atop j}\right\rangle\Gamma^{(4,1)}_{j,3}(p).$$

In other words, group operations do not affect the labels classifying the invariant subspace (upper line); they only affect the appropriate *internal* state labels (lower line). Note that states belonging to the same M but different $\vec{\lambda}$ values are not necessarily degenerated.

$$P\left|\alpha;{\vec{\lambda} \atop i}\right\rangle=\sum_j\left|\alpha;{\vec{\lambda} \atop j}\right\rangle\left\langle{\vec{\lambda} \atop j}\right|P\left|{\vec{\lambda} \atop i}\right\rangle=\sum\left|\alpha;{\vec{\lambda} \atop j}\right\rangle\Gamma^{\vec{\lambda}}_{ji}(P)\,. \tag{5.4b}$$

Referring to Fig. 2, we say that the group P_N "acts horizontally."

D. Classification of Atomic States

It can easily be shown that

$$\sum_{2J=\lambda_1-\lambda_2\geq 0}(\dim\mathfrak{D}^J)(\dim\Gamma^{\lambda_1,\lambda_2})=2^N\,. \tag{5.5}$$

This is a manifestation of a very deep and beautiful theorem.[24]

Theorem. The intersection of any SU(2) invariant subspace with any P_N invariant subspace is at most one dimensional. It is exactly one dimensional only when

$$2J=\lambda_1-\lambda_2\,,\quad N=\lambda_1+\lambda_2\,. \tag{5.6}$$

This theorem is sufficient to tell us what the sets α, α' of "all other quantum numbers" consist of. In short, a symmetrized state is specified *uniquely* by

$$\left|{J \atop M};{\vec{\lambda} \atop i}\right\rangle\,. \tag{5.7}$$

This specification is, in fact, even redundant. The labels J, $\vec{\lambda}$ are not independent but related by (5.6). This intersection theorem is illustrated in Fig. 2.

Under an arbitrary Hamiltonian (5.1), only states with the same $\vec{\lambda}$, J, M values are necessarily degenerate. But if the Hamiltonian consists of a sum of single-particle terms, all states with the same M value are, in general, degenerated. The introduction of particle-particle interactions lifts the high-M degeneracy (such as in the case of pressure broadening in a gas).

It is a simple matter to transform between the symmetrized bases (5.7) and the direct product bases (1.4). For simplicity we label the states (1.4) by $|\vec{N}\rangle$, where $\vec{N}$ is the N-component vector $(i_1, i_2, \ldots, i_N)$. We let n_1 and n_2 be the total number of one's and two's in $\vec{N}$, respectively. Then

$$n_1+n_2=N\,,\quad n_1-n_2=2M\,. \tag{5.8}$$

The states $|\vec{N}\rangle$ and $\left|{J \atop M};{\vec{\lambda} \atop i}\right\rangle$ are connected by a unitary transformation which is, in fact, also real and therefore orthogonal,

$$\left|{J \atop M};{\vec{\lambda} \atop i}\right\rangle=\sum_{(n_1,n_2)}{}'|\vec{N}\rangle\left\langle\vec{N}\right|{\vec{\lambda} \atop M,i}\rangle=\sum_{(n_1,n_2)}{}'|\vec{N}\rangle C^{\vec{N}}_{\vec{\lambda};iM}\,. \tag{5.9}$$

Here $\sum'_{(n_1,n_2)}$ indicates a summation over all distinct permutations of one's and two's with fixed M. The coefficients $C^{\vec{N}}_{\vec{\lambda};iM}$ vanish unless $2M=n_1-n_2$. Thus, the structure of C is that of a block diagonal matrix. The block-connecting states with $2M=n_1-n_2$ is a square matrix of order

$$\binom{N}{n_1}=\binom{N}{n_2}\,.$$

We illustrate what we have in mind in Fig. 3. This transformation will be used in discussing leakage from symmetrized states (Sec. VII D).

In the same manner as (5.7), Bloch states produced by rotation $R_{\theta,\varphi}$ of an $M=-J$ state are denoted

$$\left|{J \atop \theta,\varphi};{\vec{\lambda} \atop i}\right\rangle=R_{\theta,\varphi}\left|{J \atop -J};{\vec{\lambda} \atop i}\right\rangle\,. \tag{5.10}$$

VI. RADIATIVE PROPERTIES OF BLOCH STATES

A. Position of Problem

The description of the atomic states given above is now completed by considering a problem central to quantum optics, namely, the interaction of the assembly of two-level systems with the radiation field. This interaction introduces a *spatial* dependence, which we have so far neglected, and which

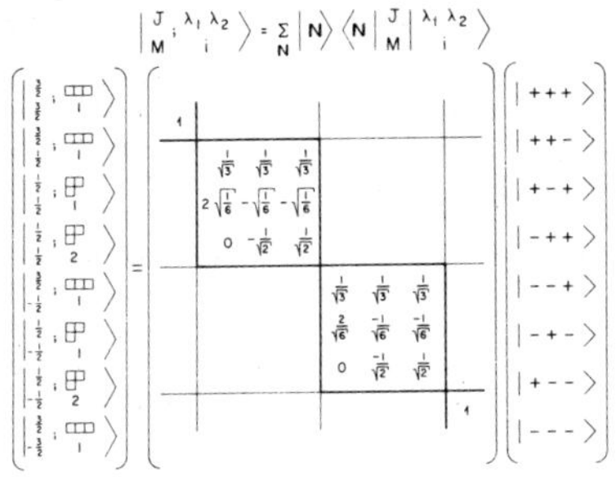

FIG. 3. Symmetrized states are linear combinations of product states. The transformation matrix partition into a direct sum of block diagonal submatrices. Each submatrix connects the direct product states with a given $M=(n_+ - n_-)$ value, with symmetrized states with exactly the same M value.

modifies somewhat the appropriate description of the atomic states.

In the Schrödinger picture the electric field operator is written

$$\vec{E}(\vec{r}) = i\sum_{\vec{k}} (\hbar\omega_k/2\epsilon_0 V)^{1/2} (\hat{e}_{\vec{k}} a^\dagger_{\vec{k}} e^{-i\vec{k}\cdot\vec{r}} - \hat{e}^*_{\vec{k}} a_{\vec{k}} e^{i\vec{k}\cdot\vec{r}}) , \quad (6.1)$$

where $\omega_k = ck$ is the frequency of the kth mode in a box of volume V. The creation and annihilation operators satisfy the commutation rules (2.3) if they belong to the same mode, and otherwise commute. The polarization vector of the kth mode is $\hat{e}_{\vec{k}}$. The total Hamiltonian consists of a field part $\mathcal{H}_F$, an atomic part $\mathcal{H}_A$, and an atom-field interaction $\mathcal{H}_{AF}$. The field Hamiltonian, neglecting the zero-point energy, is simply

$$\mathcal{H}_F = \hbar\sum_{\vec{k}} \omega_k a^\dagger_{\vec{k}} a_{\vec{k}} . \quad (6.2)$$

The atomic Hamiltonian, taking the energy reference halfway between ground and excited states, is

$$\mathcal{H}_A = \tfrac{1}{2}\hbar\sum_n \omega_n \sigma_z^n = \hbar\omega J_z , \quad (6.3)$$

the second equality being obtained by assuming that all two-level systems are degenerated ($\omega_n = \omega$, for all n). The atom-field interaction is assumed to be of electric dipole type. We write

$$\mathcal{H}_{AF} = -\sum_n \vec{d}_n \cdot \vec{E}(\vec{r}_n) , \quad (6.4)$$

where the dipole operator $\vec{d}_n$, which is Hermitian and has zero expectation value in both ground and excited states, is of the form

$$\vec{d}_n = \sigma_+^n \vec{p} + \sigma_-^n \vec{p}^* . \quad (6.5)$$

Here $\vec{p}$ is a complex c number which is assumed independent of the label n. Using (6.1) and (6.5), the interaction Hamiltonian (6.4) takes the familiar form

$$\mathcal{H}_{AF} = \hbar\sum_{\vec{k}} (g_k a_{\vec{k}} J_+^{\vec{k}} + g^*_k a^\dagger_{\vec{k}} J_-^{\vec{k}}) , \quad (6.6a)$$

where

$$g_k = i(\omega_k/2\epsilon_0\hbar V)^{1/2} \vec{p}\cdot\hat{e}^*_{\vec{k}} , \quad (6.6b)$$

$$J_+^{\vec{k}} = \sum_n \sigma_+^n e^{i\vec{k}\cdot\vec{r}_n} , \quad (6.6c)$$

$$J_-^{\vec{k}} = \sum_n \sigma_-^n e^{-i\vec{k}\cdot\vec{r}_n} . \quad (6.6d)$$

In deriving (6.6a) we have assumed that $\vec{p}\cdot\hat{e}_{\vec{k}} = \vec{p}^*\cdot\hat{e}^*_{\vec{k}} = 0$. If this selection rule does not apply, (6.6a) is obtained by neglecting energy nonconserving terms such as $a_{\vec{k}}\sigma_-^n$ and $a^\dagger_{\vec{k}}\sigma_+^n$; their contribution to transition probabilities being very small as long as $|g_k| \ll \omega$.

No exact solution is known to the dynamical problem associated with the Hamiltonian $\mathcal{H} = \mathcal{H}_F + \mathcal{H}_A + \mathcal{H}_{AF}$. The difficulty is caused by the spatial dependence in the commutation relation of $J_+^{\vec{k}}$ and $J_-^{\vec{k}'}$:

$$[J_+^{\vec{k}}, J_-^{\vec{k}'}] = \sum_n \sigma_z^n e^{i(\vec{k}-\vec{k}')\cdot\vec{r}_n} . \quad (6.7)$$

This is seen by working out, in the Heisenberg picture, the equations of motion of $a_{\vec{k}}$, $a^\dagger_{\vec{k}}$, $J_+^{\vec{k}}$, and $J_-^{\vec{k}}$. One obtains

$$\dot{a}_{\vec{k}} = -i\omega_k a_{\vec{k}} - i g^*_k J_-^{\vec{k}} , \quad (6.8a)$$

$$\dot{J}_-^{\vec{k}} = -i\omega J_-^{\vec{k}} + i\sum_{n,\vec{k}'} g_{k'} a_{\vec{k}'} \sigma_z^n e^{i(\vec{k}'-\vec{k})\cdot\vec{r}_n} . \quad (6.8b)$$

The problem can be simplified by restricting the analysis either to a single mode, or to a small volume. With a single mode k one obtains

$$\dot{J}_-^k = -i\omega J_-^k + 2 i g_k a_k J_z , \quad (6.9a)$$

$$\dot{J}_z = -i g_k a_k J_+^k + i g^*_k a^\dagger_k J_-^k . \quad (6.9b)$$

With many modes, but a pointlike medium ($r_n \simeq 0$), one has $e^{i\vec{k}\cdot\vec{r}} \simeq 1$ so that $J_\pm^{\vec{k}} \simeq J_\pm$ as defined in Eq. (1.5). The equations of motion then take the form

$$\dot{J}_- = -i\omega J_- + 2i(\sum_{\vec{k}} g_k a_{\vec{k}}) J_z , \quad (6.10a)$$

$$\dot{J}_z = -i(\sum_{\vec{k}} g_k a_{\vec{k}}) J_+ + i(\sum_{\vec{k}} g^*_k a^\dagger_{\vec{k}}) J_- . \quad (6.10b)$$

In either case Bloch states are very useful. This is obvious for the point-laser case, and will be shown presently for the single-mode case.

B. $\vec{k}$-Dependent Atomic States

The Dicke and Bloch states of the previous sections can be made $\vec{k}$ dependent by introducing phase factors in the definition of the single two-level system eigenstates $|\psi_i^n\rangle$. One defines

$$|\psi_1^{\vec{k},n}\rangle = e^{(i/2)\vec{k}\cdot\vec{r}_n} |\psi_1^n\rangle , \quad (6.11a)$$

$$|\psi_2^{\vec{k},n}\rangle = e^{-(i/2)\vec{k}\cdot\vec{r}_n} |\psi_2^n\rangle . \quad (6.11b)$$

These states are such that

$$(e^{i\vec{k}\cdot\vec{r}_n}\sigma_+^n)|\psi_2^{\vec{k},n}\rangle = |\psi_1^{\vec{k},n}\rangle , \tag{6.12a}$$

$$(e^{-i\vec{k}\cdot\vec{r}_n}\sigma_-^n)|\psi_1^{\vec{k},n}\rangle = |\psi_2^{\vec{k},n}\rangle . \tag{6.12b}$$

The new step-up and step-down operators within the parentheses are precisely those which enter in the definition of $J_\pm^{\vec{k}}$ in Eqs. (6.6c)–(6.6d). In terms of these new states and operators all the formulas of Sec. III remain valid provided one defines new operators $J_x^{\vec{k}}$ and $J_y^{\vec{k}}$ as required by Eq. (3.2). In particular, Dicke states can be defined as the simultaneous eigenstates of J_z and $(J^{\vec{k}})^2 = J_z^2 + \frac{1}{2}(J_+^{\vec{k}}J_-^{\vec{k}} + J_-^{\vec{k}}J_+^{\vec{k}})$, and Bloch states are obtained from the lowest Dicke states by application of the operator $\exp(\zeta J_+^{\vec{k}} - \zeta^* J_-^{\vec{k}})$.

These states are ideally suited for the treatment of the single-mode problem.[25] Indeed the angular momentum operator $(J^{\vec{k}})^2$ commutes with the single-mode interaction Hamiltonian of same $\vec{k}$. This means that the corresponding Dicke and Bloch states move in subspaces of constant $(J^{\vec{k}})^2$ eigenvalue. This symmetry is not preserved in the many-mode problem (6.8), as a consequence of the commutation relation (6.7). Leakage into other symmetry types results, as shown in detail for the case of the interaction with a classical field in Sec. VII.

C. Interaction with Classical Field

In certain cases, when field statistics and spontaneous emission are not important, one is allowed to treat the field classically, defining

$$\vec{E}(\vec{r}, t) = \mathrm{Re}[\vec{E}(\vec{r}, t)e^{i\omega t}] , \tag{6.13}$$

the time dependence of $\vec{E}(\vec{r}, t)$ being slow compared to ω. In this case the interaction Hamiltonian (6.4) becomes

$$\mathcal{H}_{AF} = -\tfrac{1}{2}\textstyle\sum_n [\sigma_+^n \vec{p}\cdot\vec{E}^*(\vec{r}_n, t)e^{-i\omega t} + \sigma_-^n \vec{p}^*\cdot\vec{E}(\vec{r}_n, t)e^{i\omega t}] , \tag{6.14}$$

where we have assumed that $\vec{p}\cdot\vec{E} = \vec{p}^*\cdot\vec{E}^* = 0$. If this selection rule does not apply, (6.14) implies the so-called rotating wave approximation, which consists in neglecting terms in $\sigma_+^n e^{i\omega t}$ and $\sigma_-^n e^{-i\omega t}$ on the grounds that these contribute only to small double-frequency terms in the final result. The total Hamiltonian now being $\mathcal{H}(t) = \mathcal{H}_A + \mathcal{H}_{AF}(t)$, the equations of motion of the expectation values take the form

$$\langle\dot{\sigma}_-^n\rangle = -i\omega\langle\sigma_-^n\rangle - (i/2\hbar)\vec{p}\cdot\vec{E}^*(\vec{r}_n, t)\langle\sigma_z^n\rangle e^{-i\omega t} , \tag{6.15a}$$

$$\langle\dot{\sigma}_z^n\rangle = (i/\hbar)\vec{p}\cdot\vec{E}^*(\vec{r}_n, t)\langle\sigma_+^n\rangle e^{-i\omega t} + \text{c.c.} , \tag{6.15b}$$

which are just the ordinary Bloch equations.[8]

For a pointlike medium $\vec{E}(\vec{r}_n, t) \simeq \vec{E}(t)$ for all n's, and Eqs. (6.15) can be summed over n, giving

$$\langle\dot{J}_-\rangle = -i\omega\langle J_-\rangle - (i/\hbar)\vec{p}\cdot\vec{E}^*(t)\langle J_z\rangle e^{-i\omega t} , \tag{6.16a}$$

$$\langle\dot{J}_z\rangle = (i/2\hbar)\vec{p}\cdot\vec{E}^*(t)\langle J_+\rangle e^{-i\omega t} + \text{c.c.} . \tag{6.16b}$$

These equations could have been derived directly from (6.10) by making a self-consistent-field approximation $\langle J_\mu a\rangle = \langle J_\mu\rangle\langle a\rangle$, and writing

$$\vec{E}e^{i\omega t} = 2\textstyle\sum_{\vec{k}} i\hat{e}_{\vec{k}}(\hbar\omega_k/2\epsilon_0 V)^{1/2}\langle a_{\vec{k}}^\dagger\rangle . \tag{6.17}$$

For an extended medium, assuming that the spatial dependence of $\vec{E}$ is strictly of the form $e^{-i\vec{k}\cdot\vec{r}}$, the summation of (6.15) over n gives equations similar to (6.16) but where $J_\pm$ is replaced by $J_\pm^{\vec{k}}$. These equations could also be derived from (6.9) by a self-consistent field approximation.

D. Emission Rates

Returning to the pointlike medium, it is interesting to calculate, using first-order perturbation theory, the emission probability of single photons for systems which are initially either in a Dicke state or in a Bloch state. The interaction Hamiltonian is, from (6.6a), $\mathcal{H}_{AF} = \sum_{\vec{k}}\mathcal{H}_{\vec{k}}$ with

$$\mathcal{H}_{\vec{k}} = \hbar g_k a_{\vec{k}} J_+ + \hbar g_k^* a_{\vec{k}}^\dagger J_- . \tag{6.18}$$

The transition probability from an initial atomic state i to a final atomic state f, with emission of one photon in mode $\vec{k}$, is simply

$$W_{\vec{k}} = (2\pi/\hbar)\delta(\hbar\omega - \hbar\omega_k)|\langle f, n_{\vec{k}}+1|\mathcal{H}_{\vec{k}}|i, n_{\vec{k}}\rangle|^2 , \tag{6.19}$$

where the second symbol in the state labeling indicates the photon number in the field. For a single atom initially in the upper state $|\psi_1\rangle$, and in the vacuum of photons ($n_{\vec{k}} = 0$ for all $\vec{k}$), the usual spontaneous emission intensity in the solid angle $d\Omega$ around $\vec{K}$ is obtained by summing over $\vec{k}$ in this solid angle,

$$I_0 d\Omega = \hbar\omega \sum_{\vec{k}\ \mathrm{in}\ d\Omega} W_{\vec{k}} = \frac{\omega^4|\hat{e}_{\vec{k}}^*\cdot\vec{p}|}{8\pi^2\epsilon_0 c^3}\, d\Omega . \tag{6.20}$$

For many atoms initially in the Dicke state $|M\rangle$ and making transition to the state $|M-1\rangle$, one obtains, using $\langle M-1|J_-|M\rangle = [(J+M)(J-M+1)]^{1/2}$, the spontaneous emission intensity

$$I_D = (J+M)(J-M+1)I_0 . \tag{6.21a}$$

This is the familiar result of Dicke[7] indicating superradiance for $M \simeq 0$. On the other hand, the stimulated intensity in a certain mode $\vec{k}$, which is proportional to

$$|\langle M-1|J_-|M\rangle|^2 - |\langle M+1|J_+|M\rangle|^2 = 2M , \tag{6.22a}$$

is the normal stimulated rate, simply proportional to the population inversion.

If the initial state is the Bloch state $|\theta, \varphi\rangle$, the spontaneous emission intensity is

$$I_B = \textstyle\sum_M |\langle M|J_-|\theta, \varphi\rangle|^2 I_0 = \langle\theta, \varphi|J_+J_-|\theta, \varphi\rangle I_0$$

$$= [J^2\sin^2\theta + 2J\sin^4(\tfrac{1}{2}\theta)]I_0 , \tag{6.21b}$$

TABLE II. Comparison of emission rates and dipole moments of Dicke and Bloch states of the same energy expectation value.

	Dicke state $\vert M=-J\cos\theta\rangle$	Bloch state $\vert\theta,\varphi\rangle$
Spontaneous emission	$[J^2\sin^2\theta+2J\sin^2\frac{1}{2}\theta]I_0$	$[J^2\sin^2\theta+2J\sin^4\frac{1}{2}\theta]I_0$
Stimulated emission	$-2J\cos\theta$	$-2J\cos\theta$
Dipole moment	0	$J\sin\theta\,(\vec{p}e^{i\omega t+i\varphi}+\text{c.c.})$
Classical emission	0	$[J^2\sin^2\theta]I_0$

the last equality being obtained from Eqs. (A2a) and (B6). This is also superradiant for $\theta\simeq\frac{1}{2}\pi$. The intensity can be compared to that obtained from the Dicke state having the same energy expectation value. This fact was already noted by Senitzky.[3] With $M=-J\cos\theta$ one obtains from (6.21a) $I_D=[J^2\sin^2\theta+2J\sin^2(\frac{1}{2}\theta)]I_0$. This is practically identical to (6.21b) for $J\gg1$. Similarly, the stimulated intensity is now proportional to

$$\sum_M(|\langle M|J_-|\theta,\varphi\rangle|^2-|\langle M|J_+|\theta,\varphi\rangle|^2)=-2J\cos\theta\ . \tag{6.22b}$$

This is identical to the stimulated intensity obtained from the Dicke state having the same mean energy.

Though their radiation rates are very similar, there is a considerable difference between Dicke and Bloch states of the same mean energy. This is seen by calculating the expectation value of the total dipole moment $\vec{D}$. In the Heisenberg picture, using the unperturbed Hamiltonian $\mathcal{H}=\mathcal{H}_A+\mathcal{H}_F$ (which amounts to neglecting the field reaction, and is equivalent to first-order perturbation theory), this dipole moment is

$$\vec{D}=e^{i\mathcal{H}t/\hbar}\sum_n\vec{d}_n\,e^{-i\mathcal{H}t/\hbar}$$
$$=\vec{p}\,e^{i\omega t}J_++\vec{p}^*e^{-i\omega t}J_-\ , \tag{6.23}$$

where $\vec{d}_n$ is given by (6.5). For the Dicke state $|M\rangle$ one has

$$\langle M|\vec{D}|M\rangle=0\ , \tag{6.24a}$$

whereas for the Bloch state $|\theta,\varphi\rangle$,

$$\langle\theta,\varphi|\vec{D}|\theta,\varphi\rangle=J\sin\theta(\vec{p}e^{i\omega t+i\varphi}+\vec{p}^*e^{-i\omega t-i\varphi})\ . \tag{6.24b}$$

This last result is easily obtained using the technique of Appendix B, or the relation $\langle\theta,\varphi|J_z|\theta,\varphi\rangle=-J\cos\theta$ together with the eigenvalue equations (3.10d) and (3.10e). While the Dicke state has zero expectation value for the dipole moment, the Bloch state is characterized by a macroscopic dipole which is able to radiate classically. In the wave zone, the classical radiation intensity of this dipole turns out to be

$$I_C=(J^2\sin^2\theta)I_0\ . \tag{6.21c}$$

This is practically identical to (6.21b), except for θ close to π. For the totally inverted state there is no classical emission, whereas first-order perturbation theory gives a finite intensity $2JI_0$. These results are summarized in Table II. An inspection of this table suggests that Bloch states are the closest quantum analogs to "classical" atomic states, in the same manner as Glauber states are the quantum analogs of classical fields.

VII. SEMICLASSICAL LIMIT

A. Semiclassical Theorem

A known property of field coherent states is that they can be produced by classical currents acting on the vacuum of photons.[2,4] A similar property exists for Bloch states.

An atomic system of small dimension which is initially in a Bloch state, or in particular in its ground state, and which is acted upon by classical fields, remains in a Bloch state. By a classical field is meant that the interaction Hamiltonian is of the form (6.14) or, for a point system,

$$\mathcal{H}_{AF}=-\tfrac{1}{2}\vec{p}\cdot\vec{E}^*(t)J_+e^{-i\omega t}-\tfrac{1}{2}\vec{p}^*\cdot\vec{E}(t)J_-e^{-i\omega t}\ . \tag{7.1a}$$

In the interaction picture this Hamiltonian becomes

$$\mathcal{H}_{AF}=-\tfrac{1}{2}\vec{p}\cdot\vec{E}^*(t)J_+-\tfrac{1}{2}\vec{p}^*\cdot\vec{E}(t)J_-\ . \tag{7.1b}$$

As the field $\vec{E}$ is turned on, the state vector $|(t)\rangle$, which was previously fixed in this picture and equal to $|\theta,\varphi\rangle$, changes according to

$$i\hbar\frac{\partial|(t)\rangle}{\partial t}=\mathcal{H}_{AF}|(t)\rangle\ . \tag{7.2}$$

Writing $\vec{p}^*\cdot\vec{E}=i|\vec{p}^*\cdot\vec{E}|e^{i\varphi'}$, this equation gives, for a small time increment Δt,

$$|(t+\Delta t)\rangle=[1-i(\Delta t|\vec{p}^*\cdot\vec{E}|/\hbar)$$
$$\times(J_x\sin\varphi'-J_y\cos\varphi')]\,|\theta,\varphi\rangle\ . \tag{7.3}$$

The operator within the bracket on the right-hand side produces an infinitesimal rotation by an angle $\theta'=\Delta t|\vec{p}^*\cdot\vec{E}|/\hbar$ around an axis $(\sin\varphi',-\cos\varphi',0)$. Using Eq. (A9), one has from (7.3)

$$|(t+\Delta t)\rangle=R_{\theta',\varphi'}R_{\theta,\varphi}|-J\rangle=R_{\Theta,\Phi}e^{-i\Psi J_z}|-J\rangle$$
$$=e^{i\Psi J}|\Theta,\Phi\rangle\ . \tag{7.4}$$

The new state is therefore a Bloch state, and the statement is proved. In the spirit of Sec. VI B, this also applies to an extended medium excited by a field $\vec{E}(t)e^{i\omega t-i\vec{k}\cdot\vec{r}}$. In this case however the reaction of the medium on the field will not allow, in general, the preparation of a strictly space-independent field $\vec{E}(t)$. This is related to the concepts of maximum cooperation number and cooperation time which have been exposed elsewhere.[26] It leads to the leakage into Bloch states of other symmetry types, to be treated at the end of this section.

The results concerning the correspondence between Bloch states and classical dipoles, Glauber states and classical fields, and the production of coherent states of one kind by classical excitation of the other kind have been condensed in Table III.

B. Evolution of Any Pure State

From (7.2) and (7.3) it is clear that the evolution operator, which is an infinite product of infinitesimal rotations, is itself a rotation. This rotation R can be represented by its three Euler angles (α, β, γ), by a rotation ξ about an arbitrary axis (n_1, n_2, n_3), by the product of a rotation around $\hat{z}$ by Ψ with a rotation $R_{\Theta,\Psi}$, or in any other convenient manner. These various ways of expressing the evolution operator are easily related using the techniques of Appendixes A and D. For example one simply writes

$$R = e^{-i\alpha J_z} e^{-i\beta J_y} e^{-i\gamma J_z} = e^{-i\xi(\hat{n}\cdot\vec{J})} = R_{\Theta,\Phi} e^{-i\Psi J_z}, \tag{7.5a}$$

which leads, using Eqs. (A1) and (A12), to

$$\begin{bmatrix} e^{-i\alpha/2}\cos\frac{1}{2}\beta\, e^{-i\gamma/2} & -e^{-i\alpha/2}\sin\frac{1}{2}\beta e^{i\gamma/2} \\ e^{i\alpha/2}\sin\frac{1}{2}\beta\, e^{-i\gamma/2} & e^{i\alpha/2}\cos\frac{1}{2}\beta\, e^{i\gamma/2} \end{bmatrix}$$

$$= \begin{bmatrix} \cos\frac{1}{2}\xi - in_3\sin\frac{1}{2}\xi & -i(n_1 - in_2)\sin\frac{1}{2}\xi \\ -i(n_1+in_2)\sin\frac{1}{2}\xi & \cos\frac{1}{2}\xi + in_3\sin\frac{1}{2}\xi \end{bmatrix} = (1+|T|^2)^{-1/2} \begin{bmatrix} e^{-i\Psi/2} & Te^{i\Psi/2} \\ -T^* e^{-i\Psi/2} & e^{i\Psi/2} \end{bmatrix}, \tag{7.5b}$$

where

$$T = e^{-i\Phi}\tan\tfrac{1}{2}\Theta\,. \tag{7.5c}$$

These equalities lead to the appropriate relations, between the various variables.

If we study the evolution from the ground state which is the product state $|\phi_{i_1\cdots i_N}\rangle$ with $i_n = 2$ in (1.4), we note that under the rotation operation (7.5a) the state remains a product of single-atom wave functions

$$|\psi^n(t)\rangle = \alpha^n(t)|\psi_1^n\rangle + \beta^n(t)|\psi_2^n\rangle\,, \tag{7.6}$$

where the coefficients $\alpha^n(t)$, $\beta^n(t)$ are all equal (for a point system) and determined directly from the transformation matrix (7.5b). The global atomic state is then

$$|(t)\rangle = \sum_{\text{all } \vec{N}} [\alpha(t)]^{n_1}[\beta(t)]^{n_2}|\vec{N}\rangle\,, \tag{7.7}$$

where $\vec{N} \equiv (i_1, i_2, \ldots, i_N)$ is a short-hand rotation equivalent to (1.4). It is convenient to collect together all states $|\vec{N}\rangle$ with n_1 one's and n_2 two's as follows:

$$\sum_{(n_1,n_2)}{}' |\vec{N}\rangle = \binom{N}{n_1}^{1/2} \left\{ \sum_{(n_1,n_2)}{}' |\vec{N}\rangle \Big/ \binom{N}{n_1}^{1/2} \right\}, \tag{7.8}$$

where $\sum'_{(n_1,n_2)}$ describes a summation over all distinct permutations of n_1 one's and n_2 two's. The state within the curly bracket on the right-hand side of (7.8) is fully symmetric under the permutation group and normalized to unity. Therefore, it is the Dicke state

$$\left|{J=N/2 \atop M=(n_1-n_2)/2}; {N,0 \atop 1}\right\rangle$$

introduced in Sec. V. Therefore, (7.7) becomes

$$\begin{aligned} |(t)\rangle &= \sum_{M=-J}^{J} \binom{N}{n_1}^{1/2} \alpha^{n_1}\beta^{n_2} \left|{N/2 \atop M}; {N,0 \atop 1}\right\rangle \\ &= \sum_{M=-J}^{J} \mathfrak{D}^J_{M,-J}(R) \left|{N/2 \atop M}; {N,0 \atop 1}\right\rangle \\ &= (\beta/|\beta|)^{2J} \left|{N/2 \atop 2\arctan(|\alpha|/|\beta|),\, \arg(\beta/\alpha)}; {N,0 \atop 1}\right\rangle. \end{aligned} \tag{7.9}$$

The second equality follows by definition[14] of the irreducible representation $\mathfrak{D}^J_{M,M'}(R)$ for a rotation applied to the $M' = -J$ state, and the third equality follows from the last form of (7.5a) and the definition of the Bloch state (3.8).

If the initial state is an arbitrary Dicke state $\left|{J \atop M}; {\vec{\lambda} \atop i}\right\rangle$ one has similarly,

$$|(t)\rangle = R\left|{J \atop M}; {\vec{\lambda} \atop i}\right\rangle = \sum_{M'} \mathfrak{D}^J_{M',M}(R)\left|{J \atop M'}; {\vec{\lambda} \atop i}\right\rangle, \tag{7.10}$$

TABLE III. Classical excitation and coherent states. (The single arrows indicate the direction of production of coherent states starting from classical states. The double arrows indicate states connected by the correspondence principle.)

	Atomic states	Field states
Classical states	Classical current	Classical field
Quantum states	Coherent atomic state	Coherent field state

which, for large J, is a rather complicated relation in view of Eq. (D6).

C. Evolution of Statistical Mixture of States

A real physical system will exist, in general, not in a given state, but rather in a statistical mixture of states. Upon application of a classical field it would then be necessary to deal with a summation over states of the type (7.10), which are rather unwieldy to handle except for those arising from the ground state $M=-J$, or the most excited state $M=+J$.

Fortunately, it can easily be seen, using (5.2b), that

$$\sum_{x=0}^{\lambda_2-1} \dim\Gamma^{N-x,x} \ll \dim\Gamma^{N-\lambda_2,\lambda_2}, \qquad (7.11a)$$

as long as

$$\lambda_2 \ll \tfrac{1}{3}(N+1). \qquad (7.11b)$$

The total number of states with fixed-M value lying to the left (Fig. 2) of the states with $J=|M|$ is given by the left-hand side of (7.11a). The right-hand side gives the total number of states with $J=|M|$. Therefore in a statistical mixture in which only those states with $\lambda_2 \ll \frac{1}{3}N$ are significantly populated, it is possible to neglect all states of the form (7.10), except for those with $M=-J$. In short, for fixed negative M value such that $|M| \ll \frac{1}{6}N$ it is sufficient to study those states

$$\left| {}^{|M|}_{M}\,;\, {}^{N/2+|M|,\,N/2-|M|}_{i} \right\rangle, \qquad (7.12)$$

since there are far fewer states with the same M value in all other invariant subspaces combined. The evolution of the states (7.12) is given by (7.9).

D. Evolution of Extended Systems

An extended system, initially in a product state, and acted upon by a classical field, remains in a product state. This is easy to see from the interaction Hamiltonian (6.14). The infinitesimal evolution operator, in the interaction picture, can be written

$$U(t,t+\Delta t) = 1 - \frac{i\Delta t}{\hbar}\mathcal{H}^{\rm int}_{\rm AF}(t) = \prod_{n=1}^{N}\left(1-\frac{i\Delta t}{\hbar}\mathcal{H}_n\right) = \prod_{n=1}^{N} U_n(t,t+\Delta t), \qquad (7.13a)$$

where

$$\mathcal{H}_n = -\tfrac{1}{2}\sigma_+^n \vec{p}\cdot\vec{E}^*(\vec{r}_n,t) - \tfrac{1}{2}\sigma_-^n \vec{p}^*\cdot\vec{E}(\vec{r}_n,t). \qquad (7.13b)$$

U is therefore a product of infinitesimal rotations U_n applied to each single system. Hence the state remains a product state.

Let us take, for example, the evolution from the ground state $|\phi_{22\ldots22}\rangle$. The coefficients of the single-atom wave functions in (7.6) depend now on the index n. The total atomic state cannot be written as simply as (7.7). We write

$$|(t)\rangle = \sum_{\text{all }\vec{N}} \{\alpha^n, \beta^n, \vec{N}\}\,|\vec{N}\rangle, \qquad (7.14)$$

where the notation $\{\alpha, \beta, \vec{N}\}$ means the product $\gamma^1\gamma^2\cdots\gamma^N$, where γ^n stands for α^n or β^n depending on whether the nth vector component of $\vec{N}$ is 1 or 2. We collect together all states $|\vec{N}\rangle$ with n_1 one's and n_2 two's. These correspond to a fixed M value. Any of these product states can be expanded in terms of the states $|{}^J_M;{}^{\vec{\lambda}}_i\rangle$ with fixed M value, in the spirit of (5.9). One obtains

$$|(t)\rangle = \sum_M \sum_{\vec{\lambda},i} |{}^J_M;{}^{\vec{\lambda}}_i\rangle \sum_{(n_1,n_2)}{}' (C^{\vec{N}}_{\vec{\lambda};iM})^{-1}\{\alpha^n,\beta^n,\vec{N}\}, \qquad (7.15)$$

where $\sum'_{(n_1,n_2)}$ has the same meaning as in (5.9). In the case of the point system, $\{\alpha^n, \beta^n, \vec{N}\}$ depends on n_1 and n_2 but not on which particular permutation of n_1 one's and n_2 two's is selected. As a result, the last sum on the right-hand side of (7.15) is zero in that case except for the totally symmetric state $\vec{\lambda}=(N,0)$; $i=1$. In the present case all values of $\vec{\lambda}$ and i are, in general, obtained, in spite of the fact that we started from a state with $\vec{\lambda}=(N,0)$; $i=1$. We speak of a *leakage from symmetrized states*. It is due to the fact that the atoms have different space dependences, which allow them to be distinguished, and violate (5.1). A similar difficulty will, of course, occur in a fully quantum treatment. The situation can be saved with a single mode by defining $\vec{k}$-dependent Bloch states (Sec. VI B). Similarly, the situation can be saved for classical fields if the spatial dependence is strictly of the form $e^{-i\vec{k}\cdot\vec{r}}$. In these cases, however, leakage will also occur in the presence of atomic motion.

VIII. CALCULATION OF THERMAL AVERAGES

In this last section we show how the disentangling theorem, the contraction procedure, and other group theoretical methods presented in the Appendixes, can be used to obtain statistical results beyond those considerations presented in Secs. II D, III D, and VII C. It frequently happens that thermal expectation values must be taken for operators which can be written as exponentials of other operators. Under these circumstances the Baker–Campbell–Hausdorff formulas may be used to compute these thermal averages explicitly.

To begin we consider a single column of Fig. 2. The $2J+1$ energy levels are assumed to be equally spaced by $E=\hbar\omega$. The partition function is

$$Z = \mathrm{Tr}\, e^{-\beta E J_z}, \quad \beta = 1/kT. \qquad (8.1)$$

Let us compute the thermal average of the exponential operator $e^{i(\alpha_+ J_+ + \alpha_- J_-)}$,

$$\langle e^{i(\alpha_+J_+ + \alpha_-J_-)}\rangle_J = \frac{\mathrm{Tr}\, e^{-\beta E J_z} e^{i(\alpha_+J_+ + \alpha_-J_-)}}{\mathrm{Tr}\, e^{-\beta E J_z}} . \quad (8.2)$$

This is most easily done by considering, in the spirit of Appendix A, the $J=\frac{1}{2}$, 2×2 matrix representation of the associated group. Then the 2×2 matrix whose trace is to be taken in the numerator of (8.2) is

$$\begin{bmatrix} \cos\gamma\, e^{-\beta E/2} & i\alpha_+(\sin\gamma/\gamma)e^{-\beta E/2} \\ i\alpha_-(\sin\gamma/\gamma)e^{\beta E/2} & \cos\gamma\, e^{\beta E/2} \end{bmatrix}, \quad (8.3)$$

where $\gamma^2=\alpha_+\alpha_-$. This results directly from the use of (A1). In order to simplify the calculation for higher-J values, it is useful to transform (8.3) by a unitary transformation. The matrix being non-Hermitian it cannot be diagonalized. However, it can be transformed to upper triangular form. Both the trace and the determinant being preserved by this transformation, the diagonal elements λ_+, λ_- of this upper triangular matrix are simply given by

$$\lambda_+ + \lambda_- = \cos\gamma(e^{-\beta E/2} + e^{\beta E/2}) , \quad \lambda_+\lambda_- = 1 . \quad (8.4)$$

We write

$$\lambda_+ = 1/\lambda_- \equiv \lambda = \tfrac{1}{2}\{\cos\gamma(e^{-\beta E/2}+e^{\beta E/2}) + [\cos^2\gamma(e^{\beta E/2}+e^{-\beta E/2})^2 - 4]^{1/2}\} . \quad (8.5)$$

The trace in the numerator of (8.2) is written

$$\lambda + \frac{1}{\lambda} = \sum_{m=-1/2}^{1/2} (\lambda)^{2m} = \frac{\lambda^2-\lambda^{-2}}{\lambda-\lambda^{-1}} . \quad (8.6)$$

The denominator is obtained by setting $\alpha_+=\alpha_-=0$.

To evaluate the trace for arbitrary J we merely observe that once (8.3) has been transformed to an upper triangular form all diagonal elements in the representation $\mathfrak{D}^J$ are simply powers of λ. The trace is immediately obtained by extending the summation in (8.6) from $M=-J$ to $M=+J$. The result is

$$\langle e^{i(\alpha_+J_+ + \alpha_-J_-)}\rangle_J = \frac{[\lambda^{2J+1}(\gamma) - \lambda^{-2J-1}(\gamma)]/[\lambda(\gamma)-\lambda^{-1}(\gamma)]}{[\lambda^{2J+1}(0) - \lambda^{-2J-1}(0)]/[\lambda(0)-\lambda^{-1}(0)]} . \quad (8.7)$$

Now we consider the entire tableau of Fig. 2. The traces are simply obtained by summing over columns

$$\langle e^{i(\alpha_+J_+ + \alpha_-J_-)}\rangle_N = \frac{\sum_J \dim\Gamma^{\lambda_1,\lambda_2}\, \mathrm{Tr}_J\, e^{-\beta E J_z} e^{i(\alpha_+J_+ + \alpha_-J_-)}}{\sum_J \dim\Gamma^{\lambda_1,\lambda_2}\, \mathrm{Tr}_J\, e^{-\beta E J_z}} , \quad (8.8)$$

where (λ_1, λ_2) are related to N and J by (5.8), and $\dim\Gamma^{\lambda_1,\lambda_2}$ is given by (5.2b), which can also be written

$$\dim\Gamma^{\lambda_1,\lambda_2} = \binom{N}{\frac{1}{2}N-J} - \binom{N}{\frac{1}{2}N-J-1} . \quad (8.9)$$

The summation over J in (8.8) starts from 0 or $\frac{1}{2}$, and proceeds in integer steps up to $\frac{1}{2}N$. Using (8.7) one can write, for the numerator of (8.8),

$$\sum_J \left\{\binom{N}{\frac{1}{2}N-J} - \binom{N}{\frac{1}{2}N-J-1}\right\} \sum_{M=-J}^{J} (\lambda^2)^M = \sum_{M=-N/2}^{N/2} (\lambda^2)^M \sum_{J=|M|}^{N/2} \left\{\binom{N}{\frac{1}{2}N-J} - \binom{N}{\frac{1}{2}N-J-1}\right\} . \quad (8.10)$$

The last sum in (8.10) is equal to $\binom{N}{N/2-M}$. The entire expression reduces then to $(\lambda+\lambda^{-1})^N$. Using (8.4) this gives

$$\langle e^{i(\alpha_+J_+ + \alpha_-J_-)}\rangle_N = \left(\frac{\lambda(\gamma)+\lambda^{-1}(\gamma)}{\lambda(0)+\lambda^{-1}(0)}\right)^N = (\cos\gamma)^N , \quad (8.11)$$

which is temperature independent. This result depends on the fact that the diagonal elements, in the 2×2 representation, of the operator whose trace is to be taken are equal. This causes the temperature dependence to factor in the trace (8.4).

Returning to (8.7), we note that this formula can be contracted using the procedure outlined in Table I, and writing $\alpha_\pm = c\gamma_\pm$. After some simple manipulations one obtains

$$\frac{\mathrm{Tr}\, e^{-\beta\mathcal{H}} e^{i(\gamma_+a^\dagger + \gamma_-a)}}{\mathrm{Tr}\, e^{-\beta\mathcal{H}}} = \exp[-\tfrac{1}{2}\gamma_+\gamma_- \coth(\tfrac{1}{2}\hbar\omega\beta)] . \quad (8.12)$$

This is a familiar result.[27–29] Equations such as (8.7) and (8.12) are encountered in many physical applications. One example is the calculation of the intensity of a beam scattered by atoms in thermal motion in a lattice.[29,30] Using the techniques which have been exposed here all averages of the type (8.7), (8.11), and (8.12) can be obtained very simply. In order to disentangle exponentials containing harmonic-oscillator operators, it is often more practical to disentangle first similar expressions with spin operators, and to follow this by a contraction. Examples are given in Appendix A.

IX. CONCLUSION

Starting from the direct-product representation of N two-level atoms, we have introduced two representations which have been called the Dicke and the Bloch representations. The first one is very suitable to describe problems of cooperative spontaneous emission of a radiation field, insofar as it displays strong atom-atom correlations.[7,25,26] The second one is suitable to describe the resonant interaction of a classical field and of a set of atoms. A qualitative statement on the difference between these two representations has already been given by Senitzky,[3] but this difference has seldom been taken into account in the solution of quantum-electrical problems. In the previous sections we have shown the formal properties of the two representations, and their relevance to radiation and statistical problems. The properties of the Dicke and Bloch representations are similar, respectively, to those of two well-known representations of the harmonic-oscillator Hilbert space, namely, the Fock (or energy eigenstates) and the Glauber (or annihilation-operator eigenstates) representations. Besides displaying a series of similarities, the two atomic representations contract into the corresponding harmonic-oscillator representations in the limit $N\to\infty$.

In view of their paramount importance in many quantum-electrical problems, we briefly summarize here the main results concerning Bloch states. The atomic coherent states, and the operators involved in their description, obey a number of properties: (i) The states are defined by a unitary transformation operator acting on the ground state; (ii) the states obey simple eigenvalue equations; (iii) these states are nonorthogonal and overcomplete; (iv) the angular momentum operators obey a large number of Baker–Campbell–Hausdorff formulas; (v) within a fixed Bloch subspace the statistical operators have a diagonal representation in the coherent-state representation; (vi) generating functions for normal, antinormal, and fully symmetrized orderings of powers of the operators J_+, J_z, and J_- can be constructed; (vii) minimum-uncertainty relations for noncommuting operators can be constructed within the atomic coherent states.

The relationship between the atomic and field coherent states has been effected using a group-contraction procedure. Specifically, the Lie algebra of the group U(2) is contracted to the "harmonic-oscillator algebra." In this limit the commutation properties of the four operators J_z, J_+, J_-, and J_0 go over to the commutation relations of the operators $a^\dagger a$, $a^\dagger$, a, and I, respectively. In the limit $c\to 0$ the Bloch sphere surface contracts to the phase plane of the harmonic oscillator. The matrix elements of the shift operators J_+, J_- contract to the matrix elements of the creation and annihilation operators $a^\dagger$, a. All properties of the atomic coherent states (listed above) contract immediately to the corresponding well-known properties of the field coherent states. We have also shown how, in this limit, the spherical harmonics, as well as their orthogonality and completeness relations, contract to the harmonic-oscillator eigenfunctions and their orthogonality and completeness relations. Finally, we have shown how the Baker–Campbell–Hausdorff (BCH) formulas for SU(2) contract to the BCH relations useful and familiar for the field coherent states.

As this work was being completed we became aware that Radcliffe has defined coherent spin states analogous to the atomic coherent states presented here, and that he derived their overlap and completeness properties.[31] On the other hand, Barut *et al.* have defined coherent states for a group different from ours, but contracted these states to harmonic-oscillator coherent states.[32] It should also be noted that the angular momentum coherent states defined by Arkins *et al.*[33] are based on a Hilbert space different from ours, and therefore are not appropriate to the description of N two-level atoms.

APPENDIX A: DISENTANGLING THEOREM FOR ANGULAR MOMENTUM OPERATORS

In dealing with noncommuting exponential operators it is very useful to be able to change a symmetrized exponential operator into an ordered product of exponential operators. The well-known BCH formula (2.11) is of this type. Similar expressions can be obtained for angular momentum operators. We proceed to the derivation of these expressions by first considering the 2×2 matrix representation of the rotation group's algebra,

$$J_+=\begin{pmatrix}0&1\\0&0\end{pmatrix},\quad J_z=\begin{pmatrix}\frac{1}{2}&0\\0&-\frac{1}{2}\end{pmatrix},\quad J_-=\begin{pmatrix}0&0\\1&0\end{pmatrix},$$

which is the faithful representation of smallest dimension.

By Maclaurin series expansion one finds

$$e^{w_+J_++w_-J_-+w_zJ_z}=\begin{pmatrix}\cosh K+\frac{1}{2}w_z(\sinh K)/K & w_+(\sinh K)/K\\ w_-(\sinh K)/K & \cosh K-\frac{1}{2}w_z(\sinh K)/K\end{pmatrix},\quad K=(w_+w_-+\tfrac{1}{4}w_z^2)^{1/2}. \tag{A1}$$

and similarly,

$$e^{x_+J_+}e^{(\ln x_z)J_z}e^{x_-J_-}=\begin{pmatrix}(x_z)^{1/2}+x_+x_-/(x_z)^{1/2} & x_+/(x_z)^{1/2}\\ x_-/(x_z)^{1/2} & 1/(x_z)^{1/2}\end{pmatrix}, \tag{A2}$$

$$e^{y_-J_-}e^{(\ln y_z)J_z}e^{y_+J_+}=\begin{pmatrix}(y_z)^{1/2} & y_+(y_z)^{1/2}\\ y_-(y_z)^{1/2} & 1/(y_z)^{1/2}+y_+y_-(y_z)^{1/2}\end{pmatrix}. \tag{A3}$$

Equating these three matrices element by element gives expressions for each set of coefficients in terms of the others. This procedure gives four equations for three variables, but since the J matrices are traceless, the determinant of each group operation (A1)–(A3) is unity; therefore only three of these equations are independent. The applicability of the resulting operator equation

$$e^{w_+J_+ + w_-J_- + w_zJ_z} = e^{x_+J_+}e^{(\ln x_z)J_z}e^{x_-J_-}$$
$$= e^{y_-J_-}e^{(\ln y_z)J_z}e^{y_+J_+} \quad \text{(A4)}$$

is not restricted to the 2×2 matrix representation. The algebra of infinitesimal rotation operators maps onto the rotation group, which is represented by the exponential operators. Any relation between exponential operators, i.e., between group operations, which is valid for one particular faithful representation of the group remains valid for all others. Therefore the equalities (A4) are general.

In general the parameters w, x, y appearing in (A1)–(A3) are arbitrary complex numbers. The 2×2 matrices given explicitly are then complex 2×2 matrices with determinant +1: i.e., members of the special linear group Sl(2, c). Reality restrictions on w, x, y lead to different subgroups of Sl(2, c). The condition that the parameters be real restricts consideration to the subgroup Sl(2, r) of Sl(2, c). The condition w_z real, $w_+^* = -w_-$, and similar conditions for x, y is equivalent to the restriction to the subgroup SU(2) of Sl(2, c). The BCH formula (A4) is valid for the group Sl(2, c). It is also valid for its real forms (subgroups) provided the given set of parameters (say w) is selected in such a way that the solution of Eqs. (A1)–(A4) for the other sets of parameters (say x and y) satisfies the same restrictive conditions (for instance, reality) obeyed by the given set (say w).

Using the disentangling relations, expressions for the rotation operator (3.7) are obtained:

$$R_{\theta,\varphi} = e^{\zeta J_+ - \zeta^* J_-} = e^{\tau J_+}e^{\ln(1+|\tau|^2)J_z}e^{-\tau^* J_-}$$
$$= e^{-\tau^* J_-}e^{-\ln(1+|\tau|^2)J_z}e^{\tau J_+}, \quad \text{(A5)}$$

where ζ and τ are given in (3.7b) and (3.11b). If we let $\zeta = c\alpha$, $\zeta^* = c\alpha^*$, $a^\dagger = cJ_+$, $a = cJ_-$, and $J_z = a^\dagger a - 1/2c^2$, following the contraction procedure of Sec. IV, Eq. (A5) gives, in the limit $c\to 0$,

$$T_\alpha = e^{\alpha a^\dagger - \alpha^* a} = e^{\alpha a^\dagger}e^{-|\alpha|^2/2}e^{-\alpha^* a}$$
$$= e^{-\alpha^* a}e^{|\alpha|^2/2}e^{\alpha a^\dagger} \quad \text{(A6)}$$

which is the BCH formula (2.11).

A more general expression can be obtained by contraction of Eq. (A4). If we let $w_+ = c\alpha$, $w_- = -c\beta^*$, and $w_z = 0$, we obtain

$$e^{\alpha a^\dagger - \beta^* a} = e^{\alpha a^\dagger}e^{-\alpha\beta^*/2}e^{-\beta^* a} = e^{-\beta^* a}e^{\alpha\beta^*/2}e^{\alpha a^\dagger}. \quad \text{(A7)}$$

Using this relation, together with (A6), one obtains, after some manipulations,

$$T_\alpha T_\beta = e^{(\alpha\beta^* - \alpha^*\beta)/2}T_{\alpha+\beta}, \quad \text{(A8)}$$

which describes the composition of translations T_α. The use of this equation allows derivation of Eq. (2.14a) very simply. Of course, Eq. (A8) can also be obtained by contraction of a similar equation for the composition of rotations:

$$R_{\theta',\varphi'}R_{\theta,\varphi} = R_{\Theta,\Phi}e^{-i\Psi J_z}, \quad \text{(A9)}$$

where Θ, Φ, and Ψ are to be determined. We note that the $R_{\theta,\varphi}$'s do not form a group, since we have restricted ourselves to rotations around an axis in the (x, y) plane. It is therefore necessary to allow for a rotation around the z axis on the right-hand side of (A9). This rotation simply amounts to changing the phase factor of the single-atom eigenstates $|\psi_i^n\rangle$. The angles Θ, Φ, and Ψ could be obtained by manipulating (A4) and (A5). A simpler procedure is to use the 2×2 matrix representation as in (A1)–(A3). By application of (A1) one has

$$R_{\theta,\varphi} = \frac{1}{(1+|\tau|^2)^{1/2}}\begin{pmatrix}1 & \tau\\ -\tau^* & 1\end{pmatrix}, \quad \text{(A10)}$$

$$e^{-i\Psi J_z} = \begin{pmatrix}e^{-i\Psi/2} & 0\\ 0 & e^{i\Psi/2}\end{pmatrix}. \quad \text{(A11)}$$

With $\tau' = e^{-i\varphi'}\tan\frac{1}{2}\theta'$ and $T = e^{-i\Phi}\tan\frac{1}{2}\Theta$, Eq. (A9) becomes

$$[(1+|\tau|^2)(1+|\tau'|^2)]^{1/2}\begin{pmatrix}1-\tau^*\tau' & \tau+\tau'\\ -\tau^*-\tau'^* & 1-\tau\tau'^*\end{pmatrix}$$
$$= \frac{1}{(1+|T|^2)^{1/2}}\begin{pmatrix}e^{-i\Psi/2} & Te^{i\Psi/2}\\ -T^*e^{-i\Psi/2} & e^{i\Psi/2}\end{pmatrix}, \quad \text{(A12)}$$

which determines Ψ and T in term of τ and τ'. The contraction of (A9) proceeds straightforwardly, and gives (A8). Equation (A9) can also be used to derive (3.14a) in a simple manner.

As a final case of interest let us show how to reentangle the expression

$$e^{\lambda a^\dagger a}e^{(\gamma_+ a^\dagger + \gamma_- a)}. \quad \text{(A13)}$$

The quantity is obtained from the contraction of

$$e^{\lambda J_z}e^{(\alpha_+ J_+ + \alpha_- J_-)}$$
$$= \exp\left(2\cosh\theta\sinh\tfrac{1}{2}\lambda J_z + \alpha_+\frac{\sinh\theta}{\theta}e^{\lambda/2}J_+\right.$$
$$\left. + \alpha_-\frac{\sinh\theta}{\theta}e^{-\lambda/2}J_-\right)\frac{\frac{1}{2}\Omega}{\sinh\frac{1}{2}\Omega}, \quad \text{(A14a)}$$

where $\theta^2 = \alpha_+\alpha_-$,

$$\cosh\tfrac{1}{2}\Omega = \cosh\theta\cosh\tfrac{1}{2}\lambda. \quad \text{(A14b)}$$

Making a contraction with $\alpha_\pm = c\gamma_\pm$ and following the rules of Table I one obtains

$$e^{\lambda a^\dagger a}e^{(\gamma_+ a^\dagger + \gamma_- a)}$$

$$= e^{-\lambda\gamma_+\gamma_-/4}[1 + (2/\lambda)\coth\tfrac{1}{2}\lambda - \coth^2\tfrac{1}{2}\lambda]$$

$$\times \exp[\lambda a^\dagger a + (\tfrac{1}{2}\lambda/\sinh\tfrac{1}{2}\lambda)(\gamma_+ e^{\lambda/2} a^\dagger + \gamma_- e^{-\lambda/2} a)]\ , \tag{A15}$$

which is the desired result. A similar expression has been derived in a more involved way by Weiss and Maradudin[30] in order to calculate some thermal averages which have been obtained differently in Sec. VIII.

APPENDIX B: GENERATING FUNCTIONS FOR EXPECTATION VALUES WITHIN BLOCH STATES

Using the disentangling theorem of Appendix A, together with the definition of Bloch states as rotation (3.8), or equivalently (3.12), it is easy to construct generating functions for normally ordered, antinormally ordered, and symmetrized expectation values of products of powers of the operators J_+, J_z, J_- within Bloch states.

We define the following expectation values:

$$X_N(\alpha, \beta, \gamma) = \langle \theta, \varphi | e^{\alpha J_+} e^{\beta J_z} e^{\gamma J_-} | \theta, \varphi \rangle\ , \tag{B1a}$$

$$X_A(\alpha, \beta, \gamma) = \langle \theta, \varphi | e^{\gamma J_-} e^{\beta J_z} e^{\alpha J_+} | \theta, \varphi \rangle\ , \tag{B1b}$$

$$X_S(\alpha, \beta, \gamma) = \langle \theta, \varphi | e^{\alpha J_+ + \beta J_z + \gamma J_-} | \theta, \varphi \rangle\ , \tag{B1c}$$

and will show that these functions can easily be calculated. These functions are generating functions since one has

$$\left[\left(\frac{\partial}{\partial\alpha}\right)^a \left(\frac{\partial}{\partial\beta}\right)^b \left(\frac{\partial}{\partial\gamma}\right)^c X_n\right]_{\alpha=\beta=\gamma=0} = \langle \theta, \varphi | J_+^a J_z^b J_-^c | \theta, \varphi \rangle\ , \tag{B2a}$$

$$\left[\left(\frac{\partial}{\partial\alpha}\right)^a \left(\frac{\partial}{\partial\beta}\right)^b \left(\frac{\partial}{\partial\gamma}\right)^c X_A\right]_{\alpha=\beta=\gamma=0} = \langle \theta, \varphi | J_-^c J_z^b J_+^a | \theta, \varphi \rangle\ , \tag{B2b}$$

$$\left[\left(\frac{\partial}{\partial\alpha}\right)^a \left(\frac{\partial}{\partial\beta}\right)^b \left(\frac{\partial}{\partial\gamma}\right)^c X_S\right]_{\alpha=\beta=\gamma=0} = \langle \theta, \varphi | S\{J_+^a J_z^b J_-^c\} | \theta, \varphi \rangle\ , \tag{B2c}$$

where $S\{\ \}$ means the fully symmetrized sum of products, which is equal to the sum of all distinct permutations of the factors within the brackets divided by the number $(a+b+c)!/a!b!c!$ of these permutations.

It is sufficient to compute one of the generating functions, as the other two are then given by using the disentangling theorem of Appendix A. It is simplest to do this for X_A. Using (3.12) one has

$$X_A = (1/[1 + |\tau|^2]^{2J})\langle -J | e^{(\tau^* + \gamma)J_-} e^{\beta J_z} e^{(\tau+\alpha)J_+} | -J \rangle\ . \tag{B3}$$

This expression is then put in normally ordered form (A2), in which case only the term $e^{(\ln x_z)J_z}$ contributes to the expectation value in the ground state $\langle -J | e^{(\ln x_z)J_z} | -J \rangle = 1/x_z^J$. One obtains

$$\frac{1}{(x_z)^{1/2}} = \frac{1}{(y_z)^{1/2}} + y_+ y_- (y_z)^{1/2}$$

$$= e^{-\beta/2} + (\tau + \alpha)(\tau^* + \gamma) e^{\beta/2}\ , \tag{B4}$$

which immediately gives

$$X_A = \left(\frac{e^{-\beta/2} + e^{\beta/2}(\tau+\alpha)(\tau^*+\gamma)}{1 + |\tau|^2}\right)^{2J} = [e^{-\beta/2}\cos^2(\tfrac{1}{2}\theta) + e^{\beta/2}(\sin\tfrac{1}{2}\theta\, e^{-i\varphi} + \alpha\cos\tfrac{1}{2}\theta)(\sin\tfrac{1}{2}\theta\, e^{i\varphi} + \gamma\cos\tfrac{1}{2}\theta)]^{2J}\ . \tag{B5}$$

From this expression it immediately follows that

$$X_N = \left(\frac{e^{\beta/2}|\tau|^2 + e^{-\beta/2}(\alpha\tau^* + 1)(\gamma\tau + 1)}{1 + |\tau|^2}\right)^{2J} = [e^{\beta/2}\sin^2(\tfrac{1}{2}\theta) + e^{-\beta/2}(\alpha e^{i\varphi}\sin\tfrac{1}{2}\theta + \cos\tfrac{1}{2}\theta)(\gamma e^{-i\varphi}\sin\tfrac{1}{2}\theta + \cos\tfrac{1}{2}\theta)]^{2J}\ , \tag{B6}$$

$$X_S = \left(\frac{(1 + |\tau|^2)\cosh K - (1 - |\tau|^2)\tfrac{1}{2}\beta(\sinh K)/K + (\alpha\tau^* + \gamma\tau)(\sinh K)/K)}{1 + |\tau^2|}\right)^{2J}$$

$$= \left[\cosh K - \tfrac{1}{2}\beta\frac{\sinh K}{K}\cos\theta + (\alpha e^{i\varphi} + \gamma e^{-i\varphi})\frac{\sinh K}{K}\sin\tfrac{1}{2}\theta\cos\tfrac{1}{2}\theta\right]^{2J}\ , \tag{B7a}$$

where

$$K = (\alpha\gamma + \tfrac{1}{4}\beta^2)^{1/2}\ . \tag{B7b}$$

APPENDIX C: CONTRACTION OF SPHERICAL HARMONICS TO HERMITE POLYNOMIALS

This appendix indicates additional consequences of the contraction of the angular momentum algebra (3.3) to the harmonic-oscillator algebra (2.3). The mathematical properties derived here are not directly connected to the contraction of Sec. III to Sec. II, but are of independent interest. One should note that the symbols θ and φ used here are the angular coordinates of a Schrödinger representation, and are not related to Bloch-state labels, as elsewhere in this article. The notation $|l, m\rangle$ is used for Dicke states, in order to conform with the usual spherical harmonic notation.

We first note that the eigenvalue equations

$$J^2|l, m\rangle = l(l+1)|l, m\rangle , \tag{C1a}$$

$$J_z|l, m\rangle = m|l, m\rangle \tag{C1b}$$

contract to

$$[-2n_{op} + (a^\dagger a + aa^\dagger)]|\infty, n\rangle = |\infty, n\rangle , \tag{C2a}$$

$$a^\dagger a|\infty, n\rangle = n|\infty, n\rangle , \tag{C2b}$$

respectively, where n_{op} on the left-hand side of (C2a) is the number operator. These contractions are easily performed using the rules of Table I. Equation (C2a) can be written in a slightly different form,

$$\tfrac{1}{2}(a^\dagger a + aa^\dagger)|\infty, n\rangle = (n_{op} + \tfrac{1}{2})|\infty, n\rangle , \tag{C2c}$$

where the left-hand side is now a Hamiltonian for a model physical system, while the right-hand side describes the eigenvalue spectrum for states which diagonalize n_{op}.

In the Schrödinger representation Eqs. (C1a) and (C1b) lead to wave functions $\langle q|l, m\rangle = Y_{l,m}(\theta, \varphi)$ which are the spherical harmonics, whereas (C2b) leads to wave functions $\langle q|\infty, n\rangle = \psi_n((\omega m/\hbar)^{1/2}q)$ which are related to the Hermite polynomials $H_n(x)$, where $x = (\omega m/\hbar)^{1/2}q$. It is therefore clear that the spherical harmonics $Y_{l,m}(\theta, \varphi)$ can be contracted to the Hermite polynomials $H_n(x)$, but it remains to show how the coordinates θ, φ should be contracted to x. To this effect we consider how the operators J_x, J_y, J_z, written in the coordinate representation, should be contracted according to the rules of Table I. One has

$$\lim_{c\to 0}(cJ_x) = \lim_{c\to 0}\left[ic\left(\sin\varphi \frac{\partial}{\partial\theta} + \cot\!\operatorname{an}\theta \cos\varphi \frac{\partial}{\partial\varphi}\right)\right]$$

$$= \frac{a^\dagger + a}{2} = \frac{x}{\sqrt{2}} , \tag{C3a}$$

$$\lim_{c\to 0}(cJ_y) = \lim_{c\to 0}\left[ic\left(-\cos\varphi \frac{\partial}{\partial\theta} + \cot\!\operatorname{an}\theta \sin\varphi \frac{\partial}{\partial\varphi}\right)\right]$$

$$= \frac{a^\dagger - a}{2i} = -\frac{p}{(2\hbar\omega m)^{1/2}} = \frac{i}{\sqrt{2}}\frac{\partial}{\partial x} , \tag{C3b}$$

$$\lim_{c\to 0}(2c^2 J_z) = \lim_{c\to 0}\left(-2ic^2\frac{\partial}{\partial\varphi}\right) = -1 . \tag{C3c}$$

Introducing the last equation in the previous two, one sees that cotanθ must approach zero as fast as c for the contraction to give a finite result. With $\theta = \frac{1}{2}\pi - cX$ one then obtains

$$\lim_{c\to 0}\left(-i\sin\varphi\frac{\partial}{\partial x} + \tfrac{1}{2}X\cos\varphi\right) = \frac{x}{\sqrt{2}} , \tag{C4a}$$

$$\lim_{c\to 0}\left(i\cos\varphi\frac{\partial}{\partial x} + \tfrac{1}{2}X\sin\varphi\right) = \frac{i}{\sqrt{2}}\frac{\partial}{\partial x} . \tag{C4b}$$

Though φ could be kept arbitrary, the simplest contraction procedure is clearly to take $\varphi = 0$, in which case $X = \sqrt{2}x$. The correct limiting technique is then

$$\lim_{c\to 0}\varphi = 0 , \quad \lim_{c\to 0}\theta = \tfrac{1}{2}\pi - c\sqrt{2}x , \tag{C5a}$$

together with

$$\lim_{c\to 0}(2lc^2) = 1 , \quad \lim_{c\to 0} m = -\infty , \quad l + m = n \text{ (fixed)} . \tag{C5b}$$

The angular momentum eigenfunctions are,[9] with $u = \cos\theta$,

$$Y_{l,m}(\theta, \varphi) = P_{l,m}(u)e^{im\varphi} , \tag{C6a}$$

where

$$P_{l,m}(u) = (-1)^{l+m}\frac{1}{2^l l!}\left(\frac{2l+1}{2}\right)^{1/2}\left(\frac{(l-m)!}{(l+m)!}\right)^{1/2}$$

$$\times(1-u^2)^{m/2}\frac{d^{l+m}}{du^{l+m}}(1-u^2) \tag{C6b}$$

are the associated Legendre polynomials, normalized such that

$$\int_{-1}^{1} P_{l,m}^2\, du = 1 . \tag{C6c}$$

The harmonic-oscillator eigenfunctions are[9]

$$\psi_n = \frac{(-1)^n}{(2^n n!\sqrt{\pi})^{1/2}} e^{x^2/2}\frac{d^n}{dx^n}e^{-x^2}$$

$$= \frac{1}{(2^n n!\sqrt{\pi})^{1/2}} e^{-x^2/2} H_n(x) , \tag{C7a}$$

which are normalized such that

$$\int_{-\infty}^{+\infty} \psi_n^2(x)\, dx = 1 . \tag{C7b}$$

With $\varphi = 0$, we just have to contract $P_{l,m}(u)$, with $u = xc\sqrt{2} = l^{1/2}x$, as $c \to 0$. Introducing this in (C6c), we see that the normalization will only be preserved if ψ_n is the limit of $l^{-1/4}P_{l,m}$. Indeed, we find

$$\lim_{c\to 0} l^{-1/4}P_{l,m}$$

$$= \lim_{c\to 0}(-1)^n\left(\frac{(2l)!\,l^{1/2}}{2^{2l}\,l!\,l!}\right)^{1/2}\left(\frac{1}{2^n n!\,(2lc^2)^n}\right)^{1/2}$$

$$\times[1-2c^2x^2]^{-(1/2c^2)/2}\left(\frac{d}{dx}\right)^n[1-2c^2x^2]^{1/2c^2}$$

$$= (-1)^n\pi^{-1/4}\left(\frac{1}{2^n n!}\right)^{1/2} e^{x^2/2}\left(\frac{d}{dx}\right)^n e^{-x^2} = \psi_n(x) , \tag{C8}$$

where Stirling's approximation has been used to contract the first quantity within large parentheses.

The orthogonality of the $\psi_n(x)$'s results from an orthogonality relation satisfied by the associated

Legendre polynomials

$$\int_{-1}^{1} P_{l,m}(u)P_{l,m'}(u)\,du = \delta_{mm'}\,, \qquad \text{(C9a)}$$

which immediately gives

$$\int_{-\infty}^{+\infty} \lim l^{-1/4} P_{l,m} \lim l^{-1/4} P_{l,m'}\, dx = \delta_{mm'} \qquad \text{(C9b)}$$

or

$$\int_{-\infty}^{+\infty} \psi_n(x)\psi_{n'}(x) = \delta_{nn'}\;.$$

As another example of the derivation of a property of the ψ_n's from a property of the associated Legendre polynomials, we show how the completeness relation can be obtained from the addition theorem

$$\left(\frac{2}{2l+1}\right)^{1/2} \sum_{m=-l}^{l} P_{l,m}(\cos\theta)P_{l,m}(\cos\theta') = P_{l,0}(\cos(\theta-\theta'))\;. \qquad \text{(C10a)}$$

One has, in the limit,

$$\sum_{n=0}^{\infty} \lim_{l\to\infty} l^{-1/4} P_{l,m}(\cos\theta) \lim_{l\to\infty} l^{-1/4} P_{l,m}(\cos\theta') = \lim_{l\to\infty} P_{l,0}(\cos[(x'-x)/\sqrt{l}\,]) \qquad \text{(C10b)}$$

or

$$\sum_{n=0}^{\infty} \psi_n(x)\psi_n(x') = \delta(x'-x)\;. \qquad \text{(C10c)}$$

In conclusion let us mention that all other properties of associated Legendre polynomials contract to corresponding properties of oscillator eigenfunctions. In this way, one can construct additional generating functions, recursion relations, addition theorems, etc., for the oscillator eigenfunctions from those of the associated Legendre polynomials.

APPENDIX D: RELATIONS BETWEEN BLOCH STATES, SPHERICAL HARMONICS, AND IRREDUCIBLE REPRESENTATIONS OF FULL ROTATION GROUP

In this appendix additional uses of the disentangling theorem are discussed. In particular it is shown how to derive in a simple manner the well-known $(2J+1)\times(2J+1)$ irreducible representations of the full rotation group. Properties of these representations are then related to properties of the Jacobi polynomials, of the spherical harmonics, and of the associated Legendre polynomials. This allows to derive orthogonality relations in a very simple manner. The relation between spherical harmonics and the Bloch amplitudes of Eq. (3.13) is also shown. These properties are useful to compute integrals on the sphere of Bloch-state projectors with spherical-harmonics weight functions. These integrals are then used to construct the diagonal expansion of operators into Bloch states, as in Eqs. (3.20c) and (3.20d). As a case of special interest the statistical operator of a pure Bloch state is finally derived.

1. Derivation of Irreducible Representations of Full Rotation Group

Let us consider the rotation of an object defined by the three Euler angles $(-\alpha,\ -\beta,\ -\gamma)$. The result is identical to a coordinate rotation $(\alpha,\ \beta,\ \gamma)$ and will therefore agree with the usual group theoretical expressions.[14] In the 2×2 irreducible representation one has, in the spirit of Appendix A,

$$R(-\alpha,-\beta,-\gamma) = e^{iJ_z\alpha}e^{iJ_y\beta}e^{iJ_z\gamma} = \begin{bmatrix} e^{i\alpha/2} & 0 \\ 0 & e^{-i\alpha/2}\end{bmatrix}\begin{bmatrix} \cos\frac{1}{2}\beta & \sin\frac{1}{2}\beta \\ -\sin\frac{1}{2}\beta & \cos\frac{1}{2}\beta\end{bmatrix}\begin{bmatrix} e^{i\gamma/2} & 0 \\ 0 & e^{-i\gamma/2}\end{bmatrix}$$

$$= \begin{bmatrix} (x_z)^{1/2} + x_+x_-/(x_z)^{1/2} & x_+/(x_z)^{1/2} \\ x_-/(x_z)^{1/2} & 1/(x_z)^{1/2}\end{bmatrix}\;. \qquad \text{(D1)}$$

By definition one has

$$R(-\alpha,-\beta,-\gamma)|M\rangle = \sum_{M'} \mathfrak{D}^J_{M'M}(\alpha\beta\gamma)|M'\rangle\;. \qquad \text{(D2)}$$

The occurrence of the minus signs in the left-hand side has been explained above. Using (D1) and (D2) one obtains

$$\mathfrak{D}^J_{M'M}(\alpha,\beta,\gamma) = \langle M'|R(-\alpha,-\beta,-\gamma)|M\rangle = \langle M'|e^{x_+J_+}e^{(\ln x_z)J_z}e^{x_-J_-}|M\rangle\,, \qquad \text{(D3)}$$

the last equality resulting from (A2). From (D1) one has

$$x_+ = \tan\tfrac{1}{2}\beta\, e^{i\alpha}\,,\quad x_- = -\tan\tfrac{1}{2}\beta\, e^{i\gamma}\,,$$

$$x_z = e^{i(\alpha+\gamma)}/\cos^2(\tfrac{1}{2}\beta)\;.$$

It remains then to calculate the last matrix element in (D3). This is done straightforwardly by expansion of the exponential operators containing J_+ and J_-:

$$\mathfrak{D}^J_{M'M}(\alpha,\beta,\gamma) = \sum_{\mu\nu} \frac{1}{\mu!}\,\frac{1}{\nu!}\,(x_+)^{\mu}(x_-)^{\nu}(x_z)^{M-\nu} \times\langle M'|J_+^{\mu}J_-^{\nu}|M\rangle\;. \qquad \text{(D4)}$$

Finally, using the step-up and step-down properties of J_+ and J_- one obtains

$$\mathfrak{D}^J_{M'M}(\alpha,\beta,\gamma)=\sum_\mu \frac{(-1)^{M-M'+\mu}}{\mu!\,(\mu+M-M')!}\,\frac{(J-M'+\mu)!}{(J+M'-\mu)!}\left(\frac{(J+M')!\,(J+M)!}{(J-M')!\,(J-M)!}\right)^{1/2}(\sin\tfrac{1}{2}\beta)^{2\mu+M-M'}(\cos\tfrac{1}{2}\beta)^{-M-M'}e^{i\alpha M'}e^{i\gamma M}. \tag{D5}$$

This expression is identical to the more usual form[14]

$$\mathfrak{D}^J_{M'M}(\alpha,\beta,\gamma)=\sum_\kappa (-1)^\kappa \frac{[(J+M)!\,(J-M)!\,(J+M')!\,(J-M')!]^{1/2}}{(J-M'-\kappa)!\,(J+M-\kappa)!\,\kappa!\,(\kappa+M'-M)!}(\sin\tfrac{1}{2}\beta)^{2\kappa+M'-M}(\cos\tfrac{1}{2}\beta)^{2J+M-M'-2\kappa}e^{i\alpha M'}e^{i\gamma M}. \tag{D6}$$

The identity is not easy to show, but this fact should not mask the great simplicity with which (D5) has been derived. To prove the identity, starting from (D6), one replaces $(\cos\frac{1}{2}\beta)^{2J+2M-2\kappa}$ by $(1-\sin^2(\frac{1}{2}\beta)^{J+M-\kappa}$. This expression is expanded in powers of $(\sin\frac{1}{2}\beta)^{2\nu}$, and a double sum results. The summation indices are changed to $\mu=M'-M+\nu+\kappa$, which runs from $M'-M+\kappa$ to $J+M'$, and to $\eta=M'-M+\kappa$. The summation over η is carried independently, using the relation

$$\sum_{k=0}^{p}\binom{n}{k}\binom{m}{p-k}=\binom{n+m}{p}.$$

Equation (D5) results.

The construction of the matrix elements (D5) does not give any new information, as (D6) has been known for a long time. But the procedure which has been followed throws light on mechanisms for computing representation matrix elements for other groups in an explicit form. It turns out that an exactly similar technique can be used for a large number of Lie groups. To expand on this would fall beyond the limits of the present work.[34]

2. Relations to Jacobi Polynomials, Associated Legendre Polynomials, and Spherical Harmonics

The matrix elements $\mathfrak{D}^J_{M'M}(\alpha,\beta,\gamma)$ are directly related to the Jacobi polynomials. To see this we first write (D5) or (D6) in a different way,

$$\mathfrak{D}^J_{M'M}(0,\beta,0)=\sum_{\nu=-J}^{J}\frac{(-1)^{\nu-M'}}{(\nu-M')!\,(\nu-M)!}\,\frac{(J+\nu)!}{(J-\nu)!}\left(\frac{(J-M')!\,(J+M)!}{(J+M')!\,(J-M)!}\right)^{1/2}(\sin\tfrac{1}{2}\beta)^{2\nu-M-M'}(\cos\tfrac{1}{2}\beta)^{M+M'}. \tag{D7}$$

These matrix elements are given by an elegant generating function[35,36]

$$\mathfrak{D}^J_{M',M}(0,\beta,0)=P^J_{M',M}(\cos\beta) \tag{D8}$$

with

$$P^l_{m,n}(z)=\frac{(-1)^{l-m}}{2^l(l-n)!}\left(\frac{(l-n)!\,(l+m)!}{(l+n)!\,(l-m)!}\right)^{1/2}(1+z)^{-(m+n)/2}(1-z)^{-(m-n)/2}\left(\frac{d}{dz}\right)^{l-m}[(1-z)^{l-n}(1+z)^{l+n}]. \tag{D9}$$

This is just the generating function for the Jacobi polynomials.

When $J=l$ is an integer the labels m, n can assume all integral values $-l\le m, n\le +l$. Setting $n=0$ in the generating function above yields the generating function for the associated Legendre polynomials and the spherical harmonics:

$$\mathfrak{D}^l_{m,0}(0,\beta,0)=P^l_{m,0}(\cos\beta)=\frac{(-1)^{l-m}}{2^l l!}\left(\frac{(l+m)!}{(l-m)!}\right)^{1/2}(1-z^2)^{-m/2}\times\left(\frac{d}{dz}\right)^{l-m}(1-z^2)^l \tag{D10}$$

and

$$\left(\frac{2l+1}{4\pi}\right)^{1/2}\mathfrak{D}^l_{m,0}(\varphi,\theta,-)=(-1)^m Y^l_m(\theta,\varphi). \tag{D11}$$

In short, the spherical harmonics are essentially matrix elements belonging to the $n=0$ column of the $\mathfrak{D}^l_{m,n}$ representation of the full rotation group. Moreover, the matrix element which occurs at the intersection of the $m=0$ row and the $n=0$ column is a Legendre polynomial $P^l_{0,0}(\cos\beta)$. We see that a number of special functions are associated with $\mathfrak{D}^l_{m,n}$. The orthogonality properties of all these functions are simply obtained from the well-known expression[14,37]

$$\int_0^{2\pi}d\alpha\int_0^{\pi}\sin\beta\,d\beta\int_0^{2\pi}d\gamma\,\mathfrak{D}^{l'*}_{m'n'}(\alpha,\beta,\gamma)\mathfrak{D}^l_{m,n}(\alpha,\beta,\gamma)=\frac{\int_0^{2\pi}d\alpha\int_0^{\pi}|d\cos\beta|\int_0^{2\pi}d\gamma}{\dim\mathfrak{D}^l}\delta^{l'l}\delta_{m'n}\delta_{n'n}. \tag{D12}$$

Thus the functions

$$\left(\frac{2l+1}{4\pi\times 2\pi}\right)^{1/2}\mathfrak{D}^l_{mn}(\alpha,\beta,\gamma) \tag{D13a}$$

form an orthonormal set on the parameter space (α, β, γ) with respect to the measure

$$d\alpha \, |d\cos\beta| \, d\gamma \, . \tag{D14a}$$

By setting $n' = n = 0$ in (D12), and carrying out the integral over γ, we find that the functions

$$\left(\frac{2l+1}{4\pi}\right)^{1/2} \mathfrak{D}^l_{m,0}(\varphi, \theta, -) = (-1)^m Y^l_m(\theta, \varphi) \tag{D13b}$$

are orthonormal on (θ, φ) with respect to the measure

$$d\varphi \, |d\cos\theta| = d\varphi \sin\theta \, d\theta \, . \tag{D14b}$$

Finally setting $n = n' = 0$ and $m = m' = 0$ in (D12), and carrying out the integrals over both α and γ, we find that the Legendre polynomials $P_l(\cos\theta)$,

$$\left(\frac{2l+1}{2}\right)^{1/2} \mathfrak{D}^l_{00}(-, \theta, -) = \left(\frac{2l+1}{2}\right)^{1/2} P_l(\cos\theta) \, , \tag{D13c}$$

form an orthonormal set on the interval $(-1, +1)$ with respect to the measure

$$|d\cos\theta| = \sin\theta \, d\theta \, . \tag{D14c}$$

The completeness of the functions (D13a)–(D13c) may be proved analogously starting from the completeness relation[37] for the matrix elements $\mathfrak{D}^l_{mn}$:

$$\sum_l \sum_m \sum_n \frac{2l+1}{4\pi \times 2\pi} \mathfrak{D}^l_{mn}(\alpha', \beta', \gamma') \mathfrak{D}^l_{mn}(\alpha, \beta, \gamma) = \delta(\alpha' - \alpha)\delta(\cos\beta' - \cos\beta)\delta(\gamma' - \gamma) \, . \tag{D15}$$

Equation (D15) is the dual to Eq. (D12).

3. Relations between Spherical Harmonics and Bloch Amplitudes

The spherical harmonics (D13b) and the Bloch amplitudes (3.13), respectively,

$$Y^L_M(\theta, \varphi) = (-1)^M \left(\frac{2L+1}{4\pi}\right)^{1/2} \mathfrak{D}^L_{M,0}(\varphi, \theta, -)$$

and

$$\langle J, M | \theta, \varphi \rangle \equiv \mathfrak{D}^J_{M,-J}(-\varphi, \theta, \varphi) \, ,$$

have analogous properties because they are functions drawn from different columns of the $\mathfrak{D}^J_{m',m}$ [SU(2)] representation of SU(2). Since the columns ($m = 0$ for spherical harmonics and $m = -j$ for the Bloch amplitudes) are separated by half the range of the discrete variable m, it should come as no surprise that the oscillator eigenfunctions are contracted limits of the spherical harmonics for the argument $(\frac{1}{2}\pi - c\sqrt{2}\,x)$ centered around the midpoint in the finite range of the continuous dual variable θ [Eq. (C5a)]. This was proved there in a different way.

4. Integrals over Bloch-State Projectors

The observations made so far can be used to compute explicitly integrals of the form

$$I = \int |J, \theta, \varphi\rangle Y^L_M(\theta\varphi) \langle J, \theta, \varphi| \, d\Omega \tag{D16}$$

with $d\Omega = d\varphi \sin\theta \, d\theta$. First, the Bloch states are expanded in terms of matrix elements

$$|\theta, \varphi\rangle = \sum_m |J, m\rangle \mathfrak{D}^J_{m,-J}(-\varphi, \theta, \varphi)$$

$$= e^{iJ(\gamma-\varphi)} \sum_m |J, m\rangle \mathfrak{D}^J_{m,-J}(-\varphi, \theta, \gamma) \, , \tag{D17}$$

where γ is a dummy variable, introduced for convenience. Substituting (D17) and (D13b) into (D16) one obtains

$$I = \sum_{m, m'} |J, m\rangle A^{J,L,M}_{mm'} \langle J, m'| \, , \tag{D18}$$

where

$$A^{J,L}_{mm'} = \frac{(-1)^M}{2\pi} \left(\frac{2L+1}{4\pi}\right)^{1/2} \int \mathfrak{D}^{J*}_{m,-J}(\varphi, \theta, \gamma) \mathfrak{D}^L_{M,0}(\varphi, \theta, \gamma) \mathfrak{D}^J_{m',-J}(\varphi, \theta, \gamma) \, d\mu(\varphi, \theta, \gamma) \, . \tag{D19}$$

Here $d\mu(\varphi, \theta, \gamma)$ is given by (D14a). The integral in (D19) is well known,[38–40] and can be expressed in terms of Clebsch–Gordan coefficients:

$$\int \mathfrak{D}^{j_1 *}_{a'a}(R) \mathfrak{D}^{j_2}_{b'b}(R) \mathfrak{D}^{j_3}_{c'c}(R) \, d\mu(R) = \frac{8\pi^2}{2j_1+1} \left\langle \begin{matrix} j_2 & j_3 \\ b' & c' \end{matrix} \middle| \begin{matrix} j_1 \\ a' \end{matrix} \right\rangle \left\langle \begin{matrix} j_1 \\ a \end{matrix} \middle| \begin{matrix} j_2 & j_3 \\ b & c \end{matrix} \right\rangle \, . \tag{D20}$$

Using this fact one finds

$$A^{J,L,M}_{mm'} = (-1)^M \left(\frac{4\pi(2L+1)}{(2J+1)^2}\right)^{1/2} \left\langle \begin{matrix} L & J \\ M & m' \end{matrix} \middle| \begin{matrix} J \\ m \end{matrix} \right\rangle \left\langle \begin{matrix} J & L \\ -J & 0 \end{matrix} \middle| \begin{matrix} J \\ -J \end{matrix} \right\rangle \, . \tag{D21}$$

The last Clebsch–Gordan coefficient is given explicitly by[41]

$$\left\langle \begin{matrix} J \\ -J \end{matrix} \middle| \begin{matrix} L & J \\ 0 & -J \end{matrix} \right\rangle = \left(\frac{(2J+1)!\,(2J)!}{(2J+1+L)!\,(2J-L)!}\right)^{1/2} \, . \tag{D22}$$

In particular it results that $A^{J,L,M}_{mm'} = 0$ for $L > 2J$.

This is precisely what is required to have $(2J+1)^2$ independent coefficients in the expansion (3.20d).

5. Diagonal Expansion of Operators into Bloch States

We now proceed to solve for the coefficients G_{lm} in (3.20d) in terms of the matrix elements

$\langle M|G|M'\rangle$ of (3.20a). Using (3.20a), (3.20c), (3.20d), and (D18), one obtains

$$\langle M|G|M'\rangle = \sum_{l,m} G_{lm} A_{MM'}^{J,l,m} . \qquad \text{(D23)}$$

This relation must be inverted, using the value of $A_{MM'}^{J,l,m}$ from (D21). To this effect we use the relation[41]

$$\left\langle \begin{matrix} j_1 & j_2 \\ j_2 & m_2 \end{matrix} \middle| \begin{matrix} J \\ M \end{matrix} \right\rangle = (-1)^{j_1 - J + m_2} \left(\frac{2J+1}{2j_1+1}\right)^{1/2} \left\langle \begin{matrix} J & j_2 \\ M & -m_2 \end{matrix} \middle| \begin{matrix} j_1 \\ m_1 \end{matrix} \right\rangle \qquad \text{(D24)}$$

to transform the first Clebsch–Gordan coefficient in (D21). One obtains

$$A_{MM'}^{J,l,m} = (-1)^{l-m}(-1)^{J-M'}\left(\frac{4\pi}{2J+1}\right)^{1/2} \times \left\langle \begin{matrix} J & J \\ M & -M' \end{matrix} \middle| \begin{matrix} l \\ m \end{matrix} \right\rangle \left\langle \begin{matrix} J \\ -J \end{matrix} \middle| \begin{matrix} l & J \\ 0 & -J \end{matrix} \right\rangle . \qquad \text{(D25)}$$

Introducing this in (D23), multiplying both sides by $\langle {}^{l'}_{m'} | {}^{J}_{M} \; {}^{J}_{-M'} \rangle (-1)^{J-M'}$, and summing over M and M', one obtains

$$\sum_{M,M'} (-1)^{J-M'} \left\langle \begin{matrix} l' \\ m' \end{matrix} \middle| \begin{matrix} J & J \\ M & -M' \end{matrix} \right\rangle \langle M|G|M'\rangle = (-1)^{l'-m'} G_{l'm'} \left(\frac{4\pi}{2J+1}\right)^{1/2} \left\langle \begin{matrix} J \\ -J \end{matrix} \middle| \begin{matrix} l' & J \\ 0 & -J \end{matrix} \right\rangle . \qquad \text{(D26)}$$

We made use of the relation

$$\sum_{m_1 m_2} \left\langle \begin{matrix} L \\ M \end{matrix} \middle| \begin{matrix} j & j \\ m_1 & m_2 \end{matrix} \right\rangle \left\langle \begin{matrix} j & j \\ m_1 & m_2 \end{matrix} \middle| \begin{matrix} L' \\ M' \end{matrix} \right\rangle = \delta^{LL'} \delta_{MM'} .$$

Equation (D26) gives the coefficients $G_{l'm'}$ in terms of a sum over the matrix elements $\langle M|G|M'\rangle$. These ideas can be used to generate the diagonal representation of any well-behaved operator and, in particular, of the density operator, as explained in Sec. III D. To give a simple example, let us consider the density operator of a pure Bloch state $|\theta_0, \varphi_0\rangle$,

$$\rho = |\theta_0, \varphi_0\rangle\langle\theta_0, \varphi_0| . \qquad \text{(D27)}$$

From (3.23) it is immediate that the weight function can be written

$$P(\theta, \varphi) = \delta(\theta - \theta_0; \varphi - \varphi_0) . \qquad \text{(D28)}$$

However this function is not of the form (3.28), since the summation in (3.28) is limited to the first $(2J+1)^2$ values of $\lambda = (l, m)$. The coefficients X_λ are obtained immediately from (3.27), $X_\lambda = Y_l^m(\theta_0, \varphi_0)$. The summation (3.28) gives then

$$P(\theta, \varphi) = \sum_{l=0}^{2J} \sum_{m=-l}^{l} Y_l^m(\theta_0, \varphi_0) Y_l^{-m}(\theta, \varphi) = \frac{1}{4\pi} \sum_{l=0}^{2J} (2l+1) P_l(\cos\Theta) \qquad \text{(D29)}$$

with $\cos\Theta = \cos\theta\cos\theta_0 + \sin\theta\sin\theta_0\cos(\varphi - \varphi_0)$. The two expressions (D28) and (D29) correspond to the same density matrix in view of the fact the integral (D16) vanishes for $L > 2J$.

---

*Present address: Physics Department, University of South Florida, Tampa, Fla. 33620.

[1]J. Schwinger, Phys. Rev. 91, 728 (1953).

[2]E. Schrödinger, Naturwiss. 14, 644 (1927).

[3]I. R. Senitzky, Phys. Rev. 111, 3 (1958).

[4]R. J. Glauber, Phys. Rev. 130, 2529 (1963); 131, 2766 (1963); in *Quantum Optics and Electronics*, edited by C. DeWitt, A. Blandin, and C. Cohen-Tannoudji (Gordon and Breach, New York, 1965).

[5]M. Hamermesh, *Group Theory* (Addison-Wesley, Reading, Mass., 1962), especially p. 413.

[6]H. Weyl, *The Theory of Groups in Quantum Mechanics* (Dover, New York, 1928), p. 1931.

[7]R. H. Dicke, Phys. Rev. 93, 99 (1954); in *Proceedings of the Third Quantum Electronics Conference, Paris*, 1963, edited by P. Grivet and N. Bloembergen (Columbia U. P., New York, 1964), p. 35.

[8]F. Bloch, Phys. Rev. 70, 460 (1946).

[9]A. Messiah, *Quantum Mechanics* (North-Holland, Amsterdam, 1962).

[10]H. F. Baker, Proc. London Math. Soc. 34, 347 (1902); J. E. Campbell, *ibid.* 29, 14 (1898); F. Hausdorff, Ber. Verhandl. Sachs. Akad. Wiss. Leipzig, Math.-Naturwiss. 58, 19 (1906).

[11]G. H. Weiss and A. A. Maradudin, J. Math. Phys. 3, 771 (1962).

[12]R. P. Feynman, Phys. Rev. 84, 108 (1951).

[13]R. J. Glauber, in *Physics of Quantum Electronics*, edited by P. Kelley *et al.* (Columbia U. P., New York, 1966); S. R. Klauder and E. C. G. Sudarshan, *Fundamentals of Quantum Optics* (Benjamin, New York, 1968).

[14]E. Wigner, *Group Theory and Its Applications to Quantum Mechanics* (Academic, New York, 1959).

[15]D. M. Brink and G. R. Satchler, *Angular Momentum* (Clarendon, Oxford, 1968).

[16]U. Fano and G. Racah, *Irreducible Tensorial Sets* (Academic, New York, 1959).

[17]A. R. Edmonds, *Angular Momentum in Quantum Mechanics* (Princeton U. P., Princeton, 1957).

[18]E. Callen and H. Callen, Phys. Rev. 129, 578 (1963).

[19]A. Meckler, Nuovo Cimento Suppl. 12, 1 (1959).

[20]I. E. Segal, Duke Math. J. 18, 221 (1951).

[21]E. Inönü and E. P. Wigner, Proc. Natl. Acad. Sci. (U.S.) 39, 510 (1953).

[22]E. J. Saletan, Math. Phys. 2, 1 (1961).

[23]V. Heine, *Group Theory in Quantum Mechanics* (Dover, New York, 1931).

[24]See Ref. 6, Sec. 5; also H. Weyl, *Classical Groups* (Princeton U. P., Princeton, 1946).

[25]R. Bonifacio, P. Schwendimann, and F. Haake, Phys. Rev. A 4, 302 (1971); 4, 854 (1971).

[26]F. T. Arecchi and E. Courtens, Phys. Rev. A 2, 1730 (1970).

[27]H. Ott, Ann. Physik 23, 1969 (1935).

[28]M. Born, Rept. Progr. Phys. 9, 294 (1943).

[29]A. A. Maradudin, G. H. Weiss, and E. W. Montroll,

Theory of Lattice Dynamics in the Harmonic Approximation, Solid State Physics Ser., Suppl. No. 3 (Academic, New York, 1963).

[30]G. H. Weiss and A. A. Maradudin, J. Math. Phys. 3, 771 (1962).

[31]J. M. Radcliffe, J. Phys. A 4, 313 (1971).

[32]A. O. Barut and L. Girardello, Commun. Math. Phys. 21, 41 (1971).

[33]P. W. Atkins and J. C. Dobson, Proc. Roy. Soc. (London) A321, 321 (1971).

[34]R. Gilmore, in *Proceedings of the Third Rochester Conference on Coherence and Quantum Optics*, edited by L. Mandel and E. Wolf (Plenum, New York, 1972).

[35]A. Kihlberg, V. F. Müller, and F. Halbwachs, Commun. Math. Phys. 3, 194 (1966).

[36]N. Ja. Vilenkin, *Special Functions and the Theory of Group Representations*, Transl. Math. Monographs, Vol. 22 (Am. Math. Soc., Providence, R. I., 1968), especially Chap. II.

[37]J. D. Talman, *Special Functions: A Group Theoretic Approach*, based on Lectures by E. P. Wigner (Benjamin, New York, 1968), pp. 75 and 160.

[38]Reference 36, p. 178.

[39]E. Merzbacher, *Quantum Mechanics* (Wiley, New York, 1961), p. 514.

[40]M. E. Rose, *Elementary Theory of Angular Momentum* (Wiley, New York, 1957), p. 58.

[41]Reference 9, Vol. II, pp. 1056–1059.

PHYSICAL REVIEW A

GENERAL PHYSICS

THIRD SERIES, VOL. 7, NO. 3 MARCH 1973

Phase Transition in the Dicke Model of Superradiance*

Y. K. Wang and F. T. Hioe

Institute for Fundamental Studies, Department of Physics and Astronomy, The University of Rochester, Rochester, New York 14627

(Received 10 October 1972)

A system of N two-level atoms interacting with a quantized field, the so-called Dicke model of superradiance, is studied. By making use of a set of Glauber's coherent states for the field, the free energy of the system is calculated exactly in the thermodynamic limit. The results agree precisely with those obtained by Hepp and Lieb, who studied the same model using a different method. The exhibition of a phase transition of the system is presented mathematically in an elementary manner in our approach. The generalization to the case of finitely many radiation modes is also presented.

I. INTRODUCTION

The problem of a system of N two-level atoms interacting with a radiation field has been studied by a number of authors.[1-8] In the so-called Dicke model, the atoms are considered to be at fixed positions within a linear cavity of volume V and the separations between the atoms are assumed to be large enough so that the direct interaction among them can be ignored. However, by the fact that the atoms interact with the same radiation field, the atoms cannot be treated as independent. The importance of treating the radiating atoms as a single quantum system was recognized by Dicke[1] who correctly described a coherent spontaneous radiation process of the system. An exact solution for the Hamiltonian of N identical two-level atoms interacting with a single-mode quantized radiation field at resonance was given by Tavis and Cummings.[2] One of the most significant and interesting results regarding the Dicke model was obtained very recently by Hepp and Lieb.[7] They calculated exactly the thermodynamic properties of the system, the free energy in particular, in the thermodynamic limit that N, $V\to\infty$, $N/V=$ finite and showed that for a sufficiently large value of the coupling constant between the atoms and the field, the system exhibits a second-order phase transition from normal to superradiance at a certain critical temperature. The results of Hepp and Lieb are rigorous, but the mathematics which they used in arriving at their important results is somewhat abstract.

In this paper, we study the same problem by a rather different method. Employing a set of coherent states of the field defined by Glauber,[8] we find that the partition function of the system can be expressed in terms of an integral which is readily evaluated in the thermodynamic limit. The results, which are in exact agreement with those of Hepp and Lieb, are derived, as will be seen, in a much more straightforward manner than the method used by them. The phase-transition property of the Dicke model is shown, in our approach, to be closely analogous to that of the Curie–Weiss model of ferromagnetism given by the mean-field theory.[9] Other thermodynamic quantities such as those relevant to the study of the photon statistics of the system are also given readily by our method. The generalization to the case in which the field consists of finitely many radiation modes is straightforward.

In Sec. II, the Dicke model of superradiance is briefly reviewed. The free energy of the model is evaluated exactly and analyzed in detail in Sec. III. The generalization to the multimode case is presented in Sec. IV. Finally, a conclusion and a few remarks on other possible generalizations are given in Sec. V.

II. DICKE MODEL OF SUPERRADIANCE

First let us briefly review the Dicke model of superradiance. We consider a system of N identical two level atoms, coupled through dipole interactions with an electromagnetic field in a cavity of volume V. The atoms are kept at fixed position and the dimension V is much smaller than the wavelength of the field so that all atoms see exactly the same field.

The Hamiltonian of the model in the rotating-wave approximation[3,4] is ($\hbar = c = 1$)

$$H = H_0 + H_I \ , \tag{1}$$

where

$$H_0 = H_{\rm field} + H_{\rm atoms} = \sum_s \nu_s a_s^\dagger a_s + \tfrac{1}{2}\omega \sum_{j=1}^{N} \sigma_j^z \ , \tag{2}$$

$$H_I = \frac{1}{2\sqrt{V}} \left[\left(\sum_s \lambda_s' a_s\right)\left(\sum_{j=1}^{N} \sigma_j^+\right) + \left(\sum_s \lambda_s' a_s^\dagger\right)\left(\sum_{j=1}^{N} \sigma_j^-\right)\right] , \tag{3}$$

and $a_s^\dagger$ and a_s are creation and annihilation operators for the sth mode with frequency ν_s of the electromagnetic field, ω is the energy difference between the two levels of the atoms, and λ' measures the coupling between the field and the atoms. We have also introduced the Pauli spin matrices $\vec{\sigma}_j$'s to describe the two level atoms and

$$\sigma_j^+ = \sigma_j^x + i\sigma_j^y \ , \quad \sigma_j^- = \sigma_j^x - i\sigma_j^y \ . \tag{4}$$

In the thermodynamic limit that $N \to \infty$, $V \to \infty$ such that $N/V \equiv \rho =$ finite, we have

$$H_I = \frac{\sqrt{\rho}}{2\sqrt{N}} \left[\left(\sum_s \lambda_s' a_s\right)\left(\sum_{j=1}^{N} \sigma_j^+\right) + \left(\sum_s \lambda_s' a_s^\dagger\right)\left(\sum_{j=1}^{N} \sigma_j^-\right)\right] . \tag{5}$$

For a single radiation mode of frequency ν, it is convenient to measure the energy in units of the frequency ν, and write

$$H = a^\dagger a + \sum_{j=1}^{N} \left[\tfrac{1}{2}\epsilon \sigma_j^z + (\lambda/2\sqrt{N})(a\sigma_j^+ + a^\dagger \sigma_j^-)\right] , \tag{6}$$

with

$$\epsilon = \omega/\nu \ , \quad \lambda = \lambda' \sqrt{\rho}/\nu \ , \tag{7}$$

which is the Hamiltonian of the Dicke model.

III. THERMODYNAMIC PROPERTIES

The thermodynamic functions of the Dicke model described above can be calculated from the canonical partition function $Z(N, T)$:

$$Z(N, T) = {\rm Tr}\, e^{-\beta H} \ ; \quad \beta = 1/k_B T \ . \tag{8}$$

A convenient basis to calculate the trace of the partition function is the Glauber's coherent state[8] $|\alpha\rangle$, which has the following properties:

(i) $|\alpha\rangle$ is an eigenstate of the annihilation operator a,

$$a|\alpha\rangle = \alpha|\alpha\rangle \ ; \quad \langle\alpha|a^\dagger = \langle\alpha|\alpha^* \ . \tag{9}$$

(ii) The set of all $|\alpha\rangle$'s is complete, and

$$\int \frac{d^2\alpha}{\pi} |\alpha\rangle\langle\alpha| = 1 \ . \tag{10}$$

Using $\{|\alpha\rangle\}$ for the photon field, we have

$$Z(N, T) = \sum_{s_1=\pm 1} \cdots \sum_{s_N=\pm 1} \int \frac{d^2\alpha}{\pi} \times \langle s_1 \cdots s_N | \langle\alpha| e^{-\beta H} |\alpha\rangle | s_1 \cdots s_N\rangle \ , \tag{11}$$

where $\int d^2\alpha$ means $\iint d({\rm Re}\alpha)\, d({\rm Im}\alpha)$, and the sum is taken over all the atomic states.

We shall now obtain an explicit expression for the expectation value with respect to $|\alpha\rangle$ of $e^{-\beta H}$, which turns out to be very simple for the Dicke model. Let us first write the Hamiltonian in the form

$$H = \sum_{j=1}^{N} \left[\frac{a^\dagger}{\sqrt{N}} \frac{a}{\sqrt{N}} + \frac{\epsilon}{2}\sigma_j^z + \frac{\lambda}{2}\left(\frac{a^\dagger}{\sqrt{N}}\sigma_j^- + \frac{a}{\sqrt{N}}\sigma_j^+\right)\right] . \tag{12}$$

It should be noted that H is a function of N. In the following discussion we shall write H_N for H whenever we wish to emphasize the dependence of H on N. If we write $b = a/\sqrt{N}$ and $b^\dagger = a^\dagger/\sqrt{N}$, the commutation relation of the operators b and $b^\dagger$ is given by

$$[b, b^\dagger] = (1/N)[a, a^\dagger] = 1/N \ . \tag{12a}$$

Using Eq. (12a), we shall show that in the thermodynamic limit the field operators appearing in each term of the expansion of $e^{-\beta H}$ can be replaced by the same operators arranged in the so called antinormal order, in which all the a's occur to the right-hand side of $a^\dagger$'s. It follows then that the expectation value $\langle\alpha|e^{-\beta H}|\alpha\rangle$ becomes

$$\langle\alpha|e^{-\beta H}|\alpha\rangle = \exp\left\{-\beta\left[\alpha^*\alpha + \sum_{j=1}^{N}\left(\frac{\epsilon}{2}\sigma_j^z + \frac{\lambda}{2\sqrt{N}}(\alpha^*\sigma_j^- + \alpha\sigma_j^+)\right)\right]\right\} , \tag{12b}$$

for it is known that[8] the expectation value of any operator in the antinormal order such as $(a^\dagger)^r a^s$ is simply given by substituting α for a, and α^* for $a^\dagger$, namely,

$$\langle\alpha|(a^\dagger)^r a^s|\alpha\rangle = (\alpha^*)^r \alpha^s \ .$$

Our proof showing that the field operators in $e^{-\beta H_N}$ can be replaced by the same operators arranged in the antinormal order which we shall give in the following is based upon the following two assumptions.

Assumption i: The limits as $N \to \infty$ of the field operators $a/\sqrt{N}$ and $a^\dagger/\sqrt{N}$ exist.

Assumption ii: The order of the double limit in the exponential series

$$\lim_{N\to\infty} \lim_{R\to\infty} \sum_{r=0}^{R} (-\beta H_N)^r / r!$$

can be interchanged. We hope to provide a rigorous justification of the above assumptions in a future publication.

Consider first the limits of $a/\sqrt{N}$ and $a^\dagger/\sqrt{N}$ to be different from zero. Bearing in mind the two assumptions above, let us consider the exponential series

$$e^{-\beta H_N} = \sum_r (-\beta H_N)^r / r! \tag{12c}$$

and suppose that we take sufficiently many but finite number of terms from the right-hand side of (12c). The field operators appearing in a typical term is of the form

$$b^\dagger b b b^\dagger \cdots b b^\dagger ,$$

where the number of factors is at most of the order of the highest power of H in (12c). To see how it can be arranged into the antinormal order we give a simple example

$$b^\dagger b b^\dagger b b^\dagger b = b^\dagger b^\dagger b^\dagger b b b + (3/N) b^\dagger b^\dagger b b + (1/N^2) b^\dagger b .$$

Each interchange of b and $b^\dagger$ introduces an extra term with an extra factor $1/N$ due to the commutation relation (12a). The number of interchanges required to put all the field operators into the antinormal order is clearly finite. By taking N sufficiently large, all the terms with the extra factors $1/N$, $1/N^2$, etc., can be made to drop out. Thus, in effect, the field operators appearing in each term of the expansion of $e^{-\beta H}$ in any order can be replaced by the same operators arranged in the antinormal order. Therefore, (12b) follows.

On the other hand if b and $b^\dagger$ tend to zero in the limit $N \to \infty$, the Hamiltonian becomes simply

$$H = a^\dagger a + \sum_{j=1}^{N} \frac{\epsilon}{2} \sigma_j^z .$$

The free energy per atom $f(T)$ of the system is then given by

$$-\beta f(T) = \ln[2\cosh(\tfrac{1}{2}\beta\epsilon)] .$$

Thus, the field does not contribute to the free energy and the ordering of the field operators makes no difference to the calculation.

We now proceed from Eq. (12b) and define

$$h_j = (\tfrac{1}{2}\epsilon)\sigma_j^z + (\lambda/2\sqrt{N})(\alpha^* \sigma_j^- + \alpha \sigma_j^+) . \tag{13}$$

Using Eq. (12) and the property $[h_i, h_j] = 0$, we can reduce the integrand in (11) to

$$\begin{aligned}
&\langle s_1 \cdots s_N | \langle \alpha | e^{-\beta H} | \alpha \rangle | s_1 \cdots s_N \rangle \\
&= e^{-\beta|\alpha|^2} \langle s_1 \cdots s_N | \exp\Big(-\beta \sum_{j=1}^{N} h_j\Big) | s_1 \cdots s_N \rangle \\
&= e^{-\beta|\alpha|^2} \langle s_1 \cdots s_N | \prod_{j=1}^{N} e^{-\beta h_j} | s_1 \cdots s_N \rangle \\
&= e^{-\beta|\alpha|^2} \prod_{j=1}^{N} \langle s_j | e^{-\beta h_j} | s_j \rangle .
\end{aligned} \tag{14}$$

From (14) and (11), we have

$$\begin{aligned}
Z(N, T) &= \int \frac{d^2\alpha}{\pi} \sum_{s_1=\pm 1} \cdots \sum_{s_N=\pm 1} e^{-\beta|\alpha|^2} \Big(\prod_{j=1}^{N} \langle s_j | e^{-\beta h_j} | s_j \rangle \Big) \\
&= \int \frac{d^2\alpha}{\pi} e^{-\beta|\alpha|^2} (\langle +1 | e^{-\beta h} | +1 \rangle + \langle -1 | e^{-\beta h} | -1 \rangle)^N \\
&= \int \frac{d^2\alpha}{\pi} e^{-\beta|\alpha|^2} (\mathrm{Tr}\, e^{-\beta h})^N ,
\end{aligned} \tag{15}$$

where

$$h = (\tfrac{1}{2}\epsilon)\sigma^z + (\lambda/2\sqrt{N})(\alpha^*\sigma^- + \alpha\sigma^+) . \tag{16}$$

Writing in the matrix form, the operator h becomes

$$h = \begin{pmatrix} \tfrac{1}{2}\epsilon & \lambda\alpha/\sqrt{N} \\ \lambda\alpha^*/\sqrt{N} & -\tfrac{1}{2}\epsilon \end{pmatrix} , \tag{17}$$

whose eigenvalues μ's satisfy the secular equation

$$\begin{vmatrix} \tfrac{1}{2}\epsilon - \mu & \lambda\alpha/\sqrt{N} \\ \lambda\alpha^*/\sqrt{N} & -\tfrac{1}{2}\epsilon - \mu \end{vmatrix} = 0 \tag{18}$$

or

$$\mu = \pm(\tfrac{1}{2}\epsilon)(1 + 4\lambda^2|\alpha|^2/\epsilon^2 N)^{1/2} . \tag{19}$$

Hence substituting (19) into (15), we get

$$\begin{aligned}
Z(N, T) &= \int \frac{d^2\alpha}{\pi} e^{-\beta|\alpha|^2} (e^{-\beta|\mu|} + e^{\beta|\mu|})^N \\
&= \int \frac{d^2\alpha}{\pi} e^{-\beta|\alpha|^2} \\
&\quad \times \{2\cosh[(\tfrac{1}{2}\beta\epsilon)(1 + 4\lambda^2|\alpha|^2/\epsilon^2 N)^{1/2}]\}^N .
\end{aligned} \tag{20}$$

We note that the integrand in (20) is real and depends only upon $|\alpha|$ and the integral (20) converges as $|\alpha| \to \infty$. The free energy per atom $f(T)$ is obtained from $Z(N, T)$ by the usual formula

$$f(T) = \lim_{N\to\infty} [(1/\beta N) \ln Z(N, T)] . \tag{21}$$

To evaluate (20), we write

$$\int \frac{d^2\alpha}{\pi} = \int_0^\infty r\,dr \int_0^{2\pi} \frac{d\theta}{\pi} = 2\int_0^\infty r\,dr . \tag{22}$$

Equation (20) then becomes

$$\begin{aligned}
Z(N, T) &= 2\int_0^\infty r\,dr\, e^{-\beta r^2} \\
&\quad \times \{2\cosh[(\tfrac{1}{2}\beta\epsilon)(1 + 4\lambda^2 r^2/\epsilon^2 N)]\}^N .
\end{aligned} \tag{23}$$

Let $y = r^2/N$, and write

$$Z(N, T) = N\int_0^\infty dy\, e^{-N\beta y}\, 2\cosh\{(\tfrac{1}{2}\beta\epsilon)[1 + (4\lambda^2/\epsilon^2)y]^{1/2}\} \tag{24}$$

$$= N\int_0^\infty dy \exp\{N[-\beta y + \ln(2\cosh\{(\tfrac{1}{2}\beta\epsilon)[1+(4\lambda^2/\epsilon^2)y]^{1/2}\})]\} . \quad (25)$$

By Laplace's method,[10] integral (25) is given by

$$Z(N, T) = N\frac{C}{\sqrt{N}} \max_{0\le y\le\infty} \exp\{N[-\beta y + \ln(2\cosh\{(\tfrac{1}{2}\beta\epsilon)[1+(4\lambda^2/\epsilon^2)y]^{1/2}\})]\} . \quad (26)$$

Let

$$\phi(y) = -\beta y + \ln(2\cosh\{(\tfrac{1}{2}\beta\epsilon)[1+(4\lambda^2/\epsilon^2)y]^{1/2}\}), \quad (27)$$

then

$$\phi'(y) = -\beta + (\beta\lambda^2/\epsilon)[1+(4\lambda^2/\epsilon^2)y]^{-1/2} \times \tanh\{(\tfrac{1}{2}\beta\epsilon)[1+(4\lambda^2/\epsilon^2)y]^{1/2}\} . \quad (28)$$

Putting $\phi'(y) = 0$, we get

$$(\epsilon/\lambda^2)\eta = \tanh[(\tfrac{1}{2}\beta\epsilon)\eta] , \quad (29)$$

where

$$\eta = [1+(4\lambda^2/\epsilon^2)y]^{1/2} , \quad 1 \le \eta < \infty . \quad (30)$$

We note that in the region $0 \le x < \infty$, the function $\tanh x$ is a monotonically increasing function of x as x increases and that its slope is a monotonically decreasing function of x and $\tanh x < 1$ for $x < \infty$. For $\lambda^2 < \epsilon$, Eq. (29) has no solution in the region $1 \le \eta < \infty$. Thus, for $\lambda^2 < \epsilon$, it is easy to see that the function

$$e^{-\beta y}\, 2\cosh\{(\tfrac{1}{2}\beta\epsilon)[1+(4\lambda^2/\epsilon^2)y]^{1/2}\}$$

is a monotonically decreasing function of y with the maximum of the function at $y = 0$ equal to $2\cosh\frac{1}{2}\beta\epsilon$. Thus, the free energy per particle $f(T)$ is given by

$$-\beta f(T) = \lim_{N\to\infty} (1/N)\ln Z(N, T) = \ln[2\cosh(\tfrac{1}{2}\beta\epsilon)] . \quad (31)$$

For $\lambda^2 > \epsilon$, the solution of Eq. (29) depends on the value of β. For $\beta < \beta_c$, where β_c is given by

$$(\epsilon/\lambda^2) = \tanh(\tfrac{1}{2}\beta_c\epsilon) , \quad (32)$$

Eq. (29) has no solution in the region $1 \le \eta < \infty$ and thus

$$-\beta f(T) = \ln[2\cosh(\tfrac{1}{2}\beta\epsilon)] \quad (33)$$

as in Eq. (31). For $\beta > \beta_c$, Eq. (29) has one (and only one) solution in the region $1 \le \eta < \infty$ given by

$$(\epsilon/\lambda^2)\eta_0 = \tanh[(\tfrac{1}{2}\beta\epsilon)\eta_0] . \quad (34)$$

Writing $(\epsilon/\lambda^2)\eta_0 = 2\sigma$ or $(\epsilon/\lambda^2)[1+(4\lambda^2/\epsilon^2)y_0]^{1/2} = 2\sigma$, we have

$$y_0 = \lambda^2\sigma^2 - \epsilon^2/4\lambda^2$$

and

$$-\beta f(T) = \ln(\exp(-\beta y_0)\, 2\cosh\{(\tfrac{1}{2}\beta\epsilon)[1+(4\lambda^2/\epsilon^2)y_0]^{1/2}\}) \quad (35)$$

or

$$-\beta f(T) = \ln[2\cosh(\beta\lambda^2\sigma)] - \beta\lambda^2\sigma^2 + \beta\epsilon^2/4\lambda^2 , \quad (36)$$

where

$$2\sigma = \tanh(\beta\lambda^2\sigma) \ne 0 .$$

Thus, to summarize, we have, for (i), $\lambda^2 < \epsilon$, no phase transition occurs in the system at any temperature. For (ii), $\lambda^2 > \epsilon$, there is a critical temperature T_c given by

$$\epsilon/\lambda^2 = \tanh(\tfrac{1}{2}\beta_c\epsilon) , \quad \beta_c = 1/kT_c$$

at which the system changes discontinuously from one state to another. The average number of photons in the two different states (the states at $T > T_c$ being called the normal state and the state $T < T_c$ being called the superradiant state) can be easily computed. More generally, let us consider the quantities $\langle (a^\dagger a/N)^r \rangle$ defined by

$$\left\langle \left(\frac{a^\dagger a}{N}\right)^r \right\rangle = \frac{\mathrm{Tr}(a^\dagger a/N)^r e^{-\beta H}}{\mathrm{Tr}\, e^{-\beta H}} . \quad (37)$$

It follows from our formulation that $\langle (a^\dagger a/N)^r \rangle$ is given by

$$\left\langle \left(\frac{a^\dagger a}{N}\right)^r \right\rangle = \int_0^\infty y^r dy \exp\{N[-\beta y + \ln(2\cosh\{(\tfrac{1}{2}\beta\epsilon)[1+(4\lambda^2/\epsilon^2)y]^{1/2}\})]\}$$
$$\times \left(\int_0^\infty dy \exp\{N[-\beta y + \ln(2\cosh\{(\tfrac{1}{2}\beta\epsilon)[1+(4\lambda^2/\epsilon^2)y]^{1/2}\})]\}\right)^{-1} . \quad (38)$$

By Laplace's method, it immediately follows that for (i) $\lambda^2 < \epsilon$ and (ii) $\lambda^2 > \epsilon$, $\beta < \beta_c$

$$\langle (a^\dagger a/N)^r \rangle = \delta_{r,0} . \quad (39)$$

For (iii), $\lambda^2 > \epsilon$, $\beta > \beta_c$, however, $\langle (a^\dagger a/N)^r \rangle$ is given by

$$\langle (a^\dagger a/N)^r \rangle = y_0^r = (\lambda^2\sigma^2 - \epsilon^2/4\lambda^2)^r , \quad (40)$$

where σ is the root of the equation

$$2\sigma = \tanh(\beta\lambda^2\sigma) \neq 0 \ . \tag{41}$$

The above results, which are in exact agreement with those of Hepp and Lieb, are derived, as we have seen, in a manner which is straightforward and elementary compared to the method used by Hepp and Lieb. The close analogy between the Dicke model of superradiance and the Curie-Weiss model of ferromagnetism[7] is also more apparent from our formulation.

In Sec. IV, we shall deal with the case of finitely many radiation modes.

IV. GENERALIZATION TO MULTIMODE CASE

The Hamiltonian of the Dicke model, in the case of m radiation modes of frequencies $\nu_1, \nu_2, \ldots, \nu_m$, is given by

$$H = \sum_{s=1}^{m} \nu_s a_s^\dagger a_s + \sum_{i=1}^{N} \left\{ \frac{\omega}{2}\sigma_i^z + \frac{1}{2\sqrt{N}} \times \left[\left(\sum_{s=1}^{m} \lambda_s a_s \right) \sigma_i^+ + \left(\sum_{s=1}^{m} \lambda_s a_s^\dagger \right) \sigma_i^- \right] \right\}, \tag{42}$$

where $\lambda_1, \lambda_2, \ldots, \lambda_m$ are the coupling constants. In analogy with Eq. (20), the partition function of the system is now given by

$$Z = \int \cdots \int \frac{d^2\alpha_1}{\pi} \frac{d^2\alpha_2}{\pi} \cdots \frac{d^2\alpha_m}{\pi} \exp(-\beta | \nu_1\alpha_1^2 + \nu_2\alpha_2^2 + \cdots + \nu_m\alpha_m^2 |) \times \left(2\cosh\{(\tfrac{1}{2}\beta\omega)[1 + (4/\omega^2 N) | \lambda_1\alpha_1 + \lambda_2\alpha_2 + \cdots + \lambda_m\alpha_m |^2]^{1/2}\} \right)^N , \tag{43}$$

where

$$|\nu_1\alpha_1^2 + \nu_2\alpha_2^2 + \cdots + \nu_m\alpha_m^2| = (\nu_1 x_1^2 + \nu_2 x_2^2 + \cdots + \nu_m x_m^2) + (\nu_1 y_1^2 + \nu_2 y_2^2 + \cdots + \nu_m y_m^2) ,$$
$$|\lambda_1\alpha_1 + \lambda_2\alpha_2 + \cdots + \lambda_m\alpha_m|^2 = (\lambda_1 x_1 + \lambda_2 x_2 + \cdots + \lambda_m x_m)^2 + (\lambda_1 y_1 + \lambda_2 y_2 + \cdots + \lambda_m y_m)^2 , \tag{44}$$
$$d^2\alpha_i = dx_i \, dy_i \ .$$

We now define a new set of variables $x^{(1)}, x^{(2)}, \ldots, x^{(m)}$ in place of the original variables $x_1, x_2, \ldots, x_m$ by the following relations:

$$\lambda x^{(1)} = \lambda_{11}x_1 + \lambda_{12}x_2 + \cdots + \lambda_{1m}x_m ,$$
$$\lambda x^{(2)} = \lambda_{21}x_1 + \lambda_{22}x_2 + \cdots + \lambda_{2m}x_m ,$$
$$\cdots \tag{45}$$
$$\cdots$$
$$\lambda x^{(m)} = \lambda_{m1}x_1 + \lambda_{m2}x_2 + \cdots + \lambda_{mm}x_m .$$

A new set of y variables is also defined by the same set of relations. The elements in the first row of the determinant

$$\Delta = \begin{vmatrix} \lambda_{11} & \lambda_{12} & \cdots \lambda_{1m} \\ \lambda_{21} & \lambda_{22} & \cdots \lambda_{2m} \\ \lambda_{m1} & \lambda_{m2} & \cdots \lambda_{mm} \end{vmatrix} \tag{46}$$

are chosen to be equal to the respective coupling constants $\lambda_1, \lambda_2, \ldots, \lambda_m$ so that the expression $|\lambda_1\alpha_1 + \cdots + \lambda_m\alpha_m|^2$ in Eq. (43) reduces simply to $\lambda^2(x^{(1)2} + y^{(1)2})$.

We now wish to choose the remaining elements λ_{ij} in such a way that when $\nu_1 x_1^2 + \nu_2 x_2^2 + \cdots + \nu_m x_m^2$ is expressed in terms of the new set of variables $x^{(1)}, x^{(2)}, \ldots, x^{(m)}$, the coefficients of the cross-product terms $x^{(1)}x^{(j)}$, $j = 2, 3, \ldots, m$ vanish.

Consider the solutions of (45). We have

$$x_1 = \frac{1}{\Delta} \begin{vmatrix} \lambda x^{(1)} & \lambda_{12} \cdots \lambda_{1m} \\ \lambda x^{(2)} & \lambda_{22} \cdots \lambda_{2m} \\ \lambda x^{(m)} & \lambda_{m2} \cdots \lambda_{mm} \end{vmatrix} ,$$
$$x_2 = \frac{-1}{\Delta} \begin{vmatrix} \lambda_{11} & \lambda x^{(1)} \cdots \lambda_{1m} \\ \lambda_{21} & \lambda x^{(2)} \cdots \lambda_{2m} \\ \lambda_{m1} & \lambda x^{(m)} \cdots \lambda_{mm} \end{vmatrix} , \text{ etc.} \tag{47}$$

If we substitute (47) into $\nu_1 x_1^2 + \nu_2 x_2^2 + \cdots + \nu_m x_m^2$, it is easy to see that the coefficient of $x^{(1)}x^{(j)}$, $j \neq 1$ is given by

$$(2/\Delta^2)(\nu_1 A_{11}A_{j1} + \nu_2 A_{12}A_{j2} + \cdots + \nu_m A_{1m}A_{jm}) , \tag{48}$$

where A_{kl} is the signed cofactor of λ_{kl} in the determinant Δ. If we now choose

$$A_{1j} = c\lambda_{1j} \prod_{\substack{k=1 \\ k\neq j}}^{m} \nu_k , \quad j = 1, 2, \ldots, m \tag{49}$$

where c is any arbitrary constant, then (48) becomes

$$(2/\Delta^2)\nu_1\nu_2 \cdots \nu_m c(\lambda_{11}A_{j1} + \lambda_{12}A_{j2} + \cdots + \lambda_{1m}A_{jm}) = 0 , \tag{50}$$

because the quantity inside the bracket vanishes for $j \neq 1$. It is interesting to note that the substitution defined by (49) is all that is needed for our purpose and it is not necessary for us to determine the individual λ_{ij}, because the new variables $x^{(2)}$,

$x^{(3)}, \ldots, x^{(m)}$ and $y^{(2)}, y^{(3)}, \ldots, y^{(m)}$ occurring only in the exponential factor in (43) can now be integrated independently of the variables $x^{(1)}$ and $y^{(1)}$, giving rise to a constant which is of no physical interest. Using the substitution (49), $\nu_1 x_1^2 + \nu_2 x_2^2 + \cdots + \nu_m x_m^2$ becomes

$$\nu_1 x_1^2 + \nu_2 x_2^2 + \cdots + \nu_m x_m^2 = (1/\Delta^2)(\nu_1 A_{11}^2 + \nu_2 A_{22}^2 + \cdots + \nu_m A_{1m}^2)\lambda^2 x^{(1)2} + \text{terms involving } x^{(i)2}, x^{(i)}x^{(j)}, \quad i,j \neq 1. \tag{51}$$

We have similar expression for $\nu_1 y_1^2 + \nu_2 y_2^2 + \cdots$.

The coefficient of $x^{(1)2}$ is

$$\frac{\lambda^2}{\Delta^2}\,\nu_1 A_{11}^2 + \nu_2 A_{12}^2 + \cdots + \nu_m A_{1m}^2 = \lambda^2 \frac{\nu_1\nu_2\cdots\nu_m(\lambda_1^2\nu_2\nu_3\cdots\nu_m + \lambda_2^2\nu_1\nu_3\cdots\nu_m + \cdots)}{(\lambda_1^2\nu_2\nu_3\cdots\nu_m + \lambda_2^2\nu_1\nu_3\cdots\nu_m + \cdots)^2}$$

$$= \lambda^2\left(\frac{\lambda_1^2}{\nu_1} + \frac{\lambda_2^2}{\nu_2} + \cdots + \frac{\lambda_m^2}{\nu_m}\right)^{-1}. \tag{52}$$

Thus if we define

$$\lambda^2 = \lambda_1^2/\nu_1 + \lambda_2^2/\nu_2 + \cdots + \lambda_m^2/\nu_m, \tag{53}$$

then

$$Z = C \int\int \frac{dx^{(1)}}{\sqrt{\pi}}\frac{dy^{(1)}}{\sqrt{\pi}} \exp[-\beta(x^{(1)2} + y^{(1)2})]\left(2\cosh\{(\tfrac{1}{2}\beta\omega)[1 + (4\lambda^2/\omega^2 N)(x^{(1)2} + y^{(1)2})]^{1/2}\}\right)^N$$

$$= C \int \frac{d^2\alpha}{\pi} e^{-\beta|\alpha|^2}\left(2\cosh\{(\tfrac{1}{2}\beta\epsilon)[1 + (4\lambda^2/\omega^2 N)|\alpha|^2]^{1/2}\}\right)^N \quad \text{if we let } \alpha = (x^{(1)}, y^{(1)}), \tag{54}$$

where C is a constant given by the integrals over $dx^{(i)}dy^{(i)}$, $i = 2, \ldots, m$, times the Jacobian of the variable transformation. The constant C is of no physical importance as long as m is finite, for the free energy per atom $f(T)$ is given by

$$f(T) = \lim_{N\to\infty}[-(1/\beta N)\ln Z_1], \tag{55}$$

where

$$Z_1 = \int \frac{d^2\alpha}{\pi} e^{-\beta|\alpha|^2} \times \left(2\cosh\{(\tfrac{1}{2}\beta\epsilon)[1 + (4\lambda^2/\omega^2 N)|\alpha|^2]^{1/2}\}\right)^N. \tag{56}$$

We have reduced the multimode case to the single-mode case [with λ^2 in Eq. (56) given by Eq. (53)] for which the analysis of Sec. III applies.

The generalization to the multimode case was briefly mentioned by Hepp and Lieb in their paper but no explicit formula was given by them.

V. CONCLUSION

In conclusion, we have outlined an alternative approach to the thermodynamics of the Dicke model of superradiance which is considerably simpler than the formulation of Hepp and Lieb. In our approach, the phase transition is presented mathematically in an elementary manner. We have also used the same approach to analyze the phase transition property of the Dicke model in the case of many radiation modes by reducing the multimode case to the equivalent single-mode case. Some further generalizations are under consideration: (i) to relax the very unrealistic restriction of the model that the dimension of the cavity is much smaller than the wavelength of the electromagnetic field even in the thermodynamic limit, and (ii) to generalize the model to include the effects due to the motions of the atoms. These and others will be published in a later paper.

ACKNOWLEDGMENTS

We are most grateful to Professor H. M. Nussenzveig for showing us a preprint of Hepp and Lieb's paper as well as a preprint of his review paper which provided the stimulus for the work reported in this paper, and for his many helpful comments of the manuscript. We are grateful to Professor Elliott Montroll for his interest and support.

*Research partially supported by ARPA and monitored by ONR under Contract No. N00014-67-A-0398-0005.

[1] R. H. Dicke, Phys. Rev. 93, 99 (1954).
[2] M. Tavis and F. W. Cummings, Phys. Rev. **170**, 379 (1968).
[3] M. D. Scully and W. E. Lamb, Phys. Rev. 159, A208 (1967).
[4] H. M. Nussenzveig (unpublished).
[5] R. Graham and H. Haken, Z. Phys. **237**, 31 (1970).
[6] V. DeGiorgio and M. Scully, Phys. Rev. A **2**, 1170 (1970).
[7] K. Hepp and E. H. Lieb, Ann. Phys. (N.Y.) (to be published).
[8] R. Glauber, Phys. Rev. **131**, 2766 (1963).
[9] See, for example, M. Kac, in *Fundamental Problems in Statistical Mechanics II*, edited by E. G. D. Cohen (North-Holland, Amsterdam, 1968), p. 71.
[10] See, for example, H. Jeffreys and B. S. Jeffreys, *Methods of Mathematical Physics* (Cambridge U.P., Cambridge, 1966), p. 503.

PHYSICAL REVIEW A VOLUME 8, NUMBER 5 NOVEMBER 1973

Equilibrium Statistical Mechanics of Matter Interacting with the Quantized Radiation Field

Klaus Hepp
Department of Physics, Eidgenössische Technische Hochschule, CH-8049 Zürich, Switzerland

Elliott H. Lieb*
Institut des Hautes Etudes Scientifiques, 91440-Bures-sur-Yvette, France

(Received 7 June 1973)

The thermodynamic properties of several systems of multilevel atoms interacting with a quantized radiation field are investigated. We allow a quantum-mechanical treatment of the translational degrees of freedom and do not require the rotating-wave approximation. In the finite-photon-mode case one can calculate the free energy per atom in the thermodynamic limit exactly and rigorously. In the infinite-mode case we only get upper and lower bounds, but these are sufficient to give conditions for thermodynamic stability and instability. The kind of phase transition previously found by us for the one-mode Dicke model with the rotating-wave approximation persists in the general multimode case.

I. INTRODUCTION

In a previous paper[1] we elucidated the thermodynamic properties of the Dicke maser model with one photon mode coupled to N two-level atoms in the rotating-wave approximation. We calculated the free energy per atom and the thermal expectation values of time-dependent intensive and fluctuation observables exactly in the limit $N\to\infty$. Our analysis relied heavily on the various conservation laws inherent in the model: the total "number" operator and the total atomic "spin." In the relevant part of the spectrum we effectively diagonalized the Hamiltonian in such a way that the error became negligible as $N\to\infty$. While such a procedure is necessary if one wants to study fluctuation observables, it is intuitively clear that a simpler mean field method must exist if one is content to analyze only the intensive observables. In fact our results showed that these quantities behave classically; that is to say the lack of commutativity becomes unimportant as $N\to\infty$. It is also clear that these qualitative features should continue to hold even without the above mentioned conservation laws, i.e., it should be possible to handle a much wider class of models. In this paper we show how to achieve that goal in a simple way by using coherent states and the results of a recent paper.[2] The phase transition found in Ref. 1 persists in the general multimode case.

Wang and Hioe[3] and Hioe[4] demonstrated that the Glauber coherent states of the photon field constitute a natural basis for this problem. Unfortunately, they were not able to show that their approximate evaluation of $N^{-1}\ln \mathrm{Tr}e^{-\beta H}$ is exact when $N\to\infty$. Nevertheless, it is a fact that their answer agreed precisely with our previous rigorous result[1] for the Dicke model, and we shall prove here that their results for the other models they treat are also correct. In addition, we also calculate exactly the expectation values of intensive observables for those models. Ginibre[5] has previously made use of coherent states to obtain exact results for the many-boson problem. Our methods can be easily generalized for atoms of more than two levels.

In Sec. III we make an attempt to understand the thermodynamics of the infinite-mode case which, to the best of our knowledge, was never discussed before. We cannot solve that problem exactly, but we can derive upper and lower bounds to the partition function which prove that such a system is thermodynamically stable if and only if the atoms have very repulsive cores.

In Sec. IV we return to the finite-mode Hamiltonian, but allow translational degrees of freedom (either quantum or classical) for the atoms. It is shown there that use of the Glauber coherent photon states permits an exact (as $N\to\infty$) reduction of the problem to a conventional many body problem. This, in turn, can be solved in closed form in several interesting cases.

We do not confine our attention only to the Glauber coherent photon states, but also show that the Bloch atomic coherent states[6] for the atoms may be used to advantage. The former make the radiation field classical, while the latter make the atomic variables classical. It is the Bloch states that are used in our discussion of the infinite-mode case.

In the Appendix we resolve some technical problems related to the unbounded nature of the photon operators.

II. FINITE-MODE MODELS

The Hamiltonians we shall consider are those of the form

$$H=\sum_{m=1}^{M}\nu_m a_m^* a_m+\epsilon S^z+N^{-1/2}\sum_{m=1}^{M}\sum_{n=1}^{N}[a_m(\lambda_{mn}S_n^+ +\mu_{mn}S_n^-)+\mathrm{H.c.}], \qquad (2.1)$$

where $a_1^\#,\ldots,a_M^\#$ are boson creation and annihilation operators for the photon modes having energies $\nu_m>0, \vec{S}_1,\ldots,\vec{S}_N$ are spin-$\frac{1}{2}$ operators describing N two-level atoms having energy spacing $\epsilon>0$ ($S_n^+=S_n^x+iS_n^y$ and $S^i=\sum_{n=1}^N S_n^i$, $i=x,y$ or z). The λ_{mn} (respectively, μ_{mn}) are coupling constants for the rotating (respectively, counter-rotating) wave terms. The factor $N^{-1/2}$ in (2.1) really comes from a factor $V^{-1/2}$, where V is the volume of the cavity containing the atoms. However, as we are interested in N/V fixed, the distinction merely entails a coupling constant renormalization. The interaction term in (2.1) is linear in the $a^\#$ operators, but the method we are about to describe would work equally well if one included quadratic terms (i.e., if one goes beyond the dipolar approximation).

A. Atomic Coherent State Representation

We turn now to the atomic or Bloch angular momentum coherent states[6] used in Ref. 2. With $\mathfrak{F}$ being the Fock space for all the photon modes and $\mathfrak{H}$ being the Hilbert space for the N atoms (spins) we define

$$Z=\mathrm{Tr}e^{-\beta H} \qquad (2.2)$$

where the trace is over $\mathfrak{F}\otimes\mathfrak{H}$. The bounds obtained in Ref. 2, and justified in the Appendix for boson operators, are

$$\tilde{Z}(\tfrac{1}{2})\le Z\le\tilde{Z}(\tfrac{3}{2}), \qquad (2.3)$$

where

$$\tilde{Z}(J)=2^N\,\mathrm{Tr}_{\mathfrak{F}}(4\pi)^{-N}\int d\Omega^N e^{-\beta\tilde{H}(J,\Omega^N)} \qquad (2.4)$$

and where

$$\Omega^N=(\vec{\Omega}_1,\ldots,\vec{\Omega}_N),$$
$$\vec{\Omega}_n=(\sin\theta_n\cos\varphi_n,\sin\theta_n\sin\varphi_n,\cos\theta_n). \qquad (2.5)$$

$\tilde{H}(J,\Omega^N)$ is defined from (2.1) by replacing each $\vec{S}_n$ by $J\vec{\Omega}_n$. The bounds in (2.3) for the free energy per atom do not agree as $N\to\infty$, but they are useful in proving the stability of the infinite photon mode free energy—a subject to which we shall return in Sec. III.

Further progress can be made with the Bloch-state bounds, however, if we make an assumption about H, namely that the λ_{mn} and the μ_{mn} are independent of n. This is true in the original Dicke model where one neglects the spatial variation of the radiation field. (Alternatively, the method will work if the atoms belong to a finite number of groups in each of which the coupling constants are the same.) In this case we can use the fact that the total "spin," J, is a constant of the motion, so that

$$Z=\sum_{J\ge 0}^{N/2}Y(N,J)Z(J), \qquad (2.6)$$

$$Z(J)=\mathrm{Tr}e^{-\beta H(J)}, \qquad (2.7)$$

where $H(J)$ is the Hamiltonian for a spin, $\vec{S}$, of magnitude J,

$$H(J)=\sum_{m=1}^{M}\nu_m a_m^* a_m+\epsilon S^z+N^{-1/2}\sum_{m=1}^{M}[a_m(\lambda_m S^+ +\mu_m S^-)+\mathrm{H.c.}] \qquad (2.8)$$

and

$$Y(N,J)=N!\,(2J+1)[(J+1+\tfrac{1}{2}N)!\,(\tfrac{1}{2}N-J)!]^{-1} \qquad (2.9)$$

is the number of ways to construct an angular momentum J from N spin-$\frac{1}{2}$ particles.

Using Ref. 2, one has the bounds

$$\tilde{Z}(J)\le(2J+1)^{-1}Z(J)\le\tilde{Z}(J+1), \qquad (2.10)$$

with

$$\tilde{Z}(J)=(4\pi)^{-1}\mathrm{Tr}_{\mathfrak{F}}\int d\Omega e^{-\beta\tilde{H}(J,\Omega)} \qquad (2.11)$$

and $\tilde{H}(J,\Omega)$ is (2.8) with S replaced by $J\vec{\Omega}$. Since $\tilde{H}$ is quadratic in the $a_m^\#$, it is easy to evaluate $\mathrm{Tr}_{\mathfrak{F}}$ in (2.11):

$$\tilde{Z}(J)=(4\pi)^{-1}\prod_{m=1}^{M}(1-e^{-\beta\nu_m})^{-1}\int_0^\pi\sin\theta\,d\theta\int_0^{2\pi}d\varphi\,\exp\left(-\beta\epsilon J\cos\theta+\beta N^{-1}J^2\sin^2\theta\sum_{m=1}^{M}|\lambda_m e^{i\varphi}+\mu_m e^{-i\varphi}|^2\nu_m^{-1}\right). \qquad (2.12)$$

Since $J^2\le\frac{1}{4}N^2$, we see that

$$\ln\tilde{Z}(J+1)-\ln\tilde{Z}(J)\le\beta\epsilon+\beta[(N+1)/N]\sum_{m=1}^{M}(|\lambda_m|+|\mu_m|)^2\nu_m^{-1}, \qquad (2.13)$$

which means that (2.12) gives the free energy per atom exactly to order M/N. It is also clear that for large N and fixed M, (2.12) and (2.6) can be evaluated by steepest descent, i.e., simply by maximizing with respect to θ, φ, and J. If we first maximize with respect to φ we see that *the free energy is the same as that for one mode in the rotating-wave approximation,* but with an effective coupling constant λ given by

$$\lambda^2 = \max_\varphi \sum_{m=1}^{M} |\lambda_m e^{i\varphi} + \mu_m e^{-i\varphi}|^2 \nu_m^{-1}. \tag{2.14}$$

Subsequent maximization with respect to θ and J yields the free energy found in Ref. 1. Such a result agrees with that found in Ref. 3.

Expectation values of intensive observables can be found by the foregoing procedure, but some technicalities are required. These will be discussed in the Appendix. The result is as expected: The expectation value of any polynomial in the intensive observables (defined in the Appendix) is equal to that polynomial evaluated at the maximal steepest descent points referred to above, and then averaged over the variety of maximal steepest descent points if there is more than one point.

Before turning to the photon coherent states, which we shall describe next, we remark that when there is no spatial dependence (λ_{mn} and μ_{mn} independent of n) the Bloch picture (2.12) is the more convenient to use. The main conclusion, (2.14), that even without the rotating-wave approximation there is effectively only one coupling constant, is more tedious to derive using the photon coherent states (cf. Refs. 3 and 4).

B. Photon Coherent-State Representation

For a given photon mode, $a^\#$, a Glauber coherent state is

$$|\alpha\rangle = \exp[-\tfrac{1}{2}|\alpha|^2 + \alpha a^*]|0\rangle, \tag{2.15}$$

where α is any complex number and $|0\rangle$ is the photon vacuum. The following properties are required and are justified in the Appendix:

$$\langle\alpha|\beta\rangle = \exp[-\tfrac{1}{2}|\alpha|^2 - \tfrac{1}{2}|\beta|^2 + \alpha^*\beta], \tag{2.16}$$

$$\mathrm{Tr}|\alpha\rangle\langle\beta| = \langle\beta|\alpha\rangle, \tag{2.17}$$

$$\mathrm{Tr}A = \pi^{-1}\int d\alpha\langle\alpha|A|\alpha\rangle, \tag{2.18}$$

where A is any trace-class operator and $\int d\alpha = \iint_{-\infty}^{\infty} d(\mathrm{Re}\,\alpha)d(\mathrm{Im}\,\alpha)$,

$$\langle\alpha|a|\alpha\rangle = \alpha, \tag{2.19}$$

$$\langle\alpha|a^*a|\alpha\rangle = |\alpha|^2, \tag{2.20}$$

$$I = \pi^{-1}\int d\alpha|\alpha\rangle\langle\alpha|, \tag{2.21}$$

$$a = \pi^{-1}\int d\alpha\alpha|\alpha\rangle\langle\alpha|, \tag{2.22}$$

$$a^*a = \pi^{-1}\int d\alpha(|\alpha|^2 - 1)|\alpha\rangle\langle\alpha|. \tag{2.23}$$

The same methods as in Ref. 2, but using the Glauber states instead of the Bloch states, yield upper and lower bounds for the partition function (again, see the Appendix):

$$\tilde{Z} \le Z \le \exp\Big(\beta\sum_{m=1}^{M}\nu_m\Big)\tilde{Z}, \tag{2.24}$$

where

$$\tilde{Z} = \pi^{-M}\,\mathrm{Tr}_{\mathcal{K}}\int d\alpha^M e^{-\beta\tilde{H}(\alpha^M)}, \tag{2.25}$$

and

$$\alpha^M = (\alpha_1, \ldots, \alpha_M),$$

$$\tilde{H}(\alpha^M) = \sum_{m=1}^{M}\nu_m|\alpha_m|^2 + \epsilon S^z + N^{-1/2}\times\sum_{m=1}^{M}\sum_{n=1}^{N}[\alpha_m(\lambda_{mn}S_n^+ + \mu_{mn}S_n^-) + \mathrm{H.c.}]. \tag{2.26}$$

The trace on $\mathcal{K}$ in (2.25) is now easy to compute:

$$\tilde{Z} = \pi^{-M}\int d\alpha^M \exp\Big(-\beta\sum_{m=1}^{M}\nu_m|\alpha_m|^2\Big)2^N\prod_{n=1}^{N}\cosh\times\Big[\tfrac{1}{2}\beta\Big(\epsilon^2 + 4N^{-1}\Big|\sum_{m=1}^{M}(\alpha_m\lambda_{mn} + \alpha_m^*\mu_{mn}^*)\Big|^2\Big)^{1/2}\Big]. \tag{2.27}$$

This $\tilde{Z}$ is precisely the approximate partition function derived by Wang and Hioe.[3,4] We note from (2.24) that, as long as the number of photon modes is finite, $-\beta N^{-1}\ln\tilde{Z}$ is the free energy per atom to order MN^{-1}. Another remark is that although we assumed each atom to have only two levels, the atomic trace in (2.26) can just as well be evaluated for multilevel atoms, and the interaction does not have to be linear in the atomic $S^\pm$ operators.

III. THERMODYNAMIC STABILITY OF THE INFINITE-MODE SYSTEM

In this section we return to the Hamiltonian H of (2.1) and shall show that under suitable conditions it is stable when the number of photon modes is infinite. This means that there exists a constant A such that $Z = \mathrm{Tr}e^{-\beta H} < e^{NA}$ for all N. The proof of the existence of the thermodynamic limit, $\lim_{N\to\infty}N^{-1}\ln Z$, is a more complicated question with which we shall not deal. The phrase infinite number of modes means that in any frequency interval (ν_1, ν_2) the number of modes with $\nu_1 < \nu < \nu_2$

is proportional to N, the size of the cavity. The problem we are considering here is similar to the corresponding problem for electron-phonon interactions dealt with by Gallavotti, Ginibre, and Velo.[7]

We use the bounds (2.3) and the definition (2.4). The trace over the photon modes $\mathfrak{F}$ can easily be done exactly and one finds, dropping irrelevant factors of the form $e^{\mathrm{const}\,N}$,

$$\tilde{Z}(J) \sim R_N \int d\Omega^N \exp\left(\beta J^2 \sum_{n=1}^{N} \sum_{j=1}^{N} v(\vec{\Omega}_n, \vec{\Omega}_j) - \beta\epsilon J \sum_{n=1}^{N} \cos\theta_n\right), \tag{3.1}$$

where

$$v(\vec{\Omega}_n, \vec{\Omega}_j) = N^{-1} \sum_m \nu_m^{-1} \mathrm{Re}(\lambda_{mn} S_n^+ + \mu_{mn} S_n^-) \times (\lambda_{mj}^* S_j^- + \mu_{mj}^* S_j^+), \tag{3.2}$$

$$S_n^{\pm} = \sin\theta_n e^{\pm i\varphi_n}, \tag{3.3}$$

and R_N is the ideal photon contribution to Z

$$R_N = \prod_m (1 - e^{-\beta\nu_m})^{-1}. \tag{3.4}$$

The latter is independent of the interaction and may be assumed to have a nice thermodynamic limit in a physically sensible model, e.g.,

$$\nu_m \to \nu(\vec{k}) = |\vec{k}|, \tag{3.5}$$

$\vec{k} = 2\pi N^{-1}(n_x, n_y, n_z), n_i \in Z, \vec{k} \neq 0$. We could allow several photon modes for each $\vec{k}$; such a generalization does not affect the conclusions below.

If (3.2) be inserted into (3.1), one has (dropping R_N) precisely the partition function of N classical spins of magnitude J interacting with each other via a quadratic interaction and with an external magnetic field ϵ in the z direction. This is a well understood problem. If one allows all the particles to be at the same location, i.e., λ_{mn} and μ_{mn} independent of n, then $v(\vec{\Omega}_n, \vec{\Omega}_j)$ is positive when $\vec{\Omega}_n \approx \vec{\Omega}_j$, and no matter what one assumes about the various parameters, the *lower* bound $\tilde{Z}(\frac{1}{2})$ will behave like e^{BN^2} with $B > 0$. Thus, *no stability is possible in this case* and, as we shall show later, the introduction of the quantum mechanical uncertainty principle does not save the situation.

To describe (3.2) more concretely, we use (3.5) *et seq.*; we associate a position $\vec{x}_n$ with each atomic index n, and suppose that

$$\lambda_{mn} \to \lambda(\vec{k}) e^{i\vec{k}\cdot\vec{x}_n}, \tag{3.6}$$

$$\mu_{mn} \to \mu(\vec{k}) e^{i\vec{k}\cdot\vec{x}_n}.$$

Then

$$v(\vec{\Omega}_n, \vec{\Omega}_j) \to \sum_{i=1}^{4} g^i(\vec{\Omega}_n, \vec{\Omega}_j) h^i(\vec{x}_n - \vec{x}_j), \tag{3.7}$$

where

$$h^1(\vec{x}) = N^{-1} \sum_{\vec{k}} \nu(\vec{k})^{-1} [|\lambda(\vec{k})|^2 + |\mu(\vec{k})|^2] \cos(\vec{k}\cdot\vec{x}), \tag{3.8}$$

$$g^1(\vec{\Omega}_n, \vec{\Omega}_j) = \sin\theta_n \sin\theta_j e^{i(\varphi_n - \varphi_j)}, \tag{3.9}$$

and the other pairs $g^i, h^i, r = 2, 3, 4$ are defined similarly.

Let us investigate how the generic $h(\vec{x})$ decreases as $|\vec{x}| \to \infty$. In a real system, $\lambda(\vec{k})$ and $\mu(\vec{k})$ are the Fourier transforms of functions with exponential decrease at infinity, i.e.,

$$\lambda(\vec{k}) = \int d^3x\, \psi_u^*(\vec{x}) \vec{e}(\vec{k}) \cdot \vec{p}\psi_l(\vec{x}) e^{i\vec{k}\cdot\vec{x}}, \tag{3.10}$$

where u and l refer to the upper and lower atomic states, $\vec{e}$ is a polarization vector and $\vec{p}$ is the momentum operator. Since $\lambda(\vec{0}) \neq 0$, there is an infrared problem connected with the potential (3.8). On physical grounds we do not think this difficulty is intractable but, as a somewhat painful analysis is required, we shall not try to solve this problem here. For one thing, one should certainly include the term $\vec{A}^2$ in the Hamiltonian, where $\vec{A}$ is the vector potential. This introduces a positive quadratic form in the $a^\#$ operators which can be handled by our methods, just as the photon energy, $\sum \nu_m a_m^* a_m$, was. It will certainly mitigate the k^{-1} divergence in (3.8).

From now on we shall assume $\lambda(\vec{k})$ has a zero of sufficiently high order at $\vec{k} = 0$. Then

$$h^i(\vec{x}) = \sum_{n \in Z^3} \tilde{h}(\vec{x} + \vec{n}L), \tag{3.11}$$

where L is the length of the cubic cavity and $\tilde{h}(\vec{x})$ is a smooth, real function which decreases at infinity as $|\vec{x}|^{-4}$, for example. For such a potential, it is well known[8] that the total N body potential appearing in (3.1) is stable provided the particles are not too close together, e.g., $|\vec{x}_i - \vec{x}_j| > a > 0$ for all i, j.

To generalize the above model, (2.1), we can permit the atomic coordinates $\{\vec{x}_n\}$ to be variable and replace H by $H + U(\vec{x}_1, \ldots, \vec{x}_N)$, where U is some ordinary interatomic potential, e.g., a sum of pair potentials. Now the partition function involves a configurational integral on $(\mathbf{R}^3)^N$, i.e.

$$Z \to (N!)^{-1} \int_{V^N} d^{3N}x\, Z(x) e^{-\beta U(x)}, \tag{3.12}$$

where $Z(x)$ is the partition function for each fixed set of atomic positions as defined in (2.2). The bounds (2.3) still apply and we can draw the following general conclusion: If we ignore the infrared problem as above, then the total Hamiltonian is stable if U is sufficiently repulsive whenever two atoms get close together and if U decays sufficiently fast (i.e., faster than $|x|^{-3-\epsilon}$ for $\epsilon > 0$) at large separations. Conversely, the Hamiltonian is unstable if U does not have strongly repulsive cores.

The next level of complication is to treat the atomic coordinates quantum-mechanically, i.e., $H \to H + U + T$, where T is the N-particle kinetic energy operator. Using Bloch states we can still derive the bounds (2.3), but now

$$\tilde{Z}(J) = 2^N \mathrm{Tr}_{\mathfrak{F}} \mathrm{Tr}_{\mathcal{G}} \int d\Omega^N \exp[-\beta \tilde{H}(J, \Omega^N) + U(x) + T], \tag{3.13}$$

where $\mathcal{G}$ is the Hilbert space $L^2(V^N)$ and $\tilde{H}$ is as in (2.4). Because the terms linear in $a^\#$ involve the x_n's, it is no longer possible to do the trace over $\mathfrak{F}$ in a closed form. To circumvent this difficulty, we derive upper and lower bounds to (3.13).

Lower bound. We use the variational principle $\mathrm{Tr} e^A > e^{(\psi, A\psi)}$ for any normalized ψ. With Ω^N fixed we choose $\psi \in \mathfrak{F} \otimes \mathcal{G}$ to be $\psi = \varphi \otimes \rho$. Let ρ be of the form $\rho(\vec{x}_1, \ldots, \vec{x}_N) = \prod_{n=1}^N \rho_i(\vec{x}_i)$ where the ρ_i are smooth functions with supports in V which are pair-wise disjoint. Since the supports are disjoint we can imagine the atoms to be either fermions or bosons without affecting any expectation values. With ρ given, we can calculate $(\psi, T\psi)$ and $(\psi, V\psi)$. Recalling that λ_{mn} depends upon $\vec{x}_n$, we can calculate $\langle \lambda_{mn} \rangle = (\psi, \lambda_{mn}\psi)$, and similarly $\langle \mu_{mn} \rangle$. Now we choose $\varphi \in \mathfrak{F}$ to be the ground-state eigenvector of

$$\sum \nu_m a_m^* a_m + N^{-1/2} \sum \sum \{a_m \langle \lambda_{mn} \rangle S_n^+ + \langle \mu_{mn} \rangle S_n^- + \mathrm{H.c.}\} \tag{3.14}$$

with energy

$$-N^{-1} \sum_m \nu_m^{-1} \left| \sum_n \langle \lambda_{mn} \rangle S_n^+ + \langle \mu_{mn} \rangle S_n^- \right|^2 . \tag{3.15}$$

The point of this lower bound calculation is to show that the quantum uncertainty principle does not change the previous conclusion based on classical mechanics, namely that a repulsive core is still needed in U. Without it we still have that $Z > e^{CN^2}$ for large N with $C > 0$; this can be accomplished by letting all ρ_i have support in some fixed box of size l, where l is such that the ground-state energy of (3.14) is still $O(N^2)$ and negative. Then, $(\psi, T\psi) = O(N^{5/3} l^{-2})$, and hence the ground-state energy of $H + T$ is still $O(N^2)$ and negative.

Upper bound. We can write the Hamiltonian $H + T + U$ as $H_1 + H_2$, where $H_1 = T + \frac{1}{2} \sum \nu_m a_m^* a_m$ and $H_2 = H' + U$, and where H' is as in (2.1), except that it has a factor $\frac{1}{2}$ multiplying the photon-energy term. The kinetic energy does not appear in H_2, and if U has the aforementioned repulsive core plus rapid fall-off properties then the factor $\frac{1}{2}$ in H' does not affect the previous conclusion: as an operator on $\mathfrak{F} \otimes \mathcal{G}$, H_2 is bounded below by $-BN$ for some constant B. Thus

$$Z < e^{\beta BN} \mathrm{Tr}_{\mathfrak{F}} \mathrm{Tr}_{\mathcal{G}} e^{-\beta H_1} < e^{AN}$$

for some constant A.

In summary, a proper quantum statistical treatment of the atomic center of mass motion does not affect the general conclusion above.

IV. FINITE-MODE CASE WITH TRANSLATIONAL DEGREES OF FREEDOM

In Sec. III we were obliged to use the Bloch atomic state picture for a simple, but fundamental reason: If one uses the Glauber state picture, one sees from (2.20) and (2.23) that the difference between the upper and lower bounds to the free energy (2.24) is precisely $\sum_{m=1}^M \nu_m$, and if the number of modes is infinite this sum generally diverges. On the other hand, if M is finite one can always use the Glauber state picture, and the difference of the bounds is $O(N^{-1})$ in the free energy per atom. With this in mind, we shall briefly consider the class of Hamiltonians mentioned at the end of Sec. III, but with M finite, i.e.,

$$K = H + U + T , \tag{4.1}$$

where K is the total Hamiltonian in the Hilbert space $\mathcal{K} = \mathfrak{F} \otimes \mathcal{G} \otimes \mathcal{H}$, H is given by (2.1), U is an ordinary interatomic potential and T is the N-atom kinetic energy.

Our bounds lead to the statement that to $O(1)$ in the total free energy

$$Z \equiv \mathrm{Tr}_{\mathcal{K}} e^{-\beta K} , \tag{4.2}$$

$$Z = \pi^{-M} \int d\alpha^M Z(\alpha) , \tag{4.3}$$

and

$$Z(\alpha) = \mathrm{Tr}_{\mathcal{K} \otimes \mathcal{G}} e^{-\beta K(\alpha)} , \tag{4.4}$$

where $K(\alpha)$ is K with a_m and a_m^* replaced by the c numbers α_m and α_m^*. By changing variables $\alpha_m \to \gamma_m N^{1/2}$, it is clear that one can use steepest descent (in the limit $N \to \infty$) to evaluate the integral (4.3). With $\gamma = (\gamma_1, \ldots, \gamma_M)$, we define

$$g(\gamma) = \lim_{N \to \infty} N^{-1} \ln Z(N^{1/2} \gamma) . \tag{4.5}$$

Then,

$$\lim_{N \to \infty} N^{-1} \ln Z = \max_\gamma g(\gamma) , \tag{4.6}$$

provided the set on which the maximum occurs is of dimension less than $2M$. The existence of the limit $g(\gamma)$ can be shown by customary methods[8] when U is reasonable. It is generally impossible to calculate $g(\gamma)$ exactly, but the point is that the problem has been reduced to a conventional problem of point particles with spin interacting via a spin independent potential U and whose spins are coupled to a static external field which may or may not have a spatial variation. If U is a sum of single particle potentials, $g(\gamma)$ can be evaluated in closed form and there will clearly be a phase transition for sufficiently large coupling constants λ_{mn}, μ_{mn} from $\gamma = 0$ to $\gamma \neq 0$ as β is varied, just as there is in the models considered in Sec. II. An example of this in which the translational degrees of freedom are treated classically is given in Ref. 4. If U is more complicated, this kind of phase transition in γ probably still persists.

V. CONCLUSION

In this paper we have attempted to show the usefulness of coherent states (both Glauber states for the photon field and Bloch atomic states for the atomic degrees of freedom) to calculate rigorously the free energy per atom for various models of the interaction of matter and radiation. For the finite mode case (Sec. II) the problem can be reduced to the calculation of a classical type integral which can be handled by steepest descent methods. This proves the results of Hioe and Wang, and the Glauber state picture shows that, in effect, each atom interacts with a classical radiation field whose value is determined self-consistently. Therefore, one sees that as long as the number of photon modes is held fixed and as long as the interactions of the photons and atoms has a smooth spatial dependence, the thermodynamic properties of the model are very insensitive to the details of the Hamiltonian. In particular they are insensitive to whether or not the rotating-wave approximation is used and they are insensitive to the number of atomic levels, provided that number is finite.

The Bloch picture shows clearly that for two-level atoms with or without the rotating-wave approximation, but with spatially independent coupling constants, the problem reduces to that of a single photon mode in the rotating wave approximation. Furthermore, as shown in the Appendix, expectation values of intensive observables can be calculated using the mean fields given by the steepest-descent points of the coherent-state integrals. In this context we note a similarity to those approximations in laser theory in which the photon field is treated semiclassically, and in which the atoms are treated as independent, but fully quantum-mechanical entities.[9,10]

Section III explores the question of stability of the more fundamental Hamiltonian in which there are an infinite number of photon modes. Leaving aside the infrared problem, we have established some necessary and some sufficient conditions for the thermodynamic stability. Our methods are not powerful enough, however, to prove the existence of the thermodynamic limit.

In Sec. IV, where Hamiltonians similar to those of Sec. III but with a finite number of photon modes are discussed, the existence of the thermodynamic limit can be established. The main result of Sec. IV is that the inclusion of translational degrees of freedom generally does not destroy the kind of phase transition possessed by the models of Sec. II.

ACKNOWLEDGMENT

The authors would like to thank the I.H.E.S. for its generous hospitality.

APPENDIX: SOME TECHNICAL CONSIDERATIONS AND CALCULATION OF EXPECTATION VALUES

To avoid unnecessary complications we shall consider a one-mode Hamiltonian

$$H = H_0 + R,$$
$$H_0 = a^* a, \qquad \text{(A1)}$$
$$R = A + a^* B + a B^*,$$

where the Hilbert space is $\mathcal{K} = \mathfrak{F} \otimes \mathcal{H}$, $\mathfrak{F}$ is the Fock space of the boson mode $a^\#$, $\mathcal{H}$ is a finite dimensional Hilbert space of dimension d (the spin space) and A and B are operators on $\mathcal{H}$ with A Hermitian.

Clearly R is a Kato perturbation[11] of H_0, which implies that H is self-adjoint on $D(H_0)$, the domain of H_0; H is bounded from below and has purely discrete spectrum with finite multiplicity because H_0 has these properties. Therefore $\rho = e^{-\beta H_0}$ is bounded. To prove that ρ is trace class, write $\rho = \exp[-\frac{1}{2}\beta H_0 - \beta(\frac{1}{2}H_0 + R)]$. By the Golden-Thompson inequality,[12]

$$\mathrm{Tr}\rho \leq \mathrm{Tr} e^{-\beta H_0/2} e^{-\beta(H_0/2 + R)}. \qquad \text{(A2)}$$

The first factor in (A2) is trace class and the second is bounded.

Let P_n be the projector onto the states with $\leq n$ photons, so that $P_n \to I$ strongly. Let $H_n = P_n H P_n$ and consider

$$Z_n \equiv \mathrm{Tr} P_n e^{-\beta H_n}, \qquad \text{(A3)}$$

which is a finite dimensional trace. We note that $D = \bigcup_n P_n \mathcal{K}$ is a common core for all H_n and H and, for any $\psi \in D$, $P_n H P_n \psi = \psi$ for n sufficiently large. In fact $P_n Q P_n \psi = \psi$ for n sufficiently large when Q is any polynomial in a, with coefficients which are operators on $\mathcal{K}$. Hence $H_n \to H$ in the strong resolvent sense.[13] Let $\{\psi_i, h_i\}_{i=1}^{\infty}$ (resp. $\{\psi_i^n, h_i^n\}_{i=1}^{nd}$) be the eigenvectors and eigenvalues of H (resp. $H_n | P_n \mathcal{K}$) arranged in increasing order. We can use the ψ_i^n as trial vectors for H and conclude from the mini-max principle that $h_i^n \geq h_i$ for $1 \leq i \leq nd$. Using this fact, together with the strong convergence[13] of the eigenprojections $E_n(a,b) \to E(a,b)$ for every interval (a,b) with $a,b \notin \mathrm{spec}(H)$, one can show that $h_n^i \to h^i$ and $\psi_n^i \to \psi^i$ weakly. Hence $Z_n \to Z$ by the dominated-convergence theorem.

Define the cutoff Glauber states by

$$|\alpha, n\rangle = P_n |\alpha\rangle . \tag{A4}$$

For these states the formulas (2.16)–(2.23) are correct with certain obvious modifications, i.e.,

$$\langle \alpha, n | \beta, n\rangle = K_n(\alpha,\beta) = \exp[-\tfrac{1}{2}|\alpha|^2 - \tfrac{1}{2}|\beta|^2] \times \sum_{m=0}^{n} (\alpha^*\beta)^m/m!\,, \tag{A5}$$

$$\mathrm{Tr}\,|\alpha,n\rangle\langle\beta,n| = K_n(\beta,\alpha)\,, \tag{A6}$$

$$\mathrm{Tr} P_n A P_n = \pi^{-1}\int d\alpha \langle \alpha, n | A | \alpha, n\rangle\,, \tag{A7}$$

$$\langle \alpha, n | a | \alpha, n\rangle = \alpha K_{n-1}(\alpha,\alpha)\,, \tag{A8}$$

$$\langle \alpha, n | a^* a | \alpha, n\rangle = |\alpha|^2 K_{n-1}(\alpha,\alpha)\,, \tag{A9}$$

$$P_n = \pi^{-1}\int d\alpha\, |\alpha,n\rangle\langle\alpha,n|\,, \tag{A10}$$

$$P_n a P_n = \pi^{-1}\int d\alpha\,\alpha\, |\alpha,n\rangle\langle\alpha,n|\,, \tag{A11}$$

$$P_n a^* a P_n = \pi^{-1}\int d\alpha (|\alpha|^2 - 1) |\alpha,n\rangle\langle\alpha,n|\,, \tag{A12}$$

$$\pi^{-1}\int d\alpha K_n(\beta,\alpha) K_n(\alpha,\gamma) = K_n(\beta,\gamma)\,. \tag{A13}$$

Similar formulas hold for all other polynomials in $a^\#$.

We wish to compute a lower bound to Z_n following Ref. 2. Use (A7) and the fact that $\langle\psi|e^X|\psi\rangle \geq \langle\psi|\psi\rangle \exp(\langle\psi|X|\psi\rangle/\langle\psi|\psi\rangle)$ applied to H_n. Then

$$Z_n \geq \mathrm{Tr}_{\mathcal{K}} \int d\alpha K_n(\alpha,\alpha) \times \exp\{-|\alpha|^2 K_{n-1}(\alpha,\alpha)/K_n(\alpha,\alpha) + A + [B\alpha^* K_{n-1}(\alpha,\alpha)/K_n(\alpha,\alpha) + \mathrm{H.c.}]\}\,. \tag{A14}$$

First we take $\mathrm{Tr}_{\mathcal{K}}$, and then let $n \to \infty$. By the dominated-convergence theorem, the left side of (2.24) is proved. It is equally clear that the same strategy proves the lower bounds (2.3) and (2.4) obtained from the Bloch state representation.

It is somewhat more difficult to prove the upper bounds (2.24) or (2.3). According to the method of Ref. 2 one defines

$$Z_n^m = \mathrm{Tr} P_n (1 - \beta H_n/m)^m \tag{A15}$$

and, since the trace is finite dimensional, $\lim_{n\to\infty} Z_n^m = Z_n$. The difficulty is that if one applies (A5)–(A13) directly to (A15) the dominated convergence theorem will not be clearly valid. Instead, for all $\epsilon \geq 0$ we define $H_n(\epsilon)$ as follows. For each photon operator that appears in H_n, such as $a^* a$ or $a^\#$, one introduces a convergence factor $e^{(-\epsilon|\alpha|^2)}$ in the integrands of (A11) and (A12). Using $H_n(\epsilon)$, one defines $Z_n(\epsilon)$ and $Z_n^m(\epsilon)$. Then $\lim_{\epsilon\downarrow 0} Z_n(\epsilon) = Z_n$ and $\lim_{m\to\infty} Z_n^m(\epsilon) = Z_n(\epsilon)$. Using the Golden-Thompson inequality, as shown in Ref. 2, one has

$$Z_n^m(\epsilon) \leq \pi^{-1}\int d\alpha K_n(\alpha,\alpha) \mathrm{Tr}_{\mathcal{K}}\{1 - \beta \hat{H}(\alpha,\epsilon)\}^m\,, \tag{A16}$$

where $\hat{H}(\alpha,\epsilon)$ is defined by the replacements $a^* a \to (|\alpha|^2 - 1) e^{(-\epsilon|\alpha|^2)}$, etc. Owing to the convergence factor and to the Gaussian decrease of $K_n(\alpha,\alpha)$ [cf.(A5)] one can use dominated convergence to assert that

$$Z_n(\epsilon) \leq \pi^{-1}\int d\alpha K_n(\alpha,\alpha) \mathrm{Tr}_{\mathcal{K}} e^{\{-\beta\hat{H}(\alpha,\epsilon)\}}\,. \tag{A17}$$

Again, we can let $\epsilon \downarrow 0$ and use dominated convergence to replace $Z_n(\epsilon)$ by Z_n and $\hat{H}(\alpha,\epsilon)$ by $\hat{H}(\alpha,0)$ in (A19). Now, because $\hat{H}$ has a term $|\alpha|^2$, we can let $n \to \infty$ and use dominated convergence to obtain the right-hand side of (2.24). For the Bloch state upper bound, (2.3), the same strategy works except that it is not necessary to introduce the convergence factor, i.e., one can deal directly with Z_n.

Having proved that the bounds on the partition function used in the main text can be justified for the unbounded photon operators, we turn now to the problem of evaluating expectation values of intensive observables. This latter is the algebra generated by $a^\#/N^{1/2}$ and by the atomic operators $S^i_{(N)}/N$. In particular we consider an operator Θ which is a monomial in the $a^\#/N^{1/2}$ times an operator on $\mathcal{K}$. Defining

$$\langle\Theta\rangle = \mathrm{Tr}\,\Theta e^{-\beta H}/\mathrm{Tr}\,e^{-\beta H}\,, \tag{A18}$$

we want to show that as $N \to \infty$, $\langle\Theta\rangle$ can be evaluated by simply replacing each factor in Θ by its value at the absolute maximum of the integrands of (2.11), (2.25), etc., and then *summing over those points*. [Note that in the Dicke model with the rotating-wave approximation there is a conservation law which leads to a gauge invariance. The stationary phase "points" are, in fact, curves. After integrating over those curves, one finds, for example, that $N^{-1}\langle a^*a\rangle \neq 0$ but $N^{-1}\langle a^2\rangle = 0$, as expected (cf. Ref. 1)].

We shall need two lemmas whose use will become clear later. First we give a different proof and a slightly generalized version of Griffiths's lemma.[14]

Lemma 1 (Griffiths). Let $\{g_n(x)\}$ be a sequence of convex functions on $x\in(a,b)\equiv I\subset \mathbb{R}$ with a pointwise limit $g(x)$, which, of course, is convex. Let $G_n^+(x)$[resp.$G_n^-(x)$] be the right (resp. left) derivatives of $g_n(x)$, and similarly for $G^+(x)$, $G^-(x)$. Then, for all $x\in I$,

$$\limsup_{n\to\infty} G_n^+(x) \leqslant G^+(x)\,,$$
$$\liminf_{n\to\infty} G_n^-(x) \geqslant G^-(x)\,. \tag{A19}$$

In particular, if all the $g_n(x)$ and $g(x)$ are differentiable at some point $x\in I$, then

$$\lim_{n\to\infty} dg_n(x)/dx = dg(x)/dx\,. \tag{A20}$$

Proof. Fix $x\in I$. For $y>0$ and $x\pm y\in I$,

$$g_n(x+y)\geqslant g_n(x)+yG_n^+(x)\,,$$
$$g_n(x-y)\geqslant g_n(x)-yG_n^-(x)\,.$$

Fix y and take the limit $n\to\infty$. Then

$$\limsup_{n\to\infty} G_n^+(x)\leqslant y^{-1}[g(x+y)-g(y)]$$

and similarly for $\liminf_{n\to\infty} G_n^-(x)$. Now let $y\downarrow 0$. Q.E.D.

In the following, Lemma 2, we consider a sequence of "partition functions" which, quite generally, we may write as

$$Z_n(\beta)=\int e^{-\beta n e}d\mu_n(e)\,, \tag{A21}$$

where $\{\mu_n\}$ is some sequence of nonnegative measure on $\mathbb{R}$. We assume that all $Z_n(\beta)<\infty$ for β in some open interval $I=(a,b)$. Define

$$g_n(\beta)=n^{-1}\ln Z_n(\beta)\,. \tag{A22}$$

Obviously, $g_n(\beta)$ is convex for $\beta\in I$. We also define the moments of the energy (per particle) as

$$E_n^k(\beta)=\int e^k e^{-\beta n e}d\mu_n(e)/Z_n(\beta)\,, \tag{A23}$$

where $k=0,1,2,\ldots$. It is easy to prove (by dominated convergence, or otherwise) that for $\beta\in I$, $E_n^k(\beta)$ exists and

$$E_n^k(\beta)=(-n)^{-k}Z(\beta)^{-1}d^kZ_n(\beta)/d\beta^k. \tag{A24}$$

Lemma 2. Let the sequence $\{\mu_n, Z_n(\beta), g_n(\beta)\}$ be defined as above and assume that

$$\lim_{n\to\infty} g_n(\beta)=g(\beta) \tag{A25}$$

exists. Then, for every β at which $g(\beta)$ is differentiable,

$$\lim_{n\to\infty} E_n^k(\beta)=[\lim_{n\to\infty} E_n^1(\beta)]^k=[-dg(\beta)/d\beta]^k. \tag{A26}$$

Proof. For $k=1$, the lemma is the same as Lemma 1. Let J be a compact subinterval of I. From the facts that $Z_n(\beta)<\infty$ for all $\beta\in I$, and the uniform convergence of $g_n(\beta)$ on compacts, which always holds for convex functions, we can say that if $\beta\in J$ then we can find a lower cutoff e_0 in the integral (A23) such that the error in $g_n(\beta)$ and in $E_n^k(\beta)$ is $O(e^{-cn})$ for all k less than some fixed K, and where $c>0$ is a constant independent of n and β. By adding a (trivial) constant to e we may assume, without loss of generality, that $e_0=1$, i.e., $\mu((-\infty,1))=0$. Then $E_n^k(\beta)>0$ and $e^k d\mu_n(e)$ is a non-negative measure for all k. Consider $k=2$. Replace $d\mu_n(e)$ by $e\,d\mu_n(e)$ and thereby define $Z_n^1(\beta)$ and $g_n^1(\beta)$ as in (A23) and (A24). Then $dg_n^1(\beta)/d(\beta)=-E_n^2(\beta)/E_n^1(\beta)$. Now $\exp[n(g_n^1(\beta)-g_n(\beta))]=Z_n^1(\beta)/Z_n(\beta)$, so $h_n(\beta)\equiv g_n^1(\beta)-g_n(\beta)\to 0$. Hence, $g_n^1(\beta)=h_n(\beta)+g_n(\beta)$ is a convex function having the limit $g(\beta)$, and we can apply Lemma 1 to it on int(J). This proves the $k=2$ case. Obviously, the argument can be extended inductively, i.e., one writes $E_n^k(\beta)=\prod_{j=1}^k[E_n^{k-j+1}(\beta)/E_n^{k-j}(\beta)]$. Q.E.D.

Now we return to the monomial Θ whose expectation value we wish to compute. If Θ is not self-adjoint then consider Θ to be replaced by $\Theta+\Theta^*$. Clearly there is some integer $k\geqslant 1$ such that H^k dominates Θ, i.e., $|\Theta|\leqslant\frac{1}{2}H^k+bI$ for some positive constant b. With $\beta>0$ fixed, let $e_k=\lim_{N\to\infty}N^{-k}\langle H^k\rangle$. For each $\lambda\geqslant 0$ define $H(\lambda)\equiv H+\lambda N\Theta+\lambda W$, with $W=N^{-k+1}H^k-e_{k-1}H$. Obviously, the preceding analysis for Z applies equally well to $Z(\lambda)\equiv \mathrm{Tr}e^{-\beta H(\lambda)}$, i.e., one can obtain upper and lower bounds to $Z(\lambda)$ in terms of classical integrals. Naturally, for the monomials in Θ and H^k one has to find the obvious generalizations of (2.19)–(2.23), but this can easily be done; in particular the monomial $(a^*)^p a^q$ has $(\alpha^*)^p\alpha^q$ as its leading term in (2.19)–(2.23).

The Peierls-Bogoliubov inequality

$$Z(\lambda)\geqslant Z(0)e^{-\beta\lambda\langle N\Theta+W\rangle}\,, \tag{A27}$$

can easily be proved to hold here. Then, with $g(\beta,N,\lambda)\equiv N^{-1}\ln Z(\lambda)$ and $\lambda>0$,

$$\beta\langle\Theta\rangle\geqslant\lambda^{-1}[g(\beta,N,0)-g(\beta,N,\lambda)]-\beta N^{-1}\langle W\rangle\,. \tag{A28}$$

Take the limit $N\to\infty$. In those cases (Secs. II and IV) in which our upper and lower bounds agree, they will also agree when $\lambda>0$. Thus, $\lim_{N\to\infty} g(\beta,N,\lambda)=g(\beta,\lambda)$ exists. Also, $\lim_{N\to\infty}N^{-1}\langle W\rangle=0$ by Lemma 2. Thus, for all $\lambda>0$

$$\liminf_{N\to\infty}\langle\Theta\rangle \geq \lambda^{-1}[g(\beta,\lambda)-g(\beta,0)]. \quad (A29)$$

In a parallel way, we can define $H'(\lambda)=H-\lambda N\Theta+\lambda W$ and $g'(\beta,N,\lambda)$, and obtain

$$\limsup_{N\to\infty}\langle\Theta\rangle \leq \lambda^{-1}[g(\beta,0)-g'(\beta,\lambda)]. \quad (A30)$$

To complete the demonstration one has to show that as $\lambda\downarrow 0$ the right-hand sides of (A29) and (A30) have a common limit and that this is the classical value of $\langle\Theta\rangle$ at the steepest-descent points for $\lambda=0$. When these points are isolated, as they are in the absence of the rotating-wave approximation, it is easy to see that the above is true; as $\lambda\downarrow 0$ some points approach the $\lambda=0$ points while others go off to infinity. The latter are not maximal, however. In the rotating-wave approximation, the $\lambda=0$ points are in fact curves $|\alpha|=$const. In this case, (A29) and (A30) will agree for those operators such as a^*a/N or $aS^+/N^{3/2}$ which are invariant under the gauge group. For other operators, such as a^2, the two limits will not agree because as $\lambda\downarrow 0$ the steepest-descent curves will be approached as two different points. However, in such cases it is easy to see directly that $\langle a^2\rangle=0$, and this value agrees with what one would get if one integrated over the entire curve.

This completes our discussion of the Hamiltonian (A1) and clearly no difficulty is encountered in extending it to the multimode case. However, some remarks are needed for the Hamiltonians considered in Secs. III and IV because the unbounded operators T and/or U are introduced. The additional Hilbert space, $\mathcal{G}$, is generally $L^2(V^N)$, where V is the box, but if U has a hard core it is $L^2(V^N-\text{hard-core region})$. In either case, it suffices to say that in addition to the P_n photon projection operators one can also introduce a strongly convergent sequence of projections $\{Q_n\}$ on $\mathcal{G}$ such that $D=\cup_n Q_n\mathcal{G}$ is a common core for H and all Q_nHQ_n and such that $(Q_nHQ_n-H)\varphi\to 0$ strongly for all $\varphi\in D$.

The remarks in this appendix are admittedly sketchy in places, but the interested reader can easily fill in the details required to analyze any particular model of the class we have considered in the main text.

---

*On leave from the Department of Mathematics, MIT, Cambridge, Mass. 02139, U. S. A. Work partially supported by U. S. National Science Foundation under Grant No. GP-31674X and by a Guggenheim Memorial Foundation Fellowship.

[1]K. Hepp and E. H. Lieb, Ann. Phys. (N.Y.) 76, 360 (1973).

[2]E. H. Lieb, Commun. Math. Phys. 31, 327 (1973).

[3]Y. K. Wang and F. T. Hioe, Phys. Rev. A 7, 831 (1973).

[4]F. T. Hioe (report of work prior to publication).

[5]J. Ginibre, Commun. Math. Phys. 8, 26 (1968).

[6]F. T. Arecchi, E. Courtens, R. Gilmore, and H. Thomas, Phys. Rev. A 6, 2211 (1972); J. M. Radcliffe, J. Phys. A 4, 313 (1971).

[7]G. Gallavotti, J. Ginibre, and G. Velo, Lett. al Nuovo Cimento Serie I 4, 1293 (1970).

[8]M. E. Fisher and D. Ruelle, J. Math. Phys. 7, 260 (1966); M. E. Fisher, Arch. Rat. Mech. Anal. 17, 377 (1964); D. Ruelle, *Statistical Mechanics* (Benjamin, New York, 1969).

[9]W. E. Lamb, Jr., Phys. Rev. 134, A1429 (1964).

[10]K. Hepp and E. H. Lieb, Helv. Phys. Acta (to be published).

[11]T. Kato, *Perturbation Theory for Linear Operators* (Springer, New York, 1966).

[12]S. Golden, Phys. Rev. 137, B1127 (1965); C. J. Thompson, J. Math. Phys. 6, 1812 (1965); M. B. Ruskai, Commun. Math. Phys. 26, 280 (1972); M. Breitenecker and H. R. Gruemm, Commun. Math. Phys. 26, 276 (1972).

[13]M. Reed and B. Simon, *Methods of Modern Mathematical Physics* (Academic, New York, 1972), Chap. 8.

[14]R. B. Griffiths, J. Math. Phys. 5, 1215 (1964).

PHYSICAL REVIEW A VOLUME 11, NUMBER 3 MARCH 1975

Multitime-correlation functions and the atomic coherent-state representation

L. M. Narducci* and C. M. Bowden
Quantum Physics, Physical Sciences Directorate, Redstone Arsenal, Alabama 35809

V. Bluemel, G. P. Garrazana, and R. A. Tuft
Physics Department, Worcester Polytechnic Institute, Worcester, Massachusetts 01609
(Received 25 April 1974)

We consider the evaluation of multitime-correlation functions for quantum statistical systems characterized by dynamical variables of the angular-momentum type (such as, for example, paramagnetic or nuclear spins or collections of two-level atoms), and show how the atomic coherent-state representation of Arecchi *et al.* can be used to express the multitime averages in terms of phase-space integrals. Our calculation suggests a convenient classical-quantum correspondence rule for angular-momentum degrees of freedom.

I. INTRODUCTION

The statistical behavior of irreversible processes has been the subject of considerable attention in quantum optics.[1] The starting point of many recent contributions[2,3] is the Nakajima[4]-Zwanzig[5] master equation, which provides a remarkably flexible technique for dealing with closed or open systems interacting with a collection of dynamical variables (reservoir) for which a detailed description is not required. In his recent monograph on this subject, Haake[3] emphasized the virtues and the drawbacks of such an approach and surveyed some of the applications. It is a remarkable fact that such diverse systems as lasers,[6] superconductors,[2] damped oscillators,[3] relaxing spins,[7] and collectively radiating atoms[8] can all be described by a unified technique, suitably modified to fit the situation in hand.

The above results are of interest not only because they underscore the often quoted remark that the density operator for a given system contains complete quantum statistical information, but also because they provide a powerful tool for the computation of multitime-correlation functions.[9] Such a calculation, in principle, reduces to no more than an algebraic *tour de force* once the reduced density operator is known (such indeed is the case for a number of physical problems where the existence of a small expansion parameter affords an accurate evaluation of the density operator to a low order of perturbation).

There is, however, an alternative approach based on the existence of quasiprobability functions associated with the density operator of the system.[10-14] In many ways, the use of quasiprobability distributions in quantum statistical problems is the simplest and most transparent method of attack. In fact, this quantum-classical rule of correspondence has been used with different degrees of sophistication to describe all the physical processes which are responsible for laser action. The rule of correspondence for Bose-Einstein operators is one which associates the pair of complex numbers α^* and α to normal ordered products of field operators $a^\dagger$ and a according to Glauber's diagonal coherent-state representation.

In contrast, the selection of the macroscopic classical variables associated with quantum atomic operators has been more varied and mainly dictated by convenience.[15] Haken recently derived an explicit form of the generalized Fokker-Planck equation for arbitrary quantum systems from the density matrix equation.[16] Haken's rule of correspondence for atomic operators leads, in general, to partial differential equations for the quasiprobability functions which contain derivatives of arbitrary order with respect to the classical atomic variables (more specifically with respect to the classical atomic inversion). This has not been an overwhelming difficulty because satisfactory approximation techniques have been found to reduce the differential equations that have been studied to a manageable form.

It appears desirable, however, to look for a continuous representation such that the quasiprobability function is a solution of a differential equation containing derivatives of finite order. The atomic coherent-state representation[17] (ACGT representation) appears to be an ideal candidate for certain applications in view of the recent analysis of a superradiant master equation by Narducci *et al.*[18]

In this paper we focus on the calculation of multitime-correlation functions for systems with observables satisfying an angular momentum algebra.[19] The main result of our work, which is based on the ACGT representation, is the reduction of these multitime-correlation functions to multiple integrals over the phase space of the atomic variables [Eq. (3.14)]. The formal analogy with the results obtained for the correlation functions of field operators is not fortuitous, but rather is a consequence of the analogy between the Glauber and the ACGT representations.

In Sec. II we rephrase the statement of the problem and summarize the relevant information on the multitime-correlation functions. In Sec. III we derive the explicit integral representation over the phase space of the atomic variables. For two reasons our derivation is limited to four-time correlation functions and the special cases that can be constructed trivially from them: First, the integral representations tend to become rather lengthy for an arbitrary number of operators, and second, the general expressions are not very useful in practice. In Sec. IV we discuss a few illustrations of the technique and summarize the rules of correspondence that allow the transformation of operator equations into c-number differential or integral form.

II. MULTITIME-CORRELATION FUNCTIONS FOR ARBITRARY OPERATORS

We are interested in time-ordered correlation functions of the form

$$K_n(t_1' t) = \langle A_1(t_1')A_2(t_2')\cdots A_n(t_n')B_n(t_n)\cdots B_1(t_1)\rangle$$
$$= \mathrm{tr}[B_n(t_n)\cdots B_1(t_1)\rho(t_1)A_1(t_1')\cdots A_n(t_n')], \qquad (2.1)$$

where A and B are dynamical variables for an arbitrary system and ρ is the system density operator. For simplicity, we restrict ourselves to an equal number of A and B operators (although this restriction is not necessary for our argument) and consider, for definiteness, the following time sequence:

$$t_n' \geq t_n \geq \cdots \geq t_1' \geq t_1. \qquad (2.2)$$

Other time sequences can be chosen; the one given above is adequate for our purposes. The trace in Eq. (2.1) is evaluated in the Heisenberg picture, where the operators A and B are explicitly time dependent. For most applications of interest one is usually confronted with a system comprised of a small set of relevant dynamical variables (S) and a larger set of variables (B) whose detailed evolution is not required or impossible to obtain in practice.

Procedures have been detailed in the literature[3-5,7] to express Eq. (2.1) in terms of the reduced density operator $W(t)$ of the relevant system. The explicit calculation of the multitime-correlation functions of the relevant operators in terms of the reduced density operator $W(t)$ is sufficiently involved that we cannot summarize it in a few statements. We prefer to comment on the final result and refer the reader to the original contributions by Haken and Weidlich[9] and by Haake[2] for the complete development.

For the special case of Markovian evolution the four-time correlation function is explicitly given by

$$K_2(t', t) = \mathrm{tr}_S\{V(t_2', t_2)$$
$$\times [B_2 V(t_2, t_1')[V(t_1', t_1)[B_1 W(t_1)]A_1]]A_2\}, \qquad (2.3)$$

where $W(t_1)$ is the reduced density operator at the earliest time in the chosen time sequence and V is a nonunitary time-development operator formally defined by the relation

$$W(t') = V(t', t)W(t). \qquad (2.4)$$

The trace is to be evaluated over the relevant variables. In Eq. (2.3) it must be understood that V acts only on the operators contained in the brackets $[\cdots]$ that immediately follow.[20]

It is clear that, once the effect of the evolution induced by $V(t)$ is accounted for explicitly, the calculation of an arbitrary multitime-correlation function becomes a matter of algebraic manipulations. The problem, of course, is the implementation of the formal time translation effected by the operator V.

We shall concern ourselves with this aspect of the problem in Sec. III of this paper. At this point it will be useful to analyze two special cases of Eq. (2.3). The two-time correlation function $K_1(t_1', t_1)$ follows directly from Eq. (2.3) by setting $A_2 = B_2 = 1$ and $t_2' = t_2 = t_1'$. The result is

$$K_1(t_1', t_1) = \mathrm{tr}_S\{V(t_1', t_1)[B_1 W(t_1)]A_1\}. \qquad (2.5)$$

For the common situation in which $t_1' = t_1$, we find

$$K_2(t, t) = \mathrm{tr}_S\{B_2 V(t_2, t_1)[B_1 W A_1]A_2\}. \qquad (2.6)$$

Both Eqs. (2.5) and (2.6) follow from the initial condition $V(t, t) = 1$.

III. INTEGRAL REPRESENTATION OF THE MULTITIME-CORRELATION FUNCTIONS

In the formal development summarized in Sec. II, we have not specified either the exact nature of the system operators A and B or, more crucially, how the time translation is explicitly carried out in Eq. (2.3) and Eq. (2.6). We do so at this point with the help of the ACGT representation.

We are concerned with system observables that are arbitrary functions of the angular momentum operators $J^{\pm}, J_3$ operating in a $(2J+1)$-dimensional Hilbert space. The operators $J^{\pm}, J_3$ could be, for example, the collective atomic observables describing an assembly of identical two-level systems as in recent models of superradiance.[8] The following calculation is based on the assumption that the total effective angular momentum operator J^2 is a constant of the motion during the evolution of the system.

Following the notation of ACGT, we define the atomic coherent states $|\Omega\rangle$ as follows:

$$|\Omega\rangle = \sum_{m=-J}^{J} |J, m\rangle \binom{2J}{m+J}^{1/2} \times \left(\sin\frac{\theta}{2}\right)^{J+m} \left(\cos\frac{\theta}{2}\right)^{J-m} e^{-i(J+m)\varphi}, \qquad (3.1)$$

where θ and φ are the angular parameters in a spherical coordinate system (the Bloch sphere), and where $\theta = 0$ corresponds to the south pole of the sphere.

Let $P(\Omega_1, t_1)$ be the quasiprobability density function associated with the reduced density operator $W(t_1)$; i.e., let

$$W(t_1) = \int d\Omega_1 P(\Omega_1, t_1) |\Omega_1\rangle\langle\Omega_1|, \qquad (3.2)$$

$$d\Omega_1 = \sin\theta_1\, d\theta_1\, d\varphi_1,$$

where $P(\Omega_1, t_1)$ is subject to the normalization condition

$$\int d\Omega_1 P(\Omega_1, t_1) = 1. \qquad (3.3)$$

Unlike the P function of the electromagnetic field, which may be singular, $P(\Omega, t)$ need not be; in fact, it can always be expressed as a finite superposition of at most $(2J+1)^2$ spherical harmonics. This may not always be the most convenient representation, but it eliminates the possibility of pathological behavior.

We develop an integral representation for the four-time correlation function $K_2(t', t)$ of Eq. (2.3) starting from the innermost brackets. The product $B_1 W(t_1)$ can be expressed as follows:

$$B_1 W(t_1) = \int d\Omega_1 P(\Omega_1, t_1) B_1 |\Omega_1\rangle\langle\Omega_1|. \qquad (3.4)$$

If B_1 were a destruction operator for the electromagnetic field and $|\Omega_1\rangle\langle\Omega_1|$ the diagonal projector of the field coherent states, one would find

$$B_1 |\Omega_1\rangle\langle\Omega_1| = (\text{complex number}) \times |\Omega_1\rangle\langle\Omega_1|. \qquad (3.5)$$

In our case the coherent atomic states $|\Omega_1\rangle$ are not right eigenstates of a ladder operator. Thus we look for a differential operator $\mathfrak{D}_{B_1}(\Omega_1)$ such that

$$B_1 |\Omega_1\rangle\langle\Omega_1| = \mathfrak{D}_{B_1}(\Omega_1) |\Omega_1\rangle\langle\Omega_1|. \qquad (3.6)$$

The differential operator $\mathfrak{D}_{B_1}(\Omega_1)$ can be constructed as we indicate at the end of this section and, more explicitly, in Appendix A. Using Eq. (3.6) in Eq. (3.4) we arrive at the following result:

$$B_1 W(t_1) = \int d\Omega_1 P(\Omega_1, t_1) \mathfrak{D}_{B_1}(\Omega_1)(|\Omega_1\rangle\langle\Omega_1|). \qquad (3.7)$$

The parentheses which follow $\mathfrak{D}_{B_1}$ are a reminder that $\mathfrak{D}_{B_1}$ operates on the angular variables in the expression that immediately follows.

We proceed next to produce an explicit c-number representation for the time translation operator $V(t_1', t_1)$. From Eqs. (3.7) and (2.3) we have

$$V(t_1', t_1)[B_1 W(t_1)] = \int d\Omega_1 P(\Omega_1, t_1) \mathfrak{D}_{B_1}(\Omega_1)(V(t_1', t_1)[|\Omega_1\rangle\langle\Omega_1|]). \qquad (3.8)$$

In Eq. (3.8) we have allowed V to operate directly onto the coherent-atomic-state projector in view of the fact that $\mathfrak{D}_{B_1}$ is a differential operator. We now observe that $V(t_1', t_1)[|\Omega_1\rangle\langle\Omega_1|]$ is, by definition, the reduced density operator $W(t_1')$ subject to the initial condition

$$W(t_1) = |\Omega_1\rangle\langle\Omega_1|. \qquad (3.9)$$

We can thus introduce a diagonal representation of the form (3.2) for the density operator in Eq. (3.9) under the proviso that the quasiprobability function associated with it satisfies the appropriate initial condition. Thus we let

$$V(t_1', t_1)[|\Omega_1\rangle\langle\Omega_1|] = \int d\Omega_1' P(\Omega_1', t_1' | \Omega_1, t_1) |\Omega_1'\rangle\langle\Omega_1'|, \qquad (3.10)$$

with

$$P(\Omega_1', t_1 | \Omega_1, t_1) = \delta(\Omega_1' - \Omega_1). \qquad (3.11)$$

We observe that the initial density function (3.11) need not be a distribution,[21] although it is convenient to represent it in this form in the present development.

For the next step we multiply Eq. (3.8) from the right by the operator A_1 and postulate the existence of a new differential operator such that

$$|\Omega_1'\rangle\langle\Omega_1'| A_1 = \mathfrak{D}_{A_1^\dagger}{}^*(\Omega_1') |\Omega_1'\rangle\langle\Omega_1'|, \qquad (3.12)$$

where $\mathfrak{D}_{A_1^\dagger}{}^*$ is the complex conjugate operator of $\mathfrak{D}_{A_1^\dagger}$, which in turn is defined by the following equation:

$$A_1^\dagger |\Omega_1'\rangle\langle\Omega_1'| = \mathfrak{D}_{A_1^\dagger}(\Omega_1') |\Omega_1'\rangle\langle\Omega_1'|. \qquad (3.13)$$

Proceeding in this fashion until every multiplication by the operators B and A is replaced by the differential forms $\mathfrak{D}_B$ and $\mathfrak{D}_{A^\dagger}{}^*$, respectively, and every time translation operator is represented in terms of a conditional quasiprobability density, we find the following integral representation for the correlation function:

$$K_2(t',t)=\int d\Omega_1\cdots d\Omega_2' P(\Omega_1,t_1)\mathfrak{D}_{B_1}(\Omega_1)(P(\Omega_1',t_1'|\Omega_1,t_1))\mathfrak{D}_{A_1^\dagger}{}^*(\Omega_1')(P(\Omega_2,t_2|\Omega_1',t_1'))\mathfrak{D}_{B_2}(\Omega_2)(P(\Omega_2',t_2'|\Omega_2,t_2))\langle\Omega_2'|A_2|\Omega_2'\rangle. \tag{3.14}$$

This is the main result of this section.

The generalization to arbitrary multitime-correlation functions is straightforward but of little practical use and will not be presented here. The special case of a two-time correlation function can be easily derived from the more general result of Eq. (3.14). Explicitly, we have

$$K_1(t_1',t_1)=\int d\Omega_1\, d\Omega_1' P(\Omega_1,t_1)\mathfrak{D}_{B_1}(\Omega_1)(P(\Omega_1',t_1'|\Omega_1,t_1)) \times\langle\Omega_1'|A_1|\Omega_1'\rangle. \tag{3.15}$$

Before we can discuss the integral representation for the case in which $t_1'=t_1$, we must elaborate on the construction of the differential operators and some of their properties. We consider first the basic operators $J^\pm$ and J_3. After some algebraic manipulations, one can verify the following identities:

$$J^+|\Omega\rangle\langle\Omega|\equiv\mathfrak{D}_{J^+}(\Omega)|\Omega\rangle\langle\Omega| =e^{i\varphi}\left[J\sin\theta+\cos^2\left(\frac{\theta}{2}\right)\frac{\partial}{\partial\theta}+\frac{i}{2}\cot\left(\frac{\theta}{2}\right)\frac{\partial}{\partial\varphi}\right]|\Omega\rangle\langle\Omega|, \tag{3.16}$$

$$J^-|\Omega\rangle\langle\Omega|\equiv\mathfrak{D}_{J^-}(\Omega)|\Omega\rangle\langle\Omega| =e^{-i\varphi}\left[J\sin\theta-\sin^2\left(\frac{\theta}{2}\right)\frac{\partial}{\partial\theta}-\frac{i}{2}\tan\left(\frac{\theta}{2}\right)\frac{\partial}{\partial\varphi}\right]|\Omega\rangle\langle\Omega|, \tag{3.17}$$

$$J_3|\Omega\rangle\langle\Omega|\equiv\mathfrak{D}_{J_3}(\Omega)|\Omega\rangle\langle\Omega| =\left(-J\cos\theta+\tfrac{1}{2}\sin\theta\frac{\partial}{\partial\theta}+\frac{i}{2}\frac{\partial}{\partial\varphi}\right)|\Omega\rangle\langle\Omega|. \tag{3.18}$$

Equation (3.16) will be proved in Appendix A; the others follow from the same procedure. Products of operators $B_1B_2|\Omega\rangle\langle\Omega|$, where B_1 and B_2 are any two of the three angular momentum operators $J^\pm$, J_3, can be transformed into the form $\mathfrak{D}|\Omega\rangle\langle\Omega|$ as follows:

$$\begin{aligned}B_1B_2|\Omega\rangle\langle\Omega|&=B_1\mathfrak{D}_{B_2}(\Omega)|\Omega\rangle\langle\Omega|\\&=\mathfrak{D}_{B_2}(\Omega)B_1|\Omega\rangle\langle\Omega|\\&=\mathfrak{D}_{B_2}(\Omega)\mathfrak{D}_{B_1}(\Omega)|\Omega\rangle\langle\Omega|.\end{aligned} \tag{3.19}$$

Equation (3.19) can be immediately generalized for an arbitrary number of multiplications from the left. Postmultiplication of the projector $|\Omega\rangle\langle\Omega|$ by an operator A can be handled in a similar way as shown in the following example:

$$\begin{aligned}|\Omega\rangle\langle\Omega|J^-&=[J^+|\Omega\rangle\langle\Omega|]^\dagger\\&=\mathfrak{D}_{J^+}{}^*|\Omega\rangle\langle\Omega|,\end{aligned} \tag{3.20}$$

where the differential operator $\mathfrak{D}_{J^+}{}^*$ is the complex conjugate form of the operator appearing in Eq. (3.16).

We consider finally the operator $B|\Omega\rangle\langle\Omega|A$. At first sight there seems to be some ambiguity in the form of the equivalent differential operator. We can operate with B first and arrive at the result

$$B|\Omega\rangle\langle\Omega|A=\mathfrak{D}_B\mathfrak{D}_{A^\dagger}{}^*|\Omega\rangle\langle\Omega|. \tag{3.21}$$

On the other hand, if we operate with A first we find

$$B|\Omega\rangle\langle\Omega|A=\mathfrak{D}_{A^\dagger}{}^*\mathfrak{D}_B|\Omega\rangle\langle\Omega|. \tag{3.22}$$

It is, however, a simple matter to verify that Eqs. (3.21) and (3.22) are identical for any combination of the operators of interest, $J^\pm$ and J_3. Finally, for the special case of Eq. (2.6), the correlation function can be expressed in the integral form

$$K_2(t,t)=\int d\Omega_1\, d\Omega_2 P(\Omega_1,t_1)\mathfrak{D}_{B_1}(\Omega_1)\mathfrak{D}_{A_1^\dagger}{}^*(\Omega_1) \times(P(\Omega_2,t_2|\Omega_1,t_1)\langle\Omega_2|A_2B_2|\Omega_2\rangle). \tag{3.23}$$

We can summarize our results by stating that an arbitrary multitime-correlation function for operators of the angular momentum type can be constructed in terms of the quasiprobability density, evaluated at the earliest time in the required time sequence, and the set of conditional probability densities between all consecutive pairs of times at which the operators are specified. The operators A and B are mapped into differential operators of the type $\mathfrak{D}_A$ and $\mathfrak{D}_B$, which can be constructed explicitly using the elementary relations given by Eqs. (3.16)–(3.18) and the algebraic rules discussed in this section. An example of an explicit differential equation for the system probability density is given in Appendix B.

IV. APPLICATIONS

A simple calculation based on the master equation describing the irreversible decay of a two-level system is chosen to illustrate the theory of the previous sections.

Consider the two-time correlation functions in Eqs. (3.15) and (3.23). We identify the following expectation values:

$$\langle A_1(t_1')\rangle_{\Omega_1,t_1} \equiv \int d\Omega_1' P(\Omega_1', t_1'|\Omega_1, t_1)\langle\Omega_1'|A_1|\Omega_1'\rangle, \tag{4.1}$$

$$\langle A_2B_2(t_2)\rangle_{\Omega_1,t_1} \equiv \int d\Omega_2 P(\Omega_2, t_2|\Omega_1, t_1)\langle\Omega_2|A_2B_2|\Omega_2\rangle, \tag{4.2}$$

where the notation $\langle\cdots\rangle_{\Omega_1,t_1}$ implies that the expectation values are to be calculated from the initial condition

$$W(t_1) = |\Omega_1\rangle\langle\Omega_1|.$$

It follows that the two-time correlation functions $K_1(t_1', t_1)$, $K_2(t_1, t_2, t_2, t_1)$ can be expressed in the more compact form

$$K_1(t_1', t_1) = \int d\Omega_1 P(\Omega_1, t_1)\mathfrak{D}_{B_1}(\Omega_1)\langle A_1(t_1')\rangle_{\Omega_1,t_1} \tag{4.3}$$

and

$$K_2(t_1, t_2, t_2, t_1) = \int d\Omega_1 P(\Omega_1, t_1)\mathfrak{D}_{B_1}(\Omega_1) \times \mathfrak{D}_{A_1^\dagger}^{*}(\Omega_1)\langle A_2B_2(t_2)\rangle_{\Omega_1,t_1}. \tag{4.4}$$

In general, Eqs. (4.3) and (4.4) are not simpler to evaluate than the original expressions in (3.15) and (3.23) because the calculation of the expectation values (4.1) and (4.2) requires knowledge of the conditional probability density. The special case of the irreversible relaxation of a two-level system is an exception because the one-time expectation values of the operators of interest can be calculated directly from the master equation.

Consider the irreversible master equation for the reduced atomic density operator of a two-level system.[1]

$$\frac{dW(t)}{dt} = \tfrac{1}{2}\gamma_{21}\{[J^-, W(t)J^+] + [J^-W(t), J^+]\} + \tfrac{1}{2}\gamma_{12}\{[J^+, W(t)J^-] + [J^+W(t), J^-]\} - \tfrac{1}{4}\eta W(t) + \eta J_3 W(t) J_3, \tag{4.5}$$

where γ_{12} and γ_{21} are the atomic transition rates and η is a parameter that accounts for phase-destroying effects. Equation (4.5) describes the atomic decay to thermal equilibrium with a reservoir. In view of the simple algebraic properties of the two-dimensional angular momentum operators $J^\pm$ and J_3, the equations of motion for the operators of interest can be derived directly from the master equation and integrated at once.

Consider, for example, the atomic correlation function $K_1(t_1', t_1) = \langle J_3(t_1')J_3(t_1)\rangle$. According to Eq. (4.3), we need to calculate the expectation value $\langle J_3(t_1')\rangle_{\Omega_1,t_1}$ subject to the initial condition $W(t_1) = |\Omega_1\rangle\langle\Omega_1|$. From Eq. (4.5) we arrive at the equation of motion

$$\frac{d}{dt_1'}\langle J_3(t_1')\rangle = \mathrm{tr}_S[J_3\dot{W}(t_1')] = -(\gamma_{12}+\gamma_{21})\langle J_3(t_1')\rangle - \tfrac{1}{2}(\gamma_{21}-\gamma_{12}). \tag{4.6}$$

The solution of Eq. (4.6) is

$$\langle J_3(t_1')\rangle = \left(\langle J_3(t_1)\rangle + \frac{1}{2}\,\frac{\gamma_{21}-\gamma_{12}}{\gamma_{12}+\gamma_{21}}\right)e^{-(\gamma_{12}+\gamma_{21})(t_1'-t_1)} - \frac{1}{2}\,\frac{\gamma_{21}-\gamma_{12}}{\gamma_{12}+\gamma_{21}}. \tag{4.7}$$

The initial value $\langle J_3(t_1)\rangle$ is, of course, the diagonal matrix element

$$\langle J_3(t_1)\rangle = \langle\Omega_1|J_3|\Omega_1\rangle = -\tfrac{1}{2}\cos\theta_1. \tag{4.8}$$

From the explicit expression for the operator $\mathfrak{D}_{J_3}(\Omega_1)$ given by Eq. (3.18) and from Eq. (4.8), we finally arrive at the desired result

$$\langle J_3(t_1')J_3(t_1)\rangle = \int d\Omega_1 P(\Omega_1, t_1) \times\left(-J\cos\theta_1 + \tfrac{1}{2}\sin\theta_1\frac{\partial}{\partial\theta_1} + \frac{i}{2}\frac{\partial}{\partial\varphi_1}\right) \times\langle J_3(t_1')\rangle, \tag{4.9}$$

which can be evaluated explicitly once the density function $P(\Omega_1, t_1)$ is assigned. In particular, we see that for $t_1' - t_1 \to \infty$, Eq. (4.9) reduces to

$$\langle J_3(t_1')J_3(t_1)\rangle = -\frac{1}{2}\,\frac{\gamma_{21}-\gamma_{12}}{\gamma_{12}+\gamma_{21}}\langle J_3(t_1)\rangle = \langle J_3(\infty)\rangle\langle J_3(t_1)\rangle. \tag{4.10}$$

As a second example, we calculate the correlation function

$$\langle J^+(t_1)J^+(t_2)J^-(t_2)J^-(t_1)\rangle = \int d\Omega_1 P(\Omega_1, t_1)\mathfrak{D}_{J^-}(\Omega_1) \times\mathfrak{D}_{J^-}^{*}(\Omega_1)\langle J^+J^-(t_2)\rangle_{\Omega_1,t_1}. \tag{4.11}$$

The expectation value $\langle J^+J^-(t_2)\rangle_{\Omega_1,t_1}$ in Eq. (4.11) follows directly from Eq. (4.7) and the identity

$$J^+J^- = \tfrac{1}{2} + J_3. \tag{4.12}$$

We have

$$\langle J^+J^-(t_1)\rangle_{\Omega_1,t_1} = \frac{\gamma_{12}}{\gamma_{12}+\gamma_{21}} + \left(-\tfrac{1}{2}\cos\theta_1 + \frac{1}{2}\,\frac{\gamma_{21}-\gamma_{12}}{\gamma_{12}+\gamma_{21}}\right) \times e^{-(\gamma_{12}+\gamma_{21})(t_2-t_1)}. \tag{4.13}$$

The differential operator $\mathfrak{D}_{J^-}\mathfrak{D}_{J^-}{}^*$ follows from Eq. (3.17) and has the form

$$\mathfrak{D}_{J^-}\mathfrak{D}_{J^-}{}^* = \tfrac{1}{2}(1-\cos\theta) - \frac{1}{4}\,\frac{(1-\cos\theta)^2}{\sin\theta}(2+\cos\theta)\frac{\partial}{\partial\theta} + \tfrac{1}{4}(1-\cos\theta)^2\frac{\partial^2}{\partial\theta^2} + \tfrac{1}{4}\tan^2\left(\frac{\theta}{2}\right)\frac{\partial^2}{\partial\varphi^2}\,. \quad (4.14)$$

The required correlation function can be calculated from Eqs. (4.14) and (4.13). The result is

$$\begin{aligned}\langle J^+(t_1)J^+(t_2)J^-(t_2)J^-(t_1)\rangle &= \int d\Omega_1\, P(\Omega_1, t_1)\tfrac{1}{2}(1-\cos\theta_1)\frac{\gamma_{12}}{\gamma_{12}+\gamma_{21}} \times (1-e^{-(\gamma_{12}+\gamma_{21})(t_2-t_1)}) \\ &= \langle J^+J^-(t_1)\rangle \frac{\gamma_{12}}{\gamma_{12}+\gamma_{21}}(1-e^{-(\gamma_{12}+\gamma_{21})(t_2-t_1)}).\end{aligned} \quad (4.15)$$

Notice that, as expected, the initial value of this correlation function vanishes, since $(J^\pm)^2 = 0$.

ACKNOWLEDGMENTS

One of us (LMN) wishes to express his appreciation to Professor Adriaan Walther for many long and clarifying discussions on the subject of this paper, and to Romas A. Shatas of the Quantum Physics Group, Redstone Arsenal, Alabama, for his hospitality. We are also grateful to Professor Fritz Haake for an illuminating correspondence, and to Professor J. H. Eberly for some useful suggestions in the final preparation of the manuscript.

APPENDIX A: $\mathfrak{D}$ OPERATORS

The explicit construction of the $\mathfrak{D}$ operators is a simple, but time consuming, proposition. Here we prove Eq. (3.16) of the text and write down a number of other formulas which are useful for mapping operator equations into c-number differential form.

Consider the atomic coherent-state projector

$$\Lambda(\Omega) \equiv |\Omega\rangle\langle\Omega| = \sum_{p,q=0}^{2J} \Gamma_{p,q}(\theta,\varphi), \quad (A1)$$

where we have defined

$$\Gamma_{p,q}(\theta,\varphi) = \binom{2J}{p}^{1/2}\binom{2J}{q}^{1/2} |p\rangle\langle q| \times e^{-i(p-q)\varphi}\left(\sin\frac{\theta}{2}\right)^{p+q}\left(\cos\frac{\theta}{2}\right)^{4J-(p+q)}. \quad (A2)$$

For future reference, we construct the following derivatives of the operator $\Lambda(\Omega)$:

$$\frac{\partial\Lambda}{\partial\theta} = -2J\tan\left(\frac{\theta}{2}\right)\Lambda(\Omega) + \frac{1}{\sin\theta}\sum_{p,q}(p+q)\Gamma_{p,q}(\theta,\varphi), \quad (A3)$$

$$\frac{\partial\Lambda}{\partial\varphi} = -i\sum_{p,q}(p-q)\Gamma_{p,q}(\theta,\varphi), \quad (A4)$$

and define the operators Σ_p and Σ_q as follows:

$$\Sigma_p = \sum_{p,q=0}^{2J} p\Gamma_{p,q}(\theta,\varphi), \quad (A5)$$

$$\Sigma_q = \sum_{p,q=0}^{2J} q\Gamma_{p,q}(\theta,\varphi). \quad (A6)$$

From Eqs. (A3) and (A4) we can calculate Σ_p and Σ_q in terms of the projector Λ and its first order derivatives:

$$\Sigma_p = \tfrac{1}{2}\sin\theta\frac{\partial\Lambda}{\partial\theta} + J(1-\cos\theta)\Lambda + \frac{i}{2}\frac{\partial\Lambda}{\partial\varphi}, \quad (A7)$$

$$\Sigma_q = \tfrac{1}{2}\sin\theta\frac{\partial\Lambda}{\partial\theta} + J(1-\cos\theta)\Lambda - \frac{i}{2}\frac{\partial\Lambda}{\Lambda\varphi}. \quad (A8)$$

The differential operator $\mathfrak{D}_{J^+}$ can be constructed as follows:

$$J^+\Lambda = \cot(\tfrac{1}{2}\theta)e^{i\varphi}\Sigma_p(\Omega). \quad (A9)$$

Using Eq. (A7) in Eq. (A9) yields

$$J^+\Lambda = e^{i\varphi}\left[J\sin\theta + \cos^2\left(\frac{\theta}{2}\right)\frac{\partial}{\partial\theta} + \frac{i}{2}\cot\left(\frac{\theta}{2}\right)\frac{\partial}{\partial\varphi}\right]\Lambda = \mathfrak{D}_{J^+}(\Omega)\Lambda(\Omega). \quad (A10)$$

A similar calculation leads to

$$J^-\Lambda = e^{-i\varphi}\left[J\sin\theta - \sin^2\left(\frac{\theta}{2}\right)\frac{\partial}{\partial\theta} - \frac{i}{2}\tan\left(\frac{\theta}{2}\right)\frac{\partial}{\partial\varphi}\right]\Lambda = \mathfrak{D}_{J^-}(\Omega)\Lambda(\Omega) \quad (A11)$$

and

$$J_3\Lambda = \left(-J\cos\theta + \tfrac{1}{2}\sin\theta\frac{\partial}{\partial\theta} + \frac{i}{2}\frac{\partial}{\partial\varphi}\right)\Lambda = \mathfrak{D}_{J_3}(\Omega)\Lambda(\Omega). \quad (A12)$$

Occasionally one needs to evaluate the differential operators corresponding to $\Lambda(\Omega)J^\pm$ and $\Lambda(\Omega)J_3$. These can be obtained directly from Eqs. (A10)–(A12) and the identities

$$\Lambda J^+ = \mathfrak{D}_{J^-}{}^*\Lambda\,, \quad \Lambda J^- = \mathfrak{D}_{J^+}{}^*\Lambda\,, \quad \Lambda J_3 = \mathfrak{D}_{J_3}{}^*\Lambda\,, \quad (A13)$$

where $\mathfrak{D}^*$ indicates complex conjugation. Thus, for example, we have

$$\Lambda J^+ = e^{i\varphi}\left[J\sin\theta - \sin^2\left(\frac{\theta}{2}\right)\frac{\partial}{\partial\theta} + \frac{i}{2}\tan\left(\frac{\theta}{2}\right)\frac{\partial}{\partial\varphi}\right]\Lambda\,. \quad (A14)$$

In practical calculations it is a common occurrence to have products of two angular momentum operators acting on the projector Λ. The appropriate $\mathfrak{D}$ operators in this case can be easily constructed from Eqs. (A11)–(A13). Since this calculation involves a substantial amount of algebraic manipulation, we present here the results for the most common operators:

$$\begin{aligned} J^+J^-\Lambda(\Omega) &= [J^2\sin^2\theta + \tfrac12 J(1-\cos\theta)^2]\Lambda + [J\sin\theta\cos\theta + \tfrac14\sin\theta(2-\cos\theta)]\frac{\partial\Lambda}{\partial\theta} + i(J\cos\theta + \tfrac12)\frac{\partial\Lambda}{\partial\varphi} \\ &\quad - \frac{i}{2}\sin\theta\frac{\partial^2\Lambda}{\partial\theta\partial\varphi} + \frac14\frac{\partial^2\Lambda}{\partial\varphi^2} - \tfrac14\sin^2\theta\frac{\partial^2\Lambda}{\partial\theta^2} \\ &\equiv \mathfrak{D}_{J^-}\mathfrak{D}_{J^+}\Lambda(\Omega), \end{aligned} \tag{A15}$$

$$\begin{aligned} J^-J^+\Lambda(\Omega) &= [J^2\sin^2\theta + \tfrac12 J(1+\cos\theta)^2]\Lambda + [J\sin\theta\cos\theta - \tfrac14\sin\theta(2+\cos\theta)]\frac{\partial\Lambda}{\partial\theta} + i(J\cos\theta - \tfrac12)\frac{\partial\Lambda}{\partial\varphi} \\ &\quad - \frac{i}{2}\sin\theta\frac{\partial^2\Lambda}{\partial\theta\partial\varphi} + \frac14\frac{\partial^2\Lambda}{\partial\varphi^2} - \tfrac14\sin^2\theta\frac{\partial^2\Lambda}{\partial\theta^2} \\ &\equiv \mathfrak{D}_{J^+}\mathfrak{D}_{J^-}\Lambda(\Omega), \end{aligned} \tag{A16}$$

$$\Lambda(\Omega)J^+J^- = \mathfrak{D}_{J^-}^*\mathfrak{D}_{J^+}^*\Lambda(\Omega), \tag{A17}$$

$$\Lambda(\Omega)J^-J^+ = \mathfrak{D}_{J^+}^*\mathfrak{D}_{J^-}^*\Lambda(\Omega), \tag{A18}$$

$$\begin{aligned} J^-\Lambda(\Omega)J^+ &= [J^2\sin^2\theta + \tfrac12 J(1-\cos\theta)^2]\Lambda + [-J\sin\theta(1-\cos\theta) + \tfrac14\cot\theta(1-\cos\theta)^2]\frac{\partial\Lambda}{\partial\theta} \\ &\quad + \tfrac14(1-\cos\theta)^2\frac{\partial^2\Lambda}{\partial\theta^2}\tfrac14\tan^2\left(\frac{\theta}{2}\right)\frac{\partial^2\Lambda}{\partial\varphi^2} \\ &\equiv \mathfrak{D}_{J^-}\mathfrak{D}_{J^-}^*\Lambda = \mathfrak{D}_{J^-}^*\mathfrak{D}_{J^-}\Lambda, \end{aligned} \tag{A19}$$

$$\begin{aligned} J^+\Lambda J^- &= [J^2\sin^2\theta + \tfrac12 J(1+\cos\theta)^2]\Lambda + [J\sin\theta(1+\cos\theta) + \tfrac14\cot\theta(1+\cos\theta)^2]\frac{\partial\Lambda}{\partial\theta} \\ &\quad + \tfrac14(1+\cos\theta)^2\frac{\partial^2\Lambda}{\partial\theta^2} + \tfrac14\cot^2\left(\frac{\theta}{2}\right)\frac{\partial^2\Lambda}{\partial\varphi^2} \\ &\equiv \mathfrak{D}_{J^+}\mathfrak{D}_{J^+}^*\Lambda \equiv \mathfrak{D}_{J^+}^*\mathfrak{D}_{J^+}\Lambda. \end{aligned} \tag{A20}$$

APPENDIX B: FURTHER APPLICATION OF THE $\mathfrak{D}$-OPERATOR CALCULUS

In a previous publication,[18] Narducci *et al.* derived a Fokker-Planck equation for a model of superradiant emission. The Fokker-Planck equation was constructed under the assumption that the P function associated with the reduced atomic density operator was independent of the phase angle φ. Still, the calculation required a lengthy manipulation.

The rules developed in Appendix A for the $\mathfrak{D}$-operator calculus afford a remarkable simplification of the derivation, as we proceed to show. As a matter of fact, there is no merit in assuming that the P function be independent of the angle φ (although this is probably a very sound physical condition).

The superradiant master equation derived by Bonifacio *et al.*[8] has the form

$$\frac{dW(t)}{dt} = \tfrac12\{[J^-, WJ^+] + [J^-W, J^+]\}, \tag{B1}$$

where $W(t)$ is the reduced density operator for the atomic system and J^+ and J^- are the collective polarization operators.

In terms of the diagonal representation

$$W(t) = \int d\Omega\, P(\Omega, t)\Lambda(\Omega) \tag{B2}$$

the superradiant master equation (B1) takes the form

$$\int d\Omega\frac{\partial P}{\partial t}\Lambda(\Omega) = \int d\Omega\, P(\Omega, t)\{J^-\Lambda J^+ - \tfrac12\Lambda J^+J^- - \tfrac12 J^+J^-\Lambda\}. \tag{B3}$$

Next we replace the operators in Eq. (B3) by their equivalent expressions given by Eqs. (A19), (A17), and (A15) and arrive at the following result:

$$\begin{aligned} \int d\theta\, d\varphi\frac{\partial Q}{\partial t}\Lambda &= \int d\theta\, d\varphi\, Q(\theta, \varphi, t) \\ &\quad \times\left[\left(-J\sin\theta - \frac{1-\cos\theta}{2\sin\theta}\right)\frac{\partial\Lambda}{\partial\theta} + \tfrac12(1-\cos\theta)\frac{\partial^2\Lambda}{\partial\theta^2} - \frac12\frac{\cos\theta}{1+\cos\theta}\frac{\partial^2\Lambda}{\partial\varphi^2}\right], \end{aligned} \tag{B4}$$

where the density function $Q(\theta, \varphi, t)$ is defined as

$$Q(\theta, \varphi, t) = \sin\theta\, P(\theta, \varphi, t). \tag{B5}$$

If we integrate the right-hand side of Eq. (B4) by parts, and observe that the sum of the surface terms vanishes identically, we obtain the following equation for $Q(\theta, \varphi, t)$:

$$\frac{\partial Q}{\partial t} = \frac{\partial}{\partial \theta}\left[\left(J\sin\theta + \frac{1-\cos\theta}{2\sin\theta}\right)Q\right] + \frac{\partial^2}{\partial\theta^2}\left[\frac{1-\cos\theta}{2}Q\right] - \frac{\partial^2}{\partial\varphi^2}\left[\frac{1}{2}\,\frac{\cos\theta}{1+\cos\theta}Q\right], \tag{B6}$$

which reduces to the equation discussed in Ref. 18 if Q is assumed to be independent of φ.

---

*Permanent address: Physics Department, Worcester Polytechnic Institute, Worcester, Mass. 01609.

[1] For an extensive review see H. Haken, *Handbuch der Physik* (Springer, Berlin, 1970), Vol. XXV/2.

[2] F. Haake, Phys. Rev. A 3, 1723 (1971).

[3] F. Haake, Springer Tracts Mod. Phys. 66, 98 (1973), and references therein.

[4] S. Nakajima, Prog. Theor. Phys. 20, 948 (1958).

[5] R. Zwanzig, J. Chem. Phys. 33, 1338 (1960).

[6] F. Haake, Z. Phys. 227, 179 (1969).

[7] P. N. Argyres and P. L. Kelley, Phys. Rev. 134, A98 (1964).

[8] R. Bonifacio, P. Schwendimann, and F. Haake, Phys. Rev. A 4, 302 (1971).

[9] A different procedure for calculating multitime-correlation functions without resort to the Zwanzig projector technique has been previously described by H. Haken and W. Weidlich, Z. Phys. 205, 96 (1967), and subsequently applied to problems in laser physics.

[10] R. Bonifacio and F. Haake, Z. Phys. 200, 526 (1967).

[11] M. Lax and W. Louisell, IEEE J. Quantum Electron. QE3, 47 (1967).

[12] R. Graham, F. Haake, H. Haken, and W. Weidlich, Z. Phys. 213, 21 (1968).

[13] U. Gnutzmann, Z. Phys. 225, 416 (1969).

[14] M. Lax, Phys. Rev. 172, 350 (1968).

[15] For a critical assessment of the various rules, see H. Haken, in *Laser Handbook*, edited by F. T. Arecchi and E. O. Schulz-Dubois (North-Holland, Amsterdam, 1972), p. 141.

[16] H. Haken, Z. Phys. 219, 411 (1969).

[17] F. T. Arecchi, E. Courtens, R. Gilmore, and H. Thomas, Phys. Rev. A 6, 2211 (1972). See also J. M. Radcliffe, J. Phys. A 4, 313 (1971).

[18] L. M. Narducci, C. M. Bowden, and C. A. Coulter, Phys. Rev. A 9, 829 (1974).

[19] We consider arbitrary angular momentum operators in view of the well-known formal equivalence between the algebra of the observables associated with two-level atomic systems and angular momentum operators.

[20] It may be worthwhile to point out that Eq. (2.3) is formally correct also for systems undergoing deterministic evolution. (See for instance Ref. 13, p. 417. In this case $V(t',t)$ is replaced by a unitary operator.)

[21] An illustration of this assertion may be found in Appendix D of Ref. 17.

IX. Applications in Nuclear Physics

Coherent states of the quantum-mechanical top

D. Janssen

Joint Institute for Nuclear Research
(Submitted February 25, 1976)
Yad. Fiz. **25**, 897–907 (April 1977)

A set of coherent states of the asymmetric quantum-mechanical top is constructed. The physical and mathematical properties of this set and its possible applications are investigated.

PACS numbers: 03.65.Ge

I. INTRODUCTION

1. Coherent states of the harmonic oscillator, their properties, and applications

In many problems of quantum many-body theory it is very convenient to make use of a basis of states that, while being nonorthogonal and linearly dependent, minimizes the product of the uncertainties for certain canonically conjugate operators characterizing the problem. As the simplest example let us consider the harmonic oscillator

$$H=\frac{1}{2}\hat{p}^2+\frac{1}{2}x^2,\quad \hat{p}=-i\frac{d}{dx},\quad \hbar=1, \tag{1a}$$

$$[\hat{p},x]=-i. \tag{1b}$$

The discrete spectrum of this Hamiltonian and the quantum-mechanical properties of the oscillator are related to the noncommutativity of the momentum $\hat{p}$ and coordinate x. The corresponding product of uncertainties is

$$(\Delta x)^2(\Delta\hat{p})^2\geqslant 1/4, \tag{2}$$

where

$$\Delta\hat{A}=\hat{A}-\langle\psi|\hat{A}|\psi\rangle,\quad (\Delta\hat{A})^2=\langle\psi|(\Delta\hat{A})^2|\psi\rangle$$

and ψ is an arbitrary wave function. The left-hand side of (2) is minimized by the following wave function, constructed from the eigenfunction of the Hamiltonian (1):

$$|\alpha\rangle=\exp\left\{-\frac{1}{2}|\alpha|^2\right\}\sum_{n=0}^{\infty}\frac{\alpha^n}{\sqrt{n!}}\Big|n\Big\rangle=\exp\left\{-\frac{1}{2}|\alpha|^2\right\}\exp\{\alpha b^+\}|0\rangle,$$
$$b^+=\frac{1}{\sqrt{2}}(x-i\hat{p}). \tag{3}$$

Here $|n\rangle$ is a normalized eigenfunction of the harmonic oscillator and α is an arbitrary complex number. Such a set of coherent states (3) is overcomplete and nonorthogonal. This set was first found by Schrödinger[1] and has found wide applications in the quantum-mechanical description of coherent light sources.[2] It has also been used in studies of superconductivity, superfluidity,[3] and phonons in crystals.[4] Let us point out another important property of the coherent states (3). If we write the equations of motion in the Heisenberg representation and average them over the coherent-state basis, then using (1) and (3) we obtain

$$+i\frac{d}{dt}\langle\alpha|x|\alpha\rangle=\langle\alpha|[x,H]|\alpha\rangle=+i\langle\alpha|\hat{p}|\alpha\rangle,$$

$$i\frac{d}{dt}\langle\alpha|\hat{p}|\alpha\rangle=\langle\alpha|[\hat{p},H]|\alpha\rangle=-i\langle\alpha|x|\alpha\rangle, \tag{4}$$

$$\langle\alpha|x|\alpha\rangle=A_0\sin(t-t_0),\quad \langle\alpha|p|\alpha\rangle=A_0\cos(t-t_0),$$

i.e., the mean values of the momentum and coordinate evolve in time like the corresponding classical quantities.

2. Perelomov's generalization of the concept of coherent states and the coherent spin states

After the successful use of coherent states of the type (3) numerous attempts have been made to generalize the concept of "coherent states" to the case of more complicated commutation relations than those given by (1), i.e., to the case of a more complicated algebra. In this paper we shall only use the results of Perelomov.[5] Let G be an arbitrary Lie group and T an irreducible unitary representation of G, acting on the Hilbert space $\mathscr{H}$. If $|\psi_0\rangle$ is a certain element of this space, then all elements h of the group G, satisfying

$$T(h)|\psi_0\rangle=e^{i\alpha(h)}|\psi_0\rangle,$$

constitute the stability subgroup H of the group G for the state $|\psi_0\rangle$. Let us denote

$$|\psi_{g(x)}\rangle=T(g(x))|\psi_0\rangle$$

where $g(x)$ is one of the representatives of all elements of the group G, for which the vectors $|\psi_g\rangle=T(g)|\psi_0\rangle$ differ from each other by a phase factor only. Perelomov then calls $|\psi_{g(x)}\rangle$ the set of coherent states of this group. As a special case this definition contains the harmonic oscillator coherent states (3). In the case of the simple compact group $SU(2)$, choosing $|\psi_0\rangle$ as the state with the lowest possible spin projection onto the z axis, we obtain

$$|\alpha\rangle=\sum_M\sqrt{\frac{(2S)!}{(S-M)!(S+M)!}}\alpha^{S+M}|SM\rangle(1+|\alpha|^2)^{-S}$$
$$=(1+|\alpha|^2)^{-S}\exp(\alpha\hat{S}_+)|S-S\rangle; \tag{5}$$

here $\hat{S}_+=\hat{S}_x+i\hat{S}_y$; $\hat{S}_x$, $\hat{S}_y$, and $\hat{S}_z$ are the $SU(2)$ generators and S is the total spin of the considered system. These states, called coherent spin states, were first obtained by Radcliffe[6] and have found applications in Ref. 7. It is easy to show that for $|\alpha|\ll S$ the operator $\hat{S}_+$ can be replaced by b^+ and that expressions (5) and (3) will coincide. Although Perelomov's definition of coherent states turned out to be useful for the harmonic oscillator and the $SU(2)$ group, it does not make it possible to define a unique set of coherent states for the asymmetric top.

II. DEFINITION OF COHERENT STATES FOR THE QUANTUM-MECHANICAL ASYMMETRIC TOP

1. The z-representation of coherent states

The Hamiltonian of the asymmetric top can be written as

$$H=A_xL_x^2+A_yL_y^2+A_zL_z^2, \quad A_x\neq A_y\neq A_z. \tag{6}$$

It commutes with the square of the total angular momentum $\hat{L}^2=L_x^2+L_y^2+L_z^2=\hat{J}^2$ and with the operators J_x, J_y, and J_z. Here L_i $(i=x,y,z)$ are the angular-momentum projections onto the axes of the intrinsic coordinate system and J_i are the projections onto the laboratory-system axes. The wave function corresponding to the Hamiltonian (6) can be written as

$$\Psi_a^{IM}=\sum_K C_K^{Ia}|IMK\rangle, \tag{7}$$

where $|IMK\rangle$ are the eigenstates of the operators $\hat{L}^2$, L_z, and J_z corresponding to the eigenvalues $I(I+1)$, K, and M. The set of states $|IMK\rangle$ constitutes an irreducible representation of the semi-direct product of the group $SU(2)\times SU(2)$ and an abelian group. The operators L_i, J_i, and $T_{\mu\nu}^{1/2}$ $(\mu,\nu=\pm 1/2)$ are the generators of this product and satisfy the following commutation relations

$$[L_i, L_k]=i\varepsilon_{ikl}L_l,\ [J_i, J_k]=i\varepsilon_{ikl}J_l,\ [L_i, J_k]=0, \tag{8}$$

$$[T_{\mu\nu}^{1/2}, T_{\mu'\nu'}^{1/2}]=0, \quad [L_z, T_{\mu\nu}^{1/2}]=\nu T_{\mu\nu}^{1/2}, \quad [L_\pm, T_{\mu\nu}^{1/2}]=\sqrt{(1/2\mp\nu)(3/2\pm\nu)}\,T_{\mu\nu\pm 1}^{1/2},$$
$$[J_z, T_{\mu\nu}^{1/2}]=\mu T_{\mu\nu}^{1/2}, \quad [J_\pm, T_{\mu\nu}^{1/2}]=\sqrt{(1/2\mp\mu)(3/2\pm\mu)}\,T_{\mu\pm 1/2,\nu}, \quad L_\pm=L_x\pm iL_y,$$
$$[J_\pm=J_x\pm iJ_y]. \tag{9}$$

Since the eigenstates (7) of the Hamiltonian (6) are characterized by three quantum numbers, each element of the set of coherent states must be labelled by three complex numbers, say x, y, and z. If we assume that the coherent states of a top are for $x=z=0$ constructed from $|I-I-I\rangle$ states in the same manner as the harmonic-oscillator coherent states are constructed from the states $|n\rangle$, we obtain

$$|x=0,y,z=0\rangle=\sum_{I=0,1/2,1}^{\infty}\frac{1}{\sqrt{(2I)!}}y^{2I}|I-I-I\rangle\exp\left\{-\frac{1}{2}|y|^2\right\}. \tag{10}$$

Remembering that the $SU(2)$ coherent states can be obtained by applying the operator $e^{\alpha S_+}$ to states of the lowest spin-projection value, we define the coherent states for the top as follows:

$$|xyz\rangle=\exp(-1/2\,yy^*(1+xx^*)(1+zz^*))e^{xJ_+}e^{zL_+}\sum_I\frac{y^{2I}}{\sqrt{(2I)!}}|I-I-I\rangle,$$
$$|xyz\rangle=\exp(-1/2\,yy^*(1+xx^*)(1+zz^*))e^{xJ_+}e^{zL_+}\exp\left\{y\sqrt{2\hat{I}+1}\,T_{-1/2-1/2}^{1/2}\right\}|000\rangle,$$
$$|xyz\rangle=\exp\left(-\frac{1}{2}yy^*(1+xx^*)(1+zz^*)\right)$$
$$\times\sum_{IMK}\sqrt{\frac{(2I)!}{(I+M)!(I-M)!(I+K)!(I-K)!}}x^{I+M}y^{2I}z^{I+K}|IMK\rangle; \tag{11}$$

the operator $\hat{I}$ is defined by the relation

$$\hat{I}|IMK\rangle=I|IMK\rangle. \tag{12}$$

The overlap integral for two coherent states is equal to

$$\langle xyz|x'y'z'\rangle=\exp\{y^*y'(1+x^*x')(1+z^*z')-1/2\,yy^*(1+xx^*)(1+zz^*)$$
$$-1/2\,y'y'^*(1+x'x'^*)(1+z'z'^*)\}. \tag{13}$$

Using the equation

$$\frac{1}{\pi^3}\int dx\,dz\int dy[x^m x^{*m'}y^n y^{*n'}z^l z^{*l'}\exp\{-yy^*(1+zz^*)(1+xx^*)\}$$
$$\times yy^*(yy^*(1+xx^*)(1+zz^*)-1)]=\delta_{mm'}\delta_{nn'}\delta_{ll'}\frac{m!l!(n-m)!(n-l)!}{n!}, \tag{14}$$
$$dx=d\,\mathrm{Re}(x)\,d\,\mathrm{Im}(x),\ dy=d\,\mathrm{Re}(y)\,d\,\mathrm{Im}(y),\ dz=d\,\mathrm{Re}(z)\,d\,\mathrm{Im}(z),$$

we can write the identity operator as

$$\hat{E}=\sum_{IMK}|IMK\rangle\langle IMK|$$
$$=\int dx\,dz\int dy\frac{1}{\pi^3}yy^*(yy^*(1+xx^*)(1+zz^*)-1)|xyz\rangle\langle xyz|. \tag{15}$$

The basis states $|IMK\rangle$ can be expanded in terms of coherent states

$$|IMK\rangle=\frac{1}{\pi^3}\int dx\,dz\int dy\exp\{-1/2\,yy^*(1+xx^*)(1+zz^*)\}yy$$
$$\times(yy^*(1+xx^*)(1+zz^*)-1)x^{*I+M}y^{*2I}z^{*I+K}$$
$$\times\sqrt{\frac{(2I)!}{(I-M)!(I+M)!(I-K)!(I+K)!}}|xyz\rangle. \tag{16}$$

Equations (15) and (16) show that the coherent states $|xyz\rangle$ constitute a complete set of states equivalent to the set $|IMK\rangle$.

To conclude this section let us give the formulas for the mean values of the operators (8), (9), and (12) in the coherent-state basis (these expressions are easy to verify and we shall need them below):

$$\langle xyz|L_x|xyz\rangle=\frac{z+z^*}{1+zz^*}\langle xyz|I|xyz\rangle,$$
$$\langle xyz|L_x^2|xyz\rangle=\langle xyz|L_x|xyz\rangle^2+\frac{1}{2}\langle xyz|\hat{I}|xyz\rangle,$$
$$\langle xyz|L_y|xyz\rangle=\frac{i(z-z^*)}{1+zz^*}\langle xyz|\hat{I}|xyz\rangle,$$
$$\langle xyz|L_y^2|xyz\rangle=\langle xyz|L_y|xyz\rangle^2+\frac{1}{2}\langle xyz|\hat{I}|xyz\rangle;$$
$$\langle xyz|L_z|xyz\rangle=\frac{(zz^*-1)}{1+zz^*}\langle xyz|\hat{I}|xyz\rangle, \tag{17}$$
$$\langle xyz|L_z^2|xyz\rangle=\langle xyz|L_z|xyz\rangle^2+\frac{1}{2}\langle xyz|\hat{I}|xyz\rangle,$$
$$\langle xyz|L_iL_k|xyz\rangle+\langle xyz|L_kL_i|xyz\rangle=2\langle xyz|L_k|xyz\rangle\langle xyz|L_i|xyz\rangle,$$
$$i\neq k=x,y,z,$$

$$\langle xyz|\hat{I}|xyz\rangle=\frac{1}{2}yy^*(1+xx^*)(1+zz^*). \tag{18}$$

Replacing z by x in (17) and (18), we obtain similar expressions for the operators J_x, J_y, and J_z. For the operator $T_{\mu\nu}^{1/2}$ we obtain

$$\langle xyz|T_{1/2\,1/2}^{1/2}|xyz\rangle$$
$$=(y+y^*x^*z^*)\sum\frac{(yy^*(1+xx^*)(1+zz^*))^{2I}}{(2I)!\sqrt{2I+2}}\exp\{-yy^*(1+xx^*)(1+zz^*)\},$$

$$\langle xyz|T^{1/2}_{1/2\,-1/2}|xyz\rangle$$

$$=(x^*y^*-zy)\sum_I \frac{(yy^*(1+xx^*)(1+zz^*))^{2I}}{(2I)!\sqrt{2I+2}}\exp\{-yy^*(1+xx^*)(1+zz^*)\}, \quad (19)$$

$$\langle xyz|T^{1/2}_{-1/2\,-1/2}|xyz\rangle=\langle xyz|T^{1/2}_{1/2\,1/2}|xyz\rangle^*,$$

$$\langle xyz|T^{1/2}_{-1/2\,1/2}|xyz\rangle=-\langle xyz|T^{1/2}_{1/2\,-1/2}|xyz\rangle^*.$$

The set of coherent states constructed according to Perelomov's definition can have, e.g., the following form:

$$|\mathrm{coh}\rangle \sim e^{xJ_+}e^{zL_+}e^{yT^{1/2}_{-1/2\,-1/2}}|000\rangle.$$

This set does not however satisfy the relations (17) and (18) and hence does not manifest the quasiclassical properties discussed in Sec. III.

2. Interpretation of the coherent-state parameters

Introducing the variables α, β, γ, θ, φ and the radial variable r as in

$$x=-e^{-i\alpha}\,\mathrm{tg}\frac{\beta}{2},\quad z=-e^{-i\Phi}\,\mathrm{tg}\frac{\theta}{2},\quad y=\sqrt{2r}\cos\frac{\beta}{2}\cos\frac{\theta}{2}e^{i(\alpha+\gamma+\varphi)/2}, \quad (20)$$

we can write the coherent state (11) as follows

$$|xyz\rangle=|\alpha\beta\gamma\theta\varphi r\rangle=\hat{R}_J(\alpha,\beta,\gamma)\hat{R}_L(\varphi,\theta,0)e^{-r}$$

$$\times\sum_I\frac{(2r)^I}{\sqrt{(2I)!}}|I-I-I\rangle=\sum_{IMK}D^I_{M-I}(\alpha,\beta,\gamma)D^I_{K-I}(\varphi,\theta,0)\frac{(2r)^I}{\sqrt{(2I)!}}|IMK\rangle e^{-r},$$

where

$$\hat{R}_J(\alpha,\beta,\gamma)=e^{-i\alpha\hat{J}_z}e^{-i\beta\hat{J}_y}e^{-i\gamma\hat{J}_z},$$

and $D^I_{MK}(\alpha,\beta,\gamma)$ is a Wigner D function. It follows from (17), (18), and (20) that (we put $|xyz\rangle=|\alpha\beta\gamma\theta\varphi r\rangle \equiv |\mathrm{coh}\rangle$):

$$\langle\mathrm{coh}|L_x|\mathrm{coh}\rangle=-r\cos\varphi\sin\theta,\quad \langle\mathrm{coh}|J_x|\mathrm{coh}\rangle=-r\cos\alpha\sin\beta,$$

$$\langle\mathrm{coh}|L_y|\mathrm{coh}\rangle=-r\sin\varphi\sin\theta,\quad \langle\mathrm{coh}|J_y|\mathrm{coh}\rangle=-r\sin\alpha\sin\beta,$$

$$\langle\mathrm{coh}|L_z|\mathrm{coh}\rangle=-r\cos\theta,\quad \langle\mathrm{coh}|J_z|\mathrm{coh}\rangle=-r\cos\beta,$$

$$\langle\mathrm{coh}|\hat{I}|\mathrm{coh}\rangle=r,\quad \langle\mathrm{coh}|L_x^2+L_y^2+L_z^2|\mathrm{coh}\rangle$$

$$=\langle\mathrm{coh}|J_x^2+J_y^2+J_z^2|\mathrm{coh}\rangle=r\left(r+\tfrac{3}{2}\right). \quad (21)$$

The parameter r is clearly the mean value of the angular-momentum operator in the coherent-state basis, i.e., the absolute value of the vector **J**. The angles φ and θ determine the direction of this vector in the internal coordinate frame and the angles α and β in the laboratory one. They determine the position of the internal coordinate system with respect to the l. s. up to rotations about the angular-momentum vector axis. In order to fix the position of the internal coordinate system completely, another angle is needed, namely γ, contained in the complex number y. From (19) and (20) we obtain the expressions

$$\langle\mathrm{coh}|T^{1/2}_{\mu\nu}|\mathrm{coh}\rangle=D^{1/2}_{\mu\nu}(\alpha'',\beta'',\gamma'')\sqrt{2r}\,e^{-2r}\sum_I\frac{(2r)^{2I}}{(2I)!\sqrt{2I+2}},$$

$$D^{1/2}_{\mu\nu}(\alpha'',\beta'',\gamma'')=\sum_\rho D^{1/2}_{\mu\rho}(-\alpha,\beta,-\gamma)D^{1/2}_{\rho\nu}(0,-\theta,-\varphi), \quad (22)$$

for the mean values of the operators $T^{1/2}_{\mu\nu}$ and

$$\langle\alpha\beta\gamma\theta\varphi r|\alpha'\beta'\gamma'\theta'\varphi' r'\rangle=\exp\left\{2\sqrt{rr'}\left(\cos\frac{\bar\beta}{2}\cos\frac{\bar\theta}{2}\exp[i(\bar\alpha+\bar\varphi+\bar\gamma)/2]-1\right)\right\} \quad (23)$$

for the overlap integral of two coherent states. The angles $\bar\alpha$, $\bar\beta$, $\bar\gamma$, $\bar\theta$, and $\bar\varphi$ are defined by the equations:

$$\hat{R}_J(\bar\alpha,\bar\beta,\bar\gamma)=\hat{R}_J(-\gamma,-\beta,-\alpha)\hat{R}_J(\alpha',\beta',\gamma'),$$

$$\hat{R}_L(\bar\varphi,\bar\theta,0)=\hat{R}_L(0,-\theta,-\varphi)\hat{R}_L(\varphi',\theta',0). \quad (24)$$

3. Boson representation of operators in the coherent-state basis

The boson representation of the operators $T^{1/2}_{\mu\nu}$, J_i, L_i has been considered in detail by Marshalek.[8] Here we shall obtain a representation for these operators in the coherent-state basis (11), using the differential operators d/dx, d/dy, and d/dz. This representation differs from that of Marshalek,[8] since he considered a different state space than $|xyz\rangle$. Using (8) and (16) we obtain, e.g., for the operator J_z the formula:

$$J_z|IMK\rangle=M|IMK\rangle=\int dx\,dz\int dy\,yy^*(yy^*(1+xx^*)(1+zz^*)-1)$$

$$\times\left(x^*\frac{d}{dx^*}-\frac{1}{2}y^*\frac{d}{dy^*}\right)x^{*I+M}y^{*2I}z^{*I+K}\sqrt{\frac{(2I)!}{(I+K)!(I-M)!(I+K)!(I-K)!}}$$

$$\times|xyz\rangle\exp\left(-\frac{1}{2}yy^*(1+xx^*)(1+zz^*)\right). \quad (25)$$

From (11), (14), and (25) we find:

$$J_z|IMK\rangle=\int dx\,dz\int dy\,yy^*(yy^*(1+xx^*)(1+zz^*)-1)$$

$$\times x^{*I+M}y^{*2I}z^{*I+K}\sqrt{\frac{(2I)!}{(I-M)!(I+M)!(I-K)!(I+K)!}} \quad (26)$$

$$\times\exp(-{}^1/_2yy^*(1+xx^*)(1+yy^*))\left(x\frac{d}{dx}-\frac{1}{2}y\frac{d}{dy}\right)|xyz\rangle,$$

i.e., in the coherent-state basis $|xyz\rangle$ the operator J_z acts like the operator $x(d/dx)-\frac{1}{2}y(d/dy)$. Similarly for the operators J_i, L_i, $\hat{I}$, and $T^{1/2}_{\mu\nu}$ we obtain

$$\hat{I}=y\frac{d}{dy};$$

$$L_z=z\frac{d}{dz}-\frac{1}{2}\hat{I},\quad L_+=\frac{d}{dz},\quad L_-=z\left(\hat{I}-z\frac{d}{dz}\right),$$

$$J_z=x\frac{d}{dx}-\frac{1}{2}\hat{I},\quad J_+=\frac{d}{dx},\quad J_-=x\left(\hat{I}-x\frac{d}{dx}\right), \quad (27)$$

$$T^{1/2}_{1/2\,1/2}=\left(y\frac{1}{\sqrt{2+\hat{I}}}+\frac{1}{(1+\hat{I})^2\sqrt{2+\hat{I}}}\frac{d}{dx}\frac{d}{dy}\frac{d}{dz}\right);$$

$$T^{1/2}_{-1/2\,-1/2}=xyz\frac{1}{\sqrt{2+\hat{I}}}+\frac{1}{(1+\hat{I})^2\sqrt{2+\hat{I}}}xz\frac{d}{dx}\frac{d}{dy}\frac{d}{dz}-$$

$$-\frac{1}{(1+\hat{I})\sqrt{2+\hat{I}}}\left(z\frac{d}{dz}+x\frac{d}{dx}\right)\frac{d}{dy}+\frac{1}{\sqrt{2+\hat{I}}}\frac{d}{dy},$$

$$T^{1/2}_{1/2\,-1/2}=zy\frac{1}{\sqrt{2+\hat{I}}}+\frac{z}{(1+\hat{I})^2\sqrt{2+\hat{I}}}\frac{d}{dy}\frac{d}{dz}\frac{d}{dx}-\frac{1}{(1+\hat{I})\sqrt{2+\hat{I}}}\frac{d}{dy}\frac{d}{dx},$$

$$T^{1/2}_{-1/2\,1/2}=xy\frac{1}{\sqrt{2+\hat{I}}}+\frac{x}{(1+\hat{I})^2\sqrt{2+\hat{I}}}\frac{d}{dx}\frac{d}{dz}\frac{d}{dy}-\frac{1}{(1+\hat{I})\sqrt{2+\hat{I}}}\frac{d}{dy}\frac{d}{dz}; \quad (28)$$

here $J_\pm=J_x\pm iJ_y$ and $L_\pm=L_x\pm iL_y$. It can be shown that the operators (27) and (28) satisfy the commutation relations (8) and (9) and that the following relations

$$(J_+)^+=J_-,\ (L_+)^+=L_-,\ (J_z)^+=J_z,\ (L_z)^+=L_z,$$

$$(T^{1/2}_{1/2\,1/2})^+=T^{1/2}_{-1/2\,-1/2},\quad (T^{1/2}_{1/2\,-1/2})^+=-T^{1/2}_{-1/2\,1/2}\ . \quad (29)$$

hold in the space of states $|xyz\rangle$ with the identity operator (15). Equations (27) are a generalization of the Dyson boson representation[9] for the semidirect product of the group $SU(2)\times SU(2)$ times an abelian group. If we introduce a new variable $\bar{z}=zy$ instead of z, then, using Eq. (27), we obtain

$$L_z=\frac{1}{2}\left(\bar{z}\frac{d}{d\bar{z}}-y\frac{d}{dy}\right),\quad L_+=y\frac{d}{d\bar{z}},\quad L_-=\bar{z}\frac{d}{dy}. \tag{30}$$

Making the substitutions $\bar{z}\to b^+$, $d/d\bar{z}\to b$, $y\to a^+$ and $d/dy\to a$, we see that (30) coincides with the Schwinger[10] boson representation for the angular-momentum operator.

III. PHYSICAL PROPERTIES OF THE COHERENT STATES OF A TOP

1. The product of uncertainties

Since the constructed set of states $|xyz\rangle$ is a set of coherent states of the Hamiltonian (6) it should minimize the product of uncertainties of operators, the commutation relations of which uniquely determine the discrete spectrum of the Hamiltonian (6) of the top. In this problem the noncommutativity of the different components of the angular-momentum operator plays the same role as the noncommutativity of the coordinate and momentum operators in the case of the harmonic oscillator. Let us define the operator $\Delta L_i=L_i-\langle\mathrm{coh}|L_i|\mathrm{coh}\rangle$. Using the Schwartz inequality for the product of uncertainties of the values of different components of the angular-momentum operator, we find

$$\langle\mathrm{coh}|(\Delta L_i)^2|\mathrm{coh}\rangle\langle\mathrm{coh}|(\Delta L_k)^2|\mathrm{coh}\rangle \geqslant |\langle\mathrm{coh}|\Delta L_i\Delta L_k|\mathrm{coh}\rangle|^2. \tag{31}$$

It is easy to see that

$$|\langle\mathrm{coh}|\Delta L_i\Delta L_k|\mathrm{coh}\rangle|^2=\tfrac{1}{4}|\langle\mathrm{coh}|[L_i,L_k]|\mathrm{coh}\rangle|^2 + \tfrac{1}{4}\Big|\tfrac{1}{2}\langle\mathrm{coh}|L_iL_k|\mathrm{coh}\rangle+\tfrac{1}{2}\langle\mathrm{coh}|L_kL_i|\mathrm{coh}\rangle - \langle\mathrm{coh}|L_i|\mathrm{coh}\rangle\langle\mathrm{coh}|L_k|\mathrm{coh}\rangle\Big|^2. \tag{32}$$

Using the properties (17) of the coherent states, we obtain from (31) and (32):

$$\frac{\langle\mathrm{coh}|(\Delta L_i)^2|\mathrm{coh}\rangle\langle\mathrm{coh}|(\Delta L_k)^2|\mathrm{coh}\rangle}{|\langle\mathrm{coh}|L_l|\mathrm{coh}\rangle|^2}\geqslant\frac{1}{4},\quad i\neq k\neq l. \tag{33}$$

Using the specific expressions (17) for the mean values of the operators L_i and L_i^2, then in the case when we have

$$\langle\mathrm{coh}|L_i|\mathrm{coh}\rangle=\langle\mathrm{coh}|L_k|\mathrm{coh}\rangle=0$$

and

$$\langle\mathrm{coh}|L_l|\mathrm{coh}\rangle=\langle\mathrm{coh}|\hat{I}|\mathrm{coh}\rangle,$$

Eq. (33) implies

$$\frac{\langle\mathrm{coh}|(\Delta L_i)^2|\mathrm{coh}\rangle\langle\mathrm{coh}|(\Delta L_k)^2|\mathrm{coh}\rangle}{|\langle\mathrm{coh}|L_l|\mathrm{coh}\rangle|^2}=\frac{1}{4}. \tag{34}$$

In the general case it can be shown that the state $|xyz\rangle$ minimizes the product (31) for all components of the angular-momentum vector $\mathbf{L}$ lying in the plane perpendicular to the direction of the vector $\langle xyz|\mathbf{L}|xyz\rangle$. This means that the obtained set of coherent states minimizes the fluctuations of the direction of the angular-momentum vector. It is interesting to note that for $\langle\mathrm{coh}|(\Delta\hat{I})^2|\mathrm{coh}\rangle$ we obtain

$$\langle\mathrm{coh}|(\Delta\hat{I})^2|\mathrm{coh}\rangle=\langle\mathrm{coh}|\hat{I}|\mathrm{coh}\rangle.$$

This result is analogous to the result for the harmonic oscillator, namely

$$\langle\alpha|(\Delta\hat{N})^2|\alpha\rangle=\langle\alpha|\hat{N}|\alpha\rangle$$

where $\hat{N}$ is the number-of-phonons operator.

2. The Euler equation for the asymmetric top

In the Heisenberg representation the equations of motion for the operators L_i and the Hamiltonian of the top (6) have the following form:

$$\dot{L}_i=i[H,L_i],$$

$$\dot{L}_i=\sum_{kl}(L_lL_k+L_kL_l)\varepsilon_{ikl}A_k. \tag{35}$$

Averaging this equation over a coherent state and using property (17) we obtain

$$\langle\mathrm{coh}|\dot{L}_i|\mathrm{coh}\rangle=2\sum_{kl}\langle\mathrm{coh}|L_l|\mathrm{coh}\rangle\langle\mathrm{coh}|L_k|\mathrm{coh}\rangle\varepsilon_{ikl}A_k. \tag{36}$$

Using the definition $\langle\mathrm{coh}|L_i|\mathrm{coh}\rangle=\omega_i/(2A_i)$, we can represent Eq. (36) as:

$$\begin{aligned}\dot{\omega}_x&=A_x\omega_y\omega_z\left(\frac{1}{A_y}-\frac{1}{A_z}\right),\\ \dot{\omega}_y&=A_y\omega_x\omega_z\left(\frac{1}{A_x}-\frac{1}{A_y}\right),\\ \dot{\omega}_z&=A_z\omega_x\omega_y\left(\frac{1}{A_x}-\frac{1}{A_y}\right).\end{aligned} \tag{37}$$

This is precisely the Euler equation for the top in classical mechanics.

IV. APPLICATION OF COHERENT STATES

1. Energy and transition probabilities in a nonaxial rotator

It was shown in Sec. III that the coherent states have properties close to those of classical states and hence we can assume the $\psi_{IM}=P^{IM}|xyz\rangle$ is a good approximation for the wave function of Hamiltonian (6) for $I\gg 1$. Here

$$P^{IM}=\int d\Omega\,\hat{R}_I(\Omega)\sum_K D^{I*}_{MK}(\Omega)$$

is the projection operator onto the state of total angular momentum I. It can be shown that in view of the projecting onto a state of given angular momentum and the rotational invariance of Hamiltonian (6) the mean value of Hamiltonian depends on z alone:

$$E_I(z)=\langle\Psi_{IM}|\Sigma_iA_iL_i^2|\Psi_{IM}\rangle/\langle\Psi_{IM}|\Psi_{IM}\rangle=\tfrac{1}{2}A_x(I(2I-1)\cos^2\varphi\sin^2\theta+I) + \tfrac{1}{2}A_y(I(2I-1)\sin^2\varphi\sin^2\theta+I)+\tfrac{1}{2}A_z(I(2I-1)\cos^2\theta+I),\quad z=-e^{-i\varphi}\,\mathrm{tg}\left(\frac{\theta}{2}\right) \tag{38}$$

Condition

$$\frac{dE_I(z)}{d\varphi}=0, \quad \frac{dE_I(z)}{d\theta}=0$$

implies

$$\cos^2\varphi=1, \sin^2\theta=1,$$

for $(1/A_x)>(1/A_y)$, $(1/A_x)>(1/A_z)$, i.e., the top rotates about axis x and we have

$$E_I=\frac{I}{2}(A_z+A_y)+I^2A_x, \quad \Psi_{IM}=\frac{1}{2^I}\sum_K\sqrt{\frac{(2I)!}{(I-K)!(I+K)!}}|IMK\rangle. \tag{39}$$

Assuming that the quadrupole-moment operator has the form

$$Q_{2\mu}=T^2_{\mu0}q_0+(T^2_{\mu2}+T^2_{\mu-2})q_2,$$

we obtain the following expression for the probabilities of the $E2$ transitions:

$$B(E2, I\to I')=\sum_{\mu M'}|\langle\Psi_{IM}|Q_{2\mu}|\Psi_{I'M'}\rangle|^2$$

$$=\begin{cases}\langle II20|II\rangle^2\left(\frac{q_0}{2}-\sqrt{\frac{3}{2}}q_2\right)^2, & \text{if} \quad I=I',\\ 0, & \text{if} \quad I'=I-1,\\ [(2I-3)/(2I+1)]\frac{1}{4}\left(\sqrt{\frac{3}{2}}q_0+q_2\right)^2, & \text{if} \quad I'=I-2.\end{cases} \tag{40}$$

Up to fluctuation vibrations the results (39) and (40) coincide with the results obtained previously by other authors.[11,12] These terms can be approximately taken into account using the boson representation (27) for the angular-momentum operators, as was done by Bohr and Mottelson.[11]

2. Coherent states and the forced-rotation (cranking) model

Starting from the one-particle Hamiltonian

$$H_{cr}=H_{sp}-\beta\cos\gamma Q_0-\frac{\beta}{\sqrt{2}}\sin\gamma(Q_2+Q_{-2})-\omega J_i, \quad i=x,y,z, \tag{41}$$

it is possible to calculate the total energy of the rotating nucleus.[13] The functions $E_i(\omega)$ and $f_i(\omega)$, where $E_i(\omega)$ is the energy of the nucleus rotating about the axis i with frequency ω and $f_i(\omega)$ is the mean value of the i-th component of the angular momentum, can be assumed to be known. The mean values of the other components of the angular-momentum operator are equal to zero. Using the Schwinger boson representation (30) for the components of the operator $\mathbf{L}$ in the internal coordinate system, we can write the Hamiltonian that provides the discrete spectrum of the rotating nucleus and is invariant with respect to a rotation about the principal axes through the angle π, in the form

$$\begin{aligned}H=h_0&+h_1(a^+a+b^+b)+h_{20}a^+b^+ab+h_{21}(a^+a^+bb+b^+b^+aa)\\&+h_{22}(a^+a^+aa+b^+b^+bb)+h_{30}(a^+a^+b^+bbb+b^+b^+b^+baa\\&+a^+a^+a^+abb+b^+b^+a^+aa)+h_{31}(a^+a^+a^+aaa\\&+b^+b^+b^+bbb)+\dots\end{aligned} \tag{42}$$

Averaging this Hamiltonian over the coherent states (11), we obtain the following expression for the energy of a nuclear state with a given mean value of the angular-momentum operator

$$\begin{aligned}E(y,\bar z)-h_0&+h_1(yy^*+\bar z\bar z^*)+h_{20}yy^*\bar z\bar z^*+h_{21}(y^2\bar z^{*2}+\bar z^2y^{*2})\\&+h_{22}(y^2y^{*2}+\bar z^2\bar z^{*2})+h_{30}(y^2\bar z\bar z^{*3}+\bar z^3\bar z^*y^{*2}+y^3y^*\bar z^{*2}\\&+\bar z^2yy^{*3})+h_{31}(y^3y^{*3}+\bar z^3\bar z^{*3})+\dots,\\ \langle y\bar z|L_x|y\bar z\rangle&=(\bar zy^*+y\bar z^*), \ \langle y\bar z|L_y|y\bar z\rangle=i(\bar zy^*-y\bar z^*),\\ \langle y\bar z|L_z|y\bar z\rangle&=\bar z\bar z^*-yy^*.\end{aligned} \tag{43}$$

Let us consider special cases of Eq. (43).

1.

$$\begin{aligned}y=\bar z, \ \langle L_x\rangle=\langle yy|L_x|yy\rangle=2yy^*,\\ \langle yy|L_y|yy\rangle=\langle yy|L_z|yy\rangle=0,\end{aligned} \tag{44a}$$

$$E_1=h_0+h_{10}\langle L_x\rangle+\frac{1}{4}(h_{20}+2h_{21}+2h_{22})\langle L_x\rangle^2+\frac{1}{4}(2h_{30}+h_{31})\langle L_x\rangle^3+\dots$$

2.

$$\begin{aligned}\bar z=-iy, \ \langle L_y\rangle=\langle y,-iy|L_y|y,-iy\rangle=2yy^*,\\ \langle y-iy|L_x|y-iy\rangle=\langle y-iy|L_z|y-iy\rangle=0,\end{aligned} \tag{44b}$$

$$E_2=h_0+h_{10}\langle L_y\rangle+{}^1/_4(h_{20}-2h_{21}+2h_{22})\langle L_y\rangle^2-{}^1/_4(2h_{30}-h_{31})\langle L_y\rangle^3+\dots$$

3.

$$\begin{aligned}y=0, \ \langle 0\bar z|L_z|0\bar z\rangle=\langle L_z\rangle=zz^*, \ \langle 0\bar z|L_x|0\bar z\rangle=\langle 0\bar z|L_y|0\bar z\rangle=0,\\ E_3=h_0+h_1\langle L_z\rangle+h_{22}\langle L_z\rangle^2+h_{31}\langle L_z\rangle^3+\dots\end{aligned} \tag{44c}$$

Assuming that the coherent states for sufficiently high spins provide a good approximation for the nuclear wave function, the energies $E_i(\omega)$ must coincide with the energies (44), i.e., we have

$$\begin{aligned}&E_x(\omega)=E_1, \quad f_x(\omega)=\langle L_x\rangle, \quad \langle L_y\rangle=\langle L_z\rangle=0,\\ &E_y(\omega)=E_2, \quad f_y(\omega)=\langle L_y\rangle, \quad \langle L_x\rangle=\langle L_z\rangle=0,\\ &E_z(\omega)=E_3, \quad f_z(\omega)=\langle L_z\rangle, \quad \langle L_y\rangle=\langle L_x\rangle=0\end{aligned} \tag{45}$$

for each value of ω. Knowing $E_i(\omega)$ and $f_i(\omega)$, it is thus possible to calculate the coefficients h_{nm} in the Hamiltonian (42). Diagonalizing the Hamiltonian, we obtain the complete spectrum of states of the rotating nucleus including excitations related to the nutational motion. An effective Hamiltonian of the type (42) was discussed by Mikhaĭlov *et al.*[14]

V. CONCLUSIONS

Let us point out the following interesting fact. The set of coherent states (11) contains states of both integer and half-integer spin values. Thus when investigating the quasiclassical properties of a rotating nucleus we must consider even and odd nuclei simultaneously.

After this investigation was concluded an article appeared,[15] devoted to similar questions. Their results follow from ours if we put the parameter $z=0$. Besides, they[15] did not consider questions related to the description of the motion of an asymmetric top.

The author is indebted to I. N. Mikhaĭlov for numerous discussions of problems related to the theory of coherent states, leading to the writing of this paper. The author also thanks R. V. Dzholos for reading the manuscript and making helpful comments.

[1]E. Schrödinger, Naturwiss. **14**, 664 (1926).
[2]R. I. Glauber, Phys. Rev. **130**, 2529 (1963); **131**, 2766 (1963).
[3]F. W. Cummings and I. R. Johnston, Phys. Rev. **151**, 105 (1966).
[4]P. Carruthers and K. S. Dy, Phys. Rev. **147**, 214 (1966).
[5]A. M. Perelomov, Commun. Math. Phys. **26**, 222 (1972).
[6]I. H. Radcliffe, J. Phys. A **4**, 313 (1971).

[7]R. Holtz and I. Hanus, J. Phys. A 7, No. 4, L37 (1974).
[8]E. R. Marshalek, Phys. Rev. C11, 1426 (1975).
[9]F. J. Dyson, Phys. Rev. 102, 1217 (1956).
[10]J. Schwinger, in Quantum Theory of Angular Momentum, edited by L. C. Biedenharn and N. J. Van Dam, Academic Press, New York, 1965.
[11]O. Bohr and B. Mottelson, Nuclear Structure, Vol. 2, W. A. Benjamin, New York [Russian translation, Mir, 1976].
[12]I. N. Mikhaĭlov, Preprint JINR R4-7862, Dubna, 1974.
[13]K. Neergord and V. V. Pashkevich, Preprint JINR R4-8947, Dubna, 1975.
[14]I. N. Mikhaĭlov, E. N. Nadzhakov and D. Karadzhov, Fiz. Elem. Chastits At. Yadra 4, 311 (1973) [Sov. J. Part. Nucl. 4, 129 (1973)].
[15]D. Baumik, T. Nag, and D. Dutta-Roy, J. Phys. A 8, 1868 (1975).

Translated by P. Winternitz

Volume 76B, number 1 PHYSICS LETTERS 8 May 1978

STUDIES OF THE GROUND-STATE PROPERTIES OF THE LIPKIN–MESHKOV–GLICK MODEL VIA THE ATOMIC COHERENT STATES

Robert GILMORE [1] and Da Hsuan FENG
Department of Physics and Atmospheric Science, Drexel University, Philadelphia, PA 19104, USA

Received 14 December 1977

The ground-state energy of the Lipkin–Meshkov–Glick hamiltonian was estimated by the Bogoliubov–Lieb inequalities which involve the atomic coherent states. It was found that the so-called Q-representation of the pseudospin hamiltonian provides a good estimation of the ground-state energy. Furthermore, nuclear phase transitions can be vividly described by the deviations of the pseudospin from the southpole of the so-called Bloch sphere.

The energy eigenvalue spectrum of the nuclear hamiltonian proposed by Lipkin, Meshkov and Glick [1],

$$H = \tfrac{1}{2}\epsilon \sum_{p} \sigma a^{+}_{p\sigma} a_{p\sigma} + \tfrac{1}{2} V \sum_{p,p',\sigma} a^{+}_{p\sigma} a^{+}_{p'\sigma} a_{p'-\sigma} a_{p-\sigma}$$

$$+ \tfrac{1}{2} W \sum_{p,p',\sigma} a^{+}_{p\sigma} a^{+}_{p'-\sigma} a_{p'\sigma} a_{p-\sigma} \tag{1}$$

can be obtained after transformation to a pseudospin representation by means of the identifications

$$J_z = \tfrac{1}{2} \sum_{p} \sigma a^{+}_{p\sigma} a_{p\sigma}, \quad J_+ = \sum_{p} a^{+}_{p,+1} a_{p,-1} = J_-^{\dagger}. \tag{2}$$

The resulting pseudospin hamiltonian,

$$H_s = \epsilon J_z + \tfrac{1}{2} V (J_+^2 + J_-^2) + \tfrac{1}{2} W (J_+ J_- + J_- J_+), \tag{3}$$

may then be diagonalized within each fixed-J subspace because the pseudospin operators (2) commute with the hamiltonians (1), (3). The subspace containing the ground state has maximum possible $J = N/2$.

The ground state energy E_1 can be obtained from the limit

$$E_1 = \lim_{\beta \to \infty} -\beta^{-1} \mathrm{Tr}_J \mathrm{e}^{-\beta H_s}. \tag{4}$$

The trace over the fixed-J subspace may be estimated from the inequalities [2]

$$\frac{2J+1}{4\pi} \int \mathrm{e}^{-\beta P(\hat{H}_s, \Omega)} \, \mathrm{d}\Omega \geqslant \mathrm{Tr}_J \mathrm{e}^{-\beta H_s}$$

$$\geqslant \frac{2J+1}{4\pi} \int \mathrm{e}^{-\beta Q(\hat{H}_s, \Omega)} \, \mathrm{d}\Omega. \tag{5}$$

The integrals are taken over the sphere surface, $\Omega = (\theta, \phi)$, $\mathrm{d}\Omega = \sin\theta \mathrm{d}\theta \mathrm{d}\phi$, and the functions $P(\hat{H}_s, \Omega)$, $Q(\hat{H}_s, \Omega)$ are defined over the sphere surface by means of the atomic coherent state representation [3–6].

Atomic coherent states are different from the more familiar coherent states associated with the harmonic oscillator, which have been discussed extensively by Glauber [7] and others [8], and which have been used in nuclear studies by Rowe [9] and others. The atomic coherent state $|J, \Omega\rangle$ is obtained by applying a unitary transformation $U(\Omega)$ to the state $|J, J\rangle$ in the SU(2) invariant subspace of dimension $2J + 1$:

$$U(\Omega) = \exp(\tfrac{1}{2}\theta)(\mathrm{e}^{\mathrm{i}\phi} J_- - \mathrm{e}^{-\mathrm{i}\phi} J_+),$$

$$|J, \Omega\rangle = U(\Omega) \, |J, J\rangle. \tag{6}$$

The unitary transformations $U(\Omega)$ and atomic coherent states $|J, \Omega\rangle$ exist in 1–1 correspondence with points on a sphere surface ($0 \leqslant \theta \leqslant \pi$, $0 \leqslant \phi \leqslant 2\pi$) corresponding to the Bloch sphere of quantum optics. The properties of atomic coherent states have been

[1] Permanent address: Department of Physics, University of South Florida, Tempa, FL 33620.

studied in detail. They are nonorthogonal and overcomplete. As a result of this overcompleteness every operator mapping a fixed-J subspace into itself can be expressed in "diagonal" form in the atomic coherent state representation:

$$\hat{A} = \sum_{M,M'} |J, M\rangle A_{MM'} \langle J, M'|$$

$$= \frac{2J+1}{4\pi} \int |J, \Omega\rangle P(\hat{A}, \Omega) \langle J, \Omega| \, d\Omega. \quad (7P)$$

The function $P(\hat{A}, \Omega)$ is called the "P-representative" of the operator $\hat{A}$. The "Q-representative" of $\hat{A}$ is the expectation value of $\hat{A}$ in the atomic coherent state representation

$$Q(\hat{A}, \Omega) = \langle J, \Omega|\hat{A}|J, \Omega\rangle. \quad (7Q)$$

The P- and Q-representatives of symmetrized functions of angular momentum operators are simple to construct [10]. The operator function is first expressed in terms of irreducible spherical tensor operators $\mathcal{Y}^l_m(J)$. The P-representative of $\mathcal{Y}^l_m(J)$ in the invariant subspace of dimension $2J+1$ is $(2J+1+l)!/(2J+1)!2^l \, Y^l_m(\Omega)$; its Q-representative in the same space is $(2J)!/(2J-1)! \times 2^l Y^l_m(\Omega)$. The P- and Q-representatives of H_s are

$$P(\hat{H}_s, \Omega) = \epsilon(J+1)\cos\theta + V(J+1)(J+\tfrac{3}{2})\sin^2\theta$$

$$\times (e^{2i\phi} + e^{-2i\phi})/a + W(J+1)(J+\tfrac{3}{2})\sin^2\theta,$$

$$Q(\hat{H}_s, \Omega) = \epsilon J\cos\theta + V(J)(J-\tfrac{1}{2})\sin^2\theta$$

$$\times (e^{2i\phi} + e^{-2i\phi})/2 + W(J)(J-\tfrac{1}{2})\sin^2\theta. \quad (8)$$

The trace of an operator is given by the integral of its Q-representative

$$\mathrm{Tr}_J \hat{A} = \frac{2J+1}{4\pi} \int Q(\hat{A}, \Omega) \, d\Omega. \quad (9)$$

The right-hand inequality in eq. (5) is easily obtained by combining the Bogoliubov inequality ($\langle\psi|e^X|\psi\rangle \geqslant \exp\langle\Psi|X|\Psi\rangle$ for X hermitian and $\langle\psi|\psi\rangle = 1$) with eq. (9):

$$\mathrm{Tr}\, e^{-\beta\hat{H}_s} = \frac{2J+1}{4\pi} \int \langle J, \Omega| \, e^{-\beta\hat{H}_s} |J, \Omega\rangle \, d\Omega$$

$$\geqslant \frac{2J+1}{4\pi} \int e^{-\beta Q(\hat{H}_s, \Omega)} \, d\Omega. \quad (10)$$

The left-hand inequality in eq. (5) is due to Lieb [2].

The asymptotic values of the integrals appearing in eq. (5) as $\beta \to +\infty$ are easily computed. The largest contribution to integrals of the form $\int \exp(-\beta f(\Omega))\, d\Omega$ comes from the neighborhood in which $f(\Omega)$ assumes its global minimum, at $\Omega = \Omega_m$. Then

$$\int e^{-\beta f(\Omega)} \, d\Omega = e^{-\beta f(\Omega_m)} \int e^{-\beta[f(\Omega) - f(\Omega_m)]} \, d\Omega. \quad (11)$$

The integral on the right-hand side has a power-law asymptotic dependence on β [$O(\beta^{-n/2}), n \geqslant 1$] [11], so that

$$\lim_{\beta\to\infty} -\beta^{-1} \ln \int e^{-\beta f(\Omega)} \, d\Omega = \min_{\Omega} f(\Omega) = f(\Omega_m). \quad (12)$$

This result, combined with eqs. (4) and (5), provides lower and upper bounds for the ground-state energy:

$$\min_{\Omega} P(\hat{H}_s, \Omega) \leqslant E_1 \leqslant \min_{\Omega} Q(\hat{H}_s, \Omega). \quad (13)$$

Minimization of the functions (8) over ϕ leads to the following functions of θ which must be minimized:

$$f_P(\theta)/\epsilon(J+1) = \cos\theta - \tfrac{1}{2}\tilde{V}_P \sin^2\theta,$$

$$f_Q(\theta)/\epsilon J = \cos\theta - \tfrac{1}{2}\tilde{V}_Q \sin^2\theta, \quad (14)$$

where $\tilde{V}_P = 2(|V| - W)(J+\tfrac{3}{2})/\epsilon$ and $\tilde{V}_Q = 2(|V| - W)(J-\tfrac{1}{2})/\epsilon$. The functions (12) have a global minimum at $\theta = \pi$ ($|J, -J\rangle$) if $\tilde{V} \leqslant 1$ and a global minimum at $\cos\theta = -1/\tilde{V}$ if $\tilde{V} \geqslant 1$. The upper and lower bounds on the ground-state energy are therefore

$$-\epsilon(J+1) \leqslant E_1 \leqslant -\epsilon J,$$

$$-\epsilon(J+1)\tfrac{1}{2}(\tilde{V}_P + \tilde{V}_P^{-1}) \leqslant E_1 \leqslant -\epsilon J, \quad (15)$$

$$-\epsilon(J+1)\tfrac{1}{2}(\tilde{V}_P + \tilde{V}_P^{-1}) \leqslant E_1 \leqslant -\epsilon J \tfrac{1}{2}(\tilde{V}_Q + \tilde{V}_Q^{-1}).$$

In table 1 we compare the upper and lower bounds given here with the exact ground-state eigenvalues given in ref. [1] for $N = 14, 30, 50$ for various values of NV/ϵ ($W = 0$).

The numerical results indicate that the Q-representative is a reasonably accurate upper bound on the ground-state energy. The problem of minimizing $Q(\hat{H}_s, \Omega)$ is equivalent to the problem of determining the best ground-state wave function for $\hat{H}_s$ from the class of

Volume 76B, number 1 PHYSICS LETTERS 8 May 1978

Table 1
$-E_1$ as a function of NV/ϵ for N = 14, 30, 50. The exact values are taken from ref. [1].

NV/ϵ	N = 14 exact	N = 14 bounds	N = 30 exact	N = 30 bounds	N = 50 exact	N = 50 bounds
0	7.000	7.000 8.000	15.000	15.000 16.000	25.000	25.000 26.000
0.4	7.038	7.000 8.000	15.040	15.000 16.000	25.041	25.000 26.000
0.6	7.088	7.000 8.000	15.090	15.000 16.000	25.096	25.000 26.000
0.8	7.163	7.000 8.000	15.179	15.000 16.000	25.186	25.000 26.000
1.0	7.270	7.000 8.151	15.314	15.000 16.073	25.340	25.000 26.044
2.0	8.636	8.385 11.361	18.547	18.379 21.236	31.037	30.878 33.692
5.0	17.268	17.004 24.945	38.049	37.802 45.455	64.043	63.801 71.353
∞ a)	0.942	0.929 1.388	0.973	0.967 1.173	0.984	0.980 1.102

a) The last line gives $|E_1|/V(N/2)^2$, which converges to 1 for $N \to \infty$, $V \to 0$, $NV/\epsilon \to \infty$.

variational wave functions of the form $|J, \Omega\rangle$. These considerations, together with eq. (6), suggest that it is useful to regard the ground-state wave function for $\hat{H}_s$ as a "wave-packer", or sharply peaked superposition of states $|J, \Omega\rangle$ centered at Ω_m. We can then say that the ground state "points toward" Ω_m on the Bloch sphere surface. For $\tilde{V}_P < 1$ the ground state points toward the south pole, while for $\tilde{V}_Q > 1$ it points toward some other point in the southern hemisphere. The transition from the south polar ground state (spherical closed-shell regime) to the non-south polar ground state (deformed regime) can be regarded as a second-order phase transition. Although the values of the expansion coefficients $C_m(\delta)$,

$$|\psi_1\rangle = \sum_M C_M(\delta)\, |J, M\rangle, \tag{16}$$

depend smoothly on the parameter $\delta = N(|V| - W)/\epsilon$, the value of Ω_m and the minimum values of $f_Q(\Omega)$, $f_P(\Omega)$ suffer discontinuities in their second derivatives at $\delta \approx 1$. The order parameter for this phase transition is $\theta - \pi$ and the bifurcation parameter is δ. In the limit $N \to \infty$, $|V|$, $W \to 0$ and $N(|V| - W)/\epsilon = \delta$, the upper and lower bounds on the ground-state energy per nucleon converge because $\lim_{N\to\infty} P(\hat{H}_s, \Omega_m) - Q(\hat{H}_s, \Omega_m)/N \to 0$:

$$\lim_{N\to\infty} E_1 = -\epsilon/2, \qquad \delta \leqslant 1,$$
$$= -\tfrac{1}{2}\epsilon(\delta + \delta^{-1}), \qquad \delta \geqslant 1. \tag{17}$$

The P- and Q-representatives provide a useful tool for studying the qualitative properties of nuclear hamiltonians more complicated than eq. (1), provided they can be transcribed into the pseudospin formalism. For example, eqs. (8) or (14) clearly reveal that the terms $(J_+^2 + J_-^2)$ contribute to the phase transition, and that the term $(J_+J_- + J_-J_+)$ does also if $W < 0$, but tends to suppress the transition if $W > 0$. For pseudospin hamiltonians $\hat{H}_{ps}$ involving only finite powers of angular momentum operators the ground-state energy per nucleon is given, in the $N \to \infty$ limit, by

$$\lim_{N\to\infty} E_1/N = \lim_{N\to\infty} \min_{\Omega} Q(\hat{H}_{ps}, \Omega). \tag{18}$$

References

[1] H.J. Lipkin, N. Meshkov and A.J. Glick, Nucl. Phys. 62 (1965) 188.
[2] E.H. Lieb, Commun. Math. Phys. 31 (1973) 340.
[3] J.M. Radcliffe, J. Phys. A4 (1971) 313.
[4] F.T. Arrechi, E. Courtens, R. Gilmore and H. Thomas, Phys. Rev. A6 (1972) 2211.
[5] A.M. Perelomov, Commun. Math. Phys. 26 (1972) 222.
[6] R. Gilmore, Rev. Mex. Fis. 23 (1974) 143.
[7] R. Glauber, Phys. Rev. 130 (1962) 2529.
[8] E. Schroedinger, Naturwiss. 14 (1927) 644.
[9] D.J. Rowe, Can. J. Phys. 54 (1976) 1941.
[10] R. Gilmore, J. Phys. A9 (1976) L65.
[11] R. Gilmore, J. Math. Phys. 18 (1977) 17.

Nuclear Physics **A301** (1978) 189–204;

PHASE TRANSITIONS IN NUCLEAR MATTER DESCRIBED BY PSEUDOSPIN HAMILTONIANS

R. GILMORE and D. H. FENG

Department of Physics and Atmospheric Science, Drexel University, Philadelphia, Pennsylvania 19104

Received 19 December 1977

Abstract: Upper and lower bounds on the ground-state energy per nucleon E_g/N and the free energy per nucleon $F(\beta)/N$ are constructed for nuclear systems described by pseudospin Hamiltonians. In the limit of large numbers of nucleons these bounds become equal. A simple algorithm is developed for computing E_g/N, $F(\beta)/N$ and S/N (entropy per nucleon) exactly in the $N \to \infty$ limit. The values of E_g/N and $F(\beta)/N$ are obtained by computing the minimum value of associated potential functions h_C and Φ. These potentials are constructed very simply from the pseudospin Hamiltonian. Ground-state energy phase transitions are determined by investigating how the minima of the potential h_C change as a function of changing nuclear interaction parameters. Thermodynamic phase transitions are determined by investigating how the minima of the potential Φ change as a function of changing nuclear temperature. Concise conditions are given for the occurrence of a second-order phase transition of either type. These conditions define critical values of the nuclear interaction parameters at which a ground-state energy second-order phase transition occurs and the critical temperature at which a thermodynamic second-order phase transition occurs. A "crossover theorem" relates the occurrence of a ground-state energy phase transition to a thermodynamic phase transition.

1. Introduction

We study phase transitions in nuclear systems described by pseudospin Hamiltonians in the nuclear matter limit ($N \to \infty$). Pseudospin Hamiltonians [1]) are constructed from the nuclear pseudospin operators,

$$J_3 = \tfrac{1}{2} \sum_{p,\sigma=\pm 1} \sigma a^\dagger_{p\sigma} a_{p\sigma},$$

$$J_+ = \sum_p a^\dagger_{p,+1} a_{p,-1},$$

$$J_- = \sum_p a^\dagger_{p,-1} a_{p,+1} = J^\dagger_+. \tag{1.1}$$

Here $\sigma = \pm 1$ indexes the upper or lower of two N-fold degenerate levels containing a total of N nucleons, and p indexes the various independent orthonormal degenerate states within each level. When the nuclear ground state has maximum pseudospin $J = \frac{1}{2}N$ and minimum z-component $M = -\frac{1}{2}N$, all N nucleons are in the lower shell $\sigma = -1$, all p are different, and the nucleus must assume a spherically symmetric shape. When the nuclear ground state does not have minimum M, the indices p need not all be different, and the nuclear ground-state shape need not be spherically

May 1978

symmetric. Nuclear shape phase transitions have previously been studied by computing the dependence of the ground-state energy and the gap as a function of certain nuclear interaction parameters for particular model Hamiltonians within the fixed manifold of angular momentum states with $J = \frac{1}{2}N$. Lipkin, Meshkov and Glick [1]) have shown that the nuclear pseudospin Hamiltonian

$$\mathscr{H} = \varepsilon J_3 + \tfrac{1}{2}V(J_+^2 + J_-^2) + \tfrac{1}{2}W(J_+J_- + J_-J_+) \tag{1.2}$$

describes a ground-state phase transition for the nuclear interaction parameters $V \approx 1/N\varepsilon$, $W = 0$ by computing the ground-state energy for $N = 14$, 30, and fifty for various values of NV/ε.

In the nuclear matter limit we are able to compute the ground-state energy per nucleon exactly for a large class of pseudospin Hamiltonians. We are thus able to determine the precise conditions on the nuclear interaction parameters under which a nuclear shape phase transition occurs. The class of pseudospin Hamiltonians for which the present study is valid is described in sect. 3. In sect. 4, we construct upper and lower bounds on the free energy per nucleon and the ground-state energy per nucleon. These bounds converge to the same value in the nuclear matter limit. Further, these bounds are obtained by minimizing appropriate "potentials" defined over a certain space of variables. The potentials themselves are constructed trivially from the pseudospin Hamiltonian. The physical interpretation of these potentials is presented in sect. 5.

Nuclear phase transitions are then investigated in sects. 6 and 7. We discuss two types of nuclear phase transition, the ground-state energy phase transition and the thermodynamic phase transiton. Both types of phase transition are studied by investigating how the minimum of the appropriate potential changes as a function of changing parameter (s).

A ground-state energy phase transition may occur as a function of increasing nuclear interaction parameters. In the absence of interactions, the nuclear ground state is spherically symmetric, with $J = \frac{1}{2}N$, $M = -J$. As the nuclear interaction parameters are increased, the nuclear shape becomes distorted when the ground state no longer has $M = -J$. We give necessary and sufficient conditions for the occurrence of a second-order phase transition from the spherical to the deformed regime as a function of increasing nuclear interaction parameters.

If some mechanism excites the nucleus, which then "thermalizes" before decaying completely back to the ground state, it is sensible to talk about a "nuclear temperature" [2]) and a free energy. We show that, in the limit of very high temperatures, the nucleus is spherically symmetric. As the temperature decreases, a phase transition from spherical to deformed shape occurs at a critical temperature T_c provided the nuclear ground state is deformed. When the thermodynamic phase transition is second order, we give an explicit expression for T_c. We also state a "crossover" theorem [3]) and indicate its proof: If the nuclear ground state has undergone a ground-state energy phase transition from spherical to deformed as a function of increasing nuclear interaction parameters, then the system will undergo a thermodynamic phase transition from deformed to spherical as a function of increasing nuclear temperature.

The study of phase transitions is made possible by studying the (bifurcation [4])) properties of the minima of some "potential" †. The "potential" in turn can be written down directly from the pseudospin Hamiltonian. Thus, we have reduced to an algorithm the procedure for studying phase transitions according to

$$\begin{matrix}\text{pseudospin}\\\text{Hamiltonian}\end{matrix} \rightarrow \text{potential} \rightarrow \begin{matrix}\text{free energy}\\\text{per nucleon}\end{matrix} \rightarrow \begin{matrix}\text{ground-state energy}\\\text{per nucleon}\end{matrix} \qquad (1.3)$$

This algorithm is made possible through the exploitation of atomic coherent states. Therefore, we begin our study by describing these states and their properties in sect. 2.

2. Properties of atomic coherent states

An elegant and convenient continuous-basis representation for a collection of two-level atoms in quantum optics was proposed independently by Radcliffe [5]) and by Arecchi, Courtens, Gilmore and Thomas [6]) (ACGT) in terms of what ACGT call Bloch states or atomic coherent states.

Atomic coherent states are different from, but closely analogous to, the more familiar Glauber [7]) coherent states for the electromagnetic field which have been applied in nuclear physics by Rowe and Basserman [8]) in their treatment of large amplitude heavy-ion collision problems. The atomic coherent states in the SU(2) invariant subspace J of dimension $2J+1$ are obtained by applying a unitary transformation to the state $|J, J\rangle$ of highest weight

$$\left|\begin{matrix}J\\\Omega\end{matrix}\right\rangle = \hat{U}(\Omega)\left|\begin{matrix}J\\J\end{matrix}\right\rangle,$$

$$\hat{U}(\Omega) = \exp(-\eta J_+ + \eta^* J_-), \qquad \eta = \exp(-i\phi)\tfrac{1}{2}\theta. \qquad (2.1)$$

The angle θ is measured from the north pole, and $\Omega = (\theta, \phi)$ represents a point on a sphere surface, called the Bloch sphere in quantum optics. The atomic coherent states may be expressed in terms of the more familiar angular momentum basis states $|J, M\rangle$, which obey

$$\hat{J}^2\left|\begin{matrix}J\\M\end{matrix}\right\rangle = J(J+1)\left|\begin{matrix}J\\M\end{matrix}\right\rangle,$$

$$J_3\left|\begin{matrix}J\\M\end{matrix}\right\rangle = M\left|\begin{matrix}J\\M\end{matrix}\right\rangle, \qquad (2.2)$$

† This should not be confused with the usual nuclear potential.

by

$$\left|\begin{matrix}J\\ \Omega\end{matrix}\right\rangle = \sum_{M=-J}^{+J} \left|\begin{matrix}J\\ M\end{matrix}\right\rangle\left\langle\begin{matrix}J\\ M\end{matrix}\middle|\begin{matrix}J\\ \Omega\end{matrix}\right\rangle$$
$$= \sum_{M=-J}^{+J} \binom{2J}{J\pm M}^{\frac{1}{2}} (\cos\tfrac{1}{2}\theta)^{J+M}(e^{i\phi}\sin\tfrac{1}{2}\theta)^{J-M} \left|\begin{matrix}J\\ M\end{matrix}\right\rangle. \qquad (2.3)$$

The Bloch states are non-orthogonal and overcomplete. The inner product of two Bloch states is given by

$$\left|\left\langle\begin{matrix}J\\ \Omega'\end{matrix}\middle|\begin{matrix}J\\ \Omega\end{matrix}\right\rangle\right|^2 = (\cos\tfrac{1}{2}\Theta)^{4J},$$
$$\cos\Theta = \cos\theta'\cos\theta + \sin\theta'\sin\theta\cos(\phi'-\phi). \qquad 2.4)$$

The overcompleteness allows for the (non-unique) resolution of the identity operator in the J-invariant space of dimension $2J+1$:

$$\frac{2J+1}{4\pi}\int \left|\begin{matrix}J\\ \Omega\end{matrix}\right\rangle\left\langle\begin{matrix}J\\ \Omega\end{matrix}\right| d\Omega = \sum_{M=-J}^{+J} \left|\begin{matrix}J\\ M\end{matrix}\right\rangle\left\langle\begin{matrix}J\\ M\end{matrix}\right| = I_{2J+1}. \qquad (2.5)$$

The integral is taken over the Bloch sphere, and $d\Omega = \sin\theta d\theta d\phi$.

The most important properties of the atomic coherent states required for the present work depend on their overcompleteness. Any operator $\hat{G}$ mapping the J-invariant subspace into itself can always [6]) be expressed in "diagonal form" in the atomic coherent state representation

$$\hat{G} = \frac{2J+1}{4\pi}\int \left|\begin{matrix}J\\ \Omega\end{matrix}\right\rangle P_J(\hat{G};\Omega)\left\langle\begin{matrix}J\\ \Omega\end{matrix}\right| d\Omega. \qquad (2.6P)$$

The integrand $P_J(\hat{G};\Omega)$ is called the P-representative of the operator $\hat{G}$ in the J-invariant subspace. It is a function defined over the surface of the Bloch sphere. This function depends on the operator $\hat{G}$ and the J-invariant subspace in which $\hat{G}$ acts. Another useful function associated with the operator $\hat{G}$ is its Q-representative, defined as the expectation value of $\hat{G}$ in the coherent state representation

$$Q_J(\hat{G};\Omega) = \left\langle\begin{matrix}J\\ \Omega\end{matrix}\middle| \hat{G} \middle|\begin{matrix}J\\ \Omega\end{matrix}\right\rangle. \qquad (2.6Q)$$

The Q- and P-representatives for symmetrized spherical tensor operators $\mathscr{Y}^L_M(\boldsymbol{J})$ [ref. [9])] are easily expressed in terms of the corresponding spherical harmonic functions [10])

$$Q_J(\mathscr{Y}^L_M(\boldsymbol{J});\Omega) = \frac{(2J)!}{(2J-L)!2^L} Y^L_M(\Omega), \qquad (2.7Q)$$

$$P_J(\mathscr{Y}^L_M(\boldsymbol{J});\Omega) = \frac{(2J+1+L)!}{(2J+1)!2^L} Y^L_M(\Omega). \qquad (2.7P)$$

For example, the resolution of the identity ($I_{2J+1} \approx \mathscr{Y}_0^0(\boldsymbol{J})$) given in (2.5) is the special case of (2.6*P*) and (2.7*P*) with $L = 0$.

3. Model pseudospin Hamiltonians

We treat model pseudospin Hamiltonians $\hat{\mathscr{H}}$ for N interacting nucleons having the form

$$\hat{\mathscr{H}}/N = \hat{h}_Q. \tag{3.1}$$

Here and below, $h = h(\boldsymbol{v})$ is a function of three arguments $\boldsymbol{v} = (v_3, v_+, v_-)$. The operator $\hat{h}_Q$ is obtained from the function h through the operator substitutions and symmetrization:

$$h \to \hat{h}_Q: \quad \begin{array}{l} v_3 \to J_3/N \\ v_\pm \to J_\pm/N. \end{array} \tag{3.2Q}$$

The pseudospin operators $\boldsymbol{J}$ are given in (1.1). The form (3.1) of the Hamiltonian $\hat{\mathscr{H}}$ is dictated by homogeneity requirements. That is, the binding energy of nuclear matter is essentially proportional to the number of nucleons present. The binding energy per nucleon remains bounded in the nuclear matter limit only when the pseudospin Hamiltonian has the form (3.1). The homogeneity arguments leading to this form have been discussed by Gilmore [11]). Rigorous requirements for the existence of the $N \to \infty$ limit given by Ruelle [12]) also lead to a pseudospin Hamiltonian of the form (3.1).

We further assume: (a) $h(\boldsymbol{v})$ is a polynomial of degree $k < \infty$; (b) $h_Q(\boldsymbol{J}/N)$ is hermitian and symmetrized.

For future use, it is convenient to introduce another set of substitutions into the function h. These substitutions are defined by

$$h \to h_C: \quad \begin{array}{l} v_3 \to r\cos\theta \\ v_\pm \to r\sin\theta\, e^{\pm i\phi}. \end{array} \tag{3.2c}$$

The function $h_C = h_C(r, \theta, \phi)$ will be encountered in sect. 4.

Example: For the pseudospin Hamiltonian (1.2)

$$\begin{aligned} h &= \varepsilon v_3 + \tfrac{1}{2}NV(v_+^2 + v_-^2) + NWv_+v_-, \\ \hat{h}_Q &= \varepsilon(J_3/N) + \tfrac{1}{2}NV((J_+/N)^2 + (J_-/N)^2) + \tfrac{1}{2}NW(J_+J_- + J_-J_+)/N^2 = \hat{\mathscr{H}}/N, \\ h_C &= \varepsilon r\cos\theta + \tfrac{1}{2}NVr^2\sin^2\theta(e^{2i\phi} + e^{-2i\phi}) + NWr^2\sin^2\theta. \end{aligned} \tag{3.3}$$

4. Bounds on the free energy per nucleon

For purposes of investigating phase transitions in nuclear matter, it is necessary to compute the ground-state energy per nucleon E_g/N as a function of the nuclear interaction parameters which describe the various interactions, and which appear in the Hamiltonian (3.1). Since it is almost as simple to compute the free energy per nucleon $F(\beta)/N$ as a function of "nuclear temperature" $T = 1/k_B\beta$ as it is to compute E_g/N, we will compute the former. The ground-state energy is easily obtained from the free energy in the zero-temperature limit

$$E_g/N = \lim_{T\to 0} F(\beta)/N. \tag{4.1}$$

The free energy is obtained from the partition function

$$\exp(-\beta F) = \mathrm{Tr}\exp(-\beta\hat{\mathscr{H}}). \tag{4.2}$$

The trace is carried out over the N-nucleon Hilbert space, which has dimension 2^N for N identical nucleons of pseudospin-$\frac{1}{2}$. To carry out this trace, we observe that the pseudospin operators $\boldsymbol{J}$ from which $\hat{\mathscr{H}}$ is constructed act only within each irreducible invariant subspace of dimension $2J+1$. It is therefore useful to decompose the pseudospin Hilbert space into its SU(2) invariant subspaces. Then the sum (4.2) reduces to

$$\mathrm{Tr}\exp(-\beta\hat{\mathscr{H}}) = \sum_{J=0 \text{ or } \frac{1}{2}}^{\frac{1}{2}N} Y(N, J)\mathrm{Tr}_J \exp(-\beta\hat{\mathscr{H}}). \tag{4.3}$$

In this expression, Tr_J indicates the trace is restricted to a J-invariant subspace of dimension $2J+1$, and $Y(N, J)$ is the multiplicity factor [6]) for the occurrence of the J-invariant space in the N-nucleon Hilbert space of dimension 2^N:

$$Y(N, J) = \frac{N!(2J+1)}{(\frac{1}{2}N+J+1)!(\frac{1}{2}N-J)!}. \tag{4.4}$$

The trace within each J-invariant subspace generally cannot be done in closed form. However, upper and lower bounds can be placed on these sums using the technology of atomic coherent states, as described in sect. 2. These bounds are [13])

$$\frac{2J+1}{4\pi}\int \exp(-\beta Q_J(\mathscr{H}; \Omega))\mathrm{d}\Omega \leq \mathrm{Tr}_J \exp(-\beta\hat{\mathscr{H}}) \leq \frac{2J+1}{4\pi}\int \exp(-\beta P_J(\mathscr{H}; \Omega))\mathrm{d}\Omega. \tag{4.5}$$

The P- and Q- representatives of the Hamiltonian $\mathscr{H}$ can now easily be determined by expanding $\mathscr{H}$ in terms of irreducible spherical tensor operators $\mathscr{Y}^L_M(\boldsymbol{J})$. This can easily be done by expanding h_C in terms of spherical harmonics

$$h_C = \sum_{L=1}^{k}\sum_{M=-L}^{+L} A^L_M r^L Y^L_M(\Omega), \tag{4.6C}$$

$$\hat{h}_Q = \sum_{L=1}^{k}\sum_{M=-L}^{+L} A^L_M N^{-L}\mathscr{Y}^L_M(\boldsymbol{J}). \tag{4.6Q}$$

The factor N^{-L} is present by the homogeneity requirement discussed in sect. 3, and the fact that $\mathscr{Y}^L_M(J/N) = N^{-L}\mathscr{Y}^L_M(J)$. The coordinates A^L_M in (4.6) are the nuclear interaction paramcters which describe the various pseudospin interactions. These coordinates obey $A^L_M = A^{L*}_{-M}$, have order of magnitude unity, and are independent of N on passage to the nuclear matter limit (compare the results following (1.2)). The summation terminates at finite $L = k$ by assumption (a) of sect. 3. We note in passing that h_C is a harmonic function.

Upper and lower bounds for the free energy are now obtained by combining the upper and lower bounds within each invariant subspace. These bounds are obtained from (2.7), (4.6) and (4.5). For example, the upper bound on the free energy is

$$\sum_{J=0 \text{ or } \frac{1}{2}}^{\frac{1}{2}N} Y(N,J)\frac{2J+1}{4\pi}\int \exp(-\beta Q_J(\mathscr{H};\Omega))\mathrm{d}\Omega \leq \operatorname{Tr}\exp(-\beta\hat{\mathscr{H}}) = \exp(-\beta F). \quad (4.7)$$

Such estimates are useful for finite N [ref. [14])] but we are now particularly interested in these estimates in the limit $N \to \infty$ of nuclear matter. In this limit it is useful to convert the sum over J into an integral by introducing a new parameter r,

$$0 \leqq r = J/N \leqq \tfrac{1}{2}. \quad (4.8a)$$

Then

$$\sum_{J=0 \text{ or } \frac{1}{2}}^{\frac{1}{2}N} \to N\int_0^{\frac{1}{2}} \mathrm{d}r. \quad (4.8b)$$

We note also that

$$Q_J(\hat{\mathscr{H}};\Omega) = Q_J(N\hat{h}_Q;\Omega) = NQ(\hat{h}_Q; r, \Omega), \quad (4.8c)$$

$$Y(N,J) = \mathrm{e}^{S(r)} = \mathrm{e}^{Ns(r)}, \quad (4.8d)$$

$$s(r) = S(r)/N = -\{(\tfrac{1}{2}+r)\ln(\tfrac{1}{2}+r)+(\tfrac{1}{2}-r)\ln(\tfrac{1}{2}-r)\}+\mathrm{O}\left(\frac{\ln N}{N}\right). \quad (4.8e)$$

The result (4.8e) has been obtained using Stirling's approximation [15]). Note in (4.8c) that we have absorbed the subscript on Q into its argument on passage from J to r.

By applying the results (4.8) to the inequality (4.7) we find

$$\frac{N^2}{4\pi}\int \exp[-N\beta(Q(\hat{h}_Q; r, \Omega)-\beta^{-1}s(r))]\mathrm{d}r\mathrm{d}\Omega \leq \exp(-\beta F). \quad (4.9)$$

The integral is taken throughout a sphere of radius $\frac{1}{2}$. Asymptotic values of this integral for large N can be estimated using Laplace's method [16]). Briefly, for integrals of the form

$$I(N) = \int \mathrm{e}^{-Nf(\boldsymbol{x})}\mathrm{d}\boldsymbol{x}, \quad (4.10)$$

the largest contribution to $I(N)$ comes from the neighborhood in which $f(\boldsymbol{x})$ assumes its minimum value. For large N

$$-\frac{1}{N}\ln I(N) = \min f(\boldsymbol{x})+\mathrm{O}\left(\frac{\ln N}{N}\right). \quad (4.11)$$

Estimation of the integral in (4.9) by this method leads to a rigorous upper bound on F/N [ref. [17])]

$$\min\{Q(\hat{h}_C; r, \Omega) - \beta^{-1}s(r)\} + O\left(\frac{\ln N}{N}\right) \geqq F(\beta)/N. \tag{4.12}$$

The reversed inequality is obtained by replacing $Q(\hat{h}_Q; r, \Omega)$ by $P(\hat{h}_Q; r, \Omega)$.

From (2.7)

$$\lim_{N\to\infty} P_J(\mathscr{Y}^L_M(J/N); \Omega) = \lim_{N\to\infty} Q_J(\mathscr{Y}^L_M(J/N); \Omega) = r^L Y^L_M(\Omega). \tag{4.13}$$

Since $\hat{h}_Q$ involves only a finite sum of spherical tensor operators (4.6Q), we have also

$$\lim_{N\to\infty} \min_{r,\Omega} \{P(\hat{h}_Q; r, \Omega) - Q(\hat{h}_Q; r, \Omega)\} = 0. \tag{4.14}$$

This means the difference between the upper and lower bounds on the free energy per nucleon vanishes in the $N \to \infty$ limit, so either can be used to estimate the free energy per nucleon rigorously. In fact

$$\lim_{N\to\infty} P(\hat{h}_Q; r, \Omega) = \lim_{N\to\infty} \sum_{L=1}^{k} \sum_{M=-L}^{+L} A^L_M P_J(\mathscr{Y}^L_M(J/N); \Omega)$$

$$= \sum_{L=1}^{k} \sum_{M=-L}^{+L} A^L_M r^L Y^L_M(\Omega) = h_C(r, \theta, \phi). \tag{4.15}$$

By now combining the results (4.12)–(4.15) we obtain a simple and rigorously correct expression for the free energy per nucleon in the nuclear matter limit

$$\lim_{N\to\infty} F(\beta)/N = \min_{r,\Omega} \Phi(\beta; r, \theta, \phi),$$

$$\Phi(\beta; r, \theta, \phi) = h_C(r, \theta, \phi) - \beta^{-1}s(r). \tag{4.16}$$

5. Physical interpretation

If we write the free energy per nucleon in the form

$$\lim_{N\to\infty} F(\beta)/N = \lim_{N\to\infty} (E - TS)/N, \tag{5.1}$$

then the physical interpretation of (4.16) is clear. From (4.1)

$$\lim_{N\to\infty} E_g/N = \min_{r,\theta,\phi} h_C(r, \theta, \phi). \tag{5.2}$$

Therefore $\Phi(\beta; r, \Omega)$ and $h_C(r, \theta, \phi)$ are "potential" functions whose minima give respectively the free energy per nucleon and the ground-state energy per nucleon. Similarly, when evaluated at the value r_0 which minimizes $\Phi(\beta; r, \Omega)$,

$$\lim_{N\to\infty} S/N = k_B s(r_0). \tag{5.3}$$

Therefore the logarithm of the SU(2) multiplicity factor $Y(N, J)$ determines the nuclear entropy up to a proportionality factor which is Boltzmann's constant k_B.

6. Ground-state phase transitions

The nuclear shape phase transition is determined by the ground-state energy phase transition. The existence and location of such a transition can in turn be determined by investigating the dependence of the minima of h_C on the nuclear interaction parameters A^L_M which appear in (4.6*Q*). Phase transitions occur when one local minimum either loses its stability, or else becomes metastable with respect to a deeper local minimum.

If the minima of h_C were in the interior of the sphere of radius $\frac{1}{2}$, they would be determined by setting the gradient of h_C equal to zero

$$\partial h_C/\partial\theta = 0, \tag{6.1a}$$

$$\partial h_C/\partial\phi = 0, \tag{6.1b}$$

$$\partial h_C/\partial r = 0. \tag{6.1c}$$

Since h_C is harmonic, it has no stable equilibria in the interior of this sphere [18]). As a result, the minimum of h_C must be on the sphere surface $r = \frac{1}{2}$. For $r = \frac{1}{2}$, the minimum is determined from (6.1a) and (6.1b). At the minimum $\partial h_C/\partial r < 0$.

Without detailed knowledge of h_C, it is not possible to make statements about when the location of the minimum of h_C jumps from one point on the sphere surface to another as a function of changing nuclear interaction parameters. However, it is possible to make precise statements about when a local minimum loses its stability without knowing a great deal about h_C. To do this, we observe that in the zero interaction limit. $A^1_0 = \sqrt{\frac{4}{3}\pi}\ \varepsilon$, and $A^L_M = 0$, all other L, M. The ground-state energy is then associated with the point $(r, \theta, \phi) = (\frac{1}{2}, \pi, \phi)$. We assume h_C is invariant under the transformation $\theta \to -\theta$. This means that all nuclear interaction parameters A^L_M with odd M vanish. This also guarantees that $\partial h_C/\partial\theta = 0$ at $\theta = \pi$. As a result, $\theta = \pi$ is always an equilibrium. In the absence of interactions this equilibrium is a minimum.

To determine the conditions under which the minimum at $\theta = \pi$ loses its stability as the nuclear interaction parameters A^L_M are increased from zero, we Taylor expand $h_C(\frac{1}{2}, \theta, \phi)$ about $\theta = \pi$

$$h_C(\tfrac{1}{2}, \theta, \phi) = h_C + \tfrac{1}{2}(\theta-\pi)^2 \frac{\partial^2 h_C}{\partial\theta^2} + \frac{1}{4!}(\theta-\pi)^4 \frac{\partial^4 h_C}{\partial\theta^4} + \dots \tag{6.2}$$

All derivatives are evaluated at $(\frac{1}{2}, \pi, \phi)$. The $\theta \to -\theta$ invariance of h_C excludes odd terms from this expansion. The minimum at $\theta = \pi$ is locally stable when $\partial^2 h_C/\partial\theta^2 > 0$. Loss of stability occurs when this coefficient vanishes. The phase transition is second or first order depending on the sign of the quartic coefficient (cf. fig. 1):

$$\frac{\partial^2 h_C}{\partial\theta^2} = 0 \Rightarrow \begin{matrix}\text{phase}\\ \text{transition}\end{matrix} \begin{cases} \dfrac{\partial^4 h_C}{\partial\theta^4} > 0 \Rightarrow \begin{matrix}\text{second}\\ \text{order}\end{matrix} \\[2ex] \dfrac{\partial^4 h_C}{\partial\theta^4} < 0 \Rightarrow \begin{matrix}\text{first}\\ \text{order.}\end{matrix} \end{cases} \tag{6.3}$$

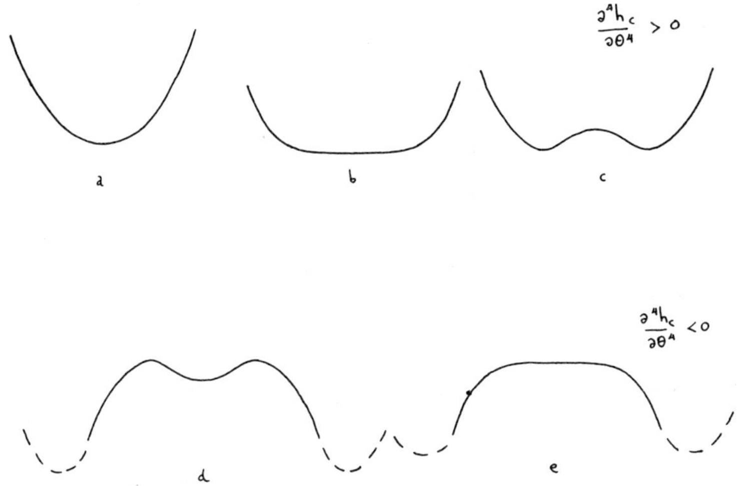

Fig. 1. Top: $\partial^4 h_C/\partial\theta^4 > 0$. The second derivative undergoes a sign change, as follows: $\partial^2 h_C/\partial\theta^2 > 0$ (a), $\partial^2 h_C/\partial\theta^2 = 0$ (b), $\partial^2 h_C/\partial\theta^2 < 0$ (c). A second-order phase transition occurs when $\partial^2 h_C/\partial\theta^2 = 0$. Bottom: $\partial^4 h_C/\partial\theta^4 < 0$, $\partial^2 h_C/\partial\theta^2 > 0$ (d), $\partial^2 h_C/\partial\theta^2 = 0$ (e). Before the second derivative undergoes a sign change, the local minimum at the center has become metastable with respect to distant minima. In order for h_C to be globally stable, the negative quartic term must be dominated at large θ by a positive term of higher power. Thus, the simplest (Ginzburg-Landau) model for a first-order phase transition involves a potential function of the form $V(x) = ax^2 - bx^4 + cx^6$, $a, b, c > 0$.

The latter condition ($\partial^4 h_C/\partial\theta^4 < 0$) guarantees that the equilibrium at $\theta = \pi$ becomes metastable with respect to some far-away minimum before it loses its local stability. Nothing but a detailed knowledge of h_C will guarantee that the second-order phase transition accompanying the loss of stability of the $\theta = \pi$ solution branch is not pre-empted by a first-order transition for smaller values of the nuclear interaction parameters.

To illustrate these ideas we consider the Lipkin-Meshkov-Glick [1]) Hamiltonian, suitably modified by the homogeneity requirement

$$h = \varepsilon v_3 + \tfrac{1}{2}V(v_+^2 + v_-^2) + W v_+ v_-,$$

$$\mathscr{H} = N\hat{h}_Q = \varepsilon J_3 + \frac{V}{2N}(J_+^2 + J_-^2) + \frac{W}{2N}(J_+J_- + J_-J_+),$$

$$h_C = \varepsilon r \cos\theta + \tfrac{1}{2}Vr^2 \sin^2\theta (e^{2i\phi} + e^{-2i\phi}) + Wr^2 \sin^2\theta. \tag{6.4}$$

The minimum of h_C occurs on the sphere boundary at $r = \frac{1}{2}$, and for $e^{2i\phi} = -1$ if $V > 0$, $e^{2i\phi} = +1$ if $V < 0$. The function of θ which remains to be minimized is

$$h_C(\tfrac{1}{2}, \theta, \phi = 0 \text{ or } \tfrac{1}{2}\pi) = \tfrac{1}{2}\varepsilon \cos\theta + \tfrac{1}{4}(W - |V|)\sin^2\theta. \tag{6.5}$$

The second and fourth derivatives on the branch $\theta = \pi$ are

$$\frac{\partial^2 h_C}{\partial \theta^2} = \tfrac{1}{2}\varepsilon + \tfrac{1}{2}(W - |V|),$$

$$\frac{\partial^4 h_C}{\partial \theta^4} = -\tfrac{1}{2}\varepsilon - 2(W - |V|). \tag{6.6}$$

The condition for the loss of stability of the solution branch $\theta = \pi$ is $\varepsilon = |V| - W$. At this point the fourth derivative has value $\frac{3}{2}\varepsilon > 0$. Therefore a second-order ground-state energy phase transition occurs when the nuclear interaction parameters V, W obey $(|V| - W) \geqq \varepsilon$.

It is possible to discuss the occurrence of ground-state energy phase transtions even without the requirement that h_C remain invariant under the transformation $\theta \to -\theta$. However, this case is somewhat more difficult, as the discussion must be non-local in nature. Nevertheless, precise statements can be made with the aid of elementry catastrophe theory [19]). To illustrate the types of problem that may arise, we consider the following pseudospin Hamiltonian

$$h = \varepsilon v_3 + A v_+ v_- + \tfrac{1}{2} B v_3 (v_+ + v_-),$$

$$\mathscr{H} = N\hat{h}_Q = \varepsilon J_3 + \frac{A}{2N}\{J_+, J_-\} + \frac{B}{4N}\{J_3, J_+ + J_-\},$$

$$h_C = \varepsilon r \cos\theta + A r^2 \sin^2\theta + \tfrac{1}{2} B r^2 \sin\theta \cos\theta (e^{i\phi} + e^{-i\phi}). \tag{6.7}$$

The potential h_C is minimized by choosing $r = \frac{1}{2}$, $e^{i\phi} = \pm 1$. The resulting potential has the form

$$h_C(\tfrac{1}{2}, \theta, 0 \text{ or } \pi) = \tfrac{1}{2}\varepsilon \cos\theta + \tfrac{1}{4} A \sin^2\theta + \tfrac{1}{4} B \sin\theta \cos\theta. \tag{6.8}$$

For $B \neq 0$ the $\theta \to -\theta$ symmetry is broken and $\theta = \pi$ is no longer a stationary point. The manifold of stationary points $\theta = \theta(A, B)$, solution of the equation $\partial h_C / \partial \theta = 0$ for θ as a function of the nuclear interaction parameters A, B, is shown in fig. 2. For some values of A, B, there is only one value of θ for which h_C has a local minimum. However, there is a cuspshaped region within the A-B plane in which the potential h_C has two local minima. In this region a small change in the values of the parameters A, B may produce a drastic change in the value of θ at which the global minimum of h_C occurs. This jump from one sheet to another corresponds to a first-order phase transition. The pleated surface shown in fig. 2 is equivalent ("diffeomorphic") to the "cusp catastrophe" [19]) manifold.

7. Thermodynamic phase transitions

We consider now another kind of phase transition. This is the thermodynamic phase transition, which occurs as a function of changing temperature for fixed values

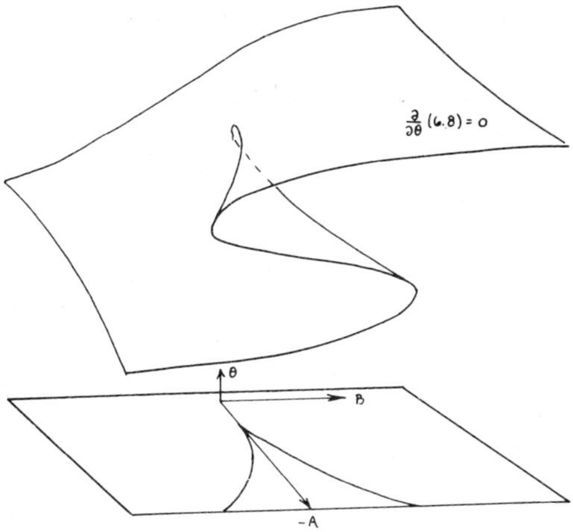

Fig. 2. The stationary values of the potential (6.8) form a pleated two-dimensional surface in the three-dimensional θ-A-B space. The cusp occurs at the point whose coordinates are $A = -\varepsilon$, $B = 0$.

of the nuclear interaction parameters. The conditions under which the thermodynamic phase transition occurs are determined from the potential $\Phi(\beta; r, \theta, \phi)$ (4.15). The stationary points of $\Phi(\beta; r, \theta, \phi)$ are determined by the zero gradient condition. Since $\Phi(\beta; r, \theta, \phi)$ is not harmonic when $T \neq 0$, the minimum can occur inside the sphere, for $r = \frac{1}{2}$. The equilibrium equations are identical to (6.1), except that (6.1c) must be replaced by

$$\frac{\partial h_C}{\partial r} = \beta^{-1} \frac{ds}{dr} = -k_B T \ln \frac{1+2r}{1-2r}. \tag{7.1}$$

The point (r, θ, ϕ) which minimizes $\Phi(\beta; r, \theta, \phi)$ at fixed temperature can easily be determined in the low and high temperature limits. In the limit $T \to 0$, r approaches $\frac{1}{2}$ exponentially like

$$r = \tfrac{1}{2} - \exp(\beta \partial h_C/\partial r) = \tfrac{1}{2} \tanh\left(-\tfrac{1}{2}\beta \frac{\partial h_C}{\partial r}\right), \tag{7.2}$$

where $\partial h_C/\partial r$ is evaluated at the point on the sphere surface $r = \frac{1}{2}$ at which h_C is minimum. To investigate the high temperature limit we recall that h_C is a polynomial function defined on a bounded domain (see sect. 3, property (a)). Therefore h_C and its derivatives are everywhere bounded. In particular, the left hand side of (7.1) is bounded, so that as $T \to \infty$ r must approach zero. This means the dominant term in (4.6) is εJ_3, with the interaction terms decreasing in importance with increasing value of L. By making T sufficiently large it is possible to make r sufficiently small so that h_C has only one local minimum, considered as a function of (θ, ϕ).

We now restrict consideration to the case where h_C is invariant under the transformation $\theta \to -\theta$. Then the unique high temperature minimum of $\Phi(\beta; r, \theta, \phi)$

occurs on the branch $\theta = \pi$, for all $T > T_c$, where T_c is some critical temperature. If we assume the nuclear interaction parameters A^L_M are sufficiently large so that the minimum of h_C occurs for $\theta \neq \pi$, then a thermodynamic phase transition must occur at some critical temperatue $T_c > 0$. At this temperature the minimum of $\Phi(\beta; r, \theta, \phi)$ jumps from the solution branch $\theta = \pi$ to a solution branch $\theta \neq \pi$. The thermodynamic phase transition is second order if the jump is local (soft jump) and is first order if the jump is non-local (hard jump).

As explained is sect. 6, nothing much can be said about non-local jumps without a detailed knowledge of h_C or equivalently $\Phi(\beta; r, \theta, \phi)$. However, local jumps, which correspond to loss of stability of the $\theta = \pi$ branch, can be determined by local (i.e. differential) methods. The function $\Phi(\beta; r, \theta, \phi)$ is expanded around $\theta = \pi$. Then $\partial^2\Phi/\partial\theta^2$ is greater than zero at high temperatures and eventually becomes negative if a second-order phase transition occurs. The condition for a second-order thermodynamic phase transition is therefore

$$\left.\frac{\partial^2 h_C}{\partial\theta^2}\right|_{r_c, \pi, \phi} = 0. \tag{7.3}$$

This equation determines r_c, which in turn determines the critical temperature through

$$\left.\frac{\partial h_C}{\partial r}\right|_{r_c, \pi, \phi} = -\beta_c^{-1} \ln \frac{1+2r_c}{1-2r_c}. \tag{7.4}$$

The transition is second or first order depending on the sign of $\partial^4 h_C/\partial\theta^4$, as indicated in (6.3), where the derivative is now evaluated at (r_c, π, ϕ).

This argument is no longer valid in the event the potential h_C lacks the symmetry $\theta \to -\theta$. This is because the high temperature minimum of Φ is no longer required to lie exactly on the line $\theta = \pi$. In this way the introduction of symmetry-breaking terms (i.e. $\{J_z, J_+\} + \{J_z, J_-\}$) into the nuclear pseudospin Hamiltonian can destroy the second-order thermodynamic phase transition.

To illustrate these ideas we again consider the modified MGL Hamiltonian (6.4). The free energy potential to be minimized, after choosing $Ve^{2i\phi} = -|V|$, is

$$\Phi(\beta; r, \theta) = \varepsilon r \cos\theta + (W - |V|)r^2 \sin^2\theta - \beta^{-1}s(r). \tag{7.5}$$

The critical temperature and stability properties are easily determined from the expansion

$$\Phi(\beta; r, \theta) = (-\varepsilon r - \beta^{-1}s(r)) + (\theta-\pi)^2(\tfrac{1}{2}\varepsilon r + (W-|V|)r^2) + (\theta-\pi)^4\left(-\frac{\varepsilon r}{4!} - \tfrac{1}{3}(W-|V|)r^2\right) + \ldots \tag{7.6}$$

The equilibrium at $\theta = \pi$ is locally stable until the coefficient of $(\theta-\pi)^2$ becomes negative. This defines r_c

$$r_c = \frac{\varepsilon}{2(|V|-W)}. \tag{7.7}$$

At this value of r_c, the coefficient of $(\theta-\pi)^4$ is $(\frac{1}{4}\varepsilon)^2/(|V|-W) > 0$, so the transition is second order. The critical temperature determined from (7.4) is given by the gap equation

$$\frac{(|V|-W)}{\varepsilon}\tanh\tfrac{1}{2}\beta_c\varepsilon = 1. \tag{7.8}$$

8. Summary and conclusions

We have exploited the properties of pseudospin Hamiltonians for the purpose of studying nuclear shape phase transitions in the nuclear matter limit. We have reduced the procedure to two stages, summarized in (1.3). In the first stage, we have developed a recipe for accurately computing E_g/N and $F(\beta)/N$. This involves computing the minimum value of the corresponding "potential" functions $h_C(r, \theta, \phi)$ and $\Phi(\beta; r, \theta, \phi)$. These potentials are constructed from the pseudospin Hamiltonian in an elementary way (cf. (3.2c) and (4.16)). In the second stage the behavior of the minima and stationary points of these potentials is studied as a function of changing nuclear interaction parameters (h_C) and temperature (Φ). This can be done in an organized way using the methods of elementary catastrophe theory [19]). Our approach is equivalent without being as complicated.

The extensive class of pseudospin Hamiltonians studied is given in (3.1). Rigorous upper and lower bounds on the free energy per nucleon (4.12) [ref. [14])] were constructed using the powerful tools of the atomic coherent state representation (sect. 2). In the $N \to \infty$ limit, the upper and lower bounds on the free energy per nucleon become equal (4.14). The problems of computing the free energy per nucleon and the ground-state energy per nucleon were reduced to the problems of minimizing potentials Φ and h_C over compact domains. The nuclear entropy is the logarithm of an SU(2) multiplicity factor multiplied by Boltzmann's constant k_B (5.3).

Phase transitions of two types were discussed: ground-state energy phase transitions and thermodynamic phase transitions. In the absence of nuclear interactions the nucleus is spherically symmetric.

A ground-state energy phase transition may occur from spherical to deformed as a function of increasing nuclear interaction parameters. A sufficient condition for the occurence of a second-order ground-state energy phase transition was given in (6.3). A necessary condition is that a first-order phase transition does not occur for smaller values of the nuclear interaction parameters.

At sufficiently high temperatures the nucleus is spherically symmetric. A thermodynamic phase transition from spherical to deformed may occur as a function of decreasing nuclear temperature. A sufficient condition for the occurrence of a second-order thermodynamic phase transition was given in (7.3). A necessary condition is that a first-order phase transition does not occur for higher values of the nuclear temperature.

The relationship between ground-state energy phase transitions and thermodynamic phase transitions is contained in the "crossover theorem", whose proof we have indicated in sect. 7. Briefly, we first increase the nuclear interaction parameters $A^L_M, L > 1$ at $T = 0$, and then for fixed A^L_M, increase T from zero. If the system undergoes a ground-state energy phase transition from spherical to deformed (increasing A^L_M, $T = 0$), then it will undergo a thermodynamic phase transition from deformed to spherical (fixed A^L_M, T increasing). The larger the nuclear interaction parameters, the higher the critical temperature (fig. 3).

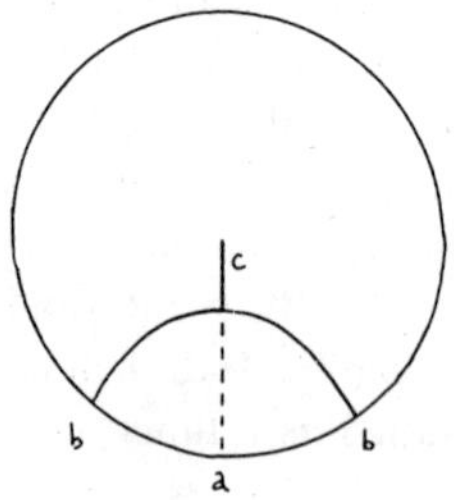

Fig. 3. The state of nuclear matter can be represented by a point in the sphere of radius $\frac{1}{2}$. In the zero temperature limit this point lies on the sphere surface. In the absence of nuclear interactions this point lies at the south polar position (a). For sufficiently large nuclear interaction parameters the ground-state energy is represented by points at $r = \frac{1}{2}$, $\theta \neq \pi$ (b, b'). The stationary point of h_C at $\theta = \pi$ is unstable. In the high temperature ($r \ll \frac{1}{2}$) limit, the point representing the nuclear state lies on the line $\theta = \pi$ (c). At some intermediate temperature T_c, determined from r_c by means of (7.3) and (7.4), the $\theta \neq \pi$ branch joins the branch $\theta = \pi$ for a thermodynamic phase transition. This is called the "crossover theorem."

Pseudospin Hamiltonians have long been used for the purpose of studying phase transitions in nuclear systems. They are particularly useful in the regime of small N ($N \leqq 50$) because the ground-state energy and gap could be computed numerically. The dependence of these quantities on the nuclear interaction parameters provided strong clues about the occurrence of a phase transition without precisely defining the phase transition condition. Unfortunately, each new pseudospin Hamiltonian had to be studied separately.

The analytic procedure described in this work and the numerical matrix diagonalization method used earlier are complementary. For small N, the numerical approach gives exact eigenvalues while the analytic approach provides only bounds on the ground-state energy. These bounds differ by a number whose order of magnitude is 1, but which is independent of N [ref. [14])]. The matrix diagonalization procedure becomes unfeasible for large N, while the analytic method gives exact and simple results for the free energy per nucleon, the ground-state energy per nucleon, and the entropy per nucleon. In adddtion, the criteria for phase transitions are precisely defined. Finally, the analytic approach provides a very pictorial representation for the nuclear (Gibbs [11-13]) state. In the atomic coherent state presentation it is a function sharply peaked about the value (r, θ, ϕ) which minimizes $\Phi(\beta; r, \theta\ \phi)$ at temperature T. For $T = 0$, the ground-state wave function is sharply peaked about the values (θ, ϕ) which minimize $h_C(r, \theta, \phi)$ on the surface of the sphere $r = \frac{1}{2}$.

We are grateful to Prof. Abraham Klein and Prof. Lorenzo Narducci for useful discussions on this subject.

References

1) H. J. Lipkin, N. Meshkov and A. J. Glick, Nucl. Phys. **62** (1965) 188
2) J. M. Blatt and V. W. Weisskopf, Theoretical nuclear physics (Wiley, NY, 1952)
3) R. Gilmore and C. M. Bowden, Classical and semiclassical treatment of the phase transition in Dicke models, in Cooperative effects in matter and radiation, ed. C. M. Bowden, D. W. Howgate and H. Robl (Plenum, NY, 1977)
4) H. Poincaré, Les méthodes nouvelles de la mécanique céleste, vol. I (Paris, 1892)
5) J. M. Radcliffe, J. of Phys. **A4** (1971) 313
6) F. T. Arecchi, E. Courtens, R. Gilmore and H. Thomas, Phys. Rev. **A6** (1972) 2211
7) R. J. Glauber, Phys. Rev. **130** (1963) 2529
8) D. J. Rowe and R. Basserman, Can. J. Phys. **54** (1976) 1941
9) D. Brink and G. R. Satchler, Angular momentum (University Press, Oxford, 1964)
10) R. Gilmore, J. of Phys. **A9** (1976) L65
11) R. Gilmore, Physica **86A** (1977) 137
12) D. Ruelle, Statistical mechanics (Benjamin, NY, 1969)
13) E. H. Lieb, Commun. Math. Phys. **31** (1973) 340
14) R. Gilmore and D. H. Feng, Phys. Lett. **B**, to be published
15) M. Abramowitz and I. A. Stegun, ed., Handbook of mathematical functions (US Government Printing Office, Washington, 1964)
16) D. V. Widder, An introduction to transform theory (Academic, NY, 1971)
17) R. Gilmore, J. Math. Phys. **18** (1977) 17
18) J. D. Jackson, Classical electrodynamics (Wiley, NY, 1962)
19) E. C. Zeeman, Scientific American **234**, no. 4 (1976) 65;
T. Poston and I. N. Stewart, Taylor expansions and catastrophes (Pitman, London, 1976)

Nuclear Physics **A319** (1979) 349–363; © *North-Holland Publishing Co., Amsterdam*

CONDENSATES AND BOSE–EINSTEIN CORRELATIONS*

G. N. FOWLER**, N. STELTE and R. M. WEINER
Physics Department, University of Marburg, W. Germany

Received 3 November 1978

Abstract: We discuss the conditions under which Bose–Einstein correlations can be used to determine the presence of condensates. One starts from the fact that any condensate is a coherent state and uses then the formalism of quantum optics. The case of final state interactions is considered explicitly in a Landau–Ginzburg type approach as used in laser physics. By applying a mean field Hartree approximation for the many mode problem we find that the second order correlation function for identical bosons $C_2(\boldsymbol{p}_1-\boldsymbol{p}_2=0)$ does effectively not depend on the phenomenological coupling constant g and thus $C_2(0)$ can be used to determine the amount of coherence and hence the existence of condensates. On the other hand $C_2(\boldsymbol{p}_1-\boldsymbol{p}_2\neq 0)$ does depend on g and therefore its measurement provides information about g. The implications of these findings for the determination of the size and lifetime of the source à la Hanbury-Brown–Twiss are discussed. Our conclusions apply both to condensates in nuclear (pion condensates, alpha and deuteron condensates) and hadronic matter.

1. Introduction

The existence of pion condensates in dense nuclear matter has for some time been the subject of interesting theoretical investigations [1,2])[†]. There is, however, so far no experimental evidence for this phenomenon and the most that can be said at present is that the evidence available is not in contradiction with the existence of this effect [1]). This applies *a fortiori* for heavy ion collisions since this domain of nuclear physics is yet in its incipient stage[††].

The difficulty in the experimental search for pion condensates lies mainly in the fact that in order to prove the existence of condensates one would have to disentangle effects which occur in the presence of condensates from effects which are present anyway ("background" effects). On the other hand given the fact that a satisfactory theory of strong interactions does not exist it is not possible to predict these "background" effects in a sufficiently reliable manner; thus the comparison of certain effects in the presence and in the absence of condensation becomes difficult.

* Work supported in part by Gesellschaft für Schwerionenforschung, Darmstadt.

** On leave from University of Exeter, England.

† Ref. [1]) contains an extensive review of this subject while ref. [2]) reviews some specific theoretical developments.

†† Possible manifestations of pion condensates in heavy ion collisions are discussed among others in refs. [3,4]).

It is the purpose of this paper to discuss a new method for the investigation of condensates in nuclear matter which apparently presents to a lesser degree the drawbacks specific for the method used so far, since it is based on general quantum mechanical considerations which are independent of the specific mechanism which gives rise to the condensation phenomenon. For this we use the fact that any condensate is a coherent state; this influences in a definite way the correlations of the corresponding quanta (fields).

Recently, by applying the methods of quantum optics, we have shown [5]) how this effect can be used to look for coherence in particle and nuclear physics. In particular the intensity correlations of identical particles (pions e.g.) constitute a measure of coherence in analogy with the Hanbury–Brown–Twiss effect in quantum optics. However, in order to apply the formalism of quantum optics to a strongly interacting system, the effect of the interaction on the correlations has to be evaluated. In the following this problem is investigated by approximating the many mode problem by a selfconsistent Hartree method. We find in a Landau–Ginzburg type approach used in laser physics that the second order correlation function at the origin (where the momenta of the two particles are equal) does not depend on the strength of the interaction g so that there the methods of quantum optics can be used to extract information about the amount of coherence [6]), i.e. about the existence of condensates. On the other hand for correlations of identical particles with unequal momenta there is a strong g-dependence. We suggest that this observation can be used to measure experimentally the coupling constant g.

It should be emphasized that our results are relevant not only for the determination of coherence but also for the determination of the size and lifetime of the source using Bose–Einstein correlations. The considerations which underly the latter determinations ignore the existence of coherence and furthermore neglect the final state interactions of emitted particles.

Our results show that such a procedure can hardly be meaningful unless a separate determination of the amount of coherence and strength of interaction is made. The formalism to be used in this case is outlined.

2. Effects of condensates in a non-interacting system

The essential ingredients of the formalism of quantum optics on which our approach is based have been summarized in ref. [5]) [a more complete exposition can be found in Glauber's lectures [7])]. In problems where one deals both with systems containing small numbers of particles as well as with macroscopic systems (condensates belong to this class), the use of the coherent state representation is appropriate since coherent states are eigenstates of the annihilation operator a_k and thus correspond to an undefined number of particles. The eigenvalues α_k are the amplitudes of the coherent fields. In a certain sense the opposite of coherent fields are chaotic fields. Indeed the density operator in the coherent state representation

$\{\alpha_k\}$ reads

$$\rho = \int \mathscr{P}(\{\alpha_k\}) |\{\alpha_k\}\rangle\langle\{\alpha_k\}| \prod_k d^2\alpha_k, \tag{2.1}$$

where $\mathscr{P}(\{\alpha_k\})$ is the distribution function of the states treated as random variables. Coherent states correspond to a well defined $\{\alpha_k\}$, i.e. to maximum "noiselessness":

$$\mathscr{P}^{\{\beta_k\}}_{\text{coherent}}(\{\alpha_k\}) = \prod_k \delta(\alpha_k - \beta_k), \tag{2.2}$$

while the maximum "noise", i.e. chaotic states[†], are described by a Gaussian distribution

$$\mathscr{P}_{\text{chaotic}}(\{\alpha_k\}) = \prod_k \frac{1}{\pi\langle n_k\rangle} e^{-|\alpha_k|^2/\langle n_k\rangle}, \tag{2.3}$$

where $\langle n_k\rangle$ is the mean occupation number in mode k. The difference between coherent and chaotic sources manifests itself quite strikingly in the correlations and the associated moments [5]) and it is this effect which we suggest be used as a method in the search of condensates.

Before explaining how this can be done, we would like, however, to point out the conditions under which the application of the quantum optics formalism is possible and useful:

(i) Since we assume that $\mathscr{P}(\alpha) > 0$ (it has the meaning of a probability), conservation constraints like energy and momentum, and quantum number conservation laws should not be significant.

(ii) It is useful to work with stationary correlation functions; for this one has to assume stationary distributions in the relevant variable. In high energy physics rapidity or pseudo-rapidity are examples of such variables since there is a (pseudo) rapidity plateau. (In quantum optics the relevant variable is time.)

For non-interacting systems the usefulness of the classification into coherent and chaotic distributions from the point of view of condensates is clearly seen from the following. We recall at first that in practice [this happens e.g. in quantum optics [7]) as well as in high energy physics [5])] one usually faces more complex situations where one has a mixture of coherent and chaotic distributions. This is defined by a convolution of (2.2) and (2.3):

$$\mathscr{P}(\{\alpha_k\}) = \int \prod_k \mathscr{P}^{\{\beta_k\}}_{\text{coherent}}(\{\gamma_k\}) \mathscr{P}_{\text{chaotic}}(\{\gamma_k - \alpha_k\}) d^2\gamma_k = \prod_k \frac{e^{-|\alpha_k - \beta_k|^2/\langle n_k\rangle}}{\pi\langle n_k\rangle}, \tag{2.4}$$

[†] A particular case of chaotic fields are fields generated by thermal sources. It should be pointed out here that distribution (2.3) leads to the usual Bose–Einstein distribution of an ideal gas, i.e. of non-interacting particles.

i.e. a Gaussian centered around the coherent field values. All correlations can then be expressed in terms of two parameters, the correlation length of the chaotic component, and the ratio

$$\kappa = \langle n_c \rangle / \langle n_{ch} \rangle, \tag{2.5}$$

where $\langle n_c \rangle$ and $\langle n_{ch} \rangle$ are the mean numbers of particles in the coherent and chaotic components respectively; they are related to the mean total number N by

$$N = \langle n_c \rangle + \langle n_{ch} \rangle. \tag{2.6}$$

Eqs. (2.4) and (2.1) define completely the physics of a coherent-chaotic mixture. Using these equations one can compute all physical quantities of interest. Among others the second and third order correlation functions are given in terms of the fields φ by

$$C_2(y_1 - y_2) = \frac{\langle I(y_1) I(y_2) \rangle}{\langle I \rangle^2} = \frac{\mathrm{Tr}\,\rho[|\varphi(y_1)|^2 |\varphi(y_2)|^2]}{(\mathrm{Tr}\,\rho|\varphi|^2)^2}, \tag{2.7}$$

$$C_3(y_1, y_2, y_3) = \frac{\langle I(y_1) I(y_2) I(y_3) \rangle}{\langle I \rangle^3}. \tag{2.8}$$

For $y_1 - y_2 = 0$ one gets (cf. the appendix)

$$C_2(0) = 2 - \eta^2, \tag{2.9}$$

where η is related to the parameter κ of (2.5) by the relation

$$\kappa = \eta/(1 - \eta) \tag{2.10}$$

Parameter η represents the amount of coherence; it varies between 0 (no coherence) and 1 (total coherence) so that $C_2(0)$ is 2 for a purely chaotic source and 1 for a purely coherent source. Any intermediate value of $C_2(0)$ means an admixture of coherent and chaotic fields; a measurement of $C_2(0)$ leads immediately to the determination of η and hence establishes whether a condensate can be present.

For the third order correlation function one finds (cf. appendix)

$$C_3(0) \equiv C_3(y_1 = y_2 = y_3) = \frac{6(1 + 3\kappa + \frac{3}{2}\kappa^2 + \frac{1}{6}\kappa^3)}{(1+\kappa)^3}. \tag{2.11}$$

3. Coherence for interacting fields

If the fields one investigates do interact, i.e. if there are final state interaction between the emitted particles as is the case e.g. for pions or α-particles in nuclear reactions, the question arises as to the parametrization to be used in order that meaningful conclusions about the presence of coherence (condensates) may be drawn.

It should be clear that in the absence of a theory of strong interactions we have to limit ourselves to a phenomenological approach. For this purpose we generalize the statistical density formalism of quantum optics in a manner closely related to the Landau–Ginzburg theory. Such an approach was suggested by Botke *et al.* [8]). However, as pointed out in ref. [5]) their formalism does not contain the important case of a mixture of a coherent and chaotic distributions in which we are interested so that we have to generalize it.

The density matrix of ref. [8]) reads:

$$\rho = Z^{-1} \int \delta\pi |\pi\rangle e^{-F(\pi)} \langle\pi|, \tag{3.1}$$

$$Z = \int \delta\pi e^{-F(\pi)}, \tag{3.2}$$

where the integrals are functional integrals, π represents the field, and F is the analogue of the Landau–Ginzburg free energy. To obtain a mixture we assume [5]) that π is a superposition of a coherent and a chaotic field,

$$\pi(y) = Y^{-\frac{1}{2}} \sum_{k=-\infty}^{\infty} f_k e^{iky} + Y^{-\frac{1}{2}}\beta; \qquad k = 2\pi n/Y;\ n = 0, 1, 2, \ldots. \tag{3.3}$$

Here $Y = 2y_{\max}$, f_k are the mode amplitudes describing the chaotic field and β^2 the intensity of the coherent field ($\beta^2 = \langle n_c \rangle$); it is related to the amount of coherence η by the relation

$$\eta = \beta^2/N. \tag{3.4}$$

We assume that the coherent component acts only in one mode which we choose as the zero mode.

The Landau–Ginzburg form for F is

$$F(\pi) = \int_0^Y dy \left[a|\pi(y)|^2 + b\left|\frac{\partial\pi(y)}{\partial y}\right|^2 + g|\pi(y)|^4 \right], \tag{3.5}$$

where a, b and g are constants. Together with β^2 they constitute the set of four parameters on which the theory depends. By comparing with the non-interacting case discussed previously it is clear that the interaction is represented by the g-term. This corresponds to the elimination of the source-field interaction in the laser case in favor of an effective field-field interaction [9]). We substitute eqs. (3.3), (3.5) into (3.1) and approximate the sums involving the $k \neq 0$ modes in a selfconsistent Hartree approximation while considering rigorously the $k = 0$ mode (this is the only mode which contains a contribution both from the coherent and chaotic components).

Essentially this amounts to writing

$$\sum_{k_1+k_2=k_3+k_4} f^*_{k_1} f^*_{k_2} f_{k_3} f_{k_4} \approx 2 \sum_{k_1,k_2} |f_{k_1}|^2 |f_{k_2}|^2 - \sum_k |f_k|^4$$

$$\approx X\left(4|f_0|^2 + 2\sum_k |f_k|^2\right) + |f_0|^4, \qquad (3.6)$$

where X is the number of "chaotic" particles in the $k \neq 0$ modes and is determined in a selfconsistent way by demanding that

$$N = \beta^2 + X_0 + X, \qquad (3.7)$$

where N is the total number of particles, and X_0 the number of "chaotic" particles in the $k = 0$ mode, and given by

$$X_0 = \frac{\exp(-d^2)}{\sqrt{\pi g}(1 - \text{Erf}(d))} - \frac{d}{\sqrt{g}}, \qquad (3.8)$$

with

$$d = \frac{1}{2\tilde{n}_0\sqrt{g}}. \qquad (3.9)$$

We get eventually in the $\mathcal{P}$-representation,

$$\rho = Z^{-1} \exp\left\{-\sum_{k\neq 0} \frac{|f_k|^2}{\tilde{n}_k(g)} - \frac{|f_0 - \beta|^2}{\tilde{n}_0} - g|f_0 - \beta|^4\right\}, \qquad (3.10)$$

with Z the normalization constant and

$$\tilde{n}_k(g) = \frac{1}{n_k^{-1} + 2gX}, \quad k \neq 0; \qquad \tilde{n}_0(g) = \frac{1}{n_0^{-1} + 4gX}, \qquad (3.11)$$

where $\tilde{n}_k(g)$ is the number of "chaotic" particles in mode $k \neq 0$ with the interaction included while[†]

$$n_k = \tilde{n}_k(g = 0) = \frac{1}{a + bk^2} \qquad (3.12)$$

is the number of "chaotic" particles in mode $k \neq 0$ in the absence of the interaction.

The different g-dependence of $\tilde{n}_{k\neq 0}$ and $\tilde{n}_0$ is due to the fact that the zero mode is treated differently (except when $\beta = 0$). Further, X is the solution of the equation

$$X = \sum_{k\neq 0} \frac{1}{a + bk^2 + 2gX}. \qquad (3.13)$$

[†] The tilde on some variables is used to distinguish the $g \neq 0$ case.

It is useful to express X in terms of the quantity $\tilde{\xi}$ which will turn out to be calculable from the observable effective correlation length:

$$\tilde{\xi}=\left(\frac{b}{a+2gX}\right)^{\frac{1}{2}}. \tag{3.14}$$

[For $g=0$ $\tilde{\xi}$ reduces to the correlation length ξ used in ref. [5]).]

In this case one has

$$X=\frac{-a+\sqrt{a^2+8gL}}{4g}, \tag{3.15}$$

with

$$L=\tfrac{1}{2}Y\tilde{\xi}^{-1}\coth\left(\tfrac{1}{2}Y\tilde{\xi}^{-1}\right)-1, \tag{3.16}$$

and in this way the constant b has been replaced by the measurable quantity $\tilde{\xi}$. Once we know ρ all physical quantities can be calculated.

Because of the computational difficulties we restrict ourselves in the present paper to the determination of the effects of the interaction in C_2 only. This is in fact the most important quantity both from the theoretical point of view (it contains already the salient features which illustrate the role of coherence and final state interaction) and from the experimental point of view (higher order correlations are much more difficult to measure).

4. Second order correlation function for an interacting system

We start from definition (2.7) of C_2 and use the expression for the density matrix ρ given in (3.10). For the fields φ we use the same expansion as in (3.3). We have

$$C_2(y_1,y_2)=\frac{\sum\limits_{p,q,r,s}\int f_p^*f_qf_r^*f_s\mathrm{e}^{i(q-p)y_1}\mathrm{e}^{i(s-r)y_2}\rho\prod\limits_k \mathrm{d}^2f_k}{\left(\sum\limits_p\int|f_p|^2\rho\prod\limits_k\mathrm{d}^2f_k\right)^2}. \tag{4.1}$$

It is clear that C_2 will depend only on the difference $\Delta y=y_1-y_2$ because, apart from the relative phase (f_0,β) which is arbitrary, ρ does not depend on the phases of the fields. The result is

$$C_2(\Delta y)=C_2(0)+\Psi_{\beta^2}(\Delta y)-1, \tag{4.2}$$

with

$$C_2(0)=2-\eta^2-\delta, \tag{4.3}$$

$$\delta=(2X_0^2-U)/N^2, \tag{4.4}$$

$$U=\frac{1+2d^2}{2g}-\frac{d\exp(-d^2)}{\sqrt{\pi}g(1-\mathrm{Erf}(d))}, \tag{4.5}$$

where X_0 and d are given by eqs. (3.8) and (3.9) respectively. The Δy dependence (and essentially also the g-dependence, as will be seen below) of C_2 is contained in the function Ψ_{β^2}:

$$\Psi_{\beta^2}(\Delta y) = \left\{\left[N - X + \frac{\mathscr{L}(\Delta y)}{n_0^{-1} + 2gX}\right] \Big/ N\right\}^2, \qquad \beta^2 \neq 0, \tag{4.6}$$

$$\Psi_{\beta^2}(\Delta y) = \left\{\left[\frac{\mathscr{L}(\Delta y) + 1}{n_0^{-1} + 2gN}\right] \Big/ N\right\}^2, \qquad \beta^2 = 0, \tag{4.7}$$

$$\mathscr{L}(\Delta y) = \tfrac{1}{2} Y \tilde{\xi}^{-1} \frac{\cosh \frac{1}{2} Y \tilde{\xi}^{-1}(1 - 2\Delta y / Y)}{\sinh \frac{1}{2} Y \tilde{\xi}^{-1}} - 1. \tag{4.8}$$

Note that $\mathscr{L}(\Delta y = 0) = L$ [cf. eq. (3.12)]. From eq. (4.8) it is now clear that $\tilde{\xi}$ can be determined by measuring the effective correlation length.

The main result of our calculations is the fact that while the interaction does not play any significant role in $C_2(0)$ (the effects are of the order of 1–2%) it plays an important part in $C_2(\Delta y \neq 0)$. This situation is illustrated in figs. 1 and 2. Thus it is seen in fig. 1b, for example, that all C_2 curves for various g coincide in the origin. This is a consequence of the parametrization of the problem in terms of the two phenomenological parameters N and $\tilde{\xi}$ which contain already the bulk of the g-dependence [cf. eqs. (3.11) and (3.12)]. The supplementary g-dependence of $C_2(\Delta y \neq 0) - C_2(0)$ is due to the explicit g-dependence of the function Ψ given by eqs. (4.6) and (4.7). We find furthermore that there is no g-dependence for $\beta^2 = 0$ (no coherence) and β^2 very large (total coherence). The first fact can easily be understood by inspection of the form of ρ [eq. (3.10)] and by recalling that in our approximation the last factor in the exponent which contains now the entire g-dependence becomes also a quadratic form; then ρ is a Gaussian and we recover the chaotic case with $g = 0$. The independence of $C_2(\Delta y)$ on g at large β^2 comes about because in this case from eq. (3.10) it follows that only terms with $f_0 \approx \beta$ contribute. In this case the fourth power term in β which contains the g-dependence will be negligible when compared with the second power term. Physically one could interpret these observations by saying that for a strongly coherent or chaotic field the final state interaction does not manage to disturb the correlation. Fig. 2 shows how one can use the second order correlation function to determine g (cf. below). Note that the curves intersect at some Δy which depends on η. This is characteristic for correlations treated by quantum optics and is different from the parametrization of coherence used e.g. in ref. [10]).

Before discussing the implications of our results we mention that they are not expected to be very satisfactory for Δy large at small β^2. This is because the results for large Δy are sensitive to the detailed treatment of the $k = 0$ mode. We have treated this mode exactly for $\beta^2 \neq 0$ and used the Hartree approximation for this as for the other modes for $\beta^2 = 0$ since there is no reason to distinguish one mode from the others for $\beta^2 = 0$.

5. Discussion and experimental implications

5.1. COHERENCE AND CONDENSATES

So far we have referred only to the effects of coherence rather than condensates it being understood that coherence implies the existence of a condensate. In any case for nuclear matter it is difficult to conceive of circumstances in which the one does not imply the other. In order to confirm the identification of a condensate an investigation of the energy dependence should reveal a characteristic threshold for the effect which in other many body systems is related to a phase transition. This can be seen from the example of the Bose–Einstein condensation for an ideal Bose gas. There the number of condensed particles which in out notation would coincide with β^2 is given by

$$n_c = N[1-(T/T_0)^{\frac{3}{2}}], \qquad T \leqq T_0,$$

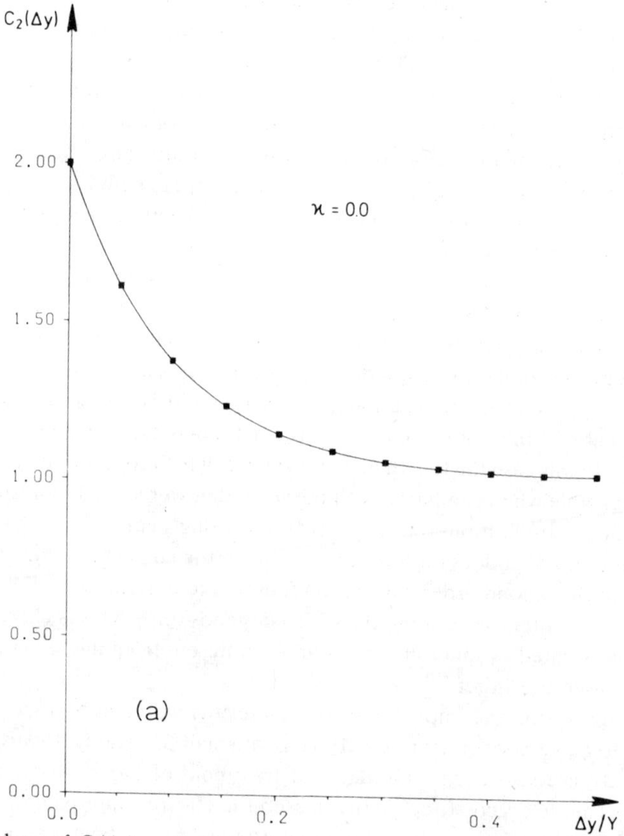

Fig. 1. Dependence of $C_2(\Delta y)$ on Δy for fixed κ and various g. The symbols are defined in fig. 1b; $Y\bar{\xi}^{-1} = 5$; $N = 3$. Note that for $\kappa = 0$ and $\kappa = 10$ the various g-curves coincide.

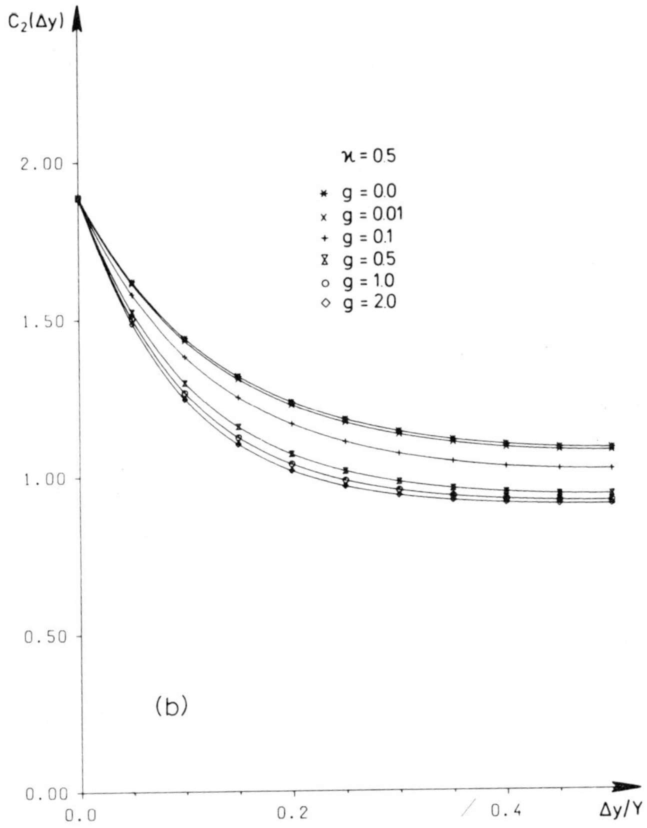

Fig. 1 (*contd*)

where T is the temperature and T_0 the condensation temperature. One should recall also that in a statistical approach T and T_0 depend on the c.m. energy and so then does n_c. Very recently Montvay and Zimanyi [4]) suggested that such a condensation process might indeed occur in heavy ion collisions.

5.2. DETERMINATION OF AMOUNT OF COHERENCE AND INTERACTION COUPLING CONSTANT

Provided conditions (i) and (ii) of sect. 2 are met, it is clear that the measurement of $C_2(0)$ provides us via eq. (4.3) with a determination of η. This applies for any system of identical bosons like mesons, alphas, deuterons, etc. The α-particles

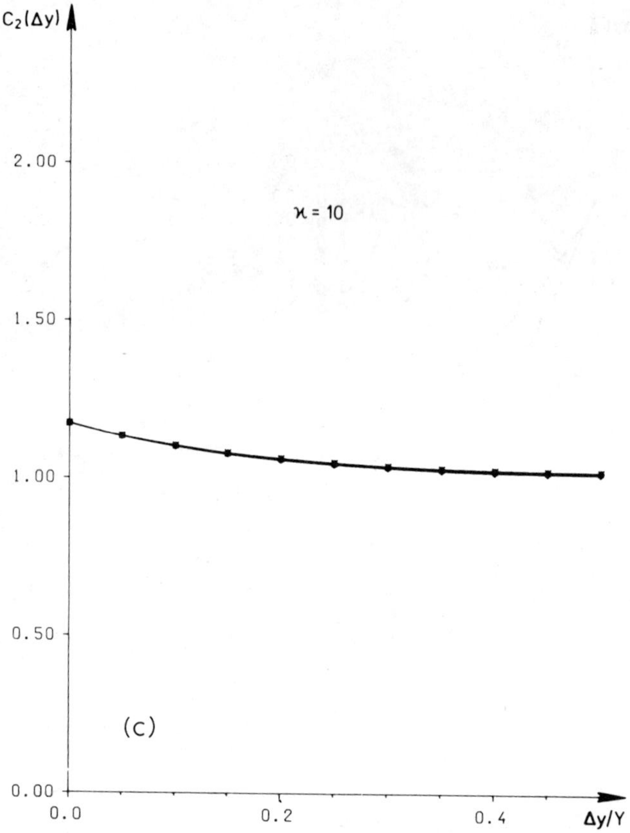

Fig. 1 (*contd*)

and deuterons have non-vanishing baryon number and this affects in principle the correlations. However, for nuclear reactions involving heavy nuclei this is not expected to be significant. This is true even more for heavy ion collisions.

Note that while $C_2(\Delta y \neq 0)$ depends on Y, $\tilde{\xi}$ and g, $C_2(0)$ does not depend on these parameters. The value of Y is given by experiment (maximum range of rapidity for example) so that we are left with two parameters g and $\tilde{\xi}$ to be determined. This can be done for example by investigating $C_2(\Delta y)$ in the same reaction at two different c.m. energies $E_{\mathrm{c.m.}}$. It is reasonable to assume that $\tilde{\xi}$ does not depend on $E_{\mathrm{c.m.}}$ while the amount of coherence does (cf. subsect. 5.1) and thus at various $E_{\mathrm{c.m.}}$ one sits on different η-curves so that both g and $\tilde{\xi}$ can be measured.

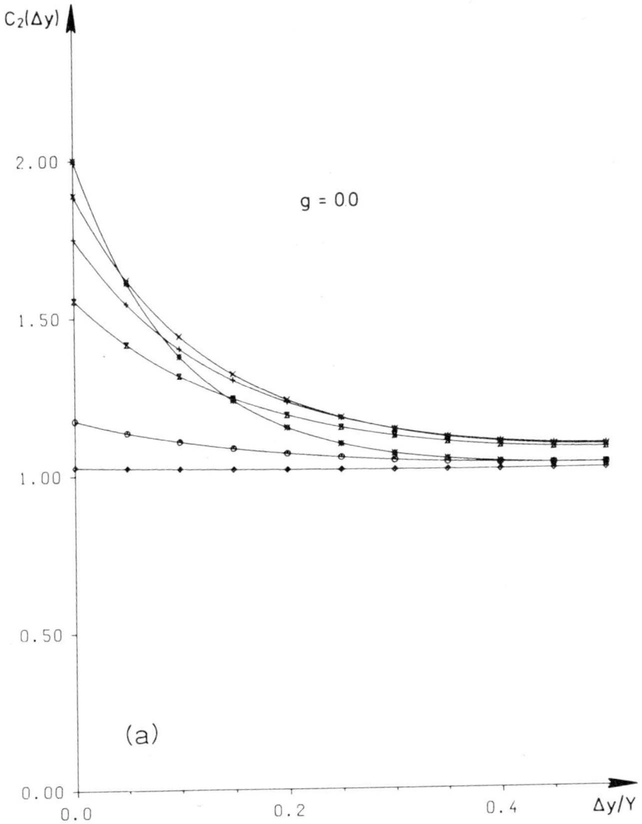

Fig. 2. Dependence of $C_2(\Delta y)$ on Δy for fixed g and various κ. The symbols are defined in fig. 2b; $Y\tilde{\xi}^{-1} = 5$; $N = 3$.

5.3. SIZE AND LIFETIME OF THE FIREBALL

The determination of these space-time characteristics of the source is usually based on the Hanbury-Brown-Twiss (HBT) effect and assumes (explicitly or implicitly) the following:

(i) The transverse and longitudinal source correlations are independent (factorization).

(ii) The emission process is stationary in the relevant variables.

(iii) Absence of coherence.

It was shown in ref. [5]) that at least in particle physics assumption (iii) probably does not hold. If there exist condensates in nuclear matter then (iii) would not hold in

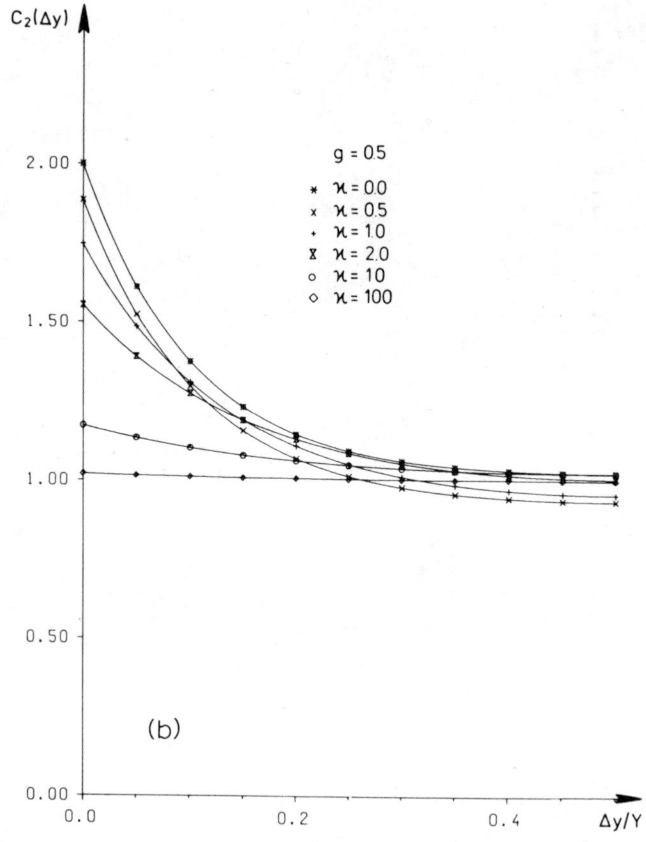

Fig. 2 (*contd*)

nuclear physics, too. The question then arises whether and how the HBT effect can still be used for the determination of the size and lifetime of the source. The answer should be clear from the results of the preceding sections. If assumptions (i) and (ii) hold the HBT effect can be used for the measurement of the space and time characteristics once the coupling g is known (g can be determined independently as discussed above). The procedure should be then the following:

Suppose that (i) and (ii) hold for the variables $q_0 = E_1 - E_2$ and $q_t = |(\boldsymbol{p}_1 - \boldsymbol{p}_2) \times (\boldsymbol{p}_1 + \boldsymbol{p}_2)| / |\boldsymbol{p}_1 + \boldsymbol{p}_2|$ where E and $\boldsymbol{p}$ are the energies and three-momenta of the two identical particles†. The conjugate variables of q_0 and q_t are the lifetime τ and the transverse radius R, respectively. Then the combined second order

† These variables are frequently used in order to parametrize the correlations [11,10]).

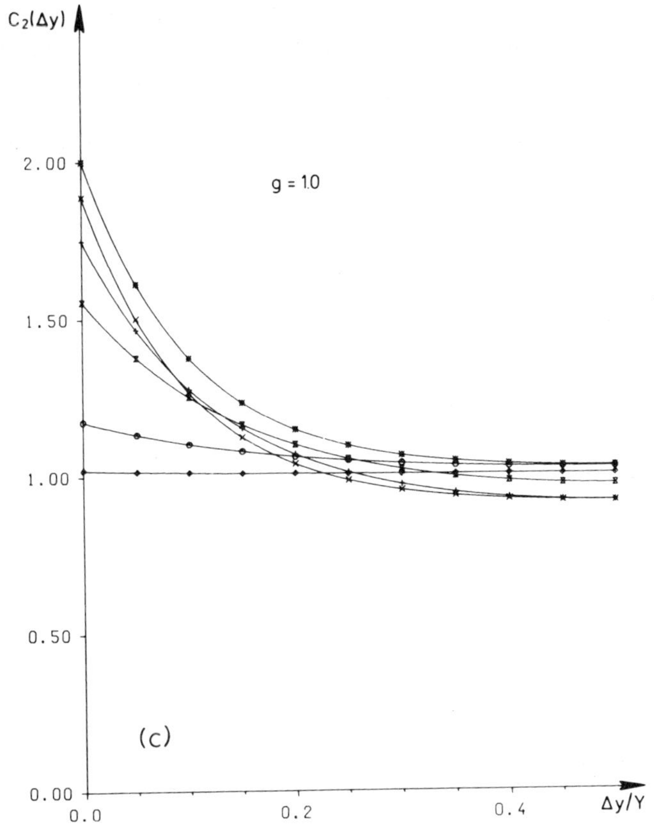

Fig. 2 (*contd*)

correlation function reads

$$C_2(q_0, q_t) = C_2(q_0)C_2(q_t),$$

where C_2 is now given by our eq. (4.2) with q_0 (q_t) taking the role of Δy, and τ and R the role of the corresponding correlation lengths $\tilde{\xi}$. The measurement of R, τ is then a measurement of $\tilde{\xi}$, i.e. the slope of C_2 at fixed β^2 (coherence) and g. As outlined previously, β^2 and g can be determined independently. (Note that in principle β would be different for the two variables, as $\tilde{\xi}$ obviously is.)

5.4. THE γ-γ CORRELATIONS VERSUS CORRELATIONS BETWEEN MESONS, ALPHAS, DEUTERONS, ETC.

Once we believe that final state interactions are not important in the determination of $C_2(0)$ the comparison of γ-γ correlations with correlations involving pairs of interacting particles like mesons, alphas, deuterons could provide us with a tool through which one could find out whether photons originate from the same fireball as hadrons. Such a comparison was not possible as long as no control of the interaction was available. If the identity of photon and hadronic sources is established the comparison of C_2 for γ-γ and hadron-hadron correlations at $g \neq 0$ provides furthermore an independent method for the measurement of the coupling g. Since the measurement of γ-γ correlations is well within the reach of present experimental possibilities[†] we can hope to find soon the answer to the above question.

Appendix

CALCULATION OF THE THIRD-ORDER CORRELATION FUNCTION FOR A NON-INTERACTING SYSTEM WITH A COHERENT-CHAOTIC MIXTURE

We use the expression for the density matrix derived in sect. 3 where we put $g = 0$. The third-order correlation function is defined by eq. (2.8). We are interested in its value for $y_1 - y_2 = y_1 - y_3 = 0$. This involves integrals of the type

$$\mathcal{I} = \int f^{2k} \mathrm{e}^{-|f-\beta|^2/\langle n\rangle} \mathrm{d}^2 f = \mathrm{e}^{-|\beta^2|} \int |f|^{2k+1} \mathrm{e}^{-|f|^2/\langle n\rangle} \mathrm{d}|f| \int \mathrm{e}^{-2|f||\beta|\cos\varphi/\langle n\rangle} \mathrm{d}(\cos\varphi)$$

$$= \frac{\mathrm{e}^{-|\beta|^2}}{\langle n\rangle} \int_0^\infty |f|^{2k+1} \mathrm{e}^{-|f|^2/\langle n\rangle} I_0\left(\frac{2|f||\beta|}{\langle n\rangle}\right) \mathrm{d}|f|,$$

which can be easily evaluated by using the relation

$$\int_0^\infty I_0(at) \mathrm{e}^{-p^2 t^2} t^{\mu-1} \mathrm{d}t = \frac{\Gamma(\frac{1}{2}\mu)}{2p\mu} \, {}_1F_1(\tfrac{1}{2}\mu; 1; a^2/4p^2).$$

Note that the same type of calculation leads to $C_2(0)$ as given by eq. (2.9).

References

1) A. B. Migdal, Rev. Mod. Phys. **50** (1978) 107
2) G. E. Brown and W. Weise, Phys. Reports **27** (1976) 1
3) V. Ruck, M. Gyulassy and W. Greiner, Z. Phys. **A277** (1976) 391;
M. Gyulassy and W. Greiner, Ann. of Phys. **109** (1977) 485
4) J. Montvay and J. Zimanyi, Hadron chemistry in heavy ion collisions, Budapest preprint, 1978
5) G. N. Fowler and R. M. Weiner, Phys. Rev. **D17** (1978) 3118 Phys. Lett. **70B** (1977) 201; Proc. VIII Int. Symp. on multiparticle dynamics, Kaysersberg, 1977, A-277
6) G. N. Fowler and R. M. Weiner, Proc. IV European Antiproton Symp., Strasbourg, France, June 26–30, 1978
7) R. Glauber, in Quantum optics and electronics, 1964 Les Houches Lectures, ed. C. DeWitt, A. Blandlin and C. Cohen-Tannoudji (Gordon and Breach, New York, 1965) p. 63
8) J. C. Botke, D. J. Scalapino and R. L. Sugar, Phys. Rev. **D9** (1974) 813
9) H. Haken, in Quantum optics, 1970, Proc. Tenth Session of the Scottish Universities Summer School in Physics (1969), p. 201
10) M. Deutschmann *et al.*, CERN/EP/PHYS 78-1 (1978) (unpublished)
11) G. J. Kopylov and M. J. Podgoretski, Yad. Fiz. **18** (1973) 656 [Sov. J. Nucl. Phys. **18** (1974) 336]
12) J. Antos *et al.*, JINR preprint P1-11033, Dubna 1977

[†] Cf. e.g. ref. [12]).

X. Applications in Elementary Particle Physics

Reprinted from ANNALS OF PHYSICS Vol. 66, No. 2, August 1971

Coherent Production of Pions*

D. HORN†

CERN, Geneva

AND

R. SILVER

California Institute of Technology, Pasadena, California 91109

Received September 22, 1970

THIS PAPER IS DEDICATED TO THE MEMORY OF AMOS DE-SHALIT

We investigate the question of independent and coherent production of pions in high energy processes. A model for coherent production is suggested and the restrictions due to momentum conservation, charge conservation, and isospin conservation are dealt with in detail. A formalism for the isospin analysis of identical pions is developed and applied to coherent states. Results and consequences are discussed. Experimental examples are compared with the theoretical concepts and ideas.

I. INTRODUCTION

In this paper we try to test theoretically the concept of coherent pion production in high energy experiments. Let us first summarize several characteristics that are known to be common to many high energy collisions. The main products of these experiments are pions whose average multiplicity grows slowly with energy. The products of the collision move with a relatively low transverse momentum k_T (perpendicular to the incoming momentum) whose average is roughly of the order of 300–400 MeV. In the Appendix we discuss experimental examples that show these characteristics. In particular, we note that many reactions show a distribution of pions that is strongly concentrated around the centre of mass (this phenomenon is sometimes referred to as pionization). It is therefore useful to describe the situa-

* Work supported in part by the U. S. Atomic Energy Commission. Prepared under contract AT(11-1)-68 for the San Francisco Operations Office, U. S. Atomic Energy Commission.

† On leave from Tel Aviv University, Tel Aviv, Israel.

tion in terms of the c.m. momenta which we shall use throughout this paper. The relativistic phase space d^3k/ω includes a convenient enhancement factor near the centre of mass. In addition to these pions, one often observes also two leading particles. These are outgoing particles that follow the trend of the incoming momenta. They can be identical with the incoming particles but may also be resonances that are emitted with low momentum transfers from the colliding particles. The relative amount of resonance emission seems to vary from one reaction to another. One may actually be tempted to explain all data in terms of resonance production. It seems, hopeless, however, to approach the problem in such a fashion. We chose therefore a different way in the present paper—motivated by looking at the problem from a different angle, and regarding pion emission as a process that is close to "classical radiation" of a pionic field.

We will define now some key concepts that we will use. We call the emission of π_i and π_j in a reaction $A + B \to \pi_i + \pi_j + \cdots$ *uncorrelated* if

$$P(\mathbf{k}_i, \mathbf{k}_j) = P(\mathbf{k}_i)\, P(\mathbf{k}_j), \tag{1}$$

where $P(\mathbf{k}_i)$ is the probability distribution of π_i in (c.m.) momentum space. $P(\mathbf{k}_i, \mathbf{k}_j)$ is the distribution for the two pions. Experimental examples that show some evidence for such behaviour are given in the Appendix. This concept can apply to pions of either the same or different charges. We expect that to the extent that (1) is satisfied, similar uncorrelated behaviour will hold also for distributions of more particles. We define now the emission of a pion as *independent* if the same $P(\mathbf{k})$ is found in all (or many) configurations of outgoing particles where this pion is tested. We assume independence also to imply that the cross section for production of n (e.g., neutral) pions of momenta $\mathbf{k}_1 \cdots \mathbf{k}_n$ (and some other fixed configuration of particles) will be proportional to

$$\frac{d^3k_1}{2\omega_1} fP(\mathbf{k}_1) \cdots \frac{d^3k_n}{2\omega_n} fP(\mathbf{k}_n).$$

$fP(\mathbf{k})$ may in principle depend on the charge of the pion but is supposed to be fixed for a given energy of incoming particles. We learn in the Appendix that this can be sometimes regarded as a crude approximation to the actual situation. If the emission is independent, one expects to find distributions of the many-particle events similar to the Poisson distribution. A state that has such characteristics and in which also the phase of the pion wave function [whose norm is $fP(\mathbf{k})$] is fixed is the *coherent* state. In the next sections we will discuss the general characteristics of this state and explain how it can be incorporated in a T matrix description of many-particle production. Since we apply the method to pions that are emitted with low energies in c.m. it can be called *coherent pionization.* We deviate perhaps from the usual picture that this name may suggest, since we allow correlations between the pions and the leading particles. We forbid only correlations among the pions in the emitted cloud. Finally, let us introduce the concept of *identical* pions. This applies when a set of uncorrelated pions is described by the same wave function in momentum space for each pion with the exception of unique but different phases and over-all magnitudes for the three different charges.

One may wonder how it is possible to talk about independent and coherent pions if there are some obvious correlations to be taken into account. The aim of this paper is to answer this question. We look for theoretical restrictions and implications with regard to these concepts. The constraints that we discuss can be divided into three general classes:

(1) four-momentum conservation (phase space);
(2) conserved quantum numbers (charge, parity, charge conjugation);
(3) isospin conservation.

Momentum conservation is taken care of by the leading particles—upon integrating on them one recovers the coherent state of pions (within our model at least). This question and related ones had already been dealt with in the literature in many different ways [1]. We develop our own version of coherent production and then contine to discuss the conserved quantum numbers and the restrictions imposed by them, a subject overlooked in many other discussions of similar models. We then develop a formalism to handle isospins of identical pions and apply it to our coherent states. The paper is divided into the following Sections:

II. — The coherent state,
III. — Emission of a coherent state,
IV. — Charge and parity,
V. — Isospin analysis of identical pions,
VI. — Isospin analysis of coherent pions,
VII. — Discussion and summary,
Appendix — Experimental examples.

We do not intend to fit experimental results or even to speculate about the exact form of the probability distribution in momentum space and its s dependence. We just study the concept of coherence to see where and when it might be applicable.

II. The Coherent State

The concept of a coherent state of bosons is quite unique in its physical interpretation and mathematical structure. This is a quantum mechanical state that is closest to a classical system in its dynamical properties [2]. A coherent state of particles can thus be regarded as a classical radiation of the corresponding field. It is used in describing electromagnetic radiation in quantum optics [2] as well as in the analysis of bremsstrahlung and the related infra-red catastrophe [3]. In the present section we apply this concept to pions. For the time being, we still disregard their quantum numbers and use only the fact that they are bosons. Modifications introduced in the following Sections can be simply implemented within the formalism of Sections II and III.

In the theory of a nonrelativistic harmonic oscillator one defines a coherent state as an eigenstate of the annihilation operator. It corresponds to a displaced ground state in configuration and momentum space and has therefore the minimal uncertainty value of $\Delta \times \Delta p$. Its time evolution looks like a classical wave packet since it retains its shape during the motion.

The definition in a relativistic theory of many bosons follows parallel lines. Let us define creation and annihilation operators satisfying

$$[a(\mathbf{k}), a^{+}(\mathbf{k}')] = \delta^{(3)}(\mathbf{k} - \mathbf{k}'). \tag{2}$$

We define the coherent state of bosons $|f\rangle$ by the equation

$$a(\mathbf{k}) \,|\, f\rangle = \frac{f(k)}{\sqrt{2\omega}} \,|\, f\rangle, \tag{3}$$

where $f(k)$ is the momentum space wave function of each boson, k is the four-momentum, and $k_0 = \omega = \sqrt{\mathbf{k}^2 + \mu^2}$. $f(k)$ is a relativistic invariant function of k and depends on some external momenta as well. The solution to Eq. (3) is given by

$$\begin{aligned} |f\rangle &= \exp\left\{\int d^3k \left[\frac{f(k)}{\sqrt{2\omega}}\, a^{+}(\mathbf{k}) - \frac{1}{2}\frac{|f(k)|^2}{2\omega}\right]\right\} |0\rangle \\ &= \exp\left\{\int d^3k \left[\frac{f(k)}{\sqrt{2\omega}}\, a^{+}(\mathbf{k}) - \frac{f^*(k)}{\sqrt{2\omega}}\, a(\mathbf{k})\right]\right\} |0\rangle \equiv S(f)\,|0\rangle, \end{aligned} \tag{4}$$

where we normalized $|f\rangle$ to $\langle f | f\rangle = 1$. The basic Eq. (3) gives the clue to the classical behaviour: The expectation value of a second quantized boson field within the state $|f\rangle$ will be given by the classical field with momentum distribution $f(k)$.

The expectation value of the four momentum operator P_μ is

$$\langle f |\, P_\mu \,| f\rangle = \langle f |\int d^3k\, k_\mu a^{+}(\mathbf{k})\, a(\mathbf{k}) \,| f\rangle = \int d^3k\, k_\mu \frac{|f(k)|^2}{2\omega}. \tag{5}$$

The expectation value of the number operator N is defined as $\bar{n}$ and given by

$$\langle f |\, N \,| f\rangle \equiv \bar{n} = \langle f |\int d^3k\, a^{+}(\mathbf{k})\, a(\mathbf{k}) \,| f\rangle = \int d^3k \frac{|f(k)|^2}{2\omega}. \tag{6}$$

Obviously, $|f\rangle$ is a combination of all n-particle states. A straightforward calculation leads to

$$|f\rangle = \sum_n \left(e^{-\bar{n}} \frac{\bar{n}^n}{n!}\right)^{1/2} |n\rangle, \tag{7}$$

thus exhibiting a Poisson distribution in the states $|n\rangle$.

In dealing with the production of a coherent state, we have to project out of it the piece that corresponds to a definite four-momentum K. We will denote this new state by $|f, K\rangle$. It is given by

$$\begin{aligned} |f, K\rangle &= \frac{1}{(2\pi)^4} \int d^4x\, e^{-iK\cdot x} S(fe^{ik\cdot x})\, |0\rangle \\ &= \frac{1}{(2\pi)^4} \int d^4x\, e^{-iK\cdot x} \exp\left\{-\frac{\bar{n}}{2} + \int d^3k \frac{f(k)}{\sqrt{2\omega}} a^+(k)\, e^{ik\cdot x}\right\} |0\rangle \end{aligned} \tag{8}$$

and obeys

$$\int d^4K\, |f, K\rangle = |f\rangle, \tag{9}$$

$$\langle f | f, K\rangle \equiv \rho_f(K) = \frac{1}{(2\pi)^4} \int d^4x\, e^{-iK\cdot x} \exp\left\{\int \frac{d^3k}{2\omega} |f|^2 (e^{ik\cdot x} - 1)\right\}, \tag{10}$$

$$\langle f, K' | f, K\rangle = \delta^{(4)}(K' - K)\, \rho_f(K). \tag{11}$$

It is instructive to decompose the quantity $\rho_f(K)$ in a series of the various n-particle contributions. This is achieved by a decomposition of the integrand in Eq. (10) in powers of e^{ikx} :

$$\begin{aligned} \rho_f(K) &= \sum_{n=0}^{\infty} \rho_f^{(n)}(K) \\ &= e^{-\bar{n}} \Big\{\delta^{(4)}(K) + |f(K)|^2\, \delta(K^2 - \mu^2)\, \theta(K_0) + \frac{1}{2!} \int d^4k\, |f(k)|^2\, |f(K-k)|^2 \\ &\quad \times \delta(k^2 - \mu^2)\, \delta[(K-k)^2 - \mu^2]\, \theta(k_0)\, \theta(K_0 - k_0) + \cdots\Big\}. \end{aligned} \tag{12}$$

Equation (12) reveals the spectrum structure that one would expect: a contribution at $K_\mu = 0$ from the vacuum component, one at $K^2 = \mu^2$ from the one-particle state and a continuum that starts from the threshold of two particles.

For the sake of further use in the next section, we list also some properties of the scalar products of two different coherent states:

$$\langle g \mid f \rangle = \exp\left\{-\frac{1}{2}\int \frac{d^3k}{2\omega}\left(|f|^2 + |g|^2 - 2g^* f\right)\right\}. \tag{13}$$

Equation (13) shows that two different coherent states are not orthogonal to each other (they are not eigenstates of Hermitian operators). Nevertheless, they do form an over-complete set [2]. The analog of Eq. (10) is

$$\begin{aligned}\langle g \mid f, K\rangle &\equiv \rho_{g,f}(K)\\ &= \frac{1}{(2\pi)^4}\int d^4x e^{-iK\cdot x} \exp\left\{-\frac{1}{2}\int \frac{d^3k}{2\omega}\left(|f|^2 + |g|^2 - 2g^* f e^{ik\cdot x}\right)\right\}.\end{aligned} \tag{14}$$

The calculation of quantities like $\rho_f(K)$ or $\rho_f^{(n)}(K)$ is not an easy matter. Thus $\rho_f^{(n)}(K)$ can be rewritten as

$$\rho_f^{(n)}(K) = \frac{e^{-\bar{n}}}{n!}\int \frac{d^3k_1}{2\omega_1}\cdots\frac{d^3k_n}{2\omega_n}\,|f(k_1)|^2\cdots|f(k_n)|^2\,\delta^{(4)}(k_1 + \cdots + k_n - K). \tag{15}$$

To simplify matters, we can define normalized distributions $\tilde{\rho}_f^{(n)}(K)$ such that

$$\rho_f^{(n)}(K) = e^{-\bar{n}}\frac{\bar{n}^n}{n!}\tilde{\rho}_f^{(n)}(K), \qquad \int d^4K\,\tilde{\rho}_f^{(n)}(K) = 1. \tag{16}$$

One can then use the central limit theorem and find that

$$\tilde{\rho}_f^{(n)}(K) \approx \frac{1}{n^2}\frac{\sqrt{\det \eta}}{4\pi^2}\exp\left\{-\frac{1}{2n}\eta_{\mu\nu}(K^\mu - n\bar{k}^\mu)(K^\nu - n\bar{k}^\nu)\right\}, \tag{17}$$

where we used

$$\begin{aligned}\bar{k}^\mu &= \frac{1}{\bar{n}}\int \frac{d^3k}{2\omega}k^\mu\,|f(k)|^2,\\ \frac{\eta_{\mu\nu}}{\bar{n}}&\int (k^\nu - \bar{k}^\nu)(k^\sigma - \bar{k}^\sigma)\,|f(k)|^2\frac{d^3k}{2\omega} = \delta_\mu{}^\sigma.\end{aligned} \tag{18}$$

This result was given by Van Hove [4] and similar expressions were analyzed in detail by Lurçat and Mazur [5]. Let us discuss here briefly the expected form for $\tilde{\rho}_f^{(n)}$ if $f(k)$ has the characteristics of the distribution functions of the pions described in the Appendix. A reasonable guess would be $\bar{k}^\mu = (\bar{\omega}; \mathbf{0})$, with $\eta_{\mu\nu}$ a diagonal matrix with elements $(\sigma_E^{-2}, \sigma_T^{-2}, \sigma_T^{-2}, \sigma_L^{-2})$, where T and L designate transverse and longitudinal directions, respectively. There is obviously a connection between these terms given by $\sigma_E{}^2 = 2\sigma_T{}^2 + \sigma_L{}^2 + \mu^2 - \bar{\omega}^2$. It then follows from (17) that

$$\tilde{\rho}_f^{(n)}(K) \approx \frac{1}{n^2}\frac{1}{4\pi^2\sigma_T{}^2\sigma_E\sigma_L}\exp\left\{-\frac{1}{n}\left[\frac{(K_0 - n\bar{\omega})^2}{2\sigma_E{}^2} + \frac{K_T{}^2}{2\sigma_T{}^2} + \frac{K_L{}^2}{2\sigma_L{}^2}\right]\right\}. \quad (19)$$

Equation (19) tells us that the over-all distribution of the coherent pions is peaked around a linearly increasing energy with an increasing width as expected from a typical random walk problem.

III. Emission of a Coherent State

In this section we discuss a formalism that describes a process in which two incoming particles (with momenta q_1 and q_2) produce two outgoing leading particles (with momenta p_1 and p_2) and n mesons of momenta $k_1 \cdots k_n$ which are part of a coherent state. For the moment we continue to ignore the quantum numbers of the pions (we will discuss this problem in the next section). We propose now the following factorized S matrix structure

$$\langle p_1 p_2 k_1 \cdots k_n \mid S \mid q_1 q_2\rangle$$
$$= i\int d^4x\, e^{ix\cdot(p_1+p_2-q_1-q_2)}\langle p_1 p_2 k_1 \cdots k_n \mid S(fe^{ikx})\, \tilde{T} \mid q_1 q_2\rangle. \quad (20)$$

To the extent that the incoming particles are not mesons of the kind appearing in the coherent cloud [or, if they are such mesons, they have momenta outside the range of $f(k)$] equation (20) can be brought into a completely factorized form

$$\langle p_1 p_2 k_1 \cdots k_n \mid S \mid q_1 q_2\rangle$$
$$= i\int d^4K\,(2\pi)^4\,\delta^{(4)}(K + p_1 + p_2 - q_1 - q_2)\langle k_1 \cdots k_n \mid f, K\rangle\langle p_1 p_2 \mid \tilde{T} \mid q_1 q_2\rangle. \quad (21)$$

$\tilde{T}$ acts only on the particles $q_1 q_2 p_1 p_2$ that form what we call the "skeleton" of the process. It can thus depend on the invariant variables:

$$\begin{aligned} s &= (q_1 + q_2)^2, & \bar{s} &= (p_1 + p_2)^2, \\ t &= (q_1 - p_1)^2, & \bar{t} &= (q_2 - p_2)^2, \\ u &= (q_1 - p_2)^2, & \bar{u} &= (q_2 - p_1)^2. \end{aligned} \quad (22)$$

The sum of all these variables is given by

$$s + t + u + \bar{s} + \bar{t} + \bar{u} = K^2 + 2\Sigma, \quad (23)$$

where

$$K = q_1 + q_2 - p_1 - p_2, \qquad \Sigma = q_1^2 + q_2^2 + p_1^2 + p_2^2.$$

Note that $f(k)$ is an invariant function of k and depends therefore on the four momenta of the skeleton. We refer to this fact by using the notation $f_{pq}(k)$. It assures the invariance of the expressions (20) and (21) under the Poincaré group.

The form (21) leads to the following result for the cross section of n meson production

$$\sigma_{2+n} = \int (dp)\, \rho^{(n)}_{f_{pq}}(q_1 + q_2 - p_1 - p_2)\, |\langle p \mid \tilde{T} \mid q\rangle|^2, \tag{24}$$

where (dp) stands for the invariant phase space element of the outgoing leading particles and the relevant flux factor. $\langle p \mid \tilde{T} \mid q\rangle$ is an abbreviation for the skeleton matrix element. Equation (24) is formally similar to the two-particle production cross section

$$\sigma_2 = \int (dp)\, \delta^{(4)}(q_1 + q_2 - p_1 - p_2)\, |\langle p \mid T \mid q\rangle|^2, \tag{25}$$

with the $\rho^{(n)}$ replacing the δ function. In other words, $\rho^{(n)}$ describes the distribution of four-momenta absorbed in the mesonic cloud. We leave it yet as an open question whether the recipe (24) can be smoothly continued to $n = 0$ to give $\sigma_2 = \tilde{\sigma}$, where

$$\begin{aligned} \tilde{\sigma} &= \int (dp)\, \rho^{(0)}_{f_{pq}}(q_1 + q_2 - p_1 - p_2)\, |\langle p \mid \tilde{T} \mid q\rangle|^2 \\ &= \int (dp)\, \delta^{(4)}(q_1 + q_2 - p_1 - p_2)\, e^{-\bar{n}_{pq}}\, |\langle p \mid \tilde{T} \mid q\rangle|^2. \end{aligned} \tag{26}$$

Equation (19) gave us the approximate form of $\tilde{\rho}^{(n)}$ which turned out to be concentrated around $K_0 = n\bar{\omega}$, $\mathbf{K} = 0$. If we assume that $\langle p \mid \tilde{T} \mid q\rangle$ is independent of K^2, then, at least until the end of phase space is reached, we can approximate (24) by

$$\begin{aligned} \sigma_{2+n} &= \int (dp)\, d^4K\, \rho^{(n)}_{f_{pq}}(K)\, \delta^{(4)}(K + p_1 + p_2 - q_1 - q_2)\, |\langle p \mid \tilde{T} \mid q\rangle|^2 \\ &\approx \int (dp)\, \delta^{(4)}(p + \bar{K} - q)\, |\langle p \mid \tilde{T} \mid q\rangle|^2 \int d^4K\, \rho^{(n)}_{f_{pq}}(K) \\ &\approx \int (dp)\, \delta^{(4)}(p_1 + p_2 - q_1 - q_2)\, |\langle p \mid \tilde{T} \mid q\rangle|^2\, e^{-\bar{n}_{pq}}\, \frac{\bar{n}^n_{pq}}{n!}, \end{aligned} \tag{27}$$

which means a Poisson distribution for the differential cross section. The last equality implies a structure in $\tilde{T}$ to which we will return below. Further, if $\bar{n}$ depends on q only then we have

$$\sigma_{2+n} = \frac{\bar{n}^n}{n!}\,\tilde{\sigma}. \tag{28}$$

This calculation makes sense only provided phase-space restrictions can be avoided. In other words, if the number of pions is smaller than the maximum allowed by energy conservation, then

$$n\bar{\omega} < \sqrt{s} - m_1 - m_2\,, \tag{29}$$

where m_1 and m_2 are the masses of p_1 and p_2. This works best for an $f(k)$ that is concentrated around the c.m. with a narrow width. For high n that violate the inequality (29) we have to expect distortions of the distribution law (28).

The skeleton can be either elastic or inelastic. By elastic skeleton we mean that the outgoing particles in the skeleton are the same as the incoming one. This does not imply $\sigma_2 = \tilde{\sigma}$; to this question we return in a minute. An inelastic skeleton can have resonances among its outgoing particles. Actually, a skeleton does not have to be a four-particle object and can also consist out of three particles or five, etc. The existence of two leading particles is, of course, suggestive of a four-point skeleton. In the case of an elastic skeleton we have some evidence that Eq. (28) cannot be continued to $n = 0$. This is pointed out in the Appendix. We have, therefore, to rely on unitarity to give us the elastic part $\langle p \mid T \mid q\rangle$ in terms of all the inelastic reactions. As a crude approximation, one may consider a model in which all inelastic reactions are described by Eq. (21) with an elastic skeleton. Unitarity leads then to the condition

$$\begin{aligned}
&i[\langle p_1p_2 \mid T^+ \mid q_1q_2\rangle - \langle p_1p_2 \mid T \mid q_1q_2\rangle] \\
&\qquad - \int \frac{d^3p_1{}'}{2E_1{}'}\frac{d^3p_2{}'}{2E_2{}'}\,(2\pi)^4\,\delta^{(4)}(p_1{}' + p_2{}' - q_1 - q_2)\langle p \mid T^+ \mid p'\rangle\langle p' \mid T \mid q\rangle \\
&\quad = \sum_{n=1}^{\infty} \int \frac{d^3p_1{}'}{2E_1{}'}\frac{d^3p_2{}'}{2E_2{}'}\,(2\pi)^4\,\rho^{(n)}_{f_{p'p},f_{p'q}}(q_1 + q_2 - p_1 - p_2) \\
&\qquad \times \langle p_1p_2 \mid \tilde{T}^+ \mid p_1{}'p_2{}'\rangle\langle p_1{}'p_2{}' \mid \tilde{T} \mid q_1q_2\rangle,
\end{aligned} \tag{30}$$

where $\rho^{(n)}_{f_{p'p},f_{p'q}}$ are the n-particle contributions to an object of the type defined in Eq. (14). The right side of Eq. (30) is analogous to Van Hove's "overlap integral" [4] that determines the t structure of the elastic amplitude. Note that an explicit t structure can and should exist in $\langle p \mid \tilde{T} \mid q\rangle$ [6] as we shall see below.

It is interesting to see how the bremsstrahlung theory [3] which actually suggests this formalism, solves the unitarity problem. The function f_{pq} is given in this case by

$$(2\pi)^{3/2} f_{pq}(k) = e_1' \frac{\epsilon \cdot p_1}{k \cdot p_1} + e_2' \frac{\epsilon \cdot p_2}{k \cdot p_2} - e_1 \frac{\epsilon \cdot q_1}{k \cdot p_1} - e_2 \frac{\epsilon \cdot q_2}{k \cdot p_2}, \tag{31}$$

where e_i are the various charges and ϵ is the photon's polarization vector. Clearly, f is peaked around $k = 0$ and the whole treatment is verified by QED only in the limit $k \to 0$ [3]. Then it turns out that indeed

$$\langle p \mid T \mid q \rangle = \langle p \mid \tilde{T} \mid q \rangle \, e^{-\bar{n}_{pq}/2}, \tag{32}$$

and unitarity is satisfied, provided $\tilde{T}$ satisfies elastic unitarity. To see this, we rewrite (30) with the use of (32):

$$i[\langle p \mid T^+ \mid q \rangle - \langle p \mid T \mid q \rangle]$$
$$= \int \frac{d^3p_1'}{2E_1'} \frac{d^3p_2'}{2E_2'} \langle p \mid \tilde{T}^+ \mid p' \rangle \langle p' \mid \tilde{T} \mid q \rangle (2\pi)^4 \int d^4x$$
$$\times \exp \left\{ -\frac{1}{2} \int \frac{d^3k}{2\omega} (\mid f_{p'p} \mid^2 + \mid f_{p'q} \mid^2 - 2 f^*_{p'p} e^{ikx} f_{p'q}) \right\} e^{ix(p_1' + p_2' - q_1 - q_2)}.$$

Using the properties of (31) and replacing here the e^{ikx} in the integrand by 1 (the limit $k \to 0$!), we find

$$i[\langle p \mid T^+ \mid q \rangle - \langle p \mid T \mid q \rangle]$$
$$= e^{-\bar{n}_{qp}/2} \int \frac{d^3p_1'}{2E_1'} \frac{d^3p_2'}{2E_2'} \langle p \mid \tilde{T}^+ \mid p' \rangle \langle p' \mid \tilde{T} \mid q \rangle (2\pi)^4 \, \delta^{(4)}(p_1' + p_2' - q_1 - q_2), \tag{33}$$

which shows that the ansatz (32) works provided $\langle p \mid \tilde{T} \mid q \rangle$ obeys by itself a unitarity equation.

There clearly are several important differences between the formalism of bremsstrahlung and the emission of mesons in high energy collision. The first is that the identification (32) cannot be made in our case. Another is that the limit $k \to 0$ is not justifiable and cannot be obtained with massive (and energetic) mesons. That can be circumvented by having a skeleton matrix element that does not vary significantly with K. A very important third difference is that we may choose f to depend on q only. In the Appendix we show characteristic distributions that depend only on k_T and k_L. These variables can be given an invariant definition in terms of $k \cdot q_1$, $k \cdot q_2$, $q_1 \cdot q_2$ and the masses involved. Therefore one needs no p dependence. This then makes it possible to go from Eq. (27) to Eq. (28) and get simple relations for the integrated cross sections.

Let us now investigate shortly some of the properties of the skeleton in the high-energy collisions. To be specific, we will usually think of an elastic skeleton although other types can also be constructed. We already hinted after Eq. (27) that it has to have some structure. Otherwise, phase space effects of the leading particles should be felt. We learn from experiment that the leading particles are indeed confined to low transverse momenta, i.e., also low momentum transfers. Such a dependence can be incorporated in $\langle p \mid \tilde{T} \mid q \rangle$. We note that an explicit t dependence will be effectively weakened for higher multiplicities. That can be seen by looking at the relevant differential phase space element

$$\frac{d^3p_1}{2E_1}\frac{d^3p_2}{2E_2}\, d^4K\, \delta^{(4)}(p_1 + p_2 + K - q_1 - q_2) \rightarrow \frac{d\varphi\, dt\, d^4K}{8q\sqrt{s}\, R}, \tag{34}$$

where φ is the azimuthal angle and q is the value of the incoming momenta $\mathbf{q}_1$ and $\mathbf{q}_2$ in the c.m. R is given by

$$R = 1 - \frac{K_0}{\sqrt{s}} + \frac{E_1 K \cos\theta_{p_1K}}{p\sqrt{s}} + \frac{E_1 K}{\sqrt{s}} \frac{\partial \cos\theta_{p_1K}}{\partial \cos\theta_{p_1q_1}} \frac{\partial \cos\theta_{p_1q_1}}{\partial p}, \tag{35}$$

where K and p are the magnitudes of $\mathbf{K}$ and $\mathbf{p}_1$ and θ_{p_1K} is the angle between them. For low values of K_0 (low multiplicities), $R \rightarrow 1$. Hence a strong t-dependent effect introduced in $\tilde{T}$ will show up for low n and get distorted for high n. This may be quite consistent with observed trends. We will not investigate further this question since in this paper we are mainly concerned with the consistency of the whole picture rather than the fine details.

The explicit construction of an example of coherent production of mesons shows that independent productions can take place. Coherence is also a statement about the phases that are not directly measurable. They will, however, be important when we discuss the isospin question in Section VI. The easiest things to measure are of course the cross sections. Their distribution, suggested by Eq. (28), will get modified by considerations of the quantum numbers of the pions to which we turn in the next Section.

IV. Charge and Parity

The coherent state must have a fixed electric charge that matches the charge of the skeleton. This is not true of simple charged coherent states of the types ($i = +$ or $-$)

$$|f_i\rangle = \exp\left\{-\frac{1}{2}\int d^3k\, \frac{|f_i|^2}{2\omega} + \int d^3k\, \frac{f_i(k)}{\sqrt{2\omega}}\, a_i^+(k)\right\} |0\rangle \qquad i = +, 0, -. \tag{36}$$

One way to deal with the problem can be to start from the state

$$|F\rangle = |f_+\rangle\,|f_0\rangle\,|f_-\rangle \tag{37}$$

and project out the required charge. An alternative is to define a state $|f_+f_-, Q\rangle$ obeying the equation

$$a_+(\mathbf{k})\,a_-(\mathbf{k})\,|f_+f_-, Q\rangle = \frac{1}{2\omega} f_+(k)\,f_-(k)\,|f_+f_-, Q\rangle \tag{38}$$

which has a definite charge Q. This is an analog of Eq. (3) and can serve as a definition of a coherent state of charged particles. The solution to Eq. (38) is

$$|f_+f_-, Q\rangle = C^{-1/2} \sum_n \frac{1}{(n+Q)!\,n!} \left(\int \frac{d^3k}{\sqrt{2\omega}} f_+(k)\,a_+{}^+(\mathbf{k})\right)^{n+Q} \times \left(\int \frac{d^3k}{\sqrt{2\omega}} f_-(k)\,a_-{}^+(\mathbf{k})\right)^n |0\rangle, \tag{39}$$

where the sum starts from $n = 0$ for positive Q and from $n = -Q$ for negative Q. The normalization constant C turns out to be

$$C = (-i)^Q\,J_Q(2ix), \tag{40}$$

where

$$x^2 = \int \frac{d^3k}{2\omega}\,|f_+(k)|^2 \int \frac{d^3k}{2\omega}\,|f_-(k)|^2. \tag{41}$$

This parameter is equal to the average value of n^2 in the resulting distribution of $(n+Q)$ positive and n negative charged states

$$x^2 = \langle n^2\rangle. \tag{42}$$

It is straightforward to show that the projection of $|F\rangle$ onto a specific charge Q does indeed contain this state. It is

$$|f, Q\rangle = |f_0\rangle|f_+f_-, Q\rangle \tag{43}$$

which we will regard as the right choice to take the place of $|f\rangle$ in Eq. (21). The distribution of charged particles that results from this state was discussed by us in Ref. [7]. It is the conditional distribution reached by multiplying two Poisson distributions (for the two different charges) and projecting out the right total charge. Thus for a cloud of charge Q the cross section for $n + Q$ positive and n negative pions is given by

$$P_n^{(Q)} = P_{n+Q}^{(-Q)} = \frac{i^Q}{J_Q(2ix)} \frac{x^{2n+Q}}{n!(n+Q)!}, \tag{44}$$

and the average is

$$\langle n \rangle = -ix \frac{J_{Q+1}(2ix)}{J_Q(2ix)}. \tag{45}$$

We will return to a discussion of the distribution in Section VII.

We come now to the question of parity conservation. If the skeleton contains two spinors among its four particles, this question can be overcome by inserting $\gamma_5 f$ instead of f in the relevant expressions. Thus (20) becomes

$$\langle p_1 p_2 k_1 \cdots k_n \mid S \mid q_1 q_2 \rangle$$
$$= i \int d^4x \, e^{ix(p_1+p_2-q_1-q_2)} \langle p_1 p_2 k_1 \cdots k_n \mid S(\gamma_5 f e^{ikx}) \tilde{T} \mid q_1 q_2 \rangle, \tag{46}$$

and this expression is invariant under parity. This little change has some severe consequences — it leads to a distinction between even and odd numbers of emitted pions. Whereas the even numbers can still be described by the arguments in Section II which lead to the basic $\tilde{\sigma}$, the odd number of pions will be described in terms of

$$\tilde{\sigma}' = \int (dp) \, e^{-\bar{n}_{pq}} \, \delta^{(4)}(p_1 + p_2 - q_1 - q_2) \, |\langle p \mid \gamma_5 \tilde{T} \mid q \rangle|^2. \tag{47}$$

The numerical differences between $\tilde{\sigma}$ and $\tilde{\sigma}'$ can of course affect the distribution. We might, however, sometimes find small deviations that will not be detected on the logarithmic scale. For instance, in the limiting case in which the incoming baryon is very fast and the outgoing very slow, one has $|\bar{u}(p_1)\, u(q_1)| \approx |\bar{u}(p_1)\, \gamma_5 u(q_1)|$.

Another way to realize the situation is to consider the case of symmetric pionic distributions, $f(\mathbf{k}) = f(-\mathbf{k})$. Then it is clear that the allowed transitions in the skeleton have to be from even j to even j and from odd j to odd j. However, in the case of an even number of pions we have also even (odd) $l \rightarrow$ even (odd) l, whereas if the number of emitted pions is odd we find even (odd) $l \rightarrow$ odd (even) l. Hence a spin transition must be involved. All this means that the completely factorized form (21) cannot be obtained and, in the best case, it breaks down into somewhat different distributions for even and odd pions.

In practice, it turns out that this problem often does not play an important role. Thus if one looks at the distribution of charged pions one sums over the neutral ones. The neutral pions are then included in an effective skeleton and the distribution of the charged ones is given by $| \gamma_5 f_+ \gamma_5 f_- , Q\rangle$ which has always either odd or even numbers of γ_5 matrices. One way to completely avoid the parity problem is simply to discuss only emission of scalar pairs of pions. This is of course farther away from independent pion production. We will return and discuss such an approach in Sections VI and VII. In future, we will refrain from using explicitly the γ_5 factor although it will be implicitly assumed.

One immediate important consequence of the parity problem is that a skeleton of spinless particles cannot emit a coherent state $| f, Q\rangle$ of pions. Moreover, one may expect that the more spins there are in the skeleton the easier it is to connect it to a pionic coherent state. It is interesting to note in this connection that $\sigma_T(\pi\pi) < \sigma_T(\pi p) < \sigma_T(pp)$ and the direction of the inequalities is also that of the number of spins involved. If independent pion production is an important inelastic mode, then this correlation may be meaningful.

Several selection rules arise for neutral systems from charge conjugation considerations. Thus a skeleton of four pions can be connected only to even number of pions, which is the same condition as that of parity conservation. A neutral system of identical pions has positive charge conjugation. Thus it cannot couple, e.g., to e^+e^- (via a photon). Similarly $\bar{p}p$ annihilation at rest is restricted by charge conjugation. Both e^+e^- and $\bar{p}p$ are different from πp and pp collisions. e^+e^- produces hadronic matter via a photon and it is unclear what can serve as the skeleton. $\bar{p}p$ collisions, although being pure hadronic reactions, have a small elastic skeleton. Hence once again it is not clear how to implement our model in this case. The model can hopefully be used in πp and pp reactions in which charge conjugation does not impose additional restrictions.

V. Isospin Analysis of Identical Pions

We develop here a formalism that enables us to deal with the isospin analysis of a system of identical pions. We limit ourselves to identical pions since this case renders itself to an elegant and simple treatment. We will give an argument at the end of this Section showing that this limitation does not actually prevent us from reaching the optimal situation (lowest isospin) in coherent states.

We start by defining a normalized momentum space distribution $\varphi(k)$ satisfying

$$\int \frac{d^3k}{2\omega} \, | \, \varphi(k)|^2 = 1. \tag{48}$$

The fact that the pions are identical is summarized in the assumption

$$f_i(k) = f_i \varphi(k), \qquad i = +, 0, -, \tag{49}$$

where the f_i are three constants. The magnitudes and phases of the f_i determine the isospin structure of a definite combination of identical pions.

Let us now define three operators

$$a_i^+ = \int d^3k \, \frac{\varphi(k)}{\sqrt{2\omega}} \, a_i^+(\mathbf{k}), \tag{50}$$

which obey the commutation relations

$$[a_i \, , a_j^+] = \delta_{ij} \, . \tag{51}$$

The isospin generators for the system of identical pions can be simply expressed in terms of these operators. They are

$$\mathbf{I} = a_i^+ \boldsymbol{\tau}_{ij} a_j \, , \tag{52}$$

where

$$\tau_x = \frac{1}{\sqrt{2}} \begin{pmatrix} 0 & 1 & 0 \\ 1 & 0 & 1 \\ 0 & 1 & 0 \end{pmatrix}, \quad \tau_y = \frac{1}{\sqrt{2}} \begin{pmatrix} 0 & -i & 0 \\ i & 0 & -i \\ 0 & i & 0 \end{pmatrix}, \quad \tau_z = \begin{pmatrix} 1 & 0 & 0 \\ 0 & 0 & 0 \\ 0 & 0 & -1 \end{pmatrix}. \tag{53}$$

Let us now define the number operators

$$N_i = a_i^+ a_i \, , \qquad N = N_+ + N_0 + N_- \, , \tag{54}$$

and the bilinear isoscalar creation and annihilation operators

$$A = a_0 a_0 - 2a_+ a_- \, , \qquad [\mathbf{I}, A] = [\mathbf{I}, A^+] = 0. \tag{55}$$

The three operators N, A and A^+ play a key role in the isospin analysis. It is interesting to note that they close on an algebra [8]

$$[N, A] = -2A, \qquad [N, A^+] = 2A^+, \qquad [A, A^+] = 4N + 6. \tag{56}$$

Their importance stems from the fact that the operator $\mathbf{I}^2$ can be written in terms of them as

$$\mathbf{I}^2 = N(N + 1) - A^+ A \tag{57}$$

as can be seen by a straightforward calculation. It follows from this equation that a state of n-identical pions will have isospin $I = n$ if and only if

$$A \mid I = n, n\rangle = 0. \tag{58}$$

It is simple to construct such a state by using the following operator (note the mixed isospin property)

$$T^+ = a_+^+ + \sqrt{2}\, a_0^+ + a_-^+. \tag{59}$$

This operator obeys

$$\tfrac{1}{2}[A, T^+] = -a_+ + \sqrt{2}\, a_0 - a_- \equiv U \qquad [U, T^+] = 0 \tag{60}$$

Because of these properties it is evident that

$$A(T^+)^n \mid 0\rangle = 0; \tag{61}$$

hence every I_z projection of the state $(T^+)^n \mid 0\rangle$ has an isospin of $I = n$. Actually, this state contains all $n + 1$ I_z projections. We can write them out explicitly and define through them the states

$$\mid I = n, I_z, n\rangle$$

$$= B^{-1/2} \sum_p \frac{n!}{(I_z + p)!\, p!(n - 2p - I_z)!} (a_+^+)^{I_z+p} (\sqrt{2}\, a_0^+)^{n-2p-I_z} (a_-^+)^p \mid 0\rangle, \tag{62}$$

where the sum is over all integer p such that the factorials can be defined. B is a normalization constant equal to

$$B = \sum_p \frac{(n!)^2\, 2^{n-2p-I_z}}{(I_z + p)!\, p!(n - 2p - I_z)!}\,. \tag{63}$$

A system of n identical pions can include, in addition to $I = n$, also all isospins of $n - 2, n - 4,...$ down to 0 or 1. Altogether, these form $\frac{1}{2}(n + 1)(n + 2)$ states, characteristic of the completely symmetric combination. We can prove that this is the case by direct construction of the isospin states. We have already seen that A^+ is a creation operator of an $I = 0$ system; indeed,

$$\mid I = 0, n = 2m\rangle = \frac{1}{\sqrt{(2m + 1)!}} (A^+)^m \mid 0\rangle. \tag{64}$$

Thus we have created the lowest isospin for even number of pions. It is now simple to construct the general state in terms of (62)

$$\mid I, I_z, n = 2m + I\rangle = D^{-1/2}(A^+)^m \mid I, I_z, n = I\rangle. \tag{65}$$

D is a normalization constant equal to

$$D = \frac{4^m m!\,\Gamma(m + I + \frac{3}{2})}{\Gamma(I + \frac{3}{2})}. \tag{66}$$

The fact that all the different states (65) form an orthonormal system can be shown by using the property (60). Simple calculation of the number of states shows that we constructed in this way all possible isospin states of identical pions.

Let us apply this formalism to the coherent state $| F\rangle$ of Eq. (37). It is defined now in terms of identical pions and satisfies

$$a_i \,| F\rangle = f_i \,| F\rangle, \qquad i = +, 0, -. \tag{67}$$

In particular,

$$A \,| F\rangle = (f_0^2 - 2f_+f_-)| F\rangle. \tag{68}$$

We see that if we choose $f_0^2 = 2f_+f_-$, we have a coherent state which contains only states with $I = n$. In other words, this choice of the parameters leads to the maximal isospin content. In our application to physics, we try to achieve the opposite goal, viz., to minimize the isospins of the coherent state since we are bound by the isospins of the skeleton. We find by using Eq. (57) that

$$\langle \mathbf{I}^2 \rangle = \langle F \,|\, \mathbf{I}^2 \,|\, F\rangle = \bar{n}(\bar{n} + 2) - | f_0^2 - 2f_+f_- |^2, \tag{69}$$

where

$$\bar{n} \equiv \langle F \,|\, N \,|\, F\rangle = | f_+ |^2 + | f_0 |^2 + | f_- |^2. \tag{70}$$

It follows from (69) that $\langle \mathbf{I}^2 \rangle$ obtains its minimal value if

$$\arg(f_0^2) = \pi + \arg(f_+f_-) \qquad \text{and} \quad | f_+ | = | f_- |. \tag{71}$$

These are also the conditions that ensure that the state $| F\rangle$ has no preferred direction in isospace, $\langle F \,|\, \mathbf{I} \,|\, F\rangle = 0$. Thus the minimal value is reached by random walk in isospace

$$\langle \mathbf{I}^2 \rangle_{\min} = 2\bar{n} \tag{72}$$

We can show quite generaly that this situation cannot be improved if we consider a coherent state $| F\rangle$ of nonidentical pions. Indeed, in this case direct computation leads to [the use of Eq. (52) is now forbidden!]

$$\langle \mathbf{I}^2 \rangle = 2\bar{n} + \left[\int \frac{d^3k}{2\omega} (| f_+(k)|^2 - | f_-(k)|^2)\right]^2 + 2\left| \int \frac{d^3k}{2\omega} [f_+^*(k) f_0(k) + f_0^*(k) f_-(k)]\right|^2. \tag{73}$$

The minimal value is once again given by Eq. (72). It is reached when the two squared terms vanish and in the limit of identical pions these conditions reduce to (71). We expect a similar situation to prevail also when we deal with the more physical state $|f, Q\rangle$ in the next section. We think, therefore, that the restriction to identical pions does not prevent us from reaching the minimal possible value of the isospin average of a coherent state.

VI. Isospin Analysis of Coherent Pions

Let us now apply the methods developed in the previous section to the coherent state of Eq. (43), namely, $|f, Q\rangle = |f_0\rangle\,|f_+f_-, Q\rangle$. We will limit ourselves in the numerical calculations to the case $Q = 0$. Other Q values can be treated similarly. The main idea is that the model (20) should not violate isospin conservation. That cannot be achieved in an absolute fashion with the emission of $|f, Q\rangle$ since this state contains all possible isospins. We may still hope that it does make sense as an approximation. That will be the case if the main isospin content of $|f, Q\rangle$ can be matched by the skeleton. Thus an elastic skeleton of πp can carry up to $I = 3$, and an elastic skeleton of pp up to $I = 2$. Hence the imposition of the necessary isospin cuts on $|f, Q = 0\rangle$ will make small difference only if its isospins are confined mainly to the above-mentioned regions.

The result of Eq. (72) may look quite pessimistic. Nevertheless, one should remember that $|F\rangle$ contains all the possible I_z projections. By limiting ourselves to $|f, Q = 0\rangle$ we can hope to do better. The calculation in this case is much more difficult than that for $|F\rangle$. The reason is that $|f, Q = 0\rangle$ is no longer an eigenstate of a_+ and a_-, separately. It is, however, an eigenstate of A:

$$A\,|f, Q\rangle = (f_0^2 - 2f_+f_-)|f, Q\rangle, \qquad a_0\,|f, Q\rangle = f_0\,|f, Q\rangle. \tag{74}$$

We limited ourselves again to identical pions (see discussion at the end of Section V). We have now the freedom to play with the phases and magnitudes of the f_i. As a matter of fact, the parameter that is of importance is

$$\xi = \frac{f_0^2}{f_+f_-}\,. \tag{75}$$

By choosing $\xi = 2$, we reach the situation of the maximal isospin state. The discussion in Section V shows us that the minimal values are reached for negative values of ξ. That, indeed, is also the case here. Before turning to the numerical evaluation, we would like still to point out that suitable choices of ξ can eliminate a particular isospin altogether from any combination of identical pions.

The reason for this result is that Eq. (62) implies

$$\langle I = n, I_z = 0, n \mid f\rangle = B^{-1/2}\langle 0 \mid f\rangle(\sqrt{2} f_0)^n \sum_p \frac{n!}{p!\,p!(n-2p)!} (2\xi)^{-p}. \quad (76)$$

Hence a suitable choice of ξ leads to $\langle I = n, I_z = 0, n \mid f\rangle = 0$. Once this is achieved, it follows from Eq. (65) that all $\langle I, I_z = 0, n \mid f\rangle = 0$. Thus the choice $\xi = -1$ eliminates $I = 2$, and the choice $\xi = -3$ eliminates $I = 3$.

Let us now turn to the question of the minimal isospin content. For simplicity, we choose f_+f_- as real and denote it by $x = f_+f_-$, in agreement with the notation in Eq. (41). We find then the following results

$$\langle \mathbf{I}^2\rangle = \frac{4xJ_1(2ix)}{iJ_0(2ix)}(1 + |f_0|^2) - 2x^2\frac{J_2(2ix)}{J_0(2ix)} - 2x^2 + 2|f_0|^2 + 4\,\mathrm{Re}(f_0{}^2x),$$
$$\langle n\rangle = \langle f \mid N \mid f\rangle = \frac{2xJ_1(2ix)}{iJ_0(2ix)} + |f_0|^2 = \langle n_{ch}\rangle + \langle n_{\pi^0}\rangle. \quad (77)$$

The results for $\langle I\rangle$ vs. $\langle n\rangle$, where $\langle I\rangle\langle I+1\rangle = \langle \mathbf{I}^2\rangle$, are plotted in Fig. 1 for several values of the parameter ξ. We see that for negative values of ξ they all lie very close to each other obeying

$$\langle \mathbf{I}^2\rangle \approx \langle n\rangle. \quad (78)$$

Thus by going from the state $|F\rangle$ to $|f, Q = 0\rangle$ we gained a factor of two in the

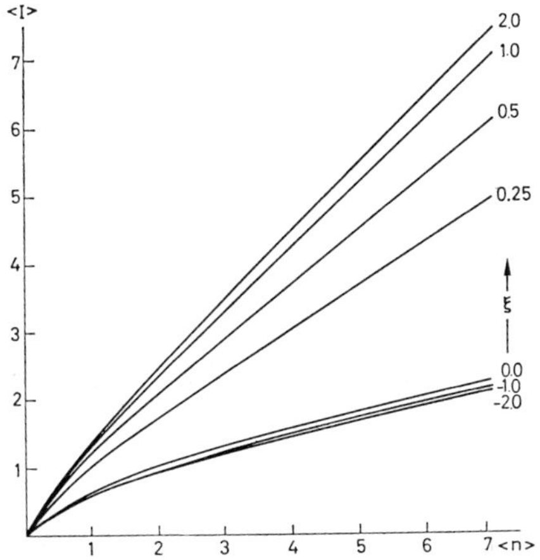

FIG. 1. Plots of $\langle I\rangle$ vs. $\langle n\rangle$ for various choices of the parameter ξ in the coherent state $|f, Q = 0\rangle$.

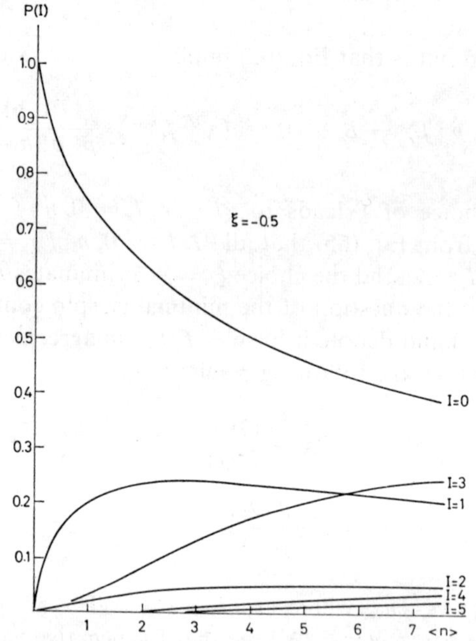

FIG. 2. The percentages of the contents of different isospins in the coherent state $|f, Q = 0\rangle$ vs. $\langle n \rangle$ for the choice $\xi = -0.5$ are plotted.

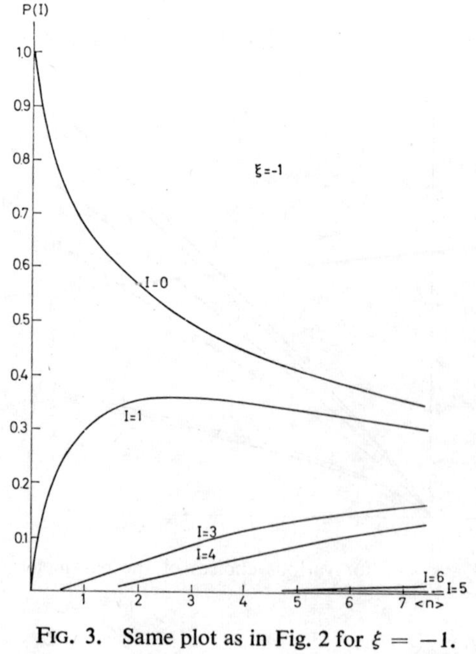

FIG. 3. Same plot as in Fig. 2 for $\xi = -1$.

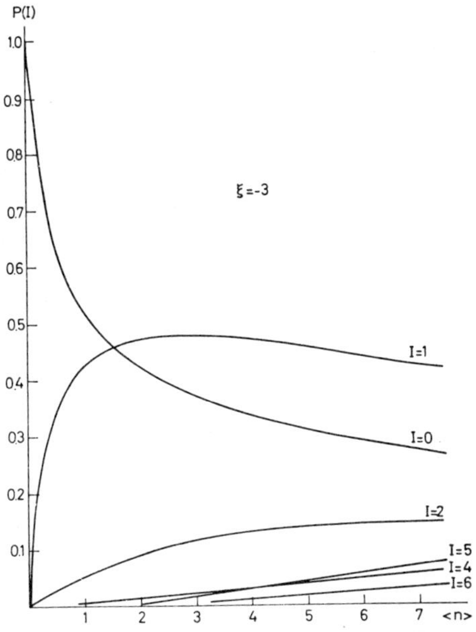

FIG. 4. Same plot as in Fig. 2 for $\xi = -3$.

minimal value of $\langle \mathbf{I}^2 \rangle$. This is of course essential in order to be able to use the coherent state $|f, Q = 0\rangle$ in a physical model.

The absolute value of ξ is the asymptotic (i.e., for large $\langle n \rangle$) ratio of the number of π^0 to the number of π^+. Therefore, we do not consider values that are too far from unity. In Figs. 2–4, we show the distribution of $\langle I \rangle$ for various choices of the parameter ξ. Figure 2 shows that for $\xi = -0.5$ all isospins higher than three are strongly quenched. Figure 3 has the choice $\xi = -1$ that eliminates $I = 2$ and Fig. 4 is drawn with $\xi = -3$ that eliminates $I = 3$. In all figures we see the important roles of low isospins for reasonable values of $\langle n \rangle$. A similar calculation leads to the distribution of the various proportions of specific isospin values in the n-pion configurations. Figure 5 shows these distributions for $\xi = -0.5$, a value that emphasizes the isospins 0, 1 and 3. Figure 6 shows the distribution for $\xi = -2$, where $I = 0$, and 1 are the important values. The relative amounts of the low isospins change slowly with ξ. If one tries to attach a coherent state to an elastic pp skeleton, one needs $I \leqslant 2$. We see from Fig. 6 that although the leading terms have low I values one still encounters sizable contributions from forbidden isospins. Hence this picture is not in very good agreement with isospin conservation. If the skeleton has higher isospins involved (like elastic πp or $pp \to p\Delta$), the situation is

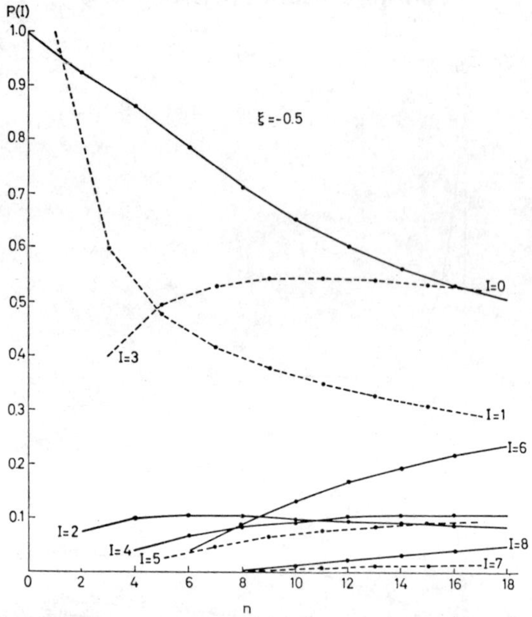

FIG. 5. The percentages of the contents of different isospins in the various n-particle states included in $|f, Q = 0\rangle$ vs. n for the choice $\xi = -0.5$ are plotted. Note the two types of curves that describe even and odd isospins for even and odd n, respectively.

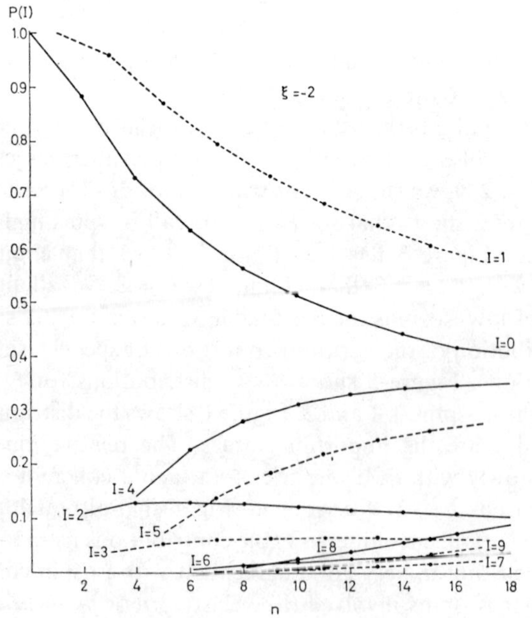

FIG. 6. Same plot as in Fig. 5 for $\xi = -2$.

better. In any case, however, the production of $|f, Q\rangle$ can only approximately satisfy the isospin restrictions.

There exists one possibility of eliminating the isospin (as well as the parity) problem altogether and that is the production of the pions in scalar isoscalar pairs. This requires of course different skeletons for even and odd pionic reactions. We will investigate here the mathematical structure of such a coherent state of identical pions. It can be called a coherent state since it can be chosen as an eigenstate of the operator A

$$A \mid g\rangle = g \mid g\rangle, \tag{79}$$

which is similar to the property (74) of $|f, Q\rangle$. The solution of Eq. (79) that is pure $I = 0$ is

$$\begin{aligned} \mid g\rangle &= \sqrt{\frac{g}{\sinh g}} \sum_m \frac{g^m}{(2m+1)!} (A^+)^m \mid 0\rangle \\ &= \sqrt{\frac{g}{\sinh g}} \sum_m \frac{g^m}{\sqrt{(2m+1)!}} \mid I = 0, n = 2m\rangle. \end{aligned} \tag{80}$$

The n-pion distribution is given by

$$P(n = 2m) = \frac{1}{\sinh g} \frac{g^{2m+1}}{(2m+1)!} \cdot \tag{81}$$

One important property of (80) is that the isoscalar state has the same amount of all the different charges

$$\langle n_{\pi^+}\rangle = \langle n_{\pi^-}\rangle = \langle n_{\pi^0}\rangle = \tfrac{1}{3}\langle n\rangle \tag{82}$$

which is easy to check. However, now the probability of finding neutral pions is correlated to that of the charged pions (in $|f, Q\rangle$ these are independent!). The probability to find r charged pairs in a state $| I = 0, n = 2m\rangle$ is

$$P(r; m) = \frac{m!\, m!}{(2m+1)!} \binom{2m-2r}{m-r} 4^r. \tag{83}$$

From Eqs. (83) and (81) one can find the probability to find r charged pairs in the coherent state $| g\rangle$. It is

$$P_r^{(g)} = \frac{(2g)^{2r}}{\sinh g} \sum_{p=0}^{\infty} g^{2p+1} \left(\frac{(p+r)!}{p!(2p+2r+1)!}\right)^2 (2p)!\,. \tag{84}$$

The average number of pions is

$$\langle n \rangle = g \coth g - 1 \tag{85}$$

and $\langle r \rangle = \frac{1}{3}\langle n \rangle$. Using such an analysis, we can calculate the expected correlation of $\langle n_{\pi^0} \rangle$ vs. r for fixed $\langle n \rangle$. These correlations are shown in Fig. 7 where $\langle n_{\pi^0} \rangle$ is plotted versus $n_{ch} = 2 + r$ in a way that can be contrasted with the Fig. 16 in the Appendix. We see that the strong correlation that exists in Fig. 7 does not look like the trend of Fig. 16. Some other aspects of this distribution will be discussed in the next section.

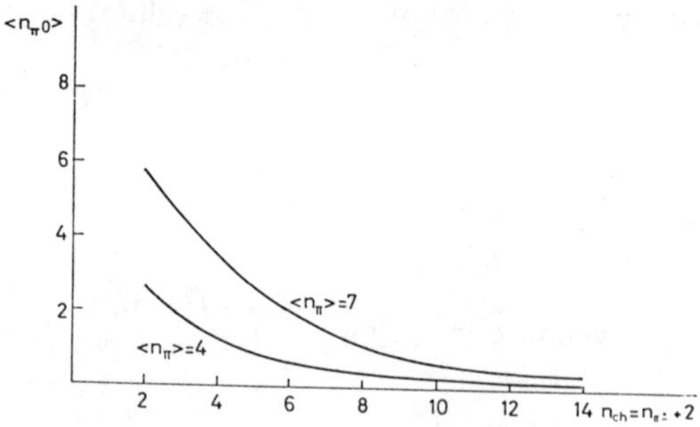

FIG. 7. The dependence of $\langle n_{\pi^0} \rangle$ on n_{π^+} is shown for two different values of $\langle n_\pi \rangle = \langle n_{\pi^0} \rangle + 2\,\langle n_{\pi^+} \rangle = 3\,\langle n_{\pi^0} \rangle$ in the coherent state $|\, g \rangle$.

VII. DISCUSSION AND SUMMARY

In Section IV we saw that imposing charge conservation on a coherent state leads to obvious restrictions of the concept of independent emissions. Nevertheless, we could solve the problem within the framework of coherent states: the projection of the coherent state $|\, F \rangle$ of Eq. (37) on a particular charge state turned out to be the coherent state $|\, f, Q \rangle$ of Eq. (43). This process cannot be continued with the isospin problem. Projecting out of $|\, f, Q \rangle$ only the allowed isospins is possible but it distorts the nature of the state. However, we learned in Section VI that optimal choices of the parameter ξ can make these distortions minimal. Thus the coherent state $|\, f, Q \rangle$ can be regarded as an approximation to an allowed state. The optimal ranges of ξ depend on the skeleton to which the relevant coherent state is coupled. The higher the isospins of the skeleton particles are, the smaller the distortions of the coherent state will be. The nature of these skeletons is still left as an open problem. We may suppose that various types of skeletons can exist and their effects may interfere with each other. The resulting picture will again be very complicated unless one particular skeleton has an overwhelming effect. Under such circumstances, one can look for regularities of the cross sections for various configurations. One may for instance think of an elastic skeleton leading to a distribution $P_n^{(0)}$ of Eq. (44). In Ref. [7], we have shown that all various $P_n^{(Q)}$ look similar. We include here again the comparison of $P_n^{(0)}$ with experiment in Fig. 8. That comparison is based on the compilation of Wang [9] of many inelastic reactions. The Fig. 8 shows also distributions suggested by Wang: W^I is a Poisson

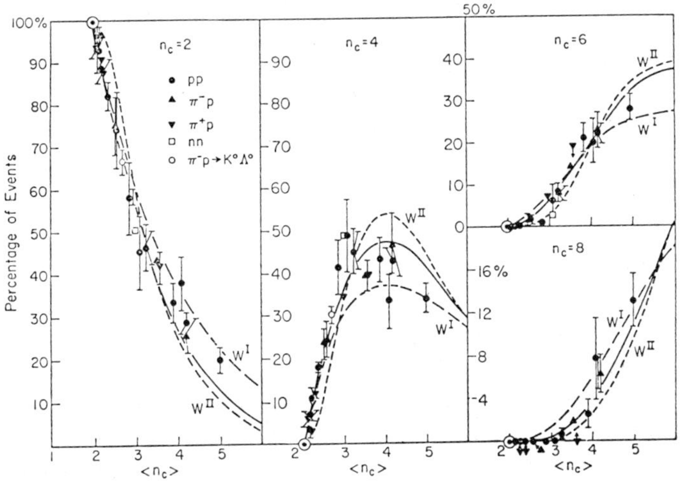

FIG. 8. Theoretical distributions are compared with Wang's compilation of inelastic production data [9]. W^I and W^{II} are taken from Wang's paper [9] and the solid curve is the result of the $P^{(0)}$ distribution following from the state $|f, Q = 0\rangle$. The figure is taken from Ref. [7].

distribution in $\frac{1}{2}(n_{ch} - 2)$ and W^{II} a hyperbolic sine distribution in $(n_{ch} - 2)$. If we replace the coherent state $|f, Q\rangle$ by the state $|g\rangle$ of Eq. (79), we find the distribution $P^{(g)}$ given by Eq. (84). Since $|g\rangle$ describes a production of pion pairs we may expect it to be somewhat similar to W^I. This is indeed the case as shown in Fig. 9, where we compare W^I, $P^{(0)}$ and $P^{(g)}$ in a fashion similar to Fig. 8. We may expect that for higher n, $P^{(0)}$ will be modified by the isospin cuts and both $P^{(0)}$ and $P^{(g)}$ will be affected by phase space limitations.

Figure 9 seems to favour the mode $|f, Q\rangle$ over $|g\rangle$. It could be expected that $P^{(g)}$ will not be a good fit since this distribution of the state $|g\rangle$ corresponds to the production of only an even number of pions. Even if another skeleton is added for the odd pions, we are still faced with the result, derived and discussed in the previous section, that a strong correlation has to be expected between $\langle n_{\pi^0}\rangle$ and n_{ch}. It is, therefore, doubtful whether this state $|g\rangle$ is actually produced in experiment. The state $|g\rangle$ has the theoretical advantage that it can be easily embedded in the formalism since its constituents are pion pairs with vacuum quantum

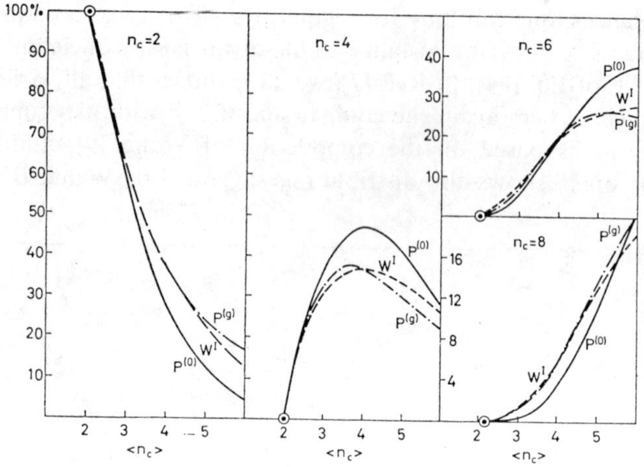

FIG. 9. Comparison of W^I, $P^{(0)}$, and $P^{(g)}$ distributions on a scale similar to Fig. 8.

numbers. Future experimental and theoretical analyses will hopefully tell us whether this mathematical simplicity can be exploited in realistic physical models.

The state $|f, Q\rangle$ is indeed closer to the concept of independent pions than the state $|g\rangle$. Throughout the paper, we have tried to follow this concept of independent production and we realize that it may look strange from a different point of view of the dynamics of multipion production. Thus the generalization of mechanisms familiar to us from two-particle productions would quite naturally lead us to expect many complicated configurations of resonances in the final states. In principle, the two different approaches may be complementary rather than contradicting. The reason is that the different production modes through resonances interfere with each other and can no longer be easily identified in the final amplitude. In such cases one usually resorts to a statistical approach. The coherent emission of pions can be regarded as a particular statistical description. We may expect, therefore, the sum of all possible production mechanisms to have a big overlap with the coherent amplitude. This leads to a new dual approach to many-particle production that should be further investigated.

Our investigation of the isospin problem answered only one necessary question: To what extent can violations of isospin inequalities be avoided. We did not give an explicit model of how the various isospins are produced. According to the picture outlined in the previous paragraph, we expect such a detailed model to be very complicated. Therefore, the correlations between various production modes of particles, that do in principle exist because of isospin conservation, do not simply follow from our description.

Our starting point was the independent production of pions and this led us to an approximate description of multipion amplitudes. There are, obviously, different statistical approaches possible. The most conventional ones use actually the conservation of isospin as their main foundation, assigning equal statistical weights to all the allowed states [10]. Their usual description does not make use of a skeleton. That is one of the differences between us and these statistical approaches. Although our model of coherent production can be regarded as a certain description of a statistical state, it does, however, incorporate dynamical assumptions and physical principles that are not of a pure statistical nature.

Finally, let us point out that our particular model for coherent production of pions — which seems to us as a natural way to approach the problem — is not necessarily the only one possible. In fact, the various papers of Ref. [1] discuss different approaches to the same problem. Our analysis in Sections IV to VI depends on the nature of the coherent state and not necessarily on the mechanism of its production (Section III). Therefore this analysis can be applied to the other theories as well. Approaching the many-pion emission problem on the basis of coherent states has some important advantages of mathematical elegance and physical simplicity, and we may hope, therefore, that it will also turn out to be useful and productive.

APPENDIX: Experimental Examples

In this Appendix, we discuss some experimental facts that can serve as a background for the theoretical discussion. In particular, we want to elucidate the concepts defined in the Introduction. We rely heavily on both published and

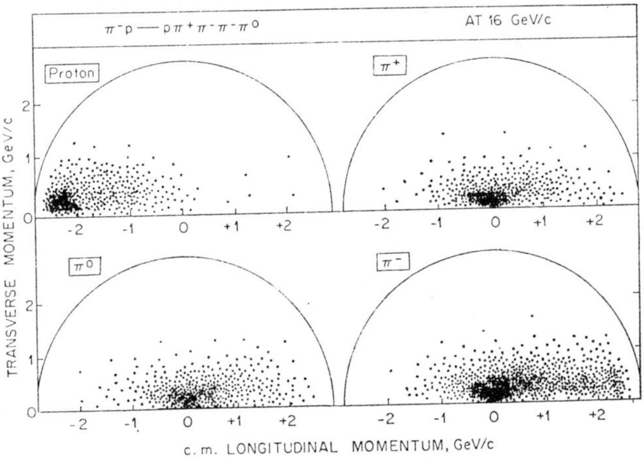

FIG. 10. Peyrou plot of $\pi^-p \to p\pi^+\pi^-\pi^-\pi^0$. Taken from Ref. [11].

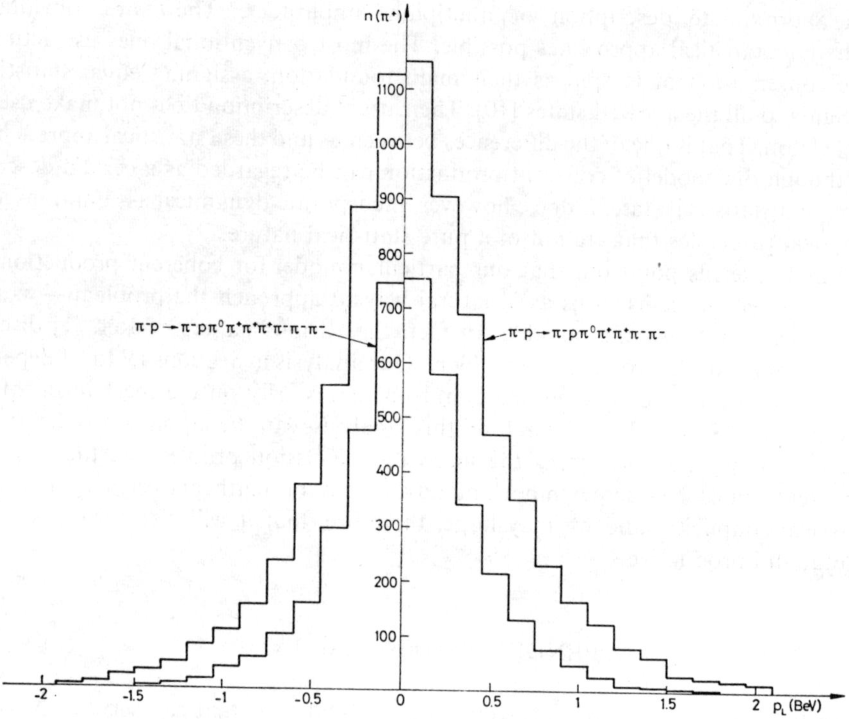

FIG. 11. Distribution of π^+ events vs. their longitudinal centre-of-mass momentum. Unpublished data of Ref. [11].

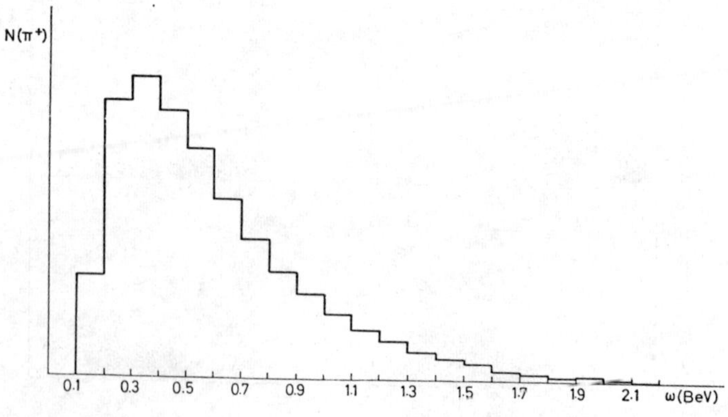

FIG. 12. Distribution of π^+ events vs. their centre-of-mass energy in the reaction $\pi^- p \to p\pi^-\pi^+\pi^+\pi^-\pi^-\pi^0$. Unpublished data of Ref. [11].

unpublished data of the ABBCCHW collaboration [11, 12] on 16 GeV $\pi^- p$ interactions.

Figure 10 shows the existence of leading particles and the concentration of the remaining pions around the centre-of-mass. This concentration is further investigated in Figs. 11 and 12. Figure 11 shows the longitudinal centre-of-mass distribution of the π^+ in two different configurations. Note the similarity between the dominant features of these two curves that would also be implied by independent production of π^+. Figure 12 gives the centre-of-mass energy distribution of the π^+ in the reaction $\pi^- p \to \pi^- p \pi^+ \pi^+ \pi^- \pi^- \pi^0$. Note the dominant peaking at energies of the order of 0.3 GeV. The strong peaking at low energies in the centre-of-mass is necessary in order for relation (29) to hold for a large range of n so that phase-space limitations do not affect the distributions characteristic of independent production.

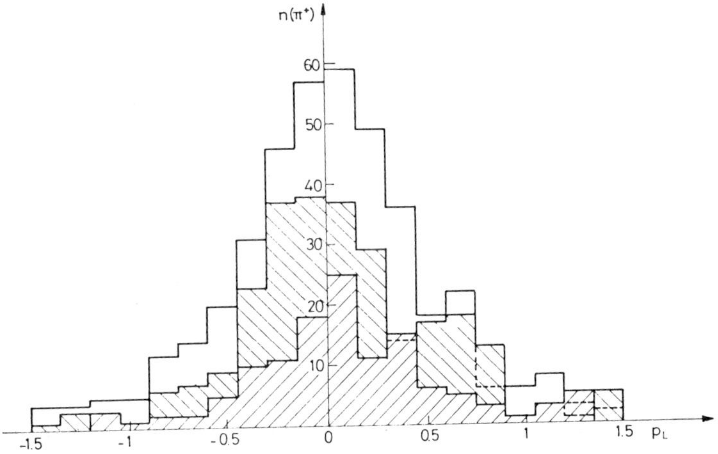

FIG. 13. Distribution of π^+ events vs. their longitudinal centre-of-mass momentum as a function of the other π^+'s longitudinal momentum in $\pi^- p \to \pi^- p \pi^+ \pi^+ \pi^- \pi^- \pi^0$. See explanation in text. Unpublished data of Ref. [11].

Let us come now to the question of uncorrelated production. Ideally, we should look at a seven-dimensional plot for two pions to check relation (1). Instead we choose to restrict ourselves to a check of the correlation between two longitudinal momentum distributions. Figure 13 shows the longitudinal centre-of-mass momentum distribution of one π^+ in $\pi^- p \to \pi^- p \pi^0 \pi^+ \pi^+ \pi^- \pi^-$ for several choices of the longitudinal momentum of the other π^+ [these choices correspond to values of (0, -0.1 GeV), (-0.2, -0.3), (-0.4, -0.5) for the three distributions shown]. The general trend is a uniform decrease of the distribution in agreement with uncorrelated production. One may argue that, since $\pi^+ \pi^+$ is an exotic channel, it should not exhibit strong correlations, whereas the $\pi^+\pi^0$ channel could be more sensitive to such effects. The truth of the matter is that a similar check for $\pi^+\pi^0$ correlations shows a far less regular behaviour than Fig. 13. Nevertheless, it is still true that the main bulk of events is concentrated around the centre of the $p_L{}^+p_L{}^0$ system. We have to conclude that uncorrelated production of $\pi^+\pi^0$ is a very crude approximation to the data.

If independent pion production can be regarded as an approximation to experiment, we may expect that the bigger the number of produced pions is the more energy is dissipated into the pionic cloud. As a consequence, the energy available for the leading particles will decrease with increasing multiplicity of pions. That this is indeed the trend of the data is seen in Fig. 14. We see that the average

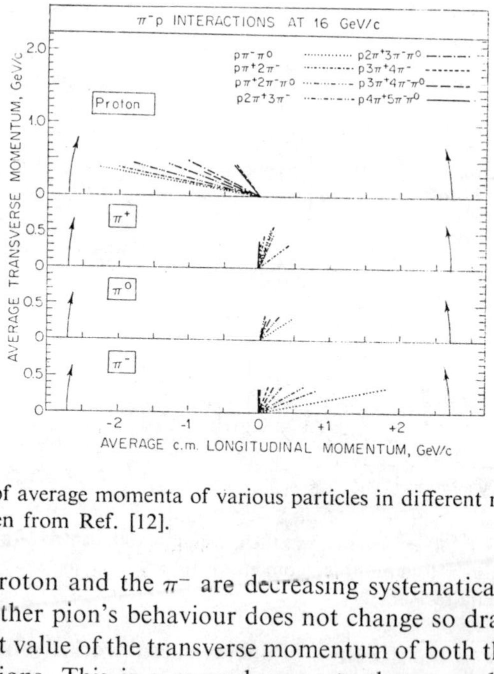

FIG. 14. Vectors of average momenta of various particles in different modes observed in the π^-p experiment. Taken from Ref. [12].

momenta of the proton and the π^- are decreasing systematically with increasing multiplicity. The other pion's behaviour does not change so drastically. Note also the almost constant value of the transverse momentum of both the leading particles and the emitted pions. This is a general property that must be common to the skeleton and the coherent state in our model.

Finally, let us discuss the cross sections distribution for the various multiplicities. Figure 15 shows the cross sections for charged nonstrange particles production in the 16 GeV π^-p experiment. The general shape is reminiscent of Poisson type distributions. For $n_{ch} = 2$, we included two entries. The dots show all inelastic processes and the cross includes also the elastic one. Using a statistical approach, one may wonder whether the elastic cross section should be included or not.

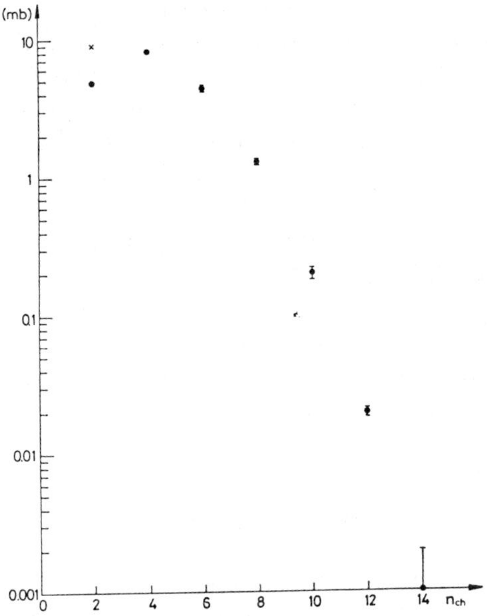

FIG. 15. Distribution of the cross section for nonstrange particle production in the $\pi^- p$ experiment. Taken from Ref. [11]. The cross represents the value obtained for $n_{ch} = 2$ if the elastic reaction is included.

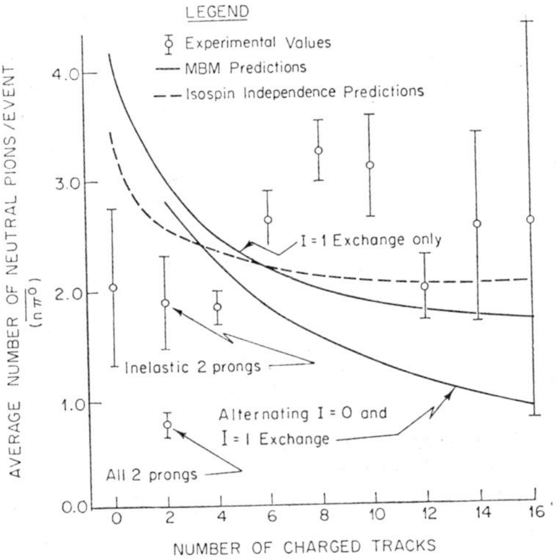

FIG. 16. The average number of π^0 produced in $\pi^- p$ at 25 GeV vs. the number of charged particles. Taken from Ref. [13].

Figure 16 (taken from a 25 GeV $\pi^- p$ experiment [13]) shows evidence against it: the elastic cross section is about equal in magnitude to all inelastic ones with $n_{ch} = 2$. In all other multiplicities, the reactions without any π^0 form a small minority of the events. Hence we may conclude that the elastic scattering has to be treated on a different footing. Note the behaviour of $\langle n_{\pi^0} \rangle$ in Fig. 16. A distribution following the state $| f, Q \rangle$ of Eq. (43) would lead to a constant $\langle n_{\pi^0} \rangle$, whereas the results of $| g \rangle$ shown in Fig. 7 do not show the trends seen in Fig. 16.

The experimental facts discussed above should be considered as a background for the ideas discussed in the paper. We are aware of the fact that different experiments show different amounts of strong resonance productions and the results shown here are not reproduced in detail in other experimental configurations. Nevertheless, the general features are similar and we may hope that they can be crudely described by the model discussed in the paper.

ACKNOWLEDGMENTS

We would like to thank all our colleagues in Caltech for their constructive criticism and advice. We are grateful to Dr. R. Stroynowski for providing us with the unpublished data of Ref. [11] that are discussed in the Appendix. One of us (D. H.) would like to thank Professor J. Prentki for his kind hospitality at the Theory Division of CERN.

REFERENCES

1. The description of the production of pions in terms of coherent states or classical radiations was suggested several times in the literature. As examples of different models based on this central idea we give the following references: H. W. LEWIS, J. R. OPPENHEIMER, AND S. A. WOUTHUYSEN, *Phys. Rev.* **73** (1948), 127; H. A. KASTRUP, *Phys. Rev.* **147** (1966), 1130, *Nucl. Phys. B* **1** (1967), 309; M. G. GUNDZIK, *Phys. Rev.* **184** (1969), 1537; P. E. HECKMAN, *Phys. Rev. D* **1** (1970), 934; G. G. ZIPFEL, JR., *Phys. Rev. Lett.* **24** (1970), 756.
2. R. J. GLAUBER, *Phys. Rev.* **130** (1963), 2529; **131** (1963), 2766. For a general review, see, e.g.; T. W. B. KIBBLE, "Cargèse Lectures in Physics," (M. Lévy, Ed.), Vol. 2, p. 299, Gordon and Breach, 1968. This paper contains also an extensive list of references.
3. T. W. B. KIBBLE, Lectures at Boulder conference (1968), *in* "Mathematical Methods in Theoretical Physics" Vol. XI D, p. 387, Gordon and Breach, 1969.
4. L. VAN HOVE, *Nuovo Cimento* **28** (1963), 798.
5. F. LURÇAT AND P. MAZUR, *Nuovo Cimento* **31** (1964), 140.
6. A. BIAŁAS, *Nuovo Cimento* **33** (1964), 972.
7. D. HORN AND R. SILVER, *Phys. Rev. D.*, **2** (1970), 2082; see also: H. A. KASTRUP, *Nucl. Phys. B* **1** (1967), 309.
8. For a discussion of such algebras and their use in other fields of physics, see, e.g.; H. J. LIPKIN, "Lie Groups for Pedestrians," Chap. 5, North-Holland, Amsterdam, 1966.
9. C. P. WANG, *Phys. Rev.* **180** (1968), 1463.

10. For a summary and discussion of such models, see, e.g., J. BARTKE, École Internationale de la Physique des Particules Élémentaires," Herceg-Novi, (1970). We note that A. DE-SHALIT, to whose memory this paper is dedicated, contributed to this field in its early stages. See, e.g., Y. YEIVIN AND A. DE-SHALIT, *Nuovo Cimento* **1** (1955), 1146.
11. M. DEUTSCHMANN *et al.*, ABBCCHW Collaboration, submitted to the Kiev Conference on Elementary Particles, 1970, and unpublished information. We would like to thank Dr. D. R. O. MORRISON and this collaboration for making their unpublished data available to us.
12. R. HONECKER *et al.*, ABBCCHW Collaboration, *Nucl. Phys. B* **13** (1969), 571.
13. J. W. ELBERT *et al.*, *Nucl. Phys. B* **19** (1970), 85.

PHYSICAL REVIEW D VOLUME 9, NUMBER 3 1 FEBRUARY 1974

Coherent states and particle production*

J. C. Botke, D. J. Scalapino, and R. L. Sugar
Department of Physics, University of California, Santa Barbara, California 93106
(Received 30 May 1973)

Hadron production at high energies is discussed in terms of coherent states. We suggest that the scattering operator and the pion field density operator take on simple forms in the coherent-state representation. In particular, we show that for a wide variety of configurations of the pion field the density operator can be written in diagonal form in the coherent-state representation. When this is possible, one obtains the statistical theory of pion production which we have discussed recently. We also construct a solvable unitary model in which the total, elastic, and inclusive cross sections go to constants at high energy, while the exclusive cross sections go to zero like a power of energy. In order to take into account the isotopic spin of the pions, we generalize the concept of coherent states and construct states with definite charge and isotopic spin.

I. INTRODUCTION

In most discussions of particle production one starts by considering matrix elements of the scattering operator between states containing definite numbers of particles. This is the natural approach when only a few particles are produced. However, at present accelerator energies the average number of particles produced in hadron-hadron collisions is already large, and it is expected to increase indefinitely with energy. Under these circumstances it may be advantageous to express the scattering operator and other operators of interest in terms of states which are not eigenfunctions of the number operator. A well-known example is the use of coherent states of photons to study the statics of the radiation field in systems such as lasers, where the average number of photons is large.[1] In this paper we shall use this approach to study the production of pions at high energies.[2]

It should be emphasized that we do not envision the pions produced in a high-energy collision as coming off in a single coherent state. Pions produced in such a state would have no correlations in their rapidity or transverse momenta aside from those imposed by energy and momentum conservation. What we do wish to suggest is that the scattering operator and the pion field density operator take on simple forms in the coherent-state representation. We show that this is the case for a wide variety of final configurations for the pion field, ranging from ones in which the pion distribution is chaotic to ones in which it is highly coherent. Although pions produced in a single coherent state do not have any correlations in momentum space, there is no problem in introducing both short- and long-range correlations in our formalism.

In order to present our ideas in the simplest possible framework, we start by considering the production of spinless, isoscalar "pions." In Sec. II the properties of the coherent states of such particles are briefly reviewed. In Sec. III the density operator for the isoscalar field is introduced, and expressions for the inclusive and semi-inclusive cross sections are given in terms of it. The density operator is then expressed in terms of coherent states and some simple examples are presented. The connection between the present approach and the statistical theory recently presented by two of us is discussed.[3]

In Sec. IV our formalism is used to construct generalizations of the models of Auerbach *et al.*,[4] which satisfy exact s-channel unitarity. We present a solvable unitary model in which the total and elastic cross sections go to constants at high energy, while the exclusive cross sections for the production of pions in the central region go to zero like a power of the energy. This model contains many of the features of the two-component models recently discussed in the literature[5]; however, in the present case the diffractive component arises naturally; it does not have to be put in by hand.

In Sec. V we confront the problems arising from the fact that the pion has isospin one. An ordinary coherent state of charged pions will not have definite charge or definite isospin. It is therefore necessary to generalize the concept of coherent states in order to construct ones with definite charge and isospin. The techniques developed here can be applied equally well to particles with larger isospin than the pion or to higher-symmetry groups.

Finally in Sec. VI we briefly discuss our results and consider possible generalizations of our work.

II. COHERENT STATES

We begin by briefly reviewing the properties of the coherent states of a spinless, isoscalar particle.[6] The four-momenta of these particles will be expressed in terms of their rapidity y and their transverse momenta, $\vec{q}_\perp$. When no confusion can arise, we shall denote the three variables $(y, \vec{q}_\perp)$ by the single symbol q. The creation and annihilation operators are normalized so that their commutation relation takes on the Lorentz-covariant form

$$[a(y,\vec{q}_\perp), a^\dagger(y',\vec{q}'_\perp)] = \delta(y-y')\delta^2(\vec{q}_\perp - \vec{q}'_\perp)\,. \tag{1}$$

The coherent states are, by definition, the eigenstates of the destruction operator. Thus,

$$a(q)|\Pi\rangle = \Pi(q)|\Pi\rangle\,, \tag{2}$$

where $\Pi(q)$ is an arbitrary complex function of q. $|\Pi\rangle$ can be written in the form

$$|\Pi\rangle = \exp\left[-\tfrac{1}{2}\int dq\,|\Pi(q)|^2\right] \times \exp\left[\int dq\,\Pi(q)a^\dagger(q)\right]|0\rangle\,. \tag{3}$$

$|0\rangle$ is the vacuum state defined by $a(q)|0\rangle = 0$ and $dq \equiv dy\,d^2q_\perp$ is the invariant phase-space volume element. It is sometimes useful to introduce the coherent-state displacement operator

$$D(\Pi) = \exp\left\{\int dq\,[\Pi(q)a^\dagger(q) - \Pi^*(q)a(q)]\right\}\,. \tag{4}$$

Notice that

$$|\Pi\rangle = D(\Pi)|0\rangle\,. \tag{5}$$

$D(\Pi)$ is a unitary operator, $D^\dagger(\Pi) = D(-\Pi) = D^{-1}(\Pi)$.

The states $|\Pi\rangle$ are normalized to unity, but they are not mutually orthogonal. In fact

$$\langle\Pi'|\Pi\rangle = \exp\left\{-\tfrac{1}{2}\int dq\,[|\Pi'(q)|^2 + |\Pi(q)|^2 - 2\Pi'^*(q)\,\Pi(q)]\right\}\,. \tag{6}$$

The average number of particles in the state $|\Pi\rangle$ is is

$$\bar{n} = \left\langle \Pi \left| \int dq\,a^\dagger(q)a(q) \right| \Pi \right\rangle = \int dq\,|\Pi(q)|^2\,. \tag{7}$$

The probability density of finding n particles with momenta $q_1, \ldots, q_n$ in this state is

$$|\langle 0|a(q_1)\cdots a(q_n)|\Pi\rangle|^2 = |\Pi(q_1)|^2 \cdots |\Pi(q_n)|^2 e^{-\bar{n}}\,, \tag{8}$$

so the probability of finding n particles of any momenta is

$$P_n = \frac{(\bar{n})^n}{n!} e^{-\bar{n}}\,. \tag{9}$$

It is sometimes convenient to introduce a complete set of states, $f_k(q)$, normalized so that

$$\int dq\,f_k^*(q)f_{k'}(q) = \delta_{k,k'}\,. \tag{10}$$

For most purposes it will not be necessary to specify the functional form of the $f_k(q)$. The creation and annihilation operators for a particular normal mode are defined by

$$a_k = \int dq\,f_k^*(q)a(q)\,, \tag{11}$$

so

$$[a_k, a_{k'}^\dagger] = \delta_{k,k'}\,. \tag{12}$$

Writing

$$\Pi(q) = \sum_k \alpha_k f_k(q)\,, \tag{13}$$

we see that the coherent state defined in Eq. (3) can be written in the form

$$|\Pi\rangle = \exp\left(-\tfrac{1}{2}\sum_k |\alpha_k|^2\right)\exp\left(\sum_k \alpha_k a_k^\dagger\right)|0\rangle \equiv |\{\alpha_k\}\rangle\,. \tag{14}$$

Although the coherent states are not linearly independent, they do form a complete set. Denoting by $\int \delta\Pi$ a functional integration over all forms of the complex field $\Pi(q)$ and by

$$\int d^2\alpha_k = \int d(\mathrm{Re}\,\alpha_k)\,d(\mathrm{Im}\,\alpha_k)$$

an integration over all complex values of α_k, we see that

$$\begin{aligned}\int \delta\Pi\,|\Pi\rangle\langle\Pi| &= \int \prod_k \frac{d^2\alpha_k}{\pi}\,|\{\alpha_k\}\rangle\langle\{\alpha_k\}| \\ &= \sum_{n_1,n_2,\ldots} \frac{(a_1^\dagger)^{n_1}}{(n_1!)^{1/2}}\frac{(a_2^\dagger)^{n_2}}{(n_2!)^{1/2}}\cdots|0\rangle \times\langle 0|\frac{(a_1)^{n_1}}{(n_1!)^{1/2}}\frac{(a_2)^{n_2}}{(n_2!)^{1/2}}\cdots \\ &= 1\,. \end{aligned} \tag{15}$$

The measure of the functional integration is defined by the first line of Eq. (15). In going from line 1 to line 2 of Eq. (15) we have made use of the fact that

$$\int \frac{d^2\alpha_k}{\pi} e^{-|\alpha_k|^2}(\alpha_k)^n(\alpha_k^*)^{n'} = \delta_{n,n'}\,n!\,. \tag{16}$$

Using Eq. (15), one can expand any vector or operator in terms of coherent states. For example,

$$|f\rangle = \int \delta\Pi \, |\Pi\rangle\langle\Pi|f\rangle \,. \tag{17}$$

However, since the coherent states are not linearly independent, the expansion is not unique.[1]

III. THE PION FIELD DENSITY OPERATOR

Most of the properties of the pions produced in a high-energy hadron-hadron collision can be expressed simply in terms of a pion field density operator. As in Sec. II we shall neglect the complications of spin and isospin. Proceeding formally we write

$$\tilde{\rho} = T|P_1P_2\rangle\langle P_1P_2|T^\dagger \tilde{Z}^{-1}, \tag{18}$$

with

$$\tilde{Z} = \mathrm{tr}[T|P_1P_2\rangle\langle P_1P_2|T^\dagger] = \langle P_1P_2|T^\dagger T|P_1P_2\rangle \,. \tag{19}$$

$|P_1P_2\rangle$ is the incident state of two hadrons with momenta P_1 and P_2, and T is related to the scattering operator by $T = 2i(1-S)$. Clearly $\tilde{Z}$ is proportional to the total cross section σ.

The inclusive cross sections for the production of a pion with momentum q is given in terms of $\tilde{\rho}$ by

$$\frac{1}{\sigma}\frac{d\sigma}{dq} = \tilde{Z}^{-1}\sum_n \langle n|a(q)T|P_1P_2\rangle\langle P_1P_2|T^\dagger a^\dagger(q)|n\rangle = \mathrm{tr}[a^\dagger(q)a(a)\tilde{\rho}] \,. \tag{20}$$

Again $dq = dy\,d^2q_\perp$. Notice that we can use closure in Eq. (20) because the total energy-momentum conservation δ function is included in the operator T. Similarly expressions for all inclusive and exclusive cross sections can be written in the form $\mathrm{tr}[O\tilde{\rho}]$.

If we restrict our interest to one type of particle, say pions, then we can perform the trace over the coordinates of all other particles at the beginning. We therefore define the pion field density operator ρ by

$$\rho = \mathrm{tr}'\tilde{\rho}, \tag{21}$$

where tr$'$ indicates a trace over the coordinates of all particles except pions. Clearly ρ satisfies the normalization condition

$$\mathrm{tr}\rho = 1 \,. \tag{22}$$

From now on tr will denote a trace over pion coordinates only. The inclusive cross section for the production of n pions is given in terms of ρ by

$$\frac{1}{\sigma}\frac{d\sigma}{dq_1\cdots dq_n} = \mathrm{tr}[a^\dagger(q_1)\cdots a^\dagger(q_n)a(q_1)\cdots a(q_n)\rho] \equiv I(q_1,\ldots,q_n), \tag{23}$$

and the semi-inclusive cross sections for the production of exactly n pions plus any number of other types of particles by

$$\frac{1}{\sigma}\sigma_n(q_1,\ldots,q_n) = \mathrm{tr}[\,|q_1,\ldots,q_n\rangle\langle q_1,\ldots,q_n|\rho] \equiv S(q_1,\ldots,q_n) \,. \tag{24}$$

Here $|q_1,\ldots,q_n\rangle$ is a state of n pions with momenta $q_1,\ldots,q_n$.

In performing the trace in Eq. (21) one integrates out the energy-momentum conservation δ functions in T, thereby introducing constraints into ρ. Since the transverse momenta of the pions are limited by the dynamics, the most important constraint is on the rapidities. In the laboratory system y must lie in the approximate range $0 \leq y \leq Y$, where Y is the rapidity of the incident projectile. There are, of course, more complicated constraints which are most important for pions produced near the edges of the rapidity plot, i.e., near 0 or Y. Since we expect most of the pions to come off in the central region, the only constraint that we shall explicitly build in is that their rapidities be restricted to the range $0 \leq y \leq Y$.

If one had a fundamental theory of strong interaction dynamics, one could of course calculate ρ. In the absence of such a theory it seems useful to attempt to find a simple phenomenological parameterization of ρ in the hopes of learning something about the underlying dynamics. One knows from the study of quantum optics that because of the overcompleteness of the coherent states, a wide class of density operators can be written in diagonal form in the coherent-state representation.[1,7] Since calculations are particularly simple for such density operators, let us start by considering them.

We write ρ in the form

$$\rho = Z^{-1}\int \delta\Pi \, |\Pi\rangle e^{-F[\Pi]}\langle\Pi| \,, \tag{25}$$

with

$$Z = \int \delta\Pi \, e^{-F[\Pi]} \,. \tag{26}$$

$F[\Pi]$ is an arbitrarily functional of Π. Because of the energy-momentum conservation constraint, the functional integration is to run only over those functions of $\Pi(y,\vec{q}_\perp)$ which vanish for y outside the range $0 \leq y \leq Y$. Recalling that $a(y,\vec{q}_\perp)|\Pi\rangle = \Pi(y,\vec{q}_\perp)|\Pi\rangle$, we see that the inclusive and semi-inclusive cross sections take on the particularly simple forms

$$I(q_1,\ldots,q_n)=Z^{-1}\int \delta\Pi\,|\Pi(q_1)|^2\cdots|\Pi(q_n)|^2 \times e^{-F[\Pi]}, \tag{27}$$

and

$$S(q_1,\ldots,q_n)=Z^{-1}\int \delta\Pi\,|\Pi(q_1)|^2\cdots|\Pi(q_n)|^2 \times e^{-F[\Pi]}\exp\left[-\int dq\,|\Pi(q)|^2\right]. \tag{28}$$

Mueller's generating function[8] $\Omega(z)$ is given by

$$\Omega(z)=1+\sum_{n=1}^{\infty}\frac{(z-1)^n}{n!}\int dq_1\cdots dq_n I(q_1,\ldots,q_n)$$

$$=Z^{-1}\int \delta\Pi\, e^{-F[\Pi]}e^{-(1-z)\int dq|\Pi(q)|^2}. \tag{29}$$

Equations (27) and (29) were the starting points for the statistical theory of particle production recently proposed by two of us.[3]

A particularly simple form for the density matrix is obtained by taking $e^{-F[\Pi]}$ to be a functional δ function. Then the density matrix is given by

$$\rho=|\Pi_0\rangle\langle\Pi_0|. \tag{30}$$

Clearly,

$$I(q_1,\ldots,q_n)=|\Pi_0(q_1)|^2\cdots|\Pi_0(q_n)|^2, \tag{31}$$

$$S(q_1,\ldots,q_n)=|\Pi_0(q_1)|^2\cdots|\Pi_0(q_n)|^2 e^{-\bar{n}}, \tag{32}$$

and

$$\Omega(z)=e^{(z-1)\bar{n}}, \tag{33}$$

with

$$\bar{n}=\int_0^Y dy\int d^2q_\perp|\Pi_0(y,q_\perp)|^2. \tag{34}$$

If $\Pi_0(y,\vec{q}_\perp)$ is independent of y for $0\leq y\leq Y$, or a slowly varying function of y in this region, then $\bar{n}$ grows like Y and the $S(q_1,\ldots,q_n)$ fall like a power of s. The inclusive cross sections are precisely what one would obtain in a multi-Regge model in which there was a single Regge trajectory with intercept one. This model is of course tremendously oversimplified as there are no correlations in either the rapidity or the transverse momentum.

In I we considered a simple form for $F[\Pi]$ based upon an analogy with the generalized Ginsburg-Landau theory of superconductivity. We wrote

$$F[\Pi]=\int_0^Y dy\left[a|\Pi(y)|^2+b|\Pi(y)|^4+c\left|\frac{\partial\Pi}{\partial y}\right|^2\right]. \tag{35}$$

For simplicity the dependence on the transverse momenta has been suppressed. Although Eq. (35) could be obtained by retaining the first few terms in the power-series expansion of a general $F[\Pi]$, it should be emphasized that it is applicable even when the fluctuations of the pion field are not small. The main justification for this form of $F[\Pi]$ is that by suitably adjusting the phenomenological parameters a, b, and c, it can describe a wide range of possible final configurations of the pion field.

To see just how wide a class of phenomena can be described by the Ginzburg-Landau functional let us consider two extreme examples. First take $a, c>0$ and $b=0$. In this case it is convenient to expand $\Pi(y)$ in terms of a complete set of normal-mode functions. Since we do not wish to restrict the values $\Pi(y)$ can take on at the boundaries,[3] the appropriate choice is

$$f_0(y)=Y^{-1/2},$$
$$f_k(y)=(\tfrac{1}{2}Y)^{-1/2}\cos\left[\frac{\pi k y}{Y}\right],\quad k=1,2,\ldots\,. \tag{36}$$

From Eq. (13) and the orthogonality of the $f_k(y)$ we see that

$$F[\Pi]=\sum_{k=0}^{\infty}|\alpha_k|^2[a+c(\pi k/Y)^2]. \tag{37}$$

Then from Eqs. (14) and (15) we find

$$\rho=\prod_{k=0}^{\infty}\rho_k, \tag{38}$$

with

$$\rho_k=(\bar{n}_k)^{-1}\int\frac{d^2\alpha_k}{\pi}|\alpha_k\rangle\langle\alpha_k|e^{-|\alpha_k|^2/\bar{n}_k} \tag{39}$$

and

$$\bar{n}_k=[a+c(\pi k/Y)^2]^{-1}. \tag{40}$$

In the occupation-number basis we have

$$\rho_k=(1+\bar{n}_k)^{-1}\sum_{m_k}|m_k\rangle\langle m_k|[\bar{n}_k/(1+\bar{n}_k)]^{m_k}, \tag{41}$$

where

$$|m_k\rangle=(m_k!)^{-1/2}(a_k^\dagger)^{m_k}|0\rangle. \tag{42}$$

Thus, each normal mode of the pion field is described by a Gaussian density operator which corresponds to random excitation characteristic of noncoherent sources.[1]

The opposite extreme, a highly coherent pion field, can be obtained by taking $a<0$, $b>0$, and $c\gg 1$. From Eqs. (26), (29), and (35) one sees that the generating function is given by

$$\Omega(z)=Z(a+1-z)/Z(a)\,. \tag{43}$$

The functional integral needed to calculate $Z(a)$ will be strongly peaked near the mean-field value Π_0 defined by

$$\frac{\delta F[\Pi_0]}{\delta \Pi_0^*}=0\,, \tag{44}$$

which gives

$$|\Pi_0|^2=-a/2b\,. \tag{45}$$

One then finds[3]

$$\Omega(z)\simeq[1+(1-z)/a]^{-1/4}\exp\{Y[(z-1)(-a/2b)+(z-1)^2/4b]\}\,. \tag{46}$$

Clearly by making $|a|$ and b large, one can come arbitrarily close to the Poisson distribution associated with a pure coherent state [Eq. (33)].

A detailed study of the models arising from the Ginzburg-Landau form for $F[\Pi]$ is given in I. The problem of performing the functional integrations reduces to the problem of solving the Schrödinger equation for the anharmonic oscillator. The point we wish to emphasize here is that by making appropriate choices for the parameters a, b, and c, this simple form for $F[\Pi]$ can describe pion field configurations ranging from chaotic to highly coherent. The present experimental data favor the range $a<0$ and $c\gg 1$. (See Ref. 3.) The $|\partial\Pi/\partial y|^2$ term in $F[\Pi]$ gives rise to nontrivial short-range correlations in rapidity. The Ginzburg-Landau form for $F[\Pi]$ can also be used to generate three-dimensional models in which the dependence on $\vec{q}_\perp$ is retained. In this case, terms which contain gradients, with respect to the transverse momentum or transverse impact parameter, will give rise to short-range correlations in $\vec{q}_\perp$.

Although a wide class of density operators can be written in diagonal form in the coherent-state representation, this is not always convenient or even possible. An example of a simple density operator that is most conveniently written in nondiagonal form is given in Sec. IV.

IV. A UNITARY MODEL

In the models that we have discussed so far the incident hadrons can be pictured as exciting the pion field via a single interaction. In fact for the diagonal density operators discussed in Sec. III, the inclusive cross sections have the same structure as a function of rapidity as one finds in the multi-Regge model.[3] Recently a rather different class of models [AABS models[4]] was presented for which the scattering operator satisfies exact s-channel unitarity. In these models the incident hadrons are pictured as propagating through the interaction region, without making significant fractional changes in their rapidities. However, in order to satisfy s-channel unitarity they must be allowed to interact an arbitrary number of times with pions being emitted and absorbed at each interaction. In the original version of these models each interaction corresponded to the exchange of a multiperipheral-like chain. In this section we shall use the coherent-state representation to discuss generalization of the AABS models.

In the AABS models the scattering operator is diagonal in the rapidity difference Y, and the relative impact parameter, $\vec{B}$, of the two incident hadrons. We write it in the form

$$S(Y,\vec{B})=\exp[i\chi(Y,\vec{B})]\,. \tag{47}$$

The Hermitian operator $\chi(Y,\vec{B})$ determines the amplitude for emitting or absorbing a given number of pions with each interaction. In general it is a functional of the pion creation and annihilation operators. If $\chi(Y,\vec{B})$ has a finite Hilbert-Schmidt norm[9] then it can always be written in the form[10]

$$\chi(Y,\vec{B})=\int\delta\Pi\, D(\Pi)[\chi(\Pi;Y,\vec{B})+\chi^*(-\Pi;Y,\vec{B})]\,, \tag{48}$$

where $D(\Pi)$ is the coherent-state displacement operator defined in Eq. (4). The Hermiticity of the operator $\chi(Y,\vec{B})$ follows from the fact that $D^\dagger(\Pi)=D(-\Pi)$.

The weight function $\chi(\Pi;Y,\vec{B})$ can be parameterized using techniques similar to those discussed in Sec. III. In particular it is not difficult to introduce short-range correlations in rapidity and transverse momentum among the pions produced in a particular interaction. However, in order to simplify our presentation we shall neglect such correlations and take $\chi(\Pi;Y,\vec{B})$ to be proportional to a functional δ function. We write

$$\chi(Y,\vec{B})=\tfrac{1}{2}\Lambda(\vec{B})[D(\Pi_0)+D(-\Pi_0)]\,. \tag{49}$$

The only requirements that we shall make on the function $\Pi_0(y,\vec{q}_\perp)$ are that it be nonzero only in the interval $0\leq y\leq Y$ and that inside this interval it be slowly varying enough so that

$$\int_0^Y dy\int d^2q_\perp|\Pi_0(y,\vec{q}_\perp)|^2=\lambda Y\,, \tag{50}$$

where λ goes to a constant at high energies. In principle Λ could also depend on Y. With our choice, the Born approximation to the elastic scattering amplitude is

$$2e^{Y}\int d^2B\langle 0|\chi(Y,B)|0\rangle = 2e^{Y(1-\lambda/2)}\int d^2B\,\Lambda(\vec{B}), \tag{51}$$

which corresponds to a fixed pole in the angular momentum plane at $l=1-\lambda$. Our choice seems to be the most natural since we are neglecting the correlation in the transverse momenta necessary to generate a moving pole.[4] One could generalize the model by allowing the incident hadrons to be diffractively excited.[11] The only change would be that Λ would become a matrix.

For any two functions Π_1 and Π_2, one easily sees that

$$D(\Pi_1)D(\Pi_2)=D(\Pi_1+\Pi_2)e^{(1/2)\int dq[\Pi_2^*\Pi_1-\Pi_1^*\Pi_2]}. \tag{52}$$

As a result,

$$[D(\Pi_0), D(-\Pi_0)]=0 \tag{53}$$

and

$$D(\Pi_0)^n D(-\Pi_0)^m = D((n-m)\Pi_0). \tag{54}$$

The scattering operator therefore has the simple series expansion

$$S(Y,\vec{B})=\sum_{n,m=0}^{\infty}\frac{(i\Lambda/2)^{n+m}}{n!\,m!}D((n-m)\Pi_0). \tag{55}$$

Since the two incident hadrons are assumed not to be pions, in order to construct the elastic scattering amplitude we need the matrix element of S in the state with no pions:

$$\langle 0|S(Y,\vec{b})|0\rangle=\sum_{n,m}\frac{(i\Lambda/2)^{n+m}}{n!\,m!}e^{-(1/2)(n-m)^2\lambda Y}$$

$$\simeq J_0(\Lambda). \tag{56}$$

In the last step in Eq. (56) we have retained only the terms of leading power in $s\sim e^{Y}$. The total cross section is given by

$$\sigma=\int d^2B\,\mathrm{Im}\{2i[1-\langle 0|S(Y,\vec{B})|0\rangle]\}$$

$$\simeq 2\int d^2B[1-J_0(\Lambda(\vec{B}))], \tag{57}$$

and the elastic cross section by

$$\sigma_{el}=\int d^2B|1-\langle 0|S(Y,\vec{B})|0\rangle|^2$$

$$\simeq\int d^2B|1-J_0(\Lambda(\vec{B}))|^2. \tag{58}$$

Clearly the total and elastic cross section go to constants at high energy for any value of the parameters λ and Λ.

In our present normalization the density operator is given by

$$\rho(Y,\vec{B})=\frac{1}{\sigma}[1-e^{i\chi(Y,\vec{B})}]|0\rangle\langle 0|[1-e^{-i\chi(Y,\vec{B})}]. \tag{59}$$

Notice that in this class of models the density operator is always separable in the impact-parameter representation. The inclusive cross section for the production of l pions is given by

$$I(q_1,\ldots,q_l)=\frac{1}{\sigma}\int d^2B\langle 0||e^{-i\chi(Y,\vec{B})}a^\dagger(q_1)\cdots a^\dagger(q_l)a(q_l)\cdots a(q_1)\,e^{i\chi(Y,\vec{B})}|0\rangle$$

$$=\frac{1}{\sigma}\prod_{i=1}^{l}|\Pi_0(q_i)|^2\int d^2B\sum_{\substack{n,m\\n',m'}}\frac{(i\Lambda/2)^{n+m}}{n!\,m!}\frac{(-i\Lambda/2)^{n'+m'}}{n'!\,m'!}[(n-m)(n'-m')]^l e^{-(1/2)[(n-m)-(n'-m')]^2\lambda Y}. \tag{60}$$

Retaining the terms of leading power in s gives

$$I(q_1,\ldots,q_n)\simeq\prod_{i=1}^{l}|\Pi_0(q_i)|^2\,2C(l)/\sigma, \tag{61}$$

where

$$C(l)=\sum_{p=1}^{\infty}p^{2l}\int d^2B\,J_p^2(\Lambda(\vec{B})). \tag{62}$$

Thus, the inclusive cross sections approach energy-independent limits at high energies. Although we have omitted the short-range correlations among the pions associated with a single χ, the correlation functions do not vanish. For example,

$$C_2(q_1,q_2)=\left[\frac{I(q_1,q_2)}{I(q_1)I(q_2)}-1\right]$$

$$=[\sigma C(2)/C(1)^2-1], \tag{63}$$

which is independent of the q_i. This long-range

correlation arises from interference effects among pions associated with different χ's. This type of long-range correlation has been discussed previously using the more familiar S-matrix Regge-pole language.[12]

Since the only particles in this model besides the pions are the two incident hadrons which are neither created nor destroyed the semi-inclusive cross sections are identical to the exclusive cross sections. We have

$$S(q_1,\ldots,q_l)=\frac{1}{\sigma}\int d^2B\,|\langle q_1,\ldots,q_l|(1-e^{i\chi(Y,\vec{B})})|0\rangle|^2$$
$$=\frac{1}{\sigma}\prod_{i=1}^{l}|\Pi_0(q_i)|^2\sum_{n,m}\int d^2B\,\frac{(i\Lambda/2)^{n+m}}{n!\,m!}(n-m)^l e^{-(1/2)(n-m)^2\lambda Y^2}\,. \tag{64}$$

Notice that all the exclusive cross sections except the elastic go to zero at high energy like a power of the energy.

The generating function takes on an interesting form in this model. From Eqs. (29), (50), and (61) we see that

$$\Omega(z)=1+\sum_{n=1}^{\infty}\frac{(z-1)^n}{n!}(\lambda Y)^n\,2C(n)/\sigma\,. \tag{65}$$

Using the identity

$$1=J_0^2(\Lambda)+2\sum_{p=1}^{\infty}J_p^2(\Lambda)\,, \tag{66}$$

we see that $\Omega(z)$ can be written in the form

$$\Omega(z)=\frac{\sigma_{el}}{\sigma}+\frac{2}{\sigma}\sum_{p=1}^{\infty}e^{(z-1)\lambda p^2Y}\int d^2B\,J_p^2(\Lambda(B))\,. \tag{67}$$

This generating function has a form recently suggested by several authors[5]; namely, it contains a diffractive component σ_{el}/σ plus a sum of multiperipheral-like components. However, it should be emphasized that in this model the diffractive component in $\Omega(z)$ has not been put in by hand. It arises naturally from interference effects among pions produced in different independent interactions between the incident hadrons. (Multiperipheralists may wish to think of these pions as being produced on different multiperipheral chains.) Thus the constant elastic cross section really does arise from the shadow cast by the inelastic channels. Notice that if Λ is small, all of the pions are produced in a single interaction. However, the important terms in the elastic amplitude are those proportional to χ^2. This is natural since it is only through terms quadratic or greater in χ that the inelastic channels can affect the elastic amplitude.

To summarize, we have constructed a simple unitary model in which the total, elastic, and inclusive cross sections approach constants at high energy, while the exclusive cross sections fall like a power of the energy. These results should be contrasted with those obtained in the solvable model of AABS,[4] which is very similar in spirit to the present model. In the AABS model χ was chosen to have the general features of the multiperipheral model. In the present notation χ_{AABS} can be written

$$\chi_{AABS}=e^{-(\alpha-1-\lambda/2)Y}f(\vec{B})\,e^{\int dq[\Pi_0(q)a^\dagger(q)+\Pi_0^*(q)a(q)]}\,. \tag{68}$$

Here α is the position of the input pole exchanged along the multiperipheral chain and λ is defined in Eq. (50). The principal difference between χ_{AABS} and the χ defined in Eq. (49) is the plus sign before $\Pi_0^*(q)a(q)$, which makes χ_{AABS} an unbounded operator. Notice that when χ_{AABS} is applied to the pion vacuum it gives rise to a coherent state proportional to $|\Pi_0\rangle$. The contributions to the elastic scattering amplitude proportional to χ_{AABS} and χ_{AABS}^2 have the energy dependence $e^{\alpha Y}$ and $e^{(2\alpha-1+\lambda)Y}$, respectively,[4] which is of course what one would expect from the multiperipheral model in the weak coupling limit. In the present model, the corresponding terms have the energy dependence $e^{(1-\lambda/2)Y}$ and e^{Y}. So, as was mentioned before, the total and elastic cross sections go to constants at high energy independent of the value of λ, which is the effective pion coupling constant. On the other hand, the input pole, which controls the energy dependence of the exclusive cross sections [see Eq. (64)], is located at $l=1-\frac{1}{2}\lambda$. This is an example of the dependence of cross sections on coupling constants recently conjectured by Harari.[13]

In the solvable AABS model the total cross section always goes to zero at high energies for $\alpha\leq 1$ because any dynamical pole that approaches $l=1$ is washed out by rapid oscillations of the S matrix. These oscillations are associated with large eigenvalues of χ_{AABS}, which is an unbounded operator. In the present model χ is a bounded operator and no such oscillations occur.

The model that we have been discussing can be made to look even more like the AABS model by replacing $\Lambda(\vec{B})$ by the quantity $\Lambda(\vec{B})e^{(\alpha-1+\lambda/2)Y}$. Then expanding the elastic amplitude in powers of χ, one finds that the first terms have energy dependence $e^{\alpha Y}$ and the Nth term $e^{Y[1+N(\alpha-1+\lambda/2)]}$ for $N\geq 2$. For $\alpha<1-\frac{1}{2}\lambda$ the total cross section goes to zero at high energies, for $\alpha>1-\frac{1}{2}\lambda$ it goes like Y^2 and the Froissart bound is saturated, and for $\alpha=1-\frac{1}{2}\lambda$ we regain the model we have discussed in detail. This generalized model provides an example of the saturation of multichain forces discussed in the second and third papers of Ref. 4.

V. ISOTOPIC SPIN

In order to present our ideas in the simplest possible setting, we have ignored internal quantum numbers in the preceding sections. We now turn to the problem of finding a set of states suitable for describing physical pions.

The creation and annihilation operators for the pions are vectors in isotopic-spin space and will be written in the form

$$\vec{a}(q)=(a_1(q),\, a_2(q),\, a_3(q))\,. \tag{69}$$

The creation operators for the physical pions are

$$a_+^\dagger(q)=\frac{-1}{\sqrt{2}}[a_1^\dagger(q)+i a_2^\dagger(q)]\,,$$
$$a_-^\dagger(q)=\frac{1}{\sqrt{2}}[a_1^\dagger(q)-i a_2^\dagger(q)]\,, \tag{70}$$
$$a_0^\dagger(q)=a_3^\dagger(q)\,.$$

The subscripts on the creation operators on the left-hand side of Eq. (70) stand for the charges of the pions.

The eigenfunctions of the annihilation operators can be written in the form

$$\vec{a}(q)\,|\vec{\Pi}\rangle=\vec{\Pi}(q)\,|\vec{\Pi}\rangle\,, \tag{71}$$

where

$$|\vec{\Pi}\rangle=e^{-(1/2)\int dq|\vec{\Pi}(q)|^2}e^{\int dq\,\vec{\Pi}(q)\cdot\vec{a}^\dagger(q)}|0\rangle\,, \tag{72}$$

and

$$\vec{\Pi}(q)=(\Pi_1(q),\, \Pi_2(q),\, \Pi_3(q)) \tag{73}$$

is an isospin vector whose components $\Pi_i(q)$ are, in general, complex. The states $|\vec{\Pi}\rangle$ are not likely to be a useful set of basis states for describing pion production since they are not eigenstates of the total charge or the total isotopic spin.

Let us begin by constructing a set of states of definite charge and isospin in which all pions have the momentum-space wave function $\Pi(q)$. We write the vector $\vec{\Pi}$ in the form

$$\vec{\Pi}(q)=\Pi(q)\hat{n}\,, \tag{74}$$

where

$$\hat{n}=(\sin\theta\cos\varphi,\; \sin\theta\sin\varphi,\; \cos\theta) \tag{75}$$

gives the direction of $\vec{\Pi}$ in isotopic-spin space. The required state with total isotopic spin I and z component of isotopic spin I_z is

$$|\Pi, I, I_z\rangle=\int d\Omega_{\hat{n}}\, Y^*_{I,I_z}(\hat{n})e^{-(1/2)\int dq|\vec{\Pi}|^2}e^{\int dq\,\vec{\Pi}\cdot\vec{a}^\dagger}|0\rangle\,. \tag{76}$$

$Y_{I,I_z}(\hat{n})=Y_{I,I_z}(\theta,\varphi)$ is the familiar spherical harmonic of angular momentum I and z component I_z. Notice that $|\Pi, I, I_z\rangle$ is not an eigenstate of the destruction operator $\vec{a}(q)$. That $|\Pi, I, I_z\rangle$ is indeed a state of isotopic spin I and z component I_z can be seen by recalling that under a rotation R in isospin space,

$$O_R a_i^\dagger O_R^{-1}=\sum_j R_{ij}a_j^\dagger\,, \tag{77}$$

so

$$\begin{aligned}O_R|\Pi, I, I_z\rangle &=\int d\Omega_{\hat{n}}\, Y^*_{I,I_z}(\hat{n})\\ &\quad\times e^{-(1/2)\int dq|\vec{\Pi}|^2}e^{\int dq\,\vec{\Pi}\cdot R\vec{a}^\dagger}|0\rangle\\ &=\int d\Omega_{\hat{n}}\, Y^*_{I,I_z}(R\hat{n}')\\ &\quad\times e^{-(1/2)\int dq|\vec{\Pi}|^2}e^{\int dq\,\vec{\Pi}'\cdot\vec{a}^\dagger}|0\rangle\\ &=\sum_{I_z'}D^I_{I_z',I_z}(R)|\Pi, I, I_z'\rangle\,,\end{aligned} \tag{78}$$

where $\hat{n}'=R^{-1}\hat{n}$ and $D^I(R)$ is the Wigner D matrix.

In order to understand the properties of these states let us consider the state with isospin zero in some detail. The states defined in Eq. (75) are not normalized to unity. In particular

$$\begin{aligned}\langle\Pi, 0, 0|\Pi, 0, 0\rangle &=e^{-c}\int d\Omega_{\hat{n}}\,d\Omega_{\hat{n}'}\,e^{c\hat{n}\cdot\hat{n}'}\\ &=(4\pi)^2 j_0(-ic)e^{-c}\equiv N\,,\end{aligned} \tag{79}$$

where

$$c=\int dq\,|\Pi(q)|^2\,, \tag{80}$$

and j_0 is the spherical Bessel function of order zero. The probability amplitude for finding n pions with momenta $q_1, \ldots, q_n$ and isospin indices $i_1, \ldots, i_n$ in the state $|\Pi, 0, 0\rangle$ is

$$\begin{aligned}&\langle 0|a_{i_1}(q_1)\cdots a_{i_n}(q_n)|\Pi, 0, 0\rangle N^{-1/2}\\ &\qquad=N^{-1/2}e^{-(1/2)c}\,\Pi(q_1)\cdots\Pi(q_n)\int d\Omega_{\hat{n}}\,\hat{n}_{i_1}\cdots\hat{n}_{i_n}\,.\end{aligned} \tag{81}$$

This amplitude clearly vanishes unless the pions are in a state of total isotopic spin zero. The probability of finding n_+ π_+'s, n_- π_-'s, and n_0 π_0's is

$$P(n_+, n_-, n_0) = \frac{N^{-1}e^{-c}c^{n_++n_-+n_0}}{n_+!\,n_-!\,n_0!\,2^{n_++n_-}} \times \left| \int d\Omega_{\hat{n}}\, e^{i(n_+-n_-)\varphi}(\sin\theta)^{n_++n_-}(\cos\theta)^{n_0}\right|^2 . \tag{82}$$

P obviously vanishes unless $n_+ = n_- \equiv \frac{1}{2}n_c$, and n_0 is even. Under these circumstances

$$P(\tfrac{1}{2}n_c, \tfrac{1}{2}n_c, n_0) = \frac{(2\pi)^2 N^{-1} c^{n_c+n_0}e^{-c}}{n_0!\,2^{n_c}} \times \left[\frac{\Gamma(\frac{1}{2}n_0+\frac{1}{2})}{\Gamma(\frac{1}{2}n_c+\frac{1}{2}n_0+\frac{3}{2})}\right]^2 . \tag{83}$$

Finally, the average number of particles is given by

$$\begin{aligned}\bar{n} &= N^{-1}\langle \Pi, 0, 0|\sum_i \int dq\, a_i^\dagger(q)a_i(q)|\Pi, 0, 0\rangle \\ &= N^{-1}e^{-c}c\int d\Omega_{\hat{n}}\,d\Omega_{\hat{n}'}\,\hat{n}\cdot\hat{n}'\,e^{c\hat{n}\cdot\hat{n}'} \\ &= icj_1(-ic)/j_0(-ic) = c\coth(c) - 1 \end{aligned} \tag{84}$$

and

$$\bar{n}_+ = \bar{n}_- = \bar{n}_0 = \tfrac{1}{3}\bar{n} . \tag{85}$$

The states $|\Pi, I, I_z\rangle$ do not form a complete set. The difficulty is that in general the real and imaginary parts of the vector $\vec{\Pi}$ need not point in the same direction in isotopic-spin space. Consider a particular normal-mode function $f_k(q)$. The most general $\vec{\Pi}$ that can be constructed from it can be written in the form

$$\vec{\Pi}_k(q) = f_k(q)[\mathrm{Re}\,\alpha_k\,\hat{n}_R + i\,\mathrm{Im}\,\alpha_k\,\hat{n}_I] , \tag{86}$$

where $\hat{n}_R$ and $\hat{n}_I$ are two independent unit vectors and α_k is any complex number. A complete set of states for the kth normal mode is

$$\begin{aligned}|f_k; I, I_z; I_R, I_I\rangle = \sum_m \langle I_R, m; I_I, I_z - m|I, I_z; I_R, I_I\rangle & \\ \times \int d\Omega_{\hat{n}_R}\,d\Omega_{\hat{n}_I} & \\ \times Y_{I_R, m}(\hat{n}_R)^*\,Y_{I_I, I_z-m}(\hat{n}_I)^* & \\ \times e^{-\int dq|\vec{\Pi}_k|^2}\,e^{\int dq\,\vec{\Pi}_k^*\cdot\vec{a}^\dagger}|0\rangle ; &\end{aligned} \tag{87}$$

$\langle I_R, m; I_I, I_z - m|I, I_z; I_R, I_I\rangle$ is the standard Clebsch-Gordan coefficient. One can check, using the techniques of Eq. (15), that the states obtained by taking direct products over all normal modes are complete. Since the problem of coupling the isospins of the various normal modes to form a definite total isotopic spin is formidable, this approach is likely to be useful only if the dynamics is such that only a few modes enter or that the isospin in each mode is zero.

In the states that we have just described, the isotopic spin is a global property. Pions which are well separated in rapidity and transverse momentum are correlated to the extent that their individual isotopic spins are coupled to form a definite total isotopic spin. In some models, such as the multi-Regge model, one assumes that there are significant correlations only among particles with small relative rapidities. It is not difficult to construct states of definite isotopic spin with this property. Let us consider an example. We write the vector $\vec{\Pi}$ in the form

$$\vec{\Pi}(y) = \Pi(y)\hat{n}(y) . \tag{88}$$

The dependence on the transverse momentum has been suppressed for simplicity. As usual we restrict y to the range $0 \le y \le Y$.

Consider the state

$$|\Pi; I_0, I_{0z}; I_Y, I_{Yz}\rangle = \int \delta\Omega \left\{\exp \int_0^Y dy\left[-c\left(\frac{d\hat{n}}{dy}\right)^2 + \vec{\Pi}(y)\cdot\vec{a}^\dagger(y)\right]\right\}|0\rangle . \tag{89}$$

$\int \delta\Omega$ indicates a functional integration over all possible forms for the unit vector $\hat{n}(y)$. (I_0, I_{0z}) is the isospin transferred to the pion field at rapidity 0 and (I_Y, I_{Yz}) is the isospin transferred from the pion field at rapidity Y. These quantities determine the boundary conditions on $\hat{n}(y)$. In order to understand the content of this state let us consider the probability amplitude for finding two pions with rapidities y_1 and y_2 and isospin indices i and j:

$$\begin{aligned}A_{ij}(y_1, y_2) &= N^{-1/2}\langle 0|a_i(y_1)a_j(y_2)|\Pi; I_0, I_{0z}; I_Y, I_{Yz}\rangle \\ &= N^{-1/2}\Pi(y_1)\Pi(y_2)\int \delta\Omega\,\hat{n}_i(y_1)\hat{n}_j(y_2) \\ &\qquad \times \exp\left[-c\int_0^Y dy\left(\frac{d\hat{n}}{dy}\right)^2\right], \\ &\equiv N^{-1/2}\Pi(y_1)\Pi(y_2)J_{ij} , \end{aligned} \tag{90}$$

with

$$N=\langle \Pi; I_0, I_{0z}; I_Y, I_{Yz} | \Pi; I_0, I_{0z}; I_Y, I_{Yz}\rangle . \qquad (91)$$

It is convenient to break up the region $0 \le y \le Y$ into N equal intervals of length Δy so that $\Delta y N = Y$. Then

$$J_{ij}=\lim_{\Delta y\to 0}(c/\Delta y\pi)^N\int\prod_{l=0}^{N} d\Omega_l\, \hat{n}_i(l_1)\hat{n}_j(l_2) \times Y^*_{I_0,I_{0z}}(\hat{n}(0))\, Y^*_{I_Y,I_{Yz}}(\hat{n}(N)) \times \exp\left[-c\sum_{l=0}^{N-1}[\hat{n}(l+1)-\hat{n}(l)]^2/\Delta y\right], \qquad (92)$$

where

$$\hat{n}(l)\equiv\hat{n}(l\,\Delta y),\qquad l=0,1,2,\ldots,N$$

and

$$\hat{n}(l_1)=\hat{n}(y_1),\qquad \hat{n}(l_2)=\hat{n}(y_2). \qquad (93)$$

The factor $(c/\Delta y\pi)^N$ defines the measure of the functional integration. The spherical harmonics $Y_{I_0,I_{0z}}$ and $Y_{I_Y,I_{Yz}}$ appear in Eq. (92) because we have insisted that a definite isospin be transferred at the boundaries $y=0$, Y. More generally one could replace them by arbitrary functions of $\hat{n}(0)$ and $\hat{n}(N)$. The forms of these functions would then determine the probability of a given isospin being transferred at the boundary.

The general integral which must be performed to evaluate J_{ij} is

$$J=(c/\Delta y\pi)\int d\Omega_l e^{-c[\hat{n}(l+1)-\hat{n}(l)]^2/\Delta y}\psi(\hat{n}(l))$$
$$=(4c/\Delta y)\sum_{I,I_z} e^{-2c/\Delta y} j_I(-2ic/\Delta y) i^I\, Y_{I,I_z}(\hat{n}(l+1)) \times\int d\Omega_l\, Y^*_{I,I_z}(\hat{n}(l))\,\psi(\hat{n}(l)). \qquad (94)$$

Using the asymptotic behavior of the spherical Bessel functions we find that to first order in Δy,

$$J\simeq\sum_{I,I_z}[1-I(I+1)\Delta y/4c]\, Y_{I,I_z}(\hat{n}(l+1)) \times\int d\Omega_l Y^*_{I,I_z}(\hat{n}(l))\,\psi(\hat{n}(l))$$
$$=[1-\vec{I}^2\Delta y/4c]\,\psi(\hat{n}(l+1))$$
$$\simeq e^{-\vec{I}^2\Delta y/4c}\,\psi(\hat{n}(l+1)), \qquad (95)$$

where $\vec{I}$ is the isotopic-spin operator. Using this result in Eq. (92) one finds for $0<y_1<y_2<Y$

$$J_{ij}=\sum_{I,I_z} e^{-y_1 I_0(I_0+1)/4c}\, g^i_{I_0,I_{0z};I,I_z} \times e^{-(y_2-y_1)I(I+1)/4c}\, g^j_{I,I_z;I_Y,I_{Yz}} \times e^{-(Y-y_2)I_Y(I_Y+1)/4c}, \qquad (96)$$

where

$$g^i_{I,I_z;I',I'_z}=\int d\Omega\, Y_{I,I_z}(\hat{n})\hat{n}_i\, Y_{I',I'_z}(\hat{n}). \qquad (97)$$

In particular for $I_0=I_Y=0$,

$$J_{ij}=\delta_{ij}(4\pi/3)e^{-|y_2-y_1|/2c}. \qquad (98)$$

Notice that for $I_0=I_Y=0$ the state $|\Pi; 0, 0; 0, 0\rangle$ has total isotopic spin zero.

The probability amplitude for finding N pions can be obtained in an analogous manner. The isospin structure is the same as in the multi-Regge model when there is only one Regge trajectory of each isospin. Secondary trajectories for each isospin will arise naturally when we introduce functional integrations over the $\Pi(y)$ as discussed in Sec. III.

VI. DISCUSSION

In this work we have constructed a general framework for describing high-energy pion production. We have seen that all of the pion cross sections can be obtained from a pion field density operator. For a wide range of configurations of the pion field, this operator can be simply expressed in the coherent-state representation.

We have considered a variety of correlations among the pions. In Sec. III we showed how to introduce short-range correlations in the rapidity and transverse momentum. In Sec. IV we presented a simple model that satisfied exact s-channel unitarity, and so how the constraints of unitarity led to long-range correlations. Finally in Sec. V we studied the correlations associated with isospin conservation. We believe that all of these correlations can and should be taken into account in a realistic theory of pion production.

In Sec. III we pointed out that the density operator could often be written in diagonal form in the coherent-state representation, and that when it could, the various pion cross section took on a particularly simple form. If one then uses the Ginsburg-Landau parameterization for $F[\Pi]$ or a more general form that leads only to short-range correlations, then the pion field can be thought of as being excited by a single interaction of the incident hadrons. As we saw in Sec. IV, multiple independent interactions lead to long-range correlations. When only short-range correlations are present, it can be shown from Eq. (29) that $\Omega(z)$ falls like a power of the energy for z near zero. Crudely speaking this follows from the fact that $\int dq|\Pi(q)|^2$ is typically of order Y. As a result, the elastic and total cross sections in such a model will always go to zero at high energies. In Sec. IV we considered a model in which the elastic and total cross sections do go to constants. In this model if one retains only the terms that contribute to leading power in s, then the density operator can be written in diagonal form. The present problem is avoided because the weight functional $e^{-F[\Pi]}$ has a singularity at $\Pi=0$ which has the form of a functional δ function. In more general unitary models in which the incident hadrons can be diffractively excited to states which subsequently decay into pions,[11] we would expect this functional δ function and its derivatives to be present in $e^{-F[\Pi]}$.

ACKNOWLEDGMENT

We would like to thank M. Kugler for a stimulating discussion.

---

*Work supported by the National Science Foundation.

[1]R. J. Glauber, Phys. Rev. 131, 2766 (1963).

[2]For earlier work in this direction see D. Horn and R. Silver, Ann. Phys. (N.Y.) 60, 509 (1971), and references contained therein. We thank M. Bander and M. Kugler for bringing this reference to our attention.

[3]D. J. Scalapino and R. L. Sugar, Phys. Rev. D 8, 2284 (1973), hereafter referred to as I.

[4]S. Auerbach, R. Aviv, R. Blankenbecler, and R. L. Sugar, Phys. Rev. Lett. 29, 522 (1972); Phys. Rev. D 6, 2216 (1972); R. L. Sugar, *ibid.* 8, 1134 (1973). These models will hereafter be denoted as AABS models.

[5]K. Wilson, Cornell Reports No. CLNS-131, 1970 (unpublished); W. R. Frazer, R. D. Peccei, S. S. Pinsky, and C.-I Tan, Phys. Rev. D 7, 2647 (1973). H. Harari and B. Rabinovici, Phys. Lett. 43B, 49 (1973); K. Fiałkowski and M. Miettinen, *ibid.* 43B, 61 (1973).

[6]For a detailed discussion of the properties of these states see Refs. 1 and 10.

[7]E. C. G. Sudarshan, Phys. Rev. Lett. 10, 277 (1963); J. R. Klauder, J. McKenna, and D. G. Currie, J. Math. Phys. 6, 734 (1965); K. E. Cahill, Phys. Rev. 180, 1244 (1969).

[8]A. H. Mueller, Phys. Rev. D 4, 150 (1971).

[9]χ is not required to have a finite Hilbert-Schmidt norm. In fact, in the models discussed in Ref. 4 it does not. Such models would have to be treated by techniques somewhat different from those we shall present here.

[10]K. E. Cahill and R. J. Glauber, Phys. Rev. 177, 1857 (1969).

[11]J. A. Skard and J. R. Fulco, Phys. Rev. D 8, 312 (1973); R. Blankenbecler, J. R. Fulco, and R. L. Sugar (unpublished).

[12]J. C. Botke, Phys. Rev. Lett. 31, 658 (1973).

[13]H. Harari, Phys. Rev. Lett. 29, 1706 (1972).

VOLUME 43, NUMBER 12 PHYSICAL REVIEW LETTERS 17 SEPTEMBER 1979

Coherent Quark-Gluon Jets

G. Curci
CERN, Geneva, Switzerland

and

M. Greco and Y. Srivastava[a]
Laboratori Nazionali di Frascati, Istituto Nazionale di Fisica Nucleare, Frascati, Italy
(Received 5 March 1979)

With use of the coherent-state formalism, general expressions are presented for arbitrary massless quark and gluon jet cross sections. Also, normalized probability distributions are given for the jet transverse momentum which take due account of correlations imposed by the momentum conservation. These results are used to discuss successfully the SPEAR data of Hanson *et al.* between 3 and 7.8 GeV.

Much attention has been directed recently to study theoretically the jet structure in quantum chromodynamics (QCD) perturbation theory.[1] In the present work, we describe the results of an approach based on the coherent-state formalism.[2,3] A simple formula is obtained for a general jet cross section in terms of an infinite number of gluons and massless quarks. We also obtain probability distributions for the jet transverse momentum which include correlations induced by the momentum conservation. A phenomenological extrapolation of our jet K_1 distribution in the nonperturbative region is then presented. This assumes that the hadronic transverse-momentum distribution follows that of the QCD radiation[4] and uses a parametrization of $\overline{\alpha}(k_\perp)$ obtained in an earlier work.[5] We find excellent agreement with the accurate SPEAR data for the two-jet process $e^+e^- \to q\overline{q} \to$ hadrons, for total energies 3–7.8 GeV.[6]

In Ref. 2, coherent states $|\overline{i}\rangle$, consisting of an indefinite number of soft gluons, corresponding to a "pure" state $|i\rangle$ were constructed to obtain ir-finite inclusive cross sections in QCD. For example, for a quark in a color-singlet potential, the inclusive cross section was found to be[2]

$$\sigma_{\text{inc}} \propto |\langle \overline{f}|S|\overline{i}\rangle|^2 = \exp\{\int d^3k\,(2k)^{-1}[j_\mu^{\,c}(k)^\dagger]\overline{g}_{\text{YM}}^{\,2}(k)\}\;|\langle f|S|i\rangle|^2, \tag{1}$$

where $j_\mu^{\,c}(k)$ is the classical color current operator and $\overline{g}_{\text{YM}}(k)$ is the effective coupling constant for pure Yang-Mills theory. The finite corrections in Eq. (1) after the cancellation of ir singularities can be written as

$$\exp\left[-\frac{C_F}{2\pi^2}\int_{\Delta\omega}^{E}\left(\frac{dk}{k}\right)\int_{m^2}^{-q^2}\frac{dk_\perp^{\,2}}{k_\perp^{\,2}}\overline{g}_{\text{YM}}^{\,2}(k)\right], \tag{2}$$

where $C_F = (N_C^{\,2}-1)/2N_C$ for SU(N_C) color, m and $\Delta\omega$ are the mass and energy loss of the quark, and q^2 is the momentum transfer.

To discuss the massless limit $m \to 0$ new coherent states have to be defined[3] which take into account the collective effect of hard and collinear gluons, i.e., the set of states degenerate with the initial and final states.[7] As discussed in Ref. 3 the mass singularities exactly cancel between soft and hard contributions provided $g_{\text{YM}}(k)$ in the above Eq. (2) is replaced by the full coupling constant $\overline{g}(k_\perp)$, since, as $m \to 0$, quark loops in addition to the gluons must be included.

Thus, for $e^+e^- \to \overline{q}q$ we have ir- and mass-singularity-free "super" inclusive cross sections given by

$$d\sigma_{\text{super}}^{(2q)} = d\sigma_0 \exp\left[-\frac{1}{\pi^2}\int_{\Delta\omega/E}^{1} dx\, P_{gq}(x)\int_{k_{\perp 1}^{\,2}}^{k_{\perp 2}^{\,2}}\frac{(d^2k_\perp)}{k_\perp^{\,2}}\overline{\alpha}(k_\perp)\right], \tag{3}$$

where the gluon distributions due to the quarks is[8]

$$P_{gq}(x) = C_F[1+(1-x)^2]/x. \tag{4}$$

In Eq. (3) $k_{\perp 1} = E\delta$, $k_{\perp 2} = Q/2 = E$, E is the quark energy, δ is the half angle of the jet cone, and $d\sigma_0$ is the "pointlike" cross section.

The ratio $d\sigma_{\text{super}}/d\sigma_0$ in Eq. (3) gives the probability of finding a fraction $\epsilon = \Delta\omega/Q$ of the total energy Q outside a pair of oppositely directed cones of half angle δ. First order expansion of Eq. (5) directly gives the result of Sterman and Weinberg[1] up to terms of order $\epsilon\ln\delta$. A more accurate kinematical analysis shows that the upper limit in x should be restricted to $1-\Delta\omega/E=1-2\epsilon$. This gives

$$I_q(\epsilon)\equiv\int_{2\epsilon}^{1-2\epsilon}dx\,P_{gq}(x)=-C_F[2\ln 2\epsilon+\tfrac{3}{2}-2\epsilon+4\epsilon^2+O(\epsilon^3)], \tag{5}$$

in agreement with Stevenson.[1] From now, we exponentiate with the right kinematics and certainly our results will remain valid for small ϵ but they may not account for all ϵ terms.

The generalization of Eq. (3) to gluon jets is obtained as follows. The probability $P_{gq}(x)$ in Eq. (4) has to be replaced by

$$P_{gg}(x)+N_fP_{qg}(x)=N_C\left[\frac{x}{1-x}+\frac{1-x}{x}+x(1-x)\right]+N_f\,\frac{x^2+(1-x)^2}{2}, \tag{6}$$

corresponding to the sum of probabilities that a gluon radiates gluons radiates gluons or $N_f(q\bar{q})$ pairs. Then Eq. (5) is replaced by

$$I_g(\epsilon)\equiv\int_{2\epsilon}^{1-2\epsilon}dx[P_{gg}(x)+N_fP_{qg}(x)]$$

$$=-N_C\left[2\ln 2\epsilon+\left(\frac{11}{6}-\frac{1}{3}\frac{N_f}{N_C}\right)-4\epsilon\left(1-\frac{N_f}{2N_C}\right)+8\epsilon^2\left(1-\frac{N_f}{2N_C}\right)+O(\epsilon^3)\right]. \tag{7}$$

The logarithmic and constant terms (in ϵ) in Eq. (7) agree with the explicit perturbative calculations of Shyzuya and Tye, Einhorn and Weeks, and Smilga and Vysotsky.[1] Finite order terms in ϵ have, as yet, not been calculated in perturbation theory. Notice that our distributions in ϵ for $I_q(\epsilon)$ and $I_g(\epsilon)$ have been obtained without any recourse to singular distributions.[8] This difference arises because we do not integrate to $x=1$.

The above results generalize for n_q and n_g quark and gluon jets, respectively, to

$$d\sigma_{\text{super}}^{(n_q,n_g)}=d\sigma_0\exp\left\{-\frac{1}{\pi}[n_qI_q(\epsilon)+n_gI_g(\epsilon)]\int_{k_{\perp1}}^{k_{\perp2}}\left(\frac{dk_\perp}{k_\perp}\right)\bar{\alpha}(k_\perp)\right\}, \tag{8}$$

where for simplicity the same kinematical limits $k_{\perp1}$ and $k_{\perp2}$ are used for all jets. The curly bracket in Eq. (8) gives the leading terms in $(\ln\delta)$. All nonleading terms in δ are included in $d\sigma_0$ and have to be calculated, if need be, for each process separately in perturbation theory. First-order expansion of Eq. (18) agrees with the results of Smilga and Vysotsky.[1]

From now on we only consider $q\bar{q}$ jets in e^+e^- annihilation. From Eq. (3), we can define a probability distribution $dP(K_\perp)$,

$$\int_0^{K_{\perp1}}\left(\frac{dP}{dK_\perp}\right)dK_\perp\equiv\exp\left[-\frac{2}{\pi}I(\epsilon)\int_{K_{\perp1}}^{Q/2}\frac{dk_\perp}{k_\perp}\bar{\alpha}(k_\perp)\right], \tag{9}$$

so that

$$\frac{dP}{dK_\perp}\simeq\frac{2I(\epsilon)}{\pi}\left[\frac{\bar{\alpha}(K_\perp)}{K_\perp}\right]\left[\frac{\bar{\alpha}(Q/2)}{\bar{\alpha}(K_\perp)}\right]^{I(\epsilon)/\pi b}, \tag{10}$$

where $b=(33-2N_f)/12\pi$ for three colors, $I(\epsilon)\equiv Iq(\epsilon)$ and we have used the asymptotic-freedom formula for $\bar{\alpha}(k_\perp)$. Thus, Eq. (10) should be valid only for $\Lambda\ll K_\perp\ll Q/2$.

So far, we have omitted any correlation in transverse momentum, since it was assumed that the individual emissions were statistically independent. We can impose transverse-momentum conservation in a manner similar to that of Greco, Pancheri-Srivastava, and Srivastava.[9] This leads us to the following results, the details of which shall be presented elsewhere. We obtain

$$\frac{d^2P}{d^2K_\perp}=\frac{1}{2\pi}\int_0^\infty x_\perp dx_\perp J_0(x_\perp K_\perp)\exp\left\{-\frac{2I(\epsilon)}{\pi}\int_0^{Q/2}\left(\frac{dk_\perp}{k_\perp}\right)\bar{\alpha}(k_\perp)[1-J_0(x_\perp k_\perp)]\right\}, \tag{11}$$

and

$$P(\epsilon,K_{\perp1})=K_{\perp1}\int_0^\infty dx_\perp J_1(x_\perp K_{\perp1})\exp\left\{-\frac{2}{\pi}I(\epsilon)\int_0^{Q/2}\left(\frac{dk_\perp}{k_\perp}\right)\bar{\alpha}(k_\perp)[1-J_0(x_\perp k_\perp)]\right\}. \tag{12}$$

VOLUME 43, NUMBER 12 PHYSICAL REVIEW LETTERS 17 SEPTEMBER 1979

Equation (12) gives the probability of finding a fraction $1-\epsilon$ of the total energy inside two opposite cones of maximum transverse momentum $K_{\perp 1}$, and replaces our earlier Eq. (3) for $(d\sigma_{\text{super}}/d\sigma_0)$. Similarly, Eq. (11), which is properly normalized to 1 for any fixed ϵ, replaces the uncorrelated emission result [Eq. (10)]. From it various moments can be derived. For example,

$$\langle K_\perp^2\rangle = (I(\epsilon)/\pi)\int_0^{Q^2/4} dk_\perp^2\, \overline{\alpha}(k_\perp), \tag{13}$$

which for large Q becomes

$$\langle K_\perp^2\rangle \simeq \text{const} + \frac{I(\epsilon)}{4\pi} Q^2 \overline{\alpha}(\tfrac{1}{2}Q)\left[1 + O\left(\frac{1}{\ln Q}\right)\right]. \tag{14}$$

The constant term can only be obtained upon a knowledge of $\overline{\alpha}(k_\perp)$ in the nonperturbative region $(K_\perp \lesssim C\Lambda)$. Using the parametrization discussed below, we find it negligible for a wide range of C values $(C \gtrsim 1)$. A comparison of quark and gluon jets then gives, for fixed ϵ and Q,

$$\frac{\langle K_\perp^2\rangle_{g\ \text{jet}}}{\langle K_\perp^2\rangle_{q\ \text{jet}}} = \frac{I_g(\epsilon)}{I_q(\epsilon)} \xrightarrow[\epsilon\to 0]{} \frac{2N_C^2}{N_C^2-1}, \tag{15}$$

which predicts gluon jets broader by a factor of two than the quark jets. Similar results have been obtained in the recent literature.[1,4]

A few remarks are in order concerning the validity of the above results, since, as is clear from Eqs. (11-12), the range of the $k_\perp$ integrals extends over a nonperturbative region $(k_\perp \lesssim C\Lambda)$ where the leading logarithmic approximation is questionable. The influence of this region is negligible for $K_\perp \gg C\Lambda$, which is the region of interest for a direct test of QCD, as can be seen by approximating $J_0(z)$ by $\theta(1-z)$ and $J_1(z)$ by $\delta(1-z)$ in Eqs. (11-12) which then reduce to the naive expressions (3) and (10). Numerical results which explicitly verify this fact, as well as giving our predictions for jet physics at the forthcoming very high-energy machines, will be presented elsewhere.

We now present an interesting result obtained by extrapolating smoothly our result (11) to the region $K_\perp \lesssim C\Lambda$. Assuming that the above $K_\perp$ distribution obtained for the radiation of a $q\overline{q}$ jet also describes the $k_\perp$ behavior of a single hadron[4] and using a simple parametrization for $\overline{\alpha}(k_\perp)$, directly suggested[5] by all the data on $R \equiv \sigma(e\overline{e} \to \text{hadrons})/\sigma(e\overline{e}\to\mu\mu)$, we find that Eq. (11) reproduces the experimental data quite well. More in detail, we limit the maximum transverse momentum allowed for a single hadron to a value $\sim Q/2\langle n\rangle$, where $\langle n\rangle$ is the average multiplicity.

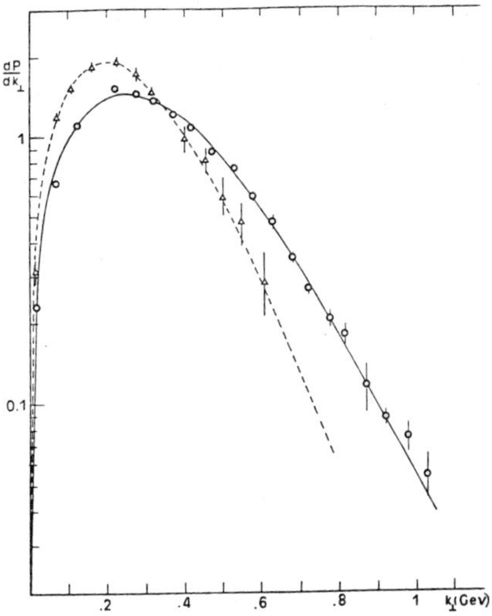

FIG. 1. Normalized $\sigma^{-1}d\sigma/dk_\perp$ vs $k_\perp$ for $Q=3$ GeV (triangles) and 7.5 GeV (circles) SPEAR single inclusive data from Ref. 6. Our results are given by the dashed line for 3 GeV and by the solid line for 7.5 GeV.

This automatically restricts the $k_\perp$ range in Eq. (11) to values $\lesssim$ 2-3 GeV for present and near future e^+e^- beam energies. For $\overline{\alpha}(k_\perp)$ we use the following parametrization:

$$\overline{\alpha}(k_\perp) = \begin{cases} \dfrac{6\pi}{(33-2N_f)\ln C}, & k_\perp \leqslant C\Lambda \\[2ex] \dfrac{6\pi}{(33-2N_f)\ln(k_\perp/\Lambda)}, & k_\perp \geqslant C\Lambda, \end{cases} \tag{16}$$

with $C \simeq 4$ and $\Lambda \simeq 0.7$ GeV. This effective $\overline{\alpha}$ has been shown[5] to bring the QCD prediction $R = \sum_s Q_i^2[1+\overline{\alpha}(S)/\pi]$ in excellent agreement with the average value $R_{\text{av}} \equiv \overline{S}^{-1}\int_{S_{\text{th}}} ds\, R_{\text{exp}}(S)$, extracted from *all* data below $\overline{S} \simeq 60\ \text{GeV}^2$.

With the above modifications in Eq. (11) our predictions are plotted in Fig. (1) for $Q=3$ and 7.5 GeV, and compared with the SLAC data[6] on $\sigma^{-1}d\sigma/dk_\perp$, suitably normalized to unity. Being unable to extract ϵ directly from the data, we have fixed $I(\epsilon)$ by requiring the same $\langle k_\perp\rangle$ as given by the data at $Q=7.5$ GeV. This leads to $2\epsilon \simeq 0.025$, which is also used for $Q=3$ GeV. The data in Fig. (1) agree very well with our expressions and thus support our extrapolation of the

QCD $K_\perp$ distribution to the nonperturbative region.

A useful analytical approximation to Eq. (11), valid for $\bar{\alpha}$ constant, is given by

$$\frac{dP}{dk_\perp} \simeq \frac{2\beta}{a\Gamma(1+\beta/2)}\left(\frac{k_\perp}{a}\right)^{\beta/2} K_{1-\beta/2}\left(\frac{2k_\perp}{a}\right), \qquad (17)$$

where $a = Q/2$, $\beta = 2\bar{\alpha}I(\epsilon)/\pi$, and $K_\nu(z)$ is the modified Bessel function. This result was derived earlier in Ref. 4. Equation (17) gives the same $\langle k_\perp\rangle$ and $\langle k_\perp^2\rangle$ as the original distribution (11).

In conclusion, we have presented a general formalism for quark and gluon jets and their $k_\perp$ distributions. A successful analysis of single-hadron $k_\perp$ distributions at SPEAR ($Q = 3-7.8$ GeV) indicates a rather smooth extrapolation of QCD for predictions to the nonperturbative region as also found[5] for the total cross section.

One of us (Y.S.) would like to thank Dr. Giulia Pancheri-Srivastava for useful discussions. This work was supported, in part, by the U. S. National Science Foundation, Washington, D. C.

---

[a]Permanent address: Northeastern University, Boston, Mass. 02115.

[1]G. Sterman and S. Weinberg, Phys. Rev. Lett. 39, 1436 (1977); H. Politzer, Nucl. Phys. B129, 301 (1977); H. Georgi and M. Machacek, Phys. Rev. Lett. 39, 1237 (1977); E. Farhi, Phys. Rev. Lett. 39, 1587 (1977); P. Stevenson, Phys. Lett. 78B, 451 (1978); K. Shizuya and S.-H. H. Tye, Phys. Rev. Lett. 41, 787 1195(E) (1978); M. Einhorn and B. Weeks, SLAC Report No. SLAC-PUB 2164, 1968 (unpublished); A. Smilga and M. Vysotsky, Institute for Theoretical and Experimental Physics (Moscow) Report No. ITEP-93, 1978 (unpublished); K. Koller and T. Walsh, Nucl. Phys. B140, 449 (1978); G. Fox and S. Wolfram, Phys. Rev. Lett. 41, 1581 (1978); K. Konishi, A. Ukawa, and G. Veneziano, Phys. Lett. 80B, 259 (1979); R. K. Ellis and R. Petronzio, Phys. Lett. 80B, 249 (1979); A. Ali, J. Körner, Z. Kunszt, J. Willrodt, G. Kramer, G. Schieroholz, and E. Pietarinen, to be published.

[2]M. Greco, F. Palumbo, G. Pancheri-Srivastava, and Y. Srivastava, Phys. Lett. 77B, 282 (1978).

[3]G. Curci and M. Greco, Phys. Lett. 79B, 406 (1978).

[4]G. Pancheri-Srivastava and Y. Srivastava, Laboratori Nazionali di Frascati Report No. LNF-78/46(P), 1978 (to be published); G. Pancheri-Srivastava and Y. Srivastava, Northeastern University (Boston) Report No. NUB 2379, 1978 (to be published).

[5]M. Greco, G. Penso, and Y. Srivastava, to be published.

[6]G. Hanson, in *Proceedings of the Thirteenth Recontre de Moriond, Les Arcs-Savoie, France, 1978*, edited by J. Tran Thanh Van (Editions Frontières, France, 1979).

[7]T. Kinoshita, J. Math. Phys. 3, 650 (1962); T. D. Lee and M. Nauenberg, Phys. Rev. B 133, 1549 (1964).

[8]G. Altarelli and G. Parisi, Nucl. Phys. B126, 298 (1977).

[9]M. Greco, G. Pancheri-Srivastava and Y. Srivastava, Nucl. Phys. B101, 234 (1975).

PHYSICAL REVIEW D VOLUME 24, NUMBER 10 15 NOVEMBER 1981

Coherent-state representation of a non-Abelian charged quantum field

K.-E. Eriksson, N. Mukunda,* and B.-S. Skagerstam[†]
Institute of Theoretical Physics, S-412 96 Göteborg, Sweden
(Received 18 November 1980)

Quasicoherent states, previously defined for bosons with SU(2) gauge charge (and no other degree of freedom), are now defined for the general (field-theoretical) case. The coherent states thus constructed form a complete basis in the Fock space. They transform according to irreducible representations of the SU(2) group and are at the same time eigenstates of isosinglet pair and isosinglet three-particle annihilation operators. Some physical applications are indicated in the contexts of multiple particle production and gluon bremsstrahlung by a quark line in e^+e^- annihilation.

I. INTRODUCTION

In a previous paper two of the present authors discussed an extension of the concept of coherent states to a situation where a non-Abelian charge is involved.[1] For one degree of freedom a construction was given of *quasi-coherent states* transforming irreducibly under isotopic spin rotations. In the present paper we will extend the results in Ref. 1 to the field-theoretical case. Also some applications are discussed at the end of the paper. We use the term "quasicoherent states" rather than "generalized coherent states" since the latter term has already been used for a similar construction in the literature.[2]

As in Ref. 1 we will only consider the rotation group (the isospin group). A more general treatment of gauge groups such as SU(3) will be discussed in a forthcoming paper. For the reader's convenience we here briefly recall the conventional treatment of coherent states[3] for an isovector boson field with N kinematical states available. Reference 1 describes simply the case $N=1$. In the present paper we keep N finite only for convenience; clearly $N\to\infty$ (field theory) can be treated in the same way.

The Fock space for the isovector boson can be generated from a vacuum state $|0\rangle$ by creation operators $a^\dagger_{i\alpha}$, where $i=1,\ldots,N$. The corresponding annihilation operators are $a_{i\alpha}$. An orthonormal basis for the Fock space is

$$|\vec{n}_1,\ldots,\vec{n}_N\rangle=\prod_{i=1}^{N}[(n_{i1}!\,n_{i2}!\,n_{i3}!)^{-1/2}\times(a^\dagger_{i1})^{n_{i1}}(a^\dagger_{i2})^{n_{i2}}(a^\dagger_{i3})^{n_{i3}}]|0\rangle\ . \tag{1}$$

We introduce in the usual way the unitary operator

$$U(\underline{x})=e^{\underline{a}^\dagger\cdot\underline{x}-\underline{x}^*\cdot\underline{a}}\ , \tag{2}$$

where the $3N$-vector $\underline{x}$ describes a one-particle state and where $\underline{x}^*\cdot\underline{a}=x^*_{i\alpha}a_{i\alpha}$. Here i runs over 1 to N and α over 1 to 3. The conventional coherent state $|\underline{x}\rangle$ is then given by[3]

$$|\underline{x}\rangle=U(\underline{x})|0\rangle=e^{-\underline{x}^*\cdot\underline{x}/2}e^{\underline{a}^\dagger\cdot\underline{x}}|0\rangle\ ,\quad a_{i\alpha}|\underline{x}\rangle=x_{i\alpha}|\underline{x}\rangle\ . \tag{3}$$

The scalar product then is

$$\langle\underline{y}|\underline{x}\rangle=e^{-(\underline{x}^*-\underline{y}^*)\cdot(\underline{x}-\underline{y})/2}\,e^{(\underline{y}^*\cdot\underline{x}-\underline{x}^*\cdot\underline{y})/2}\ . \tag{4}$$

In particular

$$\langle\underline{x}|\underline{x}\rangle=1\ . \tag{5}$$

The coherent states form an overcomplete basis with the completeness relation

$$\int d^{6N}\underline{x}\,|\underline{x}\rangle\langle\underline{x}|=1\ ,\qquad d^{6N}\underline{x}=\prod_{i=1}^{N}\prod_{\alpha=1}^{3}\left(\frac{1}{\pi}\,d\,\mathrm{Re}\{x_{i\alpha}\}\,d\,\mathrm{Im}\{x_{i\alpha}\}\right)\ . \tag{6}$$

Given a $3N$-vector $\underline{x}$ we may define a matrix $J(\underline{x})$ serving as a "tensor of inertia" in the isospin space

$$J_{\alpha\beta}(\underline{x})=x^*_{i\alpha}x_{i\beta}\quad(\alpha,\beta=1,2,3)\ . \tag{7}$$

Then

$$J(\underline{x})^\dagger=J(\underline{x})\ ,\qquad \mathrm{Tr}[J(\underline{x})]=\underline{x}^*\cdot\underline{x}\ . \tag{8}$$

The tensor of inertia (7) will turn out to be useful in the normalization of a certain basis of quasicoherent states.

In the Fock space we can define the isospin rotation operators ($\hbar=1$)

$$I_\alpha=\frac{1}{i}\epsilon_{\alpha\beta\gamma}a^\dagger_{k\beta}a_{k\gamma} \tag{9}$$

which obey

$$[I_\alpha,I_\beta]=i\epsilon_{\alpha\beta\gamma}I_\gamma\ . \tag{10}$$

Below we will give a decomposition of the state $|\underline{x}\rangle$ in terms of a complete set of states (the quasi-coherent states) which are diagonal in $\vec{I}^2$, I_3, and the isosinglet pair and three-particle annihilation operators

$$\omega_{ij}=\vec{a}_i\cdot\vec{a}_j \tag{11}$$

and

$$\Omega_{ijk}=\epsilon_{\alpha\beta\gamma}a_{i\alpha}a_{j\beta}a_{k\gamma}\ . \tag{12}$$

To obtain such a decomposition we make use of the fact that the Wigner $D^l_{\mu\nu}$ functions form a complete basis for the continuous functions on the rotation group.[4] For $R\in SO(3)$ we normalize the invariant group measure d^3R so that

$$\int_{SO(3)} d^3R=1\ . \tag{13}$$

The $D^l_{\mu\nu}$ functions satisfy (summation over repeated matrix indices is understood)

$$D^l_{\mu\nu}(R_1)D^l_{\nu\lambda}(R_2)=D^l_{\mu\lambda}(R_1R_2)\ , \tag{14}$$

$$D^l_{\mu\nu}(R)^*=D^l_{\nu\mu}(R^{\sim})\ , \tag{15}$$

$$(2l+1)\int_{SO(3)} d^3R D^{l'}_{\mu'\nu'}(R)^*D^l_{\mu\nu}(R)=\delta_{l'l}\delta_{\mu'\mu}\delta_{\nu'\nu}\ , \tag{16}$$

$$\sum_{l=0}^{\infty}(2l+1)D^l_{\mu\nu}(R)^*D^l_{\mu\nu}(R')=\delta^3(R;R')\ , \tag{17}$$

where the group δ function satisfies

$$\int_{SO(3)} d^3R' f(R')\delta^3(R;R')=f(R)\ . \tag{18}$$

Now let M be an arbitrary 3×3 matrix and let R be a 3×3 SO(3) matrix ($R=R^*$; $R_{\alpha\gamma}R_{\beta\gamma}=\delta_{\alpha\beta}$). We shall need the SO(3) decomposition of exp $[\mathrm{Tr}(RM^{\sim})]$.

Using completeness (17) and orthonormality (16) we find

$$e^{\mathrm{Tr}(RM^{\sim})}=\sum_{l=0}^{\infty}(2l+1)\phi^l_{\mu\nu}(M)D^l_{\mu\nu}(R)\ , \tag{19}$$

where

$$\begin{aligned}\phi^l_{\mu\nu}(M)&=\int_{SO(3)} d^3R\, D^l_{\mu\nu}(R)^*\, e^{M_{\alpha\beta}R_{\alpha\beta}}\\ &=D^l_{\mu\nu}\left(\frac{\partial}{\partial M^*}\right)^* I(M)\end{aligned} \tag{20}$$

with the transformation property

$$\phi^l_{\mu\nu}(RMR')=D^l_{\mu\mu'}(R)^*\phi^l_{\mu'\nu'}(M)D^l_{\nu'\nu}(R')^*\ . \tag{21}$$

In (20) $I(M)$ is the integral

$$I(M)=\int_{SO(3)} d^3R\, e^{M_{\alpha\beta}R_{\alpha\beta}} \tag{22}$$

and the functions $D^l_{\mu\nu}(M)$ are defined as homogeneous polynomials of degree l in the matrix elements $M_{\alpha\beta}$ (see Appendix A),

$$D^l_{\mu\nu}(M)=\frac{2^l}{(2l+1)!}(t^l_\mu)^*_{\alpha_1\cdots\alpha_l}(t^l_\nu)_{\beta_1\cdots\beta_l}M_{\alpha_1\beta_1}\cdots M_{\alpha_l\beta_l}\ , \tag{23}$$

where $(t^l_\mu)_{\alpha_1\cdots\alpha_l}$ are the expansion coefficients of the spherical harmonics, defined in Ref. 1 as homogeneous polyomials of an arbitrary three-vector,

$$r^lY^l_\mu(\theta,\varphi)=Y^l_\mu(\vec{r})=\frac{1}{l!}(t^l_\mu)_{\alpha_1\cdots\alpha_l}r_{\alpha_1}\cdots r_{\alpha_l}\ , \tag{24}$$

where $\vec{r}=r(\sin\theta\cos\varphi,\ \sin\theta\sin\varphi,\ \cos\theta)$. From the definition (22) it immediately follows that $I(M)$ is invariant under left-handed as well as right-handed multiplication of M by SO(3) matrices

$$I(M)=I(R_1MR_2),\quad R_1,R_2\in SO(3)\ . \tag{25}$$

Therefore $I(M)$ is a function only of the three invariants x, y, and z defined by

$$\begin{aligned}x&=\mathrm{Tr}(MM^{\sim})\ ,\\ y&=4\,\mathrm{Det}M\ ,\\ z&=\tfrac{1}{2}[\mathrm{Tr}(MM^{\sim})]^2-\mathrm{Tr}(MM^{\sim}MM^{\sim})\ .\end{aligned} \tag{26}$$

In fact $I(M)\equiv I(x,y,z)$ can be expressed as a Fourier-Mellin integral,

$$\begin{aligned}I(M)&=I(x,y,z)\\ &=\frac{1}{2\pi i}\int_{s_0-i\infty}^{s_0+i\infty} ds\, e^s(s^4-2xs^2-2ys-2z)^{-1/2}\ ,\end{aligned} \tag{27}$$

where s_0 has to be chosen in such a way that all singularities of the integrand are on the left-hand side of the integration contour. The derivation of (27) is shown in Appendix B. From (27) we get the power-series expansion

$$I(x,y,z)=\sum_{j,k,m=0}^{\infty}\frac{[2(j+k+m)-1]!!}{j!\,k!\,m!\,(1+2j+3k+4m)!}x^jy^kz^m\ . \tag{28}$$

The mathematical preliminaries have been given here because $I(M)$ and $\phi^l_{\mu\nu}(M)$ defined in terms of $I(M)$ through (22) play an important role in the coherent-state formalism to be developed in the following sections.

In Sec. II we construct three types of quasi-coherent states and derive their properties. In Sec. III we construct states which are simultaneously eigenstates of $\vec{I}^2$, I_3, and the number operator N. In Secs. IV and V we treat a few simple examples.

II. QUASICOHERENT STATES FOR SU(2) GAUGE FIELDS

Consider now an isospin rotation $R=(R_{\alpha\beta})$ $(\alpha,\beta=1,2,3)$ acting on a $3N$-vector

$$\underline{x}=(\vec{x}_1,\ldots,\vec{x}_N)\ . \tag{29}$$

The result is

$$R\underline{x}=(R\vec{x}_1,\ldots,R\vec{x}_N)\ . \tag{30}$$

From the corresponding coherent state $|R\underline{x}\rangle$ we can project out a certain irreducible representation by using the D functions,

$$|{}^{l}_{\mu\nu};\underline{x}\rangle = e^{\underline{x}^*\cdot\underline{x}/2}(2l+1)^{1/2}\int d^3R\, D^l_{\mu\nu}(R)^*\,|R\underline{x}\rangle \tag{31}$$

The effect of a rotation on such a state is easily found,

$$\begin{aligned} U(R)|{}^{l}_{\mu\nu};\underline{x}\rangle &= (2l+1)^{1/2}e^{\underline{x}^*\cdot\underline{x}/2}\int d^3R'\,D^l_{\mu\nu}(R')^*\,|RR'\underline{x}\rangle \\ &= D^l_{\mu\mu'}(R^{-1})^*(2l+1)^{1/2}e^{\underline{x}^*\cdot\underline{x}/2} \\ &\quad\times\int d^3R'\,D^l_{\mu'\nu}(RR')^*\,|RR'\underline{x}\rangle \\ &= D^l_{\mu'\mu}(R)|{}^{l}_{\mu'\nu};\underline{x}\rangle\ . \end{aligned} \tag{32}$$

The state (31) we shall call a *quasicoherent* state. It is an eigenstate of $\vec{I}^2$ and I_3 as well as of the isosinglet annihilation operators ω_{ij} and Ω_{ijk} defined by (11) and (12),

$$\begin{aligned} \vec{I}^2|{}^{l}_{\mu\nu};\underline{x}\rangle &= l(l+1)|{}^{l}_{\mu\nu};\underline{x}\rangle\ ,\\ I_3|{}^{l}_{\mu\nu};\underline{x}\rangle &= \mu|{}^{l}_{\mu\nu};\underline{x}\rangle\ , \end{aligned} \tag{33}$$

$$\begin{aligned} \omega_{ij}|{}^{l}_{\mu\nu};\underline{x}\rangle &= \vec{x}_i\cdot\vec{x}_j|{}^{l}_{\mu\nu};\underline{x}\rangle\ ,\\ \Omega_{ijk}|{}^{l}_{\mu\nu};\underline{x}\rangle &= \vec{x}_i\cdot(\vec{x}_j\times\vec{x}_k)|{}^{l}_{\mu\nu};\underline{x}\rangle\ . \end{aligned} \tag{34}$$

To get a more explicit expression for $|{}^{l}_{\mu\nu};\underline{x}\rangle$ we use (3) and (20) in (31),

$$\begin{aligned} |{}^{l}_{\mu\nu};\underline{x}\rangle &= (2l+1)^{1/2}\int d^3R\,D^l_{\mu\nu}(R)^*\,e^{\mathrm{Tr}(R(\vec{x}_i\vec{a}_i^\dagger))}|0\rangle \\ &= (2l+1)^{1/2}\,\phi^l_{\mu\nu}(\vec{a}_i^\dagger\vec{x}_i)\,|0\rangle\ , \end{aligned} \tag{35}$$

where summation of i over the N degrees of freedom is understood in $\vec{a}_i^\dagger\vec{x}_i$ which denotes the matrix (of linear combinations of creation operators) $(a^\dagger_{i\alpha}x_{i\beta})$.

Using (35) and (3) we easily get the scalar product between a coherent state and a quasicoherent state,

$$\langle\underline{y}|{}^{l}_{\mu\nu};\underline{x}\rangle = (2l+1)^{1/2}e^{-\underline{y}^*\cdot\underline{y}/2}\,\phi^l_{\mu\nu}(\vec{y}_i^*\vec{x}_i). \tag{36}$$

Like the coherent states, the quasicoherent states form an overcomplete basis. Their scalar product is easily obtained from (31) and (36) with the use of (21) and (16),

$$\begin{aligned} \langle{}^{l'}_{\mu'\nu'};\underline{y}|{}^{l}_{\mu\nu};\underline{x}\rangle &= (2l+1)\int d^3R\,D^{l'}_{\mu'\nu'}(R)\,\phi^l_{\mu\nu}(R\vec{y}_i^*\vec{x}_i)\\ &= (2l+1)\int d^3R\,D^{l'}_{\mu'\nu'}(R)\,D^l_{\mu\mu''}(R)^*\,\phi^l_{\mu''\nu}(\vec{y}_i^*\vec{x}_i) = \delta_{ll'}\,\delta_{\mu\mu'}\,\phi_{\nu\nu'}(\vec{y}_i^*\vec{x}_i)\ . \end{aligned} \tag{37}$$

Starting from (6) and using (17) and (31) we can decompose the unit operators as follows:

$$\begin{aligned} 1 &= \iint d^3R\,d^3R'\,\delta^3(R;R')\int d^{6N}\underline{x}\,|R\underline{x}\rangle\langle R'\underline{x}| \\ &= \sum_{l=0}^{\infty}(2l+1)\int d^{6N}\underline{x}\int d^3R\,D^l_{\mu\nu}(R)^*|R\underline{x}\rangle\int d^3R'\,D^l_{\mu\nu}(R')\langle R'\underline{x}| = \sum_{l=0}^{\infty}\int d^{6N}\underline{x}\,e^{-\underline{x}^*\cdot\underline{x}}\,|{}^{l}_{\mu\nu};\underline{x}\rangle\langle{}^{l}_{\mu\nu};\underline{x}|\ . \end{aligned} \tag{38}$$

We can also introduce a somewhat less overcomplete basis by choosing $\nu=0$ in $|{}^{l}_{\mu\nu};\underline{x}\rangle$. With a suitable normalization we define a *reduced* type of *quasicoherent states*,

$$\begin{aligned} |l,\mu;\underline{x}\rangle &= [\varphi_l(J(\underline{x}))]^{-1/2}|{}^{l}_{\mu 0};\underline{x}\rangle \\ &= \left(\frac{2l+1}{\varphi_l(J(\underline{x}))}\right)^{1/2}\phi^l_{\mu 0}(\vec{a}_i^\dagger\vec{x}_i)\,|0\rangle\ , \end{aligned} \tag{39}$$

where

$$\varphi_l(M) = \phi^l_{00}(M) = \phi^l_{00}(M^-) \tag{40}$$

and $J(\underline{x})$ is the tensor of inertia defined in (7).

Using (37) we get the scalar product for reduced coherent states,

$$\langle l',\mu';\underline{y}|l,\mu;\underline{x}\rangle = \delta_{ll'}\,\delta_{\mu'\mu}\,\frac{\varphi_l(\vec{y}_i^*\vec{x}_i)}{[\varphi_l(J(\underline{x}))\varphi_l(J(\underline{y}))]^{1/2}}\ , \tag{41}$$

which for $\underline{y}=\underline{x}$ reduces to the normalization

$$\langle l',\mu';\underline{x}|l,\mu;\underline{x}\rangle = \delta_{l'l}\,\delta_{\mu'\mu}\ . \tag{42}$$

The states (35) may be expressed in terms of the new states (39),

$$\left|{}^{l}_{\mu\nu};\underline{x}\right\rangle=(2l+1)\int d^3R\, D^{l}_{0\nu}(R)^*[\varphi_l(RJ(\underline{x})R^{-})]^{1/2}\left|l,\mu;R\underline{x}\right\rangle . \tag{43}$$

The completeness relation corresponding to (38) is

$$1=\sum_{l=0}^{\infty}\int d^{6N}\underline{x}\, e^{-\underline{x}^*\cdot\underline{x}}\varphi_l(J(\underline{x}))\left|l,\mu;\underline{x}\right\rangle\left\langle l,\mu;\underline{x}\right| . \tag{44}$$

The quasicoherent states (39) are closely related to the generalized coherent states as defined in Ref. 1. They are also a generalization of the states used by Botke *et al.*[5] for the case of one-particle wave functions that are parallel in the isospin space. The transformation property (32) under rotations and the eigenstate properties (33) and (34) clearly also hold for the reduced states $|l,\mu;\underline{x}\rangle$ as a special case.

A further reduction can be made. The SO(3) characters

$$X_l(R)=D^{l}_{\mu\mu}(R) \tag{45}$$

satisfy the orthonormality relation

$$\int d^3R\, X_{l'}(R)X_l(R)=\delta_{l'l} \tag{46}$$

which follows from (16). The further reduced quasicoherent states

$$\begin{aligned}\left|l;\underline{x}\right\rangle&=\left(\frac{2l+1}{\chi_l(J(\underline{x}))}\right)^{1/2}\int d^3R\, X_l(R)e^{\underline{x}^*\cdot\underline{x}/2}\left|R\underline{x}\right\rangle\\&=(2l+1)^{1/2}[\chi_l(J(\underline{x}))]^{-1/2}\int d^3R\, X_l(R)e^{\mathrm{Tr}[R(\vec{x}_i\vec{a}_i^\dagger)]}\left|0\right\rangle=(2l+1)^{1/2}[\chi_l(J(\underline{x}))]^{-1/2}\chi_l(\vec{a}_i^\dagger\vec{x}_i)\left|0\right\rangle,\quad \chi_l(M)=\phi^{l}_{\mu\mu}(M)\end{aligned} \tag{47}$$

transform as follows under rotations:

$$\begin{aligned}U(R)\left|l;\underline{x}\right\rangle&=\frac{e^{\underline{x}^*\cdot\underline{x}/2}}{[\chi_l(J(\underline{x}))]^{1/2}}\int d^3R'\, X_l(R')\left|RR'\underline{x}\right\rangle\\&=\frac{e^{\underline{x}^*\cdot\underline{x}/2}}{[\chi_l(J(\underline{x}))]^{1/2}}\int d^3R'\, X_l(RR'R^{-1})\left|RR'R^{-1}R\underline{x}\right\rangle\\&=\left|l;R\underline{x}\right\rangle .\end{aligned} \tag{48}$$

It is also an eigenstate of $\vec{I}^2$,

$$\vec{I}^2\left|l;\underline{x}\right\rangle=l(l+1)\left|l;\underline{x}\right\rangle , \tag{49}$$

but not of I_3.

The scalar product is easily found to be

$$\left\langle l';\underline{y}\middle|l;\underline{x}\right\rangle=\delta_{ll'}\frac{\chi_l(\vec{y}_i^*\vec{x}_i)}{[\chi_l(J(\underline{x}))\chi_l(J(\underline{y}))]^{1/2}} , \tag{50}$$

which for $\underline{y}=\underline{x}$ reduces to

$$\left\langle l';\underline{x}\middle|l;\underline{x}\right\rangle=\delta_{ll'} . \tag{51}$$

The completeness relation reads

$$1=\sum_{l=0}^{\infty}(2l+1)\int d^{6N}\underline{x}\, e^{-\underline{x}^*\cdot\underline{x}}\chi_l(J(\underline{x}))\left|l;\underline{x}\right\rangle\left\langle l;\underline{x}\right| . \tag{52}$$

It is also interesting to compute the scalar product between a Fock state (non-normalized)

$$\left|i_1\alpha_1;\ldots;i_n\alpha_n\right\rangle=a^\dagger_{i_1\alpha_1}\cdots a^\dagger_{i_n\alpha_n}\left|0\right\rangle \tag{53}$$

and a reduced quasicoherent state. Using (39) and (35) we then obtain

$$\begin{aligned}&\left\langle i_1\alpha_1;\ldots;i_n\alpha_n\middle|l,\mu;\underline{x}\right\rangle\\&=\frac{1}{[\varphi_l(J(\underline{x}))]^{1/2}}\left({}^{l}_{\mu 0}\middle|{}^{1}_{\alpha_1\beta_1}\cdots{}^{1}_{\alpha_n\beta_n}\right)x_{i_1\beta_1}\cdots x_{i_n\beta_n} ,\end{aligned} \tag{54}$$

where

$$\begin{aligned}&\left({}^{l}_{\mu\nu}\middle|{}^{1}_{\alpha_1\beta_1}\cdots{}^{1}_{\alpha_n\beta_n}\right)\\&=(2l+1)^{1/2}\int d^3R\, D^{l}_{\mu\nu}(R)^*\prod_{m=1}^{n}R_{\alpha_m\beta_m}\end{aligned} \tag{55}$$

is a generalized Clebsch-Gordan coefficient. The probability of having exactly n bosons in a quasicoherent state can now be expressed in terms of (54). Let Π_n be the n-particle state projector,

$$\Pi_n=\frac{1}{n!}\left|i_1\alpha_1;\ldots;i_n\alpha_n\right\rangle\left\langle i_1\alpha_1;\ldots;i_n\alpha_n\right| . \tag{56}$$

Then tho corresponding probability can be expressed as follows:

$$\begin{aligned}\left\langle l,\mu';\underline{x}\middle|\Pi_n\middle|l,\mu;\underline{x}\right\rangle&=\frac{\delta_{\mu\mu'}}{n!\,\varphi_l(J(\underline{x}))}(2l+1)\int d^3R\, x^*_{i_1\alpha_1}R_{\alpha_1\beta_1}x_{i_1\beta_1}\cdots x^*_{i_n\alpha_n}R_{\alpha_n\beta_n}x_{i_n\beta_n}D^{l}_{00}(R)^*\\&=\frac{(2l+1)^{1/2}}{n!\,\varphi_l(J(\underline{x}))}\left({}^{l}_{00}\middle|{}^{1}_{\alpha_1\beta_1}\cdots{}^{1}_{\alpha_n\beta_n}\right)J_{\alpha_1\beta_1}(\underline{x})\cdots J_{\alpha_n\beta_n}(\underline{x}) .\end{aligned} \tag{57}$$

For a coherent state the number operator

$$N=\underline{a}^\dagger\cdot\underline{a} \tag{58}$$

has the expectation value

$$\langle \underline{x}^* | N | \underline{x} \rangle = \underline{x}^* \cdot \underline{x} . \tag{59}$$

For a reduced quasicoherent state the corresponding value is independent of μ and can be obtained as the mean value

$$\langle N \rangle_{l;\underline{x}} \equiv \frac{1}{2l+1} \langle l, \mu; \underline{x} | N | l, \mu; \underline{x} \rangle = e^{\underline{x}^* \cdot \underline{x}} [\varphi_l(J(\underline{x}))]^{-1} \int\int d^3R\, d^3R'\, D^l_{\mu 0}(R') D^l_{\mu 0}(R)^* \langle R'\underline{x} | N | R\underline{x} \rangle$$

$$= e^{\underline{x}^* \cdot \underline{x}} [\varphi_l(J(\underline{x}))]^{-1} \int d^3R\, D^l_{00}(R) \langle R\underline{x} | N | \underline{x} \rangle = [\varphi_l(J(\underline{x}))]^{-1} \frac{\partial}{\partial \lambda} \int d^3R\, D^l_{00}(R) e^{\mathrm{Tr}(R J(\underline{x}))\lambda} \Big|_{\lambda=1} .$$

Thus, because of (40),

$$\langle N \rangle_{l;\underline{x}} = \frac{\partial}{\partial \lambda} \ln \varphi_l(\lambda J(\underline{x})) \Big|_{\lambda=1} . \tag{60}$$

Similarly we can determine $[(\Delta N)_{l;\underline{x}}]^2 = \langle N^2 \rangle_{l;\underline{x}} - (\langle N \rangle_{l;\underline{x}})^2$ with the result

$$[(\Delta N)_{l;\underline{x}}]^2 = \left(\frac{\partial^2}{\partial \lambda^2} + \frac{\partial}{\partial \lambda} \right) \ln \varphi_l(\lambda J(\underline{x})) \Big|_{\lambda=1} . \tag{61}$$

III. NUMBER-OPERATOR EIGENSTATES

Following the discussion in Ref. 1 we may also introduce states $|n, l\mu; x\rangle$ which are eigenstates of $\vec{I}^2$, I_3, and the number operator (58):

$$|n, l\mu; \underline{x}\rangle = \left(\frac{2l+1}{\varphi_{l;n}(J(\underline{x}))} \right)^{1/2} \int_0^{2\pi} \frac{d\psi}{2\pi} e^{-(n+l)i\psi} \int d^3R\, D^l_{\mu 0}(R)^* e^{-\underline{x}^* \cdot \underline{x}/2} |R e^{i\psi} \underline{x}\rangle . \tag{62}$$

Here we have introduced the notation [see Eqs. (3), (20), and (22)]

$$\phi^{l;n}_{\mu\nu}(M) = \int_0^{2\pi} \frac{d\psi}{2\pi} e^{-(n+l)i\psi} \phi^l_{\mu\nu}(e^{i\psi} M) = D^l_{\mu\nu}\left(\frac{\partial}{\partial M^*} \right)^* I_{n+2l}(M) \tag{63}$$

and

$$\varphi_{l;n}(M) = \phi^{l;n}_{00}(M) = D^l_{00}\left(\frac{\partial}{\partial M^*} \right)^* I_{n+2l}(M) , \tag{64}$$

where

$$I_n(M) = \int_0^{2\pi} \frac{d\psi}{2\pi} e^{-ni\psi} I(e^{i\psi} M) . \tag{65}$$

Using (28) we obtain $I_n(M)$ explicitly as the following polynomial in the invariants (26):

$$I_{2n}(M) = \frac{x^n}{(2n+1)!} \sum_{r=0}^{[n/2]} \sum_{s=0}^{[(n-2r)/3]} \frac{[2(n-r-s)-1]!!}{(n-2r-3s)!\,r!\,(2s)!} \left(\frac{z}{x^2} \right)^r \left(\frac{y^2}{x^3} \right)^s ,$$

$$I_{2n+1}(M) = \frac{x^{n-1} y}{(2n+2)!} \sum_{r=0}^{[(n-1)/2]} \sum_{s=0}^{[(n-1-2r)/3]} \frac{[2(n-r-s)-1]!!}{(n-1-2r-3s)!\,r!\,(2s+1)!} \left(\frac{z}{x^2} \right)^r \left(\frac{y^2}{x^3} \right)^s . \tag{66}$$

The number operator eigenstates (62) satisfy

$$\vec{I}^2 | n, l\mu; \underline{x}\rangle = l(l+1) | n, l\mu; \underline{x}\rangle ,$$
$$I_3 | n, l\mu; \underline{x}\rangle = \mu | n, l\mu; \underline{x}\rangle , \tag{67}$$
$$N | n, l\mu; \underline{x}\rangle = (2n+l) | n, l\mu; \underline{x}\rangle .$$

The scalar product is given by

$$\langle n', l'\mu'; \underline{y} | n, l\mu; \underline{x}\rangle = \frac{\varphi_{l;n}(\vec{y}_i^* \vec{x}_i)}{[\varphi_{l;n}(J(\underline{x})) \varphi_{l;n}(J(\underline{y}))]^{1/2}} \delta_{nn'} \delta_{ll'} \delta_{\mu\mu'} . \tag{68}$$

For $\underline{x} = \underline{y}$ this reduces to

$$\langle n', l'\mu'; \underline{x} | n, l\mu; \underline{x}\rangle = \delta_{nn'} \delta_{ll'} \delta_{\mu\mu'} . \tag{69}$$

The overlap of a state (62) with a coherent state

$|\underline{y}\rangle$ is

$$\langle \underline{y} | n, l\mu; \underline{x}\rangle = \left(\frac{2l+1}{\varphi_{l;n}(J(\underline{x}))}\right)^{1/2} e^{-\underline{y}^{*}\cdot\underline{y}/2}\,\phi_{\mu 0}^{l;n}(\vec{y}_i^{*}\vec{x}_i)\,. \tag{70}$$

A completeness relation can also be derived for the number operator eigenstates in a way similar to the derivation of (52). One obtains

$$1 = \sum_{n=0}^{\infty}\sum_{l=0}^{\infty}\int d^{6N}\underline{x}\,\varphi_{l;n}(J(\underline{x}))\,|n, l\mu;\underline{x}\rangle\langle n, l\mu;\underline{x}|\,. \tag{71}$$

IV. QUASICOHERENT STATES AND PARTICLE PRODUCTION

In discussions of multiple particle production one usually considers matrix elements of the scattering operator between Fock-space states. At high energies, when the number of produced particles is large, it may be useful instead to consider the corresponding matrix elements in terms of states which are not eigenstates of the number operator. In quantum electrodynamics (QED) it is well known (see Refs. 3 and 6) that coherent states are extremely useful for studying problems where the number of photons involved is large (or infinite). In physical processes of pion production at high energies a coherent-state basis may analogously be appropriate.[8] These states will not, however, have a definite isospin content. As suggested by Botke *et al.*[5] one could therefore generalize the concept of coherent states in order to get states transforming irreducibly under isospin transformations. Such an extension was indeed given by Botke *et al.*[5] We have now generalized their construction to a superposition of one-particle states which no longer have to be parallel in the isospin space. The quasicoherent states $|{}^{l}_{\mu\nu};\underline{x}\rangle$ (or $|l, m;\underline{x}\rangle$) which we are using constitute a complete basis in the Fock space. The model calculations of Botke *et al.*[5] could now be reconsidered in terms of quasicoherent states but we will not develop on this point further here. Here we notice that the number-operator eigenstate construction in Sec. III extends the isospin reduction of Ref. 8 to general n-pion states. It is of general interest, however, to consider in detail some properties of our quasicoherent states. For simplicity we will restrict ourselves to the isospin [SU(2)] singlet state but the extension to other representations is rather straightforward. We denote the singlet state by $|0;\underline{f}\rangle$. In the isosinglet case there is no difference between the quasicoherent states given by (35), (39), or (47). Using (35) and (40) we now get

$$|0;\underline{f}\rangle = \frac{e^{\underline{f}^{*}\cdot\underline{f}/2}}{[\varphi_0(\vec{f}_k^{*}\vec{f}_k)]^{1/2}}\int d^3R\,|R\underline{f}\rangle. \tag{72}$$

We shall now consider some specific choices of one-particle states. As our *first* example, let $\vec{n}$ denote a unit vector in isospin space and let $\underline{f} = \vec{n}f(\vec{k})$ be the one-particle state as a function of momentum. This choice of one-particle states corresponds to the construction given by Botke *et al.*[5] and to what is called identical pions in Ref. 8. The matrix J defined in (7) then takes the form

$$J_{\alpha\beta} = n_\alpha n_\beta\, c, \quad c = \int \frac{d^3k}{2\omega} f(\vec{k})^{*} f(\vec{k}), \tag{73}$$

where ω is the energy of the particle under consideration. The matrix $J_{\alpha\beta}$ given by (73) can easily be diagonalized. The group invariant $\varphi_0(J)$ can therefore be computed in closed form,

$$\varphi_0(J) = \frac{\sinh c}{c} \equiv \varphi_0(c)\,. \tag{74}$$

Using (60) we find the expectation value of the number operator

$$N_1 \equiv \langle 0;\underline{f}\,|\,N\,|\,0;\underline{f}\,\rangle = c\coth c - 1\,. \tag{75}$$

For a coherent state $|\underline{f}\rangle$ we would obtain [see Eq. (59)]

$$\langle \underline{f}\,|\,N\,|\,\underline{f}\,\rangle = c\,. \tag{76}$$

Comparison of (75) and (76) then leads to the conclusion that quasicoherent states are "less condensated" than the coherent states (see Fig. 1).

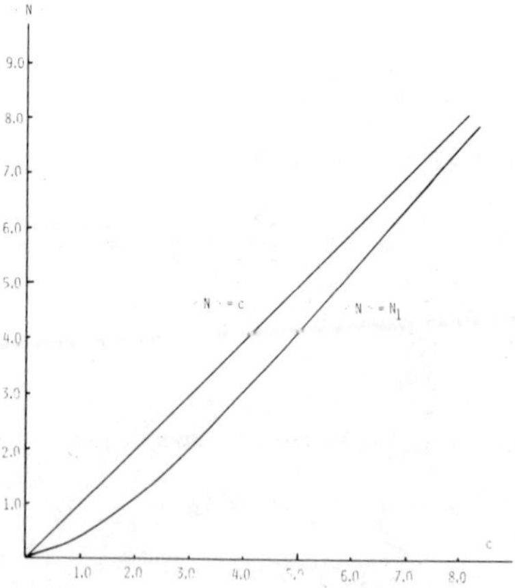

FIG. 1. The mean value of the number operator N in the state $\vec{n}f(\vec{k})$ as a function of c [Eq. (73)] for a coherent state and for a quasicoherent state. The two mean values approach the same value for large c.

This is an expected result.

The probability amplitude for finding n pions with momenta $\vec{k}_1, \ldots, \vec{k}_n$ and isospin indices $\mu_1, \ldots, \mu_n$ can be computed from the formula (57). If we let n_+, n_-, and n_0 denote the numbers of π^+, π^-, and π^0 particles, respectively, we find that $n_+ = n_-$ and that n_0 is even. The corresponding probability $P(n_+, n_-, n_0)$ is

$$P(n_+, n_-, n_0) = \frac{c^n}{\varphi_0(c)} \frac{1}{n_0!} \left[\frac{(n_0 - 1)!!}{(n+1)!!}\right]^2, \quad (77)$$

where $n = n_+ + n_- + n_0$ is the total number of pions. We notice that (74), (75), and (77) do not depend on the direction of the unit vector $\vec{n}$. This fact is obvious when we realize that $|0;R\underline{f}\rangle = |0;\underline{f}\rangle$ $R \in \mathrm{SO}(3)$, according to the definition (72).

As our *second* example of one-particle states we consider the following form of the matrix J:

$$J_{\alpha\beta} = \int \frac{d^3k}{2\omega} f_\alpha(\vec{k})^* f_\beta(\vec{k}) = c\delta_{\alpha\beta} . \quad (78)$$

In this case it is convenient to use the integral representation (27) in order to compute the group invariant $\varphi_0(J)$. The corresponding inverse Laplace transform is elementary and the result is

$$\varphi_0(J) = e^c [I_0(2c) - I_1(2c)], \quad (79)$$

where $I_n(\ )$ is the nth-order modified Bessel function. The expectation value (60) of the number operator is in this case

$$N_2 \equiv \langle 0;\underline{f} | N | 0;\underline{f}\rangle = c \frac{I_1(2c) - I_2(2c)}{I_0(2c) - I_1(2c)} . \quad (80)$$

For the corresponding coherent state $|\underline{f}\rangle$ we easily obtain

$$\langle \underline{f} | N | \underline{f} \rangle = 3c . \quad (81)$$

We find once again that the quasicoherent state $|0;\underline{f}\rangle$ contains less particles than the coherent state $|\underline{f}\rangle$ (see Fig. 2). We notice, however, that $N_2 > N_1$ (if $c \neq 0$) i.e., the orientation in isospin space of the one-particle state is essential for the physical properties of the quasicoherent states. The probability of finding n_i π^i mesons ($i = -0, +$) can be computed by integrating (57) over the momenta $\vec{k}_1, \ldots, \vec{k}_n$ ($n = n_+ + n_- + n_0$). We find that $n_+ = n_-$. The result of the computation is

$$P(n_+, n_-, n_0) = \frac{\delta_{n_+ n_-}}{\varphi_0(J)} \frac{2^{-2n_+ - 1} c^n}{(n_+!)^2 n_0!} \int_{-1}^{1} dx\,(1+x)^{2n_+} x^{n_0} , \quad (82)$$

where $\varphi_0(J)$ is given by (79). We notice that integrals of the form (82) occur in the statistical approach 9 to multiple particle production when isospin conservation is imposed.

V. GLUON BREMSSTRAHLUNG FROM A "CLASSICAL" QUARK LINE

We have constructed the analog of coherent states for [SU(2)] non-Abelian gauge fields. The corresponding construction simplifies if one considers fields carrying an Abelian charge. We refer the reader to Refs. 10 and 7 for a discussion of the one-particle and field-theoretical situations, respectively. The corresponding quasicoherent states can now be used, e.g., in a study[7] of soft emission of charged massless bosons in a scattering process. If one neglects self-interactions among the soft bosons and the quantum structure of their source, then the infrared divergences exponentiate and can be treated as in QED. Examples of the relevant Feynman diagrams are shown in Fig. 3. The emission of self-interacting bosons can, in principle, be investigated in detail by making use of functional techniques. A closed expression for the probability of soft boson emission up to a certain total energy can be written down[11] (also taking the quantum nature of the corresponding source into account). Below we will show that, under the assumptions mentioned above, the emission of soft [SU(2)] gluons from a classical quark current exponentiates in exactly the same way as in QED. Classical quark currents have been found to be useful in the study of the infrared structure of non-Abelian gauge theories[12] as well as in the study of quark and gluon jets in quantum chromodynamics (QCD),[13,14] Frautschi and Krzywicki have discussed the effect of con-

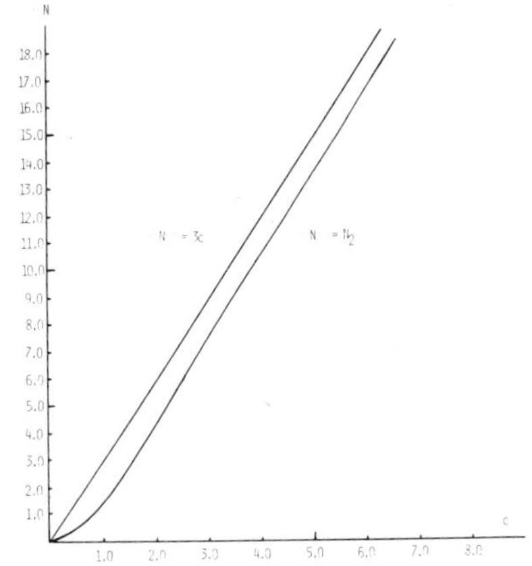

FIG. 2. The mean value of the number operator N is shown as a function of c in the state $f_\alpha(\vec{k})$ satisfying (78) for a coherent state and for a quasicoherent state. The two mean values approach the same value for large c.

finement on gluon bremsstrahlung in e^+e^- annihilation. They argued that the confinement mechanism in QCD may provide a natural infrared cutoff for the bremsstrahlung spectrum. In their analysis the gluon radiation emitted by a quark line is estimated by considering soft emission of photons from a classical current. As is well known, the infrared divergences in QED can be treated in terms of such a classical current.[6] The radiation field is then effectively described by a coherent state. In QCD the situation with regard to soft gluon emission is more complicated due to the self-coupling of the gluons and the color content of the sources. However, it has been shown[15] that properly defined transition probabilities are infrared finite order by order in a renormalized coupling constant. It is possible to develop a classical theory of Yang-Mills particles[16] interacting with non-Abelian fields.[17] The color content of a given particle can, formally, be described in terms of c numbers which after quantization are replaced by the Lie-algebra generators of the gauge group under consideration [SU(2) in our case].

Under the simplifying assumptions mentioned above, we can use the picture suggested by Frautschi and Krzywicki[14] to compute the probability for emission of soft gluons from a classical current,

$$j_\mu^\alpha(\vec{x},t)=j_\mu(\vec{x},t)I^\alpha\,, \tag{83}$$

where

$$j_0(\vec{x},t)=g\delta^3(\vec{x}-\vec{v}t)\,,\qquad \vec{j}(\vec{x},t)=g\vec{v}\,\delta^3(\vec{x}-\vec{v}t) \tag{84}$$

is the conventional form of a classical charged current ($\vec{v}$ is the velocity of the particle) and I^α denotes the classical color degrees of freedom. In general I^α will be time dependent and precess around the gluon field. For the case when the coherent radiation field generated by the "effective current" (83) is proportional to I^α, than we may consider I^α as effectively time independent. Let us consider the (perturbative) vacuum as the initial state. The final state of the soft gluons emitted will then be a coherent state. We have, however, to take color conservation into account. The final state will therefore be a quasicoherent isosinglet state

$$|0;\underline{j}\rangle=\frac{e^{\underline{j}^*\cdot\underline{j}/2}}{\varphi_0(\vec{j}_k^*\vec{j}_k)}\int d^3R\,|R\underline{j}\rangle\,, \tag{85}$$

where $\underline{j}$ is the classical current (83). Following the procedure of Ref. 7 we now compute the transition probability for the source (83) to emit soft gluons with a total energy not exceeding ΔE

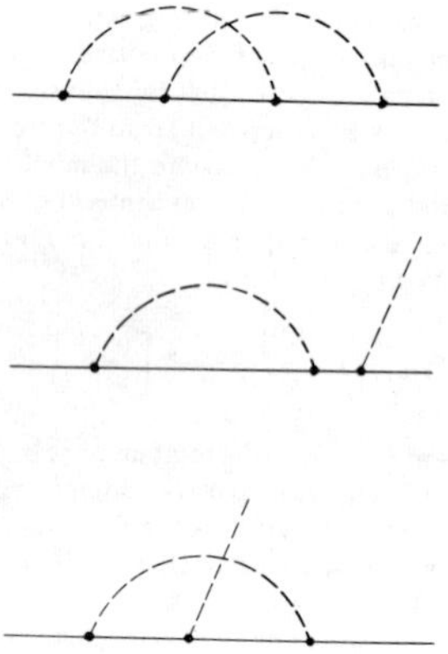

FIG. 3. Some typical Feynman diagrams contributing to the soft gluon emission from a classical quark current are exhibited. The vertices commute due to the classical nature of the current, and no gluon self-interactions are taken into account.

(other energy intervals can be treated analogously). The density operator describing the corresponding energy resolution is given by

$$\rho(\Delta E)=\int d^4p\,\theta(\Delta E-p_0)\delta^4(P-p)\,, \tag{86}$$

where P stands for the momentum operator and θ is the Heaviside step function. We observe that (85) is of the form described by the matrix (73). The transition probability $P(\Delta E)$ can then be evaluated with the result

$$P(\Delta E)=\langle 0;\underline{j}\,|\,\rho(\Delta E)\,|\,0;\underline{j}\rangle =\frac{1}{2\pi}\int_0^{\Delta E}dp_0\int_{-\infty}^{\infty}dt\,e^{-ip_0t}\frac{\varphi_0(c(t))}{\varphi_0(c)}\,, \tag{87}$$

where

$$c(t)=\vec{I}^2\int\frac{d^3k}{2\omega}e^{i\omega t}j_\mu(\vec{k})^*j_\mu(\vec{k}) \tag{88}$$

and

$$c=\vec{I}^2\int\frac{d^3k}{2\omega}j_\mu(\vec{k})^*j_\mu(\vec{k})\,. \tag{89}$$

In the analysis by Frautschi and Krzywicki, (88) and (89) are finite integrals (with an appropriate ultraviolet cutoff) due to the confinement mechanism. It is instructive to compare (87) with the corresponding result in QED (see, e.g., Ref. 7 and references cited therein). Then (88) and (89) are infrared divergent. By making use of (74) and its asymptotic form we then obtain

$$P(\Delta E)=\frac{1}{2\pi}\int_0^{\Delta E} dp_0 \int_{-\infty}^{\infty} dt\, e^{-ip_0 t} \exp\left[\bar{g}^2 \int \frac{d^3k}{2\omega}(e^{i\omega t}-1) j_\mu(\vec{k})^* j^\mu(\vec{k})\right] \quad (90)$$

which is finite and of the same form as the corresponding expression in QED.

In a more refined analysis one could (in the leading logarithm approximation) take self-interactions into account by replacing the coupling constant g in (84) by a running coupling constant as in Refs. 12 and 13.

VI. FINAL REMARKS

In Ref. 7 one of the present authors showed how quasicoherent states can be constructed in Abelian field theory where a conserved Abelian charge is present. Owing to superselection rules conventional coherent states are not appropriate as was noted by Bhaumik *et al.*[10] Here we have demonstrated a similar construction for [SU(2)] non-Abelian charges and for one-particle states (gauge bosons or pions) transforming under the adjoint representation of SU(2).

This is also a generalization of Ref. 1 which deals with the case of one single available kinematical state, and our construction also extends the work by Botke *et al.*[5]

From our presentation of the construction of quasicoherent states it is clear that our results can be extended to any compact group. Work on this extension is in progress.[18]

The construction in Sec. II of quasicoherent states can also easily be carried over to the case when the one-particle states transform according to the fundamental representation of SU(2) (appropriate for K mesons or in general two-level systems.[19,20] The corresponding complete set of states is by definition

$$|{}^{l}_{\mu\nu};\underline{x}\rangle=(2l+1)^{1/2}e^{\underline{x}^*\cdot\underline{x}/2}\int dg\, D^l_{\mu\nu}(g)^*\,|g\underline{x}\rangle\,. \quad (91)$$

The isospinor integral that corresponds to $I(M)$ in (22) and (28) for isovectors now depends only on one variable.[21] (See Appendix B.) Most of the results in Secs. II and III can now easily be carried over to the set of states given by (91).

We expect that quasicoherent states constructed in the present paper also may be a useful complete set of states with regard to physical applications. In Secs. IV and V we have indicated some properties of these states as well as some physical situations where they describe relevant properties of the system under consideration. The method can clearly be applied also to more complex situations.

With regard to the study of gluon condensates[22] and quark condensation[23] in QCD, quasicoherent states may simplify the analysis [when extended to the SU(3) case[18]]. We intend to study these questions elsewhere. Finally, it is amusing to notice that invariant integrals of the form (22) or (B9) frequently occur in the analysis of gauge-field theories in the Wilson lattice approach.[21]

ACKNOWLEDGMENTS

The authors wish to thank A. P. Balachandran for valuable discussions. One of the authors (B. S. S.) wishes also to thank R. Stora for several constructive remarks and discussions.

APPENDIX A

In this appendix we shall show how $D^l_{\mu\nu}(M)$ as homogeneous polynomials (23) are obtained.

Let

$$U(R)=e^{-i\vec{l}\cdot\vec{\theta}} \quad (A1)$$

be a unitary rotation operator. The Winger D functions are defined by

$$D^l_{\mu\nu}(R)=\langle l\mu|U(R)|l\nu\rangle\,. \quad (A2)$$

Then[1]

$$Y^l_\mu(R\vec{e})=D^l_{\mu\nu}(R)^* Y^l_\nu(\vec{e})\,, \quad (A3)$$

i.e.,

$$D^l_{\mu\nu}(R)=\frac{1}{4\pi}\int d\Omega(\vec{e})Y^l_\mu(R\vec{e})^* Y^l_\nu(\vec{e})\,. \quad (A4)$$

Since $Y^l_\mu(e)$ is a homogeneous polynomial in $\vec{e}$ by equation (24) we obtain

$$D^l_{\mu\nu}(R)=\frac{1}{(l!)^2}(t^l_\mu)^*_{\alpha_1\cdots\alpha_l}(t^l_\nu)_{\beta_1\cdots\beta_l}R_{\alpha_1\alpha_1'}\cdots R_{\alpha_l\alpha_l'}$$
$$\times\frac{1}{4\pi}\int d\Omega(\vec{e})e_{\alpha_1'}\ldots e_{\alpha_l'}e_{\beta_1}\ldots e_{\beta_l}\,. \quad (A5)$$

The integral in equation (A5) can easily be evaluated and the result is

$$D^l_{\mu\nu}(R)=\frac{2^l}{(2l+1)!}(t^l_\mu)^*_{\alpha_1\ldots\alpha_l}(t^l_\nu)_{\beta_1\ldots\beta_l}R_{\alpha_1\beta_1}\cdots R_{\alpha_l\beta_l}\,, \quad (A6)$$

which can now be extended to be valid for *any* 3×3 matrix.

APPENDIX B

In this appendix we shall derive an explicit expression for the invariant integral

$$I(M)=\int_{SO(3)} d^3R\, e^{\mathrm{Tr}(RM^\sim)} \quad (B1)$$

in terms of the three invariants

$$x = \mathrm{Tr}(MM^{\sim}), \quad y = 4\,\mathrm{Det}M,$$
$$z = \tfrac{1}{2}[\mathrm{Tr}(MM^{\sim})]^2 - \mathrm{Tr}(MM^{\sim}MM^{\sim}). \tag{B2}$$

Now $I(M)$ is an analytic function of M. We can therefore assume that M is real and *generic*, i.e., M is such that x, y, z are all nonzero. We therefore consider the case when M is a diagonal matrix:

$$M = \begin{pmatrix} m_1 & 0 & 0 \\ 0 & m_2 & 0 \\ 0 & 0 & m_3 \end{pmatrix}. \tag{B3}$$

SO(3) is the adjoint group of SU(2). Since, topologically, $\mathrm{SU}(2) \approx S^3$ we can therefore rewrite the invariant integral (B1) as follows:

$$I(M) = \pi^{-2} \int d^4u\, \delta(u^2-1) \exp\left[\tfrac{1}{2} \sum_{i=1}^{3} m_i \mathrm{Tr}(\vec{\sigma}\cdot\vec{e}_i u \vec{\sigma}\cdot\vec{e}_i u^\dagger)\right]. \tag{B4}$$

Here $\{\vec{e}_i\}$ is an orthonormal basis in R^3 and $\sigma = (\sigma_1, \sigma_2, \sigma_3)$ are the three Pauli matrices; u is a general element of SU(2), i.e., $u = u_0 \cdot 1 + i\vec{\sigma}\cdot\vec{u}$ and $u^2 = u_0^2 + \vec{u}^2$. For the δ function we use the integral representation

$$\delta(u^2-1) = \frac{1}{2\pi} \int_{-\infty}^{\infty} d\xi\, e^{i\xi(u^2-1)}. \tag{B5}$$

The u integral in (B4) is then a Gaussian integral which is easily evaluated with the result ($s = i\xi$)

$$I(M) = \frac{1}{2\pi i} \int_{s_0 - i\infty}^{s_0+i\infty} ds\, e^s f(s; m_1, m_2, m_3), \tag{B6}$$

where

$$f(s; m_1, m_2, m_3) = [s^4 - 2(m_1^2 + m_2^2 + m_3^2)s^2 - 8m_1m_2m_3 s + m_1^4 + m_2^4 + m_3^4 - 2(m_2^2m_3^2 + m_3^2m_1^2 + m_1^2m_2^2)]^{-1/2} \tag{B7}$$

and s_0 has to be chosen in such a way that all singularities of $f(s, m_1, m_2, m_3)$ are to the left of the integration contour. Inserting (B7) into (B6) and expressing the polynomials in m_1, m_2, m_3 in terms of the invariants (B2) for the matrix (B3) we finally obtain

$$I(M) = \frac{1}{2\pi i} \int_{s_0-i\infty}^{s_0+i\infty} ds\, e^s (s^4 - 2xs^2 - 2ys - 2z)^{-1/2}. \tag{B8}$$

By analyticity (B8) is true for any 3×3 matrix M. Expanding (B8) in terms of x/s^2, y/s^3, and z/s^4 and evaluating the integral term by term we obtain the expansion (28).

For *isospinor* bosons the relevant matrix is a 2×2 matrix m and the integral corresponding to $I(M)$ in (22) and (28) is

$$K(m) = \int_{\mathrm{SU}(2)} dg\, e^{\mathrm{Tr}(gm^{\sim})}. \tag{B9}$$

By invariance arguments it may be shown that $K(m)$ is a function only of Det m. One obtains

$$K(m) = \frac{I_1(2(\mathrm{Det}M)^{1/2})}{(\mathrm{Det}M)^{1/2}} = \sum_{n=0}^{\infty} \frac{(\mathrm{Det}m)^n}{n!(n+1)!}. \tag{B10}$$

---

*Permanent address: Indian Institute of Science, Bangalore 560012, India.

†Present address: CERN, Geneva, Switzerland.

[1]K.-E. Eriksson and B.-S. Skagerstam, J. Phys. A 12, 2175 (1979); 14, 545(E) (1981).

[2]J. M. Radcliffe, J. Phys. A 4, 313 (1971); A. M. Perelemov, Commun. Math. Phys. 26, 222 (1972). See also M. M. Nieto, Phys. Rev. D 22, 391 (1980); M. M. Nieto and L. M. Simmons, Jr., *ibid.* 20, 1321 (1979); 20, 1332 (1979); 20, 1342 (1979); V. P. Gutschick and M. M. Nieto, *ibid.* 22, 403 (1980); 23, 922 (1981), and references therein; F. T. Hioe, J. Math. Phys. 15, 445 (1974).

[3]A detailed discussion is given in J. R. Klauder and E. C. G. Sudarshan, *Fundamentals in Quantum Optics* (Benjamin, New York, 1968) and in J. R. Klauder, Acta Phys. Austriaca, Suppl. XVII, 1 (1977). See also B. De Facio and C. L. Hammer, J. Math. Phys. 16, 267 (1976); C. L. Hammer *et al.*, Phys. Rev. D 18, 373 (1978); 19, 667 (1979); J. R. Klauder, *ibid.* 19, 2349 (1979).

[4]See, e.g., J. D. Talman, *Special Functions—A Group Theoretic Approach* (Benjamin, New York, 1968).

[5]J. C. Botke, D. J. Scalapino, and R. L. Sugar, Phys. Rev. D 9, 813 (1974).

[6]See, e.g., T. W. B. Kibble, in *Mathematical Methods in Theoretical Physics* (Gordon and Breach, New York, 1969), Vol. XID.

[7]B.-S. Skagerstam, Phys. Rev. D 19, 2471 (1979); 22, 534 (1980).

[8]D. Horn and R. Silver, Ann. Phys. (N.Y.) 66, 509 (1971).

[9]See, e.g., D. Horn and F. Zachariasen, *Hadron Physics at Very High Energies* (Benjamin, New York, 1973), Chap. 17 and references cited therein.

[10]D. Bhaumik, K. Bhaumik, and B. Dutta-Roy, J. Phys. A 9, 1507 (1976).

[11]K.-E. Eriksson and B.-S. Skagerstam, Phys. Rev. D 18, 3858 (1978).

[12]M. Greco, F. Palumbo, G. Pancheri-Srivastava, and Y. Srivastava, Phys. Lett. 77B, 282 (1978); G. Curci and M. Greco, *ibid.* 79B, 406 (1978).

[13]G. Curci, M. Greco, and Y. Srivastava, Nucl. Phys. B159, 451 (1979).

[14]S. Frautschi and A. Krzywicki, Z. Phys. C 1, 43

(1979).

[15]Y.-P. Yao, Phys. Rev. Lett. 36, 653 (1976); T. Appelquist, J. Carazzone, H. Kluberg-Stern, and M. Roth, *ibid.* 36, 768 (1976); T. Kinoshita and A. Ukawa, Phys. Rev. D 13, 1573 (1976); 13, 1977 (1976); 15, 1596 (1977); E. Poggio and H. Quinn, *ibid.* 14, 578 (1976).

[16]S. K. Wong, Nuovo Cimento 65A, 689 (1970).

[17]A. P. Balachandran, P. Salomonson, B.-S. Skagerstam, and J.-O. Winnberg, Phys. Rev. D 15, 2308 (1977); A. Barducci, R. Casalbuoni, and L. Lusanna, Nucl. Phys. B124, 93 (1977); A. P. Balachandran, S. Borchardt, and A. Stern, Phys. Rev. D 17, 3247 (1978).

[18]K.-E. Eriksson, N. Mukunda, and B.-S. Skagerstam, in preparation.

[19]F. T. Arecchi, E. Courtens, R. Gilmore, and H. Thomas, Phys. Rev. A 6, 2211 (1972); K. Hepp and E. H. Lieb, *ibid.* 8, 2517 (1973); L. M. Narducci, C. M. Bowden, V. Bluemel, G. P. Garrazana, and R. A. Tuft, *ibid.* 11, 973 (1975).

[20]R. Holtz and I. Hanus, J. Phys. A 7, L37 (1974); D. H. Feng, L. M. Narducci, and R. Gilmore, in Proceedings of the 1978 International Meeting on Frontiers of Physics, Singapore, edited by K. K. Phua, C. K. Chew, and Y. K. Lim (unpublished). In this context see also D. Janssen, Yad. Fiz. 25, 897 (1977) [Sov. J. Nucl. Phys. 25, 479 (1977)].

[21]See, e.g., M. Creutz, Rev. Mod. Phys. 50, 561 (1978); K.-E. Eriksson, N. Svartholm, and B.-S. Skagerstam, Report No. TH-2974-CERN (unpublished) and references cited therein.

[22]P. Houston and D. Pottinger, Z. Phys. C 3, 83 (1979) and references cited therein; A. B. Migdal, Zh. Eksp. Teor. Fiz. Pis'ma Red. 28, 37 (1978) [JETP Lett. 28, 35 (1978)].

[23]J. Finger, D. Horn, and J. E. Mandula, Phys. Rev. D 20, 3253 (1979).

Volume 127B, number 3,4 PHYSICS LETTERS 28 July 1983

CORRELATIONS AND FLUCTUATIONS IN HADRONIC MULTIPLICITY DISTRIBUTIONS: THE MEANING OF KNO SCALING

P. CARRUTHERS and C.C. SHIH [1]
Theoretical Division, Los Alamos National Laboratory, Los Alamos, NM 87545, USA

Received 27 April 1983

Charged hadronic multiplicity distributions at Fermilab, ISR and collider energies are shown to be described by generalized Bose–Einstein distributions of the type encountered in quantum optics. Systematic energy-dependent deviations from KNO scaling are well described by simple cases of this formula, which in the limit of large $\bar{n}$ have a simple and explicit scaling form, valid to a few percent at the CERN collider energy. It is suggested that this framework is generic, i.e., practically independent of dynamical details. The essential feature is that the field variables of the "radiant surface" for hadronic emission are distributed as gaussian random variables (with k effective sources) as suggested by a typical central limit theorem argument. In order to bring the lower energy data into a form easily comparable with higher energy data, particularly for the crucial large multiplicity events $(n/\bar{n}) \gg 1$, it is essential to remove the two charges of the "leading" particles from the counting rules. The long tail of the KNO distribution is to be identified with coherent fluctuations analogous to those occurring in the Hanbury-Brown–Twiss effect.

Recent measurements [1–3] of charged multiplicity distributions at the CERN $\bar{p}p$ collider have confirmed that approximate Koba–Nielsen–Olesen (KNO) scaling [4] persists to extremely high energies (540 GeV CM energy). Although this effect has been investigated from many dynamical-geometrical points of view, the phenomenon itself seems much more precise than any theoretical explanation yet brought forth [5]. In addition, the detailed shape of the distribution is not satisfactorily understood, although several expressions have been proposed [6,7].

Recall that the KNO "plot" relates $\bar{n}P_n$ (n being the number of charged secondaries) to the variable $z \equiv n/\bar{n}$. On the basis of Feynman scaling it was further suggested [4] that $\bar{n}P_n$ itself is a function $\psi(n/\bar{n})$ alone. Such scaling should be expected to hold only asymptotically, although the detailed form of $\bar{n}P_n$ at various energies may give information on the underlying mechanism. In this note we restrict our attention to purely hadronic systems, which differ in important ways from hadron production in processes involving one or two initial leptons, and from $p\bar{p}$ annihilation.

We wish to propose a generic mechanism practically independent of dynamical details, modelled on the theory of partial coherence. The fundamental assumption, expressed in classical terms (or in coherent state language), is that the fields emerging from the sources behave as gaussian random variables [8–11] [‡1]. It is well known that the photoelectron counting distributions corresponding to sources whose fields are distributed as gaussian random variables satisfy KNO scaling for large mean multiplicities. Remarkably, we have found that the prototype of such distributions – the generalized Bose–Einstein distribution with k cells – describes data from Fermilab through collider energies, including scaling violations. The especially interesting large multiplicity fluctuations $n/\bar{n} \gg 1$ are then claimed to result from coherent many particle fluctuations of the type arising in intensity interferometry [14].

[1] On leave from the University of Tennessee, Knoxville, TE 37916, USA.

[‡1] Sugar and Scalapino [12] have expressed a similar point of view though their formalism is quite different. See also ref. [13].

Volume 127B, number 3,4 PHYSICS LETTERS 28 July 1983

Two particle Bose–Einstein correlations have already been observed [15,16] in a coherence "bin" of pseudorapidity $\Delta\eta \sim 1$ (η is ln tan $\theta/2$ where θ is defined relative to the collision axis). Another clue pointing in this direction is that the angular coherence expected in the coherence function $\Gamma(r, t, r't')$ is (not surprisingly) of order $\Delta\eta \sim 1$. The precise result depends slightly on the geometry of the source. For a disc source of uniform brightness the coherence extends to the first zero of the Bessel function $J_1(x)$, where [10]

$$x_0 = kbr/R \approx 3.8 , \qquad (1)$$

k is the typical momentum of the radiation, b the radius of the disc, R the distance from source to detector and $r = |r - r'|$ the separation of two quanta at the detector. A similar formula holds for a rectangular source.

Setting $k = k_\perp/\sin\theta$ and $r/R \approx \cot\theta$ we find the angle θ (measured from the longitudinal axis) corresponding to the condition (8). Typical $k_\perp b$ values are expected to be ~1,2,3 giving pseudo-rapidity of 1.4, 1.0, 0.8. Soft quanta cohere over the entire solid angle. Therefore, the arbitrary cut $|\eta| < 1.5$ commonly used by experimentalists to define the central region coincides reasonably with the width in η in which a particle produced at 90° coheres with another like particle.

From the foregoing, we expect the η distributions of soft secondaries to be spanned by a small number of basic coherence intervals $\eta_0 \sim 1$. The many particle state should exhibit extensive coherence over the whole range. In particular the central region cut used in ISR and collider data correspond to about two coherence units. The entire interval (excluding leading particles and diffractive multiplicities) is expected to require four to six coherence units η_0.

We suggest that the observed hadrons are emitted from the radiant surface of a tiny chaotic source (rather set of sources) created in the hadronic collision. Given this general framework it is not necessary to have a detailed understanding of the nature of the fields. Thus detailed dynamical models, though not necessarily wrong, miss the essence of the phenomenon. The point is that whatever the nature of the fields (which could be four velocity fields, or entropy four-currents in a statistical hydrodynamical picture) they are distributed in a chaotic, probably gaussian manner. Perhaps it is worth noting that this picture bears a qualitative resemblance to Heisenberg's original "turbulence" model [17] of particle production.

The Poisson distribution

$$P_n = \bar{n}^n e^{-\bar{n}}/n! , \qquad (2)$$

fails to describe the hadronic multiplicity distribution. LeBellac et al. [5] showed that a superposition of Poisson distributions (motivated by dynamical models) greatly improved agreement with the data. A more general superposition which is oriented towards averaging over field fluctuations, is common in quantum optics. It is possible to give simple analytic expressions for these superpositions (in the case of gaussian distributions) for counting times short, and long compared to the coherence time τ. For a single source the averaging gives the Bose–Einstein distribution

$$P_n = \bar{n}^n/(1 + \bar{n})^{n+1} . \qquad (3)$$

Now suppose we have k independent sources with identical intensity distributions whose fields are gaussian random variables. The formula is then [10]

$$P_n^{(k)} = \frac{(n + k - 1)!}{n!(k - 1)!} \left(\frac{\bar{n}/k}{1 + \bar{n}/k}\right)^n \frac{1}{(1 + \bar{n}/k)^k} \qquad (4)$$

and the variance is given by

$$\overline{n^2} - \bar{n}^2 = \bar{n}(1 + \bar{n}/k) . \qquad (5)$$

The parameter k is sometimes referred to as the number of cells. Eq. (4) is the distribution of Bose–Einstein quanta among k cells with equal a priori probability, a result apparently first obtained by Planck [18,19].

Already in 1974 Knox [20] pointed out that the Bose–Einstein formula (4) worked quite well for Fermilab data for both P_n and the variance, with $k = 6$. This suggestion does not seem to have attracted much attention, partly because the scaling behavior of (4) was not derived (see below).

We note that (4) predicts for large n, fixed $z = n/\bar{n}$ the KNO forms

$$\bar{n}P_n^{(k)} \sim \psi_k(z) = \frac{k^k}{(k - 1)!} z^{k-1} e^{-kz} . \qquad (6)$$

We shall explicitly use in our comparison

$$\psi_2(z) = 4z\, e^{-2z} , \quad \psi_4(z) = \tfrac{2}{3}(4z)^3 e^{-4z} \qquad (7)$$

The asymptotic form is an accurate rendition of (4)

(within 5%) at collider energies. At lower energies, (4) is considered to be the correct formula to which KNO scaling is but an asymptotic approximation ‡2.

Some remarks are required to explain our analysis of the data. First of all, eq. (4) is defined for all non-negative integers, whereas the experimental analysis uses $n = 2,4,6, \ldots$ when all charges are detected. (Here we shall not discuss the small $n = 0$ component and its interpretation.) When P_n^{ex} is defined as σ_n/σ_{in}, $n = 2,4,6, \ldots$ the area of the P_n^{ex} plot is 2. Second, we affirm that $\bar{n}$ in eq. (4) is the total charged multiplicity, subject to leading-particle subtractions discussed below. In order to use (4) for even integers only, we have derived suitable analytic expressions (which show a small change in normalization and in the moments from the corresponding all-integer quantities.) These are used where appropriate. When incomplete detection is the case (e.g., by a small solid angle acceptance), then odd integer events occur. Often it is appropriate to simply average over a range of multiplicities, as has been done in recent collider experiments. For uniform treatment we shall always normalize the data so that the area under the curve is unity.

We begin with the (least ambiguous) evidence at the highest available energy (540 GeV in the CM). The first result is that the central region (defined somewhat ambiguously as $|\eta| < 1.3-1.5$), which has a different form than that seen if the full nondiffractive range ($|\eta| < 5$) is included, agrees well with (4) for $k = 2$. Fig. 1 shows central region data from ISR and collider experiments for $|\eta| < 1.3-1.5$. The solid curve is eq. (7) for $k = 2$. In order to examine the especially interesting high multiplicity "tail" it is useful to use a log plot (which conceals the difference among different proposed distributions in the peak region). Fig. 2 compares UA1 data for $|\eta| < 1.5$ for $z \lesssim 4$ with the $k = 2$ formula.

Most data represent all angle (or almost all angle) integrated probabilities. For brevity of presentation and in order to exhibit our predicted deviations from KNO scaling we shall present collider data along with ISR and Fermilab results. The principal result from

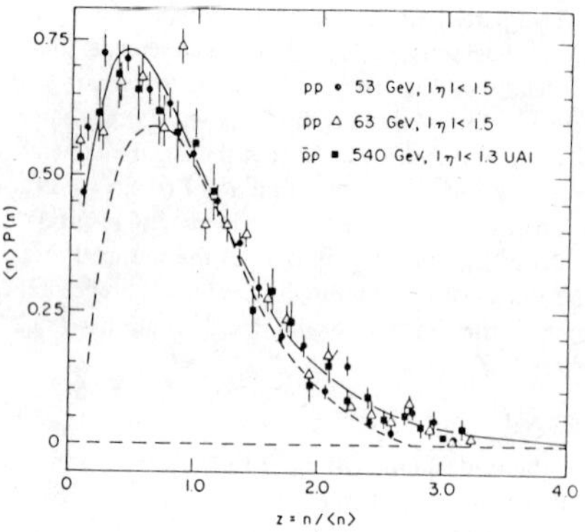

Fig. 1. Central region ISR [21] and collider data (UA1) [1] are compared with the $k = 2$ case of the generalized Bose–Einstein formula (4) using a linear vertical scale. The solid curve corresponds to eq. (6) with $k = 2$ and the dashed line is the De Groot empirical formula [6].

the "all η" collider data is that they are well described by the $k = 4$ distribution. Since $\bar{n} \sim 26-29$, the asymptotic formula (5) is reliable; indeed the shape of the curve $\psi_4(z)$ is noticeably different from $\psi_2(z)$ in the peak (near 1/2 for ψ_2 and 3/4 for ψ_4) and asymptotic $z \gg 1$ regions.

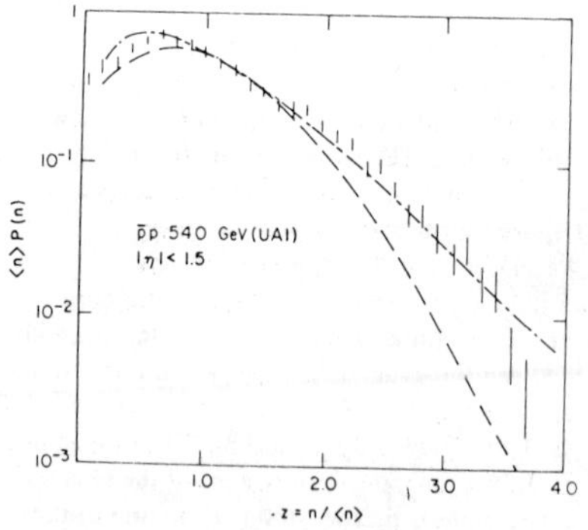

Fig. 2. Similar collider data (UA1) [3], extending to higher z, are shown using a logarithmic scale to bring out the details of the $z \gg 1$ region.

‡2 Liu and Meng have already [7] used ψ_2 and ψ_6 in data fits, on the basis of a completely different physical model based on k independent harmonic oscillators. Mathematically, however, there is a clear connection with our approach in terms of counting.

Volume 127B, number 3,4 PHYSICS LETTERS 28 July 1983

Our next point is to emphasize that the discrete $k = 4$ formula, eq. (4) also accounts for the lower energy data provided that the contribution of the leading particles is deducted from the effective charge contributing to the observed total multiplicity. This point will be stressed later in a different way when we discuss the moments of the distributions [‡3]. We therefore assume that the two incident particles (charge magnitude 2) become leading particles not associated with (4) and plot $(\bar{n} - 2) P_n$ versus $z_2 = (n-2)/(\bar{n}-2)$ for even n. The effect is only 7% in height at collider energy but is very significant at Fermilab energies. This shift brings order to the data, especially the high multiplicity Fermilab data. The moment data also strongly support this interpretation. In fig. 3 the peak region is displayed (on a linear scale) in order to exhibit the energy (more precisely $\bar{n}$) dependent scaling violations. Notice, however, the slight discrepancy near the peak for Fermilab energies, which can be fixed by using $k = 3$ at the price of hurting the large z region. One should keep in mind that at Fermilab energies the energy momentum constraints probably suppress the tail in a way not taken into account in (4). Considering various factors, however, we consider the predicted "scaling violations" to agree quite well with the predicted values of eq. (4). Fig. 4 shows the extreme sensitivity of the large z regime to the average multiplicity. Again, $k = 4$ works very well, with order of magnitude "scaling violations" for $z \gg 1$, much larger than in the peak region.

[‡3] The idea that KNO scaling is more easily recognized (alternatively, a description of violations thereof) goes back ten years [22]. If the mechanism of leading particle/diffractive excitation separates from the principal "soft" hadronization-process, then it makes sense to plot not $\bar{n}P_n$ versus $n/\bar{n}$ but $\bar{n}_\alpha P_n$ versus $n_\alpha/\bar{n}_\alpha$ where $z_\alpha \equiv n_\alpha/\bar{n}_\alpha = (n - \alpha)/(\bar{n} - \alpha)$ and α is the average number of leading particles not subject to the statistical formula (4). Offhand one would expect $1 \lesssim \alpha \lesssim 2$ where the higher value is most likely at very high energy. Early fits indeed prefer $1 < \alpha < 1.3$ at lower energy, which could indicate that at lower energy one of the incident particles participates in the hadronization by failing to qualify as a leading particle. Those fits emphasize the peak region, and we find that in the $z \gg 1$ domain, the choice $\alpha = 2$ is superior. Still, one can see that energy-momentum considerations, not considered in (4), could suppress the large z region in a way not incompatible with $\alpha \sim 1$. The possible energy dependence of α will be analyzed elsewhere, paying attention to issues too lengthy to be addressed in this note.

Another measure of the multiplicity distributions predicted by eq. (4) is given by the moments (computed over all integers)

$$\gamma_2 \equiv \langle (n - \bar{n})^2 \rangle / \bar{n}^2 = 1/k + 1/\bar{n} ,$$

$$\gamma_3 \equiv \langle (n - \bar{n})^3 \rangle / \bar{n}^3 = 2/k^2 + 3/k\bar{n} + 1/\bar{n}^2 ,$$

$$\gamma_4 \equiv \langle (n - \bar{n})^4 \rangle / \bar{n}^4 - 3\gamma_2^2$$

$$= 6/k^3 + 12/k^2\bar{n} + 7/k\bar{n}^2 + 1/\bar{n}^3 . \quad (8)$$

In order to compare with data, these formulas have been modified by restriction to even integers and use of the experimental cutoff n_c in the observed n. The highest moments deviate most from the unconstrained formula (4), due to the cutoff in n.

Eq. (8) gives an extremely simple prediction of the γ_n, which quantities are usually poorly predicted by models. In order to compare with data according to our previous philosophy, viz, that $\bar{n} - 2$ is the appropriate quantity for the nondiffractive multiplicity question, we need to instead compare with

$$\tilde{\gamma}_2 = \gamma_2 [\bar{n}/(\bar{n} - 2)]^2 , \quad \tilde{\gamma}_3 = \gamma_3 [\bar{n}/(\bar{n} - 2)]^3 ,$$

$$\tilde{\gamma}_4 = \gamma_4 [\bar{n}/(\bar{n} - 2)]^4 . \quad (9)$$

In order to condense the lengthy analysis required to compare our moment predictions with experiment we state some conclusions. A detailed analysis which eliminates the odd component of (4) confirms that the "smooth" interpretation of $\psi(z)$ is rather accurate when we use only even n rather than all n. However, the effect of a "cut" whereby $n \geqslant n_c$ contributions are not included is very significant for the higher moments.

Before discussing the all-charge data we emphasize the crucial role of eliminating leading particles from the counting rules at Fermilab and ISR energies. Experiments measuring γ_2 reveal a factor 2 difference between negative particle measurements and all-charge data, when *all* particles are counted. However, using the modified moments (9) we find that $\tilde{\gamma}_j \simeq \gamma_{\bar{j}}, j = 2,3,4$.

To complete the moment data analysis based on the foregoing, we present in table 1 a comparison with experiment. The main features to note here include the effects of (a) leading particle subtraction, (b) coincidence with negative particle correlations with sub-

Volume 127B, number 3,4 PHYSICS LETTERS 28 July 1983

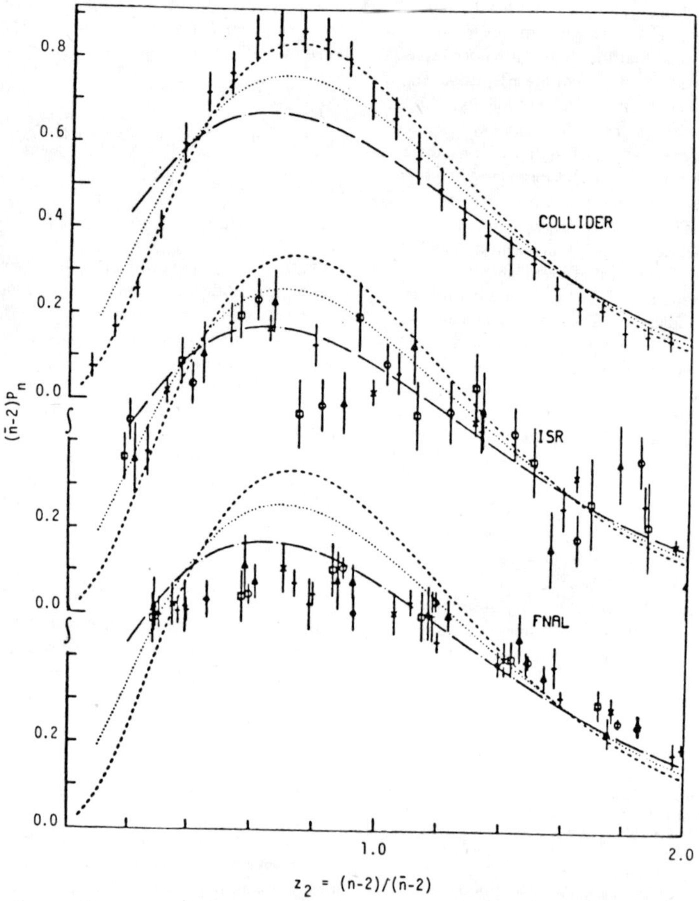

Fig. 3. From top to bottom, the plot shows collider data (UA5) [2], ISR [21] and Fermilab [23] data, compared with theoretical predictions from (4) with $k = 4$, after subtracting 2 from the total multiplicity to account for the leading particles. The dashed, dotted, and dash–dotted curves correspond, respectively, to the average multiplicities $\bar{n} = 27$, 10 and 5.

tracted all-charge correlations, (c) effect of cutoff on the moments, and (d) general agreement with the $\bar{n}$ and k dependence already surmised from the preceding analysis. The worst agreement is with ISR data, which typically exhibit the largest experimental errors. Results are not so good for γ_4, which is very sensitive.

We briefly address several issues connected with the basic formula (4) and its use. (a) First note that (4) is applied to charged particles only. However, if we assume that (4) applies to the total distribution with $n = n_{ch} + n_0$, fold with the binomial distribution, and sum over all n_0, we find again eq. (4) the same k and

Volume 127B, number 3,4 PHYSICS LETTERS 28 July 1983

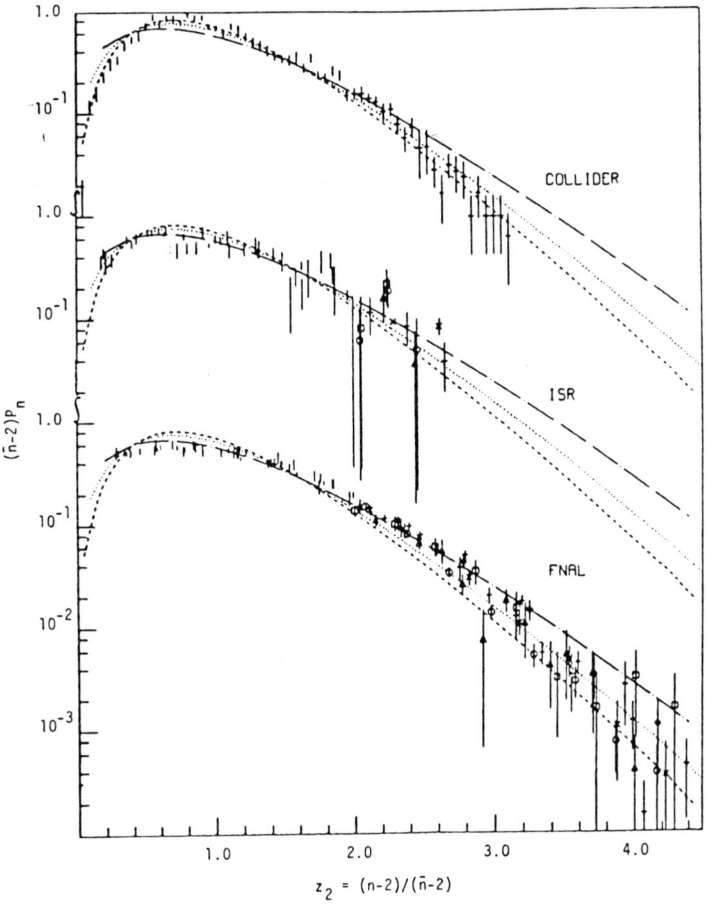

Fig. 4. Same as fig. 3 for collider data (UA1) [3], ISR [21] and Fermilab [23] data. The log scale has been used to better exhibit the region of large fluctuations. The dashed, dotted and dash–dotted curves correspond respectively to the average multiplicities $\bar{n}$ = 20, 10, 5.

with $n = n_{ch}$ alone; i.e., the distribution (4) is form invariant under elimination of the neutral component of the distribution by summation. (This remarkable property is shared by the Poisson distribution.) (b) The effect of selection of n = even on normalization and on moments was mentioned above, as was the effect of a cut in n on the moment predictions. (c) The energy–momentum constraints have not yet been included in our analysis. We expect these to be negligible at collider energy but significant in the large n tail at Fermilab energies. (d) Eq. (4) assumes k sources of *equal* strength. Until we understand the nature of these

Table 1
Comparison between experimental multiplicity moments and theoretical four-cell predictions at FNAL [23], ISR [21] and collider [1–3] energies. All values of moments are multiplied by 1×10^2.

	p(GeV/c) [FNAL]									W(GeV) [ISR]					W(GeV) [Collider]	
	102	147	147	147	205	300	303	360	405	23	31	45	53	63	540	540
$\bar{n}$	6.32	7.02	7.09	7.41	7.68	8.50	8.86	8.73	8.99	8.12	9.54	11.01	11.77	12.70	21.1	28.9
γ_2(exp) [a]	24.5	27.1	24.3	23.6	24.6	24.8	24.3	24.4	28.5	24.9	25.6	28.7	29.5	30.0	29.6	28.2
$\gamma_{\bar{2}}$(exp) [b]	52.2				45.1	42.5	40.8		45.9							
$\tilde{\gamma}_2$(exp) [c]	52	53	47	44	45	42	41	41	46	44	41	43	43	42	30	33
γ_2(cut) [d]	50	46	46	44	43	41	39	40	40	39	36	33	32	30	28	28
γ_2(all) [e]	51	46	46	45	44	41	40	40	40	42	39	36	35	34	30	29
γ_3(exp) [a]	7.9	10	7.4	6.6	8.4	8.0	8.3	6.8	11.4	9.4	9.0	12.7	12.7	12.5	12.2	15.2
$\gamma_{\bar{3}}$(exp) [b]	24.5				20.3		17.5		23.3							
$\tilde{\gamma}_3$(exp) [c]	25	27	20	17	21	18	18	15	24	22	18	23	22	21	12	19
γ_3(cut) [d]	35	31	30	29	28	25	23	25	25	18	16	13	12	10	11	14
γ_3(all) [e]	36	32	32	30	29	27	27	26	25	28	24	22	21	20	16	15

a) The experimental γ_j are calculated from P_n^{ex} using the *definitions* in eq. (8).
b) See eq. (8) for definition of experimental $\gamma_{\bar{j}}$ with negatives only.
c) See eq. (9) for definition of experimental $\tilde{\gamma}_j$ after leading particle subtraction.
d) Theoretical prediction of γ_j using eq. (8) with n = experimentally allowed even integers, and $\bar{n} \to \bar{n} - 2$.
e) Theoretical prediction of γ_j using eq. (8) with n = all even integers, and $\bar{n} \to \bar{n} - 2$.

sources it is difficult to assess this assumption.

The phenomenological success of the compound Bose–Einstein formula (4) points to an identification of the k "cells" with a set of k independent harmonic oscillator variables. Using conventional coherent state concepts [8–10], one sees that (4) corresponds to the projection onto the n quantum sector of the density matrix

$$\rho_k(\{z_i\}) = \frac{1}{\pi^k} \int \prod_{i=1}^{k} d^2 z_i\, \phi(\{z_i\}) |\{z_i\}\rangle\langle\{z_i\}| \,,$$

$$\phi_k(\{z_i\}) = \frac{1}{(\bar{n}/k)^k} \exp\left(-\sum_{i=1}^{k} |z_i|^2/(\bar{n}/k)\right) , \qquad (10)$$

i.e., $P_n^{(k)} = \langle n|\rho_k|n\rangle$. Here z_i is the complex variable parametrizing the coherent state $|z_i\rangle$. Thus in the coherent state basis the weight is gaussian in the variables z_i, corresponding to the classical description in a well-known way [8–13].

Our analysis, which has dealt with integrated multiplicity information (with the exception of the central region) can be interpreted as equivalent to a system of k oscillators, each with mean excitation $\bar{n}/k$ having the density matrix (10). This system emits a distribution of n hadronic units. These are expected to be mostly neutral, although each of the k cells may have a small charge.

In the foregoing we treated the number of cells k as a fixed parameter. However if we identify the emitting cells with clusters (or else fireballs, not necessarily the same), it is natural to regard our k as an average over distributions of these objects. We note that Lèvy [24], and Hayot and Sterman [5] have proposed a scheme, in which the average number of clusters is actually 4, possibly very slowly increasing with energy. If k were to be energy dependent, then true KNO scaling would never set in. In this regard we note that the FNAL data in the peak region are better fit by $k = 3$, but at the expense of the tail and the moment predictions. However, we have not yet taken into account the expected suppression of the tail due to the energy-momentum constraints expected to be important for large n. The fact that $k = 2$ describes the central region also points to $k = 4$ being an average. It is possible to understand this on simple kinematical grounds, if we associate the clusters with different parts of the rapidity plot (a point of view providing contact with the fireball and statistical–hydrodynamical models). Roughly speaking, the particles in the central region can emit to all angles while the fast matter (fragmentation fireballs?), say $y > 1.5$, is kept out of the central region by simple kinematic effects.

Given the above results, we need still to address the question: what fields are radiating final hadrons and what variables qualify as gaussian random variables? The decisive field variables should not carry quantum numbers to generate the observed coherence. Even gluons have two spins and eight color degrees of freedom, so that it is likely that color neutral collective coordinates describe the cluster fluctuations. The type of variable we have in mind is analogous to Landau's entropy four-current s_μ, which in the perfect relativistic fluid evolves [25] without attention to which phase of matter the medium is in. At breakup the distribution of secondaries is then determined by the entropy distribution on the emitting surface. Although expressed in a different language, a similar tacit assumption has been made [26] in recent QCD calculations in which the multiplicity is supposed to be the same as for an evolution of quarks and gluons, terminated at mass values Q_0^2 quarks and gluons, greater than those appropriate to final state hadrons. It is possible that these fluctuations are density oscillations analogous to those proposed [27] for the QCD plasma phase.

The authors are indebted to T. Goldman, Minh Duong-Van and R. Slansky for stimulating discussions.

References

[1] G. Arnison et al., UA1 Collab., Phys. Lett. 107B (1982) 320.
[2] K. Alpgard et al., UA5 Collab., Phys. Lett. 121B (1983) 209.
[3] G. Arnison et al., UA1 Collab., CERN EP/82-134, to be published.
[4] Z. Koba, H.B. Nielsen and P. Olesen, Nucl. Phys. B40 (1972) 317.
[5] S. Barshay, Phys. Lett. 42B (1972) 457; 116B (1982) 193;
K. Kikkawa and E. Ma, Phys. Rev. D9 (1974) 120;
A.J. Buras and Z. Koba, Nuovo Cimento 6 (1973) 629;
M. LeBellac, J.L. Meunier and G. Plaut, Nucl. Phys. B62 (1973) 350;
F. Hayot and G. Sterman, Phys. Lett. 121B (1983) 419.

[6] P. Slattery, Phys. Rev. D7 (1973) 2073;
E.H. deGroot, Phys. Lett. 57B (1975) 159.
[7] L. Liu and T. Meng, Inst. für Theor. Phys. preprint (Berlin, 1982).
[8] R.J. Glauber, Phys. Rev. 131 (1963) 1766.
[9] R.J. Glauber, in: Quantum optics and electronics, eds. C. DeWitt, A. Blandin and C. Cohen-Tannoudji (Gordon and Breach, New York, 1964).
[10] J. Klauder and E.C.G. Sudarshan, Quantum optics (Benjamin, New York, 1968).
[11] L. Mandel and E. Wolf, Coherence and fluctuations of light, vols. I and II (Dover, New York, 1970).
[12] R. Sugar and D. Scalapino, Phys. Rev. D8 (1973) 2284.
[13] G.N. Fowler and R.M. Weiner, Phys. Lett. 70B (1977) 201.
[14] P. Hanbury-Brown and F.R. Twiss, Nature 178 (1956) 1046; Proc. Roy. Soc. A242 (1957) 300; A243 (1957) 291.
[15] M. Goosens et al., Nuovo Cimento 48A (1978) 169.
[16] D. Drijard et al., Nucl. Phys. B155 (1979) 269.
[17] W. Heisenberg, Z. Phys. 133 (1952) 65.
[18] M. Planck, Sitz. Deutsch. Akad. Wiss. Berlin 33 (1923) 355.
[19] B. Decomps and A. Kastler, C.R. Acad. Sci. (Paris) 256 (1963) 1087.
[20] W. Knox, Phys. Rev. D10 (1974) 65.
[21] W. Thome et al., Nucl. Phys. B129 (1977) 365.
[22] R. Møller, Nucl. Phys. B74 (1974) 145.
[23] F.T. Dao et al., Phys. Rev. Lett. 24 (1972) 1627;
C. Bromberg et al., Phys. Rev. Lett. 26 (1973) 1563;
S. Barish et al., Phys. Rev. D9 (1974) 2689;
A. Firestone et al., Phys. Rev. D10 (1974) 2080;
A. Firestone et al., Phys. Rev. D14 (1976) 2902;
D. Brick et al., Phys. Rev. D25 (1982) 2794.
[24] M. Lèvy, Nucl. Phys. B59 (1973) 583.
[25] P. Carruthers and Minh Duong-Van, LA-UR-82-3412.
[26] A. Bassetto, M. Ciafaloni and G. Marchesini, Nucl. Phys. B161 (1980) 477.
[27] P. Carruthers, Phys. Rev. Lett. 50 (1983) 1179.

HADRONIC MULTIPLICITY DISTRIBUTIONS: EXAMPLE OF A UNIVERSAL STOCHASTIC MECHANISM

P. Carruthers
Theoretical Division
Los Alamos National Laboratory
Los Alamos, NM 87545

1. Introduction

The existence of approximate scaling[1] of hadronic charged multiplicity distributions when plotted in KNO form (i.e., $\bar{n}P_n$ vs. $n/\bar{n}$) continues to attract interest. Both the existence of the phenomenon, and the shape of the scaling curve $\psi(n/\bar{n}) = \bar{n}P_n$ (large n, $\bar{n}$) has been "explained" from many geometrical-dynamical points of view. Here we propose instead that these results depend on a generic framework independent of dynamical details, which context moreover occurs in many areas of science. This view imposes global constraints on any modelistic view which must be respected as are the symmetries of a theory. What one learns from the existence of scaling and the form of ψ is indeed interesting but different from traditional views.

In 1959 Mandel published a paper[2] showing that photoelectron counts from "thermal" sources (i.e., sources whose fields are distributed as gaussian random variables) follow the generalized[3] Bose-Einstein distribution (or negative binomial distribution)

$$P_n^{(k)} = \frac{(n + k - 1)!}{n!(k - 1)!} \left(\frac{\bar{n}/k}{1 + \bar{n}/k} \right)^n \frac{1}{(1 + \bar{n}/k)^k} \tag{1}$$

where k is the number of "cells." For n, $\bar{n}$ larger than k and unity the asymptotic form of (1) is ($z \equiv n/\bar{n}$)

$$\psi(z) = \frac{k^k}{(k - 1)!} z^{k-1} e^{-kz} \tag{2}$$

Recently C. C. Shih and I noticed[4] that distributions of this type do a remarkably good job of describing KNO scaling (and its violation) in purely hadronic reactions with k = 3 to 4 (where independence of the nature of the colliding hadrons appears to hold; $\bar{p}p$ annihilations are excluded here and will be mentioned below). Indeed several authors have suspected[5)-7)] that the distribution (1) is appropriate to these reactions. What we have provided is an extensive phenomenological analysis to sharpen this claim, using much more complete data, and especially to provide the proper conceptual framework to justify Eq. (1).

Before proceeding, we mention that the distribution (1) also describes photoelectron counts of coherent laser light (which would give Poisson counting) transmitted through liquids at their critical point. The conclusion is that the large fluctuations at the critical point are distributed as gaussian random variables.[8)]

Still more remarkable, we have recently shown[9)] that the probability of finding n galaxies in a Zwicky cluster[10)] follows (1) with 6 cells, after a long tail probably due to gravitational cannibalism of clusters is subtracted. Since the mechanism of cluster/galaxy formation probably originates in gravitational/hydrodynamic instabilities, we conclude that distributions of the form (1) do not depend on quantum mechanics. Indeed, it was discovered early in the construction of semi-classical pictures of photocounting that Bose-Einstein distributions result from Poisson (constant intensity) if the fields were treated as stochastic gaussian variables.[11)]

We suggest, therefore, that the essence of physical mechanisms leading to Eq. (1) is very simple: (1) There exist k independent cells; (2) The emitting field variables are distributed as gaussian random variables whereupon the magic qualities of the central limit theorem can be invoked. In the case of hadron production, the space-time surface at which hadronization occurs is occupied by emitting fields whose random behavior is characterized by highly fluctuating (gaussian) field variables.

The foregoing picture omits the possibility of a coherent admixture to the chaotic signal mentioned above. In quantum optics, this effect is quite striking.[12)-13)] Below threshold for lasing the photons are thermal and the photocounts follow Eq. (1) for k = 1 if one polarization mode is involved. Above threshold, the effect of the coherent signal is to decisively narrow the distribution: the more coherence, the closer the counting distribution is to Poisson. Although our numerical analysis is not complete, we suggest[14)] that the narrow distributions of hadron multiplicities seen in e^+e^-, $\bar{p}p$ annihilation and νN, $\bar{\nu}N$ reactions exhibit the effect of a predominantly coherent source.

In skeletal form we can exhibit the foregoing ideas in the framework of a simple harmonic oscillator. The density matrix, expressed in the coherent state basis,[15)] has a very simple form corresponding to perfect coherence, perfect incoherence, or a mixture. The coherent states $|\alpha\rangle$ is an eigenfunction of the boson destruction operator with complex eigenvalue α. In the number basis, we have

$$|\alpha\rangle = e^{-\frac{1}{2}|\alpha|^2} \sum_{n=0}^{\infty} \frac{\alpha^n}{\sqrt{n!}} |n\rangle \tag{3}$$

From this we see that the expectation value of the position operator is "classical" $\langle\alpha|x(t)|\alpha\rangle \propto \cos(\phi - \omega t)$ and that the distribution of quanta is Poisson with $\bar{n} = |\alpha|^2$.

The special class of density matrices[15)]

$$\phi = \int d^2\alpha\ \Phi(\alpha)\ |\alpha\rangle\langle\alpha| \tag{4}$$

contains the physical structure we need. For weight functions we associate number distributions as follows.[15)-17)]

1) pure coherence

$$\Phi(\alpha) = \delta(\alpha - \beta) \qquad P_n = e^{-|\beta|^2}|\beta|^{2n}/n! \tag{5}$$

2) chaotic

$$\Phi(\alpha) = \frac{e^{-|\alpha|^2/\bar{n}_\alpha}}{\pi\bar{n}_\alpha} \qquad P_n = \frac{\bar{n}_\alpha^{\,n}}{(1 + \bar{n}_\alpha)^{n+1}} \tag{6}$$

3) mixed

$$\Phi(\alpha) = \frac{e^{-|\alpha - \beta|^2/\bar{n}_\alpha}}{\pi\bar{n}_\alpha} \qquad P_n = \frac{(\bar{n}_\alpha)^n}{(1 + \bar{n}_\alpha)^{n+1}}$$

$$\times\, L_n\left(-\frac{\bar{n}_\beta}{\bar{n}_\alpha(1 + \bar{n}_\alpha)}\right) \exp\left(-\frac{\bar{n}_\beta}{1 + \bar{n}_\alpha}\right) \tag{7}$$

In Eq. (7) L_n is the LaGuerre polynomial and $\bar{n}_\beta = |\beta|^2$. The distribution (1) corresponds to a product of gaussians (6) with $\bar{n}_\alpha = \bar{n}/k$. $\bar{n}$ is $\bar{n}_\alpha + \bar{n}_\beta$ and $\bar{n}_\beta/\bar{n}_\alpha$ is commonly referred to as the signal to noise ratio.

The single cell formulas given in Eqs. (5-7) vary enormously in their scaling behavior: the Poisson does not scale; the Bose-Einstein form obeys $\bar{n}P_n \to e^{-z}$, while for the mixed case we find[14)] for large n

$$\bar{n}P_n \to \frac{2(r + 1)}{r} \exp\left(-\frac{1}{r} - \frac{r + 1}{r} z\right) I_0\left(2\sqrt{\frac{r + 1}{r}} z\right) \tag{8}$$

when the noise to signal is not too small. The definition of r is $\bar{n}_\beta/\bar{n}_\alpha$. Since Eq. (7) interpolates smoothly between (5 and 6), and since (5) does not scale, it is interesting that we have scaling for the mixed case to rather low noise values. This non-uniform behavior will be discussed elsewhere.[14)] Nevertheless, experimentalists should check as precisely as possible the accuracy of KNO scaling for the "narrow" distributions because of the possibly large Poisson component.

The basic distribution (1) has many remarkable properties, from which we stress one of particular importance in applications: its form invariance under elimination of the neutrals. If we suppose (1) to apply to $n = n_{ch} + n_0$ and eliminate n_0 by summation to obtain a fixed n_{ch}, using a binomial partition of n_{ch} and n_0 in our composition, we reproduce (1) with $n \to n_{ch}$ and the same k. This form invariance under elimination of unobserved degrees of freedom has much in common with renormalization group ideas and is apparently characteristic of the class of distributions known as "infinitely divisible."

Next we summarize some results of our analysis of purely hadronic reactions. This analysis takes into account the following. In order to compare with data (n_{ch} even), we extract the even component of (1). Secondly a cut at the observed n_{max} is necessary in the theoretical estimation of moments. Thirdly, we make a leading particle subtraction of two charge units($n_{ch} \to n_{ch} - 2$). The basis for this is (1) conceptual, that the hadronization process to first approximation should not count the charges of leading particles and (b) the excellent phenomenological simplicity that results. Two bits of evidence should be mentioned. First of all the usual moments γ_n (n = 2, 3, 4) measured separately for negatives and all charges at Fermilab become the same to 2% after this modification. Without this change the moments differ enormously, so that one might conclude that a different distribution holds for all charges and negatives. Moreover the distribution (1) describes systematic scaling violations down to FNAL energy with k = 4, although we are not sure that k = 3 is better for these energies since our calculation has not yet taken into account the suppression of the large n tail due to energy-momentum conservation.

One expression of the scaling violation is in the prediction of the moments:

$$\gamma_2 = \frac{\langle(n - \bar{n})^2\rangle}{\bar{n}^2} = \frac{1}{k} + \frac{1}{\bar{n}}$$

$$\gamma_3 = \frac{\langle(n - \bar{n})^3\rangle}{\bar{n}^3} = \frac{2}{k^2} + \frac{3}{k\bar{n}} + \frac{1}{\bar{n}^2} \qquad (9)$$

At CERN collider energy[18] $\bar{n}_{ch}$ = 28.9 including leading particles so the difference between 0.29 and 0.27 scarcely affects the (good) prediction γ_2 = 0.29 of Eq. (9) with k = 4. Figure 1 shows the prediction of (1) for k = 4 and 6 along with Slattery's early fit[19] to low energy data. These curves give an indication of the magnitude of scaling violations.

Traditionally it has been comforting to imagine that γ_2 is about 0.25 for all energies. Our point of view is that the correct (subtracted) γ_2 decreases from about 1/2 at low energy to 1/4 at high energy as predicted by Eq. (9). A partial indication is given in Table I while skeptics are invited to inspect the analysis of Refs. 4 and 14 for substantiation of this claim.

For hadronic collisions it is natural to identify k with the average number of clusters. Therefore k need not be integral. Authorities differ as to the number of clusters appropriate to phenomenological fits. However, Hayot and Sterman[20] suggest that k = 4 at collider

TABLE I

Fermilab second and third moments are given for (a) all charges, (b) negative charges, (c) modified moments (leading particle subtraction) and the prediction of Eq. (9) with $\bar{n} = n_{ch} - 2$. With the leading particle subtraction, the moments for all charges agree closely with those for negative charges and rather well with the energy dependent prediction of Eq. (9). See Ref. 4 for references to the experimental literature and more detail.

p_{lab} (GeV/c)	102	205	303	405
$\gamma_2(\mathrm{exp})^a$	24.5	24.6	24.3	28.5
$\gamma_2^-(\mathrm{exp})^b$	52.2	45.1	40.8	45.9
$\tilde{\gamma}_2(\mathrm{exp})^c$	52	45	41	46
$\gamma_2(\mathrm{th})^d$	50	43	39	40
$\gamma_3(\mathrm{exp})^a$	7.9	8.4	8.3	11.4
$\gamma_3^-(\mathrm{exp})^b$	24.5	20.3	17.5	23.3
$\tilde{\gamma}_3(\mathrm{exp})^c$	25	21	18	24
$\gamma_3(\mathrm{th})^d$	35	28	23	25

energies on the basis of leading log QCD. Hayot further suggests[21] that k varies as $(\log s)^{\frac{1}{2}}$ which would have k ranging from 3 to 5 as one goes from Fermilab to Desertron energies. These values are completely compatible with our analysis, which did not provide any prediction of k.

When k is constant, KNO scaling sets in fairly quickly (asymptopia is only 5-10% away at 540 GeV). However, a slowly increasing k will sharpen the distribution as energy increases as easily seen from the asymptotic form (2). The peak is at $1 - 1/k$ and the maximum is roughly $\psi_{max} \cong [k^2/2\pi(k-1)]^{\frac{1}{2}}$.

Another very important clue is the difference of shape for central region ($|\eta| \lesssim 1.5$) and full pseudorapidity data. The central region results[22] as well described[4] by k = 2 instead of k = 4. A possible explanation of this based on the kinematical distribution of clusters has been given in Ref. 4. It will be of great interest to see whether k increases monotonically to 4 as the rapidity bite $-\eta_0 < \eta < \eta_0$ is increased.

In closing we make some qualitative remarks on KNO distributions in lepton-induced reactions. Previously we suggested that the narrow distributions might be explained by the admixture of coherence in the hadronization process. Such a mixture was also used in the analysis of rapidity correlation by Fowler and Weiner.[23] When I mentioned this to Gunther Wolf at this meeting, he suggested a connection to the differing distributions between single jet and total e^+e^- KNO distributions in preliminary TASSO data.[24] The single jet distribution is broader, which is natural in our picture because of the loss of correlations when the single jet is selected. These correlations are necessary if only to neutralize the color of the core quark in the jet. The main difference may be due to the dominant Poisson component with $\bar{n} = 7$ (rather than 14 for the full reaction at 34 GeV).

A similar curiosity occurs in the fact[25] that the νp KNO as distribution is narrower than that in $\bar{\nu}p$. Here the only elementary distinction seems to be that ν can hit (one) d quark in the proton, while $\bar{\nu}$ has a choice of two u quarks. Whether this picture can be quantified in terms of the required coherence parameter remains to be seen.

In summary we suggest that multiplicity distributions of hadrons be viewed in the universal context of stochastic fields. For noisy fields (gaussian variables) we found that a small number of cells, probably to be identified with clusters, gives an excellent description of KNO scaling and its violation without the usual dynamical assumptions. Since the cluster concept is not sharply defined, it is not precluded that a more abstract characterization of k, for example as some topological number characterizing a submanifold of the solution space of the problem at hand, is possible.[26] With regard to the sharper distributions exemplified by e^+e^- hadronization, we have called attention to analogues in other domains of physics and the likely connection with coherence effects. In every case it appears that the (global) multiplicity distribution plotted in KNO form is insensitive to detailed dynamical models. Although rapidity correlations and other information about the phase space structure of the reaction will depend on dynamics, the integral quantities predicted in our framework impose powerful constraints which must be respected by any successful approximate dynamical model.

Acknowledgments

I am grateful to my collaborators Minh Duong-Van and C. C. Shih for many contributions to this project and to B. Hasslacher for some provocative discussions. At this meeting I am particularly grateful to J. Rushbrooke and G. Wolf for lively discussions bearing especially on the problem of multiplicity distributions. This research was supported by the U.S. Department of Energy.

References

1. Z. Koba, H. B. Nielsen and P. Olesen, Nuc. Phys. B 40, 317 (1972).
2. L. Mandel, Proc. Phys. Soc. 74, 233 (1959).
3. M. Planck, Sitzungsber. Dtsch. Akad. Wiss. Berlin 33, 355 (1923).

4. P. Carruthers and C. C..Shih, Phys. Lett. B (to be published).
5. W. J. Knox, PR D10, 65 (1974).
6. N. Suzuki, Prog. Theor. Phys. 51, 1629 (1974).
7. A. Giovannini, Nuovo Cimento 34A, 647 (1976).
8. S. H. Chen, C. C. Lai, J. Rouch, and P. Tartaglia, J. Stat. Phys. 30, 699 (1983).
9. P. Carruthers and Minh Duong-Van, LA-UR-83-1951 (to be published).
10. F. Zwicky et al, Catalog of Galaxies and Clusters of Galaxies, (California Institute of Technology, Pasadena, 1961-68, Vols. 1-6).
11. J. Klauder and E. C. G. Sudershan, "Quantum Optics" (W. A. Benjamin, N.Y., 1968).
12. C. Freed and H. A. Haus, IEEE Journal of Quantum Electron., Vol. QE-2, 190 (1966).
13. F. T. Arecchi, Phys. Rev. Lett. 15, 912 (1965).
14. P. Carruthers and C. C. Shih (to be published).
15. R. J. Glauber, Phys. Rev. 131, 109 (1963).
16. P. Carruthers and M. M. Nieto, Am. J. Phys. 33, 537 (1965) and Rev. Mod. Phys. 40, 411 (1968).
17. K. J. Glauber, in "Physics of Quantum Electronics" ed. P. L. Kelly, B. Lax and P. E. Tannenwald (McGraw-Hill, N.Y., 1966) p. 788.
18. K. Alpgard et al (UA5 Collaboration), Phys. Lett. 121B 209 (1983).
19. P. Slattery, PR D7, 2073 (1973).
20. F. Hayot and G. Sterman, Phys. Lett. 121B, 419 (1982).
21. F. Hayot, Phys. Lett. B (to be published).
22. G. Arnison et al (UA1 Collaboration), Phys. Lett. 107B, 320 (1982).
23. G. N. Fowler and R. M. Weiner, Phys. Lett. 70B, 201 (1977).
24. G. Wolf (private communication).
25. N. Schmitz (private communication).
26. B. Hasslacher (private communication).

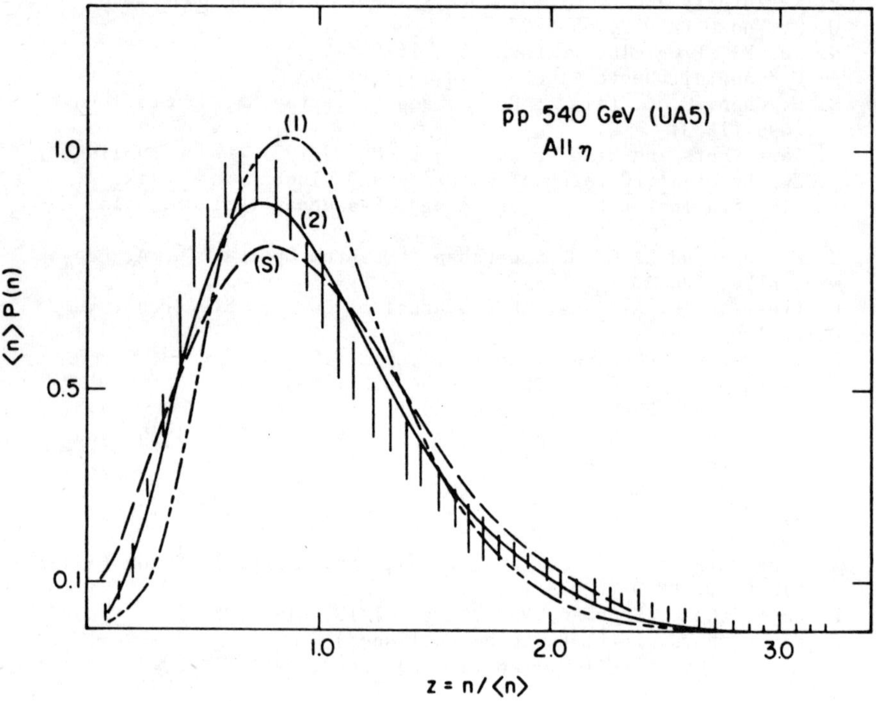

Fig. 1. UA5 data are compared with Eq. (1) for k = 6 (curve 1) and k = 4 (curve 2). The Slattery fit(s) to Fermilab data is shown to indicate the magnitude of scaling violation.

XI. Applications in Quantum Field Theory

PHYSICAL REVIEW VOLUME 140, NUMBER 4B 22 NOVEMBER 1965

Infrared Divergence in Quantum Electrodynamics*

VICTOR CHUNG
Lawrence Radiation Laboratory, University of California, Berkeley, California
(Received 15 July 1965)

The infrared divergences of quantum electrodynamics are eliminated to all orders of perturbation theory in the matrix elements by an appropriate choice of initial and final soft photon states. The condition for this cancellation restricts these states to representations of the canonical commutation rules which are unitarily inequivalent to the usual Fock representation.

I. INTRODUCTION

THE matrix element in quantum electrodynamics for the scattering from an initial state containing a finite number of electrons and photons into a similar final state contains an integral which diverges logarithmically for small momentum k. The conventional treatment of this "infrared divergence" has been to sum the cross sections over all possible final states consistent with experimental measurements. In particular, when all states with any number of soft photons with momenta below the threshold of observability are considered, the divergences cancel, and the calculated cross sections are consistent with experiment. It is therefore possible to attribute the original divergence in the matrix element to the inappropriate choice of initial and final states to represent the experimental situation. In an actual scattering experiment, an indefinite number of soft photons are emitted, so that in some sense, states which are eigenstates of the number operator are unphysical.

In this paper, we shall show that there exists a representation of the photon states for quantum electrodynamics which appears more appropriate for describing scattering than the usual Fock representation in that the matrix elements do not have infrared divergences. These states are not eigenstates of the number operator, and are parametrized in a manner similar to that used by Glauber,[1] Bargmann,[2] and others. When certain conditions of convergence are imposed, the states can be shown to form irreducible representations of the canonical commutation rules for the "in" and "out" fields which are unitarily inequivalent to the usual Fock representation. Similar results have been obtained by Shroer[3] in certain model field theories.

In the absence of known solutions to the renormalized field equations, we make no pretense to mathematical rigor. In particular, the Feynman-Dyson perturbation techniques are used throughout, and most questions of order in limiting procedures, etc., are treated heuristically.

Section II will summarize the parts of the conventional treatment of infrared divergences which we shall need. This section is based on a more complete discussion made in the article by Yennie, Frautschi, and Suura.[4] The parametrization of the states and its relationship to the usual occupation number parametrization are introduced in Sec. III. We shall make use of the algebra of states developed in Glauber's paper.[1] In Sec. IV, the cancellation of the divergences to second order is demonstrated in order to illustrate the methods used in the succeeding sections. A calculation of the matrix elements for potential scattering in Sec. V shows that the divergences indeed cancel to all orders. In Sec. VI, the structure and the physical meaning of the representations are examined. Then we show that by squaring the matrix elements and summing over the final states results are obtained in low order which agree with those obtained by Yennie *et al.*[4] by the conventional treatment. Some extensions and generalizations of our treatment are carried out in the appendices.

II. SEPARATION OF THE INFRARED FACTORS

The following exposition of the separation of the infrared parts from the matrix element can be found in the review article by Yennie *et al.*[4] We will summarize here what is relevant to our own discussion. For simplicity, we study the example of an electron scattering from a potential, although similar results can be obtained for more general situations.

Consider a process in which there are a fixed number of photons and an electron of momentum p in the initial state and a fixed number of photons with the scattered electron of momentum p' in the final state. The photons may or may not have interacted with the electron line. The complete matrix element for this process is given by

$$M(\mathbf{p},\mathbf{p}')=\sum_{n=0}^{\infty} M_n(\mathbf{p},\mathbf{p}'), \tag{1}$$

where $M_n(\mathbf{p},\mathbf{p}')$ corresponds to the sum of all diagrams in which there are n virtual photons which can be distinguished from the potential interactions in the "basic process" M_0. The real photon variables have been suppressed.

* This work was performed under the auspices of the U. S. Atomic Energy Commission.

[1] Roy J. Glauber, Phys. Rev. 131, 2766 (1963).
[2] V. Bargmann, Proc. Natl. Acad. Sci. U. S. 48, 199 (1962).
[3] B. Schroer, Fortschr. Physik 11, 1 (1963).
[4] D. R. Yennie, S. C. Frautschi, and H. Suura, Ann. Phys. (N. Y.) 13, 379 (1961).

The quantity $\rho_n(k_1,\cdots,k_n)$ is defined by the relation

$$M_n=\frac{1}{n!}\int\cdots\int\prod_{i=1}^{n}\frac{d^4k}{k_i^2-\lambda^2+i\epsilon}\rho_n(k_1,\cdots,k_n),\quad(2)$$

where λ is the photon mass which we allow to approach zero later. It has been shown that ρ_n is of the form

$$\rho_n(k_1,\cdots,k_n)=S(k_n)\rho_{n-1}(k_1,\cdots,k_{n-1})+\xi^{(1)}(k_1,\cdots,k_{n-1};k_n),\quad(3)$$

where $S(k_n)$ contains the k_n infrared divergence, and can have the form

$$S(k_n)=\frac{1}{2}\left[\frac{ie^2}{(2\pi)^4}\right]\left(\frac{2p_\mu'-k_\mu}{2p'\cdot k-k^2}-\frac{2p_\mu-k_\mu}{2p\cdot k-k^2}\right).\quad(4)$$

The remainder $\xi^{(1)}$ has no infrared divergence in k_n, and its infrared divergence in the other k's has not been made worse by the separation.

By iteration of Eq. (3), $\rho_n(k_1,\cdots,k_n)$ can be expressed as a sum over all permutations of the k's:

$$\rho_n(k_1,\cdots,k_n)=\sum_{\text{perm}}\sum_{r=0}^{n}\frac{1}{r!(n-r)!}\times\prod_{i=1}^{r}S(k_i)\xi_{n-r}(k_{r+1},\cdots,k_n).\quad(5)$$

The functions ξ_r are noninfrared and symmetrical in the k's. If we adopt the definitions

$$\alpha B(\mathbf{p},\mathbf{p}')\equiv\int\frac{d^4k\,S(k)}{k^2-\lambda^2},\quad(6a)$$

$$m_r(\mathbf{p},\mathbf{p}')\equiv\frac{1}{r!}\int\prod_{i=1}^{r}\frac{d^4k_i}{k_i^2}\xi_r(k_1,\cdots,k_n),\quad(6b)$$

then substitution of (2), (5), (6a), and (6b) into (1) results in the simple expression

$$M=\exp(\alpha B)\sum_{n=0}^{\infty}m_n.\quad(7)$$

In this expression, $m_0=\rho_0=\xi_0=M_0$. The m_n's in (7) are divergence-free, so that the whole infrared divergence has been isolated in the argument αB of the exponential. For future reference, we can write down the form of $\mathrm{Re}(\alpha B)$ which follows from (4) and (6):

$$\mathrm{Re}(\alpha B)=\frac{e^2}{4(2\pi)^3}\int\frac{d^3k}{(k^2+\lambda^2)^{1/2}}\times\left(\frac{2p_\mu'-k_\mu}{2p'\cdot k-\lambda^2}-\frac{2p_\mu-k_\mu}{2p\cdot k-\lambda^2}\right)^2.\quad(8)$$

The extraction of the infrared contribution to the matrix element for the emission of real photons has a form similar to that in Eq. (3). In this case, we let $\tilde{\rho}_n(k_1,\cdots,k_n)$ be the matrix element corresponding to the emission or absorption of n undetectable photons with momenta $k_1,\cdots,k_n$, and for some arbitrary order in the virtual photon corrections. It has been shown that

$$\tilde{\rho}_n(k_1,\cdots,k_n)=\pm\tilde{S}(k_n)\tilde{\rho}_{n-1}(k_1,\cdots,k_n)+\tilde{\xi}^{(1)}(k_1,\cdots,k_{n-1};k_n),\quad(9)$$

where $\tilde{S}(k_n)$ is the factor containing the infrared divergence and has the form

$$\tilde{S}(k)=\frac{e}{[2(2\pi)^3k_0]^{1/2}}\left[\frac{p'\cdot e}{k\cdot p'}-\frac{p\cdot e}{p\cdot k}\right],\quad(10)$$

and the (+) and (−) signs correspond to emission and absorption, respectively. Again the remainder $\tilde{\xi}^{(1)}$ is divergence-free in k_n, and the divergences in the other k's is no worse for the separation.

It has been shown that the iteration of (9) leads to the form

$$\tilde{\rho}_n(k_1,\cdots,k_n)=\sum_{\text{perm}}\sum_{r=0}^{n}(-1)^m\frac{1}{r!(n-r)!}\times\prod_{i=1}^{r}\tilde{S}(k_i)\tilde{\xi}_{n-r}(k_{r+1},\cdots,k_n),\quad(11)$$

where the functions $\tilde{\xi}$ are noninfrared and symmetrical in the k's and m corresponds to the number of absorbed photons.

III. PARAMETRIZATION OF THE PHOTON STATES

The properties of the states which we will find convenient to use have been discussed by several other authors[1,2] in different contexts from the one in which we intend to use them.

Let $\{f_i(k)\}$ be a complete and orthonormal set of functions defined on some region Ω of momentum space including $k=0$ (perhaps all of momentum space). A typical state "belonging to the ith mode" is defined by

$$|\alpha_i\rangle=\frac{\exp(\alpha_ia_i^\dagger)}{\exp(\frac{1}{2}|\alpha_i|^2)}|0\rangle=\exp(-\tfrac{1}{2}|\alpha_i|^2)\sum_n\frac{(\alpha_ia_i^\dagger)^n}{n!}|0\rangle,\quad(13)$$

where

$$a_i^\dagger=\int d^3k\,f_i(k)a^\dagger(k)\quad(14)$$

is an "in" or "out" creation operator.

In this expression, α_i is a complex number which can take on any value in the complex plane, $a^\dagger(k)$ is the photon creation operator which obeys the commutation rules

$$[a(k),a^\dagger(k')]=\delta(k-k'),$$
$$[a(k),a(k')]=[a^\dagger(k),a^\dagger(k')]=0,\quad(15)$$

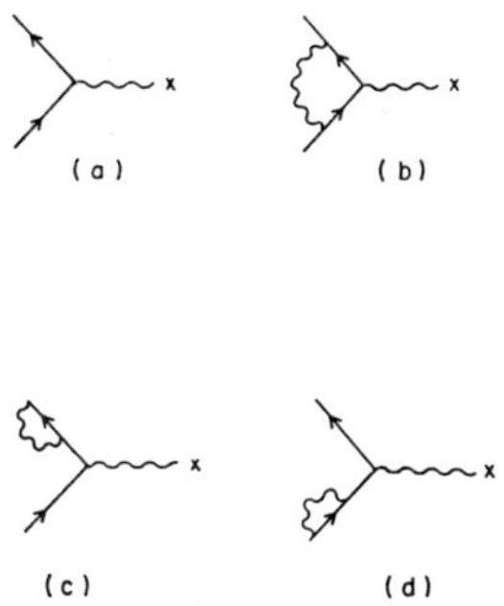

FIG. 1. Contributions to the second-order virtual-photon radiative corrections. Diagram (a) corresponds to the original uncorrected matrix element M_0.

and $|0\rangle$ is the state with no photons; $a_i^\dagger$ obeys the commutation rules

$$[a_i,a_j^\dagger]=\delta_{ij},\quad [a_i,a_j]=[a_i^\dagger,a_j^\dagger]=0. \qquad (16)$$

From the commutation rules, it is a trivial matter to show that these states are eigenfunctions of the destruction operator

$$a(k)|\alpha_i\rangle=\alpha_i f_i(k)|\alpha_i\rangle$$

or $\qquad\qquad\qquad\qquad\qquad\qquad\qquad\qquad\qquad (17)$

$$a_i|\alpha_i\rangle=\alpha_i|\alpha_i\rangle,$$

and that the mean number of "photons" is

$$\langle\alpha_i|\hat{N}|\alpha_i\rangle=\int d^3k\langle\alpha_i|a^\dagger(k)a(k)|\alpha_i\rangle=|\alpha_i|^2. \qquad (18)$$

It is sometimes useful to note that the state in Eq. (13) can be "created" by a unitary operator

$$D(\alpha_i)=\exp[\alpha_i a_i^\dagger-\alpha_i^* a_i], \qquad (19)$$

which has the following "translation" property:

$$D(\alpha_i)D(\beta_i)=\exp[\tfrac{1}{2}(\alpha_i\beta_i^*-\alpha_i^*\beta_i)]D(\alpha_i+\beta_i). \qquad (20)$$

The states defined in this manner are nonorthogonal; the overlap between two states $|\alpha_i\rangle$ and $|\beta_i\rangle$ is given by

$$|\langle\alpha_i|\beta_i\rangle|^2=\exp\{-|\alpha_i-\beta_i|^2\}. \qquad (21)$$

However, it follows from (21) that the states are normalized, i.e,

$$\langle\alpha_i|\alpha_i\rangle=1. \qquad (22)$$

Another property which these states possess is completeness. In fact, it is easy to show that

$$\frac{1}{\pi}\int d^2\alpha_i|\alpha_i\rangle\langle\alpha_i|=\sum_{n_i}|n_i\rangle\langle n_i|=I, \qquad (23)$$

where the state denoted by n_i is an eigenstate of the number of photons which have the momentum distribution described by the function $f_i(k)$, and $d^2\alpha=d(\mathrm{Re}\alpha_i)d(\mathrm{Im}\alpha_i)$ is real.

An arbitrary state of the ith mode has an expansion in terms of the n-photon states of the form

$$|\ \rangle=\sum_n c_n|n\rangle=\sum_n c_n\frac{(a_i^\dagger)^n}{(n!)^{1/2}}|0\rangle, \qquad (24)$$

where $\sum_n|c_n|^2=1$. We associate with each such state an analytic function,

$$f(z)=\sum_n c_n\frac{z^n}{(n!)^{1/2}}. \qquad (25)$$

Equation (24) may then be rewritten as

$$|f\rangle=f(a_i^\dagger)|0\rangle. \qquad (26)$$

Using (23), we can expand $|f\rangle$ in terms of the new states:

$$|f\rangle=\frac{1}{\pi}\int d^2\alpha_i|\alpha_i\rangle\langle\alpha_i|f(a_i^\dagger)|0\rangle$$
$$=\frac{1}{\pi}\int d^2\alpha_i|\alpha_i\rangle f(\alpha_i^*)\exp(-\tfrac{1}{2}|\alpha_i|^2). \qquad (27)$$

In (27) we have used the fact that the states $|\alpha_i\rangle$ are eigenstates of the destruction operator a_i:

$$a_i|\alpha_i\rangle=\alpha_i|\alpha_i\rangle. \qquad (28)$$

In a similar fashion, the adjoint state vectors $\langle g|$ can be shown to possess an analogous expansion,

$$\langle g|=\frac{1}{\pi}\int[g(\beta_i^*)]^*\langle\beta_i|\exp(-\tfrac{1}{2}|\beta_i|^2)d^2\beta_i. \qquad (29)$$

A basis for the whole electromagnetic field is a direct product of the states $|\alpha_i\rangle$ of the individual modes

$$|\ \rangle=\prod_i|\alpha_i\rangle\equiv|\{\alpha_i\}\rangle, \qquad (30)$$

and the mean number of photons in such a state is

$$\sum_i|\alpha_i|^2. \qquad (31)$$

Equations (25), (27), and (30) ensure that states containing a finite number of photons (the usual Fock representation) can be expanded in terms of the states $|\{\alpha_i\}\rangle$ which satisfy $\sum_i|\alpha_i|^2<\infty$. This will be shown in Sec. VI. However, this restriction will not be imposed in the discussion that follows, i.e., we shall allow for the possibility that there exist states in which the average number of photons is not bounded.

IV. CANCELLATION OF THE INFRARED DIVERGENCES TO SECOND ORDER

In order to illustrate the general procedure, we shall choose two particular photon states parametrized in the manner just discussed in Sec. III and calculate in lowest order the matrix element for potential scattering from one to the other. The photon is assumed to have a finite mass λ which is allowed to approach zero at the end of the calculation.

Let the initial momentum of the electron be p_i and the final momentum of the scattered electron be p_f. For the initial state, we choose

$$
\begin{aligned}
|\ \rangle_i &= \exp\left\{-\tfrac{1}{2}\sum_{l=1}^{2}\int |\tilde{S}_i^{(l)}(k)|^2 d^3k\right\} \\
&\quad \times \exp\left\{\sum_{l=1}^{2}\int d^3k\, \tilde{S}_i^{(l)}(k) e^{(l)}(k) a^{(l)\dagger}(k)\right\} |\Psi(p_i)\rangle \\
&= \exp\{-\tfrac{1}{2}\sum_{l,a}|\beta_{ia}{}^l|^2\} \\
&\quad \times \exp\left\{\sum_{l,a}\beta_{ia}{}^l \int d^3k\, f_a(k) e^{(l)}(k) a^{(l)\dagger}(k)\right\} |\Psi(p_i)\rangle,
\end{aligned}
\tag{32}
$$

where

$$
\tilde{S}_i^{(l)}(k) = \frac{e}{[2(2\pi)^3 k_0]^{1/2}} \frac{p_i \cdot e^{(l)}}{k \cdot p_i}
\tag{33}
$$

is a function which depends on the momentum of the initial electron. $|\Psi(p_i)\rangle$ is the wave function for the electron, and $e^{(l)}(k)$ are the polarization vectors. The superscript (l) is the polarization index. Since Eq. (33) is meant to define the momentum distribution only as $|k|\to 0$, the function $\tilde{S}_i(k)$ for $k\neq 0$ can be chosen in any manner which makes the integrals in (32) converge as $k\to\infty$. The second form given above exhibits the relation to the discussion in Sec. III. The coefficients $\beta_{ia}{}^l$ are the coefficients obtained in the expansion of $\tilde{S}_i{}^l(k)$ in terms of the chosen orthonormal set.

The initial state can then be expanded in lowest order to give

$$
\begin{aligned}
|\ \rangle_i &\cong \left(1-\tfrac{1}{2}\sum_{l=1}^{2}\int |\tilde{S}_i^{(l)}(k)|^2 d^3k\right) \\
&\quad \times \left(1+\sum_{l=1}^{2}\int d^3k\, \tilde{S}_i^{(l)}(k) e^{(l)}(k) a^{(l)\dagger}(k)\right) |\Psi(p_i)\rangle.
\end{aligned}
\tag{34}
$$

Similarly, the final state can be expanded to give

$$
\begin{aligned}
|\ \rangle_f &\cong \left(1-\tfrac{1}{2}\sum_{l=1}^{2}\int |\tilde{S}_f^{(l)}(k)|^2 d^3k\right) \\
&\quad \times \left(1+\sum_{l=1}^{2}\int d^3k\, \tilde{S}_f^{(l)}(k) e^{(l)}(k) a^{(l)\dagger}(k)\right) |\Psi(p_f)\rangle,
\end{aligned}
\tag{35}
$$

where

$$
\tilde{S}_f^{(l)}(k) = \frac{e}{[2(2\pi)^3 k_0]^{1/2}} \frac{p_f \cdot e^{(l)}}{k \cdot p_f}.
$$

Let the basic interaction given by the matrix element M_0 be the single-potential interaction shown in Fig. 1(a). In order to calculate the S matrix to order e^2, the contributions from all the diagrams in Fig. 1 and Fig. 2 must be summed. Diagrams (b), (c), and (d) of Fig. 1 correspond to the virtual photon radiative corrections to M_0. Diagrams (b), (c), (d), and (e) of Fig. 2 account for the possibility of emission or absorption of single real photons, while diagram (a) of Fig. 2 accounts for the possibility that the photon does not interact with the electrons at all.

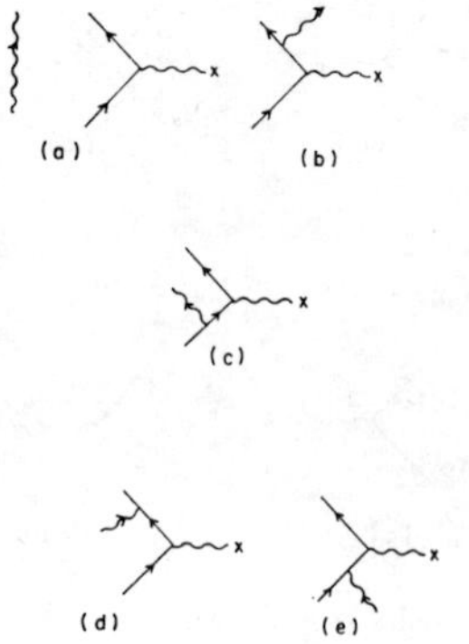

FIG. 2. Contributions to the second-order corrections due to emission or absorption of real soft photons. Diagram (a) accounts for the possibility that the photon does not interact with the electrons at all.

From the discussion in Sec. II of this paper, one knows that the diagrams in Fig. 1 will contribute

$$
M_0 + (\alpha B + \eta) M_0,
\tag{36}
$$

where η is a quantity which is not infrared divergent as $\lambda\to 0$. The contribution from diagrams (b) and (c) of Fig. 2 gives a term with a factor

$$
\sum_l \int d^3k\, \tilde{S}_f^{(l)}(k)[\tilde{S}^{(l)}(k) M_0 + \xi(k)],
\tag{37}
$$

where $\tilde{S}^{(l)}(k)$ was defined in Eq. (10). A similar contribution from diagrams (d) and (e) of Fig. 2 is

$$
-\sum_l \int d^3k\, \tilde{S}_i^{(l)}(k)[\tilde{S}^{(l)}(k) M_0 + \xi(k)].
\tag{38}
$$

The disconnected diagram (a) of Fig. 2 is given by

$$
M_0 \sum_l \int d^3k\, \tilde{S}_i^{(l)}(k) \tilde{S}_f^{(l)}(k).
\tag{39}
$$

Summing Eqs. (36) to (39) with the proper normalization given by Eq. (54), one finds the result

$$
\begin{aligned}
M_{i\to f} &\cong (1+\alpha B+\eta) M_0 \left(1-\tfrac{1}{2}\sum_l \int |\tilde{S}_i^{(l)}(k)|^2 d^3k\right) \\
&\quad \times \left(1-\tfrac{1}{2}\sum_l \int |\tilde{S}_f^{(l)}(k)|^2 d^3k\right) \\
&\quad + \left[\sum_l \int (\tilde{S}_f^{(l)} - \tilde{S}_i^{(l)}) \tilde{S}^{(l)}(k) d^3k + \bar{\eta}\right] M_0 \\
&\quad + M_0 \sum_l \int d^3k\, \tilde{S}_i^{(l)}(k) \tilde{S}_f^{(l)}(k),
\end{aligned}
\tag{40}
$$

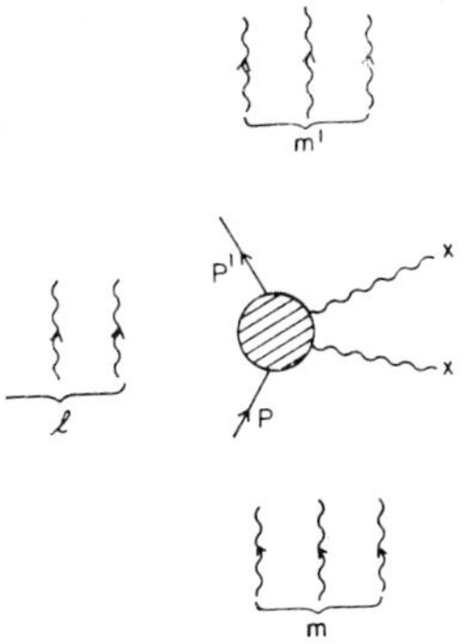

FIG. 3. Representation of l noninteracting real soft photons, m real soft photons absorbed by the electron line, and m' real soft photons emitted by the electron line.

where $\bar{\eta}$ is noninfrared divergent. Thus

$$M_{i\to f}=(1+\alpha B)M_0+(\eta+\bar{\eta})M_0+M_0\sum_l\int|\tilde{S}^{(l)}(k)|^2d^3k$$
$$-M_0\sum_l\int d^3k\{\tfrac{1}{2}|\tilde{S}_i^{(l)}(k)|^2+\tfrac{1}{2}|\tilde{S}_f^{(l)}(k)|^2-\tilde{S}_f^{(l)}(k)\tilde{S}_i^{(l)}(k)\}+O(e^3)$$
$$=\left(1+\alpha B+\tfrac{1}{2}\sum_l\int|\tilde{S}^{(l)}(k)|^2d^3k\right)M_0+(\eta+\bar{\eta})M_0+O(e^3),\qquad(41)$$

where we have used

$$\tilde{S}^{(l)}(k)\equiv\tilde{S}_f^{(l)}(k)-\tilde{S}_i^{(l)}(k).\qquad(42)$$

From (10) we have the relation

$$\tfrac{1}{2}\sum_{l=1}^{2}\int|\tilde{S}^{(l)}(k)|^2d^3k=\sum_{l=0}^{3}\frac{e^2}{4(2\pi)^3k_0}\int d^3k\left[\frac{p_\mu'\cdot e^{(l)}}{p'\cdot k}-\frac{p\cdot e^{(l)}}{p\cdot k}\right]^2(-g^{ll})$$
$$=-\frac{e^2}{4(2\pi)^3k_0}\int d^3k\left[\frac{p_\mu'}{p\cdot k}-\frac{p_\mu}{p\cdot k}\right]^2\equiv\alpha\tilde{B}.\qquad(43)$$

Thus

$$M_{i\to f}=(1+\alpha B+\alpha\tilde{B})M_0+(\eta+\bar{\eta})M_0+O(e^3).\qquad(44)$$

By comparing Eq. (43) with Eq. (8), one can see that the infrared divergence which occurs when $\lambda\to 0$ has been canceled in Eq. (44).

Note that it is the matrix element which is finite. Thus in the calculation of the cross section, there will be no need to deal with an infinite sum of divergent integrals, as must be done in the conventional treatment of infrared divergences.

V. CANCELLATION OF INFRARED DIVERGENCES TO ALL ORDERS

It is now a matter of algebra to calculate the matrix element for the transition from a state of electron momentum $\mathbf{p}$ and photon "quantum numbers" $\{\alpha_a{}^\lambda\}$ to a state of electron momentum $\mathbf{p}'$ and photon "quantum numbers" $\{\gamma_c{}^\lambda\}$, where

$$|\{\alpha_a{}^\lambda\}\rangle=\prod_a\frac{\exp\sum_\lambda\left[\alpha_a{}^\lambda\int_\Omega d^3k f_a(k)e_\mu{}^{(\lambda)}(k)a^{(\lambda)\dagger}(k)\right]}{\exp[\sum_\lambda\tfrac{1}{2}|\alpha_a{}^\lambda|^2]}|0\rangle.\qquad(45)$$

The superscripts (λ) refer to the polarization indices.

Consider all the diagrams represented by Fig. 3, in which there are m real photons absorbed by the electron line, m' real photons emitted by the electron line, and l photons which do not interact with the electron at all.

The matrix element for the process $|\{\alpha_a{}^\lambda\},p_i\rangle\to|\{\gamma_c{}^{\lambda'}\},p_f\rangle$ is then a sum over all diagrams of the kind shown in Fig. 3 for all values of m, m', and l, and with the proper factors determined by Eqs. (45).

Considerations which enter the calculation of this matrix element are expalined below:

(a) There is an infrared divergent factor $e^{\alpha B}$ due to the virtual photon corrections. [See Eq. (7)].

(b) The overlap of the l initial-state noninteracting photons with the l final-state noninteracting photons contributes a factor

$$l!\left[\sum_{\substack{\mu,a,c,\\\lambda,\lambda'}}\alpha_a{}^\lambda\gamma_c{}^{*\lambda'}\int d^3k f_c{}^*(k)f_a(k)e_\mu{}^{(\lambda)}(k)e_\mu{}^{(\lambda')}(k)\right]^l$$
$$=l![\sum_{\substack{\lambda,\lambda',\\a,c}}\alpha_a{}^\lambda\gamma_c{}^{*\lambda'}\delta_{\lambda\lambda'}\delta_{ac}]^l.\qquad(46)$$

(c) Equation (11) gives the contribution due to the interaction of m initial-state photons and m' final-state photons with the electron line

$$\tilde{\rho}_{m+m'}{}^{\{\lambda\}}(k_1,\cdots,k_{m+m'})=\sum_{\substack{\text{perm}\\k's}}\sum_{t=0}^{m+m'}(-1)^m(\prod_{i=1}^{t}\tilde{S}^{(\lambda)}(k_i))$$
$$\times\xi_{m+m'-t}{}^{\{\lambda\}}(k_{t+1},\cdots,k_{m+m'})\frac{1}{t!(m+m'-t)!}.\qquad(47)$$

(d) Contribution (c) must be integrated over the momentum distribution that is obtained from the formal expansion of the initial and final states [see Eq. (45)]:

$$\left[\left(\prod_{r=1}^{m}\sum_{\lambda,a}a_a{}^\lambda\int d^3k_r f_a(k_r)\right)\times\left(\prod_{r'=m+1}^{m+m'}\sum_{\lambda',c}\gamma_c{}^{*\lambda'}\int d^3k_{r'}\,f_c{}^*(k_{r'})\right)\times\tilde{\rho}_{m+m'}{}^{\{\lambda,\lambda'\}}(k_1,\cdots,k_{m+m'})\right].\qquad(48)$$

(e) The formal expansion of Eq. (45) also leads to the factors

$$\frac{1}{(m+l)!}\frac{1}{(m'+l)!}\exp(-\tfrac{1}{2}\sum_{a,\lambda}|\alpha_a{}^\lambda|^2)\times\exp(-\tfrac{1}{2}\sum_{c,\lambda'}|\gamma_c{}^{\lambda'}|^2). \quad (49)$$

(f) In addition to the above, there is a combinatorial factor which accounts for the number of ways that $(m+l)$ initial-state photons and $(m'+l)$ final-state photons can be distributed among m initial-state interacting photons, m' final-state interacting photons, and l noninteracting photons:

$$\frac{(m+l)!}{m!l!}\frac{(m'+l)!}{m'!l!}. \quad (50)$$

After summing over all numbers m, m', and l, we arrive at the following expression for the matrix element $\tilde{M}$:

$$\tilde{M}=e^{\alpha B}\sum_{l=0}^{\infty}\sum_{m=0}^{\infty}\sum_{m'=0}^{\infty}\left[\frac{1}{(m+l)!}\frac{1}{(m'+l)!}\right]\left[\frac{(m+l)!(m'+l)!}{m!m'!l!l!}\right][\exp(-\tfrac{1}{2}\sum_{\lambda,a}|\alpha_a{}^\lambda|^2)\exp(-\tfrac{1}{2}\sum_{\lambda,a}|\gamma_a{}^\lambda|^2)][l!][\sum_{\lambda,a}\alpha_a{}^\lambda\gamma_a{}^{*\lambda}]^l$$
$$\times\left[\left(\prod_{r=1}^{m}\sum_{\lambda,a}\alpha_a{}^\lambda\int d^3k_r\, f_a(k_r)\right)\left(\prod_{r'=m+1}^{m+m'}\sum_{\lambda',c}\gamma_c{}^{*\lambda'}\int d^3k_{r'}\, f_c{}^*(k_{r'})\right)\tilde{\rho}_{m+m'}{}^{\{\lambda,\lambda'\}}(k_1,\cdots,k_{m+m'})\right]. \quad (51)$$

Another factor corresponding to the contribution from the scattering of photons by photons could have been included explicitly, but since this term does not contribute to the cancellation of infrared divergences, nor does add to the divergences, it has not been considered in this analysis.

Making the appropriate cancellations, and combining the terms with a little bit of careful counting, we arrive at the expression

$$M=e^{\alpha B}\sum_{l=0}^{\infty}\frac{1}{l!}\sum_{m=0}^{\infty}\frac{1}{m!}\sum_{m'=0}^{\infty}\frac{1}{m'!}[\sum_{a,\lambda}\alpha_a{}^\lambda\gamma_a{}^{*\lambda}]^l\left\{\sum_{j=0}^{m}\sum_{j'=0}^{m'}\frac{m!}{j!(m-j)!}\frac{m'!}{j'!(m'-j')!}\right.$$
$$\left.\times[-\sum_{a,\lambda}\alpha_a{}^\lambda(\tilde{S}^\lambda,f_a)_\Omega]^j[\sum_{c,\lambda'}\gamma_c{}^{*\lambda'}(f_c{}^*,\tilde{S}^{\lambda'})_\Omega]^{j'}P_{m-j,m'-j'}(p_i,p_f)\right\}[\exp(-\tfrac{1}{2}\sum_{\lambda,a}|\alpha_a{}^\lambda|^2)\exp(-\tfrac{1}{2}\sum_{\lambda,a}|\gamma_a{}^\lambda|^2)], \quad (52)$$

where

$$(\tilde{S}^\lambda,f_a)_\Omega\equiv\int_\Omega d^3k\,\tilde{S}^{(\lambda)}(k)f_a(k),\quad (f_c{}^*,\tilde{S}^{\lambda'})_\Omega\equiv\int_\Omega d^3k\,\tilde{S}^{(\lambda')}(k)f_c{}^*(k),$$

and

$$P_{j,j'}\equiv(-1)^j\left(\prod_{r=1}^{j}\int d^3k_r\sum_{\lambda,a}\alpha_a{}^\lambda f_a(k_r)\right)\left(\prod_{r'=1}^{j'}\int d^3k_{r'}\sum_{\lambda',c}\gamma_c{}^{*\lambda'}f_c{}^*(k_{r'})\right)\xi_{j+j'}{}^{\{\lambda\}}(k_1,\cdots,k_{j+j'}). \quad (53)$$

Defining the "residuals" $m_{j,j'}$ by

$$m_{j,j'}(p_i,p_f)=\frac{P_{j,j'}}{j!j'!}, \quad (54)$$

and reordering the sums in (52), we can write

$$\tilde{M}=e^{\alpha B}\exp[-\tfrac{1}{2}\sum_{\lambda,a}|\alpha_a{}^\lambda|^2]\exp[-\tfrac{1}{2}\sum_{\lambda,a}|\gamma_a{}^\lambda|^2]\exp[\sum_{\lambda,a}\alpha_a{}^\lambda\gamma_a{}^{*\lambda}]$$
$$\times\exp[-\sum_{\lambda,a}\alpha_a{}^\lambda(\tilde{S}^\lambda,f_a)_\Omega]\exp[\sum_{\lambda,c}\gamma_c{}^{*\lambda}(f_c{}^*,\tilde{S}^\lambda)_\Omega]\{\sum_{m,m'=0}^{\infty}m_{m,m'}\}. \quad (55)$$

To simplify the notation further, we define the coefficients $\beta_a{}^\lambda$ by

$$\beta_a{}^\lambda\equiv(f_a{}^*,\tilde{S}^\lambda)_\Omega. \quad (56)$$

Since the function $\tilde{S}$ is real, Eq. (55) becomes

$$\tilde{M}=e^{\alpha B}\exp[-\tfrac{1}{2}\sum_{\lambda,a}|\alpha_a{}^\lambda|^2]\exp[-\tfrac{1}{2}\sum_{\lambda,a}|\gamma_a{}^\lambda|^2]\exp[\sum_{\lambda,a}\alpha_a{}^\lambda\gamma_a{}^{*\lambda}]$$
$$\times\exp[-\sum_{\lambda,a}\alpha_a{}^\lambda\beta_a{}^{*\lambda}]\exp[\sum_{\lambda,a}\beta_a{}^\lambda\gamma_a{}^{*\lambda}]\{\sum_{m,m'=0}^{\infty}m_{m,m'}(p_i,p_f)\}. \quad (57)$$

As in Sec. IV, it proves convenient to split the function $\tilde{S}^{(\lambda)}$ into two parts:

$$\tilde{S}_i^{(\lambda)}(k)=\frac{e}{[2(2\pi)^3k_0]^{1/2}}\frac{p_i\cdot e^{(\lambda)}}{k\cdot p_i},\quad \tilde{S}_f^{(\lambda)}(k)=\frac{e}{[2(2\pi)^3k_0]^{1/2}}\frac{p_f\cdot e^{(\lambda)}}{k\cdot p_f}. \tag{58}$$

We can then define the coefficients $\beta_{ia}{}^{\lambda}$ and $\beta_{fa}{}^{\lambda}$ by

$$\beta_{ia}{}^{\lambda}=(f_a{}^*,\tilde{S}_i{}^{(\lambda)})_\Omega,\quad \beta_{fa}{}^{\lambda}=(f_a{}^*,\tilde{S}_f{}^{(\lambda)})_\Omega, \tag{59}$$

so that by Eqs. (10) and (56),

$$\beta_a{}^{\lambda}=\beta_{fa}{}^{\lambda}-\beta_{ia}{}^{\lambda}. \tag{60}$$

The complex coefficients $\{\alpha_a{}^{\lambda}\}$ and $\{\gamma_c{}^{\lambda'}\}$ which specify the initial and final states of the photons may be regarded as a set of coordinates in some complex infinite-dimensional space X. It is then possible to simplify Eq. (57) by making a translation of the coordinate system in X by the amounts defined in Eq. (59):

$$\gamma_a{}^{\lambda}=\beta_{fa}{}^{\lambda}+\epsilon_{fa}{}^{\lambda},\quad \alpha_a{}^{\lambda}=\beta_{ia}{}^{\lambda}+\epsilon_{ia}{}^{\lambda}. \tag{61}$$

Thus

$$\begin{aligned}\tilde{M}=e^{\alpha B}\{\exp\sum_{\lambda,a}[-\tfrac{1}{2}|\beta_{ia}{}^{\lambda}+\epsilon_{ia}{}^{\lambda}|^2-\tfrac{1}{2}|\beta_{fa}{}^{\lambda}+\epsilon_{fa}{}^{\lambda}|^2+(\beta_{ia}{}^{\lambda}+\epsilon_{ia}{}^{\lambda})(\beta_{fa}{}^{\lambda}+\epsilon_{fa}{}^{\lambda})^*\\ -(\beta_{ia}{}^{\lambda}+\epsilon_{ia}{}^{\lambda})\beta_a{}^{*\lambda}+\beta_a{}^{\lambda}(\beta_{fa}{}^{\lambda}+\epsilon_{fa}{}^{\lambda})^*]\}\{\sum_{m,m'=0}^{\infty}m_{m,m'}\}\\ =e^{\alpha B}\{\exp\sum_{a,\lambda}[+\tfrac{1}{2}|\beta_f-\beta_i|^2-\tfrac{1}{2}|\epsilon_f-\epsilon_i|^2+i\,\mathrm{Im}(\beta_i{}^*\epsilon_i+\beta^a\epsilon_f{}^*-\beta_i\beta_f{}^*+\epsilon_i\epsilon_f{}^*)]\}\{\sum_{m,m'=0}^{\infty}m_{m,m'}\},\end{aligned} \tag{62}$$

where the mode and polarization indices have been suppressed for convenience.

By Eqs. (43), (53), (56), and (60), we have

$$\begin{aligned}\sum_{\lambda,a}\tfrac{1}{2}|\beta_{fa}{}^{\lambda}-\beta_{ia}{}^{\lambda}|^2=\sum_{\lambda,a}\tfrac{1}{2}|\beta_a{}^{\lambda}|^2=\sum_{\lambda,a}\left[\int d^3k\,\tilde{S}^{(\lambda)}(k)f_a(k)\right]\left[\int d^3k'\,\tilde{S}^{(\lambda)}(k')f_a{}^*(k')\right]\\ =\sum_{\lambda}\frac{1}{2}\int d^3k\,\tilde{S}^{(\lambda)}(k)\tilde{S}^{(\lambda)}(k)=\alpha\tilde{B}.\end{aligned} \tag{63}$$

Substituting Eq. (63) into (62), we arrive at the important result

$$\tilde{M}=\exp(\alpha B+\alpha\tilde{B})\exp(-\tfrac{1}{2}\sum_{\lambda,a}|\epsilon_{fa}{}^{\lambda}-\epsilon_{ia}{}^{\lambda}|^2)\times e^{i\phi}\{\sum_{m,m'=0}^{\infty}m_{m,m'}\}, \tag{64}$$

where ϕ is real.

The argument of the first exponential was shown in Sec. IV to be infrared divergenceless in the limit of zero photon mass. The third exponential has modulus unity, and the last sum is term by term divergence-free. If the possible states of the system are restricted by the condition

$$\sum_{\lambda,a}|\epsilon_{fa}{}^{\lambda}-\epsilon_{ia}{}^{\lambda}|^2<\infty, \tag{65}$$

the second exponential is nonzero, but less than or equal to unity. With this condition satisfied the infrared divergences have been eliminated. The interpretation of this restriction is discussed in the next section.

VI. INTERPRETATION OF THE PHOTON STATES

In the beginning of this section, we will show that Eq. (65) defines a separable Hilbert space. To do this, we study a related space $\mathfrak{F}$ which will turn out to be identical to the ordinary Fock space $\mathfrak{F}_\infty$. Translations like Eq. (61) will not change the intrinsic properties of this space. Finally, a calculation of the total cross section will relate this whole discussion to experiment.

Much of the mathematical material here will be treated heuristically, but a more rigorous formulation of the statements can be found in the papers by Bargmann.[2,5]

We will define a separable Hilbert space $\mathfrak{F}$ in the following manner: Let $\{\theta_i\}$ be an infinite sequence of complex numbers. A set of "principal vectors" $|\{\theta_i\}\rangle$ is

[5] V. Bargmann, Comm. Pure Appl. Math. 14, 187 (1961).

then defined by the equation

$$|\{\theta_i\}\rangle=\prod_i|\theta_i\rangle$$
$$=\prod_i \exp[-\tfrac{1}{2}|\theta_i|^2]\exp[\theta_i a_i^\dagger]|0\rangle$$
$$=\prod_i \exp[-\tfrac{1}{2}|\theta_i|^2]\exp\left[\theta_i\int f_i(k)a^\dagger(k)\right]|0\rangle \quad (66)$$

and the condition

$$\sum_i|\theta_i|^2<\infty\,. \quad (67)$$

The elements of $\mathfrak{F}$ are taken to be the closure of all finite linear combinations of the principal vectors.

From (66) and the commutation rules for $a^\dagger(k)$, the inner product of two elements, $|f\rangle=\sum_{j=1}^{p}\lambda_j|\{\theta_i^{(j)}\}\rangle$ and $|f'\rangle=\sum_{k=1}^{q}\mu_k|\{\theta_i^{(k)}\}\rangle$, is given by

$$\langle f|f'\rangle=\sum_{j,k}\lambda_j^*\mu_k\{\exp[\sum_i\theta_i^{*(j)}\theta_i^{(k)}]$$
$$\times\exp[-\tfrac{1}{2}\sum_i|\theta_i^{(j)}|^2]\exp[-\tfrac{1}{2}\sum_i|\theta_i^{(k)}|^2]\}\,. \quad (68)$$

In particular, the inner product of two principal vectors $|\{\theta_i^{(j)}\}\rangle$ and $|\{\theta_i^{(k)}\}\rangle$ has the property

$$|\langle\{\theta_i^{(j)}\}|\{\theta_i^{(k)}\}\rangle|^2$$
$$=|\exp[\sum_i\theta_i^{*(j)}\theta_i^{(k)}]\exp[-\tfrac{1}{2}\sum_i|\theta_i^{(j)}|^2]$$
$$\times\exp[-\tfrac{1}{2}\sum_i|\theta_i^{(k)}|^2]|^2$$
$$=\exp\{-\sum_i|\theta_i^{(j)}-\theta_i^{(k)}|^2\}\,, \quad (69)$$

so that the principal vectors are all normalized to unit length. Moreover, by Eq. (67), no two principal vectors are normal to each other.

The properties of the space could in fact have been derived by using Eq. (69) instead of Eq. (66), but we wish to retain the connection with the previous sections of this paper.

The separability of $\mathfrak{F}$ follows from the existence of a countable sequence of vectors which is dense in $\mathfrak{F}$. Let $|\{\theta_i\}\rangle$ be any principal vector. From Eq. (67), it is known that for any $\delta_N>0$ there exists an integer N such that $\sum_{i>N}|\theta_i|^2<\delta_N$. Moreover, for any $i<N$ and $\delta>0$ it is always possible to find rational numbers $\{R_i\}$ such that $|\theta_i-R_i|^2<\delta$.

Consider a principal vector $|\{\theta_i'\}\rangle$ such that $\theta_i'=0$ for $i>N$, and $\theta_i'=R_i$ for $i\leq N$. Let $\sum_i|\theta_i|^2=A^2$. Then

$$\sum_i|\theta_i'-\theta_i|^2\leqslant N\delta+\delta_N$$

and

$$|\mathrm{Im}\sum_i\theta_i^*\theta_i'|=|\mathrm{Im}\sum_i\theta_i^*\theta_i+\mathrm{Im}\sum_i\theta_i^*(\theta_i'-\theta_i)|\leqslant|\sum_i\theta_i^*(\theta_i'-\theta_i)|\leqslant(\sum_i|\theta_i|^2\sum_i|\theta_i'-\theta_i|^2)^{1/2}\leqslant A\,(N\delta+\delta_N)^{1/2}.$$

Thus,

$$\||\{\theta_i\}\rangle-|\{\theta_i'\}\rangle\|^2=2-\langle\{\theta_i\}|\{\theta_i'\}\rangle-\langle\{\theta_i'\}|\{\theta_i\}\rangle$$
$$=2-\exp[-\tfrac{1}{2}\sum_i|\theta_i|^2]\exp[-\tfrac{1}{2}\sum_i|\theta_i'|^2]\{\exp[\sum_i\theta_i^*\theta_i']+\exp[\sum_i\theta_i^{*\prime}\theta_i]\}$$
$$=2-\exp[-\tfrac{1}{2}\sum_i|\theta_i'-\theta_i|^2]\{\exp[i\,\mathrm{Im}\sum_i\theta_i^*\theta_i']+\exp[i\,\mathrm{Im}\sum_i\theta_i^{*\prime}\theta_i]\}$$
$$=2\{1-(\exp[-\tfrac{1}{2}\sum_i|\theta_i'-\theta_i|^2])(\cos[\mathrm{Im}\sum_i\theta_i^*\theta_i'])\}\leqslant(A^2+1)(N\delta+\delta_N)\,.$$

Since $|\{\theta_i\}\rangle$, δ_N, and δ were arbitrary, we have shown that any principal vector can be approximated by another principal vector belonging to a denumerable set. The denumerable set which consists of all finite sums of principal vectors like $|\{\theta_i\}\rangle$ is dense in $\mathfrak{F}$.

In the case of massless soft photons, there is no reason to restrict the photon states by Eq. (67). Let $\{\theta_i^{(0)}\}$ be a sequence of complex numbers which are not square-summable, i.e.,

$$\sum_i|\theta_i^{(0)}|^2\nless\infty\,. \quad (70)$$

Then the states defined by the complex numbers $\{\theta_i\}$, and which satisfy the condition

$$\sum_i|\theta_i-\theta_i^{(0)}|^2<\infty\,, \quad (71)$$

form a separable Hilbert space $\mathfrak{F}^{(0)}$ with all the properties of $\mathfrak{F}$, except Eq. (67). $\mathfrak{F}^{(0)}$ is unitarily inequivalent to $\mathfrak{F}$, i.e., it forms a unitarily inequivalent representation of the canonical commutation rules.

In Sec. III, we discussed the connection between the Fock states and the principal vectors for a single mode. We will now briefly study the relationship between the occupation number parametrization and the principal vector parametrization.

The states in the Fock space $\mathfrak{F}_\infty$ are specified by a set M of infinite sequences of nonnegative integers $\{m_i\}$, or "occupation numbers" of which a finite number are different from zero. An orthonormal basis of Fock space is given by

$$|u_{\{m\}}\rangle=\prod_i\frac{(a_i^\dagger)^{m_i}}{(m_i!)^{1/2}}|0\rangle\,. \quad (72)$$

An arbitrary state $|f\rangle$ of $\mathfrak{F}_\infty$ is given by

$$|f\rangle=\sum_{\{m\}\in M}\gamma_{\{m\}}|u_{\{m\}}\rangle\,, \quad (73)$$

where the complex coefficients $\gamma_{\{m\}}$ satisfy

$$\sum_{\{m\}\in M}\gamma_{\{m\}}^2<\infty\,. \quad (74)$$

At this point, it should be apparent that $\mathfrak{F}\subset\mathfrak{F}_\infty$, since

$$|\{\theta_i\}\rangle=\sum_{\{m\}\in M}\prod_i\frac{(\theta_i a_i^\dagger)^{m_i}}{m_i!}|0\rangle$$
$$=\sum_{\{m\}\in M}\prod_i\frac{(\theta_i)^{m_i}}{(m_i!)^{1/2}}|u_{\{m\}}\rangle \quad (75)$$

and

$$\sum_{\{m\}\in M}\left|\prod_i \frac{(\theta_i)^{m_i}}{(m_i!)^{1/2}}\right|^2=\prod_i\left(\sum_{m=0}^{\infty}\frac{|\theta_i|^{2m_i}}{m_i!}\right)$$
$$=\exp(\sum_i|\theta_i|^2)<\infty\,. \quad (76)$$

The scalar product of two vectors $|f\rangle$ and $|f'\rangle$ in $\mathfrak{F}_\infty$ can be obtained from (72) and (73):

$$\langle f|f'\rangle=\sum_{\{m\}\in M}\gamma_{\{m\}}{}^*\gamma_{\{m\}}{}'. \quad (77)$$

Let $Q_n\{m_i\}$ be the truncated sequence

$$Q_n\{m_i\}=(m_1,m_2,\cdots,m_n,0,0,\cdots), \quad (78)$$

and define a projection on $\mathfrak{F}_\infty$ by

$$E_n|f\rangle=E_n\sum_{\{m\}\in M}\gamma_{\{m\}}|u_{\{m\}}\rangle=\sum_{\{m\}\in M_n}\gamma_{\{m\}}|u_{\{m\}}\rangle, \quad (79)$$

where M_n is the set of all sequences of the form given by Eq. (78). Then it follows from Eq. (74) that $E_n|f\rangle$ converges strongly to $|f\rangle$ as $n\to\infty$. We will show that $E_n|f\rangle$ is contained in $\mathfrak{F}$, which implies that $\mathfrak{F}_\infty\equiv\mathfrak{F}$.

The expansion of $E_n|f\rangle$ in terms of the principal vectors follows directly from Eq. (27):

$$E_n|f\rangle=\sum_{\{m\}\in M_n}\gamma_{\{m\}}\prod_{i=1}^{n}\int d\mu_i$$
$$\times|(\theta_1,\theta_2,\cdots,\theta_n,0,0,\cdots)\rangle\frac{(\theta_i{}^*)^{m_i}}{(m_i!)^{1/2}}, \quad (80)$$

where

$$d\mu_i=\pi^{-1}\exp[-\tfrac{1}{2}|\theta_i|^2]d(\mathrm{Re}\theta_i)d(\mathrm{Im}\theta_i). \quad (81)$$

From Eq. (69), it is clear that principal vectors $|\{\theta_i\}\rangle$ which do not satisfy Eq. (67) are orthogonal to $E_n|f\rangle$. Therefore $E_n|f\rangle\in\mathfrak{F}$ and the result $\mathfrak{F}_\infty=\mathfrak{F}$ follows, i.e., the Fock space built from states with a finite number of photons, and the space of principal vectors satisfying Eq. (67), are the same space.

In order to satisfy the requirement Eq. (65) for finite matrix elements, it will be necessary not to restrict the scattering states to $\mathfrak{F}$. For if the initial state were in $\mathfrak{F}$, i.e., the $\{\alpha_a{}^\lambda\}$ of Eq. (45) satisfied the condition

$$\sum_{\lambda,a}|\alpha_a{}^\lambda|^2<\infty\,, \quad (82)$$

then the final state parametrized by the sequence of complex numbers $\{\gamma_a{}^\lambda\}$ would be given by

$$\gamma_a{}^\lambda=\alpha_a{}^\lambda+(\beta_{fa}{}^\lambda-\beta_{ia}{}^\lambda)+(\epsilon_{fa}{}^\lambda-\epsilon_{ia}{}^\lambda)\,, \quad (83)$$

where we have used Eq. (61), and the ϵ's would satisfy

$$\sum_{\lambda,a}|\epsilon_{fa}{}^\lambda-\epsilon_{ia}{}^\lambda|^2<\infty\,. \quad (84)$$

But we know from Sec. IV and Eq. (63) that

$$\sum_{\lambda,a}|\beta_{fa}{}^\lambda-\beta_{ia}{}^\lambda|^2=\tfrac{1}{2}\sum_{l=1}^{2}\int|\tilde{S}^{(l)}(k)|^2d^3k=\alpha\tilde{B}\,, \quad (85)$$

and $\tilde{B}\to\infty$ as the photon mass approaches zero. Therefore

$$\sum_{\lambda,a}|\gamma_a{}^\lambda|^2\to\infty\,, \quad (86)$$

as the photon mass approaches zero. Thus the final state cannot belong to $\mathfrak{F}$.

Nevertheless, the coefficients $\{\gamma_a{}^\lambda\}$ define a final state. It must be that the final state belongs to an inequivalent representation of the canonical commutation rules whose most outstanding feature is that the average number of photons is infinite. Not any final state will do, however, for the boundaries of this new space $\mathfrak{F}'$ are restricted by the condition Eq. (65).

One of many ways to satisfy Eq. (65) which preserves symmetry between the initial and final states is to write these states as

$$\alpha_a{}^\lambda=\epsilon_{ia}{}^\lambda+\beta_{ia}{}^\lambda+\epsilon_{0a}{}^\lambda\,,$$
$$\gamma_a{}^\lambda=\epsilon_{fa}{}^\lambda+\beta_{fa}{}^\lambda+\epsilon_{0a}{}^\lambda\,,$$

and to restrict the states by

$$\sum_{\lambda,a}|\epsilon_{ia}{}^\lambda|^2<\infty\,,\quad \sum_{\lambda,a}|\epsilon_{fa}{}^\lambda|^2<\infty\,.$$

Then we would get different theories by different choices of the sequence $\{\epsilon_{0a}{}^\lambda\}$. With such a choice, the photon states would have a dependence upon the momenta of the participating electrons.

So far we have not spoken at all about Ω, the region of momentum space on which the single photon states $\{f_a\}$ were defined. In an experimental situation, there is always a threshold below which a single photon cannot be detected. We identify Ω with what we shall call the "resolution region," i.e., all photons with momentum $k\in\Omega$ are not detectable, while those which satisfy $k\notin\Omega$ are detectable. In what follows, the nondetectable photons will be spoken of as "soft" photons, while the others will be called "hard." Furthermore, we shall indicate the resolution of the momentum space by a subscript, e.g., X_Ω.

In a practical calculation where one wants to treat, for example, the scattering of an electron with the emission of hard photons, the hard photons can be dealt with by the conventional occupation-number parametrization, while the soft photons are described in terms of the translated principal vectors. More specifically, consider the calculation of the cross section for an electron of momentum p_i scattering into a state with an electron of momentum p_f plus several hard photons. The incoming electron is associated with a photon field described by a sequence $\{\alpha_a{}^\lambda\}$ and the outgoing electron has a photon field $\{\gamma_a{}^\lambda\}$. In Eq. (64), the "basic matrix element" $m_{0,0}$ corresponds to diagrams with only the detectable real photons and those virtual photons necessary for the process to occur. The terms m_{ij} for $i, j=0$ contain the effects of the noninfrared parts of the real and virtual soft photons to higher order in the coupling constant.

To lowest order in the noninfrared photons, the squared matrix element for a particular diagram $m_{0,0}$ is from Eq. (64):

$$|\tilde{M}|^2=\exp[2(\mathrm{Re}\,\alpha B+\alpha\tilde{B}_\Omega)]$$
$$\times\exp[-\sum_{\lambda,a}|\epsilon_{fa}{}^\lambda-\epsilon_{ia}{}^\lambda|^2]|m_{0,0}|^2. \quad (87)$$

We can then sum over final states. The result (to lowest order) is independent of the initial state:

$$\sum_{\substack{\text{final}\\\text{states}}}|\tilde{M}|^2=\exp[2(\mathrm{Re}\,\alpha B+\alpha\tilde{B}_\Omega)]|m_{0,0}|^2$$
$$\times\lim_{n\to\infty}(\pi^{-1})^n\left[\int d^2\epsilon_f e^{-|\epsilon_f|^2}\right]^n$$
$$=\exp[2(\mathrm{Re}\,\alpha B+\alpha\tilde{B}_\Omega)]|m_{0,0}|^2. \quad (88)$$

The remaining exponential contains part of the effect of the choice of $\tilde{S}_i(k)$ and the region of resolution Ω, and we obtain a similar result to what Yennie *et al.* obtained (for a nonenergy-conserving potential). In fact, the "reason" why the results are the same is that, in the summation over all final states in the conventional treatment of the infrafed divergence, the main contributions came from states which were not in the usual Fock space, but were in a nonseparable space defined by Eq. (72) without any restrictions on the sequence $\{m\}$ of occupation numbers. In particular, the separable space of final states $\mathfrak{F}'$ is contained in this nonseparable space.

In the above computation, and in Sec. V, the resolution regions for the initial and final states were assumed to be the same. One can argue that the resolution region of the initial state can be made arbitrarily small, but finite, by waiting a sufficiently long time before the scattering experiment. The situation where the initial resolution region is smaller than the final state resolution region is discussed in the Appendix.

ACKNOWLEDGMENT

The author gratefully thanks Professor Stanley Mandelstam for suggesting this problem, and for his patient guidance and helpful advice.

APPENDIX A: GENERALIZATION OF THE CANCELLATION TO SEVERAL ELECTRON LINES

For simplicity, only the case of a single electron line interacting with a potential was treated in Sec. V. The generalization to several electron lines interacting with one another will be outlined here for completeness. It is possible to make an extension to positron scattering, pair production, and other processes, but we shall not do so here.

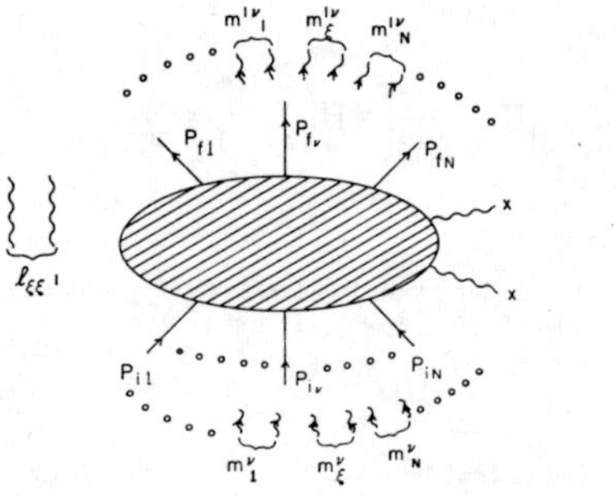

FIG. 4. Representation of l noninteracting real soft photons, $m_\xi{}^\nu$ real soft photons associated with the ξth electron and absorbed by the νth electron, and $m_\xi{}'^\nu$ real soft photons associated with the ξth electron and emitted by the νth electron.

Since the cancellation of the infrared divergences in this more complicated situation requires a proliferation of subscripts and superscripts, we drop all notation having to do with the polarization of the photons. Superscripts will now designate the electron line.

The initial state consists of N incoming electrons with momenta $p_{i1}, p_{i2}, \cdots, p_{i\nu}, \cdots, p_{iN}$, along with some photons. They scatter into a final state of N outgoing electrons with momenta $p_{f1}, \cdots, p_{f\nu}, \cdots, p_{fN}$, again with some photons. We have assumed that all resolution regions are identical.

Thus the soft-photon initial state can, for example, be conveniently written as

$$\prod_a \frac{\exp\left[\sum_\nu(\beta_{ia}{}^\nu+\epsilon_{ia}{}^\nu)\int_\Omega d^3k\, f_a(k)a^\dagger(k)\right]}{\exp[\frac{1}{2}|\sum_\nu \beta_{ia}{}^\nu+\epsilon_{ia}{}^\nu|^2]}|0\rangle, \quad (A1)$$

where

$$\beta_{ia}{}^\nu\equiv(f_a{}^*,S_i(p_{i\nu}))_\Omega. \quad (A2)$$

The final-state soft photons are described by a similar expression.

The interaction is illustrated in Fig. 4. In this diagram, our attention is focused on the νth electron line. The integer $m_\xi{}^\nu$ denotes the number of photons in the initial state which "belongs" to the ξth electron and interact with the νth electron line. Similarly, $m_\xi{}'^\nu$ is the number of photons in the final state belonging to the ξth electron which interact with the νth electron line. The number of noninteracting photons coming from the ξth incoming electron, and, becoming part of the ξ'th outgoing electron, is given by the integer $l_{\xi\xi'}$.

We arrive at an equation analogous to Eq. (64) through the following considerations.

(a) The overlap of the $l=\sum_{\xi\xi'} l_{\xi\xi'}$ initial-state noninteracting photons with the l final-state noninteracting photon contributes a factor

$$\prod_{\xi\xi'} [l_{\xi\xi'}!(\sum \alpha_a{}^{\xi}\gamma_a{}^{*\xi'})^{l_{\xi\xi'}}],$$

where

$$\alpha_a{}^{\xi}\equiv\beta_{ia}{}^{\xi}+\epsilon_{ia}{}^{\xi},\quad \gamma_a{}^{\xi}\equiv\beta_{fa}{}^{\xi}+\epsilon_{fa}{}^{\xi}.$$

Note again that the superscript ξ refers to the electron line, and not to the polarization.

(b) A contribution from the interaction of $m=\sum_{\nu,\xi}{}^{N} m_{\xi}{}^{\nu}$ initial-state photons, and the $m'=\sum_{\nu,\xi}{}^{N} m_{\xi}{}'^{\nu}$ final-state photons with the N electron lines, gives

$$\bar{\rho}(\{m_{\xi}{}^{\nu}\}\{m_{\xi}{}'^{\nu}\})=\prod_{\nu}\left[\sum_{t=0}^{m^{\nu}+m'^{\nu}}\prod_{i=1}^{t}(-1)^{m_{\nu}}\tilde{S}(k_{\nu,i})\xi_{m^{\nu}+m'^{\nu}-t}(k_{\nu,t+1},\cdots,k_{\nu,m^{\nu}+m'^{\nu}})\frac{1}{t!(m^{\nu}+m'^{\nu}-t)!}\right],$$

where $m^{\nu}=\sum_{\xi} m_{\xi}{}^{\nu}$, $m'^{\nu}=\sum_{\xi} m_{\xi}{}'^{\nu}$.

(c) Contribution (b) must be integrated over the momentum distribution that is obtained from the formal expansion of the initial and final states,

$$\left\{\prod_{\xi}\left[\left(\prod_{r=1}^{m_{\xi}}\sum_{a}\alpha_a{}^{\xi}\int d^3k_{\xi,r}\, f_a(k_{\xi,r})\right)\left(\prod_{r'=m_{\xi}+1}^{m_{\xi}+m_{\xi}'}\sum_{c}\gamma_c{}^{*\xi}\int d^3k_{\xi,r'}\, f_a{}^{*}(k_{\xi,r'})\right)\right]\right\}\{\bar{\rho}(\{m_{\xi}{}^{\nu}\},\{m_{\xi}{}'^{\nu}\};k_1,\cdots,k_m,\cdots,k_{m+m'})\},$$

where $m_{\xi}=\sum_{\nu} m_{\xi}{}^{\nu}$, $m_{\xi}'=\sum_{\nu} m_{\xi}{}'^{\nu}$.

(d) The formal expansion of Eq. (A1) also leads to the factors

$$\prod_{\xi}\left[\frac{1}{(m_{\xi}+l_{\xi})!}\frac{1}{(m_{\xi}'+l_{\xi}')!}\right]\exp[-\tfrac{1}{2}\sum_{a}|\sum_{\xi}\alpha_a{}^{\xi}|^2]\exp[-\tfrac{1}{2}\sum_{c}|\sum_{\xi}\gamma_c{}^{\xi}|^2],$$

where $l_{\xi}=\sum_{\xi'} l_{\xi\xi'}$, $l_{\xi}'=\sum_{\xi} l_{\xi\xi'}$.

(e) In addition to the above, there is a combinatorial factor which accounts for the number of ways that the $(m+l)$ initial-state photons, $(m'+l)$ final-state photons, and the l noninteracting photons, can be distributed among themselves:

$$\left[\prod_{\xi}\frac{(m_{\xi}+l_{\xi})!}{\prod_{\nu}(m_{\xi}{}^{\nu})!\prod_{\xi'}(l_{\xi\xi'})!}\right]\left[\prod_{\xi'}\frac{(m_{\xi}'+l_{\xi}')!}{\prod_{\nu}(m_{\xi}{}'^{\nu})!\prod_{\xi}(l_{\xi\xi'})!}\right].$$

(f) There is an infrared divergent factor due to virtual photon corrections (see Yennie *et al.*[4]):

$$F=\exp[\alpha B].$$

To get the matrix element $\tilde{M}$, the contributions are summed over all values of $\{m_{\xi}{}^{\nu}\}$, $\{m_{\xi}{}'^{\nu}\}$, $\{l_{\xi\xi'}\}$,

$$\tilde{M}=F\sum_{\{l_{\xi\xi'}\}=0}^{\infty}\sum_{\{m_{\xi}{}^{\nu}\}=0}^{\infty}\sum_{\{m_{\xi}{}'^{\nu}\}=0}^{\infty}\left[\prod_{\xi}\left(\frac{1}{(m_{\xi}+l_{\xi})!}\frac{1}{(m_{\xi}'+l_{\xi}')!}\right)\right]\left[\prod_{\xi}\frac{(m_{\xi}+l_{\xi})!}{\prod_{\nu}(m_{\xi}{}^{\nu})!\prod_{\xi'}(l_{\xi\xi'})!}\prod_{\xi'}\frac{(m_{\xi'}'+l_{\xi'}')!}{\prod_{\nu}(m_{\xi}{}'^{\nu})!\prod_{\xi}(l_{\xi\xi'})!}\right]$$

$$\times[\exp[-\tfrac{1}{2}\sum_{a}|\sum_{\xi}\alpha_a{}^{\xi}|^2]\exp[-\tfrac{1}{2}\sum_{a}|\sum_{\xi}\gamma_a{}^{\xi}|^2]][\prod_{\xi\xi'} l_{\xi\xi'}!(\sum_{a}\alpha_a{}^{\xi}\gamma_a{}^{*\xi'})^{l_{\xi\xi'}}]$$

$$\times\left\{\prod_{\xi}\left[\left(\prod_{r=1}^{m_{\xi}}(-1)^{m_{\xi}}\sum_{a}\alpha_a{}^{\xi}\int d^3k_{\xi,r}\, f_a(k_{\xi,r})\right)\left(\prod_{r'=m_{\xi}+1}^{m_{\xi}+m_{\xi}'}\sum_{c}\gamma_c{}^{*\xi}\int d^3k_{\xi,r'}\, f_c{}^{*}(k_{\xi,r'})\right)\right]\bar{\rho}(k_1\cdots,k_{m+m'})\right\}.$$

The last factor in the braces can be reduced by separating out the divergent terms:

$$\left\{\prod_{\nu\xi}\left[\sum_{j_{\xi}{}^{\nu}=0}^{m_{\xi}{}^{\nu}}\sum_{j_{\xi}{}'^{\nu}=0}^{m_{\xi}{}'^{\nu}}\frac{(m_{\xi}{}^{\nu})!}{(j_{\xi}{}^{\nu})!(m_{\xi}{}^{\nu}-j_{\xi}{}^{\nu})!}\frac{(m_{\xi}{}'^{\nu})!}{(j_{\xi}{}'^{\nu})!(m_{\xi}{}'^{\nu}-j_{\xi}{}'^{\nu})!}(-\sum_{a}\alpha_a{}^{\xi}(\tilde{S}_{\nu},f_a))^{j_{\xi}{}^{\nu}}(\sum_{a}\gamma_a{}^{*\xi}(f_a{}^{*},\tilde{S}_{\nu}))^{j_{\xi}{}'^{\nu}}P\right]\right\},$$

where

$$P(\{j_{\xi}{}^{\nu}\}\{j_{\xi}{}'^{\nu}\};\{p_{i\nu}\},\{p_{f\nu}\})=\left[\left(\prod_{r=1}^{j_{\xi}{}^{\nu}}\int d^3k_r\sum_{a}\alpha_a{}^{\xi}f_a(k)\right)\left(\prod_{r'=j_{\xi}{}^{\nu}+1}^{j_{\xi}{}^{\nu}+j_{\xi}{}'^{\nu}}\int d^3k_{r'}\sum_{c}\gamma_c{}^{*\xi}f_c{}^{*}(k)\right)\xi_{j_{\xi}{}^{\nu}+j_{\xi}{}'^{\nu}}(k_1,\cdots,k_{j_{\xi}{}^{\nu}+j_{\xi}{}'^{\nu}})\right\}.$$

Again, one defines a divergence-free "residual" by

$$m(j,j')=\frac{P}{j!j'!},$$

so that

$$\tilde{M}=F[\prod_{\xi\xi'}\exp[\sum_a \alpha_a{}^{\xi}\gamma_a{}^{*\xi'}]]\exp[-\sum_a|\sum_\xi \alpha_a{}^{\xi}|^2]\exp[-\sum_a|\sum_\xi \gamma_a{}^{\xi}|^2]$$

$$\times[\prod_{\nu\xi}\exp[-\sum_a \alpha_a{}^{\xi}\beta_a{}^{*\nu}]\exp[\sum_a \gamma_a{}^{*\xi}\beta_a{}^{\nu}]][\prod_{\nu\xi}\sum_{j_\xi{}^\nu=0}^{\infty}\sum_{j_\xi{}'^\nu=0}^{\infty} m(\{j_\xi{}^\nu\},\{j_\xi{}'^\nu\})].$$

Making the translation to a new coordinate system, we get

$$\alpha_a{}^{\xi}=\beta_{ia}{}^{\xi}+\epsilon_{ia}{}^{\xi},\quad \gamma_a{}^{\xi}=\beta_{fa}{}^{\xi}+\epsilon_{fa}{}^{\xi},\quad \beta_a{}^{\nu}=\beta_{fa}{}^{\nu}-\beta_{fa}{}^{\nu},$$

and with a little additional algebra, the final result is exhibited:

$$\tilde{M}=F\exp[\tfrac{1}{2}\sum_a|\sum_\xi \beta_{fa}{}^{\xi}-\beta_{ia}{}^{\xi}|^2]\exp[-\tfrac{1}{2}\sum_a|\sum_\xi \epsilon_{fa}{}^{\xi}-\epsilon_{ia}{}^{\xi}|^2]e^{i\phi}[\prod_{\nu\xi}\sum_{j_\xi{}^\nu=0}^{\infty}\sum_{j_\xi{}'^\nu=0}^{\infty} m(\{j_\xi{}^\nu\},\{j_\xi{}'^\nu\})]. \quad \text{(A3)}$$

The divergences in the factor F and the first exponential cancel (Yennie *et al.*[4]). Again the condition for finite matrix elements is

$$\sum_a|\sum_\xi \epsilon_{fa}{}^{\xi}-\epsilon_{ia}{}^{\xi}|^2<\infty. \quad \text{(A4)}$$

APPENDIX B: THE PROJECTION OF ENERGY-MOMENTUM EIGENSTATES

For completeness, the procedure for projecting out energy-momentum eigenstates will be given here. Consider a typical soft photon state described by a principal vector $|\{\alpha_a{}^{\xi}\}\rangle$:

$$|\{\alpha_a{}^{\lambda}\}\rangle=\exp[-\tfrac{1}{2}\sum_{\lambda,a}|\alpha_a{}^{\lambda}|^2]\exp\left[\sum_{\lambda,a}\alpha_a{}^{\lambda}\int_\Omega d^3k\, f_a(k)e^{(\lambda)}(k)a^{(\lambda)\dagger}(k)\right]|0\rangle$$

$$=\exp[-\tfrac{1}{2}\sum_{\lambda,a}|\alpha_a{}^{\lambda}|^2]\sum_{n=0}^{\infty}\frac{1}{n!}\left[\sum_{a,\lambda}\alpha_a{}^{\lambda}\int_\Omega d^3k_1\, f_a(k_1)e^{(\lambda)}a^{(\lambda)\dagger}(k)\right]\cdots\left[\sum_{a,\lambda}a_a{}^{\lambda}\int_\Omega d^3k_n\, f_a(k_n)e^{(\lambda)}a^{(\lambda)\dagger}(k_n)\right]|0\rangle. \quad \text{(B1)}$$

Using the formula

$$\frac{1}{2\pi}\int e^{ixy}dy=\delta(x), \quad \text{(B2)}$$

it is then póssible to project out from $|\{\alpha_a{}^{\lambda}\}\rangle$ the eigenstates of energy momentum:

$$P(E,\mathbf{K})|\{\alpha_a{}^{\lambda}\}\rangle=\exp[-\tfrac{1}{2}\sum_{\lambda,a}|\alpha_a{}^{\lambda}|^2]\sum_{n=0}^{\infty}\frac{1}{n!}$$

$$\times\left\{\int d^3k_i\cdots d^3k_n\prod_{i=1}^{n}[\sum_{a,\lambda}\alpha_a{}^{\lambda}f_a(k_i)e^{(\lambda)}(k_i)a^{(\lambda)\dagger}(k_i)]\delta(E-\sum_i\omega_i)\delta^3(\mathbf{K}-\sum_i\mathbf{k}_i)\right\}|0\rangle$$

$$=\exp[-\tfrac{1}{2}\sum_{\lambda,a}|\alpha_a{}^{\lambda}|^2]\sum_{n=0}^{\infty}\frac{1}{n!}\frac{1}{(2\pi)^4}\int dyd^3\mathbf{x}\, e^{iEy}e^{i\mathbf{K}\cdot\mathbf{x}}$$

$$\times\left\{\int d^3k_i\cdots d^3k_n\prod_i[\sum_{\lambda,a}\alpha_a{}^{\lambda}e^{-i\omega y}e^{-ik_i\cdot x}f_a(k_i)e^{(\lambda)}(k_i)a^{(\lambda)\dagger}(k_i)]\right\}|0\rangle$$

$$=\frac{1}{(2\pi)^4}\int dyd^3\mathbf{x}\exp[-\tfrac{1}{2}\sum_{\lambda,a}|\alpha_a{}^{\lambda}|^2]e^{iEy+i\mathbf{K}\cdot\mathbf{x}}\exp\left[\sum_{\lambda,a}\alpha_a{}^{\lambda}\int d^3k e^{-i\omega y}e^{-ik\cdot x}f_a(k)e^{(\lambda)}(k)a^{(\lambda)\dagger}(k)\right]. \quad \text{(B3)}$$

APPENDIX C: INITIAL AND FINAL STATES WITH DIFFERENT RESOLUTION REGIONS

Suppose that we are dealing with the situation where the resolution region Ω for the initial state is smaller than the resolution region Ω' for the final state ($\Omega\subset\Omega'$). In other words, the threshold for detecting low-energy photons is lower before the scattering experiment than afterwards. Then an infrared divergenceless matrix element in a form analogous to Eq. (64) may be obtained with very little additional complication.

Let us define a domain D of momentum space such that

$$\Omega'=\Omega\bigcup D,$$

where $\Omega\bigcup D=0$. Then we suppose that there exists a complete set of orthonormal functions $\{g_j(k)\}$ defined on D. A typical final state is now given by

$$|f\rangle=|\{\gamma_i,\gamma_{j'}\}\rangle=\frac{\exp\left[\prod_i \gamma_i\int_\Omega d^3k\, f_i(k)a^\dagger(k)\right]}{\exp[\frac{1}{2}\sum_i|\gamma_i|^2]}\times\frac{\exp\left[\prod_j \gamma_j'\int_D d^3k\, g_j(k)a^\dagger(k)\right]}{\exp[\frac{1}{2}\sum_j|\gamma_j|^2]}|0\rangle. \quad \text{(C1)}$$

The indices having to do with polarization have been suppressed. The result of the modification is that in the derivation of Eq. (55) one must make the substitutions

$$\gamma_a{}^\lambda\int_\Omega f_a e^{(\lambda)}\rightarrow\gamma_a{}^\lambda\int_\Omega f_a e^{(\lambda)}+\gamma_a{}'^\lambda\int_D g_a e^{(\lambda)},$$

so that Eq. (55) is correct only if we have on the right-hand side the additional factor

$$\exp[-\tfrac{1}{2}\sum_{\lambda,c}|\gamma_c{}'^\lambda|^2]\exp[\sum_{\lambda,c}\gamma_c{}'^{*\lambda}(g_c{}^*,\tilde{S}^{(\lambda)})_D],$$

and the divergenceless sum

$$\sum_{m,m'=0}^{\infty}{}' m_{m,m'}$$

contains integrals of the functions $\{g_c\}$ over the addition region D.

We now define new coefficients $\beta_a{}'^\lambda$ and variables $\epsilon_{fa}{}'^\lambda$ in a similar way to Eqs. (59) and (61):

$$\beta_a{}'^\lambda=(g_a{}^*,\tilde{S}^{(\lambda)})_D,\quad \gamma_a{}'^\lambda=\beta_a{}'^\lambda+\epsilon_{fa}{}'^\lambda. \quad \text{(C2)}$$

Then the additional (noninfrared-divergent) factor becomes

$$\exp[\sum_{\lambda,a}(\tfrac{1}{2}|\beta_a{}'^\lambda|^2-|\epsilon_{fa}{}'^\lambda|^2+i\,\mathrm{Im}\epsilon_{fa}{}^{*\prime\lambda}\beta_a{}'^\lambda)].$$

It is natural to define

$$\alpha\tilde{B}_D=\exp[\sum_{\lambda,a}\tfrac{1}{2}|\beta_a{}'^\lambda|^2], \quad \text{(C3)}$$

so that Eq. (64) becomes

$$\tilde{M}=\exp(\alpha B+\alpha\tilde{B}_\Omega+\alpha\tilde{B}_D)\exp[-\tfrac{1}{2}\sum_{\lambda,a}|\epsilon_{fa}{}^\lambda-\epsilon_{ia}{}^\lambda|^2]\times\exp[-\tfrac{1}{2}\sum_{\lambda,a}|\epsilon_{fa}{}'^\lambda|^2]e^{i\phi}\sum_{m,m'=0}^{\infty}m_{m,m'}. \quad \text{(C4)}$$

In this expression the infrared divergences cancel in the sum $\alpha B+\alpha B_\Omega$ of the argument of the first exponential. The term $\exp(\alpha\tilde{B}_D)$ accounts for the difference in resolution regions. The condition for finite matrix elements is now

$$\sum_{\lambda,a}|\epsilon_{fa}{}^\lambda-\epsilon_{ia}{}^\lambda|^2<\infty,\quad \sum_{\lambda,a}|\epsilon_{fa}{}'^\lambda|^2<\infty. \quad \text{(C5)}$$

PHYSICAL REVIEW D VOLUME 3, NUMBER 6 15 MARCH 1971

Coherent States and Indefinite Metric—Applications to the Free Electromagnetic and Gravitational Fields*

JAGANNATHAN GOMATAM
Physics Department, Syracuse University, Syracuse, New York 13210
(Received 16 October 1970)

The "coherent states" are constructed within the framework of an indefinite-metric linear vector space. It is found that these states have positive definite norm and are eigenstates of the annihilation operator. The theory is then applied to the free electromagnetic field and the linearized free gravitational field in the Gupta formulation. The "coherent states," restricted by the weak subsidiary condition, decompose into the usual coherent states involving only the physical degrees of freedom and an orthogonal vector of vanishing norm.

I. INTRODUCTION

THE coherent-state representation[1] has found useful applications in quantum optics and quantum field theory. But these coherent states were constructed by an infinite superposition of state vectors belonging to a Hilbert space with a positive definite metric. In view of the usefulness of the indefinite metric[2] in quantum field theory, the properties of "coherent states" (hereafter called C states)[3] in indefinite-metric vector spaces are of importance. Previous studies[4] in this direction were primarily confined to the so-called pseudo-oscillator in one dimension. We propose here a method of constructing covariant C states with an indefinite metric, with applications to the electromagnetic[5] and Einstein's linearized gravitational[6] fields in the framework of the Gupta formalism. The most important property of C states which has not been emphasized in earlier attempts[4] is the positive definiteness of the norm. The fact that the C-state vectors are amenable to direct physical interpretation, coupled with their positive definiteness despite the basic vector space being of indefinite metric, suggests the possibility that the C-state vectors may point to a way of extracting physical states out of the totality of indefinite-metric states. Imposing the Gupta subsidiary conditions[5,6] on the C-state vectors, we arrive at the remarkable result that every C-state vector decomposes into a physical coherent-state vector containing only the physical degrees of freedom and an orthogonal vector of vanishing norm.

The coherent representation of the electromagnetic field has been extensively investigated[1] in the framework of a positive-definite-metric Hilbert space. Starting with the quantization condition for the physical degrees of freedom of the vector potential, the coherent states were constructed[7,8] using standard techniques. In our investigation based on the indefinite metric, we mete out equal treatment to all four components of the electromagnetic vector potential and show that the Gupta subsidiary condition on the C states leads to only the physical degrees of freedom. The result that the C states are eigenstates of annihilation operator a_μ with eigenvalue z_μ provides a remarkable advantage, viz., Gupta's[5] weak subsidiary condition on the four-vector potential is most easily realized as a restriction on the four-dimensional complex manifold of vectors by stipulating $k^\mu z_\mu=0$, where of course $k^\mu k_\mu=0$. Thus, in our formalism, while the totality of all C states $|\{z_\mu\}\rangle$ already possess an essential ingredient of acceptable physical interpretation, viz., positive definiteness of the norm, the subclass obtained by imposing the constraint $k^\mu z_\mu=0$ satisfies the weak condition automatically. This subclass turns out to be the same (modulo vectors of vanishing norm) as the coherent states constructed by starting with the commutation relations between physical degrees of freedom only.

The insight provided by the indefinite-metric approach to coherent states raises the possibility of a similar approach to Einstein's linear model of the quantized gravitational[9] field. Here again the Gupta[6] method of quantization of the gravitational field provides the basic framework. The degrees of freedom that introduce an indefinite metric are suppressed in the

* Work supported by the National Science Foundation.

[1] The literature on coherent states is vast. A concise report of references especially on finite degrees of freedom is found in J. R. Klauder and E. C. G. Sudarshan, *Fundamentals of Quantum Optics* (Benjamin, New York, 1968) and H. Haken, *Light and Matter* (Springer, New York, 1970). For infinite degrees of freedom leading to the quantum-field-theoretic case, see S. S. Schweber, J. Math. Phys. **3**, 831 (1962) and F. Rohrlich, in *Analytic Methods in Mathematical Physics*, edited by R. P. Gilbert and R. G. Newton (Gordon and Breach, New York, 1970).

[2] K. L. Nagy, *State Vector Spaces with Indefinite Metric in Quantum Field Theory* (Noordhoff, Groningen, The Netherlands, 1966).

[3] The term "coherent states" is used in the literature where the underlying Hilbert space has a positive definite metric, with definite physical connotation. In our case although the algebraic construction resembles the standard techniques, the underlying Hilbert space has an indefinite metric. To avoid confusion, we will call our coherent states the C states.

[4] M. G. Gundzik, J. Math. Phys. **7**, 641 (1966); G. S. Agarwal, Nuovo Cimento **65B**, 266 (1970).

[5] S. N. Gupta, Proc. Phys. Soc. (London) **A63**, 681 (1950); K. Bleuler, Helv. Phys. Acta **23**, 567 (1950).

[6] S. N. Gupta, Proc. Phys. Soc. (London) **A65**, 161 (1952).

[7] V. Chung, Phys. Rev. **140**, B1110 (1965).

[8] T. W. B. Kibble, J. Math. Phys. **9**, 316 (1968).

[9] The construction of an overcomplete family of states starting with commutation relations belonging to an affine group has been considered by J. R. Klauder, in *Relativity: Proceedings of the Midwest Conference, Cincinnati, Ohio* (Plenum, New York, 1970), where models that are algebraically similar to gravitational theory are discussed. Here the Hilbert space considered is of positive definite metric. Our approach is based on an indefinite metric.

physical state by imposing subsidiary conditions on the C state. The treatment is quite analogous to the electromagnetic case.

The construction of C states with an indefinite metric is outlined in Sec. II. The theory is applied to the free electromagnetic field in Sec. III and to the linearized free gravitational field in Sec. IV. A discussion is presented in Sec. V.

II. CONSTRUCTION OF C STATES WITH INDEFINITE METRIC

The construction[4] of C states in an indefinite-metric linear vector space V is straightforward. However, we adopt a normalization different from the one used in the literature[4] with a view to exhibiting covariance explicitly in the four-dimensional case. We start with the so-called pseudo-oscillator quantization condition

$$[a,a^\dagger]=-1. \tag{2.1}$$

A realization of (2.1) in V is the analog of the usual Fock representation, however, with indefinite norm for the vectors $|n\rangle\in V$. The vacuum states are defined in the usual way:

$$a|0\rangle=0, \tag{2.2a}$$

$$\langle 0|a^\dagger=0. \tag{2.2b}$$

Note that $a^\dagger$ is the pseudoadjoint of a. We define the vectors $|n\rangle$ by

$$|n\rangle\equiv\frac{1}{(n!)^{1/2}}(a^\dagger)^n|0\rangle, \tag{2.3}$$

where n is a positive integer. Now the definition (2.3) together with (2.1) and (2.2) implies

$$a|n\rangle=-(n)^{1/2}|n-1\rangle, \tag{2.4a}$$

$$a^\dagger|n\rangle=(n+1)^{1/2}|n+1\rangle. \tag{2.4b}$$

The operator $N\equiv -a^\dagger a$ plays the role of the usual number operator:

$$(-a^\dagger a)|n\rangle=n|n\rangle. \tag{2.5}$$

Thus the Hamiltonian $H=EN$ has positive definite eigenvalues nE. The orthonormality and the completeness conditions are, respectively,

$$\langle n|m\rangle=(-1)^n\delta_{nm}, \tag{2.6}$$

$$\sum_{n=0}^{\infty}|n\rangle(-1)^n\langle n|=1, \tag{2.7}$$

where the quantity $\langle n|m\rangle$ is the inner product in V. Since

$$\||n\rangle\|^2=(-1)^n, \tag{2.8}$$

the metric in V is indefinite.

We define the C state $|z\rangle$ through

$$|z\rangle\equiv e^{|z|^2/2}\sum_{n=0}^{\infty}\frac{(-z)^n}{(n!)^{1/2}}|n\rangle. \tag{2.9}$$

Note the positive sign in front of $|z|^2$ and the negative sign in front of z. The adjoint state is

$$\langle z|\equiv e^{|z|^2/2}\sum_{n=0}^{\infty}\frac{(-z^*)^n}{(n!)^{1/2}}\langle n|. \tag{2.10}$$

The vectors $|z\rangle$ constitute a linear vector space $\mathfrak{C}$ with the inner product

$$\langle z'|z\rangle=\exp(\tfrac{1}{2}|z|^2+\tfrac{1}{2}|z'|^2-z'^*z) \tag{2.11a}$$

$$=\exp(\tfrac{1}{2}|z'-z|^2-i\,\mathrm{Im}z'^*z). \tag{2.11b}$$

For our purposes, the most significant property of the inner product (2.11) is that it leads to a positive definite norm for $|z\rangle$:

$$\||z\rangle\|^2\equiv\langle z|z\rangle=1. \tag{2.12}$$

Thus, the C states $|z\rangle$ constructed by an infinite superposition of indefinite-metric states $|n\rangle$ possess a positive definite norm. This is understandable in view of the exponential-type summation involved in the definition of C states. The previous term dominates the subsequent term, thus sequentially making the vacuum state the most important term in the summation leading to a positive norm. This property plays an important role in the construction of C states for the electromagnetic field in the Gupta formalism.

The "completeness" condition needs careful handling when the metric is indefinite. The relation reads as follows:

$$\int d\mu(z)|-z\rangle\langle z|=1, \tag{2.13}$$

where

$$d\mu(z)\equiv(1/\pi)e^{-2|z|^2}dxdy, \tag{2.14}$$

with $z=x+iy$. Note that the factor $e^{-2|z|^2}$ is introduced to secure convergence of the measure. Though this might seem artificial at this point, we have to bear in mind that the degrees of freedom leading to the indefinite metric occur in addition to those that possess positive-definite-metric vector spaces. Then the finiteness of the measure will be related to the physical dominance of the positive definiteness over negative definiteness in the exponent. The subsidiary condition will provide a precise statement of this dominance. Thus the completeness relation will find expression only in terms of physical degrees of freedom.

The states $|z\rangle$ are eigenstates of the annihilation operator with eigenvalue z,

$$a|z\rangle=z|z\rangle. \tag{2.15}$$

The C states $|z\rangle$ can be obtained from the vacuum by a pseudounitary transformation

$$|z\rangle=U(z,z^*)|0\rangle, \tag{2.16}$$

where

$$U(z,z^*)\equiv\exp[-(a^\dagger z-az^*)]. \tag{2.17}$$

That U is pseudounitary can be inferred from the fact that

$$(a^\dagger z-az^*)^\dagger=-(a^\dagger z-az^*).$$

III. FREE ELECTROMAGNETIC FIELD

A. Monochromatic Model

We will generalize the results of Sec. II to four degrees of freedom, one of which leads to an indefinite metric. The C states for the Minkowski oscillator

$$[a_\mu, a_\nu^\dagger] = g_{\mu\nu} \tag{3.1}$$

(where the signature of $g_{\mu\nu}$ is $+2$) will be first constructed. We call such states covariant C states. These results will later be generalized to the electromagnetic field. The spatial components a_1, a_2, a_3 satisfy the normal oscillator commutation relations, while a_0 obeys the pseudo-oscillator commutation condition of Sec. II. A realization of (3.1) in $V_4^{(4)}$ is governed by the following:

$$a_\mu|0,0,0,0\rangle \equiv a_\mu|0\rangle = 0\,, \tag{3.2a}$$

$$\langle 0|a_\mu^\dagger = 0\,, \tag{3.2b}$$

for all μ.

$$|n_1n_2n_3n_0\rangle \equiv |\{n_\mu\}\rangle = \frac{(a_1^\dagger)^{n_1}(a_2^\dagger)^{n_2}(a_3^\dagger)^{n_3}(a_0^\dagger)^{n_2}}{(n_1!n_2!n_3!n_0!)^{1/2}}|0\rangle, \tag{3.3}$$

$$a_1a_2a_3a_0|\{n_\mu\}\rangle = -(n_1n_2n_3n_0)^{1/2}|n_1-1,\, n_2-1,\, n_3-1,\, n_0-1\rangle, \tag{3.4}$$

$$a_1^\dagger a_2^\dagger a_3^\dagger a_0^\dagger|\{n_\mu\}\rangle = [(n_1+1)(n_2+1)(n_3+1)(n_0+1)]^{1/2} \times|n_1+1,\, n_2+1,\, n_3+1,\, n_0+1\rangle, \tag{3.5}$$

$$\langle\{n_\mu'\}|\{n_\mu\}\rangle = (-1)^{n_0}\delta_{n_0'n_0}\delta_{n_1'n_1}\delta_{n_2'n_2}\delta_{n_3'n_3}. \tag{3.6}$$

We recall that $\langle\{n_\mu'\}|\{n_\mu\}\rangle$ is the inner product in the indefinite-metric linear vector space $V_4^{(4)}$. The completeness condition is

$$\sum_{\{n_\mu\}=0}^{\infty} |\{n_\mu\}\rangle(-1)^{n_0}\langle\{n_\mu\}| = 1. \tag{3.7}$$

The covariant C states belonging to $\mathcal{C}_4^{(4)}$ are defined by

$$|\{z_\mu\}\rangle \equiv \exp(-\tfrac{1}{2}z^{\mu *}z_\mu) \times \sum_{\{n_\mu\}=0}^{\infty} \frac{(z_1)^{n_1}(z_2)^{n_2}(z_3)^{n_3}(-z_0)^{n_0}}{(n_1!n_2!n_3!n_0!)^{1/2}}|\{n_\mu\}\rangle, \tag{3.8}$$

where $z^{\mu *}z_\mu = z^{k*}z_k - |z^0|^2 \equiv |z_k|^2 - |z^0|^2 \equiv |z_\mu|^2$. The adjoint states are introduced in an obvious manner. It is easily checked, using (3.4), that

$$a_\mu|\{z_\mu\}\rangle = z_\mu|\{z_\mu\}\rangle. \tag{3.9}$$

Of course, there is no summation over μ in (3.9). The inner product in $\mathcal{C}_4^{(4)}$ is

$$\langle\{z_\mu'\}|\{z_\mu\}\rangle = \exp(-\tfrac{1}{2}|z_\mu'|^2 - \tfrac{1}{2}|z_\mu|^2 + z_\mu'^* z^\mu). \tag{3.10}$$

Note the important result that

$$\||\{z_\mu\}\rangle\|^2 = 1. \tag{3.11}$$

Therefore, the covariant C states possess positive norm.

The completeness condition on $|\{z_\mu\}|$ merits a careful study. When all z_μ are independent, we define the weight function

$$d\mu(\{z_\mu\}) \equiv (1/\pi^4)e^{-2|z^0|^2}dx_1dx_2dx_3dx_0dy_1dy_2dy_3dy_0 \tag{3.12}$$

and the completeness condition reads

$$\int d\mu(\{z_\mu\})|z_1, z_2, z_3, -z_0\rangle\langle z_1, z_2, z_3, +z_0| = 1. \tag{3.13}$$

However, in physical applications, not all z_μ are independent because of the subsidiary condition, as we will see presently. We stipulate that the states (3.8) satisfy the subsidiary condition in the Gupta[5] form,

$$k^\mu a_\mu|\{z_\mu\}\rangle = 0\,, \tag{3.14}$$

where k^μ is the null vector $k^\mu k_\mu = 0$. Since the C states are eigenstates of a_μ with eigenvalue z_μ, we can replace the left-hand side of (3.14) by $k^\mu z_\mu|\{z_\mu\}\rangle$. Thus the subsidiary condition will be satisfied if

$$k^\mu z_\mu|\{z_\mu\}\rangle = 0. \tag{3.15}$$

This implies that

$$k^\mu z_\mu = 0 \tag{3.16}$$

is necessary and sufficient to guarantee (3.14). We will denote the class of C states restricted by (3.16) by ${}_P\mathcal{C}_4^{(4)}$. Obviously, ${}_P\mathcal{C}_4^{(4)} \subset \mathcal{C}_4^{(4)}$.

There exists a pseudounitary transformation

$$U[\{z_\mu\},\{z_\mu^*\}] \equiv \exp(z^\mu a_\mu^\dagger - z^{\mu *}a_\mu) \tag{3.17}$$

such that

$$|\{z_\mu\}\rangle = U[\{z_\mu\},\{z_\mu^*\}]|0\rangle. \tag{3.18}$$

To see this we rewrite (3.8) as follows:

$$|\{z_\mu\}\rangle = \exp(-\tfrac{1}{2}|z_\mu|^2)\exp(z^\mu a_\mu^\dagger)|0\rangle \tag{3.19}$$

$$= \exp(-\tfrac{1}{2}|z_\mu|^2)\exp(z^\mu a_\mu^\dagger)\exp(-z^{\mu *}a_\mu)|0\rangle \tag{3.20}$$

and use the identity $e^{\frac{1}{2}[A,B]}e^{(A+B)} \equiv e^Ae^B$.

An interesting aspect of $|\{z_\mu\}\rangle$ is revealed by relation (3.20). With due regard to the commutation properties of a_μ, we rewrite (3.20) as

$$|\{z_\mu\}\rangle = \exp[-\tfrac{1}{2}(|z_1|^2+|z_2|^2)]\exp[-\tfrac{1}{2}(|z_3|^2-|z_0|^2)] \times\exp(z_1a_1^\dagger+z_2a_2^\dagger)\exp[-(z_1^*a_1+z_2^*a_2)] \times\exp(z_3a_3^\dagger-z_0a_0^\dagger)\exp[-(z_3^*a_3-z_0^*a_0)]|0\rangle \tag{3.21a}$$

$$= \exp[-\tfrac{1}{2}(|z_1|^2+|z_2|^2)]\exp(z_1a_1^\dagger+z_2a_2^\dagger) \times\exp[-(z_1^*a_1+z_2^*a_2)]\exp[(z_3a_3^\dagger-z_0a_0^\dagger) -(z_3^*a_3-z_0^*a_0)]|0\rangle, \tag{3.21b}$$

where we have used

$$[(z_3a_3^\dagger - z_0a_0^\dagger) - (z_3^*a_3 - z_0^*a_0)] = |z_3|^2 - |z_0|^2.$$

Note that in the calculation from (3.19) to (3.22) only the commutation properties of a_μ were used and the subsidiary condition was not used. The appearance of the factors $\exp[-\frac{1}{2}(|z_1|^2+|z_2|^2)]$ $\exp(z_1a_1^\dagger+z_2a_2^\dagger)$ $\times\exp[-(z_1^*a_1+z_2^*a_2)]$ as a unit is significant. Now we will realize the subsidiary condition (3.16) in a special frame of reference F,

$$k_1=k_2=0\,,\quad k_3=k_0\,,\quad z_3=z_0\,. \tag{3.22}$$

The frame of reference F is not a covariant frame. But since every observer can choose the frame F, the treatment remains invariant. With this choice, we have

$$\begin{aligned}|\{z_\mu\}\rangle_F=&\exp[-\tfrac{1}{2}(|z_1|^2+|z_2|^2)]\\&\times\exp(z_1a_1^\dagger+z_2a_2^\dagger)\exp[-(z_1^*a_1+z_2^*a_2)]\\&\times\exp[z_3(a_3-a_0)^\dagger-z_3^*(a_3-a_0)]|0\rangle\,.\end{aligned} \tag{3.23}$$

With $\mathfrak{a}=a_3-a_0$, we note that

$$[\mathfrak{a},\mathfrak{a}^\dagger]=0\,. \tag{3.24}$$

Thus

$$\begin{aligned}|\{z_\mu\}\rangle_F=&\exp[-\tfrac{1}{2}(|z_1|^2+|z_2|^2)]\\&\times\exp(z_1a_1^\dagger+z_2a_2^\dagger)\exp[-(z_1^*a_1+z_2^*a_2)]\\&\times\exp(z_3\mathfrak{a}^\dagger)\exp(-z_3^*\mathfrak{a})|0\rangle\end{aligned} \tag{3.25a}$$

$$\begin{aligned}=&\exp[-\tfrac{1}{2}(|z_1|^2+|z_2|^2)]\exp(z_1a_1^\dagger+z_2a_2^\dagger)\\&\times\exp[-(z_1^*a_1+z_2^*a_2)]\exp(z_3\mathfrak{a}^\dagger)|0\rangle\,.\end{aligned} \tag{3.25b}$$

The operator $\exp(z_3\mathfrak{a}^\dagger)$ generates exactly those vectors that satisfy the subsidiary condition in F because

$$\mathfrak{a}\exp(z_3\mathfrak{a}^\dagger)|0\rangle=0\,. \tag{3.26}$$

We rewrite (3.25b) as

$$|\{z_\mu\}\rangle_F=|z_1,z_2\rangle_F+|z_1,z_2,\chi(z_3)\rangle_F\,, \tag{3.27}$$

where

$$\begin{aligned}|z_1,z_2\rangle_F\equiv&\exp[-\tfrac{1}{2}(|z_1|^2+|z_2|^2)]\exp(z_1a_1^\dagger+z_2a_2^\dagger)\\&\times\exp[-(z_1^*a_1+z_2^*a_2)]|0\rangle\,,\end{aligned} \tag{3.28}$$

$$|z_1,z_2,\chi(z_3)\rangle_F\equiv|z_1,z_2\rangle_F\otimes\sum_{n=1}^{\infty}\frac{(z_3\mathfrak{a}^\dagger)^n}{n!}|0\rangle\,. \tag{3.29}$$

We see easily that

$$\||z_1,z_2\rangle_F\|^2=1\,, \tag{3.30}$$

$$\||z_1,z_2,\chi(z_3)\rangle_F\|^2=0\,, \tag{3.31}$$

$${}_F\langle z_1,z_2|z_1,z_2,\chi(z_3)\rangle_F=0\,. \tag{3.32}$$

Thus $|\{z_\mu\}\rangle_F$ decomposes into a vector of unit norm and an orthogonal vector of zero norm. Since

$$\||\{z_\mu\}\rangle_F\|^2=1\,, \tag{3.33}$$

the states $|z_1,z_2,\chi(z_3)\rangle_F$ do not make a physically relevant contribution to the C states and we can set

$$|\{z_\mu\}\rangle_F=|z_1,z_2\rangle_F\,, \tag{3.34}$$

where $|z_1,z_2\rangle_F$ are just the coherent states for the degrees of freedom a_1 and a_2 in the frame F. Then the completeness condition reads

$$\int d\mu_F(z_1,z_2)\,|z_1,z_2\rangle_F\,{}_F\langle z_1,z_2|=1\,, \tag{3.35}$$

where

$$d\mu_F(z_1,z_2)\equiv(1/\pi^2)dx_1dx_2dy_1dy_2\,.$$

B. Construction of C States for Free Electromagnetic Field

The states (3.8) can be extended to the field-theoretic case by suitably replacing each degree of freedom represented by a_μ by an infinite degree of freedom, $a_\mu(k)$, where k is the momentum four-vector. The basic equations[10] for the vector potential $A_\mu(x)$ are

$$\Box A_\mu(x)=0\,, \tag{3.36}$$

$$[A_\mu(x),A_\nu(y)]=-ig_{\mu\nu}D(x-y)\,, \tag{3.37}$$

$$\partial^\mu A_\mu^{(-)}(x)|\Psi\rangle=0\,, \tag{3.38}$$

where $A_\mu^{(-)}(x)$ is the negative-frequency (annihilation-operator) component of $A_\mu(x)$ and $|\Psi\rangle$ is an arbitrary physical state in $V_\infty^{(4)}$. We introduce the inner product

$$\langle f,g\rangle=i\int f^{\mu*}(x)\overleftrightarrow{\partial}_\nu g_\mu(x)d\sigma^\nu(x)\,, \tag{3.39}$$

where the integration is to be carried out over a space-like surface and $\overleftrightarrow{\partial}_\mu\equiv\overrightarrow{\partial}_\mu-\overleftarrow{\partial}_\mu$. If $f^\mu(x)$ and $g_\mu(x)$ are solutions of (3.36), then the inner product (3.39) is independent of time, and therefore we can choose a constant-time surface to carry out integration. Then

$$\langle f,g\rangle=i\int f^{\mu*}(x)\overleftrightarrow{\partial}_0 g_\mu(x)d^3x\,. \tag{3.40}$$

Since $f^\mu(x)$ is a solution of (3.36),

$$f^\mu(x)=\frac{1}{(2\pi)^2}\int\tilde{f}^\mu(k)e^{ikx}\delta(k^2)d^4k \tag{3.41}$$

$$=\frac{1}{(2\pi)^2}\int[\tilde{f}^\mu(k)e^{ikx}+\tilde{f}^\mu(-k)e^{-ikx}]\frac{d^3k}{2|\mathbf{k}|}\,. \tag{3.42}$$

Since $f^{\mu*}(x)=f^\mu(x)$, we obtain

$$f^\mu(x)=\frac{1}{(2\pi)^{3/2}}\times\int[\tilde{f}^\mu(\mathbf{k})e^{ikx}+\tilde{f}^{\mu*}(\mathbf{k})e^{-ikx}]\frac{d^3k}{(2|\mathbf{k}|)^{1/2}}\,, \tag{3.43}$$

where

$$\tilde{f}^\mu(\mathbf{k})\equiv(4\pi|\mathbf{k}|)^{-1/2}\tilde{f}^\mu(k)$$

and

$$\tilde{f}^{\mu*}(\mathbf{k})\equiv(4\pi|\mathbf{k}|)^{-1/2}\tilde{f}^\mu(-k)\,.$$

[10] J. Jauch and F. Rohrlich, *Theory of Photons and Electrons* (Addison-Wesley, Reading, Mass., 1959).

Thus, in terms of momentum variables,

$$\langle f,g\rangle=\frac{1}{(2\pi)}\int d^3k[\tilde{f}^{\mu*}(\mathbf{k})\tilde{g}_\mu(\mathbf{k})-\tilde{f}^\mu(\mathbf{k})\tilde{g}_\mu{}^*(\mathbf{k})]. \quad (3.44)$$

Note that

$$\langle f,g\rangle^*=\langle g,f\rangle=-\langle f,g\rangle. \quad (3.45)$$

Similarly, we can introduce

$$A_\mu(x)=\frac{1}{(2\pi)^{3/2}}\times\int[a_\mu(\mathbf{k})e^{ikx}+a_\mu{}^\dagger(\mathbf{k})e^{-ikx}]\frac{d^3k}{(2|\mathbf{k}|)^{1/2}}, \quad (3.46)$$

with corresponding inner product $\langle a,f\rangle$. The commutation relations (3.37) can be written down in terms of $a_\mu(\mathbf{k})$:

$$[a_\mu(\mathbf{k}),a_\nu{}^\dagger(\mathbf{k}')]=g_{\mu\nu}\delta(\mathbf{k}-\mathbf{k}'). \quad (3.47)$$

The subsidiary condition (3.38) reads

$$k^\mu a_\mu(\mathbf{k})|\Psi\rangle=0. \quad (3.48)$$

We are now equipped to write down the C states by replacing various inner products in the previous section by their suitably generalized version. We define the product

$$(f,g)=\frac{1}{(2\pi)}\int\tilde{f}^{\mu*}(\mathbf{k})\tilde{g}_\mu(\mathbf{k})d^3k. \quad (3.49)$$

This product is related to (3.44) through

$$\langle f,g\rangle=(f,g)-(g,f) \quad (3.50a)$$

$$=(f,g)-(f,g)^*. \quad (3.50b)$$

The commutation relations (3.47) can be expressed by

$$[(f,a),(a,g)]=(f,g). \quad (3.51)$$

Note that the product (f,f) is indefinite.

We define the states

$$|f)\equiv e^{-\frac{1}{2}(f,f)}e^{(a,f)}e^{-(f,a)}|0\rangle, \quad (3.52)$$

or, equivalently,

$$|f)\equiv e^{(a,f)}|0\rangle. \quad (3.53)$$

It is easy to check, using $(f,a)|0\rangle=0$, that

$$(f,a)|g)=(f,g)|g). \quad (3.54)$$

The inner product in $\mathfrak{C}_\infty{}^{(4)}$ is

$$\langle g|f)=e^{-\frac{1}{2}(f-g,f-g)+\frac{1}{2}(f,g)} \quad (3.55)$$

$$=e^{-\frac{1}{2}(f,f)-\frac{1}{2}(g,g)+(g,f)}. \quad (3.56)$$

We note the important result,

$$\langle f|f)=1>0. \quad (3.57)$$

The C states $|f)$ thus form a submanifold $\mathfrak{C}_\infty{}^{(4)}$ with positive definite norm. The physical states are restricted by (3.38). The C states satisfying (3.38) are given by

$$k^\mu a_\mu(\mathbf{k})|g)=0$$

or

$$\int k^\mu a_\mu(\mathbf{k})|g)d^3k=0. \quad (3.58)$$

Now, using (3.54), we rewrite the left-hand side of (3.58) to obtain

$$\frac{1}{(2\pi)}\int k^\mu g_\mu(\mathbf{k})d^3k|g)=0. \quad (3.59)$$

Thus the C states $|g)$, constructed using $g_\mu(\mathbf{k})$ such that

$$k^\mu g_\mu(\mathbf{k})=0, \quad (3.60)$$

satisfy the subsidiary condition (3.58).

As before, we can rewrite (3.52) with the help of the inner products

$$(f_1,g_1)\equiv(f,g)_1=\int\tilde{f}^{1*}(\mathbf{k})\tilde{g}_1(k)d^3k, \quad (3.61)$$

etc., to obtain

$$|f)=e^{-\frac{1}{2}[(f,f)_1+(f,f)_2]}e^{(a,f)_1+(a,f)_2}e^{-[(f,a)_1+(f,a)_2]}\times e^{(a,f)_3-(a,f)_0-[(f,a)_3-(f,a)_0]}|0\rangle. \quad (3.62)$$

We realize the subsidiary condition (3.60) in the special frame F introduced in the earlier section. Then

$$|f)_F=e^{-\frac{1}{2}[(f,f)_1+(f,f)_2]}e^{(a,f)_1+(a,f)_2}\times e^{-[(f,a)_1+(f,a)_2]}\exp(\mathfrak{A},f)_3|0\rangle,$$

where, in the frame F,

$$\tilde{f}_3(\mathbf{k})=\tilde{f}_0(\mathbf{k}) \quad\text{and}\quad (\mathfrak{A},f)_3=\int(a_3-a_0)^\dagger\tilde{f}_3(\mathbf{k})dk^3. \quad (3.63)$$

Then the states $|f)_F$ decompose into

$$|f)_F=|f_1,f_2)_F+|f_1,f_2,\chi(f_3))_F \quad (3.64)$$

in a way analogous to (3.27), with

$$|f_1,f_2)_F\equiv e^{-\frac{1}{2}[(f,f)_1+(f,f)_2]}e^{(a,f)_1+(a,f)_2}\times e^{-[(f,a)_1+(f,a)_2]}|0\rangle, \quad (3.65)$$

$$|f_1,f_2,\chi(f_3))_F\equiv e^{-\frac{1}{2}[(f,f)_1+(f,f)_2]}e^{(a,f)_1+(a,f)_2}\times e^{-[(f,a)_1+(f,a)_2]}\sum_{n=1}^{\infty}\frac{(\mathfrak{A},f)_3{}^n}{n!}|0\rangle, \quad (3.66)$$

$${}_F\langle f_1,f_2|f_1,f_2,\chi(f_3))_F=0, \quad (3.67)$$

$$\||f_1,f_2,\chi(f_3))_F\|^2=0. \quad (3.68)$$

Thus the physically relevant *coherent states*[11] are given by (3.65).

[11] *Note added in proof*. Result (3.65) is valid also in three dimensions. To see this, we introduce the coordinate system defined by the unit polarization vectors such that $e_\mu{}^{(\alpha)}e^{(\alpha')\mu}=g^{\alpha\alpha'}$. Any vector valued function $V_\mu(\mathbf{k})$ can be expanded in terms of the set $\{e_\mu{}^{(\alpha)}\}$. We have

$$V_\mu(\mathbf{k})=\sum_{\alpha=0}^{3}V^{(\alpha)}(\mathbf{k})e_\mu{}^{(\alpha)}(\mathbf{k}).$$

The commutation relations (3.67) are imposed on $a^{(\alpha)}(\mathbf{k})$. We then realize the weak subsidiary condition in the frame G defined by $k^\mu e_\mu{}^{(1)}(\mathbf{k})=0=k^\mu e_\mu{}^{(2)}(\mathbf{k})$ and $k^\mu e_\mu{}^{(3)}(\mathbf{k})=\mp k^\mu e_\mu{}^{(0)}(\mathbf{k})$. Then we have $k^\mu\tilde{g}_\mu(\mathbf{k})=k^\mu e_\mu{}^{(3)}(\mathbf{k})\tilde{g}^{(3)}(\mathbf{k})+k^\mu e_\mu{}^{(0)}(\mathbf{k})\tilde{g}^{(0)}(\mathbf{k})=0$, i.e., $k^\mu e_\mu{}^{(3)}(\mathbf{k})[\tilde{g}^{(3)}(\mathbf{k})\mp\tilde{g}^{(0)}(\mathbf{k})]=|\mathbf{k}|[\tilde{g}^{(3)}(\mathbf{k})\mp\tilde{g}^{(0)}(\mathbf{k})]=0$. Thus we choose states $|g)$ such that $\tilde{g}^{(3)}(\mathbf{k})=\pm\tilde{g}^{(0)}(\mathbf{k})$.

Now we can introduce a complete set of orthonormal Maxwell wave packets $\{h_i^{(n)}\}_{n=1,\ldots,\infty}$ with $i=1,2$ with respect to the inner product (3.61). Then

$$(f,f)_i=\sum_{n=1}^{\infty}|f_i^{(n)}|^2, \tag{3.69}$$

where

$$f_i^{(n)}\equiv(h_i^{(n)},f_i). \tag{3.70}$$

Thus

$$|f_1,f_2\rangle_F=\exp[-\tfrac{1}{2}\sum_{n=1}^{\infty}(|f_1^{(n)}|^2+|f_2^{(n)}|^2)]$$
$$\times\exp(\sum_{n=1}^{\infty}a_1^{(n)\dagger}f_1^{(n)})\exp(\sum_{n=1}^{\infty}a_2^{(n)\dagger}f_2^{(n)})|0\rangle. \tag{3.71}$$

The completeness condition now reads

$$\int|f_1,f_2\rangle_F\;{}_F\langle f_1,f_2|d\mu(f_1,f_2)=1, \tag{3.72}$$

where

$$d\mu(f_1,f_2)=\prod_{n=1}^{\infty}\frac{|f_1^{(n)}|d|f_1^{(n)}|d\theta_1^{(n)}|f_2^{(n)}|d|f_2^{(n)}|d\theta_2^{(n)}}{\pi^2}, \tag{3.73}$$

where

$$\theta_i^{(n)}=\arg f_i^{(n)}. \tag{3.74}$$

IV. LINEAR MODEL OF GRAVITATIONAL FIELD

A. Construction of C States

The coherent states with one of the degrees of freedom leading to an indefinite metric has been already considered in connection with the electromagnetic field. The construction for the linear gravitational field is completely similar. In this section, we will outline the procedure very briefly with a model where the infinite degrees of freedom arising from the momentum variable have been suppressed. The basic equations are adopted from the Gupta formalism[6] with a few minor changes in notation.

The field equations are given by

$$\Box\gamma_{\mu\nu}(x)=0, \tag{4.1}$$

$$\Box\gamma(x)=0. \tag{4.2}$$

The commutation relations for the γ's are

$$[\gamma_{\mu\nu}(x),\gamma_{\lambda\rho}(y)]=-i(g_{\mu\lambda}g_{\nu\rho}+g_{\mu\rho}g_{\nu\lambda})D(x-y), \tag{4.3}$$

$$[\gamma(x),\gamma(y)]=4iD(x-y), \tag{4.4}$$

where $D(x)$ is the usual zero-mass Green's function.[10] We express the γ's in terms of their Fourier transforms:

$$\gamma_{\mu\nu}(x)=\frac{1}{(2\pi)^{3/2}}\int[a_{\mu\nu}(\mathbf{k})e^{ikx}+a_{\mu\nu}^{\dagger}(\mathbf{k})e^{-ikx}]\frac{d^3k}{(2|\mathbf{k}|)^{1/2}}, \tag{4.5}$$

$$\gamma(x)=\frac{2}{(2\pi)^{3/2}}\int[a(\mathbf{k})e^{ikx}+a_a(\mathbf{k})e^{-ikx}]\frac{d^3k}{(2|\mathbf{k}|)^{1/2}}. \tag{4.6}$$

The quantization conditions between the γ's lead to the following commutation relations between $a_{\mu\nu}$ and a:

$$[a_{\mu\nu}(\mathbf{k}),a_{\lambda\rho}^{\dagger}(\mathbf{k}')]=(g_{\mu\lambda}g_{\nu\rho}+g_{\mu\rho}g_{\nu\lambda})\delta(\mathbf{k}-\mathbf{k}'), \tag{4.7}$$

$$[a(\mathbf{k}),a^{\dagger}(\mathbf{k}')]=-\delta(\mathbf{k}-\mathbf{k}'). \tag{4.8}$$

We note that $a_{l0}(\mathbf{k})$ and $a(\mathbf{k})$ satisfy the pseudocommutation relations. Suppressing the momentum dependence, we introduce the following new operators:

$$\begin{aligned}&b_{11}\equiv\tfrac{1}{2}(a_{11}-a_{22}), && b_{lk}\equiv a_{lk},\\ &b_{22}\equiv\tfrac{1}{2}(a_{11}+a_{22}), && b_{l0}\equiv a_{l0},\\ &b_{33}\equiv a_{33}/\sqrt{2}, && b\equiv a,\\ &b_{00}\equiv a_{00}/\sqrt{2},\end{aligned} \tag{4.9}$$

where the last three definitions are introduced for the sake of consistency of notation. We note that

$$\begin{aligned}&[b_{11},b_{11}^{\dagger}]=1,\quad [b_{22},b_{22}^{\dagger}]=1,\quad [b_{33},b_{33}^{\dagger}]=1,\\ &[b_{00},b_{00}^{\dagger}]=1,\quad [b_{12},b_{12}^{\dagger}]=1,\quad [b_{23},b_{23}^{\dagger}]=1,\\ &[b_{31},b_{31}^{\dagger}]=1,\quad [b_{10},b_{10}^{\dagger}]=-1,\\ &[b_{20},b_{20}^{\dagger}]=-1,\quad [b_{30},b_{30}^{\dagger}]=-1,\\ &[b,b^{\dagger}]=-1.\end{aligned} \tag{4.10}$$

We can introduce the occupation-number space in a way quite analogous to the electromagnetic case. We merely outline the salient features.

$$b_{\mu\nu}|0\rangle=0, \tag{4.11a}$$

$$\langle 0|b_{\mu\nu}^{\dagger}=0, \tag{4.11b}$$

$$\langle\{n_{\mu\nu}'\}|\{n_{\mu\nu}\}\rangle=(-1)^{n_{10}+n_{20}+n_{30}+n}\delta_{\{n_{\mu\nu}'\}\{n_{\mu\nu}\}}. \tag{4.12}$$

Note that the metric is indefinite.

We introduce the complex numbers $z=x+iy$ and $z_{\mu\nu}=z_{\nu\mu}=x_{\mu\nu}+iy_{\mu\nu}$ to define the C states.

$$|\{z_{\mu\nu};z\}\rangle=\exp[-\tfrac{1}{2}(\mathfrak{z},\mathfrak{z})]\exp(\mathfrak{B},\mathfrak{z})\exp[-(\mathfrak{z},\mathfrak{B})]|0\rangle, \tag{4.13}$$

where the product $(\mathfrak{B},\mathfrak{z})$ is defined by

$$\begin{aligned}(\mathfrak{B},\mathfrak{z})\equiv{}&z_{11}b_{11}^{\dagger}+z_{22}b_{22}^{\dagger}+z_{33}b_{33}^{\dagger}\\ &+z_{00}b_{00}^{\dagger}+z_{12}b_{12}^{\dagger}+z_{23}b_{23}^{\dagger}+z_{31}b_{31}^{\dagger}\\ &-(z_{10}b_{10}^{\dagger}+z_{20}b_{20}^{\dagger}+z_{30}b_{30}^{\dagger}+zb^{\dagger}).\end{aligned} \tag{4.14}$$

The following relation is useful in computations:

$$[(\mathfrak{z}',\mathfrak{B}),(\mathfrak{B},\mathfrak{z})]=(\mathfrak{z}',\mathfrak{z}). \tag{4.15}$$

Using (4.15), we can rewrite (4.13) as

$$|\{z_{\mu\nu};z\}\rangle=\exp[(\mathfrak{B},\mathfrak{z})-(\mathfrak{z},\mathfrak{B})]|0\rangle. \tag{4.16}$$

It is easily checked that

$$b_{\mu\nu}|\{z_{\mu\nu};z\}\rangle = z_{\mu\nu}|\{z_{\mu\nu};z\}\rangle, \qquad (4.17)$$

$$\langle\{z_{\mu\nu}';z'\}|\{z_{\mu\nu};z\}\rangle = \exp[-\tfrac{1}{2}(\mathfrak{z}',\mathfrak{z}')-\tfrac{1}{2}(\mathfrak{z},\mathfrak{z})+(\mathfrak{z}'\mathfrak{z})]. \qquad (4.18)$$

We arrive at the most significant result that

$$\||\{z_{\mu\nu};z\}\rangle\|^2=1. \qquad (4.19)$$

B. Subsidiary Conditions

The subsidiary conditions in the Gupta[6] formalism are imposed on the state vectors.

$$\partial^\mu \gamma_{\mu\nu}^{(-)}(x)|\Psi\rangle=0, \qquad (4.20)$$

$$[\gamma_\mu{}^{\mu(-)}(x)-\gamma^{(-)}(x)]|\Psi\rangle=0. \qquad (4.21)$$

In terms of Fourier transforms,

$$k^\mu a_{\mu\nu}|\Psi\rangle=0, \qquad (4.22)$$

$$(a_\mu{}^\mu-2a)|\Psi\rangle=0. \qquad (4.23)$$

Equivalently,

$$(k^1a_{1\nu}+k^2a_{2\nu}+k^3a_{3\nu}+k^0a_{30})|\Psi\rangle=0,$$

$$(a_{11}+a_{22}+a_{33}-a_{00}-2a)|\Psi\rangle=0.$$

In terms of the b's, we have

$$\begin{aligned}
&[k^1(b_{11}+b_{22})+k^2b_{21}+k^3b_{31}+k^0b_{01}]|\Psi\rangle=0,\\
&[k^1b_{12}+k^2(b_{22}-b_{11})+k^3b_{32}+k^0b_{02}]|\Psi\rangle=0,\\
&[k^1b_{13}+k^2b_{23}+\sqrt{2}k^3b_{33}+k^0b_{03}]|\Psi\rangle=0, \qquad (4.24)\\
&[k^1b_{10}+k^2b_{20}+k^3b_{30}+\sqrt{2}k^0b_{00}]|\Psi\rangle=0,\\
&[2b_{22}+\sqrt{2}(b_{33}-b_{00})-2b]|\Psi\rangle=0.
\end{aligned}$$

Since we want the C states to satisfy the subsidiary conditions (4.22) and (4.23), the $z_{\mu\nu}$'s and z are constrained as follows:

$$\begin{aligned}
&k^1(z_{11}+z_{22})+k^2z_{21}+k^3z_{31}+k^0z_{01}=0,\\
&k^1z_{12}+k^2(z_{22}-z_{11})+k^3z_{32}+k^0z_{02}=0,\\
&k^1z_{13}+k^2z_{23}+\sqrt{2}k^3z_{33}+k^0z_{03}=0, \qquad (4.25)\\
&k^1z_{01}+k^2z_{02}+k^3z_{03}+\sqrt{2}k^0z_{00}=0,\\
&2z_{22}+\sqrt{2}(z_{33}-z_{00})-2z=0.
\end{aligned}$$

Since the above system calls for a tedious algebraic solution, we will realize the subsidiary condition in a particular frame of reference F:

$$k_1=k_2=0, \quad k_3=k_0. \qquad (4.26)$$

This implies that

$$z_{31}=z_{01}, \quad z_{32}=z_{02}, \quad \sqrt{2}z_{33}=z_{03}, \quad z_{03}=\sqrt{2}z_{00}. \qquad (4.27)$$

The last two constraints in the system (4.27) imply that $z_{33}=z_{00}$ and hence

$$z_{22}=z. \qquad (4.28)$$

In the frame F,

$$(\mathfrak{z},\mathfrak{z})_F=|z_{11}|^2+|z_{12}|^2. \qquad (4.29)$$

We further notice that in the frame F, the operator

$$Q=z_{22}(b_{22}-b)+z_{33}(b_{33}+b_{00}-\sqrt{2}b_{30}) +z_{23}(b_{23}-b_{20})+z_{31}(b_{31}-b_{10}) \qquad (4.30)$$

commutes with $Q^\dagger$. Thus,

$$\begin{aligned}
|\{z_{\mu\nu};z\}\rangle_F
&=\exp[-\tfrac{1}{2}(|z_{11}|^2+|z_{12}|^2)]\exp(z_{11}b_{11}^\dagger+z_{12}b_{12}^\dagger)\\
&\quad\times\exp[-(z_{11}^*b_{11}+z_{12}^*b_{12})]e^{Q^\dagger}e^{-Q}|0\rangle \qquad (4.31)\\
&=\exp[-\tfrac{1}{2}(|z_{11}|^2+|z_{12}|^2)]\exp(z_{11}b_{11}^\dagger+z_{12}b_{12}^\dagger)\\
&\quad\times\exp[-(z_{11}^*b_{11}+z_{12}^*b_{12})]\left[1+\sum_{r=1}^{\infty}\frac{(Q^\dagger)^r}{r!}\right]|0\rangle \qquad (4.32)\\
&=|z_{11},z_{12}\rangle_F|z_{11},z_{12},\chi(z_{22},z_{33},z_{32},z_{31}\rangle_F, \qquad (4.33)
\end{aligned}$$

where

$$|z_{11},z_{12}\rangle_F \equiv\exp[-\tfrac{1}{2}(|z_{11}|^2+|z_{12}|^2)]\exp(z_{11}b_{11}^\dagger+z_{12}b_{12}^\dagger) \times\exp[-(z_{11}^*b_{11}+z_{12}^*b_{12})] \qquad (4.34)$$

and

$$|z_{11},z_{12},\chi(z_{22},z_{33},z_{32},z_{31})\rangle_F =|z_{11},z_{12}\rangle_F\otimes\sum_{r=1}^{\infty}\frac{(Q^\dagger)^r}{r!}|0\rangle. \qquad (4.35)$$

Notice that

$$\||z_{11},z_{12}\rangle_F\|^2=1, \qquad (4.36)$$

$${}_F\langle z_{11},z_{12}|z_{11},z_{12},\chi(z_{22},z_{33},z_{32},z_{31})\rangle_F=0, \qquad (4.37)$$

and

$$\||z_{11},z_{12},\chi(z_{22},z_{33},z_{32},z_{31})\rangle_F\|^2=0. \qquad (4.38)$$

Thus only b_{11} and b_{12} gravitons contribute to the coherent states.

V. DISCUSSION

We have dealt with the construction of C states for zero-mass boson systems wherein the basic field variables are further constrained by subsidiary conditions. We have arrived at a remarkable result that the C-state vectors have positive definite norm. We have shown that in a special frame F (chosen for the sake of simplicity of algebraic treatment), the C-state vectors decompose into the usual coherent-state vectors involving only the physical degrees of freedom and an orthogonal null vector. Since the frame F can be chosen by every observer, relativistic invariance is preserved.

ACKNOWLEDGMENTS

Sincere thanks are due to Professor F. Rohrlich and Professor Guy Johnson for useful discussions and critical comments.

Volume 53B, number 2 PHYSICS LETTERS 25 November 1974

EXTENDED PARTICLES AND SOLITONS

K. CAHILL

Centre de Physique Théorique de l'Ecole Polytechnique, 75230 Paris Cedex 05, France

Received 21 October 1974

A variational method is introduced for approximating the eigenstates of the Hamiltonian in relativistic quantum field theory. The method has three ingredients: the coherent states, the solutions of the corresponding classical field theory, and the generator-coordinate technique of nuclear physics. It provides a quantum-mechanical interpretation for solitons, some of which resemble extended particles.

It has been pointed out by Glauber [1], and more recently by Hepp [2], that the coherent states lend themselves to the comparison of a quantum field theory with its classical counterpart. In this note it is suggested that a variational method based upon these states may be used to draw from the solutions to a given classical field theory information about the spectrum of the corresponding quantum theory.

The method breaks translational invariance and other symmetries of the Hamiltonian in the first approximation, which is the tree approximation. These symmetries are restored by the second and higher approximations. The method is similar to one invented by Wheeler that is called the method of generator coordinates in the context of nuclear physics [3]. It is non-perturbative and non-singular.

Recently various solutions to certain classical field theories have been discovered that share some of the features of extended hadrons [4, 5]. The present method provides a quantum-mechanical interpretation for these solutions, which are called solitons.

It will be simplest to consider a theory of a single scalar Hermitian field $\varphi(x)$ described by a Hamiltonian of the form

$$H = :\int d^3x\{\tfrac{1}{2}\pi(x)^2 + \tfrac{1}{2}[\nabla\varphi(x)]^2 + V[\varphi(x)]\}: . \quad (1)$$

The field $\varphi(x)$ and its conjugate momentum $\pi(x)$ will be assumed to obey the canonical equal-time commutation relation

$$[\varphi(t, \boldsymbol{x}), \pi(t, \boldsymbol{x}')] = \mathrm{i}\delta^{(3)}(\boldsymbol{x}-\boldsymbol{x}') . \quad (2)$$

The colons in the definition of H denote normal ordering with respect to the annihilation and creation operators which are defined as

$$a(\boldsymbol{k}) = \int d^3\boldsymbol{x} \exp(-\mathrm{i}\boldsymbol{k}\cdot\boldsymbol{x})[k^0\varphi(0,\boldsymbol{x}) + \mathrm{i}\pi(0,\boldsymbol{x})] , \quad (3)$$

and as $a^\dagger(\boldsymbol{k}) = [a(\boldsymbol{k})]^\dagger$. It is convenient to put $k^0 = \sqrt{\boldsymbol{k}^2}$.

For any two real functions $q(\boldsymbol{x})$ and $p(\boldsymbol{x})$ the coherent state $|q,p\rangle$ is defined as

$$|q,p\rangle = \exp\{\mathrm{i}\int d^3\boldsymbol{x}[p(\boldsymbol{x})\varphi(0,\boldsymbol{x}) - q(\boldsymbol{x})\pi(0,\boldsymbol{x})]\}|0\rangle , \quad (4)$$

where $|0\rangle$ is the state for which $a(\boldsymbol{k})|0\rangle = 0$ for all $\boldsymbol{k}$. Since q and p are real, the exponential is an isometric operator and the coherent state is normalized to unity.

The coherent state $|q,p\rangle$ is an eigenstate of the annihilation operator $a(\boldsymbol{k})$ for every $\boldsymbol{k}$, $a(\boldsymbol{k})|q,p\rangle = \alpha(\boldsymbol{k})|q,p\rangle$, with an eigenvalue $\alpha(\boldsymbol{k})$ that is given by the integral (3) with $\varphi(0,\boldsymbol{x})$ and $\pi(0,\boldsymbol{x})$ replaced by $q(\boldsymbol{x})$ and $p(\boldsymbol{x})$, respectively. This property entails that the expectation value of the Hamiltonian (1) in the coherent state $|q,p\rangle$ is given by the classical expression

$$\langle q,p|H|q,p\rangle = \int d^3\boldsymbol{x}\{\tfrac{1}{2}p(\boldsymbol{x})^2 + \tfrac{1}{2}[\nabla q(\boldsymbol{x})]^2 + V[q(\boldsymbol{x})]\} . \quad (5)$$

It is easy to identify the coherent states $|q,p\rangle$ for which this energy is stationary under small variations of the functions $q(\boldsymbol{x})$ and $p(\boldsymbol{x})$. The requirement

$$0 = \delta\langle q,p|H|q,p\rangle , \quad (6)$$

implies that the function $p(\boldsymbol{x})$ vanish identically and that the function $q(\boldsymbol{x})$ be a solution of the static classical field equation

$$\nabla^2 q(x) = \partial V[q(x)]/\partial q(x) . \tag{7}$$

This equation usually has several solutions corresponding to the relative minima of the energy (5). The solution q_o corresponding to the absolute minimum is a constant, equal to the value q_o that minimizes $V(q)$. The state $|q_o\rangle \equiv |q_o, 0\rangle$ associated with it and given by eq. (4) with $q(x) = q_o$ and $p(x) = 0$ will be interpreted as an approximation to the physical vacuum. It will be convenient to adopt the convention $V(q_o) = 0$ so that $\langle q_o|H|q_o\rangle = 0$.

Among the solutions to the field equation (7) that are not translationally invariant, those with finite energy as given by eq. (5) will be interpreted as particles. For each such solution $q(x)$, the state $|q\rangle \equiv |q,0\rangle$, defined by eq. (4) with $p(x) = 0$, will be considered as an approximation to the state that represents one physical particle of type q at rest. The first approximation to the mass of the particle is $m(q) = \langle q|H|q\rangle$, which is the classical value (5) and also that of the tree approximation [2,6].

The solutions $q(x)$ are highly degenerate. For every solution $q(x)$ of eq. (7) there is another solution $q_y(x) = q(x-y)$ for each displacement vector y. The corresponding displaced state $|q_y\rangle = \exp(-\mathrm{i}P\cdot y)|q\rangle$, where P is the momentum operator, has the same energy (5) as the state $|q\rangle$ since the Hamiltonian (1) is translationally invariant.

It is therefore possible to obtain a better approximation to the eigenstate of the Hamiltonian H corresponding to the solution $q(x)$ by forming a superposition of the states $|q_y\rangle$ with an arbitrary weight function $f(y)$

$$|q,f\rangle = \int \mathrm{d}^3y f(y) \exp(-\mathrm{i}P\cdot y)|q\rangle , \tag{8}$$

and by demanding that the energy $\langle q,f|H|q,f\rangle / \langle q,f|q,f\rangle$ be stationary with respect to small variations of the weight function $f(y)$. The function $\exp(\mathrm{i}p\cdot y)$ satisfies this criterion; for each p it projects the state $|q\rangle$ onto the momentum p subspace. Thus an improved approximation to the state of one physical particle at rest is provided by the translationally invariant superposition

$$|q, \mathbf{0}\rangle = \int \mathrm{d}^3y \exp(-\mathrm{i}P\cdot y)|q\rangle . \tag{9}$$

The energy of this state

$$m(q;\mathbf{0}) = \langle q;\mathbf{0}|H|q;\mathbf{0}\rangle / \langle q;\mathbf{0}|H|q;\mathbf{0}\rangle , \tag{10}$$

will be interpreted as a second approximation to the mass of the particle associated with the classical solution $q(x)$.

The invariance of the Hamiltonian under rotations leads to an analogous projection procedure. The same is true for any internal symmetry. In general, the projection of an approximate eigenstate onto a suitably invariant subspace restores the symmetry and improves the approximation.

Suppose that the Lorentz transformation L is represented by the unitary operator $U(L)$, so that the four-momentum operators p^μ transform as $U(L)^{-1} P^\mu U(L) = L^\mu_\nu P^\nu$. Then since $P|q;\mathbf{0}\rangle = 0$, it follows that the expectation value of P^μ in the boosted state $U(L)|q;\mathbf{0}\rangle$ is $p^\mu = L^\mu_0\, m(q;\mathbf{0})$, whence the desired relation $p^2 = m(q;\mathbf{0})^2$.

This method has been applied to the theory described by the Hamiltonian

$$H = :\int \mathrm{d}x\, [\tfrac{1}{2}\pi(x)^2 + \tfrac{1}{2}\varphi'(x)^2 - \mu^2\varphi(x)^2 + \tfrac{1}{2}\lambda^2\varphi(x)^4 + \tfrac{1}{2}\lambda^{-2}\mu^4] : , \tag{11}$$

in two-dimensional space time. The vacuum solutions are degenerate, $q_o = \pm\lambda^{-1}\mu$, as are the soliton solutions, $q_\pm = \pm\lambda^{-1}\mu \tanh \mu x$. For $\lambda \ll \mu$ the formula (10) yields

$$m(q_\pm;0) = (4/3)(\mu/\lambda)^2\mu - [\pi^3/90\zeta(3)]\mu + O(\lambda^2) , \tag{12}$$

in which the first term is $m(q_\pm)$ as given by eq. (5) and $\zeta(3) \approx 1.20$. This result is similar but unequal to that obtained by Dashen, Hasslacher, and Neveu [5] for the case of weak coupling.

It is possible to improve the accuracy of the present method, at the expense of additional computational complexity, by first projecting the coherent state $|q,p\rangle$ onto a subspace with the appropriate momentum and angular-momentum quantum numbers and by then seeking the functions q and p that minimize the energy of the projected state.

It is a pleasure to recall helpful conversations with R. Balian, J. Bros, S. Coleman, A. Neveu, and R. Stora, as well as the very valuable advice of C.G. Callan, Jr. and J. Goldstone.

References

[1] R.J. Glauber, Phys. Rev. Letters 10 (1963) 84; Phys. Rev. 130 (1963) 2529; 131 (1963) 2766;
K.E. Cahill and R.J. Glauber, Phys. Rev. 177 (1969) 1857, 1882.
[2] K. Hepp, Commun. Math. Phys. 35 (1974) 265.
[3] J.A. Wheeler, Phys. Rev. 52 (1937) 1083, 1107;
J.J. Griffen and J.A. Wheeler, Phys. Rev. 108 (1957) 311;
C.W. Wong, UCLA report (sept. 1974).
[4] A.C. Scott, F.Y.F. Chu and D.W. McLaughlin, Proc. I.E.E.E. 61 (1973) 1443;
L.D. Fadeev, Max-Planck-Institut report MPI-PAE/Pth 16 (1974);
L.D. Fadeev and L.A. Takhtajan, Dubna report E2-7998 (1974);
A.M. Polyakov, Pisma v. ZhETF (June 1974);
G. 't Hooft, Nucl. Phys. B79 (1974) 276 and CERN report TH. 1902;
H. Van Dam, M. Veltman and T.T. Wu (to appear).
[5] R. Dashen, B. Hasslacher and A. Neveu, Princeton I.A.S. reports (to appear).
[6] S. Coleman and E. Weinberg, Phys. Rev. D7 (1973) 1888.

PHYSICAL REVIEW D VOLUME 18, NUMBER 10 15 NOVEMBER 1978

Classical source emitting self-interacting bosons

Karl-Erik Eriksson and Bo-Sture Skagerstam*
Institute of Theoretical Physics, Fack, S-402 20 Göteborg 5, Sweden
(Received 30 November 1977)

A general expression is derived for the S-matrix elements for a self-interacting boson field in interaction with a classical current. The probability for the emission of boson radiation up to a certain total energy is determined, and it is shown how infrared divergences are canceled or reduced. The formal extension by functional methods to a quantum current is indicated.

I. INTRODUCTION

The aim of this note is to derive the state of bosons radiated from a classical source in the case of boson self-interaction and to obtain the result in a form which is as closed as possible. By use of functional methods the result may then be generalized to a quantum source. An important application is the interaction of the soft part of a Yang-Mills field with a source that may be treated as semiclassical (recoil neglected).[1]

We shall denote the various states (in some basis) of a boson by $i, j, k, \ldots$, which may take on an infinity of values $1, 2, \ldots$. The annihilation and creation operators satisfy

$$[a_i, a_j] = 0, \quad [a_i^\dagger, a_j^\dagger] = 0, \qquad (1)$$
$$[a_i, a_j^\dagger] = \delta_{ij}.$$

A unique normalized vacuum state is assumed to exist,

$$a_i |0\rangle = 0 \quad \text{for all } i, \quad \langle 0|0\rangle = 1. \qquad (2)$$

We introduce one-boson wave functions $\underline{f}$, where

$$\underline{f} = (f_1, f_2, \ldots), \quad \sum_i |f_i|^2 < \infty, \qquad (3)$$

and a scalar product between two such wave functions $\underline{f}$ and $\underline{g}$,

$$\underline{f}^* \cdot \underline{g} = \sum_i f_i^* g_i. \qquad (4)$$

Corresponding to a wave function $\underline{f}$ we have an annihilation operator and a creation operator

$$a_{\underline{f}} = \underline{f}^* \cdot \underline{a}, \qquad (5)$$
$$a_{\underline{f}}^\dagger = \underline{a}^\dagger \cdot \underline{f}.$$

The nonvanishing commutator for such operators is

$$[a_{\underline{f}}, a_{\underline{g}}^\dagger] = \underline{f}^* \cdot \underline{g}. \qquad (6)$$

To the wave function $\underline{f}$ corresponds a unitary operator

$$U(\underline{f}) = e^{a_{\underline{f}}^\dagger - a_{\underline{f}}} = e^{-\underline{f}^* \cdot \underline{f}/2} e^{a_{\underline{f}}^\dagger} e^{-a_{\underline{f}}}, \qquad (7)$$
$$U(\underline{f})^\dagger a_i U(\underline{f}) = a_i + f_i,$$

and a coherent state

$$|f\rangle = U(\underline{f})|0\rangle, \qquad (8)$$
$$a_i |f\rangle = f_i |f\rangle, \quad \langle f|f\rangle = 1,$$

which is thus a normalized eigenstate of the annihilation operator. The coherent states have the scalar product

$$\langle f|g\rangle = e^{(\underline{f}^* \cdot \underline{g} - \underline{g}^* \cdot \underline{f})/2} e^{-(\underline{f} - \underline{g})^* \cdot (\underline{f} - \underline{g})/2}. \qquad (9)$$

Completeness is expressed through

$$1 = \int [d\underline{f}] |f\rangle\langle f|, \quad [d\underline{f}] = \prod_{i=1}^{\infty} \frac{1}{\pi} d(\text{Re} f_i) d(\text{Im} f_i) \qquad (10)$$

The Hamiltonian is assumed to be

$$H = H_0 + H_1 + H_2. \qquad (11)$$

Here, H_0 is the free Hamiltonian

$$H_0 = \underline{a}^\dagger \cdot \underline{\Omega} \cdot \underline{a}, \qquad (12)$$

where the matrix Ω may be assumed to be already diagonalized with eigenvalue ω_i for the ith mode,

$$[H_0, a_i] = -\omega_i a_i. \qquad (13)$$

Furthermore, in (11)

$$H_1 = i(\underline{a}^\dagger \cdot \underline{s} - \underline{s}^* \cdot \underline{a}) \qquad (14)$$

is the interaction with the external source and, finally,

$$H_2 = V(\underline{a}^\dagger; \underline{a}) \qquad (15)$$

gives the self-interaction.

It is convenient to start in the interaction picture, where

$$a_j(t) = a_j(0) e^{i\omega_j t}, \qquad (16)$$

and where a state vector $|\,\rangle_t$ satisfies a Schrödinger equation

$$i\frac{d}{dt}|\rangle_t=[H_1(t)+H_2(t)]|\rangle_t\,, \tag{17}$$

with

$$H_1(t)=i[\underline{a}^\dagger(t)\cdot\underline{s}(t)-\underline{s}^*(t)\cdot\underline{a}(t)]\,, \tag{18}$$

$$H_2(t)=V(\underline{a}^\dagger(t);\underline{a}(t))\,. \tag{19}$$

II. THE SOURCE-FREE CASE

For the case of no external source $\underline{s}(t)=0$, H_1 vanishes and the scattering operator is

$$\mathcal{S}=T\exp\left(-i\int_{-\infty}^{\infty}dt\,V(\underline{a}^\dagger(t);\underline{a}(t))\right)\,. \tag{20}$$

If we use Wick's theorem to expand this in terms of normal-ordered products and take the matrix elements between coherent states

$$\langle f|\mathcal{S}|g\rangle\,,$$

then all annihilation (creation) operators may be replaced by $\underline{g}$ ($\underline{f}^*$). The matrix element turns out to be a sum over products of factors corresponding to disconnected graphs. The combinatorics is such that

$$\langle f|\mathcal{S}|g\rangle=e^{-iW(\underline{f}^*;\underline{g})}\langle f|g\rangle\,, \tag{21}$$

where $iW(\underline{f}^*;\underline{g})$ is the sum over all connected graphs with all ingoing (outgoing) particles in the state $\underline{g}$ ($\underline{f}^*$).

We shall now use unitarity of $\mathcal{S}$ to derive a condition for W. In matrix form, unitarity reads

$$\langle f|g\rangle=\int[d\underline{h}]\langle h|\mathcal{S}|f\rangle^*\langle h|\mathcal{S}|g\rangle\,, \tag{22}$$

Here we have used the decomposition (10) of unity. Inserting (21) into (22) we get, using (9),

$$\langle f|g\rangle=\int[d\underline{h}]\langle f|h\rangle\langle h|g\rangle\exp[iW(\underline{h}^*;\underline{f})^*-iW(\underline{h}^*;\underline{g})]$$

$$=\exp[-\tfrac{1}{2}(\underline{f}^*\cdot\underline{f}+\underline{g}^*\cdot\underline{g})]\exp\left[iW\left(\frac{\delta}{\delta\underline{\alpha}};\underline{f}\right)^*-iW\left(\frac{\delta}{\delta\underline{\beta}};\underline{g}\right)\right]\int[d\underline{h}]e^{-\underline{h}^*\cdot\underline{h}}e^{(\underline{f}^*+\underline{\alpha}^*)\cdot\underline{h}+\underline{h}^*\cdot(\underline{g}+\underline{\beta})}\Big|_{\underline{\alpha}=\underline{\beta}=0}\,.$$

The Gaussian integral on the right-hand side reduces to $e^{(\underline{f}^*+\underline{\alpha}^*)\cdot(\underline{g}+\underline{\beta})}$, and thus

$$\exp[iW(\delta/\delta\underline{\alpha};\underline{f})^*-iW(\delta/\delta\underline{\beta};\underline{g})]\exp(\alpha^*\cdot\underline{g}+f^*\cdot\underline{\beta}+\underline{\alpha}^*\cdot\underline{\beta})\Big|_{\underline{\alpha}=\underline{\beta}=0}=1$$

or

$$\exp[iW(\underline{g}^*+\delta/\delta\underline{\alpha};\underline{f})^*]\exp[-iW(\underline{f}^*+\underline{\alpha}^*;\underline{g})]\Big|_{\underline{\alpha}=0}=1\,, \tag{23}$$

which thus is an expression for the unitarity of the scattering operator.

In the case of a vanishing external field the result is thus given by (21) where W satisfies (23) and expresses the sum over connected graphs.

III. A NONZERO CLASSICAL SOURCE

For a *nonzero* $\underline{s}(t)$, it is convenient to introduce the operator

$$U_1(t)=e^{i\varphi(t)}e^{\underline{a}^\dagger\cdot\underline{S}(t)-\underline{S}^*(t)\cdot\underline{a}}=e^{i\varphi(t)}U(\underline{S}(t))\,, \tag{24}$$

where[2]

$$\varphi(t)=\frac{1}{2i}\int_{-\infty}^{t}dt'\int_{-\infty}^{t'}dt''\sum_j[s_j(t')s_j^*(t'')e^{-i\omega_j(t'-t'')}-s_j^*(t')s_j(t'')e^{i\omega_j(t'-t'')}]\,,$$

$$S_j(t)=\int_{-\infty}^{t}dt'\,s_j(t')e^{-i\omega_j t'}\,. \tag{25}$$

The operator (24) satisfies the equation

$$i\frac{d}{dt}U_1(t)=H_1(t)U_1(t),\quad U_1(-\infty)=1\,, \tag{26}$$

and is thus the time-evolution operator in the case of no self-interaction. It is useful for introducing a new "intermediate picture" between the interaction picture and the Heisenberg picture,

$$|\rangle_t=U_1(t)|\rangle_t'\,. \tag{27}$$

From the Schrödinger equation for $|\rangle_t$, (17), and the equation (26) for $U_1(t)$, it then follows that $|\rangle_t'$ satisfies the following Schrödinger equation:

$$i\frac{d}{dt}|\rangle_t'=H_2'(t)|\rangle_t'\,, \tag{28}$$

where, using (19) and (7),

$$H_2'(t)=U_1(t)^{-1}H_2(t)U_1(t)$$
$$=V([a_j^\dagger+S_j^*(t)]e^{-i\omega_j t};[a_k+S_k(t)]e^{i\omega_k t})\,. \tag{29}$$

From (28) one obtains the time evolution

$$|\rangle_t^{\prime}=T\exp\left(-i\int^t dt' H_2^{\prime}(t')\right)|\rangle_{t_0}^{\prime}, \tag{30}$$

or by (27) in the interaction picture

$$|\rangle_t=U_1(t)T\exp\left(-i\int_{t_0}^t dt' H_2^{\prime}(t')\;U_1(t_0)^{-1}\right)|\rangle_{t_0}. \tag{31}$$

In the limit $t_0\to-\infty$, $t\to+\infty$ with the boundary condition of (26), we then have $|\rangle_{+\infty}=\mathcal{S}|\rangle_{-\infty}$, with

$$\mathcal{S}=U_1(\infty)T\exp\left(-i\int_{-\infty}^{\infty} dt H_2^{\prime}(t)\right). \tag{32}$$

Let us now put $t=\infty$ in (25) and define

$$\varphi=\varphi(\infty)=\frac{1}{2i}\int_{-\infty}^{\infty}dt\int_{-\infty}^{t}dt'\sum_j[s_j(t)s_j^*(t')e^{-i\omega_j(t-t')}-s_j^*(t)s_j(t')e^{i\omega_j(t-t')}],$$
$$S_j=S_j(\infty)=\int_{-\infty}^{\infty}dt s_j(t)e^{-i\omega_j t}, \tag{33}$$

and use this in (32). Then from (24) and (29), we get

$$\mathcal{S}=e^{i\varphi}e^{-\underline{S}^*\cdot\underline{S}/2}e^{\underline{a}^\dagger\cdot\underline{S}}e^{-\underline{S}^*\cdot\underline{a}}\;T\exp\left(-i\int_{-\infty}^{\infty}dt V([a_j^\dagger+S_j^*(t)]e^{-i\omega_j t};[a_k+S_k(t)]e^{i\omega_k t})\right). \tag{34}$$

In infrared-divergent cases, φ and $\underline{S}$ in (33) are divergent. With integrals cut off for large positive and negative times ($\int_{-T}^{T}$), we then have

$$\varphi,\underline{S}^*\cdot\underline{S}\sim\ln T. \tag{35}$$

One may then keep T large but finite and take the limit $T\to\infty$ in transition probabilities.

To go a step towards normal ordering, we commute the factor $e^{\underline{S}^*\cdot\underline{a}}$ through the T product, obtaining

$$\mathcal{S}=e^{i\varphi}e^{-\underline{S}^*\cdot\underline{S}/2}e^{\underline{a}^\dagger\cdot\underline{S}}T\exp\left(-i\int_{-\infty}^{\infty}dt\,V([a^\dagger+S_j^*(t)-S_j]e^{-i\omega_j t};[a_k+S_k(t)]e^{i\omega_k t})\right)e^{-\underline{S}^*\cdot\underline{a}}. \tag{36}$$

We then take the matrix element of (36) between coherent states, and use (8) and the expression (21) for the matrix element of (20)

$$\langle f|\mathcal{S}|g\rangle=e^{i\varphi}e^{-\underline{S}^*\cdot\underline{S}/2}e^{\underline{f}^*\cdot\underline{S}-\underline{S}^*\cdot\underline{g}}\langle f|g\rangle\exp\{iW([f_j^*+S_j^*(t)-S_j^*]e^{-i\omega_j t};[g_k+S_k(t)]e^{i\omega_k t})\}. \tag{37}$$

The W functional in (37) is now written in terms of time-dependent wave functions. This is because $\underline{S}(t)$ introduces an unknown time dependence. In the case $\underline{s}(t)=0$, then also $\underline{S}(t)=0$ and the time dependence in $W(f_je^{-i\omega_j t},g_ke^{i\omega_k t})$ is completely known.

With the vacuum as the initial state (37) gives

$$\langle f|\mathcal{S}|0\rangle=e^{i\varphi}e^{-\underline{S}^*\cdot\underline{S}/2}e^{\underline{f}^*\cdot\underline{S}}e^{-\underline{f}^*\cdot\underline{f}/2}\exp\{iW([f_j^*+S_j^*(t)-S_j^*]e^{-i\omega_j t};S_k(t)e^{i\omega_k t})\}. \tag{38}$$

We shall now use the projection operator

$$\Pi(\Delta E)=\frac{1}{2\pi}\int_0^{\Delta E}dE\int_{-\infty}^{\infty}d\tau\,e^{i(H_0-E)\tau} \tag{39}$$

to project out radiation in the energy interval 0 to ΔE. The total transition probability for the source to create radiation in this energy interval is then

$$\begin{aligned}P(\Delta E)&=\langle 0|\mathcal{S}^\dagger\Pi(\Delta E)\mathcal{S}|0\rangle\\&=\int\int[d\underline{f}][d\underline{g}]\langle f|\mathcal{S}|0\rangle^*\langle g|\mathcal{S}|0\rangle\langle f|\Pi(\Delta E)|g\rangle\\&=e^{-\underline{S}^*\cdot\underline{S}}\int\int[d\underline{f}][d\underline{g}]e^{-(\underline{f}^*\cdot\underline{f}+\underline{g}^*\cdot\underline{g})/2}e^{\underline{S}^*\cdot\underline{f}+\underline{g}^*\cdot\underline{S}}\langle f|\Pi(\Delta E)|g\rangle\\&\quad\times\exp\{iW([f_j^*+S_j^*(t)-S_j^*]e^{-i\omega_j t};S_k(t)e^{i\omega_k t})^*-iW([g_j^*+S_j^*(t)-S_j^*]e^{-i\omega_j t};S_k(t)e^{i\omega_k t})\}.\end{aligned} \tag{40}$$

The matrix element of $\Pi(\Delta E)$ entering (40) is

$$\begin{aligned}\langle f|\Pi(\Delta E)|g\rangle&=\frac{1}{2\pi}\int_0^{\Delta E}dE\int_{-\infty}^{\infty}d\tau\,e^{-iE\tau}\langle f|g_\tau\rangle\\&=\frac{1}{2\pi}\int_0^{\Delta E}dE\int_{-\infty}^{\infty}d\tau\,e^{-iE\tau}e^{-(\underline{f}^*\cdot\underline{f}+\underline{g}^*\cdot\underline{g})/2}e^{\underline{f}^*\cdot\underline{g}_\tau},\end{aligned} \tag{41}$$

with

$$(g_\tau)_j = g_j e^{i\omega_j \tau}. \tag{42}$$

We now insert (41) into (40) and use again the method of taking out the W's from the integral by putting them in functional form, as we did in the derivation of (23):

$$P(\Delta E)=\frac{1}{2\pi}\int_0^{\Delta E} dE \int_{-\infty}^{\infty} d\tau\, e^{-iE\tau} e^{-\underline{S}^*\cdot \underline{S}} \exp\{iW([\delta/\delta\alpha_j + S_j^*(t) - S_j^*]e^{-i\omega_j t}; S_k(t)e^{i\omega_k t})^*$$

$$-iW([\delta/\delta\beta_j + S_j^*(t) - S_j^*]e^{-i\omega_j t}; S_k(t)e^{i\omega_k t})\}$$

$$\times \int\int [d\underline{f}][d\underline{g}] e^{-\underline{f}^*\cdot\underline{f} - \underline{g}^*\cdot\underline{g}} e^{(\underline{S}^*+\underline{\alpha}^*)\cdot(\underline{S}+\underline{\beta}) + \underline{f}^*\cdot\underline{g}_\tau}\Big|_{\underline{\alpha}=\underline{\beta}=0}.$$

The Gaussian integration here is straightforward and gives the result $\exp(\sum_j (S_j^* + \alpha_j^*)(S_j + \beta_j)e^{i\omega_j\tau})$. This gives the result

$$P(\Delta E)=\frac{1}{2\pi}\int_0^{\Delta E} de \int_{-\infty}^{\infty} d\tau\, e^{-iE\tau}\exp\Big(\sum_m S_m^* S_m (e^{i\omega_m \tau} - 1)\Big)$$

$$\times \exp\{iW([S_j^*(t) + S_j^*(e^{-i\omega_j\tau} - 1) + \delta/\delta\alpha_j]e^{-i\omega_j t}; S_k(t)e^{i\omega_k t})^*\}$$

$$\times \exp\{iW([S_j^*(t) + S_j^*(e^{i\omega_j \tau} - 1) + \alpha_j^* e^{i\omega_j\tau}]e^{-i\omega_j t}; S_k(t)e^{i\omega_k t})\}\big|_{\underline{\alpha}=0}. \tag{43}$$

If $\underline{s}(t)$ and then also $\underline{S}(t)$ and $\underline{S}$ are such as to cause infrared divergences in the phase φ in (25) and in the norm of $\underline{S}$ as indicated in (35), then the factor

$$e^{i\omega_m \tau} - 1 \tag{44}$$

helps to remove such divergences (or to make them less severe) in the probability.

The double integral in (43) is common to theories with or without self interaction (Abelian and non-Abelian gauge theories). It is well known from quantum electrodynamics, where it can be explicitly evaluated.

IV. RADIATION FROM A QUANTUM SOURCE

The source may be changed into a quantum source $\underline{j}(t)$ by means of a functional operator

$$T\exp\Big[\int_{-\infty}^{\infty} dt\, \underline{j}(t)\cdot\frac{\delta}{\delta \underline{s}(t)}\Big] \tag{45}$$

operating on (38) [or (37) which is a more general case], after which $\underline{s}(t)$ is put equal to zero. Then (43) is replaced by

$$P(\Delta E) = \Big\langle \beta \Big| T\exp\Big[\int_{-\infty}^{\infty} dt\, \underline{j}(t)\cdot\frac{\delta}{\delta \underline{s}^{(A)}(t)}\Big]\Big|\alpha\Big\rangle^* \Big\langle \beta \Big| T\exp\Big[\int_{-\infty}^{\infty} dt\, \underline{j}(t)\cdot\frac{\delta}{\delta \underline{s}^{(B)}(t)}\Big]\Big|\alpha\Big\rangle e^{i(\varphi^{(A)} - \varphi^{(B)})}$$

$$\times \frac{1}{2\pi}\int_0^{\Delta E} dE\int_{-\infty}^{\infty} d\tau e^{-iE\tau}\langle S^{(A)} | S_\tau^{(B)}\rangle$$

$$\times [\exp\{iW([S_j^{(A)*}(t) + S_j^{(B)*}e^{-i\omega_j\tau} - S_j^{(A)*} + \delta/\delta\alpha_j]e^{-i\omega_j t}; S_k^{(A)}(t)e^{i\omega_k t})^*\}$$

$$\times \exp\{-iW([S_j^{(B)*}(t) + S_j^{(A)*}e^{i\omega_j\tau} - S_j^{(B)*} + \alpha_j e^{i\omega_j\tau}]e^{-i\omega_j t}; S_k^{(B)}(t)e^{i\omega_k t})\}]_{\alpha=0}\big|_{s^{(A)}=s^{(B)}=0}. \tag{46}$$

Here $|\alpha\rangle$ ($|\beta\rangle$) is the initial (final)state of the source. The φ's, $\underline{S}(t)$'s, and $\underline{S}$'s are defined in terms of $\underline{s}^{(A)}(t)$ and $\underline{s}^{(B)}(t)$ through equations analogous to (25) and (33).

The general method outlined here might be of some use in the study of the infrared behavior of the quanta of non-Abelian gauge fields. Formally, self-interaction effects and effects of the quantum nature of the current have been separated out. This separation is only formal, however, since in (46) the problem of ordering $\underline{j}(t)$ still remains. Also the W's of the last exponent remain to be computed in perturbation theory.

ACKNOWLEDGMENTS

This work arose out of collaboration with A. Din, H. Rubinstein, and G. Peressutti, and we thank them all for critical and stimulating discussions. Professor Rubinstein has provided us with valuable information on the development of the field. We also thank A. P. Balachandran who for five years has been encouraging one of us (K.E.E.) to solve the infrared problem in non-Abelian gauge field theory and we apologize for making a contribution which is still far from a solution. The work of B-S. S. was supported by the Swedish Natural Science Research Council under Contract No. 8244-008.

---

*Present address: Physics Department, Syracuse University, Syracuse, New York 13210.

[1]For a general treatment of the infrared-divergence problem see T. Kinoshita, J. Math. Phys. 3, 650 (1962); and T. D. Lee and M. Nauenberg, Phys. Rev. 133, B1549 (1964). A special case of non-Abelian gauge field theory is the theory of gravitation. This is comparatively simple, however, since soft gravitons carry negligible gravitational charge. See S. Weinberg, Phys. Rev. 138, B988 (1965); 140, B516 (1965). Some works dealing with the infrared problem of non-Abelian gauge theories are the following: E. C. Poggio and H. R. Quinn, Phys. Rev. D 14, 578 (1976); Y. P. Yao, Phys. Rev. Lett. 36, 653 (1976); T. Appelquist, J. Carazzone, H. Kluberg-Stern, and M. Roth, *ibid.* 36, 768 (1976); 36, 1161(E) (1976); J. Cornwall and G. Tiktopoulos, *ibid.* 35, 338 (1975); Phys. Rev. D 13, 3370 (1976); E. C. Poggio, H. R. Quinn, and J. B. Zuber, *ibid.* 15, 1630 (1977); T. Kinoshita and A. Ukawa, *ibid.* 15, 1596 (1977); E. C. Poggio, Phys. Lett. 68B, 347 (1977).

[2]Here we adopt a method which was applied to the study of the detailed time evolution of the soft electromagnetic bremsstrahlung field in K. E. Eriksson, Phys. Scr. 1, 3 (1970). Gravitons have been similarly treated in C. Alvegård, K. E. Eriksson, and C. Högfors, Phys. Scr. 17, 95 (1978). For the Coulomb field, see J. D. Dollard, J. Math. Phys. 5, 729 (1964). Some works on the infrared problem in quantum electrodynamics which are relevant to the present problem but which appeared later than K.E.E.'s 1970 paper and which, therefore, are not referred to therein are the following: P. Kulish and L. Faddéev, Teor. Mat. Fiz. 4, 153 (1970) [Theor. Math. Phys. 4, 745 (1970)]; D. Zwanziger, Phys. Rev. D 7, 1082 (1973). G. Grammer and D. R. Yennie, *ibid.* 8, 4332 (1973).

PHYSICAL REVIEW D VOLUME 18, NUMBER 2 15 JULY 1978

Path-integral representation for the S matrix

C. L. Hammer and J. E. Shrauner*
Ames Laboratory—USDOE, Iowa State University, Ames, Iowa 50011

B. De Facio
Physics Department, University of Missouri-Columbia, Columbia, Missouri 65211

(Received 16 January 1978; revised manuscript received 13 March 1978)

We present a formulation of quantum field theory as a path-integral representation for elements of the U or S matrix in the coherent-state basis. These matrix elements are shown to serve as generating functionals for all the usual S-matrix elements between states of definite particle number. The coherent-state formalism for general bilinear quantum field theories is described and then incorporated into the construction of a path integral such that the formulation is independent of any canonical formalism. We discuss the relationship of this formulation to the usual path-integral formulation and show in what sense they are equivalent for canonical theories. We also discuss how this formulation is more general than the usual one in that it is well defined even for theories having no canonical form and for which a Lagrangian action and the usual path-integral formulation may not apply. Applications of this formulation to specific calculations are exhibited for the cases of a quantized field interacting with a given external source and for the renormalization of the simple quantum field theory model of a scalar meson field interacting with a nonrelativistic nucleon field.

I. INTRODUCTION

We consider the formulation of a quantum field theory that combines the path-integral representation with the coherent-state representation for elements of the U or S matrices. This formulation is useful both for proving fundamental properties of quantum field theories and for performing specific calculations.

The path-integral formulation of quantum field theory has been very prominent recently in elementary particle physics, and there exist many fine reviews of the subject in the literature.[1] Some of the earliest work in this subject was by Feynman, who developed a path-integral representation for quantum amplitudes and applied them to integrating out (the degrees of freedom of) the electromagnetic field in quantum electrodynamics.[2]

Coherent-state techniques have also been rising to prominence in general quantum field theory and elementary particle physics as well as in quantum optics. The application of coherent-state techniques to problems involving large quantum numbers in quantum optics is by now rather well known.[3] Recently coherent states have been used as the foundation for statistical theories of correlations of many-particle production processes at high energies.[4] These statistical theories and other phenomenological theories correspond to detailed microscopic field theories at a stage at which some subset of the interacting field variables have been integrated out as mentioned above. Also Bolsterli[5] has used coherent-state methods, without path integrals or lattice limits, to study covariant theories of mesons with static sources. He has shown these methods to be powerful.

The power of combining the path-integral formulation and the coherent-state representation has been shown in the recent solution by Dente[6] of the long-standing problem of demonstrating that quantum electrodynamics has a classical limit. In this work it was shown by integrating out the electromagnetic field in transition matrix elements between nonvacuum coherent states that QED in its usual form has the expected classical limit without the necessity of imposing any artificial alteration of the theory as was done in the "absorber theory" of Feynman and Wheeler.[7]

The coherent-state formalism for general bilinear quantum field theories as described by DeFacio and Hammer[8] is incorporated in an essential way into our path-integral formulation. The formalism of bilinear field theories, including their quantization, derives totally from their free-field equations of motion and associated currents and does not depend on a canonical quantization procedure.[8,9,10,11] This generalization makes our path-integral formulation in the basis of coherent states of general bilinear field theories applicable even for such theories for which no canonical form exists.

Earlier work[12] uniting the coherent-state formalism and path-integral formalism has been done by Klauder and Schweber, who adapted the lattice-integration methods of Feynman, much the same as we do here. More recently fundamental work in this area has been done by Faddeev, Berezin, Klauder, and others.[13] Thus, the basic mathematical details in this area have been, with the exceptions of a few possible lacunae, rather thoroughly developed. In the present work we assemble the formulation with modest innovations as described above, emphasizing the extended generality and the potential for modeling and treating important phenomenological problems. We start on a program of explicit calculations using this formalism.

A path-integral representation for U- and S-matrix elements in the coherent-state basis is thus constructed. Because of the properties of the coherent-state matrix elements of field operators, the quantum field theory problem is reduced to a quantum mechanics problem. The path-integral formulation then reduces the quantum mechanics problem to a consideration of an effective classical action problem. Thus a convenient and natural formulation is obtained for the treatment of transitions between initial and final states, i.e., end points of the functional path integral that are nontrivial, nonvacuum states such as classical condensates, solitons,[14] vexictons,[15] vortices,[16] etc. This, combined with the fact that this formulation is in terms of the U- and S-matrix elements, means that we have a powerful method for approaching spontaneous symmetry breaking in quantum field theory in a way not previously used for this purpose.

In this paper we describe in Sec. II the general formalism of the coherent-state basis for a general bilinear quantum field theory. In Sec. III we show how the S-matrix element in the coherent-state basis of a given field may serve as a generating functional for all the usual S-matrix elements between states of the occupation-number basis for that field. We also discuss there the time dependence of the operators of a general bilinear quantum field theory upon which is based not only the asymptotic condition but also the extension to the coherent-state formalism of a method of quantization that is independent of any postulates of canonical quantization or locality conditions.[8,9,10,11] In Sec. IV we construct a path-integral representation of U- and S-matrix elements in the coherent-state basis for general bilinear quantum field theories. In Sec. V we demonstrate the application of this formulation to the calculation of a specific problem with the prototype of a quantized field interacting with a given external current source. In Sec. VI we extend application to the renormalization of the simple scalar meson model of interacting fields. In Sec. VII we discuss the relation to our formulation of quantum field theory to the usual path-integral representation. Specifically, we show that our path integral is equivalent to the usual one for the case of canonical field theories. However, our formulation is more general in that it remains well defined even for theories that have no canonical form and for which the Lagrangian and its action, and therefore, possibly, the usual path-integral formulation, does not exist.

II. THE COHERENT-STATE BASIS

For a field theory of the usual class[8] in which there exists, as a consequence of the field equations of motion, a conserved current $J_\mu(\psi_1, \psi_2)$ that is bilinear in ψ_1 and ψ_2, which are any pair of solutions of the field equations, the timelike current component $J_0(\psi_1, \psi_2)$ defines a bilinear structure that can be used to construct invariant inner products as

$$(\phi(t), \chi(t)) \equiv \int d^3x\, J_0(\phi(\vec{x}, t), \chi(\vec{x}, t)). \tag{2.1}$$

This inner product can be formed with any pair of suitably well-behaved functions. It will be independent of time t whenever ϕ and χ each satisfy the field equations from which the conserved current J_μ is defined.

Instead of the usual canonical commutation relations (CCR's) we adopt the more general form[8,17] required by the symmetries of the equations of motion:

$$[\psi, S] = s\psi\,, \tag{2.2}$$

where S is a symmetry operator

$$S = \int d^3x\, J_0(\psi, s\psi)\,.$$

This expression, coupled with the locality assumption for independent fields

$$[\psi_i(\vec{x}, t), \psi_j(\vec{y}, t)]_\pm = 0\,, \tag{2.3}$$

can be used to obtain the remaining equal-time commutators. They are not necessarily CCR's.[8,17]

Of particular interest for our use is the operator

$$\Sigma_\phi(t) \equiv z^{*-1/2}(\psi(t), \phi(t))\eta_A - z^{-1/2}\bar{\eta}_A(\phi(t), \psi(t)), \tag{2.4}$$

where $\phi(\vec{x}, t)$ is any suitably well-behaved c-number function; $\psi(\vec{x}, t)$ is an operator-valued solution of the field equations, and z is a renormalization constant. For bosons $\eta_A = \bar{\eta}_A = \eta_B = 1$, and for fermions $\eta_A = \eta_F$ and $\bar{\eta}_A = \bar{\eta}_F$ are a pair of anticommuting c-number quantities for which

$$\{\eta_F, \bar{\eta}_F\} = \{\eta_F, \eta_F\} = \{\bar{\eta}_F, \bar{\eta}_F\} = \{\eta_F, \psi\} = \{\bar{\eta}_F, \psi\} = 0\,.$$

The commutator of the operator $\Sigma_\phi(t)$ with the field operator follows from the form of Eq. (2.2) as

$$[\psi(\vec{x}, t), \Sigma_\phi(t)] = z^{*-1/2}\eta_A \phi(\vec{x}, t)\,. \tag{2.5}$$

This commutation relation implies that the operator Σ_ϕ generates transformations of the field operator ψ of the form

$$e^{-\Sigma_\phi(t)}\psi(\vec{x},t)e^{\Sigma_\phi(t)}=\psi(\vec{x},t)+z^{*-1/2}\eta_A\phi(\vec{x},t)\,. \tag{2.6}$$

This means that $\exp[\Sigma_\phi(t)]$ generates coherent states of the field $\psi(\vec{x},t)$ as

$$e^{\Sigma_\phi(t)}|\Omega\rangle\equiv|\phi(t)\rangle\,,\qquad \langle\Omega|e^{-\Sigma_\phi(t)}\equiv\langle\phi(t)|\,, \tag{2.7}$$

where $|\Omega\rangle$ is the Fock-space vacuum state. These coherent states are eigenstates of the field operator $\psi(\vec{x},t)$ as

$$\langle\phi(t)|\psi(\vec{x},t)|\phi(t)\rangle=z^{*-1/2}\eta_A\phi(\vec{x},t)\,, \tag{2.8}$$

or if $\psi_+(\vec{x},t)$ is the pure annihilation-operator part of $\psi(\vec{x},t)$,

$$\psi_+(\vec{x},t)|\phi(t)\rangle=z^{*-1/2}\eta_A\phi(\vec{x},t)|\phi(t)\rangle\,. \tag{2.9}$$

The function $\phi(\vec{x},t)$ is arbitrary, with the condition that it is well behaved with respect to the inner products that it enters for the specific theory under consideration in a given case. This *coherent-state wave function* $\phi(\vec{x},t)$ can also be used as a variational parameter in our formulation of quantum field theory in the coherent-state basis.

Using the Baker-Campbell-Hausdorf theorem we obtain the important relations

$$\begin{aligned}e^{\Sigma_\phi(t)}=\exp[z^{*-1/2}(\psi(t),\phi(t))\eta_A]\\ \times\exp[-z^{-1/2}\bar{\eta}_A(\phi(t),\psi(t))]\\ \times\exp[-\tfrac{1}{2}|z|^{-1/2}(\phi(t),\phi(t))]\,,\end{aligned} \tag{2.10}$$

$$\begin{aligned}e^{\Sigma_\phi(t)}e^{\Sigma_\chi(t)}=e^{\Sigma_{\phi+\chi}(t)}\\ \times\exp\{\tfrac{1}{2}|z|^{-1}[(\chi(t),\phi(t))-(\phi(t),\chi(t))]\}\,,\end{aligned} \tag{2.11}$$

and therefore

$$\begin{aligned}\langle\phi(t)|\chi(t)\rangle=\exp\{-\tfrac{1}{2}|z|^{-1}[(\chi(t),\chi(t))+(\phi(t),\phi(t))\\ -2(\phi(t),\chi(t))]\}\,,\end{aligned} \tag{2.12}$$

where $\phi(\vec{x},t)$ and $\chi(\vec{x},t)$ are arbitrary coherent-state wave functions and $\psi(\vec{x},t)$ is the quantized field operator. The coherent states $|\phi(t)\rangle$ satisfy the completeness relation

$$\int\left[\frac{D^2\phi}{\pi|z|}\right]|\phi(t)\rangle\langle\phi(t)|=1\,. \tag{2.13}$$

The differential element for our functional integrations on the coherent-state wave-function space is written in the forms

$$\begin{aligned}\left[\frac{D^2\phi}{\pi|z|}\right]&=\left[\frac{D\,\mathrm{Re}\phi}{(\pi|z|)^{1/2}}\right]\left[\frac{D\,\mathrm{Im}\phi}{(\pi|z|)^{1/2}}\right]\\ &=\left[\frac{D\phi}{(2\pi z)^{1/2}}\right]\left[\frac{D\phi^*}{(2\pi z^*)^{1/2}}\right].\end{aligned} \tag{2.14}$$

Here we have adopted the notation for any constant b,

$$[Db\phi]=\prod_k(bd\beta_k)=(bd\beta_1)(bd\beta_2)\cdots\,,$$

where β_k are the expansion coefficients of the field ϕ expanded in normal modes.

All the relations for coherent states mentioned above apply for any operator-valued field ψ satisfying canonical equal-time commutation or anticommutation relations and their associated conserved currents.[9,10] However, in what follows we will always be using complete sets of states that are coherent states of free fields and so the inner products involved in the generators and matrix elements of such states will always be based on the conserved currents of the free fields; i.e., the usual type of inner products. In this case the commutation relation Eq. (2.5) on which our coherent-state formalism is based is well defined and determined even if no canonical commutation relation or locality (or quasilocality) condition exists.[8,9,10] Quite generally, all operators of a quantized field theory may be expressed in bilinear form such as we use for the Σ_ϕ operator.[8,11]

III. S-MATRIX ELEMENT IN COHERENT-STATE BASIS AS GENERATING FUNCTIONAL FOR USUAL S-MATRIX ELEMENTS

The S-matrix element in the coherent-state basis of a given field in a given theory may serve as a generating functional for all the usual S-matrix elements between states of definite numbers of quanta of that field. The mechanism of this generation is the variation with respect to the projection of the coherent-state wave function ϕ onto the single-particle modes of the in-states of this field. The reason that all the usual particle S-matrix elements may be obtained from a coherent-state S-matrix element is because a coherent state is comprised of a definite combination of all the n-particle Fock states of all modes of the field. This may be shown by projecting the in-field operator $\psi_{\rm in}$ and the coherent-state wave function ϕ onto the positive-energy free-particle wave function of the kth mode $\hat{\phi}_k(x)$ as

$$\psi_{k,\rm in}\equiv(\hat{\phi}_k(t),\psi_{\rm in}(t))\,,\quad\text{independent of }t$$

and

$$\beta_k(t)\equiv(\hat{\phi}_k(t),\phi(t))\,,$$

where the $\hat{\phi}_k(x,t)$'s are assumed to form a complete orthonormal set with $(\hat{\phi}_k, \hat{\phi}_q) = \delta^3_{k,q}$. So then the generator of the coherent in state $|\phi(t)\rangle_{in}$ is expanded as

$$\Sigma_{\phi\,in}(t) = (\psi_{in}(t), \phi(t))\eta_A - \bar{\eta}_A(\phi(t), \psi_{in}(t))$$
$$= \sum_k [\psi^\dagger_{k,in}\eta_A\beta_k(t) - \bar{\eta}_A\psi_{k,in}\beta^*_k(t)], \quad (3.1)$$

from which follows

$$|\phi(t)\rangle_{in} = e^{\Sigma_{\phi\,in}(t)}|\Omega\rangle_{in}$$
$$= \prod_k e^{\psi^\dagger_{k,in}\eta_A\beta_k(t)} e^{-(1/2)|\beta_k(t)|^2}|\Omega\rangle_{in}$$
$$= \prod_k \left[\sum_{n_k} \frac{[\eta_A\beta_k(t)]^{n_k}}{(n_k!)^{1/2}}|n_k\rangle_{in}\right] e^{-(1/2)|\beta_k(t)|^2}. \quad (3.2)$$

Thus the general particle in-state of the Fock basis is

$$|l_1 l_2 \cdots l_k \cdots\rangle_{in} = \prod_k (\partial/\partial\beta_k)^{l_k}(l_k!)^{-1/2} \times e^{(1/2)|\beta_k|^2}|\phi\rangle_{in}\Big|_{\phi=0=\beta_k}, \quad (3.3)$$

where the t dependence has been suppressed, as it goes away after the $\phi \to 0, \beta_k \to 0$ limit. This implies that the usual S-matrix element between such states of definite particle number is

$$_{out}\langle l_1 \cdots l_k \cdots | n_1 \cdots n_q \cdots\rangle_{in}$$
$$= \prod_{k,q} (\partial/\partial\beta'^*_k)^{l_k}(\partial/\partial\beta_q)^{n_q}(l_k!n_q!)^{-1/2} \times e^{1/2(|\beta'_k|^2+|\beta_q|^2)}\,{}_{out}\langle\phi'|\phi\rangle_{in}\Big|_{\substack{\phi=0\\ \phi'=0}}, \quad (3.4)$$

with $\beta'_k(t) \equiv (\hat{\phi}_k(t), \phi'(t))$. These relations are direct consequences of the definitions of the coherent states.

The asymptotic in/out limits of our coherent states require some discussion. The usual asymptotic condition is given as the weak operator limit relating (matrix elements of) interacting unrenormalized Heisenberg fields $\psi(\vec{x}, t)$ with the free asymptotic in- or out-fields $\psi_{in,out}(\vec{x}, t)$, each in terms of their particle-mode projections as

$$\text{wk-lim}_{t\to-\infty(+\infty)} [(\hat{\phi}_k(t), \psi(t)) - z^{1/2}(\hat{\phi}_k(t), \psi_{in,out}(t))] = 0, \quad (3.5)$$

where z is a wave-function renormalization constant. The term $(\hat{\phi}_k(t), \psi_{in,out}(t))$ of this relation is independent of the time t since both $\hat{\phi}_k(\vec{x}, t)$ and $\psi_{in}(\vec{x}, t)$ satisfy the free-field equation which determines the conserved current that is used to construct these inner products. The same inner product based on the free-field conserved current is used in both terms of this relation.

The generator of the coherent in/out states, $\Sigma_{\phi\,in/out}$, is given by Eq. (3.1) in terms of this free-field-based inner product and these same $\hat{\phi}_k$'s. If we now use this same free-field-based inner product in

$$\Sigma_\phi(t) = z^{*-1/2}(\psi(t), \phi(t))\eta_A - z^{-1/2}\bar{\eta}_A(\phi(t), \psi(t)), \quad (3.6)$$

then it is clear that

$$\text{wk-lim}_{t\to-\infty\,(+\infty)} [\Sigma_\phi(t) - \Sigma_{\phi\,in\,(out)}(t)] = 0. \quad (3.7)$$

Justification of the use of the free-field inner product in the $\Sigma_\phi(t)$ defined in Eq. (3.6) will be more apparent after a discussion of the t dependence of these operators, which also leads us into our U-matrix treatment in Sec. IV.

The time dependence of the Heisenberg field operators is described as

$$\psi(t) = e^{iHt}\psi(0)e^{-iHt}.$$

We can also define a set of fields $\psi_0(t)$ that propagate as free fields with the Hamiltonian $H_0 = H - V$ and coincide with the interacting fields of the Heisenberg and Dirac pictures all at time $t = 0$ as

$$\psi_0(t) = e^{iH_0t}\psi(0)e^{-iH_0t}$$
$$= U_c(t, 0)\psi(0),$$

where $U_c(t, 0)$ is the c-number time translation operator for the free field ψ_0. For example $U_c(t, 0) = \exp(it\nabla^2/2m)$ in the case of a Hamiltonian theory of nonrelativistic particles of mass m. Then

$$e^{-iH_0t}(\phi(0), \psi(0))e^{iH_0t} = (\psi(0), U_c^{-1}(t, 0)\psi(0))$$
$$= (U_c(t, 0)\phi(0), \psi(0)).$$

For the purposes of the rest of the present paper we shall restrict our considerations to coherent-state wave functions that satisfy the free-field equations, so that $U_c(t, 0)\phi(0) = \phi(t)$ [and incidentally $\beta_k(t)$ defined above becomes time independent], and so

$$e^{iHt}e^{-iH_0t}(\phi(0), \psi(0))e^{iH_0t}e^{-iHt} = (\phi(t), \psi(t))$$

and (3.8)

$$\Sigma_\phi(t) = e^{iHt}e^{-iH_0t}\Sigma_\phi(0)e^{iH_0t}e^{-iHt}.$$

Thus the coherent state evolves in time as

$$|\phi(t)\rangle = e^{iHt}e^{-iH_0t}|\phi(0)\rangle$$
$$= e^{\Sigma_\phi(t)}|\Omega(t)\rangle, \quad (3.9)$$

where

$$|\Omega(t)\rangle = e^{iHt}e^{-iH_0t}|\Omega\rangle .$$

The asymptotic in/out states are given as

$$\underset{t\to-\infty\,(+\infty)}{\text{wk-lim}}\,[e^{iHt}e^{-iH_0t}|\phi(0)\rangle - |\phi(t)\rangle_{\text{in}}] = 0 , \qquad (3.10)$$

and the S-matrix element in the coherent-state basis is given by

$$\underset{\substack{t\to-\infty\\ t'\to+\infty}}{\text{wk-lim}}\,[\langle\phi'(0)|U(t',t)|\phi(0)\rangle - {}_{\text{out}}\langle\phi'(t')|\phi(t)\rangle_{\text{in}}] = 0 , \qquad (3.11)$$

where

$$U(t',t) = e^{iH_0t'}e^{-iH(t'-t)}e^{-iH_0t}$$

is the familiar time evolution operator.[18] The time $t = 0$ is when our free and interacting Heisenberg fields coincide and when the Heisenberg and Dirac pictures coincide.

This justifies the use of the free-field inner products in constructing the generator $\Sigma_\phi(t)$ in Eq. (3.6), because all our coherent states can be viewed as being built at time $t = 0$ when the free and interacting Heisenberg fields coincide and then they evolve in time to other values of t according to the above descriptions. Finally, in terms of the above limits and the Heisenberg field operators, the S-matrix element Eq. (3.4) is

$$\begin{aligned}
{}_{\text{out}}\langle l_1\cdots l_k\cdots|n_1\cdots n_q\cdots\rangle_{\text{in}}
&= \lim_{\substack{t\to-\infty\\ t'\to\infty}} \prod_{k,q} (\partial/\partial\beta_k'^*)^{l_k}(\partial/\partial\beta_q)^{n_q}(l_k!n_q!)^{-1/2} \\
&\quad\times e^{+(1/2)|z|^{-1}(|\beta_k'|^2+|\beta_q|^2)} \\
&\quad\times \langle\phi'(t')|\phi(t)\rangle\big|_{\substack{\phi=0\\ \phi'=0}} . \qquad (3.12)
\end{aligned}$$

IV. PATH-INTEGRAL REPRESENTATION OF U AND S MATRICES IN COHERENT-STATE BASIS

The interaction Hamiltonian in the Dirac picture is defined as

$$V(t) = e^{iH_0t}Ve^{-iH_0t} . \qquad (4.1)$$

Then the U operator above in the limit that $t' - t = \epsilon$ becomes infinitesimally small is

$$U(t+\epsilon, t) = e^{-i\epsilon V(t)} . \qquad (4.2)$$

This expression is valid to lowest order in ϵ even if V should depend explicitly on time.

The field operators of which H, H_0, and V are constructed are assumed to be separated into pure annihilation and creation parts as $\psi = \psi_+ + \psi_-$. A normal ordering is defined so that in matrix elements between coherent states the annihilation parts act to the right on their eigenstates as $\psi_+(\vec{x},t)|\phi(0)\rangle = \phi(\vec{x},t)|\phi(0)\rangle$, and the creation parts ψ_- act similarly to the left on their eigenstates. This definition of the normal ordering clearly is the same as the conventional normal ordering. A result is that in the infinitesimal-time-interval U-matrix element the exponential operator can be replaced by a c-number functional as

$$\begin{aligned}
&\langle\phi'(0)|e^{-i\epsilon V(\psi_-(t),\psi_+(t))}|\phi(0)\rangle \\
&\quad = \langle\phi'(0)|\phi(0)\rangle e^{-i\epsilon V(z^{-1/2}\phi'(t')^*,\, z^{*-1/2}\phi(t))} . \qquad (4.3)
\end{aligned}$$

We consider $\eta_A = \bar{\eta}_A = 1$ here and in the following. Because our matrix element is first order in the infinitesimal ϵ we have some liberty as to the convention for specifying its time arguments. For example, it might be convenient to label them as $V(\phi'(t+\epsilon)^*, \phi(t))$ in the exponent functional. We shall also make use of the choice $V(\phi'(t)^*, \phi(t))$.

To obtain the finite-time-interval matrix elements we start by inserting a complete set of intermediate coherent states as

$$\begin{aligned}
&\langle\phi'(0)|U(t',t)|\phi(0)\rangle \\
&\quad = \int \left[\frac{D^2\phi_1(0)}{\pi|z|}\right]\langle\phi'(0)|U(t',t_1)|\phi_1(0)\rangle \\
&\qquad\times\langle\phi_1(0)|U(t_1,t)|\phi(0)\rangle , \qquad (4.4)
\end{aligned}$$

and we do this for each $t_l = t + l\epsilon$, $l = 1, 2, \ldots, N$, with $t' - t = (N+1)\epsilon$. After integrating over all the sets of coherent intermediate states we let $N\to\infty$ and $\epsilon\to 0$. This gives

$$\begin{aligned}
\langle\phi'(0)|U(t',t)|\phi(0)\rangle = \lim_{\substack{N\to\infty\\ \epsilon\to 0}} \int &\langle\phi'(0)|\phi_N(0)\rangle\left[\frac{D^2\xi_N}{\pi}\right]e^{-i\epsilon V(z^{-1/2}\phi'^*(t'),\,\xi_N(t_N))} \\
&\times\langle\phi_N(0)|\phi_{N-1}(0)\rangle\left[\frac{D^2\xi_{N-1}}{\pi}\right]e^{-i\epsilon V(\xi_N^*(t_N),\,\xi_{N-1}(t_{N-1}))}\cdots \\
&\times\langle\phi_1(0)|\phi(0)\rangle e^{-i\epsilon V(\xi_1^*(t_1),\,\phi(t)z^{*-1/2})} , \qquad (4.5)
\end{aligned}$$

where the integration variables are defined as $\xi_l \equiv z^{*-1/2}\phi_l$. Altogether a path integral is built up in which each succeeding set of intermediate states may be labeled by the l associated with t_l. A representative integration is

$$\begin{aligned}
&\int\left[\frac{D^2\xi_l(0)}{\pi}\right]\langle\phi_{l+1}(0)|\phi_l(0)\rangle e^{-i\epsilon V(\xi_{l+1}^*(t_{l+1}),\,\xi_l(t_l))} \\
&\qquad\times e^{-i\epsilon V(\xi_l^*(t_l),\,\xi_{l-1}(t_{l-1}))}\langle\phi_l(0)|\phi_{l-1}(0)\rangle . \qquad (4.6)
\end{aligned}$$

This results in the product of $\langle\phi_{l+1}(0)|\phi_{l-1}(0)\rangle$, which just describes the propagation of the kinematical weight factor, times a dynamical part from the V's. Clearly this feature will propagate through the whole lattice giving for the matrix element for the finite interval $t'-t$

$$\langle\phi'(0)|U(t',t)|\phi(0)\rangle=\langle\phi'(0)|\phi(0)\rangle \times \exp\left[-i\int_t^{t'} dt''\mathcal{V}(t'')\right], \quad (4.7)$$

where $\mathcal{V}(t)$ is an effective interaction potential functional of the fields ϕ' and ϕ that describes the results of the path integration.

V. INTEGRATION OF EXTERNAL CURRENT PROBLEM

The problem of a quantized field interacting with a given external source current is the prototype for calculations with interacting quantum field theories. Our treatment in subsequent articles of more complex interacting theories will refer to this simple prototype.

We consider a system of a quantized field $\psi(\vec{x},t)$ interacting with a given external source current $j(\vec{x},t)$ through the interaction Hamiltonian

$$V=\int d^3x\, \psi(\vec{x},t)j(\vec{x},t)+\text{H.c.} \quad (5.1)$$

The normal-ordered U-matrix element for an infinitesimal time interval in the coherent-state basis in the Dirac interaction picture is

$$\langle\phi'(0)|U(t+\epsilon,t)|\phi(0)\rangle$$
$$=\exp\left\{-i\epsilon\int d^3x[z^{*-1/2}\phi'^*(\vec{x},t+\epsilon)j(\vec{x},t)+z^{-1/2}j^*(\vec{x},t+\epsilon)\phi(\vec{x},t)]\right\}\langle\phi'(0)|\phi(0)\rangle. \quad (5.2)$$

Note that the kinematic weight factor is evaluated at $t=0$. The functional integrations over complete sets of coherent states at each of the intermediate lattice times which comprise the path integral are in this case Gaussian. That this is true independently of the structure of the inner product in the above expression for $U(t+\epsilon,t)$ is readily seen by reexpressing it in terms of expansions on a complete set of eigenfunctions $\hat{\phi}_k(\vec{x},t)$ of the free Hamiltonian. Defining $\beta_k'(t)=(\hat{\phi}_k(t),\phi'(t))$ and $\beta_k(t)=(\hat{\phi}_k(t),\phi(t))$ as before and similarly

$$\sum_k \beta_k^*(t)j_k(t)\equiv\int d^3x\,\phi^*(\vec{x},t)j(\vec{x},t),$$

we have

$$\langle\phi'(0)|U(t+\epsilon,t)|\phi(0)\rangle=\exp\left\{-i\epsilon\sum_k[z^{*-1/2}\beta_k'(t+\epsilon)^*j_k(t)+z^{-1/2}j_k(t+\epsilon)^*\beta_k(t)]\right\}$$
$$\times\exp\left(-\tfrac{1}{2}|z|^{-1}\sum_k[|\beta_k'|^2+|\beta_k|^2-2\beta_k'^*\beta_k]\right) \quad (5.3)$$

and

$$\left[\frac{D^2\phi_l(0)}{\pi|z|}\right]=\prod_k\left(\frac{d\,\mathrm{Re}\beta_{lk}}{(\pi z^*)^{1/2}}\right)\left(\frac{d\,\mathrm{Im}\beta_{lk}}{(\pi z)^{1/2}}\right) \quad (5.4)$$

for the integration labeled by the time $t_l=t+l\epsilon$, with $l=0,1,2,\ldots,N$, $\epsilon(N+1)=(t'-t)$. In this decomposition it is thus clear that all our integrals are Gaussian. After integrating N intermediate stages the finite-lattice path integral is

$$\langle\phi'(0)|U(t',t)|\phi(0)\rangle=\prod_k\left\{\exp\left[\frac{-1}{2|z|}(|\beta_k'|^2+|\beta_k|^2-2\beta_k'^*\beta_k)\right]\right.$$
$$\times\exp\left[-i\sum_{l=1}^{N+1}[z^{*-1/2}\beta_k'^*(t_l)j_k(t_{l-1})+z^{-1/2}j_k^*(t_l)\beta_k(t_{l-1})]\right]$$
$$\left.\times\exp\left[-\epsilon^2\sum_{l=1}\sum_{r=1}^{l}j_k^*(t_{l+1})j_k(t_r)e^{-i\omega_k(t_l-t_r)}\right]\right\}. \quad (5.5)$$

The basic integral is seen by setting $N=1$ in this expression. In terms of the original inner products this may be written in the continuous limit $N\to\infty$, $\epsilon\to dt''$ as

$$\langle\phi'(0)|U(t',t)|\phi(0)\rangle=\langle\phi'(0)|\phi(0)\rangle\exp\left\{-i\int_t^{t'}d^3x\,dt''[z^{*-1/2}\phi'^*(t'')j(t'')+z^{-1/2}j^*(t'')\phi(t'')]\right\}$$
$$\times\exp\left[-\int_t^{t'}dt''d^3x\int_t^{t''}dt'''d^3y\,j(\vec{x},t'')G(\vec{x},t'';\vec{y},t''')j(\vec{y},t''')\right], \quad (5.6)$$

where

$$G(\vec{x},t';\vec{y},t) \equiv \sum_k \hat{\phi}_k(\vec{x},t')\hat{\phi}_k^*(\vec{y},t). \quad (5.7)$$

As expected, the kinematic factor $\langle\phi'(0)\,|\,\phi(0)\rangle$ appears in the finite-interval U-matrix element in the same way as in the matrix element for the infinitesimal interval. The exponential factor containing the double time integral is common to all U-matrix elements of this theory, including all the particle-Fock-states matrix elements that can be generated from a matrix element between coherent states. In particular this factor by itself is clearly the Fock-space vacuum-vacuum U-matrix element. All these observations also apply equally well, after taking the asymptotic limits, to the corresponding S-matrix elements.

Taking the asymptotic limit and adapting our previous prescription Eq. (3.12) for deriving the usual particle S-matrix elements we obtain the well-known result for the vacuum-vacuum matrix element,

$$_{\text{out}}\langle\Omega\,|\,\Omega\rangle_{\text{in}} = {}_{\text{in}}\langle\Omega\,|S|\,\Omega\rangle_{\text{in}} = \lim_{\substack{t\to-\infty\\ t'\to\infty}} \langle\phi'(0)\,|U(t',t)|\,\phi(0)\rangle\,|_{\phi'=0=\phi}$$

$$= \lim_{\substack{t\to-\infty\\ t'\to\infty}} \exp\left[-\int_t^{t'} dt'' \int_t^{t''} dt''' \sum_k j_k^*(t'')j_k(t''')e^{-i\omega_k(t''-t''')}\right].$$

VI. SCALAR MESON MODEL

A simple quantum field theory that we use to illustrate our formulation and methods is the model of a neutral Hermitian scalar meson field interacting with very massive nonrelativistic nucleons.[19] This model is characterized by the Hamiltonian $H = H_0 + V$ with the free part

$$H_0 = \sum_q m\psi_q^\dagger\psi_q + \sum_k \omega_k a_k^\dagger a_k, \quad (6.1)$$

in which the energies of the nucleons are taken to be independent of their momenta $\vec{q}$, the meson energies are $\omega_k = (k^2+\mu^2)^{1/2}$, and the nucleon and meson creation and annihilation operators, respectively, satisfy the equal-time commutation relations

$$\{\psi_{q'},\psi_q^\dagger\} = \delta^3(\vec{q},\vec{q}'), \qquad [a_{k'},a_k^\dagger] = \delta^3(\vec{k},\vec{k}'). \quad (6.2)$$

Here m and μ are the physical masses of the nucleons and the mesons.

The interaction is given by

$$V = \lambda\sum_{q,k}(\psi_{q+k}^\dagger\psi_q a_k g_k + \text{H.c.}) - \delta m\sum_q \psi_q^\dagger\psi_q, \quad (6.3)$$

where λ is the coupling constant, $g_k = f(k^2)/(2\omega_k)^{1/2}$ with $f(k^2)$ a form factor, and δm is a mass-renormalization counterterm.

A. Renormalization

We shall first consider S-matrix elements for transitions between initial and final states that are coherent states of the meson field and that have a single nucleon in a state of definite momentum since it is this sector that contains all the essential renormalization aspects of the scalar meson model. Thus the S-matrix element of interest is

$$_{\text{out}}\langle p',\phi'\,|p,\phi\rangle_{\text{in}} = \lim_{\substack{t\to-\infty\\ t'\to\infty}} |z_p|^{-1}\langle\phi'(0)\,|\psi_{p'}e^{iH_0t'}e^{-iH(t'-t)} \times e^{-iH_0t}\psi_p^\dagger\,|\,\phi(0)\rangle, \quad (6.4)$$

where z_p is the nucleon wave-function renormalization constant.

Following the analysis of Sec. IV, we find the U-matrix element for the infinitesimal interval $t_{j+1} - t_j = \epsilon$ for the interaction potential of Eq. (6.3) to first order in ϵ is

$$M_{j+1,j} = \langle\phi_{j+1}(0)\,|\psi_{p_{j+1}}e^{-i\epsilon V(t_j)}\psi_{p_j}^\dagger\,|\,\phi_j(0)\rangle$$

$$= \langle\phi_{j+1}(0)\,|\,\phi_j(0)\rangle\Big\{\delta^3(\vec{p}_{j+1},\vec{p}_j)(1+i\epsilon\delta m) - i\epsilon\lambda\sum_{q,k}[\delta^3(\vec{q},\vec{p}_j)\delta^3(\vec{q}-\vec{k},\vec{p}_{j+1})g_k\beta_{j,k}e^{-i\omega_k t_j}$$

$$+ \delta^3(\vec{q},\vec{p}_{j+1})\delta^3(\vec{q}-\vec{k},\vec{p}_j)g_k^*\beta_{j,k}^* e^{i\omega_k t_j}]\Big\}, \quad (6.5)$$

with $\psi_{p_j}\,|\,\phi_l(0)\rangle = 0$ for all l and p_j, and where $\beta_{j,k} = (\hat{\phi}_k(0),\phi_j(0))$, ϕ_j is the coherent-state wave function in the jth lattice stage, and $\hat{\phi}_k$ is the kth normal-mode wave function as before. The reason that only one nucleon operator enters into the intermediate-state calculation is determined by the fermion-number superselection rule and the fact that our nonrelativistic model contains no antinucleons. The momentum δ functions, which arise from the anticommutation rules of the nucleon operators, Eq. (6.2), are now replaced by their familiar Fourier integral representation. Then the integration over the set of intermediate states at the first lattice point $t_1 = t + \epsilon$ is

$$M_{2,0} = \int \left[\frac{D^2\phi_1}{\pi}\right] d^3p_1 \langle\phi_2(0)\,|\,\phi_1(0)\langle\phi_1(0)\,|\,\phi(0)\rangle$$
$$\times \int d^3x_1 e^{i\vec{x}_1\cdot(\vec{p}_1-\vec{p}_2)} \int d^3x\, e^{i\vec{x}\cdot(\vec{p}-\vec{p}_1)} e^{+2i\epsilon\delta m} \exp\Big\{-i\epsilon\lambda \sum_k [g_k(\beta_{1k}e^{i\vec{k}\cdot\vec{x}_1 - i\omega t_1} + \beta_k e^{i\vec{k}\cdot\vec{x}-i\omega_k t})$$
$$+ g_k^*(\beta_{2k}^* e^{-i\vec{k}\cdot\vec{x}_1 + i\omega_k t_1} + \beta_{1k}^* e^{i\vec{k}\cdot\vec{x}+i\omega_k t})]\Big\}. \tag{6.6}$$

Summation over the momenta $\vec{p}_1$ of the intermediate nucleon states gives $\delta^3(\vec{x} - \vec{x}_1)$, which in turn saturates the d^3x_1 integration. Thus the first lattice integration gives

$$M_{2,0} = \langle\phi_2(0)\,|\,\phi(0) e^{2i\epsilon\delta m} \int d^3x\, e^{i\vec{x}\cdot(\vec{p}-\vec{p}_2)} \exp\Big\{-i\epsilon\lambda \sum [g_k^*\beta_{2k}^* e^{-i\cdot\vec{k}\cdot\vec{x}}(e^{-i\omega_k t_1} + e^{-i\omega_k t})$$
$$+ g_k\beta_k e^{i\vec{k}\cdot\vec{x}}(e^{-i\omega_k t_1} + e^{-i\omega_k t}) - i\epsilon\lambda |g_k|^2 e^{-i\omega_k(t_1 - t)}]\Big\}. \tag{6.7}$$

After N such lattice integrations, and letting $\epsilon \to 0$ and $N \to \infty$ keeping $(N+1)\epsilon = t' - t$ fixed we get the continuum limit

$$\langle p', \phi'(0)\,|U(t',t)|\,p, \phi(0)\rangle = \langle\phi'(0)\,|\,\phi(0)\rangle \int d^3x\, e^{i\vec{x}\cdot(\vec{p}-\vec{p}')} \exp\left(i\delta m \int_t^{t'} dt''\right)$$
$$\times \exp\Big\{-i\lambda \int_t^{t'} dt'' \sum_k [\beta_k'^* g_k^* e^{-i\vec{k}\cdot\vec{x} + i\omega_k t''} + \beta_k g_k e^{i\vec{k}\cdot\vec{x} - i\omega_k t''}]$$
$$-\lambda^2 \int_t^{t'} dt'' \int_t^{t''} dt''' \sum_k |g_k|^2 e^{-i\omega_k(t''-t''')}\Big\}. \tag{6.8}$$

The one-nucleon S-matrix element is obtained from this by setting $\phi = \phi' = 0$. Thus

$${}_{\text{out}}\langle p'\,|\,p\rangle_{\text{in}} = \lim_{\substack{t\to-\infty \\ t'\to\infty}} \delta^3(\vec{p} - \vec{p}')\,|z_p|^{-1} \exp\left[i\delta m \int_t^{t'} dt'' - \lambda^2 \int_t^{t'} dt'' \int_t^{t''} dt''' \sum_k |g_k|^2 e^{-i\omega_k(t''-t''')}\right]$$
$$\equiv \delta^3(\vec{p}' - \vec{p})\,. \tag{6.9}$$

Then equating logarithms of coefficients of $\delta^3(\vec{p} - \vec{p}')$ gives

$$\lim_{t'-t\to\infty} [\ln|z_p| - i\delta m(t'-t)] = \lim_{t'-t\to\infty} (-\lambda^2) \int_t^{t'} dt'' \int_t^{t''} dt''' \sum_k |g_k|^2 e^{i\omega_k(t''-t''')}$$
$$= \lim_{t'-t\to\infty} (-\lambda^2) \sum_k |g_k|^2 \left[\frac{(t'-t)}{i\omega_k} + \frac{1}{\omega_k^2}(e^{-i\omega_k(t'-t)} - 1)\right]. \tag{6.10}$$

The Riemann-Lebesque lemma can be invoked to drop the oscillating terms. This gives the well-known[19] results

$$\delta m = -\lambda^2 \sum_k |g_k|^2 \omega_k^{-1}, \tag{6.11}$$

$$\ln|z_p| = \lambda^2 \sum_k |g_k|^2 \omega_k^{-2}. \tag{6.12}$$

B. Meson-nucleon and meson-meson scattering

The S-matrix elements for scattering in the one-nucleon sector, as obtained from Eq. (3.12), are

$${}_{\text{out}}\langle q_1'q_2'\cdots q_{l_{\text{n}}}'; p'\,|\,q_1q_2\cdots q_l; p\rangle_{\text{in}} = \delta_{m,l}\,\delta^3(\vec{p}' - \vec{p}) \prod_{i=1}^{l} \delta^3(\vec{q}_i' - \vec{q}_i)\,, \tag{6.13}$$

where the renormalization conditions, Eqs. (6.1) and (6.12), have been applied, and where one notes that the $\delta(\omega_k)$ terms which arise are zero for the case of massive mesons. This expresses the well-known results that in this model there is no nontrivial meson-nucleon or meson-meson scattering.

C. N-N scattering and effective N-N potential

To consider nucleon-nucleon scattering with our formulation we must calculate S-matrix elements between initial and final states that are coherent states of the meson field and have two nucleons in states of definite momentum,

$$_{\rm out}\langle p',q',\phi'|p,q,\phi\rangle_{\rm in}=\lim_{\substack{t\to-\infty\\ t'\to\infty}}|z_p|^{-2}\langle\phi'(0)|\psi_{p'}\psi_{q'}e^{iH_0t'}e^{-iH(t'-t)}e^{iH_0t}\psi_p^\dagger\psi_q^\dagger|\phi(0)\rangle. \tag{6.14}$$

The elementary U-matrix element for the infinitesimal time interval, $t_j-t_i=\epsilon$, in which we are interested is

$$N_{j,i}=\langle\phi_j(0)|\psi_{p_j}\psi_{q_j}e^{-i\epsilon V(t_j)}\psi_{p_i}^\dagger\psi_{q_i}^\dagger|\phi_i(0)\rangle. \tag{6.15}$$

As before in the single-nucleon matrix element $M_{j,i}$ the way the nucleon operators enter $N_{j,i}$ is fixed by the fermion number superselection rule and the fact that there are no antinucleons in our nonrelativistic model. The anticommutation rules for the nucleon operators, Eq. (6.2), give rise to momentum δ functions in $N_{j,i}$ which we replace by their Fourier integral representations to give

$$N_{j,i}=\tfrac{1}{2}\langle\phi_j(0)|\phi_i(0)\rangle\int d^3x d^3y[e^{i\vec{x}\cdot(\vec{q}_i-\vec{p}_j)+i\vec{y}\cdot(\vec{q}_j-\vec{p}_i)}-e^{i\vec{x}\cdot(\vec{q}_i-\vec{q}_j)+i\vec{y}\cdot(\vec{p}_j-\vec{p}_i)}]$$
$$\times\left[1+2i\epsilon\,\delta m-i\epsilon\lambda\sum_k(e^{i\vec{k}\cdot\vec{x}}+e^{-i\vec{k}\cdot\vec{y}})g_k\beta_{ik}e^{-i\omega_kt_i}-i\epsilon\lambda\sum_k(e^{i\vec{k}\cdot\vec{x}}+e^{i\vec{k}\cdot\vec{y}})g_k^*\beta_{jk}^*e^{i\omega_kt_i}\right]. \tag{6.16}$$

The form of these expressions for $N_{j,i}$ with respect to its dependence on the variables of functional integration, ϕ_j or β_{jk} and ϕ_i or β_{ik}, is the same as in the earlier case for $M_{j,i}$. The differences occur only because of the more complicated double δ functions; $(e^{i\vec{k}\cdot\vec{x}}+e^{-i\vec{k}\cdot\vec{y}})$ replaces $e^{i\vec{k}\cdot\vec{x}}$, and $2\delta m$ replaces δm. Thus although the algebra is slightly more tedious, the same description propagates through at every succeeding lattice integration. The complete S-matrix element becomes

$$_{\rm out}\langle p',q',\phi'|p,q,\phi\rangle_{\rm in}=\lim_{t'-t\to\infty}\langle\phi'(0)|\phi(0)\rangle\int d^3xd^3y[e^{i\vec{x}\cdot(\vec{q}-\vec{p}')+i\vec{y}\cdot(\vec{q}'-\vec{p})}-e^{i\vec{x}\cdot(\vec{q}-\vec{q}')+i\vec{y}\cdot(\vec{p}'-\vec{p})}]$$
$$\times\exp\Big\{-i\lambda\int_t^{t'}dt''\sum_k[\beta_k^*g_k^*(e^{i\vec{k}\cdot\vec{x}}+e^{-i\vec{k}\cdot\vec{y}})e^{i\omega_kt''}+\beta_kg_k(e^{-i\vec{k}\cdot\vec{x}}+e^{i\vec{k}\cdot\vec{y}})e^{-i\omega_kt''}]$$
$$+2\lambda^2\sum_k|g_k|^2e^{i\vec{k}\cdot(\vec{x}+\vec{y})}\int_t^{t'}dt''\int_t^{t''}dt'''e^{-i\omega_k(t''-t''')}\Big\}, \tag{6.17}$$

where we have already cancelled $2\ln|z_p|-2i\delta m(t'-t)$ out of the double time integral using the renormalization identities defined in the single-nucleon matrix element.

The exact nucleon-nucleon scattering matrix element is obtained by setting $\phi=0=\phi'$. To find the static potential between a pair of nucleons we expand to lowest order in the interaction, i.e., to order λ^2, giving

$$_{\rm out}\langle p'q'|p,q\rangle_{\rm in}\cong\tfrac{1}{2}\Big[\delta^3(\vec{q}'-\vec{p})\,\delta^3(\vec{p}'-\vec{q})-\delta^3(\vec{q}-\vec{q}')\,\delta^3(\vec{p}-\vec{p}')$$
$$+\delta^3(\vec{p}+\vec{q}-\vec{p}'-\vec{q}')2\lambda^2T\left(\frac{|g_{q-p'}|^2}{i\omega_{q-p'}}-\frac{|g_{q-q'}|^2}{i\omega_{q-q'}}\right)\Big], \tag{6.18}$$

where we retained only the leading asymptotic behavior

$$\lim_{\substack{t\to-\infty \\ t'\to\infty}} \int_t^{t'} dt'' \int_t^{t''} dt''' e^{-i\omega_k(t''-t''')} = \lim_{t'-t\to} \frac{t'-t}{i\omega_k}$$

$$\equiv \frac{T}{i\omega_k}. \qquad (6.19)$$

We can identify T with the energy δ function

$$T = \int_{-\infty}^{\infty} dt = 2\pi\delta(2m - 2m),$$

and we can identify the static potential between a pair of nucleons as

$$V(\vec{x} - \vec{y}) = \lambda^2 \int d^3k e^{i\vec{k}\cdot(\vec{x}-\vec{y})} \frac{|g_k|^2}{\omega_k}. \qquad (6.20)$$

All of these results are well known, having been previously obtained by more familiar methods.[19]

VII. RELATION OF THIS FORMULATION TO SOME OTHER PATH-INTEGRAL REPRESENTATIONS OF QUANTUM AMPLITUDES

Our construction of the path integral for the U matrix emphasizes its composition of distinct kinematical and dynamical parts. The dynamical part is, of course, evident in the interaction exponential factor of the differential element of the path integral for the infinitesimal time interval. The kinematical part appears in the differential path element as the measure with which the dynamical effects are integrated. This kinematical measure has the coherent-state overlap factor $\langle\phi_{j+1}(0)|\phi_j(0)\rangle$ as a weight functional times $D^2\phi_j$.

It is also of interest to point out the relation of our path-integral formulation to the usual one. The variable of functional integration over the jth complete set of intermediate states, which we take as coherent states of the free field, is the coherent-state wave function ϕ_j which is associated with the time $t_j = t + j\epsilon$. By reabsorbing the time dependence back into the free-field coherent states the jth differential element of the path integral may be written

$$\langle\phi_{j+1}(0)|\phi_j(0)\rangle \exp[-i\epsilon V(z^{-1/2}\phi^*_{j+1}(t_{j+1}), \phi_j(t_j)z^{*-1/2})]D^2\phi_j$$
$$= \langle\phi_{j+1}(t_j+\epsilon)|\phi_j(t_j)\rangle \exp[-i\epsilon H(z^{-1/2}\phi^*_{j+1}(t_j+\epsilon), \phi_j(t_j)z^{*-1/2})]D^2\phi_j. \qquad (7.1)$$

The kinematical weight functional now has the form

$$\exp\{|z|^{-1}[-\tfrac{1}{2}(\phi_{j+1}(t_j+\epsilon), \phi_{j+1}(t_j+\epsilon)) - \tfrac{1}{2}(\phi_j(t_j), \phi_j(t_j)) + (\phi_{j+1}(t_j+\epsilon), \phi_j(t_j))]\}. \qquad (7.2)$$

The two-time inner product is the obvious generalization of the inner product defined earlier. For canonical systems of fields there inner products are always of the form

$$(\phi_j(t_j), \phi_i(t_i)) = \int d^3x\, J_0(\phi_j(\vec{x}, t_j), \phi_i(\vec{x}, t_i))$$

$$= -i\int d^3x\, \pi_j(t_j)\phi_i(t_i), \qquad (7.3)$$

where $\pi(\vec{x}, t)$ is the field canonically conjugate to $\phi(\vec{x}, t)$ in some Lagrangian $\pi = \partial\mathcal{L}/\partial\dot{\phi}$. The kinematical weight factor is now, to order ϵ, the exponential of

$$i\epsilon\int d^3x[\tfrac{1}{2}\pi_j(\vec{x}, t_j)\dot{\phi}_j(\vec{x}, t_j) - \tfrac{1}{2}\dot{\pi}_j(\vec{x}, t_j)\phi_j(\vec{x}, t_j)]$$

$$= i\epsilon\int d^3x\Big\{\pi_j(\vec{x}, t_j)\dot{\phi}_j(\vec{x}, t_j)$$
$$- \frac{1}{2}\frac{d}{dt_j}[\pi_j(\vec{x}, t_j)\phi_j(\vec{x}, t_j)]\Big\}. \qquad (7.4)$$

The sum of this with the term $-i\epsilon H = -i\epsilon\int d^3x\,\mathcal{H}(\phi_j(\vec{x}, t_j), \pi_j(\vec{x}, t_j))$, we define to be the action

$$i\epsilon\int d^3x\,\mathcal{L}(\phi_j(\vec{x}, t)),$$

with

$$\mathcal{L}(\phi(\vec{x}, t)) = \pi(\vec{x}, t)\dot{\phi}(\vec{x}, t) - \mathcal{H}(\phi(\vec{x}, t), \pi(\vec{x}, t))$$
$$- \frac{1}{2}\frac{d}{dt}[\pi(\vec{x}, t)\phi(\vec{x}, t)]. \qquad (7.5)$$

This differs from the usual expression only by the total time derivative which is inconsequential for the determination of the equations of motion.

Thus the path integral in our formulation is equivalent to the usual path-integral representation for quantum field theories that are of canonical form. However, our formulation is in fact more general and can be applied to theories which have no canonical form and no Lagrangian action with which to formulate the usual path-integral representation.[20,17] We simply do not have to assume the canonical form for the bilinear current J_μ.

The operators of our formulation are based on

the bilinear conserved currents which are determined solely from the field equations, rather than on a canonical quantization procedure.[8,9,10,17] These bilinear operators are essentially self-adjoint on a dense domain spanned by a set of coherent states and have been shown to be general.[11] These operators describe the whole q-number theory, ensure its covariance, and give conditions on equal-time commutation relations, if any exist in the theory; all independently of any separate postulation of a locality or a quasilocality condition.

Examples of such noncanonical, non-Lagrangian quantum field theories are ultralocal models which can be obtained by omitting spatial gradient terms from the Hamiltonians of certain covariant theories. These models are interesting because, although they are noncovariant, they are exactly solvable.[8,21] Interacting ultralocal systems are not continuously connected to the corresponding free systems in the limit as the interaction coupling constant is turned off. The dynamics of these systems can be well defined. However, $\pi = i\dot{\phi}$ does not share a dense domain in common with the Hamiltonian, and thus is not well behaved. Since no canonical momentum operator exists for the interacting theory no canonical form exists and the Legendre transformation to the Lagrangian does not exist.[22] Therefore the path-integral representation in the usual form with the Lagrangian action does not exist. On the other hand our formulation of a path-integral representation for the U-matrix does exist.

Another, less exotic, example occurs for field theories which contain ghost fields. Such fields are introduced into a theory whenever, in addition to the usual field equations, auxiliary conditions must also be satisfied by one or more of the fields. The ghost fields "act as Lagrange multipliers which enforce these constraints. Without the use of these ghost fields, a Lagrangian cannot be constructed which generates the field equations and the auxiliary conditions. In other words, because the physical fields are not linearily independent the theory is not canonical. These extra fields "clutter up" the usual path-integral formalism because they must be treated as dummy variables which are integrated out. The extra integrations can be avoided in the formalism presented here because it is unnecessary to postulate the ghost fields in the first instance. The particular example of a system of a Dirac field minimally coupled to a massive vector field, as well as the ultralocal models, is treated in Ref. 8.

In the usual path-integral formulation of a quantum field theory the generating functional of the vacuum Green's functions is integrated with the field interacting with an auxiliary external source current. This source current is purely auxiliary serving as a variational parameter for generating the Green's functions; after which it is set to zero. However, the calculations are done before this external source is set to zero. This source usually violates some of the conservation laws, so that the calculation is deprived of some of the symmetries that could otherwise facilitate the calculation. The symmetries are recovered only after setting the external sources to zero. Also, this treatment is biased in favor of the consideration of transitions between initial and final states that are the Fock vacuum or states near it in Fock space.

However, a totally different perspective can now be taken for this analysis because we have shown within the Lehmann-Symanzik-Zimmermann framework, that for canonical theories the usual generating functional, with the external source set to zero, *is* the S matrix in the coherent-state representation. Consequently, Eq. (3.4) applies, and the S-matrix elements can be obtained directly by using the coherent-state wave functions as variational parameters.

On the other hand, one need not take this approach when considering problems with nonvacuum ground states such as condensates of solitons, superfluids, etc., since it is the coherent-state S-matrix element itself which is then of interest.

ACKNOWLEDGMENTS

One of us (J.E.S.) wishes to thank Ames Laboratory and the U. S. Energy Research and Development Administration for support as a visiting faculty participant while most of this work was done. Part of this work was also done while J.E.S. was at the Aspen Center for Physics and the hospitality shown him there is gratefully acknowledged. We also wish to thank P. Carruthers, L. M. Simmons, D. E. Soper, and T. A. Weber for useful conversations. This work was supported by the U. S. Department of Energy, Division of Basic Energy Sciences.

---

*Permanent address: Physics Department, Washington University, St. Louis, Missouri 63130.

[1]E. Abers and B. Lee, Phys. Rep. 9C, 1, (1973); J. Iliopoulos, C. Itzykson, and A. Martin, Rev. Mod. Phys. 47, 165 (1975); H. M. Fried, *Functional Methods and Models in Quantum Field Theory* (M. I. T. Press, Cambridge, 1972). R. J. Finkelstein, J. S. Kvitky, and J. O. Mouton, Phys. Rev. D 4, 2220 (1971).

[2]R. P. Feynman, Rev. Mod. Phys. 20, 367 (1948); Phys. Rev. 80, 440 (1950); R. P. Feynman and A. R. Hibbs, *Quantum Mechanics and Path Integrals* (McGraw-Hill, New York, 1965). Another early program on functional methods and path integrals was by K. Symanzik, J. Math. Phys. 1, 249 (1960). Our approach follows closely that of Feynman.

[3]R. J. Glauber, Phys. Rev. 131, 2766 (1963); J. R. Klauder and E. C. G. Sudarshan, *Fundamentals of Quantum Optics* (Benjamin, New York, 1968); P. A. Carruthers and M. M. Nieto, Rev. Mod. Phys. 40, 441 (1968).

[4]D. J. Scalapino and R. L. Sugar, Phys. Rev. D 8, 2284 (1973); J. C. Botke, D. J. Scalapino, and R. L. Sugar, *ibid.* 10, 1604 (1974); H. Minakata, Prog. Theor. Phys. 56, 1217 (1976).

[5]M. Bolsterli, Phys. Rev. D 13, 1722 (1976); 16, 1749 (1977).

[6]G. C. Dente, Phys. Rev. D 12, 1733 (1975); Phys. Rev. D (to be published).

[7]R. P. Feynman and J. A. Wheeler, Rev. Mod. Phys. 17, 157 (1945); 21, 425 (1949).

[8]B. DeFacio and C. L. Hammer, J. Math. Phys. 17, 267 (1976).

[9]Y. Takahashi and H. Umezawa, Prog. Theor. Phys. 9, 14 (1953); Nucl. Phys. 51, 193 (1964); Y. Takahashi, *An Introduction to Field Quantization* (Pergamon, London, 1969).

[10]C. L. Hammer and R. H. Tucker, J. Math. Phys. 12, 1327 (1971); Phys. Rev. D 3, 2448 (1971).

[11]J. R. Klauder, J. Math. Phys. 11, 609 (1970).

[12]J. R. Klauder, Ann. Phys. (N.Y.) 11, 123 (1960); S. S. Schweber, J. Math Phys. 3, 831 (1962).

[13]L. D. Faddeev, Theor. Math. Phys. 1, (1969); in *Proceedings of the Conference on Methods in Field Theory*, edited by R. Balian and J. Zinn-Justin (North-Holland, Amsterdam, 1978), p. 1.; F. A. Berezin, Theor. Math. Phys. 6, 194 (1971); J. R. Klauder, Bell Laboratory report, 1977 (unpublished).

[14]P. Vinciarelli, Nucl. Phys. B89, 463 (1975); B89, 493 (1975).

[15]Z. F. Ezawa and H. C. Tze, Nucl. Phys. B96, 264 (1975).

[16]H. Matsumoto, N. J. Papastamatiou, and H. Umezawa, Nucl. Phys. B97, 90 (1975).

[17]These commutation relationships also follow as a special case of the affine field commutators discussed by J. R. Klauder which were designed to replace CCR's for the case where the conjugate momentum is ill defined, J. R. Klauder, J. Math. Phys. 18, 1711 (1977).

[18]S. Gasiorowicz, *Elementary Particle Physics* (Wiley, New York, 1966), p. 316; S. S. Schweber, *An Introduction to Relativistic Quantum Field Theory* (Harper and Row, New York 1961), p. 113.

[19]S. S. Schweber, Ref. 18, Sec. 12a, pp. 339–351.

[20]Recently Klauder has developed a noncanonical formulation of quantum field theory in which the classical action is "augmented" by a term which changes the functional integration without changing the classical theory. It does effectively change the basic measure in the functional integration. J. R. Klauder, Phys. Rev. D 15, 2830 (1977); 14, 1952 (1976); J. Math. Phys. 18, 1711 (1977).

[21]J. R. Klauder, Commun. Math. Phys. 18, 307 (1970); Acta Phys. Austriaca Suppl. 8, 227 (1971); in Proceedings of the Symposium on Basic Quotations in Elementary Particle Physics, Munich, 1971 (unpublished) in *L Lectures in Theoretical Physics*, edited by W. Brittin (Colorado Assoc. Universities Press, Boulder, 1973), Vol. XIVB; Ann. Phys. (N. Y.) 79, 111 (1973).

[22]G. C. Hegerfeldt and J. R. Klauder, Nuovo Cimento 10A, 723 (1972).

Volume 77B, number 3 PHYSICS LETTERS 14 August 1978

COHERENT STATE APPROACH TO THE INFRA-RED BEHAVIOUR OF NON-ABELIAN GAUGE THEORIES

M. GRECO, F. PALUMBO, G. PANCHERI-SRIVASTAVA [1] and Y. SRIVASTAVA [1]
INFN, Laboratori Nazionali di Frascati, Rome, Italy

Received 31 January 1978

The infrared behaviour of matrix elements between coherent states of definite color is studied in non-abelian gauge theories. The matrix elements are shown to be finite to the lowest non-trivial order, and to all orders if the soft meson formula for real gluons holds to all orders. Factorization occurs in the fixed angle regime.

In this letter we study the infrared (IR) behaviour of non-abelian gauge theories (NAGT) using the coherent state formalism. This method has been applied [1] in QED providing one with a definition of matrix elements which are (i) free from IR divergences, (ii) factorizable and directly comparable with the experimental cross sections. This approach is exactly equivalent to the standard one of summing cross sections for real and virtual photon emission [1,2].

Our extension to NAGT is based on the (assumed) validity, to all orders, of the so-called soft meson formula [3]. This formula has been proved to all orders in the leading logarithms for virtual gluons [4].

It has also been checked for single-gluon emission in different processes [5], and for double-gluon emission both in quark scattering in an external colorless potential [6] and in quark–quark scattering [7].

In all cases the leading divergences cancel out in the inclusive cross sections after color average over both initial and final states, in agreement with the Lee and Nauenberg theorem [8]. Due to the color average, however, nothing can be said about cross sections for states of definite color. The coherent state formalism, on the other hand, is such as to provide one with matrix elements which are free from IR divergences for any initial or final color states. This formalism, therefore, turns out to be not equivalent to the standard one of summing cross sections for real and virtual gluon emission, in contrast to QED.

The factorization properties of the new matrix elements are also different from QED. In fact, factorization occurs only in the fixed angle regime (see below for definitions), as also found for the inclusive color averaged cross section [5–7].

We shall not discuss in this letter the connection between our results and physically measurable quantities.

For given initial and final states $|\mathrm{i}\rangle$ and $|\mathrm{f}\rangle$, we define the corresponding coherent states $|\tilde{\mathrm{i}}\rangle$ and $|\tilde{\mathrm{f}}\rangle$:

$$|\tilde{\mathrm{i}}\rangle = \mathrm{e}^{\mathrm{i}\Lambda_\mathrm{i}}|\mathrm{i}\rangle, \quad |\tilde{\mathrm{f}}\rangle = \mathrm{e}^{\mathrm{i}\Lambda_\mathrm{f}}|\mathrm{f}\rangle\,, \tag{1}$$

with

$$\Lambda = \frac{1}{(2\pi)^4}\int \mathrm{d}^4k\, j^c_\mu(k)\, A^c_\mu(-k). \tag{2}$$

In eq. (2) $A^c_\mu(x)$ is the quantized gluon field of color index c, and $j^c_\mu(k)$ is the "classical" current

[1] Physics Department, Northeastern University, Boston, MA, USA. Work supported in part by a grant from the National Science Foundation, USA.

$$j_\mu^c(k) = \frac{\mathrm{i}}{(2\pi)^{3/2}}\,\bar{g}(k) \sum_\alpha \frac{\eta_\alpha p_{\alpha\mu}}{(p_\alpha k)}\, t^c(\alpha) \equiv \sum_\alpha j_{\alpha\mu}(k) t^c(\alpha), \tag{3}$$

where α is the label of particles in initial and final states, $\eta_\alpha = +1$ for incoming particles and outgoing antiparticles and $\eta_\alpha = -1$ otherwise. The operators $t^c(\alpha)$ are the appropriate generators of the color group for the αth particle ($t^c(\alpha)$ must be replaced by $t^{c*}(\alpha)$ for outgoing particles) and $\bar{g}(k)$ is the effective coupling constant for pure Yang–Mills theory, which appears in the soft meson formula for virtual gluons [4] (see discussion later).

Let us define new S-matrix elements:

$$\bar{M} = \langle \tilde{\mathrm{f}} | S | \tilde{\mathrm{i}} \rangle = \langle \mathrm{f} | \mathrm{e}^{-\mathrm{i}\Lambda_\mathrm{f}} S\, \mathrm{e}^{\mathrm{i}\Lambda_\mathrm{i}} | \mathrm{i} \rangle \equiv \langle \mathrm{f} | \bar{S} | \mathrm{i} \rangle. \tag{4}$$

In terms of a perturbative expansion in $\bar{g}(k)$,

$$\bar{M} = \sum_{n=0}^{\infty} \bar{M}_n, \tag{5}$$

with

$$\bar{M}_n = \sum_{k=0}^{n} \sum_{l=0}^{k} \langle \mathrm{f} | \frac{(-\mathrm{i}\Lambda_\mathrm{f})^{k-l}}{(k-l)!} S_{n_0,l} \frac{(\mathrm{i}\Lambda_\mathrm{i})^{n-k}}{(n-k)!} | \mathrm{i} \rangle, \tag{6}$$

where n_0 corresponds to the lowest-order non-vanishing transition amplitude. $\bar{M}_n$ and $S_{n_0,l}$ are of order n_0 in the bare coupling constant but are of order n and l, respectively, in the effective coupling constant $\bar{g}(k)$. This follows from eq. (3) and the results of ref. [4].

Let us concentrate on $\bar{M}_2$, the first radiatively corrected matrix element. One of the three terms in $\bar{M}_2$ is given by

$$\langle \mathrm{f} | S_{n_0,1} (\mathrm{i}\Lambda_\mathrm{i}) | \mathrm{i} \rangle = \frac{\mathrm{i}}{(2\pi)^4} \int \mathrm{d}^4 k \sum_{\substack{\alpha \in \{\mathrm{i}\} \\ \sigma}} j_{\alpha\mu}(k) \langle \mathrm{f} | S_{n_0,1} A_\mu^c(-k) | \mathrm{i}, \lambda_\alpha \to \sigma \rangle \langle \sigma | t^c | \lambda_\alpha \rangle, \tag{7}$$

where λ_α is the color index of the αth particle in the initial state i.

The matrix element appearing in eq. (7) is evaluated by using the assumed soft meson formula for the emission of a soft gauge vector from any on-shell process:

$$\langle \mathrm{f} | T_\nu^a(k) | \mathrm{i} \rangle = \sum_{\substack{\gamma \in \{\mathrm{i}\} \\ \lambda}} \bar{g}(k) \eta_\gamma \frac{p_{\gamma\nu}}{(p_\gamma \cdot k)} \langle \lambda | t^a | \lambda_\gamma \rangle \langle \mathrm{f} | S | \mathrm{i}, \lambda_\gamma \to \lambda \rangle + \sum_{\gamma \in \{\mathrm{f}\}} \bar{g}(k) \eta_\gamma \frac{p_{\gamma\nu}}{(p_\gamma \cdot k)} \langle \lambda_\gamma | t^\alpha | \lambda \rangle \langle \mathrm{f}, \lambda_\gamma \to \lambda | S | \mathrm{i} \rangle. \tag{8}$$

Then eq. (7) becomes

$$\begin{aligned} \langle \mathrm{f} | S_{n_0,1} (\mathrm{i}\Lambda_\mathrm{i}) | \mathrm{i} \rangle = \int \mathrm{d}^4 k\, \delta(k^2) \theta(k_0) \Bigg\{ \sum_{\alpha \in \{\mathrm{i}\},\sigma} j_{\alpha\mu}(k) \langle \sigma | t^c | \lambda_\alpha \rangle \Bigg\} \\ \times \Bigg\{ \sum_{\alpha' \in \{\mathrm{i}\},\sigma'} j_{\alpha'}^{\mu}(-k) \langle \sigma' | t^c | \lambda_{\alpha'}(\sigma) \rangle \langle \mathrm{f} | S_{n_0,0} | \mathrm{i}, \begin{bmatrix} \lambda_\alpha \to \sigma \\ \lambda_{\alpha'}(\sigma) \to \sigma' \end{bmatrix} \rangle \\ + \sum_{\beta' \in \{\mathrm{f}\},\rho'} j_{\beta'}^{\mu}(-k) \langle \lambda_{\beta'} | t^c | \rho' \rangle \langle \mathrm{f}, \lambda_{\beta'} \to \rho' | S_{n_0,0} | \mathrm{i}, \lambda_\alpha \to \sigma \rangle \Bigg\}, \end{aligned} \tag{9}$$

where in the state $|\lambda_{\alpha'}(\sigma)\rangle$ the color index is $\lambda_{\alpha'}$ for $\alpha' \neq \alpha$ and σ otherwise. A similar evaluation of the other

Volume 77B, number 3 PHYSICS LETTERS 14 August 1978

terms in $\bar{M}_2$ leads to the result:

$$\bar{M}_2 = \langle \mathrm{f}|S_{n_0,2}|\mathrm{i}\rangle + \langle \mathrm{f}|R_{n_0,2}|\mathrm{i}\rangle , \tag{10}$$

with

$$\begin{aligned}\langle \mathrm{f}|R_{n_0,2}|\mathrm{i}\rangle = \tfrac{1}{2}\int \mathrm{d}^4k\,\delta(k^2)\theta(k_0)\Bigg\{ & \sum_{\alpha,\alpha'\in\{\mathrm{i}\},\sigma,\sigma'} j_{\alpha\mu}(k) j^{\mu}_{\alpha'}(-k)\langle\sigma|t^c|\lambda_\alpha\rangle\langle\sigma'|t^c|\lambda_{\alpha'}(\sigma)\rangle\langle \mathrm{f}|S_{n_0,0}|\mathrm{i},\begin{bmatrix}\lambda_\alpha\to\sigma\\ \lambda_{\alpha'}(\sigma)\to\sigma'\end{bmatrix}\rangle \\
&+\sum_{\alpha\in\{\mathrm{i}\},\beta'\in\{\mathrm{f}\},\sigma,\rho'} j_{\alpha\mu}(k) j^{\mu}_{\beta'}(-k)\langle\sigma|t^c|\lambda_\alpha\rangle\langle\lambda_{\beta'}|t^c|\rho'\rangle\langle \mathrm{f},\lambda_{\beta'}\to\rho'|S_{n_0,0}|\mathrm{i},\lambda_\alpha\to\sigma\rangle \\
&+\sum_{\beta\in\{\mathrm{f}\},\alpha'\in\{\mathrm{i}\},\rho,\sigma'} j_{\beta\mu}(k) j^{\mu}_{\alpha'}(-k)\langle\sigma'|t^c|\lambda_{\alpha'}\rangle\langle\lambda_\beta|t^c|\rho\rangle\langle \mathrm{f},\lambda_\beta\to\rho|S_{n_0,0}|\mathrm{i},\lambda_{\alpha'}\to\sigma'\rangle \\
&+\sum_{\beta,\beta'\in\{\mathrm{f}\},\rho,\rho'} j_{\beta\mu}(k) j^{\mu}_{\beta'}(-k)\langle\lambda_\beta|t^c|\rho\rangle\langle\lambda_{\beta'}(\rho)|t^c|\rho'\rangle\langle \mathrm{f},\begin{bmatrix}\lambda_\beta\to\rho\\ \lambda_{\beta'}(\rho)\to\rho'\end{bmatrix}|S_{n_0,0}|\mathrm{i}\rangle \ .\end{aligned} \tag{11}$$

In shorthand notation

$$\langle \mathrm{f}|R_{n_0,2}|\mathrm{i}\rangle = \tfrac{1}{2}\int \mathrm{d}^4k\,\delta(k^2)\theta(k_0)\langle \mathrm{f}|\{S_{n_0,0}\, j^c_{\mathrm{in}\,\mu}(k) j^{c\mu}_{\mathrm{in}}(-k) + 2 j^c_{\mathrm{out}\,\mu}(k) S_{n_0,0}\, j^{c\mu}_{\mathrm{in}}(-k) + j^c_{\mathrm{out}\,\mu}(k) j^{c\mu}_{\mathrm{out}}(-k) S_{n_0,0}\}|\mathrm{i}\rangle . \tag{12}$$

On the other hand, for the pure virtual gluon contribution $\langle \mathrm{f}|S_{n_0,2}|\mathrm{i}\rangle$ one gets the same result, but for the replacement of $\delta(k^2)\theta(k_0)$ by $(-\mathrm{i}/2\pi)(1/k^2)$.

One concludes therefore that $\bar{M}_2$ is finite and has the following form:

$$\bar{M}_2 = \tfrac{1}{2}\int \mathrm{d}^4k \left[\delta(k^2)\theta(k_0) - \frac{\mathrm{i}}{2\pi}\frac{1}{k^2}\right]\langle \mathrm{f}|\{ j^c_{\mathrm{out}\,\mu}(k) j^{c\mu}_{\mathrm{out}}(-k)\bar{S}_0 + 2 j^c_{\mathrm{out}\,\mu}(k)\bar{S}_0\, j^{c\mu}_{\mathrm{in}}(-k) + \bar{S}_0\, j^c_{\mathrm{in}\,\mu}(k) j^{c\mu}_{\mathrm{in}}(-k)\}|\mathrm{i}\rangle . \tag{13}$$

By finite we mean that the standard infrared singularities have been eliminated. It is always possible however that the complete $\bar{g}(k)$ is such that $\bar{M}$ is still not convergent.

We are now going to relate the matrix element $R_{n_0,2}$ to the probability of emission of one real gluon. From eq. (8), rewritten in short as

$$\langle \mathrm{f}|T^c_\mu(k)|\mathrm{i}\rangle = \langle \mathrm{f}|j^c_{\mathrm{out}\,\mu}(k) S_{n_0,0}|\mathrm{i}\rangle + \langle \mathrm{f}|S_{n_0,0}\, j^c_{\mathrm{in}\,\mu}(k)|\mathrm{i}\rangle , \tag{14}$$

we have

$$\begin{aligned}&\int \mathrm{d}^4k\,\delta(k^2)\theta(k_0)\langle \mathrm{f}|T^c_\mu(k)|\mathrm{i}\rangle\langle \mathrm{f}|T^{c\mu}(k)|\mathrm{i}\rangle^* \\
&\quad = \int \mathrm{d}^4k\,\delta(k^2)\theta(k_0)\{\langle \mathrm{f}|j^c_{\mathrm{out}}(k) S_{n_0,0}|\mathrm{i}\rangle\langle \mathrm{i}|S^+_{n_0,0}\, j^c_{\mathrm{out}}(-k)|\mathrm{f}\rangle + \langle \mathrm{f}|j^c_{\mathrm{out}}(k) S_{n_0,0}|\mathrm{i}\rangle\langle \mathrm{i}|j^c_{\mathrm{in}}(-k) S^+_{n_0,0}|\mathrm{f}\rangle \\
&\quad + \langle \mathrm{f}|S_{n_0,0}\, j^c_{\mathrm{in}}(k)|\mathrm{i}\rangle\langle \mathrm{i}|S^+_{n_0,0}\, j^c_{\mathrm{out}}(-k)|\mathrm{f}\rangle + \langle \mathrm{f}|S_{n_0,0}\, j^c_{\mathrm{in}}(k)|\mathrm{i}\rangle\langle \mathrm{i}|j^c_{\mathrm{in}}(-k) S^+_{n_0,0}|\mathrm{f}\rangle\} .\end{aligned} \tag{15}$$

On the other hand, from eq. (5) we have

$$\begin{aligned}|\bar{M}|^2 = |\bar{M}_0 + \bar{M}_2 + \ldots|^2 = |\langle \mathrm{f}|S_{n_0,0}|\mathrm{i}\rangle|^2 \\
+ \{\langle \mathrm{f}|S_{n_0,0}|\mathrm{i}\rangle\langle \mathrm{f}|S_{n_0,2}|\mathrm{i}\rangle^* + \langle \mathrm{f}|S_{n_0,0}|\mathrm{i}\rangle\langle \mathrm{f}|R_{n_0,2}|\mathrm{i}\rangle^* + \mathrm{c.c.}\} + \text{higher orders.}\end{aligned} \tag{16}$$

From eqs. (12) and (16) it is clear that the term $R_{n_0,2}$ in $\bar{M}$ is equivalent to the single gluon bremsstrahlung only upon average over color of the external particles.

Volume 77B, number 3 PHYSICS LETTERS 14 August 1978

A recursion formula is conjectured for the matrix element $\bar{M}$,

$$\langle \mathrm{f}|\bar{S}|\mathrm{i}\rangle_{n+2} = \tfrac{1}{2}\int \mathrm{d}^4k \left[\delta(k^2)\theta(k^0) - \frac{\mathrm{i}}{2\pi}\frac{1}{k^2}\right]\langle \mathrm{f}|\{j^c_{\mathrm{out}\,\mu}(k)j^{c\mu}_{\mathrm{out}}(-k)\bar{S}_n + 2j^c_{\mathrm{out}\,\mu}(k)\bar{S}_n j^{c\mu}_{\mathrm{in}}(-k) + \bar{S}_n j^c_{\mathrm{in}\,\mu}(k)j^{c\mu}_{\mathrm{in}}(-k)\}|\mathrm{i}\rangle, \tag{17}$$

which is based on the extension of the soft-meson formula for virtual gluon to $\bar{S}$. Eq. (13) is the first step towards eq. (17), which proves the finiteness of $\bar{M}$. As for the lowest-order calculation, the equivalence of the present calculation to the nth order inclusive cross section holds only after average over initial and final color states, in contrast to the abelian case.

The virtual gluon part of eq. (17) has been derived in refs. [3] and [4] in the form of a differential equation.

In the fixed angle regime, when all the invariant squared energies and momentum transfers become large, the r.h.s. of eq. (17) gets factorized and $\bar{M}$ exponentiates as in the abelian case. One finds

$$\langle \mathrm{f}|\bar{S}|\mathrm{i}\rangle_{n+2} = \tfrac{1}{2}B\left(\sum_\alpha C_\alpha\right)\langle \mathrm{f}|\bar{S}|\mathrm{i}\rangle_n, \tag{18}$$

where C_α is the eigenvalue of the Casimir operator for the αth particle, the sum runs over both the initial and final particles, and

$$B = B_{\alpha\beta} = \int \mathrm{d}^4k \left[\delta(k^2)\theta(k^0) - \frac{\mathrm{i}}{2\pi}\frac{1}{k^2}\right]\frac{(p_\alpha p_\beta)}{(p_\alpha k)(p_\beta k)}\frac{\bar{g}^2(k)}{(2\pi)^3}. \tag{19}$$

Notice that in the same regime, exponentiation occurs for the real bremsstrahlung only for the color averaged cross sections. Details of this result shall be presented elsewhere.

Finally, we compare our formalism with some perturbative calculations, which in particular allows us to justify the introduction of $\bar{g}(k)$ in eq. (3) [‡1]. More precisely, let us consider massive quark scattering in an external colourless potential. Then eqs. (17)–(19) give

$$\langle \mathrm{f}|\bar{S}|\mathrm{i}\rangle_{n+2} = \left\{\tfrac{1}{2}C_{\mathrm{F}}M(q^2)\int_0^{\ln\Delta} \mathrm{d}(\ln k)\,\bar{g}^2(k)\right\}\langle \mathrm{f}|\bar{S}|\mathrm{i}\rangle_n, \tag{20}$$

where C_{F} is the eigenvalue of the Casimir operator for the fundamental representation, Δ is the fractional energy loss and $M(q^2) = (1/2\pi^2)\ln(q^2/m^2)$ $(q^2 \gg m^2)$. This leads to

$$\mathrm{d}\sigma \propto |\langle \mathrm{f}|\bar{S}|\mathrm{i}\rangle|^2 = \exp\left\{C_{\mathrm{F}}M(q^2)\int_0^{\ln\Delta} \mathrm{d}(\ln k)\bar{g}^2(k)\right\}|\langle \mathrm{f}|\bar{S}|\mathrm{i}\rangle_0|^2. \tag{21}$$

Comparison with the calculation of the inclusive cross section by Frenkel et al. [6] then gives, for the leading terms,

$$\bar{g}^2(t) = g^2\left(1 - \frac{11}{24\pi^2}C_{\mathrm{YM}}g^2t + \ldots\right) = g^2\left(1 + \frac{11}{24\pi^2}C_{\mathrm{YM}}g^2t\right)^{-1}, \tag{22}$$

where C_{YM} is the eigenvalue of the Casimir operator for the adjoint representation and $t \sim \ln k$.

Summarizing, we have constructed a coherent state formalism for NACT, showing that matrix elements between coherent states of definite color are finite and factorized in the fixed angle regime.

Two of us (M.G. and F.P.) are grateful to A. Ukawa for very useful discussions.

‡1 In this context see also ref. [9].

Volume 77B, number 3 PHYSICS LETTERS 14 August 1978

References

[1] V. Chung, Phys. Rev. B140 (1965) 1110;
M. Greco and G. Rossi, Nuovo Cimento 50 (1967) 168.
[2] T. Kibble, J. Math. Phys. 9 (1968) 315; Phys. Rev. 173 (1968) 1527; 174 (1968) 1882; 175 (1968) 1624;
P. Kulish and L. Faddeev, Teor. Math. Fiz. 4 (1970) 153; Engl. transl. Theor. Math. Phys. (USSR) 4 (1970) 745;
J. Storrow, Nuovo Cimento 54A (1968) 15;
S. Schweber, in: Cargese Summer School Lectures, eds. M. Levy and J. Sucher (Gordon and Breach, New York, 1972);
D. Zwanziger, Phys. Rev. D11 (1975) 3481, 3504;
M. Greco, G. Pancheri-Srivastava and Y. Srivastava, Nucl. Phys. B101 (1975) 234.
[3] J.M. Cornwall and G. Tiktopoulos, Phys. Rev. D13 (1976) 3370; D15 (1977) 2937.
[4] T. Kinoshita and A. Ukawa, Phys. Rev. D15 (1977) 1596; D16 (1977) 332.
[5] Y.P. Yao, Phys. Rev. Lett. 36 (1976) 653;
T. Appelquist, J. Carazzone, H. Kluberg-Stern and M. Roth, Phys. Rev. Lett. 36 (1976) 768;
L. Tyburski, Phys. Rev. Lett. 37 (1976) 319.
[6] J. Frenkel, R. Meuldermans, I. Mohammad and J.C. Taylor, Phys. Lett. 64B (1976) 211; Nucl. Phys. B121 (1977) 58.
[7] M. Greco, F. Palumbo, G. Pancheri-Srivastava and Y. Srivastava, to be published.
[8] T.D. Lee and N. Nauenberg, Phys. Rev. B133 (1964) 1549.
[9] E. Poggio, Phys. Rev. Lett. 36 (1976) 1511.

Volume 78B, number 4 PHYSICS LETTERS 9 October 1978

CREATION AND ANNIHILATION OPERATORS FOR NIELSEN–OLESEN VORTICES IN THE COHERENT STATE APPROXIMATION

D.E.L. POTTINGER
School of Mathematics, Trinity College, Dublin 2, Ireland

Received 7 June 1978

Using the standard techniques of canonical quantization, we construct approximate expressions for the creation and annihilation operators for Nielsen–Olesen vortices. The forms for the creation and annihilation operators are appropriate to situations where large-scale vortex condensation takes place. In particular, a phenomenon such as this is basic in the quark confinement schemes of Mandelstam and 't Hooft.

In this letter we wish to present approximate expressions for the creation and annihilation operators of Nielsen–Olesen vortices. We employ coherent states, so that the approximation involved should be valid whenever we encounter situations involving macroscopic (large-scale) vortex condensation [1]. Such schemes have recently been discussed (in the nonabelian framework) by 't Hooft [2] and Mandelstam [3] in the context of a possible mechanism for quark confinement. A more mathematical but less physical approach to the same problem has recently been given by Ezawa [4].

As an illustrative example of our approach we consider the Higgs model in 3 + 1 dimensions. This system is well known to have vortex solutions [5], the Nielsen–Olesen gauge solitons. The Lagrange density for the Higgs model, in the Coulomb gauge, is

$$\begin{aligned}\mathcal{L}(A_\mu, \phi, \phi^*, C) = &-\tfrac{1}{4}(\partial_\mu A_\nu - \partial_\nu A_\mu)^2 \\ &+\tfrac{1}{2}(\partial_\mu + ieA_\mu)\phi^*(\partial^\mu - ieA^\mu)\phi \\ &-\tfrac{1}{4}\lambda(\phi^*\phi - f)^2 - C\partial_i A^i ,\end{aligned} \tag{1}$$

where we have introduced the Lagrange multiplier C in order to implement the gauge constraint.

By performing the transformations

$$\begin{aligned}\phi(x) &= [f + \rho(x)] \exp[-i\chi(x)] , \\ A_\mu(x) &= B_\mu(x) - e^{-1}\partial_\mu\chi(x) ,\end{aligned} \tag{2}$$

we may write the Lagrange density of eq. (1) in the following form:

$$\begin{aligned}\mathcal{L}(B_\mu, \rho, \chi, C) = &-\tfrac{1}{4}(\partial_\mu B_\nu - \partial_\nu B_\mu)^2 \\ &+\tfrac{1}{2}e^2(\rho + f)^2 B_\mu B^\mu - \tfrac{1}{4}\lambda\rho^2(\rho + 2f)^2 \\ &+\tfrac{1}{2}\partial_\mu\rho\partial^\mu\rho - C\partial_i(B^i - e^{-1}\partial^i\chi) .\end{aligned} \tag{3}$$

When the gauge field A_μ lies in the sector of trivial topology, the excitation spectrum of eq. (3) consists of

(i) massive vector mesons of mass m_v (= ef in the tree approximation), and

(ii) massive scalar mesons of mass m_s (= $\sqrt{2\lambda}\, f$ in the tree approximation).

The Coulomb gauge constraint implies that

$$\nabla^2\chi = -e\partial_k B^k . \tag{4}$$

Hence, up to a global gauge transformation, the χ field is determined by the vector field B_μ.

The hamiltonian density $\mathcal{H}$ corresponding to the Lagrange density of eq. (3) is

$$\begin{aligned}\mathcal{H} = &\tfrac{1}{2}E^2 + \tfrac{1}{2}(\text{curl}\, \boldsymbol{B})^2 + \tfrac{1}{2}e^2(\rho + f)^2[B_0^2 + \boldsymbol{B}^2] + \tfrac{1}{2}\pi^2 \\ &+\tfrac{1}{2}(\nabla\rho)^2 + \tfrac{1}{4}\lambda\rho^2(\rho + 2f)^2 + \text{surface term},\end{aligned} \tag{5}$$

where

$$B_0 = e^{-2}(\rho + f)^{-2}\text{div}\, \boldsymbol{E} , \tag{6}$$

Volume 78B, number 4 PHYSICS LETTERS 9 October 1978

$$E_i = \partial_0 B_i - \partial_i B_0 \,, \tag{7}$$

also

$$\pi = \partial_0 \rho \,. \tag{8}$$

The independent field degrees of freedom are $\boldsymbol{E}$, $\boldsymbol{B}$, π and ρ. Canonical quantization is achieved by implementing the following set of equal-time commutation relations

$$[B_i(\boldsymbol{x}, t), E_j(\boldsymbol{x}', t)] = \mathrm{i}\delta_{ij}\delta(\boldsymbol{x}-\boldsymbol{x}')\,,$$
$$[\rho(\boldsymbol{x}, t), \pi(\boldsymbol{x}', t)] = \mathrm{i}\delta(\boldsymbol{x}-\boldsymbol{x}')\,. \tag{9}$$

In an attempt at exploring the collective nonlinear phenomena of the coupled set of operator field equations derived from eqs. (5) and (9), we try a variational approach. In other words, we guess a trial state (which in this case will be a functional of a set of c-number parametrizing fields) and compute the expectation value of the energy operator (eq. (5) with eq. (9)) in this state. To obtain a "best fit" for the set of parametrizing functions we impose the requirement of stationarity of the energy expectation value with respect to variations in these functions (minimisation of the energy).

In the present paper we use, as a trial state, a coherent state [6]. This particular approximation scheme has the net effect of reducing an intractable set of coupled nonlinear operator equations to a classical problem of interacting fields where nontrivial solutions are already known to exist (the Nielsen–Olesen vortex solutions). Hence, by a suitable choice of parametrizing functions we obtain a candidate for the creation (also annihilation) operator for a Nielsen–Olesen vortex.

This particular realization of the topological charge raising operator should be useful in describing situations where large-scale vortex condensation takes place. We now write down our coherent trial state.

$$|\mathrm{trial}\rangle = \exp\left[-\mathrm{i}\int \mathrm{d}^3x\,(\boldsymbol{Q}\cdot\boldsymbol{E} - \boldsymbol{P}\cdot\boldsymbol{B})\right] \times \exp\left[-\mathrm{i}\int \mathrm{d}^3x\,(u\pi - v\rho)\right]|0\rangle\,. \tag{10}$$

Here the free-field Fock vacuum $|0\rangle$ is defined with respect to the set of asymptotic fields of eq. (3). The expectation value of the normal-ordered hamiltonian operator (eq. (5) and eq. (9)) in the state given by eq. (10) is

$$\langle\mathrm{trial}|:\mathcal{H}:|\mathrm{trial}\rangle \equiv E(\boldsymbol{P}, \boldsymbol{Q}, u, v)$$
$$= \tfrac{1}{2}\boldsymbol{P}^2 + \tfrac{1}{2}(\mathrm{curl}\,\boldsymbol{Q})^2 + \tfrac{1}{2}e^2(u+f)^2[Q_0^2 + \boldsymbol{Q}^2] \tag{11}$$
$$+ \tfrac{1}{2}v^2 + \tfrac{1}{2}(\nabla u)^2 + \tfrac{1}{4}\lambda u^2(u+2f)^2 + \text{surface term}\,,$$

where $Q_0 = e^{-2}(u+f)^{-2}\mathrm{div}\,\boldsymbol{P}$. To evaluate the "best fit" we require that $E(\boldsymbol{P}, \boldsymbol{Q}, u, v)$ be stationary with respect to variations in the independent c-number fields $\boldsymbol{P}$, $\boldsymbol{Q}$, u and v. This leads to the following set of field equations

$$v = 0\,, \tag{12}$$

$$\boldsymbol{P} = \mathrm{grad}\,[e^2(u+f)^2 Q_0]\,, \tag{13}$$

$$-\nabla^2 Q_0 + e^2(u+f)^2 Q_0 = 0\,, \tag{14}$$

$$-\nabla^2\boldsymbol{Q} + \boldsymbol{\nabla}(\boldsymbol{\nabla}\cdot\boldsymbol{Q}) + e^2(u+f)^2\boldsymbol{Q} + \text{surface term} = 0\,, \tag{15}$$

$$-\nabla^2 u + \lambda u(u+f)(u+2f) + e^2(u+f)[Q_0^2 + \boldsymbol{Q}^2] = 0\,. \tag{16}$$

These equations are known to have vortex solutions. To see this put $Q_0 = 0$ and $\partial_i Q_i = 0$ and also write $Q_r = Q_z = 0$ with $Q_\theta = Q - n/er$. We then obtain the axially symmetric equations of ref. [5]

$$-\frac{\mathrm{d}}{\mathrm{d}r}\left(\frac{1}{r}\frac{\mathrm{d}}{\mathrm{d}r}(rQ)\right) + e(u+f)^2\left(eQ - \frac{n}{r}\right) = 0\,, \tag{17}$$

$$-\frac{1}{r}\frac{\mathrm{d}}{\mathrm{d}r}\left(r\frac{\mathrm{d}u}{\mathrm{d}r}\right) + \lambda u(u+f)(u+zf) + (u+f)(eQ - n/r)^2 = 0\,. \tag{18}$$

The integer n measures the amount of topological charge (the quantised magnetic flux) carried by the vortex since

$$\oint Q_\mu \mathrm{d}x^\mu = 2\pi n/e\,. \tag{19}$$

In conclusion the operator

$$\exp\left[-\mathrm{i}\int \mathrm{d}^3x(\boldsymbol{Q}\cdot\boldsymbol{E} - \boldsymbol{P}\cdot\boldsymbol{B})\right]\exp\left[-\mathrm{i}\int \mathrm{d}^3x(u\pi - v\rho)\right] \tag{20}$$

creates a vortex (gauge soliton) of magnetic flux (topological charge) n, localized around $r = 0$, from the translationally invariant Fock space vacuum (which has $n = 0$) when Q, P, u and v satisfy eqs. (12)–(18).

Mathematically speaking, the unitary operator of eq. (20) intertwines between the physically and mathematically inequivalent representations of the algebra of observables [7] (i.e. causes transitions between the different topological charge sectors of the theory).

The method outlined here can be extended in a straightforward way to include the nonabelian Higgs model. This development together with some quantitative applications of the coherent state formalism will be presented in a forthcoming publication.

The author wishes to thank the National Science Council of Ireland for financial support.

References

[1] J.S. Langer, Phys. Rev. 167 (1968) 183.
[2] G. 't Hooft, On the phase transition towards permanent quark confinement, Utrecht preprint (1977).
[3] S. Mandelstam, Confinement in nonabelian gauge theories, Invited talk at the Washington Meeting of the APS (April 1977).
[4] Z.F. Ezawa, Quantum vortex operators and superselection rules in the Higgs model, Max Planck preprint (Jan. 1978).
[5] H.B. Nielsen and P. Olesen, Nucl. Phys. B61 (1973) 45.
[6] J.R. Klauder and E.C.G. Sudarshan, Fundamentals of quantum optics (Benjamin, 1968) Ch. 7.
[7] K. Cahill, Phys. Lett. 53B (1974) 174.
[8] J.A. Swieca, Solitons and confinement, Fortschritte der Physik, Heft 5 (1977) Band 25.

PHYSICAL REVIEW D VOLUME 27, NUMBER 4 15 FEBRUARY 1983

Vacuum instability and the critical value of the coupling parameter in scalar QED

Guang-jiong Ni
*Fudan University, Shanghai, People's Republic of China
and Department of Physics, University of Illinois at Urbana-Champaign,
1110 W. Green Street, Urbana, Illinois 61801**

Yi-ping Wang
Shanghai Science and Technology University, People's Republic of China
(Received 3 May 1982; revised manuscript received 2 August 1982)

Using the coherent state as a trial wave function describing the particle-antiparticle condensate if the coupling parameter $\alpha \equiv e^2/4\pi$ exceeds some critical value α_c, we find $\alpha_c = 2.543$ for scalar QED.

I. INTRODUCTION

It is well known that the Dirac equation for an electron moving around a nucleus with charge number Z can be solved exactly under the condition

$$Z\alpha \equiv Z\frac{e^2}{4\pi\hbar c} \equiv Z\frac{e^2}{4\pi} < 1 \; . \qquad (1.1)$$

In other words, there is a critical value of the product of α with Z,

$$(Z\alpha)_c = 1 \; , \qquad (1.2)$$

above which the motion of the electron becomes unstable, with a complex energy eigenvalue as a consequence of the Hamiltonian turning non-Hermitian. Actually, this is a symptom stemming from the restricted applicability of quantum mechanics and can be cured to some extent.[1]

For the Klein-Gordon equation, an earlier investigation showed that the motion of a charged particle in a nonsingular potential will become unstable once this potential is too deep and broad.[2] In the case of the Coulomb potential for a pointlike nucleus with charge number Z, there exists a critical value[3]

$$(Z\alpha)_c = \tfrac{1}{2} \; , \qquad (1.3)$$

which is comparable to that for the Dirac equation as given by Eq. (1.2).

In recent years the instability problem for non-Abelian gauge fields was initiated by Mandula *et al.*[4] It was found that the classical Coulomb solution for a point external source Q in a Yang-Mills field equation becomes unstable when the product of Q with α (here $\alpha \equiv g^2/4\pi$, g is the coupling constant in the Yang-Mills field) exceeds the following critical value:

$$(Q\alpha)_c = \tfrac{3}{2} \; . \qquad (1.4)$$

This problem is related intimately to that in massless scalar electrodynamics. The instability of the Coulomb solution implies that there is another solution with lower energy which screens the external source.

Instead of a pointlike external source, a constant field, say, a constant magnetic field along a fixed space and isospin direction will make the vacuum unstable. The corresponding unstable modes have been examined by many authors,[5] with speculation that the gluon condensation would occur even in the absence of the external field.[6]

While the above investigations were carried out mainly with classical field equations, there is a very interesting paper by Finger, Horn, and Mandula (denoted as FHM in the following) putting the problem in the quantum version.[7] They argued that even in the absence of any external source or field, a particle-antiparticle pair may be created in vacuum once the coupling constant α exceeds a critical value:

$$\alpha_c \equiv \frac{e_c^{\,2}}{4\pi} = \tfrac{3}{2} \;\; (\text{QED}) \; , \qquad (1.5)$$

$$\alpha_c \equiv \frac{g_c^{\,2}}{4\pi} = \tfrac{9}{8} \;\; (\text{QCD}) \; . \qquad (1.6)$$

While working within a limited Fock-space approximation, FHM neglected the multiparticle interactions in their calculation and this simplification was reflected in the final result shown in Eq. (1.5), indicating no difference between scalar and spinor QED.

As there is a common feeling shared by many physicists that the problem of vacuum instability of a gauge field must be linked with various important questions such as charge screening, phase transitions, confinement, and mass generation as well, we wish to improve the FHM scheme by taking the many-body effects into account. Instead of using a particle-antiparticle state

$$|\psi\rangle = \int d^3p\, \psi(\vec{p}) a^\dagger_{\vec{p}} b^\dagger_{-\vec{p}} |0\rangle , \tag{1.7}$$

we will use a coherent state of many-particle configurations to calculate the critical values of scalar QED with a result somewhat larger than that in Eq. (1.5).[8]

The organization of this paper is as follows. In Sec. II we first discuss the problem of massless scalar QED by using relativistic coherent states.[9] A momentum distribution function $f(\vec{p})$ is introduced similar to $\psi(\vec{p})$ in Eq. (1.7), but can only be calculated numerically by choosing two types of trial function $f(\vec{p})$, to get the critical values $\alpha_c = 2.543$ and 2.705. In Sec. III we point out the addition of a mass term will not affect the above critical values. In Sec. VI we give some further discussion. An appendix gives the necessary formulas used in the calculations.

II. MASSLESS SCALAR QED

A. Lagrangian and quantization

We begin to discuss first the massless scalar QED, the mass term will be added in Sec. III:

$$\mathscr{L} = (\partial^\mu + ieA^\mu)\phi^*(\partial_\mu - ieA_\mu)\phi - \tfrac{1}{4}F_{\mu\nu}F^{\mu\nu} . \tag{2.1}$$

In the first stage, the calculation is carried out in the same manner as FHM's. The canonical momenta are defined as

$$\Pi = \frac{\partial \mathscr{L}}{\partial \dot{\phi}^*} = (\partial_0 - ieA_0)\phi \tag{2.2}$$

and

$$\Pi^* = \frac{\partial \mathscr{L}}{\partial \dot{\phi}} = (\partial_0 + ieA_0)\phi^* . \tag{2.3}$$

Then in the Coulomb gauge

$$\vec{\nabla}\cdot\vec{A} = 0 \tag{2.4}$$

the Hamiltonian reads

$$\mathscr{H} = \Pi^*\dot{\phi} + \Pi\dot{\phi}^* - \vec{E}^t\cdot\vec{A}^t - \mathscr{L} \tag{2.5}$$

$$= \Pi^*\Pi + (\vec{\nabla} - ie\vec{A})\phi^*\cdot(\vec{\nabla} + ie\vec{A})\phi + \tfrac{1}{2}[(\vec{E}^t)^2 + (\vec{B})^2 + (\vec{\nabla}\chi)^2] . \tag{2.6}$$

Following FHM, we confine ourselves to discussing the states containing no physical photons. So the transverse electric field potential $\vec{A}^t$ together with $\vec{E}^t$ and $\vec{B}$ can be neglected:

$$\mathscr{H} = \Pi^*\Pi + \vec{\nabla}\phi^*\cdot\vec{\nabla}\phi + \tfrac{1}{2}(\vec{\nabla}\chi)^2 . \tag{2.7}$$

Here the scalar potential χ is determined by charge density ρ as follows:

$$\nabla^2\chi = -e\rho = ie(\Pi^*\phi - \Pi\phi^*) . \tag{2.8}$$

The canonical quantization procedure is also done in the usual form:

$$\phi(\vec{x},0) = \frac{1}{(2\pi)^{3/2}} \int \frac{d^3k}{(2|k|)^{1/2}} (e^{i\vec{k}\cdot\vec{x}} a_{\vec{k}} + e^{-i\vec{k}\cdot\vec{x}} b^\dagger_{\vec{k}}) ,$$

$$\phi^*(\vec{x},0) = \frac{1}{(2\pi)^{3/2}} \int \frac{d^3k}{(2|k|)^{1/2}} (e^{-i\vec{k}\cdot\vec{x}} a^\dagger_{\vec{k}} + e^{i\vec{k}\cdot\vec{x}} b_{\vec{k}}) , \tag{2.9}$$

$$\Pi(\vec{x},0) = \frac{-i}{(2\pi)^{3/2}} \int \frac{d^3k}{\sqrt{2}} \sqrt{|k|} (e^{i\vec{k}\cdot\vec{x}} a_{\vec{k}} - e^{-i\vec{k}\cdot\vec{x}} b^\dagger_{\vec{k}}) ,$$

$$\Pi^*(\vec{x},0) = \frac{i}{(2\pi)^{3/2}} \int \frac{d^3k}{\sqrt{2}} \sqrt{|k|} (e^{-i\vec{k}\cdot\vec{x}} a^\dagger_{\vec{k}} - e^{i\vec{k}\cdot\vec{x}} b_{\vec{k}}) .$$

Here the annihilation and creation operators a, $a^\dagger$, b, and $b^\dagger$ of particles and antiparticles with charge $+1$ and -1 obey the following commutation relations:

$$[a_{\vec{k}}, a^\dagger_{\vec{k}'}] = [b_{\vec{k}}, b^\dagger_{\vec{k}'}] = \delta^3(\vec{k} - \vec{k}') . \tag{2.10}$$

B. The expectation value of the Hamiltonian in a coherent state

We will evaluate the energy of a new neutral state which may form from the naive vacuum $|0\rangle$ by creating some particle-antiparticle pairs in it. For this purpose we integrate the Hamiltonian density to get

$$H=\int d^3x = H_0 + H_I \ , \tag{2.11}$$

$$H_0=\int d^3x\,|\vec{k}|\,(a^\dagger_{\vec{k}}a_{\vec{k}}+b^\dagger_{\vec{k}}b_{\vec{k}}) \ , \tag{2.12}$$

$$H_I=\tfrac{1}{2}\int d^3x(\vec{\nabla}\chi)^2$$
$$=\tfrac{1}{2}\int\int d^3x\,d^3x' {:}\rho(\vec{x})\frac{e^2}{4\pi|\vec{x}-\vec{x}'|}\rho(\vec{x}'){:} \ . \tag{2.13}$$

Substituting Eqs. (2.8) and (2.9) into Eq. (2.13) we get H_I as

$$\begin{aligned}H_I=\frac{e^2}{8(2\pi)^3}\int d^3k\,d^3k'\,d^3k''\frac{1}{k^2}{:}\Bigg\{&\Bigg[\left[\frac{|\vec{k}'|}{|\vec{k}+\vec{k}'|}\right]^{1/2}+\left[\frac{|\vec{k}+\vec{k}'|}{|\vec{k}'|}\right]^{1/2}\Bigg](a^\dagger_{-\vec{k}'}a_{-\vec{k}-\vec{k}'}-b^\dagger_{\vec{k}+\vec{k}'}b_{\vec{k}'})\\ &+\Bigg[\left[\frac{|\vec{k}'|}{|\vec{k}+\vec{k}'|}\right]^{1/2}-\left[\frac{|\vec{k}+\vec{k}'|}{|\vec{k}'|}\right]^{1/2}\Bigg](a^\dagger_{-\vec{k}'}b^\dagger_{\vec{k}+\vec{k}'}-a_{-\vec{k}-\vec{k}'}b_{\vec{k}'})\Bigg\}\\ \times\Bigg\{&\Bigg[\left[\frac{|\vec{k}''|}{|\vec{k}-\vec{k}''|}\right]^{1/2}+\left[\frac{|\vec{k}-\vec{k}''|}{|\vec{k}''|}\right]^{1/2}\Bigg](a^\dagger_{-\vec{k}''}a_{\vec{k}-\vec{k}''}-b^\dagger_{-\vec{k}+\vec{k}''}b_{\vec{k}''})\\ &+\Bigg[\left[\frac{|\vec{k}''|}{|\vec{k}-\vec{k}''|}\right]^{1/2}-\left[\frac{|\vec{k}-\vec{k}''|}{|\vec{k}''|}\right]^{1/2}\Bigg](a^\dagger_{-\vec{k}''}b^\dagger_{-\vec{k}+\vec{k}''}-a_{\vec{k}-\vec{k}''}b_{\vec{k}''})\Bigg\}{:}\ .\end{aligned} \tag{2.14}$$

Obviously, the normal-product order ensured the expectation value of H in the naive vacuum equals zero:

$$E_{\text{naive}}\equiv\langle 0|{:}H{:}|0\rangle=0 \ . \tag{2.15}$$

The Coulomb interactions H_I are composed of 16 types of terms corresponding to the Feynman diagrams as shown in Fig. 1. Since in the FHM scheme a pair of particle-antiparticle states as

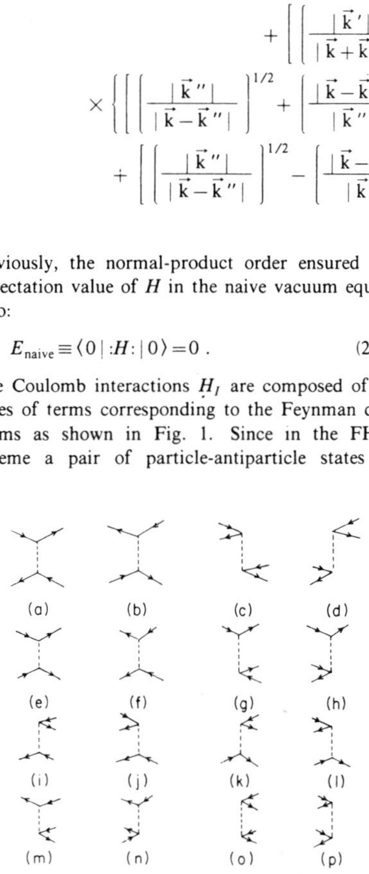

FIG. 1. The Feynman diagrams corresponding to 16 types of terms in H_I, Eq. (2.14). Read from left to right with the dashed lines denoting the Coulomb photon exchange.

shown by (1.7) is assumed, only the first four diagrams contribute to

$$E_\psi=\langle\psi|{:}H{:}|\psi\rangle \ . \tag{2.16}$$

Now we try to consider a many-particle state within which the interactions H_I will display in full. The relativistic coherent state suggested by Skagerstam[9] (see Appendix) can serve this purpose and is denoted by $|qfg\rangle$. Here q is the charge number of the state, $f(\vec{k})$ and $g(\vec{k})$ are the momentum distribution functions of particles and antiparticles, respectively. The neutrality of the state demands $q=0$ and the symmetry between particles and antiparticles implies $f(\vec{k})=g(\vec{k})$. Thus, we will choose $|0ff\rangle$ as a candidate of a new vacuum,

$$|\tilde{0}\rangle=|0ff\rangle \ , \tag{2.17}$$

and begin to calculate the expectation value of energy in this state:

$$E=\langle 0ff|{:}H{:}|0ff\rangle=E_0+E_1 \ , \tag{2.18}$$

$$E_0=\langle 0ff|{:}H_0{:}|0ff\rangle \ , \tag{2.19}$$

$$E_1=\langle 0ff|{:}H_I{:}|0ff\rangle \ . \tag{2.20}$$

The properties of a coherent state, especially formulas (A9) and (A10), allow us to turn the operators (q numbers) into c numbers directly without an appearance of divergence. The results are

$$E_0=2\frac{I_1(z)}{I_0(z)}\int d^3k\,|\vec{k}|[f(|\vec{k}|)]^2\,, \tag{2.21}$$

$$E_1=-\frac{2e^2}{(2\pi)^3}\frac{1}{z}\frac{I_1(z)}{I_0(z)}\int d^3k\frac{1}{|\vec{k}|^2}\int d^3k'\int d^3k''\left[\frac{|\vec{k}+\vec{k}'|}{|\vec{k}'|}\right]^{1/2}\left[\frac{|\vec{k}''|}{|\vec{k}-\vec{k}''|}\right]^{1/2}$$
$$\times f(|\vec{k}'|)f(\vec{k}''|)f(|\vec{k}+\vec{k}'|)f(|\vec{k}-\vec{k}''|)\,. \tag{2.22}$$

Here and in the following, we restrict further the function $f(\vec{k})$ to being a real function only depending on the magnitude of momentum $\vec{k}$. $I_n(z)$ is the modified Bessel function of rank n with argument

$$z\equiv 2\|f\|^2\equiv 2\int d^3k[f(|\vec{k}|)]^2\,. \tag{2.23}$$

We see that the kinetic energies of particles contribute a positive term E_0 while the Coulomb interactions among them contribute a negative one, E_1. The latter can be understood by counting the numbers of attracting pairs and that of repulsing ones. Therefore, a critical value of e, say e_c, exists above which the naive vacuum $|0\rangle$ will turn automatically into a new one, $|\tilde{0}\rangle=|0ff\rangle$:

$$\alpha\equiv\frac{e_c{}^2}{4\pi}=\frac{(2\pi)^2\int d^3p[f(|\vec{p}|)]^2\int d^3k\,|\vec{k}|[f(|\vec{k}|)]^2}{\int d^3k\frac{1}{|\vec{k}|^2}\int d^3k'\int d^3k''\left[\frac{|\vec{k}+\vec{k}'||\vec{k}''|}{|\vec{k}'||\vec{k}-\vec{k}''|}\right]^{1/2}f(|\vec{k}'|)f(|\vec{k}''|)f(|\vec{k}+\vec{k}'|)f(|\vec{k}-\vec{k}''|)}\,. \tag{2.24}$$

C. The critical value of the coupling parameter

We are able to establish an equation for $f(|\vec{k}|)$ by taking the variation

$$\frac{\delta E}{\delta f(|\vec{p}|)}=0 \tag{2.25}$$

and find it in a formidable appearance:

$$\Bigg\{8\left[1-\frac{1}{z}\frac{I_1(z)}{I_0(z)}-\left[\frac{I_1(z)}{I_0(z)}\right]^2\right]\int d^3k\,|\vec{k}|[f(|k|)]^2+\frac{4I_1(z)}{I_0(z)}|\vec{p}|$$
$$+\frac{4e^2}{(2\pi)^3}\left[-\frac{2}{z}+\frac{4}{z^2}\frac{I_1(z)}{I_0(z)}+\frac{2}{z}\left[\frac{I_1(z)}{I_0(z)}\right]^2\right]$$
$$\times\int\frac{d^3k}{|\vec{k}|^2}\int d^3k'\int d^3k''\left[\frac{|\vec{k}+|\vec{k}'|}{|\vec{k}'|}\right]^{1/2}\left[\frac{|\vec{k}''|}{|\vec{k}-\vec{k}''|}\right]^{1/2}$$
$$\times f(|\vec{k}'|)f(|\vec{k}''|)f(|\vec{k}+\vec{k}'|)f(\vec{k}-\vec{k}''|)\Bigg\}f(|\vec{p}|)$$
$$-\frac{2e^2}{(2\pi)^3}\left[\frac{2}{z}-\frac{I_1(z)}{I_0(z)}\right]\int\frac{d^3k}{|\vec{k}|^2}\int d^3k'\left[\frac{|\vec{k}+\vec{k}'|}{|\vec{k}+\vec{p}|}\right]^{1/2}\left[\left[\frac{|\vec{p}|}{|\vec{k}+\vec{p}|}\right]^{1/2}+\left[\frac{|\vec{k}+\vec{p}|}{|\vec{p}|}\right]^{1/2}\right]$$
$$\times f(|\vec{k}'|)f(|\vec{k}+\vec{k}'|)f(|\vec{k}+\vec{p}|)=0\,. \tag{2.26}$$

But in view of the original Lagrangian, Eq. (2.1), containing no restoring term like ϕ^4, we may anticipate that there should be one solution in Eq. (2.26) which describes a collapsing state with the averaged number of particles (and antiparticles):

$$\langle N\rangle=2z\frac{I_1(z)}{I_0(z)}\underset{z\to\infty}{\sim}2z\to\infty\,. \tag{2.27}$$

So we resort to making an ansatz of $f(|\vec{k}|)=f(k)$:

$$f^{(1)}(k)=A\frac{1}{(|\vec{k}|)^{1/2}}e^{-\beta k^2} . \tag{2.28}$$

Then by direct integration of Eqs. (2.21) and (2.22), we find

$$E=\frac{1}{\sqrt{\beta}}\frac{\sqrt{2\pi}}{4}B\left[\frac{1}{B}-\frac{e^2}{4\pi}\right]\langle N\rangle , \tag{2.29}$$

where

$$B=\frac{1}{\sqrt{\pi}}\int_0^\infty \frac{1}{\xi^2}[\mathrm{erf}(\xi)]^2 e^{-2\xi^2}d\xi=0.3932 \tag{2.30}$$

is evaluated by expanding the error function

$$\mathrm{erf}(\xi)=\frac{2}{\sqrt{\pi}}\int_0^\xi e^{-\eta^2}d\eta \tag{2.31}$$

into a series and integrating term by term. Now it is evident from Eq. (2.29) that the energy of coherent state E is proportional to $\langle N\rangle$, the averaged number of particles in it, and the sign of E depends on the magnitude of the coupling parameter $e^2/4\pi$ $(=\alpha)$ with a critical value

$$\alpha_c^{(1)}=\frac{e_c^2}{4\pi}=\frac{1}{B}=2.543 . \tag{2.32}$$

If $\alpha<\alpha_c^{(1)}$, the condensation $(\langle N\rangle>0)$ will cost a certain amount of energy, so only fluctuations with small amplitude can exist. In other words, the naive vacuum $|0\rangle$ is stable. But if $\alpha>\alpha_c^{(1)}$, the formation of a coherent state with $\langle N\rangle>0$ will be a favorable process in energy, so the naive vacuum becomes unstable against collapsing as shown in Fig. 2.

As a check of accuracy in our approximation, a second ansatz is made as follows:

$$f^{(2)}(k)=A'(|k|)^{1/2}e^{-\beta' k^2} . \tag{2.33}$$

Then

$$E=\frac{1}{\sqrt{\beta'}}\frac{3\sqrt{2\pi}}{8}B'\left[\frac{1}{B'}-\frac{e^2}{4\pi}\right]\langle N\rangle , \tag{2.34}$$

with

$$B'=\frac{1}{6\pi}+\frac{2}{3\pi}\int_0^\infty d\xi\left[\frac{1}{\xi}+2\xi\right]\mathrm{erf}(\xi)e^{-3\xi^2}$$
$$+\frac{1}{6\sqrt{\pi}}\int_0^\infty d\xi[\mathrm{erf}(\xi)]^2\left[\frac{1}{\xi^2}+4+4\xi^2\right]e^{-2\xi^2} . \tag{2.35}$$

Thus

$$\alpha_c^{(2)}=\frac{1}{B'}=2.705 . \tag{2.36}$$

We see the discrepancy between these two values

$$\frac{\alpha_c^{(2)}-\alpha_c^{(1)}}{\alpha_c^{(1)}}\simeq 6.37\% \tag{2.37}$$

is not too bad in view of the crudeness of our approximations.

III. SCALAR QED WITH MASS

We have to answer the following question: Will incorporation of a mass term into the massless L affect the values of α? The answer is no. Actually, in the case of scalar QED, if we add a mass term $-m^2\phi^*\phi$ in Eq. (2.1) and adopt the ansatz (2.33), one can expand the expectation value of energy in the vicinity of $z=0$ $(z=2\|f\|^2)$:

$$E=\frac{m^2}{2}\left[\frac{\pi\beta'}{2}\right]^{1/2}z$$
$$+\frac{3}{8}\left[\frac{\pi}{2\beta'}\right]^{1/2}\left[\left[1-\frac{e^2}{4\pi}B'\right]\right.$$
$$\left.+\tfrac{2}{3}\beta' m^2\right]z^2+\cdots . \tag{3.1}$$

It can be seen that if $\beta'\to 0$ (this implies the condensation occurs over the whole momentum range), we still have

$$\alpha_c^{(2)}=\frac{1}{B'}=2.705 . \tag{3.2}$$

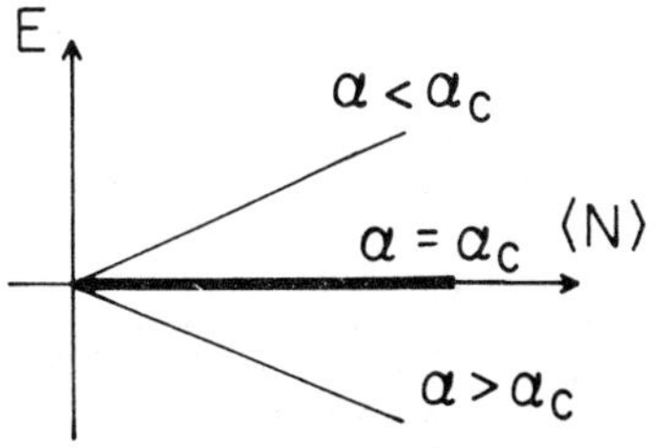

FIG. 2. The energy E of the coherent state as a function of $\langle N\rangle$, the average number of particles, with three different coupling parameters.

IV. DISCUSSION

Recently, Hey and Mandula also carried out a coherent-state variational calculation in massless scalar QED.[10] They used a trial state in analogy with the BCS type as follows:

$$|C\rangle=\exp\left[\int d^3k \tanh\theta(\vec{k})a^\dagger_{\vec{k}}b^\dagger_{-\vec{k}}\right]|0\rangle\ . \tag{4.1}$$

Through a numerical analysis they found that a phase transition from perturbative vacuum to condensate phase occurs in the region between $\alpha=3$ and $\alpha=3.5$. It seems higher than our prediction $\alpha_c=2.543$.

First of all, both these values are larger than that of FHM,[7] Eq. (1.5), and are quite understandable. The repulsive forces between particles of the same kind tend to prevent the whole system from collapsing, so the critical value of α which signals the appearance of one Cooper pair with negative energy is often lower than that which signals a transition to condensate phase.

Second, it seems to us that while a BCS type of trial state is a suitable candidate describing a condensate phase of a fermion system,[8,11] it may be too restrictive for a boson system. Actually, the momenta of particle and of antiparticle are just opposite within a pair as in Eq. (4.1) and so are less flexible than that in the coherent state proposed by Skagerstam[9] as shown in the Appendix of this paper.

The other advantage of this kind of coherent state we have taken is the simplicity in calculation. There is no explicit ultraviolet divergence and thereby no need of introducing a counterterm such as $\lambda(\phi^*\phi)^2$ which in turn brings no direct contact with the work of Coleman and Weinberg.[12] In the terminology of Ref. 10, we are also working in the constrained version of scalar QED, i.e., the vacuum expectation value of the scalar field vanishes: $\langle\phi\rangle=0$.

ACKNOWLEDGMENTS

The authors would like to thank Professors Cai Jian-hua, Hao Bai-lin, Hu Yao-guang, Ruan Tunan, Su Ru-keng, Tao Rui-bao, and Yin Pong-cheng for helpful discussions. One of the authors (G-j Ni) is also grateful to Professor R. O. Simmons and the Physics Department, University of Illinois at Urbana-Champaign for hospitality and to Professors S-J Chang, J. B. Kogut, M. Stone, M. Wortis, J. A. Wright, and H. W. Wyld for discussions during which this work was in its final stage. This work was supported by Fudan University, Shanghai Science and Technology University, the University of Illinois Research Board, the Fudan University-University of Illinois at Urbana-Champaign exchange program, and NSF Contract No. PHY82-01948.

APPENDIX: THE RELATIVISTIC COHERENT STATE

We simply cite some formulas[9] which are used in this paper. Define

$$|f\rangle=\int d^3k\, f(\vec{k})a^\dagger_{\vec{k}}|0\rangle \equiv(a^\dagger,f)|0\rangle\ , \tag{A1}$$

$$||f||^2=\langle f|f\rangle =\int d^3k\, f^*(\vec{k})f(\vec{k})\ . \tag{A2}$$

Then a coherent state of a scalar field with net charge q is represented by

$$|qfg\rangle=N_qF_q((a^\dagger,f)(b^\dagger,g)) \times\frac{1}{||f||^q}(a^\dagger,f)^q|0\rangle\ , \tag{A3}$$

where

$$F_q(y)=y^{-q/2}I_q(2y^{1/2}) \tag{A4}$$

and $I_q(z)$ is the modified Bessel function of rank q:

$$I_q(z)=\sum_{k=0}^{\infty}\frac{1}{k!\Gamma(k+q+1)}\left[\frac{z}{2}\right]^{q+2k}\ , \tag{A5}$$

$$N_q=\left[\frac{||f||^q||g||^q}{I_q(2||f||\ ||g||)}\right]^{1/2}\ . \tag{A6}$$

The average number of particles in a coherent state is

$$\langle N\rangle=\int\langle a^\dagger_{\vec{k}}a_{\vec{k}}+b^\dagger_{\vec{k}}b_{\vec{k}}\rangle d^3k =2||f||\ ||g||)\frac{I_q'(2||f||\ ||g||)}{I_q(2||f||\ ||g||)}\ , \tag{A7}$$

while the charge number of this state is

$$Q=\int\langle a^\dagger_{\vec{k}}a_{\vec{a}}-b^\dagger_{\vec{k}}b_{\vec{k}}\rangle d^3k=q\ . \tag{A8}$$

This coherent state is an eigenstate of a pair of annihilation operators

$$a_{\vec{k}}b_{\vec{k}}|qfg\rangle=f(\vec{k})g(\vec{k})|qfg\rangle\ . \tag{A9}$$

In this paper $q=0$ and $f=q$, so we further have

$$\langle 0ff \mid a^{\dagger}_{\vec{k}_1} a^{\dagger}_{\vec{k}_2} a_{\vec{k}_3} a_{\vec{k}_4} \mid 0ff \rangle = f^*(\vec{k}_1) f^*(\vec{k}_2) f(\vec{k}_3) f(\vec{k}_4) \frac{I_2(z)}{I_0(z)}, \tag{A10}$$

$$\langle 0ff \mid a^{\dagger}_{\vec{k}_1} b^{\dagger}_{\vec{k}_2} a^{\dagger}_{\vec{k}_3} a_{\vec{k}_4} \mid 0ff \rangle = f^*(\vec{k}_1) f^*(\vec{k}_2) f^*(\vec{k}_3) f(\vec{k}_4) \frac{I_1(z)}{I_0(z)},$$

where

$$z = 2\|f\|^2 . \tag{A11}$$

---

*Permanent and present address.

[1]K. M. Case, Phys. Rev. 80, 797 (1950); Dai Xian-xi, Huang Fa-yang, and Ni Guang-jiong, in *Proceedings of the Guangzhou Conference on Theoretical Particle Physics, Guangzhou, China, 1980* (Academica Sinica, Beijing, 1980), Vol. II, p. 1373.

[2]L. I. Schiff, H. Snyder, and J. Weinberg, Phys. Rev. 57, 315 (1940).

[3]A. J. G. Hey and J. E. Mandula, Phys. Rev. D 19, 1856 (1979).

[4]J. E. Mandula, Phys. Rev. D 14, 3497 (1976); Phys. Lett. 67B, 175 (1977); 69B, 495 (1977); M. Magg, *ibid.* 78B, 481 (1978).

[5]N. K. Nielsen and P. Olesen, Nucl. Phys. B144, 376 (1978); Shau-Jin Chang and N. Weiss, Phys. Rev. D 20, 869 (1979); P. Sikivie, *ibid.* 20, 877 (1979).

[6]J. Ambjørn and P. Olesen, Nucl. Phys. B170, 60 (1980); B170, 265 (1980); S.-J. Chang and G-J. Ni, Phys. Rev. D 26, 864 (1982).

[7]J. Finger, D. Horn, and J. E. Mandula, Phys. Rev. D 20, 3253 (1979).

[8]This paper is an enlarged version of part of a short note in Chinese (Kexue Tongbao): Ni Guang-jiong and Wang Yi-ping, Sci. Bull. 26, 853 (1981).

[9]B. S. Skagerstam, Phys. Rev. D 19, 2471 (1979); 22, 534(E) (1980).

[10]A. J. G. Hey and J. E. Mandula, Nucl. Phys. B198, 237 (1982).

[11]J. R. Finger and J. E. Mandula, Nucl. Phys. B199, 168 (1982).

[12]S. Coleman and E. Weinberg, Phys. Rev. D 7, 1888 (1973).